# BRIEF CONTENTS

iii

# Organic Chemistry

## STRUCTURE AND REACTIVITY

### FIFTH EDITION

Seyhan N. Eğe

THE UNIVERSITY OF MICHIGAN

Houghton Mifflin Company     Boston     New York

Publisher: Charles Hartford
Executive Editor: Richard Stratton
Development Editor: Rita Lombard
Editorial Associate: Rosemary Mack
Project Editor: Andrea Cava
Senior Production/Design Coordinator: Jill Haber
Senior Designer: Henry Rachlin
Senior Manufacturing Coordinator: Priscilla J. Bailey
Senior Marketing Manager: Katherine Greig
Marketing Associate: Alexandra Shaw

Cover illustration: Kelly Chichak

Cover image: Cover image of a single-walled carbon nanotube wrapped in an amylose coil, based on research by Professor J. Fraser Stoddart (A. Star, D. W. Steurman, J. R. Heath, and J. F. Stoddart, *Angewandte Chemie. International Edition 2002, 41,* 2508–2512).

About the cover: Single-walled carbon nanotubes consist of small test-tube-like structures about 1 nm in diameter and some micrometers in length made up essentially of rolled-up sheets of graphene, the carbon structure of graphite (Figure 10.7, p. 365 in the text). They have high strength and elasticity and thermal and electric conductivities. Much research is being done on applications for these new materials.

One area that has been difficult to examine is possible biological applications of these nanotubes because they are not soluble in water. Professor J. Fraser Stoddart of the Department of Chemistry and Biochemistry at the University of California, Los Angeles, and his coworkers discovered that the rod-shaped nanotubes could be made soluble in water if they were wrapped in the coiled structures that a fraction of starch known as amylose assumes when it complexes with molecules of iodine to give the familiar starch-iodine blue color. Figure 23.7, page 1004 in the text, shows how the carbon nanotube displaces iodine molecules from the coiled amylose. The interior of the amylose coil is hydrophobic enough to provide a hospitable environment for the carbon rod. The exterior of the coil hydrogen bonds to the surrounding water molecules to bring the whole complex into solution. The image on the cover of this book shows a single-walled carbon nanotube (the blue rod showing the six-membered rings characteristic of graphite) wrapped in a coil of amylose (shown in red, violet, and silver).

Once the nanotubes are made water soluble by wrapping in amylose, they dissolve to the extent of 3 grams per liter. The nanotubes in water solution can be separated from the other carbon materials that are the side products of the production of the nanotubes but that do not have the shapes that allow them also to complex with amylose. The water solutions of the nanotubes are quite stable as long as nothing is done to destroy the amylose coil. Our saliva contains $\alpha$-amylase, an enzyme that digests starch. Professor Stoddart reports that the nanotube solutions "are stable for weeks provided you do not spit on them!" Addition of saliva (or solutions of starch-splitting enzymes) precipitate the nanotubes as the enzyme breaks amylose down into smaller and smaller carbohydrate fragments, finally resulting in the formation of glucose. Once the three-dimensional structure of amylose is destroyed, it can no longer surround and solubilize the nanotubes.

A list of acknowledgments appears following the index. This list is considered an extension of the copyright page.

Printed in the U.S.A.

Library of Congress Control Number: 2003102314

ISBN: 0-618-31809-7

123456789-DOW-07 06 05 04 03

# CONTENTS

## 3  Reactions of Organic Compounds as Acids and Bases   78

# 4   Reaction Pathways   115

# 5   Alkanes and Cycloalkanes   144

# 6  Stereochemistry  191

# 7 Nucleophilic Substitution and Elimination Reactions    233

## 10 The Chemistry of Aromatic Compounds. Electrophilic Aromatic Substitution   351

## 11 Nuclear Magnetic Resonance Spectroscopy   389

## 15  Carboxylic Acids and Their Derivatives. Acyl-Transfer Reactions   594

## 16  Structural Effects in Acidity and Basicity Revisited. Enolization   653

## 17 Enols and Enolate Anions as Nucleophiles. Alkylation and Condensation Reactions 692

# 22   The Chemistry of Heterocyclic Compounds   931

# 23 Structure and Reactivity in Biological Macromolecules   970

# Complete Reaction Mechanisms as Featured in
## VISUALIZING THE REACTION

**xxiii**

# PREFACE

The Fifth Edition of *Organic Chemistry: Structure and Reactivity* incorporates some major changes, continuing the evolution of the textbook that began with the Fourth Edition. Unchanged is its primary pedagogical aim: to engage students actively in developing a practical understanding of the causes of chemical change rather than trying to master organic chemistry through memorization. I help students to think as practicing chemists do in predicting reactivity from structure by encouraging them to learn new ways of thinking—of analyzing problems, sorting facts, reasoning by analogy, looking for patterns—and that gives them skills for all the rest of their work.

## Emphasis on Process

The Fifth Edition is therefore unchanged from previous editions in its emphasis on process, on an understanding of reactivity rather than on an encyclopedic knowledge of reactions. Students suffer from an overload of different things to remember and from not learning to think their way toward predicting the outcome of a reaction that they have never seen before. A thorough understanding, with an emphasis on the mechanism of a few major reactions, enables students to apply the principles they have learned and gives them the confidence to tackle new situations. Once students master the fundamental thought processes of chemistry, teachers find that they can introduce their own favorite reactions to their students. To achieve this end, in writing the Fifth Edition, I continued to choose or reject topics on the basis of whether they contribute to an overall mechanistic understanding of the subject at a level that is appropriate for the students for whom the book is written.

## Integration of Biological Examples

Who are the students in organic chemistry? We know that most of them are not going to become chemists. They are mostly heading toward the biological and medical sciences or will become chemical engineers. As the basis of the biological sciences becomes more and more chemical, there are calls for students in modern biology also to be firmly grounded in chemistry. Those heading for chemistry and chemical engineering, on their part, need to appreciate and understand the biological applications of chemistry as well as the use of biological methods, such as genetic engineering, for the production of chemicals.

To achieve a greater integration of the applications of the concepts of organic chemistry to biological systems into earlier chapters, I decided to give this edition of the book a spiral approach to various topics. The first decision from which others flowed was to recognize that the design of syntheses of organic compounds is an excellent exercise in the analysis of structures and the application of reactions. However, beyond a certain level of sophistication, there is no reason to familiarize students who will not become chemists with certain reagents or processes. A separate chapter on synthesis (Chapter 21) has been added to the book to allow instructors who wish to do so to revisit certain concepts or to introduce reactions that are primarily useful for syntheses. Thus, for example, students meet the concept of protecting groups first in Chapter 15 in the context of acyl-transfer reaction and peptide synthesis. In Chapter 21, protecting groups for alcohols and carbonyl groups are introduced, and those for amines are expanded. All of these protecting groups are used in syntheses. Later in Chapter 21, solid phase syntheses are introduced and protecting groups are once more applied. Then in a new chapter on structure and reactivity in biological macromolecules (Chapter 23), protecting groups are used again in the synthesis of oligonucleotides.

Other examples of this multilayered approach to topics, which gives instructors more flexibility in how deeply they wish to go into a subject, may be found in Chapters 15, 17, and 18. Acyl-transfer reactions, including their application to peptide syntheses, are treated in Chapter 15, but the kinds of transformations of acid derivatives that are primarily useful in organic syntheses, such as hydride reductions or reactions with organometallic reagents, are postponed to Chapter 21. Chapter 17 is devoted to the reactions of enolate ions and introduces aldol and Claisen condensations. These reactions are reviewed and applied in the context of biological and biosynthetic pathways. Applications of these condensation reactions to more sophisticated organic syntheses, including the construction of ring systems, are revisited in Chapter 21. Chapter 18, the chemistry of polyenes, includes an introduction to the basics of the Diels-Alder reaction as an example of 1,4-addition to a diene, but the finer points of regioselectivity and stereoselectivity are reserved for Chapter 21, and the frontier molecular orbital treatment of the reaction for Chapter 25, the chapter on concerted reactions.

## Organization and Coverage

The first six chapters of the Fifth Edition continue to lead students quickly to the concept that the structures of organic compounds determine their chemical reactivity. In these chapters

- The emphasis on acidity and basicity to introduce chemical reactivity (a hallmark of this book since its first edition) continues, as does the emphasis on the use of the curved-arrow notation for the representation of reaction mechanisms.

- Proton and carbon nuclear magnetic resonance spectroscopy are used in Chapter 5 in a rudimentary way as tools for the investigation of questions of connectivity, molecular symmetry, and equivalence of atoms and groups.

- The basics of structure, nomenclature, stereochemistry, and reaction mechanisms are presented, equipping students to understand the chemistry presented to them in the rest of the book.

- Extensive cross-referencing allows instructors considerable flexibility in choosing the order of subjects to follow.

A major change in the order of topics in the Fifth Edition is the moving of aromatic chemistry from Chapter 20 in the Fourth Edition to Chapter 10 in this edition. In this move the chapter was shortened to concentrate on the concepts of aromaticity and electrophilic aromatic substitution. The first 10 chapters of the book now form a coherent introduction to organic chemistry with a consistent emphasis on carbocations as reactive intermediates, and form the basis of a first-term course.

Nuclear magnetic resonance spectroscopy is now the subject of Chapter 11, with ultraviolet, infrared, and mass spectral methods found in Chapter 12. These topics are taught at the University of Michigan in the laboratory courses that accompany the lecture courses.

The interconversions of alcohols and carbonyl compounds (Chapters 13 and 14), acyl-transfer reactions (Chapter 15), and the chemistry of enolates (Chapters 16 and 17) constitute the core of a second term where emphasis shifts to carbanions (and free radicals, if Chapter 19 is included) as reactive intermediates. The use of biological examples to review and enrich the fundamental concepts and mechanisms of these chapters accomplishes the goal of introducing students to some of the most important properties and reactions of amino acids, peptides, monosaccharides, and lipids, even in courses that never get to the "biochemistry" chapters usually found at the back of organic chemistry textbooks.

A new chapter, Chapter 23, brings together the chemistry of carbohydrates, proteins, and nucleic acids, examining the structures of their simple units and the different levels of complexity that result as these units bind to each other to give larger and larger molecules. The chapter includes an introduction to the techniques that made the Human Genome Project possible, as well as to the synthesis of oligonucleotides. This chapter plays the same role in the book as Chapter 21, the synthesis chapter. Instructors who wish to go beyond the introduction to structure and reactivity of biological molecules found in the earlier chapters will find several additional layers of complexity in Chapter 23.

## Features of the Fifth Edition

The features that were introduced in previous editions as aids to learning are retained in this edition. I have already discussed the emphasis on process rather than memorization in the book, as well as the greater integration of biological examples in the text. A description of additional features follows.

### Visualizing the Reaction and the Pedagogical Use of Color

As in all previous editions, complete mechanisms are given for each type of reaction. These mechanisms are highlighted in "Visualizing the Reaction," a boxed feature set apart from the text. Students must practice developing their powers of imagination and following a process with the "inner eye." These complete mechanisms feature the judicious use of color to enhance the process of visualizing. Acidic or electrophilic sites are highlighted as red atoms, blue shading signifies basic or nucleophilic species, and gray shading emphasizes leaving groups. Color is used to indicate whether the reactive species on one side of an equation are converted into new reactive species after reaction to show students the reversibility of many reactions, especially acid-base reactions. Color also is used throughout the book to stress, with consistency, various important structural features. For example, in sections where students are just learning to see stereochemistry, green and red shading highlights stereochemical relationships.

### Structure and Reactivity on the Web: "One Small Step"

This textbook prepares students to extend the understanding gained from the specific examples they have studied to new reactions and concepts. It takes advantage of the growing capacity of students to think this way in a feature called "One Small Step." A few of the "One Small Steps" appear in the textbook at the point at which an extension of a concept just studied will allow a student to move a little further in understanding. Most of the "One Small Steps," however, will appear on the web site associated with the book available through http://chemistry.college.hmco.com, with notes in the margins to indicate that such a resource is available. In many cases, after the statement of the problem, hints for the solution of the problem will also appear on the web. Finally, students will be able to access answers to the problems, some of which will contain animation, as appropriate, on the web. The answers to the "One Small Step" problems that appear in the textbook will also be available to instructors in the *Instructor's Manual*.

Some of the material that has been removed from the textbook in an effort to make it more mechanistically cohesive is still available to instructors who wish to use it as enrichment through the "One Small Steps" on the web. In addition, totally new reagents and reactions also appear there. The web format allows us to continue to update and renew the types of problems and thought-provoking questions that accompany the book.

## Sections on the Art of Solving Problems

In working with my students, I have become convinced that encouraging them to analyze problems systematically is the single most important factor in increasing their overall intellectual skills. The "Art of Solving Problems" sections, unique among organic chemistry texts, offer students a systematic, questioning approach to solving organic chemistry problems. These sections do not simply provide a way for students to learn to plug data into a prelearned formula; rather, students learn to reason their way to a solution. Some examples of such sections are listed below.

- In Chapter 1, students are introduced to the idea that the solution to a problem in chemistry requires a step-by-step analysis of the problem. This analysis takes the form of questions that students pose to themselves in a systematic way.

- In Chapter 4, students are shown how to reason backward in solving problems involving simple syntheses.

- In Chapter 7, students are led through the types of questions that help them to predict the product of a reaction and to transform a given starting material into a desired product. These are complex questions with many types of answers, depending on the particular problem being solved. Not all of the questions are directly applicable to the problem under consideration, but they represent steps in the processes of deciding how to use the data given in the problem.

- The same method of questioning is applied in Chapter 9 to writing mechanisms. To reinforce this practice for students, in most chapters a problem is worked out using the same set of questions.

- Chapter 11 introduces the application of the same systematic method of questioning to the solution of spectroscopic problems.

The *Study Guide* further reinforces the questioning approach used in the book by applying it to solving some of the problems in and at the end of chapters.

## Problems

We try as much as possible to create examination questions using examples from the recent research literature. We tell the students that the problem they are solving is related to real chemistry being performed by chemists doing cutting-edge research in interesting and important problems. That same philosophy is behind the problems in my textbook. Except for a few routine drill problems, all of the problems reflect the chemistry that is very much alive in the research literature today. Many of the problems include spectra or spectral data. Many others illustrate the relationships between chemistry and biology or medicine. In addition to the problems in the textbook, students have access to supplemental problems that appear in the *Study Guide* and for which there are no answers. Answers to these problems are available to instructors in the *Instructor's Manual.*

## In-Text Summaries

Two forms of summary appear in the textbook itself. Each chapter (except Chapter 23) contains either

- An end-of-chapter summary that offers a concise review of the major concepts covered in the chapter, or

- End-of-chapter tables that summarize the reactions that appear in the chapter. Organized so as to remind students of how the reactions proceed, the summary tables are not made up of general reactions to be memorized, but instead take students briefly through the stages of the reaction again, reminding them of the types of reagents needed, reactive intermediates involved, and the stereochemistry of the reaction. These tables are particularly helpful to students when they are used together with the concept maps in the *Study Guide.*

# Study Guide

Roberta W. Kleinman of Lock Haven University and Peggy Zitek of the University of Michigan are my coauthors for the *Study Guide.* Both of them have given me invaluable help. I especially want to acknowledge Roberta Kleinman and Agnes Soderbeck, of the University of Michigan, who, with their considerable skill at the computer, are both responsible for transforming the material into camera-ready copy. As in the previous editions, the Fifth Edition contains detailed solutions to every problem in the text as well as explanations of the reasoning processes behind the answers for many problems. The answers have been expanded to include many three-dimensional molecular representations. Some problems in this edition are worked out using the questions developed in the "Art of Problem Solving" sections of the text to reinforce students' understanding of this approach. In addition, the *Study Guide* contains the supplemental problems for each chapter that were described earlier.

## Concept Maps

The concept maps in the *Study Guide* were conceived as a practical way to organize and summarize the material presented in the book, and they present major ideas in outline form. Notes in the margins of the textbook alert students that the concept maps are available. Students are encouraged to examine the maps, and then create their own, because the process of creating a concept map requires them to give up a purely linear way of thinking about a subject and to explore interrelationships.

## Workbook Exercises

The *Study Guide* also includes workbook exercises for Chapters 1–11, 14, 15, and 17. Notes in the margins of the textbook alert students to the presence of the exercises in the *Study Guide.* The exercises are intended to encourage students to think about and review concepts learned in earlier chapters in a different context before starting new work. No answers are provided for the workbook exercises. Students are encouraged to do these exercises with other students and to talk about the issues they raise.

# Additional Supplements

## Test Bank

My colleagues and I firmly believe that the real curriculum in a course is defined not by the syllabus but by the examinations for the course. To that end, we spend considerable effort in making sure our examinations truly reflect the values that we wish students to gain from our courses. Building upon the foundation laid in previous editions by my colleague Brian Coppola of University of Michigan, Steven R. Boone and Scott E. McKay, both of Central Missouri State University, have greatly enhanced the strength of this edition's *Test Bank*. Two kinds of questions may be found there. Some of them lend themselves to multiple-choice answers. Others more closely resemble the types of questions on our examinations, showing how we use the research literature of organic chemistry to design examination questions that encourage students to join in the reasoning processes by practicing elements in their research. More may be found on this in the *Instructor's Manual*.

## Instructor's Manual

Since the University of Michigan put its chemistry curriculum into practice 10 years ago, we have been asked on many different occasions to share our ideas on how a first-year course can be successfully based on the content of organic chemistry. We are, of course, not alone in such a course. The *Instructor's Manual* that accompanies this textbook, written with my colleague Brian Coppola, shares some of the ways in which we conduct our first-year courses. We also have invited Paul Scudder of New College to share his perspectives on the subject. The *Instructor's Manual* contains the following features:

- Teaching and learning methods we and Paul Scudder have found useful in taking full advantage of the pedagogical possibilities built into the textbook

- Suggestions on how to design examinations for a course based on the book

- Suggestions for demonstrations that may be used with some of the early chapters of the book

- Answers to the supplemental problems that appear in the *Study Guide*

- Answers to the "One Small Step" problems that appear in the textbook

# Software

Several exciting multimedia products provide new teaching and learning tools to support the Fifth Edition.

## *HMClass Prep* with *HMTesting* CD-ROM Package

This package includes both *HMClass Prep* and *HMTesting* on one CD-ROM (ISBN 0-618-31814-3). It allows an instructor to access both lecture aids and testing software in one place. These components cannot be ordered separately.

- *HMClass Prep* includes everything instructors will need to develop their lectures—PowerPoint slides with line art and spectra from the textbook, concept maps from the *Study Guide,* as well as *Instructor's Manual* and Microsoft® Word files of the printed *Test Bank*.

● *HMTesting* combines a flexible test-editing program with a comprehensive grade-book function for easy administration and tracking. It enables instructors to administer tests via network server or the web. The *HMTesting* database contains a wealth of questions and can produce multiple-choice as well as essay tests. Questions can be customized based on the chapter being covered, the question format, level of difficulty, and specific topics. *HMTesting* provides for the utmost security in accessing both test questions and grades.

The "One Small Step" feature of the textbook, designed to help students apply their conceptual understanding to new reactions, is supplemented by further examples on the Houghton Mifflin web site (http://chemistry.college.hmco.com). The web site contains additional "One Small Step" problems with hints, graphics, and animations to illustrate the answers. Students may work with these examples directly on the web, or instructors may print them out for use in the classroom discussion.

In addition, there are other available resources:

● *ChemOffice Ltd.* includes the introductory student version of *ChemDraw* and *Chem3D,* CambridgeSoft's premiere chemical drawing and modeling programs (cross-platform CD-ROM WIN/MAC).

● *Darling Molecular Visions Kit* is a flexible modeling kit that helps students to visualize the organic structures and reactions they are learning. The unique flexible properties of the *Darling Molecular Visions Kit* provide the degree of freedom necessary to study the spatial relationships of atoms, types of atomic bonding, geometric relationships, interatomic distances, and molecular strain in various molecular conformations.

● *The Chemistry Tutor,* version 2.0 CD-ROM, is an interactive introduction to topics in organic chemistry with a review of topics in general chemistry.

## Acknowledgments

Many people have contributed to the Fifth Edition of my book. I am grateful to all of my colleagues at the University of Michigan as a department that is supportive of curricular innovation. In particular, the organic chemists have thought creatively about what constitutes the right organic chemistry course for students who will not become chemists, and even for those who will become chemists as the boundaries between chemistry and other fields continue to blur. Their thinking has helped to clarify mine. Of all my colleagues at the University of Michigan, Brian Coppola continues to be a major contributor to the evolution of my thinking about how organic chemistry could be taught and, hence, to the development of the book. He is a superb and creative teacher, constantly seeking ways to make students see the linguistic and mechanistic logic and unity of reaction that seem so disparate to their eyes. We conduct an ongoing informal seminar, along with anyone else who wants to join the argument, about the endlessly fascinating subject of how students learn or, unfortunately, do not learn chemistry. He is a vigorous critic of my ideas and, as such, serves me as consultant. Some of his ideas and innovations have inevitably found their way into the book. Among his contributions are the workbook exercises and essay on how to learn organic chemistry in the *Study Guide,* the concept of the "One Small Step" feature, as well as his work on the *Instructor's Manual.*

Suggestions and corrections from colleagues and students who have used the book are particularly valuable. I owe special thanks to Richard Lawton of the University of Michigan and to Hernando A. Trujillo of Grinnell College, who supplied me with a list of notes and suggestions arising from his use of the third and fourth editions of the text. Dr Alex Aisen, formerly of the Department of Radiology at the University of Michigan, supplied me with information on magnetic resonance imaging and the photograph that appears on page 435 of the text. Brian Eklov took all of the carbon and proton nuclear magnetic resonance spectra that now appear in the book. Frank Parker and James Windak helped with the spectra illustrating Fourier transform nuclear magnetic resonance and infrared methods.

I very much appreciate the helpful comments of the following reviewers: Ed Blackburn, University of Alberta; Robert Coleman, Ohio State University; Jim Dailey, University of Pennsylvania; Loretta T. Dorn, Fort Hays University; Steve Hardinger, University of California, Los Angeles; Dalila Kovacs, Michigan State University; Paul J. Kropp, University of North Carolina; Kenneth Laili, Kent State University; Jeff Moore, University of Illinois; Erach R. Talaty, Wichita State University. I would also like to thank Patricia Pieper at Anoka Ramsey Community College for her help in reviewing the text for accuracy.

Roberta Kleinman and Peggy Zitek contributed substantially to the text as well as to the *Study Guide.* All of the three-dimensional figures in the text were developed by Roberta Kleinman, who combines artistic talent and a knowledge of computer graphics with an interest in how students visualize and learn, and her skills have contributed to the look of the book. She joins me in struggling to see things as the student sees them and not as we, with years of experience, know them to be. She is also a major contributor to the setting up and maintenance of the part of the web site that carries the "One Small Step" features. I owe a great deal to her critical eye. Peggy Zitek brings a sharp eye and a critical mind to the thankless task of proofreading. She also prepared the index for this edition. I value the help of both of these good friends.

Charles Hartford, Editor-in-Chief and Publisher, and Richard Stratton, Executive Editor, have guided the book through this revision. I am grateful for their support and encouragement. Rita Lombard, Development Editor, Andrea Cava, Project Editor, Charlotte Miller, Art Editor, and Henry Rachlin, Senior Designer, competently and patiently coped with the countless details necessary to make a chemistry book in four colors technically accurate as well as aesthetically pleasing. Without their expertise, the production process would have been much more traumatic. I thank them for their help. I also wish to thank Katherine Greig, Senior Marketing Manager, and Alexandra Shaw, Marketing Associate, for their tireless efforts in promoting this textbook.

Finally, no acknowledgments would be complete without saying how much I appreciate the warm support I receive from my family and my friends. My thanks go to all of them for their patience and care. In particular, I dedicate this edition of the book to the memory of my parents, Ragip and Nezahet Nurettin Eğe.

Seyhan N. Eğe

# Features of *Organic Chemistry,* FIFTH EDITION

## GREATER INTEGRATION OF BIOLOGICAL EXAMPLES

The Fifth Edition takes a multilayered approach to the introduction of important concepts, integrating biological examples where possible.

For example, students meet the concept of protecting groups first in Chapter 15 in the context of acyl-transfer reactions and peptide synthesis.

Then, in Chapter 21, a new chapter on **synthesis,** protecting groups are applied to alcohols and carbonyl groups while those for amines are expanded. All of these protecting groups are used in syntheses, including those in a new section on the application of solid-phase synthesis to small molecules.

---

630    **Chapter 15   Carboxylic Acids and Their Derivatives. Acyl-Transfer Reactions**

Alanine also can react with another molecule of itself to give alanylalanine, and glycine can react with itself to give glycylglycine. (Remember that while we write equations showing single molecules of each reagent, in reality, billions and billions of them are randomly colliding with each other in the reaction mixture.) We should expect to get a mixture of four peptides. The complexity of the mixture, of course, increases greatly as the number of amino acids in the peptide increases.

**Problem 15.34**    Using the structures of the two peptides shown on page 629 as guides, write structures for alanylalanine and glycylglycine.

To prepare a peptide of a particular structure, the amino acids must be added onto the chain in a precise order. The desired reaction is a nucleophilic substitution by an amino group at the carbonyl group of a carboxylic acid, but there are nucleophilic groups such as other amino groups, thiols, and alcohols on the side chains of amino acids that are also expected to react with carboxylic acids. The structures of the amino acids usually found in peptides and proteins may be found on pages 983 and 985. The reactions to form the backbone of a peptide chain must involve the amino group at carbon 2 and not any other nucleophilic functional group.

Another complication is that a free carboxylic acid group does not react readily with amines to give amides. Usually an acid chloride or an acid anhydride is used so that a better leaving group than a hydroxyl group is present when an acid is converted into an amide (p. 628).

Thus there are two major aspects to peptide synthesis. First, the reaction must be directed to the desired part of the molecule. This means that other nucleophilic functional groups must be hidden, protected, so that they do not react. For example, if the amino group in alanine were protected so that it was no longer a nucleophile, then only one peptide could be formed between alanine and the amino group of glycine.

The equation above shows the amino group in alanine masked by a **protecting group** that reduces its nucleophilicity to the point that it will not react with the carboxylic acid group of glycine (nor with the carboxylic acid group of another alanine molecule). A protecting group converts a reactive functional group into a different group that is inert to the conditions of the reaction to be carried out. Another requirement for a protecting group is that it must be easily removable once the job is done so that the original functional group can be restored.

The second aspect of peptide synthesis requires that the carboxylic acid group of an amino acid must be made reactive enough to form an amide bond under conditions that do not destroy peptide bonds and other functional groups that may be present in the molecule. The equation above shows a second protecting group on the carboxylic acid group of glycine to prevent its activation so that glycine does not react with other molecules of itself.

---

Chapter 23, a new chapter on **structure and reactivity in biological macromolecules,** sees protecting groups used again in the synthesis of oligonucleotides.

---

**21.2   Protecting Groups in Synthesis**

A.  Acetals and Ketals

Most naturally occurring compounds with interesting biological properties have a variety of functional groups in them. One functional group may be adversely affected by or interfere with a reaction that a chemist wishes to carry out at another one. For example, it is not possible to prepare a Grignard reagent from an alkyl halide containing an alcohol function, as we saw on page 842. And yet it may be useful to make such an organometallic reagent in a complex synthesis (p. 844). To do this, chemists use protecting groups. A protecting group converts a reactive functional group into a different group that is inert to the conditions of some reaction (or reactions) that is to be carried out as part of a synthetic pathway. For example, a hy-

B.  Ethers

Even some ethers that are not acetals or ketals are easily cleaved. Benzyl ethers are often used as protecting groups for alcohols. The benzylic ether bond is cleaved by hydrogenation reactions. We saw an example of such a cleavage of a bond between oxygen and a benzyl group by hydrogenation in the removal of the carbobenzyloxy protecting group in peptide syntheses (p. 633).

Silyl ethers, in which there is an oxygen–silicon bond, are usually cleaved by fluoride ion. The high bond energy of the silicon–fluorine bond (Table 2.4, p. 64) serves as a thermodynamic driving force for this cleavage. Silyl ethers are prepared by the reaction of an alcohol and a silyl halide in the presence of a base to aid in the deprotonation of the alcohol. For example, the *tert*-butyldimethylsilyl group (abbreviated TBDMS) is used to protect the hydroxyl group of 3-butyn-1-ol so that the terminal alkyne can be selectively deprotonated. The carbanion resulting from that deprotonation adds to acetaldehyde.

C.  Protecting Groups for Amines

The synthesis of peptides, discussed in Section 15.8B (p. 631), required the use of protecting groups on the amino groups of amino acids to be able to prepare a peptide with the amino acids in the order we desired and not have random reactions between amino groups and carboxylic acid groups. We needed to protect the amino and carboxylic acid groups that we did not want to have react and activate the carboxylic acid group that was to be part of the new peptide bond. The two amine protecting groups we used in Section 15.8B were the *tert*-butoxycarbonyl group (Boc) and benzyloxycarbonyl group (Cbz). These groups are also used as protecting groups in syntheses that do not involve peptides. For example, the following steps are part of the synthesis of a polyamine used to make a complexing agent for metals to be used in connection with radiation therapies.

---

D.  Synthesis of Oligonucleotides

The synthesis of oligonucleotides presents many of the same challenges that the synthesis of a peptide (p. 629) does. Each nucleotide unit has multiple reactive sites, hydroxyl and amino groups. The nucleotides have to be strung together in a precise order if the compound that results is going to be the carrier of genetic information. Phosphate groups that are not very reactive at one stage of the synthesis have to be activated at another stage in order to create the phosphate ester linkage between two nucleotide units. These phosphate bonds must be between the 5'-hydroxyl group of one nucleotide unit and the 3'-hydroxyl unit of another one.

In a classic experiment in 1968, Hai Gobind Khorana, who shared the Nobel Prize for Medicine that year, demonstrated that oligonucleotides could indeed be synthesized. Just as it was for peptides, syntheses of large oligonucleotides required solid-state syntheses and automation of the steps. Khorana quickly extended his work to synthesis on a solid support. We will look at such a synthesis of a dinucleotide as a reminder of ideas about protecting groups and activating reagents first encountered in Section 15.8.

Khorana started his synthesis with the same resin that Merrifield used for peptide synthesis (p. 901). He modified it by a series of standard organic reactions to create a triphenylmethyl chloride (trityl chloride) substituted on one of the phenyl rings with a *p*-methoxy group (methoxytrityl chloride) attached to the polymer support.

# TOOLS FOR EFFECTIVE LEARNING

## Visualizing the Reaction and the Pedagogical Use of Color

Complete mechanisms–highlighted in "Visualizing the Reaction" features–are given for each type of reaction. Color is used to enhance the process of visualizing. Acidic or electrophilic sites are highlighted as red atoms; blue shading signifies basic or nucleophilic species; and gray shading emphasizes leaving groups. Color is used to indicate whether the reactive species on one side of an equation are converted into new reactive species after the reaction in order to show students the reversibility of many reactions, especially acid–base reactions.

In the above reaction, iodide ion is the nucleophile, the species with electrons to donate. It reacts with 1-bromobutane at the carbon atom bearing the partial positive charge. The carbon atom is an **electrophilic center,** a site of electron deficiency, and bromide ion is the leaving group.

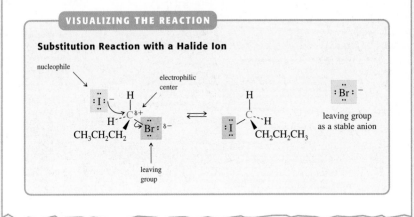

**VISUALIZING THE REACTION**

**Substitution Reaction with a Halide Ion**

## Structure and Reactivity on the Web: "One Small Step"

"One Small Step" helps students extend their understanding of a specific example in the textbook to new concepts and reactions.

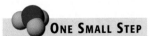

**ONE SMALL STEP**

The addition of metal hydrides to multiple bonds is a quite general reaction for elements from groups 13 and 14 of the periodic table.

**PROBLEM:** Predict the products of the following reactions:

$$2 \; CH_3\overset{\overset{\displaystyle CH_3}{|}}{C}{=}CH_2 + AlH_3 \longrightarrow ?$$

$$(CH_3CH_2)_2GaH +$$

$$H_2C{=}CH(CH_2)_7CH_3 \longrightarrow ?$$

**Hint:** Use the chemistry of diborane to guide you in finding the answers to these problems.

$$CH_3CH_2CH{=}CH_2 \xrightarrow{BH_3} (CH_3CH_2CH_2CH_2)_3B + (CH_3CH_2\overset{\overset{\displaystyle CH_3}{|}}{CH}{-})_3B$$

1-butene     tri(*n*-butyl)borane     tri(*sec*-butyl)borane

$$\Big\downarrow \begin{array}{l} H_2O_2 \\ NaOH \\ H_2O \end{array}$$

$$CH_3CH_2CH_2CH_2OH + CH_3CH_2\overset{\overset{\displaystyle}{}}{\underset{\underset{\displaystyle OH}{|}}{CH}}CH_3$$

*n*-butyl alcohol     *sec*-butyl alcohol
93%        7%

Note that the chief product of the reaction sequence, *n*-butyl alcohol, looks as if an anti-Markovnikov addition of water to the double bond has occurred.

$$CH_3CH_2CH{=}CH_2 \longrightarrow \longrightarrow CH_3CH_2\underset{\underset{\displaystyle H}{|}}{CH}{-}\underset{\underset{\displaystyle OH}{|}}{CH_2}$$

     H—OH

anti-Markovnikov addition of H—OH to an alkene

In contrast, in the hydration reaction (p. 284), alcohols are produced by a Markovnikov addition of water to the double bond.

Most of the "One Small Steps" appear on a text-specific web site available through http://chemistry.college.hmco.com with icons in the margins of the textbook to indicate that such a resource is available.

On the Web: ONE SMALL STEP

The interactive nature of the web format allows students to check hints for solutions after seeing the statement of the problem. Students will also be able to access answers to the problems, some of which will contain animations as appropriate.

# End-of-Chapter Summaries

Two forms of summary appear in the textbook. Each chapter (except Chapter 23) contains either:

An end-of-chapter summary that offers a concise review of the major concepts covered in the chapter.

or

Tables that summarize the reactions that appear in the chapter. Organized to remind students of how the reactions proceed, these tables take students briefly through the stages of the reaction again, reinforcing the types of reagents needed, the reactive intermediates involved, and the stereochemistry of the reaction.

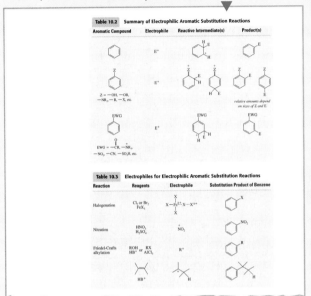

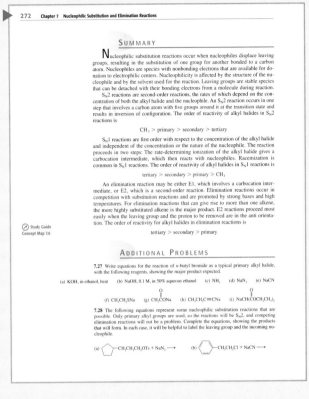

Chapter 7 Nucleophilic Substitution and Elimination Reactions summary text (page 272):

SUMMARY

Nucleophilic substitution reactions occur when nucleophiles displace leaving groups, resulting in the substitution of one group for another bonded to a carbon atom. Nucleophiles are species with nonbonding electrons that are available for donation to electrophilic centers. Nucleophilicity is affected by the structure of the nucleophile and by the solvent used for the reaction. Leaving groups are stable species that can be detached with their bonding electrons from a molecule during reaction.

$S_N2$ reactions are second-order reactions, the rates of which depend on the concentration of both the alkyl halide and the nucleophile. An $S_N2$ reaction occurs in one step that involves a carbon atom with five groups around it at the transition state and results in inversion of configuration. The order of reactivity of alkyl halides in $S_N2$ reactions is

$$CH_3 > \text{primary} > \text{secondary} > \text{tertiary}$$

$S_N1$ reactions are first order with respect to the concentration of the alkyl halide and independent of the concentration or the nature of the nucleophile. The reaction proceeds in two steps: The rate-determining ionization of the alkyl halide gives a carbocation intermediate, which then reacts with nucleophiles. Racemization is common in $S_N1$ reactions. The order of reactivity of alkyl halides in $S_N1$ reactions is

$$\text{tertiary} > \text{secondary} > \text{primary} > CH_3$$

An elimination reaction may be either E1, which involves a carbocation intermediate, or E2, which is a second-order reaction. Elimination reactions occur in competition with substitution reactions and are promoted by strong bases and high temperatures. For elimination reactions that can give rise to more than one alkene, the more highly substituted alkene is the major product. E2 reactions proceed most easily when the leaving group and the proton to be removed are in the anti orientation. The order of reactivity for alkyl halides in elimination reactions is

$$\text{tertiary} > \text{secondary} > \text{primary}$$

Study Guide
Concept Map 7.6

ADDITIONAL PROBLEMS

7.27 Write equations for the reaction of n-butyl bromide as a typical primary alkyl halide, with the following reagents, showing the major product expected.

(a) KOH, in ethanol, heat    (b) NaOH, 0.1 M, in 50% aqueous ethanol    (c) $NH_3$    (d) $NaN_3$    (e) NaCN

(f) $CH_3CH_2SNa$    (g) $CH_3CONa$    (h) $CH_3CH_2C\equiv CNa$    (i) $NaCH(COCH_2CH_3)_2$

7.28 The following equations represent some nucleophilic substitution reactions that are possible. Only primary alkyl groups are used, so the reactions will be $S_N2$, and competing elimination reactions will not be a problem. Complete the equations, showing the products that will form. In each case, it will be helpful to label the leaving group and the incoming nucleophile.

(a) —CH$_2$CH$_2$CH$_2$OTs + NaN$_3$ →    (b) —CH$_2$CH$_2$Cl + NaCN →

## HMClass Prep with HMTesting CD-ROM

NEW TO THIS EDITION

This package includes both **HMClass Prep** and **HMTesting** on one CD-ROM. It allows an instructor to access both lecture aids and testing software in one place. These components cannot be ordered separately.

▶ **HMClass Prep** includes everything instructors will need to develop their lectures—PowerPoint slides with line art and spectra from the textbook, concept maps from the *Study Guide,* as well as *Instructor's Manual* and Microsoft® Word files of the printed *Test Bank.*

▶ **HMTesting** combines a flexible test-editing program with a comprehensive gradebook function for easy administration and tracking. It enables instructors to administer tests via network server or the web. The **HMTesting** database contains a wealth of questions and can produce multiple-choice as well as essay tests. Questions can be customized based on the chapter being covered, the question format, level of difficulty, and specific topics. **HMTesting** provides for the utmost security in accessing both test questions and grades.

# The Art of Problem Solving

The Art of Solving Problems sections, unique among organic chemistry texts, offer students a step-by-step approach to solving organic chemistry problems. Students learn to reason their way to a solution rather than simply plugging data into a prelearned formula.

Through a series of systematic questions, students are first introduced to this problem-solving approach in Chapter 1.

Chapter 4 then shows students how to reason backward in solving problems involving, in this example, simple synthesis.

PROBLEM: Write resonance contributors for the ion having the following connectivity: $CH_2CHO^-$. Evaluate each resonance contributor in terms of the rules given in Section 1.6A and decide which are major contributors and which minor.

Question: Which atoms are present, and how many electrons are available?

Answer:

$$
\begin{aligned}
&\text{2C (Group 14)} &&= 2 \times 4 \text{ electrons} &&= 8 \text{ electrons} \\
&\text{3H (Group 1)} &&= 3 \times 1 \text{ electron} &&= 3 \text{ electrons} \\
&\text{1O (Group 16)} &&= 1 \times 6 \text{ electrons} &&= 6 \text{ electrons} \\
&\text{1 negative charge} &&= 1 \text{ electron} &&= \underline{1 \text{ electron}} \\
& && && 18 \text{ electrons}
\end{aligned}
$$

Question: What is a Lewis structure for this ion?

Answer: Start with the connectivity given in the statement of the problem. Bond the carbon and oxygen atoms together, and put in the hydrogen atoms; then use the electrons that are left to complete octets around the carbon and oxygen atoms.

The process of working backwards that is illustrated above is important. It is a good idea to practice it verbally, reasoning aloud to yourself as you work the problems, until this type of analysis becomes second nature.

**Problem 4.18** How could each of the following transformations be carried out? Some of them may require more than one reaction. In each case, show as much of your reasoning process as possible.

(a) $CH_3CH_2CH{=}CH_2 \xrightarrow{?} CH_3CH_2\underset{\underset{SH}{|}}{C}HCH_3$     (b) $CH_3CH{=}CHCH_3 \xrightarrow{?} CH_3CH_2\underset{\underset{Cl}{|}}{C}HCH_3$

# Problems

Except for a few routine drill problems, all of the problems in the textbook reflect chemistry that is very much alive in research literature today. Many of the problems include spectra or spectral data. Many others illustrate the relationships between chemistry and biology or medicine.

**Problem 13.1** The boiling point of 1,2-ethanediol is 197.2 °C. Why is it a good compound to use as an antifreeze?

**Problem 13.2** Would you expect cholesterol to be particularly soluble in water? Explain your answer.

**Problem 13.3** Cholesterol has more than one functional group. To what other class of compounds does it belong? Write equations using cholesterol to illustrate three reactions that are typical of that other functional group class.

**17.39** The following reactions were carried out in the synthesis of large-ring compounds that capture and hold metal ions. Provide structural formulas for Compounds A and B. What structural features of A and B do the infrared absorption bands point to?

$CH_3CH_2O\overset{\overset{O}{\|}}{C}CH_2\overset{\overset{O}{\|}}{C}NHCH_2 {-}\hspace{-2pt}\bigcirc\hspace{-2pt}{-}NO_2 \xrightarrow[\underset{\overset{\|}{O}}{CH_3COH}]{Br_2} \quad A$

$C_{12}H_{14}N_2O_5Br$

$\bar{\nu}_{max}$ 1750, 1640 cm$^{-1}$

$H_2NCH_2\overset{\overset{O}{\|}}{C}OCH_2CH_3 \xrightarrow[\text{dimethylformamide}]{} \xrightarrow{(CH_3CH_2)_3N \text{ (base)}} \quad B$

$C_{16}H_{12}N_3O_7$

$\bar{\nu}_{max}$ 1740, 1670 cm$^{-1}$

# An Introduction to Structure and Bonding in Organic Compounds

## A Look Ahead

Organic chemistry was born in 1828 when Friedrich Wöhler attempted to synthesize ammonium cyanate, $NH_4CNO$, and instead obtained urea,

$$\underset{NH_2\overset{\displaystyle O}{\overset{\|}{C}}NH_2}{}$$

Wöhler, who studied to be a doctor of medicine before he decided to become a chemist, discovered that the compound he had made was identical with urea recovered from urine. Up to that time, scientists had thought that the compounds present in living plants and animals could not be synthesized in the laboratory from inorganic reagents. Wöhler recognized the importance of his experiment and wrote to a friend, "I must tell you that I can make urea without the use of kidneys, either man or dog. Ammonium cyanate is urea."

Wöhler's discovery was important because it gave impetus to a long series of experiments in which chemists probed the nature of the chemical substances that exist in living organisms and in petroleum and coal, which are formed from the remains of plants and animals that lived long ago. As early chemists struggled to isolate and purify the components of plants, animals, coal, and petroleum, they quickly realized that the chemistry of carbon was associated with life in a special way that distinguished that element from the others. Compounds containing carbon were called *organic compounds* to reflect their origin in living systems and to distinguish them from the inorganic compounds, the acids, bases, and salts derived from the other elements in the periodic table.

Organic chemistry, the chemistry of the compounds of carbon, is central to many disciplines. Life processes are supported by the chemical reactions of complex organic compounds such as enzymes, hormones, proteins, carbohydrates, lipids, and nucleic acids. Chemists have created millions of organic compounds that did not exist in nature originally as they have attempted to create new materials and new medications. Industrial chemists have developed synthetic rubber, called neoprene, and synthetic silks, such as rayon and nylon, that improve on the properties of the natural substances and offset shortages of natural supplies. Food additives, dyes, artificial flavorings, artificial sweeteners, preservatives, and pesticides, most of them organic compounds, are regularly in the headlines. Crude petroleum is converted by organic reactions into fuels that supply energy for heat, transportation, and industry. Petroleum is also used in the synthesis of giant molecules engineered to have useful properties. The names of these materials—Teflon, Orlon, Acrilan, polystyrene, polypropylene, polyurethane—have become household words.

The progress of chemistry as a science depends on experimental manipulations of substances in the laboratory, leading to the observation of new phenomena. In

buckyball

*a form of carbon*

(p. 364)

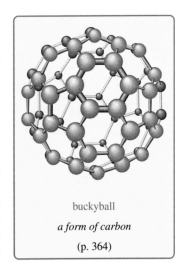

testosterone

*a male sex hormone*

(p. 537)

nicotine

*from tobacco*

(p. 805)

CH_3  OH  CH_3

$CH_3C$ 〈ring〉 $CCH_3$

CH_3        CH_3

CH_3

BHT

*a food preservative*

(p. 794)

CH_2OH

HO — 〈ring〉 O

HO — — OH

HO

glucose

*a simple sugar*

(p. 566)

thinking about these phenomena, chemists arrive at ideas about the nature of the chemical substances under investigation. These ideas are tested by new experiments and new observations. The range of manipulations and observations available to chemists has expanded enormously in recent years. The power that chemists have to transform chemical compounds into new ones, sometimes useful and sometimes harmful, also has increased.

Chemists have refined the way they visualize and think about the submicroscopic units called *atoms* and *molecules*. They have created models that help them picture and understand experimental facts about chemical substances. Scientists are constantly creating and refining models that explain the physical world. Some models are pictorial; others are more abstract and mathematical. Some models are widely adopted because they are useful; others are soon modified or discarded on the basis of new observations. In this first chapter we will examine some of the models chemists currently find most useful for thinking about experimental observations of structures and properties of organic compounds.

## 1.1    How to Study Organic Chemistry

Learning organic chemistry is like learning a new language, a language that is both verbal and pictorial. Organic chemistry is a highly organized discipline, based on the premise that the structure of a compound determines its reactivity. Organic chemistry is, therefore, a study of the relationship between the structures of molecules and their reactions. As you develop an understanding of this relationship, you will be able to make predictions about molecules and reactions that are new to you. You will learn to dissect complex structures and to distinguish the pieces you recognize. You will be able to reason by analogy from systems and reactions that you have already learned to new systems and reactions that resemble the earlier ones in important ways.

Your pencil is an indispensable tool in your studies. While you are training your eye to look carefully, you should be training your hand to draw. Professional organic chemists cannot talk to each other without drawing structures. To learn organic chemistry, you must draw and redraw the structures of compounds and write out equations as you are studying. The ways structures are illustrated in this book and drawn by your instructor on the chalkboard represent the result of years of evolution in thinking about organic compounds. Different kinds of pictures represent different degrees of precision. You must train your hand to produce the correct pictures; *precision in drawing leads to precision in thinking.* The correct representation of a structure often will give you insight into the solution to a problem.

Organic chemists are concerned about the shapes of molecules. You should acquire molecular models and examine the structures of carbon compounds in three dimensions. Then you will learn to translate three-dimensional structures to the two dimensions of a page. Chemists have developed specific ways of representing three-dimensionality on a flat surface in order to convey the maximum amount of information clearly and consistently. If the rules are not followed exactly, the resulting pictures are meaningless.

To study effectively, you should read the assignments before attending lectures. Chemists are problem solvers. You too will spend most of your time learning to solve problems. *Work all the problems,* no matter how simple they seem, in writing and in full detail. There is no other way to develop the skills you will need to work with the more and more complex structures and concepts you will encounter as you go through this book. The important reactions and ideas will come up again and again in different problems throughout the book. If you find that you cannot do a problem, go back to the text and review the reactions and ideas that apply to it; then try the problem once more. This method of studying will ensure that you spend the most time working with the ideas that you find difficult. As you go back and forth between the problems and the text, you will gradually learn the most important facts without making a specific effort to memorize them. You also will learn a way of thinking, of looking for patterns, and of recognizing qualitative similarities between seemingly unrelated facts. As an aid to your study, this book has sections on the art of solving problems. They outline the thinking processes that are necessary for the successful solution of the most common types of problems in organic chemistry. The development of such skills will be one of the most important results of your study of organic chemistry, because an ability to think this way can be applied to problems in every area of life.

Because organic chemistry is a language, a good way to study the subject is to talk about it. Scientists solve important problems by cooperating with each other. They share data, they argue about theories, and they help each other examine facts from different points of view. Much of the content of this book consists of descriptions—"stories"—about how chemists reason about facts. Study with friends, and tell each other how you understand the major concepts in each chapter and how you think out the answers to problems. By saying out loud the words that are used in describing structures and reactions, you reinforce your understanding of them. You also will find that there are many ways in which you can arrive at correct answers. The object of your studying should not be to "learn the correct answer" to any particular problem but to develop confidence in your ability to figure out the answers to problems you have never seen. Working with other students in learning to think about structure and reactivity is a good way for you to understand both chemistry and the scientific process.

The answers to the problems in the text are found in the *Study Guide* for this book. In addition, the *Study Guide* contains concept maps, which are summaries of important ideas presented in outline form. Notes have been placed in the margin to alert you that these summaries are available. After looking at some of the concept maps, you may want to practice making your own (which need not look exactly like the ones in the *Study Guide*) to summarize material presented in lectures or to reorganize your notes. Concept maps may be helpful to you as an alternative way of organizing ideas and presenting relationships. The margins of the text also will alert you to the presence of workbook exercises in the *Study Guide*. These exercises give you a chance to review earlier material or prepare for new concepts by working on a few problems for which there are no answers in the *Study Guide*. The workbook exercises, therefore, are particularly good ones to discuss with your study partners.

Gradually, you will learn the names of a large number of organic compounds in order to be able to communicate your thoughts about them. All compounds have formal names that can be assigned by the application of definite rules agreed on by chemists. Some compounds also have trivial names by which they have been known historically. These names still appear on the sides of tanker cars, on labels of reagent bottles, and in articles in scientific journals. To be literate in organic chemistry, you must be able to recognize both formal and trivial names.

The successful mastery of organic chemistry requires a lot of hard work and consistent studying. It is not a subject that can be crammed. Many students make the mistake of trying to memorize the text. An understanding of the basis of chemical transformations is what is really needed. It is true that facts must be learned, but you will be overwhelmed by them unless you develop an ability to see relationships.

In this book you will find features that we call "One Small Step." These are identified in the margins at different places in the book. These features will allow you to apply the concepts that you are learning to new and unfamiliar compounds and reactions. Some of the "One Small Step" features are right in the book; others are on the Web site that supplements this book. Working with these problems will be another opportunity to talk chemistry with your friends as you deepen your understanding of the subject matter.

Although the study of organic chemistry requires a lot of concentrated work, many students find that it is also fun. The thinking that goes into such study is related to the thinking used in solving puzzles. For example, solving problems in organic chemistry requires you to recognize patterns and fill in the missing pieces, much as you do when you put together a jigsaw puzzle. You also learn to be precise in thinking about qualitative concepts, just as you are already precise in quantitative ways in mathematics. You will experience the power of your mind to analyze an unfamiliar problem and to arrive at a correct picture of the disparate facts that must be brought together for a solution. You will come to trust in your ability to think correctly to predict experimental outcomes. Such self-knowledge is exhilarating.

## 1.2   Ionic and Covalent Compounds

Compounds are divided broadly into two bonding models, ionic and covalent. **Ionic compounds** are composed of ions, structural units that may be single atoms or groups of atoms, bearing positive or negative charges. In **covalent compounds,** the structural units are molecules having no net charge. Ionic compounds are usually crystalline solids with high melting points. Many of these compounds dissolve in water to form solutions that conduct electricity. Common table salt, NaCl, is a typical example of an ionic compound in which the ions, $Na^+$ and $Cl^-$, consist of single atoms. Magnesium sulfate, $MgSO_4$, commonly known as Epsom salts, and sodium bicarbonate, $NaHCO_3$, which is baking soda, are other familiar ionic compounds. In these compounds, the negatively charged ions (the sulfate anion, $SO_4^{2-}$, and the bicarbonate anion, $HCO_3^-$) are composed of atoms held together by covalent bonds. In other ionic compounds, such as ammonium chloride, $NH_4Cl$, the positively charged ion (the ammonium ion, $NH_4^+$) contains covalent bonds.

Covalent compounds may be gases, liquids, or solids. Methane, $CH_4$, is the principal component of natural gas. Carbon tetrachloride, $CCl_4$, is a typical covalent liquid that was once commonly used in dry cleaning. *para*-Dichlorobenzene, $C_6H_4Cl_2$, which is the main constituent of mothballs, is a covalent solid. Methane, carbon tetrachloride, and *para*-dichlorobenzene do not dissolve in water to any great extent. Such compounds are **nonpolar covalent compounds.**

Other covalent compounds, such as ethanol, $CH_3CH_2OH$ (a liquid), and glucose, $C_6H_{12}O_6$ (a solid), are quite soluble in water and form solutions that do not conduct electricity. Such compounds ionize only slightly in aqueous solutions. Water, ethanol, and glucose are examples of **polar covalent compounds.** In the rest of this chapter we will look more closely at ideas about the nature of chemical bonding and at the various factors that influence the different physical properties of compounds, such as boiling point, melting point, and solubility.

## 1.3   Ionic Bonding

**The forces that act between charged particles are called electrostatic forces.** Most of the chemistry that we will discuss can be understood on the basis of these electrostatic forces, which are summarized simply in the two statements that follow: **Particles that have opposite charges—positive and negative—are strongly attracted to each other. Particles that have the same charge—both positive or both negative—repel each other.** A corollary of these statements is that energy must be exerted to separate a negatively charged particle from a positively charged one.

We experience the forces of attraction and repulsion when we play with magnets. To understand better the preceding statements about electrostatics, recall how it felt when you brought the north pole of a magnet near the south pole of another one (attraction) or what you felt as you tried to bring the north poles of two magnets close to each other (repulsion). The first happens easily; in fact, the magnets will snap together with no effort on your part. In the second case, you experience in your muscles the work that has to be done to bring the two magnets close together. In other words, you have to use up energy to make it happen. You also have to use energy to separate the north pole of a magnet from the south pole of another magnet. In playing with magnets, we are dealing with electromagnetic forces instead of electrostatic ones, but the experience of attraction and repulsion in the case of magnets can stand in for the other, which is not as easy to feel directly.

Crystalline sodium chloride consists of an arrangement of positively charged sodium ions and negatively charged chloride ions arranged alternately in a three-dimensional array called a **crystal lattice.** Each sodium ion is surrounded by six chloride ions, and each chloride ion by six sodium ions. The ions are held in place by strong electrostatic forces between the positively and negatively charged ions. **Ionic bonding** consists of electrostatic attractions between ions of opposite charge. The individual ion is a sphere bearing a symmetrical distribution of charge. For this reason, there is no particular direction to bonding in ionic compounds. In the solid state, there are no individual molecules of sodium chloride composed of one $Na^+$ and one $Cl^-$.

Sodium chloride has a high melting point, 801 °C, and a very high boiling point, 1413 °C. These physical properties are an indication of the strength of the electrostatic forces holding the ions together. Large amounts of energy must be applied to the sodium chloride crystal to overcome the electrostatic forces that hold the ions in place in the crystal lattice and allow them to move past each other in liquid sodium chloride. Even more energy is necessary to further separate the ions in going from the liquid to the vapor state. In the vapor state, the positively charged sodium ion, $Na^+$, and the negatively charged chloride ion, $Cl^-$, come together to give a molecule of NaCl, which is isolated from other molecules of NaCl by large spaces. In each of these molecules, a sodium ion is closely associated with a chloride ion. Much energy, indicated by the high temperature to which sodium chloride has to be heated before it vaporizes, is needed to separate individual molecules of NaCl from the crystal lattice, which is like a gigantic extended molecule of sodium chloride.

*Study Guide*
Concept Map 1.1 and 1.2

## 1.4    Covalent Bonding

### A. Lewis Structures

Early in this century, Gilbert N. Lewis at the University of California at Berkeley proposed that the **covalent bond** be represented as the sharing of a pair of electrons between two atoms. He also proposed that, with a few exceptions, stable molecules or ions have eight electrons, or four pairs of electrons, in the outermost shell, the valence shell, of each atom. This stable configuration of electrons is called an **octet.** An atom having a filled valence shell is said to have a **closed shell configuration.** Lewis's suggestions for drawing structures of covalent compounds have proved enormously useful to organic chemists.

The **Lewis structure** of a covalent molecule shows all the electrons in the valence shell of each atom; the bonds between atoms are shown as shared pairs of electrons. The periodic table in this book shows two systems for numbering the groups, one consisting of Arabic numerals and the other of Roman numerals. The total number of electrons in the valence shell of each atom can be determined from its group number, shown in Roman numerals. For the Arabic numerals, for elements in Groups 13 to 18, the number of valence electrons equals the group number minus ten. The shared electrons are called the **bonding electrons** and also may be represented by a line or lines between the two atoms. The valence electrons that are not being shared are the **nonbonding electrons:** they are shown by dots oriented in a square around the symbol of the atom. The construction of the Lewis structure for a covalent compound, hydrogen chloride, is shown below.

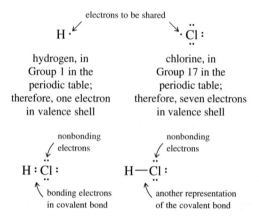

The hydrogen atom in hydrogen chloride shares two electrons, which is all its valence shell can hold. The chlorine atom has eight electrons around it in a stable octet.

Table 1.1 includes some more examples of Lewis structures of compounds and ions written first in the abbreviated form known as the **condensed formula** and then as both the full Lewis structure showing all the valence electrons and the Lewis structure with the covalent bonds represented by lines and only the nonbonding electrons shown as dots.

In all the structures in Table 1.1, the valence electrons are arranged so that the atoms of most elements except hydrogen share eight electrons. (An exception is found in Section 1.4D, p. 13.) The restriction that a hydrogen atom usually cannot share more than two electrons means that we cannot draw structures in which a hydrogen atom has more than one covalent bond. In constructing the water molecule, for example, we see from the periodic table that the oxygen atom has six valence electrons and the hydrogen atoms one each. A total of eight dots representing the electrons can be placed around the oxygen atom. The hydrogen atoms can each share two of these electrons with the oxygen atom, which then has two pairs of unshared, or nonbonding, electrons.

When drawing Lewis structures, we must be careful to keep track of the number of electrons available to form bonds and the location of the electrons. The hy-

| | | Lewis Structures | |
|---|---|---|---|

**Table 1.1**  **Condensed Formulas and Lewis Structures of Some Ions and Compounds**

| Name | Condensed Formula | Lewis Structures | |
|---|---|---|---|
| water | $H_2O$ | H:O:H | H—O—H |
| hydronium ion | $H_3O^+$ | H:O:H (+) over O, H below | H—O—H (+) over O, H below |
| ammonia | $NH_3$ | H:N:H, H below | H—N—H, H below |
| methane | $CH_4$ | H above, H:C:H, H below | H above, H—C—H, H below |
| methanol | $CH_3OH$ | H above, H:C:O:H, H below | H above, H—C—O—H, H below |
| methoxide anion | $CH_3O^-$ | H above, H:C:O:$^-$, H below | H above, H—C—O:$^-$, H below |
| hydroxylamine | $H_2NOH$ | H:N:O:H, H below | H—N—O—H, H below |

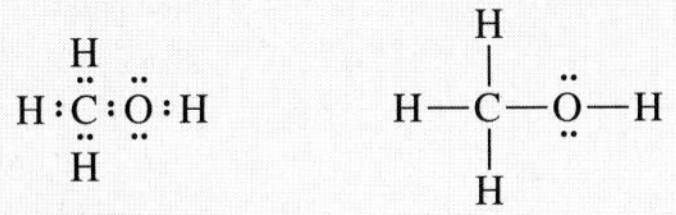

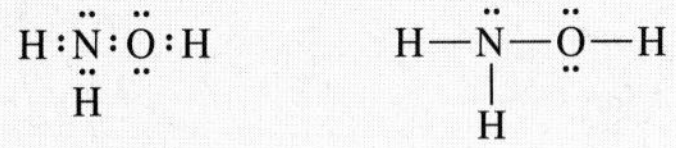

dronium ion, for example, is formed when a water molecule accepts a proton (a hydrogen atom without its electron) from an acid. Eight electrons are available to bond one oxygen atom with three hydrogen atoms.

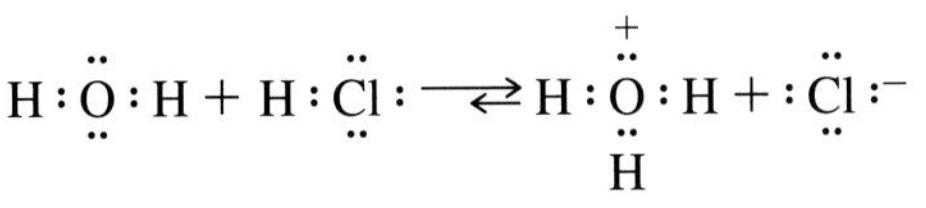

$$\text{H:O:H} + \text{H:Cl:} \;\rightleftharpoons\; \overset{+}{\text{H:O:H}} + \text{:Cl:}^-$$
$$\qquad\qquad\qquad\qquad\quad\text{H}$$

| water | hydrogen chloride | hydronium ion | chloride ion |
|---|---|---|---|

In the equation above, two uncharged species, water and hydrogen chloride, react to give a positively charged hydronium ion and a negatively charged chloride ion. The sum of the charges on one side of the equation, zero in this case, equals the sum of the charges, also zero, on the other side. Charges as well as numbers of atoms must always be balanced in writing equations.

## B. Connectivity

The Lewis structures shown in Table 1.1 illustrate the concept of **connectivity** in covalent molecules. Connectivity means just what it sounds like: the structure of the molecule is described in terms of the atoms that are connected to each other by means of covalent bonds. The connectivity in a molecule is determined by looking at the structure and describing which atom is bonded to which others. For compounds containing only single covalent bonds, the connectivity is the same as the backbone of the Lewis structure. For example, by looking at the condensed formula for methanol, $CH_3OH$, we can see that the connectivity in methanol is one carbon atom bonded to three hydrogen atoms and an oxygen atom, which is also bonded to a hydrogen atom.

$$CH_3OH$$

condensed formula for
methanol indicating
connectivity

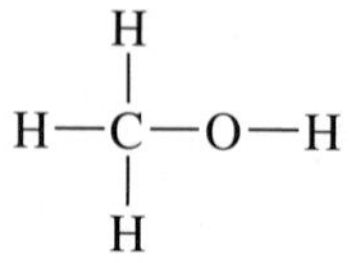

connectivity in methanol
shown in detail; one
carbon atom bonded to
three hydrogen atoms and
an oxygen atom; oxygen
atom also bonded to one
hydrogen atom

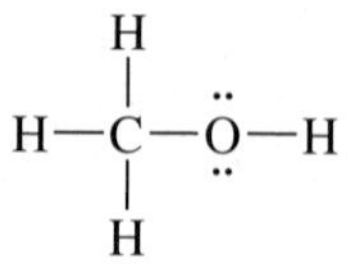

Lewis structure
for methanol

For compounds that have more than one carbon atom, it is important to be able to interpret the connectivities shown in condensed formulas. For example, 2-methylbutane can be written in three different ways, each one making the connectivity clearer.

$$CH_3CH_2CH(CH_3)CH_3$$

condensed formula for
2-methylbutane

$$CH_3CH_2CHCH_3$$

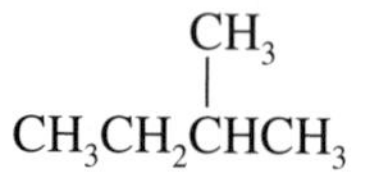

another version of the
condensed formula for
2-methylbutane

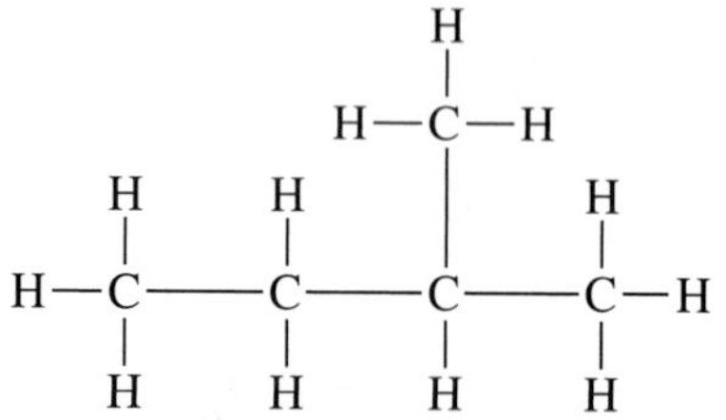

Lewis structure for
2-methylbutane

Carbon in both methanol and 2-methylbutane can be represented as an atom that can be bonded by four single bonds to other atoms.

a carbon atom

In methanol these other atoms are hydrogen and oxygen. In 2-methylbutane they are carbon and hydrogen. If we dissect 2-methylbutane into its carbon atoms, we discover five of them, each with the possibility of forming four bonds.

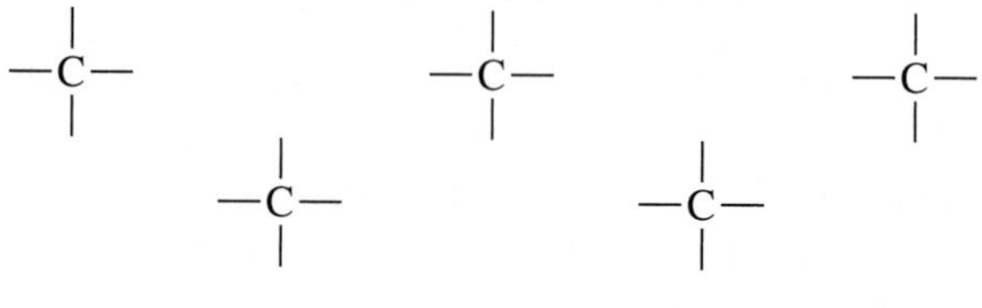

We also discover that we can now reconnect the five carbon atoms in more than just the single way in which they were connected in 2-methylbutane. In fact, there are a total of three ways.

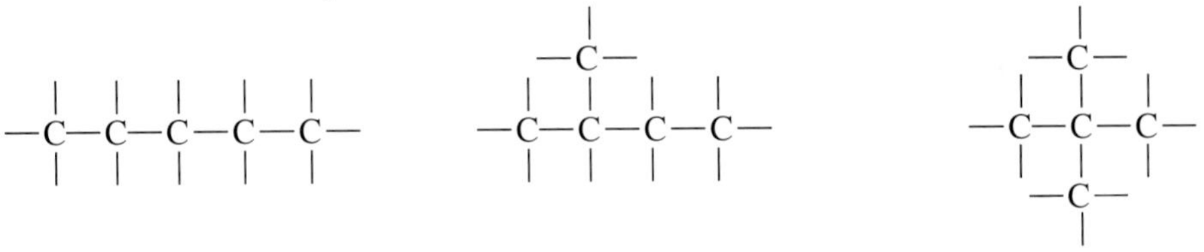

```
                              |                          |
                            —C—                        —C—
  |   |   |   |   |         |   |   |   |          |   |   |
—C—C—C—C—C—             —C—C—C—C—              —C—C—C—
  |   |   |   |   |         |   |   |   |          |   |   |
                                                        —C—
                                                          |
```

| five carbons connected | four carbons connected in a row | three carbons connected in a row |
| to each other in a row | with one carbon attached to | with two carbons attached to |
|  | the second carbon in the row | the middle carbon in the row |

Other ways of drawing these connections may seem different but are not really. For example,

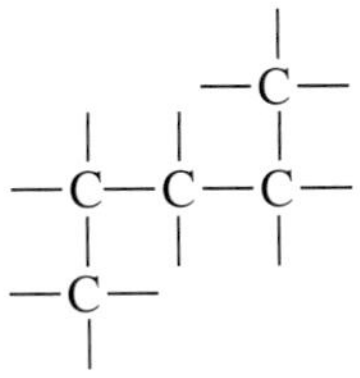

```
                  |
                —C—
          |   |   |
        —C—C—C—
          |   |   |
        —C—
          |
```

looks different from the connections shown above but is really the same as the first structure. It has five carbons connected to each other in a row. Two of the carbon atoms are twisted out of a straight line, but this does not change the connectivity of the molecule.

This is an excellent time to start working with molecular models. With them, you can actually connect carbon atoms and see how there are three, and only three, ways in which you can put five carbon atoms together. You also will discover that the molecule does not really look like the pictures we draw on the page but has angles around the carbon atoms that give the backbone of the molecule a zigzag appearance. We will discuss this in greater depth later (Sections 1.8 and 5.8).

Once the carbon atoms are connected to each other, bonding sites are left over for other atoms. Note that no matter how we connected the five carbon atoms, twelve open sites are available for bonding to other atoms. This will always be true as long as (1) the carbon atoms are forming only single bonds and (2) there are no rings in the compound. If these compounds are **hydrocarbons,** compounds containing only carbon and hydrogen, they will all have the molecular formula $C_5H_{12}$. More generally, compounds containing just carbon atoms and all the hydrogen atoms that they can hold have the molecular formula $C_nH_{2n+2}$ ($C_5H_{2\times5+2} = C_5H_{12}$). Such compounds are said to be **saturated hydrocarbons.** Our exploration above showed us that there are three compounds having the same molecular formula, $C_5H_{12}$, but different connectivities. Compounds with the same molecular formula but different structural formulas are called **isomers.** Isomers will be discussed further in Section 1.7 (p. 21).

A molecular formula does not give us information about connectivity, but it does tell us one very important thing: whether the structural formula for the compound can contain rings or atoms bonded to each other by more than one bond. If the formula has $2n + 2$ hydrogen atoms for every $n$ carbon atoms, the structure will not have any rings or more than one bond between the same two atoms in it. The condensed and Lewis structures for the three compounds with the molecular formula $C_5H_{12}$ are shown on the next page.

$CH_3CH_2CH_2CH_2CH_3$

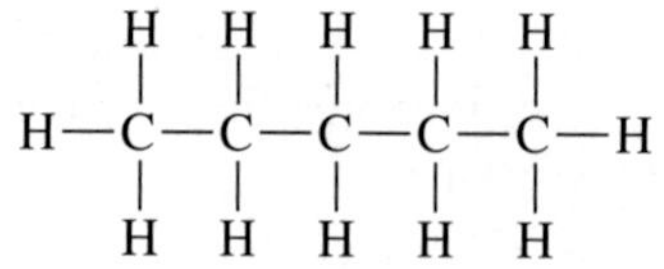

H—C—C—C—C—C—H (with H above and below each carbon)

condensed formula for pentane    Lewis structure for pentane

*connectivity:    5 carbon atoms in a row*

$CH_3CH_2CH(CH_3)CH_3$

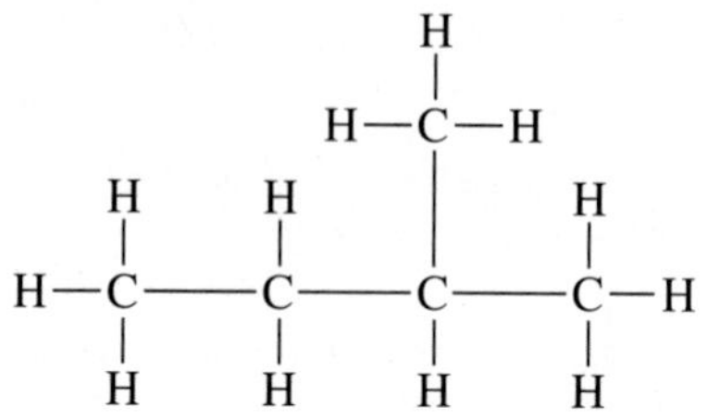

Lewis structure for 2-methylbutane

condensed formula for 2-methylbutane

*connectivity:    4 carbon atoms in a row, 1 carbon attached to the second carbon atom in the row*

$CH_3C(CH_3)_2CH_3$

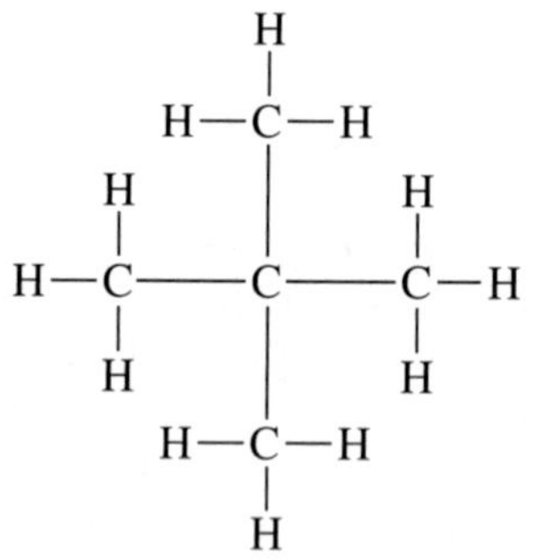

condensed formula for 2,2-dimethylpropane    Lewis structure for 2,2-dimethylpropane

*connectivity:    3 carbon atoms in a row, 2 carbon atoms attached to the second carbon atom in the row*

Carbon atoms are also found bonded to atoms other than hydrogen, atoms such as oxygen, the halogens, and nitrogen. Saturated compounds of carbon, hydrogen, and oxygen also have $2n + 2$ hydrogen atoms for $n$ carbon atoms. For example, the formula $C_4H_{10}O$ tells us that compounds having this molecular formula will not have any rings or multiple bonds in them because the formula contains enough hydrogen atoms to fill any vacant bond sites left over from putting together the four carbon atoms and one oxygen atom units.

4 carbon atoms and 1 oxygen atom that can be bonded
together in a variety of ways

Two of the ways in which these atoms can be bonded together are shown below.

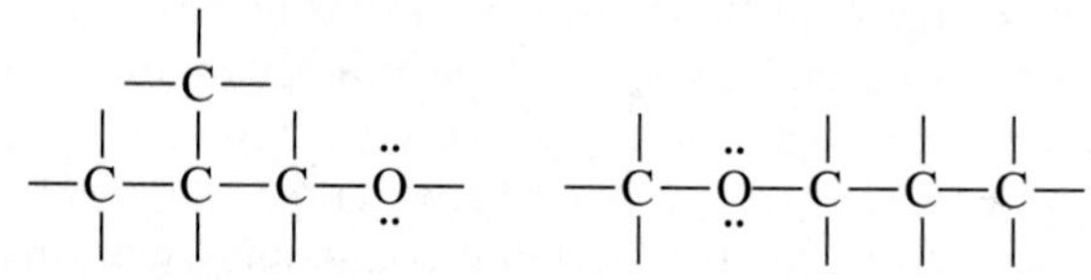

Both structural formulas have ten vacant bonding sites to be occupied by hydrogen atoms. The completed structural formulas are

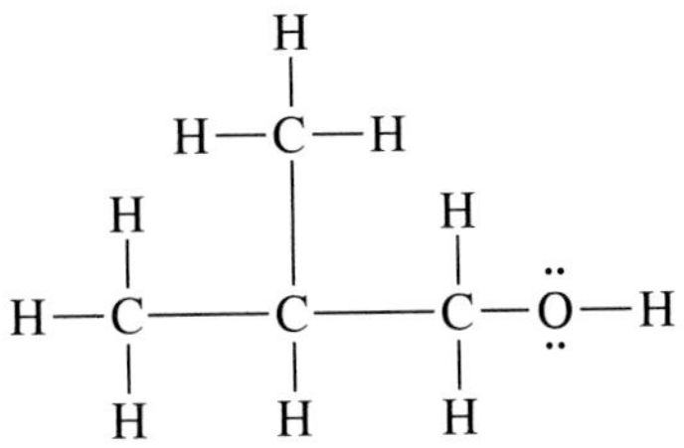

$$
\begin{array}{c}
\text{H} \\
| \\
\text{H}-\text{C}-\text{H} \\
\end{array}
$$

Lewis structure for
2-methyl-1-propanol

condensed formula for
2-methyl-1-propanol

$CH_3CH(CH_3)CH_2OH$

*connectivity:* *3 carbon atoms in a row with 1 oxygen atom*
*at the end of the row and a carbon atom on the second*
*carbon atom of the carbon chain*

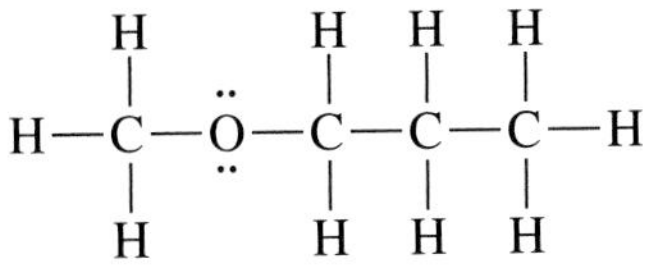

Lewis structure for
methyl propyl ether

condensed formula for
methyl propyl ether

$CH_3OCH_2CH_2CH_3$

*connectivity:* *1 carbon atom connected to an oxygen atom,*
*which is connected to a row of 3 carbon atoms*

**Problem 1.1**    Besides the two compounds shown above, there are five more ways in which four carbon atoms, one oxygen atom, and ten hydrogen atoms can be connected. Find them. Molecular models will be very helpful in this task. Write Lewis structures and condensed formulas for each compound. Describe the connectivity of each compound in a way that makes clear how it differs from the other compounds with the same molecular formula.

Halogens in most organic compounds bond only once to other atoms; therefore, they occupy the same slot as a hydrogen atom does in a structural formula. For example, compounds with the molecular formula $C_4H_9Br$ will be saturated—that is, they will not contain any rings or multiple bonds. So will those with the molecular formula $C_4H_8Cl_2$. In both cases, if we mentally substitute H for Br or Cl, we get $C_4H_{10}$ (or $C_nH_{2n+2}$) as the formula.

**Problem 1.2**    There are four compounds with the molecular formula $C_4H_9Br$. Starting with four carbon atoms, $-\overset{|}{\underset{|}{C}}-$, and one bromine atom, $-\overset{..}{\underset{..}{Br}}:$, construct them. Write Lewis structures and condensed formulas for them and describe the connectivity of each one. Molecular models will be helpful.

Nitrogen normally bonds to three other atoms. A compound containing nitrogen, therefore, has to contain one more hydrogen atom than a compound containing only carbon or carbon with oxygen atoms. For example, $CH_3OH$ has the molecular formula $CH_4O$, but $CH_3NH_2$ has the molecular formula $CH_5N$.

**Problem 1.3**    The molecular formula $C_3H_9N$ corresponds to four structural formulas. Find them, starting with three carbon atoms, $-\overset{|}{\underset{|}{C}}-$, and one nitrogen atom, $-\overset{|}{\underset{|}{N}}:$. What are the connectivities that you find? Write Lewis structures and condensed formulas for each compound.

Chemists are very interested in the connectivities of molecules. Chemical reactions usually result in changes in connectivity, and a close look at the structural formulas representing a molecule before and after it undergoes a reaction tells chemists what type of reaction has taken place and, therefore, what reagent to use. The first workbook exercises in the *Study Guide* give you practice in describing connectivities and making predictions about reactions the way chemists do.

**Workbook Exercises**

**Problem 1.4** Which of the following compounds have connectivities with all singly bonded atoms and no rings in their structural formulas?

(a) $C_2F_4$ (the monomer tetrafluoroethene, used to make Teflon)
(b) $C_{14}H_9Cl_5$ (the formula for DDT, an insecticide)
(c) $C_2HBrClF_3$ (the formula for halothane, an anesthetic)

(d) $C_{10}H_{16}$ (the formula for adamantane, an interesting molecule that resembles diamond in its structure)
(e) $C_{20}H_{42}$ (the formula for icosane, found in Vaseline)
(f) $C_2H_3Cl$ (the formula for vinyl chloride, used to make plastics)
(g) $CHCl_3$ (the formula for chloroform, used in old movies to knock people out)
(h) $C_4H_7Cl$ (the formula for methallyl chloride, a fumigant)

## C. Formal Charges

Organic chemists are not satisfied with the simple statement that an ion, such as the hydronium ion, $H_3O^+$, is positively charged. They find it useful to locate the charge on a particular atom in the ion. In the equation on p. 7, the hydronium ion is shown with the positive charge next to the oxygen atom. The decision as to where to put the charge is made by calculating the formal charge for each atom in an ion or a molecule.

The **formal charge** for an atom may be calculated using this formula:

$$\text{Formal charge} = (\text{number of valence electrons})$$
$$- (\text{number of nonbonding electrons})$$
$$- \tfrac{1}{2}(\text{number of bonding electrons})$$

oxygen has six bonding electrons, two nonbonding electrons

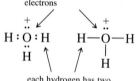

each hydrogen has two bonding electrons, no nonbonding electrons

To locate the formal charge in the hydronium ion, all the electrons in the valence shell of each atom are counted. Each hydrogen atom in the hydronium ion has no formal charge because each has one valence electron, no nonbonding electrons, and two bonding electrons.

$$\text{Formal charge for hydrogen} = 1 - 0 - \tfrac{1}{2}(2) = 0$$

In the hydronium ion, oxygen shares six electrons with the hydrogen atoms and therefore has six bonding electrons. It has two nonbonding electrons. Because it is in Group 16 of the periodic table, it has sixteen minus ten, or six, valence electrons.

$$\text{Formal charge for oxygen} = 6 - 2 - \tfrac{1}{2}(6) = 6 - 5 = +1$$

Calculating a formal charge is essentially the same as asking whether an atom in a molecule or ion has in its valence shell more electrons or fewer electrons than are necessary to balance its nuclear charge, the number of protons in the nucleus. An uncharged oxygen atom has to have six electrons in its valence shell. In the hydronium ion, oxygen bonds with three hydrogen atoms, so only five electrons effectively belong to oxygen, which is one fewer than it needs; therefore, it bears a formal charge of +1. With a little practice you will find this way of thinking about formal charge much easier than using the formula. Memorizing the formula has two disadvantages: Not only is it possible to forget a formula, it is also possible to remember it incorrectly. Understanding the basis for the formula frees you from having to use it.

In the case of the hydronium ion, the formal charge for oxygen also represents the charge on the ion because no other atoms in the ion are charged. *The sum of the formal charges on the atoms in an uncharged molecule is zero. For an ion, the sum of the formal charges on different atoms should add up to the charge on the ion.*

**Problem 1.5**    Draw Lewis structures for the species the connectivities of which are represented by the following condensed formulas. Identify any atoms bearing formal charges.

(a) $CCl_4$      (b) $CH_3Br$      (c) $CH_3OH_2^+$      (d) $NH_2^-$      (e) $CH_3NH_3^+$      (f) $H_2NNH_2$      (g) $PH_3$      (h) $H_2S$

(i) $CH_3CH_2OH$      (j) $HOCH_2CH_2OH$      (k) $CH_3OCH_3^+$
                                                              |
                                                             $CH_3$

**Problem 1.6**    Decide whether the central atom in each of the following formulas is uncharged, positively charged, or negatively charged. All nonbonding electrons are shown.

(a) $CH_3$—$\overset{\displaystyle CH_3}{\underset{\displaystyle CH_3}{N}}$—$CH_3$      (b) $:\overset{..}{Br}$—$\overset{..}{C}$—$\overset{..}{Br}:$      (c) $CH_3$—$\overset{\displaystyle H}{\overset{|}{\underset{..}{O}}}$—$CH_3$

(d) $CH_3$—$\overset{..}{N}$—$H$      (e) $:\overset{..}{Cl}$—$\overset{\displaystyle :\overset{..}{Cl}:}{\underset{}{C}}$—$\overset{..}{Cl}:$      (f) $CH_3$—$\overset{\displaystyle CH_3}{\underset{\displaystyle CH_3}{C}}$—$CH_3$

## D. Molecules with Open Shells

Sometimes there are not enough electrons in a system to provide an octet around the central atom. Boron trifluoride, $BF_3$, is such a molecule. Boron is in Group 13 of the periodic table and has only three electrons in its valence shell. Adding these three to the twenty-one electrons contributed by the three fluorine atoms gives twenty-four electrons available for bonding. If we put octets around the fluorine atoms, the boron atom ends up with only six electrons. This leaves the boron atom with an **open shell.** It is electron-deficient, meaning that it can accept another pair of electrons to complete an octet. As a result, boron trifluoride reacts with compounds such as ammonia that have nonbonding electrons.

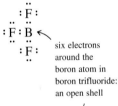

six electrons around the boron atom in boron trifluoride: an open shell

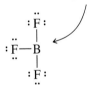

$$:\overset{\displaystyle :\overset{..}{F}:}{\underset{\displaystyle :\overset{..}{F}:}{\overset{..}{F}}}\!\!-\!\!B \;+\; :\overset{\displaystyle H}{\underset{\displaystyle H}{N}}\!\!-\!\!H \;\rightleftarrows\; :\overset{\displaystyle :\overset{..}{F}:}{\underset{\displaystyle :\overset{..}{F}:}{\overset{..}{F}}}\!\!-\!\!\overset{-}{B}\!\!-\!\!\overset{\displaystyle H}{\underset{\displaystyle H}{\overset{+}{N}}}\!\!-\!\!H$$

boron trifluoride    ammonia      compound of boron trifluoride and ammonia

**Problem 1.7**    In the equation above, formal charges are shown in the formula for the reaction product of boron trifluoride and ammonia. Make sure that you understand how they were obtained.

**Problem 1.8**    Of the species shown in Table 1.1 (p. 7), pick out the ones that would react with $BF_3$. Write equations for the reactions. Be sure to include any formal charges that result. *Hint:* Are all pairs of nonbonding electrons equally available for forming a bond to the boron atom? Think about whether electrons (which are negatively charged) will be donated most easily by a species with no charge, with a negative charge, or with a positive charge. Remember the rules of electrostatics (p. 5).

## 1.5 Multiple Bonds

The compounds discussed earlier in this chapter all have carbon atoms bonded by single bonds to other atoms. Carbon, however, is not limited to forming single bonds. In some compounds carbon atoms are held to other atoms by double or triple covalent bonds, which are also called **multiple bonds.**

For example, ethylene, $C_2H_4$, has the connectivity $CH_2CH_2$. The molecular formula of ethylene has two fewer hydrogen atoms than would be needed to form the corresponding saturated compound, which would be ethane, $C_2H_6$. The molecular formula of ethylene corresponds to the general formula $C_nH_{2n}$. Such a compound, which is missing two hydrogen atoms when compared with the general formula $C_nH_{2n+2}$, is said to contain a **unit of unsaturation.** A unit of unsaturation in the molecular formula always appears in the structural formula as a double bond or a ring. If the ethylene molecule is constructed from two carbon units and four hydrogen atoms, we find that one of the bonding sites on each carbon remains vacant and can be satisfied only by a second bond between the two carbon atoms.

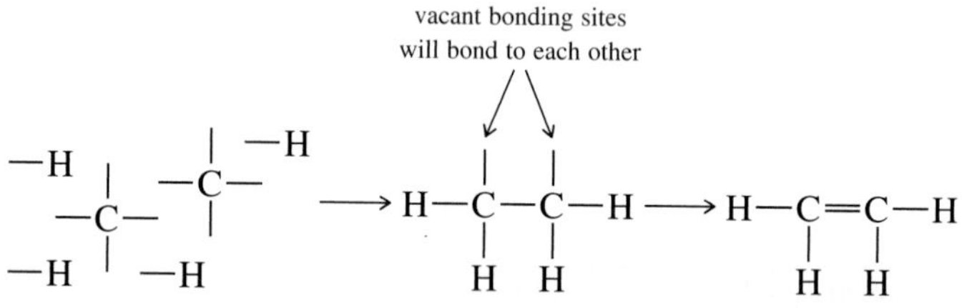

No other way of connecting the two carbon atoms and four hydrogen atoms gives octets around each carbon atom. Another way of looking at this is that the carbon atoms in ethylene share two pairs of electrons in a double bond.

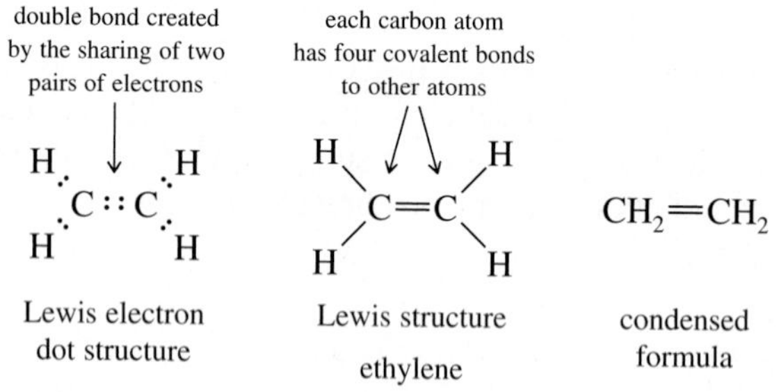

Acetylene, $C_2H_2$, has the connectivity CHCH. It is even more unsaturated than ethylene is and, in fact, is missing four hydrogen atoms when compared with ethane, $C_2H_6$. Each two missing hydrogen atoms correspond to one unit of unsaturation, so acetylene has two units of unsaturation. Two units of unsaturation appear in its structural formula as a triple bond. If we construct acetylene from two carbon atoms and two hydrogen atoms, we find four vacant bond sites that need to be bonded together to put octets around the carbon atoms.

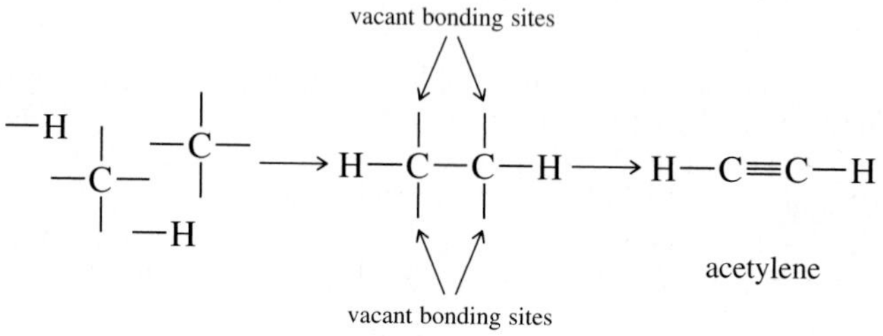

The same structural formula is arrived at by writing a Lewis electron dot structure.

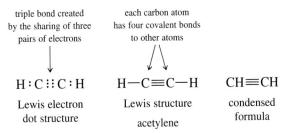

$H:C::C:H$     $H-C\equiv C-H$     $CH\equiv CH$

Lewis electron     Lewis structure     condensed
dot structure        acetylene        formula

Many important organic compounds have double bonds between carbon and oxygen. Formaldehyde, for example, has the formula and the connectivity $CH_2O$ (or $H_2CO$). It has two fewer hydrogen atoms than the corresponding saturated compound, $CH_4O(C_nH_{2n+2}O)$. $CH_2O$ corresponds to $C_nH_{2n}O$; therefore, it has one unit of unsaturation. It must contain a double bond (or a ring, but when there is a ring, either the connectivity or the question clearly indicates that). Attempts to join one carbon, one oxygen, and two hydrogen atoms give only one way in which we get octets around the carbon and oxygen atoms if we exclude structures that have charges on some atoms.

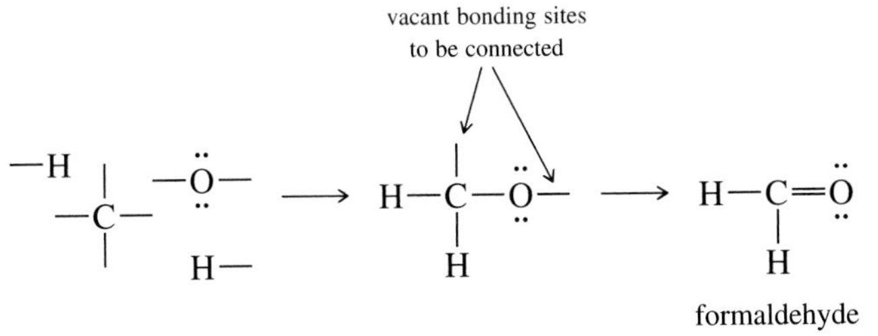

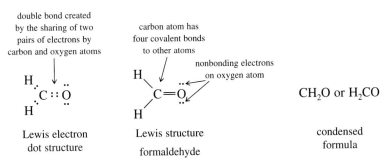

We can arrive at the same structural formula by working with Lewis structures.

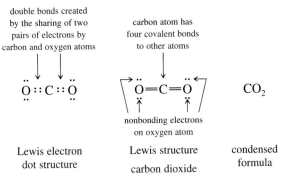

$CH_2O$ or $H_2CO$

Lewis electron     Lewis structure     condensed
dot structure      formaldehyde      formula

Carbon dioxide, $CO_2$, connectivity OCO, has even fewer hydrogen atoms than formaldehyde does and has two units of unsaturation, which appear as two double bonds.

double bonds created
by the sharing of two
pairs of electrons by
carbon and oxygen atoms

carbon atom has
four covalent bonds
to other atoms

$O::C::O$     $O=C=O$     $CO_2$

nonbonding electrons
on oxygen atom

Lewis electron     Lewis structure     condensed
dot structure     carbon dioxide     formula

Note that although the traditional condensed formula for carbon dioxide is written as $CO_2$, the actual structure of carbon dioxide has the carbon atom between two oxygens. Compounds in which oxygen atoms are bonded to each other are usually highly reactive and have names containing words such as "peroxy" or "peroxide." Hydrogen peroxide, $H_2O_2$, with connectivity HOOH, is an example of such a compound that owes its reactivity to the presence of an oxygen–oxygen single bond and therefore can be used as a disinfectant and as a bleach. Unless the name of a compound gives the indication that the compound has a peroxide structure, it is best not to write structural formulas that contain oxygen–oxygen bonds.

Important compounds known as cyanides or nitriles contain triple bonds between carbon and nitrogen. Hydrogen cyanide, an extremely toxic gas, is an example of such a compound. A saturated compound containing one carbon and one nitrogen atom would have the molecular formula $CH_5N$ (p. 11). Hydrogen cyanide, HCN, is missing four hydrogen atoms and therefore has two units of unsaturation. They appear as a triple bond between the carbon and nitrogen atoms.

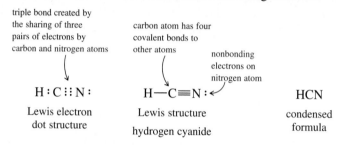

triple bond created by the sharing of three pairs of electrons by carbon and nitrogen atoms

carbon atom has four covalent bonds to other atoms

nonbonding electrons on nitrogen atom

$$H:C::N:$$

Lewis electron dot structure

$$H—C≡N:$$

Lewis structure
hydrogen cyanide

HCN

condensed formula

**Problem 1.9**   Draw Lewis structures for the compounds represented by the following formulas.

(a) $CH_3\overset{O}{\overset{\|}{C}}CH_3$    (b) $CH_3C≡CH$    (c) $HC\overset{O}{\overset{\|}{N}}H_2$    (d) $HC\overset{O}{\overset{\|}{O}}H$

(e) $CH_3\overset{O}{\overset{\|}{C}}OCH_3$    (f) $HON=O$    (g) $CH_3N=NCH_3$    (h) $CH_2=CHCl$

**Problem 1.10**    Many of the compounds in Problem 1.4 (p. 12) have units of unsaturation in them. Decide how many units of unsaturation there are in each of these compounds.

structure determined experimentally for the carbonate ion, $CO_3^{2-}$

Lewis structures for the carbonate ion, $CO_3^{2-}$

## 1.6  Resonance

### A. Resonance Contributors

Some covalent molecules and ions cannot be represented satisfactorily by a single Lewis structure. The carbonate ion, $CO_3^{2-}$, is an example of such a species. The carbonate ion in calcite, $CaCO_3$, is planar, with bond angles of 120° and three equivalent carbon–oxygen bonds 1.29 Å ($1.29 \times 10^{-8}$ cm) long. However, if there is to be an octet around each atom, the Lewis structure of $CO_3^{2-}$ must be drawn with a double bond between carbon and one of the oxygen atoms. Two of the oxygen atoms have formal charges of −1, giving a total charge of −2 for the ion. This Lewis structure for the carbonate ion suggests that one of the oxygen atoms is different from the other two. The experimental evidence, however, indicates that all three oxygen atoms are equivalent. For example, the distances from the carbon atom to each oxygen atom are equal. As far as we can tell experimentally, each oxygen atom has some negative charge and an equal probability of reacting with an acid to pick up a proton. The Lewis structure does not depict the experimental reality adequately.

The experimental observations for the carbonate ion are better represented by a picture in which the electrons are equally distributed among all three oxygen atoms. Three structures may be drawn for the carbonate ion, each having the same connectivity and each differing only in the location of pairs of electrons. The individual structures are **resonance contributors** to the structure of the carbonate ion; the carbonate ion is pictured as a **resonance hybrid** of these contributors. These structures differ only in the arrangement of electrons, not in the positions of the atoms.

*resonance contributors to the*
*structure of the carbonate ion*

The actual properties of the carbonate ion cannot be represented by any one of the Lewis structures taken alone. The experimental facts are rationalized by drawing the three resonance contributors. These three taken together indicate that each oxygen atom bears two-thirds of the charge on an electron and that the carbon–oxygen bonds are all the same length, a length between that typical of a carbon–oxygen single bond (1.43 Å) and that typical of a carbon–oxygen double bond (1.22 Å).

Resonance contributors are significant for many other ions. The nitrate ion is another example of a species for which a single Lewis structure is not satisfactory.

$NO_3^-$

*resonance contributors to the*
*structure of the nitrate ion*

Again, these structures have the same connectivity and differ from each other only in the location of pairs of electrons. In each resonance contributor in the nitrate ion, two oxygen atoms bear a formal charge of $-1$, and the nitrogen atom has a formal charge of $+1$. The sum of these formal charges is the charge on the anion. Experimentally, the three oxygen atoms in the nitrate ion are equivalent.

Structures for uncharged molecules also may have resonance contributors. Nitromethane is an example of a molecule for which we can draw more than one Lewis structure.

$CH_3NO_2$

*resonance contributors for nitromethane*

In nitromethane, the formal charge of $+1$ on the nitrogen atom and the formal charge of $-1$ on the oxygen atom cancel each other, so the molecule as a whole is not charged.

A double-headed arrow, ↔ , is used between the resonance contributors of nitromethane to indicate their relationship. This symbol does *not* mean that the two forms are in equilibrium with each other. No reaction is implied by the double-headed resonance arrow. There is only one structure for nitromethane, which is a hybrid of the two Lewis structures we are able to draw. The symbol for equilibrium is two arrows pointing in opposite directions, showing a reversible chemical reaction, for example,

$$H_2CO_3 \;+\; H_2O \;\rightleftharpoons\; H_3O^+ \;+\; HCO_3^-$$

Here there are differences in connectivity in the species on the left and the right of the arrows; a reaction has taken place.

Resonance is an example of a model that was developed to deal with experimental observations that could not be explained in terms of a simpler model, such as a single Lewis structure for a molecular species. It is important to remember that the individual representations of resonance contributors have no reality. The compound, such as nitromethane, for which resonance contributors are written does not exist as a mixture of different forms. The actual molecular species has properties suggested by all the resonance contributors taken together. For example, in nitromethane, the nitrogen atom bears a positive charge, each oxygen atom bears part of a negative charge, and both nitrogen–oxygen bonds are the same length.

The resonance contributors shown for the carbonate ion are equivalent to each other, as are the ones shown for the nitrate ion and nitromethane on page 17. This is not always the case. For example, three resonance contributors can be written for the formate ion.

major contributors            minor contributor

*resonance contributors for the formate ion*

Two of them are equivalent, with a double bond to one oxygen atom and a single bond to the other one. Each atom (except hydrogen) has an octet of electrons around it. The third structure is different. The carbon atom has only six electrons around it and bears a positive charge. There is no double bond, and both oxygen atoms are negatively charged. This third structure has a higher energy and is less stable than the other two because it has fewer bonds and a separation of charge. A **separation of charge,** by which one atom becomes positively charged while another one becomes negatively charged, can be achieved only through an expenditure of energy. The resonance contributors with no separation of charge, with the maximum number of covalent bonds, and with octets around each atom (except hydrogen) contribute the most to the experimentally observed properties of the species being represented. These resonance contributors are thus more important and are known as **major contributors.** Those that have fewer covalent bonds and a separation of charge have less effect on the properties of the species and are often called **minor contributors.**

Minor resonance contributors in which there is a separation of charge are frequently used in order to explain chemical reactivity. For example, carbon dioxide reacts readily with hydroxide ion, $OH^-$. The ease with which this reaction occurs can be rationalized by writing a resonance contributor for carbon dioxide in which the carbon atom has only six electrons around it and bears a positive charge.

$$:\ddot{O}=C=\ddot{O}: \longleftrightarrow :\ddot{O}=C\overset{+}{-}\ddot{O}:^- \longleftrightarrow {}^-:\ddot{O}-C\overset{+}{=}\ddot{O}:$$

*resonance contributors for carbon dioxide, with a separation of charge and six electrons around a positively charged carbon*

$$:\ddot{O}=C\overset{+}{-}\ddot{O}:^- \longrightarrow :\ddot{O}=C-\ddot{O}:$$

$$^-:\ddot{O}-H \qquad :\ddot{O}-H$$

covalent bond formed by the sharing of a pair of electrons from the negatively charged hydroxide ion with the positively charged carbon atom

bicarbonate anion

These are the rules for writing resonance contributors:

1. Resonance contributors have the same connectivity. Only nonbonding electrons and electrons in multiple bonds change locations from one resonance contributor to another. The electrons in single covalent bonds are not involved.

2. The nuclei of atoms in different resonance contributors are in the same positions.

3. All resonance contributors must have the same numbers of paired and unpaired electrons.

4. Resonance contributors in which atoms of elements from the second period all have eight electrons around them are more important than those in which such atoms have fewer than eight electrons. Similarly, resonance contributors with a greater number of covalent bonds are more important than those with a smaller number. For atoms of elements from periods beyond the second period, such as sulfur and phosphorus, it is possible to write structures with ten or more electrons around a central atom.

5. Resonance contributors in which there is little or no separation of charge are more important than those with a large separation of charge. (Remember that it takes energy to separate opposite charges.) Resonance contributors containing like charges (both positive or both negative) on adjacent atoms are not favored. (Like charges repel each other.)

6. When structures with a separation of charge are written, the more important resonance contributor has the negative charge on the more electronegative atom. (A review of the concept of electronegativity is found on p. 26.)

The concept of resonance was developed by Linus Pauling of the California Institute of Technology. In 1954, Pauling was awarded the Nobel Prize for his research into the nature of the chemical bond and his application of this knowledge to determination of the structures of complex substances. Resonance is best understood in the context of different examples. The rationalization of the reactivity of carbon dioxide with the hydroxide ion illustrates the way the idea is most often used. Based on the experimental facts known about a compound, chemists draw resonance contributors for the molecule to explain reactivity. In the case of carbon dioxide, resonance contributors are used to explain the reactivity of the carbon atom toward reagents, such as hydroxide ion, that have pairs of electrons to share with atoms having open shells. Section 2.10 (p. 67) will introduce another important application of the concept of resonance, the idea of stabilization of a species by resonance. You will gradually develop an intuitive understanding of the concept as it is applied to many situations throughout this book.

⊘ Study Guide
Concept Map 1.3

## B. The Art of Solving Problems

When chemists solve a problem, they usually do so by systematically analyzing it and approaching the answer in steps. When they get very good at it, they are not always conscious of all the steps in the process because they do them rapidly and automatically. Only when they are faced with a difficult and unfamiliar problem do they slow down and become conscious of all the different things they do in order to reach a solution.

For students just starting the study of organic chemistry, all problems are unfamiliar. Therefore, this book has sections on the skills used in solving problems. These sections ask the kinds of questions and give the kinds of answers chemists do when they have to solve similar problems. Asking yourself these questions and finding the answers to them for different problems will help you develop the skills you need for the successful study of organic chemistry.

PROBLEM: Write resonance contributors for the ion having the following connectivity: $CH_2CHO^-$. Evaluate each resonance contributor in terms of the rules given in Section 1.6A and decide which are major contributors and which minor.

**Question:** Which atoms are present, and how many electrons are available?

**Answer:**

$$
\begin{array}{llll}
2C\ (Group\ 14) & = 2 \times 4\ electrons & = & 8\ electrons \\
3H\ (Group\ 1) & = 3 \times 1\ electron & = & 3\ electrons \\
1O\ (Group\ 16) & = 1 \times 6\ electrons & = & 6\ electrons \\
1\ negative\ charge = & 1\ electron & = & \underline{1\ electron} \\
& & & 18\ electrons
\end{array}
$$

**Question:** What is a Lewis structure for this ion?

**Answer:** Start with the connectivity given in the statement of the problem. Bond the carbon and oxygen atoms together, and put in the hydrogen atoms; then use the electrons that are left to complete octets around the carbon and oxygen atoms.

Stage 1.    The connectivity given shows a carbon atom bonded to two hydrogen atoms and a second carbon atom. The second carbon atom is bonded to a hydrogen atom and an oxygen atom.

*uses 10 electrons*

Stage 2.        *uses 18 electrons*

Stage 3.    Simplify the structure, and locate any formal charge.

Oxygen has six valence electrons; in this structure it has six nonbonding electrons and two bonding electrons, one of which effectively belongs to oxygen. Therefore, oxygen has seven electrons in its valence shell, one more than is necessary to balance its nuclear charge; thus it has a formal charge of $-1$. The same result can be obtained by applying the formula for formal charge.

$$\text{Formal charge} = 6 - 6 - \tfrac{1}{2}(2) = -1$$

Note that the task for this part of the problem was made easier because the original problem statement showed the connectivity of the atoms in the compound. It is important to use all the information given in a problem.

**Question:** What resonance contributors are possible?

**Answer:** Explore ways of moving nonbonding electrons and electrons in the double bond.

1                2

There seem to be two resonance contributors.

**Question:** Which one is the more important (major) contributor?

**Answer:** (Review the rules on p. 19 if necessary.) Both contributors have the same number of covalent bonds and eight electrons around each carbon and oxygen atom and the same separation of charge. Contributor 1 is the major contributor because the negative charge is on the oxygen atom rather than on the carbon atom, as it is in contributor 2, and oxygen is the more electronegative element. ■

**Problem 1.11**    Write resonance contributors for the following ions and molecules. Names and connectivities are given, followed by the more conventional condensed formula in parentheses when it is different from the connectivity. For species that contain multiple oxygen atoms and another atom such as C, N, S, or P, assume that the oxygen atoms are each bonded to that other atom and not to each other. Include formal charges where applicable. Evaluate each resonance contributor you write in terms of the rules given in Section 1.6A, and decide which are major contributors and which are minor contributors.

(a) bicarbonate ion, $HOCO_2^-$ ($HCO_3^-$)
(b) formaldehyde, $H_2CO$        (c) nitrite ion, $NO_2^-$

(d) isocyanate ion, $NCO^-$      (e) nitric acid, $HONO_2$ ($HNO_3$)
(f) ozone, $O_3$ (experimental evidence shows the molecule is not cyclic)
(g) nitronium ion, $NO_2^+$        (h) chlorate ion, $ClO_3^-$

**Problem 1.12**    The reaction between hydroxide ion and carbon dioxide is similar in some ways to the one shown between ammonia and boron trifluoride on p. 13. What similarities do you see?

**Problem 1.13**    On the basis of the resonance contributors you wrote for formaldehyde, $H_2CO$ (part b of Problem 1.11), and your answer to Problem 1.12, predict whether ammonia will react with formaldehyde.

## 1.7    Isomers

Carbon atoms form strong single and multiple bonds among themselves and with oxygen and nitrogen. The compounds of carbon, therefore, have wide structural variety. The versatility of carbon allows for the creation of the complex structures that are important in living organisms. As the early organic chemists determined the molecular formulas for the compounds they had recovered from natural sources, they discovered that it is possible for two or more compounds with very different properties to have the same molecular formula. In 1830, the Swedish chemist Jakob

Berzelius named such compounds "isomeric bodies" from the Greek words *isos,* meaning "equal," and *meros,* meaning "part." We call them **isomers.**

Isomers are compounds that have identical molecular formulas but differ in the ways in which the atoms are bonded to each other. Isomers may be constitutional isomers or stereoisomers. **Constitutional isomers** differ in the order and the way in which the atoms are bonded together in their molecules. In other words, they differ in their connectivities (p. 7). **Stereoisomers** have the same connectivities; they differ only in the arrangement of their atoms in space. They can be distinguished only by exploring their structures in three dimensions. Stereoisomerism will be discussed in Chapter 6.

Two structural formulas may be written for the molecular formula $C_4H_{10}$, indicating two different ways in which four carbon atoms and ten hydrogen atoms can be connected to each other.

$$CH_3CH_2CH_2CH_3$$

butane
bp $-0.6\ °C$

$$CH_3CHCH_3 \ (CH_3)$$

2-methylpropane
bp $-10.2\ °C$

Butane and 2-methylpropane are constitutional isomers of each other; they have the same molecular formula but different structural formulas. Their structures differ in connectivity. Because their structures differ, the compounds have different physical properties, for example, different boiling points.

Similarly, the molecular formula $C_2H_6O$ gives rise to two different connectivities.

$$CH_3CH_2OH$$

ethanol
bp $78\ °C$

$$CH_3OCH_3$$

dimethyl ether
bp $-24\ °C$

Ethanol and dimethyl ether differ from each other more than do butane and 2-methylpropane, which contain only carbon–carbon and carbon–hydrogen bonds. For example, the oxygen atom is bonded to a carbon atom and a hydrogen atom in ethanol, whereas it is bonded only to carbon atoms in dimethyl ether. This difference in structure is reflected in the large difference in boiling point between ethanol and dimethyl ether. Some of the ways in which structure affects physical properties are discussed starting on p. 26.

As the number of atoms in a molecule increases, the number of possible structures increases rapidly. For example, the molecular formula $C_3H_6O$ indicates one unit of unsaturation ($C_3H_6O$ contains two fewer hydrogen atoms than $C_3H_8O$). Nine constitutional isomers (each of which contains either a double bond or a ring) result, of which seven are shown on the next page.

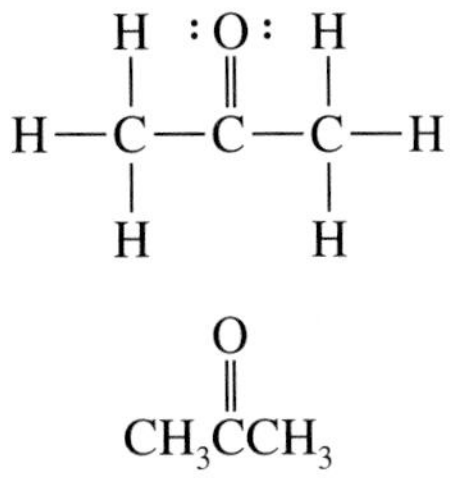

$CH_3CCH_3$
acetone
bp 56.5 °C

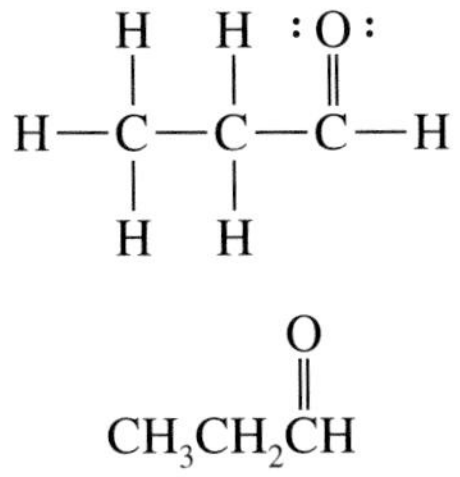

$CH_3CH_2CH$
propanal
bp 48.8 °C

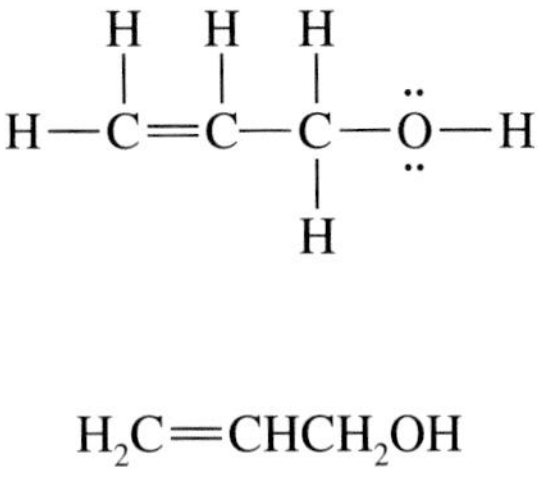

$H_2C=CHCH_2OH$
2-propen-1-ol
allyl alcohol
bp 97 °C

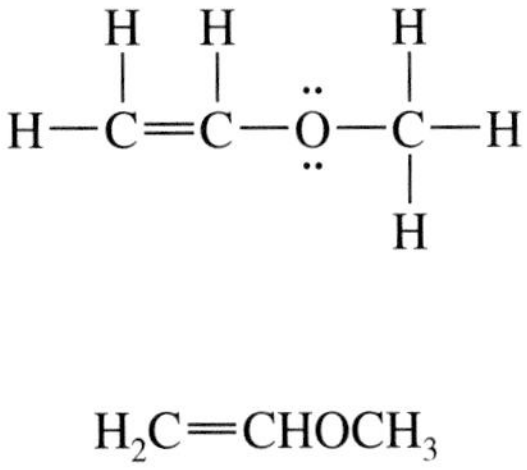

$H_2C=CHOCH_3$
methyl vinyl ether
bp 8 °C

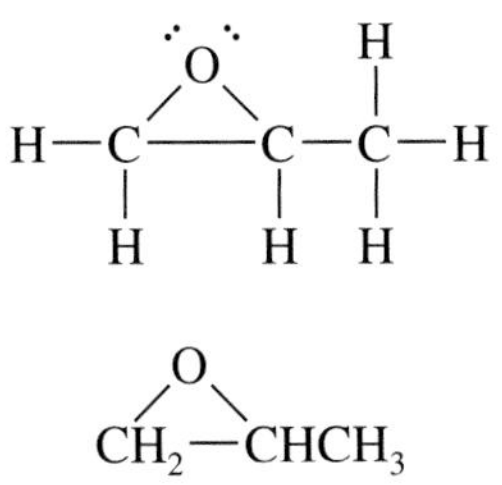

$CH_2-CHCH_3$
2-methyloxirane
bp 35 °C

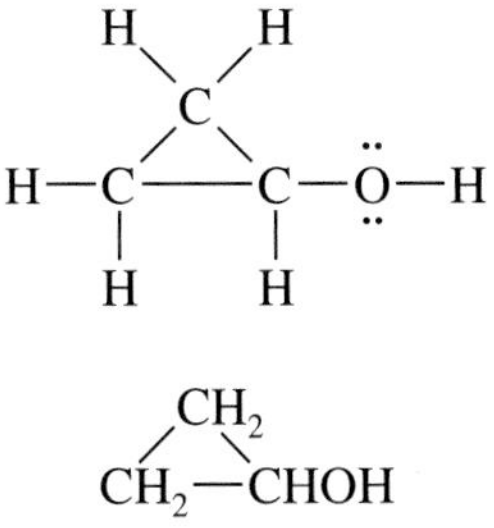

$CH_2-CHOH$
cyclopropanol
bp 100 °C

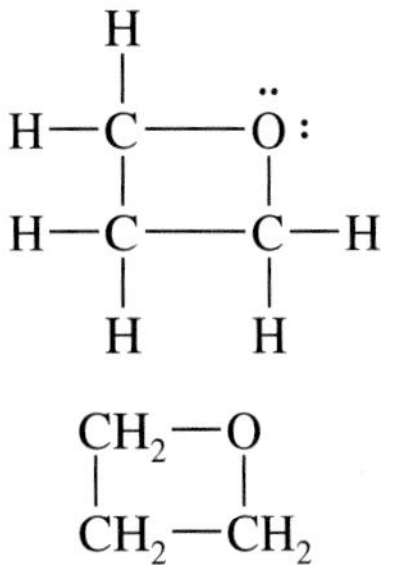

$CH_2-CH_2$
oxetane
bp 47.9 °C

Two more structural formulas can be written for $C_3H_6O$. Both of these have a hydroxyl group, —OH, on one of the carbon atoms of a carbon–carbon double bond. These unstable species, called enols, exist in equilibrium with acetone and propanal.

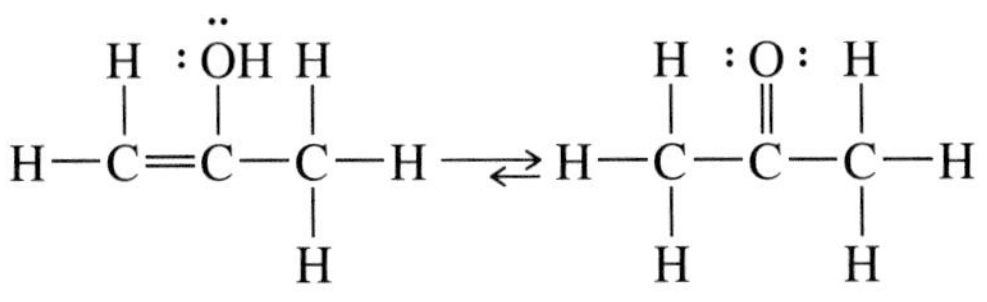

enol of acetone        acetone

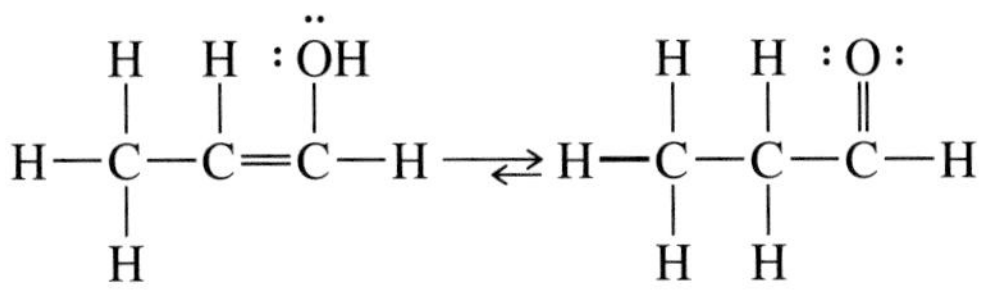

enol of propanal       propanal

We will see later (in Chapters 16 and 17, for example) that such species are important in the reactions of compounds such as acetone and propanal, even though they cannot be isolated and are usually not counted as constitutional isomers.

The formulas shown above demonstrate the large variety of ways in which carbon atoms bond among themselves and with atoms of another element, oxygen in this case. In determining the number of isomers for a given molecular formula, we must consider structures with single and multiple bonds and cyclic structures.

Study Guide
Concept Map 1.4

**Problem 1.14**    Draw structural formulas for the constitutional isomers of $C_3H_7Cl$, $C_3H_8O$, $C_4H_8Cl_2$, and $C_2H_4O$.

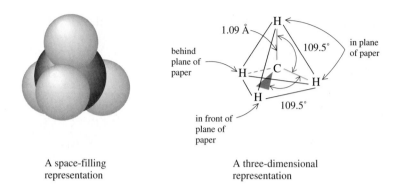

*Figure 1.1*
Three-dimensional representations of methane.

A space-filling representation

A three-dimensional representation

## 1.8  Shapes of Covalent Molecules

### A. Tetrahedral Molecules

The covalent bond, in contrast to the ionic bond (p. 5), has direction in space. If we have more than two atoms covalently bound to each other, we must decide how to arrange them in three dimensions. In 1916, when Lewis postulated that four pairs of electrons form an octet around a central atom, he also suggested that the pairs of electrons are located at the corners of a tetrahedron, as far from each other as possible.

About twenty years after Lewis made his suggestion, experimental values for the distance and angles between atoms in simple covalent molecules were determined using a technique called electron diffraction. In this section we will consider some of the experimental observations about the shapes of covalent compounds. In Chapter 2 we will study a theory of covalent bonding that explains the molecular geometries that have been observed.

Chemists use two parameters, bond lengths and bond angles, to describe the three-dimensional structure of a molecule. A **bond length** is the average distance between the nuclei of the atoms that are covalently bound together. A **bond angle** is the angle formed by the intersection of two covalent bonds at the atom common to both. Electron diffraction experiments have shown that the four hydrogens bonded to the carbon atom in methane, $CH_4$, lie at the corners of a regular tetrahedron, with the carbon atom itself at the center of the tetrahedron. A three-dimensional representation of methane, showing carbon–hydrogen bond lengths of 1.09 Å and H—C—H bond angles of 109.5°, the tetrahedral angle, is shown in Figure 1.1.

Figure 1.1 shows a space-filling model and the molecule in perspective. In the perspective drawing, the carbon atom and two of the hydrogen atoms are shown to be lying in the plane of the paper by the use of ordinary solid lines as bonds. The solid wedge used to attach one of the hydrogen atoms to the carbon atom indicates that this hydrogen atom is coming out of the plane of the paper toward the viewer. The dashed bond to the fourth hydrogen atom indicates that the atom is behind the plane of the paper. You should look at molecular models of tetrahedral carbon atoms so that you can visualize these directional relationships clearly and draw them.

Ethane, $C_2H_6$, has carbon–hydrogen bond lengths and H—C—H bond angles similar to those in methane. It also has a carbon–carbon bond that is 1.53 Å long, as shown in Figure 1.2.

The ammonium ion, $NH_4^+$, has the same shape as the methane molecule. The experimentally determined shape of the ammonia molecule is pyramidal, with bond angles that are close to the tetrahedral angle. The experimental methods for determining structure show only the locations of the nuclei of atoms; therefore, in the

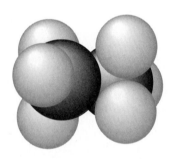

A space-filling representation

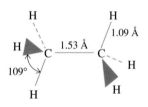

A three-dimensional representation

*Figure 1.2*
Three-dimensional representations of ethane.

ammonia molecule, the nonbonding electrons that occupy the fourth corner of the tetrahedron cannot be seen. Similarly, in the water molecule, pairs of nonbonding electrons occupy two corners of the tetrahedron. The experimentally determined shape of the water molecule is bent, with a bond angle of 104.5°.

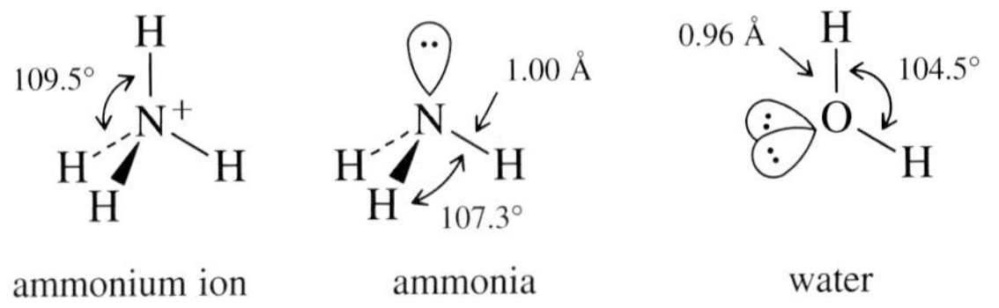

representations of the ammonium ion, ammonia, and water

In fact, whenever there are four pairs of electrons, either in single bonds or as non-bonding electrons, around a central atom from the second row of the periodic table, the bond angles for the species are close to the tetrahedral angle, 109.5°. This generalization allows us to predict the shapes of many organic molecules and ions.

Bond lengths depend on the identities of the atoms being bonded. For example, a carbon–hydrogen bond is longer than a nitrogen–hydrogen bond, which is longer than an oxygen–hydrogen bond. The length of a carbon–carbon single bond in ethane, however, does not vary much from that of carbon–carbon single bonds in other organic molecules. Similarly, the carbon–hydrogen bond length in methane can be considered representative of carbon–hydrogen bond lengths in other organic compounds having such bonds at tetrahedral carbon atoms.

**Problem 1.15** Draw three-dimensional representations of the following compounds.

(a) $CCl_4$     (b) $CHCl_3$     (c) $CH_3Br$     (d) $CH_3CH_2Cl$     (e) $CH_3OH$
(f) $CH_3NH_2$     (g) $CH_3OCH_3$     (h) $CH_3SCH_3$

For the cases for which you have the necessary information, show the approximate bond lengths and bond angles you expect to find in these compounds.

## B. Planar Molecules

Not all carbon compounds contain only tetrahedral carbon atoms in them. Organic molecules containing double or triple bonds have characteristic shapes. Ethylene, $CH_2{=}CH_2$, is a flat molecule with bond angles of approximately 120° around each carbon atom. The length of the carbon–carbon double bond is 1.34 Å, considerably shorter than the carbon–carbon single bond in ethane. The carbon–hydrogen bonds are also a little shorter than the ones in methane and ethane. Two representations of ethylene, one in which the molecule is entirely in the plane of the paper and the other in which the molecule is being viewed from one edge, are shown below.

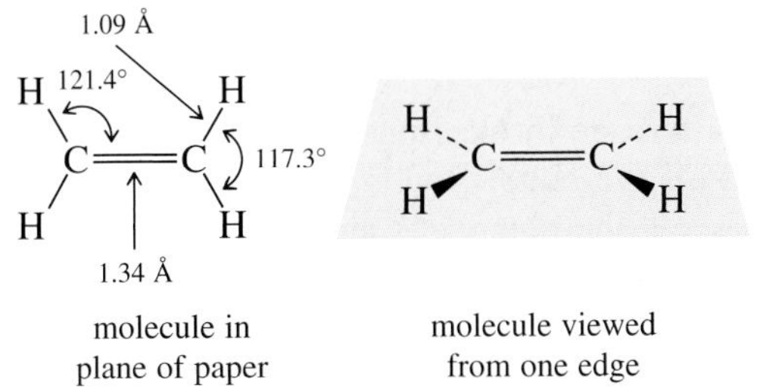

molecule in plane of paper

molecule viewed from one edge

ethylene

## C. Linear Molecules

The triple bond in acetylene, $HC{\equiv}CH$, is 1.20 Å, shorter than the carbon–carbon bond in ethane or ethylene. The carbon–hydrogen bond in acetylene is only 1.06 Å, which again is shorter than that type of bond in ethane and ethylene. Acetylene is linear, with bond angles of 180°.

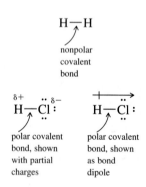

acetylene

The bond lengths in ethylene and acetylene are very close to the bond lengths for double and triple carbon–carbon bonds in other organic compounds. The bond angles are even more general. Whenever carbon is bonded by a double bond to one other atom, bond angles of approximately 120° are observed. If carbon is bonded by a triple bond to another atom, the portion of the molecule containing the triple bond is linear.

**Problem 1.16**    Draw three-dimensional representations of the following compounds. Give approximate bond lengths and bond angles wherever you can.

(a) $CH_2{=}CHCl$    (b) $CH_3\overset{\overset{\displaystyle CH_3}{|}}{C}{=}CH_2$    (c) $CH_3C{\equiv}CH$    (d) $H_2C{=}O$

(e) $CHCl{=}CHCl$    (Two different arrangements of the atoms are possible.)

## 1.9    The Polarity of Covalent Molecules

### A. Polar Covalent Bonds

Electrons in covalent bonds are shared equally when the two bonded atoms are the same but unequally when different elements participate in bonding. For example, in the hydrogen molecule, where electrons are shared equally by two hydrogen atoms, there is a **nonpolar covalent bond.** In hydrogen chloride, the electrons of the covalent bond are drawn closer to the chlorine atom, which is the more electronegative of the two atoms. A covalent bond in which there is unequal sharing of electrons is a **polar covalent bond.** The presence of such a bond is shown by writing partial positive and partial negative charges, $\delta+$ and $\delta-$, over the bonded atoms, pointing out a permanent and directional distortion of the electrons in the bond. The polarized covalent bond also may be shown as a **bond dipole.** The dipole has a negative pole and a positive pole and is represented by the symbol $\longmapsto$, with the point of the arrow drawn toward the more electronegative atom.

H—H

nonpolar
covalent
bond

$\overset{\delta+}{H}{-}\overset{\displaystyle\cdot\cdot}{\underset{\displaystyle\cdot\cdot}{Cl}}\!:$    $\overset{\displaystyle\longmapsto}{H}{-}\overset{\displaystyle\cdot\cdot}{\underset{\displaystyle\cdot\cdot}{Cl}}\!:$

polar covalent    polar covalent
bond, shown    bond, shown
with partial    as bond
charges    dipole

The **polarity** in a bond arises from the different electronegativities of the two atoms participating in the bond. The **electronegativity** of an element was defined by Pauling, who developed the concept, as "the power of an atom in a molecule to attract electrons to itself." The most electronegative elements are in the upper right-hand corner of the periodic table; electronegativity increases as one moves up in a group or to the right in any period. The electronegativities of some elements of interest in organic chemistry are shown in Table 1.2, arranged according to the groups in the periodic table.

The greater the difference in electronegativity between the bonded atoms, the greater is the polarity of the bond. Bonds formed between the metals at the left-hand side of the table and the nonmetals at the extreme right are so polar that we call them ionic. Covalent bonds of all gradations of polarity form between nonmetals and between many metals and nonmetals. For example, the electronegativities of carbon and hydrogen are close enough that carbon–hydrogen bonds do not have much polarity. Bonds of high polarity are those between hydrogen and oxygen, fluorine, chlorine, or nitrogen. Carbon–halogen, carbon–oxygen, and

**Table 1.2   Electronegativity Values for Some Elements**

| 1 | 2 | 13 | 14 | 15 | 16 | 17 |
|---|---|----|----|----|----|----|
| H<br>2.1 | | | | | | |
| Li<br>1.0 | | B<br>2.0 | C<br>2.5 | N<br>3.0 | O<br>3.5 | F<br>4.0 |
| Na<br>0.9 | Mg<br>1.2 | Al<br>1.5 | Si<br>1.8 | P<br>2.1 | S<br>2.5 | Cl<br>3.0 |
| K<br>0.8 | | | | | | Br<br>2.8 |
| | | | | | | I<br>2.4 |

carbon–nitrogen bonds are also polar. Bond polarities contribute significantly to the physical and chemical properties of molecules, and we will refer to them often as we consider chemical reactivity.

**Problem 1.17**   Write $\delta+$ and $\delta-$ to predict the direction of polarization of the covalent bonds indicated by the arrows in the following compounds.

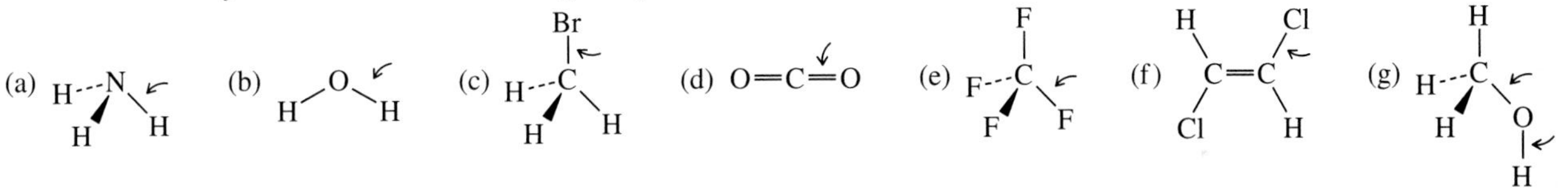

(a) $H\text{---}N$ with H, H     (b) $H$—$O$—$H$     (c) $H\text{---}C$ with Br, H, H     (d) $O{=}C{=}O$     (e) $F\text{---}C$ with F, F, F     (f) $C{=}C$ with H, Cl, Cl, H     (g) $H\text{---}C$ with H, H, O—H

## B. Dipole Moments of Covalent Molecules

The overall polarity of a molecule is measured by its **dipole moment, $\mu$.** For diatomic molecules, those containing two atoms, the bond dipole is also the dipole moment. The dipole moment results from the separation of the centers of positive and negative charge and is given a unit, D, the debye, that is derived from the magnitude of the overall charge and the distance separating the centers of charge. Diatomic molecules in which both atoms are the same have no dipole moment. For the hydrogen halides, the more electronegative the halogen, the larger is the dipole moment for the molecule.

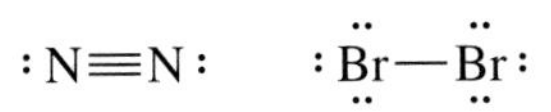

$$:N{\equiv}N: \qquad :\ddot{B}r{-}\ddot{B}r:$$

no dipole moment for diatomic molecules in
which both atoms are the same

$$H{-}\ddot{F}: \quad H{-}\ddot{C}l: \quad H{-}\ddot{B}r: \quad H{-}\ddot{I}:$$

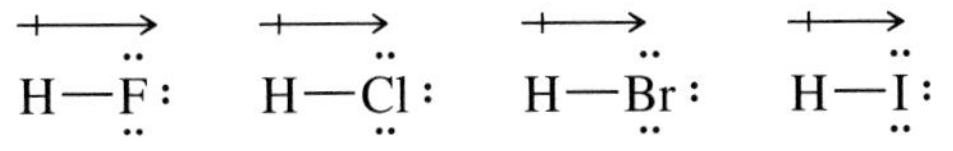

$\mu$, 1.98 D     $\mu$, 1.03 D     $\mu$, 0.78 D     $\mu$, 0.38 D

decreasing dipole moment for hydrogen halides
with decreasing electronegativity of the halogen atom

The overall dipole moment of a molecule containing more than two atoms is the vector sum of the individual bond dipole moments. A molecule may contain polar bonds but have no overall dipole moment if the shape of the molecule is such that the individual bond moments cancel out. In such a case, the average position of the partial positive charges coincides with the average position of the partial negative charges. This is the case for carbon dioxide, $CO_2$, and carbon tetrachloride, $CCl_4$.

carbon dioxide        carbon tetrachloride
μ, 0 D                        μ, 0 D

*cancellation of individual bond moments*
*in symmetrical polyatomic molecules*

In carbon dioxide, bond moments of equal magnitude point in exactly opposite directions. The net result is that the molecule as a whole has no dipole moment, even though the individual bond moments are large. The absence of a measurable dipole moment is one of the reasons chemists have assigned a linear structure to carbon dioxide. The cancellation of the bond moments in carbon tetrachloride is harder to visualize. Because of the symmetrical nature of the tetrahedron, the vector sum of the bond moments of any three carbon–chlorine bonds is exactly equal to and opposite in direction to the bond moment for the fourth carbon–chlorine bond. The overall result is that the molecule as a whole has no permanent dipole, although the individual carbon–chlorine bonds are polarized.

In most molecules, the vector sum of the individual bond moments is not zero, and there is a dipole moment. Water, for example, has a dipole moment with the negative pole at the oxygen atom and the positive pole between the two hydrogen atoms. Similarly, the dipole moment in chloromethane, $CH_3Cl$, has the negative pole at the chlorine atom. Thus the dipole moment of a molecule is related to the polarities of its individual bonds and to its molecular geometry.

water                    chloromethane
μ, 1.84 D               μ, 1.86 D

*molecules with dipole moments resulting from*
*vector sums of individual bond moments*

Study Guide
Concept Map 1.2

**Problem 1.18**    Draw three-dimensional representations for the following compounds. For each compound, predict whether it will have a dipole moment and, if so, the direction of the moment.

(a) $CHCl_3$    (b) $CH_3OH$    (c) $ICl$    (d) $NH_3$    (e) $NH_4^+$    (f) $CH_2Br_2$    (g) $CH_3OCH_3$    (h) $CH_2{=}CCl_2$

## 1.10    Nonbonding Interactions Between Molecules

Covalent compounds may be gases, liquids, or solids. Compounds that have low molecular weights and no permanent dipole moments, such as methane and carbon

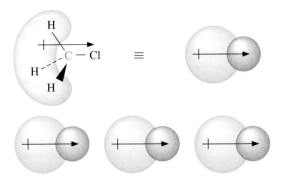

*Figure 1.3*
Schematic representation of dipole–dipole interactions in chloromethane.

dioxide, are gases. The forces that act between such molecules are very weak, so condensation to the liquid or solid phase takes place only at low temperatures or high pressures. The forces that act between molecules are called **intermolecular forces** or **intermolecular nonbonding interactions.** These interactions increase significantly as the molecular weight, and hence the size of the molecules, increases. They also increase with increasing polarity of molecules. Three types of intermolecular forces are important: (1) dipole–dipole interactions, (2) van der Waals interactions, and (3) hydrogen bonding.

In the following sections, a number of organic compounds with a variety of structures are introduced. These compounds appear many times in later chapters, and you will gradually become familiar with their structures, their names, and their reactions. For the moment, you should concentrate only on bond polarities and the interactions that are possible between these kinds of molecules. As always, use the problems as a guide to what you should learn from this chapter.

## A. Dipole–Dipole Interactions

Molecules with dipole moments tend to orient themselves in the liquid and solid phases so that the negative end of one molecule is facing the positive end of another one. The interactions of the permanent dipoles in different molecules are called **dipole–dipole interactions.** Chloromethane has a dipole moment of 1.86 D, which lies along the carbon–chlorine bond. Chloromethane molecules orient themselves so that the positive end of one dipole is pointed toward the negative end of another dipole, as depicted in the schematic representation in Figure 1.3.

An ordinary covalent bond has bond energy in the range from 30 to 130 kcal/mol (p. 64). Dipole–dipole interactions are much weaker, approximately 1 to 3 kcal/mol.

## B. Van der Waals Interactions

Intermolecular forces act to attract even nonpolar molecules to each other, as demonstrated by the physical properties of three nonpolar compounds: methane, $CH_4$; hexane, $C_6H_{14}$; and icosane, $C_{20}H_{42}$.

$$CH_4 \qquad\qquad CH_3CH_2CH_2CH_2CH_2CH_3$$

| methane | hexane |
|---|---|
| molecular weight 16 | molecular weight 86 |
| bp −162 °C | bp 69 °C |
| gas at room temperature | liquid at room temperature |

$$CH_3CH_2CH_2CH_2CH_2CH_2CH_2CH_2CH_2CH_2CH_2CH_2CH_2CH_2CH_2CH_2CH_2CH_2CH_2CH_3$$

icosane
molecular weight 282
mp 36 °C
solid at room temperature

The forces of attraction between the small molecules of methane are so weak that methane exists as a gas at room temperature. Molecules of hexane are larger than those of methane, and the attractive forces between hexane molecules are increased enough that hexane is a liquid. The still larger molecules of icosane attract each other so strongly that the compound is a solid at room temperature.

The weak forces of attraction that exist between nonpolar molecules are called **van der Waals interactions.** These forces are the result of the constant motion of electrons within bonds and molecules, giving rise to effects known as *London dispersion forces.* The motion of electrons creates small distortions in the distribution of charge in nonpolar molecules. A small and momentary dipole results. This small dipole in one molecule can then create a dipole with the opposite orientation, an **induced dipole,** in a second molecule. Although the induced dipoles are constantly changing, the net result is a slight attraction between molecules. As the number of carbon and hydrogen atoms increases, the additive effect of these weak intermolecular forces becomes more significant, as evidenced by the increase in boiling and melting points from methane to hexane to icosane.

Van der Waals interactions can act only through the parts of different molecules that are within a certain distance of each other. The three-dimensional shapes of molecules, therefore, determine to some extent the intermolecular interactions between molecules (Figure 1.4). For example, the isomers butane and 2-methylpropane, both with the molecular formula $C_4H_{10}$, have different boiling points.

2-Methylpropane is a more compact molecule than butane. If you build models of the two compounds, you can see that 2-methylpropane is almost spherical, whereas butane is elongated. The molecules of butane have a greater surface area for interaction with each other than do the molecules of 2-methylpropane. The stronger interactions that are possible for butane are reflected in its boiling point, which is higher than the boiling point of 2-methylpropane.

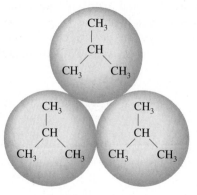

Interactions among
spherical molecules

2-methylpropane
bp −10.2 °C

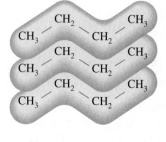

Interactions among
elongated molecules

butane
bp −0.6 °C

*Figure 1.4*
A comparison of
intermolecular interactions for
2-methylpropane and butane.

## C. Hydrogen Bonding

When a hydrogen atom is covalently bonded to a strongly electronegative atom, such as oxygen, fluorine, or nitrogen, the bond is very polar. A hydrogen atom in this situation has a large affinity for nonbonding electrons on other oxygen, fluorine, or nitrogen atoms. The strong interaction that results is called a **hydrogen bond.** A hydrogen bond is a particularly strong dipole–dipole interaction.

**Significant formation of hydrogen bonds is seen only with hydrogen atoms covalently bonded to oxygen, fluorine, and nitrogen.** Even though chlorine is as electronegative as nitrogen, chlorine atoms are not significantly involved in hydrogen bonding. Chlorine, in the third period of the periodic table, has larger atoms than those of fluorine, oxygen, or nitrogen. The partial negative charge on a chlorine atom in a covalent bond is therefore more diffuse and does not attract the hydrogen atom as strongly. Hydrogen atoms bonded to carbon are not usually involved in the formation of hydrogen bonds.

Because water is a major constituent of our surroundings and of living organisms, an understanding of hydrogen bonding is important. The strength of hydrogen bonds is reflected in the physical properties of compounds in which such bonding occurs. Hydrogen bonding, for example, is responsible for the form of ice crystals, resulting in a lower density for ice than for liquid water. (Ice floats in water, a property that has profound consequences in nature. Imagine a lake freezing from the bottom up!) It is also responsible for the unexpectedly high boiling point of water. For example, water, $H_2O$, with a molecular weight of 18, boils at 100 °C, but the much larger and heavier hydrogen sulfide, $H_2S$, with a molecular weight of 34, boils at −62 °C and is a gas at room temperature. In converting liquid water into water vapor, energy must be supplied to break the hydrogen bonds that exist between the molecules of water in the liquid phase so that they can escape into the vapor phase, where they are far enough from each other that little or no hydrogen bonding exists. Figure 1.5 (p. 32) shows the progressive loss of hydrogen bonding as we go from the solid to the liquid to the vapor phase of water.

The attractive forces of hydrogen bonding are usually indicated by a dashed line rather than the solid line used for a covalent bond. The strength of a hydrogen bond involving an oxygen, fluorine, or nitrogen atom ranges from 3 to 10 kcal/mol, making hydrogen bonds the strongest known type of intermolecular interaction. Strong as a hydrogen bond is, though, it is still much weaker than a chemical bond (p. 64), whether ionic or covalent. Hydrogen bonds are interactions *between* molecules and should not be confused with covalent bonds to hydrogen *within* a molecule.

strong hydrogen bond *between* molecules

a covalent bond to hydrogen *within* the molecule, *not* a hydrogen bond

water
molecular weight 18
bp 100 °C

The highly organized structure of ice results from complete hydrogen bonding between molecules of water. Ice is less dense than water near the freezing point. Very few molecules are in the vapor phase.

Some of the hydrogen bonds are broken when ice melts; molecules of water can move closer to one another; some more can escape into the vapor phase.

If enough energy is supplied, more hydrogen bonds are broken; the molecules move even farther apart and occupy a much larger volume in the vapor phase.

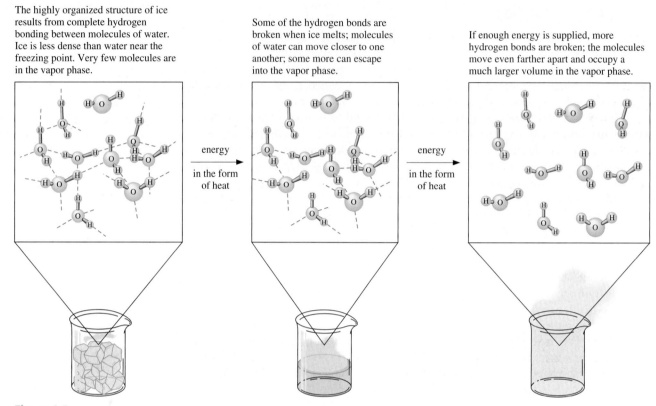

energy in the form of heat

energy in the form of heat

**Figure 1.5**
The solid, liquid, and vapor phases of water.

The effect of hydrogen bonding is demonstrated dramatically by the difference in the boiling points of ethanol, $CH_3CH_2OH$, and its isomer, dimethyl ether, $CH_3OCH_3$. Ethanol is an alcohol. An alcohol resembles a water molecule in which one of the hydrogen atoms has been replaced by a nonpolar group containing carbon and hydrogen atoms. The polar part of the molecule is the hydroxyl group, —OH.

ethanol
molecular weight 46
bp 78 °C

dimethyl ether
molecular weight 46
bp −24 °C

The hydrogen bonding that takes place between molecules of ethanol contributes to a much higher boiling point for this compound than that of dimethyl ether, in which the molecules are held together only by dipole–dipole interactions. In an ether, both hydrogen atoms that are present in water have been replaced by nonpolar groups containing carbon and hydrogen atoms. As a result, an ether does not have a polar hydroxyl group.

**Problem 1.19**    For each set, predict which compound will have the highest boiling point. Indicate, with drawings, the reasoning behind your conclusions.

(a) $CH_3CH_2CH_2CH_2CH_3$,    $CH_3CH_2CH_2CH_2OH$,    $CH_3CH_2OCH_2CH_3$

(b) $CH_3CH_3$,    $CH_3F$,    $CH_3OH$        (c) $CH_3CH_2CH_3$,    $CH_3SH$,    $CH_3OH$

Hydrogen bonding is important in determining the solubility of organic compounds in water. Molecules that can participate in the formation of hydrogen bonds with water will dissolve in water if the nonpolar part of the molecule (the part made up of carbon and hydrogen alone) is not too large. For example, the solubilities in water of three alcohols—ethanol, 1-butanol, and 1-hexanol—vary with the size of the nonpolar portion of their molecules.

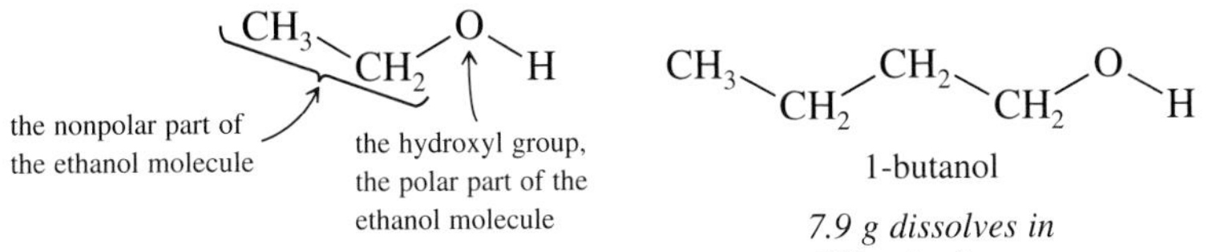

the nonpolar part of
the ethanol molecule

the hydroxyl group,
the polar part of the
ethanol molecule

*ethanol mixes with water
in all proportions*

1-butanol

*7.9 g dissolves in
100 mL of water*

the nonpolar part of
1-hexanol is large in comparison with the polar part

1-hexanol

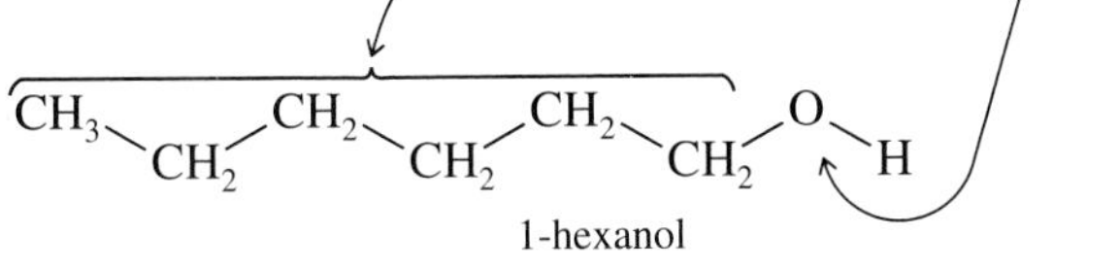

*0.59 g dissolves in
100 mL of water*

Ethanol, in which the polar part of the molecule, the hydroxyl group, is large in proportion to the whole molecule, is completely miscible with water. The hydroxyl group is a much less significant portion of the 1-hexanol molecule, so that compound has low water solubility. The structure of 1-butanol is intermediate between those of ethanol and 1-hexanol, and so is its solubility in water.

The interaction between a dissolved species and the molecules of a solvent is known as **solvation.** Hydrogen bonding is a particularly effective interaction between solute and solvent. The solvation of ethanol by water is shown below.

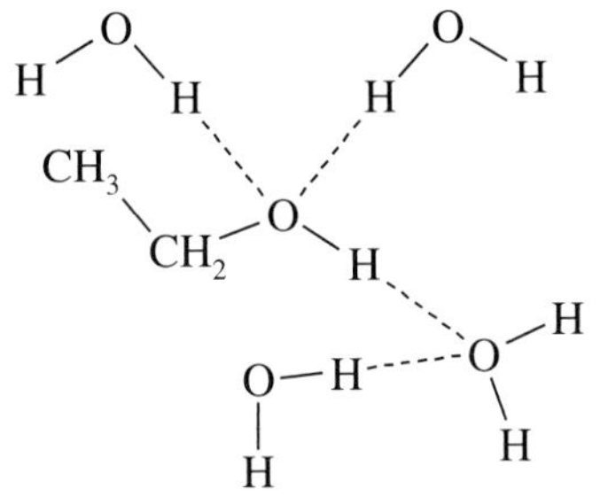

*solvation of ethanol by water*

Solvation is especially important in stabilizing ionic species. For example, when magnesium bromide, $MgBr_2$, is dissolved in water, the magnesium ions and the bromide ions are solvated by water molecules. The **dielectric constant, $\varepsilon$,** of a solvent measures the ability of the solvent to separate ionic charges and, therefore, to dissolve ionic compounds. Water has a high dielectric constant, 78.5 at 25 °C. (In contrast, the dielectric constant of carbon tetrachloride is only 2.2.)

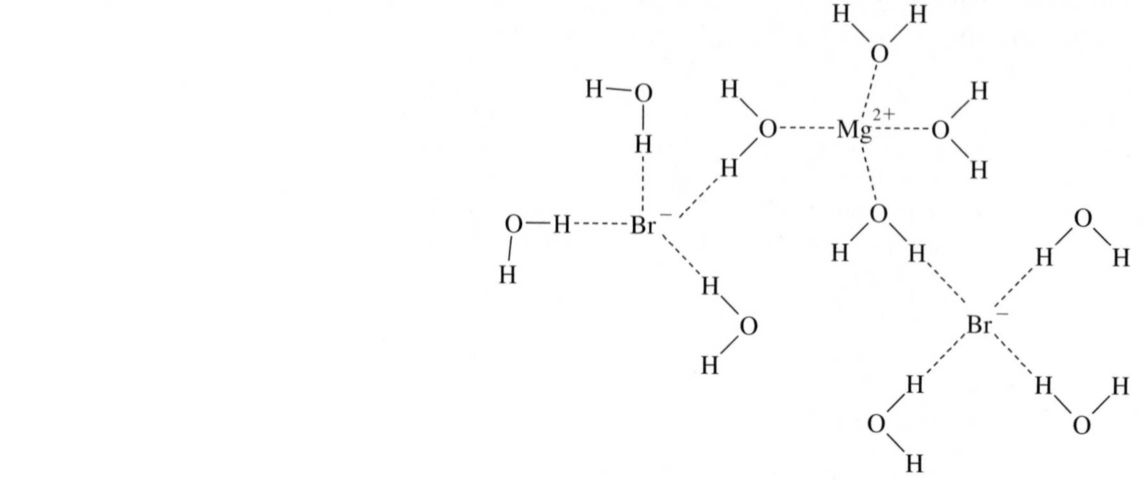

*solvation of magnesium bromide by water*

The interaction between a polar molecule and an ion is a strong one known as **ion–dipole interaction.** The hydroxyl group in water and alcohols, having the large polarity of the oxygen–hydrogen bond and a high concentration of positive charge on the small hydrogen atom, is particularly effective at stabilizing anions.

Compounds that contain oxygen or nitrogen atoms but lack hydrogen atoms bonded to these electronegative elements may participate in hydrogen bonding as **hydrogen-bond acceptors** even though they cannot function as **hydrogen-bond donors.** The relative solubilities in water of diethyl ether, $CH_3CH_2OCH_2CH_3$, and pentane, $CH_3CH_2CH_2CH_2CH_3$, demonstrate this fact.

diethyl ether

*7.5 g dissolves in*
*100 mL of water*

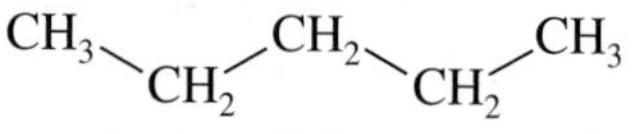

pentane

*0.036 g dissolves in*
*100 mL of water*

Diethyl ether is almost as soluble in water as 1-butanol (p. 33) and much more so than pentane, which is a completely nonpolar compound containing only carbon and hydrogen atoms.

---

**Problem 1.20**    For each pair of compounds, predict which one will be more soluble in water. Indicate, with drawings, the reasons for your conclusions.

(a) $CH_3CH_2Cl$    or    $CH_3CH_2OH$    (b) $CH_3CH_2CH_2OH$    or    $CH_3CH_2CH_2SH$

(c) $CH_3CH_2CH_2CH_2CH_2OH$    or    $HOCH_2CH_2CH_2CH_2CH_2OH$

(d) $CH_3CH_2\overset{\overset{O}{\|}}{C}OH$    or    $CH_3\overset{\overset{O}{\|}}{C}OCH_3$    (e) $CH_3CH_2CH_2Cl$    or    $CH_3CH_2CH_2NH_2$

The hydrogen bond plays an important role in chemistry and biochemistry. It is especially important in interactions between different parts of large molecules where many hydrogen bonds are possible. The shape of an enzyme that catalyzes chemical processes in the human body, for example, is determined to a great extent by hydrogen bonding between distant parts of the large molecule. The two strands of the double helix of deoxyribonucleic acid are held together by a precise pattern of hydrogen bonding believed to be responsible for the transmission of the genetic code. You will learn much more about this when we look at the structures and the chemistry of proteins, carbohydrates, and other giant molecules found in living systems.

⊘ **Study Guide**
**Concept Map 1.6**

## SUMMARY

Compounds are classified as ionic compounds or covalent compounds. An ionic bond consists of electrostatic forces holding together positively and negatively charged ions. A covalent bond arises from the sharing of a pair of electrons by two atoms.

Covalent compounds may be represented by Lewis structures, in which lines are used to represent covalent bonds and dots are used to show nonbonding electrons. Structures are drawn so as to have eight electrons, an octet, around each atom except hydrogen, which can have only two electrons. A few other atoms, such as boron and aluminum from Group 13 of the periodic table, which often has an open shell, or sulfur and phosphorus from the third period, which are, therefore, not limited by the octet rule, are also exceptions. Double or triple bonds may be drawn between atoms in order to create structures in which there are octets around multivalent atoms such as carbon, oxygen, and nitrogen. In Lewis structures, any of the atoms may bear a formal charge, that is, may have in its valence shell more electrons or fewer electrons than are necessary to balance its nuclear charge.

Compounds that have $2n + 2$ hydrogen (or equivalent) atoms for every $n$ carbon atoms are said to be saturated. A saturated compound contains only single bonds between its atoms. Each pair of hydrogen atoms "missing" from a molecular formula means that the compound has a unit of unsaturation. A unit of unsaturation may be a ring or a double bond. Examining the molecular formula of a compound to detect units of unsaturation is one step toward determining what kind of structural formula is possible for the compound.

Not all compounds can be represented satisfactorily by a single Lewis structure. The physical and chemical properties of some species are best accounted for by a hybrid of several Lewis structures, called resonance contributors, that differ from each other in the location of electrons.

Covalent bonds have direction in space. Shapes of covalent molecules are defined by the bond lengths and bond angles. Molecules in which there are four pairs of bonding and/or nonbonding electrons around a central atom have bond angles that approach the tetrahedral angle, 109.5°. Molecules in which the central atom is bonded to three other atoms and there are no nonbonding electrons are planar and have bond angles of 120°. Molecules in which the central atom is bonded to two other atoms are linear with bond angles of 180°.

Covalent bonds may be nonpolar or polar. Nonpolar covalent bonds occur when the atoms that are bonded have similar electronegativities. Polar covalent bonds result from bonding between atoms of differing electronegativities.

A molecule has a dipole moment, $\mu$, when there is a separation of the centers of positive and negative charge in the molecule. Molecules containing only nonpolar covalent bonds are nonpolar and have very small permanent dipole moments or none at all. Molecules containing polar covalent bonds have significant dipole moments unless the shape of the molecule causes the individual bond moments to cancel one another out.

Physical properties such as boiling points, melting points, and solubilities are determined to a large extent by intermolecular nonbonding interactions. Interactions between polar molecules are dipole–dipole interactions, of which hydrogen bonding is the strongest form. Compounds containing hydrogen bonded to oxygen or nitrogen participate in hydrogen bonding both as hydrogen-bond donors and as hydrogen-bond acceptors. Compounds containing oxygen or nitrogen to which no hydrogen is bonded are hydrogen-bond acceptors. The degree of hydrogen bonding possible for a compound strongly influences its boiling point and solubility in water. Interactions between nonpolar molecules are called van der Waals interactions. These interactions depend on fluctuating induced dipoles and therefore on the sizes and shapes of covalent molecules.

## ADDITIONAL PROBLEMS

**1.21** For each of the following compounds, tell whether its bonds are ionic, covalent, or of both kinds. Of the covalent bonds, show which ones have polarity.

(a) $MgF_2$    (b) $SiF_4$    (c) $NaH$    (d) $ClF$    (e) $SCl_2$    (f) $OF_2$
(g) $SiH_4$    (h) $PH_3$    (i) $NaOCH_3$    (j) $CH_3Na$    (k) $Na_2CO_3$
(l) $BrCN$

**1.22** Draw Lewis structures for the following species using lines for covalent bonds and showing any nonbonding electrons that are present. Show formal charges where relevant.

(a) $NF_3$    (b) $AlCl_3$    (c) $CH_3SCH_3$    (d) $CH_3NH_2$    (e) $CH_3CHClCH_3$
(f) $OH^-$    (g) $CH_3CH_2CH_2OH$    (h) $H_2O_2$    (i) $CH_3NHOH$
(j) $CH_3SH$    (k) $SiH_4$    (l) $H_2SO_4$    (m) $HNO_3$

**1.23** Draw each of the following compounds and ions in three dimensions, showing in each case the geometry you would expect to find around the atom that is shaded with color.

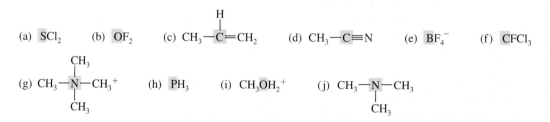

**1.24** Which of the following molecules will have a dipole moment? Show your reasoning by drawing a three-dimensional representation of each molecule showing the direction of the dipole.

(a) $CH_2{=}CHCl$    (b) $CH_3SCH_3$    (c) $CH_3C{\equiv}CCH_3$    (d) $FCBr_3$
(e) $CH_3CH_2OH$    (f) $HC{\equiv}CCl$    (g) $Cl_2C{=}CCl_2$

**1.25** For which of the following compounds will hydrogen bonding among its molecules be important?

(a) $CH_2F_2$    (b) $CH_3CH_2CH_2OH$    (c) $CH_3CH_2OCH_3$    (d) $HOCH_2CH_2CH_2CH_2OH$

(e) $CH_3CH_2CH_2NH_2$    (f) $CH_3CH_2\overset{\displaystyle O}{\overset{\|}{C}}NH_2$    (g) $CH_3CH_2CH_2SH$

**1.26** Which of the following compounds will participate in hydrogen bonding with water? For each compound, indicate whether it will be a hydrogen-bond donor, hydrogen-bond acceptor, or both. Illustrate your reasoning with drawings.

(a) $CH_3CH_2OCH_3$    (b) $CH_3CH_2\underset{\displaystyle CH_2CH_3}{N}CH_2CH_3$    (c) $CH_3\overset{\displaystyle O}{\overset{\|}{C}}NHCH_3$    (d) $CH_3C\equiv N$    (e) $CH_3CH_2\overset{\displaystyle O}{\overset{\|}{C}}OH$

(f) $CH_3CH_2CH_2Cl$    (g) $CH_3NHOH$    (h) $\underset{CH_3}{\overset{CH_3}{>}}C=C\underset{Cl}{\overset{H}{<}}$    (i) $CH_3\overset{\displaystyle O}{\overset{\|}{S}}CH_3$

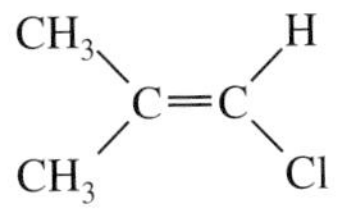

**1.27** Researchers are devoting significant effort to discovering organic compounds that can serve as conductors of electricity and/or as components of electric circuits. Compound A, shown below, has properties that may be useful as a rectifier (a device that controls which way electrons flow through it).

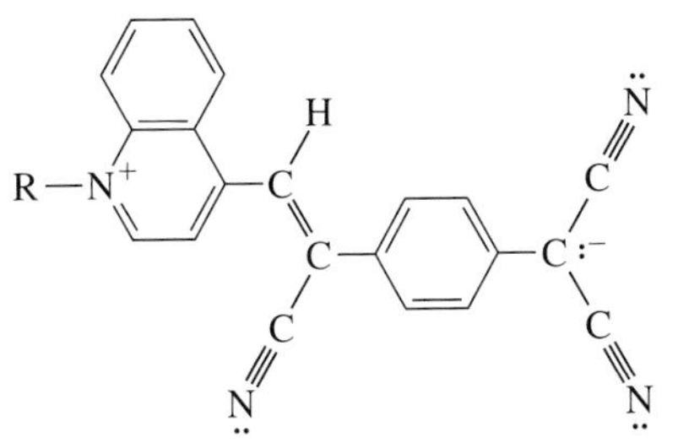

Compound A, where R is a long saturated hydrocarbon chain

(a) Draw a resonance contributor of Compound A that has no separation of charge.
(b) Draw another resonance contributor of Compound A with a separation of charge and filled shells on all atoms.

**1.28** Creatinine is an important end product in the metabolism of proteins. Blood tests for creatinine and urea are used to find out whether kidneys are functioning properly. A new and much more convenient blood test for creatinine has been reported. It depends on three precise intermolecular interactions between creatinine (called the guest molecule) and another molecule specifically designed to hold it (called the host molecule). The two molecules are shown below.

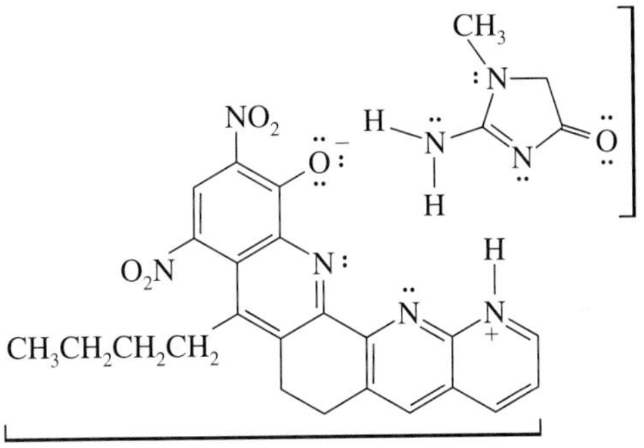

creatinine, the guest molecule

the host molecule

(a)   What are the important intermolecular interactions between the host molecule and the guest molecule? Show these interactions pictorially using the structural formulas given on the previous page.

(b)   In discussing the interactions between the host molecule and creatinine, the authors who reported this work drew two structural representations of species with the same molecular formula as creatinine (shown again in the middle below).

Structure A                    creatinine                    Structure B

What is the relationship of Structure A to creatinine? What is the relationship of Structure B to creatinine?

**1.29** The bond length that is expected for a boron–fluorine single bond is 1.52 Å. The actual bond length observed for boron trifluoride is 1.30 Å. What model do you propose to account for this experimental observation?

**1.30** Acetronitrile, $CH_3C{\equiv}N$, has a large dipole moment (greater than 3 D).

(a)   Draw a Lewis structure for acetonitrile.
(b)   Predict the direction of the dipole moment for acetonitrile.
(c)   The large dipole moment for acetonitrile has been interpreted to mean that the molecule has a major resonance contributor that reflects a separation of charge. Draw such a resonance contributor for acetonitrile.

**1.31** Nitric oxide, NO, has recently been discovered to have many important biological functions in the immune system, in the transmission of nerve impulses, and in the treatment of heart disease. Nitric oxide is a highly reactive compound.

(a)   Draw a Lewis structure and three resonance contributors for nitric oxide.
(b)   Which of the structures that you drew is the major resonance contributor and why? Which one is second best?
(c)   Why is nitric oxide such a highly reactive compound?

**1.32**

(a)   Carbon monoxide, CO, has a much smaller dipole moment than expected. This observation has puzzled chemists and led to much argument about the nature of the bonding in the molecule. The small dipole moment for CO can be rationalized by concluding that there are three important resonance contributors for the molecule. One of them has no formal charges; the other two are polarized in opposite directions. Write these resonance contributors and analyze each one to see whether (1) each atom has an octet around it and (2) the formal charges are in accord with the relative electronegativities of carbon and oxygen.

(b)   Carbon monoxide is highly toxic because it binds tightly to iron in hemoglobin and thus prevents that molecule from binding to and carrying oxygen in the blood. The carbon atom in carbon monoxide binds to iron(II), $Fe^{2+}$. What does this experimental fact indicate about the relative importance of the various resonance contributors to the structure of carbon monoxide?

**1.33** Acetone (propanone) has a boiling point of 56 °C, whereas 1,1,1-trifluoropropanone has a boiling point of 22 °C.

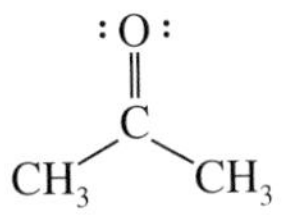

acetone (propanone)     1,1,1-trifluoropropanone
bp 56 °C                bp 22 °C

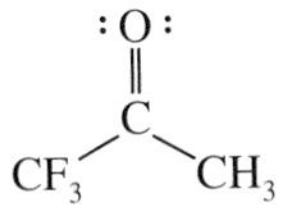

(a) Which of the three major types of intermolecular interactions, hydrogen bonding, dipole–dipole interactions, or van der Waals forces, is the most important in determining the boiling points of these compounds?

(b) How do you rationalize the difference in boiling points for acetone and 1,1,1-trifluoropropanone? Draw a structural formula for each that illustrates your reasoning.

**1.34** Amides are compounds containing this group of atoms.

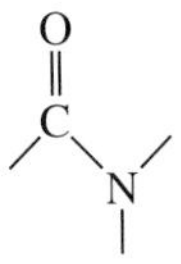

In protein chains, amino acids are held together by amide bonds. The properties of an amide bond are best rationalized on the basis of resonance contributors for an amide. Write three resonance contributors for the amide group and discuss their relative importance.

**1.35** Sulfur dioxide, $SO_2$, has the connectivity OSO. The S—O bond length is $1.43 \pm 0.01$ Å, which is shorter than the expected S—O single-bond length of 1.70 Å and the expected S=O double-bond length of 1.49 Å. The experimental O—S—O bond angle is $121 \pm 5°$. Sulfur dioxide has a significant dipole moment. Write Lewis structures for sulfur dioxide that are in accord with these experimental facts.

**1.36** Compounds of the general structure shown below have been shown to complex selectively with carbonate anion, $CO_3^{2-}$, by intermolecular hydrogen bonding between two distinct pairs of atoms. Draw a Lewis structure for carbonate ion and show, with dashed lines, how it would hydrogen-bond to the compound shown below. An intramolecular hydrogen bond is also important. Show that as well.

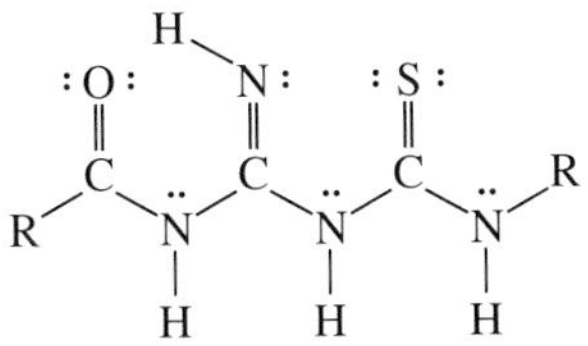

R is a general symbol for groups containing carbon

# Covalent Bonding and Chemical Reactivity

### A Look Ahead

The complete three-dimensional structures of covalent compounds can be described in terms of bond lengths and bond angles that are experimentally determined. For example, two carbon atoms are bonded to each other and to three hydrogen atoms each in ethane, $C_2H_6$, a molecule in which the bond angles around the carbon atoms are 109.5°. In ethylene, $C_2H_4$, the two carbon atoms are held together by a double bond. All the atoms in ethylene lie in a plane, with bond angles close to 120°. The two carbon atoms in acetylene, $C_2H_2$, share a triple bond, and each carbon is bonded to one hydrogen atom in this linear molecule. In general, two carbon atoms can be bonded together in three different ways: with a single bond, a double bond, or a triple bond.

single bond
ethane

double bond
ethylene

triple bond
acetylene

The double bond in ethylene and the triple bond in acetylene are functional groups. A **functional group** is a structural unit consisting of an atom or a group of atoms that serves as a site of chemical reactivity in a molecule.

How can the electronic structure of carbon be used to explain the different kinds of bonding and the different shapes that are observed for various organic molecules? Scientists have put forth many ideas about the nature of chemical bonding and the reasons for chemical reactivity. In this chapter we will examine two systems of ideas, called **molecular orbital theory** and **orbital hybridization,** that are currently used by organic chemists to rationalize the facts known about the compounds of carbon. We will also learn to recognize important functional groups.

**Workbook Exercises**

## 2.1 Introduction

Chemists' ideas about electrons in molecules and how they participate in bonding are derived from quantum mechanics. In the mathematical equations of quantum mechanics, electrons are treated as if they have the properties of both waves and particles. In 1923, the French physicist Louis de Broglie first introduced the idea that the motion of electrons can be described by equations similar to those associated with waves. This idea was further developed independently by Erwin Schrödinger of Austria and Werner Heisenberg of Germany. In the Schrödinger equation, the motion of the electron is related to a set of allowed energy values by mathematical expressions known as **wave equations.** Each energy level allowed for an electron corresponds to a particular solution of the wave equation, called a **wave function.**

Nobel Prizes in Physics were awarded to de Broglie (1929), Heisenberg (1932), and Schrödinger (1933) in recognition of the importance of quantum mechanics to our understanding of the nature of chemical bonding. This way of looking at electrons and bonding is a highly sophisticated and mathematical model; in contrast, the Lewis structure for a covalent compound is a simple and pictorial one. Both systems, however, are human creations; we impose them on the facts of nature as explanations and aids to prediction.

## 2.2 Atomic Orbitals

### A. The Atomic Orbitals of Hydrogen. A Review

According to the prevailing model of atomic structure, the exact location of an electron in an atom cannot be determined. Theories of atomic structure deal with the probability of finding an electron at a given distance and direction from the nucleus. This probability is determined by the wave function for the electron, which is called an **atomic orbital.** An orbital is usually represented by a picture that shows the boundary of the space surrounding the nucleus of an atom within which there is some finite probability (such as 90%) of finding the electron.

The hydrogen atom has only a single proton and a single electron. The two atomic orbitals of lowest energy in the hydrogen atom are the $1s$ and $2s$ orbitals, which are spherically symmetrical (Figure 2.1). There is equal probability of finding the electron at a given distance in all directions from the nucleus. The $2s$ orbital is larger than the $1s$ orbital. An electron in the $2s$ orbital is, on average, farther away from the nucleus than is one in the $1s$ orbital. The electron in the $2s$ orbital is less attracted by the nucleus and is, therefore, in a higher energy level.

The three $2p$ atomic orbitals of the hydrogen atom are oriented at right angles to each other and are usually given the designation of the coordinate axes $2p_x$, $2p_y$, and $2p_z$. Note that a $p$ orbital does *not* have spherical symmetry but does have symmetry about the axis along which it lies (see Figure 2.1). The probability of finding an electron in a $2p$ atomic orbital is greatest in its two lobes, which are on opposite sides of the nucleus. There is zero probability of finding an electron at the nucleus in a $p$ orbital. A region where the probability of finding an electron is zero is called a **node.** A nodal plane passes through the nucleus for $p$ orbitals. The $2s$ orbital also has a nodal region, the surface of a sphere buried within the boundary surface seen in Figure 2.1.

A more precise mathematical way of describing a $p$ atomic orbital is to assign different mathematical signs to the two lobes of the orbital to indicate that the wave function defining the orbital changes sign as it goes through the nodal plane. Unfortunately, positive and negative signs also mean positive and negative charges in the language of chemistry, so pictures with many such signs in them could be confusing. Therefore, in this book the two lobes of a $p$ orbital will be shown with different colored shading to indicate the change in sign that takes place across the nodal plane (Figure 2.2, p. 42).

It is of no importance which lobe of a $p$ orbital is seen as positive and which negative, or which gray and which colored. Such an assignment is purely arbitrary. The change in sign within such an orbital is important because it determines how the orbital will interact with other orbitals.

The characteristics of the $s$ and $p$ atomic orbitals of the hydrogen atom are used as rough approximations of the properties of similar orbitals for elements such as carbon, oxygen, nitrogen, and the halogens. The relative energies and the exact sizes

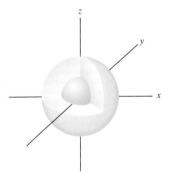

The $1s$ and $2s$ orbitals in three dimensions

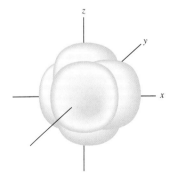

The three $p$ orbitals pointing along the three axes

*Figure 2.1*

**The $s$ and $p$ atomic orbitals of the hydrogen atom.**

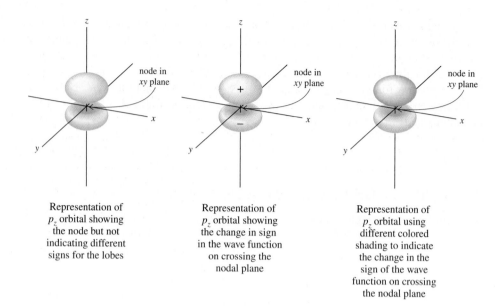

Representation of $p_z$ orbital showing the node but not indicating different signs for the lobes

Representation of $p_z$ orbital showing the change in sign in the wave function on crossing the nodal plane

Representation of $p_z$ orbital using different colored shading to indicate the change in the sign of the wave function on crossing the nodal plane

*Figure 2.2*
Different ways in which a *p* orbital may be represented.

of atomic orbitals change with the charge on the nucleus of an atom and with the number of electrons it has. Nevertheless, chemists use the picture developed for the hydrogen atom when making predictions about bonding in more complex atoms.

## B. Orbitals of Atoms in the Second Row of the Periodic Table

One of the complications that arises whenever there is more than one electron in an atom is the question of **electron spin.** Electrons can have one of two possible spin orientations, symbolized by arrows pointing up or down, ↑ or ↓. Three rules govern the assignment of the most stable electronic configuration to an atom.

1. The Aufbau Principle states that electrons fill atomic orbitals in order according to increasing energy. This means that the 1*s* orbital is filled first, then the 2*s*, then the 2*p*.

**Table 2.1    Assignment of Electrons to Orbitals**

| Element | Electronic Configuration | Electrons in Orbitals | | | | |
|---|---|---|---|---|---|---|
| | | 1s | 2s | 2p | | |
| Hydrogen | $1s^1$ | ↑ | | | | |
| Helium | $1s^2$ | ↑↓ | | | | |
| Lithium | $1s^2 2s^1$ | ↑↓ | ↑ | | | |
| Beryllium | $1s^2 2s^2$ | ↑↓ | ↑↓ | | | |
| Boron | $1s^2 2s^2 2p^1$ | ↑↓ | ↑↓ | ↑ | | |
| Carbon | $1s^2 2s^2 2p^2$ | ↑↓ | ↑↓ | ↑ | ↑ | |
| Nitrogen | $1s^2 2s^2 2p^3$ | ↑↓ | ↑↓ | ↑ | ↑ | ↑ |
| Oxygen | $1s^2 2s^2 2p^4$ | ↑↓ | ↑↓ | ↑↓ | ↑ | ↑ |
| Fluorine | $1s^2 2s^2 2p^5$ | ↑↓ | ↑↓ | ↑↓ | ↑↓ | ↑ |
| Neon | $1s^2 2s^2 2p^6$ | ↑↓ | ↑↓ | ↑↓ | ↑↓ | ↑↓ |

2. The Pauli Exclusion Principle states that two electrons in the same orbital must have opposing spins.

3. Hund's Rule states that when orbitals of equal energy are available, one electron must be assigned to each of those orbitals before any orbital receives two.

Table 2.1 shows the assignment of electrons to the orbitals of the first ten elements in the periodic table. Note that the three $2p$ orbitals are of equal energy. They are said to be **degenerate.**

**Problem 2.1**    The orbitals that are in the next higher energy level above the $2p$ orbitals are the $3s$ orbitals, followed by the $3p$ orbitals. Assign electronic configurations to the elements sodium, magnesium, aluminum, silicon, phosphorus, sulfur, chlorine, and argon.

## 2.3  Overlap of Atomic Orbitals. The Formation of Molecular Orbitals

### A. The Hydrogen Molecule

The Lewis structure of a molecule shows a covalent bond arising from the sharing of a pair of electrons, usually one from each atom. The orbital picture of the covalent bond shows a molecular orbital formed by the overlap of two atomic orbitals, each containing an electron. The directional properties of atomic orbitals are important in determining the extent of overlap that is possible and the geometry of the molecule that results when an atom forms more than one covalent bond.

The hydrogen molecule, arising from the combination of two hydrogen atoms, is the simplest molecule possible. An orbital picture of bonding requires that we imagine two hydrogen atoms, each with one electron in a $1s$ orbital, approaching each other. If the orbitals have the same mathematical sign, they can interact to reinforce each other. They are said to be **in phase.** The interaction of two orbitals of the same mathematical sign results in the formation of a **bonding molecular orbital,** shown schematically in Figure 2.3. The two electrons, one from each hydrogen atom, occupy the bonding molecular orbital. There is an increase in electron density between the two nuclei.

The hydrogen molecule is more stable than the individual hydrogen atoms because the electron of each atom is attracted to the positively charged nucleus of the other atom as well as to its own nucleus. The interaction shown in Figure 2.3 results in a lowering of the overall energy, $E$, of the system. The bonding molecular orbital

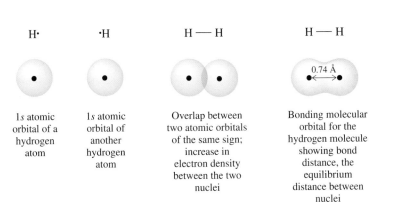

H·    ·H    H — H    H — H

| $1s$ atomic orbital of a hydrogen atom | $1s$ atomic orbital of another hydrogen atom | Overlap between two atomic orbitals of the same sign; increase in electron density between the two nuclei | Bonding molecular orbital for the hydrogen molecule showing bond distance, the equilibrium distance between nuclei |

0.74 Å

*Figure 2.3*
A schematic representation of the formation of a bonding molecular orbital in hydrogen by the overlap of two $1s$ atomic orbitals.

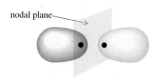

nodal plane

Antibonding molecular orbital for the hydrogen molecule; no overlap between the two atomic orbitals and a nodal plane between the two nuclei

1s atomic orbital of a hydrogen atom

1s atomic orbital of another hydrogen atom; this orbital of opposite sign to the first one

*Figure 2.4*

A schematic representation of the antibonding interaction between 1s atomic orbitals of two hydrogen atoms.

is lower in energy than the two separate atomic orbitals. Note that the bonding molecular orbital of hydrogen is symmetrical about an axis connecting the two nuclei. A bonding molecular orbital with cylindrical symmetry about an internuclear axis is called a **σ molecular orbital** (σ is read "sigma").

The hydrogen molecule is most stable when the nuclei of its atoms are a certain distance apart. If the nuclei are any closer, they repel each other too strongly. If they are farther apart, the atomic orbitals do not overlap enough to form a good covalent bond. The equilibrium distance that allows for the most overlap without excessive nuclear repulsion is called the **bond distance,** which is 0.74 Å, or 0.074 nm (a nanometer is $10^{-9}$ m), for the hydrogen molecule (see Figure 2.3).

There is another possible combination of the two 1s atomic orbitals of hydrogen, one in which the orbitals are of opposite mathematical sign and are said to be **out of phase** with each other. In this combination, there is a node, a region of no electron density, between the two nuclei (Figure 2.4). The lack of electron density between the two nuclei results in repulsion between them. This interaction gives rise to an **antibonding molecular orbital,** a molecular orbital that is of higher energy than the two separate atomic orbitals. The antibonding molecular orbital corresponding to the bonding molecular orbital is called the **σ\* molecular orbital** (σ\* is read "sigma star").

When atomic orbitals are combined to give molecular orbitals, the number of molecular orbitals formed equals the number of atomic orbitals used. Thus the combination of two atomic orbitals gives rise to two molecular orbitals, a bonding molecular orbital and an antibonding molecular orbital. The relationship between the bonding and antibonding molecular orbitals of the hydrogen molecule is shown schematically in Figure 2.5.

The two electrons of the hydrogen molecule are usually in the bonding molecular orbital (see Figure 2.5). Molecular orbitals, like atomic orbitals, can hold only two electrons of opposing spin. When a bonding molecular orbital is occupied by two electrons, a covalent bond results. The bond between the hydrogen atoms in a molecule of hydrogen is called a **σ bond.** Electrons in an antibonding molecular orbital do not contribute to bonding between atoms and, in fact, serve to destabilize a molecule.

A hydrogen molecule is more stable than two separate hydrogen atoms by about 104 kcal/mol. This is the amount of energy required to break the bond in the

*Figure 2.5*

The relative energies and shapes of the σ and σ\* molecular orbitals for the hydrogen molecule.

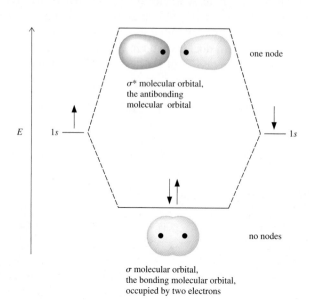

one node

σ\* molecular orbital, the antibonding molecular orbital

E    1s    1s

σ molecular orbital, the bonding molecular orbital, occupied by two electrons

no nodes

hydrogen molecule and separate the two hydrogen atoms. It is known as the **bond dissociation energy** (p. 62) for hydrogen.

## B. The Concept of Antibonding Molecular Orbitals

The model for bonding that creates bonding molecular orbitals from the interaction of atomic orbitals, discussed in the last section, also postulates the existence of an unoccupied higher-energy orbital, the antibonding molecular orbital. What is the usefulness of the concept that such an orbital exists? At the simplest level, we can say that the antibonding molecular orbital is part of a model that allows us to explain why the hydrogen molecule ion, $H_2^+$, exists and the diatomic helium molecule, $He_2$, does not.

The simple Lewis electron dot model for covalent bonding, which defines a bond as a shared pair of electrons, does not allow for the existence of a molecule with a one-electron bond. Yet $H_2^+$ has been detected experimentally, particularly in environments such as the sun and the stars. The picture developed for the hydrogen molecule in Figure 2.5 can help us to rationalize its existence as the result of bonding between a hydrogen atom and a hydrogen ion, a proton.

The orbital diagram in Figure 2.6 shows us that placing a single electron in the bonding molecular orbital, the $\sigma$ orbital, arising from interaction of the $1s$ orbitals of a hydrogen atom with one electron, and proton, the nucleus of a hydrogen atom minus its electron, results in a net lowering of energy, hence a bond.

If we draw a similar orbital diagram for $He_2$, however, we find a different situation (Figure 2.7). Each helium atom has two electrons. The bonding molecular orbital, $\sigma$, can hold only two electrons; therefore, the other two must go into the antibonding molecular orbital, the $\sigma^*$ orbital. Note that the $\sigma^*$ orbital is higher in energy, higher above the level of the $1s$ orbitals of the isolated helium atoms than the $\sigma$ orbital is below them. In other words, there is no gain in stability, no net lowering of energy, when two helium atoms are brought together.

Antibonding molecular orbitals are also useful in describing the interaction of electromagnetic radiation with molecules. This topic will be the subject of Section 2.11.

**Problem 2.2**     Predict, by creating an orbital diagram, whether the $He_2^+$ molecule ion exists or not.

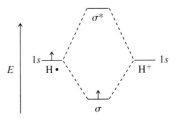

*Figure 2.6*
Orbital diagram showing bonding in the $H_2^+$ molecule ion.

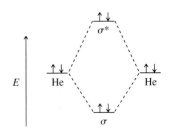

*Figure 2.7*
Orbital diagram showing no net bonding when two helium atoms interact.

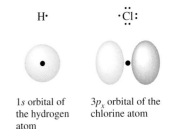

$1s$ orbital of the hydrogen atom        $3p_x$ orbital of the chlorine atom

Bonding molecular orbital of hydrogen chloride, a $\sigma$ orbital

*Figure 2.8*
A schematic representation of the formation of a bonding molecular orbital in hydrogen chloride by the overlap of the $1s$ orbital of hydrogen and the $3p_x$ orbital of chlorine.

## C. Sigma Bonds Involving $p$ Orbitals

Different types of atomic orbitals can overlap to create $\sigma$ bonds. For example, the covalent bond in a molecule of hydrogen chloride is thought to arise from the overlap of a $1s$ orbital of a hydrogen atom with a $3p$ orbital of a chlorine atom. All the $p$ orbitals are equivalent, so any one of them may be used for the bond. The bonding molecular orbital that is formed when the spherically symmetrical $1s$ orbital of the hydrogen atom approaches the $3p_x$ orbital of the chlorine atom along the $x$ axis is shown schematically in Figure 2.8.

The molecular orbital in hydrogen chloride is shaped differently from the atomic orbitals that go into its formation. In particular, the lobe of the $3p_x$ orbital that is not involved in the bonding is greatly contracted. The greatest electron density in the molecular orbital is observed between the two nuclei. The orbital is cylindrically symmetrical with respect to the axis between the two nuclei and is, therefore, a $\sigma$ orbital. The presence of two electrons of opposing spin in the orbital

**Study Guide**
Concept Map 2.1

gives rise to a single bond, also called a $\sigma$ bond, between the hydrogen and chlorine atoms in hydrogen chloride.

In general, a $\sigma$ bond is formed by the overlap of two $s$ atomic orbitals, an $s$ and a $p$ atomic orbital, or two $p$ atomic orbitals. The bonding molecular orbitals that are formed by the overlap of atomic orbitals are called **$\sigma$ orbitals** when they are symmetrical around the axis joining the two atomic nuclei. All $\sigma$ bonds have electron density concentrated between the nuclei of the bonded atoms.

In summary, the overlap of atomic orbitals gives rise to bonding (and antibonding) molecular orbitals. Bonds are formed when bonding molecular orbitals are occupied by pairs of electrons. For the sake of simplicity, we will ignore the antibonding molecular orbitals. In all the compounds we will discuss, the bonding molecular orbitals are filled with electrons, so distinctions between the formation of a bonding orbital and the formation of a bond are not always made.

**Problem 2.3** Draw the atomic orbitals involved and show the bonding molecular orbitals of hydrogen fluoride (HF) and the fluorine molecule ($F_2$).

**Problem 2.4** The following drawings show atomic orbitals approaching each other along the $x$ axis for each atom. In each case, decide whether good overlap and bonding will occur for those orientations of the orbitals.

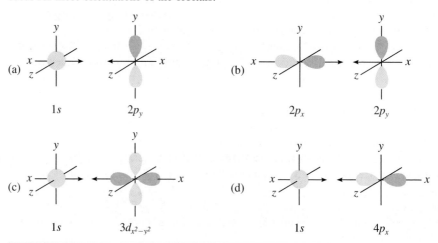

## 2.4 Hybrid Orbitals

### A. Tetrahedral Carbon Atoms

The simplest compound of carbon and hydrogen, methane, $CH_4$, is a symmetrical molecule with four carbon–hydrogen bonds of equal length directed toward the corners of a regular tetrahedron (Figure 1.1, p. 24). A similar tetrahedral arrangement of bonds has been shown experimentally to occur whenever carbon is bonded to four other atoms.

Methane has eight bonding electrons, four from the carbon atom and one from each of the four hydrogen atoms. In the language of molecular orbital theory, methane has four bonding molecular orbitals, each of which holds two of the bonding electrons. The hydrogen atoms (or the electron pairs) repel each other and, therefore, move as far as possible from each other in a tetrahedral arrangement. In this model, more than one molecular orbital contributes to each carbon–hydrogen bond.

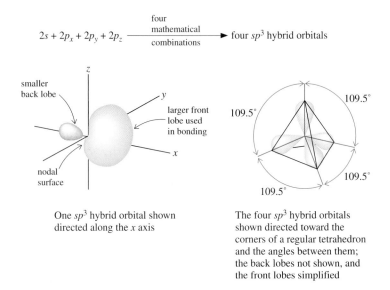

$2s + 2p_x + 2p_y + 2p_z \xrightarrow{\text{four mathematical combinations}}$ four $sp^3$ hybrid orbitals

One $sp^3$ hybrid orbital shown directed along the $x$ axis

The four $sp^3$ hybrid orbitals shown directed toward the corners of a regular tetrahedron and the angles between them; the back lobes not shown, and the front lobes simplified

**Figure 2.9**

Hybridization of the $2s$, $2p_x$, $2p_y$, and $2p_z$ atomic orbitals of carbon to produce $sp^3$ hybrid orbitals.

Chemists also explain the bonding at a carbon atom that is bonded to four other atoms in a tetrahedral geometry by using the concept of **orbital hybridization.** According to this idea, overlap of hybrid atomic orbitals instead of the pure atomic orbitals of atoms forms molecular orbitals. **Hybrid orbitals** are mathematical combinations of atomic orbitals. For a tetrahedral carbon atom, the wave functions for the four atomic orbitals—$2s$, $2p_x$, $2p_y$, and $2p_z$—are combined to create four new hybrid orbitals, shown in Figure 2.9.

The hybrid atomic orbitals are called *$sp^3$ **hybrid orbitals*** to show that they arise from a mathematical combination of one $s$ orbital and three $p$ orbitals, indicated by the superscript 3 on the $p$. *The number of hybrid orbitals generated is always equal to the number of atomic orbitals combined.*

Methane has four identical $\sigma$ bonds; therefore, four atomic orbitals of the central carbon atom must be mixed. Understanding how chemists decide which hybrid orbitals to use is important. They have a problem they are trying to solve. They *know* the connectivity and the shape of the molecule, the structure of which they are trying to rationalize. They *choose* the combination of atomic orbitals that will give them that result. The mathematical combination of one $s$ and three $p$ orbitals was chosen deliberately to create four hybrid orbitals directed to the corners of a tetrahedron because that is the shape of methane. Like the pictures of pure atomic orbitals, the pictures of hybrid orbitals represent regions in space within the boundaries of which there is some finite probability of finding an electron.

Note that the overall shape of a single $sp^3$ hybrid orbital resembles that of a $p$ orbital in that it has a node at the nucleus of the carbon atom. An $sp^3$ orbital, however, does not have two equal lobes, as a pure $p$ orbital does. The larger lobe of an $sp^3$ hybrid orbital extends farther into space from the nucleus of the carbon atom than any of the atomic orbitals do. Thus the hybrid orbitals of carbon are better able to overlap with orbitals of other atoms and to form stronger covalent bonds than the pure atomic orbitals.

According to the orbital hybridization picture, each of the four carbon–hydrogen bonds in methane is formed by the overlap of a $1s$ orbital of a hydrogen atom and an $sp^3$ hybrid orbital of a carbon atom (Figure 2.10). In this representation, the eight bonding electrons of methane are localized in four equivalent carbon–hydrogen bonds directed to the corners of a tetrahedron. Methane is pictured as having

Schematic drawing of the overlap between the $1s$ orbital of the hydrogen atoms and the $sp^3$ hybrid orbitals of the carbon atom, forming four bonds directed toward the corners of a tetrahedron

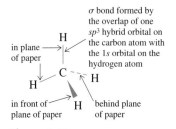

**Figure 2.10**

Formation of methane by the overlap of hybrid $sp^3$ orbitals of a carbon atom with $1s$ orbitals of four hydrogen atoms.

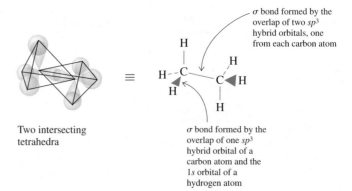

Two intersecting
tetrahedra

$\sigma$ bond formed by the
overlap of two $sp^3$
hybrid orbitals, one
from each carbon atom

$\sigma$ bond formed by the
overlap of one $sp^3$
hybrid orbital of a
carbon atom and the
$1s$ orbital of a
hydrogen atom

*Figure 2.11*
Orbital picture and structural
formula of the ethane
molecule.

four equivalent $\sigma$ bonds with an angle of 109.5°, the tetrahedral angle, between any two.

Organic chemists find the model of orbital hybridization to be useful in accounting for many experimental observations and thus use this way of thinking about bonding in organic compounds. We will use this language in describing the structure and bonding of a number of representative organic compounds in the remainder of this chapter.

A single bond between two carbon atoms is pictured as being formed by the overlap of an $sp^3$ hybrid orbital from each one. For example, ethane, $C_2H_6$, has this type of single bond. Tetrahedral geometry is maintained around each carbon atom. In Figure 2.11, the orbital picture of the bonding in ethane is shown, along with a three-dimensional structural formula showing the geometry of the molecule.

Methane and ethane are the simplest members of a family of compounds known as **alkanes.** Alkanes are a subclass of a much larger group of organic compounds called **hydrocarbons.** Hydrocarbons contain only the elements carbon and hydrogen. The alkane family is characterized by the presence of tetrahedral carbon atoms bonded to hydrogen atoms or to other tetrahedral carbon atoms. All alkanes have the molecular formula $C_nH_{2n+2}$ and are therefore **saturated hydrocarbons** (p. 9).

A group derived from an alkane by removal of one of its hydrogen atoms is known as an **alkyl group.** For example, $CH_3$—, from methane, $CH_4$, is the methyl group; $CH_3CH_2$—, from ethane, $CH_3CH_3$, is the ethyl group. The systematic naming of alkanes and alkyl groups is covered in Chapter 5.

**Problem 2.5**    Chloroform, $CHCl_3$, is another compound that contains a tetrahedral carbon atom. Sketch an orbital picture for it showing how the bonding in the molecule arises. Also draw a three-dimensional representation of the molecule.

**Problem 2.6**    Ethyl fluoride, $C_2H_5F$, is like ethane, except that one hydrogen atom has been replaced by a fluorine atom. Draw an orbital picture and a three-dimensional representation of ethyl fluoride.

### B. $sp^3$–Hybridized Atoms Other Than Carbon

The concept of $sp^3$ hybridization used to describe the bonding of tetrahedral carbon atoms also can be used to describe bonding in ammonia and related organic compounds. Like methane, ammonia has bond angles close to 109.5° (p. 24). Nitrogen, the central atom in ammonia, is bonded to three hydrogen atoms and has a pair of nonbonding electrons. Nitrogen has two electrons in the $2s$ orbital and one electron

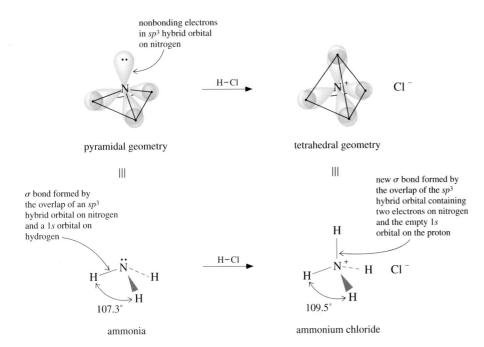

pyramidal geometry

tetrahedral geometry

*Figure 2.12*
Orbital pictures and three-dimensional structural formulas of ammonia and ammonium chloride.

in each $2p$ orbital, for a total of five electrons in its valence shell. Chemists explain the bonding in ammonia by postulating the hybridization of one $2s$ and three $2p$ orbitals to create the four $sp^3$ hybrid orbitals of nitrogen. One of the $sp^3$ hybrid orbitals of nitrogen is filled with a pair of electrons. The other three electrons of the valence shell of the nitrogen atom, along with three electrons from three hydrogen atoms, are in bonding molecular orbitals formed by the overlap of the $sp^3$ hybrid orbitals of the nitrogen atom with the $1s$ orbitals of the hydrogen atoms (Figure 2.12).

When ammonia reacts with an acid such as hydrogen chloride, the nonbonding electrons of nitrogen form a covalent bond with a proton, which has no electrons in the $1s$ orbital. The resulting ammonium ion, $NH_4^+$, contains four equal nitrogen–hydrogen bonds directed to the corners of a tetrahedron (see Figure 2.12). The positively charged ammonium ion is held by an ionic bond to the negatively charged chloride ion, which is also formed in the reaction.

**Problem 2.7** What is wrong with a picture of the covalent bonding in ammonia that locates the nonbonding electrons of the molecule in the $2s$ orbital of nitrogen and postulates that the three nitrogen–hydrogen single bonds are formed by the overlap of $2p$ atomic orbitals of nitrogen with $1s$ atomic orbitals of hydrogen?

There are many organic compounds that contain a nitrogen atom having tetrahedral geometry with respect to its bonds and a pair of nonbonding electrons. One class of such compounds that reacts with acids in the way shown in Figure 2.12 is the **amines,** organic relatives of ammonia in which one or more of the hydrogen atoms of ammonia have been replaced by a group that contains carbon. The simplest amine is methylamine, $CH_3NH_2$. The carbon–nitrogen bond in this molecule is formed by the overlap of an $sp^3$ hybrid orbital of a carbon atom with an $sp^3$ hybrid orbital of a nitrogen atom. The C—N—H bond angle in methylamine is 107°, which is close to the bond angles in ammonia. The carbon–nitrogen bond length is 1.47 Å, which is shorter than the 1.54 Å of a carbon–carbon $\sigma$ bond (Figure 2.13, p. 50). The portion of the methylamine molecule consisting of the nitrogen and two

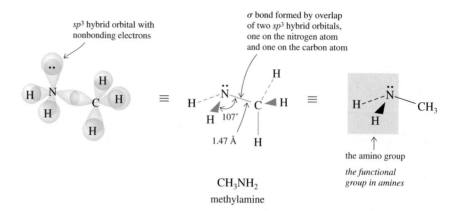

**Figure 2.13**
Orbital picture and three-dimensional structural formula of methylamine.

hydrogen atoms is called the **amino group.** The amino group is a **functional group,** a structural unit that serves as the site of chemical reactivity that is typical of amines.

Another element that is often found in organic compounds is oxygen. The H—O—H bond angle in water has been measured as 104.5°, somewhat smaller than the tetrahedral angle, 109.5°. Oxygen has six valence electrons, two in the $2s$ orbital and four in the three $2p$ orbitals. After hybridization, two of the $sp^3$ hybrid orbitals of oxygen are filled with pairs of nonbonding electrons. The $\sigma$ bonds in water are formed when the two unfilled $sp^3$ hybrid orbitals of the oxygen atom overlap with the $1s$ orbitals of two hydrogen atoms. Just as ammonia reacts with an acid to give an ammonium ion (see Figure 2.12), water reacts with an acid to give a hydronium ion. A hybrid orbital of the oxygen atom that contains nonbonding electrons overlaps with the empty $1s$ orbital of a proton. The molecular orbital pictures for water and the hydronium ion are shown in Figure 2.14.

The fact that the bond angle in the water molecule deviates from the tetrahedral angle is explained by postulating that the two pairs of nonbonding electrons on oxygen occupy more space and repel each other more strongly than do the two pairs of electrons in the oxygen–hydrogen bonds because bonding electrons are constrained in space by their interaction with two nuclei.

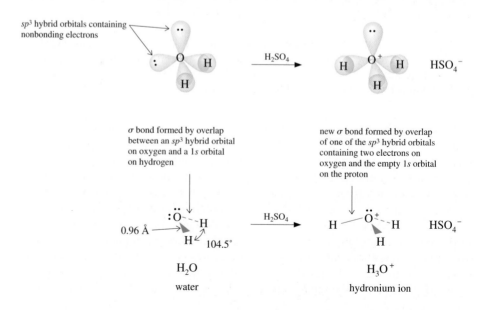

**Figure 2.14**
Orbital pictures and three-dimensional structural formulas for the water molecule and the hydronium ion.

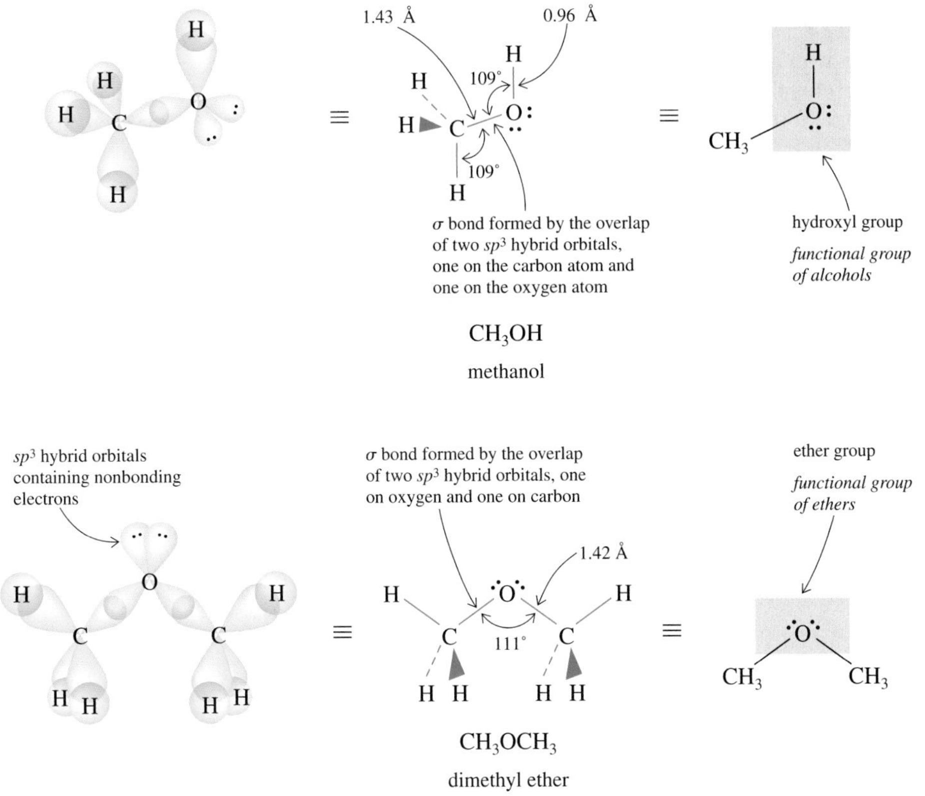

**Figure 2.15**
Orbital pictures and three-dimensional structural formulas of methanol and dimethyl ether.

**Alcohols** are organic relatives of water. In alcohol molecules, there is the same geometry around the oxygen atom as in the water molecule. In alcohols, one of the hydrogen atoms of water has been replaced by a group containing one or more tetrahedral carbon atoms. The functional group in an alcohol is the **hydroxyl group,** —OH. The structure of a typical alcohol, methanol, $CH_3OH$, is shown in Figure 2.15. The carbon–oxygen bond in this compound is formed by the overlap of an $sp^3$ hybrid orbital from each atom.

**Ethers** are also organic relatives of water, in which both hydrogens have been replaced by groups containing carbon atoms. The structure of dimethyl ether is shown in Figure 2.15. Ethers, like water and alcohols, contain an $sp^3$-hybridized oxygen atom with two pairs of nonbonding electrons occupying two of its hybrid orbitals. The oxygen atom and its nonbonding electrons are the functional group in ethers.

In this section we have examined some compounds that contain atoms of carbon, nitrogen, and oxygen that are covalently bonded to other atoms by single bonds with bond angles that are close to tetrahedral. The compounds fall into four functional group classes. Alkanes contain only tetrahedral carbon atoms bonded to hydrogen atoms. Amines are organic relatives of ammonia. Alcohols and ethers are related to water. The orbital pictures and three-dimensional structures of simple representatives of each of these functional group classes are also characteristic of their more complex members.

The reactions of organic compounds are directly related to their structures. Thus, in many cases, we can predict the chemistry of a complex molecule from what we know about the reactivity of a simple one of the same functional group class. The ability to predict reactivity from an inspection of the structure of a compound is important in the study of organic chemistry. By solving the following problems, you can take the first steps in developing this skill.

**Problem 2.8** Figure 2.12 illustrates the change in bonding that takes place when ammonia reacts with an acid. The nonbonding electrons on amines also interact with the proton from an acid in the same way. Draw structural formulas showing what happens when methylamine (Figure 2.13) reacts with hydrogen chloride.

**Problem 2.9** Alcohols and ethers react with acids the same way water does because of the presence of nonbonding electrons on the oxygen atom. Using Figure 2.14 as your guide, write structural formulas that show what happens when methanol and dimethyl ether (Figure 2.15) react with sulfuric acid.

**Problem 2.10** Write three-dimensional structural formulas for the following species. In each case, indicate which types of orbitals overlap to form each bond to the atom that is underlined. For some of the species, you may find it helpful to write Lewis structures first.

(a) $H_2N-\underline{N}H_2$    (b) $\underline{B}F_4^-$    (c) $CH_3-\overset{\overset{\displaystyle CH_3}{|}}{\underset{\underset{\displaystyle CH_3}{|}}{\underline{N}}}\overset{+}{-}CH_3$    (d) $CH_3-\overset{\overset{\displaystyle CH_3}{|}}{\underline{O}}\overset{+}{-}CH_3$

**Problem 2.11** Identify each of the following compounds as belonging to one of these functional group classes: alkanes, amines, alcohols, or ethers.

(a) $CH_3CH_2CH_2CH_2CH_2OH$    (b) $CH_3CH_2CH_2OCH_3$    (c) $CH_3CH_2CH_2CH_3$

(d) $CH_3CH_2NH_2$    (e) $CH_3\underset{\underset{\displaystyle OH}{|}}{C}HCH_2CH_3$    (f) $CH_3CH_2\overset{\overset{\displaystyle H}{|}}{N}CH_2CH_3$

**Problem 2.12** Draw a structural formula illustrating the important electronic features of the functional group in each of the compounds in Problem 2.11. Predict which compounds will react with sulfuric acid (see Problems 2.8 and 2.9).

## 2.5 The Orbital Picture for Compounds Containing Trigonal Planar Carbon Atoms

### A. Covalent Bonding in Alkenes

Not all compounds of carbon contain tetrahedral carbon atoms. Ethylene, $CH_2=CH_2$, for example, is a flat molecule in which all six atoms lie in the same plane (p. 25). Experiments show that the H—C—H and H—C—C bond angles are close to 120° and that the carbon–carbon bond length is 1.34 Å, shorter than the carbon–carbon bond in ethane. Ethylene is the simplest member of the family of hydrocarbons called **alkenes.** The functional group of the alkenes is the carbon–carbon double bond. In ethylene, each carbon atom is bonded to three other atoms, all lying in a plane. Such a carbon atom is known as a **trigonal planar carbon atom.** Other atoms, such as boron in boron trifluoride, $BF_3$, are also trigonal planar.

Note that here, as in the case of methane, chemists start with their knowledge of the connectivity and the geometry of ethylene to decide what type of hybrid orbitals to create. There are three atoms bonded to each carbon atom and no nonbonding electrons to be accommodated. Therefore, only three hybrid orbitals are needed to form the carbon and hydrogen framework of the molecule. The molecule is planar with bond angles of about 120°. Three new $sp^2$ **hybrid orbitals** with this geometry are created when two of the $2p$ orbitals of carbon are combined with the $2s$ orbital.

$\overset{\diagdown}{\diagup}C=C\overset{\diagup}{\diagdown}$

carbon–carbon double bond

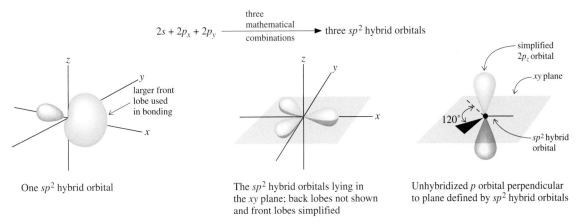

$$2s + 2p_x + 2p_y \xrightarrow[\text{combinations}]{\substack{\text{three} \\ \text{mathematical}}} \text{three } sp^2 \text{ hybrid orbitals}$$

larger front lobe used in bonding

One $sp^2$ hybrid orbital

The $sp^2$ hybrid orbitals lying in the $xy$ plane; back lobes not shown and front lobes simplified

simplified $2p_z$ orbital

$xy$ plane

$sp^2$ hybrid orbital

Unhybridized $p$ orbital perpendicular to plane defined by $sp^2$ hybrid orbitals

*Figure 2.16*
Formation of the $sp^2$ hybrid orbitals of a trigonal carbon atom.

The shape of an $sp^2$ orbital is similar to that of an $sp^3$ hybrid orbital, but the spatial orientation is quite different. The three $sp^2$ hybrid orbitals lie in a plane, are directed to the corners of an equilateral triangle, and have angles of 120° between them. The third $2p$ orbital of an $sp^2$-hybridized carbon is unhybridized. This orbital retains its shape as an atomic orbital and is perpendicular to the plane defined by the three $sp^2$ hybrid orbitals (Figure 2.16).

The skeleton of ethylene is made up of the $\sigma$ bonds between the various atoms. One carbon–carbon bond in ethylene is formed by the overlap of $sp^2$ hybrid orbitals from each carbon atom. The carbon–hydrogen bonds are created by the overlap of $sp^2$ hybrid orbitals of the carbon atoms with $1s$ orbitals of the hydrogen atoms. Each carbon atom contributes three of its four valence electrons to these bonds (Figure 2.17).

The fourth valence electron of each carbon atom is in the unhybridized $2p$ orbital. Two $p$ atomic orbitals on adjacent atoms can interact to create two new molecular orbitals, the $\pi$ molecular orbitals ($\pi$ is read "pi"). Two $p$ orbitals next to each other can be oriented so that lobes of the same sign (same color) are on the same

overlap of two $sp^2$ hybrid orbitals

overlap of $1s$ orbital of hydrogen with $sp^2$ hybrid orbital of carbon

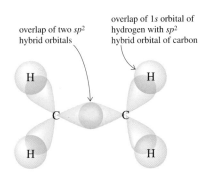

Schematic drawing of the overlap between the $1s$ orbitals of the hydrogen atoms and the $sp^2$ hybrid orbitals of the carbon atoms forming the C—H $\sigma$ bonds of ethylene. The C—C $\sigma$ bond is formed by overlap of two $sp^2$ hybrid orbitals.

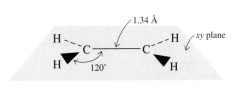

1.34 Å

$xy$ plane

120°

$\sigma$ bond skeleton of ethylene lying in $xy$ plane.

*Figure 2.17*
Skeleton of $\sigma$ bonds in ethylene.

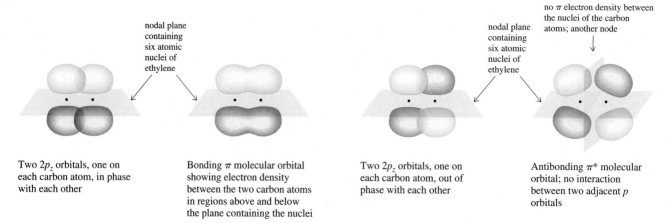

nodal plane containing six atomic nuclei of ethylene

nodal plane containing six atomic nuclei of ethylene

no $\pi$ electron density between the nuclei of the carbon atoms; another node

Two $2p_z$ orbitals, one on each carbon atom, in phase with each other

Bonding $\pi$ molecular orbital showing electron density between the two carbon atoms in regions above and below the plane containing the nuclei

Two $2p_z$ orbitals, one on each carbon atom, out of phase with each other

Antibonding $\pi^*$ molecular orbital; no interaction between two adjacent $p$ orbitals

**Figure 2.18**
Molecular orbital picture of the $\pi$ bond in ethylene.

side of the nodal plane. In this case, the lobes that bear the same mathematical sign interact to reinforce each other. The two orbitals are in phase and combine to form a **bonding $\pi$ molecular orbital.** If, on the other hand, the $p$ orbitals are located so that lobes of opposite sign (differing colors) are on the same side of the nodal plane, they are out of phase and no bonding takes place. In this situation, there is a node (a region of no electron density) between the nuclei of the two atoms, and an **antibonding $\pi^*$ molecular orbital** ($\pi^*$ is read "pi star") results.

When the two carbon atoms in ethylene form a $\sigma$ bond, their $p$ atomic orbitals are close enough that their parallel lobes interact as shown in Figure 2.18. Two electrons, one from each carbon atom, are in the bonding $\pi$ molecular orbital, creating a $\pi$ bond between the two carbon atoms. Thus the double bond in ethylene and other alkenes consists of a $\sigma$ bond and a $\pi$ bond.

Each $p$ atomic orbital has two lobes and a node at the nucleus. The $\pi$ molecular orbital, because it is created by the side-to-side overlap of $p$ orbitals, also has two lobes and a nodal plane. Thus in a $\pi$ bond there is some finite probability of finding electrons in lobes above and below the plane of the molecule. Figure 2.18 shows that a $\pi$ bond does not have the cylindrical symmetry of a $\sigma$ bond.

## B. Covalent Bonding in Carbonyl Compounds. Aldehydes and Ketones

A number of organic compounds have $\pi$ bonds between carbon and oxygen atoms. An important structural unit in such compounds is the carbonyl group, which contains a carbon–oxygen double bond.

$$\begin{array}{c} \diagdown \\ \diagup \end{array} C{=}O$$

Compounds in which there is a carbonyl group are divided into different functional group classes depending on other groups or atoms that are bonded to the carbon atom of the carbonyl group. In **aldehydes,** for example, the functional group is the carbonyl group bonded to at least one hydrogen atom. The functional group of **ketones** is the carbonyl group bonded to two carbon atoms. The bonding in acetaldehyde, an aldehyde, and in acetone, a ketone, is shown in Figure 2.19.

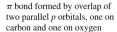

π bond formed by overlap of
two parallel *p* orbitals, one on
carbon and one on oxygen

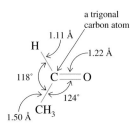

a trigonal
carbon atom

1.11 Å

1.22 Å

118°

124°

1.50 Å

acetaldehyde

*an aldehyde*

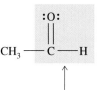

Carbonyl group bonded
to hydrogen

*the functional group
in aldehydes*

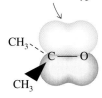

π bond formed by overlap of
two parallel *p* orbitals, one on
carbon and one on oxygen

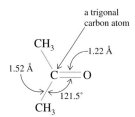

a trigonal
carbon atom

$CH_3$

1.22 Å

1.52 Å

121.5°

$CH_3$

acetone

*a ketone*

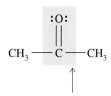

Carbonyl group bonded
to two carbons

*the functional group
in ketones*

*Figure 2.19*
Bonding in acetaldehyde and
acetone.

The carbon atom of the carbonyl group is bonded to three other atoms, all lying in a plane, and therefore is a trigonal planar carbon atom. The bond angles of the carbonyl group are approximately 120°. The carbon atom of the carbonyl group is $sp^2$-hybridized. Its three hybrid orbitals form the skeleton of σ bonds for the carbonyl group. The σ bond between carbon and oxygen is formed by overlap of one of these hybrid orbitals with an $sp^2$ hybrid orbital of oxygen.

Just as with ethylene, we can identify a $2p$ orbital on the carbon atom of the carbonyl group that is perpendicular to the plane defined by the three $sp^2$ hybrid orbitals. Overlap between this orbital and a $2p$ orbital of oxygen results in the π bond between carbon and oxygen. The π bond in a carbonyl compound has the same shape and symmetry as the π bond in ethylene. In the absence of information to the contrary, nonbonding electrons are also placed in hybrid orbitals, $sp^2$ orbitals in the case of the oxygen atom of the carbonyl group.

**Problem 2.13**   Construct the skeleton of σ bonds for acetaldehyde (Figure 2.19) showing the different orbitals that must overlap to form the bonds. In addition, show how the π bond is formed.

## C. Carboxylic Acids and Esters

An oxygen atom attached by a single bond to the carbon atom of a carbonyl group is characteristic of several classes of organic compounds, two of which are carboxylic acids and esters. In **carboxylic acids,** the functional group is the **carboxyl group,** in which the carbonyl group is bonded to a hydroxyl group. In **esters,** the

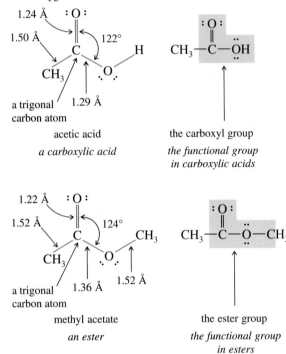

**Figure 2.20**
Structures of acetic acid and methyl acetate.

hydrogen atom on the hydroxyl group of a carboxylic acid has been replaced by a group containing carbon. In other words, both functional groups, carboxyl and ester, contain a carbonyl group bonded to an $sp^3$-hybridized oxygen atom. Figure 2.20 shows the structures of a carboxylic acid and an ester.

**Problem 2.14** Identify the hybridization of all the atoms and the origin of all the bonds in methyl acetate (Figure 2.20).

**Problem 2.15** The functional group classes you have already learned to recognize are alkanes, alkenes, amines, alcohols, ethers, aldehydes, ketones, carboxylic acids, and esters. To which functional group class does each of the following compounds belong?

(a) $CH_3CHCH_3$
      |
      OH

(b) $\underset{CH_3}{\overset{CH_3}{\diagdown}} C = C \underset{H}{\overset{CH_3}{\diagup}}$

(c) $CH_3CH_2\overset{\overset{O}{\|}}{C}H$

(d) $CH_3CH_2\overset{\overset{O}{\|}}{C}CH_2CH_3$

(e) $CH_3CH_2CH_2NH_2$

(f) $CH_3CHCH_2CH_3$
      |
      CH_3

(g) $CH_3CH_2CH_2\overset{\overset{O}{\|}}{C}OH$

(h) $CH_3CH_2O\overset{}{C}HCH_3$
              |
              CH_3

(i) $CH_3CH_2\overset{\overset{O}{\|}}{C}OCH_2CH_3$

**Problem 2.16** Some compounds contain more than one functional group. Identify the functional groups that are present in the following naturally occurring compounds.

(a) CH₃CHCOH  (b) CH₃CHCOH  (c) CH₃C—COH  (d) HOCH₂CHCH

$$\text{(a) }CH_3\overset{\displaystyle OH}{\underset{\displaystyle OH}{C}}HCOH \qquad \text{(b) }CH_3\underset{NH_2}{CH}COH \qquad \text{(c) }CH_3C{-}COH \qquad \text{(d) }HOCH_2\underset{OH}{CH}CH$$

lactic acid          alanine          pyruvic acid          glyceraldehyde

(e) CH₃CH₂CH₂CH₂CH₂CH₂CH₂CH₂CH=CHCH₂CH₂CH₂CH₂CH₂CH₂CH₂COH

oleic acid

**Problem 2.17**   What is the hybridization of the atoms shown in color in each of the following compounds?

$$\text{(a) }CH_3{-}\overset{:\ddot{O}:}{\underset{\displaystyle H}{C}}{-}\overset{}{N}{-}H \qquad \text{(b) }CH_3CH_2\overset{:\ddot{O}:}{C}{-}\ddot{O}{-}\overset{H}{C}{=}\overset{H}{C}{-}H \qquad \text{(c) }CH_3{-}\overset{:\ddot{O}CH_3}{\underset{:\ddot{O}CH_3}{C}}{-}CH_3$$

$$\text{(d) }H{-}\overset{H}{C}{=}\overset{H}{C}{-}\overset{H}{\underset{H}{C}}{-}H \qquad \text{(e) }CH_3{-}\ddot{N}{=}\overset{CH_3}{C}{-}CH_3$$

## 2.6   The Orbital Picture for Linear Molecules

### A. Bonding in Alkynes

**Alkynes** are a class of hydrocarbons in which the functional group is a carbon–carbon triple bond. The simplest member of this class is acetylene, HC≡CH. Experimental measurements show that the four atoms in acetylene lie in a straight line with an H—C—C bond angle of 180° (p. 26). The molecule is linear.

Each carbon atom in acetylene is bonded to only two other atoms. Therefore, we need only two hybrid orbitals on each carbon atom to create the carbon–carbon and carbon–hydrogen bonds that define the connectivity in this molecule. A carbon atom with only two hybrid orbitals is said to be *sp*-hybridized. Two **sp hybrid orbitals** are formed by a combination of one 2s orbital and one 2p orbital. These hybrid orbitals of carbon point away from each other along a straight line. The two carbon atoms of acetylene are joined by the overlap of one *sp* hybrid orbital from each atom. The carbon–hydrogen σ bond results from the overlap of the *sp* hybrid orbital of carbon with the 1s orbital of hydrogen (Figure 2.21, p. 58).

Two of the valence electrons of each carbon atom in acetylene are used to form the σ bonds. Each carbon atom also has two unhybridized 2p orbitals at right angles to each other, with one electron in each. As is the case for ethylene, the p orbitals of the carbon atoms overlap to form bonding π molecular orbitals. Because there are two sets of p orbitals, two sets of π molecular orbitals are formed. Each bonding π molecular orbital has two electrons in it, giving rise to a π bond. Thus the triple bond in acetylene (and other alkynes) consists of a σ bond and two π bonds. Two views of the π bonds are shown in Figure 2.22 (p. 58). A side view of the molecule shows that there is electron density above and below the line defined by the nuclei,

—C≡C—

carbon–carbon triple bond

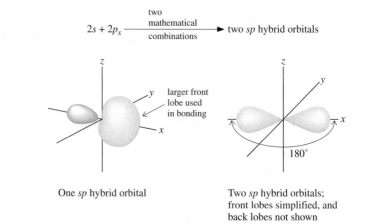

$$2s + 2p_x \xrightarrow[\text{combinations}]{\text{two mathematical}} \text{two } sp \text{ hybrid orbitals}$$

One *sp* hybrid orbital

Two *sp* hybrid orbitals; front lobes simplified, and back lobes not shown

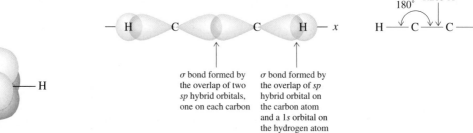

σ bond formed by the overlap of two *sp* hybrid orbitals, one on each carbon

σ bond formed by the overlap of *sp* hybrid orbital on the carbon atom and a 1*s* orbital on the hydrogen atom

*Figure 2.21*
**Orbital hybridization and the σ bond skeleton of acetylene.**

Two perpendicular *p* orbitals on each carbon atom, each one parallel to one on the other carbon atom

Side view of the two π bonds perpendicular to each other

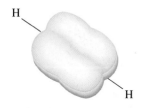

View of the two π bonds showing toroidal (doughnut-shaped) symmetry

*Figure 2.22*
**The π bonds in acetylene.**

(➔) **Study Guide**
**Concept Maps 2.2, 2.3**

as well as in front and behind it. A view from the end of the molecule shows that there is electron density all around the axis of the molecule. There is a nodal axis along the line between the nuclei.

## B. Bonding in Nitriles

**Nitriles** are organic compounds that contain a triple bond between a carbon and a nitrogen atom. The functional group in nitriles is the cyano group, —C≡N. The carbon atom of the cyano group is bonded to nitrogen and to one other atom. A typical nitrile is acetonitrile, $CH_3C≡N$ (Figure 2.23).

The carbon atom and the nitrogen atom in the cyano group are *sp*-hybridized, and the carbon–nitrogen triple bond is formed in the same way that the carbon–carbon triple bond in acetylene is. The cyano triple bond consists of a σ bond, formed by the overlap of one *sp* hybrid orbital from carbon and one from nitrogen, and two π bonds, formed by the overlap of two *p* orbitals on each atom. One of the *sp* orbitals of carbon overlaps with the $sp^3$ orbital of the tetrahedral carbon atom to which it is bonded. The nonbonding electrons of nitrogen occupy its second *sp* orbital.

We now have had an overview of many of the major functional group classes. They are summarized in a table that appears inside the back cover of this book.

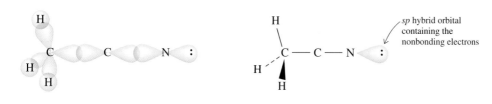

The orbital picture of acetonitrile showing the formation of the σ bonds

σ bond skeleton of acetonitrile

*sp* hybrid orbital containing the nonbonding electrons

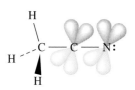

*p* orbitals used to form π bonds

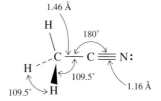

1.46 Å

180°

109.5°

109.5°

1.16 Å

*Figure 2.23*
The orbital picture and structure of acetonitrile.

**Problem 2.18**   Identify the functional group class to which each of the following compounds belongs.

(a) $CH_3CH_2C{\equiv}CH$

(b) $CH_3C{=}CHCH_2CH_2CH_3$
      $\quad\ \ |$
      $\quad\ CH_3$

(c) $CH_3CH_2\overset{\displaystyle O}{\overset{\displaystyle \|}{C}}CH_2CH_3$

(d) $CH_3CH_2CH_2\overset{\displaystyle O}{\overset{\displaystyle \|}{C}}OCH_3$

(e) $CH_3CH_2CH_2C{\equiv}N$

(f) $CH_3CH_2CH_2\overset{\displaystyle O}{\overset{\displaystyle \|}{C}}H$

(g) $CH_3CH_2CHCOH$
      $\qquad\quad |$
      $\qquad\ CH_3$
      (with $\overset{\displaystyle O}{\overset{\displaystyle \|}{\phantom{C}}}$)

(h) $CH_3CH_2CHCH_2CH_3$
      $\qquad\quad |$
      $\qquad\ NH_2$

**Problem 2.19**   How many σ and how many π bonds are present in each of the following molecules?

(a) $H{-}\overset{\displaystyle |}{\underset{\displaystyle H}{C}}{=}O$

(b) $CH_3CH_2OH$

(c) $H{-}C{\equiv}C{-}C{\equiv}C{-}H$

(d) $H{-}\overset{\displaystyle |}{\underset{\displaystyle H}{C}}{=}\overset{\displaystyle |}{\underset{\displaystyle H}{C}}{-}C{\equiv}N$

(e) $CH_3{-}\overset{\displaystyle O}{\overset{\displaystyle \|}{C}}{-}OCH_3$

(f) $CH_3CH_2CH_3$

**Problem 2.20**   Draw a detailed picture of the orbitals involved in the bonding in hydrogen cyanide, HCN.

**Problem 2.21**   Draw a detailed picture of the orbitals involved in the bonding in propyne, $CH_3C{\equiv}CH$.

## 2.7   Covalent Bond Lengths and Their Relation to Orbital Hybridization

### A. Bond Lengths in Hydrocarbons

Bond lengths and bond angles for a number of typical covalent compounds are given in Chapter 1 and in the earlier sections of this chapter. The values are remarkably constant for each particular kind of bond and are determined not only by the atoms involved but also by the hybridization of orbitals of those atoms.

Electrons in $2s$ atomic orbitals are closer to the nucleus than are electrons in $2p$ atomic orbitals. Orbital hybridization affects bond length because a hybrid orbital with a greater percentage of $s$ character does not extend as far from the nucleus as does a hybrid orbital with less $s$ character. The percentage of $s$ character is the ratio of the number of $s$ orbitals to the total number of orbitals used in hybridization. For example, in ethane the $sp^3$-hybridized carbon atoms have one-fourth, or 25%, $s$ character and three-fourths, or 75%, $p$ character. For an $sp$-hybridized carbon atom, such as is found in acetylene, the hybrid orbitals have 50% $s$ character.

The more $s$ character a hybrid orbital has, the closer to the nucleus and the more tightly held are its electrons. An orbital with a large percentage of $s$ character, therefore, forms a shorter $\sigma$ bond than one with a small percentage of $s$ character. The lengths of carbon–carbon and carbon–hydrogen bonds in propane, propene, and propyne are compared in Table 2.2.

The carbon–carbon single bond between $sp^3$-hybridized carbon atoms in propane is longer (1.53 Å) than the carbon–carbon single bond between an $sp^3$-hybridized carbon atom and an $sp^2$-hybridized carbon atom in propene (1.50 Å). The comparable bond between $sp^3$-hybridized and $sp$-hybridized carbon atoms in propyne is the shortest of all (1.46 Å). A similar trend is seen for carbon–hydrogen bonds as the hybridization of the carbon atom changes: $sp^3$, 1.10 Å; $sp^2$, 1.08 Å; and $sp$, 1.06 Å.

This series of compounds also demonstrates the effect of $\pi$ bonding between two carbon atoms. The carbon–carbon double bond in propene is shorter (1.34 Å) than the carbon–carbon single bond in propane (1.53 Å). Propyne, with two $\pi$ bonds between two of the carbon atoms, has the shortest carbon–carbon bond length of all (1.21 Å). Carbon atoms are held closer together by multiple bonds than by single bonds. The shorter bonds are also stronger bonds. This relationship is presented in detail in Section 2.8 (p. 62).

These data are given here to illustrate trends and principles. You should not attempt to memorize these numbers but should understand why they have the relative values that they do.

### B. Orbital Hybridization and the Electronegativities of Carbon Atoms

Chemists have rationalized many experimental observations of the properties and reactivity of compounds containing carbon atoms with tetrahedral, trigonal planar, or linear geometries by suggesting that these carbon atoms have different electronegativities. This argument is based directly on the differing percentages of $s$ character in the hybrid orbitals used in bonding by different types of carbon atoms. An electron in an $s$ orbital is, on average, closer to the nucleus than an electron in a $p$ orbital. The hybrid orbitals of the carbon atoms in acetylene, with 50% $s$ character, hold the electrons of

| | | Bond | Orbitals Used |
|---|---|---|---|
| **Compound** | **Bond** | **Length, Å** | **for Bonding** |
| 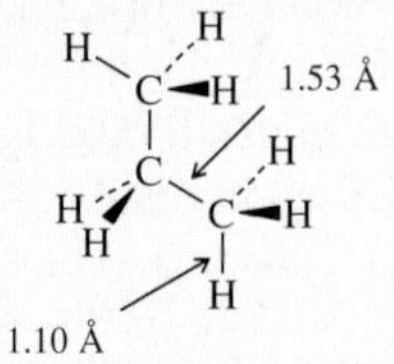 propane | C—C | 1.53 | C, $sp^3$; C, $sp^3$ |
| | C—H | 1.10 | C, $sp^3$; H, $s$ |
| 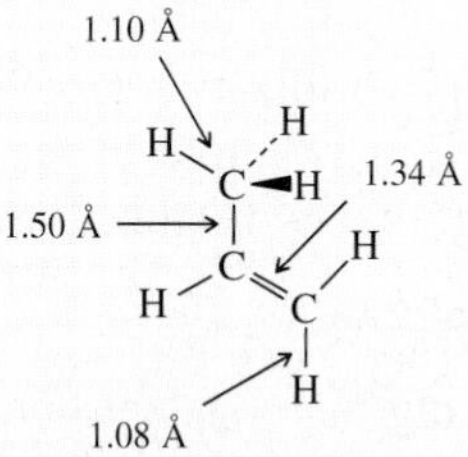 propene | C—C | 1.50 | C, $sp^3$; C, $sp^2$ |
| | C=C | 1.34 | C, $sp^2$; C, $sp^2$ |
| | C—H | 1.10 | C, $sp^3$; H, $s$ |
| | C—H | 1.08 | C, $sp^2$; H, $s$ |
| 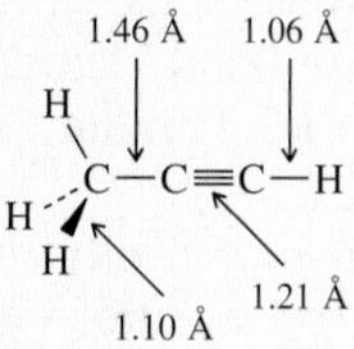 propyne | C—C | 1.46 | C, $sp^3$; C, $sp$ |
| | C≡C | 1.21 | C, $sp$; C, $sp$ |
| | C—H | 1.10 | C, $sp^3$; H, $s$ |
| | C—H | 1.06 | C, $sp$; H, $s$ |

**Table 2.2   Orbital Hybridization and Bond Lengths in Propane, Propene, and Propyne**

the bond more tightly to the nucleus of the carbon atom and are thus more electronegative than the carbon atoms in ethane, whose hybrid orbitals have only 25% $s$ character. This kind of reasoning is used many times in this book to explain the different chemical properties of compounds composed of different types of carbon atoms.

## C. Bond Lengths in Organic Compounds Containing Oxygen and Nitrogen

Like carbon–carbon bonds, carbon–oxygen and carbon–nitrogen bonds have lengths related to orbital hybridization and the presence of multiple bonds. Carbon–nitrogen bonds are shorter than comparable carbon–carbon bonds and longer than comparable carbon–oxygen bonds. These differences reflect the decrease in atomic size that occurs from left to right in the second row of the periodic table.

The trends for carbon–carbon bonds are also observed for carbon–nitrogen and carbon–oxygen bonds. The carbon–nitrogen single bond in methylamine is longer (1.47 Å) than the carbon–nitrogen triple bond in acetonitrile (1.16 Å). Similarly, a carbon–oxygen single bond in either methanol or dimethyl ether (1.43 Å) is longer than the carbon–oxygen double bond (1.22 Å) in acetaldehyde. Three different carbon–oxygen bond lengths are seen in methyl acetate (Figure 2.20, p. 56).

It is worthwhile to pay this much attention to the structural details of these few compounds because the types of bonding observed in them are representative of the bonding found in a wide range of organic compounds. The bond lengths for propane are typical of those in all compounds where two or more tetrahedral carbon atoms are bonded to each other. Whenever we see a double or a triple bond between carbon atoms, we can assume that it has properties similar to the bond in propene or propyne, respectively. Similarly, the properties of the functional groups in the simple amines, alcohols, ethers, and carbonyl compounds analyzed here in detail closely resemble the properties of those functional groups in more complex compounds.

## 2.8   Covalent Bond Strengths

### A. Bond Dissociation Energies

Bond strengths, as well as bond lengths, are related to orbital hybridization (p. 47). One measure of bond strength is bond dissociation energy. For example, it takes 104 kcal/mol to break the bond in a hydrogen molecule to get two hydrogen atoms.

$$\text{H—H(g)} \longrightarrow 2\,\text{H}\cdot\text{(g)} \qquad \Delta H° = 104 \text{ kcal/mol}$$

This reaction is **endothermic;** energy is consumed in breaking the covalent bond. The enthalpy (the heat of reaction) of this endothermic reaction is called the **bond dissociation energy, *DH,*** for the hydrogen molecule.

A covalent bond can be broken in two ways. Both electrons of the covalent bond can go to one of the two atoms sharing the bond. Such a bond cleavage is called a **heterolytic cleavage.** Acid–base reactions (Chapter 3) are familiar examples of reactions in which bonds are broken heterolytically.

The covalent bond in the hydrogen molecule is broken homolytically in the reaction shown above. In a **homolytic cleavage** of a bond, one electron of the covalent bond being broken goes to each fragment of the molecule. The **standard bond dissociation energy, *DH°,*** is defined as the change in enthalpy for a reaction in which one specific covalent bond in a molecule is broken homolytically while the reactants and the products are in the standard state (in the gas phase at 1 atm and 25 °C).

Bond dissociation energies have been measured for a variety of bonds involving carbon. Bond dissociation energies for some carbon–hydrogen bonds are shown in Table 2.3 in both kilojoules per mole and kilocalories per mole; kilocalories per mole will be used throughout the rest of this book. The values for bond dissociation energies for carbon–hydrogen bonds in ethylene and acetylene show the extra strength of such bonds at $sp^2$- and $sp$-hybridized carbon atoms as compared with similar bonds at $sp^3$-hybridized carbon atoms. Also, bond dissociation energies may be correlated with bond lengths of carbon–hydrogen bonds in propane, propene, and propyne (p. 60).

| CH$_3$CH$_2$CH$_2$——H | CH$_3$CH=CH——H | CH$_3$C≡C—H |
|---|---|---|
| carbon–hydrogen bond at $sp^3$-hybridized carbon atom; 1.10 Å | carbon–hydrogen bond at $sp^2$-hybridized carbon atom; 1.08 Å | carbon–hydrogen bond at $sp$-hybridized carbon atom; 1.06 Å |
| $DH° = 98$ kcal/mol | $DH° = 108$ kcal/mol | $DH° = 125$ kcal/mol |
| *longest carbon–hydrogen bond; weakest bond* | | *shortest carbon–hydrogen bond; strongest bond* |

Bond dissociation energies for some other types of bonds are also given in Table 2.3. The carbon–carbon triple bond in acetylene is shorter (1.21 Å) and

**Table 2.3  Standard Bond Dissociation Energies for Representative Bonds**

| Bond | $DH°$ (kcal/mol) | $DH°$ (kJ/mol) | Bond | $DH°$ (kcal/mol) | $DH°$ (kJ/mol) |
|---|---|---|---|---|---|
| $CH_3-H$ | 104 | 435 | $HC\equiv C-H$ | 125 | 523 |
| $CH_3CH_2-H$ | 98 | 410 | ⬡—H | 110 | 460 |
| $CH_3CH_2CH_2-H$ | 98 | 410 | | | |
| $CH_3\overset{\underset{\|}{CH_3}}{CH}-H$ | 95 | 397 | ⬡—$CH_2-H$ | 85 | 356 |
| | | | $CH_2=CHCH_2-H$ | 89 | 372 |
| $CH_3\overset{\underset{\|}{CH_3}}{CHCH_2}-H$ | 98 | 410 | $CH_3-CH_3$ | 88 | 368 |
| | | | $CH_2=CH_2$ | 145 | 607 |
| $CH_3\overset{\underset{\|}{\underset{CH_3}{\|}}}{\overset{CH_3}{C}}-H$ | 91 | 381 | $(CH_3)_2C=O$ | 176 | 736 |
| $CH_2=CH-H$ | 108 | 452 | $HC\equiv CH$ | 190 | 795 |

stronger than the carbon–carbon double bond in ethylene (1.34 Å), which in turn is shorter and stronger than the carbon–carbon single bond in ethane (1.54 Å). Note, however, that a double bond is not twice as strong as a single bond, nor is a triple bond three times as strong. The values for the multiple bonds represent the energy necessary to break both the $\sigma$ and the $\pi$ bonds in these compounds. Most important reactions of alkenes and alkynes break the $\pi$ bond but leave the $\sigma$ bond intact. Examples of two such reactions will be shown in the next sections. Such reactions require the input of less energy than the values given in Table 2.3.

## B. Average Bond Energies

By making a variety of experimental measurements on large numbers of organic and inorganic compounds, chemists have arrived at values for average bond energies for different types of covalent bonds. These bond energies differ from bond dissociation energies in that they do not refer to a specific bond in a particular molecule. Instead, they represent average values for types of bonds, taken from data for many different molecules. Average bond energies are useful in estimating the changes in energy that occur in chemical reactions, which always involve the breaking and forming of bonds. Table 2.4 (p. 64) gives average bond energies for single bonds between the elements shown and for some multiple bonds.

Using average bond energies, we can predict whether a reaction is exothermic or endothermic, that is, whether it has a negative or positive enthalpy of reaction. We do this by calculating the energy consumed in breaking bonds and the energy released in forming new bonds in the reaction. If more energy is released than is consumed, the reaction is **exothermic.** The following problem illustrates how this is done.

**Table 2.4** Average Bond Energies [kJ/mol (*above*) and kcal/mol (*below*)]

**Single Bonds**

| | H | C | N | O | F | Cl | Br | I | Si |
|---|---|---|---|---|---|---|---|---|---|
| **H** | 435<br>104 | 414<br>99 | 389<br>93 | 464<br>111 | 565<br>135 | 431<br>103 | 364<br>87 | 297<br>71 | 318<br>76 |
| **C** | | 347<br>83 | 305<br>73 | 360<br>86 | 485<br>116 | 339<br>81 | 285<br>68 | 218<br>52 | 301<br>72 |
| **N** | | | 163<br>39 | 222<br>53 | 272<br>65 | 192<br>46 | | | |
| **O** | | | | 197<br>47 | 188<br>45 | 218<br>52 | 201<br>48 | 234<br>56 | 452<br>108 |
| **F** | | | | | 155<br>37 | | | | 565<br>135 |
| **Cl** | | | | | | 243<br>58 | | | 381<br>91 |
| **Br** | | | | | | | 192<br>46 | | 310<br>74 |
| **I** | | | | | | | | 151<br>36 | 234<br>56 |
| **Si** | | | | | | | | | 222<br>53 |

**Multiple Bonds**

| | | | |
|---|---|---|---|
| C=C | 611<br>146 | C≡C | 837<br>200 |
| C=N | 615<br>147 | C≡N | 891<br>213 |
| C=O (aldehydes) | 736<br>176 | | |
| C=O (ketones) | 749<br>179 | | |

**PROBLEM:** Ethylene, an alkene, reacts with hydrogen to give ethane, an alkane. Calculate the enthalpy of the reaction, $\Delta H_r$.

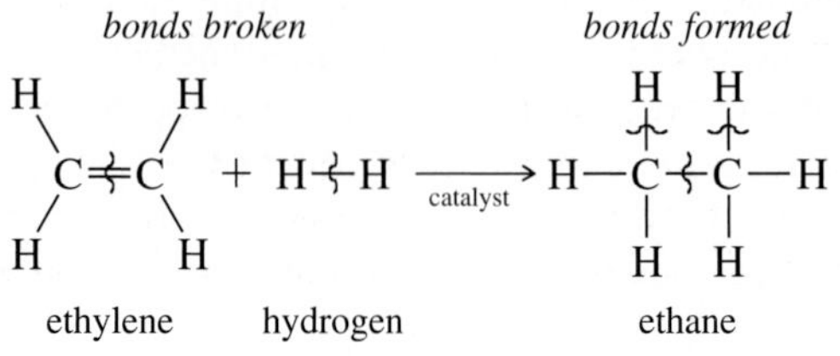

*bonds broken*        *bonds formed*

ethylene    hydrogen        ethane

**Solution:** The carbon–carbon double bond and the hydrogen–hydrogen single bond are broken in the reaction, requiring an input of energy. Therefore, the enthalpy changes for the bonds being broken are given as positive numbers. The formation of a carbon–carbon single bond and two carbon–hydrogen single bonds releases energy; the enthalpy changes for the bonds being formed are negative numbers.

| Bonds broken: | | Bonds formed: | |
|---|---|---|---|
| C=C | + 146 kcal/mol | C—C | −83 kcal/mol |
| H—H | + 104 kcal/mol | 2 × C—H | 2 × −99 kcal/mol |
| | + 250 kcal/mol | | −281 kcal/mol |

Enthalpy of reaction, $\Delta H_r$ = +250 kcal/mol − 281 kcal/mol = −31 kcal/mol

The sum of changes on both sides of the reaction equation indicates that more energy is released than is consumed, so the reaction is exothermic. ∎

⊘ **Study Guide**
**Concept Map 2.4**

**Problem 2.22**   Calculate $\Delta H_r$ for each of the following reactions using the average bond energies given in Table 2.4.

(a) $CH_3$—H + Cl—Cl ⟶ $CH_3$—Cl + H—Cl

(b) $H_2C$=O + H—C≡N ⟶ $H_2C\begin{smallmatrix}\diagup C≡N \\ \diagdown O—H\end{smallmatrix}$

(c) $H_2C$=$CH_2$ + H—Br ⟶ $H_2C$—$CH_2$ with H and Br substituents

(d) $CH_3CH$—$CH_2$ (with OH and H) ⟶ $CH_3CH$=$CH_2$ + H—OH

**Problem 2.23**   The carbon–chlorine bond in chloroethene, $CH_2CHCl$, is shorter and stronger than the carbon–chlorine bond in chloroethane, $CH_3CH_2Cl$. The relevant data are given below.

| | | C—Cl bond length | C—Cl bond dissociation energy |
|---|---|---|---|
| chloroethane | $CH_3CH_2Cl$ | 1.78 Å | 80 kcal/mol |
| chloroethene | $CH_2CHCl$ | 1.72 Å | 90 kcal/mol |

(a) Draw a Lewis structure for chloroethane and one for chloroethene.
(b) Chemists rationalize the shorter length of the chloroethene C—Cl bond in two ways. One of these involves resonance. How would you use resonance to explain the experimental data? Include Lewis structures in your explanation.
(c) Draw three-dimensional orbital pictures for your chloroethene resonance contributors using lines (——), dashed lines (------), and wedges (◀) for σ bonds and the orbitals that overlap for π bonds. You may depict any nonbonding electron pairs *not* involved in resonance as dots around the appropriate atom(s).
(d) In a few words, describe the other rationalization useful for explaining the difference in bond length and bond dissociation energy between chloroethane and chloroethene.

# 2.9   Effects of Bonding on Chemical Reactivity

Functional groups of organic compounds are the sites of chemical reactions. Among the hydrocarbons, ethane, a representative of the alkane family, and ethylene, a typical member of the alkene family, behave quite differently in the presence of bromine. When ethane, $CH_3CH_3$, is treated with bromine in carbon tetrachloride (a solvent) in the dark at room temperature, no reaction takes place. The solution remains reddish brown, the color of bromine.

$$\underset{\substack{\text{ethane}\\ \textit{an alkane}}}{H-\underset{\underset{H}{|}}{\overset{\overset{H}{|}}{C}}-\underset{\underset{H}{|}}{\overset{\overset{H}{|}}{C}}-H} + \underset{\substack{\text{reddish}\\ \text{brown}}}{Br_2} \xrightarrow[\substack{\text{tetrachloride}\\ \text{dark}\\ 25\ °C}]{\text{carbon}} \text{no reaction}$$

$sp^3$

When ethylene, $H_2C=CH_2$, is treated with bromine under the same conditions, the red color disappears rapidly, indicating that a chemical transformation has occurred.

$$\underset{\substack{\text{ethylene}\\ \textit{an alkene}}}{\underset{H}{\overset{H}{\diagdown}}C=C\underset{H}{\overset{H}{\diagup}}} + \underset{\substack{\text{reddish}\\ \text{brown}}}{Br_2} \xrightarrow[\substack{\text{tetrachloride}\\ \text{dark}\\ 25\ °C}]{\text{carbon}} \underset{\substack{\text{1,2-dibromoethane}\\ \text{colorless}\\ \textit{an alkyl halide}}}{H-\underset{\underset{Br}{|}}{\overset{\overset{H}{|}}{C}}-\underset{\underset{Br}{|}}{\overset{\overset{H}{|}}{C}}-H}$$

$sp^2$     $sp^3$

*an addition reaction*

The product of the reaction of bromine with ethylene, 1,2-dibromoethane, is colorless, so the progress of the reaction is easily seen. The reaction of bromine with ethylene is an example of an addition reaction. In an **addition reaction,** the product contains all the elements of the two reacting species. The reaction of an alkene with hydrogen to give an alkane (p. 64) is another example of an addition reaction.

In ethane, all the bonding electrons are in $\sigma$ bonds involving $sp^3$-hybridized carbon atoms. Electrons in $\sigma$ bonds are not readily involved in chemical reactions. The electrons in the $\pi$ bond of ethylene, in contrast, react easily with bromine. The addition product 1,2-dibromoethane has no $\pi$ bond, and the $sp^2$ carbon atoms of ethylene have become rehybridized into $sp^3$ carbon atoms in this product. In general, electrons in $\pi$ bonds (called $\pi$ electrons) are involved more readily in chemical reactions than are electrons in $\sigma$ bonds. The $\pi$ molecular orbitals are at a higher energy level than the $\sigma$ molecular orbitals are. We can visualize $\pi$ electrons as being more loosely held than are electrons in a $\sigma$ bond and thus more available to reagents that are seeking electrons.

Acetylene, $HC\equiv CH$, the simplest member of the alkyne family, also undergoes addition of bromine in the dark. Bromine adds successively to each of the two $\pi$ bonds of the alkyne.

$$\underset{\substack{\text{acetylene}\\ \textit{an alkyne}}}{H-C\equiv C-H} \xrightarrow{Br_2} \underset{\substack{\text{1,2-dibromoethene}\\ \textit{an alkene}}}{\underset{Br}{\overset{H}{\diagdown}}C=C\underset{H}{\overset{Br}{\diagup}}} \xrightarrow{Br_2} \underset{\substack{\text{1,1,2,2-tetrabromoethane}\\ \textit{an alkyl halide}}}{H-\underset{\underset{Br}{|}}{\overset{\overset{Br}{|}}{C}}-\underset{\underset{Br}{|}}{\overset{\overset{Br}{|}}{C}}-H}$$

$sp$    $sp^2$    $sp^3$

*addition reactions*

At the end of the first stage of the reaction, the linear carbon atoms in acetylene have been converted to the trigonal planar carbon atoms of an alkene, 1,2-dibromoethene. The addition of another molecule of bromine to the $\pi$ bond of this alkene converts it to 1,1,2,2-tetrabromoethane, which has only tetrahedral carbon atoms.

The addition of bromine is a reaction that is typical of many compounds having $\pi$ electrons in carbon–carbon double or triple bonds. The final products formed by such addition reactions of bromine are members of a class of organic compounds called **alkyl halides.** Alkyl halides are compounds in which a halogen atom is bonded to a tetrahedral carbon atom. The details of the addition reactions of bromine to $\pi$ bonds are presented in Section 8.8.

Not all hydrocarbons that contain $\pi$ bonds react with bromine at room temperature in the dark. Benzene has the molecular formula $C_6H_6$, which indicates that the carbon atoms in benzene must be linked to each other by multiple bonds. Benzene is the simplest member of a class of hydrocarbons known as **aromatic hydrocarbons.** These hydrocarbons are pictured as having ring structures in which double bonds alternate with single bonds. Yet benzene does not react with bromine under the conditions that ethylene does.

$$+ \; Br_2 \xrightarrow[\substack{\text{carbon}\\\text{tetrachloride}\\\text{dark}\\25\ °C}]{} \text{no reaction}$$

benzene

*an aromatic hydrocarbon*

How can this significant difference in reactivity between ethylene and benzene be explained? At first glance, we see that both molecules contain $\pi$ bonds and $\pi$ electrons, but the bonds of benzene must somehow be different from those of ethylene if the addition reaction with bromine does not occur. The lack of reactivity of benzene and other aromatic hydrocarbons in addition reactions is explored in detail in Chapter 10. The molecular orbital picture of benzene is related to ideas already presented; therefore, it can be examined now.

**Problem 2.24**   Write equations showing the addition of bromine to the following compounds. How many moles of bromine will react with one mole of each compound before the reaction stops?

(a) $CH_3CH_2CH{=}CHCH_3$     (b) $CH_3C{\equiv}CCH_3$
(c) $HC{\equiv}CCH_2C{\equiv}CH$
(d) $CH_3CH{=}CHCH_2CH{=}CHCH_3$

**Problem 2.25**   Different classes of compounds containing important functional groups have been introduced throughout this chapter. Review the chapter now, making a list of the different functional group classes and including the structural features that are typical of each one.

## 2.10   The Structure of Benzene

The experimental technique of x-ray crystallography provides convincing evidence that the structure of benzene is symmetrical. The six carbon atoms lie at the corners of a regular hexagon. Each carbon atom is bonded to two other carbon atoms and a hydrogen atom, all of which lie in a plane. Each carbon–carbon bond in benzene is 1.39 Å long, and each carbon–hydrogen bond is 1.09 Å long. All H—C—C and C—C—C bond angles are 120°, as expected for a regular hexagon. Benzene, in fact, is usually represented by a hexagon from which the symbols for the carbon and hydrogen atoms are omitted.

benzene

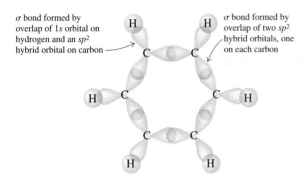

σ bond formed by overlap of 1s orbital on hydrogen and an sp² hybrid orbital on carbon

σ bond formed by overlap of two sp² hybrid orbitals, one on each carbon

The orbital picture of the σ bonds in benzene

1.39 Å    120°    120°    1.09 Å    A regular hexagon

The experimentally determined bond lengths and bond angles of benzene

*Figure 2.24*
**The σ bond skeleton of benzene.**

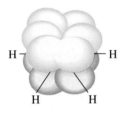

Side view showing the six parallel *p* orbitals in phase

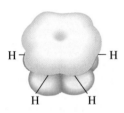

Lowest energy bonding π molecular orbital showing delocalization

*Figure 2.25*
**Molecular orbital picture of the π bonds in benzene.**

The carbon atoms in benzene are trigonal planar. They are bonded to three other atoms in a plane with bond angles of 120°. Such carbon atoms are $sp^2$-hybridized. Thus the σ bonds in benzene result from the overlap of $sp^2$ hybrid orbitals between adjacent carbon atoms and the overlap of $sp^2$ orbitals of carbon atoms with 1s orbitals of hydrogens (Figure 2.24).

Three valence electrons from each carbon atom are involved in the σ bonding in benzene. Each carbon atom also has a 2p orbital, which has a lobe above and below the plane of the benzene ring, and an electron in that orbital. All six p orbitals are perpendicular to the plane of the ring, an orientation that allows for overlap of the orbitals to form π bonds between adjacent carbon atoms. The overlap of six p orbitals in benzene gives rise to six π molecular orbitals. Three of them are bonding molecular orbitals, and three are antibonding molecular orbitals. The six electrons occupy the three bonding π orbitals of benzene (pictures and relative energies of these orbitals are shown on p. 354). The lowest-energy bonding molecular orbital in benzene is shown in Figure 2.25. This molecular orbital has two circular lobes, one above and one below the plane of the ring. The π electrons in benzene are completely **delocalized** over the entire ring. Thus the molecule is highly symmetrical with all carbon–carbon bonds of equal length (1.39 Å), longer than that of a double bond (1.34 Å) but shorter than that of a single bond (1.53 Å).

The structural formula for benzene is written with three double bonds in the ring. Such a representation is called the **Kekulé formula** for benzene, after the German chemist August Kekulé, who first proposed that structure for benzene in 1865. Even that long ago, however, Kekulé recognized that a structure that depicts benzene with single and double bonds is not a satisfactory representation of the properties of the compound. He suggested that the structure of benzene should be represented by two formulas in which the single and double bonds have different positions.

Kekulé formulas for benzene

Kekulé thought that these structural formulas corresponded to different species that were in dynamic equilibrium with each other. It is now recognized that structures that differ from each other only in the location of electrons cannot be distinguished. Such structures are called resonance contributors (p. 16) to the true structure of the compound, which has properties best represented by all the resonance contributors taken together. The correct structure for the compound is said to be a resonance hybrid of the various resonance contributors that can be written for it.

Each Kekulé formula for benzene suggests that benzene has three double bonds and three single bonds in the six-membered ring. Yet we saw in Section 2.9 that benzene does not react with bromine in the same way as do compounds such as ethylene that have normal carbon–carbon double bonds. Thus neither Kekulé formula is an accurate picture of the real benzene molecule. The resonance hybrid, the two Kekulé formulas taken together, suggests that the double bonds in benzene are not fixed in position and that the $\pi$ electrons are delocalized over the entire ring. As a result, the reactivity of benzene is unlike that of compounds containing localized double bonds.

Aromatic hydrocarbons represent a large class of organic compounds the chemistry of which is both interesting and often different from that of alkanes and alkenes. The chemistry of such compounds is described in detail in Chapter 10, but aromatic rings appear in many compounds in all parts of the book. Recognition of the special stability of this type of hydrocarbon is therefore important.

**Problem 2.26**     Before the structure of benzene was finally determined experimentally, one of the many structures that had been proposed for the hydrocarbon with the molecular formula $C_6H_6$ was called Dewar benzene (shown at right). What would be the hybridization of each carbon atom in the structure shown? How would you expect this compound to behave if a solution of bromine were added to it?

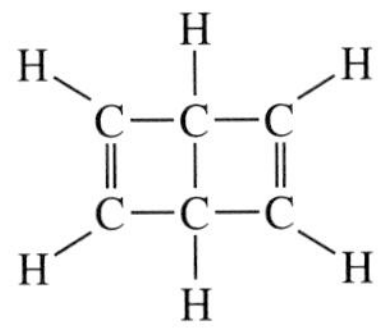

Dewar benzene

## 2.11 Electronic Transitions Between Bonding and Antibonding Molecular Orbitals

### A. Transitions Between Bonding and Antibonding Molecular Orbitals in Hydrogen

Atoms and molecules are said to exist in their **ground state** when their electrons are in the lowest-energy orbital available to them. For example, for the hydrogen atom, the ground state is $1s^1$, a single electron in the $1s$ orbital. For the hydrogen molecule, as shown in Figure 2.5 (p. 44), it is $\sigma^2$, two electrons in the bonding $\sigma$ molecular orbital. Both atoms and molecules, however, have many other states that lie above the ground state and are unoccupied. They have no electrons in them in the ground state. If energy corresponding to the gap in energy between an occupied orbital and an unoccupied orbital is supplied to the atom or molecule, an electron will move from a lower to a higher energy level. An atom or molecule, after absorption of such a unit of energy represented by $h\nu$, is said to be in the **excited state.** Return to the ground state occurs with loss of energy, very often with the emission of light. Fluorescent lights in classrooms and offices, yellow sodium lamps in parking lots, and the multicolored "neon" lights of storefronts and theaters all use electrical energy to create light by exciting atoms to higher energy levels with subsequent return to the ground state and

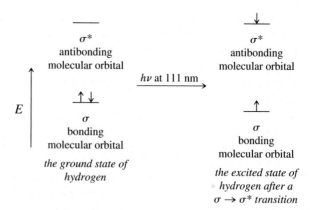

*Figure 2.26*

**The ground state of hydrogen and the excited state of hydrogen after absorption of ultraviolet radiation. The $\sigma \rightarrow \sigma^*$ transition.**

emission of the color of light that is characteristic of the gap in energy between the orbital occupied by the electron in the excited state and that of the ground state.

The antibonding molecular orbital shown in Figure 2.5 for the hydrogen molecule is an energy level that is unoccupied in the ground state but could become occupied if enough energy were put into the molecule that one of the electrons in the bonding molecular orbital moves to the higher energy level. Such a transformation takes place when the hydrogen molecule absorbs ultraviolet radiation at 111 nm, a region of the spectrum known as the vacuum ultraviolet. The energy corresponding to light at this wavelength is 262 kcal/mol. The shorter the wavelength of light, the higher the energy it provides (see Table 11.1, p. 395).

Figure 2.26 shows the distribution of electrons for the excited state that results when one electron is moved from the bonding $\sigma$ molecular orbital to the antibonding $\sigma^*$ molecular orbital in what is known as a $\sigma \rightarrow \sigma^*$ transition. The electronic transition shown in Figure 2.26 weakens the bond between the two hydrogen atoms because one electron is now in the antibonding molecular orbital, which has a node between the two nuclei (see Figure 2.5) and thus no electron density there. That electron does not contribute to the covalent bond but, in fact, subtracts from it. The molecule returns to the ground state when the electron in the antibonding molecular orbital falls back to the bonding molecular orbital with the loss of energy.

Electronic transitions between bonding and antibonding molecular orbitals take place when light of a wavelength corresponding to the energy gap between those orbitals is absorbed by a molecule. This absorption is recorded by ultraviolet-visible spectrophotometers and serves to help determine the structure of compounds or detect them in mixtures, as we will see in Chapter 12. The excited state of a molecule also can undergo chemical reactions and photochemical reactions, some of which are discussed in Chapter 25. In this section we will continue to explore the kinds of electronic transitions possible for some functional group classes.

## B. Electronic Transitions in Compounds with $\sigma$ Bonds and Nonbonding Electrons

Compounds that contain only single bonds, such as saturated hydrocarbons, have absorption in the vacuum ultraviolet, just as hydrogen does. For these compounds too, this absorption is attributed to the $\sigma \rightarrow \sigma^*$ transition, which is observed at 125 nm for methane, for example. If compounds contain nonbonding electrons (represented by $n$), the molecule also has energy levels occupied by those electrons, designated as $n$ orbitals. These orbitals are higher in energy than $\sigma$ but lower than $\sigma^*$ orbitals. Figure 2.27 shows the relative energies of the types of orbitals we have studied in this chapter. Looking at the ordering of the energy levels in Figure 2.27, we can predict that the wavelength of light absorbed for an $n \rightarrow \sigma^*$ transition will be longer than that for a $\sigma \rightarrow \sigma^*$ transition for the same compound. The $n$ orbital is higher in energy than the $\sigma$ orbital; therefore, it is closer to the $\sigma^*$ orbital. The gap in energy between the $n$ and $\sigma^*$ orbitals is smaller than it is for the $\sigma$ and $\sigma^*$ orbitals. For example, ammonia has two absorption bands, at 194 nm for the $n \rightarrow \sigma^*$ absorption and at 152 nm for the $\sigma \rightarrow \sigma^*$ absorption. Note again that a longer wavelength of light corresponds to a lower-energy transition, whereas a shorter wavelength of light gives rise to a higher-energy transition. The $n \rightarrow \sigma^*$ transition at 194 nm corresponds to a smaller energy gap between orbitals than the $\sigma \rightarrow \sigma^*$ transition at 152 nm. This kind of a relationship, discussed further on page 395, is one of the fundamental ideas in quantum mechanics.

Another interesting fact is that the energy of the $n$ orbital is related to the electronegativity of the atom bearing the nonbonding electrons. This is not surprising. A promotion of an $n$ electron to a higher energy level is moving it further from the nucleus. The more electronegative the atom, the harder it is to do this. Thus the energy of the $n$ orbital becomes higher as the electronegativity of the atom holding the $n$ electrons decreases and the wavelength of the absorption for the $n \rightarrow \sigma^*$ transition increases. As an example of this, the $n \rightarrow \sigma^*$ absorption for $H_2O$ comes at 167 nm, whereas that for $NH_3$ (nitrogen is less electronegative than oxygen) is at 194 nm.

*Figure 2.27*
Relative energies of molecular orbitals.

**Problem 2.27**    Draw two energy diagrams showing the relative energy levels for the $\sigma$, $n$, and $\sigma^*$ orbitals for $H_2O$ and $NH_3$. You may assume that the gap between the $\sigma$ and $\sigma^*$ orbitals is larger for $H_2O$ than it is for $NH_3$.

**Problem 2.28**    Absorption bands have been recorded at 173 nm for Compound A, 204 nm for Compound B, and 258 nm for Compound C. These compounds are all halomethanes, $CH_3X$, where X can be Cl, Br, or I. Assign structures to Compounds A, B, and C.

## C. Electronic Transitions in Compounds with Double Bonds

Compounds that contain double bonds such as alkenes or carbonyl compounds have bonding $\pi$ molecular orbitals and the corresponding antibonding $\pi^*$ molecular orbitals. Transitions between these orbitals are also possible, given the right amount of energy. For ethylene, absorption of ultraviolet radiation at 165 nm results in a $\pi \rightarrow \pi^*$ transition (Figure 2.28), where the molecule goes from the ground state, in which both $\pi$ electrons are in the bonding molecular orbital, to the excited state, where one electron has been promoted to the antibonding molecular orbital.

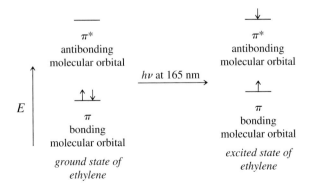

*Figure 2.28*
The ground state of ethylene and the excited state of ethylene after absorption of ultraviolet radiation. The $\pi \rightarrow \pi^*$ transition.

The bonding $\pi$ molecular orbital has electron density between the two carbon atoms. An antibonding $\pi^*$ molecular orbital not only has a node in the plane of the alkene but also has a nodal plane bisecting the bond between the two carbon atoms (see Figure 2.18, p. 54). If one of the electrons of a $\pi$ orbital is promoted to a $\pi^*$ orbital, electron density between the two carbon atoms decreases, and the two ends of the bond can now rotate relative to each other, something that is not possible for the $\pi$ bond in the ground state.

**Problem 2.29**   Prove to yourself by using molecular models that you cannot rotate the two carbon atoms of ethylene relative to each other. Compare this with what happens when you try to rotate the two carbon atoms of ethane relative to each other. Explain the difference using concepts about the nature of $\sigma$ and $\pi$ molecular orbitals.

One biological consequence of a $\pi \rightarrow \pi^*$ transition in alkenes is related to the reaction that takes place in the retina of our eye as the primary event of vision. Light arrives at the retina and is absorbed by retinal, a compound containing multiple double bonds. As a result, a rotation relative to each other of the two carbon atoms of one of the double bonds takes place. The molecule changes shape, which sets off a whole other chain of events that eventually registers in our brain as sight. More can be found on this on page 758.

Benzene (Section 2.10) also has double bonds. The ultraviolet spectrum of benzene is interesting for its complexity. The energy gap between the highest occupied bonding molecular orbital and the lowest unoccupied antibonding molecular orbital in benzene with its interacting double bonds is smaller than it is for the double bond in ethylene. Therefore, the $\pi \rightarrow \pi^*$ transition for benzene comes at a longer wavelength, 203 nm, than the corresponding transition in ethylene. In addition, it has a number of still longer-wavelength bands that are the result of the extensive delocalization of its $\pi$ electrons. In general, such spectra (see Figure 12.8, p. 451, for toluene, for example) are typical of aromatic compounds.

Chemists use observations about the transitions between different kinds of molecular energy levels to get information about molecular structures in various forms of spectroscopy. Ultraviolet-visible spectroscopy, which deals with the electronic transitions we have been discussing in this section, as well as other forms of spectroscopy, will be the main subject of Chapters 11 and 12.

## SUMMARY

When two atomic orbitals interact, two molecular orbitals result. In-phase interactions give bonding molecular orbitals, whereas out-of-phase interactions give antibonding molecular orbitals. A covalent bond results when a bonding molecular orbital is occupied by a pair of electrons.

Both $\sigma$ orbitals and $\sigma$ bonds have cylindrical symmetry around the axis connecting the two nuclei. Sigma bonds can arise from the overlap of $s$ atomic orbitals either with other $s$ orbitals or with $p$ orbitals or from the overlap of two $p$ orbitals. Both $\pi$ orbitals and $\pi$ bonds have a nodal plane and arise from the sideways overlap of $p$ orbitals.

Organic chemists find the concept of hybrid orbitals useful for rationalizing the different shapes of covalent molecules and the different kinds of bonds found in them. Hybrid orbitals are created by mathematical combinations of the wave functions for $s$ and $p$ orbitals. When carbon is bonded by single bonds to four other atoms, it is said to be $sp^3$-hybridized. The $sp^3$ hybrid orbitals are directed to the corners of a tetrahedron, with angles of 109° between them. Such a carbon is also called a tetrahedral carbon atom.

When a central atom is bonded to three other atoms by single or multiple bonds, it is said to be an $sp^2$-hybridized atom. The $sp^2$ hybrid orbitals are directed to the corners of an equilateral triangle, with angles of 120° between them. Such an atom is also called a trigonal planar atom.

A central atom bonded to two other atoms by single or multiple bonds is an $sp$-hybridized atom. The $sp$ hybrid orbitals of an atom point away from each other at an angle of 180°. The portion of a molecule containing $sp$ hybrid orbitals is linear.

In the orbital hybridization picture of bonding, $\sigma$ bonds are formed by the overlap of hybrid orbitals with other hybrid orbitals or with $s$ or $p$ atomic orbitals.

A carbon atom that is $sp^2$-hybridized has an unhybridized $p$ orbital, which can form a $\pi$ bond by overlapping side to side with a $p$ orbital of an adjacent atom (carbon, oxygen, or nitrogen) to which the carbon is joined by a $\sigma$ bond. The $\pi$ bond and $\sigma$ bond together constitute a double bond. An $sp$-hybridized carbon atom has two unhybridized $p$ orbitals and forms a triple bond, consisting of a $\sigma$ and two $\pi$ bonds, with another carbon or a nitrogen atom.

The strength of a covalent bond is related to its length, which in turn is related to the sizes of the atoms bonded, their hybridization, and whether the bond is a single, double, or triple bond. The shorter a bond, the stronger it is.

Bond strengths may be expressed as bond dissociation energies, *DH* (the energy necessary to break a bond so that one electron of the covalent bond goes to each fragment of the molecule) or as average bond energies (determined from many experimental measurements to be the average covalent bond energy for a variety of typical molecules). Average bond energies are useful for estimating enthalpy changes in reactions.

Carbon atoms bond among themselves and to atoms of other elements with single, double, and triple bonds, giving rise to a variety of functional groups. A functional group is a structural unit consisting of an atom or group of atoms that serves as a site of reactivity in a molecule. Organic compounds that contain the same functional group are classified together into families of compounds. The names of the functional group classes met so far and the structure of the functional group in each are given in a table that appears inside the back cover of this book.

## ADDITIONAL PROBLEMS

**2.30** Draw Lewis structures for each of the following species and indicate the hybridization and geometry of the central atom.

(a) $BeH_2$  (b) $CO_2$  (c) $CH_3^+$  (d) $CF_4$  (e) $Cl_2CO$  (f) $CCl_3^-$  (g) $NF_3$  (h) $BF_4^-$

**2.31** Boron forms a compound with hydrogen that has the molecular formula $B_2H_6$. The nature of the bond that holds the two boron atoms together has been a source of much research and discussion.

(a) Can $B_2H_6$ be represented by a conventional Lewis structure? Explain your answer in a few words or with structural formulas.

(b) The $B_2H_6^{2-}$ anion has been synthesized at Ohio State University. This anion is exactly analogous in the number of electrons available for bonding to a compound of carbon and hydrogen that has no net charge. The $B_2H_6^{2-}$ anion has exactly the same three-dimensional structure as this carbon–hydrogen compound. Which carbon–hydrogen compound would be a good analogy for $B_2H_6^{2-}$?

(c) Draw a Lewis structure for $B_2H_6^{2-}$. (All atoms have closed shells.)

(d) Draw a three-dimensional picture for $B_2H_6^{2-}$.

**2.32** Rank each of the following groups of compounds according to lengths of the indicated bonds.

(a) $CH_3C{\equiv}CH$, $CH_2{=}CHCH_3$, $CH_3CH_2{-}CH_3$,

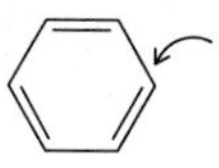

(b) $CH_3{-}OH$, $CH_2{=}O$, $HC{-}OCH_3$ (with $\overset{O}{\|}$)   (c) $CH_3C{\equiv}N$, $CH_3{-}NH_2$, $CH_2{=}NCH_3$

**2.33** Interesting compounds isolated from nature often have more than one functional group. Identify the functional groups found in each of the natural products shown below.

(a)

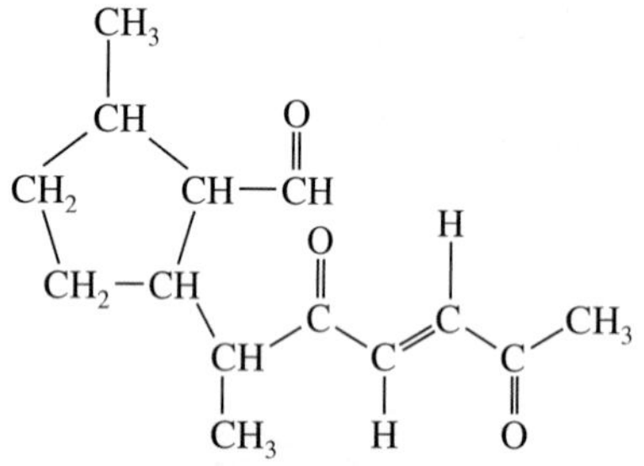

*a defensive toxin produced
by the whirligig beetle*

(b)

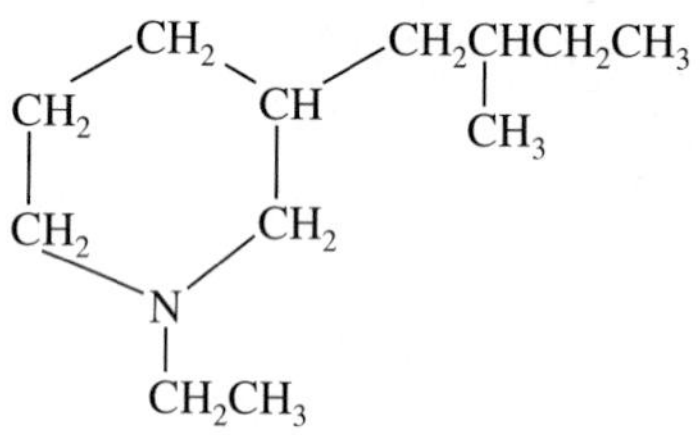

*a substance produced by a beetle
to help it float on water*

(c)

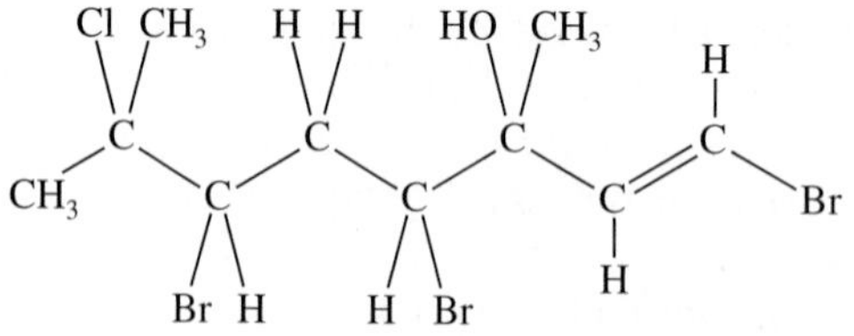

*a metabolite of a seaweed
that has strong antimicrobial activity*

(d)

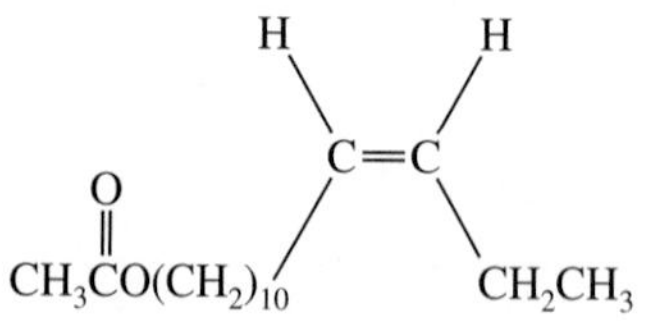

*used by the red-banded leaf roller to
communicate with others of its species*

(e)

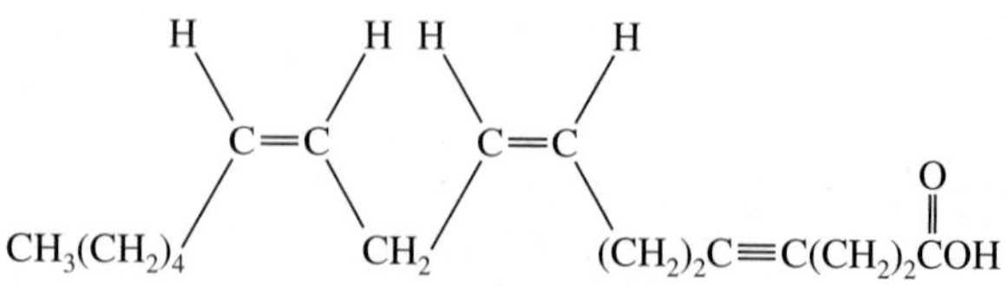

*a metabolite of a seaweed
that is poisonous to fish*

**2.34** For each of the following compounds, assign hybridization to each carbon, oxygen, and nitrogen atom. Show which types of orbitals overlap to create the different kinds of bonds in the molecules.

(a) $CH_2$=O  (b) $CH_3NHOH$  (c) $CH_3CH$=$CHCH_3$  (d) $CH_3C$≡$CCH_3$

**2.35** The compound shown below is the oxime of formaldehyde. The oxime functional group is found in some natural products, including antibiotics.

$$H_2C=N-OH$$

formaldehyde oxime

(a)  What is the hybridization at the carbon, the nitrogen, and the oxygen atoms in the oxime?
(b)  Draw a three-dimensional representation of the compound. Use lines, dashes, and wedges to represent $\sigma$ bonds and to show the directions of nonbonding electrons. Show the $p$ orbitals that overlap to give any $\pi$ bonds present in the molecule.

**2.36** The formation of a bond between nonbonding electrons of nitrogen and oxygen atoms and the empty $1s$ orbital of a proton was shown in Figures 2.12 and 2.14. Referring to those figures, predict, by writing an equation, what would happen if each of the following compounds were put into sulfuric acid.

(a) $CH_3C$≡N:   (b) $CH_3\overset{\overset{\displaystyle :O:}{\|}}{C}CH_3$   (c) $CH_3\overset{\overset{\displaystyle CH_3}{|}}{C}$=$NCH_3$   (d) $CH_3\overset{\overset{\displaystyle :O:}{\|\cdot\cdot}}{C}OCH_3$

**2.37** Predict, by completing the equations, what would happen if the following reagents were mixed. (A review of Sections 2.8, 2.9, and 2.10 might be helpful.)

(a) $CH_3CH$=$CHCH_3 + Br_2 \xrightarrow[\substack{\text{carbon}\\ \text{tetrachloride}\\ \text{dark, 25 °C}}]{}$

(b) $CH_3CH_2CH$=$CH_2 + H_2 \xrightarrow[\text{catalyst}]{}$

(c) $CH_3CH_2CH_2CH_3 + Br_2 \xrightarrow[\substack{\text{carbon}\\ \text{tetrachloride}\\ \text{dark, 25 °C}}]{}$

(d)  ⬡—$CH$=$CH_2 + Br_2 \xrightarrow[\substack{\text{carbon}\\ \text{tetrachloride}\\ \text{dark, 25 °C}}]{}$

(e) $CH_3C$≡$CH + Br_2$ (excess) $\xrightarrow[\substack{\text{carbon}\\ \text{tetrachloride}\\ \text{dark, 25 °C}}]{}$

(f) $CH_3C$≡$CH + H_2$ (excess) $\xrightarrow[\text{catalyst}]{}$

**2.38** Calculate $\Delta H_r$ for each of the following reactions using the average bond energies given in Table 2.4 (p. 64).

(a) $H_2C$=$CH_2 + Br-Br \longrightarrow H_2\overset{\overset{\displaystyle \,}{|}}{C}\underset{\underset{\displaystyle Br}{|}}{\,}-\overset{\overset{\displaystyle \,}{|}}{C}\underset{\underset{\displaystyle Br}{|}}{H_2}$

(b) $CH_3-\overset{\overset{\displaystyle CH_3}{|}}{\underset{\underset{\displaystyle CH_3}{|}}{C}}-OH + H-Cl \longrightarrow CH_3-\overset{\overset{\displaystyle CH_3}{|}}{\underset{\underset{\displaystyle CH_3}{|}}{C}}-Cl + H-OH$

(c) $CH_3\overset{\overset{\displaystyle O}{\|}}{C}CH_3 + H-OH \longrightarrow CH_3\overset{\overset{\displaystyle O-H}{|}}{\underset{\underset{\displaystyle OH}{|}}{C}}CH_3$

**2.39** Allene has the structural formula shown at the right. What is the hybridization of each of the carbon atoms? What are the orbitals involved in the formation of the $\sigma$ bond skeleton and of the $\pi$ bonds in this molecule? Illustrate your answers with drawings. (*Hint:* Figures 2.18 and 2.22 may be helpful.)

$CH_2$=$C$=$CH_2$

allene

**2.40** Boron trifluoride, $BF_3$, is a gas, bp $-99.9$ °C. It is sold commercially as boron trifluoride etherate, a compound that is formed when boron trifluoride is dissolved in diethyl ether, $CH_3CH_2OCH_2CH_3$. Diethyl ether boils at 35 °C; boron trifluoride etherate boils at 126 °C.

(a) Show the hybridization of the central atoms, boron and oxygen, in boron trifluoride and in diethyl ether by drawing three-dimensional structural formulas for the compounds, emphasizing the bond angles at the central atoms.

(b) Why does boron trifluoride react so readily with diethyl ether? Answer this by showing the interaction of the orbitals involved in the reaction.

(c) Why does boron trifluoride etherate have a boiling point so much higher than those of the compounds that combined to form it? A few words and a pictorial representation would be helpful.

**2.41** One of the widely prescribed medications for individuals who are infected with the HIV virus is the compound known as AZT (azidothymidine), shown on p. 277. The therapeutic value has been, in part, correlated with the azido group ($N_3$ subunit) that is part of the AZT structure.

(a) Azidomethane ($CH_3N_3$; connectivity is $H_3CNNN$, an uncharged molecule) is one of the simplest organic compounds containing an azido group. Draw three resonance contributors for $CH_3N_3$. All the atoms in this uncharged molecule have closed shell configurations in all three resonance contributors. Which of these three resonance contributors makes the least contribution to the overall structure of azidomethane? Why?

(b) What is the anticipated N—N—N bond angle and the hybridization predicted for the center nitrogen atom?

**2.42** Isothiocyanates are versatile reagents in syntheses. There are three important resonance forms for methyl isocyanate ($CH_3NCS$) in which all the atoms have closed shells and formal charges of $-1, 0$, or $+1$.

(a) Draw these three resonance contributors.

(b) Which of the three forms that you drew is the most important contributor to the actual structure of methyl isocyanate? Why do you think so?

(c) Draw a picture of the most important resonance contributor of methyl isocyanate showing $\sigma$ bonds as lines, dashed lines, or wedges and indicating which $p$ orbitals interact to give $\pi$ bonds. In order to do this, you will have to decide on the hybridization of each of the atoms in methyl isocyanate. Your choice of hybridizations needs to accommodate all three resonance contributors. (*Hint:* A review of Problem 2.39 will be helpful.)

**2.43** A tear gas often used to control crowds in demonstrations contains the compound *o*-chlorobenzylidene malononitrile, also called CS. CS has the structure shown below and owes its potency to the fact that the $\pi$ orbitals of the carbon–carbon double bond and those of the carbon–nitrogen triple bonds interact, conveying the electronegativity of the nitrogen atoms to the carbon, three atoms away.

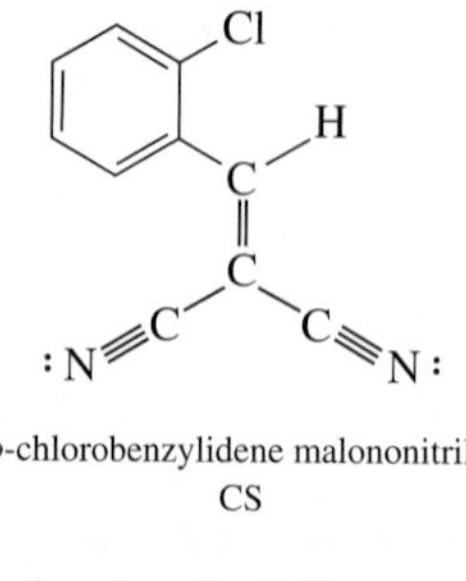

*o*-chlorobenzylidene malononitrile,
CS

(a) Draw resonance contributors showing the influence of the nitrogen atoms on the carbon three atoms away.

(b) Use the abbreviated structure below to create an orbital picture for the portion of CS shown. Sigma bonds should be shown using lines, dashed lines, or wedges to indicate directionality. Show the $p$ orbitals involved in the $\pi$ bonds. You do not need to show the directionality of the lone pairs.

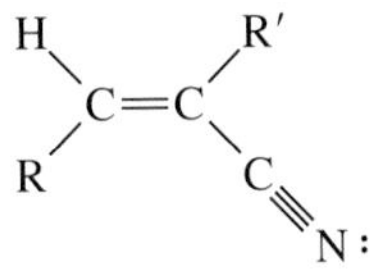

**2.44**

(a) Sections 2.5 and 2.6 put forth the ideas that a carbon atom that is trigonal planar is $sp^2$-hybridized and that a carbon atom that is bonded to two other groups in a linear molecule is $sp$-hybridized. Similar reasoning can be applied to other elements. For example, water also can be viewed as a planar triangular molecule. What would the hybridization of the oxygen atom be in this model of water? Which orbitals would the two pairs of nonbonding electrons occupy?

(b) Photoelectron emission spectroscopy, a form of spectroscopy that measures how easily electrons can be removed from various energy levels in a molecule, was used in an experiment that indicated that the two pairs of nonbonding electrons in water occupy different energy levels. Which model for the water molecule better fits this experimental fact: the one in which the oxygen atom is $sp^3$-hybridized or the one in which it is $sp^2$-hybridized?

(c) Would any other model for the bonding in water account for the observed experimental fact?

(d) Lithium hydroxide, LiOH, is a linear covalent molecule. Propose an orbital picture of the bonding in lithium hydroxide. What kind of hybridization does the oxygen atom have here?

**2.45** The unusual cationic molecule $[HeH]^+$ has a bond energy of 43 kcal/mol.

(a) Using Figure 2.6 (p. 45) as a model, draw an energy diagram showing the atomic orbitals of hydrogen and helium atoms and the molecular orbitals of the $[HeH]^+$ molecule. (*Hint:* The $1s$ orbital for helium is slightly lower in energy than the $1s$ orbital for hydrogen.) Place the appropriate numbers of electrons in the atomic orbitals and show the bonding electrons.

(b) Would you expect $[HeH]^-$ to be stable? Explain why or why not.

(c) Draw the Lewis structure for $[HeH]^+$ showing clearly which atom has the formal charge of $+1$.

# 3 Reactions of Organic Compounds as Acids and Bases

## A Look Ahead

The first step in most chemical reactions is the interaction of a pair of nonbonding electrons in one molecule or ion with a center of electron deficiency (which may be a vacant orbital or a partial positive charge) at an atom in another species. For example, the nonbonding electrons of a molecule of ammonia are donated to a vacant orbital of boron trifluoride (p. 13) to give a new covalent bond. The movement of the pair of electrons from the nitrogen to the boron atom is symbolized by a curved arrow.

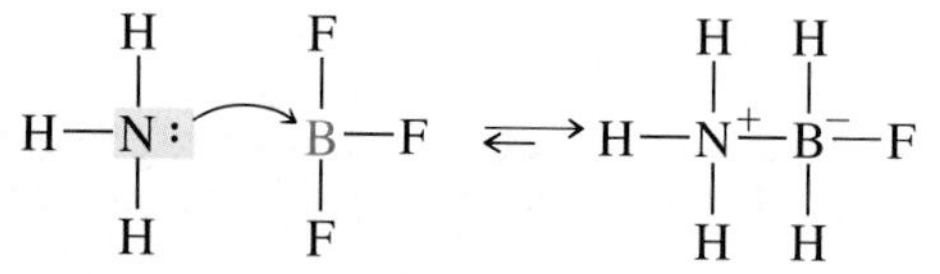

*curved arrow symbolizing
the donation of a pair of
electrons from the nitrogen
atom to the boron atom*

An understanding of the generality of this way of looking at organic reactions is so fundamental that we will explore it in this chapter using acid–base reactions as examples. Your goal in this chapter is to learn to see the sites of electron density and electron deficiency in the structures of compounds. These sites make the molecules vulnerable to a variety of chemical reagents. A species with electrons to donate is a base. An electron-deficient species that accepts an electron pair is an acid.

You should not attempt, at this point, to memorize structures or names of compounds; you will become familiar with them gradually as you study and write reactions. In this chapter you should concentrate on learning to predict acid–base reactivity from the structural features that are apparent in the formulas and on depicting reaction mechanisms correctly using the curved-arrow notation.

**Workbook Exercises**

## 3.1  Brønsted and Lewis Acids and Bases

### A. Brønsted-Lowry Theory of Acids and Bases

What are acids and bases? Two typical acid–base reactions are shown on the next page.

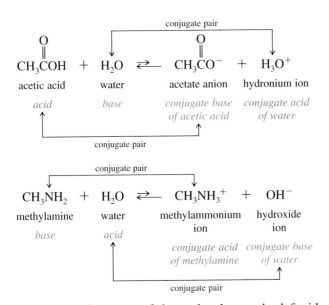

In each reaction, a proton from one of the molecules on the left side of the equation is transferred to the other molecule on the same side of the equation. In the **Brønsted-Lowry theory** of acids and bases, an **acid** is defined as a **proton donor** and a **base** as a **proton acceptor.** An alternative and more accurate way to define Brønsted acids and bases is to say that an acid is a substance from which a proton can be removed and a base is a substance that can remove a proton from an acid. In the first equation, therefore, acetic acid is the acid (the substance from which a proton was removed), and water is the base (the substance that removed the proton from acetic acid). Acetate ion, which is formed by the loss of a proton from acetic acid, is a base. It can remove a proton from the hydronium ion to become acetic acid again. A pair of species that can be interconverted by the loss and gain of a proton is called a **conjugate acid–base pair.** The conjugate acid of any base will have an additional hydrogen and an increase in positive charge (or a decrease in negative charge). The conjugate base of an acid will have one hydrogen fewer and will have an increase in negative charge (or a decrease in positive charge). Brønsted acids are also called **protic acids**—that is, acids that react via the transfer of a proton. Note that water can be an acid or a base. It can gain a proton to become a hydronium ion, $H_3O^+$, its conjugate acid, or lose a proton to become the hydroxide ion, $OH^-$, its conjugate base.

⊘ Study Guide
Concept Map 3.1

**Problem 3.1**    Identify the acid and base in each of the following conjugate acid–base pairs. Draw Lewis structures for each species and locate any formal charge.

(a) $(CH_3)_2O$, $(CH_3)_2OH^+$       (b) $H_2SO_4$, $HSO_4^-$       (c) $NH_2^-$, $NH_3$
(d) $CH_3OH_2^+$, $CH_3OH$       (e) $H_2CO$, $H_2COH^+$       (f) $CH_3OH$, $CH_3O^-$

Acidity and basicity are described in terms of equilibria. A review of the quantitative treatment of acid–base equilibria is given later in this chapter. First, we will concentrate on achieving a qualitative understanding of the phenomena.

A useful way to picture acid–base reactions is to imagine two bases competing for the same proton, which is transferred from one species to the other in the reaction. For example, in the reaction of acetic acid with water, acetate ion and water are the bases in competition for the proton. Their relative strengths as bases determine where the equilibrium lies. Methylamine and hydroxide ion are the bases that compete for the proton in an equilibrium process when methylamine reacts with water. An amine (p. 49) has basic properties similar to those of ammonia and reacts with acids to give a substituted ammonium ion. In both these cases, the conjugate base on the right side of the equilibrium arrows is stronger than the base on the left side. The reactants are only partially converted to products. In fact, most of the molecules present at equilibrium for the first equation are those of acetic acid and water, with only a few acetate and hydronium ions being formed. Similarly, there are more methylamine and water molecules present at equilibrium than methylammonium ions and hydroxide ions. *At equilibrium, the weaker acid and the weaker base are the major species found in the reaction mixture.*

In contrast, in the reaction of hydrogen chloride with dimethylamine, chloride ion, the conjugate base of the strong acid hydrogen chloride, is a weak base and does not compete effectively with dimethylamine for the proton. *Strong acids have weak conjugate bases, while weak acids have strong conjugate bases.* The reaction is written with the longer arrow pointing to the right and only a short one pointing to the left, indicating that most of the reactant molecules have been converted into product.

conjugate pair

$$(CH_3)_2NH \;+\; HCl \;\rightleftharpoons\; (CH_3)_2NH_2^{+} \;+\; Cl^{-}$$

dimethylamine (*base*) — hydrogen chloride (*acid*) — dimethylammonium ion (*conjugate acid of dimethylamine*) — chloride ion (*conjugate base of hydrogen chloride*)

conjugate pair

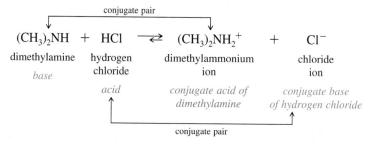

At equilibrium, very few dimethylamine and hydrogen chloride molecules remain. They have been converted to dimethylammonium and chloride ions.

In discussing equilibria, we must not forget that in reality we are talking about the dynamic interactions of very large numbers of molecules. The smallest amount of material that we can weigh in the laboratory contains billions and billions of molecules. For example, if we were to work with a single drop of acetic acid, depending on the size of the dropper, it would weigh approximately 10–50 mg. A mole of acetic acid weighs 60 g and contains $6.022 \times 10^{23}$ molecules. A millimole ($10^{-3}$ mole) of acetic acid weighs 60 mg and contains $6.022 \times 10^{20}$ molecules, still an enormous number. Even though we write balanced equations showing one or, at most, a few molecules of each reactant and product, the molecular formulas used in the equations represent large numbers of molecules.

In writing equations, the formulas for the species that chemists mix together in the laboratory are put on the left side of the reaction arrows and are called *reactants*. The species that form in the reaction appear on the right side of the reaction arrows and are called *products*, even though in many reactions more of the reactants than products may be present at equilibrium and even though the "products" also react with each other to give the "reactants" in the reverse reaction of the equilibrium. Reactants are also often called *starting materials*, perhaps a clearer term indicating their experimental origin.

## B. Lewis Theory of Acids and Bases

The proton acts as an acid because it has an empty 1*s* orbital that can accept a pair of electrons from some species that is serving as a base. This view of an acid–base reaction is shown in Figures 2.12 and 2.14 (pp. 49 and 50) and leads to another definition of acids and bases. The **Lewis theory** of acids and bases defines an **acid** as an **electron-pair acceptor** and a **base** as an **electron-pair donor.** Thus a proton is only one of a large number of species that may function as a Lewis acid. Any molecule or ion may be an acid if it has an empty orbital to accept a pair of electrons. Any molecule or ion with a pair of electrons to donate can be a base. For example, dimethyl ether acts as a Lewis base toward boron trichloride, a Lewis acid.

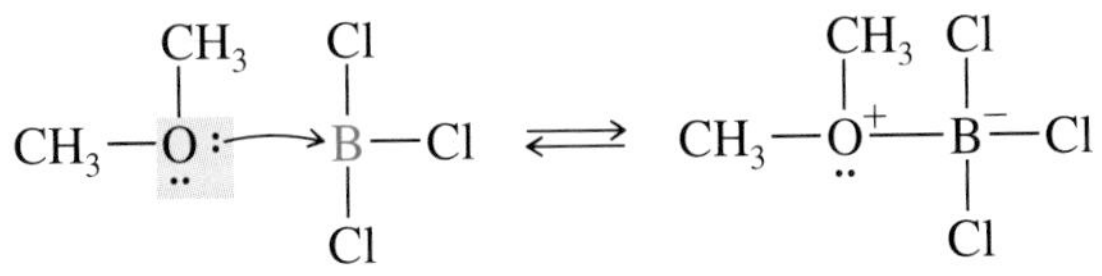

| dimethyl ether | boron trichloride | complex of dimethyl ether and boron trichloride |
| Lewis base | Lewis acid | |

Ethanol may donate a pair of electrons to a metal ion such as zinc(II).

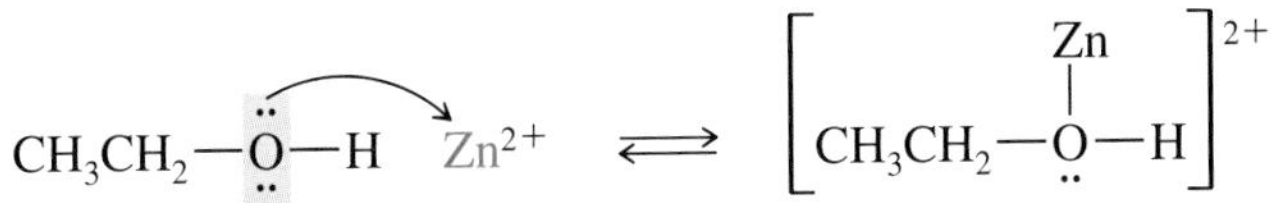

| ethanol | zinc cation | complex of zinc cation with ethanol |
| Lewis base | Lewis acid | |

In both these cases, a species with a vacant orbital accepts a pair of electrons from a donor species. Lewis acids are known as **aprotic acids,** compounds that react with bases by accepting pairs of electrons, not by donating protons.

⊘ **Study Guide**
**Concept Map 3.2**

**Problem 3.2** Write Lewis structures for the ions or molecules involved in the following reactions, and point out which ones are Lewis acids and which are Lewis bases.

(a) $Cu^{2+} + 4\,NH_3 \rightleftharpoons Cu(NH_3)_4^{2+}$

(b)

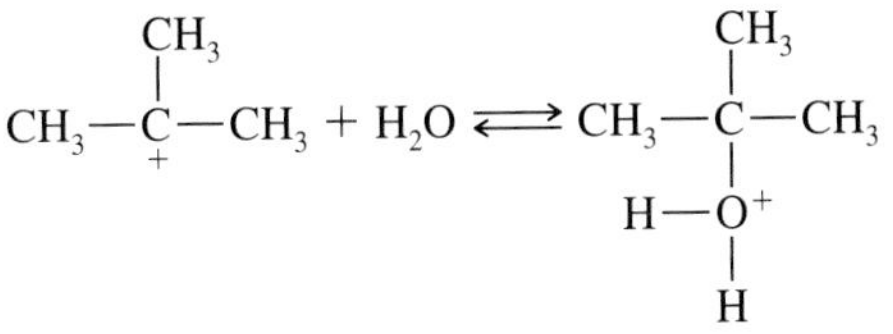

(c)

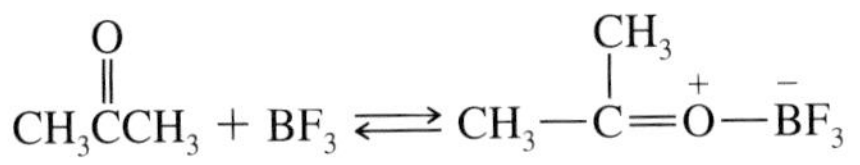

(d) 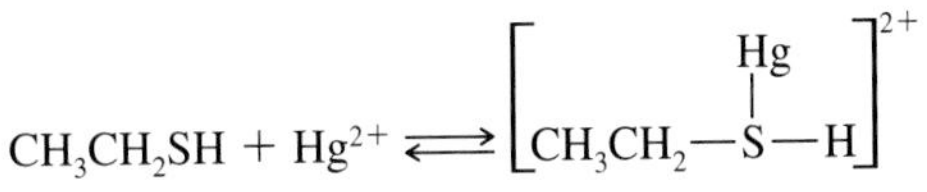 $CH_3CH_2SH + Hg^{2+} \rightleftharpoons \left[CH_3CH_2-\overset{Hg}{\underset{|}{S}}-H\right]^{2+}$

## 3.2 Reactions of Organic Compounds as Bases

Organic compounds with nonbonding electrons on nitrogen, oxygen, sulfur, or phosphorus can react as Lewis bases or Brønsted bases. They react with Lewis acids or Brønsted acids. We have already considered the reactions of nitrogen bases, or amines, with acids (pp. 49 and 79). Just as water is protonated to give a hydronium ion, organic oxygen compounds are protonated to give **oxonium ions.** The reaction of diethyl ether with concentrated hydriodic acid is typical of that of an oxygen base with a protic acid.

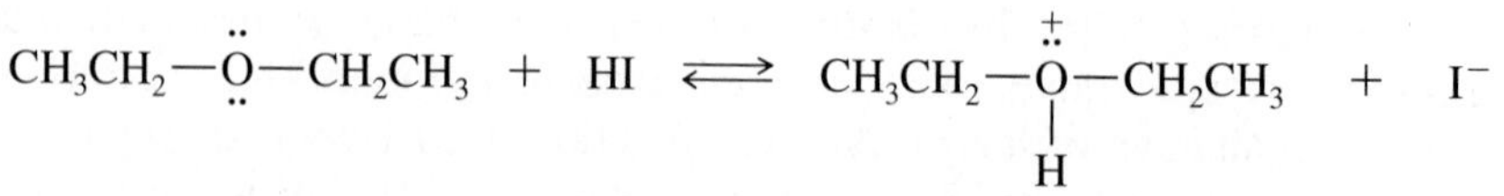

$$CH_3CH_2-\overset{..}{\underset{..}{O}}-CH_2CH_3 \;+\; HI \;\rightleftharpoons\; CH_3CH_2-\overset{+}{\underset{\underset{H}{|}}{\overset{..}{O}}}-CH_2CH_3 \;+\; I^-$$

| diethyl ether | hydriodic acid | diethyloxonium ion | iodide ion |
|---|---|---|---|
| *base* | *acid* | *conjugate acid of diethyl ether* | *conjugate base of hydriodic acid* |

Alcohols such as ethanol are protonated by acids.

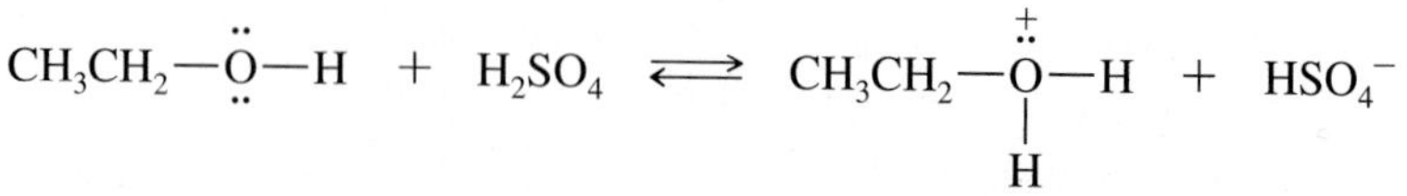

$$CH_3CH_2-\overset{..}{\underset{..}{O}}-H \;+\; H_2SO_4 \;\rightleftharpoons\; CH_3CH_2-\overset{+}{\underset{\underset{H}{|}}{\overset{..}{O}}}-H \;+\; HSO_4^-$$

| ethanol | sulfuric acid | ethyloxonium ion | hydrogen sulfate anion |
|---|---|---|---|
| *base* | *acid* | *conjugate acid of ethanol* | *conjugate base of sulfuric acid* |

A ketone is another type of oxygen compound that can behave as a base (p. 54). For example, acetone donates electrons to boron trifluoride, a Lewis acid.

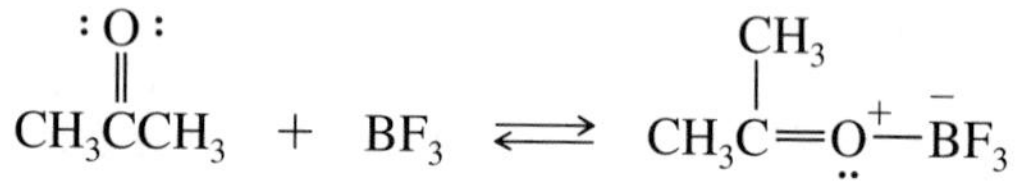

$$\overset{\overset{\displaystyle :\!\overset{..}{O}\!:}{\|}}{CH_3CCH_3} \;+\; BF_3 \;\rightleftharpoons\; \overset{\overset{\displaystyle CH_3}{|}}{CH_3C}=\overset{+}{\underset{..}{O}}-\overset{-}{BF_3}$$

| acetone | boron trifluoride | complex of boron trifluoride with acetone |
|---|---|---|
| *base* | *acid* | |

A **thiol** is a typical organic sulfur compound, formed when an organic group replaces one of the hydrogen atoms in hydrogen sulfide, $H_2S$. Ethanethiol reacts with the strong protic acid sulfuric acid.

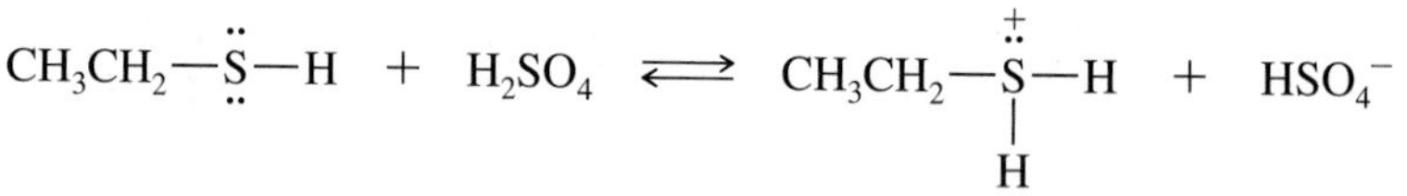

$$CH_3CH_2-\overset{..}{\underset{..}{S}}-H \;+\; H_2SO_4 \;\rightleftharpoons\; CH_3CH_2-\overset{+}{\underset{\underset{H}{|}}{\overset{..}{S}}}-H \;+\; HSO_4^-$$

| ethanethiol | sulfuric acid | conjugate acid of ethanethiol | hydrogen sulfate anion |
|---|---|---|---|
| *base* | *acid* | | *conjugate base of sulfuric acid* |

Compounds with $\pi$ bonds, such as alkenes and alkynes, are also bases. They too react with strong acids.

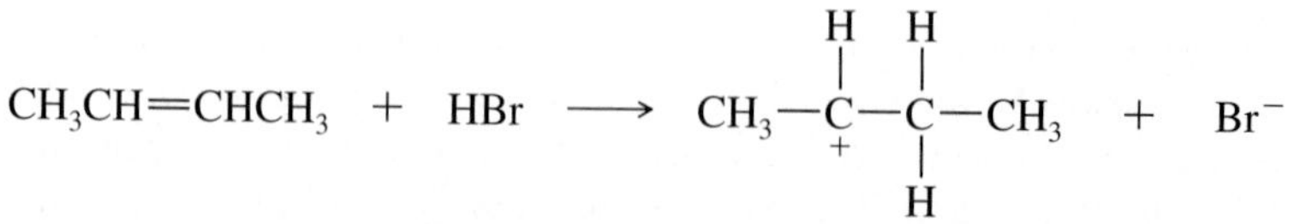

$$CH_3CH=CHCH_3 \;+\; HBr \;\longrightarrow\; \overset{\overset{\displaystyle H \;\; H}{| \;\; |}}{CH_3-\underset{\underset{H}{|}}{\overset{+}{C}}-\underset{}{C}-CH_3} \;+\; Br^-$$

| 2-butene | hydrogen bromide | sec-butyl cation a carbocation | bromide ion |
|---|---|---|---|
| *base* | *acid* | *conjugate acid of 2-butene* | *conjugate base of hydrogen bromide* |

When 2-butene reacts with the strong acid hydrogen bromide, a positively charged carbon atom, a **carbocation,** is formed. The $sp^2$-hybridized cationic carbon atom has an empty $p$ orbital (see Problem 2.30c) and is, therefore, a Lewis acid. In a subsequent reaction, the cation can combine with bromide anion, a Lewis base, to give 2-bromobutane.

$$\overset{..}{\underset{..}{:}}\!\!Br\!:^- \;+\; CH_3-\overset{\overset{H}{|}}{\underset{\underset{H}{|}}{C}}-\overset{\overset{H}{|}}{\underset{+}{C}}-CH_3 \;\longrightarrow\; CH_3-\overset{\overset{H}{|}}{\underset{\underset{Br}{|}}{C}}-\overset{\overset{H}{|}}{\underset{\underset{H}{|}}{C}}-CH_3$$

|  |  |  |
|---|---|---|
| bromide ion | *sec*-butyl cation | 2-bromobutane |
| *base* | *acid* | |

Generally, nonbonding electrons on nitrogen, oxygen, and sulfur are more loosely held and are protonated more easily than the electrons of $\pi$ bonds (see Figure 2.27).

In a Brønsted acid–base reaction, a proton is transferred from the acid to the base to give products that are the conjugate base and conjugate acid of the reactants. In a Lewis acid–base reaction, a pair of electrons is donated from the Lewis base to the Lewis acid to give a single product in which those electrons are shared in a covalent bond.

---

**Problem 3.3** Write equations showing the reactions you would expect for the following compounds with $AlCl_3$ (a Lewis acid) and with $H_2SO_4$ (a strong protic acid).

(a) $CH_3NHCH_3$
dimethylamine

(b) $\overset{\overset{O}{\|}}{HCH}$
formaldehyde

(c) $HC\equiv CH$
acetylene

(d) $CH_3CH_2\overset{\overset{CH_2CH_3}{|}}{P}CH_2CH_3$
triethylphosphine

(e) $CH_3CH_2OH$
ethanol

(f) $CH_3CH_2SCH_2CH_3$
diethyl sulfide

---

Some organic compounds have more than one atom with nonbonding electrons; more than one site in such a molecule can react with acids. For example, acetamide has nonbonding electrons on both an oxygen and a nitrogen atom, and either may be protonated.

$$\overset{\overset{:O:}{\|}}{CH_3CNH_2} \;+\; H_2SO_4 \;\rightleftharpoons\; \overset{\overset{+}{\overset{:O-H}{\|}}}{CH_3CNH_2} \;\text{or}\; \overset{\overset{:O:}{\|}}{CH_3CNH_3}\!\!^+ \;+\; HSO_4^-$$

| acetamide | sulfuric acid | conjugate acids of acetamide | hydrogen sulfate anion |
|---|---|---|---|
| *base* | | | |

$$\overset{\overset{:O:}{\|}}{CH_3COH} \;+\; H_2SO_4 \;\rightleftharpoons\; \overset{\overset{+}{\overset{:O-H}{\|}}}{CH_3COH} \;\text{or}\; \overset{\overset{:O:}{\|}}{CH_3COH_2}\!\!^+ \;+\; HSO_4^-$$

| acetic acid | sulfuric acid | conjugate acids of acetic acid | hydrogen sulfate anion |
|---|---|---|---|
| *base* | | | |

Note that neither compound is protonated twice. Once a molecule has been protonated and bears a positive charge, the availability of the other electrons in the system is greatly reduced. In principle, it is possible to protonate the compound again, but only under very strongly acidic conditions. Normally, the reaction stops when one proton is added to the molecule.

Both acetamide and acetic acid are more readily protonated at the oxygen atom of the carbonyl group than at their other basic site. This experimental observation can be explained by writing resonance contributors for each cation. The form of acetamide in which the oxygen atom is protonated, for example, has resonance contributors with the positive charge distributed among an oxygen, a carbon, and a nitrogen atom.

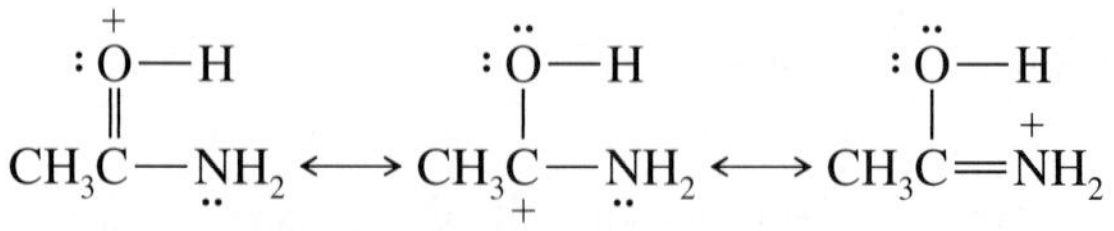

$$\overset{+}{:}\!\overset{\cdot\cdot}{\text{O}}\!-\!\text{H} \qquad :\overset{\cdot\cdot}{\text{O}}\!-\!\text{H} \qquad :\overset{\cdot\cdot}{\text{O}}\!-\!\text{H}$$

$$\text{CH}_3\text{C}\!-\!\text{NH}_2 \longleftrightarrow \text{CH}_3\overset{}{\text{C}}\!-\!\text{NH}_2 \longleftrightarrow \text{CH}_3\text{C}\!=\!\overset{+}{\text{NH}}_2$$

*resonance contributors for the cation
resulting from the protonation of acetamide
at the oxygen atom; the cation is
stabilized by delocalization of charge*

The spreading of charge over two or more atoms is called **delocalization of charge.** Cations (or anions) for which delocalization of charge is possible are more stable than species in which charge is localized. The cation shown above can be compared with the cation that results if the nitrogen atom in acetamide is protonated.

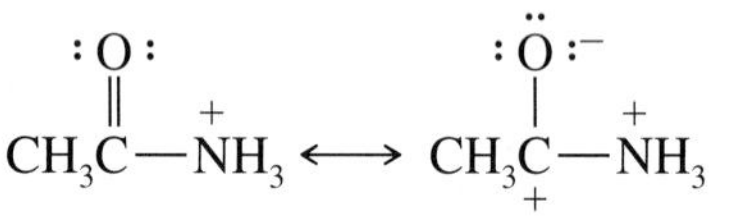

$$:\text{O}: \qquad\qquad :\overset{\cdot\cdot}{\text{O}}:^{-}$$

$$\text{CH}_3\text{C}\!-\!\overset{+}{\text{NH}}_3 \longleftrightarrow \text{CH}_3\overset{+}{\text{C}}\!-\!\text{NH}_3$$

*highly unfavorable
resonance contributor;
two positive charges
adjacent to each other
and a negative charge
is separated from a
positive charge*

*no resonance contributor that delocalizes
the positive charge at the nitrogen atom*

This cation from acetamide in which the nitrogen has been protonated has a high concentration of positive charge at a single atom and is, therefore, less stable than the one in which the oxygen has been protonated.

The protonation of the nonbonding electrons on the oxygen atom of a carbonyl or a hydroxyl group is an important first step in the reactions under acidic conditions such as acetamide, acetic acid, and alcohols. The conjugate acids of these compounds are usually more reactive toward Lewis bases than the unprotonated forms are. Consequently, acids are often used as catalysts to promote reactions of organic compounds.

**Problem 3.4**    The following compounds all have more than one site where they can react with a strong protic acid such as sulfuric acid. For each compound, draw Lewis structures and write equations showing the structures of the different products you would expect to obtain from protonation reactions.

(a) $\overset{\text{O}}{\overset{\|}{\text{CH}_3\text{COCH}_2\text{CH}_3}}$

ethyl acetate

(b) $\text{HOCH}_2\text{CH}_2\overset{\overset{\text{CH}_3}{|}}{\text{N}}\text{CH}_3$

2-(*N,N*-dimethylamino) ethanol

(c) $\text{CH}_2\!=\!\text{CHCH}_2\text{OH}$

2-propen-1-ol

(d) $\text{CH}_3\text{N}\!=\!\text{C}\!=\!\text{O}$

methyl isocyanate

**Problem 3.5**   Give the structural formula of the conjugate acid of each of the following compounds.

(a)
$$CH_3CH_2 \quad H$$
$$C=C$$
$$H \quad CH_2CH_3$$

(b) $CH_3CH_2OCH_2CH_3$

(c) $CH_3CH_2NCH_2CH_3$ with H on N

(d) $CH_3CH_2CH_2SCH_3$

(e)
$$O$$
$$\|$$
$$CH_3CH_2CH$$

(f) $CH_3CH_2CH_2OH$

(g)
$$O$$
$$\|$$
$$CH_3CH_2CCH_2CH_3$$

## 3.3   The Use of Curved Arrows in Writing Mechanisms

Almost all chemical transformations involve a change in the connectivity of atoms. Bonds are broken; new bonds are formed. Chemists are interested not only in the reactants and the products of a reaction but also in the details, especially the timing, of the bond-breaking and bond-forming processes. In order to study these details, chemists examine the progress of reactions using many experimental techniques. Modern laser technology, especially, is giving chemists a way to look at reactions that take place very fast, in $10^{-13}$ seconds, for example. Chemists study the rate of a chemical reaction, how fast it goes, and see how the rate and the products vary when experimental conditions are changed. They explore changes in stereochemistry that occur during the reaction. From observations such as these, they postulate the details of the pathway that the reactants follow as they are converted into the products. Such a detailed reaction pathway is called the **mechanism** of the reaction.

Mechanisms serve as an aid to understanding the chemical reactivity of different types of organic compounds. Subsequent sections of this book will contain many mechanisms proposed for reactions. In this section we explore the symbolism widely used by organic chemists to trace the details of the changes that occur as reactants are converted to products. The symbolism is a powerful tool with which chemists organize their knowledge of a large number of reactions into a relatively few types of transformations. Understanding reaction mechanisms and practice in drawing them will enable you to learn, recall, and predict chemical reactions.

Any strong acid, when put into water, transfers protons to the water to give hydronium ions, the real acids in aqueous solutions. The overall equation is familiar. For example, gaseous hydrogen chloride dissolves in water to give the solution we call hydrochloric acid. Hydrochloric acid is really a solution containing hydronium ions and chloride ions.

$$HCl(g) \;+\; H_2O \;\rightleftarrows\; H_3O^+ \;+\; Cl^-$$

| hydrogen chloride | water | hydronium ion | chloride ion |
|---|---|---|---|
| *acid* | *base* | *conjugate acid* | *conjugate base* |

In this reaction, a bond is broken between hydrogen and chlorine, and a new bond is formed between hydrogen and the oxygen in water. There are at least three possible ways in which we can imagine this process as taking place.

1. Bond-breaking occurs first, followed by bond-forming.

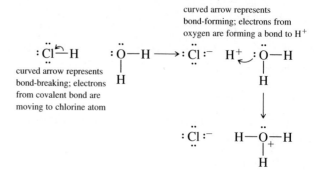

curved arrow represents bond-forming; electrons from oxygen are forming a bond to H⁺

curved arrow represents bond-breaking; electrons from covalent bond are moving to chlorine atom

2. Bond-forming takes place first, and then bond-breaking occurs.

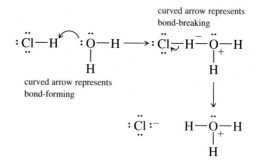

curved arrow represents bond-breaking

curved arrow represents bond-forming

3. Bond-breaking and bond-forming take place simultaneously.

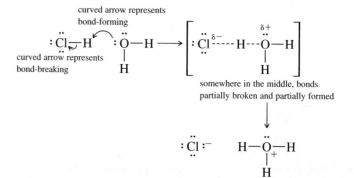

curved arrow represents bond-forming

curved arrow represents bond-breaking

somewhere in the middle, bonds partially broken and partially formed

If we examine these three possible processes critically, we find problems with Process 1 and Process 2. In the first process we create the hydrogen cation, usually called a **proton.** The problem with this is that a naked proton, without the electron cloud that makes it a hydrogen atom, is very small ($\sim 10^{-5} \times$ the size of an atom). The positive charge concentrated in such a small volume attracts electrons strongly. A proton, $H^+$, is unstable in the presence of other species that have electrons to donate. It is always transferred from one base to another.

The equation for Process 2 shows a species with a hydrogen atom with two covalent bonds (or four electrons) to it and a negative formal charge. Such a species violates the rule that hydrogen can accept only two electrons into its bonding molecular orbital (p. 44). We can therefore dismiss this as a reasonable process for hydrogen. (Note that for atoms with room in their valence shells for larger numbers of electrons, such a process might very well be a good one.)

The third equation, representing a process in which bond-breaking and bond-forming occur at roughly the same rate, seems the most reasonable one for this reaction. In the equation, in brackets, we see a species that represents the halfway point at which the proton is losing its bond to the chlorine atom, which is beginning to look like a chloride ion. Meanwhile, the proton is taking electrons away from the oxygen atom of the water molecule. The oxygen atom is beginning to acquire a positive charge. This transitional stage is usually not shown when we write a mechanism. For example, the transfer of the proton from hydrogen chloride to water is shown below.

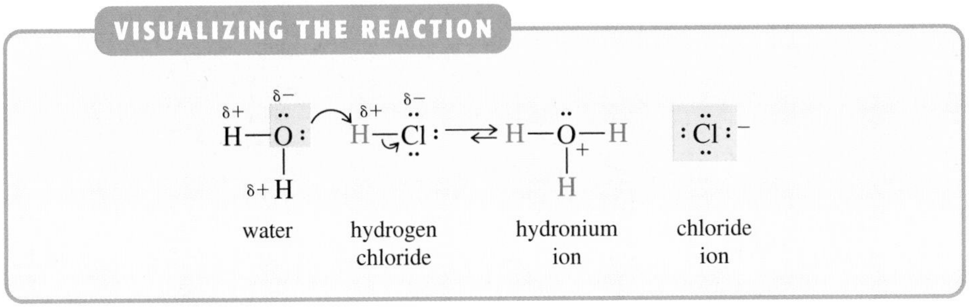

**VISUALIZING THE REACTION**

| water | hydrogen chloride | hydronium ion | chloride ion |

The curved arrows in the equation above show the electronic changes that take place as covalent bonds are formed and broken during the reaction. Such arrows are part of a symbolic language used by many chemists. This kind of symbolism has several important features. First, the structures of both reactants are written to show reactive electrons and sites of electron deficiency. A **curved arrow** is drawn from the site of electron availability, such as a pair of nonbonding electrons, to the site of electron deficiency, such as an atom with a partial positive charge. This arrow depicts the start of the bonding process. It represents the motion of an electron pair rather than the movement of atoms. If the transfer of electrons would result in too many electrons around an atom, such as two covalent bonds to a hydrogen atom or a violation of the octet rule for a second period atom, another curved arrow is drawn to symbolize the simultaneous release of bonding electrons to some atom or group of atoms, which then detaches itself. Chloride ion is the species formed by this process in the equation above.

Let us look more closely at another reaction, the reaction of ethanol with sulfuric acid.

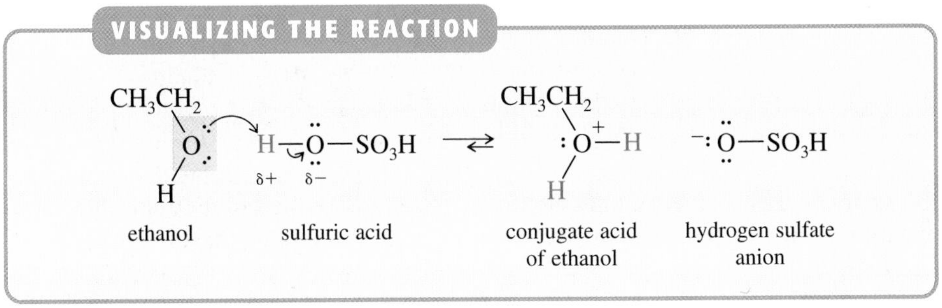

**VISUALIZING THE REACTION**

| ethanol | sulfuric acid | conjugate acid of ethanol | hydrogen sulfate anion |

The nonbonding electrons on the oxygen atom are shown moving toward the partially positively charged hydrogen atom of sulfuric acid. As the proton is transferred to the oxygen atom of ethanol, the hydrogen sulfate ion is detached, taking with it the pair of electrons that bonded it to the proton.

The $\pi$ electrons of a multiple bond may also serve as the site of electron availability in a reaction.

**VISUALIZING THE REACTION**

**Protonation of an Alkene**

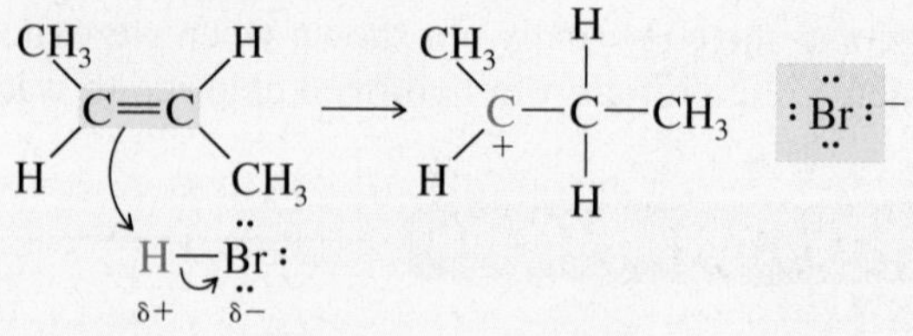

2-butene    hydrobromic    *sec*-butyl    bromide
acid    cation    ion

The reaction of 2-butene with hydrobromic acid starts with the bonding of the $\pi$ electrons with the positively polarized hydrogen atom of the hydrogen halide. The carbon atom that loses its share of the electrons in the $\pi$ bond becomes positively charged. Bromide ion is the other product of the reaction.

Whenever a mechanism is written for a reaction, it is important to make sure that the net charge is the same on both sides of the equation. Thus, in all the previous examples within this chapter, two neutral reactants with a net charge of zero give products with one positive and one negative charge, also a net charge of zero.

The site of electron deficiency may be an empty orbital on a positively charged atom. The reaction of the *sec*-butyl cation with bromide ion illustrates this kind of Lewis acid–base reaction.

**VISUALIZING THE REACTION**

**Reaction of a Carbocation with an Anion**

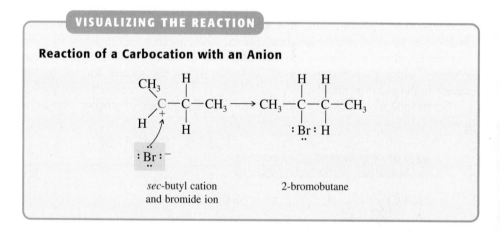

*sec*-butyl cation    2-bromobutane
and bromide ion

The curved arrow shows the transfer of electrons from the bromide ion, the site of electron availability, to the positively charged carbon atom, the site of electron deficiency.

The equations and the description given in this section provide an introduction to a language that will be used frequently in the rest of this book. It is important that you begin to practice it right away. Observe carefully the way that the curved arrows are drawn, especially where they start and where they end. These arrows, which represent the movement of an electron pair as bonds are formed and broken, must be used with precision. The tail of the arrow is associated with the pair of electrons that is being used to form a new bond or is being transferred to an atom in the breaking of a bond. The head of the arrow points between the two atoms where the new bond is to be formed or to the atom to which electrons are being transferred in the breaking of a bond.

**Problem 3.6**    Rewrite the equations in your answers for Problems 3.3 and 3.4 to show how electrons move in the reactions.

**Problem 3.7**    In the following equations, reactants and arrows showing the transfer of electrons are indicated. Supply the products of the reactions.

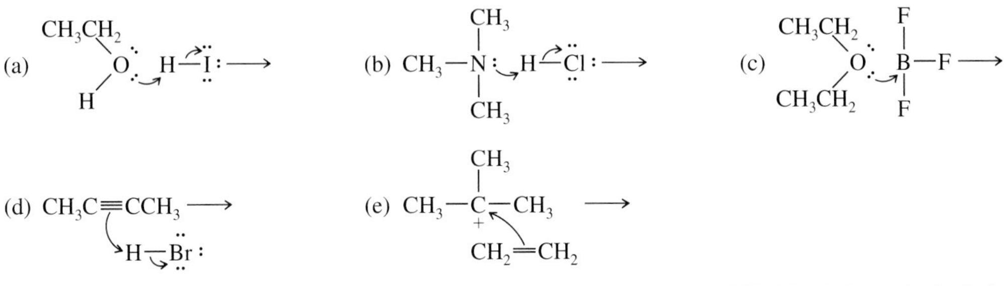

(a)
$$CH_3CH_2 \overset{\displaystyle}{\underset{H}{\diagdown}} \ddot{O} \cdot H \overset{\frown}{-} \ddot{I}: \longrightarrow$$

(b) $CH_3 - \overset{\displaystyle CH_3}{\underset{\displaystyle CH_3}{\overset{|}{N}}} : \, H \overset{\frown}{\diagdown} \ddot{C}l: \longrightarrow$

(c)
$$CH_3CH_2 \overset{\displaystyle}{\underset{CH_3CH_2}{\diagdown}} \ddot{O} \cdot \overset{\displaystyle F}{\underset{F}{\overset{|}{B}}} - F \longrightarrow$$

(d) $CH_3C \equiv CCH_3 \longrightarrow$
$$\overset{\displaystyle}{\underset{\displaystyle \diagdown H \overset{\frown}{\diagdown} \ddot{B}r:}{}}$$

(e) $CH_3 - \overset{\displaystyle CH_3}{\underset{\displaystyle \overset{+}{C}H_2 = CH_2}{\overset{|}{C}}} - CH_3 \longrightarrow$

# 3.4    Carbon, Nitrogen, Oxygen, Sulfur, and Halogen Acids

## A. Relative Acidities and Basicities Across a Period in the Periodic Table

The reactions of an organic compound as an acid depend on the ease with which it can lose a proton to a base. The acidity of a hydrogen atom bound to a central atom increases the further to the right the central atom is in any period of the periodic table. Ethane, in which the hydrogen atoms are bonded to carbon, is a very weak acid. The hydrogen atoms on the nitrogen atom of methylamine are more acidic, while the hydrogen atom bonded to the oxygen atom of methanol is even more acidic.

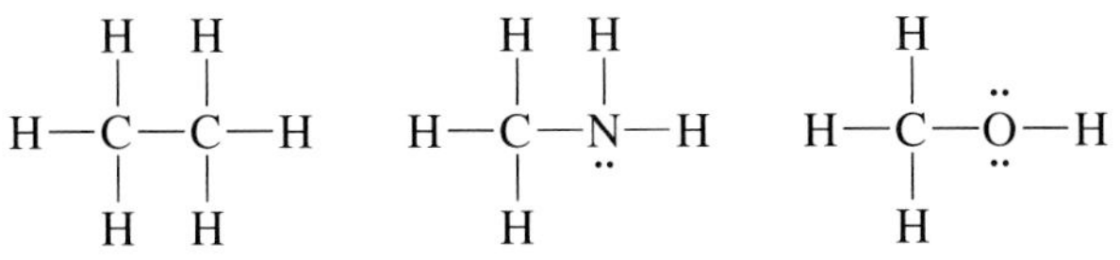

$$\underset{\text{ethane}}{H - \overset{\displaystyle H \quad H}{\underset{\displaystyle H \quad H}{\overset{|\quad\ |}{\underset{|\quad\ |}{C - C}}}} - H} \qquad \underset{\text{methylamine}}{H - \overset{\displaystyle H \quad H}{\underset{\displaystyle H}{\overset{|\quad\ |}{\underset{|}{C - \ddot{N}}}}} - H} \qquad \underset{\text{methanol}}{H - \overset{\displaystyle H}{\underset{\displaystyle H}{\overset{|}{\underset{|}{C - \ddot{O}}}}} - H}$$

*increasing acidity of hydrogen bonded to carbon, nitrogen, and oxygen*

The conjugate bases of ethane, methylamine, and methanol are shown below.

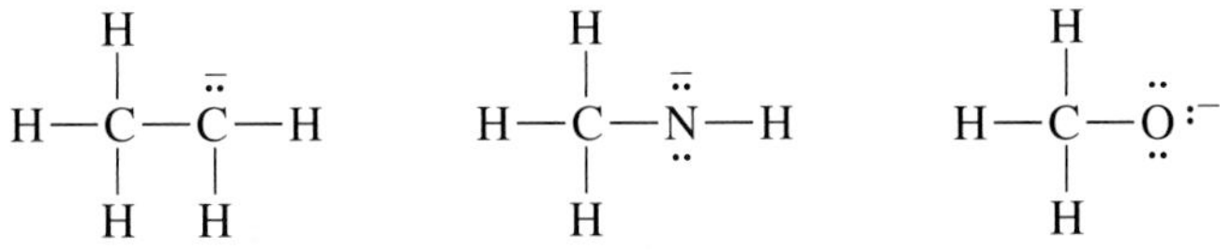

$$H - \overset{\displaystyle H}{\underset{\displaystyle H \quad H}{\overset{|}{\underset{|\quad\ |}{C - \overset{..}{\underset{..}{C}}}}}} - H \qquad H - \overset{\displaystyle H}{\underset{\displaystyle H}{\overset{|}{\underset{|}{C - \overset{..}{\underset{..}{N}}}}}} - H \qquad H - \overset{\displaystyle H}{\underset{\displaystyle H}{\overset{|}{\underset{|}{C - \overset{..}{\underset{..}{O}}}}}} :^-$$

ethyl anion                    methylamide                    methoxide
                                   anion                            anion

*a carbanion*

*conjugate base of ethane;*          *conjugate base*          *conjugate base of*
*strongest base*                     *of methylamine*          *methanol; weakest base*

The weakest acid in the series, ethane, has the strongest conjugate base, and the strongest acid, methanol, has the weakest conjugate base.

The conjugate base of ethane is ethyl anion, a carbanion. **Carbanions** are very strong bases. The ethyl anion, for example, will deprotonate methylamine.

| ethyl anion | methylamine | ethane | methylamide anion |
|---|---|---|---|
| *base* | *acid* | *conjugate acid of ethyl anion* | *conjugate base of methylamine* |

The anion of methylamine, in turn, will deprotonate methanol.

| methylamide anion | methanol | methylamine | methoxide anion |
|---|---|---|---|
| *base* | *acid* | *conjugate acid of methylamide anion* | *conjugate base of methanol* |

The relative basicities of the ethyl anion, the methylamide anion, and the methoxide anion are related to the electronegativities of the carbon, nitrogen, and oxygen atoms. Carbon is less electronegative than nitrogen, which means that it attracts and holds the electrons around it less firmly than nitrogen does. For this reason, an ethyl anion is less stable relative to its conjugate acid, ethane, than methylamide anion is relative to methylamine. Similarly, because oxygen is more electronegative than nitrogen is, methoxide anion is more stable relative to methanol than methylamide is relative to methylamine. Any factor that stabilizes an anion lowers its basicity, making it less likely to be protonated to form its conjugate acid.

Assertions about acidity that depend on the concept of electronegativity can only be made about species containing atoms from the same period of the periodic table. Comparisons based on the electronegativity of atoms in different periods do not work. Atomic size becomes an important factor in comparing elements from different periods, as we will see in the next section.

**Problem 3.8**  Arrange the species in each of the following sets in order of decreasing basicity.

(a) $CH_3NHCH_3$, $CH_3OCH_3$  (b) $NH_2^-$, $CH_3^-$, $F^-$, $OH^-$  (c) $CH_3\bar{N}CH_3$, $CH_3O^-$

## B. Relative Acidities and Basicities Within Groups in the Periodic Table

In the reaction between methanol and hydroxide ion, equilibrium is reached when the concentrations of hydroxide ion and methoxide ion are roughly equal, because the acidities of methanol and water are similar. The substitution of a methyl group for a hydrogen atom in the water molecule does not greatly affect the acidity of the hydroxyl group.

**VISUALIZING THE REACTION**

**A Proton Transfer Reaction**

site of electron deficiency

$$H-\overset{\overset{\displaystyle H}{|}}{\underset{\underset{\displaystyle H}{|}}{C}}-\overset{\delta-}{\ddot{O}}\overset{\delta+}{-}H \quad \overset{-}{:}\ddot{O}-H \rightleftharpoons H-\overset{\overset{\displaystyle H}{|}}{\underset{\underset{\displaystyle H}{|}}{C}}-\ddot{\ddot{O}}:^{-} \quad H-\ddot{O}-H$$

methanol          hydroxide              methoxide          water
                      ion                        ion

In the reaction of methanethiol with hydroxide ion, the equilibrium lies much more to the right.

**VISUALIZING THE REACTION**

**Deprotonation of a Thiol**

$$H-\overset{\overset{\displaystyle H}{|}}{\underset{\underset{\displaystyle H}{|}}{C}}-\ddot{S}-H \quad \overset{-}{:}\ddot{O}-H \rightleftharpoons H-\overset{\overset{\displaystyle H}{|}}{\underset{\underset{\displaystyle H}{|}}{C}}-\ddot{S}:^{-} \quad H-\ddot{O}-H$$

methanethiol        hydroxide          methanethiolate          water
                        ion                    anion

A hydrogen atom bonded to a sulfur atom is more acidic than one bonded to oxygen and is transferred more completely to hydroxide ion. Another way of describing the same phenomenon is to say that the methanethiolate anion is a weaker base than hydroxide ion (or methoxide ion). Sulfur is below oxygen in the periodic table and is a larger atom. Therefore, a negative charge on sulfur is more spread out—less concentrated at any one point in space.

The negative charge on a sulfur anion is more diffused than the negative charge on oxygen. The sulfur anion is, therefore, more stable relative to its conjugate acid than the oxygen anion is. The sulfur acid is more likely to lose its proton to the base, hydroxide ion. In other words, methanethiol is a stronger acid than methanol, and methanethiolate is a weaker base than methoxide ion. A more stable anion corresponds to a weaker base, one that is less likely to be protonated to form its conjugate acid. The relationships discussed above can be shown pictorially by drawing two diagrams showing the relative energies of the species involved in the two equations (Figure 3.1, p. 92).

The more stable a species is, the lower in energy it is. It appears lower on an energy diagram than a less stable species, which has higher energy. The energy diagram for the reaction of methanol with hydroxide ion shows little difference in energy between the reactants and the products in that reaction. There is a much larger difference in energy between the reactants and the products when methanethiol and hydroxide ion react. The products are much farther "downhill" in this case.

One of the reactants, hydroxide ion, and one of the products, water, are the same for both reactions; therefore, we can use these energy diagrams to tell us something

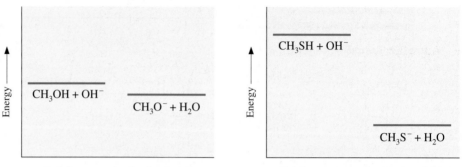

**Figure 3.1**
Energy diagrams for the reactions of methanol and methanethiol with hydroxide ion.

about how each acid compares with its own conjugate base. Methanethiolate anion is much more stable (lower in energy) relative to its conjugate acid, methanethiol, than methoxide ion is relative to methanol. Stabilization of the conjugate base results in increased acidity for the conjugate acid. Therefore, at equilibrium, there is more methanethiolate ion and water than there is methanethiol and hydroxide ion.

In summary, the basicity of an anion is lower the further down the atom is in any group in the periodic table; the acidity of the corresponding conjugate acid increases in the same order.

$$R-\ddot{O}:^{-} \ > \ R-\ddot{S}:^{-}$$

*stronger base     weaker base*

$$R-\ddot{O}-H < R-\ddot{S}-H$$

*weaker acid     stronger acid*

The same trend is characteristic of the halogen acids, which become stronger from hydrogen fluoride to hydrogen iodide.

$$HF < HCl < HBr < HI$$

*weakest acid          strongest acid*

$$F^{-} > Cl^{-} > Br^{-} > I^{-}$$

*strongest base          weakest base*

⊘ **Study Guide**
**Concept Map 3.3**

In general, then, within any group in the periodic table, the acidity of a protic acid increases with the size of the atom to which the proton is bonded.

**Problem 3.9**   Which is the stronger acid, $H_2O$ or $H_2S$?

**Problem 3.10**   Write an equation showing the reaction you expect between ethanethiol, $CH_3CH_2SH$, and methoxide anion, $CH_3O^{-}$.

## C. Organic Cations as Acids

The protonated forms of organic compounds, which are organic cations, are also acids. The conjugate base of methylammonium ion is methylamine; methylamine is a stable, neutral molecule; therefore, methylammonium ion loses a proton relatively easily.

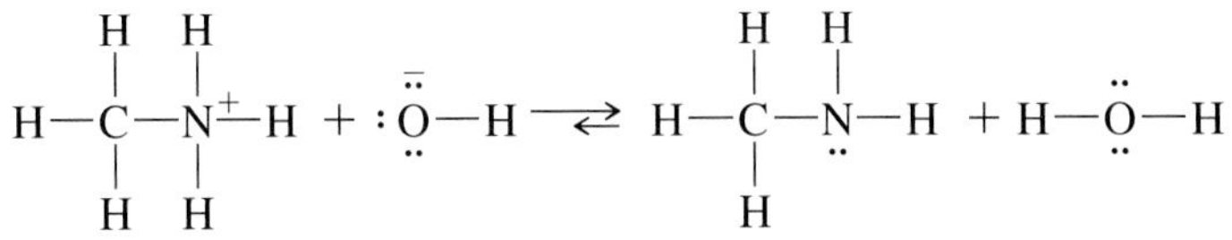

$$H\overset{\overset{\displaystyle H}{|}}{\underset{\underset{\displaystyle H}{|}}{C}}\overset{\overset{\displaystyle H}{|}}{\underset{\underset{\displaystyle H}{|}}{N^{+}}}H \; + \; :\!\overset{..}{\underset{..}{O}}\!-\!H \; \rightleftharpoons \; H\overset{\overset{\displaystyle H}{|}}{\underset{\underset{\displaystyle H}{|}}{C}}\overset{\overset{\displaystyle H}{|}}{\underset{}{\overset{..}{N}}}\!-\!H \; + \; H\!-\!\overset{..}{\underset{..}{O}}\!-\!H$$

| methylammonium ion | hydroxide ion | methylamine | water |
|---|---|---|---|
| | | *conjugate base of methylammonium ion* | *conjugate acid of hydroxide ion* |
| *acid* | *base* | | |

Protonation reactions of alcohols, ethers, and ketones yield oxonium ions, which are acidic. Methyloxonium ion, the conjugate acid of methanol, has an acidity similar to that of the hydronium ion.

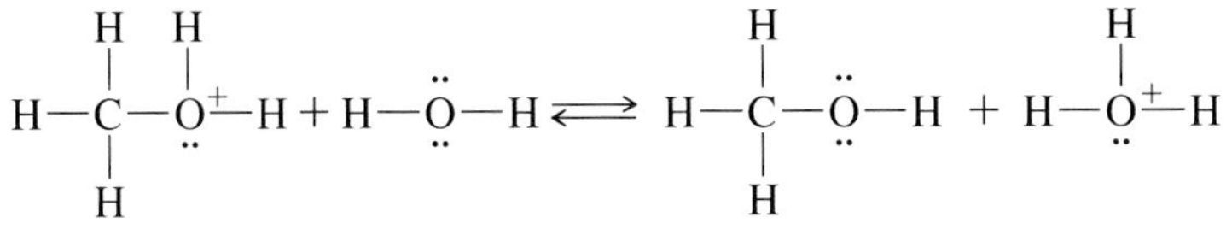

$$H\overset{\overset{\displaystyle H}{|}}{\underset{\underset{\displaystyle H}{|}}{C}}\overset{\overset{\displaystyle H}{|}}{\underset{..}{O^{+}}}H + H\!-\!\overset{..}{\underset{..}{O}}\!-\!H \rightleftharpoons H\overset{\overset{\displaystyle H}{|}}{\underset{\underset{\displaystyle H}{|}}{C}}\!-\!\overset{..}{O}\!-\!H + H\!-\!\overset{\overset{\displaystyle H}{|}}{\underset{..}{O^{+}}}H$$

| methyloxonium ion | water | methanol | hydronium ion |
|---|---|---|---|
| | *base* | *conjugate base of methyloxonium ion* | |
| *acid* | | | *conjugate acid of water* |

Carbocations are strong Brønsted acids as well as Lewis acids (p. 79). The hydrogen atoms on the carbon atoms adjacent to the cationic carbon are acidic and easily lost to bases, even weak bases such as water.

$$\overset{+}{CH_3CHCH_3} \; + \; H_2O \; \rightleftharpoons \; CH_3CH{=}CH_2 \; + \; H_3O^{+}$$

| isopropyl cation | water | propene | hydronium ion |
|---|---|---|---|
| *acid* | *base* | *conjugate base of isopropyl cation* | *conjugate acid of water* |

**Problem 3.11**   (a) Water is a Lewis base as well as a Brønsted base. What other reaction besides the one shown above is possible between a carbocation and water? Write a mechanism for the reaction using the curved-arrow notation.
(b) The product of the reaction you wrote in part (a) is itself an acid. How will it react with excess water?

**Problem 3.12**   Arrange the species in each of the following sets in order of increasing acidity.

(a) $CH_3CH_2OH,$   $CH_3CH_2CH_3,$   $CH_3CH_2NH_2,$   $CH_3CH_2SH$

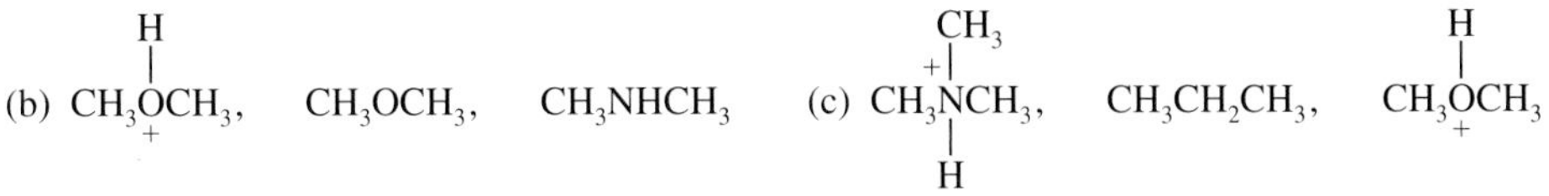

(b) $CH_3\overset{\overset{\displaystyle H}{|}}{\underset{\underset{\displaystyle +}{}}{O}}CH_3,$   $CH_3OCH_3,$   $CH_3NHCH_3$   (c) $CH_3\overset{\overset{\displaystyle CH_3}{+|}}{\underset{\underset{\displaystyle H}{|}}{N}}CH_3,$   $CH_3CH_2CH_3,$   $CH_3\overset{\overset{\displaystyle H}{|}}{\underset{\underset{\displaystyle +}{}}{O}}CH_3$

**Problem 3.13**   Give the structural formula for the conjugate base of each of the following species.

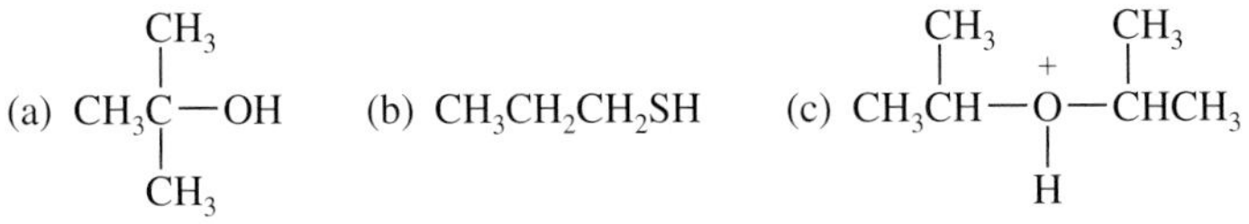

(a) $CH_3\overset{\overset{\displaystyle CH_3}{|}}{\underset{\underset{\displaystyle CH_3}{|}}{C}}{-}OH$   (b) $CH_3CH_2CH_2SH$   (c) $CH_3\overset{\overset{\displaystyle CH_3}{|}}{\underset{\underset{\displaystyle H}{|}}{CH}}{-}\overset{+}{O}{-}\overset{\overset{\displaystyle CH_3}{|}}{CH}CH_3$

(d) $CH_3-\!\!\!\bigcirc\!\!\!-\overset{\overset{H}{|}}{\underset{\underset{H}{|}}{\overset{+}{N}}}\!\!-H$   (e) $\bigcirc\!\!\!-OH$   (f) $CH_3CH_2\underset{\underset{F}{|}}{CH}\overset{\overset{O}{\|}}{C}OH$

## 3.5   Equilibria in Acid–Base Reactions

### A. Acidity Constants and pKₐ

Measuring the acidities of organic compounds involves the measurement of equilibrium constants. The **equilibrium constant, $K_{eq}$,** for any reaction is defined as the product of the concentrations of the products of the reaction divided by the product of the concentrations of the reactants at equilibrium. The result is a number that remains constant for that reaction under a given set of conditions, such as solvent and temperature.

When acetic acid is dissolved in water, certain concentrations of acetic acid, water, acetate ions, and hydronium ions are present at equilibrium. Acetate ion is more basic than water, so the equilibrium lies on the side of the undissociated acetic acid.

$$CH_3\overset{\overset{O}{\|}}{C}OH \;+\; H_2O \;\rightleftarrows\; CH_3\overset{\overset{O}{\|}}{C}O^- \;+\; H_3O^+$$

acetic acid      water      acetate anion      hydronium ion

The equilibrium constant for this reaction, $K_{eq}$, is written as

$$K_{eq} = \frac{[CH_3CO_2^-][H_3O^+]}{[CH_3CO_2H][H_2O]}$$

For dilute solutions, in which the concentration of water is large and almost constant, another expression, the **acidity constant, $K_a$,** is used. For acetic acid,

$$K_{eq}[H_2O] = K_a = \frac{[CH_3CO_2^-][H_3O^+]}{[CH_3CO_2H]} = 1.75 \times 10^{-5}$$

The acidity constants for different acids have magnitudes ranging from $10^{14}$ to $10^{-50}$. Such large and small numbers are most conveniently expressed as logarithms. Just as pH is $-\log [H_3O^+]$, so p$K_a$ is defined as $-\log K_a$. The p$K_a$ for acetic acid is 4.76, a value that is typical of many organic acids. A strong inorganic acid such as sulfuric acid has $K_a \sim 10^9$ and p$K_a \sim -9$. The p$K_a$ of hydronium ion is $-1.7$; that of ammonium ion is 9.4. Hydrocarbons, such as methane, are extraordinarily weak acids, with p$K_a$ values of approximately 50.

Inside the front cover of this book is a table of p$K_a$ values for a variety of acids, shown with their conjugate bases. Note that the strongest acids, with the weakest conjugate bases, are at the top of the table with negative or small positive p$K_a$ values. *The weaker the acid, the stronger is its conjugate base and the larger is its p$K_a$ value.* The p$K_a$ values for the very weak acids at the bottom of the table can be measured only indirectly, so different values for the same compound can be found in different tables. These p$K_a$ values are rough indications of relative acidities. Note that p$K_a$ values are only given for protic acids. As you use the table, you will acquire some sense of which acids are strong and which are weak. Do not attempt to memorize p$K_a$ values; refer to the table if the solution of a problem requires it.

**Problem 3.14**   The equation for the reaction of acetic acid with water is given on the preceding page. Write the equation for the reaction of sulfuric acid, $H_2SO_4$, with water. Sulfuric acid has a $K_a$ of $\sim 10^9$, while acetic acid has a $K_a$ of $\sim 10^{-5}$. What are the relative lengths of the reaction arrows for the reaction of sulfuric acid with water? Using Figure 3.1 (p. 92) as your guide, draw energy diagrams for the reaction of acetic acid with water and of sulfuric acid with water.

**Problem 3.15**   This is formic acid:

$$\overset{\overset{\textstyle O}{\|}}{HCOH}$$

Its $K_a$ is $1.99 \times 10^{-4}$. What is its $pK_a$? Is it a stronger or weaker acid than acetic acid, with a $pK_a$ of 4.76?

**Problem 3.16**   The imidazole ring is present in the amino acid histidine. This amino acid plays an important role in proton-transfer reactions in the human body. The protonated form of imidazole (imidazolinium ion) has a $pK_a$ of 7.0. What is $K_a$ for this acid?

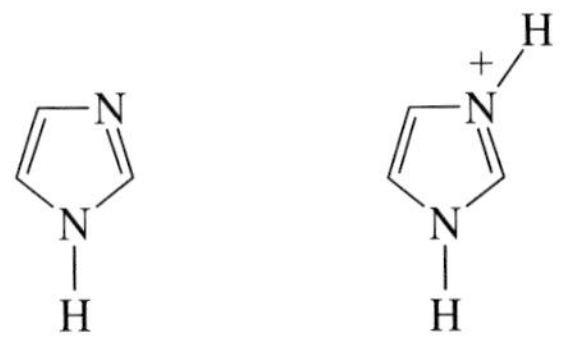

imidazole

imidazolinium ion

*conjugate acid of imidazole*

# B. Equilibrium, Free Energy, Enthalpy, and Entropy

The equilibrium constant for a reaction is related to the change in energy for the reaction. We have already seen this when we compared the reactions of methanol and methanethiol with hydroxide ion (p. 91). The equilibrium for the reaction of methanethiol with hydroxide ion in which there is a decrease in energy lies to the right, in the direction in which more product molecules than reactant molecules are present at equilibrium.

There are many different kinds of energy, such as heat energy, electrical energy, kinetic energy, and potential energy. The kind of energy that is related to the equilibrium constant is called the **free energy of a reaction.** The equation showing the relationship between the standard free energy of a reaction and the equilibrium constant of that reaction is

$$\Delta G^\circ = -RT \ln K_{eq}$$

or
$$\Delta G^\circ = -2.303 RT \log K_{eq}$$

$\Delta G^\circ$ is the change in free energy for the reaction when the reactants and the products are in their standard states (1 M solution for solutions; 1 atm of pressure for gases); $R$ is the gas constant, $1.987 \times 10^{-3}$ kcal/deg·mol; $T$ is the absolute temperature; and $K_{eq}$ is the equilibrium constant for the reaction. If there is a decrease in free energy during the reaction, that is, if $\Delta G^\circ$ is negative, the equilibrium constant is greater than 1. In other words, in that case the reaction proceeds so that more products than reactants are present at equilibrium. If $\Delta G^\circ$ is positive, the reverse reaction is favored. For example, chloroacetate anion and acetate anion compete for a proton in the following reaction:

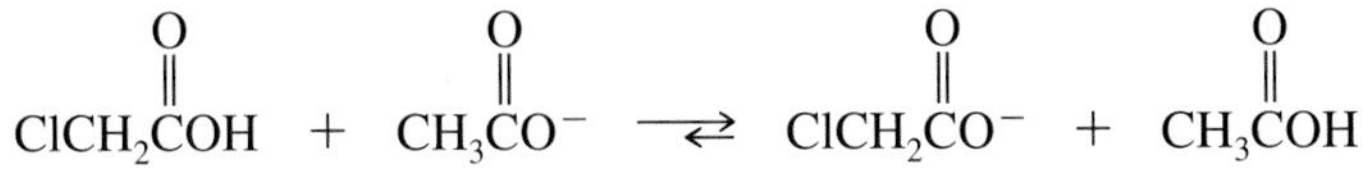

$$\underset{\substack{\text{chloroacetic}\\\text{acid}}}{\overset{\overset{\textstyle O}{\|}}{ClCH_2COH}} + \underset{\substack{\text{acetate}\\\text{anion}}}{\overset{\overset{\textstyle O}{\|}}{CH_3CO^-}} \;\rightleftharpoons\; \underset{\substack{\text{chloroacetate}\\\text{anion}}}{\overset{\overset{\textstyle O}{\|}}{ClCH_2CO^-}} + \underset{\substack{\text{acetic}\\\text{acid}}}{\overset{\overset{\textstyle O}{\|}}{CH_3COH}}$$

*acid*                    *base*                 *conjugate base of chloroacetic acid*        *conjugate acid of acetate anion*

$\Delta G°$ for this reaction at 298 K has been determined to be $-2.59$ kcal/mol. There is a *decrease* in free energy during the reaction; therefore, the equilibrium lies to the right. Acetate anion takes a proton away from chloroacetic acid; it is a stronger base than chloroacetate anion. Therefore, at equilibrium, there is more chloroacetate anion present than acetate anion.

The equilibrium constant for the above reaction can be calculated from $\Delta G°$ and is also related to the acidity constants for chloroacetic acid and acetic acid by this equation:

$$K_{eq} = \frac{K_a \text{ of } ClCH_2CO_2H}{K_a \text{ of } CH_3CO_2H}$$

**Problem 3.17**   Using the discussion of the $K_a$ for acetic acid (p. 94) as a model, write an equation for the ionization of chloroacetic acid, and derive the expression for the $K_a$ for the acid. Use the expressions for the $K_a$ of acetic acid and for that of chloroacetic acid to prove for yourself the relationship between $K_{eq}$ and $K_a$ for the two acids given above.

**Problem 3.18**   Calculate $K_{eq}$ for the reaction of chloroacetic acid and acetate anion and the $pK_a$ for chloroacetic acid from the data given above. $K_a$ for acetic acid is $1.75 \times 10^{-5}$.

The change in free energy, $\Delta G$, of the system determines whether the reaction will take place as shown in the equation for the reaction. $\Delta G$ has two components. One of them, $\Delta H$, is the change in **enthalpy,** also called the **heat of the reaction.** In Section 2.8, we used tables of bond energies to find out whether a reaction was endothermic or exothermic. In doing this, we were determining $\Delta H$, the change in enthalpy for that reaction.

The other component of $\Delta G$ is the change in entropy during the reaction, $\Delta S$. **Entropy** is a measure of disorder or randomness in the system. Anything that restricts the freedom of motion of molecules or within molecules leads to a decrease in entropy. For example, a reaction in which two molecules combine to give a single molecule results in a decrease in entropy. Examples of such reactions were seen in Chapter 2. One is given below:

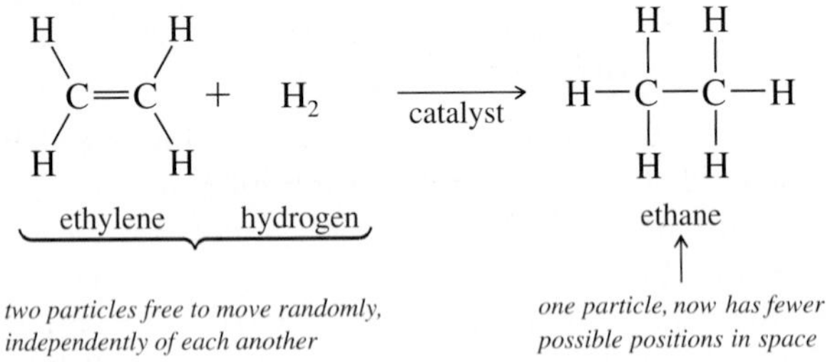

|                    |   |   |
| ethylene     hydrogen |   | ethane |
| *two particles free to move randomly, independently of each another* |   | *one particle, now has fewer possible positions in space* |

There is a decrease in entropy in going from reactants to products, $-\Delta S$, for this reaction.

A process in which a molecule dissociates to give two or more fragments gives an increase in entropy because each piece is free to move in ways that were not possible when they were bound together. For example, a change in entropy accompanies the dissolving of a solid in a liquid. An example is the solution of a crystal of sodium chloride.

$$\text{NaCl (s)} \xrightarrow[H_2O]{} \text{Na}^+ \text{ (aq) } + \text{Cl}^- \text{ (aq)}$$

| | |
| *sodium ions and chloride ions held firmly in place in crystal lattice* | *sodium ions and chloride ions free to move around in solution* |

In the crystal, all the sodium ions and the chloride ions are constrained to their positions in the crystal lattice. At ordinary temperatures, they vibrate in position but do not move from one part of the crystal to another. When the crystal dissolves, each ion becomes free to move throughout the entire volume of the solution. We know this because if we put crystalline salt or sugar in water and let it stand, eventually the entire volume of the solution will taste salty or sweet. Stirring speeds the process by which the dissolved particles move through the solution, but diffusion occurs without stirring by the random motion of the ions or molecules. This greatly increased possibility for motion results in an increase in the entropy of the system, $+\Delta S$. Part of the change in entropy comes from changes in the entropy of the solvent. The loss of hydrogen bonds among water molecules leads to an increase in entropy for water. On the other hand, the ordering of water molecules around the positively and negatively charged ions results in a decrease in entropy. The sum of all the changes—for the crystal, for the ions, and for the solvent molecules—is the overall change in entropy for the system.

The relationship among the changes in free energy, enthalpy, and entropy under standard conditions (p. 95) is summed up in the following equation:

$$\Delta G^\circ = \Delta H^\circ - T\Delta S^\circ$$

Most reactions, however, are not run under standard conditions. The relationship shown below holds for any set of conditions. The difference is indicated by removing the small superscript from the symbols for free energy, enthalpy, and entropy. For reactions that are not at standard conditions,

$$\Delta G = \Delta H - T\Delta S$$

If $\Delta G$ has a negative value for a particular reaction, that is, if there is a decrease in free energy during the reaction, the reactants will be converted to the products as shown in the equation for the reaction. If $\Delta G$ is positive, the reaction shown in the equation will not proceed; instead, the reverse reaction will occur. The species shown as the products on the right side of the equation will be converted to the reactants on the left side of the equation. If $\Delta G$ is zero, the reaction must be at equilibrium. Reactants and products are being converted to each other at the same rates, so no net change in the concentrations of the different species can be observed.

Negative values for $\Delta H$ and positive values for $\Delta S$ contribute to making $\Delta G$ negative. Temperature also plays a role, because small changes in entropy can be magnified when multiplied by the absolute temperature, especially at high temperatures. Exothermic reactions, reactions in which energy is lost to the surroundings as heat (p. 63), will occur if there is no large decrease in entropy. For many organic reactions, entropy changes tend to be small, and predictions about reactions can be based on the values of the enthalpy of reaction alone. However, although we have an intuitive feeling that exothermic reactions are favored and endothermic ones are not, this is not totally justified by the facts.

Study Guide
Concept Map 3.4

## C. Using the Table of p$K_a$ Values to Predict Acid–Base Reactions

Reactions in which a proton is transferred are fast and reversible. They are often the first reactions to occur whenever a strong enough acid is brought into the presence of a base.

The table of p$K_a$ values can be used to predict the extent to which acid–base reactions will take place. In general, an acid will transfer a proton to the conjugate base of any acid that is below it in the table. If the p$K_a$ values for the two acids differ by only one or two units, both acids will be present in substantial quantities at equilibrium. If there is a large difference in acidity, the transfer of the proton will be nearly complete. For example, suppose we wanted to know whether we could use ammonia, $NH_3$, to prepare free trimethylamine, $(CH_3)_3N$, from trimethylammonium chloride, $(CH_3)_3NH^+Cl^-$. The proton donor in this reaction is the trimethylammonium ion, $(CH_3)_3NH^+$, and the proton acceptor is ammonia.

$$CH_3\text{—}\overset{\overset{\displaystyle CH_3}{|}}{\underset{\underset{\displaystyle CH_3}{|}}{\overset{+}{N}}}\text{—}H \;+\; H\text{—}\overset{\overset{\displaystyle \cdot\cdot}{}}{\underset{\underset{\displaystyle H}{|}}{N}}\text{—}H \;\rightleftharpoons\; CH_3\text{—}\overset{\overset{\displaystyle CH_3}{|}}{\underset{\underset{\displaystyle CH_3}{|}}{N}}\!: \;+\; H\text{—}\overset{\overset{\displaystyle H}{|}}{\underset{\underset{\displaystyle H}{|}}{\overset{+}{N}}}\text{—}H$$

| trimethylammonium ion | ammonia | trimethylamine | ammonium ion |
|---|---|---|---|
| $pK_a$ 9.8 | *base* | *conjugate base of trimethylammonium ion* | $pK_a$ 9.4 |
| *acid* | | | *conjugate acid of ammonia* |

We look up the two acids, trimethylammonium ion and ammonium ion, in the $pK_a$ table and find that their $pK_a$ values are very close to each other, 9.4 for the ammonium ion and 9.8 for the trimethylammonium ion. The larger $pK_a$ value for the trimethylammonium ion indicates that trimethylamine is a stronger base than ammonia. In a competition for a proton, trimethylamine is more likely to retain the proton than to release it to ammonia. The species shown in the equation above are in equilibrium, but the arrow pointing to the left is larger than the one pointing to the right. This reaction could not be used to generate trimethylamine quantitatively from trimethylammonium ion.

To remove the proton from trimethylammonium ion, we must find a stronger base than the amine. A search of the table of $pK_a$ values reveals that hydroxide ion, the conjugate base of water with $pK_a$ 15.7, would be a good reagent to use.

$$CH_3\text{—}\overset{\overset{\displaystyle CH_3}{|}}{\underset{\underset{\displaystyle CH_3}{|}}{\overset{+}{N}}}\text{—}H \;+\; {}^-\!:\!\overset{\cdot\cdot}{\underset{\cdot\cdot}{O}}\text{—}H \;\rightleftharpoons\; CH_3\text{—}\overset{\overset{\displaystyle CH_3}{|}}{\underset{\underset{\displaystyle CH_3}{|}}{N}}\!: \;+\; H\text{—}\overset{\cdot\cdot}{\underset{\cdot\cdot}{O}}\text{—}H$$

| trimethylammonium ion | hydroxide ion | trimethylamine | water |
|---|---|---|---|
| $pK_a$ 9.8 | *base* | *conjugate base of trimethylammonium ion* | $pK_a$ 15.7 |
| *acid* | | | *conjugate acid of hydroxide ion* |

The difference in the $pK_a$ values of the two acids trimethylammonium ion and water is large, indicating that hydroxide ion is a much stronger base than trimethylamine. In the competition between the two bases, the proton is transferred essentially completely from the trimethylammonium ion to the hydroxide ion, freeing the weaker base, trimethylamine. As we have seen before, at equilibrium, the weaker acid and the weaker base are the major species found in the reaction mixture.

The table of $pK_a$ values does not include all possible acids and bases. What happens when we need to make a prediction about a compound that does not appear in the table? For instance, we might wish to know whether propyne, $CH_3C\equiv CH$, will be deprotonated by amide anion, $NH_2^-$.

$$CH_3C\equiv C\text{—}H \;+\; {}^-\!:\!\overset{\overset{\displaystyle H}{|}}{\underset{\underset{\displaystyle \cdot\cdot}{}}{N}}\text{—}H \;\rightleftharpoons\; CH_3C\equiv C\!:^- \;+\; H\text{—}\overset{\overset{\displaystyle H}{|}}{\underset{\underset{\displaystyle \cdot\cdot}{}}{N}}\text{—}H$$

| propyne | amide anion | methylacetylide anion | ammonia |
|---|---|---|---|
| $pK_a \sim 26$ | *base* | | $pK_a \sim 36$ |
| *acid* | | *conjugate base of propyne* | *conjugate acid of amide anion* |

Propyne does not appear in the table of $pK_a$ values. It has two kinds of hydrogen atoms, three bonded to the $sp^3$-hybridized carbon atom of the methyl group and one on the $sp$-hybridized carbon atom of the triple bond. The best analogy for the $pK_a$ of the hydrogen atoms on the methyl group is the $pK_a$ of methane, $pK_a$ 49. All other methyl groups that are identified as acids in the $pK_a$ table are next to groups such as carbonyl and nitro groups, which strongly affect the acidity of hydrogen on the carbon atoms next to them (p. 668). This is not true of the triple bond. The hydrogen atoms on the methyl group in propyne will, therefore, have $pK_a \sim 49$. Acetylene, HC≡CH, which resembles propyne, is listed with $pK_a$ 26. We can assign the same $pK_a$ value to propyne, reasoning that the acidity of propyne, like that of acetylene, is derived from the hydrogen atom bonded to the $sp$-hybridized carbon atom (p. 57) of the triple bond. (Because there are no other kinds of hydrogen atoms in acetylene, the acidity must come from that hydrogen.) We also reason that the substitution of an alkyl group for a hydrogen atom in acetylene will not dramatically change the acidity of the hydrogen atom on the triple bond. In general, functional groups are affected only in subtle ways by the particular alkyl group attached to them. Therefore, we can generalize further that a series of alkynes such as

$$CH_3C≡CH \qquad CH_3CH_2C≡CH \qquad CH_3CH_2CH_2C≡CH$$

propyne         1-butyne         1-pentyne

will have similar $pK_a$ values and similar reactivity. Propyne, therefore, has two sites of acidity, with the hydrogen on the $sp$-hybridized carbon atom being much more acidic ($\sim 10^{13}$ times so) than the one on the methyl group. It will behave as an acid with $pK_a \sim 26$. The process of reasoning by analogy to make judgments and estimates about unknown situations is one you should cultivate during your study of organic chemistry. Once we conclude that the $pK_a$ of propyne is approximately 26, we can complete the equation by showing that the alkyne will be deprotonated by amide anion, the conjugate base of ammonia, which has $pK_a$ 36.

The table of $pK_a$ values can be used to make predictions about a large number of acid–base reactions. Developing familiarity with the table and skill in using it will help you learn chemistry.

**Problem 3.19** From the table of $pK_a$ values, select an oxonium ion, an ammonium ion, a carboxylic acid, an alcohol, a thiol, an amine, and an alkane, and arrange them in order of increasing acidity. This series summarizes the trends that were discussed in Section 3.4 and will be useful in problems that follow.

**Problem 3.20** Complete the following equations. Use the table of $pK_a$ values to decide where equilibrium will lie for each. Show the direction of the equilibrium by the relative sizes of the reaction arrows you draw.

(a) $CH_3CH_2OH + KOH \longrightarrow$

(b) $CH_3CH_2\overset{+}{N}H_2 \ Cl^- + NaOH \longrightarrow$
    $\quad\ \ |$
    $\quad\ CH_2CH_3$

(c) $CH_2FC\overset{O}{\overset{||}{C}}OH + NaHCO_3 \longrightarrow$

(d) $CH_3-\!\!\left\langle \bigcirc \right\rangle\!\!-OH + NaHCO_3 \longrightarrow$

(e) $\left\langle \bigcirc \right\rangle\!\!-NH_2 + HCl \longrightarrow$

(f) $CH_3CH_2CH_2SH + CH_3CH_2ONa \longrightarrow$

(g) $CH_3-\!\!\left\langle \bigcirc \right\rangle\!\!-\overset{O}{\overset{||}{C}}OH + NaCN \longrightarrow$

(h) $CH_3\overset{O}{\overset{||}{C}}CH_2\overset{O}{\overset{||}{C}}CH_3 + CH_3ONa \longrightarrow$

## D. $pK_a$ Values, the Equilibrium Constant, and Free Energy Changes in Reactions

In previous sections we used $pK_a$ values to predict the extent of a given acid–base reaction and also to choose a particular base that would deprotonate an acid we wished to convert to its conjugate base. We also learned to predict the direction of the equilibrium by looking at the relative acidities and basicities of the two acids and bases involved in the proton transfer reaction. The equilibrium constant is related to the acidity constants (p. 94) and, therefore, to the $pK_a$ values of the two acids involved in the reaction. It is useful to be able to estimate the equilibrium constant of a proposed reaction within a power of ten. This can be done easily by looking at the difference in the $pK_a$ values of the two acids involved in the reaction. The two reactions of the trimethylammonium ion discussed in the last section will serve as examples.

When trimethylammonium ion is mixed with ammonia, the proton is reversibly transferred between the trimethylamine and ammonia.

$$\underset{\substack{\text{trimethylammonium}\\\text{ion}\\ pK_a\ 9.8\\ \textit{reactant acid}}}{CH_3\!-\!\overset{\displaystyle CH_3}{\underset{\displaystyle CH_3}{\overset{+}{N}}}\!-\!H} \;+\; \underset{\substack{\text{ammonia}}}{H\!-\!\overset{\displaystyle \cdot\cdot}{\underset{\displaystyle H}{N}}\!-\!H} \;\;\rightleftharpoons\;\; \underset{\substack{\text{trimethylamine}}}{CH_3\!-\!\overset{\displaystyle CH_3}{\underset{\displaystyle CH_3}{N}}\!:} \;+\; \underset{\substack{\text{ammonium ion}\\ pK_a\ 9.4\\ \textit{product acid}}}{H\!-\!\overset{\displaystyle H}{\underset{\displaystyle H}{\overset{+}{N}}}\!-\!H}$$

The equilibrium constant for the reaction of trimethylammonium ion (abbreviated as $TMAH^+$ below) with ammonia is related to the acidity constants of the two acids in the equation above (see Problem 3.17, p. 96).

$$K_{eq} = \frac{K_{a_{TMAH^+}}}{K_{a_{NH_4^+}}} \quad \text{or more generally} \quad K_{eq} = \frac{K_{a_{\text{reactant acid}}}}{K_{a_{\text{product acid}}}}$$

if

$$K_{eq} = \frac{K_{a_{TMAH^+}}}{K_{a_{NH_4^+}}}$$

then

$$\log K_{eq} = \log K_{a_{TMAH^+}} - \log K_{a_{NH_4^+}}$$

and

$$-\log K_{eq} = -\log K_{a_{TMAH^+}} - (-\log K_{a_{NH_4^+}})$$

Therefore,

$$pK_{eq} = pK_{a_{TMAH^+}} - pK_{a_{NH_4^+}}$$

or generally,

$$pK_{eq} = pK_{a_{\text{reactant acid}}} - pK_{a_{\text{product acid}}}$$

Substituting the $pK_a$ values of trimethylammonium ion (9.8) and ammonium ion (9.4) into the equation, we get numerical values for $pK_{eq}$ and for $K_{eq}$.

$$pK_{eq} = 9.8 - 9.4 = 0.4$$

$$pK_{eq} = -\log K_{eq}$$

Therefore,

$$K_{eq} = \text{antilog of } -0.4 = 10^{-0.4}$$

$$K_{eq} = \frac{[(CH_3)_3N][NH_4^+]}{[(CH_3)_3NH^+][NH_3]} = 0.4$$

(It is purely coincidence that $10^{-0.4}$ happens to be 0.4. The next example will show this.)

$K_{eq}$ is less than 1, telling us that the reactants, the species that appear in the denominator, are favored over the products for the reaction system shown in the previous equation. This is the same conclusion we had reached in the previous section for this reaction.

The reaction of trimethylammonium ion with hydroxide ion has a different outcome.

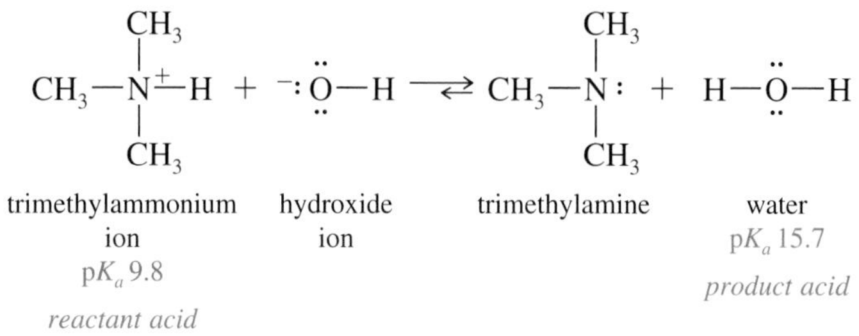

$$CH_3-\overset{\overset{\displaystyle CH_3}{|}}{\underset{\underset{\displaystyle CH_3}{|}}{N}}\!\overset{+}{-}H \;+\; {^-}\!:\!\ddot{O}-H \;\rightleftarrows\; CH_3-\overset{\overset{\displaystyle CH_3}{|}}{\underset{\underset{\displaystyle CH_3}{|}}{N}}\!: \;+\; H-\ddot{O}-H$$

| trimethylammonium ion | hydroxide ion | trimethylamine | water |
|---|---|---|---|
| $pK_a$ 9.8 | | | $pK_a$ 15.7 |
| *reactant acid* | | | *product acid* |

Here, use of the equation relating $pK_{eq}$ to $pK_a$ tells us that the reaction to the right is favored by about six orders of magnitude.

$$pK_{eq} = pK_{a_{TMAH^+}} - pK_{a_{H_2O}}$$

$$pK_{eq} = [9.8 \text{ (or } \sim 10)] - [15.7 \text{ (or } \sim 16)] = -5.9 \text{ (or } \sim -6)$$

$$K_{eq} = \frac{[(CH_3)_3N][H_2O]}{[(CH_3)_3NH^+][OH^-]} = 10^6 \text{ or } 1,000,000$$

The product of concentrations of the species on the right side of the equation is a million times greater than the product of the concentrations of the species on the left side of the equation. Most of the trimethylammonium ions have been converted to trimethylamine molecules.

Equilibrium constants are related to the change in standard free energy for the reaction (p. 95). At room temperature, 298 K, the equation for the relationship between standard free energy and the equilibrium constant becomes

$$\Delta G° = 1.4 \text{ kcal/mol} \times pK_{eq}$$

**Problem 3.21**   Verify the equation shown above by using the equation for the relationship between the standard free energy change for a reaction and its equilibrium constant and the value for $R$ given on page 95.

The change in standard free energy for the reaction of the trimethylammonium ion and ammonia is

$$\Delta G° = 1.4 \text{ kcal/mol} \times 0.4$$
$$= 0.56 \text{ kcal/mol}$$

This is a slight increase in free energy. The reaction of trimethylammonium ion with hydroxide ion, in contrast, has a decrease in free energy.

$$\Delta G° = 1.4 \times (-5.9) \text{ kcal/mol}$$
$$= -8.3 \text{ kcal/mol}$$

These energy relationships can be summarized in energy diagrams (Figure 3.2, p.102).

The reaction of trimethylammonium ion with ammonia can be described as a slightly "uphill" or "almost level" reaction. A small climb in energy occurs between the reactants and the products. Such a reaction has an equilibrium constant that is less than 1, and the reactant side of the equation is favored over the product side. Because

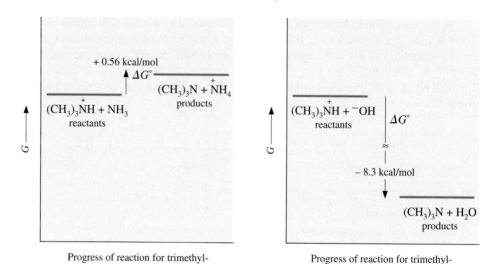

**Figure 3.2**
Comparison of free energy changes for the reaction of trimethylammonium ion with ammonia and with hydroxide ion.

the difference in energy is small, the amount of trimethylamine formed in the reaction mixture can be increased by increasing the amount of ammonia used in the deprotonation reaction. This does not change the equilibrium constant. Instead, it changes the concentrations of all the species present at equilibrium as they shift to accommodate the increase in the concentration of one of the participants in the reaction.

increases     increases
↓                ↓

$$K_{eq} = \frac{[(CH_3)_3N][NH_4^+]}{[(CH_3)_3NH^+][NH_3]} = 0.4$$

↑                ↑
decreases    increase in this quantity
                 causes changes in all other
                 quantities so that the
                 quotient remains 0.4

The response of the system consisting of trimethylammonium ion, ammonia, trimethylamine, and ammonium ion to an increase in the concentration of ammonia is an example of **Le Châtelier's Principle.** The principle states that when a stress is placed on a system at equilibrium, the equilibrium shifts in a way that reduces the stress. Thus an increase in the amount of one of the reactants, ammonia, shifts the equilibrium in the direction that uses up the reactants, ammonia molecules and trimethylammonium ions, and produces more products, trimethylamine molecules and ammonium ions.

The products of the reaction of trimethylammonium ion with hydroxide ion are considerably "downhill" from the reactants. The equilibrium constant for such a reaction is much greater than 1, and the products are favored over the reactants at equilibrium.

**Problem 3.22**     A number of acid–base reactions were given in Section 3.4. Rewrite each equation leaving out the arrows between the reactants and the products. Use the $pK_a$ table to assign $pK_a$ values to the acids that appear in each equation. For each equation, determine $pK_{eq}$ and whether there is a decrease or an increase in free energy. Make your own judgment in each case about whether the reaction goes essentially to completion as written or whether reactants and products are present together in significant quantities.

## E. The Importance of Solvation in Acidity

The $pK_a$ of an acid varies somewhat depending on the solvent in which it is dissolved and the temperature at which the determination is made. The importance of the solvent has become much clearer since the development of mass spectrometers (Chapter 12), instruments that allow chemists to study reactions in the gas phase.

For example, the proton-transfer reaction of acetic acid with water in the gas phase has been examined.

$$CH_3\overset{\overset{\displaystyle O}{\|}}{C}OH(g) \; + \; H_2O(g) \; \rightleftharpoons \; CH_3\overset{\overset{\displaystyle O}{\|}}{C}O^-(g) \; + \; H_3\overset{+}{O}(g)$$

| acetic acid in the gas phase $pK_a$ 130 | water in the gas phase | acetate anion in the gas phase | hydronium ion in the gas phase |

The $pK_a$ of acetic acid under these conditions is 130, indicating that it is enormously difficult to separate the negatively charged acetate ion from the positively charged hydronium ion unless solvent molecules are present to stabilize the charged particles. Clearly, the interaction of the solvent with the undissociated acid and with anions and cations that form on dissociation contributes significantly to the stabilization of the various species and, therefore, to the position of the equilibrium.

Such interactions are particularly important in water, which participates in extensive hydrogen bonding (p. 31). When an acid is dissolved in water, the bonding between the solvent molecules is disrupted, and new hydrogen bonds are formed between water molecules and the solute particles. Large changes in the entropy and the enthalpy of the system occur. Experimentation has shown that such interactions may be the most important factor in determining the relative acidities of closely related compounds.

Section 3.6 will discuss the relative acidities of a series of acids and offer rationalizations for the trends observed. The language used here to talk about acidity is used by chemists to correlate facts and predict reactivity in all areas of organic chemistry.

## 3.6 The Effects of Structural Changes on Acidity

### A. The Resonance Effect

Acetic acid is a typical member of the family of organic compounds known as the **carboxylic acids** (p. 55). Carboxylic acids, in which a hydroxyl group is bonded to the carbon of a carbonyl group, are much stronger acids than are alcohols, in which the hydroxyl group is bonded to an $sp^3$-hybridized carbon atom. A simple experimental observation illustrates the difference in acidity. When a carboxylic acid is put into a solution containing bicarbonate ion, bubbles of carbon dioxide are seen.

$$\overset{\text{carboxyl group}}{CH_3\overset{\overset{\displaystyle O}{\|}}{C}OH} \; + \; HCO_3^- \; \rightleftharpoons \; \overset{\text{carboxylate anion}}{CH_3\overset{\overset{\displaystyle O}{\|}}{C}O^-} \; + \; H_2CO_3 \; \rightleftharpoons \; H_2O \; + \; CO_2 \uparrow$$

| acetic acid $pK_a$ 4.8 | bicarbonate anion | acetate anion | carbonic acid $pK_a$ 6.5 | water | carbon dioxide |
| acid | base | conjugate base of acetic acid | conjugate acid of bicarbonate anion | | bubbles |

When an alcohol is placed in a bicarbonate solution, there is no such evidence of reaction. Ethanol is not a strong enough acid to protonate bicarbonate ion.

$$CH_3CH_2OH \quad + \quad HCO_3^- \quad \longrightarrow \quad \text{no bubbles}$$

ethanol        bicarbonate<br>
$pK_a$ 17        anion

Acetic acid, $pK_a$ 4.8, on the other hand, is a stronger acid than carbonic acid, $H_2CO_3$, $pK_a$ 6.5, and protonates bicarbonate anion, the conjugate base of carbonic acid. The stronger acid, acetic acid, transfers a proton to the conjugate base of the weaker acid, carbonic acid. Carbonic acid is formed by dissolving carbon dioxide in water; in the absence of external pressure (such as exists in capped bottles of soft drink), bubbles of carbon dioxide can be seen whenever reactions lead to carbonic acid.

The difference in the acidities of ethanol and acetic acid can be rationalized by comparing the relative basicities of their conjugate bases. The conjugate base of acetic acid, acetate anion, is a weaker base than ethoxide anion, the conjugate base of ethanol. In the acetate anion, the negative charge is distributed equally between the two oxygen atoms. This delocalization of charge (p. 84) by resonance stabilizes the anion. An anion in which the charge is delocalized to two oxygen atoms is a weaker base than one in which the charge is concentrated on a single oxygen atom, as it is in the ethoxide anion.

$$CH_3CH_2 \overset{\cdot\cdot}{\underset{\cdot\cdot}{O}}{:}^- \qquad\qquad CH_3 \overset{\overset{\displaystyle :O:}{\|}}{-}\overset{\cdot\cdot}{\underset{\cdot\cdot}{C}}{-}\overset{\cdot\cdot}{\underset{\cdot\cdot}{O}}{:}^- \longleftrightarrow CH_3 \overset{\overset{\displaystyle :\overset{\cdot\cdot}{O}:^-}{|}}{-}C{=}\overset{\cdot\cdot}{\underset{\cdot\cdot}{O}}$$

ethoxide anion            resonance contributors for the<br>
                        acetate anion

*charge localized on*      *charge delocalized to two oxygen atoms*<br>
*one oxygen atom*

*less stable*                      *more stable*<br>
*strong base*                     *weak base*

In summary, one factor that affects the basicity of an anion and therefore the acidity of its conjugate acid is whether there can be delocalization of charge over several atoms by resonance. The resonance effect is used to account for part of the observed difference of about twelve units between the $pK_a$ of ethanol and that of acetic acid. It is a strong and important effect.

## B. Inductive and Field Effects

Other structural features of acids also affect their acidity. For example, one of the hydrogen atoms on the carbon atom attached to the carboxyl group of acetic acid can be replaced by another group or atom. The carbon atom adjacent to a carboxyl group is called the **α-carbon** of the acid, and the hydrogen atoms bonded to it are the **α-hydrogen atoms**.

the α-hydrogen        H    O    the carboxyl group<br>
atoms               |    ‖<br>
          H—C—C—OH<br>
              |<br>
              H<br>
                the α-carbon<br>
                atom

| Table 3.1 | The Acidity of Substituted Carboxylic Acids Determined in Water at 25 °C | | | | | |
|---|---|---|---|---|---|
| Name | Structure | $pK_a$ | Name | Structure | $pK_a$ |
| acetic acid | $\underset{\underset{\text{H}}{\mid}}{\overset{\overset{\text{O}}{\parallel}}{\text{CH}_2\text{COH}}}$ | 4.76 | chloroacetic acid | $\underset{\underset{\text{Cl}}{\mid}}{\overset{\overset{\text{O}}{\parallel}}{\text{CH}_2\text{COH}}}$ | 2.86 |
| propanoic acid | $\underset{\underset{\text{CH}_3}{\mid}}{\overset{\overset{\text{O}}{\parallel}}{\text{CH}_2\text{COH}}}$ | 4.87 | bromoacetic acid | $\underset{\underset{\text{Br}}{\mid}}{\overset{\overset{\text{O}}{\parallel}}{\text{CH}_2\text{COH}}}$ | 2.90 |
| fluoroacetic acid | $\underset{\underset{\text{F}}{\mid}}{\overset{\overset{\text{O}}{\parallel}}{\text{CH}_2\text{COH}}}$ | 2.59 | iodoacetic acid | $\underset{\underset{\text{I}}{\mid}}{\overset{\overset{\text{O}}{\parallel}}{\text{CH}_2\text{COH}}}$ | 3.17 |

In the series of compounds shown in Table 3.1, an $\alpha$-hydrogen atom in acetic acid has been replaced in turn by a methyl group, a fluorine atom, a chlorine atom, a bromine atom, and an iodine atom. The $pK_a$ values given indicate that acidity is decreased slightly by substitution of a methyl group but increased by substitution of a halogen atom. Substitution by a fluorine atom at the $\alpha$-carbon produces the greatest increase in acidity. The acidity decreases as the substituted atom changes from fluorine to chlorine to bromine to iodine.

The $pK_a$ values for another series of acids illustrate the effect of a halogen atom that has been substituted farther and farther away from the carboxyl group (Table 3.2). Butanoic acid has an acidity very close to that of propanoic acid (see Table 3.1).

$$\overset{\overset{\text{O}}{\parallel}}{\text{CH}_3\text{CH}_2\text{CH}_2\text{COH}} \qquad \overset{\overset{\text{O}}{\parallel}}{\text{CH}_3\text{CH}_2\text{COH}}$$

$$\underset{pK_a\,4.82}{\text{butanoic acid}} \qquad \underset{pK_a\,4.87}{\text{propanoic acid}}$$

Acidity does not change significantly with the size of the alkyl group attached to the carboxyl group. A chlorine atom on the $\alpha$-carbon atom increases the acidity of butanoic acid, but the effect falls off rapidly as the chlorine atom is moved farther from the carboxyl group in 3-chlorobutanoic acid and still farther in 4-chlorobutanoic acid.

| Table 3.2 | The Effect on Acidity (Determined in Water at 25 °C) of the Distance Between the Substituent and the Carboxyl Group | | | | |
|---|---|---|---|---|---|
| Name | Structure | $pK_a$ | Name | Structure | $pK_a$ |
| butanoic acid | $\overset{\overset{\text{O}}{\parallel}}{\text{CH}_3\text{CH}_2\text{CH}_2\text{COH}}$ | 4.82 | 3-chlorobutanoic acid | $\underset{\underset{\text{Cl}}{\mid}}{\overset{\overset{\text{O}}{\parallel}}{\text{CH}_3\text{CHCH}_2\text{COH}}}$ | 4.05 |
| 2-chlorobutanoic acid | $\underset{\underset{\text{Cl}}{\mid}}{\overset{\overset{\text{O}}{\parallel}}{\text{CH}_3\text{CH}_2\text{CHCOH}}}$ | 2.86 | 4-chlorobutanoic acid | $\underset{\underset{\text{Cl}}{\mid}}{\overset{\overset{\text{O}}{\parallel}}{\text{CH}_2\text{CH}_2\text{CH}_2\text{COH}}}$ | 4.52 |

These data suggest that the acidity of an acid increases when electronegative atoms are introduced into the molecule. The greater the electronegativity of the substituent, the greater is the increase in acidity (see Table 3.1). This increase in acidity also depends on the distance between the substituent and the carboxyl group (see Table 3.2).

Chemists explain these facts in two ways. An electronegative substituent, such as a chlorine atom, draws the electrons of the carbon–chlorine bond toward the chlorine atom, creating an electron deficiency in that carbon atom and in neighboring carbon atoms. This effect results in a withdrawal of electron density from the carboxyl group, which acts to stabilize the carboxylate anion. The carboxylate anion becomes a weaker base; consequently, the corresponding conjugate acid is stronger.

$$
\begin{array}{cc}
\underset{\displaystyle\downarrow}{\overset{\displaystyle O}{\underset{\displaystyle Cl}{CH_2 \leftarrow \overset{\|}{C} \leftarrow O^-}}} & \overset{\displaystyle O}{\underset{\displaystyle H}{CH_2 - \overset{\|}{C} - O^-}}
\end{array}
$$

chloroacetate anion  acetate anion

*stabilized by the
inductive effect,
a weaker base
than acetate anion*

The transmission of the effect of an electron-withdrawing (or electron-donating) group through $\sigma$ bonds is called an **inductive effect.** It diminishes rapidly as the number of bonds between the substituent and the carboxyl group increases.

The **field effect** arises in the bond dipole moments of a molecule and is transmitted not through the bonds but through the environment around the molecule. For example, chloroacetate ion is stabilized relative to acetate ion by the bond dipole of the carbon–chlorine bond.

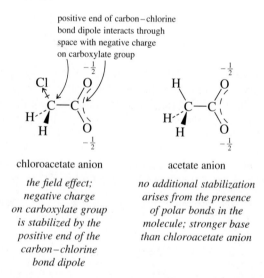

positive end of carbon–chlorine
bond dipole interacts through
space with negative charge
on carboxylate group

chloroacetate anion  acetate anion

*the field effect;
negative charge
on carboxylate group
is stabilized by the
positive end of the
carbon–chlorine
bond dipole*

*no additional stabilization
arises from the presence
of polar bonds in the
molecule; stronger base
than chloroacetate anion*

It is theorized that the positive end of the dipole of the carbon–chlorine bond in chloroacetate anion interacts directly through space with the negative charge on the carboxylate anion to reduce charge density at the carboxylate group. The effect of the carbon–chlorine bond dipole on the carboxylate anion decreases as the distance between the chlorine atom and the carboxylate group increases (see Table 3.2).

The inductive effect and the field effect operate in the same direction. Many ingenious experiments have been devised to measure these effects separately. The most successful experiments indicate that the transmission of the effect through space by interaction of bond dipoles is more important than the transmission of the effect through bonds. Customarily, inductive effect is used to refer to the combination of both these effects.

**Problem 3.23**    Predict whether $CF_3CH_2OH$ is more or less acidic than $CH_3CH_2OH$. Why?

**Problem 3.24**    Look up the $pK_a$ values for chloroacetic acid, $ClCH_2COH$, dichloroacetic acid, $Cl_2CHCOH$, and trichloroacetic acid, $Cl_3CCOH$, in the table of $pK_a$ values. How do you explain the data? Use structural formulas and energy diagrams to illustrate your answer.

Other substituents that increase the acidity of carboxylic acids are the hydroxyl, cyano, and nitro groups (Table 3.3). In hydroxyacetic acid, the electronegative oxygen atom substituted on the $\alpha$-carbon atom reduces electron density at the carboxylate anion, making it a weaker base in comparison with the unsubstituted acetate ion. Consequently, hydroxyacetic acid is a stronger acid than acetic acid.

The cyano and nitro groups are strongly electron-withdrawing. Note that nitroacetic acid is a stronger acid than fluoroacetic acid (see Table 3.1) and cyanoacetic acid is slightly stronger than fluoroacetic acid. The powerful electron-withdrawing effects of these groups are rationalized by pointing to their electronic structures. In each case, the atom attached to the $\alpha$-carbon atom of the acid either has a positive formal charge or has a positive charge in an important resonance contributor. The presence of this partial positive charge in the substituent contributes to the stabilization of the carboxylate anion.

**Problem 3.25**    Draw Lewis structures for cyanoacetic acid and nitroacetic acid (see Table 3.3). Draw resonance contributors for them. Explain in each case why the substituent increases the acidity of the acid over that of acetic acid, and also explain why nitroacetic acid is stronger than cyanoacetic acid.

In summary, an electron-withdrawing atom or group substituted on the alkyl chain of a carboxylic acid stabilizes the corresponding carboxylate anion, thereby increasing the acidity of the carboxylic acid. Important electron-withdrawing groups are the halogens, hydroxyl and ether groups, the cyano group, and the nitro group.

**Table 3.3    The Effects of Electron-Withdrawing Groups on the Acidity of Carboxylic Acids**

| Name | Structure | $pK_a$ | Name | Structure | $pK_a$ |
|------|-----------|--------|------|-----------|--------|
| acetic acid | $CH_2COH$ $H$ | 4.76 | cyanoacetic acid | $CH_2COH$ $CN$ | 2.46 |
| hydroxyacetic acid | $CH_2COH$ $OH$ | 3.83 | nitroacetic acid | $CH_2COH$ $NO_2$ | 1.68 |

**Problem 3.26**    In Section 3.6A (p. 103) the argument was made that the difference in acidity between ethanol and acetic acid is due to resonance delocalization of the negative charge on the acetate ion and the lack of such stabilization in the ethoxide ion. Chemists also rationalize the difference in acidity between the two compounds using inductive effects. To what kind of carbon atom is the oxygen atom bonded in ethoxide ion and in acetate ion? On what would such an argument be based? A review of Sections 1.9 and 2.7B will be helpful.

In water, propanoic acid is slightly weaker than acetic acid. The nature of the inductive effect of an alkyl group is debated by chemists. Carbocations, introduced on page 83, are electron-deficient. They have an open shell. They are stabilized by alkyl groups (p. 129), which suggests that alkyl groups are electron-releasing when they are bonded to the cationic carbon atom. They also increase the basicity of amines (below), again suggesting that they are electron-releasing. On the other hand, though **tert**-butyl alcohol ($pK_a$ 19) is a weaker acid than ethanol ($pK_a$ 17) in water, it is a stronger acid in the gas phase. This experimental observation suggests that alkyl groups can stabilize anions as well as cations and that solvation plays an important role in determining relative acidities. Thus a word of caution is necessary. The relative acidities on which the generalizations presented in this chapter are based were determined in water. In the gas phase, reversals in the order of related compounds are often seen.

The models presented above, which rationalize acidity solely in terms of the electronegativity of the substituents and of the polarization of bonds within molecules, are too simple. Such models are retained, however, because they are enormously useful in rationalizing a wide range of experimental observations and in making predictions about reactivity in many different systems. Resonance effects and inductive effects will be invoked frequently as we try to understand how the structures of organic compounds determine their reactivity.

## 3.7    Amines

Amines (p. 49) are the most important organic bases as well as being weak acids (p. 89). The basicities of amines are determined by the relative availability of the nonbonding electrons on the nitrogen atom to a proton donor or Lewis acid and by the stabilization of the positively charged nitrogen atom by solvation or, in some special cases, by resonance. On page 94, the basicities of various species were related to the $pK_a$ values of their conjugate acids. For an amine, too, the larger the $pK_a$ of its conjugate acid, the weaker the acid and the stronger the corresponding base.

Amines in which alkyl groups are substituted on the nitrogen atom have basicities similar to that of ammonia. Their conjugate acids are alkylammonium ions with $pK_a$ values of approximately 10.8 to 9.5. Methylamine, dimethylamine, and trimethylamine are protonated by water just as ammonia is. All amines with low molecular weights, in which the nitrogen atom is a significant portion of the molecule, dissolve in water to give solutions that turn red litmus paper blue.

$$NH_3 \;+\; H_2O \;\rightleftharpoons\; NH_4^+ \;+\; OH^-$$

ammonia    $pK_a$ 15.7        ammonium ion    $pK_a$ 9.2

$$CH_3NH_2 \;+\; H_2O \;\rightleftharpoons\; CH_3NH_3^+ \;+\; OH^-$$

methylamine    $pK_a$ 15.7        methylammonium ion    $pK_a$ 10.6

$$(CH_3)_2NH \;+\; H_2O \;\rightleftharpoons\; (CH_3)_2NH_2^+ \;+\; OH^-$$

dimethylamine    $pK_a$ 15.7        dimethylammonium ion    $pK_a$ 10.7

$$(CH_3)_3N \;+\; H_2O \;\rightleftharpoons\; (CH_3)_3NH^+ \;+\; OH^-$$

trimethylamine    $pK_a$ 15.7        trimethylammonium ion    $pK_a$ 9.8

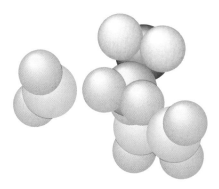

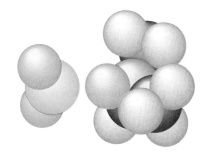

Methylammonium ion
hydrogen bonding with
three water molecules

Trimethylammonium ion
hydrogen bonding with
only one water molecule

*Figure 3.3*
A comparison of
the solvation of
trimethylammonium and
methylammonium ions.

The one methyl group on the nitrogen atom in methylamine and the two methyl groups in dimethylamine increase basicity. Trimethylamine is slightly more basic than ammonia but is a weaker base than dimethylamine. The three alkyl groups around the nitrogen atom interfere with protonation and stabilization of the cation by solvation, making the amine less basic (Figure 3.3).

⊘ **Study Guide**
**Concept Map 3.5**

## SUMMARY

Many organic reactions can be classified as acid–base reactions. Two ways of defining acids and bases are useful. According to the Brønsted-Lowry definition, an acid is a proton donor, and a base, a proton acceptor. A Brønsted acid is also known as a protic acid. According to the Lewis definition, an acid is an electron-pair acceptor, and a base, an electron-pair donor. In general, any species, including a proton, that can accept a pair of electrons is classified as a Lewis acid. Important Lewis acids include compounds such as boron trifluoride, $BF_3$, and aluminum chloride, $AlCl_3$. Such acids are known as aprotic acids.

Atoms bearing nonbonding electrons, such as oxygen, nitrogen, and sulfur, are the basic sites in molecules or ions. Protic acids transfer a proton to such atoms, and Lewis acids coordinate with them. The electrons in bonds (usually $\pi$ bonds) also react with protic acids or Lewis acids.

The basicity of a species depends on the position in the periodic table of the atom bearing the nonbonding electrons. Basicity decreases for a series of atoms that go from left to right across any period or down any group. An anion containing a given atom is always more basic than an uncharged molecule containing the same atom (for example, $OH^-$ is more basic than $H_2O$).

The acidities of protic acids are quantified as their $pK_a$ values, which are derived from equilibrium constants. Equilibrium constants, $K_{eq}$, in turn, reflect the change in standard free energy, $\Delta G°$, for the reaction.

The loss of a proton converts a protic acid into its conjugate base. Similarly, a base that gains a proton is converted into its conjugate acid. A strong acid has a weak conjugate base, and a weak acid has a strong conjugate base. Bases can be stabilized by delocalization of electron density through resonance or inductive effects. These stabilized bases are weaker (have stronger conjugate acids) than bases in which such effects are not as important.

# ADDITIONAL PROBLEMS

**3.27** Complete the following equations. Use the table of $pK_a$ values to decide where the equilibrium will lie by determining $K_{eq}$ and whether there is a decrease or increase in free energy for each reaction.

(a) [piperidinium structure with N$^+$, H, H and Cl$^-$] + NaOH $\longrightarrow$    (b) $CH_3CH_2\overset{\overset{H}{|}}{\underset{+}{O}}CH_2CH_3 + H_2O \longrightarrow$

(c) $CH_3\overset{O}{\overset{||}{C}}OH + CH_3CH_2SNa \longrightarrow$    (d) $CCl_3\overset{O}{\overset{||}{C}}OH + NaHCO_3 \longrightarrow$

**3.28** Arrange the following compounds in order of increasing acidity of the hydrogen atoms.

$$AlH_3, \qquad H_2S, \qquad HCl, \qquad NaH$$

**3.29** Arrange the following compounds in order of increasing acidity.

(a) $CF_3\overset{O}{\overset{||}{C}}OH, \quad CCl_3\overset{O}{\overset{||}{C}}OH, \quad CH_3\overset{O}{\overset{||}{C}}OH$    (b) $CH_3CH_2OH, \quad FCH_2CH_2OH, \quad ClCH_2CH_2OH$

(c) $CH_3CH_2SH, \quad CH_3CH_2OH, \quad CH_3CH_2NH_2$    (d) $CH_3\overset{O}{\overset{||}{C}}OH, \quad CH_3\overset{+OH}{\overset{||}{C}}OH, \quad CH_3\overset{O}{\overset{||}{C}}OCH_3$

(e) $CH_3\overset{O}{\overset{||}{C}}OH, \quad CH_3\overset{O}{\overset{||}{C}}NH_2, \quad CH_3\overset{O}{\overset{||}{C}}OCH_3$

**3.30** Arrange each group of species in order of increasing basicity.

(a) $Cl_3C:^-, \quad CH_3\overset{\overset{CH_3}{|}}{N}CH_3, \quad CH_3\overset{..}{\underset{..}{O}}CH_3$    (b) $\overset{..}{N}F_3, \quad \overset{..}{N}H_3, \quad \overset{..}{N}H_2\overset{..}{\underset{..}{O}}H$

(c) $\overset{..}{N}H_3, \quad :\overset{..}{N}H_2^-, \quad CH_3NH_2$    (d) $CH_3\overset{..}{\underset{..}{S}}CH_3, \quad CH_3\overset{\overset{CH_3}{|}}{\underset{..}{P}}CH_3, \quad CH_3\overset{\overset{CH_3}{|}}{\underset{\underset{CH_3}{|}}{Si}}CH_3$

**3.31** For each acid–base reaction given below, label acids, bases, conjugate acids, and conjugate bases, and tell whether Brønsted acids or Lewis acids are present.

(a) $F^- + BF_3 \longrightarrow BF_4^-$    (b) $Ag^+ + 2 NH_3 \longrightarrow Ag(NH_3)_2^+$

(c) $Al(H_2O)_6^{3+} + OH^- \longrightarrow Al(OH)(H_2O)_5^{2+} + H_2O$

(d) $CH_3CH_2\overset{O}{\overset{||}{C}}OH + OH^- \longrightarrow CH_3CH_2\overset{O}{\overset{||}{C}}O^- + H_2O$

(e) $CH_3CH_2\overset{\overset{CH_2CH_3}{|}}{N}CH_2CH_3 + CF_3\overset{O}{\overset{||}{C}}OH \longrightarrow CH_3CH_2\overset{\overset{CH_2CH_3}{|}}{\underset{+}{N}}HCH_2CH_3 + CF_3\overset{O}{\overset{||}{C}}O^-$

**3.32** The conjugate acids of the following amines have the $pK_a$ values shown. Explain the observed trend.

$$CH_3CH_2CH_2CH_2NH_2 \qquad CH_3OCH_2CH_2CH_2NH_2 \qquad CH_3OCHCH_2NH_2 \qquad N{\equiv}CCH_2CH_2NH_2$$
$$\qquad\qquad\qquad\qquad\qquad\qquad\qquad\qquad\qquad\qquad\qquad\qquad OCH_3$$

| | | | | |
|---|---|---|---|---|
| $pK_a$ of conjugate acid | 10.60 | 9.92 | 8.54 | 7.80 |

**3.33** Complete the following equations.

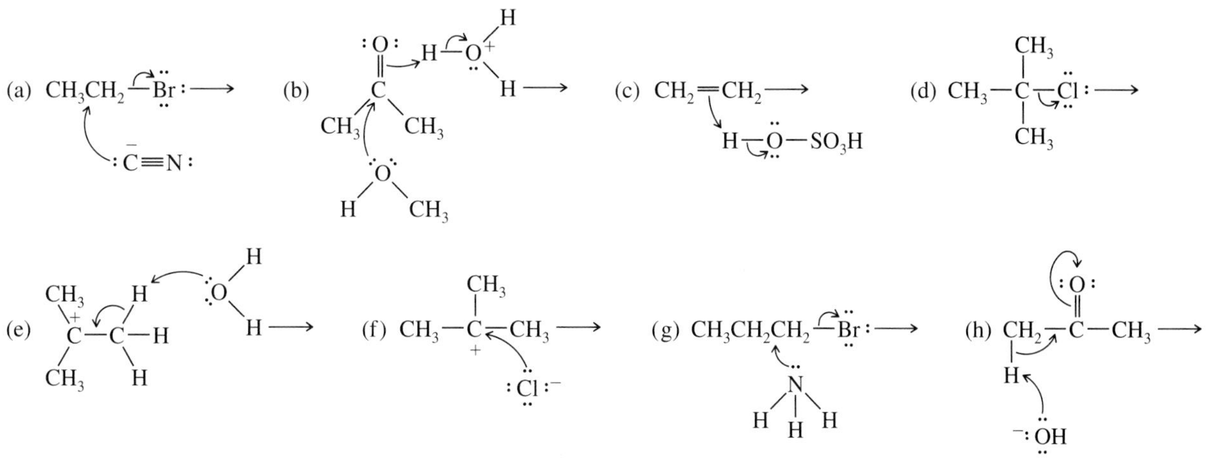

(a) $CH_3CH_2{-}Br: \longrightarrow$

$\qquad\qquad :C{\equiv}N:$

(b) [structure with C=O, CH₃, CH₃, O–CH₃, H groups] $\longrightarrow$

(c) $CH_2{=}CH_2 \longrightarrow$
$\qquad\qquad H{-}O{-}SO_3H$

(d) $CH_3{-}\underset{CH_3}{\overset{CH_3}{C}}{-}Cl: \longrightarrow$

(e) [structure with CH₃, CH₃, H, C–C–H, O, H groups] $\longrightarrow$

(f) $CH_3{-}\underset{+}{\overset{CH_3}{C}}{-}CH_3 \longrightarrow$
$\qquad\qquad :Cl:^-$

(g) $CH_3CH_2CH_2{-}Br: \longrightarrow$
$\qquad\qquad\qquad \underset{H\ H\ H}{N}$

(h) $CH_2{-}\overset{:O:}{C}{-}CH_3 \longrightarrow$
$\qquad H \qquad ^-:OH$

**3.34** The following steps were proposed in a mechanism for reactions observed in a study of the mutagenicity of compounds. Supply the curved arrows necessary to complete the reactions shown below.

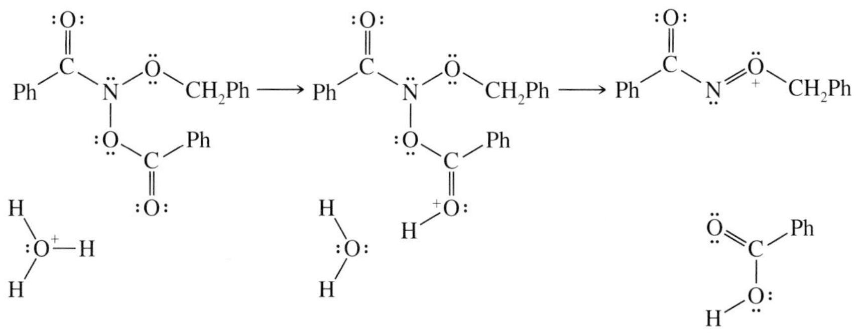

[mechanism structures: Ph–C(=O)–N–O–CH₂Ph with O–C(=O)–Ph and HO⁺H₂ → Ph–C(=O)–N–O–CH₂Ph with O–C(=O)⁺ → Ph–C(=O)–N⁺=O–CH₂Ph and O=C(Ph)–O–H]

**3.35** For the following compounds, $pK_a$ values are given. Calculate the acidity constants.

(a) $CHCl_2\overset{O}{\underset{\|}{C}}OH,$ $pK_a$ 1.3
(b) $CH_3NH_3{}^+,$ $pK_a$ 10.4
(c) $CCl_3CH_2OH,$ $pK_a$ 12.2

**3.36** The following compounds have the acidity constants shown. What are their $pK_a$ values?

(a) $CH_3CH_2OH,$ $K_a = 10^{-17}$
(b) $CH_3CH_2SH,$ $K_a = 3.16 \times 10^{-11}$
(c) $CH_3CH_2{-}\overset{H}{\underset{+}{O}}{-}CH_2CH_3,$ $K_a = 3.98 \times 10^3$

**3.37** The compound of sodium with hydrogen, sodium hydride, is ionic, NaH.

(a) What is the Lewis electron dot structure of hydride ion?
(b) When sodium hydride is placed in water, hydride ion is converted to hydrogen, $H_2$, and

the resulting solution has pH > 7 and turns red litmus paper blue. Write a balanced equation showing the reaction of sodium hydride with water.

(c)  Is hydride ion an acid or a base? What is the relationship of hydrogen, $H_2$, to hydride ion?

(d)  Hydride ion reacts with an alcohol, such as ethanol, in exactly the same way as it does with water, generating hydrogen as one of the products. Complete the following equation.

$$CH_3CH_2OH + Na^+H^- \longrightarrow$$

**3.38** The following are partial equations representing acid–base reactions. Looking carefully at the connectivity of the atoms shown in the equation and at the conjugate acid–base pairs found in the p$K_a$ table, construct structural formulas for the missing species designated by letters. Indicate the direction of the equilibrium in each case by the relative sizes of the reaction arrows. For each reaction, draw an energy diagram indicating whether the reactants or the products are lower in energy.

(a)

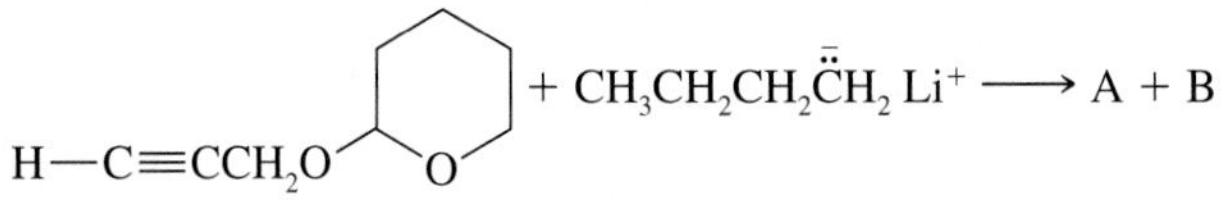

$$H-C\equiv CCH_2O \quad + \; CH_3CH_2CH_2\ddot{C}H_2 \; Li^+ \longrightarrow A + B$$

(b) 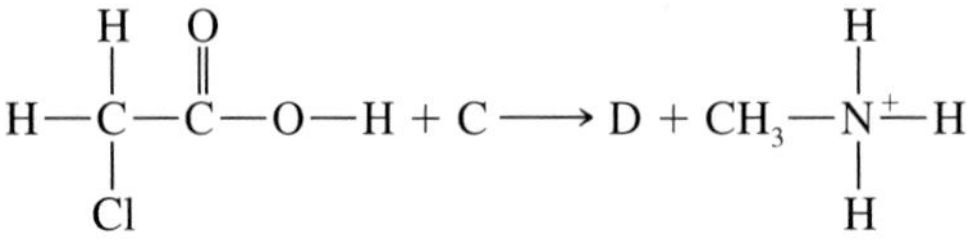 
$$H-\underset{\underset{Cl}{|}}{\overset{\overset{H}{|}}{C}}-\overset{\overset{O}{\|}}{C}-O-H + C \longrightarrow D + CH_3-\underset{\underset{H}{|}}{\overset{\overset{H}{|}}{N}}\overset{+}{-}H$$

(c) 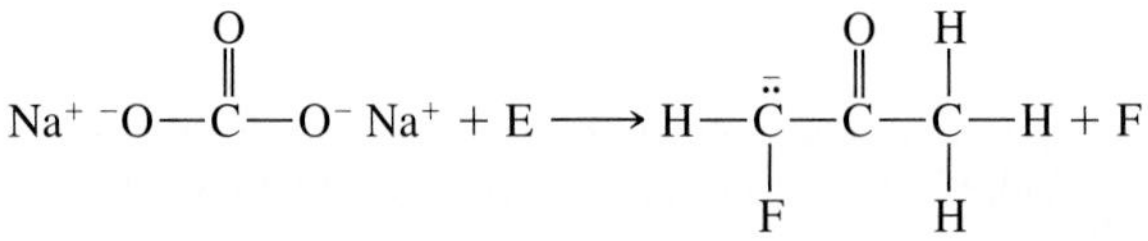 
$$Na^+ \; {}^-O-\overset{\overset{O}{\|}}{C}-O^- \; Na^+ + E \longrightarrow H-\underset{\underset{F}{|}}{\overset{\overset{..}{}}{C}}-\overset{\overset{O}{\|}}{C}-\underset{\underset{H}{|}}{\overset{\overset{H}{|}}{C}}-H + F$$

(d) 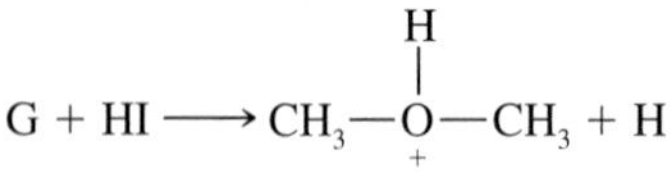 
$$G + HI \longrightarrow CH_3-\underset{+}{\overset{\overset{H}{|}}{O}}-CH_3 + H$$

(e)

$$\underset{OH}{\overset{\overset{\overset{O}{\|}}{COH}}{}} \quad + Na^+ \; {}^-O-\overset{\overset{O}{\|}}{C}-OH \longrightarrow I + J$$

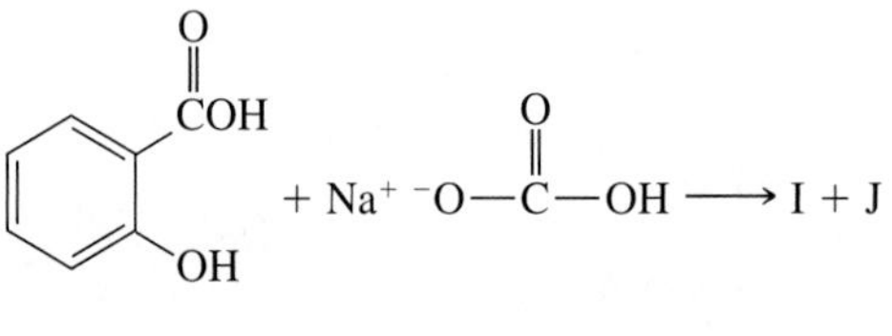

**3.39** Trifluoromethanesulfonic acid, also called triflic acid, has the formula $CF_3SO_3H$. It is a strong acid.

(a)  Write the Lewis structure of triflic acid.

(b)  Would you expect it to be a stronger acid or a weaker acid than benzenesulfonic acid,

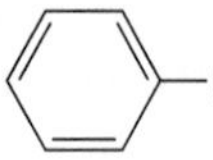

$$\text{⬡}-SO_3H? \text{ Why?}$$

(c)  The following reaction was carried out.

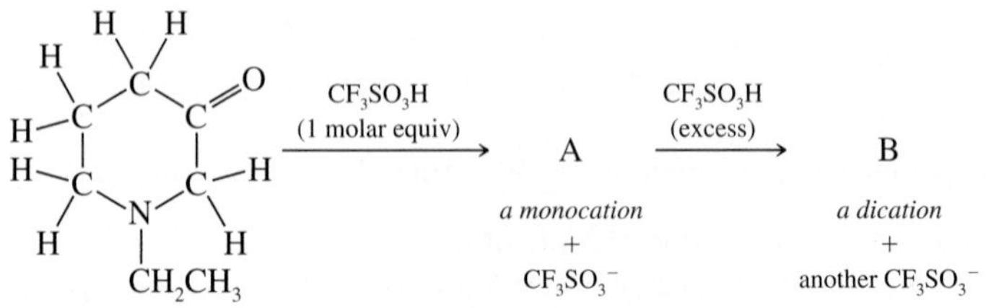

$$\underset{CH_2CH_3}{\text{[cyclic structure]}} \xrightarrow[\text{(1 molar equiv)}]{CF_3SO_3H} \; A \; \xrightarrow[\text{(excess)}]{CF_3SO_3H} \; B$$

$$\begin{array}{ccc} & a\ monocation & a\ dication \\ & + & + \\ & CF_3SO_3{}^- & \text{another } CF_3SO_3{}^- \end{array}$$

Provide structural formulas for the cations A and B. Think carefully about which site in the starting material is the most basic.

**3.40** When strong acids or strong bases are dissolved in a solvent such as water, a phenomenon known as the leveling effect is observed. Differences in acidity for strong acids such as sulfuric acid and hydrochloric acid cannot be measured in water. They appear to have the same acidity in that solvent. What is happening? What is the strongest acid that can exist in aqueous solution? What is the strongest base? Write equations illustrating your answers.

**3.41** Proteins are built up from units called amino acids. The simplest amino acid is glycine, the structure of which can be written in two ways:

(a) Consult the table of $pK_a$ values and decide which is the better representation of the structure of the compound. (*Hint:* Which is more basic, an amino group or a carboxylate anion?)
(b) Glycine is a crystalline solid that decomposes at 233 °C as it melts. It has a solubility of 26 g in 100 mL of water. Do these facts fit the structure that you have chosen for glycine? Explain.

**3.42** The acidity constant for any acid, HA, in water is

$$K_a = \frac{[A^-][H_3O^+]}{[HA]}$$

This expression for $K_a$ can be converted into another one showing the relationship between the $pK_a$ of an acid and the pH of the solution:

$$pH = pK_a + \log \frac{[A^-]}{[HA]}$$

This expression, sometimes called the Henderson-Hasselbalch equation, is useful in calculating the relative amounts of dissociated and undissociated acid present at any pH when the $pK_a$ of the acid is known.

(a) Derive the Henderson-Hasselbalch equation from the expression for $K_a$.
(b) What does the Henderson-Hasselbalch equation tell you about the concentrations of HA and $A^-$ when $pK_a = pH$?
(c) In Problem 3.16 you learned that the imidazole ring in the amino acid histidine is important in proton-transfer reactions in the human body. The protonated form of imidazole has $pK_a$ 7.0. The pH of body fluids is ~6.5. What can you say in a qualitative way about the relative quantities of protonated and unprotonated imidazole rings present in the human body?

**3.43** When treated with strong acid, hydroxamic acids can form three different monoprotonated species. Draw the structure of the conjugate acid resulting from the most favorable protonation of the following hydroxamic acid. Explain your choice:

a hydroxamic acid

**3.44** Vitamin C, ascorbic acid, has the structure shown below. The most acidic proton in ascorbic acid has a p$K_a$ of 4.10; the next most acidic one has a p$K_a$ of 11.99.

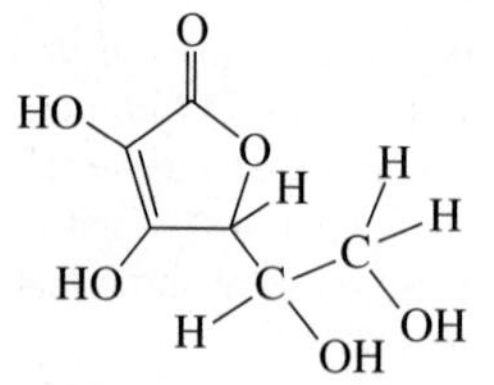

vitamin C, ascorbic acid

(a) Draw the structure of the conjugate base that results from the first ionization of ascorbic acid, along with other structures that rationalize your choice of which proton to remove first.

(b) Physiologic pH, the pH of body fluids, is ~6.5. Draw the structure of ascorbic acid as it exists in the body.

# Reaction Pathways

## A Look Ahead

In Chapter 3 you learned to make predictions about one important class of reactions, proton-transfer reactions between Brønsted acids and bases. You also learned to recognize Lewis acids, which react by accepting pairs of electrons, and Lewis bases, which are electron-pair donors.

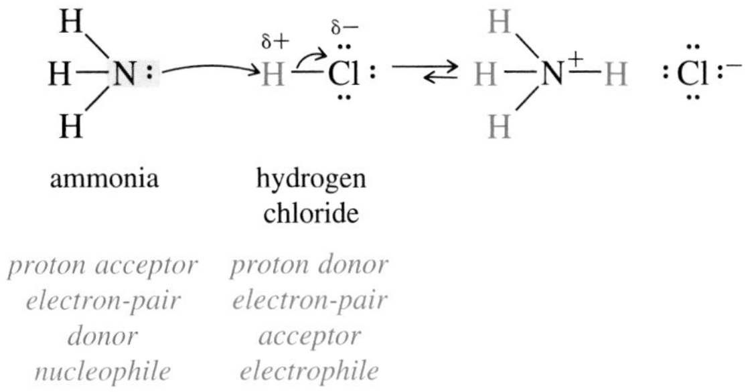

ammonia     hydrogen
chloride

*proton acceptor*    *proton donor*
*electron-pair*      *electron-pair*
*donor*           *acceptor*
*nucleophile*       *electrophile*

Hydrogen chloride, an acid, is also an electrophile, a species that is "electron-seeking" and can accept a pair of electrons. Ammonia, a base, is a nucleophile, a species "seeking a nucleus" to which to donate a pair of electrons.

In this chapter we will examine two important classes of reactions that result from interactions of nucleophiles and electrophiles. In nucleophilic substitution reactions, a nucleophile displaces a group already present in another molecule. The group that is displaced is called a **leaving group.**

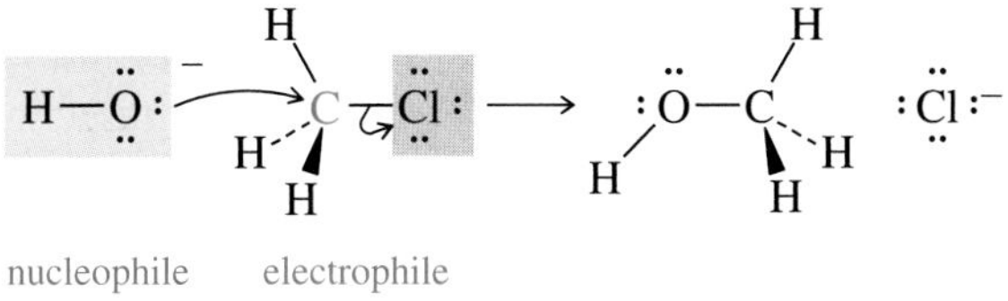

nucleophile      electrophile

*a nucleophilic substitution reaction*

In electrophilic addition reactions, an acid adds to a double bond.

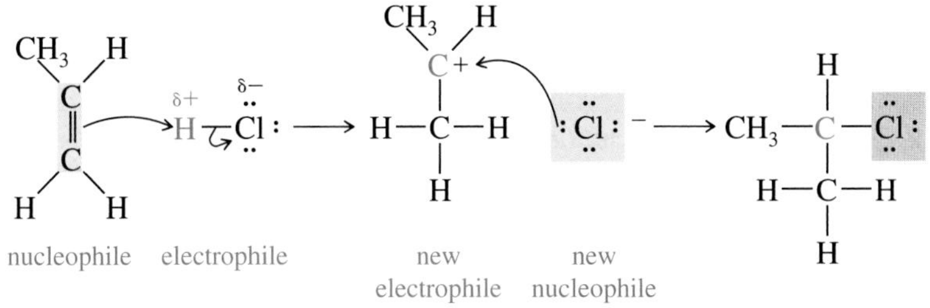

nucleophile    electrophile       new         new
electrophile    nucleophile

*an electrophilic addition reaction*

**115**

Workbook Exercises

We will study in depth the two reactions shown at the bottom of page 115 and will apply the ideas they illustrate about chemical reactivity to many other reactions.

## 4.1 Introduction. Electrophiles and Nucleophiles

Two acid–base reactions are shown below.

VISUALIZING THE REACTION

**Acid–Base Reactions**

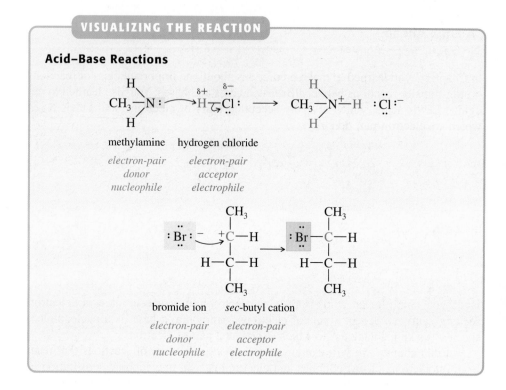

The acids in these reactions are examples of **electrophiles,** "electron-seeking" reagents that have room in their orbitals to accept a pair of electrons. Electrophiles are shown in red in the equations above. Hydrogen chloride is an electrophile because the electrons of the covalent bond between hydrogen and chlorine are pulled closer to the more electronegative atom, chlorine. Hydrogen has a partial positive charge and is therefore attractive to the nonbonding electrons of methylamine. Carbon in the *sec*-butyl cation is an electrophile because it has only six electrons in its outermost shell, leaving room in its orbitals to accept a pair of electrons from the bromide ion.

The species that react with electrophiles are called **nucleophiles,** reagents "seeking a nucleus" to which to donate a pair of electrons. In the preceding examples, the nucleophiles, methylamine and bromide ion, are identified by blue shaded boxes. Most Lewis and Brønsted bases are nucleophiles.

In predicting whether a reaction between two species is likely, we must consider three factors. First, we examine the structures of the two species to see whether there are sites of electron deficiency and electron availability. In other words, there must be an electron acceptor and an electron donor, an electrophile and a nucleophile, for each reaction. In most cases, the identities of the electrophile and the nucleophile are easy to determine, although sometimes they are not obvious. However, in every reaction we will be looking for structural features that allow for interactions between orbitals with easily available electrons and orbitals with room to accept these electrons. Only then is there the possibility for the formation of a new bond.

The second factor we must consider is the position of the equilibrium between the reactants and the products of the reaction in question. For example, $pK_a$ values are used to make predictions about acid–base reactions (p. 97). Equilibrium is determined by thermodynamic considerations, the overall enthalpy and entropy changes that occur during a reaction (p. 95).

Finally, we need to consider the mechanism of the reaction, the details of the bond-breaking and bond-making that must take place to get from reactants to products (p. 85). A reaction may have an equilibrium constant that favors the products but may occur too slowly to be practical. The rate of a reaction (p. 121) depends on the reaction pathway, the mechanism of the reaction. The field of kinetics deals with the factors that affect the rate of a reaction. In the next two sections we will examine two reactions that represent important classes of reactions. We will identify electrophiles and nucleophiles and examine energy considerations, equilibria, and mechanisms for the two reactions.

⊘ Study Guide
Concept Map 4.1

---

**Problem 4.1**    Decide which of the following compounds and ions are electrophiles and which are nucleophiles. Writing Lewis structures may be helpful.

(a) $CH_3O^-$    (b) $PH_3$    (c) $Cu^{2+}$    (d) HBr    (e) $CH_3Cl$    (f) $CH_3NH_2$    (g) $H^-$    (h) $B(CH_3)_3$

(i) $HONH_2$    (j) $HC\equiv CH$    (k) $AlCl_3$    (l) $\overset{H}{\underset{H}{>}}C=O$    (m) $I^-$    (n) $CH_3CH_2S^-$    (o) $Hg^{2+}$

---

## 4.2   The Reaction of Chloromethane with Hydroxide Ion

### A. A Nucleophilic Substitution Reaction

Chloromethane reacts with sodium hydroxide in solution in water to give methanol and sodium chloride.

$$CH_3Cl + Na^+OH^- \xrightarrow{H_2O} CH_3OH + Na^+Cl^-$$

The polarization of the carbon–chlorine bond in chloromethane results from the differing electronegativities of carbon and chlorine. The unequal sharing of the electrons in this covalent bond creates a site of electron deficiency at the carbon atom. Chloromethane is, therefore, an electrophile, or more accurately, it contains an electrophilic center. Hydroxide ion is a good nucleophile. It has pairs of electrons to share and a negative charge, making it a site of electron density.

The reaction starts with the interaction of the nucleophile with the electrophile. Just as we did for the proton-transfer reaction (p. 86), we can ask questions about the timing of bond-breaking and bond-forming in this reaction. The options are the same as they were for proton transfer.

1. Bond-breaking comes first; then a new bond forms.

bond-breaking                    bond-forming

2. Bond-forming comes first; then the carbon–halogen bond breaks.

$$\text{bond-forming} \qquad\qquad \text{bond-breaking}$$

3. Bond-breaking and bond-forming are more or less simultaneous.

*bond-breaking and bond-forming*
*take place at the same time*

The first option gives rise to an open shell on carbon with a positive charge on the carbon atom. Some carbon cations have been detected experimentally (p. 129). The methyl cation, which would be formed in the case shown in Option 1, is not among them. It is too unstable to form under these conditions. In Chapter 7 we will see that in some substitution reactions, bond-breaking comes before bond-forming when the cations that form are more stable than the methyl cation (see Section 7.3).

The second option puts ten electrons around the carbon atom, thereby violating the octet rule. Chemists favor the third option, in which bond-breaking and bond-forming processes occur more or less at the same time for chloromethane reacting with hydroxide ion. This reaction is similar to the reaction of hydrogen chloride with water in that both reactions may be seen as displacement of chloride ion by an incoming nucleophile.

Just as the transfer of a proton from one base to another is called a **proton-transfer** or a **protonation reaction,** the reaction of chloromethane with a nucleophile can be seen as the transfer of an alkyl group or an **alkylation reaction.**

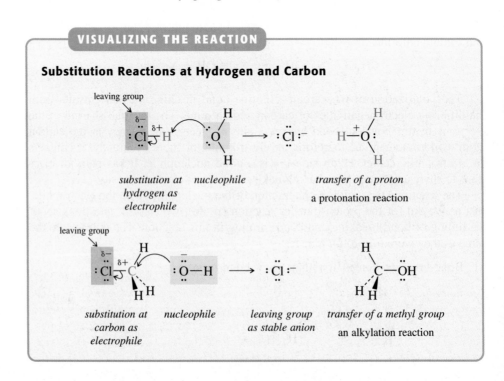

**VISUALIZING THE REACTION**

**Substitution Reactions at Hydrogen and Carbon**

*substitution at*    *nucleophile*        *transfer of a proton*
*hydrogen as*                 a protonation reaction
*electrophile*

*substitution at*    *nucleophile*    *leaving group*    *transfer of a methyl group*
*carbon as*                 *as stable anion*     an alkylation reaction
*electrophile*

Carbon can be involved in only four covalent bonds, so the carbon–chlorine bond breaks as the carbon–oxygen bond forms. The hydroxide ion, the nucleophile in this reaction, displaces chloride ion. The chloride ion is called a **leaving group,** a stable species that can be detached from a molecule in a bond-breaking step. The overall reaction is called a **nucleophilic substitution reaction** or a **nucleophilic displacement reaction.**

To balance charges, sodium ion, $Na^+$, appears on both sides of the overall reaction equation shown at the beginning of this section. Such ions are called **spectator ions** or **counter ions.** They appear in the balanced equation for a reaction because without them, the reagents used as starting materials in the reactions and the products obtained would not be complete. They are not shown when writing mechanisms because they do not participate directly in the bond-breaking and bond-forming processes.

**Problem 4.2**   Predict, by using curved arrows to show the movement of electrons, what will happen when each pair of reagents shown below is mixed. Identify the electrophile and the nucleophile in each case.

(a) $CH_3Br + Na^+OH^- \longrightarrow$     (b) $CH_3I + Na^+OH^- \longrightarrow$
(c) $CH_3CH_2Cl + Na^+OH^- \longrightarrow$     (d) $CH_3I + NH_3 \longrightarrow$

**Problem 4.3**   Sodium ion, $Na^+$, is not a good electrophile. How would you rationalize this fact? Why are mercury(II), $Hg^{2+}$, and copper(II), $Cu^{2+}$, better electrophiles than sodium ion?

**Problem 4.4**   Bleach solution is prepared by bubbling chlorine gas, $Cl_2$, into a solution of sodium hydroxide. The nucleophile, hydroxide ion, reacts with the electrophile, chlorine, to give hypochlorous acid. The acid is then deprotonated by excess base to give sodium hypochlorite, which is in equilibrium with free chlorine in chlorine bleaches.

$$Cl_2 \;+\; Na^+OH^- \longrightarrow HOCl \;+\; Na^+Cl^-$$
electrophile     nucleophile     hypochlorous
acid

$$HOCl \;+\; Na^+OH^- \longrightarrow Na^+OCl^- + H_2O$$
hypochlorous     base     sodium
acid     hypochlorite

Other household cleaners contain ammonia, $NH_3$. Bleach bottles contain warnings not to mix bleach with other household cleaners, especially ammonia. This is because chlorine reacts with ammonia to give a highly toxic compound, chloramine, $H_2N—Cl$. This compound is formed as a result of a nucleophilic attack of ammonia on chlorine, and the subsequent deprotonation of the resulting compound by hydroxide ion in the bleach solution. This reaction is exactly analogous to the reaction of chlorine with hydroxide ion shown in this problem.

(a) Write the overall equation for the formation of chloramine from chlorine and ammonia.
(b) Use curved arrows to show electronic changes that take place in the conversion of chlorine and ammonia to chloramine.

## B. A Biological Nucleophilic Substitution Reaction

Nucleophilic substitutions are an important class of reactions for chemical transformations both in the laboratory and in biological systems. Biological alkylation reactions, many of them involving the transfer of a methyl group, constitute important steps in the synthesis of compounds essential to the functioning of living organisms. An example is the formation of the amino acid methionine from homocysteine.

**VISUALIZING THE REACTION**

**A Biological Nucleophilic Substitution. A Methyl Transfer Reaction**

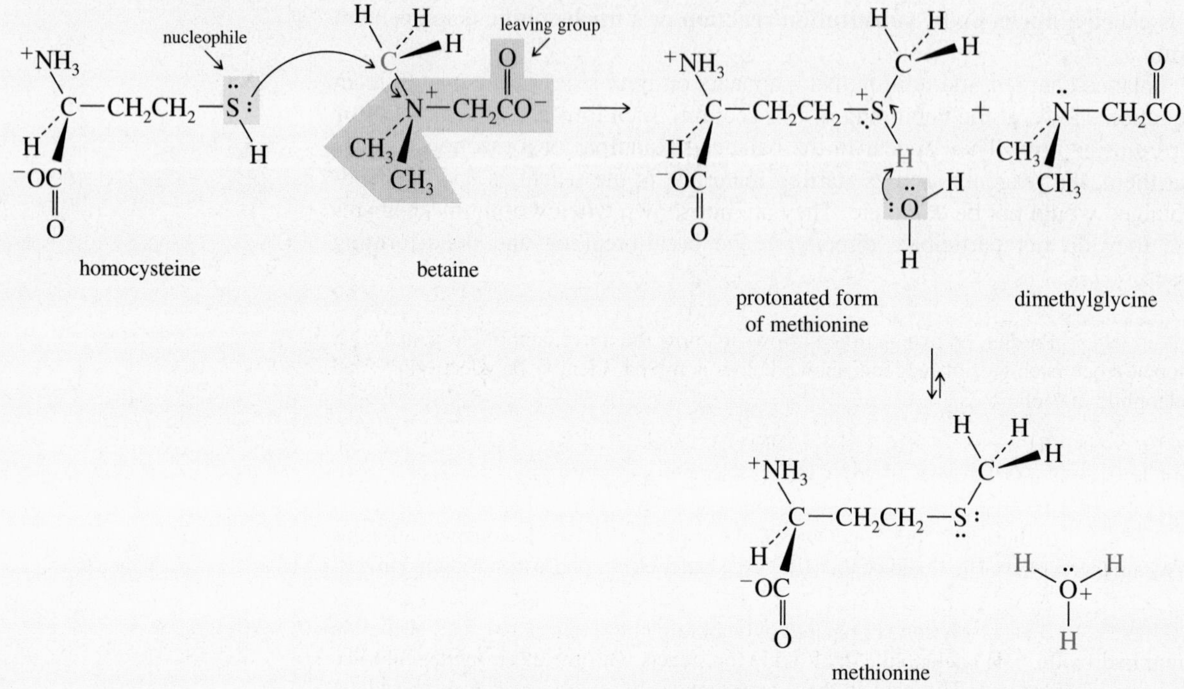

homocysteine

betaine

protonated form
of methionine

dimethylglycine

methionine

In this reaction, the sulfur atom on homocysteine is the nucleophile. The methyl groups on the positively charged nitrogen atom of betaine are electrophilic. The leaving group is an amine in this reaction, providing an example of how a leaving group need not be a halide ion. We will meet many different kinds of leaving groups as we continue our study of substitution reactions.

A transfer of a methyl group from betaine to homocysteine results in the formation of methionine, an important amino acid. In the body, such reactions take place in the liver and are catalyzed by an enzyme called betaine homocysteine methyl transferase. Nucleophilic substitution reactions are an important class of organic reactions and will be covered in greater detail in Chapter 7.

(↗) **Study Guide**
**Concept Map 4.2**

## C. Energy Changes in the Reaction. Equilibrium

The extent of the reaction of chloromethane with hydroxide ion to give methanol and chloride ion is indicated by the equilibrium constant for the reaction, which can be calculated from the change in standard free energy that occurs when the reaction takes place (p. 95). The change in standard free energy for the reaction is the difference between the standard free energies of formation for the products and those for the reactants. Values for the standard free energies of formation for many covalent and ionic species can be found in handbooks of chemistry and physics. The values for the reaction we are considering follow. The standard conditions for these values are defined as 1 M solutions in water at a temperature of 298 K. These values are used to calculate the change in standard free energy for the reaction.

$$CH_3Cl(aq) + OH^-(aq) \longrightarrow CH_3OH(aq) + Cl^-(aq)$$

$\Delta G_f^{\circ}$, kcal/mol
$\qquad\qquad -12.3 \qquad\qquad -37.6 \qquad\qquad\qquad -41.9 \qquad\qquad -31.4$

$\qquad\qquad \Delta G_f^{\circ}(\text{reactants}) = -49.9\,\text{kcal/mol} \qquad \Delta G_f^{\circ}(\text{products}) = -73.3\,\text{kcal/mol}$

$$\Delta G_r^{\circ}(\text{reaction}) = \Delta G_f^{\circ}(\text{products}) - \Delta G_f^{\circ}(\text{reactants})$$
$$= -73.3\,\text{kcal/mol} - (-49.9\,\text{kcal/mol})$$
$$= -23.4\,\text{kcal/mol}$$

There is a decrease in the standard free energy of the system for the reaction as it is written above; $\Delta G_r^{\circ}$ for the reaction has a negative value. This means that the reaction will go as written; that is, the products are favored thermodynamically over the reactants.

The equilibrium constant for the reaction can be calculated from the value for $\Delta G_r^{\circ}$.

$$\Delta G_r^{\circ} = -2.303RT \log K_{eq}$$

$$\log K_{eq} = \frac{-\Delta G_r^{\circ}}{2.303RT}$$

$$= \frac{-(-23.4\,\text{kcal/mol})}{2.303 \times 1.987 \times 10^{-3}\,\text{kcal/deg·mol} \times 298\,\text{K}}$$

$$= \frac{23.4\,\text{kcal/mol}}{1.36\,\text{kcal/mol}}$$

$$\log K_{eq} = 17.2$$

$$K_{eq} = 1.61 \times 10^{17}$$

The value for the equilibrium constant is very large; this reaction is essentially complete at equilibrium. No detectable amount of the reactant chloromethane remains. The equation relating the equilibrium constant to $\Delta G_r^{\circ}$ contains a temperature term because equilibrium constants depend on temperature. The one calculated above is for the reaction at room temperature, 25 °C (298 K).

Note that in talking about equilibrium, we are concerned only with the difference in energy between the initial state of the system, the reactants, and the final state of the system when the products have formed. Nothing has been said about how fast the reaction will take place. That is the subject of the next section.

**Study Guide**
**Concept Map 4.3**

## D. The Rate of the Reaction. Kinetics

Questions about how fast a chemical reaction goes are in the realm of kinetics. **Kinetics** deals with the rate of a chemical reaction and the factors that influence that rate.

The rate of the reaction of chloromethane with hydroxide ion has been studied in water solution at 38 °C, with the initial concentration of chloromethane at 0.003 M and that of the hydroxide ion at 0.01 M. The rate was studied by measuring the decrease in the concentration of hydroxide ion or the increase in the concentration of chloride ion as the reaction progressed.

$$CH_3Cl \quad + \quad OH^- \quad \longrightarrow \quad CH_3OH \quad + \quad Cl^-$$

chloromethane $\qquad$ hydroxide ion $\qquad$ methanol $\qquad$ chloride ion

*concentration* $\qquad\qquad\qquad\qquad$ *concentration*
*decreases as the* $\qquad\qquad\qquad\qquad$ *increases as the*
*reaction progresses* $\qquad\qquad\qquad$ *reaction progresses*

From these quantities, the amount of chloromethane that had reacted at any given time was determined.

The rate of the reaction was found to be dependent on the concentrations of both the reactants. The rate equation is

$$\text{Rate} = k_r[CH_3Cl][OH^-]$$

where $k_r$ is the rate constant for the reaction. The reaction is a **second-order reaction.** The **order** of a chemical reaction is defined as the sum of the exponents of the concentration terms that appear in the rate equation. In this rate equation, the concentration of chloromethane and the concentration of hydroxide ion each appear once; each exponent is 1. (Of course, exponents that are ones are not generally shown in written equations.) Therefore, the order of the reaction is $1 + 1$, or 2. The reaction is first-order with respect to chloromethane, first-order with respect to hydroxide ion, and second-order overall. The order of a reaction is always determined experimentally and cannot be predicted from its equation. The rate constant, $k_r$, is a constant for any given temperature. On page 125 we will discuss the effect of temperature on $k_r$.

For the reaction of chloromethane with hydroxide ion at 38 °C (311 K), the value for $k_r$ is $3.55 \times 10^{-5}$ L/mol·s. Therefore, for the initial reaction conditions where the concentration of chloromethane is 0.003 M and that of hydroxide ion is 0.01 M, we have

$$\text{Initial rate} = 3.55 \times 10^{-5} \text{ L/mol} \cdot \text{s} \times 3 \times 10^{-3} \text{ mol/L} \times 1 \times 10^{-2} \text{ mol/L}$$
$$= 1.07 \times 10^{-9} \text{ mol/L} \cdot \text{s}$$

The reaction is very slow. Only a very small percentage of the chloromethane molecules in this reaction mixture are converted to methanol in a day. Thus, even though the equilibrium is highly favorable for the formation of methanol from chloromethane (p. 121), the reaction proceeds extremely slowly. We will examine why this is so in the next three parts of this section.

⊘ **Study Guide**
**Concept Map 4.4**

**Problem 4.5**    What would be the initial rate of the reaction of chloromethane with hydroxide ion at 311 K if the concentration of hydroxide ion were increased to 0.05 M? What would happen to the rate if the initial concentration of chloromethane were decreased to 0.001 M?

## E. The Transition State

The mechanism for the reaction of chloromethane with hydroxide ion given on page 118 shows the two species coming together in a way that allows for bonding to start between the oxygen atom of the hydroxide ion and the carbon atom of chloromethane. The two reagents collide with each other and must do so with an orientation that allows bonding to take place. The hydroxide ion could collide with the chlorine end of the chloromethane molecule, and no reaction would take place. Instead, repulsion would occur between the negatively charged hydroxide ion and the partially negatively charged chlorine atom. The two species would bounce off one another, with no chemical change taking place.

*approach of hydroxide ion that may lead to reaction with chloromethane*

*attraction between negatively charged hydroxide ion and carbon atom with partial positive charge*

*approach of hydroxide ion to chloromethane that will not lead to chemical reaction*

*repulsion between negatively charged hydroxide ion and chlorine atom with partial negative charge*

Therefore, not all encounters between potentially reactive species produce the chemical reaction.

The reaction of chloromethane with hydroxide ion involves the breaking of a bond between carbon and chlorine and the forming of a new bond between oxygen and carbon. At some point during the reaction, the carbon atom is partially bonded to both the hydroxide ion and the chlorine atom. This state, lying between the reactants and the products, is known as the **transition state.** The transition state is exactly what it sounds like, a transitory state that the system passes through on its way from reactants to products. It has no lifetime but does have geometry and a charge distribution that is represented by a molecular complex called the **activated complex.** This molecular complex exists at the transition state in equilibrium with the reactants and is represented in equations enclosed in square brackets with a symbol called a *double dagger* ($\ddagger$) as a superscript.

*reactants*

*the activated complex the molecular configuration at the transition state*

*products*

The transition state is a high-energy state. Energy has been put into the molecule to partially break the carbon–chlorine bond, but the energy from the formation of the full carbon–oxygen bond has not yet been gained.

Once the activated complex is formed, it is converted very rapidly to product, at some absolute rate, $k_0$. At 25 °C, the **absolute rate constant,** $k_0$, is $6.2 \times 10^{12}$/s. This number is of the same magnitude as the frequency of the vibration of a covalent bond in a molecule (see Section 12.2A for a description of some vibrational motions of covalent bonds). Thus, once the reactants acquire enough energy to become an activated complex, bonds break and new bonds form during the time required for a molecular vibration.

## F. Free Energy of Activation

The energy relationships between the reactants, the transition state, and the products for the reaction of chloromethane and hydroxide ion can be represented on an **energy diagram** (Figure 4.1, p. 124). The *y* axis represents energy, in this case, the free energy of the system. The *x* axis is called the **reaction coordinate** and represents the changes that must take place in bond lengths and bond angles within

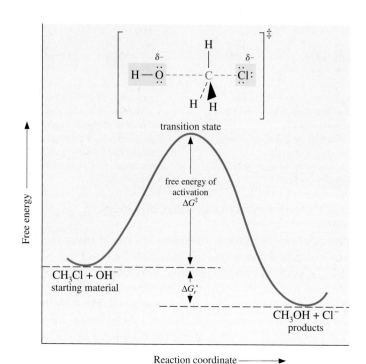

**Figure 4.1**
Relative energy levels for reactants, transition state, and products for the reaction of chloromethane with hydroxide ion.

molecules as reactants are converted to products. Moving along the $x$ axis from left to right corresponds to the progress of the reaction from reactants to products.

In going from the reactants to the products, the system has to go over an energy barrier. The height of this energy barrier is the **free energy of activation,** $\Delta G^{\ddagger}$, for the reaction, which is the energy that the reactants attain in reaching the transition state. The free energy of activation is a quantity that has to be determined experimentally for each reaction; it cannot be predicted.

The rate constant, $k_r$, for a reaction is related to the free energy of activation by the following equation:

$$k_r = k_0 e^{-(\Delta G^{\ddagger}/RT)}$$

in which $e$ is 2.718, the base of natural logarithms, and $k_0$ is the absolute rate constant mentioned at the end of the last subsection. The negative exponential relationship between the rate constant and $\Delta G^{\ddagger}$ is very important. *As $\Delta G^{\ddagger}$ gets larger, $e^{-(\Delta G^{\ddagger}/RT)}$ must get smaller; therefore, the observed rate of the reaction decreases. And a smaller $\Delta G^{\ddagger}$ means an increased rate of reaction.* Small differences in energies of activation produce very large differences in rates. For example, a doubling of the free energy of activation from 10 to 20 kcal/mol would decrease the rate to a ten-millionth of its original value.

Another feature of Figure 4.1 is of interest. The activated complex is shown with unequal distances between the carbon atom undergoing substitution and the incoming hydroxide ion and the departing chloride ion. The transition state is closer in structure to the reactants, chloromethane and hydroxide ion, than to the products, methanol and chloride ion. In assigning this structure to the transition state, we are applying the **Hammond Postulate,** which states that the geometry of the transition state most closely resembles the side to which it is closer in free energy. In Figure 4.1, the transition state is closer in free energy to the reactants than to the products; therefore, we postulate that the activated complex looks more like the reactants than the products.

⊘ **Study Guide**
**Concept Map 4.5**

## G. The Effect of Temperature on the Rate of the Reaction

The rate of a reaction is different at different temperatures. For example, the rate constant, $k_r$, for the reaction of chloromethane with hydroxide ion has been measured at several temperatures. The results are shown in Table 4.1. The rate of the reaction increases by a factor of about two to four for each rise in temperature of $10°$, according to the data in the table.

We are already familiar with the equation that relates the rate constant with temperature:

$$k_r = k_0 e^{-(\Delta G^{\ddagger}/RT)}$$

Temperature, $T$, appears in the denominator of the exponent; therefore, an increase in $T$ increases the value of $e^{-(\Delta G^{\ddagger}/RT)}$ and also the value of $k_r$. We are assuming that $\Delta G^{\ddagger}$ does not vary much with temperature.

What physical model accounts for the dependence of rate on temperature? At any temperature, the molecules in a reaction mixture have a wide distribution of energies, as shown in Figure 4.2 for two temperatures. Most of the molecules have energies close to some average value, but some have much lower energies or much higher ones. The molecules are moving rapidly and collide frequently with one another. As a result of these collisions, energy is transferred from one molecule to another. The kinetic energy of the molecules is also converted to other forms of energy, such as the vibrational energy of bonds within a molecule. Therefore, the amount of energy available to a given molecule is constantly changing.

An increase in temperature increases the kinetic energy of the molecules and thus the number of molecules with higher kinetic energies. If $\Delta G_1^{\ddagger}$ is the free energy of activation for the reaction, more molecules will have enough energy to get to the top of the energy barrier in the reaction pathway (see Figure 4.1) at the higher temperature, $T_2$, than at the lower temperature, $T_1$. Thus an increase in temperature increases the rate of a reaction.

If we compare a reaction with a low free energy of activation ($\Delta G_1^{\ddagger}$ in Figure 4.2) to one with a high free energy of activation ($\Delta G_2^{\ddagger}$), we see that at either temperature more molecules will have enough energy to get over the lower energy barrier. *The rate of the reaction with the lower free energy of activation will be greater than the rate of reaction with the higher free energy of activation.*

In summary, the rate of a reaction is affected by four factors:

1. Concentration of reagents

2. Energy of activation

**Table 4.1**   **The Effect of Temperature on the Rate of the Reaction of Chloromethane with Hydroxide Ion**

| T, K | $k_r$, L/mol·s |
|------|------|
| 311 | $3.55 \times 10^{-5}$ |
| 322 | $1.46 \times 10^{-4}$ |
| 329 | $3.00 \times 10^{-4}$ |
| 333 | $4.88 \times 10^{-4}$ |
| 340 | $9.97 \times 10^{-4}$ |

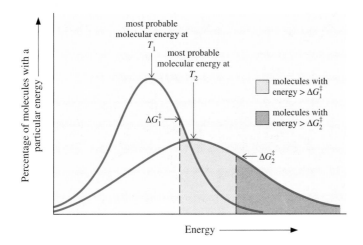

*Figure 4.2*
Distribution of energy among molecules at temperatures $T_1$ and $T_2$.

3. Temperature

4. Spatial effects that are due to the sizes and shapes of the colliding molecules and that determine which orientations of collisions lead to reactions

**Study Guide**
**Concept Map 4.6**

In the rate expression that appears on page 122, all these factors except concentration are included in the experimental rate constant, $k_r$.

**Problem 4.6**    Bromomethane, $CH_3Br$, also reacts with hydroxide ion in the way that chloromethane does. The second-order rate constant, $k_r$, for the reaction at 308 K is $5.3 \times 10^{-4}$ L/mol·s. What is the initial rate of the reaction in a water solution that is 0.001 M in bromomethane and 0.01 M in hydroxide ion?

## 4.3   Addition of Hydrogen Bromide to Propene

### A. An Electrophilic Addition Reaction

The $\pi$ electrons in alkenes are protonated by strong acids (p. 82). The vulnerability of a carbon–carbon double bond to acids is illustrated by the reaction of hydrogen bromide and propene.

$$CH_3CH{=}CH_2 \ + \ HBr \xrightarrow[\substack{18\ h \\ \text{absence} \\ \text{of oxygen}}]{25\ °C} \underset{\substack{Br}}{CH_3CHCH_3}$$

propene         hydrogen          2-bromopropane
                bromide           isopropyl bromide
                                  95%

When the reaction is carried out in the gas phase in the absence of oxygen (air) and with highly purified reagents for 18 hours at room temperature, a 95% yield of 2-bromopropane, also called isopropyl bromide, is obtained.

The electrons of the $\pi$ bond in an alkene are available for reaction. The alkene, therefore, is the nucleophile in this reaction. The hydrogen bromide molecule is po-

**VISUALIZING THE REACTION**

**Protonation of an Alkene**

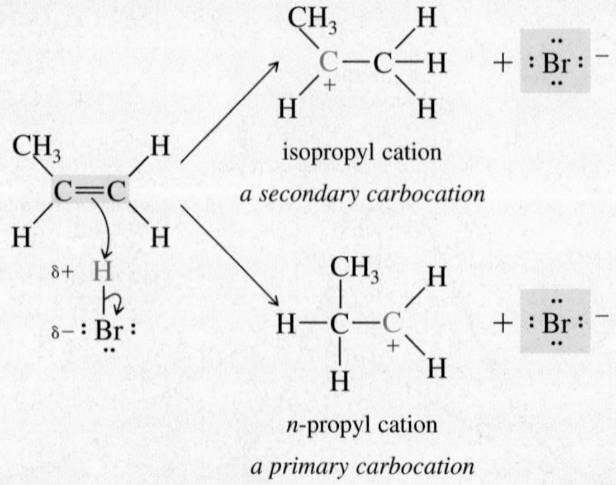

isopropyl cation

*a secondary carbocation*

*n*-propyl cation

*a primary carbocation*

larized; the hydrogen atom has a partial positive charge, and the electronegative bromide atom has a partial negative charge. The hydrogen atom of hydrogen bromide is thus the electrophile. The approach of the polarized hydrogen bromide molecule to the $\pi$ bond of propene initiates the reaction. The mechanism of the reaction shows the donation of electrons from the nucleophile, specifically, the electrons of the $\pi$ bond of propene, to the electrophile, the proton from hydrogen bromide.

A bond can form between the proton and the double bond of propene in two ways. The proton can bond to the carbon atom at the end of the propene molecule, leaving the middle carbon atom deficient in electrons. This creates a **secondary carbocation,** a positively charged species in which two alkyl groups are attached to the electron-deficient carbon atom. If bonding occurs between the proton and the middle carbon atom in the chain, the end carbon atom will have a positive charge, creating a **primary carbocation,** a species in which only one alkyl group is attached to the electron-deficient carbon atom.

Carbocations are Lewis acids. They are electron-seeking reagents, electrophiles, and react with nucleophiles that are present in a reaction mixture. A negatively charged bromide ion, which is one of the nucleophiles in the reaction mixture we are considering, combines with an isopropyl cation to give the observed product.

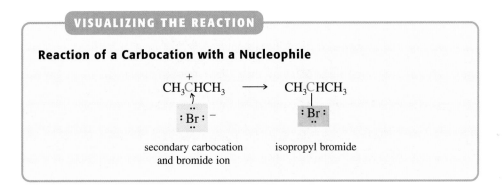

VISUALIZING THE REACTION

**Reaction of a Carbocation with a Nucleophile**

secondary carbocation
and bromide ion

isopropyl bromide

No *n*-propyl bromide is obtained, suggesting that little of the primary carbocation is formed in this reaction mixture.

The overall reaction, the addition of hydrogen bromide to the double bond, is an example of a large and important class of reactions known as **electrophilic addition reactions.** Many more examples of this type of reaction will be seen in Chapters 8 and 9.

*Study Guide*
Concept Map 4.7

**Problem 4.7**  What other nucleophile might be found in the reaction mixture for the reaction of propene with hydrogen bromide? Write a mechanism for the reaction of isopropyl cation with this nucleophile. (*Hint:* A review of Sections 3.2 and 3.3 may be helpful.)

## B. Markovnikov's Rule. Regioselectivity in a Reaction

Propene and hydrogen bromide are both unsymmetrical molecules. The reaction between them could form 1-bromopropane (*n*-propyl bromide) as well as 2-bromopropane, but under the conditions described earlier (p. 126), only the latter compound is produced.

$$CH_3CH\!=\!CH_2 + H\!-\!Br$$

$$\begin{array}{cc} CH_3CH\!-\!CH_2 & CH_3CH\!-\!CH_2 \\ \;\;|\qquad\; | & \;\;|\qquad\; | \\ \;\;H\qquad Br & \;\;Br\qquad H \end{array}$$

1-bromopropane          2-bromopropane
n-propyl bromide        isopropyl bromide

*not observed*            *the only product*

Of the two possible ways in which these reagents can react, one reaction pathway leading to one product is favored. Such an observed preference in the direction in which molecules react with each other is called **regioselectivity.** Such selectivity is useful if we wish to synthesize the compound, in this case isopropyl bromide, that is favored in a reaction. On the other hand, it works against us if we want to synthesize the unfavored product of the reaction.

The regioselectivity of the addition of protic acids to alkenes has been recognized for a long time. In 1869, the Russian chemist Vladimir Markovnikov summarized his experimental observations in a rule that bears his name. Markovnikov said, "When a hydrocarbon of unsymmetrical structure combines with a halogen hydracid, the halogen adds itself to the less hydrogenated carbon atom, i.e., to the carbon atom that is more under the influence of other carbon atoms." In propene, one carbon atom of the double bond is bonded to two hydrogen atoms and the other carbon atom is bonded to a single hydrogen atom. As we saw, the bromine atom from hydrogen bromide bonds to the carbon atom of the double bond with the single hydrogen atom instead of the one with two hydrogens to give rise to isopropyl bromide, which is the product that Markovnikov's Rule calls for.

**Study Guide**
**Concept Map 4.8**

**Problem 4.8**    Predict the products of the following reactions.

$$\overset{\displaystyle CH_3}{\underset{\displaystyle |}{}}$$

(a)  $CH_3C\!=\!CH_2 + HBr \longrightarrow$    (b)  $CH_3CH_2CH\!=\!CH_2 + HBr \longrightarrow$

(c)  $CH_3CH_2CH\!=\!CHCH_3 + HBr \longrightarrow$

## C. Relative Stabilities of Carbocations

Carbocations are **reactive intermediates** that are formed as the reaction progresses from the reactants to the products. They are less stable than the reactants and the products and have only a short lifetime in the reaction mixture. Nevertheless, there is good experimental evidence for the existence of such intermediates, including physical measurements that have been made on some solutions. The cationic carbon atom, bonded to three other atoms, is trigonal planar and $sp^2$-hybridized. An empty $p$ orbital is perpendicular to the plane containing the carbon atom and the three atoms to which it is bonded (Figure 4.3).

Why are secondary carbocations formed more easily than primary cations in the reaction mixture we have been studying? This observation is explained by postulating that the methyl group is slightly electron-releasing in its effect on neighboring atoms. A methyl group is more **polarizable** (p. 253) than a hydrogen atom; it has a larger density of electrons that can be drawn toward a site of electron deficiency. This electron-releasing effect is particularly noticeable when the $sp^3$-hybridized carbon atom of an alkyl group is bonded to a more electronegative $sp^2$- or $sp$-hybridized

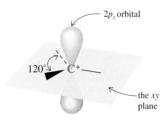

*Figure 4.3*
The geometry at a carbocation.

carbon atom (p. 60). The electrons around the carbon atom of the alkyl group are drawn toward the more electronegative carbon atom to which it is bonded. In the case of a carbocation, this effect results in the delocalization of some of the positive charge to the alkyl group, which stabilizes the ion.

In a primary carbocation, one alkyl group and two hydrogen atoms are bonded to the carbon atom bearing the positive charge. In a secondary carbocation, there are two alkyl groups and one hydrogen atom on the electron-deficient carbon atom. A secondary carbocation is more stable than a primary carbocation because the electron deficiency in the secondary carbocation is partially offset by the electron-releasing effect of two methyl groups. Only one alkyl group is present to stabilize the electron-deficient carbon atom of a primary carbocation.

$$CH_3\overset{\delta+}{CH_2} \rightarrow \underset{H}{\overset{H}{C}}{}^{\delta+}$$

*n*-propyl cation

*a primary carbocation*
*delocalization of charge*
*to one alkyl group*

$$\overset{\delta+}{CH_3} \rightarrow \underset{\delta+}{\overset{H}{C}} \leftarrow \overset{\delta+}{CH_3}$$

isopropyl cation

*a secondary carbocation*
*delocalization of*
*charge to two alkyl groups*

**Problem 4.9**    Would you expect the methyl cation, $CH_3{}^+$, to be more stable or less stable than the *n*-propyl cation, $CH_3CH_2CH_2{}^+$?

**Problem 4.10**    Would a tertiary carbocation, $(CH_3)_3C^+$, be more stable or less stable than the isopropyl cation, $(CH_3)_2CH^+$, which is a secondary carbocation? List the four cations given in this problem and Problem 4.9 in order of decreasing stability.

The regioselectivity observed for the addition of hydrogen bromide to propene can be rationalized by saying that the more stable of the two possible cationic intermediates is formed in the reaction. A modern version of Markovnikov's Rule is as follows: *For the addition of an unsymmetrical electrophilic reagent to a double bond, the major product is that isomer that results from the formation of the more stable cationic intermediate.* Markovnikov himself recognized the importance of the "carbon that is under the influence of other carbon atoms" as the site at which the halogen would attach itself. The stability of a cationic intermediate increases as the number of carbon atoms bonded to the positively charged carbon atom increases. Thus a tertiary carbocation is more stable than a secondary one, which in turn is more stable than a primary one. The methyl cation is the least stable of all alkyl carbocations.

$$CH_3-\underset{\underset{+}{|}}{\overset{\overset{CH_3}{|}}{C}}-CH_3 \qquad CH_3-\underset{\underset{+}{|}}{\overset{\overset{CH_3}{|}}{C}}-H \qquad CH_3CH_2-\underset{\underset{+}{|}}{\overset{\overset{H}{|}}{C}}-H \qquad H-\underset{\underset{+}{|}}{\overset{\overset{H}{|}}{C}}-H$$

*tert*-butyl cation        isopropyl cation        *n*-propyl cation        methyl cation

*most stable*                                                                *least stable*

In recent years, carbocations have been generated by the ionization of alkyl halides in solvents that are highly polar but do not react with carbocations as nucleophiles. In these solvents, the formation of a carbocation can be an exothermic process. For example, in sulfuryl chlorofluoride, $SO_2ClF$, to which antimony pentafluoride, $SbF_5$, has been added, the formation of *tert*-butyl ion from *tert*-butyl chloride at $-55\ °C$ has an enthalpy of ionization, $\Delta H_i$, of $-24.8$ kcal/mol. The ionization of isopropyl chloride is less exothermic under the same conditions, with a $\Delta H_i$ value of

−15.3 kcal/mol. The ionization is helped by the interaction of the Lewis acid, $SbF_5$, with chloride ion, a Lewis base.

$$RCl + SbF_5 \xrightarrow[\substack{-55\,°C}]{SO_2ClF} R^+SbF_5Cl^-$$

**Study Guide**
**Concept Map 4.9**

These experiments have shown that a tertiary carbocation is more easily formed than a secondary carbocation. Primary carbocations have never been observed in these experiments.

**Problem 4.11**  A study of carbocations has shown that the 1-fluoroethyl cation is 10.3 kcal/mol more stable than the ethyl cation, whereas the 1-cyanoethyl cation is 18.4 kcal/mol less stable than the ethyl cation.

$$\overset{+}{CH_3CHF} \qquad \overset{+}{CH_3CH_2} \qquad \overset{+}{CH_3CHCN}$$

*most stable*                 *least stable*

Use structural formulas to show how chemists rationalize these observations.

## D. Equilibria in the Addition of Hydrogen Bromide to Propene

The reaction of propene with hydrogen bromide can give two products, isopropyl bromide and *n*-propyl bromide. The two products do not differ much in energy. The change in standard free energy (p. 121) in going from the reactants to either one of the products is calculated as follows:

$$CH_3CH{=}CH_2(g) \quad + \quad HBr(g) \longrightarrow CH_3CH_2CH_2Br(g)$$

$\Delta G_f^\circ$    +14.99 kcal/mol     −12.73 kcal/mol     −5.37 kcal/mol

$$\Delta G_r^\circ = -5.37 - (14.99 - 12.73)$$
$$= -7.63 \text{ kcal/mol}$$

$$CH_3CH{=}CH_2(g) \quad + \quad HBr(g) \longrightarrow \underset{\underset{Br}{|}}{CH_3CHCH_3}$$

$\Delta G_f^\circ$    +14.99 kcal/mol     −12.73 kcal/mol     −6.51 kcal/mol

$$\Delta G_r^\circ = -6.51 - (14.99 - 12.73)$$
$$= -8.77 \text{ kcal/mol}$$

The change in standard free energy for each reaction is a negative value, indicating that both reactions will proceed as written. Note that isopropyl bromide is more stable than *n*-propyl bromide; its $\Delta G^\circ$ value is lower by 1.14 kcal/mol.

Equilibrium constants for the two reactions can be calculated using the expression relating free energy changes to the equilibrium constant (p. 121). The equilibrium constant for the formation of *n*-propyl bromide from propene and hydrogen bromide is $3.89 \times 10^5$. The equilibrium constant for the formation of isopropyl bromide is $2.63 \times 10^6$. The ratio of the two equilibrium constants is 6.76, indicating that the equilibrium mixture should contain 87% isopropyl bromide and 13% *n*-propyl bromide. In actual experiments, however, isopropyl bromide is the only product. The composition of the product of this reaction is *not* determined by the relative stabilities of isopropyl bromide and *n*-propyl bromide.

**Problem 4.12**  Carry out the calculation of the equilibrium constants for the formation of isopropyl bromide and *n*-propyl bromide.

## E. The Energy Diagram for the Addition of Hydrogen Bromide to Propene

The reaction of propene with hydrogen bromide proceeds by means of a carbocation, a reactive intermediate that is higher in energy than the reactants or the product (p. 128). These energy relationships can be represented on an energy diagram (Figure 4.4). In going from the reactants to the reactive intermediate, the system has to go over an energy barrier that is greater than the difference in energy between propene and isopropyl cation. This high-energy state is the transition state for the formation of the secondary carbocation, and the energy barrier represents the free energy of activation for the reaction, $\Delta G^{\ddagger}$. Note that the peaks labeled as transition states in Figure 4.4 represent the highest points from the reactants to the reactive intermediate and from the intermediate to the product. The molecular configuration at the transition state for the formation of the secondary carbocation is shown in detail below.

*starting materials*          *activated complex at*          *intermediates:*
                              *the transition state*          *isopropyl cation*
                                                              *and bromide ion*

The two molecules, propene and hydrogen bromide, must first approach each other in the critical orientation with the positive end of the hydrogen bromide molecule pointing toward the $\pi$-electron cloud of propene. A flow of electrons starts

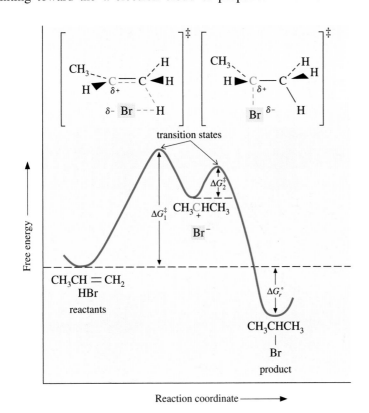

**Figure 4.4**
Relative energy levels for reactants, reactive intermediate, and product in the conversion of propene and hydrogen bromide to isopropyl bromide.

from the $\pi$ bond to the hydrogen atom. As this bonding starts, the bond between the hydrogen and bromine atoms becomes weaker as the bonding electrons move closer to bromine, which becomes more like a bromide ion. At the same time, one of the carbon atoms involved in the double bond develops a partial positive charge and begins to look like a carbocation. The hybridization of the carbon atom that is bonding to the hydrogen atom starts to change from $sp^2$ to $sp^3$.

The energy diagram in Figure 4.4 indicates that the formation of the secondary carbocation and bromide ion from propene and hydrogen bromide is an endothermic reaction. The molecules of the reagents must absorb energy from their surroundings in order to get to that state. The free energy of activation, the energy necessary to get to the transition state, is even greater than the energy difference between the starting materials and the intermediate. The isopropyl cation is converted to isopropyl bromide by a highly exothermic reaction. This reaction also has a transition state and a free energy of activation, though a small one.

According to the Hammond Postulate (p. 124), we expect both transition states to be closer in structure to the carbocation intermediate than they are to the reactants for the first transition state or to the products for the second one.

## F. The Basis for Markovnikov's Rule

Two different carbocations are possible from the protonation of propene. The product observed for the overall reaction comes from only one of these, the secondary carbocation. Two reaction pathways are in competition in this addition reaction. What determines which pathway leads to the major product of the reaction?

Isopropyl bromide is more stable than *n*-propyl bromide by about 1 kcal/mol (p. 130). Indirect measurements show that the isopropyl cation is more stable than the *n*-propyl cation by about 16 kcal/mol. The energy diagram in Figure 4.5 shows the two different pathways that the addition of hydrogen bromide to propene can take. The large difference in the stabilities of the secondary and primary carbocations is reflected in the large difference in the energies of activation for the formation of the two species. The reaction pathway of lower energy proceeds via the more stable secondary carbocation. The rate at which this secondary carbocation is formed is much greater than the rate at which the primary carbocation is formed.

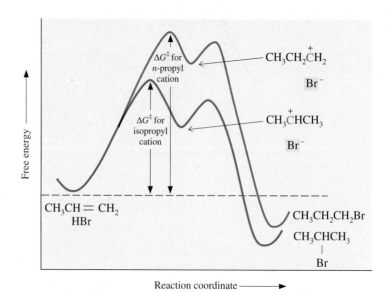

*Figure 4.5*
Comparison of the reaction pathways leading from propene and hydrogen bromide to the two possible products.

Therefore, the product of the reaction is the one from the intermediate the formation of which has the lower energy of activation, even though overall energy considerations (p. 130) allow for the formation of the other product. The products obtained in this reaction are determined by the relative rates at which they are formed and not by their relative stabilities. Such a reaction is said to be *under kinetic (rate) control*.

⊘ **Study Guide**
**Concept Map 4.10**

**Problem 4.13**   The reaction of hydrogen iodide with propene in the gas phase at 238.5 °C has been shown to be a second-order reaction.

(a) Write an equation for this reaction.
(b) Write a rate equation for the reaction.

**Problem 4.14**   At 238.5 °C when propene and hydrogen iodide are both present at an initial pressure of 45 mm Hg, the second-order rate constant is determined to be $1.66 \times 10^{-6}$/atm·s. What is the overall rate of the reaction under these conditions? (760 mm Hg = 1 atm.)

**Problem 4.15**   The energy of activation for the formation of an *n*-propyl cation from propene and hydrogen iodide is about 16 kcal/mol higher than that for the formation of an isopropyl cation (see Problem 4.14). Compare the value of the exponential term $e^{-\Delta G^{\ddagger}/RT}$ for these two cations, assuming a temperature of 490 K and values of $\Delta G^{\ddagger}$ of 22 kcal/mol for the formation of the isopropyl cation and 38 kcal/mol for the formation of the *n*-propyl cation. How much faster is the reaction that produces the isopropyl cation than the one leading to the *n*-propyl cation?

## G. Kinetics of a Reaction with Two Steps. The Rate–Determining Step

The reaction of chloromethane with hydroxide ion occurs in one step (p. 118). No reactive intermediate is formed in the reaction, and the energy diagram for the reaction (see Figure 4.1) shows a single transition state (and therefore, a single free energy of activation) on the path between the reactants and the products. The rate for this reaction is determined by the height of the energy barrier for this single step.

The energy diagram for the reaction of propene with hydrogen bromide (see Figure 4.4) has two transition states and two free energies of activation. The reaction occurs in two consecutive steps. These two consecutive steps of a single overall reaction should not be confused with the two competing reactions for the addition of hydrogen bromide to propene illustrated in Figure 4.5.

So far we have focused on the conversion of propene to isopropyl cation and have not considered the reaction leading from isopropyl cation to isopropyl bromide, the product obtained from the reaction. Isopropyl cation reacts with bromide ion in the second step of the reaction, which has a small energy of activation, goes through a second transition state (see Figure 4.4), and gives isopropyl bromide in an exothermic reaction. Because the energy of activation for this second step is so small in comparison with the energy of activation for the first step, the rate constant for the reaction of isopropyl cation with bromide ion is large. Once the molecules make it to the top of the first hill, they also can acquire enough energy to make it over the second hill. Thus, although there are actually two different rate constants involved in the overall reaction leading to the formation of isopropyl bromide, experimental measurements are made of only one of them, $k_1$, the rate constant for the first step.

$$CH_3CH{=}CH_2 + HBr \xrightarrow[\text{(slow step)}]{k_1} CH_3\overset{+}{C}HCH_3 + Br^- \xrightarrow[\text{(fast step)}]{k_2} \underset{\underset{Br}{|}}{CH_3CHCH_3}$$

The slow step of a reaction sequence, the step with the transition state of highest energy, is known as the **rate-determining step** of the reaction. The overall rate of a reaction cannot be any faster than the rate-determining step. For the reaction of propene and hydrogen bromide, the formation of the high-energy reactive intermediate, isopropyl cation, is the rate-determining step. Once the cation is formed, it reacts relatively rapidly with bromide ion to give the product.

## 4.4   Chemical Transformations

### A. Writing Equations for Organic Reactions

The equations in previous sections of this chapter are attempts to represent accurately the reagents and reaction conditions used in the organic reactions being discussed. The exact details are given because it is important for you to become familiar with the nature of experiments in organic chemistry. It is not necessary, however, for you to memorize all the details that appear in equations. The important reagents will appear many times in problems. If you use the problems as your guide, as you were advised to do in Section 1.1, your attention will be focused on the most important reagents. In other words, you need to learn only the detail that is necessary to answer the problems.

Two types of equations are written for organic reactions. The first type is represented in this chapter by the equations for the reaction of chloromethane with hydroxide ion on page 117 and of hydrogen bromide with propene on page 126. In these equations, all the reactants are shown on the left side of the arrow, and reaction conditions are shown below the arrow. Another way of showing organic reactions is the type of equation in which some reagents are written above the reaction arrow. These are often inorganic reagents, especially acids and bases. For example, equations for the two reactions we have been studying also can be written like this:

$$CH_3Cl \xrightarrow[H_2O]{Na^+OH^-} CH_3OH$$

$$CH_3CH{=}CH_2 \xrightarrow[\substack{25\,°C \\ 18\,h \\ \text{absence of} \\ \text{oxygen}}]{HBr} CH_3\underset{\underset{Br}{|}}{C}HCH_3$$

In this type of equation, the emphasis is on the transformation of the organic compounds, and no attempt is made to show a complete and balanced equation. For example, in the top equation, chloride ion, one of the products of the reaction, is not shown. To help you distinguish between the chief reagent and other reaction conditions, the important reagents will appear over the arrow, and any catalysts, solvents, and other conditions, such as temperature and reaction time, will appear below the arrow. Thus, for the second equation, hydrogen bromide is the important reagent and is written over the arrow. It adds to the double bond in propene. The conditions that are necessary for the reaction to give isopropyl bromide with the best yield are shown under the arrow. You are not required to memorize these conditions.

### B. Thinking about Reactions

Each reaction has three components, which can be placed on the corners of a triangle. They are the starting materials, the products, and the reagents that are necessary to make the transformation. Two such triangles, which we will call **concept triangles,** are shown in Figure 4.6. They represent the two reactions we have studied so far.

In the center of each triangle is the name for the type of reaction, the important concept that connects the three corners of the triangle. For any reaction, we can

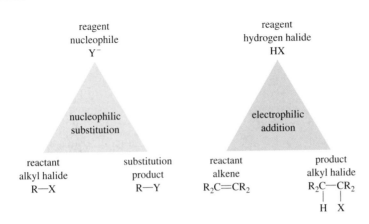

*Figure 4.6*
Concept triangles for nucleophilic substitution and electrophilic addition reactions.

complete the equation if we have information about two of the three components of the reaction represented on the corners of a triangle and if we know the type of reaction that ties the components of the reaction together. Once we understand the concept of "nucleophilic substitution," for example, we can fill in any one of the missing corners in a series of equations. Two examples are shown in Figure 4.7.

In the first triangle in Figure 4.7, we see that the connectivity in the reactant has been changed in going to the product by the substitution of $N_3$, an azido group, for Br. This analysis suggests that a nucleophilic substitution reaction has occurred, the concept that we need to complete the equation. A nucleophile, in this case $N_3^-$, will be used to complete the equation.

$$CH_3Br + N_3^- \longrightarrow CH_3N_3 + Br^-$$
reactant    nucleophile    product

In practice, because we cannot find any bottles containing just azide ions, $N_3^-$, in the laboratory, we add a spectator ion, $Na^+$, to each side of the equation.

$$CH_3Br + Na^+N_3^- \longrightarrow CH_3N_3 + Na^+Br^-$$

For the second triangle, our thinking is somewhat different. We recognize the reagent, $I^-$, iodide ion, as a nucleophile and see that it is bonded to a carbon atom in the product. This suggests that a nucleophilic substitution reaction has occurred. Once we arrive at that concept, the structure of the reactant can be determined. It must have the same connectivity as the product, except that a leaving group, such as another halogen atom, must be connected where the iodine atom is in the product.

$$CH_3CH_2Br + I^- \longrightarrow CH_3CH_2I + Br^-$$
reactant    nucleophile    product

Once again we complete the equation using a spectator ion, $K^+$.

$$CH_3CH_2Br + K^+I^- \longrightarrow CH_3CH_2I + K^+Br^-$$

## C. Choosing Reagents for Simple Syntheses

Organic chemists are interested in synthesizing organic compounds, transforming simple, readily available reagents into more complex compounds with interesting physical and chemical properties. Syntheses are carried out to make biologically active compounds that were originally isolated from natural sources or to make similar compounds that will have even more useful properties. Much synthetic work is

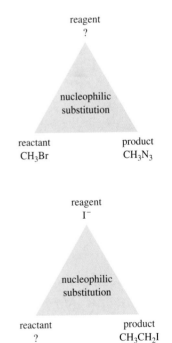

*Figure 4.7*
Concept triangles with missing components.

done to produce medicinal products, to find medications having the maximum effectiveness and the minimum number of undesirable side effects.

The first three sections of this chapter covered two important types of reactions, nucleophilic substitution and electrophilic addition reactions. Now we will look at other examples of these types, as we have already done in Problems 4.2 (p. 119) and 4.8 (p. 128). We also will begin to put reactions together to transform one organic compound into another.

Other simple alkyl halides undergo a nucleophilic substitution reaction in the same way that chloromethane does. Many other anions besides hydroxide ion can serve as nucleophiles. Neutral compounds such as ammonia or amines are also nucleophilic (Problem 4.1, p. 117). Other alkenes will add hydrogen halides in the way that propene does. Halogens also add to double bonds (p. 66). Table 4.2 lists some nucleophiles that react readily with alkyl halides as well as some electrophiles, both acids and halogens, that add to alkenes. The following problem provides practice in writing equations with these reagents.

**Problem 4.16**    Complete the following equations, showing the major product(s) expected in each case. Identify the electrophile and the nucleophile for each. Write a mechanism for the reactions in parts a–d that follow.

(a) $CH_3CH_2Br + NaI \longrightarrow$

(b) $CH_3\underset{\underset{Br}{|}}{C}HCH_3 + NaCN \longrightarrow$

(c) $CH_3CH_2CH{=}CH_2 + HCl \longrightarrow$

(d) $CH_3CH_2\underset{\underset{Br}{|}}{C}HCH_3 + CH_3SNa \longrightarrow$

(e) $CH_3CH{=}CH_2 + Cl_2 \longrightarrow$

In planning a synthesis, organic chemists analyze the structures of the desired product and the reagents that are available and propose reactions that will convert the starting material to the products. For example, how could the following transformation be carried out?

$$CH_3Cl \xrightarrow{?} CH_3CN$$

$$\text{chloromethane} \qquad \text{acetonitrile}$$

An inspection of the two structures reveals that the only change in connectivity has been the substitution of a cyano group (—CN) for a chlorine atom. Sodium cyanide, a source of the nucleophilic cyanide ion, is the reagent needed to make this transformation.

$$CH_3Cl \xrightarrow{NaCN} CH_3CN$$

**Table 4.2**    **Nucleophilic and Electrophilic Reagents**

| Nucleophiles | Electrophiles |
|---|---|
| $HO^-$, $CH_3O^-$, $CH_3CH_2O^-$ | $HCl$, $HBr$, $HI$, $H_3O^+$ |
| $HS^-$, $CH_3S^-$, $CH_3CH_2S^-$ | $H_2SO_4$ ($HOSO_3H$) |
| $CN^-$, $N_3^-$ | $Cl_2$, $Br_2$ |
| $I^-$, $Br^-$ | |
| $NH_3$, $CH_3NH_2$ | |
| $PH_3$ | |

In thinking this way we are applying the concept triangle for nucleophilic substitution shown in Figure 4.6 (p. 135).

What about the following transformation?

$$CH_3CH=CHCH_3 \xrightarrow{?} CH_3CH_2CHCH_3$$
$$\underset{I}{|}$$

2-butene     2-iodobutane

The product has an extra hydrogen atom and an iodine atom attached to the carbon atoms that had a $\pi$ bond between them in the starting material. The $\pi$ bond is a source of electrons, a nucleophile; therefore, an electrophilic reagent incorporating hydrogen and iodine, HI, is needed. Here we use the concept of electrophilic addition to tie reactant, product, and reagent together.

$$CH_3CH=CHCH_3 \xrightarrow{HI} CH_3CHCHCH_3$$
$$\overset{}{\underset{H\ \ I}{|\ \ |}}$$

The starting material is symmetrical; the carbon atoms on both sides of the double bond have the same substituents. Therefore, the reaction will show no regioselectivity.

---

**Problem 4.17**    What reagent will bring about the chemical change shown in each of the following equations? Identify each reagent as an electrophile or a nucleophile, and explain why you chose it. Constructing a concept triangle for each reaction will be helpful.

(a) $CH_3CH_2I \xrightarrow{?} CH_3CH_2OCH_3$

(b) $CH_3CH_2CH=CH_2 \xrightarrow{?} CH_3CH_2CHCH_3$
$$\underset{OSO_3H}{|}$$

(c) $CH_3CH_2CH=CH_2 \xrightarrow{?} CH_3CH_2CHCH_2$
$$\underset{Br\ \ Br}{|\ \ |}$$

(d) $CH_3CH_2CH_2Br \xrightarrow{?} CH_3CH_2CH_2\overset{+}{N}H_3Br^-$

---

Sometimes a chemical transformation requires more than one reaction. For example, how could the following change be carried out?

$$CH_3CH_2CH=CH_2 \xrightarrow{?} CH_3CH_2CHCH_3$$
$$\underset{SCH_3}{|}$$

1-butene     sec-butyl methyl thioether

Unlike the previous case that required the addition of HI to a double bond, there is no reagent that will carry out this change in one step. Hydroiodic acid, HI, as well as the other acids listed in Table 4.2 as electrophiles, all have negative $pK_a$ values. In other words, they are very strong acids. Only such strong acids protonate the $\pi$ bond.

In this case, it looks as if $H—SCH_3$ is being added to the double bond, but that compound is not acidic enough ($pK_a \sim 10.5$ from the table inside the front cover) to serve as an electrophile toward a double bond. The $—SCH_3$ group, however, appears in Table 4.2 as a nucleophile. If a compound that has a halogen atom where the $—SCH_3$ group is found in the product were available, it could be transformed into the desired product, the thioether. The problem thus requires two reactions.

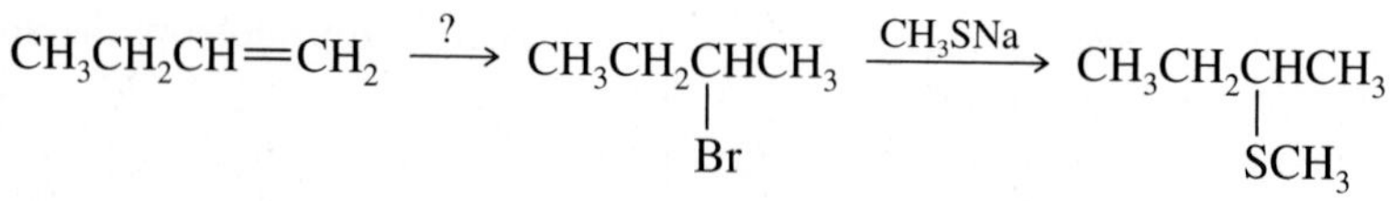

$$CH_3CH_2CH=CH_2 \xrightarrow{?} CH_3CH_2\underset{\underset{Br}{|}}{C}HCH_3 \xrightarrow{CH_3SNa} CH_3CH_2\underset{\underset{SCH_3}{|}}{C}HCH_3$$

The electrophilic reagent HBr will convert the alkene to the alkyl bromide in the first step of this sequence of reactions.

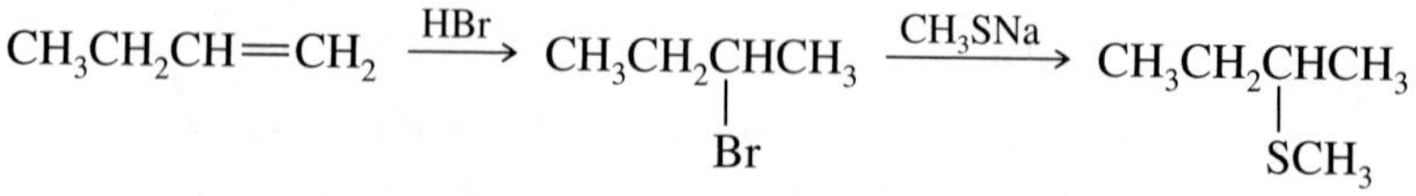

$$CH_3CH_2CH=CH_2 \xrightarrow{HBr} CH_3CH_2\underset{\underset{Br}{|}}{C}HCH_3 \xrightarrow{CH_3SNa} CH_3CH_2\underset{\underset{SCH_3}{|}}{C}HCH_3$$

When solving synthesis problems that require more than one step, you should work backwards. Another example will illustrate this thinking process.

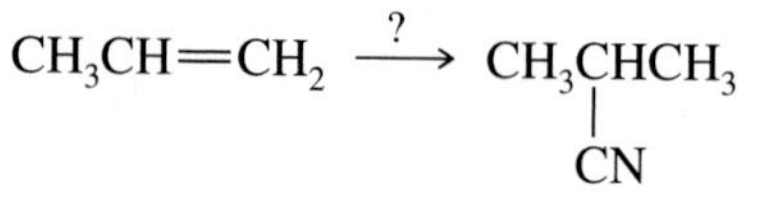

$$CH_3CH=CH_2 \xrightarrow{?} CH_3\underset{\underset{CN}{|}}{C}HCH_3$$

propene                     isopropyl cyanide

A check of the p$K_a$ table shows that the acid (HCN) that would have to be added to the double bond for a one-step reaction is too weak (p$K_a$ 9.1) to serve as a good electrophile toward a double bond. A direct electrophilic addition reaction is not possible. The structure of the product must be analyzed in order to decide where and what type of reaction is likely.

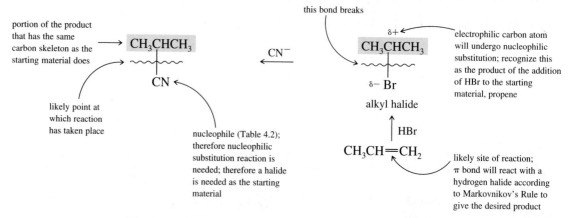

The process of working backwards that is illustrated above is important. It is a good idea to practice it verbally, reasoning aloud to yourself as you work the problems, until this type of analysis becomes second nature.

---

**Problem 4.18**    How could each of the following transformations be carried out? Some of them may require more than one reaction. In each case, show as much of your reasoning process as possible.

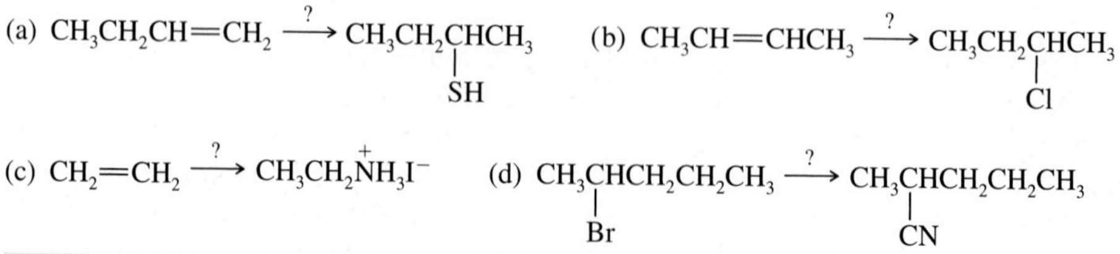

(a) $CH_3CH_2CH=CH_2 \xrightarrow{?} CH_3CH_2\underset{\underset{SH}{|}}{C}HCH_3$       (b) $CH_3CH=CHCH_3 \xrightarrow{?} CH_3CH_2\underset{\underset{Cl}{|}}{C}HCH_3$

(c) $CH_2=CH_2 \xrightarrow{?} CH_3CH_2\overset{+}{N}H_3I^-$    (d) $CH_3\underset{\underset{Br}{|}}{C}HCH_2CH_2CH_3 \xrightarrow{?} CH_3\underset{\underset{CN}{|}}{C}HCH_2CH_2CH_3$

## SUMMARY

Reagents in chemical reactions can be classified as electrophiles or nucleophiles. Electrophiles are electron-deficient species that are seeking electrons. Nucleophiles are electron-rich species that are seeking a nucleus to donate electrons to.

One example of a nucleophilic substitution reaction is when an alkyl halide, the electrophile, reacts with hydroxide ion, the nucleophile, with the loss of a halide ion.

$$H-\overset{..}{\underset{..}{O}}:^{-} \quad \overset{\delta+}{C}H_3 \overset{\delta-}{\overset{..}{\underset{..}{Cl}}}: \longrightarrow H\overset{..}{\underset{..}{O}}-CH_3 \quad :\overset{..}{\underset{..}{Cl}}:^{-}$$

*nucleophile*     *electrophile*

One example of an electrophilic addition reaction is when an alkene, acting as the nucleophile, reacts with a strong acid, acting as the electrophile.

$$CH_3CH=CH_2 \quad \overset{\delta+}{H}\overset{\delta-}{\overset{..}{\underset{..}{Br}}}: \longrightarrow CH_3\overset{+}{C}HCH_3 \longrightarrow CH_3CHCH_3$$

$$:\overset{..}{\underset{..}{Br}}:^{-} \qquad :\overset{..}{\underset{..}{Br}}:$$

*nucleophile*        *electrophile*

Chemical reactions occur when there is a decrease in free energy in going from reactants to products. A decrease in free energy corresponds to an equilibrium constant that favors the products over the reactants.

For a reaction to be practical, it must proceed at a reasonable rate. The rate of a reaction is determined by the free energy of activation for the rate-determining step, that is, the energy necessary to reach the transition state on the way to a reactive intermediate or a product. An increase in temperature increases the number of molecules having energy equal to or above the free energy of activation and, therefore, increases the rate of a reaction.

A nucleophilic substitution reaction may be a one-step reaction with only one free energy of activation and one transition state. Electrophilic addition to an alkene is invariably a two-step reaction. The first step gives a reactive intermediate, a carbocation; then the second step leads to the product by a combination of the cation with an anion. The overall reaction has two transition states and two free energies of activation. The step with the transition state of highest energy is the rate-determining step.

When a reaction gives as the major product one of several possible products, the reaction is said to be regioselective. The addition of an acid to an unsymmetrical alkene is a regioselective reaction that follows Markovnikov's Rule. The observed regioselectivity is often determined by the relative stabilities of the different reactive intermediates that lead to the products. For the addition of hydrogen bromide to propene, the more stable secondary carbocation is favored over the primary carbocation as the reactive intermediate that leads to the product.

## ADDITIONAL PROBLEMS

**4.19** Complete the following equations, showing the major product(s) expected in each case. Identify each reagent as an electrophile or a nucleophile. Write the full mechanism for the reactions in parts a–d.

(a) $CH_3CH{=}CH_2 + H_2SO_4 \longrightarrow$    (b) $CH_3CH_2CH_2Br + PH_3 \longrightarrow$    (c) $CH_3CH_2I + CH_3CH_2SNa \longrightarrow$

(d) 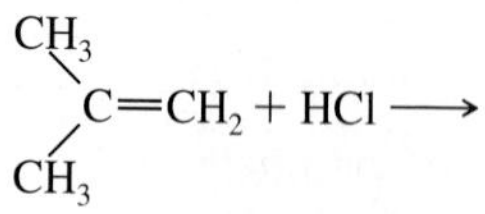 $\underset{CH_3}{\overset{CH_3}{\diagdown}}C{=}CH_2 + HCl \longrightarrow$    (e) $CH_3CH_2CH{=}CHCH_3 + Br_2 \longrightarrow$

**4.20** Supply a reagent that will give the chemical change shown in each of the equations below. Identify the reagents as electrophiles or nucleophiles and explain why you chose them. Constructing a concept triangle for each reaction may be helpful.

(a) $CH_3Br \overset{?}{\longrightarrow} CH_3SH$    (b) $CH_3CH_2Br \overset{?}{\longrightarrow} CH_3CH_2CN$    (c) 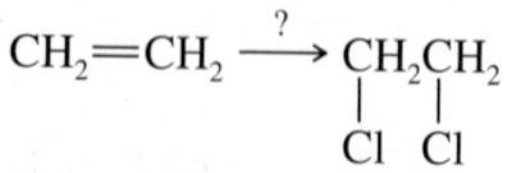 $CH_2{=}CH_2 \overset{?}{\longrightarrow} \underset{\underset{Cl}{|}}{CH_2}\underset{\underset{Cl}{|}}{CH_2}$

(d) $CH_3CH_2CH_2Br \overset{?}{\longrightarrow} CH_3CH_2CH_2OCH_2CH_3$    (e) 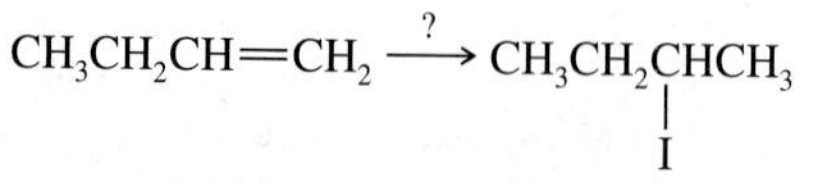 $CH_3CH_2CH{=}CH_2 \overset{?}{\longrightarrow} CH_3CH_2\underset{\underset{I}{|}}{CH}CH_3$    (f) 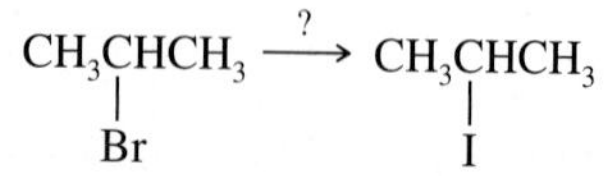 $CH_3\underset{\underset{Br}{|}}{CH}CH_3 \overset{?}{\longrightarrow} CH_3\underset{\underset{I}{|}}{CH}CH_3$

**4.21** How could each of the following transformations be carried out? Some of these transformations may require more than one reaction. In each case, show as much of your reasoning process as possible.

(a) $CH_2{=}CH_2 \overset{?}{\longrightarrow} CH_3CH_2OCH_3$    (b) 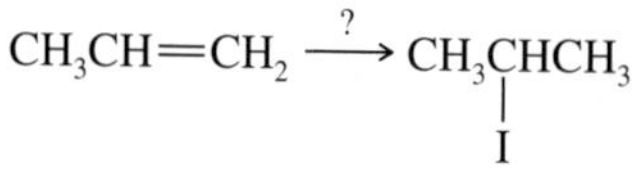 $CH_3CH{=}CH_2 \overset{?}{\longrightarrow} CH_3\underset{\underset{I}{|}}{CH}CH_3$

(c) 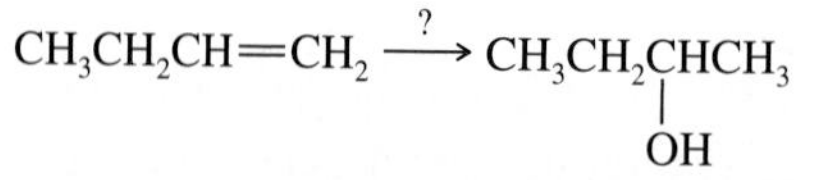 $CH_3CH_2CH{=}CH_2 \overset{?}{\longrightarrow} CH_3CH_2\underset{\underset{OH}{|}}{CH}CH_3$    (d) 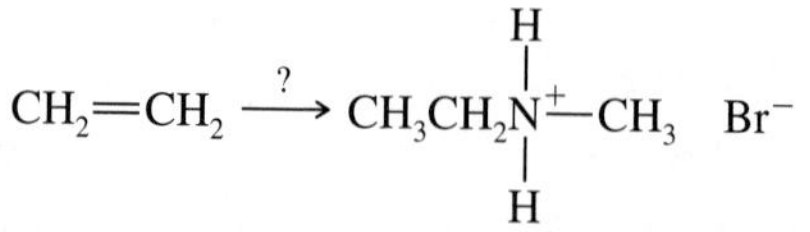 $CH_2{=}CH_2 \overset{?}{\longrightarrow} CH_3CH_2\underset{\underset{H}{|}}{\overset{\overset{H}{|}}{N^+}}CH_3 \quad Br^-$

**4.22** For the following reactions, most of which are as yet unfamiliar, complete the Lewis structures for the pertinent parts of the molecules and identify electrophiles and nucleophiles. Draw arrows showing the movements of electrons that take place to convert the reactants to the products. For example, for

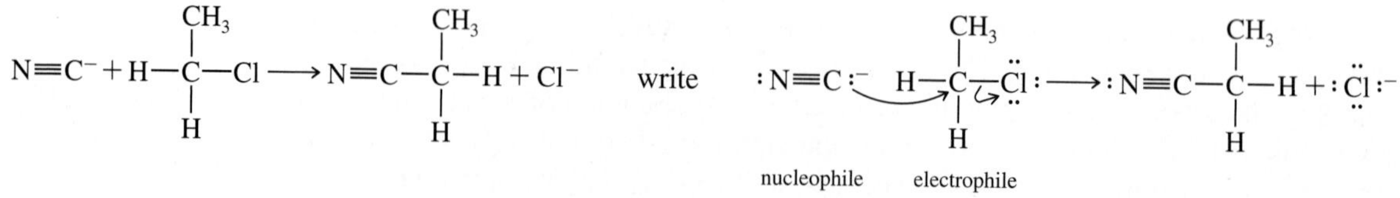

$N{\equiv}C^- + H{-}\underset{\underset{H}{|}}{\overset{\overset{CH_3}{|}}{C}}{-}Cl \longrightarrow N{\equiv}C{-}\underset{\underset{H}{|}}{\overset{\overset{CH_3}{|}}{C}}{-}H + Cl^-$    write    $:N{\equiv}C:^- \quad H{\overset{\overset{CH_3}{|}}{\underset{\underset{H}{|}}{C}}}{\overset{..}{\underset{..}{Cl}}}: \longrightarrow :N{\equiv}C{-}\underset{\underset{H}{|}}{\overset{\overset{CH_3}{|}}{C}}{-}H + :\overset{..}{\underset{..}{Cl}}:^-$

nucleophile        electrophile

(a) $I^- + CH_3Cl \longrightarrow ICH_3 + Cl^-$    (b) 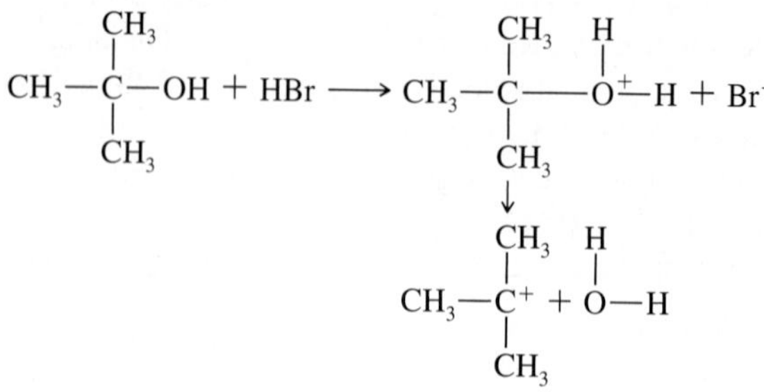 $CH_3{-}\underset{\underset{CH_3}{|}}{\overset{\overset{CH_3}{|}}{C}}{-}OH + HBr \longrightarrow CH_3{-}\underset{\underset{CH_3}{|}}{\overset{\overset{CH_3 \quad H}{|}}{C}}{-}\overset{\overset{|}{}}{O}{\overset{+}{\underset{-}{}}}H + Br^-$

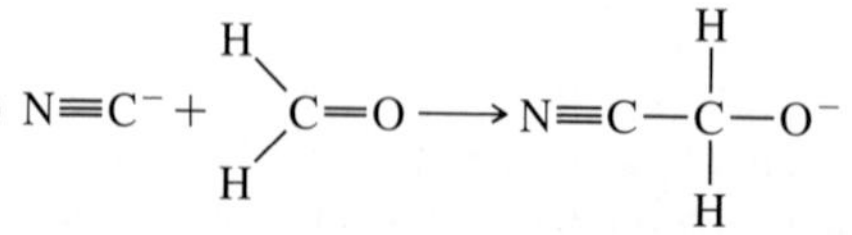

$\downarrow$

$CH_3{-}\underset{\underset{CH_3}{|}}{\overset{\overset{CH_3 \quad H}{|}}{C}}{}^+ + \overset{\overset{|}{}}{O}{-}H$

(c) $N{\equiv}C^- + \underset{\underset{H}{\diagup}}{\overset{\overset{H}{\diagdown}}{}}C{=}O \longrightarrow N{\equiv}C{-}\underset{\underset{H}{|}}{\overset{\overset{H}{|}}{C}}{-}O^-$    (d) 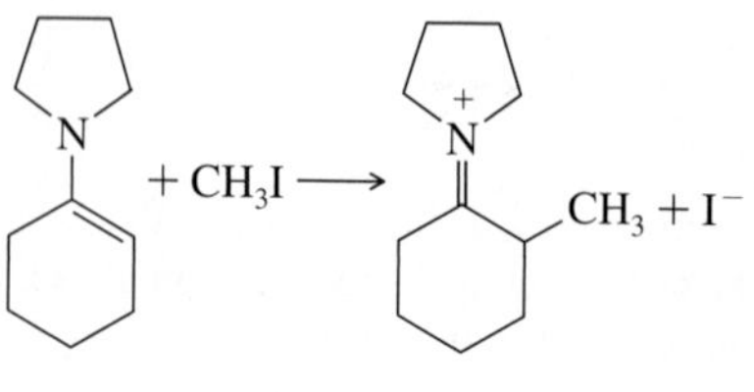 $\underset{\text{(cyclohexene with pyrrolidine N)}}{} + CH_3I \longrightarrow \underset{\text{(product)}}{} CH_3 + I^-$

(e) 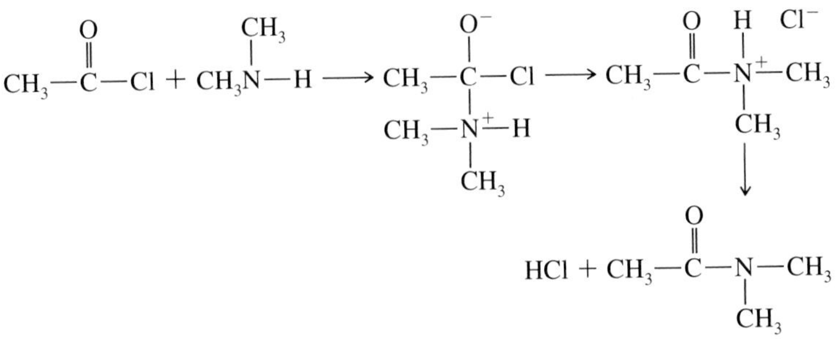 
$$CH_3-\overset{\overset{\displaystyle O}{\|}}{C}-Cl \;+\; CH_3\overset{\displaystyle N}{\underset{\displaystyle |}{N}}-H \longrightarrow CH_3-\overset{\overset{\displaystyle O^-}{|}}{\underset{\underset{\displaystyle CH_3}{\overset{\displaystyle |}{\underset{\displaystyle CH_3-\overset{+}{N}-H}{|}}}}{C}}-Cl \longrightarrow CH_3-\overset{\overset{\displaystyle O}{\|}}{C}-\overset{\overset{\displaystyle H}{|}}{\underset{\underset{\displaystyle CH_3}{\displaystyle |}}{\overset{+}{N}}}-CH_3 \;\; Cl^-$$

$$HCl \;+\; CH_3-\overset{\overset{\displaystyle O}{\|}}{C}-\underset{\underset{\displaystyle CH_3}{\displaystyle |}}{N}-CH_3$$

(f) 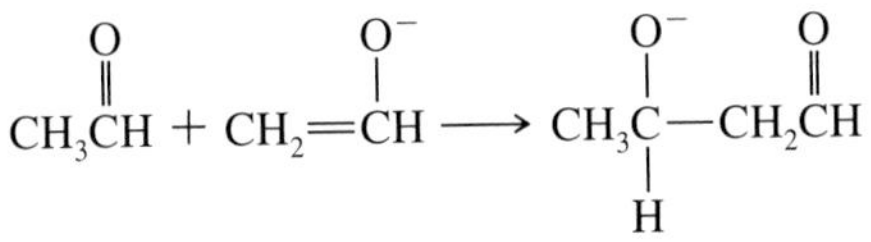 
$$CH_3\overset{\overset{\displaystyle O}{\|}}{C}H \;+\; CH_2{=}\overset{\overset{\displaystyle O^-}{|}}{C}H \longrightarrow CH_3\overset{\overset{\displaystyle O^-}{|}}{\underset{\underset{\displaystyle H}{\displaystyle |}}{C}}-CH_2\overset{\overset{\displaystyle O}{\|}}{C}H$$

(g) 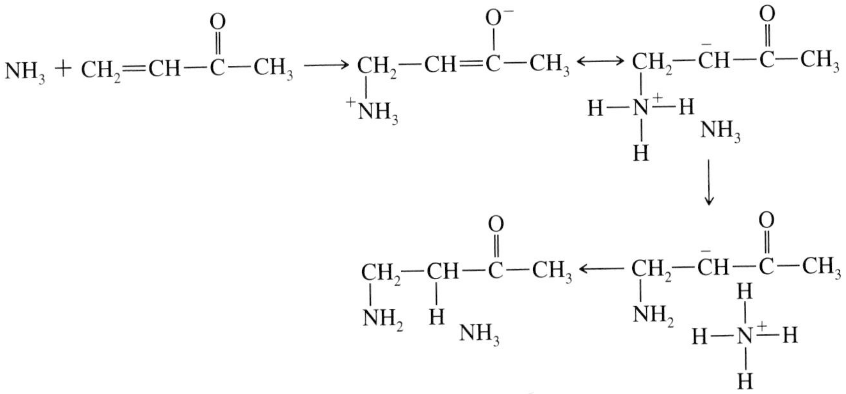 
$$NH_3 \;+\; CH_2{=}CH-\overset{\overset{\displaystyle O}{\|}}{C}-CH_3 \longrightarrow \underset{\underset{\displaystyle {}^+NH_3}{\displaystyle |}}{CH_2}-CH{=}\overset{\overset{\displaystyle O^-}{|}}{C}-CH_3 \longleftrightarrow CH_2-\overset{\overset{\underset{\displaystyle H-\overset{+}{N}-H}{\displaystyle |}}{\displaystyle |}}{CH}-\overset{\overset{\displaystyle O}{\|}}{C}-CH_3 \\ \underset{\displaystyle NH_3}{}$$

$$\underset{\underset{\displaystyle NH_2 \;\; H}{\displaystyle |\;\;\;\;|}}{CH_2}-CH-\overset{\overset{\displaystyle O}{\|}}{C}-CH_3 \longleftarrow \underset{\underset{\displaystyle \underset{\displaystyle H}{\overset{\displaystyle H-\overset{+}{N}-H}{|}} }{\displaystyle NH_2}}{CH_2-CH}-\overset{\overset{\displaystyle O}{\|}}{\underset{\underset{\displaystyle H}{\displaystyle |}}{C}}-CH_3 \\ \underset{\displaystyle NH_3}{}$$

**4.23** When HBr adds to the double bond in methyl vinyl ketone as shown below, the product appears to show anti-Markovnikov regioselectivity. Use words and structural formulas to rationalize this experimental observation.

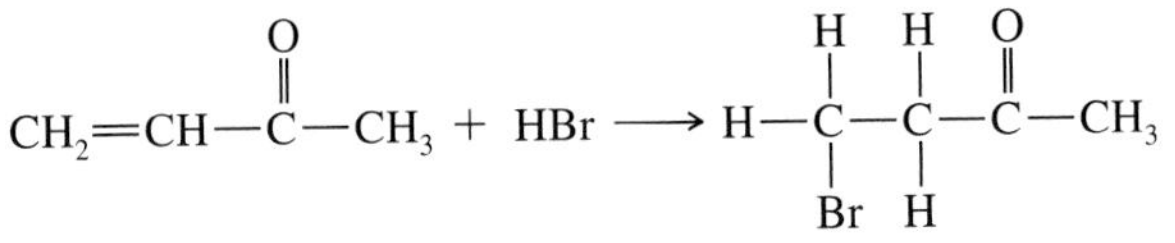

$$CH_2{=}CH-\overset{\overset{\displaystyle O}{\|}}{C}-CH_3 \;+\; HBr \longrightarrow H-\overset{\overset{\displaystyle H}{|}}{\underset{\underset{\displaystyle Br}{\displaystyle |}}{C}}-\overset{\overset{\displaystyle H}{|}}{\underset{\underset{\displaystyle H}{\displaystyle |}}{C}}-\overset{\overset{\displaystyle O}{\|}}{C}-CH_3$$

major product

**4.24** The following energy diagram is for a reaction in which the starting material, Compound A, is converted into several products, of which we are interested in the pathways that lead to two, Compounds F and G. The table gives the values for the energies of different points in the energy diagram relative to that of Compound A.

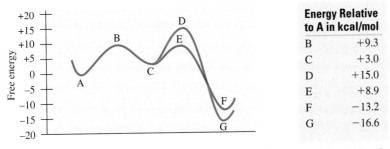

| | Energy Relative to A in kcal/mol |
|---|---|
| B | +9.3 |
| C | +3.0 |
| D | +15.0 |
| E | +8.9 |
| F | −13.2 |
| G | −16.6 |

Answer the following questions by referring to the energy diagram and the table of energy values. Use the letters on the energy diagram in your answers.

(a) Which point(s) on the diagram correspond to transition state(s)?
(b) What is the rate-determining step in the conversion of A to G?
(c) Assuming that the formations of F and G are reversible and that the reaction is allowed to come to equilibrium, which product will be the major one?
(d) Which product forms the fastest?
(e) What is the $\Delta G^{\ddagger}$ for the C to F step?
(f) What is the $\Delta G_r^{\circ}$ for the formation of the most stable product?

**4.25** The initial rate of the reaction of chloromethane with hydroxide ion was calculated on page 122. Does the rate of the reaction remain the same during the course of the reaction? Explain.

**4.26** Refer back to Problem 4.6 (p. 126) and calculate the rate of the reaction of bromomethane with hydroxide ion when the reaction is halfway complete.

**4.27** What value of $\Delta G^{\circ}$ is necessary if a reaction is to go 99% to completion at 298 K?

**4.28**

(a) The thermodynamic data shown have been determined for the following reaction.

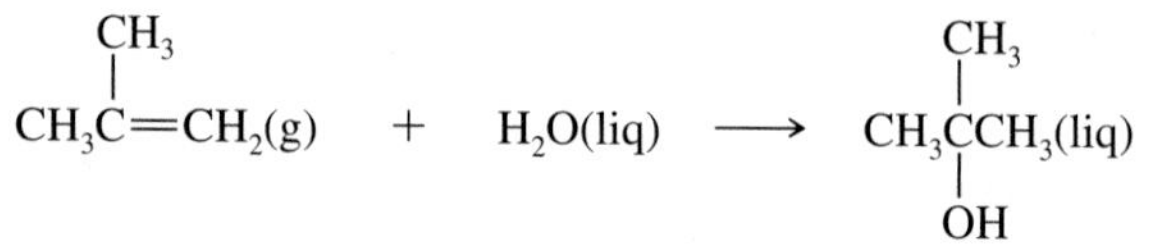

$$\underset{\substack{CH_3 \\ |}}{CH_3C}{=}CH_2(g) \quad + \quad H_2O(liq) \quad \longrightarrow \quad \underset{\substack{| \\ OH}}{\underset{\substack{CH_3 \\ |}}{CH_3CCH_3}}(liq)$$

| | | | |
|---|---|---|---|
| $S_f^{\circ}$ (at 298 K) | 70.17 cal/deg·mol | 16.72 cal/deg·mol | 46.30 cal/deg·mol |
| $\Delta H_f^{\circ}$ (at 298 K) | −4.04 kcal/mol | −68.32 kcal/mol | −85.87 kcal/mol |

What is the equilibrium constant for this reaction at 298 K? A review of pages 94–97 may be helpful.

(b) Water alone does not add to the double bond, but when a solution of sulfuric acid in water is used, the alkene is converted into an alcohol. Propose an explanation for this observation based on the $pK_a$ values for the species involved. What is the electrophilic species in a dilute solution of sulfuric acid in water? What is the nucleophile? Why won't pure water add to the double bond?

**4.29** A convenient way to convert a carboxylic acid to its methyl ester is to use diazomethane. A typical reaction is shown below.

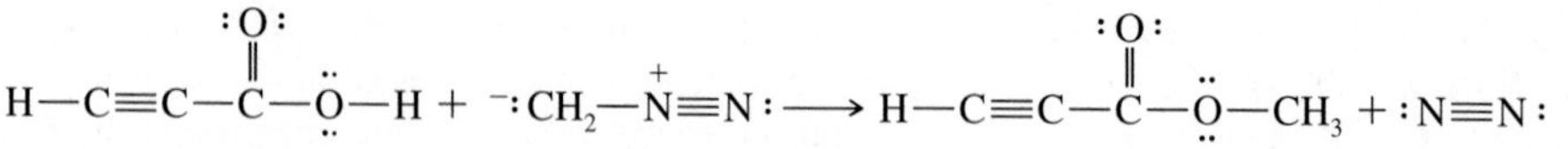

$$H{-}C{\equiv}C{-}\overset{\overset{\ddots O \ddots}{\|}}{C}{-}\overset{\ddots}{\underset{\ddots}{O}}{-}H + {}^{-}{:}CH_2{-}\overset{+}{N}{\equiv}N{:} \longrightarrow H{-}C{\equiv}C{-}\overset{\overset{\ddots O \ddots}{\|}}{C}{-}\overset{\ddots}{\underset{\ddots}{O}}{-}CH_3 + {:}N{\equiv}N{:}$$

propynoic acid          diazomethane
$pK_a$ 3

(a) In the first part of this reaction, diazomethane deprotonates the acid. The $pK_a$ of the conjugate acid of diazomethane is 8. Draw the mechanism for the deprotonation reaction using the curved arrow symbolism and give the structural formulas for the conjugate acid–base pair resulting from the deprotonation reaction.
(b) In the second part of the reaction, a substitution reaction takes place between the conjugate acid and the conjugate base formed in part a to give the products shown in the equation above. In this substitution reaction, nitrogen, $N_2$, rather than a halide ion, leaves. Show the mechanism for the substitution reaction that gives the products shown above.
(c) Draw an energy diagram that describes the progress of the substitution reaction from part b. You should include labels for each axis, the free energies of reaction and activa-

tion, and structural formulas for the activated complex at the transition state(s) and of reactive intermediates (if any).

**4.30** Chemists have studied the electrophilic addition reaction of trifluoroacetic acid,

$$\underset{\text{CF}_3\text{COH},}{\overset{\text{O}}{\parallel}}$$ to the alkenes shown below:

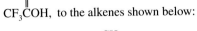

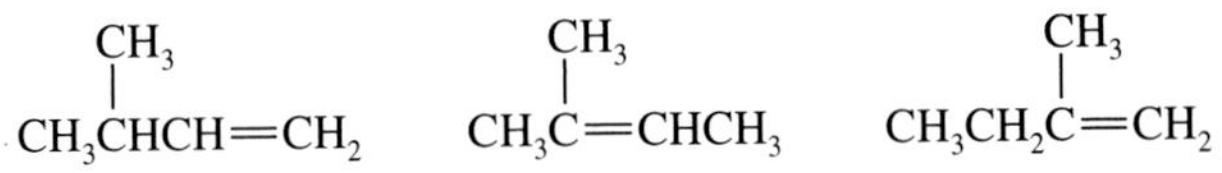

$$\underset{\text{3-methyl-1-butene}}{\overset{\overset{\text{CH}_3}{|}}{\text{CH}_3\text{CHCH}=\text{CH}_2}} \qquad \underset{\text{2-methyl-2-butene}}{\overset{\overset{\text{CH}_3}{|}}{\text{CH}_3\text{C}=\text{CHCH}_3}} \qquad \underset{\text{2-methyl-1-butene}}{\overset{\overset{\text{CH}_3}{|}}{\text{CH}_3\text{CH}_2\text{C}=\text{CH}_2}}$$

(a) Using the curved-arrow notation, give the complete mechanism for the electrophilic addition of trifluoroacetic acid to 2-methyl-2-butene.

(b) It was found that 2-methyl-2-butene and 2-methyl-1-butene react with trifluoroacetic acid at approximately the same rate, but both react about $6 \times 10^4$ times faster than 3-methyl-1-butene does. Use structural formulas and words to rationalize these experimental observations.

# Alkanes and Cycloalkanes

## A Look Ahead

Alkanes are saturated hydrocarbons with the general formula $C_nH_{2n+2}$. The structures of methane, $CH_4$, and ethane, $C_2H_6$, the two smallest members of the family of alkanes, were examined thoroughly on page 24. Removal of a hydrogen atom from an alkane gives rise to an alkyl group (p. 48). Cycloalkanes are hydrocarbons with the general formula $C_nH_{2n}$ and therefore have one unit of unsaturation. In cycloalkanes, the carbon atoms form a ring.

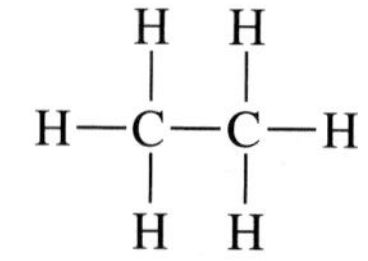

$C_2H_6$
ethane

*an alkane*

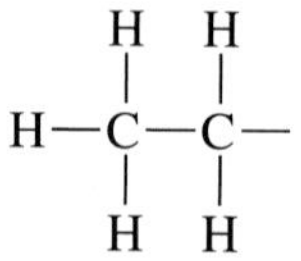

$C_2H_5$
ethyl group

*an alkyl group*

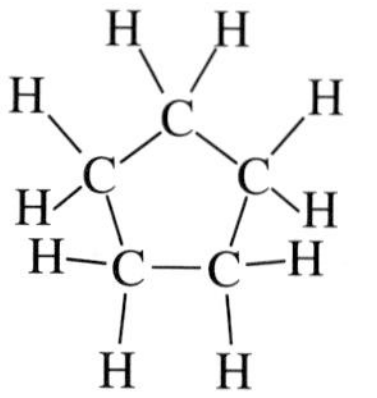

$C_5H_{10}$
cyclopentane

*a cycloalkane*

Many organic compounds have alkyl or cycloalkyl groups as part of their structures. Therefore, in this chapter, alkanes and cycloalkanes are used to further examine ideas about structure and isomerism and to introduce systematic ways of naming organic compounds.

## 5.1  Isomerism and Physical Properties

Alkanes have only $\sigma$ bonds (p. 48) and are described as **saturated hydrocarbons,** meaning that all valence electrons of carbon are involved in single bonds. Compounds that differ from each other in their molecular formulas by the unit —$CH_2$— are called members of a **homologous series.** Thus methane and ethane belong to a homologous series that has many other members, all of which are saturated hydrocarbons. All alkanes have similar chemical properties, but their physical properties vary with molecular weight and the shape of the molecules. Table 5.1 shows the boiling and melting points of some representative alkanes.

The boiling points of the alkanes increase steadily with molecular weight from methane to butane; these first four alkanes are all gases at room temperature. In compounds with four or more carbon atoms, several different arrangements of the carbon chains are possible. The boiling points of different compounds having the same molecular formula and functional groups depend on the compactness of the molecular shape. Compounds with a long carbon chain tend to have higher boiling points than those with a more spherical shape (p. 30).

**Table 5.1 Boiling Points and Melting Points for Some Alkanes**

| Molecular Formula | Name | Molecular Weight | bp, °C | mp, °C |
|---|---|---|---|---|
| $CH_4$ | methane | 16 | − 164 | − 182.5 |
| $C_2H_6$ | ethane | 30 | − 88.6 | − 183.3 |
| $C_3H_8$ | propane | 44 | − 42.1 | − 189.7 |
| $C_4H_{10}$ | butane | 58 | − 0.6 | − 138.4 |
| $C_4H_{10}$ | 2-methylpropane | 58 | − 10.2 | − 138.3 |
| $C_5H_{12}$ | pentane | 72 | 36.1 | − 129.7 |
| $C_5H_{12}$ | 2-methylbutane | 72 | 27.9 | − 159.9 |
| $C_5H_{12}$ | 2,2-dimethylpropane | 72 | 9.5 | − 16.6 |
| $C_6H_{14}$ | hexane | 86 | 68.9 | − 93.5 |
| $C_7H_{16}$ | heptane | 100 | 98.4 | − 90.6 |
| $C_8H_{18}$ | octane | 114 | 125.7 | − 56.8 |
| $C_9H_{20}$ | nonane | 128 | 150.8 | − 51.0 |
| $C_{10}H_{22}$ | decane | 142 | 174.1 | − 29.7 |
| $C_{20}H_{42}$ | icosane | 282 | 343 | 36.8 |

On the other hand, compact, spherical molecules pack better in crystal lattices, so compounds such as these have higher melting points. These trends are illustrated by the different compounds that have the molecular formula $C_5H_{12}$, which are shown below.

| pentane | 2-methylbutane | 2,2-dimethylpropane |
|---|---|---|
| bp 36.1 °C | bp 27.9 °C | bp 9.5 °C |
| mp −129.7 °C | mp −159.9 °C | mp −16.6 °C |
| *straight-chain alkane* | *branched-chain alkanes* | |

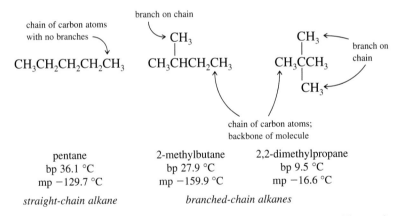

Pentane has the most elongated molecular shape and the highest boiling point of the three compounds. 2,2-Dimethylpropane has the most compact and spherical shape, the lowest boiling point, and by far the highest melting point of the three.

Pentane, 2-methylbutane, and 2,2-dimethylpropane all have the molecular formula $C_5H_{12}$ but have different connectivities and are constitutional isomers (p. 22). Pentane is a **straight-chain hydrocarbon;** 2-methylbutane and 2,2-dimethylpropane are **branched-chain hydrocarbons.** In pentane, all the carbon atoms form one continuous chain. In the other two compounds, some of the carbon atoms form the "backbone" of the molecule, but others branch off it. Constitutional isomers have different physical properties, as shown above for the isomers of $C_5H_{12}$.

The higher the number of carbon atoms in an alkane (and therefore the higher its molecular weight), the higher are its boiling and melting points. The actual values are given in Table 5.1 for the straight-chain hydrocarbons $CH_4$ through $C_{10}H_{22}$

and $C_{20}H_{42}$. If a linear alkane contains approximately eighteen carbon atoms, it is a solid at room temperature. Icosane, $C_{20}H_{42}$, has a melting point of 36.8 °C, which is very close to body temperature.

## 5.2    Methane

Methane, $CH_4$, is the simplest hydrocarbon. The details of the structure of methane were given on pages 24 and 27. The molecule is tetrahedral and has four equivalent carbon–hydrogen bonds. It may be represented in three dimensions, as a Lewis structure, or by a condensed formula (Figure 5.1).

Any one of the hydrogen atoms in methane may be replaced by another atom or group to give a new compound. Chloromethane, for example, is a compound in which one of the hydrogen atoms in methane has been substituted by a chlorine atom. Chloromethane is an alkyl halide; the hydrocarbon portion of the molecule is called a methyl group. Methanol, an alcohol, is another example of a compound containing a methyl group.

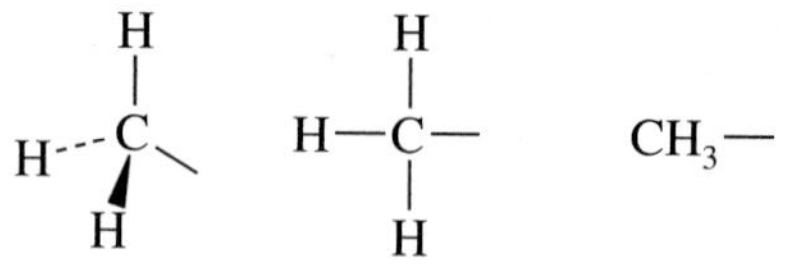

the methyl group

*an alkyl group*

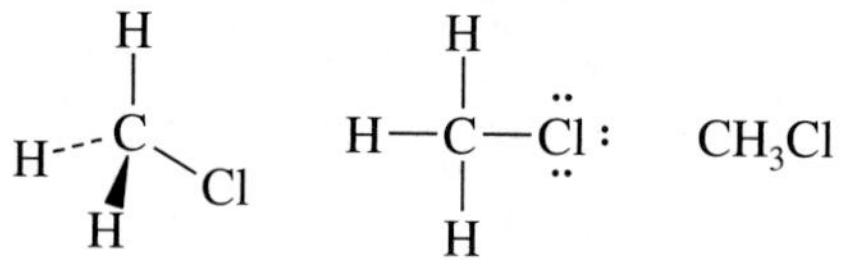

chloromethane
or methyl chloride

*an alkyl halide*

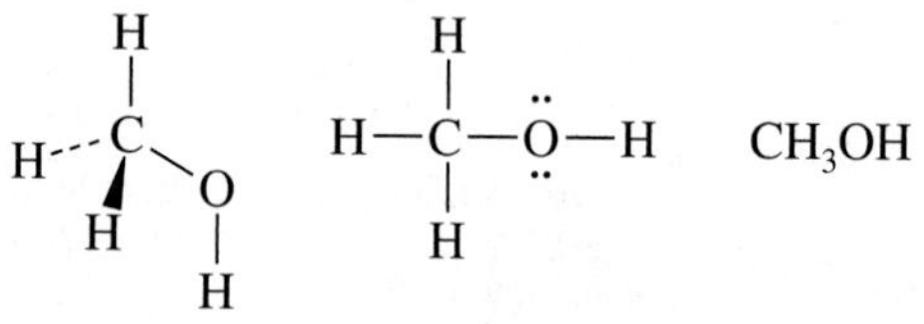

methanol
or methyl alcohol

*an alcohol*

The first name given for each compound above is its systematic name; the second is another acceptable and widely used name. The **methyl group** is the alkyl group derived from a particular alkane, methane. The name of an alkyl group is obtained by changing the **ane** ending of the name of an alkane to **yl**. Thus the methyl group is clearly related to methane. An alkane is sometimes represented in tables or equations by the symbol RH; the corresponding alkyl group is symbolized by R. An introduction to the rules for the systematic naming of organic compounds will be given in Section 5.10. In this and the next three sections, you should concentrate on the structures and the names of the first four alkanes and the alkyl groups derived from them.

It is important to recognize that the chlorine atom in chloromethane or the hydroxyl group in methanol could have replaced any one of the hydrogen atoms in

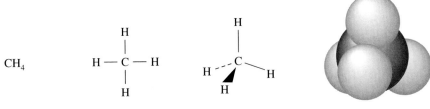

CH₄

condensed formula | Lewis structure | a three-dimensional representation | a space-filling representation

*Figure 5.1*
Different representations of methane.

methane; in each case, the same compound would have been formed. The structural formulas below, for example, represent a single methanol molecule tumbling in space, not three different molecular species.

*different orientations in space of a
molecule of methanol*

All the hydrogen atoms in methane are equivalent. Substitution of a particular group for any one of them results in the same compound.

Similarly, only one compound results if two of the hydrogen atoms of methane are replaced. For example, a molecule of dichloromethane may be represented in a number of ways.

*perspective formulas*

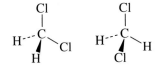

$CH_2Cl_2$

*Lewis structures*      *condensed formula*

*dichloromethane*

The representation a chemist chooses to use depends on the particular molecular features he or she is trying to emphasize.

**Problem 5.1**    Draw at least two different three-dimensional representations for each of the following compounds. (You may wish to review Sections 1.8 and 2.4 before attempting this.)

(a) $CH_2ClBr$     (b) $CH_3SH$     (c) $CH_3OCH_3$     (d) $CHCl_3$     (e) $CH_3NH_2$

## 5.3 Ethane

### A. The Ethyl Group

Ethane, $C_2H_6$, follows methane in the series of alkanes. The structure of ethane (considered in detail on pages 24 and 48 may be represented by a structural formula showing the molecule in three dimensions, by a Lewis structure, or by a condensed formula.

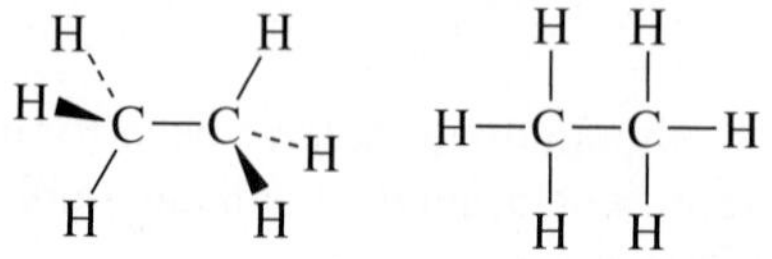

$CH_3CH_3$

*perspective formula* *Lewis structure* *condensed formula*

ethane

All the hydrogen atoms in ethane are equivalent. Removal of any one of them gives the **ethyl group,** and substitution of another atom or group for any one of them results in the same compound.

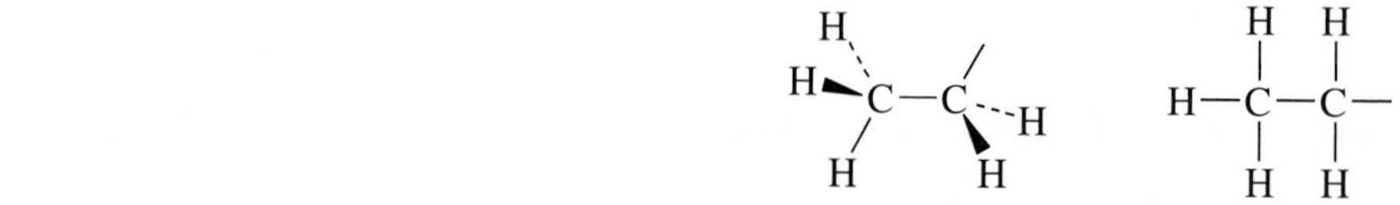

$CH_3CH_2—$

the ethyl group

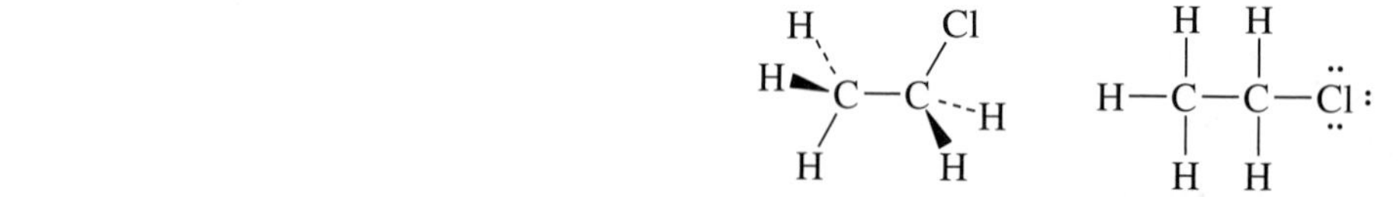

$CH_3CH_2Cl$

chloroethane
or ethyl chloride

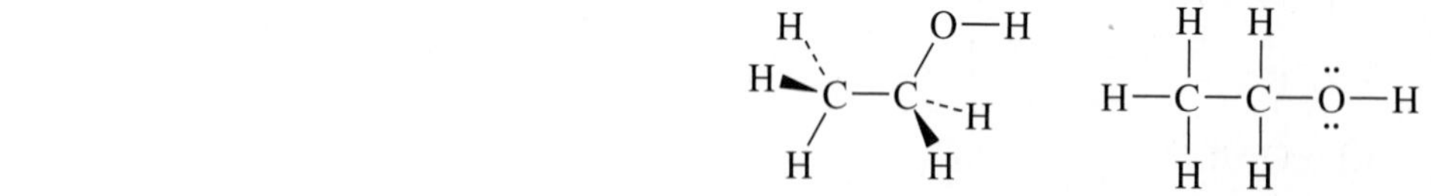

$CH_3CH_2OH$

ethanol
or ethyl alcohol

*perspective formula* *Lewis structure* *condensed formula*

The three-dimensional representations of chloroethane shown below are drawn in different ways to represent views of the molecule from different points in space. All four of these structural formulas represent the same molecule.

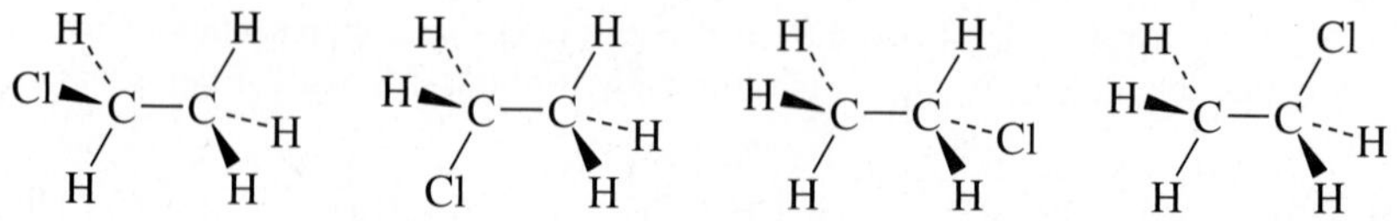

*representations of chloroethane, viewed from several different perspectives*

## B. Isomerism in Disubstituted Ethanes

The replacement of any two hydrogen atoms of methane by two other atoms gives the same compound (p. 147). An inspection of the structure of chloroethane, however, reveals that the five hydrogen atoms are not equivalent to each other. There are two hydrogen atoms on the carbon atom already bearing a chlorine atom and three on the other carbon atom. If one of the remaining hydrogen atoms is replaced by a chlorine atom, the new compound, $C_2H_4Cl_2$, could have one of two possible structures.

| | | |
|---|---|---|
| (structure) | (structure) 1,2-dichloroethane | $ClCH_2CH_2Cl$ |
| bp 83 °C | mp −35 °C | $\mu = 1.42$ D |
| (structure) | (structure) 1,1-dichloroethane | $CH_3CHCl_2$ |
| bp 57 °C | mp −97 °C | $\mu = 1.95$ D |

*isomers of $C_2H_4Cl_2$*

1,2-Dichloroethane is derived from chloroethane by replacement of one of the three hydrogen atoms on the originally unsubstituted carbon atom. 1,1-Dichloroethane is derived from chloroethane by replacement of a hydrogen atom on the same carbon atom as the first chlorine atom. The two compounds are another example of constitutional isomers, compounds having the same molecular formula but different connectivities. They have different physical properties, including boiling and melting points and dipole moments. Their names reflect the difference in their structures. The name 1,1-dichloroethane indicates that both chlorine atoms are on the same carbon atom. 1,2-Dichloroethane identifies the positions of the chlorine atoms on adjacent carbon atoms.

**Problem 5.2**    How many different connectivities are possible for the molecular formula $C_2H_3Cl_3$? How about $C_2H_2Cl_4$? Draw condensed formulas, Lewis structures, and three-dimensional representations of these compounds.

## 5.4  Nuclear Magnetic Resonance as a Tool for the Study of Molecular Structures

The equivalence or nonequivalence of atoms or groups in a structure is observable experimentally, most notably by nuclear magnetic resonance spectroscopy. This technique will be discussed more fully in Chapter 11. In this section we will use the nuclear magnetic resonance spectra of 1,2-dichloroethane and 1,1-dichloroethane to explore how many different kinds of carbon atoms and hydrogen atoms exist in their structures.

Figure 5.2 on the next page shows the carbon and proton (hydrogen) nuclear magnetic resonance spectra of 1,2-dichloroethane. Each spectrum has only one peak

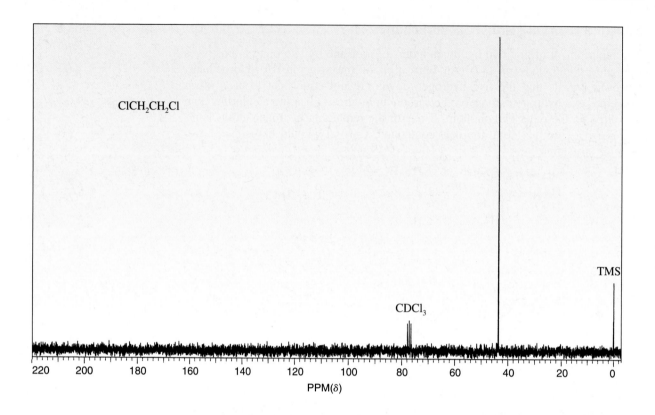

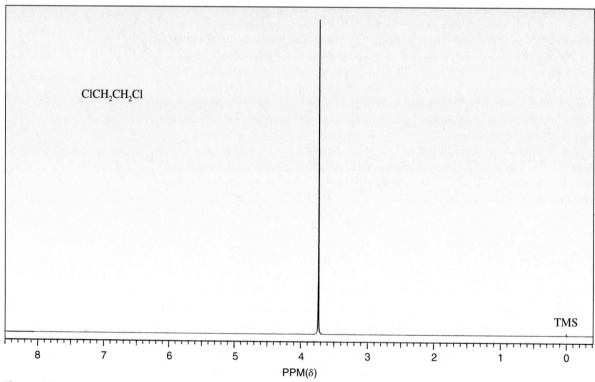

*Figure 5.2*

Nuclear magnetic resonance spectra of 1,2-dichloroethane: *(top)* carbon spectrum; *(bottom)* proton spectrum.

(other than the ones labeled $CDCl_3$, the solvent, and TMS, the peak for the standard reference compound), indicating that there is only one kind of carbon atom and one kind of hydrogen atom in the compound. In other words, the two carbon atoms are equivalent to each other. The four hydrogen atoms also are equivalent to each other. We see this when we describe the connectivity of the molecule. Each carbon atom is bonded to two hydrogen atoms and one chlorine atom and to another carbon atom also bonded to two hydrogen atoms and one chlorine atom. The molecule is completely symmetrical, and the spectra reflect that symmetry.

The carbon and proton nuclear magnetic resonance spectra of 1,1-dichloroethane appear in Figure 5.3 on page 152. Each spectrum has two peaks, indicating that there are two kinds of carbon atoms and two kinds of hydrogen atoms in the compound. A check of the structural formula for the compounds confirms that this is true. We have the carbon that (besides the other carbon atom) is bonded only to hydrogen atoms and the one that has one hydrogen and two chlorine atoms bonded to it. The one that has chlorine atoms bonded to it appears farther to the left in the spectrum. In other words, the environment of the carbon atom, the electronegativity of the groups attached to it, determines where the peak for that carbon atom appears in the spectrum. So not only do we find out how many different kinds of carbon atoms we have, we also find out something about the other atoms that are attached to the carbon.

The proton nuclear magnetic resonance spectrum of 1,1-dichloroethane (Figure 5.3) gives us similar information. It tells us that there are two kinds of hydrogen atoms in the compound and something about the environment of those hydrogen atoms. The smaller peak represents the hydrogen atom on the carbon atom with the two chlorine atoms on it. The electronegativity of the chlorine atoms influences the position of the peak for the hydrogen atom as well as that for the carbon atom to which the chlorine and hydrogen atoms are attached.

The proton nuclear magnetic resonance spectrum has two other useful features. One is hidden under the broad peaks of the spectrum of 1,1-dichloroethane and may be seen in the more detailed spectrum of the compound shown in Figure 11.22 (p. 420). In Chapter 11, we will discover how to make use of the features that appear in the spectra of compounds that contain nonequivalent hydrogen atoms when the spectra are taken at higher resolution. The feature that we will focus on now is the step-like curves over the peaks in the proton nuclear magnetic resonance spectrum of 1,1-dichloroethane (Figure 5.3). These curves represent integrations of the areas under the peaks. From the relative heights of the steps (13 and 44 mm), we can tell the relative numbers of hydrogen atoms represented by each peak. The numbers in this case tell us that we have a ratio of $13:44$, or about $1:3$ for the relative numbers of hydrogen atoms in each environment, one hydrogen on the carbon atom that also bears the chlorine atoms and three on the other carbon atom.

A comparison of the spectra of Figures 5.2 and 5.3 shows us the effect of bonding to one chlorine atom or two chlorine atoms on a carbon atom or on the hydrogen atoms bonded to that carbon atom. In Figure 5.2, the peak for the carbon atom in 1,2-dichloroethane with one chlorine atom on it appears at 44 ppm (from the scale at the bottom of the spectrum). The peak for the carbon atom in 1,1-dichloromethane having two chlorine atoms on it (Figure 5.3) appears farther to the left at 69 ppm. The three peaks at 77 ppm that appear in both spectra come from the carbon atom in the solvent, deuteriochloroform, $CDCl_3$, and show that having three chlorine atoms on a carbon moves the peak even farther to the left. (The presence of three peaks is a result of interactions between the nucleus of the deuterium atom and that of the carbon atom; more on this in Chapter 11.)

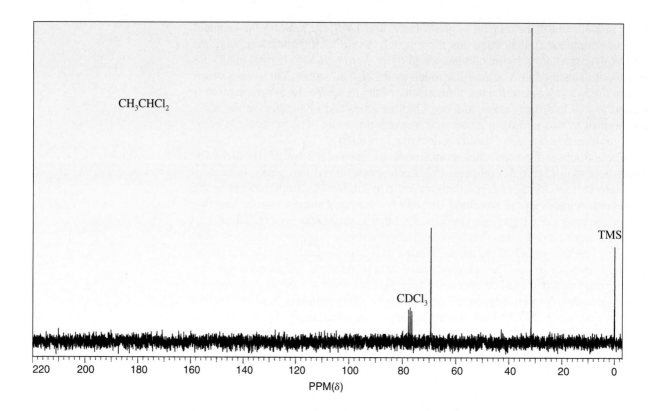

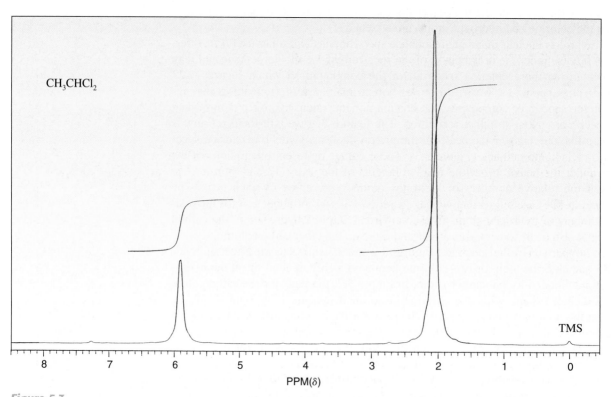

*Figure 5.3*
Nuclear magnetic resonance spectra of 1,1-dichloroethane: *(top)* carbon spectrum;
*(bottom)* proton spectrum.

The same shift to the left is seen for the position of the peak for the hydrogen atoms on the carbon atoms bearing chlorine from 1,2-dichloroethane at 3.75 ppm (Figure 5.2) to 1,1-dichloroethane at 5.92 ppm (Figure 5.3). The values, read off the bottom of the spectra, are called the **chemical shifts** for the carbon and hydrogen atoms. They represent how far their peaks have shifted from the standard position represented in these spectra by a small peak for tetramethylsilane (TMS) at 0 ppm. For example, the peak for the hydrogen atoms on the carbon atom with no chlorines on it in 1,1-dichloroethane is at 2.05 ppm, much closer to the peak for TMS. We will learn much more about chemical shifts and their use in determining the structure of a compound in Chapter 11.

In summary, nuclear magnetic resonance spectroscopy gives us three useful pieces of information about the connectivity of a molecule:

1. The number of different signals in the carbon or proton nuclear magnetic resonance spectrum of the compound gives us information about the symmetry of the molecule and the equivalence or nonequivalence of the different atoms there.

2. The chemical shifts of the signals tell us about the environment in which the carbon and hydrogen atoms exist, especially whether any strongly electronegative atoms are close by.

3. The integration of the proton nuclear magnetic resonance spectrum tells us the relative numbers of hydrogen atoms at each chemical shift.

We will use these techniques and these ideas to explore further the concepts of equivalence and nonequivalence of atoms and groups in molecules.

**Problem 5.3**     The carbon and proton nuclear magnetic resonance spectra of one of the $C_2H_3Cl_3$ compounds you explored in Problem 5.2 are given in Figure 5.4 (figure continues on p. 154). What is the structure of the compound?

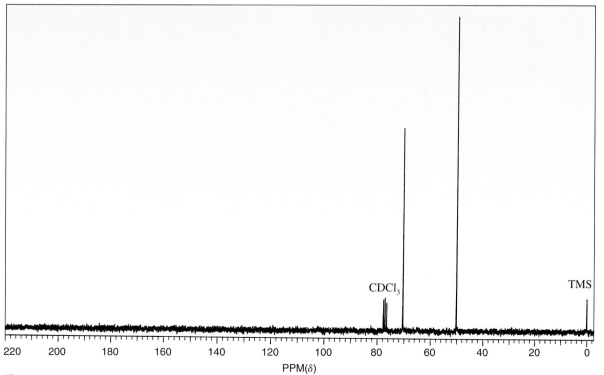

*Figure 5.4*

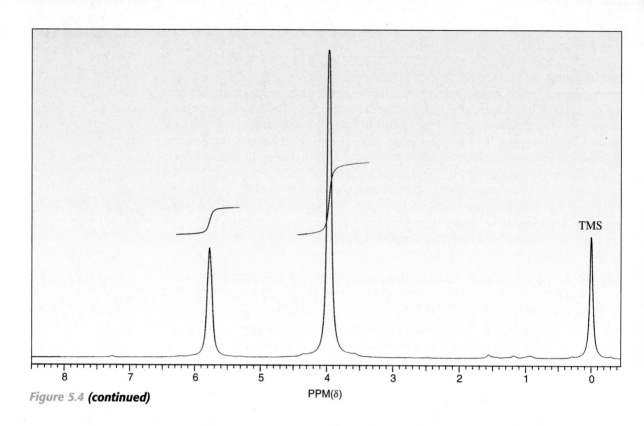

*Figure 5.4 (continued)*

PPM(δ)

**Problem 5.4**    Figure 5.5 shows the carbon nuclear magnetic resonance spectra of $C_2H_2Cl_4$ compounds, Compound A and Compound B, also among your answers to Problem 5.2. Each compound has only one peak in its proton nuclear magnetic resonance spectrum. The peak in the spectrum of Compound A is at 4.30 ppm, while that for Compound B is at 5.95 ppm. Draw structural formulas for Compounds A and B.

Compound A

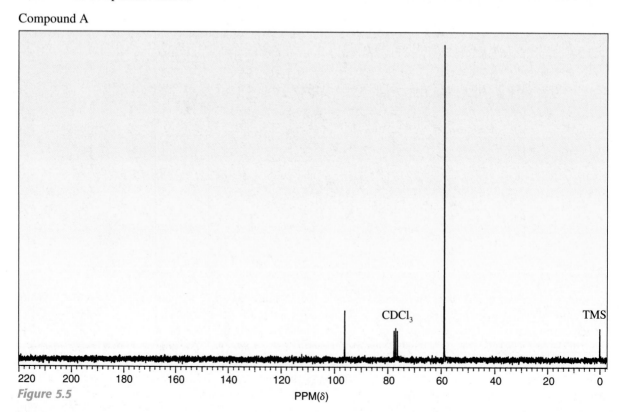

*Figure 5.5*

PPM(δ)

Compound B

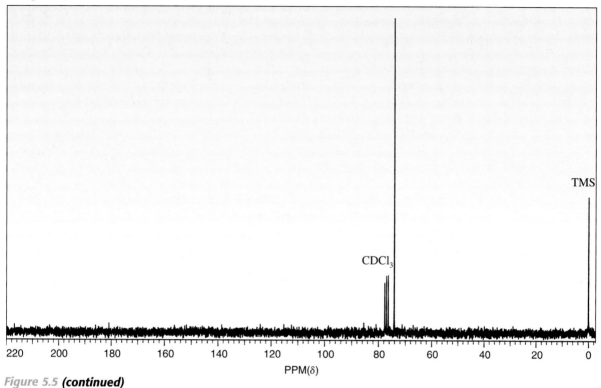

Figure 5.5 *(continued)*

## 5.5 Propane

The three carbon atoms and eight hydrogen atoms in propane, $C_3H_8$, can be connected in only one way. However, propane is the first of the alkane series to contain two different kinds of carbon and hydrogen atoms.

*perspective formula*      *Lewis structure*      *condensed formula*

propane

The two end carbon atoms in propane that are each attached to only one other carbon atom are called **primary carbon atoms.** The hydrogen atoms on these primary carbon atoms are informally called **primary hydrogen atoms.** There are two equivalent primary carbon atoms and six equivalent primary hydrogen atoms in propane. The middle carbon atom in propane is bonded to two carbon atoms and is a **secondary carbon atom,** bearing two hydrogen atoms usually called **secondary hydrogen atoms.**

Two different propyl groups may be created from propane. Removal of a primary hydrogen atom gives the **propyl,** or **n-propyl, group** (*n* stands for "normal"), a primary alkyl group.

propyl, or *n*-propyl, group

*a primary alkyl group*

$CH_3CH_2CH_2$—

$CH_3CH_2CH_2Br$

1-bromopropane
or *n*-propyl bromide
bp 71 °C

*a primary alkyl halide*

$CH_3CH_2CH_2OH$

1-propanol
or *n*-propyl alcohol
bp 97.4 °C

*a primary alcohol*

If one of the secondary hydrogen atoms is removed, a secondary alkyl group known as the **1-methylethyl,** or **isopropyl, group** is formed.

1-methylethyl, or isopropyl, group

*a secondary alkyl group*

$CH_3CHCH_3$

$CH_3CHCH_3$
|
Br

2-bromopropane
or isopropyl bromide
bp 59 °C

*a secondary alkyl halide*

$CH_3CHCH_3$
|
OH

2-propanol
or isopropyl alcohol
bp 82.4 °C

*a secondary alcohol*

1-Bromopropane and 2-bromopropane are constitutional isomers. They have different physical properties, for example, different boiling points. 1-Bromopropane is classified as a **primary alkyl halide** because the halogen is bonded to a primary carbon atom; 2-bromopropane is a **secondary alkyl halide** because the halogen is bonded to a secondary carbon atom. The two compounds have some chemical properties that are similar and some that are different. These properties are treated in more detail in Chapter 7. Likewise, 1-propanol and 2-propanol are isomeric alcohols. They share some chemical properties but differ in that the first is a **primary alcohol** and the other is a **secondary alcohol** (Chapter 13). The kind of carbon atom to which a functional group is bonded often makes a difference in reactivity. Therefore, it is important that you learn to distinguish isomers and to classify them according to their structural features. For example, distinctions were made among primary, secondary, and tertiary carbocations in Chapter 4 (p. 129). Differences in the stabilities of primary and secondary carbocations determine the products formed when hydrogen bromide adds to propene, for example.

**Problem 5.5**    The carbon nuclear magnetic resonance spectra of 1-bromopropane and 2-bromopropane are given in Figure 5.6.

(a) Which spectrum corresponds to which compound?
(b) Predict how many peaks there will be in the proton nuclear magnetic resonance spectrum of 1-bromopropane. Make a rough sketch of the spectrum you expect, showing the relative positions of the different kinds of hydrogens in the spectrum and the relative heights of the integration curves.
(c) Do the same thing for 2-bromopropane.

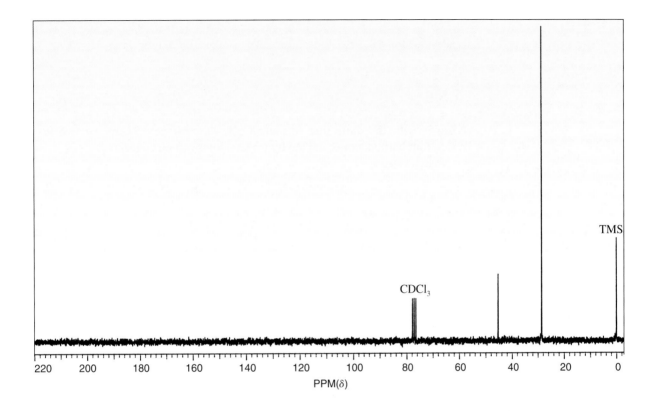

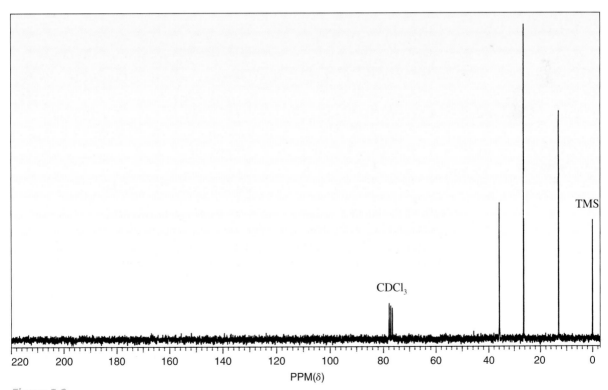

*Figure 5.6*

**Problem 5.6**   How many constitutional isomers of $C_3H_6Br_2$ are possible? Draw them as condensed formulas, Lewis structures, and three-dimensional formulas.

## 5.6   Butanes

Alkanes with the molecular formula $C_4H_{10}$ are butanes. There are two ways in which four carbon atoms and ten hydrogen atoms can be connected.

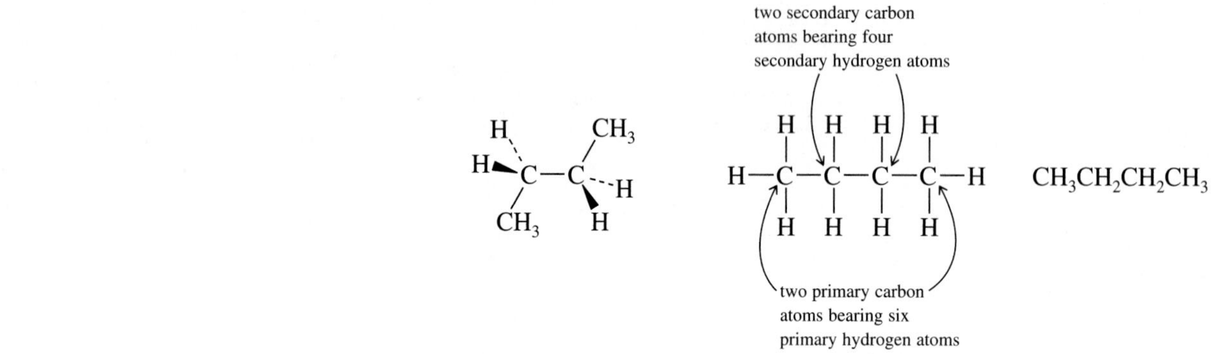

two secondary carbon
atoms bearing four
secondary hydrogen atoms

two primary carbon
atoms bearing six
primary hydrogen atoms

$CH_3CH_2CH_2CH_3$

butane or *n*-butane
bp −0.6 °C

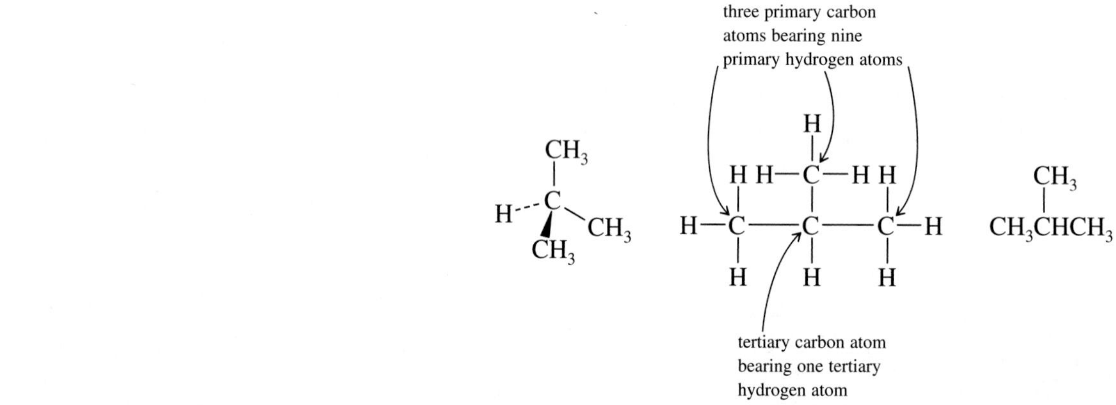

three primary carbon
atoms bearing nine
primary hydrogen atoms

tertiary carbon atom
bearing one tertiary
hydrogen atom

$CH_3CHCH_3$

2-methylpropane or isobutane
bp −10.2 °C

One of the butanes is the straight-chain isomer with two equivalent primary carbon atoms and two equivalent secondary ones. The other isomer of $C_4H_{10}$ is a branched-chain alkane. In this isomer, the three carbon atoms that are attached to only one other carbon atom are, by definition, primary carbon atoms. All the hydrogen atoms on these three carbon atoms are equivalent and are also primary. The fourth carbon atom, which is bonded to three other carbon atoms, is called a **tertiary carbon atom,** and the hydrogen atom on it is usually referred to as a **tertiary hydrogen atom.**

The straight-chain isomer of $C_4H_{10}$ is called butane or normal or *n*-butane. The systematic name for the branched-chain isomer is 2-methylpropane, indicating that the compound has a three-carbon chain, propane, on which a methyl group is substituted on the second carbon atom. This compound is also called isobutane.

Four different alkyl groups with the molecular formula $C_4H_9$ can be created by removing a different type of hydrogen atom from butane or 2-methylpropane.

Removal of a primary hydrogen atom from butane creates a primary alkyl group, the **butyl,** or **_n_-butyl, group.**

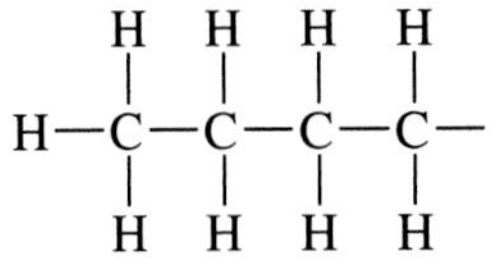

$$H-\underset{\underset{H}{|}}{\overset{\overset{H}{|}}{C}}-\underset{\underset{H}{|}}{\overset{\overset{H}{|}}{C}}-\underset{\underset{H}{|}}{\overset{\overset{H}{|}}{C}}-\underset{\underset{H}{|}}{\overset{\overset{H}{|}}{C}}- \qquad CH_3CH_2CH_2CH_2-$$

butyl, or _n_-butyl, group

_a primary alkyl group_

$$CH_3CH_2CH_2CH_2Cl \qquad CH_3CH_2CH_2CH_2OH$$

1-chlorobutane
or _n_-butyl chloride
bp 78.4 °C

1-butanol
or _n_-butyl alcohol
bp 117 °C

_a primary alkyl halide_       _a primary alcohol_

Removal of one of the secondary hydrogen atoms of butane gives the **1-methyl-propyl,** or _sec_**-butyl, group.**

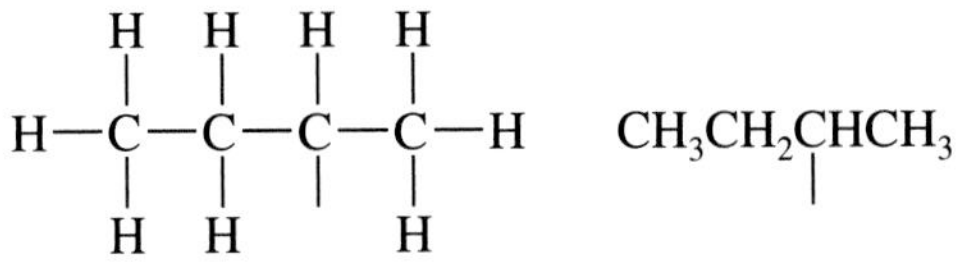

$$H-\underset{\underset{H}{|}}{\overset{\overset{H}{|}}{C}}-\underset{\underset{H}{|}}{\overset{\overset{H}{|}}{C}}-\underset{\underset{H}{|}}{\overset{\overset{H}{|}}{C}}-\underset{\underset{H}{|}}{\overset{\overset{H}{|}}{C}}-H \qquad CH_3CH_2\underset{|}{CH}CH_3$$

1-methylpropyl, or _sec_-butyl, group

_a secondary alkyl group_

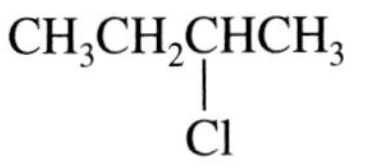

$$CH_3CH_2\underset{\underset{Cl}{|}}{CH}CH_3 \qquad CH_3CH_2\underset{\underset{OH}{|}}{CH}CH_3$$

2-chlorobutane
or _sec_-butyl chloride
bp 68.3 °C

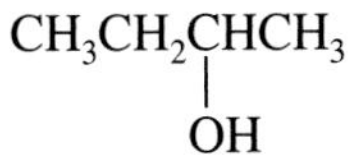

2-butanol
or _sec_-butyl alcohol
bp 99.5 °C

_a secondary alkyl halide_       _a secondary alcohol_

Two different alkyl groups can be formed from 2-methylpropane. A primary alkyl group, the **2-methylpropyl,** or **isobutyl, group,** results from the removal of any one of the primary hydrogen atoms.

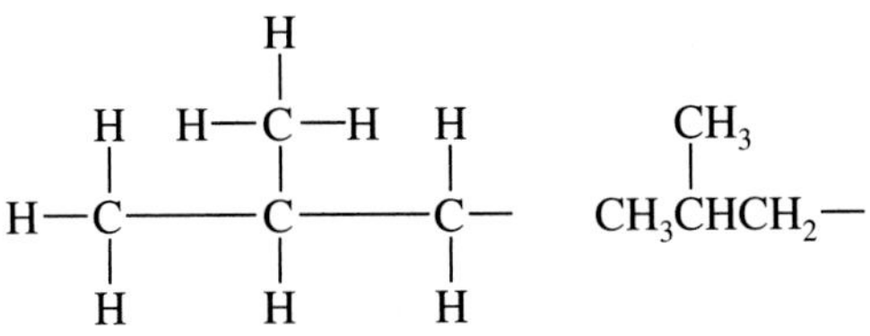

$$H-\underset{\underset{H}{|}}{\overset{\overset{H}{|}}{C}}-\underset{\underset{H}{|}}{\overset{\overset{H-\overset{\overset{H}{|}}{C}-H}{|}}{C}}-\underset{\underset{H}{|}}{\overset{\overset{H}{|}}{C}}- \qquad \underset{\underset{|}{CH_3CHCH_2-}}{\overset{CH_3}{}}$$

2-methylpropyl, or isobutyl, group

_a primary alkyl group_

$$\underset{CH_3CHCH_2Cl}{\overset{CH_3}{|}} \qquad \underset{CH_3CHCH_2OH}{\overset{CH_3}{|}}$$

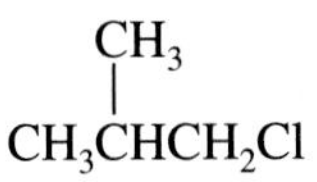

1-chloro-2-methylpropane
or isobutyl chloride
bp 68.9 °C

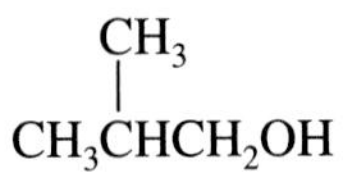

2-methyl-1-propanol
or isobutyl alcohol
bp 108 °C

_a primary alkyl halide_       _a primary alcohol_

Removal of the tertiary hydrogen atom gives a tertiary alkyl group, usually called the **tertiary butyl** (abbreviated *tert*-butyl) **group.**

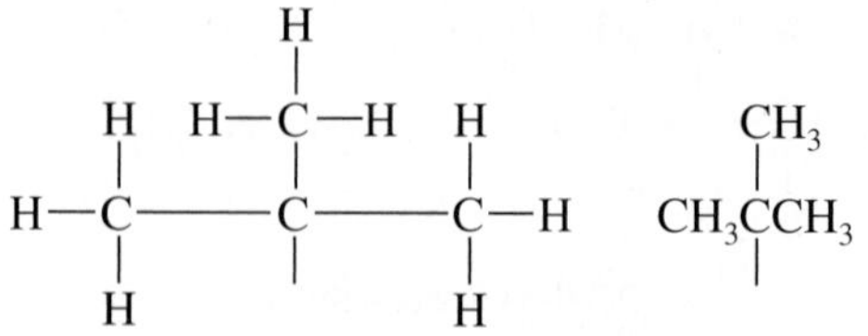

$$\underset{\begin{array}{c}\text{1,1-dimethylethyl or } tert\text{-butyl group}\\ \textit{a tertiary alkyl group}\end{array}}{\text{H---C} \quad \quad \text{C} \quad \quad \text{C---H}} \qquad \underset{}{CH_3CCH_3}$$

1,1-dimethylethyl or *tert*-butyl group
*a tertiary alkyl group*

$$\underset{\begin{array}{c}\text{2-chloro-2-methylpropane}\\ tert\text{-butyl chloride}\\ \text{bp 52 °C}\\ \textit{a tertiary alkyl halide}\end{array}}{CH_3CCH_3} \qquad \underset{\begin{array}{c}\text{2-methyl-2-propanol}\\ tert\text{-butyl alcohol}\\ \text{bp 82 °C}\\ \textit{a tertiary alcohol}\end{array}}{CH_3CCH_3}$$

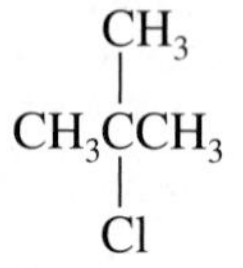

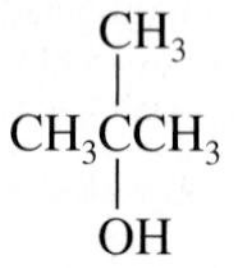

The four alkyl halides with the molecular formula $C_4H_9Cl$ shown above are constitutional isomers of one another, as are the four alcohols with the molecular formula $C_4H_{10}O$. Each constitutional isomer has distinctive physical properties; for example, note the different boiling points for the isomeric compounds shown in this section. Thus various isomers may be separated from each other by methods such as fractional distillation of liquids, recrystallization of solids, and chromatography.

You should learn the structures and the names of the alkyl groups derived from methane, ethane, propane, and the two butanes because they are used in naming more complicated molecules.

**Problem 5.7**    Many aromatic hydrocarbons have alkyl groups on benzene rings. Identify the alkyl group(s) on the benzene ring in each of the following compounds.

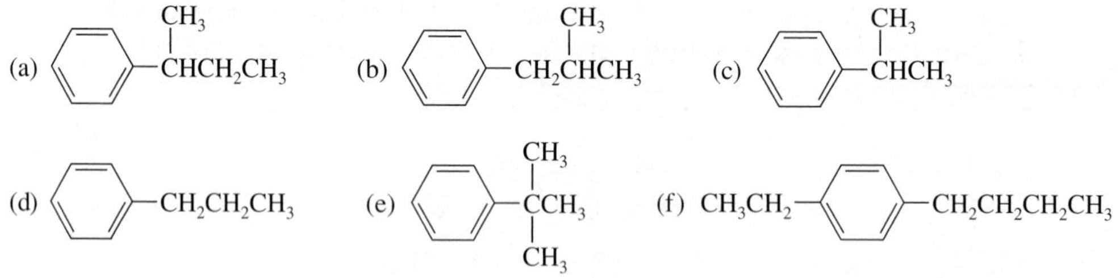

(a) $\text{C}_6\text{H}_5$—CHCH₂CH₃ with CH₃ substituent    (b) $\text{C}_6\text{H}_5$—CH₂CHCH₃ with CH₃ substituent    (c) $\text{C}_6\text{H}_5$—CHCH₃ with CH₃ substituent

(d) $\text{C}_6\text{H}_5$—CH₂CH₂CH₃    (e) $\text{C}_6\text{H}_5$—C(CH₃)₃    (f) CH₃CH₂—C₆H₄—CH₂CH₂CH₂CH₃

**Problem 5.8**    Decide whether or not each of the following sets of structural formulas represents constitutional isomers.

(a) $CH_3CH{=}CHCH_2OH, \quad CH_3CH{=}CHOCH_3$    (b) $ClCH_2CH_2CH_2CH_3, \quad CH_3CH_2CH_2CH_2Cl$

(c) $CH_3CH_2CH_2CH_2NH_2, \quad CH_3CH_2NHCH_2CH_3$    (d) $CH_3CH_2CH_2\overset{O}{\overset{\|}{C}}H, \quad CH_3CH_2\overset{O}{\overset{\|}{C}}CH_3$

(e) $ClCH_2CH_2CH_2Cl, \quad CH_3\overset{Cl}{\underset{Cl}{C}}CH_3$    (f) $CH_2{=}CHCH_2CH_2OH, \quad CH_3CH_2\overset{O}{\overset{\|}{C}}CH_3$    (g) $CH_3\overset{}{\underset{OH}{C}}HCH_2CH_3, \quad CH_3CH_2\overset{}{\underset{OH}{C}}HCH_3$

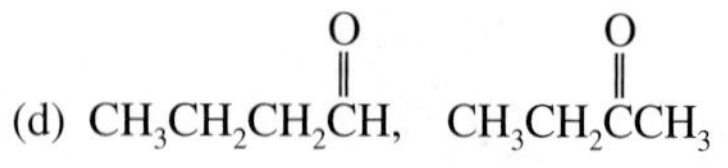

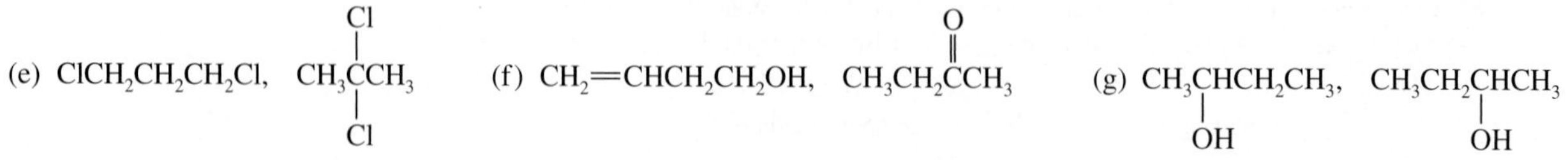

## 5.7   A Closer Look at Equivalence of Groups and Atoms in Molecular Structures

### A. Nuclear Magnetic Resonance Spectra and Equivalence of Atoms and Groups in Molecular Structures

In Section 5.4 we saw how chemists use instruments such as nuclear magnetic resonance spectrometers to explore molecular structures. To interpret the spectra obtained, they analyze the molecular symmetry found in the compounds with which they work and categorize which groups of atoms are equivalent to each other and which are not.

The carbon and proton nuclear magnetic resonance spectra of 1-bromo-2-methylpropane (isobutyl bromide) are given in Figure 5.7 on the next page. The compound, shown below, has three groups of equivalent carbon atoms and three groups of equivalent hydrogen atoms, labeled $a$, $b$, and $c$.

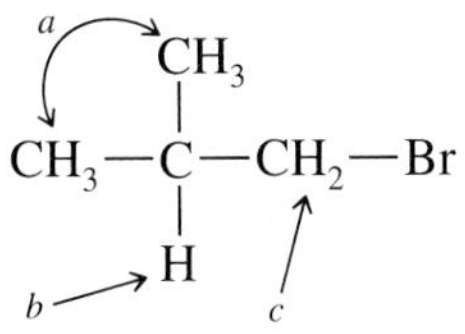

$$a \quad CH_3$$
$$CH_3-C-CH_2-Br \qquad (CH_3)_2CHCH_2Br$$
$$b \quad H \quad c$$

1-bromo-2-methylpropane
isobutyl bromide

The ratio of the numbers of hydrogen atoms in groups $a$, $b$, and $c$ is $6:1:2$, from the integration of the proton magnetic resonance spectrum (30 mm to 5 mm to 10 mm), as well as from an inspection of the structural formula. In other words, all six hydrogen atoms labeled $a$ are equivalent to each other but different from those labeled $b$ and $c$. Similarly, the hydrogen labeled $b$ is different from those labeled $a$ and $c$. The two hydrogens labeled $c$ are equivalent to each other but distinguishable from those labeled $a$ and $b$. The same is true of the carbon atoms, the $a$ carbon atoms, the $b$ carbon atom, and the $c$ carbon atom, represented by the three peaks in the carbon atom nuclear magnetic resonance spectrum of the compound.

Study Guide
Concept Map 5.1

### Problem 5.9

(a) Use the proton magnetic resonance spectrum in Figure 5.7 to assign chemical shifts to the $a$ hydrogens, the $b$ hydrogen, and the $c$ hydrogens.
(b) Do the same thing for the $a$, $b$, and $c$ carbons, using the carbon nuclear magnetic resonance spectrum.

### B. Chemical Substitution as a Way to Explore Equivalence of Atoms and Groups in Molecular Structures

Another way to explore the equivalence of the hydrogen atoms is by substitution, a criterion that we have already used. We ask whether the substitution of another atom for any one of those hydrogen atoms will give the same connectivity. Let us try the exercise of substituting a chlorine atom for one of the hydrogen atoms in 1-bromo-2-methylpropane.

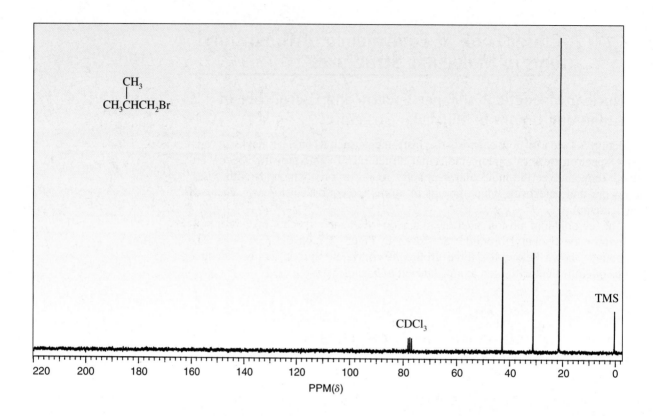

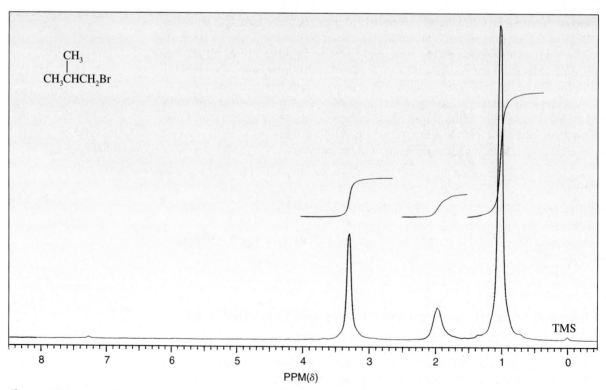

*Figure 5.7*
Nuclear magnetic resonance spectra of 1-bromo-2-methylpropane: *(top)* carbon spectrum;
*(bottom)* proton spectrum.

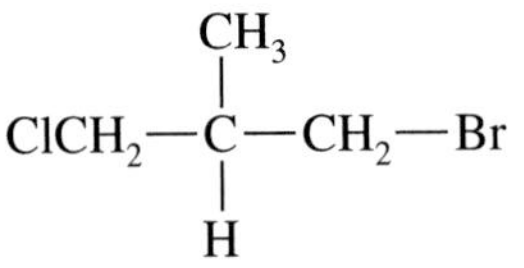

$$CH_3$$
$$ClCH_2-C-CH_2-Br$$
$$H$$

substitution of Cl for
one of the *a* group of
hydrogens

$$CH_2Cl$$
$$CH_3-C-CH_2-Br$$
$$H$$

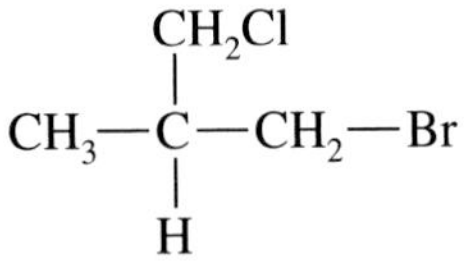

substitution of Cl for
another one of the *a*
group of hydrogens

$$CH_3$$
$$CH_3-C-CH_2-Br$$
$$Cl$$

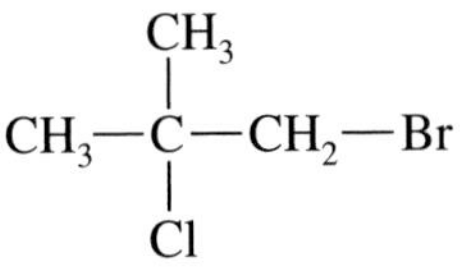

substitution of Cl for
the *b* hydrogen

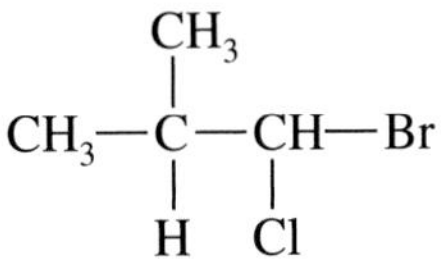

$$CH_3$$
$$CH_3-C-CH-Br$$
$$H \quad Cl$$

substitution of Cl for
one of the *c* group of
hydrogens

In the four structural formulas above, it clearly makes a big difference whether a chlorine atom is substituted for an *a*, *b*, or *c* hydrogen atom. It is harder to tell whether the first two structures shown above are identical or not. The right angle in the structural formula between the two groups on the tertiary carbon makes the two methyl groups look different, but, in fact, they are equivalent because in reality the carbon atom has tetrahedral bond angles. The two methyl groups, therefore, have identical spatial relationships to the tertiary hydrogen and to the other carbon atom in the molecule. Substitution of any one of the hydrogen atoms in group *a* by chlorine gives rise to the same connectivity. Experimenting with molecular models will convince you of this fact.

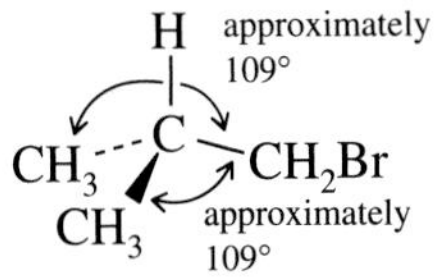

H approximately 109°

$$CH_3 \cdots C \rightarrow CH_2Br$$
$$CH_3 \text{ approximately } 109°$$

the spatial relationship between
the two methyl groups bearing the
*a* group of hydrogen atoms

### Problem 5.10

(a) How many peaks would you expect in the carbon nuclear magnetic resonance spectrum of each of the $C_4H_8BrCl$ compounds shown above?
(b) How about the number of peaks their proton magnetic spectra would exhibit?

## C. The Concept of Neighboring Atoms in Analyzing Molecular Structures

Another way to analyze whether groups of atoms are equivalent is to look at how a group of atoms is related to the other atoms in the molecule. Chemists speak of neighbors to the atom in which they are interested. Neighbors may be close—separated by one, two, or three bonds—or they may be more distant. In searching for neighbors, we look for atoms that are not equivalent, that are distinguishable from each other. For example, for 1-bromo-2-methylpropane, we find the following relationships: All the *a* hydrogens are equivalent. Therefore, they do not count as neighbors to each other. The closest hydrogen atom that is a neighbor to *a* hydrogens is the *b* hydrogen, which is three bonds away. We say that the *b* hydrogen is a *three-bond neighbor* to the *a* hydrogens and also to the *c* hydrogens. The *a* hydrogens are *four-bond neighbors* of the *c* hydrogens.

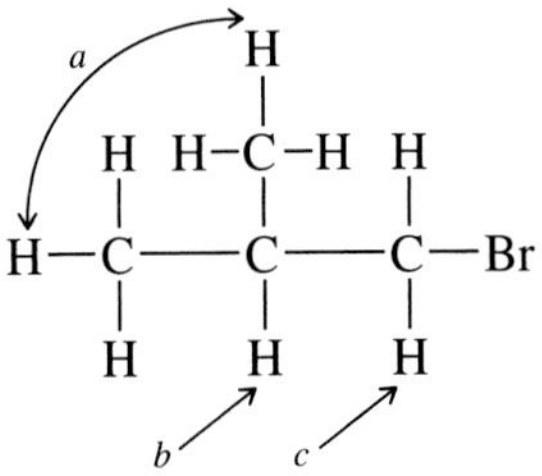

The carbon atoms and the bromine atom may also be used in our description of neighbors. For example, all the hydrogen atoms are one-bond neighbors to one carbon atom. The *a* hydrogen atoms are four-bond neighbors to the bromine atom.

The structural formulas that result from substitution of a chlorine atom for one of the hydrogen atoms in group *a* may now be analyzed in a different way. The first

$$a \quad H$$
$$H \quad H-C-H \quad H$$
$$H-C-C-C-Br$$
$$H \quad H \quad H$$
$$b \quad c$$

Substituting a chlorine atom for either one of the *c* hydrogens in 1-bromo-2-methylpropane gives a compound with the same connectivity. To explore whether having the same connectivity necessarily means that two structures are the same, carry out the following experiment.

**PROBLEM:** Build two molecular models of 1-bromo-2-methylpropane. Replace one of the *c* hydrogen atoms with a chlorine atom in one of the models. In the other model, replace the other *c* hydrogen atom with chlorine. Are the two compounds you made identical?

**Hint:** Can you take one molecular model and put it on top of the other one and have all parts coincide exactly?

**PROBLEM:** Repeat the exercise by replacing an *a* hydrogen from one methyl group by chlorine in one model and an *a* hydrogen from the other methyl group in the other model. Are the two compounds you made identical to each other?

thing we notice is that once the chlorine atom is in place, we have *four* different groups of hydrogen atoms rather than three.

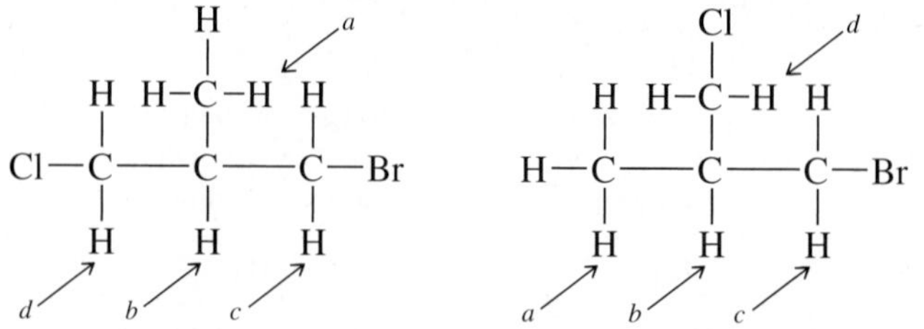

The hydrogen atoms in group *a* are equivalent to each other but are no longer equivalent to the hydrogens now labeled as group *d*. The *d* hydrogen atoms are four-bond neighbors to those in group *a* and group *c*; they are three-bond neighbors to the *b* hydrogen. They are one-bond neighbors to one carbon atom, two-bond neighbors to another carbon atom, and a four-bond neighbor to the bromine atom. Because these descriptions are true for either structural formula, both therefore represent the same connectivity.

**Problem 5.11**    Three $C_5H_{11}Br$ isomers are shown below. For each one, identify the equivalent groups of carbon atoms and hydrogen atoms. Find all hydrogen atoms that are three-bond neighbors of each other.

| | | |
|---|---|---|
| Compound A | 1-bromo-2,2-dimethylpropane | $CH_3C(CH_3)_2CH_2Br$ |
| Compound B | 1-bromopentane | $CH_3CH_2CH_2CH_2CH_2Br$ |
| Compound C | 3-bromopentane | $CH_3CH_2CHBrCH_2CH_3$ |

## 5.8   Conformation

### A. Conformations of Ethane

The different three-dimensional pictures of chloroethane shown on page 148 are drawn so that the carbon–chlorine bond bisects the angle made by two of the hydrogen atoms on the adjacent carbon atom. The molecule, however, is not frozen in that form. The carbon atoms are constantly in motion, rotating relative to each other around the single bond that joins them, giving rise to different orientations in space of the hydrogen atoms. The **conformations** of a molecule are different arrangements in space of the atoms within the molecule due to rotations around single bonds. In ethane, the conformation in which the carbon–hydrogen bonds on the two carbon atoms are as far apart as possible is called the **staggered conformation.** This is a low-energy conformation. As the two carbon atoms rotate with respect to each other, at some point the carbon–hydrogen bonds on one carbon atom line up paralleling the similar bonds on the other carbon atom so that they eclipse them if viewed from one end of the molecule. This **eclipsed conformation** corresponds to a maximum energy.

The relative positions of the hydrogen atoms on the two carbon atoms are best seen in a Newman projection, named for Melvin Newman of Ohio State University. A **Newman projection** is a view of a molecule down the axis of a carbon–carbon bond. The carbon atom toward the front is represented by a dot, and the one toward the rear, by a circle. The atoms or groups on the carbon atoms are shown as being bonded to the dot or the circle. Newman projections of the staggered and eclipsed conformations of ethane are shown on page 165, with their equivalent perspective formulas.

viewed from this point
for Newman projection

staggered conformation of ethane

carbon atom
in the rear

60° dihedral angle
between hydrogen
atoms on adjacent
carbon atoms

carbon atom
in the front

eclipsed conformation of ethane

hydrogen atom
on the rear carbon
atom directly behind
a hydrogen atom on
the front carbon atom

0° dihedral angle
between hydrogen
atoms on adjacent
carbon atoms

*perspective
formulas*

*Newman
projections*

A Newman projection allows us to visualize the spatial relationship between atoms bonded to one carbon atom and those bonded to an adjacent carbon atom. This spatial relationship can be described in terms of the **dihedral angle, $\theta$,** which is defined as the angle between the atoms on adjacent carbon atoms as those atoms appear in the projection formula. For example, in the staggered conformation of ethane, the dihedral angles between the hydrogen atoms on the front carbon atom and those on the back carbon atom are 60°. In the eclipsed conformation, the dihedral angles are 0°.

The difference in energy between the staggered and eclipsed conformations of ethane is approximately 3 kcal/mol. The reason for this energy difference in ethane has been the subject of some debate. It is now believed to arise from unfavorable interactions between orbitals for the carbon–hydrogen bonds when the bonds are closer together in the eclipsed conformation and favorable interactions between bonding and antibonding orbitals of the same bonds when they are staggered relative to each other (Figure 5.8).

The Newman projections of ethane in Figure 5.8 show the rear carbon atom rotating relative to the front one around the carbon–carbon bond. A rotation of 60° converts a staggered conformation into an eclipsed one. Another 60° rotation (or a total rotation of 120°) returns the molecule to a staggered conformation. At room temperature, molecules have enough kinetic energy to get over energy barriers up to approximately 20 kcal/mol. As Figure 5.8 illustrates, the energy required for rotation

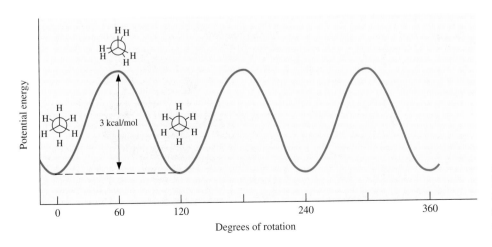

3 kcal/mol

Potential energy

Degrees of rotation

*Figure 5.8*
Energy diagram showing the energy difference between the staggered and eclipsed conformations of ethane.

around the carbon–carbon bond is only 3 kcal/mol. As a result, rotation around the carbon–carbon bond in ethane molecules is constant. Yet the molecules spend most of their time in the energy valleys represented by the staggered conformations.

**Problem 5.12**    Draw perspective formulas and Newman projections for the different conformations of chloroethane.

## B. Conformations of Butane

The three-dimensional representation of butane on page 158 shows the first and last carbon atoms of the chain as far apart from each other as possible. The hydrogen atoms on the second and third carbon atoms are staggered. This arrangement, in which there is a dihedral angle of 180° between the methyl groups, is the most stable one and is called the **anti conformation** of butane. There are other conformations also. Shown on the facing page are perspective formulas and Newman projections for the conformations of butane that result from a rotation of 60° around the bond between the second and third carbon atoms in the chain.

Starting with the anti conformation, rotation of the front carbon atom relative to the rear one gives different conformations of butane. Besides the anti conformation, A, there are two other staggered conformations of butane, C and E; these are called **gauche conformations.** In C and E, the methyl groups have a dihedral angle of 60° with respect to each other in the Newman projection (Figure 5.9). When the methyl groups are that close, nonbonding interactions (p. 28) known as **van der Waals repulsions** occur. Figure 5.9 (right side) shows that atoms have space-filling properties and exert an influence that is not apparent from looking at molecular formulas. The effective sizes of atoms in molecules are expressed in terms of van der Waals radii for those atoms. The van der Waals radius of a hydrogen atom is 1.2 Å. Nonbonding interactions between molecules or between different parts of the same molecule result in van der Waals attractions (p. 30), as long as the interacting atoms do not get too close to each other. At distances shorter than the van der Waals radii of the atoms, repulsion occurs. In the gauche conformation of butane, the hydrogen atoms of the two methyl groups are close enough that van der Waals repulsions result. Consequently, the gauche conformations are less stable than the anti form by approximately 0.9 kcal/mol.

The different conformations of butane are isomeric. The isomeric species corresponding to the valleys in the energy diagrams such as Figure 5.8 and Figure 5.10 (pp. 165 and 168) are called **conformers.** Conformers differ from constitutional

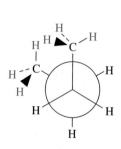

Newman projection

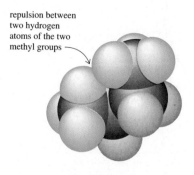

repulsion between two hydrogen atoms of the two methyl groups

source of van der Waals repulsion between gauche methyl groups

*Figure 5.9*
Repulsion between the methyl groups in the gauche conformation of butane.

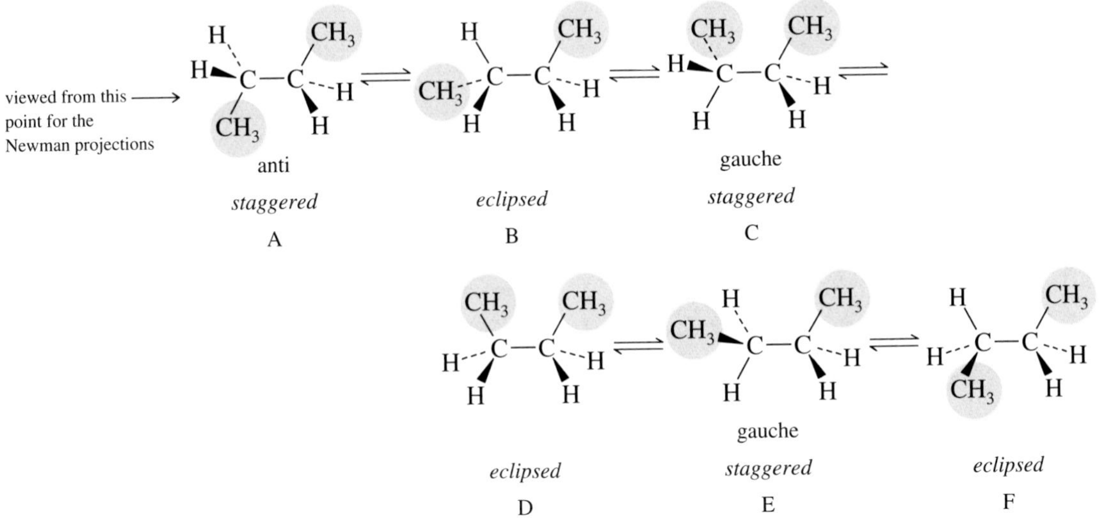

viewed from this
point for the
Newman projections

anti

staggered
A

eclipsed
B

gauche

staggered
C

eclipsed
D

gauche

staggered
E

eclipsed
F

**perspective formulas**

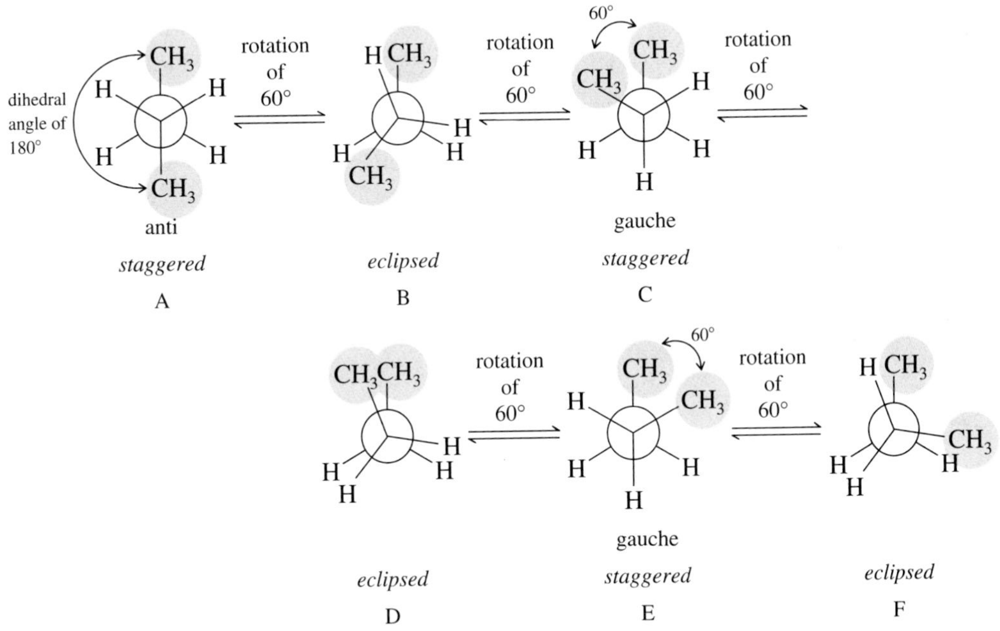

dihedral
angle of
180°

anti

staggered
A

eclipsed
B

gauche

staggered
C

eclipsed
D

gauche

staggered
E

eclipsed
F

**Newman projections**

isomers in that they have the same connectivities. They differ only in the spatial positions of atoms relative to each other. These spatial relationships are dynamic, that is, changing continuously. Usually conformers are being converted so rapidly from one into the other that they cannot be isolated as individual species but are in equilibrium with each other.

At room temperature, about 25% of butane molecules are in gauche conformations and 75% in the anti conformation. The energy barrier to rotation around the central bond in butane is small enough, approximately 3.8 kcal/mol, that the different conformers cannot be separated from each other at room temperature; they are

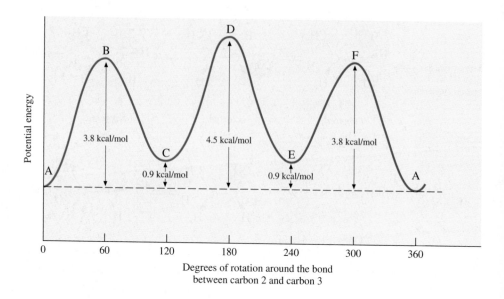

*Figure 5.10*
Energy diagram for the
conformations of butane.

rapidly interconverted. The physical properties actually observed for butane repre-sent an average of all the properties that would be expected for all of the different forms present at any one time.

The energy relationships of the different conformations of butane are shown in Figure 5.10. The interconversions of the anti and gauche conformers require rota-tion through high-energy eclipsed conformations. The highest-energy conformation is D, in which the two methyl groups eclipse each other.

For larger alkanes, too, the most stable conformations are generally staggered, with the largest groups anti to each other. The gauche conformations are also pres-ent and are only slightly less stable than the anti ones. Thus, even though the struc-tural formulas of alkanes are usually written in a straight line for convenience, it is important to keep in mind the zigzag nature of hydrocarbon chains, as illustrated below for octane, $CH_3CH_2CH_2CH_2CH_2CH_2CH_2CH_3$.

*the most stable conformation of octane;*
*largest substituents at any one carbon–carbon bond anti to each other*

⊘ **Study Guide**
**Concept Map 5.2**

**Problem 5.13**     Draw possible conformations of 1-chloropropane, $CH_3CH_2CH_2Cl$, around the bond connecting carbon atom 1 to carbon atom 2 as perspective formulas and as Newman projections. Draw an energy diagram showing the relative energies of the different conformers.

**Problem 5.14**     Show the conformers of 1,2-dichloroethane (p. 149) as perspective for-mulas and as Newman projections. Predict, from looking at your pictures, whether or not each conformer has a dipole moment. (You may wish to review pages 26 to 28.) 1,2-Dichloroethane has a dipole moment of 1.42 D. What does this tell you about the con-formational composition of 1,2-dichloroethane at room temperature?

## 5.9  Line Structures for Organic Compounds

Chemists, in drawing complex organic structures, especially those containing rings, often use a simplified way of representing them. In this type of representation, known as **line structures,** or **skeletal structural formulas,** each junction of two straight lines or the end of a straight line represents a carbon atom and the number of hydrogen atoms necessary to give the carbon four bonds, for example:

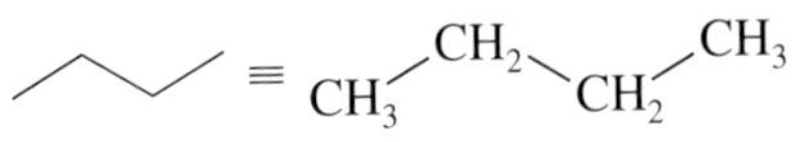

butane 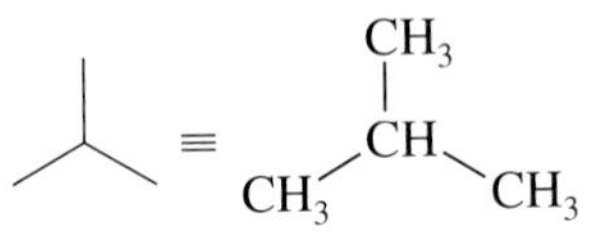                2-methylpropane 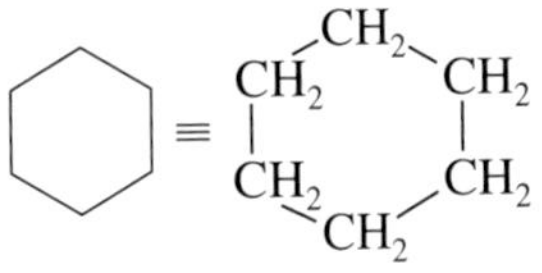                cyclohexane

Note that the end of a line is equivalent to $CH_3$. The intersection of two lines is $CH_2$, and the intersection of three lines is CH. Four lines meeting at a point represent a carbon atom with no hydrogen atoms, a **quaternary carbon atom.**

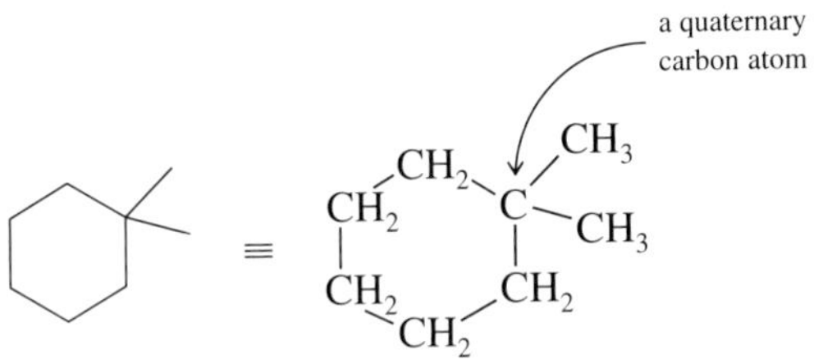

Functional groups are represented by their usual symbols. A carbon–carbon double bond is shown as two parallel lines, a triple bond as three.

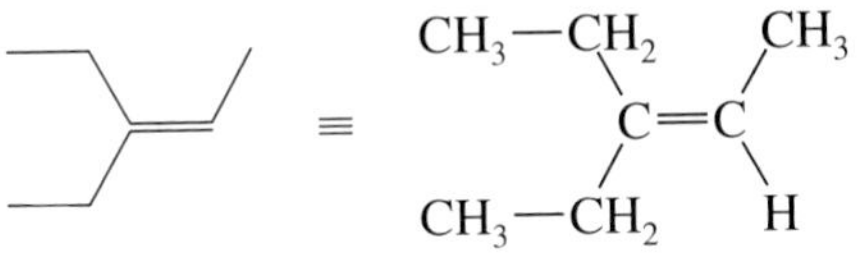

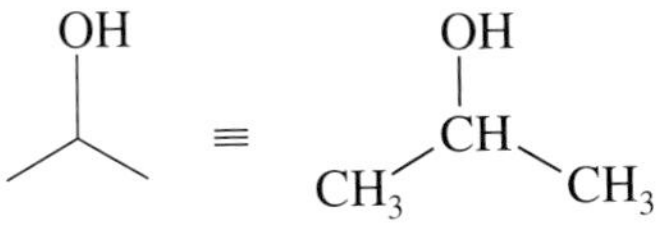

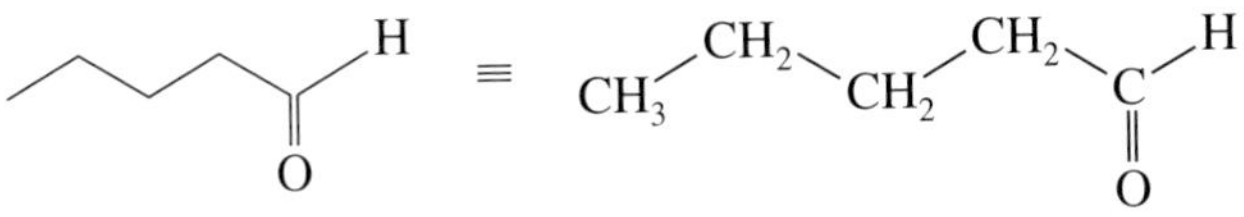

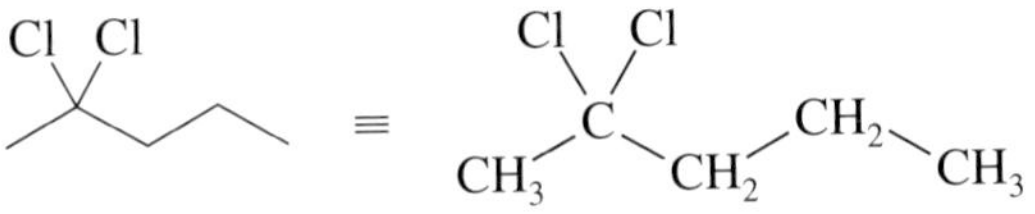

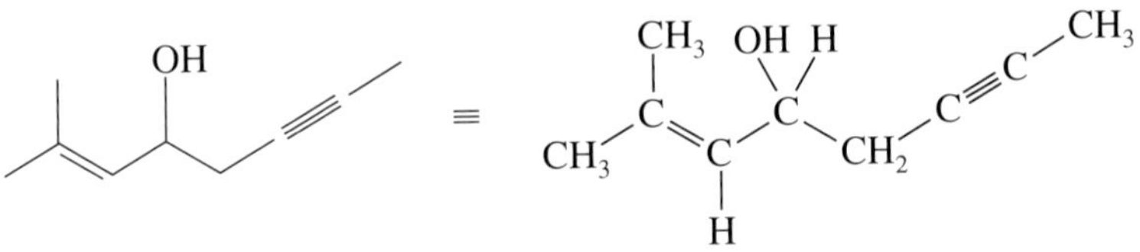

If necessary, nonbonding electrons and formal charges may also be shown on line structures.

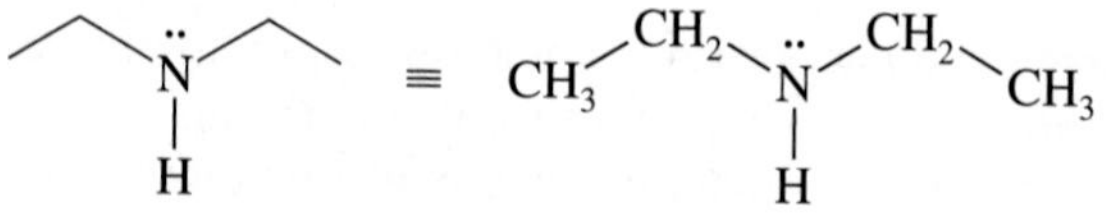

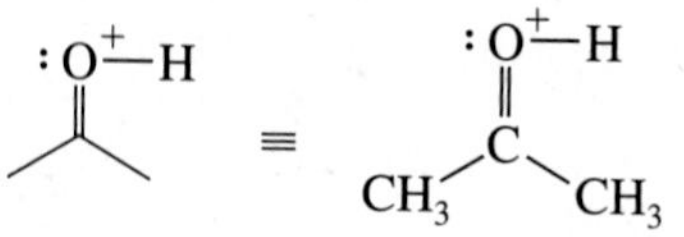

Wedges and dashed lines may be used with line structures to indicate the orientation of groups in space.

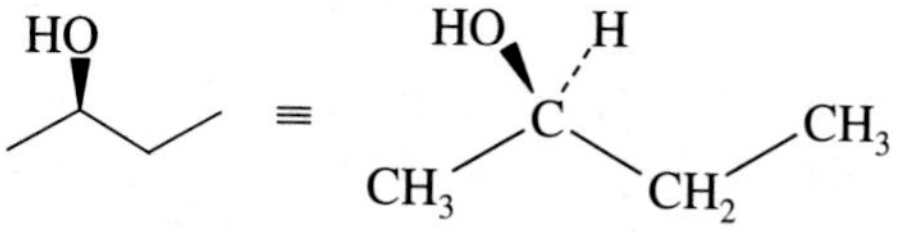

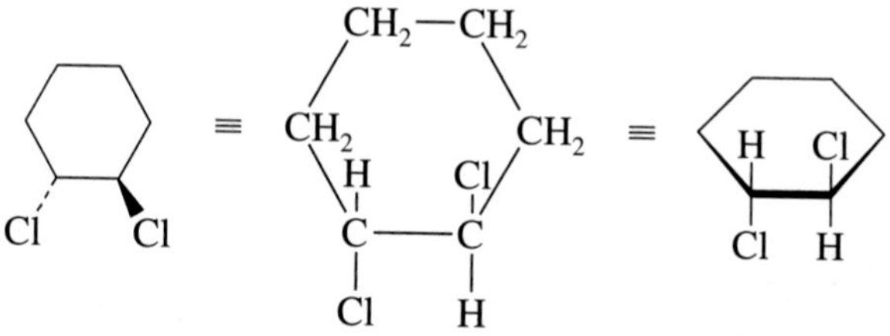

In this book, line structures will be used for drawing cyclic molecules and some complex molecules. Most of the time, structural formulas showing all the carbon and hydrogen atoms will be used.

⊘ **Study Guide**
**Concept Map 5.3**

**Problem 5.15**   Write line structural formulas for the following compounds.

(a) CH₃CH₂CH₂CH=CH₂

(b)

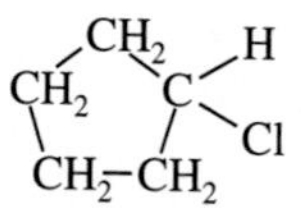

(c)

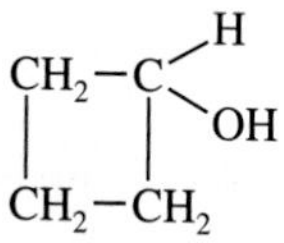

(d) 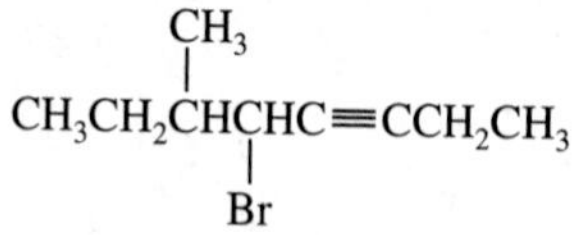 CH₃CH₂CHCHC≡CCH₂CH₃
$\quad\quad\quad$ with CH₃ above and Br below

(e) 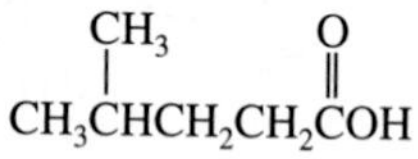 CH₃CHCH₂CH₂COH with CH₃ and O above

**Problem 5.16**   Write condensed structural formulas for the following compounds.

(a)

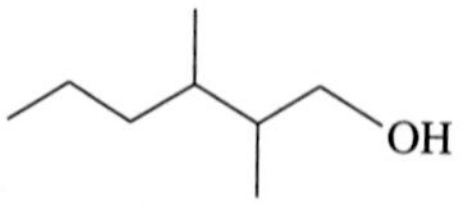

(b)

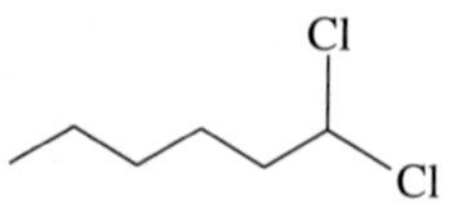

(c)

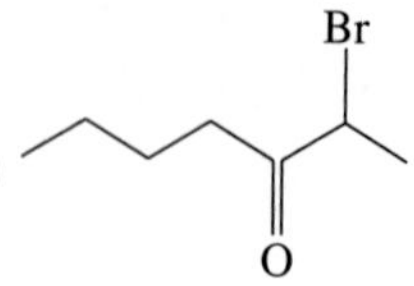

(d)

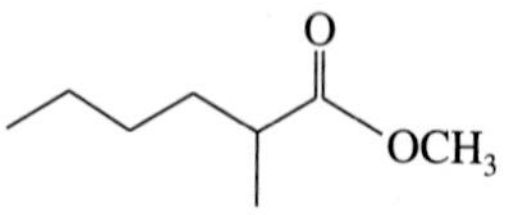

(e)

## 5.10  Nomenclature

### A. Introduction

Historically, compounds were given names that reflected their origin or their properties. For example, the painkiller morphine was named for Morpheus, the Greek god of dreams. Cholesterol, the chief component of gallstones, got its name from the Greek words for bile and solid. This way of naming compounds is descriptive but does not lead to any systematic procedure for assigning names to new and related compounds.

By the end of the nineteenth century, the number of organic compounds that were being synthesized or isolated from natural sources was growing rapidly. Under these conditions, it eventually became impossible for chemists to learn the names randomly assigned to compounds by their discoverers, especially when the names showed no correlation to the structures. Since 1892, chemists from all over the world have met periodically to decide on systematic rules for naming organic compounds. These rules, which are constantly evolving, are called the *International Union of Pure and Applied Chemistry* (abbreviated as *IUPAC*) *rules*. The IUPAC system of nomenclature was developed so that each organic compound would have a unique name that would allow the structural formula to be written from it.

Ideally, a chemist would have to learn only the systematic names of compounds. In the real world, however, there are three complications. First, even the IUPAC rules allow for some variations in the naming of compounds. Second, the correct IUPAC names for some compounds are so complicated and cumbersome that everybody continues to use the common, unsystematic names for them. These names are usually derived from the natural origins of compounds or their molecular shapes or sometimes even the whimsy of their creators. The structural formulas and systematic names of morphine and cholesterol are shown below, not so you can memorize them but so you will have some idea why the IUPAC names are not generally used to refer to these compounds.

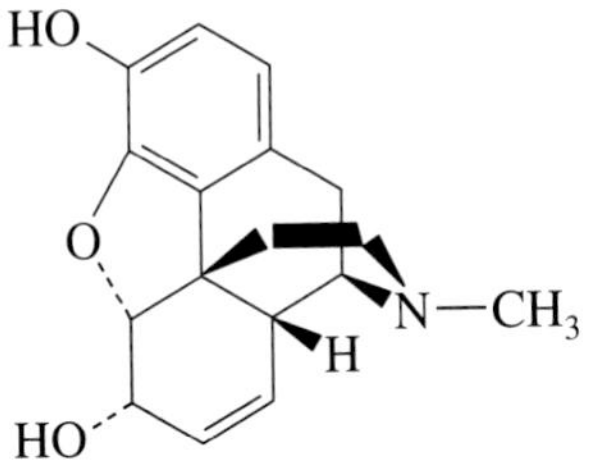

(5α, 6α)-7,8-didehydro-4,5-
epoxy-17-methylmorphinan-
3,6-diol;
morphine

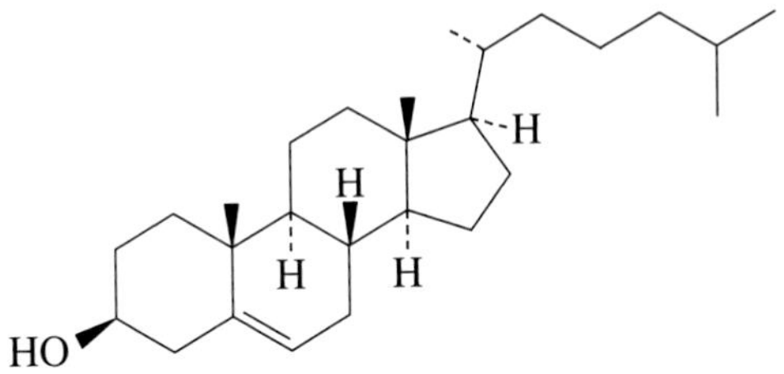

(3β)-cholest-5-en-3-ol;
cholesterol

Finally, many compounds had names before the IUPAC rules came into being, and these names are still used, especially by chemical supply houses and by industry. To be literate in the laboratory, you must be able to recognize many common names. Therefore, while the emphasis in this book is on the systematic names of compounds, the common names of important compounds are also given and are used in cases where they are the ones used overwhelmingly by chemists in their daily work.

### B. Nomenclature of Alkanes

The names of the first four alkanes are methane, ethane, propane, and butane. The systematic nomenclature of the other members of the series is based on a prefix indicating the number of carbon atoms in the chain, followed by the suffix **ane**. The

**Table 5.2**    **The Names of Some Straight-Chain Alkanes**

| Molecular Formula | Name |
|---|---|
| $CH_4$ | methane |
| $C_2H_6$ | ethane |
| $C_3H_8$ | propane |
| $C_4H_{10}$ | butane |
| $C_5H_{12}$ | pentane |
| $C_6H_{14}$ | hexane |
| $C_7H_{16}$ | heptane |
| $C_8H_{18}$ | octane |
| $C_9H_{20}$ | nonane |
| $C_{10}H_{22}$ | decane |
| $C_{11}H_{24}$ | undecane |
| $C_{12}H_{26}$ | dodecane |
| $C_{16}H_{34}$ | hexadecane |
| $C_{18}H_{38}$ | octadecane |
| $C_{20}H_{42}$ | icosane |

prefixes come from Greek or Latin words for the numbers. The names of the first twelve straight-chain alkanes, along with those of three larger alkanes that form the basis for names of biologically important compounds, are shown in Table 5.2. These names are used in naming all other types of organic compounds derived from alkanes, so you should learn them.

For a compound that is not a straight-chain alkane, the name is derived from the name of the alkane that corresponds to the longest continuous chain of carbon atoms in the molecule. For example, the branched-chain compounds with the molecular formula $C_5H_{12}$ (p. 145) are named 2-methylbutane and 2,2-dimethyl-propane. The longest chain in the first one is four carbons long; therefore, the compound is classified as a substituted butane. The longest chain in the second one is three carbons long, so the compound is a substituted propane.

Alkanes with the molecular formula $C_6H_{14}$ exist in five isomeric forms. The straight-chain isomer is called hexane. Two of the branched-chain isomers have five carbon atoms in the longest continuous chain and are named as substituted pentanes.

$$\underset{1}{CH_3}\underset{2}{CH_2}\underset{3}{CH_2}\underset{4}{CH_2}\underset{5}{CH_2}\underset{6}{CH_3}$$

hexane

*straight-chain isomer*
*of* $C_6H_{14}$

$$\begin{array}{cc} CH_3 & CH_3 \\ | & | \\ \underset{1}{CH_3}\underset{2}{CH}\underset{3}{CH_2}\underset{4}{CH_2}\underset{5}{CH_3} & \underset{1}{CH_3}\underset{2}{CH_2}\underset{3}{CH}\underset{4}{CH_2}\underset{5}{CH_3} \\ \text{2-methylpentane} & \text{3-methylpentane} \end{array}$$

*two of the branched-chain isomers of* $C_6H_{14}$

The other two isomers of $C_6H_{14}$ have only four carbon atoms in the longest continuous chain and are, therefore, substituted butanes.

$$\begin{array}{cc} CH_3 & CH_3 \quad CH_3 \\ | & | \quad\quad | \\ \underset{1}{CH_3}-\underset{2}{C}-\underset{3}{CH_2}\underset{4}{CH_3} & \underset{1}{CH_3}\underset{2}{CH}-\underset{3}{CH}\underset{4}{CH_3} \\ \quad | & \\ \quad CH_3 & \\ \text{2,2-dimethylbutane} & \text{2,3-dimethylbutane} \end{array}$$

*two more of the branched-chain isomers of* $C_6H_{14}$

After the longest continuous chain of carbon atoms in the molecule is identified, the alkyl groups attached to that backbone are named. Each one is then assigned a number indicating its position on the chain. The chain is numbered so that the substituents have the lowest possible numbers. The name of the compound is then created by listing the names of the substituents along with their positions on the chain. For example, the substituents in the branched-chain $C_6H_{14}$ compounds are all methyl groups. The first isomer is called 2-methylpentane to indicate that the methyl group is on the second carbon atom from the end of the five-carbon chain. (The name 4-methylpentane, which is arrived at by numbering the chain in the opposite direction, is incorrect because it violates the rule that substituents must be given the lowest possible numbers.) The second isomer has the methyl group attached to the third carbon atom of the chain and is named 3-methylpentane.

When there is more than one substituent on a chain, the lowest possible number is the number of the first position in the chain at which a choice has to be made, for example:

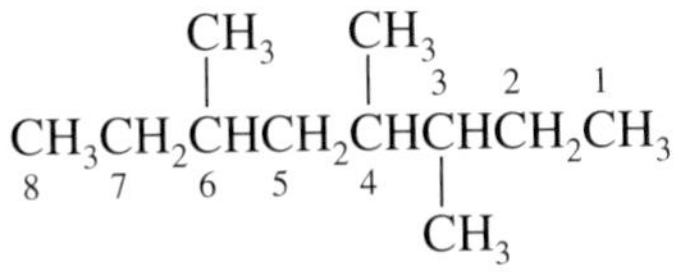

$$\underset{8\quad7\quad6\quad5\quad\;4}{CH_3CH_2CHCH_2CHCHCH_2CH_3}$$

with CH₃ groups at positions 6, 5 (labeled 3 2 1) and CH₃ at position 4.

CH₃   CH₃
 |     |  3  2  1
CH₃CH₂CHCH₂CHCHCH₂CH₃
 8   7  6  5   4 |
              CH₃

3,4,6-trimethyloctane

*not 3,5,6-trimethyloctane*
*because 4 is a lower number than 5*

If there is more than one substituent of the same kind, the name of the alkyl group is given the prefix **di** (for two), **tri** (for three), or **tetra** (for four). As before, each substituent is also given a number indicating its position on the chain. These rules are illustrated by the names of the last three of the compounds shown above. The name 2,2-dimethylbutane indicates that the compound has a chain of four carbon atoms with two methyl groups attached to the second carbon of that chain. Note that the number 2 is repeated for each methyl group so that there is no doubt about the placement of the groups. The name of the isomeric compound 2,3-dimethyl-butane differs from 2,2-dimethylbutane in only one number. The name 3,4,6-trimethyl-octane also has a number for each methyl group on the chain. Note that the numbers are separated by commas and joined to the name by a hyphen.

An application of these rules to somewhat more complicated examples may be helpful. A $C_{12}H_{26}$ hydrocarbon is shown below.

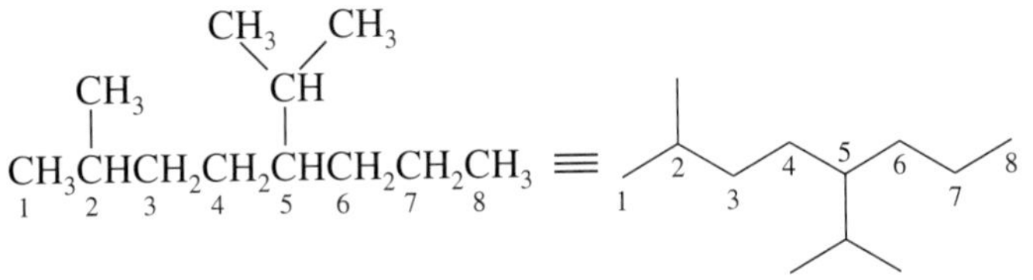

CH₃  CH₃
  \  /
CH₃   CH
 |     |
CH₃CHCH₂CH₂CHCH₂CH₂CH₃  ≡
 1  2  3  4  5  6  7  8

5-isopropyl-2-methyloctane

The longest continuous chain contains eight carbon atoms. The compound is, therefore, named as a substituted octane. The alkyl substituents on the chain are identified as the methyl group (on carbon atom 2) and the isopropyl group (on carbon atom 5). The compound is given the name 5-isopropyl-2-methyloctane. When there are several different kinds of substituents on the chain, they are listed in alphabetical order. Prefixes such as di or tri are not considered when substituents are alphabetized. Neither are italicized prefixes that are followed by hyphens; for example, *tert*-butyl is treated as a butyl group.

Another example further illustrates the rules.

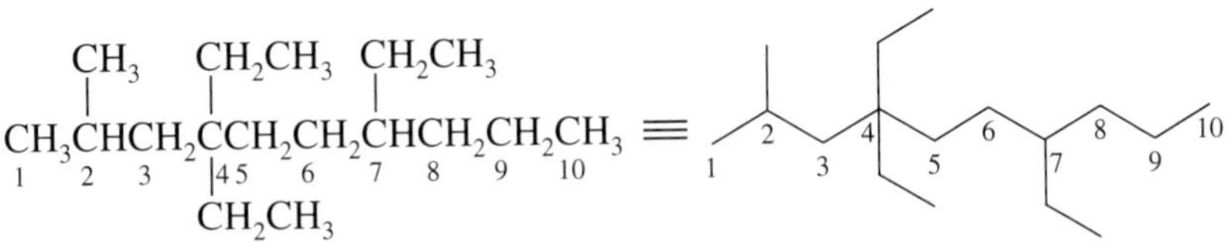

CH₃   CH₂CH₃ CH₂CH₃
 |      |      |
CH₃CHCH₂CCH₂CH₂CHCH₂CH₂CH₃  ≡
 1  2  3 |4 5  6  7  8  9  10
        CH₂CH₃

4,4,7-triethyl-2-methyldecane

In this compound, the longest continuous chain contains ten carbon atoms, and therefore, this is a substituted decane. The backbone of the molecule has three ethyl groups (two at carbon atom 4 and one at carbon atom 7) and a methyl group (at carbon atom 2) attached. The name is 4,4,7-triethyl-2-methyldecane; the ethyl groups are listed before the methyl group because the prefix tri is ignored in alphabetizing. Note again the use of commas to separate a series of numbers and the use of hyphens to join numbers to the substituents. The name of the last substituent listed is merged with that assigned to the backbone.

This list summarizes the rules for the nomenclature of alkanes:

1. Locate the longest continuous straight chain of carbon atoms in the molecule. The name of the straight-chain alkane with the same number of carbon atoms becomes the last part of the name of the compound.

2. Find and name all of the alkyl groups that are branches off the backbone of the molecule. Assign each one a position on the chain, numbering the chain so that the substituents have the lowest possible numbers.

3. If there are several substituents of the same kind, indicate how many by using the prefix di, tri, or tetra, and use a number to assign a position on the chain to each one.

4. Construct the name of the compound by listing all the substituents in alphabetical order, ignoring the prefixes di, tri, and tetra and italicized prefixes such as *n-, sec-,* and *tert-.*

5. Use commas to separate numbers that are grouped together. Separate numbers from names of groups by hyphens. Merge the name of the last substituent with the name of the straight chain alkane that is the basis of the name of the compound.

One last example will provide a review of these points.

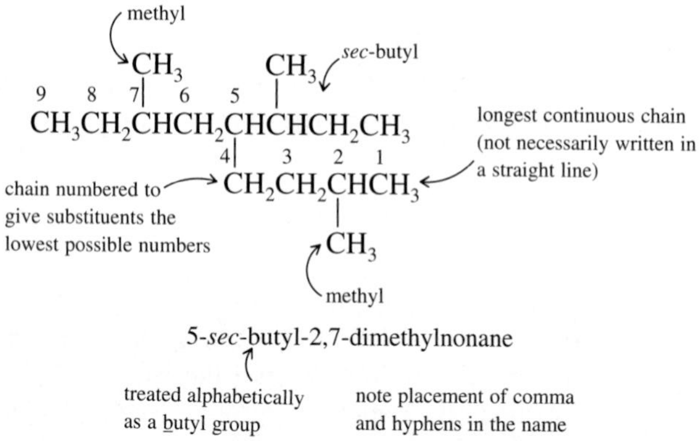

5-*sec*-butyl-2,7-dimethylnonane

---

**Problem 5.17**   Name the following compounds.

(a) (b) (c) (d) (e) (f)

---

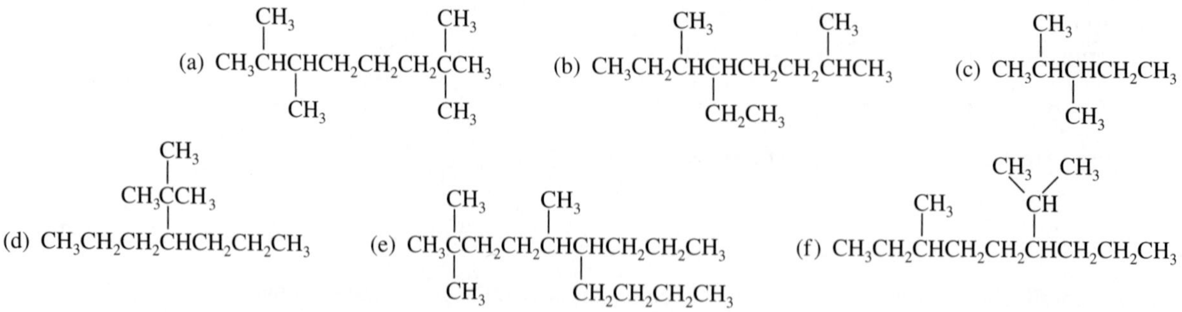

**Problem 5.18**   Draw structural formulas for the following compounds.

(a) 2,2-dimethylheptane   (b) 6-isobutyl-2-methyldecane
(c) 5,5-diisopropyl-2,8-dimethylnonane
(d) 6-ethyl-2,2,4-trimethyldodecane

**Problem 5.19**   Draw structural formulas for the nine alkanes having the molecular formula $C_7H_{16}$. Identify primary, secondary, and tertiary carbon atoms in each one. Show which hydrogen atoms are equivalent. Name each compound.

## C. Nomenclature of Alkyl Halides and Alcohols

The systematic names of alkyl halides are assigned in the same way as the names of branched-chain alkanes. The prefixes **fluoro, chloro, bromo,** and **iodo** are used to indicate the presence of halogens, for example.

$$\underset{\substack{|\\F}}{\overset{\substack{CH_3\\|}}{\underset{1\quad 2\quad 3\quad 4\quad 5}{CH_3CHCHCH_2CH_3}}}$$

3-fluoro-2-methylpentane

$$\underset{\substack{|\\Cl}}{\overset{\substack{Cl\\|}}{\underset{4\quad 3\quad 2\quad 1}{CH_3CH_2CCH_3}}}$$

2,2-dichlorobutane

$$\underset{\substack{|\quad\;|\\Br\;\;Br}}{\overset{\substack{CH_3\;CH_3\\|\quad\;|}}{\underset{1\quad 2\quad 3\quad 4\;5\quad 6}{CH_3CH_2C\!-\!\!-\!CCH_2CH_3}}}$$

3,4-dibromo-3,4-dimethylhexane

$$\underset{\substack{|\\I}}{\overset{\substack{CH_2CH_3\\|}}{\underset{1\quad 2\quad 3\;4\quad 5\quad 6\quad 7}{CH_3CH_2CCH_2CH_2CH_2CH_3}}}$$

3-ethyl-3-iodoheptane

Some organic compounds are named by changing the suffix of the name of the hydrocarbon chain, instead of adding a prefix to it as is done for an alkyl or halogen substituent. Alcohols are usually named by changing the **e** ending of the name of the alkane to **ol** and using a number to indicate the position of the hydroxyl group, which is not specifically named. An alcohol owes its characteristic properties and reactivity to the hydroxyl group, so in naming an alcohol, the carbon chain is numbered so that the carbon atom bearing the hydroxyl group has the lowest possible number.

$$\overset{\substack{CH_3\\|}}{CH_3CHCH_2CH_2CH_2OH}$$

4-methyl-1-pentanol

$$\underset{\substack{|\\CH_3}}{\overset{\substack{CH_3\\|}}{CH_3CCH_2OH}}$$

2,2-dimethyl-1-propanol

$$HOCH_2CH_2CH_2OH$$

1,3-propanediol

$$ClCH_2CH_2CH_2CH_2OH$$

4-chloro-1-butanol

---

**Problem 5.20**    Name the following compounds.

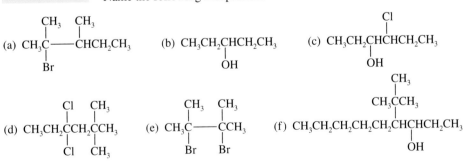

---

**Problem 5.21**    Write structural formulas for the following compounds.

(a) 2-iodooctane     (b) 3-hexanol     (c) 2,2,2-trifluoroethanol
(d) 1-chloro-4,4-dimethyl-2-pentanol     (e) 1,2,3-propanetriol     (f) 1-pentanol
(g) 2,3-dichloro-3-ethylheptane

**Problem 5.22**    There are seventeen constitutional isomers with the molecular formula $C_6H_{13}Cl$. Write structural formulas for them. Name each one, and tell whether it is a primary, secondary, or tertiary halide. Reviewing Section 5.6 (p. 158) may help you remember the structural features of alkyl halides.

**Problem 5.23**    Name all the isomeric compounds having the molecular formula $C_4H_8Cl_2$ (Problem 1.14, p. 23).

## D. The Phenyl Group

There is one other group that you should learn at this stage. The **phenyl group** is formed by the removal of one of the six equivalent hydrogen atoms on benzene. Groups such as the phenyl group, derived from aromatic hydrocarbons (p. 67), are known as **aryl groups.** An aromatic hydrocarbon is sometimes represented by the symbol ArH, and an aryl group by Ar.

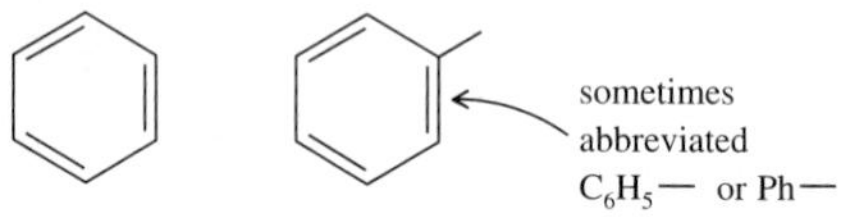

sometimes abbreviated $C_6H_5$— or Ph—

benzene        phenyl group

The phenyl group is used in much the same way as alkyl groups are used in naming compounds. When the hydrocarbon chain attached to the benzene ring is complex or contains more than five carbon atoms, phenyl is used as a part of the name of the compound. However, benzene rings on which small alkyl groups are substituted are usually named as benzene derivatives.

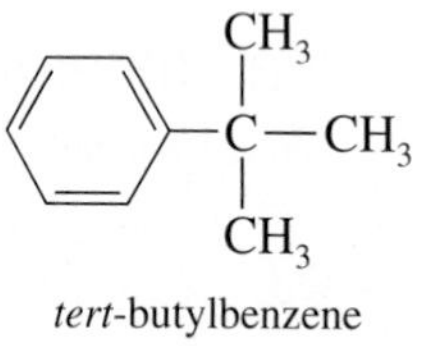

*tert*-butylbenzene 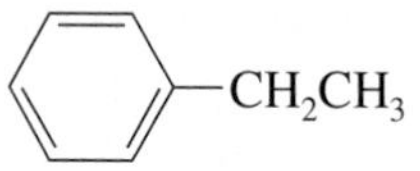            ethylbenzene

*simple alkyl substituents; therefore, compounds
named as substituted benzene*

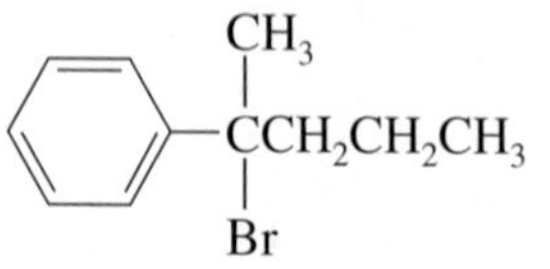

2-bromo-2-phenylpentane 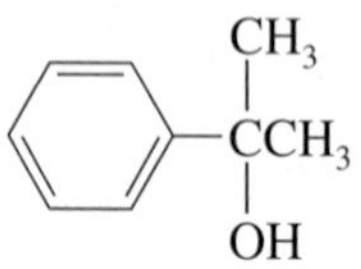        2-phenyl-2-propanol

*complex substituents containing other groups that also need to be named;
therefore, compounds named as alkanes with phenyl substituents*

**Problem 5.24**    Name the compounds in Problem 5.7 (p. 160). (*Hint:* Positions on a benzene ring may be numbered too.)

**Problem 5.25**   Name the following compounds.

(a) 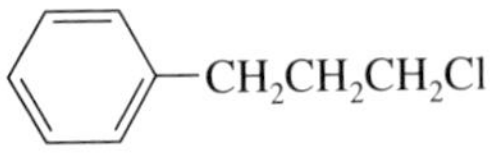 [ring]—$CH_2CH_2CH_2Cl$

(b) 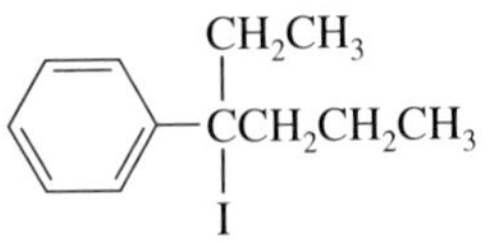 [ring]—$\overset{\displaystyle CH_2CH_3}{\underset{\displaystyle I}{C}}CH_2CH_2CH_3$

(c) [ring]—$\overset{\displaystyle CH_3}{CHCH_3}$  with $CH_3$

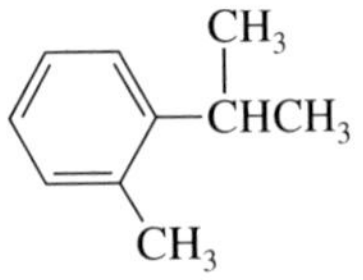

(d) 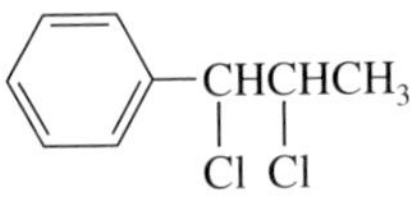 [ring]—$\underset{\displaystyle Cl\ \ Cl}{CHCHCH_3}$

(e) 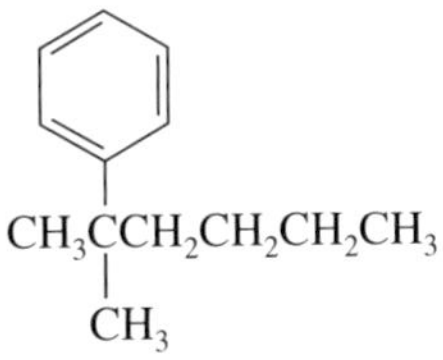 $CH_3\overset{\displaystyle [ring]}{\underset{\displaystyle CH_3}{C}}CH_2CH_2CH_2CH_3$

(f) 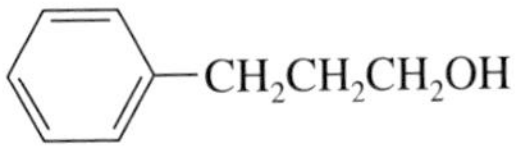 [ring]—$CH_2CH_2CH_2OH$

**Problem 5.26**   Draw structural formulas for the following compounds.

(a) 1-phenyl-1-pentanol       (b) 3-bromo-3-methyl-1-phenylbutane
(c) 4-methyl-2,6-diphenylheptane       (d) 4-*tert*-butyl-2-phenyloctane

# 5.11  Cycloalkanes

## A. Cyclopropane and Cyclobutane. Ring Strain

**Cycloalkanes** are hydrocarbons that have the general formula $C_nH_{2n}$ and in which some or all of the carbon atoms form a ring. A cycloalkane is named by adding the prefix **cyclo** to the name of the alkane having the same number of carbon atoms as are in the ring. Alkanes are divided into two classes: the cycloalkanes, which contain a ring, and the **open-chain alkanes,** which are sometimes called **acyclic compounds** to distinguish them from the cyclic ones.

The three carbon atoms in cyclopropane, $C_3H_6$, the smallest cycloalkane, define a plane. Cyclopropane has internal carbon–carbon bond angles of 60° and external carbon–carbon–hydrogen and hydrogen–carbon–hydrogen bond angles of 116° and 118°, respectively. All the bonds are eclipsed in cyclopropane.

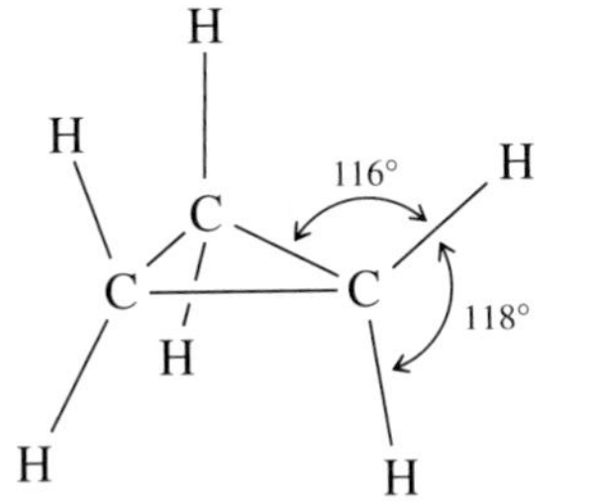

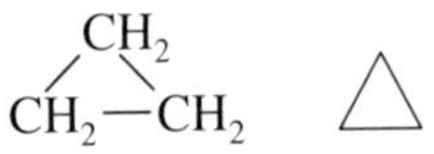

cyclopropane 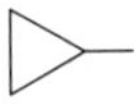          cyclopropyl
group

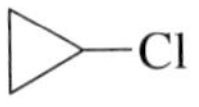

chlorocyclopropane
cyclopropyl chloride

Cyclobutane, $C_4H_8$, is not planar. One of the atoms in the ring is bent out of the plane of the other three by about 20°. This causes the expected internal bond angles of 90° to be reduced to 88° and also minimizes the eclipsing of the hydrogen atoms on adjacent carbon atoms.

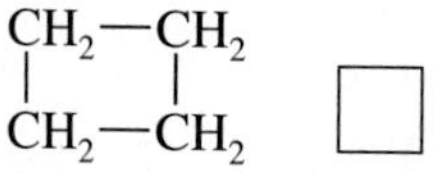

cyclobutane 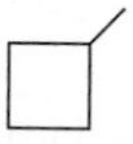          cyclobutyl          bromocyclobutane
                        group           cyclobutyl bromide

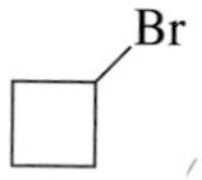

Cycloalkanes are usually symbolized in equations by regular polygons the corners of which correspond to the number of carbon atoms in the ring. Thus cyclopropane is represented by a triangle and cyclobutane by a square. It is understood that each corner of a polygon represents a carbon atom and two hydrogen atoms unless another substituent is shown bonded to that position. Cycloalkyl groups are derived from cycloalkanes, just as alkyl groups are derived from alkanes. Thus cyclopropane gives a cyclopropyl group, and cyclobutane gives a cyclobutyl group.

The bond angles in cyclopropane and cyclobutane are quite different from the normal tetrahedral bond angle of 109.5°. Cyclopropane and cyclobutane are unstable compared with the larger cycloalkanes such as cyclopentane and cyclohexane, which have tetrahedral bond angles. The two small cycloalkanes are said to have **ring strain,** to be destabilized by the deformation of their bond angles and by the nonbonding repulsions between the electrons in covalent bonds on adjacent atoms. This results in weaker carbon–carbon bonds. For example, the normal bond dissociation energy (p. 63) for a carbon–carbon single bond is about 88 kcal/mol. That for a carbon–carbon bond in cyclopropane is 65 kcal/mol. The strained state of the small ring compounds is demonstrated by their reactivity, which is greater than is expected of alkanes. For example, compounds with double bonds, such as ethylene, react easily with hydrogen gas in the presence of metal catalysts (p. 64) to form alkanes.

$$CH_2{=}CH_2 \xrightarrow[\text{Ni} \quad 20\,°C]{H_2} CH_3CH_3$$

ethylene                ethane

Cyclopropane, even though it is not an alkene, also reacts with hydrogen. In the presence of a nickel catalyst, the ring opens, and propane is formed.

$$\triangle \xrightarrow[\text{Ni} \quad 120\,°C]{H_2} CH_3CH_2CH_3$$

cyclopropane                propane

A higher temperature is required for this reaction than for the one with an alkene, and a still higher temperature is necessary to cleave a cyclobutane ring.

$$\text{cyclobutane} \quad \xrightarrow[\substack{Ni \\ 200\ °C}]{H_2} \quad CH_3CH_2CH_2CH_3 \quad \text{butane}$$

The larger rings in cyclopentane and cyclohexane do not open when treated with hydrogen in the presence of a catalyst under these conditions.

$$\text{cyclopentane} \quad \text{or} \quad \text{cyclohexane} \quad \xrightarrow[\substack{Ni \\ 200\ °C}]{H_2} \quad \text{no reaction}$$

The reactivity of compounds containing a three-membered ring raises questions about the exact nature of the bonding in such compounds. The H—C—H bond angles in cyclopropane, for example, appear to fit better with $sp^2$-hybridized (p. 52) than with $sp^3$-hybridized (p. 47) carbon atoms. The reactivity also suggests some unsaturated character to the ring. The small C—C—C bond angles in the molecule make overlap between orbitals on adjacent carbon atoms difficult, and certainly weaken the carbon–carbon bonds.

## B. Cyclopentane

The bond angles in cyclopentane are very close to the tetrahedral angle and also close to the internal bond angle of 108° for a regular pentagon. Cyclopentane might be expected to be stable in a planar form except that the hydrogen atoms on adjacent carbon atoms are eclipsed in this form. Cyclopentane is most stable when one of the carbon atoms is out of the plane of the other four (see Figure 5.11 on the next page). This shape is called the **envelope form** of cyclopentane. The carbon atom that is out of the plane moves around the ring in a phenomenon known as **pseudorotation** so that the eclipsing of the hydrogen atoms at various points in the ring is relieved at least part of the time. Note, however, that the structure of the ring does not allow complete rotation around the carbon–carbon single bonds. The type of conformational isomerism seen for ethane or butane is not possible for a cyclic alkane. Substitution of another group for a hydrogen atom on cyclopentane gives rise to cyclopentyl compounds.

**Problem 5.27** Name the following compounds.

(a) [structure: cyclobutane with CH₃ and OH] (b) [structure: cyclopropane—CH₂CH₃] (c) [structure: cyclopentane with CH₂CH₃ and CH₃] (d) [structure: cyclopentane—I]

**Problem 5.28** Draw structures for the following compounds.

(a) *tert*-butylcyclopentane    (b) cyclopropanol    (c) 1-methyl-1-cyclopentanol
(d) isopropylcyclobutane    (e) phenylcyclopentane (or cyclopentylbenzene)

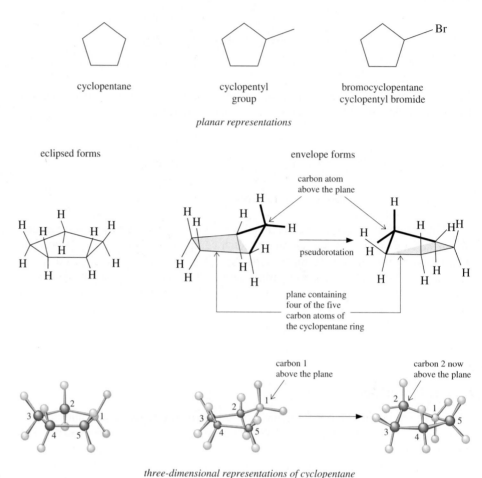

**Figure 5.11**
Planar and envelope
conformations of cyclopentane.

## C. Cyclohexane. Conformation in Cyclohexane and in Cyclohexanes with One Substituent

Cyclohexane, $C_6H_{12}$, was known to be an unstrained molecule in the late nineteenth century. This fact was puzzling because a planar regular hexagon with internal bond angles of 120°, which is larger than the tetrahedral angle, would be expected to exhibit some ring strain. In 1890, the German chemist Ulrich Sachse pointed out that cyclohexane need not be a planar molecule, that puckering of the ring (having one or more of the carbon atoms out of the plane) would allow for tetrahedral bond angles. He suggested that cyclohexane should exist in two forms, now called the **chair form** and the **boat form.** At that time, these forms could not be observed experimentally, so Sachse's ideas were not accepted until about 35 years later when experimental evidence was found for puckered six-membered rings in some more complex compounds. Then, in 1943, the Norwegian chemist Odd Hassel used electron diffraction to show that the chair form of cyclohexane was the predominant conformer in the gas phase. For this work Hassel shared the Nobel Prize in Chemistry in 1969 with Sir Derek Barton.

In the chair form of cyclohexane, four of the carbon atoms in the ring are in a plane, a fifth carbon atom is above the plane, and a sixth is below it. The plane is shown in blue in the structures on the facing page.

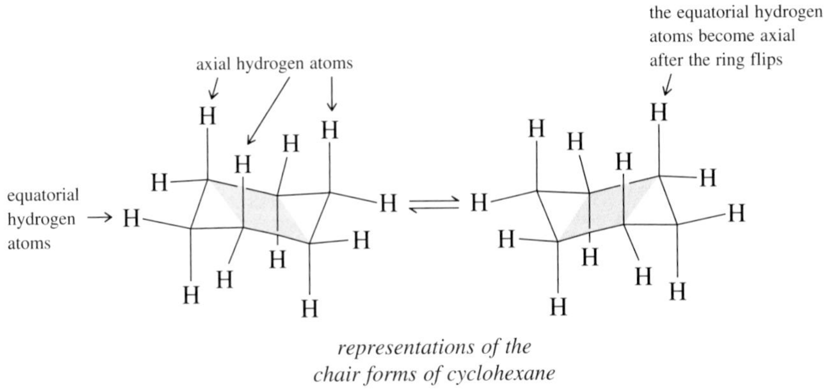

*representations of the
chair forms of cyclohexane*

Partial rotation around the carbon–carbon bonds in cyclohexane results in the conversion of one chair form to another. The energy barrier between the two forms is 10.8 kcal/mol, low enough that at room temperature the ring is constantly undergoing a process called **ring reversal** (or **ring inversion**) (approximately $10^5$ times a second) from one conformation to the other.

All the carbon–hydrogen bonds on adjacent carbon atoms in any chair form of cyclohexane are staggered, as shown by the Newman projection.

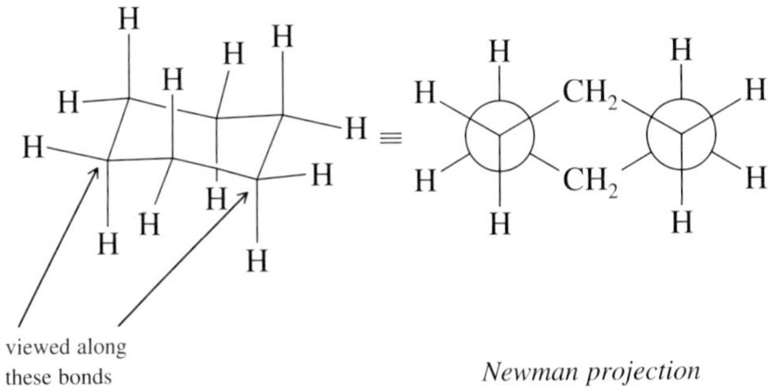

*Newman projection*

There are two types of hydrogen atoms in the chair form of cyclohexane. Six of the hydrogen atoms lie in a ring outside of the carbon skeleton and roughly in the plane of the molecule. These hydrogen atoms are called the **equatorial hydrogen atoms.** The bonds to the other six hydrogen atoms are parallel to an axis that goes through the center of the cyclohexane ring. Of these **axial hydrogen atoms,** three are above the plane of the carbon ring and three are below. The rapid reversal of the ring at room temperature converts the hydrogen atoms of one type into the other. You should use a molecular model to demonstrate to yourself how this works. The easiest way to convert a model of cyclohexane from one chair form to another is to move the carbon atom that is below the plane of the other four up, and the one that is above the plane down, as demonstrated in Figure 5.12.

*Figure 5.12*
The interconversion of axial and equatorial hydrogen atoms by ring reversal in a model of cyclohexane.

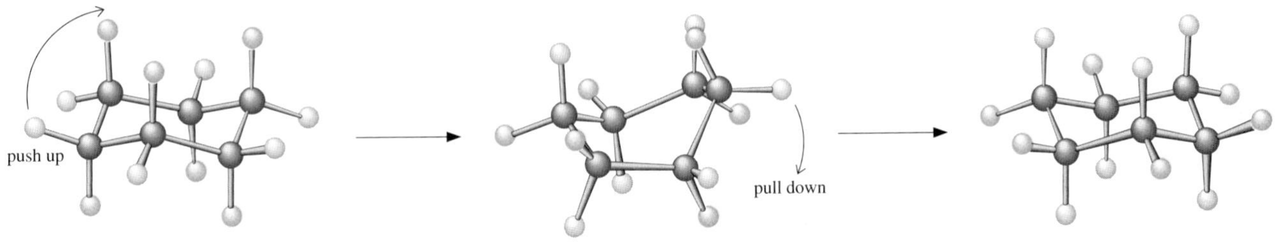

Besides the chair forms, there are other conformations of cyclohexane that have higher energies (are less stable), such as the boat form and the twist form.

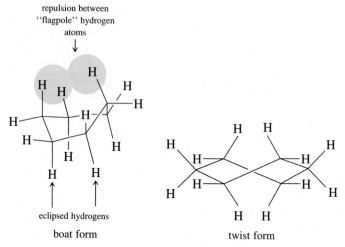

*high-energy conformations of cyclohexane*

In the **boat form,** two of the carbon atoms lie above the plane defined by the other four carbon atoms. Two of the hydrogen atoms on these two carbon atoms (called the "flagpole" positions) are brought close enough that they repel each other. In addition, the carbon–hydrogen bonds on the carbon atoms in the plane are all eclipsed. The **twist form** is a little more stable than the boat form because some of these interactions are lessened. The eclipsed bonds move so they are out of alignment with one another, but the staggered bonds move closer. At room temperature, most cyclohexane molecules exist in the most stable conformation, the chair form. The twist form occurs as an intermediate stage in the conversion of one chair form to another.

If a substituent has replaced one of the hydrogen atoms of cyclohexane, the chair conformations are no longer equivalent. For example, molecules of methylcyclohexane are mostly in the form in which the methyl group is in an equatorial position (Figure 5.13). This form is more stable by approximately 1.7 kcal/mol than the chair form with the axial methyl group, which is created by the ring reversal. When

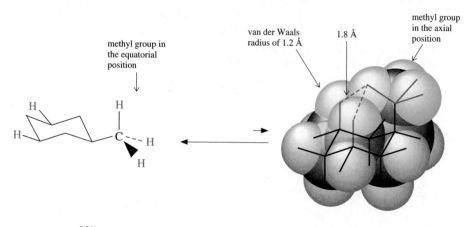

*Figure 5.13*
Chair conformations of
methylcyclohexane.

95%

5%

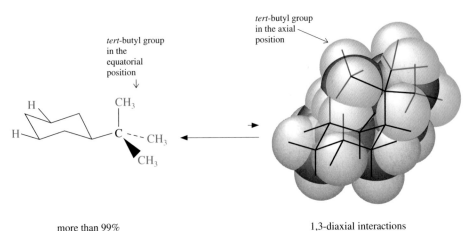

**Figure 5.14**
Chair conformations of *tert*-butylcyclohexane.

the methyl group is in an axial position, it is close to the two axial hydrogen atoms that are on the same side of the plane of the ring. The distance between a hydrogen atom of the methyl group and either one of these two axial hydrogen atoms is 1.8 Å, which is shorter than 2.4 Å, the sum of the van der Waals radii of two hydrogen atoms (p. 166). The steric interaction, the repulsion between the methyl group and each of these two hydrogen atoms, is called a **1,3-diaxial interaction.** The 1,3-diaxial interaction serves to make the conformer less stable than one in which the methyl group is equatorial.

A methyl group in the axial position on a cyclohexane ring is gauche (p. 166) to the largest substituent on each of the carbon atoms adjacent to the one to which it is bonded. These substituents happen to be other carbon atoms of the ring.

If a methyl group is equatorial on the ring, it is anti to the ring atoms that are the largest substituents on adjacent carbon atoms. This analysis also predicts a lower stability for the conformer of cyclohexane with a substituent in the axial position.

In general, the larger the substituent, the greater the repulsive interactions when it occupies an axial position. A *tert*-butyl group, for example, is so bulky that the conformation in which it is in the equatorial position is overwhelmingly favored (Figure 5.14). The difference in energy between the two chair forms of *tert*-butylcyclohexane is 5.6 kcal/mol.

*Study Guide*
Concept Map 5.4

**Problem 5.29**   Draw structural formulas for chlorocyclohexane and cyclohexanol in the planar form (using a regular hexagon to represent the ring, p. 169) and in the two chair conformations.

## 5.12   The Chemical Properties of Alkanes

Alkanes have structures in which all the valence electrons of carbon are involved in $\sigma$ bonds with carbon and hydrogen atoms. Saturated hydrocarbons are inert toward most reagents. Their reactions generally take place at high temperatures or in the presence of catalysts that promote the cleavage of single bonds.

Alkanes are important constituents of fuels, and their combustion in air provides much of the energy consumed in the modern world. Methane is the chief constituent of natural gas. Once ignited, it reacts with oxygen to evolve heat.

$$CH_4 + 2\,O_2 \longrightarrow CO_2 + 2\,H_2O \qquad \Delta H = -210.8\ \text{kcal/mol}$$
methane

The equation represents the complete combustion of methane and assumes that sufficient oxygen is available to convert the methane to carbon dioxide and water. Often, however, incomplete combustion takes place, forming some carbon monoxide, which is highly toxic, and carbon, which gives the flame of a Bunsen burner its yellow color when the air supply is limited.

Petroleum is a mixture of a large variety of hydrocarbons, including alkanes and aromatic hydrocarbons (p. 67). The hydrocarbons are separated into fractions according to their boiling points. Propane and butane, for example, are gases that are liquefied under pressure and used as fuel that can be transported in small tanks such as in cigarette lighters. The hydrocarbons with boiling points up to 25 °C (see Table 5.1) make up the fraction of petroleum known as gas and liquefied gas.

The next and largest fraction obtained from petroleum is gasoline, consisting of hydrocarbons with four to twelve carbon atoms and a boiling-point range of 20–200 °C. Hydrocarbons at the upper end of this range are converted into more useful mixtures of lower-molecular-weight alkanes by heating in the presence of hydrogen gas and catalysts in a process known as **hydrocracking.** For example, undecane, $C_{11}H_{24}$, has a boiling point of 196 °C. Undecane is cracked to give a mixture of lower-weight alkanes in which butanes, pentanes, hexanes, and heptanes are the major constituents.

$$C_{11}H_{24} \ \xrightarrow[\substack{\text{catalyst} \\ 275\ °C}]{H_2} \ C_4H_{10} \ + \ C_5H_{12} \ + \ C_6H_{14} \ + \ C_7H_{16}$$

undecane     butanes    pentanes    hexanes    heptanes
bp 196 °C    bp~0 °C    bp~30 °C    bp~68 °C    bp~98 °C

The boiling-point range of these compounds is approximately 0–100 °C; thus they are more easily volatilized than undecane and can be used more efficiently in internal combustion engines.

Kerosene, consisting of hydrocarbons containing nine to sixteen carbon atoms and having a boiling range of 175–275 °C, is a higher-boiling fraction of petroleum. Gas oil and diesel oil come next, boiling at 200–400 °C and consisting of hydrocarbons with fifteen to twenty-five carbon atoms. Lubricating oil contains alkanes of even higher molecular weight, ones with twenty to seventy carbon atoms in the chain.

Compared with other functional group classes, alkanes have limited chemical reactivity. An important type of reaction of alkanes is their reaction with halogens at high temperatures or under the influence of light (light is symbolized in equations by $h\nu$ under the reaction arrow). Such reactions are complex and give mixtures of products depending on the ratios of reactants that are used. For example, methane

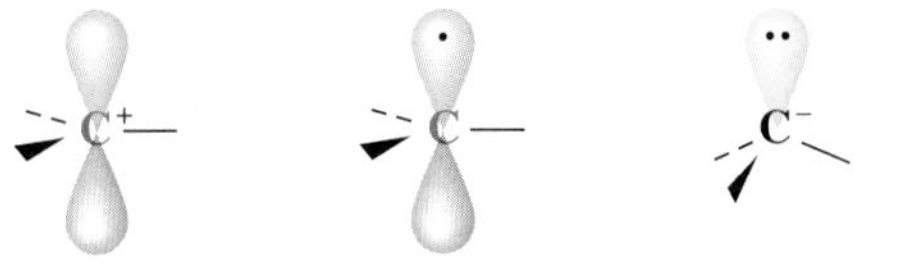

*Figure 5.15*
A carbocation, a free radical,
and a carbanion.

reacts with an equal volume of chlorine in the presence of light to give a mixture of alkyl halides.

$$CH_4 \ + \ Cl_2 \ \xrightarrow{h\nu} \ CH_3Cl \ + \ CH_2Cl_2 \ + \ CHCl_3 \ + \ CCl_4 \ + \ Cl_3CCCl_3$$

| methane | chlorine | chloromethane | dichloromethane | trichloro-methane | tetrachloro-methane | hexachloroethane |

methylene chloride     chloroform     carbon tetrachloride

*1 volume*  *1 volume*      *80%*      *10%*      *small amounts*

If six times as much methane as chlorine is used, the product is almost completely chloromethane.

When a large excess of cyclopentane is heated with chlorine at 250 °C, chloro-cyclopentane is formed, along with small amounts of dichlorocyclopentanes.

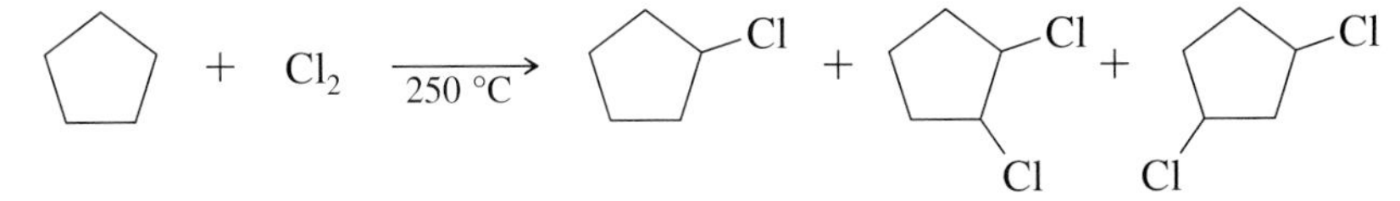

cyclopentane    chlorine      chlorocyclopentane    1,2-dichloro-cyclopentane    1,3-dichloro-cyclopentane

                         *95%*          *4%*          *1%*

Halogenation reactions usually give mixtures of products. They are used to prepare alkyl halides in industry but not often in the laboratory.

The reactive intermediates in these halogenation reactions are free radicals. The detailed mechanism for these reaction may be found on pages 769–774. A **free radical** has an unpaired electron occupying a *p* orbital on an $sp^2$-hybridized carbon atom and no charge. It is shown in Figure 5.15 with two other reactive intermediates we have learned about, a carbocation and a carbanion. The chemistry of free radicals will be the subject of Chapter 19.

---

**Problem 5.30**    The carbon atom in a free radical has only seven electrons around it, which is not a complete octet, and is, therefore, electron-deficient. Using your knowledge of the relative stabilities of primary, secondary, and tertiary carbocations (p. 129), predict the relative stabilities of these radicals.

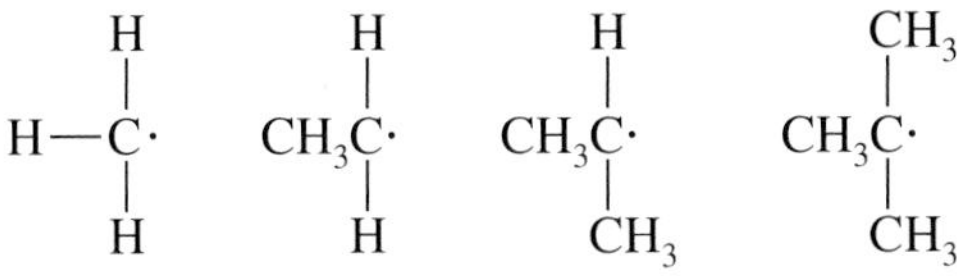

---

Halogenated hydrocarbons are important in our everyday lives. Besides dichloromethane, chloroform, and carbon tetrachloride, several halogenated ethenes and ethanes, shown on page 186, are important solvents for dry cleaning, for the cleaning of metal parts, for coatings, and for reactions. Vinyl chloride is the starting material for making the important polymer PVC, poly(vinyl chloride) (p. 782).

Cl—C=C—Cl
|          |
Cl        Cl
1,1,2,2-tetrachloroethene
perchloroethylene

Cl—C=C—H
|          |
Cl        Cl
1,1,2-trichloroethene
trichloroethylene

Cl—C=C—H
|          |
H         H
chloroethene
vinyl chloride

Cl  Cl
|    |
Cl—C—C—H
|    |
H    H
1,1,2-trichloroethane

Cl  H
|    |
Cl—C—C—H
|    |
Cl   H
1,1,1-trichloroethane

Cl  Cl
|    |
Cl—C—C—Cl
|    |
H    H
1,1,2,2-tetrachloroethane

Most halogenated hydrocarbons have some degree of toxicity and must be handled with care.

1,1,2-Trichloro-1,2,2-trifluoroethane is a member of a class of halogenated hydrocarbons known collectively as **chlorofluorocarbons,** or **CFCs.** Other widely used members of this group are trichlorofluoromethane and dichlorodifluoromethane.

Cl  Cl
|    |
Cl—C—C—F
|    |
F    F
1,1,2-trichloro-1,2,2-trifluoroethane
Freon 113

Cl
|
Cl—C—F
|
Cl
trichlorofluoromethane
Freon 11

Cl
|
Cl—C—F
|
F
dichlorodifluoromethane
Freon 12

Chlorofluorocarbons were developed because of their stability and relative lack of toxicity (in contrast to compounds such as carbon tetrachloride, which are quite toxic). These properties make them suitable for use in home appliances. The low boiling points of the chlorofluoromethanes allow them to vaporize easily and yet to condense into the liquid state upon compression. This property is valuable in the cooling coils of refrigerators and air conditioners, where these compounds are mostly used. Other compounds containing bromine, called **halons,** are highly effective as fire extinguishers.

F
|
F—C—Br
|
F
bromotrifluoromethane
Halon 1301

F
|
F—C—Br
|
Cl
bromochlorodifluoromethane
Halon 1211

Cl  F
|    |
Cl—C——C—F
|    |
H    F
2,2-dichloro-1,1,1-trifluoroethane
HCFC-123

For the last 25 years, there has been increasing concern that chlorofluorocarbons and halons, because of their stability, survive in the atmosphere long enough to reach the ozone layer and there decompose to give chlorine and bromine atoms that catalyze the destruction of ozone. The story of how this happens is told in Section 19.1E. Compounds retaining some of the hydrogen atoms on carbon also have been developed as substitutes for chlorofluorocarbons to try to get around this problem. One such compound, known as a **hydrochlorofluorocarbon,** is 2,2-dichloro-1,1,1-trifluoroethane, given the short designation of HCFC-123. The presence of hydrogen in the molecule causes it to decompose lower down in the atmosphere so that it does not survive to reach the ozone layer. Recently, however, discovery of liver damage in workers exposed to relatively high levels of the compound has raised concerns about the possible toxicity of the compound.

The alkyl portions of molecules that contain other functional groups usually remain unchanged in reactions that transform the functional groups. You should keep in mind this unreactivity of alkyl groups as you study the following chapters.

## SUMMARY

Alkanes and cycloalkanes are organic compounds that have only carbon–hydrogen and carbon–carbon single bonds. The different ways in which the carbon atoms can be bonded together in alkanes give rise to constitutional isomers, which are straight-chain or branched-chain compounds. The carbon atoms in alkanes are classified according to the number of other carbon atoms they are bonded to: primary (bonded to one other carbon atom), secondary (bonded to two other carbon atoms), tertiary (bonded to three other carbon atoms), and quaternary (bonded to four other carbon atoms). Hydrogen atoms are also classified as primary, secondary, and tertiary according to the type of carbon atom to which they are bonded.

The carbon chains in alkanes assume different conformations through rotations around the carbon–carbon single bonds. Staggered conformations are more stable than eclipsed conformations because, in the latter, bonds on adjacent carbon atoms are aligned with each other and, therefore, repel each other. Anti conformations are staggered conformations in which the largest groups are as far apart as they can be. Anti conformations are more stable than other staggered conformations, such as the gauche conformation in which larger groups are close enough to repel each other.

The most stable conformation of cyclohexane is the chair form, which has six equatorial hydrogen atoms and six axial ones. A substituted cyclohexane is most stable when the substituent is in an equatorial position. A cyclohexane with a substituent in the axial position is destabilized by 1,3-diaxial interactions between the substituent and the two hydrogens also in axial positions on the same face of the ring.

Small ring compounds such as cyclopropane and cyclobutane are unstable in comparison to cyclopentane and cyclohexane because of ring strain. Such strain is due to bond angles that are smaller than the tetrahedral angle and to the eclipsing of bonds on adjacent atoms.

Alkanes are named according to the IUPAC rules of nomenclature. The names of alkanes are used as the basis for the names of other classes of organic compounds.

Alkanes have a low reactivity compared to that of compounds containing functional groups. Combustion and halogenation, both of which require the input of considerable energy, are the most important reactions of alkanes.

## ADDITIONAL PROBLEMS

**5.31** Name the following compounds.

(a)

(b) $CH_3\underset{\underset{CH_3}{|}}{\overset{\overset{CH_3}{|}}{C}}CH_2CH_2\underset{\underset{CH_3}{|}}{\overset{\overset{CH_3}{|}}{C}}CH_3$

(c)

(d) $CH_3\underset{\underset{Cl}{|}}{\overset{\overset{CH_3}{|}}{C}}CH_2CH_2CH_3$

(e) $CH_3\overset{\overset{CH_3}{|}}{C}H$

(f) 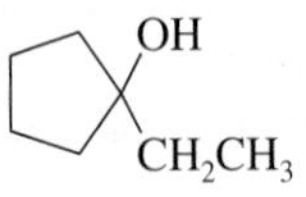 [structure: cyclopentane with OH and CH$_2$CH$_3$ substituents]

(g) 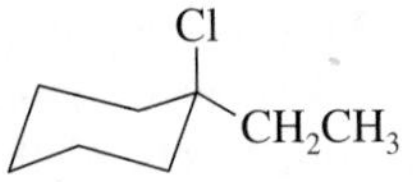 [structure: cyclohexane with Cl and CH$_2$CH$_3$ substituents]

(h) 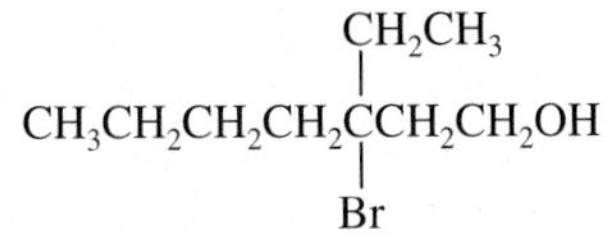 $CH_3CH_2CH_2CH_2CCH_2CH_2OH$ with $CH_2CH_3$ above and $Br$ below the central carbon

**5.32** 2-Bromo-2-chloro-1,1,1-trifluoroethane (bp 50.2 °C) is used as an inhalation anesthetic. What is the structure of this compound?

**5.33** Draw a structural formula for each of the following compounds.

(a) 2-methyl-2-hexanol          (b) 1-methyl-1-chlorocyclobutane          (c) iodocyclopropane          (d) 4-*tert*-butylheptane
(e) 5-cyclopropylnonane          (f) isobutylcyclopentane          (g) 3-phenylheptane          (h) 1-bromo-1-methylcyclohexane
(i) 7-chloro-2,7-dimethyl-2-octanol

**5.34** Three of the isomers having the molecular formula $C_3H_8O_2$ are liquids with the following boiling points: Compound A has a boiling point of 45 °C, Compound B has a bp of 124 °C, and Compound C has a bp of 215 °C. All have good solubility in water. Compounds that have an —OH group on a carbon atom that also has another —OH or —OR group on it are unstable. Such isomers should be excluded from among the structures you consider for Compounds A, B, and C. Propose Lewis structures for Compounds A, B, and C that explain the trend in their boiling points. In working out your answers, remember that ethanol, $CH_3CH_2OH$, boils at 78 °C.

**5.35** For each molecular formula, identify the number of units of unsaturation and draw structural formulas for three different constitutional isomers. Use examples that contain different structural features such as different functional groups, multiple bonds, and rings.

(a) $C_5H_9Br$          (b) $C_5H_{10}O$          (c) $C_5H_8O$
(d) An isomer of $C_5H_9Br$ does not react easily with bromine in carbon tetrachloride. What is a possible structure for it?
(e) An isomer of $C_5H_{10}O$ is treated with hydrogen in the presence of a catalyst and gives 3-methyl-1-butanol. What structures are possible for the original compound?

**5.36** The chair conformation of fluorocyclohexane in which the fluorine atom occupies the equatorial position is 0.2 kcal/mol lower in energy than the conformation in which the fluorine occupies the axial position.

(a) Draw the two chair conformations of fluorocyclohexane and indicate by the length of arrows between the two, which conformer predominates at equilibrium.
(b) Is the equilibrium constant for fluorocyclohexane (equatorial) $\rightleftarrows$ fluorocyclohexane (axial) smaller than 1, exactly 1, or greater than 1?
(c) Consider the structure of bromocyclohexane and the difference in energy between the two chair conformations of this compound. Predict whether the difference in energy for these two conformers will be smaller than, larger than, or the same as the difference in energy for the two conformers of fluorocyclohexane. Explain the reasons for your prediction.

**5.37** Predict the products of each of the following reactions. Analysis of each problem to identify acids and bases or electrophiles and nucleophiles will be helpful in most cases.

(a) $CH_3CH_2NH_2 + HCl \longrightarrow$          (b) $CH_3CH_2C{\equiv}CH + NaNH_2 \longrightarrow$          (c) $(CH_3CH_2CH_2)_2NH + CH_3CH_2Br \longrightarrow$

(d) $CH_3CH_2CH{=}CHCH_2CH_3 + HBr \longrightarrow$          (e) $CH_3CH_2CH_2SH + NaOH \longrightarrow$          (f) 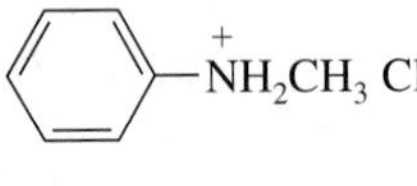 [phenyl]—$\overset{+}{N}H_2CH_3$ $Cl^- + NaOH \longrightarrow$

(g) $CH_3CH_2CH{=}CHCH_3 + Br_2 \longrightarrow$     (h) 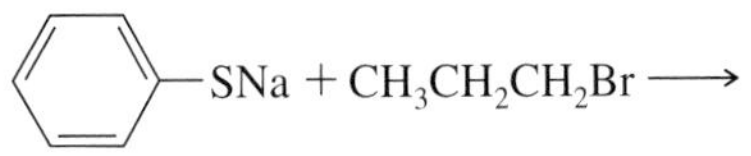 ⬡$-SNa + CH_3CH_2CH_2Br \longrightarrow$

(i) 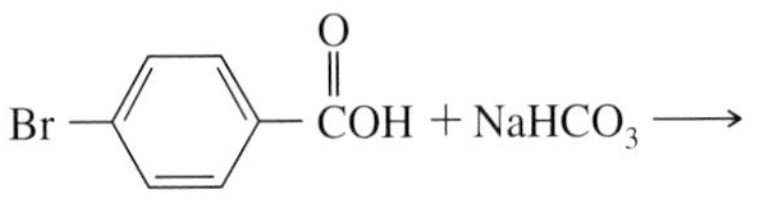  $Br-$⬡$-\overset{\overset{\textstyle O}{\|}}{C}OH + NaHCO_3 \longrightarrow$     (j) $CH_3CH_3 + Cl_2 \xrightarrow{h\nu}$
                                                    1 volume   1 volume

**5.38** Which reagents could be used to make each of the following transformations? More than one step may be necessary in some cases. A good way to start each problem is to analyze the change in connectivity that occurs in going from the starting material or reagent to the product. Looking for possible nucleophiles and electrophiles is also a good idea.

(a) 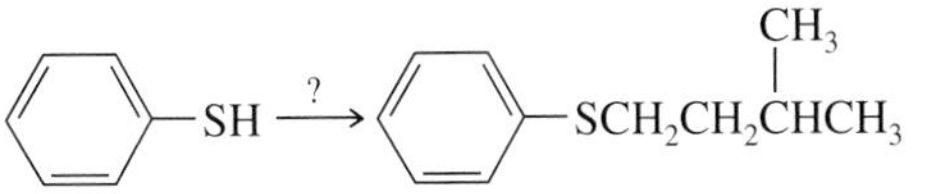 ⬡$-SH \xrightarrow{?}$ ⬡$-SCH_2CH_2\overset{\overset{\textstyle CH_3}{|}}{C}HCH_3$     (b) $? \xrightarrow{CH_3NH_2} CH_3CH_2\overset{+}{N}H_2CH_3Br^-$

(c) $CH_2{=}CH_2 \xrightarrow{?} CH_3CH_2OCH_3$     (d) $CH_3CH_2CH_2CH_2NH_2 \xrightarrow{?} CH_3CH_2CH_2CH_2\overset{+}{N}H_3Cl^-$

(e) $? \xrightarrow[\text{(excess)}]{Br_2} CH_3CH_2\overset{\overset{\textstyle Br}{|}}{\underset{\underset{\textstyle Br}{|}}{C}}{-}\overset{\overset{\textstyle Br}{|}}{\underset{\underset{\textstyle Br}{|}}{C}}H$     (f) $CH_3CH_2CH_2CH_2OH \xrightarrow{?} CH_3CH_2CH_2CH_2OCH_3$

(g) $CH_3\overset{\overset{\textstyle }{}}{\underset{\underset{\textstyle OH}{|}}{C}}HCH_2CH{=}CH_2 \xrightarrow{?} CH_3\underset{\underset{\textstyle OH}{|}}{C}HCH_2CH_2CH_3$     (h) 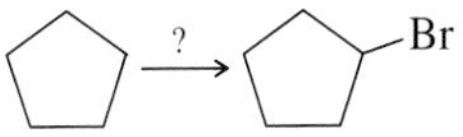 ⬠ $\xrightarrow{?}$ ⬠$-Br$

**5.39** Studies of 1,2-ethanediol ($HOCH_2CH_2OH$) show that the conformer in which the hydroxyl groups are gauche to each other is 2.3 kcal/mol more stable than the conformer in which the hydroxyl groups are anti to each other. For 1,2-dimethoxyethane ($CH_3OCH_2CH_2OCH_3$), the anti conformer is favored.

(a) Draw perspective formulas and Newman projections for the gauche and anti conformers of 1,2-ethanediol and of 1,2-dimethoxyethane.
(b) Why is the gauche form of 1,2-ethanediol more stable than the anti form? Draw a picture of the factor that is responsible for the stabilization of the gauche form.
(c) The three staggered conformations of a related molecule, 1,2-propanediol, are given below.

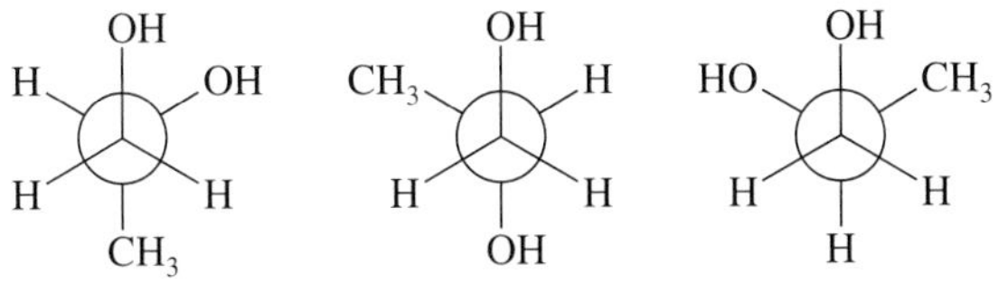

      **A**           **B**          **C**

Given the data on 1,2-ethanediol, rank the three conformers of 1,2-propanediol according to their relative stabilities, going from most to least stable.
(d) Draw an energy diagram that represents the bond rotation necessary to go from Conformer A to Conformer B. Draw the Newman projection for the species that exists at the energy maximum in going from A to B.

**5.40** Different conformations of cyclic as well as acyclic compounds have been studied by computer modeling as well as by experimental methods.

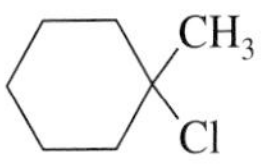

(a) Name the compound shown at right.
(b) Draw the two chair conformations of the compound.

(c)   Computer modeling has shown that the chair form in which the methyl group is equatorial and the chloro group is axial is 1.42 kcal/mol *more* stable than the other chair form. Why is this so? Briefly explain using words and a picture.

**5.41** The equatorial conformer of ethynylcyclohexane is more stable than its axial form by −0.41 kcal/mol. The difference in energy for ethylcyclohexane favors the equatorial form by −1.75 kcal/mol.

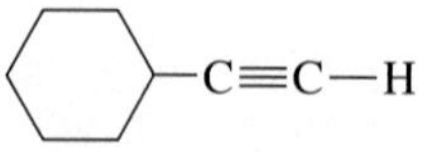

ethynylcyclohexane 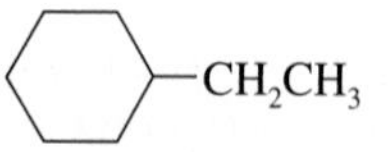          ethylcyclohexane

Show the equilibrium between the equatorial and the axial forms for ethynylcyclohexane and for ethylcyclohexane by drawing the chair forms and using arrows of different lengths to indicate the position of equilibrium for each compound. Why is there a smaller difference in energy between the axial and the equatorial forms of ethynylcyclohexane than between the axial and equatorial forms of ethylcyclohexane?

**5.42** The carbon and proton nuclear magnetic resonance spectra of 1-bromo-2-methylpropane were given in Figure 5.7 (p. 162). Make rough sketches of the carbon and proton nuclear magnetic resonance spectra that you would expect for the three other isomers with molecular formula $C_4H_9Br$.

**5.43** (a) Draw the structure of 2-methyl-1-butanol and circle the carbon atom in the molecule that would give the peak farthest to the left in the carbon nuclear magnetic resonance spectrum.
(b) How many peaks are expected for the carbon nuclear magnetic resonance spectrum of 2-methyl-1-butanol?
(c) How many peaks are expected for the proton nuclear magnetic resonance spectrum of 2-methyl-1 butanol?
(d) Draw Newman projections for the staggered conformation looking down the carbon-1−carbon-2 bond in 2-methyl-1-butanol and identify the least stable and the most stable conformations. Draw the perspective formula for the most stable conformer.
(e) Draw the Newman projection for the highest energy eclipsed conformation and the perspective formula corresponding to this conformation.

# Stereochemistry

## A Look Ahead

Stereochemistry is chemistry in three-dimensional space. It is of immense importance, especially in the study of complex molecules that are biologically important, such as proteins, carbohydrates, and nucleic acids (Chapter 23). Much of the rest of this book will be concerned with the stereochemistry of chemical reactions, that is, how the reactions actually proceed in three dimensions. First, we must look more closely at the structures of organic molecules.

Chapter 5 explored the question of the equivalency of hydrogen atoms. Replacing different types of hydrogen atoms on a molecule with halogen atoms gives rise to constitutional isomers. Carrying this exercise a little further uncovers a more subtle kind of isomerism called **stereoisomerism.** Stereoisomers have the same connectivity and differ from each other only in the way their atoms are oriented in space. They cannot be distinguished by looking at their condensed structural formulas, which are identical.

Chapter 5 introduced one aspect of stereochemistry, the dynamic interconversions molecules undergo as the atoms within them move relative to one another by rotation around single bonds. Such motions give rise to one type of stereoisomer, called conformational isomers, or conformers (p. 166). This chapter explores another type of stereoisomerism called **configurational isomerism. Configurational isomers** differ from each other only in the arrangement of their atoms in space and cannot be converted from one into another by rotations about single bonds within the molecules. Configurational isomers may be enantiomers or diastereomers. **Enantiomers** are stereoisomers that are mirror images of one another. All stereoisomers that are not enantiomers are called **diastereomers.** Molecular models will help you to see this more subtle kind of isomerism, and you should use them as you read this chapter.

**Workbook Exercises**

⊘ **Study Guide**
**Concept Map 6.1**

## 6.1    Enantiomers

The substitution of a chlorine atom for one of the secondary hydrogen atoms of butane has already been shown (on p. 159) to give rise to 2-chlorobutane. Butane exists as a single isomer, and at first glance the two hydrogen atoms on the secondary carbon atom may appear to be equivalent. However, if first one and then the other of these two hydrogen atoms are replaced by a chlorine atom (Figure 6.1, p. 192), the two molecular species that are formed are not identical. It should be clear to you that Structure A cannot be made to coincide perfectly with Structure B. If you were to try to pick A up out of the page and place it over B, the methyl and ethyl groups and the central carbon atom would coincide, but the positions of the chlorine and hydrogen atoms would be reversed. Molecular models will help you see this. Thus Structure A and Structure B represent two molecular species that differ from each

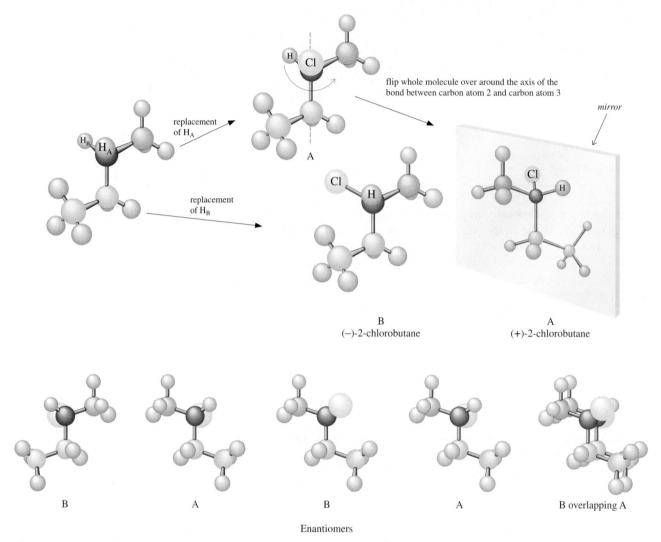

**Figure 6.1**
Creation of nonsuperimposable mirror-image isomers by the replacement of one or the other of two secondary hydrogen atoms in butane by chlorine. Structures A and B are enantiomers.

other only in the orientation of the atoms in space. They are **stereoisomers** of 2-chlorobutane. If Structure A is first flipped over by rotating its parts around the axis of the bond between carbon atom 2 and carbon atom 3 and then placed next to Structure B, the relationship between the two isomers becomes more clear. Structures A and B are related as an object is related to its mirror image. One looks like the reflection of the other. Thus the two molecular species represented by these structural formulas are mirror-image isomers of each other.

A mirror image may be drawn of any molecular structure. Isomerism exists only when mirror images are not superimposable on each other. Structure A in Figure 6.1 cannot be picked up and placed over Structure B so that all points coincide, no matter how the two structures are rotated in space. Stereoisomers that are nonsuperimposable mirror images of each other are called **enantiomers.** Thus the stereoisomers represented by structural formulas A and B in Figure 6.1 are enantiomers of 2-chlorobutane.

CH$_3$—C◂H     H►C—CH$_3$
         Cl   Cl
CH$_3$CH$_2$     CH$_2$CH$_3$
      A              B
(+)-2-chlorobutane     (−)-2-chlorobutane

*enantiomers*

Enantiomers have identical physical properties with one exception—they rotate plane-polarized light in opposite directions. Compounds that rotate the plane of polarized light are said to have **optical activity,** or to be **optically active** (p. 199). The two enantiomers of 2-chlorobutane rotate the plane of polarized light in opposite directions, as indicated by the plus and minus signs in their names. (−)-2-Chlorobutane rotates the plane of polarized light counterclockwise, and (+)-2-chlorobutane rotates it an equal amount clockwise. All other physical properties of enantiomers are identical. Therefore, they cannot be separated from each other by physical means such as distillation or crystallization. Enantiomers are one type of configurational isomer (p. 191).

## 6.2  Chirality

A molecule or any other object that cannot be superimposed on its mirror image is said to be **chiral.** Thus 2-chlorobutane (p. 192) is chiral. The concept of chirality is important in the study of the chemistry of biological systems. Many of the compounds that occur in living organisms, such as carbohydrates and proteins, are chiral. Living organisms, including human beings, are chiral. Our internal organs are arranged asymmetrically, and we have distinctive twists and whorls in the way our hair grows. Our left and right hands are nonsuperimposable mirror images of each other (Figure 6.2). In fact, the word *chiral* comes from the Greek word *cheir,* which means "hand."

   Many naturally occurring compounds exist as one of two possible enantiomers. Biological systems can distinguish between one enantiomer and its mirror image. For example, all amino acids that are present in proteins exhibit a certain spatial

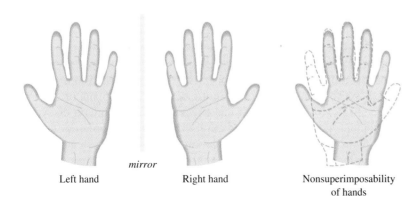

*mirror*

Left hand          Right hand          Nonsuperimposability
                                          of hands

*Figure 6.2*
**Hands as models of chiral objects that have nonsuperimposable mirror images.**

orientation of the groups around a central carbon atom. The natural enantiomer of the amino acid alanine and its mirror image are shown below.

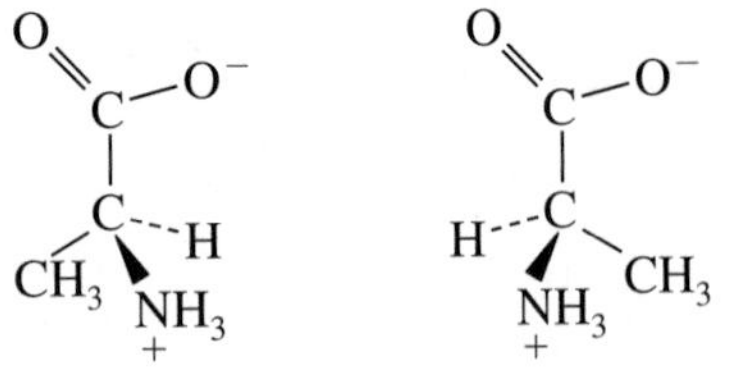

(+)-alanine      (−)-alanine

*the natural enantiomer*    *the mirror image*

The chirality of the human body, which is reflected at the most fundamental level in the stereochemistry of enzyme systems, requires that chemical reactions take place with a particular orientation in space. You will have some idea of what is involved if you try to put your left glove on your right hand or your left foot into your right shoe. Thus one enantiomer of a compound may be a hormone or may be active as a medication, and its mirror image may be biologically inactive. While enantiomers have the same physical and chemical properties when they are in achiral environments, they have different properties in chiral environments, such as biological systems. The sensitivity of various biological systems to stereochemistry is one of the reasons chemists are concerned both with the stereoisomerism of organic molecules and with the way their stereochemistry is determined and changed by chemical reactions.

An easy way to test for chirality is to look for a **plane of symmetry.** If an object or a molecule can be divided by a plane into two equal halves that are mirror images of each other, the plane is a plane of symmetry, and the object or molecule is not chiral. For example, butane in the anti conformation has a plane of symmetry that goes through all four carbon atoms and two primary hydrogen atoms. The plane bisects the angles between the pairs of secondary hydrogen atoms. Two planes of symmetry can be found in the eclipsed conformation of butane (Figure 6.3). 2-Chlorobutane does not have a plane of symmetry. No conformation of 2-chlorobutane can be divided by a plane into two halves that are mirror images of each other (see Figure 6.3).

As a consequence of its symmetry, a molecule of butane is superimposable on its mirror image. In Figure 6.4 (p. 196), representation C is the mirror image of representation D, but these structural formulas do not represent different molecular species. If D is flipped over by a rotation around the axis of the bond between carbon atom 2 and carbon atom 3, it becomes identical with C; it could be lifted out of the page and fitted exactly over C. In other words, D is a superimposable mirror image of C. An object that is superimposable on its mirror image has a plane of symmetry and is **achiral.**

Some achiral molecules and common objects are shown in Figure 6.5 (p. 196). Bromochloromethane has a plane of symmetry that bisects the bromine atom, the chlorine atom, the carbon atom, and the angle between the two hydrogen atoms. (Remember that individual atoms are spherically symmetrical even though the symbols chemists use to represent them are not.) Acetone has a plane of symmetry that bisects the central carbon atom and the oxygen atom. The plane of symmetry of methylcyclobutane goes through the methyl group, the carbon atom to which it is bound, the hydrogen atom on that carbon atom, and the carbon atom and two hydrogen atoms on the opposite side of the ring. A pencil has many planes of symmetry; one is shown. The plane of symmetry for the mug bisects the handle of the mug. An object or molecule that has a plane of symmetry is achiral.

## ONE SMALL STEP

In thinking about symmetry, it is important to think about the relationship between conformation and configuration.

**PROBLEM:** The two gauche forms of butane (p. 167) are enantiomers of each other. Why don't we have two optically active forms of the compound?

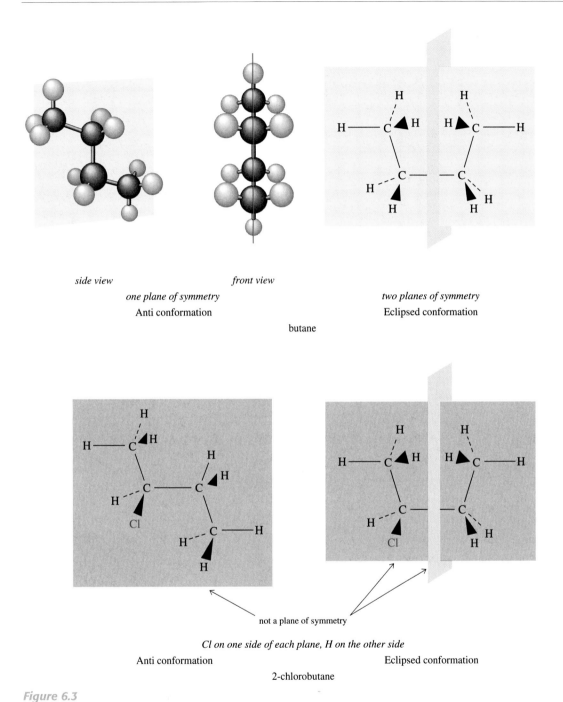

*side view*        *front view*

*one plane of symmetry*

Anti conformation

*two planes of symmetry*

Eclipsed conformation

butane

*not a plane of symmetry*

*Cl on one side of each plane, H on the other side*

Anti conformation                Eclipsed conformation

2-chlorobutane

*Figure 6.3*
**Planes of symmetry in different conformations of butane; lack of symmetry in 2-chlorobutane.**

There are, however, complex molecules that do not have a plane of symmetry and yet are achiral because they possess other kinds of symmetry. For the molecules we are concerned with, the test as to whether or not a plane of symmetry is present is sufficient to distinguish between chiral and achiral systems.

⊘ **Study Guide**
**Concept Maps 6.2 and 6.3**

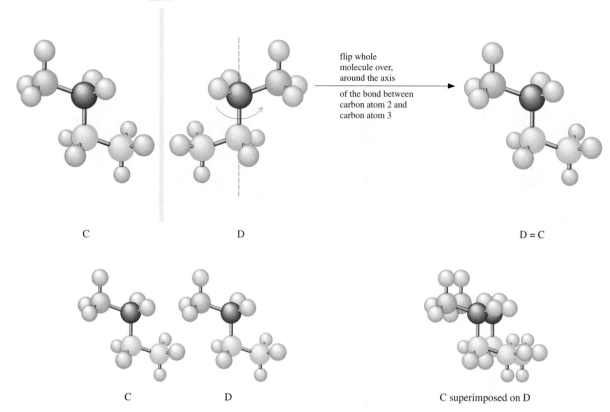

*mirror*

flip whole
molecule over,
around the axis

of the bond between
carbon atom 2 and
carbon atom 3

C

D

D = C

C

D

C superimposed on D

*Figure 6.4*
Two representations of butane;
each is a superimposable
mirror image of the other.

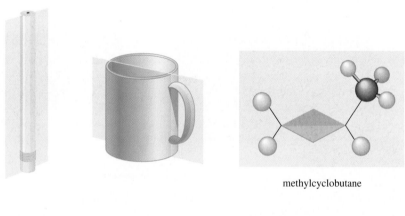

methylcyclobutane

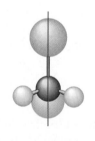

bromochloromethane
*side view*

bromochloromethane
*front view*

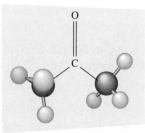

acetone

*Figure 6.5*
Some achiral molecules and
common objects, with planes
of symmetry.

**Problem 6.1**

(a) Suppose the mug in Figure 6.5 had the word *MOTHER* on one side. Would it be chiral or achiral? What if it had the word *MOM?*
(b) Would a spool with thread on it be chiral or achiral?
(c) What about an empty spool?

**Problem 6.2**    Of the following common objects, which ones are chiral and which achiral?

(a) a screw    (b) a nail    (c) a hammer    (d) a spade
(e) socks    (f) mittens    (g) a stocking cap (with no decoration on it)    (h) a shirt    (i) a tee shirt with *GO BLUE* written on it

# 6.3    Stereocenters

2-Chlorobutane (p. 193) and alanine (p. 194) are examples of chiral molecules, and they have one thing in common: One carbon atom in each of them has four different substituents. A tetrahedral atom that is bonded to four different substituents constitutes a **stereogenic center,** one that gives rise to stereoisomerism. The term *stereogenic center* has been shortened to the word **stereocenter,** which is now widely used by chemists and will be used in this book. Such a center is also sometimes called a *chiral center* or an *asymmetric carbon atom*, but chemists no longer use this last term.

In Section 6.8 we will see that molecules containing more than one stereocenter may or may not exhibit chirality depending on whether they have a plane of symmetry and are superimposable on their mirror images. In this and the next few sections we are only concerned with molecules containing a single stereocenter. Such molecules do not have a plane of symmetry and are not superimposable on their mirror images. They are, therefore, chiral. Another way of looking at such molecules is to say that four different groups may be arranged around a carbon atom in two (and only two) different ways. Such arrangements may be considered to be left-handed or right-handed. You should work with models to convince yourself that any four groups around a carbon atom can only have two different arrangements and that these two different forms are nonsuperimposable mirror images.

The groups around the stereocenter do not have to be very different in order to qualify as different. For example, the stereocenter in 2-butanol has a hydrogen atom, a hydroxyl group, a methyl group, and an ethyl group as the four different substituents. Even though the methyl and ethyl groups are both alkyl groups, they are different.

enantiomers of 2-butanol

Isotopes of the same element are different, as are large alkyl groups that differ from each other in subtle ways at some distance from the stereocenter. To discover whether a carbon atom is a stereocenter involves searching carefully in all directions, moving away from the potential stereocenter until some difference is found or until it is established that at least two of the groups are the same.

point at which difference
between two alkyl groups is found

$$CH_3CHCH_2CH_2 \overset{\displaystyle CH_3}{\underset{\displaystyle Cl}{\overset{\displaystyle |}{\,}}} \overset{\displaystyle CH_2CH_2CH_3}{\underset{\displaystyle }{\overset{\displaystyle |}{C}}} \cdots H$$

stereocenter

*one enantiomer of 5-chloro-2-methyloctane*

**Problem 6.3**    Determine whether each of the following compounds contains a stereocenter. If so, draw the two enantiomers of the compound. If not, draw the compound in such
a way that you can identify the plane of symmetry.

(a) $CH_3CH_2CH_2Br$    (b) $CH_3CHCH_2Cl$    (c) $CH_3CH_2CHCH_2Cl$
                                 $\quad\quad\;\; |$                          $\quad\quad\quad\overset{CH_3}{\overset{|}{\phantom{x}}}$
                                 $\quad\quad OH$

(d) [cyclohexane with CH₃ and Br]    (e) $CH_3CHCH_2CH_2CH_3$    (f) $CH_3CHCH_2CHCH_3$
                                                $\quad\quad |$                       $\quad\quad\overset{CH_3}{\overset{|}{\phantom{x}}}\quad\quad\overset{|}{\phantom{x}}$
                                                $\quad\quad Br$                      $\quad\quad\quad\quad\quad\quad OH$

## 6.4  Plane-Polarized Light and Optical Activity

### A. Plane-Polarized Light

Light is generally characterized as a wave with associated oscillating electrical and
magnetic fields. Thus, for a beam of light, one vector describes the electrical field
strength and another the magnetic field strength. These vectors are in planes that are
perpendicular to each other and to the direction of the propagation of the wave (Figure 6.6). In ordinary light, there are a great number of these electromagnetic waves;
the planes of the electrical and magnetic vectors of each individual wave are randomly oriented with respect to those of the other waves. Such light is said to be **unpolarized.**

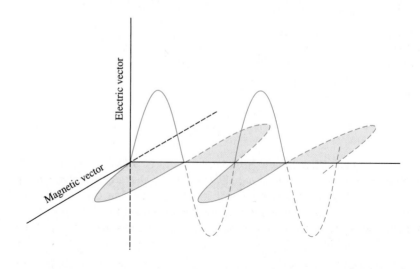

*Figure 6.6*
Electric and magnetic field
strength vectors of a light
wave.

Certain crystalline substances—calcite is a good example—only allow the passage of light waves with their electrical vectors in one single plane. Prisms made of such crystals transmit light that is said to be **plane-polarized.**

## B. The Experimental Determination of Optical Activity

When plane-polarized light interacts with molecules, the direction of the plane of polarization is changed. If the molecules are achiral, the interactions that take place with large numbers of molecules in many random orientations do not result in an overall change in the plane of polarization. A twist given to the plane of polarization by a molecule in one orientation is canceled out by an equal twist caused by a molecule in another orientation that looks like the mirror image of the first one. The result is that no rotation of the plane of polarization is observed. If the molecules are chiral, then no molecular orientation is the exact mirror image of any other. Therefore, the changes that take place in the plane of polarized light as it interacts with the molecules are not averaged out but instead add up so that rotation of the plane of polarization is observed. The direction of this rotation may be counterclockwise, in which case the species is said to be **levorotatory.** If it is clockwise, the species is **dextrorotatory.** Whether an optically active compound rotates the plane of polarization clockwise or counterclockwise is indicated by the sign of rotation $(+)$ or $(-)$ at the beginning of the name of the compound.

The instrument that is used to measure optical activity is called a **polarimeter.** It has a source of light and two prisms—one is used to produce the plane-polarized light and the other to detect any rotation in the plane of polarization. A tube containing a solution of the compound to be investigated is placed between the prisms. Finally, there is a viewing device with a scale on it so that the angle by which the plane of polarized light has been rotated can be measured (Figure 6.7).

The light that goes through the polarizer prism emerges as plane-polarized light. If the axis of the analyzer prism is aligned with that of the polarizer prism and there is no optically active material in the sample tube, the light will go through to the scale unchanged. If the sample tube contains an optically active compound, however, the plane of polarization will be rotated. The analyzer prism is then rotated until its axis coincides with the new plane of polarization, and light is once more transmitted. The angle by which the analyzer prism has to be rotated in order to allow the transmission of light is the angle of rotation for the optically active compound (Figure 6.8, p. 200).

How much the plane of polarization is rotated depends on several things. First, it depends on the particular optically active compound on which the measurement is being made. The sign of rotation for a compound is an experimental measurement. There are rules for predicting the direction and extent of the rotation for some compounds, but ultimately the optical rotation for a compound must be determined experimentally. Once the optical rotation for one enantiomer is known, however, it is certain that the other isomer will have the same degree of rotation but in the opposite direction.

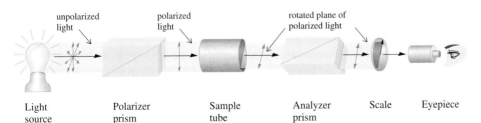

| unpolarized light | polarized light | rotated plane of polarized light |
|---|---|---|

| Light source | Polarizer prism | Sample tube | Analyzer prism | Scale | Eyepiece |
|---|---|---|---|---|---|

*Figure 6.7*
Sketch of the main parts of a polarimeter.

Electric vectors
of a beam of
unpolarized
light

Direction of the
electric vector of
plane polarized light
as it leaves the
polarizer

Plane of polarization
rotated counterclockwise;
angle of rotation = $-\alpha$

Plane of polarization
rotated clockwise;
angle of rotation = $+\alpha$

**Figure 6.8**
Schematic representation of
the rotation of the plane of
polarized light.

Second, because the total rotation depends on the number of interactions between the light beam and the molecules, the concentration of the solution and the
path length of the beam of polarized light through the solution must be considered.
The particular conformations of the molecules being examined and any interactions
between the molecules and the solvent are important, so temperature and solvent are
also recorded.

Finally, optical rotation changes with the wavelength of the light used in the
measurement, so a complete report must also include that information. For purposes
of comparing different compounds, the specific rotation of a compound is calculated using this formula

$$[\alpha]_D^T = \frac{\alpha}{lc} \text{ (concentration; solvent)}$$

The symbol $[\alpha]_D^T$ means the specific rotation, calculated from data measured at a
temperature of $T$ °C using the light from the yellow line (the so-called D line) in the
visible spectrum of sodium. This wavelength of 589 nm corresponds to the yellow
color seen when a sodium salt is heated in the flame of a Bunsen burner. In the nineteenth century, it was the only readily available source of light of a single wavelength. It is still used today so that contemporary results can be compared with
those of earlier days. The measured angle of rotation is $\alpha$. The length of the sample
tube, $l$, is expressed in decimeters, and a standard tube is 10 cm or 1 dm long. The
concentration of the sample, $c$, is given in grams per milliliter of solution. In most
cases, the concentration of the solution and the solvent used are reported in parentheses after the value for the specific rotation.

The optical rotation of a chiral compound is a specific physical property of
the compound and is determined and reported just as other physical properties are,
for example, boiling and melting points. Different chiral compounds have widely
varying specific rotations. The specific rotations of some compounds isolated from
natural sources illustrate this. In ethanol, cholesterol has $[\alpha]_D^{20}$ $-3.15°$, and for
nicotine $[\alpha]_D^{20}$ is $+163.2°$. The specific rotation for cane sugar (sucrose) in water is
$[\alpha]_D^{20}$ $+66.4°$.

Calculating specific rotation with the above formula is illustrated by the following problem.

**PROBLEM:** A solution (100 mL) of 16.5 g of the levorotatory form of camphor in ethanol has an optical rotation of $-7.29°$ at 20 °C, using a 10-cm sample tube and a sodium lamp for the measurement. What is the specific rotation of this compound?

**Solution**

$$[\alpha]_D^{20} = \frac{\alpha}{1 \text{ dm} \times 16.5 \text{ g}/100 \text{ mL}}$$

$$= \frac{\alpha \times 100}{1 \times 16.5} = \frac{-7.29° \times 100}{1 \times 16.5}$$

Thus the specific rotation for camphor is

$$[\alpha]_D^{20} = -44.2° \ (c \ 0.165; \text{ ethanol}) \ \blacksquare$$

🗗 **Study Guide**
**Concept Map 6.4**

**Problem 6.4**    The optical rotation of sugar is used in industry as a quick way to check on the concentration of sugar solutions. If 20 g of cane sugar dissolved in water to make up 100 mL of solution and placed in a tube 40 cm long rotates the plane of polarized light $+53.2°$ at 20 °C, what is the concentration of another solution of sugar, measured at the same temperature and in the same polarimeter tube, if the optical rotation is $+13.3°$?

**Problem 6.5**    Menthol has an optical rotation of $+2.46°$ when the measurement is made with a sodium lamp at 20 °C

on a solution containing 5 g of menthol in 100 mL of ethanol solution using a sample tube 10 cm long. What is the specific rotation of menthol?

**Problem 6.6**    When the analyzer prism is rotated 90° clockwise, it arrives at the same position as if it were rotated counterclockwise 270° ($-270°$). If the optical rotation were being determined for the first time, how would it be possible to establish whether the rotation should be reported as $+90°$ or $-270°$?

## 6.5  The Formation of Stereoisomers in Chemical Reactions. Racemic Mixtures

### A. The Addition of Hydrogen Bromide to 1–Butene

Addition of hydrogen bromide to 1-butene produces 2-bromobutane.

$$CH_3CH_2CH{=}CH_2 \xrightarrow{HBr}$$

(−)-2-bromobutane    (+)-2-bromobutane

An examination of the mechanism of this reaction shows that it should give equal amounts of the two stereoisomers. The reaction proceeds by way of a planar carbocation intermediate, a symmetrical achiral species. The bromide ion reacts with equal probability at either face of this cation. The transition state corresponding to the reaction of the cation with a bromide ion that approaches from the right is the enantiomer of the other transition state, which corresponds to the reaction of the cation with a bromide ion approaching from the left. The two transition states leading from the cation to the two different products are equal in energy. The two processes have the same energy of activation and, therefore, the same rate. (−)-2-Bromobutane and (+)-2-bromobutane are formed in equal amounts when hydrogen bromide adds to 1-butene. *Equal amounts of enantiomeric products are always formed when two achiral reagents react to give a chiral product.* The two

**VISUALIZING THE REACTION**

**Formation of Enantiomers**

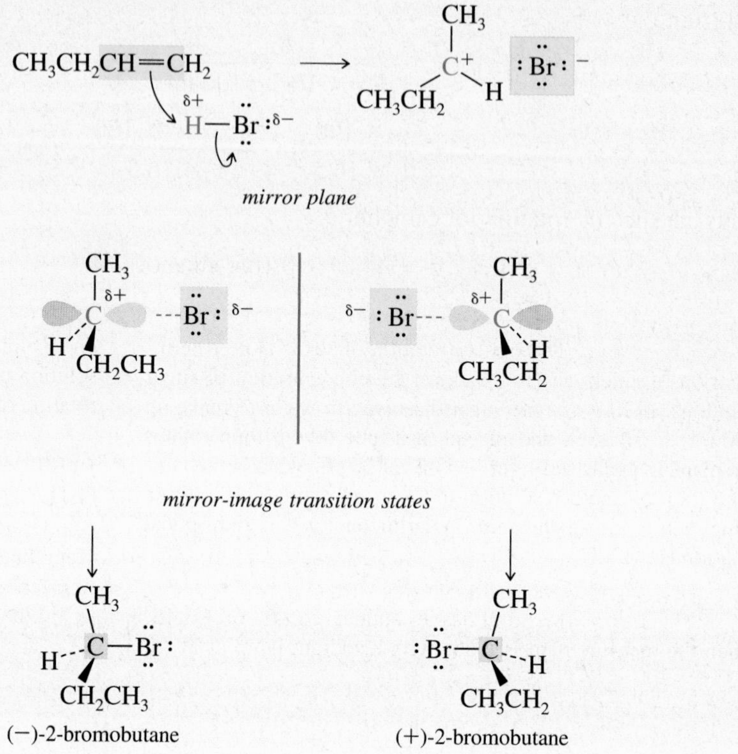

*mirror plane*

*mirror-image transition states*

(−)-2-bromobutane                (+)-2-bromobutane

enantiomers have the same physical properties, except for optical rotation, and cannot be separated from each other by ordinary physical methods, such as distillation.

The equations written in this section show the formation of both enantiomers of 2-bromobutane in order to increase your awareness of what is happening stereochemically in such a reaction. Normally, both enantiomers are not shown as products of a chemical reaction.

## B. Racemic Mixtures and Enantiomeric Excess

The addition of hydrogen bromide to 1-butene results in the formation of a mixture consisting of equal amounts of (−)-2-bromobutane and (+)-2-bromobutane. Such a mixture is known as a racemic mixture. A **racemic mixture** contains equal numbers of molecules of two enantiomers and shows no optical rotation. Because a racemic mixture contains equal concentrations of molecules of opposite optical activity, plane-polarized light that is given a twist to the right by an encounter with a molecule of one enantiomer is twisted back by an encounter with a molecule of the mirror-image isomer. The plane of polarization remains unchanged as the light passes through the mixture. A racemic mixture is indicated by the use of the symbol (±) at the beginning of the name of the compound. For example, (−)-2-bromobutane has $[\alpha]_D^{22}$ −23.1° and (+)-2-bromobutane has $[\alpha]_D^{22}$ +23.1° when the determinations are done on the pure liquids. The mixture of equal quantities of the two enantiomers of 2-bromobutane obtained as the product of the addition of hydrogen bromide to 1-butene has $[\alpha]_D^{22}$ 0° and thus is not optically active. It is designated as ($\pm$)-2-bromobutane.

If one enantiomer of a pair is present to a greater extent, the mixture will show an optical rotation corresponding to the percentage of the species that is present in excess. The percentage of the enantiomer that is present in excess is known as the **enantiomeric excess.** It is calculated from a formula involving the rotation observed for a mixture and the known optical rotation of the pure enantiomer:

$$\frac{\text{measured specific rotation of mixture}}{\text{specific rotation for the pure enantiomer}} \times 100 = \% \text{ enantiomeric excess}$$

**PROBLEM:**  A sample of 2-bromobutane has $[\alpha]_D^{22}$ +11.55°. The specific rotation of (+)-2-bromobutane at 22 °C is +23.1°. What is the enantiomeric excess of (+)-2-bromobutane in this sample?

**Solution**

$$\frac{\text{measured specific rotation of mixture}}{\text{specific rotation of pure (+)-2-bromobutane}} \times 100 = \frac{+11.55}{+23.1} \times 100$$

$$= 50\% \text{ enantiomeric excess of (+)-2-bromobutane}$$

The sample of 2-bromobutane described in the problem above is dextrorotatory because it has an excess of the dextrorotatory enantiomer. The exact composition of the mixture can be calculated from the enantiomeric excess. Leaving out the excess dextrorotatory molecules, the rest of the mixture has no optical rotation because it consists of equal numbers of dextrorotatory and levorotatory molecules. The optical rotations they cause cancel one another out. ■

**PROBLEM:**  If a mixture of 2-bromobutanes has an enantiomeric excess of 50% of (+)-2-bromobutane, what is the stereoisomeric composition of the mixture?

**Solution:**  Of the total mixture, $(100 - 50)\%$ consists of equal numbers of dextrorotatory and levorotatory molecules. Thus half of 50%, or 25%, of these molecules are dextrorotatory, and 25% of them are levorotatory. Therefore, the mixture is 25% ($-$)-2-bromobutane and 75% (50% enantiomeric excess +25%) (+)-2-bromobutane. ■

A racemic mixture cannot be separated into its components by ordinary physical methods because the physical properties of the two components, except for the direction of rotation of a plane of polarized light, are identical. For example, (+)-2-bromobutane and ($-$)-2-bromobutane both boil at 91 °C. A separation of enantiomers must, therefore, always involve the use of chiral reagents, which will interact differently with molecules of differing chirality. Enzymes in biological systems are such reagents. One way of separating enantiomers is to use such living organisms to metabolize one form and leave the other form untouched.

A more general method for separating enantiomers, known as the resolution of a racemic mixture, involves the formation of compounds from the enantiomers by reaction of the mixture with a chiral reagent. Understanding this method requires that we first look at molecules containing more than one stereocenter. For this reason, we will postpone the discussion of this topic until Section 6.11.

*Study Guide*
*Concept Map 6.5*

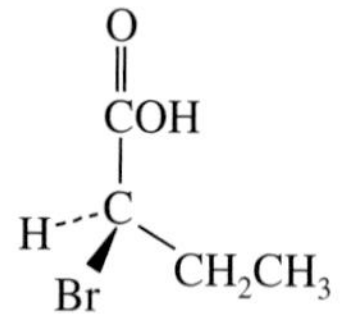

O
‖
COH
|
H---C
Br    CH₂CH₃

**Problem 6.7**    (+)-2-Bromobutanoic acid, shown at left, has $[\alpha]_D^{25}$ +39.5° ($c$ 9.1; ether). A sample of 2-bromobutanoic acid having $[\alpha]_D^{25}$ −14.70° was recovered by resolution of a racemic mixture of the acid.

(a) What is the enantiomeric excess of the sample of acid recovered from the racemic mixture?

(b) Draw the correct stereochemical formula for (−)-2-bromobutanoic acid, and say what percentage of the mixture each enantiomer is.

## 6.6    The Discovery of Molecular Dissymmetry

Optical activity was known early in the nineteenth century to be a property of crystals, such as quartz, that are demonstrably dissymmetric (without symmetry) in appearance. In Paris in 1848, Louis Pasteur noticed that crystals of the sodium ammonium salt of (+)-tartaric acid, a by-product of winemaking, had dissymmetric crystals that could not be superimposed on their mirror images. He thought that this crystalline dissymmetry might indicate a similar lack of symmetry in the molecules of (+)-tartaric acid. Another form of tartaric acid, known as paratartaric acid, did not rotate the plane of polarized light. Pasteur studied the sodium ammonium salt of paratartaric acid, expecting to find that its crystals were symmetrical. Instead, he saw that some of the crystals had a right-handed appearance and others a left-handed one. The two crystalline forms were mirror images of each other.

Pasteur separated the two crystalline forms of sodium ammonium paratartrate with tweezers as he looked through a microscope. A solution of the right-handed crystals in water rotated the plane of polarized light to the right, exactly the way the (+)-tartaric acid salt did. The left-handed crystals gave a solution with an optical rotation of equal magnitude but opposite sign. A mixture of equal weights of the two crystal forms, when dissolved in water, gave a solution with no optical rotation. With these experiments, Pasteur demonstrated that optical activity was the result of a molecular property that survived even when the crystal form was destroyed by dissolving it. He saw molecular dissymmetry as the cause of the phenomenon of optical activity, and he recognized that there were two dissimilar molecular forms of optically active tartaric acid.

In further experiments in 1854, Pasteur showed that the microorganism *Penicillum glaucum* consumed (+)-tartaric acid but not (−)-tartaric acid. His work with optically active compounds led Pasteur to say, "Life is dominated by dissymmetrical actions. I can even foresee that all living species are primordially in their structure, in their external forms, functions of cosmic dissymmetry."

Although the phenomenon of molecular dissymmetry was recognized in the 1840s, there was no clear picture of how it came about until 1874. Up to that time, chemists were still struggling with ways to represent molecular structures and had not yet made clear distinctions between constitutional isomers and stereoisomers. The idea that groups around a carbon atom are arranged in a tetrahedron was suggested in 1874 by the Dutch chemist J. H. van't Hoff, who was twenty-two years old at the time. He recognized that it was necessary to think of structures in three dimensions in order to solve the problems of isomerism that were being discovered in the laboratory. A carbon atom with four different substituents arranged tetrahedrally around it would account for the two isomers observed experimentally for compounds with a single stereocenter. The right-handed and left-handed arrangements that are possible for four groups around a tetrahedral carbon atom could be used to explain the phenomenon of optical activity.

The French chemist J. A. Le Bel started with the work of Pasteur and also arrived at the idea that a carbon atom with four different substituents around it is the basis for optical activity in organic molecules. He published his ideas in 1874, the same year as van't Hoff did. Le Bel emphasized the lack of symmetry, in particular, the lack of a plane of symmetry, at the molecular level as a necessary condition for optical activity. He hit on the idea of a tetrahedral carbon atom by exploring the number of isomers that are formed as one, two, and then three different groups are substituted on a carbon that originally had four identical groups on it. The earlier discussion (Section 6.1) about creating isomers by substituting the hydrogen atoms on butane follows very closely his way of thinking about stereoisomerism. Le Bel's approach differed from van't Hoff's in that van't Hoff took a tetrahedral carbon atom as his starting point.

Le Bel's ideas remained more abstract than van't Hoff's. For example, van't Hoff created molecular models of tetrahedral carbon atoms and sent them to leading chemists of the time in an effort to gain acceptance of his ideas. He drew structural formulas in perspective to make his ideas clear. He made predictions about optical activity or the lack of it for compounds yet to be investigated, predictions that were found to be correct when the experimental work was done. In spite of this, his ideas were not generally accepted for a number of years. The opposition got some support from conflicting experimental results. In those days, many organic compounds were isolated from natural sources and contained small amounts of optically active compounds as impurities. Because of this contamination, confusing data were obtained that showed the presence of optical activity for compounds that did not have stereocenters. But van't Hoff himself supervised much of the experimental work that proved that pure compounds were not optically active unless molecular dissymmetry was present. In 1901, he received the first Nobel Prize in Chemistry for his work in other areas.

By the end of the nineteenth century, the tetrahedral carbon atom was accepted as the basis of structural organic chemistry. Much of the research being done by that time on sugars, which contain several stereocenters, would not have been possible without this basis. The experimental results confirmed the correctness of van't Hoff's ideas. His predictions about the number of isomers that are possible and the kinds of compounds that should have optical activity were demonstrated to be true.

## 6.7   Configuration. Representation and Nomenclature of Stereoisomers

### A. Configuration of Stereoisomers

The actual orientation in space of the groups around a stereocenter is called the **absolute configuration** of the compound. The determination of this configuration for a particular compound is not a trivial matter. Later chapters will discuss reactions that take place with known stereochemical results. Experimental evidence from many such reactions has been accumulated over the years, so the stereochemical relationships among many series of compounds are known. The ultimate determination of exactly how these molecules look depended on the development of sophisticated x-ray diffraction techniques for determining the structures of crystals. The solution of this problem, known as the determination of the absolute configuration of stereoisomers, was completed in 1951 (see Section 6.7C, p. 210).

*The configuration of a compound is unchanged unless at least one bond at the stereocenter is broken.* Thus configuration must not be confused with conformation, which changes continuously at room temperature as a result of rotation about single bonds and the flipping of rings in molecules. A stereoisomer has a single configuration but may exist in a number of conformations, depending on the solvent and the temperature. For example, four representations of $(-)$-2-chlorobutane are shown below; three of them represent different conformations of the molecule and one of them represents a rotation of the whole molecule in space. In all of these, the configuration of the molecule at the stereocenter remains unchanged.

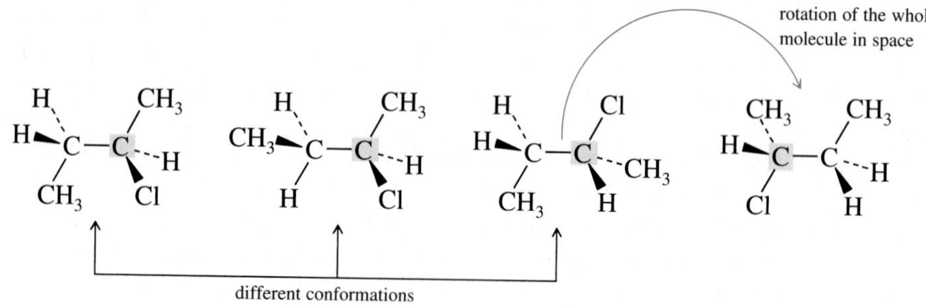

rotation of the whole molecule in space

different conformations

$(-)$-2-chlorobutane

Chemical reactions that involve breaking a bond at a stereocenter often result in a change in the configuration of a chiral compound. In the next chapter we will examine such reactions.

## B.  Nomenclature of Stereoisomers

The rules of nomenclature described in Section 5.10 are not adequate for naming stereoisomers. For example, unless you memorize the structures for $(+)$-2-chlorobutane and $(-)$-2-chlorobutane (p. 193), there is no way for you to draw a unique structure for each isomer from the names alone. There is no simple correlation between the sign of rotation and the structure. The sign of rotation by itself does not tell what the configuration of the compound is.

To solve this problem, another set of rules, the **Cahn-Ingold-Prelog Rules,** were created. The configuration at a stereocenter is described as being **R**, from the Latin *rectus* (or "right-handed"), or **S**, from the Latin *sinister* (or "left-handed"), depending on the order in which the different substituents are arranged around the stereocenter. The rules that are applied to determine configuration are as follows:

1. Each group attached to the stereocenter is assigned a priority. The higher the **atomic number** of the atom bonded directly to the stereocenter, the higher is the priority of the substituent; for example:

$$Cl > O > N > C > H$$

Among isotopes, which have identical atomic numbers, the one with the higher atomic weight takes priority. Thus tritium, the isotope of hydrogen with an atomic weight of 3, has higher priority than deuterium, with an atomic weight of 2. Hydrogen, with an atomic weight of 1, has the lowest priority, not only of these three isotopes but in any case.

$$T > D > H$$

2. If two identical atoms are attached to the stereocenter, the next atoms in both chains are investigated, moving away from the stereocenter until some difference is found. A priority assignment is made at the first point at which atoms of different priorities are found.

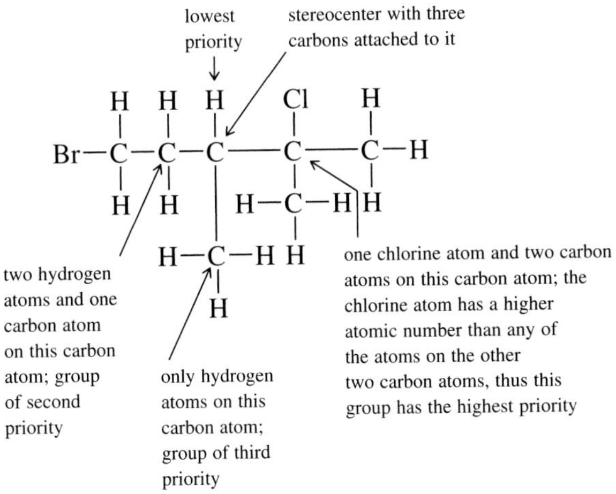

lowest priority

stereocenter with three carbons attached to it

two hydrogen atoms and one carbon atom on this carbon atom; group of second priority

only hydrogen atoms on this carbon atom; group of third priority

one chlorine atom and two carbon atoms on this carbon atom; the chlorine atom has a higher atomic number than any of the atoms on the other two carbon atoms, thus this group has the highest priority

Note that there is a bromine atom, an atom of higher atomic number than any other atom in the above molecule, at the end of the chain. This bromine atom does not influence the assignment of priorities because it is beyond the point of difference.

3. A double bond is counted as two single bonds for both of the atoms involved.

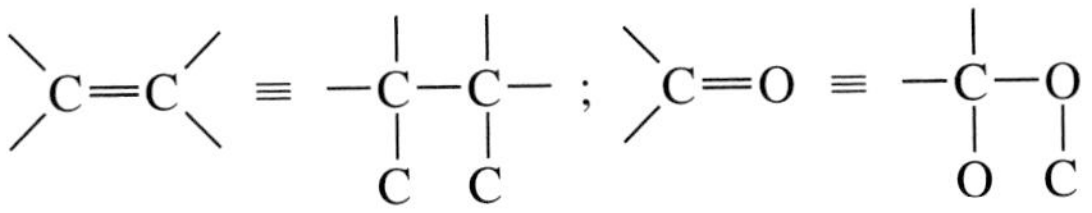

The same principle is extended to triple bonds.

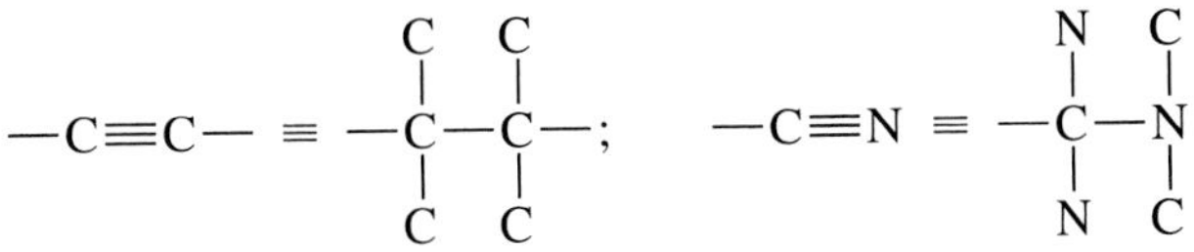

For example,

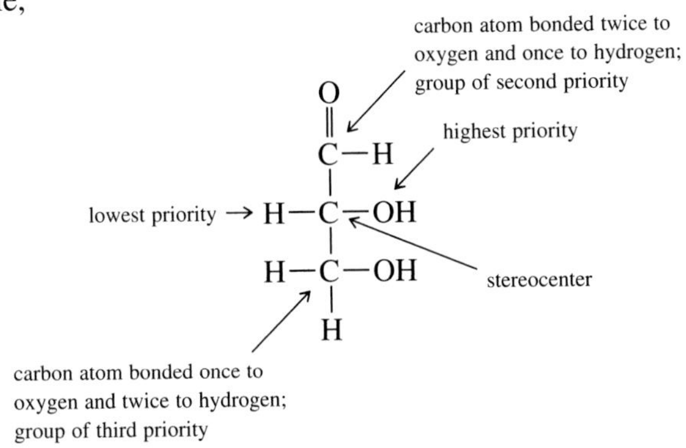

carbon atom bonded twice to oxygen and once to hydrogen; group of second priority

highest priority

lowest priority →

stereocenter

carbon atom bonded once to oxygen and twice to hydrogen; group of third priority

4. After priorities have been assigned, the molecule is viewed with the substituent of lowest priority away from the viewer. If you can trace a clockwise path from the group of highest (or first) priority to the one of second priority and then to the one of third priority, the stereocenter is assigned the *R* configuration. If the arrangement of the groups in order of decreasing priority follows a counterclockwise path, the configuration is *S*.

The Cahn-Ingold-Prelog Rules can be clarified by applying them to assign configurations to stereoisomers of bromochloroiodomethane.

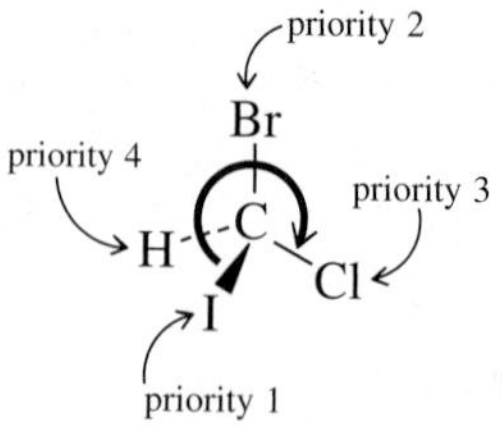

*clockwise path*
*from priority 1 to 2 to 3;*
*therefore, this is*
*(R)-bromochloroiodomethane*

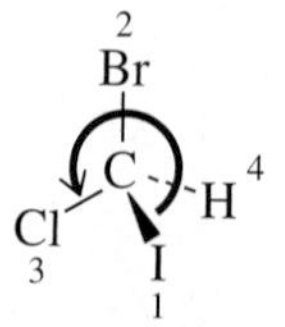

*counterclockwise path*
*from 1 to 2 to 3; therefore, this*
*is (S)-bromochloroiodomethane*

These rules can also be clarified by applying them to assign configurations to stereoisomers of 2-chlorobutane.

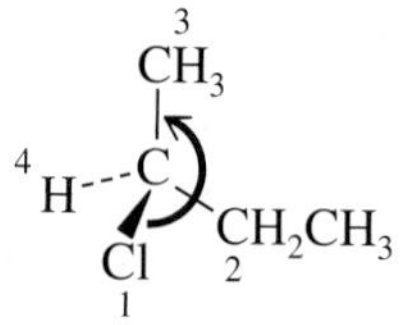

*counterclockwise path*
*from 1 to 2 to 3; therefore,*
*this is (S)-(+)-2-chlorobutane*

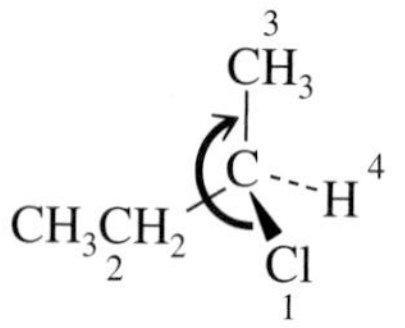

*clockwise path*
*from 1 to 2 to 3; therefore,*
*this is (R)-(−)-2-chlorobutane*

Note that the molecules are drawn so that the group of lowest priority is projecting away from the viewer. The assignment of priorities in bromochloroiodomethane is straightforward: The priorities are determined by looking up the atomic numbers for the atoms attached to the stereocenter. For 2-chlorobutane, two of the atoms bonded to the stereocenter are carbons. The ethyl group takes priority over the methyl group because the second carbon atom in the ethyl group has priority over any one of the three hydrogen atoms in the methyl group.

(+)-Alanine and (−)-lactic acid, two naturally occurring compounds, are assigned the configurations below.

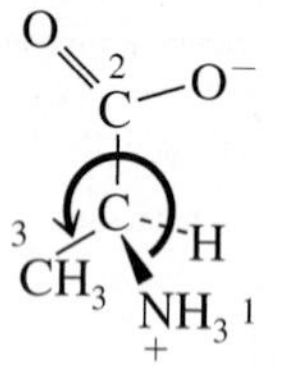

*counterclockwise order*
*of groups by priority*

(S)-(+)-alanine

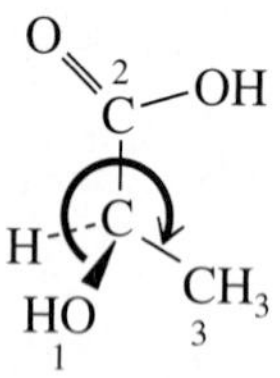

*clockwise order of*
*groups by priority*

(R)-(−)-lactic acid

In alanine, the nitrogen atom takes priority over the methyl group and the carboxylic acid group. The carboxylic acid group, with oxygen bonded to its carbon atom, is of higher priority than the methyl group, in which only hydrogens are bonded to the carbon atom. Thus the enantiomer of alanine that has a positive sign of rotation is assigned the *S* configuration. Similar reasoning is used to assign the *R* configuration to the form of lactic acid that is levorotatory.

Note that the designation of a compound as *R* or *S* has nothing to do with the sign of rotation. The Cahn-Ingold-Prelog Rules can be applied to any three-dimensional representation of a stereocenter to determine whether it is *R* or *S*. To find out whether a particular picture of a molecule corresponds to the dextrorotatory or levorotatory form requires much experimental work and cannot be determined just by looking at the structural formula. If the sign of rotation of one enantiomer is known, however, the other enantiomer will have the opposite sign of rotation. For example, knowing that (S)-(+)-alanine and (R)-(−)-lactic acid have the structures shown on page 208 means that the following must be the configurations of (R)-(−)-alanine and (S)-(+)-lactic acid.

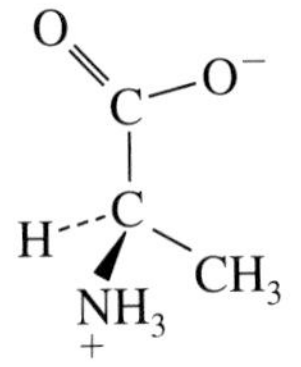

(R)-(−)-alanine
enantiomer of
(S)-(+)-alanine

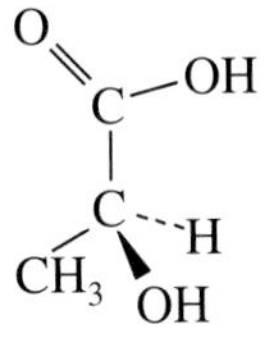

(S)-(+)-lactic acid
enantiomer of
(R)-(−)-lactic acid

The use of R or S in the name of a compound assigns a particular configuration, a specific orientation in space, to the atoms in the molecule. Thus the Cahn-Ingold-Prelog Rules enable chemists to describe the stereochemistry of a compound without having to draw a three-dimensional picture of the molecule. For example, the name (R)-2-bromo-1-propanol fully describes the orientation of the groups around the stereocenter in that compound.

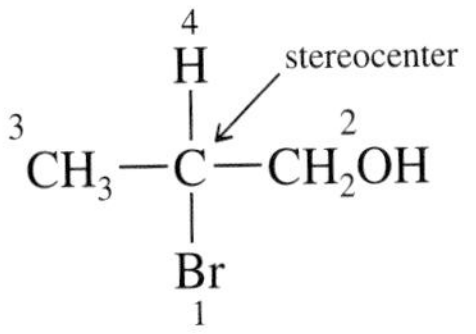

2-bromo-1-propanol

*(no stereochemistry shown)*
*priorities assigned to the groups*
*at the stereocenter*

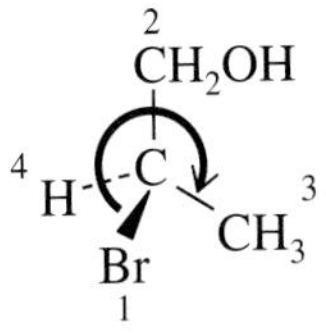

(R)-2-bromo-1-propanol

*one possible three-dimensional*
*picture showing stereochemistry*

The Cahn-Ingold-Prelog Rules fulfill the requirement for a good system of nomenclature: Each name corresponds to a unique structure that can be reproduced from the name alone.

**On the Web: ONE SMALL STEP**

**Problem 6.8** Name each of the following compounds, including an assignment of configuration.

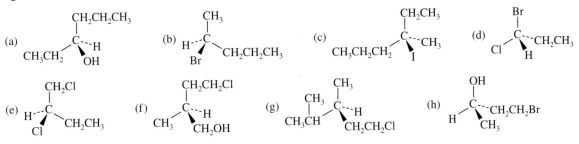

**Problem 6.9** Draw three-dimensional structural formulas for the following compounds.

(a) (R)-1-bromo-1-chloroethane    (b) (S)-1-chloro-2-propanol    (c) (S)-2,3-dimethylpentane    (d) (R)-2-butanol
(e) (R)-ethanol-1-d (CH₃CHDOH)    (f) (S)-1-bromo-3-chloro-2-methylpropane

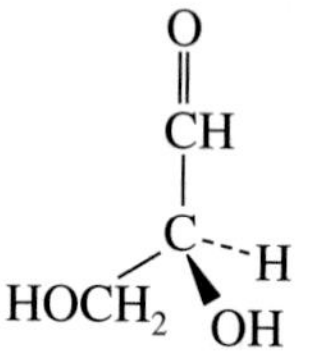

(S)-glyceraldehyde

(R)-glyceraldehyde

(R)-(+)-glyceraldehyde

## C. Relative and Absolute Configurations

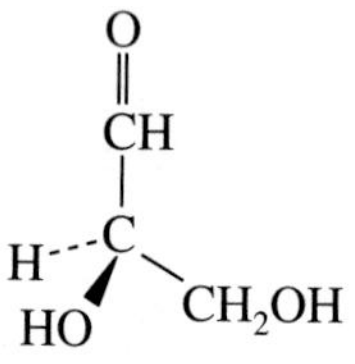

The phenomenon of the optical activity of natural products, especially sugars, was known by chemists early in the nineteenth century. For over a century, however, the problem of determining absolute configuration (p. 205) stumped scientists who were working with optically active compounds. For example, the simplest optically active sugar is 2,3-dihydroxypropanal, usually called glyceraldehyde. It has a single stereocenter and exists as two enantiomers, (+)-glyceraldehyde and (−)-glyceraldehyde. The configurations of the two enantiomers are shown in the margin. There is no way of knowing just by looking at these structures whether the *R* or the *S* isomer corresponds to (+)-glyceraldehyde. In other words, the actual arrangement of the atoms around the stereocenter, the absolute configuration of a compound, cannot be determined from the sign of the optical rotation.

Chemists recognized that they could relate the configurations of various optically active compounds to each other by chemical interconversions, even when they did not know absolute configurations. Some of the ways in which this was done in the 1930s and 1940s are described in Chapter 7. Thus chemists came to know the relative configurations of large numbers of compounds. In 1951, the absolute configuration of (+)-tartaric acid was determined by the Dutch chemist J. M. Bijvoet and his colleagues using a sophisticated modification of x-ray diffraction. Their work was done in the laboratory named for van't Hoff, who had recognized that the existence of optically active enantiomers could be explained by postulating a tetrahedral carbon atom as the stereocenter (p. 205).

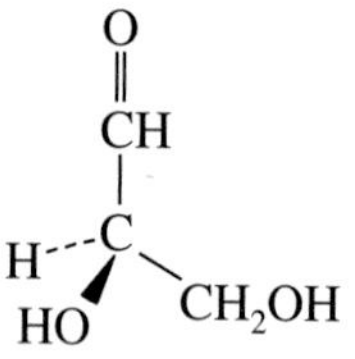

It was only after Bijvoet's determination that chemists were able to assign actual three-dimensional structures to the chiral compounds that they worked with. For example, it is now known that (+)-glyceraldehyde has the *R* configuration. The story of how this came about is part of the history of the determination of the structure of glucose (Chapter 23).

(*R*)-(+)-Glyceraldehyde, in turn, has served as a reference compound for the assignment of configuration to many compounds. If the name of a compound includes both the sign of rotation and the designation *R* or *S*, then the actual three-dimensional arrangement of its atoms has been determined by careful experiments. In other words, the absolute configuration of that particular compound is known.

## 6.8   Diastereomers

### A. Compounds with More Than One Stereocenter

Replacement of one of the secondary hydrogen atoms in butane with a chlorine atom creates one of the enantiomeric 2-chlorobutanes (Section 6.1, p. 191). Replacement of either of the hydrogen atoms on carbon atom 3 in each of the enantiomers of 2-chlorobutane with a bromine atom is shown below and on the facing page.

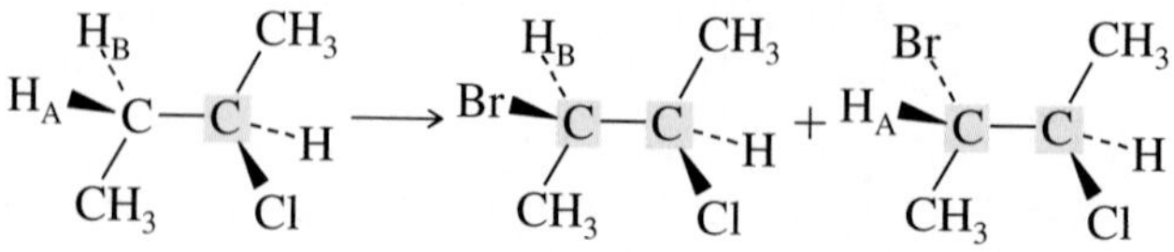

(*R*)-2-chlorobutane → (2*R*,3*R*)-2-bromo-3-chlorobutane + (2*S*,3*R*)-2-bromo-3-chlorobutane

*from replacement of H_A*    *from replacement of H_B*

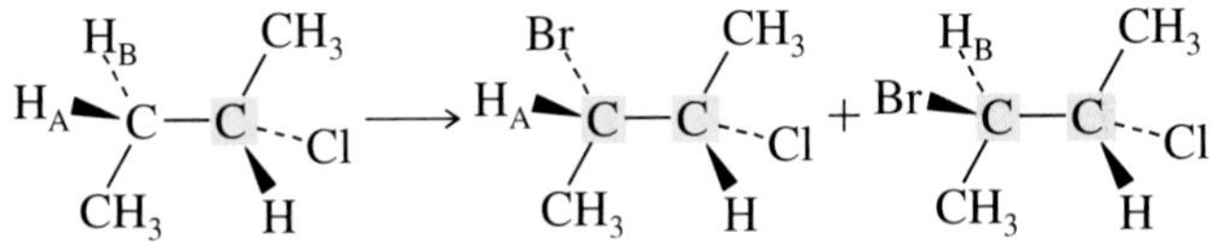

(S)-2-chlorobutane → (2S,3S)-2-bromo-3-chlorobutane + (2R,3S)-2-bromo-3-chlorobutane

*enantiomer of (2R,3R)-2-bromo-3-chlorobutane*     *enantiomer of (2S,3R)-2-bromo-3-chlorobutane*

Note that the original stereocenter is left undisturbed as one or the other of the two hydrogen atoms on carbon atom 3 is replaced by a bromine atom. Four compounds result. *The maximum number of stereoisomers for a compound with n stereocenters is $2^n$.* A compound with two stereocenters will have at most $2^2$, or 4, stereoisomers.

A close examination of the four 2-bromo-3-chlorobutanes reveals that they are two pairs of enantiomers. **Diastereomers** are stereoisomers that are not enantiomers. Therefore, each of the four is an enantiomer of one of the other three compounds and a diastereomer of the other two.

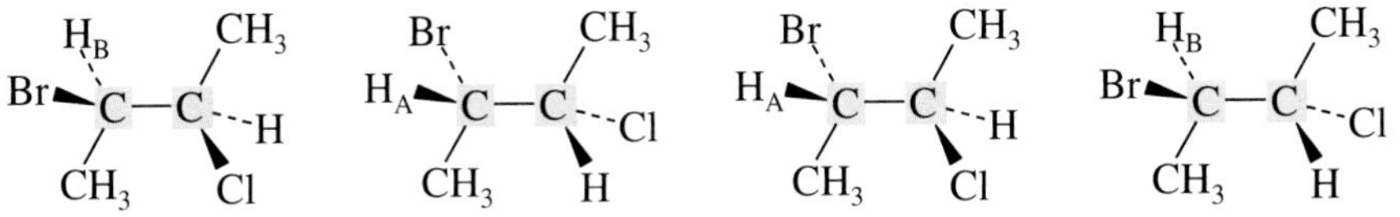

(2R,3R)-2-bromo-3-chlorobutane    (2S,3S)-2-bromo-3-chlorobutane    (2S,3R)-2-bromo-3-chlorobutane    (2R,3S)-2-bromo-3-chlorobutane

*enantiomers of each other*        *enantiomers of each other*

*diastereomers of the compounds to the right*     *diastereomers of the compounds to the left*

The stereochemical relationships are apparent not only from the perspective formulas but also from the names of the compounds. The enantiomers have the opposite configuration at each of the stereocenters. The diastereomers have the same configuration at one of the two centers and the opposite configuration at the other. You should examine these structures and their names until you convince yourself of these relationships.

Unlike enantiomers (p. 193), diastereomers have different physical properties, such as boiling points, melting points, dipole moments, and solubilities. Diastereomers can be separated from each other by the usual means of purification, such as fractional distillation. For example, a mixture of the four 2-bromo-3-chlorobutanes shown above could be separated into two racemic mixtures, each one consisting of a pair of enantiomers.

The examples in this section illustrate the stereochemical relationships that arise when two different stereocenters are present in a molecule. In the next section we will explore the stereochemistry of a compound that contains two stereocenters that have the same substituents.

**Problem 6.10** Draw and name all the stereoisomers that are possible for each of the following compounds. Identify the enantiomers and diastereomers.

(a) $CH_3CH_2CHCH_2CH_2CH_3$
         |
        $CH_3$

(b) $CH_3CH_2CHCH_2CHCH_3$
         |     |
        $CH_3$  Br

(c) $CH_3CH_2CHCHCH_3$
         |  |
       Br Br

**Problem 6.11** Draw and name all the stereoisomers that are possible for 3-bromo-2-butanol and 2-chloro-3-methylheptane. Identify the enantiomers and diastereomers.

## B. Compounds Containing Two Stereocenters with Identical Substituents. Meso Forms

The substitution of a bromine atom on the enantiomeric 2-chlorobutanes (described in Section 6.8A) can be done with a chlorine atom instead.

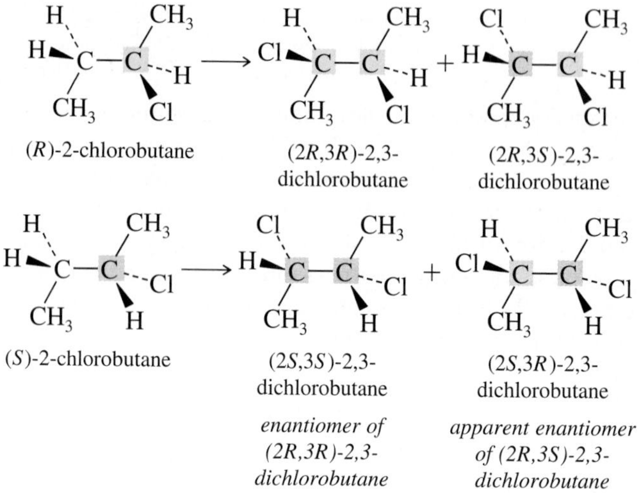

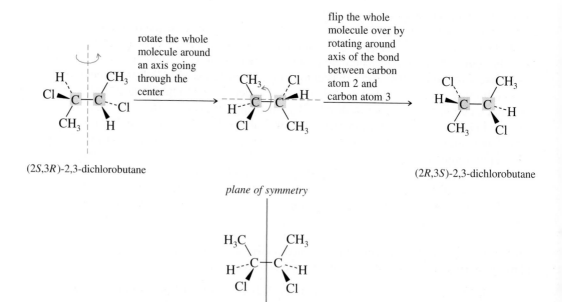

(R)-2-chlorobutane    (2R,3R)-2,3-dichlorobutane    (2R,3S)-2,3-dichlorobutane

(S)-2-chlorobutane    (2S,3S)-2,3-dichlorobutane    (2S,3R)-2,3-dichlorobutane

*enantiomer of (2R,3R)-2,3-dichlorobutane*    *apparent enantiomer of (2R,3S)-2,3-dichlorobutane*

It appears that four stereoisomers are formed, that is, two pairs of enantiomers. But a closer examination of the structures labeled (2R,3S)-2,3-dichlorobutane and (2S,3R)-2,3-dichlorobutane reveals that they are representations of the same molecule. These two structures are superimposable mirror images of each other. Two rotations of one of the structures in space make it identical with the other one.

rotate the whole molecule around an axis going through the center

flip the whole molecule over by rotating around axis of the bond between carbon atom 2 and carbon atom 3

(2S,3R)-2,3-dichlorobutane

(2R,3S)-2,3-dichlorobutane

*plane of symmetry*

eclipsed conformation of (2R,3S)-2,3-dichlorobutane

The eclipsed conformation (2R,3S)-2,3-dichlorobutane has a plane of symmetry because the two stereocenters have identical substituents. One half of the molecule is the mirror image of the other half. The molecule is not chiral and has no optical activity. It is known as a meso form. A **meso form** of a compound has stereocenters but no net chirality because of the existence of internal symmetry.

2,3-Dichlorobutane has three stereoisomers. They are the enantiomers (2R,3R)-2,3-dichlorobutane and (2S,3S)-2,3-dichlorobutane and the meso form *meso*-2,3-dichlorobutane, which is named (2R,3S)-2,3-dichlorobutane. In meso compounds, the stereocenter with the lower number is designated as having the R configuration. The meso form is a diastereomer of either one of the enantiomers. In this case, because of symmetry in the compound, we find fewer than the $2^n$ or 4 stereoisomers expected of a compound having two stereocenters.

Study Guide
Concept Map 6.6

**Problem 6.12**    Draw perspective formulas for all stereoisomers possible for 2,3-butanediol, 2,3-pentanediol, and 2,4-pentanediol. Identify the enantiomers and diastereomers.

**Problem 6.13**    The following unusual amino acid, isostatine, is found in dolastatin 10, a cytotoxic peptide isolated from the Indian Ocean sea hare. Point to each stereocenter in isostatine with an arrow and assign configuration R or S to it.

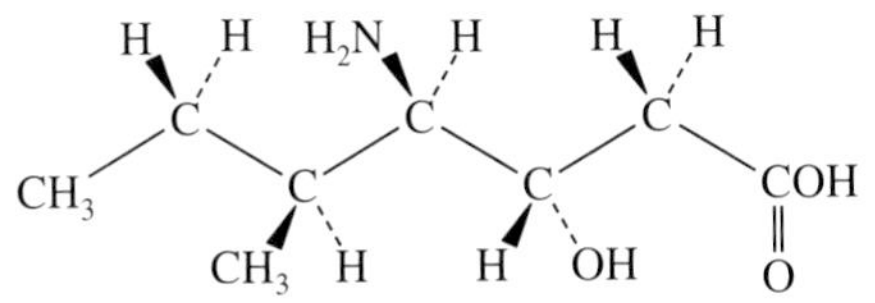

**Problem 6.14**    How many stereoisomers are possible for a compound with three different stereocenters? 2,3,4-Tribromohexane is such a compound. Draw all possible stereoisomers for the compound, and identify the enantiomers and diastereomers. One stereoisomer is shown below to demonstrate how they can be drawn.

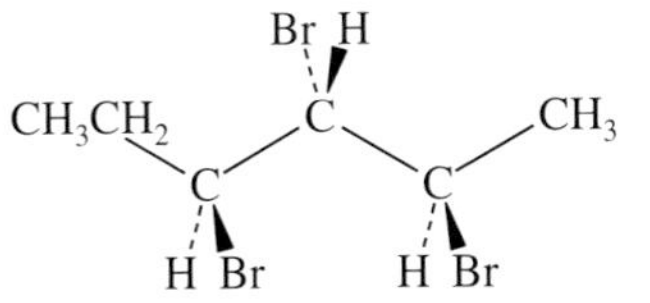

(2R,3R,4S)-2,3,4-tribromohexane

## 6.9    Stereoisomerism in Cyclic Compounds

### A. Cis and Trans Compounds

The addition of bromine to the $\pi$ bond in cyclopentene, a typical reaction of alkenes (Section 2.9), gives 1,2-dibromocyclopentanes in which the two bromine atoms are on opposite sides of the plane of the ring.

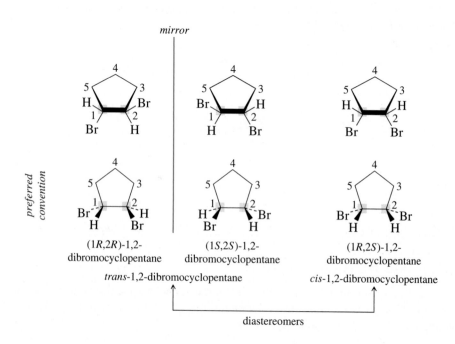

Two stereoisomers, enantiomers of each other, are formed in equal amounts in the reaction, resulting in a racemic mixture.

The rigidity of the ring and the lack of free rotation around the carbon–carbon bonds (p. 179) give rise to stereoisomerism. Two kinds of 1,2-dibromocyclopentanes are possible. The two substituents may be on opposite sides of the plane of the ring, or **trans** to each other. Alternatively, they may be on the same side of the ring, or **cis** to each other (Figure 6.9).

Note how the stereochemistry on the ring is indicated. Two conventions are shown. In one, the bottom of the ring is drawn with heavier lines, and wedged ring bonds are used to indicate that the lower edge of the ring projects out at us. One bromine atom is above and one below that projecting plane in each compound. Simpler and less ambiguous stereochemical formulas use wedges and dashed lines to indicate which bonds are above and which below the plane of the ring. The second way of drawing structures is preferred. In using these structural formulas, however,

*Figure 6.9*

Different stereoisomers of 1,2-dibromocyclopentane.

you must remember that hydrogen atoms are also present at the carbon atoms bearing the bromine atoms. When the bromine atom is below the plane of the ring, the hydrogen atom is above, and vice versa.

Each of the *trans*-1,2-dibromocyclopentanes has two stereocenters. Configuration is assigned to stereocenters in rings exactly the same way as it is to other stereocenters. For example, the isomer shown in the margin is (1*R*,2*R*)-1,2-dibromocyclopentane. At carbon atom 1, the bromine atom has the highest priority. The side of the ring going to the carbon atom bearing a bromine atom is next. The other side of the ring is third priority. The hydrogen atom, which is sticking up, has fourth priority. Your eye travels counterclockwise from the group of priority 1 to that of 2 to that of 3; the stereocenter appears to be *S* but is really *R* because we are viewing the molecule from the wrong face with the group of lowest priority pointing toward us. If you were to go behind the page, you would be looking at the stereocenter from the correct angle and would then have a clockwise movement of your eye from the group of priority 1 to that of 2 to that of 3.

The hydrogen atom on the stereocenter at carbon atom 2 points away from us. Your eye travels clockwise from the group of priority 1 to that of 2 to that of 3. The stereocenter, therefore, has the *R* configuration.

The other *trans*-1,2-dibromocyclopentane has the *S* configuration at each stereocenter. Thus the two isomers are mirror-image isomers, or enantiomers, of each other. The molecules of the isomers are not superimposable on each other and are, therefore, truly different species.

*cis*-1,2-Dibromocyclopentane is a stereoisomer of the two trans compounds but is not a mirror image of either one. Having the *R* configuration at one stereocenter and the *S* configuration at the other one, it is a diastereomer (p. 211) of the trans compounds and a meso compound (p. 213). A plane of symmetry bisects carbon atom 4, the two hydrogen atoms on it, and the carbon–carbon bond between carbon atoms 1 and 2. *cis*-1,2-Dibromocyclopentane is not chiral and has no optical activity (Figure 6.10).

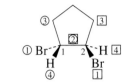

(1*R*,2*R*)-1,2-dibromocyclopentane

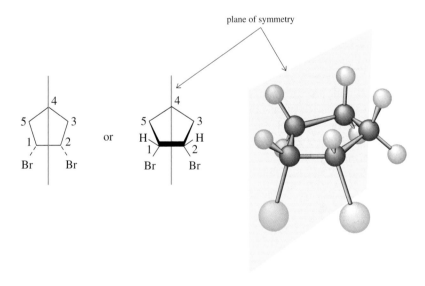

*cis*-1,2-dibromocyclopentane

*achiral*

*a meso compound*

**Figure 6.10**
The symmetry of *cis*-1,2-dibromocyclopentane.

**Problem 6.15** How many stereoisomers are possible for 1,3-dimethylcyclopentane? Draw the structural formulas, name the compounds, and discuss the stereochemical relationships among them.

**Problem 6.16** How many stereoisomers are possible for 1-bromo-2-methylcyclopentane? Draw structural formulas, and give them names that designate configurations at the stereocenters. You may want to use molecular models.

**Problem 6.17** Kainic acid is an important neurotoxin that is used widely in research into diseases such as epilepsy and Alzheimer's disease. It has the structure shown at right.

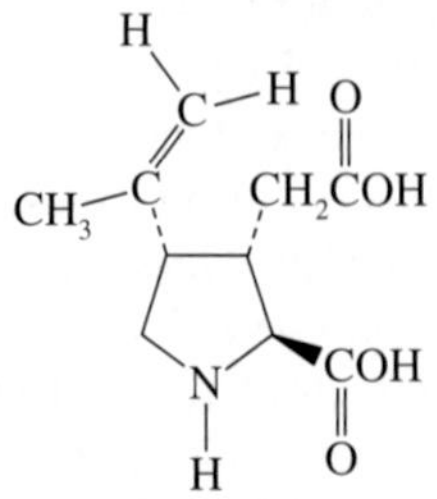

(a) Identify each source of stereoisomerism in kainic acid, and assign configuration *R*, *S*, *E*, or *Z* as appropriate.
(b) Draw the structural formula for a stereoisomer of kainic acid. What is the relationship between the structure you drew and kainic acid?

## B. Configuration and Conformation of Disubstituted Cyclohexanes

For a cyclohexane with a single substituent on the ring, the preferred conformation is the chair form in which the substituent is in the equatorial position (p. 183). When there are two substituents on the ring, they may be either cis or trans to each other. For example, there are three stereoisomers of 1,2-dimethylcyclohexane, a pair of enantiomeric trans isomers and a cis isomer.

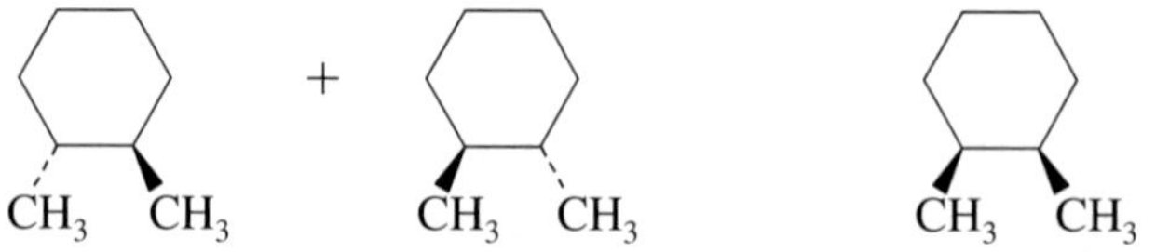

CH₃ CH₃ + CH₃ CH₃ CH₃ CH₃

(1*R*,2*R*)-1-2-
dimethylcyclohexane

(1*S*,2*S*)-1-2-
dimethylcyclohexane

*trans*-1,2-dimethylcyclohexane

(1*S*,2*R*)-1-2-
dimethylcyclohexane

*cis*-1,2-dimethylcyclohexane

In the trans isomers, the two methyl groups are on opposite sides of the ring. In the cis isomer, they are on the same side of the ring.

To convert a drawing of the planar form of a cyclohexane compound with more than one substituent on it to a chair conformation, care must be taken to make sure that the two structural formulas correspond to the same stereoisomer. The process is illustrated below for (1*S*,2*S*)-1,2-dimethylcyclohexane.

(1*S*,2*S*)-1,2-Dimethylcyclohexane has a methyl group above the plane of the ring at carbon atom 1 and below the plane of the ring at carbon atom 2. We draw any chair conformation of cyclohexane, for example:

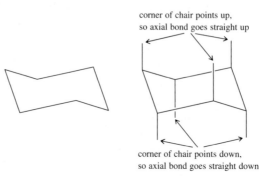

corner of chair points up,
so axial bond goes straight up

corner of chair points down,
so axial bond goes straight down

We then fill in all the axial bonds to remind ourselves which bonds are above and which below the plane of the ring. The axial bonds always point in the same direction as the point of the chair to which they are attached. There will always be three axial bonds pointing up and three pointing down. The equatorial bonds point slightly up at the carbon atoms that have axial bonds pointing down and slightly down at the carbon atoms that have axial bonds pointing up.

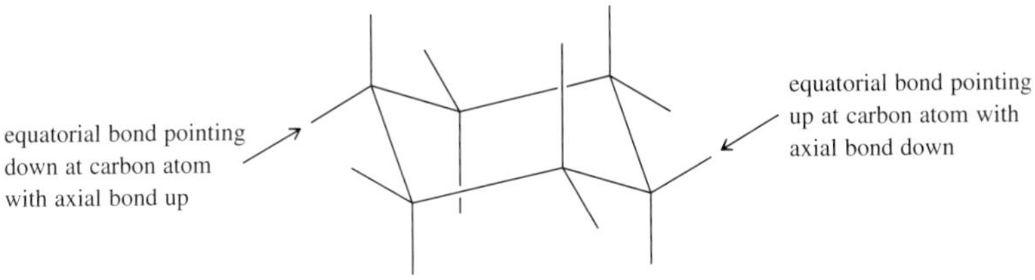

It does not matter which of these bonds we use to attach the methyl groups as long as one is up and the other is down on the next carbon atom in a counterclockwise direction from the point of first attachment. *Either equatorial or axial positions may be used to make the first attachment.* Whether the group is above or below the plane of the ring is the important factor, *not* whether it is axial or equatorial. Note that an axial or an equatorial bond may point either above or below the average plane of the ring. It is easiest to see the direction of the bonds in the axial positions, so we will use them for our first drawing.

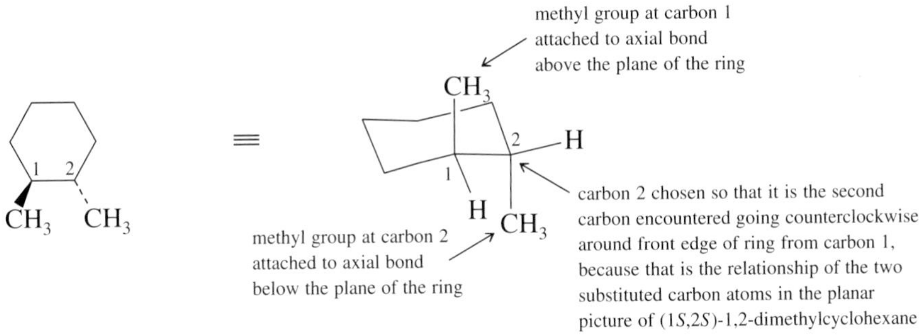

Note that we use wedges and dashed lines to indicate which bonds are up and which bonds are down when we use a planar hexagon to show the cyclohexane ring. If we use a chair form of the cyclohexane ring, the bonds on that drawing already show stereochemistry by whether they are axial or equatorial. Therefore, no wedges or dashed lines are needed, and none are used with drawings of chair forms.

The diaxial conformer of (1*S*,2*S*)-1,2-dimethylcyclohexane that we drew above is in equilibrium with the other chair conformer in which the two methyl groups occupy equatorial positions on the ring (Section 5.11C).

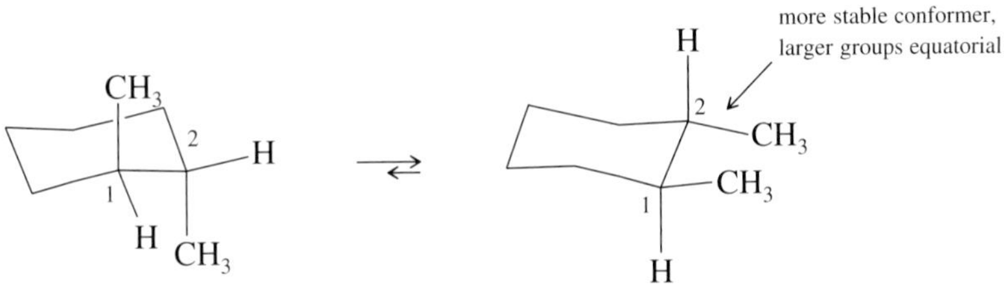

diaxial form of
(1*S*,2*S*)-1,2-dimethylcyclohexane

diequatorial form of
(1*S*,2*S*)-1,2-dimethylcyclohexane

Note that the diequatorial form retains the spatial relationship between the two methyl groups found in the diaxial form. The methyl group at carbon atom 1 is above the plane of the ring; the methyl group at carbon atom 2 is below the plane of the ring. If the two are conformers of the same stereoisomer, the methyl groups must take these positions. Changes in conformation do not change the configuration at stereocenters (p. 206). Each enantiomer of *trans*-1,2-dimethylcyclohexane exists mainly in two conformations. The diequatorial form, which has the large substituents in equatorial positions, predominates in the equilibrium mixture at room temperature.

In *cis*-1,2-dimethylcyclohexane, one methyl group is axial, and the other one is equatorial. Reversal of the ring puts the methyl group that was originally equatorial into the axial position, and the one that was originally axial becomes equatorial.

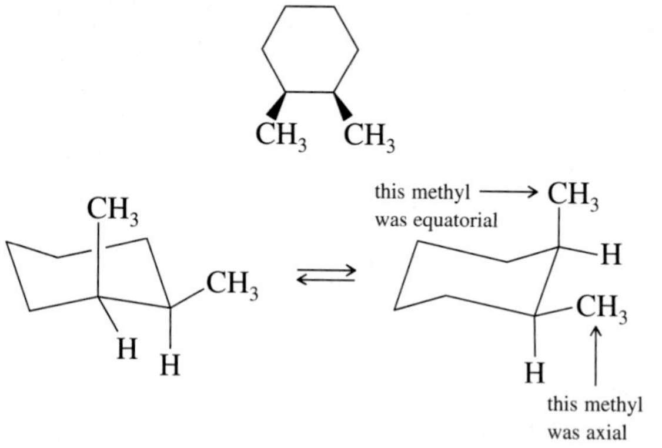

*cis*-1,2-dimethylcyclohexane

These two conformers have equal energy and are present in equal amounts. Although each conformer is chiral, the rapid interconversion between the two at room temperature means that they cannot be separated. The compound is not optically active. *cis*-1,2-Dimethylcyclohexane is a diastereomer of each of the *trans*-1,2-dimethylcyclohexanes.

For 1,3-disubstituted cyclohexanes, the cis isomer is the more stable one. *cis*-1,3-Dimethylcyclohexane has two conformations, one in which both alkyl groups are equatorial and one in which both are axial (Figure 6.11). The diequatorial conformer of *cis*-1,3-dimethylcyclohexane is more stable than the diaxial one by about 5.4 kcal/mol, largely because of 1,3-diaxial interactions (p. 183) of the two methyl

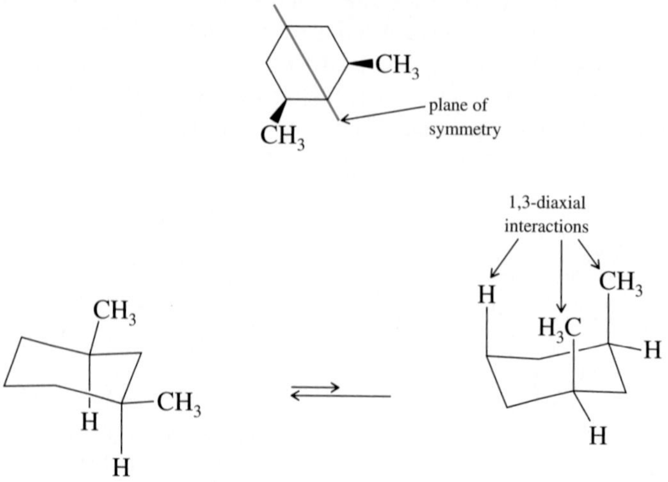

*Figure 6.11*
Conformers and symmetry of
*cis*-1,3-dimethylcyclohexane.

*cis*-1,3-dimethylcyclohexane

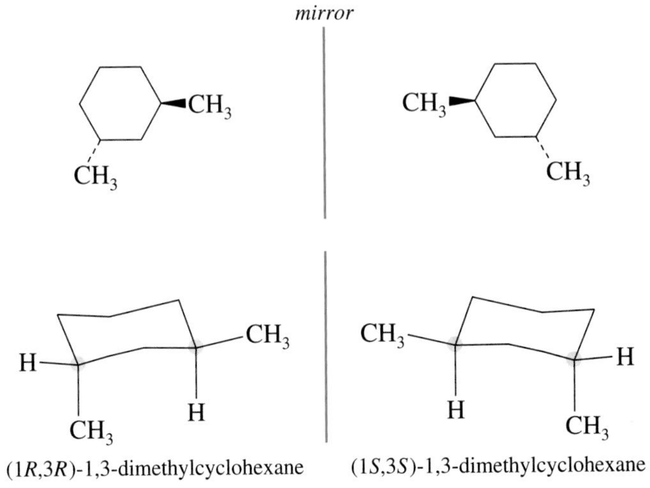

Figure 6.12
The enantiomeric *trans*-1,3-dimethylcyclohexanes.

groups. *cis*-1,3-Dimethylcyclohexane has a plane of symmetry and is a meso compound.

*trans*-1,3-Dimethylcyclohexane exists as two enantiomers. Each enantiomer has one axial and one equatorial methyl group (Figure 6.12).

Neither *cis*- nor *trans*-1,4-dimethylcyclohexane has stereocenters, and both are optically inactive. Starting with either carbon atom 1 or carbon atom 4, an investigation of the substituents on that carbon atom turns up a hydrogen atom, a methyl group, and then two branches of the ring that are identical going in either direction. The molecules as a whole have planes of symmetry bisecting carbon atoms 1 and 4 and the methyl group and hydrogen atom on each one.

*trans*-1,4-Dimethylcyclohexane has two conformers, one in which the two methyl groups are equatorial and one in which they are axial (Figure 6.13). The form with two equatorial substituents is more stable than the other by about 3.6 kcal/mol.

In *cis*-1,4-dimethylcyclohexane, one methyl group is axial and the other equatorial (Figure 6.14, p. 220). Ring reversal gives a form that is identical with the starting structure.

In these examples, the two substituents on the ring were identical, giving rise to symmetry in *cis*-1,2 and *cis*-1,3 compounds and reducing the number of stereoisomers. If two different stereocenters are present, then there will be four stereoisomers. For example, 1-bromo-3-methylcyclohexane has four stereoisomers, two enantiomeric cis isomers and two enantiomeric trans isomers (Figure 6.15, p. 220).

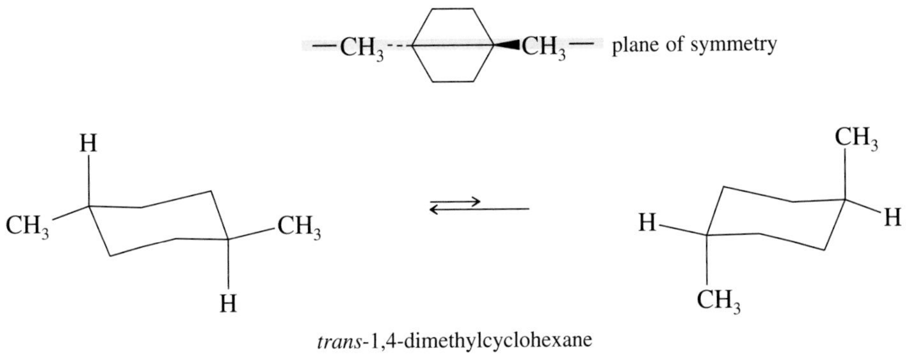

*trans*-1,4-dimethylcyclohexane

Figure 6.13
Conformers and symmetry of *trans*-1,4-dimethylcyclohexane.

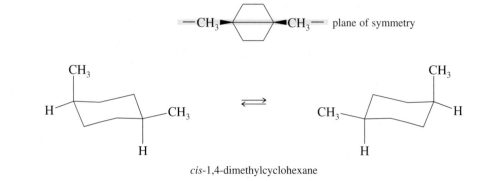

**Figure 6.14**
The symmetry of *cis*-1,4-dimethylcyclohexane.

*cis*-1,4-dimethylcyclohexane

*mirror*                                                             *mirror*

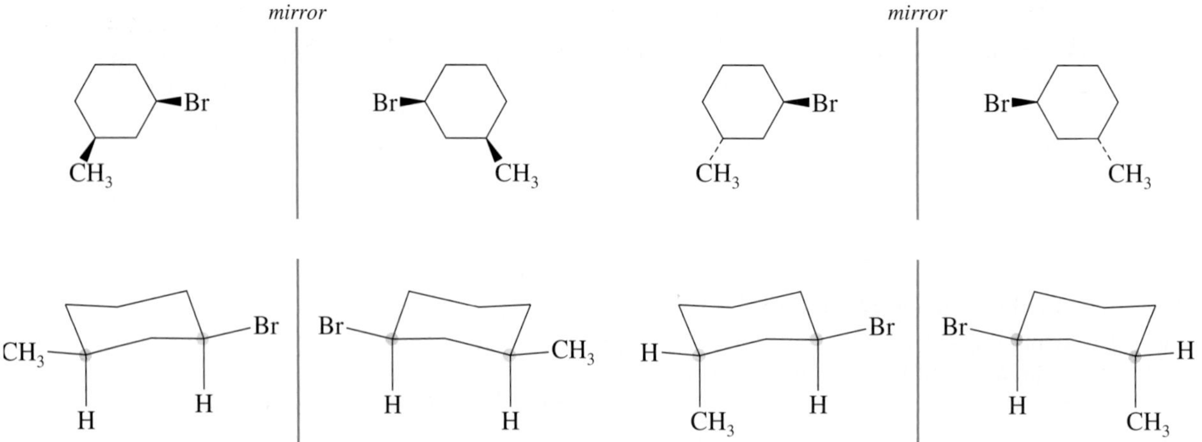

(1*R*,3*S*)-1-bromo-3-methylcyclohexane

(1*S*,3*R*)-1-bromo-3-methylcyclohexane

(1*R*,3*R*)-1-bromo-3-methylcyclohexane

(1*S*,3*S*)-1-bromo-3-methylcyclohexane

*cis*-1-bromo-3-methylcyclohexanes

*trans*-1-bromo-3-methylcyclohexanes

**Figure 6.15**
The stereoisomers of 1-bromo-3-methylcyclohexane.

**Problem 6.18**    How many different configurational isomers are possible for 1-bromo-2-methylcyclohexane? Illustrate by drawing planar formulas and chair conformations. Do the same for 1-bromo-4-methylcyclohexane.

**Problem 6.19**    L-Vancosamine is an essential constituent of vancomycin, an antibiotic that is the last line of defense against severe bacterial infections.

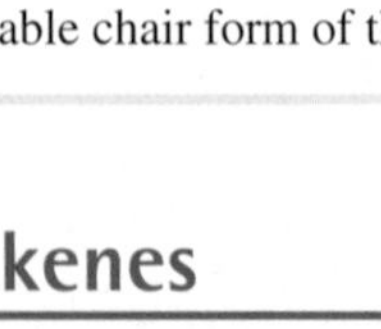

L-vancosamine

(a) Assign configuration to the stereocenters in L-vancosamine.
(b) Draw the most stable chair form of the compound.

## 6.10    Stereoisomerism in Alkenes

### A. The Origin of Stereoisomerism in Alkenes

The bonding between two carbon atoms involved in a double bond has been described (p. 54) as consisting of a $\sigma$ bond and a $\pi$ bond, which is created by overlap of the *p* orbitals. The necessity that the *p* orbitals in the $\pi$ bond must overlap im-

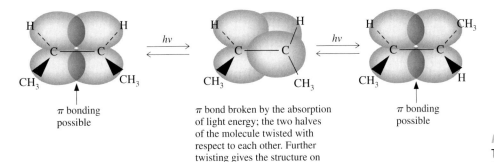

π bonding
possible

π bond broken by the absorption
of light energy; the two halves
of the molecule twisted with
respect to each other. Further
twisting gives the structure on
the right.

π bonding
possible

*Figure 6.16*
The interconversion of *cis*-2-
butene and *trans*-2-butene.

parts a certain rigidity to the double bond. Rotation of the carbons on the two ends of the double bond with respect to each other does not take place unless enough energy is supplied to break the π bond. A carbon–carbon single bond consisting of a σ bond with cylindrical symmetry (p. 44), however, does allow free rotation (p. 164) of the bonded atoms with respect to each other.

The rigidity of the double bond gives rise to the possibility of stereoisomerism. For example, 2-butene can have two methyl groups on the same side of the molecule (or cis to each other) or on opposite sides (or trans to each other). Each isomer is converted to the other when enough energy is supplied, for example, by absorption of ultraviolet radiation (p. 72) or being heated to temperatures around 300 °C. The conversion takes place because the π bond breaks when energy is absorbed, and the two halves of the molecule can then rotate with respect to each other before the π bond forms again (Figure 6.16).

The interconversion of stereoisomers having a double bond requires much more energy (~65 kcal/mol) than does the interconversion of conformational isomers by rotations around single bonds (less than 10 kcal/mol; see p. 165). Conformational isomers cannot be separated and isolated as individual molecular species at ordinary temperatures (where ~20 kcal/mol of kinetic energy is available to molecules), but the stereoisomers created by different spatial arrangements of groups around double bonds can. Double-bond stereoisomers are also configurational isomers (p. 191). They are stereoisomers but not mirror-image isomers and, therefore, are diastereomers (p. 211). Alkenes, because of the planarity of the double bond, have a plane of symmetry and are not chiral. An alkene, therefore, does not show optical activity unless there is a stereocenter elsewhere in the molecule.

Even alkenes with relatively simple structures show several kinds of isomerism. For example, there are four different noncyclic structures that can be drawn for $C_4H_8$.

$$CH_3CH_2 \quad H$$
$$C{=}C$$
$$H \qquad H$$

identical substituents
at one end of double
bond; no stereoisomerism
possible at a double bond

$$CH_3 \quad H$$
$$C{=}C$$
$$CH_3 \quad H$$

also identical
in this case

1-butene
bp −5 °C

2-methylpropene
bp −6 °C

different substituents
at each end of the
double bond;
stereoisomerism
possible

$$CH_3 \quad H$$
$$C{=}C$$
$$H \qquad CH_3$$

*trans*-2-butene
bp 2.5 °C

$$H \qquad H$$
$$C{=}C$$
$$CH_3 \quad CH_3$$

*cis*-2-butene
bp 1 °C

*isomeric structures for alkenes with the
molecular formula $C_4H_8$*

1-Butene and 2-methylpropene are constitutional isomers of the 2-butenes. They have different connectivities. They do not have stereoisomers. Stereoisomerism is not possible for an alkene unless two different substituents are present on the carbon at each end of the double bond. *cis*-2-Butene and *trans*-2-butene are stereoisomers of each other. They have the same connectivity and differ from each other only in the spatial arrangement of the methyl groups and the hydrogen atoms. This similarity is reflected in their names, which are identical except for the designation for spatial orientation. *cis*-2-Butene and *trans*-2-butene, being diastereomers (p. 211), have different physical properties, for example, their different boiling points.

## B. Nomenclature of Stereoisomeric Alkenes

When an alkene has stereoisomers, the complete name of the compound should include a definition of the stereochemistry, if it is known. In simple alkenes such as the 2-butenes, the designations cis and trans are used with no ambiguity. In the cis isomer, similar groups are on the same side of the double bond; in the trans isomer, they are on opposite sides of the double bond. In some simple cases, such nomenclature is used with no confusion.

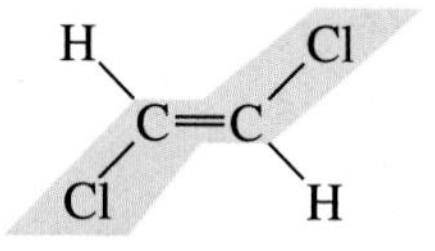

*trans*-1,2-dichloroethene 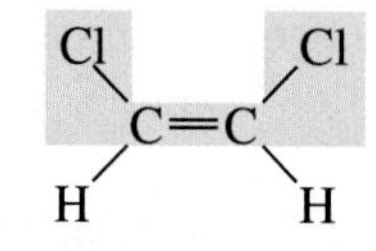    *cis*-1,2-dichloroethene 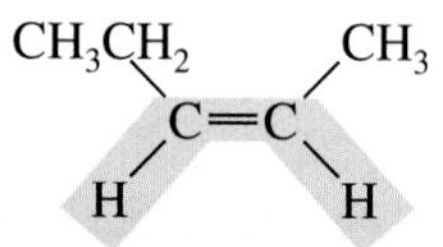    *cis*-2-pentene 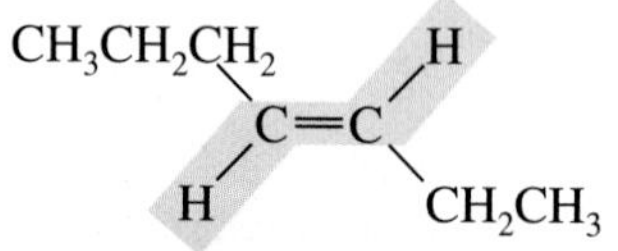    *trans*-3-heptene

The stereochemistry of more highly substituted alkenes is harder to define as cis or trans. The Cahn-Ingold-Prelog Rules (p. 206) are used systematically to assign priorities to the substituents on a double bond. The double bond is assigned a **Z** (for *zusammen*, the German word for "together") configuration if the two groups of higher priority at each end of the double bond are on the same side of the molecule. If the two groups of higher priority are on the opposite sides of the double bond, the configuration is denoted by an **E** (for *entgegen*, the German word for "opposite").

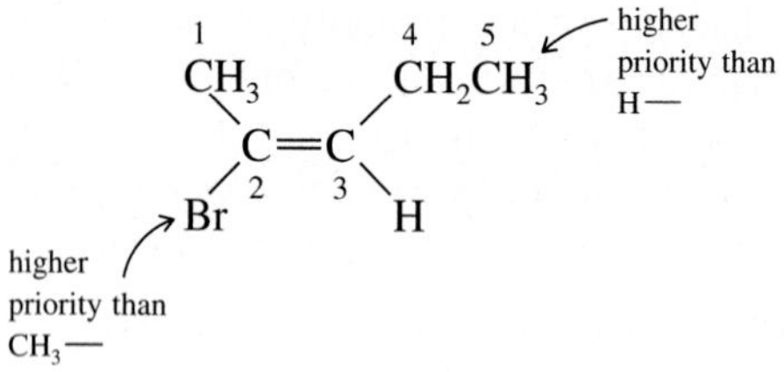

*groups of higher priority are on opposite sides of the double bond; therefore this is*
*(E)-2-bromo-2-pentene*

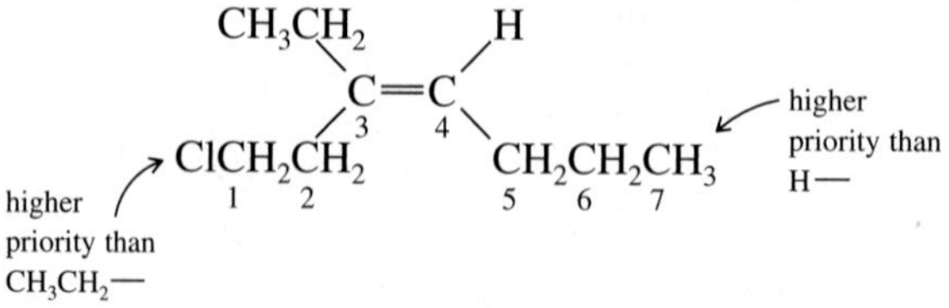

*groups of higher priority are on the same side of the double bond; therefore this is*
*(Z)-1-chloro-3-ethyl-3-heptene*

CH₃
|
higher     →   CH₃CH        CH₂CH₂CH₃
priority than   1    2
CH₃—                \ 3   4 /
                    C=C
                   /        \
                CH₃       CH₂CH₂CH₂CH₃       ← higher
                           5   6   7   8        priority than
                                                CH₃CH₂CH₂—

*groups of higher priority are on opposite*
*sides of the double bond; therefore this*
*is (E)-2,3-dimethyl-4-propyl-3-octene*

To avoid any possible confusion, this system of assigning configuration to stereoisomers of alkenes is used in this book. The terms *cis* and *trans* are reserved for descriptions of the spatial relationship between groups. For example, in (*E*)-2-bromo-2-pentene shown on previous page, the methyl and the ethyl group on the carbon atoms of the double bond are cis to each other. The molecule, however, has the *E* configuration.

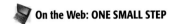

**On the Web: ONE SMALL STEP**

**Study Guide**
Concept Map 6.7

---

**Problem 6.20**   Assign configuration to each of the following alkenes.

(a)
CH₃CH₂        CH₃
        \    /
         C=C
        /    \
       H      CH₂CH₃

3-methyl-3-hexene

(b)
Br           CH₂Br
  \         /
   C=C
  /         \
Cl           CH₃

1,3-dibromo-1-chloro-
2-methylpropene

(c)
ClCH₂CH₂        O
           \    ‖
            C=C  COH
           /    \
       CH₃CH₂    H

5-chloro-3-ethyl-
2-pentenoic acid

(d)
CH₃CH₂            H
        \        /
         C=C    C
        /    \ ‖‖
       H      CH₂CH₃

3-ethyl-3-hexen-1-yne

---

# 6.11   The Resolution of a Racemic Mixture

## A. Separation of Mixtures

Mixtures of organic compounds are separated by means of a variety of laboratory techniques. Compounds with different boiling points may be separated by distillation, for example. If compounds have different solubilities in solvents, they may be separable by extraction or recrystallization. An acidic compound can be removed from a mixture by washing the mixture with a dilute base, such as a solution of sodium bicarbonate. Chromatographic techniques will separate not only compounds belonging to different functional group classes but also members of the same functional group class with differing molecular structures. Diastereomers, such as cis and trans isomers of alkenes or cyclic compounds, also can be separated chromatographically.

What about compounds that differ from each other only in the configuration *R* or *S* of stereocenters? Compounds that have only one stereocenter exist as two mirror-image isomers, or enantiomers. Enantiomers have identical physical properties except for the direction in which they rotate the plane of polarized light (p. 199). Enantiomers are indistinguishable with respect to the polarity of their molecules and have identical solubilities in ordinary solvents, so they cannot be separated by recrystallization or chromatography in the usual way. They have identical boiling points, so distillation cannot be used to separate them.

When a chiral compound containing a single stereocenter is prepared in the laboratory, starting with achiral reagents, both enantiomers are formed in equal amounts. The product is a racemic mixture. An example of such a reaction is the addition of hydrogen bromide to 1-butene (p. 201). Once a racemic mixture is formed, the two enantiomers in the mixture are inseparable by the techniques normally used to separate constitutional isomers and diastereomers.

As discussed earlier (p. 211), compounds that have two different stereocenters have four stereoisomers, which are two sets of enantiomers. The members of one pair of enantiomers are diastereomers of the compounds in the other pair of enantiomers. The diastereomeric pairs have physical properties sufficiently different that they can be separated by usual techniques such as recrystallization, distillation, and chromatography.

## B. Resolution of Racemic Mixtures

The separation of a mixture of enantiomers is called the **resolution of a racemic mixture.** Louis Pasteur performed a very unusual resolution using a pair of tweezers on the crystals of sodium ammonium tartrate to pick out left-handed and right-handed crystals under a microscope (p. 204). Living organisms have enzyme systems that are highly sensitive to the stereochemistry of the compounds with which they interact; these systems perform resolutions when they metabolize one enantiomer and reject another one. Yeast cells, for example, if allowed to ferment in a medium containing sugar and a racemic amino acid, consume the enantiomer of the amino acid that is found naturally in proteins and leave the other. Most vertebrates, including human beings, also separate enantiomers biologically. For example, if racemic alanine (p. 194) is fed to a human being, the $S$ isomer is metabolized and the $R$ isomer is excreted in the urine. Researchers have observed that in most biological resolutions, only one enantiomer, the one that cannot be used by the living organism, is recovered.

A common way to separate enantiomers is to convert them into compounds that are diastereomers, which have different physical properties and, as a result, are separable. Reactions of enantiomers with achiral reagents cause no change in their relationship as mirror-image isomers of each other. Only when the reagent itself is chiral is new stereochemistry introduced into the molecule and, with it, the possibility of a physical separation. The most widely used technique for the resolution of mixtures of enantiomers depends on the reaction of a chiral acid with a racemic base (or a chiral base with a racemic acid) to give a mixture of diastereomeric salts. The salts are separated by recrystallization, and the individual enantiomers, as well as the chiral reagent, are recovered by further acid–base reactions.

The next section contains a description of an actual resolution that will help to clarify these ideas.

## C. The Process of Resolution

Many synthetic processes give rise to racemic mixtures (p. 201). For biologically active compounds, the two enantiomers often have different physiological properties and effects. In this section we are going to examine in detail how a racemic mixture of a medication, amphetamine, is separated into its two enantiomers as an example of the general principles discussed in Section 6.11B.

Amphetamine, also known as benzedrine, is 2-amino-1-phenylpropane.

2-amino-1-phenylpropane
or amphetamine
or benzedrine

Amphetamine has one stereocenter and, therefore, two stereoisomers. Each enantiomer has distinctive physiological properties. The dextrorotatory isomer has the *S* configuration and is the form that stimulates the central nervous system. It is sold as its sulfuric acid salt under names such as dextroamphetamine sulfate. The levorotatory isomer acts on the sympathetic nervous system.

Amphetamine is synthesized as a racemic mixture, which must be resolved to isolate the *S* isomer. (+)-Tartaric acid is used as the chiral reagent for the resolution. The mixture of the enantiomers of amphetamine and a molar equivalent of (+)-tartaric acid are dissolved in ethanol. A proton is transferred from one of the carboxylic acid groups of (+)-tartaric acid to the amine group of amphetamine. Two salts are formed, one from each enantiomer of the amphetamine.

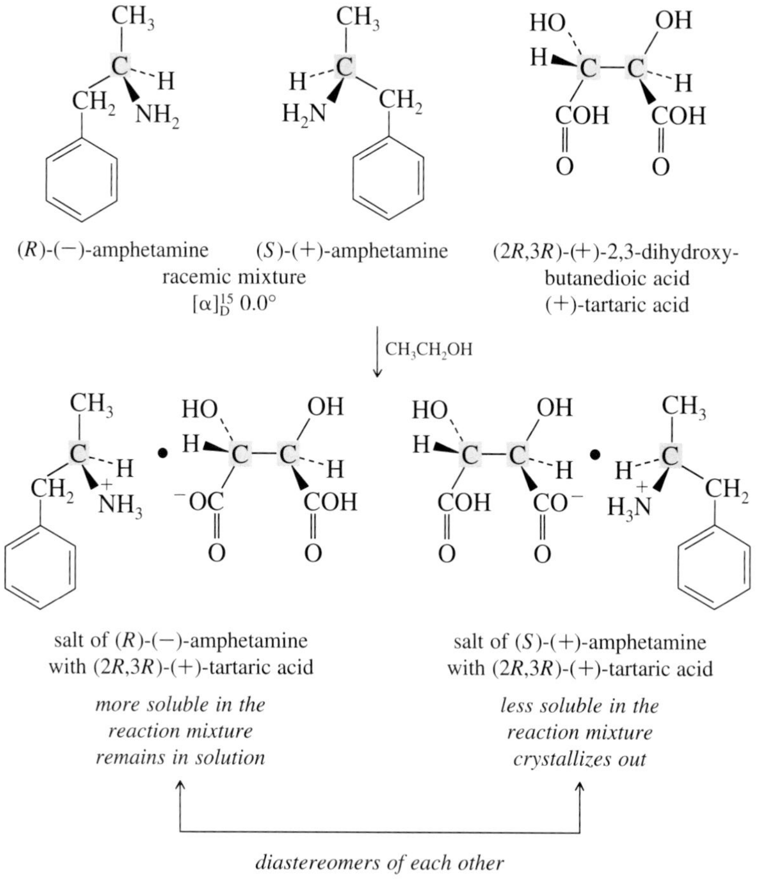

| (R)-(−)-amphetamine | (S)-(+)-amphetamine | (2R,3R)-(+)-2,3-dihydroxy- |
|---|---|---|
| racemic mixture | | butanedioic acid |
| $[\alpha]_D^{15}\ 0.0°$ | | (+)-tartaric acid |

$CH_3CH_2OH$

| salt of (R)-(−)-amphetamine | salt of (S)-(+)-amphetamine |
|---|---|
| with (2R,3R)-(+)-tartaric acid | with (2R,3R)-(+)-tartaric acid |
| *more soluble in the* | *less soluble in the* |
| *reaction mixture* | *reaction mixture* |
| *remains in solution* | *crystallizes out* |

*diastereomers of each other*

The products are diastereomers of each other. If you examine closely the structural formulas written for the salts, you will see that the two compounds are *not* mirror-image isomers of each other. The amine portions of the molecules remain enantiomeric, but attached to each one is the ion of an acid, and it is identical in both cases. The two salts have three stereocenters but do not have the opposite configuration at each of these points. One salt could be called the *R,R,R* salt and the other the *S,R,R* salt. The two compounds are thus stereoisomers but not enantiomers. By definition, all stereoisomers that are not enantiomers are diastereomers of each other (p. 211).

The diastereomeric salts have different physical properties. The salt derived from the *S* amine is less soluble in the reaction mixture than the other salt is and crystallizes out of solution. The free amine is recovered by treating this salt with potassium hydroxide.

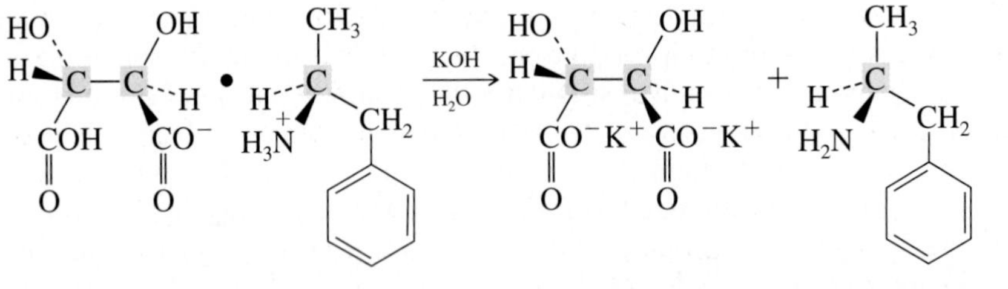

salt of (S)-(+)-amphetamine
with (2R,3R)-(+)-tartaric acid

*soluble in water
remains in solution*

*insoluble in water
extracted out of
solution and distilled
$[\alpha]_D^{15}$ +40.2°
(c 8.7; benzene)*

The strong base hydroxide ion removes a proton from the organic ammonium ion and regenerates the free water-insoluble amine, which is separated from the water-soluble, nonvolatile dipotassium salt of (+)-tartaric acid by extraction and distillation. The purified amine has an optical rotation of $[\alpha]_D^{15}$ +40.2° (c 8.7; benzene).

Crystals of the salt of the *R* amine with (+)-tartaric acid are also recovered from the solution. These are contaminated by the crystals of the salt of the *S* amine, so further purification is necessary. The *R* amine is recovered in the same way the *S* amine is and then is treated with (−)-tartaric acid, which is also available from natural sources.

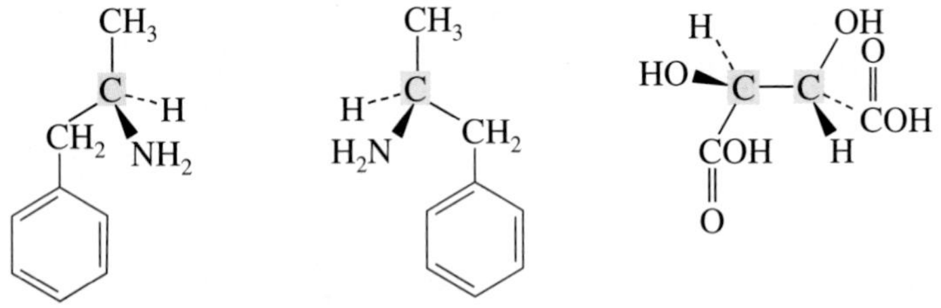

(R)-(−)-amphetamine
*most of mixture*

(S)-(+)-amphetamine
*present as impurity*

(2S,3S)-(−)-2,3-dihydroxy-
butanedioic acid
(−)-tartaric acid

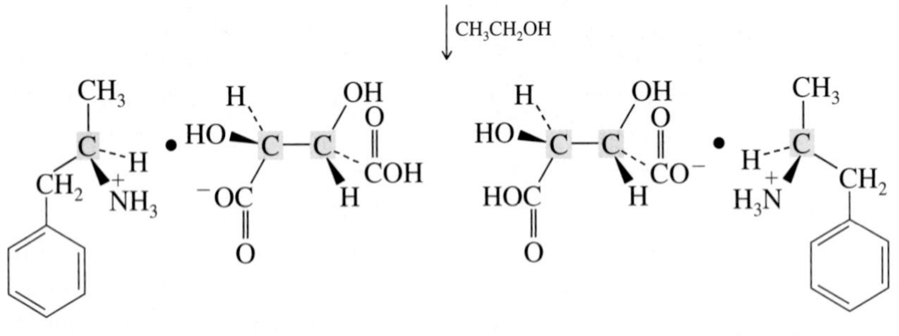

$\downarrow$ CH₃CH₂OH

salt of (R)-(−)-amphetamine
with (2S,3S)-(−)-tartaric acid

*less soluble in the
reaction mixture
crystallizes out*

salt of (S)-(+)-amphetamine
with (2S,3S)-(−)-tartaric acid

*more soluble in the
reaction mixture
remains in solution*

Here the salt of the *R* amine with (−)-tartaric acid is the less soluble one and crystallizes out of the solution. The *R* amine is recovered from the salt using exactly the same method used for recovering the *S* amine and, after being distilled, has $[\alpha]_D^{15}$ −40.1° (c 8.83; benzene). The two enantiomeric amines, having been separated from a racemic mixture, are seen to have optical rotations that are equal in magnitude (within experimental error) but opposite in sign.

The process described above is the resolution of a racemic mixture. The chiral reagent used, tartaric acid in this case, is called the **resolving agent.** A large number of resolving agents are available; many of them are acids and bases from natural sources or new ones created by modifications of naturally occurring chiral compounds. Some synthetic compounds that are easily resolved are then used as resolving agents themselves.

Resolutions of racemic mixtures are not easy procedures. They require a great deal of patience and skill, and some good luck, too. In many cases, they depend on taking advantage of the properties of organic compounds as acids and bases. The most important step in any resolution, however, is the conversion of the mixture of enantiomers into a mixture of diastereomers that have physical properties that allow for their separation.

The separation of diastereomers need not be done so that isolable compounds end up in different flasks. For example, if a chromatography column were to contain a chiral adsorbent, one enantiomer would interact more strongly than the other with the material in the column, and a separation would take place. Such methods are already being used and will certainly be used more and more in the future as new chiral adsorbents and complexing agents are developed.

(↗) **Study Guide**
**Concept Map 6.8**

**Problem 6.21**    Write an equation showing how (R)-(−)-amphetamine can be recovered from the salt it forms with (−)-tartaric acid.

**Problem 6.22**    Write structural formulas showing the details for the resolution of a racemic mixture of 1-amino-1 phenylethane using (+)-tartaric acid. The separation takes place in methanol as the solvent. The salt of (S)-(−)-1-amino-1-phenylethane with (+)-tartaric acid is the one with the lower solubility.

The difference in physiological properties exhibited by the enantiomers of amphetamine (p. 225) is turning out to be general. Synthetic medications in the past have been sold as racemic mixtures, sometimes with tragic results (see Problem 6.32). The Food and Drug Administration is now requiring that more and more of them be prepared, tested, and sold as single enantiomers. Resolution of a racemic mixture is one way to obtain one enantiomer out of a racemic mixture. Because the process is cumbersome and results in a waste of half the product, however, chemists are increasingly turning to methods that use chiral reagents and chiral catalysts to obtain enantiomerically pure compounds in the first place.

## SUMMARY

Stereochemistry deals with the spatial properties of compounds and chemical reactions. Compounds that differ from each other only in how their atoms are arranged in space are known as stereoisomers. Stereoisomers are classified into conformational isomers, which are interconvertible by rotations around single bonds at room temperature, and configurational isomers. Configurational isomers can be either enantiomers, which are related to each other as nonsuperimposable mirror images, or diastereomers, which include all other stereoisomers. An important structural feature of many stereoisomers is the stereocenter, a tetrahedral atom bonded to four different substituents.

Enantiomers have identical physical properties except that they rotate a plane of polarized light in opposite directions. Compounds that rotate plane-polarized light are said to have optical activity. A mixture of equal amounts of two enantiomers has no optical activity and is called a racemic mixture. To separate the enantiomers in a racemic mixture, a resolution must be performed.

Diastereomers have different physical properties and may be separated from each other by ordinary physical means. Diastereomers may have optical activity if they contain stereocenters; however, if they have a plane of symmetry, they are meso compounds and are not optically active. Compounds that contain $n$ stereocenters have $2^n$ stereoisomers unless symmetry reduces this number. Diastereomers such as cis and trans isomers of alkenes do not have optical activity.

Configurational isomers are named using the Cahn-Ingold-Prelog Rules. Priorities are assigned to substituents on carbon atoms that are stereocenters or are involved in double bonds. Depending on the spatial relationships of the groups, stereocenters are assigned $R$ or $S$ configurations and double bonds, $E$ or $Z$ configurations. These configurations can be assigned by looking at structural formulas of compounds. Both the sign of optical rotation, $(+)$ or $(-)$, and the actual arrangement in space of the atoms in a compound that shows a particular rotation have to be determined experimentally. The spatial arrangement of the atoms is called the absolute configuration of the compound.

## ADDITIONAL PROBLEMS

**6.23** Name the following compounds.

(a)  (b)  (c)  (d)

(e)  (f)  (g)  (h)

**6.24** Which of the following compounds have stereocenters? Draw three-dimensional pictures of those that do, showing any enantiomers and diastereomers.

(a)   (b) $ClCH_2CHCH_2CH_2Cl$
                                        $\quad\quad\quad\;\; | $
                                        $\quad\quad\quad\; Cl$
   (c)   (d) $Br-\bigcirc-Br$   (e) $CH_3CH_2CCH_2OH$ with $CH_3$ top and $OH$ bottom

(f) $CH_3CHCH_2CH_3$ with $CH_3$ above   (g)   (h) $CH_3C-CHCH_2CH_3$ with $CH_3$ top, $HO$ and $OH$ below   (i)

**6.25** For each of the following pairs of structural formulas, tell whether the two represent identical molecular species, conformers of the same species, constitutional isomers, enantiomers, or diastereomers.

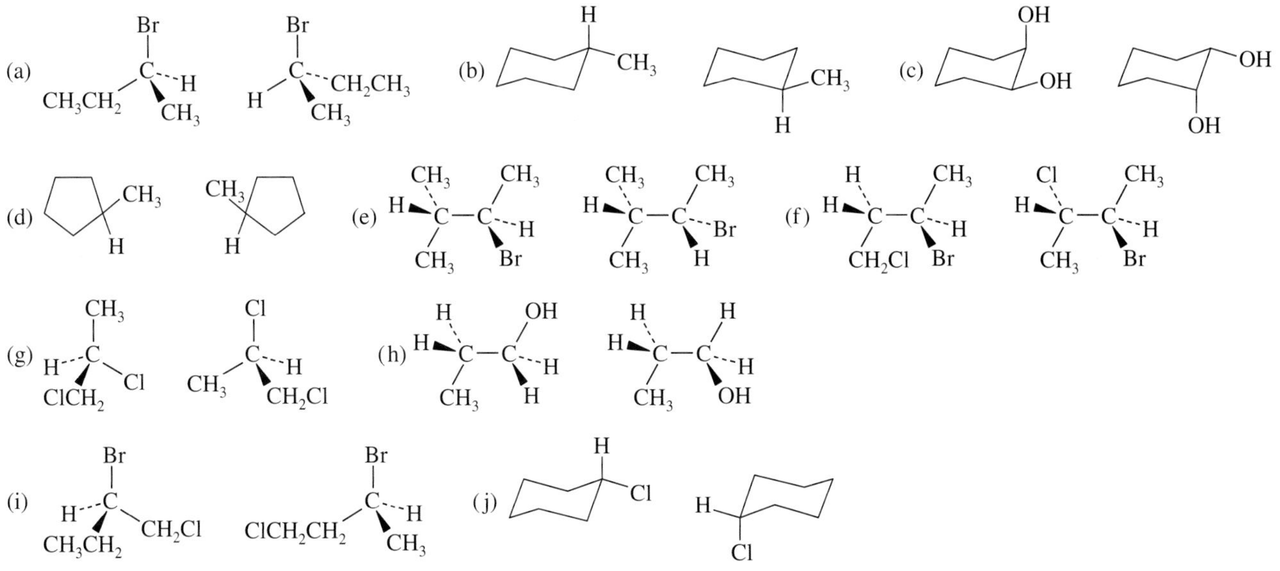

(a) [structure: Br, C, H, CH₃CH₂, CH₃]   [structure: Br, C, CH₂CH₃, H, CH₃]   (b) [structure with H, CH₃]   [structure with CH₃, H]   (c) [structure with OH, OH]   [structure with OH, OH, OH]

(d) [cyclopentane with CH₃, H]   [cyclopentane with CH₃, H]   (e) [structure: CH₃, CH₃, H, C, C, H, CH₃, Br]   [structure: CH₃, CH₃, H, C, C, Br, CH₃, H]   (f) [structure: H, CH₃, H, C, C, H, CH₂Cl, Br]   [structure: Cl, CH₃, H, C, C, H, CH₃, Br]

(g) [structure: CH₃, H, C, Cl, ClCH₂]   [structure: Cl, CH₃, C, H, CH₂Cl]   (h) [structure: H, OH, H, C, C, H, CH₃, H]   [structure: H, H, H, C, C, H, CH₃, OH]

(i) [structure: Br, H, C, CH₂Cl, CH₃CH₂]   [structure: Br, C, H, ClCH₂CH₂, CH₃]   (j) [structure with H, Cl]   [structure with H, Cl]

**6.26** The compound shown below with no stereochemistry indicated is an intermediate in the synthesis of a neurotoxin isolated from a marine sponge.

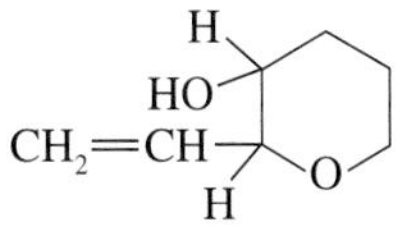

[structure: H, HO, CH₂=CH, O, H on a tetrahydropyran ring]

(a) The hydroxyl group and the side chain with the double bond in it are trans to each other in this intermediate. Draw one possible stereoisomer for this compound using the flat hexagon for the ring.
(b) Draw the two chair conformations of the compound you have drawn in part (a). Which way does the equilibrium lie between the two forms that you have drawn?

**6.27** Chiral acetic acid in which two of the hydrogen atoms on the methyl group have been replaced by deuterium, D, and tritium, T (p. 206), has been synthesized for studies of the stereochemistry of biological reactions. Write three-dimensional formulas for (R)- and (S)-DHTCCOH.
$$\overset{\|}{\underset{O}{}}$$

**6.28** Addition of HBr to optically active (R)-(+)-6-bromo-1-heptene, shown below, gives two 2,6-dibromoheptane stereoisomers. One is optically active and the other is not.

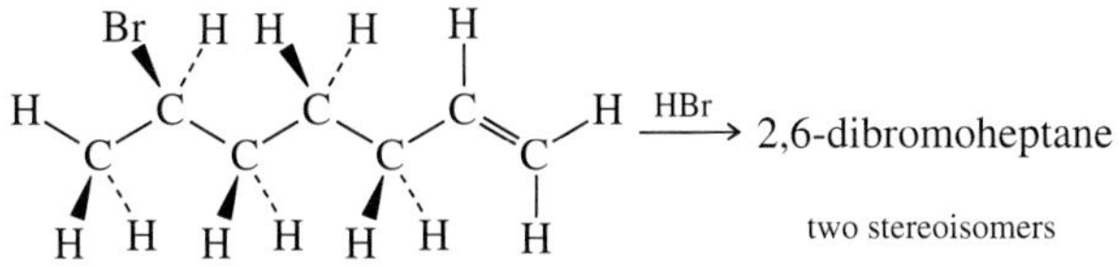

[structure: Br, H, H, H, H, H, C, C, C, C, C, H, H, H, H, H, H, H, H, with HBr arrow → 2,6-dibromoheptane, two stereoisomers]

(R)-(+)-6-bromo-1-heptene

Using the curved-arrow convention, show the mechanism for the addition of HBr to the compound in a way that clearly indicates how the two stereoisomers are formed. Label any stereocenters in the two products, and indicate which product is optically active and which is not.

**6.29** The females of the *Culex quinquefasciaties* species of mosquitoes decide where to lay their eggs by sensing a pheromone (a chemical used in communication) given off by previously laid mature eggs. This pheromone is (5R,6S)-6-acetoxy-5-hexadecanolide, shown below with the numbering but no stereochemistry.

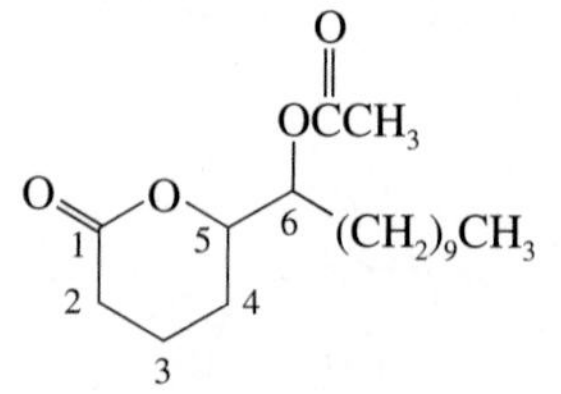

(5R,6S)-6-acetoxy-5-hexadecanolide
with no stereochemistry shown

(a)   Redraw the structure to show the correct stereochemistry for the compound.

(b)   The pheromone is synthesized very efficiently from (Z)-5-hexadecenoic acid, shown below with no stereochemistry. Redraw it in the correct stereochemical form.

$$\underset{1}{\overset{O}{\|}} \underset{}{\overset{2\ \ 3\ \ 4\ \ 5}{\text{HOCCH}_2\text{CH}_2\text{CH}_2\text{CH}}}{=}\overset{6}{\text{CH}}(\text{CH}_2)_9\overset{16}{\text{CH}_3}$$

**6.30** Fruit sugar, or fructose, has $[\alpha]_D^{20}$ −92° (c 0.02; water). Calculate the rotation that would be observed for 100 mL of a solution made with 1 g of fructose in water and measured in a tube 5 cm long at 20 °C using a sodium lamp.

**6.31** (+)-2-Butanol has $[\alpha]_D^{20}$ +13.90° when the measurement is made on the pure liquid. A sample of 2-butanol was found to have an optical rotation of −3.5°. What is the stereoisomeric composition of this mixture?

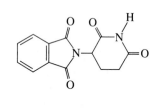

**6.32** Thalidomide, prescribed in countries other than the United States as an antidepressant in the 1960s, has the structure shown at left. It was used as a racemic mixture and given to pregnant women to treat morning sickness, resulting in the birth of severely deformed infants. It has since been shown that the R enantiomer is the antidepressant, while the S enantiomer is mutagenic, resulting in the deformities. Draw stereochemically accurate representations of (R)- and (S)-thalidomide.

**6.33** A widely prescribed analgesic (painkiller) is Darvon. The medically active form of Darvon is the propanoic ester of (2S,3R)-4-dimethylamino-1,2-diphenyl-3-methyl-2-butanol, which is sold as the hydrochloric acid salt of the amine. The structural formula of the amino-alcohol used in the preparation of Darvon is shown below. Identify the stereocenters, write structural formulas for all possible stereoisomers, and pick out the medicinally useful form of the compound.

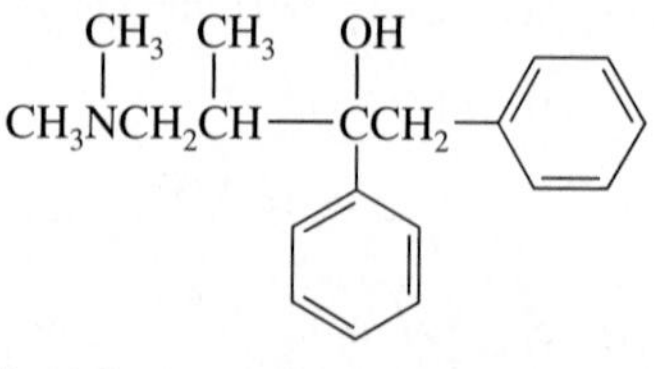

4-dimethylamino-1,2-diphenyl-3-methyl-2-butanol

Darvon alcohol

**6.34** The compound decalin has two cyclohexane rings fused together. The hydrogen atoms at the ring junction are on the same side of the rings in *cis*-decalin and on opposite sides of the rings in *trans*-decalin, as shown on the next page. Decide which stereoisomer of decalin is more stable. (*Hint:* Think of the one ring as substituents on the other ring. It may help you to construct the ring systems with models.)

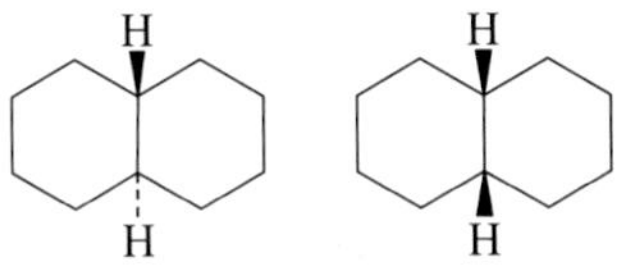

*trans*-decalin          *cis*-decalin

**6.35** The following transformation has been carried out.

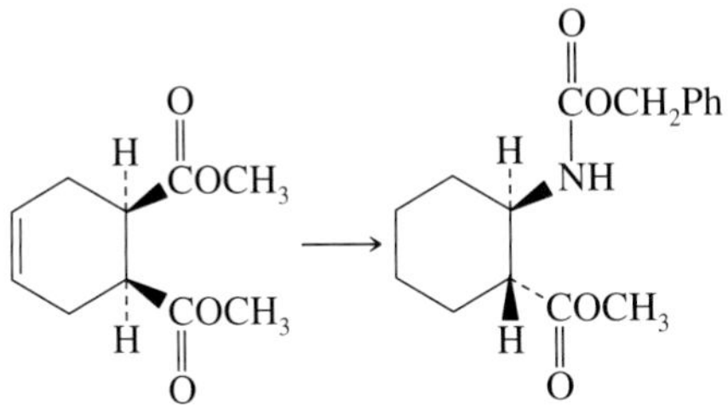

Compound A          Compound B

For each of these compounds, answer the following questions.

(a) How many enantiomers does each compound have?
(b) Is the compound optically active?
(c) What is the stereochemical configuration of the upper stereocenter in the compound as drawn above?
(d) How many diastereomers does this compound have?
(e) Draw the two chair conformations of Compound B, and decide which is the more stable one.

**6.36** Both the cis and the trans isomers of 3-methoxycyclohexanol are found as components of a group of antibiotics called macrolides. Both 3-methoxycyclohexanols are chiral. However, one diastereomer reacts with base, and then with methyl iodide to give an optically inactive 1,3-dimethoxycyclohexane. The other diastereomer gives a racemic mixture of optically active 1,3-dimethoxycyclohexanes.

(a) Draw a good three-dimensional representation of a stereoisomer of 3-methoxycyclohexanol that would give optically inactive 1,3-dimethoxycyclohexane when treated with base and methyl iodide.
(b) Draw a good three-dimensional representation of a stereoisomer of 3-methoxycyclohexanol that would give optically active 1,3-dimethoxycyclohexane when treated with base and methyl iodide.
(c) Draw the possible conformers of *cis*- and of *trans*-dimethoxycyclohexane. Which one is the most stable conformer of those possible for these two compounds?

**6.37** (+)-Sandresolide A is a natural product isolated from a Caribbean coral.

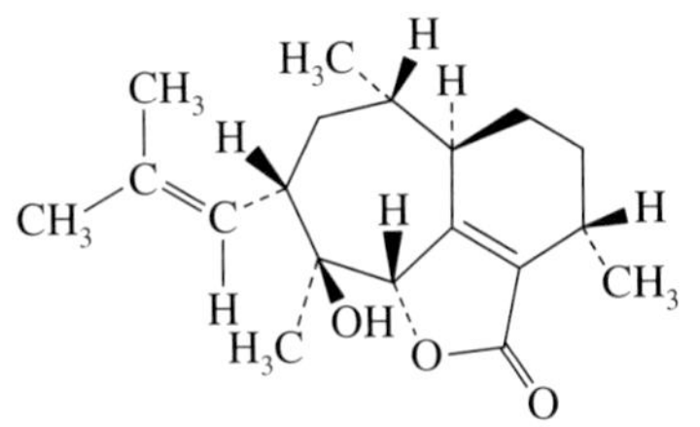

(+)-sandresolide A

(a) Identify each stereocenter in (+)-sandresolide A and label it as *R* or *S*.
(b) Is this stereoisomer of sandresolide A dextrorotatory or levorotatory?

(c)  If *one* stereocenter of (+)-sandresolide A were to be inverted, that is, changed from *R* or *S* or vice versa, the resulting compound would have what relationship to (+)-sandresolide A? Would the resulting compound be dextrorotatory or levorotatory, or would it be impossible to tell?

**6.38** The following structures represent only a few of the possible constitutional isomers of $C_4H_8O_2$.

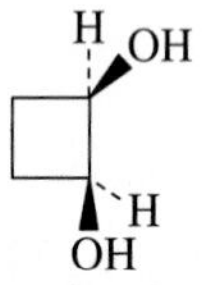

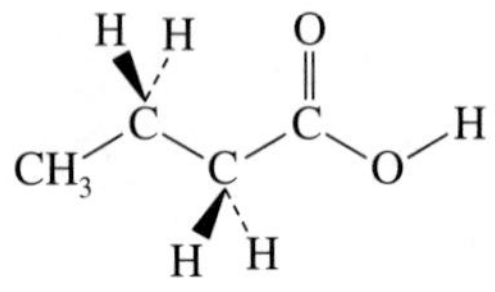

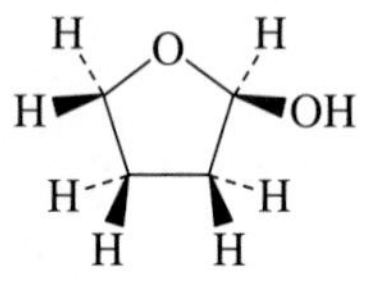

The following statements describe the properties of some of these compounds. Use the letters given with the structures to identify the compound (or compounds) that have the properties described in each case.

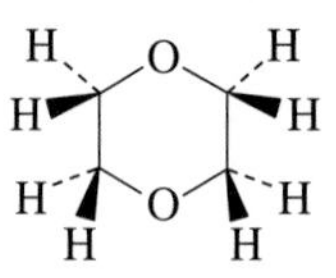

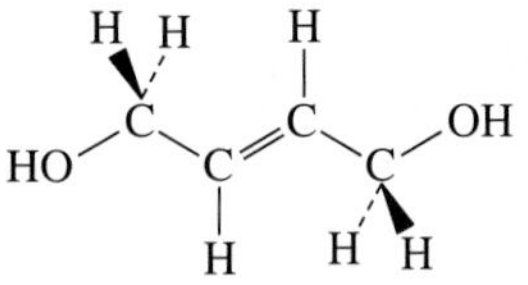

(a)  Is chiral.    (b)  Is meso.    (c)  Is a hydrogen bond donor.
(d)  Has at least one diastereomer.    (e)  Has a $pK_a$ of less than 5.0.
(f)  Has three different peaks in its proton nuclear magnetic resonance spectrum and two different peaks in its carbon nuclear magnetic resonance spectrum.
(g)  Has four different absorption bands in its carbon nuclear magnetic resonance spectrum.

# Nucleophilic Substitution and Elimination Reactions

## A Look Ahead

The displacement of a halide ion by a nucleophile is known as a **nucleophilic substitution reaction.** We examined this type of reaction in Section 4.2. Alkyl halides are representative of a much larger set of compounds in which a carbon atom is bonded to an atom or a group, known as a leaving group, that can be displaced in a substitution reaction. The presence of a leaving group (symbolized below as LG) on a carbon atom makes possible two different kinds of reactions involving $\sigma$ bonds at or next to the carbon atom. The two sites of attack are the carbon atom bearing the leaving group and the hydrogen atom on a carbon atom adjacent to the first one. Attack at the carbon atom results in substitution; attack at the hydrogen atom gives an elimination reaction.

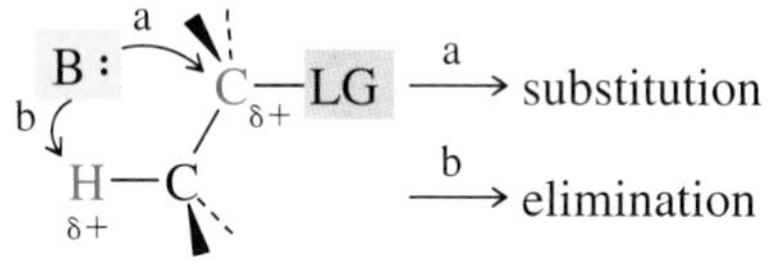

The electronegativity of the atom in the leaving group to which the carbon atom is bonded polarizes that bond, giving the carbon atom and any hydrogen atoms close to it a partial positive character. The carbon atom bearing a partial positive charge is an electrophilic center. A Lewis base that is a good nucleophile will react at that carbon atom, displacing the leaving group. For example,

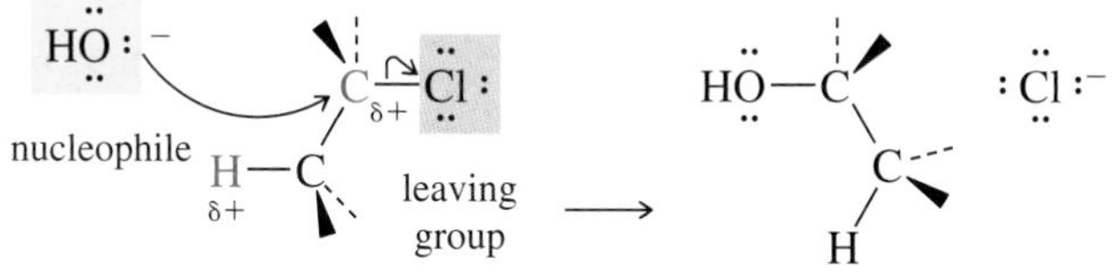

a nucleophilic substitution reaction

A Lewis base that is also a strong Brønsted base may remove a proton from the carbon adjacent to the electrophilic center as well. A double bond is formed by an internal displacement of the leaving group by the electrons that are released as the proton is removed from the carbon atom to which it is bonded. This reaction is called an **elimination reaction** and is the reversal of an addition reaction such as the addition of a hydrogen halide to a double bond (p. 126).

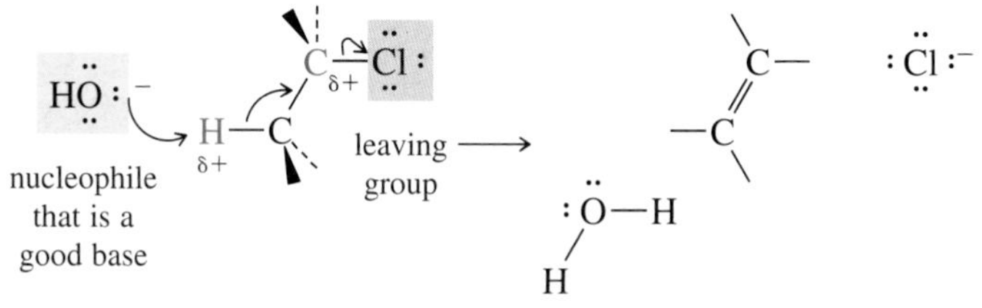

Substitution and elimination reactions have tremendous potential for transforming organic molecules; as a result, they have been studied intensively. Early work on substitution reactions of alkyl halides was done by Sir Christopher Ingold and Edward Hughes and their collaborators at University College, London, in the 1930s and 1940s. Their ideas have influenced the thinking of organic chemists about the mechanisms of many different reactions and have inspired much experimental work and continuing vigorous debate.

The kinds of substituents that can be introduced into organic molecules depend on the availability of starting materials with suitable leaving groups and of nucleophiles that will give substitution reactions with them. Elimination and substitution reactions almost always occur in competition with each other. Factors such as the nature of the reagents and the solvents used play a part in determining which reaction is the major one in a given case. In this chapter, we will explore the detailed mechanisms for these transformations so that you will understand the limitations on them and how to control them.

**Workbook Exercises**

## 7.1   Synthesis. Nucleophilic Substitution Reactions in Chemical Transformations

A new bond is formed whenever a nucleophile substitutes at an electrophilic center bearing a leaving group. New carbon–oxygen, carbon–sulfur, carbon–nitrogen, and carbon–carbon bonds are created in this way. Nucleophilic reagents that can be used to make bonds of each of these types are shown in Table 7.1. Most of these reagents are familiar from Section 4.4C.

A simple nucleophilic substitution reaction involves substitution of one halide ion for another. For example, iodide ion in acetone as the solvent will react with 1-bromobutane to substitute for bromide ion to give 1-iodobutane.

$$CH_3CH_2CH_2CH_2Br + Na^+ + I^- \xrightarrow[\substack{\text{acetone} \\ 25\,°C}]{} CH_3CH_2CH_2CH_2I + Na^+Br^- \downarrow$$

| 1-bromobutane | *fast reaction* | 1-iodobutane |
| *n*-butyl bromide | | *n*-butyl iodide |

The reaction is reversible. In a solvent such as ethanol, treatment of 1-iodobutane with sodium bromide gives 1-bromobutane. However, sodium bromide, which is not as soluble as sodium iodide is in acetone, precipitates out of solution, drawing the forward reaction toward completion when the reaction is carried out in acetone. Any product that leaves the phase in which the reaction is taking place helps to shift the equilibrium toward the products. Chemists often choose reaction conditions, such as the solvent above, so that a product will be removed from the equilibrium.

**Table 7.1   Common Nucleophilic Reagents for Creating New Bonds to Carbon**

| Carbon–Oxygen Bonds | Carbon–Sulfur Bonds | Carbon–Nitrogen Bonds | Carbon–Carbon Bonds |
|---|---|---|---|
| $HO^-$ | $HS^-$ | $NH_3$ | $CN^-$ |
| $RO^-$ | $RS^-$ | $RNH_2$ | $RC{\equiv}C^-$ |
| $ArO^-$ | $ArS^-$ | $N_3^-$ | $\underset{}{(CH_3CH_2OC)_2CH^-}$ with $\overset{O}{\overset{\|}{}}$ |

In the above reaction, iodide ion is the nucleophile, the species with electrons to donate. It reacts with 1-bromobutane at the carbon atom bearing the partial positive charge. The carbon atom is an **electrophilic center,** a site of electron deficiency, and bromide ion is the leaving group.

---

### VISUALIZING THE REACTION

**Substitution Reaction with a Halide Ion**

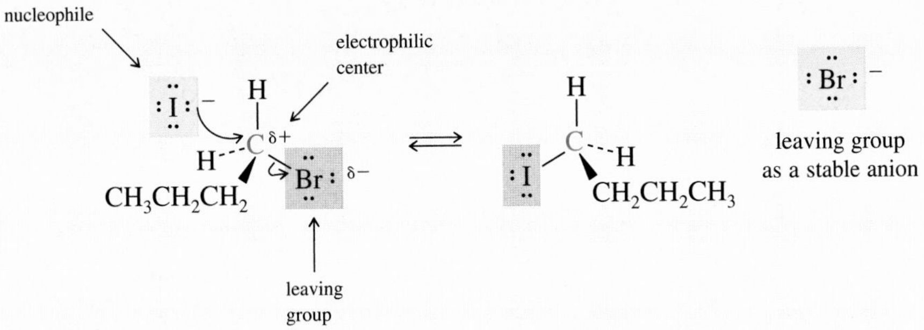

---

Ethers are formed when oxygen anions such as alkoxide or aryloxide anions are used as nucleophiles. Deprotonation of phenol with sodium hydroxide, for example, gives the phenolate ion, a good nucleophile, which will react with 1-iodopropane to give phenyl propyl ether.

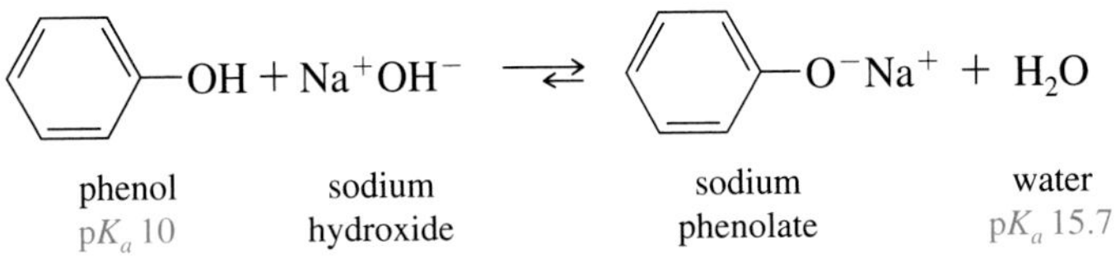

| phenol | sodium | sodium | water |
| $pK_a$ 10 | hydroxide | phenolate | $pK_a$ 15.7 |

---

### VISUALIZING THE REACTION

**Substitution Reaction with an Oxygen Anion**

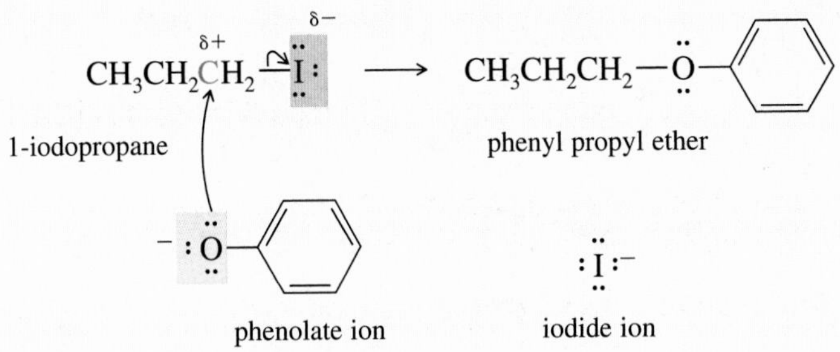

---

Some of the most useful nucleophilic substitution reactions are those that create new carbon–carbon bonds. These reactions require reagents in which carbon atoms are nucleophilic centers—that is, carbanions. For example, a carbanion is generated when an alkyne with a triple bond at the end of the carbon chain is deprotonated by a strong base (p. 98).

$$CH_3CH_2CH_2C \equiv CH + Na^+NH_2^- \xrightleftharpoons[NH_3 \text{(liq)}]{} CH_3CH_2CH_2C \equiv C:^- Na^+ + NH_3$$

| 1-pentyne | sodium | conjugate base of 1-pentyne | ammonia |
| $pK_a \sim 26$ | amide | | $pK_a$ 36 |

Such carbanions react with alkyl halides to give more highly substituted alkynes.

---

**VISUALIZING THE REACTION**

**Substitution Reaction Using a Carbanion**

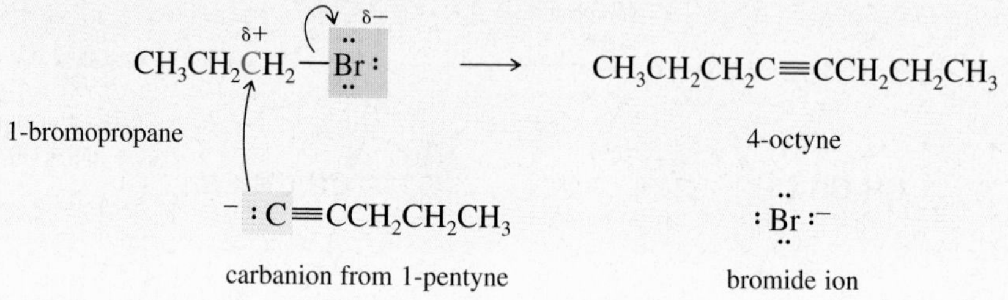

| 1-bromopropane | | 4-octyne |

carbanion from 1-pentyne          bromide ion

---

Azide ion is a good nucleophile and is often used to introduce carbon–nitrogen bonds.

---

**VISUALIZING THE REACTION**

**Substitution Reaction Using Azide Ion**

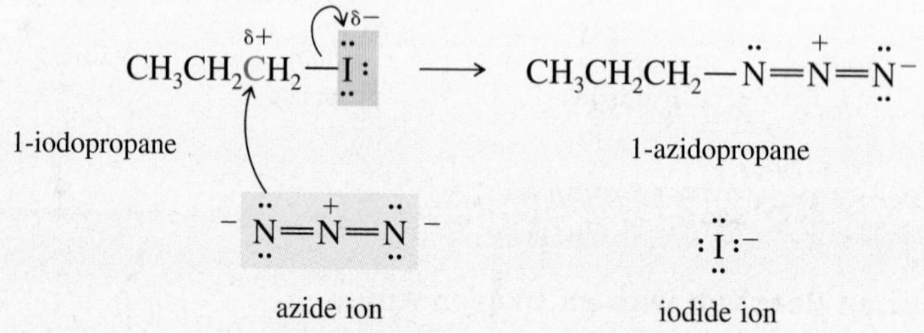

1-iodopropane          1-azidopropane

azide ion          iodide ion

---

Amines also displace leaving groups to give more highly substituted amines.

---

**VISUALIZING THE REACTION**

**Substitution Reaction Using an Amine**

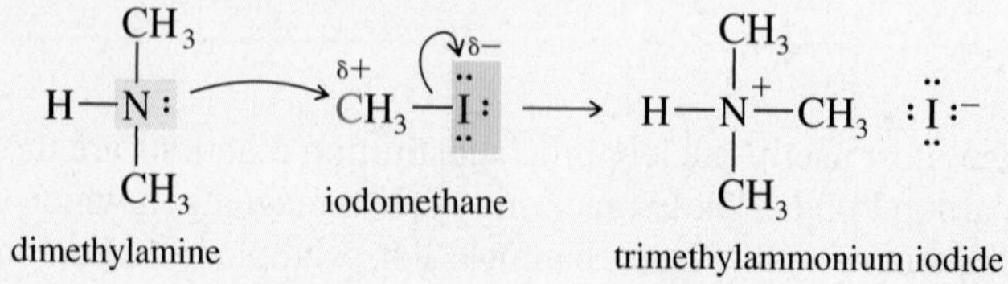

dimethylamine          iodomethane          trimethylammonium iodide

## 7.2 The Art of Solving Problems

In Section 4.4B (p. 134) you learned ways to analyze the structures of reactants and products in order to decide how to transform one into the other. This section offers some further guidelines on how to approach the solution of problems in organic chemistry.

Two categories of problems are (1) predicting the products of a reaction, given the reagents, and (2) deciding which reagents are necessary to convert a given starting material into a desired product. You can solve such problems by systematically asking yourself a series of questions that will help you analyze the important features of a reaction. In this section, two problems are analyzed in detail in this way to illustrate the necessary thought processes.

PROBLEM: Predict the product(s) of the following reaction.

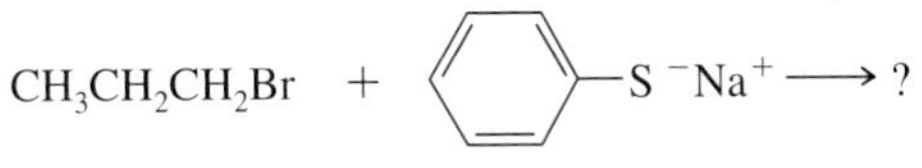

$$CH_3CH_2CH_2Br \ + \ \langle \!\!\!\bigcirc\!\!\!\rangle \!\!-\!\!S^-Na^+ \longrightarrow \ ?$$

### Solution

1. To what functional group classes do the reactants belong?

   One reactant is an alkyl halide, the other is a sulfur anion (remember that sodium ions are usually spectator ions, there to balance charge but not to participate in the reaction).

2. Does either reactant have a leaving group?

   Yes, the bromine atom in 1-bromopropane is a leaving group.

3. Are any of the reactants acids, bases, nucleophiles, or electrophiles?

   The sulfur anion is a nucleophile (see Table 7.1). The alkyl halide contains an electrophilic carbon atom.

4. What is the most likely first step for the reaction? Most common reactions can be classified as either protonation–deprotonation reactions or reactions of a nucleophile with an electrophile.

   No strong acids ($pK_a < 1$) or bases ($pK_a$ of conjugate acid $> \sim 13$) are present in this reaction mixture, so a protonation–deprotonation reaction is not probable. Reaction of the nucleophilic sulfur atom with the electrophilic carbon atom is most likely.

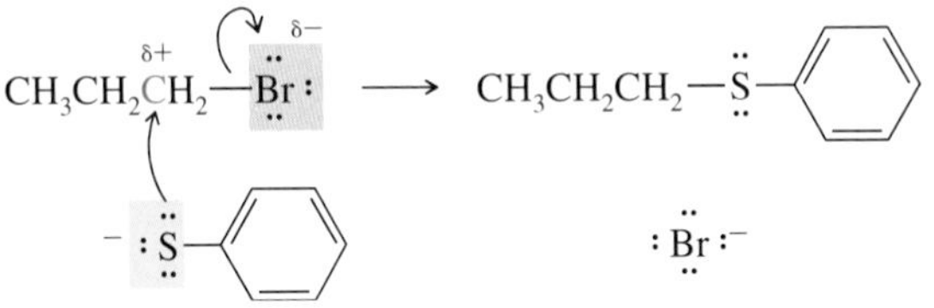

5. What are the properties of the species present in the reaction mixture after this first step? Is any further reaction likely?

   All species formed are stable. No further reaction will occur. The complete equation is

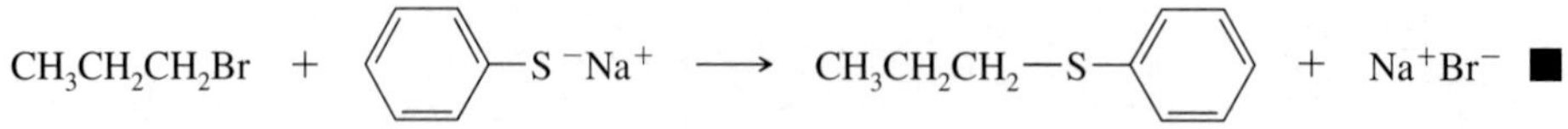

$$CH_3CH_2CH_2Br \ + \ \langle\ \rangle-S^-Na^+ \ \longrightarrow \ CH_3CH_2CH_2-S-\langle\ \rangle \ + \ Na^+Br^- \ ■$$

---

**PROBLEM:** How would you synthesize $CH_3CH_2C\equiv CH$ from $CH_3CH_2I$?

**Solution**

1. What are the connectivities of the two compounds? How many carbon atoms does each contain? Are there any rings? What are the positions of branches and functional groups on the carbon skeletons?

   The starting material has a two-carbon chain, an ethyl group, with an iodine atom bonded to one carbon. The product has a four-carbon chain in which the two carbons that have been added to the ethyl group are bonded to each other by a triple bond.

2. How do the functional groups change in going from starting material to product? Does the starting material have a good leaving group?

   The iodine atom in the starting material is a leaving group and has been replaced in the product by an alkyne functional group.

3. Is it possible to dissect the structures of the starting material and product to see which bonds must be broken and which formed?

$$CH_3CH_2{\dashv}I \qquad CH_3CH_2{\dashv}C\equiv CH$$

   bond to be broken      new bond to be formed

$$CH_3CH_2- \qquad -C\equiv CH$$

   the pieces that have to be brought together

4. New bonds are created when an electrophile reacts with a nucleophile. Do we recognize any part of the product molecule as coming from a good nucleophile or an electrophilic addition?

   The ethyl group is attached to a leaving group in the starting material and so must serve as the electrophilic center. A check of Table 7.1 reminds us that the $-C\equiv CH$ group is related to the nucleophile $HC\equiv C:^-$, the conjugate base of an alkyne.

5. What type of compound would be a good precursor to the product?

   A displacement of the iodine atom by the nucleophile will give us the product we want.

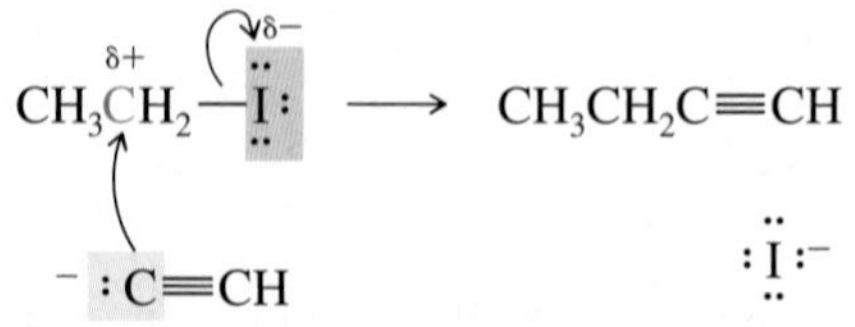

$$\overset{\delta+}{CH_3CH_2}-\overset{\curvearrowleft\,^{\delta-}}{\ddot{I}:} \ \longrightarrow \ CH_3CH_2C\equiv CH$$

$$^-:C\equiv CH \qquad\qquad\qquad :\ddot{I}:^-$$

6. After this last step, do we see how to get from starting material to product? If not, we need to analyze the structure obtained in step 5 by applying questions 4 and 5 to it. ■

These steps are a restatement of the way of thinking about problems that was introduced in Section 4.4. It will be helpful to you to ask yourself these questions in a systematic way for each problem you work on until this way of thinking becomes familiar and easy. Only consistent practice will develop the problem-solving skills you need to succeed in organic chemistry.

**Problem 7.1**   Use the curved-arrow notation to show how each of the following substitution reactions will take place. Also show what the products will be.

(a) $CH_3CHCH_2CH_2I$  +  benzene ring—$O^-Na^+$ $\xrightarrow{\text{acetone}}$

with $CH_3$ on the first carbon

(b) $CH_3I$  +  piperidine ring (N–H) $\longrightarrow$

(excess)

(c) $CH_3CH_2CH_2Br$  +  $CH_3S^-Na^+$ $\longrightarrow$

**Problem 7.2**   The following incomplete reaction sequences give a starting material and a product that can be prepared from it. Show the reagents that would be necessary to carry out each transformation and any major products that would be formed in the intermediate stages. There may be more than one correct way to complete a sequence.

(a) $CH_3CH{=}CH_2 \longrightarrow CH_3CHCH_3$ with $SCH_2CH_3$

(b) $CH_3CH_2CH_2CH_2CH_2Br \longrightarrow CH_3CH_2CH_2CH_2CH_2CH(COCH_2CH_3)_2$

(c) $HC(=O)$—aromatic ring with I and OCH$_3$—OH $\longrightarrow$ $HC(=O)$—aromatic ring with I and OCH$_3$—OCH$_2$CH$_2$CH$_3$

## 7.3  Substitution Reactions

### A. Experimental Evidence. Rate of Reaction

Alkyl halides differ in the timing of the breaking of the bond between the carbon and halogen atoms in substitution reactions. We discussed the different possibilities for the timing of bond-forming and bond-breaking processes in Section 4.2A.

An investigation of the rate of a reaction such as the reaction of chloromethane with hydroxide ion (studied in Chapter 4) shows that it depends on the concentrations of both reactants. An increase in the concentration of either reactant causes a proportional increase in the rate of the reaction. This is also true of the reactions in Section 7.1. For example, for the reaction of n-butyl bromide with iodide ion (p. 234), the rate expression is

$$R = k[nBuBr][I^-]$$

where Bu is butyl, $R$ is the rate of the reaction, and $k$ is the second-order rate constant for this particular reaction at a given temperature and in a given solvent.

For the reaction of n-butyl bromide and iodide ion, a transition state that involves both species is consistent with the second-order rate expression determined experimentally.

reactants

activated complex
at the transition state

products

iodide ion and
n-butyl bromide

bromide ion and
n-butyl iodide

Iodide ion, a nucleophile, approaches *n*-butyl bromide. The electrons on the iodide ion are drawn toward the electrophilic carbon atom, while the bond to the bromide atom loosens and lengthens. The lowest-energy way for the iodide ion to approach the alkyl halide molecule, and the least crowded, is from the side opposite the departing bromide ion. In this way, the negative charge on the iodide ion and the developing negative charge on the bromide ion can be kept as far apart as possible. At the transition state, partial bonding occurs between carbon and iodine. The carbon atom also remains partially bonded to the bromine atom, which is beginning to become a bromide ion. In the activated complex at the transition state, there are five groups around the carbon atom. The activated complex can revert to starting material or proceed to products.

The reaction of *n*-butyl bromide with iodide ion belongs to a class of reactions known as **bimolecular nucleophilic substitution reactions.** *Bimolecular* refers to the fact that *two* species undergo bonding changes in the same transition state of such a reaction. The reaction is initiated by the approach of a nucleophile and is therefore a nucleophilic substitution reaction. Chemists refer to these reactions as $S_N2$ **reactions,** where the letters and number stand for substitution, nucleophilic, and bimolecular, respectively. An $S_N2$ reaction takes place in one step. The reaction of hydroxide ion and chloromethane, discussed in Chapter 4, is another $S_N2$ reaction.

Not all halides react in the same way that *n*-butyl bromide does. An experiment to explore the effect of structure on reactivity in $S_N2$ reactions used radioactive bromide ion to see how fast it was incorporated into different alkyl bromides in acetone at 25 °C. The relative rates of reaction for different alkyl groups are given in Table 7.2. Methyl bromide reacts the most quickly. Even ethyl bromide, in which a methyl group has replaced one of the hydrogen atoms of methyl bromide, reacts much

**Table 7.2** **Relative Rates for the Reaction of Alkyl Bromides, RBr, with Radioactive Bromide Ion in Acetone at 25 °C**

$$Br^{*-} + R\text{---}Br \underset{\underset{25\ °C}{acetone}}{\rightleftharpoons} R\text{---}Br^* + Br^-$$

| R | $CH_3-$ | $CH_3CH_2-$ | $CH_3CH_2CH_2-$ | $\underset{}{CH_3\overset{CH_3}{\underset{|}{CH}}-}$ | $CH_3\overset{CH_3}{\underset{\underset{CH_3}{|}}{\overset{|}{C}}-$ |
|---|---|---|---|---|---|
| | methyl | ethyl | n-propyl | isopropyl | tert-butyl |
| Relative rates | 100 | 1.31 | 0.81 | 0.015 | 0.004 |
| Nature of R | | primary | primary | secondary | tertiary |

more slowly than methyl bromide. The reaction is very slow for the tertiary halide *tert*-butyl bromide.

The halogen in a *tert*-butyl halide is not displaced unless the reaction is carried out in a solvent such as water or ethanol, which have high dielectric constants (p. 33). Such solvents, which are called **protic solvents** or **ionizing solvents,** promote ionization because they are able to solvate (p. 34) both the cation and the anion that form in the process. Once ionization occurs, a variety of reactions is possible. For example, when *tert*-butyl chloride is placed in a mixture of 80% ethanol and 20% water in the presence of potassium hydroxide, a complex mixture of products is formed.

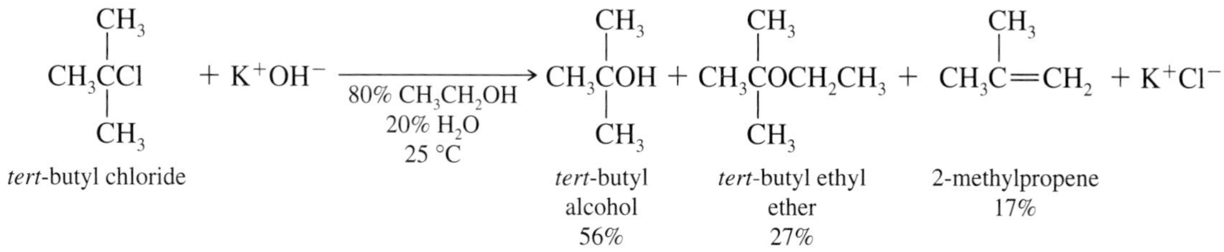

The rate of the reaction is proportional only to the concentration of alkyl halide in the reaction mixture. The reaction is said to be first order in alkyl halide concentration.

$$R = k[tert\text{-BuCl}]$$

Only one species undergoes changes in bonding in the transition state for the rate-determining step, so this reaction is classified as a **unimolecular reaction.** The rate-determining step (p. 133) must be the ionization of the alkyl halide.

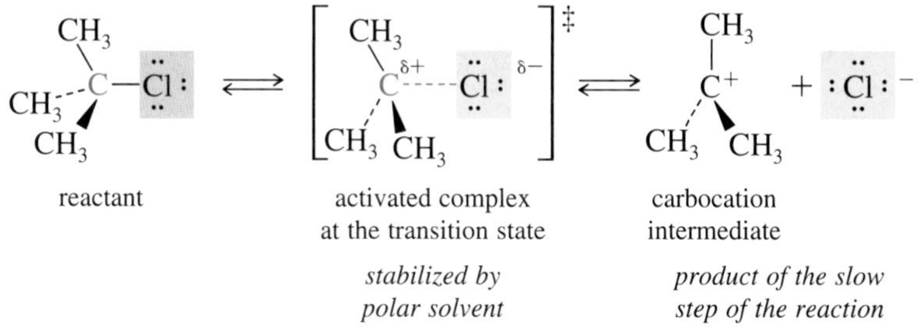

The reaction takes place in ethanol and water, both good ionizing solvents. Energy is released by the electrostatic interactions of the negative ends of the dipoles of the oxygen–hydrogen bonds with the carbocation and the positive ends of those dipoles with the chloride ion.

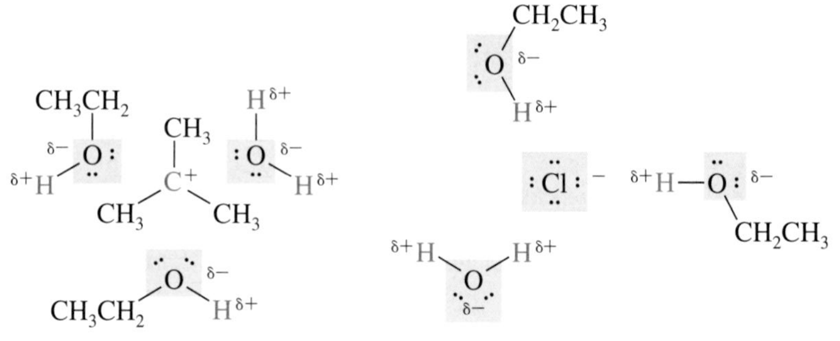

*solvation of* tert-*butyl cation and chloride ion by ethanol and water*

This energy from solvation helps to offset the energy that is necessary to break the carbon–chlorine bond. Ethanol and water, because of their polar oxygen–hydrogen bonds, stabilize chloride anions as well as carbocations. The importance of solvent in stabilizing ionic species has already been emphasized earlier in this book (pp. 34, 103, and 129).

For a tertiary halide in a good ionizing solvent, breaking of the carbon–halogen bond occurs before any bond-forming to a nucleophile takes place. The products seen in the equation for the reaction of *tert*-butyl chloride in ethanol and water come from reactions of the carbocation that formed in the rate-determining step with the solvents and with hydroxide ion.

---

**VISUALIZING THE REACTION**

**The S$_N$1 Reaction**

These reactions are discussed further on page 245. The substitution reactions of *tert*-butyl chloride are called **S$_N$1 reactions,** where the sequence of letters and numbers stand for substitution, nucleophilic, and unimolecular, respectively. An S$_N$1 reaction takes place in two steps: first, the ionization of the halide, and second, the reaction of the carbocation with nucleophiles.

---

**Problem 7.3**    The rate constant, $k$, for the reaction

$$CH_3CH_2Br + OH^- \xrightarrow[\substack{80\% \text{ ethanol} \\ 55\,°C}]{} CH_3CH_2OH + Br^-$$

is $1.7 \times 10^{-3}$ L/mol·s. What is the initial rate of the reaction when 0.05 M ethyl bromide is allowed to react with 0.07 M sodium hydroxide? What are the units in which the rate of the reaction, $R$, is expressed? What would the initial rate of the reaction be if 0.1 M ethyl bromide were used instead?

**Problem 7.4**    Suppose 0.06 M *n*-butyl bromide is allowed to react with 0.02 M potassium iodide in acetone at 25 °C. The rate constant, $k$, for the reaction at that temperature is $1.09 \times 10^{-1}$ L/mol·min. What is the initial rate at which *n*-butyl bromide disappears from the reaction mixture under these conditions?

**Problem 7.5**    The rate constant, $k$, for the reaction of *tert*-butyl chloride in 70% aqueous ethanol at 25 °C has been determined to be 0.145/h. If the initial concentration of *tert*-butyl chloride is 0.0824 M, what is the initial rate of the reaction? What units are used to express this rate? Will this rate remain the same as the reaction progresses? What will it be when half of the *tert*-butyl chloride initially present has reacted?

## B. Competing Mechanisms for Substitution Reactions

Why is the substitution reaction of halide ions in acetone slower for a secondary alkyl halide than for a primary one, and why does the reaction go hardly at all with a tertiary halide? To explain rate data, chemists refer to the structure of the activated complex at the transition state on the way from the reactants to the products of the reaction. The energy diagram for a second-order reaction shows the reactants being converted to products by way of one transition state, with no energy minimum representing a reactive intermediate (see Figure 4.1, p. 124).

As we go from a primary alkyl halide to a tertiary alkyl halide, the transition states for the bimolecular reaction become progressively more crowded as the number of alkyl groups on the carbon atom undergoing substitution increases. A transition state with five groups crowded around the central carbon atom is a high-energy transition state. Only a few molecules in a reaction system will have enough energy to react by that pathway. When the solvent does not help the process of ionization, no lower-energy pathway is available for substitution at the tertiary carbon atom, and the reaction is very slow. The structure of the alkyl halide, especially the number of alkyl groups bonded to the carbon atom undergoing substitution, determines how high the energy of the transition state for the reaction will be. Reactions that can go through transition states of relatively low energy will proceed faster than those that must reach high-energy states.

Three energy diagrams illustrate these ideas (Figure 7.1). These energy diagrams represent the displacement of bromide ion by radioactive bromide ion in a primary, a secondary, and a tertiary alkyl halide. The activated complex at the transition state

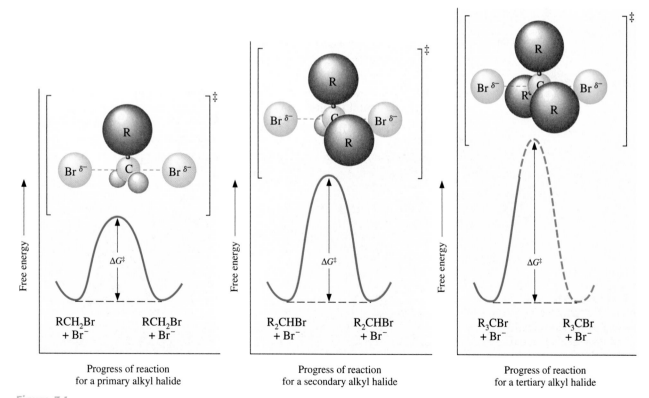

**Figure 7.1**

Comparison of the free energies of activation for bimolecular substitution reactions at primary, secondary, and tertiary carbon atoms.

becomes progressively more crowded as we substitute alkyl groups (R groups) for hydrogen atoms. The reaction of a tertiary halide by this mechanism has a significantly higher free energy of activation than does the reaction of a primary halide and is, therefore, much slower than the reaction of the primary halide. In fact, very few tertiary halide molecules make it to the top of the hill, as indicated in the energy diagram by a dashed line for the completion of the reaction.

Because the $S_N2$ reaction is so hindered for a tertiary halide, another reaction pathway, one that goes through the formation of a cation, is favored if reaction conditions permit. In an ionizing solvent, the reaction of a tertiary halide has two steps. The rate-determining step (p. 241) is the one giving rise to the high-energy reactive intermediate, the carbocation. The structural feature that is important for the activated complex at the transition state is the development of a separation of charge between the carbon atom and the departing leaving group. The carbon atom begins to have cationic character.

According to the Hammond Postulate (p. 124), we expect the activated complex to resemble the carbocation intermediate. Any factor that stabilizes the developing charge also stabilizes the activated complex and, therefore, lowers the free energy of activation necessary to reach the transition state. The greater the number of alkyl groups on the carbon atom that is developing the positive charge, the more stable the developing carbocation and the lower the energy of activation will be. Solvents with high dielectric constants, such as water ($\epsilon$ 78.5) and ethanol ($\epsilon$ 24.3), also help to stabilize the activated complex and lower the energy of activation. This increases the rate of the reaction. Therefore, tertiary alkyl halides, which give rise to tertiary cations, react faster under these reaction conditions than secondary ones, which react faster than primary halides do.

Even though the formation of a carbocation is a high-energy process, for a tertiary halide, the $S_N2$ reaction is even more unfavorable. The reverse is true for a primary halide. The $S_N2$ reaction has a lower energy of activation than the formation of a primary carbocation. These comparisons are shown schematically in Figure 7.2. The primary carbocation is so unstable that it forms hardly at all, even under the best ionizing conditions. Figure 7.2 shows this by an interrupted line for the progress of an $S_N1$ reaction for the primary halide.

In summary, if a nucleophile can approach the carbon atom bearing the leaving group and start to bond with it before the carbon–halogen bond breaks, then the alkyl halide will undergo a bimolecular substitution reaction, $S_N2$. $S_N2$ reactions oc-

*Figure 7.2*
Comparison of the free energies of activation of $S_N1$ and $S_N2$ reactions for primary and tertiary alkyl halides. Dashed lines represent the path the reaction could take if the lower-energy path were not available. The product in each reaction would be the same regardless of the path taken.

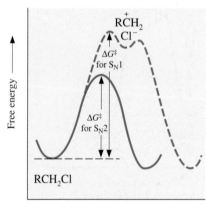

Progress of reaction
for a primary alkyl halide

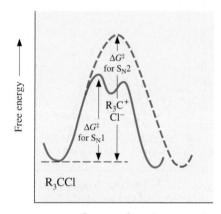

Progress of reaction
for a tertiary alkyl halide

cur with methyl, primary, and secondary halides. All the reactions shown in Section 7.1 are $S_N2$ reactions. If, on the other hand, the carbon atom bearing the leaving group is a tertiary halide, then the nucleophile cannot get close enough to the electrophilic carbon atom for bonding to start. No reaction occurs unless the halide is present in an ionizing solvent. In such a case, the tertiary halide ionizes to give a stable tertiary carbocation. The substitution products come from the cation in $S_N1$ reactions. The reactions of carbocations are described in the next section.

Experimental evidence indicates that secondary alkyl halides react with good nucleophiles by $S_N2$ reactions. If the secondary halide is present in a good ionizing solvent and no good nucleophile is present, solvolysis reactions (discussed in the next section) by an $S_N1$ mechanism are seen.

## C. Reactions of Carbocations

Once the *tert*-butyl cation is formed (p. 241), it has two reaction pathways open to it: reaction with a nucleophile or loss of a proton. Both these reactions are fast compared to the ionization that is the rate-determining step. The carbocation is a Lewis acid, which means it is an electrophile (p. 116) that lacks a pair of electrons and can accept them to attain a stable octet. When the *tert*-butyl cation is formed in ethanol, the cation and the alcohol, which is a nucleophile, react. The oxonium ion formed is a strong acid and loses a proton to the solvent.

**VISUALIZING THE REACTION**

**Reaction of a Carbocation with a Nucleophilic Solvent**

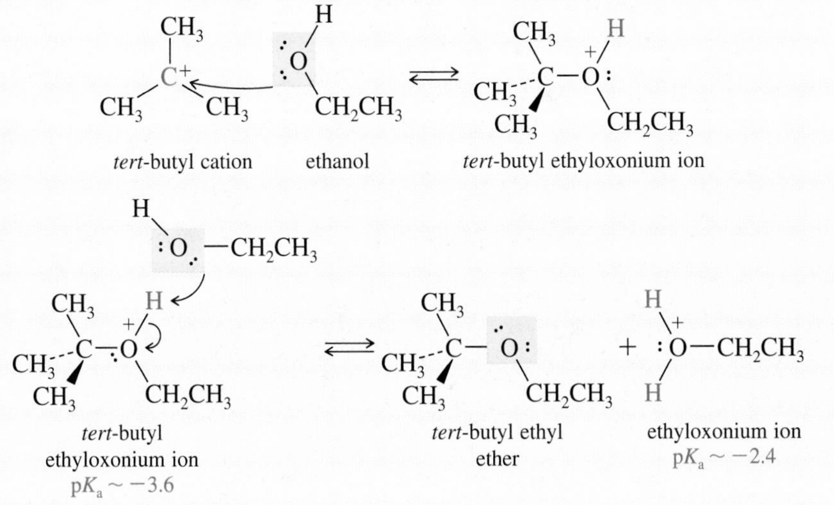

In the above reaction, the substitution product is derived from the solvent. A substitution reaction in which the solvent acts as the nucleophile is called a **solvolysis reaction.**

When a number of possible bases are found in the reaction mixture, as there are in this case, the symbol :B is often used in equations to represent any one of them. The last step of the solvolysis reaction can be rewritten using this symbol, as shown on the next page.

**VISUALIZING THE REACTION**

tert-butyl ethyloxonium ion          tert-butyl ethyl ether          conjugate acid of the base

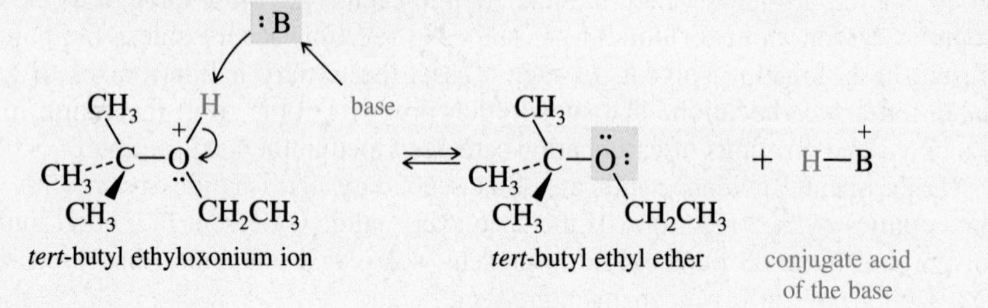

**Problem 7.6**    Of all the species present in the reaction mixture for the first equation on page 241, make a list of those that can act as bases.

The *tert*-butyl cation reacts rapidly with the other nucleophiles that are present in the reaction mixture (water and hydroxide ion) as well as with ethanol.

**VISUALIZING THE REACTION**

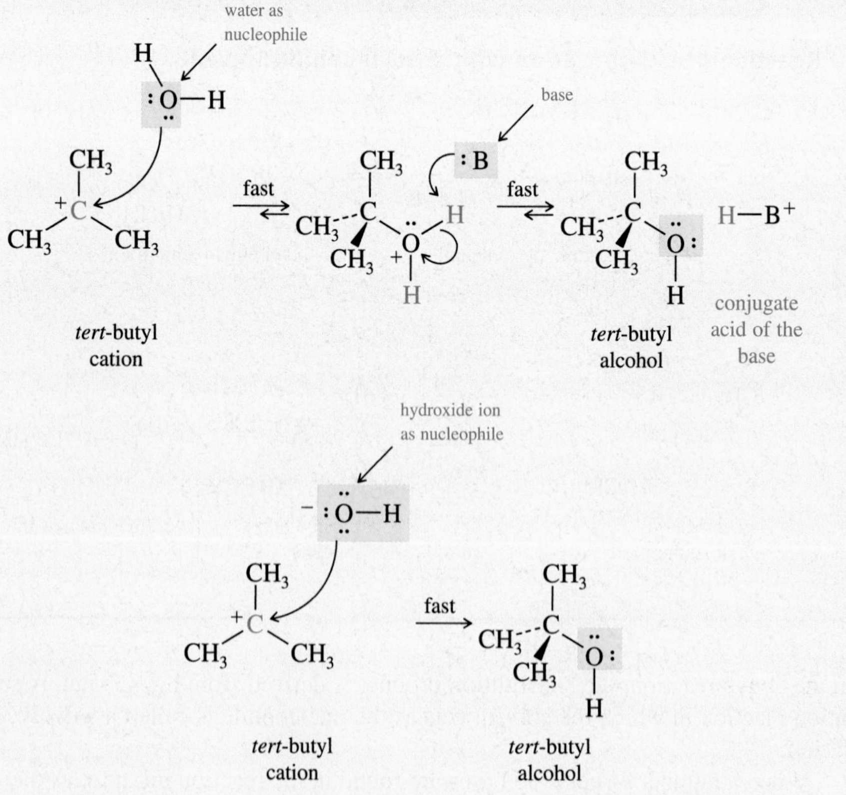

The intermediate tertiary carbocation is also a Brønsted acid (p. 79). The cationic carbon atom gives a partial positive character to the hydrogen atoms on carbon atoms adjacent to it. These hydrogen atoms are acidic enough to be removed by any of the bases present in the reaction mixture to give the minor product (p. 241), the product of the reaction called an E1 (unimolecular elimination) reaction. Elimination reactions are discussed in Section 7.7.

**VISUALIZING THE REACTION**

**Deprotonation of a Carbocation. The E1 Reaction**

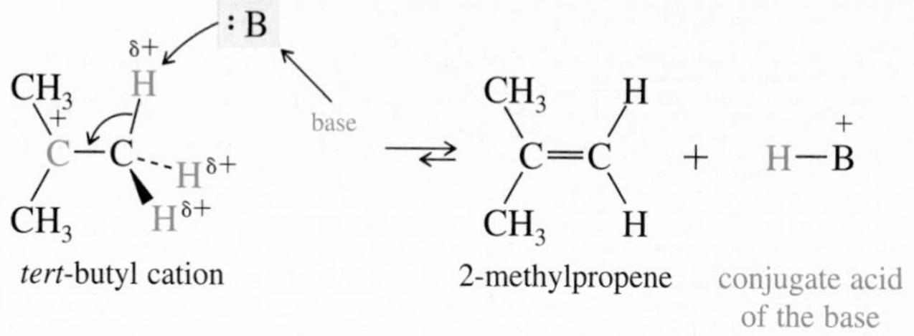

*tert*-butyl cation        2-methylpropene    conjugate acid of the base

When a carbocation is the intermediate for a substitution reaction, an alkene is one of the products because of the ease with which protons are lost from that intermediate.

## D. Stereochemistry of S_N2 Reactions

When (*S*)-(+)-2-bromobutane reacts with iodide ion in acetone, the 2-iodobutane that is formed has the *R* configuration. The iodine atom becomes attached to the stereocenter on the side opposite to the position that was occupied by the bromine atom. This result provides evidence in favor of the transition state for the $S_N2$ reaction that is an activated complex involving a carbon atom with five groups around it (p. 240).

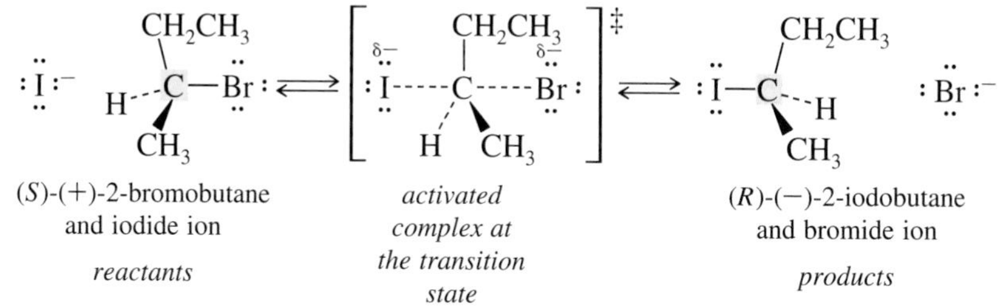

(*S*)-(+)-2-bromobutane    *activated*        (*R*)-(−)-2-iodobutane
and iodide ion             *complex at*       and bromide ion
                           *the transition*
*reactants*                *state*            *products*

An inversion of configuration has taken place at the stereocenter. Another example may make this a little clearer. If bromide ion is used to displace the bromine in (*S*)-(+)-2-bromobutane, the original chiral compound is converted into its enantiomer, (*R*)-(−)-2-bromobutane.

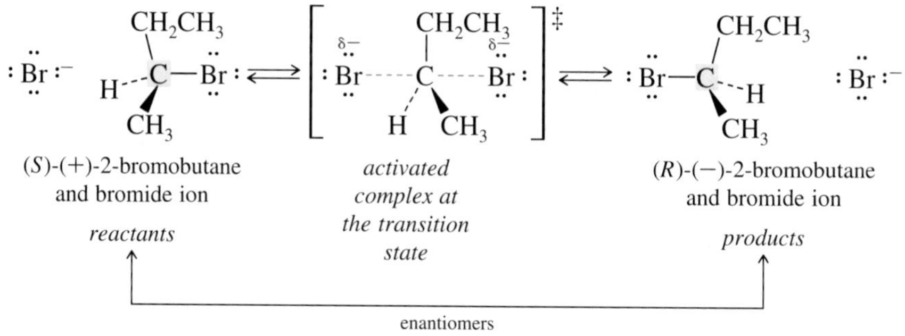

(*S*)-(+)-2-bromobutane    *activated*        (*R*)-(−)-2-bromobutane
and bromide ion            *complex at*       and bromide ion
                           *the transition*
*reactants*                *state*            *products*

enantiomers

Remember that the assignment of $S$ and $R$ can be made just by looking at the three-dimensional pictures. The use of ($+$) or ($-$) indicates whether the compound actually rotates the plane of polarized light to the right or to the left, an experimental determination. Optical rotation cannot be assigned by looking at the structure. Once the sign of rotation is determined for one enantiomer, the other must have optical rotation of the opposite sign (p. 199).

**Problem 7.7** What will happen to the optical rotation observed for pure $(S)$-($+$)-2-bromobutane after it has been exposed to a bromide ion in acetone over a period of time?

**Inversion of configuration** is the conversion of a stereocenter from one configuration to the opposite one. It was discovered in 1893 by Paul Walden at the University of Rostock. This stereochemical transformation, known as the **Walden inversion,** can take place only if bonds are broken and re-formed at the stereocenter. If the species that undergoes inversion has the same groups on it as before inversion, the product is the enantiomer of the starting compound. If the groups are different, an inversion has taken place if the point of attachment of the new group is on the opposite side of the stereocenter from that of the leaving group. In the iodide displacement reaction, iodide ion displaces bromide ion with inversion of configuration because its point of attachment to the stereocenter is clearly on the opposite side of the molecule from the original position of the bromine atom.

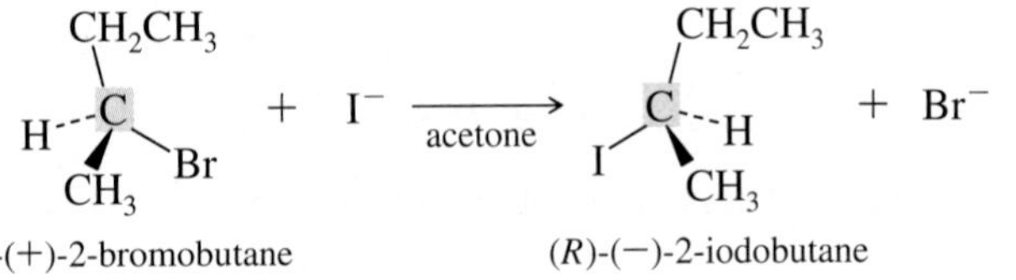

$(S)$-($+$)-2-bromobutane                $(R)$-($-$)-2-iodobutane

*inversion of configuration*

Study Guide
Concept Map 7.1

All experimental evidence shows that when a bimolecular nucleophilic substitution reaction takes place, there is a complete inversion of configuration at the carbon atom undergoing the substitution.

**Problem 7.8** In one of the classic experiments used to explore the stereochemistry of $S_N2$ reactions, two different rates were measured for the reaction shown below. Radioactive iodide ion was used to measure the rate at which iodine was substituted in $(S)$-($+$)-2-iodooctane. The rate at which the starting alkyl iodide lost its optical activity, the **rate of racemization,** also was measured. The rate of racemization is twice the rate of the substitution reaction. Explain these observations.

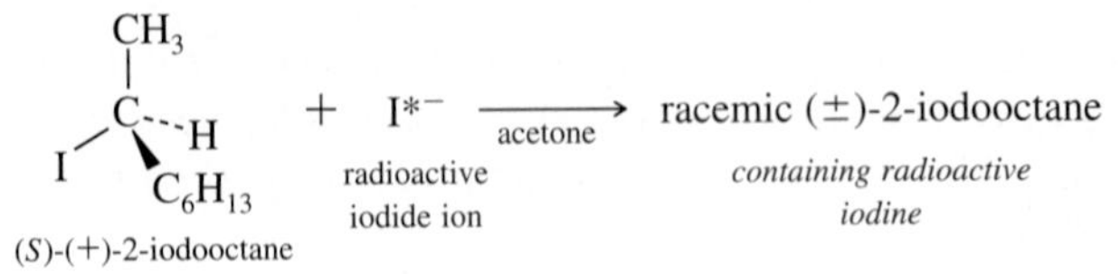

$(S)$-($+$)-2-iodooctane            radioactive            racemic ($\pm$)-2-iodooctane
                                     iodide ion             *containing radioactive*
                                                                    *iodine*

## E. Stereochemistry of $S_N1$ Reactions

What kind of stereochemistry will characterize a reaction that goes through a carbocation intermediate? The simplest tertiary alkyl halide that is chiral is 3-bromo-3-methylhexane. The rate-determining step in the hydrolysis of $(S)$-3-bromo-3-methylhexane is expected to give a planar carbocation solvated on both sides by water molecules (Figure 7.3).

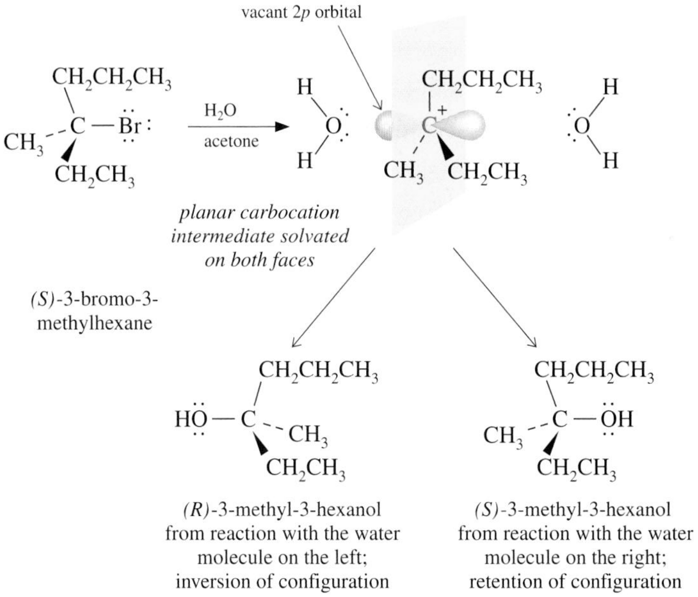

*Figure 7.3*
The hydrolysis of (S)-3-bromo-3-methylhexane.

Two of the water molecules are in a position to donate an electron pair to the empty *p* orbital. If reaction takes place with the water molecule on the right, the alcohol (S)-3-methyl-3-hexanol is formed, with retention of configuration. If the other water molecule reacts, the alcohol with the inverted configuration, (R)-3-methyl-3-hexanol, is formed. When a symmetrical intermediate, such as the planar carbocation, is formed in a reaction at the stereocenter, the chirality of the starting material is lost. In this case, the products are enantiomers. If they are formed in equal amounts, there will be no optical activity detected in the product, which will be a racemic mixture of alcohols.

Experiments with chiral compounds that can undergo $S_N1$ reactions show that a great deal of racemization does take place. The reactions of (S)-(−)-1-chloro-1-phenylethane illustrate this best.

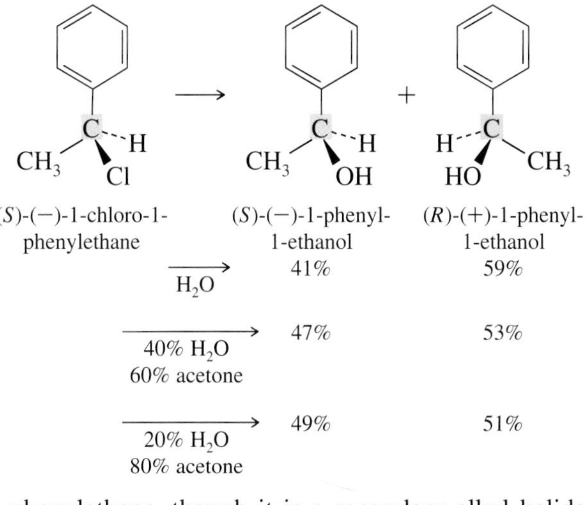

1-Chloro-1-phenylethane, though it is a secondary alkyl halide, ionizes easily because the carbocation that forms has high stability. The halogen is bonded to an $sp^3$-hybridized carbon atom that is bonded to an aromatic ring. Such halides, called **benzylic halides** (see Problem 7.9), ionize to give **benzylic carbocations.** The

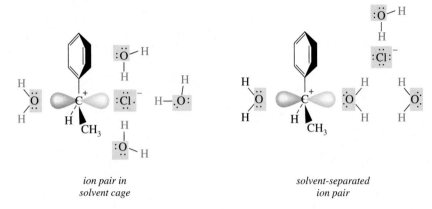

*Figure 7.4*
**Stages of the ionization of 1-chloro-1-phenylethane.**

*ion pair in solvent cage*

*solvent-separated ion pair*

stability of benzylic carbocations can be rationalized by writing resonance contributors in which the positive charge is delocalized to the aromatic ring.

*resonance contributors of the 1-phenylethyl cation*

An examination of the experimental facts shows that 1-chloro-1-phenylethane undergoes the $S_N1$ reaction with extensive racemization but with a slight excess of inversion. For example, when the reaction is carried out in water, an excess of approximately 18% of the inverted product appears. As the percentage of water in the reaction mixture decreases to 40% and then to 20% and is replaced by an increasing amount of acetone, the observed product becomes more completely racemized.

We can interpret these facts to mean that the carbocation that is originally formed is not completely free to react with water on both sides. One side of the carbocation is shielded by the departing chloride ion for some time. In fact, the carbocation and the chloride ion form an **ion pair,** held together by a **cage of solvent molecules.** At this stage, only the water molecules on the side of the cation opposite to the departing chloride ion can react with the cation, giving rise to inversion. If the cation is stable enough to survive this stage, the ions can diffuse farther apart, and water molecules can get close to the other side of the cation (Figure 7.4). Products with retention as well as inversion of configuration result.

If the solvent is pure water, the concentration of water molecules is high enough that some of the reaction takes place while the cation is still shielded by chloride ion in the front. As the proportion of water molecules in the solvent decreases from 100% to 40% to 20%, an increasing number of the cations can survive long enough without reacting with water that the ions can diffuse apart. Reaction with water can then take place on both sides of the cation.

⊘ **Study Guide**
**Concept Map 7.2**

benzyl chloride

**Problem 7.9**    Benzyl chloride has the structure shown at left. Write an equation for the ionization of benzyl chloride, and draw resonance contributions showing the delocalization of charge in the benzylic cation.

**Problem 7.10**    Write detailed mechanisms showing the conversion of benzyl chloride to benzyl alcohol by both $S_N1$ and $S_N2$ pathways.

In summary, reaction of a nucleophile with the alkyl halide before ionization or with an ion pair gives inversion of configuration. If the nucleophile is the solvent, reaction of the cation with a solvent molecule, either in a solvent-separated ion pair or as the free cation, gives extensive racemization. With careful experiments, chemists have shown that most secondary alkyl compounds undergo nucleophilic substitution reactions with almost complete inversion of configuration, even in solvents that encourage ionization. Therefore, reaction of these secondary alkyl compounds must occur through nucleophilic attack on either the alkyl halide or the ion pair.

## 7.4 Nucleophilicity

The **nucleophilicities** of different Lewis bases are determined by measuring the rates of substitution reactions. For example, the relative nucleophilicities of various ions and neutral molecules have been determined using the rates of their substitution reactions with methyl bromide in water.

$$SH^- \geq CN^- > I^- > OH^- > N_3^- > Br^- > \langle\!\!\langle \bigcirc \rangle\!\!\rangle\!-\!O^- > CH_3\overset{\overset{\displaystyle O}{\|}}{C}O^- \geq Cl^- > F^- > NO_3^- > H_2O$$

The hydrosulfide anion, $SH^-$, is a better nucleophile than hydroxide ion, $OH^-$, in water because it reacts faster with the electrophilic center in methyl bromide. That is, the second of these two reactions is faster than the first.

$$\overset{..}{\underset{..}{HO}}:^- \;+\; \overset{\delta+ \;\; \delta-}{CH_3Br} \;\xrightarrow{H_2O}\; \overset{..}{HO}CH_3 \;+\; Br^-$$

$$\overset{..}{\underset{..}{HS}}:^- \;+\; \overset{\delta+ \;\; \delta-}{CH_3Br} \;\xrightarrow{H_2O}\; \overset{..}{HS}CH_3 \;+\; Br^-$$

An examination of the relative nucleophilicities given above shows several trends.

1. For nucleophiles of the same period in the periodic table, nucleophilicity decreases with increasing electronegativity. For example:

$$\overset{..}{\underset{..}{HO}}:^- \qquad > \qquad :\!\overset{..}{\underset{..}{F}}\!:^-$$

*more nucleophilic*   *less nucleophilic*

Oxygen is less electronegative than fluorine is, meaning that it holds the electrons around it less firmly than fluorine does. Thus the oxygen atom in hydroxide ion is better able to donate its electrons than a fluoride ion is. Consequently, the hydroxide ion, which has nonbonding electrons on oxygen, is more nucleophilic than the fluoride ion. This reasoning applies to other elements in the same period as oxygen and fluorine.

$$\underset{\underset{\displaystyle H}{|}}{\overset{\overset{\displaystyle H}{|}}{H-N}}:\qquad \underset{\displaystyle H}{\overset{\displaystyle H}{\diagdown}}\!\overset{..}{\underset{..}{O}}:\qquad H-\overset{..}{\underset{..}{F}}:$$

*most nucleophilic*        *least nucleophilic*

$$\underset{\underset{\displaystyle R}{|}}{\overset{\overset{\displaystyle R}{|}}{R-C}}:^-\qquad \underset{\displaystyle R}{\overset{\displaystyle |}{R-N}}:^-\quad R-\overset{..}{\underset{..}{O}}:^-\qquad :\!\overset{..}{\underset{..}{F}}\!:^-$$

*most nucleophilic*                    *least nucleophilic*

2. Anions are more powerful nucleophiles than their uncharged conjugate acids. For example:

$$HO\!:^- \quad > \quad \overset{H}{\underset{H}{\diagdown}}O\!:$$

*more nucleophilic    less nucleophilic*

The electrons on an atom bearing a negative charge are more loosely held and therefore more easily donated than are the electrons in an uncharged molecule. Thus alkoxide anions are better nucleophiles than alcohols, and amide anions are better nucleophiles than amines.

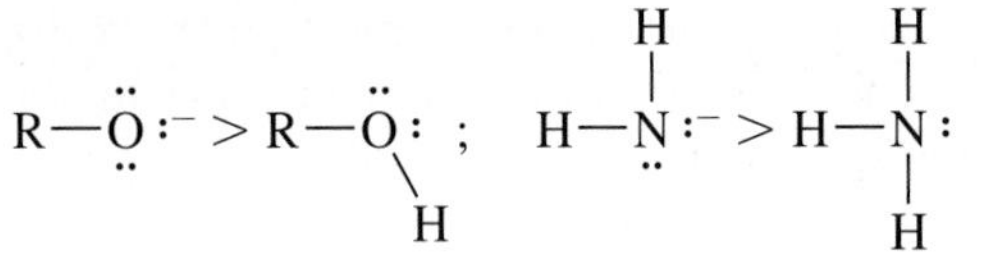

$$R\!-\!\ddot{O}\!:^- > R\!-\!\underset{H}{\ddot{O}}\!: \quad ; \quad H\!-\!\overset{H}{\underset{}{\ddot{N}}}\!:^- > H\!-\!\overset{H}{\underset{H}{N}}\!:$$

*anions more nucleophilic than the corresponding conjugate acids*

3. For anions from a given group in the periodic table, nucleophilicity increases going down the group. For example:

$$HS\!:^- \quad > \quad HO\!:^-$$

*more nucleophilic    less nucleophilic*

$$:\!\ddot{I}\!:^- \qquad :\!\ddot{Br}\!:^- \qquad :\!\ddot{Cl}\!:^- \qquad :\!\ddot{F}\!:^-$$

*most nucleophilic                    least nucleophilic*

These relative nucleophilicities can be partially attributed to the effect of solvent. The experiments that led to the conclusions given above were carried out in water, a protic solvent that hydrogen bonds. Hydrogen bonding decreases the availability of electrons and, therefore, the nucleophilicity of an anion in solution. The effect of solvent will be discussed in greater detail in the next section.

The reactivity of an uncharged nucleophile is not affected much by changes in the solvent. As with anions, the nucleophilicity of uncharged molecules increases the farther down in a given group of the periodic table the nucleophilic center appears. For example, if the reaction of an amine is compared to that of a phosphine, the phosphine reacts about 1000 times faster under the same conditions. That is, the second of these two reactions is 1000 times faster than the first:

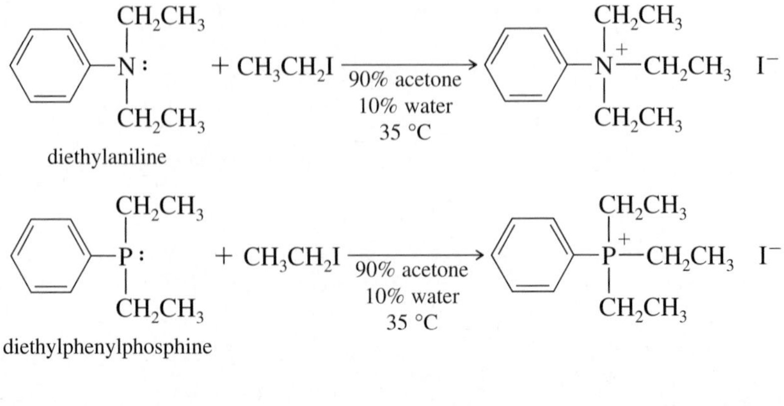

diethylaniline

diethylphenylphosphine

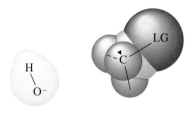 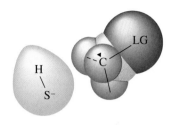

Smaller, more electronegative atom does not start
bonding easily to electrophile; higher $\Delta G^{\ddagger}$;
slower reaction

Larger atom with more loosely held electron
cloud bonds more easily to electrophile;
lower $\Delta G^{\ddagger}$; faster reaction

**Figure 7.5**
Difference in polarizability
between small and large
atoms.

The phosphorus atom is larger than the nitrogen atom. The outermost electrons on the phosphorus are farther away from the nucleus and more loosely held. When phosphorus is close to the center of a partial positive charge, its orbitals containing nonbonding electrons are easily distorted in the direction of the charge, allowing the bonding process to start more easily. The ease with which the electron cloud on an atom can be distorted is called the **polarizability** of the atom. Atoms become larger going down a group in the periodic table and generally become more polarizable and more nucleophilic.

We can picture the effect of polarizability as shown in Figure 7.5. A small highly electronegative atom holds its electrons tightly. It is hard for such an atom to start to bond to the electrophilic carbon atom, and the energy of activation required for such a reaction is high, making the reaction proceed relatively slowly. A large atom with lower electronegativity does not hold the outermost electrons tightly to its nucleus. These electrons are more easily drawn toward the partial positive charge on the carbon atom, and bonding starts easily. The energy of activation required for such a reaction is lower than in the first case, and the reaction is faster. We measure relative nucleophilicities by measuring relative rates of reactions. Polarizabilities affect these rates by affecting how easy it is to get to the transition state in a substitution reaction. Polarizability is important in the process of bonding to a carbon atom that is an electrophilic center bearing a partial positive charge.

**Problem 7.11**    For each pair of reagents, decide which species is more nucleophilic.

(a) ⟨benzene⟩—$S^-$  or  ⟨benzene⟩—$O^-$    (b) $OH^-$  or  $CH_3\overset{\overset{O}{\|}}{C}O^-$    (c) $OH^-$  or  $NO_3^-$

**Problem 7.12**    Draw energy diagrams that compare the energy of activation for the reaction of hydroxide ion ($OH^-$) with methyl bromide ($CH_3Br$) with the energy of activation for the reaction of acetate ion, $CH_3\overset{\overset{O}{\|}}{C}O^-$, with the same halide. Explain why you chose to draw the energy diagrams as you did.

# 7.5    Solvent Effects

The nucleophilicity of a species is relative, not absolute; it depends on the other reagents in the reaction, the other ions present in the solution, the solvent, and the degree of solvation of the ions. The importance of solvent in determining nucleophilicity has become much clearer in recent years as more reactions are studied in the gas phase, where relative nucleophilicities are quite different from those given on page

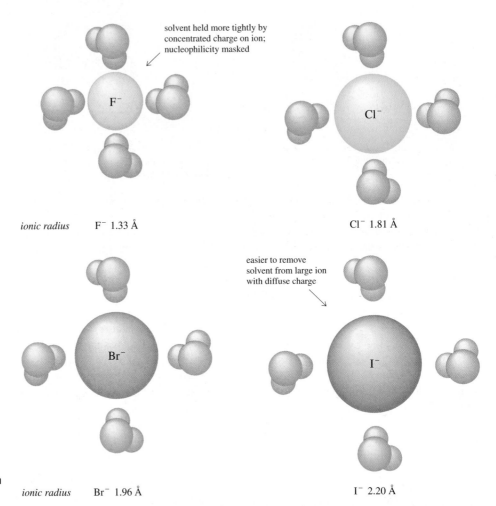

solvent held more tightly by
concentrated charge on ion;
nucleophilicity masked

F⁻

*ionic radius*    F⁻ 1.33 Å

Cl⁻

Cl⁻ 1.81 Å

easier to remove
solvent from large ion
with diffuse charge

Br⁻

I⁻

*Figure 7.6*
Effect of the size of an anion
on its interaction with water, a
protic solvent.

*ionic radius*    Br⁻ 1.96 Å

I⁻ 2.20 Å

251. For example, in the gas phase, fluoride ion is a much better nucleophile than bromide ion. The relative nucleophilicities given for the halide ions were determined in solvents such as water, ethanol, and methanol. These solvents have hydroxyl groups and participate in hydrogen bonding. The smaller the anion, the more concentrated is its negative charge, and the more strongly solvated it is by a solvent such as ethanol (p. 241). Thus chloride ion will be more strongly hydrogen-bonded than bromide ion in such a solution. Hydrogen bonding diminishes the availability of the nonbonding electrons of the anion and decreases its nucleophilicity (Figure 7.6). In hydroxylic solvents, chloride ion is a weaker nucleophile than bromide ion.

Some polar solvents lack a functional group that can serve as a proton donor in hydrogen bonding. Such polar but aprotic solvents are acetone, acetonitrile, dimethylformamide, dimethyl sulfoxide, and hexamethylphosphoric triamide.

$$CH_3\overset{\overset{\displaystyle O}{\|}}{C}CH_3 \qquad CH_3C\equiv N \qquad H\overset{\overset{\displaystyle O}{\|}}{C}\underset{\underset{\displaystyle CH_3}{|}}{N}CH_3 \qquad CH_3\overset{\overset{\displaystyle O}{\|}}{S}CH_3 \qquad CH_3\underset{\underset{\displaystyle CH_3NCH_3}{|}}{N}\overset{\overset{\displaystyle CH_3\ \ \ O\ \ \ CH_3}{\ \ \ \ \|\ \ \ }}{P}NCH_3$$

| acetone | acetonitrile | dimethylformamide | dimethyl sulfoxide | hexamethylphosphoric triamide |
|---------|--------------|-------------------|--------------------|-------------------------------|
| bp 56.5 °C | bp 81.6 °C | bp 153 °C | bp 189 °C | bp 232 °C |
| $\mu$ 2.88 | $\mu$ 3.92 | $\mu$ 3.82 | $\mu$ 3.96 | $\mu$ 4.30 |
| $\varepsilon$ 20.7 | $\varepsilon$ 36.2 | $\varepsilon$ 36.7 | $\varepsilon$ 49 | $\varepsilon$ 30 |

All these solvents have large dipole moments because of the presence of the nitrile group (the carbon–nitrogen triple bond) or of a polar group such as carbon, sulfur, or phosphorus doubly bonded to oxygen. These solvents also have high dielectric constants (p. 33) and can be used as solvents for organic reactions involving ionic reagents or reaction intermediates. When ionic compounds are dissolved in these solvents, the anions are not solvated as strongly as they would be in hydroxylic solvents, so they are freer to participate in nucleophilic substitution reactions. Rates of nucleophilic substitution reactions in these solvents are higher than they are in alcohols.

⊘ **Study Guide**
Concept Map 7.3

**Problem 7.13**    The rate of the reaction of methyl iodide with chloride ion was measured in a series of solvents. The structures of the solvents and the relative rates are shown below the reaction equation. How would you explain these observations?

$$CH_3I \;+\; Cl^- \longrightarrow CH_3Cl \;+\; I^-$$

| Solvent | $CH_3OH$ | $\overset{\displaystyle O}{\overset{\|}{HCNH_2}}$ | $\overset{\displaystyle O}{\overset{\|}{HCN(CH_3)_2}}$ | $\overset{\displaystyle O}{\overset{\|}{CH_3CN(CH_3)_2}}$ |
|---|---|---|---|---|
| Relative rate | 1 | $1.25 \times 10^1$ | $1.2 \times 10^6$ | $7.4 \times 10^6$ |

**Problem 7.14**

(a) Hydroxylamine, $H_2NOH$, has two sites of potential basicity and nucleophilicity. Predict what the product will be for each of the following reactions.

1. $H_2NOH \xrightarrow{\ HCl\ }$    2. $H_2NOH \xrightarrow{\ CH_3I\ (1\ equivalent)\ }$

(b) The conjugate acid of hydroxylamine has $pK_a$ 5.97. What does this tell you about the structure of the conjugate acid? To which acid in the table of $pK_a$ values is it comparable? How do you explain the difference in the acidities of the two species?

# 7.6  Leaving Groups

## A. Leaving Groups

In Chapter 4, a leaving group was defined as a stable species that can be detached from a molecule with its bonding electrons during a displacement reaction (p. 115). What makes a good leaving group? For example, *tert*-butyl chloride ionizes easily in water to give *tert*-butyl cation and chloride ion, but *tert*-butyl alcohol does not ionize under the same conditions to give the cation and hydroxide ion. The difference between *tert*-butyl chloride and *tert*-butyl alcohol lies in the leaving group, which is chloride ion in the one case and hydroxide ion in the other.

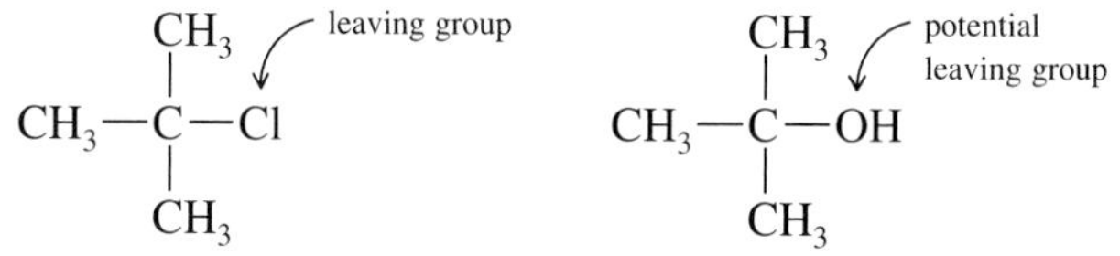

*tert*-butyl chloride            *tert*-butyl alcohol

Hydroxide ion resembles chloride ion in that it is a stable species with an octet of electrons and a negative charge on an electronegative atom. It differs from chloride ion in being the conjugate base of water, a much weaker acid than hydrochloric acid, the conjugate acid of chloride ion.

In Chapter 3, acidity was discussed in terms of a competition between different conjugate bases for the same proton. The more successful a species was in this competition, the stronger a base it was and the weaker its conjugate acid. Strong bases with weak conjugate acids appear in the lower part of the table of p$K_a$ values found inside the front cover of this book. The table can be used to make predictions about how good an ion or molecule will be as a leaving group. *The same factors that make a species a weak base also make it a good leaving group.* Among the halogens, covalently bonded iodine (which becomes iodide ion, the weakest base) is the best leaving group; fluorine (which becomes fluoride ion, the strongest base) is the poorest.

$$-F \ <-Cl<-Br< \ -I$$

| | |
|---|---|
| *poorest leaving group* | *best leaving group* |

For example, in aqueous ethanol, *tert*-butyl iodide reacts 2.5 times faster than *tert*-butyl bromide, which in turn reacts 44 times faster than *tert*-butyl chloride. That is, the following reaction is fastest when X = I and slowest when X = Cl.

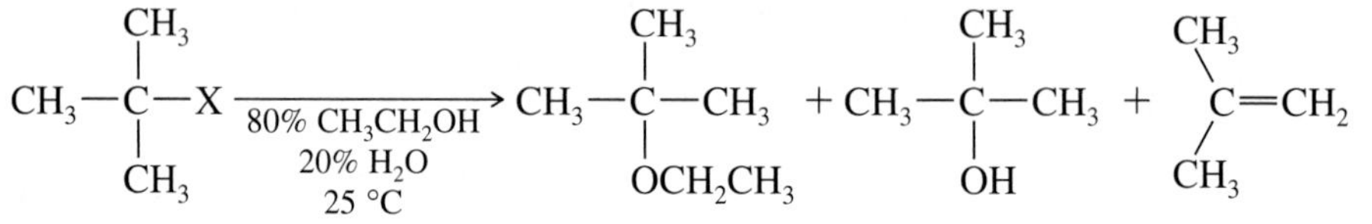

$$CH_3-\underset{\underset{CH_3}{|}}{\overset{\overset{CH_3}{|}}{C}}-X \xrightarrow[\substack{20\%\ H_2O \\ 25\ °C}]{80\%\ CH_3CH_2OH} CH_3-\underset{\underset{OCH_2CH_3}{|}}{\overset{\overset{CH_3}{|}}{C}}-CH_3 \ + CH_3-\underset{\underset{OH}{|}}{\overset{\overset{CH_3}{|}}{C}}-CH_3 \ + \underset{\underset{CH_3}{\diagup}}{\overset{CH_3}{\diagdown}}C=CH_2$$

*tert*-butyl halide        *tert*-butyl ethyl ether    *tert*-butyl alcohol    2-methylpropene

$$\sim 85\%$$

Generally, the conjugate bases of acids in the top half of the p$K_a$ table, with p$K_a$ values less than 8, are reasonably good leaving groups. Highly basic species generally are very poor leaving groups.

## B. Making Leaving Groups Better

Certain kinds of species are good leaving groups, and reaction conditions can be tailored in order to make poor leaving groups into better ones. To bring about the transformation of *tert*-butyl alcohol into *tert*-butyl bromide, for example, it might appear that a hydroxyl group has to be replaced by a bromine atom.

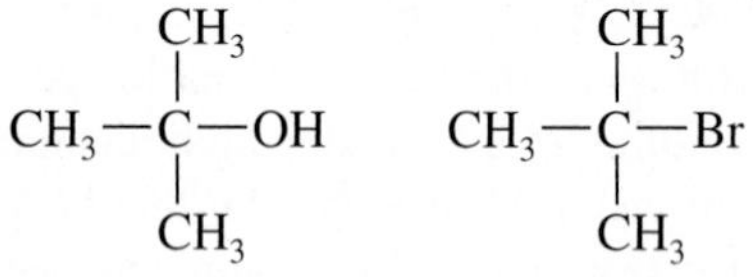

$$CH_3-\underset{\underset{CH_3}{|}}{\overset{\overset{CH_3}{|}}{C}}-OH \qquad CH_3-\underset{\underset{CH_3}{|}}{\overset{\overset{CH_3}{|}}{C}}-Br$$

*tert*-butyl alcohol      *tert*-butyl bromide

The reaction appears to be a substitution achieved by addition of bromide ion to the reaction mixture to give displacement of hydroxide ion. In fact, nothing happens when sodium bromide is added to *tert*-butyl alcohol.

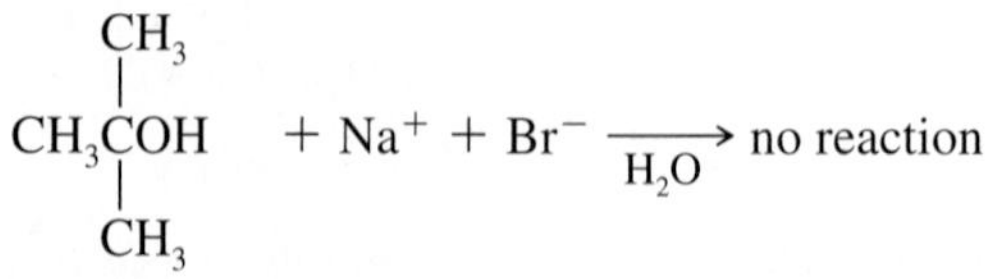

$$CH_3\underset{\underset{CH_3}{|}}{\overset{\overset{CH_3}{|}}{C}}OH \quad + Na^+ + Br^- \xrightarrow[H_2O]{} \text{no reaction}$$

*tert*-butyl alcohol

Hydroxide ion is not a good leaving group, so the substitution reaction cannot take place.

If the reaction conditions are changed by using hydrobromic acid instead of sodium bromide, *tert*-butyl bromide is easily formed.

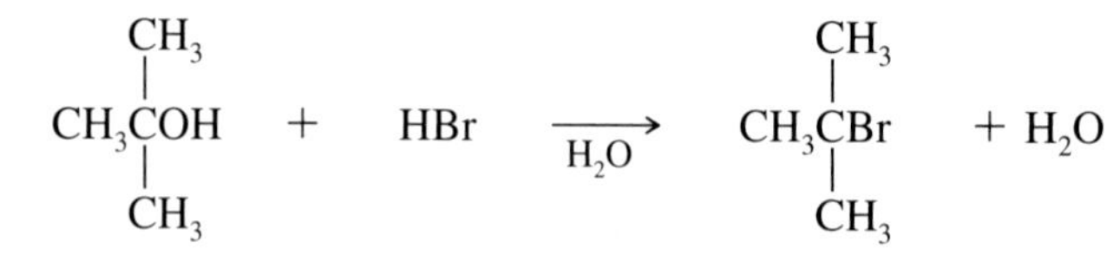

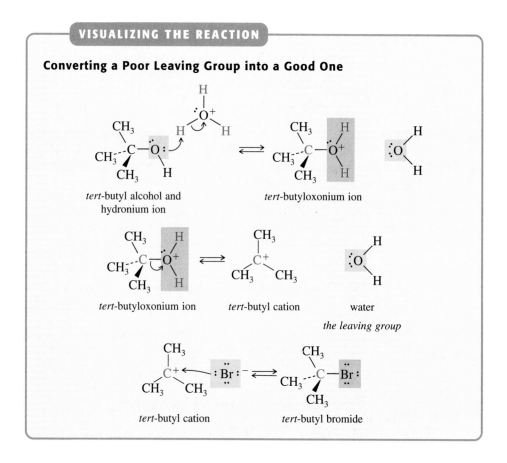

The acid protonates the unshared electrons on the oxygen atom of the alcohol to give a new leaving group, water, which is the conjugate base of the strong acid hydronium ion ($pK_a$ −1.7).  Water is a much better leaving group than hydroxide ion is, and ionization can take place. The *tert*-butyl cation can then combine with the nucleophile, bromide ion.

**VISUALIZING THE REACTION**

**Converting a Poor Leaving Group into a Good One**

*tert*-butyl alcohol and hydronium ion

*tert*-butyloxonium ion

*tert*-butyloxonium ion

*tert*-butyl cation

water

*the leaving group*

*tert*-butyl cation

*tert*-butyl bromide

**Problem 7.15**    Where did the hydronium ion in the first step of the mechanism given above come from?

**Problem 7.16**    What other nucleophiles are present in the reaction mixture created by combining *tert*-butyl alcohol, hydrobromic acid, and water? What other reactions are possible for the tertiary carbocation?

A hydroxyl group is frequently protonated in order to convert it into water, a good leaving group. However, a protonated alcohol is an ionic intermediate that cannot be isolated and put into a bottle. Furthermore, the oxonium ion (the protonated alcohol) is a strong acid and cannot be an intermediate when nucleophiles that are stronger bases than alcohols are to be used in the displacement reactions. For example, suppose we want to convert 1-butanol into pentanenitrile.

$$CH_3CH_2CH_2CH_2OH \xrightarrow{\ ?\ } CH_3CH_2CH_2CH_2C{\equiv}N$$

1-butanol                                    pentanenitrile

A check of the $pK_a$ table shows us that the nucleophile we need for this displacement, cyanide ion, $CN^-$, is a fairly strong base. Its conjugate acid, hydrocyanic acid, HCN, has $pK_a$ 9.1. This means that we cannot use HCN to protonate the alcohol directly, the way we used HBr ($pK_a \sim -8$). The equilibrium favors the reverse reaction.

$$CH_3CH_2CH_2CH_2OH \ + \ HCN \ \rightleftarrows \ CH_3CH_2CH_2CH_2\overset{\overset{\textstyle H}{\textstyle |}}{O}{-}H \ + \ CN^-$$

1-butanol              hydrocyanic acid              an oxonium ion          cyanide ion
                         $pK_a$ 9.1                 conjugate acid of
                        weak acid                    1-butanol
                      *highly toxic*                  $pK_a \sim -2$
                                                     strong acid

Even if we were to form the oxonium ion by using a strong acid with a weakly nucleophilic conjugate base—sulfuric acid, for example—to protonate the alcohol, we cannot use the oxonium ion with a basic nucleophile such as cyanide ion. Amines and azide ion are other important nucleophiles that are incompatible with oxonium ions.

**Problem 7.17**    Use the $pK_a$ table to predict the products of the reaction of methylamine, $CH_3NH_2$, with the conjugate acid of 1-butanol.

**Problem 7.18**    Use the $pK_a$ table to make a list of all the nucleophiles appearing in Table 7.1 (p. xxx) that cannot be used with oxonium ions. The conjugate acid of azide ion, hydrazoic acid, $HN_3$, has $pK_a$ 4.7.

**Problem 7.19**    When ethanol is heated with a small amount of sulfuric acid, diethyl ether is formed and can be distilled out of the reaction mixture. This is, in fact, how the ether once used for anesthesia was prepared.

$$2\ CH_3CH_2OH \xrightarrow[\Delta]{H_2SO_4} CH_3CH_2OCH_2CH_3 + H_2O$$

Write a detailed mechanism for this reaction, applying all the ideas developed so far about acids, bases, conversion of poor leaving groups into good leaving groups, and displacement of leaving groups by nucleophiles.

**Problem 7.20**    When diethyl ether is heated with hydrogen iodide, iodoethane is the product.

$$CH_3CH_2OCH_2CH_3 + 2\ HI \xrightarrow{\Delta} 2\ CH_3CH_2I + H_2O$$

Write a detailed mechanism for this reaction.

## C. Sulfonate Esters

Alcohols are readily available compounds, so it is important to be able to convert them into stable isolable compounds with good leaving groups. A successful strategy is to convert the alcohol into the ester of the relatively strong acid *p*-toluenesulfonic acid, $pK_a \sim -0.6$.

*p*-toluenesulfonic acid                              *p*-toluenesulfonyl group

TsOH                                                    Ts—

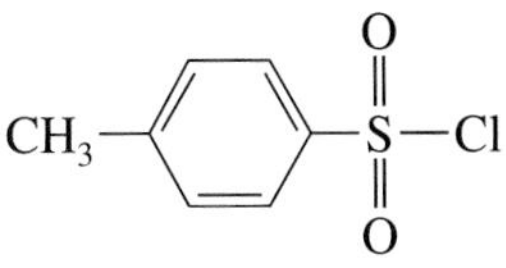

p-toluenesulfonyl chloride
tosyl chloride
TsCl

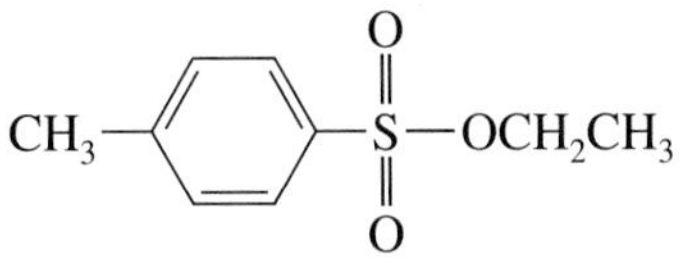

ethyl p-toluenesulfonate
ethyl tosylate
TsOCH$_2$CH$_3$

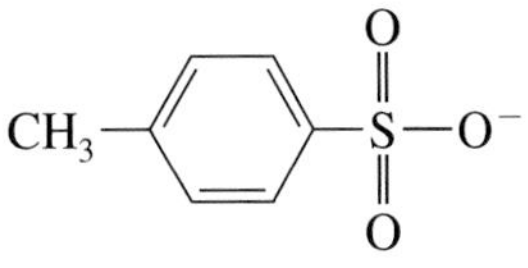

p-toluenesulfonate anion
tosylate
TsO$^-$

The p-toluenesulfonyl group has a large structure and a long name. Because it is used so often in organic reactions, it has been given an abbreviation for the sake of convenience. The name p-toluenesulfonyl has been condensed by taking only the underlined sections to form the shorter name **tosyl.** In structural formulas, the symbol **Ts** is used, as shown above.

The alcohol 1-butanol reacts with tosyl chloride in pyridine as a solvent to give a tosylate.

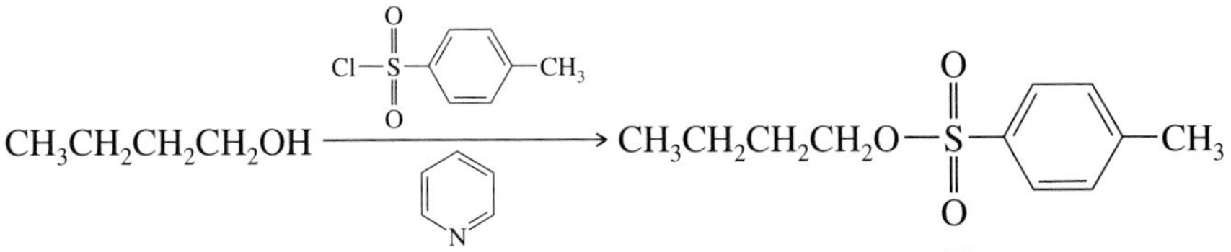

CH$_3$CH$_2$CH$_2$CH$_2$OH

1-butanol

*leaving group OH$^-$*

CH$_3$CH$_2$CH$_2$CH$_2$O—S

or

CH$_3$CH$_2$CH$_2$CH$_2$OTs

butyl tosylate

*leaving group TsO$^-$*

A basic solvent, pyridine, deprotenates the oxonium ion that forms in the reaction. In the reaction, the poor leaving group, the hydroxide ion, is converted into a good leaving group, the tosylate anion.

## VISUALIZING THE REACTION

### Formation of a Tosylate

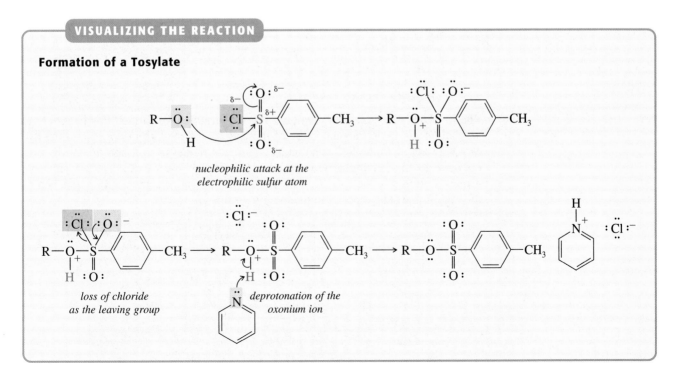

*nucleophilic attack at the electrophilic sulfur atom*

*loss of chloride as the leaving group*

*deprotonation of the oxonium ion*

Once butyl tosylate is prepared, it will react with cyanide ion to give pentanenitrile (p. 258).

$$CH_3CH_2CH_2CH_2OTs + K^+CN^- \xrightarrow{\text{ethanol}} CH_3CH_2CH_2CH_2CN + TsO^-K^+$$

butyl tosylate      potassium cyanide      pentanenitrile      potassium tosylate

Note that the alcohol is converted to a tosylate without breaking the carbon–oxygen bond in the alcohol. Thus the tosylate of a chiral alcohol can be prepared without affecting stereochemistry. For example, if (*R*)-(−)-2-butanol is converted into its tosylate, then the tosylate also has the *R* configuration because no bonds to the stereocenter are broken in this reaction (and nothing is done to any group that changes its priority in the rules of nomenclature of chiral compounds). The configuration of the molecule is unchanged; that is, the reaction takes place with **retention of configuration.** The tosylate anion is a good leaving group, and a nucleophilic substitution reaction with cyanide ion gives 2-methylbutanenitrile with the *S* configuration. The second reaction goes by an $S_N2$ mechanism with inversion of configuration.

(*R*)-(−)-2-butanol      (*R*)-2-butyl tosylate      (*S*)-2-methylbutanenitrile

retention of configuration      inversion of configuration

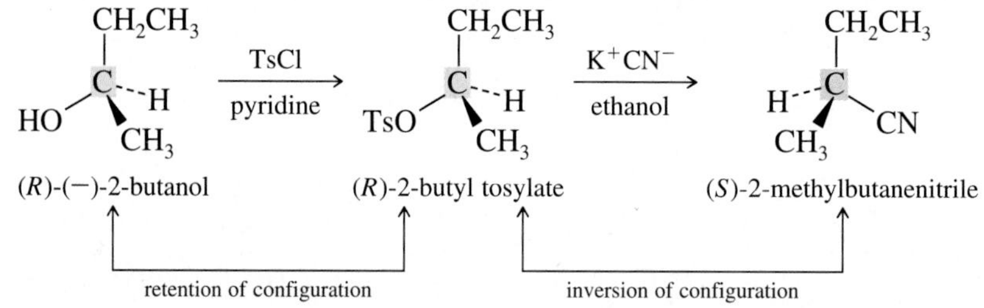

In another example, *cis*-3-methyl-1-cyclopentanol is converted to *trans*-1-azido-3-methylcyclopentane by way of the tosylate.

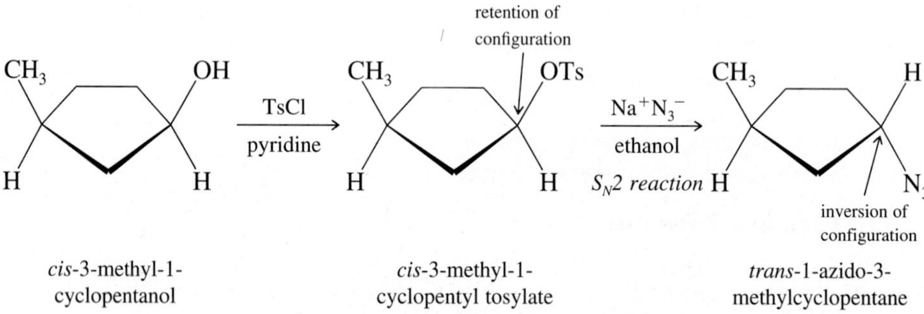

*cis*-3-methyl-1-cyclopentanol      *cis*-3-methyl-1-cyclopentyl tosylate      *trans*-1-azido-3-methylcyclopentane

The tosylate is formed with retention of configuration. We can tell this because the tosylate group is on the same side of the ring as the methyl group, just as the hydroxyl group is. In the $S_N2$ reaction, the azide ion moves in from the side of the ring opposite to the leaving group, the tosylate ion, giving a carbon atom with five groups around it at the transition state. The product has the azido group attached to the ring on the side opposite to the methyl group.

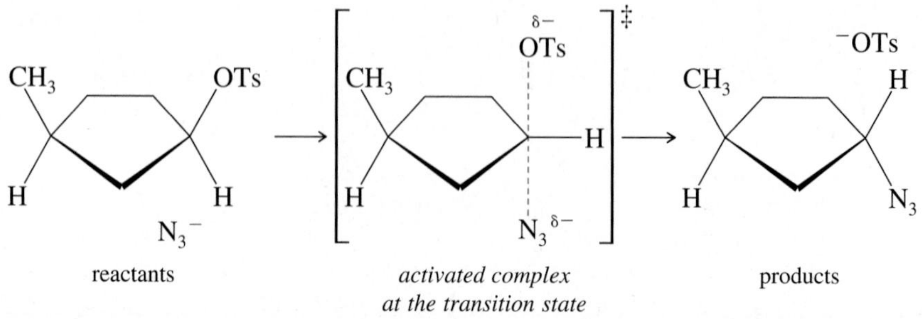

reactants      *activated complex at the transition state*      products

Another way of showing the stereochemistry of the reactions described on the previous page is seen below.

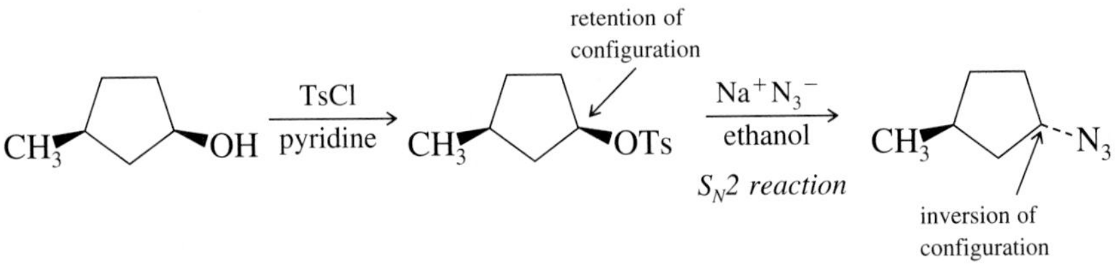

cis-3-methyl-1
cyclopentanol

cis-3-methyl-1
cyclopentyl tosylate

trans-1-azido-3-
methylcyclopentane

In the first two structures, the methyl group and the hydroxyl or tosylate groups are both above the plane of the ring. After the $S_N2$ reaction, the azido group is below the plane of the ring, while the stereochemistry at the carbon atom bearing the methyl group is untouched.

Researchers have developed many other reagents to convert alcohols to compounds containing good leaving groups. Most of them are related to p-toluenesulfonic acid. One that is frequently used is shown below with its abbreviations.

methanesulfonyl chloride
mesyl chloride

mesyl chloride

mesylate ion

Study Guide
Concept Map 7.4

**Problem 7.21** The following tosylate was prepared as an intermediate for the synthesis of compounds used in a study of palladium-catalyzed cyclizations of enantiomerically pure acyclic compounds. The compound was prepared from the S isomer of a chiral diol. Explain the regioselectivity of the tosylation reaction and draw a stereochemically correct structural formula for the product.

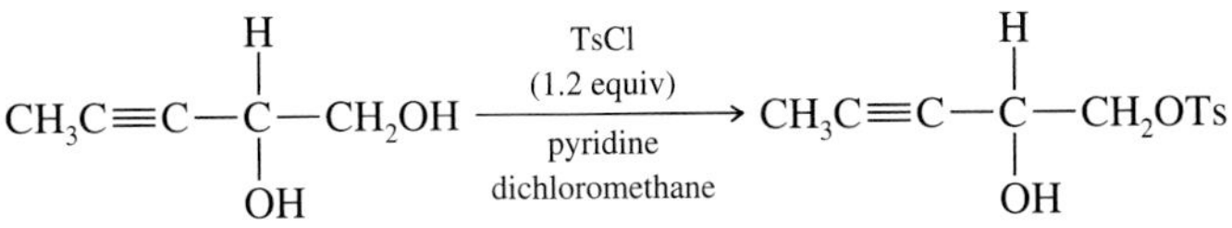

**Problem 7.22** Mesyl chloride, MsCl, was used in the synthesis of gleosporone, an inhibitor of the germination of the spores of fungi. Provide structural formulas for the products of the reactions shown below.

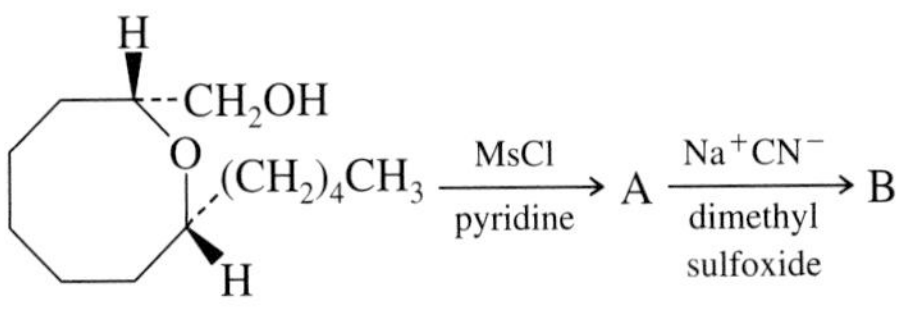

## D. Biological Leaving Groups

Alkyl transfer reactions play important roles in biological processes. The leaving groups in these reactions are not halide ions. Instead, a number of groups serve that role. On page 120 we saw how a methyl group was transferred from a nitrogen bearing four substituents and a positive charge; in that case, an amine was the leaving group. For hydroxyl groups, phosphoric acid esters are the biological counterparts of tosylates. Often two or three phosphorus atoms are bound together by oxygen atoms in compounds that are derived from the anhydrides of phosphoric acid (p. 680). In the example shown below, adenosine triphosphate (ATP, p. 680) serves to alkylate methionine, which thus acquires a good leaving group and itself becomes a methyl transfer agent, S-adenosylmethionine, abbreviated as SAM. S-Adenosylmethionine transfers methyl groups in the biological synthesis of many important compounds. The one shown below is the synthesis of creatine, which is essential for the functioning of muscles.

Leaving groups thus may take many forms. What is essential is that the group allow for electrophilic character at the atom to which it is attached and that the species formed in the displacement reaction be a stable ion or molecule.

## VISUALIZING THE REACTION

### Biological Leaving Groups

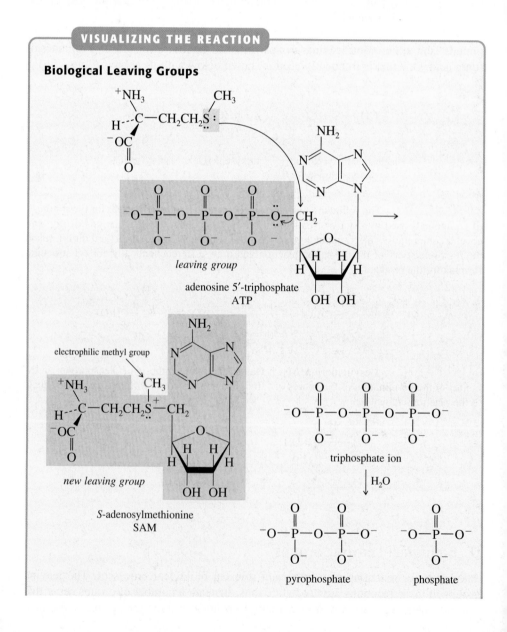

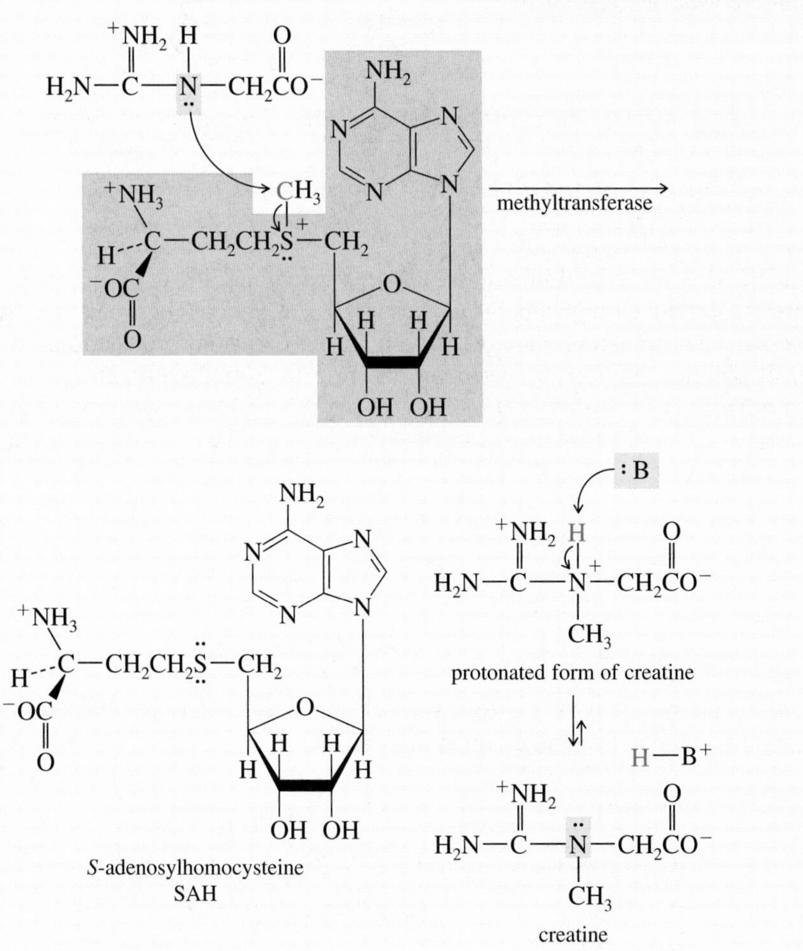

The mechanism scheme shows structures with labels:

$^+NH_2$ H O
$H_2N-C-N-CH_2CO^-$

$NH_2$ (adenine)

$^+NH_3$
$H \cdots C-CH_2CH_2\overset{..}{S}-CH_2$
$^-OC$
O
H H H H
OH OH

→ methyltransferase →

$CH_3$
$^+NH_3$
$H \cdots C-CH_2CH_2\overset{+}{S}-CH_2$
$^-OC$
O
H H H H
OH OH

S-adenosylhomocysteine
SAH

: B

$^+NH_2$ H O
$H_2N-C-\overset{+}{N}-CH_2CO^-$
$CH_3$
protonated form of creatine

↕ H—B$^+$

$^+NH_2$ O
$H_2N-C-\overset{..}{N}-CH_2CO^-$
$CH_3$
creatine

---

## ONE SMALL STEP

Leaving groups are stable ions or molecules that are displaced in nucleophilic substitution reactions.

**PROBLEM:** For each of the following sets of reagents, identify the potential leaving group, and complete the reaction.

(a) 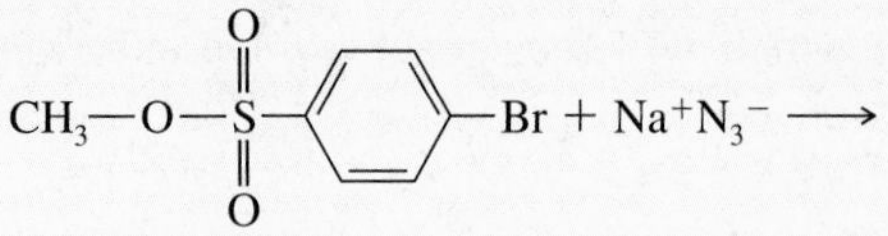 $CH_3-O-\overset{\overset{O}{\|}}{\underset{\underset{O}{\|}}{S}}-C_6H_4-Br + Na^+N_3^- \longrightarrow$

(b) $CH_3-\overset{\overset{CH_3}{|}}{\underset{\underset{CH_3}{|}}{O^+}}$   $BF_4^- + CH_3CH_2SNa \longrightarrow$

(c) $CH_3CH_2CH_2CH_2-\overset{+}{N}\equiv N$   $Cl^- \xrightarrow{H_2O}$

**Hint:** Identifying the nucleophile and the electrophilic carbon that will react with each other is an important first step for solving this problem.

On the Web: ONE SMALL STEP

## 7.7   Elimination Reactions

### A. Elimination Reactions in Competition with Substitution Reactions

An **elimination reaction** results when a proton and a leaving group are removed from adjacent carbon atoms, giving rise to a $\pi$ bond between the two carbon atoms. An elimination reaction is the reverse of an addition reaction. For example, electrophilic addition of hydrogen chloride to 2-methylpropene gives *tert*-butyl chloride.

$$\underset{\substack{\text{2-methylpropene}}}{\overset{\substack{\text{CH}_3 \\ |}}{\text{CH}_3\text{C}}\text{=}\text{CH}_2} + \text{HCl} \longrightarrow \underset{\substack{\text{tert-butyl chloride}}}{\overset{\substack{\text{CH}_3 \\ |}}{\text{CH}_3\text{—}\overset{\displaystyle |}{\underset{\displaystyle |}{\text{C}}}\text{—CH}_3}}$$

$$\overset{\text{Cl}}{}$$

*an electrophilic addition reaction*

One of the products of the treatment of *tert*-butyl chloride with base (p. 241) is 2-methylpropene from an elimination reaction.

$$\underset{\substack{\text{tert-butyl chloride}}}{\overset{\substack{\text{CH}_3 \\ |}}{\text{CH}_3\text{—}\overset{\displaystyle |}{\underset{\substack{| \\ \text{Cl}}}{\text{C}}}\text{—CH}_3}} + \text{K}^+\text{OH}^- \longrightarrow \underset{\substack{\text{2-methylpropene}}}{\overset{\substack{\text{CH}_3 \\ |}}{\text{CH}_3\text{C}}\text{=}\text{CH}_2} + \text{K}^+\text{Cl}^- + \text{H}_2\text{O}$$

*an elimination reaction*

Any alkyl halide (except a methyl or benzyl halide) has three competing routes of reaction with bases or nucleophiles open to it. They are

1. Bimolecular displacement of the leaving group ($S_N2$)

2. Bimolecular elimination (E2)

3. Ionization to give a carbocation, which will then react to give substitution or elimination ($S_N1$ and E1; see Section 7.3C)

Which route will be taken depends on the relative heights of the energy barriers on each reaction pathway. Each molecule explores these different pathways and takes the one that has the lowest energy barrier under the conditions of the reaction. For example, a secondary alkyl halide reacts more slowly with a nucleophile in an $S_N2$ reaction than a primary halide does because the activated complex at the transition state is more crowded (Figure 7.1, p. 243). The $S_N2$ reaction for a tertiary halide is even slower. Such changes in the structure of the halides allow the elimination reaction to become more competitive. The three equations on the next page show that it is highly impractical to try to carry out nucleophilic substitution reactions on secondary and tertiary halides with a strongly basic reagent because the chief product in each case comes from an elimination reaction.

$$CH_3CH_2Br \xrightarrow[\substack{\text{ethanol} \\ 55\,°C}]{CH_3CH_2O^- Na^+} CH_3CH_2OCH_2CH_3 + CH_2{=}CH_2$$

ethyl bromide      diethyl ether    ethylene

99%      1%

*a primary alkyl*      *$S_N2$ product*      *E2 product*
*halide*

1° halide, low $\Delta G^{\ddagger}$ for the $S_N2$ reaction, $S_N2$ reaction favored

$$CH_3\underset{\underset{\displaystyle Br}{|}}{C}HCH_3 \xrightarrow[\substack{\text{ethanol} \\ 55\,°C}]{CH_3CH_2O^- Na^+} CH_3\underset{\underset{\displaystyle OCH_2CH_3}{|}}{C}HCH_3 + CH_3CH{=}CH_2$$

isopropyl bromide      ethyl isopropyl    propene
ether    79%
21%

*a secondary alkyl*      *$S_N2$ product*      *E2 product*
*halide*

2° halide, higher $\Delta G^{\ddagger}$ for the $S_N2$ reaction, E2 reaction becomes competitive

$$CH_3\underset{\underset{\displaystyle Br}{|}}{\overset{\overset{\displaystyle CH_3}{|}}{C}}CH_3 \xrightarrow[\substack{\text{ethanol} \\ 55\,°C}]{CH_3CH_2O^- Na^+} CH_3\overset{\overset{\displaystyle CH_3}{|}}{C}{=}CH_2$$

*tert*-butyl bromide      2-methylpropene
100%

*a tertiary alkyl*      *E1 + E2 product*
*halide*

3° halide, $\Delta G^{\ddagger}$ for the $S_N2$ reaction so high that only E products seen

The change in the nature of the activated complex at the transition state, and hence $\Delta G^{\ddagger}$ for a reaction, can come from the base as well as from the alkyl halide. The use of a sterically hindered base, such as the *tert*-butoxide ion, also favors the elimination reaction over the substitution reaction. For example, when the primary alkyl halide 1-bromooctadecane is treated with sodium methoxide, methyl octadecyl ether is the major product.

$$CH_3(CH_2)_{15}CH_2CH_2Br \xrightarrow[\substack{\text{methyl alcohol} \\ 65\,°C \\ 12\,h}]{CH_3O^- Na^+ \,(1\,M)} CH_3(CH_2)_{15}CH_2CH_2OCH_3 + CH_3(CH_2)_{15}CH{=}CH_2$$

methyl octadecyl ether    1-octadecene
96%    1%

*$S_N2$ product*      *E2 product*

When potassium *tert*-butoxide is used as the base, the major product is the alkene.

$$CH_3(CH_2)_{15}CH_2CH_2Br \xrightarrow[\substack{\textit{tert}\text{-butyl alcohol} \\ 80\,°C \\ 20\,h}]{CH_3\underset{\underset{\displaystyle CH_3}{|}}{\overset{\overset{\displaystyle CH_3}{|}}{C}}O^- K^+ \,(1\,M)} CH_3(CH_2)_{15}CH_2CH_2O\underset{\underset{\displaystyle CH_3}{|}}{\overset{\overset{\displaystyle CH_3}{|}}{C}}CH_3 + CH_3(CH_2)_{15}CH{=}CH_2$$

*tert*-butyl octadecyl ether    1-octadecene
12%    85%

*$S_N2$ product*      *E2 product*

This effect is primarily due to the size of the base. The rate of the elimination reaction, in which a proton is removed from the surface of the molecule, is increased relative to the rate of the substitution reaction, in which the bulky *tert*-butoxide ion has to form a bond to an electrophilic center surrounded by other groups.

Temperature is also a factor. Elimination reactions are more likely at higher temperatures as molecules acquire enough energy to climb energy barriers that were too high to cross at lower temperatures.

How can we decide whether a substitution or an elimination reaction will take place?

It is useful to analyze such cases by answering four questions for each one:

1. Is the carbon atom bearing the leaving group primary, secondary, or tertiary?

2. Is there a good nucleophile?

3. Is the nucleophile also a strong base?

4. Is the solvent protic and, therefore, an ionizing solvent?

If we have a methyl or primary halide and a good nucleophile, the predominant reaction will be substitution by the $S_N2$ mechanism. If we have a tertiary halide (or another halide that gives a stable cation) and an ionizing solvent, we will get a mixture of substitution and elimination products, by way of an intermediate carbocation via the $S_N1$ and E1 mechanisms. If we have a secondary halide, the major product will depend on the conditions.

Imagine a secondary alkyl halide — for example, isopropyl bromide — reacting under several different conditions and predict what the products will be.

1. $CH_3CHCH_3$  $\xrightarrow[\substack{\text{ethanol} \\ 25\ °C}]{Na^+N_3^-}$ ?  |  2. $CH_3CHCH_3$  $\xrightarrow[\substack{80\% \text{ ethanol} \\ 20\% \text{ water}}]{}$ ?  |  3. $CH_3CHCH_3$  $\xrightarrow[\substack{tert\text{-butyl alcohol} \\ \Delta}]{\substack{CH_3 \\ | \\ CH_3CO^-K^+ \\ | \\ CH_3}}$ ?
   $|$                              $|$                              $|$
   $Br$                            $Br$                            $Br$

$S_N2$, $S_N1$, and E reactions are all potentially available to the alkyl halide, but the most probable reaction is chosen by identifying the lowest-energy pathway.

In the first case, the azide ion, a good nucleophile that is not a strong base, is present (see Table 7.1). A direct displacement of the halide ion in an $S_N2$ reaction is most probable.

In the second case, no nucleophile that appears in Table 7.1 is present. Water and ethanol are only weakly nucleophilic. A mixture of water and ethanol acting as a solvent will stabilize ions. Any reaction that takes place will probably involve a carbocation intermediate. Both $S_N1$ and E1 products are expected.

In the third case, the *tert*-butoxide ion, a nucleophile that is a strong and sterically hindered base, is present. The reaction mixture is heated. Elimination, the E2 reaction, is most probable under these conditions, with the major product being an alkene.

---

**Problem 7.23**  Draw detailed mechanisms using the curved-arrow convention to show how the products predicted for equations 1, 2, and 3 above will be formed. The mechanism for an E2 reaction was given in A Look Ahead on page 233.

## B. The Mechanism and Regioselectivity of the Elimination Reaction

An elimination may result from the formation of a cation, in which case it is known as a **unimolecular elimination** or **E1 reaction.** Such reactions always accompany $S_N1$ reactions. The mechanism of the E1 reaction was shown on page 247.

Alternatively, the elimination may occur by removal of a proton and simultaneous loss of a leaving group in a **bimolecular elimination** or **E2 reaction.** The reaction of *sec*-butyl bromide with concentrated potassium hydroxide is an example of such a reaction.

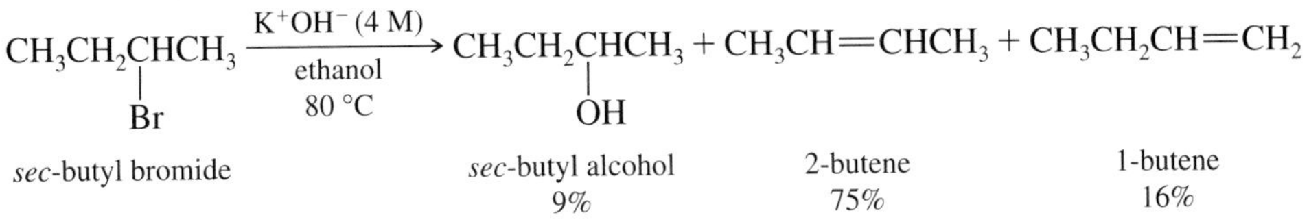
CH3CH2CHCH3
Br
K+OH− (4 M)
ethanol
80 °C
CH3CH2CHCH3 + CH3CH=CHCH3 + CH3CH2CH=CH2
OH
sec-butyl bromide
sec-butyl alcohol
9%
2-butene
75%
1-butene
16%

The reaction gives rise to a mixture of alkenes in which the more highly substituted alkene is formed in the larger amount. The more highly substituted alkene is defined as the one with fewer hydrogen atoms on the carbons involved in the double bond. 2-Butene has two alkyl groups and two hydrogen atoms on the double bond, whereas 1-butene has one alkyl group and three hydrogen atoms on its double bond. Thus 2-butene has a more highly substituted double bond than 1-butene does. The more highly substituted an alkene is, the more stable it is. A further discussion of the relative stabilities of alkenes is given on page 282.

The regioselectivity of an E2 reaction is determined by the relative stabilities of the transition states. The molecular complex at the transition state for the formation of an alkene resembles the alkene in structure and is stabilized by the same factors that stabilize the alkene. The base attacks 2-bromobutane at two different sites to give the alkenes observed as products.

**VISUALIZING THE REACTION**

### The E2 Reaction

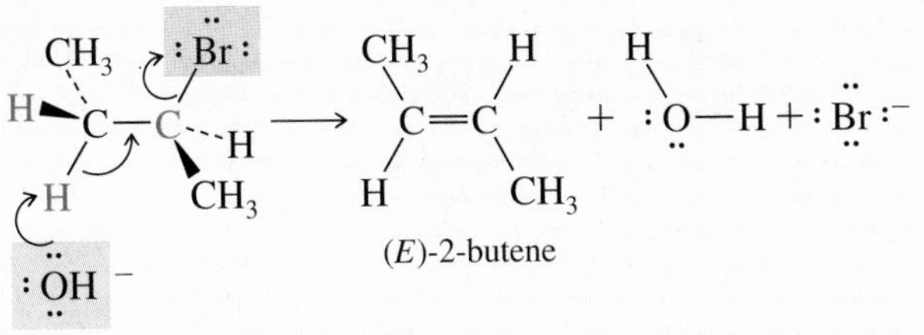
CH3
:Br:
H
C
C
H
H
CH3
:OH −
CH3
H
C=C
H
CH3
H
:O—H + :Br:−
(E)-2-butene

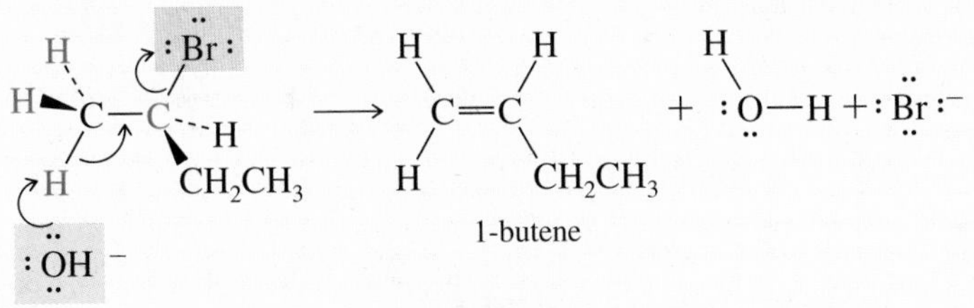
H
:Br:
H
C
C
H
H
CH2CH3
:OH −
H
H
C=C
H
CH2CH3
H
:O—H + :Br:−
1-butene

The activated complex at the transition state leading to 2-butene resembles that alkene and is, therefore, lower in energy than the activated complex at the transition state leading to 1-butene.

*activated complex at the transition state for the formation of 2-butene; lower-energy transition state resembling the more highly substituted alkene product*

*activated complex at the transition state for the formation of 1-butene; higher-energy transition state resembling the less highly substituted alkene product*

2-butene

*major product*

1-butene

The reaction goes chiefly by the pathway leading through the transition state of lower energy (p. 132) to give the more highly substituted alkene, which is the more stable one (Figure 7.7).

This generalization is usually true if the leaving group is a halide ion and the base is a hydroxide or ethoxide ion. Much experimental work has been done to show that the nature and size of the leaving group, the size of the base used to remove the proton, and especially the stereochemistry of the compound undergoing elimination can affect the exact proportions of the alkenes formed as products.

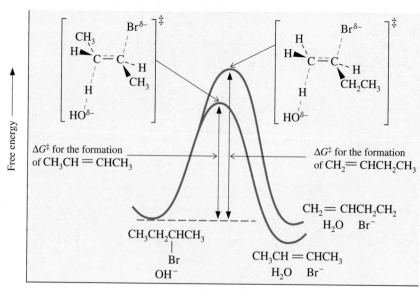

**Figure 7.7**

A comparison of the relative transition state energies for the formation of 2-butene and 1-butene from *sec*-butyl bromide.

**Problem 7.24**   Is another stereochemistry possible for the 2-butene that is formed in the elimination reaction of *sec*-butyl bromide? Draw any other conformations of the halide that will give 2-butene as the product.

**Problem 7.25**   Write the structure(s) of the product(s) of the elimination reactions that you would expect in the following cases. If more than one product can be formed, predict which one will be the major product.

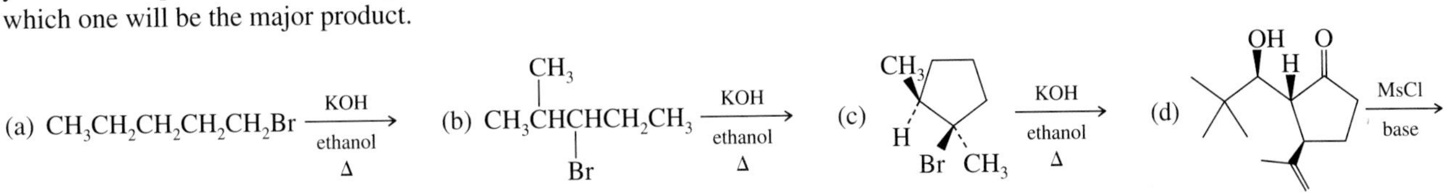

(a) $CH_3CH_2CH_2CH_2CH_2Br$ $\xrightarrow[\Delta]{\text{KOH} \atop \text{ethanol}}$

(b) $CH_3\overset{\overset{\displaystyle CH_3}{|}}{C}HCHCH_2CH_3$ $\underset{\underset{\displaystyle Br}{|}}{}$ $\xrightarrow[\Delta]{\text{KOH} \atop \text{ethanol}}$

(c) $\xrightarrow[\Delta]{\text{KOH} \atop \text{ethanol}}$

(d) $\xrightarrow[\text{base}]{\text{MsCl}}$

## C. Stereochemistry of Elimination Reactions

A definite conformation for the species undergoing an E2 reaction has been suggested. There is evidence that E2 elimination proceeds best when the hydrogen to be removed and the leaving group are in an anti conformation with respect to each other. For example, (2*R*,3*R*)-3-phenyl-2-butyl tosylate gives exclusively (*E*)-2-phenyl-2-butene, and its diastereomer, (2*S*,3*R*)-3-phenyl-2-butyl tosylate gives (*Z*)-2-phenyl-2-butene.

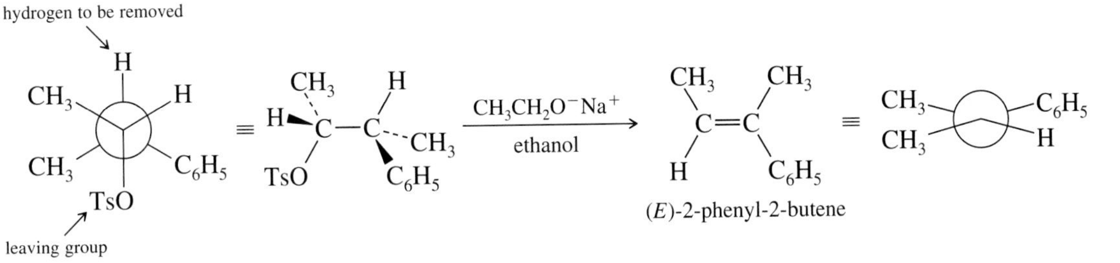

(2R,3R)-3-phenyl-2-butyl tosylate        (*E*)-2-phenyl-2-butene

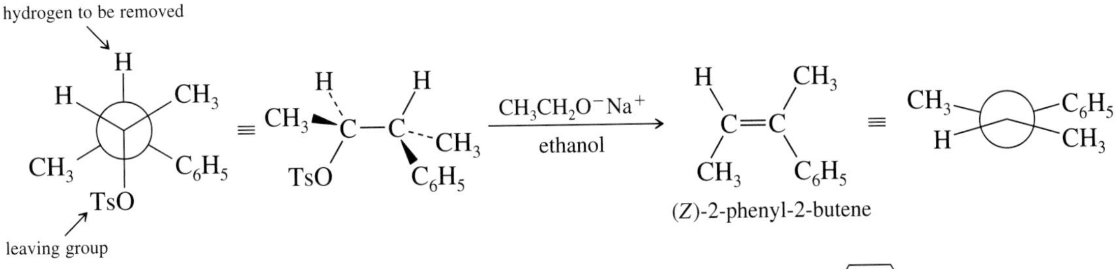

(2S,3R)-3-phenyl-2-butyl tosylate        (*Z*)-2-phenyl-2-butene

$C_6H_5-$ ≡

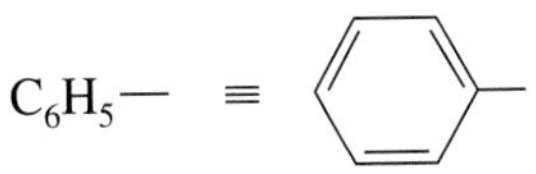

   In an E2 reaction, the stereochemistry of the alkene that is formed is governed by the conformation that the starting material must adopt in the transition state so that the leaving group and the proton being removed by the base are anti to each other. There is no such stereochemical requirement for an E1 reaction, because the leaving group is lost and a carbocation has been formed before the proton is removed.

**Problem 7.26**   Compound A reacts with potassium *tert*-butoxide to give a mixture of diastereomeric alkenes, Compounds B and C, as shown on the next page.

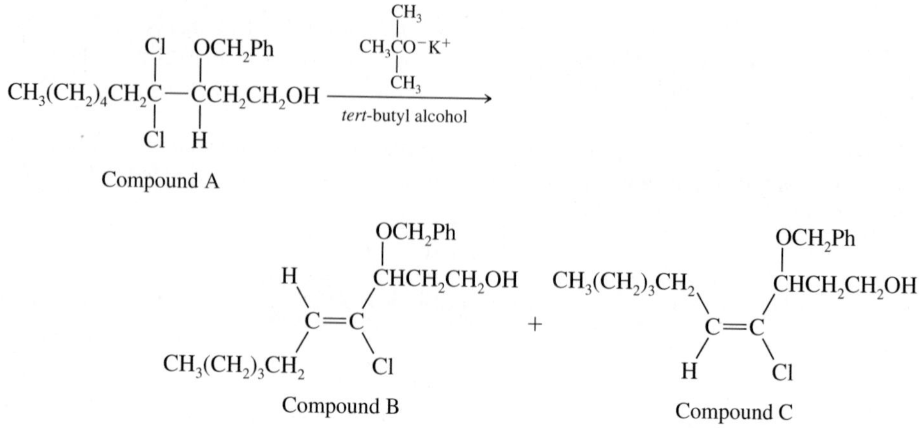

Compound A

Compound B                                  Compound C

(a) Assign configuration to Compounds B and C.
(b) Draw perspective formulas and Newman projections for the conformations of Compound A that give rise to Compounds B and C. Show the mechanism, using the curved-arrow convention, for the formation of the alkenes.
(c) Predict which alkene is the major product and explain why you chose it.

An anti conformation for the leaving group and the proton that is lost is also necessary for elimination reactions of certain cyclic compounds. For example, *trans*-2-methylcyclohexyl tosylate gives only 3-methylcyclohexene when treated with potassium *tert*-butoxide.

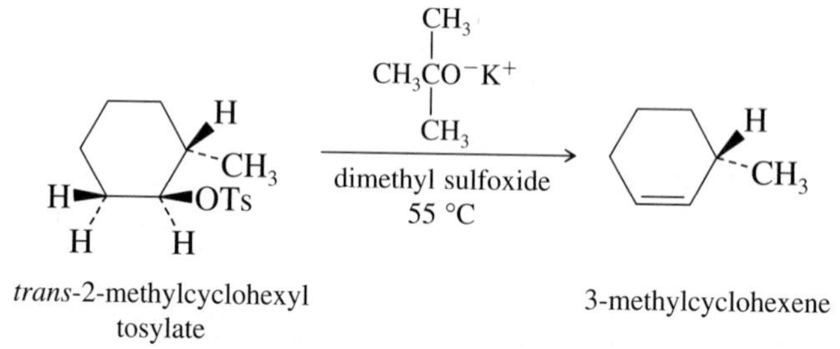

*trans*-2-methylcyclohexyl                        3-methylcyclohexene
tosylate

In the more stable diequatorial conformation of *trans*-2-methylcyclohexyl tosylate, no hydrogen atom is anti to the leaving group. If the cyclohexane ring flips, one hydrogen atom becomes anti to the tosylate group. The only observed product is the alkene that would be formed by removal of that proton, suggesting that the molecule reacts in this conformation. There is always a small amount of the diaxial conformation in equilibrium with the more stable conformation at room temperature.

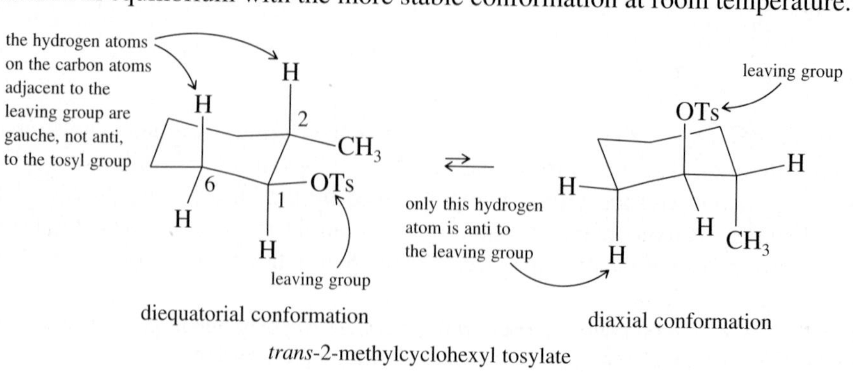

the hydrogen atoms on the carbon atoms adjacent to the leaving group are gauche, not anti, to the tosyl group

only this hydrogen atom is anti to the leaving group

leaving group

diequatorial conformation                          diaxial conformation

*trans*-2-methylcyclohexyl tosylate

Although the anti orientation between the leaving group and the hydrogen atom appears to be the most favorable one for elimination reactions, it is not the only one possible. In cyclopentyl compounds especially, the elimination of a hydrogen and a leaving group that are trans to each other is easier, but the elimination of groups that are cis to each other also takes place. *cis*-2-Phenylcyclopentyl tosylate and *trans*-2-phenylcyclopentyl tosylate both give 1-phenylcyclopentene as the product.

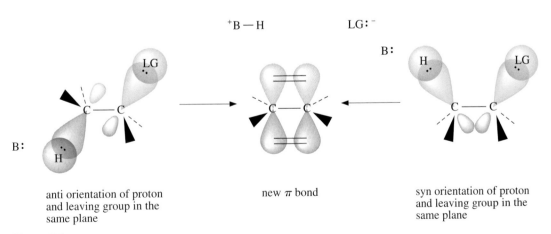

*anti elimination*                                    *syn elimination*

The first reaction, in which the groups that must be eliminated to give the more stable alkene are trans to each other, is nine times faster than the second reaction, in which the groups eliminated are cis to each other. An elimination in which the groups that are lost are cis to each other is called a **syn elimination.** In an **anti elimination** reaction, the groups that are lost are trans to each other on a ring or anti to each other in one conformation of an acyclic compound.

In order to form the new $\pi$ bond in an elimination reaction, the orbital that contains the electrons that are being released by the removal of the proton at one carbon atom must overlap with the orbital being vacated as the leaving group departs with the pair of electrons that bonded it to the adjacent carbon atom. This necessity governs the stereochemistry of the E2 reaction. The $p$ orbitals that form the $\pi$ bond have to be in the same plane for sideways overlap to take place (p. 54). Therefore, the proton and the leaving group that are lost in the E2 reaction also must be in the same plane, whether they are syn or anti to each other, as shown in Figure 7.8.

⊘ **Study Guide**
**Concept Map 7.5**

**+B—H**                          **LG:⁻**

anti orientation of proton
and leaving group in the
same plane

new $\pi$ bond

syn orientation of proton
and leaving group in the
same plane

*Figure 7.8*
Orientation that is necessary for the departing proton and the leaving group so that the developing $p$ orbitals overlap smoothly to give the new $\pi$ bond.

## SUMMARY

Nucleophilic substitution reactions occur when nucleophiles displace leaving groups, resulting in the substitution of one group for another bonded to a carbon atom. Nucleophiles are species with nonbonding electrons that are available for donation to electrophilic centers. Nucleophilicity is affected by the structure of the nucleophile and by the solvent used for the reaction. Leaving groups are stable species that can be detached with their bonding electrons from a molecule during reaction.

$S_N2$ reactions are second-order reactions, the rates of which depend on the concentration of both the alkyl halide and the nucleophile. An $S_N2$ reaction occurs in one step that involves a carbon atom with five groups around it at the transition state and results in inversion of configuration. The order of reactivity of alkyl halides in $S_N2$ reactions is

$$CH_3 > \text{primary} > \text{secondary} > \text{tertiary}$$

$S_N1$ reactions are first order with respect to the concentration of the alkyl halide and independent of the concentration or the nature of the nucleophile. The reaction proceeds in two steps: The rate-determining ionization of the alkyl halide gives a carbocation intermediate, which then reacts with nucleophiles. Racemization is common in $S_N1$ reactions. The order of reactivity of alkyl halides in $S_N1$ reactions is

$$\text{tertiary} > \text{secondary} > \text{primary} > CH_3$$

An elimination reaction may be either E1, which involves a carbocation intermediate, or E2, which is a second-order reaction. Elimination reactions occur in competition with substitution reactions and are promoted by strong bases and high temperatures. For elimination reactions that can give rise to more than one alkene, the more highly substituted alkene is the major product. E2 reactions proceed most easily when the leaving group and the proton to be removed are in the anti orientation. The order of reactivity for alkyl halides in elimination reactions is

$$\text{tertiary} > \text{secondary} > \text{primary}$$

(7) **Study Guide**
**Concept Map 7.6**

## ADDITIONAL PROBLEMS

**7.27** Write equations for the reaction of *n*-butyl bromide as a typical primary alkyl halide, with the following reagents, showing the major product expected.

(a) KOH, in ethanol, heat     (b) NaOH, 0.1 M, in 50% aqueous ethanol     (c) $NH_3$    (d) $NaN_3$    (e) NaCN

(f) $CH_3CH_2SNa$    (g) $CH_3\overset{\displaystyle O}{\overset{\|}{C}}ONa$    (h) $CH_3CH_2C{\equiv}CNa$    (i) $NaCH(\overset{\displaystyle O}{\overset{\|}{C}}OCH_2CH_3)_2$

**7.28** The following equations represent some nucleophilic substitution reactions that are possible. Only primary alkyl groups are used, so the reactions will be $S_N2$, and competing elimination reactions will not be a problem. Complete the equations, showing the products that will form. In each case, it will be helpful to label the leaving group and the incoming nucleophile.

(a) ⬠—$CH_2CH_2CH_2OTs + NaN_3 \longrightarrow$      (b) ⬡—$CH_2CH_2Cl + NaCN \longrightarrow$

(c) $CH_3CH_2CH_2CH_2Br + CH_3CH_2SNa \longrightarrow$      (d) $CH_3CH_2CH_2Cl + CH_3C\equiv CNa \longrightarrow$

(e) $CH_3\overset{\overset{\displaystyle CH_3}{|}}{C}HCH_2CH_2Br + \underset{\text{(excess)}}{NH_3} \longrightarrow$      (f) $CH_3I + NaCH(\overset{\overset{\displaystyle O}{\|}}{C}OCH_2CH_3)_2 \longrightarrow$

**7.29** Decide which species in each of the following pairs is more nucleophilic.

(a) $SH^-$   or   $Cl^-$      (b) $(CH_3)_3B$   or   $(CH_3)_3P$      (c) $CH_3NH^-$   or   $CH_3NH_2$

(d) $CH_3SCH_3$   or   $CH_3OCH_3$

**7.30** The following compounds undergo elimination reactions when heated with base. Complete the following equations, showing the products from the elimination reactions that will occur if each is heated with ethanolic potassium hydroxide. Wherever you can, indicate which product will be the major one and which the minor.

(a) 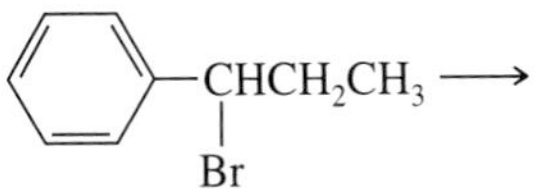 Ph$-\overset{\overset{\displaystyle}{|}}{\underset{\underset{\displaystyle Br}{|}}{C}}HCH_2CH_3 \longrightarrow$    (b) $CH_3CH_2\overset{\overset{\displaystyle CH_3}{|}}{\underset{\underset{\displaystyle Br}{|}}{C}}CH_2CH_2CH_3 \longrightarrow$    (c) 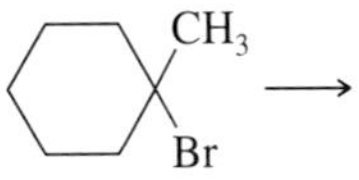 (cyclohexyl with $CH_3$ and $Br$) $\longrightarrow$    (d) $CH_3CH_2CH_2\overset{\overset{\displaystyle}{|}}{\underset{\underset{\displaystyle Br}{|}}{C}}H\overset{\overset{\displaystyle O}{\|}}{C}OH \longrightarrow$

**7.31** Complete the following reaction sequences, showing the major product you expect for each stage. When you know what the stereochemistry of a product will be, show it in your answer.

(a) 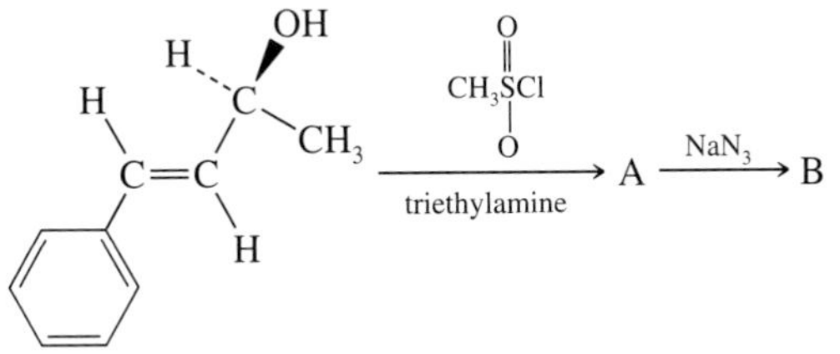 (alkene structure) $\xrightarrow[\text{triethylamine}]{CH_3SCl / O} A \xrightarrow{NaN_3} B$

(b) Ph$-(CH_2)_8CH_2Br \xrightarrow{CH_3NH_2 \text{ (excess)}} C$

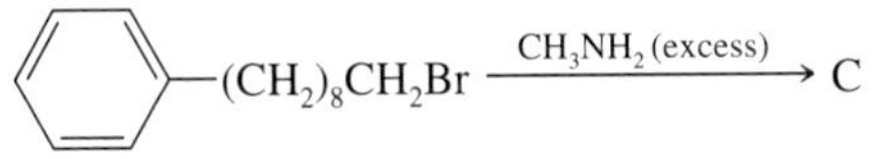

(c) $CH_3\overset{\overset{\displaystyle O}{\|}}{C}CH_2OH \xrightarrow{D} CH_3\overset{\overset{\displaystyle O}{\|}}{C}CH_2OTs \xrightarrow{Na^+ \, ^{18}F^-} E$

(d) 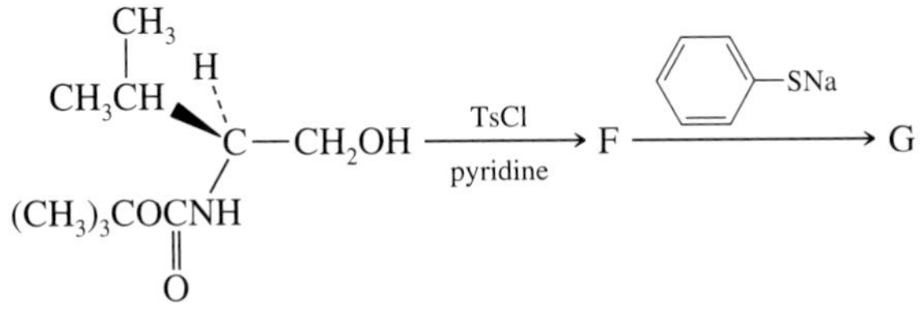 (structure with $CH_3CH$, $CH_3$, $(CH_3)_3COCNH$) $C-CH_2OH \xrightarrow[\text{pyridine}]{TsCl} F \xrightarrow{Ph-SNa} G$

(e) (dibromo-bis(bromomethyl)benzene) $\xrightarrow[\text{dimethylformamide}]{Ph-CH_2ONa \, \text{(excess)}} H$

(f) Ph$-OH \xrightarrow{I}$ Ph$-ONa \xrightarrow[\text{(1 equiv)}]{BrCH_2(CH_2)_6CH_2Br} J$

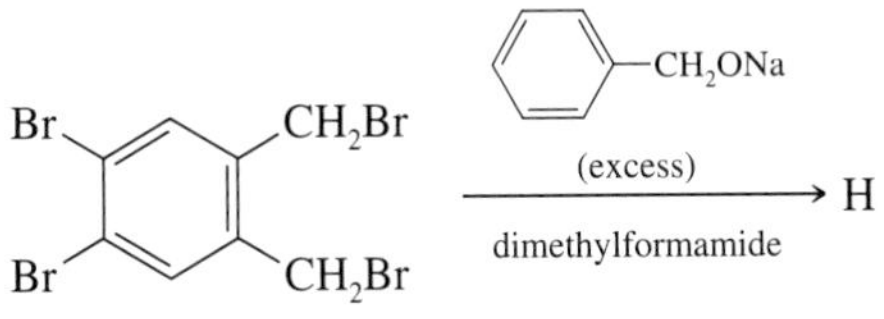

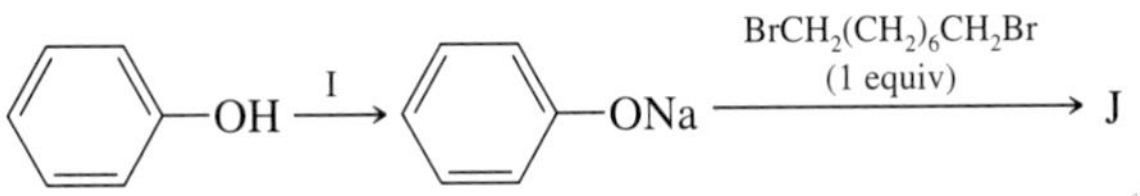

**7.32** More practice in recognizing reactions follows.

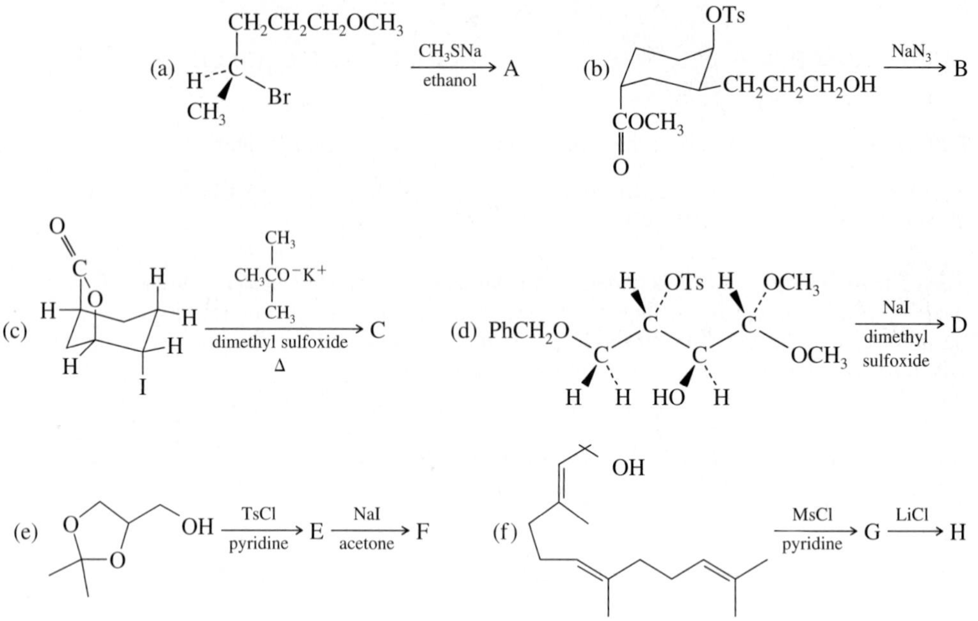

(a)
$$CH_2CH_2CH_2OCH_3$$
H⋯C—Br
CH_3
$\xrightarrow[\text{ethanol}]{CH_3SNa}$ A

(b)
OTs
—CH_2CH_2CH_2OH
COCH_3 ‖ O
$\xrightarrow{NaN_3}$ B

(c)
O ‖ C—O
H—, H—, H—, H
I
$\xrightarrow[\text{dimethyl sulfoxide}]{CH_3\overset{CH_3}{\underset{CH_3}{C}}O^-K^+} \atop \Delta$ C

(d) PhCH_2O—C—C—C—C—OCH_3
H OTs H OCH_3
H H HO H
$\xrightarrow[\text{dimethyl sulfoxide}]{NaI}$ D

(e)
O, O
OH
$\xrightarrow[\text{pyridine}]{TsCl}$ E $\xrightarrow[\text{acetone}]{NaI}$ F

(f)
OH
$\xrightarrow[\text{pyridine}]{MsCl}$ G $\xrightarrow{LiCl}$ H

**7.33** For each of the following pairs of reactions indicate which one will be faster and explain briefly why you think so.

(a) 1. $CH_3CH_2CH_2Br + Na^+CN^- \xrightarrow[\text{sulfoxide}]{\text{dimethyl}} CH_3CH_2CH_2CN + Na^+Br^-$

2. $CH_3\underset{Br}{CH}CH_3 + Na^+CN^- \xrightarrow[\text{sulfoxide}]{\text{dimethyl}} CH_3\underset{CN}{CH}CH_3 + Na^+Br^-$

(b) 1. $CH_3CH_2CH_2Br + CH_3O^-Na^+ \xrightarrow{H_2O} CH_3CH_2CH_2OCH_3 + Na^+Br^-$

2. $CH_3CH_2CH_2Br + CH_3S^-Na^+ \xrightarrow{H_2O} CH_3CH_2CH_2SCH_3 + Na^+Br^-$

(c) 1. $CH_3\underset{CH_3}{\overset{CH_3}{C}}-Br + H_2O \longrightarrow CH_3\underset{CH_3}{\overset{CH_3}{C}}-OH + HBr$

2. $CH_3CH_2CH_2CH_2Br + H_2O \longrightarrow CH_3CH_2CH_2CH_2OH + HBr$

(d) 1. $CH_3CH_2CH=CH_2 + HCl \longrightarrow CH_3CH_2\underset{Cl}{CH}CH_3$

2. $CH_3\overset{CH_3}{C}=CH_2 + HCl \longrightarrow CH_3\underset{Cl}{\overset{CH_3}{C}}CH_3$

**7.34** The following transformation was carried out in synthesizing an enantiomerically pure natural product. Show how you would do this transformation.

$$HOCH_2\diagdown \underset{CH_2\quad OH}{C}\diagup CH_3 \quad \xrightarrow{?} \quad NCCH_2\diagdown \underset{CH_2\quad OH}{C}\diagup CH_3$$

**7.35** The following incomplete reaction sequences give only a starting material and a product that can be prepared from it. Show the reagents that would be necessary to carry out each transformation and any major products that would be formed in the intermediate stages. There may be more than one correct way to complete a sequence.

(a) $\bigcirc-CH_2OH \longrightarrow \bigcirc-CH_2SCH_2CH_3$ (b) $CH_3CH_2CH_2CH_2CH_2OH \longrightarrow CH_3CH_2CH_2CH_2CH_2I$

(c) $\underset{HO}{\overset{CH_3}{\underset{CH_2CH_3}{C}}}\diagdown H \longrightarrow \underset{CH_3CH_2}{\overset{CH_3}{\underset{}{C}}}\diagup\overset{}{\underset{NH_2}{H}}$   (d) $\bigcirc-CH_2CH_2CH=CH_2 \longrightarrow \bigcirc-CH_2CH_2\underset{N_3}{CHCH_3}$

(e) $CH_3\overset{O}{\overset{\|}{C}}OCH_2\overset{CH_3}{\underset{}{C}}=CHCH_2Cl \longrightarrow CH_3\overset{O}{\overset{\|}{C}}OCH_2\overset{CH_3}{\underset{}{C}}=CHCH_2NHCHCH_2OH$ with $\overset{CH=CH_2}{\underset{}{}}$

**7.36** Answer the questions below for this $S_N2$ reaction:

$$CH_3CH_2CH_2CH_2Br + NaOH \xrightarrow[ethanol]{} CH_3CH_2CH_2CH_2OH + NaBr$$

(a) What is the rate expression for the reaction?
(b) Draw an energy diagram for the reaction. Label all parts. You may assume that the products are lower in energy than the reactants.
(c) What will be the effect on the rate of the reaction of doubling the concentration of *n*-butyl bromide?
(d) What will be the effect on the rate of the reaction of halving the concentration of sodium hydroxide?
(e) Will the rate of the reaction change significantly if the solvent is changed to 80% ethanol, 20% water?

**7.37** Answer the questions below for this $S_N1$ reaction:

$$\underset{Br}{\overset{C_6H_5}{\underset{}{C_6H_5CCH_3}}} \xrightarrow[CH_3CH_2OH]{} \underset{OCH_2CH_3}{\overset{C_6H_5}{\underset{}{C_6H_5CCH_3}}} + HBr$$

1-bromo-1,1-diphenylethane

(a) What is the rate expression for the reaction?
(b) Draw an energy diagram for the reaction. Label all parts. You may assume that the products are lower in energy than the reactants.
(c) What will be the effect on the rate of the reaction of doubling the initial concentration of 1-bromo-1,1-diphenylethane?
(d) Will the rate of the reaction change significantly if some water is added to the solvent, which is ethanol?

**7.38** When 1-chloro-2-butene reacts in 50% aqueous acetone at 47 °C, the product is a mixture of two alcohols.

$$CH_3CH=CHCH_2Cl \xrightarrow[\substack{50\% H_2O \\ 50\% \text{ acetone} \\ 47\text{ }°C}]{} CH_3CH=CHCH_2OH + CH_3\underset{OH}{CHCH}=CH_2$$

1-chloro-2-butene         2-buten-1-ol      3-buten-2-ol
                                  56%             44%

The reaction goes through a resonance-stabilized cation, called an **allylic cation.** What is the structure of the cation? Explain why it has high stability. Write a detailed mechanism that accounts for the observed experimental facts.

**7.39** *cis*-2-Phenylcyclohexyl tosylate reacts with potassium *tert*-butoxide in *tert*-butyl alcohol at 50 °C to give exclusively 1-phenylcyclohexene. *trans*-2-Phenylcyclohexyl tosylate does not give any alkene under the same conditions. Draw structures for the two compounds, and explain these observations.

**7.40** The following alkyl chlorides undergo first-order nucleophilic substitution by water (solvolysis) with dramatically different rates. The first-order reaction rates were measured for the following alkyl chlorides using water-dioxane mixtures as the solvent system.

$$CH_3CH_2OCH_2Cl \quad > \quad CH_3CH_2Ch_2CH_2CL \quad > \quad CH_3OCH_2CH_2Cl$$

| | chloroethoxymethane | 1-chlorobutane | 1-chloro-2-methoxyethane |
|---|---|---|---|
| Relative rates | $1 \times 10^9$ | 1 | 0.2 |

(a) Write the mechanism for the rate-determining step of the reaction of chloroethoxymethane.

(b) What determines the relative rates of substitution of the compounds shown above? Use structural formulas and words to rationalize the experimental data.

**7.41** When an alkyl halide reacts with a thiocyanate ion, $SCN^-$, reaction takes place at the sulfur atom. With the cyanate ion, $OCN^-$, reaction takes place at the nitrogen.

$$RX + SCN^- \longrightarrow RSCN + X^-$$

thiocyanate ion   alkyl thiocyanate

$$RX + OCN^- \longrightarrow RNCO + X^-$$

cyanate ion   alkyl isocyanate

Write structures for the thiocyanate and cyanate ions and their alkyl derivatives, and explain the difference in the reactivity of the ions. (*Hint:* You might find it useful to review Section 7.4.)

**7.42** The specific rotation of (*S*)-2-bromopentane in ethanol is $[\alpha]$ +31°.

Experiment A: When sodium bromide is added to a solution of (*S*)-2-bromopentane, the initially observed rotation ($\alpha_{obs}$ = +20°) gradually decreases until the solution is optically inactive.

Experiment B: When the concentration of the added sodium bromide is increased, the *rate* of formation of the optically inactive solution is observed to increase.

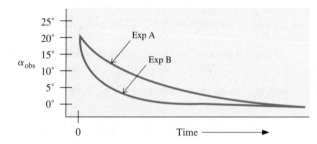

Analysis of the solution at any point along the way indicates that only 2-bromopentane and sodium bromide are present.

(a) Draw (*S*)-2-bromopentane and the structure of the reaction product of (*S*)-2-bromopentane with sodium bromide. What is the relationship between these two compounds?

(b) Briefly describe the significance of Experiment B. Why does the rate of this reaction increase with an increase in the concentration of sodium bromide?

(c) Which of the following statements describes why the solution becomes optically inactive during the course of these experiments?

1. All the starting material is irreversibly converted to the product.
2. One-half of the starting material is irreversibly converted to the product.
3. The starting material is reversibly converted to the product.

Justify your choice by explaining how this statement can be used to rationalize the observed optical inactivity. An energy diagram should be part of your answer.

**7.43** The two enantiomers of a chiral compound generally show different biological activity. Recently a group of chemists in Copenhagen prepared the enantiomer of 3′-azido-2′-deoxythymidine (AZT), which is used in the treatment of AIDS. They wanted to see whether the enantiomer of AZT would be an even better treatment of the disease.

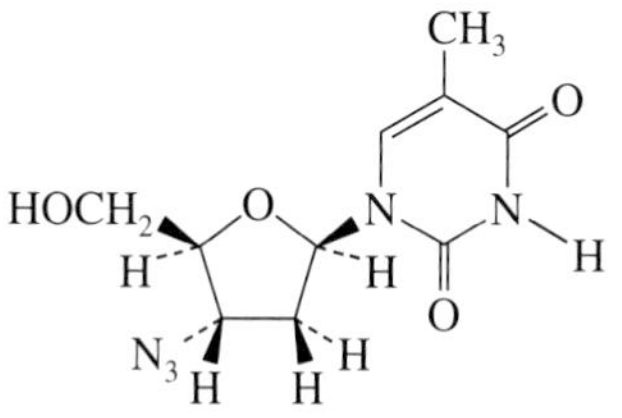

(+)-3′-azido-2′-deoxythymidine
(+)-AZT
$[\alpha]_D^{20}$ +56° (c 0.01; methanol)

(a) Draw (−)-AZT, the enantiomer of (+)-AZT.
(b) Assign configuration to all of the stereocenters in (+)-AZT.
(c) What is the specific rotation expected for (−)-AZT?
(d) What is the total number of stereoisomers possible for a compound having the same connectivity as AZT?

**7.44** As part of research into the synthesis of a new natural product, fulvanin 1, isolated from two species of marine sponges, Compound A was synthesized.

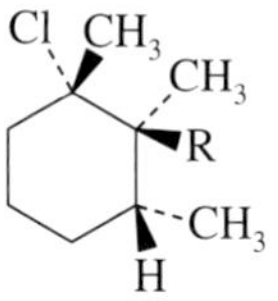

Compound A        where R is        $-CH_2CH_2$

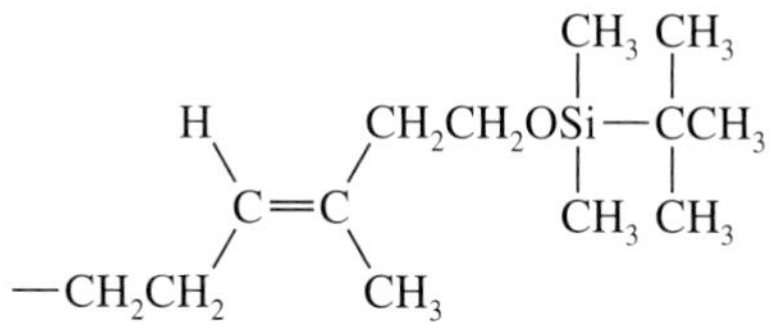

(a) Compound A is converted by treatment with base to a mixture of two alkenes in an E2 elimination.

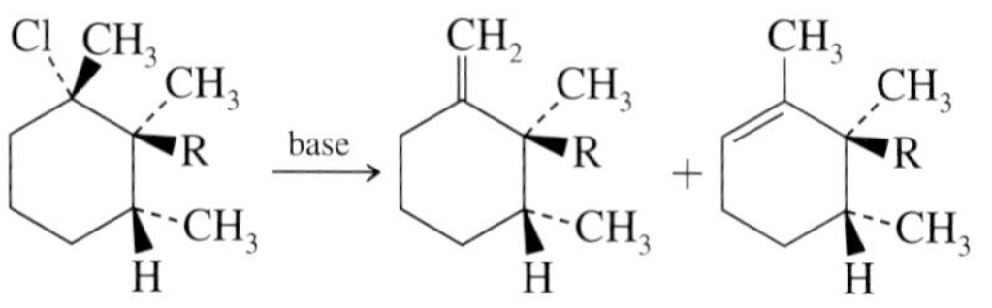

Compound A        alkene B        alkene C

Three times as much of one of the alkenes is formed as of the other. Which alkene is the major one? Why?

(b) Draw the two chair forms for Compound A, being careful to show the stereoisomer given above. You may use the shorthand R in your drawings. Indicate the direction of the equilibrium between the two forms.

(c) Using the curved-arrow convention and the correct chair form of Compound A, show the mechanism for the formation of alkene C from Compound A. You may use B: and

$HB^+$ as bases and acids as needed and show the alkene in the planar formed used above.

**7.45** When the following compound is treated with an excess of tetrabutylammonium bromide and allowed to come to equilibrium, 44% of the mixture has the *S* configuration at carbon 2′ and 56% has the *R* configuration at the same carbon.

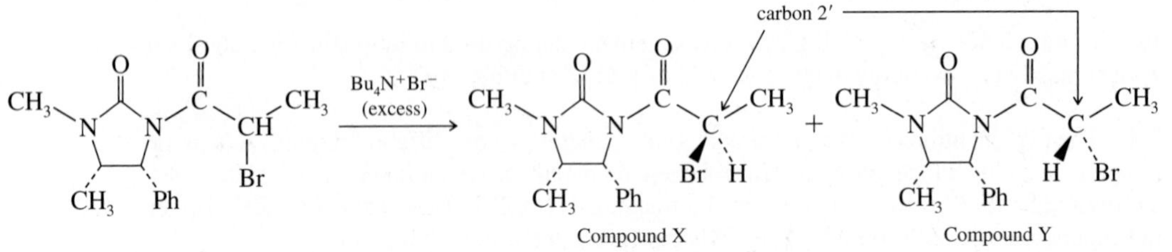

Compound X                    Compound Y

Which compound forms 56% of the mixture, and why are Compounds X and Y not formed in equal amounts at equilibrium?

**7.46** Recently, chemists have discovered that oleamide, the amide of oleic acid, the chief fatty acid in olive oil and in our bodies, is the compound that induces sleep. The following reactions were carried out in the study that determined the structure of the sleep-inducing compound.

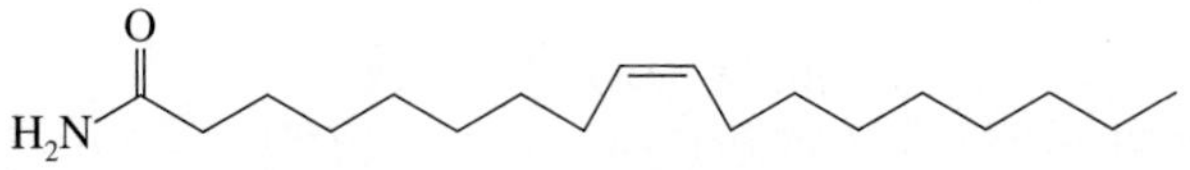

oleamide

(a)  Assign configuration to the double bond in oleamide according to the Cahn-Ingold-Prelog Rules.
(b)  Fill in the structures for the missing starting materials or products as indicated.

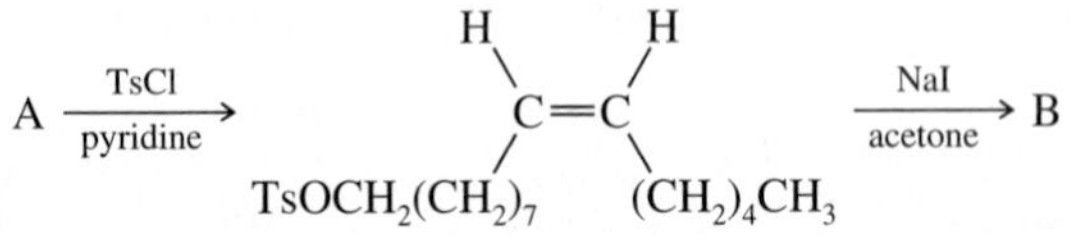

(c)  Treatment of the alcohol shown below with mesyl chloride in base gives a mesylate that very rapidly undergoes an E2 reaction to give an alkene in which the $\pi$ orbitals of the newly formed double bond can interact with the $\pi$ orbitals of the carbonyl group. The new double bond has the *E* configuration. Draw structural formulas for the species indicated.

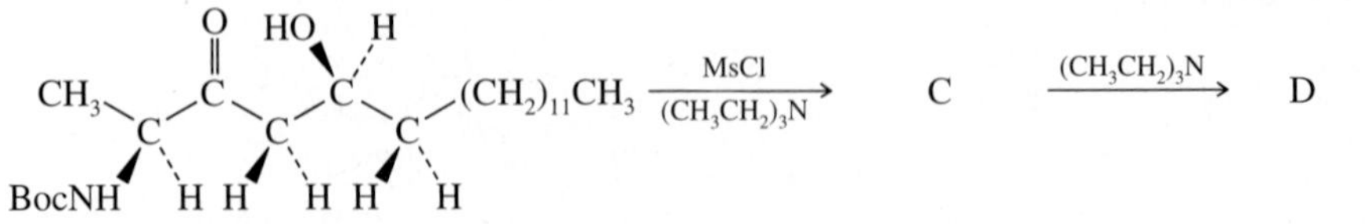

Boc is a ''protecting group'' that keeps the           mesylate that              E2 product
nitrogen atom from reacting with MsCl.                  easily undergoes
                                                        E2 reaction

# Alkenes

## A Look Ahead

The most important reaction of alkenes is electrophilic addition to the double bond. You are familiar with an example of this type of reaction from Chapter 4, the addition of hydrogen bromide to propene.

The functional group in alkenes is the double bond.

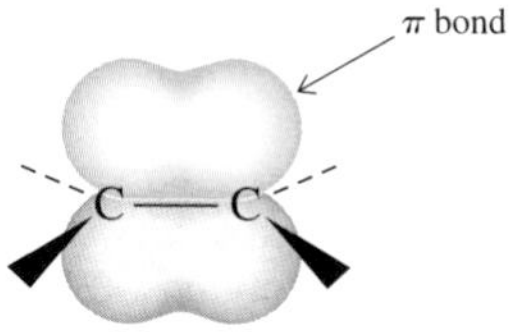

π bond

The electrons of the π bond are donated to electrophilic reagents, which form bonds to the carbon atoms of the double bond. An acid such as hydrogen bromide is an example of one type of electrophilic reagent that adds to the double bond.

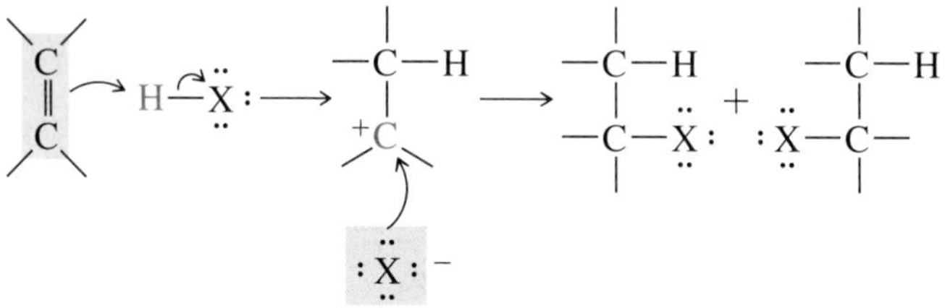

In the reaction shown above, the electrons of the π bond are donated to an electrophile, creating a deficiency of electrons on the carbon skeleton, which then accepts a pair of electrons from a nucleophile. The π bond is broken, and two new σ bonds are formed. The movement of electrons "out" from the π bond and the subsequent movement of electrons "in" from a nucleophile are features common to all the reactions discussed in this chapter. Look for this common theme as you study the reactions that may appear at first to be quite different from each other.

In this chapter we will study the addition reactions of other electrophiles: Lewis acids, such as diborane, and oxidizing agents, such as the halogens, osmium tetroxide, and ozone. We will examine the mechanisms by which they react with alkenes and the stereochemistry of those reactions.

**Workbook Exercises**

## 8.1   Structure and Isomerism in Alkenes

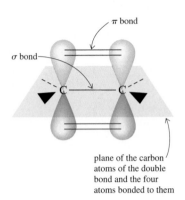

*Figure 8.1*
The orbitals and stereochemistry of a double bond.

Alkenes are hydrocarbons that contain carbon–carbon double bonds. The carbon atoms joined by the double bond, and the four atoms bonded to them lie in a plane, with bond angles of approximately 120° around the two carbon atoms. Such trigonal planar carbon atoms are described as being $sp^2$-hybridized (p. 52). The double bond in an alkene consists of a $\sigma$ bond resulting from the overlap of $sp^2$ hybrid orbitals on two carbon atoms and a $\pi$ bond resulting from the side-to-side overlap of the remaining $p$ orbital on each carbon (Figure 8.1).

Alkenes form a homologous series with the molecular formula $C_nH_{2n}$ and are, thus, unsaturated hydrocarbons (p. 14). The smallest members of the series—ethene, usually called ethylene ($C_2H_4$), propene ($C_3H_6$), and butenes ($C_4H_8$)—are gases at room temperature. Larger alkenes up to $C_{20}H_{40}$ are liquids. As with alkanes, the boiling points and melting points of alkenes increase with molecular weight but show some variations that depend on the shape of the molecule. Alkenes with the same molecular formula may be isomers of one another if the position or the stereochemistry of the double bond differ. The different noncyclic isomers of $C_4H_8$—1-butene, 2-methylpropene, (*E*)-2-butene, and (*Z*)-2-butene—were shown on page 221.

**Problem 8.1**    Write structural formulas for the six noncyclic isomers of $C_5H_{10}$. Show any stereoisomerism that exists for these compounds. (*Hint:* A review of Section 6.10 will be helpful.)

## 8.2   Nomenclature of Alkenes

### A.  IUPAC Names for Alkenes

The systematic name of an alkene is derived from the name of the alkane corresponding to the longest continuous chain of carbon atoms that contains the double bond. The **ane** ending of the name of the alkane is changed to **ene** for the alkene. Thus the IUPAC name ethene comes from ethane. An alkene with five carbon atoms in a continuous chain is named as a pentene; one with six carbon atoms in the longest continuous chain is a hexene.

$$CH_2{=}CH_2 \qquad \overset{5}{C}H_3\overset{4}{C}H_2\overset{3}{C}H_2\overset{2}{C}H{=}\overset{1}{C}H \qquad \overset{6}{C}H_3\overset{5}{C}H_2\overset{4}{C}H{=}\overset{3}{C}H\overset{2}{C}H_2\overset{1}{C}H_3$$

ethene
ethylene

1-pentene

3-hexene

The chain is numbered so as to give the carbon atoms joined by the double bond the lowest possible numbers. The position of the double bond is indicated by the number of the *first* of the two carbon atoms. Thus the examples shown above are 1-pentene, because the double bond is between carbon atoms 1 and 2 of a five-carbon chain, and 3-hexene, because the double bond is between carbon atoms 3 and 4 of a six-carbon chain. If other substituents are present, they too are named and their positions are indicated by numbers.

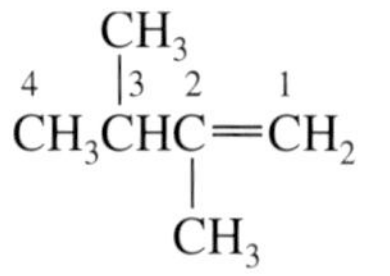

$$\underset{\text{2,3-dimethyl-1-butene}}{\overset{\displaystyle CH_3}{\underset{\displaystyle CH_3}{CH_3CHC=CH_2}}}$$

2,3-dimethyl-1-butene

$$\overset{4}{CH_3}\overset{3}{CH}=\overset{2}{CH}\overset{1}{CH_2}OH$$

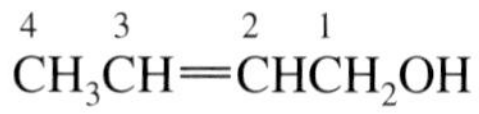

2-buten-1-ol

$$\underset{6\quad5\quad4}{ClCH_2CH_2CH_2}\overset{CH_3}{\underset{3}{\diagup}}C=C\overset{CH_3}{\underset{2}{\diagup}}\overset{CH_3}{\underset{1}{}}$$

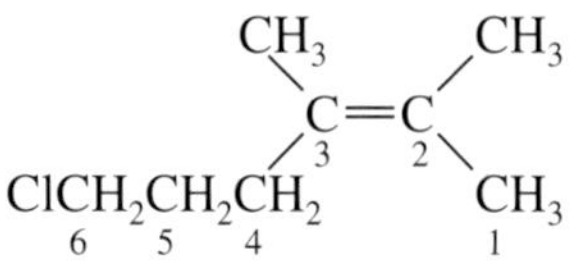

6-chloro-2,3-dimethyl-2-hexene

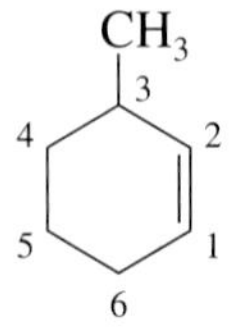

3-methylcyclohexene

*the first carbon of the double bond is understood to be carbon 1 in the cyclic case*

The nomenclature of stereoisomeric alkenes was discussed on pages 222 to 223. You may want to review those pages before you do the next two problems.

**Problem 8.2** Name all the isomers of $C_5H_{10}$ for which you drew structural formulas in Problem 8.1.

**Problem 8.3** Draw a structural formula for each of the following compounds (a–f), showing the correct stereochemistry when it is designated.

(a) 1-phenylcyclohexene
(b) 3,3-dimethylcyclopentene
(c) (Z)-2-phenyl-2-butene
(d) (E)-1,3-dichloro-2-methyl-2-pentene
(e) (E)-2-penten-1-ol
(f) (Z)-3-methyl-3-heptene

## B. Vinyl and Allyl Groups

Two unsaturated groups have common names that are used as substituents or to indicate structural features in molecules. The **vinyl group** is formed when a hydrogen atom is removed from ethylene.

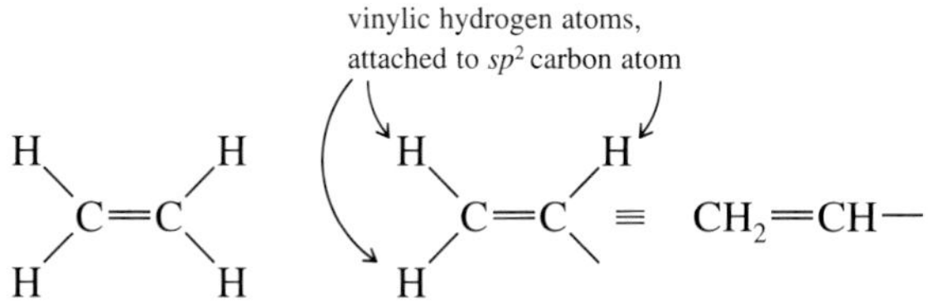

ethylene    the vinyl group

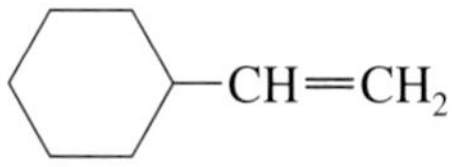

vinylic hydrogen atoms, attached to $sp^2$ carbon atom

$$\equiv \quad CH_2=CH-$$

cyclohexylethene
vinylcyclohexane

$$H_2C=CHCl$$

chloroethene
vinyl chloride

*a vinylic halide*

Hydrogen atoms attached to the $sp^2$-hybridized carbon atoms of a double bond are described as **vinylic hydrogen atoms.** Similarly, an organic halide with a halogen atom bonded directly to an $sp^2$-hybridized carbon atom of a double bond is known as a **vinylic halide.**

Removing a hydrogen atom from the $sp^3$-hybridized carbon atom of propene gives an **allyl group.**

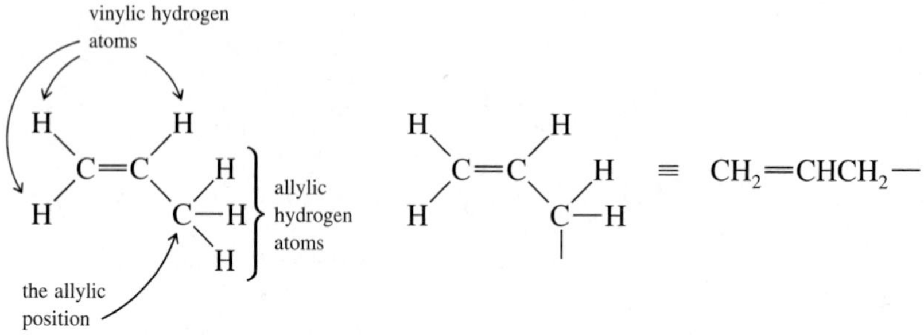


propene

an allyl group

$CH_2{=}CHCH_2Cl$

3-chloropropene
allyl chloride

*an allylic halide*

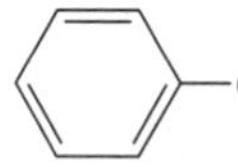


3-phenylpropene
allylbenzene

In general, $sp^3$-hybridized carbon atoms adjacent to a double bond are known as the **allylic positions** in the molecule. The hydrogen atoms on a $sp^3$-hybridized carbon atom are **allylic hydrogen atoms.** Replacement of one of the allylic hydrogen atoms by a halogen atom gives an **allylic halide.** Hydrogen atoms (or halogen atoms) in a vinylic position are much less reactive than hydrogen atoms (or halogen atoms) in an allylic position (pp. 775 and 1075), so the kind of structural distinctions described here are important.

## 8.3 Relative Stabilities of Alkenes

Alkenes react with hydrogen in the presence of a metal catalyst to give alkanes (pp. 298–301). This hydrogenation reaction is exothermic. The convention is to write the heat of an exothermic reaction as a negative quantity to emphasize that the total internal energy of the system decreases during the reaction. Some of the energy stored in the molecules on the left-hand side of the reaction equation has been released to the environment as heat. Experimental heats of hydrogenation of isomeric alkenes can be used to measure their relative stabilities.

1-Butene and ($Z$)- and ($E$)-2-butene each add one equivalent of hydrogen to become butane.

$$\underset{\text{1-butene}}{CH_3CH_2CH{=}CH_2} + H_2 \xrightarrow[\text{catalyst}]{} \underset{\text{butane}}{CH_3CH_2CH_2CH_3} \qquad \Delta H° = -30.3 \text{ kcal/mol}$$

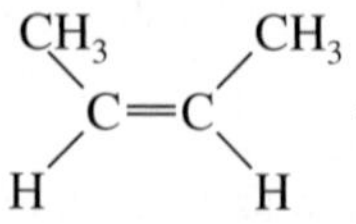

$$+ H_2 \xrightarrow[\text{catalyst}]{} \underset{\text{butane}}{CH_3CH_2CH_2CH_3} \qquad \Delta H° = -28.6 \text{ kcal/mol}$$

($Z$)-2-butene

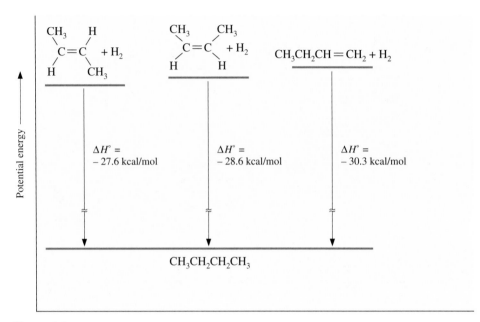

*Figure 8.2*
Difference in energy level of butane from that of hydrogen plus 1-butene, (Z)-2-butene, and (E)-2-butene, as reflected by heats of hydrogenation.

$$CH_3CH=CHCH_3 + H_2 \xrightarrow[\text{catalyst}]{} CH_3CH_2CH_2CH_3 \qquad \Delta H^\circ = -27.6 \text{ kcal/mol}$$

(E)-2-butene                                   butane

Each reaction has the same product, butane, and an identical reagent, hydrogen. The differences in the heats of hydrogenation, the energy evolved as the alkene is converted to the alkane, must reflect the differences in energy in the alkenes. These differences are shown schematically in Figure 8.2. Less heat is evolved when (E)-2-butene is hydrogenated than when (Z)-2-butene undergoes the same reaction. Thus the energy of the trans compound must be closer before the reaction to that of butane. (E)-2-Butene is at a lower energy level than (Z)-2-butene, or in other words, it is more stable than (Z)-2-butene. The same argument demonstrates that 1-butene is less stable than either of the 2-butenes. More heat is evolved as 1-butene is transformed into butane than for either of the other two cases. 1-Butene starts at a higher energy level than either of the 2-butenes and has farther to drop to reach the energy level of butane.

A **terminal alkene,** in which the double bond is at the end of a chain, is less stable than an **internal alkene,** with the double bond somewhere in the middle of the chain. A trans alkene, in which the larger substituents on the double bond are farther apart from each other, is more stable than the corresponding cis alkene in which the larger substituents are closer together.

An important factor governing stability seems to be the number of substituents on the double bond. This is illustrated in the following reactions by the experimental heats of hydrogenation for three more alkenes, 3-methyl-1-butene, 2-methyl-1-butene, and 2-methyl-2-butene, all of which give the same alkane when hydrogenated.

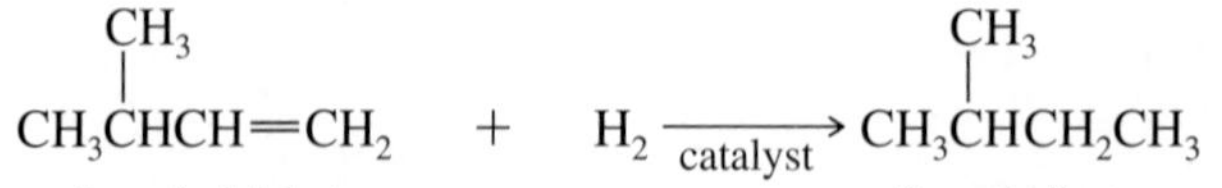

$$\underset{\substack{\text{3-methyl-1-butene}}}{\overset{\overset{\displaystyle CH_3}{|}}{CH_3CHCH=CH_2}} \quad + \quad H_2 \xrightarrow{\text{catalyst}} \underset{\substack{\text{2-methylbutane}}}{\overset{\overset{\displaystyle CH_3}{|}}{CH_3CHCH_2CH_3}} \qquad \Delta H° = -30.3 \text{ kcal/mol}$$

*terminal alkene;*
*one substituent on double bond;*
*least stable*

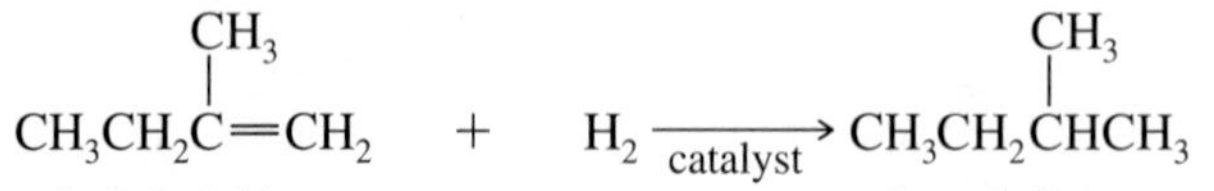

$$\underset{\substack{\text{2-methyl-1-butene}}}{\overset{\overset{\displaystyle CH_3}{|}}{CH_3CH_2C=CH_2}} \quad + \quad H_2 \xrightarrow{\text{catalyst}} \underset{\substack{\text{2-methylbutane}}}{\overset{\overset{\displaystyle CH_3}{|}}{CH_3CH_2CHCH_3}} \qquad \Delta H° = -28.5 \text{ kcal/mol}$$

*terminal alkene;*
*two substituents on double bond*

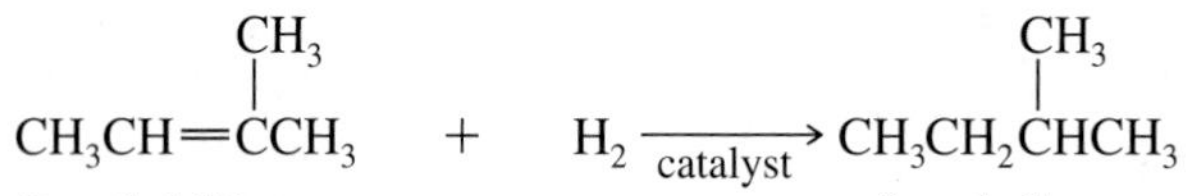

$$\underset{\substack{\text{2-methyl-2-butene}}}{\overset{\overset{\displaystyle CH_3}{|}}{CH_3CH=CCH_3}} \quad + \quad H_2 \xrightarrow{\text{catalyst}} \underset{\substack{\text{2-methylbutane}}}{\overset{\overset{\displaystyle CH_3}{|}}{CH_3CH_2CHCH_3}} \qquad \Delta H° = -26.9 \text{ kcal/mol}$$

*internal alkene;*
*three substituents on double bond;*
*most stable*

The product in each case is the same compound, 2-methylbutane. One of the reactants, hydrogen, is the same for each reaction. The differences in the heats of hydrogenation must arise from the differences in energy of the other reactants, the three different alkenes.

In summary, the position of the double bond and the number of alkyl substituents on it seem to be more important in predicting the relative stability of alkenes than the nature of the alkyl groups on carbon atoms of the double bond. Internal alkenes are more stable than terminal alkenes. The stability of an alkene increases with the number of alkyl groups that are substituted on carbon atoms of the double bond. Alkenes in which bulky substituents are trans to each other are more stable than the corresponding cis alkenes.

**Problem 8.4**    Are *Z* and *E* isomers possible for any of the methylbutenes shown above and on the previous page?

**Problem 8.5**    Draw an energy diagram like Figure 8.2 to illustrate the relative stabilities of the methylbutenes in a schematic way.

## 8.4    Electrophilic Addition of Acids to Alkenes

### A. Hydration and Dehydration Reactions of Alkenes

The addition of hydrogen bromide to propene was examined in detail in Chapter 4 as an example of an electrophilic addition to a double bond. Other acids, acting as electrophiles, also react with alkenes. For example, water adds to alkenes in the presence of acids to form alcohols. Water ($pK_a$ 15.7) is not acidic enough to protonate the double bond, so a stronger acid, such as sulfuric acid, is used to create hydronium ions ($pK_a$ −1.7), which react with the alkene. The addition of water is known as a **hydration reaction.** The hydration reactions of simple alkenes that are available from petroleum as by-products in the manufacture of gasoline are of commercial importance. The hydration of 2-methylpropene is one example.

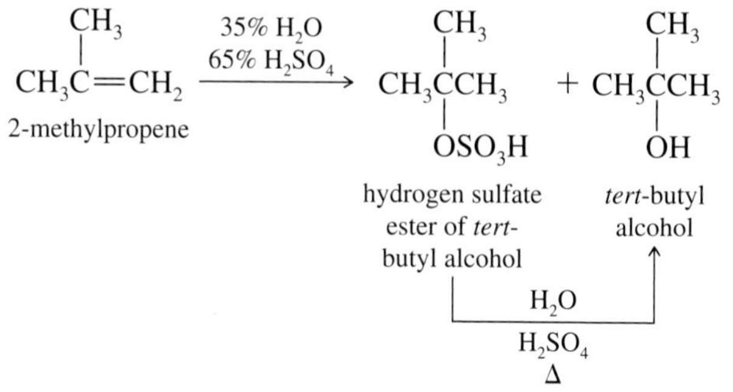

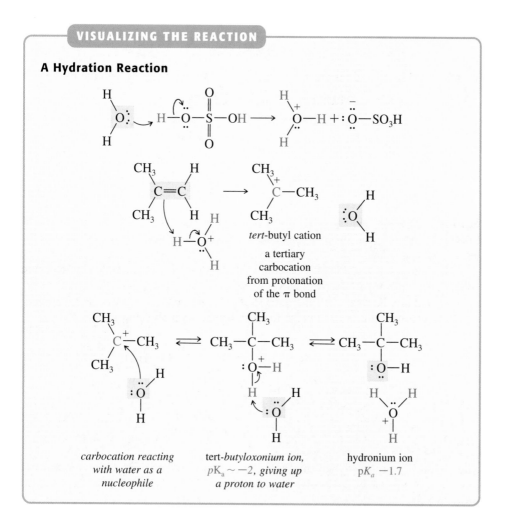

The alkene reacts with hydronium ion to give a tertiary carbocation. The nucleophiles present in the reaction mixture are water and hydrogen sulfate ion. Some of the carbocations combine with water to form *tert*-butyl alcohol; some combine with hydrogen sulfate ion to form an ester. In the industrial process, the sulfate ester is decomposed by heating with dilute aqueous acid, and the alcohol is isolated as the product.

**VISUALIZING THE REACTION**

**A Hydration Reaction**

*tert*-butyl cation
a tertiary
carbocation
from protonation
of the π bond

carbocation reacting
with water as a
nucleophile

tert-*butyloxonium ion,*
pK$_a$ ~ −2, *giving up
a proton to water*

hydronium ion
pK$_a$ −1.7

The overall process results in Markovnikov addition of water to an alkene.

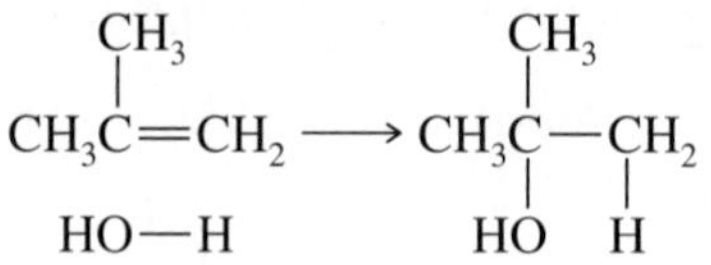

$$CH_3C{=}CH_2 \longrightarrow CH_3\overset{\displaystyle CH_3}{\underset{\displaystyle HO\ \ H}{C{-}CH_2}}$$

Markovnikov addition of H—OH to an alkene

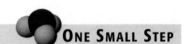

## ONE SMALL STEP

Your understanding of the mechanism of the addition of water to a double bond can be used to make predictions about the reactions of compounds containing hydroxyl groups. Two important categories of such compounds are alcohols and carboxylic acids.

**PROBLEM:** A great deal of research is going into developing alternative fuels as we become more concerned about the environmental effects of the fuels we now use. Methyl *tert*-butyl ether is a compound that improves the octane rating of gasoline and can, therefore, be substituted for lead, an environmental toxin. It is synthesized from 2-methylpropene and methanol in a reaction catalyzed by a small amount of strong acid such as sulfuric acid. Write a detailed mechanism for this reaction. Your mechanism must account for the fact that the reaction takes place with very high regioselectivity and that the isomeric ether, methyl isobutyl ether, is not formed in significant amounts.

**Hint:** What is the strongest acid that is present when a small amount of sulfuric acid is put in an excess of methanol?

**PROBLEM:** Predict what will happen if 2-methylpropene and acetic acid react in the presence of a trace of sulfuric acid. What is the mechanism of this reaction? How is it similar to that of the reaction in the first problem in this small step?

The reactivity of an alkene toward acids is highly dependent on the stability of the carbocation that it forms on protonation. For example, for reactions in aqueous acid, 2,3-dimethyl-2-butene, 2-methyl-2-butene, and 2-methylpropene (which give tertiary carbocations on protonation) react about 10,000 times faster than (*E*)-2-butene and propene (which give secondary carbocations) do.

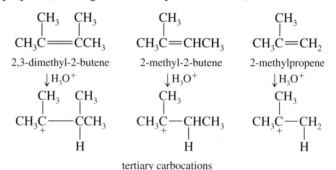

tertiary carbocations

That is, the above reactions are approximately 10,000 times faster than the following:

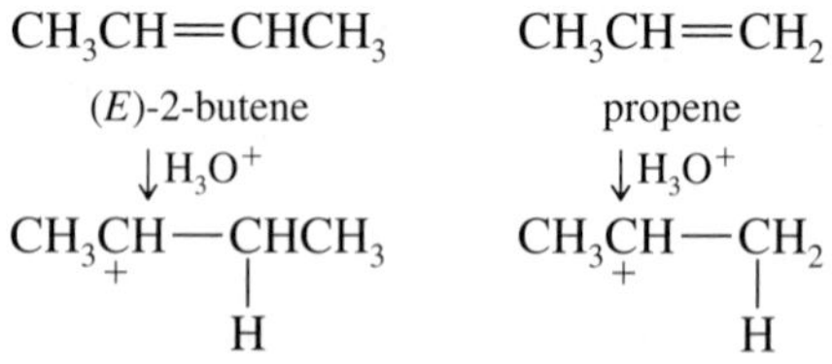

$$CH_3CH{=}CHCH_3 \qquad CH_3CH{=}CH_2$$

(*E*)-2-butene                    propene

$$\downarrow H_3O^+ \qquad\qquad \downarrow H_3O^+$$

$$\underset{+}{CH_3CH}{-}\underset{H}{CHCH_3} \qquad \underset{+}{CH_3CH}{-}\underset{H}{CH_2}$$

secondary carbocations

Such experimental evidence reinforces the idea that alkyl groups are electron-releasing when bonded to $sp^2$-hybridized carbon atoms and stabilize carbocation intermediates (p. 129).

The hydration reaction is a reversible one. Alcohols, when heated in acid, are converted to alkenes in an elimination reaction. The reaction goes through a series of steps that are the reverse of the ones shown for the hydration reaction. Cyclopentanol, for example, gives cyclopentene when heated with phosphoric acid.

$$\text{cyclopentanol} \xrightarrow[\Delta]{H_3PO_4} \text{cyclopentene} + H_2O$$

cyclopentanol
bp 140 °C

cyclopentene
bp 45 °C
distilled out of
the reaction mixture
90%

The alcohol is protonated by the acid, giving an oxonium ion. The oxonium ion loses a molecule of water to give a carbocation. The carbocation loses a proton to water, a base.

**VISUALIZING THE REACTION**

**Dehydration. An Elimination Reaction**

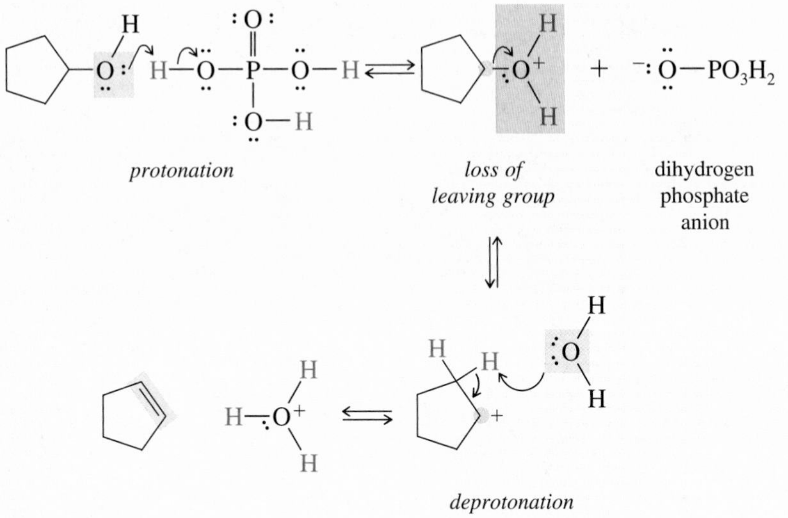

*protonation*     *loss of leaving group*     *dihydrogen phosphate anion*

*deprotonation*

The hydration reaction, addition of water to an alkene, occurs when an alkene is treated with dilute acid. The reverse reaction, the **dehydration of an alcohol,** occurs when an alcohol is heated with concentrated acid. If the alkene that is formed is removed during the reaction by distillation, a good yield is obtained.

**Problem 8.6**   Predict the products of the following reactions.

(a) 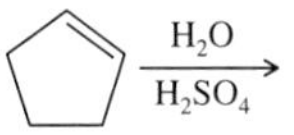 $\xrightarrow[\text{H}_2\text{SO}_4]{\text{H}_2\text{O}}$

(b) 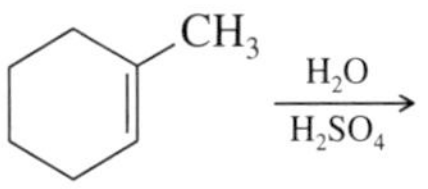 $\xrightarrow[\text{H}_2\text{SO}_4]{\text{H}_2\text{O}}$

(c) $CH_3CH_2CH_2CH_2CH{=}CH_2 \xrightarrow[\text{H}_2\text{SO}_4]{\text{H}_2\text{O}}$

(d) $CH_3\overset{\overset{\displaystyle CH_3}{|}}{\underset{\underset{\displaystyle OH}{|}}{C}}CH_3 \xrightarrow[\Delta]{\text{H}_3\text{PO}_4}$

(e) $CH_3\overset{\overset{\displaystyle CH_3}{|}}{C}{=}CHCH_2CH_3 \xrightarrow[\text{H}_2\text{SO}_4]{\text{H}_2\text{O}}$

(f) 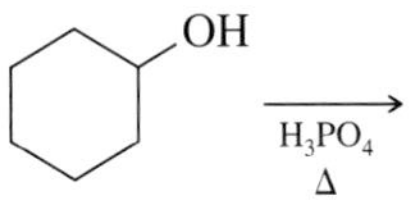 $\xrightarrow[\Delta]{\text{H}_3\text{PO}_4}$ (cyclohexanol, OH)

## B. Reaction of Carbocations with Alkenes. Polymerization

Alkenes are electron-rich reagents; the electrons of the $\pi$ bond are available for reaction with electrophiles. Whenever carbocations are formed in reaction mixtures containing alkenes, some reaction between these electron-deficient intermediates and the electron-rich alkenes can be expected. Under some conditions, this reaction may be a major one and leads to the formation of polymers. For example, protonation of 2-methylpropene gives rise to the *tert*-butyl cation, which adds to the double bond of the alkene to give a new carbocation. The reaction shown does not have to stop with the combination of two alkene units. Further additions of 2-methylpropene can occur. In fact, with longer heating, complex mixtures containing high-molecular-weight alkenes are formed.

## VISUALIZING THE REACTION

**Polymerization of an Alkene**

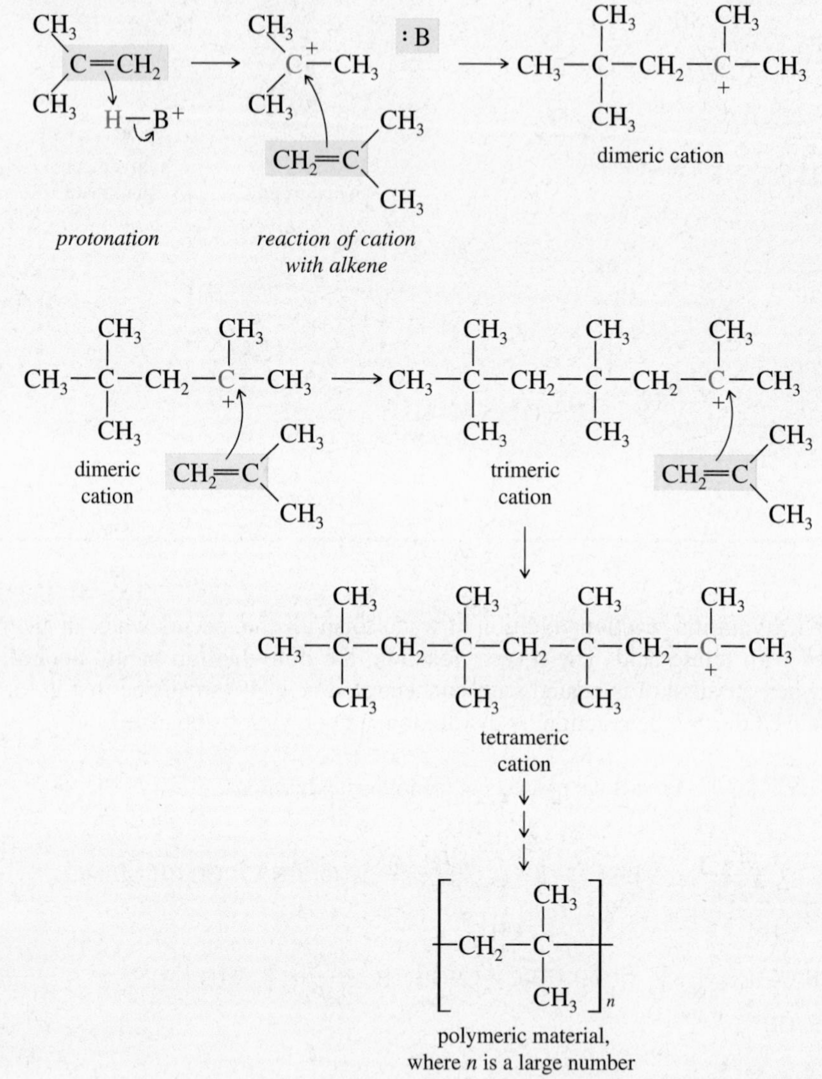

*protonation*                *reaction of cation
with alkene*

*dimeric cation*

*dimeric cation*          *trimeric cation*

*tetrameric cation*

$$\left[ -CH_2-\overset{\overset{\displaystyle CH_3}{|}}{\underset{\underset{\displaystyle CH_3}{|}}{C}}- \right]_n$$

polymeric material,
where *n* is a large number

In these reactions, 2-methylpropene is the **monomer,** a low-molecular-weight unit that adds to itself in a repetitious fashion to give a molecule having a higher molecular weight. Two units of the monomer combine to give a **dimer;** three form a **trimer.** A large molecule consisting of many units of monomer bonded together is called a **polymer.** The synthesis of polymeric materials with useful properties is an important area of organic chemistry that we will consider in Chapter 24.

## 8.5 Rearrangements of Carbocations

### A. Shifts of Hydrogen Atoms

At the end of the nineteenth century, Georg Wagner, who was working at the University of Warsaw, investigated the reactions of some naturally occurring alkenes and alcohols with acidic reagents. He found that some of the products had carbon skeletons that differed from those of the starting materials. These reactions were further investigated by Hans Meerwein in Germany in the 1920s, and he confirmed that the bonding between carbon atoms was being rearranged.

Chemists have also observed unexpected products in some reactions of simple alkenes. For example, 3-methyl-1-butene reacts with hydrogen chloride to give 2-chloro-2-methylbutane, in which the chlorine atom is bound to a tertiary carbon atom, as well as the expected product, 2-chloro-3-methylbutane.

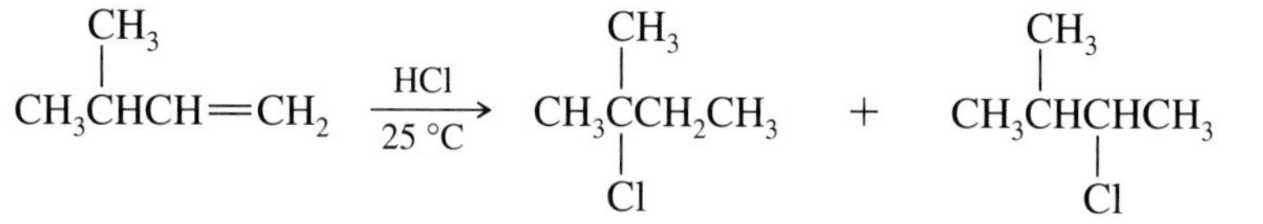

$$
\underset{\substack{\text{3-methyl-1-butene}}}{\overset{\overset{\displaystyle CH_3}{|}}{CH_3CHCH{=}CH_2}} \xrightarrow[25\,°C]{HCl} \underset{\substack{\text{2-chloro-2-methylbutane}\\ \text{~50\%}}}{\overset{\overset{\displaystyle CH_3}{|}}{\underset{\underset{\displaystyle Cl}{|}}{CH_3CCH_2CH_3}}} + \underset{\substack{\text{2-chloro-3-methylbutane}\\ \text{~50\%}}}{\overset{\overset{\displaystyle CH_3}{|}}{\underset{\underset{\displaystyle Cl}{|}}{CH_3CHCHCH_3}}}
$$

Frank C. Whitmore, who did much early work on the chemistry of alkenes at Pennsylvania State University, suggested that the secondary carbocation that formed as an intermediate in this reaction rearranged to a more stable tertiary carbocation. The two products result from the combination of chloride ion with a secondary or tertiary carbocation intermediate.

**VISUALIZING THE REACTION**

**1,2-Hydride Shift**

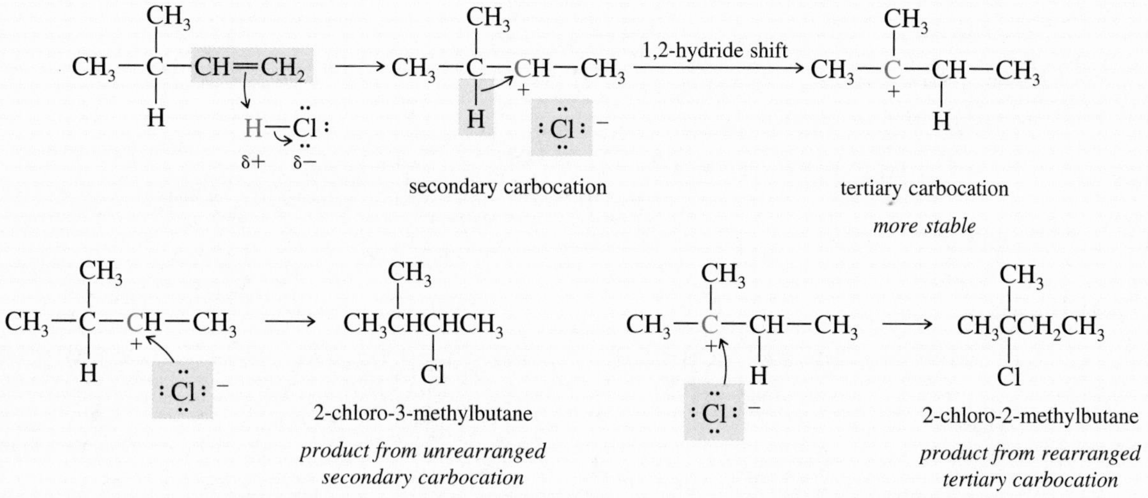

secondary carbocation

tertiary carbocation
*more stable*

2-chloro-3-methylbutane

*product from unrearranged
secondary carbocation*

2-chloro-2-methylbutane

*product from rearranged
tertiary carbocation*

The rearrangement shown on the previous page results when a pair of electrons binding a hydrogen atom to a carbon atom moves to an adjacent carbon atom that has a deficiency of electrons. A hydrogen atom with a pair of electrons is a **hydride ion,** so the rearrangement is called a **hydride shift.** When it takes place between adjacent carbon atoms, it is a **1,2-hydride shift.** The loss of the hydrogen atom and its bonding electrons from one carbon atom leaves a new cationic center in the molecule. In the example, a secondary carbocation is transformed into a more stable tertiary carbocation. Some of the secondary carbocations react with chloride ions without rearranging, so products derived from both secondary and tertiary carbocations are seen.

## B.  Shifts of Carbon Atoms

Rearrangements are not limited to hydride shifts. Any pair of electrons can move to an adjacent site of electron deficiency. The group attached to the pair of electrons goes along. For example, carbon atoms can migrate within carbocations to give rearrangements of carbon skeletons. Another reaction studied by Frank C. Whitmore, shown below, illustrates this process.

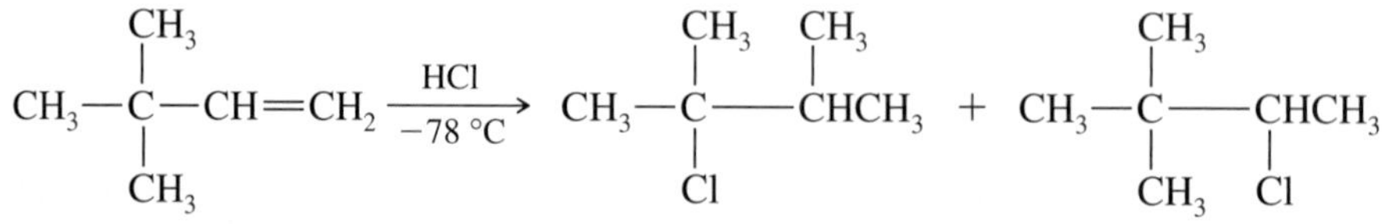

3,3-dimethyl-1-butene     2-chloro-2,3-dimethylbutane     3-chloro-2,2-dimethylbutane
                                           61%                             37%

3,3-Dimethyl-1-butene reacts with hydrogen chloride to give a secondary and a tertiary alkyl halide. The following mechanism has been proposed for the formation of these compounds.

---

**VISUALIZING THE REACTION**

**1,2-Alkyl Shift**

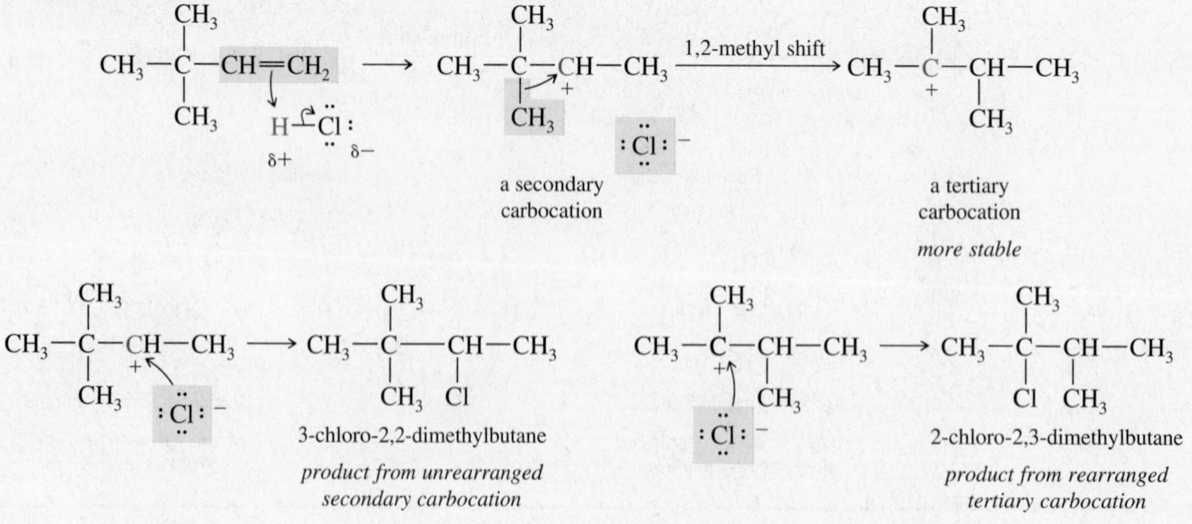

a secondary carbocation

a tertiary carbocation

*more stable*

3-chloro-2,2-dimethylbutane

*product from unrearranged secondary carbocation*

2-chloro-2,3-dimethylbutane

*product from rearranged tertiary carbocation*

A secondary carbocation forms when 3,3-dimethyl-1-butene reacts with hydrogen chloride. The shift of a methyl group, along with the pair of electrons that binds it, from one carbon atom to the adjacent cationic carbon atom creates a new tertiary carbocation. This rearrangement is called a **1,2-methyl shift.** More generally, alkyl or aryl groups of all sorts may participate in such rearrangements. Note that a 1,2-hydride shift does not change the carbon skeleton of the molecule, but a 1,2-alkyl shift does.

Carbocation intermediates may be formed by the loss of a water molecule from an alcohol in acid, as well as by the addition of a proton to an alkene. For example, the dehydration of cyclopentanol by phosphoric acid occurs by way of a carbocation (p. 287). Rearrangements occur in the conversion of some alcohols to alkenes, which is evidence for the existence of carbocation intermediates in these reactions. The dehydration of 3,3-dimethyl-2-butanol is an example. The unrearranged alkene, 3,3-dimethyl-1-butene, is formed in very small amounts. The major products, 2,3-dimethyl-2-butene and 2,3-dimethyl-1-butene, are formed by rearrangement.

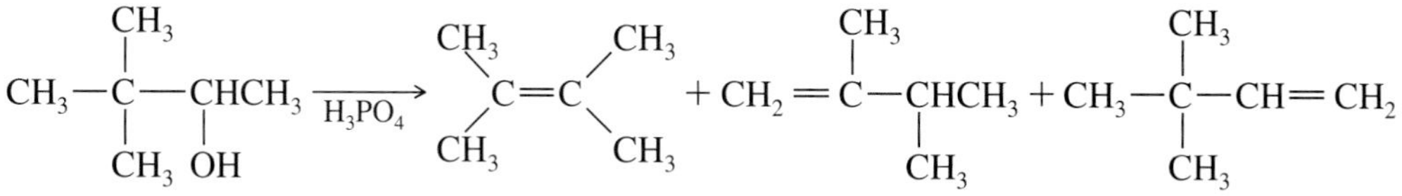

| 3,3-dimethyl-2-butanol | 2,3-dimethyl-2-butene 61% | 2,3-dimethyl-1-butene 31% | 3,3-dimethyl-1-butene 3% |

The above equation represents a fundamental reaction of carbocations. Carbocations are strong acids and lose a proton easily, even to a weak base. Most reactions that proceed through a carbocation intermediate give rise to some alkene as a product (p. 247).

Interconversion of carbocations is a common phenomenon. The alkene mixture seen when 3,3-dimethyl-2-butanol is dehydrated is formed whenever any one of those three product alkenes isolated in the pure state is subjected to the acid conditions of the dehydration reaction. Such a product composition represents, therefore, an equilibrium mixture of secondary and tertiary carbocations.

In summary, two ways of forming alkyl cations are by the protonation of an alkene or by loss of water from a protonated alcohol. Carbocations react as Lewis acids with electron-rich species or as Brønsted acids to lose a proton to a base. If their structure permits, they also rearrange to other carbocations in equilibrium reactions.

(↗) Study Guide
Concept Maps 8.1 and 8.2

**Problem 8.7** The following products were obtained in the reaction shown below. Write a mechanism showing how these products arise.

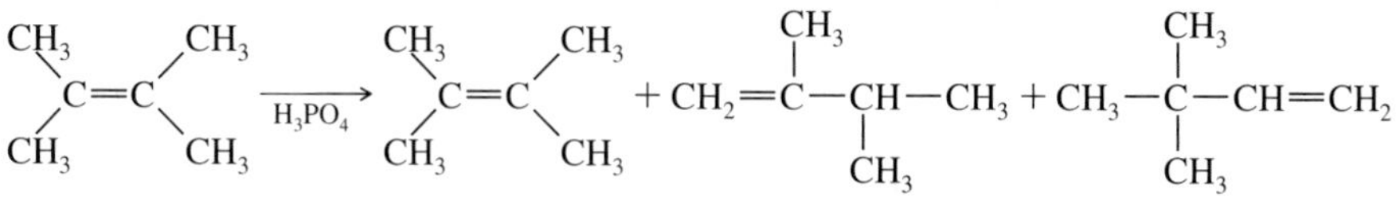

**Problem 8.8** The following reaction was observed. Write a mechanism that accounts for the product that was obtained.

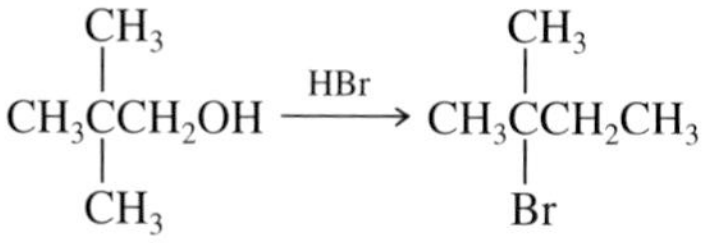

## 8.6    Addition of Diborane to Alkenes

### A. Diborane as an Electrophile. The Hydroboration Reaction

Diborane is an interesting compound that has only six pairs of bonding electrons. The question of what holds the two boron atoms together has been controversial and has led to much lively discussion and research.

Diborane exists as $B_2H_6$, a dimer of $BH_3$, in the absence of Lewis bases. It dissociates easily in the presence of ethers, which have nonbonding electrons, to give ether-borane complexes.

$$B_2H_6 \;+\; 2 \; \underset{\text{tetrahydrofuran}}{\underset{\text{\textit{a cyclic ether}}}{\text{O}}} \;\longrightarrow\; 2 \; \underset{\text{borane-tetrahydrofuran}}{\underset{\text{complex}}{\text{H}-\overset{\overset{+}{\text{O}}}{\underset{\text{H}}{\text{B}}}-\text{H}}}$$

diborane    tetrahydrofuran                borane-tetrahydrofuran
                *a cyclic ether*                    complex

Ethers are used as solvents for reactions of diborane, so the reagent exists in the complexed form. Therefore, the formula $BH_3$ will be used in this book to represent the reactive species.

---

**Problem 8.9**    Write the Lewis structure for $BH_3$. What shape do you expect the molecule to have? What kind of hybridization does the boron atom have in $BH_3$? Draw an orbital picture of the bonding in the molecule.

---

$BH_3$ is a Lewis acid and, therefore, an electrophile. It reacts with the $\pi$ electrons of an alkene with an orientation that gives rise to a partial positive charge at the carbon atom that will lead to the more stable carbocation.

$$CH_3CH_2CH{=}CH_2 \;\longrightarrow\; \overset{\delta+}{CH_3CH_2CH}{=\!=}CH_2$$
$$\underset{\text{H}}{\overset{\downarrow}{\text{H}-\text{B}-\text{H}}} \qquad\qquad \underset{\underset{\delta-}{\text{H}}}{{}^{\delta-}\text{H}-\text{B}-\text{H}^{\delta-}}$$

1-butene and borane                    partial bonding between
                                        1-butene and borane

At the same time, negative charge develops at the boron atom. The hydrogen atoms bonded to the boron acquire partial negative character, and one of them is transferred as a hydride ion to the carbon atom that has the partial positive charge. The addition of hydrogen and boron to the double bond is known as **hydroboration.** The reaction may be represented as a single step.

**Addition of Diborane to an Alkene. Hydroboration**

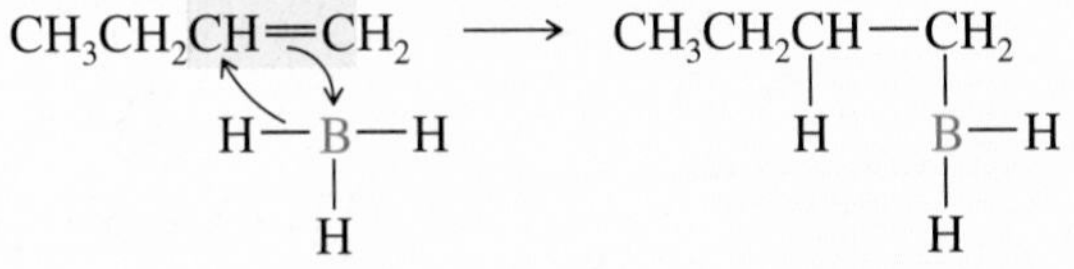

$$CH_3CH_2CH\!\!=\!\!CH_2 \longrightarrow CH_3CH_2CH\!-\!CH_2$$

1-butene and borane

*n*-butylborane

*an organoborane*

The reaction is said to have a **four-center transition state,** with four atoms (two carbon, one boron, and one hydrogen) undergoing changes in bonding at the same time. The hydroboration reaction follows the modern sense of Markovnikov's Rule (p. 132) in that the electrophile, the boron atom, is attached to the less highly substituted carbon atom, creating partial positive character at the carbon atom better able to stabilize the charge. Hydrogen, with a partial negative charge, is the nucleophile in this reaction and adds to the more highly substituted carbon atom.

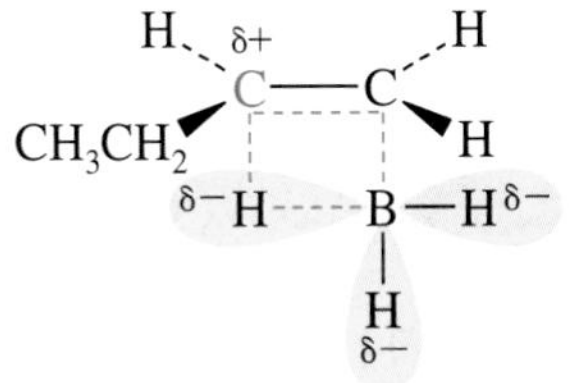

The product from the addition of borane to an alkene is known as an organoborane. Such a compound, for example, *n*-butylborane, which is shown below, is still a Lewis acid in which the boron atom can accept another pair of electrons from an alkene and can transfer a hydride ion to it. The reaction proceeds until all three hydrogen atoms that were originally bonded to boron have been replaced by alkyl groups.

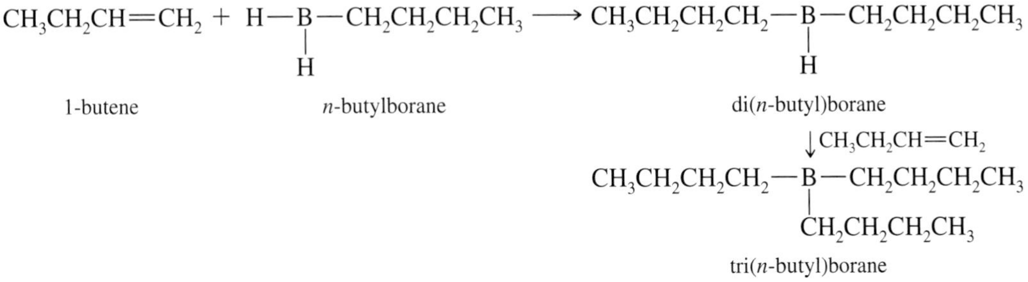

*four-center transition state for the reaction of 1-butene with borane*

$$CH_3CH_2CH\!\!=\!\!CH_2 + H\!-\!\underset{\underset{H}{|}}{B}\!-\!CH_2CH_2CH_2CH_3 \longrightarrow CH_3CH_2CH_2CH_2\!-\!\underset{\underset{H}{|}}{B}\!-\!CH_2CH_2CH_2CH_3$$

1-butene        *n*-butylborane

di(*n*-butyl)borane

$$\downarrow CH_3CH_2CH\!\!=\!\!CH_2$$

$$CH_3CH_2CH_2CH_2\!-\!\underset{\underset{CH_2CH_2CH_2CH_3}{|}}{B}\!-\!CH_2CH_2CH_2CH_3$$

tri(*n*-butyl)borane

**Problem 8.10**    Write equations for the reactions of cyclopentene, 1-decene, and 3,3-dimethyl-1-butene with diborane.

## B. The Hydroboration–Oxidation Reaction

The **hydroboration–oxidation reaction,** which was discovered in 1956 by Herbert C. Brown at Purdue University, has made it possible to synthesize alcohols in high yields under mild conditions with high regioselectivity (p. 127). In this reaction, diborane adds to an alkene to give organoboranes. Oxidation of the carbon–boron bond in the organoboranes with hydrogen peroxide in basic solution gives alcohols, as illustrated by the overall reaction for 1-butene shown on the next page.

### ONE SMALL STEP

The addition of metal hydrides to multiple bonds is a quite general reaction for elements from groups 13 and 14 of the periodic table.

**PROBLEM:** Predict the products of the following reactions:

$$2 \ CH_3\overset{\overset{\displaystyle CH_3}{|}}{C}{=}CH_2 + AlH_3 \longrightarrow \ ?$$

$$(CH_3CH_2)_2GaH \ +$$

$$H_2C{=}CH(CH_2)_7CH_3 \longrightarrow \ ?$$

**Hint:** Use the chemistry of diborane to guide you in finding the answers to these problems.

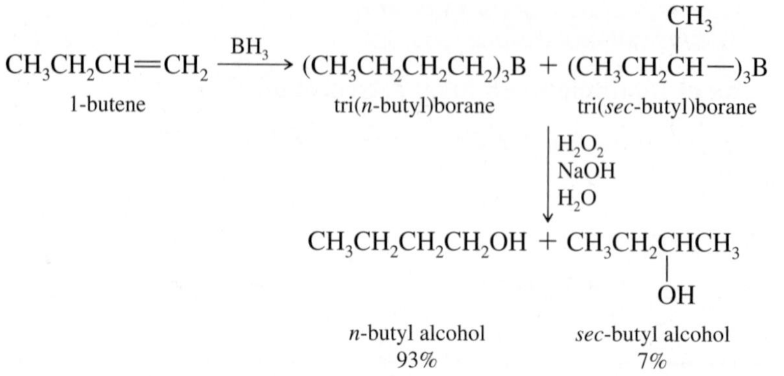

$$CH_3CH_2CH{=}CH_2 \ \xrightarrow{\ BH_3\ } \ (CH_3CH_2CH_2CH_2)_3B \ + \ (CH_3CH_2\overset{\overset{\displaystyle CH_3}{|}}{CH}{-})_3B$$

1-butene                              tri(*n*-butyl)borane              tri(*sec*-butyl)borane

$$\Big\downarrow \begin{array}{l} H_2O_2 \\ NaOH \\ H_2O \end{array}$$

$$CH_3CH_2CH_2CH_2OH \ + \ CH_3CH_2\overset{\overset{\displaystyle }{}}{\underset{\underset{\displaystyle OH}{|}}{CH}}CH_3$$

*n*-butyl alcohol          *sec*-butyl alcohol
93%                          7%

Note that the chief product of the reaction sequence, *n*-butyl alcohol, looks as if an anti-Markovnikov addition of water to the double bond has occurred.

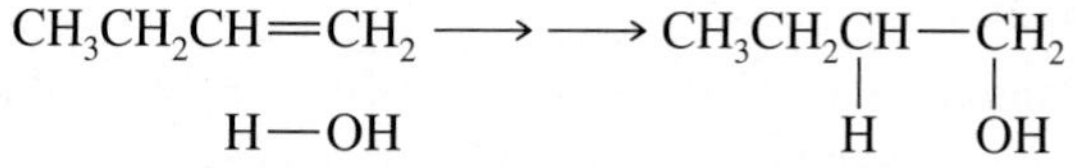

$$CH_3CH_2CH{=}CH_2 \ \longrightarrow \longrightarrow \ CH_3CH_2\underset{\underset{\displaystyle H}{|}}{CH}{-}\underset{\underset{\displaystyle OH}{|}}{CH_2}$$

$$\underset{\displaystyle H{-}OH}{}$$

anti-Markovnikov addition of H—OH to an alkene

In contrast, in the hydration reaction (p. 284), alcohols are produced by a Markovnikov addition of water to the double bond.

## C. Regioselectivity in Hydroboration Reactions

In the unsymmetrical terminal alkene 1-butene, the boron atom adds chiefly to the less highly substituted of the doubly bonded carbon atoms. In the case of internal alkenes in which one of the doubly bonded carbon atoms has two substituents and the other has one, the boron atom also attaches itself to the less substituted carbon atom. For example, 2-methyl-2-butene reacts with diborane to give chiefly the organoborane in which the boron atom is attached to the less highly substituted carbon atom.

$$CH_3\overset{\overset{\displaystyle CH_3}{|}}{C}{=}CHCH_3 \ \xrightarrow[\text{diglyme}]{\ BH_3\ } \ CH_3\overset{\overset{\displaystyle CH_3}{|}}{\underset{\underset{\displaystyle H}{|}}{C}}{-}\overset{\overset{\displaystyle H}{|}}{\underset{\underset{\displaystyle BH_2}{|}}{C}}CH_3 \ + \ CH_3\overset{\overset{\displaystyle CH_3}{|}}{\underset{\underset{\displaystyle H_2B}{|}}{C}}{-}\overset{\overset{\displaystyle H}{|}}{\underset{\underset{\displaystyle H}{|}}{C}}CH_3$$

2-methyl-2-butene                        98%                        2%

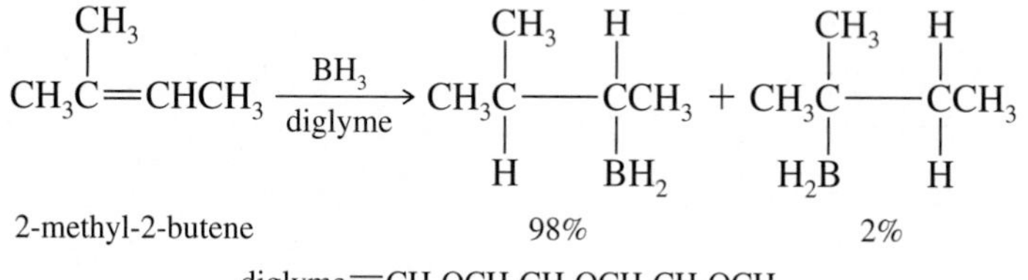

diglyme ≡ CH_3OCH_2CH_2OCH_2CH_2OCH_3

*an ether*

For the more or less symmetrically disubstituted alkenes, however, two products are possible. They are formed in roughly equal amounts even when the bulkiness of one of the substituents would seem to favor addition of the boron atom to the other carbon atom, as illustrated by the reaction of (*E*)-4-methyl-2-pentene.

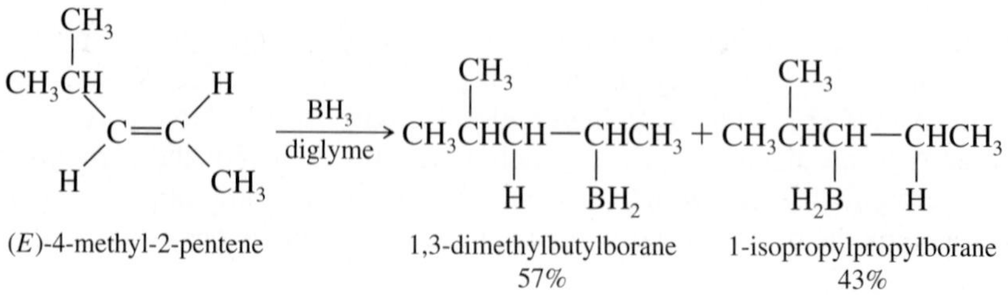

$$\underset{(E)\text{-4-methyl-2-pentene}}{\overset{\overset{\displaystyle CH_3}{|}}{\underset{\displaystyle }{CH_3CH}}\underset{\displaystyle H}{\diagdown}{\underset{\displaystyle }{C}}{=}\underset{\displaystyle }{C}\overset{\displaystyle H}{\diagup}\underset{\displaystyle CH_3}{\diagdown}} \ \xrightarrow[\text{diglyme}]{\ BH_3\ } \ \underset{\text{1,3-dimethylbutylborane}}{CH_3\overset{\overset{\displaystyle CH_3}{|}}{CH}CH{-}\overset{}{\underset{\underset{\displaystyle BH_2}{|}}{\underset{\underset{\displaystyle H}{|}}{CH}}}CH_3} \ + \ \underset{\text{1-isopropylpropylborane}}{CH_3\overset{\overset{\displaystyle CH_3}{|}}{CH}CH{-}\overset{}{\underset{\underset{\displaystyle H}{|}}{\underset{\underset{\displaystyle H_2B}{|}}{CH}}}CH_3}$$

(*E*)-4-methyl-2-pentene          1,3-dimethylbutylborane          1-isopropylpropylborane
57%                          43%

Herbert C. Brown reasoned that a bulkier reagent than borane might show greater selectivity in the way it added to a double bond, because steric interactions between substituents on the double bond and on the reagent would increase and become more important. One of the reagents he developed for greater selectivity is the organoborane 9-borabicyclo[3.3.1]nonane (9-BBN), which is formed when borane adds twice to the cyclic diene 1,5-cyclooctadiene, to give a bicyclic compound. In a bicyclic compound, two or more atoms are shared by two different rings (p. 745).

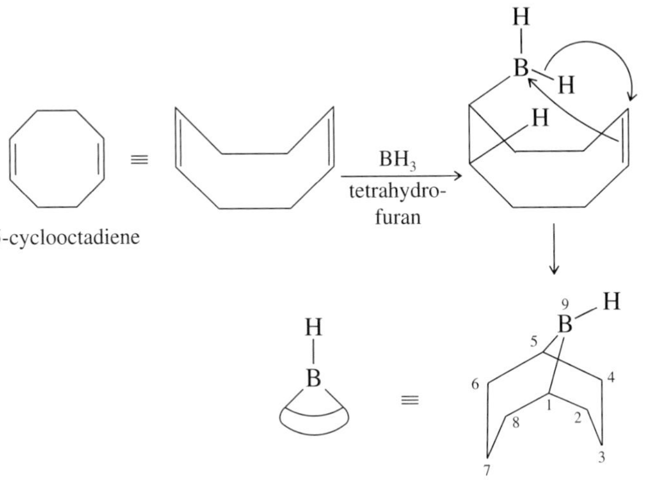

1,5-cyclooctadiene

a symbol
frequently used
for 9-BBN

9-borabicyclo[3.3.1]nonane
or 9-BBN

9-BBN is a reasonably stable compound that can be purchased in a bottle and handled with minor precautions in the air, in contrast to diborane, which is a gas and reacts violently with air. This organoborane is thus a more convenient reagent for hydroboration reactions. The bulkiness of substituents on the boron atom gives the reagent high selectivity in its reaction with substituted alkenes. With (Z)-4-methyl-2-pentene, for example, only the product in which the boron atom has bonded to the carbon atom that has the less bulky of the two substituents is formed.

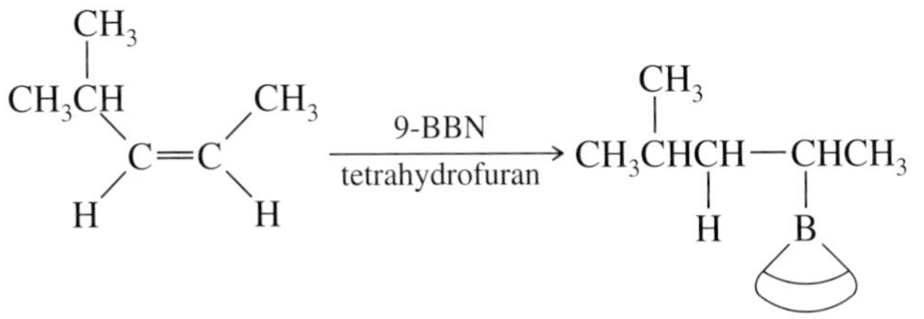

(Z)-4-methyl-2-pentene

99.8%

This contrasts with the equal distribution of products for the reaction of diborane with the E isomer shown on the previous page.

**Problem 8.11**    Write an equation showing the chief product from the reaction of 9-BBN with (a) 2-methyl-1-pentene, (b) styrene (phenylethene), (c) (Z)-4,4-dimethyl-2-pentene, (d) 2,3-dimethyl-2-butene, and (e) (Z)-3-hexene.

# D. Stereoselective Reactions. Stereochemistry of Hydroboration Reactions

The transition state shown on page 293 for the addition of borane to an alkene predicts that the boron atom and the hydrogen atom will be bonded to the same side of the molecule. An addition reaction in which the incoming groups are added to the same side of the molecule is called a **syn addition.** Depending on the structure of the starting reagent, a syn addition may give products with cis (see p. 314) or trans (see below) stereochemistry. In the hydroboration reaction, syn addition of boron and hydrogen to the double bond is thought to occur.

There is no easy way to determine the stereochemistry of organoboranes. The alcohols formed from the oxidation of the organoborane intermediates show overall syn addition of water to the double bond in the cases where stereochemistry can be determined. For example, 1-methylcyclohexene gives *trans*-2-methylcyclohexanol in the hydroboration reaction.

| 1-methylcyclohexene and 9-borabicyclo[3.3.1]nonane | organoborane from syn addition | *trans*-2-methylcyclohexanol 98% |

Careful experiments with several different systems have led chemists to conclude that the overall syn addition of water is the result of syn addition of the boron hydride to the double bond followed by oxidation of the carbon–boron bond with retention of configuration. We will examine the details of the oxidation process in the next section.

A reaction in which one of several possible stereochemical results predominates is called a **stereoselective reaction.** Alcohols are formed stereoselectively in the hydroboration–oxidation reaction. *trans*-2-Methylcyclohexanol is the major product obtained above. The other possible stereoisomer, *cis*-2-methylcyclohexanol, is not found.

# E. Oxidation of Organoboranes

The oxidation of alkylboranes with hydrogen peroxide in basic solution gives alcohols. Both the carbon atom and the boron atom in the organoborane are oxidized, and hydrogen peroxide is reduced to water. The reaction requires the addition of 6 M sodium hydroxide and 30% hydrogen peroxide directly to the hydroboration reaction mixture, so that the conversion of an alkene to an alcohol takes place in one reaction vessel without the isolation of the organoborane.

The reaction is believed to start with the formation of the hydroperoxide anion, $HOO^-$, in base.

$$HOOH \quad + \quad Na^+OH^- \longrightarrow Na^+OOH^- \quad + \quad H_2O$$

hydrogen peroxide

*more acidic than*
*$H_2O$ because of the*
*inductive effect*
*of the second*
*oxygen atom*

This nucleophilic anion then reacts with the boron atom. The important step is the migration of an alkyl group to oxygen, a rearrangement that takes place with retention of configuration.

---

**VISUALIZING THE REACTION**

**Oxidation of an Organoborane**

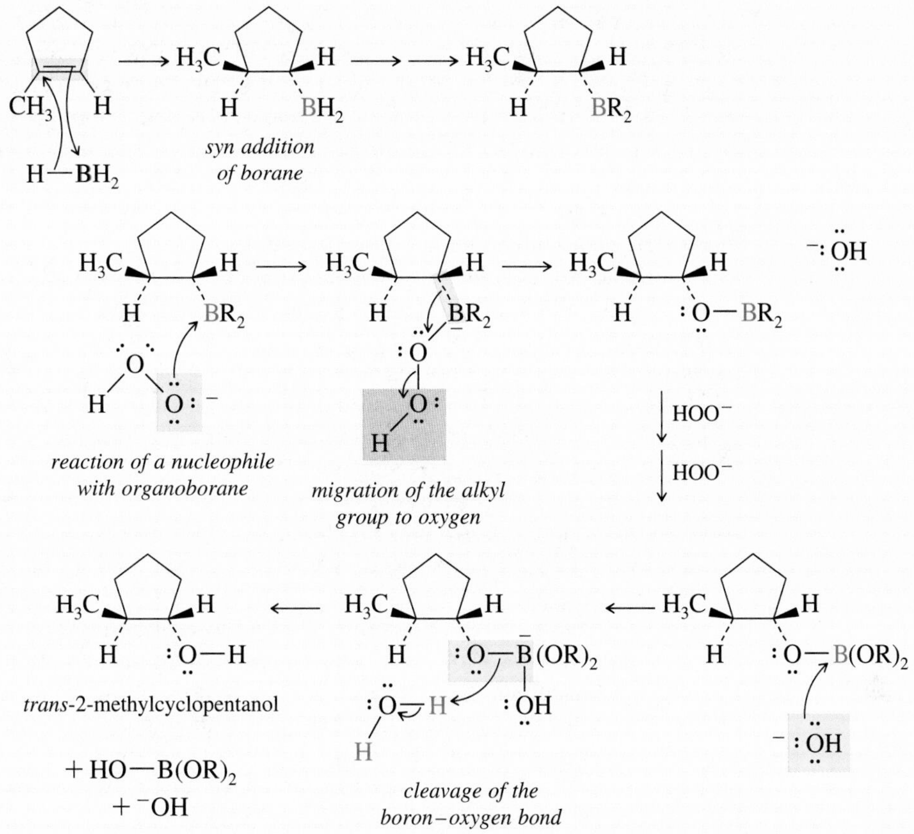

*syn addition of borane*

*reaction of a nucleophile with organoborane*

*migration of the alkyl group to oxygen*

*trans-2-methylcyclopentanol*

$+ HO—B(OR)_2$
$+ {}^-OH$

*cleavage of the boron–oxygen bond*

---

The hydroxyl group in *trans*-2-methylcyclopentanol ends up on the side of the molecule where the boron atom was originally attached in the organoborane. Boron is oxidized to boric acid, present as the sodium salt, sodium borate.

The hydroboration–oxidation sequence provides a very mild and convenient way to prepare alcohols from alkenes, with water adding to the double bond with high regioselectivity and stereoselectivity. The rearrangements and polymerization reactions typical of acid-catalyzed hydrations of alkenes do not take place under these conditions. The importance of this type of reaction was recognized in 1979 when Herbert C. Brown was honored with the Nobel Prize for his work with organoboranes.

*(↗) Study Guide*
*Concept Map 8.3*

**Problem 8.12** For the organoboranes formed in the reactions in Problems 8.10 and 8.11, show the products you would expect from peroxide oxidation in basic solution.

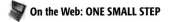 **On the Web: ONE SMALL STEP**

## 8.7　Addition of Hydrogen to Alkenes. Catalytic Hydrogenation Reactions

### A. Heterogeneous Catalysis

Hydrogen adds to a double bond in the presence of a metallic catalyst in a **hydrogenation reaction.** (A hydrogenation reaction is one type of reduction reaction. Oxidation–reduction reactions are discussed more fully in Section 13.7.) A hydrogenation reaction is usually carried out in some inert solvent. The alkene is stirred in the presence of the solid catalyst while hydrogen gas is introduced into the reaction mixture. The amount of hydrogen used up in the reaction can be monitored by watching the drop in pressure or the decrease in the volume of the gas in the system. Hydrogen gas does not add to double bonds unless a specially prepared metal surface is present.

The presence of the catalyst as a separate solid phase makes the reaction mixture a **heterogeneous** one, in contrast to reactions in which all reagents are present in a single **homogeneous** phase. Metals that are commonly used as **hydrogenation catalysts** are palladium, platinum, and nickel. Palladium and platinum catalysts are usually prepared by the reduction of a salt of the metal by hydrogen, very often in the presence of a larger amount of an inert material that serves to dilute and support the catalyst. An example is palladium chloride on carbon, which when reduced by hydrogen gives a solid that consists of palladium metal in a finely divided state supported on powdered carbon. Reduction of platinum oxide gives a platinum catalyst. A form of nickel called **Raney nickel** is made by using sodium hydroxide to dissolve out the aluminum in a nickel–aluminum alloy. The reaction of aluminum with sodium hydroxide gives hydrogen gas, which is adsorbed on finely divided nickel from the alloy.

Exactly how a catalyst functions in a hydrogenation reaction is a matter of debate and may vary with the particular catalyst, the relative amounts of alkene and hydrogen gas used, and the temperature. The important factor in catalysis is the nature and extent of the available metal surface. The alkene is believed to be adsorbed onto the surface of the catalyst, forming bonds to the metal atoms. The $\pi$ bond is broken at this stage. This reaction may be viewed as a Lewis acid–base interaction between the metal atoms, which have room in their orbitals for electrons, and the $\pi$ electrons of the multiple bond (Figure 8.3). In some cases, the hydrogen molecule may also be adsorbed to the surface of the metal near the organic molecule, with some loosening of the hydrogen–hydrogen bond. Reaction occurs so that a hydrogen atom is bonded to each of the carbon atoms.

The addition of hydrogen to an alkene to give an alkane is an exothermic process (p. 282). In the absence of a catalyst, however, the reaction has a high energy of activation and takes place at a negligible rate. The catalyst changes the nature of the transition state for the reaction and thereby lowers the energy of activation. The reaction can then proceed at a reasonable rate under practical temperature and pressure. Figure 8.4 is an energy diagram showing how the catalyzed reaction compares with the uncatalyzed reaction.

Because the uncatalyzed hydrogenation reaction has a very high energy of activation, no reaction takes place unless a catalyst is used. By showing the formation of an intermediate on the pathway from reactants to product, Figure 8.4 indicates that the mechanism of the catalyzed reaction is different from that of the uncatalyzed reaction. The interaction between the alkene and the surface of the catalyst, depicted

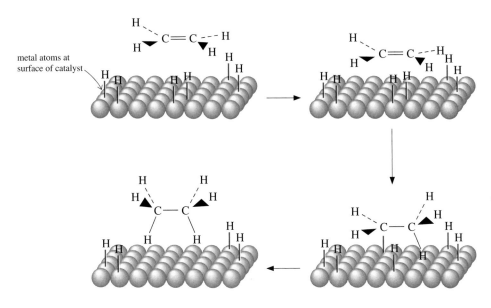

metal atoms at surface of catalyst

**Figure 8.3**
A schematic representation of the interaction leading to the transfer of hydrogen atoms from the surface of a metallic catalyst to the carbon atoms of an alkene.

in Figure 8.3, changes the nature of the bonding in the alkene and creates a species that shows greater reactivity toward hydrogen. The change in the energy of activation that accompanies catalysis occurs because the nature of the transition state is changed by the presence of the catalyst. Note that the catalyst also lowers the energy of activation for the reverse reaction, the dehydrogenation of an alkane to an alkene. Such reactions do take place in the presence of hydrogenation catalysts if no external hydrogen gas is added to the reaction system.

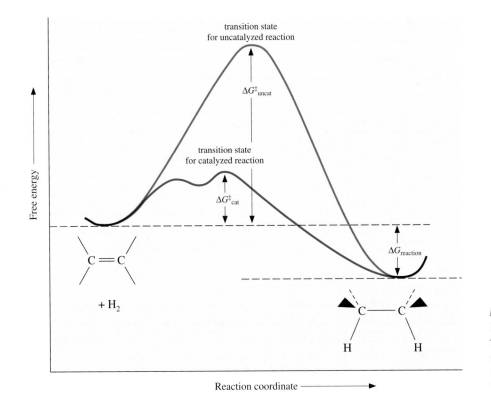

**Figure 8.4**
Energy diagram comparing the free energy of activation for a catalyzed hydrogenation reaction with that for an uncatalyzed reaction.

## B. Some Hydrogenation Reactions of Alkenes

Alkenes are converted to alkanes by catalytic hydrogenation. Some typical examples are the conversion of 2-methyl-2-pentene to 2-methylpentane and 3-methylcyclopentene to methylcyclopentane.

$$
\underset{\substack{\text{2-methyl-2-pentene} \\ \text{25 °C}}}{\overset{\overset{\displaystyle CH_3}{|}}{CH_3C\!\!=\!\!CHCH_2CH_3}} \xrightarrow[\text{Pt}]{H_2} \underset{\text{2-methylpentane}}{\overset{\overset{\displaystyle CH_3}{|}}{CH_3CHCH_2CH_2CH_3}}
$$

3-methylcyclopentene $\xrightarrow[\substack{\text{Pt} \\ \text{25 °C}}]{H_2}$ methylcyclopentane

Hydrogenation reactions are generally efficient reactions that give the products in high yield and are important in industrial processes. For example, hydrogenation of a mixture of trimethylpentenes obtained from the acid-catalyzed dimerization of 2-methylpropene (p. 288) gives 2,2,4-trimethylpentane, known commercially as isooctane and used as a component of high-octane gasoline.

$$
\underset{\substack{\text{2,4,4-trimethyl-1-pentene}}}{\overset{\overset{\displaystyle CH_3\ CH_3}{|\quad\ |}}{\underset{\underset{\displaystyle CH_3}{|}}{CH_3CCH_2C\!\!=\!\!CH_2}}} \;+\; \underset{\substack{\text{2,4,4-trimethyl-2-pentene}}}{\overset{\overset{\displaystyle CH_3\ \ CH_3}{|\quad\ |}}{\underset{\underset{\displaystyle CH_3}{|}}{CH_3CCH\!\!=\!\!CCH_3}}} \xrightarrow[\text{acetic acid}]{H_2 \atop Pt} \underset{\substack{\text{2,2,4-trimethylpentane} \\ \text{isooctane}}}{\overset{\overset{\displaystyle CH_3\ CH_3}{|\quad\ |}}{\underset{\underset{\displaystyle CH_3}{|}}{CH_3CCH_2CHCH_3}}}
$$

Another important industrial application of hydrogenation is the conversion of vegetable oils, which are mixtures of esters of unsaturated acids, into solid fats, which are esters of saturated acids (p. 638). Oleic acid, the acid found in olive oil, is (Z)-9-octadecenoic acid. Hydrogenation of oleic acid converts it to stearic acid, also called octadecanoic acid, which is found in butter and beef fat.

$$
\underset{\substack{\text{(Z)-9-octadecenoic acid} \\ \text{oleic acid} \\ \text{mp 14 °C}}}{\overset{CH_3(CH_2)_7}{\underset{H}{}}\!\!C\!\!=\!\!C\!\!\overset{(CH_2)_7COH}{\underset{H}{}}} \xrightarrow[\text{Ni}]{H_2} \underset{\substack{\text{octadecanoic acid} \\ \text{stearic acid} \\ \text{mp 69 °C}}}{CH_3(CH_2)_{16}\overset{\displaystyle O}{\overset{\|}{C}}OH}
$$

*liquid at room temperature,*          *solid at room temperature,*
*a component of vegetable oil*          *a component of fats*

Oils and fats are similar in structure except for the presence of double bonds in the acid components of oils, which lowers their melting points and makes them liquids at room temperature. Because fats are more stable toward oxidation by air and more convenient to handle and store, hydrogenation is used to convert oils to fats such as margarine or vegetable shortening.

Recent concerns with the amount of saturated fat in our diets have led to attempts to retain some of the unsaturation typical of oils in the production of margarines. Softer margarines with lower melting points and higher percentages of unsaturated fats have become available. This has, in turn, led to a new concern. The hydrogenation reaction on the surface of a catalyst is a reversible process. New double bonds can be created in the hydrogenation process, especially if the amount of hydrogen being used is being limited in order to keep some of the unsaturation in the molecule. Chemists are finding new fatty acids in margarines in which the position of the double bonds has been shifted or the cis stereochemistry of the double bond typical of the natural oil has been converted to the thermodynamically more stable trans form. In other words, the unsaturation left in the margarine does not necessarily result from the natural unsaturated fatty acids found in oils. The full nutritional effects of the unnatural fatty acids are not known yet, but *trans*-fatty acids have been implicated in a higher risk of heart disease and breast cancer. The chemistry of fats and oils is further explored in Chapter 15.

⊘ **Study Guide**
**Concept Map 8.4**

**Problem 8.13**  Complete the following equations.

(a) $CH_3\overset{\displaystyle CH_3}{\underset{\displaystyle CH_3}{C}}CH{=}CH_2 \xrightarrow[Ni]{H_2} A$

(b) $CH_3CH_2CH{=}CHCH_2CH_3 \xrightarrow{H_2}{Pt} B$

(c) $CH_3CH_2\overset{\displaystyle CH_3}{\underset{\displaystyle OH}{C}}CH_2CH_3 \xrightarrow[\Delta]{H_2SO_4} C + D \xrightarrow{H_2}{Pt} E$

(d) $\square\!\!<^{CH_3} \xrightarrow{H_2}{Pd/C} F$

(e) ⬡ $\xrightarrow[Pd/C]{H_2 \text{ (excess)}} G$

# 8.8  Addition of Bromine to Alkenes

## A. Introduction

The addition of acids, diborane, and elemental hydrogen to the double bond proceeds by attack of the $\pi$ bond on some electrophilic center (a proton, a boron atom, the surface of a metallic catalyst) and results in the addition of the elements of each reagent, H—X, H—BR$_2$, H—H, across the double bond.

The next series of reagents that we will examine is oxidizing agents. **Oxidizing agents** are reagents that seek electrons and are, therefore, also classified as electrophiles. In the process of gaining electrons, oxidizing reagents become reduced. Oxidation–reduction reactions will be examined in greater detail in Chapter 13. The oxidizing agents that we will examine now are the halogens, ozone, peroxyacids, and the oxygen compounds of transition metals such as osmium (Os, atomic number 76). The transition metals, which appear in the middle of the periodic table, have variable valences, gaining and losing electrons with relative ease. This property makes oxides of transition metals good oxidizing agents. They are able to accept electrons from nucleophiles (such as $\pi$ bonds) with ease. In doing so, they go to a lower valence state and are said to be reduced.

All the reagents that are discussed in the next two sections have another feature in common: They react with the double bond to give a cyclic reactive intermediate or product. For these reagents, the movement of electrons "out" from the $\pi$ bond appears to proceed simultaneously with the donation of a pair of electrons back "in" from the reagent to the carbon skeleton of the alkene, resulting in the formation of two new $\sigma$ bonds in a ring. Look for the formation of a cyclic intermediate as a unifying feature of all of the reactions that follow.

## B. Bromine as an Electrophile

Halogens are electrophiles that react by accepting electrons to form halide anions. Bromine therefore adds to carbon–carbon double bonds (pp. 66 and 214). A bromine molecule is normally symmetrical, but as it approaches the $\pi$ electrons of an alkene, the distribution of the electrons in the covalent bond changes. One of the bromine atoms becomes more positive, the other more negative. This polarization of the bond in the bromine molecule enables bonding to take place between it and the alkene.

A solution of bromine in carbon tetrachloride has the reddish brown color typical of elemental bromine. When such a solution is added to an alkene, the color of bromine rapidly disappears. This reaction serves as a test for the presence of carbon–carbon multiple bonds and distinguishes alkenes and alkynes, which have $\pi$ bonds, from alkanes, which have none. Aromatic compounds such as benzene, in which electrons are present in especially stable orbitals, do not react with bromine under these conditions (p. 67).

The addition of bromine to ethylene appears to proceed in two steps. Evidence is provided by the following experiments. If bromination is carried out in the presence of species that can act as nucleophiles, such as negatively charged ions, or in water as the solvent, mixtures of products are obtained. When bromide ion or chloride ion is the nucleophile, the products are alkyl halides. If water is the nucleophile, the product has a hydroxyl group and a halogen atom on adjacent carbon atoms. Such compounds are known as **halohydrins.**

$$\underset{\text{ethylene}}{CH_2{=}CH_2} \;+\; \underset{\text{bromine}}{Br_2} \;\xrightarrow[\text{H}_2\text{O}]{\text{NaCl (saturated)}}\; \underset{\substack{\text{1,2-dibromoethane}\\54\%}}{BrCH_2CH_2Br} \;+\; \underset{\substack{\text{1-bromo-2-chloro-}\\\text{ethane}\\46\%}}{BrCH_2CH_2Cl}$$

$$\underset{\text{ethylene}}{CH_2{=}CH_2} \;+\; \underset{\text{bromine}}{Br_2} \;\xrightarrow[\text{0 °C}]{\text{H}_2\text{O}}\; \underset{\substack{\text{2-bromoethanol}\\54\%}}{BrCH_2CH_2OH} \;+\; \underset{\substack{\text{1,2-dibromoethane}\\37\%}}{BrCH_2CH_2Br}$$

*a halohydrin*

Note that, in the second reaction, water is both the solvent and the nucleophile and is present in a high concentration. Therefore, the halohydrin is the major product.

These facts support the proposal that reaction of bromine with the double bond gives an intermediate cation that then reacts with any nucleophilic species present.

## C. The Bromonium Ion

If the addition of bromine to cyclopentene (p. 214) proceeds by nucleophilic attack of the double bond on bromine with the resulting formation of a planar carbocation as an intermediate, some *cis*-1,2-dibromopentane would be expected as a product because the second step, attack of bromide ion, can take place from either side of the planar carbocation.

*trans*-1,2-dibromo-cyclopentane

*racemic mixture only product*

*cis*-1,2-dibromo-cyclopentane

*not observed*

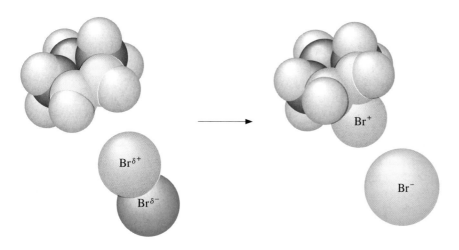

*Figure 8.5*
Formation of the cyclopentylbromonium ion.

In fact, only the trans isomer is obtained from the reaction. This observation led to the idea that the intermediate cation is a cyclic **bromonium ion.** The bromine atom is so large in comparison to carbon that the electron cloud of bromine overlaps with *p* orbitals of both of the doubly bonded carbon atoms that were originally part of the double bond (Figure 8.5).

As shown in the Visualizing the Reaction feature below, the opening of a bromonium ion is a nucleophilic substitution reaction. The nucleophile is the bromide ion, and the leaving group is the positively charged bromine atom of the bromonium ion. The reaction produces inversion of configuration at the carbon atom attacked by the bromide ion and retention of configuration at the carbon atom that winds up with the bromine atom of the bromonium ion. The overall addition to the double bond results in a racemic mixture with trans stereochemistry, as seen for 1,2-dibromocyclopentane.

**VISUALIZING THE REACTION**

**Addition of Bromine to an Alkene**

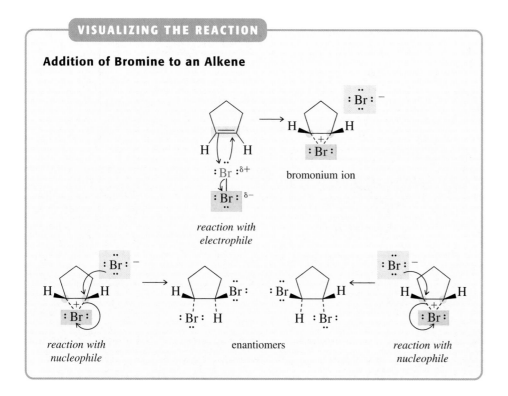

## D. Stereochemistry of the Addition of Bromine to Alkenes

The addition of bromine to the double bond in cyclopentene is 100% stereoselective (p. 296), leading only to *trans*-1,2-dibromocyclopentane as a product. The bromonium ion is symmetrical in the case of cyclopentene, so nucleophilic attack at either one of the carbon atoms is equally likely. A racemic mixture of the two enantiomeric *trans*-1,2-dibromocyclopentanes is formed.

An addition reaction in which the two components that add to the double bond end up trans to each other is said to proceed with anti stereochemistry and is an **anti addition reaction.** An anti addition contrasts with a syn addition, an example of which is the addition of diborane to a $\pi$ bond (p. 296). The words **syn** and **anti** refer to the mechanism of a reaction; the words **cis** and **trans** are used to describe the stereochemistry of the product of the reaction.

The origin of the term *anti* is illustrated clearly by the addition of bromine to an acyclic alkene. The reaction of (*Z*)-2-pentene with bromine gives (2*S*,3*S*)-2,3-dibromopentane and (2*R*,3*R*)-2,3-dibromopentane. The products from the reaction of (*E*)-2-pentene with bromine are (2*S*,3*R*)-2,3-dibromopentane and (2*R*,3*S*)-2,3-dibromopentane.

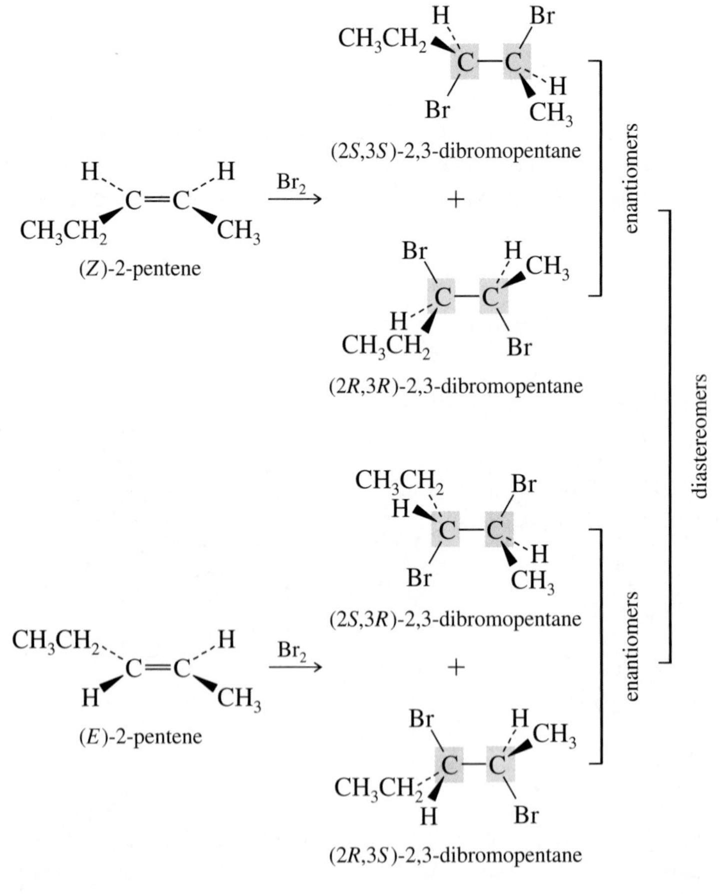

Each of these stereoisomers is shown in a conformation that illustrates the anti relationship between the bromine atoms. (*Z*)-2-Pentene gives rise to one set of enantiomeric 2,3-dibromopentanes; (*E*)-2-pentene gives rise to another set. The addition of bromine to either of these two alkenes proceeds with high stereoselectivity.

There are four stereoisomers of 2,3-dibromopentane, which has two different stereocenters. They exist as the pairs of enantiomers shown on the previous page. (2*S*,3*S*)-2,3-Dibromopentane (or its enantiomer) is also a diastereomer of (2*S*,3*R*)-2,3-dibromopentane (or its enantiomer). The presence of two stereocenters gives rise to $2^2$, or 4, stereoisomers (p. 211), unless some symmetry is present to reduce the number observed. For example, $2^2$, or 4, stereoisomers would be expected for the 1,2-dibromocyclopentanes, but the symmetry of *cis*-1,2-dibromocyclopentane reduces to three the number of isomers actually observed (p. 213).

The two pentenes and bromine are achiral. The bromonium ion that results from either pentene is chiral and is, therefore, formed as a racemic mixture (p. 201). The two forms arise from addition of bromine to each face of the planar double bond.

---

**VISUALIZING THE REACTION**

**Formation of Enantiomeric Bromonium Ions from (*Z*)-2-Pentene**

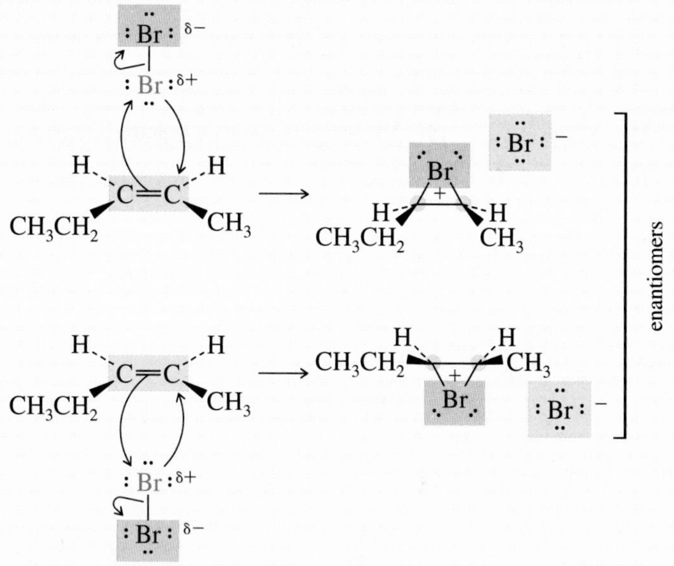

---

The enantiomeric bromonium ions are opened by attack of a bromide ion to give enantiomeric 2,3-dibromopentanes. If the bromonium ion were symmetrical, attack at either carbon atom would be equally probable. In this case, the two carbon atoms of the bromonium ion are not identical. The mechanism on page 306 shows the bromide ion attacking the less hindered of the two carbon atoms, carbon 2, of each bromonium ion.

**VISUALIZING THE REACTION**

**Reaction of Enantiomeric Bromonium Ions with Bromide Ion**

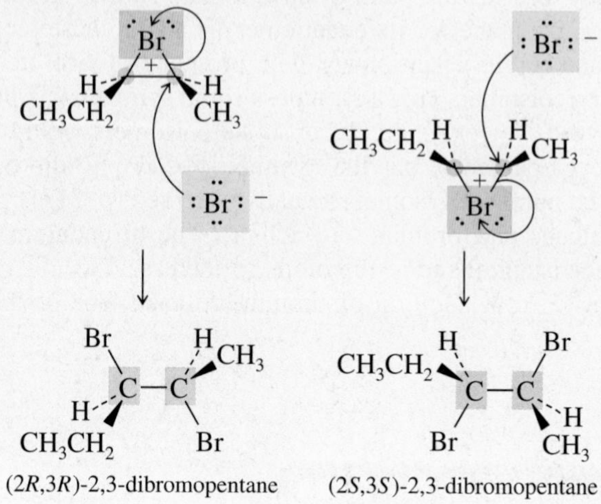

(2R,3R)-2,3-dibromopentane          (2S,3S)-2,3-dibromopentane

Attack at carbon atom 3 of each of the enantiomeric bromonium ions derived from (Z)-2-pentene will give the same mixture of enantiomeric 2,3-dibromopentanes. A similar sequence of reactions results in the enantiomers obtained from (E)-2-pentene.

**Problem 8.14**   Prove to yourself that the two statements just above are true.

**Problem 8.15**   Complete the following equations.

(a) (E)-CH$_3$CH$_2$CH=CHCH$_2$CH$_3$ $\xrightarrow[\text{tetrachloride}]{\text{carbon}}$ $\overset{\text{Br}_2}{}$

(b) (Z)-CH$_3$CH$_2$CH=CHCH$_2$CH$_3$ $\xrightarrow[\text{tetrachloride}]{\text{carbon}}$ $\overset{\text{Br}_2}{}$

(c) 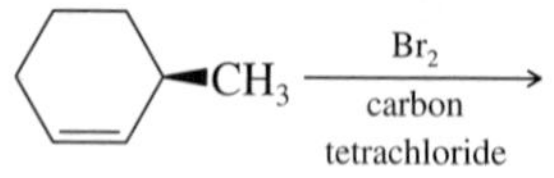 $\xrightarrow[\text{tetrachloride}]{\text{carbon}}$ $\overset{\text{Br}_2}{}$

(d) 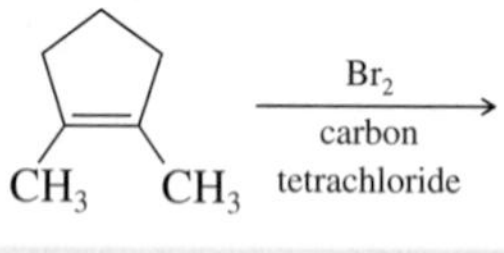 $\xrightarrow[\text{tetrachloride}]{\text{carbon}}$ $\overset{\text{Br}_2}{}$

## E. The Bromonium Ion and Nucleophiles. Halohydrins

The addition of bromine to ethylene in the presence of high concentrations of chloride ions gives 1-bromo-2-chloroethane, as well as 1,2-dibromoethane (p. 302). Chloride ion does not add to the double bond unless bromine is also present, and no 1,2-dichloroethane is obtained. Similarly, the use of water instead of an unreactive solvent for the addition reaction results in the formation of 2-bromoethanol, a halohydrin (p. 302). These facts provide additional experimental evidence for a bromonium ion as the intermediate. The intermediate bromonium ion is attacked by a nucleophile—bromide ion, chloride ion, or water—to give the final products observed.

**VISUALIZING THE REACTION**

## Reactions of Bromonium Ions with Nucleophiles

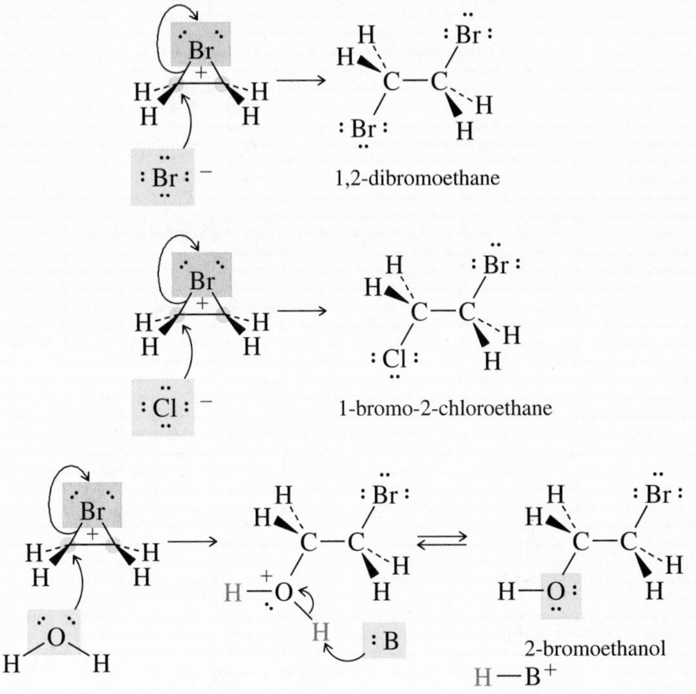

1,2-dibromoethane

1-bromo-2-chloroethane

2-bromoethanol

Halohydrins are also formed by reaction of alkenes with hypochlorous acid, HOCl, or hypobromous acid, HOBr, prepared by the acidification of solutions of sodium or calcium hypohalites. Sodium hypochlorite is the chief constituent of laundry bleach and gives a solution of hypochlorous acid on treatment with cold dilute nitric acid.

$$\text{NaOCl} + \text{HNO}_3 \xrightarrow[\text{cold}]{\text{H}_2\text{O}} \text{HOCl} + \text{NaNO}_3$$

sodium           hypochlorous
hypochlorite          acid

Hypochlorous acid adds to an alkene, such as cyclohexene, to give a trans chlorohydrin.

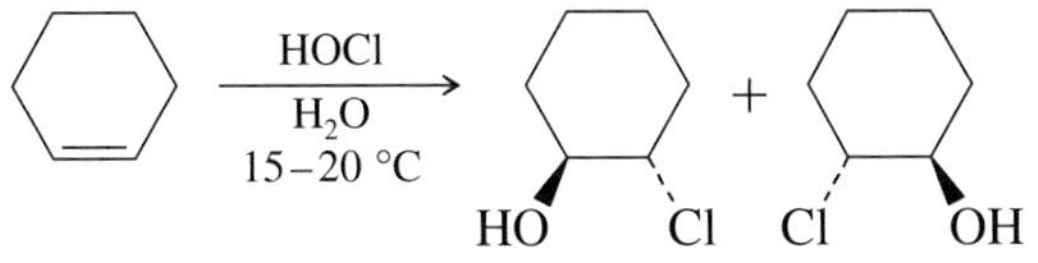

cyclohexene                     *trans*-2-chlorocyclohexanol;
chlorohydrin of cyclohexene
70%

*racemic mixture*

The reaction is believed to involve a chloronium ion intermediate that results from an attack of the electrons of the $\pi$ bond on chlorine, which is present in solutions of hypochlorous acid.

Ethylene and cyclohexene are symmetrical molecules; therefore, the same halohydrin is obtained no matter which carbon atom the halogen atom bonds to and which carbon the hydroxyl group bonds to. The reactions of unsymmetrical alkenes have also been studied. For example, 2-methylpropene reacts with hypochlorous acid to give 1-chloro-2-methyl-2-propanol.

$$\underset{\text{2-methylpropene}}{\overset{\overset{\displaystyle CH_3}{|}}{CH_3C{=}CH_2}} \quad \xrightarrow[\text{H}_2\text{O}]{\text{HOCl}} \quad \underset{\substack{\text{1-chloro-2-methyl-} \\ \text{2-propanol}}}{\overset{\overset{\displaystyle CH_3}{|}}{CH_3\underset{\underset{\displaystyle HO}{|}}{C}{-}\underset{\underset{\displaystyle Cl}{|}}{CH_2}}}$$

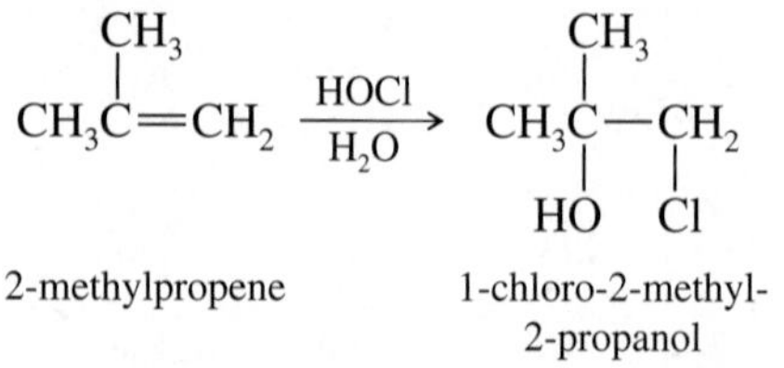

Similarly, when styrene (phenylethene) is treated with a solution of chlorine in water, 2-chloro-1-phenylethanol is observed.

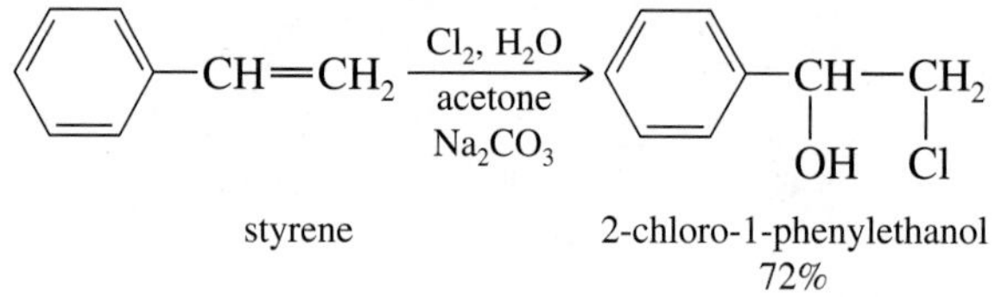

styrene → 2-chloro-1-phenylethanol 72%

Reagents: Cl₂, H₂O, acetone, Na₂CO₃

The products observed for these reactions can be explained by proposing that the intermediate chloronium ion is unsymmetrical, with considerable cationic character at the more highly substituted carbon atom.

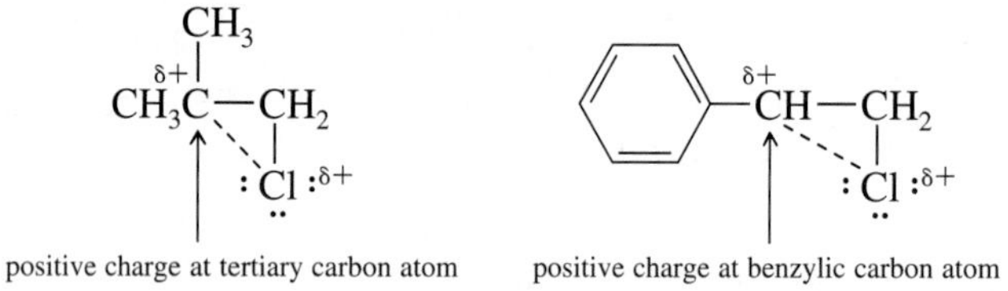

positive charge at tertiary carbon atom      positive charge at benzylic carbon atom

*unsymmetrical chloronium ion intermediates*

The nucleophile, water in these examples, reacts with the relatively stable unsymmetrical intermediate, and the product obtained has the hydroxyl group on the carbon atom that had the partial positive charge in the intermediate.

Why couldn't the intermediate in such reactions be an ordinary carbocation with no bridging to the halogen atom? Evidence for a halonium ion intermediate comes from the stereochemistry that results from reactions in which halohydrins are formed. An example involving a bromonium ion intermediate is as follows. A brominated amide, *N*-bromosuccinimide, is often used as the source of electrophilic bromine.

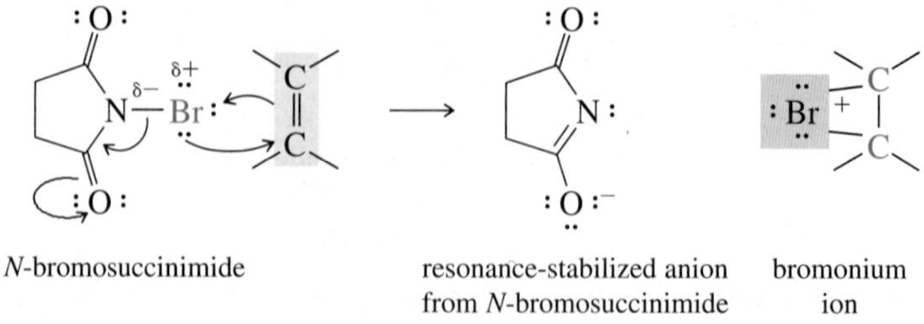

*N*-bromosuccinimide      resonance-stabilized anion      bromonium
                          from *N*-bromosuccinimide        ion

For example, when treated with *N*-bromosuccinimide in aqueous dimethyl sulfoxide, (*E*)-1-phenyl-1-propene gives a racemic mixture of (1*R*,2*S*)- and (1*S*,2*R*)-2-bromo-1-phenyl-1-propanol.

(E)-1-phenyl-1-propene

(1R,2S)-2-bromo-
1-phenyl-1-propanol

(1S,2R)-2-bromo-
1-phenyl-1-propanol

$C_6H_5-$  ≡

The stereochemistry of the products shows clearly that anti addition to the double bond has taken place. The intermediate, therefore, must be more like a bromonium ion than like a planar benzylic carbocation, which would be able to react with water on either side.

**Problem 8.16**   Write a detailed mechanism showing an unsymmetrical bromonium ion and one showing a planar carbocation as the intermediate for the formation of the halohydrins from (E)-1-phenyl-1-propene. Prove to yourself that the stereochemistry that is observed experimentally would not be possible with a planar carbocation as the intermediate.

**Problem 8.17**   2-Methyl-1-phenyl-1-propene, when it is treated with N-bromosuccinimide and water in dimethyl sulfoxide, gives a mixture of two halohydrins in roughly equal amounts. What are the structures of the products? What accounts for the lack of regioselectivity in this case?

**On the Web: ONE SMALL STEP**

A proton is small in size and has no nonbonding electrons, so it does not usually form bridged cations the way bromine does. Carbocations formed when acids add to alkenes therefore give more rearranged products (p. 289) and show less stereoselectivity than bromonium ions do (p. 304).

**Study Guide**
Concept Map 8.5

# 8.9   Reactions of Alkenes with Oxygen Electrophiles

## A. Ozone and the Ozone Layer

**Ozone,** $O_3$, is formed from oxygen, $O_2$, in the presence of ultraviolet light or an electric discharge. It is also produced when compounds in the exhausts from gasoline engines interact with oxygen in the presence of sunlight.

In the high reaches of the atmosphere, the stratosphere, ozone is constantly being formed from oxygen by the absorption of ultraviolet radiation from the sun. In an equilibrium process, ozone itself absorbs another wavelength of light from the sun and breaks down into oxygen.

$$O_2 \xrightarrow[h\nu]{} 2\,O$$

oxygen   140 nm   oxygen
molecule            atoms

$$O + O_2 \longrightarrow O_3 \quad \text{formation of ozone}$$

$$O_3 \xrightarrow[260\,nm]{h\nu} O + O_2$$

$$O + O_3 \longrightarrow O_2 + O_2$$

breakdown of ozone

These two processes, when undisturbed, maintain a steady concentration of ozone in the stratosphere. This is important because ozone absorbs sunlight in the same region of the spectrum as DNA does. When DNA absorbs radiation at approximately 280 nm, changes occur that bind adjacent thymine bases in DNA together (p. 1087), thereby giving rise to changes in the genetic code and destroying the cell. For this reason, ultraviolet radiation can be used to kill bacteria and sterilize equipment.

In the last twenty-five years, scientists have become increasingly concerned as measurements of ozone in the stratosphere have shown an overall net decrease, especially over the polar regions. For example, 70% of the ozone over Antarctica was lost during a six-week period in September and October of 1989, a time period corresponding to the coming of spring and the increase of sunlight on that continent. Careful monitoring of ozone over the Arctic in the winter of 1997 showed that ozone was being lost over significant areas of the northern hemisphere as well. Such a decrease in the ozone layer is expected to lead to increased damage to all living organisms because of increased ultraviolet damage to DNA. In fact, scientists have invented a gauge to measure how much of the harmful portion of the ultraviolet radiation of the sun reaches the earth by measuring how much DNA in a standard sample undergoes the thymine dimerization reaction. For human beings, the concern is that additional ultraviolet radiation will increase the rates of skin cancer. The effects on plants and animals, and on the vast array of organisms from very small to very large that live in the oceans, is not known yet. The chemistry of how human activity impacts on the ozone layer is discussed in Section 19.1E.

While ozone in the stratosphere is beneficial, ozone at ground level in smog is not. Ozone is a highly reactive species. It causes irritation of mucous membranes in the eyes and lungs. It is also especially damaging to anything made of rubber, which is a polymer containing many double bonds (p. 1073). The reaction of ozone with alkenes is the subject of the next section.

## B.  Reaction of Alkenes with Ozone. Ozonolysis

When ozone, $O_3$, reacts with an alkene, the carbon–carbon double bond is broken and a carbon–oxygen double bond is formed on each fragment of the molecule. The overall reaction is called an **ozonolysis reaction,** meaning a cleavage (*lysis* means "to loosen") of bonds by ozone. For example, 2-methylpropene is converted to acetone and formaldehyde by ozone.

cleavage of double bond; formation of carbonyl groups

*an ozonolysis reaction*

The ozone molecule, which cannot be symbolized satisfactorily by a single Lewis structure, is represented by a set of resonance contributors.

*resonance contributors of ozone*

Ozone is an electrophile that adds to an alkene to give an unstable cyclic compound called a **molozonide.** The molozonide decomposes to a charged intermediate and a carbonyl compound held close together in a solvent cage. If the carbonyl fragment is an aldehyde, the two pieces recombine to form an **ozonide.** These reactions are illustrated for 2-methylpropene, which was studied extensively by Rudolf Criegee in the middle of the twentieth century to establish the mechanism shown below.

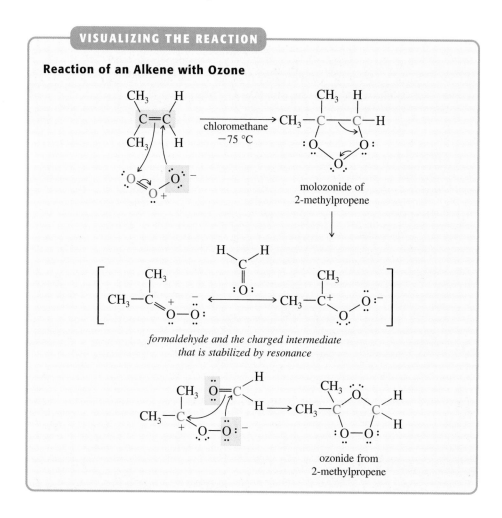

**VISUALIZING THE REACTION**

**Reaction of an Alkene with Ozone**

molozonide of
2-methylpropene

*formaldehyde and the charged intermediate
that is stabilized by resonance*

ozonide from
2-methylpropene

Ozonides and peroxides are unstable compounds that break up in water to give carbonyl compounds. The products obtained from an ozonolysis reaction depend on the conditions used. For example, 1-octene gives formic acid and heptanoic acid when an oxidizing agent such as hydrogen peroxide is present. Such conditions are called an **oxidative work-up** of the reaction mixture.

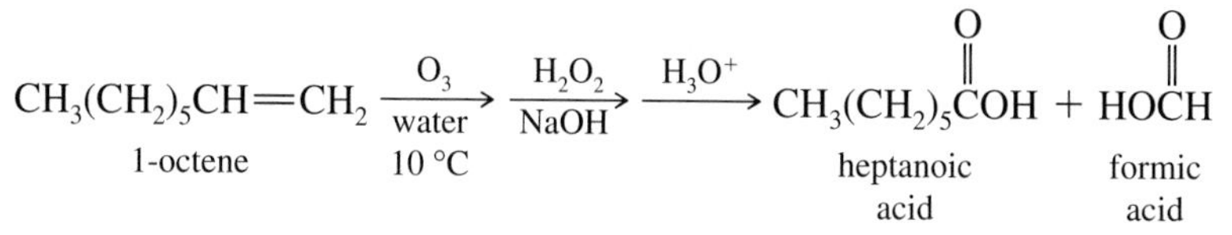

$$CH_3(CH_2)_5CH{=}CH_2 \xrightarrow[\substack{water \\ 10\ °C}]{O_3} \xrightarrow[NaOH]{H_2O_2} \xrightarrow{H_3O^+} CH_3(CH_2)_5\overset{\overset{\displaystyle O}{\|}}{C}OH + HO\overset{\overset{\displaystyle O}{\|}}{C}H$$

1-octene                             heptanoic     formic
                                       acid         acid

*oxidative work-up of ozonolysis mixture*

If a reducing agent such as zinc or dimethyl sulfide is used, in what is called a **reductive work-up,** the products are aldehydes.

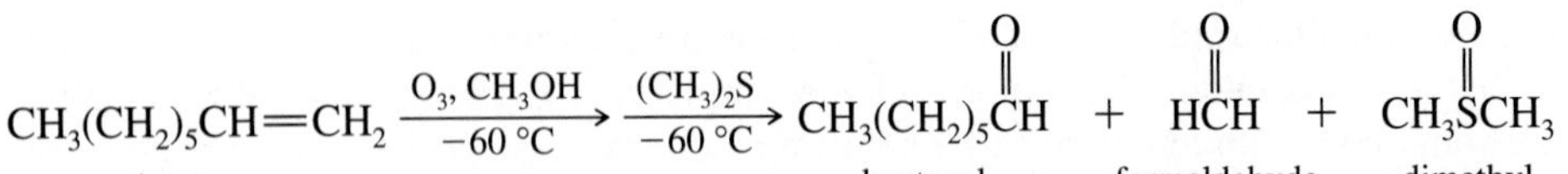

$$CH_3(CH_2)_5CH{=}CH_2 \xrightarrow[-60\ °C]{O_3,\ CH_3OH} \xrightarrow[-60\ °C]{(CH_3)_2S} CH_3(CH_2)_5\overset{\overset{\displaystyle O}{\|}}{C}H \ + \ H\overset{\overset{\displaystyle O}{\|}}{C}H \ + \ CH_3\overset{\overset{\displaystyle O}{\|}}{S}CH_3$$

1-octene

heptanal          formaldehyde          dimethyl
                                        sulfoxide

*oxidation*
*product of*
*dimethyl*
*sulfide*

*reductive work-up of ozonolysis mixture*

Dimethyl sulfide acts as a reducing agent because it reacts with oxygen to form a stable compound, dimethyl sulfoxide, and thereby prevents oxidation of the aldehyde products to carboxylic acids.

For the reaction of ozone with cyclohexene, different products are also obtained by changing the reaction conditions. Cyclohexene is a cyclic alkene. Thus its reaction with ozone, when followed by treatment with dimethyl sulfide to decompose the ozonide, gives a compound in which two aldehyde groups are still attached to each other by a chain of carbon atoms that was the rest of the ring.

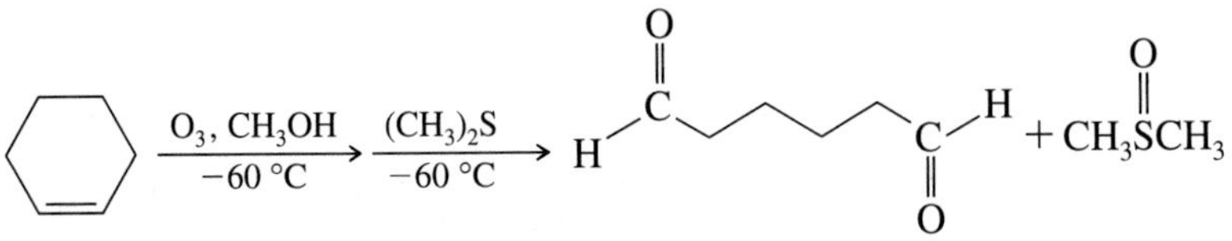

cyclohexene          *reductive*          1,6-hexanedial          dimethyl
                     *work-up*            62%                     sulfoxide

*dialdehyde from*
*cleavage of the double*
*bond in cyclohexene*
*by ozone*

If hydrogen peroxide is used after the addition of ozone, the corresponding dicarboxylic acid is formed.

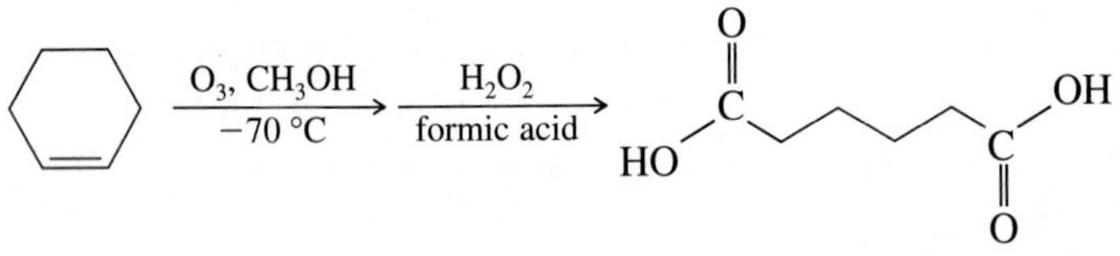

cyclohexene          *oxidative*          1,6-hexanedioic acid
                     *work-up*            adipic acid
                                          85%

Thus, if a reducing agent is not used in the decomposition of an ozonide, the aldehyde products are oxidized to carboxylic acids. In fact, ozone is used to synthesize carboxylic acids by cleavage of multiple bonds (p. 609).

If there are two organic groups as substituents on one of the doubly bonded carbon atoms in the alkene, a product of the reaction with ozone is a ketone. 2-Methylpropene, for example, gives acetone on treatment with ozone (p. 310). A ketone, in contrast to an aldehyde, cannot be easily oxidized. In order to add any more oxygen atoms to the carbonyl group, a carbon–carbon single bond would have to be broken.

Reactions that result in the cleavage of bonds are used to convert large molecules into smaller, more easily identifiable fragments. Such reactions are known as **degradation reactions.** Degradation reactions have been particularly important in determining the structures of complex molecules isolated from natural sources. Ozonolysis is a degradation reaction. The compounds formed as a result of ozonolysis—aldehydes, ketones, and carboxylic acids—have smaller molecules than the starting alkene as well as reactive functional groups. They are, therefore, more easily identified than the starting alkene. Knowing what the fragments are makes it possible to reconstruct the original molecule. For example, a chemist who identified heptanal and formaldehyde as ozonolysis products from an alkene of unknown structure would be able to conclude that the alkene was 1-octene.

$$CH_3(CH_2)_5 \diagdown C=O \qquad O=C \diagup{}^{H}_{H} \xleftarrow{\text{ozonolysis}} CH_3(CH_2)_5 \diagdown{}^{}_{H} C=C \diagup{}^{H}_{H}$$

$\qquad$ heptanal $\qquad\qquad$ formaldehyde $\qquad\qquad\qquad$ 1-octene

The structure of the alkene is easily arrived at by mentally removing the oxygen atoms and joining the carbon atoms of the two carbonyl groups with a double bond.

**On the Web: ONE SMALL STEP**

**Problem 8.18** Complete the following equations.

(a) $CH_3\overset{\overset{\displaystyle CH_3}{|}}{C}=CHCH_3 \xrightarrow{O_3} \xrightarrow[\text{water}]{Zn}$

(b) $CH_3\overset{\overset{\displaystyle CH_3}{|}}{C}HCH=\overset{\overset{\displaystyle CH_2CH_3}{|}}{C}CH_2CH_3 \xrightarrow{O_3, CH_3OH} \xrightarrow{(CH_3)_2S}$

(c) $CH_3\overset{\overset{\displaystyle CH_3}{|}}{C}HCH=\overset{\overset{\displaystyle CH_2CH_3}{|}}{C}CH_2CH_3 \xrightarrow[\text{water}]{O_3} \xrightarrow{H_2O_2, NaOH} \xrightarrow{H_3O^+}$

(d)  $\xrightarrow{O_3, CH_3OH} \xrightarrow{(CH_3)_2S}$

(e) $\xrightarrow[\substack{\text{dichloromethane} \\ 0\,°C}]{O_3} \xrightarrow{H_2O_2, NaOH}$

**Problem 8.19** In early research into the structures of branched hydrocarbons, chemists determined the structures of a number of alkenes by ozonolysis. What products would you expect to get from the ozonolysis of a mixture of alkenes consisting of 90% 3-ethyl-5,5-dimethyl-2-hexene and 10% 4-ethyl-2,2-dimethyl-3-hexene? The reaction mixture was treated with zinc and water.

## C. Oxidation of Alkenes with Osmium Tetroxide

The oxidizing agent osmium tetroxide, $OsO_4$, reacts with alkenes to give *cis*-diols. Osmium tetroxide is expensive and highly toxic, so it is often used in catalytic amounts along with another oxidizing agent that keeps recycling the osmium back to the tetroxide stage. For example, cyclohexene is converted to *cis*-cyclohexane-1,2-diol in this way. Osmium tetroxide is shown as being present in catalytic amounts. It is reoxidized by *N*-methylmorpholine-*N*-oxide.

*N*-methylmorpholine-*N*-oxide

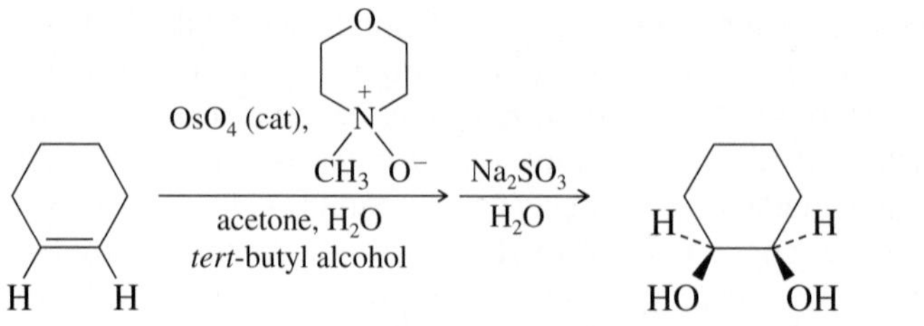

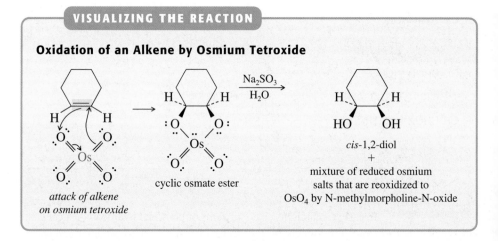

cyclohexene

*cis*-cyclohexane-1,2-diol
91%

The reaction takes place in two steps. The osmium tetroxide adds to one face of the double bond in a syn addition (p. 296) to give a cyclic osmate ester, which is then cleaved by the addition of water and a reducing agent. Sodium sulfite, $Na_2SO_3$, is the reducing agent used in the example shown above, but others such as sodium bisulfite, $NaHSO_3$, sodium thiosulfate, $Na_2S_2O_3$, and sodium metabisulfite, $Na_2S_2O_5$, are also used. The oxygen atoms that end up in the diol come from the osmium tetroxide and not from the water used in the cleavage of the osmate ester. This was shown in other experiments when water containing oxygen-18 was used in the work-up of the reaction, and no isotopic oxygen was found in the diols that formed.

The formation of the cyclic osmate ester is the important step of the reaction and the one that determines the stereochemistry of the reaction.

**VISUALIZING THE REACTION**

**Oxidation of an Alkene by Osmium Tetroxide**

attack of alkene
on osmium tetroxide

cyclic osmate ester

*cis*-1,2-diol
+
mixture of reduced osmium salts that are reoxidized to $OsO_4$ by N-methylmorpholine-N-oxide

The structure of this type of cyclic osmate ester was captured by x-ray crystallography in research into the reactions of buckyball (p. 364; see also Figure 10.6, p. 365).

Other oxidizing agents, such as peroxides, are also used in osmium tetroxide oxidations. For example, (*E*)-4-octene is converted into a racemic mixture of (4*R*,5*R*)-octane-1,2-diol and (4*S*,5*S*)-octane-1,2-diol by a catalytic amount of osmium tetroxide, recycled by *tert*-butyl peroxide.

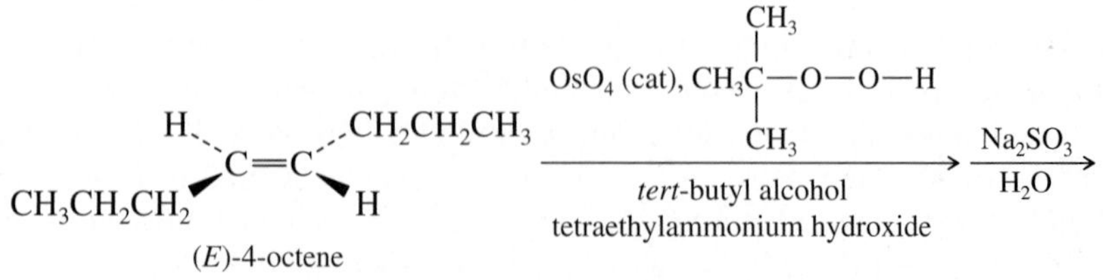

(*E*)-4-octene

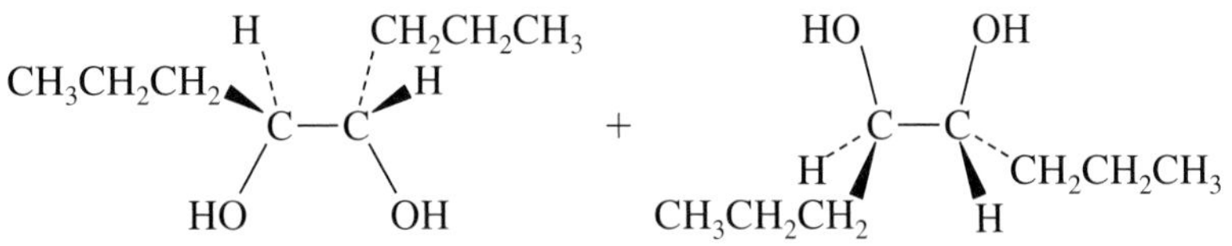

(4S,5S)-octane-4,5-diol                    (4R,5R)-octane-4,5-diol

**Problem 8.20**    In the reaction shown above, the products were drawn in the eclipsed conformation to emphasize the stereochemistry that results from the syn addition of the osmium tetroxide. Redraw the *R*,*R*-isomer in a more stable conformation.

**Problem 8.21**    (*Z*)-4-Octene also has been converted to an octane-4,5-diol under the same conditions used for the *E*-isomer. Show this reaction, and explore the stereochemistry of the product(s) that form(s) in this case.

The stereoselectivity of the osmium tetroxide reaction can be further increased if the reaction is carried out in the presence of chiral amines that coordinate with the osmium atom. The structure in Figure 10.6 (p. 364) shows two 4-*tert*-butylpyridine molecules coordinated with osmium; this amine is not chiral and would not, therefore, change the stereochemical course of the reaction. The transition state corresponding to the approach of the top face of the alkene to osmium tetroxide is the mirror image of the transition state resulting from the approach of the bottom face of the alkene to the reagent. The enantiomeric transition states are of equal energy, so the two enantiomeric products form in equal amounts, giving rise to racemic mixtures (p. 201). For example, when (*E*)-1,2-diphenylethene (stilbene) is oxidized with osmium tetroxide in the presence of pyridine, a racemic mixture of (1*R*,2*R*)- and (1*S*,2*S*)-1,2-diphenylethane-1,2-diols forms.

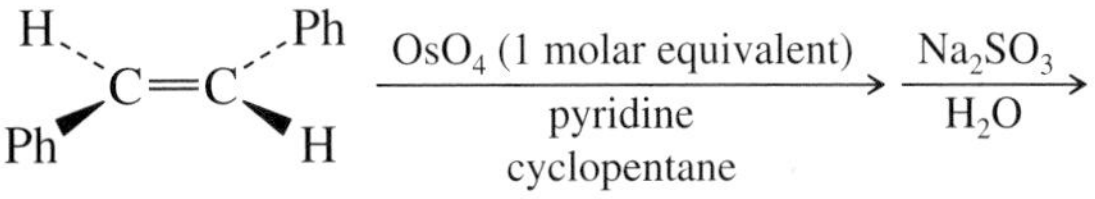

(1S,2S)-1,2-diphenylethane-1,2-diol    (1R,2R)-1,2-diphenylethane-1,2-diol

Note that this is an example where a full equivalent of osmium tetroxide is used. There is no other oxidizing agent.

K. Barry Sharpless shared the Nobel Prize in Chemistry in 2001 for his exploration of oxidation reactions catalyzed by chiral reagents to give product mixtures that are richer in one enantiomer than the other. In particular, he used compounds related to the alkaloid quinine (p. 805). When the chiral amine complexes with osmium tetroxide, the reagent as a whole becomes chiral. The transition state for the approach of one face of the alkene to the reagent is now the diastereomer of the transition state resulting from approach of the other face of the alkene. The two transition states are of different energies; one is favored over the other and forms faster. Most of the product comes from that transition state. For example, if

(*E*)-1,2-diphenylethene is oxidized with osmium tetroxide in the presence of a quinine derivative, the product is predominantly the *S,S* product.

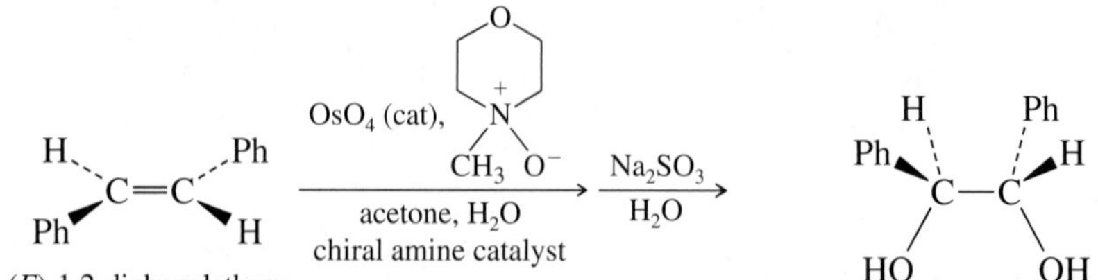

(*E*)-1,2-diphenylethene

(1*S*,2*S*)-1,2-diphenylethane-1,2-diol
in 80% enantiomeric excess

An 80% enantiomeric excess means that 90% of the product is the enantiomer that is shown and 10% is the other enantiomer (p. 202). Professor Sharpless shared the Nobel Prize with two other chemists, William S. Knowles and Ryoji Nyori, both of whom also had developed chiral catalysts, in their case, for catalytic hydrogenation reactions (p. 298). This work is important because the synthesis of stereochemically pure compounds for use especially as medications is becoming more and more essential.

**Problem 8.22**   Complete the following equations, predicting the products indicated by letters and showing stereochemistry wherever possible.

(a) 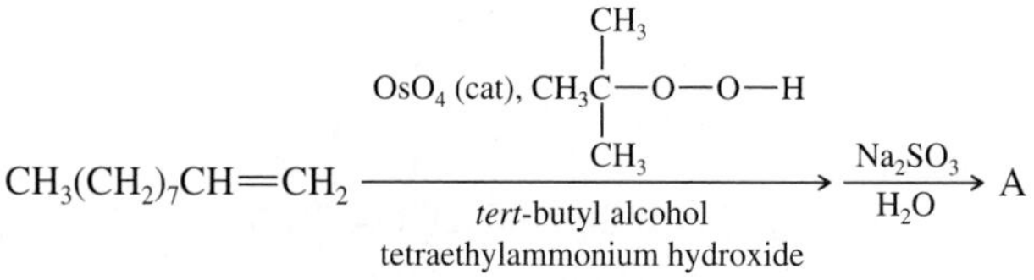  $CH_3(CH_2)_7CH{=}CH_2$

(b)

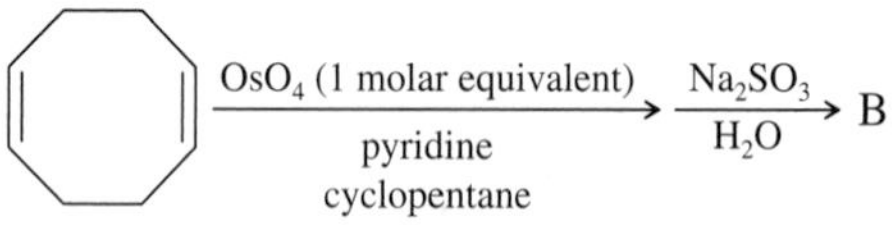

(c)

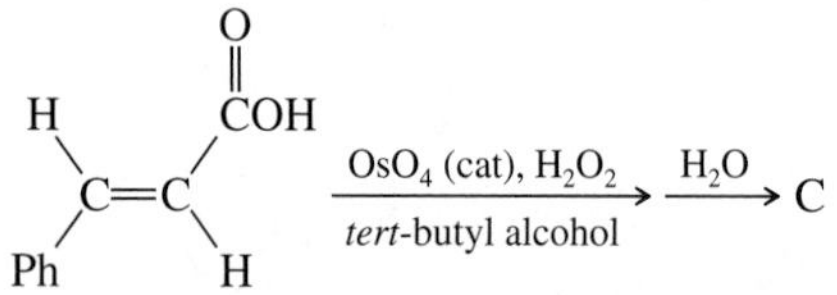

(*Hint:* In this case, what is the oxidizing agent used to reoxidize the catalytic $OsO_4$?)

## D.  Oxidation of Alkenes with Peroxyacids

Carboxylic acids have the general formula

$$\overset{\overset{\displaystyle O}{\|}}{RCOH}$$

and may be regarded as being derived from water, HOH, by the replacement of one of the hydrogen atoms by an acyl group,

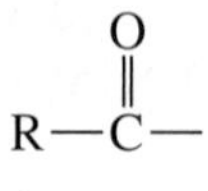

$$\overset{\overset{\displaystyle O}{\|}}{R{-}C{-}}$$

Carboxylic acids have an acidic hydrogen atom and act as Brønsted acids. They are *not* oxidizing agents.

**Peroxycarboxylic acids** have the general formula

$$R-\overset{\overset{\displaystyle O}{\|}}{C}-O-O-H$$

and may be viewed as being related to the oxidizing agent hydrogen peroxide, HOOH, in the same way that carboxylic acids are related to water. Replacement of one of the hydrogen atoms in hydrogen peroxide by an acyl group gives the formula for a peroxycarboxylic acid. Peroxyformic acid, for example, is an unstable compound that can be made by mixing formic acid with hydrogen peroxide.

$$\underset{\text{formic acid}}{\overset{\overset{\displaystyle O}{\|}}{HCOH}} + \underset{\substack{\text{hydrogen}\\\text{peroxide}}}{HOOH} \rightleftharpoons \underset{\substack{\text{peroxyformic}\\\text{acid}}}{\overset{\overset{\displaystyle O}{\|}}{HCOOH}} + \underset{\text{water}}{H_2O}$$

Peroxyacids are oxidizing agents and react with alkenes to give three-membered cyclic ethers known as **epoxides** or **oxiranes.** These small ring compounds are quite reactive; the ring opens easily, especially in acidic solutions. These reactions are considered in more detail in Section 13.5. This section examines the formation of oxiranes from alkenes using *m*-chloroperoxybenzoic acid as the reagent.

*m*-Chloroperoxybenzoic acid is sold commercially and is stable enough that it can be stored and used over a period of time. Its reactions are carried out in organic solvents, such as ethers or halogenated hydrocarbons, so the solubility of starting materials and products is not a problem. The oxiranes that are formed are stable in the reaction mixture and are isolated in reasonably good yields.

The reaction of *m*-chloroperoxybenzoic acid with alkenes is stereoselective. Stereoisomeric alkenes give the corresponding oxiranes with virtually no change in stereochemistry. For example, (*Z*)-2-butene is oxidized to *cis*-2,3-dimethyloxirane, and (*E*)-2-butene gives *trans*-2,3-dimethyloxirane with well over 99% stereochemical purity.

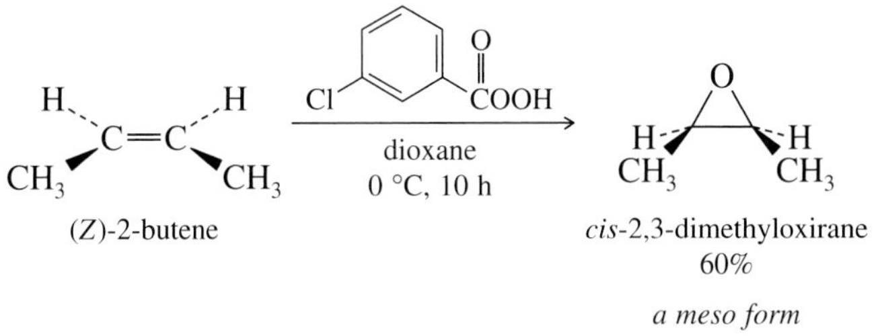

(*Z*)-2-butene

*cis*-2,3-dimethyloxirane
60%

*a meso form*

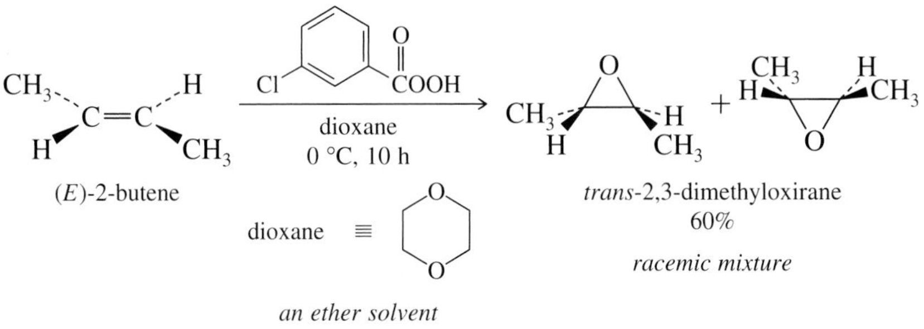

(*E*)-2-butene

dioxane ≡

*an ether solvent*

*trans*-2,3-dimethyloxirane
60%

*racemic mixture*

*cis*-2,3-Dimethyloxirane is not optically active because it is a meso form; it has a plane of symmetry bisecting the oxygen atom and the carbon–carbon bond of the oxirane ring. *trans*-2,3-Dimethyloxirane is formed as a racemic mixture; the two enantiomers result because there is an equal probability of attack from above and below the plane of the double bond.

An oxygen atom is transferred from the peroxyacid to the double bond, giving the cyclic ether. The other product of the reaction is *m*-chlorobenzoic acid, no longer a peroxyacid. The mechanism of the reaction involves a single-step transfer of the peroxy oxygen atom to the double bond.

## VISUALIZING THE REACTION

**Reaction of a Peroxyacid with an Alkene**

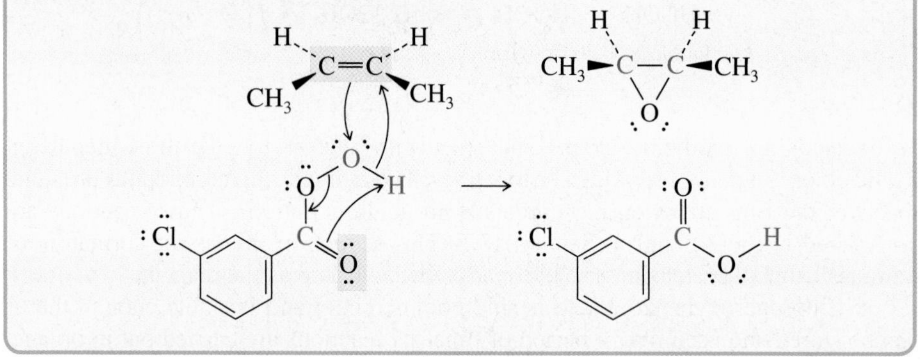

The peroxyacid behaves as an electrophile in the reaction shown above. Its electrophilic character is revealed in the selectivity with which it reacts with double bonds having different degrees of substitution. The more alkyl substituents on a double bond, the more readily it is attacked by the peroxyacid. This fact suggests that higher electron density in a $\pi$ bond favors the reaction with peroxyacids. 1,2-Dimethyl-1,4-cyclohexadiene, for example, reacts selectively with one molar equivalent of *m*-chloroperoxybenzoic acid at the more highly substituted of the two double bonds.

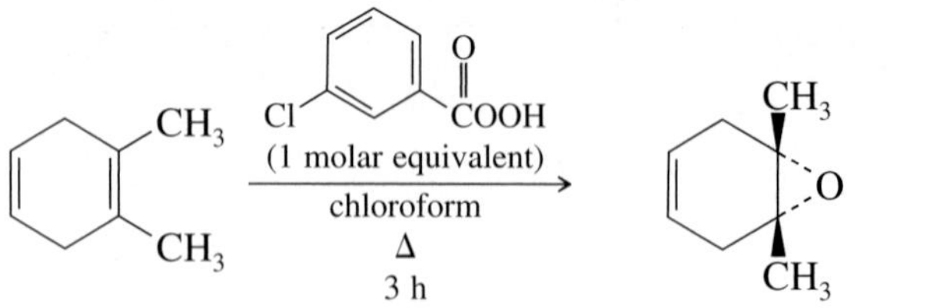

1,2-dimethyl-1,4-cyclohexadiene

1,2-dimethyl-1,2-epoxy-4-cyclohexene

Study Guide
Concept Map 8.6

This observed result demonstrates the effect that alkyl groups have in increasing the availability of electrons in the functionalities to which they are attached.

**Problem 8.23**    Predict the major product of each of the following reactions. Write a structure showing the stereochemistry when it is pertinent.

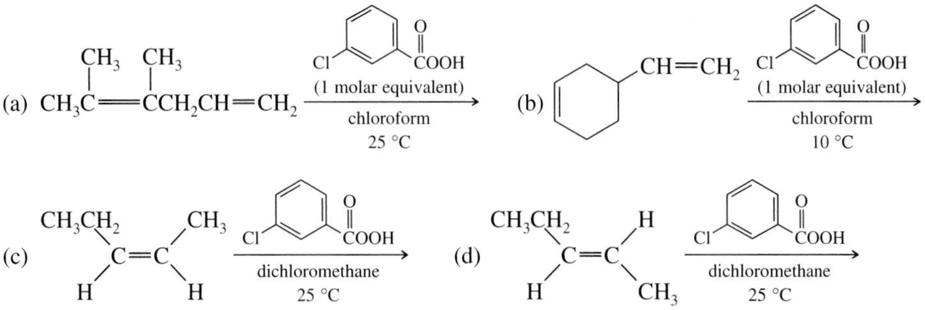

(a)

(b)

(c)

(d)

## E. The Art of Solving Problems

**PROBLEM:** *meso*-2,3-Butanediol is prepared from (Z)-2-butene. Suggest a reagent that can be used for this transformation.

### Solution

1. What are the connectivities of the two compounds? How many carbon atoms does each contain? Are there any rings? What are the positions of branches and functional groups on the chains?

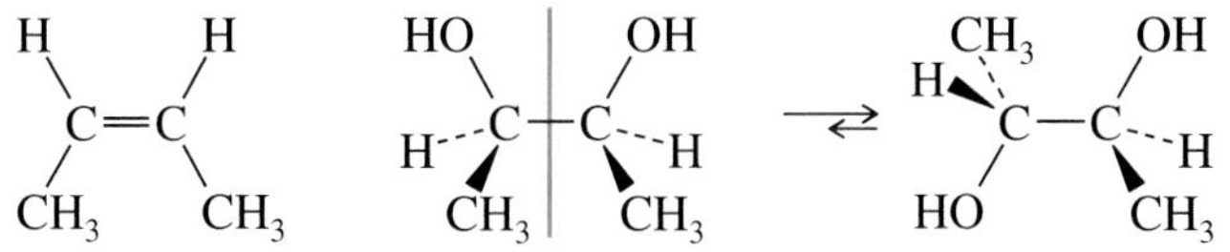

The carbon skeletons of the two compounds are identical. The two carbon atoms that are joined by a double bond in the starting material have hydroxyl groups on them in the product. The methyl groups in the starting alkene and in the eclipsed conformation of the resulting diol are on the same side of the two carbon atoms to which they are bonded.

2. How do the functional groups change in going from starting material to product? Does the starting material have a good leaving group?

The double bond disappears, and the two carbon atoms of the double bond end up with hydroxyl groups on them. There is no leaving group present in the starting material.

3. Is it possible to dissect the structures of the starting material and product to see which bonds must have been broken and which formed?

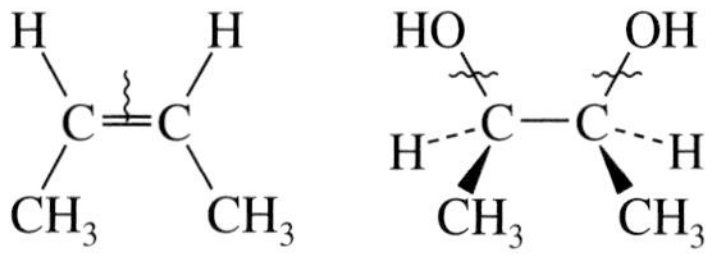

4. New bonds are created when an electrophile reacts with a nucleophile. Do we recognize any part of the product molecule as coming from a good nucleophile or an electrophilic addition?

A diol results from reaction of an alkene with an oxidizing agent (an electrophile) such as osmium tetroxide. The reaction of osmium tetroxide is a syn addition and therefore will give the desired stereochemistry.

5. What type of compound would be a good precursor to the product?

No other precursor is needed. The starting material is converted directly to the product.

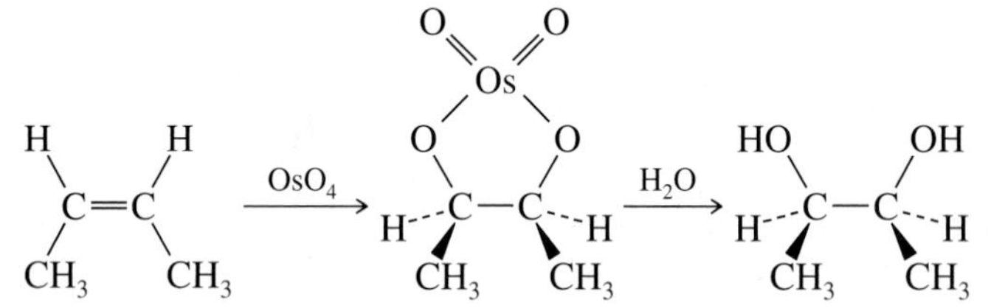

**Problem 8.24**    How would you synthesize racemic 2,3-butanediol from (*E*)-2-butene?

## ADDITIONAL PROBLEMS

**8.25** Name each of the following compounds, assigning the correct configuration to any for which stereochemistry is shown.

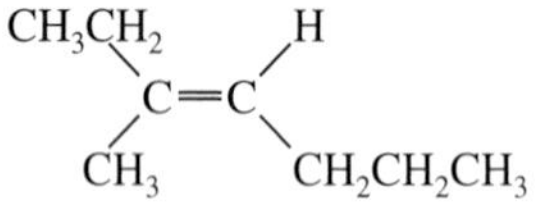

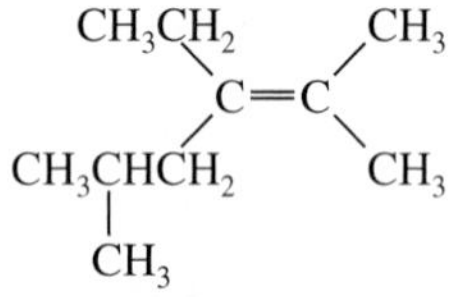

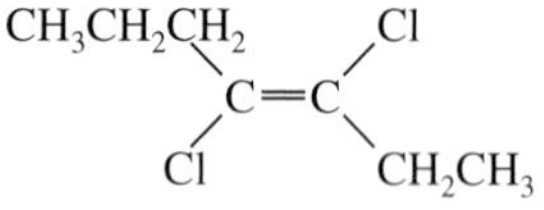

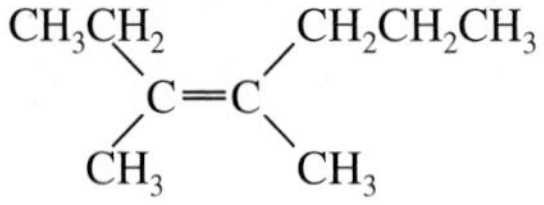

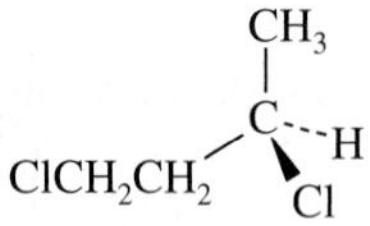

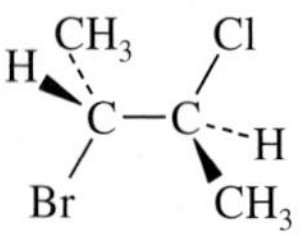

**8.26** Draw a structural formula for each of the following compounds. Be sure to show stereochemistry when it is indicated.

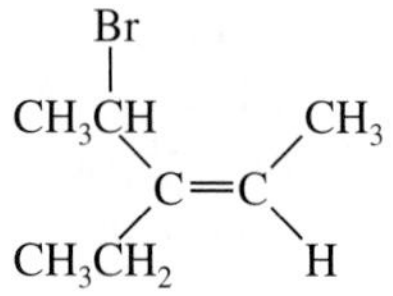

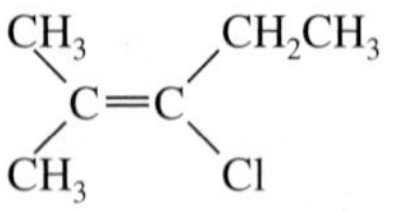

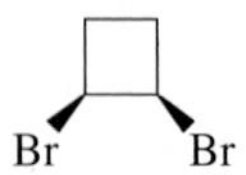

(a) (*Z*)-5-chloro-2-pentene    (b) *trans*-1,2-dimethylcyclopropane    (c) *meso*-2,3-dibromobutane
(d) 3-chloro-2-methyl-1-pentene    (e) *cis*-1,2-dichlorocyclopentane    (f) 1,2-dimethylcyclohexene

**8.27** Draw structural formulas for the stereoisomers of the following compounds.

(a) 3-hexene    (b) 1,2-dimethylcyclopropane    (c) 1-bromo-2-methylcyclopentane    (d) 2-chloropentane
(e) 3,4-dimethylhexane    (f) 2-bromo-3-methylpentane    (g) 5-bromo-2-hexene

## Table 8.1 Summary of the Reactions of Alkenes

### Electrophilic Addition Reactions

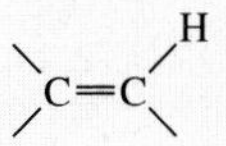

| Alkene | Electrophile | Intermediate | Reagent in Second Step | Product(s) | Stereochemistry |
|---|---|---|---|---|---|
| $\text{C=C}$ (with H) | HX | carbocation $-\overset{H}{\underset{+}{C}}-\overset{H}{\underset{H}{C}}-$ | $X^-$ | $-\overset{H}{\underset{X}{C}}-\overset{H}{\underset{H}{C}}-$ | |
| | $H_3O^+$ | | $H_2O$ | 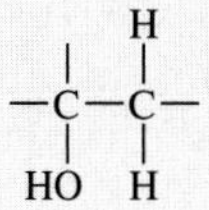 $-\overset{H}{\underset{HO}{C}}-\overset{H}{\underset{H}{C}}-$ | |
| | $X_2$ | 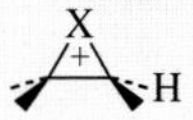 halonium ion (X, H) | $X^-$, ROH | 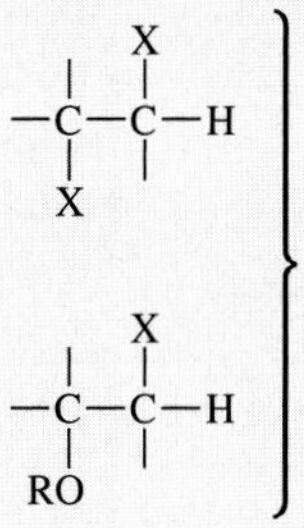 $-\overset{X}{\underset{X}{C}}-\overset{}{\underset{}{C}}-H$  and  $-\overset{X}{\underset{RO}{C}}-\overset{}{\underset{}{C}}-H$ | anti addition |
| | $BH_3$ | $-\overset{H}{\underset{H}{C}}-\overset{H}{\underset{BR_2}{C}}-$ | $H_2O_2$, $OH^-$ | 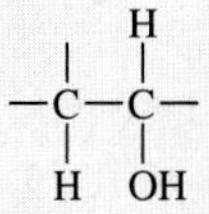 $-\overset{H}{\underset{H}{C}}-\overset{H}{\underset{OH}{C}}-$ | syn addition |

### Oxidation Reactions

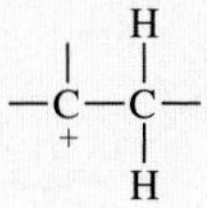

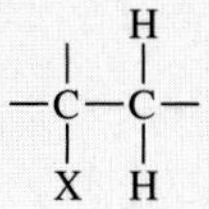

| Alkene | Electrophile | Intermediate | Reagent in Second Step | Product(s) | Stereochemistry |
|---|---|---|---|---|---|
| | $O_3$ | 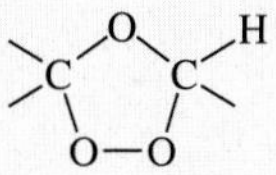 ozonide (C–O–O–C, O) | Zn, $H_2O$ or $CH_3SCH_3$ (reductive work-up) | 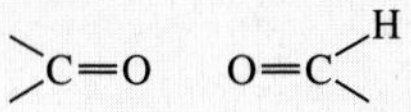 $\text{C=O}$   $O=\text{C}$ (with H) | |
| | | | $H_2O_2$ (oxidative work-up) | 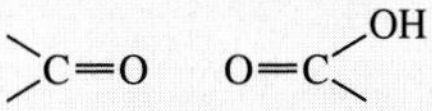 $\text{C=O}$   $O=\text{C}$ (with OH) | |
| | $OsO_4$ | 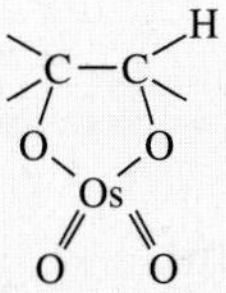 cyclic osmate ester | $Na_2SO_3$ $H_2O$ | 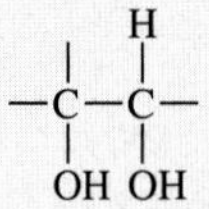 $-\overset{H}{\underset{OH}{C}}-\overset{}{\underset{OH}{C}}-$ | syn addition |
| | 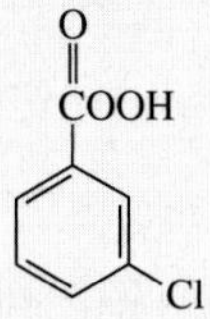 *m*-chloroperoxybenzoic acid (COOH, Cl) | none observed | none |  epoxide (O, H) | syn addition |

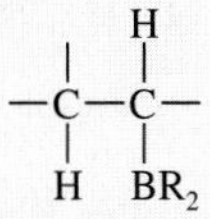

### Reduction Reaction

| Alkene | Electrophile | Intermediate | Reagent in Second Step | Product(s) | Stereochemistry |
|---|---|---|---|---|---|
| | $H_2$, $PtO_2$ | none observed | none | 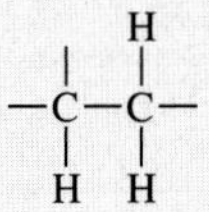 $-\overset{H}{\underset{H}{C}}-\overset{H}{\underset{H}{C}}-$ | syn addition |

**8.28** A report in *Nature* contains the astonishing news that female moths and female elephants use the same ester as a sex pheromone (a chemical used in communication) to signal their readiness to mate. The ester has the structure shown below.

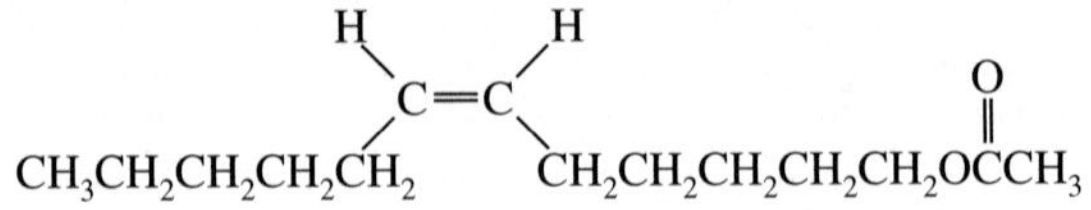

The alcohol corresponding to this ester has the structure shown below. What is the correct IUPAC name for this alcohol?

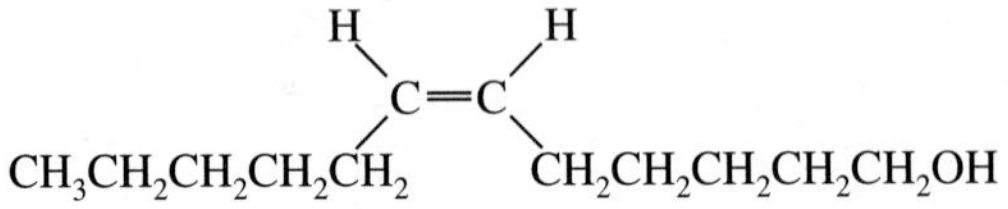

**8.29** Write equations predicting the reaction of (*E*)-3-methyl-2-pentene with each of the following reagents. In cases where you know what the stereochemistry of the product(s) should be, illustrate it with a three-dimensional representation.

(a) $H_2$, Pt    (b) $H_2O$, $H_2SO_4$    (c) HBr    (d) 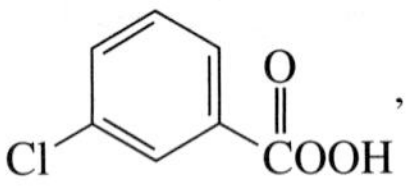 , chloroform

(e) (1) $O_3$, $CH_3OH$; (2) $(CH_3)_2S$    (f) (1) $OsO_4$, 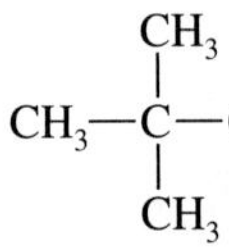 ; (2) $Na_2SO_3$, water

(g) (1) $O_3$, water; (2) $H_2O_2$, NaOH; (3) $H_3O^+$    (h) $Br_2$, carbon tetrachloride    (i) $Br_2$, $H_2O$

**8.30** Complete the following equations, showing the stereochemistry wherever it is known.

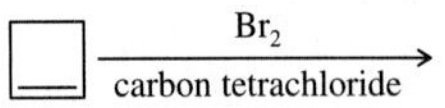

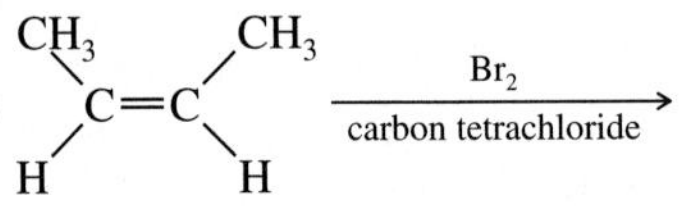

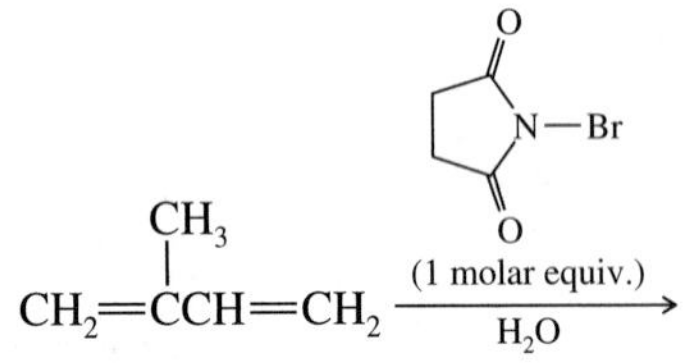

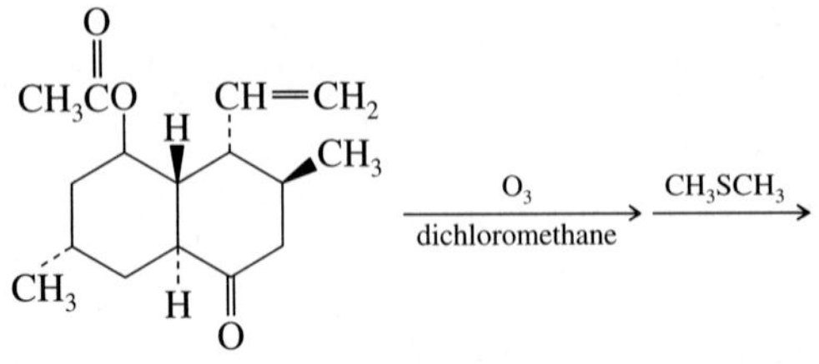

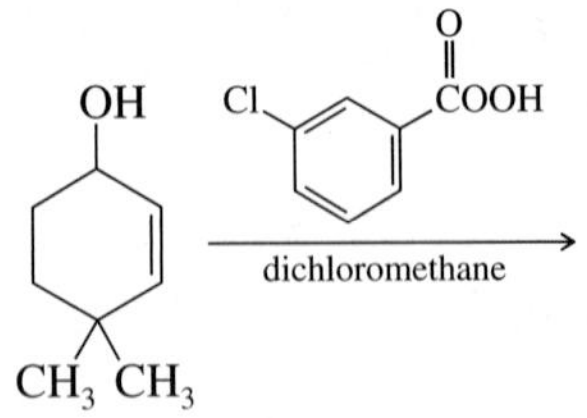

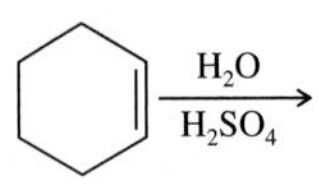

**8.31** Supply structural formulas for the intermediates and products designated by capital letters. Show the stereochemistry of the product(s) when known.

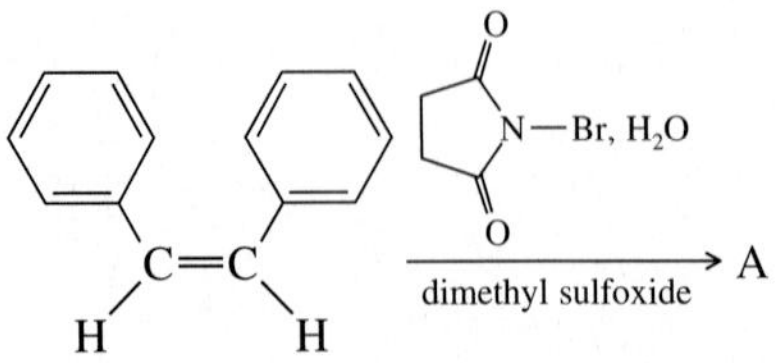

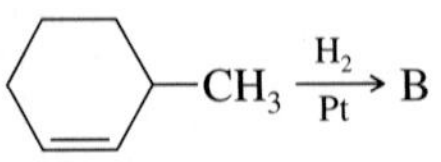

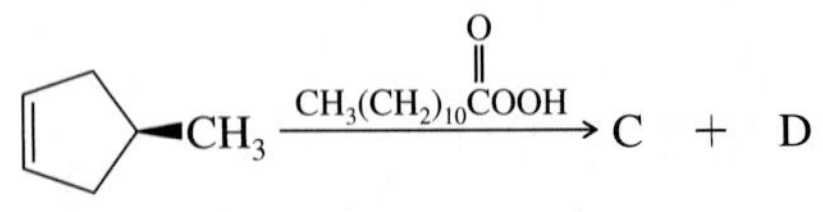

(d)

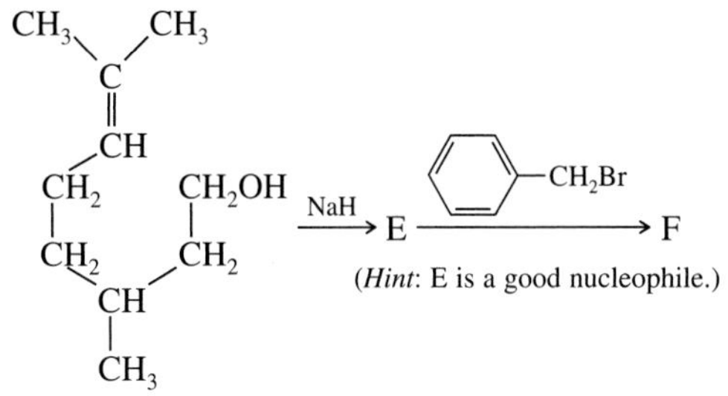

$$\xrightarrow{\text{NaH}} E \xrightarrow{\text{—CH}_2\text{Br}} F$$

(*Hint*: E is a good nucleophile.)

(e)

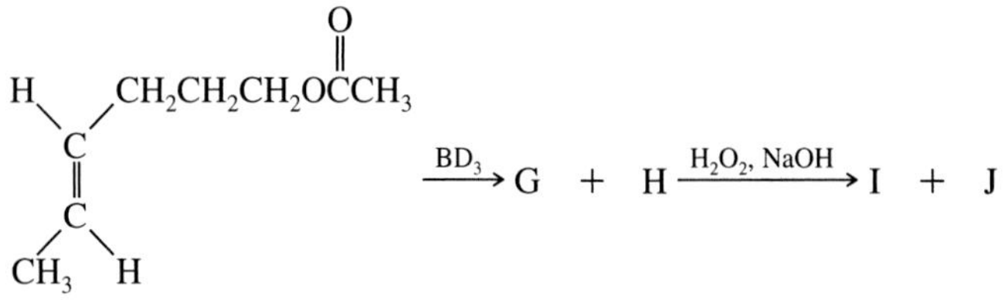

$$\xrightarrow{\text{BD}_3} G + H \xrightarrow{\text{H}_2\text{O}_2,\ \text{NaOH}} I + J$$

(f)

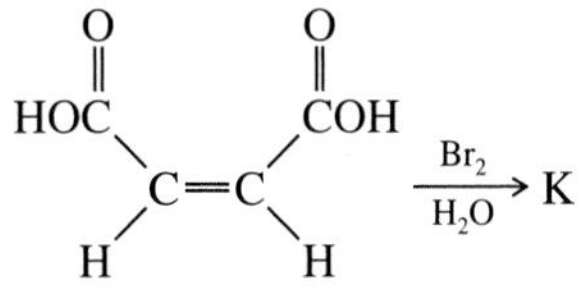

$$\xrightarrow[\text{H}_2\text{O}]{\text{Br}_2} K$$

(g)

$$\xrightarrow[\substack{\text{tetrahydrofuran}\\ \text{pyridine}}]{\text{OsO}_4} \xrightarrow[\substack{\text{H}_2\text{O}\\ \text{pyridine}}]{\text{NaHSO}_3} L$$

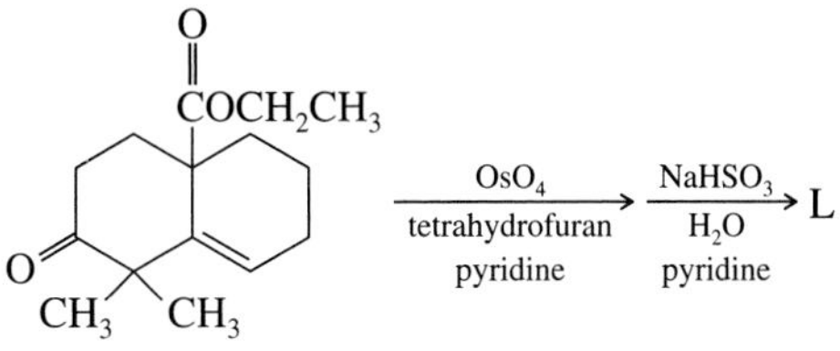

**8.32** More practice in recognizing reactions follows. Show the stereochemistry of the product(s) when known.

(a)

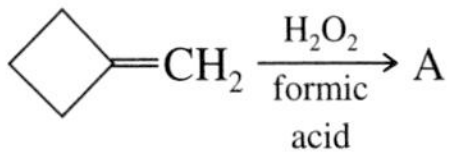

$$\xrightarrow[\substack{\text{formic}\\ \text{acid}}]{\text{H}_2\text{O}_2} A$$

(b)

$$\xrightarrow[\substack{\text{chloroform}\\ -20\ ^\circ\text{C}}]{\text{O}_3} \xrightarrow{\text{H}_2\text{O}} B$$

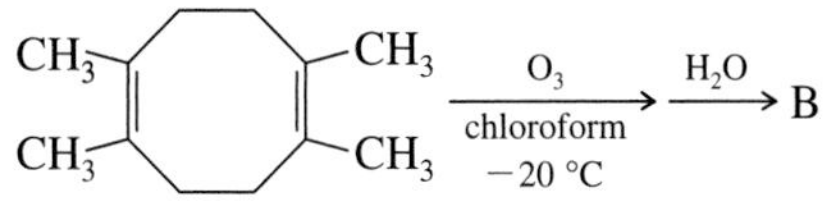

(c)

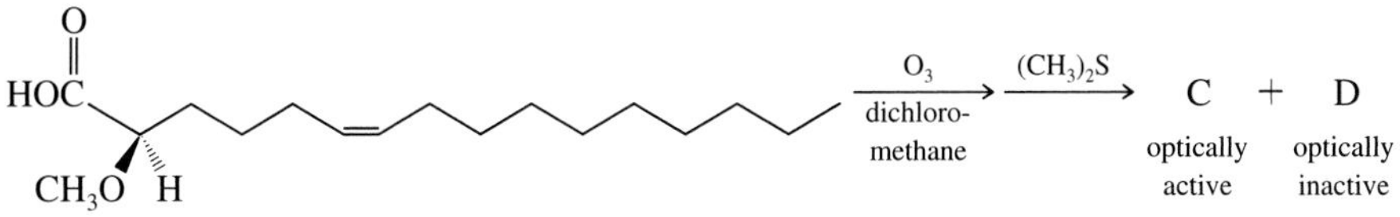

$$\xrightarrow[\substack{\text{dichloro-}\\ \text{methane}}]{\text{O}_3} \xrightarrow{(\text{CH}_3)_2\text{S}} \underset{\substack{\text{optically}\\ \text{active}}}{C} + \underset{\substack{\text{optically}\\ \text{inactive}}}{D}$$

(d)

$$\xrightarrow[\substack{\text{pyridine}\\ \textit{tert}\text{-butyl alcohol}}]{\substack{\text{OsO}_4\ (\text{cat}),\\ \text{CH}_3\quad\text{O}^-}} \xrightarrow[\text{H}_2\text{O}]{\text{Na}_2\text{SO}_3} E$$

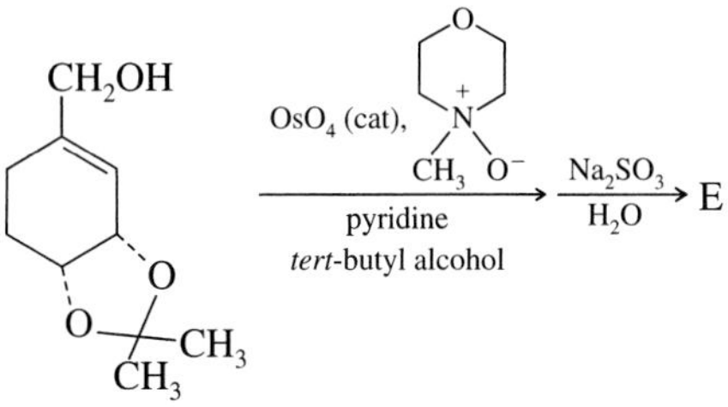

(*Hint:* Which side of the ring is less crowded as the alkene reacts with OsO$_4$?)

(e) 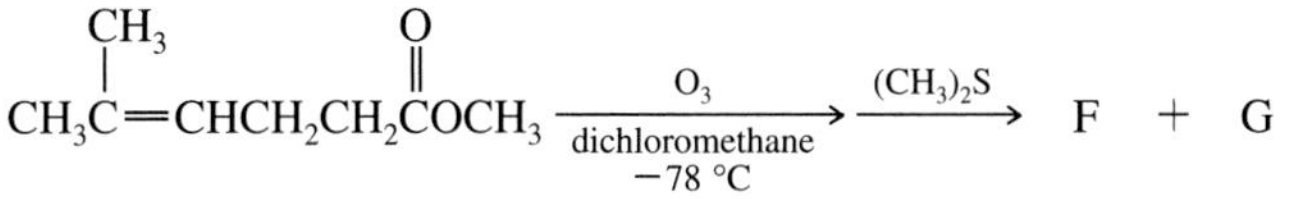 $\text{CH}_3\text{C}{=}\text{CHCH}_2\text{CH}_2\text{COCH}_3$ 

$$\xrightarrow[\substack{\text{dichloromethane}\\ -78\ ^\circ\text{C}}]{\text{O}_3} \xrightarrow{(\text{CH}_3)_2\text{S}} F + G$$

**8.33** The following are examples of reactions used in the synthesis and study of natural products and their analogues. Supply precise formulas, including stereochemistry when it is known, for the starting materials, reagents, or major products as indicated by the letters.

(a) 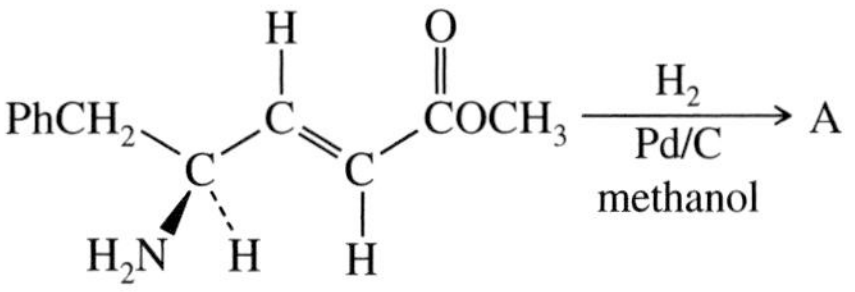 PhCH$_2$

$$\xrightarrow[\substack{\text{Pd/C}\\ \text{methanol}}]{\text{H}_2} A$$

(b)

$$\xrightarrow[\substack{\text{K}_2\text{CO}_3\\ \text{dimethylformamide}}]{B}$$

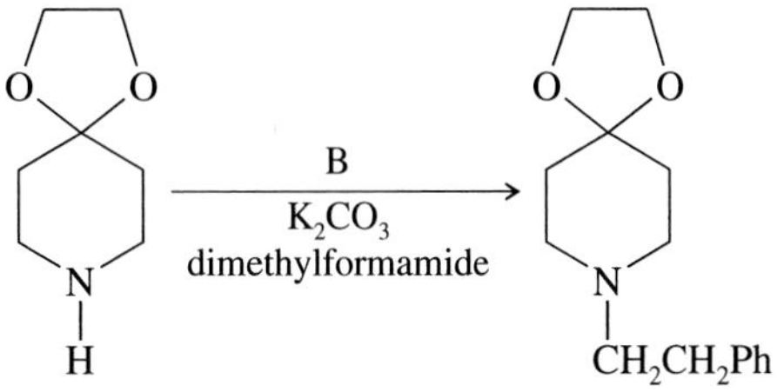

(c) $(CH_3)_3COC$ [structure with O, benzene ring, $CH=CH_2$] $\xrightarrow[\text{tetrahydrofuran}]{\text{H-B}}$ $\xrightarrow[\text{H}_2\text{O}]{\text{H}_2\text{O}_2, \text{NaOH}}$ C

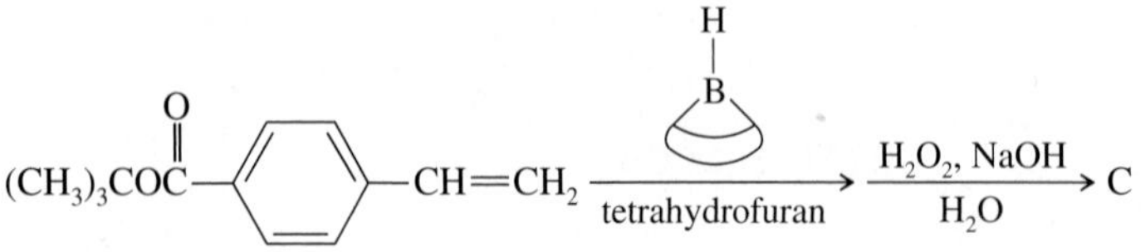

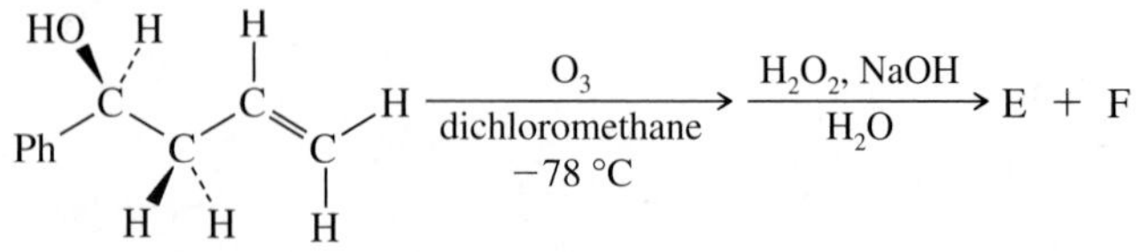

(d) 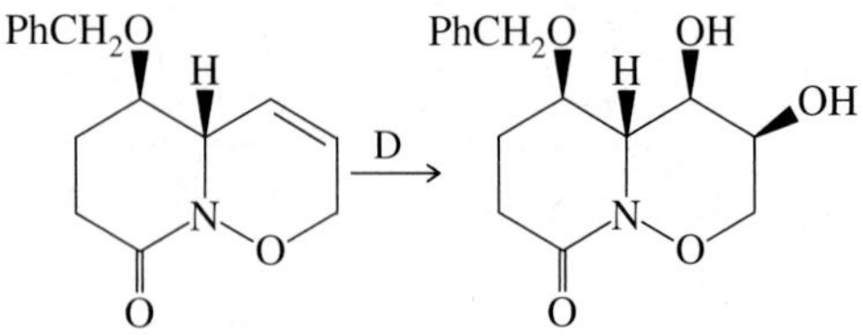 [bicyclic structure with PhCH$_2$O, H, N, O, O] $\xrightarrow{D}$ [bicyclic structure with PhCH$_2$O, H, OH, OH, N, O, O]

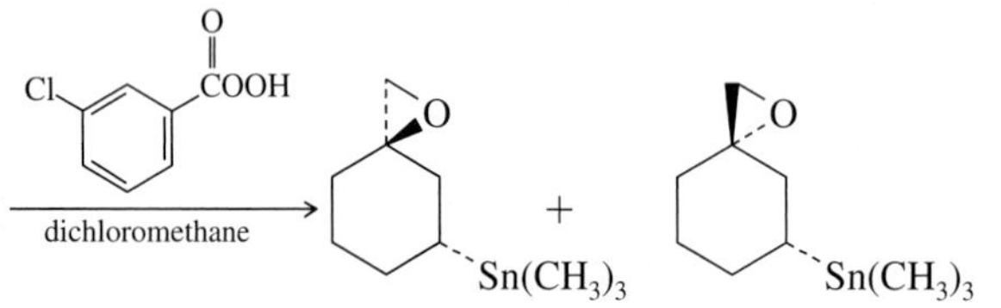

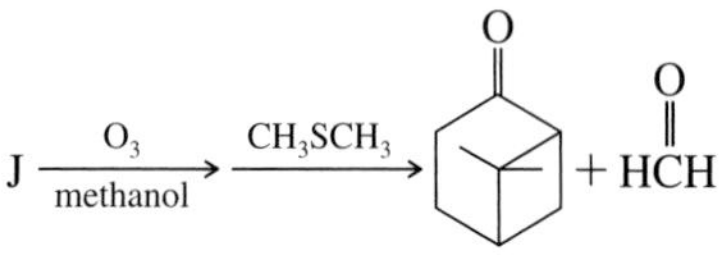

(e) [structure: HO, H, H, Ph, C, C, C, H] $\xrightarrow[\substack{\text{dichloromethane} \\ -78\,°C}]{O_3}$ $\xrightarrow[\text{H}_2\text{O}]{\text{H}_2\text{O}_2, \text{NaOH}}$ E + F

(f) G $\xrightarrow[\substack{\text{carbon} \\ \text{tetrachloride}}]{\text{Br}_2}$ [structure: $C_6H_5$, Br, H, C, C, H, Br, CH$_3$] + [structure: H, Br, $C_6H_5$, C, C, CH$_3$, Br, H]

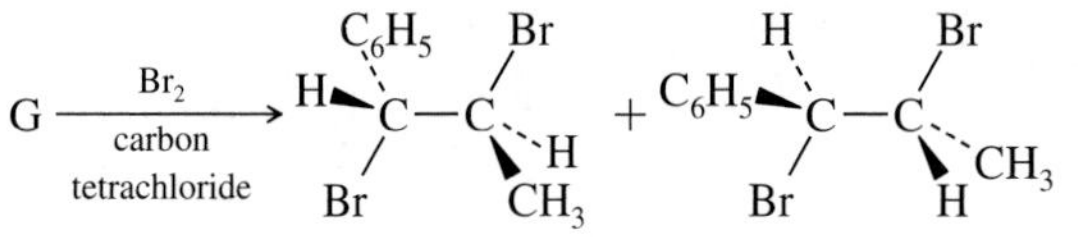

(g) $CH_3(CH_2)_6CH=CH_2 \xrightarrow{\text{H}} CH_3(CH_2)_6\underset{\underset{\text{Br}}{|}}{C}HCH_3$

(h) I $\xrightarrow[\text{dichloromethane}]{\text{[Cl-benzene-COOH structure]}}$ [cyclohexane with O epoxide, Sn(CH$_3$)$_3$] + [cyclohexane with O epoxide, Sn(CH$_3$)$_3$]

(i) J $\xrightarrow[\text{methanol}]{O_3}$ $\xrightarrow{\text{CH}_3\text{SCH}_3}$ [bicyclic ketone structure] + HCH [with O]

(j) [structure: H, C=C, H, H, H, PhCH$_2$O, N, H, OCH$_2$Ph] $\xrightarrow{K}$ [structure: HO, H, OH, H, C, C, H, H, PhCH$_2$O, N, OCH$_2$Ph]

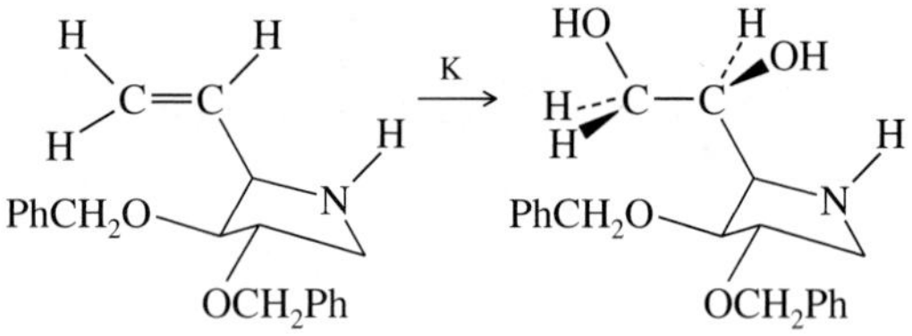

**8.34** When 2,2-dimethylcyclohexanol is treated with acid, 1,2-dimethylcyclohexene and isopropylidenecyclopentane are the products obtained. Draw a detailed mechanism that explains this result.

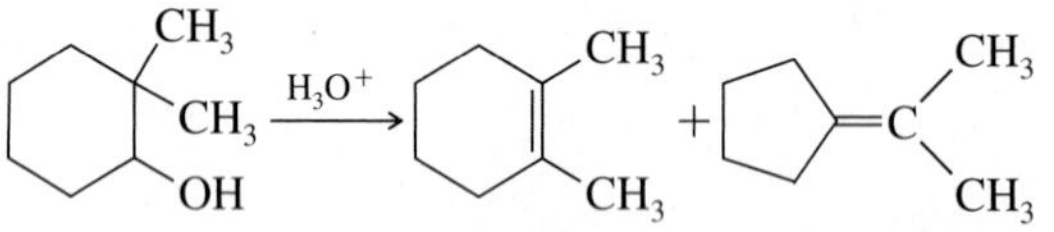

[reaction scheme] [cyclohexane with CH$_3$, CH$_3$, OH] $\xrightarrow{\text{H}_3\text{O}^+}$ [cyclohexene with CH$_3$, CH$_3$] + [cyclopentane =C with CH$_3$, CH$_3$]

2,2-dimethyl-        1,2-dimethyl-        isopropylidene-
cyclohexanol        cyclohexene        cyclopentane

**8.35** Research into the constituent of black tea that is responsible for its aroma involved the following set of reactions.

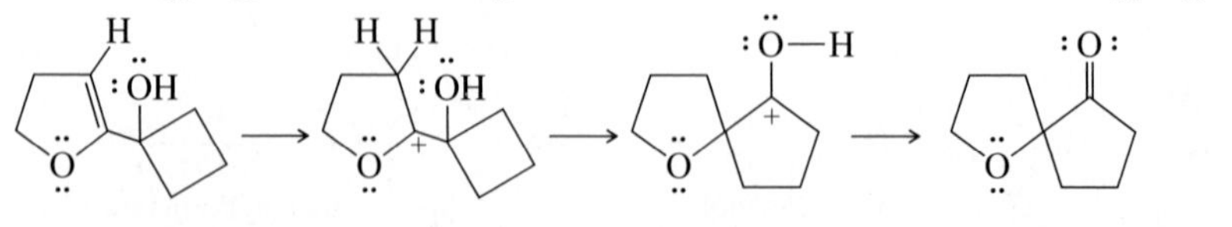

[reaction scheme with structures A, B, C, D]

H—B$^+$        : B        :$\overset{..}{O}$—H        H—B$^+$

A        B        C        D

(a) Provide curved arrows that depict the mechanism of the transformations shown here.

(b) Cations B and C have special stability. What do they have in common? Draw a picture using either structure that illustrates the source of this stability.

**8.36** 2,3-Dimethyl-2,3-butanediol has the common name pinacol. When it is treated with a strong acid, it forms 3,3-dimethyl-2-butanone, commonly known as pinacolone. The product results from the loss of water and a molecular rearrangement known as the **pinacol-pinacolone** rearrangement. Look closely at the structure of pinacolone and decide what type of rearrangement has occurred. Why does it happen? Write a mechanism for the formation of pinacolone from pinacol.

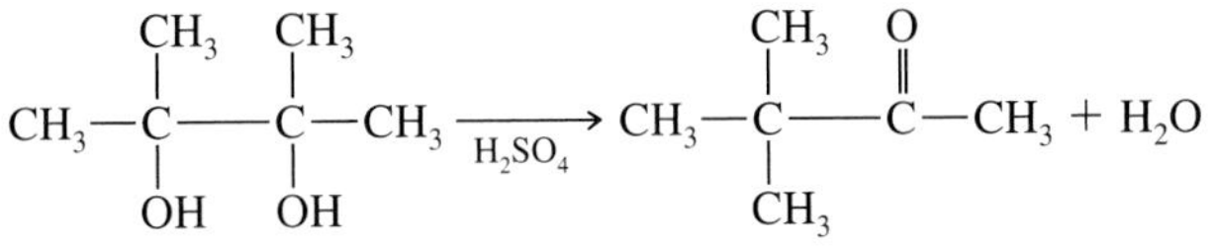

pinacol                    pinacolone

**8.37** Compound A, which is a degradation product of the antibiotic vermiculine (p. 626), has the following structure:

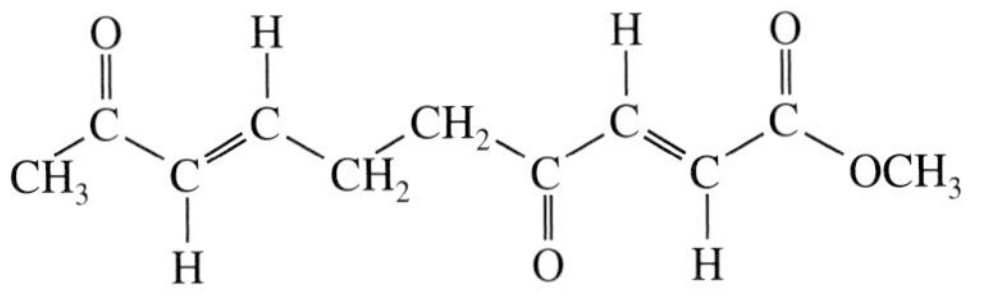

Compound A

The structure was confirmed by converting A to Compound B, $C_{11}H_{18}O_4$, which was also prepared by ozonolysis of Compound C, $C_{11}H_{18}O_2$.

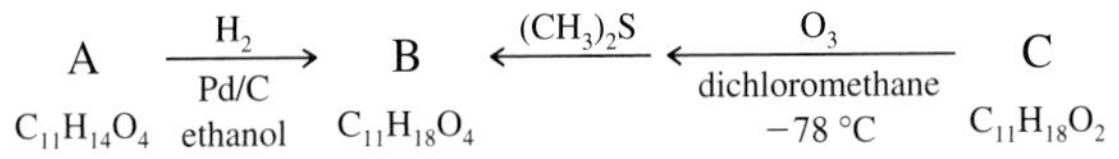

$$A \xrightarrow[\substack{Pd/C \\ C_{11}H_{14}O_4 \quad ethanol}]{H_2} B \xleftarrow[C_{11}H_{18}O_4]{(CH_3)_2S} \xleftarrow[\substack{dichloromethane \\ -78\,°C}]{O_3} C$$
$$\phantom{xxxxxxxxxxxxxxxxxxxxxxxxxxxxxxxxxxxxxxxxxxxxxxx} C_{11}H_{18}O_2$$

Assign structures to Compounds B and C.

**8.38** The following reactions were carried out during research into the stereochemistry of the hydroboration reaction.

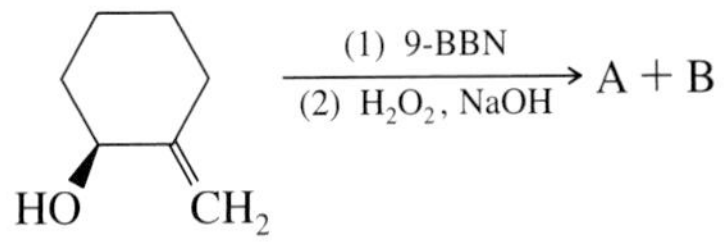

Compounds A and B are stereoisomers of each other and are formed in roughly equal amounts. Compound A is thermodynamically more stable than Compound B.

(a) Draw structural formulas for Compounds A and B.

(b) Draw a structural formula for the most stable conformer of Compound A.

**8.39** When the tosylate of 3-methyl-3-phenyl-2-butanol is heated, 2-methyl-3-phenyl-2-butene is formed. In order to determine the mechanism of this reaction, the compound was prepared with a radioactive carbon-14 ($^{14}$C) label at the tosylate carbon. The results of the experiment with the labeled compound are shown on the following page.

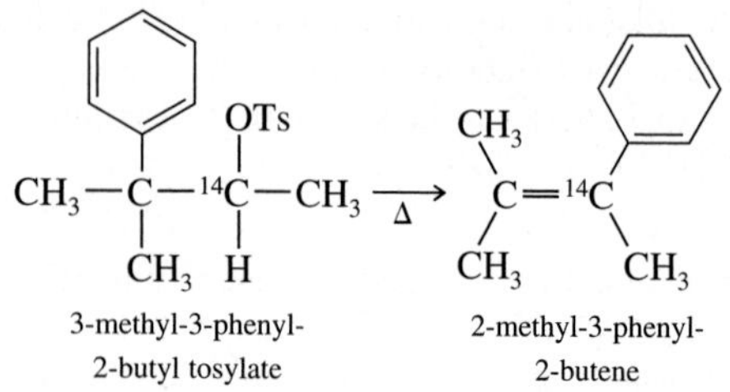

3-methyl-3-phenyl-
2-butyl tosylate

2-methyl-3-phenyl-
2-butene

Draw out a complete stepwise mechanism showing each step and intermediate in the conversion of the tosylate to the alkene. (*Hint:* Describe out loud the changes in connectivity that occur in the reaction.)

**8.40** The following reaction was carried out in the synthesis of a compound to be used in the treatment of congenital heart failure.

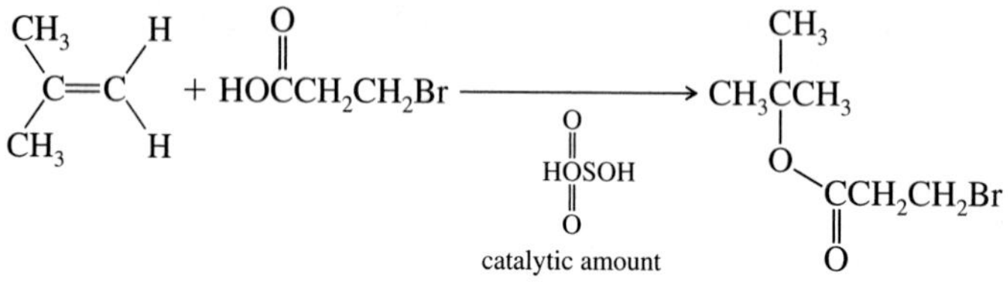

(a) Write a detailed mechanism for the reaction using the curved-arrow convention. The reaction is very slow unless a catalytic amount of sulfuric acid is added, so think carefully about the role of the inorganic acid.
(b) How many distinct peaks will there be in the carbon nuclear magnetic resonance spectrum of the product of the reaction shown above?
(c) How many distinct peaks will there be in the proton magnetic resonance spectrum of the compound?

**8.41** The following reaction was used in the synthesis of (+)-asteriscanolide, a natural product.

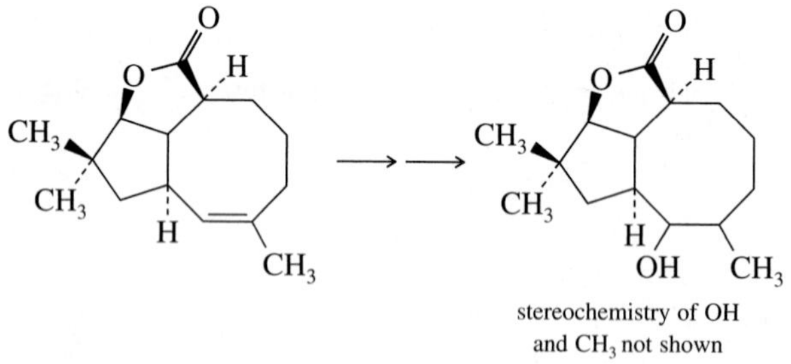

stereochemistry of OH
and CH$_3$ not shown

(a) What reagents would you use for this transformation?
(b) Two approaches to the double bond by the reagents are possible. Reaction at one face of the double bond rather than the other is highly favored. Draw a structural formula that clearly shows what you think the stereochemistry of the final product will be.

**8.42** When 1-methylcyclohexene reacts with *N*-bromosuccinimide, a source of electrophilic bromine atoms (p. 308), in the presence of a source of fluoride ions, anti addition of bromine and fluorine atoms occurs to give 2-bromo-1-fluoro-1-methylcyclohexane.

(a) Draw an accurate three-dimensional representation of the two chair conformations for one of the enantiomers of the product of the reaction described above.

(b) What are the configurations of the two stereocenters in the isomer you have drawn? Give the correct name for this isomer.

(c) Draw a mechanism for the reaction that is consistent with its regioselectivity.

**8.43** When the unsaturated alcohol shown below is treated with iodine in the presence of base, a cyclic ether containing iodine is formed.

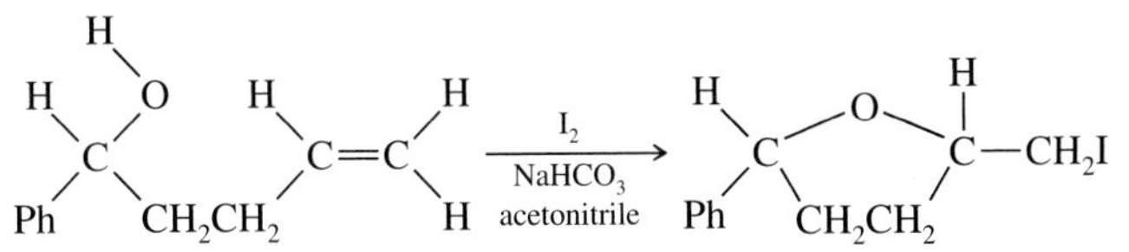

1-phenyl-4-penten-1-ol

The formation of the ether can be rationalized by proposing the same kind of reactive intermediate seen when bromine adds to a double bond. Using the curved-arrow convention, propose a mechanism for the formation of the ether. You may use $HB^+$ and $B:$ as general acids and bases as necessary.

**8.44** Aziridinium ions contain a positively charged nitrogen in a three-membered ring. The following reactions were observed for the aziridinium ion shown below.

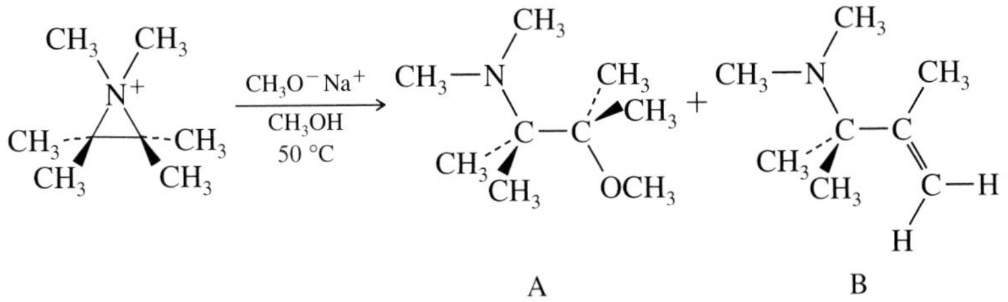

A
67%

B
33%

The kinetics of the reactions shows that both Compound A and Compound B are formed in second-order reactions. If the aziridinium ion is dissolved in methanol *without* sodium methoxide, only the first product, Compound A, is formed.

(a) Draw curved-arrow mechanisms for the two reactions of methoxide ion with the aziridinium ion shown above.

(b) Why is Compound A the only product formed when the aziridinium ion is put in methanol, while both Compound A and Compound B form when sodium methoxide is the reagent?

(c) The effect of structure on the ring-opening reaction of aziridinium ion was studied. The following results were found.

$k_2$, M$^{-1}$s$^{-1}$    relative rate ($k_{rel}$)

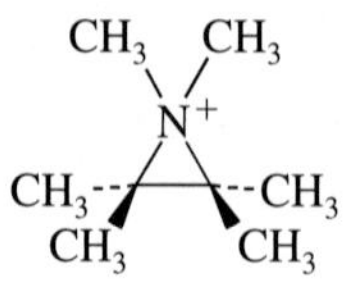

2.27 × 10$^{-2}$    1

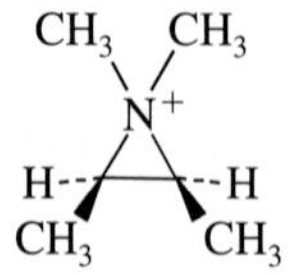

0.541    24

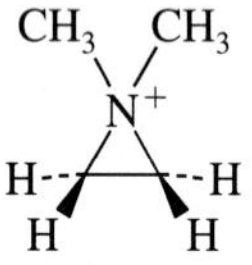

12.9    568

How would you explain this trend in the rates of the reaction? Pictures of activated complexes at the transition state and energy diagrams will be helpful.

**8.45** Recent studies have been directed toward the synthesis of analogs of 1-(1-phenylcyclohexyl)piperidine, a compound that has medical use as an intravenous anesthetic but has found its way onto the streets as PCP or angel dust.

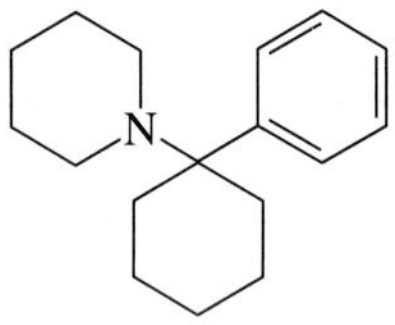

1-(1-phenylcyclohexyl)piperidine

The type of reaction investigated is shown with stereoisomeric 4-*tert*-butyl-1-phenylcyclohexanols.

or    CCl$_3$COH, Na$^+$N$_3^-$    +    +

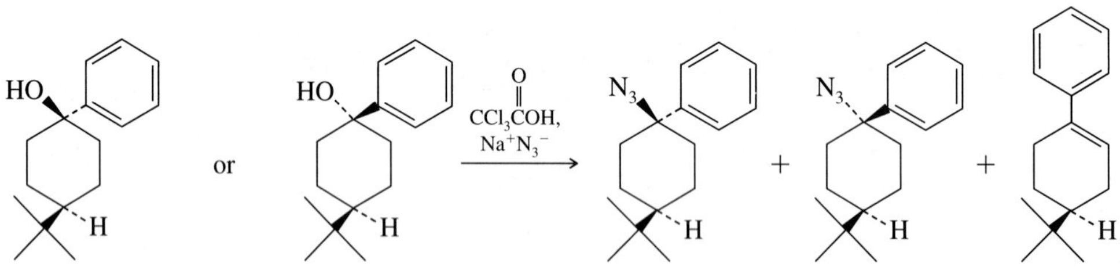

36%    46%    18%

(a) Draw the two chair conformers for 4-*tert*-butyl-1-phenylcyclohexanol in which the *tert*-butyl group is trans to the phenyl group. Which one is more stable?

(b) In a few words, explain why one of the chair forms you drew in part (a) is a higher-energy conformer than the other one.

(c) Use the curved-arrow convention to draw a mechanism that rationalizes the fact that the two starting materials, the 4-*tert*-butyl-1-phenylcyclohexanols of different stereochemistry, give the same mixture of products.

# Alkynes

## A Look Ahead

Alkynes undergo addition reactions at the triple bond.

π bonds

C — C

An alkyne

The π bonds in alkynes react with the same electrophilic reagents that the π bonds in alkenes react with. Alkynes, however, are able to add two molar equivalents of a reagent such as hydrogen bromide, hydrogen, or bromine.

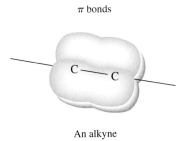

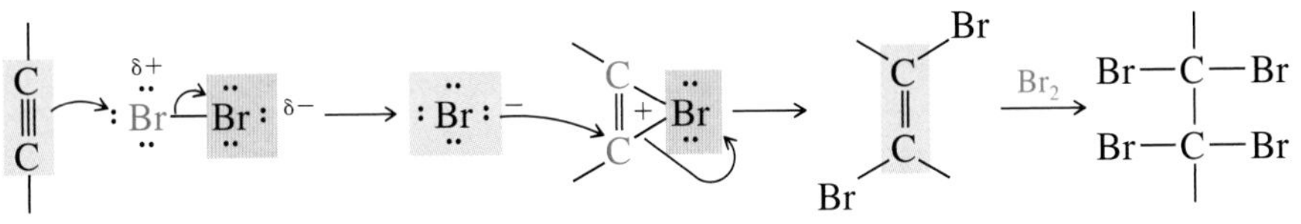

Alkynes differ from alkenes in that the hydrogen atom bonded to an *sp*-hybridized atom in an alkyne is acidic enough to be removed by a strong base, such as amide ion. The conjugate base of the alkyne is a good nucleophile and is useful in creating carbon–carbon bonds.

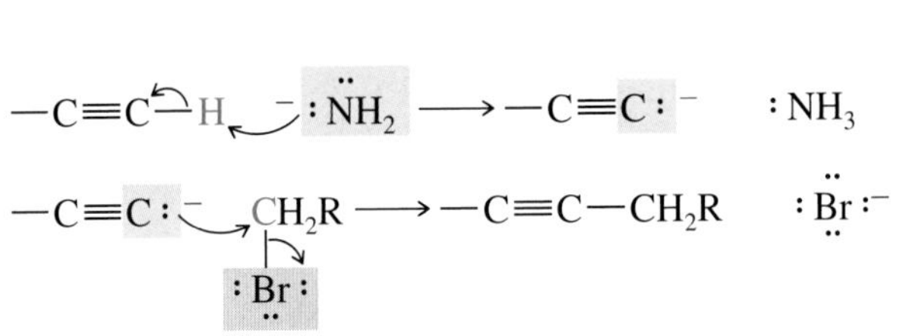

$$-C{\equiv}C{-}H \quad :NH_2 \longrightarrow -C{\equiv}C:^- \quad :NH_3$$

$$-C{\equiv}C:^- \quad CH_2R \longrightarrow -C{\equiv}C{-}CH_2R \quad :Br:^-$$
$$:Br:$$

   In this chapter you will study alkynes and their reactions, which in some ways are similar to and in other ways differ from the reactions of alkenes. You will also learn how to construct more complex molecules from simpler ones.

**Workbook Exercises**

## 9.1    Structure and Isomerism of Alkynes

Alkynes contain a carbon–carbon triple bond consisting of a $\sigma$ bond and two $\pi$ bonds between two $sp$-hybridized carbon atoms (p. 57). Because an alkyne group is linear, the only isomerism shown for those with the same formula is due to the position of the triple bond in the carbon chain.

$$CH_3CH_2CH_2C\equiv CH \qquad\qquad CH_3CH_2C\equiv CCH_3$$

1-pentyne
bp 40 °C

a terminal alkyne

2-pentyne
bp 56 °C

an internal alkyne

In a **terminal alkyne,** the triple bond is at the end of the carbon chain. A terminal alkyne has a hydrogen atom bonded to an $sp$-hybridized carbon atom, a structural feature that is important in its chemistry (p. 331). In an **internal alkyne,** the triple bond is at least one carbon atom away from the end of the carbon chain. There is no hydrogen atom bonded to either of the carbon atoms joined by the triple bond in an internal alkyne.

## 9.2    Nomenclature of Alkynes

To name an alkyne, the **ane** ending of the name of the alkane corresponding to the longest continuous chain of carbon atoms that contains the triple bond is changed to **yne.** Thus the IUPAC name for the simplest alkyne is **ethyne** (commonly known as **acetylene**). For more complex alkynes, the chain is numbered so that the carbon atoms involved in the triple bond have the lowest possible numbers. The position of the triple bond is indicated by the lower of the two numbers for the triply bonded carbon atoms. If substituents are present on the chain, they are named and their positions also indicated by numbers.

$$HC\equiv CH \qquad \overset{6\ \ 5\ \ \ 4\ \ \ \ 3\ 2\ \ \ 1}{CH_3CH_2C\equiv CCH_2CH_3} \qquad \overset{5\ \ \ \ 4|3\ \ \ \ 2\ 1}{CH_3CC\equiv CCH_3}$$

ethyne
acetylene

3-hexyne

4,4-dimethyl-2-pentyne

---

**Problem 9.1**    Name each of the following compounds.

(a) $CH_3\underset{\underset{Cl}{|}}{\overset{\overset{Cl}{|}}{C}}CH_2C\equiv C\underset{}{\overset{\overset{CH_3}{|}}{C}}HCH_3$

(b) $CH_3CH_2\underset{\underset{CH_3}{|}}{\overset{\overset{CH_3}{|}}{C}}C\equiv CH$

(c) $CH_3CH_2\underset{\underset{Br}{|}}{C}HC\equiv CCH_3$

(d) $CH_3\underset{\underset{CH_3CH_2}{|}}{C}HCH_2C\equiv C\underset{\underset{CH_3}{|}}{C}HCH_3$

---

**Problem 9.2**    Draw a structural formula for each of the following compounds.

(a) 2-chloro-2-methyl-3-heptyne    (b) 5-phenyl-2-octyne    (c) 2,2-dimethyl-3-hexyne

# 9.3 Alkynes as Acids

## A. The Preparation of Sodium Acetylide, a Source of Nucleophilic Carbon

An inspection of the $pK_a$ table shows that terminal alkynes have $pK_a$ values of ~26. The proton on the $sp$-hybridized carbon atom of a triple bond is more easily removed than one on the $sp^2$-hybridized carbon atom of a double bond ($pK_a$ ~36). The proton on an $sp^3$-hybridized carbon atom is the hardest to remove ($pK_a$ ~49). Chemists rationalize the trend in acidity found experimentally by pointing to the relative electronegativities of $sp$-, $sp^2$-, and $sp^3$-hybridized carbon atoms (p. 60). The negative charge on the conjugate base of an alkyne is better stabilized by the greater electronegativity of an $sp$-hybridized carbon atom than is the charge on an $sp^2$- or $sp^3$-hybridized carbon atom. This means that the conjugate base of an alkyne can be easily prepared by deprotonation of a terminal alkyne (p. 98) by a strong base such as an amide anion, the conjugate base of ammonia ($pK_a$ 36). Acetylene, for example, reacts with sodium amide, prepared by putting sodium metal in liquid ammonia, to form sodium acetylide.

$$2\ Na\ +\ 2\ NH_3 \longrightarrow 2\ Na^+NH_2^-\ +\ H_2\uparrow$$

sodium   ammonia      sodium    hydrogen
metal                   amide

$$HC\equiv CH + Na^+:\ddot{N}H_2^- \xrightarrow[-33\,°C]{NH_3(liq)} HC\equiv CNa + NH_3$$

acetylene      sodium           sodium
acid         amide          acetylide
             base

The anion from a terminal alkyne is a very strong nucleophile and base and must be protected from water or other acids that will protonate it.

$$HC\equiv C:^- Na^+ + H-\ddot{O}-H \rightleftarrows HC\equiv CH + Na^+\ ^-:\ddot{O}-H$$

base             acid
            $pK_a$ 15.7         $pK_a$ 26

In other words, water is a much stronger acid than a terminal alkyne.

## B. Reaction of Nucleophilic Acetylide Anions with Alkyl Halides

A nucleophile reacts at an electrophilic center that is an atom bearing a good leaving group to create a new $\sigma$ bond (pp. 234–236). Reagents that contain nucleophilic carbon atoms are used in nucleophilic substitution reactions to create new carbon–carbon bonds. Carbon–carbon bonds are formed by reactions between a carbon atom that is a nucleophilic center and a carbon atom that is an electrophilic center.

The reaction of nucleophilic acetylide anions with alkyl halides is a widely used way of synthesizing compounds that contain terminal triple bonds, for example:

$$CH_3CH_2CH_2CH_2Br + HC\equiv C:^-Na^+ \xrightarrow{NH_3(liq)} CH_3CH_2CH_2CH_2C\equiv CH + Na^+Br^-$$

1-bromobutane     sodium acetylide          1-hexyne        sodium bromide

The acetylide anion is a strong base, so only primary alkyl halides can be used in this reaction. With secondary or tertiary alkyl halides, E2 reactions (p. 264) lead to alkenes as the major products.

A terminal alkyne can be converted into its anion and used to make a larger internal alkyne. (*Z*)-9-Tricosene is a substance that enables the female housefly to attract the male housefly. (Chemical substances used by insects and animals to communicate with each other are known as pheromones; see p. 747.) 9-Tricosyne, an intermediate in the synthesis of (*Z*)-9-tricosene, is made from 1-pentadecyne and 1-chlorooctane in two steps.

$$CH_3(CH_2)_{12}C\equiv CH + LiNH_2 \xrightarrow[\text{tetrahydrofuran}]{} CH_3(CH_2)_{12}C\equiv C:^-Li^+ \ + \ NH_3$$

| 1-pentadecyne | lithium amide | hexamethylphosphoric triamide | conjugate base of 1-pentadecyne | ammonia |

$$CH_3(CH_2)_{12}C\equiv C:^-Li^+ + CH_3(CH_2)_6CH_2Cl \longrightarrow CH_3(CH_2)_{12}C\equiv CCH_2(CH_2)_6CH_3$$

| 1-chlorooctane | 9-tricosyne 54% |

**Study Guide**
**Concept Map 9.1**

This synthesis is an example of the general method by which a terminal alkyne is converted into its anion, which is then used to make an internal alkyne. The conversion of 9-tricosyne to (*Z*)-9-tricosene is shown on page 340.

---

**Problem 9.3**    Complete the following equations by supplying structural formulas for the products and reagents symbolized by the letters A–G.

(a) $CH_3CH_2CH_2C\equiv CH \xrightarrow[\text{NH}_3\text{(liq)}]{\text{NaNH}_2} A \xrightarrow{CH_3CH_2CH_2Br} B$

(b) $C \xrightarrow[\text{NH}_3\text{(liq)}]{D} CH_3\overset{\overset{\displaystyle CH_3}{|}}{C}HCH_2CH_2C\equiv CH$

(c) $CH_3CH_2CH_2C\equiv CH \xrightarrow[\text{NH}_3\text{(liq)}]{E} F \xrightarrow{(CH_3)_2SO_4} G$

[*Hint:* $(CH_3)_2SO_4$ is dimethyl sulfate with connectivity $CH_3OSO_2OCH_3$. What is the leaving group in $(CH_3)_2SO_4$?]

---

## 9.4   Electrophilic Addition Reactions of Alkynes

### A. Addition of Hydrogen Halides

The ionic addition of hydrogen bromide to an alkyne is similar to its addition to an alkene, except that two equivalents of hydrogen bromide are added before a saturated compound is formed. The overall reaction is regioselective, and the product is the one that is predicted by Markovnikov's Rule. 1-Hexyne, for example, adds hydrogen bromide twice to give 2,2-dibromohexane.

$$CH_3CH_2CH_2CH_2C\equiv CH \xrightarrow[15\,°C]{HBr} CH_3CH_2CH_2CH_2\overset{\overset{\textstyle }{}}{C}=CH_2 \xrightarrow{HBr} CH_3CH_2CH_2CH_2\overset{\overset{\displaystyle Br}{|}}{\underset{\underset{\displaystyle Br}{|}}{C}}CH_3$$

| 1-hexyne | 2-bromohexene | 2,2-dibromohexane |

The reaction goes by way of a vinyl cation formed by electrophilic attack on the triple bond.

**Addition of Hydrogen Bromide to an Alkyne**

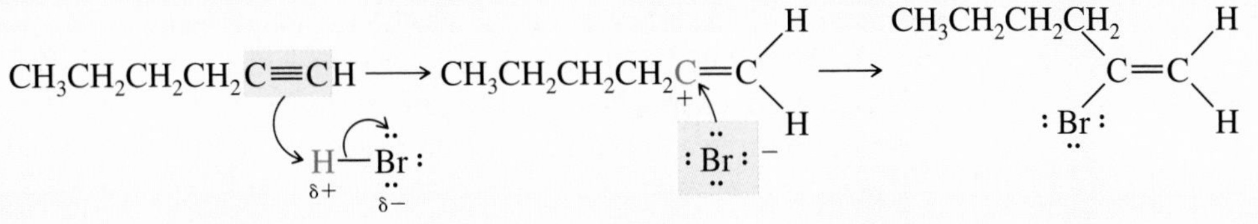

$$CH_3CH_2CH_2CH_2C{\equiv}CH \longrightarrow CH_3CH_2CH_2CH_2\overset{+}{C}{=}C\overset{H}{\underset{H}{<}} \longrightarrow \overset{CH_3CH_2CH_2CH_2}{\underset{:\ddot{B}r:}{>}}C{=}C\overset{H}{\underset{H}{<}}$$

$$H\overset{\frown}{-}\ddot{B}r:$$
$$\underset{\delta+}{\phantom{H}}\underset{\delta-}{\phantom{Br}}$$

1-hexyne and hydrogen          vinyl cation          2-bromo-1-hexene
bromide

*an unstable carbocation*

The electron-deficient carbon atom of the vinyl cation is bonded to two other atoms and may be considered to be *sp*-hybridized. It is more electronegative (p. 60) than the *sp²*-hybridized carbon atom found in an alkyl cation and thus less able to bear a positive charge. Vinyl cations are intermediate in energy between methyl and ethyl (or *n*-propyl) cations (p. 129) and less stable than secondary or tertiary cations. However, an alkyne is a higher-energy molecule than an alkene, so the energy of activation in going from an alkyne to a vinyl cation is not much larger than the energy of activation for the formation of an alkyl cation from an alkene. A $\pi$ bond in an alkyne is, therefore, about as reactive toward acids as the $\pi$ bond in an alkene is.

2-Bromo-1-hexene is less reactive toward electrophilic addition than a similar alkene without a halogen substituent would be. The halogen atom on the double bond is electron-withdrawing and decreases the availability of the $\pi$ electrons for reaction with a second molecule of hydrogen bromide. The orientation of the second addition reaction is also interesting. Two intermediates are possible, as seen below.

**Addition of Hydrogen Bromide to a Vinyl Halide**

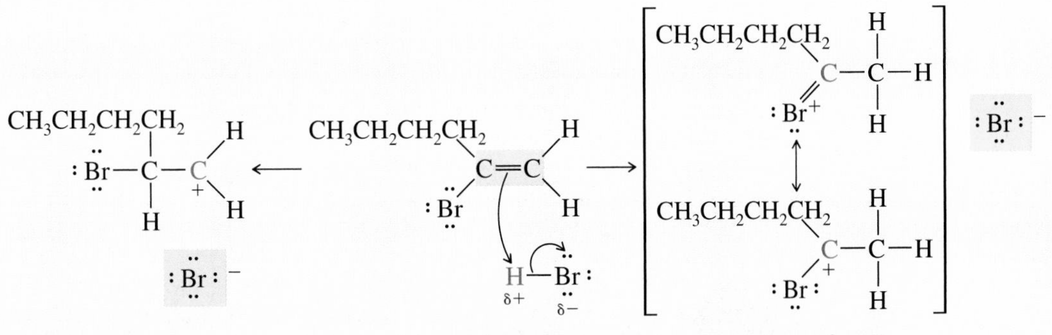

*a primary carbocation*

*does not lead
to product*

2-bromo-1-hexene
and
hydrogen bromide

*a secondary carbocation
stabilized further by
delocalization of charge
to bromine atom*

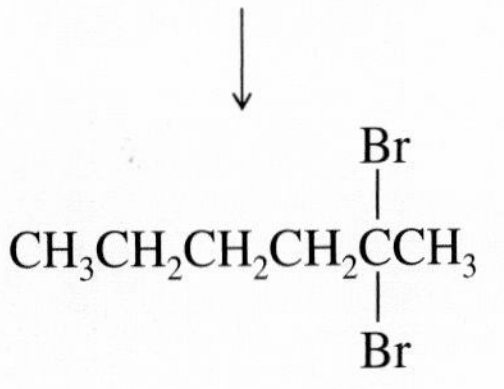

$$CH_3CH_2CH_2CH_2\underset{\underset{Br}{|}}{\overset{\overset{Br}{|}}{C}}CH_3$$

2,2-dibromohexane

One of the intermediates is a primary carbocation, and the other is a secondary carbocation with a bromine atom at the cationic center. The halogen, though electron-withdrawing in its inductive effect, does stabilize the carbocation by some delocalization of charge to the bromine atom by resonance. The product of the addition of hydrogen bromide to 2-bromo-1-hexene is the product derived from the reaction of bromide ion with the more stable of the two possible intermediates.

Resonance stabilization of a cation by a halogen atom is even more clearly present in the addition of hydrogen chloride to acetylene.

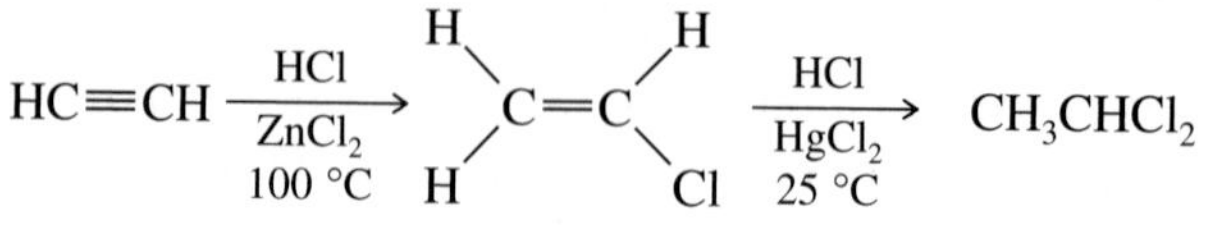

$$HC\equiv CH \xrightarrow[\substack{ZnCl_2 \\ 100\ ^\circ C}]{HCl} \underset{\substack{\\ \text{vinyl chloride}}}{\overset{\substack{H \\ C=C \\ H \quad\quad Cl}}{}} \xrightarrow[\substack{HgCl_2 \\ 25\ ^\circ C}]{HCl} \underset{\text{1,1-dichloroethane}}{CH_3CHCl_2}$$

acetylene

## VISUALIZING THE REACTION

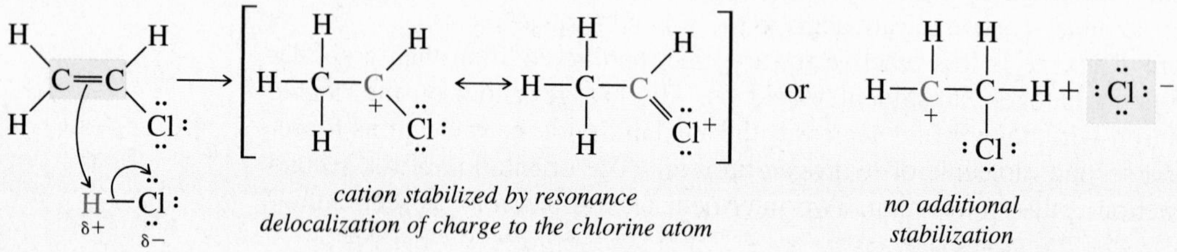

*cation stabilized by resonance delocalization of charge to the chlorine atom*

*or*

*no additional stabilization of this cation; does not lead to product*

Both possible intermediates are primary cations, and in one case a chlorine atom with its electron-withdrawing inductive effect would be expected to further destabilize the cation. The observed product can be rationalized by predicting resonance stabilization of one of the cations by the chlorine atom. Metal salts such as zinc(II) chloride and mercury(II) chloride, which act as Lewis acids, are often used to catalyze the reactions of hydrogen chloride with alkynes and vinyl halides.

## ONE SMALL STEP

Vinyl esters are important intermediates in the preparation of polymers used in making latex paint. The vinyl esters that are most useful have a branched structure in the acid portion of the ester. Acetylene is used in the synthesis of these esters.

**PROBLEM:** Predict the product of the following reaction.

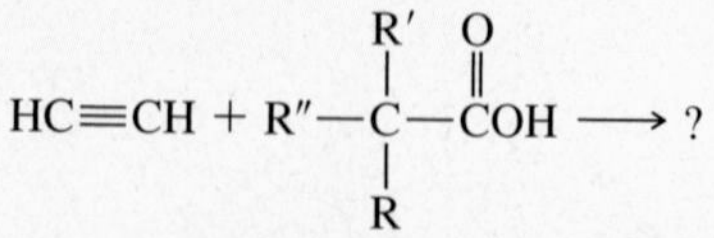

$$HC\equiv CH + R''-\underset{\underset{R}{|}}{\overset{\overset{R'}{|}}{C}}-\overset{\overset{O}{\|}}{C}OH \longrightarrow ?$$

**Hint:** Identify the nucleophiles and the electrophiles involved in the reaction.

**Problem 9.4**   Show the major product for each of the following reactions.

(a) 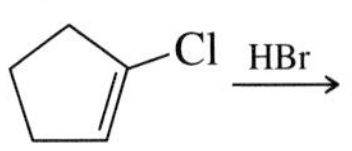 [cyclopentene with Cl substituent] $\xrightarrow{\text{HBr}}$   (b) $CH_3C\equiv CCH_3 \xrightarrow{\text{HBr (excess)}}$   (c) $CH_3CH_2CH_2CH=CH_2 \xrightarrow{\text{HCl}}$   (d) 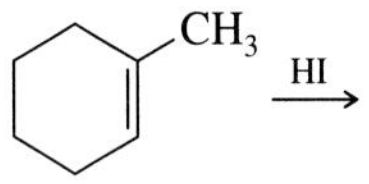 [cyclohexene with CH$_3$ substituent] $\xrightarrow{\text{HI}}$

## B. Addition of Water to Alkynes

Water adds to acetylene in the presence of acid and mercury and iron salts to give acetaldehyde. This reaction is important industrially.

$$HC\equiv CH \xrightarrow[\substack{H_2SO_4 \\ HgSO_4 \\ Fe_2(SO_4)_3}]{H_2O} CH_3\overset{\displaystyle O}{\overset{\|}{C}}H$$

acetylene                     acetaldehyde

All other alkynes give ketones. For example, 1-hexyne is converted into 2-hexanone.

$$CH_3CH_2CH_2CH_2C\equiv CH \xrightarrow[\substack{H_2SO_4 \\ HgSO_4}]{H_2O} CH_3CH_2CH_2CH_2\overset{\displaystyle O}{\overset{\|}{C}}CH_3$$

1-hexyne                           2-hexanone
                                        80%

The hydration of an alkyne proceeds by electrophilic attack on the $\pi$ electrons of the triple bond. Mercury(II) ion, $Hg^{2+}$, is a Lewis acid and serves as a catalyst in the hydration reaction. Alkynes that are substituted on both sides of the triple bond, such as 2-pentyne, are more reactive and can be hydrated without the use of a catalyst.

$$CH_3C\equiv CCH_2CH_3 \xrightarrow[\substack{H_2SO_4 \\ 0\,°C,\ 10\ min}]{H_2O} CH_3CH_2\overset{\displaystyle O}{\overset{\|}{C}}CH_2CH_3 + CH_3CH_2CH_2\overset{\displaystyle O}{\overset{\|}{C}}CH_3$$

2-pentyne                          3-pentanone          2-pentanone
                                       ~50%                 ~50%

In summary, hydration of an alkyne gives rise to a ketone, except in the case of acetylene, where acetaldehyde is the product.

**Problem 9.5**   9-Undecynoic acid, shown below, is hydrated in 80% sulfuric acid. Write an equation for the reaction.

$$CH_3C\equiv C(CH_2)_7\overset{\displaystyle O}{\overset{\|}{C}}OH$$

## C. The Art of Solving Problems

Some of the mechanisms presented in this book, such as the ones for the $S_N2$ and $S_N1$ reactions, are supported by large amounts of experimental data, including careful studies of kinetics and stereochemistry. For many of the mechanisms presented in the Visualizing the Reaction features, however, all the details are not rigorously supported by experimental data. These mechanisms represent attempts made by organic chemists to rationalize the transformations of the reactants into the products under the conditions of the reaction. Organic chemists do this by reasoning by analogy from other better-known and more thoroughly researched reactions. The kinds of questions chemists ask when proposing the mechanism for a reaction are a combination of those used in solving synthesis and transformation problems. The following problem illustrates how to figure out a mechanism.

---

**PROBLEM:** Write a detailed mechanism for the addition of water to 2-butyne.

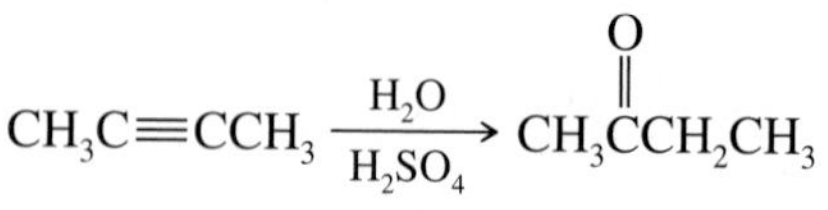

$$CH_3C{\equiv}CCH_3 \xrightarrow[H_2SO_4]{H_2O} CH_3\overset{\displaystyle O}{\overset{\|}{C}}CH_2CH_3$$

### Solution

1. How have connectivities changed in going from reactant to product? How many carbon atoms does each contain? What bonds must be broken and formed to transform reagent into product?

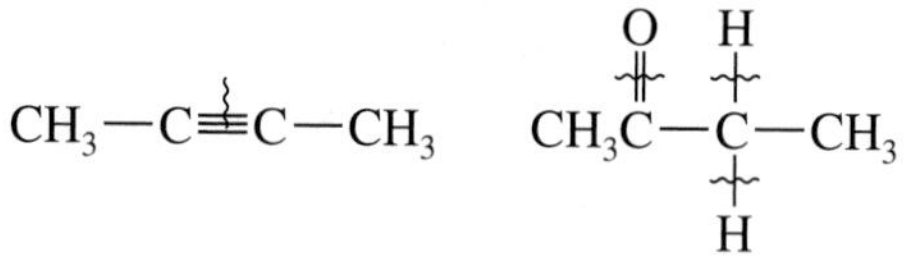

$$CH_3-C{\equiv}C-CH_3 \qquad CH_3\overset{O}{\underset{\underset{H}{|}}{C}}-\overset{H}{\underset{}{C}}-CH_3$$

The reactant and the product have the same number of carbon atoms arranged in the same way. Both $\pi$ bonds in the alkyne must be broken. A double bond between carbon and oxygen and two carbon–hydrogen bonds must be formed.

2. What reagents are present? Are they good acids, bases, nucleophiles, or electrophiles?

$H_2SO_4$ is a strong acid. In the presence of a strong acid, $H_2O$ behaves as a weak base. It is also a nucleophile. The mixture of water and sulfuric acid contains the hydronium ion, $H_3O^+$. The organic reagent is an alkyne. The triple bond has a cloud of $\pi$ electrons, which are nucleophilic.

3. What is the most likely first step for the reaction: protonation or deprotonation, ionization, attack by a nucleophile, or attack by an electrophile?

The $\pi$ bond will be protonated.

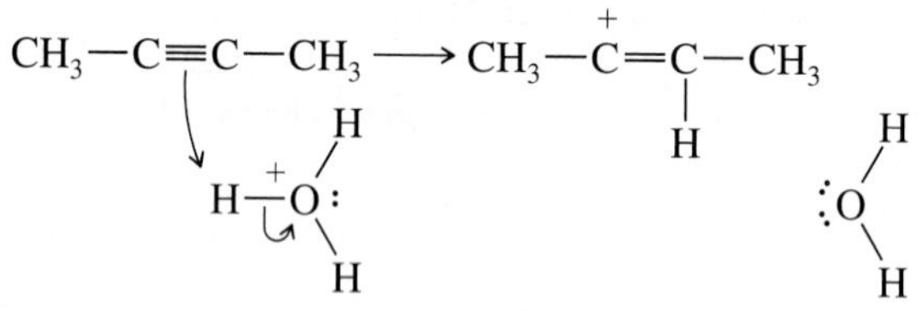

$$CH_3-C{\equiv}C-CH_3 \longrightarrow CH_3-\overset{+}{C}{=}C-CH_3$$

4. What are the properties of the species present in the reaction mixture after the first step? What is likely to happen next?

The vinyl cation is an electrophile. It will react with the nucleophile, water.

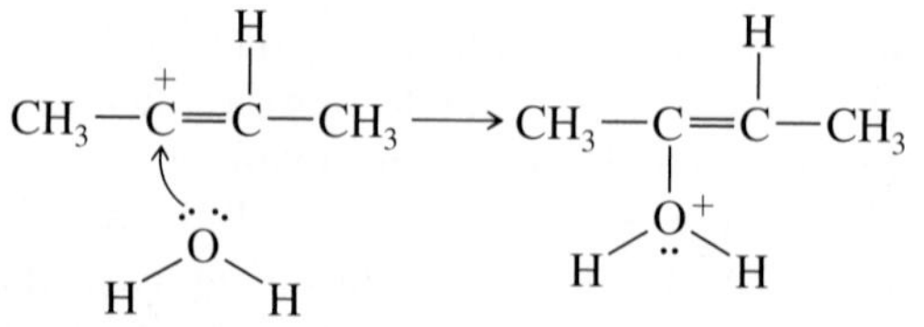

$$CH_3-\overset{+}{C}{=}\overset{\overset{H}{|}}{C}-CH_3 \longrightarrow CH_3-C{=}\overset{\overset{H}{|}}{C}-CH_3$$

4. (repeated) What are the properties of the species present in the reaction mixture now? What is likely to happen next?

The species resulting from the preceding step is an oxonium ion, an acid. It will be deprotonated by the solvent.

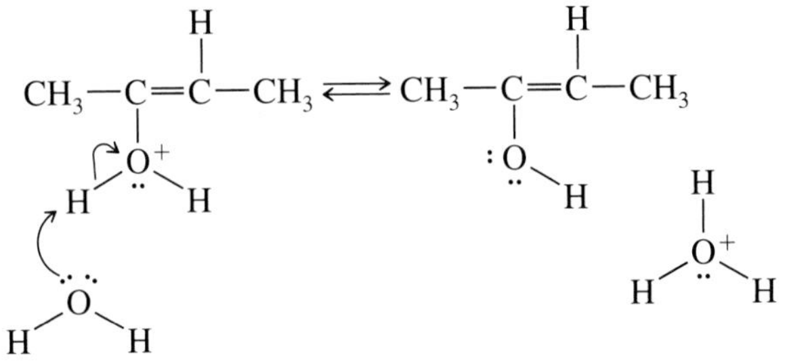

Comparing this species to the starting material and to the product, as shown below, reveals that it has a carbon–oxygen bond and a carbon–hydrogen bond at the right locations—on the carbon atoms that were originally part of the triple bond. This intermediate is called an **enol,** indicating that it contains a hydroxyl group (ol) on a double bond (ene).

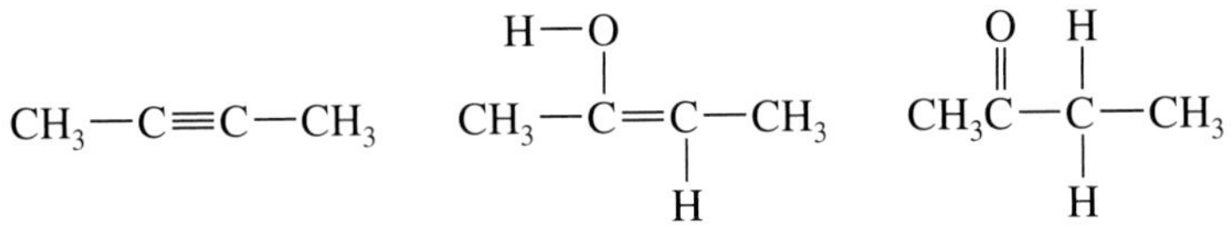

an enol

To get from the enol to the product, a hydrogen must be added at the double bond, a hydrogen must be removed from oxygen, and the double bond must be shifted to the oxygen.

4. (repeated) What are the properties of the species present in the reaction mixture now? What is likely to happen next?

The enol will be protonated in the acidic solution. Two sites of protonation are available.

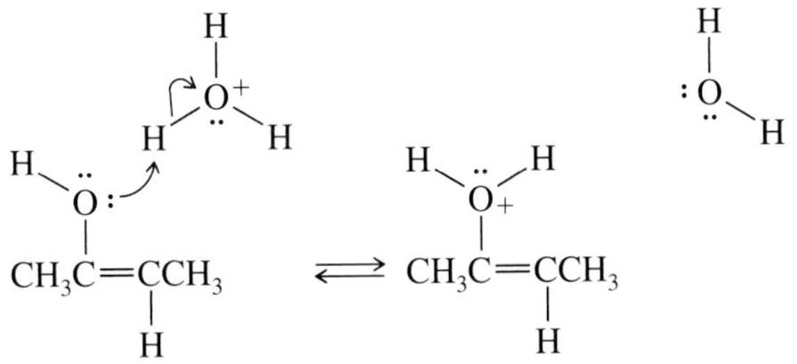

gain of a proton at oxygen

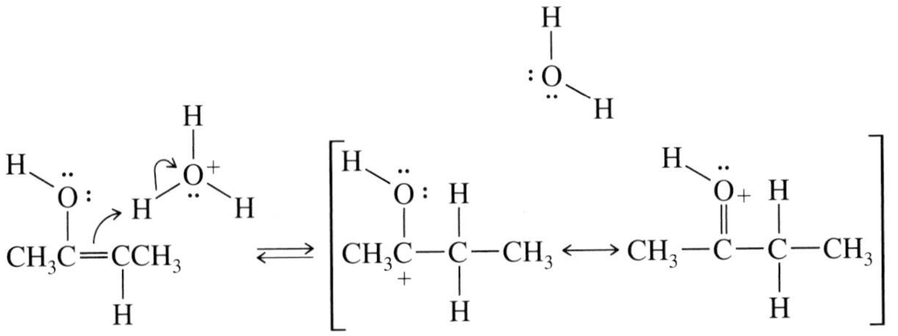

gain of a proton at carbon

Protonation at the oxygen is the reverse of the deprotonation reaction of the preceding step. It occurs but does not further the overall reaction that gives the desired product. Protonation of the double bond in an enol is particularly easy because the positive charge in the resulting carbocation is adjacent to the oxygen atom and stabilized by resonance. In one of the resonance contributors for this cation, carbon and oxygen each have an octet of electrons, and the positive charge is delocalized to the oxygen atom. A close inspection of this resonance-stabilized cation reveals that it is a protonated ketone. Loss of the proton to a base gives 2-butanone.

The complete correct answer is as follows:

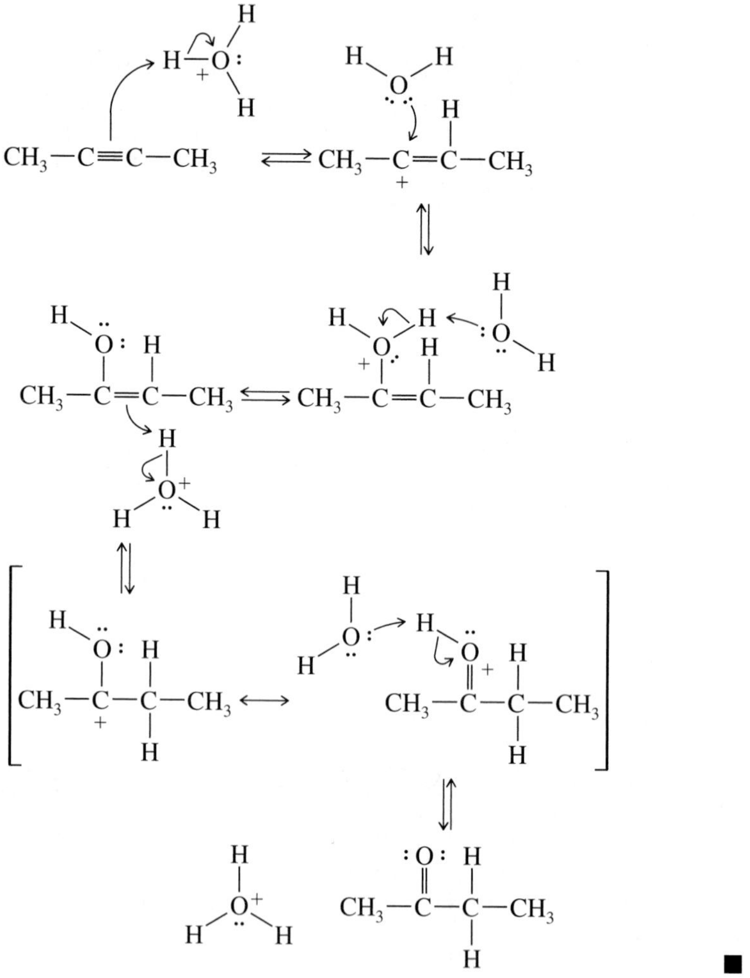

## D. Tautomerism

The conversion of an enol to a ketone by protonation at the carbon atom of the double bond and deprotonation at the oxygen atom is known as **tautomerization.** The ketone and its enol form are examples of **tautomers,** readily interconvertible constitutional isomers that exist in equilibrium with each other. Isomers that differ from each other only in the location of a hydrogen atom and a double bond are **proton tautomers.** Proton tautomers are isomers in which a hydrogen atom and a double bond switch locations between a carbon atom and a **heteroatom,** which is an atom other than carbon, such as oxygen or nitrogen. Tautomers differ from each other in the locations of atoms as well as of electrons. Therefore, they are not resonance contributors, which are different representations of the same structure (p. 16).

$$CH_3 \quad CH_2CH_3 \atop \underset{\text{H}-\text{O} \quad \text{H}}{C=C} \quad \rightleftharpoons \quad CH_3-\underset{\underset{\text{O}}{\|}}{\overset{\overset{\text{H}}{|}}{C}}-CHCH_2CH_3$$

enol form of                                keto form of
2-pentanone                               2-pentanone

*tautomers of each other*

$$CH_3-\underset{\underset{\text{O}-\text{H}}{|}}{C}=NH \quad \rightleftharpoons \quad CH_3-\underset{\underset{\text{H}}{|}}{\overset{\overset{\text{O}}{\|}}{C}}-NH$$

enol form of                keto form of
acetamide                   acetamide

*tautomers of each other*

When keto-enol tautomerization occurs, the keto form, the one in which a carbonyl group is present, is usually the more stable and predominates at equilibrium.

⊘ **Study Guide**
Concept Map 9.2

💻 **On the Web: ONE SMALL STEP**

**Problem 9.6**   Write a detailed mechanism for the conversion of 2-pentyne to 3-pentanone.

**Problem 9.7**   Write a detailed mechanism for the formation of acetaldehyde from acetylene.

## 9.5   Reduction of Alkynes

### A. Catalytic Hydrogenation of Alkynes

Addition of 1 molar equivalent of hydrogen to an alkyne gives an alkene. The catalysts that are used in such hydrogenation reactions are active toward alkynes but not alkenes, so the reaction stops when 1 equivalent of hydrogen has reacted with the triple bond. One such catalyst, known as a **poisoned catalyst,** is palladium on barium sulfate or calcium carbonate to which a small amount of lead or the organic base quinoline has been added to make the catalyst less reactive. For example, phenylethyne can be converted to phenylethene (styrene) by the use of such a catalyst.

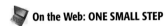

$$\text{phenylethyne} \xrightarrow[\substack{\text{Pd/CaCO}_3 \\ \text{quinoline} \\ 25\ °C}]{\text{H}_2(1-2\ \text{atm})} \text{phenylethene}$$

phenylethyne                                       phenylethene
styrene

quinoline  ≡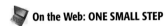

The reaction stops when 1 equivalent of hydrogen has been used up. The alkene is not reduced to an alkane on this catalyst, even though such a double bond would normally undergo that reaction (p. 300, for example). Note that the aromatic ring is not affected under these conditions. High temperatures (~100 °C) and high pressures (up to 100 atm) are usually required to add hydrogen to aromatic rings.

Various experiments with different metallic catalysts have shown that catalytic hydrogenation of internal alkynes gives predominantly cis alkenes. For example, 9-tricosyne (p. 332) is converted to (Z)-9-tricosene, the sex pheromone of the housefly, in this way.

$$CH_3(CH_2)_{12}C{\equiv}CCH_2(CH_2)_6CH_3 \xrightarrow[\substack{quinoline \\ hexane}]{\substack{H_2 \\ Pd/BaSO_4}}$$

$$\underset{\substack{(Z)\text{-9-tricosene} \\ 84\%}}{\overset{\displaystyle \underset{CH_3(CH_2)_{12}}{\overset{H}{\diagdown}} C{=}C \overset{H}{\diagup} CH_2(CH_2)_6CH_3}{}}$$

9-tricosyne

The reaction stops cleanly when exactly 1 equivalent of hydrogen has reacted. The stereochemistry that results from catalytic reductions of alkynes indicates that syn addition occurs. Hydrogen atoms are added at the surface of the catalyst to the same side of the alkyne bond.

When the catalysts that are used to hydrogenate alkenes (pp. 298–301) are used, alkynes add 2 equivalents of hydrogen to give alkanes. For example, tricosane, another pheromone, is prepared by hydrogenating 10-tricosyne over platinum.

$$CH_3(CH_2)_{11}C{\equiv}CCH_2(CH_2)_7CH_3 \xrightarrow[\substack{PtO_2 \\ hexane}]{H_2 \ (excess)} CH_3(CH_2)_{21}CH_3$$

10-tricosyne                                                                 tricosane

**Problem 9.8**    10-Tricosyne has been prepared starting with 1-tetradecyne. How would you synthesize 10-tricosyne?

**Problem 9.9**    The equation above showing the hydrogenation of 10-tricosyne indicates that an excess of hydrogen is used. What would happen if only 1 molar equivalent of hydrogen had been used under these conditions? A review of page 80 may be helpful in thinking your way through this problem.

## B. Reduction of Alkynes to Alkenes by Dissolving Metals

Alkynes are also reduced by sodium or lithium metal in liquid ammonia. In contrast to catalytic hydrogenation, this reaction produces trans alkenes. For example, 3-octyne is reduced to (E)-3-octene with sodium metal in ammonia.

$$CH_3CH_2C{\equiv}CCH_2CH_2CH_2CH_3 \xrightarrow[\substack{NH_3(liq) \\ -33\,°C}]{Na}$$

$$\underset{\substack{(E)\text{-3-octene} \\ 95\%}}{\overset{\displaystyle \underset{H}{\overset{CH_3CH_2}{\diagdown}} C{=}C \overset{H}{\diagup} \underset{CH_2CH_2CH_2CH_3}{\diagdown}}{}}$$

3-octyne

This reaction occurs by a transfer of an electron from sodium metal to the alkyne. The intermediate formed is a **radical anion,** a species that bears a negative charge and has an unpaired electron. Such a strongly basic species is protonated by ammonia. The radical that is formed is reduced again by sodium metal and then protonated to give the most stable alkene (p. 282), having the E configuration.

**VISUALIZING THE REACTION**

## Reduction of an Alkyne by Sodium in Ammonia

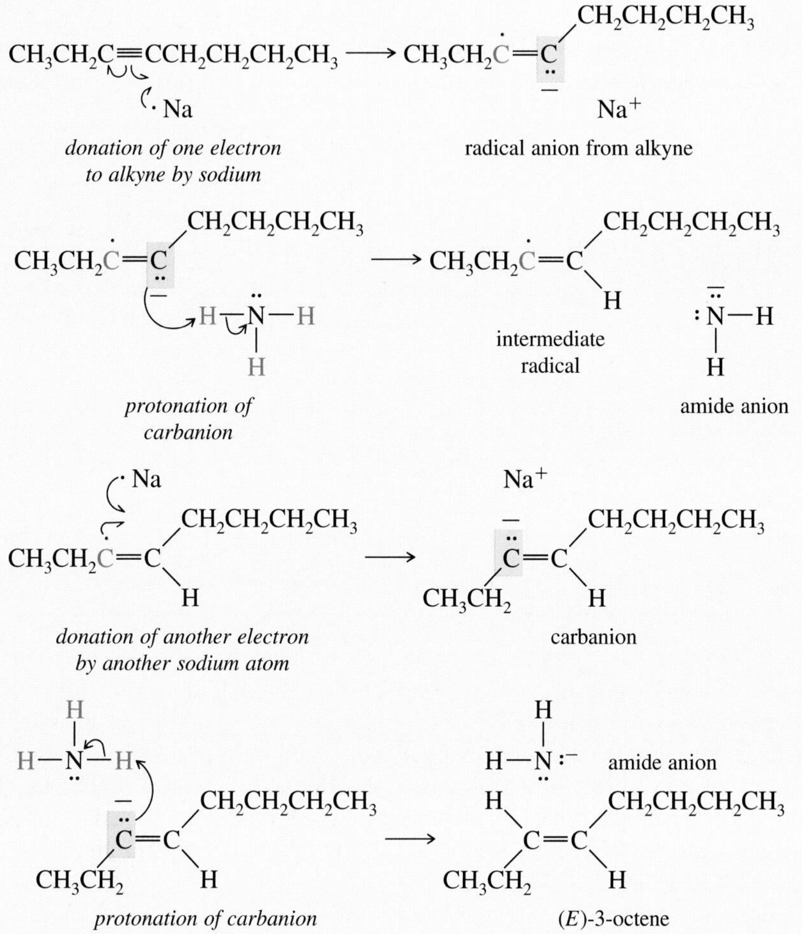

$CH_3CH_2C{\equiv}CCH_2CH_2CH_3 \longrightarrow CH_3CH_2\overset{\displaystyle .}{C}{=}\overset{\displaystyle CH_2CH_2CH_2CH_3}{\underset{\underset{Na^+}{-}}{\ddot{C}}}$

*donation of one electron*            radical anion from alkyne
*to alkyne by sodium*

$CH_3CH_2\overset{\displaystyle .}{C}{=}\overset{\displaystyle CH_2CH_2CH_2CH_3}{\ddot{C}} \longrightarrow CH_3CH_2\overset{\displaystyle .}{C}{=}C\overset{CH_2CH_2CH_2CH_3}{\underset{H}{}}$        $:\overset{\overline{\phantom{.}}}{\underset{H}{\ddot{N}}}{-}H$

$H{-}\overset{\cdot\cdot}{N}{-}H$
        $|$
        $H$

*protonation of*            intermediate            amide anion
*carbanion*            radical

$CH_3CH_2C{=}C\overset{CH_2CH_2CH_2CH_3}{\underset{H}{}}$  · Na  $\longrightarrow$  $Na^+$  $\overset{\underset{\ddot{C}}{-}}{}{=}C\overset{CH_2CH_2CH_2CH_3}{\underset{H}{}}$
                $CH_3CH_2$

*donation of another electron*            carbanion
*by another sodium atom*

$H{-}\overset{\overset{\textstyle H}{|}}{\underset{\cdot\cdot}{N}}{-}H$
    $\overset{-}{\ddot{C}}{=}C\overset{CH_2CH_2CH_2CH_3}{\underset{H}{}}$
$CH_3CH_2$

$\longrightarrow$

$H{-}\overset{\overset{\textstyle H}{|}}{\underset{\cdot\cdot}{N}}:^-$  amide anion

$\overset{H}{\underset{CH_3CH_2}{}}C{=}C\overset{CH_2CH_2CH_2CH_3}{\underset{H}{}}$

*protonation of carbanion*            (*E*)-3-octene

In the above mechanism, a new type of arrow, called a **fishhook,** ⌒➤ , is used. A fishhook represents the motion of a single electron, such as the donation of one electron to the alkyne by sodium. By contrast, a regular curved arrow, ⌒➤ , is used to indicate the movements of pairs of electrons. The protonation of a carbanion, for example, is represented with regular arrows.

In contrast to the result of the preceding reaction, when 3-octyne is hydrogenated on a metal surface, the product is predominantly (*Z*)-3-octene.

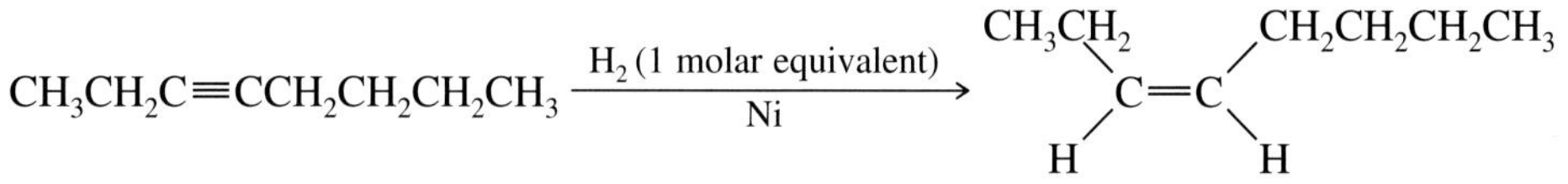

$CH_3CH_2C{\equiv}CCH_2CH_2CH_3 \xrightarrow[\text{Ni}]{\text{H}_2\,(1\text{ molar equivalent})} \overset{CH_3CH_2}{\underset{H}{}}C{=}C\overset{CH_2CH_2CH_2CH_3}{\underset{H}{}}$

3-octyne            (*Z*)-3-octene
            98% this isomer

⊘ **Study Guide**
**Concept Map 9.3**

The equations in this section demonstrate that alkynes can be converted with stereoselectivity (p. 296) to either cis or trans alkenes, depending on the reagents used.

---

**Problem 9.10**   Complete the following equations. Be sure to show the stereochemistry of the product where it is pertinent.

(a) $CH_3CH_2CH_2CH_2C{\equiv}CH \xrightarrow[\substack{Pd/BaSO_4 \\ quinoline}]{H_2}$

(b) $CH_3CH_2CH_2CH_2C{\equiv}CCH_3 \xrightarrow[\substack{Pd/BaSO_4 \\ quinoline}]{H_2}$

(c) $CH_3CH_2CH_2CH_2C{\equiv}CCH_3 \xrightarrow[\substack{NH_3 \,(liq)}]{Na}$

(d) $CH_3CH_2CH_2CH_2C{\equiv}CCH_3 \xrightarrow[\substack{Pt \\ acetic\ acid}]{H_2\ (excess)}$

---

**Problem 9.11**   When 2-butyne in the gas phase is treated with deuterium gas ($D_2$), with palladium on an alumina support as the catalyst, one alkene is obtained in 99% yield. Predict the structure of the alkene.

---

## 9.6   Planning Syntheses

Organic chemists are constantly developing new methods for synthesizing compounds having industrial, medicinal, or biological importance. For example, modifications of natural hormones and antibiotics are synthesized in attempts to understand how the natural substances function and to create more effective ones. The design of a synthesis of a complex molecule having several functional groups and a specific stereochemistry is an intellectual exercise that appeals to chemists. The actual synthesis of such a compound in the laboratory is also challenging and requires skill and care.

To design a synthesis, a chemist has to know what reactions can be used to give different types of functional groups. The reactions that create carbon–carbon bonds are among the most important. You know how to synthesize an alkyne with the triple bond in any position (pp. 331–332) and how to convert an alkyne to an alkene with either a cis or trans double bond (pp. 339–342). Other useful reactions you have learned include nucleophilic substitution reactions of halides and electrophilic addition reactions of alkenes. This is a good time to review these types of transformations in Table 7.1 (p. 234) and Table 8.1 (p. 321).

The thinking that goes into the design of syntheses (pp. 135, 238) is reviewed in the two problems that follow.

---

**PROBLEM:** How would you synthesize (*E*)- and (*Z*)-3-heptene from acetylene and any other chemicals?

**Solution**

1. What are the connectivities of the starting material and the products? How many carbon atoms does each contain? Are there any rings? What are the functional groups and their positions on the carbon skeletons?

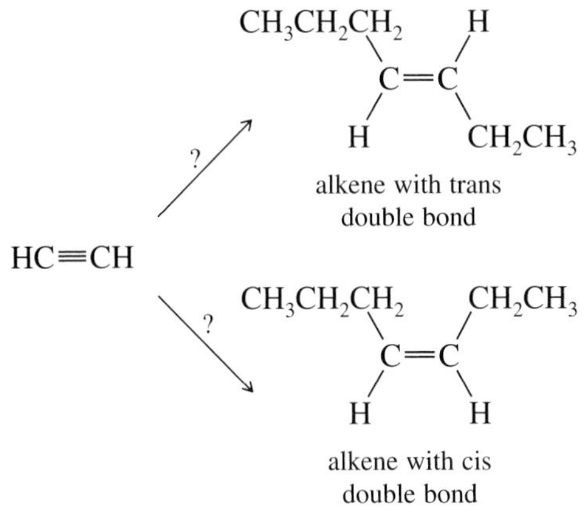

$$CH_3CH_2CH_2 \quad \diagdown \qquad / H$$
$$C=C$$
$$/ \qquad \diagdown$$
$$H \qquad CH_2CH_3$$

alkene with trans
double bond

?

HC≡CH

?

$$CH_3CH_2CH_2 \quad \diagdown \qquad / CH_2CH_3$$
$$C=C$$
$$/ \qquad \diagdown$$
$$H \qquad H$$

alkene with cis
double bond

The starting material, acetylene, has only two carbon atoms. The products each have seven carbon atoms and a double bond starting at the third carbon atom of the chain.

2. How do the functional groups change in going from starting material to product? Does the starting material have a good leaving group?

   Alkyl groups have been added to both ends of the alkyne. The triple bond has been converted to a double bond.

3. Is it possible to dissect the structures of the starting material and product to see which bonds must be broken and which formed?

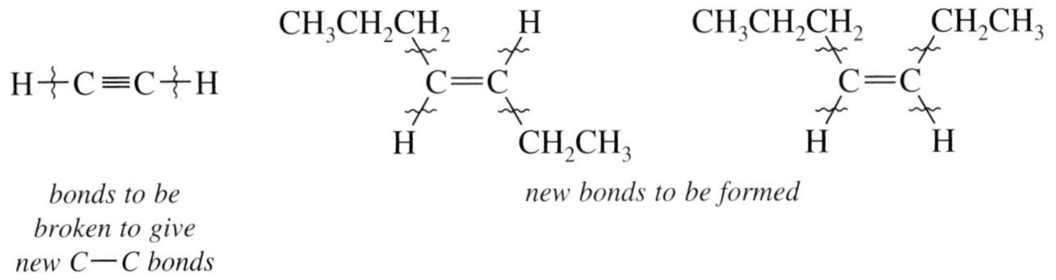

$$H \!+\! C \!\equiv\! C \!+\! H$$

$$CH_3CH_2CH_2 \quad H \qquad\qquad CH_3CH_2CH_2 \quad CH_2CH_3$$
$$C=C \qquad\qquad\qquad C=C$$
$$H \quad CH_2CH_3 \qquad\qquad H \qquad H$$

*bonds to be*
*broken to give*
*new C—C bonds*

*new bonds to be formed*

The pieces we need are

$$CH_3CH_2CH_2— \qquad —C\equiv C— \qquad —CH_2CH_3 \qquad —H \qquad —H$$

4. New bonds are created when an electrophile reacts with a nucleophile. Do we recognize any part of the product molecule as coming from a good nucleophile or an electrophilic addition?

   The alkyne $CH_3CH_2CH_2C\equiv CCH_2CH_3$ can be converted into either the cis or the trans alkene. The carbon–carbon bonds in the alkyne must be the result of nucleophilic substitution reactions. Acetylene can be converted into a good nucleophile. The alkyl groups, attached to leaving groups, must supply the electrophiles. The alkyne can be put together from these reagents:

$$CH_3CH_2CH_2Br \qquad HC\equiv CH \qquad BrCH_2CH_3$$

5. What type of compound would be a good precursor to the product?

   The alkyne $CH_3CH_2CH_2C\equiv CCH_2CH_3$ is the precursor.

6. After this step, do we see how to get from starting material to product? If not, we need to analyze the structure obtained in step 5 by applying questions 4 and 5 to it.

The complete synthesis is as follows:

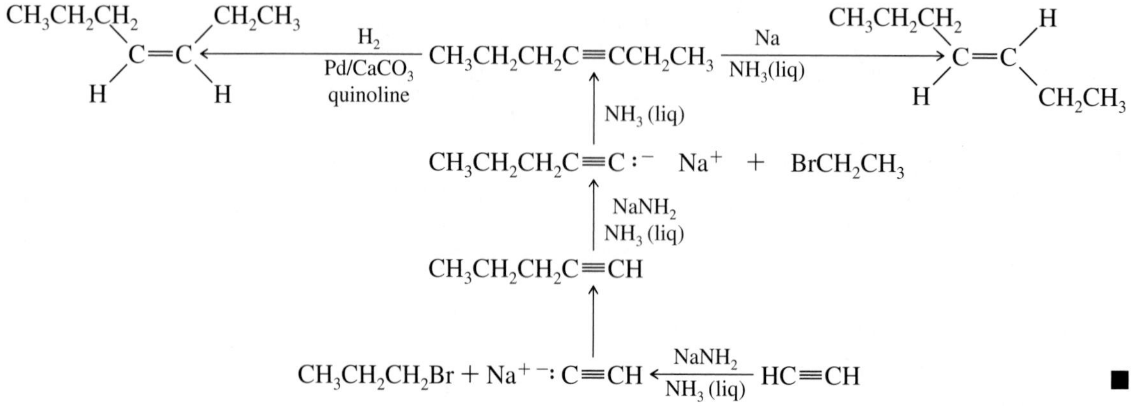

$$
\begin{array}{c}
\underset{H}{\overset{CH_3CH_2CH_2}{>}} C = C \underset{H}{\overset{CH_2CH_3}{<}}
\xleftarrow[\substack{Pd/CaCO_3 \\ quinoline}]{H_2}
CH_3CH_2CH_2C \equiv CCH_2CH_3
\xrightarrow[NH_3(liq)]{Na}
\underset{H}{\overset{CH_3CH_2CH_2}{>}} C = C \underset{CH_2CH_3}{\overset{H}{<}}
\end{array}
$$

$$\big\uparrow NH_3 \text{ (liq)}$$

$$CH_3CH_2CH_2C \equiv C:^- \ Na^+ \ + \ BrCH_2CH_3$$

$$\big\uparrow \substack{NaNH_2 \\ NH_3 \text{ (liq)}}$$

$$CH_3CH_2CH_2C \equiv CH$$

$$CH_3CH_2CH_2Br + Na^{+\,-}{:}C \equiv CH \xleftarrow[NH_3 \text{ (liq)}]{NaNH_2} HC \equiv CH \qquad \blacksquare$$

---

**PROBLEM:** Synthesize 5-methyl-1-hexanol from acetylene, using any other reagents you need.

**Solution**

1. What are the connectivities of the starting material and the products? How many carbon atoms does each contain? Are there any rings? What are the functional groups and their positions on the carbon skeletons?

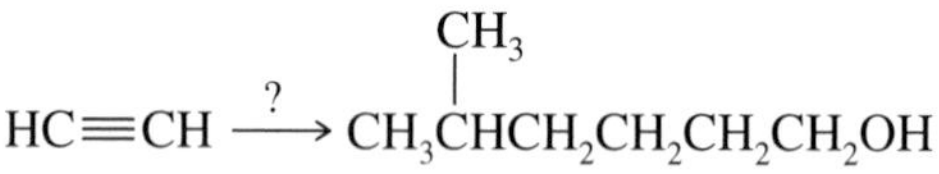

$$HC \equiv CH \xrightarrow{\ ?\ } \underset{}{\overset{CH_3}{\underset{|}{CH_3CHCH_2CH_2CH_2CH_2OH}}}$$

alkyne     primary alcohol

Acetylene has only two carbon atoms. The product has a chain of six carbon atoms with a hydroxyl group on the first carbon and a methyl group on the fifth carbon.

2. How do the functional groups change in going from starting material to product? Does the starting material have a good leaving group?

The triple bond disappears. The rest of the chain must have been added to the alkyne, probably on one side.

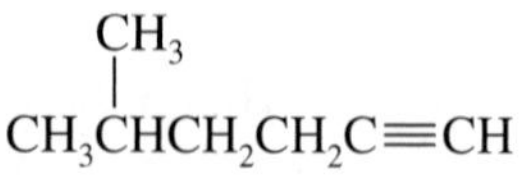

$$\overset{CH_3}{\underset{|}{CH_3CHCH_2CH_2C \equiv CH}}$$

3. Is it possible to dissect the structures of the starting material and product to see which bonds must be broken and which formed?

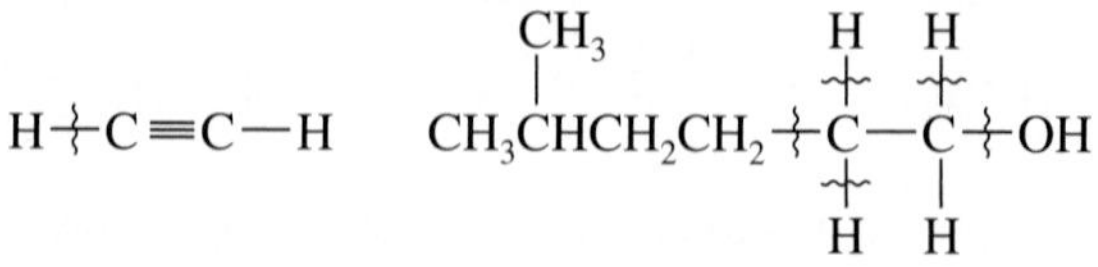

$$H \!+\! C \equiv C \!-\! H \qquad \overset{CH_3}{\underset{|}{CH_3CHCH_2CH_2}} \!+\! \overset{H}{\underset{H}{\overset{|}{\underset{|}{C}}}} \!-\! \overset{H}{\underset{H}{\overset{|}{\underset{|}{C}}}} \!+\! OH$$

*bond broken*      *new bonds formed*

The pieces we need are

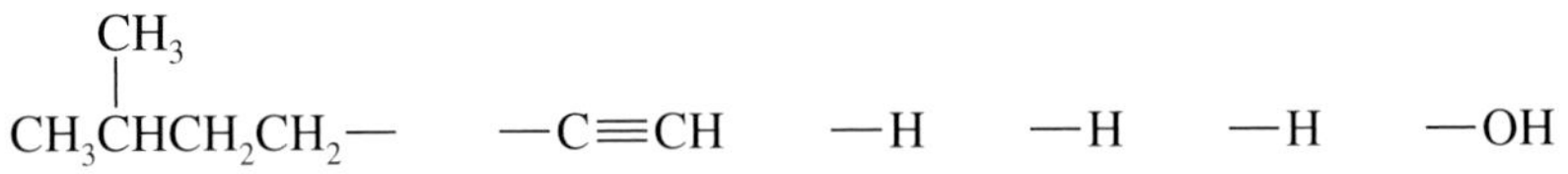

$$\underset{\overset{|}{CH_3}}{CH_3CHCH_2CH_2-} \qquad -C\equiv CH \qquad -H \qquad -H \qquad -H \qquad -OH$$

4. New bonds are created when an electrophile reacts with a nucleophile. Do we recognize any part of the product molecule as coming from a good nucleophile or an electrophilic addition?

   The carbon–carbon bond must have resulted from a nucleophilic substitution reaction. Acetylene can be converted into a good nucleophile. An alkyl group attached to a leaving group must be the electrophile. The other pieces that have been added look like hydrogen, H—H, and water, H—OH. This molecule can be put together from the following reagents.

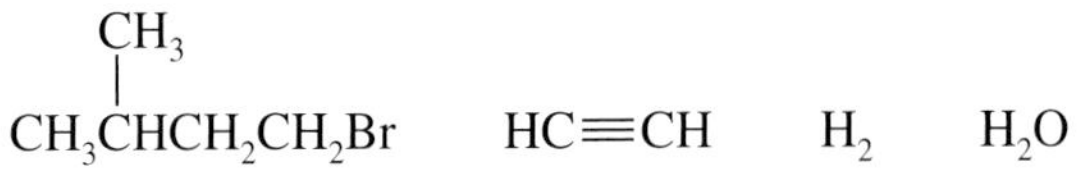

$$\underset{\overset{|}{CH_3}}{CH_3CHCH_2CH_2Br} \qquad HC\equiv CH \qquad H_2 \qquad H_2O$$

5. What type of compound would be a good precursor to the product?

   Anti-Markovnikov addition of water to a double bond will give a primary alcohol. If we had a terminal alkene having the carbon connectivity of the product, we could make the alcohol.

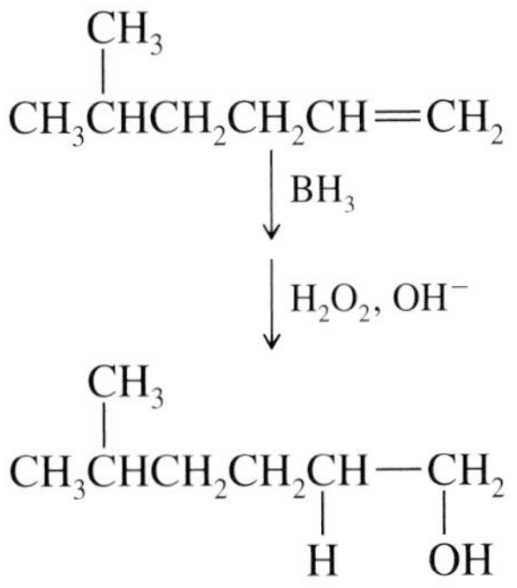

$$\underset{\overset{|}{CH_3}}{CH_3CHCH_2CH_2CH=CH_2}$$

$$\downarrow BH_3$$

$$\downarrow H_2O_2, OH^-$$

$$\underset{\overset{|}{CH_3}}{CH_3CHCH_2CH_2\underset{\overset{|}{H}}{CH}-\underset{\overset{|}{OH}}{CH_2}}$$

6. After this step, do we see how to get from starting material to product? If not, we need to analyze the structure obtained in step 5 by applying questions 4 and 5 to it.

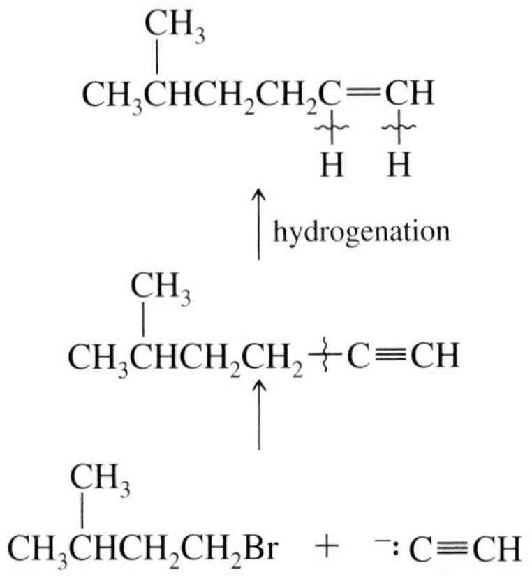

$$\underset{\overset{|}{CH_3}}{CH_3CHCH_2CH_2\underset{\overset{|}{H}}{C}=\underset{\overset{|}{H}}{CH}}$$

$$\uparrow \text{hydrogenation}$$

$$\underset{\overset{|}{CH_3}}{CH_3CHCH_2CH_2 + C\equiv CH}$$

$$\uparrow$$

$$\underset{\overset{|}{CH_3}}{CH_3CHCH_2CH_2Br} \quad + \quad {}^-{:}C\equiv CH$$

The complete synthesis is as follows:

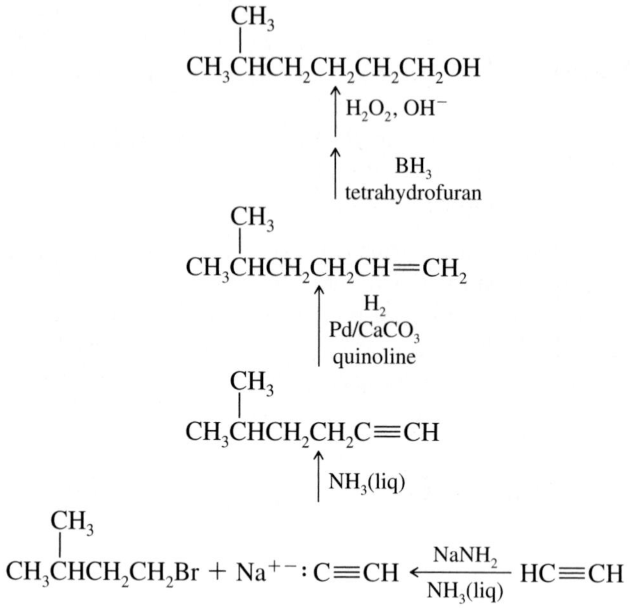

$$CH_3$$
$$CH_3CHCH_2CH_2CH_2CH_2OH$$

$\uparrow$ $H_2O_2$, OH$^-$

$\uparrow$ BH$_3$
tetrahydrofuran

$$CH_3$$
$$CH_3CHCH_2CH_2CH=CH_2$$

$\uparrow$ H$_2$
Pd/CaCO$_3$
quinoline

$$CH_3$$
$$CH_3CHCH_2CH_2C\equiv CH$$

$\uparrow$ NH$_3$(liq)

$$CH_3$$
$$CH_3CHCH_2CH_2Br + Na^{+-}:C\equiv CH \xleftarrow[NH_3(liq)]{NaNH_2} HC\equiv CH$$    ■

**Problem 9.12**   Show how you would carry out each of the following transformations.

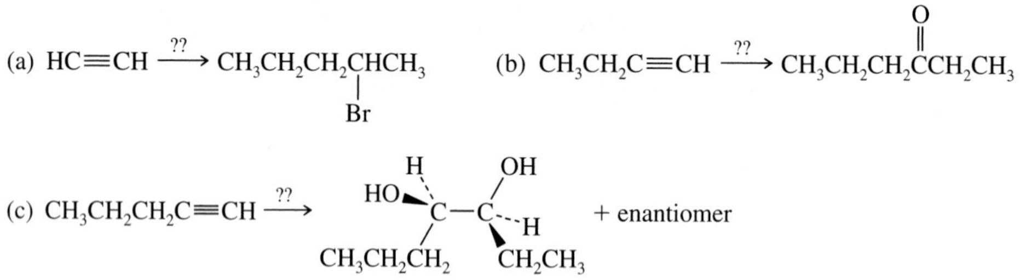

(a) HC$\equiv$CH $\xrightarrow{??}$ CH$_3$CH$_2$CH$_2$CHCH$_3$
                                   |
                                   Br

(b) CH$_3$CH$_2$C$\equiv$CH $\xrightarrow{??}$ CH$_3$CH$_2$CH$_2$CCH$_2$CH$_3$
                                                              ‖
                                                              O

(c) CH$_3$CH$_2$CH$_2$C$\equiv$CH $\xrightarrow{??}$ 

$$\begin{array}{ccc} & H & OH \\ HO & \diagdown C - C \diagup & H \\ & CH_3CH_2CH_2 & CH_2CH_3 \end{array}$$  + enantiomer

# ADDITIONAL PROBLEMS

**9.13** Name the following compounds, specifying the stereochemistry where indicated.

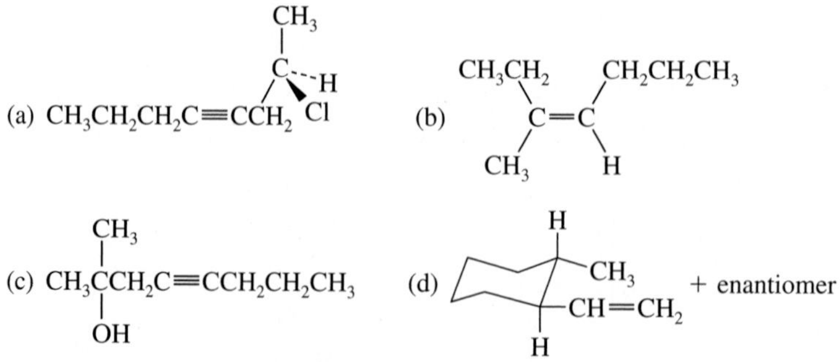

(a) CH$_3$CH$_2$CH$_2$C$\equiv$CCH$_2$Cl
        with CH$_3$ and H and C above

(b) 
$$\begin{array}{cc} CH_3CH_2 & CH_2CH_2CH_3 \\ C=C & \\ CH_3 & H \end{array}$$

(c) CH$_3$CCH$_2$C$\equiv$CCH$_2$CH$_2$CH$_3$
         |
         CH$_3$ (top), OH (bottom)

(d) structure with H, CH$_3$, CH=CH$_2$, H   + enantiomer

**9.14** Draw a structural formula for each of the following compounds.

(a) (*E*)-2-chloro-2-methyl-4-octene    (b) (*S*)-4-bromo-1-heptyne    (c) *cis*-3-methylcyclopentanol    (d) 3-hexyn-1-ol

### Table 9.1    Summary of the Reactions of Alkynes

#### Electrophilic Addition Reactions

| Alkyne | Electrophile | Intermediates | Product |
|---|---|---|---|
| RC≡CR'<br>R may be H | $X_2$ | 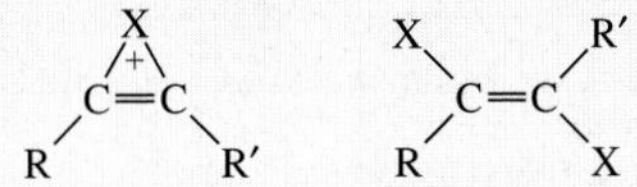 | 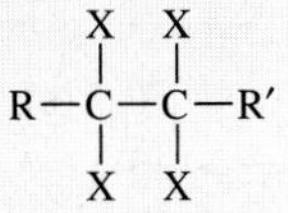 |
|  | HX | 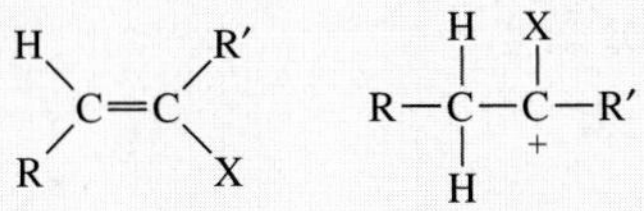 | 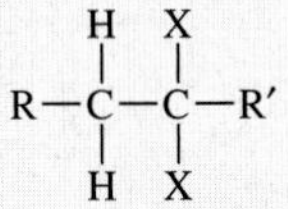 |
|  | $H_3O^+$ | 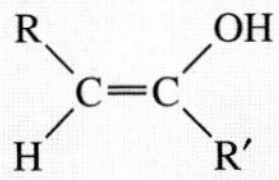 | 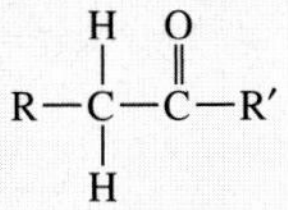 |

#### Reduction Reactions

| | | | |
|---|---|---|---|
|  | $H_2$, $PtO_2$ | | 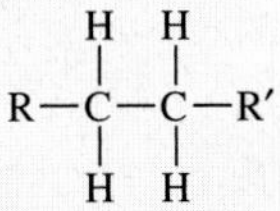 |
|  | $H_2$, Pd, $BaSO_4$<br>quinoline | | 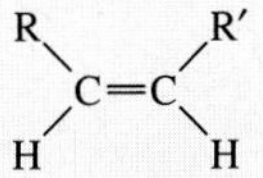 <br>(syn addition) |
|  | Na, $NH_3$ | 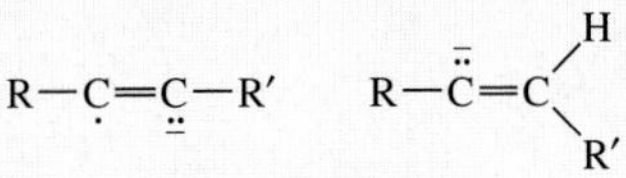 | 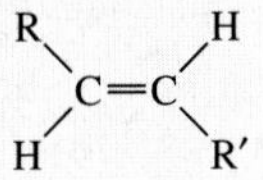 <br>(anti addition) |

| Terminal Alkyne | Reagent | Intermediate | Reagent | Product |
|---|---|---|---|---|
| RC≡CH | $NaNH_2$, $NH_3$(liq)<br>strong base | RC≡C:⁻<br>nucleophile | $R'CH_2X$<br>electrophile | $RC≡CCH_2R'$ |

**9.15** Write equations predicting the reaction of 2-pentyne with each of the following reagents. In cases where you know what the stereochemistry of the product(s) should be, use appropriate conventions to illustrate it.

(a) $H_2$, Pd/$CaCO_3$, quinoline    (b) $H_2$ (excess), Pt    (c) Na, $NH_3$(liq)    (d) product of part (a) +

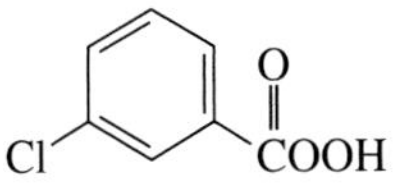

(e) product of part (c) + 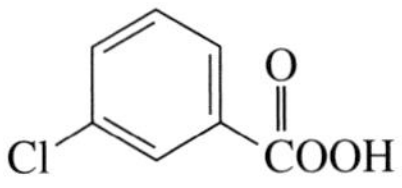    (f) HBr (1 molar equivalent)    (g) HBr (2 molar equivalents)

(h) $Br_2$ (1 molar equivalent)    (i) $Br_2$ (2 molar equivalents)    (j) $H_2O$, $H_2SO_4$, $HgSO_4$

**9.16** Supply structural formulas for the reagents, intermediates, and products designated by capital letters. Show the stereochemistry of the product(s) wherever it is known.

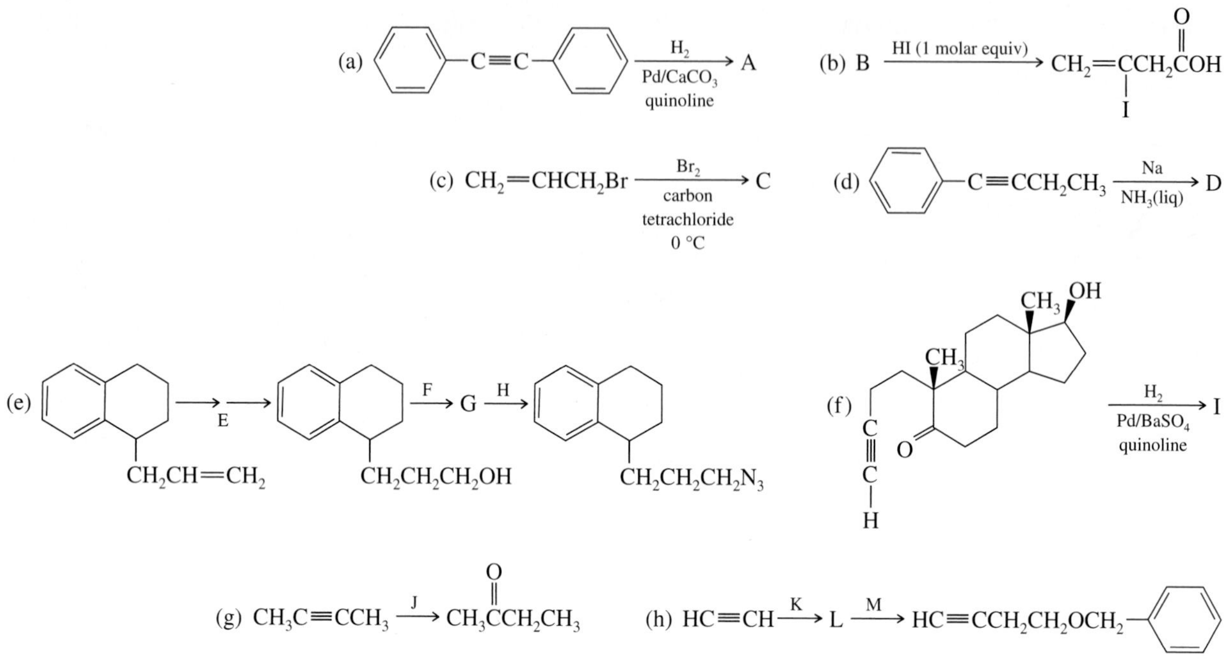

(a) [phenyl]—C≡C—[phenyl] $\xrightarrow[\substack{Pd/CaCO_3 \\ quinoline}]{H_2}$ A

(b) B $\xrightarrow{HI\ (1\ molar\ equiv)}$ $CH_2=CCH_2\overset{\displaystyle O}{\overset{\|}{C}}OH$ with I

(c) $CH_2=CHCH_2Br$ $\xrightarrow[\substack{carbon \\ tetrachloride \\ 0\ °C}]{Br_2}$ C

(d) [phenyl]—C≡CCH₂CH₃ $\xrightarrow[NH_3(liq)]{Na}$ D

(e) [bicyclic with CH₂CH=CH₂] $\xrightarrow{E}$ [bicyclic with CH₂CH₂CH₂OH] $\xrightarrow{F}$ G $\xrightarrow{H}$ [bicyclic with CH₂CH₂CH₂N₃]

(f) [steroid structure with CH₃, OH, CH₃, and C≡CH group] $\xrightarrow[\substack{Pd/BaSO_4 \\ quinoline}]{H_2}$ I

(g) $CH_3C\equiv CCH_3$ $\xrightarrow{J}$ $CH_3\overset{\displaystyle O}{\overset{\|}{C}}CH_2CH_3$

(h) $HC\equiv CH$ $\xrightarrow{K}$ L $\xrightarrow{M}$ $HC\equiv CCH_2CH_2OCH_2$—[phenyl]

**9.17** How would you carry out each of the following transformations? For some, more than one step may be necessary.

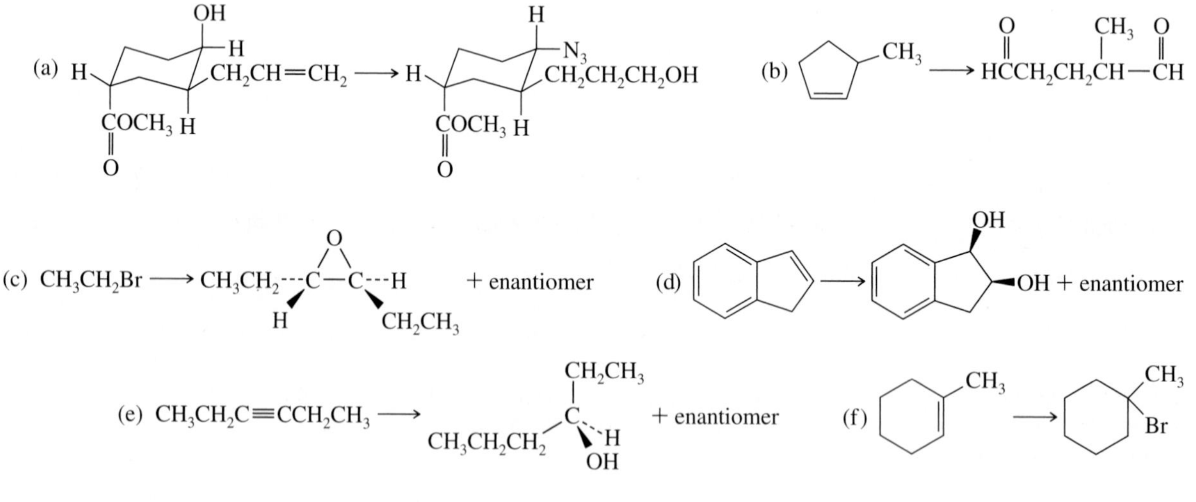

(a) [cyclohexane with OH, H, CH₂CH=CH₂, COCH₃ groups] $\longrightarrow$ [cyclohexane with H, N₃, CH₂CH₂CH₂OH, COCH₃ groups]

(b) [cyclopentene with CH₃] $\longrightarrow$ $H\overset{\displaystyle O}{\overset{\|}{C}}CH_2CH_2\overset{\displaystyle CH_3}{\underset{}{\overset{|}{CH}}}-\overset{\displaystyle O}{\overset{\|}{CH}}$

(c) $CH_3CH_2Br \longrightarrow CH_3CH_2\cdots\overset{\displaystyle O}{\overset{}{C}}-C\cdots H$ with H and CH₂CH₃ + enantiomer

(d) [indene structure] $\longrightarrow$ [indane structure with OH, OH] + enantiomer

(e) $CH_3CH_2C\equiv CCH_2CH_3 \longrightarrow$ $CH_3CH_2CH_2\overset{\displaystyle CH_2CH_3}{\underset{OH}{\overset{|}{C}}}H$ + enantiomer

(f) [cyclohexene with CH₃] $\longrightarrow$ [cyclohexane with CH₃, Br]

**9.18** 2-Methylheptadecane is a sex pheromone for certain species of insects. It has been synthesized using 1-undecyne as the starting material. How would you carry out this synthesis?

**9.19** Trichlorin A is a natural product with significant toxicity against breast cancer cells. Some steps in a partial synthesis of the compound are shown on the next page.

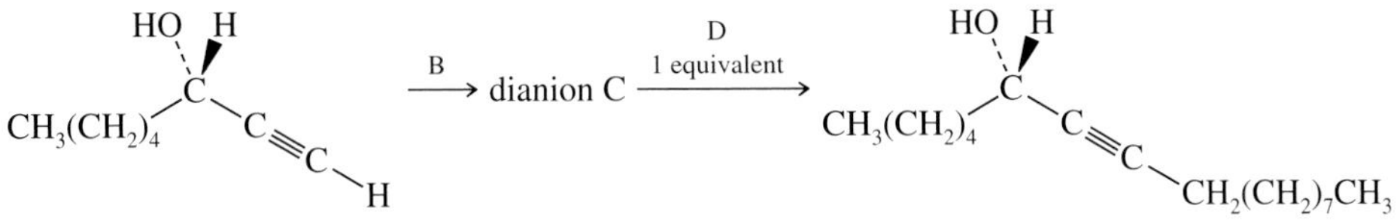

Compound A

*Hint*: Which are the two most acidic protons in Compound A? Assign approximate p$K_a$ values to them.

(a) What reagents are necessary for the transformation shown?
(b) The dianion produced in part (a) reacts selectively at one of its nucleophilic sites. What are the two nucleophiles created in the first step of this reaction, and why is the reaction so selective?
(c) What is the IUPAC name, including stereochemistry, for Compound A in part (a)?

**9.20** Synthesis of a compound isolated from a marine sponge involved the following set of reactions. Provide structural formulas for the reagents or products indicated by letters.

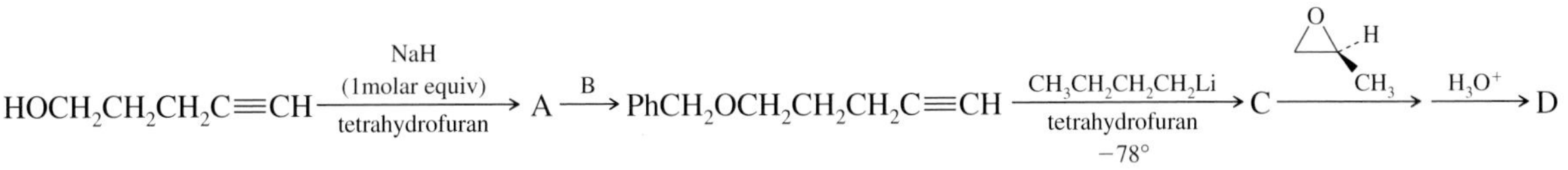

**9.21** The sex attractant of the fall armyworm moth is produced by the female. This material, which attracts the male moth at very low concentrations, is called spodoptol and is a primary alcohol with the molecular formula $C_{14}H_{28}O$. Hydrogenation of this alcohol gives 1-tetradecanol, $C_{14}H_{30}O$. Ozonolysis of spodoptol is known to give two aldehydes, Aldehyde A, $C_5H_{10}O$, and Aldehyde B, $C_9H_{18}O_2$, indicating the position of the double bond in spodoptol. At this point the structure of the sex attractant is known, except for the stereochemistry of the double bond. So little of the material is isolated from the female moths that the compound had to be synthesized before stereochemistry could be determined.

(a) Draw the structural formula for Compounds (or intermediates) D, E, and F formed in the synthesis of spodoptol.

$$HOCH_2CH_2CH_2CH_2CH_2CH_2CH_2CH_2C{\equiv}CH \xrightarrow[NH_3(liq)]{Na^+NH_2^-(excess)}$$

$$\underset{\text{dianion}}{D} \xrightarrow[(2)\ H_3O^+]{(1)\ CH_3CH_2CH_2CH_2Br} E \xrightarrow[\text{poisoned catalyst}]{H_2} \underset{\text{spodoptol}}{F}$$

(b) As a check on the stereochemistry of the double bond, Compound E, prepared above, was also reduced with sodium metal in liquid ammonia. The alkene that resulted attracted no male moths, indicating that this was not the stereochemistry found in the natural product. What is the structure of the alkene formed in the sodium metal reduction of E?
(c) Assign structures to Aldehyde A and Aldehyde B.

$$\text{spodoptol} \xrightarrow[(2)\ CH_3SCH_3]{(1)\ O_3} \underset{C_5H_{10}O}{\text{Aldehyde A}} + \underset{C_9H_{18}O_2}{\text{Aldehyde B}}$$

**9.22** Disparlure, the sex pheromone of the gypsy moth (p. 748) is the epoxide of (*Z*)-2-methyl-7-octadecene. The starting material for a synthesis of disparlure was 1-dodecyne.

(a) Outline a synthesis for disparlure.
(b) How would you modify the synthesis of part (a) to end up with the isomer of disparlure in which the alkyl groups are trans to each other?

**9.23** The following reactions were carried out in the synthesis of (−)-callystatin, a potent cytotoxic natural product isolated from a sponge. Supply structural formulas for the reagents or products designated by capital letters. Show stereochemistry wherever it is known.

(a)

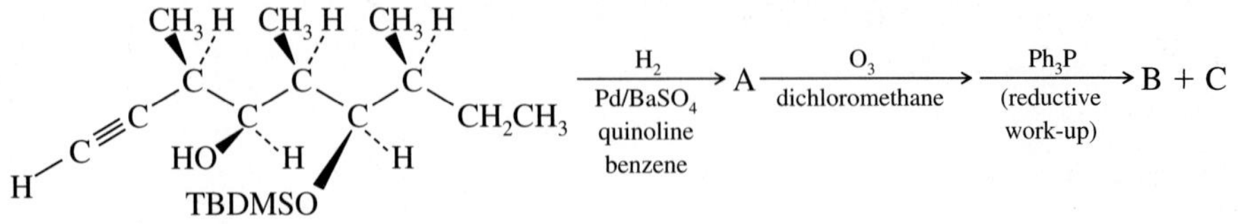

$$\xrightarrow[\substack{\text{Pd/BaSO}_4 \\ \text{quinoline} \\ \text{benzene}}]{\text{H}_2} \text{A} \xrightarrow[\text{dichloromethane}]{\text{O}_3} \xrightarrow[\substack{\text{(reductive} \\ \text{work-up)}}]{\text{Ph}_3\text{P}} \text{B} + \text{C}$$

TBDMS- is a group that protects alcohols from reactions (p. 847). It does not itself react with the reagents shown.

**9.24** A vinyl ether reacts with an alcohol in the presence of a trace of acid as a catalyst. Predict what the product of the reaction will be by evaluating the relative stabilities of possible intermediates.

$$\text{CH}_3\text{OCH}{=}\text{CH}_2 + \text{CH}_3\text{OH} \xrightarrow{\text{HCl}}$$

**9.25** The kinetics of the addition of hydrogen chloride in the gas phase to 2-methylpropene and to ethyl vinyl ether, $\text{CH}_2{=}\text{CHOCH}_2\text{CH}_3$, has been studied. The reaction of 2-methylpropene with hydrogen chloride has an experimental energy of activation $E_a$ (analogous to $\Delta G^{\ddagger}$) of 28.8 kcal/mol. For the reaction of ethyl vinyl ether with hydrogen chloride, the energy of activation is 14.7 kcal/mol. How do you account for the large difference in energy of activation for the two reactions? Write equations and draw energy diagrams to illustrate your answer.

**9.26**

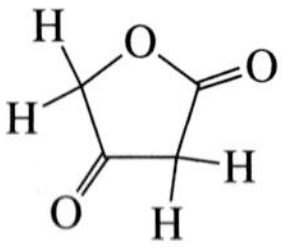

Compound X

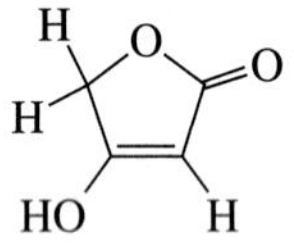

Compound Y
tetronic acid

(a) The dicarbonyl Compound X, shown in the margin, exists primarily in its enol form Y, in which form it is known as tetronic acid. Write a mechanism using the curved-arrow convention that shows how X could be converted to Y. You may use B: and HB$^+$ to indicate bases and acids.

(b) Which is the most acidic proton in tetronic acid? Show this by writing the conjugate base of the acid and using structural formulas that show how chemists rationalize the acidity of the compound.

# The Chemistry of Aromatic Compounds. Electrophilic Aromatic Substitution

10

## A Look Ahead

Aromatic hydrocarbons have multiple $\pi$ bonds and yet are stable toward the electrophilic addition reactions that alkenes undergo. Aromatic hydrocarbons undergo electrophilic substitution reactions instead, an example being the reaction of benzene with bromine in the presence of catalysts.

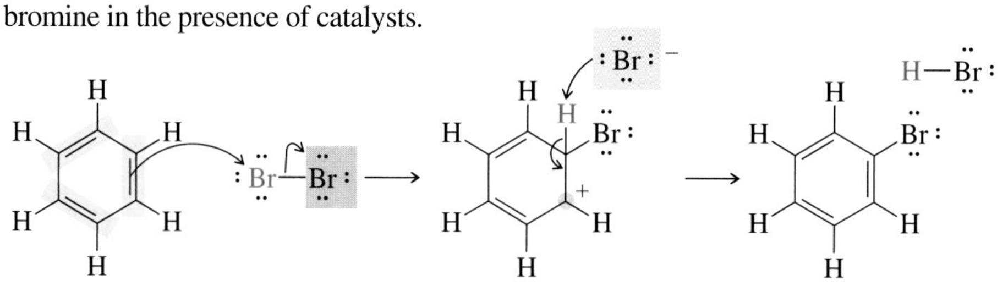

*reaction with the electrophile*     *deprotonation of cationic intermediate*     *substituted hydrocarbon*

Many electrophiles react with aromatic hydrocarbons, so electrophilic aromatic substitution reactions are among the most important reactions for transforming organic compounds. The structure of the compound undergoing the substitution determines its reactivity and the regioselectivity of the reaction. This chapter will look at some of the different electrophiles that react with aromatic compounds and show how to predict the major product of an electrophilic aromatic substitution reaction.

Chemists have long debated the question of what makes a compound aromatic. This chapter also will examine the current model for aromaticity.

**Workbook Exercises**

## 10.1 Aromaticity

### A. Introduction. Resonance Stabilization of Benzene

Vanilla and oil of wintergreen are representatives of a class of flavoring agents isolated from plants and containing chemical substances called *aromatic* because of their characteristic fragrances.

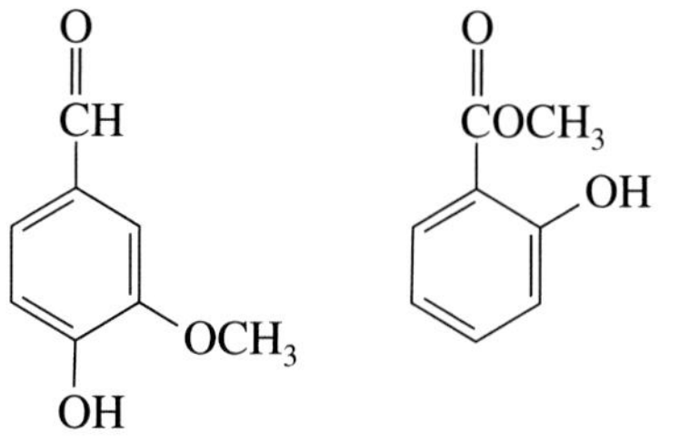

vanilla

*constituent of vanilla*

methyl salicylate

*constituent of oil of wintergreen*

351

When the structures of a number of these compounds were determined, they were all found to contain substituted benzene rings. In time, benzene and its derivatives were called **aromatic compounds** to distinguish them from saturated acyclic and cyclic hydrocarbons and simple unsaturated compounds. Still later, the term was redefined to include compounds that do not have a benzene ring but have chemical properties resembling those of benzene. Such compounds are said to have **aromaticity.**

Chemists are still arguing about what aromaticity means. At first the term was used to indicate that a compound had special stability, usually defined as resonance energy, and that it resisted the addition reactions expected of a molecule with that degree of unsaturation (p. 67). Benzene and substituted benzenes do not react with a large number of reagents that do react with other functional groups. This unreactivity has been demonstrated many times in earlier chapters in reactions that transform substituents on aromatic rings but leave the rings untouched. For example, an alkene usually can be hydrogenated using hydrogen gas at atmospheric pressure and at room temperature over a variety of catalysts such as palladium, platinum, or nickel (p. 298). To add hydrogen to an aromatic nucleus, however, special catalysts and high temperatures and pressures have to be used. For instance, o-xylene is hydrogenated to a mixture of *cis-* and *trans-*1,2-dimethylcyclohexane at 100 °C and at 1000 psi (~66 atm).

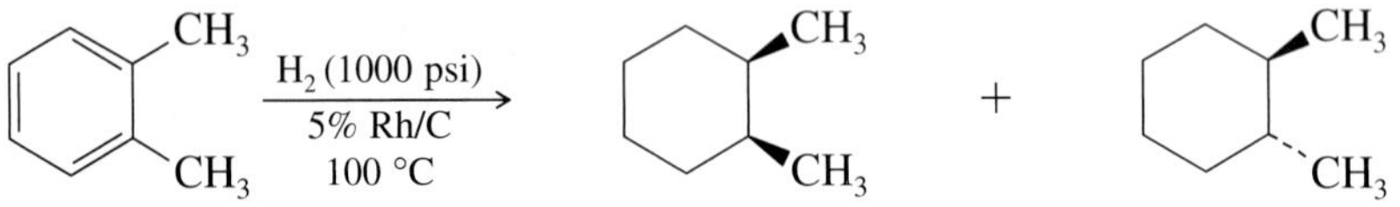

| | | |
|---|---|---|
| *o*-xylene | 9 : 1 | |
| | *cis*-1,3-dimethylcyclohexane | *trans*-1,3-dimethylcyclohexane |

The experimental heat of hydrogenation (p. 282) of benzene is −49.8 kcal/mol. This is 36 kcal/mol less than the heat of hydrogenation expected for "1,3,5-cyclohexatriene," a hypothetical six-membered ring containing three localized double bonds, shown below as containing alternating short double bonds and long single bonds. This value is calculated as three times the value of the heat of hydrogenation of cyclohexene.

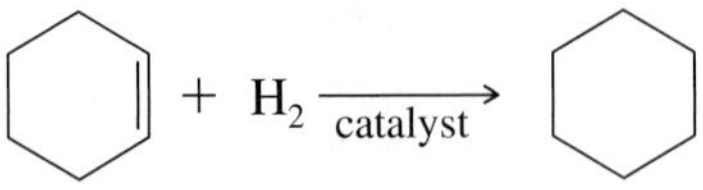

cyclohexene + H$_2$ $\xrightarrow{\text{catalyst}}$ cyclohexane    $\Delta H° = -28.6$ kcal/mol    experimental

"1,3,5-cyclohexatriene" + 3 H$_2$ $\xrightarrow{\text{catalyst}}$ cyclohexane    $\Delta H° = -85.8$ kcal/mol    calculated

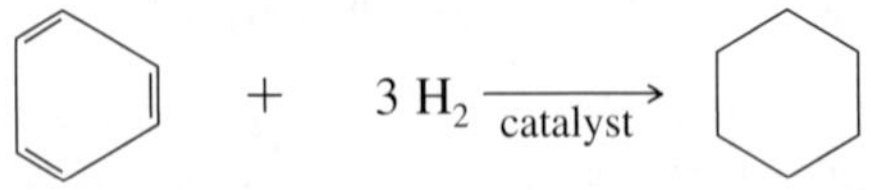

benzene + 3 H$_2$ $\xrightarrow[\substack{\text{heat} \\ \text{pressure}}]{\text{catalyst}}$ cyclohexane    $\Delta H° = -49.8$ kcal/mol    experimental

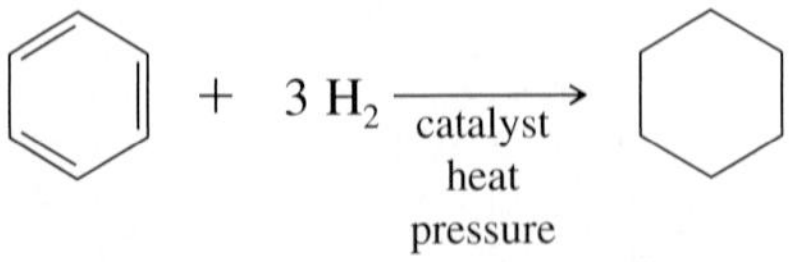

Benzene thus is more stable by 36 kcal/mol than expected for a compound containing three double bonds in a six-membered ring. The value 36 kcal/mol is called the **empirical resonance energy** of benzene. It reflects the degree of stabilization of the

aromatic ring that is attributed to the delocalization of $\pi$ electrons. This resonance stabilization of benzene and other aromatic compounds is responsible for the lower reactivity of their $\pi$ bonds in comparison with that of the $\pi$ bonds of alkenes and alkynes.

## B. The Aromatic Sextet

In 1926, Sir Robert Robinson, a British chemist who was among the first to suggest and use the ideas that form the basis for the mechanisms chemists write for organic reactions, recognized that the compounds that had been classified as aromatic always had six electrons, in either $\pi$ or nonbonding orbitals, in a planar ring. Benzene is such a compound. Robinson suggested that there was a special stability associated with what he called an **aromatic sextet.** Some other compounds that exhibit the experimentally observed properties that have been assigned to aromaticity, such as resistance to addition reactions, are shown below.

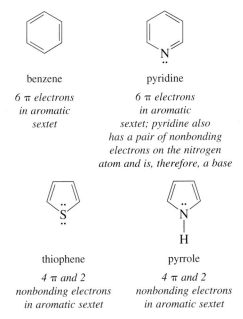

benzene

*6 π electrons*
*in aromatic*
*sextet*

pyridine

*6 π electrons*
*in aromatic*
*sextet; pyridine also*
*has a pair of nonbonding*
*electrons on the nitrogen*
*atom and is, therefore, a base*

thiophene

*4 π and 2*
*nonbonding electrons*
*in aromatic sextet*

pyrrole

*4 π and 2*
*nonbonding electrons*
*in aromatic sextet*

As the theory of molecular bonding developed further, it became clear that a sextet was indeed a significant number, corresponding to the number of electrons that fill the three bonding molecular orbitals of benzene (p. 68). Fewer electrons would leave unpaired electrons in the bonding orbitals and create a species with radical character. Any more electrons would have to go in antibonding molecular orbitals and would create an unstable, high-energy species. Benzene has a filled shell of molecular orbitals, just as an element such as neon has a filled shell of atomic orbitals (Figure 10.1, p. 354).

In 1931, German chemist Erich Hückel pointed out that there were other numbers of electrons that corresponded to filled shells for smaller or larger rings. He proposed what has since become known as **Hückel's Rule:** Any conjugated monocyclic polyene that is planar and has $(4n + 2)$ $\pi$ and/or nonbonding electrons, with $n = 0, 1, 2$, etc., will exhibit the special stability associated with aromaticity. Since then, researchers have tried to synthesize compounds that will test Hückel's prediction.

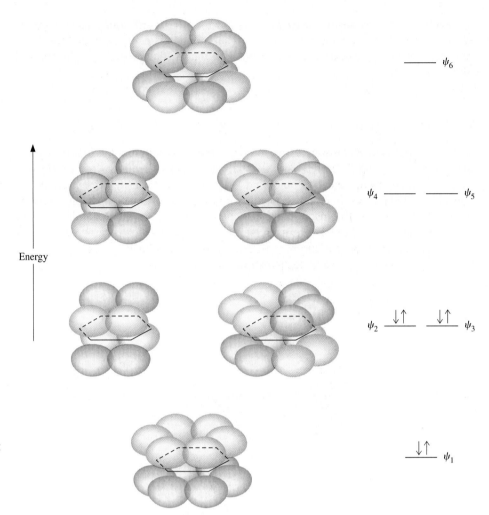

Energy

$\psi_6$

$\psi_4$ —— —— $\psi_5$

$\psi_2$ ↓↑ ↓↑ $\psi_3$

↓↑ $\psi_1$

**Figure 10.1**
Combinations of the six 2*p* atomic orbitals of benzene that form the six molecular π orbitals, with the relative energies of the three bonding and three antibonding orbitals.

Benzene is the ideal aromatic compound. It has $4n + 2$, or 6, π electrons with $n = 1$. The six carbon atoms in benzene are $sp^2$-hybridized, so their regular bond angles of 120° create a perfect hexagon. The molecule is planar and totally symmetrical. Because all the ring atoms are carbon, no differences in polarity are introduced by the presence of other elements in the ring. The electrons are, therefore, delocalized evenly over the entire system.

Cyclobutadiene, a conjugated cyclic polyene with four carbon atoms in the ring, is unstable, and cyclooctatetraene, with an eight-membered ring, behaves as a polyene.

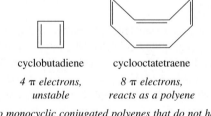

cyclobutadiene      cyclooctatetraene

*4 π electrons,*      *8 π electrons,*
*unstable*      *reacts as a polyene*

*two monocyclic conjugated polyenes that do not have*
*(4n + 2) π electrons and are not aromatic*

Cyclobutadiene has 4 $\pi$ electrons and thus does not meet the requirement of Hückel's Rule that an aromatic compound have $(4n + 2)$ $\pi$ electrons. The compound has been isolated only when frozen in an argon matrix at 4 K, where it has been detected by infrared spectroscopy. Cyclooctatetraene is an ordinary polyene in its reactivity. It is not planar (a regular octagon would have to have bond angles of 135°), so there cannot be continuous conjugation of the double bonds around the ring. It, too, lacks the proper number of electrons to be aromatic, according to Hückel's Rule.

Larger rings with conjugated cis double bonds cannot be planar because the bond angles required for such geometric figures are even larger than the 135° of a regular octagon. However, trans double bonds are possible for large rings. Such a configuration puts hydrogen atoms inside the ring, where they interfere with each other and prevent planarity. Only for rings consisting of eighteen or more carbon atoms is this steric hindrance removed. None of these larger ring compounds, called **annulenes,** has the stability of benzene, although some of them have some of the properties associated with aromaticity, such as high resonance energy. [18]Annulene is shown below.

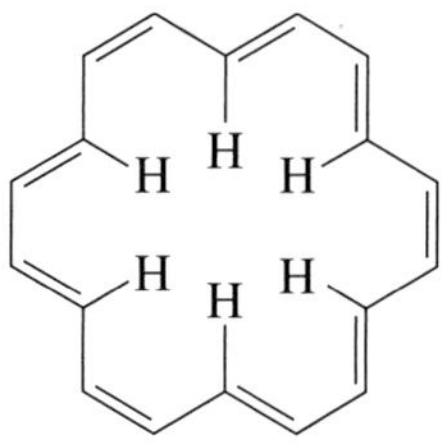

[18]annulene

*planar*

*[18]annulene has $(4n+2)$ $\pi$ electrons with n $= 4$,*
*but lacks the stability of benzene*

---

**Problem 10.1**  Decide whether or not each of the following species is aromatic.

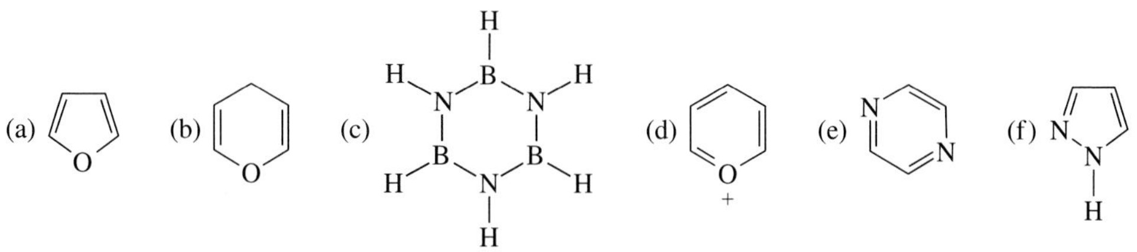

[*Hint:* Draw an orbital picture of c.]

---

## C. Detection of Aromaticity by Nuclear Magnetic Resonance Spectroscopy

Aromatic compounds have distinctive peaks in their nuclear magnetic resonance spectra. A comparison of the proton magnetic resonance spectra of benzene and 1,3,5,7-cyclooctatetraene (Figure 10.2, p. 356) shows clearly that the hydrogen

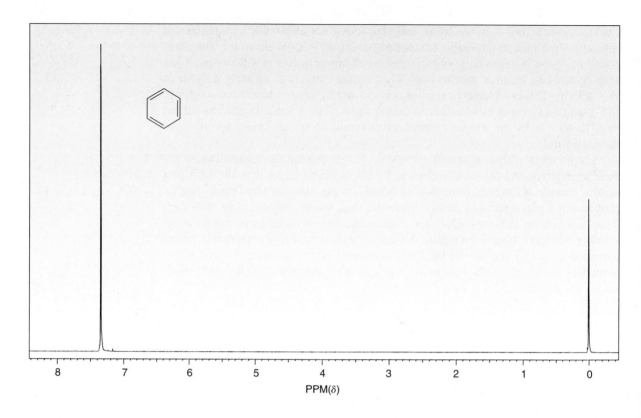

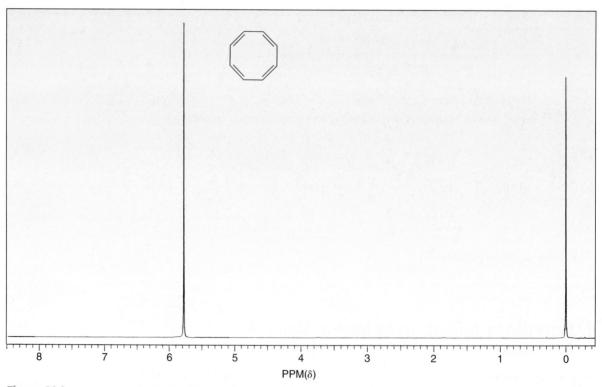

*Figure 10.2*
Proton magnetic resonance spectra for benzene and 1,3,5,7-cyclooctatetraene.

atoms on the benzene ring absorb at a different frequency than do those on the cyclooctatetraene, which behaves chemically as an alkene. The observed difference is attributed to the delocalization of electrons around the benzene ring (p. 354), which creates what is known as a ring current. How this affects the hydrogen atoms on the benzene ring is the subject of Section 11.4D (p. 407). A modern criterion of aromaticity for an unsaturated cyclic compound then becomes the observation of ring current in the nuclear magnetic resonance spectrum of the compound. The spectrum of benzene shows it; that of cyclooctatetraene supports the earlier conclusion, which is based on chemical evidence as well as on theory, that the double bonds in this cyclic polyene are localized. The hydrogens in cyclooctatetraene absorb at the same frequency as those in cyclohexene, for example (Figure 11.11, p. 408). The compound is, therefore, not aromatic.

**Problem 10.2** The proton magnetic resonance spectrum of thiophene (p. 353) is shown in Figure 10.3. In comparison, the hydrogen atoms on the $sp^2$-hybridized carbon atoms of methyl vinyl sulfide, $CH_3SCH{=}CH_2$, absorb at $\delta$ 4.95, 5.18, and 6.43. What conclusions can you draw from these data?

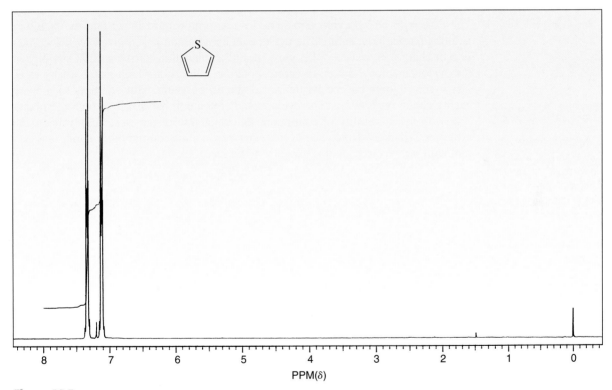

*Figure 10.3*

## D. Aromaticity of Hydrocarbon Anions and Cations

Ions, as well as neutral molecules, can be said to have aromaticity. For example, cyclopentadiene ($pK_a$ 15) is a strong acid compared with other types of alkenes and can be deprotonated by *tert*-butoxide anion.

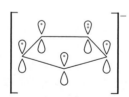

cyclopentadiene

$pK_a$ *15*

cyclopentadienyl anion

$4n + 2 = 6,$
n = *1*

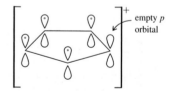

*resonance contributors for the*
*cyclopentadienyl anion*

The ease with which cyclopentadiene loses a proton must mean that the resulting anion is particularly stable. The cyclopentadienyl anion is stabilized by delocalization of charge. However, if this were the full explanation for the special stability of the cyclopentadienyl anion, the corresponding cation, for which just as many resonance contributors can be written, also should be stable. But attempts to prepare such a cation have failed. The cyclopentadienyl anion has six electrons in a planar ring with the possibility of continuous delocalization of the electrons. It meets the criteria of Hückel's Rule. The cyclopentadienyl cation, on the other hand, is a system with 4 $\pi$ electrons and is expected to be unstable.

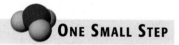

**ONE SMALL STEP**

Cyclopentadienyl anion reacts with iron(II) chloride to give a stable compound in which an iron atom is sandwiched between two cyclopentadiene rings. This compound, bis(cyclopentadienyl)iron, is also called ferrocene. Ferrocene undergoes reactions that are typical of aromatic systems.

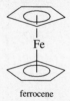

ferrocene

*all C—H bonds on*
*ferrocene are equivalent*

PROBLEM:  How do you explain the stability of ferrocene?

Hint:  A similarly stable compound is not formed with iron(III). Looking at the periodic table may give you a clue.

*p orbitals of cyclopentadienyl anion*
*showing the presence*
*of 6 $\pi$ electrons*
*(4n + 2, n = 1)*

*p orbitals of cyclopentadienyl cation*
*showing the presence*
*of only 4 $\pi$ electrons*

empty *p* orbital

Cycloheptatriene is a compound with 6 $\pi$ electrons in a ring. It is not aromatic because there cannot be continuous delocalization of the $\pi$ electrons around the ring as long as one of the carbon atoms is an $sp^3$-hybridized carbon atom with no unhybridized $p$ orbital to overlap with the $p$ orbitals on the other carbon atoms. Cycloheptatriene adds bromine and then easily loses hydrogen bromide to give a crystalline compound. This compound is insoluble in nonpolar organic solvents such as ether but dissolves readily in cold water and gives an instant precipitate of silver bromide when silver nitrate is added to its solution. The compound melts at 203 °C with decomposition. All these properties suggest that the monobromo compound ionizes as shown on the following page.

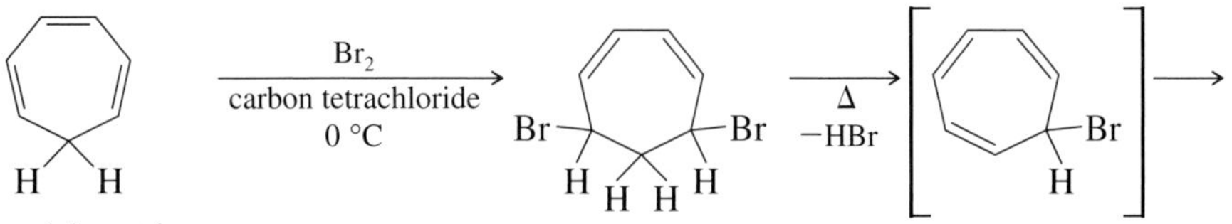

1,3,5-cycloheptatriene

dibromide from
1,3,5-cycloheptatriene
shown as the
1,6-addition product
100%

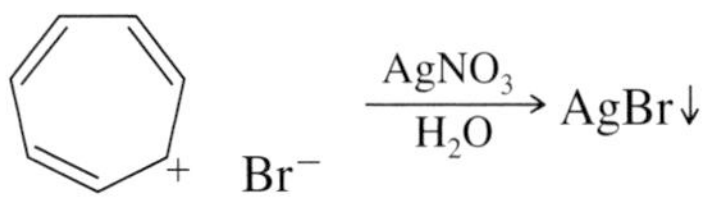

cycloheptatrienyl
bromide
tropylium bromide

*mp 203 °C with decomposition,
insoluble in nonpolar organic
solvents, soluble in water*

In other words, the compound is the salt of a stable carbocation, known as the **tropylium ion,** with bromide ion. The aromaticity possible for the cation stabilizes it so much that the normal tendency for covalent bonds to form between carbon and bromine is overcome.

⟲ **Study Guide**
**Concept Map 10.1**

**Problem 10.3**   Write a mechanism showing how tropylium bromide is formed from the dibromide of 1,3,5-cycloheptatriene.

**Problem 10.4**   Draw a representation of the tropylium ion showing the skeleton of $\sigma$ bonds for the molecule and the $p$ orbitals used for delocalization of charge. Include all of the available $\pi$ electrons.

## 10.2 Kekulé Structures and Nomenclature for Aromatic Compounds

Michael Faraday isolated benzene in 1825 from the residues of gaseous fractions of coal. The structural formula that chemists now use for benzene was first written in 1865 by the German chemist August Kekulé after a dream in which he saw rows of atoms twisting with a snakelike motion until finally, as he said: "One of the snakes had seized hold of its own tail, and the form whirled mockingly before my eyes. As if by a flash of lightning, I awoke, and this time also I spent the rest of the night in working out the consequences of the hypothesis." Kekulé recognized that the chemical properties of benzene were not compatible with the alternating double and single bonds indicated by his structural formula, and by 1872, he arrived at the idea that the double bonds were not localized. He wrote two structures, the ones now known as resonance contributors for benzene (p. 68), to show the changing nature of the double bonds, a remarkable feat of chemical imagination for his time, when electrons had not yet been discovered.

Along with benzene, other hydrocarbons that are unsaturated in their molecular formulas but stable toward reagents that add to double bonds have been isolated from various sources, chiefly from fractions of coal tar. Some of these compounds are shown with their names on the next page.

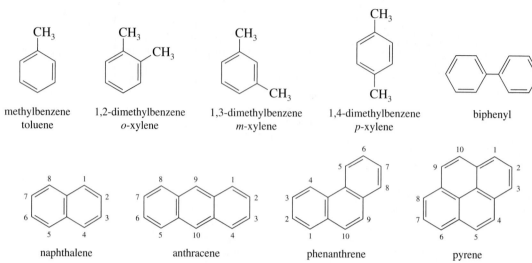

| | | | | |
|---|---|---|---|---|
| methylbenzene toluene | 1,2-dimethylbenzene *o*-xylene | 1,3-dimethylbenzene *m*-xylene | 1,4-dimethylbenzene *p*-xylene | biphenyl |

| | | | |
|---|---|---|---|
| naphthalene | anthracene | phenanthrene | pyrene |

Benzene, other aromatic hydrocarbons, and their alkyl derivatives as a class are known as **arenes.** Besides hydrocarbons, coal tar also contains aromatic compounds containing nitrogen, sulfur, or oxygen. Among the following, those having elements other than carbon in a ring—pyridine, quinoline, indole, and thiophene—are also known as **heterocyclic compounds.**

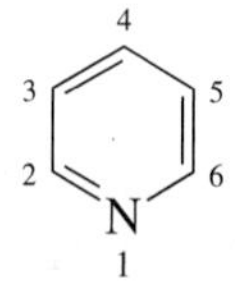

| | | | |
|---|---|---|---|
| pyridine | quinoline | indole | thiophene |

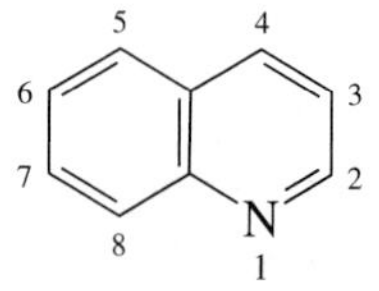

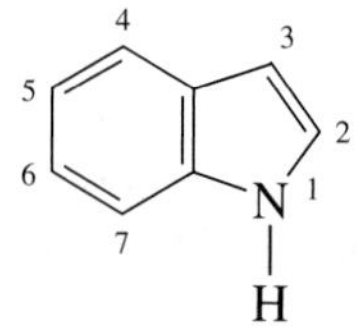

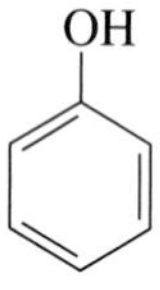

| | | | | |
|---|---|---|---|---|
| phenol | *o*-cresol | *p*-cresol | 1-naphthol α-naphthol | 2-naphthol β-naphthol |

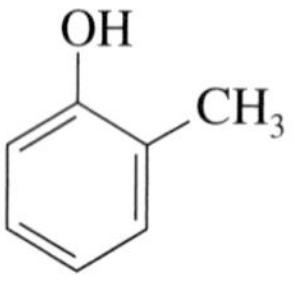

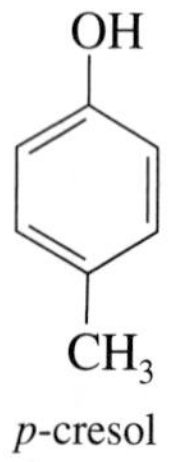

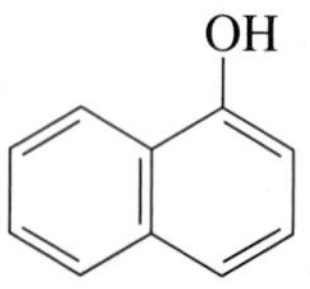

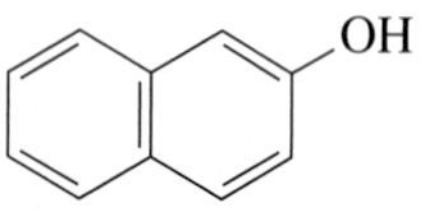

Other aromatic compounds are named as substitution products of benzene and the compounds shown above. The name of the parent compound is given a prefix with the name of each substituent and numbers indicating relative positions. For benzene derivatives, ***ortho*** (*o*), ***meta*** (*m*), and ***para*** (*p*) are also used to designate 1,2, 1,3, and 1,4 relationships between two substituents. The following examples illustrate these rules.

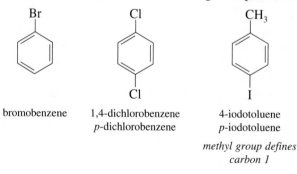

| | | |
|---|---|---|
| bromobenzene | 1,4-dichlorobenzene *p*-dichlorobenzene | 4-iodotoluene *p*-iodotoluene |

*methyl group defines carbon 1*

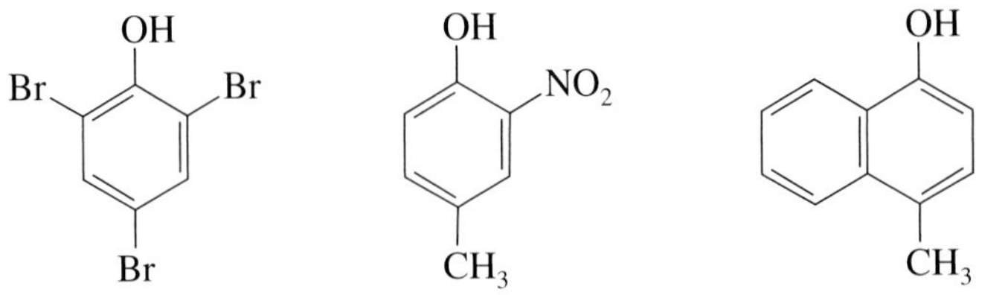

2,4,6-tribromophenol

*hydroxyl group defines carbon 1*

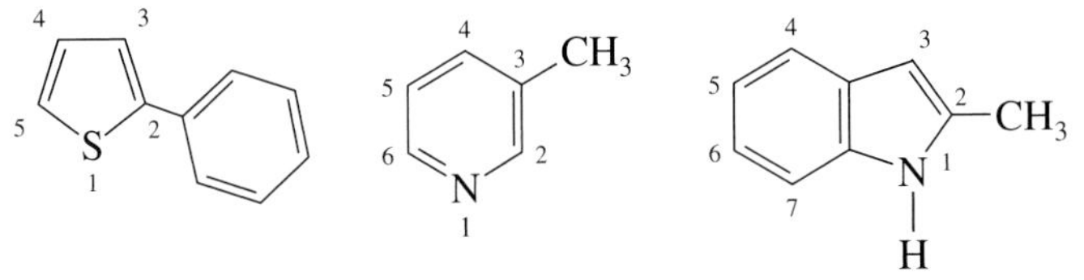

4-methyl-2-nitrophenol

4-methyl-1-naphthol

2-phenylthiophene

3-methylpyridine

2-methylindole

Some of the compounds that are shown on page 360 contain more than one aromatic ring. In biphenyl, for example, the two rings are joined by a single bond and are conjugated with each other. In other cases, such as naphthalene, anthracene, phenanthrene, and pyrene, the aromatic rings share at least one side and are said to contain **fused-ring systems** or to be **polycyclic aromatic hydrocarbons.**

The fused-ring aromatic hydrocarbons do not have completely identical resonance contributors, as benzene does. For example, the Kekulé structures, the structural formulas with localized double and single bonds, that can be written as resonance contributors for naphthalene are not equivalent.

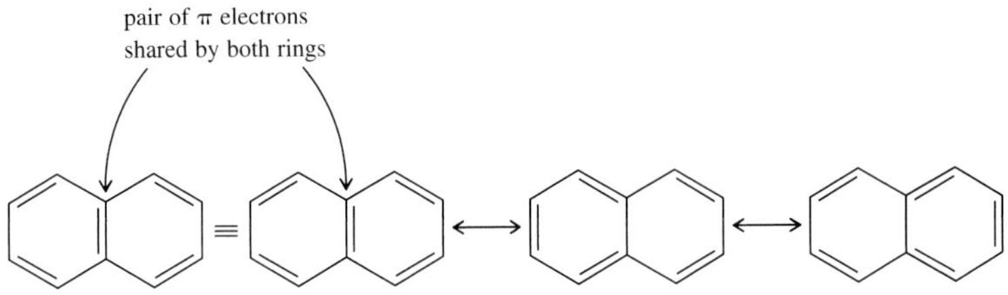

pair of π electrons
shared by both rings

*resonance contributors for naphthalene*

This nonequivalence of the resonance contributors is reflected in the chemical properties of polycyclic aromatic hydrocarbons. The difference in reactivity between rings in such compounds is clearly illustrated by anthracene and phenanthrene. The center ring in anthracene is highly reactive. When bromine is added to anthracene, it adds to the central ring in a reaction that allows the rings on either side to remain aromatic.

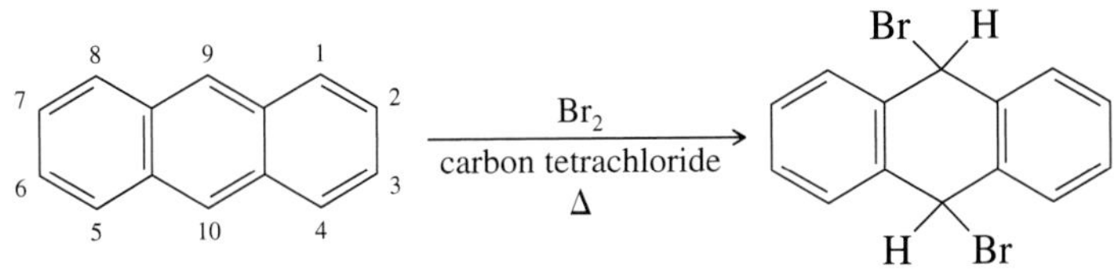

$\xrightarrow[\Delta]{\substack{Br_2 \\ \text{carbon tetrachloride}}}$

anthracene

9,10-dibromo-9,10-dihydro-anthracene

Phenanthrene also has a central ring that is more vulnerable to chemical reagents than the other two rings are. Most of the resonance contributors that can be written for phenanthrene have a double bond localized between the two carbon atoms usually numbered 9 and 10 in this system. Phenanthrene adds bromine at these positions as though it were an alkene (p. 302).

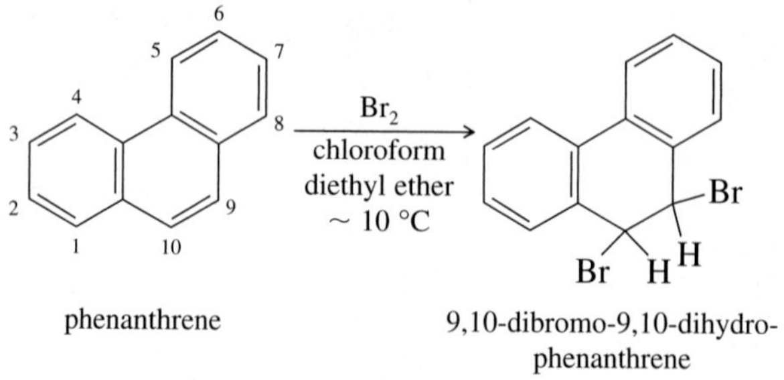

phenanthrene                    9,10-dibromo-9,10-dihydro-
                                      phenanthrene

Thus, although it is possible to talk in theoretical terms of molecular energy levels and the delocalization of electrons over entire molecules, Kekulé structures, with their alternating double- and single-bond character, are reminders of the experimental facts of the chemistry of aromatic compounds. Indeed, chemists have been unable to synthesize compounds for which electron delocalization over large molecules is possible but for which Kekulé structures cannot be written. For these reasons, Kekulé structures are used for benzene in this book, not the convention of a circle in a hexagon that is sometimes used to symbolize benzene.

**Problem 10.5**    Name the following compounds.

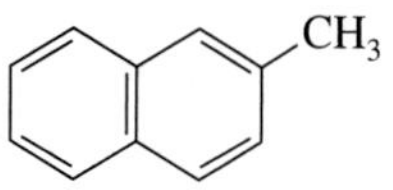

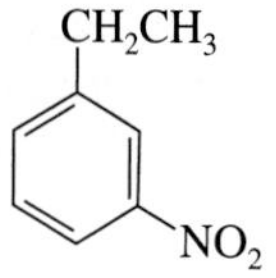

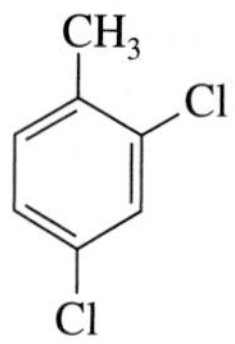

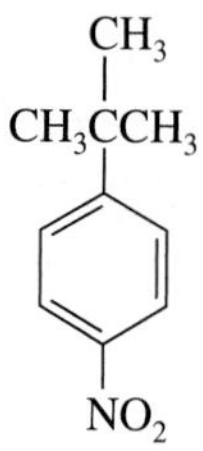

**Problem 10.6**    Draw a structural formula for each of the following compounds.

(a) 2,4-dinitrophenol      (b) 2,4,6-trimethylpyridine      (c) 8-hydroxyquinoline      (d) 2,4,6-trinitrotoluene (TNT)

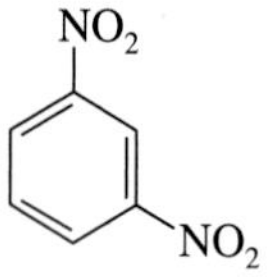

(e) 4-bromo-2-methylnaphthalene      (f) 4,4'-dibromobiphenyl      (g) 8-methyl-1-naphthol      (h) 3-ethylthiophene

(i) 2,4-dibromo-1-pentylbenzene

**Problem 10.7**    Draw resonance contributors for phenanthrene. Prove to yourself that the 9,10 bond in phenanthrene would be expected to behave more like a double bond than any of the other bonds in the molecule.

## 10.3   $C_{60}$. Buckminsterfullerene

### A. The Discovery and the Structure of $C_{60}$

In 1985, Harold Kroto of the University of Sussex and Richard Smalley of Rice University reported the discovery of stable structures consisting only of carbon atoms. The discovery came during a series of experiments in which the researchers were investigating how gaseous carbon is converted into particles—ultimately of soot, for example—in the kinds of processes that take place in stars. The most abundant and stable of the new structures was a $C_{60}$ molecule, which instantly cap-

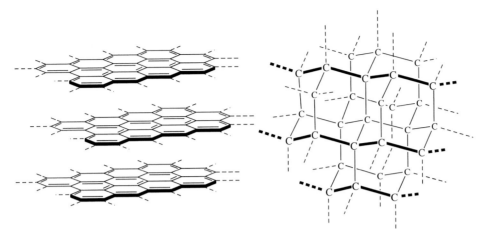

*Figure 10.4*
The three-dimensional
structures of graphite (left) and
diamond (right).

tured the imagination of chemists and has since been the subject of intensive research. The species is a new form of carbon, adding to the two other forms known,
graphite and diamond (Figure 10.4). The new form of carbon, C$_{60}$, is different from
graphite and diamond in that it consists of discrete molecules rather than extended
networks of carbon atoms.

In graphite, carbon atoms are bonded to each other in layers resembling huge
polycyclic aromatic hydrocarbons known as graphene. The layers are held to each
other by van der Waals interactions (p. 29). As a result, they slip past each other under pressure. Graphite is used as a dry lubricant because of this property. The "lead"
in pencils is really graphite combined with a binder to make it hard enough to with-
stand the pressure on the point. Different amounts of binder create pencils of different hardness. When we write with a pencil, we are laying down a trace of graphite
on the page.

Diamond, on the other hand, consists of carbon atoms that are covalently
bonded to each other by $\sigma$ bonds in a three-dimensional network that gives extraor-
dinary rigidity and hardness to the structure. Diamonds are among the hardest materials known and are used industrially to cut other materials.

When soot is extracted with benzene, a red solution results. Chromatography of
the solution on alumina results in the isolation of two species in a ratio of ~ 5 : 1.
The more abundant species is C$_{60}$, which has a molecular weight of 720 amu, as determined by mass spectrometry, an instrumental technique that allows us to determine molecular weights with very small samples of material (p. 467). The other
species, which has a molecular weight of 840 amu, is C$_{70}$. When C$_{60}$ is isolated
from soot, it is a mustard-colored solid. C$_{60}$ is soluble in common organic solvents,
especially solvents such as benzene and toluene, with which it forms magenta-
colored solutions. The red color of the benzene extract of soot comes from C$_{70}$.
Both C$_{60}$ and C$_{70}$ have very high melting points. Although they are soluble in organic solvents, they dissolve only slowly, indicating that the molecules pack together very well in their crystals.

The $^{13}$C nuclear magnetic resonance spectrum of C$_{60}$ shows a single peak at
142.68 ppm in the aromatic region (p. 403) of the spectrum, indicating that all the
carbon atoms in C$_{60}$ are equivalent to each other. This is possible only with a highly
symmetrical structure, also indicated by the high melting point and the close pack-
ing of the crystal structure of C$_{60}$. The structure proposed for C$_{60}$ is that of the familiar soccer ball, known technically as a truncated icosahedron (Figure 10.5). In

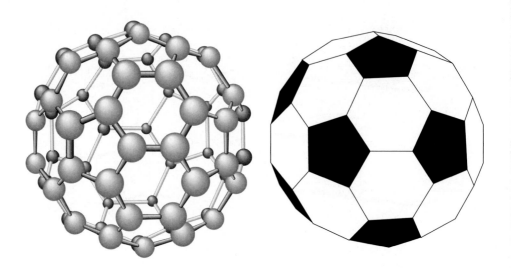

*Figure 10.5*
Buckminsterfullerene and a soccer ball.

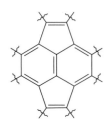

this structure, every carbon atom is at the juncture of two six-membered rings and one five-membered ring. The six-membered rings each contain three double bonds. The important unit of structure in $C_{60}$ may be symbolized by the structural fragment shown in the margin. The $C_{60}$ species has been given the name buckminsterfullerene to honor Buckminster Fuller, inventor of the architectural structures known as geodesic domes. More familiarly, $C_{60}$ is called "buckyball." The name **fullerenes** has been chosen for the class of structures consisting of five- and six-membered rings. For example, $C_{70}$ is also a fullerene.

Chemists have learned how to make $C_{60}$ easily by laser vaporization of graphite and how to purify it. This has enabled them to carry out chemical studies on $C_{60}$.

## B. The Chemistry of $C_{60}$

The studies done so far with $C_{60}$ indicate that it is not aromatic in the sense that benzene is. For instance, its $^{13}C$ nuclear magnetic resonance band at 142.68 ppm is more typical of a strained aromatic structure than it is of a structure such as naphthalene, which absorbs at 133.7 ppm. Electrophiles add to the double bonds in $C_{60}$. For example, it reacts with bromine.

$$C_{60} + Br_2 \xrightarrow{25\ °C} C_{60}Br_2 + C_{60}Br_4$$

Because of the size and the symmetry of $C_{60}$, it is hard to obtain structural information about the exact location of a reaction. This problem was solved by the formation of crystalline products that could be examined by x-ray crystallography. The first of these products was an osmium tetroxide adduct.

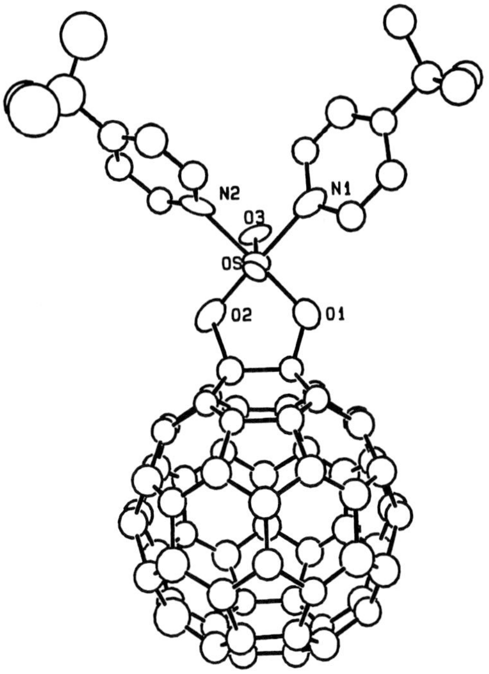

N2
N1
O3
OS
O2
O1

Reprinted with permission from *Science*, Vol. 253, p. 312. Copyright © 1991 by American Association for the Advancement of Science.

*Figure 10.6*
Picture from the x-ray crystallographic structure determination of the $C_{60}$-osmium tetroxide adduct with two 4-*tert*-butylpyridine molecules complexed with the osmium atom.

X-ray crystallography shows very clearly that the osmium tetroxide adds to a double bond that is shared by two six-membered rings. The x-ray crystallographic picture also confirms that $C_{60}$ indeed is a truncated icosahedron. The structure shown in Figure 10.6 has been called "bunny ball," for obvious reasons.

No chemist can look at the structure of $C_{60}$ without imagining some other chemical species encapsulated in the cavity inside the ball. It turns out to be relatively easy to capture metal ions inside the cage. When graphite is soaked in a solution of the metal salt and then dried and vaporized, some of the $C_{60}$ cages formed have metal ions in them. For example, when graphite is treated with a solution of lanthanum chloride in water before drying and vaporizing, a species corresponding to the inclusion of a lanthanum ion in $C_{60}$ is detected in the mass spectrum. The compound is symbolized as La@$C_{60}$, in which "@" indicates a species that is fully surrounded by a fullerene cage. Interesting arguments are in progress as to what interactions an ion in a fullerene cage would have with the $\pi$ electrons of the cage.

The discovery of $C_{60}$ in experiments with a purpose other than making a new form of carbon is a good example of how fundamental scientific research produces unexpected results with unforeseen consequences. A whole new field of research is now under active investigation, with many imaginative ideas on how a stable cage of carbon atoms can be used to create new chemistry and new materials. For example, Professor Smalley and many other scientists are exploring the properties of carbon nanotubes, which are related to the fullerenes. The very small tubes look like test tubes made up of a rolled-up sheet of graphene, which is a single layer of graphite (see Figure 10.4). The bottom of the tube is closed with six pentagons (Figure 10.7). The tubes are about 11 Å in diameter and in the range of microns in length. They form very strong fibers and are impermeable to other chemicals. Multiple nanotubes associate with each other to form "nanotube ropes." They also can

*Figure 10.7*
A view of the start of the formation of a single-wall nanotube, consisting of 310 carbon atoms, with a nickel atom attached to its open edge. The nickel atom keeps the tube open by scooting around its open edge and helps ensure that the carbon atoms rearrange to give mainly hexagons in the walls of the tube.

Reprinted with permission from "Crystalline Ropes of Metallic Carbon Nanotubes," Andreas Thess et al., *Science*, Vol. 273, July 26, 1996. Copyright © 1996 American Association for the Advancement of Science.

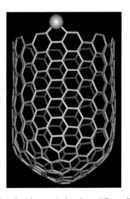

form in multiple layers, one tube within another. The possibilities for new kinds of materials made of these nanotubes seem almost endless. The significance of the work on fullerenes was recognized by the 1996 Nobel Prize in Chemistry awarded to Harold Kroto and Richard Smalley, along with their colleague Robert Curl.

## 10.4  Electrophilic Aromatic Substitution Reactions

### A. Experimental Observations from Halogenation and Nitration Reactions

Benzene is rich in electrons yet inert toward electrophilic addition reactions. Addition reactions would break up the aromatic sextet that gives benzene its high stability. Benzene and other aromatic hydrocarbons will react, however, with electrophilic reagents in reactions in which an incoming group substitutes for one of the hydrogen atoms on the ring. Such reactions are called **electrophilic aromatic substitution reactions.**

For example, benzene reacts with bromine in the presence of a catalyst, such as iron, or a Lewis acid, such as aluminum chloride, to give chiefly bromobenzene. Small amounts of $o$-dibromobenzene and $p$-dibromobenzene are also formed; the other product of the reaction is hydrogen bromide.

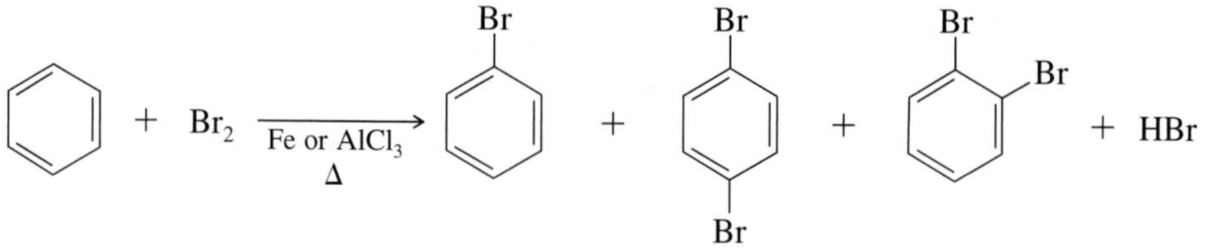

benzene                    bromobenzene  $p$-dibromobenzene  $o$-dibromobenzene
                             *major*          *minor*            *trace*

In a similar fashion, treating benzene with nitric acid in concentrated sulfuric acid converts it to nitrobenzene.

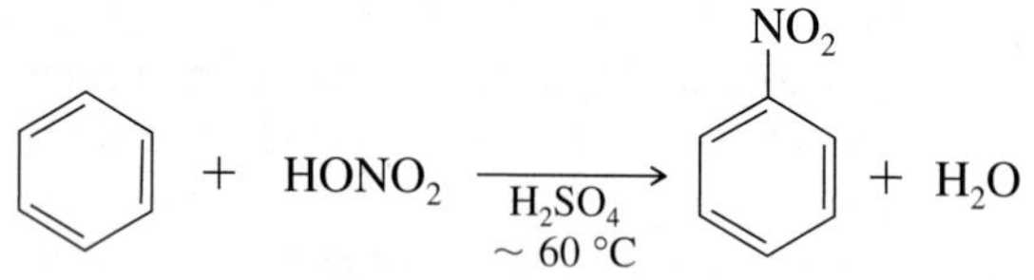

benzene        nitric acid              nitrobenzene

Aromatic compounds vary in their susceptibility toward electrophilic substitution reactions. For example, liquid bromine, a Lewis acid catalyst, and temperatures of approximately 80 °C are necessary to form bromobenzene from benzene. Phenol reacts with a dilute solution of bromine in acetic acid at 30 °C.

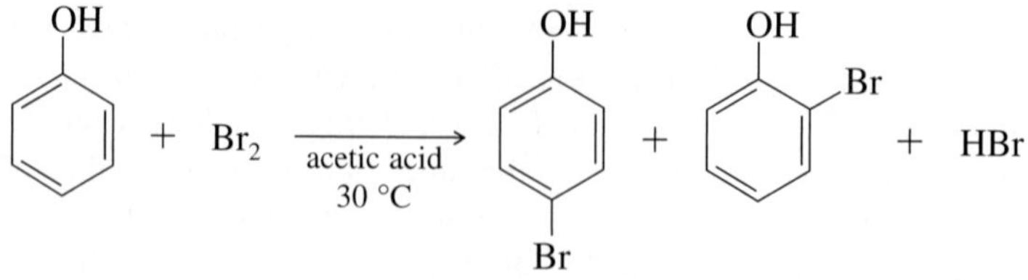

phenol                  $p$-bromophenol   $o$-bromophenol
                              88%               12%

*p*-Bromophenol is the major product of the reaction, along with some *o*-bromophenol. If phenol is treated with bromine in water, 2,4,6-tribromophenol is formed instantly. In water, phenol is in equilibrium with the phenolate ion, which bears a negative charge. The phenolate ion is much more reactive toward electrophilic attack than phenol itself is.

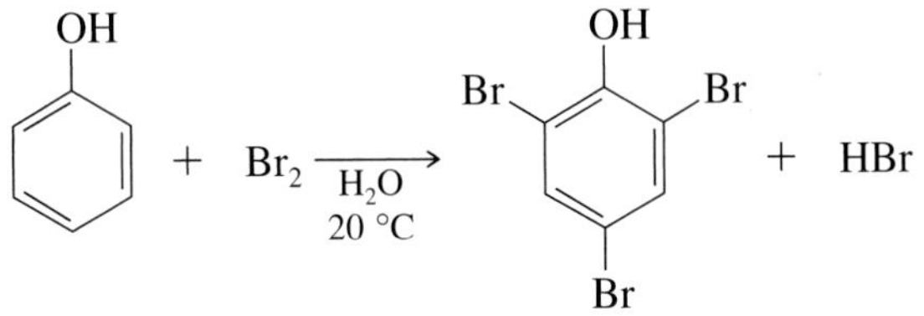

phenol                2,4,6-tribromophenol

2,4,6-Tribromophenol is much less soluble in water than phenol is and precipitates from the solution. The reaction is so rapid that it is sometimes used as a test for phenols in qualitative analysis.

**Problem 10.8**   Examine the equilibria that exist between phenol and its conjugate base in water and in acetic acid. Which solution contains a higher concentration of phenolate ion?

**Problem 10.9**   The phenolate ion is more susceptible to attack by electrophiles than phenol is. This is so because electrophilic attack goes through a positively charged reactive intermediate, which is stabilized by resonance. The general mechanism for the reaction with bromine was shown in the A Look Ahead section at the beginning of this chapter. Explore the reactive intermediates that form when phenol and phenolate ion are attacked in the para position by bromine. Which one is better stabilized by resonance? Why? What effect will this have on the relative rates of the reactions of phenol and phenolate ion with bromine?

Some substituents make it harder to introduce a second group onto an aromatic ring. For example, to introduce a second nitro group onto nitrobenzene, fuming nitric acid, which is a more powerful reagent than ordinary concentrated nitric acid, and a temperature of 100 °C must be used.

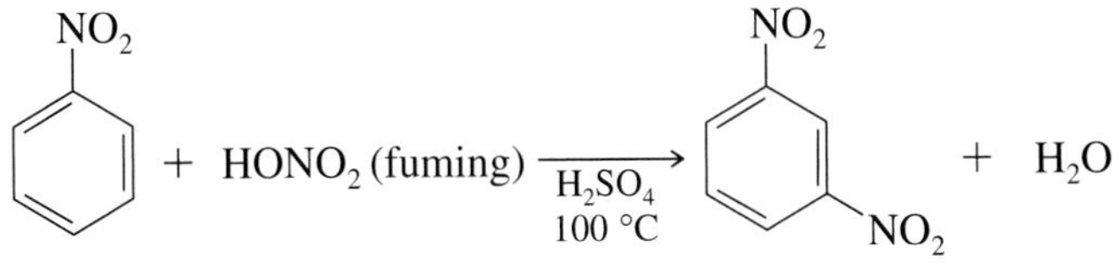

nitrobenzene    nitric acid                  *m*-dinitrobenzene
                                                       88%

These conditions are more severe than those used to substitute the first nitro group on the benzene ring (see above, p. 366). The second nitro group takes a position meta to the first one.

On the other hand, phenol can be nitrated with dilute nitric acid at room temperature. The products are *o*-nitrophenol and *p*-nitrophenol.

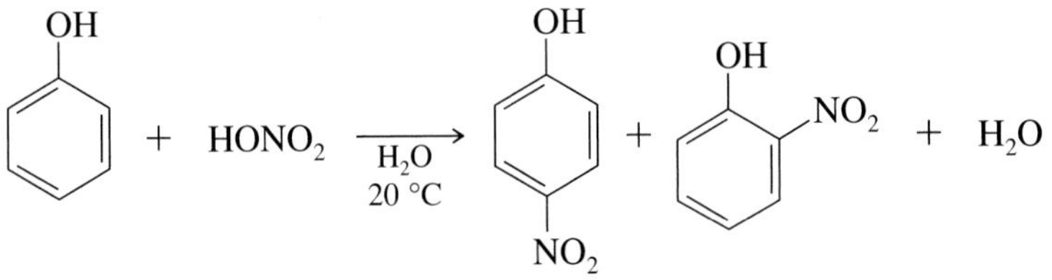

phenol        nitric acid        *p*-nitrophenol        *o*-nitrophenol    + H$_2$O
                                        60%                   40%

The experiments summarized in the preceding equations lead to two important observations. The first is that a group already present on the aromatic ring may make it easier or harder to introduce a second substituent onto the ring. A group that makes it easier to introduce new substituents is said to activate the ring toward electrophilic substitution or to be **ring-activating.** The hydroxyl group in phenol, for example, is a ring-activating substituent. Groups that make the introduction of a second substituent harder are said to deactivate the ring toward electrophilic substitution or to be **ring-deactivating.** The nitro group is such a ring-deactivating substituent.

The second observation is that the position that a second substituent takes on the ring is influenced by the group that is already there. When phenol undergoes electrophilic substitution, the new group takes up a position ortho or para to the hydroxyl group. The hydroxyl group is said to direct the incoming substituent to those positions or to be **ortho,para-directing.** The nitro group, on the other hand, directs the new substituent to the meta position and is said to be **meta-directing.**

Substituents on the aromatic ring generally fall into one or the other of these categories. Some are ortho,para-directing; others are meta-directing. For example, an alkyl group on the aromatic ring is predominantly ortho,para-directing, as shown by the nitration of ethylbenzene.

| ethylbenzene | nitric acid | o-nitroethylbenzene 45% | p-nitroethylbenzene 48% | m-nitroethylbenzene 7% |

The reaction gives a mixture, the chief components of which are o-nitroethylbenzene and p-nitroethylbenzene in approximately equal amounts. Very little meta-substituted product is formed. When methyl benzoate is nitrated, on the other hand, methyl m-nitrobenzoate is formed as the major product.

methyl benzoate    nitric acid    methyl m-nitrobenzoate ~ 85%

Thus the ester function on the aromatic ring is meta-directing. This reaction, however, does not require the harsh conditions that are necessary to introduce a second nitro group into nitrobenzene.

The results of many experiments are summarized in Table 10.1, in which different substituents on the aromatic ring are classified as ortho,para-directing or meta-directing. Most of the ortho,para-directing groups, except for the halogens, activate the ring toward further substitution. This effect is strong only for the hydroxyl group and the unprotonated amino group. An examination of the structures of the groups that are ortho,para-directing shows that, except for alkyl groups, they have at least one pair of nonbonding electrons available on the atom directly bonded to the

**Table 10.1** **Substituents on Aromatic Rings Classified as Ortho,Para- and Meta-Directing Groups**

| Ortho,Para-Directing | | Meta-Directing | |
|---|---|---|---|
| $-N(CH_3)_2$ | very strongly activating | $-\overset{+}{N}(CH_3)_3$ | very strongly deactivating |
| $-NH_2$ | | $-NO_2$ | |
| $-OH$ | | $-C\equiv N$ | |
| $-OCH_3$ | | $-SO_3H$ | |
| $-NH\overset{O}{\overset{\|}{C}}CH_3$ | | $-\overset{O}{\overset{\|}{C}}H$ | |
| $-O\overset{O}{\overset{\|}{C}}CH_3$ | | $-\overset{O}{\overset{\|}{C}}CH_3$ | |
| $-R$ | | $-\overset{O}{\overset{\|}{C}}OH$ | |
| $-Cl, Br, I$ | mildly deactivating | $-\overset{O}{\overset{\|}{C}}OCH_3$ | |
| | | $-\overset{O}{\overset{\|}{C}}NH_2$ | |
| | | $-NH_3^+$ | |

aromatic ring. All the meta-directing groups either bear a positive charge on the atom directly bonded to the aromatic ring or have a structure that can be polarized to put partial positive charge there. The meta-directing groups at the top of the list are strongly ring-deactivating. Experimentally, the others, even though they are meta-directing, allow substitution reactions under relatively mild conditions. In other words, they do not deactivate the ring strongly. Tables 10.2 and 10.3 (p. 384) summarize the different kinds of electrophiles that react with aromatic compounds, and the regioselectivity of their reactions.

The next section examines the mechanism proposed to explain the trends just summarized.

**Study Guide**
Concept Map 10.2

**Problem 10.10** Complete the following equations.

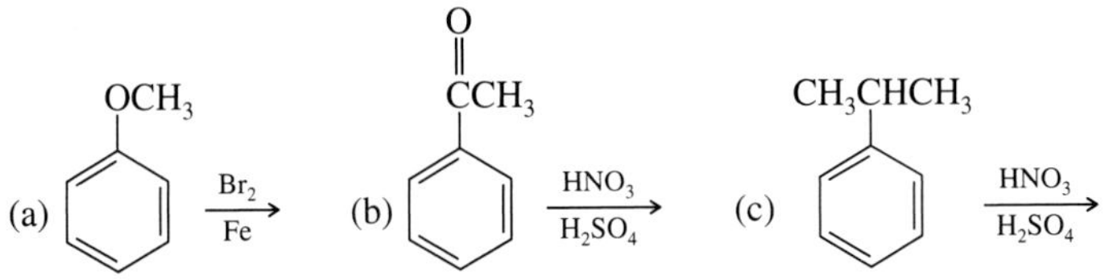

(a) [OCH₃ benzene] $\xrightarrow[Fe]{Br_2}$

(b) [benzene with $\overset{O}{\overset{\|}{C}}CH_3$] $\xrightarrow[H_2SO_4]{HNO_3}$

(c) [benzene with $CH_3CHCH_3$] $\xrightarrow[H_2SO_4]{HNO_3}$

## B. The Mechanism of Electrophilic Aromatic Substitution

Bromine and strong acids are examples of reagents that introduce substituents on the aromatic ring; they are electrophilic, electron-seeking reagents. Unless the ring is strongly activated, a Lewis acid also must be used with bromine. For example, a piece of iron is added to the reaction vessel in the bromination of benzene (p. 366). Iron reacts with bromine to give ferric bromide, which is a Lewis acid. The Lewis acid coordinates with the nonbonding electrons of the bromine molecule to polarize the bromine–bromine bond, making one of the bromine atoms more electrophilic and the other one a better leaving group.

$$3\ Br_2\ +\ 2\ Fe\ \longrightarrow\ 2\ FeBr_3$$

  bromine        iron            ferric bromide

*a Lewis acid*

$$Br\!-\!Br\text{----}\overset{\displaystyle Br}{\underset{\displaystyle Br}{Fe}}\!-\!Br\ \longrightarrow\ \overset{\delta+}{Br}\text{----}\overset{\delta-}{Br}\!-\!\overset{\displaystyle Br}{\underset{\displaystyle Br}{Fe}}\!-\!Br$$

*a complex of bromine*
*with a Lewis acid*

The $\pi$ electrons of benzene are attracted to the bromine atom at which positive charge is developing. A covalent bond forms between a carbon atom in benzene and that bromine atom, as the other bromine atom becomes part of a complex with the Lewis acid. The resulting species is a cation that loses a proton readily to restore the aromatic ring. Bromide ion (or $FeBr_4^-$) is a base strong enough to remove a proton from such an intermediate.

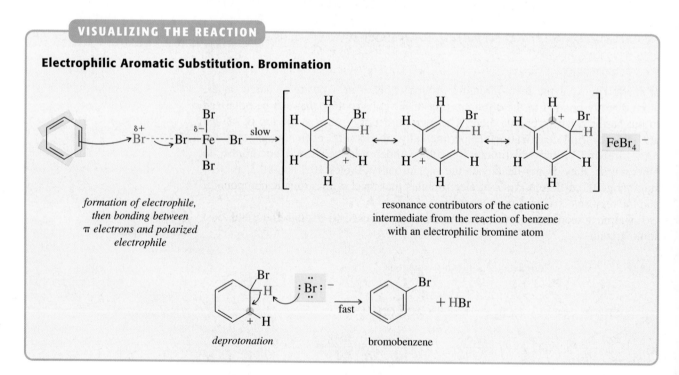

### VISUALIZING THE REACTION

**Electrophilic Aromatic Substitution. Bromination**

*formation of electrophile, then bonding between π electrons and polarized electrophile*

resonance contributors of the cationic intermediate from the reaction of benzene with an electrophilic bromine atom

*deprotonation*                    bromobenzene

The equations above represent the essential steps in all electrophilic substitution reactions.

1. The substituting reagent is polarized or ionized in such a way as to create an electron-deficient species.

2. A cationic intermediate forms by bonding between the electrons of the aromatic ring acting as the nucleophile and the electrophilic reagent.

3. A proton is lost from the carbocation to restore the aromatic ring, on which a substituent has replaced a hydrogen atom.

*The rate-determining step in most electrophilic aromatic substitution reactions is the formation of the carbocation intermediate.* Going from benzene and bromine to the transition state involves not only breaking bonds but also disrupting the aromaticity in the benzene ring. The energy barrier for this reaction is high (Figure 10.8).

The loss of a proton from the intermediate is a fast reaction; this is so because the aromatic sextet is regenerated. It is possible to test whether the loss of the proton is the rate-determining step by substituting deuterium atoms for the hydrogen atoms on the aromatic ring. A deuterium atom is chemically similar to a hydrogen atom because it, too, has one proton and one electron. A deuterium atom, however, has twice the mass of a hydrogen atom. The breaking of a covalent bond results from an increase in the vibrational energy of the bond to a level represented by the bond dissociation energy (p. 62). The vibrational energy, in turn, depends on the masses of the atoms involved in the bond (p. 454). The vibrational energy of a bond is lower when the atoms joined by the bond have greater mass. Therefore, a C—D bond has a lower ground-state vibrational energy than does a C—H bond. For this reason, a reaction involving the breaking of a bond between carbon and deuterium has a higher energy of activation and is slower than a reaction involving the breaking of a carbon–hydrogen bond. If the removal of the proton from the cationic intermediate were the rate-determining step, a decrease in the rate of the reaction should be seen if deuteriobenzene, $C_6D_6$, is used in an electrophilic substitution reaction. The experiment has been tried, and no difference in the rate of nitration has been observed for $C_6H_6$ and $C_6D_6$ with a variety of reaction conditions. These experimental results confirm that the rate-determining step does not involve the breaking of the carbon–hydrogen bond.

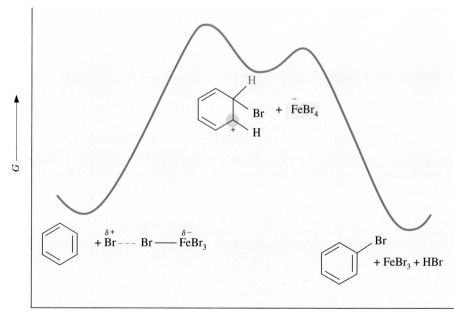

*Figure 10.8*
An energy diagram for the bromination of benzene.

In the nitration reaction, the electrophile is the nitronium ion, $NO_2^+$, which is created by protonation of nitric acid by the strong acid sulfuric acid, followed by the loss of water.

**VISUALIZING THE REACTION**

**Formation of the Nitronium Ion**

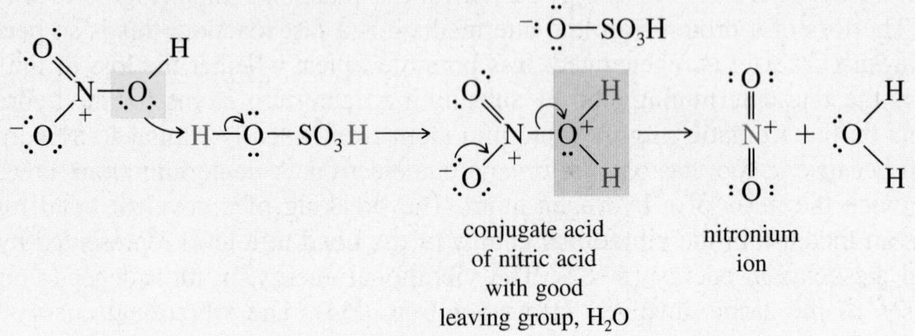

conjugate acid
of nitric acid
with good
leaving group, $H_2O$

nitronium
ion

Nitronium ion reacts with the $\pi$ electrons of the aromatic ring to give a cationic intermediate, which loses a proton to recreate the aromatic sextet.

**VISUALIZING THE REACTION**

**The Nitration Reaction**

benzene          nitronium          carbocation                    nitrobenzene
                 ion                intermediate

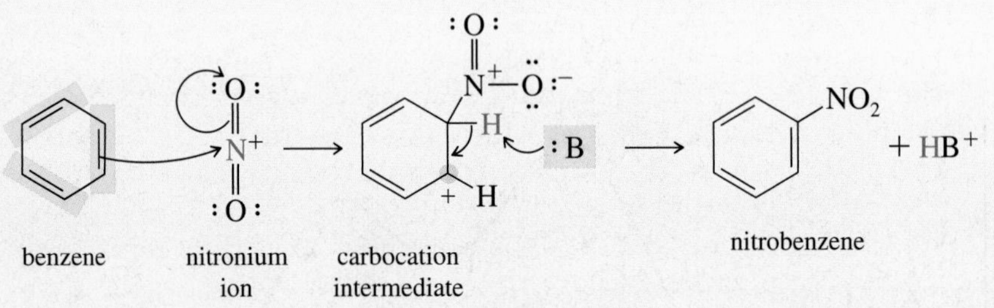

The reactivity of the aromatic ring toward acids is demonstrated by using benzene in which deuterium has been substituted for one of the hydrogen atoms. In aqueous sulfuric acid at room temperature, the deuterium atom is exchanged for a hydrogen atom.

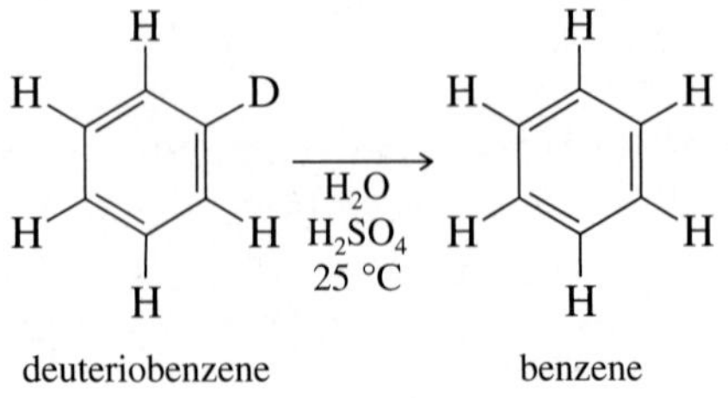

deuteriobenzene          benzene

**Acid-Catalyzed Exchange of Hydrogen Atoms**

This observation means that in an acid medium the aromatic ring reacts constantly with protons. The protonation and deprotonation of the ring are usually not detected because the starting material and the product of the reaction are the same. The reaction can be detected, however, if the ring is labeled with isotopes of hydrogen. The reactive intermediate for this exchange reaction is a cation, arising from the reaction of a proton with the $\pi$ electrons of the aromatic ring.

Study Guide
Concept Map 10.3

## C. Orientation in Electrophilic Aromatic Substitution on Compounds Having a Ring–Activating Substituent

How does the mechanism proposed in the preceding section account for the different orientations observed for substituents when the aromatic ring already has a group on it? This is a question regarding the regioselectivity of a reaction, a question that has come up again and again in many different contexts, for example, the addition of hydrogen bromide to propene (p. 126), the formation of alkenes in elimination reactions (p. 267), and the addition of diborane to double bonds (p. 294). In each case, the answer came from a consideration of the transition states that would lead to the different products and a selection of the transition state of lowest energy, or greatest stability, as the one that would give rise to the major product. According to the Hammond Postulate (p. 124), the transition states are assumed to resemble the reactive intermediates to which they are closest in energy. The reactive intermediates are therefore used as models for the structures of the activated complexes at the transition states. The relative energies of different transition states have been assumed to be parallel to the relative energies of the possible reactive intermediates.

The rationalization of the regioselectivity seen for electrophilic aromatic substitution is similar. A consideration of the relative energies of transition states that would lead to the different products expected from a given substitution reaction should reveal which ones are the most stable and, therefore, most likely to yield products.

When phenol reacts with bromine, the bromine atom can become bonded to a ring carbon at one of three possible positions: ortho, meta, and para to the hydroxyl group. The cationic intermediates that would result from bonding of the bromine atom to these positions, with their resonance contributors, are shown on the next page.

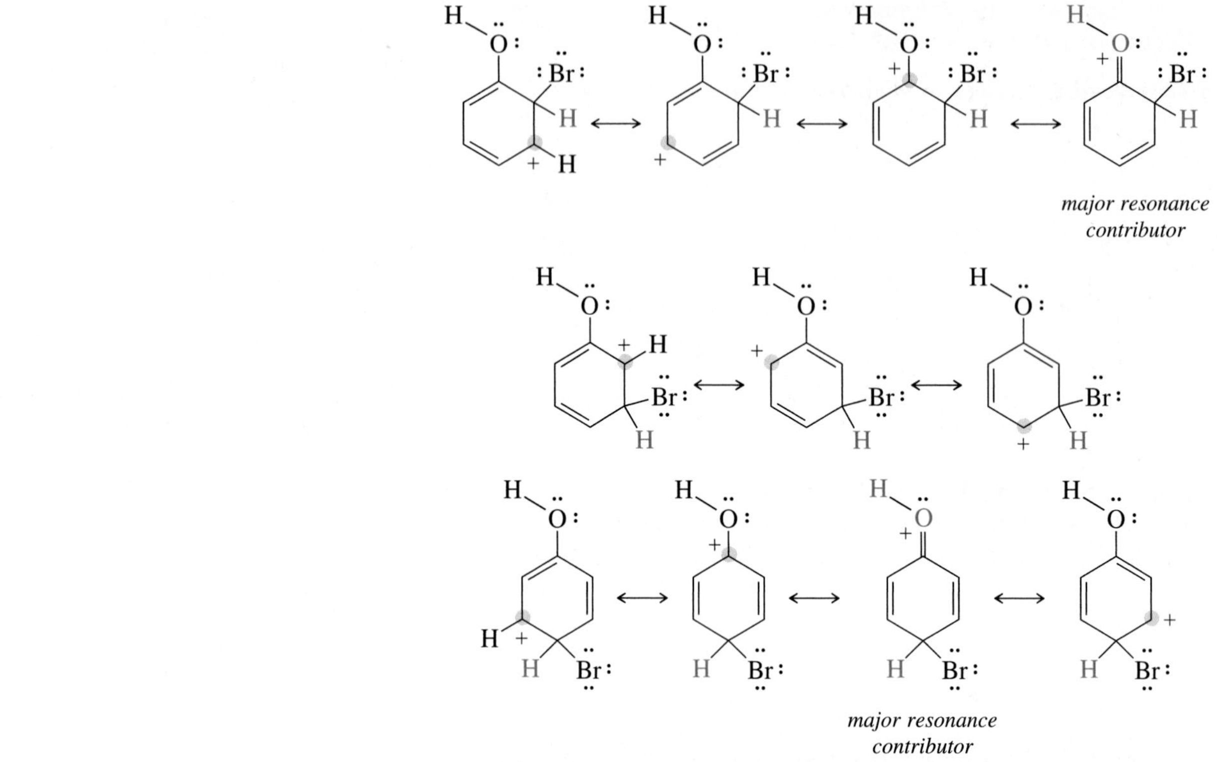

*major resonance contributor*

*major resonance contributor*

*resonance contributors for the cationic intermediates that would result from bonding of a bromine atom to the ortho, meta, or para positions of phenol*

The intermediates that result from the bonding of the bromine atom to the ortho and para positions of the ring have more resonance contributors than does the intermediate that would give rise to *m*-bromophenol. Also, for the intermediates arising from attack at the ortho and para positions, one of the resonance contributors has eight electrons around each atom, which is a particularly stable situation. The reasons just mentioned lead to the conclusion that the transition states for ortho and para substitution on a phenol ring are of lower energy than the one for meta substitution. The energies of activation for the formation of *o*-bromophenol and *p*-bromophenol are lower than that for the formation of *m*-bromophenol. Most of the reactant molecules will be channeled through the intermediates leading to those products.

The conclusions arrived at for phenol will be true for any substituent in which there is a pair of nonbonding electrons on the atom bonded to the aromatic ring. An alkyl group also stabilizes the intermediates resulting from ortho and para substitution more than the intermediate from meta substitution.

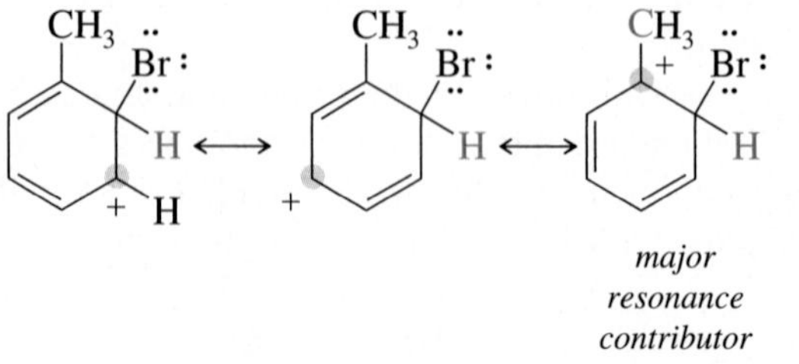

*major resonance contributor*

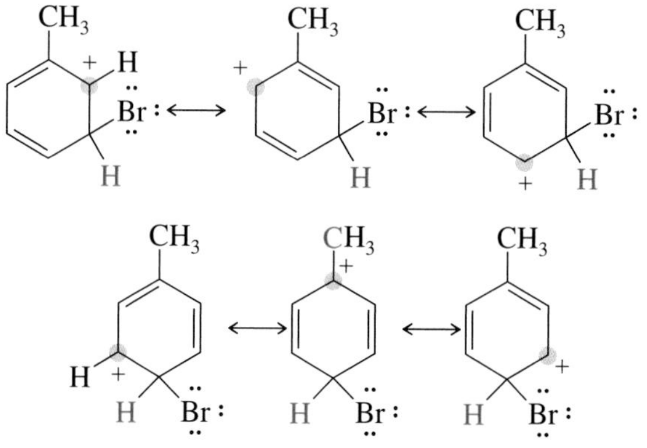

*resonance contributors for the cationic intermediates that would
result from bonding of a bromine atom to the ortho, meta,
or para positions of toluene*

For a second substituent on an aromatic ring, the orientation that gives rise to the most stable reactive intermediate is favored.

## D. Orientation in Electrophilic Aromatic Substitution on Compounds Having a Ring-Deactivating Substituent

When a strongly electron-withdrawing substituent is on the aromatic ring, it is harder to form the carbocation intermediate. The cation is destabilized by the charge distribution in the polar electron-withdrawing group. The intermediate formed by reaction at the meta position is the most stable one because it is least destabilized. This conclusion is illustrated by the nitration of nitrobenzene.

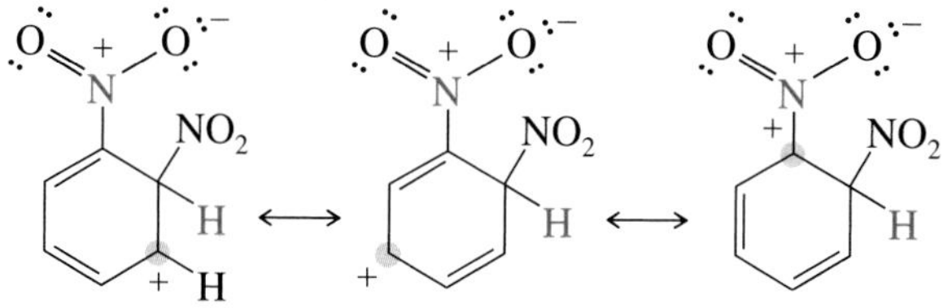

*unfavorable resonance contributor,
two adjacent positive charges*

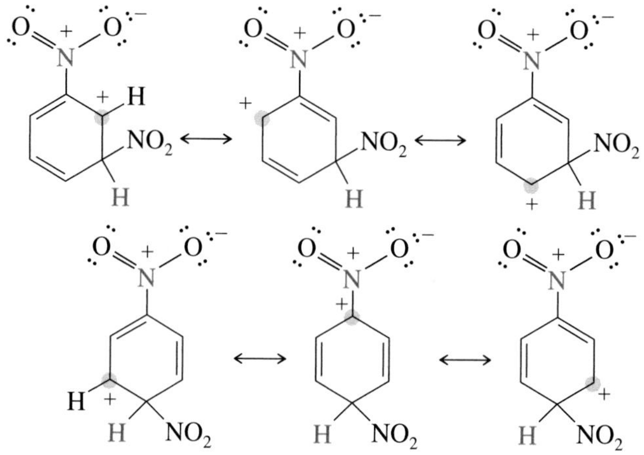

*unfavorable resonance contributor,
two adjacent positive charges*

*resonance contributors for the cationic intermediates
that would result from the bonding of a nitronium ion
to the ortho, meta, and para positions of nitrobenzene*

None of the intermediates shown on the previous page has any special stabilization other than that due to delocalization of charge in general. Among the resonance contributors for the intermediates resulting from bonding of a nitronium ion at the ortho and para positions of nitrobenzene, however, one is a particularly unfavorable one, having a juxtaposition of two positive charges. As a result, the transition state arising from an initial approach of the nitronium ion to the meta position is the most stable among those for the three possible intermediates, and substitution on nitrobenzene takes place mostly at the meta position.

The substitution of a second group on a halobenzene is an interesting case. The halogens are ring-deactivating; their presence makes it harder for electrophilic aromatic substitution to take place. This would be expected from their electronegativity and thus their negative inductive effect. Yet the halogens are ortho,para-directing. As a positive charge develops on the aromatic ring during substitution, the non-bonding electrons of a halogen are drawn toward the charge and help to stabilize it. Thus, just as a halogen atom stabilizes a carbocation intermediate in the addition of hydrogen halides to an alkyne (p. 333), it is possible for halogen atoms to stabilize the carbocation intermediate in an electrophilic substitution reaction if the incoming substituent takes an ortho or para position.

For example, the intermediate leading to *p*-dibromobenzene is favored over the one leading to *m*-dibromobenzene.

*delocalization of charge*
*to bromine atom*

*resonance contributors for the intermediate for* p-*dibromobenzene*

*no resonance contributor in which charge*
*can be delocalized to bromine atom*

*resonance contributors for the intermediate for* m-*dibromobenzene*

The positive charge in the carbocation intermediate for *p*-dibromobenzene can be delocalized to a bromine atom. Such a resonance contributor has eight electrons around each atom and is thus a major resonance contributor. No such delocalization is possible for the intermediate that would give rise to *m*-dibromobenzene. Only a very small amount of the meta isomer (1.8%) is formed in the reaction of bromobenzene with bromine.

## E. Steric Effects in Electrophilic Aromatic Substitution

The arguments presented on page 374 to rationalize the fact that the hydroxyl group in phenol or the methyl group in toluene is ortho,para-directing did not indicate any distinction between the ortho and para positions on the basis of the energies of the transition state. If these two were really indistinguishable, twice as much ortho as para isomer should be formed because there are two ortho positions on the ring and only one para position. Statistically, it would be twice as likely that collision would occur with the reagents correctly oriented for reaction at the ortho position. The experimental observations prove otherwise (p. 366). Actually, the ratio of ortho to para product can vary greatly, but substitution at the para position is generally favored over what is predicted statistically. The group already on the aromatic ring has some bulk and hinders the approach of the incoming reagent to the positions closest to it, the ortho positions. This idea has been tested by systematically increasing the bulk of a substituent and measuring the change in the composition of the product mixture formed. Toluene, ethylbenzene, isopropylbenzene, and *tert*-butylbenzene are shown below with numbers that indicate the percentage of substitution that takes place at each position on the ring in a nitration reaction.

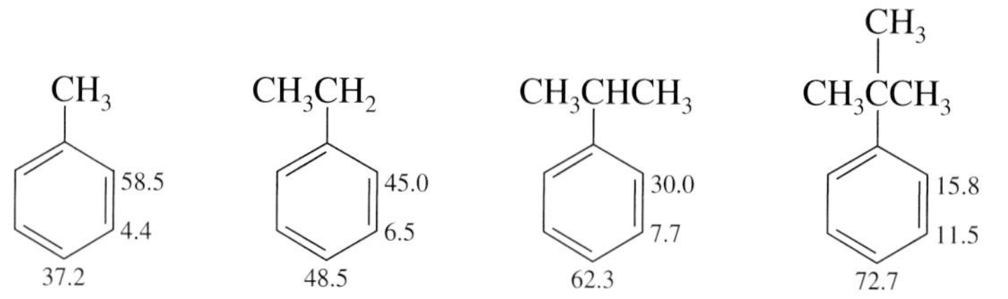

CH₃
$$\text{CH}_3$$

CH₃            CH₃CH₂            CH₃CHCH₃            CH₃CCH₃

| | 58.5 | | 45.0 | | 30.0 | | 15.8 |

4.4              6.5              7.7              11.5

37.2              48.5              62.3              72.7

toluene        ethylbenzene        isopropylbenzene        *tert*-butylbenzene

*percentage of ortho, meta, and para isomers*
*in the nitration of a series of alkylbenzenes*

From these experimental facts, it is quite clear that the proportion of ortho isomer in the product mixture becomes smaller as the alkyl group on the aromatic ring becomes larger.

There is also some evidence that the size of the incoming substituent affects the ratio of ortho to para isomers formed as products. For example, when bromobenzene reacts with chlorine, 42% of the product is *o*-bromochlorobenzene and 53% is the para isomer. If bromine, a larger atom, is the incoming group, only 13% *o*-dibromobenzene is formed, and 85% of the product mixture is *p*-dibromobenzene.

These data are a reminder that electronic factors are not the only ones to consider in thinking about a reaction. Steric interactions of the reagents are also important in determining relative energies of transition states and the resulting reaction pathways.

(✎) **Study Guide**
Concept Map 10.4

**Problem 10.11**   The trimethylammonium ion, $(CH_3)_3\overset{+}{N}-$, is a strongly deactivating group that is also meta-directing. Explain this fact. (*Hint:* A picture is worth a thousand words!)

**Problem 10.12**   How would you rationalize the experimental result given for the following reaction? (*Hint:* What are the reacting species when *N,N*-dimethylaniline is dissolved in a mixture of concentrated nitric and sulfuric acids?)

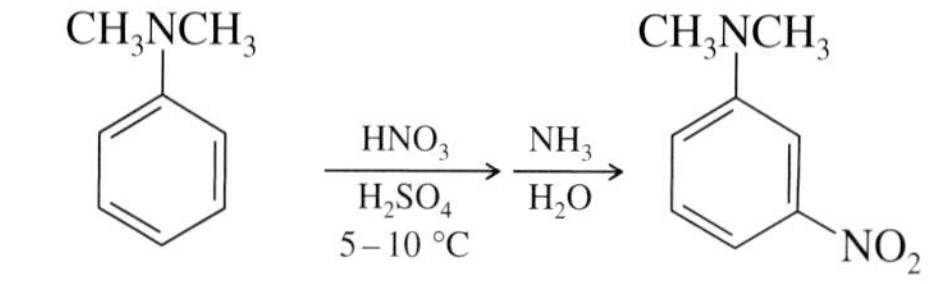

CH₃NCH₃

⬡  →  CH₃NCH₃

HNO₃      NH₃
─────  ────→
H₂SO₄      H₂O
5–10 °C

⬡
     NO₂

*N,N*-dimethylaniline        ~ 60%

**Problem 10.13**   *p*-Nitrophenol is shaken with 1 equivalent of D₂O in the presence of the strong acid perchloric acid, HClO₄, at 100 °C for 100 hours. The product has two deuterium atoms in it. One of these is lost instantly when the product is treated with ordinary water, H₂O. Propose a structure for the original product and explain the difference in the ease with which the two deuterium atoms can be removed from the molecule.

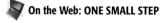 **On the Web: ONE SMALL STEP**

## 10.5 Aromatic Substitution Reactions with Carbocations as Electrophiles

### A. Friedel–Crafts Alkylation Reactions

In the reactions described in Section 10.4A, the electrophilic reagent was bromine or a nitronium ion. When carbon is cationic, it is also an electrophile; therefore, reactions that give rise to carbocations, when carried out in the presence of an aromatic ring, create carbon–carbon bonds. Three common types of reactions give rise to carbocations: the protonation of an alkene, the protonation of an alcohol with the subsequent loss of water, and the ionization of an alkyl halide. All these are used to substitute carbon side chains on aromatic rings in the **Friedel-Crafts reactions,** named for Charles Friedel, a French chemist, and James Crafts, an American chemist, who developed these methods.

For example, sulfuric acid catalyzes the reaction between the alkene cyclohexene and benzene, which gives mostly cyclohexylbenzene and some *p*-dicyclohexylbenzene, even when an excess of benzene is used.

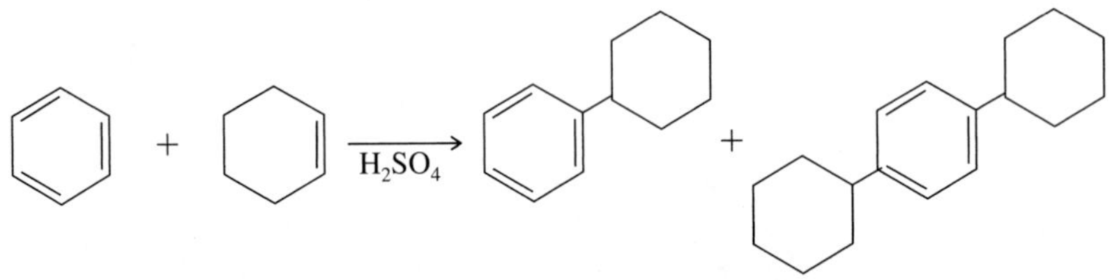

| benzene | cyclohexene | cyclohexylbenzene | *p*-dicyclohexylbenzene |
|---|---|---|---|
| *(used in excess)* | | *major* | *minor* |

This reaction, an example of a **Friedel-Crafts alkylation,** illustrates one of the difficulties of such reactions. The product of the reaction, an alkylbenzene, is more reactive toward electrophilic aromatic substitution than is benzene itself. The alkyl group is ring-activating and ortho,para-directing. As a result, as soon as any quantity of mono-alkylated benzene accumulates in the reaction mixture, it starts to undergo further alkylation to give di- and even trisubstituted products.

The mechanism for the alkylation reaction of benzene with cyclohexene is straightforward and follows the steps outlined earlier (pp. 370–371). First, a good electrophile is created by protonation of cyclohexene by sulfuric acid, the catalyst in the reaction. A carbon–carbon bond forms when the electrophile reacts with the $\pi$ electrons of the aromatic ring. A proton is lost from the cationic intermediate that is formed.

---

**VISUALIZING THE REACTION**

**A Friedel-Crafts Alkylation**

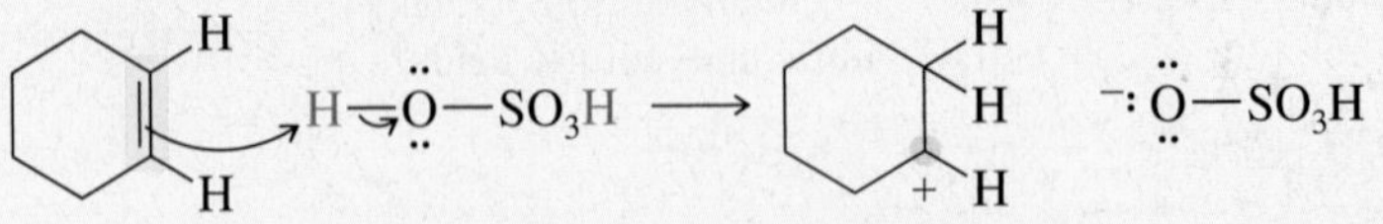

cyclohexyl cation

*the electrophile*

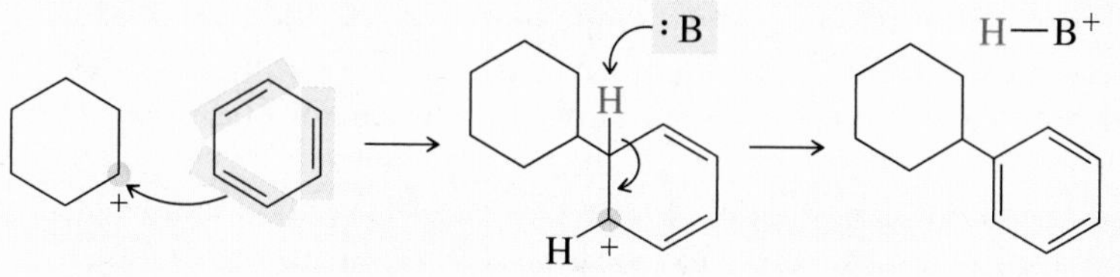

| reaction of the electrophile with the π electrons | the cationic intermediate losing a proton | |

Because carbocations are the electrophiles in the Friedel-Crafts alkylation, the types of compounds that can be synthesized by this method are limited. For example, it is difficult to put a primary alkyl chain on an aromatic ring. When benzene reacts with *n*-propyl chloride with aluminum chloride as a catalyst, a mixture of two alkylbenzenes is formed. The composition of the mixture depends on the temperature at which the reaction is run.

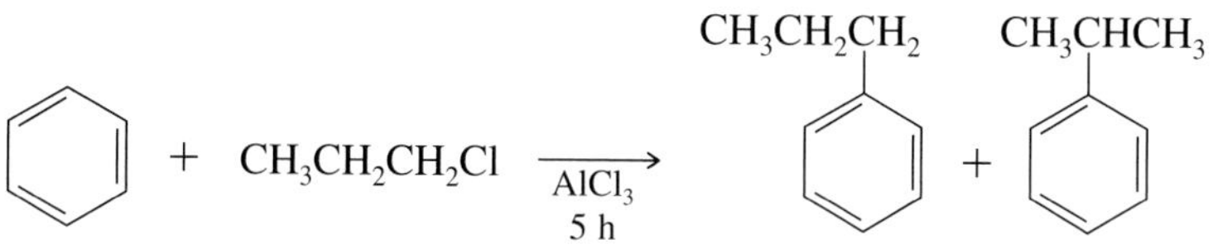

| benzene | *n*-propyl chloride | | propylbenzene | isopropylbenzene |
|---------|---------------------|---|---------------|------------------|
| | | at − 6 °C | 60% | 40% |
| | | at 35 °C | 40% | 60% |

Aluminum chloride, which is a Lewis acid, complexes with nonbonding electrons of the chlorine atom in *n*-propyl chloride, polarizing the carbon–chlorine bond. The intermediate formed must have enough cationic character at the primary carbon atom that some rearrangement to the more stable secondary carbocation takes place by a 1,2-hydride shift (p. 289). The carbocation intermediate is seen as being tightly complexed with the aluminum halide in an ion pair.

---

**VISUALIZING THE REACTION**

**Formation of the Electrophile in a Friedel-Crafts Alkylation**

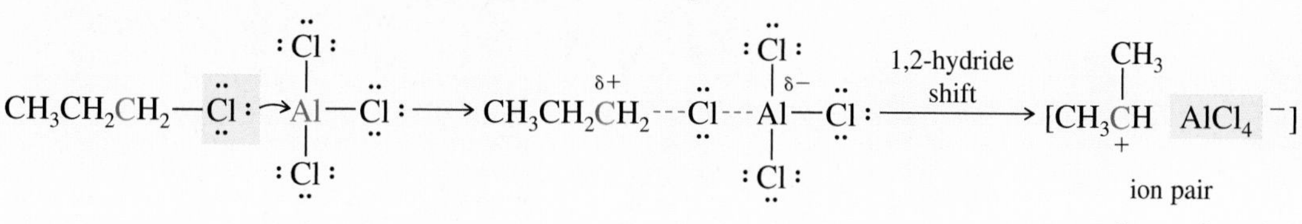

The ratios of the two products from benzene and *n*-propyl chloride are evidence that free primary carbocations are not easily formed, even with a powerful Lewis acid such as aluminum chloride to complex with the chloride ion. At the lower temperature, most of the product comes from reaction at the first carbon of the propyl chain. Only at the higher temperature, where there is a better chance of breaking the carbon–chlorine bond, does the rearranged product derived from the thermodynamically more stable cation become the major one. Clearly, the reaction is not a good one for the synthesis of either propylbenzene or isopropylbenzene.

There is no problem obtaining pure products if the alkyl halide used is one that ionizes to form the most stable carbocation directly. Thus *tert*-butyl chloride reacts with ethylbenzene to give *p-tert*-butylethylbenzene quantitatively.

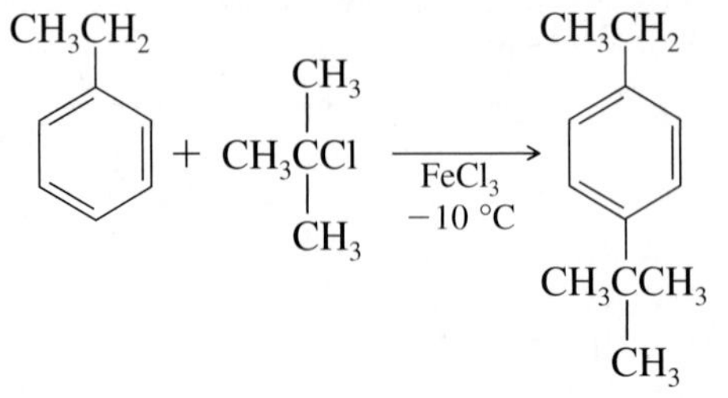

| ethylbenzene | *tert*-butyl chloride | *p-tert*-butyl-ethylbenzene |

Carbocations are created by the reaction of alcohols with protic acids or with Lewis acids. Alcohols, therefore, are used as reagents in Friedel-Crafts alkylations. In these reactions, the stability of the carbocation formed as the electrophile is again important. Isopropyl alcohol reacts with benzene to give isopropylbenzene.

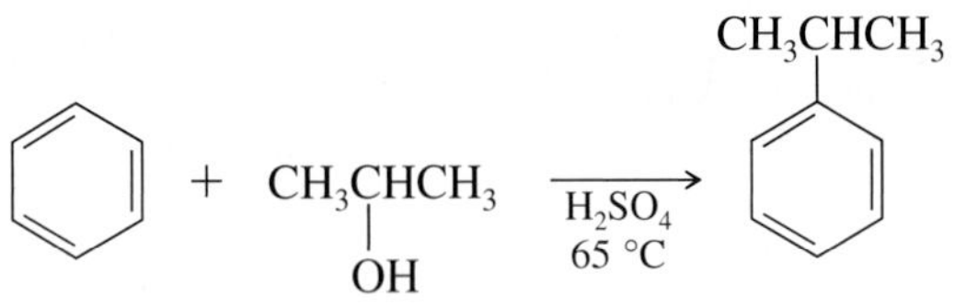

| benzene | isopropyl alcohol | isopropylbenzene cumene 65% |

**Study Guide**
Concept Map 10.5

**Problem 10.14**    Complete the following equations.

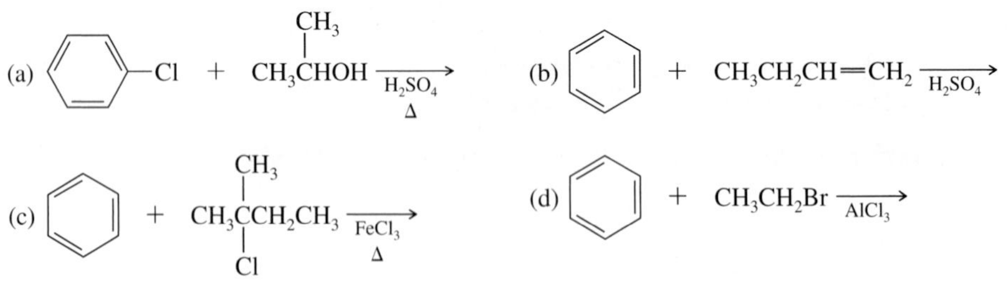

(a)

(b)

(c)

(d)

**Problem 10.15**    The same compound is formed when either of the two alcohols shown below is treated with boron trifluoride etherate. Propose a structure for the product and write a mechanism for its formation.

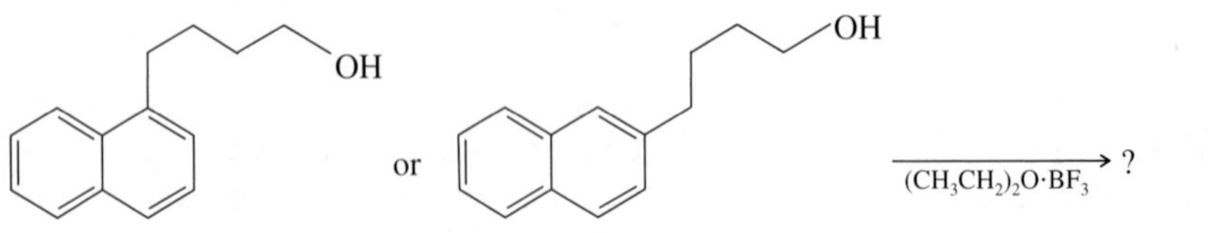

## B. The Art of Solving Problems

**PROBLEM:**  How would you carry out the following conversion?

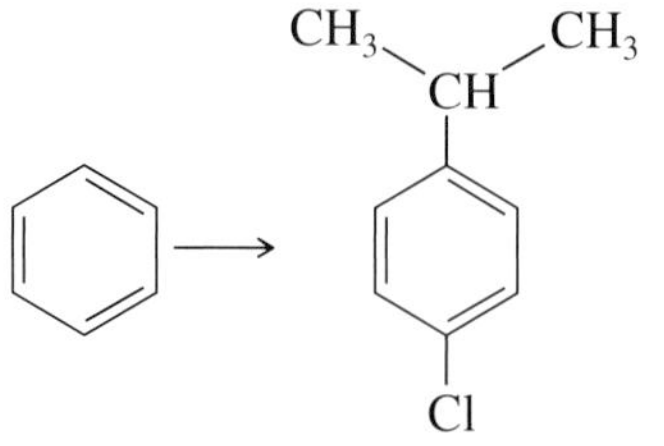

## Solution

1. What are the connectivities of the two compounds? How many carbon atoms does each contain? Are there any rings? What are the positions of branches and functional groups on the carbon skeleton?

   The starting material is benzene, an aromatic hydrocarbon. The product has an alkyl group, an isopropyl group, and a chlorine atom para to it.

2. How do the functional groups change in going from starting material to product? Does the starting material have a good leaving group?

   The aromatic hydrocarbon has undergone substitution at two positions.

3. Is it possible to dissect the structures of the starting material and product to see which bonds must be broken and which formed?

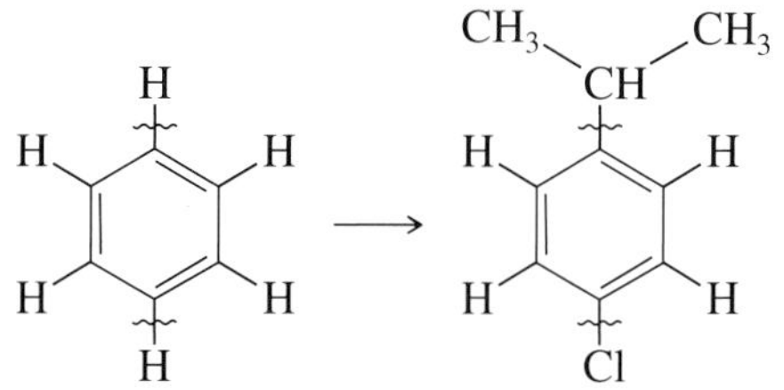

   *bonds to be broken*          *bonds to be formed*

4. New bonds are created when an electrophile reacts with a nucleophile. Do we recognize any part of the product molecule coming from a good nucleophile or an electrophilic addition?

   Both groups can be introduced by electrophilic aromatic substitution reactions. The alkyl group can be introduced by a Friedel-Crafts alkylation reaction and the halogen atom by the use of chlorine and a Lewis acid.

5. What type of compound would be a good precursor to the product?

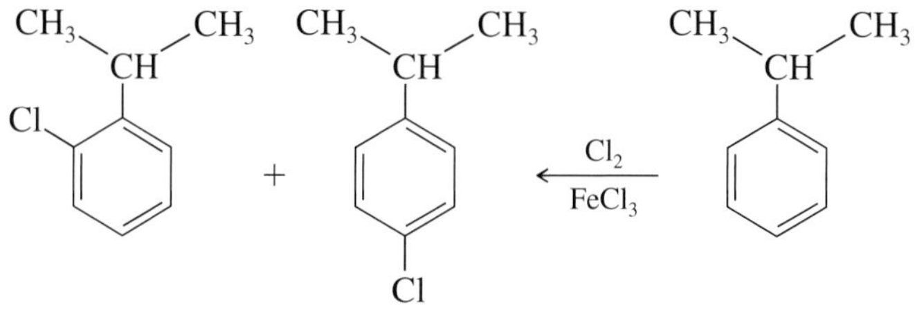

   *minor*                    *major*

   We assume that we can separate the ortho and para isomers in the laboratory.

6. After this last step, do we see how to get from starting material to product? If not, we need to analyze the structure obtained in step 5 by applying questions 4 and 5 to it.

We need to make the isopropylbenzene from benzene.

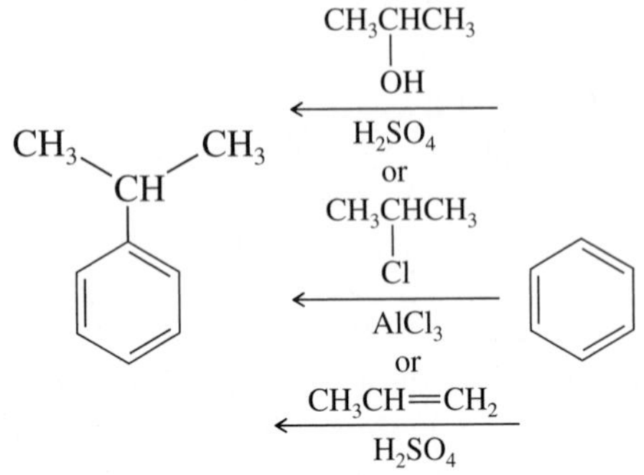

We alkylate first because that activates the ring toward further reaction, making the second step easier. ∎

**ONE SMALL STEP**

The examples used in the preceding sections of this chapter do not by any means exhaust the reagents that can act as electrophiles in electrophilic aromatic substitution reactions. Some others are given below:

$$HONO, \quad HOCl, \quad ICl,$$
$$SCl_2 \text{ with } AlCl_3,$$
$$SO_3 \text{ with } H_2SO_4$$

**PROBLEM:**   Predict how each one of the reagents shown above would react with methoxybenzene.

**Problem 10.16**    How would you synthesize the following compounds from benzene or toluene and any other organic reagents containing not more than three carbon atoms? You may assume that ortho and para isomers are easily separable in the laboratory.

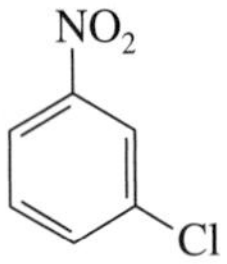

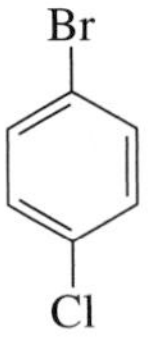

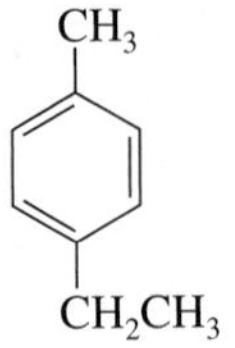

## 10.6  Polychlorinated Aromatic Hydrocarbons

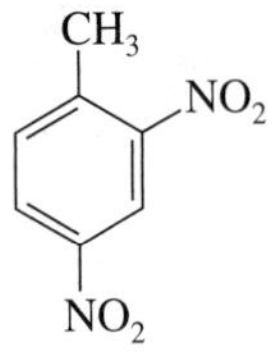

Polychlorinated aromatic hydrocarbons, especially polychlorinated biphenyl (p. 360), are widely used in industry, about 1 billion kilograms per year being produced worldwide in the 1990s. Typical polychlorinated biphenyls (PCBs) are shown below.

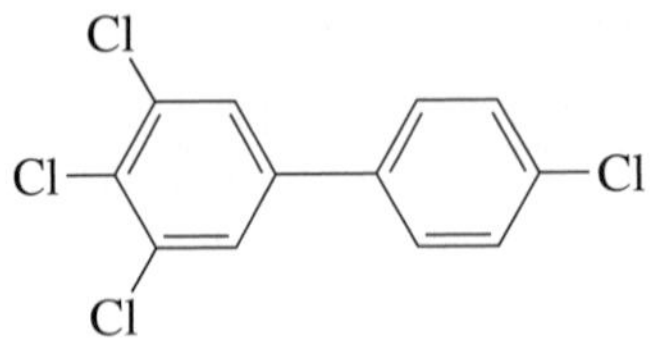

3,4,4',5-tetrachlorobiphenyl 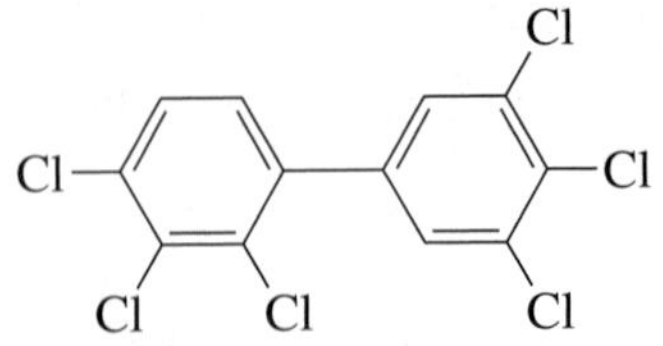          2,3,3',4,4',5'-hexachlorobiphenyl

Commercial PCBs contain around 132 different compounds. They do not burn easily, so they have been used extensively in the electrical industry in capacitors and transformers. They also have been used in lubricating oils, pesticides, adhesives, plastics, inks, paints, and sealants. They are very stable in the environment. Careless disposal of the compounds in the past has resulted in their spread, especially in lakes and oceans, where they are carried throughout the ecosystem and concentrated in higher levels of the food chain. PCBs have significant toxicity. When fed to experimental animals, they have a variety of effects, including suppression of the immune system and disruption of the endocrine system. Recently, there has been much interest and concern as to whether PCBs and other compounds with similar structures might not act as mimics of estrogens and thus disrupt both male and female sexual functions.

Two compounds derived from polychlorinated phenols have been used extensively as herbicides, substances that kill plants. These compounds, (2,4,5-trichlorophenoxy)acetic acid (abbreviated as 2,4,5-T) and (2,4-dichlorophenoxy)acetic acid (abbreviated as 2,4-D), mimic a plant hormone that controls the growth of cells. An application of the herbicide to a plant causes unchecked growth, which soon kills the plant.

(2,4,5-trichlorophenoxy)acetic acid  
2,4,5-T

(2,4-dichlorophenoxy)acetic acid  
2,4-D

These compounds are components of Agent Orange, which was used as a defoliant during the Vietnam War.

A polychlorinated compound that is found widely distributed in the environment is one of the most toxic organic chemicals known, 2,3,7,8-tetrachlorodibenzo[b,e][1,4]dioxin, usually just called dioxin. Small amounts of dioxin occur as a contaminant of 2,4,5-T.

dioxin

Dioxin is so toxic that it has an $LD_{50}$ value of 0.6 $\mu g/kg$ ($\mu g$ is a microgram, $10^{-6}$ g) for guinea pigs. This means that a dose that small will kill half the guinea pigs to which it is fed. Dioxin is also highly carcinogenic. The reasons for this high degree of toxicity are not well understood but are being studied intensively in many laboratories. By comparison, sodium cyanide has an $LD_{50}$ (oral) value of 15 mg/kg in rats, while strychnine, often used as a poison for pests, has an $LD_{50}$ value of 0.96 mg/kg. On the other hand, dioxin is only $10^{-5}$ times as toxic as botulism toxin.

**Problem 10.17**     The herbicide 2,4-D is made by mixing 2,4-dichlorophenol and chloroacetic acid with slightly more than 2 equivalents of sodium hydroxide in water solution and then acidifying the solution with hydrochloric acid. Write an equation showing all the species that are formed when the phenol and the acid are dissolved in base. Identify all the nucleophiles and the leaving groups present in the solution. 2,4-D is formed in 87% yield. How do you explain the selectivity of the reaction?

**Table 10.2 Summary of Electrophilic Aromatic Substitution Reactions**

| Aromatic Compound | Electrophile | Reactive Intermediate(s) | Product(s) |
|---|---|---|---|
|  | $E^+$ |  |  |
|  Z = —OH, —OR, —$NR_2$, —R, —X, etc. | $E^+$ |  |  *relative amounts depend on sizes of* Z *and* E |
|  EWG = $-\overset{\text{O}}{\overset{\|}{\text{C}}}\text{R}$, $-\overset{+}{\text{N}}\text{R}_3$, —$NO_2$, —CN, —$SO_3H$, etc. | $E^+$ |  |  |

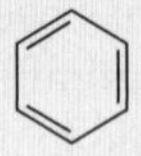

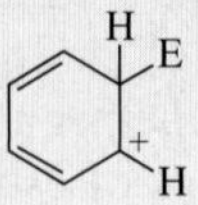

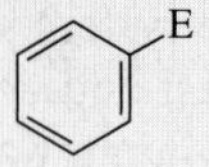

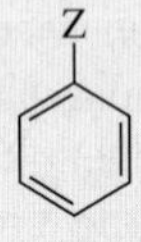

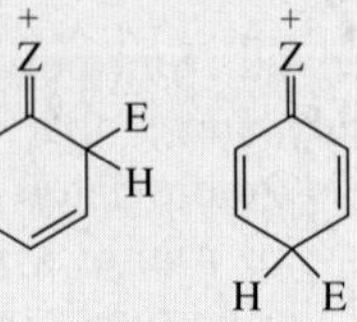

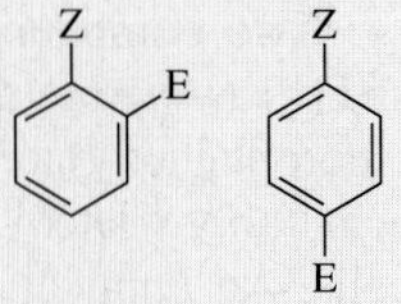

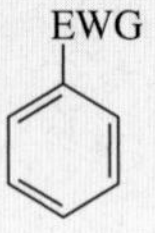

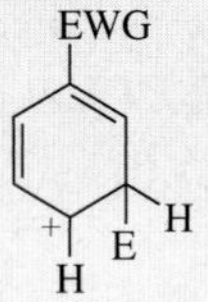

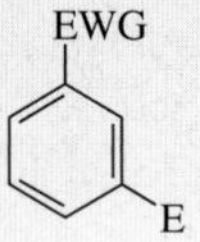

**Table 10.3 Electrophiles for Electrophilic Aromatic Substitution Reactions**

| Reaction | Reagents | Electrophile | Substitution Product of Benzene |
|---|---|---|---|
| Halogenation | $Cl_2$ or $Br_2$ $FeX_3$ | $\text{X}_3\text{Fe}^{\delta-}\text{---X---X}^{\delta+}$ |  |
| Nitration | $HNO_3$ $H_2SO_4$ | $\overset{+}{N}O_2$ |  |
| Friedel-Crafts alkylation | ROH $HB^+$ or RX $AlCl_3$ | $R^+$ |  |
| |  $HB^+$ |  |  |

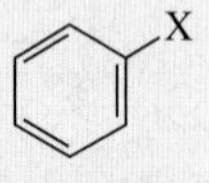

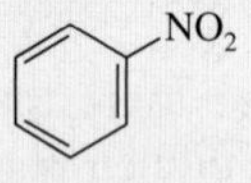

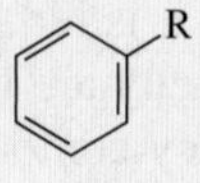

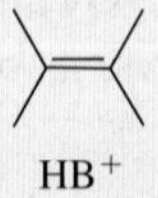

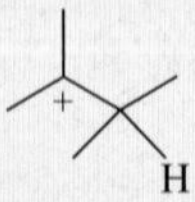

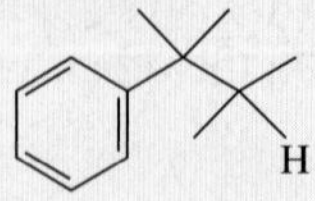

## ADDITIONAL PROBLEMS

**10.18** Name the following compounds.

(a)

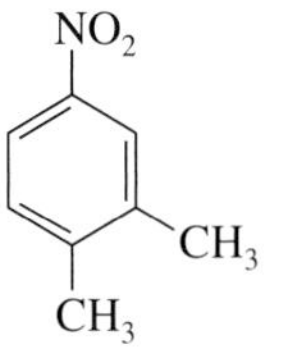

(b)

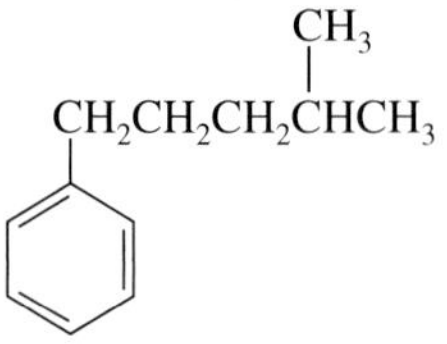

(c)

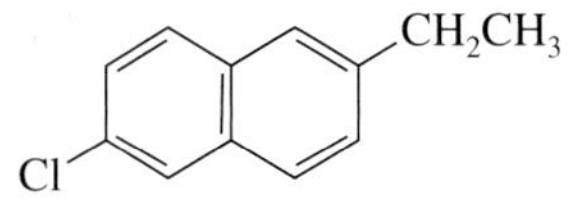

(d)

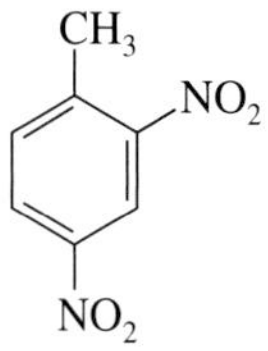

(e)

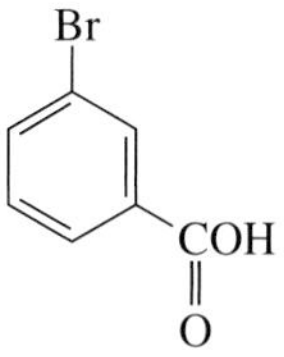

(f)

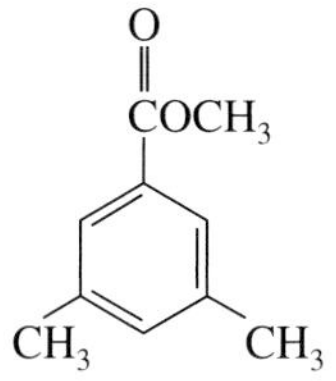

(g)

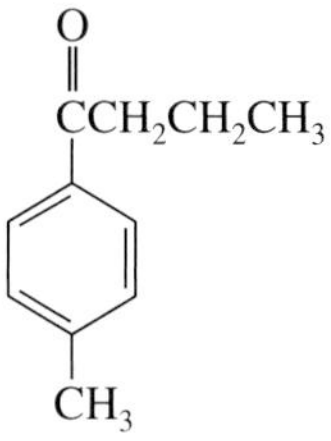

(h)

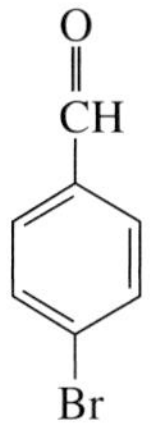

**10.19** Write structural formulas for the following compounds.

(a) ethyl *m*-chlorobenzoate

(b) 1,4-dimethylnaphthalene

(c) 1-phenyl-3-methyl-1-hexanone

(d) 1-phenyl-1-chloropropane

(e) *m*-bromonitrobenzene

(f) *p*-chlorobromobenzene

**10.20** Predict the major products of the following reactions.

(a) 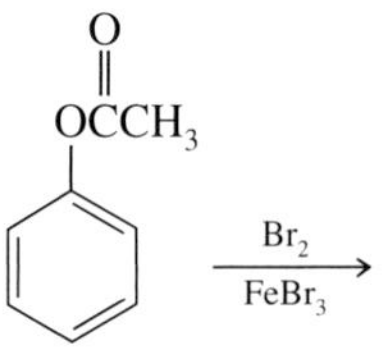 $\xrightarrow[\text{FeBr}_3]{\text{Br}_2}$

(b) 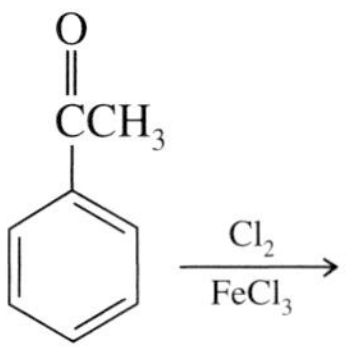 $\xrightarrow[\text{FeCl}_3]{\text{Cl}_2}$

(c) 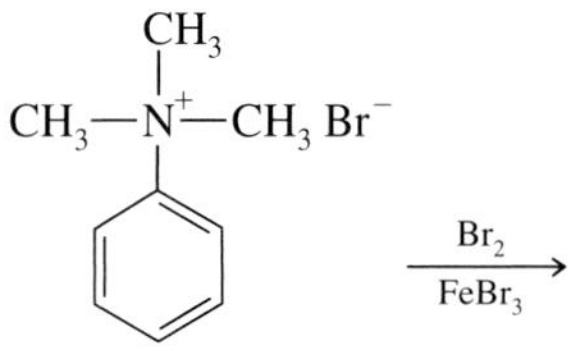 $\xrightarrow[\text{FeBr}_3]{\text{Br}_2}$

(d) 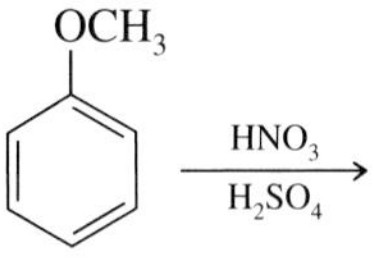 $\xrightarrow[\text{H}_2\text{SO}_4]{\text{HNO}_3}$

(e) 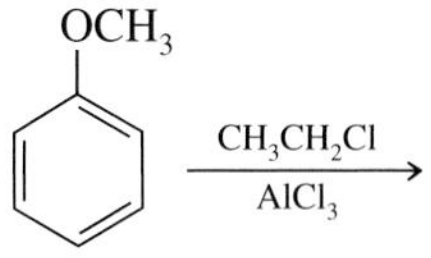 $\xrightarrow[\text{AlCl}_3]{\text{CH}_3\text{CH}_2\text{Cl}}$

(f) 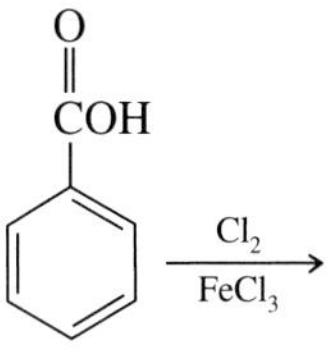 $\xrightarrow[\text{FeCl}_3]{\text{Cl}_2}$

**10.21** Give structures for compounds designated by letters in the following equations.

(a) 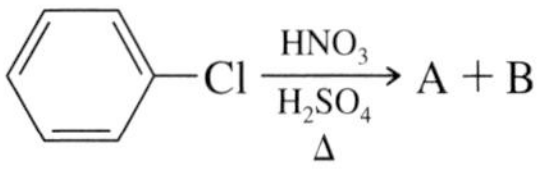 $\xrightarrow[\substack{\text{H}_2\text{SO}_4 \\ \Delta}]{\text{HNO}_3}$ A + B

(b) $\xrightarrow{\text{AlCl}_3}$ C + D

(c) $\xrightarrow{\text{HF}}$ E + F

(d) $\xrightarrow{\text{H}_3\text{PO}_4}$ G

(e) $\xrightarrow[\substack{\text{FeBr}_3 \\ \Delta}]{\text{Br}_2}$ H

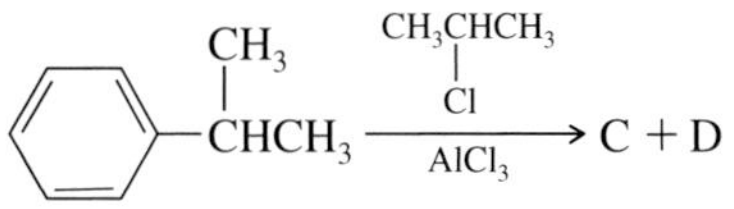

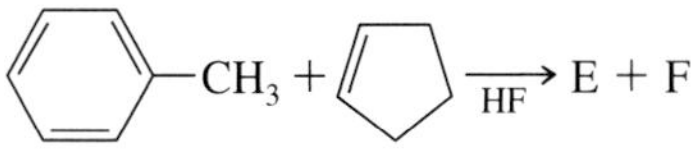

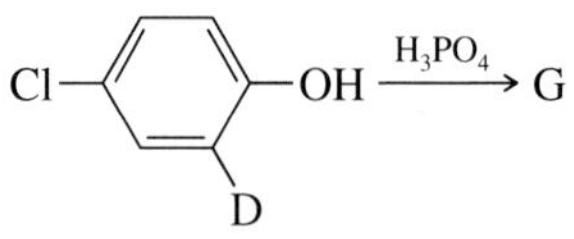

(f) $\xrightarrow{\text{I}}$

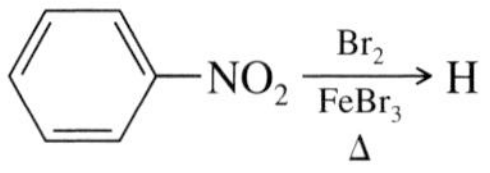

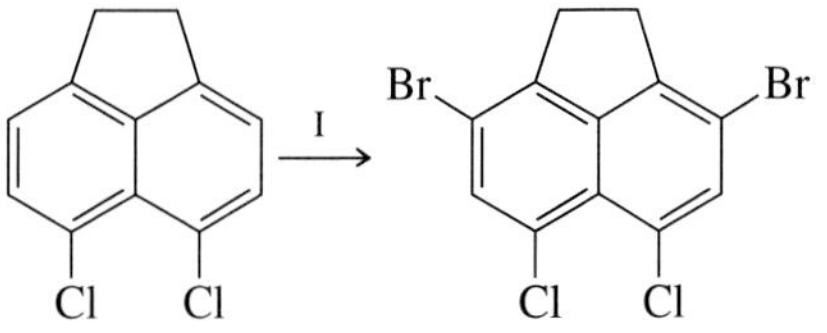

(g) 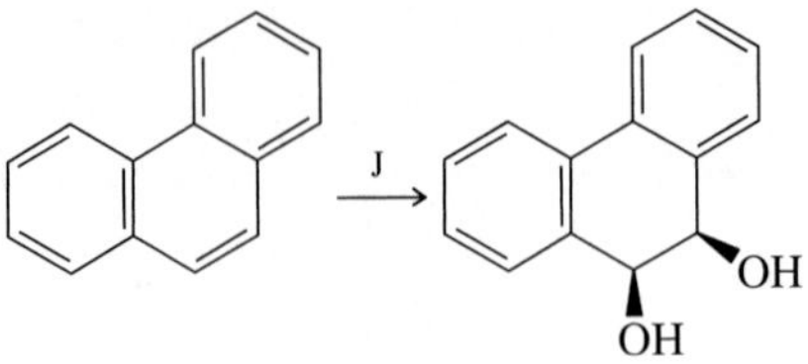 [structure of phenanthrene] → J → [structure of dihydrophenanthrene diol with OH, OH]

(h) 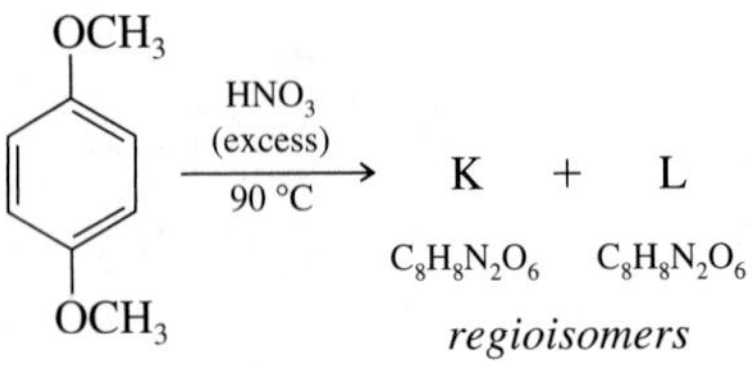 [structure of 1,4-dimethoxybenzene with OCH₃ groups] $\xrightarrow[90\ °C]{HNO_3\ (excess)}$ K + L

$C_8H_8N_2O_6$    $C_8H_8N_2O_6$
*regioisomers*

(i) 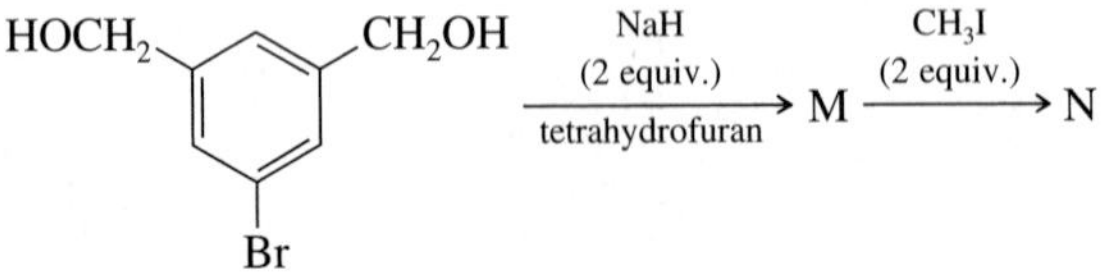 [structure: benzene with HOCH₂, CH₂OH, and Br substituents] $\xrightarrow[tetrahydrofuran]{\substack{NaH \\ (2\ equiv.)}}$ M $\xrightarrow{\substack{CH_3I \\ (2\ equiv.)}}$ N

**10.22** Propose syntheses for the following compounds starting with benzene or toluene.

(a) 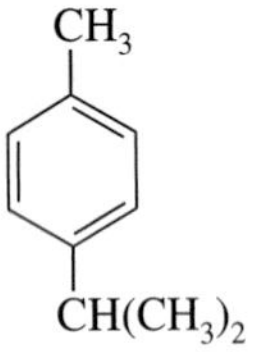 [structure: benzene with CH₃ and CH(CH₃)₂]

(b)  [structure: benzene with Br and Br]

(c) 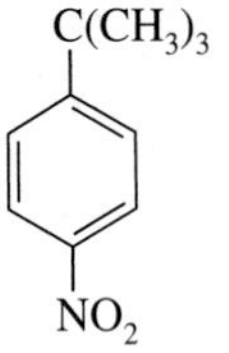 [structure: benzene with C(CH₃)₃ and NO₂]

(d) [structure: benzene with CH₃, D, and D]

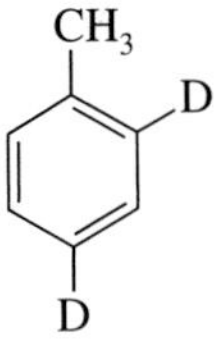

(e) 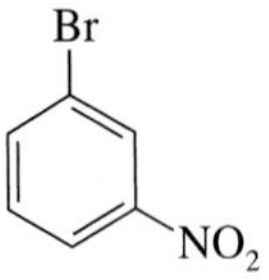 [structure: benzene with Br and NO₂]

**10.23** During research into phenol derivatives that have activity against hormones involved in blood clotting, the following synthesis was carried out in two steps. The first is a Friedel-Crafts alkylation reaction starting with an alcohol. Provide structural formulas for the reagents that would be necessary for the transformation of the starting material into the final product, as well as for the structure of the product of the first step.

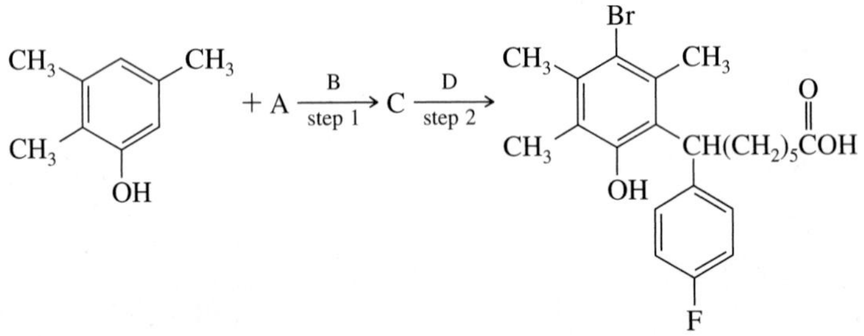

[structure: 3,5-dimethylphenol type compound with CH₃, CH₃, CH₃, OH] + A $\xrightarrow[step\ 1]{B}$ C $\xrightarrow[step\ 2]{D}$ [product structure with Br, CH₃, CH₃, CH₃, OH, CH(CH₂)₅COH, and F-labeled phenyl ring]

**10.24** When naphthalene (p. 360) is treated with bromine and aluminum bromide, the product is 99% 1-bromonaphthalene and only 1% 2-bromonaphthalene.

(a   Write the equation for this reaction.
(b)   How would you rationalize the regioselectivity of this reaction?

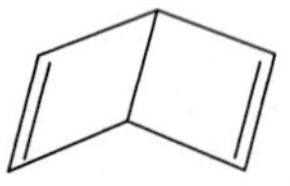

Dewar benzene

**10.25** Dewar benzene (Problem 2.26, p. 69) has actually been synthesized and shown to have reactivity that differs from that of benzene. The following reactions have been carried out with Dewar benzene; predict what the products will be.

(a) Br₂(excess)    (b) OsO₄(excess), then H₂S    (c)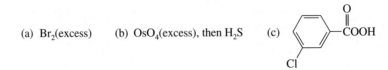

(d) When Dewar benzene is treated with either a Lewis acid, AlCl$_3$, or a protic acid, H$_2$SO$_4$, benzene is formed. Suggest a mechanism for the conversion of Dewar benzene to benzene in sulfuric acid.

**10.26** Melatonin, *N*-acetyl-5-methoxytryptamine, is a neurohormone produced by the pineal gland that is involved in the regulation of sleep and wake cycles. Compounds related to melatonin were synthesized in order to better study the physiologic role of the hormone. Propose reagents or products represented by letters in the equations shown below.

(a)

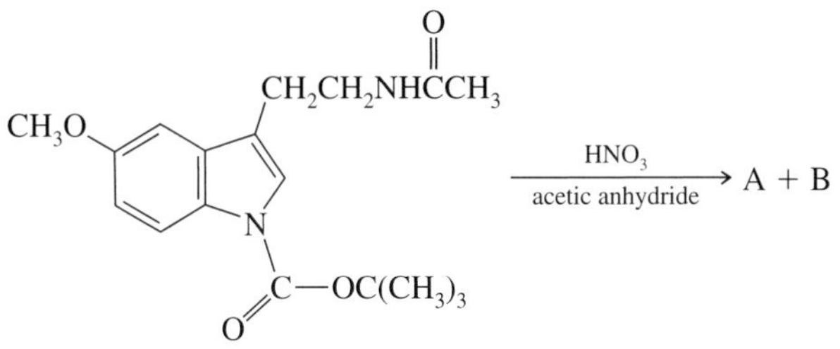

$$\xrightarrow[\text{acetic anhydride}]{\text{HNO}_3} \text{A} + \text{B}$$

(b)

$$\xrightarrow[\text{(a deprotonation step)}]{\text{C}} \text{D} \xrightarrow{\text{E}}$$

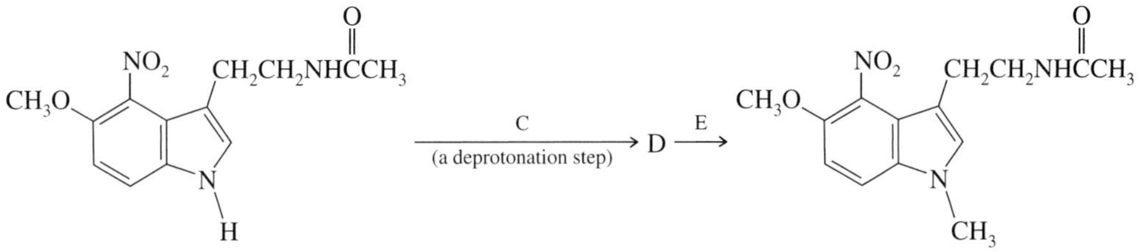

**10.27** When allyl alcohol is treated with hydrogen fluoride in the presence of benzene, two products are formed: 3-phenyl-1-propene and 1,2-diphenylpropane. Write equations showing the mechanisms for the formation of these products.

**10.28** When a piperidone, Compound A, is treated with trifluoromethanesulfonic acid, a very strong acid, in the presence of benzene, Compound B is formed. Propose a detailed mechanism for this conversion using the curved-arrow convention. The first two steps of this mechanism were part of Problem 3.39 (p. 112).

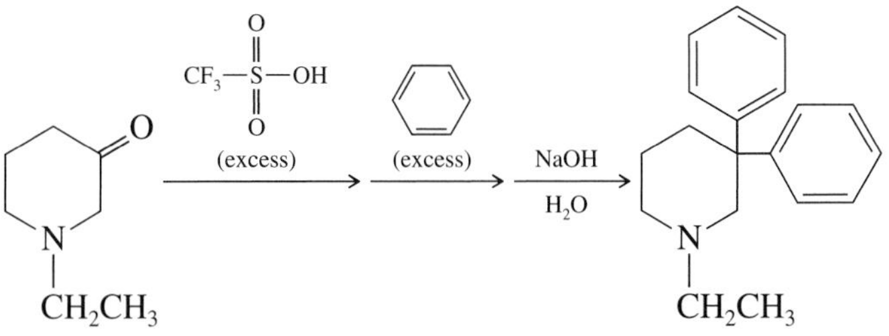

$$\xrightarrow[\text{(excess)}]{} \xrightarrow[\text{(excess)}]{} \xrightarrow[\text{H}_2\text{O}]{\text{NaOH}}$$

Compound A                                    Compound B

**10.29** An intensely blue hydrocarbon, called azulene, has the structure shown at right. Predict whether it has aromaticity.

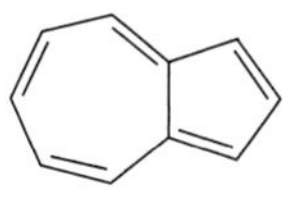

azulene

**10.30** Experimentally, it has been found that the cyclononatetraenyl anion and the cyclooctatetraenyl dianon can be prepared and are reasonably stable species. The reaction of cyclononatetraene with the carbanion generated when dimethyl sulfoxide, CH$_3$SOCH$_3$, is treated with sodium hydride gives cyclononatetraenyl anion. Cyclooctatetraenyl dianion is formed when cyclooctatetraene reacts with 2 equivalents of potassium metal in tetrahydrofuran. Write equations for the reactions described here, and explain the source of the stability of these hydrocarbon anions. A review of Section 9.5B will be helpful.

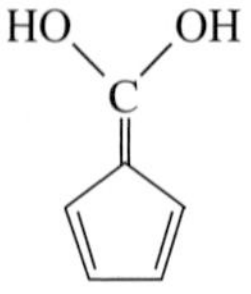

HO    OH
\    /
 C
 ‖

fulvene-6,6-diol
$pK_a$ 1.3

**10.31** The enediol fulvene-6,6-diol, shown at left, has very high acidity, $pK_a$ 1.3. Usually enols (compounds having a hydroxy group on a double bond) have $pK_a \sim 10$. How would chemists rationalize the unexpected acidity of this compound?

**10.32** Heptafulvenes are compounds in which a carbon atom of the cycloheptatriene ring is doubly bonded to a carbon outside the ring. The parent compound, shown below, is highly unstable, but substitution of two cyano groups for the hydrogen atoms on the double bond outside the ring gives a stable compound. How would you rationalize these experimental observations? (*Hint:* Writing resonance contributors for both compounds will be helpful.)

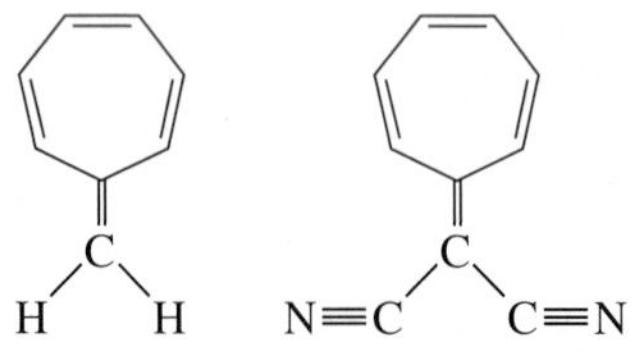

heptafulvene            a dicyanoheptafulvene

*unstable*                      *stable*

**10.33** Alkyl-substituted aromatic compounds undergo rearrangements of their carbon skeletons during Friedel-Crafts reactions. One rearrangement was studied using $^{13}C$ nuclear magnetic resonance spectroscopy. $^{13}C$-labeled tetrahydrophenanthrene is converted into an "isotopomer," an isomer that differs only in the location of the isotopic label, upon heating with aluminum chloride containing a little water. At equilibrium there are equal amounts of the two compounds.

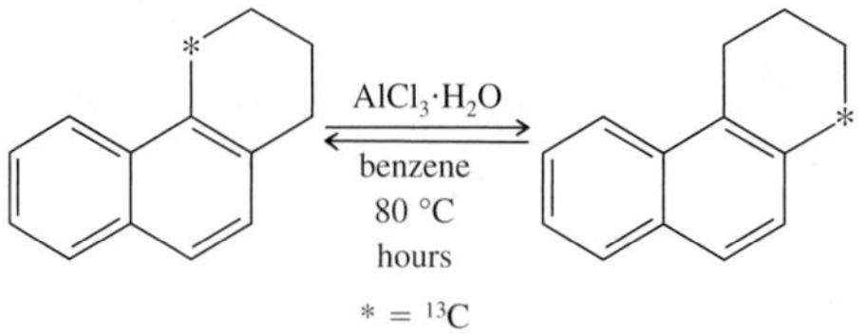

$AlCl_3 \cdot H_2O$
benzene
80 °C
hours

$* = {}^{13}C$

Propose a mechanism for this interconversion. Water in the presence of aluminum chloride is a source of protons. Why?

**10.34** The last step in a recent synthesis of corannulene, which was first synthesized by Richard Lawton at the University of Michigan, involves heating the compound shown below with palladium in the absence of hydrogen. Propose a structure for corannulene. (*Hint:* Corannulene is like a fragment of buckminsterfullerene [p. 364].)

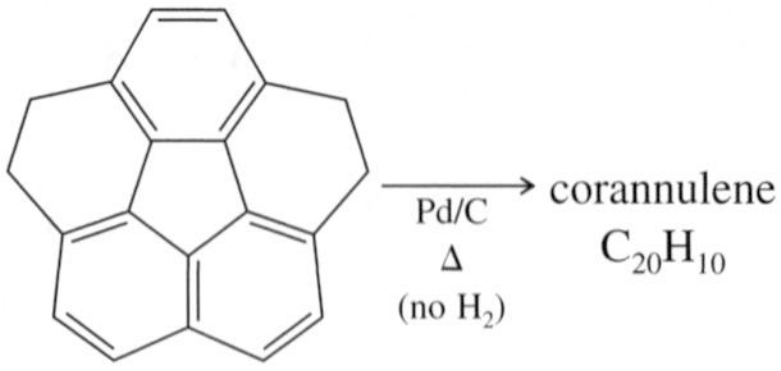

Pd/C
Δ
(no $H_2$)

corannulene
$C_{20}H_{10}$

**10.35** When chlorobuckminsterfullerene, $C_{60}Cl_n$, is treated with benzene and aluminum chloride, a compound that has broad absorption at $\delta$ 7.2 (p. 406) in its proton nuclear magnetic resonance spectrum is formed. Write an equation for the reaction of $C_{60}Cl_n$ under these reaction conditions that accounts for the spectral data.

# Nuclear Magnetic Resonance Spectroscopy

**11**

## A Look Ahead

Chemists gather information about the structure of molecules by making physical measurements on compounds. The dipole moment of a compound can be used to deduce the shape and symmetry of its molecules. Electron diffraction is a method for obtaining information about bond lengths and bond angles (Chapter 1). Ultraviolet spectroscopy probes electronic transitions between bonding and antibonding molecular orbitals (Chapter 2). Nuclear magnetic resonance spectroscopy gives us information about symmetry and connectivity (Chapter 5), and this is the technique that we will explore further in this chapter. The next chapter (Chapter 12) will extend our understanding of ultraviolet and visible spectroscopy and introduce two very important methods of structure determination. One is infrared spectroscopy, which gives us information about the functional groups that are present in a molecule. The other is mass spectrometry, which allows us to accurately determine the molecular weight and molecular formula of a compound, as well as, often, its connectivity.

The nuclei of atoms such as hydrogen, H, and an isotope of carbon, $^{13}C$, behave like small magnets. If a sample of a compound containing these elements is placed in a strong magnetic field, slightly more than half the nuclei align themselves with the field. Nuclei in this state absorb radiation in the radiofrequency range of the spectrum (p. 395) and are raised to a higher-energy state, in which they are aligned against the external magnetic field. The record of these transitions is a nuclear magnetic resonance spectrum. We have already seen examples of such spectra in Chapter 5, where we used them as tools for exploring concepts about symmetry and equivalence of carbon and hydrogen atoms in molecular structures.

The exact amount of energy necessary to cause a nucleus to undergo a transition from a lower-energy state to a higher-energy one depends on the strength of the magnetic field and on the environment in the molecule, as we saw in Chapter 5. A carbon atom bonded to an electronegative atom such as chlorine, for example, experiences the external magnetic field in a different way than one that is not. The same is true for the hydrogen atoms bonded to those carbon atoms (see Figure 5.3, p. 152, for example). Therefore, nuclear magnetic resonance spectra have a number of peaks corresponding to the different types of carbon atoms or hydrogen atoms in the molecule. Together the two spectra enable us to determine the connectivity of a molecule. This chapter will describe the different kinds of information that chemists obtain from nuclear magnetic resonance spectra and how to use such spectra in determining the structure of organic compounds.

**Workbook Exercises**

## 11.1   The Experimental Observations

### A. Chemical Shifts and Integration

The nuclei of $^{13}C$ (the isotope of carbon having 6 protons and 7 neutrons in its nucleus) and of hydrogen atoms absorb radiofrequency radiation when placed in a magnetic field. The record of these absorptions of energy are known as **nuclear magnetic resonance spectra.** A stable isotope of carbon, $^{13}C$, has a natural abundance of about 1%; that is, approximately 1% of all the carbon atoms in any sample of a compound containing carbon are $^{13}C$ atoms. Because these atoms occur randomly throughout molecules of a given compound, the $^{13}C$ magnetic resonance spectrum is representative of a compound.

Figure 11.1 shows the $^{13}C$ and the proton magnetic resonance spectra of methyl acetate. Both the spectra have a signal at the far right, which is the absorption band for the carbon atoms (for the $^{13}C$ spectrum) or the hydrogen atoms (for the proton spectrum) of tetramethylsilane, $(CH_3)_4Si$. Tetramethylsilane, usually abbreviated TMS, is added to the solution of a compound for which nuclear magnetic resonance spectra are being run to give a reference point. The divisions on the scale of the spectra are called $\delta$ (delta) and represent chemical shift values (pp. 153, 399). The signal for TMS appears at $\delta$ 0. The $^{13}C$ spectrum ($^{13}C$ NMR) of methyl acetate has four other signals at $\delta$ 21, $\delta$ 52, and $\delta$ 172 and three closely spaced peaks centered at $\delta$ 77, which is the carbon atom of the solvent deuteriochloroform, $CDCl_3$. There are two signals, besides the one for TMS, in the proton spectrum ($^1H$ NMR) of methyl acetate at $\delta$ 2.07 and $\delta$ 3.67.

Methyl acetate, the $^{13}C$ spectrum tells us, has three different kinds of carbon atoms. The proton nuclear magnetic resonance spectrum of methyl acetate indicates the presence of two different kinds of hydrogen atoms.

methyl acetate

The carbon atoms are the $a$, $b$, and $c$ carbon atoms. The hydrogen atoms are those on carbon $a$, the $a$ hydrogens of the acetyl group, and those on carbon $b$, the $b$ hydrogens of the methoxy group. There are no hydrogens on carbon $c$. The three $a$ hydrogens of the acetyl group are all equivalent to each other, as are the three $b$ hydrogens of the methoxy group. They are said to be **chemical-shift equivalent.** In other words, each group gives rise to a single peak in the proton magnetic resonance spectrum. Whether hydrogen atoms are chemical-shift equivalent can be determined by mentally substituting another species for each one of them in turn and seeing if the same or a different compound results, the same process we used in Section 5.7B.

**Problem 11.1**   Prove to yourself that replacement of any one of the $a$ hydrogen atoms with a methyl group gives the same compound but one that is different from the compound obtained by replacing any one of the $b$ hydrogen atoms with a methyl group.

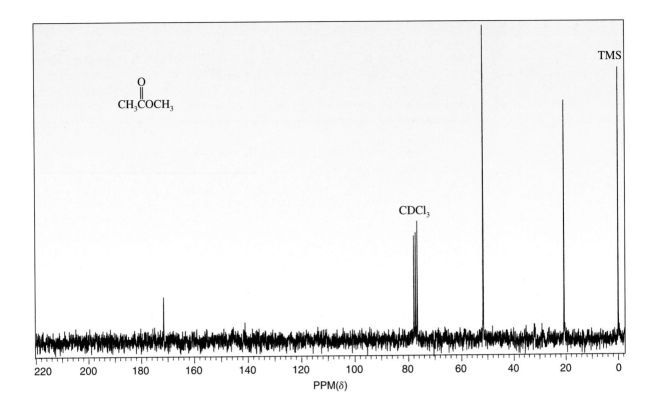

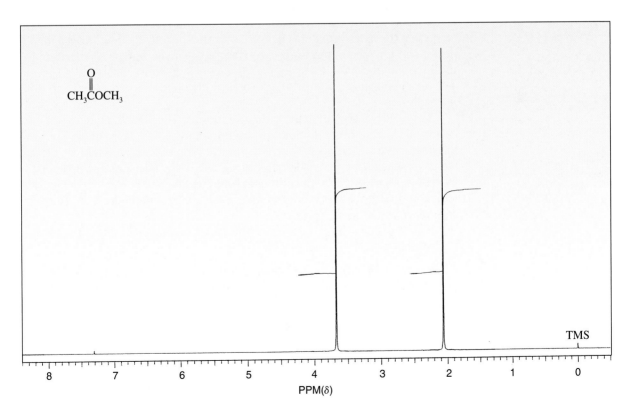

*Figure 11.1*
$^{13}$C and proton nuclear magnetic resonance spectra of methyl acetate.

As we have seen already in Chapter 5, the chemical shifts of the different carbon and hydrogen atoms are related to their environments. In the $^{13}$C spectrum, the carbon atom that is farthest from the signal for TMS at $\delta$ 172 is the one for the carbon of the carbonyl group, an $sp^2$-hybridized carbon atom doubly bonded to the electronegative oxygen atom. The next peak to the right of it at $\delta$ 52 is the carbon atom of the methoxy group, also directly bonded to oxygen. The carbon atom that is least affected, the one closest to the TMS signal at $\delta$ 21, is the carbon atom bonded to the carbonyl group.

A similar pattern of shifts is seen for the hydrogen atoms bonded to the different carbon atoms. The hydrogens of the methoxy group appear farther from TMS at $\delta$ 3.67 than do the ones on the acetyl group, at $\delta$ 2.07.

The proton magnetic resonance spectrum of methyl acetate shows another feature that we have already met in Chapter 5. The step-like tracing over the two peaks of the spectrum is known as the **integration,** a measurement of the area under each absorption peak, which in turn is proportional to the relative number of hydrogen atoms giving rise to the signal. An examination of the spectrum of methyl acetate shows that the two steps are of the same height. In other words, the areas under the two peaks are equal; therefore, there are equal numbers of protons of the two types in methyl acetate. Note that the integration does *not* give the absolute number of protons of each kind but only their relative abundance, as was discussed in Section 5.4 (p. 151).

## B. Interactions Between Neighboring Atoms

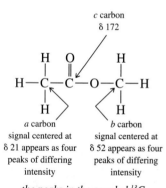

*c* carbon
$\delta$ 172

*a* carbon
signal centered at
$\delta$ 21 appears as four
peaks of differing
intensity

*b* carbon
signal centered at
$\delta$ 52 appears as four
peaks of differing
intensity

*the peaks in the coupled $^{13}$C magnetic resonance spectrum of methyl acetate*

Not all nuclear magnetic resonance spectra are as simple as the ones shown in Figure 11.1. Another version of the $^{13}$C magnetic resonance spectrum of methyl acetate is given in Figure 11.2. This spectrum, in which the interaction between the carbon atoms and the hydrogen atoms bonded to them is recorded, now shows a number of peaks, still grouped in three areas. A closer comparison of the $^{13}$C spectrum in Figure 11.1 with the "coupled spectrum" in Figure 11.2 shows that while the peak at $\delta$ 172 remains unchanged, the other two each now appear as four peaks of differing intensities.

The bands belonging to the carbons of the acetyl and methoxy groups in methyl acetate are said to be *split* in the spectrum in Figure 11.2. This splitting arises from the interaction of the nuclei of the hydrogen atoms with the nuclei of the carbon atoms to which they are bonded, the interaction of the nuclei of one-bond neighbors by a process known as **spin–spin coupling.** Section 11.5A examines how such interactions create the observed patterns, which appear in proton magnetic resonance spectra as well as in $^{13}$C spectra.

Four essential pieces of information about a compound are obtained from its nuclear magnetic resonance spectra:

1. The number of different groupings of peaks shows how many different kinds of carbon or hydrogen atoms are present in the compound (Sections 11.1A and 11.4).

2. The $\delta$ values (chemical shifts) for the carbon and hydrogen atoms give information about the environment of those atoms (Section 11.4).

3. The integration of the proton magnetic resonance spectrum reveals the relative numbers of hydrogen atoms of each type that are present in the molecule (Section 11.1A). The $^{13}$C magnetic resonance spectrum is not integrated because factors other than numbers of carbon atoms enter into the relative intensities of the peaks (p. 403).

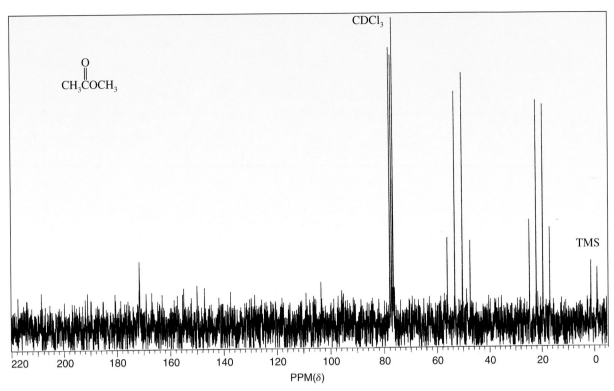

*Figure 11.2*
Proton-coupled $^{13}$C nuclear magnetic resonance spectra of methyl acetate.

4. The splitting patterns that appear in the spectrum tell us about the numbers of nonequivalent neighboring nuclei that have nuclear spin and can, therefore, interact with each other (Section 11.5).

**Problem 11.2**   For each of the following compounds, pick out the groups of carbon and hydrogen atoms that are chemical-shift equivalent to each other. How many one-bond neighbors does each group of carbon atoms in (c) and (d) have? How many three-bond neighbors does each group of hydrogen atoms in (b) and (d) have? Remember that equivalent hydrogen atoms within a group do not count as neighbors.

(a) $CH_3CH_2OCH_2CH_3$     (b) $CH_3CH_2CH_2Br$     (c) H—⟨benzene ring with H, H, H, H⟩—$OCH_2CH_3$

(d) $CH_3\overset{\overset{\displaystyle CH_3}{|}}{C}HCH_2Cl$     (e) $CH_3CH_2\overset{\overset{\displaystyle O}{||}}{C}CH_2CH_3$     (f) $CH_3CH_2CH_2CH_2CH_3$

**Problem 11.3**   The two $^{13}$C spectra shown in Figure 11.3 (p. 394) are of the following compounds. Decide which spectrum belongs to which compound, and assign chemical shift values to the carbon atoms in each compound by comparing these spectra to the ones you have seen already and reasoning by analogy.

(a) ⟨cyclopentane ring with O⟩     (b) $CH_3CH_2\overset{\overset{\displaystyle O}{||}}{C}\overset{\overset{\displaystyle O}{||}}{O}CCH_2CH_3$

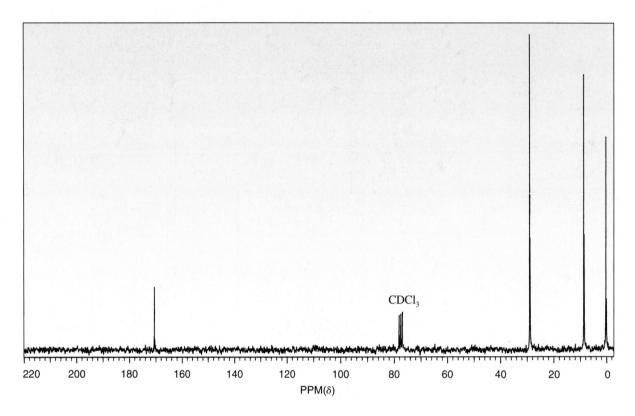

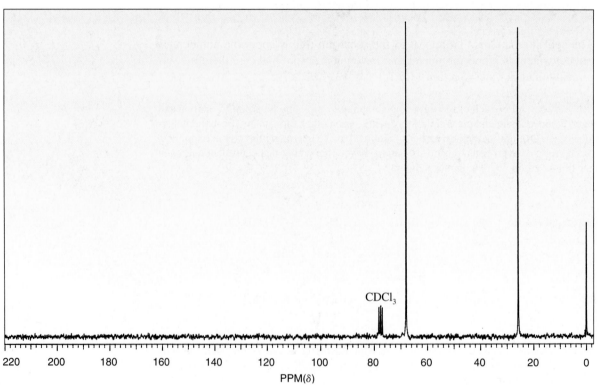

*Figure 11.3*

## 11.2   The Electromagnetic Spectrum and Absorption Spectroscopy

Molecules have several different kinds of energy levels and, therefore, absorb radiation in several regions of the electromagnetic spectrum. For example, molecules absorb ultraviolet radiation in making electronic transitions between bonding or nonbonding molecular orbitals and antibonding molecular orbitals (Section 2.11, p. 69). In doing so, they go from their ground state to an excited state. The molecule in the excited state may return to the ground state by emitting energy, either as heat or light, or it may undergo chemical reactions in the excited state.

The difference in energy between two molecular energy levels is proportional to the frequency of the light absorbed:

$$\Delta E = h\nu$$

where $h$ is Planck's constant, $6.624 \times 10^{-27}$ erg $\cdot$ s, and $\nu$ is the frequency of light in cycles per second (hertz, or Hz). The frequency of light is related to its wavelength, $\lambda$:

$$\nu = \frac{c}{\lambda}$$

where $c$ is the velocity of light in a vacuum, $2.998 \times 10^{10}$ cm/s.

The electromagnetic spectrum has a tremendous range of energy. At one end of the spectrum are cosmic rays, which have energies of approximately $10^9$ kcal/mol (1 erg/photon = $1.439 \times 10^{13}$ kcal/mol of photons). They have very short wavelengths ($10^{-12}$ cm) and high frequencies ($10^{22}$ Hz). Medically useful x-rays have wavelengths of $10^{-8}$ to $10^{-6}$ cm and energies in the range of $10^3$ kcal/mol. On the low-energy end of the electromagnetic spectrum are radiowaves, which have long wavelengths ($10^3$ to $10^6$ cm) and low frequencies ($10^7$ to $10^4$ Hz).

The regions of the electromagnetic spectrum that are used for investigating molecular structures are summarized in Table 11.1. Transitions between electronic energy levels occur when a molecule absorbs radiation in the **ultraviolet and visible regions** of the spectrum. A molecule goes from one vibrational energy level to another on absorbing **infrared radiation.** Absorption of **microwaves** causes changes in rotational energy levels. Finally, transitions between energy levels associated with

**Table 11.1   Regions of the Electromagnetic Spectrum Used in Spectroscopy**

| Region of Spectrum | Molecular Change | Energy, kcal/mol | Frequency, Hz | Wavelength, cm |
|---|---|---|---|---|
| ultraviolet | electronic transitions (Chapters 2 and 12) | 100 | $10^{15}$ | $2 \times 10^{-5}$ to $4 \times 10^{-5}$ |
| visible | electronic transitions | 50 | $5 \times 10^{14}$ | $4 \times 10^{-5}$ to $8 \times 10^{-5}$ |
| infrared | molecular vibrations (Chapter 12) | 5 | $10^{13}$ to $10^{14}$ | $10^{-4}$ to $10^{-2}$ |
| microwave | molecular rotations | $3 \times 10^{-3}$ | $3 \times 10^{10}$ | 1 |
| radiofrequency | orientation of spin of nucleus in magnetic field (Chapters 5 and 11) | $10^{-7}$ | $10^6$ | $5 \times 10^5$ |

certain nuclei such as those of $^{13}$C or hydrogen occur on the absorption of **radiofrequency waves.** Table 11.1 shows the approximate energy associated with each of these types of transitions and also indicates the chapters in which ultraviolet and infrared spectroscopy are discussed.

## 11.3   The Origin of Nuclear Magnetic Resonance Spectra

Just as electrons in an atom are assigned spin quantum numbers, nuclei of atoms with an odd mass number have **nuclear spin** and are given a **spin quantum number, *I*,** of $\frac{1}{2}$. Thus the nuclei of the common isotope of hydrogen, $^1$H, and the stable isotope of carbon, $^{13}$C, have a spin number, *I*, of $\frac{1}{2}$. Nuclei of atoms that have an even mass number but an odd atomic number also possess nuclear spin. Deuterium, $^2$H, and the common isotope of nitrogen, $^{14}$N, belong in this category, and their nuclei have a spin number, *I*, of 1. Nuclei of atoms that have an even mass number and an even atomic number have no net spin. The nuclei of the most abundant isotopes of carbon, $^{12}$C, and oxygen, $^{16}$O, fall in this category; their spin number, *I*, is 0.

*Only atoms with nuclear spin give rise to nuclear magnetic resonance.* The nucleus is a charged particle, so a spinning nucleus creates a magnetic field. Such a nucleus behaves like a small magnet; it has a **magnetic moment** and is affected by an external magnetic field. When atoms containing such nuclei are placed in a strong magnetic field, $(2I + 1)$ energy levels, where *I* is the spin number of the nuclei, become available to the nuclei. For example, when a sample of a compound containing hydrogen is put in an external magnetic field, the hydrogen exists in two energy states. Slightly more than half the nuclei line up so that their magnetic moments are aligned with the external magnetic field. This situation is the lower-energy state for the nuclei. The difference between this lower-energy state and the higher-energy one corresponds to the energy of radiation in the radiofrequency range of the electromagnetic spectrum. If an external magnetic field of 47,000 gauss (or G) is used, the hydrogen atoms in the sample absorb radiation having a frequency of 200,000,000 cycles/s, or hertz (Hz), or 200 MHz. In the same magnetic field, $^{13}$C nuclei absorb at 50,000,000 Hz, or 50 MHz. Nuclei are raised to the higher-energy level, in which their magnetic moments are opposed to the external magnetic field. The exact amount of energy required for the transition depends on the strength of the magnetic field experienced by the nuclei (Figure 11.4). The instrument that records these transitions as a spectrum is shown in Figure 11.5.

Once a nucleus is in the higher-energy level, it returns to the lower one by various processes that involve losing energy to its surroundings. These processes are known as **relaxation** and are especially important to the biological and medical applications of nuclear magnetic resonance spectroscopy (p. 435).

Not all the nuclei in a molecule of an organic compound experience the same external magnetic field. The electrons around a nucleus **shield** it from the effects of the magnetic field. Thus atoms in different electronic environments in a molecule experience an external magnetic field to different degrees. Slightly different amounts of energy are, therefore, needed to promote the nuclei of different types of atoms from the lower-energy level to the higher-energy level.

Modern Fourier-transform spectrometers use short pulses of radiofrequency radiation to excite all nuclei of a given type ($^{13}$C, for example) at once. The data are collected as an exponentially decaying sine wave (shown in Figure 11.6, p. 398) containing information about the frequencies at which the atoms in the sample ab-

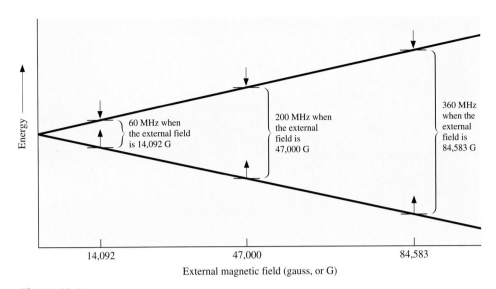

*Figure 11.4*
Dependence of the difference in energy between lower and higher nuclear spin levels of the hydrogen atom on the strength of the external magnetic field: ↑ and ↓ represent the magnetic moments of nuclei aligned with and opposed to the external field, respectively.

sorb energy. A computer stores the data and converts them into a spectrum. Figure 11.6 contains the original data for 4-methyl-2-pentanone and the spectrum derived from them. The spectrum has five peaks in it. (The three closely spaced peaks at δ 77 come from the solvent $CDCl_3$.) These peaks are the absorptions for the five different kinds of carbon atoms in 4-methyl-2-pentanone.

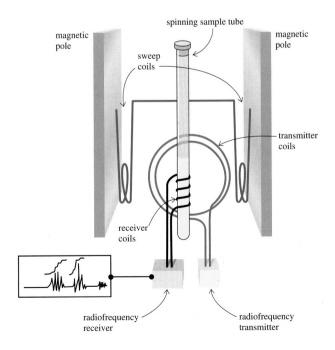

*Figure 11.5*
Schematic diagram of a typical nuclear magnetic resonance spectrometer.

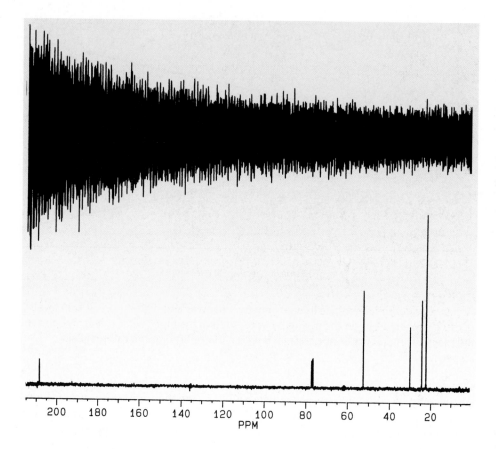

**Figure 11.6**
Free-induction decay data and proton-decoupled $^{13}C$ nuclear magnetic resonance spectrum for 4-methyl-2-pentanone, recorded on a Bruker WM-360 nuclear magnetic resonance spectrometer.

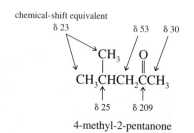

4-methyl-2-pentanone

The magnetic field in nuclear magnetic resonance spectrometers must be highly homogeneous if small differences in energy are to be observed. One way to average out the field experienced by all the molecules in a sample is to spin the tube containing the sample inside a cavity in the magnet. In practice, then, nuclear magnetic resonance spectra usually are taken of liquid samples or of solids in solution. The ideal solvent for proton magnetic resonance spectroscopy is one such as carbon tetrachloride that does not contain any hydrogen atoms. But carbon tetrachloride is not polar enough to dissolve many organic compounds, so other solvents in which hydrogen atoms have been replaced by deuterium, which does not absorb in the same region of the spectrum as hydrogen does, are used. Among these, deuteriochloroform, $CDCl_3$, is the most generally useful. Because it is difficult to measure the absolute value of an applied magnetic field, all nuclear magnetic resonance spectra are measured with reference to a standard compound, usually tetramethylsilane (p. 390).

**Problem 11.4** Predict whether or not each of the following isotopes has a nuclear magnetic moment.

(a) $^{28}Si$ (b) $^{15}N$ (c) $^{19}F$ (d) $^{31}P$ (e) $^{11}B$ (f) $^{32}S$

# 11.4 Chemical Shift

## A. The Origin of the Chemical Shift

The absorption of radiation of a given radiofrequency by the nucleus of an atom depends on the effective magnetic field that it experiences. This, in turn, depends on the environment of the atom in the molecule as well as on the strength of the external magnetic field (see Figure 11.4). The carbon and hydrogen atoms in tetramethylsilane, $(CH_3)_4Si$, the commonly used reference compound, have a higher electron density around them and are said to be more **shielded** than most other such atoms in organic compounds. This is so because carbon and hydrogen are more electronegative (2.5 and 2.1) than silicon (1.8), so electron density flows from silicon to the methyl groups. Atoms bonded to electronegative elements such as oxygen or the halogens (or to carbon atoms bonded to such atoms) have less electron density around them and are said to be **deshielded** relative to the methyl groups in tetramethylsilane (TMS). The effect is cumulative; a carbon atom bearing two chlorine atoms is more deshielded than one bearing a single halogen, as we saw in Figures 5.2 and 5.3 (pp. 150 and 152). The carbon bearing a single chlorine atom in 1,2-dichloroethane has a band at $\delta$ 44, and that with two chlorine atoms on it in 1,2-dichloroethane absorbs at $\delta$ 69. The same shift can be seen for the hydrogen atoms on those carbons ($\delta$ 3.75 and $\delta$ 5.92).

The effect of electronegativity also can be seen in a series of compounds that differ only in the type of halogen substituent. The carbon atoms bearing the halogen atom and the hydrogen atoms on those carbon atoms become progressively more deshielded in bromomethane, chloromethane, and fluoromethane as the electronegativity of the halogen increases. The deshielding is reflected in larger $\delta$ values.

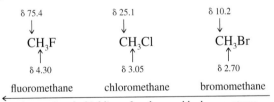

$\delta$ 75.4      $\delta$ 25.1      $\delta$ 10.2
     $\downarrow$          $\downarrow$          $\downarrow$
    $CH_3F$      $CH_3Cl$      $CH_3Br$
     $\uparrow$          $\uparrow$          $\uparrow$
   $\delta$ 4.30      $\delta$ 3.05      $\delta$ 2.70

fluoromethane    chloromethane    bromomethane

*increasing deshielding of carbon and hydrogen atoms
with increasing electronegativity of the halogen*

The $^{13}C$ and proton magnetic resonance spectra of methyl acetate were given in Figure 11.1. The spectra are shown again in Figure 11.7 (p. 400) with a difference. These spectra also show the actual frequencies at which the different nuclear transitions take place.

The difference between the position of the absorption band for a given type of atom and the position of the peak for TMS is called the **chemical shift** for that atom. The difference in energy between the peak for TMS and those for the atoms in methyl acetate can be read off the spectra in either hertz or delta units. For example, the first absorption band for the carbon atom of the acetyl group occurs at frequency, $\nu$, 1040 Hz, the second one at 2595 Hz, and the third one at 8600 Hz **downfield**

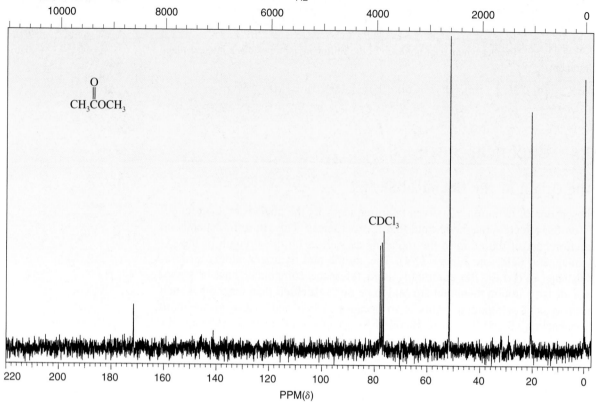

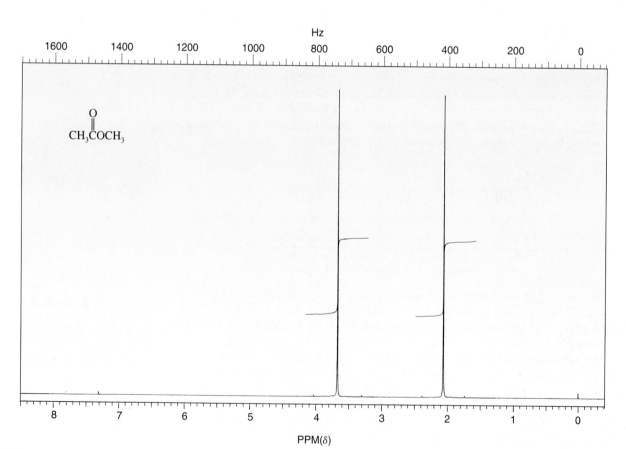

from the signal for TMS (to the left of TMS). These chemical shift values depend on the strength of the magnetic field in the instrument and, therefore, on the frequency, $\nu$, of the radiation used to make the measurements. The fundamental measurement is the frequency in hertz at which absorption of energy takes place, and this is converted to $\delta$, which is independent of instrumentation and has units of parts per million (ppm):

$$\delta = \frac{\nu_{\text{sample}} - \nu_{\text{TMS}}}{\nu_{\text{applied of the instrument}}} \times 10^6 \text{ ppm}$$

For the instrument used to take these spectra, a radiofrequency of 50 MHz is used to measure $^{13}$C resonances, so

$$\delta = \frac{\nu_{\text{sample}} - \nu_{\text{TMS}}}{50 \times 10^6} \times 10^6 \text{ ppm}$$

Thus, for the band appearing at 1040 Hz downfield from TMS,

$$\delta = \frac{1040 - 0}{50 \times 10^6} \times 10^6 \text{ ppm}$$

$$= 20.8 \text{ ppm}$$

Note that the differences in chemical shifts that are being measured in a nuclear magnetic resonance spectrum are very small, on the order of parts per million in the frequency of the radiation being used to give rise to the nuclear transition. The entire range in chemical shifts for carbon atoms is approximately 250 ppm. Note also that $\delta$ values increase from right to left, which is opposite to what is usually seen on graphs.

A radiofrequency of 200 MHz is used to take the proton nuclear magnetic resonance spectra in this book. The proton resonances for methyl acetate are at 413 and 734 Hz downfield from TMS. Applying the same formula as above but changing the applied frequency from 50 to 200 MHz gives us chemical shifts of $\delta$ 2.07 and $\delta$ 3.67. The range of chemical shifts for hydrogen atoms is much smaller than it is for carbon, only about 20 ppm.

## B. Chemical Shifts for Carbon Atoms

The range of chemical shifts for carbon atoms (~250 ppm) means that different carbon atoms, even in complex molecules, are likely to have different chemical shifts; $^{13}$C nuclear magnetic resonance spectroscopy has become, therefore, a powerful tool for structure determination in organic chemistry. For example, the spectrum of 2-octanol is given in Figure 11.8. The eight different carbon atoms of 2-octanol all have different absorption bands in this spectrum. The most highly shielded carbon (carbon $a$, $\delta$ 14) is in the methyl group farthest away from the hydroxyl group. The most highly deshielded carbon atom (carbon $h$, $\delta$ 68) is, of course, the one bonded to the oxygen atom. The others fall somewhere in between.

The spectrum of methyl benzoate (Figure 11.9) has several features in common with that of methyl acetate (Figure 11.1, p. 391), namely, the carbon atom of the

*Figure 11.7* **(shown at left)**
$^{13}$C and proton nuclear magnetic resonance spectra of methyl acetate showing the frequencies at which the nuclear transitions take place.

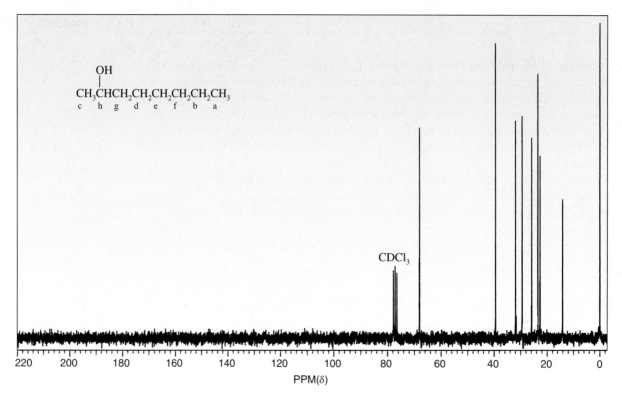

*Figure 11.8*
$^{13}$C nuclear magnetic resonance spectrum of 2-octanol.

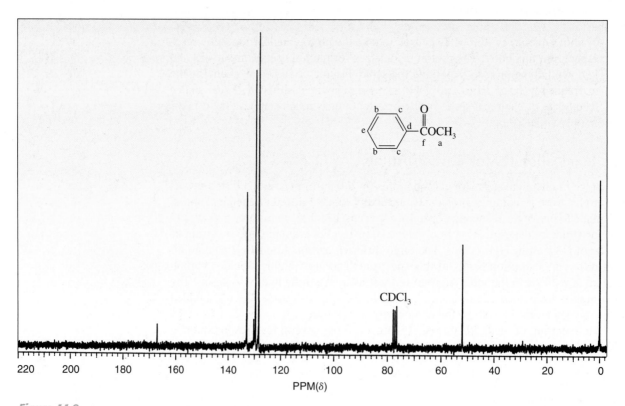

*Figure 11.9*
$^{13}$C nuclear magnetic resonance spectrum of methyl benzoate.

methoxy group ($\delta$ 52) and the carbon atom of the carbonyl group ($\delta$ 167). In addition, it has four types of carbon atoms in the phenyl group attached to the carbonyl group. These $sp^2$-hybridized carbon atoms usually appear between $\delta$ 115 and $\delta$ 150, depending on whether they are part of double bonds or aromatic rings.

Two other features of this spectrum are noteworthy. One is the difference between the position of the peak for the carbonyl carbon in the esters methyl acetate and methyl benzoate and that for the carbonyl group of the ketone shown in Figure 11.6. The carbon for the ester carbonyl groups appears at around $\delta$ 170, while that for the ketone appears at $\delta$ 209. Thus the carbon atom of the carbonyl group of an ester is more shielded than that of the carbonyl group of a ketone. This observed difference in chemical shifts of the two carbons is reflected in the chemistry of ketones and esters (p. 596). The carbonyl group of a ketone is more electrophilic than the carbonyl group of an ester.

Note too that $^{13}$C nuclear magnetic resonance spectra do not have integration curves. The intensity of a peak is not directly related to the number of carbon atoms undergoing that energy transition. In general, carbon atoms that have hydrogen atoms bonded to them give rise to more intense peaks than those that have no hydrogen atoms on them. This can be seen clearly in the spectrum of methyl benzoate (see Figure 11.9). The two peaks of lowest intensity at $\delta$ 130 and $\delta$ 167 belong to carbon atom $d$ of the aromatic ring and to carbon atom $f$ of the carbonyl group, respectively. Neither of them bears a hydrogen atom.

Typical chemical shifts observed for different types of carbon atoms are given in Table 11.2. As the first three entries in the table show, a more highly substituted

**Table 11.2    Chemical Shifts for Carbon Atoms in $^{13}$C Nuclear Magnetic Resonance Spectra**

| Type of Carbon Atom | $\delta^*$ | Type of Carbon Atom | $\delta^*$ |
|---|---|---|---|
| $RCH_2CH_3$ | 13–16 | $RCH{=}CH_2$ | 115–120 |
| $RCH_2CH_3$ | 16–25 | $RCH{=}CH_2$ | 125–140 |
| $R_3CH$ | 25–38 | $RC{\equiv}N$ | 117–125 |
| | | $ArH$ | 125–150 |
| $CH_3\overset{O}{\overset{\|}{C}}R$ | ~30 | | |
| | | $R\overset{O}{\overset{\|}{C}}OR'$ | 170–175 |
| $CH_3\overset{O}{\overset{\|}{C}}OR$ | ~20 | | |
| $RCH_2Cl$ | 40–45 | $R\overset{O}{\overset{\|}{C}}OH$ | 177–185 |
| $RCH_2Br$ | 28–35 | | |
| $RCH_2NH_2$ | 37–45 | $R\overset{O}{\overset{\|}{C}}H$ | 190–200 |
| $RCH_2OH$ | 50–64 | | |
| $RC{\equiv}CH$ | 67–70 | $R\overset{O}{\overset{\|}{C}}R'$ | 205–220 |
| $RC{\equiv}CH$ | 74–85 | | |

* The chemical shifts are given in parts per million (ppm) relative to tetramethylsilane at $\delta$ 0.00 and are for the carbon atoms shown in boldface in the formulas.

carbon atom usually has a higher chemical shift than a less substituted one with a similar environment. The numbers in the table are, as usual, to be applied with caution and only after taking into account all the available information.

**Problem 11.5**    2-Methylpentane, 3-methylpentane, and 2,3-dimethylbutane are isomeric hydrocarbons; $^{13}C$ nuclear magnetic resonance spectral data for the compounds are given below.

Isomer 1:   $\delta$ 19.5, 34.0
Isomer 2:   $\delta$ 11.4, 18.8, 29.3, 36.4
Isomer 3:   $\delta$ 14.3, 20.6, 22.6, 27.9, 41.6

Draw a structural formula for each compound, and decide which set of data corresponds to which compound.

**Problem 11.6**    Three $^{13}C$ nuclear magnetic resonance spectra are given in Figure 11.10. Decide which spectrum belongs to each of the following compounds. Assign the peaks in the spectra to specific carbon atoms.

(a)     (b) $CH_3CCH_2CH_3$     (c)   H

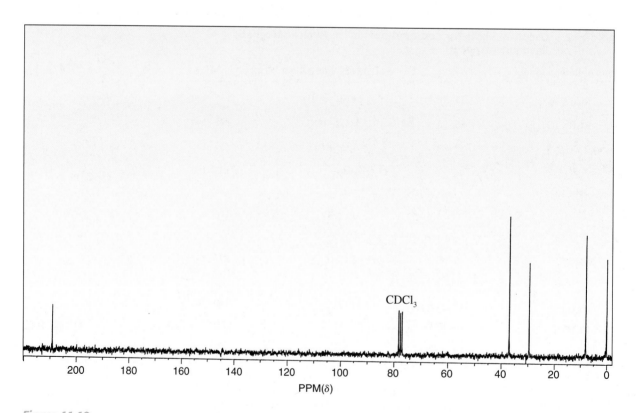

*Figure 11.10*

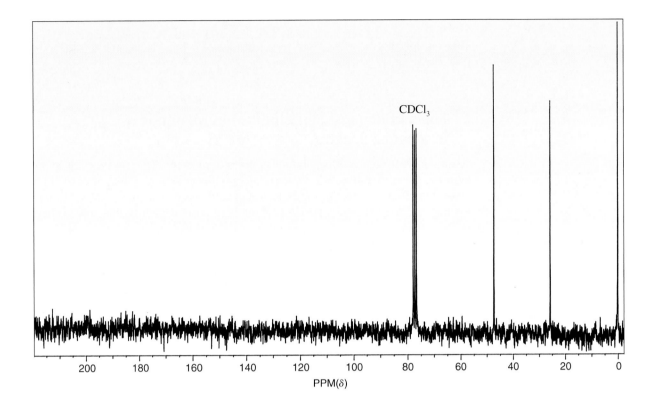

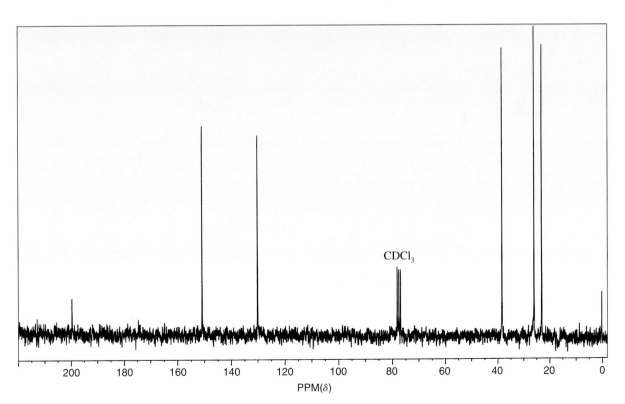

Figure 11.10 *(Continued)*

**Problem 11.7**    Compound A, $C_5H_8O$, has peaks in its $^{13}C$ nuclear magnetic resonance spectrum at $\delta$ 23.4, 38.2, and 219.6. Assign a structure to this compound.

## C. Typical Chemical Shifts in Proton Magnetic Resonance Spectra

A summary of the chemical shifts usually observed for different types of hydrogen atoms is given in Table 11.3. We have already discussed the effect of deshielding of hydrogen atoms by neighboring electronegative atoms (Section 11.4A). In the following sections we will discuss two other important factors that affect chemical shifts, the presence of $\pi$ bonds and the bonding of hydrogen directly to electronegative elements such as oxygen.

An important trend that appears from an examination of Table 11.3 is that the hydrogen atom of a methine group, a tertiary carbon atom with a hydrogen, is more deshielded than those of a methylene group, a secondary carbon with two hydrogen atoms on it, which in turn are more deshielded than the hydrogen atoms of a methyl group. The effects of substituents are cumulative, so the observed chemical shift of a hydrogen atom may be different from the value given in the table if several factors influence it at the same time. It is important to pay attention to all the information available when interpreting nuclear magnetic spectra, including the molecular formula of the compound, the integration of the spectrum, and the appearance of spin–spin coupling (p. 414), as well as the chemical shift values.

**Table 11.3    Typical Chemical Shifts for Types of Hydrogen Atoms Seen in Proton Magnetic Resonance Spectra**

| Type of Hydrogen Atom | $\delta^*$ | Type of Hydrogen Atom | $\delta^*$ | Type of Hydrogen Atom | $\delta^*$ |
|---|---|---|---|---|---|
| $RCH_3$ | 0.9–1.0 | $RNHCH_3$ | 2–3 | $RNH_2$ | 1–3 |
| $RCH_2R$ | 1.2–1.7 | $RCH_2X$ (X=Cl,Br,I) | 2.6–4.3 | $ArNH_2$ | 3–5 |
| $R_3CH$ | 1.5–2.0 | $ROCH_3$, $RCOCH_3$ (C=O) | 3.8 | $RCNHR$ (C=O) | 5–9 |
| $R_2C{=}CCH_3$ ($R'$) | 1.5–1.8 | $RCH{=}CH_2$ | 3.8–6.5 | $ROH$ | 1–5 |
| | | $ArH$ | 6.0–8.8 | $ArOH$ | 4–7 |
| $RCCH_3$ (C=O) | 2.0–2.3 | $RCH$ (C=O) | 9.5–10 | $RCOH$ (C=O) | 10–13 |
| $ArCH_3$ | 2.3 | | | | |
| $RC{\equiv}CH$ | 2.3–3.0 | | | | |

* The chemical shifts are given in parts per million (ppm) relative to tetramethylsilane at $\delta$ 0.00 and are for the hydrogen atoms shown in boldface in the formulas. The values for hydrogen atoms on oxygen and nitrogen are highly dependent on solvent, concentration, and temperature.

## D. Special Effects Seen in Proton Magnetic Resonance Spectra of Compounds with $\pi$ Bonds

The hydrogen atoms that are on $sp^2$-hybridized carbon atoms involved in double bonds and aromatic rings and the hydrogen atom on the carbonyl group of an aldehyde are much more deshielded than would be expected just from the effect of the electronegativity of the atoms to which they are bonded. For example, the hydrogen atoms on the double bond in cyclohexene absorb at $\delta$ 5.70; those on the aromatic ring of toluene absorb at $\delta$ 7.2 (Figure 11.11, p. 408). In the spectrum for benzaldehyde, the signal for the hydrogen atom on the carbonyl group appears at $\delta$ 10.00 (Figure 11.12, p. 409). This peak appears far downfield from the range normally recorded for proton magnetic spectra, so the spectrum in Figure 11.12 has an expanded scale.

The hydrogen atom bonded to the $sp$-hybridized carbon atom of a terminal alkyne, on the other hand, absorbs at $\delta$ 2.5, far upfield from a vinylic hydrogen atom. The hydrogen atom bonded to an $sp$-hybridized carbon atom should, in fact, be more deshielded than a hydrogen atom on an $sp^2$-hybridized carbon atom if electronegativity (p. 60) is the only factor that is operative.

To explain the effects described here, it has been proposed that the motion of electrons in a multiple bond sets up an induced magnetic field around the molecule containing such a bond. For certain orientations of the molecule, the induced magnetic field may either reinforce the external magnetic field or oppose it. Any hydrogen atoms that fall in the region of space around the molecule where the induced magnetic lines of force reinforce the external field appear to be deshielded in the spectrum. In other words, a lower external field strength is necessary for those nuclei to absorb energy and be promoted to the higher-energy level. Hydrogen atoms that fall in the region of space in which the induced magnetic field opposes the external field are shielded. The external magnetic field must be increased before absorption of energy takes place. Thus molecules with multiple bonds have deshielding and shielding regions about them. These regions are most easily seen for benzene. In Figure 11.13 (p. 409), the flow of electrons around the aromatic ring, called the **ring current,** the induced magnetic lines of force that result from this flow of charge, and the shielding and deshielding regions around the benzene molecule are shown.

One of the most dramatic demonstrations of the ring-current effect in a conjugated cyclic polyene can be seen in the proton magnetic resonance spectrum of [18]annulene.

[18]annulene

The spectrum of this compound has two peaks, one at $\delta$ 8.9 and the other at $\delta$ −1.8 (1.8 ppm to the right of, or upfield from, TMS). These peaks have relative intensities of 2:1. Therefore, the low-field signal is assigned to the hydrogen atoms around the outside of the ring, which lie in the deshielding region around the molecule. The

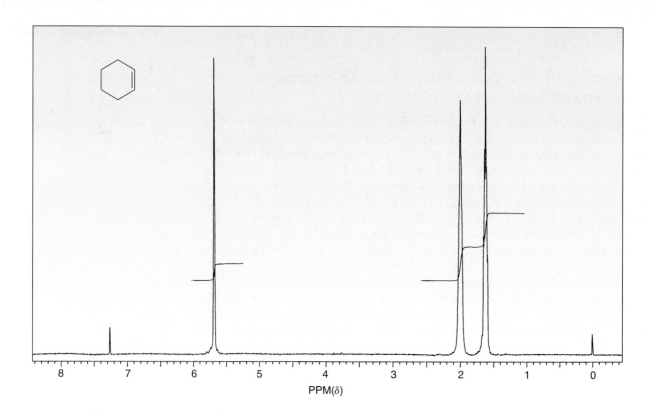

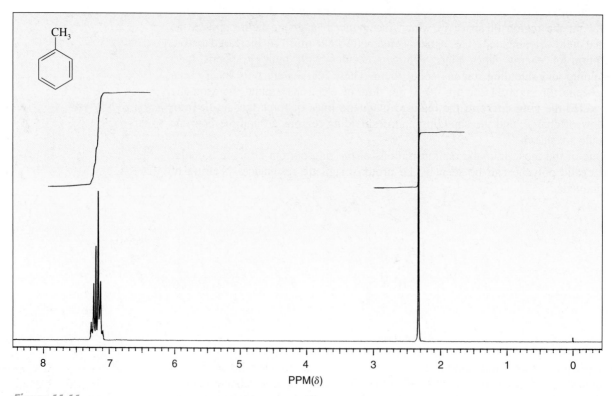

*Figure 11.11*
Proton magnetic resonance spectra for cyclohexene and toluene.

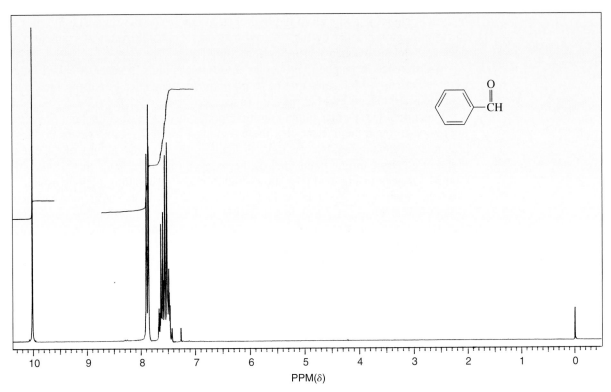

*Figure 11.12*
The proton magnetic resonance spectrum of benzaldehyde.

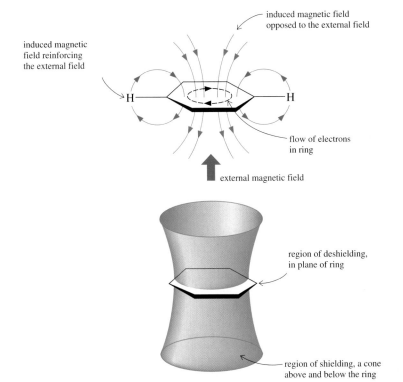

*Figure 11.13*
Simplified representations of the ring current and induced magnetic lines of force and the shielding and deshielding regions around the aromatic ring of benzene.

high-field signal is assigned to the hydrogen atoms inside the ring. They lie in the shielding region created by the induced magnetic field. [18]Annulene satisfies the nuclear magnetic resonance criterion for aromaticity.

A difference in magnetic properties at different points in space is called **magnetic anisotropy.** Molecules that have $\pi$ bonds are magnetically anisotropic. Thus benzene is a molecule that has magnetic anisotropy.

Functional groups that contain multiple bonds also have shielding and deshielding regions around them. Basically, these regions arise in the same way as do those around an aromatic ring, but it is more difficult to give a simple picture of the circulation of electrons in the bonds and of the induced magnetic fields that result. The conical surfaces in Figure 11.14 show shielding and deshielding regions around a carbonyl group, a double bond, and a triple bond. The hydrogen atom on a carbonyl group of an aldehyde or the hydrogen atoms on the carbons of a double bond lie in the plane of the respective molecules and in the deshielding regions. In an alkyne, the absence of a nodal plane and the circular symmetry of the $\pi$ electrons around the linear axis of the molecule (p. 58) means that the major electronic current is around the axis. In this case, the shielding regions lie along the axis and the deshielding regions around it. A hydrogen atom on a triply bonded carbon atom is therefore more shielded than one on a doubly bonded carbon but deshielded relative to one on a tetrahedral carbon.

### E. Chemical Shifts of Hydrogen Atoms of Hydroxyl Groups

In general, the chemical shifts of hydrogen atoms bonded to oxygen are quite variable because they depend on the extent to which hydrogen bonding is taking place, and this, in turn, depends on the concentration of the solution. The proton magnetic resonance spectrum of methanol with no solvent (Figure 11.15) shows the typical chemical shift for the hydrogen atom on a hydroxyl group. The spectrum shows two peaks, at $\delta$ 3.40 for the three hydrogen atoms of the methyl group and at $\delta$ 4.85 for the hydrogen atom of the hydroxyl group. The hydrogen atom of the hydroxyl

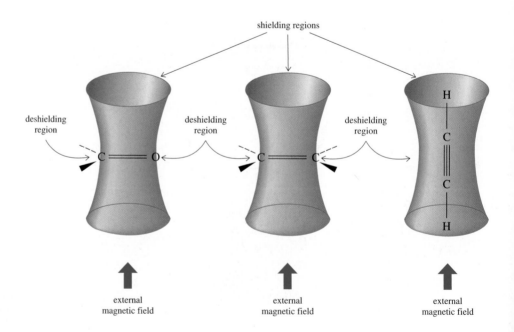

*Figure 11.14*
Shielding and deshielding regions around functional groups containing multiple bonds.

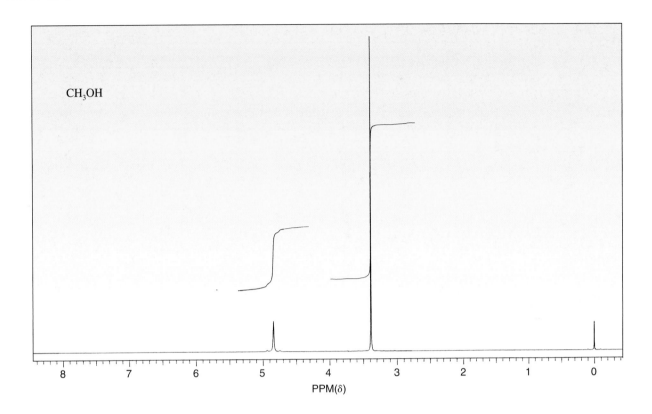

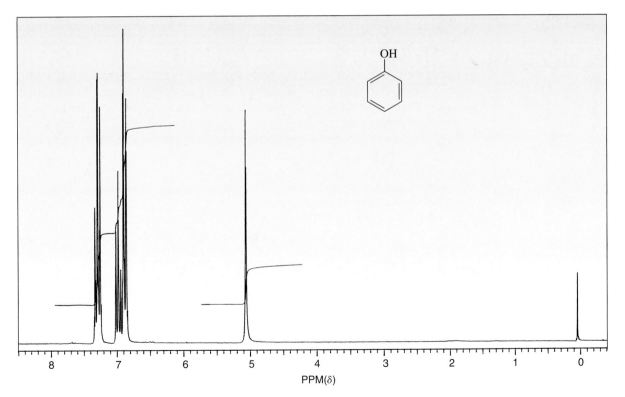

*Figure 11.15*
Proton magnetic resonance spectra for methanol and phenol.

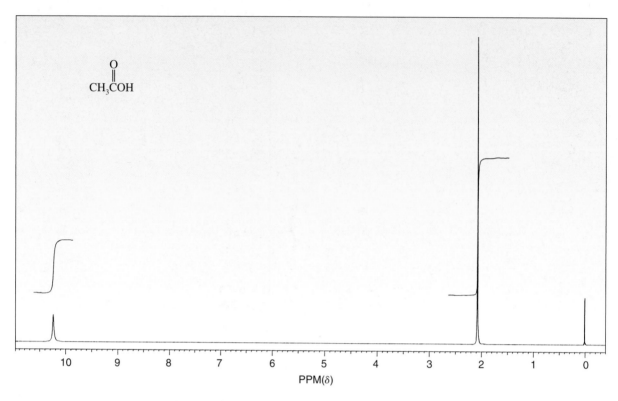

*Figure 11.16*
Proton magnetic resonance spectrum of acetic acid.

group in a phenol, in which the hydroxyl group is on an aromatic ring, is usually more deshielded than that of an alcohol. In Figure 11.15, the peak for the hydrogen atom of the hydroxyl group in phenol appears at $\delta$ 5.10, and those for the aromatic hydrogen atoms appear at $\delta$ 6.8–7.4.

The proton on the carboxyl group of an acid is the most deshielded of all hydrogen atoms bonded to oxygen. The spectrum of acetic acid, which, like that for benzaldehyde, has an expanded scale, provides an example (Figure 11.16). The spectrum has two singlets, at $\delta$ 2.10 for the hydrogen atoms of the methyl group and at $\delta$ 10.25 for the acidic hydrogen atom.

**Problem 11.8**    If an alcohol is diluted with $CDCl_3$, one of the peaks in the proton magnetic resonance spectrum moves upfield as dilution increases. Explain what is happening. Why does hydrogen bonding further deshield a hydrogen atom?

**Problem 11.9**    Using the chemical-shift values given in Table 11.3, assign the absorption bands between $\delta$ 1.0 and $\delta$ 3.0 in the spectrum of cyclohexene (Figure 11.11). Remember that values given in such a table are approximate and that you have to reason by analogy in making assignments for the spectrum of a particular compound.

**Problem 11.10**    Molecular formulas and proton magnetic resonance spectra for Compounds B–D are given in Figure 11.17 (pp. 413 and 414). Assign structures to the compounds. (*Hint:* Calculating the units of unsaturation for a compound is a good way to start solving this kind of problem.)

Compound B, $C_8H_{10}O_2$

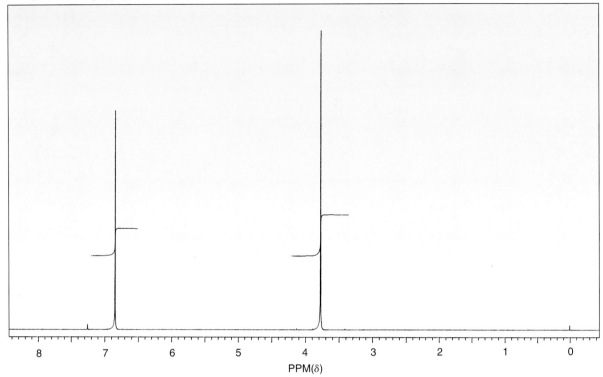

Compound C, $C_7H_7Cl$

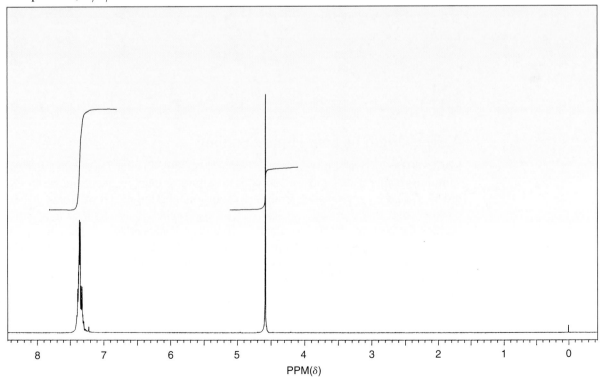

*Figure 11.17*

Compound D, $C_8H_8O$

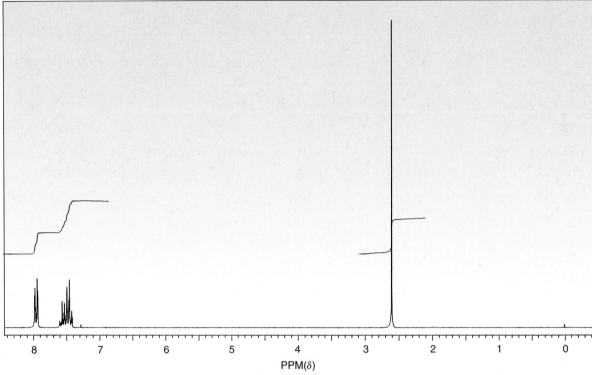

PPM($\delta$)

*Figure 11.17 **(Continued)***

## 11.5 Spin–Spin Coupling

### A. The Origin of Spin–Spin Coupling

Many of the spectra shown in earlier sections of this chapter have peaks that are split into smaller peaks, most notably the spectrum of methyl acetate shown in Figure 11.2. These patterns arise from interactions among the nuclei of atoms that are close enough that their nuclear spins affect each other. This effect is transmitted through the electrons of the covalent bonds and so is dependent on the number and types of bonds separating the atoms. For example, each hydrogen atom in the molecule has one of two possible orientations in a magnetic field. It can be oriented with the field or against the field. At any given time, about half the hydrogen atoms in a sample will have their spin aligned with the external field and half against it. As a result, the net magnetic field experienced by a neighboring atom, the carbon atom, for example, will be slightly weaker or slightly stronger than the external field supplied by the instrument, and that carbon atom, if bonded to one hydrogen atom, will absorb energy at one of two slightly different frequencies.

This interaction between the nuclei of different atoms is called **spin–spin coupling.** The interaction occurs through the intervening bonds. The strength of the interaction thus depends on factors such as the number of bonds between the interacting nuclei, the types of those bonds, and the stereochemical relationship between the atoms. The measure of the interaction is the **coupling constant, $J$,** given in hertz.

Two $^{13}C$ magnetic resonance spectra of 1,1,2-trichloroethane are shown in Figure 11.18 (p. 416). The top spectrum represents an experiment in which the nuclei of the hydrogen atoms and those of the carbon atoms to which they have been bonded have been **decoupled.** This is done by using strong radiation of a second radiofrequency, one that is absorbed by hydrogen atoms in that magnetic field, to excite all the hydrogen atoms in the molecule. Because hydrogen nuclei are constantly flipping to the higher-energy spin state and back again, they do not spend enough time in either energy level to affect the magnetic field felt by the neighboring carbon nuclei. The spectrum that results is **proton-decoupled.** In this spectrum we see the two singlets corresponding to the two nonequivalent carbon atoms in the molecule, at $\delta$ 50 for the carbon with one chlorine atom on it and $\delta$ 71 for the carbon atom bearing two chlorine atoms.

In taking the bottom spectrum shown in Figure 11.18, the carbon and hydrogen nuclei were allowed to interact with each other. The coupling constants, $J$, for carbon atoms and hydrogen atoms bonded to them can range from 100 to 320 Hz and, for compounds with many carbon atoms, can lead to spectra of great complexity, with much overlapping of peaks. For this reason, most $^{13}C$ spectra are proton-decoupled.

An examination of the two spectra with coupling between the carbon and the hydrogen nuclei shows that the peak at $\delta$ 50 has been split into three peaks, centered at $\delta$ 50. These three peaks have relative intensities that are roughly 1:2:1 (42:81:40 mm). Such a three-peak pattern, with relative intensities of 1:2:1, is called a **triplet.** The peak at $\delta$ 71 has been split into two peaks, equally spaced on either side of $\delta$ 71 and of roughly equal intensity (48:50 mm, or 1:1). A pattern that has two peaks of equal intensity is called a **doublet.**

the appearance of the $^{13}C$ magnetic resonance spectrum of
1,1,2-trichloroethane showing coupling between the carbon and
hydrogen atoms bonded to them; one-bond neighbor coupling

To better understand the patterns we see, let us examine the interactions that are possible between the magnetic moments of the nuclei of the hydrogen atoms and an external magnetic field. Figure 11.19 shows the different ways in which one hydrogen atom and two hydrogen atoms can orient themselves with respect to an external field. From this diagram, we see, as described earlier in this section, that one hydrogen atom has two possible orientations with respect to the external field, so carbon atoms attached to a single hydrogen experience the external field in two slightly different ways, with equal probability. The absorption of energy that took place at one frequency corresponding to $\delta$ 71 in the decoupled spectrum now is split into two absorptions, one at higher frequency than the original and one at lower. The signal appears as a doublet. The strength of the coupling between the

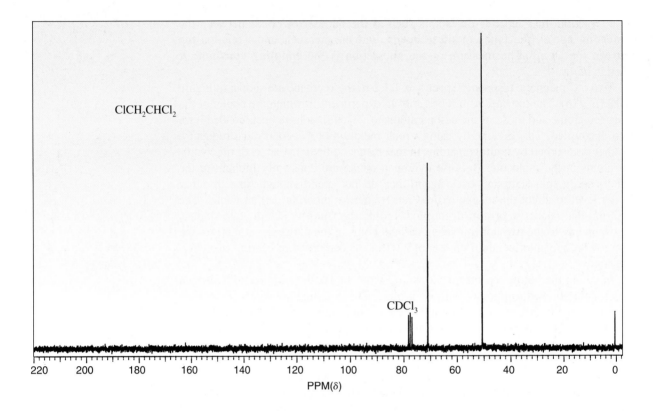

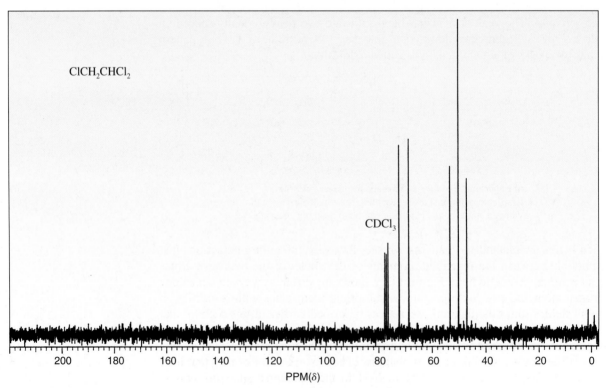

*Figure 11.18*
Proton-decoupled and coupled $^{13}$C magnetic resonance spectra of 1,1,2-trichloroethane.

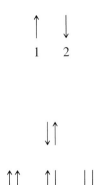

2 possible spin orientations of the nucleus of a single hydrogen atom, representing different interactions with the external field, occurring with equal probability

4 possible spin orientations of the nuclei of two hydrogen atoms, representing three different interactions with the external field, occurring with a probability of 1:2:1

*Figure 11.19*
The interactions possible between the magnetic moments of the nuclei of the hydrogen atoms in 1,1,2-trichloroethane and an external magnetic field.

carbon and the hydrogen nuclei is given by the distance between the two peaks of the doublet. In this particular case, for the carbon atom with two chlorine atoms on it, measurement gives a separation of 2.5 mm for the two peaks, which translates from the scale of the spectrum and a conversion from parts per million (the units of $\delta$) back to hertz (p. 401) to a $J$ value of 172 Hz for this interaction.

Two nuclei have three different ways in which they can be oriented with respect to the external magnetic field. Because individual spin orientations are indistinguishable, both cases in which one nucleus is against the field and one with the field are equal in energy. The carbon atom at $\delta$ 50 may thus experience three slightly different magnetic fields. The probability that any one of these three situations will occur is 1:2:1 (see Figure 11.19). Thus three peaks, a triplet, having those relative intensities arise. The coupling constant, $J$, for the interaction between this carbon atom (the one with one chlorine atom on it) and the hydrogens on it is 155 Hz (calculated from a distance of 2.25 mm between the peaks).

The proton magnetic resonance spectrum of 1,1,2-trichloroethane is given in Figure 11.20 (p. 418). In this spectrum we see the effect of coupling between the nuclei of the hydrogen atoms in the molecule. The triplet centered at $\delta$ 5.78 represents a single hydrogen (integration 16 mm), while the doublet at $\delta$ 3.98 represents two hydrogens (integration 32 mm). Therefore, we must assign those peaks as shown in Figure 11.21 (p. 418).

The coupling that we see in the proton spectrum is the result of coupling between the hydrogen atoms that are three-bond neighbors of each other on adjacent carbon atoms. Note that because these nuclei are farther apart from each other than was the case with the one-bond neighbors carbon and hydrogen (see Figure 11.18), the coupling constant is much smaller, about 7 Hz for hydrogens that are three-bond neighbors on $sp^3$-hybridized carbon atoms.

The patterns here have exactly the same origin as they did for the spin–spin coupling patterns we saw for the carbon spectra. The one proton at $\delta$ 5.78 feels the three slightly different fields caused by the different orientations that the two neighboring hydrogen atoms can have and so absorbs energy at three different frequencies to give rise to a triplet. The two protons at $\delta$ 3.98 are chemical-shift equivalent and so do not couple with each other but do feel the effect that the two different orientations of their neighbor on the other carbon has. The signal for these two protons, therefore, appears as a doublet.

**ONE SMALL STEP**

The nucleus of the deuterium atom contains 1 proton and 1 neutron and has spin number, $I = 1$. The number of spin states permitted for a nucleus is $(2I + 1)$.

**PROBLEM:**
(a) How many equally probable energy levels are available to the deuterium atom?
(b) All the $^{13}C$ nuclear magnetic resonance spectra shown have three peaks labeled $CDCl_3$ at $\delta$ 77. Describe why the carbon atom in $CDCl_3$ appears as a set of three lines of equal intensity.
(c) The three signals for $CDCl_3$ appear even in the "decoupled" $^{13}C$ spectra. Why is this so?

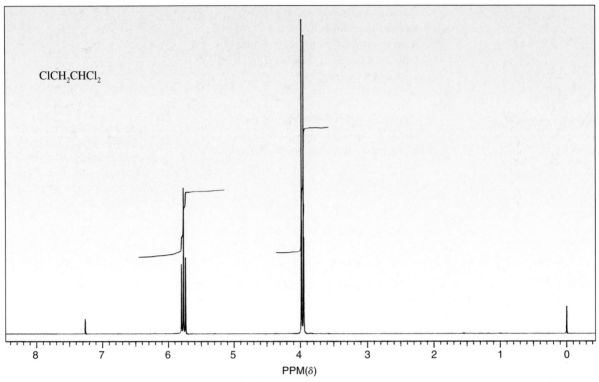

*Figure 11.20*

The proton magnetic resonance spectrum of 1,1,2-trichloroethane.

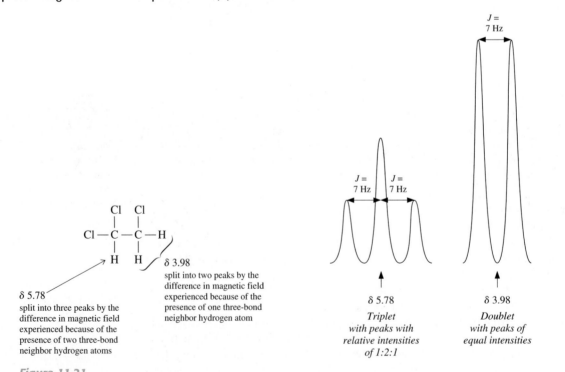

*Figure 11.21*

A representation of the coupling between three-bond neighbor hydrogen atoms on adjacent carbon atoms in 1,1,2-trichloroethane.

**Table 11.4 Splitting Patterns That Result from the Presence of N Neighboring Hydrogen Atoms**

| N | Multiplet (N + 1) | Relative Intensities of Peaks |
|---|---|---|
| 0 | singlet (1) | 1 |
| 1 | doublet (2) | 1:1 |
| 2 | triplet (3) | 1:2:1 |
| 3 | quartet (4) | 1:3:3:1 |
| 4 | quintet (5) | 1:4:6:4:1 |
| 5 | sextet (6) | 1:5:10:10:5:1 |
| 6 | septet (7) | 1:6:15:20:15:6:1 |

In the case of the $^{13}C$ spectrum of 1,1,2-trichloroethane, we were looking at the coupling of the carbon atom to the hydrogen atom directly on it. Each carbon coupled in a different way with its own hydrogen atom or atoms. In the proton spectrum, the distances between the small peaks of the patterns are the same because both patterns arise from the same interaction. The coupling is between the hydrogen on carbon 1 and the hydrogens on carbon 2, so whatever the strength of that interaction is, it is the same for both, giving rise to one coupling constant, J. Thus the bands arising from hydrogen atoms that are coupled to each other always can be identified. In practice, very often proton magnetic resonance spectra also will offer another clue to which hydrogen atoms are coupled to each other and, therefore, near neighbors in molecular structures. A close look at Figure 11.20 will show you that the triplet and the doublet are not symmetrical. The inner peaks of the triplet and of the doublet, the ones that are facing each other, are slightly more intense than the outer peaks. Chemists say that the **multiplets** (a generic word for peaks that show splitting) are "leaning" toward each other. In this case we have only these two groups of hydrogens in the molecule, so it is not hard to decide what is coupled to what. In more complex spectra, however, this phenomenon of leaning is often helpful in assigning structure.

For hydrogen atoms that have differing chemical shifts and are on adjacent carbon atoms, the splitting patterns seen are determined by the number of hydrogen atoms on the neighboring carbon atoms. The number of peaks seen is N + 1, where N equals the number of neighboring hydrogen atoms that are chemical-shift equivalent or that have equivalent coupling constants. The same patterns also result for carbon atoms that are one-bond neighbors to hydrogen atoms. The patterns that are to be expected and the relative intensities of the peaks within each pattern are given in Table 11.4.

**Problem 11.11** The proton magnetic resonance spectrum of 1,1-dichloroethane is given in Figure 11.22 on page 420. The chemical shifts for the hydrogen atoms were assigned with Figure 5.3 in Section 5.4 (p. 152). Analyze the spin–spin coupling in this spectrum. Make sure that you understand why the hydrogen atom on the carbon bearing two chlorine atoms appears as a quartet by demonstrating to yourself the number of different orientations possible for the nuclei of the three hydrogen atoms of a methyl group. What gives rise to the relative intensities seen for the four peaks of the quartet?

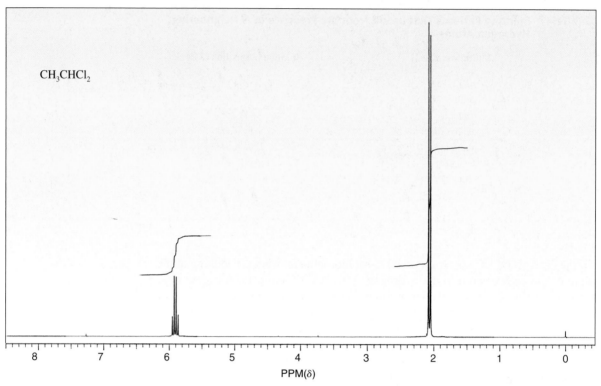

CH₃CHCl₂

*Figure 11.22*

Looking at other spectra that show the effects of spin–spin coupling will help to clarify them. Figure 11.23 shows the spectra of 2-bromopropane and 1-bromopropane. The spectrum of 2-bromopropane indicates the presence of two types of hydrogen atoms, in a ratio of 1:6. Those on the two methyl groups absorb at $\delta$ 1.70, and the one on the carbon atom bearing the bromine atom absorbs at $\delta$ 4.30. The signal for the methyl hydrogen atoms, which are all equivalent to each other, appears as a doublet because of the presence of one hydrogen atom on the adjacent carbon atom. It is very important to understand that the methyl signal in this spectrum is a doublet *not* because there are two methyl groups but because of spin–spin coupling. The influence of the six equivalent hydrogen atoms of the methyl groups, in turn, splits the absorption band for the hydrogen atom on the secondary carbon atom into 6 + 1, or 7, peaks. The intensities of the outer peaks of the septet are very weak; they do not appear unless that portion of the spectrum is enhanced. They can only be seen in the insert in the spectrum, which was recorded at a higher sensitivity. The distance between adjacent peaks in the septet is equal to the distance between the two peaks of the doublet. The value, read off the spectrum, is about 7 Hz, the usual coupling constant for hydrogen atoms on neighboring $sp^3$-hybridized carbon atoms.

doublet, split by
one hydrogen atom on
neighboring carbon atom ⟶ $\delta$ 1.70

CH₃—C—Br
with CH₃ and H

$\delta$ 4.30

septet, split by
six hydrogen atoms on
neighboring carbon atoms

2-bromopropane

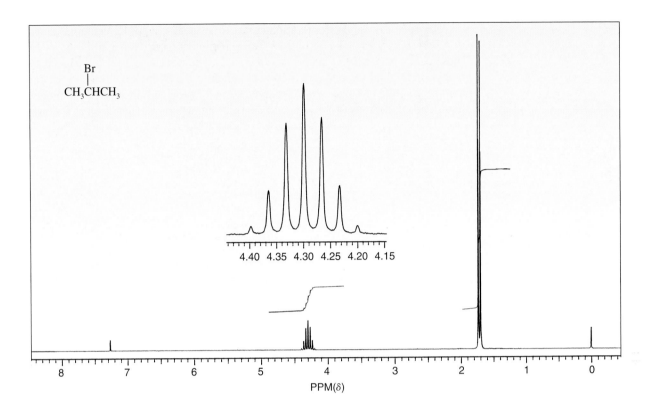

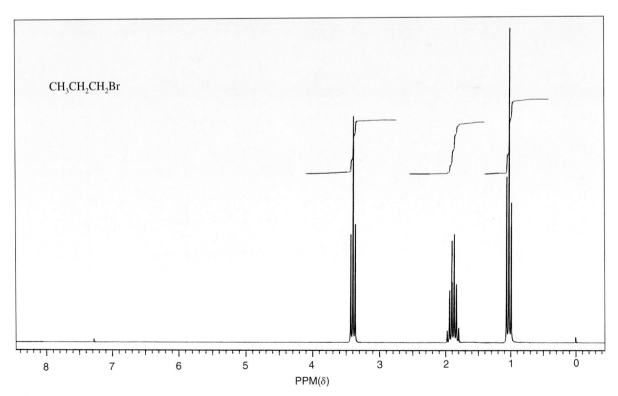

*Figure 11.23*
Proton magnetic resonance spectra of 2-bromopropane and 1-bromopropane.

The spectrum of 1-bromopropane has three peaks, a triplet centered at $\delta$ 1.00 ($J = 7$ Hz), a sextet at $\delta$ 1.88 ($J = 7$ Hz), and another triplet at $\delta$ 3.40 Hz ($J = 7$ Hz). These peaks have relative intensities of 3:2:2. The peak at $\delta$ 1.00 is assigned to the methyl group, and the one at $\delta$ 3.40 is assigned to the hydrogen atoms on the carbon atom bearing the bromine atom. The hydrogen atoms on carbon atom 2 fall in between these. The interactions that give rise to the splitting patterns are analyzed as follows:

triplet, split by
two hydrogen atoms on
neighboring carbon atom

triplet, split by
two hydrogen atoms on
neighboring carbon atom

$$Br-\underset{\substack{\delta\ 3.40}}{\overset{1}{CH_2}}-\underset{\substack{\delta\ 1.88}}{\overset{2}{CH_2}}-\underset{\substack{\delta\ 1.00}}{\overset{3}{CH_3}}$$

sextet, split by
five hydrogen atoms with the
same coupling constant on
neighboring carbon atoms

1-bromopropane

The peak for the hydrogen atoms on carbon 2 appears as a sextet because the coupling constants for the interactions of these hydrogen atoms with those on carbons 1 and 3 happen to be of the same magnitude. Thus the hydrogen atoms on carbon 2 appear to be coupled with five equivalent hydrogen atoms. If the hydrogen atoms on carbon 2 and those on carbons 1 and 3 had different coupling constants, a more complex pattern would have resulted.

**Problem 11.12**    $^{13}C$ nuclear magnetic resonance spectra, both decoupled and coupled, of a $C_4H_{10}O$ isomer are given in Figure 11.24. Assign a structure to the compound, clearly describing your reasoning.

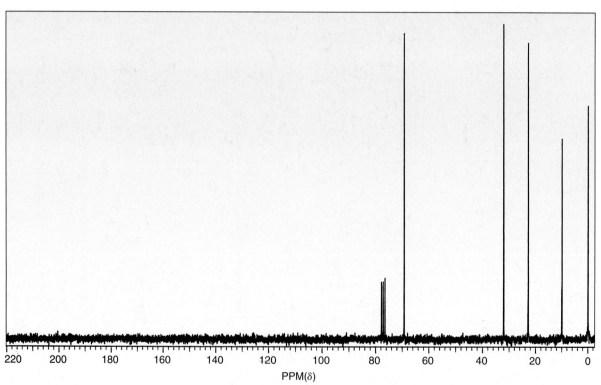

*Figure 11.24*

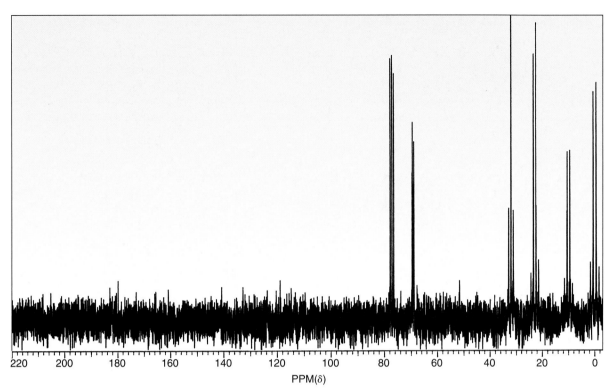

*Figure 11.24* **(Continued)**

## B. The Art of Solving Problems

In earlier sections of this book we have encountered exercises in solving problems involving choosing reagents for chemical transformations and those involving working our way through a reaction mechanism. In working with the next problem we will learn the kinds of questions to ask ourselves in solving a spectroscopy problem.

---

**PROBLEM:** Compound E, $C_{10}H_{12}O$, gives the proton magnetic resonance spectrum shown in Figure 11.25 (p. 424). Assign a structure to the compound, and show how the spectral data support your assignment.

### Solution

1. What different kinds of data do I have that are related to the structure of the compound? Do I know its molecular formula? If I do, how many units of unsaturation does the molecule have?

   The molecular formula is $C_{10}H_{12}O$. The formula for a saturated compound with 10 carbon atoms is $C_{10}H_{22}O$ ($C_nH_{2n+2}O$). Compound E has 10 fewer hydrogen atoms than necessary for saturation; therefore, it has 5 units of unsaturation. These may be in rings, double or triple bonds between carbon atoms, or a carbonyl group (because the compound contains oxygen). Such a high number of units of saturation usually means that the compound contains an aromatic ring,

   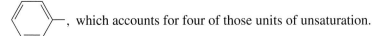, which accounts for four of those units of unsaturation.

Compound E, $C_{10}H_{12}O$

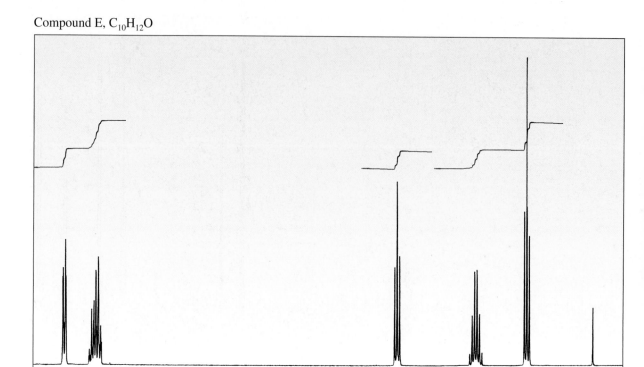

*Figure 11.25*
Proton magnetic resonance spectra of Compound E.

2. What kinds of spectral data do I have?

I have the proton magnetic resonance spectrum. The spectrum gives me three different kinds of information:

(a) The different types of hydrogen atoms in E.

There appear to be four (or five?) different kinds of hydrogen atoms in Compound E. To find out what kinds of hydrogen atoms they are, I have to consult Table 11.3 (p. 406) to see what the chemical shifts from the spectrum indicate. Because the peaks appear as groups of multiplets, the chemical shift values for each type of hydrogen atom corresponds to the $\delta$ value for the center of the multiplet. Therefore, the chemical shifts, as read from the spectrum, are $\delta$ 1.05, 1.80, 2.95, 7.50, and 7.98. A comparison of these values with those in Table 11.3, remembering that deshielding effects are cumulative, suggests the following types of assignments:

| Chemical Shifts | Assignment | Value in Table 11.3 | |
|---|---|---|---|
| $\delta$ 1.05 | $CH_3-$ | 0.9–1.0 | $(-CH_2-\overset{\overset{O}{\|\|}}{C}-R$ would be more |
| $\delta$ 1.80 | $-CH_2-$ | 1.2–1.7 | |
| $\delta$ 2.95 | $-CH_2-\overset{\overset{O}{\|\|}}{C}-$ | 2.0–2.3 for $CH_3-\overset{\overset{O}{\|\|}}{C}-R$ | deshielded than $CH_3-\overset{\overset{O}{\|\|}}{C}-R$, just as $-CH_2-$ is more deshielded than $CH_3-$.) |
| $\delta$ 7.50, 7.98 | ArH | 6.0–8.8 | |

The chemical shift values are consistent with what we learned from the molecular formula. They confirm that we probably have an aromatic ring and a carbonyl group, accounting for all five units of unsaturation.

(b) The number of hydrogen atoms of each type in E.

To determine this, we have to look at the integration of the spectrum and measure the heights of each step in the integration curves over the peaks. This is easiest to do if we actually extend the lower level of each step to the right and measure the height between the lower level and upper level with a ruler. The values are $\delta$ 1.05, 7.5 mm; $\delta$ 1.80, 5 mm; $\delta$ 2.95, 5 mm; $\delta$ 7.50, 7.5 mm; and $\delta$ 7.98, 5 mm.

These values correspond to the relative numbers of hydrogen atoms, not the actual numbers. In the case of Compound E, we have the additional information that the total number of hydrogen atoms in the molecule is 12. So we find some common denominator to divide by so that the total number of hydrogen atoms will equal the number in the molecular formula. That common denominator is 2.5. The chemical shift values can now be written with the number of hydrogen atoms they represent: $\delta$ 1.05 (3H), $\delta$ 1.80 (2H), $\delta$ 2.95 (2H), $\delta$ 7.50 (3H), and $\delta$ 7.98 (2H) for a total of 12 hydrogens in the molecule.

(c) The number of near neighbors each type of hydrogen atom has.

This information comes from the splitting patterns we see in the spectrum. The number of smaller peaks in each grouping is one more than the number of neighboring hydrogen atoms causing the splitting (see Table 11.4, p. 419). Therefore, the triplet at $\delta$ 1.05 indicates that those hydrogen atoms have two neighboring hydrogen atoms. The sextet at $\delta$ 1.80 means that those hydrogen atoms have five near neighbors. The triplet at $\delta$ 2.95 shows the presence of two near neighbors of those hydrogens. A combination of this information with the chemical shift values and the integration of the spectrum allows us to create the fragment of the molecule shown in the margin: The more complex multiplets at $\delta$ 7.50 and 7.98 indicate the possibility of different kinds of interactions between the nuclei of hydrogen atoms on aromatic rings. Some of these will be discussed in the next section.

3. Do we now have enough analyzed data to assign a structure to the compound?

We do. The fragment we arrived at by the end of question 2 has the formula $C_4H_7O$. Subtracting this from the molecular formula gives us $C_6H_5$, which is the formula for a phenyl ring and includes the five aryl hydrogen atoms that had not yet been included in the structure.

$C_{10}H_{12}O$

The two hydrogen atoms that are closest to the carbonyl group are more deshielded than the other aromatic hydrogen atoms. The structure we have assigned fits all the data we have for the compound. ∎

## C. Spin–Spin Coupling in Compounds with Multiple Bonds

Not all hydrogen atoms have a coupling constant of 7 Hz. It is easiest to see different types of coupling in the spectra of alkenes. The spectrum of vinyl acetate is shown in Figure 11.26. Besides the singlet for the hydrogen atoms of the acetyl group, the spectrum of vinyl acetate has complex peaks for three other hydrogen atoms. These peaks are centered at $\delta$ 4.57, 4.88, and 7.27. Note that the vinyl hydrogen atom on the carbon atom that is also bonded to an oxygen atom is deshielded by that oxygen atom so that its chemical shift, 7.27, lies outside the values given in Table 11.3 (p. 406) for vinyl hydrogen atoms. Each peak is split into a doublet, and the peaks of each doublet are split again into another doublet. This pattern is called a **doublet of doublets.** None of these groupings of four peaks is a quartet because the four peaks are not evenly spaced and do not have relative intensities of 1 : 3 : 3 : 1.

The hydrogen atoms on the terminal carbon atom of the double bond in vinyl acetate are distinct from each other. One ($\delta$ 4.57) is trans to the oxygen atom and cis to the hydrogen atom on the other carbon atom of the double bond, and the other ($\delta$ 4.88) is cis to the oxygen and trans to the hydrogen. Three coupling constants are seen in the spectrum. The hydrogen atoms that are trans to each other across the

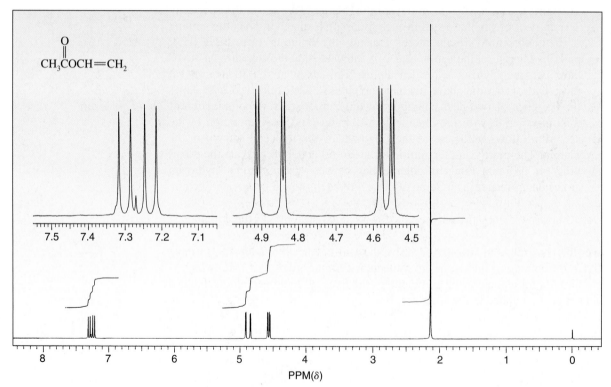

*Figure 11.26*
Proton magnetic resonance spectrum of vinyl acetate.

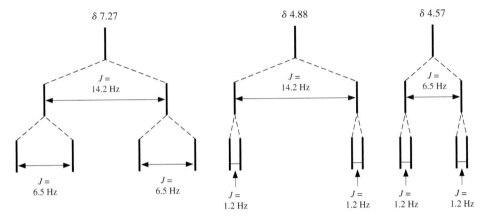

**Figure 11.27**
Analysis of the spin–spin coupling seen in the proton magnetic resonance spectrum of vinyl acetate.

double bond have the strongest interaction, $J = 14.2$ Hz. The coupling constant for the hydrogen atoms that are cis to each other is 6.5 Hz. The two hydrogen atoms on the same carbon atom have a small coupling constant, $J = 1.2$ Hz. $J_{trans}$ values are always larger than $J_{cis}$ values for hydrogen atoms on doubly bonded carbons.

The splitting patterns seen in the spectrum of vinyl acetate can be duplicated by constructing a diagram. The diagram shows the lines that will result if the peak for a hydrogen atom is split by interaction with two other hydrogen atoms with two different coupling constants. If a ruler or graph paper is used so that the distances in the diagram are really proportional to the coupling constants, the patterns seen in the spectrum emerge (Figure 11.27).

Several different coupling constants are also seen in the spectrum of an aromatic compound, represented here by 2-bromo-4-nitrotoluene (Figure 11.28, p. 428). There are four different kinds of hydrogen atoms in 2-bromo-4-nitrotoluene. The methyl group at $\delta$ 2.50 is clearly separated from the aromatic hydrogen atoms, which absorb at $\delta$ 7.40, 8.04, and 8.38. In general, very little coupling is seen between hydrogen atoms on the side chains of aromatic rings and the hydrogen atoms on the ring. The signal farthest downfield is assigned to the hydrogen atom that is between the nitro group and the bromine atom on the ring, labeled $H_C$ in the structural formula. It is split into a doublet by $H_B$, with a very small coupling constant, $J = 2.2$ Hz. The signal at highest field is assigned to the hydrogen atom next to the methyl group, $H_A$. It is split into a doublet by $H_B$, with $J = 8.6$ Hz. Note that coupling between hydrogen atoms next to each other is stronger than coupling between the more distant hydrogens. No coupling is seen between hydrogen atoms $H_A$ and $H_C$. The hydrogen atom labeled $H_B$ gives rise to a doublet of doublets. This hydrogen atom interacts with $H_A$ with one coupling constant, $J = 8.6$ Hz, and with $H_C$ with another coupling constant, $J = 2.2$ Hz.

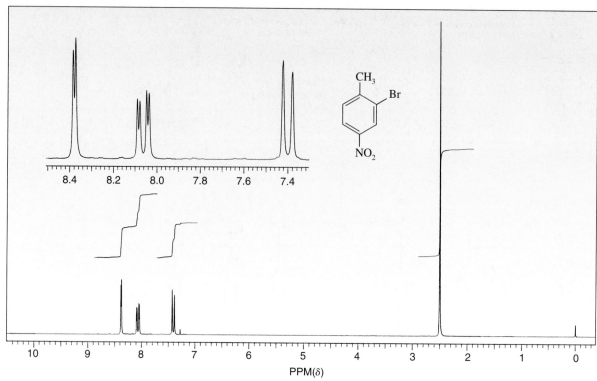

*Figure 11.28*
Proton magnetic resonance spectrum of 2-bromo-4-nitrotoluene.

δ 2.50
singlet

δ 7.40
doublet
J = 8.6 Hz H_A

δ 8.04
doublet of doublets H_B
J = 8.6, 2.2 Hz

CH₃

Br

δ 8.38
H_C doublet
J = 2.2 Hz

NO₂

2-bromo-4-nitrotoluene

Note that the proton magnetic resonance spectra of compounds that contain multiple bonds show coupling between hydrogen atoms that have as many as four or five bonds between them. In general, coupling constants for these interactions are small but do give rise to complexity in the spectra.

A list of coupling constants for carbon and hydrogen atoms in various environments is given in Table 11.5.

**Problem 11.13**   Use the coupling constants given for 2-bromo-4-nitrotoluene (p. 427) to construct a diagram like the one in Figure 11.27, showing the pattern of lines you predict for the aryl hydrogen atoms.

**Problem 11.14**   The signal for the hydrogen atoms on carbon atom 2 of 1-bromopropane (p. 422) may be regarded

as being split into a quartet by three hydrogen atoms on carbon 3, with each one of those four peaks further split into a triplet by the hydrogen atoms on carbon 1. Assume that the coupling constant for both types of interactions is 7 Hz, and construct a diagram like the one in Figure 11.27 to determine the number of lines you expect to see in the pattern that results.

**Table 11.5** Coupling Constants Seen in Nuclear Magnetic Resonance Spectra

| Atoms Coupling* | J, Hz | Atoms Coupling* | J, Hz |
|---|---|---|---|
| (geminal) C(**H**)(**H**) | 12–15 | (vinyl, cis/branched) **H**C=C—C**H** | 0–3 |
| **CH**—**CH** | 6–8 | **CH**—C≡C—**H** | 2–3 |
| **CH**—C(=O)—**H** | 2–3 | (ortho aromatic) **H**...**H** | 6–10 |
| (cis) **H**C=C**H** | 12–18 | (meta aromatic) **H**...**H** | 1–3 |
| (trans) **H**C=C**H** | 6–12 | (para aromatic) **H**—...—**H** | 0–1 |
| (geminal vinyl) C=C(**H**)(**H**) | 0–2 | $sp^3$ C—**H** | 125–130 |
| 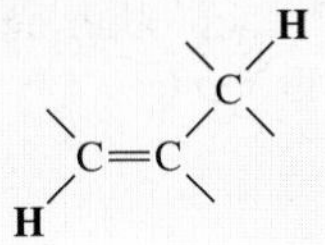 (allylic) **H**C=C—C**H** | 0–3 | $sp^2$ C—**H** | 155–170 |
| | | $sp$ C—**H** | 245–250 |

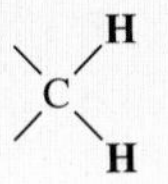

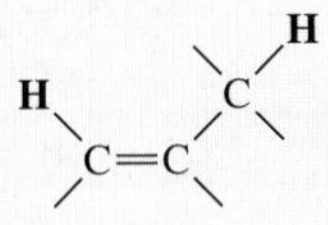

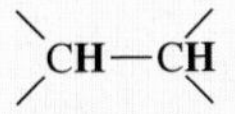

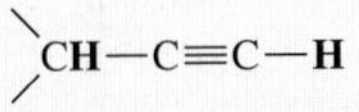

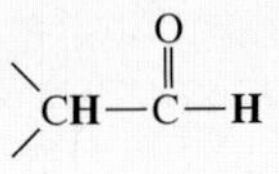

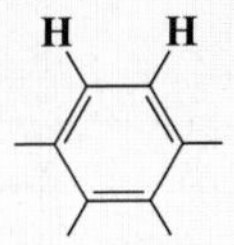

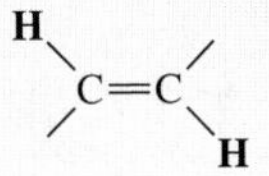

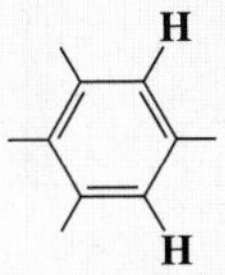

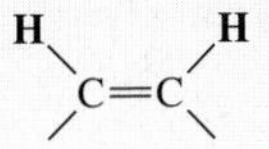

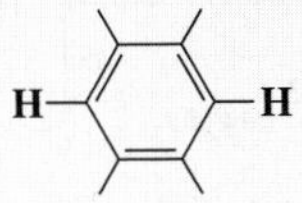

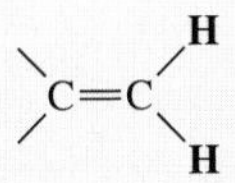

* The atoms that are coupling are shown in boldface.

## D. More Complex Splitting Patterns

The spin–spin splitting patterns such as are seen in the spectra used as examples so far arise only when the difference in chemical shifts between the hydrogen atoms that are interacting is large compared with their coupling constant, *J*. For example, in the spectrum of 1,1,2-trichloroethane (Figure 11.20, p. 418), two sets of signals are seen centered at δ 3.98 and δ 5.98. The difference in chemical shift between the two peaks is 1.80 ppm, which corresponds to 360 Hz (p. 401). This difference in chemical shift is large in comparison with the coupling constant, *J*, of 7 Hz for the interaction between hydrogen atoms that are three-bond neighbors on adjacent $sp^3$-hybridized carbon atoms. In such a case, the splitting pattern that is seen can be predicted from Table 11.4 (p. 419). When hydrogen atoms of similar chemical shift interact, the patterns become much more complex. The spectrum for 1-bromopentane in Figure 11.29 (p. 430) demonstrates this. In the spectrum of 1-bromopentane, the absorption band for the hydrogen atoms on carbon atom 1 (δ 3.42) is a recognizable triplet. These hydrogen atoms are deshielded by the bromine atom and have a chemical shift quite different from those for the rest of the hydrogen atoms in the molecule. The signal for the

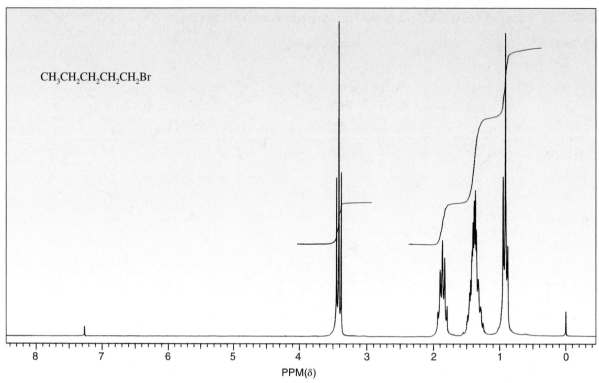

CH₃CH₂CH₂CH₂CH₂Br

*Figure 11.29*
Proton magnetic resonance spectrum of 1-bromopentane.

hydrogen atoms on carbon atom 2 appears as a distorted quintet at δ 1.90. Those on 3 and 4 are similar in chemical shift but not identical. They couple with each other and also with the hydrogen atoms on carbon 2 or with those of the methyl group.

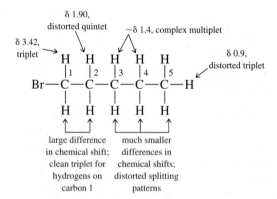

The signals for the hydrogen atoms on carbons 3 and 4 appear as a complex pattern called a multiplet (see p. 419). The signal for the methyl group is a distorted triplet. This distortion of the triplet arises because the chemical shift for the hydrogen atoms of the methyl group is too close to that of the methylene hydrogen atoms. The appearance of the spectrum of 1-bromopentane is typical of that for substituted alkanes in which most of the hydrogen atoms have similar chemical shifts. The closer the chemical shifts of the interacting hydrogen atoms, the more distorted are the patterns in the spectrum.

As seen in Figure 11.30, 1,4-substituted or *para*-disubstituted (p. 360) aromatic compounds in which the two substituents differ from each other in electronic effects

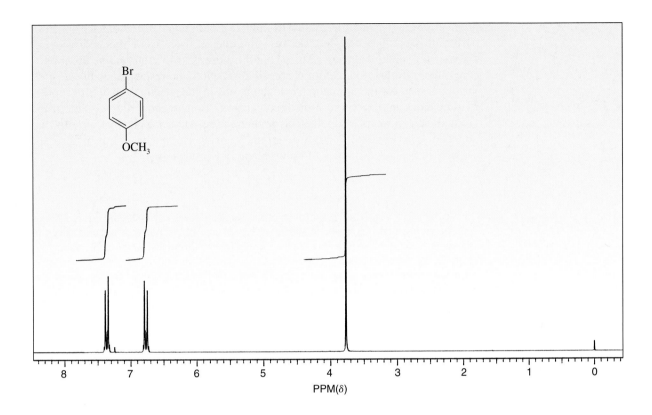

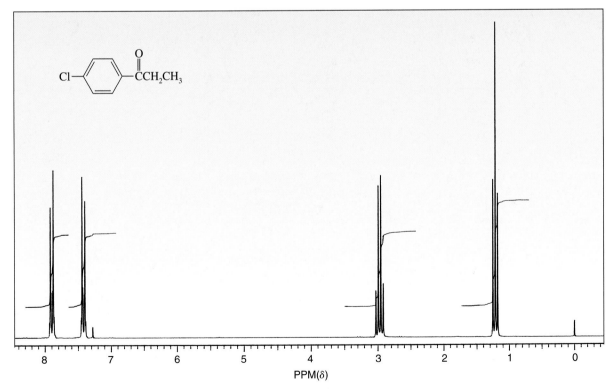

*Figure 11.30*
Proton magnetic resonance spectra of 1-bromo-4-methoxybenzene and 1-(4-chlorophenyl)-1-propanone.

typically show symmetrical patterns in their proton magnetic resonance spectra. For 1-bromo-4-methoxybenzene, because the chemical shifts of the hydrogen atoms next to the methoxy group ($\delta$ 6.79) and of those next to the bromine atom ($\delta$ 7.40) are quite different, the pattern is spread apart. Note the resemblance of the splitting patterns seen for the aromatic hydrogen atoms in the spectra of 1-(4-chlorophenyl)-1-propanone and 1-bromo-4-methoxybenzene. Both spectra show a pattern that is roughly a symmetrical doublet of doublets. The pattern typical of an ethyl group bonded to a deshielding atom or group, such as a carbonyl group, is also shown clearly in the spectrum of 1-(4-chlorophenyl)-1-propanone.

**Problem 11.15**   The proton magnetic resonance spectrum in Figure 11.31 belongs to one of the isomeric methylbenzonitriles. The $^{13}C$ nuclear magnetic resonance spectrum of the compound has six distinct bands. Methylbenzonitriles have the molecular formula $C_8H_7N$ and have a methyl group and a cyano group substituted on a benzene ring. Draw structural formulas for all possible isomers of methylbenzonitrile, and decide which one has the spectrum shown here.

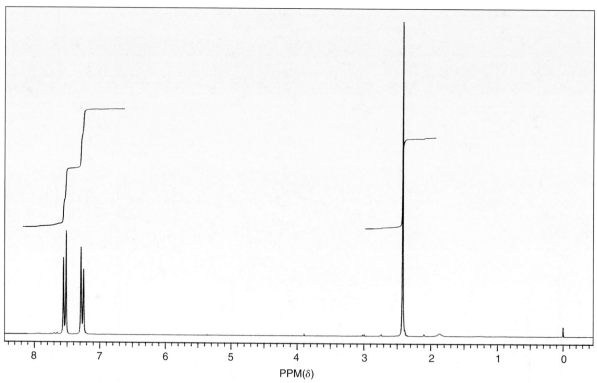

*Figure 11.31*

**Problem 11.16**   The proton magnetic resonance spectra given in Figure 11.32 belong to $C_4H_9Br$ isomers. Assign structures to Compounds F and G.

Compound F

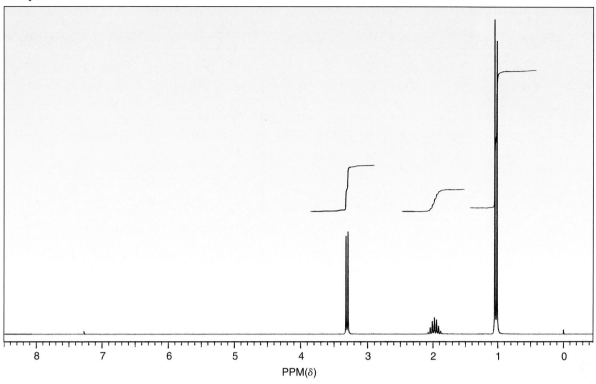

Compound G

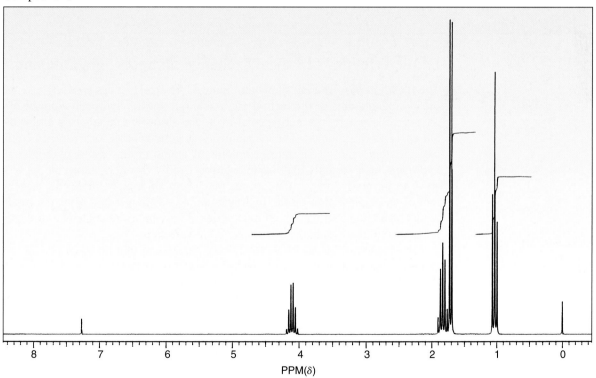

*Figure 11.32*

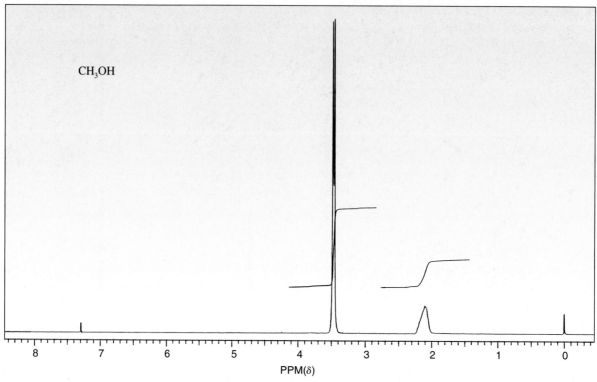

*Figure 11.33*
Proton magnetic resonance spectrum of highly purified methanol.

## E. The Effect of Chemical Exchange on Spin–Spin Coupling

The spectrum of methanol, shown in Figure 11.15 (p. 411), has two singlets, one for the hydrogen atoms of the methyl group and one for the hydrogen atom bonded to oxygen. These two different types of hydrogen atoms are on adjacent atoms but do not appear to be coupled. This phenomenon is quite general for hydrogen atoms on oxygen atoms and on the nitrogen atoms of amino groups. Such hydrogen atoms hydrogen bond strongly and are transferred rapidly from one molecule to another. As a result, no hydrogen atom spends much time on a given molecule, and the effect of the nuclear spin on the hydrogen atom is averaged out. Thus the hydrogen atoms on an adjacent carbon atom experience only an average change of the external magnetic field, and no spin–spin coupling appears.

If a sample of an alcohol is carefully purified so that no trace of acid, which catalyzes the exchange reaction, remains, coupling between the hydrogen atom of the hydroxyl group and hydrogen atoms on an adjacent carbon atom can be observed. The spectrum of a highly purified sample of methanol appears in Figure 11.33. In this spectrum, the band for the hydrogen atom of the hydroxyl group is a multiplet, and that for the methyl group is clearly a doublet. The coupling constant for the interaction between the two types of hydrogen atoms is approximately 5 Hz.

Deuterium oxide, $D_2O$, also known as "heavy water," contains deuterium, the isotope of hydrogen having one neutron in addition to one proton in its nucleus. Compounds containing hydrogen atoms bonded to oxygen or nitrogen atoms (and some carbon atoms, p. 674) exchange their hydrogen atoms for deuterium by way of hydrogen bonding. Deuterium is not detectable in a proton magnetic resonance spectrum. Such an exchange may be used to identify readily exchangeable hydrogen atoms in a compound.

**Problem 11.17**  The exchange of hydrogen atoms for deuterium atoms is an acid–base equilibrium reaction. Write a mechanism showing how the following transformation takes place.

$$ROH \ + \ DOD \ \rightleftharpoons \ ROD \ + \ HOD$$

## 11.6  Medical Applications of Nuclear Magnetic Resonance

Proton nuclear magnetic resonance spectroscopy has been used since the early 1980s to obtain images from living beings. Figure 11.34 shows an image of the brain of a patient with multiple sclerosis obtained by a technique called **magnetic resonance imaging (MRI).** The white regions near the center of the brain corre-

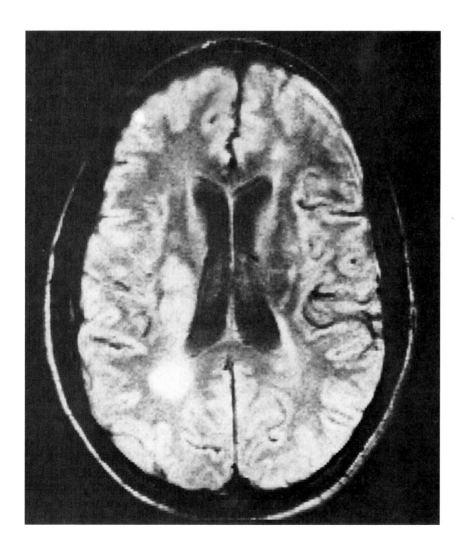

*Figure 11.34*
Image of the brain of a patient with multiple sclerosis, obtained by the technique of magnetic resonance imaging. (Courtesy of Dr. Alex Aisen.)

spond to the lesions produced by the disease. Signals are collected from many different locations in the brain in the form of a free-induction decay signal (Figure 11.6, p. 398) containing many frequencies. A computer then performs a Fourier transformation and translates these frequencies into an image. Such an image is essentially a map showing the density of the protons in water molecules and lipid molecules (p. 638) contained in the tissues of the brain.

Two factors contribute to forming this image. One, of course, is the number of protons in the substances in the head. For example, the bones of the skull do not show up in the image because they do not contain much water or other substances containing hydrogen. The most important factor for distinguishing between healthy and diseased tissues are the **relaxation times.** Relaxation is the process by which a nucleus that has been raised to the higher-energy level returns to the lower-energy level. Two types of relaxation occur. In one, the energy of a nucleus at the upper level is dissipated to the atoms of adjacent molecules. This is called spin–lattice relaxation, and $T_1$ is the time it takes for a nucleus to return to the lower-energy level by this route. Another type of relaxation occurs when a nucleus at the upper level transfers its energy to the nucleus of a nearby atom. This type of relaxation is called spin–spin relaxation. The time it takes for a nucleus to lose energy by spin–spin relaxation is called $T_2$. $T_1$ and $T_2$ for protons in pure water are both about 3 seconds; these times are much shorter (from a few hundred milliseconds to 2 seconds) for water in biological systems.

The principal factor in determining $T_1$ and $T_2$ is the rate at which molecules tumble in solution. Small molecules can change their orientation relative to each other easily and thus have inefficient relaxation processes and relatively large values for $T_1$ and $T_2$. Large molecules, such as cholesterol or other lipids, tumble more slowly because of their greater inertia and the greater friction between themselves and their neighbors. They undergo relaxation more efficiently and have smaller values of $T_1$ and $T_2$. The ordering of water molecules in normal cells is different from that in tumor cells, for example, which also contributes to a difference in $T_1$ and $T_2$ values for normal and diseased tissues.

The timing of the radiofrequency signal used to excite nuclei to the higher-energy level is used to differentiate between tissues with components that have different values of $T_1$ and $T_2$. If, for example, the interval between radiofrequency pulses is short compared to the $T_1$ of a sample, the pulse arrives before the nuclei have had a chance to return to the lower-energy level. No absorption of energy can take place; therefore, no signal is detected from that portion of the tissue. The lesion from multiple sclerosis produces edema, an abnormal accumulation of fluid with a high water content. That region has a large $T_2$ compared to adjoining tissues. A technique known as $T_2$-weighting combines different timings and orientations of the radiofrequency pulse to make areas with large $T_2$ values look bright. Magnetic resonance imaging, therefore, generates detailed images based on the chemical composition and the ordering of molecules in different types of tissues. Physicians use this technique to get images of the spinal cord, the brain, and the liver without the use of ionizing radiation (such as x-rays) and without the injection of contrasting agents, which sometimes cause allergic reactions or paralysis when used in the central nervous system.

The nuclear magnetic resonance of phosphorus, $^{31}P$, is used in another area of biomedical research. Compounds that are esters of phosphoric acid play a central role in cellular metabolism. Adenosine triphosphate (ATP) and adenosine diphosphate (ADP) are particularly important to the supply and use of chemical energy in cells. The role of phosphate esters of glucose in the metabolism of glucose is described on page 682. Phosphate esters are present in high enough concentration

(~0.5 mM) that they can be detected in intact tissues by $^{31}$P nuclear magnetic resonance. In addition, phosphorus-31 nuclear magnetic resonance is used to study DNA (p. 1016), which also contains phosphate groups.

Of particular interest is the use of phosphorus-31 nuclear magnetic resonance to follow cellular metabolism during exercise, as a way of identifying normal metabolic pathways so that this knowledge can be applied to diagnose muscle diseases. Knowledge of what went on in muscle used to be obtained only by doing muscle biopsies, removal of small amounts of muscle for testing. Athletes were understandably reluctant to provide many samples. Nuclear magnetic resonance provides a noninvasive way to follow cellular metabolism. For example, during strenuous exercise, lactic acid accumulates in muscle tissue (p. 720), producing a decrease in pH. This process can be followed using phosphorus-31 nuclear magnetic resonance because the chemical shift of the phosphorus atom in phosphate ion depends on the degree of protonation of the ion.

high pH
appears at lower field
in phosphorus-31 nuclear
magnetic resonance spectrum

low pH
appears at higher field
in phosphorus-31 nuclear
magnetic resonance spectrum

In addition, different chemical shifts are seen for $^{31}$P depending on the nature of the organic group bound to the phosphate. Therefore, it is possible to follow the decrease of compounds being metabolized and the buildup of the products of metabolism in muscle tissue. Such studies are beginning to lead to a better understanding of the enzymatic processes that are important in metabolism. Faulty metabolic pathways that lead to disease and to the failure of muscles can thus be identified. As an interesting sidelight, similar $^{31}$P nuclear magnetic resonance monitoring of the muscles in the forearms of world-class distance runners provides evidence that they may be better able than most to produce the ATP (p. 681) that fuels their athletic feats.

## SUMMARY

When a sample of an organic compound is placed in a strong magnetic field, slightly more than half of the nuclei of atoms such as hydrogen and $^{13}$C align themselves with the external field. In this state, the nuclei are in the lower of the two energy levels available to them; they absorb energy in the radiofrequency range in going to a higher-energy level in which they are aligned against the external magnetic field. The exact amount of energy necessary for this transition depends on the strength of the external magnetic field and on the environment of the hydrogen and carbon atoms in the compound being investigated. A record of these transitions is a nuclear magnetic resonance spectrum, which can be a proton or a $^{13}$C nuclear magnetic resonance spectrum.

A magnetic resonance spectrum gives four different kinds of information. The chemical shifts ($\delta$ values) tell how many different kinds of carbon (Table 11.2,

p. 403) or hydrogen atoms (Table 11.3, p. 406) are present in the compound and something about their environments. Proton magnetic resonance spectra are integrated, which reveals the relative numbers of the different hydrogen atoms in the compound. The splitting patterns (Table 11.4, p. 419) tell which atoms are close enough that their nuclei can interact. The magnitudes of the coupling constants (Table 11.5, p. 429) give information about the relative positions of the hydrogen atoms.

In $^{13}C$ nuclear magnetic resonance spectra, coupling between the nucleus of a carbon atom and the nuclei of the hydrogen atoms on it gives a splitting pattern that can be used to identify methyl, methylene, and methine carbon atoms, as well as those that have no hydrogen atoms on them.

Nuclear magnetic resonance is also important in biological and medical fields. Proton magnetic resonance can be used to create images of tissues that distinguish diseased from normal tissues. Phosphorus-31 nuclear magnetic resonance is used to follow cellular metabolism.

# ADDITIONAL PROBLEMS

**11.18** Molecular formulas and nuclear magnetic resonance spectra for Compounds H—K are given in Figure 11.35. All spectra show spin–spin coupling. Assign structures to the compounds.

Compound H, $C_4H_8O_2$

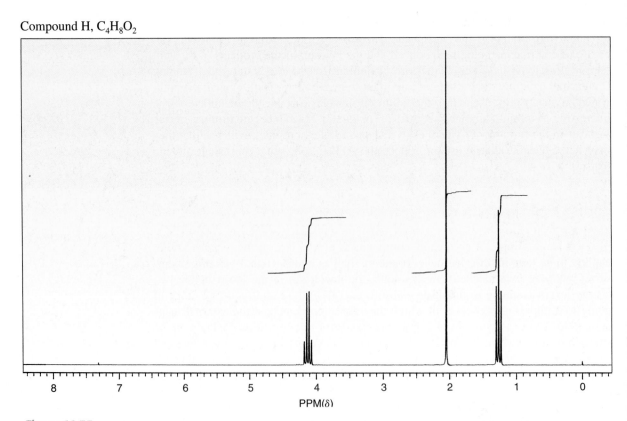

*Figure 11.35*

Compound I, C₅H₈O₂

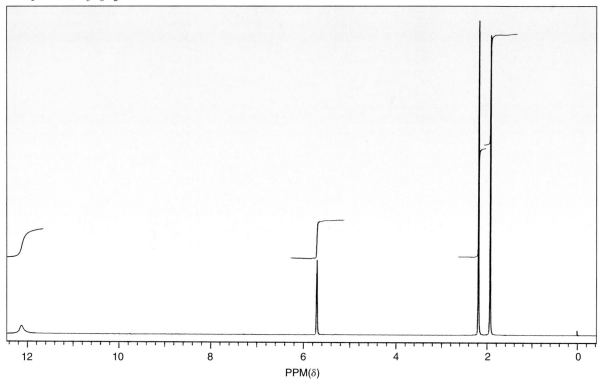

Compound J, C₇H₁₂O₃

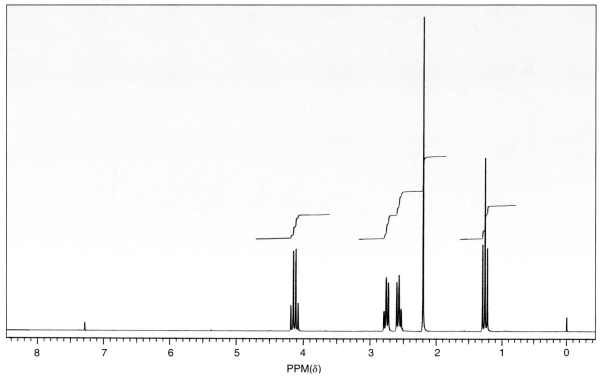

*Figure 11.35 (Continued)*

Compound K, C₄H₈O

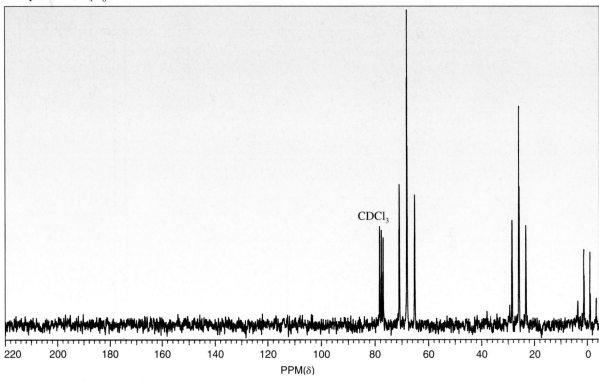

Figure 11.35 *(Continued)*

**11.19** The proton magnetic resonance spectrum of *sec*-butylbenzene is given in Figure 11.36. Assign all the peaks in the spectrum, and analyze the splitting patterns.

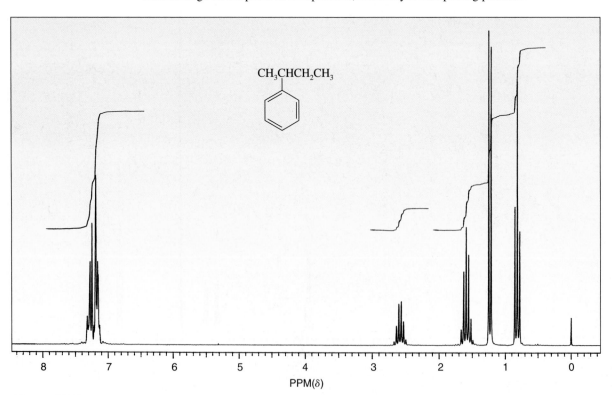

Figure 11.36

**11.20** The proton magnetic resonance spectrum of 2,4-dibromoaniline appears in Figure 11.37. Assign all the peaks in the spectrum, and analyze the splitting patterns.

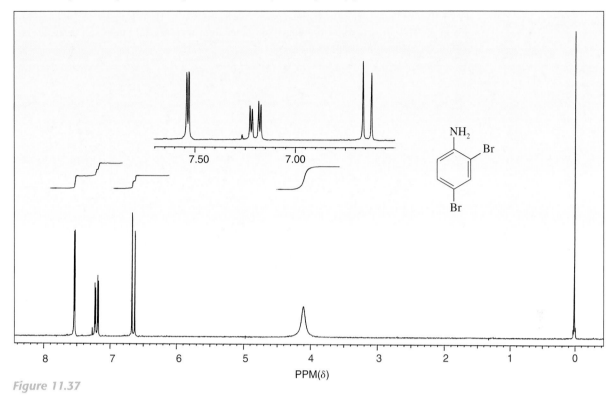

*Figure 11.37*

**11.21** A $^{13}C$ nuclear magnetic resonance spectrum of acetic acid showing spin–spin coupling is given in Figure 11.38, with an expansion of the signal at $\delta$ 178 also shown. Analyze the spin–spin coupling that is seen in the spectrum, commenting on the origins of each of the multiplets.

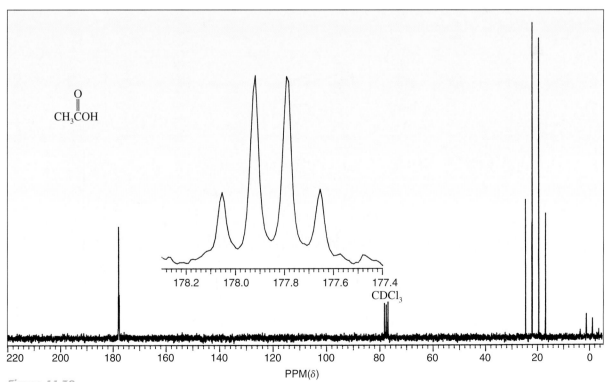

*Figure 11.38*

**11.22** Compound L has the molecular formula $C_4H_8Cl_2$ and shows $^{13}C$ nuclear magnetic resonance absorption bands at $\delta$ 30.0 and 44.1. Assign a structure to Compound L.

**11.23** The $^{13}C$ nuclear magnetic resonance spectra of two $C_4H_9Br$ isomers are given in Figure 11.39. Assign structures to them. What other information would help you be really certain of your structural assignments?

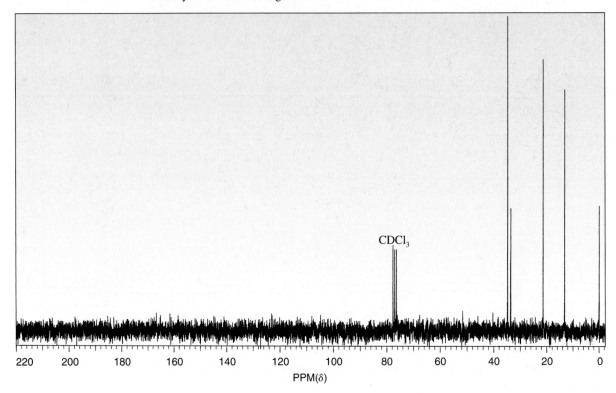

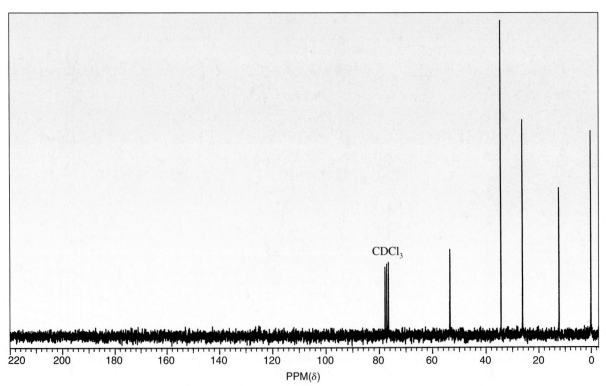

*Figure 11.39*

**442**

# Ultraviolet-Visible and Infrared Spectroscopy. Mass Spectrometry

# 12

## A Look Ahead

In Chapter 11, we learned how nuclear magnetic resonance spectroscopy allowed us to assign structures to compounds. In this chapter we will extend our understanding of spectroscopic methods such as infrared and ultraviolet and visible spectroscopy. When chemists take the infrared spectrum of a compound, they are investigating the changes in the vibrational motions of the atoms in its molecules. The energy associated with each vibrational transition depends on the masses of the atoms that are bonded together and on the strength of the bond. The infrared spectrum is, therefore, a source of information about the types of bonds the compound contains and is most useful in identifying the presence of specific functional groups, such as hydroxyl and carbonyl groups. We will also learn about the structural features that give rise to useful spectra in the ultraviolet and visible regions of the electromagnetic radiation spectrum.

One other technique for structure determination is mass spectrometry, which differs from the other types of spectroscopy discussed so far in that a mass spectrum is not a record of the energy absorbed by a molecule in going from one energy level to another. A mass spectrum is a record of the exact masses of a series of ions that are formed by fragmentation of a molecule on collision with a high-energy particle, usually an electron. The collision knocks an electron out of the molecule, giving rise to a radical-cation (a species with a positive charge and an unpaired electron) called a molecular ion. The molecular ion is unstable and breaks up to give a number of other species, which are separated by the mass spectrometer so that their exact masses and relative numbers can be recorded. This record can be used to recreate the connectivity of the molecule.

## 12.1 Ultraviolet Spectroscopy

### A. Transitions Between Electronic Energy Levels

Absorption of energy corresponding to the ultraviolet and visible regions of the electromagnetic spectrum results in transitions between electronic energy levels in molecules (Section 2.11, p. 69). A simple alkene such as ethylene, for example, absorbs radiation at a wavelength of 165 nm. A nanometer, abbreviated nm, is $10^{-7}$ cm ($10^{-9}$ m or 10 Å). The energy of the transition can be calculated from the relationships

$$\Delta E = h\nu \quad \text{or} \quad \Delta E = \frac{hc}{\lambda}$$

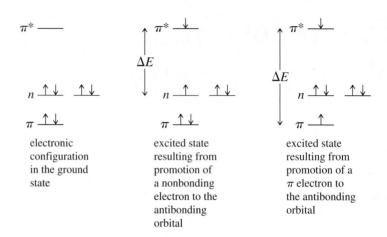

*Figure 12.1*
Electronic configurations and relative energy levels of the $\pi$ and nonbonding ($n$) electrons in formaldehyde for the ground state and for the $n \rightarrow \pi^*$ and $\pi \rightarrow \pi^*$ excited states.

where $h$ is Planck's constant, $6.624 \times 10^{-27}$ erg·s, and $c$ is the velocity of light, $2.998 \times 10^{10}$ cm/s. Radiation having a wavelength of 165 nm corresponds to $1.20 \times 10^{-11}$ erg/molecule, $7.24 \times 10^{12}$ erg/mol, or 173 kcal/mol. Therefore, the energy absorbed in the ultraviolet and visible regions of the spectrum is of the same order as covalent bond energies (Table 2.4, p. 64) and often causes chemical changes. The branch of chemistry that deals with the transformations that organic compounds undergo when they absorb ultraviolet or visible radiation is known as **photochemistry** (Chapter 25). One well-known photochemical reaction is the primary event in vision, the isomerization of (11Z)-retinal to (11E)-retinal on the absorption of light, described on page 758.

Molecular orbital theory is used to picture what happens when light is absorbed by a molecule. As we have already seen in Chapter 2, Figure 2.28 (p. 71), for example, illustrates the $\pi \rightarrow \pi^*$ transition for an alkene. A carbonyl group has a $\pi$ bond and also undergoes $\pi \rightarrow \pi^*$ transitions. In addition, the carbonyl group has two pairs of nonbonding, $n$, electrons on the oxygen atom. The energy levels for the $\pi$, $n$, and $\pi^*$ orbitals for the simplest carbonyl compound, formaldehyde, $H_2C=O$, are shown in Figure 12.1. The nonbonding electrons are more easily promoted to the $\pi^*$ orbital of the carbonyl group than are the $\pi$ electrons. Thus formaldehyde absorbs ultraviolet radiation in two regions. The lower-energy transition at 295 nm corresponds to the promotion of one of the nonbonding electrons to the antibonding $\pi$ orbital, that is, an $n \rightarrow \pi^*$ transition. The transition corresponding to the promotion of an electron from the bonding $\pi$ orbital to an antibonding $\pi$ orbital is the $\pi \rightarrow \pi^*$ transition, requires more energy, and comes at 185 nm.

**Problem 12.1**  The $n \rightarrow \pi^*$ transition for cyclohexanone comes at 290 nm. Would you expect the $n \rightarrow \pi^*$ transition for cyclohexanethione to be at a shorter or longer wavelength? A review of Section 2.11B (pp. 70-71) will be helpful.

cyclohexanone          cyclohexanethione

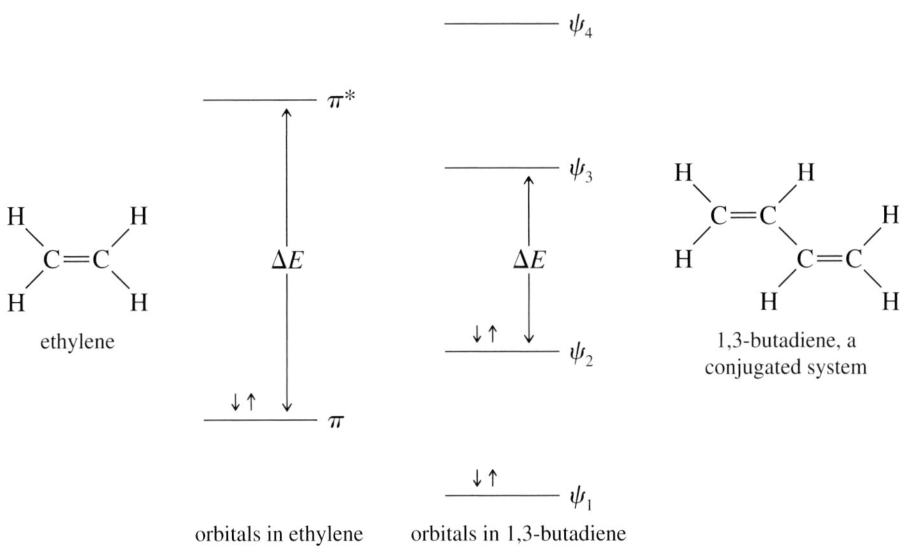

$\psi_4$

$\pi^*$

$\psi_3$

$\Delta E$     $\Delta E$

ethylene

$\psi_2$

$\pi$

$\psi_1$

orbitals in ethylene     orbitals in 1,3-butadiene

1,3-butadiene, a
conjugated system

*Figure 12.2*
A comparison of the electronic configuration and energy levels for a simple alkene with those for a conjugated alkene.

Typical ultraviolet and visible spectrophotometers do not record at wavelengths below 200 nm. The ultraviolet spectra of most organic compounds become interesting when there are two or more multiple bonds in **conjugation.** Conjugated multiple bonds are those for which $p$ orbitals on adjacent carbon atoms are separated only by a single $\sigma$ bond so that they can interact over the entire system. The Kekulé formula for benzene, for example, has conjugated double bonds (p. 68), as does 1,3-butadiene, shown in Figure 12.2. The extended interaction between the conjugated $\pi$ orbitals decreases the gap in energy between bonding and antibonding molecular orbitals (Figure 12.2). This lowers the amount of energy necessary for a $\pi \rightarrow \pi^*$ transition and increases the intensity of the absorption of radiation. If the conjugated system is extended enough, absorption takes place in the visible region of the spectrum. The carotene in vegetables such as winter squash, carrots, and red peppers appears red-orange to us because it is absorbing the violet-blue portion of the visible spectrum, allowing most of the red-orange portion to reach our eyes (p. 449).

## B. Chromophores

The ultraviolet spectrum of a compound shows only the presence or absence of certain portions of the molecule that undergo $\pi \rightarrow \pi^*$ or $n \rightarrow \pi^*$ transitions. These distinctive groupings that absorb ultraviolet or visible radiation are usually conjugated double bonds, double bonds conjugated with carbonyl groups, or aromatic rings and are known as **chromophores.**

Ultraviolet spectroscopy reveals the presence of chromophores in complicated molecules but tells nothing about the large differences in structure that may exist. For example, cholesta-4-en-3-one and 4-methyl-3-penten-2-one have essentially the same ultraviolet spectrum, with a $\pi \rightarrow \pi^*$ transition around 240 nm and an $n \rightarrow \pi^*$ transition around 310 nm.

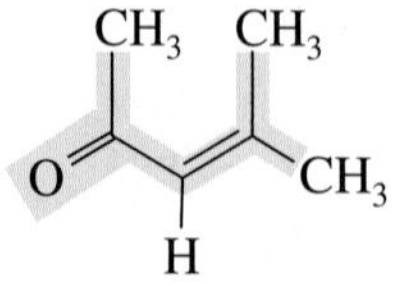

4-methyl-3-penten-2-one 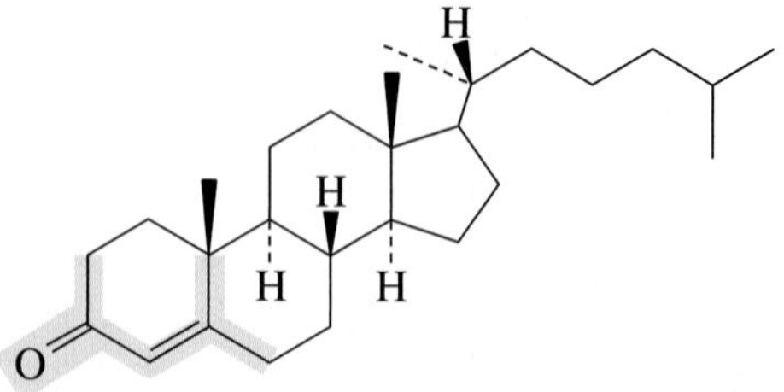                    cholesta-4-en-3-one

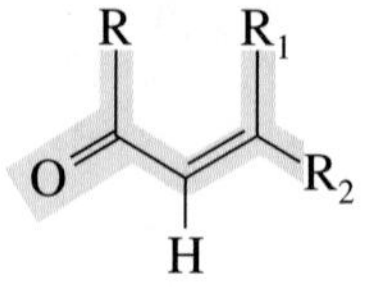

the chromophore
common to both
compounds

Both compounds have a carbonyl group conjugated with a double bond. The carbonyl group is part of an alkyl ketone, and two alkyl groups are substituted on the carbon atom at the end of the double bond that is farther from the carbonyl group. These things taken together contribute to the distinctive wavelength and intensity of the absorption of radiation, which is quite similar for both compounds. Only the chromophore in the large steroid molecule interacts with ultraviolet radiation in a way that results in absorption of energy in the region of the spectrum that an ultraviolet spectrophotometer scans. The rest of the molecule is invisible as far as this spectroscopic technique is concerned. This simplifying quality of ultraviolet spectroscopy has been extraordinarily useful in identifying compounds that have similar chromophores, regardless of the complexities and differences in the rest of their molecules. The same quality makes it useful as an analytical tool when the compound sought has an intense and distinctive absorption spectrum.

## C. The Absorption Spectrum

An ultraviolet or visible spectrum of a compound is obtained by comparing the radiation absorbed by a solution of the compound with the radiation absorbed by a similar thickness of pure solvent. A sketch of an ultraviolet spectrophotometer is shown in Figure 12.3.

The two critical aspects of an ultraviolet spectrum are the wavelength at which absorption takes place and the intensity of the absorption. The ultraviolet absorption spectrum of 1,3-pentadiene shown in Figure 12.4 is typical of a conjugated diene. The absorption appears as a broad band. Unless the radiation supplied to the compound is all of a single wavelength, molecular transitions will take place not only between one electronic energy level and a higher one but also between the closely spaced vibrational energy levels belonging to each electronic level. The infrared spectrum of a compound is a record of the transitions between those vibrational levels (Section 12.2). The ultraviolet spectrum is really the envelope for a series of closely spaced transitions clustered around a major electronic transition.

The wavelength corresponding to the energy of the electronic transition is read off the spectrum, which plots **molar absorptivity, $\epsilon$,** a measure of the intensity of

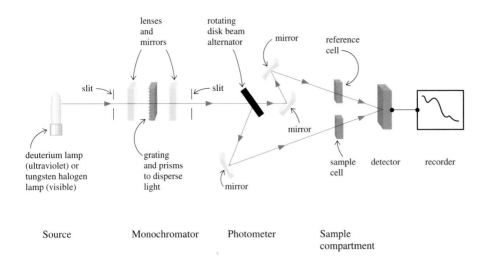

the absorption of radiation characteristic of the compound, against wavelength (or wavenumber). The position of maximum absorption is recorded as $\lambda_{max}$, 224 nm for 1,3-pentadiene. Because the solvent can affect the spectrum (an alcohol interacts with the nonbonding electrons on a carbonyl group, for example), the solvent is always specified, too. The complete designation of the wavelength of maximum absorption for the sample in Figure 12.4 is

$$\lambda_{max}^{heptane}\ 224\ nm$$

The molar absorptivity, $\epsilon$, for 1,3-pentadiene is determined from the absorbance measured for the solution of the compound in heptane. The absorbance is related to the number of molecules of 1,3-pentadiene that were in the path of the light. This depends on the concentration of the solution and on the thickness of the cell through

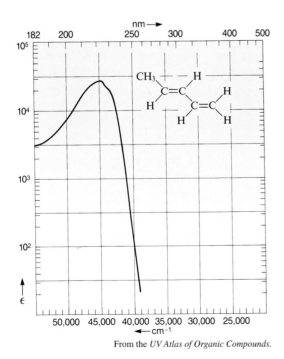

From the *UV Atlas of Organic Compounds.*

*Figure 12.4*
Ultraviolet spectrum of 1,3-
pentadiene in heptane.

which the radiation passes. The thicker the cell, the longer the path length for the beam, and the more molecules that will be encountered by the radiation on its way through, which means a greater potential for absorption of energy. The molar absorptivity is related to the absorbance, $A$, by this equation:

$$\epsilon = \frac{A}{bc}$$

Here $A$ is the experimentally determined absorbance, $b$ is the path length in centimeters, and $c$ is the concentration in moles per liter. For 1,3-pentadiene, $\epsilon$ is 26,000 L/mol · cm.

Information about the ultraviolet spectrum of a compound is thus given in two parts: the wavelength or wavelengths of maximum absorption and the molar absorptivity for the compound at those wavelengths. For the ultraviolet spectrum of 1,3-pentadiene, the complete information is $\lambda_{max}^{heptane}$ 224 nm ($\epsilon$ 26,000). Note that the units of $\epsilon$ are usually not given.

## D. The Relationship Between Structure and the Wavelength of Maximum Absorption

The wavelength of maximum absorption, $\lambda_{max}$, is dependent on the exact structure of the chromophore. For example, 2,5-dimethyl-2,4-hexadiene has the ultraviolet spectrum shown in Figure 12.5. Note the general similarity of this spectrum to that of 1,3-pentadiene (Figure 12.4). The difference consists of the shift of the position of maximum absorption to a longer wavelength, $\lambda_{max}^{heptane}$ 243 nm ($\epsilon$ 24,500). It takes less energy for the $\pi \rightarrow \pi^*$ transition to take place in this more highly substituted diene. This phenomenon is quite general; the position of $\lambda_{max}$ is dependent on the extent of conjugation and the degree of substitution on the conjugated system.

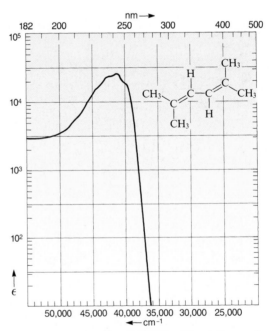

*Figure 12.5*
Ultraviolet spectrum of 2,5-dimethyl-2,4-hexadiene.

From the *UV Atlas of Organic Compounds*.

For ketones, the $\pi \rightarrow \pi^*$ transition gives rise to a more intense absorption band than the $n \rightarrow \pi^*$ transition does. Acetone has two absorption bands: $\lambda_{max}^{hexane}$ 189 nm ($\epsilon$ 900) and 279 nm ($\epsilon$ 15). The shorter-wavelength, higher-energy absorption is the $\pi \rightarrow \pi^*$ transition; the absorption at 279 nm corresponds to the $n \rightarrow \pi^*$ transition. Neither band is intense in a simple alkyl ketone.

The effect of conjugation on the spectrum of a ketone is shown in the spectra of 3-penten-2-one and 4-methyl-3-penten-2-one (Figure 12.6). For 3-penten-2-one, two bands are seen at $\lambda_{max}^{ethanol}$ 220 ($\epsilon$ 13,000) and 311 nm ($\epsilon$ 35). The chromophore in this compound is different from that in acetone, and this $\pi \rightarrow \pi^*$ transition occurs over the entire conjugated system and not just in any one $\pi$ bond. The absorption of this conjugated system at 220 nm has a high intensity. When substitution is increased on the carbon–carbon double bond of the conjugated system, the wavelength of the major absorption increases. For 4-methyl-3-penten-2-one, the absorption bands have $\lambda_{max}^{ethanol}$ 236 ($\epsilon$ 12,600) and 314 nm ($\epsilon$ 58).

When a system of really extended conjugation is present, such as in vitamin A or carotene, $\lambda_{max}$ moves into the region of the spectrum from approximately 380 to 780 nm, classified as visible. The intensity of the absorption also increases. The major absorption bands in $\beta$-carotene (p. 757), for example, occur at $\lambda_{max}^{hexane}$ 425 ($\epsilon$ 103,000), 450 ($\epsilon$ 145,000), and 477 nm ($\epsilon$ 130,000). This absorption in the blue-violet region of the visible spectrum is responsible for the yellow-orange color of carotene.

You should be aware of the following important points:

1. Ultraviolet spectra reveal the presence of distinctive groupings of atoms, usually involving conjugated systems, called chromophores.

2. In general, the more extended the conjugation, the longer the wavelength at which absorption takes place and the greater the intensity of the absorption.

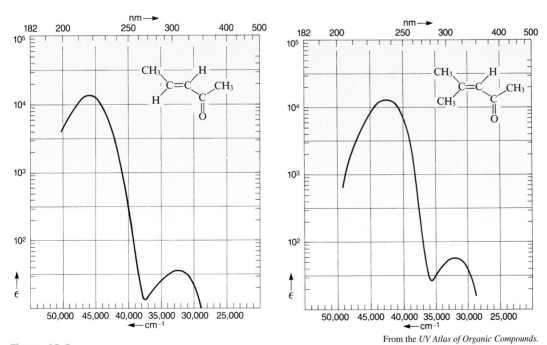

From the *UV Atlas of Organic Compounds*.

*Figure 12.6*
Ultraviolet spectra of 3-penten-2-one and 4-methyl-3-penten-2-one in ethanol.

3. Added substitution, even of alkyl groups, on the conjugated system also increases the wavelength of the absorption.

4. As the wavelength of maximum absorption increases, it moves into the visible region of the spectrum (Table 11.1, p. 395). Compounds with absorption in the visible region of the spectrum are colored.

At a more advanced level, it is possible to be much more precise about how conjugation and substitution affect absorption of energy in the ultraviolet region of the electromagnetic spectrum. However, it is sufficient for you to appreciate and be able to apply the principles listed above.

⊘ **Study Guide**
**Concept Map 12.1**

**Problem 12.2**    Which of the following compounds will absorb at the longest wavelength?

O    O    H    H
||    ||    \    /
C    C    C=C
HO      C=C    OH    HO      C=C    H    H        C=C    H
  H    C        H    C=C    OH            H    C    OH
      ||        H    H    C            H        ||
      O                  ||                    O
                        O

**Problem 12.3**    The ultraviolet spectrum of 1-acetyl-2-methyl-1-cyclohexene in ethanol is shown in Figure 12.7. Identify the electronic transitions that are responsible for the absorption bands in the spectrum. Determine $\lambda_{max}$ and $\epsilon$ for the compound and present the information about the ultraviolet absorption of 1-acetyl-2-methyl-1-cyclohexene in the standard way.

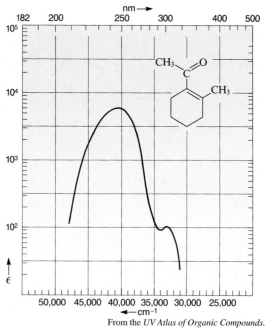

From the *UV Atlas of Organic Compounds*.

*Figure 12.7*

## E. Ultraviolet Spectroscopy of Aromatic Compounds

The aromatic ring is itself a chromophore. Benzene has major absorption bands in the ultraviolet region of the electromagnetic spectrum at 180 ($\epsilon$ 60,000), 200 ($\epsilon$ 8000), and 254 nm ($\epsilon$ 212), with other bands at 234, 239, 243, 249, 261, and 268 nm. Substitution of an alkyl group on the aromatic ring slightly increases the wavelengths of maximum absorptions. Toluene absorbs at $\lambda_{max}^{hexane}$ 189 ($\epsilon$ 55,000), 208 ($\epsilon$ 7900), and 262 nm ($\epsilon$ 260) (Figure 12.8). The chromophore is an aromatic ring with one alkyl substituent on it.

For polycyclic hydrocarbons, the absorption bands in ultraviolet spectra shift to longer wavelengths and become more intense. For example, naphthalene has $\lambda_{max}^{methanol}$ 311 nm ($\epsilon$ 239) for the band corresponding to the absorption at 254 nm ($\epsilon$ 212) for benzene.

The interaction of the nonbonding electrons of an oxygen atom with the aromatic ring shifts the absorption bands observed for benzene to higher values for phenols and aromatic ethers. For this effect to appear, the oxygen atom must, of course, be directly bonded to the aromatic ring. A comparison of the spectrum of methoxybenzene (Figure 12.8) with that of toluene demonstrates this phenomenon. Methoxybenzene has $\lambda_{max}^{isooctane}$ 220 ($\epsilon$ 8100), 271 ($\epsilon$ 2200), and 278 nm ($\epsilon$ 2250).

The conversion of a phenol into a phenolate anion results in a shift of the absorption maxima in the ultraviolet spectrum to even longer wavelengths. This effect reflects the greater electron density in the aromatic ring that is possible when the phenol is deprotonated. Both absorption bands for phenol (211 and 270 nm) move to longer wavelengths (235 and 287 nm) and increase in intensity for the phenolate anion. This change is typical of the spectra of phenols and can be used as a diagnostic test that the aromatic compound containing oxygen is indeed a phenol.

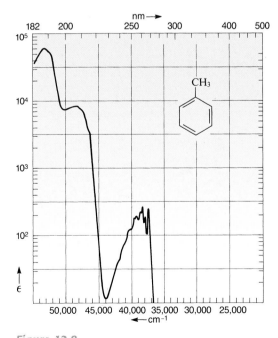 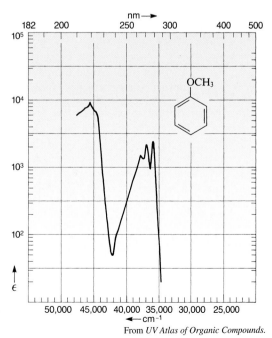

From *UV Atlas of Organic Compounds.*

*Figure 12.8*
Ultraviolet spectra of toluene in hexane and methoxybenzene in isooctane.

For amines in which the nitrogen atom is bonded directly to an aromatic ring, interaction between the nonbonding electrons of the nitrogen atom and the $\pi$ electrons of the ring is possible (p. 658). Just as occurs with phenols and aryl ethers, the chromophore of the aromatic ring changes, and the absorption bands in the ultraviolet spectrum appear at longer wavelength and become more intense. Protonation destroys the interaction responsible for this shift.

**Problem 12.4**   The ultraviolet spectrum shown in Figure 12.9 belongs to Compound A, $C_{10}H_{15}N$, and was taken in heptane.

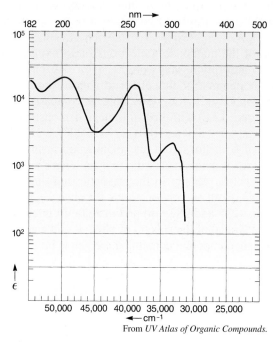

From *UV Atlas of Organic Compounds.*

*Figure 12.9*

By comparing the spectrum to the other spectra in this section, distinguish between the following possible structures for Compound A.

$$CH_3CH_2 \underset{\underset{\displaystyle \text{(phenyl)}}{N}}{} CH_2CH_3 \qquad \underset{\displaystyle \text{(phenyl)}}{CH_2CH_2}\overset{\displaystyle CH_3}{\underset{}{N}}-CH_3$$

What is the chromophore?

**Problem 12.5**   Compound B, $C_7H_8O$, has $\lambda_{max}^{methanol}$ 214 ($\epsilon$ 6030) and 273 nm ($\epsilon$ 1820). When Compound B is put into methanol with some potassium hydroxide in it, the ultraviolet spectrum has bands at 238 ($\epsilon$ 5510) and 282 nm ($\epsilon$ 1990). What is a possible structure for Compound B?

## 12.2  Infrared Spectroscopy

### A. Molecular Vibrations and Absorption Frequencies in the Infrared Region

The atoms within a molecule are constantly in motion, distorting the chemical bonds. Such motions are called **molecular vibrations.** One type of vibration possible for a molecule produces changes in bond length; such vibrations are called **stretching vibrations** (Figure 12.10). Other vibrations result in changes in bond angles and are called **bending vibrations** (Figure 12.11). Any given molecule has a number of energy levels corresponding to the different vibrational states possible for that molecule. The spacing between these energy levels corresponds to the energy of radiation in the infrared region of the electromagnetic spectrum (Table 11.1, p. 395). The differences in vibrational energy levels in the region of the infrared that is most useful to chemists correspond to radiation having frequencies, $\bar{\nu}$, wavenumbers of 4000 to 666 $cm^{-1}$ ($\bar{\nu} = 1/\lambda$ with $\lambda$ given in cm), or wavlengths of 2.5 to 15.0 $\mu m$ (a micrometer, $\mu m$, is $10^{-4}$ cm, or $10^{-6}$ m).

When radiation having energy that is the same as the difference in energy between molecular vibrational energy levels impinges on a molecule, the radiation is absorbed and the amplitude of molecular vibration increases. The molecule moves from one vibrational energy level to a higher one. The energy that is absorbed is ultimately returned to the environment as heat. This energy is roughly 1 to 10 kcal/mol, not enough to break chemical bonds or cause chemical reactions. The fact that the radiation is absorbed is recorded as a tracing by an instrument called an **infrared spectrophotometer** (Figure 12.12, p. 454).

The infrared spectra shown in this book were recorded on a type of instrument called a Fourier-transform infrared spectrophotometer. Such spectra are called FT-IR spectra for short. With such a spectrophotometer, the spectrum is obtained in

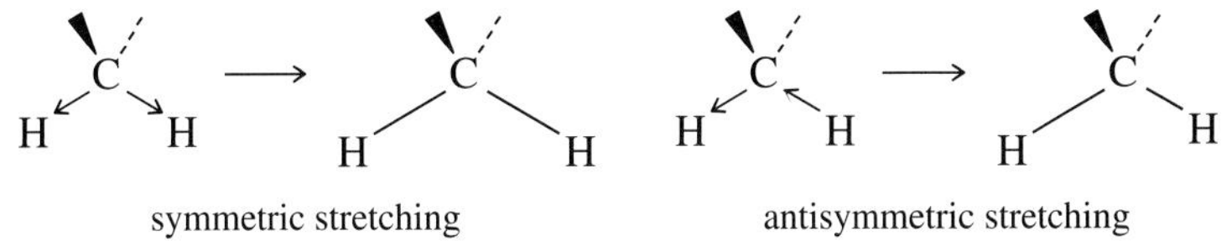

symmetric stretching        antisymmetric stretching

*Figure 12.10*
Typical stretching vibrations of a methylene group.

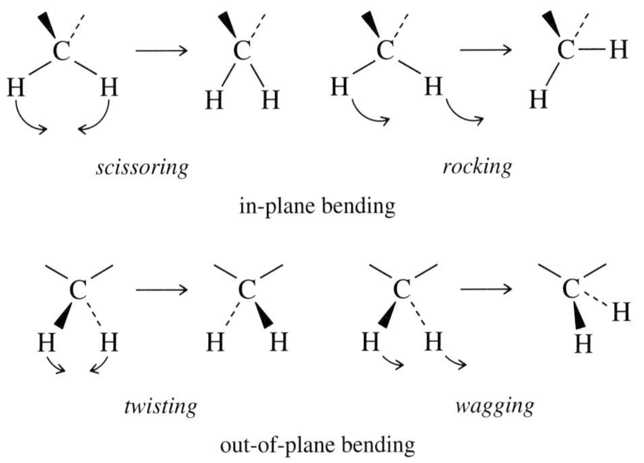

*scissoring*        *rocking*
in-plane bending

*twisting*        *wagging*
out-of-plane bending

*Figure 12.11*
Different kinds of bending vibrations for a methylene group.

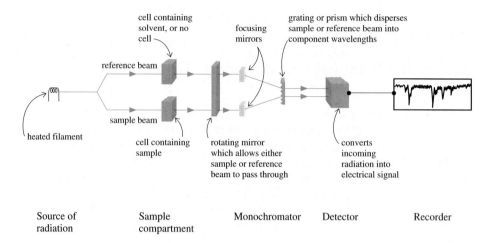

cell containing
solvent, or no
cell

focusing
mirrors

grating or prism which disperses
sample or reference beam into
component wavelengths

reference beam

heated filament

sample beam

cell containing
sample

rotating mirror
which allows either
sample or reference
beam to pass through

converts
incoming
radiation into
electrical signal

Source of
radiation

Sample
compartment

Monochromator

Detector

Recorder

**Figure 12.12**
Schematic diagram
of a typical infrared
spectrophotometer.

a second or less in the form of an interferogram, which is a plot of the sums of the cosine waves of all the frequencies present in the source of infrared radiation as modified by passage through the sample. A computer stores these signals in its memory, carries out a mathematical operation known as a Fourier transformation on them, corrects for the frequencies generated by the source of the infrared radiation, and plots the FT-IR spectrum (Figure 12.13).

Even a simple molecule can have a large number of possible molecular vibrations, and an infrared spectrum usually has many absorption bands. Infrared spectra are useful to chemists because different functional groups absorb at different frequencies, corresponding to certain vibrations typical of that portion of a molecule. Tables that list the infrared absorption frequencies for different types of functional groups are available. Table 12.1 on page 456 is an abbreviated listing of such group frequencies. The frequencies at which functional groups absorb energy are related to the types of bonds present. Figure 12.14 (p. 456) makes this clearer by summarizing some trends that can be seen by carefully examining Table 12.1.

The **stretching frequency** of a bond is related to the masses of the two atoms involved in the bond and to the strength of the bond.

$$\bar{\nu} = \frac{1}{2\pi c}\sqrt{\frac{f(m_1 + m_2)}{m_1 m_2}}$$

Here $\bar{\nu}$ is the frequency in $cm^{-1}$, $c$ is the velocity of light, $m_1$ and $m_2$ are the masses of the two atoms in grams, and $f$ is the force constant in dyne/cm. The force constant for a single bond is approximately $5 \times 10^5$ dyne/cm. The force constant for a double bond is $10 \times 10^5$ dyne/cm, about twice that for a single bond; that for a triple bond is $15 \times 10^5$ dyne/cm. Table 12.1 and Figure 12.14 show, for example, that triple bonds absorb at higher frequencies ($2260-2100$ $cm^{-1}$) than do double bonds ($1800-1390$ $cm^{-1}$), which in turn absorb at higher frequencies than do single bonds ($1360-1030$ $cm^{-1}$). It takes more energy to stretch a stronger bond.

The stretching vibrations for bonds to hydrogen, the lightest of the elements, also occur at high frequencies ($3650-2500$ $cm^{-1}$). A bond between hydrogen and an $sp$-hybridized carbon is shorter and absorbs at a higher frequency ($3300$ $cm^{-1}$) than does one between hydrogen and an $sp^2$-hybridized carbon ($3080-3020$ $cm^{-1}$). The carbon–hydrogen bonds of alkanes absorb at the lowest frequencies ($2960-2850$ $cm^{-1}$).

The major frequencies that are typical of functional groups usually appear between 4000 and 1400 $cm^{-1}$. The portion of an infrared spectrum between 1400 and

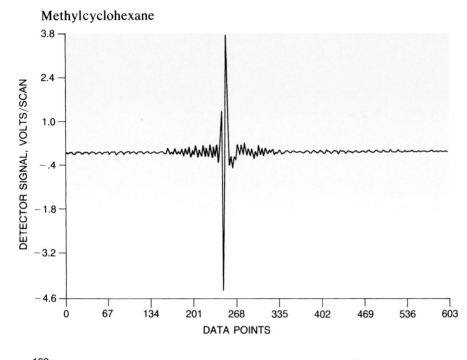

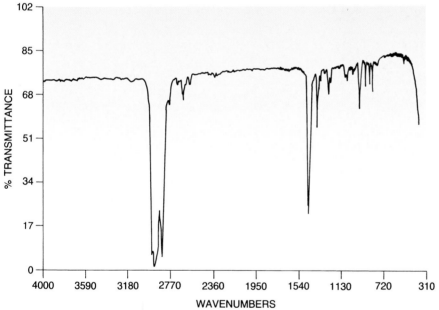

*Figure 12.13*
Interferogram and spectrum
for methylcyclohexane
obtained on a Nicolet 60-SX
Fourier-transform infrared
spectrophotometer.

200 cm$^{-1}$ is called the **fingerprint region.** It is more difficult to make specific assignments for bands in that region because they are dependent on the structure of the molecule as a whole. This region is immensely useful in making a positive identification of a compound. If spectra of two species show the same bands, with the same relative intensities, in this region as well as in the higher-frequency region, this is considered to be proof of the identity of the two species.

Not all bands in an infrared spectrum have the same intensity, and this too is useful in identifying functional groups. In general, during a vibration that corresponds to a change in the dipole of a molecule, the molecule absorbs radiation

## Table 12.1    Characteristic Infrared Absorption Frequencies

| Bond Type | Stretching, cm⁻¹ | Bending, cm⁻¹ |
|---|---|---|
| C—H alkanes | 2960–2850 (*s*) | 1470–1350 (*s*) |
| C—H alkenes | 3080–3020 (*m*) | 1000–675 (*s*) |
| C—H aromatic | 3100–3000 (*v*) | 870–675 (*v*) |
| C—H aldehyde | 2900, 2700 (*m*, 2 bands) | |
| C—H alkyne | 3300 (*s*) | |
| C≡C alkyne | 2260–2100 (*v*) | |
| C≡N nitrile | 2260–2220 (*v*) | |
| C=C alkene | 1680–1620 (*v*) | |
| C=C aromatic | 1600–1450 (*v*) | |
| C=O ketone | 1725–1705 (*s*) | |
| C=O aldehyde | 1740–1720 (*s*) | |
| C=O α, β-unsaturated ketone | 1685–1665 (*s*) | |
| C=O aryl ketone | 1700–1680 (*s*) | |
| C=O ester | 1750–1735 (*s*) | |
| C=O acid | 1725–1700 (*s*) | |
| C=O amide | 1690–1650 (*s*) | |
| O—H alcohols (not hydrogen bonded) | 3650–3590 (*v*) | |
| O—H alcohols (hydrogen bonded) | 3600–3200 (*s*, broad) | 1620–1590 (*v*) |
| O—H acids | 3000–2500 (*s*, broad) | 1655–1510 (*s*) |
| N—H amines | 3500–3300 (*m*) | |
| N—H amides | 3500–3350 (*m*) | |
| C—O alcohols, ethers, esters | 1300–1000 (*s*) | |
| C—N amines, alkyl | 1220–1020 (*w*) | |
| C—N amines, aromatic | 1360–1250 (*s*) | |
| NO₂ nitro | 1560–1515 (*s*) | |
|  | 1385–1345 (*s*) | |

*s* = strong absorption     *w* = weak absorption
*m* = medium absorption     *v* = variable absorption

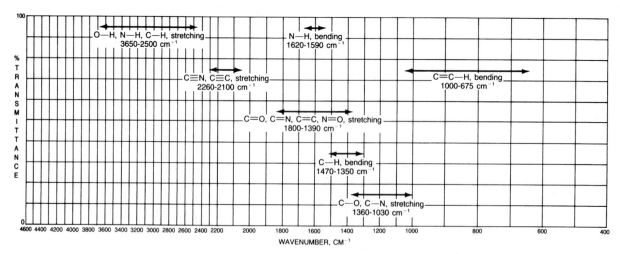

**Figure 12.14**
Infrared regions for different types of bonds.

strongly. For a vibration in which there is a small or no change in the dipole of the molecule, an absorption band may not be seen. For example, stretching of a carbonyl group gives a strong absorption, and stretching of a carbon–carbon double bond, a weak one.

| | | | |
|---|---|---|---|
| polar bond in carbonyl group | *increase in dipole moment on stretching of bond in carbonyl group* | nonpolar carbon–carbon double bond | *small or no change in dipole moment on stretching of carbon–carbon double bond* |
| | intense absorption | | absorption weak or absent |

Variations such as these are why the interpretation of infrared spectra is an art requiring careful observation of relative intensities and appearances of bands rather than a mere reading of numbers from tables. The examples of spectra of compounds with different functional groups presented in the next section will make the important points of this discussion clearer.

## B. Using Infrared Spectroscopy to Study Chemical Transformations. Infrared Spectra of Alkynes, Alkenes, and Alcohols ,

The characteristic infrared absorption frequencies of major functional groups can be used to document the transformation of one chemical species into another. For example, the terminal alkyne 1-hexyne can be converted by hydrogenation (p. 339) to 1-hexene, which, in turn, gives 1-hexanol on hydroboration (p. 293). Infrared spectra for these three compounds are shown in Figure 12.15 (p. 458). The absorption bands typical of a terminal alkyne are the triple-bond stretching frequency in the region of 2260–2100 cm$^{-1}$ and the stretching frequency for the terminal carbon–hydrogen bond, the strong sharp band at 3300 cm$^{-1}$. These two features taken together identify a terminal alkyne. Both these bands are missing in the spectrum of 1-hexene. The important band in the spectrum of the alkene comes from the stretching frequency of the carbon–carbon double bond at 1680–1640 cm$^{-1}$. In addition, the band for the carbon–hydrogen stretching frequency for the vinyl hydrogens appears near 3100 cm$^{-1}$. In the spectrum of 1-hexanol, the most important band arises from the hydroxyl (O—H) stretching frequency. Most of the time, this is a strong broad band at 3600–3200 cm$^{-1}$, indicating the presence of a hydrogen-bonded hydroxyl group. There is also a strong band around 1300–1080 cm$^{-1}$, which is attributable to carbon–oxygen single-bond stretching vibrations.

In all three spectra in Figure 12.15, other bands appear; these are characteristic of carbon–hydrogen bonds involving $sp^3$-hybridized carbon. Among these bands are those for carbon–hydrogen stretching frequencies around 2900 cm$^{-1}$ and those for the bending vibrations for such bonds around 1400 cm$^{-1}$.

The bands that have been singled out in this description are the most important in these spectra. Many of the other bands cannot be assigned to any one simple vibration of the molecule. They are the result of combinations of vibrations or are overtones of other bands. These bands are, however, characteristic of each molecule and can be used for its identification.

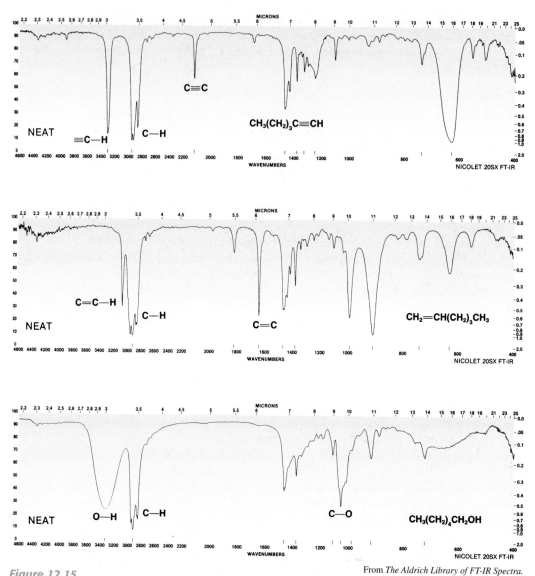

*Figure 12.15*
Infrared spectra of 1-hexyne, 1-hexene, and 1-hexanol.

From *The Aldrich Library of FT-IR Spectra.*

The three infrared spectra shown in Figure 12.15 also illustrate the usefulness of such spectra in keeping track of chemical transformations. Once a chemist's eye is trained to recognize the characteristic absorption bands, he or she can use infrared spectra to "see" what is happening in a reaction mixture by noting the disappearance of bands that characterize the functional group in the starting material and the appearance of bands that characterize the functional group being created in the product. To get information from all kinds of spectra, you must learn to look beyond the tracings and numbers produced by spectrometers in order to see the structural features that cause them.

**Problem 12.6**  The infrared spectra of Compounds C–F appear in Figure 12.16. The compounds may be alcohols, alkenes, or alkynes. Assign the correct functional group class to each compound. Point out the bands that you used in making each assignment.

## Compound C

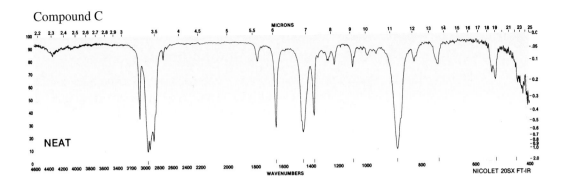

NEAT

NICOLET 20SX FT-IR

## Compound D

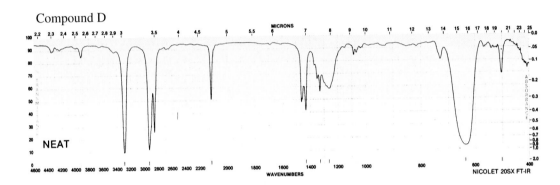

NEAT

NICOLET 20SX FT-IR

## Compound E

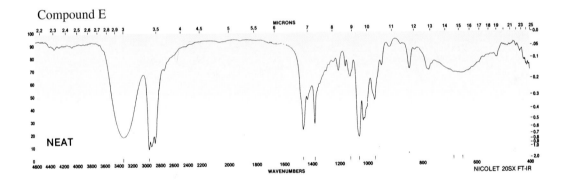

NEAT

NICOLET 20SX FT-IR

## Compound F

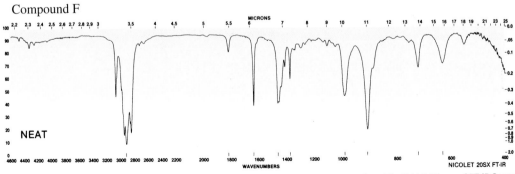

NEAT

NICOLET 20SX FT-IR

From *The Aldrich Library of FT-IR Spectra.*

*Figure 12.16*

459

## C. Infrared Spectra of Aldehydes and Ketones

The band for the stretching frequency of a carbonyl group is usually one of the strongest in the spectrum. The spectra of butanal and 2-butanone are given in Figure 12.17. Butanal has a strong absorption due to the carbonyl group at 1725 cm$^{-1}$. Another absorption that is typical of aldehydes is the carbon–hydrogen stretching frequency for the carbon–hydrogen bond on the carbonyl group. Two bands are actually present for the hydrogen on the carbonyl group; one is visible around 2700 cm$^{-1}$, but the other is buried in the strong absorption band for alkane-type hydrogens around 2900 cm$^{-1}$. The spectrum of butanal from 1400 to 400 cm$^{-1}$ is characterized by bands for the stretching and bending frequencies for the alkane portion of the molecule. The carbonyl band in 2-butanone appears at 1710 cm$^{-1}$. Note that the band for the carbon–hydrogen stretching frequency seen around 2700 cm$^{-1}$ in the spectrum of the aldehyde is missing from the spectrum of the ketone.

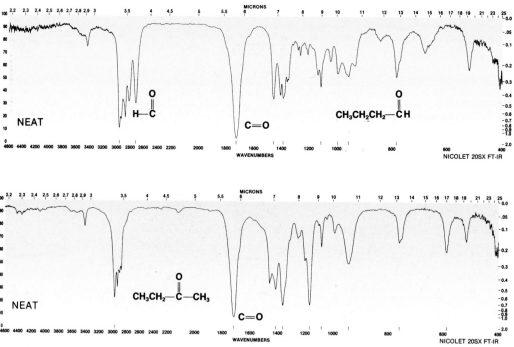

*Figure 12.17*
Infrared spectra of butanal and 2-butanone.

From *The Aldrich Library of FT-IR Spectra.*

**Problem 12.7** The infrared spectra and molecular formulas of Compounds G–J, all of which contain oxygen, are given in Figure 12.18. Decide whether each compound is an alcohol, an aldehyde, or a ketone. Propose a structure for each compound that is compatible with both the spectrum and the molecular formula. Calculating units of unsaturation is a good way to start solving such problems.

Compound G, $C_4H_{10}O$

Compound H, $C_6H_{12}O$

Compound I, $C_6H_{12}O$

Compound J, $C_4H_8O$

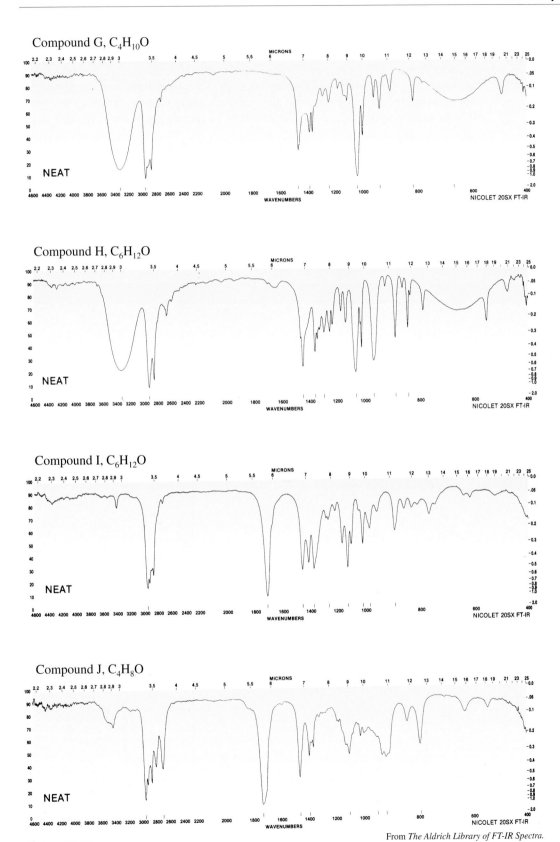

From *The Aldrich Library of FT-IR Spectra.*

*Figure 12.18*

Note that even when given its molecular formula, it is usually not possible to assign a unique structure to a compound using the infrared spectrum alone. The answers to Problem 12.7 given in the *Study Guide* all give two or more possible structures for each one of the unknown compounds. Chemists, therefore, combine data from several spectroscopic methods to solve structural problems. Problem 12.8 will be your first opportunity to combine infrared and nuclear magnetic resonance information to make a structural assignment.

**Problem 12.8**    Infrared and proton magnetic resonance spectra for Compound K, $C_5H_{10}O$, are given in Figure 12.19. The $^{13}C$ nuclear magnetic resonance spectrum of Compound K has bands at $\delta$ 18.15, 27.43, 41.65, and 212.47. Assign a structure to Compound K, and explain your reasoning in detail.

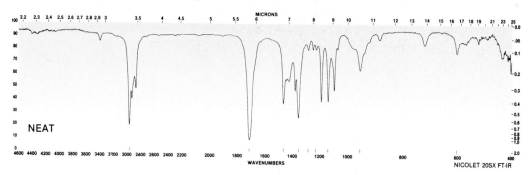

From *The Aldrich Library of FT-IR Spectra.*

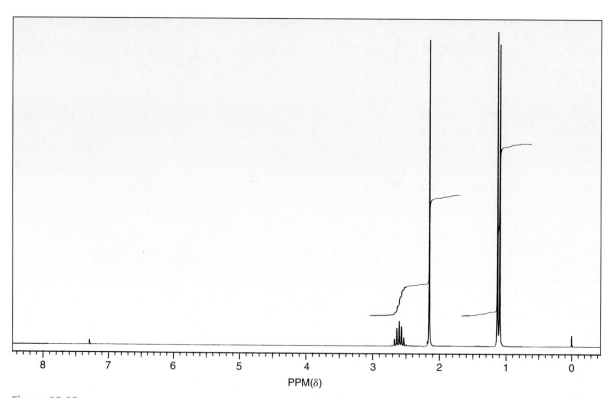

*Figure 12.19*

# D. Infrared Spectra of Conjugated Carbonyl Compounds

When a carbonyl group in an aldehyde or ketone is separated from a carbon–carbon double bond (or an aromatic ring) by only a single bond, the carbonyl group is said to be conjugated with the double bond (or the ring).

carbonyl group conjugated
with a double bond

carbonyl group conjugated
with an aromatic ring

A ketone or aldehyde in which the carbonyl group is conjugated has a lower carbonyl stretching frequency than does a ketone or aldehyde in which there is no such conjugation. This fact is rationalized by writing resonance contributors, of which one shows single-bond character for the carbon–oxygen bond.

Single bonds are easier to stretch than double bonds and, therefore, absorb at lower frequencies (p. 454). For example, comparing the infrared spectrum of 2-butanone (Figure 12.17), in which the carbonyl group is not conjugated with the double bond, with that of 2-cyclohexen-1-one, in which it is, shows clearly the shift of the carbonyl band to a lower frequency (Figure 12.20). In 2-butanone, the carbonyl stretching frequency is 1710 cm$^{-1}$; in 2-cyclohexen-1-one, it is 1675 cm$^{-1}$. Note,

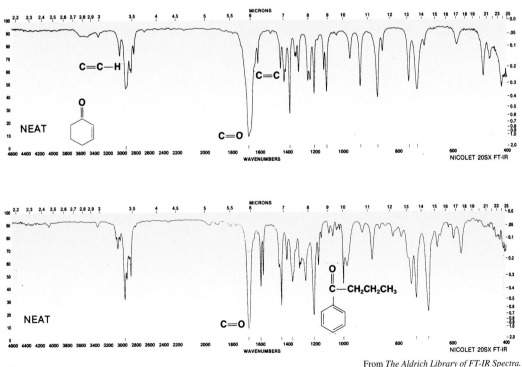

From *The Aldrich Library of FT-IR Spectra.*

*Figure 12.20*
The infrared spectra of 2-cyclohexen-1-one and 1-phenyl-1-butanone.

too, in the spectrum of 2-cyclohexen-1-one the band for the carbon–carbon double-bond stretching frequency at 1630 $cm^{-1}$ and the band for the stretching frequency at 3080 $cm^{-1}$ for the hydrogen atoms on the double bond (p. 457).

A similar comparison may be made between the infrared spectrum of 2-butanone and that of 1-phenyl-1-butanone (Figure 12.20). The carbonyl stretching frequency for 1-phenyl-1-butanone is at 1695 $cm^{-1}$, indicating that in this compound the carbonyl group is conjugated with an unsaturated system. Note the strong, sharp bands between 1600 and 1450 $cm^{-1}$ and between 850 and 700 $cm^{-1}$ in its spectrum in Figure 12.20. These bands indicate the presence of aromatic rings.

### E. Infrared Spectra of Carboxylic Acids and Esters

The infrared spectra of carboxylic acids display two important features, which are illustrated in the spectrum of hexanoic acid (Figure 12.21). First, because of the very strong hydrogen bonding between the carboxyl groups of acid molecules, a strong and broad absorption band appears from 3300 $cm^{-1}$ to as low as 2500 $cm^{-1}$ in the region of the spectrum for stretching frequencies for the oxygen–hydrogen single bond. The stretching frequencies for the carbon–hydrogen bonds of carboxylic acids are usually buried within this band. Second, the stretching frequency for the carbonyl group of a carboxylic acid appears in one of two regions. Acids in which the carbonyl group is bonded to a tetrahedral carbon atom absorb around 1725–1700 $cm^{-1}$. For acids in which the carbonyl group is conjugated with a double bond or with an aromatic ring, of which benzoic acid is an example, the carbonyl stretching frequency is around 1710–1680 $cm^{-1}$.

The important absorption bands for esters lie in the carbonyl stretching region and the carbon–oxygen single-bond stretching region of the spectrum. The carbonyl stretching frequency for alkyl esters is around 1750–1735 $cm^{-1}$. If the carbonyl group of the ester is conjugated, absorption occurs at 1730–1715 $cm^{-1}$. The carbon–oxygen single-bond stretching frequency varies with the nature of the group attached to the oxygen atom. In all esters, two bands are seen in the region between 1300 and 1000 $cm^{-1}$, the same region in which a similar absorption is seen for alcohols (p. 457). Typical spectra of esters are those of methyl hexanoate and ethyl benzoate (Figure 12.22).

Note that the infrared spectra of esters do not show a band for the stretching frequency for the hydroxyl group. Also note the greater complexity of the spectrum of the aromatic ester compared to that of the alkyl acid (Figure 12.21) and ester. The several sharp bands from 1600 to 1450 $cm^{-1}$ and the strong bands between 870 and 675 $cm^{-1}$ are typical of compounds that contain aromatic rings.

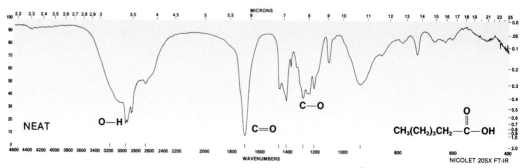

*Figure 12.21*

Infrared spectrum of hexanoic acid.

From *The Aldrich Library of FT-IR Spectra.*

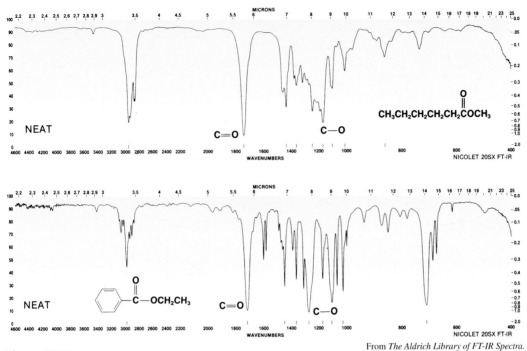

From *The Aldrich Library of FT-IR Spectra.*

*Figure 12.22*

Infrared spectra of methyl hexanoate and ethyl benzoate.

(→) **Study Guide**
**Concept Map 12.2**

**Problem 12.9**   Spectra numbered 1–5 are given in Figure 12.23 (pp. 465 and 466).
Match each spectrum with one of the following compounds.

(a) $CH_3CH_2CH_2CH_2CH_2CH_2CH_2CH_2OH$   (1-octanol)   (b) (3-methylcyclohexanone)

(c) $CH_3CH_2CH_2CH_2\overset{O}{\overset{\|}{C}}OCH_3$   (methyl pentanoate)   (d) $CH_2{=}CHCH_2\overset{O}{\overset{\|}{C}}OH$   (3-butenoic acid)

(e) $CH_2{=}CHCH_2CH_2CH_2CH_2CH_2CH_2CH_2CH_2\overset{O}{\overset{\|}{C}}H$   (10-undecenal)

**Spectrum 1**

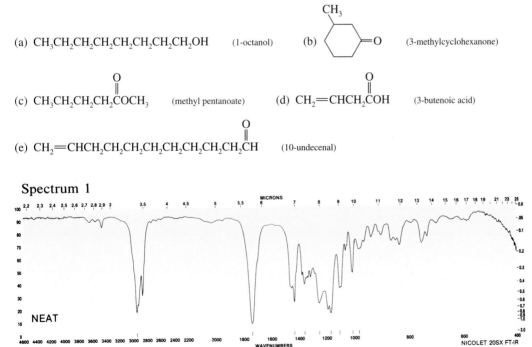

From *The Aldrich Library of FT-IR Spectra.*

*Figure 12.23*

## Spectrum 2

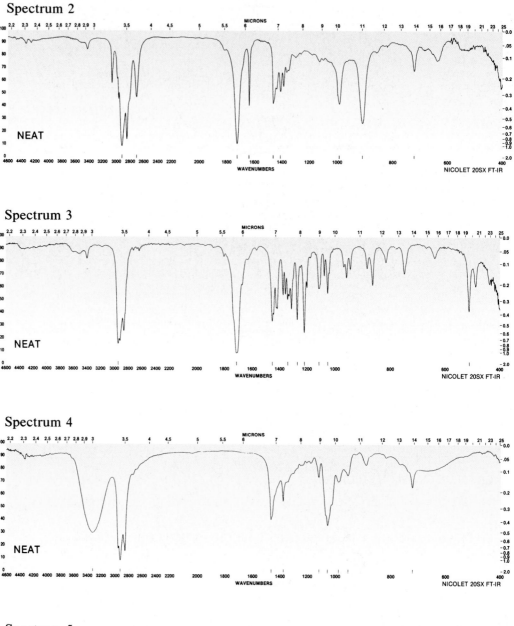

## Spectrum 3

## Spectrum 4

## Spectrum 5

**Figure 12.23** *(Continued)*

From *The Aldrich Library of FT-IR Spectra.*

## 12.3 Mass Spectrometry

### A. The Mass Spectrum

A **mass spectrum** is obtained by injecting a very small sample (a millionth of a gram, $10^{-6}$ g, will do) of a compound into a mass spectrometer (Figure 12.24), where it is subjected to bombardment by a high-energy electron beam. The first reaction that occurs in the mass spectrometer is the formation of a **molecular ion** by loss of an electron from the molecule. This electron may come from any bond or from nonbonding electrons in the molecule. Its loss gives rise to a **radical-cation,** a reactive intermediate with an unpaired electron and a positive charge. A nonbonding electron is in a higher-energy molecular orbital than an electron in a $\pi$ bond, which in turn is in a higher-energy molecular orbital than an electron in a $\sigma$ bond (Figure 2.27, p. 71). Therefore, a nonbonding electron is removed more easily than an electron of a $\pi$ bond, which is lost more easily than an electron of a $\sigma$ bond.

Fragmentation of the molecular ion gives other radical-cations, carbocations, and neutral molecules. The positively charged species are separated according to their **mass-to-charge ratios, $m/z$,** by a variety of techniques. Most ions created in a mass spectrometer have a single positive charge; therefore, the separation is essentially done according to mass. One way this is done is that the ions of different masses are accelerated in an electric field and are then deflected by different amounts as they travel through a magnetic field, as shown in Figure 12.24. Thus ions of different masses impinge separately on an electron multiplier and are recorded on a chart. The calibration of the instrument allows not only the masses of the ions to be recorded but the relative number of each kind.

The mass spectra of acetone and acetaldehyde, plotted from the original recordings as bar graphs, are shown in Figure 12.25 (p. 468). Vertical lines appear in the spectra above numbers on the horizontal axis that correspond to the masses of the ions that are formed when molecules of these compounds are bombarded with a high-energy beam of electrons. The first reaction in both cases is the loss of a nonbonding electron from the oxygen atom, to give the molecular ions, which are radical-cations.

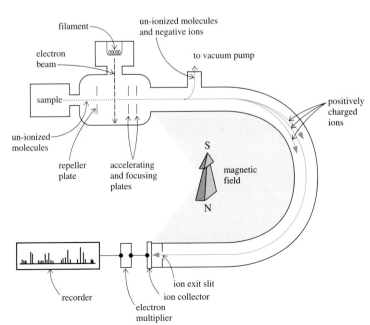

*Figure 12.24*
Schematic diagram of a typical mass spectrometer.

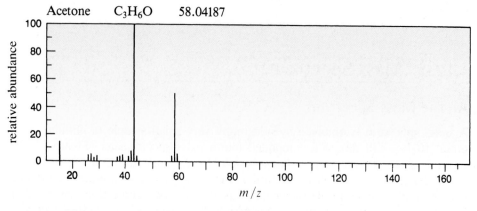

Acetone    $C_3H_6O$    58.04187

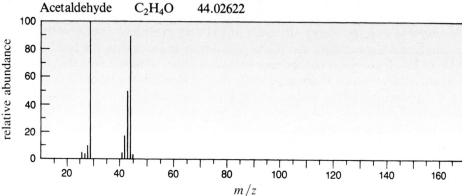

Acetaldehyde    $C_2H_4O$    44.02622

**Figure 12.25**

**Mass spectra of acetone and acetaldehyde.**

Adapted from *Registry of Mass Spectral Data*, Vol. 1, by E. Stenhagen, S. Abrahamsson, and F. W. McLafferty. Copyright © 1974 by John Wiley & Sons, Inc. Reprinted by permission of John Wiley & Sons, Inc.

The radical-cations are created in the gas phase with a large amount of excess energy. They are unstable species and fragment in a series of unimolecular reactions. An inspection of the mass spectrum of acetone indicates that the most abundant ion has m/z 43. The most abundant ion in the mass spectrum of a compound is called the **base peak.** The bar graphs are designed so that the base peak always has an intensity of 100%. The abundance of the other ions is shown relative to that of the base peak. Thus the molecular ion of acetone, m/z 58, has an abundance of approximately 50% of the ion with m/z 43.

The base peak in the spectrum of acetone corresponds to the loss of a fragment with a mass of 15 units (58 − 43 = 15). A methyl group, $CH_3$, has a mass of 15. Thus the molecular ion of acetone fragments primarily by the loss of a methyl group.

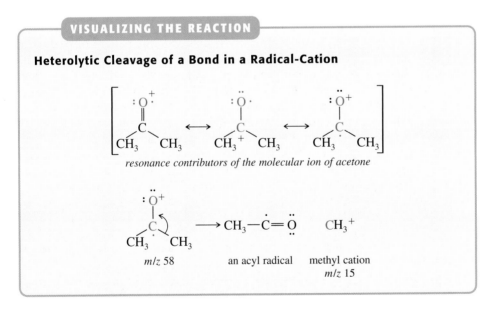

The reaction is shown as a homolytic cleavage (represented by fishhooks, p. 341) of one of the bonds between the carbonyl group and a methyl group. As a consequence, an acylium ion and a methyl radical are formed. The methyl radical is not a charged particle and thus does not appear in the mass spectrum. There is, however, a small peak at *m/z* 15, which corresponds to the formation of a small amount of methyl cations by an alternative cleavage.

The relative abundances of the acylium cation (*m/z* 43, 100%) and methyl cation (*m/z* 15, 15%) reflect the relative stabilities of the two species. In any case, it is important that the reactions are taking place in the gas phase in species that have much excess energy. Note, too, that in any fragmentation the numbers of unpaired electrons and charges present in the original species must be preserved.

Besides the molecular ion, *m/z* 44, the mass spectrum of acetaldehyde shows important peaks at *m/z* 43 and *m/z* 29 (base peak). These peaks may be assigned to the ions that form on the loss of a hydrogen atom (44 − 43 = 1) and a methyl radical (44 − 29 = 15) from the molecular ion.

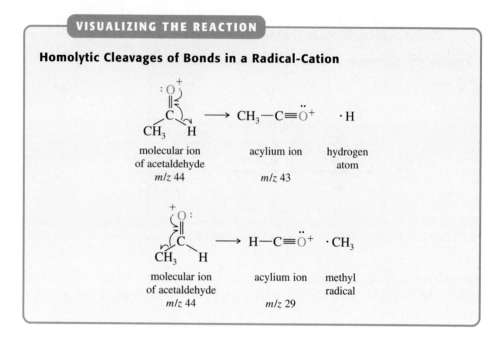

**VISUALIZING THE REACTION**

**Homolytic Cleavages of Bonds in a Radical-Cation**

The mass spectrum is a record of the cationic species that are formed when a beam of electrons is used to create radical-cations called molecular ions, which then fragment into smaller ions. The masses of these ions may be used in making judgments about the structure of the compound that gives rise to them. The mass spectrum of a compound of any complexity is unique in the pattern of peaks and the relative intensities of those peaks. As such, it can serve as a fingerprint for the compound. Because a very small sample of a compound is sufficient to obtain a mass spectrum, mass spectrometry is a powerful tool for identifying minute quantities of compounds. Mass spectrometers are often connected to gas-phase chromatographs, which separate mixtures efficiently, and computers, which automatically identify the components of mixtures by comparing mass spectral information with stored patterns for known compounds. Such techniques make it possible to obtain information about compounds present at trace levels in complex mixtures, such as physiological fluids or discharges from industrial plants.

**Problem 12.10**    The mass spectrum in Figure 12.26 is that of an aldehyde. Assign a structure to the aldehyde, and write equations showing the formation of the molecular ion and the ion giving rise to the base peak.

*Figure 12.26*

Adapted from *Registry of Mass Spectral Data,* Vol. 1, by E. Stenhagen, S. Abrahamsson, and F. W. McLafferty. Copyright © 1974 by John Wiley & Sons, Inc. Reprinted by permission of John Wiley & Sons, Inc.

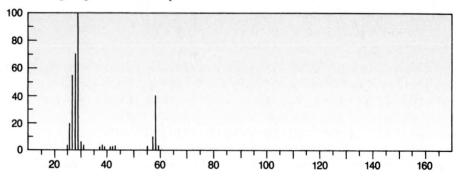

**Problem 12.11** The mass spectrum of a carboxylic acid is shown in Figure 12.27. Assign a structure to the acid, and write equations showing the formation of any ions having an abundance greater than 40%.

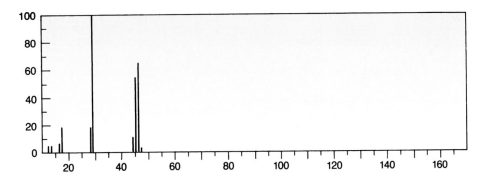

*Figure 12.27*

Adapted from *Registry of Mass Spectral Data,* Vol. 1, by E. Stenhagen, S. Abrahamsson, and F. W. McLafferty. Copyright © 1974 by John Wiley & Sons, Inc. Reprinted by permission of John Wiley & Sons, Inc.

**Problem 12.12** Two mass spectra are shown in Figure 12.28. One is that of 2-pentanone; the other is of 3-pentanone. Which is which? Write equations showing how you made your decision.

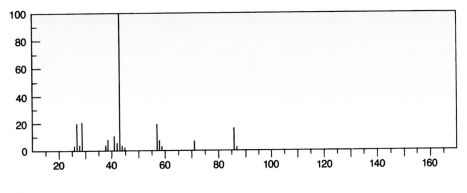

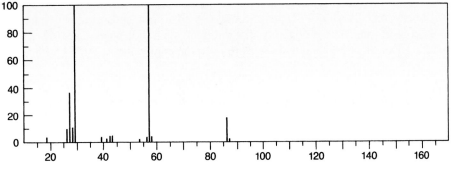

*Figure 12.28*

Adapted from *Registry of Mass Spectral Data,* Vol. 1, by E. Stenhagen, S. Abrahamsson, and F. W. McLafferty. Copyright © 1974 by John Wiley & Sons, Inc. Reprinted by permission of John Wiley & Sons, Inc.

**Problem 12.13** The mass spectrum of 2,2-dimethylpropanal is given in Figure 12.29 (p. 472). Assign a structure to the base peak in the spectrum, and write an equation for the formation of that species from the molecular ion. What would be the comparable species from the fragmentation of the molecular ion of acetaldehyde? Is such a species evident in the mass spectrum of acetaldehyde (Figure 12.25, p. 468)? How do you explain the difference in the reactions of the molecular ions for these two compounds?

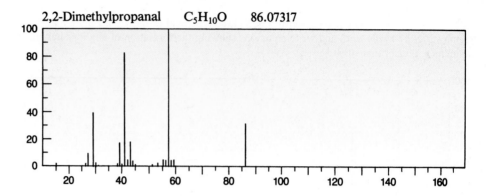

**Figure 12.29**

Adapted from *Registry of Mass Spectral Data*, Vol. 1, by E. Stenhagen, S. Abrahamsson, and F. W. McLafferty. Copyright © 1974 by John Wiley & Sons, Inc. Reprinted by permission of John Wiley & Sons, Inc.

## B. The Molecular Ion

In the mass spectra of acetone and acetaldehyde (Figure 12.25, p. 468), there are small peaks next to the peaks assigned to the molecular ions, at $m/z$ 59 and $m/z$ 45. These peaks arise from the presence of isotopes of carbon, hydrogen, and oxygen in these compounds. The common isotopes of elements that are of importance in organic chemistry, their atomic weights rounded off to the nearest mass unit, and their abundance relative to that of the most abundant isotope of that element (shown as 100%) are given in Table 12.2. Thus any compound that contains a number of carbon, hydrogen, and oxygen atoms will have some peaks in its mass spectrum at mass values one or two (or more) units above the mass calculated from its molecular formula using the atomic weights of the most common isotopes. For example, a compound containing ten carbon atoms is expected to have an ion having an intensity of about 11% (10 × 1.08%) of that of the peak for its molecular ion (M) at a mass of (M + 1). This peak would correspond to all the species in which one of the carbon atoms in the molecule is $^{13}C$ instead of $^{12}C$. The intensity of this peak, relative to the intensity of the molecular ion, would increase with increasing numbers of hydrogen and nitrogen atoms. Large numbers of oxygen atoms would increase the intensity of the (M + 2) peak, arising from molecular species in which $^{18}O$ instead of $^{16}O$ is found. The (M + 2) peak would also arise from species in which both $^{13}C$ and $^2H$ are found. The probability that there are such molecules increases with the total number of carbon and hydrogen atoms in the compound.

**Table 12.2    Natural Isotopic Abundance of Some Elements**

| | | Most Common Isotopes | | |
| Element | Mass | % | Mass | % |
| --- | --- | --- | --- | --- |
| H | 1 | 100 | 2 | 0.016 |
| C | 12 | 100 | 13 | 1.08 |
| N | 14 | 100 | 15 | 0.36 |
| O | 16 | 100 | 18 | 0.20 |
| F | 19 | 100 | — | — |
| Cl | 35 | 100 | 37 | 32.5 |
| Br | 79 | 100 | 81 | 98.0 |
| I | 127 | 100 | — | — |

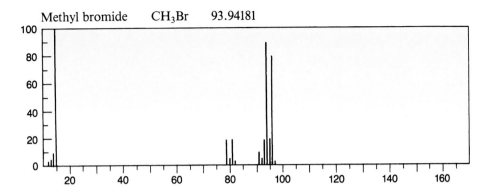

*Figure 12.30*

**Mass spectrum of methyl bromide.**

Adapted from *Registry of Mass Spectral Data*, Vol. 1, by E. Stenhagen, S. Abrahamsson, and F. W. McLafferty. Copyright © 1974 by John Wiley & Sons, Inc. Reprinted by permission of John Wiley & Sons, Inc.

The most distinctive (M + 2) peaks in mass spectra arise when the halogens bromine or chlorine are present in a molecule. Bromine has two isotopes, $^{79}$Br and $^{81}$Br, of nearly equal abundance. Thus, in a sample of a compound containing a bromine atom, half the molecules contain $^{79}$Br and half $^{81}$Br. Every species containing bromine will appear as a pair of peaks separated by two mass units. The spectrum of methyl bromide is shown in Figure 12.30. The molecular ion for $CH_3{}^{79}$Br appears at *m/z* 94 and that for $CH_3{}^{81}$Br at *m/z* 96. These two peaks have roughly the same intensity, reflecting the almost equal abundance of the two isotopes of bromine in nature. Peaks corresponding to $^{79}$Br$^+$ and $^{81}$Br$^+$ are seen at *m/z* 79 and *m/z* 81. The base peak in the spectrum is at *m/z* 15, corresponding to the methyl cation.

### VISUALIZING THE REACTION

**Isotopic Ions from Methyl Bromide**

$$CH_3 \underset{}{\overset{\cdot\cdot}{79}Br} \cdot^+ \longrightarrow \cdot CH_3 \qquad ^{79}Br :^+$$

*m/z* 94         *m/z* 79

$$CH_3 - {}^{79}Br : \xrightarrow{-e^-} CH_3 - {}^{79}Br \cdot^+$$

*m/z* 94

$$CH_3 \underset{}{\overset{\cdot\cdot}{81}Br} \cdot^+ \longrightarrow \cdot CH_3 \qquad ^{81}Br :^+$$

*m/z* 96         *m/z* 81

$$CH_3 - {}^{81}Br : \xrightarrow{-e^-} CH_3 - {}^{81}Br \cdot^+$$

*m/z* 96

$$CH_3 \overset{\cdot\cdot}{Br} \cdot^+ \longrightarrow CH_3{}^+ \qquad : Br \cdot$$

*m/z* 94 or *m/z* 96    *m/z* 15   bromine atom

*not charged, does not appear in the mass spectrum*

It is important to understand that the mass spectrum shows ions having different isotopic composition as individual peaks. The mass spectrum of a compound contains peaks corresponding to individual species containing all possible combinations of isotopes. The mass of a species as determined by mass spectrometry is *not* the same as the molecular weight calculated by using the average atomic weights found in the periodic table.

Compounds containing a chlorine atom show an (M + 2) peak having an intensity about a third of that of the peak for the molecular ion. The spectrum of methyl

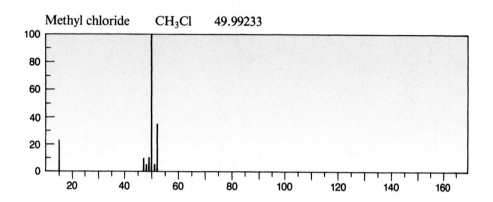

**Figure 12.31**

**Mass spectrum of methyl chloride.**

Adapted from *Registry of Mass Spectral Data,* Vol. 1, by E. Stenhagen, S. Abrahamsson, and F. W. McLafferty. Copyright © 1974 by John Wiley & Sons, Inc. Reprinted by permission of John Wiley & Sons, Inc.

chloride shows this feature (Figure 12.31). The molecular ion of methyl chloride with the composition $CH_3{}^{35}Cl$ has its peak at *m/z* 50. A peak having an intensity of about a third of the one at *m/z* 50 appears at *m/z* 52 and corresponds to the molecular ion for the composition $CH_3{}^{37}Cl$. The chief fragmentation seen for these molecular ions is the loss of a chlorine atom giving rise to a methyl cation, *m/z* 15.

**Problem 12.14**    Write equations showing the formation of the molecular ions for methyl chloride and their fragmentation.

Many of the spectra used as examples in this chapter give the molecular weight of the compound to five decimal places. This is the accuracy with which masses can be determined in modern high-resolution mass spectrometers. At that accuracy, the mass of the molecular ion can be used to determine the molecular formula of the compound. For example, acetone, $C_3H_6O$ ($M^{\overset{\cdot}{+}}$ 58.04187), can be distinguished

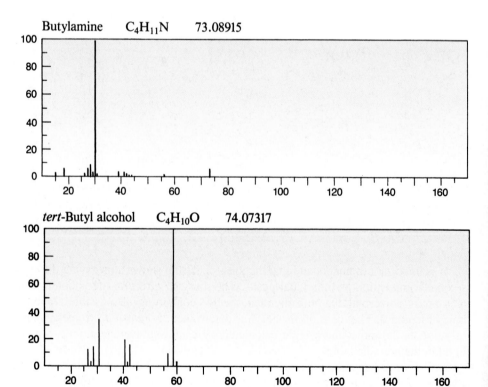

**Figure 12.32**

**Mass spectra of butylamine and *tert*-butyl alcohol.**

Adapted from *Registry of Mass Spectral Data,* Vol. 1, by E. Stenhagen, S. Abrahamsson, and F. W. McLafferty. Copyright © 1974 by John Wiley & Sons, Inc. Reprinted by permission of John Wiley & Sons, Inc.

from butane, $C_4H_{10}$ ($M^{\overset{+}{\cdot}}$ 58.07825), even though both have a mass of 58 amu, rounded off to the nearest mass unit. In fact, the elemental composition of each ion in a mass spectrum is also determined in the same way.

Not all compounds give molecular ions that are easily detectable in a mass spectrometer. The molecular ions formed by some compounds are so unstable that they fragment before they reach the detector. Alcohols and amines are among the compounds with very weak molecular ion peaks. The molecular ion from butyl-amine, for example, is so unstable that only a very small number of these ions survive to be recorded by the detector; careful inspection of the spectrum (Figure 12.32) is necessary to see the peak ($m/z$ 73). The molecular ion for *tert*-butyl alcohol does not appear at all on its spectrum (Figure 12.32).

⊙ **Study Guide**
Concept Map 12.3

**Problem 12.15**    The mass spectrum of a haloalkane is given in Figure 12.33. Assign a structure to the compound.

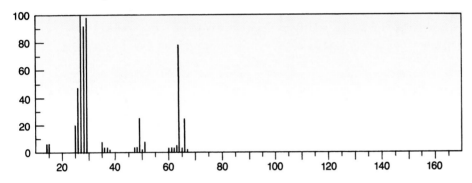

*Figure 12.33*

Adapted from *Registry of Mass Spectral Data*, Vol. 1, by E. Stenhagen, S. Abrahamsson, and F. W. McLafferty. Copyright © 1974 by John Wiley & Sons, Inc. Reprinted by permission of John Wiley & Sons, Inc.

**Problem 12.16**    The compound for which a mass spectrum appears in Figure 12.34 contains only one carbon atom and two types of halogen. After consulting Table 12.2 (p. 472), assign a structure to the compound, and account for the peaks that appear in the spectrum.

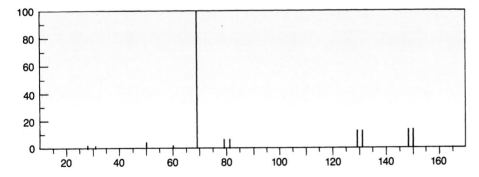

*Figure 12.34*

Adapted from *Registry of Mass Spectral Data*, Vol. 1, by E. Stenhagen, S. Abrahamsson, and F. W. McLafferty. Copyright © 1974 by John Wiley & Sons, Inc. Reprinted by permission of John Wiley & Sons, Inc.

## C. Important Fragmentation Pathways

Molecular ions may fragment unimolecularly by either heterolytic or homolytic cleavage of bonds adjacent to the site of the radical-cation formed by the loss of an electron. In the preceding two subsections, examples were shown of how the loss of a nonbonding electron from an oxygen atom or a halogen atom creates a radical-cation. The molecular ion then loses different radical species and is converted to the fragment ions. Other types of reactions are also possible, but this section will concentrate on the simpler fragmentation pathways.

The most abundant ions recorded in a mass spectrum are those formed either because the ion itself is very stable (such as the acylium ion, $m/z$ 43, in the mass spectrum of acetone, p. 469) or because the uncharged fragment that is lost is stable. For example, the methyl cation ($m/z$ 15) in the spectrum of methyl bromide (p. 473), though not very stable itself, is formed by the loss of a stable bromine atom from the molecular ion.

The fragmentation of an amine or an alcohol is determined by the groups adjacent to the nitrogen or the oxygen atom. For example, in the mass spectra of butylamine and *tert*-butyl alcohol (Figure 12.32, p. 474), the base peak in each case can be rationalized as arising from the homolytic cleavage of a bond one removed ($\alpha$-cleavage) from the radical-cation centered on the oxygen or the nitrogen atom.

**VISUALIZING THE REACTION**

**$\alpha$-Cleavage Reaction of Radical-Cations**

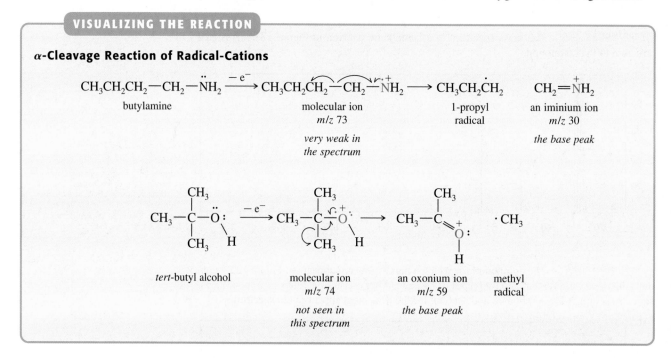

Stable cations, oxonium or iminium ions, are formed in these fragmentations. The molecular ion peak is usually weak or nonexistent in the spectra of alcohols and amines because of the ease with which the fragmentations occur.

**Problem 12.17**     The mass spectra of three isomeric amines, *sec*-butylamine, isobutylamine, and *tert*-butylamine, are given in Figure 12.35. Decide which spectrum belongs to which amine, and write equations justifying your conclusions.

*Figure 12.35*

Adapted from *Registry of Mass Spectral Data,* Vol. 1, by E. Stenhagen, S. Abrahamsson, and F. W. McLafferty. Copyright © 1974 by John Wiley & Sons, Inc. Reprinted by permission of John Wiley & Sons, Inc.

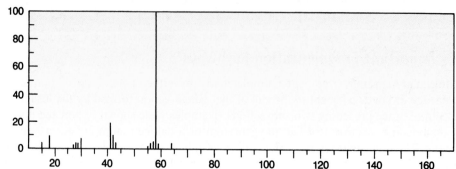

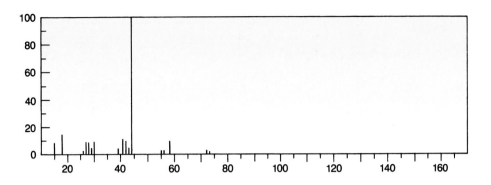

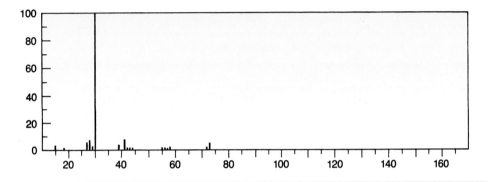

**Figure 12.35** **(Continued)**

Adapted from *Registry of Mass Spectral Data*, Vol. 1, by E. Stenhagen, S. Abrahamsson, and F. W. McLafferty. Copyright © 1974 by John Wiley & Sons, Inc. Reprinted by permission of John Wiley & Sons, Inc.

The cleavage of bonds adjacent to carbonyl groups and the loss of halogens as atoms or ions have already been discussed with examples on pages 469 and 473.

The mass spectra of alkanes and alkenes are complicated because rearrangements of intermediate cations and radicals occur when the molecule is of any size. The molecular ion of alkenes arises from the loss of an electron from the $\pi$ bond. Then an allylic bond in the molecular ion often cleaves to give an alkyl radical and an allylic cation. The ion arising from such processes is seen in the spectrum of 2-methyl-2-pentene (Figure 12.36). The formation of the molecular ion and the cleavage of the allylic bond in 2-methyl-2-pentene are shown on page 478.

**2-Methyl-2-pentene**   $C_6H_{12}$   84.09390

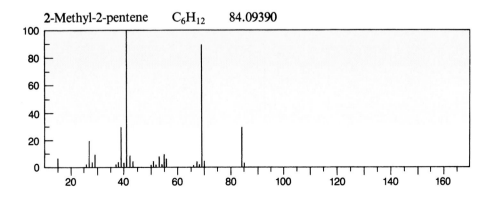

**Figure 12.36**

Mass spectrum of 2-methyl-2-pentene.

Adapted from *Registry of Mass Spectral Data*, Vol. 1, by E. Stenhagen, S. Abrahamsson, and F. W. McLafferty. Copyright © 1974 by John Wiley & Sons, Inc. Reprinted by permission of John Wiley & Sons, Inc.

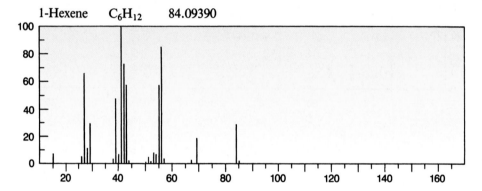

**VISUALIZING THE REACTION**

**α-Cleavage Reaction of the Radical-Cation from an Alkene**

2-methyl-2-pentene — molecular ion *m/z* 84 — an allylic cation *m/z* 69 — methyl radical

---

**Problem 12.18**    The mass spectrum of 1-hexene is shown in Figure 12.37. Write equations showing how the ion giving rise to the base peak is formed.

1-Hexene    $C_6H_{12}$    84.09390

*Figure 12.37*

Adapted from *Registry of Mass Spectral Data*, Vol. 1, by E. Stenhagen, S. Abrahamsson, and F. W. McLafferty. Copyright © 1974 by John Wiley & Sons, Inc. Reprinted by permission of John Wiley & Sons, Inc.

Another category of fragmentation reaction that is important is cleavage at benzylic bonds. The mass spectrum of toluene illustrates such fragmentation (Figure 12.38). The mass spectrum of toluene shows the peak for the molecular ion at *m/z* 92 and the base peak at *m/z* 91, which arises from the loss of a hydrogen atom from the molecular ion.

**VISUALIZING THE REACTION**

**α-Cleavage Reaction of the Radical-Cation from an Arene**

toluene — molecular ion *m/z* 92 — cycloheptatrienyl cation tropylium ion $C_7H_7^+$ *m/z* 91

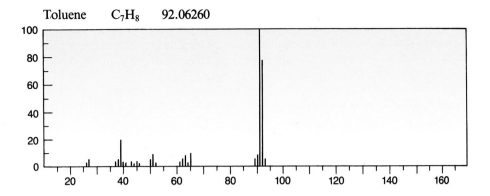

Figure 12.38

**Mass spectrum of toluene.**

Adapted from *Registry of Mass Spectral Data*, Vol. 1, by E. Stenhagen, S. Abrahamsson, and F. W. McLafferty. Copyright © 1974 by John Wiley & Sons, Inc. Reprinted by permission of John Wiley & Sons, Inc.

The cation formed on the loss of the hydrogen atom is believed to have the stable ring-expanded structure of the cycloheptatrienyl cation (p. 359) rather than that of a typical benzyl cation.

**Problem 12.19** Mass spectra for the isomeric compounds 2,2-dimethyl-1-phenylpropane and 2-methyl-3-phenylbutane are shown in Figure 12.39. Decide which spectrum belongs to which compound, and write equations showing the formation of the ions giving rise to the base peaks.

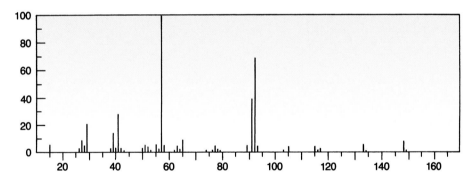

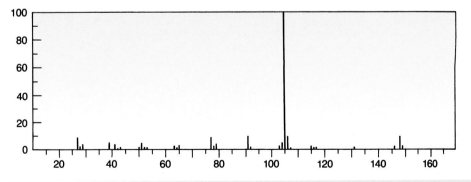

Figure 12.39

Adapted from *Registry of Mass Spectral Data*, Vol. 1, by E. Stenhagen, S. Abrahamsson, and F. W. McLafferty. Copyright © 1974 by John Wiley & Sons, Inc. Reprinted by permission of John Wiley & Sons, Inc.

## D. Rearrangements of Molecular Ions

One of the most common and most important rearrangements observed in the mass spectrometer is the transfer of a hydrogen atom from one part of a chain to a radical site at another part of the chain, usually via a six-membered cyclic transition state. The reaction is particularly easy to see with carbonyl compounds but is also one of the mechanisms by which rearrangements of alkenes occur. Such rearrangements are postulated to account for radical-cations that give rise to peaks in the mass spectra of 2-methyl-1-pentene and 2-hexanone (Figure 12.40).

The base peak in the spectrum of 2-methyl-1-pentene is at $m/z$ 56 and results from the loss of a fragment having a mass of 28 units ($84 - 56 = 28$). Such a fragment could be ethylene, $C_2H_4$.

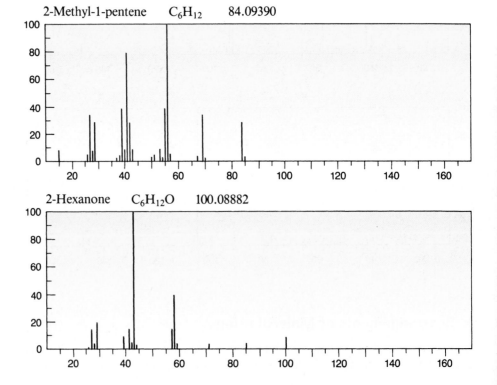

**VISUALIZING THE REACTION**

**Rearrangement of a Radical-Cation from an Alkene**

**Figure 12.40**

Mass spectra of 2-methyl-1-pentene and 2-hexanone.

Adapted from *Registry of Mass Spectral Data*, Vol. 1, by E. Stenhagen, S. Abrahamsson, and F. W. McLafferty. Copyright © 1974 by John Wiley & Sons, Inc. Reprinted by permission of John Wiley & Sons, Inc.

A hydrogen atom is transferred from one carbon atom to another carbon atom, a radical site at another part of the chain, by means of a six-membered ring transition state. The new radical cation fragments to give a stable alkene and another radical cation of lower mass.

The base peak in the spectrum of 2-hexanone, *m/z* 43, arises from cleavage of a bond next to the carbonyl group. A significant peak appearing at *m/z* 58 arises from a migration of a hydrogen atom from the γ-carbon atom to the carbonyl group. The radical cation with *m/z* 58 may be seen as being a tautomer of the radical cation of acetone and will undergo further fragmentations typical of acetone. This type of re-arrangement is seen for a wide variety of compounds containing carbon–oxygen double bonds and is known as the **McLafferty rearrangement** for Fred W. McLaf-ferty of Cornell University, who first discovered how general it is.

**VISUALIZING THE REACTION**

**The McLafferty Rearrangement**

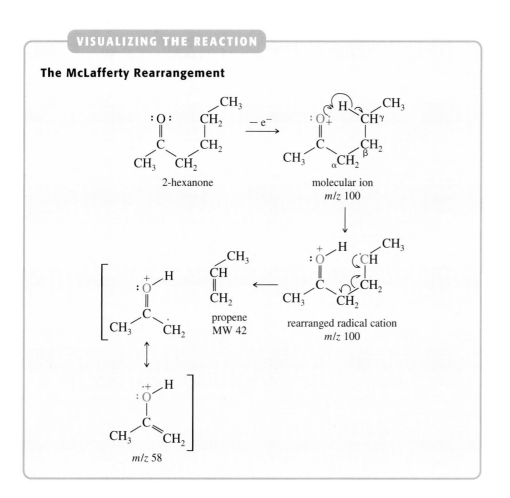

**Problem 12.20**   The mass spectrum of 3-methyl-2-pentanone is shown in Figure 12.41 (p. 482). Write equations accounting for formation of the ions with *m/z* 29, 43, 56, and 72. (*Hint:* A review of the equations in Section 12.3A may be helpful.)

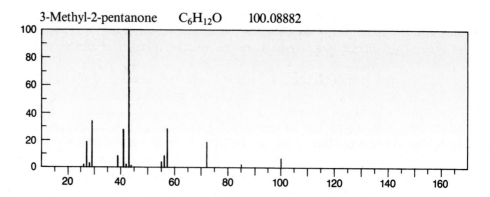

*Figure 12.41*

Adapted from *Registry of Mass Spectral Data,* Vol. 1, by E. Stenhagen, S. Abrahamsson, and F. W. McLafferty. Copyright © 1974 by John Wiley & Sons, Inc. Reprinted by permission of John Wiley & Sons, Inc.

**Problem 12.21**     Two of the important peaks in the mass spectrum of 1-hexene (Figure 12.37, p. 478), at *m/z* 56 and 42, can be explained by writing mechanisms of the kind shown on p. 480 for 2-methyl-1-pentene. See whether you can use rearrangement and fragmentation reactions to arrive at structures for these radical-cations. (*Hint:* Either of the carbon atoms originally involved in the double bond may have radical character or cation character.)

# SUMMARY

A molecule absorbs electromagnetic radiation that corresponds to transitions between fixed energy levels for that species. A spectrum is a record of the change in the absorption of energy by the compound plotted against the wavelength or the frequency of the radiation used.

Ultraviolet spectroscopy is used to detect and identify compounds having conjugated systems. The distinctive grouping of conjugated bonds in a molecule that absorbs ultraviolet or visible radiation is known as a chromophore. Each chromophore absorbs radiation at a distinctive wavelength and intensity. The wavelength at which a chromophore absorbs energy is related to the energy gap between the bonding and antibonding molecular orbitals for the conjugated system. An increase in conjugation or of alkyl substitution on the multiple bonds brings the bonding and antibonding molecular orbitals closer in energy, which decreases the energy necessary to promote an electron from a bonding to an antibonding orbital. This, in turn, increases $\lambda_{max}$ and, usually, the intensity (given as $\epsilon$, the molar absorptivity) of the absorption band.

Transitions occur between vibrational energy levels in a molecule on absorption of infrared radiation, usually at a frequency from 4000 to 600 $cm^{-1}$. Chemical bonds undergo various stretching, bending, and twisting vibrations. Vibrations that give rise to a change in the dipole moment result in absorption of radiation.

An infrared spectrum is generated by the molecule as a whole, but certain absorption bands are typical of particular groups of atoms, such as doubly bonded carbons, a carbonyl group, or a hydroxyl group. Typical group frequencies are compiled in tables (Table 12.1, p. 456, for example) that can be consulted in the interpretation of spectra. The group frequency is sensitive to the environment around the functional group. For example, the absorption band for a carbonyl group is at a different frequency for a ketone, an ester, and a conjugated carbonyl compound. Such variations in absorption allow chemists to use infrared spectra to assign structures to organic compounds. The frequencies that are useful in detecting the presence of functional groups are between 4000 and 1400 $cm^{-1}$. The portion of an infrared spectrum from 1400 to 600 $cm^{-1}$ is called the fingerprint region. Here it is not possible to assign all the bands, but the number of bands and their relative intensities are typical of the compound and can be used to identify it.

A mass spectrum is a record of the relative abundances of a series of ions of different masses that form when a sample of a compound is bombarded by high-energy electrons in a mass spectrometer. This bombardment causes an electron to be lost from a molecule of the compound, resulting in the formation of a radical-cation known as the molecular ion. The molecular ion fragments by the loss of radicals or neutral molecules to give positively charged ions. The mass spectrometer separates these cations according to their mass-to-charge ratios, $m/z$, and records their masses and their relative abundances. The most abundant ion in the spectrum is recorded as the base peak with an abundance of 100%.

The masses recorded by a mass spectrometer are exact masses; therefore, molecules of different isotopic composition give different molecular ions. For example, methyl bromide has two molecular ions of about equal abundance, one arising from $CH_3{}^{79}Br$ and the other from $CH_3{}^{81}Br$.

The fragmentation patterns seen for different types of compounds are typical of their structures. Fragmentations occur so as to produce stable cations or by the loss of stable neutral fragments. Rearrangements also occur, such as McLafferty rearrangement, in which a hydrogen atom from the $\gamma$-carbon atom is transferred to the oxygen atom in a radical-cation derived from a carbonyl compound.

Mass spectra can be obtained from very small samples. Computer techniques that allow comparison of fragmentation patterns with those of known compounds make mass spectrometry a powerful tool for identifying organic compounds.

## ADDITIONAL PROBLEMS

**12.22** The infrared spectrum of Compound L is given in Figure 12.42. Compound L has five bands in its $^{13}C$ nuclear magnetic resonance spectrum at $\delta$ 14.05, 22.56, 28.02, 32.47, and 62.76. Its proton magnetic resonance spectrum has the following bands: $\delta$ 0.90 (3H, t), 1.35 (4H, m), 1.55 (2H, m), 2.90 (1H, s), and 3.60 (2H, m). (Chemists use this shorthand way of describing nuclear magnetic resonance spectra. For each chemical shift value, the number of hydrogen atoms, the multiplicity of the band, and sometimes, the coupling constant, $J$, are given in parentheses; s = singlet, d = doublet, m = multiplet, q = quartet, and t = triplet.) Assign a structure to Compound L, showing how each piece of spectral data supports your conclusions.

Compound L

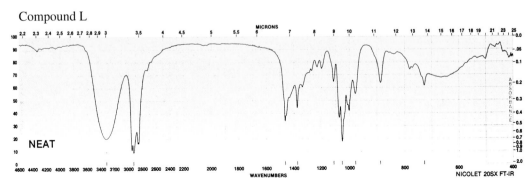

*Figure 12.42*

From *The Aldrich Library of FT-IR Spectra.*

**12.23** Compound M has the infrared spectrum shown in Figure 12.43. Its proton magnetic resonance spectrum has peaks at $\delta$ 0.95 (6H, d), 2.20 (1H, m), 2.30 (2H, m), and 9.75 (1H, t). Peaks in its $^{13}C$ nuclear magnetic resonance spectrum appear at $\delta$ 22.57, 23.50, 52.59, and 202.65. Assign a structure to Compound M that is supported by all the spectral data.

Compound M

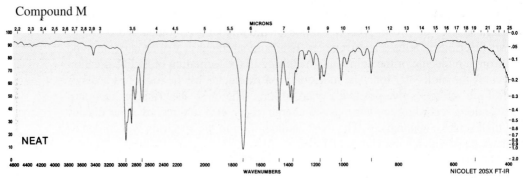

*Figure 12.43*

From *The Aldrich Library of FT-IR Spectra.*

**12.24** Three samples, labeled Compounds N, O, and P, are known to be unsaturated carboxylic acids. Ultraviolet spectra were taken of the three samples with the following results: Compound N, $\lambda_{max}^{hexane}$ 302 nm ($\epsilon$ 36,500); Compound O, 208 nm ($\epsilon$ 12,000); Compound P, 261 nm ($\epsilon$ 25,000). Which structural formula given below belongs to Compounds N, O, and P?

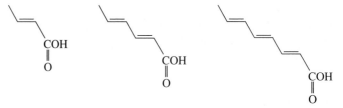

**12.25** A compound with the molecular formula $C_4H_6O$ has $\lambda_{max}^{ethanol}$ 219 nm ($\epsilon$ 16,600) and 318 nm ($\epsilon$ 30).

(a) How many units of unsaturation does the compound have?
(b) Draw some structures compatible with its molecular formula and its ultraviolet spectrum.
(c) What are some structures that would not be compatible with the ultraviolet spectrum, even though they fit the molecular formula?

**12.26** (+)-17-Methyltestosterone is a steroid related to the male sex hormones. Its infrared spectrum and structural formula appear in Figure 12.44. Assign as many of the absorption bands as you can to specific vibrational transitions in the molecule.

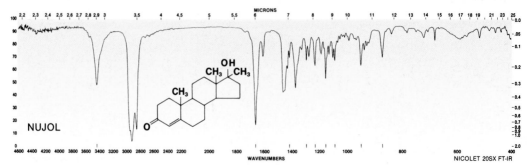

*Figure 12.44*

From *The Aldrich Library of FT-IR Spectra.*

**12.27** Compound Q, $C_7H_7BrO$, has $\lambda_{max}^{methanol}$ 227 ($\epsilon$ 14,200), 281 ($\epsilon$ 1580), and 288 nm ($\epsilon$ 1280). The infrared spectrum of Compound Q has no absorption bands of any intensity between 4000 and 3200 $cm^{-1}$ and between 2000 and 1600 $cm^{-1}$. Assign a structure to Compound Q that fits these data.

**12.28** Compound R contains only carbon, hydrogen, and a halogen. The mass spectrum is shown in Figure 12.45. Assign a structure to the compound.

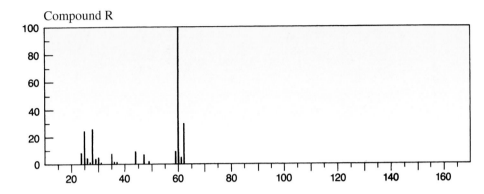

*Figure 12.45*
Adapted from *Registry of Mass Spectral Data*, Vol. 1, by E. Stenhagen, S. Abrahamsson, and F. W. McLafferty. Copyright © 1974 by John Wiley & Sons, Inc. Reprinted by permission of John Wiley & Sons, Inc.

**12.29** Compound S has a strong band in its infrared spectrum at 1685 $cm^{-1}$. Its proton magnetic resonance spectrum is shown in Figure 12.46. Assign a structure to Compound S.

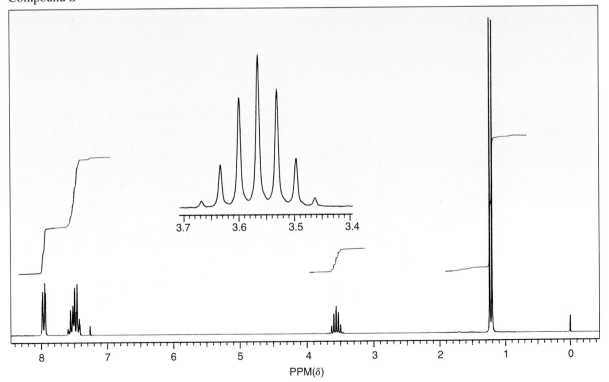

PPM($\delta$)

*Figure 12.46*

**12.30** The molecular ion for Compound T does *not* appear in its mass spectrum. The peaks for the important fragments from the molecular ion, however, are clearly seen in Figure 12.47. Assign a structure to Compound T.

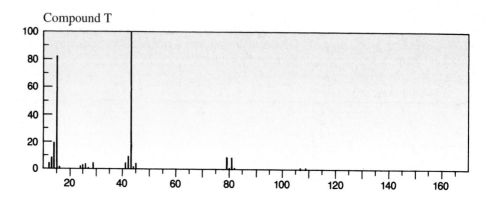

**Figure 12.47**

Adapted from *Registry of Mass Spectral Data,* Vol. 1, by E. Stenhagen, S. Abrahamsson, and F. W. McLafferty. Copyright © 1974 by John Wiley & Sons, Inc. Reprinted by permission of John Wiley & Sons, Inc.

**12.31** The infrared spectra in Figure 12.48 are of Compounds U–X, each of which has two major functional groups. Use the spectra to identify the functional groups that are present.

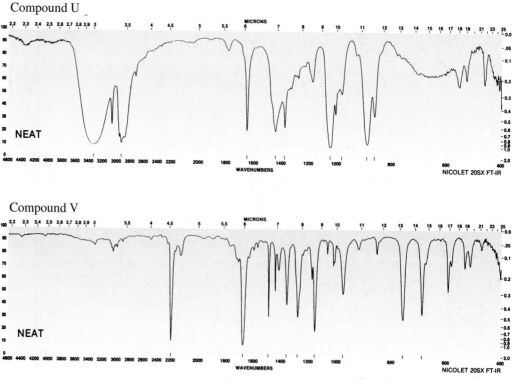

**Figure 12.48**

From *The Aldrich Library of FT-IR Spectra.*

Compound W

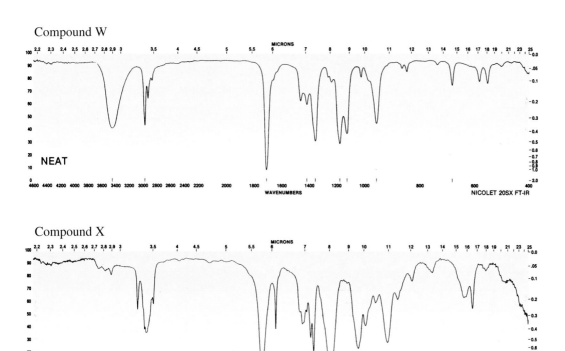

Compound X

From *The Aldrich Library of FT-IR Spectra*.

*Figure 12.48* **(Continued)**

**12.32** Two isomeric compounds have the structural formulas shown in the margin. One has $\lambda_{max}^{ethanol}$ 254 ($\epsilon$ 219), 259 ($\epsilon$ 239), 262 ($\epsilon$ 234), 265 ($\epsilon$ 186), and 268 ($\epsilon$ 173) nm. The other has $\lambda_{max}^{ethanol}$ 246 ($\epsilon$ 17780), 284 ($\epsilon$ 977), and 293 ($\epsilon$ 691) nm. Which ultraviolet spectrum corresponds to which structure?

**12.33** The base peak in the mass spectrum of pentanal, shown in Figure 12.49, is the result of a rearrangement. Write an equation that shows how the ion represented by the base peak comes into being.

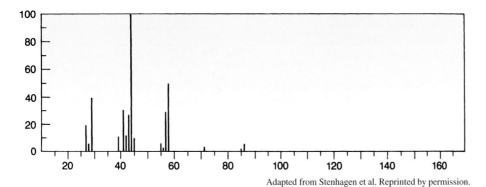

*Figure 12.49*

Adapted from Stenhagen et al. Reprinted by permission.

**12.34** Compound Y absorbs at 1703 and 2730 cm$^{-1}$ in the infrared. Its molecular ion has $m/z$ 120, and there are two base peaks, at $m/z$ 119 and 91, in its mass spectrum. Its $^{13}$C nuclear magnetic resonance spectrum has bands at 21.6 (q), 129.6 (d), 134.4 (s), 145.3 (s), and

191.4 (d) ppm. The proton magnetic resonance spectrum of Compound Y is shown in Figure 12.50. Assign a structure to the compound. Analyze all the spectral data, and write equations that account for the presence of the two base peaks in the mass spectrum.

Compound Y

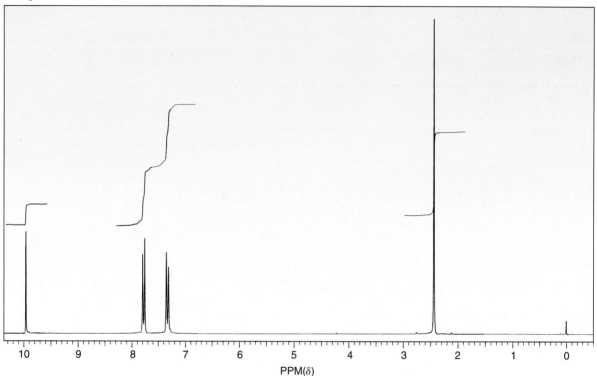

PPM($\delta$)

*Figure 12.50*

**12.35** When $C_{60}$ (p. 364) is treated with an excess of benzene in the presence of aluminum chloride, a compound is formed with a molecular weight of 1656 amu. The compound shows bands for aromatic C—H bonds in its infrared spectrum. The proton magnetic resonance spectrum has peaks at $\delta$ 7.4 and 4.5 in a ratio of 5:1.

(a) Use the molecular weight of the compound to determine the molecular formula of the compound that forms from $C_{60}$ and benzene.
(b) What do the nuclear magnetic resonance spectral data suggest about the structure of the product?
(c) Water in the presence of aluminum chloride can serve as a source of protons. Suggest a mechanism for the reaction of $C_{60}$ with benzene. You do not have to draw out the structure of $C_{60}$; simply use $C_{60}$ in equations.

# Alcohols, Diols, and Ethers

## A Look Ahead

The functional group in alcohols is the hydroxyl group.

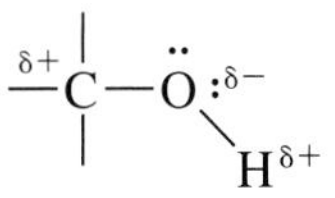

$$\overset{\displaystyle |}{\underset{\displaystyle |}{\overset{\delta+}{C}}}-\overset{\displaystyle ..}{O}\!:^{\delta-}$$
$$\diagdown \text{H}^{\delta+}$$

Alcohols resemble water in their acidity and basicity and in their ability to hydrogen-bond. Alcohols are easily converted by substitution reactions to alkyl halides when the hydroxyl group is first converted into a good leaving group. Primary alcohols are oxidized to aldehydes, and secondary alcohols to ketones. Deprotonation of alcohols gives alkoxide ions, which are both good nucleophiles and strong bases. Alkoxide ions react with primary and secondary alkyl halides to give ethers.

Ethers have two hydrocarbon groups bonded to oxygen.

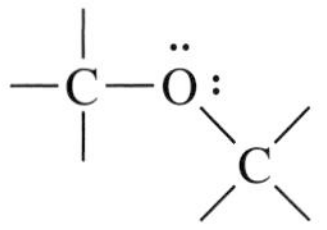

$$-\overset{\displaystyle |}{\underset{\displaystyle |}{C}}-\overset{..}{O}\!:$$

Ethers serve as good solvents. The carbon–oxygen bond can be broken by some strong acids, but ethers are otherwise not very reactive. Oxiranes, in which the oxygen atom is part of a small strained ring, are an exception. They are easily opened in acids and by nucleophiles.

$$:\!O\!:$$

In this chapter you will study the reactions of alcohols and ethers.

## 13.1  Structure and Nomenclature of Alcohols and Ethers

### A. Some Typical Alcohols and Ethers and Their Properties

**Alcohols** are compounds that contain a hydroxyl group attached to a tetrahedral carbon atom. The presence of this hydroxyl group greatly influences the physical and chemical properties of alcohols. The polarity of the oxygen–hydrogen bond leads to hydrogen bonding between alcohol molecules, which results in the high boiling points characteristic of alcohols. The water solubility of low-molecular-weight alcohols and of larger compounds with several hydroxyl groups also can be attributed to hydrogen bonding (p. 33). Their properties make alcohols good solvents for a variety of reactions (pp. 243–245).

The simplest alcohol is methanol, or methyl alcohol, a compound that is highly toxic and can cause blindness and death. Methanol can be generated by heating wood and is thus also called **wood alcohol.** Ethanol, or ethyl alcohol, is the familiar **grain alcohol.** Some alcohols have more than one hydroxyl group. One of these is 1,2-ethanediol, also known as **ethylene glycol,** which is used as automobile antifreeze. Other alcohols are large, complicated molecules of which the hydroxyl group is a small but important part. The major constituent of most gallstones, cholesterol, which also has been implicated as a possible risk factor in arterial and heart disease, is such an alcohol.

$$CH_3OH \qquad CH_3CH_2OH \qquad HOCH_2CH_2OH$$

| | | |
|---|---|---|
| methanol | ethanol | 1,2-ethanediol |
| methyl alcohol | ethyl alcohol | ethylene glycol |
| bp 64.7 °C | bp 78.5 °C | bp 197.2 °C |

cholesterol
mp 148.5 °C

**Problem 13.1**    The boiling point of 1,2-ethanediol is 197.2 °C. Why is it a good compound to use as an antifreeze?

**Problem 13.2**    Would you expect cholesterol to be particularly soluble in water? Explain your answer.

**Problem 13.3**    Cholesterol has more than one functional group. To what other class of compounds does it belong? Write equations using cholesterol to illustrate three reactions that are typical of that other functional group class.

Ethers act only as hydrogen-bond acceptors (p. 34) and therefore have much lower boiling points than do alcohols of comparable molecular weight. Because ethers can donate their nonbonding electrons, they complex with Lewis acids such as diborane (p. 292) and serve as good solvents for them. Some common ethers are diethyl ether, tetrahydrofuran, and diglyme. Diethyl ether was formerly used as an anesthetic. The name diglyme is a contraction of its more descriptive name diethylene glycol dimethyl ether.

$$CH_3CH_2OCH_2CH_3$$

| | |
|---|---|
| diethyl ether | tetrahydrofuran |
| bp 34.5 °C | bp 67 °C |

$$CH_3OCH_2CH_2OCH_2CH_2OCH_3$$

diethylene glycol dimethyl ether
diglyme
bp 162 °C

**Problem 13.4**    For each set of experimental facts presented on the next page, give an explanation for the observed trend, based on the relationships between structure and physical properties. Whenever possible, use pictures as well as words in your answers.

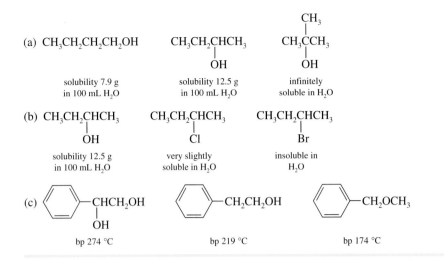

## B. Nomenclature of Alcohols

Simple alcohols can be named by using the name of the alkyl group they contain with the word *alcohol.* Methyl alcohol, ethyl alcohol, isopropyl alcohol, *tert*-butyl alcohol, and allyl alcohol are common names that are created this way. But this system becomes cumbersome when the alkyl groups are large and complicated or when other substituents are also present in the molecule.

Alcohols are named systematically according to the IUPAC rules by substituting the ending **ol** for the final **e** of the systematic name of the hydrocarbon portion of the molecule. The location of the hydroxyl group on the hydrocarbon chain is indicated by a number, usually placed at the beginning of the name. Below are some examples of IUPAC nomenclature for simple alcohols. Common names are also shown for some compounds.

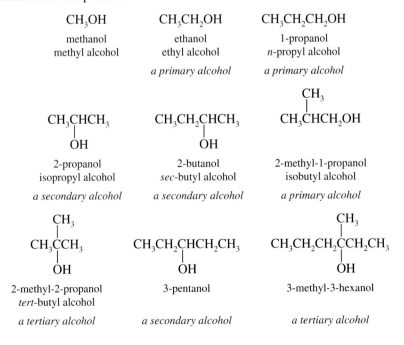

The reactivity of alcohols, like that of alkyl halides, varies according to whether the functional group is on a primary, secondary, or tertiary carbon atom (pp. 243 and 516). Therefore, alcohols are designated as primary, secondary, or tertiary alcohols, as shown on the preceding page.

The alcohol functional group is considered to be more important than double or triple bonds or halogen substituents and is consequently given the lower number in naming compounds with several functional groups.

$$\underset{\text{ClCH}_2\text{CH}_2\overset{\displaystyle \overset{\text{CH}_3}{|}}{\text{CH}}\text{CH}_2\text{OH}}{}$$

4-chloro-2-methyl-1-butanol

*a primary alcohol*

$$\text{CH}_3\overset{\displaystyle \overset{\text{CH}_3}{|}}{\text{CH}}\text{CHCH}_2\text{CH}_2\text{CH}_2\text{OH}$$
$$\underset{\text{OH}}{|}$$

4-methyl-1,5-hexanediol

*a primary and secondary alcohol*

HOCH₂C≡CCH₂OH

2-butyne-1,4-diol

*presence of a triple bond indicated by changing the name of the backbone from butane to butyne*

*a primary alcohol*

CH₂=CHCH₂OH

2-propen-1-ol
allyl alcohol

*presence of a double bond indicated by changing the name of the backbone from propane to propene*

*a primary alcohol, also an allylic alcohol*

*note location of numbers for the hydroxyl groups in these two names; necessary for clarity*

## C.  Nomenclature of Ethers

**Ethers** have two hydrocarbon groups bonded to an oxygen atom. There are two ways of naming ethers. For simple ethers, the name is obtained by naming each of the organic groups on the oxygen atom and adding the word *ether*, for example:

$$\text{CH}_3\overset{\displaystyle \overset{\text{CH}_3}{|}}{\text{CH}}\text{OCH}\overset{\displaystyle \overset{\text{CH}_3}{|}}{}\text{CH}_3$$

diisopropyl ether

CH₃OCH₂CH₂CH₂CH₃

butyl methyl ether

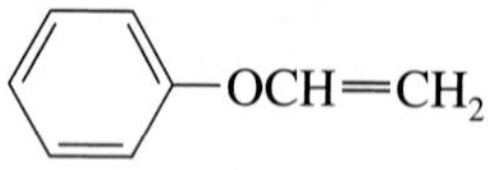

⟨ ⟩—OCH=CH₂

phenyl vinyl ether

⟨ ⟩—OCH₂CH₂CH₂CH₂CH₃

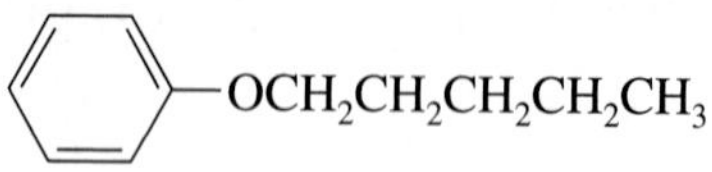

pentyl phenyl ether

For more complex molecules, the simplest organic group along with the oxygen atom is named as an **alkoxy** or **aryloxy group** and considered to be a substituent on a more complex chain. Important alkoxy or aryloxy groups are the following:

CH₃O—    CH₃CH₂O—    CH₃CH₂CH₂O—    ⟨ ⟩—O—    ⟨ ⟩—CH₂O—

methoxy        ethoxy            propoxy            phenoxy 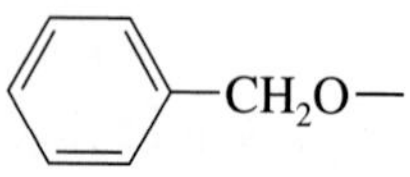        benzyloxy

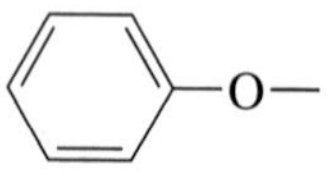

Some examples of this nomenclature are shown on the next page.

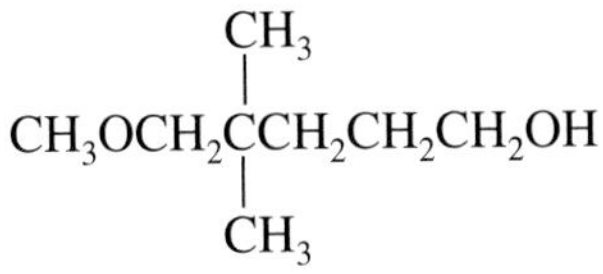

$$CH_3OCH_2\overset{\overset{\displaystyle CH_3}{|}}{\underset{\underset{\displaystyle CH_3}{|}}{C}}CH_2CH_2CH_2OH \qquad CH_3OCH_2CH_2OCH_2CH_2CH_3$$

5-methoxy-4,4-dimethyl-1-pentanol        1-methoxy-2-propoxyethane

$$CH_3CH_2O\!-\!\!\bigcirc\!\!-\!OCH_2CH_3 \qquad CH_3OCH_2\overset{\overset{\displaystyle OCH_3}{|}}{C}HCH_2OCH_3$$

1,4-diethoxybenzene 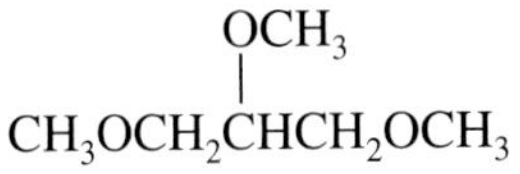        1,2,3-trimethoxypropane

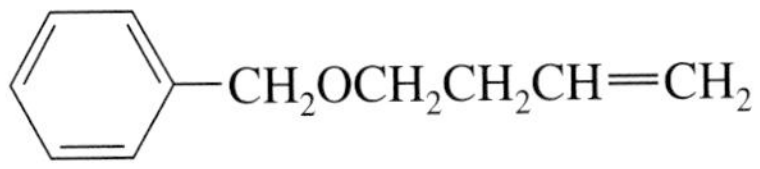

$$\bigcirc\!-\!CH_2OCH_2CH_2CH\!=\!CH_2 \qquad CH_3CH_2CH_2\overset{\overset{\displaystyle O\!-\!\bigcirc}{|}}{C}HCH_2CH_2CH_3$$

4-benzyloxy-1-butene 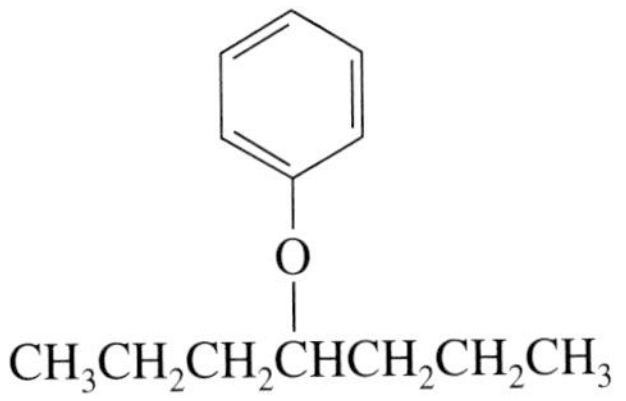        4-phenoxyheptane

**Problem 13.5** Name the following compounds, including stereochemistry if shown, according to IUPAC rules.

(a) 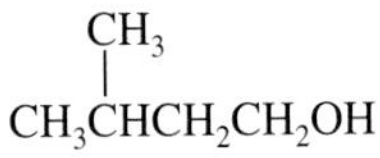 $CH_3\overset{\overset{\displaystyle CH_3}{|}}{C}HCH_2CH_2OH$

(b) 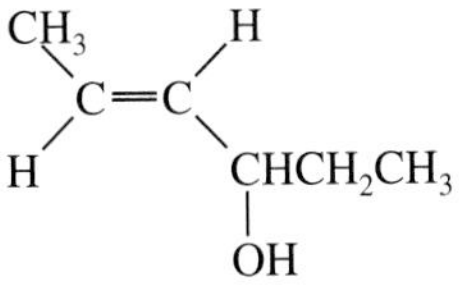 $\overset{\displaystyle CH_3}{\underset{\displaystyle H}{}}C\!=\!C\overset{\displaystyle H}{\underset{\underset{\underset{\displaystyle OH}{|}}{\displaystyle CHCH_2CH_3}}{}}$

(c) $CH_3CH_2C\!\equiv\!CCH_2OH$

(d)  $CH_3\overset{\overset{}{}}{\underset{\underset{\displaystyle Cl}{|}}{C}}H\overset{}{\underset{\underset{\displaystyle Cl}{|}}{C}}HCH_2CH_2CH_2OH$

(e) $CH_3CH_2OCH_2CH_2CH_2Cl$

(f) 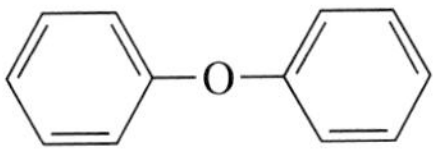 $\bigcirc\!-\!O\!-\!\bigcirc$

(g) 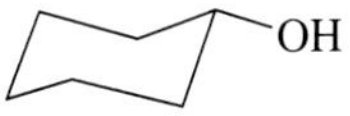 a cyclohexane ring with —OH

(h) 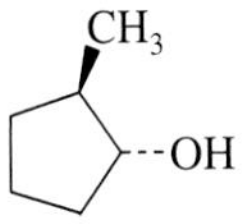 a cyclopentane ring with $CH_3$ and ---OH

(i) 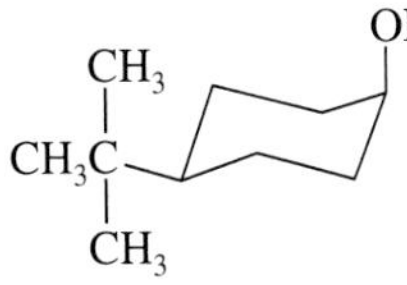 $CH_3\overset{\overset{\displaystyle CH_3}{|}}{\underset{\underset{\displaystyle CH_3}{|}}{C}}$ cyclohexane ring with OH

(j) 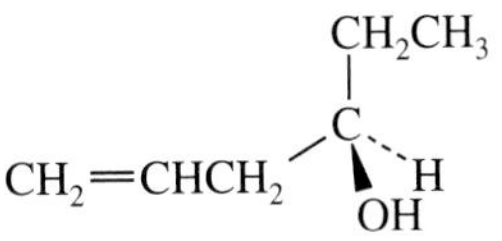 $CH_2\!=\!CHCH_2\overset{\overset{\displaystyle CH_2CH_3}{|}}{C}\overset{\displaystyle H}{\underset{\displaystyle OH}{}}$

(k) $CH_3OCH_2CH_2CH_2OCH_3$

(l) 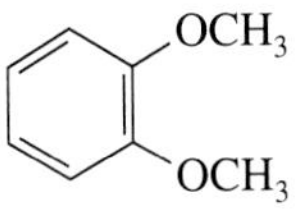 benzene ring with $OCH_3$ and $OCH_3$

## 13.2 Preparation of Alcohols

An alcohol is formed when water adds to the double bond in an alkene (p. 284). For example, propene is converted to isopropyl alcohol in dilute acid.

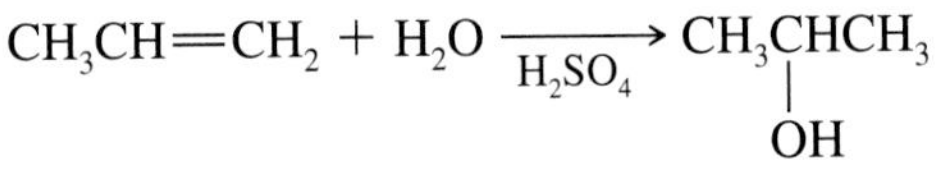

$$CH_3CH\!=\!CH_2 + H_2O \xrightarrow[H_2SO_4]{} CH_3\overset{}{\underset{\underset{\displaystyle OH}{|}}{C}}HCH_3$$

propene        isopropyl alcohol

The alcohol that results is from Markovnikov addition of water to the double bond. The reaction is initiated by the reaction of the electrophile, hydronium ion, with the double bond.

This reaction is the basis of an important industrial process for the production of simple alcohols but is not useful for the preparation of more complex alcohols. The acidic reagent can attack other functional groups in the molecule and cause unwanted side reactions. Alkenes also isomerize in acids via carbocation intermediates (p. 289). Attempts to add water to such alkenes with aqueous acid give rise to alcohols with rearranged carbon skeletons. For example, the Markovnikov addition product is not obtained from the reaction of 3-methyl-1-butene with 60% aqueous sulfuric acid. The only alcohol recovered is 2-methyl-2-butanol.

$$\underset{\text{3-methyl-1-butene}}{\overset{\overset{\displaystyle CH_3}{|}}{CH_3CHCH=CH_2}} \xrightarrow[\text{60\% }H_2SO_4]{H_2O} \underset{\text{2-methyl-2-butanol}}{\overset{\overset{\displaystyle CH_3}{|}}{\underset{\underset{\displaystyle OH}{|}}{CH_3CCH_2CH_3}}}$$

**Problem 13.6**

(a) What would be the structure of the Markovnikov addition product for the reaction of 3-methyl-1-butene with dilute acid?

(b) Write a detailed mechanism that accounts for the observed product.

**Problem 13.7**    3,3-Dimethyl-1-butene is isomerized by acid (p. 290). Predict, by writing a mechanism, what the structures of hydration products would be if a reaction of this alkene with aqueous acid were attempted.

Hydroboration followed by oxidation (p. 293) converts alkenes to alcohols, with an overall anti-Markovnikov addition of water to the double bond. The reaction does not involve carbocation intermediates, so no rearrangements are observed. 4-Penten-1-ol, for example, is converted into 1,5-pentanediol in high yield by hydroboration–oxidation.

$$\underset{\text{4-penten-1-ol}}{H_2C=CHCH_2CH_2CH_2OH} \xrightarrow[\text{tetrahydrofuran}]{\text{9-BBN}} \xrightarrow[H_2O]{H_2O_2,\ NaOH} \underset{\substack{\text{1,5-pentanediol} \\ 98\%}}{HOCH_2(CH_2)_3CH_2OH}$$

**Problem 13.8**    If 4-penten-1-ol were treated with aqueous acid, what products would be obtained?

Alcohols are also formed when a leaving group is displaced from an alkyl halide or alkyl tosylate by hydroxide ion or by water in nucleophilic substitution reactions. Elimination reactions often accompany these reactions, so they are not useful for the preparation of high-purity alcohols in good yield (p. 264). Also, because alkyl halides or tosylates are often prepared from the corresponding alcohol in the first place, they are not appropriate starting materials in an independent synthesis of an alcohol.

Two important methods for preparing alcohols are reduction of aldehydes or ketones and the Grignard reaction. Each of these involves the addition of a nucleophile to a carbonyl group. These reactions are discussed on pages 545 and 552 as part of the chemistry of aldehydes and ketones.

⌐⌐ **Study Guide**
**Concept Map 13.1**

## 13.3 Converting Alcohols to Alkyl Halides

### A. Reactions of Alcohols with Hydrogen Halides

Alkyl halides, which participate in many different substitution reactions (pp. 234 and following), may be prepared from alcohols in different ways. The hydroxyl group in an alcohol can be converted into a good leaving group (p. 256) so that it can be displaced by an incoming halide ion. For example, dissolving gaseous hydrogen bromide in 1-heptanol converts the alcohol to 1-bromoheptane.

$$CH_3(CH_2)_5CH_2OH + HBr \longrightarrow CH_3(CH_2)_5CH_2Br + H_2O$$

|   |   |   |   |
|---|---|---|---|
| 1-heptanol | hydrogen bromide | 1-bromoheptane 90% | water |

Alcohols have basic properties because of the presence of nonbonding electrons on the oxygen atom; these electrons can be donated to a proton. The protonation of the hydroxyl group creates a new leaving group, a water molecule. If this reaction is carried out on a primary alcohol, the water is displaced by halide ion in an $S_N2$ reaction, giving the alkyl halide.

---

**VISUALIZING THE REACTION**

**Conversion of an Alcohol to an Alkyl Halide by Hydrogen Bromide**

$$CH_3(CH_2)_5CH_2-\overset{H}{\underset{base}{O}}: \longrightarrow CH_3(CH_2)_5CH_2-\overset{H}{\underset{H}{O^+}} \xrightarrow[\text{of 1-heptanol}]{\text{conjugate acid}} CH_3(CH_2)_5CH_2-\overset{\cdot\cdot}{\underset{\cdot\cdot}{Br}}: \quad :\overset{H}{\underset{H}{O}}$$

$$H-\overset{\cdot\cdot}{\underset{\cdot\cdot}{Br}}: \qquad :\overset{\cdot\cdot}{\underset{\cdot\cdot}{Br}}:^-$$

acid

---

In secondary alcohols or other alcohols in which $S_N2$ reactions are hindered, rearrangements are observed, suggesting carbocations as intermediates. For example, when 3-pentanol is treated with hydrogen bromide, 3-bromopentane and 2-bromopentane are obtained in a mixture of varying composition depending on whether aqueous or gaseous hydrogen bromide is used as the reagent.

$$CH_3CH_2CHCH_2CH_3 \xrightarrow[\Delta]{HBr} CH_3CH_2CHCH_2CH_3 + CH_3CH_2CH_2CHCH_3$$

|   |   |   |
|---|---|---|
| OH | Br | Br |
| 3-pentanol | 3-bromopentane | 2-bromopentane |

**Problem 13.9**    Write the full mechanism for the conversion of 3-pentanol to the two bromopentanes in 48% aqueous hydrobromic acid.

---

## ONE SMALL STEP

Carbocations equilibrate to give mixtures that reflect their relative stabilities.

**PROBLEM:** Whenever either 2-chloropentane or 3-chloropentane is allowed to stand in concentrated hydrochloric acid in which zinc chloride has been dissolved, a mixture of two compounds consisting of 66% 2-chloropentane and 34% 3-chloropentane is obtained. Write a mechanism for this interconversion, starting with 3-chloropentane. What does the composition of the equilibrium mixture tell you about the relative stabilities of 2-pentyl and 3-pentyl cations?

**Hint:**   What is the role of zinc chloride?

**Hint:**   What are the relative numbers of 2-pentyl and 3-pentyl cations that can form?

Another way to convert the hydroxyl group in an alcohol into a good leaving group is to make the tosylate ester (p. 259), from which the tosylate group can be displaced by halide ion in a nucleophilic substitution reaction. In this way, 3-pentanol can be converted to 3-bromopentane free of 2-bromopentane; this result suggests that a carbocation intermediate is not formed under these conditions.

$$CH_3CH_2\underset{\underset{OH}{|}}{C}HCH_2CH_3 \ + \ TsCl \ \xrightarrow{\text{pyridine}} \ CH_3CH_2\underset{\underset{OTs}{|}}{C}HCH_2CH_3 \ \xrightarrow[\text{dimethyl sulfoxide}]{Na^+Br^-} \ CH_3CH_2\underset{\underset{Br}{|}}{C}HCH_2CH_3$$

| 3-pentanol | *p*-toluenesulfonyl chloride | 3-pentyl tosylate | 3-bromopentane 85% |

Tertiary alcohols react readily with hydrogen halides to give alkyl halides. For example, *tert*-butyl alcohol is converted to *tert*-butyl chloride by shaking it with cold concentrated hydrochloric acid for a few minutes.

$$CH_3-\underset{\underset{CH_3}{|}}{\overset{\overset{CH_3}{|}}{C}}-OH \ \xrightarrow[25\ °C]{HCl} \ CH_3-\underset{\underset{CH_3}{|}}{\overset{\overset{CH_3}{|}}{C}}-Cl$$

| *tert*-butyl alcohol | *tert*-butyl chloride |

This is an $S_N1$ reaction. As long as the temperature is kept low, not much of the product from the competing E1 reaction is obtained. Even if 2-methylpropene forms under these conditions, it also reacts with hydrochloric acid to give *tert*-butyl chloride.

**Problem 13.10** Write a mechanism for the conversion of *tert*-butyl alcohol to *tert*-butyl chloride using hydrochloric acid. How would 2-methylpropene be formed? How would that alkene react with hydrochloric acid?

## B. Reactions with Thionyl Chloride and Phosphorus Halides

Other reagents also can be used to avoid carbocation intermediates and minimize rearrangements in the conversion of alcohols to alkyl halides. One of these is thionyl chloride, $SOCl_2$. 3-Pentanol will react with thionyl chloride in pyridine to give 3-chloropentane, and no 2-chloropentane.

$$CH_3CH_2\underset{\underset{OH}{|}}{C}HCH_2CH_3 \quad SOCl_2 + \ \overset{\bigcirc}{\underset{N}{}} \ \longrightarrow \ CH_3CH_2\underset{\underset{Cl}{|}}{C}HCH_2CH_3 \ + \ \overset{\bigcirc}{\underset{\overset{N^+}{\underset{H}{|}}}{}} Cl^- \ + \ SO_2\uparrow$$

| 3-pentanol | thionyl chloride | pyridine | 3-chloropentane | pyridine hydrochloride | sulfur dioxide |

However, this reaction is often accompanied by elimination reactions and does not give high yields.

This reaction also starts with the conversion of the alcohol to a compound with a good leaving group, an intermediate chlorosulfite ester, which decomposes to give the alkyl halide. In the presence of organic bases such as pyridine, inversion of configuration is seen for chiral compounds.

## VISUALIZING THE REACTION

### Conversion of an Alcohol to an Alkyl Halide by Thionyl Chloride

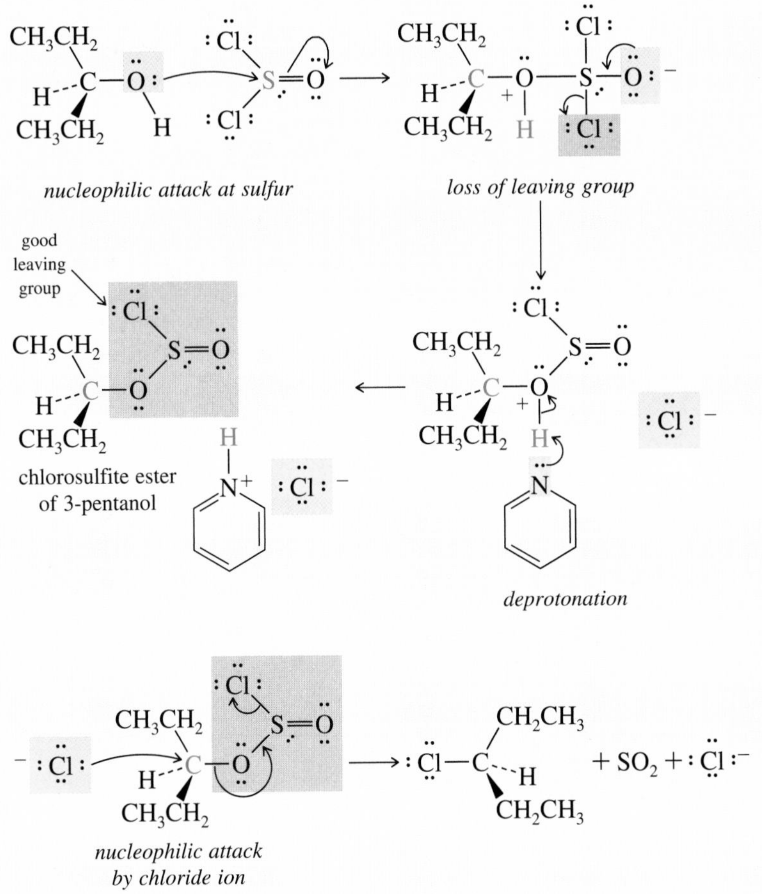

*nucleophilic attack at sulfur*

*loss of leaving group*

good leaving group

chlorosulfite ester of 3-pentanol

*deprotonation*

*nucleophilic attack by chloride ion*

$+ SO_2 + :\ddot{\underset{..}{Cl}}:^-$

---

**Problem 13.11** When 2-ethyl-1-butanol is treated with zinc chloride in concentrated hydrochloric acid, a mixture of chloroalkanes forms, including chiefly 2-ethyl-1-chlorobutane, 3-chlorohexane, 2-chlorohexane, and 3-chloro-3-methylpentane. When 2-ethyl-1-butanol is treated with thionyl chloride in pyridine, only 1-chloro-2-ethylbutane is formed. Write detailed mechanisms that account for these observations.

---

Another reagent that is used to substitute a halogen for a hydroxyl group with a minimum of rearrangement is phosphorus tribromide. For example, it is used to convert the primary alcohol isobutyl alcohol into isobutyl bromide.

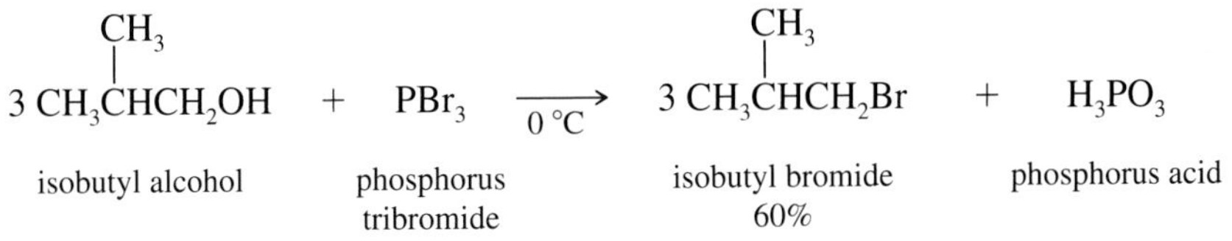

$$3\ CH_3CHCH_2OH \quad + \quad PBr_3 \quad \xrightarrow{0\ °C} \quad 3\ CH_3CHCH_2Br \quad + \quad H_3PO_3$$

(with $CH_3$ on each)

isobutyl alcohol     phosphorus tribromide     isobutyl bromide 60%     phosphorus acid

**On the Web: ONE SMALL STEP**

**Study Guide**
**Concept Map 13.2**

The mechanism suggested for the conversion of an alcohol to an alkyl halide using phosphorus tribromide resembles the one proposed for the conversion using thionyl chloride. Its first step is the formation of a phosphorus–oxygen bond between the alcohol and phosphorus tribromide, giving a phosphorus ester that has a good leaving group.

**Problem 13.12**    When optically active 2-octanol is treated with phosphorus tribromide, 2-bromooctane is formed with almost complete inversion of configuration. The other product of the reaction is phosphorous acid, $H_3PO_3$. Write a detailed mechanism for the reaction that accounts for these facts.

**Problem 13.13**    Phosphorous acid has the structure

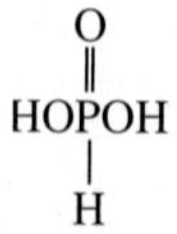

$$HOPOH$$

with $\overset{O}{\overset{\|}{\underset{H}{\phantom{.}}}}$ instead of the structure $P(OH)_3$, which might be expected from the mechanism for the formation of an alkyl halide from an alcohol in Problem 13.12. Write Lewis formulas for the two structural formulas shown above, and propose a mechanism for the conversion of $P(OH)_3$ to the correct structure. Remember that the system is acidic.

**Problem 13.14**    Complete the following equations.

(a) 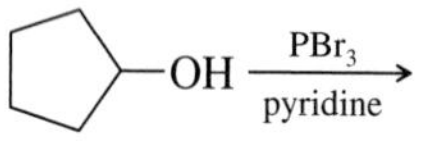 $\underset{\text{pyridine}}{\overset{PBr_3}{\longrightarrow}}$

(b) 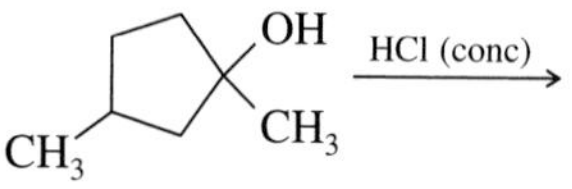 $\overset{HCl\ (conc)}{\longrightarrow}$

(c) $CH_3(CH_2)_9CH_2OH \xrightarrow{\overset{SOCl_2}{\Delta}}$

(d) $CH_3(CH_2)_8CH_2OH \xrightarrow{\overset{HBr}{\Delta}}$

(e) $CH_3SCH_2CH_2OH \xrightarrow[\text{chloroform}]{SOCl_2}$

(f) $CH_3CH_2OCH_2CH_2OH \xrightarrow[\text{pyridine}]{PBr_3}$

(g) 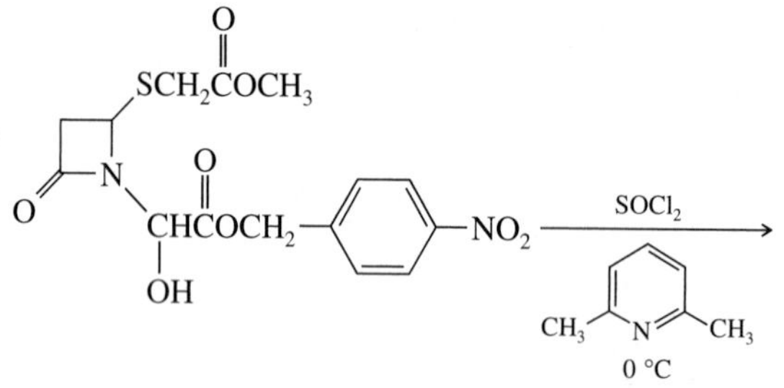 $\xrightarrow[\text{0 °C}]{SOCl_2}$

# 13.4    Reactions of Alkoxide Anions

## A. Preparation of Alkoxide Anions

Alcohols have acidic properties because of the presence of the hydrogen atom bonded to the oxygen atom in the hydroxyl group. This hydrogen atom is approximately as acidic as the hydrogen atom in water (p. 90). Thus the concentrations of hydroxide ion and methoxide anion are roughly equal when sodium hydroxide is dissolved in methanol.

$$CH_3OH\ +\ OH^-\ \rightleftharpoons\ CH_3O^-\ +\ H_2O$$

| methanol | hydroxide ion | methoxide anion | water |
|---|---|---|---|
| *acid* | *base* | *conjugate base of methanol* | *conjugate acid of hydroxide ion* |
| $pK_a\ {\sim}15.5$ | | | $pK_a\ {\sim}15.7$ |

The comparable acidities of water and alcohols make it impossible to achieve high concentrations of alkoxide anions from alcohols by using hydroxide ions. Sodium hydride, which is a source of the strongly basic hydride ion, is also used frequently to deprotonate alcohols to make alkoxide ions. Hydrogen gas is evolved, so the reaction goes to completion. For example, when 2-methyl-1,3-propanediol is treated with 1 equivalent of sodium hydride, the monoalkoxide ion is formed.

$$
\underset{\substack{\text{2-methyl-1,3-}\\\text{propanediol}}}{\overset{\overset{\displaystyle CH_3}{\underset{|}{}}}{HOCH_2CHCH_2OH}} \;+\; \underset{\substack{\text{sodium hydride}\\\text{1 equivalent}}}{NaH} \;\longrightarrow\; \underset{\substack{\text{conjugate base of}\\\text{2-methyl-1,3-propanediol}}}{\overset{\overset{\displaystyle CH_3}{\underset{|}{}}}{HOCH_2CHCH_2O^-\,Na^+}} \;+\; \underset{\substack{\text{conjugate acid}\\\text{of hydride ion}}}{H_2\uparrow}
$$

*acid*
*pK$_a$ ~17*

*base*

*pK$_a$ 35*

## B. Reactions of Alkoxide Anions with Alkyl Halides. The Williamson Synthesis of Ethers

Alkoxide ions as good nucleophiles react at the electrophilic center of primary alkyl halides in an S$_N$2 reaction (p. 235) to give ethers. This way of making ethers is known as the **Williamson synthesis.** For example, the alkoxide ion from 2-methyl-1,3-propanediol (prepared above) reacts with benzyl bromide to give a benzyl ether.

$$
\underset{\substack{\text{alkoxide ion from}\\\text{2-methyl-1,3-propanediol}}}{\overset{\overset{\displaystyle CH_3}{\underset{|}{}}}{HOCH_2CHCH_2O^-\,Na^+}} \;+\; \underset{\text{benzyl bromide}}{\text{⟨benzene⟩}-CH_2Br} \;\longrightarrow\; \underset{\substack{\text{monobenzyl ether of}\\\text{2-methyl-1,3-propanediol}}}{\overset{\overset{\displaystyle CH_3}{\underset{|}{}}}{HOCH_2CHCH_2OCH_2-\text{⟨benzene⟩}}} \;+\; NaBr
$$

For these reactions, the alkoxide ion and the alkyl halide must be chosen carefully. Alkoxide ions are strong bases and will cause E2 reactions to occur if used with secondary or tertiary halides (p. 265). The next problem provides practice in thinking about the competition between S$_N$2 and E2 reactions.

---

**Problem 13.15** Synthesize each of the following ethers by the Williamson method, choosing the alkoxide anion and alkyl halide that will give the best yield. Show the preparation of the alkoxide anions.

(a) $CH_3\overset{\overset{\displaystyle CH_3}{\underset{|}{}}}{CH}OCH_2CH_2CH_3$

(b) $CH_3\overset{\overset{\displaystyle CH_3}{\underset{|}{\phantom{C}}}}{\underset{\underset{\displaystyle CH_3}{|}}{C}}OCH_2-\text{⟨benzene⟩}$

(c) $CH_3CH_2OCH_2CH{=}CH_2$

(d) $CH_3OCH_2CH_2CH_2CH_3$

(e) $CH_3CH_2OCH_2CH_2CH_2CH_2CH_2CH_3$

(f) $\text{⟨cyclohexane⟩}-OCH_2CH_3$

## C. Intramolecular Reactions of Alkoxide Anions. The Preparation of Cyclic Ethers

If 4-chloro-1-butanol is treated with sodium hydroxide, a cyclic ether called tetrahydrofuran is formed in high yield.

$$ClCH_2CH_2CH_2CH_2OH \xrightarrow[\text{H}_2\text{O}]{\text{NaOH}}$$

4-chloro-1-butanol        tetrahydrofuran
95%

The formation of this cyclic ether is an **intramolecular reaction,** a reaction that occurs between two functional groups in the same molecule. Note that the intramolecular reaction is favored over an **intermolecular reaction,** a reaction between two different molecules. No 1,4-butanediol from the displacement of chloride ion by hydroxide ion is obtained.

The first and fastest step of the reaction is the deprotonation of the alcohol by the base. The resulting alkoxide ion then displaces chloride ion. The reacting centers in 4-chloro-1-butanol are held in close proximity by the intervening chain of four carbon atoms so that collision and reaction between them is more probable than reaction between the same reactive groups on separate molecules. Note that hydroxide ion can be used as the base for this reaction because each alkoxide ion that is formed reacts very rapidly with the electrophilic center at the other end of the molecule, and the overall reaction goes to completion. Hydroxide ion cannot be used to make alkoxide ions for intermolecular Williamson syntheses (p. 499).

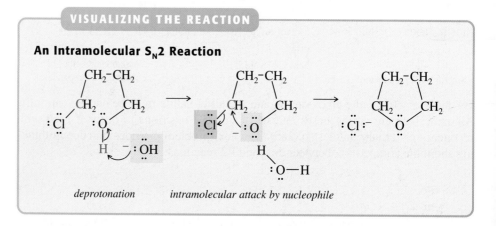

### VISUALIZING THE REACTION

**An Intramolecular S$_N$2 Reaction**

deprotonation      *intramolecular attack by nucleophile*

It is particularly easy to form rings of five and six carbon atoms (p. 179). A three-membered ring also forms easily in an intramolecular reaction. For example, when the trans halohydrin from cyclohexene (p. 307) is treated with base, an oxirane is the product.

$$\xrightarrow[\substack{\text{H}_2\text{O} \\ 25\,°\text{C} \\ 1\text{ h}}]{\text{NaOH}}$$

*trans*-2-chlorocyclohexanol        cyclohexene oxide
70%

The reaction occurs easily because in one conformation of the molecule the nucle-ophilic alkoxide ion is in position to displace halide ion in an intramolecular $S_N2$ reaction.

---

**VISUALIZING THE REACTION**

**Stereochemistry of the Formation of an Oxirane by an Intramolecular $S_N2$ Reaction**

| chair conformation with substituents diequatorial | chair conformation with substituents axial | leaving group | |

---

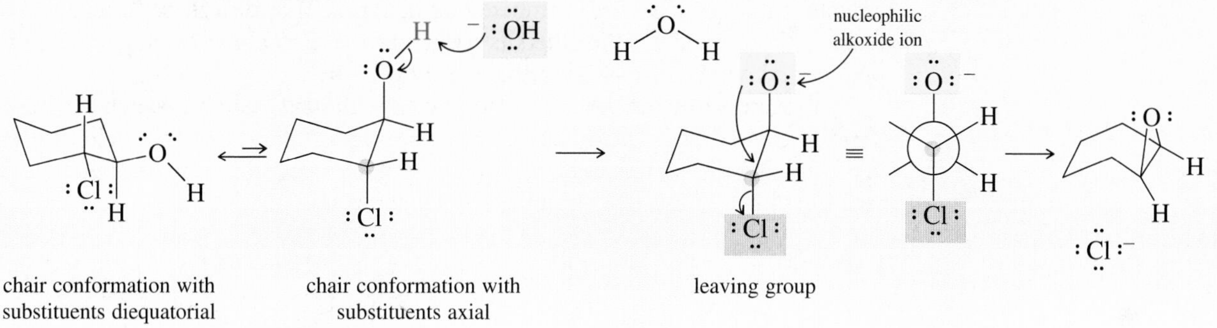

In this conformation, the alkoxide ion and the leaving group are anti to each other, as shown in the Newman projection for that portion of the cyclohexane ring. Halo-hydrins that cannot achieve an anti orientation between the nucleophile and the leaving group do not give oxiranes.

(⤢) **Study Guide**
**Concept Map 13.3**

**Problem 13.16** Write a detailed mechanism for the following reaction. Be sure to show the conformation of the molecule in the transition state.

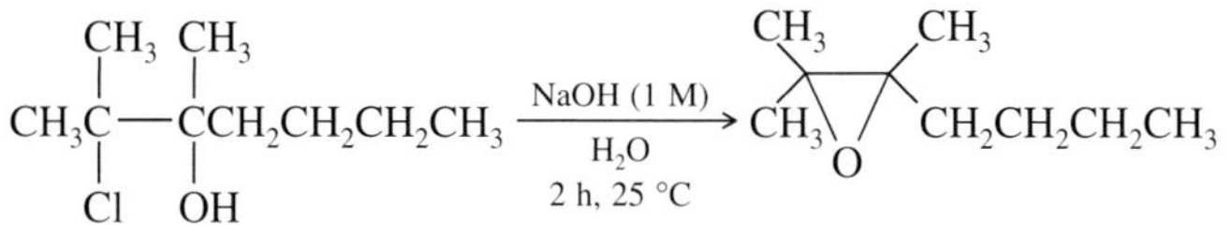

2-chloro-2,3-dimethyl-3-heptanol

2,2,3-trimethyl-3-butyloxirane

**Problem 13.17** Give structures for the products of the following reactions. Show stereochemistry if it is known.

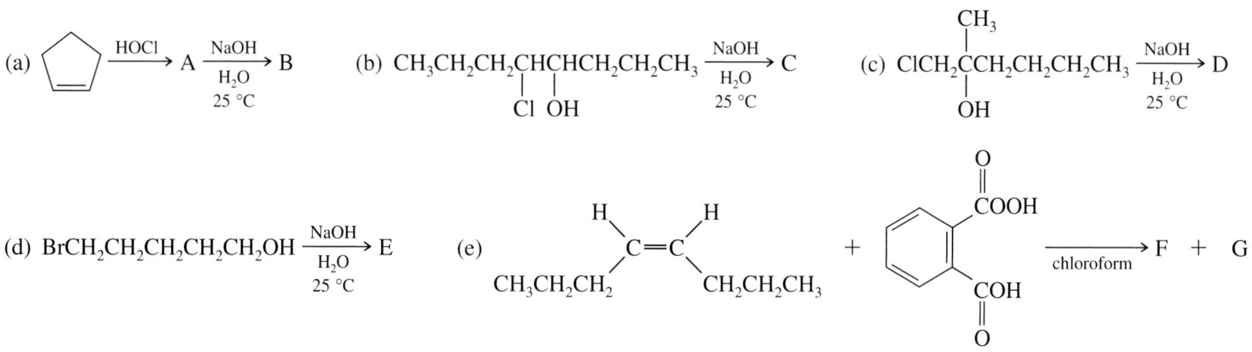

(a)

(b) $CH_3CH_2CH_2CHCHCH_2CH_2CH_3 \xrightarrow[\substack{H_2O \\ 25\ °C}]{NaOH} C$
    with Cl OH substituents

(c) $ClCH_2CCH_2CH_2CH_2CH_3 \xrightarrow[\substack{H_2O \\ 25\ °C}]{NaOH} D$
    with $CH_3$ and OH substituents

(d) $BrCH_2CH_2CH_2CH_2CH_2OH \xrightarrow[\substack{H_2O \\ 25\ °C}]{NaOH} E$

(e)
$+$
$\xrightarrow{\text{chloroform}} F\ +\ G$

## 13.5 Ring–Opening Reactions of Oxiranes

### A. Acid-Catalyzed Ring–Opening Reactions of Oxiranes

Oxiranes are compounds of industrial and biological importance because of their high reactivity. The ring of an oxirane opens readily. This is true whether electrophilic or nucleophilic reagents are used. The high reactivity of the oxirane ring, compared with that of an acyclic ether, which is not easily cleaved, is due to the strained nature of the three-membered oxirane ring. The difference in reactivity between oxiranes and acyclic ethers is particularly noticeable in reactions with nucleophiles and will be discussed further (Section 13.5B).

An example of the reaction of oxiranes with acids is what happens when the 2,3-dimethyloxiranes (p. 317) react with 48% aqueous hydrobromic acid. Both *cis*- and *trans*-2,3-dimethyloxirane give a pair of enantiomeric 3-bromo-2-butanols. The reaction of *cis*-2,3-dimethyloxirane gives one racemic mixture.

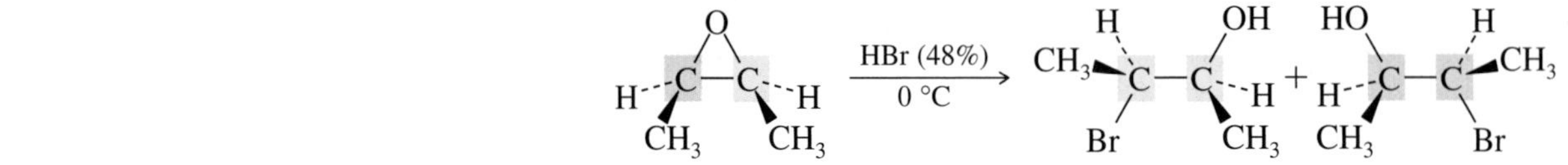

*cis*-2,3-dimethyloxirane → (2*S*,3*S*)-3-bromo-2-butanol + (2*R*,3*R*)-3-bromo-2-butanol
*racemic mixture*
85%

*trans*-2,3-Dimethyloxirane, which exists as a racemic mixture, gives another pair of enantiomeric 3-bromo-2-butanols.

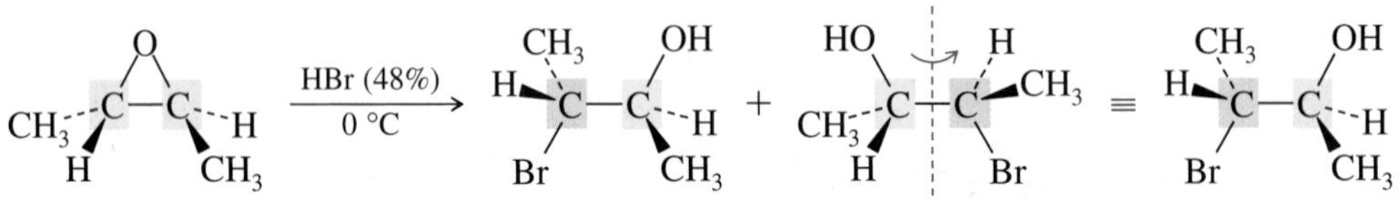

*trans*-2,3-dimethyloxirane
*one enantiomer* → (2*S*,3*R*)-3-bromo-2-butanol

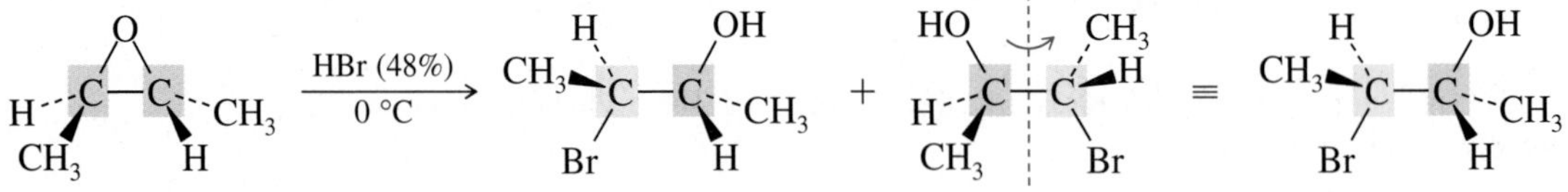

*trans*-2,3-dimethyloxirane
*the other enantiomer* → (2*R*,3*S*)-3-bromo-2-butanol

The 3-bromo-2-butanols obtained from *cis*-2,3-dimethyloxirane are diastereomers of those obtained from the trans oxirane.

This reaction starts with the protonation of the oxygen atom in the oxirane ring. The protonated oxirane is an oxonium ion and is electronically similar to the bromonium ion suggested as an intermediate in the addition of bromine to a double bond (p. 302). The oxonium ion intermediate is attacked at either carbon atom in an $S_N2$ reaction by the nucleophilic halide ion, as shown on the next page for *cis*-2,3-dimethyl oxirane. There is an inversion of configuration at the carbon atom that is attacked and retention of configuration at the other carbon atom.

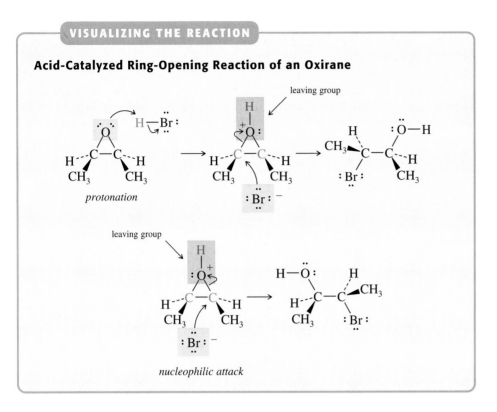

VISUALIZING THE REACTION

**Acid-Catalyzed Ring-Opening Reaction of an Oxirane**

*protonation*

*nucleophilic attack*

The reactions of the 2,3-dimethyloxiranes are examples of stereoselectivity in chemical transformations. Two starting materials with differing stereochemistry are converted into products that are also stereochemically different from each other.

**Problem 13.18**   When *cis*-2,3-dimethyloxirane is treated with water containing a trace of perchloric acid, HClO₄, a racemic mixture of 2,3-butanediols is formed. *trans*-2,3-Dimethyloxirane gives *meso*-2,3-butanediol under the same conditions. Write mechanisms for these reactions, showing all the stereochemistry.

**Problem 13.19**   When one of the two enantiomeric *trans*-2,3-dimethyloxiranes, (2R,3R)-(+)-2,3-dimethyloxirane, was treated with methanol containing a trace of sulfuric acid, there

was obtained a 57% yield of a single, optically active 3-methoxy-2-butanol.

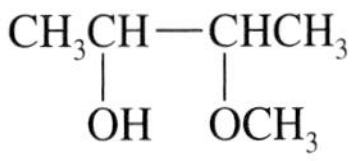

$$CH_3CH-CHCH_3$$
$$\ \ \ \ \ |\ \ \ \ \ \ \ |$$
$$\ \ \ OH\ \ \ OCH_3$$

Draw the correct structure for the oxirane and write a stereochemically correct mechanism showing how the 3-methoxy-2-butanol is formed. Assign configurations to the stereocenters in the product and give the correct name for the compound.

The 2,3-dimethyloxiranes have symmetrical structures. Regioselectivity is possible only in ring-opening reactions of unsymmetrical oxiranes. 2,2,3-Trimethyloxirane reacts with methanol in the presence of sulfuric acid to give 2-methoxy-2-methyl-3-butanol as the major product.

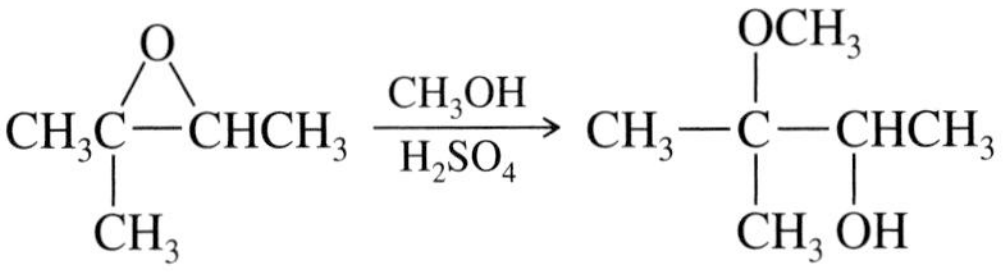

$$CH_3\overset{\overset{\displaystyle O}{\diagup\ \ \diagdown}}{C}-CHCH_3 \xrightarrow[\text{H}_2\text{SO}_4]{\text{CH}_3\text{OH}} CH_3-\overset{\overset{\displaystyle OCH_3}{|}}{\underset{\underset{\displaystyle CH_3}{|}}{C}}-\overset{}{\underset{\underset{\displaystyle OH}{|}}{C}}HCH_3$$
$$\ \ \ \ \ \ |$$
$$\ \ \ \ CH_3$$

2,2,3-trimethyloxirane

2-methoxy-2-methyl-
3-butanol
76%

In this reaction mixture, no good nucleophile is present. The protonated oxygen atom of the oxirane ring is a good leaving group, and the carbon–oxygen bonds in the strained ring may start to cleave before the approach of the weak nucleophile. The reaction, therefore, resembles an $S_N1$ reaction. In an unsymmetrically substituted oxirane, one carbocationic intermediate will be favored over the other. For 2,2,3-trimethyloxirane, the tertiary carbocation rather than the secondary one is the favored intermediate. The nucleophile in this reaction, methanol, attacks at the tertiary carbon atom. This mechanism resembles the one proposed for reactions of unsymmetrical halonium ions (p. 308).

---

**VISUALIZING THE REACTION**

**Acid-Catalyzed Ring-Opening Reaction of an Unsymmetrical Oxirane**

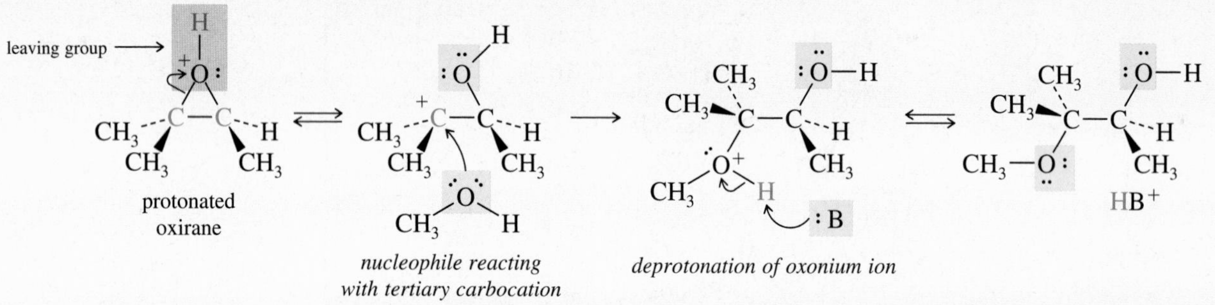

leaving group ⟶

protonated oxirane

*nucleophile reacting with tertiary carbocation*

*deprotonation of oxonium ion*

---

**Problem 13.20**   Tetrahydrofuran is a stable ether, but the ring will open when the molecule is heated with acid.

$$\text{THF} \xrightarrow[\Delta]{\text{HCl}} ClCH_2CH_2CH_2CH_2OH$$

Propose a mechanism for this reaction.

---

⬤ **ONE SMALL STEP**

Ethers are cleaved when treated with reagents such as $BBr_3$ or $AlCl_3$. For example:

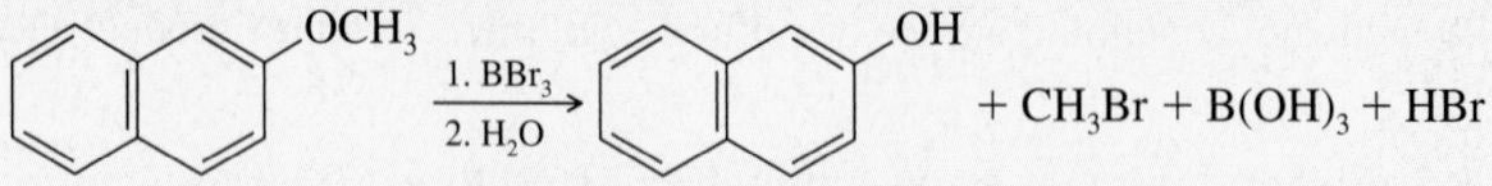

$$\text{naphthalene-}OCH_3 \xrightarrow[\text{2. }H_2O]{\text{1. }BBr_3} \text{naphthalene-}OH + CH_3Br + B(OH)_3 + HBr$$

**PROBLEM:**   Propose a mechanism for the reaction shown above.

**Hint:**   Think in terms of electrophiles (Lewis acids), nucleophiles (Lewis bases), and leaving groups.

---

## B. Ring-Opening Reactions of Oxiranes with Nucleophiles

The ring of an oxirane is opened easily by nucleophilic reagents even without protonation of the oxygen atom. In this, an oxirane differs from an acyclic ether, which is unreactive toward bases and nucleophiles. For example, 2,2,3-trimethyloxirane reacts with sodium methoxide to give 3-methoxy-2-methyl-2-butanol.

$$CH_3\overset{\displaystyle O}{\overset{\diagup \diagdown}{C}}-CHCH_3 \quad \xrightarrow[CH_3OH]{CH_3ONa} \quad CH_3\overset{\displaystyle CH_3}{\underset{\displaystyle OH}{\overset{|}{C}}}-\underset{\displaystyle OCH_3}{\overset{|}{C}}HCH_3$$

2,2,3-trimethyloxirane

3-methoxy-2-methyl-2-butanol
53%

By contrast, an acyclic ether such as diethyl ether is unreactive toward methoxide ion.

$$CH_3CH_2OCH_2CH_3 \xrightarrow[CH_3OH]{CH_3ONa} \text{no reaction}$$

In the absence of a strong acid, no cationic intermediate for the reaction can form. An $S_N2$ reaction occurs at the less highly substituted carbon atom of the oxirane ring. The leaving group is an alkoxide ion, normally a very poor leaving group (p. 255), and one that does not form in reactions of an acyclic ether. The ring-opening reaction of the oxirane is completed by protonation of the alkoxide ion by the solvent. The stereochemistry of the ring-opening reaction is similar to that of the bromonium ion (p. 303). The incoming nucleophile and the hydroxyl group that forms are anti to each other.

**VISUALIZING THE REACTION**

**Opening of an Oxirane by a Nucleophile**

Why is it so easy to displace an alkoxide ion by opening the ring of the cyclic ether, the oxirane, under conditions that will not touch an acyclic ether? The three-membered oxirane ring is strained, just as a cyclopropane ring is (p. 178). This means that the reactants are at a higher energy level for an $S_N2$ reaction on an oxirane than they are for an acyclic ether. It is easier to reach the transition state from a reactant of higher energy than it is from one of lower energy (Figure 13.1, p. 506).

Two factors help the reaction of the oxirane. Because the oxirane is at higher energy level than the acyclic ether, the difference in energy between the starting material and the transition state is smaller. Also, the strain energy of the oxirane ring is relieved by the lengthening of one of the carbon–oxygen bonds and the increase in internal bond angle in the activated complex for the opening of the ring. There is no comparable stabilization of the transition state relative to the starting material in the acyclic ether. Raising the point from which the reaction starts and lowering the energy of the transition state relative to the reactants together decrease the height of the energy barrier between reactants and products for the oxirane in comparison to an acyclic ether. A lower $\Delta G^{\ddagger}$ translates into a faster reaction for the oxirane than for the acyclic ether. In fact, the $\Delta G^{\ddagger}$ for the acyclic ether is so high that no exchange of alkoxide ions is observed.

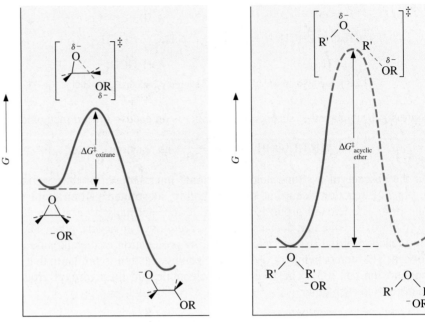

**Figure 13.1**
Energy diagrams showing how the relative energies of an oxirane and an acyclic ether affect their relative reactivities toward nucleophiles.

Progress of reaction for oxirane          Progress of reaction for acyclic ether

The reactions of oxiranes with nucleophiles are important in syntheses and biological processes. Proteins and nucleic acids are rich in nucleophiles such as amino groups. Oxiranes, especially low-molecular-weight oxiranes such as ethylene oxide, are toxic because they react with biological nucleophiles and thus disrupt the normal functioning of cells. Ethylene oxide is used to sterilize surgical instruments because it is toxic to bacteria. Higher-molecular-weight oxiranes and their reactions with biological nucleophiles are implicated in the carcinogenicity of compounds found in smoke (Section 13.6).

In syntheses, the reactions of oxiranes with carbon nucleophiles (p. 554), such as anions derived from alkynes, are particularly important. However, azide ions, amines, and sulfur and oxygen nucleophiles also react easily with oxiranes. These reactions always give rise to compounds having a group derived from a nucleophile on one carbon atom and a hydroxyl group anti to it on an adjacent carbon atom. The following problem illustrates some of these reactions.

 **On the Web: ONE SMALL STEP**

---

**Problem 13.21**    Suggest reagents for the following transformations. Remember that protic solvents are present to protonate the alkoxide ions that are formed in the nucleophilic attack on the oxirane.

(e) 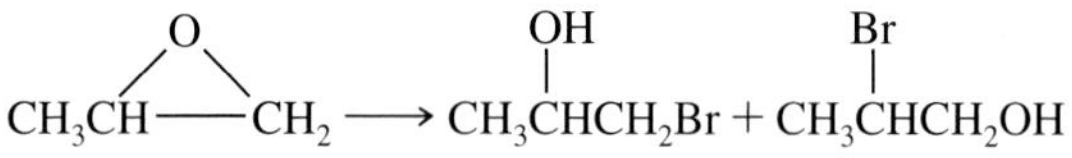 $CH_3CH\overset{O}{\overbrace{\phantom{xx}}}CH_2 \longrightarrow CH_3\underset{\underset{OH}{|}}{C}HCH_2Br + CH_3\underset{\underset{Br}{|}}{C}HCH_2OH$

(f)  $CH_2\overset{O}{\overbrace{\phantom{xx}}}CH_2 \longrightarrow Na^{+-}OCH_2CH_2C\equiv CCH_2CH_3 \xrightarrow{H_3O^+} HOCH_2CH_2C\equiv CCH_2CH_3$

## C. Stereoselective Formation of *trans*-1,2-Diols from Oxiranes

The oxirane from cyclohexene gives a racemic mixture of *trans*-cyclohexane-1,2-diols when treated with water.

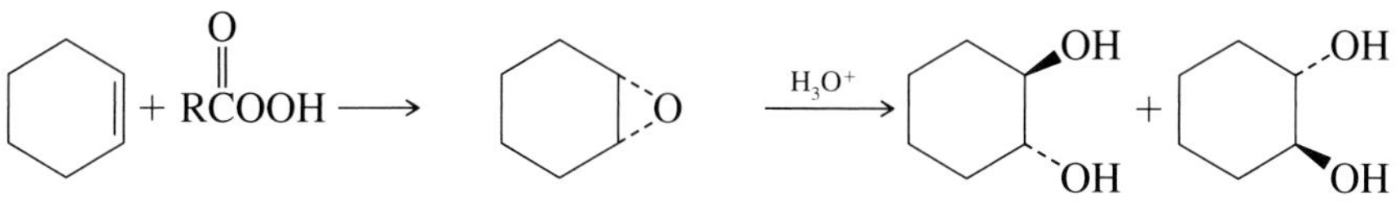

cyclohexane oxide
1,2-epoxycyclohexane

*trans*-cyclohexane-1,2-diols
racemic mixture

The ring-opening reaction starts with protonation of the oxygen atom of the oxirane to give a reactive oxonium ion. Nucleophilic attack by a water molecule on either one of the carbon atoms of the oxirane ring opens the ring to give the trans orientation of the two hydroxyl groups. The reaction is stereoselective, producing only the trans diol.

**VISUALIZING THE REACTION**

**Stereochemistry of the Opening of the Oxirane Ring in Cyclohexene Oxide**

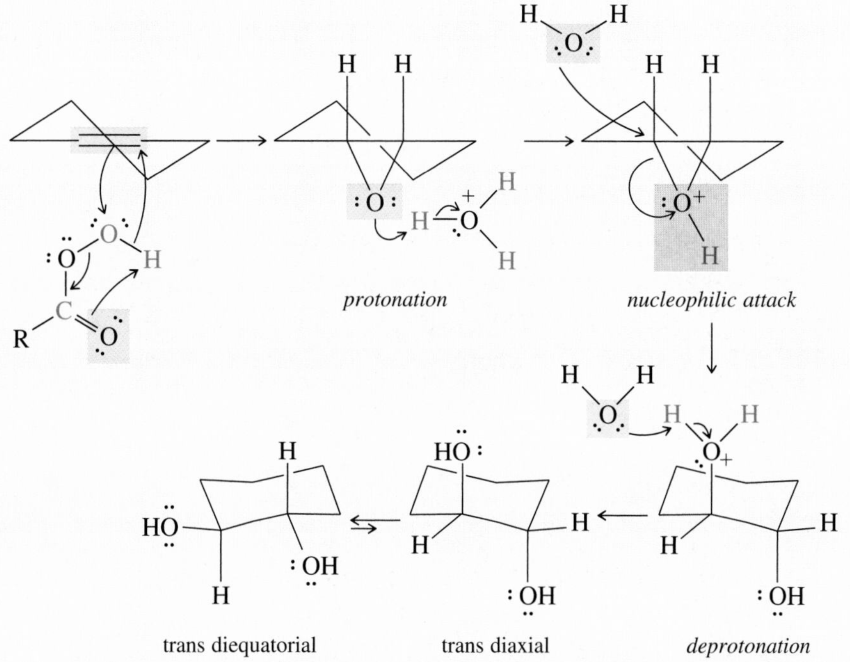

*protonation*

*nucleophilic attack*

trans diequatorial

trans diaxial

*deprotonation*

Study Guide Concept Map 13.4

As the preceding conformational formulas demonstrate, the opening of the three-membered oxirane ring first gives the two hydroxyl groups in a trans diaxial orientation. The more stable conformation for the diol is the diequatorial one, which predominates at equilibrium. Attack by the nucleophile, water, on the protonated oxirane gives rise to enantiomeric *trans*-1,2-cyclohexanediols.

You can think of the above sequence of reactions as being complementary to the oxidation of cyclohexene with osmium tetroxide to give *cis*-cyclohexane-1,2-diol (p. 314).

cyclohexene

*cis*-cyclohexane-1,2-diol
91%

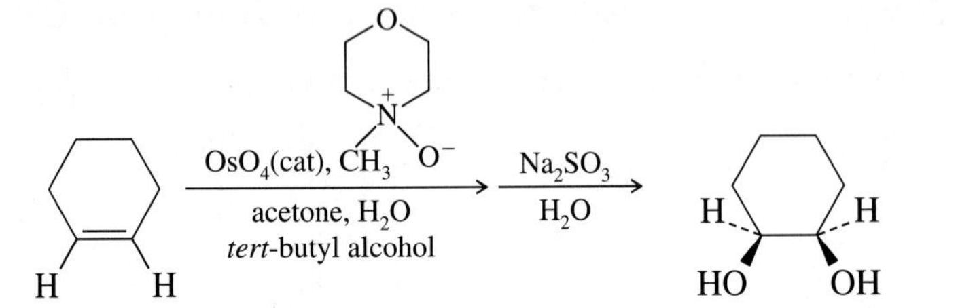

The stereochemistry of the cis diol is determined by the cyclic intermediate formed when osmium tetroxide reacts with the alkene.

*cis*-cyclohexane-1,2-diol

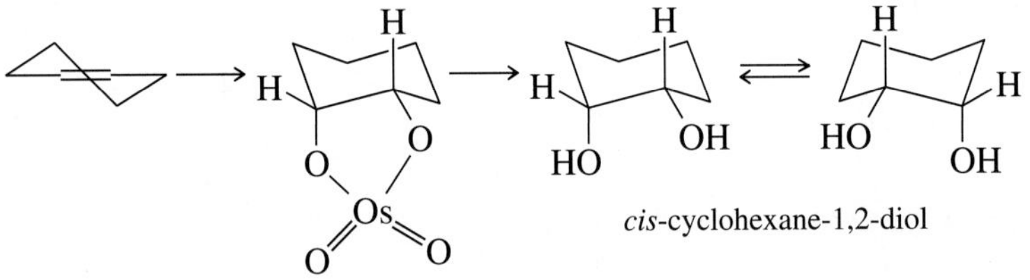

**Problem 13.22** Because of the bulk of the *tert*-butyl group, *tert*-butylcyclohexane exists mainly in one conformation (p. 183). *trans*-4-*tert*-Butylcyclohexene oxide, when treated with methanethiol, $CH_3SH$, in ethanol in the presence of sodium ethoxide, gives *trans*-2-methylthio-*trans*-5-*tert*-butylcyclohexanol. Write a mechanism for this reaction showing the conformation you expect the product to have.

**Problem 13.23**

(a) Predict the products of the following series of reactions, including the stereochemistry.

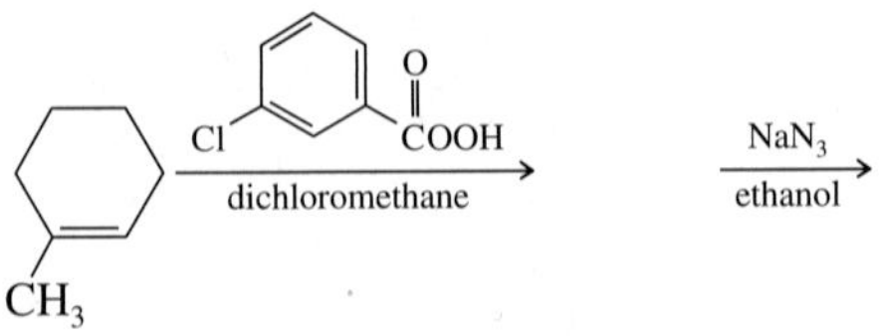

(b) How would you carry out the following transformation?

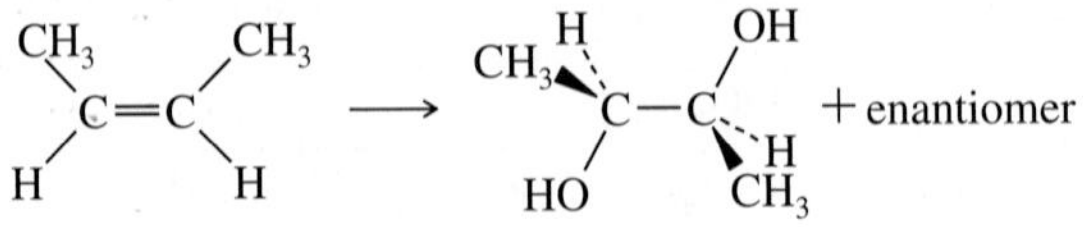

# 13.6  The Biological Reactivity of Oxiranes. Arene Oxides

The human body converts nonpolar compounds, especially aromatic compounds, into oxygenated derivatives containing hydroxyl groups in order to detoxify and excrete them. The totally nonpolar hydrocarbon is given polar functional groups that make it more soluble in the physiologic solvent, water. The hydroxyl group on the hydrocarbon is used by the body to link the aromatic residue to other functionalities that increase the solubility of the molecule in water still further. For example, sulfate esters of a phenol may be formed, or the hydroxyl group may be linked to glucuronic acid (p. 972) as an acetal (p. 558). These functions serve to hold the hydrocarbon residue in solution and aid in its transport out of the body.

The liver is the body's major organ for detoxification. Drugs, pollutants, food additives, and even some natural components of food are chemical substances that cannot be usefully incorporated into the structure of the body and must be excreted somehow. Many substances are transformed in oxidation reactions catalyzed by enzymes found in abundance in the liver. Most important among these enzymes are a group of highly colored proteins known as **cytochromes.** These enzymes, which contain iron, are found in the microsomal fraction of liver cells. Their function is to convert the oxygen from the air into a highly reactive form that can attack systems normally resistant to oxidation by air.

When aromatic hydrocarbons undergo these oxidation reactions, the products are unstable oxiranes known as **arene oxides.** Naphthalene, for example, is converted by microsomal enzymes from the liver of the rat into naphthalene-1,2-oxide.

naphthalene        naphthalene-1,2-oxide

The reactive oxide is opened to a *trans*-1,2-diol by another enzyme. Experiments have shown that the oxygen atom in the benzylic hydroxyl group comes from molecular oxygen and the other one comes from water.

naphthalene-1,2-oxide        *trans*-1,2-dihydro-
1,2-dihydroxynaphthalene

The exact mechanism by which arene oxides react is of great interest because there is evidence that such oxides are responsible for the carcinogenicity of many aromatic hydrocarbons. It is ironic that the same oxidation processes that detoxify harmful chemical substances and rid the body of them also may create metabolites that are far more carcinogenic than the original hydrocarbons themselves.

Suspicion that substances in the residues from the burning of coal and wood cause cancer first arose in 1775 when a British doctor noticed that chimney sweeps were more likely to have cancer of the scrotum than men in other occupations. The harmful compounds, found widely in the environment in cigarette smoke, automobile exhausts, and industrial emissions, are polycyclic aromatic hydrocarbons. The one on which most interest has centered is benzo[a]pyrene. About 40 of its oxygenated metabolites have been isolated and characterized. One of them in particular has been shown to be a potent mutagen. Compounds that are mutagens are suspected of being carcinogens as well because they disrupt the genetic processes in cells. A cancer cell is a cell that has been transformed so that the processes that normally control growth and reproduction no longer operate.

The reactions that are thought to be involved in the conversion of benzo[a]pyrene into its mutagenic metabolite are very much like those shown for the biological conversion of naphthalene to *trans*-1,2-dihydro-1,2-dihydroxynaphthalene. A cytochrome catalyzes the conversion of benzo[a]pyrene into an arene oxide, which adds water enzymatically to give a trans diol. Another epoxidation takes place at the double bond in the same ring as the diol function, and two stereoisomeric diol-epoxides are formed (Figure 13.2). The new oxirane ring can have two orientations with respect to the hydroxyl groups that are already present. The 7,8-diol-9,10-epoxide of benzo[a]pyrene with the oxirane ring trans to the benzylic hydroxyl group, called the anti isomer, is more potent as a mutagen than is the other isomer.

Benzo[a]pyrene-7,8-diol-9,10-epoxide disrupts the genetic mechanism of the cell in several ways. The chemistry that is relevant to this discussion involves the reactivity of oxiranes toward nucleophiles. There are many nucleophilic sites on a DNA molecule. One such site is an amino group on a heterocyclic ring in deoxyguanosine (p. 988), which is part of the backbone of DNA (p. 1016).

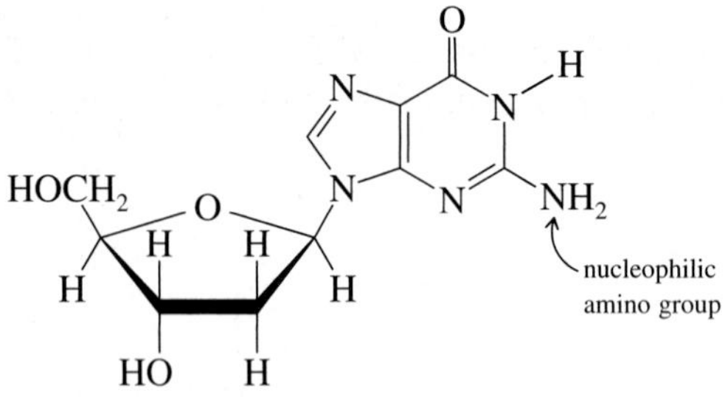

deoxyguanosine

*a component of DNA*

Reaction of the benzo[a]pyrene-7,8-diol-9,10-epoxide with DNA, degradation of the giant molecule, and purification of the fractions containing the hydrocarbon residue lead to the isolation of the following compound.

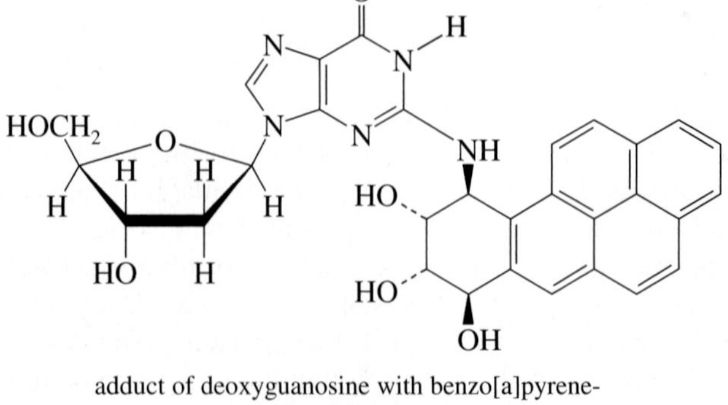

adduct of deoxyguanosine with benzo[a]pyrene-
7,8-diol-9,10-epoxide

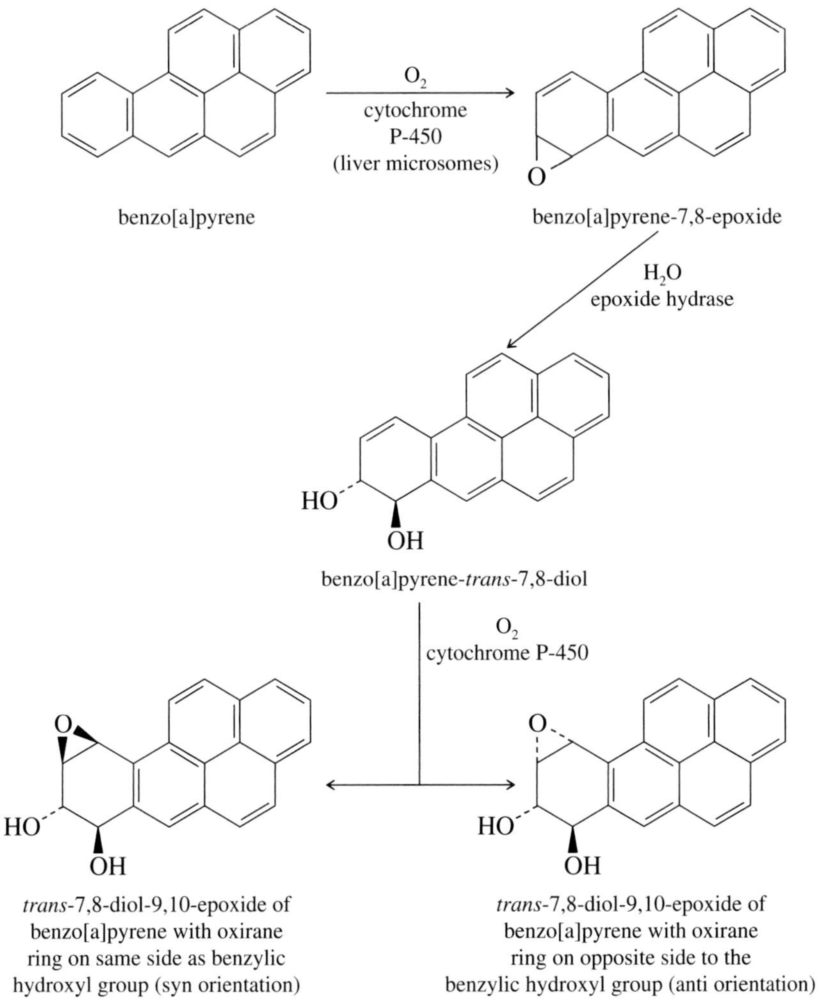

*Figure 13.2*
The conversion of benzo[a]pyrene into its mutagenic metabolite.

The amino group in deoxyguanosine reacts by a nucleophilic opening of the oxirane ring in benzo[a]pyrene-7,8-diol-9,10-epoxide. The same adduct is obtained when benzo[a]pyrene is incubated with mouse embryo cells, strongly indicating that the diol-epoxide is the metabolic intermediate responsible for the reaction with DNA.

If the aromatic hydrocarbon being metabolized contains an alkyl side chain, biological oxidations resembling the oxidation reaction discussed on page 785 take place at the benzylic position. Benzene, with no alkyl side chain, is detoxified slowly by the body and is a cumulative poison because it cannot be excreted rapidly. It affects bone marrow and causes aplastic anemia and leukemia. In contrast, toluene is much less toxic. The major pathway for its detoxification by the body is oxidation of the side chain to benzyl alcohol.

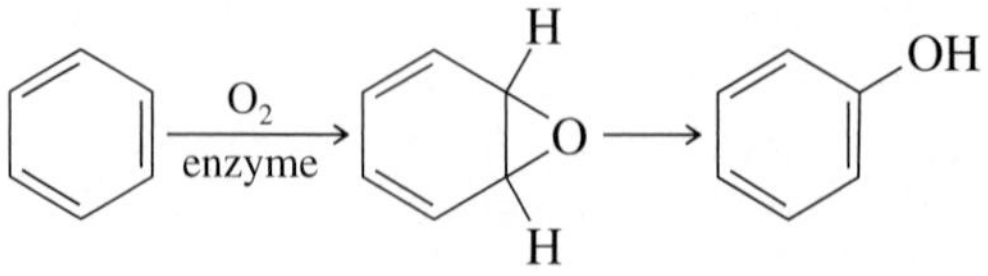

*major oxidative route for the
detoxification of benzene, slow*

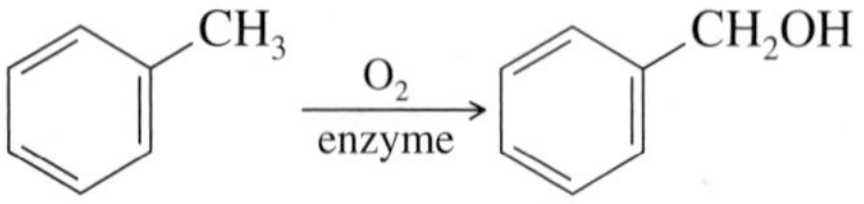

*major oxidative route for the
detoxification of toluene, fast*

A realization of these facts has led to the substitution of toluene for benzene in the laboratory whenever possible.

**Problem 13.24**  Two experiments provided part of the evidence for the formation of an arene oxide as an intermediate in the metabolism of naphthalene to *trans*-1,2-dihydro-1,2-dihydroxynaphthalene. In one experiment, naphthalene was incubated with air containing $^{18}O_2$ in ordinary water, $H_2{}^{16}O$. In the other, naphthalene was incubated with ordinary air in $H_2{}^{18}O$. Referring to the equations on page 509, write mechanisms that predict the isotopic composition of the trans diol that was formed in each experiment.

## 13.7 Oxidation Reactions of Alcohols

### A. Oxidation–Reduction of Organic Compounds

In Chapter 3 we explored the properties of molecules from the two opposite poles of acidity and basicity. Oxidation and reduction reactions are another pair of polar opposites. For many metals and their ions, the processes of oxidation and reduction can be seen easily. When copper metal is put into a solution of silver nitrate, crystals of silver metal grow on the surface of the copper and the solution turns the blue color that is typical of water solutions of copper(II) ion. Copper metal loses electrons to silver ions and is oxidized, and silver ions are reduced to silver metal by the gain of electrons.

$$Cu + 2 Ag^+ \xrightarrow{\text{water}} Cu^{2+} + 2 Ag$$

| | | | | |
|---|---|---|---|---|
| *shiny red metal* | *colorless* | | *blue* | *silver-gray crystals* |

For organic compounds, however, changes in oxidation states are generally not observable in this way. Yet it is of great importance in organizing reaction types to be able to decide whether a given compound has been oxidized or reduced. You are already familiar with several oxidation (pp. 313 and 316, for example) and reduction (pp. 298 and 340) reactions. In being oxidized, carbon "loses" electrons by forming bonds with elements that are more electronegative than it is, elements such as oxygen, nitrogen, or the halogens. In being reduced, carbon "gains" electrons by giving up bonds to more electronegative elements and forming bonds with hydrogen atoms instead.

In deciding whether an organic compound is being oxidized or reduced, the first thing to look for is changes in the number of bonds to hydrogen or to oxygen (or other electronegative elements) at the carbon atoms undergoing reaction. Loss of bonds to a hydrogen atom and/or gain of bonds to an oxygen atom at a carbon atom is an oxidation, for example:

$$CH_3-\underset{\underset{H}{|}}{\overset{\overset{OH}{|}}{C}}-H \longrightarrow CH_3-\overset{\overset{O}{\|}}{C}-H \longrightarrow CH_3-\overset{\overset{O}{\|}}{C}-O-H$$

| *loss of a bond* | *loss of a bond* |
| *to hydrogen,* | *to hydrogen,* |
| *gain of a bond* | *gain of a bond* |
| *to oxygen;* | *to oxygen;* |
| *therefore, oxidation* | *therefore, oxidation* |

The changes that ethanol must undergo to be converted first to acetaldehyde and then to acetic acid are oxidations; these reactions require the use of an oxidizing agent. The reverse reactions must be reduction reactions, carried out by some reducing agent. Formally, they involve the loss of bonds to oxygen or the gain of bonds to hydrogen.

$$CH_3-\overset{\overset{O}{\|}}{C}-O-H \longrightarrow CH_3-\overset{\overset{O}{\|}}{C}-H \longrightarrow CH_3-\underset{\underset{H}{|}}{\overset{\overset{OH}{|}}{C}}-H$$

| *loss of a bond* | *loss of a bond* |
| *to oxygen,* | *to oxygen,* |
| *gain of a bond* | *gain of a bond* |
| *to hydrogen;* | *to hydrogen;* |
| *therefore, reduction* | *therefore, reduction* |

Note that only one of the carbon atoms in this series of compounds undergoes oxidation or reduction. The methyl group is unchanged in these reactions.

A similar reasoning process allows us to conclude that the carbon atoms in an alkyne are more highly oxidized than the carbon atoms in an alkene, which in turn are more oxidized than those in an alkane.

$$H-C\equiv C-H \longrightarrow H-\underset{}{\overset{\overset{H}{|}}{C}}=\underset{}{\overset{\overset{H}{|}}{C}}-H \longrightarrow H-\underset{\underset{H}{|}}{\overset{\overset{H}{|}}{C}}-\underset{\underset{H}{|}}{\overset{\overset{H}{|}}{C}}-H$$

| *gain of two* | *gain of two* |
| *bonds to hydrogen;* | *bonds to hydrogen;* |
| *therefore, reduction* | *therefore, reduction* |

A reducing agent is necessary to convert an alkyne to an alkene and then to an alkane. This reduction would be carried out by adding hydrogen to the multiple bonds (pp. 339 and 300). However, it is not necessary to know the reagents that are used for the transformations that are shown in this section. Your goal here is to learn how to recognize which reactions involve oxidation and which reduction. The ability to distinguish between these two types of reactions is important.

Giving oxidation numbers to various substituents allows us to be quantitative about the oxidation states of carbon. A carbon–carbon bond contributes nothing to the oxidation state of a carbon atom because the bonding electrons are shared equally by the two carbon atoms. Atoms that are more electronegative than carbon pull bonding electrons toward themselves and are given negative oxidation numbers. A halogen or a hydroxyl group is assigned a $-1$ oxidation number, while a doubly bonded oxygen atom is in a $-2$ oxidation state. Each hydrogen atom is in a +1 oxidation state. Using these numbers, we can assign oxidation states to the carbon atoms of the compounds shown earlier in this section, for example:

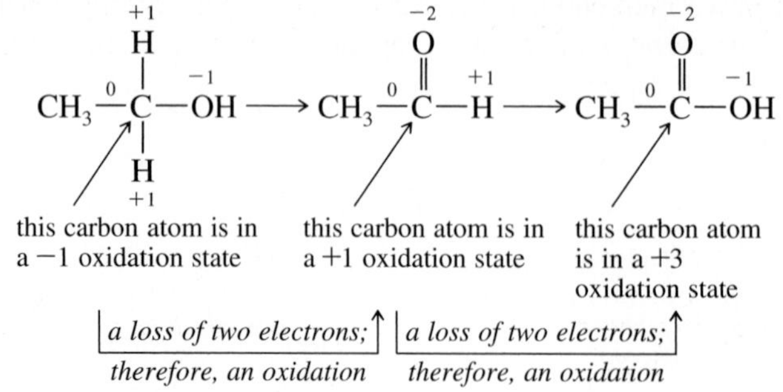

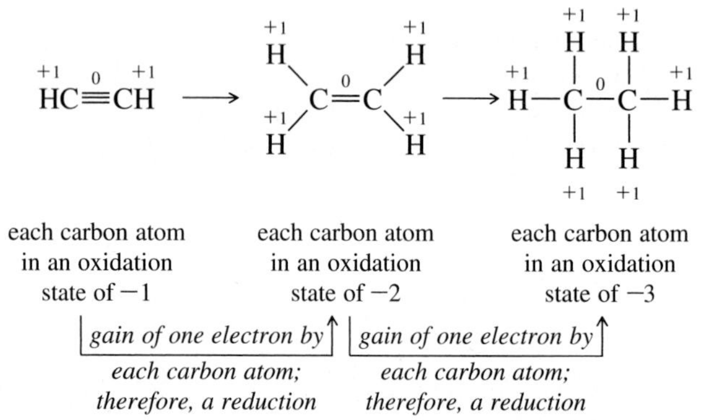

this carbon atom is in        this carbon atom is in    this carbon atom
a −1 oxidation state          a +1 oxidation state      is in a +3
                                                        oxidation state

*a loss of two electrons;↑*  *a loss of two electrons;↑*
*therefore, an oxidation*    *therefore, an oxidation*

each carbon atom          each carbon atom          each carbon atom
in an oxidation           in an oxidation           in an oxidation
state of −1               state of −2               state of −3

*gain of one electron by↑*  *gain of one electron by↑*
*each carbon atom;*         *each carbon atom;*
*therefore, a reduction*    *therefore, a reduction*

Both the qualitative approach presented earlier and this quantitative one lead to the same conclusions about oxidation and reduction.

You should work to develop skill in deciding when processes involve oxidation–reduction. If you are able to recognize the different oxidation states of carbon, you will be able to make predictions about the kinds of reagents needed for the transformations you wish to achieve.

(↗) **Study Guide**
**Concept Map 13.5**

**Problem 13.25**    For each of the following transformations, decide whether the starting material has undergone oxidation or reduction or neither. Specify whether an oxidizing or reducing agent would be needed to carry out the change that is shown.

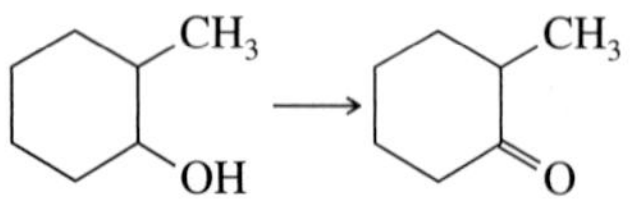

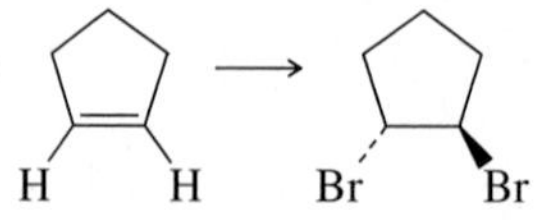

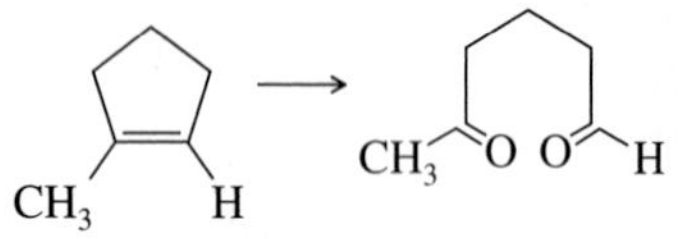

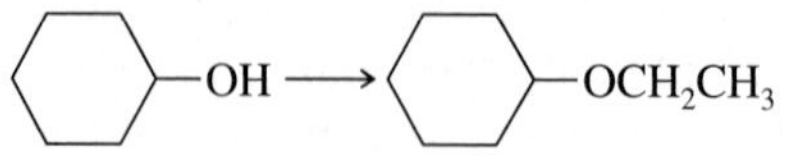

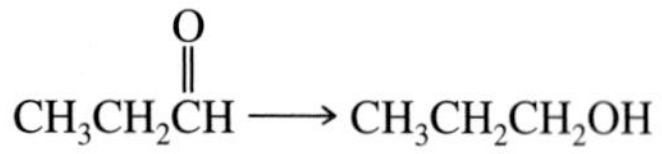

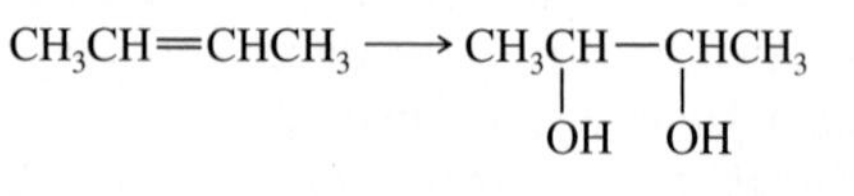

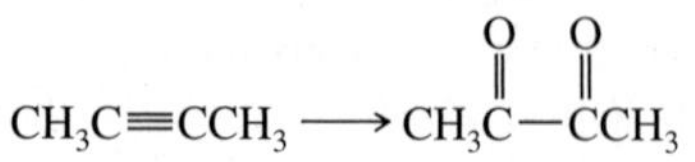

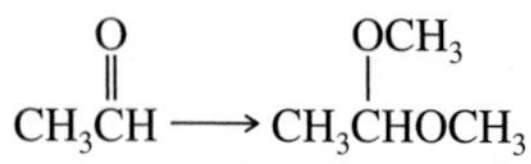

**Problem 13.26**   Methane is converted to carbon tetrachloride by successive replacement of hydrogen atoms by chlorine atoms. Methane represents the most highly reduced state of carbon, and carbon tetrachloride, the most highly oxidized. Calculate oxidation numbers for the carbon atom in methane, in carbon tetrachloride, and in the intermediate alkyl halides.

## B. Oxidation of Alcohols to Carbonyl Compounds

Primary and secondary alcohols are easily oxidized to carbonyl compounds. Tertiary alcohols are not. The structural feature necessary for this oxidation to take place is the presence of a hydrogen atom and a hydroxyl group on the same carbon atom.

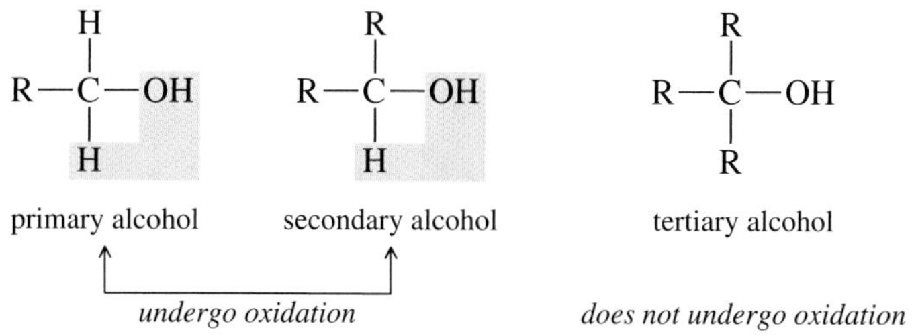

The general mechanism for the oxidation reaction is similar to that of the E2 reaction (p. 267) except that a carbon–oxygen double bond is being created in place of a carbon–carbon double bond.

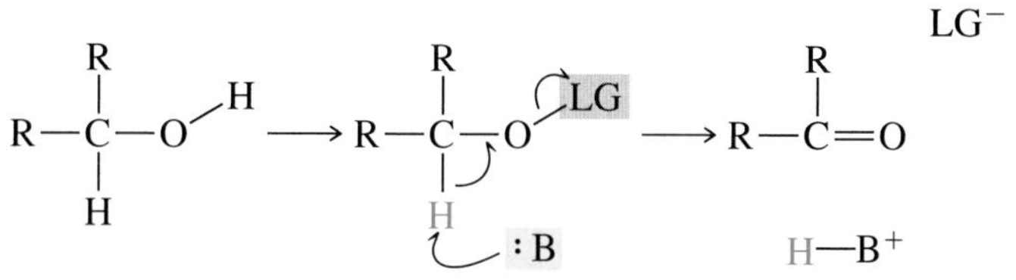

The reactions that follow will fit this pattern. The first steps of the oxidation reaction result in replacement of the hydrogen atom of the hydroxyl group and formation of a leaving group. The oxidation step involves removal of the hydrogen atom on the carbon atom undergoing oxidation with simultaneous loss of the leaving group and formation of the carbon–oxygen double bond.

## C. Chromium(VI) Reagents as Oxidizing Agents

Primary and secondary alcohols are easily oxidized by reagents containing chromium(VI), chromium in the $+6$ oxidation state. Tertiary alcohols are not. This difference in reactivity has been used as the basis of a qualitative test for primary and secondary alcohols. The alcohol is dissolved in acetone, and Jones reagent, a solution of chromium trioxide, $CrO_3$, dissolved in water and sulfuric acid, is added drop by drop. Primary and secondary alcohols give an instant blue-green color.

In aqueous acidic solutions, the chromium(VI) species that reacts with alcohols is the acid chromate ion, $HCrO_4^-$, which is yellow-orange in color. It is reduced to the blue-green ion Cr(III), chromium in the $+3$ oxidation state, by the alcohol, which in turn is oxidized. Primary alcohols give carboxylic acids by way of aldehydes; secondary alcohols give ketones under these conditions. For example:

$$3 \ CH_3CH_2CH_2CH_2OH \ + \quad 2 \ HCrO_4^- \quad + \ 4 \ H_2SO_4 \longrightarrow 3 \ CH_3CH_2CH_2\overset{\overset{\displaystyle O}{\|}}{C}H \ + \quad 2 \ Cr^{3+} \quad + \ 4SO_4^{2-} + 8 \ H_2O$$

|  1-butanol | acid chromate ion | | butanal | chromium(III) ion | |
| *primary alcohol* | *yellow-orange* | | *aldehyde* | *blue-green* | |

The aldehyde is easily further converted to a carboxylic acid. This type of reactivity for aldehydes was first seen in the ozonolysis reaction, where special conditions, the reductive work-up, had to be used to retain an aldehyde as a product; otherwise, it was converted by oxidation to the carboxylic acid (p. 311). The same thing happens here. The aldehyde reacts in aqueous solution with acid chromate to give the corresponding carboxylic acid.

$$3 \ CH_3CH_2CH_2\overset{\overset{\displaystyle O}{\|}}{C}H \ + \quad 2 \ HCrO_4^- \quad + \ 4 \ H_2SO_4 \longrightarrow 3 \ CH_3CH_2CH_2\overset{\overset{\displaystyle O}{\|}}{C}OH \ + \quad 2 \ Cr^{3+} \quad + \ 4 \ SO_4^{2-} + 5 \ H_2O$$

| butanal | acid chromate ion | | butanoic acid | chromium(III) ion | |
| *aldehyde* | *yellow-orange* | | *carboxylic acid* | *blue-green* | |

A secondary alcohol, such as *sec*-butyl alcohol, is oxidized to a ketone, which usually does not undergo further oxidation, as we also saw in the ozonolysis reaction (p. 312). For a ketone to be oxidized, carbon–carbon bonds would have to be broken because there are no more carbon–hydrogen bonds available on the carbonyl carbon atom. Ketones, therefore, survive under conditions in which aldehydes are readily oxidized to carboxylic acids.

$$3 \ CH_3CH_2\overset{\displaystyle CHCH_3}{\underset{\displaystyle OH}{|}} \ + \quad 2 \ HCrO_4^- \quad + \ 4 \ H_2SO_4 \longrightarrow 3 \ CH_3CH_2\overset{\overset{\displaystyle O}{\|}}{C}CH_3 \ + \quad 2 \ Cr^{3+} \quad + \ 4 \ SO_4^{2-} + 8 \ H_2O$$

| *sec*-butyl alcohol | acid chromate ion | 2-butanone | chromium(III) ion |
| *a secondary alcohol* | *yellow-orange* | | *blue-green* |

The tertiary alcohol *tert*-butyl alcohol is not oxidized by Jones reagent. The carbon atom bearing the hydroxyl group in this alcohol does not have a hydrogen atom on it. Here too, carbon–carbon bonds would have to be broken to allow oxidation to take place.

$$CH_3\overset{\displaystyle CH_3}{\underset{\displaystyle OH}{\overset{|}{\underset{|}{C}}}}CH_3 \quad + \quad HCrO_4^- \quad + \quad H_2SO_4 \longrightarrow \text{no oxidation}$$

| *tert*-butyl alcohol | acid chromate ion | *no change* |
| *a tertiary alcohol* | *yellow-orange* | *in color* |

Chromium(VI), present in the acid chromate ion, is the oxidizing agent. It is reduced to chromium(III) in several steps, with chromium(IV) and chromium(V) present at some stages of the reaction. It has been suggested that the reaction involves, first, the formation of a complex between the alcohol and the chromium atom, followed by a series of deprotonation and protonation steps leading to the loss of water from the complex and the formation of the chromate ester of the alcohol in much the same way as a carboxylic acid ester is formed (pp. 623–624).

The oxidation states of carbon and chromium do not change during the formation of the chromate ester. The removal of the hydrogen atom from the carbon atom that originally bore the hydroxyl group, along with the loss of the chromium-containing leaving group, is the oxidation–reduction step of the reaction.

Chromium(IV) is formed in this step and is ultimately reduced to chromium(III). The exact details of all the processes involved are quite complex. It is sufficient if you understand that a hydroxyl group and a hydrogen atom on the same carbon atom are necessary for the oxidation reaction that proceeds through the formation of the chromate ester of the alcohol.

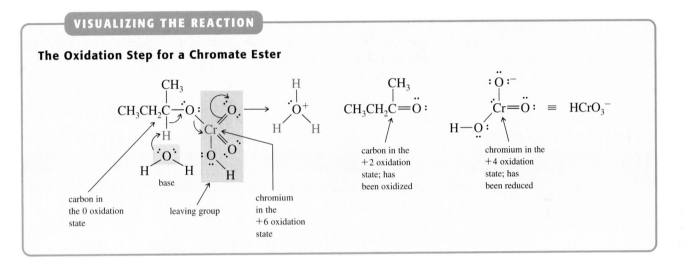

**VISUALIZING THE REACTION**

**The Oxidation Step for a Chromate Ester**

carbon in the 0 oxidation state

base

leaving group

chromium in the +6 oxidation state

carbon in the +2 oxidation state; has been oxidized

chromium in the +4 oxidation state; has been reduced

**Problem 13.27** Write the structures of the following alcohols, and predict whether each one will give a positive test with Jones reagent, the chromium trioxide reagent described in this section. For each alcohol you expect to give a positive test, write an equation for the reaction.

(a) 4-methyl-2-pentanol     (b) 2-*tert*-butylcyclohexanol     (c) cholesterol (p. 490)
(d) 1-ethylcyclopentanol     (e) 1-pentanol     (f) 2,3,3-trimethyl-2-butanol

## D. Syntheses of Aldehydes and Ketones. Selective Oxidations with Pyridinium Chlorochromate

A convenient reagent for the conversion of alcohols to aldehydes and ketones is made by dissolving chromium trioxide in hydrochloric acid and adding the basic solvent pyridine to the solution.

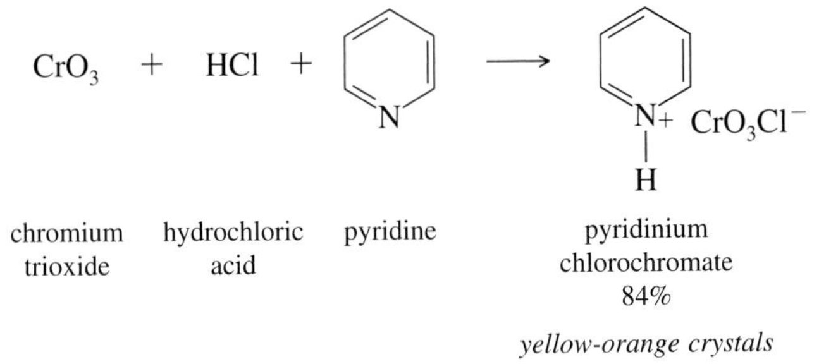

$$CrO_3 \ + \ HCl \ + \ \text{[pyridine]} \longrightarrow \text{[pyridinium chlorochromate]} \ CrO_3Cl^-$$

chromium trioxide     hydrochloric acid     pyridine

pyridinium chlorochromate 84%

*yellow-orange crystals*

This reagent is called pyridinium chlorochromate and is used in organic solvents such as dichloromethane to oxidize primary and secondary alcohols. For example, 1-decanol is oxidized to decanal with a high yield.

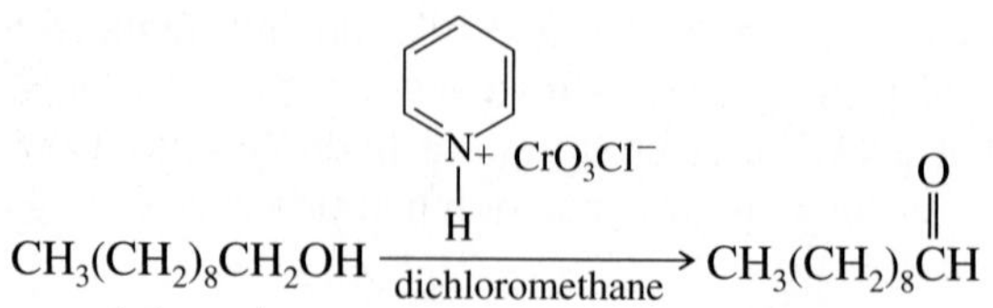

$$CH_3(CH_2)_8CH_2OH \xrightarrow[\text{dichloromethane}]{} CH_3(CH_2)_8\overset{\overset{\displaystyle O}{\|}}{C}H$$

1-decanol

decanal
92%

The oxidation stops at the aldehyde stage because acid chromate ion is not present to form a chromate ester with the aldehyde, which is the necessary intermediate for further oxidation to the carboxylic acid by chromium(VI).

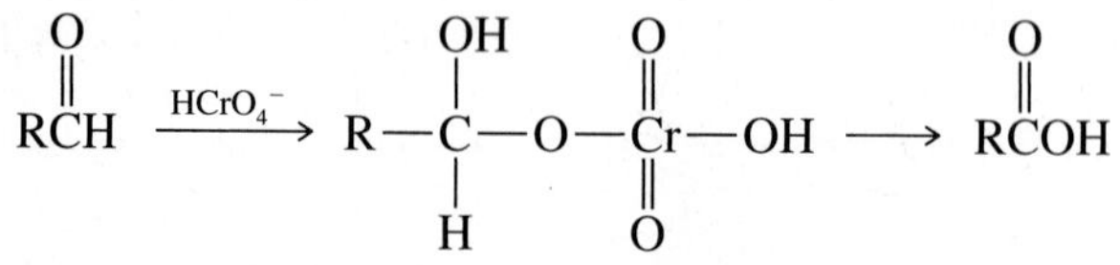

$$R\overset{\overset{\displaystyle O}{\|}}{C}H \xrightarrow{HCrO_4^-} R-\overset{\overset{\displaystyle OH}{|}}{\underset{\underset{\displaystyle H}{|}}{C}}-O-\overset{\overset{\displaystyle O}{\|}}{Cr}-OH \longrightarrow R\overset{\overset{\displaystyle O}{\|}}{C}OH$$

aldehyde      chromate ester from aldehyde      carboxylic acid

Similarly, 4-*tert*-butylcyclohexanol is oxidized to 4-*tert*-butylcyclohexanone.

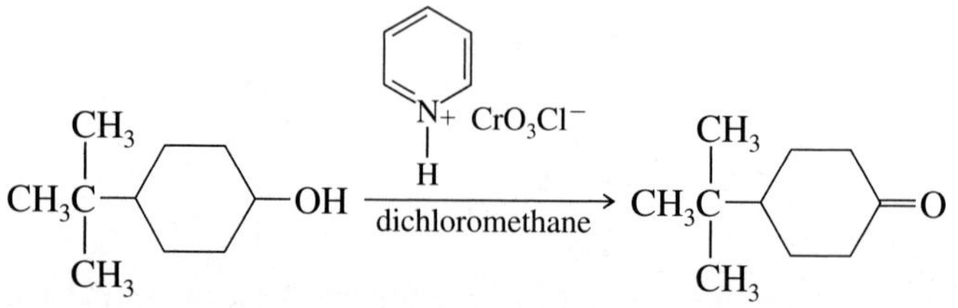

4-*tert*-butylcyclohexanol        4-*tert*-butylcyclohexanone

    Pyridinium chlorochromate is also useful for oxidizing alcohols containing other functional groups that may react with aqueous acid, such as double bonds. For example, the unsaturated alcohol citronellol, which is one of the fragrant compounds in rose oil and geranium oil, is oxidized by this reagent to the corresponding aldehyde in good yield.

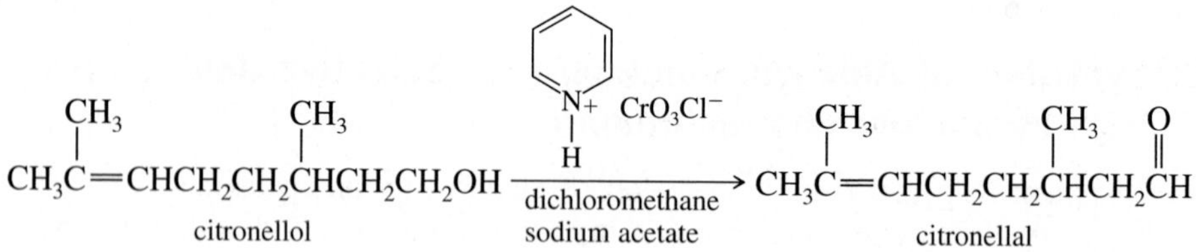

citronellol      citronellal

The double bond is unaffected by this reaction.

**Problem 13.28**      Write an equation showing the product you would expect if the oxidation of citronellol were carried out with chromic acid in the presence of water. Can you foresee any complications from having a strong acid such as sulfuric acid in the reaction mixture?

**Problem 13.29**      Complete the following equations.

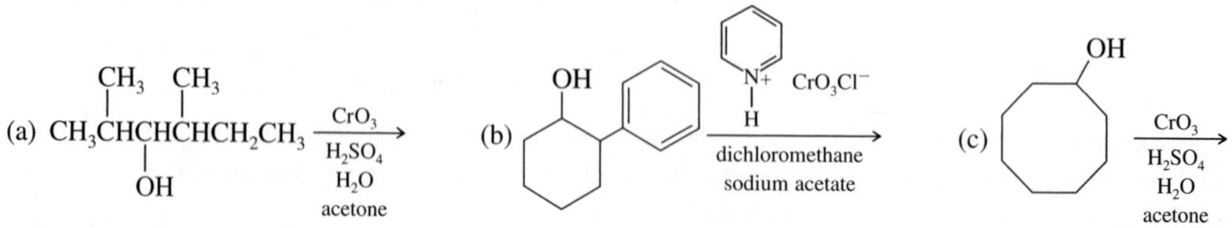

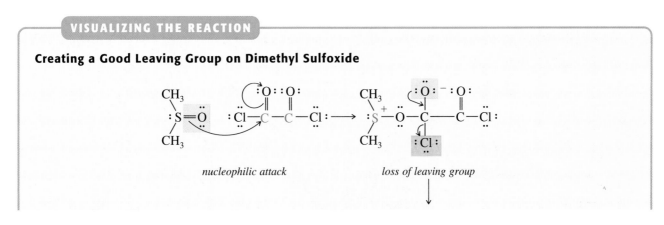

(d), (e), (f) reaction schemes

## E. The Swern Oxidation

While Cr(VI) compounds are efficient oxidizing agents for primary and secondary alcohols, they have an important disadvantage. Compounds containing chromium, especially Cr(VI), are toxic, and some are carcinogenic. Such reagents have to be used and disposed of with great care. A number of other methods for the oxidation of alcohols have been developed, many of which use dimethyl sulfoxide as the oxidizing agent. The Swern oxidation is one of the most widely used of these methods. It uses dimethyl sulfoxide [$(CH_3)_2SO$], dioxalyl chloride [$(COCl)_2$], and a base such as triethylamine [$(CH_3CH_2)_3N$] at low temperatures to achieve oxidation. For example, the primary alcohol 1-decanol is converted into decanal under these conditions.

$$CH_3(CH_2)_8CH_2OH \xrightarrow[\substack{\text{dichloromethane} \\ -60\,°C}]{\substack{CH_3SCH_3,\ ClC—CCl}} \xrightarrow[\substack{\text{dichloromethane} \\ -60\,°C}]{(CH_3CH_2)_3N} CH_3(CH_2)_8CH$$

1-decanol, decanal ~95%

Secondary alcohols are converted into ketones. 3,3-Dimethyl-2-butanol gives *tert*-butyl methyl ketone by the Swern oxidation.

$$\underset{\substack{| \\ CH_3 \\ | \\ OH}}{CH_3C}—CHCH_3 \xrightarrow[\substack{\text{dichloromethane} \\ -60\,°C}]{\substack{CH_3SCH_3,\ ClC—CCl}} \xrightarrow[\substack{\text{dichloromethane} \\ -60\,°C}]{(CH_3CH_2)_3N} \underset{\substack{| \\ CH_3}}{CH_3C}—CCH_3$$

3,3-dimethyl-2-butanol, *tert*-butyl methyl ketone 76%

The way in which dimethyl sulfoxide oxidizes an alcohol combines elements of many different reactions that we have studied. First, the dimethyl sulfoxide reacts as a nucleophile with dioxalyl chloride to give a species in which the oxygen atom of dimethyl sulfoxide has been converted into a good leaving group. These steps are similar to those of the reaction of an alcohol with thionyl chloride (p. 496).

### VISUALIZING THE REACTION

**Creating a Good Leaving Group on Dimethyl Sulfoxide**

*nucleophilic attack*            *loss of leaving group*

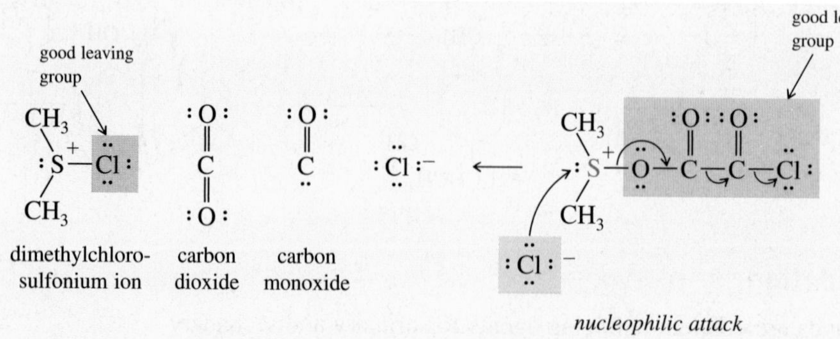

good leaving group

dimethylchloro-sulfonium ion    carbon dioxide    carbon monoxide

nucleophilic attack

good leaving group

The dimethylchlorosulfonium ion that is the product of the first part of the reaction reacts with the alcohol to give a new sulfonium ion.

## VISUALIZING THE REACTION

### Reaction of an Alcohol with the Dimethylchlorosulfonium Ion

good leaving group

nucleophilic attack          deprotonation

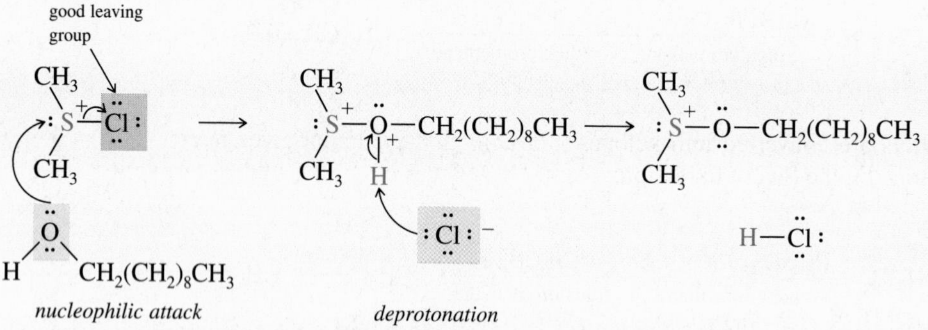

The new sulfonium ion reacts with base in the second stage of the reaction. The most acidic proton in the sulfonium ion is located next to the positively charged sulfur atom. The negative charge on the resulting carbanion is stabilized by the adjacent positive charge.

## VISUALIZING THE REACTION

### The Oxidation Step of the Swern Oxidation

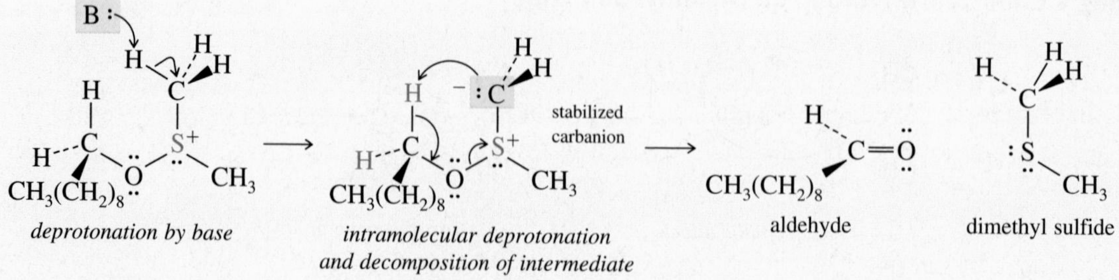

deprotonation by base

stabilized carbanion

intramolecular deprotonation and decomposition of intermediate

aldehyde          dimethyl sulfide

Once the carbanion forms on the carbon next to the sulfur atom, it removes a proton from the carbon adjacent to the oxygen atom derived from the alcohol, creating a flow of electrons toward the positively charged sulfur atom. The stable products of the reaction are dimethyl sulfide and the oxidation product of the alcohol.

⊘ **Study Guide**
**Concept Map 13.6**

**Problem 13.30**   Predict the products of the following reactions.

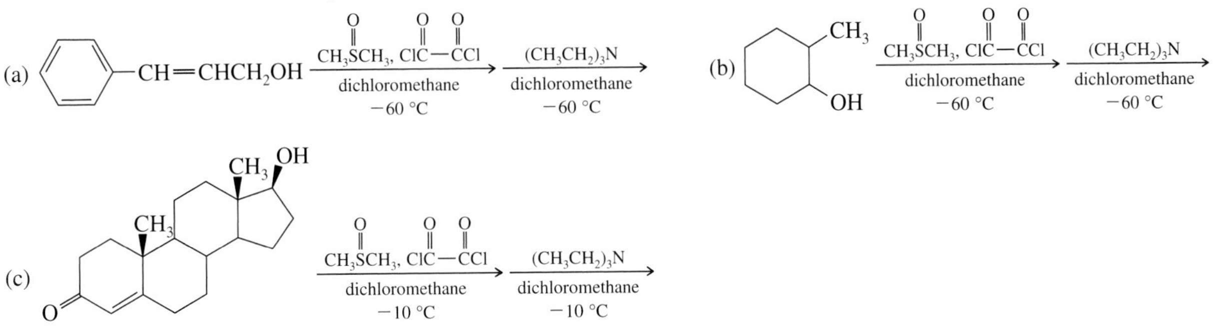

(a) 
$$\text{C}_6\text{H}_5-\text{CH}=\text{CHCH}_2\text{OH} \xrightarrow[\substack{\text{dichloromethane} \\ -60\ °C}]{\text{CH}_3\text{SCH}_3,\ \text{ClC}-\text{CCl}} \xrightarrow[\substack{\text{dichloromethane} \\ -60\ °C}]{(\text{CH}_3\text{CH}_2)_3\text{N}}$$

(b) 
$$\xrightarrow[\substack{\text{dichloromethane} \\ -60\ °C}]{\text{CH}_3\text{SCH}_3,\ \text{ClC}-\text{CCl}} \xrightarrow[\substack{\text{dichloromethane} \\ -60\ °C}]{(\text{CH}_3\text{CH}_2)_3\text{N}}$$

(c) 
$$\xrightarrow[\substack{\text{dichloromethane} \\ -10\ °C}]{\text{CH}_3\text{SCH}_3,\ \text{ClC}-\text{CCl}} \xrightarrow[\substack{\text{dichloromethane} \\ -10\ °C}]{(\text{CH}_3\text{CH}_2)_3\text{N}}$$

**Problem 13.31**   Equations for two of the experiments that were carried out to determine the mechanism of the Swern oxidation are shown below. How do the experimental results support the mechanism shown earlier in this section? Your answers should include partial mechanisms that explain the observed products as well as a verbal explanation.

Experiment 1.

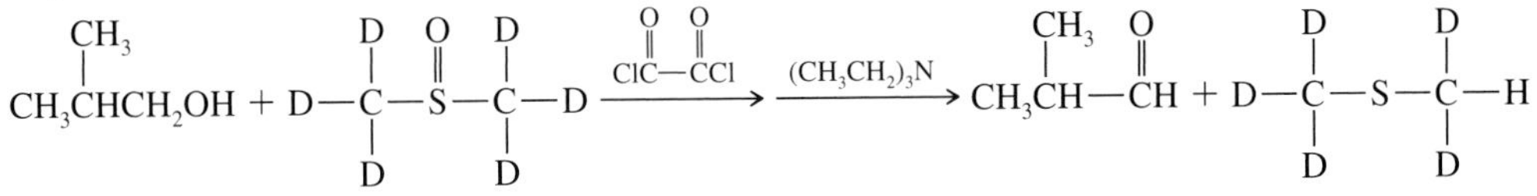

$$\text{CH}_3\text{CHCH}_2\text{OH} + \text{D}-\overset{\text{D}}{\underset{\text{D}}{\text{C}}}-\text{S}-\overset{\text{D}}{\underset{\text{D}}{\text{C}}}-\text{D} \xrightarrow[]{\text{ClC}-\text{CCl}} \xrightarrow{(\text{CH}_3\text{CH}_2)_3\text{N}} \text{CH}_3\text{CH}-\text{CH} + \text{D}-\overset{\text{D}}{\underset{\text{D}}{\text{C}}}-\text{S}-\overset{\text{D}}{\underset{\text{D}}{\text{C}}}-\text{H}$$

Experiment 2.

$$\text{Ph}-\overset{+}{\text{S}}-\text{OCH}_3 + \text{CH}_3\text{O}^-\ \text{Na}^+ \xrightarrow[\text{methanol}]{} \text{Ph}-\overset{\text{O}}{\underset{}{\text{S}}}-\text{Ph} + \text{CH}_3\text{OCH}_3$$

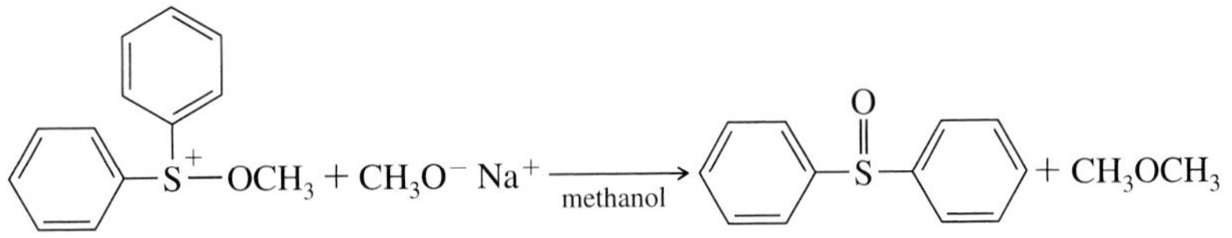

# F. The Art of Solving Problems

**PROBLEM:** How could we carry out the following transformation?

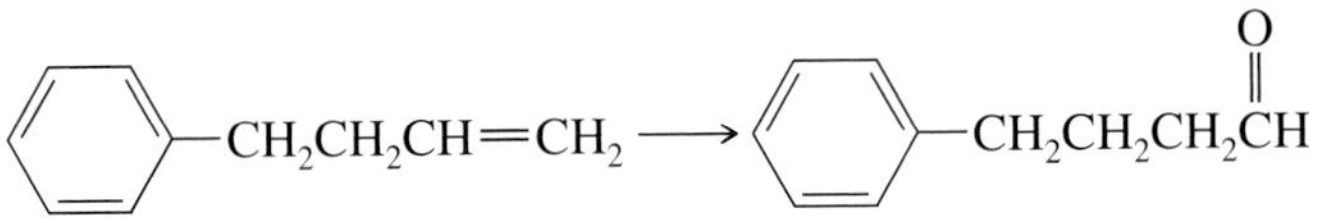

$$\text{C}_6\text{H}_5-\text{CH}_2\text{CH}_2\text{CH}=\text{CH}_2 \longrightarrow \text{C}_6\text{H}_5-\text{CH}_2\text{CH}_2\text{CH}_2\overset{\text{O}}{\overset{\|}{\text{CH}}}$$

## Solution

1. What are the connectivities of the two compounds? How many carbon atoms does each contain? Are there any rings? What are the positions of branches and functional groups on the carbon skeleton?

   Both compounds have a straight chain of four carbon atoms with an aryl group at one end. The starting material has a carbon–carbon double bond at the end of the chain; the product is an aldehyde.

2. How do the functional groups change in going from starting material to product? Does the starting material have a good leaving group?

An aldehyde functional group is formed at the first carbon atom of the double bond. There is no good leaving group in the starting material.

3. Is it possible to dissect the structures of the starting material and product to see which bonds must be broken and which formed?

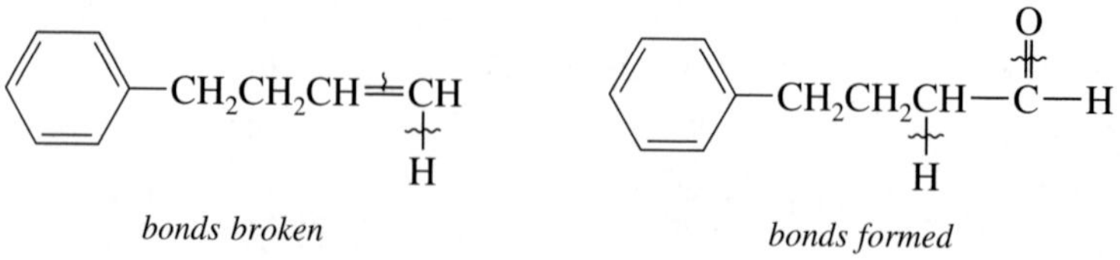

*bonds broken*                    *bonds formed*

A $\pi$ bond and a carbon–hydrogen bond must be broken. Carbon–hydrogen and carbon–oxygen bonds are formed.

4. New bonds are created when an electrophile reacts with a nucleophile. Do we recognize any part of the product molecule as coming from a good nucleophile or an electrophilic addition?

The formation of bonds to a hydrogen and an oxygen on adjacent carbon atoms looks like an addition of H—OH to the double bond, in an anti-Markovnikov fashion.

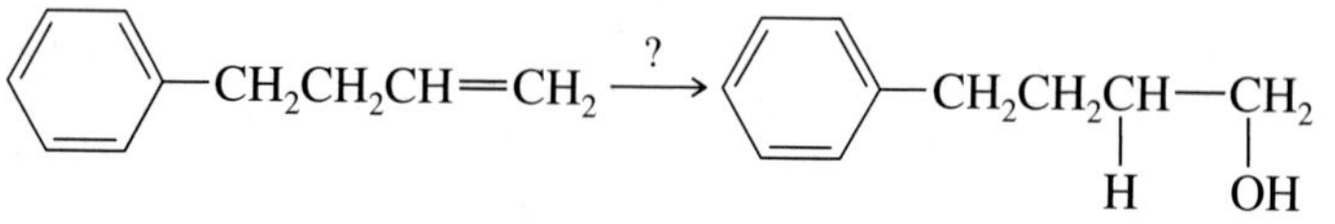

5. What type of compound would be a good precursor to the product?

An aldehyde is formed by the oxidation of a primary alcohol.

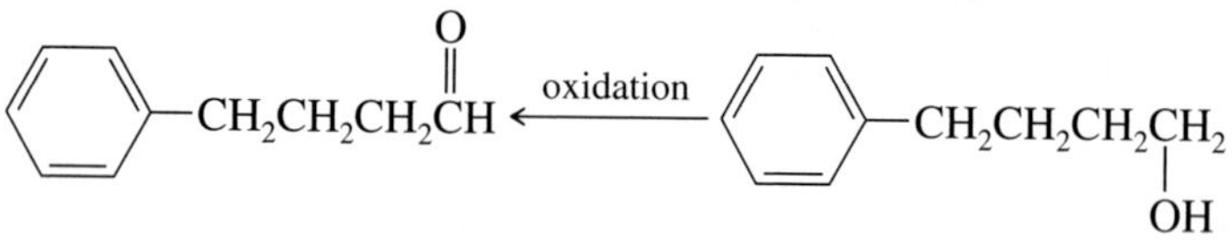

6. After this step, do we see how to get from starting material to product? If not, we need to analyze the structure obtained in step 5 by applying questions 4 and 5 to it.

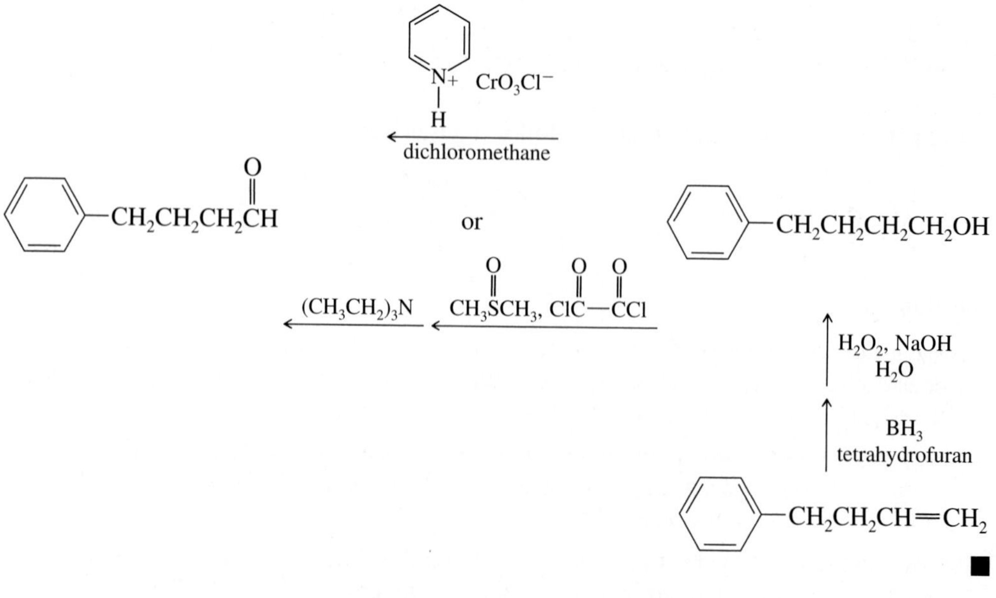

**Problem 13.32**    How would you carry out each of the following transformations?

(a)

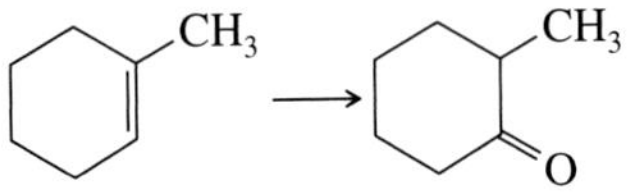

(b) $CH_3CH_2CH{=}CH_2 \longrightarrow CH_3CH_2\underset{\underset{NH_2}{|}}{C}HCH_3$

(c)

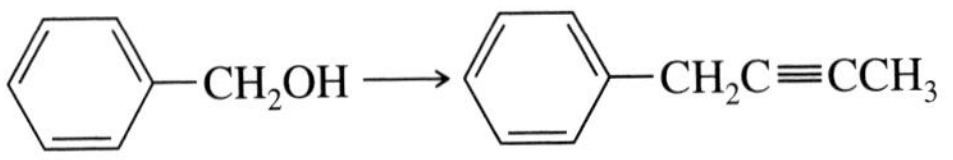

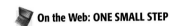
On the Web: **ONE SMALL STEP**

**Table 13.1    Summary of the Reactions of Alcohols**

| Alcohol | Reagent | Intermediates | Product |
|---|---|---|---|
| | | **Conversion to Alkyl Halides** | |
| ROH | HX | $ROH_2^+ X^-$ | RX |
| | SOCl₂ | R—O—S(=O)(Cl) Cl⁻ | RCl |
| | PBr₃ | 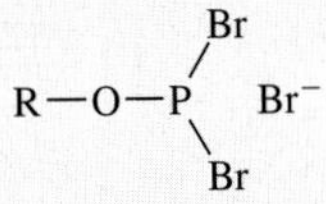 R—O—P(Br)(Br) Br⁻ | RBr |
| | | **Conversion to Ethers** | |
| ROH | NaH | $RO^- + R'X$ (primary or secondary) | ROR′ |
| 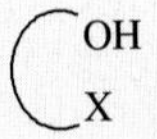 —OH —X | NaOH | 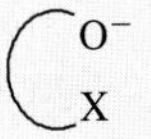 —O⁻ —X |  (five- or six-membered ring) |
| 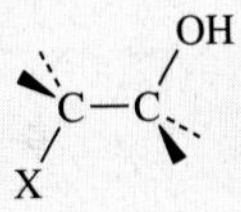 X—C—C—OH | NaOH | 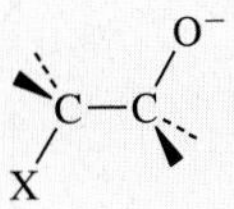 X—C—C—O⁻ |  (epoxide) |

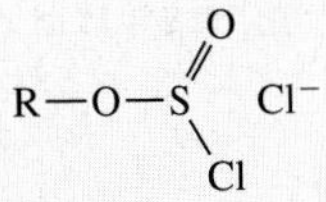

(continued)

**Table 13.1**    **(Continued)**

| Alcohol | Reagent | Intermediates | Product |
|---|---|---|---|
| | | **Oxidation Reactions** | |

$RCH_2OH$

$CrO_3$, $H_2O$, $H_3O^+$

$$RCH{-}O{-}\overset{\overset{O}{\|}}{\underset{\underset{O}{\|}}{Cr}}{-}OH$$
(with H on RCH)

$$\overset{O}{\overset{\|}{R}CH},\ \overset{O}{\overset{\|}{R}C}OH$$

$CH_3SCH_3$, $ClC{-}CCl$, $(CH_3CH_2)_3N$

or

$$RCH{-}O{-}\overset{+}{S}{-}CH_3$$
(with H below and $CH_3$ above)

$$\overset{O}{\overset{\|}{R}CH}$$

(pyridinium) $CrO_3Cl^-$

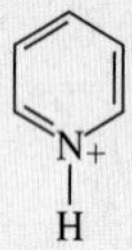

$$RCH{-}O{-}\overset{\overset{O}{\|}}{Cr}{-}OH$$
(with H and O)

$$\underset{H}{\overset{R}{R{C}}}{-}OH$$

$CH_3SCH_3$, $ClC{-}CCl$, $(CH_3CH_2)_3N$

$$\underset{H}{\overset{R}{R{C}}}{-}O{-}\overset{+}{S}{-}CH_3$$
(with $CH_3$)

or

$CrO_3$, $H_2O$, $H_3O^+$

or

$$\overset{O}{\overset{\|}{R}CR}$$

$$\underset{H}{\overset{R}{R{C}}}{-}O{-}\overset{\overset{O}{\|}}{Cr}{-}OH$$

(pyridinium) $CrO_3Cl^-$

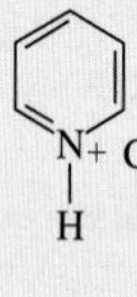

$$\underset{R}{\overset{R}{R{C}}}{-}OH$$

$CH_3SCH_3$, $ClC{-}CCl$, $(CH_3CH_2)_3N$

or

$CrO_3$, $H_2O$, $H_3O^+$

or

(pyridinium) $CrO_3Cl^-$

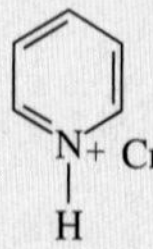

not oxidized

**Table 13.2** Summary of the Cleavage Reactions of Oxiranes

| Oxiranes | Reagent | Intermediate | Product |
|---|---|---|---|
| | | **Acids** | |

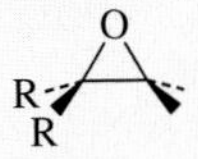

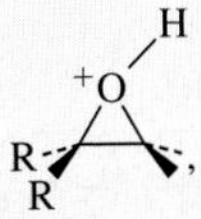

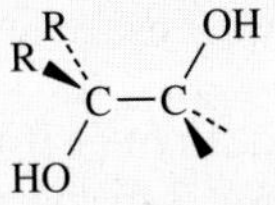

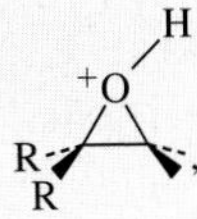

**Nucleophiles**

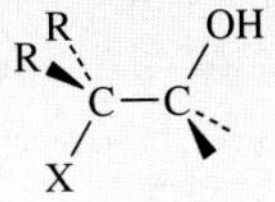

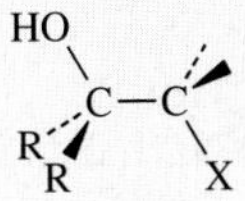

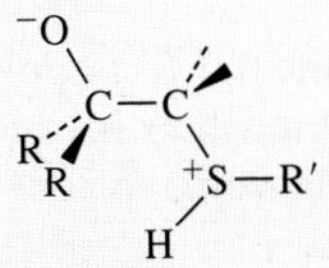

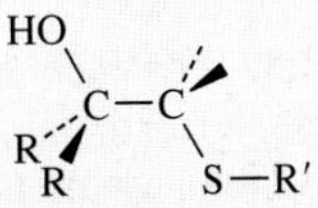

Study Guide
Concept Map 13.7

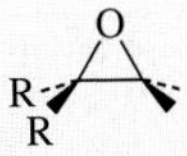

## ADDITIONAL PROBLEMS

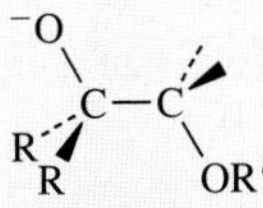

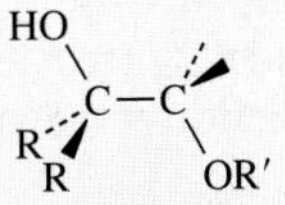

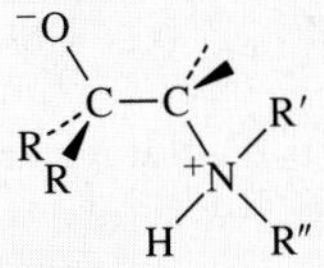

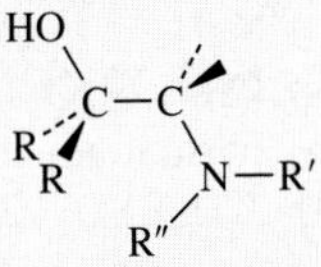

**13.33** Name each of the following compounds according to the IUPAC rules, including stereochemistry when shown.

(a) $HOCH_2CH_2CH_2OH$

(b) $CH_3OCH_2CH_2OCH_3$

(c) $CH_3CH_2OCH_2\overset{\underset{\displaystyle CH_3}{|}}{\underset{\underset{\displaystyle CH_3}{|}}{C}}CH_2OH$

(d) $CH_3OCH_2CH_2Br$

(e) 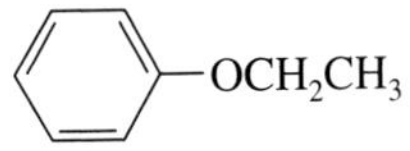 $-OCH_2CH_3$

(f) $CH_3CH_2OCH_2CH_2OH$

(g)

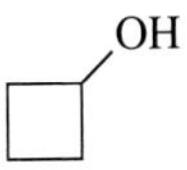

(h)

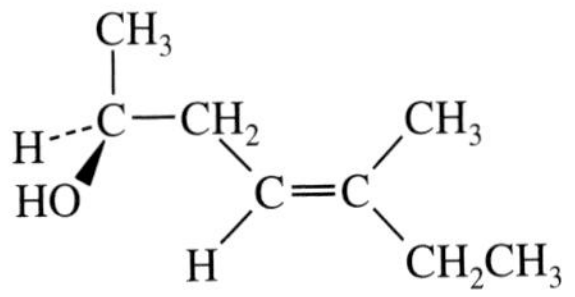

(i) 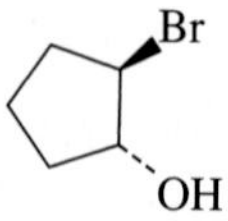 [structure: cyclopentane ring with Br and OH]

(j)  [structure: $C_6H_5$—$CH_2OCH_2$—$C_6H_5$]

(k) 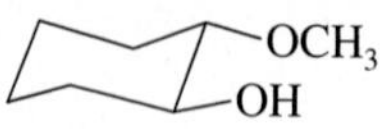 [structure: cyclohexane ring with $OCH_3$ and $OH$]

(l) 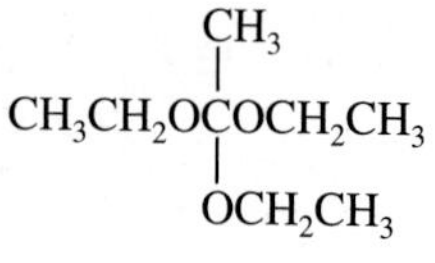 $CH_3CH_2OCOCH_2CH_3$ with $CH_3$ and $OCH_2CH_3$ substituents

**13.34** Draw structures for the following compounds. When necessary, show stereochemistry by the use of appropriate conventions.

(a) 1-chloro-1-ethoxyethene
(b) (*E*)-1-methoxy-2-propoxyethene
(c) 1,3-dichloro-2-propanol
(d) cyclobutylmethanol
(e) (*R*)-3-methyl-5-hexen-3-ol
(f) (*S*)-2-chloro-1-propanol
(g) (*S*)-3-methyl-1-pentyn-3-ol
(h) 2-nitroethanol
(i) (*R*)-5,5-dimethyl-3-heptanol
(j) (3*S*,4*R*)-4-methyl-3-hexanol

**13.35** Write equations for the reactions of 1-propanol with the following reagents under the conditions shown.

(a) HBr, $\Delta$      (b) PBr$_3$      (c) SOCl$_2$, pyridine
(d) NaH, then $CH_3CH_2CH_2CH_2Br$      (e) $H_3O^+$, cold
(f) $H_2SO_4$, $\Delta$      (g) $CrO_3$, $H_2O$, $H_3O^+$
(h) pyridinium chlorochromate, dichloromethane
(i) $(CH_3)_2SO$, $(COCl)_2$, then $(CH_3CH_2)_3N$

**13.36** Write equations for the reactions of 2-hexanol with the following reagents under the conditions shown.

(a) HBr, $\Delta$      (b) PBr$_3$      (c) SOCl$_2$, pyridine
(d) NaH      (e) ZnCl$_2$, HCl      (f) $H_2SO_4$, $\Delta$
(g) $CrO_3$, $H_2O$, $H_3O^+$
(h) pyridinium chlorochromate, dichloromethane
(i) $(CH_3)_2SO$, $(COCl)_2$, then $(CH_3CH_2)_3N$

**13.37** For each set of experimental facts presented below, give an explanation for the observed trend, based on the relationships between structure and physical properties. Whenever possible, use pictures as well as words in your answers.

(a) 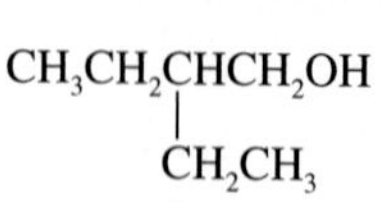 $CH_3CH_2CHCH_2OH$ with $CH_2CH_3$
solubility 0.63 g
in 100 mL $H_2O$

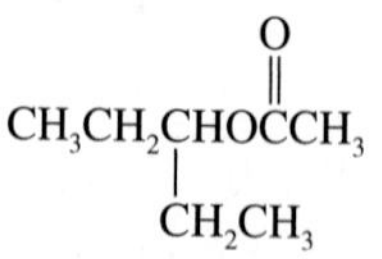

$CH_3CH_2CHOCCH_3$ (O double bond) with $CH_2CH_3$
solubility 0.06 g
in 100 mL $H_2O$

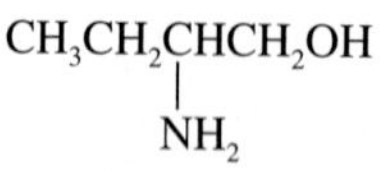

$CH_3CH_2CHCH_2OH$ with $NH_2$
infinitely soluble
in $H_2O$

(b) $CH_3CH_2OH$
pK$_a$ 15.9

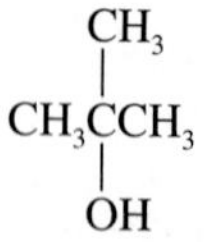

$CH_3CCH_3$ with $CH_3$ and $OH$
pK$_a$ ~ 18

$CF_3CH_2OH$
pK$_a$ 12.4

(c) $CH_3CH_2CH_2CH_2OH$
bp 118 °C
solubility 7.9 g
in 100 mL $H_2O$

$CH_3CH_2CH_2CH_2CH_2CH_2OH$
bp 157 °C
solubility 0.59 g
in 100 mL $H_2O$

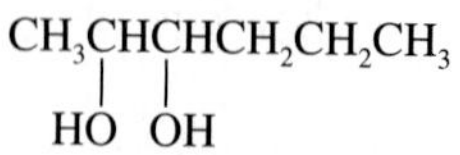

$CH_3CHCHCH_2CH_2CH_3$ with HO OH
bp 207 °C
infinitely soluble
in $H_2O$

(d) $CH_3CH_2CH_2CH_2OH$
solubility 7.9 g
in 100 mL $H_2O$

$CH_3CH_2CH_2CH_2SH$
slightly soluble
in $H_2O$

**13.38** Complete the following equations.

(a) $\underset{\underset{\displaystyle CH_3}{|}}{CH_3CHCH_2OH}$ $\xrightarrow[\text{pyridine}]{SOCl_2}$

(b) $\underset{\underset{\displaystyle OH}{|}}{\overset{\overset{\displaystyle CH_3}{|}}{CH_3CCH_2CH_3}}$ $\xrightarrow{\text{HCl (conc)}}$

(c) $\underset{\underset{\displaystyle OH}{|}}{CH_3CH_2CHCH_3}$ $\xrightarrow{PBr_3}$

(d) $-CH_2CH_3$ $\xrightarrow[\text{diglyme}]{BH_3}$ $\xrightarrow[H_2O]{H_2O_2,\ NaOH}$

(e) $HOCH_2(CH_2)_4CH_2OH$ $\xrightarrow{\text{HBr(g), excess}}$

(f) $-OH$ $\xrightarrow{HI}$

(g) $\underset{\underset{\displaystyle HO\ \ CH_3}{|\ \ \ \ |}}{\overset{\overset{\displaystyle CH_3\ \ CH_3}{|\ \ \ \ |}}{CH_3CHCHCH_2OCCH_3}}$ $\xrightarrow[\text{dichloromethane}]{}$

(h) $\xrightarrow[]{CH_3SCH_3,\ ClC-CCl}$ $(CH_3CH_2)_3N$ $\xrightarrow{}$

**13.39** More practice in recognizing reactions follows.

(a) $\xrightarrow[\substack{H_2O \\ \text{acetone}}]{CrO_3,\ H_2SO_4}$

(b) $\underset{\underset{\displaystyle CH_2CH_3}{|}}{HOCH_2CHCH_2CH_3}$ $\xrightarrow[\text{pyridine}]{PBr_3}$

(c) $\xrightarrow[\substack{H_2O \\ \text{acetone}}]{CrO_3,\ H_2SO_4}$

(d) $-CH_2OH$ $\xrightarrow[\substack{\text{pyridine} \\ \Delta}]{SOCl_2}$

(e) $\underset{\underset{\displaystyle OH}{|}}{\overset{\overset{\displaystyle CH_3}{|}}{BrCH_2CCH=CH_2}}$ $\xrightarrow[H_2O]{NaOH}$

(f) $\xrightarrow[\text{dichloromethane}]{}$

(g) $\underset{\underset{\displaystyle H}{\diagdown}}{CH_3}\overset{O}{\diagup}$ $\xrightarrow[\text{ethanol}]{}$

(h) $\underset{\underset{\displaystyle CH_3}{|}}{CH_3C}=C-C\equiv\underset{\underset{\displaystyle OH}{|}}{\overset{\overset{\displaystyle CH_3}{|}}{CCH_3}}$ $\xrightarrow[\substack{Pd/BaSO_4 \\ \text{quinoline}}]{H_2}$

**13.40** Find the products or reagents corresponding to each of the capital letters. Show stereochemistry by appropriate conventions. When a racemic mixture forms, show the structure for one enantiomer and write "and enantiomer" below it.

(a) $-CH_2CH_2CH=CH_2$ $\xrightarrow[\text{tetrahydrofuran}]{BH_3}$ A $\xrightarrow[H_2O]{H_2O_2,\ NaOH}$ B $\xrightarrow[\text{dichloromethane}]{}$ C

(b) $\blacktriangleleft CH_3$ $\xrightarrow{\text{9-BBN}}$ D + E; $\quad$ D $\xrightarrow[H_2O]{H_2O_2,\ NaOH}$ F; $\quad$ E $\xrightarrow[H_2O]{H_2O_2,\ NaOH}$ G
$\phantom{xxxxxxxxxxxxx}$ *major* $\phantom{x}$ *minor*

(c) [cyclohexenyl-benzene] $\xrightarrow{\text{9-BBN}}$ H $\xrightarrow[\text{H}_2\text{O}]{\text{H}_2\text{O}_2,\ \text{NaOH}}$ I $\xrightarrow[\text{H}_2\text{O}]{\text{CrO}_3,\ \text{H}_2\text{SO}_4}$ K

(d) [cyclopentyl]—$\text{CH}_2\text{CH}_2\text{Br}$ $\xrightarrow[\text{NH}_3(\text{l})]{\text{HC}\equiv\text{C}:^-\ \text{Na}^+}$ M $\xrightarrow[\text{HgSO}_4]{\text{H}_2\text{SO}_4,\ \text{H}_2\text{O}}$ N

(c) ... $\downarrow$ $\begin{array}{c}\text{H}_2\text{O}\\\text{H}_2\text{SO}_4\end{array}$  L

... $\downarrow$ $\begin{array}{c}\text{PBr}_3\\\text{pyridine}\end{array}$  J

(d) ... $\downarrow$ $\begin{array}{c}\text{KOH (4 M)}\\\text{ethanol}\\\Delta\end{array}$  O $\xrightarrow[\text{carbon tetrachloride}]{\text{Br}_2}$ P;  O $\xrightarrow{\text{HBr}}$ Q $\xrightarrow[\text{ethanol}]{\text{NaCN}}$ R

(e) $\xrightarrow{\text{S}}$

**13.41** Supply the reagents that are necessary and the intermediate compounds that will form in the following transformations. There may be more than one good way to carry out each synthesis.

(a) $\text{CH}_3\overset{\text{CH}_3}{\underset{|}{\text{CH}}}\text{CH}_2\text{CH}=\text{CH}_2 \longrightarrow \text{CH}_3\overset{\text{CH}_3}{\underset{|}{\text{CH}}}\text{CH}_2\overset{\text{O}}{\overset{\|}{\text{C}}}\text{CH}_3$

(b) $\text{CH}_3(\text{CH}_2)_4\text{CH}=\text{CH}_2 \longrightarrow \text{CH}_3(\text{CH}_2)_4\text{CH}_2\overset{\text{O}}{\overset{\|}{\text{C}}}\text{H}$

(c) [phenyl]—$\text{CH}_2\text{CH}_2\text{CH}=\text{CH}_2 \longrightarrow$ [phenyl]—$\text{CH}_2\text{CH}_2\text{CH}_2\text{CH}_2\text{O}$—[phenyl]

(d) $\longrightarrow$

(e) $\longrightarrow$

(f) [phenyl]—$\text{CH}=\text{CH}_2 \longrightarrow$ [phenyl]—$\overset{}{\underset{\underset{\text{OH}}{|}}{\text{CH}}}\text{CH}_2\text{N(CH}_3)_2$

**13.42** The following sequence of reactions is a good way to convert a 1-alkylcyclohexene to a 3-alkylcyclohexene. Draw structural formulas showing the stereochemistry for each step of the reaction sequence, and show why it gives the desired product.

[cyclohexene-R] $\xrightarrow{\text{BH}_3}$ $\xrightarrow[\text{H}_2\text{O}]{\text{H}_2\text{O}_2,\ \text{NaOH}}$ [cyclohexane-R-OH] $\xrightarrow[\text{pyridine}]{\text{TsCl}}$ [cyclohexane-R-OTs] $\xrightarrow[\text{\textit{tert}-butyl alcohol}]{\overset{\text{CH}_3}{\underset{\text{CH}_3}{\text{CH}_3\text{C}-\text{O}^-\ \text{K}^+}}}$ [cyclohexene-R]

**13.43** An important component of the visual pigment in the eye is retinal, the aldehyde structurally related to vitamin A. Retinal, as it exists in the retina before absorption of light, has the *Z* configuration at the double bond at carbon 11. Absorption of light converts it to the *E* configuration at that double bond.

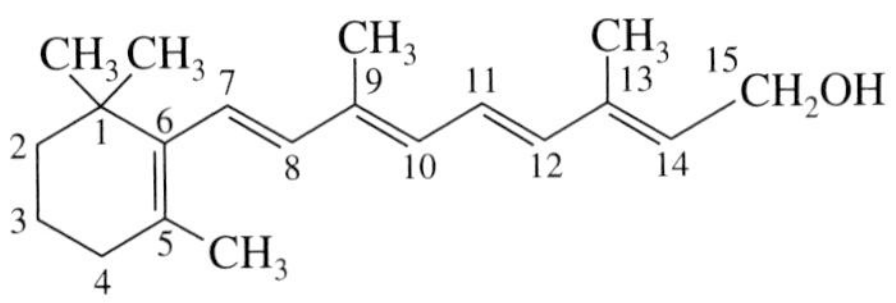

vitamin A

(a)  Draw the structure of (11*E*)-retinal.
(b)  Draw the structure of (11*Z*)-retinal.
(c)  What reagent would you use to oxidize vitamin A to retinal?

**13.44** The following reactions were carried out in research into antifungal agents isolated from natural sources. Fill in the structural formulas for the compounds indicated by letters.

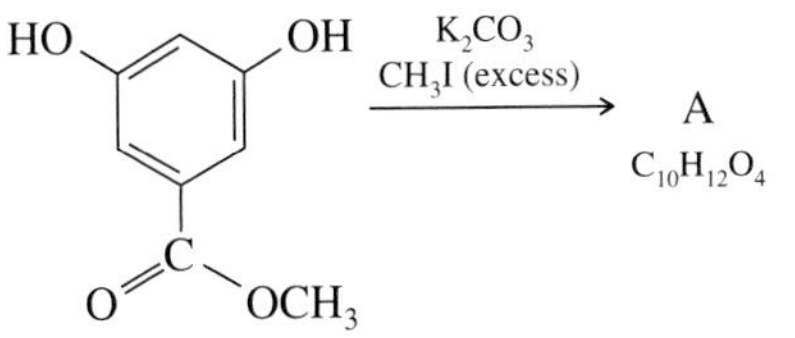

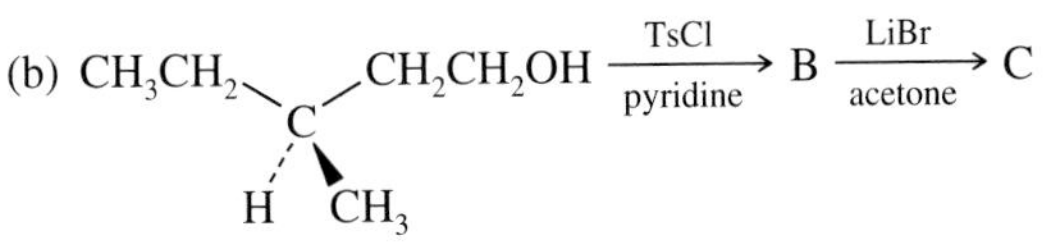

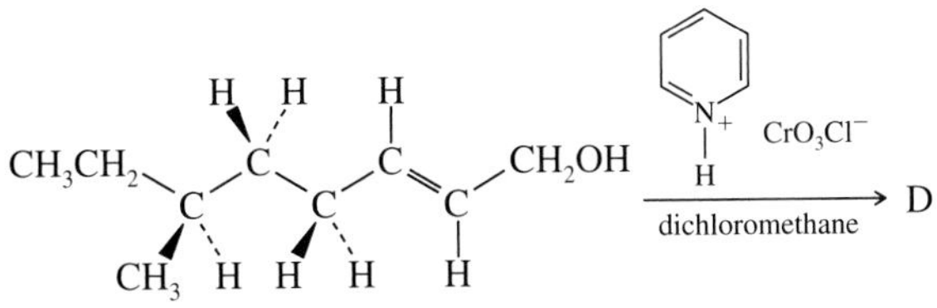

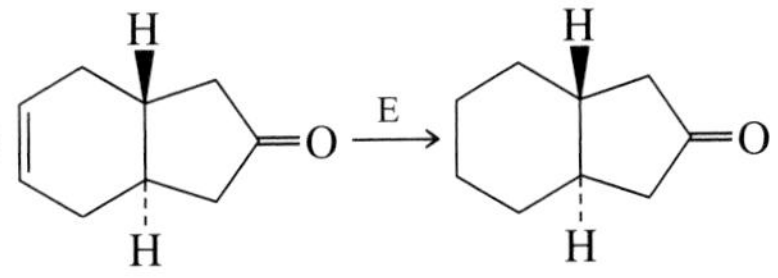

Draw a conformationally correct structural formula for Compound F. The bonds of the five-membered ring in Compound F may be seen as substituents on the six-membered ring.

**13.45** The following transformation has been observed. Assign a structure to Compound A and suggest a mechanism for the conversion of A to the final product.

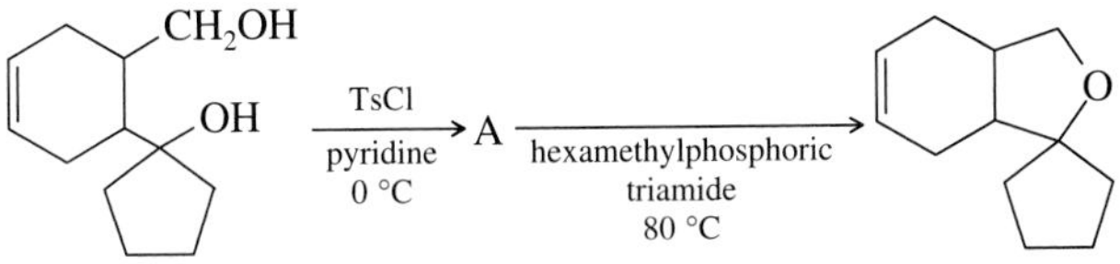

**13.46** An intermediate in the synthesis of cerulein, an antibiotic, was prepared in the following way. Fill in structural formulas for the missing compounds.

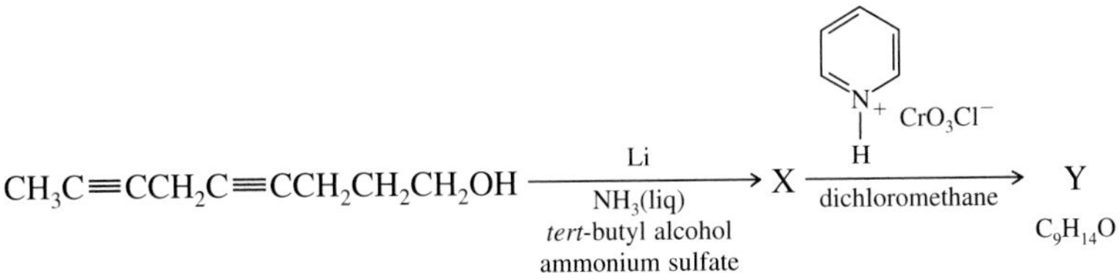

**13.47** 2-Phenyloxirane is treated with sulfuric acid in methanol and, in a separate reaction, with sodium methoxide in methanol. Predict what the major product in each reaction will be and write detailed mechanisms to rationalize your predictions.

**13.48** Chemists are always interested in new ways to construct five-membered ring compounds because such rings are found in many important natural products. The following reactions were carried out during research into such syntheses. Supply the structural formula for any missing reagents, starting materials or products indicated by letters. Show stereochemistry by a proper convention whenever it is known.

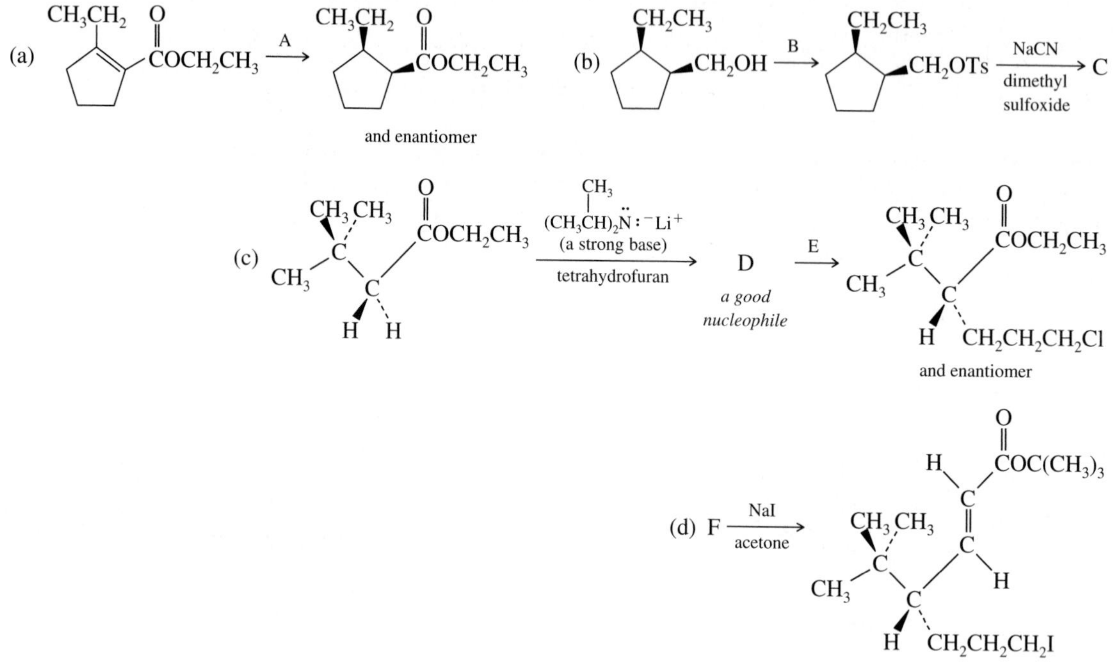

**13.49** Penaresidin A was isolated from an Okinawan marine sponge. It has potent biological activity. It has the structure shown below.

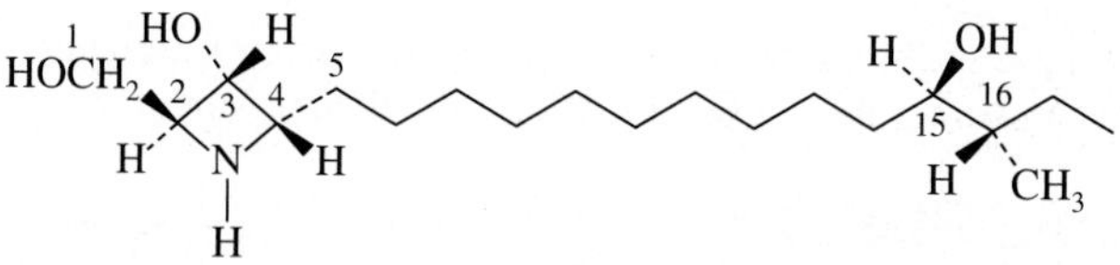

penaresidin A

(a)  Identify and assign configuration to the stereocenters in penaresidin A.
(b)  Two steps in the synthesis of the compound follow. What are the products of these steps?

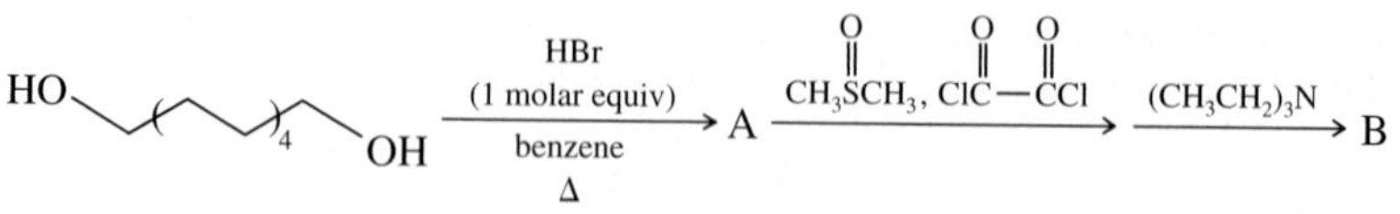

1,10-decanediol

**13.50** Microorganisms reduce carbonyl compounds to alcohols stereoselectively. 5-Chloro-2-pentanone is reduced to (*S*)-5-chloro-2-pentanol in this way. Distillation of this alcohol over sodium hydride gives a chiral cyclic ether. Write stereochemically correct structural formulas for the alcohol and the cyclic ether.

**13.51** 2-Octyn-1-ol is needed as an intermediate in the synthesis of certain polyunsaturated acids that are related to prostaglandins (p. 788), which are implicated in the clotting of blood, inflammation, and allergic responses. 2-Octyn-1-ol is prepared by treating 2-propyn-1-ol with 2 equivalents of lithium amide and then treating the resulting intermediate with 1 equivalent of 1-bromopentane. The reaction mixture is then treated with dilute acid to convert organic ions into uncharged species.

(a) Provide structures for the intermediate, A, and for 2-octyn-1-ol.

$$HC\equiv CCH_2OH + 2\ Li^+NH_2^- \xrightarrow[NH_3(liq)]{} A \xrightarrow{CH_3(CH_2)_3CH_2Br} \xrightarrow[H_2O]{H_3O^+} \text{2-octyn-1-ol}$$

(b) Another product could have been formed from the reaction of the intermediate with 1-bromopentane. What is its structure?
(c) Explain briefly why 2-octyn-1-ol and not the alternative product is the major one.
(d) What would have been the structure of the intermediate formed if *only* 1 equivalent of lithium amide had been used in part (a)?

**13.52** The following reactions were carried out in the synthesis of a natural product that inhibits the activity of an important class of enzymes.

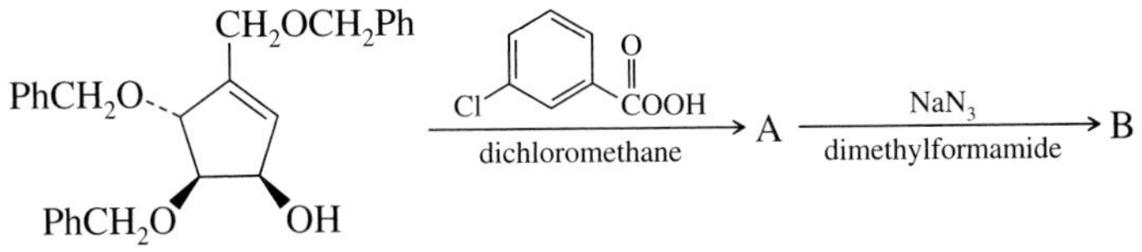

$$\xrightarrow[\text{dichloromethane}]{} A \xrightarrow[\text{dimethylformamide}]{NaN_3} B$$

Assign structures to Compounds A and B, paying special attention to stereochemistry and regioselectivity.

**13.53** The stereochemically pure alcohol (S)-(−)-2-methyl-1-butanol is converted to the corresponding aldehyde by oxidation. The aldehyde is dextrorotatory.

(a) What is the structure of (S)-(−)-2-methyl-1-butanol?
(b) Write an equation showing the reagent that you would use to convert this alcohol to the aldehyde, and draw a stereochemically correct structure for the aldehyde.

**13.54** Hydroboration of 1-methylcyclopentene using (2S,5S)-2,5-dimethyl-1-boracyclopentane followed by oxidation with hydrogen peroxide and sodium hydroxide gives an optically active methylcyclopentanol. This methylcyclopentanol can be oxidized to a 2-methylcyclopentanone that has $[\alpha]_D$ −26°.

(a) Boracyclopentane has the following structure:

Draw the structural formula of (2S,5S)-2,5-dimethyl-1-boracyclopentane, indicating clearly how you assigned configuration to the stereocenters.
(b) (S)-2-Methylcyclopentanone has $[\alpha]_D$ + 26°. Draw the structural formula for (S)-(+)-2-methylcyclopentanone. What is the structure of (−)-2-methylcyclopentanone, the product of the sequence of reactions described above?
(c) What is the structure of the optically active methylcyclopentanol prepared from 1-methylcyclopentene in the first step of the synthesis described above?

**13.55** Epoxide hydrolases are enzymes that are found widely in nature, in mammals, plants, and microorganisms. They open epoxides with the addition of water. These enzymes act with high stereo- and regioselectivity.

(a) When racemic styrene oxide is treated with the epoxide hydrolase from the fungus *Aspergillus niger*, the *R* enantiomer is opened to give (*R*)-1-phenylethane-1,2-diol, and the *S* enantiomer of styrene oxide is recovered unchanged. The equation for this reaction is shown below without the stereochemistry.

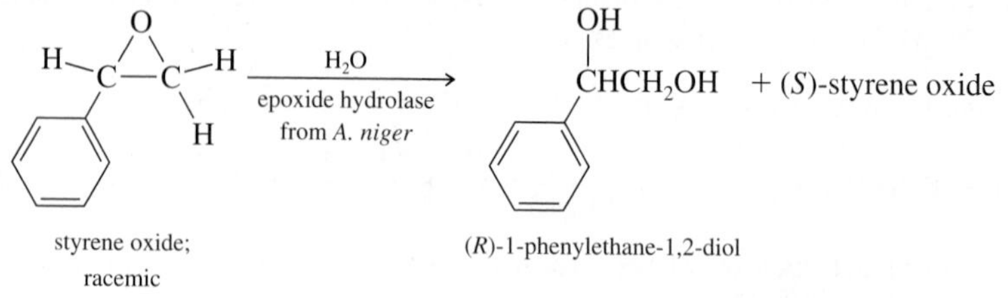

styrene oxide;
racemic

(*R*)-1-phenylethane-1,2-diol

Draw a stereochemically correct structural formula for (*R*)-styrene oxide, and use the curved-arrow convention to show how it is converted to (*R*)-1-phenylethane-1,2-diol. Also show (*R*)-1-phenylethane-1,2-diol in stereochemical detail. $H_2O$, $HB^+$, and $B:$ are available as reagents for this reaction.

(b) When the epoxide hydrolase from another fungus, *Beauveria sulfurescens*, is used in the same reaction, racemic styrene oxide is again converted to (*R*)-1-phenylethane-1,2-diol, but this time (*R*)-styrene oxide is recovered unchanged. Draw the stereochemically correct structural formula for the reactive intermediate that may be used to rationalize the stereochemical outcome of the reaction catalyzed by this microorganism.

**13.56** Consider the kinetics of the formation of oxirane from 2-chloroethanol and the formation of tetrahydrofuran from 4-chlorobutanol. These reactions were found to have the following heats and entropies of activation at 30 °C. Which reaction do you expect to be faster? Why? (*Hint:* $\Delta G^{\ddagger} = \Delta H^{\ddagger} - T\Delta S^{\ddagger}$)

at 30 °C

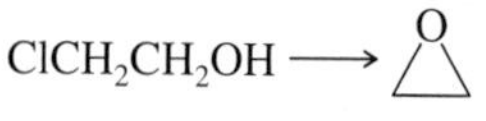

$ClCH_2CH_2OH \longrightarrow$  $\Delta H^{\ddagger} = 23.2$ kcal/mol
$\Delta S^{\ddagger} = 9.9 \times 10^{-3}$ kcal/mol · deg

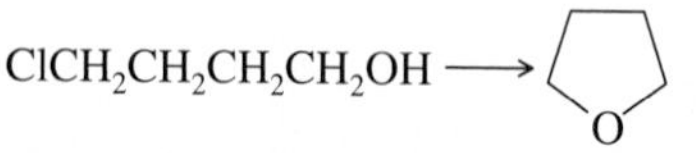

$ClCH_2CH_2CH_2CH_2OH \longrightarrow$  $\Delta H^{\ddagger} = 19.8$ kcal/mol
$\Delta S^{\ddagger} = -5.0 \times 10^{-3}$ kcal/mol · deg

**13.57** Compound A has the molecular formula $C_7H_{16}O$. Oxidizing it with pyridinium chlorochromate in dichloromethane gives Compound B, $C_7H_{14}O$. The infrared spectra of Compounds A and B are given in Figure 13.3. The proton magnetic resonance spectrum of A has peaks at $\delta$ 0.9 (3H, t), 1.3 (10H, m), 2.2 (1H, s), and 3.6 (2H, t). Its $^{13}C$ nuclear magnetic resonance spectrum has peaks at 14.2, 23.1, 26.4, 29.7, 32.4, 33.2, and 62.2 ppm. The proton magnetic resonance spectrum of B has peaks at $\delta$ 0.9 (3H, t), 1.3 (8H, m), 2.4 (2H, m), and 9.8 (1H, t). The most distinctive peak in its $^{13}C$ spectrum is at 202.2 ppm. Assign structures to the compounds that are compatible with their spectra. What structures for Compound A are ruled out by the nuclear magnetic resonance data?

**13.58** Compound C, $C_5H_{12}O$, has a strong band at 3400 cm$^{-1}$ in its infrared spectrum. When it is dissolved in acetone and a solution of chromium trioxide in sulfuric acid and water is added, the solution remains orange. On the other hand, Compound D, also $C_5H_{12}O$, with a band at 3400 cm$^{-1}$ in its infrared spectrum, turns the solution of chromium trioxide green under the same conditions. The proton magnetic resonance spectrum of C has a sharp, strong singlet at $\delta$ 1.2 and two multiplets at $\delta$ 0.9 and 1.5. Its $^{13}C$ nuclear magnetic resonance spectrum has peaks at 8.8, 28.9, 36.8, and 70.6 ppm. The proton magnetic resonance spectrum of D has a complex multiplet at $\delta$ 0.9–1.4, a doublet at $\delta$ 1.2, and a multiplet at $\delta$ 3.7. Peaks appear in its $^{13}C$ nuclear magnetic resonance spectrum at 14.3, 19.4, 23.6, 41.9, and 67.3 ppm. Assign structures to the two compounds that are compatible with all the information given.

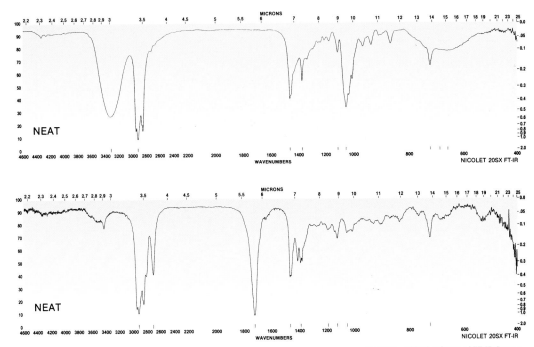

From *The Aldrich Library of FT-IR Spectra.*

*Figure 13.3*

**13.59** Compound E, $C_9H_{12}O$, has strong bands in its infrared spectrum at 3338 and 1031 cm$^{-1}$. Peaks appear at δ 32.02 (t), 34.13 (t), 61.95 (t), 125.74 (d), 128.29 (d), 128.33 (d), and 141.78 (s) in its $^{13}C$ nuclear magnetic resonance spectrum. The proton magnetic resonance spectrum of E has peaks at δ 1.8 (2H, quintet), 2.6 (1H, s), 2.8 (2H, t), 3.6 (2H, m), and 7.2 (5H, m). The mass spectrum of E is shown in Figure 13.4. Assign a structure to Compound E. Analyze the splitting patterns that appear in the nuclear magnetic resonance spectra in terms of your structure. Write an equation showing the formation of the ion that gives rise to the base peak in its mass spectrum.

**13.60** Compounds F and G are isomers with molecular ions at *m/z* 122. Their proton magnetic resonance spectra appear in Figure 13.5, p. 534. Compound F has strong bands at 3329 and 1046 cm$^{-1}$ in its infrared spectrum and a base peak at *m/z* 91 in its mass spectrum. The important bands in the infrared spectrum of Compound G are at 1245 and 1049 cm$^{-1}$, and the base peak for G is at *m/z* 94 in its mass spectrum. Assign structures to Compounds F and G. Show how the spectral data support your assignments. Write equations for the formation of the ion responsible for the base peak for each compound. (*Hint:* To figure out how the ion from G forms, decide on the structure of the neutral fragment that is lost.)

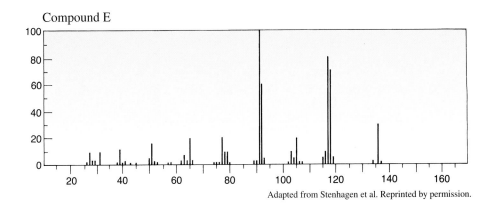

Compound E

Adapted from Stenhagen et al. Reprinted by permission.

*Figure 13.4*

Compound F

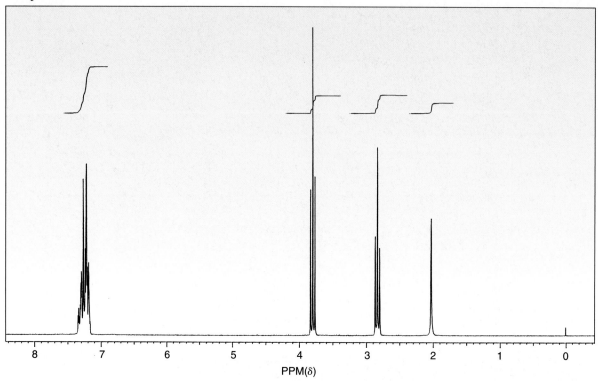

Compound G

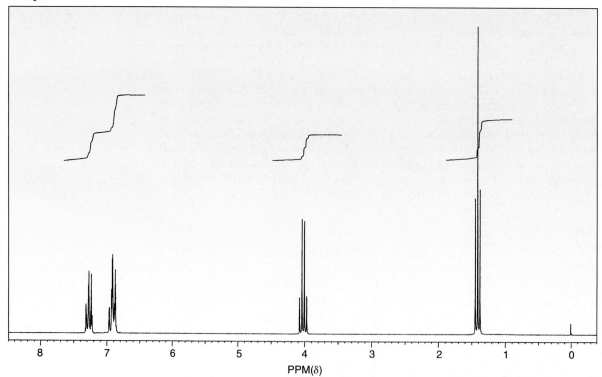

*Figure 13.5*

# Aldehydes and Ketones. Addition Reactions at Electrophilic Carbon Atoms

## A Look Ahead

Aldehydes and ketones contain the carbonyl group, in which a carbon atom is doubly bonded to an oxygen atom. The carbonyl group is highly polarized, with a very electrophilic carbon atom.

$$\delta-:\!\overset{\displaystyle :O:}{\underset{\displaystyle \delta+ C}{\|}} \quad \text{very electrophilic carbon atom}$$

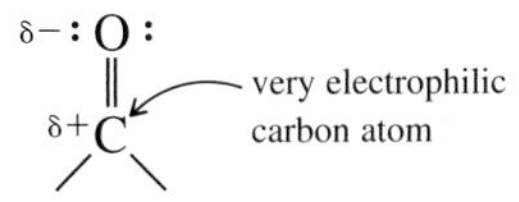

Many reactions of aldehydes and ketones start with a nucleophilic attack at the carbon atom of the carbonyl group by a nucleophile such as a cyanide ion:

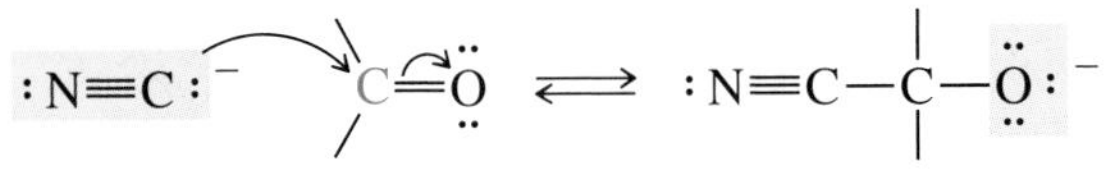

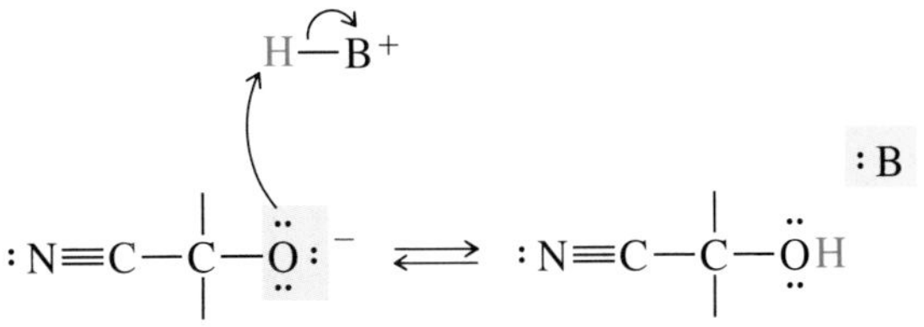

$$:N\equiv C:^- \quad \overset{}{\underset{}{C = \ddot{O}}} \quad \rightleftharpoons \quad :N\equiv C - \overset{|}{\underset{|}{C}} - \ddot{O}:^-$$

nucleophile

$$H - B^+$$

$$:B$$

$$:N\equiv C - \overset{|}{\underset{|}{C}} - \ddot{O}:^- \quad \rightleftharpoons \quad :N\equiv C - \overset{|}{\underset{|}{C}} - \ddot{O}H$$

addition compound

As a result of the nucleophilic attack at the carbonyl group, the $\pi$ electrons of the carbon–oxygen double bond become localized on the oxygen atom. The alkoxide ion that forms is protonated in a subsequent step. The overall reaction is the addition of the elements of H—CN to the carbon–oxygen double bond.

Many nucleophiles will react with the carbonyl group, and depending on the reagents, the reaction conditions, and the nature of the intermediate formed, further reactions are possible before the final product is obtained. But the most important step in all these reactions is the bonding between a nucleophile and the carbon atom of the carbonyl group.

Workbook Exercises

## 14.1 Introduction to Carbonyl Compounds

### A. The Carbonyl Group

The carbonyl group, a carbon atom doubly bonded to an oxygen atom, determines most of the physical and chemical properties of aldehydes and ketones. Except for

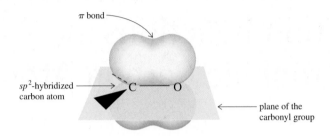

**Figure 14.1**
The bonding and geometry of
the carbonyl group.

different representations of the
polarity of the carbonyl group

$\mu = 2.88$ D

resonance contributors for acetone

formaldehyde, in aldehydes, the carbonyl group is bonded to an organic group on one side and a hydrogen atom on the other. In ketones, it is bonded to two organic groups.

The carbon atom of the carbonyl group is $sp^2$-hybridized; therefore, the carbonyl group and the two atoms bonded to it lie in a plane (Figure 14.1). Oxygen is more electronegative than carbon, so the carbonyl group has a permanent dipole moment with the negative end of the dipole at the oxygen atom. For example, acetone, a typical ketone, has a dipole moment of 2.88 D. The high polarity of the carbonyl group is rationalized by pointing to a resonance contributor in which the carbon atom bears a positive formal charge and the oxygen atom a negative one.

The polarity of the carbonyl group makes low-molecular-weight aldehydes and ketones miscible with water. The oxygen atom acts as an acceptor of hydrogen bonds, and the carbonyl compound dissolves in water as long as the hydrocarbon portion of the molecule does not contain more than four or five carbon atoms. The kinds of interactions that occur between water molecules and acetone, an example of a low-molecular-weight carbonyl compound, are shown below.

**Problem 14.1**    The boiling points of an alcohol, a ketone, and an ether, all of which contain four carbon atoms and one oxygen atom, are given below. How would you rationalize the trends shown? Use words and pictures in your answer.

| OH | | O | |
|----|--|---|--|
| $CH_3CHCH_2CH_3$ | | $CH_3CCH_2CH_3$ | $CH_3CH_2OCH_2CH_3$ |
| 2-butanol | | 2-butanone | diethyl ether |
| bp 99.5 °C | | bp 79.6 °C | bp 37 °C |

## B. The Occurrence of Aldehydes and Ketones in Nature

Many compounds of biological interest are aldehydes or ketones. The most abundant natural aldehyde is glucose, a carbohydrate that is metabolized in the human body to produce energy. Many of the steroid hormones are ketones, including testosterone, the hormone that controls the development of male sex characteristics; progesterone, the hormone secreted at the time of ovulation in females; and cortisone, a hormone of the adrenal glands that is used medicinally to relieve inflammation. Another hormone from the adrenal glands, aldosterone, is an aldehyde as well as a ketone. Aldosterone is important in regulating the concentration of sodium ions in the body and thus water retention. Prednisone is an example of a synthetic drug; it is designed to be a substitute for cortisone in relieving the symptoms of rheumatoid arthritis with fewer undesirable side effects.

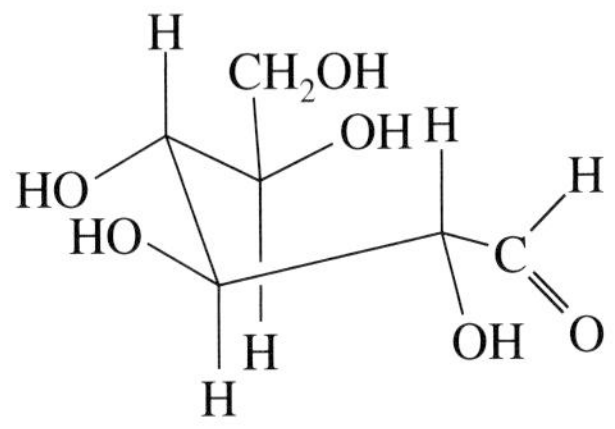

glucose

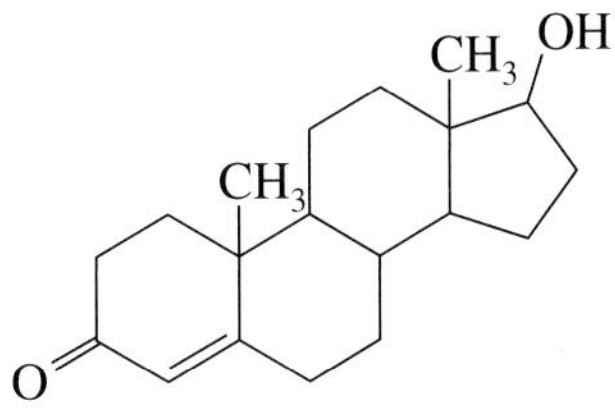

testosterone

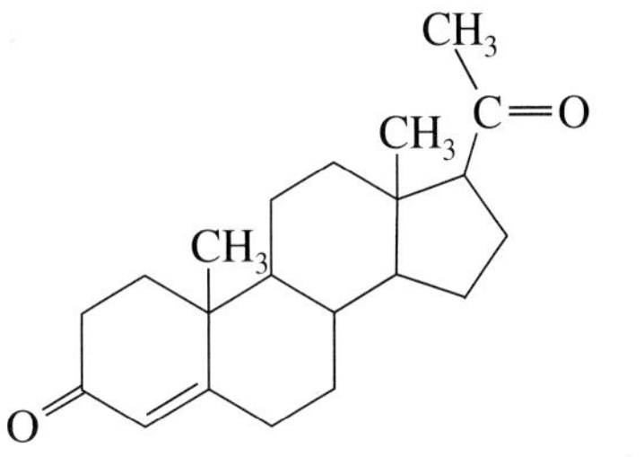

progesterone

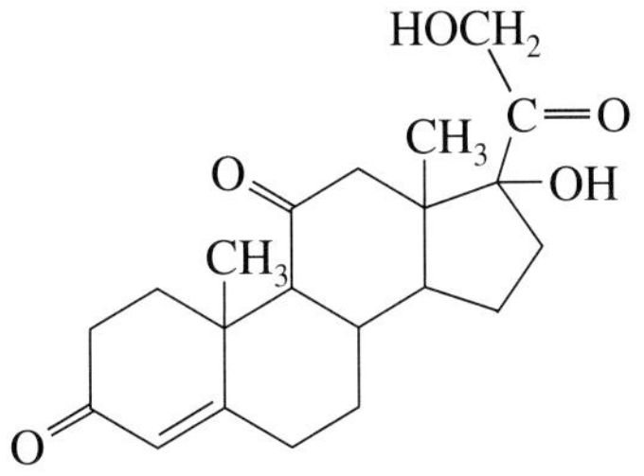

cortisone

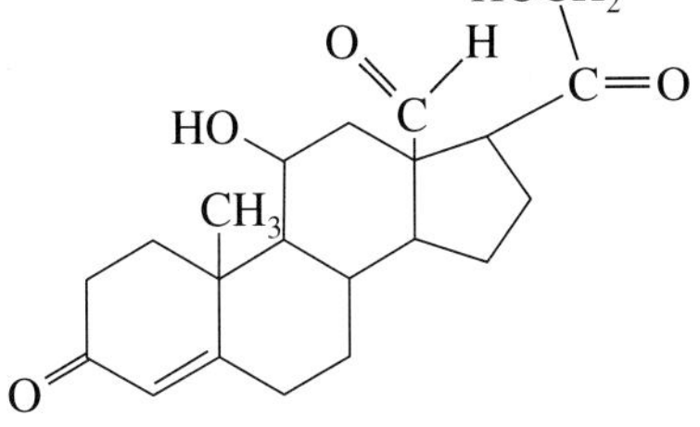

aldosterone

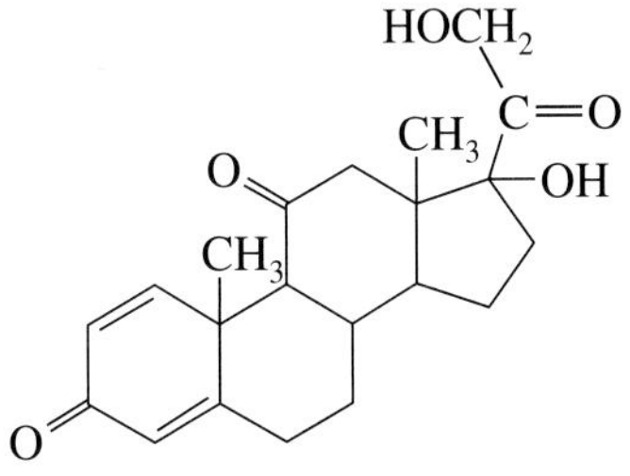

prednisone

Careful inspection of the structural formulas of the steroids shown above may lead you to wonder how molecules that are so similar to each other in structure can perform such different functions in the body. What makes testosterone a male hormone and progesterone a female one, for example, and the function of cortisone something else entirely? Contemplating such questions will give you an appreciation of the finely discriminating systems that exist in living organisms. Scientists have been unable to explain fully how the structures of hormones enable them to perform their physiologic functions. Organic chemists and biochemists constantly refine ways of looking at such compounds in order to understand their interactions with large molecules, such as the proteins known as hormone receptors that occur at the surface of cells or inside them. The binding of the hormone to the receptor triggers a series of reactions that ultimately result in the physiologic changes associated with that hormone. Interactions between hormones and receptors are described using the same concepts that chemists use in describing the reactivity of simpler compounds. Among the ideas chemists use as they try to understand the multiple reactions in the human body are polarity of bonds, hydrogen bonding, ionic interactions, van der Waals forces between nonpolar portions of molecules, transfer of protons from one basic site to another one, and nucleophilic centers reacting with electrophilic sites.

## 14.2  Nomenclature of Carbonyl Compounds

### A. Nomenclature of Aldehydes and Ketones

Simple aldehydes are named systematically by changing the **e** at the end of the name of the hydrocarbon having the same number of carbon atoms in the longest chain to **al**. The aldehyde function is at the first carbon atom of the chain. Other substituents are named using prefixes and numbered to indicate their positions relative to the aldehyde group. The common names of the two simplest aldehydes, formaldehyde and acetaldehyde, are almost always used rather than the systematic names.

$$
\underset{\substack{\text{methanal}\\\text{formaldehyde}}}{\text{HCH}}
\qquad
\underset{\substack{\text{ethanal}\\\text{acetaldehyde}}}{\text{CH}_3\text{CH}}
\qquad
\underset{\text{propanal}}{\text{CH}_3\text{CH}_2\text{CH}}
\qquad
\underset{\text{pentanal}}{\text{CH}_3\text{CH}_2\text{CH}_2\text{CH}_2\text{CH}}
\qquad
\underset{\text{2-chloro-5-methylhexanal}}{\text{CH}_3\text{CHCH}_2\text{CH}_2\text{CHCH}}
$$

If the aldehyde group is a substituent on a ring, the suffix **-carbaldehyde** is used in the name.

cyclohexanecarbaldehyde      *trans*-2-methylcyclo-
                             pentanecarbaldehyde

Note that in these cases the aldehyde group is considered to be a substituent. The carbon atom of the carbonyl group is not counted in assigning a name to the hydrocarbon portion of the molecule.

Benzene on which an aldehyde group is substituted is usually called benzaldehyde. A second substituent on the aromatic ring of benzaldehyde can have one of three positions in relation to the aldehyde group. A second substituent at carbon atom 2 of the ring is designated as ortho (*o*-) to the carbonyl group. One at carbon atom 3 is meta (*m*-), and one at carbon atom 4 is para (*p*-) (p. 360).

benzaldehyde    *o*-bromobenzaldehyde    *m*-nitrobenzaldehyde    *p*-ethylbenzaldehyde

Ketones are named according to the IUPAC rules by substituting the ending **one** for the final **e** of the name of the hydrocarbon having the same number of carbon atoms in the longest chain and using a number to indicate the position of the carbonyl group. Other substituents are named using prefixes, and their positions are indicated by numbers. In a few cases, it is useful to name the organic groups that are present as substituents on the carbonyl group, which is then designated by the word **ketone**. A few ketones, such as acetone, are generally known by their common names.

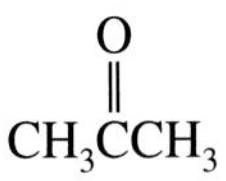

O
||
$CH_3CCH_3$

propanone
acetone

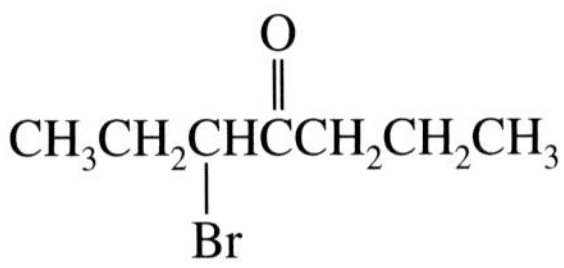

O
||
$CH_3CH_2CHCCH_2CH_2CH_3$
|
Br

3-bromo-4-heptanone

O
||
$CH_3CH_2CCH_2CH_3$

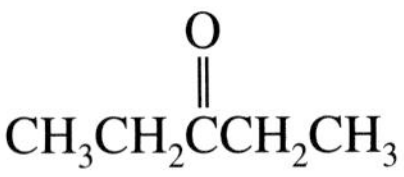

3-pentanone

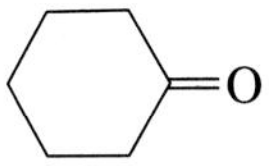

cyclohexanone

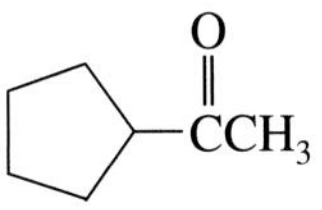

O
||
—$CCH_3$

cyclopentyl methyl
ketone

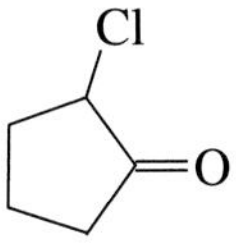

Cl

=O

2-chlorocyclopentanone

Ketones having a phenyl group adjacent to the carbonyl group have common names ending in **phenone.** Two of these common names that are widely used by chemists are acetophenone and benzophenone. Other ketones containing aromatic rings are generally named systematically or as substituted ketones.

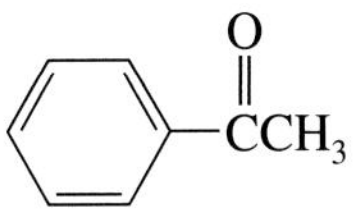

O
||
—$CCH_3$

acetophenone

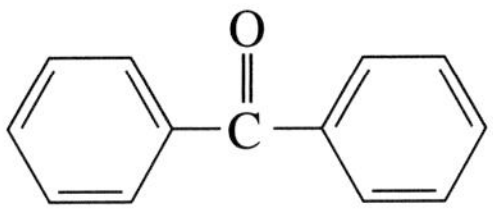

O
||
—C—

benzophenone

O
||
—$CCH_2CH_3$

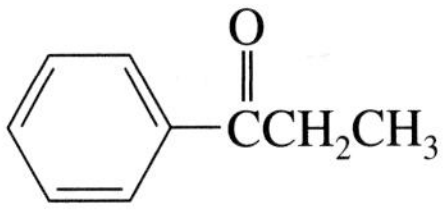

1-phenyl-1-propanone
ethyl phenyl ketone
propiophenone

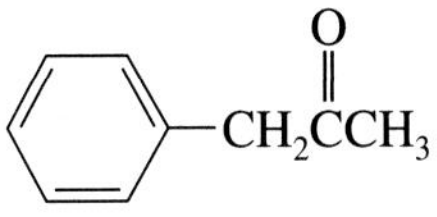

O
||
—$CH_2CCH_3$

1-phenyl-2-propanone
benzyl methyl ketone

## B. Nomenclature of Polyfunctional Compounds

When a compound contains more than one major functional group, the suffix for only one of them can be used as the ending of the name. The IUPAC rules establish priorities that specify which suffix is used. The priorities among the common functional groups are given in Table 14.1. The presence of double or triple bonds in a

**Table 14.1  Decreasing Order of Precedence of Principal Functional Groups Used as Suffixes in Nomenclature**

| Chemical Structure | Functional Group | Chemical Structure | Functional Group |
|---|---|---|---|
| O<br>\|\|<br>—COH | carboxylic acid | O<br>\|\|<br>—CH | aldehyde |
| O<br>\|\|<br>—SOH<br>\|\|<br>O | sulfonic acid | O<br>\|\|<br>—C— | ketone |
| O<br>\|\|<br>—COR | ester | —OH | alcohol |
| O<br>\|\|<br>—CNH$_2$ | amide | —NH$_2$ | amine |
| —C≡N | nitrile | —OR | ether |

molecule is indicated by naming the chain (or ring) as an alkene or alkyne. Halogen substituents and alkoxyl groups are always named using prefixes. An aldehyde or a ketone function takes precedence over an alkene or an alcohol function as the suffix used in naming the compound. These examples illustrate these priorities of nomenclature:

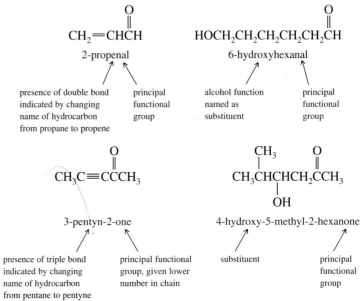

When it is necessary to name the carbonyl group as a substituent, the prefix **oxo** with a number to indicate the position is used. The prefix oxo is used to indicate either an aldehyde or a ketone. This type of nomenclature is needed when a functional group of higher priority than an aldehyde or ketone, such as a carboxylic acid or ester, is present.

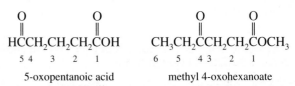

Greek letters are used to designate positions on a carbon chain in relation to a carbonyl group for describing positions of special reactivity or relationships between functional groups. For example, the carbon atoms adjacent to the carbonyl group are called the α-carbon atoms and the hydrogen atoms on them, the α-hydrogen atoms.

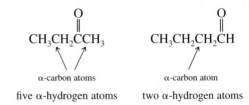

The positions alpha to a carbonyl group are significant because of the high reactivity of the α-hydrogen atoms (p. 669). The four positions farther from the carbonyl group than the α-carbon are designated, in order, by the Greek letters β (beta), γ (gamma), δ (delta), and ε (epsilon). This nomenclature is also used with substituted carboxylic acids.

$$-CH_2CH_2CH_2CH_2CH_2\underset{\alpha}{-}\overset{\overset{O}{\|}}{C}-$$
$$\underset{\epsilon}{\phantom{-}}\underset{\delta}{\phantom{CH_2}}\underset{\gamma}{\phantom{CH_2}}\underset{\beta}{\phantom{CH_2}}\underset{\alpha}{\phantom{CH_2}}$$

$$\underset{\gamma}{CH_3}\overset{OH}{\underset{\beta}{\overset{|}{C}H}}CH_2\overset{O}{\underset{\alpha}{\overset{\|}{C}H}}$$

3-hydroxybutanal

*a β-hydroxyaldehyde*

$$CH_3\overset{O}{\overset{\|}{C}}CH_2Br$$
$$\phantom{CH_3}\underset{\alpha}{\phantom{C}}\underset{\alpha}{\phantom{CH_2}}$$

1-bromo-2-propanone

*an α-bromoketone*

$$CH_3\overset{O}{\overset{\|}{C}}CH_2CH_2\overset{O}{\overset{\|}{C}}OH$$
$$\phantom{CH_3}\underset{\delta}{\phantom{C}}\underset{\gamma}{\phantom{CH_2}}\underset{\beta}{\phantom{CH_2}}\underset{\alpha}{\phantom{C}}$$

4-oxopentanoic acid

*a γ-ketoacid*

**Problem 14.2** Name the following compounds according to the IUPAC rules, including stereochemistry wherever it is shown.

(a) $CH_3CH_2\overset{\overset{CH_3}{|}}{C}H\overset{O}{\overset{\|}{C}H}$ with OH on the CH

(b) Cl—⟨benzene ring⟩—$\overset{O}{\overset{\|}{C}H}$

(c) $CH_3CH_2C{\equiv}CCH_2CH_3$ with C=O

(d) structure with phenyl, C=C, phenyl, C=O

(e) $\underset{CH_3}{\overset{CH_3CH_2CH_2}{}}C{=}C\underset{H}{\overset{CH_2CH_2\overset{O}{\overset{\|}{C}}CH_3}{}}$

(f) cyclohexanone with Br, Br and O

(g) $CH_3\overset{O}{\overset{\|}{C}}CH_2\overset{O}{\overset{\|}{C}}CH_3$

(h) $Cl_3C\overset{O}{\overset{\|}{C}}H$

(i) cyclopentane with CHO and CH₂CH₂CH₃

**Problem 14.3** Draw structural formulas for the following compounds.

(a) 3-methylcyclobutanone (b) 4,4-dimethyl-2,5-cyclohexadienone (c) phenyl cyclopentyl ketone

(d) (*E*)-4-heptenal (e) *p*-nitrobenzaldehyde (f) 1-hydroxy-3-pentanone (g) hexanedial

(h) 2-hydroxy-1,2-diphenylethanone (i) 1,4-cyclohexanedione

# 14.3 Preparation of Aldehydes and Ketones

## A. Oxidation of Alcohols

Aldehydes and ketones are often prepared by the oxidation of alcohols (p. 515). Primary alcohols are oxidized to aldehydes either by chromium(VI) reagents such as pyridinium chlorochromate (p. 517) or by dimethyl sulfoxide in the presence of dioxalyl chloride followed by base (p. 519). Both these reaction conditions avoid the complication that aldehydes are easily further oxidized to carboxylic acids (pp. 312 and 516). For example, 5-methyl-4-hexenol is oxidized to 5-methyl-4-hexenal by pyridinium chlorochromate.

$$\underset{CH_3C=CHCH_2CH_2CH_2OH}{\overset{CH_3}{\overset{|}{\phantom{C}}}} \xrightarrow[\text{dichloromethane}]{\text{pyridinium } H\ CrO_3Cl^-} \underset{CH_3C=CHCH_2CH_2CH}{\overset{CH_3}{\overset{|}{\phantom{C}}}}\overset{O}{\overset{\|}{\phantom{C}}}$$

5-methyl-4-hexen-1-ol

*primary alcohol*

5-methyl-4-hexenal
75%

*aldehyde*

Either set of reaction conditions also can be used to oxidize a secondary alcohol to a ketone. For example, the secondary alcohol function in the steroid cholestanol is converted to a ketone in a Swern oxidation.

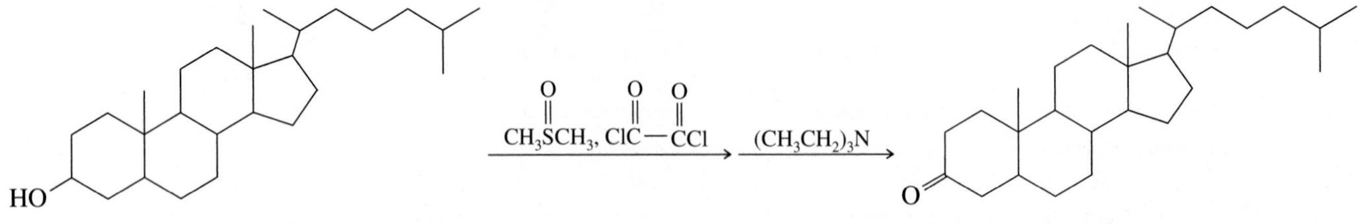

$$CH_3SCH_3, \; ClC\!\overset{O}{\overset{\|}{-}}\!CCl \quad \xrightarrow{\;(CH_3CH_2)_3N\;}$$

cholestanol

*secondary alcohol*

cholestanone
96%

*ketone*

Ozonolysis is also used to prepare aldehydes and ketones. Ozonolysis is most useful as a method of preparing an aldehyde or a ketone when the second fragment from the ozonolysis is a low-molecular-weight compound that is easily separated from the desired product of the reaction. The cleavage of 1-octene into heptanal and formaldehyde (p. 312) is a case where the desired aldehyde is easily separated from the other product. Cyclic alkenes are frequently cleaved by ozonolysis to make di-carbonyl compounds; the preparation of 1,6-hexanedial from cyclohexene (p. 312) is an example.

Aldehydes can be prepared by reduction of esters or nitriles. These reactions will be studied in Chapter 21. The reverse of the oxidation of an alcohol to an alde-hyde or a ketone is a reduction reaction (p. 513), in which an aldehyde or a ketone is converted to an alcohol.

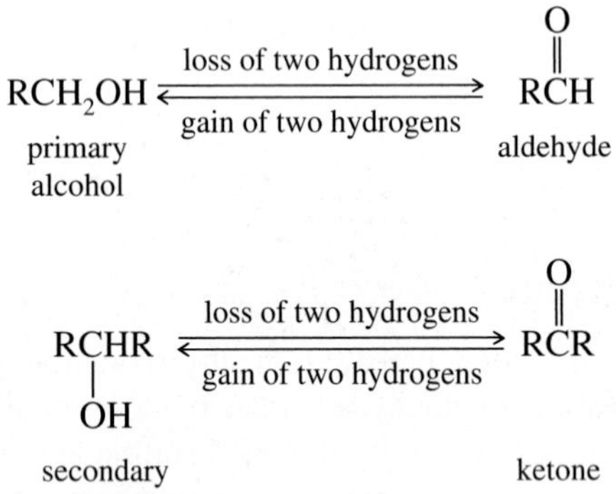

$$RCH_2OH \underset{\text{gain of two hydrogens}}{\overset{\text{loss of two hydrogens}}{\rightleftarrows}} \overset{O}{\overset{\|}{R C H}}$$

primary
alcohol

aldehyde

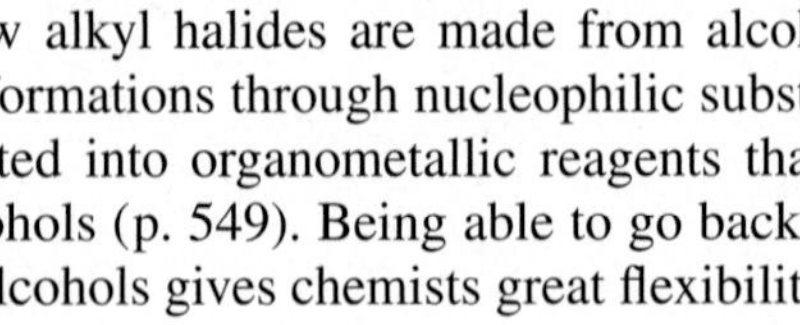

$$\underset{\underset{OH}{|}}{RCHR} \underset{\text{gain of two hydrogens}}{\overset{\text{loss of two hydrogens}}{\rightleftarrows}} \overset{O}{\overset{\|}{R C R}}$$

secondary
alcohol

ketone

Chapter 13 showed how alkyl halides are made from alcohols. Alkyl halides un-dergo a variety of transformations through nucleophilic substitution reactions. Alkyl halides are also converted into organometallic reagents that react with aldehydes and ketones to give alcohols (p. 549). Being able to go back and forth between car-bonyl compounds and alcohols gives chemists great flexibility in planning syntheses of organic compounds.

---

**Problem 14.4**    Provide structural formulas for starting materials, reagents, or products in the following reactions.

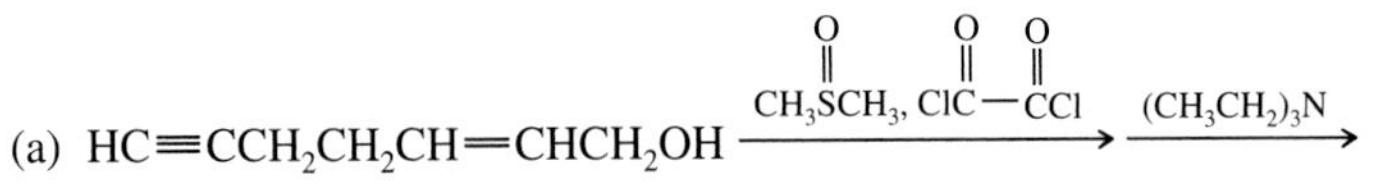

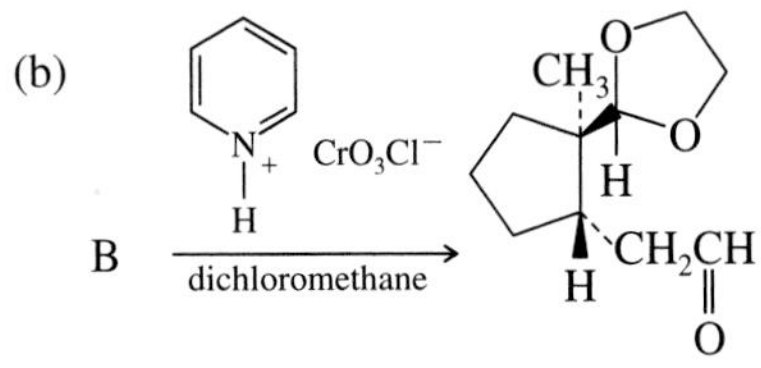

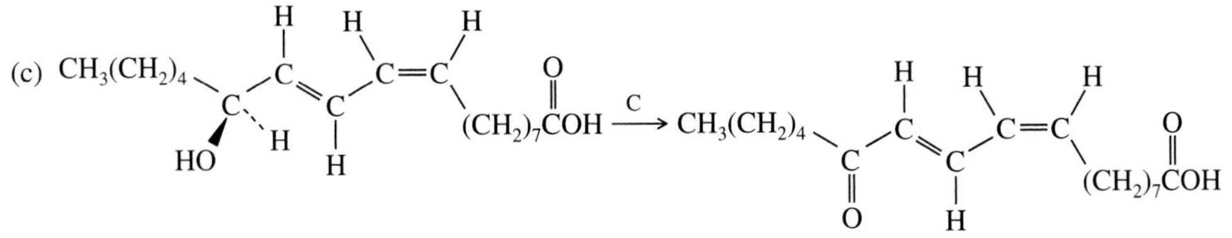

(a) $HC{\equiv}CCH_2CH_2CH{=}CHCH_2OH$ $\xrightarrow[\text{CH}_3\text{SCH}_3,\ \text{ClC}-\text{CCl}]{}$ $\xrightarrow{(CH_3CH_2)_3N}$ A

(b)

(c)

# B. Aromatic Ketones. The Friedel–Crafts Acylation Reaction

In Chapter 10 we saw the Friedel-Crafts alkylation reaction in which alkenes, alkyl halides, and alcohols were used to generate carbocationic intermediates that served as electrophiles in electrophilic aromatic substitution reactions (p. 378). It is also possible to use acid chlorides or acid anhydrides (p. 597) with aluminum chloride as a Lewis acid to generate cations known as acylium ions that also react as electrophiles with aromatic rings. For example, benzene reacts with acetic anhydride in the presence of aluminum chloride to give acetophenone.

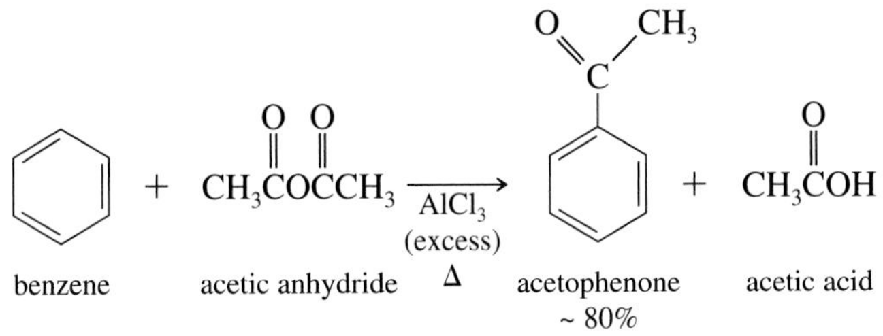

The product is a ketone. In effect, an acyl group has been substituted on the aromatic ring, so the reaction is called a **Friedel-Crafts acylation.**

An example in which an acid chloride is used is the acylation of bromobenzene with acetyl chloride.

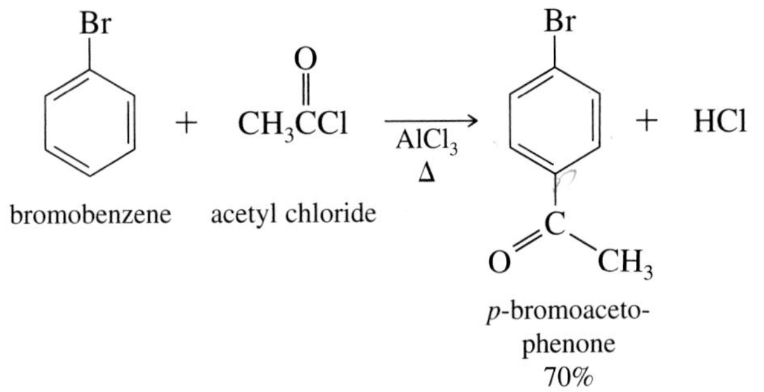

Both acid chlorides and acid anhydrides have a good leaving group (p. 597). Aluminum chloride can complex with either of these types of compounds to help remove the leaving group, creating an acylium cation.

## Formation of the Electrophile in a Friedel-Crafts Acylation

$$CH_3\overset{\overset{\displaystyle :O:}{\|}}{C}-\overset{..}{\underset{..}{Cl}}: \rightsquigarrow AlCl_3 \longrightarrow CH_3\overset{\overset{\displaystyle :O:}{\|}}{C}\overset{\curvearrowright}{-}\overset{..}{\underset{..}{Cl}}{}^{+}-\overset{-}{Al}Cl_3 \longrightarrow [CH_3\overset{+}{C}=\overset{..}{\underset{..}{O}} \quad :\overset{..}{\underset{..}{Cl}}-AlCl_3]^{-}$$

complex                                    acylium ion
ion pair

$$CH_3\overset{\overset{\displaystyle :O:}{\|}}{C}-\underset{\underset{AlCl_3}{\curvearrowright}}{\overset{..}{O}}-\overset{\overset{\displaystyle :O:}{\|}}{C}CH_3 \longrightarrow CH_3\overset{\overset{\displaystyle :O:}{\|}}{C}\overset{\curvearrowright}{-}\underset{\underset{{}^{-}AlCl_3}{\overset{+}{}}}{\overset{..}{O}}-\overset{\overset{\displaystyle :O:}{\|}}{C}CH_3 \longrightarrow [CH_3\overset{+}{C}=\overset{..}{\underset{..}{O}} \quad CH_3-\overset{\overset{\displaystyle :O:}{\|}}{C}-\overset{..}{\underset{..}{O}}-AlCl_3]^{-}$$

complex                                    acylium ion
ion pair

The acylium ion is particularly stable because the positive charge can be delocalized to the oxygen atom.

$$\left[CH_3\overset{+}{C}=\overset{..}{\underset{..}{O}} \longleftrightarrow CH_3C\equiv\overset{+}{O}:\right]$$

*particularly good
resonance contributor,
8 electrons around
each atom*

*resonance contributors of the acylium ion*

Acylium ions do not undergo rearrangement. Thus the acid chloride from a long straight-chain carboxylic acid can be used to put a straight carbon chain on an aromatic ring as an acyl group. The acid chloride of propanoic acid has been used to put a straight chain of three carbon atoms on the benzene ring.

$$\text{benzene} + CH_3CH_2\overset{\overset{\displaystyle O}{\|}}{C}Cl \xrightarrow{AlCl_3} \text{1-phenyl-1-propanone}$$

benzene    propanoyl chloride           1-phenyl-1-propanone
                                                    65%

Friedel-Crafts alkylation reactions do not usually give unbranched alkyl chains on aromatic rings because of rearrangements of carbocation intermediates (p. 379). Later in this chapter (p. 578) we will see how to reduce carbonyl groups to methylene groups. This allows us to use Friedel-Crafts acylation followed by such a reduction of the carbonyl group to introduce unbranched alkyl groups onto an aromatic ring. Another advantage of Friedel-Crafts acylation is that the carbonyl group attached to the ring deactivates it toward further substitution. Side products resulting from multiple substitution reactions are not observed with Friedel-Crafts acylations as they are with Friedel-Crafts alkylation reactions.

Ⓐ **Study Guide**
**Concept Map 14.1**

 **On the Web: ONE SMALL STEP**

**Problem 14.5**    Complete the following equations.

(a) [benzene] + CH₂ with CH₂—CCl (top, O double bond) and CH₂—CCl (bottom, O double bond) →AlCl₃ (excess)→

2 equivalents    1 equivalent

(b) [benzene]—CH₂CCl (O double bond) + [benzene] →AlCl₃→

(c) [benzene]—OCH₃ + CH₃COCCH₃ (two O double bonds) →AlCl₃→

# 14.4 Addition of the Nucleophile Hydride Ion. Reduction of Aldehydes and Ketones to Alcohols

Nucleophilic addition of the hydride ion to the electrophilic carbon atom of the carbonyl group results in the reduction of aldehydes to primary alcohols and ketones to secondary alcohols. These reductions are carried out using **sodium borohydride** or **lithium aluminum hydride.** In both these reagents, the anion is a source of hydride ion, which is very basic and a powerful nucleophile. Lithium aluminum hydride reacts violently with water or alcohols to generate hydrogen gas.

$$\text{LiAlH}_4 \;+\; 4\,\text{H}_2\text{O} \xrightarrow{\text{fast}} \text{Li}^+\,\text{Al(OH)}_4^- + 4\,\text{H}_2 \uparrow$$

lithium aluminum     water
hydride

Sodium borohydride reacts in the same way as lithium aluminum hydride does, though much more slowly.

$$\text{NaBH}_4 \;+\; 4\,\text{CH}_3\text{OH} \xrightarrow{\text{slow}} \text{Na}^+\,\text{B(OCH}_3)_4^- + 4\,\text{H}_2 \uparrow$$

sodium       methanol
borohydride

Because sodium borohydride reacts slowly with water, it is easier to handle than lithium aluminum hydride and may be used with water or an alcohol as the solvent. The reactions of this reagent with carbonyl compounds are fast enough to compete with its decomposition reaction. The reduction of butanal to *n*-butyl alcohol and of 2-butanone to 2-butanol by sodium borohydride in water are typical reactions of this reagent.

$$4\,\text{CH}_3\text{CH}_2\text{CH}_2\overset{\text{O}}{\overset{\|}{\text{CH}}} + \text{NaBH}_4 \xrightarrow{\text{H}_2\text{O}} 4\,\text{CH}_3\text{CH}_2\text{CH}_2\text{CH}_2\text{OH} + \text{Na}^+ + \text{BO}_3{}^{3-}$$

butanal                                           *n*-butyl alcohol
                                                       85%

$$4\,\text{CH}_3\text{CH}_2\overset{\text{O}}{\overset{\|}{\text{C}}}\text{CH}_3 + \text{NaBH}_4 \xrightarrow{\text{H}_2\text{O}} 4\,\text{CH}_3\underset{\text{OH}}{\text{CHCH}_2}\text{CH}_3 + \text{Na}^+ + \text{BO}_3{}^{3-}$$

2-butanone                               2-butanol
                                            87%

The mechanism for the reduction of 2-butanone with sodium borohydride is shown below. A hydride ion, a nucleophile, is transferred to the electrophilic carbon atom of the carbonyl group. The alkoxide ion that forms is protonated by the solvent. The carbonyl group, having a trigonal planar carbon atom, is symmetrical. It is equally probable that a hydride ion will be donated to the carbonyl group from either side of the molecule, giving rise to equal numbers of molecules of the two enantiomers of 2-butanol. The alcohol recovered as the product of the reaction is a racemic mixture and not optically active.

**VISUALIZING THE REACTION**

**Reduction of a Carbonyl Group**

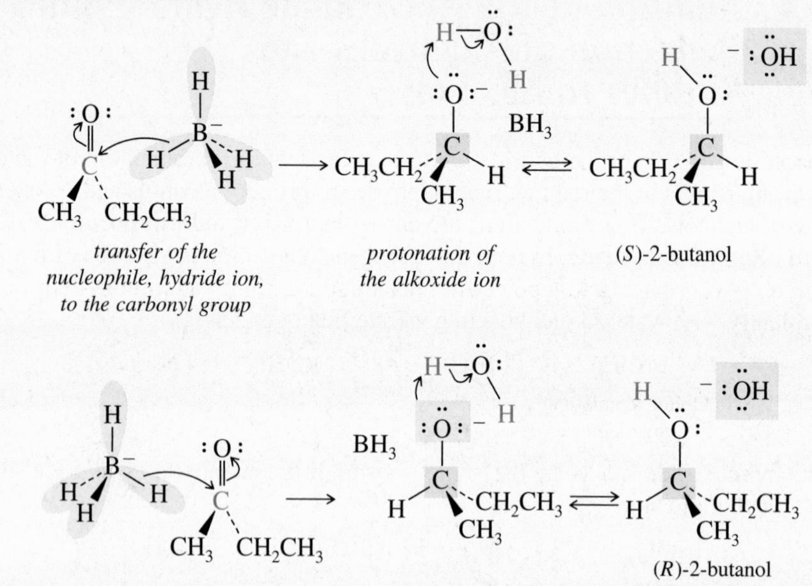

transfer of the
nucleophile, hydride ion,
to the carbonyl group

protonation of
the alkoxide ion

(S)-2-butanol

(R)-2-butanol

For the sake of simplicity, the transfer of only one of the four hydride ions possible from sodium borohydride is shown in the above mechanism. Actually, each molecule of sodium borohydride or lithium aluminum hydride can reduce four carbonyl groups. The reduction of an aldehyde to a primary alcohol by lithium aluminum hydride in diethyl ether proceeds in two steps. The reduction step gives an alkoxide anion, which complexes with the metal ions present. To recover the alcohol, the complex is treated with water and dilute acid. For example, heptanal is converted to 1-heptanol.

A similar sequence of reactions gives a secondary alcohol from 4-methyl-2-pentanone.

$$4 \ CH_3CHCH_2CCH_3 \xrightarrow[\text{diethyl ether}]{\text{LiAlH}_4} \left[ CH_3CHCH_2C\overset{\overset{\displaystyle CH_3}{|}}{\underset{\underset{\displaystyle H}{|}}{C}}-O- \right]_4 Al^-Li^+ \xrightarrow{H_3O^+} 4 \ CH_3CHCH_2CHCH_3$$

4-methyl-2-pentanone                                                          4-methyl-2-pentanol
                                                              85%

Lithium aluminum hydride is used in dry ether solvents because of its reactivity with water and alcohols. It is a much more reactive donor of hydride ions than sodium borohydride is, and it reacts with carboxylic acids and esters (p. 855) as well as with aldehydes and ketones.

Sodium borohydride and lithium aluminum hydride do not reduce isolated carbon–carbon double bonds. Thus it is possible to selectively reduce a carbonyl group in a molecule containing a double bond. An example of such a selective reduction is the synthesis of 6-methyl-5-hepten-2-ol, a pheromone (pp. 340 and 747) of the ambrosia beetle.

$$CH_3C{=}CHCH_2CH_2CCH_3 \xrightarrow[\text{ethanol}]{\text{NaBH}_4} CH_3C{=}CHCH_2CH_2CHCH_3$$

6-methyl-5-hepten-2-one                              6-methyl-5-hepten-2-ol
                                                    95%

Reductions of double bonds adjacent to carbonyl groups do sometimes occur with metal hydride reagents but are dependent on conditions such as solvent, temperature, the presence of excess hydride, and the presence of other metal ions. Researchers have found, for example, that the use of cerium(III) chloride greatly increases the selectivity of the reduction reaction of sodium borohydride. 2-Cyclohexenone, when reduced with sodium borohydride alone, gives a mixture of unsaturated and saturated alcohols.

2-cyclohexenone                          2-cyclohexenol   cyclohexanol
                                         51%         49%

In the presence of cerium(III) chloride, the reduction gives exclusively the product in which the double bond is untouched.

2-cyclohexenone                              2-cyclohexenol
                                              > 99%

Lithium aluminum hydride usually gives the unsaturated alcohol.

Study Guide
Concept Map 14.2

$$CH_3CH{=}CHCH \xrightarrow[\text{diethyl ether}]{\text{LiAlH}_4} \xrightarrow{H_3O^+} CH_3CH{=}CHCH_2OH$$

**Problem 14.6**    Complete the following equations.

(a)   CH$_3$O—⟨benzene ring⟩—$\overset{\overset{\displaystyle O}{\|}}{C}$H $\xrightarrow[\text{methanol}]{\text{NaBH}_4}$

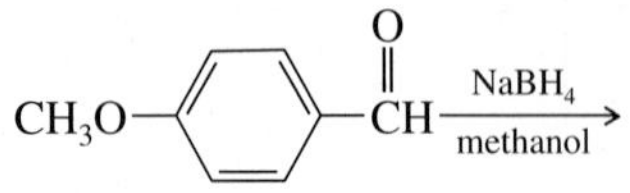

(b) CH$_3$(CH$_2$)$_5$$\overset{\overset{\displaystyle CH_3}{|}}{C}$HCH$_2$$\overset{\overset{\displaystyle O}{\|}}{C}$H $\xrightarrow[\text{diethyl ether}]{\text{LiAlH}_4}$ $\xrightarrow{\text{H}_3\text{O}^+}$

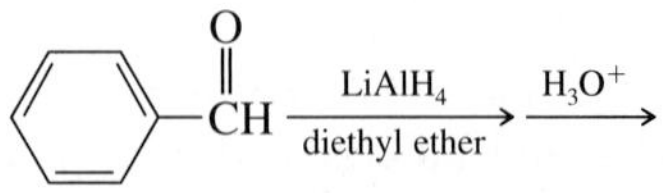

(c) ⟨benzene ring⟩—CH$_2$$\overset{\overset{\displaystyle O}{\|}}{C}$CH$_3$ $\xrightarrow[\text{diethyl ether}]{\text{LiAlH}_4}$ $\xrightarrow{\text{H}_3\text{O}^+}$

(d) ⟨benzene ring⟩—$\overset{\overset{\displaystyle O}{\|}}{C}$H $\xrightarrow[\text{diethyl ether}]{\text{LiAlH}_4}$ $\xrightarrow{\text{H}_3\text{O}^+}$

(e) CH$_3$$\overset{\overset{\displaystyle O}{\|}}{C}$CH$_2$$\overset{\overset{\displaystyle CH_3}{|}}{C}$HCH$_3$ $\xrightarrow[\text{diethyl ether}]{\text{LiAlH}_4}$ $\xrightarrow{\text{H}_3\text{O}^+}$

(f) CH$_3$CH=CH$\overset{\overset{\displaystyle O}{\|}}{C}$CH$_3$ $\xrightarrow[\substack{\text{CeCl}_3 \cdot 7\,\text{H}_2\text{O} \\ \text{methanol}}]{\text{NaBH}_4}$

## ONE SMALL STEP

Lithium aluminum hydride is a source of nucleophilic hydride ion for reactions other than those shown above. For example, alkyl halides are converted to alkanes when treated with lithium aluminum hydride in tetrahydrofuran. The following equation illustrates the reaction.

$$CH_3(CH_2)_6CH_2Br \xrightarrow[\substack{\text{tetrahydrofuran} \\ 25\,°C \\ 30\,\text{min}}]{\text{LiAlH}_4} CH_3(CH_2)_6CH_3$$

1-bromooctane                                          octane
                                                        96%

Rate studies on this reaction have shown the following orders of reactivity.

$$CH_3CH_2CH_2CH_2Br \ggg CH_3\overset{\overset{\displaystyle CH_3}{|}}{\underset{\underset{\displaystyle CH_3}{|}}{C}}CH_2Br$$

$$CH_3CH_2CH_2CH_2I \gg CH_3CH_2CH_2CH_2Cl$$

**PROBLEM:**   Propose a mechanism for the reaction, and discuss how the rate data support your mechanism.

## 14.5    Addition of the Nucleophile Cyanide Ion to the Carbonyl Group. Cyanohydrin Formation

Another nucleophilic addition reaction of aldehydes and ketones is the addition of the elements of hydrogen cyanide. The reagents generally used are sodium or potassium cyanide followed in a second step by acid. This type of reaction is illustrated below for acetone.

$$CH_3\overset{\overset{\displaystyle O}{\|}}{C}CH_3 \xrightarrow[\text{H}_2\text{O}]{\text{NaCN}} \xrightarrow[\text{H}_2\text{O}]{\text{H}_2\text{SO}_4} CH_3\overset{\overset{\displaystyle OH}{|}}{\underset{\underset{\displaystyle CN}{|}}{C}}CH_3$$

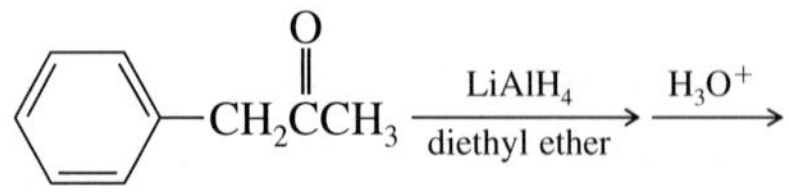

acetone                                      acetone
                                          cyanohydrin

The products of such a reaction are called **cyanohydrins** because they include a hydroxyl and a cyano group in the same molecule. The reaction is initiated when the cyanide ion attacks the electrophilic carbon of the carbonyl group. Then the alkoxide ion that is generated is protonated.

**Formation of a Cyanohydrin**

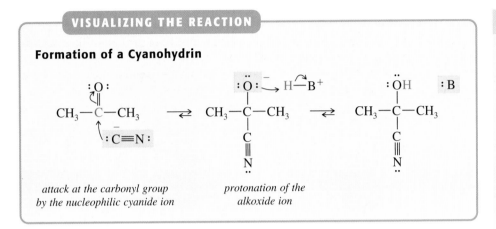

attack at the carbonyl group
by the nucleophilic cyanide ion

protonation of the
alkoxide ion

The addition of hydrogen cyanide to an aldehyde or ketone increases the number of carbon atoms in the molecule by one and introduces two functional groups that can undergo further reactions. An important reaction of the cyano group is the addition of water to the carbon–nitrogen triple bond (p. 614). An example of the formation and further transformation of a cyanohydrin is the conversion of acetaldehyde to lactic acid, which is one of the products of the metabolism of glucose in muscle tissue and also gives sour milk its acid taste.

**ONE SMALL STEP**

An important set of reactions used to convert a sugar to one containing one more carbon atom is known as the Kiliani-Fischer synthesis. The first step involves the reaction of sodium cyanide in the presence of sodium carbonate and water with the aldehyde group of a sugar to generate a cyanohydrin. Complete the following equation showing the first step of this synthesis with the simplest sugar, (R)-glyceraldehyde.

What are the stereochemical implications of this reaction?

$$\underset{\text{acetaldehyde}}{CH_3CH}\overset{O}{\overset{\|}{}} \xrightarrow[\text{H}_2\text{O}]{\text{NaCN}} \xrightarrow[\text{H}_2\text{O}]{\text{H}_2\text{SO}_4} \underset{\substack{\text{acetaldehyde}\\\text{cyanohydrin}}}{CH_3CHCN}\overset{OH}{\overset{|}{}} \xrightarrow[\Delta]{\text{H}_3\text{O}^+} \underset{\substack{\text{2-hydroxypropanamide}\\\text{amide of lactic acid}}}{CH_3CHCNH_2}\overset{O}{\overset{\|}{}}_{\underset{OH}{|}} \xrightarrow[\Delta]{\text{H}_3\text{O}^+} \underset{\substack{\text{2-hydroxypropanoic}\\\text{acid}\\\text{lactic acid}}}{CH_3CHCOH}\overset{O}{\overset{\|}{}}_{\underset{OH}{|}} + NH_4^+$$

**Problem 14.7**   The lactic acid present in muscle tissue is (R)-(−)-lactic acid, and that found in sour milk is (S)-(+)-lactic acid. What is the stereochemistry of the addition of hydrogen cyanide to acetaldehyde? What is the stereochemistry of the lactic acid formed by the hydrolysis of the cyano group?

## 14.6   Addition of Other Carbon Nucleophiles to the Carbonyl Group

### A. Preparation of Organometallic Reagents

Reactions in which carbon–carbon bonds are formed are important in synthesizing new organic compounds. A carbon–carbon bond forms when a reagent containing a carbon that is a nucleophilic center reacts with one containing a carbon that is an electrophilic center. For example, the reactions of alkyl halides with carbanions derived from terminal alkynes are used to create new carbon–carbon bonds (p. 236). In this section we will look at other ways to create carbon nucleophiles, and in the next sections we will examine their reactions with compounds containing an electrophilic carbon center.

One way of creating nucleophiles is to deprotonate an organic compound using a strong base. Acetylide anions (p. 331) are examples of nucleophiles prepared in this way. Another way of preparing carbon nucleophiles is by the reaction of organic halides with metals such as lithium or magnesium to give **organometallic reagents.** Alkyl, aryl, and vinyl halides all react with metals. The organometallic reagents prepared in this way with magnesium are named **Grignard reagents,** after Victor Grignard, a French chemist who developed them at the beginning of this century.

The reaction of methyl iodide with magnesium metal in ether gives methylmagnesium iodide, a methyl Grignard reagent.

$$\text{CH}_3\text{I} \ + \quad \text{Mg} \quad \xrightarrow[\text{diethyl ether}]{} \quad \text{CH}_3\text{MgI}$$

| methyl iodide | magnesium metal | methylmagnesium iodide |

*a methyl Grignard reagent*

Grignard reagents are also formed from aryl and vinyl halides. Phenylmagnesium bromide, for example, is a useful Grignard reagent.

bromobenzene                    phenylmagnesium bromide

*a phenyl Grignard reagent*

The organometallic compound is formed by transfer of electrons in an oxidation–reduction reaction from the magnesium atom to the more electronegative carbon atom. As a result of this electron transfer, the carbon atom that is an electrophilic center in the alkyl halide becomes a nucleophilic center in the Grignard reagent.

electrophilic center                    nucleophilic center

In all organometallic compounds, the carbon atom bonded to the metal reacts as a Lewis base and is a strong nucleophile. All organometallic reagents must be prepared in the absence of water or alcohols, which are acidic enough to protonate carbanions.

water          phenylmagnesium bromide          benzene

*acid*                *base*                *the conjugate acid of phenyl anion*

Therefore, *Grignard reagents cannot be prepared from compounds that contain acidic hydrogen atoms.* An organic halide that has a hydroxyl, carboxyl, thiol, or amino group elsewhere in the molecule cannot be used to make a Grignard reagent. Carbonyl and nitro groups in the reagent are also incompatible with the formation of Grignard reagents, because such groups react with organometallic reagents.

The basicity of the carbon atom bound to magnesium in the Grignard reagent can be utilized to make Grignard reagents from terminal alkynes. The relatively acidic hydrogen atom of a terminal alkyne is removed by an alkyl Grignard reagent.

$$CH_3(CH_2)_3C \equiv CH + CH_3CH_2MgBr \xrightarrow{\text{diethyl ether}} CH_3(CH_2)_3C \equiv CMgBr + CH_3CH_3 \uparrow$$

| 1-hexyne | ethylmagnesium bromide | 1-hexynylmagnesium bromide | ethane |
|---|---|---|---|
| *acid* *pK$_a$ ~26* | *base* | *from the conjugate base of 1-hexyne* | *the conjugate acid of ethyl anion* *pK$_a$ ~49* |

In a competition between the two bases, the proton is transferred from the alkyne to the ethyl group. Ethane gas evolves from the reaction mixture, pulling the reaction to completion.

When lithium metal reacts with an organic halide, an **organolithium reagent** is formed. The reaction is illustrated below for isopropyl bromide and bromobenzene.

$$\underset{\text{CH}_3}{\overset{\text{CH}_3}{|}}CH_3CHBr + 2\,Li \xrightarrow{\text{pentane}} \underset{\text{CH}_3}{\overset{\text{CH}_3}{|}}CH_3CHLi + Li^+Br^-$$

isopropyl bromide   lithium metal   isopropyllithium

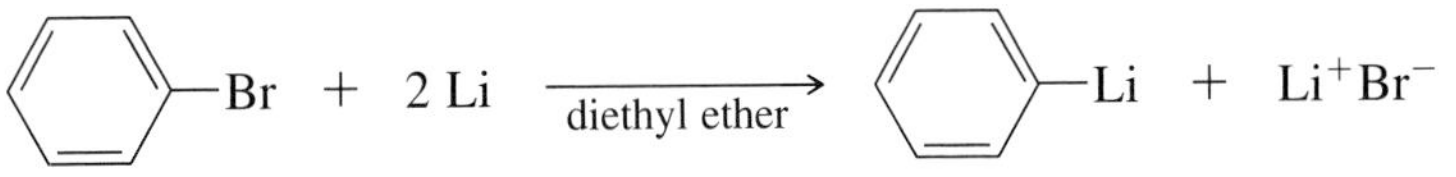

bromobenzene   lithium metal   phenyllithium

Organolithium reagents are also used as a source of basic and nucleophilic carbon atoms.

**Problem 14.8**   Complete the following equations.

(a) $CH_3CH_2CH_2CH_2Br \xrightarrow[\text{tetrahydrofuran}]{Li}$

(b) 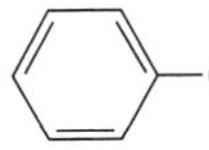 ⟨benzene ring⟩—$CH_2CH_2Br \xrightarrow[\text{diethyl ether}]{Mg}$

(c) $CH_2 = CHCl \xrightarrow[\Delta]{Mg \atop \text{tetrahydrofuran}}$

(d) 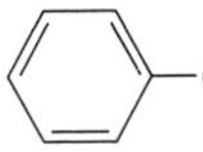 ⟨benzene ring⟩—$C \equiv CH \xrightarrow[\text{diethyl ether}]{CH_3CH_2MgBr}$

(e) 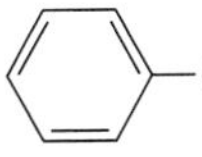 ⟨benzene ring⟩—$CH_2Br \xrightarrow[\text{tetrahydrofuran}]{Li}$

(f) $CH_3CH_2CH_2Li + NH_3 \longrightarrow$

(g) $CH_3CH_2MgI + CH_3OH \longrightarrow$

(h) 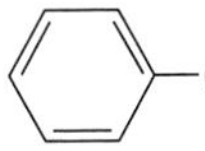 ⟨benzene ring⟩—$C \equiv CH \xrightarrow[\text{NH}_3\text{(liq)}]{NaNH_2}$

## B. Reactions of Organometallic Reagents with Aldehydes or Ketones as Electrophiles

All the organometallic reagents prepared in Section 14.6A add to aldehydes or ketones to produce alcohols. The reactions take place in two steps. The first involves attack by the nucleophilic carbon atom of the organometallic reagent on the electrophilic carbon atom of the carbonyl group and gives an alkoxide anion complexed with a metal ion. In the second step, this complex is broken up with water, and usually some dilute acid, to produce the alcohol.

---

**VISUALIZING THE REACTION**

### Addition of an Organometallic Reagent to a Carbonyl Compound

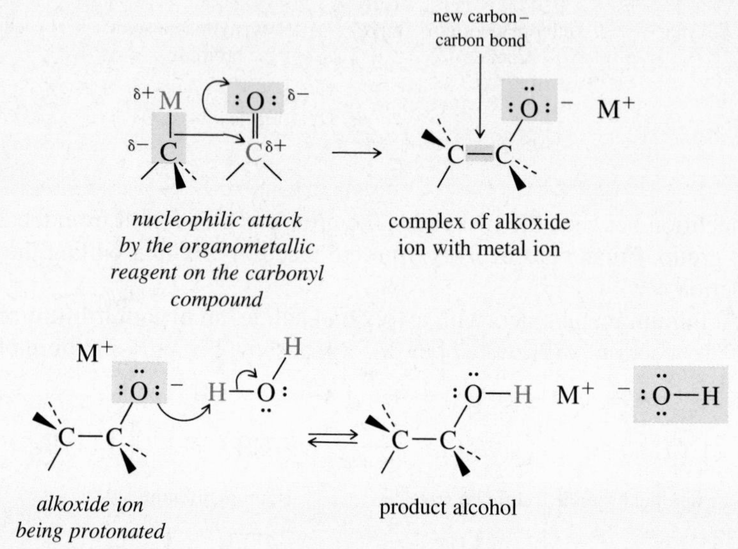

nucleophilic attack
by the organometallic
reagent on the carbonyl
compound

complex of alkoxide
ion with metal ion

alkoxide ion
being protonated

product alcohol

---

Because it is possible to vary the structure of the aldehyde or ketone as well as that of the organometallic reagent, many different alcohols may be prepared by such reactions. Historically, Grignard reagents have been used extensively to synthesize alcohols by means of the **Grignard reaction.** In each of the three reactions shown below, the Grignard reagent is prepared, an aldehyde or a ketone is added to it to give an alkoxide ion complexed with magnesium ion, and the alcohol is formed by protonation of the alkoxide ion. In the last equation, a very weak acid, ammonium chloride, is used because a stronger acid would cause dehydration of the tertiary alcohol.

$$CH_3(CH_2)_3C{\equiv}CH + CH_3CH_2MgBr \xrightarrow{\text{diethyl ether}} CH_3(CH_2)_3C{\equiv}CMgBr + CH_3CH_3\uparrow \xrightarrow{\overset{O}{\overset{\|}{HCH}}\text{ (formaldehyde)}}$$

1-hexyne

1-hexynylmagnesium
bromide

$$CH_3(CH_2)_3C{\equiv}CCH_2OMgBr \xrightarrow{H_3O^+} CH_3(CH_2)_3C{\equiv}CCH_2OH$$

2-heptyn-1-ol
82%

$$\underset{\substack{|\\ \text{CH}_3}}{CH_3CHBr} + Mg \xrightarrow{\text{diethyl ether}} \underset{\substack{|\\ \text{CH}_3}}{CH_3CHMgBr} \xrightarrow{\overset{O}{\overset{\|}{CH_3CH}}\text{ (acetaldehyde)}} \underset{\substack{|\\ \text{OMgBr}}}{\overset{\substack{\text{CH}_3\\ |}}{CH_3CHCHCH_3}} \xrightarrow{H_3O^+} \underset{\substack{|\\ \text{OH}}}{\overset{\substack{\text{CH}_3\\ |}}{CH_3CHCHCH_3}}$$

isopropyl bromide

isopropylmagnesium
bromide

3-methyl-2-butanol
50%

*racemic mixture*

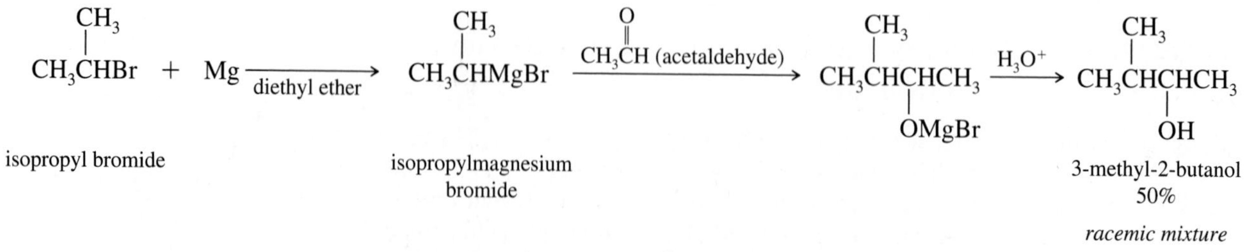

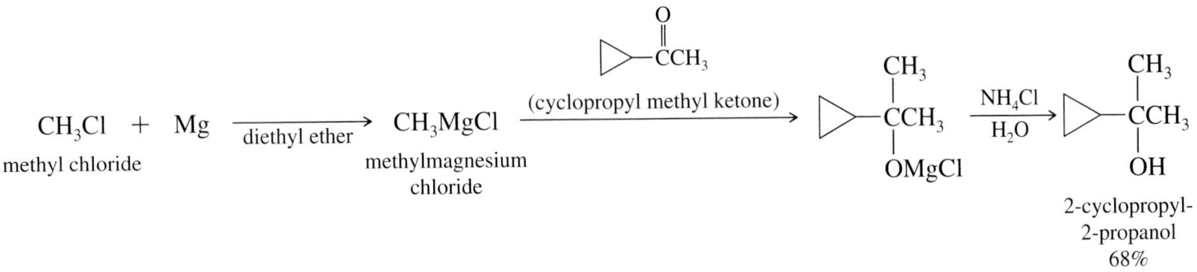

The reaction scheme:

$$CH_3Cl + Mg \xrightarrow{\text{diethyl ether}} CH_3MgCl$$

methyl chloride → methylmagnesium chloride

(cyclopropyl methyl ketone) leads to:

2-cyclopropyl-2-propanol 68%

The type of alcohol formed depends on the type of carbonyl compound that is used. Formaldehyde produces a primary alcohol that has one more carbon atom in it than the alkyl halide used to make the Grignard reagent. Other aldehydes such as acetaldehyde produce secondary alcohols. A ketone, such as cyclopropyl methyl ketone, produces a tertiary alcohol.

A secondary or tertiary alcohol formed in a Grignard reaction may contain stereocenters, depending on the nature of the organic groups bonded to the carbon attached to the hydroxyl group. As in the reduction of a carbonyl group by sodium borohydride (p. 546), the organometallic reagent attacks either face of the carbonyl group with equal probability. A racemic mixture of product alcohol results, unless, of course, chirality was already present in the Grignard reagent or existed elsewhere in the carbonyl compound; in such a case, a mixture of diastereomers is formed.

**On the Web: ONE SMALL STEP**

Organolithium reagents also add to aldehydes or ketones to form alcohols. The reactions parallel those of Grignard reagents. In many cases, organolithium reagents give better yields than Grignard reagents do. In fact, organolithium reagents form alcohols with some sterically hindered ketones that do not react with Grignard reagents. Some representative reactions of organolithium reagents are shown below.

$$CH_3CH_2CH_2CH_2CH_2CH_2Br + 2\,Li \xrightarrow[\text{0 °C}]{\text{tetrahydrofuran}} CH_3CH_2CH_2CH_2CH_2CH_2Li + Li^+Br^-$$

1-bromohexane → 1-hexyllithium

↓ HCH (formaldehyde)

$$CH_3CH_2CH_2CH_2CH_2CH_2CH_2OH \xleftarrow{H_3O^+} CH_3CH_2CH_2CH_2CH_2CH_2CH_2O^-Li^+$$

1-heptanol
72%

*a primary alcohol*

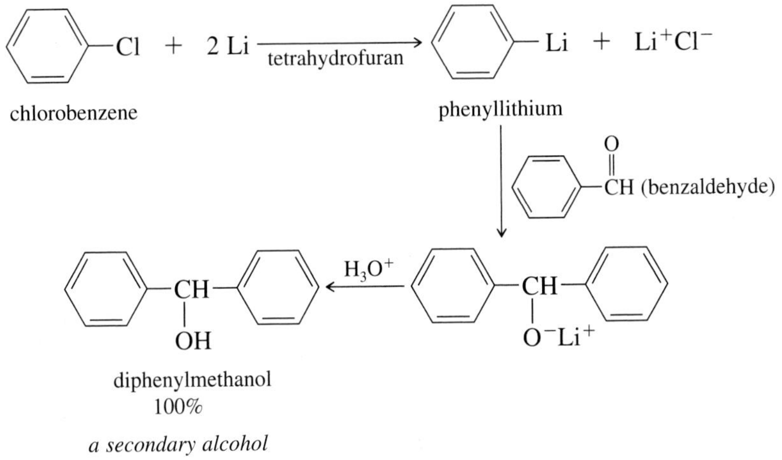

chlorobenzene $+ 2\,Li \xrightarrow{\text{tetrahydrofuran}}$ phenyllithium $+ Li^+Cl^-$

↓ CH (benzaldehyde)

diphenylmethanol
100%

*a secondary alcohol*

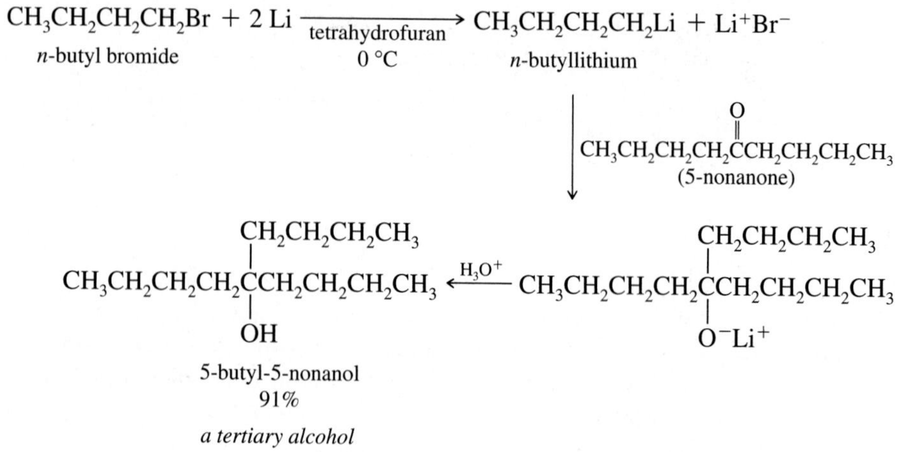

$$CH_3CH_2CH_2CH_2Br + 2\ Li \xrightarrow[\substack{tetrahydrofuran \\ 0\ °C}]{} CH_3CH_2CH_2CH_2Li + Li^+Br^-$$

*n*-butyl bromide            *n*-butyllithium

$$CH_3CH_2CH_2CH_2\overset{\overset{\displaystyle O}{\|}}{C}CH_2CH_2CH_2CH_3$$
(5-nonanone)

$$CH_3CH_2CH_2CH_2\overset{\overset{\displaystyle CH_2CH_2CH_2CH_3}{|}}{\underset{\underset{\displaystyle OH}{|}}{C}}CH_2CH_2CH_2CH_3 \xleftarrow{H_3O^+} CH_3CH_2CH_2CH_2\overset{\overset{\displaystyle CH_2CH_2CH_2CH_3}{|}}{\underset{\underset{\displaystyle O^-Li^+}{|}}{C}}CH_2CH_2CH_2CH_3$$

5-butyl-5-nonanol
91%

*a tertiary alcohol*

In each case, the organometallic reagent adds to the carbonyl group, and the resulting alkoxide anion is protonated by dilute acid to recover the alcohol. Formaldehyde yields a primary alcohol, other aldehydes give secondary alcohols, and ketones give tertiary alcohols.

Sodium acetylide, prepared from sodium amide and acetylene in liquid ammonia, also adds to carbonyl groups. It reacts with benzaldehyde to give an acetylenic secondary alcohol and with cyclohexanone to form a tertiary alcohol.

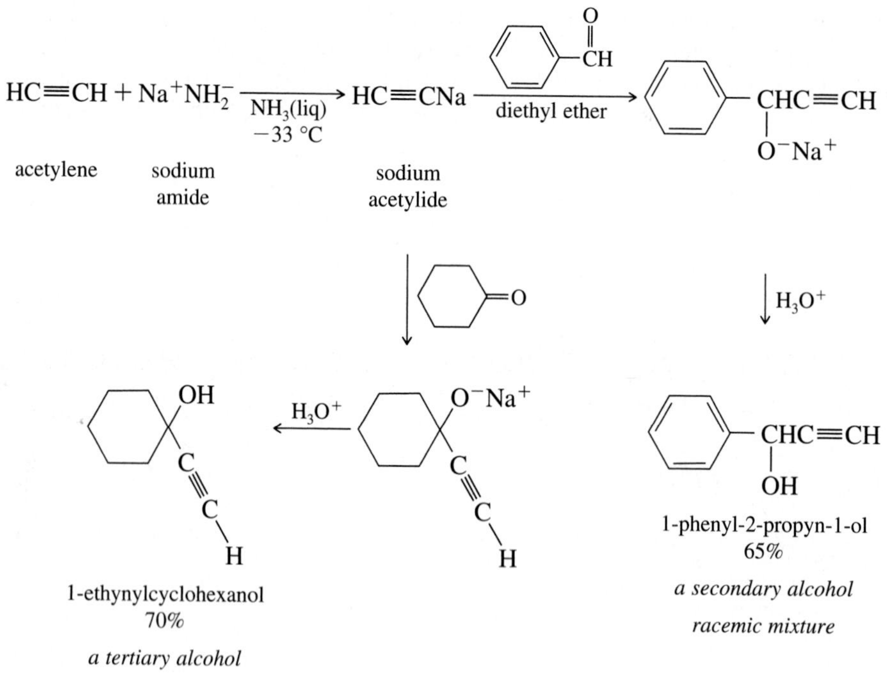

$$HC{\equiv}CH + Na^+NH_2^- \xrightarrow[\substack{NH_3(liq) \\ -33\ °C}]{} HC{\equiv}CNa \xrightarrow[diethyl\ ether]{} $$

acetylene    sodium            sodium
           amide            acetylide

1-phenyl-2-propyn-1-ol
65%

*a secondary alcohol*

*racemic mixture*

1-ethynylcyclohexanol
70%

*a tertiary alcohol*

## C. Reaction of Organometallic Reagents with Oxirane as an Electrophile

The three-membered ring of an oxirane is strained enough that it will open under nucleophilic attack (p. 504) by an organometallic reagent. In this respect, the reaction of the oxirane mirrors that of a carbonyl group: The nucleophilic carbon atom

**Table 14.2** **Alcohols from Reactions of Organometallic Reagents with Aldehydes, Ketones, and Oxirane**

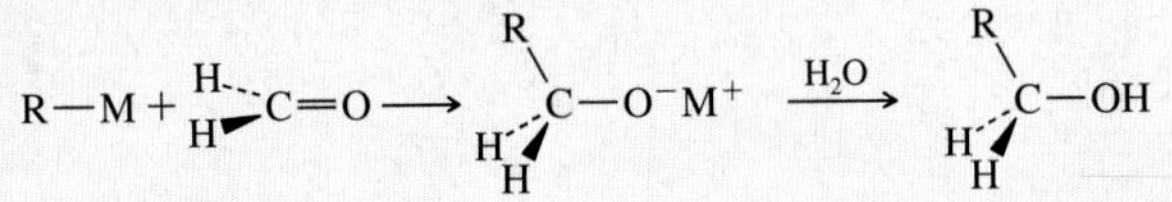

formaldehyde

a primary alcohol with one more carbon atom than the organometallic reagent had

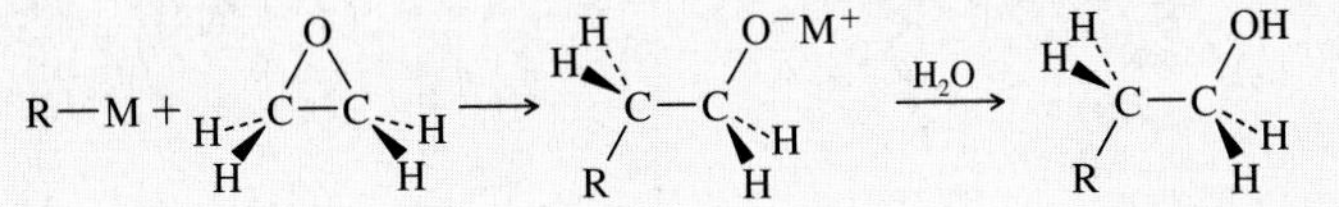

oxirane

a primary alcohol with two more carbon atoms than the organometallic reagent had

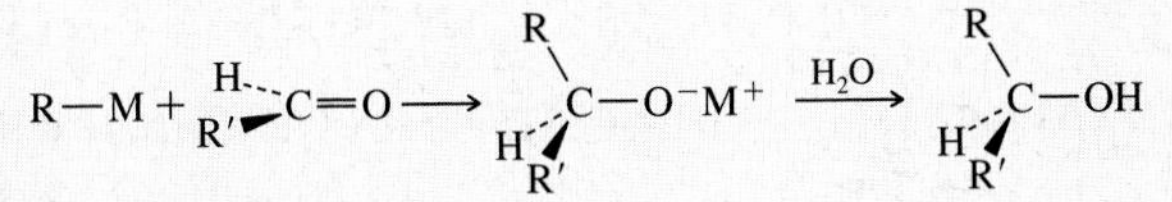

aldehyde

a secondary alcohol

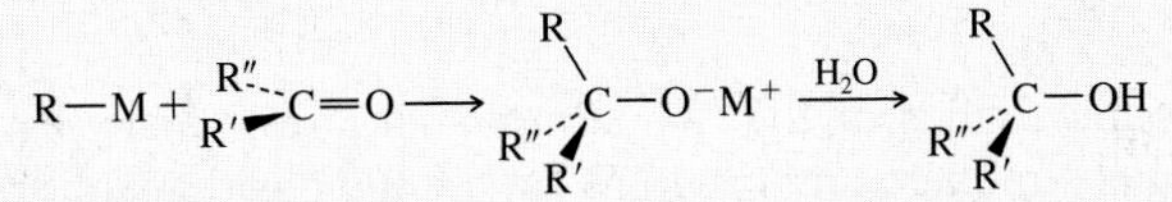

ketone

a tertiary alcohol

of the organometallic reagent bonds to one of the carbon atoms of the oxirane ring, and the metal ion complexes with the alkoxide ion that forms when the ring opens. Because of this similarity, the reaction of oxirane is discussed here where it can be compared with the reactions of aldehydes and ketones with carbon nucleophiles. *n*-Butylmagnesium bromide, for example, reacts with oxirane to give 1-hexanol, a primary alcohol; the carbon chain in the product is two carbon atoms longer than the chain in the starting organometallic reagent.

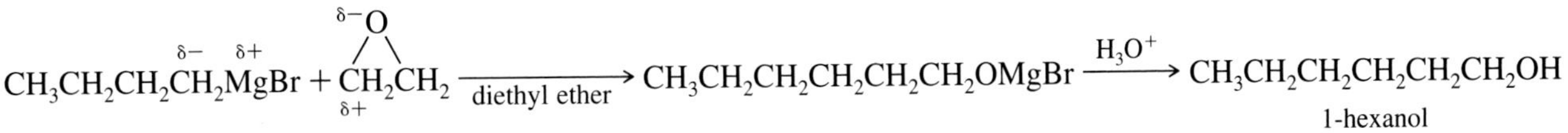

$$CH_3CH_2CH_2CH_2MgBr + \overset{\delta-}{CH_2}\overset{O}{\diagdown}CH_2 \xrightarrow{\text{diethyl ether}} CH_3CH_2CH_2CH_2CH_2CH_2OMgBr \xrightarrow{H_3O^+} CH_3CH_2CH_2CH_2CH_2CH_2OH$$

*n*-butylmagnesium bromide

oxirane

1-hexanol
60%

The structural changes that occur when organometallic reagents react with formaldehyde, other aldehydes, ketones, and oxiranes to produce alcohols are summarized in Table 14.2.

**Problem 14.9** Complete the following equations, showing the structures of all the intermediates and final products in the reactions.

(a) 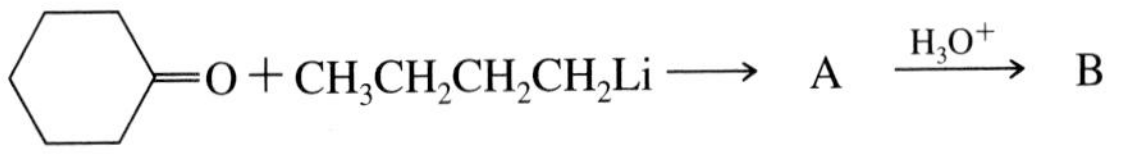 $\bigcirc{=}O + CH_3CH_2CH_2CH_2Li \longrightarrow$ A $\xrightarrow{H_3O^+}$ B

(b) 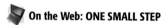 —Br $\xrightarrow[\text{diethyl ether}]{\text{Mg}}$ C $\xrightarrow{\overset{\text{O}}{\triangle}\ \text{CH}_2\text{CH}_2}$ D $\xrightarrow{\text{H}_3\text{O}^+}$ E

(c) —Br $\xrightarrow[\text{tetrahydrofuran}]{\text{Li}}$ F $\xrightarrow{\overset{\overset{\text{O}}{\|}}{\text{CH}_3\text{CH}_2\text{CH}_2\text{CH}}}$ G $\xrightarrow{\text{H}_3\text{O}^+}$ H

**Study Guide**
Concept Maps 14.3 and 14.4

**On the Web: ONE SMALL STEP**

(d) $\text{CH}_3\text{CH}_2\text{C}\equiv\text{CH} \xrightarrow[\text{NH}_3\,(\text{liq})]{\text{NaNH}_2}$ I $\xrightarrow[\text{diethyl ether}]{\text{(cyclopentanone)} =\text{O}}$ J $\xrightarrow{\text{H}_3\text{O}^+}$ K

## D. The Art of Solving Problems

**PROBLEM:** How would you carry out the following transformation?

$$\text{HOCH}_2\text{CH}_2\text{CH}_2\text{CH}_2\text{CH}=\text{CH}_2 \xrightarrow{?} \text{CH}_2=\text{CH}-\overset{\overset{\text{H}}{|}}{\underset{\overset{|}{\text{OH}}}{\text{C}}}\text{CH}_2\text{CH}_2\text{CH}_2\text{CH}_2\text{CH}=\text{CH}_2$$

### Solution

1. What are the connectivities of the two compounds? How many carbon atoms does each contain? Are there any rings? What are the positions of branches and functional groups on the carbon skeletons?

   The starting material is a six-carbon chain with a primary alcohol function at one end and a carbon–carbon double bond at the other. The product has a nine-carbon chain with a secondary alcohol function at the third carbon atom and carbon–carbon double bonds at both ends of the chain.

2. How do the functional groups change in going from starting material to product? Does the starting material have a good leaving group?

   The primary alcohol function has disappeared. A new double bond has been introduced into the molecule as part of a three-carbon addition. There is no good leaving group in the starting material, but the primary alcohol may be converted into one.

3. Is it possible to dissect the structures of the starting material and product to see which bonds must be broken and which formed?

$$\text{HO}\!+\!\text{CH}_2\text{CH}_2\text{CH}_2\text{CH}_2\text{CH}=\text{CH}_2 \qquad \text{CH}_2=\text{CH}-\overset{\overset{\text{H}}{|}}{\underset{\overset{|}{\text{OH}}}{\text{C}}}\!\!+\!\!\text{CH}_2\text{CH}_2\text{CH}_2\text{CH}_2\text{CH}=\text{CH}_2$$

bond broken                           bond formed

4. New bonds are created when an electrophile reacts with a nucleophile. Do we recognize any part of the product molecule as coming from a good nucleophile or an electrophilic addition?

   The product is a secondary alcohol. Secondary alcohols are formed by the addition of a nucleophilic organometallic reagent to the electrophilic carbonyl group of an aldehyde.

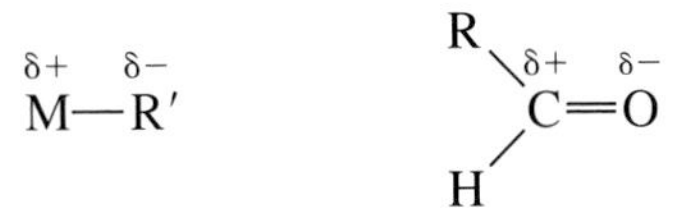

$$\overset{\delta+ \quad \delta-}{M-R'} \qquad \overset{R}{\underset{H}{\diagdown}}\overset{\delta+ \quad \delta-}{C=O}$$

organometallic        aldehyde with
reagent          electrophilic carbonyl
group

5. What type of compound would be a good precursor to the product?

An aldehyde and a Grignard or an organolithium reagent are needed. The position of the secondary alcohol function in the product suggests that a three-carbon aldehyde has reacted with an organometallic reagent derived from the rest of the molecule.

$$CH_2=CH-\overset{\overset{\displaystyle H}{|}}{C}=O \qquad BrMgCH_2CH_2CH_2CH_2CH=CH_2$$

precursors to the product

6. After this last step, do we see how to get from starting material to product? If not, we need to analyze the structures obtained in step 5 by applying questions 4 and 5 to them.

The Grignard reagent can be made from the corresponding alkyl halide, which we can make from the primary alcohol. The complete synthesis is shown below.

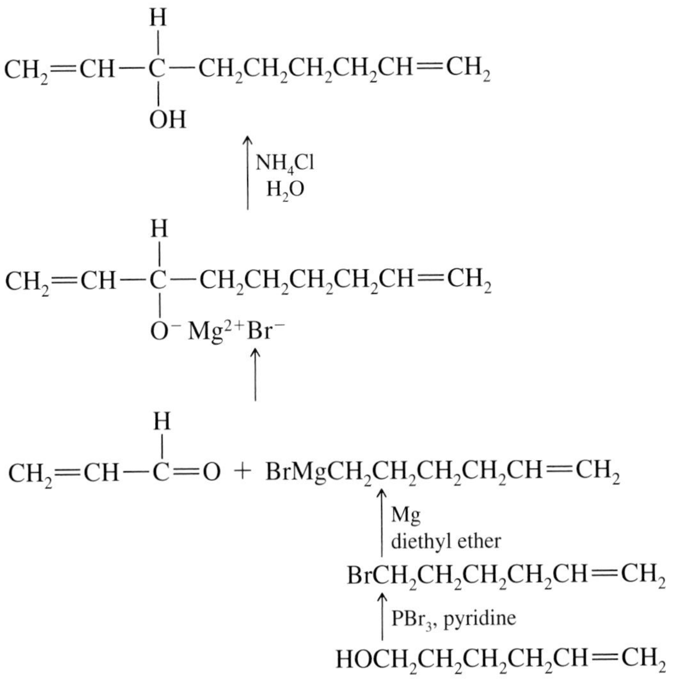

$$CH_2=CH-\overset{\overset{\displaystyle H}{|}}{\underset{\underset{\displaystyle OH}{|}}{C}}-CH_2CH_2CH_2CH_2CH=CH_2$$

$$\uparrow \begin{array}{l}NH_4Cl \\ H_2O\end{array}$$

$$CH_2=CH-\overset{\overset{\displaystyle H}{|}}{\underset{\underset{\displaystyle O^-\ Mg^{2+}Br^-}{|}}{C}}-CH_2CH_2CH_2CH_2CH=CH_2$$

$$\uparrow$$

$$CH_2=CH-\overset{\overset{\displaystyle H}{|}}{C}=O \ + \ BrMgCH_2CH_2CH_2CH_2CH=CH_2$$

$$\uparrow \begin{array}{l}Mg \\ diethyl\ ether\end{array}$$

$$BrCH_2CH_2CH_2CH_2CH=CH_2$$

$$\uparrow PBr_3, pyridine$$

$$HOCH_2CH_2CH_2CH_2CH=CH_2 \qquad \blacksquare$$

If one of the starting materials had not been given to us in the problem, our analysis of the problem would have led to two sets of precursors. An alternative dissection of the secondary alcohol gives us the second pair of reagents that would give the same alcohol.

$$CH_2=CH-\overset{\overset{\displaystyle H}{|}}{C}-CH_2CH_2CH_2CH_2CH=CH_2 \longleftarrow CH_2=CHMgBr + O=\overset{\overset{\displaystyle H}{|}}{C}CH_2CH_2CH_2CH_2CH=CH_2$$
$$\underset{OH}{|}$$

The following problem provides practice in devising syntheses for alcohols.

**Problem 14.10**    Dissect the following alcohols and propose reagents that would lead to their formation.

(a) $CH_3\overset{\overset{\displaystyle CH_3}{|}}{C}=CH\overset{\overset{\displaystyle CH_3}{|}}{C}CH_2CH_2CH_2CH_3$
$\phantom{xxx}\underset{OH}{|}$

(b) $CH_3\overset{\overset{\displaystyle CH_3}{|}}{C}=CHCH_2CH_2\overset{\overset{\displaystyle CH_3}{|}}{C}=CHCH_2CH_2\overset{}{C}HC\equiv CH$
$\phantom{xxxxxxxxxxxxxxxxxxx}\underset{OH}{|}$

(c) ⬡—$CH_2CH_2OH$

## 14.7 Reactions of Aldehydes and Ketones with Alcohols. The Formation of Acetals and Ketals

### A. Hydrates, Acetals, and Ketals

When carbonyl compounds are dissolved in water, they exist in equilibrium with low concentrations of their **hydrates,** which are compounds in which water has been added to the carbonyl group.

$$CH_3\overset{\overset{\displaystyle O}{\|}}{C}H + H_2O \rightleftarrows CH_3-\overset{\overset{\displaystyle OH}{|}}{\underset{\underset{\displaystyle OH}{|}}{C}}-H$$

acetaldehyde    water    hydrate of<br>acetaldehyde

Alcohols react with aldehydes in an equilibrium process, just as water does, to yield compounds known as **hemiacetals.** In a hemiacetal, a hydroxyl group and an alkoxyl group are bonded to the same carbon atom. Hemiacetals, after losing a molecule of water, react with yet another molecule of alcohol to give **acetals.** In an acetal, two alkoxyl groups are bonded to the same carbon atom. When acetaldehyde is placed in an excess of methanol, the aldehyde, its hemiacetal, and its acetal are all present at equilibrium.

$$CH_3\overset{\overset{\displaystyle O}{\|}}{C}H + CH_3OH \rightleftarrows CH_3\overset{\overset{\displaystyle OH}{|}}{\underset{\underset{\displaystyle OCH_3}{|}}{C}}H + CH_3\overset{\overset{\displaystyle OCH_3}{|}}{\underset{\underset{\displaystyle OCH_3}{|}}{C}}H + H_2O$$

acetaldehyde    methanol    1-methoxy-    1,1-dimethoxyethane<br>1-ethanol        *an acetal*<br>*a hemiacetal*

Ketones give **hemiketals** and **ketals** with alcohols. Hemiketals and ketals do not form as spontaneously as hemiacetals and acetals do. The amount of the hemiketal in equilibrium with acetone dissolved in methanol, for example, is very small.

$$\underset{\text{acetone}}{CH_3\overset{\overset{\displaystyle O}{\|}}{C}CH_3} + \underset{\text{methanol}}{CH_3OH} \;\rightleftharpoons\; \underset{\substack{\text{2-methoxy-2-propanol}\\\text{0.3\% at equilibrium}}}{CH_3\overset{\overset{\displaystyle OH}{|}}{\underset{\underset{\displaystyle OCH_3}{|}}{C}}CH_3}$$

*a hemiketal*

Hemiacetals and hemiketals are generally unstable compounds unless the alcohol and carbonyl groups are part of the same molecule, in which case equilibrium favors the formation of stable cyclic hemiacetals and hemiketals containing five- or six-membered rings. The most important examples of these structures are found in carbohydrates. We will examine these in detail in Section 14.8. Simpler hydroxy-aldehydes also form stable cyclic hemiacetals. 4-Hydroxybutanal, for example, forms a hemiacetal that is a five-membered ring.

4-hydroxybutanal          2-hydroxytetrahydrofuran
11% at equilibrium          89% at equilibrium

**Problem 14.11**    5-Hydroxypentanal gives a six-membered cyclic hemiacetal. Write an equation for the formation of the hemiacetal, showing the six-membered ring in the chair form. How many stereoisomers can the hemiacetal have?

If an aldehyde or ketone is placed in an excess of an alcohol in the presence of an acid, an acetal or ketal is formed. Acetone and methanol, in a reaction that is catalyzed by an arylsulfonic acid, are converted into the dimethyl ketal of acetone, 2,2-dimethoxypropane.

$$\underset{\text{acetone}}{CH_3\overset{\overset{\displaystyle O}{\|}}{C}CH_3} \;+\; \underset{\text{methanol}}{2\,CH_3OH} \;\xrightarrow{ArSO_3H}\; \underset{\text{2,2-dimethoxypropane}}{CH_3\overset{\overset{\displaystyle OCH_3}{|}}{\underset{\underset{\displaystyle OCH_3}{|}}{C}}CH_3} \;+\; H_2O$$

*a ketal*

An acetal or ketal structurally resembles a diether and as such is stable to bases and nucleophilic reagents. Acetals and ketals, however, are sensitive to dilute acid and are easily converted to the original carbonyl compound and alcohol, as shown for the hydrolysis of 2,2-dimethoxypropane.

$$\underset{\text{2,2-dimethoxypropane}}{CH_3\overset{\overset{\displaystyle OCH_3}{|}}{\underset{\underset{\displaystyle OCH_3}{|}}{C}}CH_3} \;+\; H_2O \;\xrightarrow{HCl}\; \underset{\text{acetone}}{CH_3\overset{\overset{\displaystyle O}{\|}}{C}CH_3} \;+\; \underset{\text{methanol}}{2\,CH_3OH}$$

In order to get good yields of acetals and ketals, the water formed in the equilibrium reaction must be removed.

The conversion of an acetal or ketal to the original carbonyl compound and alcohol is an example of a hydrolysis reaction. A **hydrolysis reaction** is one in which a $\sigma$ bond is cleaved by the addition of the elements of water to the fragments formed in the cleavage. Note that **hydrolysis,** in which *lysis* (meaning "to loosen") takes place, contrasts with **hydration,** in which water is added to a multiple bond, but no fragmentation of the molecule occurs (p. 284).

## B. Mechanism of Acetal or Ketal Formation and Hydrolysis

The formation of a hemiacetal may be viewed as an addition reaction to the carbon–oxygen double bond.

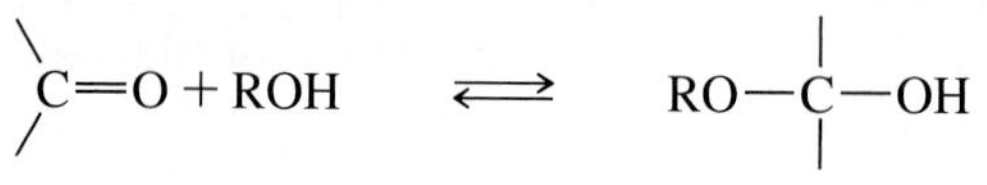

$$\begin{array}{c} \diagdown \\ \diagup \end{array} C{=}O + ROH \quad \rightleftarrows \quad RO{-}\overset{\textstyle |}{\underset{\textstyle |}{C}}{-}OH$$

addition reaction

Going from the hemiacetal to the acetal is a substitution reaction.

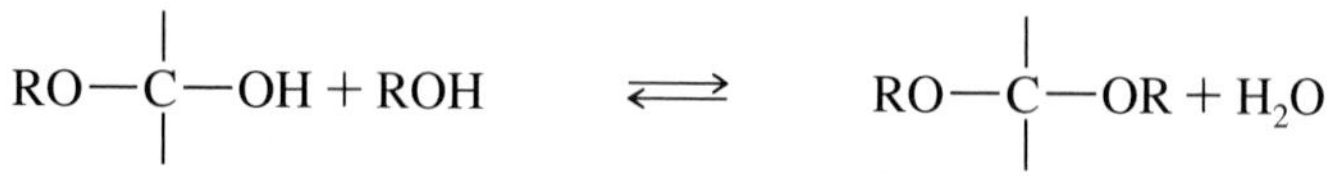

$$RO{-}\overset{\textstyle |}{\underset{\textstyle |}{C}}{-}OH + ROH \quad \rightleftarrows \quad RO{-}\overset{\textstyle |}{\underset{\textstyle |}{C}}{-}OR + H_2O$$

substitution reaction

Like other substitution reactions in which a hydroxyl group is replaced, this reaction goes best in the presence of an acid catalyst that converts the poor leaving group into a good one (p. 256).

In the reaction of an aldehyde or a ketone with an alcohol, the oxygen atom of the alcohol serves as the nucleophile that attacks the carbon atom of the carbonyl group. The acid catalyst protonates the oxygen atom of the carbonyl group, increasing the electrophilicity of the carbonyl carbon atom. Deprotonation of the oxonium ion that results from the reaction of the alcohol with the protonated carbonyl compound gives the hemiacetal or hemiketal. The mechanism is illustrated below with the acid-catalyzed reaction of acetaldehyde with methanol.

---

**VISUALIZING THE REACTION**

**Formation of a Hemiacetal**

protonation of the carbonyl group

nucleophilic attack on the carbonyl group

deprotonation

a hemiacetal

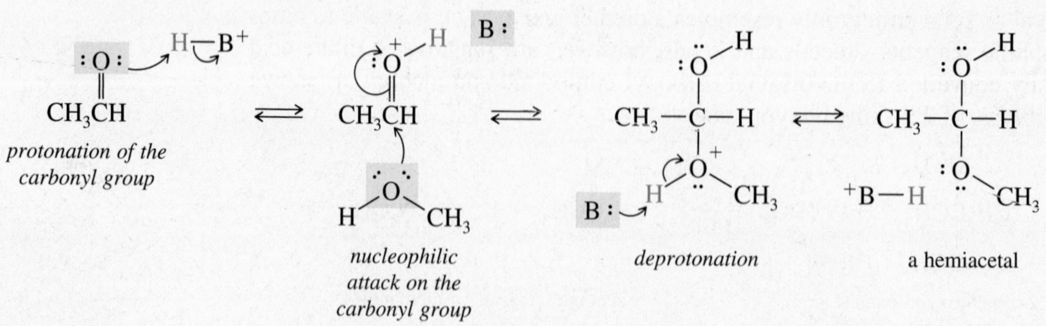

The hemiacetal of acetaldehyde is converted to the acetal in an $S_N1$ reaction, which starts with the protonation of the hemiacetal. Both oxygen atoms are reversibly protonated in multiple equilibrium reactions, but only protonation of the hydroxyl group leads to further reaction by loss of water as a leaving group. Protonation of the methoxy group is simply a reversal of the last step in the formation of the hemiacetal.

The reaction proceeds via an $S_N1$ mechanism because the carbocation that forms is particularly stable. The positive charge can be delocalized to an oxygen atom in a resonance contributor in which carbon and oxygen atoms both have octets of electrons. The carbocation reacts with methanol as the nucleophile, and deprotonation of the new intermediate gives the dimethyl acetal of acetaldehyde. Continuous removal from the reaction mixture of the water that is formed draws the equilibrium toward the product. In practice, these reactions are often carried out in the presence of solvents such as benzene or toluene that codistill with water, carrying it out of the reaction mixture.

### VISUALIZING THE REACTION

**Conversion of a Hemiacetal to an Acetal**

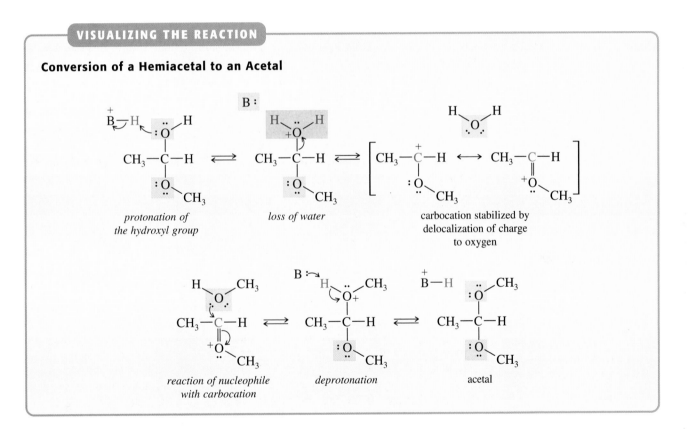

*protonation of the hydroxyl group*     *loss of water*     carbocation stabilized by delocalization of charge to oxygen

*reaction of nucleophile with carbocation*     *deprotonation*     acetal

The hydrolysis of the acetal is a reversal of all the steps in its formation. If the acetal is put in an excess of water with a trace of acid, an ether oxygen atom is protonated, and methanol is lost. Water adds to the carbocation intermediate, the hemiacetal is obtained, and the reverse process continues until aldehyde and alcohol are regenerated.

**Problem 14.12**    When acetals or ketals are hydrolyzed in water containing $^{18}O$, the resulting alcohols contain very little $^{18}O$. Write a mechanism for the hydrolysis of 2,2-dimethoxypropane in $H_2{}^{18}O$ with a trace of hydrochloric acid added, and show where the $^{18}O$ will turn up in the products.

**Problem 14.13**    Acetals, though they resemble ethers in structure, are cleaved by acids much more easily than ethers are. Vinyl ethers, also called **enol ethers,** are hydrolyzed easily with dilute acid, just as acetals are, to give aldehydes or ketones as products. Propose a mechanism for the following reaction.

$$CH_2=CH-OCH_2CH_3 \xrightarrow[\text{HClO}_4(0.1\ M)]{\text{H}_2\text{O}} \overset{\displaystyle O}{\overset{\|}{CH_3CH}} + CH_3CH_2OH$$

(A review of Problem 9.24 may be helpful.) Why does this hydrolysis reaction occur so easily?

## C. Cyclic Acetals or Ketals

Because acetals or ketals can be hydrolyzed under very mild conditions to regenerate the original carbonyl compound, they are often prepared in order to protect carbonyl groups in multifunctional compounds while reactions are carried out elsewhere in the molecule. The use of such protecting groups in syntheses will be explored in Chapter 21. The acetal function is stable to bases, nucleophiles, and reducing agents, so it can be used with a wide variety of reagents as long as acidic conditions are avoided. The acetals or ketals of simple aldehydes and ketones with low-molecular-weight alcohols are relatively easy to form, but the reaction becomes difficult if the ketone or alcohol is large and sterically hindered.

Ethylene glycol is used to make cyclic acetals or ketals in which both alcohol groups that react with the carbonyl group are on the same molecule. The reaction is shown below for cyclohexanone.

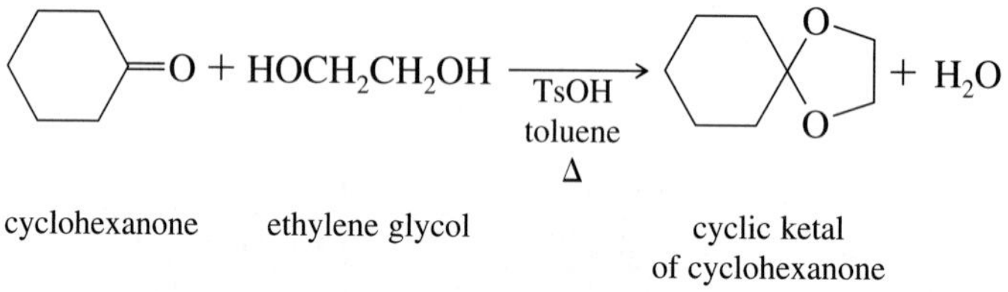

$$\text{(cyclohexanone)} =O + HOCH_2CH_2OH \xrightarrow[\substack{\text{toluene}\\\Delta}]{\text{TsOH}} \text{(cyclic ketal)} + H_2O$$

| cyclohexanone | ethylene glycol | | cyclic ketal of cyclohexanone |

Cyclic acetals and ketals are formed much more easily than their acyclic counterparts for reasons of entropy. In the reaction shown above, two reactants, cyclohexanone and ethylene glycol, form two products, the cyclic ketal and water. The number of molecules neither increases nor decreases, and hence there is no large change in entropy. If, on the other hand, we had made the ketal of cyclohexanone with methanol, a decrease in the number of molecules would have taken place in going from starting material to products.

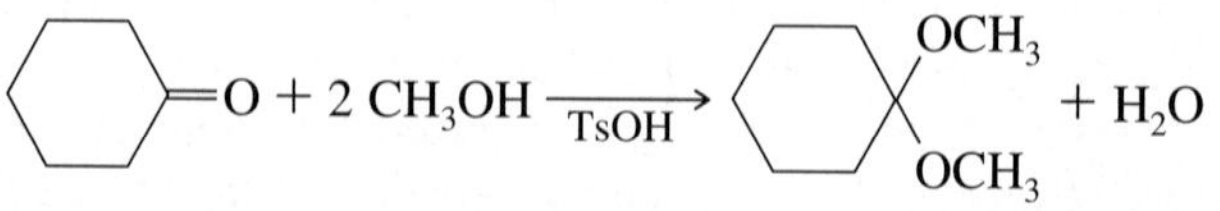

$$\text{(cyclohexanone)} =O + 2\ CH_3OH \xrightarrow{\text{TsOH}} \overset{OCH_3}{\underset{OCH_3}{\text{(ketal)}}} + H_2O$$

Three molecules of reagents (one cyclohexanone and two methanols) are converted into two molecules (the ketal and water) of product. There is a decrease in possibilities for independent position and movement and hence a decrease in entropy in going from reactants to products for the case of the acyclic ketal. A decrease in entropy leads to a less favorable free energy change for the reaction (p. 97) and, therefore, a smaller equilibrium constant for the reaction.

Entropy also affects the ease with which an intramolecular reaction takes place in another way. The hemiketal from the reaction of cyclohexanone with ethylene glycol is more likely to form another carbon–oxygen bond and hence the ketal than in the case of an acyclic ketal because the first is an intramolecular reaction with the reagent needed for the second step held close to the reactive site. Such collisions between reactants are much more probable than when separate molecules have to find each other.

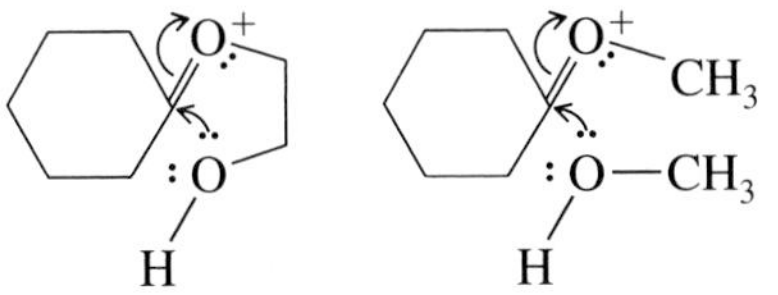

intramolecular  
reaction,  
more probable

intermolecular  
reaction,  
less probable

The cyclic ketal has a five-membered ring. Five- and six-membered rings are favored and form whenever possible in intramolecular reactions. As the number of atoms in the chain between the point of first attachment and the free reactive end increases beyond three or four, entropy begins work against ring formation. Longer chains have many more possible conformations, and the probability that the free end will find the reactive site decreases.

⤤ **Study Guide**  
**Concept Map 14.5**

**Problem 14.14**    Predict the products of the following reactions.

(a)

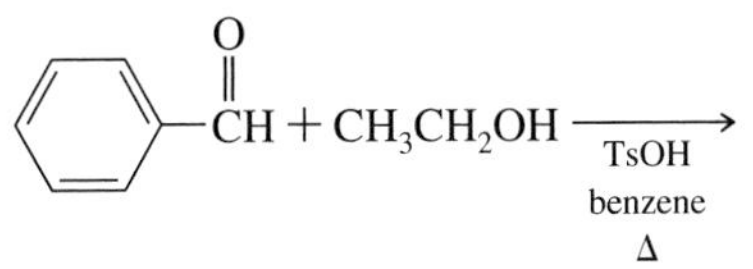

(b) $CH_3CH_2CH_2CH$ (=O) $+ CH_3OH \xrightarrow[\substack{benzene \\ \Delta}]{TsOH}$

(c) 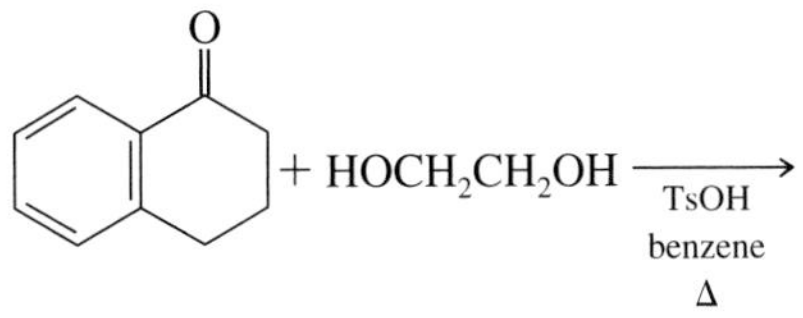 $+ HOCH_2CH_2OH \xrightarrow[\substack{benzene \\ \Delta}]{TsOH}$

**Problem 14.15**    The mechanism for the formation of a cyclic ketal is exactly the same as that shown for the formation of the acetal on page 561. The only difference is that the two alcohol groups that are part of two different molecules of methanol in the mechanism shown there are now both part of the same molecule of ethylene glycol. Write a detailed mechanism for the formation of the cyclic ketal of cyclohexanone shown on the previous page using the Visualizing the Reaction on page 561 as your model.

## 14.8   Biological Hemiacetals and Acetals. Carbohydrates

### A. Equilibrium Between Hydroxyaldehydes and Their Cyclic Hemiacetals

Glucose, the sugar from which we derive much of our energy and which is also the building block for more complex carbohydrates such as starch and cellulose, has the formula $C_6H_{12}O_6$ and is (2R, 3S, 4R, 5R)-(+)-2,3,4,5,6-pentahydroxyhexanal.

D-(+)-glucose
(2*R*, 3*S*, 4*R*, 5*R*)-(+)-2,3,4,5,6-pentahydroxyhexanal

Glucose is dextrorotatory (p. 199) and is called D-glucose to describe its stereochemical relationship to a series of other sugars. The story behind this nomenclature is in Chapter 23. In this section we will explore the equilibrium between glucose and its hemiacetals as an example and a further exploration of the chemistry of hydroxyaldehydes.

Glucose has hydroxy groups three and four carbons away from the carbonyl group and therefore forms cyclic hemiacetals (p. 562; also Problem 14.11) in equilibrium with the open-chain form shown above. Evidence that the equilibrium lies in the direction of the hemiacetal forms can be seen in the spectra of glucose. The infrared spectrum of glucose is given in Figure 14.2. The proton nuclear magnetic resonance spectra of glucose taken 5 minutes and 2 weeks after solution in $D_2O$ are shown in Figure 14.3. These spectra are special for a reason. The infrared spectrum does not show absorption in the region we would expect for the carbonyl group of an aldehyde ($\sim$1740–1720 cm$^{-1}$). Similarly, the band typical of the hydrogen atom of an aldehyde ($\sim\delta$ 9.7) does not appear in the proton magnetic resonance spectra of the compound. The picture of glucose as a pentahydroxyaldehyde does not fit modern spectroscopic evidence. The proton magnetic resonance spectra show that the composition of a solution of glucose changes over time, as indicated by the change in the relative numbers of the protons giving rise to the doublets at $\delta$ 4.65 and 5.25. The origins of these peaks are discussed on page 568.

Even before spectroscopic data were obtained for glucose, chemists had evidence that an open-chain aldehyde structure was not the best way to depict glucose and that there was more than one form of glucose. Early chemists got their information from polarimetry, the determination of the optical rotation (p. 200) of glucose. When a freshly prepared solution of glucose that has been recrystallized from methanol and has a melting point of 147°C is put into the tube of a polarimeter, an

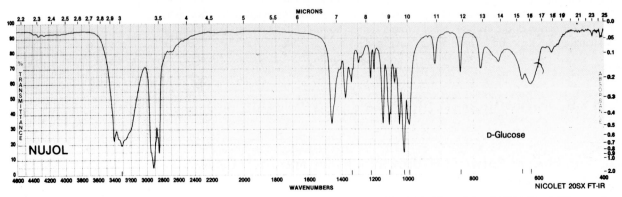

Reprinted with permission of Aldrich Chemical Company, Inc.

*Figure 14.2*
Infrared spectrum of α-D-glucose.

D-Glucose in D$_2$O, after five minutes in solution

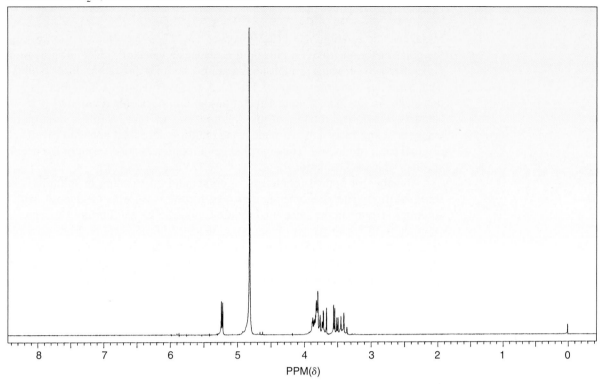

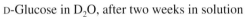
PPM($\delta$)

D-Glucose in D$_2$O, after two weeks in solution

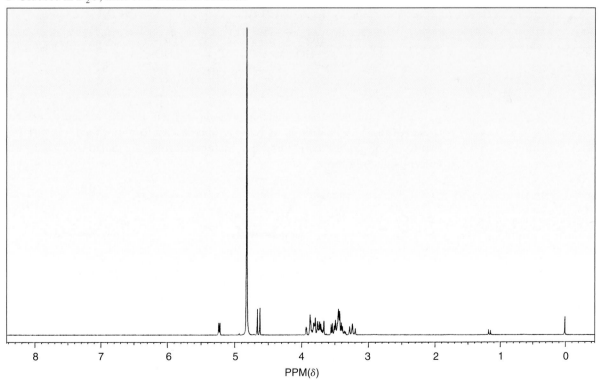
PPM($\delta$)

*Figure 14.3*
Proton magnetic resonance spectra of D-glucose: *top,* taken 5 minutes after solution in
D$_2$O; *bottom,* taken 2 weeks later.

initial specific rotation of $+112.2°$ is observed. When the solution is left to stand in the polarimeter, the rotation falls until it reaches a value of $+52.7°$. If glucose is recrystallized from water at high temperatures, another crystalline form is obtained. This form has a melting point of 150°C. A freshly prepared solution of these crystals placed in the tube of a polarimeter has an initial specific rotation of $+18.7°$. On standing, the optical rotation of this solution rises to $+52.7°$. Either solution can be evaporated and recrystallized under the condition described above to give back the original form of glucose. Thus the change in rotation is not a result of the decomposition of glucose in solution.

A change of optical rotation for a compound on standing in solution is called **mutarotation.** The phenomenon observed for the two forms of glucose, with both solutions arriving at the same final rotation, suggests an equilibrium between stereochemically different forms of the compound. Multiple equilibria are possible for glucose in aqueous solution. For example, the aldehyde group can react with water to give a hydrate. Different conformers of glucose in equilibrium with each other bring the hydroxyl groups at carbon 4 or 5 into position for intramolecular reaction with the carbonyl group to give cyclic hemiacetals. This reaction converts the carbonyl carbon into a new stereocenter, making four cyclic hemiacetals possible, two with five-membered rings and two with six-membered rings (Figure 14.4). These multiple equilibria are catalyzed by acids and bases.

A six-membered ring that includes an oxygen atom is related to the heterocyclic compound **pyran.** A saccharide in its six-membered ring cyclic form is called a **pyranose.** The names commonly used by carbohydrate chemists for the six-membered cyclic forms of glucose are $\alpha$-D-glucopyranose and $\beta$-D-glucopyranose.

|  |  |  |
|---|---|---|
| pyran | $\alpha$-D-glucopyranose | $\beta$-D-glucopyranose |

A five-membered ring that includes an oxygen atom is related to the heterocycle **furan.** A saccharide in its five-membered ring form is called a **furanose.** The two forms of glucose containing five-membered rings are called $\alpha$-D-glucofuranose and $\beta$-D-glucofuranose.

|  |  |  |
|---|---|---|
| furan | $\alpha$-D-glucofuranose | $\beta$-D-glucofuranose |

All four of the cyclic forms of glucose are in equilibrium with each other, but the pyranose forms predominate in water almost completely. Furanose forms constitute less than 0.5% at equilibrium.

**On the Web: ONE SMALL STEP**

When the pyranose ring forms, a new stereocenter is created at carbon 1. The orientation of the hydroxyl group on carbon 1 can be axial or equatorial. The form in which it is axial is called **$\alpha$-glucose,** which crystallizes at ordinary temperatures and has an initial rotation of $+112.2°$. The form in which the hydroxyl group at carbon 1 is equatorial is called **$\beta$-glucose.** It crystallizes out of water at high temperatures and has an initial rotation of $+18.7°$. In aqueous solution, each of these two

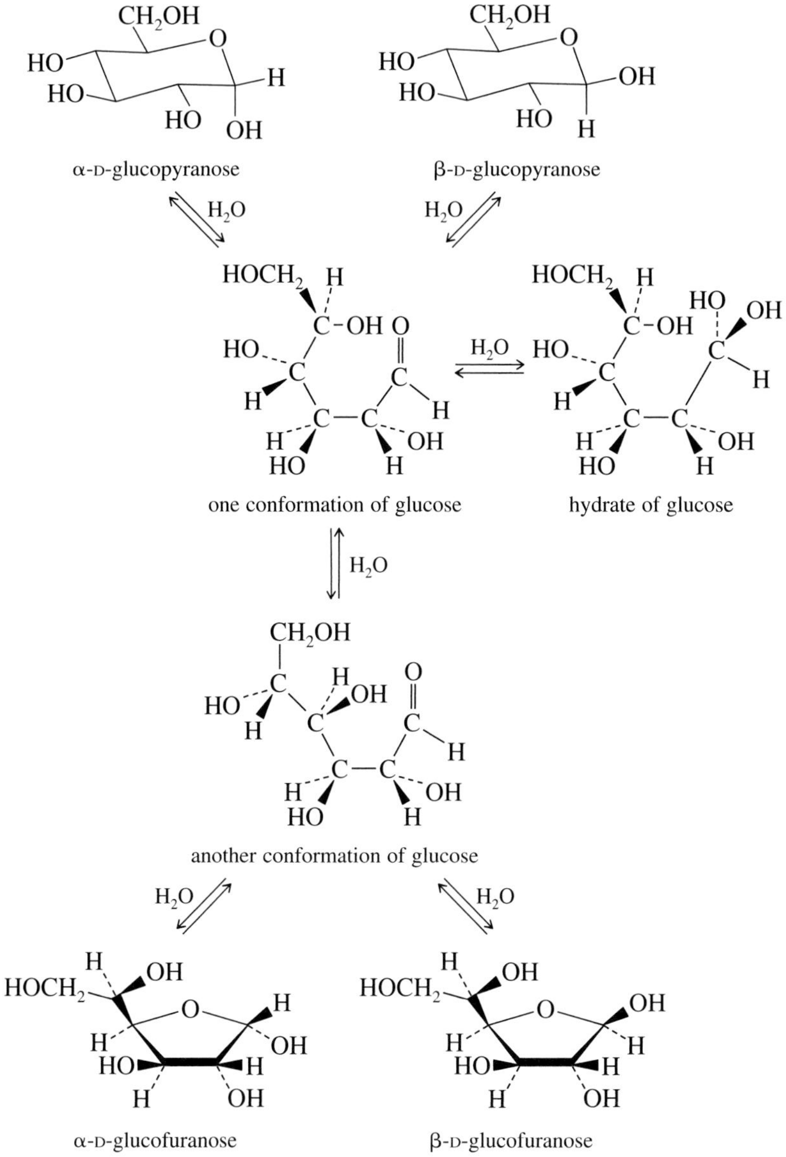

CH$_2$OH

HO
HO
HO   OH
H

α-D-glucopyranose

CH$_2$OH

HO
HO
HO   H
OH

β-D-glucopyranose

HOCH$_2$  H
C–OH O
HO
C   C
H   H
H–C–C   OH
HO   H

one conformation of glucose

HOCH$_2$  H
C–OH
HO   OH
HO
C   C
H   H
H–C–C   OH
HO   H

hydrate of glucose

CH$_2$OH
C   H
HO   OH
H   C   O
C
C–C   H
H   OH
HO   H

another conformation of glucose

H   OH
HOCH$_2$   O   H
H   OH
HO   H
H   OH

α-D-glucofuranose

H   OH
HOCH$_2$   O   OH
H   H
HO   H
H   OH

β-D-glucofuranose

Figure 14.4
The mutarotation of glucose.

forms is in equilibrium with the open-chain form, which has the free aldehyde group.

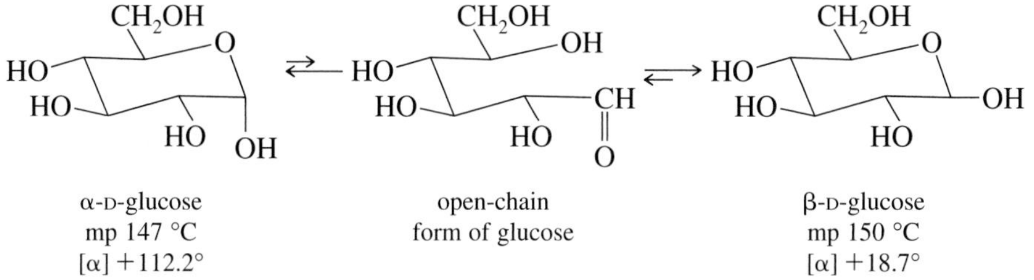

CH$_2$OH
HO
HO
HO   OH

α-D-glucose
mp 147 °C
[α] +112.2°

CH$_2$OH
HO   OH
HO
HO   CH
O

open-chain
form of glucose

CH$_2$OH
HO
HO
HO   OH

β-D-glucose
mp 150 °C
[α] +18.7°

The small concentration of this open-chain form in solution is responsible for the reactions of glucose that are typical of aldehydes. The equilibrium that exists among all forms is responsible for the change in rotation from the initial values of $+112.2°$ for $\alpha$-glucose and $+18.7°$ for $\beta$-glucose to the intermediate value of $+52.7°$ for the equilibrium mixture. This value corresponds to a mixture consisting of 38% $\alpha$-glucose and 62% $\beta$-glucose. $\beta$-Glucose has all the large substituents in the equatorial positions in the chair conformation of the six-membered ring. It is more stable in water solution than $\alpha$-glucose, in which the hydroxyl group at carbon 1 is axial, although the $\alpha$-pyranose is most easily obtained in crystalline form.

$\alpha$-Glucopyranose and $\beta$-glucopyranose are stereoisomers that differ from each other at one stereocenter, carbon 1. Therefore, they are diastereomers. Diastereomers that differ from each other in stereochemistry at the carbon atom of a potential carbonyl group in a cyclic hemiacetal are called **anomers.** $\alpha$-Glucopyranose and $\beta$-glucopyranose are, therefore, best defined as anomers of each other. Carbon 1 is the anomeric carbon atom, bearing an anomeric hydroxyl group. The hydrogen atom that is on the carbon atom of the carbonyl group in the aldehyde form of glucose is now the anomeric hydrogen atom.

Because it is on a carbon atom bonded to two oxygen atoms, the anomeric hydrogen atom is more deshielded than the other hydrogen atoms bonded to carbon in glucose. The proton magnetic resonance spectra (Figure 14.3, p. 565) were taken of a sample of $\alpha$-D-glucopyranose dissolved in $D_2O$. In this solvent, all the hydrogen atoms bonded to oxygens are exchanged for deuterium (p. 434). The strong singlet at $\delta$ 4.8 comes from DOH, resulting from this exchange. The other peaks come from hydrogens bonded to carbon atoms. Two stand out—the doublets centered at $\delta$ 4.65 and 5.25, which correspond to the anomeric hydrogen atoms for $\beta$-D-glucopyranose and $\alpha$-D-glucopyranose, respectively. When the spectrum is taken within 5 mnutes of the solution of $\alpha$-D-glucopyranose in $D_2O$, only traces of $\beta$-D-glucopyranose ($\delta$ 4.65) are seen. After the solution is allowed to stand for 2 weeks and mutarotation occurs, peaks for both anomers are clearly visible in the spectrum, with $\beta$-D-glucopyranose now being the major isomer in solution.

$\delta$ 5.25 split, by H at carbon 2
$J = \sim 3$ Hz for hydrogens
that are gauche to each other

crystalline $\alpha$-D-glucopyranose          $\alpha$-D-glucopyranose          $\beta$-D-glucopyranose

$\delta$ 4.65, split by H at carbon 2
$J = \sim 8$ Hz for hydrogens
that are anti to each other

The anomeric hydrogen atom in $\alpha$-D-glucopyranose occupies an equatorial position and is gauche to the hydrogen atom at carbon 2. The coupling constant for such an interaction is small, $\sim 3$ Hz. The anomeric hydrogen atom in $\beta$-D-glucopyranose is axial and is anti to the hydrogen atom at carbon 2. This gives a stronger interaction between the two, resulting in the larger coupling constant, $\sim 8$ Hz. The study of the nuclear magnetic resonance spectroscopy of carbohydrates was important in learning how the geometry of interacting protons influences coupling constants.

## B. Glycosides. Equilibria Among the Acetals of Glucose

When glucose is treated with methanol containing hydrogen chloride, acetals are formed. In these compounds, the hemiacetal function is converted into the cyclic monomethyl acetal. The structures of the acetals obtained depend on reaction conditions. At low temperatures and after short reaction times, the principal products are the five-membered ring methyl glucofuranosides.

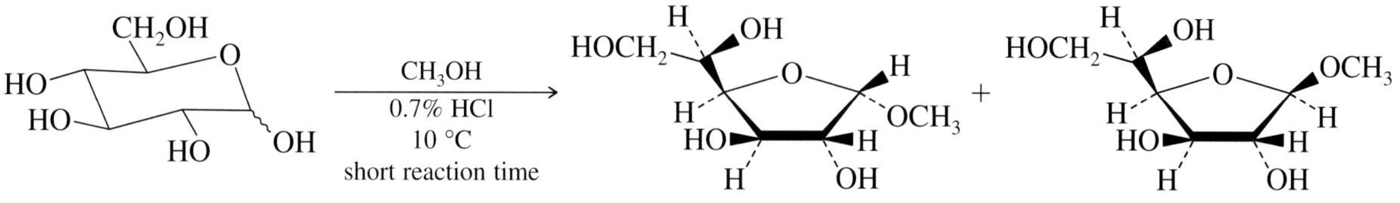

methyl α-D-glucofuranoside          methyl β-D-glucofuranoside

Prolonged heating with methanol and hydrogen chloride, on the other hand, gives mainly the methyl glucopyranosides.

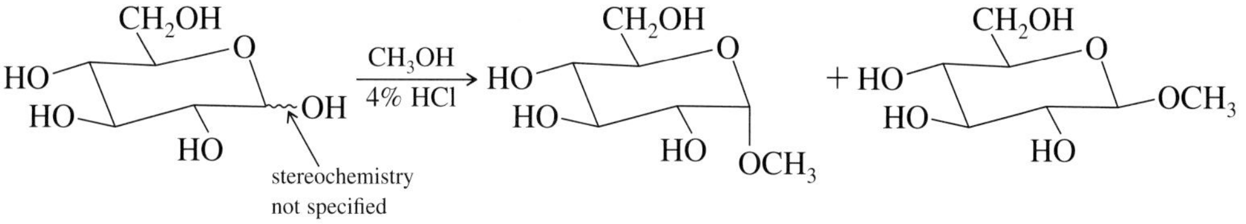

α- or β-
D-glucopyranose

methyl α-D-
glucopyranoside
mp 166 °C
[α] +158°

*major product*

methyl β-D-
glucopyranoside
mp 105 °C
[α] −34°

*minor product*

At equilibrium, 66% of the mixture consists of the α-isomer, the product with the methoxy group axial, and ~33% of the β isomer, with the methoxy group equatorial. This is true whether the starting compound is α- or β-glucopyranose. This stereochemistry is an exception to the rule that larger groups occupy equatorial positions in the chair form of six-membered rings. Chemists believe that the replacement of a —$CH_2$— group in cyclohexane by the oxygen atom in the pyranose ring is responsible for this effect, which we will discuss in the next section.

The formation of two different types of products depending on reaction conditions is familiar. Five-membered rings form faster than six-membered ones do. Under the conditions of the reaction, however, the ring can reopen, so equilibrium among all the different forms is possible. At higher temperatures and over longer periods of time, equilibrium is established. The methyl D-glucopyranosides are the thermodynamic products, with methyl α-D-glucopyranoside being the thermodynamically most stable.

As a class, carbohydrates with a full acetal linkage are known as **glycosides.** The compounds shown above are called methyl glucosides or methyl glucopyranosides to reflect their cyclic nature.

**Problem 14.16**   Write a mechanism for the formation of methyl α-D-glucopyranoside from α-D-glucopyranose. How does the mechanism that you propose explain the fact that both α- and β-glucopyranoses give the same mixture of methyl glucopyranosides? (*Hint:* Reviewing Section 14.7B, p. 560, may be helpful.)

Methyl glucopyranosides do not undergo mutarotation or show any aldehyde reactions. For example, they cannot be oxidized easily to carboxylic acids. As an acetal, the carbonyl group is effectively protected. The glucopyranosides are stable in basic solutions. They are hydrolyzed easily in acidic solutions to give an equilibrium mixture of $\alpha$-D-glucopyranose and $\beta$-D-glucopyranose.

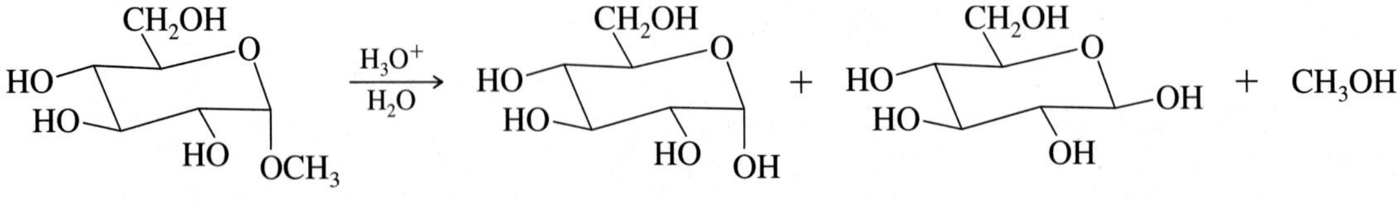

| methyl $\alpha$-D-glucopyranoside | $\alpha$-D-glucopyranose | $\beta$-D-glucopyranose |

Enzymes, which are stereoselective biological catalysts, can be used to hydrolyze glucosides selectively. The enzyme maltase will cleave only $\alpha$-glucopyranosides, and the enzyme emulsin cleaves only $\beta$-glucopyranosides.

**Problem 14.17**    Give a detailed mechanism for the hydrolysis of methyl $\alpha$-D-glucopyranoside in dilute acid.

## ONE SMALL STEP

When methyl $\alpha$-D-glucopyranoside is treated with dimethyl sulfate in the presence of aqueous sodium hydroxide, the alcohol functions are converted to methyl ether groups.

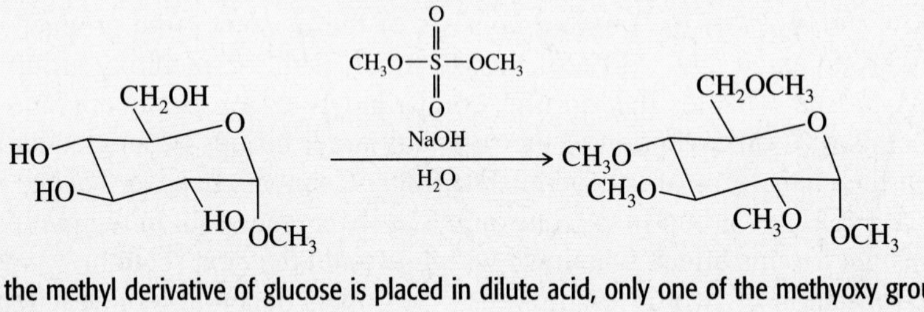

When the methyl derivative of glucose is placed in dilute acid, only one of the methyoxy groups is affected.

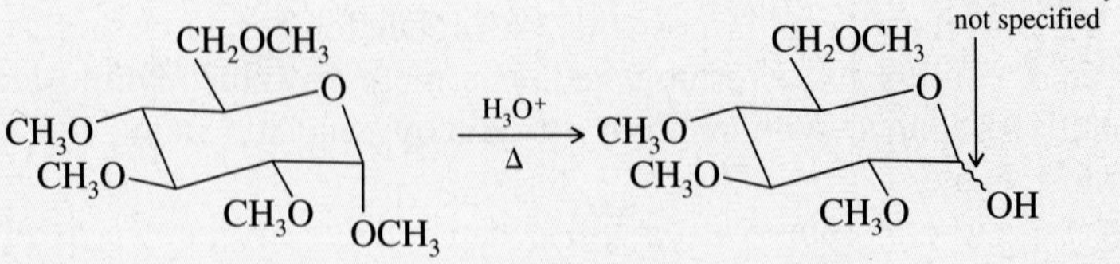

(a) In the discussion of the Williamson synthesis of ethers (p. 498), we emphasized the need to use strong bases to deprotonate alcohols to alkoxide ions and specifically showed why sodium hydroxide was not a good choice. Why can it be used in the case of glucose?

(b) Write a mechanism for the reaction of the fully methylated ether of glucose with dilute acid that demonstrates why only the anomeric methoxy group is removed.

(c) No stereochemistry at the anomeric carbon is specified for the product of acid hydrolysis of the fully methylated ether of glucose. Why? What is happening?

The glycosidic linkage occurs widely in nature. Such bonding between the anomeric carbon of one sugar unit and a hydroxyl group of another is the way in which complex carbohydrates such as starch (p. 1002) and cellulose (p. 1004) are formed. In plants, sugars are also found bonded to any of a large number of alcohols and phenols, resulting in a variety of natural products that have medicinal and other practical uses. The hydroxyl compounds that are bonded to sugars in glycosides are called **aglycons.**

An example of a glycoside that has been used in medicine for some time is salicin, found in the bark of the willow tree. Salicin is the $\beta$-glycoside of o-(hydroxymethyl)phenol.

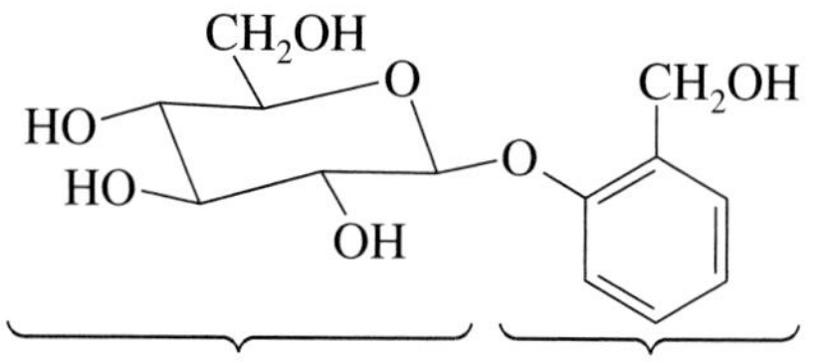

glucose       o-(hydroxymethyl)phenol

*a sugar*            *an aglycon*

*salicin, a naturally occurring glycoside*

Willow bark preparations have been known since the time of the ancient Greeks as pain relievers. They were usually used externally because the willow juice is so bitter. Chemists sought to isolate the compound that gave willow its analgesic properties. The active component, salicin, was finally isolated from other plant sources and converted to salicylic acid. Salicylic acid has valuable medicinal properties, but it cannot be taken internally. In 1899, salicylic acid was converted into its acetyl derivative, acetylsalicylic acid, which is now commonly known as aspirin.

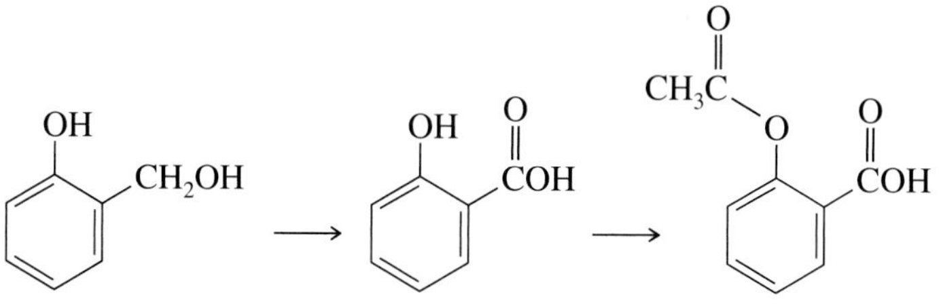

o-(hydroxymethyl)phenol        o-hydroxybenzoic        o-acetoxybenzoic acid
aglycon of salicin                    acid                acetylsalicylic acid
                                salicylic acid                 aspirin

Glycosidic linkages are also found in compounds such as digitoxin, called a **cardiac glycoside** (p. 990) because it affects the action of the heart.

**Problem 14.18**    Indican is the $\beta$-glucoside (p. 569) of D-glucose with 3-hydroxyindole, shown below. It is a precursor in plants of the blue pigment indigo. Draw the structure of indican.

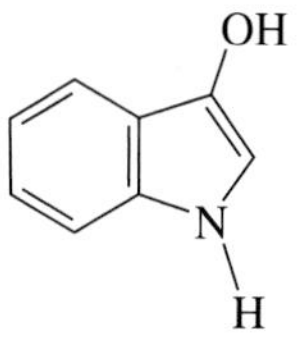

3-hydroxyindole

## C. The Anomeric Effect

Chemists have determined experimentally that many reactions that introduce oxygen or halogen atoms at the anomeric carbon atom of a pyranose ring give as the major product the stereoisomer in which the new group occupies the axial position. This is the case when the methyl D-glucopyranosides are prepared (p. 569). When equilibrium is achieved, methyl α-D-glucopyranoside is the major constituent of the product mixture, indicating that it is thermodynamically the most stable species. This preference of electronegative substituents for the axial position at the anomeric carbon atom is called the **anomeric effect.**

If we compare a pyranose to cyclohexane, we find that an oxygen atom occupies one corner of the chair form of the ring, replacing a methylene group. In place of the hydrogen atoms on the methylene group, the oxygen atom has pairs of nonbonding electrons, which can interact with the nonbonding electrons of a substituent on the adjacent carbon atom.

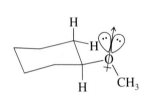

methoxy group in the equatorial position in cyclohexane

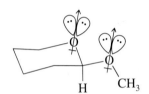

methoxy group in the equatorial position on the carbon atom adjacent to the oxygen atom in tetrahydropyran

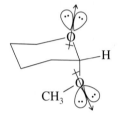

methoxy group in the axial position on the carbon atom adjacent to the oxygen atom in tetrahydropyran

The nonbonding electrons on the oxygen atom of the ring and the nonbonding electrons of the methoxy group (or a halogen) repel each other when the substituent at carbon 1 is in the equatorial position. They are farther apart when the substituent is in the axial position.

Another way of looking at this is to draw Newman projections.

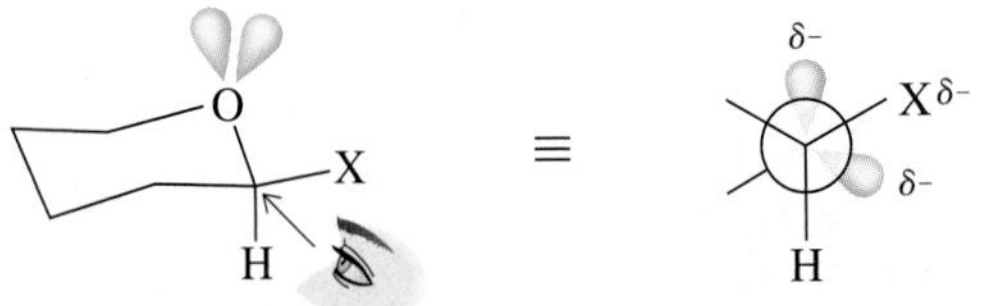

Electronegative group gauche to both lobes bearing nonbonding electrons

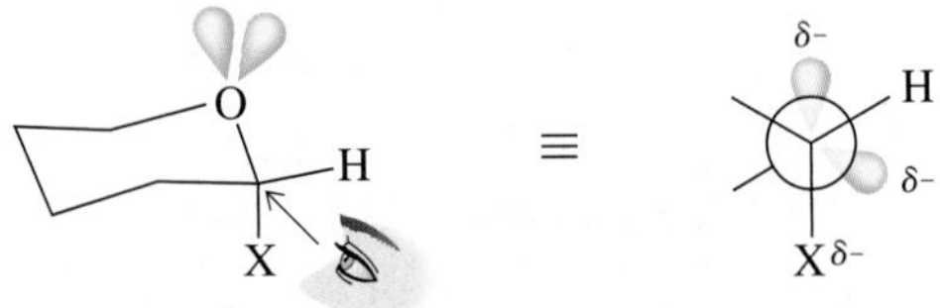

Electronegative group gauche to one but anti to another lobe bearing nonbonding electrons

The picture at the right in each case is a Newman projection looking down the bond between the carbon atom bearing the electronegative substituent (which is usually oxygen or a halogen in carbohydrate chemistry) and the oxygen atom of the ring. The rear atom of the Newman projection is the oxygen atom, showing a bond to the rest of the ring and two lobes symbolizing nonbonding electron pairs. The front carbon atom has the electronegative group gauche to the two lobes when the substituent is equatorial but anti to one lobe and gauche to the other when the substituent is axial. Less repulsion exists between the dipoles when the electronegative substituent is axial.

The exact position of equilibrium for α- and β-isomers depends on the nature of the substituents as well as the stereochemistry of other substituents on the ring. Solvent effects are important as well, with the anomeric effect being more visible in nonpolar solvents than in water. For example, the preferred conformation of 2-methoxytetrahydropyran changes with solvent. In carbon tetrachloride, 17% of 2-methoxytetrahydropyran has the methoxy group in the equatorial position, compared to 48% in $D_2O$.

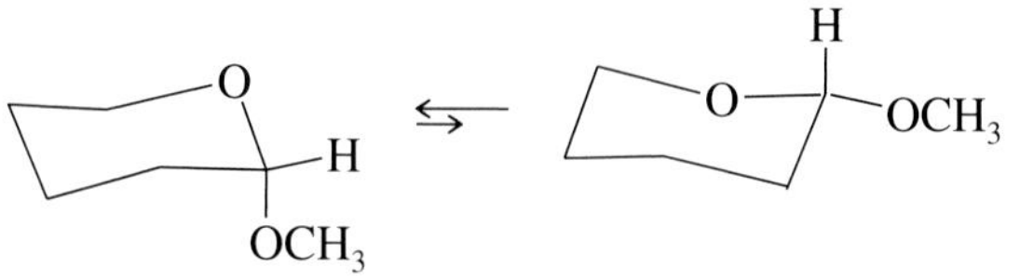

$$\mu \sim 0 \text{ D} \qquad\qquad \mu\ 1.84 \text{ D}$$

|  | | |
|---|---|---|
| in $CCl_4$ | 83% | 17% |
| in $D_2O$ | 52% | 48% |

The form in which the methoxy group is axial is much less polar than the form in which the group is equatorial. The more polar solvent, water ($D_2O$ in this case), stabilizes the more polar conformer. Hydrogen bonding is also an important factor. Especially in carbohydrates, which contain many hydroxyl groups, hydrogen-bonding interactions between the hydroxyl groups within a molecule and with the solvent are important in determining the position of equilibrium for different conformational isomers as well as for the formation of stereoisomers at the anomeric carbon atom.

**Problem 14.19**   The equilibrium between α- and β-D-glucopyranose lies on the side of the β-isomer that constitutes 62% of the mixture in water. This seems to be a case where the anomeric effect is not as important as it is for the methyl D-glucopyranosides (p. 569). What factor could be responsible for the difference in stability between the α- and β-pyranoses in aqueous solution that would not be as much a factor in the case of the α- and β-pyranosides?

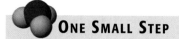

**ONE SMALL STEP**

Proton nuclear magnetic resonance is a tool that is used extensively to study the conformation of compounds.

**PROBLEM:**   Proton nuclear magnetic resonance shows that the conformation in which all the substituents are axial is favored in deuteriochloroform for the methyl pyranoside shown below. The conformation in which all the substituents are equatorial is favored in deuterium oxide.

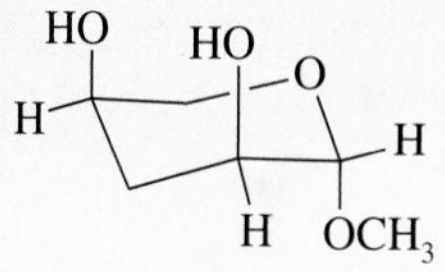

major conformer in CDCl₃

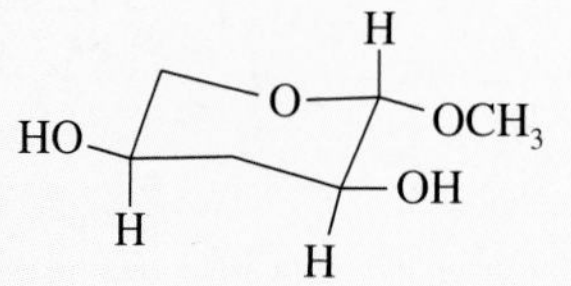

major conformer in $D_2O$

How would you rationalize this experimental observation?

## 14.9 Addition Reactions of Nucleophiles Related to Ammonia

### A. Reactions of Carbonyl Compounds with Amines

Amines and other compounds related to ammonia make up an important class of reagents that behave as nucleophiles toward the carbonyl group. Some typical compounds with this kind of reactivity are shown below.

$RNH_2$

alkylamine

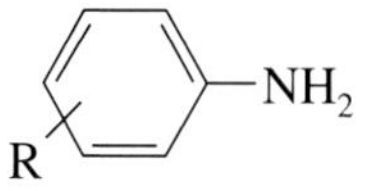

$R-\!\!\!\bigcirc\!\!\!-NH_2$

arylamine

*may have other substituents on the ring*

$HONH_2$

hydroxylamine

$H_2NNH_2$

hydrazine

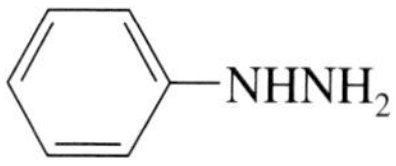

$\bigcirc\!\!\!-NHNH_2$

phenylhydrazine

$O_2N-\!\!\!\bigcirc\!\!\!-NHNH_2$

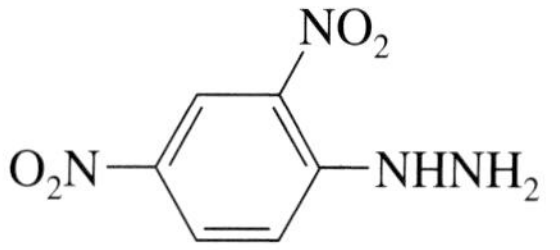

2,4-dinitrophenylhydrazine

All these compounds react with aldehydes and ketones to give products in which there is a double bond between the carbon atom of the original carbonyl group and the nitrogen atom. Water is the other product of the reaction. For example, aldehydes react readily with amines to give compounds known as **imines** or **Schiff bases.** Usually, heating the aldehyde and the amine together while distilling off the water that forms is all that is necessary, as illustrated below by the formation of *N*-methylbenzaldimine.

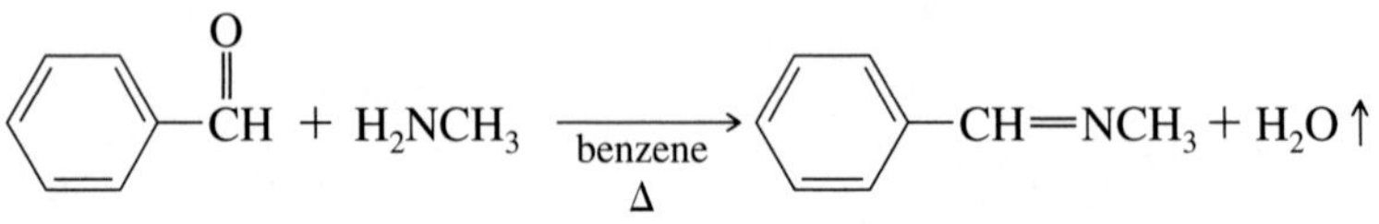

benzaldehyde    methylamine                    *N*-methylbenzaldimine
                                                    90%

This reaction may be viewed as an addition to the carbon–oxygen double bond followed by an elimination reaction to create a new double bond.

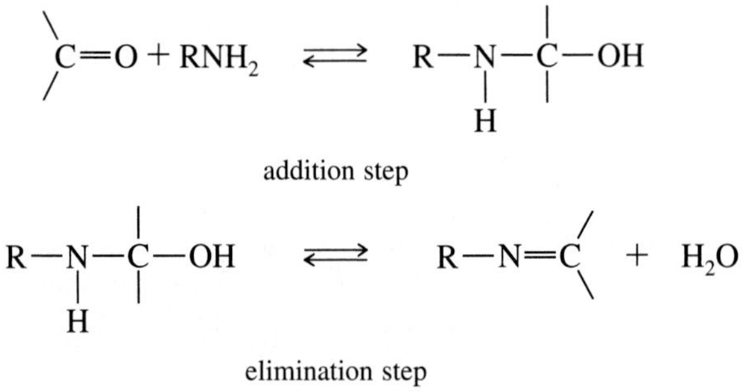

$$\text{\textbackslash C=O} + RNH_2 \quad \rightleftharpoons \quad R-N-C-OH$$

addition step

$$R-N-C-OH \quad \rightleftharpoons \quad R-N=C \quad + \quad H_2O$$

elimination step

All the types of compounds shown on page 573 react with carbonyl compounds in the same way.

Ketones, like aldehydes, form imines, although generally not as readily as aldehydes do. A typical reaction is that of acetone with propylamine. The reaction is catalyzed by hydrochloric acid, and the acid is then neutralized by sodium hydroxide.

$$CH_3CCH_3 + CH_3CH_2CH_2NH_2 \xrightarrow[HCl]{NaOH} CH_3C{=}NCH_2CH_2CH_3 + H_2O$$

acetone        propylamine                    imine of acetone and propylamine
                                                    67%

The reactions shown above are reversible; imines are converted back to carbonyl compounds and amines by reaction with water in hydrolysis reactions. The imines derived from aromatic carbonyl compounds and amines are more stable than those derived from alkyl components.

The other amine derivatives react in ways similar to amines. Cyclohexanone, for example, reacts with hydroxylamine (available in the laboratory as its hydrochloride salt) to form a compound called an **oxime.**

$$HONH_3^+Cl^- + CH_3CO^-Na^+ \longrightarrow HONH_2 + CH_3COH + Na^+Cl^-$$

hydroxylamine        sodium          hydroxylamine
hydrochloride        acetate

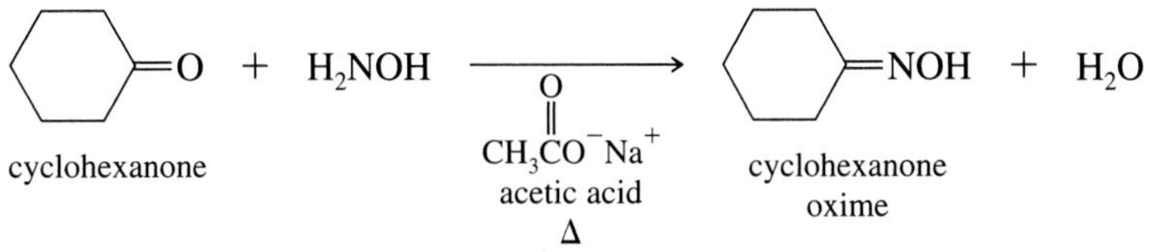

cyclohexanone + $H_2NOH$ → cyclohexanone oxime + $H_2O$

(reagent: $CH_3CO^- Na^+$ / acetic acid, $\Delta$)

Phenylhydrazine and 2,4-dinitrophenylhydrazine are used to prepare compounds from aldehydes and ketones known as **phenylhydrazones** and **2,4-dinitrophenyl-hydrazones.**

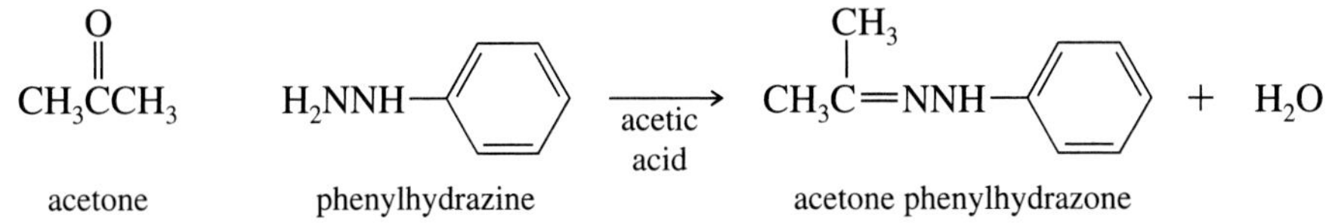

acetone + phenylhydrazine → acetone phenylhydrazone + $H_2O$ (acetic acid)

2-methylpropanal → 2-methylpropanal 2,4-dinitrophenylhydrazone ($H_2SO_4$)

---

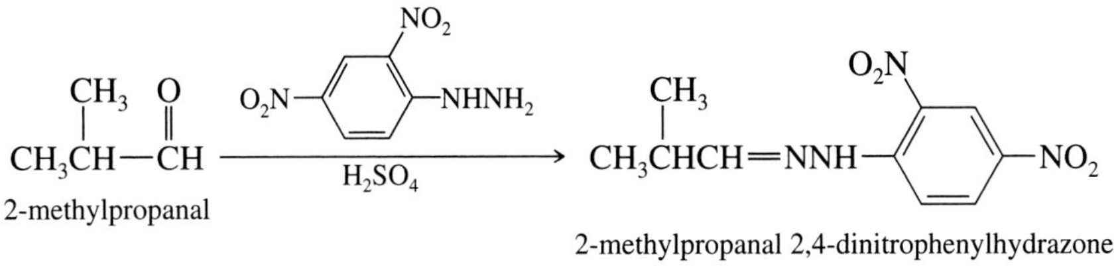

**Problem 14.20**   Complete the following equations by writing structures for the expected products.

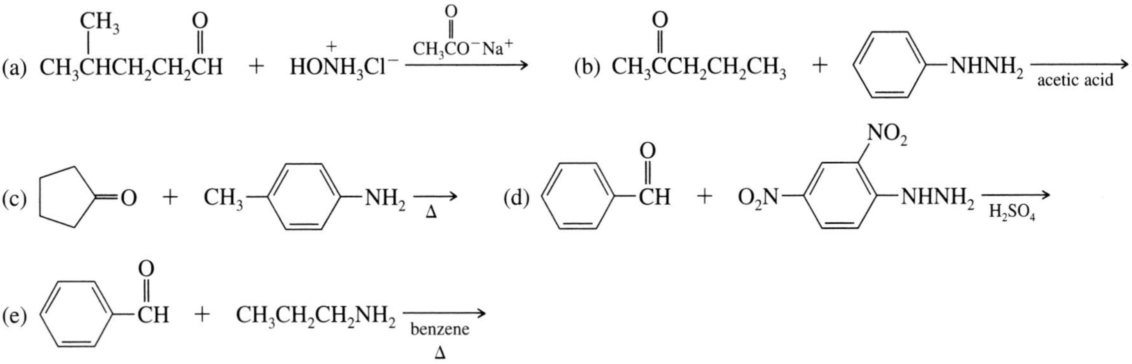

(a) $CH_3CHCH_2CH_2CH$ (with $CH_3$ branch) $O$ + $HONH_3Cl^-$ $\xrightarrow{CH_3CO^-Na^+}$

(b) $CH_3CCH_2CH_2CH_3$ $O$ + ⬡—$NHNH_2$ $\xrightarrow{\text{acetic acid}}$

(c) ⬠=O + $CH_3$—⬡—$NH_2$ $\xrightarrow{\Delta}$

(d) ⬡—CH ($O$) + $O_2N$—⬡($NO_2$)—$NHNH_2$ $\xrightarrow{H_2SO_4}$

(e) ⬡—CH ($O$) + $CH_3CH_2CH_2NH_2$ $\xrightarrow[\Delta]{\text{benzene}}$

---

## B. A Mechanism for the Reaction of Compounds Related to Ammonia with Aldehydes and Ketones

The reaction of cyclohexanone with hydroxylamine can be used as a model for the general reaction of aldehydes and ketones with such nitrogen compounds. Sodium acetate is used as a base to remove a proton from the nitrogen atom in hydroxylamine hydrochloride to form the weak acid acetic acid. Sodium acetate and acetic acid together form a buffer system, so the acidity of the reaction mixture remains close to pH 5. It is important that the solution not be too acidic because the unprotonated nitrogen atom in hydroxylamine is the nucleophile that reacts with the electrophilic carbon atom of the carbonyl group. On the other hand, acid is necessary in order to assist in the loss of water in the final stages of the reaction, so the solution must be somewhat acidic. The rate of the reaction is greatest at pH ~5.

The reaction of hydroxylamine with cyclohexanone takes place in two steps. In the first step, illustrated on page 576, the nucleophile attacks the carbonyl group.

**VISUALIZING THE REACTION**

**Addition of the Nucleophile to the Carbonyl Group**

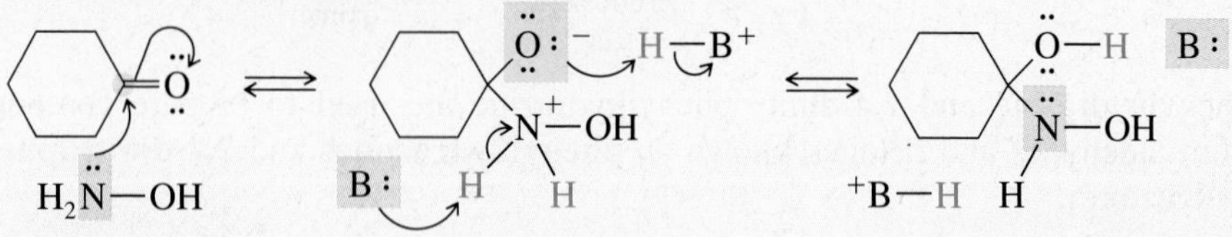

*nucleophilic attack at the carbonyl group*  *protonation and deprotonation steps*  *addition compound*

The intermediate that forms loses a proton from the positively charged nitrogen atom and is protonated at the oxygen atom. The addition compound with a hydroxyl group and an amino group on the same carbon atom loses water easily, and a carbon–nitrogen double bond forms. The loss of water is catalyzed by acid and is the rate-determining step for the reaction at moderate acidity.

**VISUALIZING THE REACTION**

**Elimination of Water from the Addition Compound**

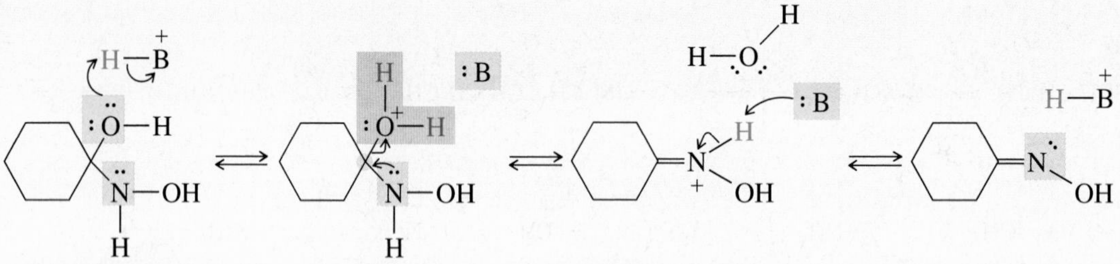

*protonation of hydroxyl group*  *loss of water*  *deprotonation*  *an oxime*

The hydroxyl group is converted into a good leaving group by protonation, and the water molecule is displaced by the nonbonding electrons on the nitrogen atom. Removal of a proton from the nitrogen atom gives the oxime.

All the processes shown above for the formation of cyclohexanone oxime may be applied to reactions of other compounds that are related to ammonia with aldehydes and ketones.

**Problem 14.21**   Suggest a mechanism for the formation of *N*-methylbenzaldimine (p. 574).

**Problem 14.22**   The steps shown for the mechanism for the formation of cyclohexanone oxime above are all reversible. Using that mechanism as a guide, propose a mechanism for the reaction of *N*-methylbenzaldimine with water to give benzaldehyde and methylamine.

⊘ **Study Guide**
**Concept Map 14.6**

**On the Web: ONE SMALL STEP**

## C. The Biological Importance of Imines

Imines have biological importance. An imine linkage between the aldehyde derived from vitamin A and the protein opsin in the retina of the eye plays an important role in the chemistry of vision (p. 758). Vitamins are also called coenzymes, meaning that they are essential to the functioning of many enzymes, which are large proteins that catalyze chemical changes in cells. An important vitamin, $B_6$, serves as a coenzyme in its aldehyde form by forming an imine with an amino group in an enzyme. The coenzyme, bound to the enzyme, is involved in **transamination reactions**—the transfer of amino groups from one amino acid to another—which are important in the metabolism and the biosynthesis of amino acids (p. 984). The main reactions in the process are shown below.

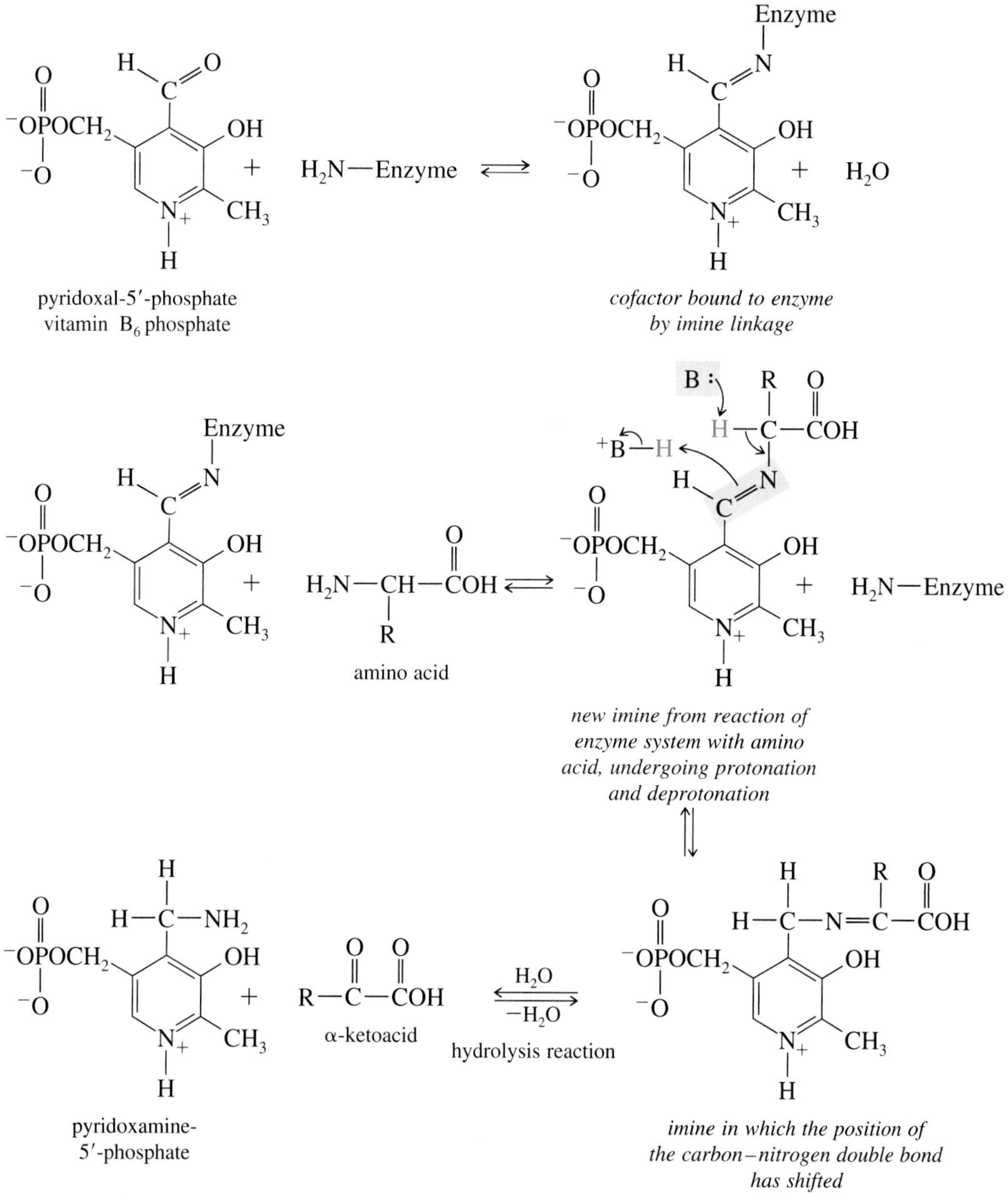
pyridoxal-5′-phosphate
vitamin B6 phosphate
H2N—Enzyme
H2O
cofactor bound to enzyme by imine linkage
amino acid
H2N—Enzyme
new imine from reaction of enzyme system with amino acid, undergoing protonation and deprotonation
pyridoxamine-5′-phosphate
α-ketoacid
H2O
−H2O
hydrolysis reaction
imine in which the position of the carbon–nitrogen double bond has shifted

As a result of the series of reactions shown above, the amino group is removed from an amino acid, which is converted to an $\alpha$-ketoacid. Pyridoxal is converted to pyridoxamine, another form of vitamin $B_6$, which reacts with another $\alpha$-ketoacid in a series of reactions similar to the ones shown above to transfer an amino group to it, creating a new amino acid and regenerating pyridoxal.

**Problem 14.23** Pyridoxamine transfers an amino group to an $\alpha$-ketoacid by forming an imine, undergoing protonation and deprotonation steps that isomerize the imine to a new imine and hydrolysis of this new imine to pyridoxal and an amino acid. Using the reactions shown on page 577 as a guide, write the steps for the conversion of 2-oxopropanoic acid to the amino acid alanine.

## 14.10 Reduction of Carbonyl Groups to Methylene Groups

### A. The Wolff–Kishner Reduction

Carbonyl groups are converted to methylene groups by the **Wolff-Kishner reduction**. In the Wolff-Kishner reduction, the hydrazone of an aldehyde or ketone is prepared and decomposed under basic conditions. An example is the reduction of 1-phenyl-1-propanone (p. 544) to give propylbenzene as a way of introducing unbranched alkyl chains on aromatic rings.

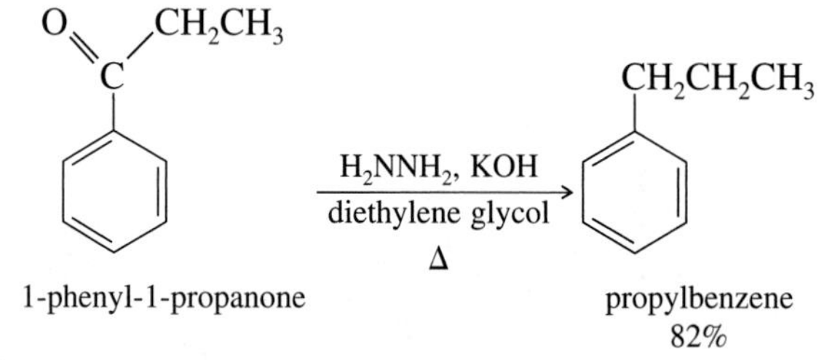

1-phenyl-1-propanone → propylbenzene 82%

$$\text{H}_2\text{NNH}_2, \text{KOH} \quad \text{diethylene glycol} \quad \Delta$$

Another example is the reduction of 2-octanone to octane.

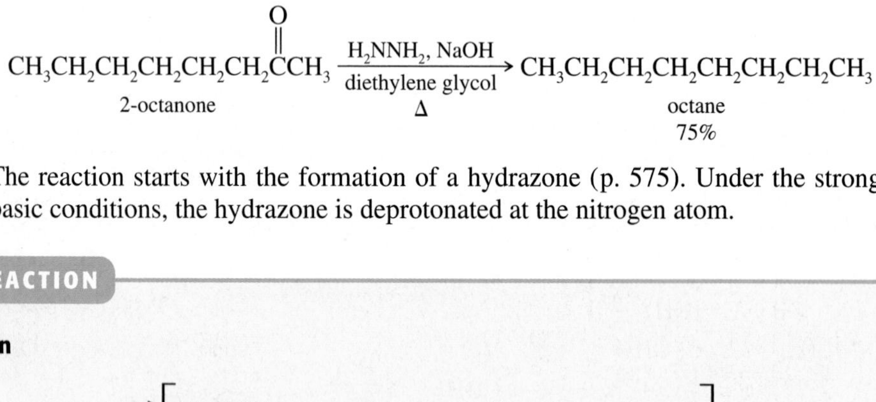

$$\text{CH}_3\text{CH}_2\text{CH}_2\text{CH}_2\text{CH}_2\text{CH}_2\overset{\displaystyle O}{\overset{\|}{\text{C}}}\text{CH}_3 \xrightarrow[\substack{\text{diethylene glycol} \\ \Delta}]{\text{H}_2\text{NNH}_2, \text{NaOH}} \text{CH}_3\text{CH}_2\text{CH}_2\text{CH}_2\text{CH}_2\text{CH}_2\text{CH}_2\text{CH}_3$$

2-octanone → octane 75%

The reaction starts with the formation of a hydrazone (p. 575). Under the strongly basic conditions, the hydrazone is deprotonated at the nitrogen atom.

---

**VISUALIZING THE REACTION**

**The Wolff-Kishner Reduction**

deprotonation of hydrazone

resonance-stabilized anion

---

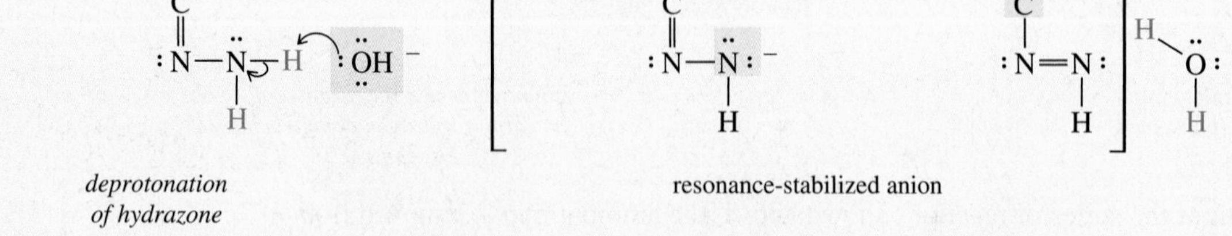

Such a deprotonation is possible because there is delocalization of the charge to give anionic character to the carbon atom that was part of the original carbonyl group. A negatively charged carbon atom is more basic than a negatively charged nitrogen atom because carbon is less electronegative than nitrogen (p. 90). The anion accepts a proton at the carbon atom and loses another one from the nitrogen atom.

**VISUALIZING THE REACTION**

| protonation at carbon atom | deprotonation at nitrogen atom | anion with nitrogen as leaving group |

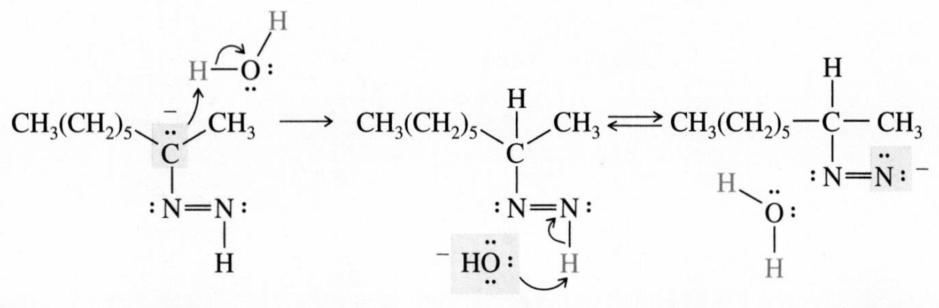

For this new anion, the very stable nitrogen molecule is a good leaving group. Usually the loss of a stable molecule as a leaving group creates a cation. In this case, the species as a whole is negatively charged, so the loss of the neutral molecule, nitrogen, leaves behind a carbanion. Protonation of this carbanion completes the reduction.

**VISUALIZING THE REACTION**

| loss of nitrogen | protonation of carbanion |

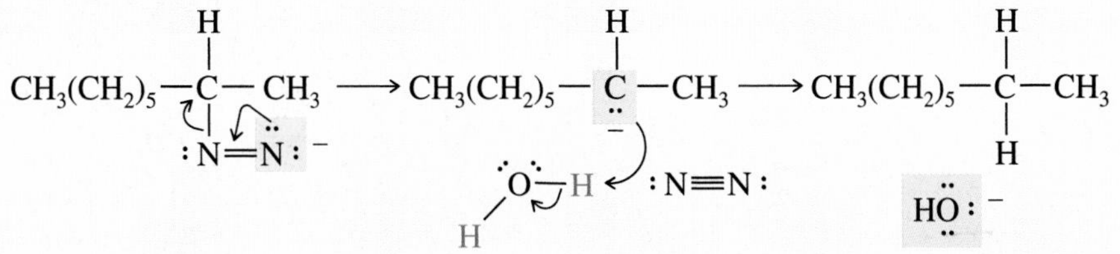

## B. Raney Nickel Reduction of Thioacetals and Thioketals

A way of converting a carbonyl group to a methylene group without the use of strong base is to make the thioacetal or thioketal of the carbonyl compound and then to cleave the carbon–sulfur bonds with hydrogen that is present in Raney nickel (p. 298). Because thiols are more nucleophilic than alcohols (p. 251), 1,2-ethanedithiol also can be used to make cyclic acetals or ketals. The reaction is catalyzed by a solution of the Lewis acid boron trifluoride in diethyl ether. The steroid diketone cholestan-3,6-dione is reduced to cholestane by this method.

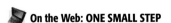 **On the Web: ONE SMALL STEP**

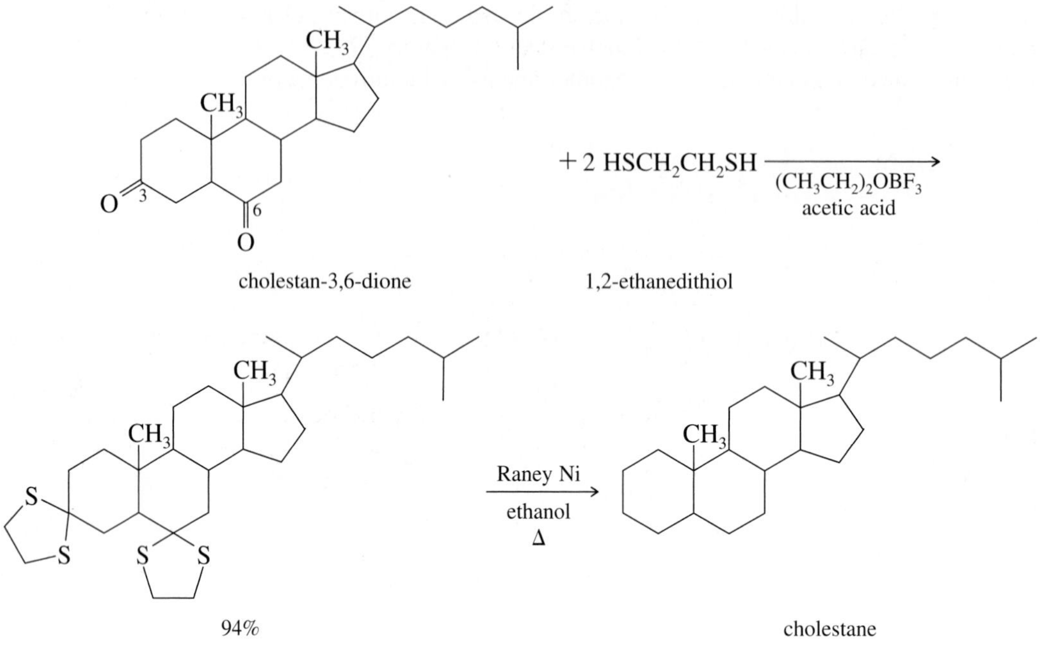

$+ 2 \ HSCH_2CH_2SH \xrightarrow[\substack{(CH_3CH_2)_2OBF_3 \\ \text{acetic acid}}]{}$

cholestan-3,6-dione                    1,2-ethanedithiol

$\xrightarrow[\substack{\text{ethanol} \\ \Delta}]{\text{Raney Ni}}$

94%                                              cholestane

⊘ **Study Guide**
**Concept Map 14.7**

As seen in the structures above, cholestan-3,6-dione is converted into cyclic dithio-ketals at both ketone groups, and the sulfur atoms are replaced by hydrogen atoms to yield the hydrocarbon cholestane. This way of simplifying the structure of a steroid can be used to prove that a series of reactions did not change the carbon skeleton of the molecule or to establish the stereochemistry at various ring junctions.

---

**Problem 14.24**   Predict what the products of the following reactions will be.

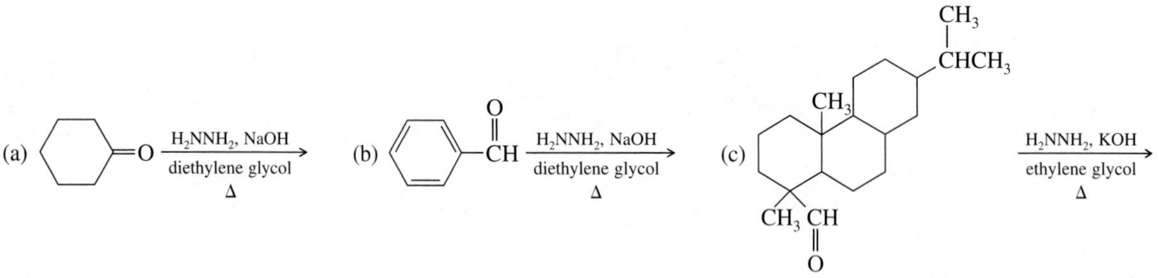

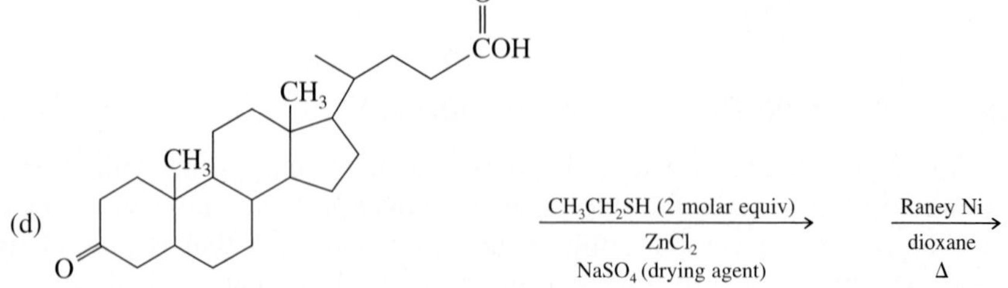

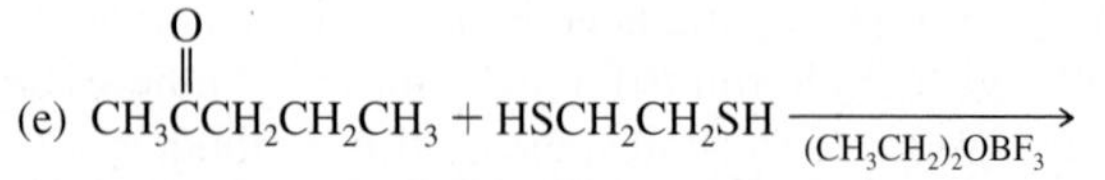

(a) $\xrightarrow[\substack{\text{diethylene glycol} \\ \Delta}]{H_2NNH_2, \ NaOH}$

(b) $\xrightarrow[\substack{\text{diethylene glycol} \\ \Delta}]{H_2NNH_2, \ NaOH}$

(c) $\xrightarrow[\substack{\text{ethylene glycol} \\ \Delta}]{H_2NNH_2, \ KOH}$

(d) $\xrightarrow[\substack{ZnCl_2 \\ NaSO_4 \ \text{(drying agent)}}]{CH_3CH_2SH \ \text{(2 molar equiv)}} \xrightarrow[\substack{\text{dioxane} \\ \Delta}]{\text{Raney Ni}}$

(e) $CH_3\overset{\overset{\displaystyle O}{\|}}{C}CH_2CH_2CH_3 + HSCH_2CH_2SH \xrightarrow[(CH_3CH_2)_2OBF_3]{}$

(f)

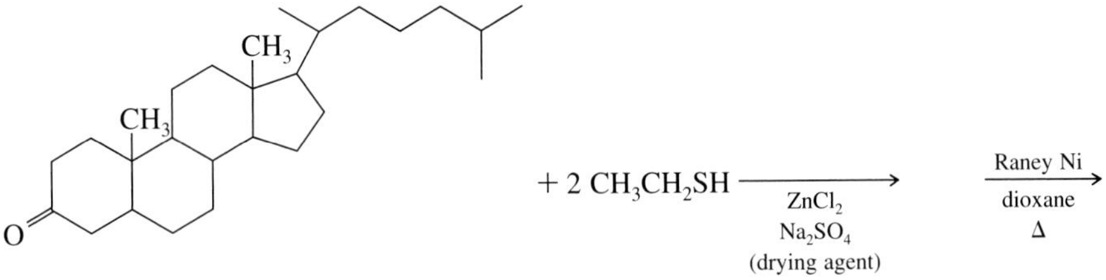

$+ 2\ CH_3CH_2SH \xrightarrow[\substack{Na_2SO_4 \\ (drying\ agent)}]{ZnCl_2} \xrightarrow[\substack{dioxane \\ \Delta}]{Raney\ Ni}$

## Table 14.3  Summary of Ways to Prepare Aldehydes and Ketones

| Starting Material | Reagent | Intermediate | Reagent in Second Step | Product |
|---|---|---|---|---|
| | | **Oxidation of Alcohols** | | |
| $RCH_2OH$ | 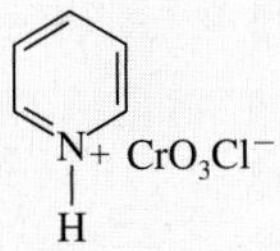 pyridinium $CrO_3Cl^-$ | $RCH_2OCrO^-$ (with C=O) | | $RCH$ (aldehyde) |
| | $(CH_3)_2SO$, $(COCl)_2$ | $RCH_2OS(CH_3)_2^+$ | $(CH_3CH_2)_3N$ | |
| $\underset{H}{RCOH}$ with R′ | pyridinium $CrO_3Cl^-$ | $R'RCOCrO^-$ (with C=O) | | $RCR'$ (ketone) |
| | $(CH_3)_2SO$, $(COCl)_2$ | $\underset{H}{R'RCOS(CH_3)_2^+}$ | $(CH_3CH_2)_3N$ | |
| | | **Friedel-Crafts Acylation Reaction** | | |
| 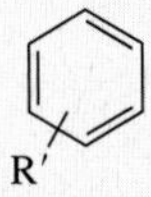 $R'$—C$_6$H$_5$ | $RCCl$, $AlCl_3$ | $RC\equiv\overset{+}{O}$ | | 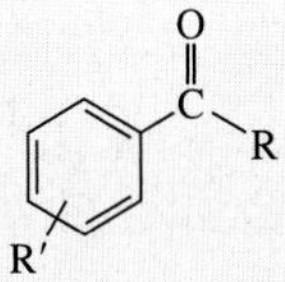 $R'$—C$_6$H$_4$—CO—R |
| 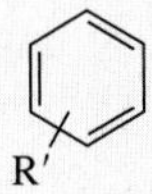 $R'$—C$_6$H$_5$ | $RCOCR$, $AlCl_3$ | $RC\equiv\overset{+}{O}$ | | 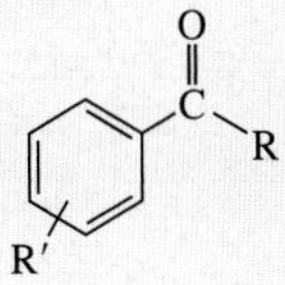 $R'$—C$_6$H$_4$—CO—R |
| | | **Ozonolysis of Alkenes** | | |
| 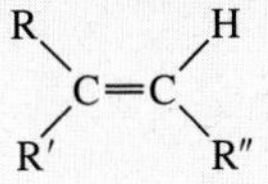 $\underset{R'}{\overset{R}{C}}=\underset{R''}{\overset{H}{C}}$ | $O_3$ | 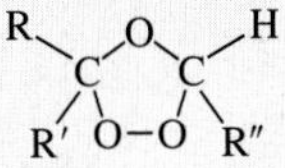 molozonide (R, R′, R″, H) | Zn, H$_2$O or CH$_3$SCH$_3$ | 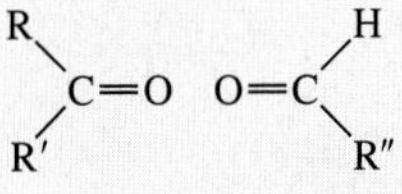 $\underset{R'}{\overset{R}{C}}=O \quad O=\underset{R''}{\overset{H}{C}}$ |
| | | | $H_2O_2$ | 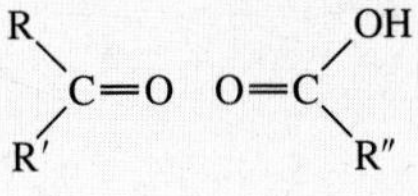 $\underset{R'}{\overset{R}{C}}=O \quad O=\underset{R''}{\overset{OH}{C}}$ |

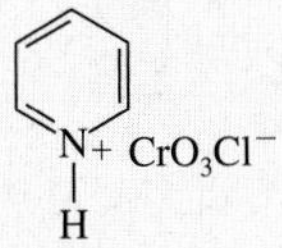

**Table 14.4** **Summary of Reactions of Aldehydes and Ketones**

| Starting Material | Reagent | Intermediate | Reagent in Second Step | Product |
|---|---|---|---|---|
| | | **Nucleophilic Addition Reactions** | | |
| $R{-}C(=O){-}R''$<br>R, R″ may be H | $NaBH_4$<br>$LiAlH_4$ | $R{-}C(O^-)(H){-}R''$ | $H_2O$ | $R{-}C(OH)(H){-}R''$ |
| | NaCN | $R{-}C(O^-)(R''){-}C{\equiv}N$ | $H_3O^+$ | $R{-}C(OH)(R''){-}C{\equiv}N$ |
| | R′MgX, R′Li | $R{-}C(O^-)(R''){-}R'$ | $H_3O^+$ | $R{-}C(OH)(R''){-}R'$ |
| (See also Table 14.2, p. 555.) | | | | |
| $R{-}C(=O){-}R''$<br>R, R″ may be H | R′OH | $R{-}C(O^-)(\overset{+}{H}OR'){-}R''$ | | $R{-}C(OH)(OR'){-}R''$ |
| | R′OH, $HB^+$ | $R{-}C(OH)(OR'){-}R''$ | R′OH, $HB^+$ | $R{-}C(OR')(OR'){-}R''$ |
|  | | <br>and<br> | | <br>and<br> |

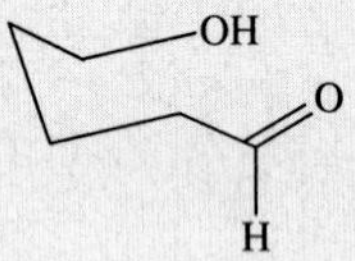

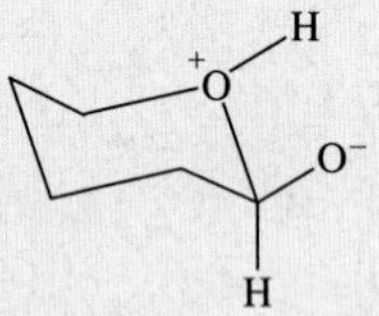

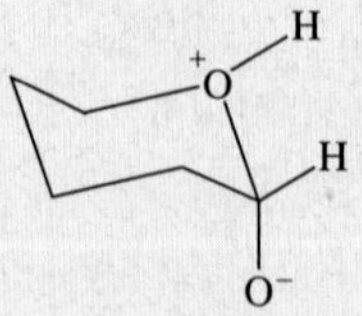

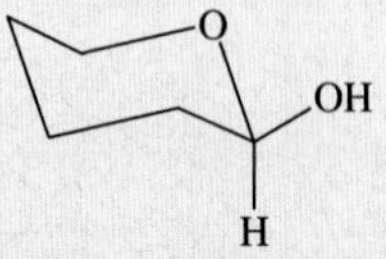

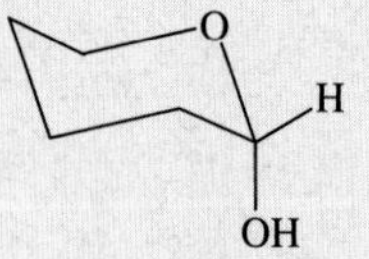

**Table 14.4**    **(Continued)**

| Starting Material | Reagent | Intermediate | Reagent in Second Step | Product |
|---|---|---|---|---|

**Nucleophilic Addition Reactions**

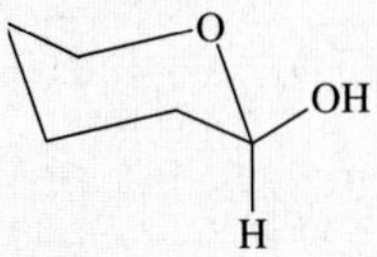

$HB^+$ — intermediate with $\overset{+}{O}H_2$ — $ROH$ — products (OR, two isomers shown with "and")

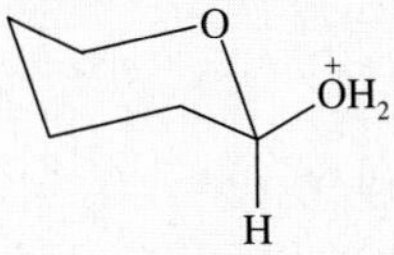

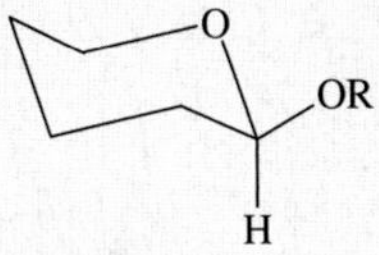

**Nucleophilic Addition Followed by Elimination of Water**

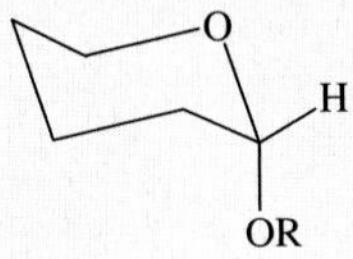

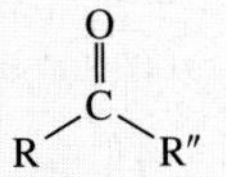

Starting material (R, R″ may be H), reagent $H_2NR'$, intermediate, product.

$H_2NOH$

$H_2NNH$— (aryl $R'$), $HB^+$

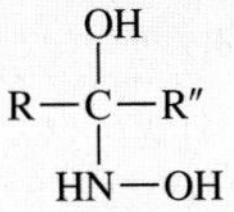

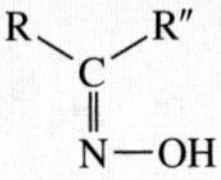

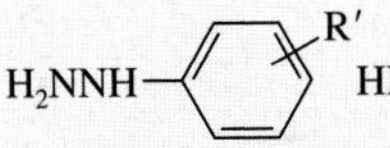

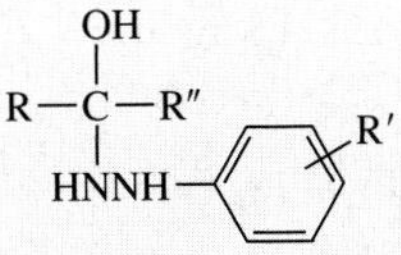

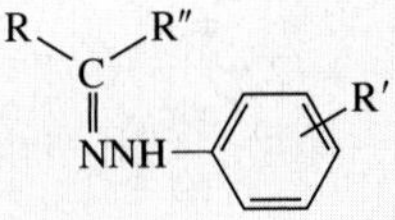

**Reduction to Methylene Groups**

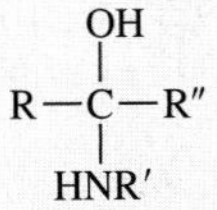

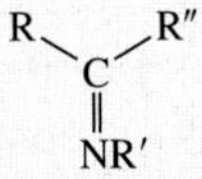

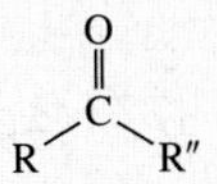

Starting material (R, R″ may be H), reagent $H_2NNH_2$, $OH^-$, intermediate, $OH^-$, product.

$HSCH_2CH_2SH$, $BF_3$ — intermediate dithiolane — Raney Ni ($H_2$) — product.

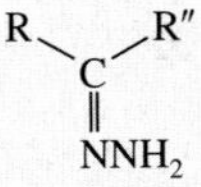

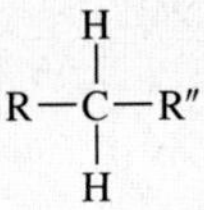

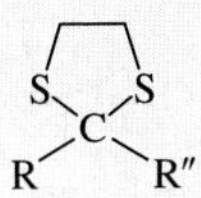

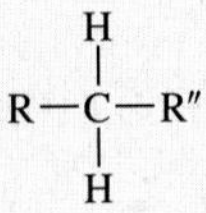

# ADDITIONAL PROBLEMS

**14.25** Name the following compounds according to the IUPAC rules, specifying stereo-chemistry if it is indicated in the structure.

(a), 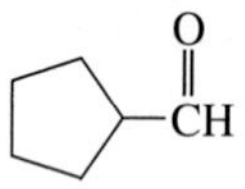 (b), 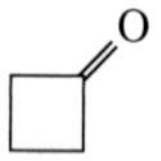 (c), 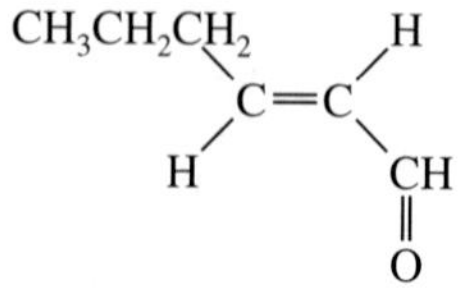 (d), 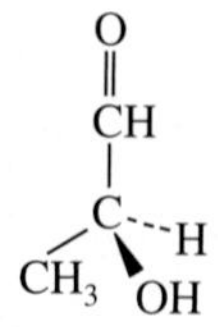 (e)

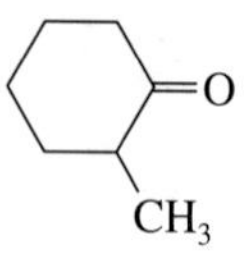

(f) 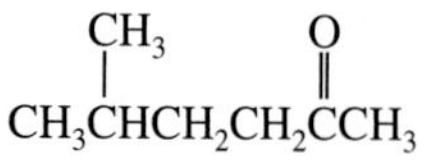 $CH_3CHCH_2CH_2CCH_3$   (g) 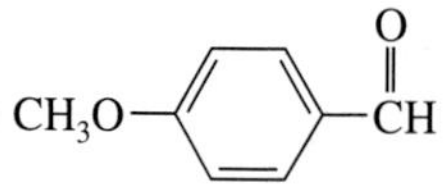   (h) 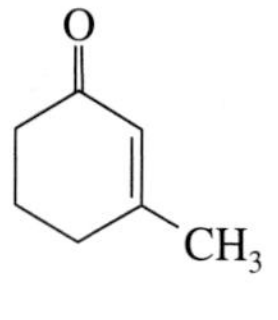   (i)

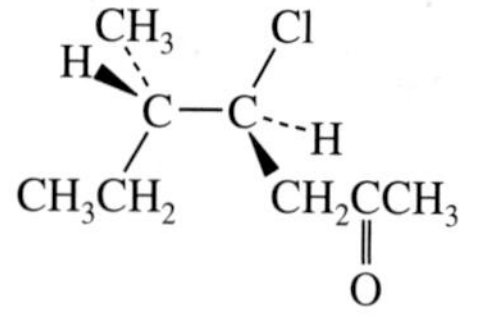

**14.26** Draw structural formulas for the following compounds.

(a) 3,3-dimethylcyclopentanone     (b) 1-bromo-2-hexanone     (c) *m*-chlorobenzaldehyde     (d) *p*-methylacetophenone
(e) 3,5-hexadien-2-one     (f) 2-methylpentanal     (g) (*R*)-2-chlorocyclobutanone     (h) (*S*)-3-bromobutanal

**14.27** Using butanal as a typical aldehyde, write equations showing the reactions that you predict will occur if the following reagents are used. Write "no reaction" if you expect none.

(a) $NaBH_4$, $H_2O$     (b) $O_2N$—⟨ ⟩—$NHNH_2$, $H_2SO_4$     (c) 1. $CH_3CH_2CH_2CH_2Li$; 2. $H_3O^+$

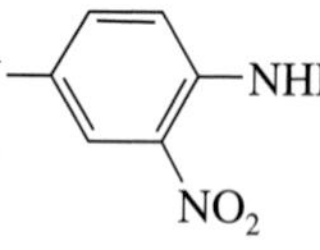

   NO₂

(d) 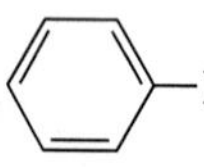 ⟨ ⟩—$NH_2$, Δ     (e) NaCN, then $H_2SO_4$(aq)     (f) $CrO_3$, $H_2O$, $H_3O^+$     (g) $HONH_3{}^+Cl^-$, $CH_3CO^-Na^+$

(h) (1) $CH_3MgI$, diethyl ether; (2) $H_3O^+$     (i) (1) $LiAlH_4$, diethyl ether; (2) $H_3O^+$     (j) $H_2NNH_2$, KOH, diethylene glycol, Δ

**14.28** Using 2-pentanone as a typical ketone, write equations showing the reactions that you predict will occur if the following reagents are used. Write "no reaction" if you expect none.

(a) $NaBH_4$, $H_2O$     (b) $O_2N$—⟨ ⟩—$NHNH_2$, $H_2SO_4$     (c)  ⟨ ⟩—$NH_2$, Δ

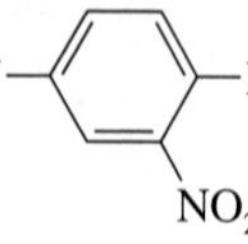

   NO₂

(d) 1. 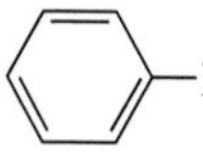 ⟨ ⟩—Li, tetrahydrofuran; 2. $NH_4Cl$, $H_2O$     (e) NaCN, then $H_2SO_4$(aq)     (f) $HONH_3{}^+Cl^-$, $CH_3CO^-Na^+$

(g) (1) $CH_3CH_2MgBr$, diethyl ether; (2) $H_3O^+$     (h) (1) $CH_3C≡CNa$; (2) $NH_4Cl$, $H_2O$     (i) $H_2NNH_2$, KOH, diethylene glycol, Δ

**14.29** Complete each of the following equations, showing the structures of the products that will be formed.

(a) [cyclohexanone] =O + [phenyl]—NHNH₂ $\xrightarrow{\text{acetic acid}}$

(b) [2,6-dimethylcyclohexane-1,4-dione structure] $\xrightarrow[\substack{\text{TsOH} \\ \text{benzene} \\ \Delta}]{\substack{\text{HOCH}_2\text{CH}_2\text{OH} \\ \text{(1 equiv)}}}$

(c) $CH_2=CHCH\overset{O}{\underset{\phantom{O}}{}}$ $\xrightarrow[\text{diethyl ether}]{\substack{\text{Li-phenyl}}}$ $\xrightarrow{H_3O^+}$

(d) [cyclohexene-CH₂OH] $\xrightarrow[\text{dichloromethane}]{\substack{\text{pyridinium} \\ \text{CrO}_3\text{Cl}^-}}$

(e) [methylcyclopentene] + $Br_2$ $\xrightarrow{\text{carbon tetrachloride}}$

(f) [methylcyclopentene] + $KMnO_4$ $\xrightarrow[H_2O]{\text{NaOH}}$

(g) [methylcyclopentene] + [3-chloro-benzoic acid, COOH] $\xrightarrow{\text{dichloromethane}}$

(h) [methylcyclopentene] $\xrightarrow[\text{tetrahydrofuran}]{\text{BH}_3}$ $\xrightarrow{\text{H}_2\text{O}_2, \text{NaOH}}$

(i) [phenyl-C(=O)-CH₃] $\xrightarrow[\text{tetrahydrofuran}]{\text{CH}_3\text{CH}_2\text{CH}_2\text{CH}_2\text{Li}}$ $\xrightarrow[\text{H}_2\text{O}]{\text{NH}_4\text{Cl}}$

(j) [cyclopentyl-CH=O] + $O_2N$—[benzene ring with NO₂]—NHNH₂ $\xrightarrow[\substack{\text{ethanol}}]{\text{H}_2\text{SO}_4}$

(k) [phenyl-C(=O)-CH₃] $\xrightarrow[\substack{\text{diethylene glycol} \\ \Delta}]{\text{H}_2\text{NNH}_2, \text{KOH}}$

**14.30** Here is more practice in recognizing reactions.

(a) [bicyclic ketone with CH₃ groups] + $HONH_3^+$ $Cl^-$ $\xrightarrow[\text{ethanol}]{\text{CH}_3\text{CO}^- \text{Na}^+}$

(b) [furan-CH=O] + [structure with CH₂OH, H₂N, H, phenyl] $\xrightarrow[\Delta]{\text{benzene}}$

(c) $CH_3CH_2CH_2C{\equiv}CH + NaNH_2$ $\xrightarrow{\text{NH}_3\text{(liq)}}$

(d) $CH_3CH_2CH_2C{\equiv}CH + CH_3CH_2MgBr$ $\xrightarrow{\text{diethyl ether}}$

(e) [cyclohexanone] =O $\xrightarrow[\text{diethyl ether}]{\text{CH}_3\text{CH}_2\text{MgBr}}$ $\xrightarrow[\text{H}_2\text{O}]{\text{NH}_4\text{Cl}}$

(f) $CH_3$—[benzene ring]—$\overset{O}{\underset{\phantom{O}}{}}CCH_2CH_3 + HOCH_2CH_2OH$ $\xrightarrow[\text{benzene, } \Delta]{\text{TsOH}}$

(g) [cyclohexanone] =O + $HSCH_2CH_2OH$ $\xrightarrow{(\text{CH}_3\text{CH}_2)_2\text{OBF}_3}$

(h) $CH_3CH_2C{\equiv}C\overset{OTHP}{\underset{\phantom{O}}{C}}HCH_2CH_2CH{=}CHC\overset{O}{\underset{\phantom{O}}{}}CH_3 + CH_3ONH_3^+$ $Cl^-$ $\xrightarrow{\text{base}}$

(i) $HOCH_2CH_2\overset{CH_3}{\underset{CH_3}{C}}CH_2CH{=}CHC\overset{O}{\underset{\phantom{O}}{}}OCH_3$ $\xrightarrow[\text{dichloromethane}]{\substack{\text{pyridinium} \\ \text{CrO}_3\text{Cl}^-}}$

(j) [phenyl-C(=O)-CH₂CH₂CH₃] $\xrightarrow[\substack{\text{BF}_3 \\ \Delta}]{\text{HSCH}_2\text{CH}_2\text{SH}}$ $\xrightarrow[\substack{\text{ethanol} \\ \Delta}]{\text{Raney Ni}}$

**14.31** Predict the structures of the products or intermediates designated by letters in the following equations. Show stereochemistry when it is known.

(a)

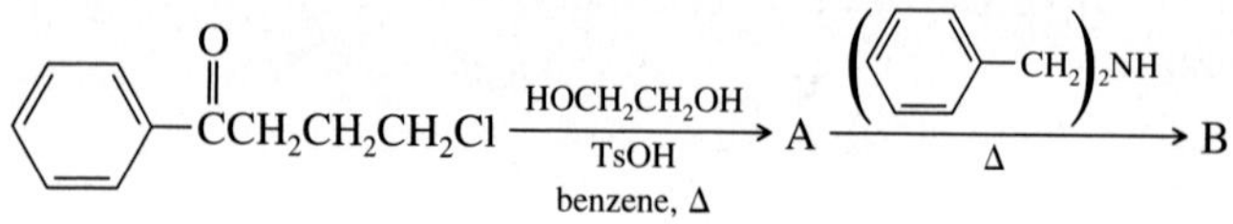

(b)

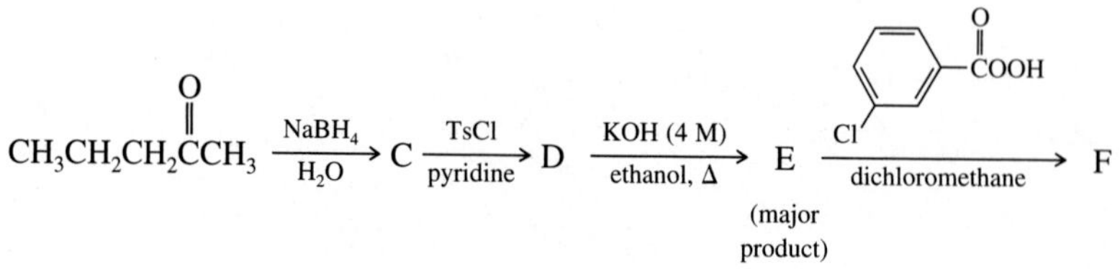

(c)

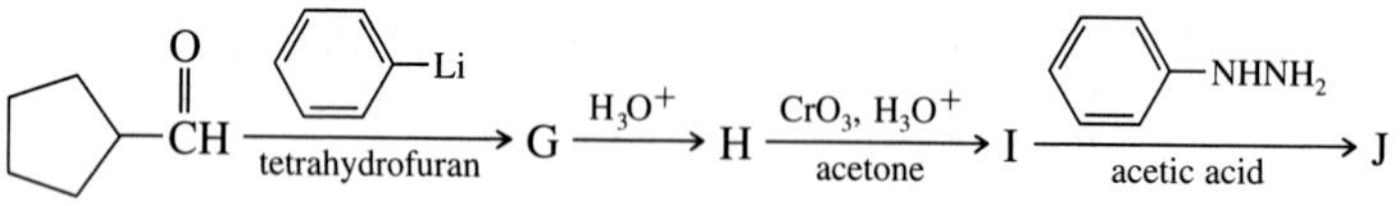

(d) 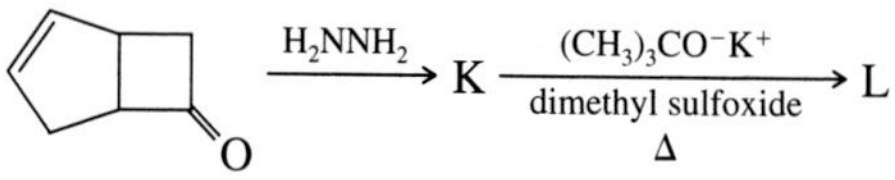    (e)

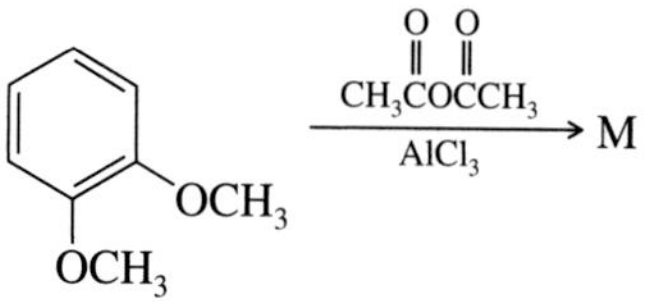

(f)

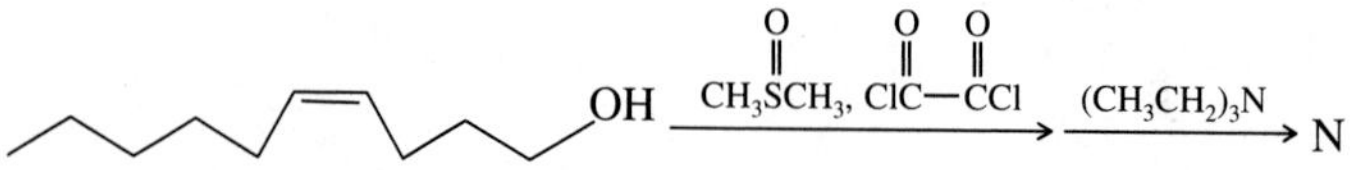

**14.32** For each product or intermediate designated by a letter or letters in the following equations, give the structure.

(a)

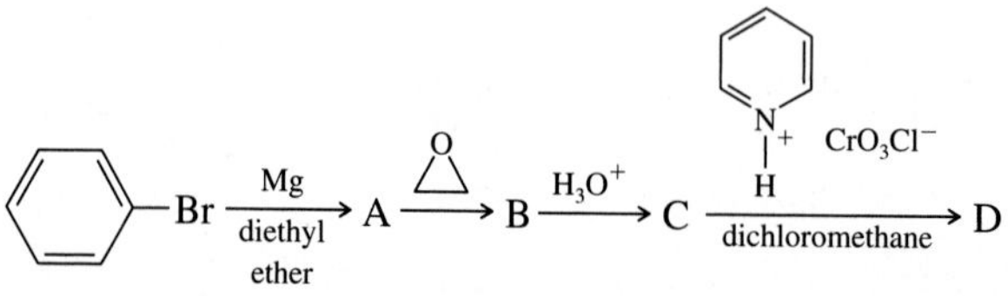

(b)

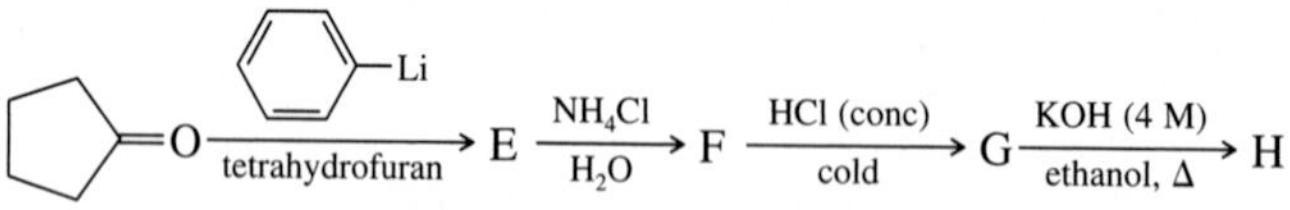

(c)

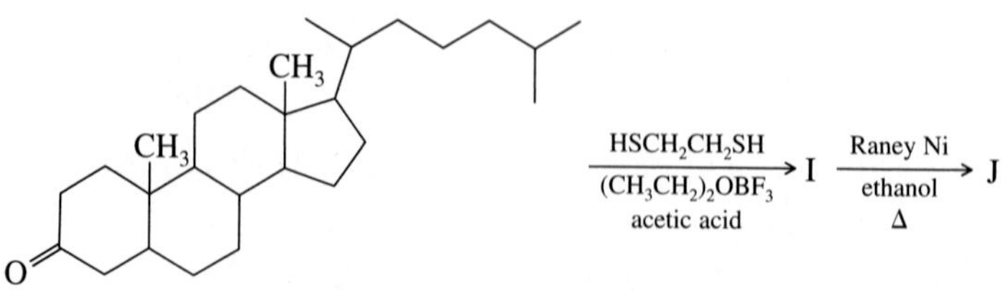

(d)

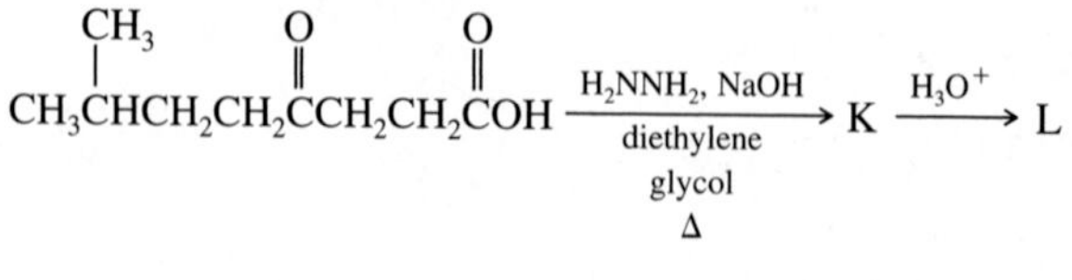

(e) 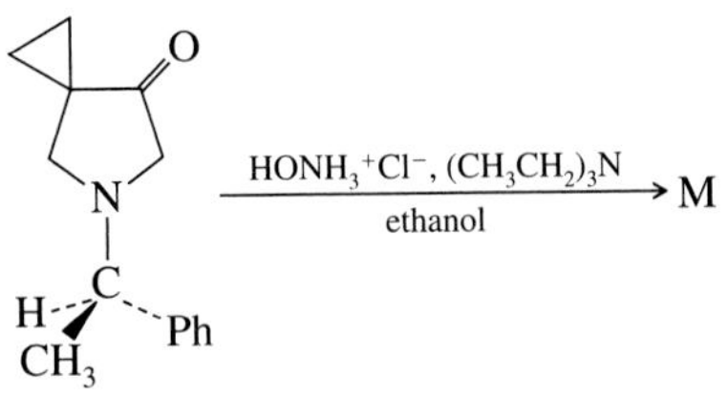 $\xrightarrow{\text{HONH}_3^+\text{Cl}^-, \ (\text{CH}_3\text{CH}_2)_3\text{N}}$ M  (f) 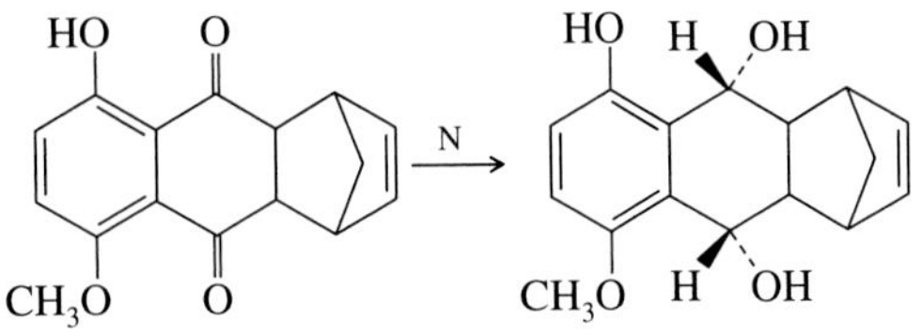 $\xrightarrow{\text{N}}$

(g) 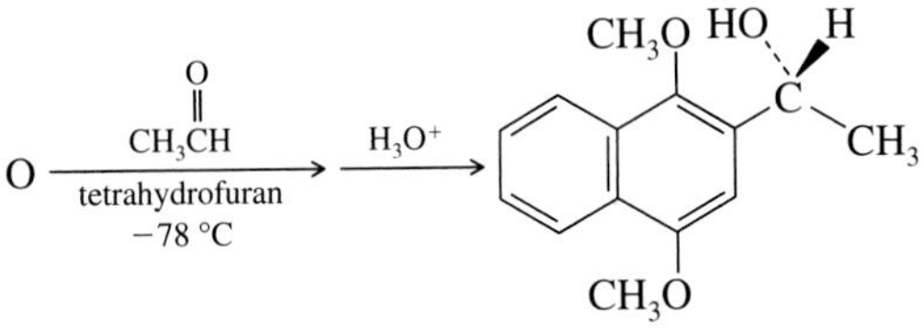 O $\xrightarrow[\substack{\text{tetrahydrofuran} \\ -78\,°\text{C}}]{\substack{\text{O} \\ \| \\ \text{CH}_3\text{CH}}}$ $\xrightarrow{\text{H}_3\text{O}^+}$

racemic mixture

(h) 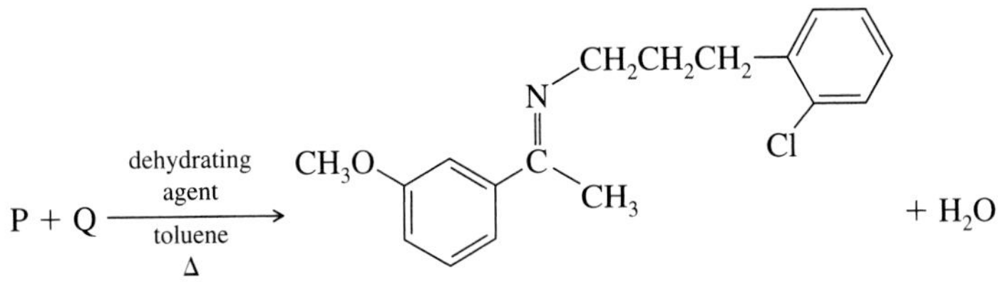 P + Q $\xrightarrow[\substack{\text{toluene} \\ \Delta}]{\substack{\text{dehydrating} \\ \text{agent}}}$ ... + H$_2$O

(i) 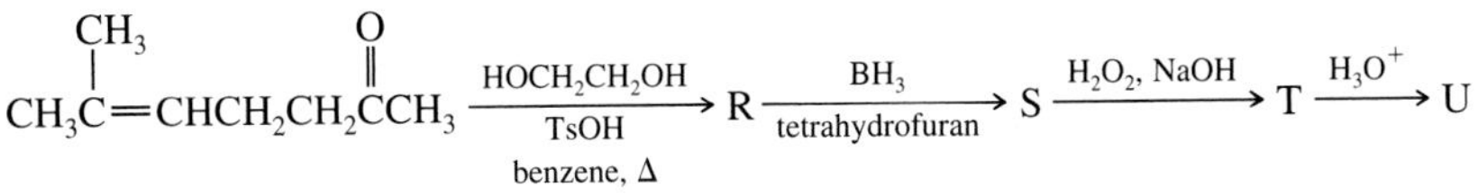 $\text{CH}_3\overset{\text{CH}_3}{\underset{}{\text{C}}}=\text{CHCH}_2\text{CH}_2\overset{\text{O}}{\overset{\|}{\text{C}}}\text{CH}_3$ $\xrightarrow[\substack{\text{TsOH} \\ \text{benzene, } \Delta}]{\text{HOCH}_2\text{CH}_2\text{OH}}$ R $\xrightarrow[\text{tetrahydrofuran}]{\text{BH}_3}$ S $\xrightarrow{\text{H}_2\text{O}_2, \ \text{NaOH}}$ T $\xrightarrow{\text{H}_3\text{O}^+}$ U

**14.33** Plan a synthesis for each of the following compounds. You have bromobenzene, any organic reagents containing three or fewer carbon atoms, and any inorganic reagents you need. There may be more than one acceptable route for a given product. Try to find the shortest route.

(a) 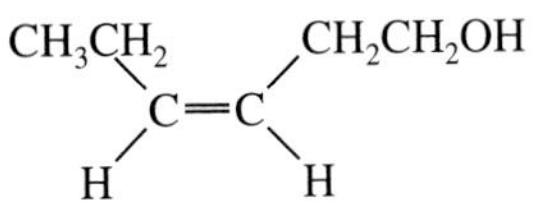  (b) 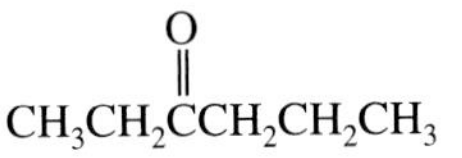 $\text{CH}_3\text{CH}_2\overset{\text{O}}{\overset{\|}{\text{C}}}\text{CH}_2\text{CH}_2\text{CH}_3$  (c) $\text{CH}_3\overset{\text{CH}_3}{\underset{}{\text{C}}}=\text{CHCH}_2\text{CH}_2\text{CH}_3$  (d) 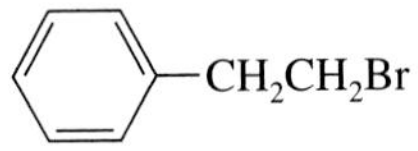 —CH$_2$CH$_2$Br

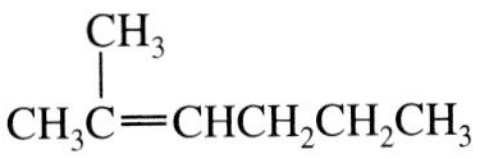

(e) 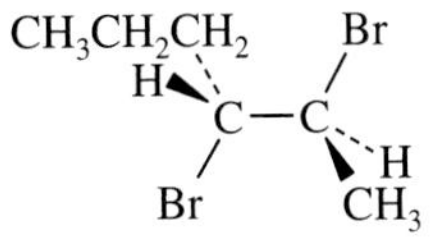  (f) 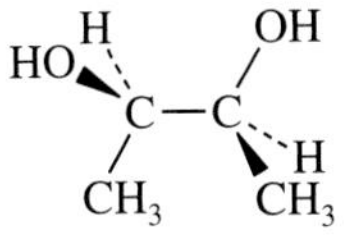  (g) 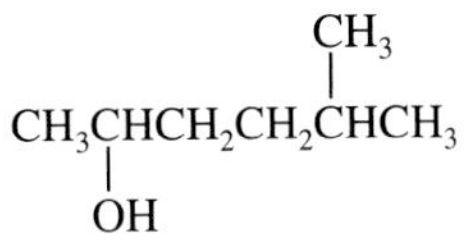 $\text{CH}_3\text{CHCH}_2\text{CH}_2\overset{\text{CH}_3}{\underset{}{\text{CHCH}_3}}$ (h) 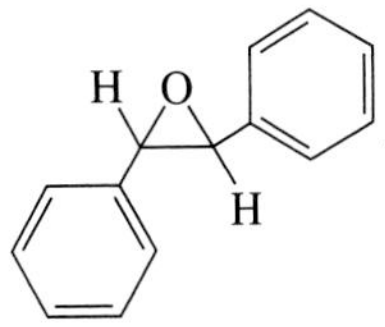  + enantiomer

**14.34** 2,2-Dimethoxypropane is converted into 2,2-dibutoxypropane when it is heated with 1-butanol and a trace of acid. Some experimental facts are summarized in the following equation.

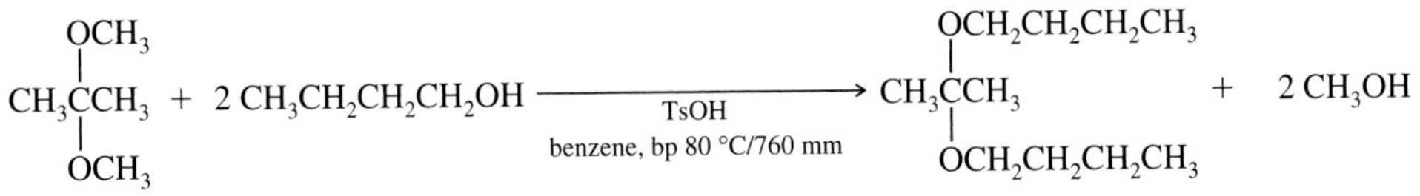

$$\text{CH}_3\overset{\text{OCH}_3}{\underset{\text{OCH}_3}{\text{CCH}_3}} + 2\,\text{CH}_3\text{CH}_2\text{CH}_2\text{CH}_2\text{OH} \xrightarrow[\text{benzene, bp } 80\,°\text{C/760 mm}]{\text{TsOH}} \text{CH}_3\overset{\text{OCH}_2\text{CH}_2\text{CH}_2\text{CH}_3}{\underset{\text{OCH}_2\text{CH}_2\text{CH}_2\text{CH}_3}{\text{CCH}_3}} + 2\,\text{CH}_3\text{OH}$$

bp 83 °C/760 mm    bp 118 °C/760 mm    bp 90 °C/20 mm    bp 65° C/760 mm
75%

Write a mechanism for the reaction and suggest some practical measures that could be taken to ensure a good yield of 2,2-dibutoxypropane.

**14.35** Hydrates (p. 558) are usually present in only small amounts in equilibrium with the corresponding carbonyl compound but are important intermediates in reactions such as the following, in which an exchange of oxygen atoms occurs in acetone.

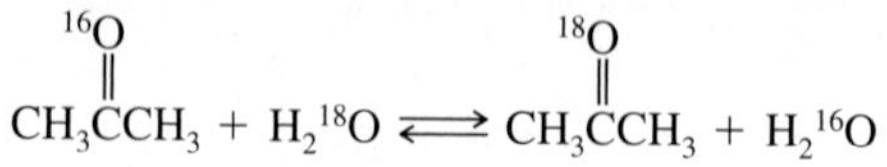

$$CH_3\overset{^{16}O}{\overset{\|}{C}}CH_3 + H_2{}^{18}O \rightleftharpoons CH_3\overset{^{18}O}{\overset{\|}{C}}CH_3 + H_2{}^{16}O$$

The exchange is detected by using water containing $^{18}O$, an isotope of the more abundant $^{16}O$. Propose a mechanism for the observed exchange of oxygen atoms between the carbonyl group and water.

**14.36** When glucose is dissolved in $H_2{}^{18}O$, one oxygen atom in glucose slowly exchanges with the labeled water, so the glucose eventually contains one $^{18}O$. Starting with a cyclic structure of glucose, write a mechanism that explains this experimental observation.

**14.37** The pyranoside of the sugar L-fucose is important in cell–cell recognition. Much work is being done on the synthesis of inhibitors of enzymes that cleave such pyranosides. Draw a mechanism using the curved-arrow convention for the acid-catalyzed conversion of one pyranose form of L-fucose to its anomer as shown below.

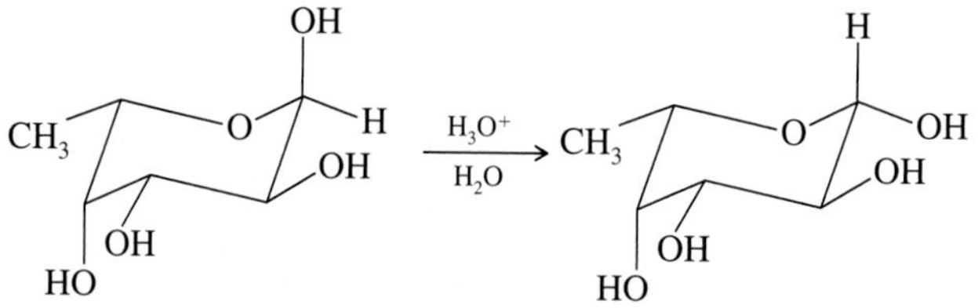

one anomer of
L-fucose

the other anomer of
L-fucose

**14.38** Specify the reagents that are necessary to carry out the following transformations.

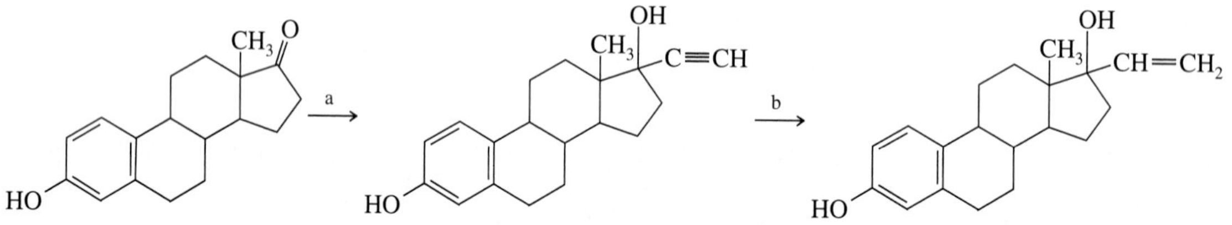

estrone

a female sex hormone

a synthetic modification
of estrone that is as active
as the natural hormone

a synthetic hormone
that is half as active
as estrone

**14.39** When 4-*tert*-butylcyclohexanone is reduced with sodium borohydride, the product is 86% *trans*-4-*tert*-butylcyclohexanol and 14% *cis*-4-*tert*-butylcyclohexanol. When 3,3,5-trimethylcyclohexanone is reduced with sodium borohydride, the product is a mixture, with 48% in which the hydroxyl group is cis to the methyl group at carbon 5 and 52% in which the hydroxyl group is trans to that methyl group. Draw structures showing the correct conformations for the starting materials and the products of these reactions. Offer an explanation for the facts observed experimentally.

**14.40** The following reaction was observed in an investigation into the synthesis of alkaloids using chiral amino acids as starting material.

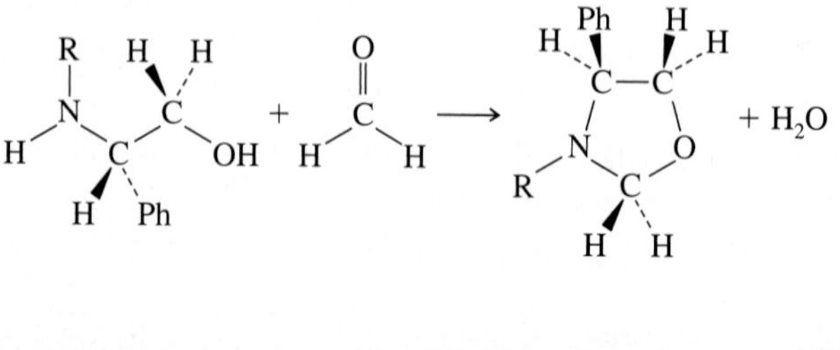

The reaction starts by nucleophilic attack of the amino group on the carbonyl group of formaldehyde and continues after that by way of a resonance-stabilized carbocation. Propose a mechanism for this reaction using the curved-arrow convention and $HB^+$ and BC: as generic acids and bases as necessary.

**14.41** Supply reagents for each of the following transformations. More than one step may be necessary for some.

(a)

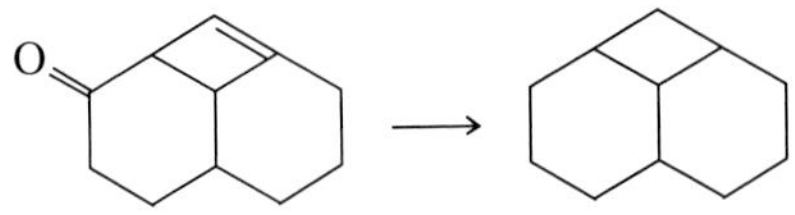

(b)

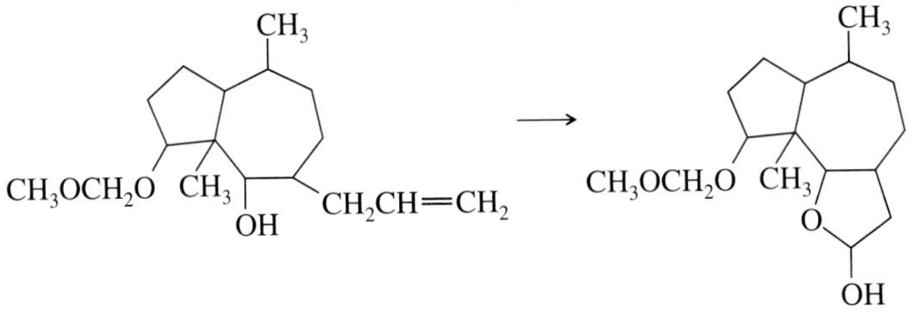

(c)

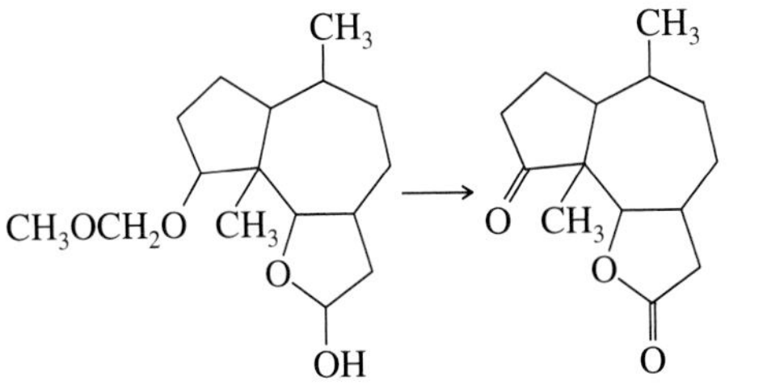

(d)

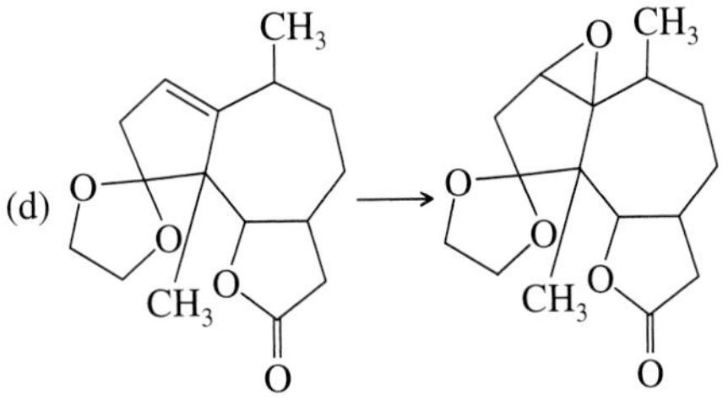

**14.42** The following sequence of reactions has been carried out in a synthesis of pentalenene, a natural product related to an antibacterial and antifungal agent. What reagents would you use to accomplish the transformations shown?

(a)

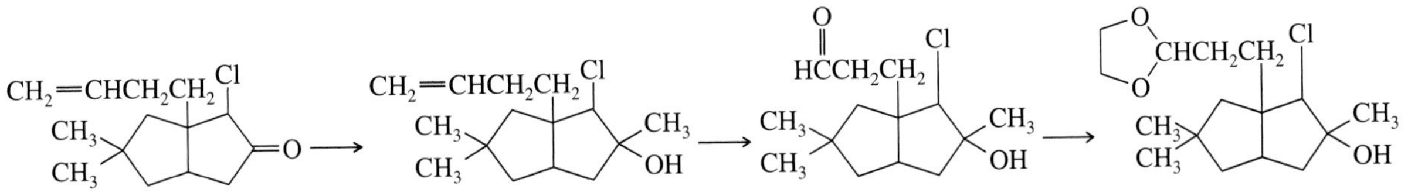

(b)

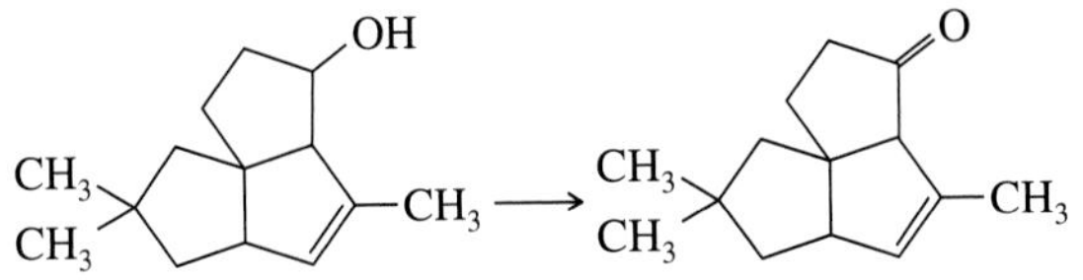

**14.43** A synthesis of natural products that have antileukemic and cytotoxic properties involves the following transformation. How would you carry it out?

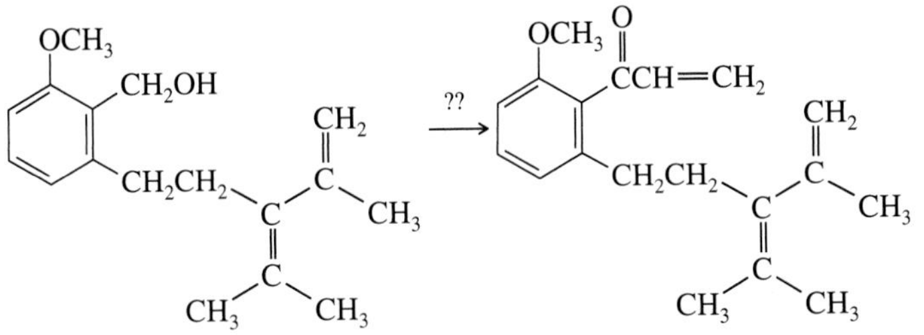

**14.44** The following transformation was carried out in three steps in a synthesis of compounds to be used in a study of reaction mechanisms. A "step" is defined as arrival at an isolable compound, so each step may have more than one substep. Propose a synthesis for Compound B starting from Compound A. Show clearly the reagents you would use for each step and the product of that step.

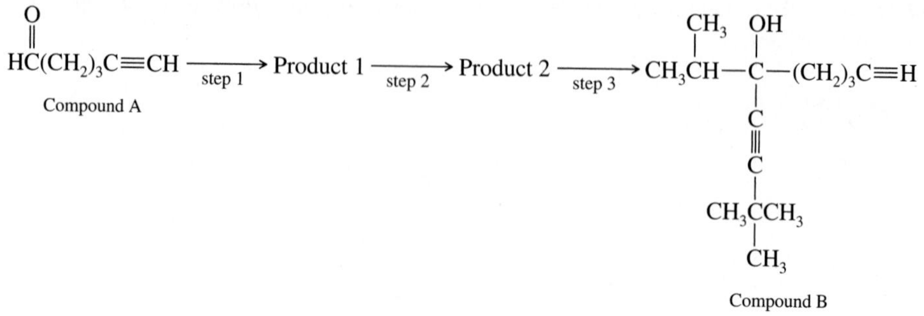

$$HC(CH_2)_3C\equiv CH \xrightarrow[\text{step 1}]{} \text{Product 1} \xrightarrow[\text{step 2}]{} \text{Product 2} \xrightarrow[\text{step 3}]{} CH_3CH-\underset{\displaystyle C}{\overset{\displaystyle CH_3}{\underset{\displaystyle \|}{\overset{\displaystyle |}{C}}}}-(CH_2)_3C\equiv H$$

Compound A

Compound B

**14.45** The following reactions were carried out in the synthesis of sugar analogs. Fill in the structural formulas for the compounds indicated by letters.

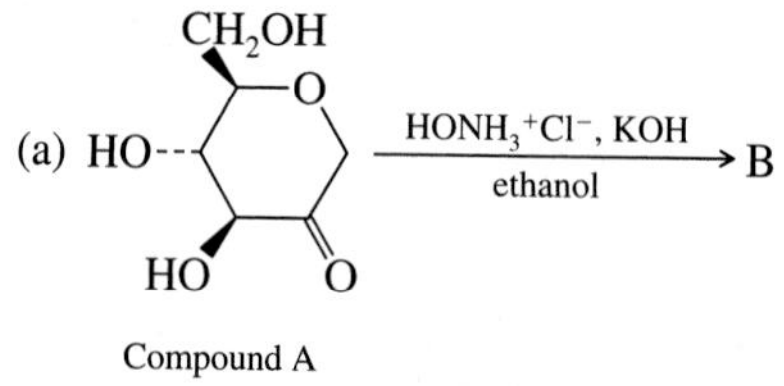

(a) 

$$\text{Compound A} \xrightarrow[\text{ethanol}]{HONH_3{}^+Cl^-,\ KOH} B$$

Compound A

(b) Compound A $\xrightarrow[\text{ethanol}]{NaBH_4}$ C + D

Compounds C and D are diastereomers. Compound C is the major product by 4:1 and has the same stereochemistry as glucose (p. 566) at the new stereocenter.

(c) Draw the two chair conformations of the starting material, Compound A, putting the carbonyl group at the right-hand corner of the ring pointing either up or down. Which conformation would be more stable? Does your drawing explain why Compound C is formed as the major product of this reduction?

**14.46** The following synthesis was carried out during a study of the mechanism by which vitamin $B_{12}$ functions in the human body.

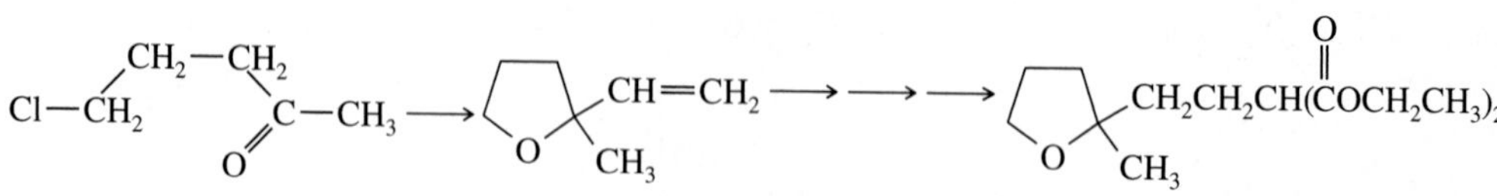

Provide a reagent and a mechanism for the first step. How would you carry out the rest of the transformation? (*Hint:* A review of Table 7.1, p. 234, may be helpful with respect to the last step of the synthesis.)

**14.47** Vitamin $D_3$ has two important physiologic roles. It controls the metabolism of calcium, including the absorption of calcium from the intestines and the resorption of calcium from bones. It is also involved in the processes of cell division and differentiation, processes that are critical in whether a cell develops normally or becomes malignant. Physicians would like to try vitamin $D_3$ in treatment of cancer but cannot because of the effect that it also has on bones. Chemists have been working to synthesize compounds that have the ability to affect cell proliferation without affecting calcium metabolism. One part of such a research project is shown on the next page. Fill in the structures of the missing reagents and products.

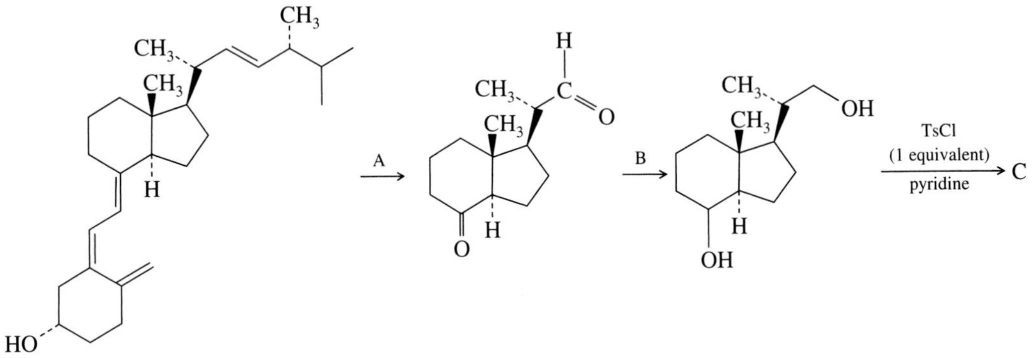

**14.48** The following transformation was observed. Propose a stepwise mechanism for it using the curved-arrow convention to show how the starting material is converted to product.

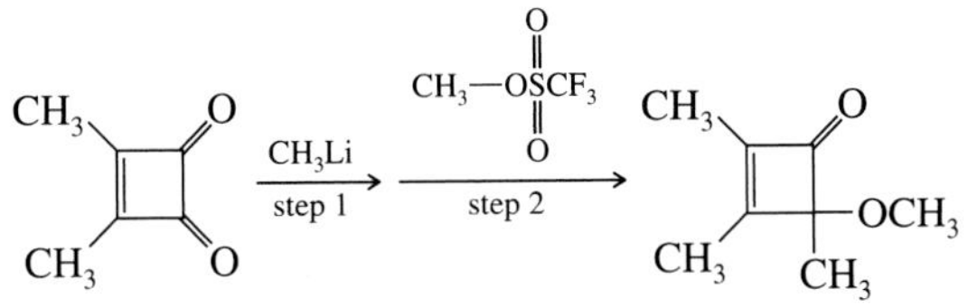

**14.49** Research into the synthesis of carbasugars, sugar analogs that are missing the oxygen atom in the ring, involved the following reaction.

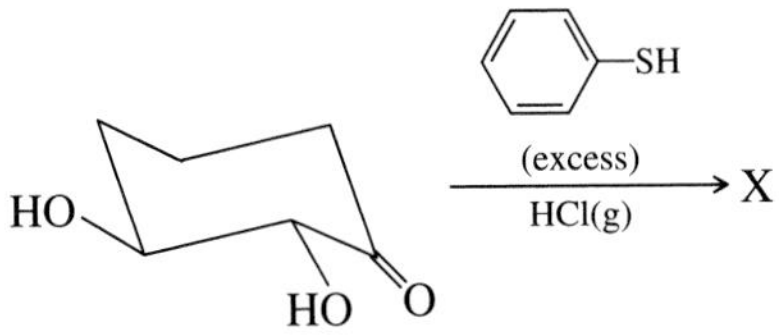

Compound X had an absorption band in its infrared spectrum at 3453 cm$^{-1}$ but no significant absorption from 1800–1600 cm$^{-1}$. Its proton nuclear magnetic resonance spectrum had peaks at $\delta$ 7.75–7.72 (10H), 4.21 (1H), 3.43 (1H), and 2.07–0.94 (6H). Two of the hydrogens in the product absorb at varying frequencies and so are not listed above. What is the structure of Compound X?

**14.50** Compounds A and B have the molecular formula $C_6H_{12}O$. They both show absorption in their infrared spectra at 1717 cm$^{-1}$. Proton magnetic resonance spectra for the compounds are given in Figure 14.5 (p. 592). $^{13}C$ nuclear magnetic resonance data for the compounds are

A: $\delta$ 13.81, 22.34, 26.05, 29.75, 43.50, 208.88

B: $\delta$ 24.69, 26.37, 44.30, 214.21

Assign structures to Compounds A and B, and show how the spectral data given above support your assignments.

**14.51** The proton magnetic resonance spectrum of Compound C, $C_{10}H_{12}O$, is shown in Figure 14.6 (p. 593). The infrared spectrum of Compound C has a strong band at 1717 cm$^{-1}$. Assign a structure to the compound.

**14.52** The proton magnetic resonance spectrum of Compound D is shown in Figure 14.7 (p. 593). The compound absorbs strongly in the infrared at 1676 cm$^{-1}$. In its mass spectrum, the peak due to the molecular ion is at $m/z$ 150, and the base peak is at $m/z$ 135. Assign a structure to Compound D, showing how you used each piece of spectral information.

Compound A

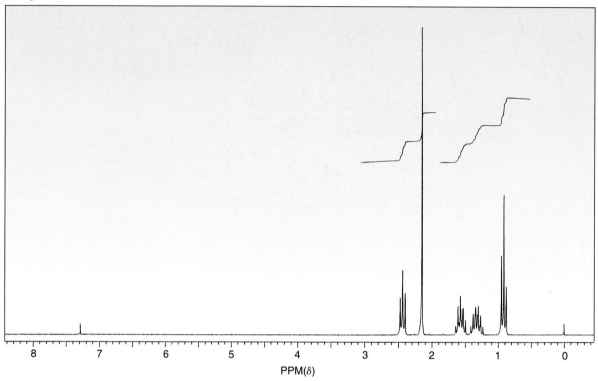

Compound B

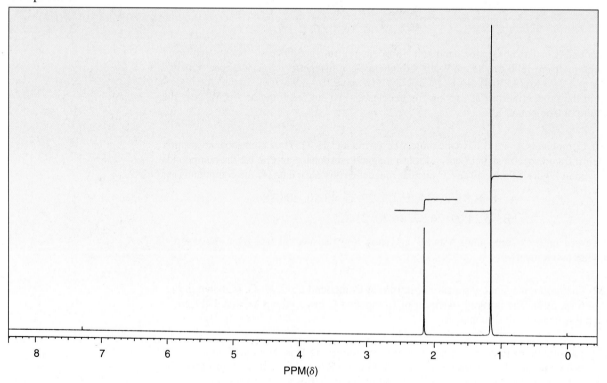

*Figure 14.5*

Compound C, C₁₀H₁₂O

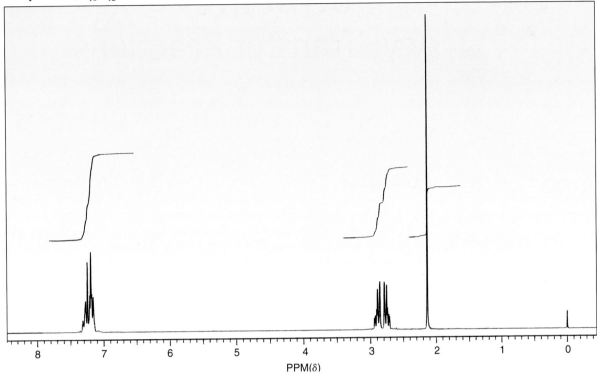

*Figure 14.6*

Compound D

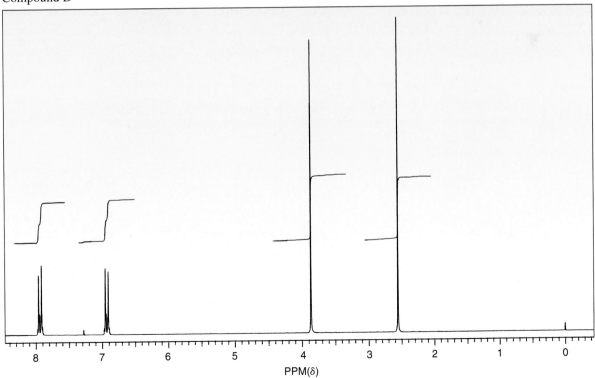

*Figure 14.7*

# 15

# Carboxylic Acids and Their Derivatives. Acyl-Transfer Reactions

## A Look Ahead

Carboxylic acids and their derivatives are compounds in which a carbonyl group is bonded to an atom that has at least one pair of nonbonding electrons on it. Acetic acid and its derivatives are examples.

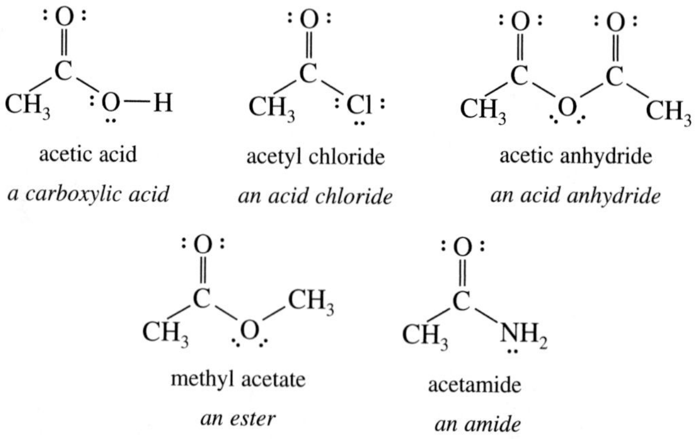

acetic acid

*a carboxylic acid*

acetyl chloride

*an acid chloride*

acetic anhydride

*an acid anhydride*

methyl acetate

*an ester*

acetamide

*an amide*

Carboxylic acids are strong organic acids. Also, the carbon atom of the carbonyl group is electrophilic and reacts with nucleophiles.

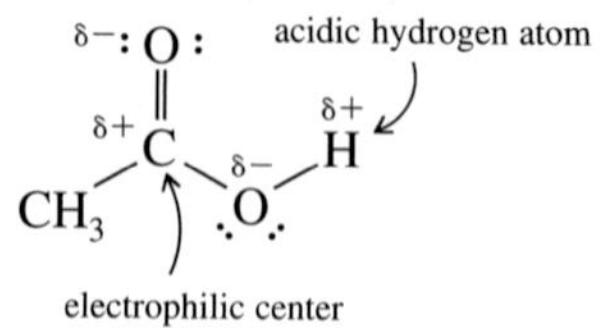

An acid derivative may be thought of as having been created from a carboxylic acid by replacement of the hydroxyl group of the carboxyl group by another atom or group. This group either is a good leaving group or may be converted to a good leaving group by protonation. Acids and acid derivatives, therefore, undergo nucleophilic substitution reactions, an example of which is the reaction of acetyl chloride with ammonia.

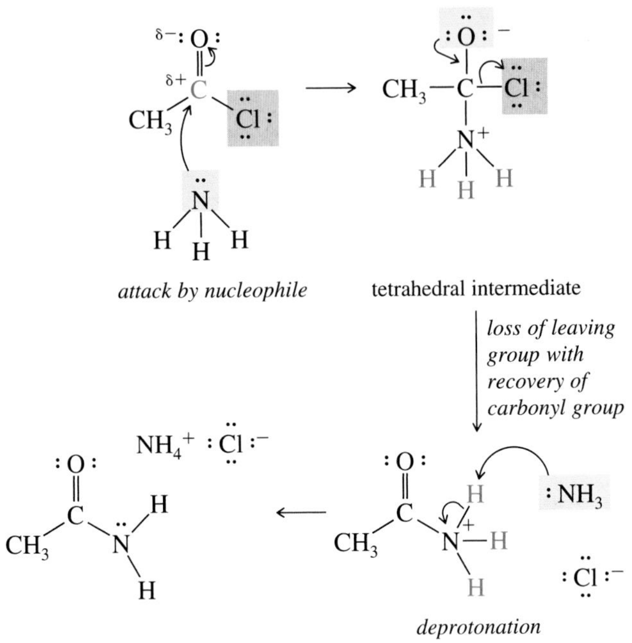

*attack by nucleophile*     tetrahedral intermediate

*loss of leaving group with recovery of carbonyl group*

*deprotonation*

Most nucleophilic substitution reactions of acids and acid derivatives have two steps.

1. Nucleophilic attack on the carbon atom of the carbonyl group, with formation of a tetrahedral intermediate.

2. Loss of a leaving group, with the recovery of the carbonyl group.

In the reaction shown above, the acetyl group, an acyl group, is transferred from a chlorine atom to a nitrogen atom.

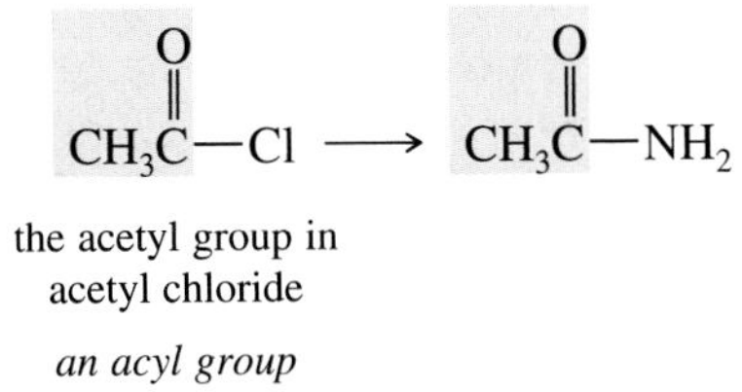

the acetyl group in
acetyl chloride

*an acyl group*

These important reactions of acid derivatives are called **acylation,** or **acyl-transfer,** reactions.

The reactions of acid derivatives differ from those of aldehydes and ketones, which do not have good leaving groups bonded to the carbonyl group. The first step of the reaction with nucleophiles is the same for acid derivatives as it is for aldehydes and ketones (p. 549, for example). Unlike aldehydes and ketones, however, acid derivatives undergo nucleophilic substitution rather than nucleophilic addition.

This chapter will emphasize the interconversions of the different acid derivatives through acyl-transfer reactions, which have biological importance too in the chemistry of amino acids and peptides.

**Workbook Exercises**

## 15.1 Properties of the Functional Groups in Carboxylic Acids and Their Derivatives

### A. The Functional Groups in Carboxylic Acids and Their Derivatives

Carboxylic acids are organic compounds that contain the carboxyl group, a functional group in which a hydroxyl group is directly bonded to the carbon atom of a carbonyl group. Interaction between the carbonyl group and the hydroxyl group affects the properties of both. A comparison of the resonance contributors possible for a carbonyl group and for a carboxyl group shows why this is the case.

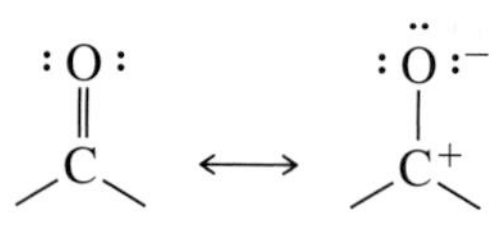

*resonance contributors for a carbonyl group*

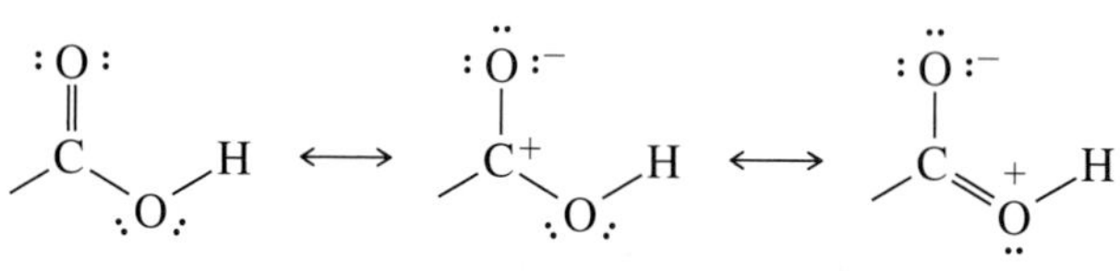

*resonance contributors for a carboxyl group*

The hydroxyl group of a carboxylic acid is unlike the hydroxyl group of an alcohol. The drain of electrons away from the hydroxyl group by the carbonyl group increases the positive character of the hydrogen atom and stabilizes the carboxylate anion. The hydrogen atom of the hydroxyl group of a carboxyl group is much more easily lost as a proton than is the hydrogen atom of the hydroxyl group of an alcohol. The acidity of carboxylic acids is discussed further in Chapter 16.

One of the resonance contributors for the carbonyl group of an aldehyde or a ketone has an open shell and a positive charge on the carbon atom. This resonance contributor is used to rationalize the electrophilic character of the carbon atom of the carbonyl group and its reactions with a variety of nucleophiles such as alcohols, amines, and organometallic reagents (Chapter 14).

In carboxylic acids and their derivatives, the functional group is stabilized by resonance more than the carbonyl group in aldehydes and ketones is. As seen above, the carboxyl group, in addition to the resonance contributor in which the carbon atom of the carbonyl group has an open shell, has a resonance contributor in which the carbon atom and the oxygen atoms have octets of electrons around them, and the positive charge is delocalized from the carbon atom to the oxygen atom. The carbonyl group in a carboxyl function is stabilized by resonance relative to the carbonyl group in an aldehyde or ketone and is, therefore, less reactive. The difference in energy between the starting material and the tetrahedral intermediate is larger for the more stable carboxyl group than for the carbonyl group of an aldehyde, for example. For this reason, many reagents that react easily with the carbonyl group of aldehydes or ketones react more slowly or only in the presence of catalysts when attacking the carbonyl group of a carboxylic acid derivative. This idea was illustrated in the comparison of the reactivity of an oxirane with that of an acyclic ether in Figure 13.1 (p. 506).

**Problem 15.1**    Compare the formation of tetrahedral intermediates from the reaction of an amine with an aldehyde or a ketone and with an acid derivative such as an ester.

| amine | aldehyde or ketone | ester |

Draw a mechanism for the formation of the intermediate and an energy diagram showing the relative energy levels of the starting material and the transition state on the way to the tetrahedral intermediate in each case.

In an **acid chloride,** the hydroxyl group of a carboxylic acid has been replaced by a chlorine atom. In an **acid anhydride,** the anion corresponding to a carboxylic acid has taken the place of the original hydroxyl group. In an **ester,** an alkoxyl group replaces the hydroxyl group. In an **amide,** an amino group is the replacement.

|  acid chloride  |  acid anhydride  |  ester  |  amide  |

In each acid derivative, the atom bonded directly to the carbonyl group has at least one pair of nonbonding electrons on it and can therefore interact with the carbonyl group in the same way that the hydroxyl group does in carboxylic acids. Also, each of the shaded groups above is a good leaving group or may be converted into a good leaving group by protonation. These structural features are important in the chemistry of acids and acid derivatives.

**Problem 15.2**    (a) Write structural formulas for propanoic acid, $CH_3CH_2CO_2H$, and its acid chloride, acid anhydride, ethyl ester, and amide.

(b) Write equations for the reactions that you would expect between propanoic acid and concentrated sulfuric acid. Repeat the process for ethyl propanoate and propanamide. (Reviewing Section 3.2 may be helpful.)

(c) Encircle any good leaving groups that you see in the structural formulas you have written in parts (a) and (b).

Acid derivatives themselves vary in their reactivity in nucleophilic substitution reactions. Acid chlorides, for example, react quickly with cold water (p. 613), but esters and amides must be heated with aqueous acid or base in order to hydrolyze them (pp. 613 and 614). The reactivity of acid derivatives toward nucleophilic substitution reactions at the carbonyl group follows this order:

$$\underset{RCCl}{\overset{O}{\parallel}} > \underset{RCOCR}{\overset{O\ \ O}{\parallel\ \ \parallel}} > \underset{RCOR'}{\overset{O}{\parallel}} \geq \underset{RCOH}{\overset{O}{\parallel}} > \underset{RCNH_2}{\overset{O}{\parallel}} > \underset{RCO^-}{\overset{O}{\parallel}}$$

Reactivity decreases as the effect of resonance stabilization increases. The relative stabilities of acids and acid derivatives are shown schematically in Figure 15.1 (p. 598).

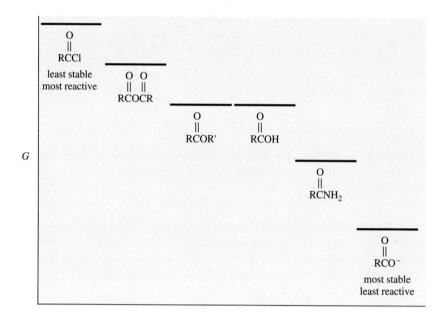

*Figure 15.1*
Relative stabilities of acid derivatives.

The acid chloride, which is the least stable of the series, is the most reactive. The chlorine atom, unlike an oxygen or nitrogen atom, is much larger than carbon and its orbitals do not overlap as well with the orbitals of the carbonyl group in resonance. For this reason, the positive charge is more localized on the carbon atom of the carbonyl group, and the group as a whole is less stable for an acid chloride than it is for the other acid derivatives (Figure 15.2).

Of all the acid derivatives, the carboxylate ion, in which the negative charge is delocalized over two oxygen atoms, is the most stabilized by resonance. With its overall negative charge, it is also least likely to be approached by a nucleophile. It is the least reactive toward nucleophiles and is formed whenever any of the other acid derivatives reacts with water in the presence of base.

Resonance stabilization is lost when the carbonyl group is converted to a tetrahedral intermediate. The energy of activation (p. 123) for the conversion of an acid chloride, which is higher on an energy diagram as a reactant, to a tetrahedral intermediate is lower than that for the conversion of the more stable amide, for example, to a similar intermediate. The ease with which a given acid derivative can be converted to the other acid derivatives by nucleophilic substitution reactions decreases

*Figure 15.2*
Comparison of the possibility of resonance stabilization in a carboxylate ion and in an acid chloride.

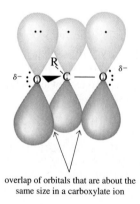

overlap of orbitals that are about the same size in a carboxylate ion

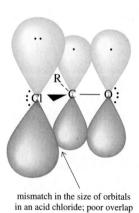

mismatch in the size of orbitals in an acid chloride; poor overlap

from left to right in the reactivity series shown earlier. That is, an acid chloride can be rather easily converted to any of the other compounds in the series; an amide, on the other hand, can be hydrolyzed to a carboxylic acid or a carboxylate anion but is not easily transformed into an ester, an acid anhydride, or an acid chloride. These relationships are summarized in Table 15.2 (p. 643).

⊘ Study Guide
Concept Map 15.1

**Problem 15.3** Write resonance contributors for propanoic acid and its acid chloride, acid anhydride, ethyl ester, and amide. Show in each case how the stability of the functional group is affected by the presence of an atom having a pair of nonbonding electrons adjacent to the carbonyl group.

**Problem 15.4** Propanamide is much less basic than propylamine. In fact, the protons on the nitrogen atom are about

as acidic as those in water ($pK_a \sim 15$), while those on the nitrogen atom of propylamine have $pK_a \sim 36$.

$$CH_3CH_2\overset{\overset{\displaystyle O}{\|}}{C}NH_2 \qquad CH_3CH_2CH_2NH_2$$

How would you explain these facts? (You may want to review the factors affecting acidity and basicity in Sections 3.6 and 3.7.)

## B. Physical Properties of Low-Molecular-Weight Acids and Acid Derivatives

Carboxylic acids of low molecular weight have boiling points that are relatively high, and they are very soluble in water. Molecular weight determinations indicate that carboxylic acids exist as dimers even in the vapor state. All these data suggest that the carboxyl group participates both as a donor and an acceptor in extensive hydrogen bonding, as illustrated below for acetic acid in the vapor state, in the liquid state, and in solution in water.

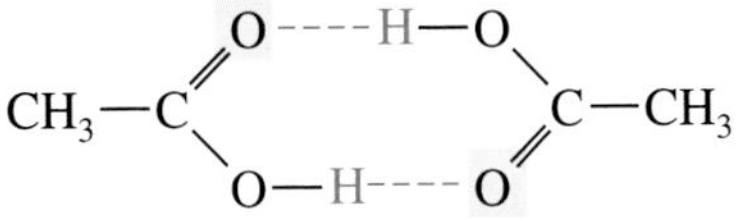

*dimer of acetic acid
held together by hydrogen
bonding in the vapor state*

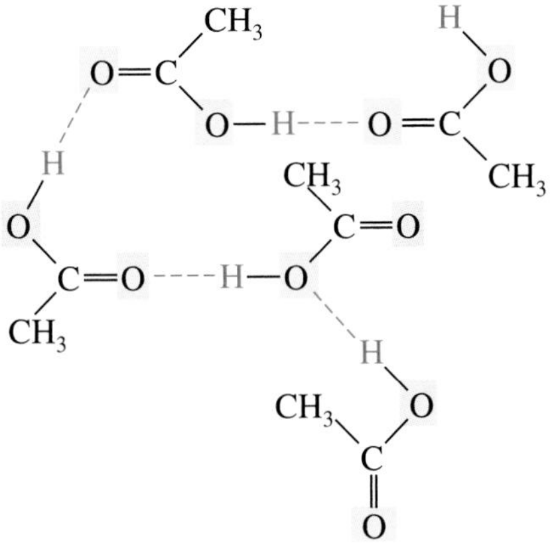

*network of hydrogen bonding
between molecules of acetic
acid in the liquid state*

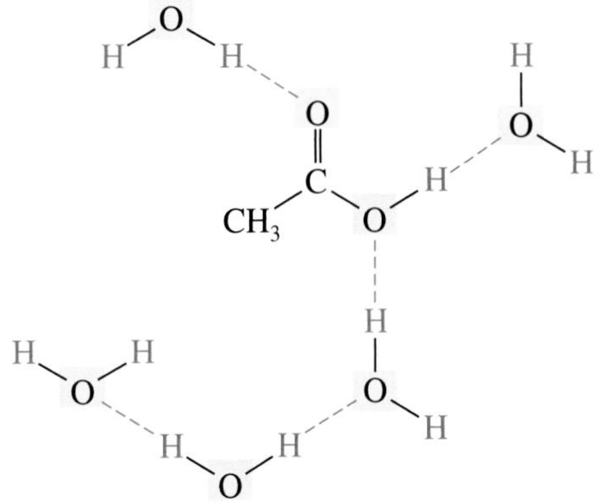

*acetic acid, hydrogen bonded
to water molecules in aqueous
solution*

Carboxylic acids with no other functional group and fewer than ten carbon atoms in the chain are liquids at room temperature. Acetic acid has a particularly high melting point, 16.7 °C, for a compound with such a low molecular weight and is known as **glacial acetic acid** in its pure state. It is a liquid at room temperature but freezes easily in an ice bath, a phenomenon that has practical importance in the laboratory. Oxalic acid and the larger dicarboxylic acids, as well as the aromatic carboxylic acids, are all solids at room temperature.

$$
\begin{array}{cccc}
\underset{\displaystyle\text{HCOH}}{\overset{\displaystyle\text{O}}{\|}} &
\underset{\displaystyle\text{CH}_3\text{COH}}{\overset{\displaystyle\text{O}}{\|}} &
\underset{\displaystyle\text{CH}_3\text{CH}_2\text{CH}_2\text{COH}}{\overset{\displaystyle\text{O}}{\|}} &
\underset{\displaystyle\text{CH}_3\text{CH}_2\text{CH}_2\text{CH}_2\text{COH}}{\overset{\displaystyle\text{O}}{\|}}
\end{array}
$$

|  |  |  |  |
|---|---|---|---|
| formic acid | acetic acid | butanoic acid | pentanoic acid |
| bp 100.5 °C | bp 118.2 °C | bp 163 °C | bp 186.4 °C |
| mp 8.4 °C | mp 16.7 °C | mp −7.9 °C | mp −34.5 °C |
| miscible with water | miscible with water | miscible with water | solubility 3.7 g in 100 g of water |

$$
\underset{\displaystyle\text{CH}_3(\text{CH}_2)_8\text{COH}}{\overset{\displaystyle\text{O}}{\|}} \qquad
\underset{\displaystyle\text{HOC}-\text{COH}}{\overset{\displaystyle\text{O}\quad\text{O}}{\|\quad\|}} \qquad
\text{benzoic acid structure } -\overset{\displaystyle\text{O}}{\overset{\|}{\text{C}}}\text{OH}
$$

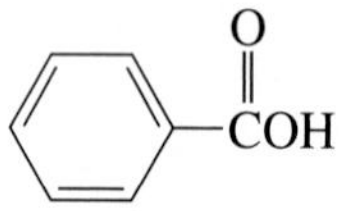

|  |  |  |
|---|---|---|
| decanoic acid | oxalic acid | benzoic acid |
| bp 270.0 °C |  |  |
| mp 31.3 °C | mp 187 °C | mp 122 °C |
| solubility 0.015 g in 100 g of water | solubility 9.0 g in 100 g of water | solubility 0.29 g in 100 g of water |

Monocarboxylic acids and dicarboxylic acids of low molecular weight are soluble in water. When two compounds dissolve in each other in all proportions, they are said to be **miscible.** The lower-molecular-weight liquid carboxylic acids shown above are miscible with water. When the hydrocarbon portion of the molecule has more than about five carbon atoms for each carboxyl group, solubility decreases. The high-molecular-weight carboxylic acids are almost insoluble in water.

Carboxylic acids that have low solubility in water, such as benzoic acid, are converted to water-soluble salts by reaction with aqueous base (p. 103). Protonation of the carboxylate anion by a strong acid regenerates the water-insoluble acid. These properties of carboxylic acids are useful in separating them from reaction mixtures containing neutral and basic compounds.

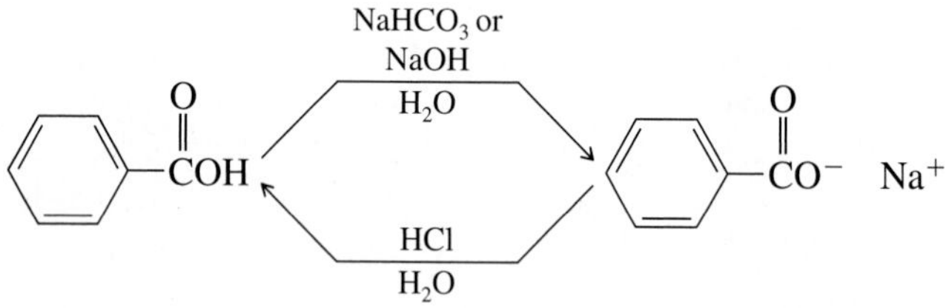

benzoic acid

*covalent,*
*insoluble in water*

sodium benzoate

*ionic,*
*soluble in water*

The importance of hydrogen bonding to the physical properties and solubility in water of carboxylic acids is demonstrated by comparing acetic acid with two of its derivatives, an ester and an amide.

$$
\underset{\displaystyle\text{CH}_3\text{COH}}{\overset{\displaystyle\text{O}}{\|}} \qquad
\underset{\displaystyle\text{CH}_3\text{COCH}_2\text{CH}_3}{\overset{\displaystyle\text{O}}{\|}} \qquad
\underset{\displaystyle\text{CH}_3\text{CNH}_2}{\overset{\displaystyle\text{O}}{\|}}
$$

|  |  |  |
|---|---|---|
| acetic acid | ethyl acetate | acetamide |
| bp 118 °C | bp 77 °C | mp 82 °C |
| miscible with water | solubility 8.6 g in 100 g of water | solubility 97.5 g in 100 g of water |

Acetic acid boils at 118 °C and is fully miscible with water, but its ethyl ester has a boiling point of 77 °C and a solubility of 8.6 g in 100 g of water. Ethyl acetate cannot hydrogen-bond to itself in the liquid state. In water, it can serve only as a hydrogen-bond acceptor at its oxygen atoms. Therefore, it has a low boiling point and relatively low solubility in water. Acetamide, on the other hand, is a solid (mp 82 °C) with a very high solubility in water. The hydrogen atoms on the nitrogen atom of an amide participate strongly in hydrogen bonding, a fact of crucial importance to the structure of proteins, which are polyamides (p. 615).

**Problem 15.5**    Predict which compound in each of the following series will have the highest solubility in water and which will have the lowest.

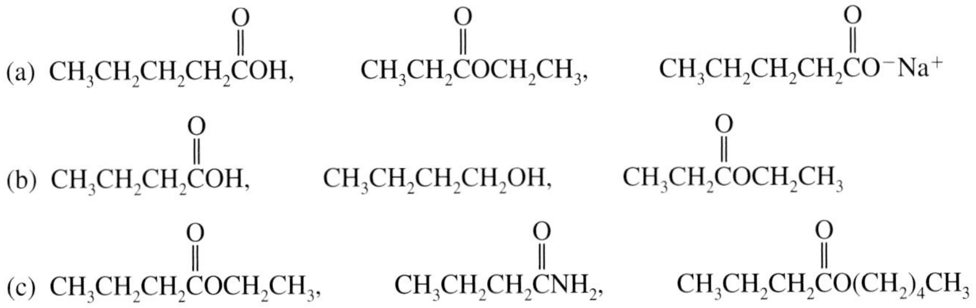

(a)  $CH_3CH_2CH_2CH_2COH$,    $CH_3CH_2COCH_2CH_3$,    $CH_3CH_2CH_2CH_2CO^-Na^+$

(b)  $CH_3CH_2CH_2COH$,    $CH_3CH_2CH_2CH_2OH$,    $CH_3CH_2COCH_2CH_3$

(c)  $CH_3CH_2CH_2COCH_2CH_3$,    $CH_3CH_2CH_2CNH_2$,    $CH_3CH_2CH_2CO(CH_2)_4CH_3$

# 15.2 Nomenclature of Carboxylic Acids and Their Derivatives

## A. Naming Carboxylic Acids

The systematic name of an alkyl carboxylic acid is derived by replacing the **e** at the end of the name of the hydrocarbon having the same number of carbon atoms in the chain with **-oic acid.** The carboxyl function is always assumed to be the first carbon atom of the chain. The presence of other substituents is indicated by assigning a name and a position number to each one. The two smallest carboxylic acids, formic acid (from *formica,* Latin for "ant") and acetic acid (from *acetum,* Latin for "vinegar"), are usually known by their common names.

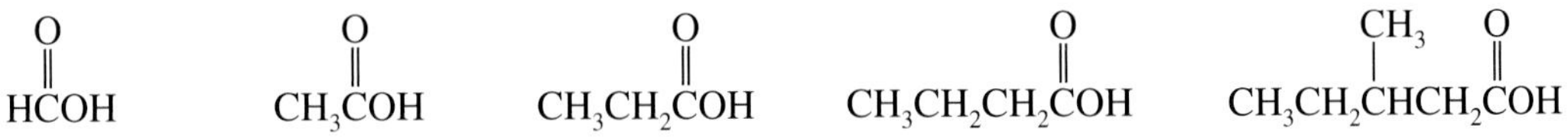

$HCOH$

methanoic acid
formic acid

$CH_3COH$

ethanoic acid
acetic acid

$CH_3CH_2COH$

propanoic acid
propionic acid

$CH_3CH_2CH_2COH$

butanoic acid
butyric acid

$CH_3CH_2CHCH_2COH$ (with $CH_3$)

3-methylpentanoic acid

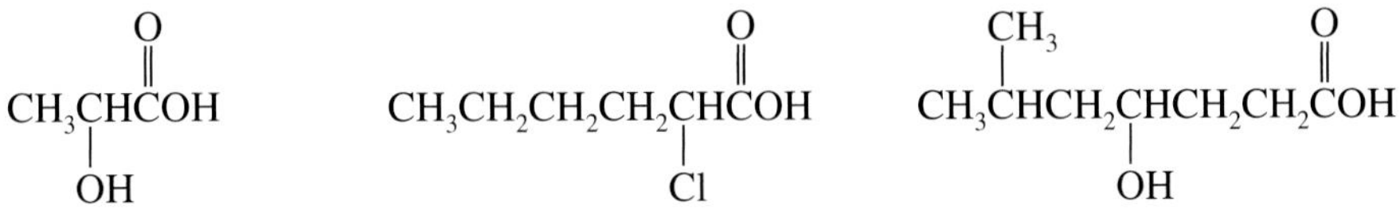

$CH_3CHCOH$ (with $OH$)

2-hydroxypropanoic acid
lactic acid

$CH_3CH_2CH_2CH_2CHCOH$ (with $Cl$)

2-chlorohexanoic acid

$CH_3CHCH_2CHCH_2CH_2COH$ (with $CH_3$ and $OH$)

4-hydroxy-6-methylheptanoic acid

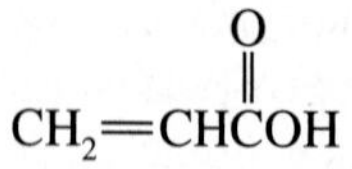

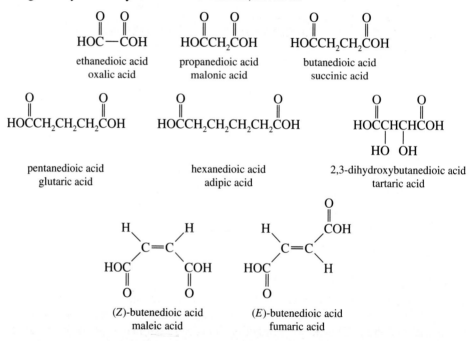

$$CH_2{=}CHCOH \quad\quad CH_3\overset{O}{\underset{}{C}}{-}\overset{O}{\underset{}{C}}OH \quad\quad CH_3CHCO^-$$

propenoic acid
acrylic acid

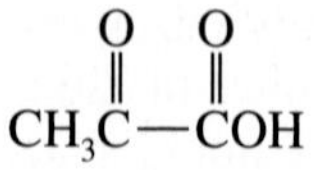

2-oxopropanoic acid
pyruvic acid

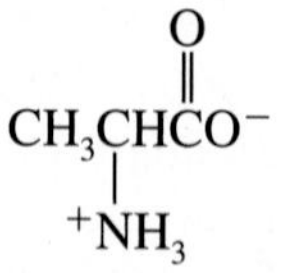

2-aminopropanoic acid
alanine

Biologically important carboxylic acids are also usually known by their common names; the hydroxyacid lactic acid, the $\alpha$-ketoacid pyruvic acid, and the amino acid alanine are shown above.

The presence of two carboxyl groups is indicated in a systematic name by using **-dioic acid** after the full name (including the final **e**) of the hydrocarbon having the same number of carbon atoms in the chain. The dicarboxylic acids shown below are generally known by their common names, however.

HOC—COH

ethanedioic acid
oxalic acid

HOCCH₂COH

propanedioic acid
malonic acid

HOCCH₂CH₂COH

butanedioic acid
succinic acid

HOCCH₂CH₂CH₂COH

pentanedioic acid
glutaric acid

HOCCH₂CH₂CH₂CH₂COH

hexanedioic acid
adipic acid

HOCCHCHCOH

HO  OH

2,3-dihydroxybutanedioic acid
tartaric acid

(Z)-butenedioic acid
maleic acid

(E)-butenedioic acid
fumaric acid

In aromatic carboxylic acids, the carboxyl group is attached to an aromatic ring. Benzoic acid, which is the simplest unsubstituted aromatic acid, and some other examples are shown below.

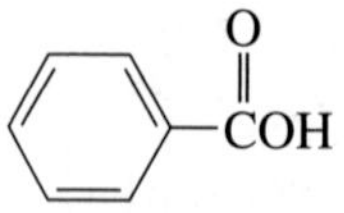

benzoic acid

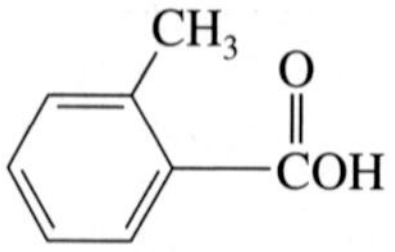

o-methylbenzoic acid
o-toluic acid

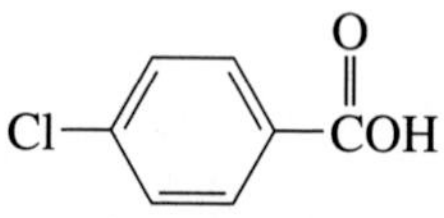

p-chlorobenzoic acid

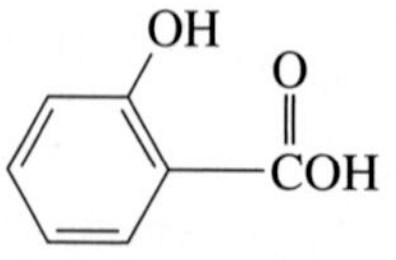

o-hydroxybenzoic acid
salicylic acid

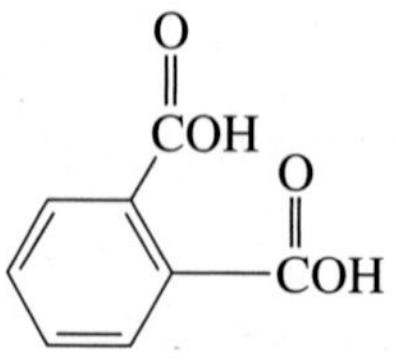

phthalic acid

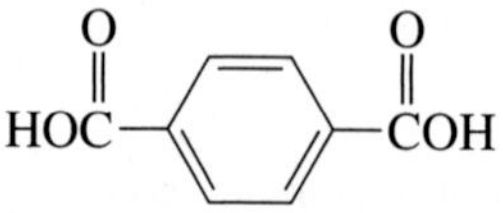

terephthalic acid

When the carboxyl group is attached to a cycloalkane, **-carboxylic acid** is added to the name of the hydrocarbon constituting the rest of the molecule.

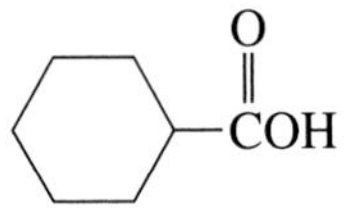

cyclohexanecarboxylic
acid

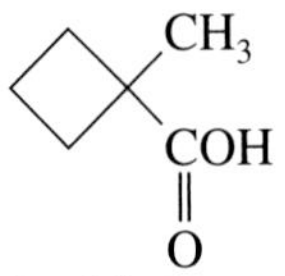

1-methylcyclobutanecarboxylic
acid

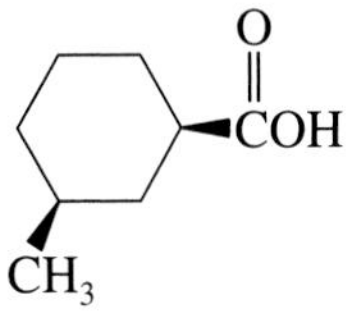

*cis*-3-methylcyclohexanecarboxylic
acid

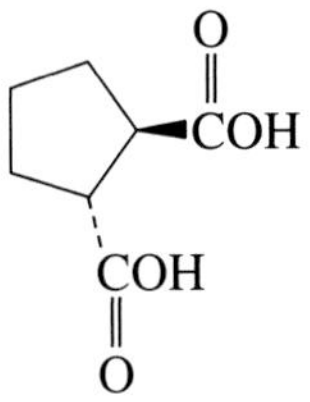

*trans*-1,2-cyclopentanedicarboxylic
acid

**Problem 15.6**    Some of the carboxylic acids shown in this section have more than one stereoisomer. Identify the compounds for which this is true, and draw all possible stereoisomers, naming each one correctly.

**Problem 15.7**    Name the following compounds, including an indication of the stereochemistry where appropriate.

(a)  $CH_3(CH_2)_8CH_2COH$

(b)

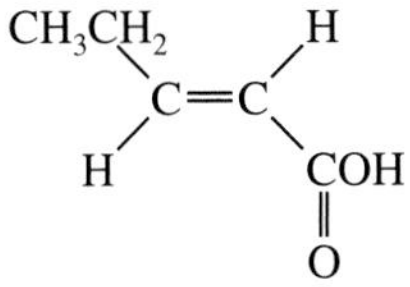

(c)

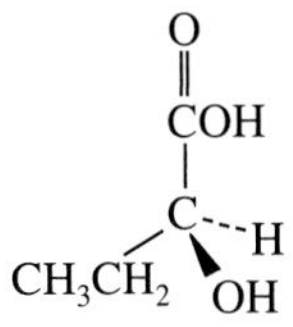

(d)

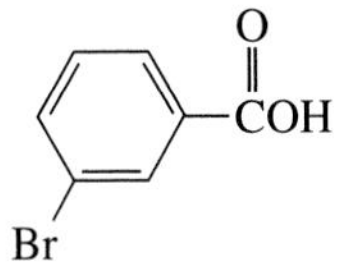

(e) 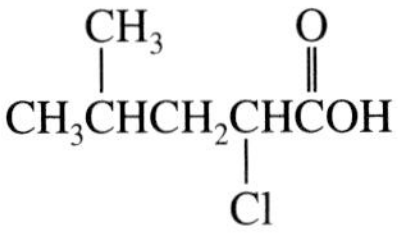 $CH_3CHCH_2CHCOH$

(f) 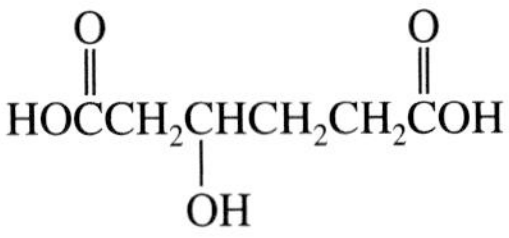 $HOCCH_2CHCH_2CH_2COH$

# B. Naming Acyl Groups, Acid Chlorides, and Anhydrides

The group obtained from a carboxylic acid by the removal of the hydroxyl portion is known as an **acyl group.** The name of an acyl group is created by changing the **ic** at the end of the name of the carboxylic acid to **yl.** This applies to common as well as systematic names of acids. When **-carboxylic acid** is used in the name of a compound, this ending is changed to **-carbonyl** for the corresponding acyl group. Some important acyl groups are shown below with their related acids.

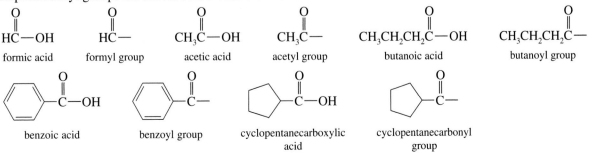

$HC—OH$        $HC—$        $CH_3C—OH$        $CH_3C—$        $CH_3CH_2CH_2C—OH$        $CH_3CH_2CH_2C—$

formic acid        formyl group        acetic acid        acetyl group        butanoic acid        butanoyl group

benzoic acid        benzoyl group        cyclopentanecarboxylic
acid        cyclopentanecarbonyl
group

Acid chlorides are named systematically as acyl chlorides.

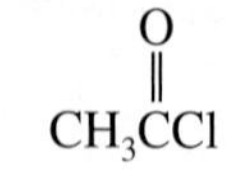

$$CH_3\overset{\displaystyle O}{\overset{\|}{C}}Cl \qquad CH_3CH_2CH_2CH_2\overset{\displaystyle O}{\overset{\|}{C}}Cl \qquad Cl\overset{\displaystyle O}{\overset{\|}{C}}CH_2CH_2\overset{\displaystyle O}{\overset{\|}{C}}Cl$$

acetyl chloride 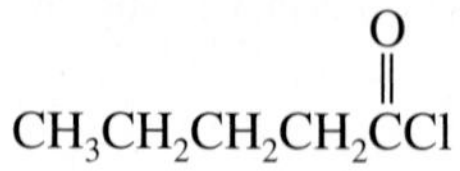          pentanoyl chloride           butanedioyl dichloride
                                                      succinyl dichloride

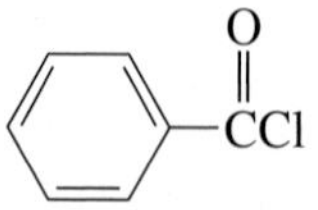

benzoyl chloride 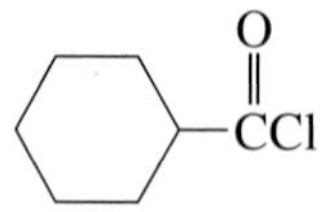          cyclohexanecarbonyl chloride 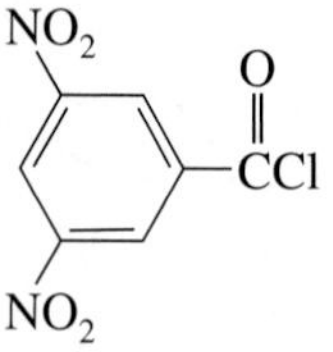          3,5-dinitrobenzoyl chloride

Important acid anhydrides are the anhydride of acetic acid and some cyclic anhydrides formed from dicarboxylic acids. An acid anhydride is named by substituting **anhydride** for **acid** in the name of the acid from which it is derived. Cyclic anhydrides containing five- or six-membered rings are stable and easily formed. Of the aromatic dicarboxylic acids, only the ones with the carboxylic groups in adjacent positions on the aromatic ring form cyclic anhydrides.

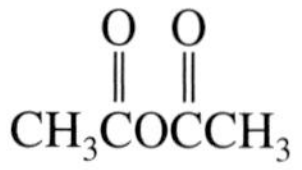

$$CH_3\overset{\displaystyle O}{\overset{\|}{C}}O\overset{\displaystyle O}{\overset{\|}{C}}CH_3$$

acetic anhydride 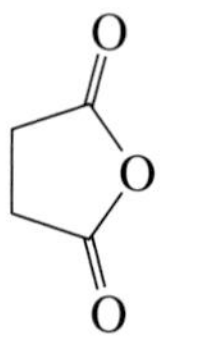          succinic anhydride 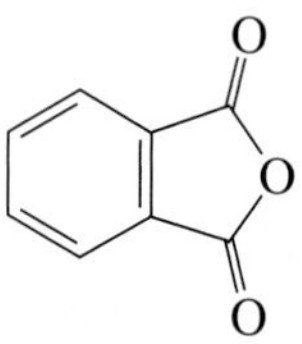          phthalic anhydride 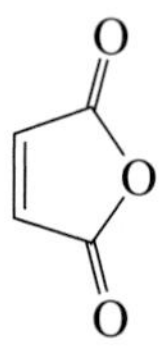          maleic anhydride

## C. Naming Salts and Esters

Salts and esters of carboxylic acids are named in the same way. The name of the cation (in the case of a salt) or the name of the organic group attached to the oxygen of the carboxyl group (in the case of an ester) precedes the name of the acid. The **-ic acid** part of the name of the acid is converted to **-ate.**

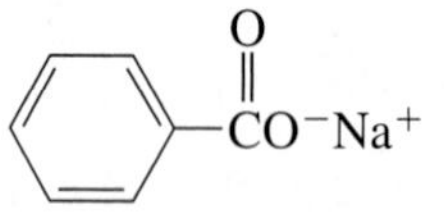

$$-\overset{\displaystyle O}{\overset{\|}{C}}O^-Na^+ \qquad (CH_3CH_2\overset{\displaystyle O}{\overset{\|}{C}}O^-)_2Ca^{2+}$$

sodium benzoate 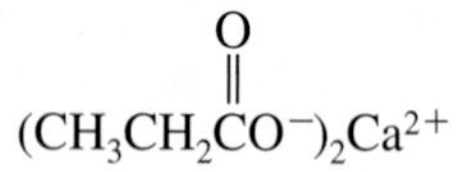          calcium propanoate
                         calcium propionate

*both salts used to retard spoilage in foods*

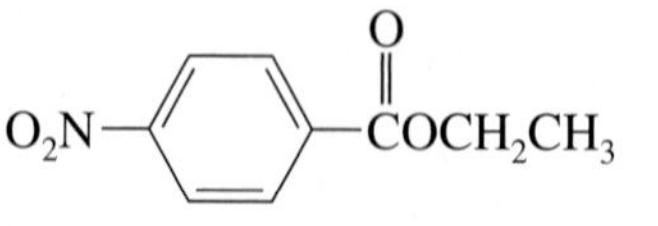

$$O_2N-\overset{\displaystyle O}{\overset{\|}{C}}OCH_2CH_3 \qquad CH_3O\overset{\displaystyle O}{\overset{\|}{C}}CH_2CH_2\overset{\displaystyle O}{\overset{\|}{C}}OCH_3$$

ethyl *p*-nitrobenzoate           dimethyl succinate

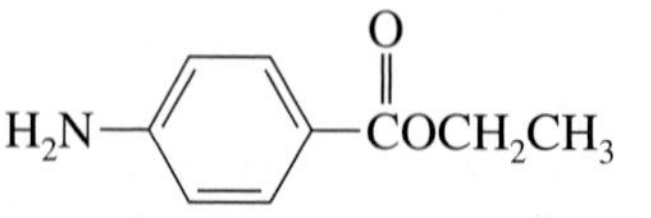

$$H_2N-\overset{\displaystyle O}{\overset{\|}{C}}OCH_2CH_3 \qquad -\overset{\displaystyle O}{\overset{\|}{C}}O\overset{\displaystyle CH_3}{\overset{|}{C}}HCH_3$$

ethyl *p*-aminobenzoate          isopropyl cyclohexanecarboxylate

*a local anesthetic*

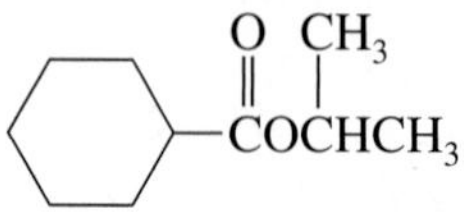

**Problem 15.8**    Name the following compounds, including an indication of the stereo-chemistry where appropriate.

(a) $CH_3CH_2CH_2CH_2CH_2\overset{\displaystyle O}{\overset{\|}{C}}Cl$

(b) $CH_3CH_2CH_2\overset{\displaystyle O}{\overset{\|}{C}}O\overset{\displaystyle O}{\overset{\|}{C}}CH_2CH_2CH_3$

(c) 
$$\underset{\underset{H}{|}}{\overset{\underset{CH_3}{|}}{C}}=\underset{\underset{H}{|}}{\overset{\overset{\displaystyle CH_2CH_2\overset{\displaystyle O}{\overset{\|}{C}}OCH_3}{|}}{C}}$$

(d) $Cl-\!\!\!\left\langle\!\!\bigcirc\!\!\right\rangle\!\!-\overset{\displaystyle O}{\overset{\|}{C}}OCH_2\overset{\overset{\displaystyle CH_3}{|}}{C}HCH_3$

## D. Naming Amides, Imides, and Nitriles

The names of amides are formed by replacing **-oic acid** (or **-ic acid** for common names) by **-amide** or **-carboxylic acid** by **-carboxamide**.

$\overset{\displaystyle O}{\overset{\|}{H C}}NH_2$     $CH_3\overset{\displaystyle O}{\overset{\|}{C}}NH_2$     $CH_3CH_2CH_2CH_2\overset{\displaystyle O}{\overset{\|}{C}}NH_2$     $\left\langle\!\!\bigcirc\!\!\right\rangle\!\!-\overset{\displaystyle O}{\overset{\|}{C}}NH_2$     $\left\langle\!\!\bigcirc\!\!\right\rangle\!\!-\overset{\displaystyle O}{\overset{\|}{C}}NH_2$

methanamide     ethanamide                 pentanamide               benzamide        cyclohexanecarboxamide
formamide       acetamide

If the nitrogen atom of the amide has any alkyl groups as substituents, the name of the amide is prefixed by the capital letter *N*-, to indicate substitution on nitrogen, followed by the name(s) of the alkyl group(s).

$\overset{\displaystyle O}{\overset{\|}{H C}}-\overset{\overset{\displaystyle CH_3}{|}}{N}CH_3$          $O_2N-\!\!\!\left\langle\!\!\bigcirc\!\!\right\rangle\!\!-\overset{\displaystyle O}{\overset{\|}{C}}-\overset{\overset{\displaystyle CH_3}{|}}{N}CH_2CH_3$

*N,N*-dimethylformamide     *N*-ethyl-*N*-methyl-*p*-nitrobenzamide

*abbreviated DMF*

If the substituent on the nitrogen atom of an amide is a phenyl group, the ending for the name of the carboxylic acid is changed to **-anilide.**

$CH_3\overset{\displaystyle O}{\overset{\|}{C}}NH-\!\!\!\left\langle\!\!\bigcirc\!\!\right\rangle$          $\left\langle\!\!\bigcirc\!\!\right\rangle\!\!-\overset{\displaystyle O}{\overset{\|}{C}}NH-\!\!\!\left\langle\!\!\bigcirc\!\!\right\rangle$

acetanilide                 benzanilide

Some dicarboxylic acids form cyclic amides in which two acyl groups are bonded to the nitrogen atom. The suffix **-imide** is given to such compounds, easily produced when five- or six-membered rings may form.

phthalimide     succinimide     *N*-bromosuccinimide

*abbreviated NBS*

Nitriles, compounds containing the cyano group, $—C\equiv N$, are considered to be acid derivatives because they can be hydrolyzed to form amides and carboxylic acids (p. 614). In the systematic nomenclature of these compounds, the suffix **-nitrile** is added to the name of the hydrocarbon containing the same number of carbon atoms, counting the carbon atom of the cyano group.

$$\underset{\text{pentanenitrile}}{\overset{5\quad4\quad3\quad2\quad1}{CH_3CH_2CH_2CH_2C\equiv N}}$$

$$\underset{\text{3-methyl-2-butenenitrile}}{\overset{\overset{\textstyle CH_3}{|}}{\underset{4\quad\ \ \ 3\quad2\quad1}{CH_3C=CHC\equiv N}}}$$

The nitriles related to acetic acid and benzoic acid are called acetonitrile and benzonitrile.

$$\underset{\text{acetonitrile}}{CH_3C\equiv N}$$

benzonitrile

When the cyano group is on a cycloalkane, the suffix **-carbonitrile** is used with the name of the hydrocarbon.

cyclohexanecarbonitrile

*trans*-2-methylcyclo-
propanecarbonitrile

If other functional groups are present, the cyano group is treated as a substituent, and its presence is indicated by the prefix **cyano-**. Note that when the suffix **-nitrile** is used, the carbon atom bonded to the nitrogen is included in the count of carbon atoms that determines the name. When **cyano-** (or **-carbonitrile**) is used, that carbon atom is part of a substituent and is not included in the numbering of the rest of the chain (or ring).

$$\underset{\text{4-cyanobutanoic acid}}{\overset{\overset{\textstyle O}{\|}}{HOCCH_2CH_2CH_2C\equiv N}}$$

ethyl *p*-cyanobenzoate

---

**Problem 15.9**    Write a structural formula for each of the following compounds.

(a) (*Z*)-4-heptenoic acid    (b) *trans*-2-methylcyclobutanecarboxylic acid    (c) (*R*)-2-bromopentanoic acid
(d) octanedioic acid    (e) dimethyl propanedioate (dimethyl malonate)    (f) benzoic anhydride
(g) *N,N*-dimethylbenzamide    (h) pentanedioyl dichloride (glutaryl dichloride)
(i) disodium ethanedioate (sodium oxalate)    (j) methyl 3-nitrobenzoate

---

**Problem 15.10**    Name the following compounds, including an indication of the stereochemistry where appropriate.

(a) $\underset{}{CH_3CH_2\overset{\overset{\textstyle O}{\|}}{C}CH_2CH_2CH_2\overset{\overset{\textstyle O}{\|}}{C}OCH_2CH_3}$    (b) $CH_3(CH_2)_8\overset{\overset{\textstyle O}{\|}}{C}OH$    (c)

(d) $CH_3(CH_2)_{16}\overset{\overset{O}{\|}}{C}O^-Na^+$ (*a soap*)

(e) $CH_3\overset{\overset{O}{\|}}{C}NHCH_3$

(f) $CH_3-\!\!\left\langle\!\bigcirc\!\right\rangle\!\!-\overset{\overset{O}{\|}}{C}NH_2$

(g) [cyclopentane ring with $\overset{\overset{O}{\|}}{C}OH$ and ---Br]

(h) $\overset{CH_3}{\underset{H}{\phantom{}}}\!C\!=\!C\!\overset{H}{\underset{\overset{\overset{O}{\|}}{C}OH}{}}$

(i) $H\text{-}\!\!\overset{\overset{\overset{O}{\|}}{CH_2COH}}{\underset{OH}{C}}\!\!\text{-}CH_3$

(j) $CH_3CH_2CH_2CH_2CH_2CH_2C\!\equiv\!N$

(k) $CH_3\overset{CH_3}{\underset{|}{CH}}CH_2\overset{\overset{O}{\|}}{C}O\overset{\overset{O}{\|}}{C}CH_2\overset{CH_3}{\underset{|}{CH}}CH_3$

(l) $CH_3CH_2CH_2\overset{\overset{O}{\|}}{C}NH-\!\!\left\langle\!\bigcirc\!\right\rangle$

# 15.3 Preparation of Carboxylic Acids

## A. Carboxylic Acids as Products of Oxidation Reactions

Carboxylic acids are formed as products of certain oxidation reactions. You have already studied many of these reactions in earlier chapters.

Primary alcohols and aldehydes are oxidized to carboxylic acids containing the same number of carbon atoms (p. 516). Oxidations of alcohols with acidic chromic acid solutions often give esters, the result of the oxidation of the hemiacetal formed from the starting material alcohol with the product aldehyde (pp. 518 and 558). For this reason, base-catalyzed oxidation with potassium permanganate followed by acidification is preferred when the acid is the desired product. 2-Ethyl-1-hexanol and the corresponding aldehyde have both been converted to 2-ethylhexanoic acid by this method.

$$CH_3CH_2CH_2CH_2\overset{\overset{CH_3CH_2}{|}}{CH}CH_2OH$$
2-ethyl-1-hexanol

or

$$CH_3CH_2CH_2CH_2\overset{\overset{O}{\|}}{CHCH}\!\!\underset{CH_3CH_2}{}$$
2-ethylhexanal

$\xrightarrow[H_2O]{\overset{KMnO_4}{NaOH}}$ $CH_3CH_2CH_2CH_2\underset{\overset{|}{CH_3CH_2}}{CH}\overset{\overset{O}{\|}}{C}O^-Na^+$ $\xrightarrow[\substack{SO_2 \\ \text{(reducing} \\ \text{agent for} \\ \text{excess } MnO_4^-)}]{H_3O^+}$ $CH_3CH_2CH_2CH_2\underset{\overset{|}{CH_3CH_2}}{CH}\overset{\overset{O}{\|}}{C}OH$

2-ethylhexanoic
acid
~75%

In basic solution, the acid is formed as its salt and therefore is not reactive toward unused alcohol in the reaction mixture. The free carboxylic acid is generated by adding sulfuric acid after the oxidation is completed (p. 600).

A very mild oxidizing agent for aldehydes is moist silver oxide. Heptanal is oxidized to heptanoic acid in very high yield with this reagent.

$$CH_3(CH_2)_5\overset{\overset{\displaystyle O}{\|}}{C}H \xrightarrow[\substack{H_2O \\ 95\,°C}]{\substack{Ag_2O \\ NaOH}} Ag\downarrow + CH_3(CH_2)_5\overset{\overset{\displaystyle O}{\|}}{C}O^-Na^+ \xrightarrow{H_3O^+} CH_3(CH_2)_5\overset{\overset{\displaystyle O}{\|}}{C}OH$$

heptanal                                                     heptanoic acid
                                                                 97%

In this reaction, silver(I) is reduced to metallic silver. If the reaction is carried out in a clean test tube or flask, a silver mirror deposits on the glass. The reaction is therefore also used as a test to distinguish aldehydes, by the ease with which they are oxidized, from ketones. The reagent is known as **Tollens reagent** and the test as the **Tollens test** or the **silver mirror test.**

In compounds containing other functional groups that would be sensitive to stronger oxidizing agents, such as potassium permanganate (p. 607), an aldehyde function can be successfully oxidized to a carboxyl group using silver oxide. For example, the unsaturated aldehyde 9,12-octadecadiynal is converted to the corresponding acid.

$$CH_3(CH_2)_4C\equiv CCH_2C\equiv C(CH_2)_7\overset{\overset{\displaystyle O}{\|}}{C}H \xrightarrow[\substack{ethanol \\ N_2\ atmosphere}]{\substack{Ag_2O \\ NaOH}} CH_3(CH_2)_4C\equiv CCH_2C\equiv C(CH_2)_7\overset{\overset{\displaystyle O}{\|}}{C}O^-Na^+$$

9,12-octadecadiynal

$$\Big\downarrow H_3O^+$$

$$CH_3(CH_2)_4C\equiv CCH_2C\equiv C(CH_2)_7\overset{\overset{\displaystyle O}{\|}}{C}OH$$

9,12-octadecadiynoic acid
78%

The reaction shown above is carried out in an atmosphere of nitrogen gas because these unsaturated compounds are sensitive to oxidation, even by the oxygen in air. (The reasons for the great sensitivity of such polyunsaturated compounds to oxygen are discussed on page 786.) The oxidation reaction with silver oxide is highly selective. The aldehyde is converted to a carboxyl group, and the triple bonds are untouched.

Tollens reagent also reacts with the aldehyde function in glucose (p. 564) to give silver metal. Glucose is, in fact, the reducing agent used to make silvered glass objects. Sugars such as glucose that reduce metal ions are known as **reducing sugars.** The reaction is evidence that the carbonyl group in the sugar is present in a hemiacetal linkage in equilibrium with the open-chain aldehyde form. If the carbonyl group is tied up in an acetal linkage, as it is in methyl glucopyranoside (p. 569), the sugar does not reduce metal ions and is known as a **nonreducing sugar.**

Reducing sugars also reduce Benedict's reagent, which is a solution in aqueous base of copper(II) sulfate and sodium citrate (used to complex with copper ion). The overall reactions that carbohydrates undergo in these systems are complex because of the multiple reactions that take place under basic conditions. The important part of each reaction is that the reducing sugar reduces a metal ion to a lower oxidation state that can be detected visually. In Benedict's reagent, it is the conversion of the blue copper(II) ion to an orange-yellow or orange-red precipitate. This reaction is the traditional one used in the past to test for the presence of reducing sugars in urine in suspected cases of diabetes.

Bromine also oxidizes sugars to carboxylic acids. Glucose, for example, gives gluconic acid.

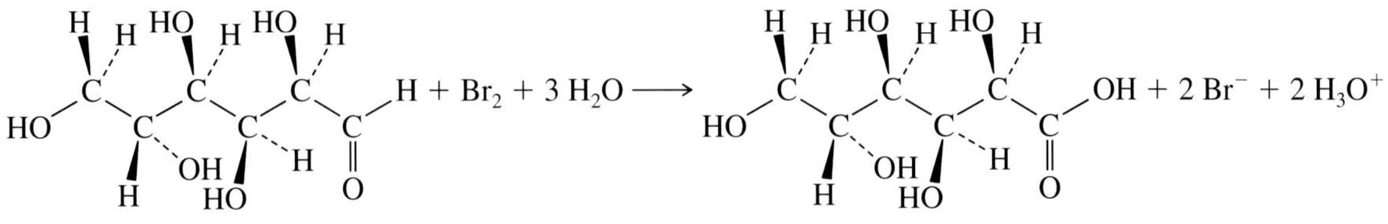

glucose

gluconic acid

In acidic solution, the hydroxyacid exists as a lactone, a cyclic ester, as we shall see later in this chapter (p. 624). The lactone also may be seen as the product of the oxidation of the cyclic hemiacetal of glucose.

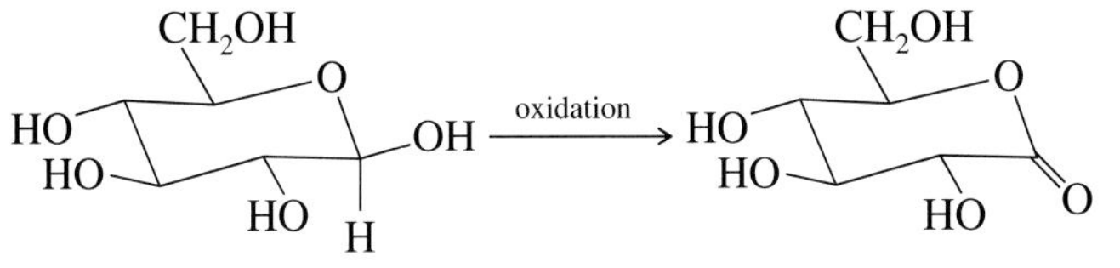

β-D-glucopyranose

lactone of gluconic acid

Ozonolysis is one of the most effective ways to cleave a carbon–carbon double bond in order to produce acids by way of aldehyde intermediates (p. 512). The conversion of 1-tridecene to dodecanoic acid is an example. In this reaction, silver oxide oxidizes the aldehydes that are the products of the breakdown of 1-tridecene ozonide.

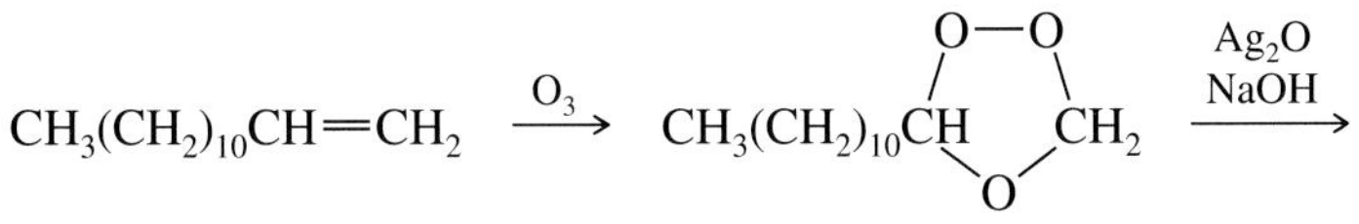

1-tridecene

1-tridecene ozonide

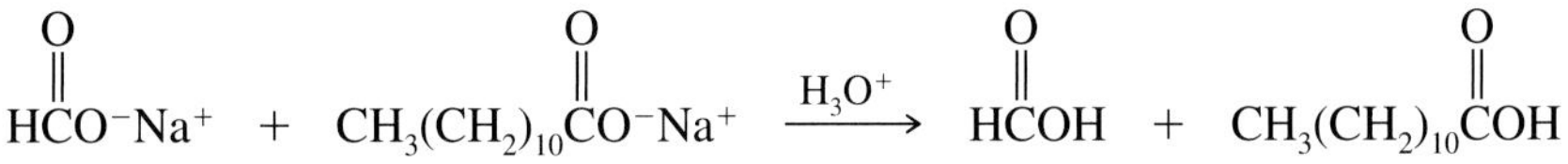

| sodium formate | sodium dodecanoate | formic acid | dodecanoic acid 94% |

Oxidative cleavage reactions give rise to carboxylic acids having fewer carbon atoms than the starting alkene does, unless, of course, a cyclic alkene is the starting material.

# B. Reactions of Organometallic Reagents with Carbon Dioxide

Organometallic reagents react with carbon dioxide to give salts of carboxylic acids. The salt is treated with a strong mineral acid to recover the carboxylic acid. The Grignard reaction is used to prepare acids that have one more carbon atom than the alkyl or aryl halide used to make the organomagnesium reagent. The conversion of 2-chlorobutane to 2-methylbutanoic acid illustrates this synthesis.

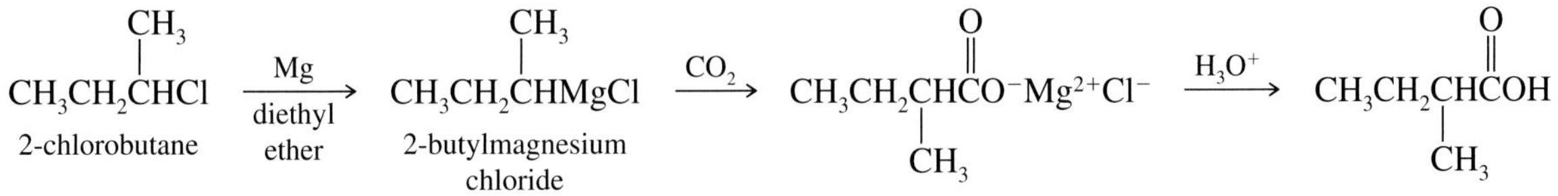

2-chlorobutane  diethyl ether  2-butylmagnesium chloride

2-methylbutanoic acid ~77%

Aromatic as well as alkyl Grignard reagents undergo these reactions. For example, 1-bromonaphthalene is converted to 1-naphthoic acid.

**Study Guide**
**Concept Map 15.2**

1-bromonaphthalene → (Mg, diethyl ether) → 1-naphthylmagnesium bromide → (CO$_2$) → → (H$_3$O$^+$) → 1-naphthoic acid

---

**Problem 15.11**    Assign structures to all compounds symbolized by a letter in the following equations.

(a) CH$_3$—⟨C$_6$H$_4$⟩—Br $\xrightarrow[\text{diethyl ether}]{\text{Mg}}$ A $\xrightarrow{\text{CO}_2}$ B $\xrightarrow{\text{H}_3\text{O}^+}$ C

(b) CH$_3$CH$_2$CH$_2$CHCH$_2$CH$_2$OH $\xrightarrow[\substack{\text{NaOH} \\ \text{H}_2\text{O} \\ \Delta}]{\text{KMnO}_4}$ D $\xrightarrow{\text{H}_3\text{O}^+}$ E
with CH$_3$ substituent

(c) ⟨cyclohexene⟩ $\xrightarrow[\text{chloroform}]{\text{O}_3}$ $\xrightarrow[\substack{\text{NaOH} \\ \text{H}_2\text{O} \\ \Delta}]{\text{Ag}_2\text{O}}$ F $\xrightarrow{\text{H}_3\text{O}^+}$ G → H

(d) ⟨C$_6$H$_5$⟩—CH$_2$CH$_2$CH(=O) $\xrightarrow[\substack{\text{NaOH} \\ \text{H}_2\text{O} \\ \Delta}]{\text{Ag}_2\text{O}}$ I $\xrightarrow{\text{H}_3\text{O}^+}$ J

(e) CH$_3$(CH$_2$)$_9$CH=CH$_2$ $\xrightarrow[\text{chloroform}]{\text{O}_3}$ $\xrightarrow[\text{NaOH}]{\text{H}_2\text{O}_2}$ $\xrightarrow{\text{H}_3\text{O}^+}$ K + L

---

## 15.4   Converting Carboxylic Acids to Acid Chlorides and Acid Anhydrides

### A. Preparation of Acid Chlorides

Nucleophilic substitution reactions occur with difficulty on carboxylic acids themselves (p. 597). The carbonyl group of the acid has to be activated—that is, it must be made more reactive before the acid becomes a good reagent for transferring acyl groups. Such reagents are called **acyl-transfer agents** or **acylating agents.** The activation of a carboxylic acid usually involves converting the hydroxyl group of the carboxyl group into a good leaving group, just as is done with alcohols.

Phosphorus halides and thionyl chloride, the same reagents that convert alcohols into alkyl halides (p. 496), change carboxylic acids into acid halides. For example, when glacial acetic acid is heated with phosphorus trichloride, it is converted to acetyl chloride.

$$3\ \text{CH}_3\overset{\text{O}}{\overset{\|}{\text{C}}}\text{OH}\ +\ \text{PCl}_3\ \xrightarrow{\Delta}\ 3\ \text{CH}_3\overset{\text{O}}{\overset{\|}{\text{C}}}\text{Cl}\ +\ \text{H}_3\text{PO}_3$$

acetic acid    phosphorus trichloride    acetyl chloride 67%    phosphorous acid

*distilled out of reaction mixture*

Thionyl chloride is particularly useful for the preparation of acid halides because it has a low boiling point, 79 °C, and the other products of the reaction, sulfur dioxide and hydrogen chloride, can be easily removed from the reaction mixture because they are gases. Acid chlorides of higher molecular weight can be purified sufficiently for most uses by heating to expel those gases and then distilling out the excess thionyl chloride. Benzoyl chloride is prepared in this way.

$$\underset{\substack{\text{benzoic acid} \\ \text{mp 122 °C}}}{\text{C}_6\text{H}_5\text{COH}} + \underset{\substack{\text{thionyl} \\ \text{chloride} \\ \text{bp 79 °C}}}{\text{SOCl}_2} \longrightarrow \underset{\substack{\text{benzoyl chloride} \\ \text{bp 198 °C} \\ 91\%}}{\text{C}_6\text{H}_5\text{CCl}} + \text{SO}_2\uparrow + \text{HCl}\uparrow$$

Acid chlorides are attacked by water or other nucleophiles (pp. 613 and 628), so in most cases they are prepared just before they are used in reactions with alcohols or amines. Note, therefore, that the water solutions of hydrohalic acids, such as HCl and HBr, used to convert alcohols into alkyl halides cannot be used to make acid halides.

## B. Preparation of Carboxylic Acid Anhydrides

As the name indicates, carboxylic acid anhydrides are compounds that are derived from carboxylic acids by the loss of water. Many of them, especially cyclic anhydrides formed from dicarboxylic acids, can actually be prepared by heating the acid and driving off the water. Acetic anhydride, a compound of great commercial importance, is prepared industrially by heating acetic acid to 800 °C.

$$\underset{\text{acetic acid}}{2 \text{ CH}_3\text{COH}} \xrightarrow[\substack{\text{quartz tube} \\ \text{porcelain chips} \\ \Delta \\ 800 \text{ °C}}]{} \underset{\text{acetic anhydride}}{\text{CH}_3\text{COCCH}_3} + \text{H}_2\text{O}\uparrow$$

Acetic anhydride is used as a dehydrating agent. Other acid anhydrides that have boiling points higher than that of acetic acid can be prepared by heating the required acid with acetic anhydride and distilling out acetic acid as it forms. For example, dodecanoic anhydride is prepared by heating dodecanoic acid with an excess of acetic anhydride.

$$\underset{\substack{\text{dodecanoic acid} \\ \text{bp 225°/100 mm} \\ \text{mp 44 °C}}}{2 \text{ CH}_3(\text{CH}_2)_{10}\text{COH}} + \underset{\substack{\text{acetic anhydride} \\ \text{bp 138 °C}}}{\text{CH}_3\text{COCCH}_3} \xrightarrow[\substack{\Delta \\ 6-8 \text{ h}}]{} \underset{\substack{\text{dodecanoic} \\ \text{anhydride} \\ \text{mp 42 °C}}}{\begin{array}{c}\text{CH}_3(\text{CH}_2)_{10}\text{C} \\ \diagdown \\ \diagup \\ \text{CH}_3(\text{CH}_2)_{10}\text{C}\end{array}\text{O}} + \underset{\substack{\text{acetic acid} \\ \text{bp 118 °C}}}{2 \text{ CH}_3\text{COH}}$$

As mentioned earlier, cyclic anhydrides, especially those having a five- or six-membered anhydride ring, form easily. Consider, for example, the preparation of maleic anhydride from maleic acid.

## ONE SMALL STEP

Acid chlorides may be transformed into other acid halides by nucleophilic substitution reactions.

**PROBLEM:** Propose a reagent and a mechanism for the following transformation.

$$\text{CH}_3\overset{\text{O}}{\overset{\|}{\text{C}}}-\text{Cl} \longrightarrow \text{CH}_3\overset{\text{O}}{\overset{\|}{\text{C}}}-\text{I}$$

**Hint:** How would you carry out this transformation if you were working with alkyl halides instead of acyl halides?

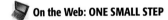

**On the Web: ONE SMALL STEP**

$$ \text{maleic acid} \quad \xrightarrow[\Delta]{\text{1,1,2,2-tetrachloroethane}} \quad \text{maleic anhydride} \quad + \quad H_2O $$

maleic acid

maleic
anhydride
89%

*distilled out
of the reaction
mixture with
the solvent*

With all the above preparations, some method to remove water from the reaction mixture must be used in order to obtain a high yield of the acid anhydride. Otherwise, the reverse reaction, the hydrolysis of an anhydride by water (p. 613), will take place. In the preparation of acetic anhydride, the water is vaporized because of the high temperature and distills out of the reaction mixture. The water formed when dodecanoic acid is converted into its anhydride reacts with the readily available acetic anhydride that is present in excess and is thus prevented from hydrolyzing the dodecanoic anhydride. In the preparation of maleic anhydride, 1,1,2,2-tetrachloroethane, a solvent that codistills with water, is used to remove the water from the reaction mixture at relatively low temperatures.

Another way of making an acid anhydride is to carry out a nucleophilic substitution reaction on an acid chloride using the salt of an acid. This reaction is illustrated by the preparation of 2-methylpropanoic anhydride.

$$ CH_3CHCCl \quad + \quad Na^+{}^-OCCHCH_3 \quad \longrightarrow \quad CH_3CHCOCCHCH_3 \quad + \quad Na^+Cl^- $$

2-methylpropanoyl
chloride

sodium 2-methyl-
propanoate

2-methylpropanoic
anhydride

**Problem 15.12**    Complete the following equations.

(a) (phenyl)—$CH_2CH_2COH \xrightarrow[\Delta]{SOCl_2}$

(b) $CH_3CH_2CH_2COH \xrightarrow{PCl_3}$

(c) $CH_3CH_2CH_2CCl + Na^+{}^-OCCH_2CH_2CH_3 \longrightarrow$

(d) $HOCCH_2CH_2COH \xrightarrow{CH_3COCCH_3}$

(e) $CH_3CH_2COH \xrightarrow[\substack{\Delta \\ 650\ °C}]{clay}$

(f) $HOC(CH_2)_5COH \xrightarrow[\Delta]{excess\ SOCl_2}$

**Problem 15.13**    The following amides were needed for a study on the stereochemistry of Michael reactions (p. 710). They were each made from an acid chloride and an amine (Section 15.7). Which acid chloride and which amines would be needed to prepare these amides?

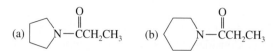

(a) (pyrrolidine)$N-CCH_2CH_3$    (b) (piperidine)$N-CCH_2CH_3$

## 15.5 Hydrolysis Reactions. Acylation of Water

### A. Hydrolysis Reactions

All acid derivatives react with water to give carboxylic acids. Low-molecular-weight acid chlorides and acid anhydrides, such as acetyl chloride and acetic anhydride, react rapidly with water.

$$\underset{\text{acetyl chloride}}{CH_3\overset{O}{\overset{\|}{C}}Cl} + H_2O \longrightarrow \underset{\text{acetic acid}}{CH_3\overset{O}{\overset{\|}{C}}OH} + HCl$$

$$\underset{\text{acetic anhydride}}{CH_3\overset{O}{\overset{\|}{C}}O\overset{O}{\overset{\|}{C}}CH_3} + H_2O \longrightarrow \underset{\text{acetic acid}}{2\ CH_3\overset{O}{\overset{\|}{C}}OH}$$

Each of these compounds has a good leaving group, chloride ion for the acid chloride and acetate ion for the anhydride. Water is the nucleophile in these reactions.

**Problem 15.14**  Using the mechanism on page 595 as a guide, write a detailed mechanism for the reaction of acetyl chloride with water.

Esters are much less reactive toward nucleophilic substitution than either acid chlorides or acid anhydrides (p. 597). An ester has to be heated with water and an acid or a base as a catalyst to be hydrolyzed to a carboxylic acid. For example, when ethyl acetate is heated with water in the presence of acid, it is converted to acetic acid and ethanol.

$$\underset{\text{ethyl acetate}}{CH_3\overset{O}{\overset{\|}{C}}OCH_2CH_3} + H_2O \underset{\underset{\Delta}{H_2SO_4}}{\rightleftarrows} \underset{\text{acetic acid}}{CH_3\overset{O}{\overset{\|}{C}}OH} + \underset{\text{ethanol}}{CH_3CH_2OH}$$

This type of reaction is reversible. Acids react with alcohols to give esters (p. 622). Using a large excess of water pushes the hydrolysis reaction toward completion.

Hydrolysis of an ester using a base gives the salt of a carboxylic acid. The acid itself is not formed until the reaction mixture is acidified with a stronger acid such as hydrochloric or sulfuric acid. An example is the hydrolysis of methyl 3-methylfuroate to the corresponding acid.

methyl 3-methylfuroate          methanol     sodium 3-methylfuroate          3-methylfuroic acid
                                                                                          ~90%

This reaction goes essentially to completion because one of the reactants necessary for the reverse reaction, the carboxylic acid, is removed from the reaction mixture as its salt.

Nitriles (prepared by nucleophilic substitution reactions of alkyl halides with cyanide ion) also can be hydrolyzed to carboxylic acids. Thus alkyl halides can be converted to carboxylic acids by these two steps. The conversion of benzyl chloride to phenylacetonitrile and then to phenylacetic acid, shown below, is an example. Amides are formed as intermediates in the hydrolysis reaction and may be isolated under certain conditions. For example, phenylacetonitrile is converted to phenyl-acetamide if treated with hydrochloric acid for 1 hour at 40 °C but to phenylacetic acid if heated to boiling with aqueous sulfuric acid for 3 hours.

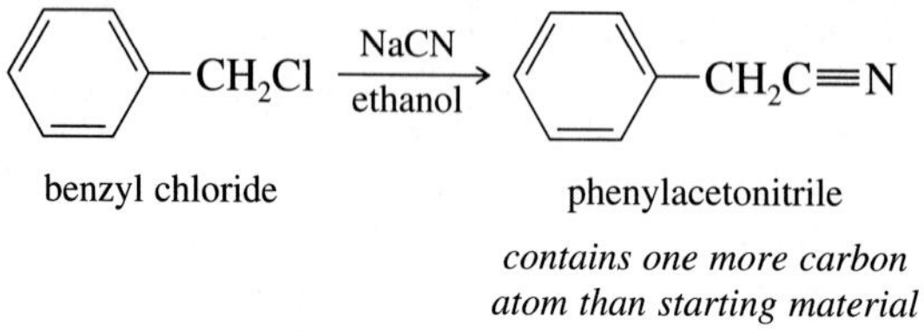

benzyl chloride                    phenylacetonitrile

*contains one more carbon*
*atom than starting material*

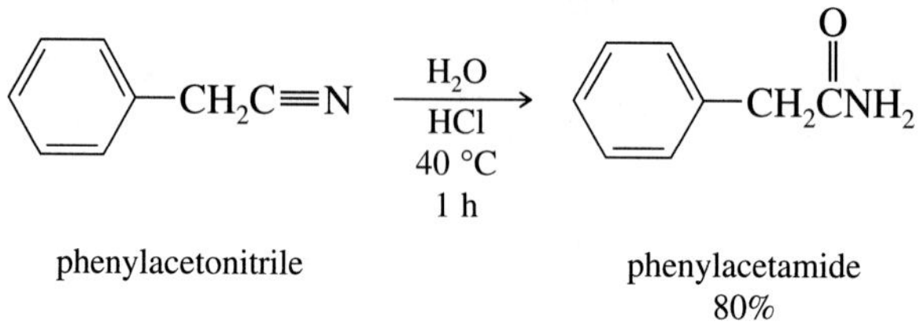

phenylacetonitrile                          phenylacetamide
                                                80%

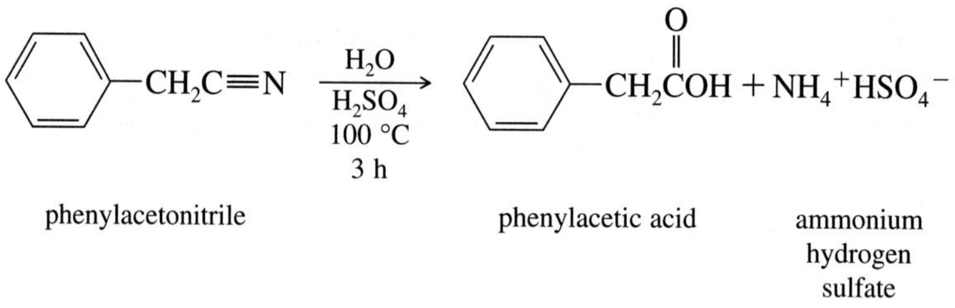

phenylacetonitrile                 phenylacetic acid        ammonium
                                                           hydrogen
                                                           sulfate

Nitriles or amides are also hydrolyzed in base. For example, chloroacetic acid in the form of its sodium salt is converted to sodium cyanoacetate. The nitrile function is then hydrolyzed in base, and malonic acid is recovered by careful acidification of the solution containing its sodium salt.

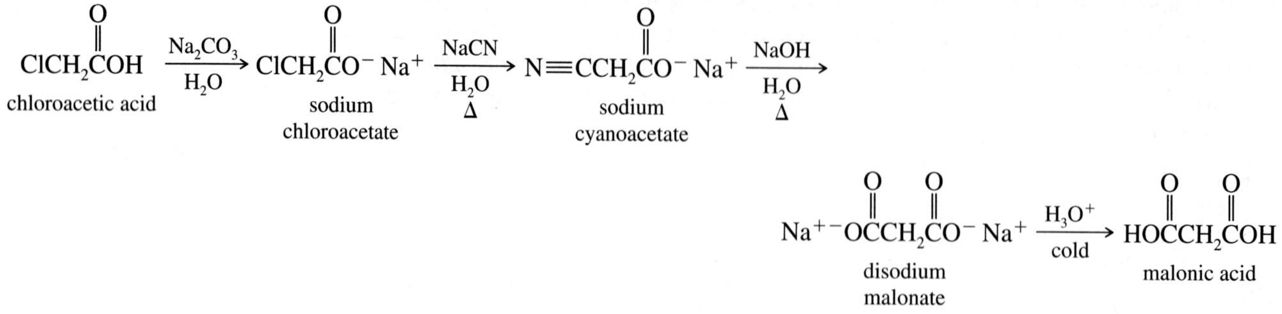

chloroacetic acid        sodium              sodium
                         chloroacetate       cyanoacetate

                         disodium            malonic acid
                         malonate

**Problem 15.15**    In the above sequence of reactions, why is chloroacetic acid converted to its salt before sodium cyanide is added to the reaction mixture? (*Hint:* The table of p$K_a$ values inside the front cover of the book may be helpful.)

## B. The Biological Importance of the Hydrolysis of Amides. The Cleavage of Peptides

The hydrolysis of amides to amines and carboxylic acids is one of the most important types of chemical reactions. Proteins are large molecules held together chiefly by amide groups known as **peptide linkages.** A large part of Chapter 23 is devoted to a study of the peptide bond and the structure of proteins. Digestion breaks down proteins into smaller units by the hydrolytic cleavage of amide bonds. The smallest units resulting from such hydrolysis of proteins are the amino acids; the amino group of one amino acid forms an amide bond with the carboxylic acid of another one. Later in this chapter (Section 15.8, p. 629) we shall look at reactions that create peptide bonds. Molecules made up of a few amino acids held together by amide bonds are called **peptides.** Glycylglycylglycine is a simple peptide made up of three units of the amino acid glycine. Hydrolysis of the peptide with aqueous acid gives the free amino acid units.

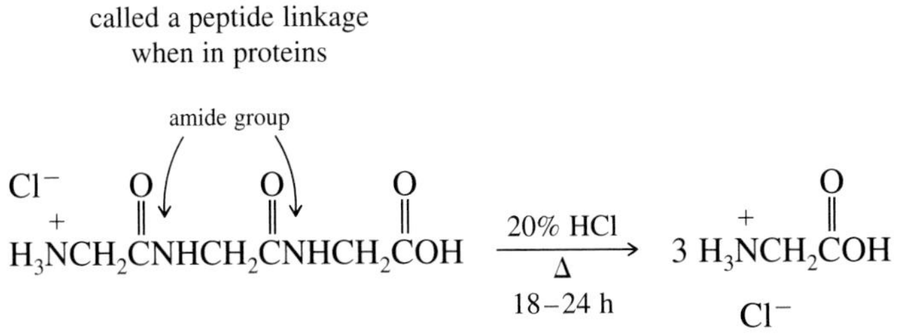

called a peptide linkage
when in proteins

amide group

$$Cl^- \quad + \quad H_3NCH_2CNHCH_2CNHCH_2COH \xrightarrow[\substack{\Delta \\ 18-24\ h}]{20\%\ HCl} 3\ H_3NCH_2COH$$

glycylglycylglycine

*a peptide made up of three units of glycine*

glycine hydrochloride

In the body, the cleavage of the peptide linkages occurs under very mild conditions and is catalyzed by enzymes (pp. 994 and 1012).

**⊘ Study Guide**
**Concept Map 15.3**

## C. Mechanisms of Hydrolysis Reactions

The nucleophilic substitution reaction involving acetyl chloride was shown on page 595 as a two-step reaction with the formation of a tetrahedral intermediate instead of as a simple $S_N2$ reaction. Evidence for the formation of such an intermediate comes from studies of the hydrolysis of isotopically labeled esters. Ethyl benzoate labeled with $^{18}O$ in the carbonyl group is allowed to react with water in the presence of acid or base. The reaction is stopped before all the ester has been hydrolyzed, and it is found that some of the unreacted ester molecules no longer contain $^{18}O$.

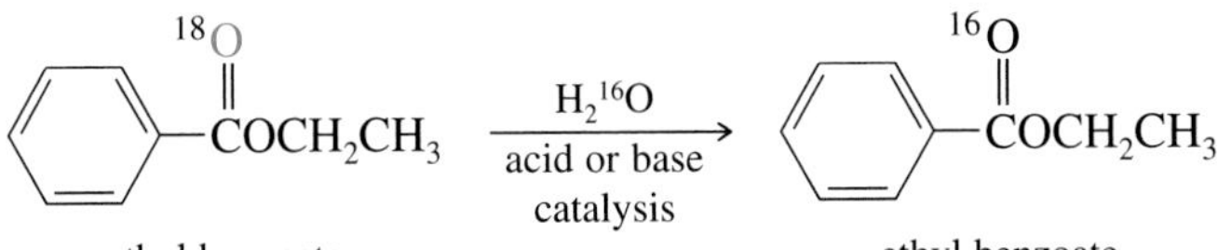

ethyl benzoate

*labeled with $^{18}O$*

ethyl benzoate

*recovered from the
reaction mixture; has
lost its label*

This observation can be explained only by assuming the reversible formation of a symmetrical intermediate in which $^{18}O$ and $^{16}O$ are both bonded to the carbon atom of the carbonyl group. The formation of such an intermediate for the acid-catalyzed hydrolysis reaction is shown below.

**VISUALIZING THE REACTION**

**Formation of a Symmetrical Intermediate in the Acid-Catalyzed Hydrolysis of an Ester**

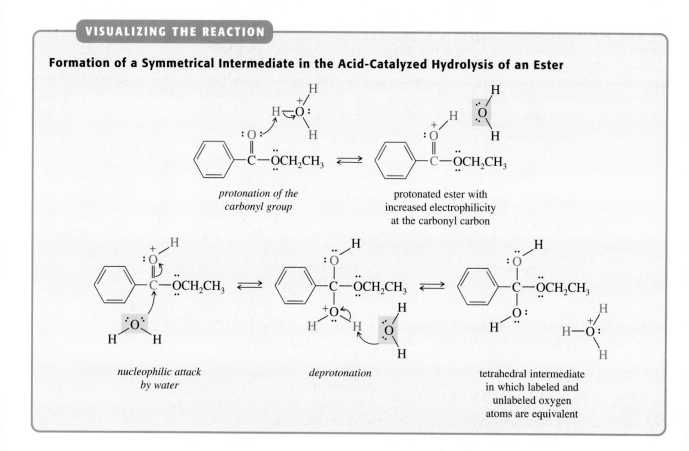

In aqueous acid, the most abundant nucleophile is the water molecule itself, which is a weak nucleophile. Therefore, the uncatalyzed hydrolysis of an ester in pure water is slow. The reversible protonation of the carbonyl group is the first step in the hydrolysis of an ester in acidic solution. The electrophilicity of the carbonyl group is increased by protonation, and it is attacked by water. Loss of a proton from the resulting intermediate gives a symmetrical tetrahedral intermediate. All these steps are reversible, so some of the tetrahedral intermediate is converted to the ester having the unlabeled oxygen atom.

In the hydrolysis reaction, the tetrahedral intermediate is cleaved to give a carboxylic acid and an alcohol.

**VISUALIZING THE REACTION**

**Acid-Catalyzed Hydrolysis of an Ester**

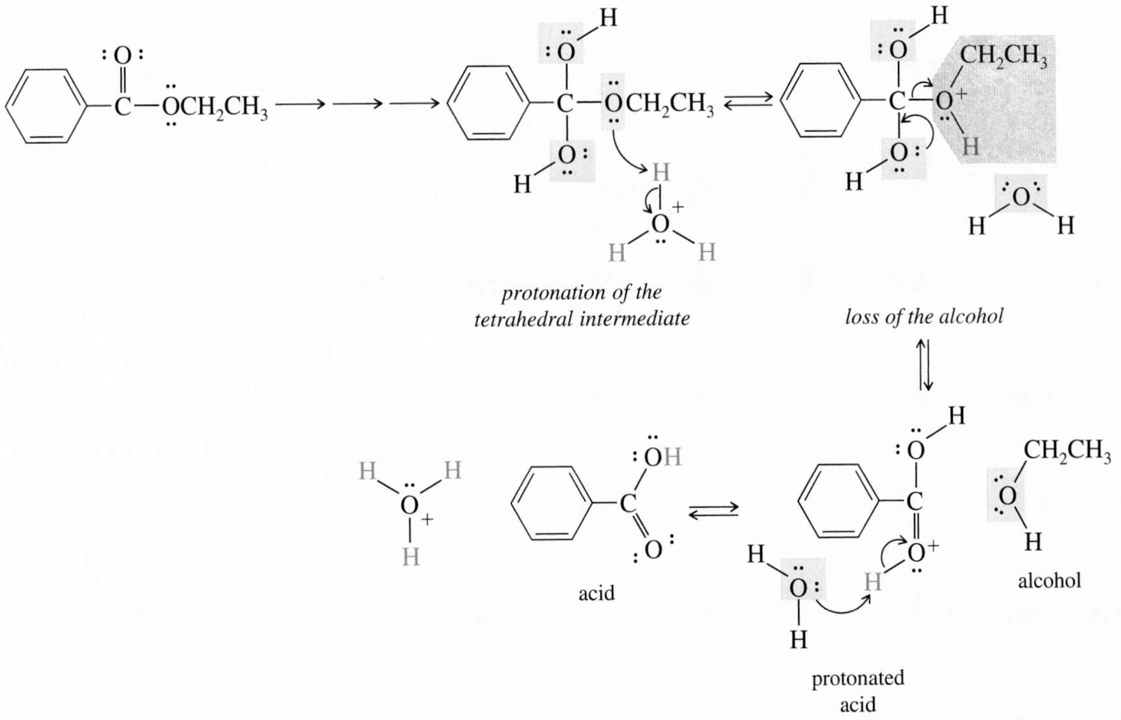

protonation of the
tetrahedral intermediate

loss of the alcohol

acid

protonated
acid

The tetrahedral intermediate has three sites at which it can be protonated. Protonation at either of the hydroxyl groups is simply a reversal of the preceding step and could lead, by loss of water, back to the protonated ester. Only when the oxygen atom of the alkoxyl group is protonated is a further reaction possible. The protonated alkoxyl group is a good leaving group, and the bond between the alkoxyl group and the carbonyl group is cleaved.

**Problem 15.16**   Write a mechanism that shows how the symmetrical tetrahedral intermediate shown in the preceding mechanisms is converted into ethyl benzoate that has no label in the carbonyl group.

Further evidence for the mechanism shown above comes from other studies using isotopic tracers. When pentyl acetate is allowed to stand in water labeled with $^{18}O$ in the presence of sodium hydroxide, none of the isotopic oxygen shows up in the alcohol recovered from the reaction mixture.

$$CH_3\overset{O}{\overset{\|}{C}}OCH_2CH_2CH_2CH_2CH_3 + Na^{+\,18}OH^- \xrightarrow[H_2^{\,18}O]{} CH_3CH_2CH_2CH_2CH_2OH + CH_3\overset{^{18}O}{\overset{\|}{C}}{-}^{18}O^-\,Na^+$$

pentyl acetate

1-pentanol
containing no $^{18}O$

sodium acetate
containing $^{18}O$
distributed between
the two oxygen atoms

This experimental evidence excludes a mechanism that involves an $S_N2$ reaction between hydroxide ion and the first carbon of the pentyl group.

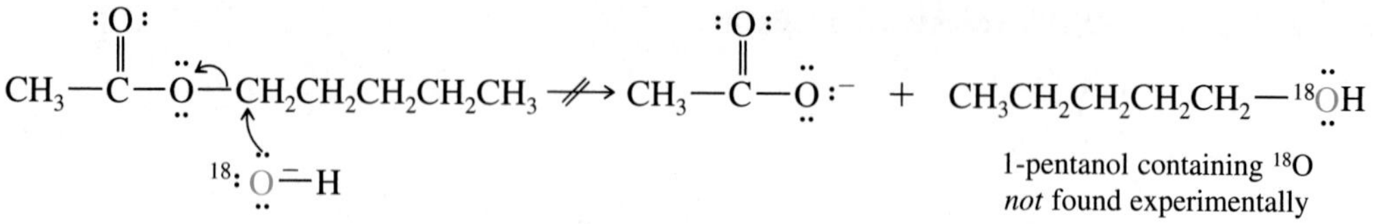

$$CH_3-\overset{\overset{\displaystyle :O:}{\|}}{C}-\ddot{O}-CH_2CH_2CH_2CH_2CH_3 \not\rightarrow CH_3-\overset{\overset{\displaystyle :O:}{\|}}{C}-\ddot{O}:^- + CH_3CH_2CH_2CH_2CH_2-{}^{18}\ddot{O}H$$

$${}^{18}:\!\overset{..}{O}\!=\!H$$

1-pentanol containing $^{18}O$
*not* found experimentally

The observed results are consistent with an attack by hydroxide ion containing $^{18}O$ at the carbonyl group, followed by cleavage of the bond between the carbonyl group and the alkoxyl group.

---

**Problem 15.17**    $^{18}O$ is shown above as being present as either one of the two oxygen atoms in acetate ion after the hydrolysis of pentyl acetate in $H_2{}^{18}O$. Why is this so?

---

In aqueous base, the best nucleophile present is the hydroxide ion. It attacks the carbonyl group directly to give a tetrahedral intermediate, which breaks down into an acid and an alkoxide ion. An alkoxide ion is not a good leaving group for an $S_N2$ reaction. The bond energy of a carbonyl group is so high (~179 kcal/mol), however, that there is a powerful driving force for the expulsion of the alkoxide ion and the formation of the carbonyl group in this type of reaction. The last step of the reaction goes essentially to completion because alkoxide ion (or hydroxide ion) is a much stronger base than acetate ion is.

---

**VISUALIZING THE REACTION**

**Basic Hydrolysis of an Ester**

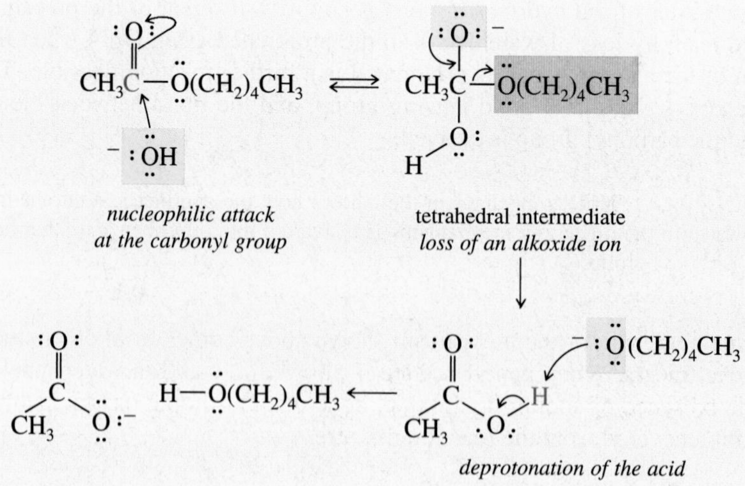

nucleophilic attack
at the carbonyl group

tetrahedral intermediate
loss of an alkoxide ion

deprotonation of the acid

---

⊘ **Study Guide**
**Concept Map 15.4**

The steps shown for the acid-catalyzed and base-catalyzed hydrolysis of esters apply to the nucleophilic substitution reactions of all acid derivatives.

On the Web: ONE SMALL STEP

**Problem 15.18** Write a detailed mechanism for the acid-catalyzed hydrolysis of acetamide to acetic acid and ammonium ion. (*Hint:* A close look at the mechanism for the acid-catalyzed hydrolysis of an ester on pages 615–617 will be helpful.)

**Problem 15.19** In sodium hydroxide solution, acetamide is hydrolyzed to sodium acetate and ammonia. Write a detailed mechanism for this reaction.

**Problem 15.20** How would you carry out the following transformations?

(a) 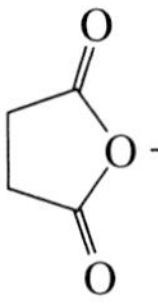 [structure of succinic anhydride] $\longrightarrow$ $Na^+ {}^-OCCH_2CH_2CO^- {}^+Na$

(b) $CH_3CH_2CH_2CH \overset{O}{\underset{}{\parallel}} \longrightarrow CH_3CH_2CH_2CHCOH$ with $CH_3$ branch

**Problem 15.21** Write the mechanism for the acid-catalyzed conversion of acetonitrile, $CH_3CN$, to acetamide, $CH_3CONH_2$. The mechanism for the addition of water to the carbon–nitrogen bond is very similar to that for the addition of water to the carbon–carbon triple bond (The Art of Solving Problems, 335).

**Problem 15.22** Using acetonitrile, $CH_3CN$, write a mechanism for the basic hydrolysis of a nitrile to an amide.

## D. The Art of Solving Problems

**PROBLEM:** How would you carry out the following transformation?

$$CH_3CHCH_2CH_2OH \longrightarrow CH_3CHCH_2CH_2COH$$

with $CH_3$ branches and an $O$ (carbonyl) on the product acid.

### Solution

1. What are the connectivities of the two compounds? How many carbon atoms does each contain? Are there any rings? What are the positions of branches and functional groups on the chains?

   The main chain of the carbon skeleton of the starting material is four carbons long, with a methyl group on the third carbon and a hydroxyl group on the first carbon. The main chain of the product is five carbons long, with a branch at the fourth carbon atom. The chain in the reagent has been lengthened by one carbon atom, which appears as a carboxylic acid group.

2. How do the functional groups change in going from starting material to product? Does the starting material have a good leaving group?

   A hydroxyl group has been converted into a carboxylic acid group.

$$-OH \longrightarrow -COH$$

with $O$ carbonyl.

   The starting material does not have a good leaving group, but the hydroxyl group can be converted into one.

3. Is it possible to dissect the structures of the starting material and product to see which bonds must be broken and which formed?

$$\underset{\overset{|}{CH_3}}{CH_3}CHCH_2CH_2 \overset{+}{\underset{\uparrow}{}} \overset{\overset{O}{\parallel}}{COH} \longleftarrow \underset{\overset{|}{CH_3}}{CH_3}CHCH_2CH_2 \overset{+}{\underset{\uparrow}{}} OH$$

new bond formed        bond to be broken

4. New bonds are created when an electrophile reacts with a nucleophile. Do we recognize any part of the product molecule as coming from a good nucleophile or an electrophilic addition?

   A carboxylic acid can be prepared from a nitrile.

$$\overset{\overset{O}{\parallel}}{-COH} \longleftarrow -C\equiv N$$

   The cyano group comes from cyanide ion, a good nucleophile.

5. What type of compound would be a good precursor to the product? The corresponding cyano compound would be a good precursor.

$$\underset{\overset{|}{CH_3}}{CH_3}CHCH_2CH_2 - \overset{\overset{O}{\parallel}}{COH} \longleftarrow \underset{\overset{|}{CH_3}}{CH_3}CHCH_2CH_2 - C\equiv N$$

   The best precursor for the cyano compound would have a good leaving group where the $-C\equiv N$ group is.

$$\underset{\overset{|}{CH_3}}{CH_3}CHCH_2CH_2 - X \quad \overset{\text{leaving group}}{\swarrow}$$

6. After this step, do we see how to get from starting material to product? If not, we need to analyze the structure obtained in step 5 by applying questions 4 and 5 to it.

   The precursor with a good leaving group can be obtained directly from the starting primary alcohol, so a complete synthetic pathway is as follows:

$$\underset{\overset{|}{CH_3}}{CH_3}CHCH_2CH_2\overset{\overset{O}{\parallel}}{COH} \underset{\Delta}{\overset{H_3O^+}{\longleftarrow}} \underset{\overset{|}{CH_3}}{CH_3}CHCH_2CH_2C\equiv N$$

$$\uparrow NaC\equiv N$$

$$\underset{\overset{|}{CH_3}}{CH_3}CHCH_2CH_2OH \underset{\text{pyridine}}{\overset{TsCl}{\longrightarrow}} \underset{\overset{|}{CH_3}}{CH_3}CHCH_2CH_2OTs$$

   There is at least one other set of answers to questions 4 through 6.

4. New bonds are created when an electrophile reacts with a nucleophile. Do we recognize any part of the product molecule as coming from a good nucleophile or an electrophilic addition?

   A carboxylic acid having one more carbon atom than the starting material can also be prepared by the reaction of an organometallic reagent with carbon dioxide.

$$\overset{O}{\underset{\|}{RCOH}} \longleftarrow \overset{O}{\underset{\|}{RCOMgX}} \longleftarrow RMgX + CO_2$$

The Grignard reagent is a good nucleophile, and the carbon atom of carbon dioxide is electrophilic.

5. What type of compound would be a good precursor to the product?

A good precursor would be an alkyl halide, from which a Grignard reagent can be prepared.

$$\overset{CH_3}{\underset{|}{CH_3CHCH_2CH_2Br}}$$

6. After this last step, do we see how to get from starting material to product? If not, we need to analyze the structure obtained in step 5 by applying questions 4 and 5 to it.

Another complete synthesis is as follows:

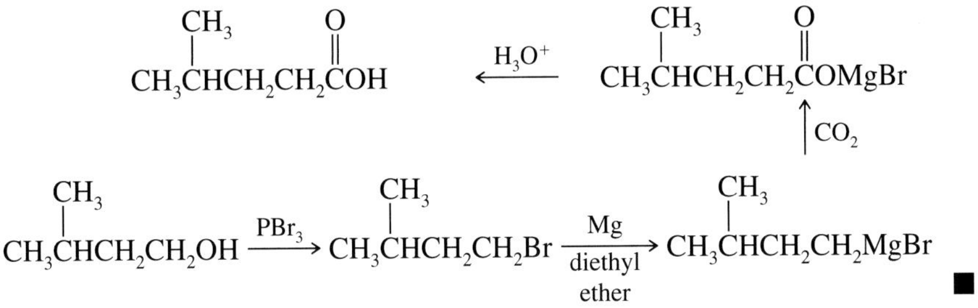

$$\overset{CH_3}{\underset{|}{CH_3CHCH_2CH_2}}\overset{O}{\underset{\|}{COH}} \xleftarrow{H_3O^+} \overset{CH_3}{\underset{|}{CH_3CHCH_2CH_2}}\overset{O}{\underset{\|}{COMgBr}}$$

$$\uparrow CO_2$$

$$\overset{CH_3}{\underset{|}{CH_3CHCH_2CH_2OH}} \xrightarrow{PBr_3} \overset{CH_3}{\underset{|}{CH_3CHCH_2CH_2Br}} \xrightarrow[\substack{\text{diethyl} \\ \text{ether}}]{Mg} \overset{CH_3}{\underset{|}{CH_3CHCH_2CH_2MgBr}} \quad ■$$

## 15.6 Esterification. Acylation of Alcohols

### A. Acid Chlorides and Acid Anhydrides as Acylating Agents

Nucleophilic substitution reactions of acid chlorides and acid anhydrides with alcohols give esters. For example, cyclohexanol is converted to a 3,5-dinitrobenzoic ester by reaction with 3,5-dinitrobenzoyl chloride.

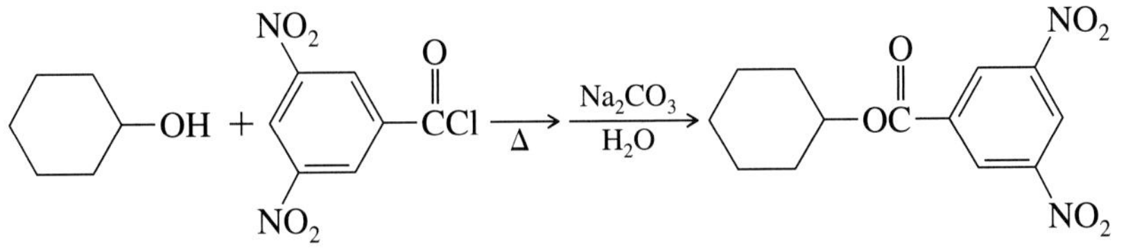

cyclohexanol
bp 160 °C

3,5-dinitrobenzoyl
chloride

cyclohexyl 3,5-dinitrobenzoate
mp 112 °C

**Problem 15.23** The **Schotten-Baumann reaction** is another way of converting an alcohol into its ester with benzoyl chloride. The alcohol and the acid chloride are mixed in 10% aqueous sodium hydroxide. Write an equation for this reaction using *n*-butyl alcohol. Show any other products that will be formed under these conditions.

Acid anhydrides are generally less reactive than acid chlorides, but do react readily with alcohols. Acetic anhydride, for example, is often used to make derivatives of natural products such as cholesterol and sugars.

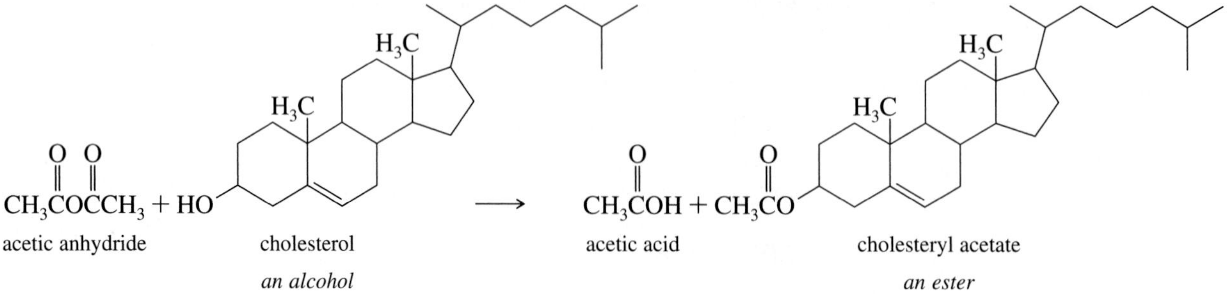

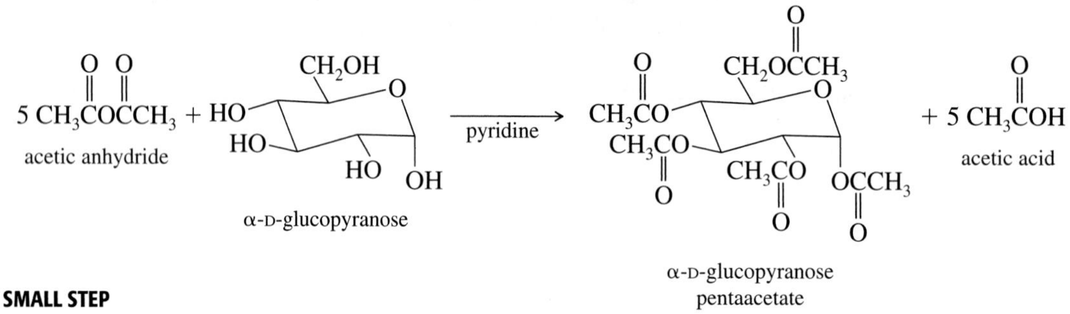

$$CH_3COCCH_3 + HO\text{—} \quad \longrightarrow \quad CH_3COH + CH_3CO\text{—}$$

acetic anhydride     cholesterol     acetic acid     cholesteryl acetate

*an alcohol*     *an ester*

$$5\ CH_3COCCH_3 + HO\text{—} \xrightarrow{\text{pyridine}} \quad + 5\ CH_3COH$$

acetic anhydride     α-D-glucopyranose     acetic acid

α-D-glucopyranose
pentaacetate

**On the Web: ONE SMALL STEP**

**Problem 15.24**    Complete the following equations.

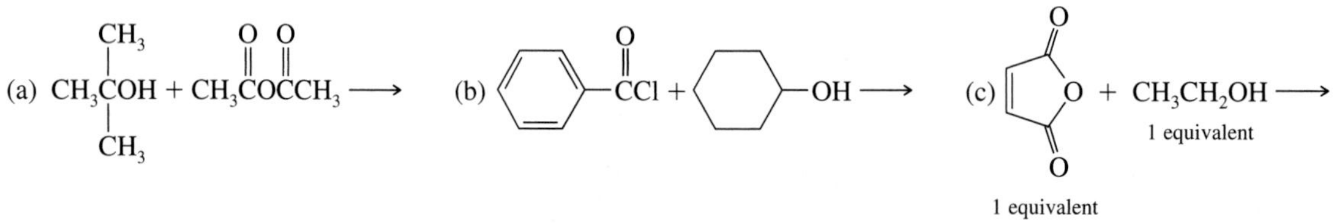

(a) $CH_3COH + CH_3COCCH_3 \longrightarrow$

(b)    benzoyl chloride + cyclohexanol $\longrightarrow$

(c)    maleic anhydride + $CH_3CH_2OH \longrightarrow$
   1 equivalent

1 equivalent

## B. Esterification

Carboxylic acids react with alcohols in the presence of a strong acid, such as dry hydrogen chloride, concentrated sulfuric acid, or *p*-toluenesulfonic acid, to give esters. The reaction reaches equilibrium. For example, when 1 mole of acetic acid and 1 mole of ethanol are mixed, 0.667 mole of ester and 0.667 mole of water are produced, and 0.333 mole of acid and 0.333 mole of alcohol remain at equilibrium.

$$CH_3COH + CH_3CH_2OH \underset{H_2SO_4}{\rightleftharpoons} CH_3COCH_2CH_3 + H_2O$$

| | | | | |
|---|---|---|---|---|
| At start: | 1 mole | 1 mole | 0 mole | 0 mole |
| At equilibrium: | 0.333 mole | 0.333 mole | 0.667 mole | 0.667 mole |

The yield of ester can be increased either by removing one of the products of the reaction as it is formed or by increasing the concentration of one of the reactants. In practice, either water is distilled out of the reaction mixture or a greatly increased amount of alcohol, which is relatively inexpensive if it is methanol or ethanol, is used. In some particularly tough cases, both measures are taken. For example, adipic acid is converted into its diester by heating it with ethanol in the presence of sulfuric acid and toluene. Toluene, ethanol, and water form a constant-boiling mixture that allows for the removal of water from the reaction mixture as it is formed.

$$HOC(CH_2)_4COH + 2 CH_3CH_2OH \xrightarrow[\substack{H_2SO_4 \\ \Delta}]{} CH_3CH_2OC(CH_2)_4COCH_2CH_3 + 2 H_2O$$

| adipic acid | ethanol | toluene | diethyl adipate |
|---|---|---|---|
| 1 mole | 3 moles | Δ | 96% |

**Problem 15.25** Assign a structure to each compound indicated by a letter in the following equations.

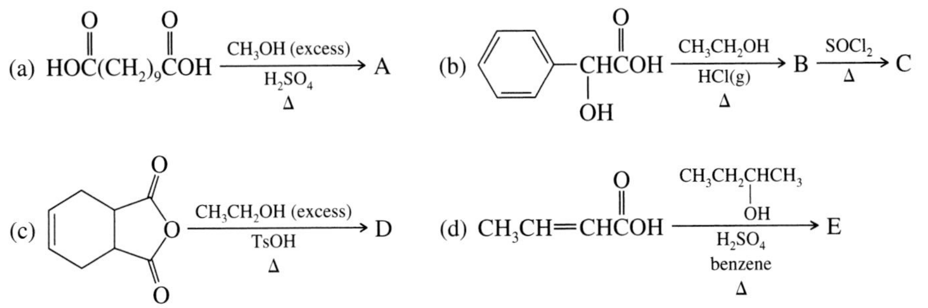

(a) $HOC(CH_2)_9COH \xrightarrow[\substack{H_2SO_4 \\ \Delta}]{CH_3OH \text{ (excess)}} A$

(b) a benzene ring—$CHCOH$ (with OH below the CH) $\xrightarrow[\substack{HCl(g) \\ \Delta}]{CH_3CH_2OH} B \xrightarrow[\Delta]{SOCl_2} C$

(c) (bicyclic anhydride) $\xrightarrow[\substack{TsOH \\ \Delta}]{CH_3CH_2OH \text{ (excess)}} D$

(d) $CH_3CH=CHCOH \xrightarrow[\substack{H_2SO_4 \\ benzene \\ \Delta}]{CH_3CH_2CHCH_3 \text{ (OH)}} E$

**Problem 15.26** Methyl benzoate is prepared by heating 10 g of benzoic acid with 25 mL of methanol and 3 mL of concentrated sulfuric acid. (a) Write an equation for the equilibrium reaction. (b) How would you remove excess methanol, sulfuric acid, and unreacted benzoic acid from the reaction mixture in order to isolate pure methyl benzoate? (*Hint:* Think about the physical and chemical properties of each reagent.)

## C. The Mechanism of the Esterification Reaction

The steps in the mechanism for the formation of an ester from an acid and an alcohol are the reverse of the steps for the acid-catalyzed hydrolysis of an ester (pp. 615–617). As the equation for the equilibrium reaction above indicates, the reaction can go in either direction depending on the conditions used. A carboxylic acid does not react with an alcohol unless a strong acid is used as a catalyst. Protonation makes the carbonyl group more electrophilic and enables it to react with the alcohol, which is a weak nucleophile.

**VISUALIZING THE REACTION**

**Esterification**

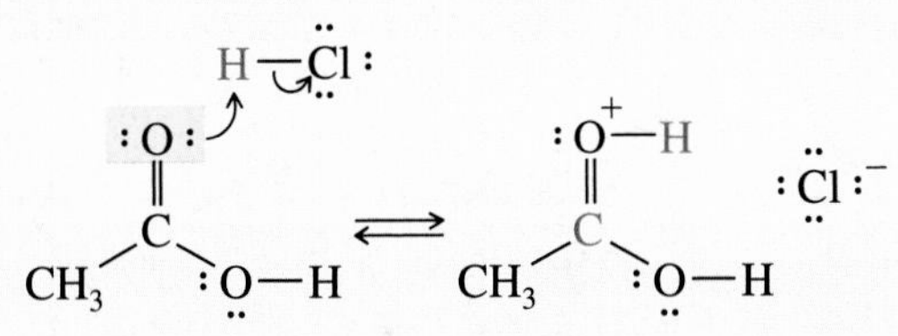

protonation of the carbonyl group

protonated acetic acid with increased electrophilicity at the carbon atom of the carbonyl group

*nucleophilic attack at the carbonyl group*

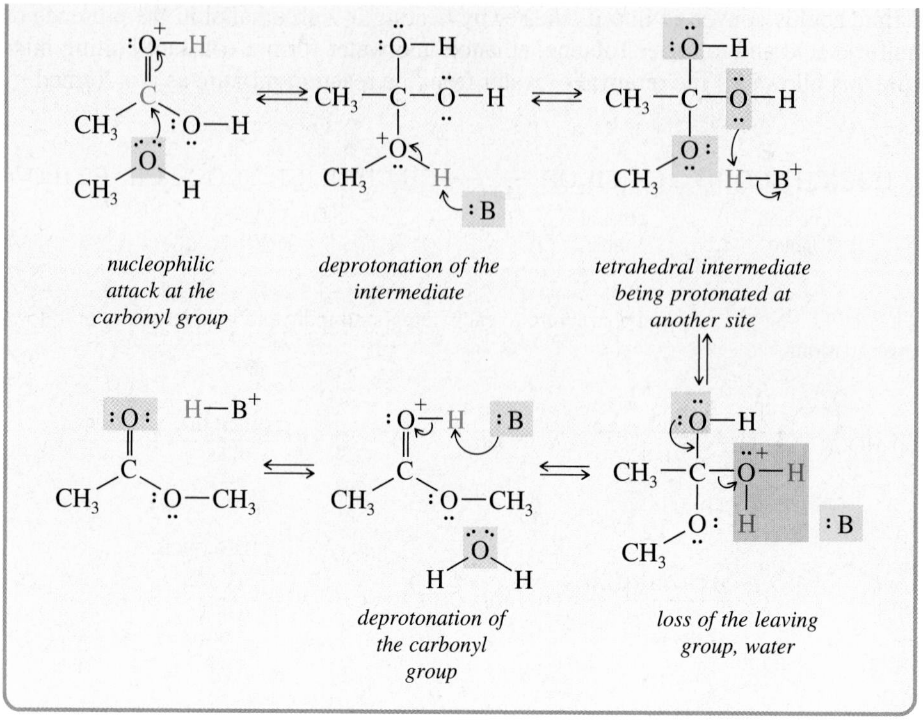

*deprotonation of the intermediate*

*tetrahedral intermediate being protonated at another site*

*deprotonation of the carbonyl group*

*loss of the leaving group, water*

Note that the above reaction has the same two steps described earlier (p. 595). The carbon atom of the carbonyl group is attacked by a nucleophile with formation of a tetrahedral intermediate, and the carbonyl group is regenerated by the loss of a leaving group. Because this reaction is carried out in strongly acidic solution, protonation and deprotonation steps start and finish it.

## D. Lactones

**Lactones** are cyclic esters formed when a carboxyl and a hydroxyl group are present within the same molecule. In fact, the formation of a five- or six-membered ring is so highly favored that reactions leading to a compound in which a carboxyl group and a hydroxyl group are separated by three or four carbon atoms usually give a lactone directly as the product. An example of such a reaction is the reduction of 4-oxopentanoic acid with sodium borohydride. Sodium borohydride reduces the ketone function but not the carboxylic acid (p. 547). The product of the reaction after acidification is the lactone of 4-hydroxypentanoic acid.

4-oxopentanoic acid

sodium salt of 4-hydroxypentanoic acid

lactone of 4-hydroxypentanoic acid

81%

*a γ-lactone*

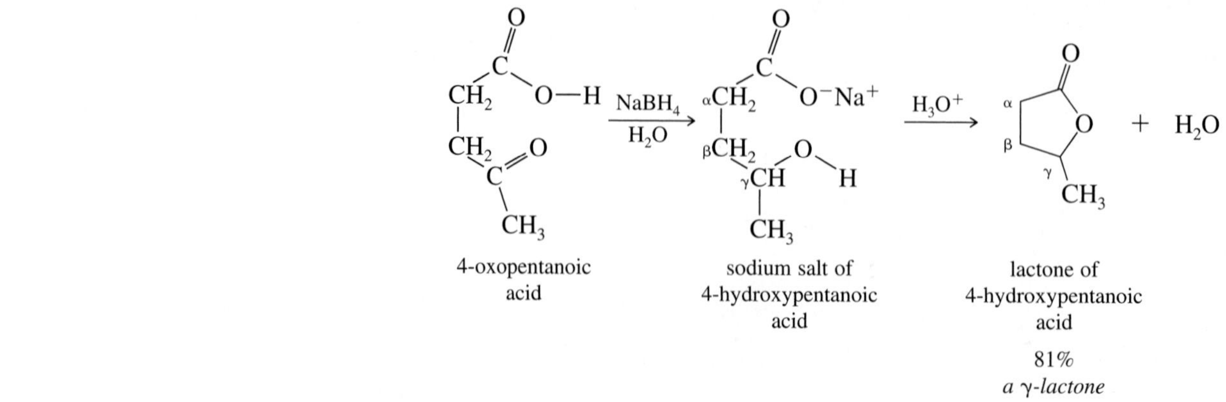

A lactone having a five-membered ring is called a **γ-lactone** because it is formed when an intramolecular reaction takes place between a carboxyl group and the hydroxyl group on the γ-carbon atom of the acid (p. 541). A six-membered or **δ-lactone** is formed when ethyl 5-acetoxypentanoate, for example, synthesized by a nucleophilic substitution reaction of ethyl 5-chloropentanoate, is hydrolyzed.

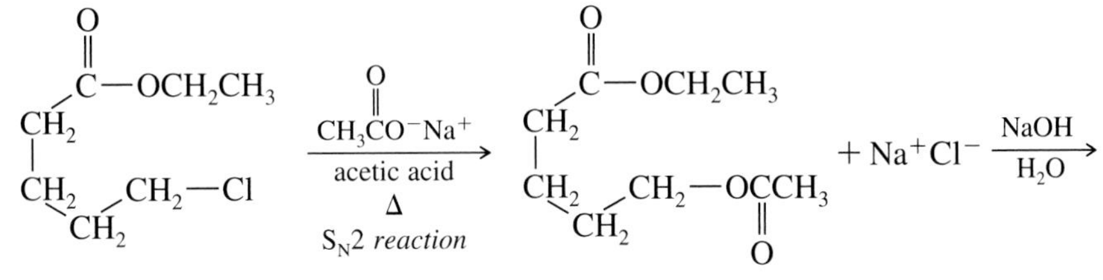

ethyl 5-chloropentanoate $\xrightarrow[\substack{\text{acetic acid} \\ \Delta \\ S_N2 \text{ reaction}}]{CH_3CO^-Na^+}$ ethyl 5-acetoxypentanoate $+ Na^+Cl^- \xrightarrow[H_2O]{NaOH}$

$CH_3CO^-Na^+ + CH_3CH_2OH +$ sodium 5-hydroxypentanoate
*both ester functions hydrolyzed*
$\xrightarrow{H_3O^+}$ lactone of 5-hydroxypentanoic acid 50%

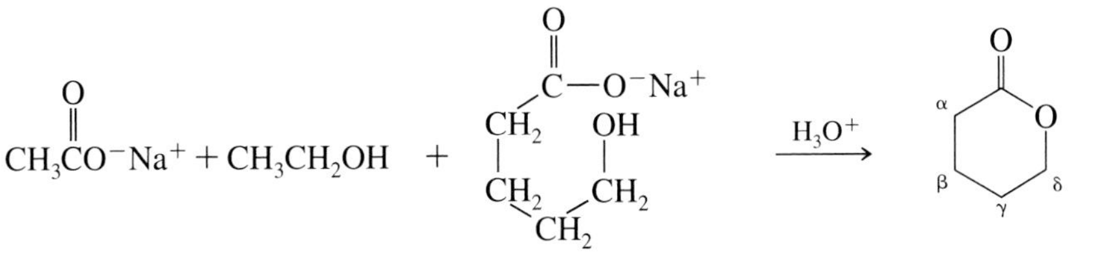

*a δ-lactone*

Large-ring lactones are found in nature. Some of them, as well as the large-ring ketones muscone, isolated from the scent glands of the musk deer, and civetone, from the civet cat, are highly valued by perfume manufacturers. In perfumes, these substances enhance and fix scents derived from other sources. The structures of four of these large-ring compounds, two ketones and two lactones, are shown below.

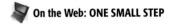
**On the Web: ONE SMALL STEP**

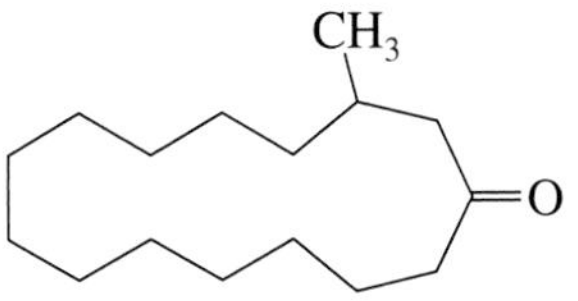

muscone
3-methylcyclopentadecanone
from the musk deer
*15-membered ring*

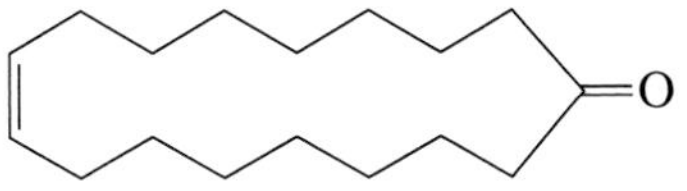

civetone
(*Z*)-9-cycloheptadecenone
from the civet cat
*17-membered ring*

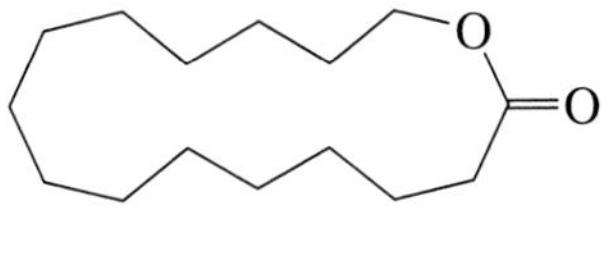

lactone of 15-hydroxypentadecanoic acid from angelica oil, from the roots of *Angelica archangelica*
*16-membered ring*

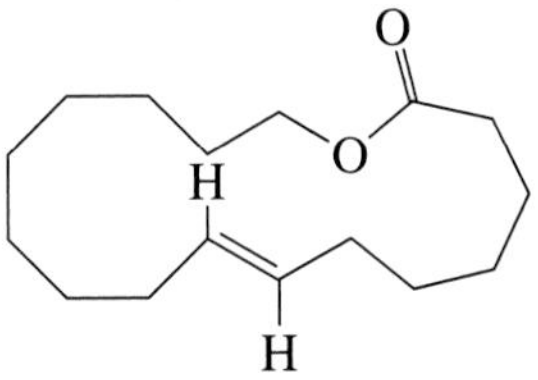

lactone of (*E*)-16-hydroxy-7-hexadecenoic acid; musk ambrette from the oil from the seeds of *Hibiscus abelmoschus*
*17-membered ring*

The rings in these lactones and ketones contain 15 to 17 atoms. The odor of cyclic ketones has been found to correlate with ring size; the musky odors desirable for perfumes are characteristic of compounds having 14 to 17 members in the ring. Compounds with larger rings have much less odor—at least, to the human nose.

Many other large-ring lactones that have been isolated from natural sources are interesting because they have antibiotic or antitumor activity. These lactones are known as **macrolides.** Two macrolides are shown below.

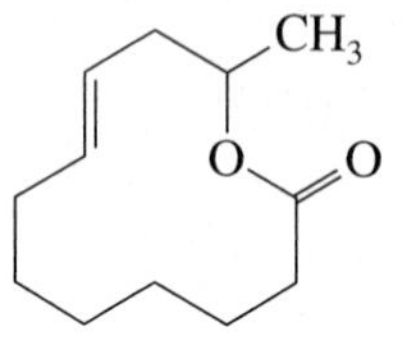

recifeiolide
from the fungus
*Cephalosporium recifei*

a 12-membered ring
lactone natural product

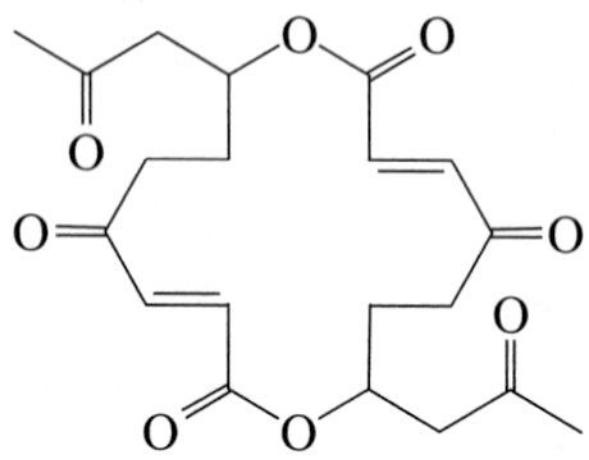

vermiculine
from the microorganism
*Penicillium vermiculatum*

a 16-membered ring
dilactone

**Problem 15.27**   Complete the following equations.

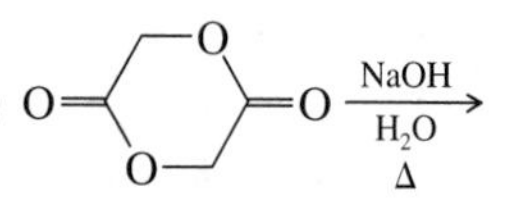

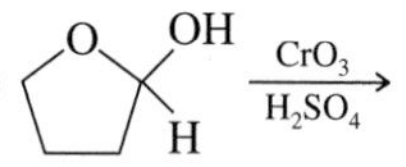

## E. Transesterification

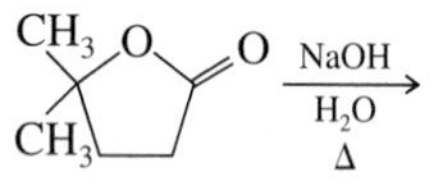

**Transesterification** is the conversion of an ester into another ester by heating it with an excess of either an alcohol or a carboxylic acid in the presence of an acidic or a basic catalyst. In essence, it is the transfer of an acyl group from one alcohol function to another. In an equilibrium reaction, either the alcohol or the acid portion of the original ester is freed. Such a reaction is carried out when there are practical reasons why an ordinary hydrolysis reaction would not work. For example, 2-chloro-2-phenylacetic acid is freed from its ethyl ester by a transesterification reaction.

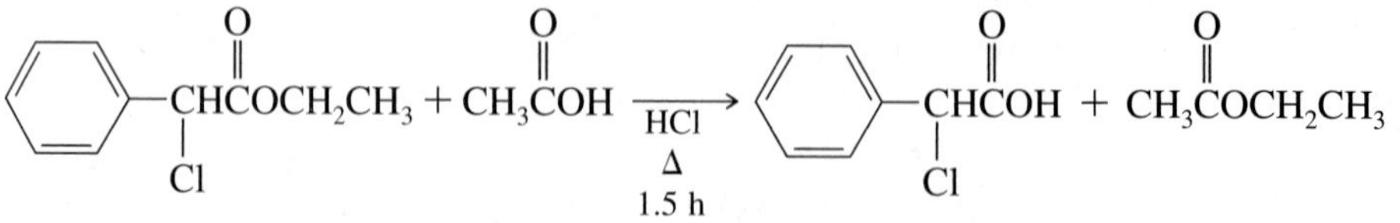

ethyl 2-chloro-2-phenylacetate
1 mole

acetic acid
7 moles
bp 118 °C

2-chloro-2-phenylacetic
acid
mp 78 °C

ethyl acetate

bp 77 °C

The presence of the chlorine substituent on the carboxylic acid makes prolonged heating in aqueous base unwise; a nucleophilic substitution at that carbon atom as well as the hydrolysis reaction may take place. Even long heating in dilute acid may lead to some substitution at the benzylic position. Concentrated hydrochloric acid is used to catalyze the transesterification reaction, in which the ethoxy group is transferred from 2-chloro-2-phenylacetic acid to acetic acid. Ethyl acetate and the excess of acetic acid are easily separated from the solid 2-chloro-2-phenylacetic acid by distillation.

In another example, methyl acrylate is converted into *n*-butyl acrylate by heating it with *n*-butyl alcohol in the presence of *p*-toluenesulfonic acid as the catalyst.

$$CH_2{=}CHCOCH_3 + CH_3CH_2CH_2CH_2OH \xrightarrow[\Delta]{TsOH} CH_2{=}CHCOCH_2CH_2CH_2CH_3 + CH_3OH$$

| methyl acrylate | *n*-butyl alcohol | | *n*-butyl acrylate | methanol |
|---|---|---|---|---|
| bp 81 °C | bp 117 °C | | bp 145 °C | bp 65 °C |

The difference in the boiling points of the alcohols allows the equilibrium to be shifted toward the higher-molecular-weight ester by distilling the methanol out of the reaction mixture.

⊘ **Study Guide**
Concept Map 15.5

**Problem 15.28**    An intermediate in the synthesis of the vitamin biotin is prepared by the following sequence of reactions.

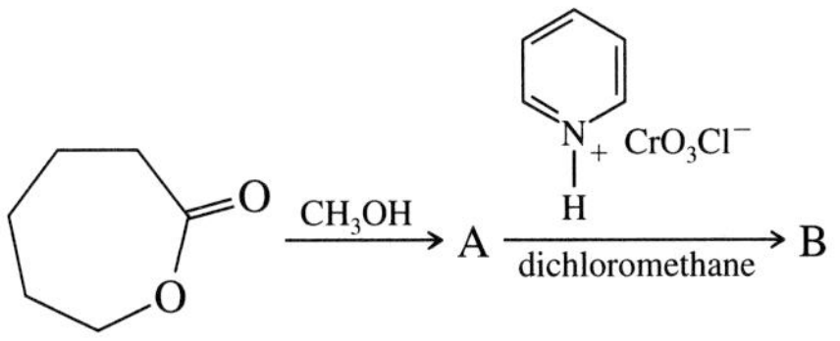

Assign structures to Compounds A and B. Biotin has the following structure.

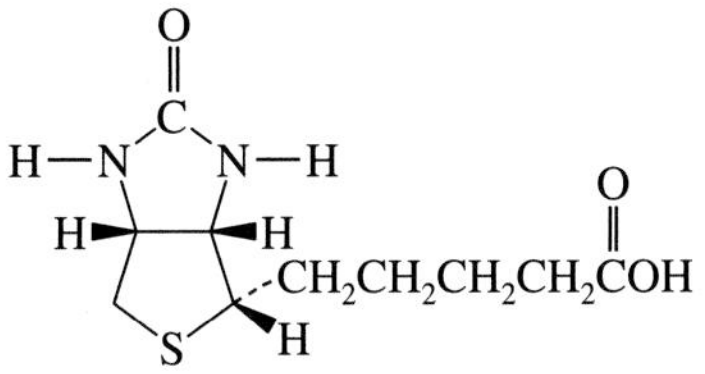

**Problem 15.29**    Coconut oil consists mainly of triglycerides (p. 639) of octanoic, dodecanoic, and tetradecanoic acids and is used to synthesize the ethyl esters of those acids. Write an equation showing how this could be done.

**Problem 15.30**    Write a detailed mechanism for the following transesterification reaction.

$$CH_3COCH_2CH_3 + CH_3OH \underset{H_2SO_4}{\rightleftharpoons} CH_3COCH_3 + CH_3CH_2OH$$

**Problem 15.31**    The ester functions in glucopyranose pentaacetate (p. 622) undergo the typical ester reactions. For example, the acetyl groups can be transesterified with sodium methoxide in methanol to regenerate glucose. Write an equation for this reation, and then write the mechanism for the reaction using the structure below to represent glucose pentaacetate.

$$R{-}O{-}\overset{\overset{\textstyle O}{\|}}{C}{-}CH_3$$

## 15.7  Formation of Amides. Acylation of Ammonia and Amines

Acid chlorides and acid anhydrides transfer acyl groups to ammonia or amines to give amides. For example, *p*-toluidine is converted to *N*-acet-*p*-toluidide by a reaction with acetic anhydride.

$$CH_3-\!\!\!\bigcirc\!\!\!-NH_2 \ + \ CH_3\overset{O}{\overset{\|}{C}}O\overset{O}{\overset{\|}{C}}CH_3 \ \longrightarrow \ CH_3-\!\!\!\bigcirc\!\!\!-NH\overset{O}{\overset{\|}{C}}CH_3 \ + \ CH_3\overset{O}{\overset{\|}{C}}OH$$

| *p*-toluidine | acetic anhydride | *N*-acet-*p*-toluidide | acetic acid |

*N,N*-Disubstituted amides are formed when acid chlorides or anhydrides react with amines having two substituents on the nitrogen atom. An example is the conversion of cyclohexanecarboxylic acid to *N,N*-dimethylcyclohexanecarboxamide; as is usual, the acid is converted into the acid chloride, which then reacts with the amine.

$$\bigcirc\!\!\!-\overset{O}{\overset{\|}{C}}OH \ + \ SOCl_2 \ \longrightarrow \ \bigcirc\!\!\!-\overset{O}{\overset{\|}{C}}Cl \ + \ SO_2\!\uparrow \ + \ HCl\!\uparrow$$

| cyclohexanecarboxylic acid | thionyl chloride | cyclohexanecarbonyl chloride |

$$\Big\downarrow \ \begin{matrix}2 \ (CH_3)_2NH \\ \text{benzene}\end{matrix}$$

$$\bigcirc\!\!\!-\overset{O}{\overset{\|}{C}}-\overset{CH_3}{\overset{|}{N}}CH_3 \ + \ CH_3\overset{CH_3}{\overset{|}{\underset{+}{N}}}H_2 \ \ Cl^-$$

| *N,N*-dimethylcyclohexanecarboxamide | dimethylammonium chloride |
| 86% | |

Esters react with ammonia or amines to give amides and alcohols. For example, ethyl lactate is converted into lactamide and ethanol by treatment with liquid ammonia.

$$\underset{\underset{\text{OH}}{|}}{CH_3CH}\overset{O}{\overset{\|}{C}}OCH_2CH_3 + NH_3(\text{liq}) \xrightarrow{-70\,°C} \underset{\underset{\text{OH}}{|}}{CH_3CH}\overset{O}{\overset{\|}{C}}NH_2 + CH_3CH_2OH$$

| ethyl lactate | ammonia | lactamide | ethanol |
| | | 70% | |

Another example demonstrates clearly the differing reactivity toward nucleophilic substitution reactions of a carbonyl group and a carbon–halogen bond. Ethyl chloroacetate, when treated with aqueous ammonia at low temperatures and for short periods of time, undergoes substitution quite selectively at the carbonyl group.

**On the Web: ONE SMALL STEP**

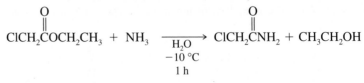

$$ClCH_2\overset{O}{\overset{\|}{C}}OCH_2CH_3 \ + \ NH_3 \ \xrightarrow[\substack{-10\,°C \\ 1\,h}]{H_2O} \ ClCH_2\overset{O}{\overset{\|}{C}}NH_2 \ + \ CH_3CH_2OH$$

| ethyl chloroacetate | ammonia | chloroacetamide | ethanol |

**Study Guide**
Concept Map 15.6

**Problem 15.32**     Suggest a mechanism for the reaction of ethyl acetate with ammonia. What is the nucleophile, and what is the leaving group?

**Problem 15.33**     Assign structures to the compounds designated by letters in the following equations.

(a) $CH_3CH_2OCCH=CHCOCH_2CH_3$ $\xrightarrow[\substack{NH_4Cl \\ H_2O}]{NH_3 \text{ (excess)}}$ A

(b) 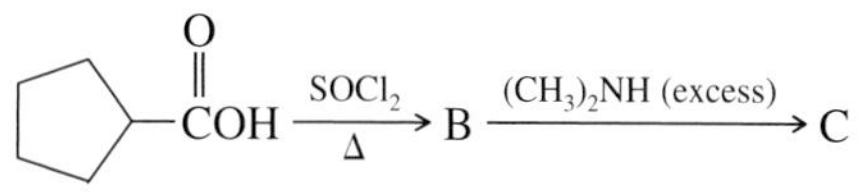 [cyclopentyl]$-COH$ $\xrightarrow[\Delta]{SOCl_2}$ B $\xrightarrow{(CH_3)_2NH \text{ (excess)}}$ C

(c) $CH_3CH_2OCCH_2CH_2CH_2CH_2COCH_2CH_3$ $\xrightarrow{H_2NNH_2 \text{ (excess)}}$ D

(d) 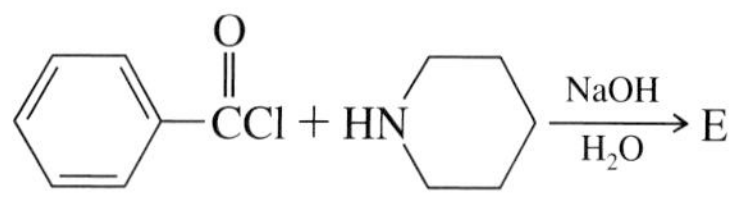 [phenyl]$-CCl + HN$[piperidyl] $\xrightarrow[H_2O]{NaOH}$ E

# 15.8   Acyl–Transfer Reactions in Biological Systems

## A. Acyl–Transfer Reactions in the Synthesis of Peptides in the Laboratory. Strategy

In nature, amino acids are linked together by amide bonds not only in large molecules known as proteins but also in smaller molecules, many of which have hormonal activity. Such small molecules, containing only a few amino acid residues, are known as peptides (p. 615). Peptide synthesis is an active area of research because of the biological activity of these compounds.

The synthesis of peptide requires that amide bonds, also called peptide linkages, be formed between the amino group of one amino acid and the carboxylic acid group of another one. For example, if we wish to synthesize the peptide alanylglycine, we need to create an amide bond between the carboxylic acid group of the amino acid alanine and the amino group of the amino acid glycine. (See Problem 3.41 on page 113 as a reminder of why the structures of the amino acids are written as they are.)

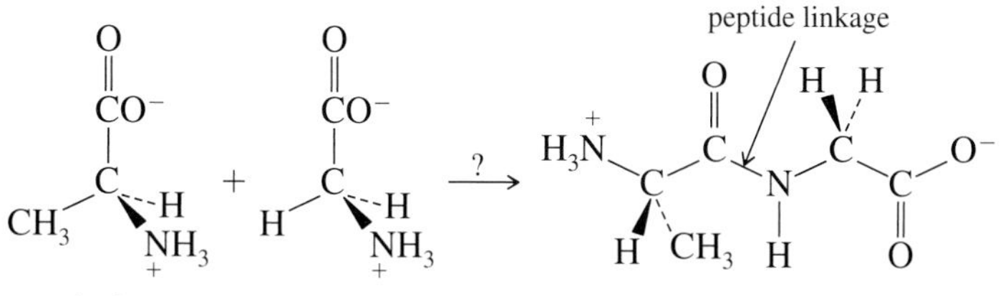

alanine          glycine                    alanylglycine

The problem with this is immediately apparent. If we mix alanine and glycine together as they are, there is nothing to stop the carboxylic acid group of glycine from reacting with the amino group of alanine to give glycylalanine, a peptide isomeric with the one we want.

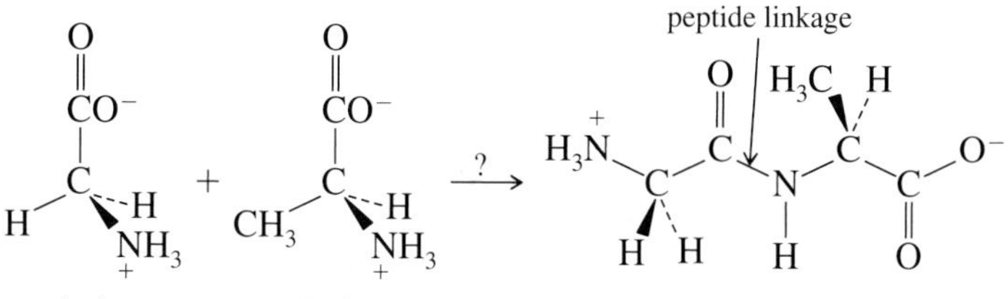

glycine          alanine                    glycylalanine

Alanine also can react with another molecule of itself to give alanylalanine, and glycine can react with itself to give glycylglycine. (Remember that while we write equations showing single molecules of each reagent, in reality, billions and billions of them are randomly colliding with each other in the reaction mixture.) We should expect to get a mixture of four peptides. The complexity of the mixture, of course, increases greatly as the number of amino acids in the peptide increases.

**Problem 15.34** Using the structures of the two peptides shown on page 629 as guides, write structures for alanylalanine and glycylglycine.

To prepare a peptide of a particular structure, the amino acids must be added onto the chain in a precise order. The desired reaction is a nucleophilic substitution by an amino group at the carbonyl group of a carboxylic acid, but there are nucleophilic groups such as other amino groups, thiols, and alcohols on the side chains of amino acids that are also expected to react with carboxylic acids. The structures of the amino acids usually found in peptides and proteins may be found on pages 983 and 985. The reactions to form the backbone of a peptide chain must involve the amino group at carbon 2 and not any other nucleophilic functional group.

Another complication is that a free carboxylic acid group does not react readily with amines to give amides. Usually an acid chloride or an acid anhydride is used so that a better leaving group than a hydroxyl group is present when an acid is converted into an amide (p. 628).

Thus there are two major aspects to peptide synthesis. First, the reaction must be directed to the desired part of the molecule. This means that other nucleophilic functional groups must be hidden, protected, so that they do not react. For example, if the amino group in alanine were protected so that it was no longer a nucleophile, then only one peptide could be formed between alanine and the amino group of glycine.

*PG— is a generic protecting group that masks the nucleophilicity of the amino group in alanine and the reactivity of the carboxyl group in glycine*

The equation above shows the amino group in alanine masked by a **protecting group** that reduces its nucleophilicity to the point that it will not react with the carboxylic acid group of glycine (nor with the carboxylic acid group of another alanine molecule). A protecting group converts a reactive functional group into a different group that is inert to the conditions of the reaction to be carried out. Another requirement for a protecting group is that it must be easily removable once the job is done so that the original functional group can be restored.

The second aspect of peptide synthesis requires that the carboxylic acid group of an amino acid must be made reactive enough to form an amide bond under conditions that do not destroy peptide bonds and other functional groups that may be present in the molecule. The equation above shows a second protecting group on the carboxylic acid group of glycine to prevent its activation so that glycine does not react with other molecules of itself.

The problem that faces peptide chemists is complex. The next section will discuss acyl-transfer reactions that are used to protect and deprotect the $\alpha$-amino group (the amino group on carbon 2) in amino acids. The act of protecting different types of functional groups in a protein synthesis is the subject of lively research, and many different and highly selective protecting groups are being developed. In principle, the arguments concerning the $\alpha$-amino group extend to the problems of protecting all functional groups. Section 21.2 will explore the use of a variety of protecting groups in other types of synthesis.

## B. Acyl–Transfer Reactions in the Protection of Amino Groups in Peptide Synthesis

The most important factor to consider when choosing a protecting group for the $\alpha$-amino group is the ease with which the protecting group can be removed later in the synthesis. The nitrogen atom of an amine can be made less nucleophilic by converting it to an amide, for example, but an ordinary amide would not be a suitable protecting group in a peptide synthesis because there would be no way to remove the protecting group except by hydrolysis, which would also cleave peptide bonds.

A protecting group that can be cleaved under anhydrous acid conditions that do not affect peptide bonds is a benzyl or *tert*-butyl carbamate. The carbamate is prepared by reaction of the amino acid with a half ester of carbonic acid, the other half of which has a good leaving group on it (p. 829). Benzyl chloroformate, for example, gives benzyl carbamates with amino acids.

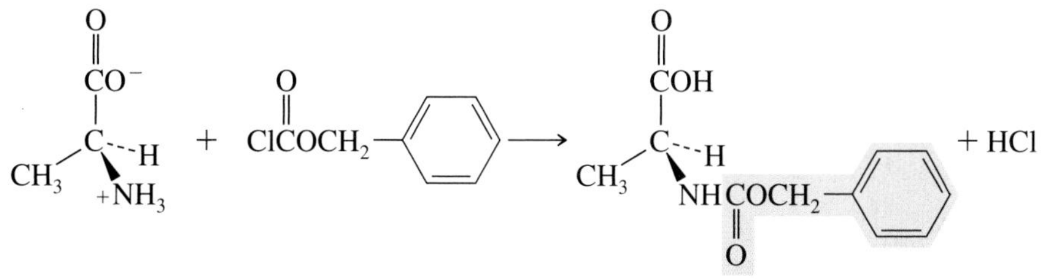

| alanine | benzyl chloroformate | N-benzyloxycarbonylalanine<br>carbobenzoxyalanine<br>Cbz-Ala |

The N-benzyloxycarbonyl protecting group is abbreviated as Cbz.

*tert*-Butyl chloroformate is not very stable, so *tert*-butyl azidoformate, in which the azide ion is the leaving group for the substitution reaction, is more commonly used as a reagent for protecting amino groups.

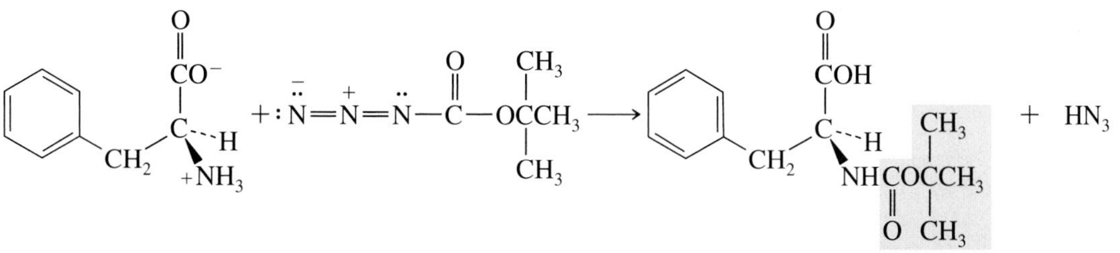

| phenylalanine | *tert*-butyl azidoformate | N-*tert*-butoxycarbonylphenylalanine<br>Boc-Phe |

The *tert*-butoxycarbonyl protecting group is abbreviated as Boc.

The benzyl and *tert*-butyl groups make good protecting groups because they both yield stable carbocations or radicals. Under acidic conditions, in the absence of water, they undergo S$_N$1 reactions to give carbamic acids, which lose carbon dioxide and regenerate the amino group. The *tert*-butoxycarbonyl group comes off when treated with an acid such as dry trifluoroacetic acid in dichloromethane or dry hydrogen chloride in ether.

**VISUALIZING THE REACTION**

**Removal of the *tert*-Butoxycarbonyl Protecting Group**

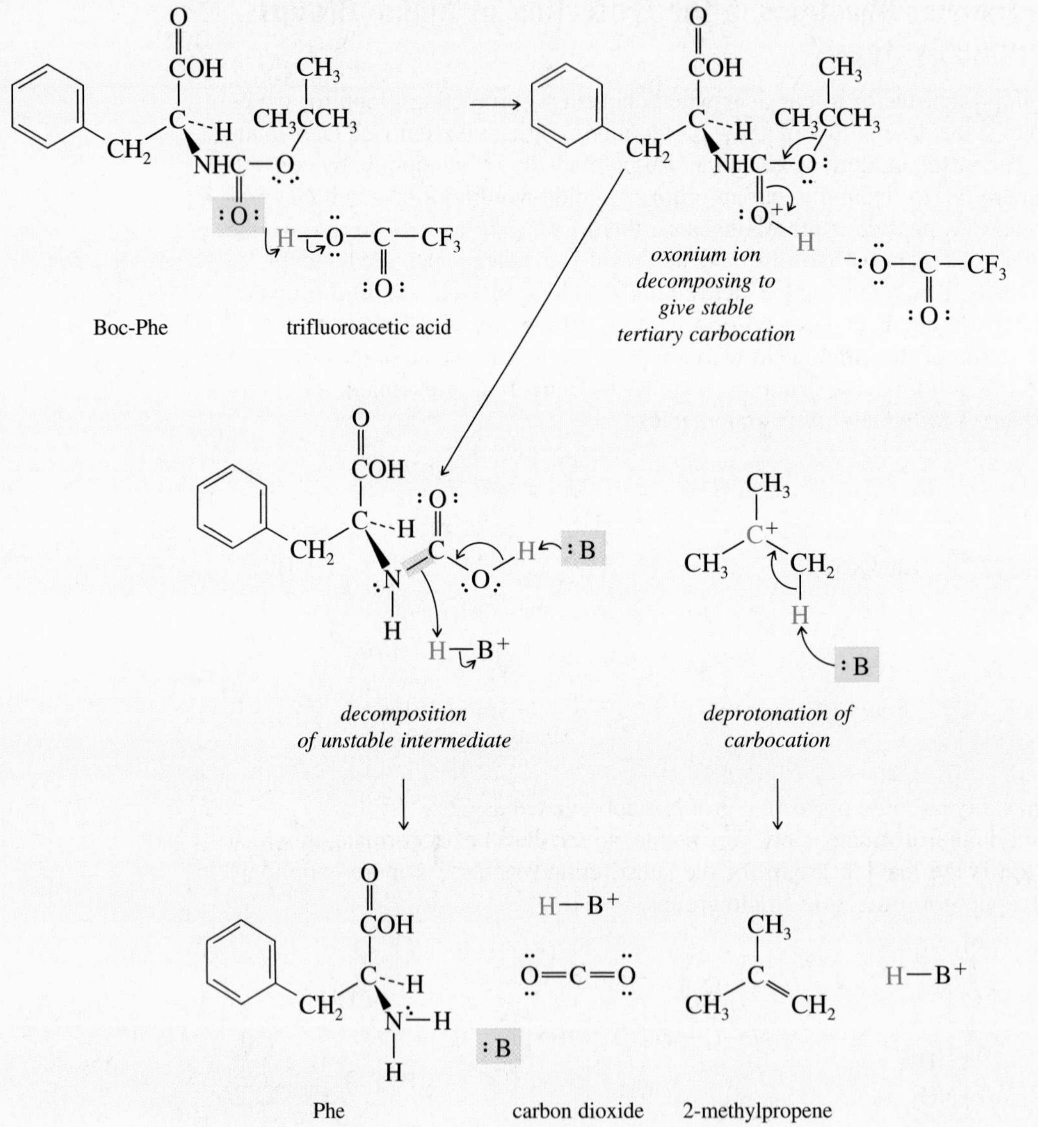

The products, other than the amino acid, are the gases carbon dioxide and 2-methylpropene. The alkene comes from the *tert*-butyl cation.

The benzyloxycarbonyl group is removed by a stronger acid, such as dry hydrogen bromide dissolved in trifluoroacetic or acetic acid.

**VISUALIZING THE REACTION**

## Removal of the Benzyloxycarbonyl Protecting Group

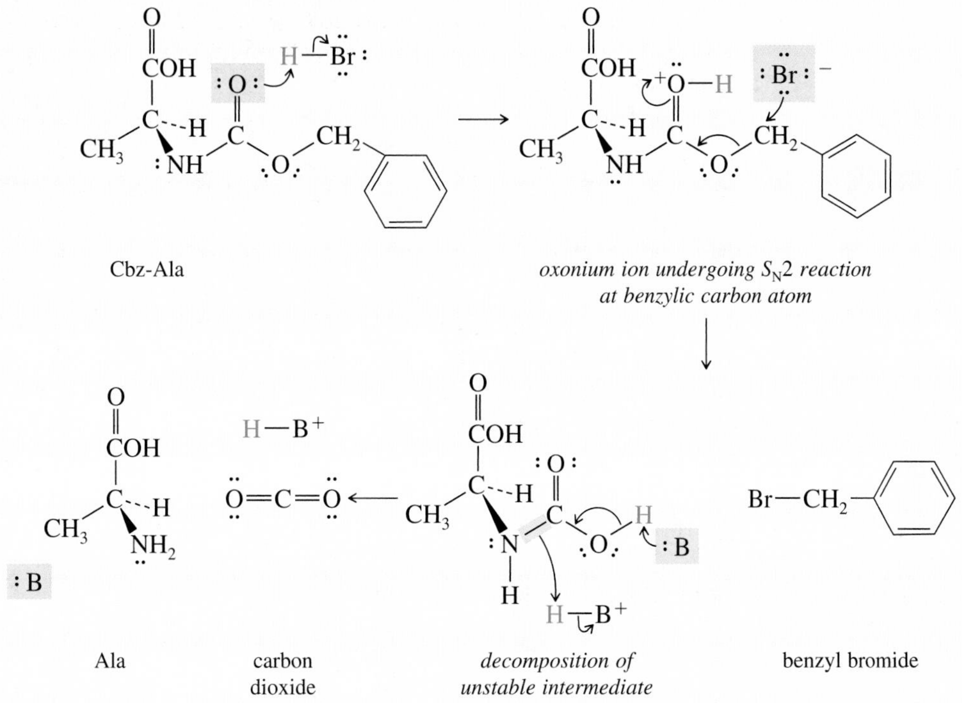

Cbz-Ala

*oxonium ion undergoing $S_N2$ reaction
at benzylic carbon atom*

Ala          carbon
            dioxide

*decomposition of
unstable intermediate*

benzyl bromide

Alternatively, the bond between oxygen and the benzyl group is cleaved by hydrogenation.

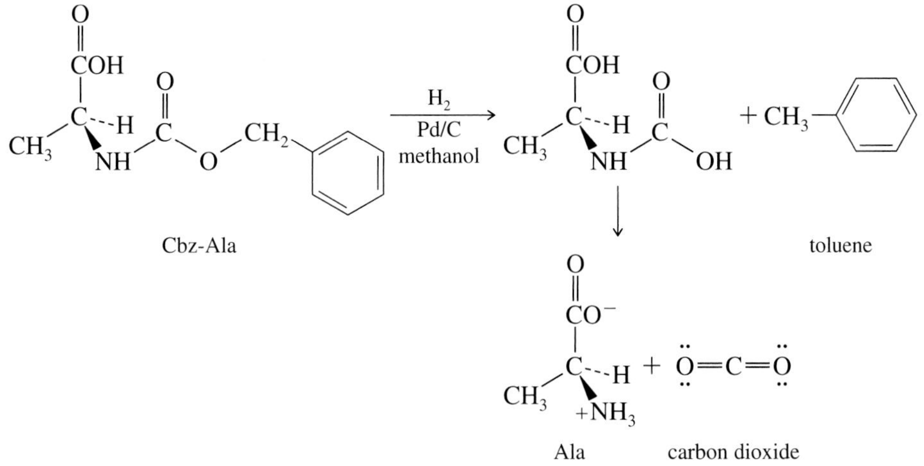

Cbz-Ala

$H_2$
Pd/C
methanol

+ $CH_3$—

toluene

Ala          carbon dioxide

The peptide bonds are stable under either of these sets of conditions. As carbamates, the amino groups are no longer basic and nucleophilic, and thus they are protected from reaction until the protecting group is removed at the appropriate time in the synthesis.

The carboxyl group of the amino acid is usually protected as a relatively unreactive ester, such as the methyl or ethyl ester.

⊘ **Study Guide
Concept Map 15.7**

## C. Acyl–Transfer Reactions in the Activation of the Carboxyl Group and Formation of Peptide Bonds

For a peptide bond to be formed, it is necessary that the hydroxyl group present in the carboxylic acid function be converted into a good leaving group. This is done in one of two ways. In the presence of dicyclohexylcarbodiimide, an acid reacts with an amine to give an amide.

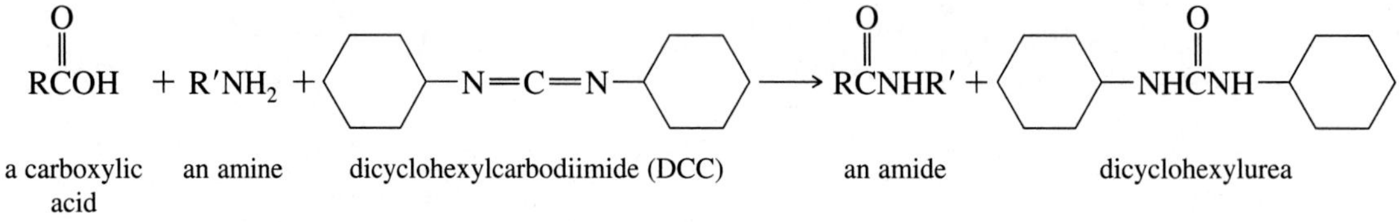

a carboxylic   an amine   dicyclohexylcarbodiimide (DCC)   an amide   dicyclohexylurea
acid

Dicyclohexylcarbodiimide (often abbreviated as DCC) converts the carboxylic acid into an intermediate that has the properties of an acid anhydride. The hydroxyl group of the carboxylic acid is converted into a good leaving group, so reaction with an amine to give an amide takes place.

**VISUALIZING THE REACTION**

### Activation of a Carboxylic Acid by Dicyclohexylcarbodiimide

*nucleophilic attack on carbodiimide; protonation at nitrogen*

*deprotonation resulting in intermediate with properties like those of an anhydride*

*nucleophilic attack at carbonyl group; formation of tetrahedral intermediate*

amide

dicyclohexylurea

*regeneration of carbonyl group; protonation of nitrogen*

*deprotonation*

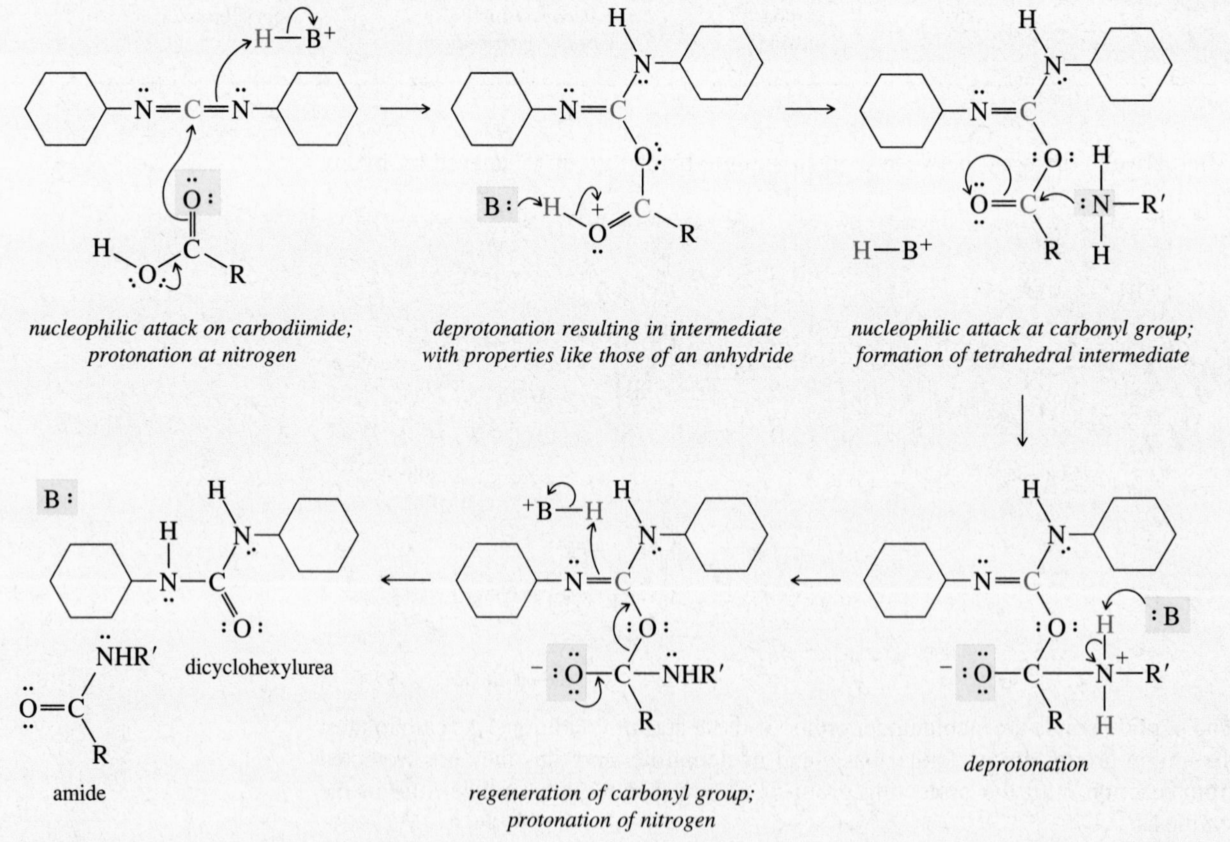

The hydroxyl group also can be converted into a good leaving group by the preparation of an active ester of the amino acid. The *p*-nitrophenyl ester is often used. For example, the *p*-nitrophenyl ester of glycine, protected at the amino group by the benzyloxycarbonyl group, is prepared. The activated carboxyl group then reacts with the amino acid valine.

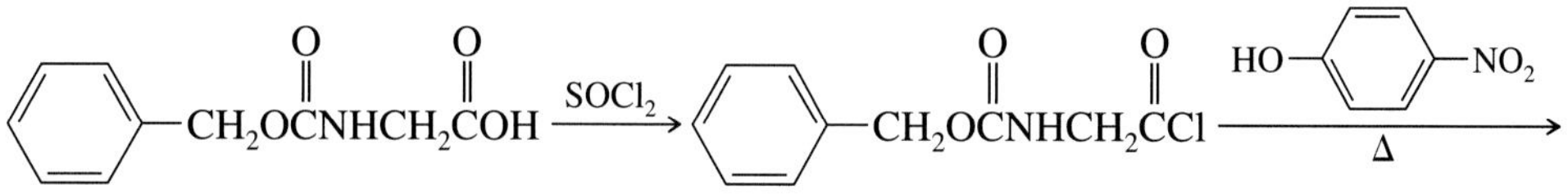

Cbz-Gly

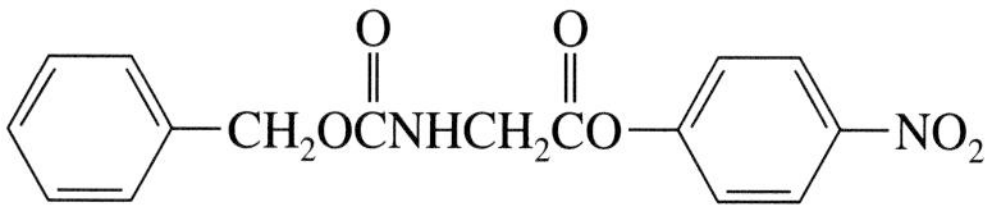

*p*-nitrophenyl ester of
*N*-benzyloxycarbonylglycine

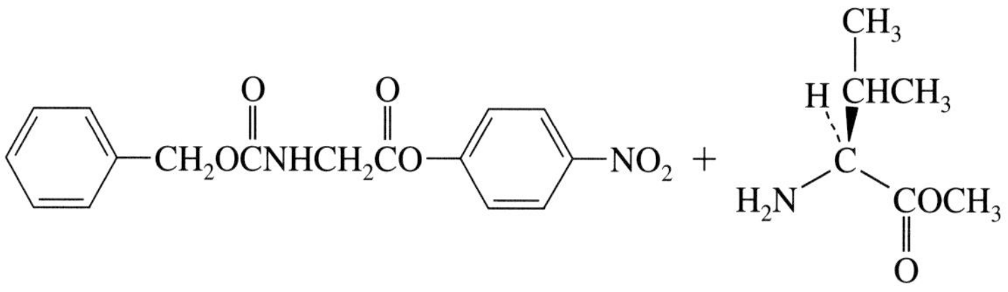

valine

*protected at its carboxyl
group as the methyl ester*

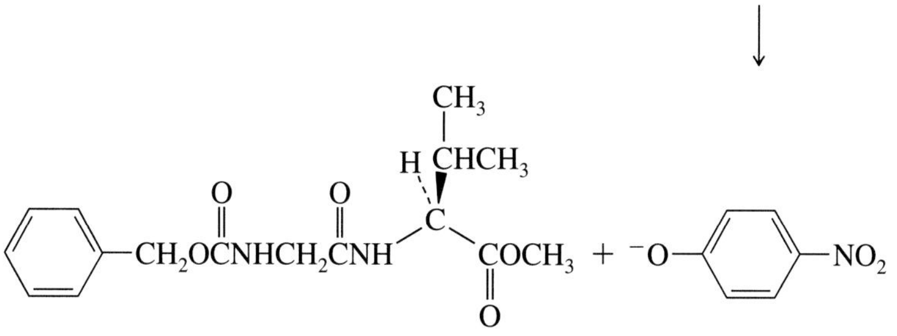

Cbz-Gly-Val-OMe
75%

*p*-nitrophenolate anion

*good leaving group,
stabilized by resonance
and electron-withdrawing
effect of nitro group*

The conditions for the formation of the *p*-nitrophenyl ester are rather harsh and might not be suitable for activating the carboxyl group of a larger peptide. Other types of esters, formed in the presence of dicyclohexylcarbodiimide, are used in many peptide syntheses. Esters of *N*-hydroxysuccinimide are often used. Resonance stabilization of the *N*-oxysuccinimide anion makes it a good leaving group.

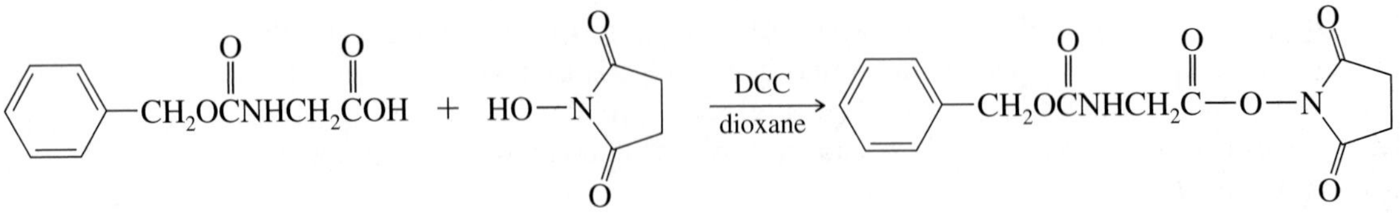

Cbz-Gly               N-hydroxysuccinimide                                    Cbz-Gly-Su

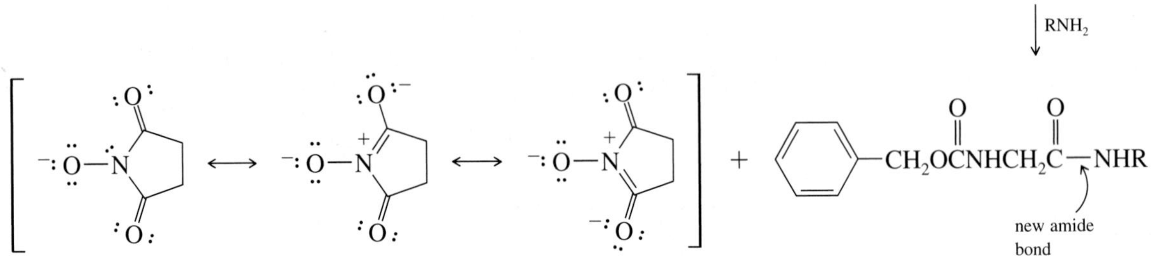

*negative charge on anion stabilized by the electron-withdrawing effect of the adjacent nitrogen atom*

**Study Guide**
**Concept Map 15.8**

The *N*-oxysuccinimide leaving group is abbreviated Su.

**Problem 15.35**   Write structural formulas for the products and intermediates represented by letters in the following equations.

(a)

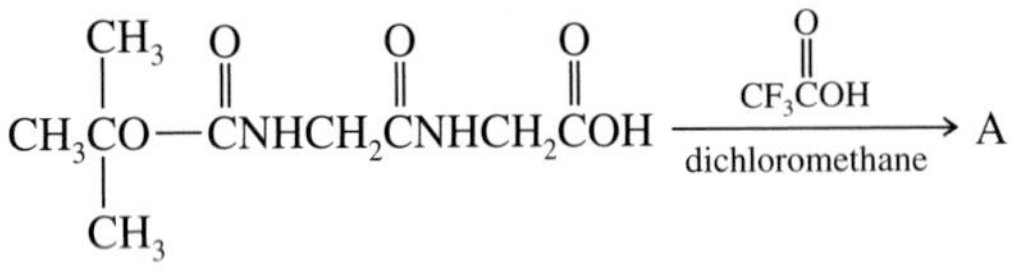

(b)

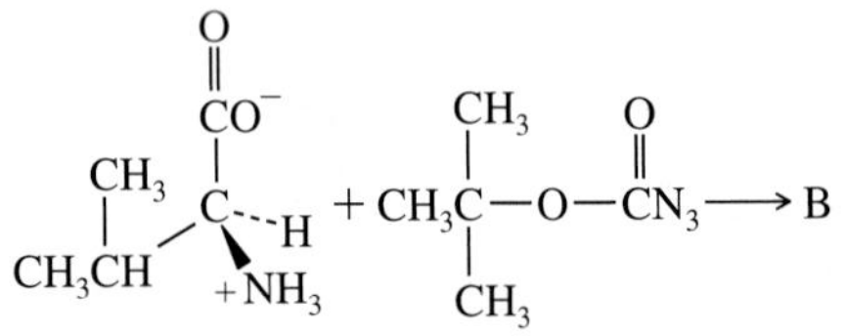

(c)

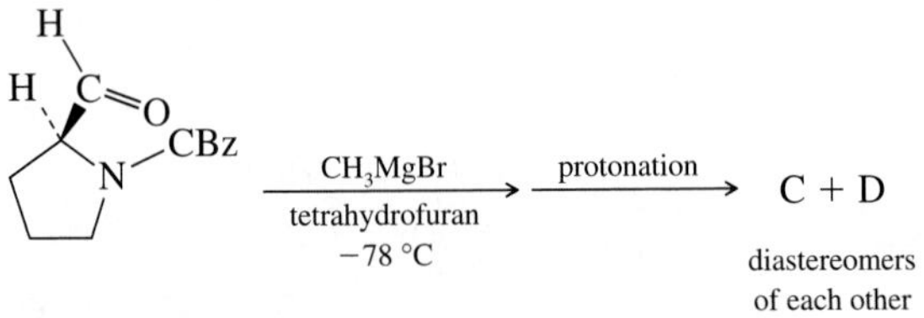

(d)

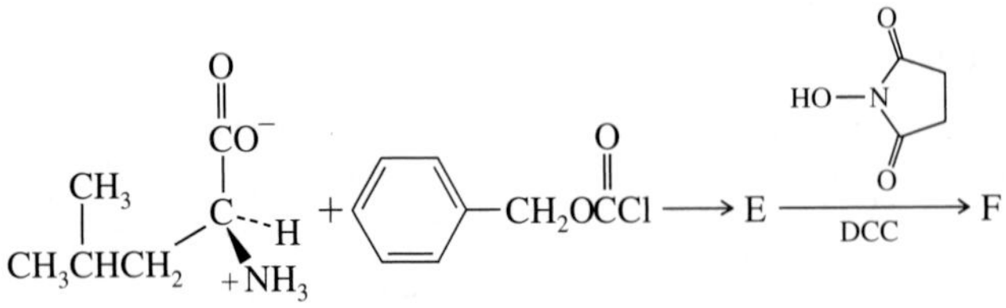

(e)

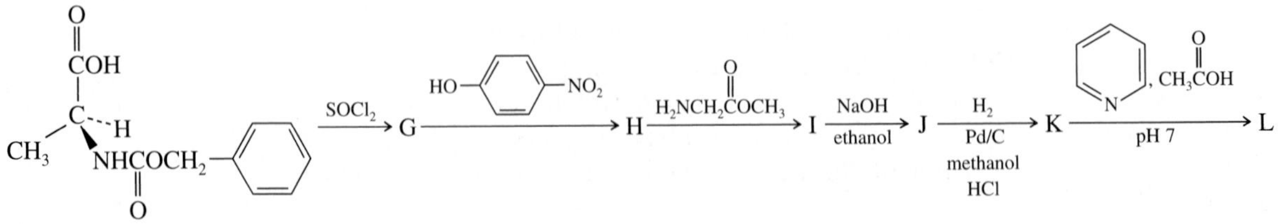

## D.  Biological Transesterification Reactions

The process of transesterification goes on all the time in the human body, usually involving thioesters. **Thioesters** are compounds in which the oxygen atom of the alkoxyl group of an ester has been replaced by a sulfur atom. An important thioester that participates in many physiologic processes is acetyl coenzyme A.

$$ROPOCH_2C\overset{CH_3}{\underset{CH_3}{\overset{|}{\underset{|}{C}}}}\overset{OH}{\overset{|}{CH}}-\overset{O}{\overset{\parallel}{C}}NHCH_2CH_2\overset{O}{\overset{\parallel}{C}}NHCH_2CH_2\overset{O}{\overset{\parallel}{S}}CCH_3$$

acetyl coenzyme A

$$usually\ abbreviated\ CoA-\overset{O}{\overset{\parallel}{S}}CCH_3\ or\ acetyl\ CoA$$

The vitamin present in coenzyme A is pantothenic acid, one of the B vitamins. It is linked by an amide bond at one end to 2-aminoethanethiol and by a carbon–oxygen bond at the other end to a phosphate group of a nucleotide, adenosine 3′,5′-diphosphate. (The structures of nucleotides are discussed on page 987.)

$$HOCH_2C\overset{CH_3}{\underset{CH_3}{\overset{|}{\underset{|}{C}}}}\overset{OH}{\overset{|}{CH}}-\overset{O}{\overset{\parallel}{C}}NHCH_2CH_2\overset{O}{\overset{\parallel}{C}}OH \qquad H_2NCH_2CH_2SH$$

pantothenic acid

*a B vitamin*

2-aminoethanethiol

$$ROPOCH_2C\overset{CH_3}{\underset{CH_3}{\overset{|}{\underset{|}{C}}}}\overset{OH}{\overset{|}{CH}}-\overset{O}{\overset{\parallel}{C}}NHCH_2CH_2\overset{O}{\overset{\parallel}{C}}NHCH_2CH_2SH$$

nucleotide

coenzyme A

*abbreviated* CoA—SH

A thioester is more reactive than the corresponding oxyester. There is less resonance interaction between the larger sulfur atom and the adjacent carbonyl and, therefore, less resonance stabilization of the carbonyl group of a thioester than an oxygen ester. The acyl group is therefore more reactive toward nucleophilic attack. Thus the acyl group of a thioester is easily transferred to another sulfur, oxygen, or nitrogen atom.

Acetyl coenzyme A is a good acyl-transfer agent. For example, a reaction that is important to the smooth functioning of the nervous system is the synthesis of acetylcholine at the nerve synapses as it is needed (p. 818). Acetyl coenzyme A participates in what is essentially a transesterification reaction.

$$CH_3\overset{CH_3}{\underset{CH_3}{\overset{|+}{\underset{|}{N}}}}CH_2CH_2OH \ + \ CoA-\overset{O}{\overset{\parallel}{S}}CCH_3 \xrightarrow[\text{acetylase}]{\text{choline}} CH_3\overset{CH_3}{\underset{CH_3}{\overset{|+}{\underset{|}{N}}}}CH_2CH_2O\overset{O}{\overset{\parallel}{C}}CH_3 + CoA-SH$$

2-(trimethylammonium)ethanol

choline

acetyl CoA

acetylcholine

*essential in the transmission of nerve impulses*

coenzyme A

The thioester acetyl coenzyme A is converted into an oxyester by the alcohol function of choline. Coenzyme A is the leaving group. In the body, the process is catalyzed by an enzyme called, appropriately, choline acetylase.

**On the Web: ONE SMALL STEP**

**Problem 15.36**    Write a complete mechanism for the acyl-transfer reaction shown on page 637. You may assume that the enzyme is a good source of protons, HB$^+$, and of bases to remove protons, B:.

**Problem 15.37**    The vitamin pantothenic acid is sold over the counter as calcium pantothenate. Write the structural formula for calcium pantothenate.

## 15.9   Lipids, Fats, Oils, and Waxes

**Lipids** are substances that are found in living organisms and are insoluble in water but soluble in organic solvents. A lipid contains a long hydrocarbon chain as part of its structure but can have a variety of functional groups. The fat-soluble vitamins A (p. 758), D (p. 1082), E (p. 795), and K (p. 795) are classified as lipids, as are cholesterol (p. 490), its esters, and other similar steroids (p. 537). Other lipids, however, are fats in body tissues and the bloodstream, oils isolated from plant sources (such as sunflower seed oil, safflower seed oil, peanut oil, and olive oil), or waxes, which can have either animal or plant sources. All these compounds are esters.

**Fats** and **oils** are esters of glycerol (1,2,3-propanetriol) and long-chain carboxylic acids, usually having twelve or more carbon atoms (p. 639). Carboxylic acids became known as fatty acids because they were originally isolated from fats. The acids found in fats are predominantly saturated; those in oils tend to be unsaturated, with one or more double bonds in the chain. The following are the important fatty acids.

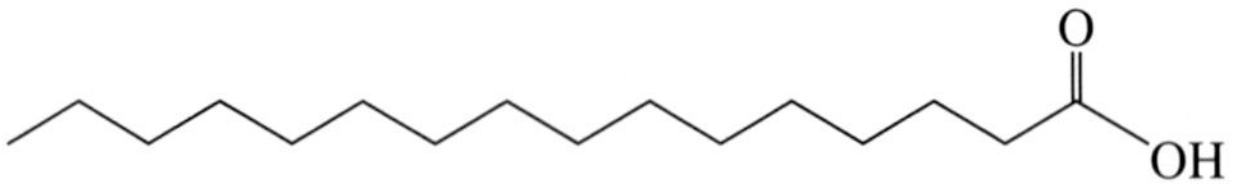

palmitic acid (hexadecanoic acid)
mp 62.9 °C

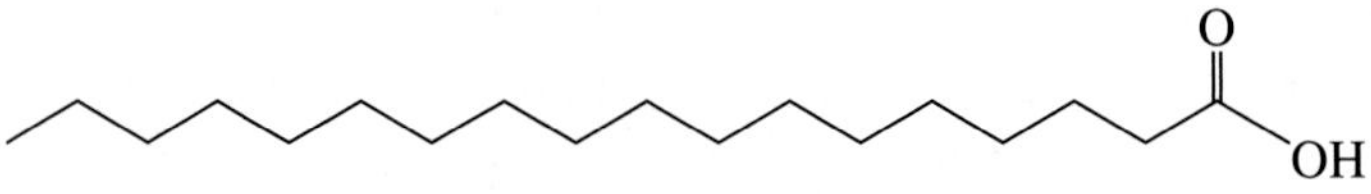

stearic acid (octadecanoic acid)
mp 69.6 °C

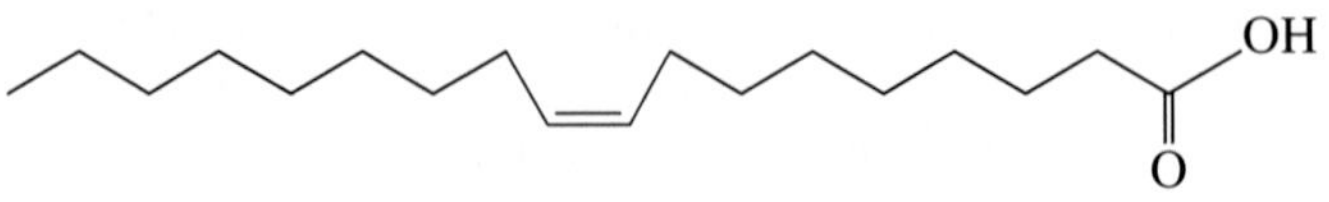

oleic acid [(Z)-9-octadecenoic acid]
mp 13 °C

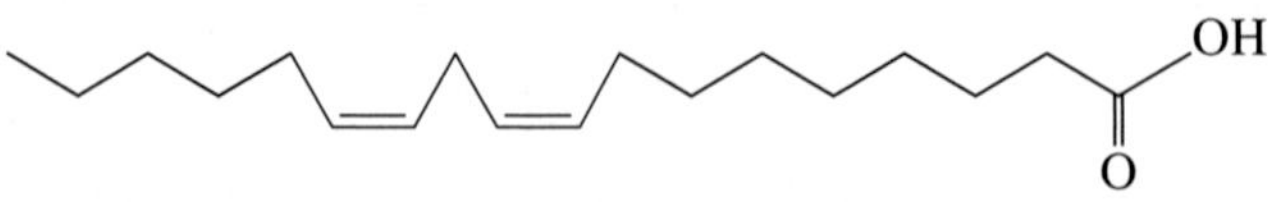

linoleic acid [(9Z, 12Z)-9,12-octadecadienoic acid]
mp −5 °C

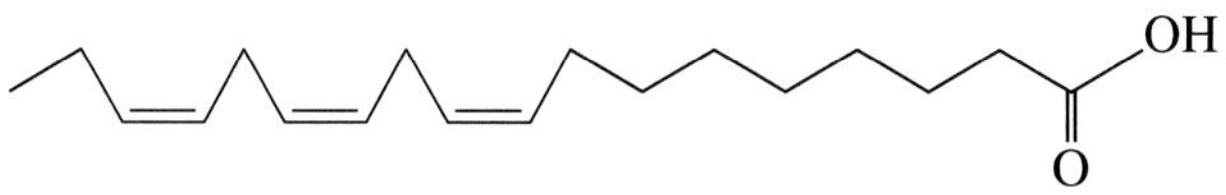

linolenic acid [(9Z, 12Z, 15Z)-9,12,15-octadecatrienoic acid]
mp −16 °C

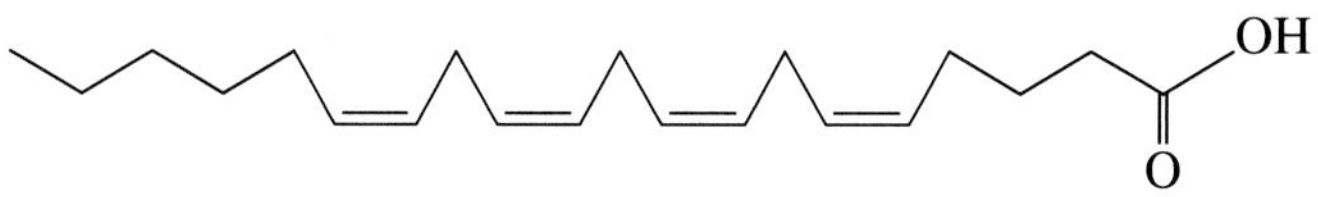

arachidonic acid [(5Z, 8Z, 11Z, 14Z)-5,8,11,14-icosatetraenoic acid]

The long-chain acids, palmitic acid and stearic acid, occur extensively in nature as the **saturated fatty acids** in solid fats. Note that they are solids at room temperature and at body temperature (37 °C). The melting points of long-chain fatty acids decrease with the introduction of double bonds into the chain. Oleic acid melts below body temperature but will solidify in a refrigerator. It is the chief fatty acid in olive oil and also in human body fat. As unsaturation increases in linoleic acid, linolenic acid, and arachidonic acid, these **polyunsaturated fatty acids** have lower and lower melting points. The melting point of arachidonic acid is so low that it has not yet been determined accurately. Polyunsaturated fatty acids are found as constituents of oils from the seeds of plants, such as safflower seed oil, sesame oil, and peanut oil. Hydrogenation of the double bonds in oils (p. 300) converts the liquid oils into fats.

The properties of fatty acids are reflected in the properties of their esters. Medical researchers are interested in the esters formed from cholesterol and fatty acids. Esters of cholesterol with saturated fatty acids are solids and have been implicated in the formation of solid deposits on the walls of blood vessels, causing diseases of the heart and arteries. Physicians have recommended the substitution of vegetable oils containing unsaturated fatty acids for animal fats and margarines, which are rich in saturated fatty acids.

The glycerol esters of fatty acids are known as **triglycerides.** A typical triglyceride is shown below.

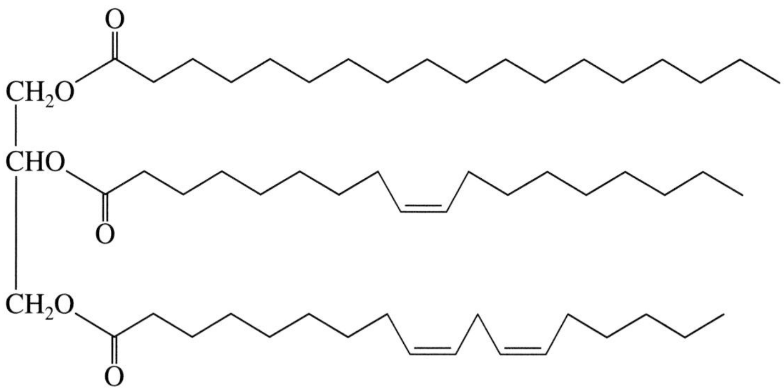

the ester of glycerol with
stearic acid, oleic acid, and linoleic acid

*a typical triglyceride*

In nature, two or three different fatty acids are usually found in a single triglyceride.

As a result of a growing consciousness about the connections between nutrition and heart disease, many people monitor the level of cholesterol in their blood. High levels of this component of the blood have been correlated with higher risks of heart attacks and strokes, which together account for more than 50% of deaths in the

United States. Researchers have discovered, however, that some of the cholesterol and blood fat can be tied up with blood proteins in **high-density lipoproteins,** or **HDLs.** High-density lipoproteins appear to have some protective value against heart disease, as opposed to **low-density lipoproteins,** or **LDLs,** which do not have such properties. Regular exercise is especially helpful in raising the ratio of high-density lipoproteins to low-density lipoproteins in the blood.

**Waxes** are esters formed from long-chain fatty acids, usually containing 24 to 28 carbon atoms, and long-chain (16 to 36 carbon atoms) primary alcohols or alcohols of the steroid series. Waxes are solids used for lubricating and protecting surfaces. In plants, for example, leaves and fruits are coated with a waterproof layer of such waxes.

$$CH_3(CH_2)_{15}OC(CH_2)_{16}CH_3$$

cetyl stearate

*common in bacteria*

$$CH_3(CH_2)_{15}OC(CH_2)_{14}CH_3$$

cetyl palmitate

*chief wax ester of spermaceti from the sperm whale*

$$CH_3(CH_2)_{29}OC(CH_2)_{14}CH_3$$

triacontyl palmitate

*a wax ester typical of those found in beeswax*

**Problem 15.38**    9,12-Octadecadiynoic acid (p. 608) is a synthetic precursor to one of the polyunsaturated fatty acids. Which one? How can 9,12-octadecadiynoic acid be transformed into the natural product?

## 15.10 Surface–Active Compounds. Soaps

Hydrolysis of esters using a basic solution (p. 618) is called a **saponification reaction** because such a reaction is used to make soap. Pioneer women, for example, made soap by heating animal fat saved from their cooking together with wood ashes, a source of potassium hydroxide. The conversion of a typical fat to glycerol and soap is shown below.

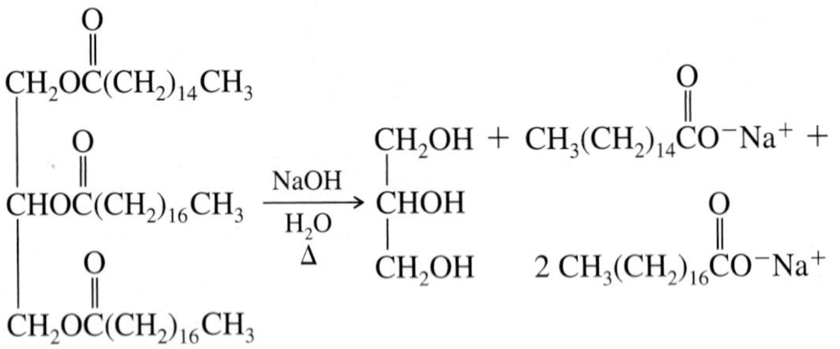

$$CH_2OC(CH_2)_{14}CH_3$$
$$CHOC(CH_2)_{16}CH_3 \xrightarrow[\substack{H_2O \\ \Delta}]{NaOH}$$
$$CH_2OC(CH_2)_{16}CH_3$$

a triglyceride

*from fat*

$$CH_2OH + CH_3(CH_2)_{14}CO^-Na^+ +$$
$$CHOH$$
$$CH_2OH \quad 2\,CH_3(CH_2)_{16}CO^-Na^+$$

glycerol                soap

The salt of a carboxylic acid is usually more soluble in water than the parent acid itself, but an interesting phenomenon arises when the hydrocarbon portion of the acid is very large in comparison with the carboxyl group. The ionic portion of the molecule interacts well with water and dissolves in it, but the rest of the long molecule will not go into solution. The hydrocarbon portions of adjacent molecules are more attracted to each other by van der Waals forces than they are to the polar water molecules. They are in fact hydrophobic, or water-repelling, in their behavior. A long-chain acid salt thus has two domains: a **hydrophilic** head, the ionic carboxylate

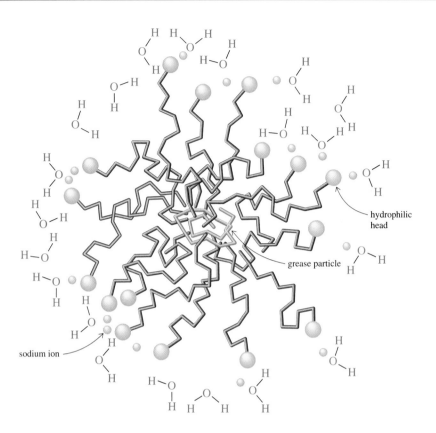

*Figure 15.3*
A cross section of a sodium
stearate micelle solubilizing a
particle of grease.

group that is soluble in water, and a **hydrophobic** tail, the long hydrocarbon portion that is repelled by water molecules and attracted to other hydrocarbon residues.

The structure of such a compound allows for a particular orientation of its molecules at the surface of water; the heads are in water and the hydrocarbon tails stick up into the air. The concentration of molecules at the surface of the water has the effect of lowering the surface tension of water. Compounds that behave in this way are said to be **surface-active compounds** or **surfactants.** Soaps are one category of surface-active compound. All good surfactants have structures with a hydrophilic head and a long hydrophobic tail. The property of surfactancy is exhibited by acids with twelve or more carbon atoms in the hydrocarbon portion of the molecule.

At a certain critical concentration of the surfactant, the surface layer is broken up into smaller units, clusters of ions called micelles. **Micelles** are particles in which the long hydrocarbon tails, repelled by the water molecules and attracted to each other, make up the interior and the negatively charged heads coat the surface and interact with the surrounding water molecules and positive ions. Figure 15.3 depicts a spherical micelle. Such micelles are formed when sodium octadecanoate (sodium stearate), which is a common ingredient of soap, is put into water.

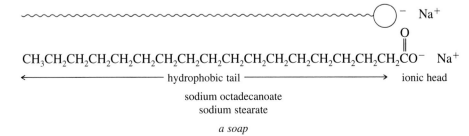

The opalescent appearance of a soap solution, which is colloidal, is evidence that the particles present are larger than individual molecules. The particles in a colloidal solution are large enough that light is scattered instead of being transmitted as is the case with a true solution, which appears clear to the eye.

Grease, in its chemical composition, resembles the hydrocarbon tails in the micelle. Rubbing a greasy spot with a soap solution causes the grease to break up into small enough particles to be trapped inside the micelles. The particles are held in solution by the hydrocarbon portion of the soap but are kept in colloidal suspension by the interaction of the ionic surface of the micelles with the surrounding water. The grease is said to be **emulsified,** held in suspension in a medium in which it would not normally be soluble.

Soaps are usually the sodium salts of long-chain carboxylic acids. Ordinary soaps have disadvantages in hard water. Hard water has dissolved calcium and magnesium ions in it, so when soap is used in such water, the calcium and magnesium salts of the carboxylic acids in the soap precipitate out. This precipitate is the scum seen in hard water and causes the ring around the bathtub. A great number of surfactants with more soluble calcium and magnesium salts have been developed. An example of a simple one prepared from nut oils is sodium dodecanyl sulfate (sodium lauryl sulfate), the sodium salt of the ester of 1-dodecanol with sulfuric acid.

$$CH_3CH_2CH_2CH_2CH_2CH_2CH_2CH_2CH_2CH_2CH_2CH_2O\overset{\overset{\displaystyle O}{\|}}{\underset{\underset{\displaystyle O}{\|}}{S}}O^- \ Na^+$$

$$\xleftarrow{\hspace{3cm}} \text{hydrophobic tail} \xrightarrow{\hspace{3cm}}$$

ionic head

sodium dodecanyl sulfate
sodium lauryl sulfate

*a surfactant used in commercial
detergents*

In sodium dodecanyl sulfate, a sulfuric acid analog instead of a carboxylic acid group provides the ionic portion of the molecule.

Molecules that have hydrophilic heads and long-chain hydrophobic tails made up of the hydrocarbon portions of saturated and unsaturated fatty acids also play an important part in the structure of cell membranes, which provide the boundaries between the outer and inner worlds of the cell. The functioning of these molecules in membranes will be described in Section 20.4D.

**Problem 15.39**    Which of the following compounds are surfactants?

(a) $\langle\text{C}_6\text{H}_5\rangle$—$\overset{\overset{\displaystyle O}{\|}}{C}O^-Na^+$

(b) $CH_3(CH_2)_{12}CH_2O\overset{\overset{\displaystyle O}{\|}}{\underset{\underset{\displaystyle O}{\|}}{S}}O^-Na^+$

(c) $CH_3(CH_2)_{10}CH_2\overset{+}{N}(CH_3)_3Cl^-$

(d) $\langle\text{C}_6\text{H}_5\rangle$—$\overset{\overset{\displaystyle O}{\|}}{\underset{\underset{\displaystyle O}{\|}}{S}}O^-Na^+$

(e) $CH_3(CH_2)_{10}CH_2$—$\langle\text{C}_6\text{H}_4\rangle$—$\overset{\overset{\displaystyle O}{\|}}{\underset{\underset{\displaystyle O}{\|}}{S}}O^-Na^+$

(f) $CH_3(CH_2)_{15}CH_2\overset{\overset{\displaystyle O}{\|}}{C}O^- \ ^+HN(CH_2CH_2OH)_3$

(g) $Na^+ {}^-O\overset{\overset{\displaystyle O}{\|}}{C}(CH_2)_{12}\overset{\overset{\displaystyle O}{\|}}{C}O^-Na^+$

## Table 15.1 Summary of the Reactions Used to Prepare Acids

| Starting Material | Reagent | Intermediate | Reagent in Second Step | Product |
|---|---|---|---|---|
| RCH₂OH | CrO₃, H₃O⁺ | RCHO | CrO₃, H₃O⁺ | RCOOH, with RCOOCH₂R as a side product |
|  | MnO₄⁻, OH⁻ | RCO⁻ | H₃O⁺ | RCOH |
| RCHO | MnO₄⁻, OH⁻ | RCO⁻ | H₃O⁺ | RCOH |
|  | Ag₂O, H₂O | RCO⁻ | H₃O⁺ | RCOH |
| R(H)C=C< (alkene) | O₃ | ozonide (R,H–C–O–O–C<) | H₂O₂ or Ag₂O, H₂O | RCOH |
| RMgX (from RX) | CO₂ | RCO⁻ | H₃O⁺ | RCOH |
| R—X | C≡N⁻ | RC≡N | H₃O⁺, Δ or OH⁻, H₂O, Δ then H₃O⁺ | RCOH |

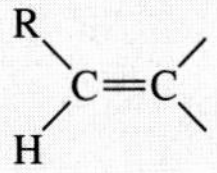

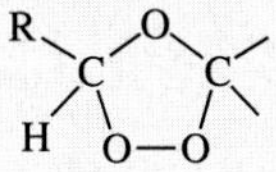

## Table 15.2 Reagents for Interconverting Acids and Acid Derivatives

| From ↓ \ To make → | RCCl | RCOCR | RCOR′ | RCOH | RCNR₂ | RCO⁻ |
|---|---|---|---|---|---|---|
| RCCl | — | RCO⁻ | R′OH | H₂O | R₂NH | H₂O, OH⁻ |
| RCOCR | — | — | R′OH | H₂O | R₂NH | H₂O, OH⁻ |
| RCOR″ | — | — | R′OH, Δ HB⁺ or B: | H₂O, H₃O⁺ Δ | R₂NH | H₂O, OH⁻ Δ |
| RCOH | SOCl₂ or PCl₃ | Δ, −H₂O (cyclic anhydride) | R′OH HB⁺, Δ | — | R₂NH, Δ | OH⁻ |
| RCNR₂ | — | — | — | H₂O, H₃O⁺ Δ | — | H₂O, OH⁻ Δ |
| RCO⁻ | — | RCCl | — | H₃O⁺ | — | — |

**Table 15.3    Some Reactions Used in the Preparation of Lactones**

| Starting Material | Reagent | Intermediate | Reagent in Second Step | Product |
|---|---|---|---|---|
| 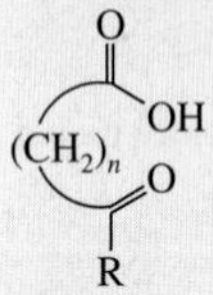 (CH$_2$)$_n$ with O=C–OH, C=O, R; $n$ = 2 or 3 | NaBH$_4$, H$_2$O | (CH$_2$)$_n$ with C=O–O$^-$ Na$^+$, OH, R H | H$_3$O$^+$ | (CH$_2$)$_n$ with O=C–O, R H |
| 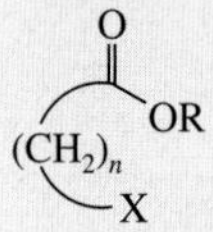 (CH$_2$)$_n$ with O=C–OR, X; $n$ = 3 or 4 | R'CO$^-$ (O=) | (CH$_2$)$_n$ with O=C–OR, O=C–OCR' | OH$^-$, H$_2$O, Δ then H$_3$O$^+$ | (CH$_2$)$_n$ with O=C–O |

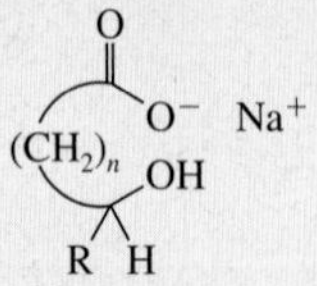

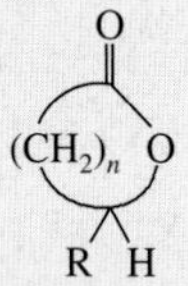

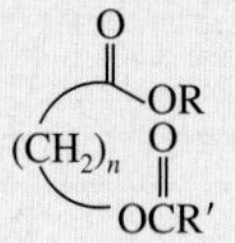

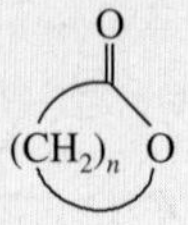

## ADDITIONAL PROBLEMS

**15.40** Name the following compounds, specifying the stereochemistry where appropriate.

(a) 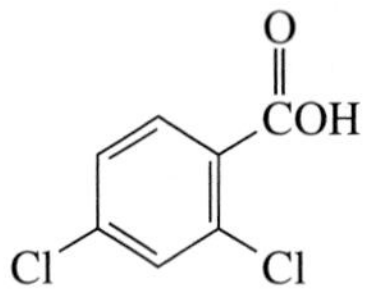 [benzene ring with COH (C=O), 2,4-dichloro substituents]

(b) CH$_3$CH$_2$CHCH$_2$CH$_2$COH (C=O), with Br substituent

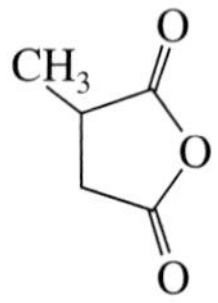

(c) [five-membered lactone ring with CH$_3$ and =O]

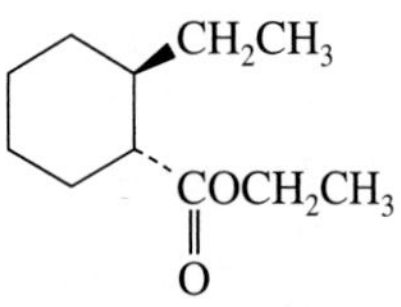

(d) CH$_3$CH$_2$CH$_2$CHC≡N, with CH$_3$ substituent

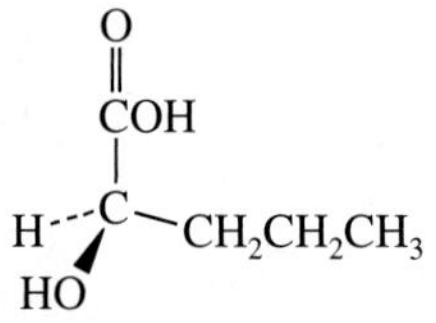

(e) [cyclohexane with CH$_2$CH$_3$ and COCH$_2$CH$_3$ (C=O)]

(f) CH$_3$CH$_2$CH$_2$CH$_2$CNCH$_3$ (C=O), with CH$_3$ on N

(g) H⋯C with COH (C=O), CH$_2$CH$_2$CH$_3$, HO

(h) CH$_3$CH$_2$CH$_2$CHCCl (C=O), with Cl substituent

(i) 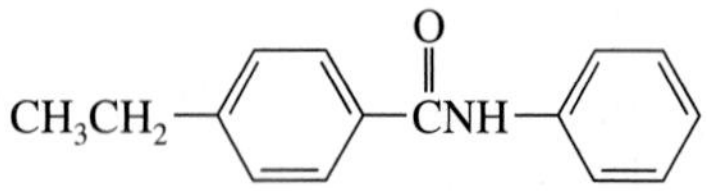 CH$_3$CH$_2$–[benzene ring]–CNH (C=O)–[benzene ring]

(j) 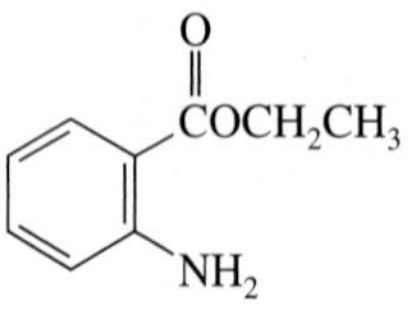 [benzene ring with COCH$_2$CH$_3$ (C=O) and NH$_2$]

(k) 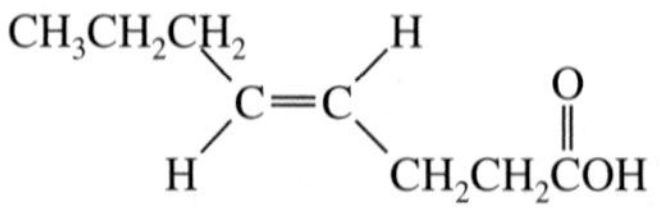 CH$_3$CH$_2$CH$_2$ / H  C=C  H / CH$_2$CH$_2$COH (C=O)

(l) 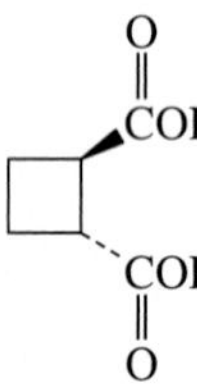 [cyclobutane with COH (C=O) and COH (C=O)]

**15.41** Draw structural formulas for the following compounds.

(a) methyl (*S*)-3-bromopentanoate   (b) *cis*-3-hydroxycyclopentanecarboxylic acid   (c) 2,4-dibromobenzamide
(d) ethyl (*Z*)-3-hexenoate   (e) heptanenitrile   (f) *N,N*-diethylhexanamide   (g) diethyl 2,2-dimethylpentanedioate
(h) methyl *p*-nitrobenzoate   (i) *p*-bromobenzoic anhydride   (j) *p*-methoxybenzoyl chloride

**15.42** Give structural formulas for all of the organic intermediates and products indicated by letters in the following equations.

(a) $CH_3O$—⟨benzene ring⟩—$NH_2$ $\xrightarrow[\substack{\text{acetic acid} \\ H_2O \\ 0-5\ °C}]{CH_3COCCH_3 \text{ (O O)}}$ A

(b) ⟨benzene ring⟩—$\overset{O}{\overset{\|}{C}}OH$ $\xrightarrow[\Delta]{PBr_3}$ B

(c) $CH_2 \overset{CH_2COH}{\underset{CH_2COH}{\big\langle}}$ (O,O) $\xrightarrow[\Delta]{CH_3COCCH_3 \text{ (O O)}}$ C

(d) ⟨3,4-dimethoxyphenyl⟩$\overset{H\ H}{\underset{Ar\ H}{C-C}}NH_2$ $\xrightarrow[\substack{\text{base} \\ \text{dichloromethane}}]{CH_3CCl \text{ (O)}}$ D

(e) ⟨benzene ring⟩$\overset{O}{\underset{CH_2CH_3}{CH\overset{\|}{C}NH_2}}$ $\xrightarrow[\substack{H_2SO_4 \\ \Delta}]{H_2O}$ E

(f) ⟨2-methylphenyl⟩—$CN$ $\xrightarrow[\substack{H_2SO_4 \\ \Delta \\ 5\ h}]{H_2O}$ F

(g) $CH_3CH_2\overset{O}{\overset{\|}{C}}Cl$ $\xrightarrow{NH_3 \text{ (excess)}}$ G

(h) ⟨benzene ring⟩—$\overset{O}{\overset{\|}{C}}OCH_2CH_3$ $\xrightarrow{H_2NOH}$ H

(i) ⟨macrocyclic lactone with $CH_3$⟩ $\xrightarrow[\text{dichloromethane}]{\substack{Cl \\ \text{⟨3-chlorophenyl⟩}—COOH \text{ (O)}}}$ I

(j) ⟨methylenedioxyphenyl⟩—$\overset{O}{\overset{\|}{CH}}$ $\xrightarrow[\substack{H_2O \\ \Delta}]{KMnO_4}$ J $\xrightarrow{H_3O^+}$ K

(k) ⟨2-bromo-1,3,5-trimethylphenyl⟩ $\xrightarrow[\substack{\text{diethyl} \\ \text{ether}}]{Mg}$ L $\xrightarrow{CO_2}$ M $\xrightarrow{H_3O^+}$ N

**15.43** This problem provides more practice in recognizing reactions.

(a) $CH_3\overset{O}{\overset{\|}{C}}OCH_2\overset{CH_3}{\underset{\big|}{C}}=CHCH_2NHCHCH_2OH$ (with $CH=CH_2$) $\xrightarrow[\text{base}]{\text{⟨benzene⟩}\overset{O}{\overset{\|}{C}}Cl \text{ (excess)}}$ A

(b) $CH_3(CH_2)_{10}\overset{O}{\overset{\|}{C}}OCH_2(CH_2)_{10}CH_3 + CH_3OH$ $\xrightarrow[\Delta]{H_2SO_4}$ B + C

(c) $CH_3(CH_2)_{13}O$—⟨phenyl⟩$\overset{H}{\underset{\|}{C}}=\overset{\|}{\underset{H}{C}}\overset{O}{\overset{\|}{C}}OCH_3$ $\xrightarrow[CH_3OH]{KOH}$ $\xrightarrow[\text{(protonation)}]{H_3O^+}$ D + E

(d) ⟨cyclohexenyl⟩$\overset{O}{\overset{\|}{CH}}$ $\xrightarrow[H_2O]{Ag_2O}$ F $\xrightarrow{H_3O^+}$ G

(e) $HO\overset{O}{\overset{\|}{C}}(CH_2)_4\overset{O}{\overset{\|}{C}}OH$ $\xrightarrow{SOCl_2 \text{ (excess)}}$ H $\xrightarrow[\substack{CH_3 \\ \text{⟨phenyl⟩}—NCH_3}]{\substack{CH_3 \\ CH_3COH \text{ (excess)} \\ \big| \\ CH_3}}$ I

(f) $O_2N$—⟨benzene ring⟩$\overset{O}{\overset{\|}{C}}OCH_3$ $\xrightarrow[\Delta]{NaOH \\ H_2O}$ J + K

(g) ⟨azocane ring with N—H and —CH_2OH, CH_2—phenyl⟩ $\xrightarrow[\substack{NaHCO_3 \\ \text{dichloromethane}}]{\text{⟨benzene⟩}\overset{O}{\overset{\|}{C}}Cl \text{ (1 equivalent)}}$ L + M (major product, minor product)

(h) $CCl_3CH_2CH_2CH_2CH_2OH$ $\xrightarrow[\substack{H_2O \\ \Delta}]{KMnO_4}$ N $\xrightarrow{H_3O^+}$ O

(i)

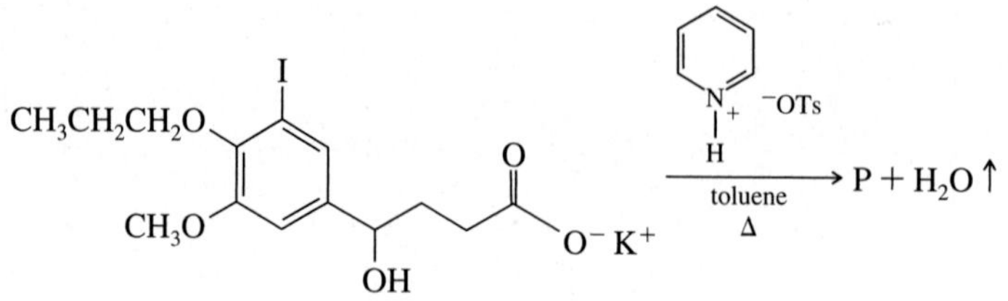

$$CH_3CH_2CH_2O\text{-, } CH_3O\text{-, } I, OH \text{ substituted benzene with } COO^- K^+ \text{ side chain} \xrightarrow[\text{toluene, } \Delta]{\text{pyridinium}^+ \, ^-OTs} P + H_2O \uparrow$$

**15.44** Assign structures to all the organic intermediates and products designated by letters in the equations below.

(a) 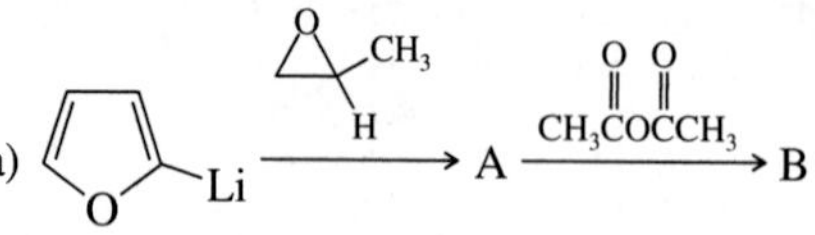 furan-Li $\xrightarrow{\text{epoxide with } CH_3, H}$ A $\xrightarrow{CH_3COCCH_3 \text{ (anhydride)}}$ B

(*Hint:* What regioselectivity do you expect for the reaction of an unsymmetrical oxirane with a nucleophile?)

(b) 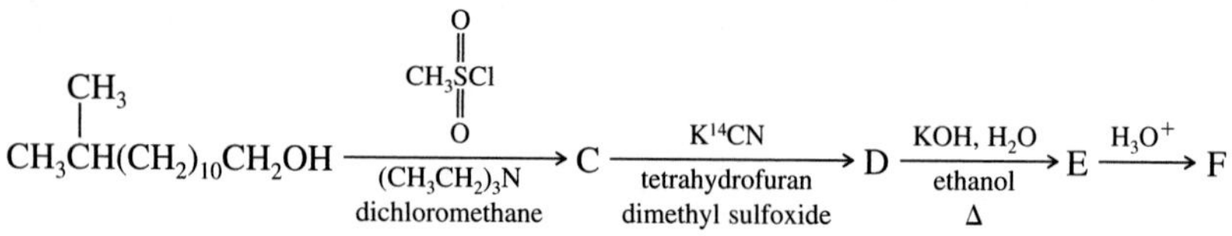 $CH_3CH(CH_3)(CH_2)_{10}CH_2OH \xrightarrow[\text{(CH}_3\text{CH}_2)_3\text{N} \atop \text{dichloromethane}]{CH_3SO_2Cl} C \xrightarrow[\text{tetrahydrofuran} \atop \text{dimethyl sulfoxide}]{K^{14}CN} D \xrightarrow[\text{ethanol} \atop \Delta]{KOH, H_2O} E \xrightarrow{H_3O^+} F$

(c) dimethyl cyclopentanone with COCH₃ $\xrightarrow[\text{methanol}]{NaBH_4} G \xrightarrow[\text{pyridine}]{SOCl_2} H \xrightarrow{\text{(DBU)}} I \xrightarrow{NaOH, H_2O} J$

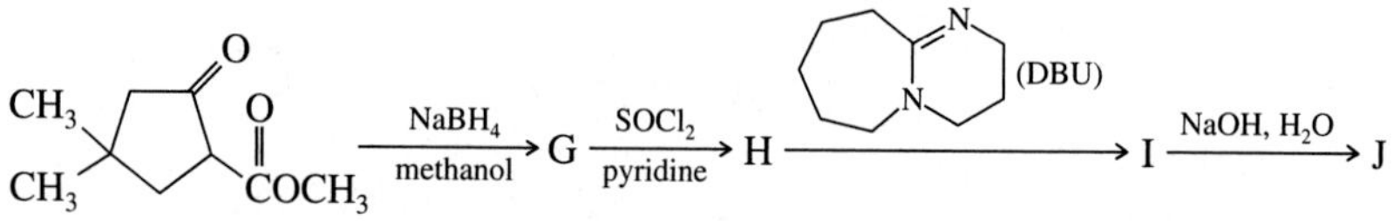

(*Hint:* DBU, 1,8-diazabicyclo[5.4.0] undec-7-ene, is a strong base but not a good nucleophile.)

(d) 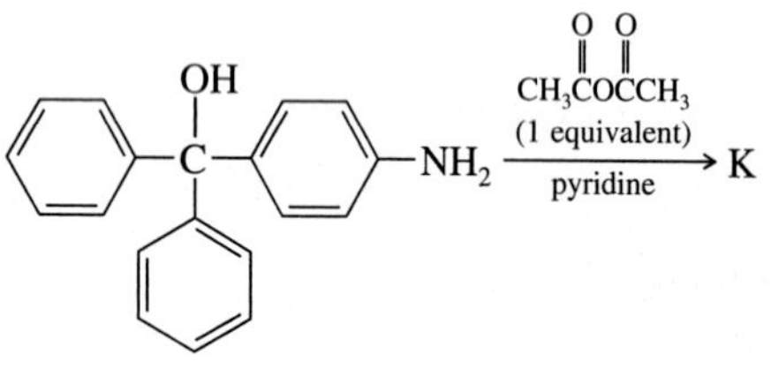 triphenylmethanol-type structure with OH, $-NH_2$ $\xrightarrow[\text{pyridine}]{CH_3COCCH_3 \atop \text{(1 equivalent)}} K$

(e) 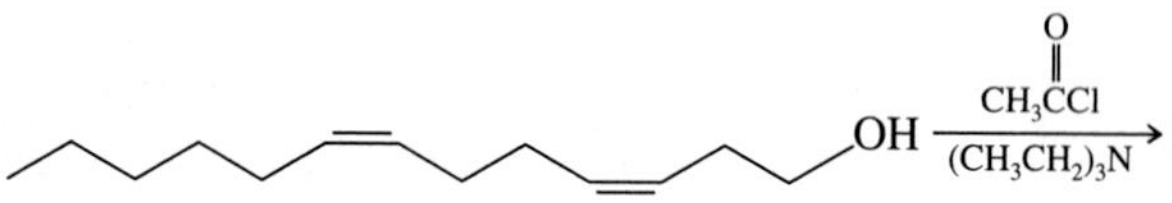 long chain diene with OH $\xrightarrow[\text{(CH}_3\text{CH}_2)_3\text{N}]{CH_3CCl \, (O)} L$

(f) 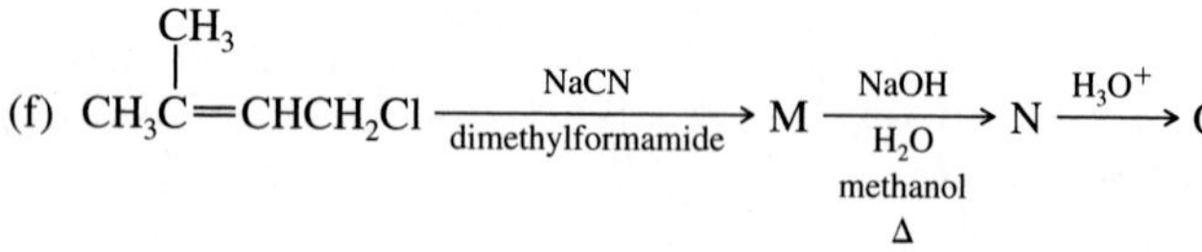 $CH_3C(CH_3)=CHCH_2Cl \xrightarrow[\text{dimethylformamide}]{NaCN} M \xrightarrow[\text{H}_2\text{O} \atop \text{methanol} \, \Delta]{NaOH} N \xrightarrow{H_3O^+} O$

**15.45** Sarin was the nerve gas used in a terrorist attack of a Tokyo subway station in 1995. The compound acts by inhibiting the enzyme cholinesterase, necessary to the deactivation of acetylcholine (p. 637) when its role in the transmission of a nerve impulse is over. Interference with the process of deactivation results in paralysis and death. Sarin is easily detoxified by treatment with base.

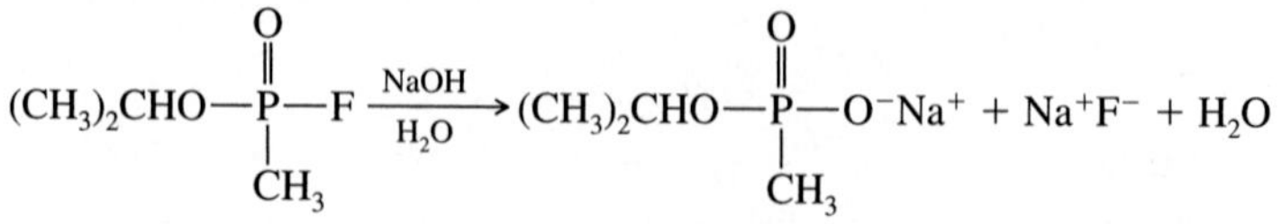

$$(CH_3)_2CHO-\overset{\overset{\textstyle O}{\|}}{\underset{\underset{\textstyle CH_3}{\,}}{P}}-F \xrightarrow[\text{H}_2\text{O}]{NaOH} (CH_3)_2CHO-\overset{\overset{\textstyle O}{\|}}{\underset{\underset{\textstyle CH_3}{\,}}{P}}-O^-Na^+ + Na^+F^- + H_2O$$

Sarin

*highly toxic*

nontoxic degradation
product of Sarin

Propose a mechanism using the curved-arrow convention for the reaction of Sarin with hydroxide ion. Assume that the phosphorus–oxygen double bond is analogous to a carbonyl group.

**15.46** When tributylphosphine, $Bu_3P$, is added in catalytic amounts (5–20 mol %) to acetic anhydride, it greatly increases the rate at which the acid anhydride reacts with an alcohol to give an ester. This effect is postulated to be the result of an acyl-transfer reaction between the strongly nucleophilic tributylphosphine and acetic anhydride, which results in a new acyl-transfer agent that is more reactive than acetic anhydride itself.

(a) What is the structure of this new acyl-transfer agent, and why is it more reactive than acetic anhydride?
(b) Use the curved-arrow convention to show how the acyl-transfer agent created in part (a) would react with an alcohol of your choice in the presence of a base, such as triethylamine.

**15.47** The synthesis of a novel peptide involved the following steps. Supply structural formulas for the reagents and products indicated by letters.

(a)

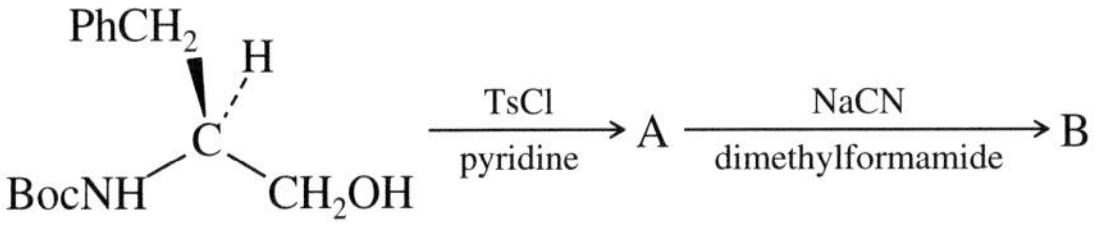

$$\underset{\underset{CH_2OH}{BocNH}}{\overset{\overset{PhCH_2\ \ H}{|}}{C}} \quad \xrightarrow[\text{pyridine}]{\text{TsCl}} A \xrightarrow[\text{dimethylformamide}]{\text{NaCN}} B$$

(b)

$$\underset{\underset{CH_2COCH_3}{BocNH}}{\overset{\overset{PhCH_2\ \ H}{|}}{C}} \quad \xrightarrow[\text{H}_2\text{O}]{\text{LiOH}} C \xrightarrow[\text{(protonation)}]{\text{H}_3\text{O}^+} D$$

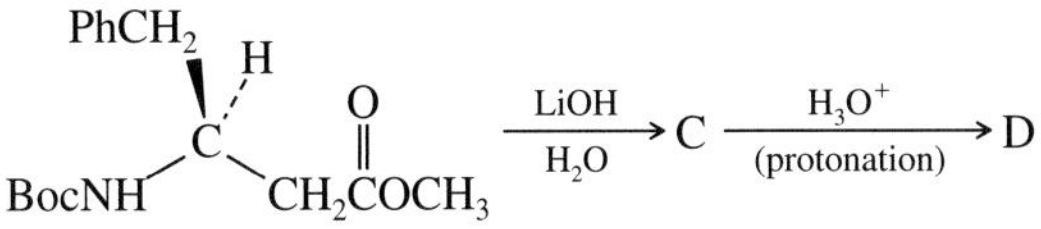

**15.48** In the synthesis of a compound that acts against a blood factor that promotes clotting, the following reactions were carried out. Suggest a structure for Compound A and a mechanism for the conversion of Compound A to the final product.

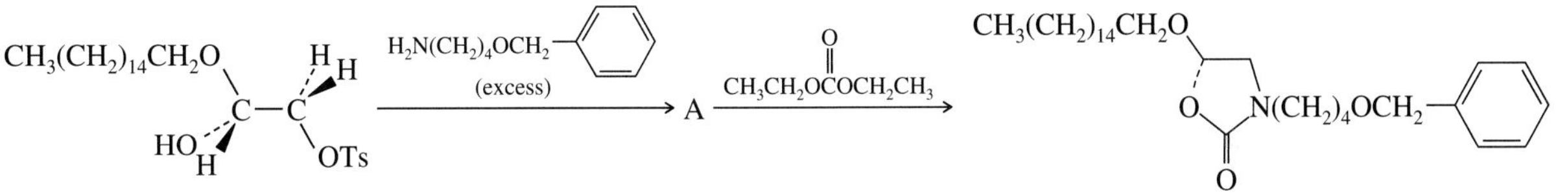

**15.49** The synthesis of a compound that mimics peptides involves the following steps. Supply the structure of reagent A and a mechanism for the formation of the *tert*-butyl ester of the carboxylic acid under the conditions shown.

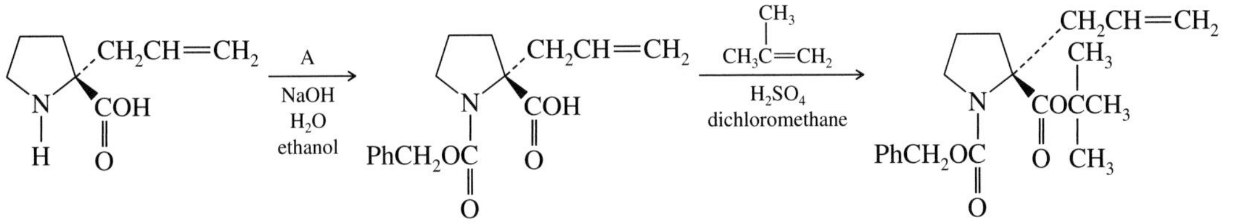

**15.50** Coriolic acid, isolated from the hearts of cows, is of interest because of its ability to transport calcium ions through membranes. It was synthesized by the following sequence of steps. Supply structural formulas for the compounds designated by letters.

(a) 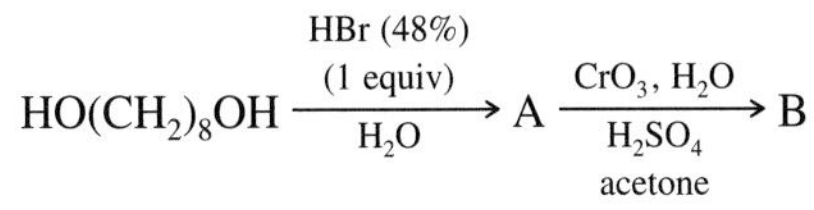 $HO(CH_2)_8OH \xrightarrow[\substack{\text{(1 equiv)}\\ \text{H}_2\text{O}}]{\text{HBr (48\%)}} A \xrightarrow[\substack{\text{H}_2\text{SO}_4\\ \text{acetone}}]{\text{CrO}_3,\ \text{H}_2\text{O}} B$

(b) $HC\equiv C-\underset{\underset{H}{|}}{\overset{\overset{H}{|}}{C}}-CH_2OH \xrightarrow[\substack{\text{NH}_3\text{(liq)}\\ \text{tetrahydrofuran}}]{\substack{\text{Li}^+\text{NH}_2^-\\ \text{(2 equiv)}}} C$

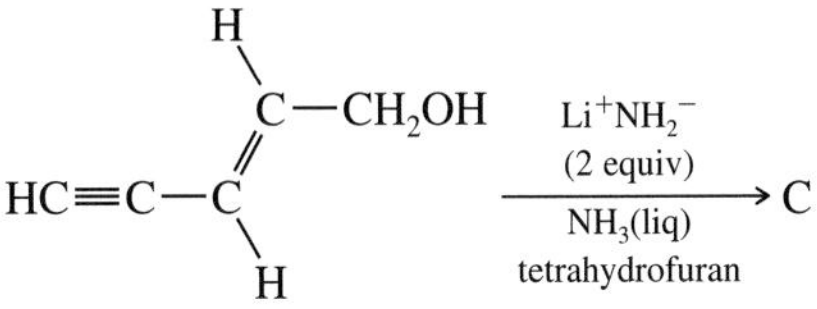

(c)

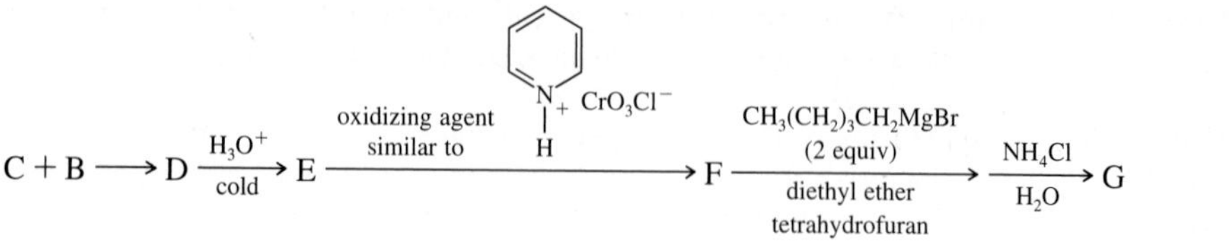
C + B → D
$H_3O^+$
cold
E
oxidizing agent
similar to
$CrO_3Cl^-$
F
$CH_3(CH_2)_3CH_2MgBr$
(2 equiv)
diethyl ether
tetrahydrofuran
$NH_4Cl$
$H_2O$
G

(d)

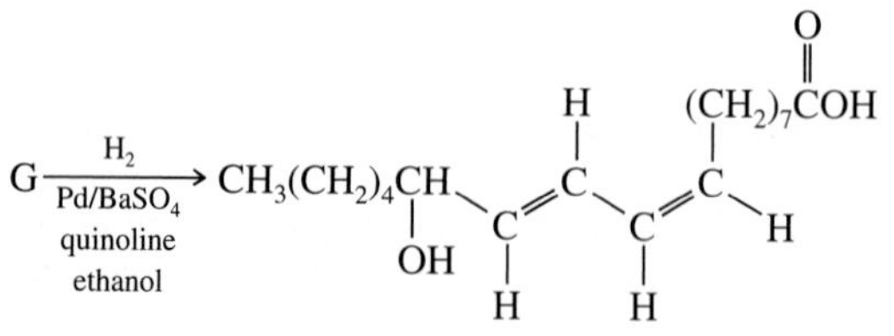
G
$H_2$
Pd/BaSO$_4$
quinoline
ethanol
$CH_3(CH_2)_4CH$
OH
$(CH_2)_7COH$

coriolic acid

**15.51** Chemists more and more are using enzymes to help them resolve racemic mixtures and to convert achiral compounds into chiral ones. Achiral dimethyl 3-hydroxypentanedioate reacts with ammonia in the presence of an enzyme, a lipase from the microorganism *Candida antarctica,* selectively at one of the ester groups to give the *S* enantiomer of the monomethylester-monoamide of 3-hydroxypentanedioic acid. Reaction with ammonia in the absence of the enzyme gives a racemic mixture of the monoester-monamide.

(a) Draw a structure for the starting material ester, and show how it reacts with ammonia to give the chiral product.
(b) When the product from the reaction in part a is esterified with (*S*)-3,3,3-trifluoro-2-methoxy-2-phenylpropanoyl chloride, the ester that forms has one signal in its $^{19}F$ nuclear magnetic resonance spectrum. Predict what the spectrum looks like when the racemic half ester–half amide of 3-hydroxypentanedioic acid is esterified with (*S*)-3,3,3-trifluoro-2-methoxy-2-phenylpropanoyl chloride.

**15.52** In research into the stereochemistry of nucleophilic reactions, the following reaction was observed.

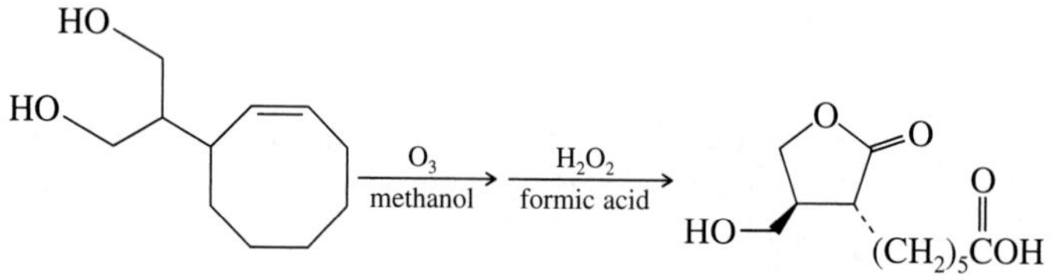
HO
HO
$O_3$
methanol
$H_2O_2$
formic acid
O
O
O
HO
$(CH_2)_5COH$

major stereoisomer formed

Show, by drawing structural formulas for any intermediate compounds formed, how the starting material is converted to the product.

**15.53** The following reaction was observed in the synthesis of an alkaloid.

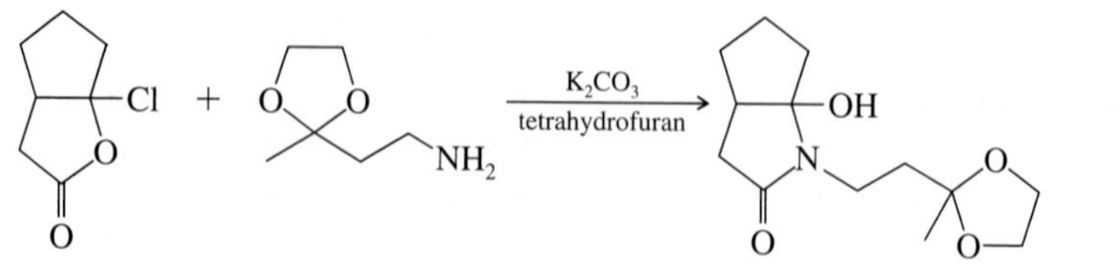
Cl
O
O
O
O
$NH_2$
$K_2CO_3$
tetrahydrofuran
OH
N
O
O
O

71%

Propose a mechanism for the reaction.

**15.54** The following reactions were carried out in the synthesis of amino acid analogues. Supply structural formulas for the products indicated by letters.

(a)

(b)

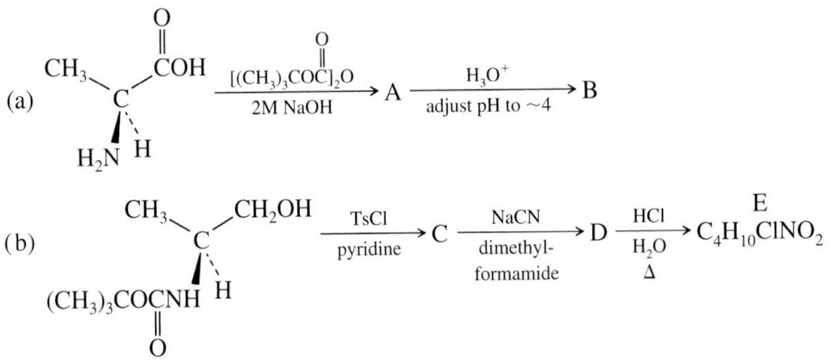

**15.55** The sweetener aspartame was discovered by accident when a researcher synthesizing part of the gastric hormone gastrin noticed that the methyl ester of L-($\alpha$)-aspartyl-L-phenylalanine was intensely sweet. The structure of aspartame is shown below.

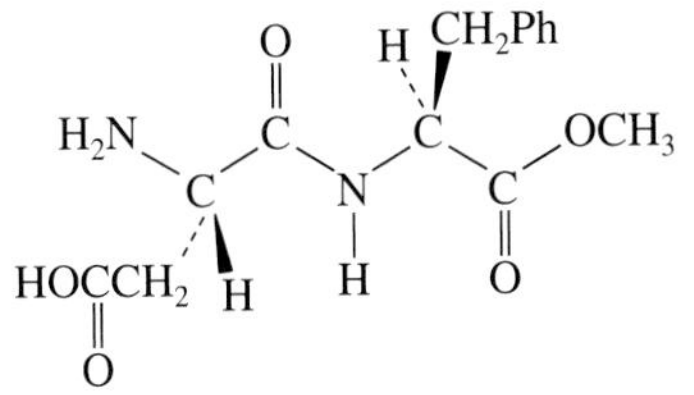

aspartame
L-($\alpha$)-aspartyl-L-phenylalanine methyl ester

The researcher and his colleagues went on the synthesize a number of analogs of aspartame to try to pinpoint the source of its sweetness. Some of the steps in the synthesis of a stereoisomer of aspartame are shown below. Fill in the missing reagents and the products indicated by letters, paying careful attention to stereochemistry.

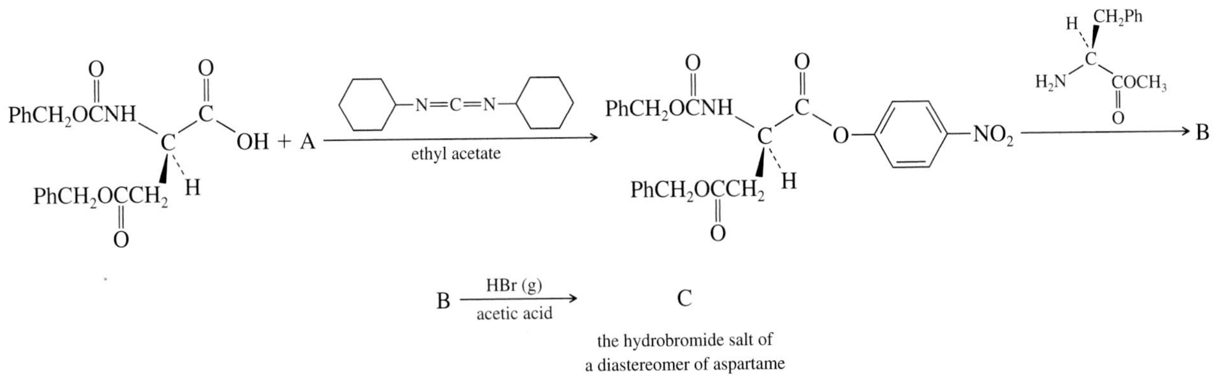

the hydrobromide salt of
a diastereomer of aspartame

**15.56** The proton magnetic resonance spectrum of *N,N*-dimethylformamide (Figure 15.4) shows two different chemical shifts for the hydrogen atoms on the methyl groups. The phenomenon observed is not due to spin–spin coupling. How do you know this? What does this observation reveal about the nature of the bond between the nitrogen atom and the carbonyl group? Draw structural formulas that account for the observation. (*Hint:* Writing resonance contributors for the compound may be helpful.)

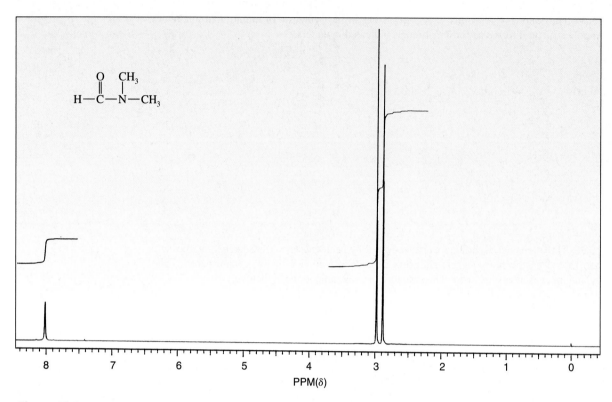

*Figure 15.4*

**15.57** Compound A, $C_9H_{10}O_2$, has sharp singlets in its proton magnetic resonance spectrum at $\delta$ 2.2 (3H), 5.2 (2H), and 7.4 (5H). The important bands in its infrared spectrum are at 1743, 1229, and 1027 $cm^{-1}$. Its $^{13}C$ nuclear magnetic resonance spectrum has peaks at 20.7, 66.1, 128.1, 128.4, 136.2, and 170.5 ppm. Assign a structure to Compound A, showing how the spectroscopic data support it.

**15.58** Compound B, $C_9H_{10}O_3$, has important bands in its infrared spectrum at 3300–2600 and 1719 $cm^{-1}$. Its proton magnetic spectrum has peaks at $\delta$ 3.56 (2H, s), 3.78 (3H, s), 6.85 (2H, m), 7.18 (2H, m), and 11.80 (1H, s). In its $^{13}C$ nuclear magnetic resonance spectrum, there are peaks at $\delta$ 40.14, 55.19, 114.00, 125.24, 130.35, 158.75, and 178.50. Assign a structure to Compound B that is compatible with the spectral data given above.

**15.59** Compound C, $C_5H_8O_2$, has industrial importance, being widely used in polymerization reactions. The proton magnetic resonance spectrum of C is shown in Figure 15.5 (p. 651). In its infrared spectrum, absorption bands appear at 1731, 1635, 1279, 1207, 1069, and 989 $cm^{-1}$. Its $^{13}C$ nuclear magnetic resonance spectrum has peaks at $\delta$ 18.33, 51.77, 125.33, 136.28, and 167.85. What is the structure of C? Analyze all the spectral data, being as precise as possible in assigning the absorptions appearing in the spectra.

Compound C

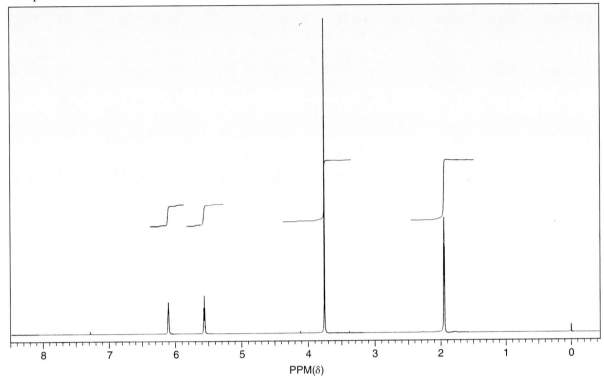

*Figure 15.5*

**15.60** Compound D, $C_8H_{14}O_4$, has absorption bands in its infrared spectrum at 1736, 1160, and 1032 cm$^{-1}$. It has four peaks in its $^{13}C$ nuclear magnetic resonance spectrum at $\delta$ 14.19, 29.24, 60.61, and 172.20. Its proton magnetic resonance spectrum has three peaks, $\delta$ 1.26 (3H, t), 2.62 (2H, s), and 4.15 (2H, q). Show how each part of the spectral data supports the assignment of a structure to Compound D.

**15.61** The proton magnetic resonance spectra of Compounds E and F are shown in Figure 15.6. Compound E absorbs in the infrared at 1743, 1243, and 1031 cm$^{-1}$. It has peaks in its $^{13}C$ nuclear magnetic resonance spectrum at $\delta$ 13.70, 19.19, 20.95, 30.75, 64.33, and 171.05. The peaks in the $^{13}C$ nuclear magnetic resonance spectrum of Compound F are at $\delta$ 13.68, 18.49, 36.02, 51.34, and 174.01. It absorbs at 1741, 1198, and 1097 cm$^{-1}$ in the infrared. Assign structures to Compounds E and F, and show how the spectral data support your assignments.

Compound E

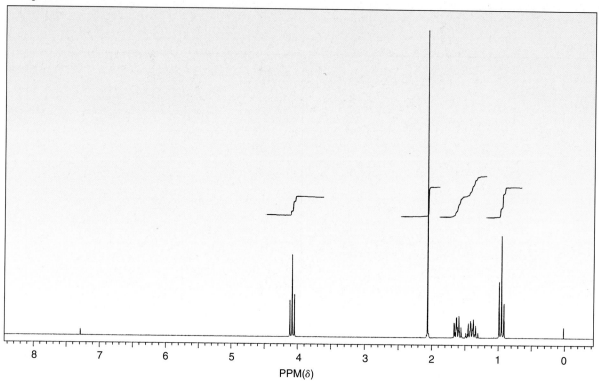

Compound F

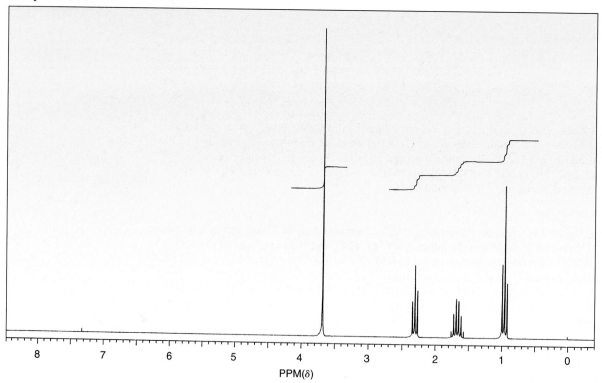

*Figure 15.6*

652

# Structural Effects in Acidity and Basicity Revisited. Enolization

## A Look Ahead

The reactivity of compounds as acids and bases was first explored in Chapter 3 and has formed the underpinning for an understanding of much of the chemistry discussed since that chapter. For example, we need to know the $pK_a$ of protic acids in order to decide whether they will add to multiple bonds (Chapters 4 and 8). Whether a group will be a good leaving group or not depends to a considerable extent on the $pK_a$ of its conjugate acid (Chapters 7, 13, and 15). And while nucleophilicity does not always parallel basicity, it is still useful to consider the two properties together (Chapters 7, 14, and 15). In addition, in many reactions, protonation and deprotonation steps are important when we look at detailed mechanisms.

The concepts that enable us to make predictions about the acidity and basicity of the compounds we considered in Chapter 3, that of stabilization of conjugate bases of acids by resonance and inductive effects, will be extended in this chapter to a whole series of new classes of compounds. Two major themes will dominate the chapters that follow this one. They are (1) the chemistry of compounds with more than one functional group, many of them of biological importance and (2) the chemistry of enolates, carbon bases that form when the hydrogen atoms adjacent to groups such as carbonyl groups, sulfonyl groups, nitro groups, and cyano groups, the $\alpha$-hydrogens, are removed as protons. They are important in syntheses and in biological transformations. This chapter will serve as an introduction to some of the concepts that will be important in understanding the chemistry that follows while building on the knowledge that you have been using in your earlier work.

## 16.1 Resonance and Inductive Effects in Acidity and Basicity. A Review in the Context of Aromatic Compounds

### A. Acidity of Aromatic Carboxylic Acids

Carboxylic acids ($pK_a$ ~5) are much weaker acids than are the strong inorganic acids, such as hydrogen chloride ($pK_a$ −7) and sulfuric acid ($pK_a$ −9). They are, however, strong for organic acids. The strength of a carboxylic acid depends on the other groups that are substituted on the hydrocarbon portion of the molecule. The effect of substitution on the acidity of alkyl carboxylic acids is discussed extensively in Chapter 3 and is summarized in Tables 3.1 and 3.2 (p. 105), and 3.3 (p. 107). Inductive effects are used to explain the observed differences in acidity for substituted acids (p. 104).

Aromatic carboxylic acids are stronger acids than alkyl carboxylic acids. Benzoic acid, for example, has a $pK_a$ of 4.2 and is a slightly stronger acid than acetic acid, $pK_a$ 4.8. In benzoic acid, the carboxyl group is bonded to an $sp^2$-hybridized carbon atom, which is more electronegative than an $sp^3$-hybridized carbon atom (p. 60).

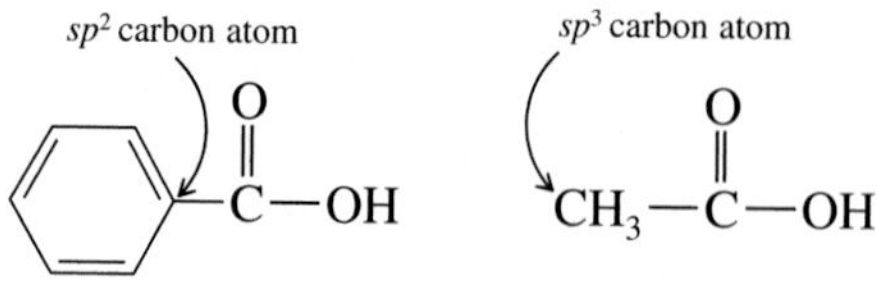

benzoic acid
$pK_a$ 4.2

*stronger acid than acetic acid, carboxyl group bonded to a more electronegative carbon atom*

acetic acid
$pK_a$ 4.8

The greater electronegativity of an $sp^2$-hybridized carbon atom is used to explain the apparent electron-withdrawing inductive effect that the phenyl group demonstrates in stabilizing the carboxylate anion.

Just as substituents on acetic acid (Table 3.1, p. 105) change its acidity, so do substituents on the aromatic ring in benzoic acid. For example, all the fluorobenzoic acids are more acidic than benzoic acid itself.

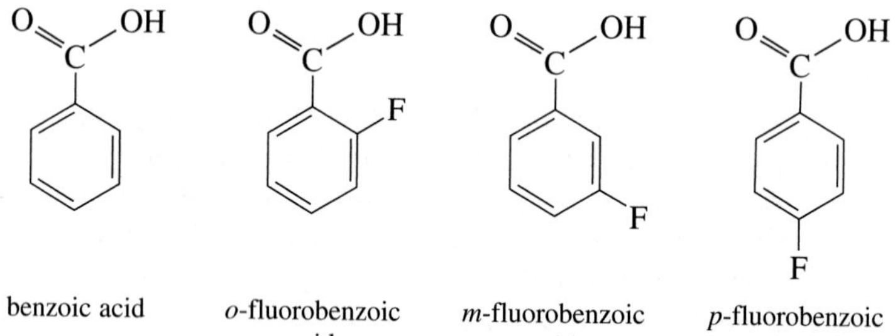

| benzoic acid | *o*-fluorobenzoic acid | *m*-fluorobenzoic acid | *p*-fluorobenzoic acid |
|---|---|---|---|
| $pK_a$ 4.2 | $pK_a$ 3.3 | $pK_a$ 3.9 | $pK_a$ 4.1 |

*o*-Fluorobenzoic acid, in which the fluorine atom is closest to the carboxyl group, is the strongest acid of the four; *p*-fluorobenzoic acid, in which the fluorine atom is farthest from the carboxyl group, is only slightly more acidic than benzoic acid. The electronegative fluorine atom exerts its effect through bonds and through space, withdrawing electronic density from the vicinity of the carboxyl group and thus stabilizing the conjugate base.

For some substituents, the inductive effect is not sufficient to explain the experimental observations. *p*-Nitrobenzoic acid is a stronger acid than *m*-nitrobenzoic acid. This observation contrasts with the observed acidities for the fluorobenzoic acids, for which acid strength decreases as the electron-withdrawing group is moved farther away from the carboxyl group.

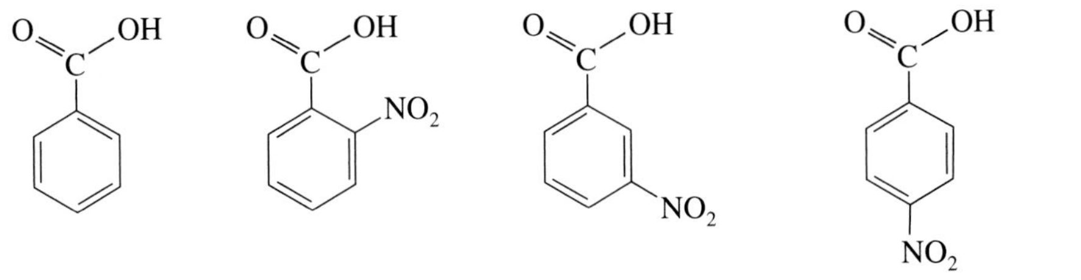

| benzoic acid | o-nitrobenzoic acid | m-nitrobenzoic acid | p-nitrobenzoic acid |
|---|---|---|---|
| $pK_a$ 4.2 | $pK_a$ 2.2 | $pK_a$ 3.5 | $pK_a$ 3.4 |
| | *strong inductive effect and resonance effect* | *inductive effect* | *weak inductive effect and strong resonance effect* |

Resonance contributors having a positive charge at the carbon atom bearing the carboxylate anion can be written for the ortho and para isomers of nitrobenzoic acid, whereas the nitro group in the meta position exerts primarily an inductive effect. The carboxylate anion in each case is stabilized by combinations of electron-withdrawing effects. All the nitrobenzoic acids are stronger than benzoic acid. In o-nitrobenzoic acid, the strongest acid of the three isomers, the inductive effect is strong, and a resonance effect is also present. In p-nitrobenzoic acid, a resonance effect operates, but the inductive effect is weak.

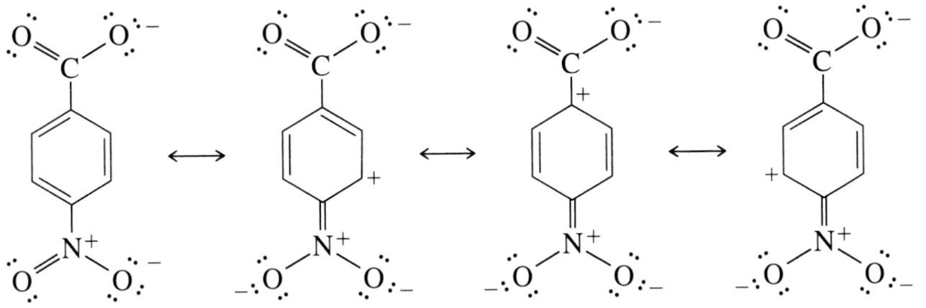

*this resonance contributor puts a positive charge on the carbon atom to which the carboxylate group is bound*

The fact that m-nitrobenzoic acid is a weaker acid than p-nitrobenzoic acid reconfirms that when the resonance effect operates, it is more important than the inductive effect.

**Problem 16.1** Draw resonance contributors for the anions of o- and m-nitrobenzoic acid. Point out the contributor that is important in explaining the acidity of the ortho compound, and prove to yourself that no such effect is present in the meta isomer.

## B. Acidity of Phenols

**Phenols** are compounds in which a hydroxyl group is bonded directly to an $sp^2$-hybridized carbon atom of an aromatic ring. A typical phenol is much more acidic than an alcohol but less so than a carboxylic acid. Phenol has $pK_a$ 10.0; the $pK_a$ of benzyl alcohol is approximately 16, and that of benzoic acid is 4.2. As was emphasized earlier (p. 104), the alkoxide anion formed on the removal of a proton from an alcohol is strongly basic because the negative charge is localized on the oxygen atom. A carboxylate anion is much less basic than an alkoxide anion because the negative charge on the former is delocalized to two oxygen atoms. The anion of a phenol, a phenolate anion, also has delocalization of charge. Resonance contributors can be written that have the negative charge at the ortho and para positions of the aromatic ring.

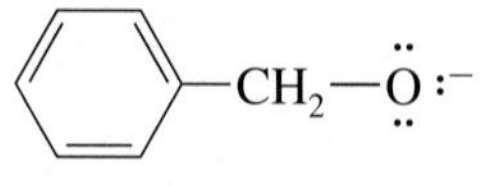

benzylate anion

*no delocalization of charge*
*most basic*

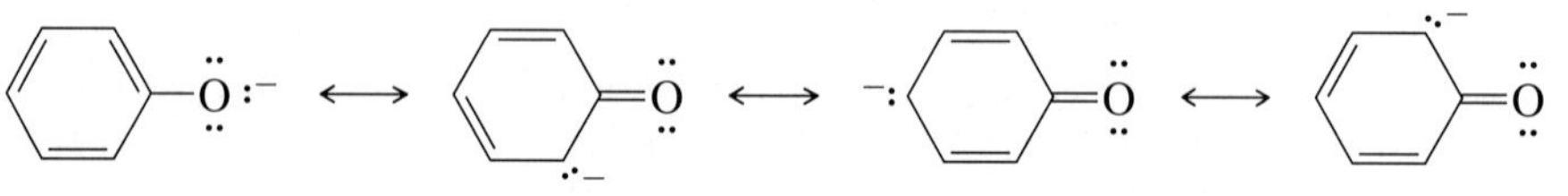

phenolate anion

*delocalization of charge to aromatic ring*

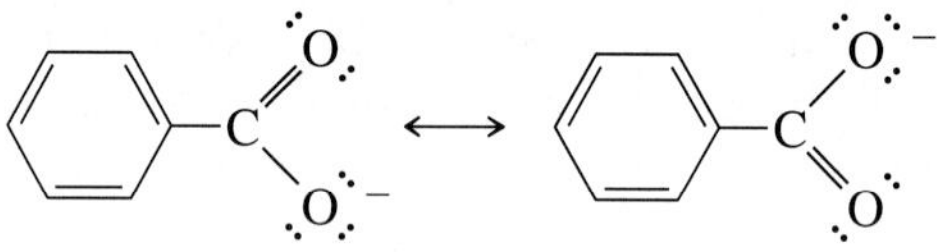

benzoate anion

*delocalization of charge to 2 oxygen atoms*
*least basic*

The benzoate anion has the greatest stability and the lowest basicity of the three anions shown above. The two resonance contributors for the benzoate anion are equivalent, and both have the negative charge on the most electronegative of the atoms present. These factors stabilize the anion.

There are more resonance contributors for the phenolate anion than there are for the benzoate anion. However, the four contributors for the phenolate anion are not all equivalent. In three of them the negative charge is on a carbon atom instead of on the more electronegative oxygen atom. Thus the phenolate anion is stabilized more than the benzylate anion, in which no delocalization is possible, but not as much as the benzoate anion is.

The acidity of phenols has practical consequences in the laboratory. For example, a phenol can be separated from a carboxylic acid by taking advantage of the difference in acidity. Benzoic acid, a solid carboxylic acid, is insoluble in cold water but dissolves in a solution of sodium bicarbonate (p. 600). Its dissolving is accompanied by the evolution of bubbles of carbon dioxide, indicating that carbonic acid is being formed (p. 103). A phenol, such as 2-naphthol, is also insoluble in water and does not react with aqueous sodium bicarbonate because it is less acidic than carbonic acid. It is not a strong enough acid to protonate the bicarbonate anion. A phenol is a stronger acid than water, however, and easily protonates hydroxide ion. Thus 2-naphthol will dissolve in a dilute solution of sodium hydroxide in water.

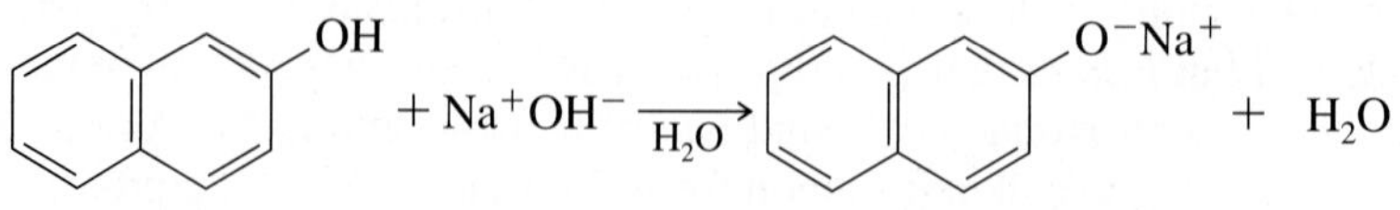

2-naphthol
$pK_a \sim 10$

*covalent compound*
*insoluble in water*

sodium 2-naphtholate

*ionic compound*
*soluble in water*

water
$pK_a$ 15.7

The acidity of phenols is affected by substituents on the aromatic ring.

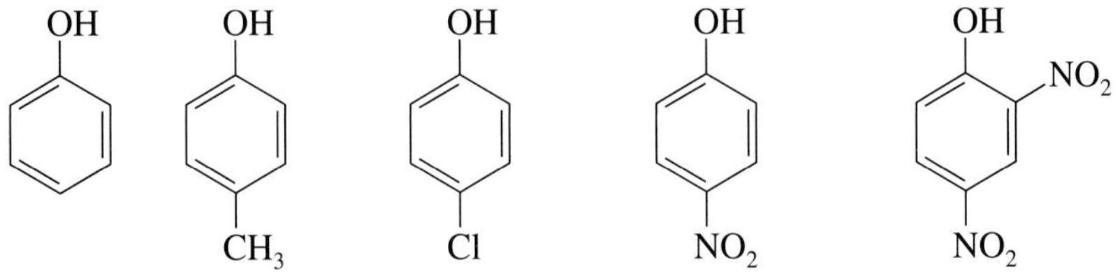

| phenol | *p*-cresol | *p*-chlorophenol | *p*-nitrophenol | 2,4-dinitrophenol |
|---|---|---|---|---|
| p$K_a$ 10.0 | p$K_a$ 10.2 | p$K_a$ 9.38 | p$K_a$ 7.15 | p$K_a$ 4.02 |

Note that the effects of the above substituents on phenols parallel the effects that the same substituents have on the acidity of carboxylic acids (pp. 654–655).

### Problem 16.2

(a) The p$K_a$ of *m*-nitrophenol is 8.39, and that of *p*-nitrophenol is 7.15. How would you rationalize these facts? The p$K_a$ of phenol is 10.0.

(b) 2,4,6-Trinitrophenol, also called picric acid, has a p$K_a$ value of 0.25. Is picric acid soluble in sodium bicarbonate solution? Explain your answer.

### Problem 16.3 For the following pair of compounds, predict which is the stronger acid. Draw structural representations that support your choice.

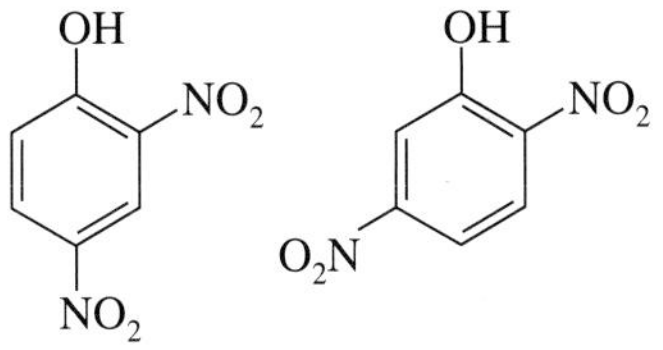

## C. Basicity of Aromatic Amines

Alkyl amines are somewhat more basic than ammonia (p. 108). Their basicity depends on the number of alkyl groups bonded to the nitrogen atom; tertiary amines are less basic than primary and secondary amines. The presence of three organic groups around the nitrogen atom gives rise to steric hindrance that makes both protonation of the amine and solvation of the cation that results more difficult. These effects lead to a decrease in the basicity of a tertiary amine in comparison with that of secondary and primary amines.

### Problem 16.4 For the reaction

$$R_3\overset{+}{N} : \overset{-}{BR'}_3 \rightleftharpoons R_3N : + BR'_3$$

the following equilibrium constants, called dissociation constants, were measured:

| $H_3\overset{+}{N} : \overset{-}{B}(CH_3)_3$ | $CH_3\overset{+}{N}H_2 : \overset{-}{B}(CH_3)_3$ | $(CH_3)_2\overset{+}{N}H : \overset{-}{B}(CH_3)_3$ | $(CH_3)_3\overset{+}{N} : \overset{-}{B}(CH_3)_3$ |
|---|---|---|---|
| $K_{diss}$ 4.6 | $K_{diss}$ 0.0350 | $K_{diss}$ 0.0214 | $K_{diss}$ 0.477 |

(a) Write the expression for the dissociation constant for the reaction.
(b) Give a brief rationalization of the observed experimental facts.

**Problem 16.5** The following $pK_a$ values were determined for the conjugate acids of the amines shown. How would you rationalize the trend observed?

$CH_3(CH_2)_3NH_2$         $CH_3O(CH_2)_3NH_2$         $CH_3OCHCH_2NH_2$         $N{\equiv}CCH_2CH_2NH_2$         $N{\equiv}CCH_2NH_2$
                                                    (with $CH_3$ above $CHCH_2NH_2$ carbon)

$pK_a$ 10.60              $pK_a$ 9.92              $pK_a$ 8.54              $pK_a$ 7.8              $pK_a$ 5.34

Aryl amines are much weaker bases than ammonia. For example, the conjugate acid of aniline is slightly more acidic ($pK_a$ 4.6) than acetic acid ($pK_a$ 4.8) and much more acidic than methylammonium ion ($pK_a$ 10.6).

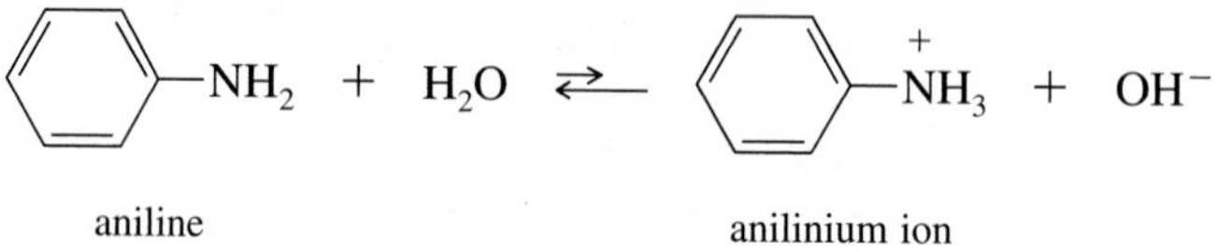

aniline

*base*

anilinium ion

$pK_a$ 4.6

*conjugate acid*

The nonbonding electrons on the nitrogen atom in aniline are said to be less available for reactions with acids for two reasons. (1) The nitrogen is bonded to an $sp^2$-hybridized carbon atom of the aromatic ring (p. 68), which is more electronegative (p. 60) than the $sp^3$-hybridized carbon atom of methylamine. (2) The nonbonding electrons can be delocalized to the aromatic ring. Resonance contributors for aniline indicate that it has decreased electron density at the nitrogen atom and increased electron density in the ring.

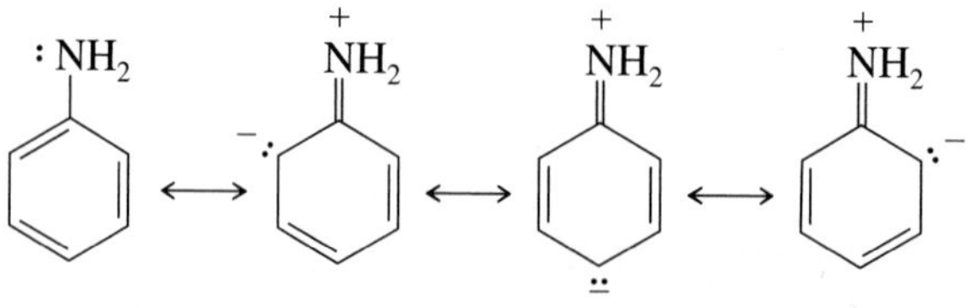

*resonance contributors for aniline*

The basicity of aromatic amines is affected by substituents on the aromatic ring. For example, the conjugate acid of *p*-nitroaniline has a $pK_a$ value of 1.0, indicating that that amine is a very weak base. The low basicity of an aryl amine with a nitro group substituted on the ring is rationalized by showing resonance contributors in which there is further delocalization of the nonbonding electrons of the amino group.

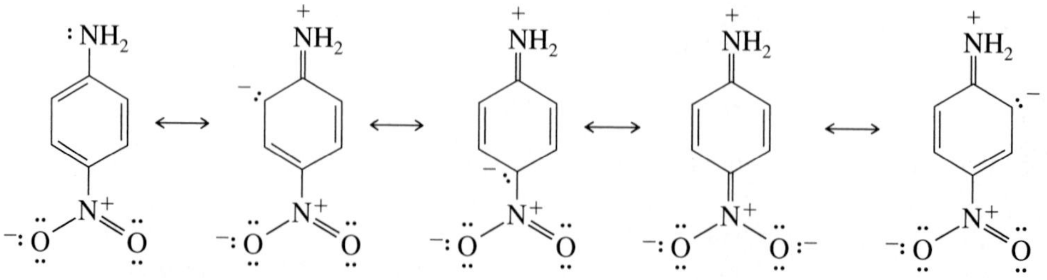

However, the conjugate acid of *p*-anisidine, also called 4-methoxyaniline, has a $pK_a$ value of 5.34. Although an oxygen substituent on an aromatic ring has an electron-withdrawing inductive effect, it has an electron-donating resonance effect. The methoxy group is, therefore, expected to increase electron density near the amino group and to stabilize the conjugate acid of the amine.

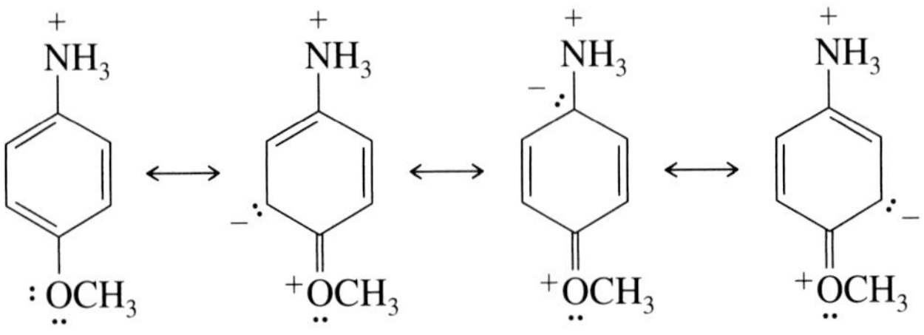

An amine that has low solubility in water, such as aniline, is converted to a water-soluble salt on protonation in dilute acid. Treatment of the conjugate acid of the amine with a strong base results in the regeneration of the water-insoluble amine.

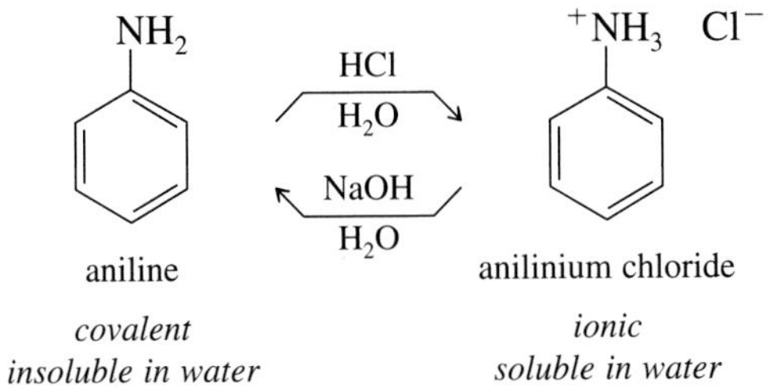

This easy interconversion of an amine and its conjugate acid is useful in separating amines from compounds that are not basic enough to be protonated to give ionic conjugate acids in dilute hydrochloride acid.

**Problem 16.6**    The interconversion of benzoic acid and its conjugate base was shown on page 600. Use the chemistry shown there and in the equations above to devise a way to separate a mixture of aniline, benzoic acid, and benzanilide (p. 605). It is best to dissolve all the organic compounds in some common solvent such as diethyl ether or dichloromethane before attempting to extract them into water as their salts.

**Problem 16.7**    The following $pK_a$ values were measured for the conjugate acids of amines substituted by cyano groups. Rationalize the trend that is apparent in the data.

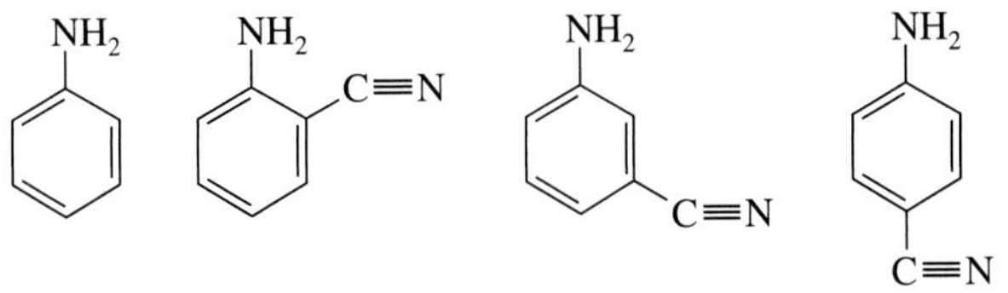

| $pK_a$ | 4.60 | 0.95 | 2.75 | 1.74 |

# 16.2 Acidity and Basicity in Polyfunctional Compounds

## A. Acidity in Dicarboxylic Acids

The $pK_a$ table inside the front cover of this book indicates the acidities of many different types of hydrogen atoms, including those on oxygen, nitrogen, sulfur, and carbon. Any compound that has more than one hydrogen atom on different atoms is thus potentially a **polyprotic acid,** an acid that can lose more than one proton in acid–base reactions. Such acid–base reactions have already been encountered in problems earlier in the book (Problems 9.19 and 13.51, for example). A more common use of the term *polyprotic acid* refers to compounds such as $H_2SO_4$, which has two hydrogen atoms on oxygen that can be lost in ionization reactions. For example:

$$H_2SO_4 \; + \; H_2O \; \rightleftharpoons \; H_3O^+ \; + \; HSO_4^-$$

| sulfuric acid | water | hydronium ion | hydrogen sulfate |
| $pK_a$ −9 | | $pK_a$ −1.7 | ion |
| *acid* | *base* | *conjugate acid* | *conjugate base* |

$$HSO_4^- \; + \; H_2O \rightleftharpoons \; H_3O^+ \; + \; SO_4^{2-}$$

| hydrogen sulfate ion | water | hydronium ion | sulfate ion |
| $pK_a$ 1.9 | | $pK_a$ −1.7 | |
| *acid* | *base* | *conjugate acid* | *conjugate base* |

The first ionization of sulfuric acid goes nearly to completion because of the large difference in $pK_a$ between the acid and hydronium ion, the conjugate acid of water. Hydrogen sulfate ion is a very weak base because of resonance stabilization of the ion.

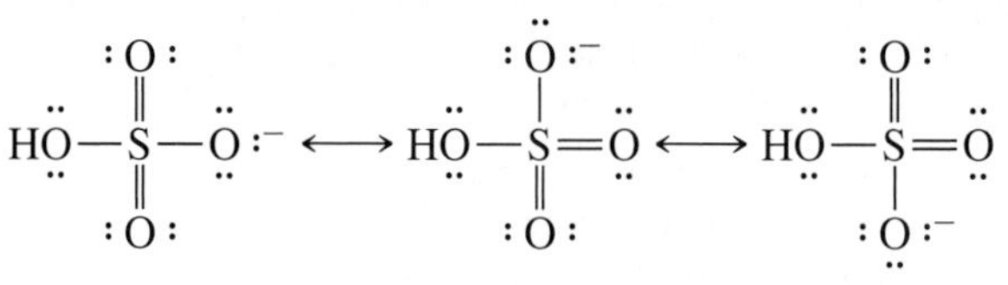

resonance stabilization of the hydrogen sulfate anion

Once the anion has a negative charge on it, removal of a second proton becomes much more difficult. To do so, we have to separate a positively charged particle from one that is becoming doubly negatively charged, which requires the input of more energy; remember playing with magnets (p. 5).

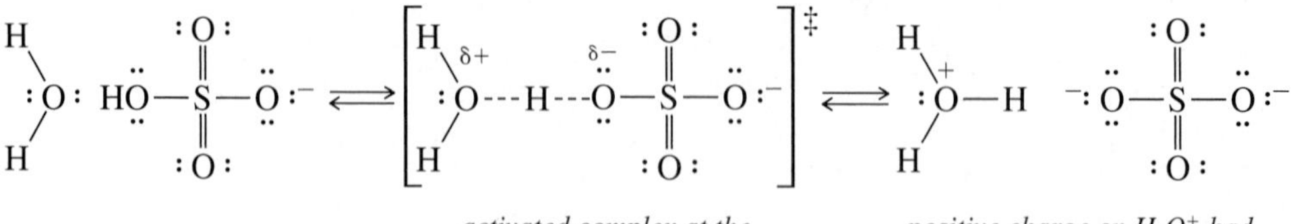

*activated complex at the transition state for the deprotonation of the hydrogen sulfate anion*

*positive charge on $H_3O^+$ had to be separated from the double negative charge of $SO_4^{2-}$*

Many organic acids, including many that are biologically important, are polyprotic acids in the sense that they contain two or more carboxylic acid groups and, just as is the case for sulfuric acid, have different $pK_a$ values for the protons that are removed in successive ionizations. Some of the polyprotic acids found in nature are shown on the next page.

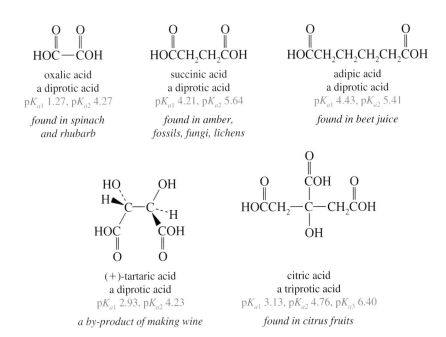

oxalic acid
a diprotic acid
$pK_{a1}$ 1.27, $pK_{a2}$ 4.27

*found in spinach
and rhubarb*

succinic acid
a diprotic acid
$pK_{a1}$ 4.21, $pK_{a2}$ 5.64

*found in amber,
fossils, fungi, lichens*

adipic acid
a diprotic acid
$pK_{a1}$ 4.43, $pK_{a2}$ 5.41

*found in beet juice*

(+)-tartaric acid
a diprotic acid
$pK_{a1}$ 2.93, $pK_{a2}$ 4.23

*a by-product of making wine*

citric acid
a triprotic acid
$pK_{a1}$ 3.13, $pK_{a2}$ 4.76, $pK_{a3}$ 6.40

*found in citrus fruits*

If we compare the acidities of these acids to that of acetic acid, $pK_a$ 4.8, we see that all of them are stronger acids in their first ionization and, in some cases, even in their second ionization. Oxalic acid, for example, is a strong acid, $pK_{a1}$ 1.27, as would be expected from the stabilization of its anion by the electron-withdrawing inductive effect of the second carboxylic acid group in the molecule.

This inductive effect gets weaker as the distance between the second carboxylic acid group and the carboxylate ion resulting from the first ionization of the acid increases. For example, succinic acid, with two carbon atoms between the carboxyl groups, is a stronger acid in its first ionization ($pK_{a1}$ 4.21) than adipic acid, with four carbon atoms between the two groups ($pK_{a1}$ 4.43). Tartaric acid ($pK_{a1}$ 2.93) and citric acid ($pK_{a1}$ 3.13) are stronger yet because they have the inductive effects of the electronegative hydroxyl groups acting to stabilize anions in addition to the effects of the other carboxyl groups.

The $pK_a$ values for the second ionization for each of the acids shown earlier is larger than the first one (and the third $pK_a$ value for citric acid is the largest of all). Just as it was for the hydrogen sulfate anion, it is harder to remove a proton from a negatively charged ion than from an uncharged molecule. A comparison of succinic acid ($pK_{a2}$ 5.64) and adipic acid ($pK_{a2}$ 5.41) shows that this effect also diminishes with the distance between the two carboxylic acid groups, as we would expect it to. Oxalic acid is unusual in that even in its second ionization it is a stronger acid than acetic acid, even though its two carboxylic acid groups are right next to each other. Chemists believe that the doubly charged anion from oxalic acid is well stabilized by solvation because of the high concentration of charge on the ion.

Chemists are interested in which species, the carboxylic acid, its monoanion, or its dianion, exists in different solutions at different acidities. For example, succinic acid exists in three forms, and the relative concentrations of those forms may be changed by making the solution more or less acidic. At low pH and high concentrations of hydronium ion, the acid is fully protonated. As base is added to the solution, and pH rises, one of the protons is removed, and the monoanion becomes the major species in solution. As the solution becomes even more basic, the second proton is removed, and the dianion is the major species in the solution.

*inductive effect of the
second carboxylic acid
group stabilizing the
conjugate base of oxalic acid*

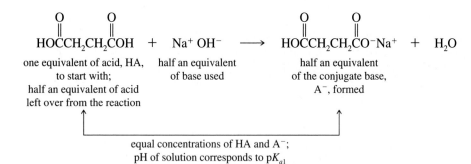

fully protonated acid                          monoanion                          dianion

*form that exists at low pH*          *form that exists at*          *form that exists*
                                      *intermediate pH values*          *at high pH*

The points of transition between the different species can be found by application of the Henderson-Hasselbalch equation (Problem 3.42, p. 113). According to this equation, $pK_a$ is equal to pH when the concentration of an acid and its conjugate base are equal.

$$pH = pK_a + \log \frac{[A^-]}{[HA]}$$

when          $[HA] = [A^-]$,   $\log \frac{[A^-]}{[HA]} = 0$,   and   $pH = pK_a$

If we dissolve succinic acid in water and add enough strong base to convert half the acid to its conjugate base and measure the pH of the solution at that point, the pH will correspond to $pK_{a1}$ for succinic acid.

HOCCH₂CH₂COH   +   Na⁺ OH⁻   ⟶   HOCCH₂CH₂CO⁻Na⁺   +   H₂O

one equivalent of acid, HA,          half an equivalent          half an equivalent
   to start with;          of base used          of the conjugate base,
half an equivalent of acid                          A⁻, formed
left over from the reaction

equal concentrations of HA and A⁻;
pH of solution corresponds to $pK_{a1}$

The relationships described among the three forms in which succinic acid can exist at different pH values are most easily seen when the relative amounts of the three species are plotted against the pH of the solution (Figure 16.1). At low pH, succinic acid predominates. At pH 4.21, $pK_{a1}$ for succinic acid, there are equal amounts of the acid and the monoanion in solution. At pH values between 4.21 and 5.64, the monoanion is the predominant species in solution but gives way to its conjugate base, the dianion, at pH values higher than 5.64, corresponding to $pK_{a2}$ for the acid. Important points to be learned from Figure 16.1 are that the relative amounts of the three species shift constantly with pH and that only at extremely low or high pH values will there be nearly complete protonation or deprotonation of the acid. At all intermediate pH values, all three species are in equilibrium with each other.

**Problem 16.8**    Which carboxylic acid group in citric acid is likely to have the lowest $pK_a$ value? Why?

**Problem 16.9**    Using Figure 16.1 as a model, construct a similar graph for citric acid, showing the relative abundances of the acid and the different anions that result from its deprotonation at different pH values.

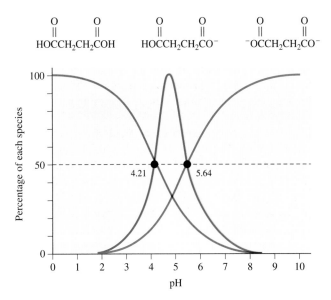

*Figure 16.1*
The relative amounts of the different acid–base forms of succinic acid as pH varies.

## B. Amino Acids and Peptides as Polyprotic Acids

Amino acids found in proteins from animals and plants are all $\alpha$-aminocarboxylic acids. The acidity and basicity of amino acids, derived from the presence of both a carboxyl and an amino group, are important properties. Chemists, biochemists, and physicians are especially interested in the properties of amino acids and the peptides and proteins derived from them (p. 983) at **physiologic pH,** the pH that exists in the cells and fluids of living organisms, which is about 6–7.

Amino acids have relatively high melting points, usually decomposing above 200 °C. They have high solubility in water and low solubility in nonpolar solvents. They have large dipole moments, and their solutions in water have high dielectric constants. All these facts suggest that the units in the crystal lattice and in solution are charged species.

An examination of the relative basicities of a carboxylate anion and an amino group indicate that an amino acid should exist as a dipolar ion, also called a **zwitterion,** in which the amino group is protonated and the carboxyl group exists as a carboxylate anion (Problem 3.41, p. 113). In other words, the acidic group in an amino acid is a substituted ammonium ion and the basic group is the carboxylate anion.

The acid–base reactions of an amino acid are shown below.

From this equation it is clear that at very low pH an amino acid exists in a form in which it is protonated at both the amino group and the carboxyl group (species 1). At such a pH, it bears a net positive charge and is a diprotic acid. At high pH, the amino acid bears a net negative charge and has two basic sites at which it can be protonated (species 3). Note that the equation indicates that it will be protonated first at the amino group and then at the carboxylate anion, which agrees with the relative basicities of the two groups. At some intermediate pH, the amino acid exists primarily as a zwitterion with no net charge (species 2). The pH at which this occurs is known as the **isoelectric point, p***I*, for the amino acid. At this pH, the amino acid is stationary in an electric field; it migrates neither to the negative pole nor to the positive pole because the charges on it are balanced. In contrast, at low pH, the amino acid bears a positive charge and migrates to the negative pole of an electric field, and at high pH, it migrates to the positive pole.

The behavior of a fully protonated amino acid as a diprotic acid can be seen clearly if a titration is carried out. For the typical amino acid, two $pK_a$ values are determined. When half of an equivalent of base is added to the amino acid alanine ($R = CH_3$) in its fully protonated form, the pH of the solution is 2.34. At this point, the concentration of the zwitterionic form (species 2) equals the concentration of the diprotic acid (species 1). According to the Henderson-Hasselbalch equation (p. 113), $pK_a$ is equal to pH when the concentration of an acid and its conjugate base are equal. Therefore, the $pK_a$ of the carboxylic acid group in alanine must be 2.34.

When an equivalent of base has been added, the zwitterion predominates. The pH at this point is 6.02, the isoelectric point of alanine. Finally, addition of another half of an equivalent of base gives a pH of 9.69, which corresponds to the $pK_a$ of the ammonium ion derived from the amino acid. At this point, equal concentrations of the zwitterion and its conjugate base (species 3 above) exist.

Again, the relationships described above are most easily seen when the concentrations of the three species mentioned above are plotted against the pH of the solution (Figure 16.2). At low pH, the predominant species in solution is the fully protonated form of alanine. As the pH is raised, the percentage of alanine in this form decreases, and the percentage of alanine that has been deprotonated at the carboxylic acid group increases. The two concentrations are equal to each other at pH 2.34, the $pK_a$ of the acid. As pH increases, the concentration of the zwitterion increases. It is the predominant form at pH 6.02, the isoelectric point, and is in equilibrium with small amounts of the fully protonated and fully deprotonated forms. At

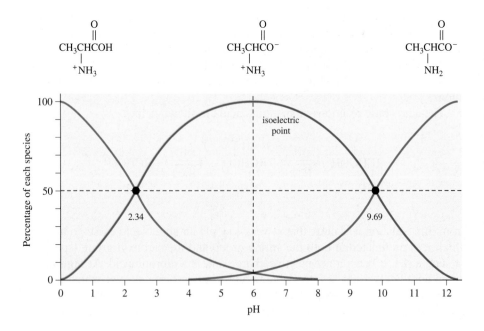

*Figure 16.2*
The relative amounts of the different acid–base forms of alanine as pH varies.

pH 9.69, the $pK_a$ of the protonated amino group in the amino acid, the concentrations of the zwitterion and the fully deprotonated form of alanine are equal. At pH values higher than 9.69, the fully deprotonated form of the amino acid becomes the predominant species.

The p$I$ is the arithmetic mean of the $pK_a$ values for the two acid groups in the compound.

$$pI = \tfrac{1}{2}(pK_{a_1} + pK_{a_2})$$

An isoelectric point of approximately 6 is typical for the amino acids that do not have additional acidic or basic groups on their side chains.

The $pK_a$ value of 2.34 for the carboxylic acid group of alanine indicates that it is a stronger acid than a typical aliphatic carboxylic acid such as acetic acid, which has $pK_a$ 4.76. The protonated amino group with its positive charge has an electron-withdrawing effect that strengthens the acid (p. 106). The amino group is also affected by the presence of the adjacent carboxylic acid group. It is less basic and its conjugate acid more acidic than is usually the case for an aliphatic amine. The conjugate acid of methylamine, for example, has $pK_a$ 10.6, compared to $pK_a$ 9.69 for the protonated amino group in alanine.

Amino acids that have additional amino or carboxyl groups in their molecules have isoelectric points that reflect their tendency either to be further protonated or to lose an additional proton. The amino acid lysine, for example, has three $pK_a$ values: 2.18 for the carboxylic acid group, 8.95 for the $\alpha$-amino group, and 10.53 for the amino group on carbon 6, the $\epsilon$-amino group.

form that predominates at physiologic pH

form that predominates at pH 9.74

lysine at low pH

lysine at high pH

The isoelectric point for lysine is 9.74. Note that the $\epsilon$-amino group in lysine is more basic than the $\alpha$-amino group, and thus the zwitterion for lysine is shown with the protonated amino group at carbon 6. A plot of the change in concentration of the various forms of lysine against pH has three crossing points because the fully protonated form of lysine has three sites from which it can lose a proton and three $pK_a$ values (Figure 16.3, p. 666). The isoelectric point of lysine indicates that lysine would bear a net positive charge at pH 7, or roughly the pH of cells. This is very important for the structure and function of proteins, as we will see in Section 23.8B. Only if the pH of the solution were raised to 9.74 would most of the lysine molecules be converted to the zwitterionic form and have no net charge.

Glutamic acid, on the other hand, has an isoelectric point of 3.2. The carboxylic acid group on carbon 4 is like a normal aliphatic carboxylic acid group and has $pK_a$ 4.25. The other carboxylic acid group has $pK_a$ 2.19, and the protonated amino group has $pK_a$ 9.67.

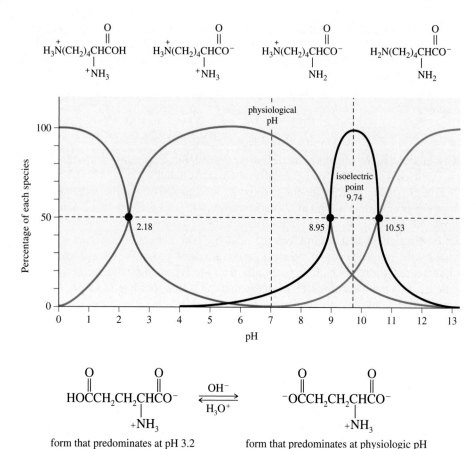

**Figure 16.3**
The relative amounts of the different acid–base forms of lysine as pH varies.

At physiologic pH, glutamic acid bears a net negative charge. Only when the pH is lowered to 3.2 does it exist as the zwitterion.

**Problem 16.10**   Construct a plot that shows the concentrations of the various acid–base forms of glutamic acid at pH values from 0 to 12.

## 16.3   Carbon Acids

### A. Carbanions. Stabilization by the Inductive Effect

**Carbanions** are reactive intermediates in which a carbon atom has a pair of non-bonding electrons and bears a negative charge. They are generated in two ways. One method is to remove a proton from an organic compound using a base. For example, the hydrogen atom on the $sp$-hybridized carbon atom of an alkyne can be removed to give a carbanionic intermediate because the electronegativity of such a carbon atom stabilizes the anion sufficiently that it is practical to generate such intermediates (Section 9.3, p. 331). The other common method of generating carbanions is by reaction of organic halides with metals to give organometallic reagents (Section 14.6A, p. 549). Prepared this way, Grignard reagents and organolithium reagents act as sources of groups containing strongly basic and nucleophilic carbon atoms. The reactions of carbanions both as bases and as nucleophiles were explored in Sections 9.3B and 14.6. Such reagents have great versatility because they are capable of forming new carbon–carbon bonds; they will be explored extensively in Chapters 17 and 21.

While terminal alkynes can be deprotonated to give carbanions, most carbon–hydrogen bonds are not acidic enough to be broken easily in this way unless there are electron-withdrawing groups on the carbon atom to help stabilize the negative charge on the anion. Methane, for example, is an extremely weak acid and will not react with hydroxide ion, but trichloromethane (chloroform) does react to give low concentrations of trichloromethyl anion.

$$\begin{array}{c} \text{H} \\ | \\ \text{H}-\text{C}-\text{H} \\ | \\ \text{H} \end{array} + \quad {}^-\!:\!\ddot{\text{O}}-\text{H} \quad \longrightarrow \quad \text{no reaction}$$

methane   hydroxide
$pK_a$ ~50    ion

$$\begin{array}{c} \text{Cl} \\ | \\ \text{Cl}-\text{C}-\text{H} \\ | \\ \text{Cl} \end{array} + \quad {}^-\!:\!\ddot{\text{O}}-\text{H} \quad \rightleftharpoons \quad \begin{array}{c} \text{Cl} \\ | \\ \text{Cl}-\text{C}\!:\!^- \\ | \\ \text{Cl} \end{array} + \quad \text{H}-\ddot{\text{O}}-\text{H}$$

trichloromethane hydroxide  trichloromethyl   water
$pK_a$ ~25    ion    anion    $pK_a$ 15.7

*acid*    *base*   *conjugate base*  *conjugate acid*
          *of trichloromethane* *of hydroxide ion*

The substitution of electron-withdrawing chlorine atoms for three of the hydrogen atoms in methane makes the remaining hydrogen atom acidic enough to react with hydroxide ion. Trichloromethane is a stronger acid than methane because the negative charge on the trichloromethyl anion can be stabilized by the inductive effect due to the chlorine atoms.

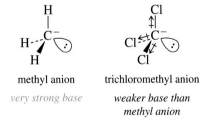

methyl anion    trichloromethyl anion

*very strong base*  *weaker base than*
        *methyl anion*

**Problem 16.11**  Would you expect triiodomethane (iodoform), $CHI_3$, to be a stronger or weaker acid than trichloromethane, $CHCl_3$? Explain using words and structural formulas.

The groups that are most effective in stabilizing carbanions are those such as carbonyl groups, nitro groups, sulfonyl groups, and cyano groups. Not only do these groups have strongly electron-withdrawing inductive effects, but they are also capable of resonance stabilization of the anion. The acidity of hydrogen atoms adjacent to such groups will be explored in the next section.

## B. Electron–Withdrawing Groups and the Acidity of Hydrogen Atoms Adjacent to Them

An examination of the $pK_a$ table inside the front cover of this book reveals an interesting set of patterns. For example:

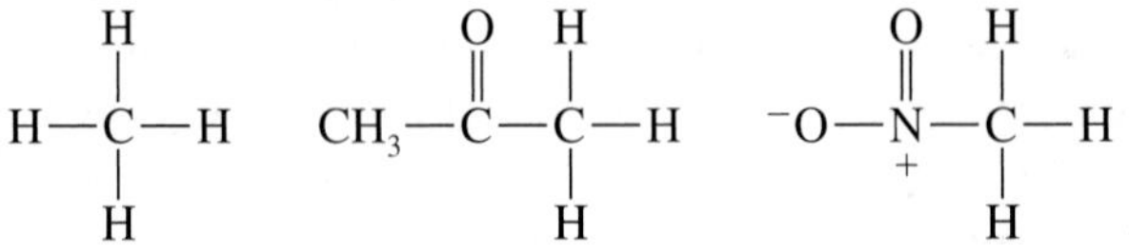

methane
$pK_a$ 49

acetone
$pK_a$ 19

nitromethane
$pK_a$ 10.2

or

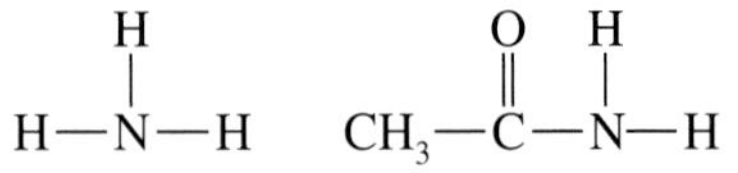

ammonia
$pK_a$ 36

acetamide
$pK_a$ 15.0

or

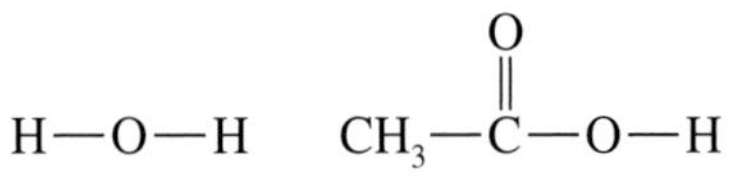

water
$pK_a$ 15.7

acetic acid
$pK_a$ 4.8

In the first set, the acidity of a proton on carbon increases markedly as we substitute first a carbonyl group then a nitro group for one of the hydrogen atoms on methane. In the second set, substitution of a carbonyl group for a hydrogen atom on ammonia greatly increases the acidity of the protons on nitrogen. Finally, we get a similar result by performing a similar substitution for one of the hydrogen atoms on oxygen in water. Introduction of an electron-withdrawing group adjacent to it significantly lowers the $pK_a$ of that proton. This means that the compound is a stronger acid (and its conjugate base, a weaker base) than the original unsubstituted compound. A weaker base means a base that is more highly stabilized in comparison with the original one (p. 90).

Groups such as the carbonyl group and the nitro group stabilize anions adjacent to them in two ways. One is the inductive effect. Both groups have positive character and serve to stabilize an anionic center adjacent to them by withdrawing electron density through bonds and through space (the field effect, considered here as part of the inductive effect, p. 106).

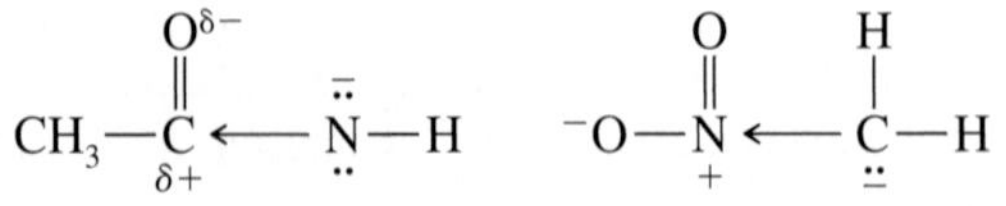

*carbonyl group stabilizing
the acetamide anion by the
inductive effect*

*nitro group stabilizing
carbanion by the
inductive effect*

Both groups also can stabilize the anions by resonance delocalization of the charge on the anion.

$$CH_3-\overset{\overset{\displaystyle :O:}{\|}}{C}-\overset{..}{\underset{..}{N}}-H \longleftrightarrow CH_3-\overset{\overset{\displaystyle :\overset{..}{O}:^-}{|}}{C}=\overset{..}{N}-H$$

resonance contributors for the acetamide anion

$$^-:\overset{..}{\underset{..}{O}}-\overset{\overset{\displaystyle :O:}{\|}}{\underset{+}{N}}-\overset{\overset{\displaystyle H}{|}}{\underset{..}{C}}-H \longleftrightarrow \overset{..}{O}=\overset{\overset{\displaystyle ^-:\overset{..}{O}:}{|}}{\underset{+}{N}}-\overset{\overset{\displaystyle H}{|}}{\underset{..}{C}}-H \longleftrightarrow ^-:\overset{..}{\underset{..}{O}}-\overset{\overset{\displaystyle ^-:\overset{..}{O}:}{|}}{\underset{+}{N}}=\overset{\overset{\displaystyle H}{|}}{C}-H$$

resonance contributors for the carbanion

In each case we can write resonance contributors that delocalize the negative charge from nitrogen or the carbon atom to an oxygen atom, which is more electronegative than the original anionic site. The acetamide anion is less basic than the amide anion, obtained by deprotonating ammonia, where the negative charge must remain on the nitrogen atom. Similarly, the anion from nitromethane is a much weaker base than the anion from methane, where the negative charge is localized on a single atom, the carbon atom.

Stabilization by resonance is always much more important than stabilization by the inductive effect. We will see many more examples where we can rationalize the stability of ions by writing resonance contributors.

**Problem 16.12**   For each of the following compounds, identify the most acidic proton(s) by writing a mechanism, using the curved-arrow convention, for the reaction with hydroxide ion. What factors are important in the stabilization of each organic anion?

(a) $HOCH_2\overset{\overset{\displaystyle O}{\|}}{C}OH$    (b) $CH_3CH_2CH_2NO_2$    (c) $HSCH_2CH_2CH_2OH$    (d) $Br\overset{\overset{\displaystyle }{}}{\underset{\underset{\displaystyle Br}{|}}{C}}H\overset{\overset{\displaystyle O}{\|}}{C}CH_3$

## 16.4   Enolization

### A. Enols and Enolates

Enols are tautomers of ketones or aldehydes and exist in equilibrium with them, as we saw when we looked at the addition of water to alkynes (Sections 9.4B, 9.4C, and 9.4D). The process of converting a carbonyl compound into its enol is called **enolization.** Enolization occurs in neutral, basic, or acidic solutions. The base-initiated enolization of acetone is shown on the next page. Acetone has a p$K_a$ of 19. A hydrogen atom can be removed from the carbon atom adjacent to the carbonyl group by a base to give a carbanion. The carbanion is stabilized by the delocalization of electrons to the oxygen atom of the carbonyl group. Such an anion is called an **enolate anion.**

## ONE SMALL STEP

Sulfonamides, an example of which is shown below, are more acidic than carboxylic acid amides. For example, sulfonamides that have a proton on the nitrogen atom dissolve easily in sodium hydroxide solution, whereas carboxylic acid amides do not.

**PROBLEM:** Rationalize the experimental observation that benzenesulfonamide is more acidic than benzamide.

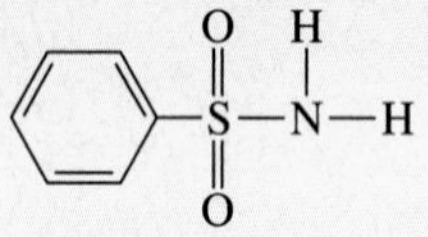

benzenesulfonamide

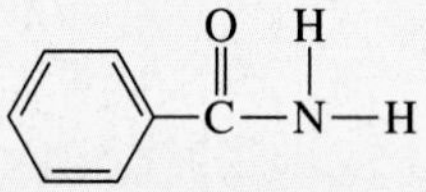

benzamide

**PROBLEM:** Predict which of the following compounds is more acidic, and explain your answer.

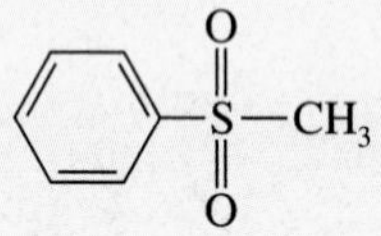

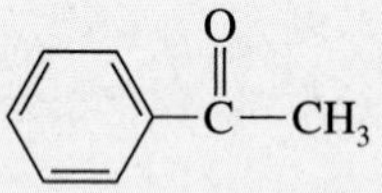

**PROBLEM:** How about these?

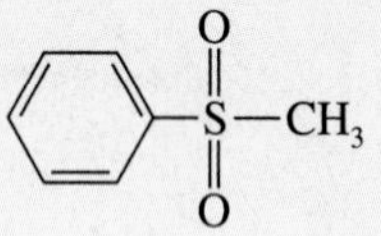

and

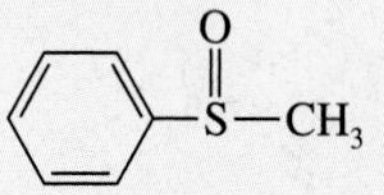

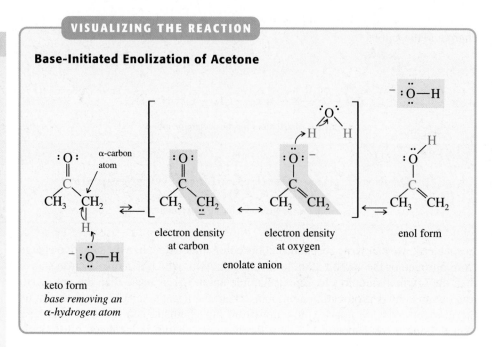

**VISUALIZING THE REACTION**

### Base-Initiated Enolization of Acetone

α-carbon atom

keto form
*base removing an*
*α-hydrogen atom*

electron density at carbon

electron density at oxygen

enolate anion

enol form

The enolate anion is protonated at the oxygen atom to give an enol. Acids also promote enolization, as demonstrated by the acid-catalyzed enolization of acetone.

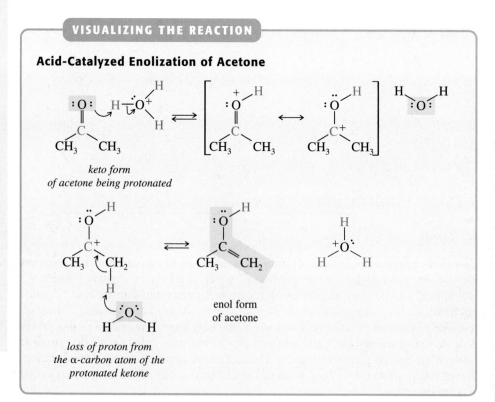

**VISUALIZING THE REACTION**

### Acid-Catalyzed Enolization of Acetone

*keto form*
*of acetone being protonated*

*loss of proton from*
*the α-carbon atom of the*
*protonated ketone*

enol form
of acetone

The ketone, protonated at the carbonyl group in acidic solution, loses a proton from the $\alpha$-carbon atom to give an enol.

The percentage of enol present at equilibrium with the keto form depends on the structure of the compound, as well as on other factors such as the solvent. For a compound with a single carbonyl group, such as acetone, the equilibrium lies far on the side of the ketone.

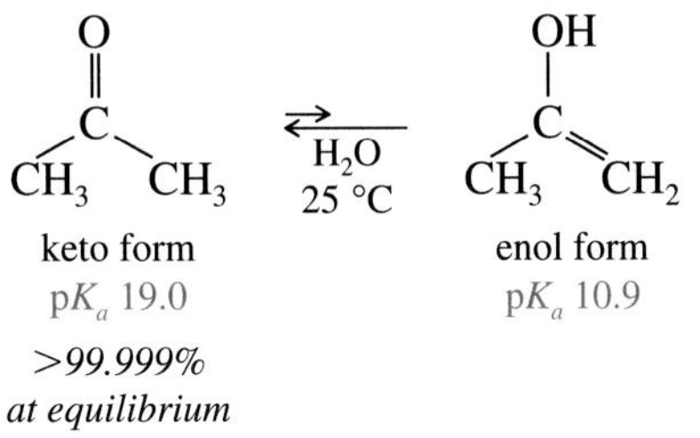

keto form
pK$_a$ 19.0

>99.999%
at equilibrium

enol form
pK$_a$ 10.9

**Problem 16.13**   Decide whether each set of structural formulas shown below represents resonance contributors or tautomers.

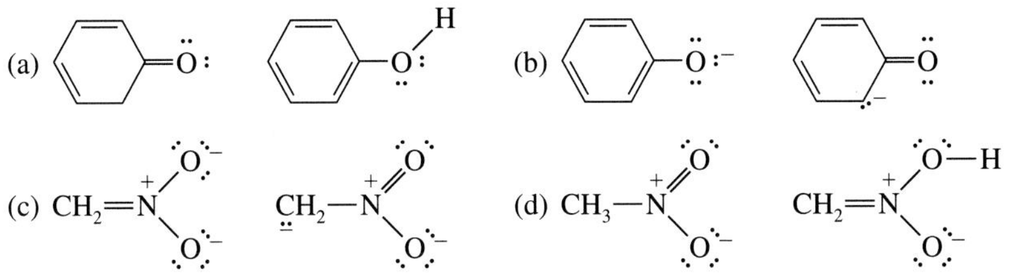

(a)

(b)

(c)

(d)

Study Guide
Concept Map 16.1

**On the Web: ONE SMALL STEP**

## B. Active Methylene Compounds

A methylene group that is alpha to two carbonyl groups is said to be an **active methylene group.** For compounds that contain such groups, 1,3-dicarbonyl compounds such as 2,4-pentanedione, the enol form is a significant part of the equilibrium mixture. The extent of enolization can be determined by taking the proton magnetic resonance spectrum of the compound. Figure 16.4 on the next page shows the spectrum of 2,4-pentanedione taken in deuteriochloroform. The relative amounts of keto and enol forms can be determined by the integration of the peaks for the methyl groups in the two forms of the dione.

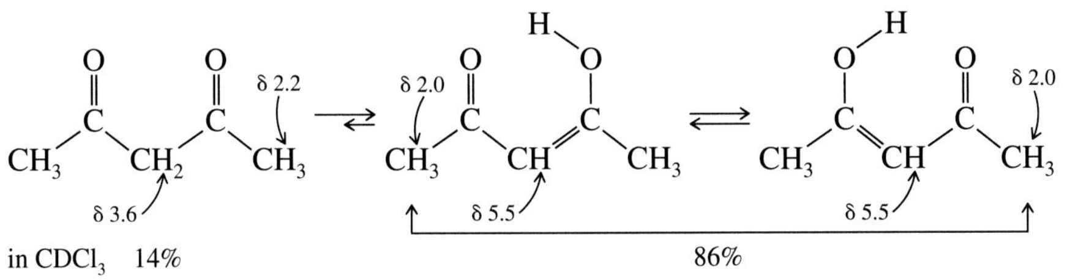

δ 2.2      δ 2.0      δ 2.0

δ 3.6      δ 5.5      δ 5.5

in CDCl$_3$   14%                    86%

For 2,4-pentanedione, the most acidic hydrogen atom (pK$_a$ 9.0) is one of those bonded to the active methylene group. Removal of a proton results in the formation of an enolate ion stabilized by delocalization of the negative charge to the two different oxygen atoms of the carbonyl groups. Protonation of the enolate ion at the negatively charged oxygen atom gives rise to the enol.

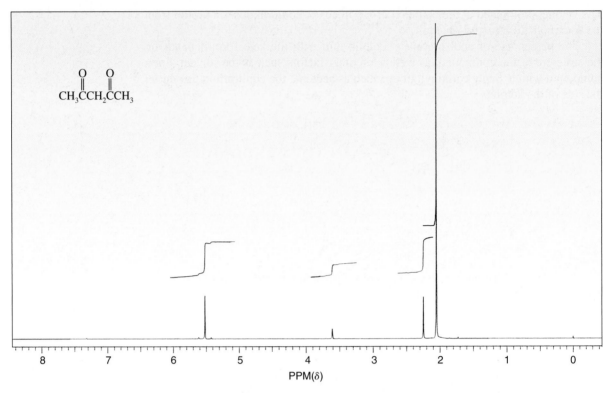

*Figure 16.4*
Proton magnetic resonance spectrum for 2,4-pentanedione in deuteriochloroform.

**VISUALIZING THE REACTION**

**Base-Catalyzed Enolization Reaction of a Dione**

The hydrogen atoms of an active methylene group are easily removed by bases such as alkoxide anions (conjugate acid $pK_a \sim 17$) because the conjugate base, the enolate ion, is highly stabilized by resonance. Acetone ($pK_a$ 19) is much less acidic than 2,4-pentanedione. Removal of a proton from acetone is more difficult because the resulting ion is stabilized by delocalization of charge to only one oxygen atom. The conjugate base of acetone is not stabilized relative to its acid as much as the conjugate base of 2,4-pentanedione is.

Active methylene groups need not be between two ketone functions. Ethyl 3-oxobutanoate (ethyl acetoacetate, $pK_a$ 11.0) also has an active methylene group, as do ethyl nitroacetate ($pK_a$ 5.8) and ethyl cyanoacetate ($pK_a \sim 9$). All these esters are considerably more acidic than ethyl acetate ($pK_a$ 23).

active methylene group

$CH_3COCH_2CH_3$
ethyl acetate
$pK_a$ 23

no active methylene group

$CH_3CCH_2COCH_2CH_3$
ethyl acetoacetate
$pK_a$ 11.0

active methylene group

$^-O-NCH_2COCH_2CH_3$
ethyl nitroacetate
$pK_a$ 5.8

active methylene group

$N\equiv CCH_2COCH_2CH_3$
ethyl cyanoacetate
$pK_a \sim 9$

The $pK_a$ values given above show that the carbonyl group of an ester does not contribute as much to the stabilization of an enolate anion as the carbonyl group of a ketone does. Ethyl acetoacetate is a weaker acid than 2,4-pentanedione, and ethyl acetate is a weaker acid than acetone. The differences in polarity and electrophilicity between carbonyl groups in ketones (and aldehydes) and those in acid derivatives have been thoroughly discussed elsewhere in this book, especially in connection with the reactivity of acid derivatives (p. 596). The presence of unshared pairs of electrons on the oxygen atom of the alkoxyl group of an ester diminishes the effectiveness of the carbonyl group of the ester in delocalizing charge. On the other hand, a nitro group is even more effective than a carbonyl group in delocalizing charge (p. 107). A cyano group is not quite as effective as a ketone in stabilizing charge but is better than an ester group.

On the Web: ONE SMALL STEP

**Problem 16.14**    Write mechanisms showing the base-catalyzed enolization of (a) ethyl acetate, (b) ethyl acetoacetate, (c) ethyl nitroacetate, and (d) ethyl cyanoacetate. Be sure to indicate all major resonance contributors for each enolate ion.

**Problem 16.15**    For each of the pairs of compounds below and on the following page, predict which one will be more extensively enolized.

(a)    or

(b)    or

$$\text{(c)} \quad \underset{\substack{| \\ CH_3}}{CH_3\overset{O}{\overset{\|}{C}}NCH_3} \quad \text{or} \quad CH_3\overset{O}{\overset{\|}{C}}OCH_2CH_3 \qquad \text{(d)} \quad CH_3O\overset{O}{\overset{\|}{C}}CH_2CH_2\overset{O}{\overset{\|}{C}}OCH_3 \quad \text{or} \quad CH_3O\overset{O}{\overset{\|}{C}}CH_2\overset{O}{\overset{\|}{C}}OCH_3$$

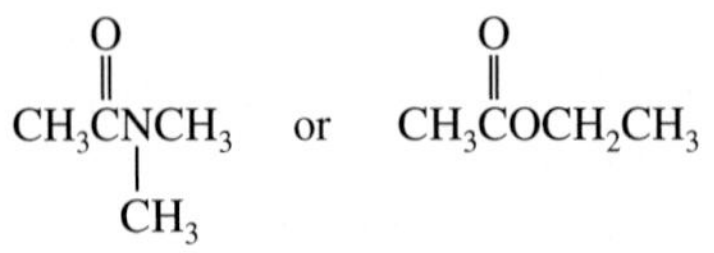

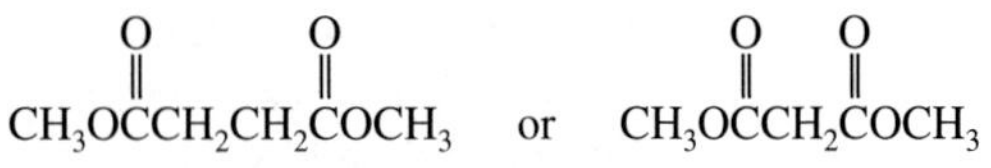

## 16.5  Enols and Enolates as Intermediates in the Exchange of Protons in Carbon Acids

### A. Detection of Proton Exchange Reactions by the Use of Deuterium

All the mechanisms that we have written for enolization reactions show the different steps to be reversible. This means that an enolate ion that is formed by deprotonation of a carbonyl compound will be protonated by proton sources present in the reaction mixture. In other words, the enolate ion is a nucleophile and reacts with suitable protic acids acting as electrophiles.

The reaction of an enolate ion with an acid is not detectable experimentally unless an isotope of hydrogen is used to distinguish the atom that has been introduced into the molecule from the hydrogen atoms already there. Chemists use deuterium oxide, $D_2O$, to look for protonation reactions. For example, if acetone is put in deuterium oxide with a trace of sulfuric acid, its $\alpha$-hydrogens are exchanged for deuterium atoms. The rate of the exchange reaction depends only on the concentration of the ketone. Enolization of the ketone is the rate-determining step.

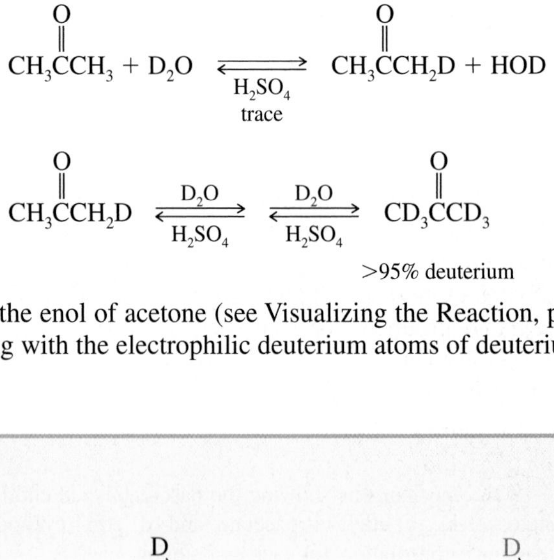

$$CH_3\overset{O}{\overset{\|}{C}}CH_3 + D_2O \underset{\substack{H_2SO_4 \\ \text{trace}}}{\rightleftharpoons} CH_3\overset{O}{\overset{\|}{C}}CH_2D + HOD$$

$$CH_3\overset{O}{\overset{\|}{C}}CH_2D \underset{H_2SO_4}{\overset{D_2O}{\rightleftharpoons}} \underset{H_2SO_4}{\overset{D_2O}{\rightleftharpoons}} CD_3\overset{O}{\overset{\|}{C}}CD_3$$

$$>95\% \text{ deuterium}$$

In this reaction, the enol of acetone (see Visualizing the Reaction, p. 670) is the nucleophile reacting with the electrophilic deuterium atoms of deuterium oxide.

---

**VISUALIZING THE REACTION**

**The Reaction of an Enol with an Acid**

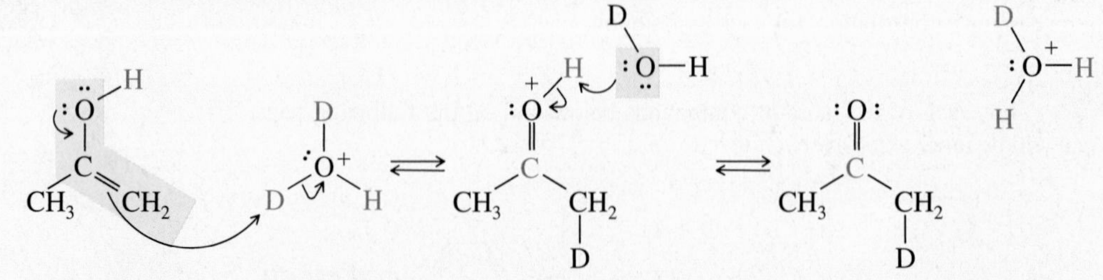

enol form of acetone        electrophile
nucleophile

If the reaction time is extended and fresh deuterium oxide is periodically added to the reaction mixture, eventually all the enolizable hydrogen atoms in a carbonyl compound are replaced by deuterium.

The incorporation of deuterium into the molecule is easily detected experimentally in two ways. For example, when deuterium oxide, $D_2O$, is added to the deuteriochloroform solution of 2,4-pentanedione, the peaks arising in the proton magnetic resonance spectrum from the active methylene hydrogens (see Figure 16.4; $\delta$ 3.6 for the keto form and $\delta$ 5.5 for the enol form) disappear. They are replaced by a peak at $\delta$ 4.7, which comes from HOD, $D_2O$ that has exchanged one of its deuterium atoms for a hydrogen. Chemists use this technique to identify protons that are easily exchangeable with the protons in water, including protons on hydroxyl, carboxyl, amino, and amide groups, as well as those that are readily enolizable.

The mass spectrum of a compound with deuterium substituted for hydrogen also will be different, with $m/z$ for each ion that contains deuterium increasing by one unit for each deuterium atom that has replaced a hydrogen atom. A comparison of the original mass spectrum with that of the deuterated compound will tell us not only how many deuterium atoms have been incorporated into the molecule but also which parts of the molecule were affected.

**Problem 16.16** 3-Methyl-2,4-pentanedione exchanges one hydrogen for deuterium if it is allowed to stand in deuterium oxide at room temperature for several days. Write an equation and a mechanism for this reaction. How do you explain the selectivity of the reaction?

**Problem 16.17** Predict the products of the following sequence of reactions by providing structural formulas for the intermediates or products indicated by letters.

## B. Racemization of Carbonyl Compounds with Stereocenters at the α-Carbon Atom

A careful look at the process of enolization (see Visualizing the Reaction on page 670, for example) shows that the $sp^3$-hybridized tetrahedral carbon atom $\alpha$ to the carbonyl group becomes an $sp^2$-hybridized trigonal planar carbon atom in the enolate ion. If such a carbon atom is a stereocenter in the carbonyl compound, it loses its stereochemistry upon enolization. The now planar $\alpha$-carbon atom can be reprotonated from either the top or the bottom of the plane, so we expect to see both configurations of the newly reconstituted stereocenter form over time. If we start with an optically active compound, it gradually loses its optical activity; in other words, it racemizes. This is exactly what is observed experimentally. When ($S$)-1-phenyl-2-methyl-1-butanone is allowed to stand in acetic acid, it loses its optical activity as its enantiomer builds up in the reaction mixture.

($S$)-1-phenyl-2-methyl-1-butanone

*chiral*

enol

*achiral*

($R$)-1-phenyl-2-methyl-1-butanone

*chiral*

**Problem 16.18**    When (R)-1-phenyl-2-methyl-1-butanone is treated with NaOD in D$_2$O, the rate of deuterium incorporation is equal to the rate of racemization, the rate at which the system loses optical activity. Show with structural formulas why this is so.

(S)-1-Phenyl-2-methyl-1-butanone has one stereocenter, so inversion of configuration at that carbon atom converts it into its enantiomer. If the compound contains more than one stereocenter, the process by which one stereocenter is inverted is called **epimerization.** Epimers are diastereomers that differ in stereochemistry at one stereocenter. For example, the ant lactone (−)-iridomyrmecin, used by ants to defend themselves against insects that prey on them, is converted to its isomer by treatment with base.

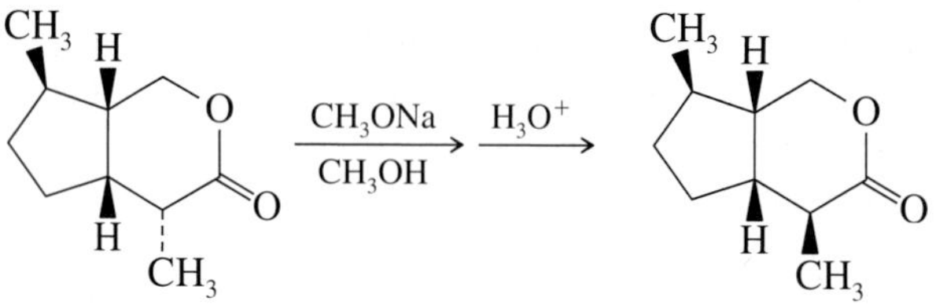

(−)-iridomyrmecin

(−)-isoiridomyrmecin

*an epimer of
(−)-iridomyrmecin;
a diastereomer that differs
in configuration at one
of four stereocenters*

an epimerization reaction

In this epimerization, the methyl group α to the carbonyl group moves to the same side of the six-membered ring as the hydrogen atoms at the junction of the two rings and away from the larger substituents, the bonds to the carbon atoms of the five-membered ring.

**Problem 16.19**    Write a detailed mechanism for the epimerization of (−)-iridomyrmecin.

## C. Racemization of Amino Acids. A Technique for Dating Fossils

The amino acids present in living organisms have the stereochemical configuration shown below for three examples.

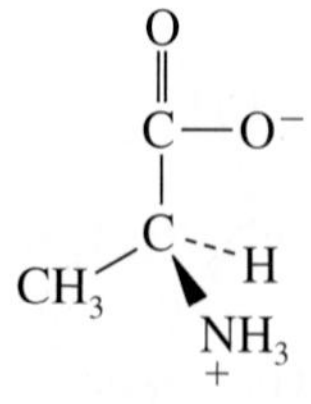

(S)-alanine

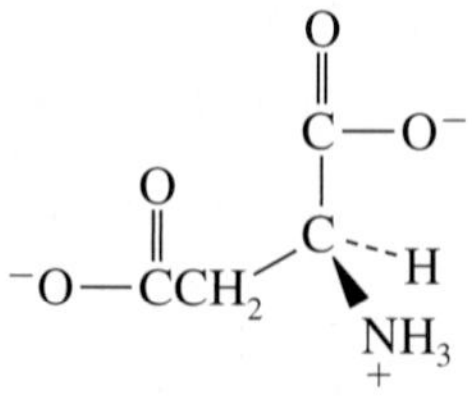

(S)-aspartic acid

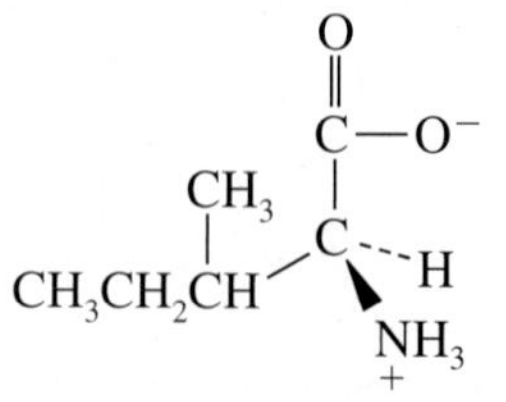

(S)-isoleucine

After the death of the organism, the amino acids in the tissues undergo slow racemization. The rate of this process depends on the organism, its environment, and the temperature of its surroundings. Materials that are relatively stable in structure such as shells of mollusks, or egg shells, or teeth, especially when they are found in environmentally constant circumstances, such as in the deep sea or in caves, can be analyzed for the ratio of natural to unnatural amino acids present in them. If we know the rate of racemization for that amino acid, we can use the ratio to tell us when the organism being analyzed died.

Interestingly, some parts of our bodies are already more "dead" than others. For example, the layer of our teeth known as dentin does not incorporate any more protein after the teeth are fully grown. Therefore, analysis of the ratio of unnatural to natural aspartic acid in dentin can serve as a way of determining the age of a person (or a corpse). The ratio of *R/S* aspartic acid in a 60-year-old tooth is about 0.05. The lens of the eye and the white matter of the brain happen to be two other tissues that can be used to determine the age of a body.

## 16.6   The Biological Importance of Enolization Reactions

### A. Enolization in Glycolysis

Enzyme-catalyzed enolization reactions are important in many biochemical processes. In this section, we will examine two reactions that are used in the metabolism of glucose, the primary source of energy in our bodies. Some of the other steps along the metabolic pathway for glucose, known as glycolysis, will appear in Chapter 17 (p. 719)

An important early step in the metabolism of glucose is its conversion to fructose. This reaction, which is achieved enzymatically in the body, also takes place if glucose is treated with dilute base in what is known as the **Lobry deBruyn–Alberda van Ekenstein rearrangement,** after the two Dutch chemists who discovered it at the end of the nineteenth century.

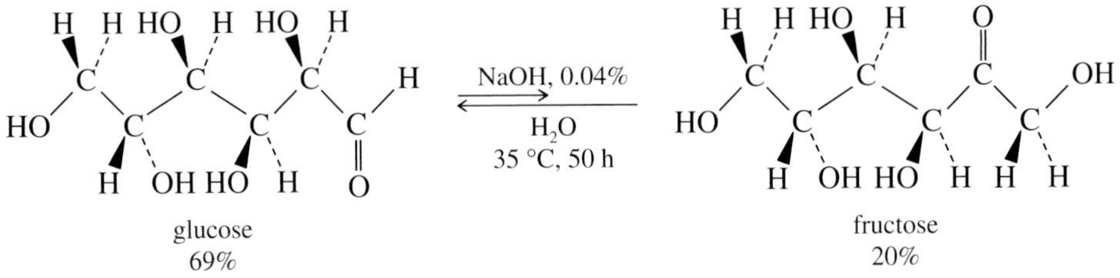

The transformation starts with an enolization to give an enediol intermediate. The two protons on the enolic hydroxyl groups are acidic. Removal of the proton from the hydroxyl group at carbon 2 yields the enolate ion that is converted to fructose by protonation at carbon 1.

**VISUALIZING THE REACTION**

## Conversion of Glucose to Fructose

glucose

*undergoing deprotonation at
carbon 2 and protonation at
the carbonyl group*

enediol intermediate

*undergoing deprotonation at
the hydroxyl group at carbon 2
and protonation at carbon 1*

fructose

---

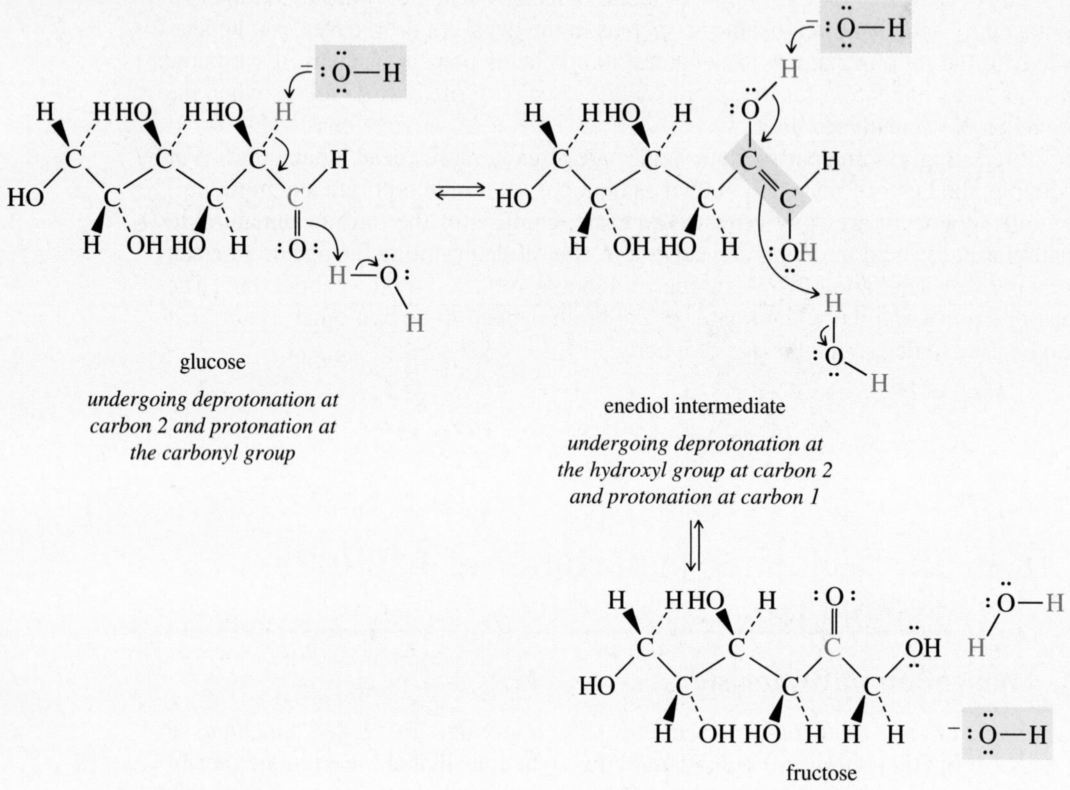

**Problem 16.20** Mannose is the epimer of glucose at carbon 2. About 1% mannose is formed in the base-catalyzed conversion of glucose to fructose shown above.

(a) Draw the structural formula for mannose.
(b) Show how the enediol intermediate formed in the conversion of glucose to fructose can be transformed to mannose.

**Problem 16.21** (a) When the reaction described above was carried out in $D_2O$, it was found that mannose (Problem 16.20) and glucose had deuterium attached to carbon 2, but in fructose deuterium was attached to carbon 1. Show how this evidence supports the mechanism proposed for the reaction.
(b) The mannose obtained as a product of the experiment in part (a) had 1.4 deuterium atoms attached to carbon per molecule. The fructose obtained had an average of 1.7 deuterium atoms attached to carbon. Referring to the mechanism, explain how fructose could have more than one deuterium atom per molecule. What does the deuterium content of mannose suggest?

---

At a later stage of glycolysis, after fructose has been broken down into two three-carbon fragments (p. 720), one of these, glyceraldehyde 3-phosphate, is converted to 2-phosphoglyceric acid. 2-Phosphoglyceric acid by an enolization reaction is transformed into phosphoenolpyruvate, a high-energy compound that provides the driving force for the formation of adenosine triphosphate (ATP) from adenosine diphosphate (ADP), as we shall see in the next section.

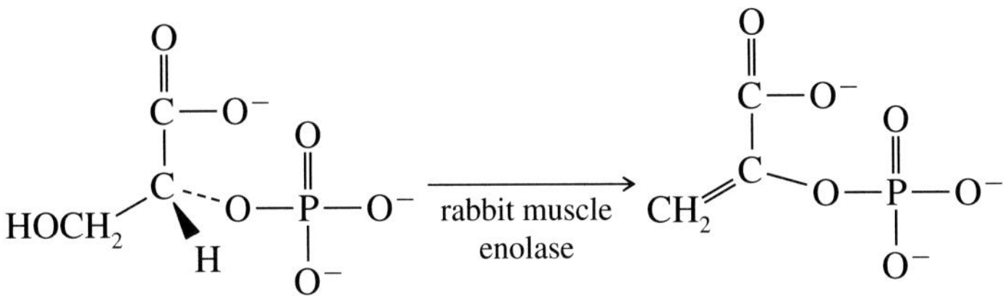

2-phosphoglyceric acid → (rabbit muscle enolase) → phosphoenolpyruvate

**VISUALIZING THE REACTION**

**Formation of Phosphoenolpyruvate from 2-Phosphoglyceric Acid**

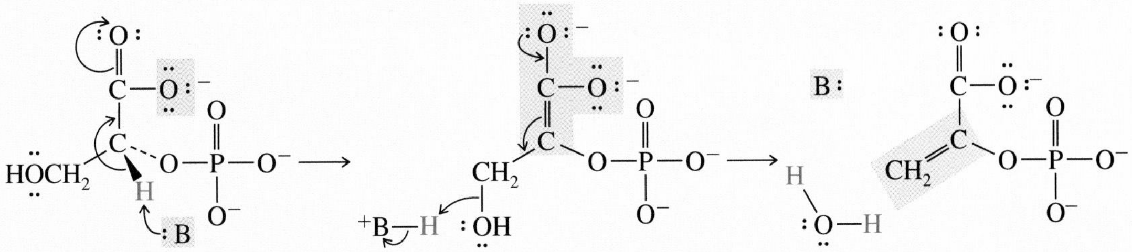

*deprotonation of the α-carbon of 2-phosphoglyceric acid*

*ejection of the hydroxy group*

phosphoenolpyruvate

A word needs to be said about the mechanism shown above. The deprotonation of the α-carbon atom is shown taking place on an anion, the form of 2-phosphoglyceric acid that exists at physiologic pH. This is possible only because the reaction is taking place in the active site of an enzyme (see p. 1012, for example). The enzyme that catalyzes the reaction shown above complexes with magnesium(II) ion, $Mg^{2+}$, so that much of the charge on the anionic substrate for the enzyme is neutralized. To carry out such a deprotonation reaction outside the active site of an enzyme would require very strong base. A base weak enough to survive in aqueous solution would not deprotonate the α-carbon of 2-phosphoglyceric acid in a test tube.

The precise placement of acidic and basic, electrophilic and nucleophilic functional groups in the active site of an enzyme minimizes the activation energy and maximizes the rate of a reaction. Reactions that are usually difficult, such as the loss of a hydroxyl group as a leaving group, also are aided by the enzyme, though such reactions are not uncommon when accompanied by the formation of a carbonyl group (pp. 703 and 706).

## B. The Enol–Keto Equilibrium as a Driving Force in Biological Reactions

Biochemical processes in a living organism are in a state of balance. Reactions that degrade nutrients and the constituents of cells to use their parts and to generate energy (catabolism) are balanced against those that synthesize biological molecules from simple compounds in processes that use up energy (anabolism). This balancing act is mediated largely by the compound adenosine triphosphate (ATP). Processes that produce energy create ATP from adenosine diphosphate (ADP) and phosphate ions. Processes that consume energy result in the conversion of ATP to ADP, with the phosphate that is lost going to create intermediates in important metabolic changes.

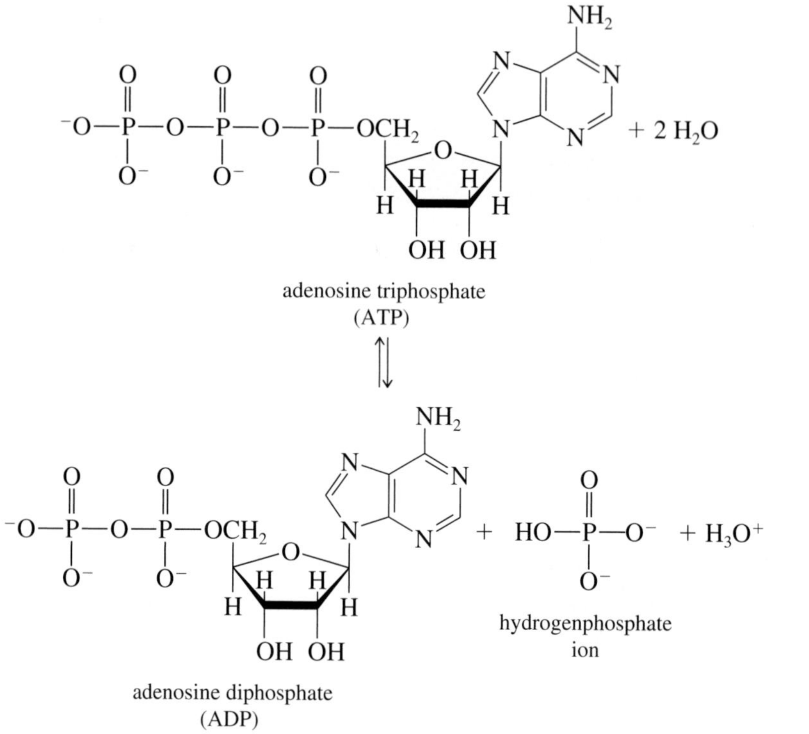

adenosine triphosphate
(ATP)

hydrogenphosphate
ion

adenosine diphosphate
(ADP)

The bonds that are being formed in the synthesis of ATP are phosphoric acid anhydride bonds. There are three forms in which phosphoric acid appears in biological systems. Two of these may be regarded as anhydrides of orthophosphoric acid, $H_3PO_4$.

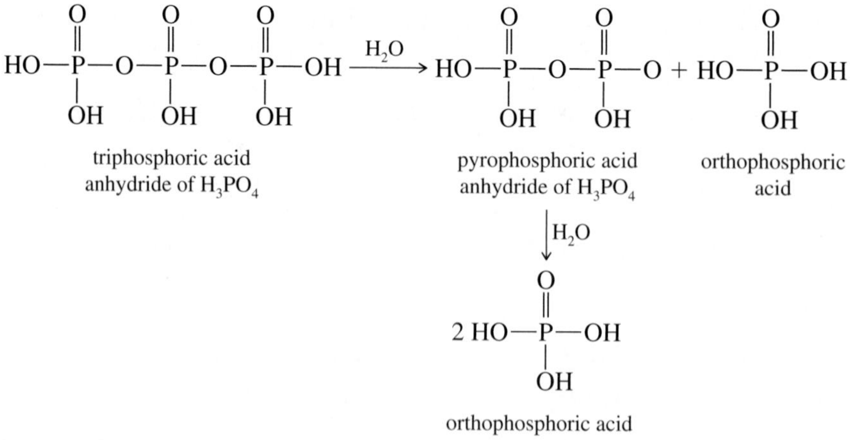

triphosphoric acid
anhydride of $H_3PO_4$

pyrophosphoric acid
anhydride of $H_3PO_4$

orthophosphoric
acid

orthophosphoric acid

All three forms of phosphoric acid are found in nucleotides, the phosphoric acid esters of nucleosides, which are the building blocks of ribonucleic and deoxyribonucleic acids (Section 23.4). Adenosine triphosphate is the triphosphate ester of one of these nucleosides, adenosine (p. 987). It is formed extensively in living systems. Its presence is so synonymous with life on earth that when experiments are designed to search for life in outer space, it is one of the key compounds sought.

ATP is not stored in cells but is constantly being synthesized and used up at the astonishing rate of about 3 mol/h (~1.5 kg/h) for a resting person. About ten times as much is needed for vigorous exercise. In fact, there is some evidence (p. 437) that world-class distance runners are genetically more capable of generating ATP in their muscles during exercise than individuals who do not achieve such athletic prominence.

Because ATP itself is a storehouse of energy, reactions that create it in the body must themselves have considerably negative $\Delta G_r^0$ values. One of the most important of the reactions that regenerate ATP involves phosphoenolpyruvate, seen as one of the products of the metabolism of glucose on page 679. This particularly reactive phosphate ester of an enol transfers a phosphate group to ADP to regenerate ATP. At this stage, some of the energy generated by the breakdown of glucose in the body is returned to storage in the triphosphate group.

The reaction is believed to take place in two steps. The first step is an enzyme-catalyzed transfer of the phosphate group from phosphoenolpyruvate to ADP.

**VISUALIZING THE REACTION**

**Biosynthesis of ATP from ADP**

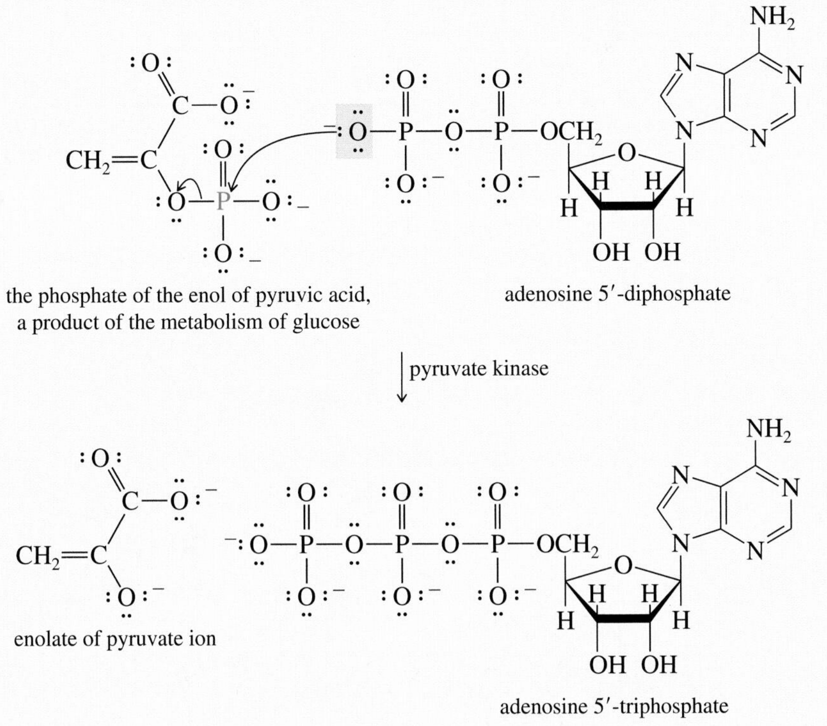

the phosphate of the enol of pyruvic acid, a product of the metabolism of glucose

adenosine 5′-diphosphate

pyruvate kinase

enolate of pyruvate ion

adenosine 5′-triphosphate

The second step of the reaction is the tautomerization of the enolate anion to the much more stable keto form of pyruvate ion.

---

### VISUALIZING THE REACTION

**Enol-Keto Tautomerization of the Enolate of Pyruvate Ion**

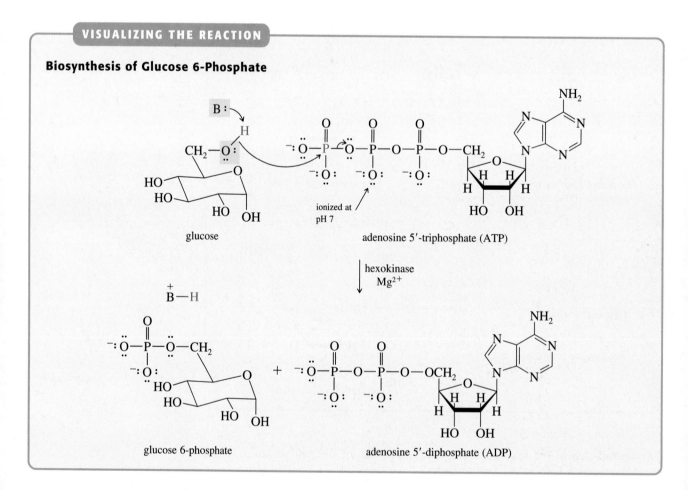

enolate of pyruvate ion                    pyruvate ion

---

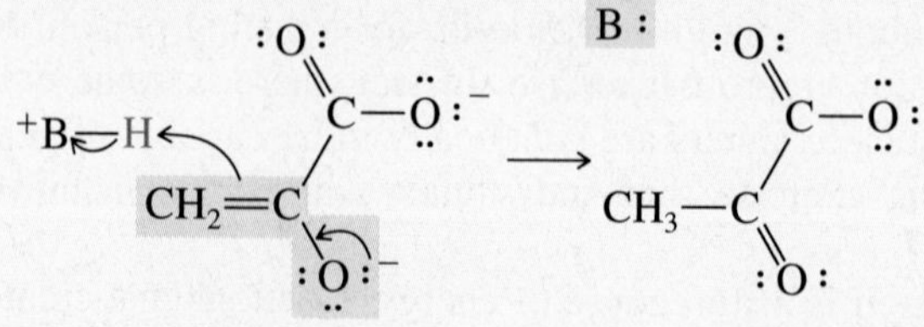

This last step contributes much of the driving force for the biosynthesis of ATP, which occurs with a decrease in free energy, $\Delta G_r^0$, of 7.5 kcal/mol.

Adenosine triphosphate stores chemical energy in the new phosphoric anhydride bond created above and makes it available to specific cell processes. It gives up energy by transferring a phosphate group to another molecule, which is converted into a reactive form as its phosphate ester. A specific example that starts off the process of glycolysis, the metabolism of glucose in the body, is the conversion of glucose to glucose 6-phosphate.

### VISUALIZING THE REACTION

**Biosynthesis of Glucose 6-Phosphate**

glucose

adenosine 5′-triphosphate (ATP)

hexokinase
$Mg^{2+}$

glucose 6-phosphate

adenosine 5′-diphosphate (ADP)

The pyrophosphate group is a leaving group in many biological substitution reactions (p. 752, for example). It has the same function here. The reaction of glucose with adenosine triphosphate is a nucleophilic substitution reaction. The reaction is catalyzed by an enzyme, hexokinase, and magnesium ion is necessary. The magnesium ion forms a complex with the two terminal phosphate groups of adenosine triphosphate, masking the negative charge on the ion at physiologic pH and making it easier for the substitution reaction to take place.

The story of the later stages of glycolysis, which also involve enolate chemistry, will be told in the next chapter.

**Table 16.1  Formation of Enols and Enolate Anions**

| Compound That Enolizes | Conditions for Enolization | Enol or Enolate Anion |
|---|---|---|
| 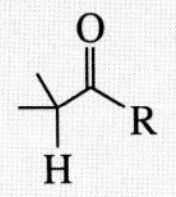 | $H{-}B^+$ | 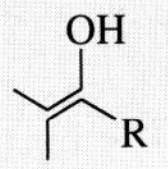 |
| 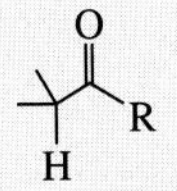 | $B:$ | 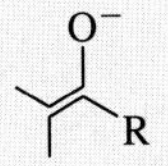 |
| 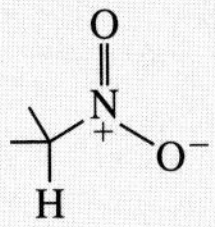 | $B:$ | 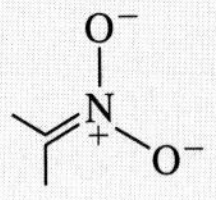 |
| 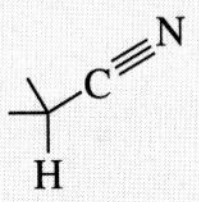 | $B:$ | 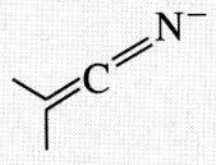 |
| 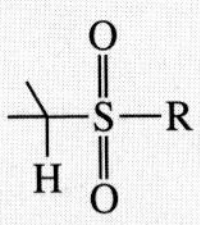 | $B:$ | 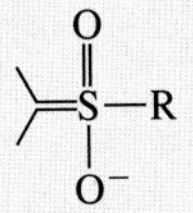 |
| 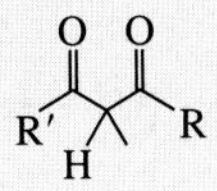 | $B:$ | 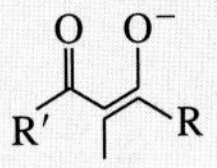 |

## ADDITIONAL PROBLEMS

**16.22** For each of the following sets of compounds, decide which one is the most acidic.

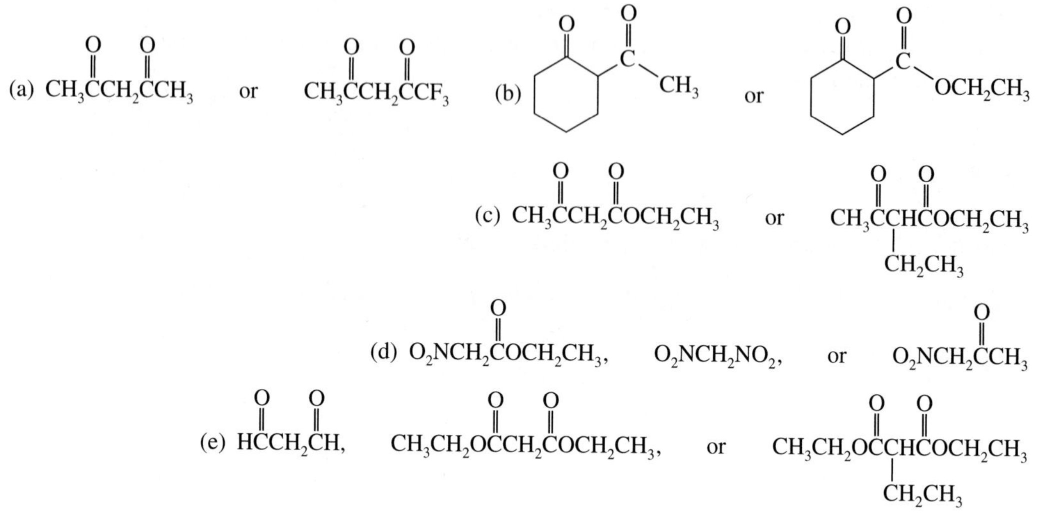

(a) $CH_3CCH_2CCH_3$    or    $CH_3CCH_2CCF_3$

(b) or

(c) $CH_3CCH_2COCH_2CH_3$    or    $CH_3CCHCOCH_2CH_3$
                                                     $CH_2CH_3$

(d) $O_2NCH_2COCH_2CH_3$,    $O_2NCH_2NO_2$,    or    $O_2NCH_2CCH_3$

(e) $HCCH_2CH$,    $CH_3CH_2OCCH_2COCH_2CH_3$,    or    $CH_3CH_2OCCHCOCH_2CH_3$
                                                                                $CH_2CH_3$

**16.23** Nitromethane has $pK_a$ 10.2, while that for dinitromethane is reported to be 3.6 and that for trinitromethane to be lower than 1. How would you rationalize these experimental data?

**16.24** The $pK_a$ for 2,4-pentanedione is 9.0. The $pK_a$ for 1,1,1-trifluoro-2,4-pentanedione is reported to be 4.7. Use structural formulas and words to rationalize this observation.

**16.25**

(a) Identify the two most acidic protons in 2-(phenylsulfonyl)ethanol, shown below, by showing the structure of the dianion that will form when it is treated with 2 equivalents of butyllithium.

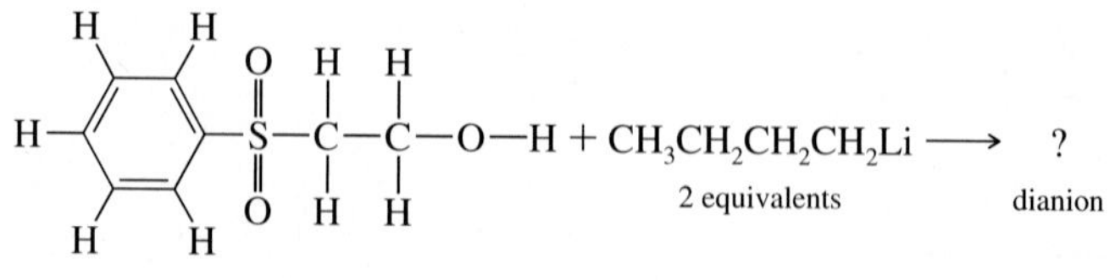

                    2-(phenylsulfonyl)ethanol

(b) The dianion contains two nucleophilic sites. Predict how the dianion will react with an electrophile such as benzyl bromide.

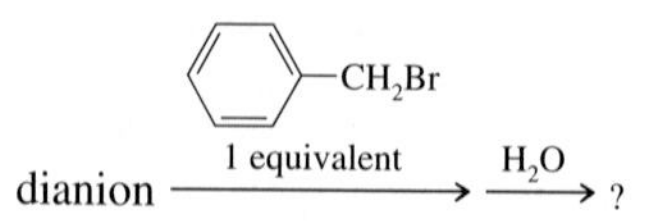

dianion $\xrightarrow[\text{1 equivalent}]{} \xrightarrow{H_2O}$ ?

**16.26** Thalidomide (Problem 6.32) racemizes at physiologic pH by an enolization reaction. Show how the racemization happens with the stereoisomer shown on the next page.

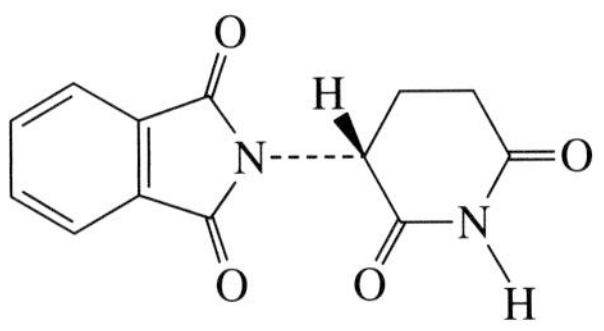

thalidomide

**16.27** The following $pK_a$ values were measured for the conjugate acids of various amines. Rationalize the trends that are apparent in the data.

(a)

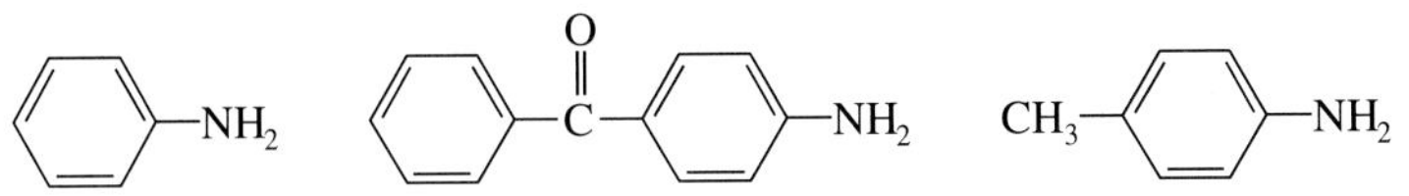

$pK_a$   4.60          2.17          5.10

(b)    $(CH_3CH_2)_2NH$   $CH_3CH_2NH_2$     (c)

$pK_a$      10.6          10.5          5.10      $pK_a$   4.60          0.8

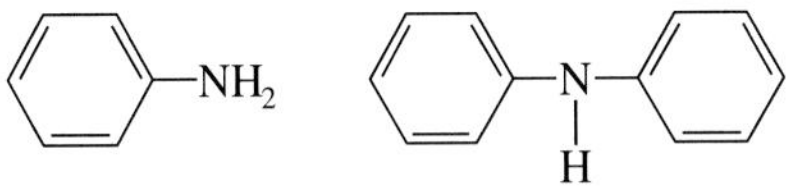

**16.28** The naturally occurring amino acid sarcosine, shown below, has $pK_a$ values 2.21 and 10.12.

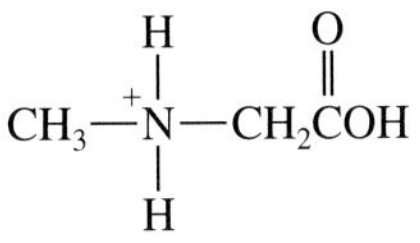

sarcosine

(a) Assign $pK_a$ values to the acidic protons in sarcosine.
(b) Construct a plot of the abundances of different species against pH to identify the approximate pH at which sarcosine has the structure shown above and the structure it would have at physiologic pH, ~6.2.
(c) What species would be most abundant at pH 13?

**16.29** Ammonium ions, $NH_4^+$, are toxic. The conversion of ammonia into urea keeps this toxin out of the blood stream. The amino acid arginine is important in the "urea cycle" by which ammonium ions are biologically converted into urea molecules. When arginine is protonated, the proton is attached at a very specific site.

(a) Explore the structure of arginine, given below, and decide which conjugate acid of arginine would be the most stable. (*Hint:* The species that is formed is stabilized by resonance.)

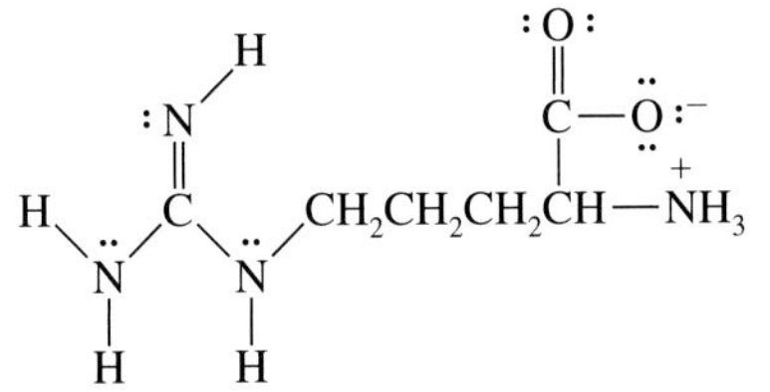

arginine

(b) Arginine is closely related to another amino acid, lysine, shown on the next page. Is lysine more or less likely to be protonated (more or less basic) than arginine? Why?

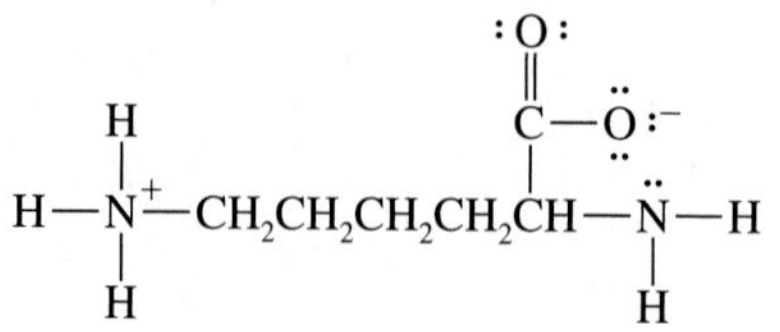

$$\text{H}-\overset{+}{\underset{\text{H}}{\overset{\text{H}}{\text{N}}}}-\text{CH}_2\text{CH}_2\text{CH}_2\text{CH}_2\overset{\overset{\text{:O:}}{\overset{\|}{\text{C}-\overset{..}{\underset{..}{\text{O}}}:^-}}}{\underset{\underset{\text{H}}{|}}{\text{CH}}}-\overset{|}{\underset{\text{H}}{\text{N}}}-\text{H}$$

lysine

**16.30** An investigation of the relative basicity of N, O, and S bases has determined that diphenylsulfinamide is protonated on the oxygen atom.

(a) Complete the equation for the protonation of diphenylsulfinamide by trifluoromethane-sulfonic acid. What is the approximate p$K_a$ of trifluoromethanesulfonic acid?

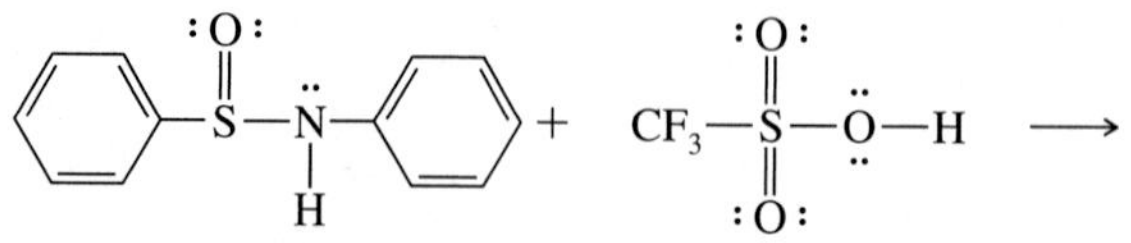

diphenylsulfinamide     trifluoromethanesulfonic acid

(b) Use words and structural formulas to provide a rationalization for the primary site of protonation of the sulfinamide. (*Hint:* It is useful to look at the conjugate acids that would result from protonation at the other basic sites in the molecule.)

**16.31**

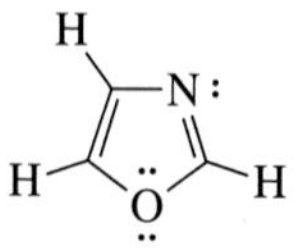

oxazole

(a) Oxazole, shown at left, has three sites where it may be deprotonated. The one proton on the carbon atom between the oxygen and the nitrogen atoms has been found to have p$K_a$ 26. Write an equation for the deprotonation of oxazole with butyllithium.
(b) There are two likely sites of protonation on oxazole. Draw these two conjugate acids, and decide which one is the more stable acid and, therefore, the one that forms.

**16.32** Researchers have found that the regioselectivity of an acyl-transfer reaction depends on the pH of the solution. Their observations are summarized below. How would you rationalize these observations?

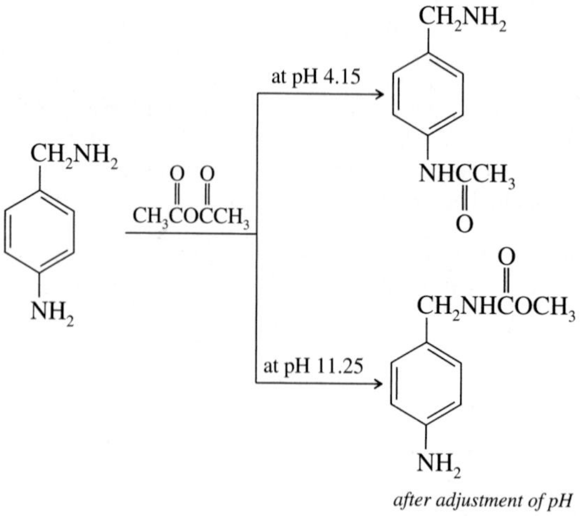

at pH 4.15

at pH 11.25

*after adjustment of pH to isolate the amine*

**16.33** The two compounds shown on the next page have different p$K_a$ values. Resonance stabilization of the conjugate base has been used to rationalize this difference for the more acidic substance. Decide which compound is the more acidic one, and draw the resonance contributor that best represents the stabilization of the conjugate base of that acid.

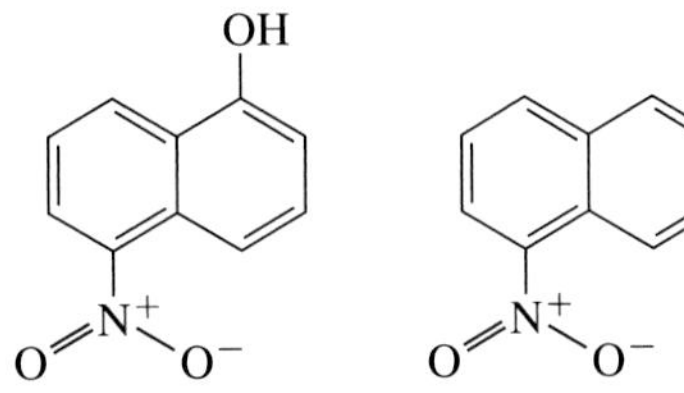

**16.34**

(a) Lithium diisopropylamide is a strong Brønsted base used in many organic reactions, as we will see below and in Chapter 17. It is prepared by deprotonating diisopropylamine, shown below, with one of the following three reagents. Use the p$K_a$ table to decide on an approximate p$K_a$ value for diisopropylamine, and complete the equation for the deprotonation reaction.

Reagents: (1) $CH_3\overset{O}{\overset{\|}{C}}OLi$, (2) $CH_3CH_2CH_2CH_2Li$, (3) $CH_3OLi$

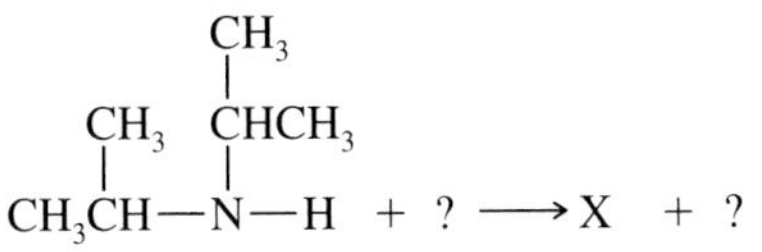

p$K_a$ = ?
diisopropylamine

lithium diisopropylamide
*conjugate base of diisopropyl amine*

(b) Predict what will happen if lithium diisopropylamide is added to benzyl propanoate.

**16.35** Thymine, one of the bases in deoxyribonucleic acid (DNA), exists in two tautomeric forms, keto and enol. The form that exists in DNA is the keto form. Identify the most acidic protons on the keto form shown at right, and draw the enol form of the compound.

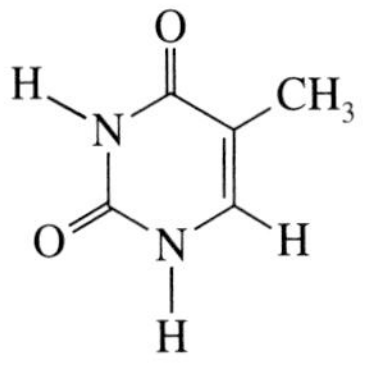

keto form of thymine

**16.36** The enzyme cytochrome C has an isoelectric point of 10.6. Will cytochrome C have a net positive charge, a net negative charge, or no net charge at physiologic pH, ~6?

**16.37** Nicotine has the structure shown at right. It has two sites of protonation. Identify them, and assign p$K_a$ values to the conjugate acids of these basic sites. Assign configuration to the stereocenter in nicotine. Draw the structural formula for the form in which nicotine will exist at physiologic pH, ~6, in the body.

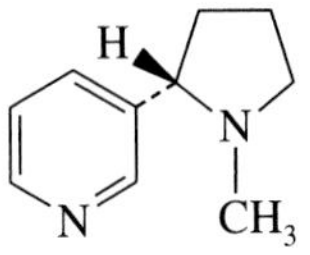

nicotine

**16.38** The energy diagram for the interconversion of the enol form of acetone with the ketone form of acetone, represented by the equation below, is shown in Figure 16.5 (p. 688). Answer the following questions using the equation and the energy diagram as the basis of your answers.

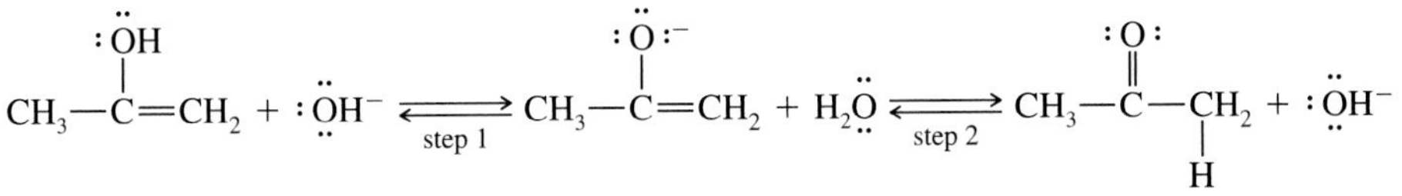

enol form of acetone | enolate ion | keto form of acetone

(a) The difference in which two energy levels, represented by letters on the energy diagram, corresponds to the free energy change, $\Delta G_r^\circ$, for the conversion of the enol to the ketone?
(b) Will $\Delta G_r^\circ$ for the reaction be >0, 0, or <0?
(c) Will the equilibrium constant, $K_{eq}$, for the conversion of the enol to the ketone be >1, 0, or <1?

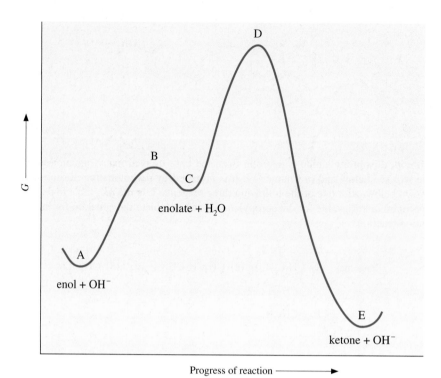

*Figure 16.5*
Energy diagram for the interconversion of acetone and its enol.

(d) The difference in which two levels represented by letters on the energy diagram corresponds to the free energy of activation, $\Delta G^{\ddagger}$, for the deprotonation of the enol to the enolate anion?

(e) Is protonation of the enolate ion at the oxygen atom faster or slower than protonation at carbon?

(f) Which point on the energy diagram represents the transition state for the deprotonation of the ketone to the enolate ion?

(g) Which is the rate-determining step for the conversion of the enol to the ketone?

**16.39** The conversion of Compound A to its diastereomer B is an example of epimerization (p. 676). Use the curved-arrow convention to write a detailed mechanism for this epimerization.

**16.40** In a study of the stereochemistry and conformation of steroids, the following reaction was observed.

The researchers point to the ease of epimerization of α-azido ketones as the reason for the presence of the two epimeric products in roughly equal amounts. Show a mechanism, using the curved-arrow convention, for the epimerization reaction. A partial structure for one of the α-azido ketones is shown at right. You may use H—B$^+$ and B: as acid and base catalysts as needed.

**16.41** Compound C has $\lambda_{max}$ 225 ($\epsilon$ 8900) and 280 nm ($\epsilon$ 1600) in 0.1 M hydrochloric acid. In 1 M sodium hydroxide solution, $\lambda_{max}$ 244 ($\epsilon$ 11,700) and 298 nm ($\epsilon$ 2600) are observed. The infrared spectrum of Compound C has a broad absorption band at 3400 cm$^{-1}$. The mass and proton magnetic resonance spectra of C are shown in Figure 16.6. Its $^{13}$C nuclear mag-

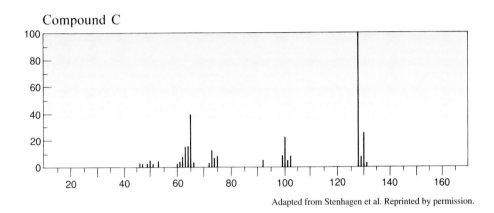

Compound C

Adapted from Stenhagen et al. Reprinted by permission.

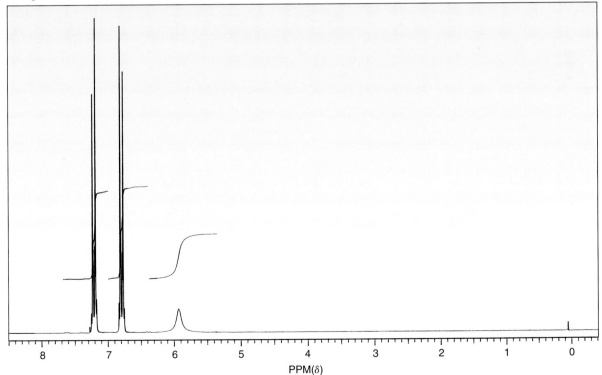

Compound C

PPM(δ)

*Figure 16.6*

netic resonance spectrum has peaks at 116.9 (d), 126.2 (s), 129.7 (d), and 153.6 (s) ppm. Assign a structure to Compound C. (*Hint:* The molecular ion is responsible for the base peak.) Discuss the different spectral data, showing how each piece supports your structural assignment.

**16.42** The ultraviolet spectra of acids and bases are often sensitive to pH, which is why such substances can be used as indicators (p. 826, for example). The spectra of aromatic phenols and amines, which have nonbonding electrons on atoms directly bonded to the aromatic ring, especially, show large changes on change of pH. Figure 16.7 shows the spectra of aniline at two different pH values, in aqueous sulfuric acid and in a buffer at pH 8.0. Which spectrum corresponds to which solution? A rereading of Section 12.1E will be helpful.

**16.43** The proton magnetic resonance spectrum of Compound D is shown in Figure 16.8 (p. 691). After a sample of the compound is shaken with $D_2O$, the band at $\delta$ 3.6 disappears. Compound D has bands in its infrared spectrum at 3480, 3376, and 1621 cm$^{-1}$. Its $^{13}C$ nuclear magnetic resonance spectrum has peaks at 13.0, 23.9, 115.4, 118.6, 126.7, 128.0, 128.3, and 144.3 ppm. When its ultraviolet spectrum is taken in 2 M hydrochloric acid, $\lambda_{max}$ is 259 nm; at pH 9.2, $\lambda_{max}$ is 280 nm. Assign a structure to Compound D, and show how the various pieces of spectral data support your assignment. Do the data allow you to make a definite assignment of structure?

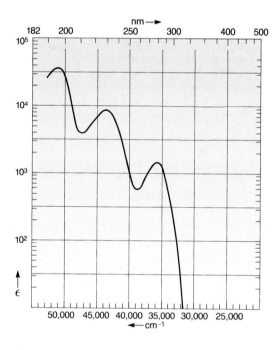

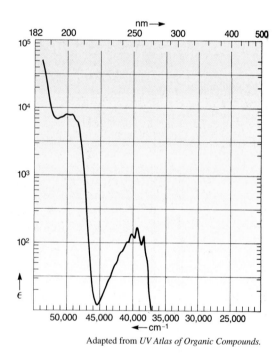

Adapted from *UV Atlas of Organic Compounds.*

*Figure 16.7*

Compound D

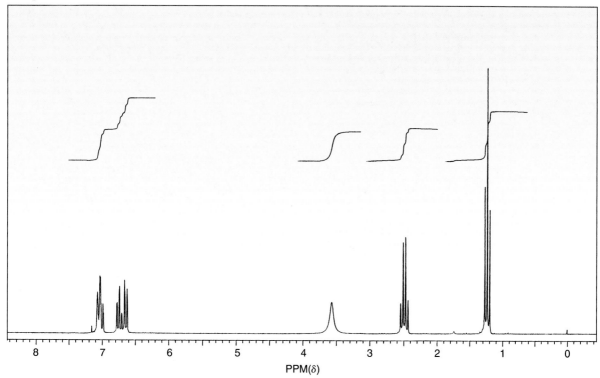

*Figure 16.8*

# 17

**Enols and Enolate Anions as Nucleophiles. Alkylation and Condensation Reactions**

## A Look Ahead

Enols and enolate anions (Section 16.4A) are nucleophiles and react with electrophiles. The features that are common to many of the reactions we will study in this chapter are summarized below.

### Step 1. Enolization

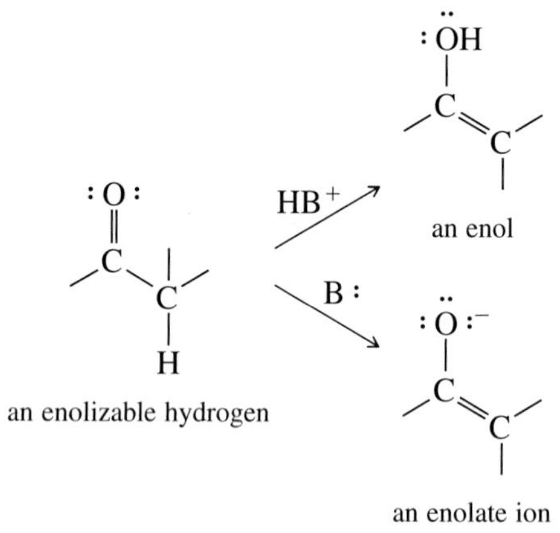

an enolizable hydrogen

an enol

an enolate ion

### Step 2. Reaction with the electrophile

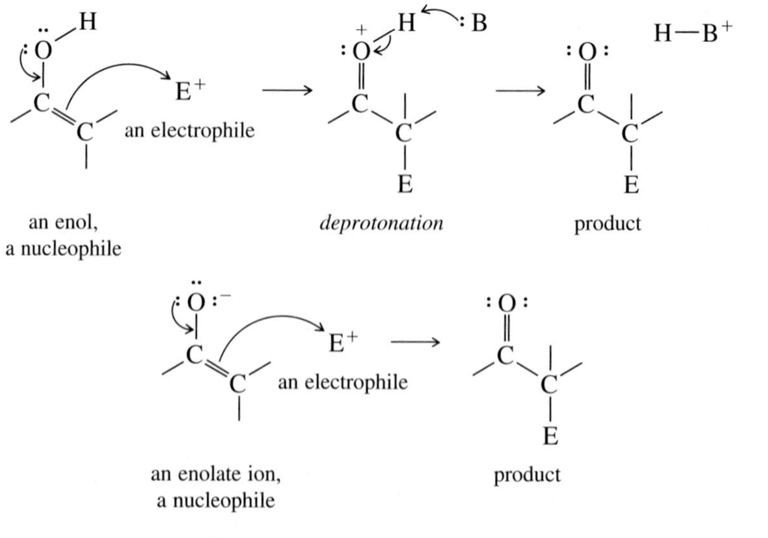

an enol,
a nucleophile

*deprotonation*

product

an enolate ion,
a nucleophile

product

The electrophile may be a halogen, an alkyl halide, or a carbonyl compound.
Enolate ions also add to double bonds that are conjugated with a carbonyl group.

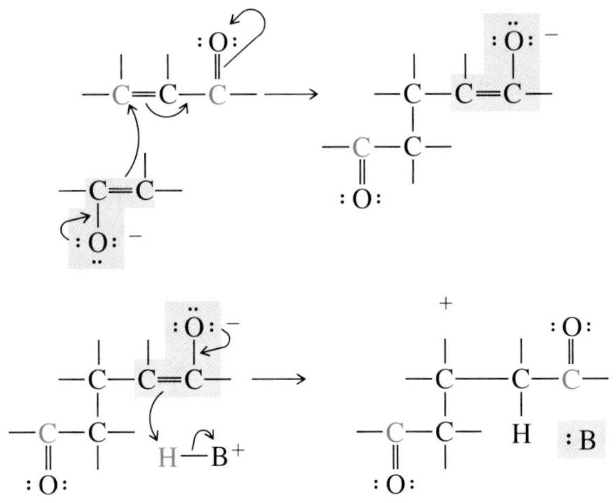

When enolate anions react as nucleophiles with electrophilic carbon atoms, carbon–carbon bonds are formed. Such reactions are, therefore, important in syntheses, including many biosynthetic pathways.

As you study the later sections of this chapter, it may appear that there are many details to learn. It will be helpful if you keep in mind the fundamental similarity of all the different reactions.

**Workbook Exercises**

⊘ **Study Guide**
**Concept Map 17.1**

## 17.1  Reactions of Enols and Enolates with Halogens as Electrophiles

Enols and enolates are nucleophiles and, therefore, react with electrophiles. For example, the enol of acetone reacts with bromine to give, chiefly, 1-bromo-2-propanone.

$$\underset{\substack{\text{acetone}}}{\overset{O}{\overset{\parallel}{CH_3CCH_3}}} + \underset{\substack{\text{bromine}}}{Br_2} \xrightarrow[65\ ^\circ C]{H_2O} \underset{\substack{\text{1-bromo-2-}\\\text{propanone}\\\sim 50\%}}{\overset{O}{\overset{\parallel}{CH_3CCH_2Br}}} + \underset{\substack{\text{1,1-dibromo-}\\\text{2-propanone}}}{\overset{O}{\overset{\parallel}{CH_3CCHBr_2}}} + \underset{\substack{\text{1,3-dibromo-}\\\text{2-propanone}}}{\overset{O}{\overset{\parallel}{BrCH_2CCH_2Br}}} + \underset{\substack{\text{hydrobromic}\\\text{acid}}}{HBr}$$

small amounts

In this reaction, bromine is the electrophile that reacts with the nucleophilic enol. Deprotonation of the resulting haloketone gives the final product.

## VISUALIZING THE REACTION

**Reaction of an Enol with a Halogen**

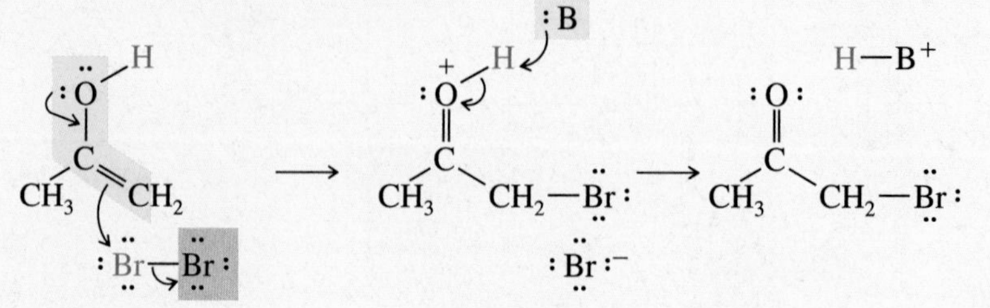

*reaction of the enol*
*with the electrophile, bromine*

*deprotonation*
*of the protonated*
*haloketone*

### ONE SMALL STEP

An α-hydrogen atom in a carboxylic acid also can be replaced by a halogen atom to give an α-haloacid. This reaction, which is known as the **Hell-Volhard-Zelinsky** reaction, works best if the acid is first converted by phosphorus tribromide in catalytic amounts into its acid halide.

**PROBLEM:** Write a mechanism for the conversion of an acyl bromide to an α-bromoacyl bromide, and explain why the halogenation works better with the acid halide than with the acid itself.

The rate of the halogenation reaction is equal to the rate at which deuterium is introduced into acetone (p. 674) and depends only on the concentration of the ketone. These facts suggest that the formation of the enol from the ketone is the rate-determining step for both these reactions and that once the enol is formed, it reacts quickly with any electrophile that may be present.

### Problem 17.1

(a) 3-Methyl-2,4-pentanedione reacts with bromine in water to give a product containing only one bromine atom. Write an equation for this reaction.
(b) If 3-methyl-2,4-pentanedione is dissolved in water and allowed to stand for a day, the initial rate at which this solution reacts with bromine is very fast. After awhile the bromine color disappears much more slowly but steadily. Explain these experimental observations.

### ONE SMALL STEP

When a methyl ketone is treated with halogen in the presence of base, the following reaction, using acetone as an example, is observed.

$$\underset{\substack{\text{acetone}}}{CH_3\overset{O}{\overset{\|}{C}}CH_3} + \underset{\substack{\text{iodine}\\\text{excess}}}{I_2} + \underset{\substack{\text{sodium}\\\text{hydroxide}}}{NaOH} \xrightarrow{H_2O} \underset{\substack{\text{triiodomethane}\\\text{iodoform}}}{CHI_3} + \underset{\substack{\text{sodium}\\\text{acetate}}}{CH_3\overset{O}{\overset{\|}{C}}O^-Na^+}$$

**PROBLEM:** Propose a mechanism for the reaction shown above.

**Hint:** Why do all the halogen atoms substitute on one side of the ketone? How is the carbon–carbon bond broken to give iodoform and a carboxylate anion? In other words, why does something that is normally not a good leaving group behave as one in this case?

## 17.2   Reactions of Enolate Anions with Alkyl Halides as Electrophiles

### A. Alkylation of Ketones

Enolate anions react with alkyl halides in nucleophilic substitution reactions known as **alkylation reactions** (p. 118). For example, cyclohexanone can be deprotonated to give an enolate ion that reacts with allyl bromide to give a product with a new carbon–carbon bond at the $\alpha$-carbon atom.

cyclohexanone   sodium amide                 enolate ion of cyclohexanone                ammonia

major resonance          allyl bromide          2-allylcyclohexanone
contributor of                                        ~ 60%
the enolate ion of
cyclohexanone

In the equation above, both resonance contributors of the enolate anion are shown in the deprotonation step. These resonance contributors remind us that the enolate ion has electron density and negative charge at two sites. It is basic and nucleophilic at the $\alpha$-carbon atom and at the oxygen atom of the carbonyl group; it also can be protonated (see Visualizing the Reaction, p. 674) or can react with other electrophiles at either of these sites. An anion that has the capacity to react at either of two positions is called an **ambident anion** or an **ambident nucleophile.** Whether such an anion reacts at the carbon atom or at the oxygen atom depends on reaction conditions, such as the nature of the cation associated with the anion, the solvent, and the nature of the electrophile present. The yield of 2-allylcyclohexanone, for example, is only about 60%. Clearly, the starting material undergoes other reactions than the one desired, perhaps including alkylation at the oxygen atom, which was not investigated. Generally, carbon electrophiles favor alkylation at the carbon atom. In the next section we will see a silicon electrophile that reacts exclusively at the oxygen atom.

In writing equations and mechanisms, chemists use the major resonance contributor for a species. For example, we write the carbonyl group as $\overset{\backslash}{C}=\overset{..}{\underset{..}{O}}$, even as we remember that it has a resonance contributor, $\overset{\backslash}{\underset{/}{C}}{}^+\!\!-\overset{..}{\underset{..}{O}}{}^{:-}$, that we use to rationalize the electrophilicity of the carbon atom of the carbonyl group. The resonance contributor in which the negative charge resides on the more electronegative oxygen atom is the major one for an enolate ion and so was used in completing the equation for the reaction of cyclohexanone with allyl bromide. We will use it also in writing the mechanism of the reaction.

---

**VISUALIZING THE REACTION**

**Alkylation of an Enolate Ion**

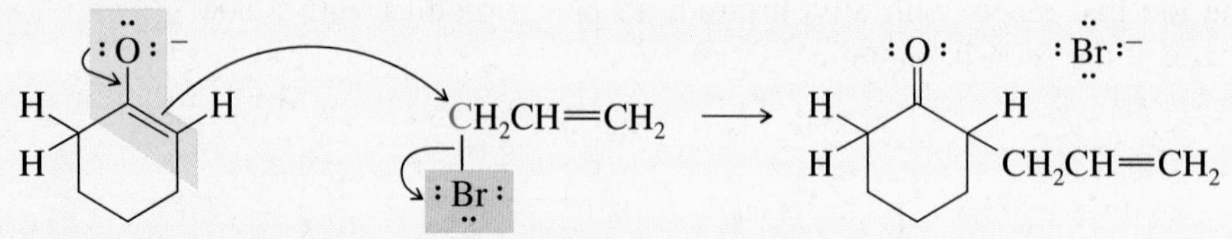

major resonance
contributor of the enolate
anion from cyclohexanone,
a nucleophile

electrophile

---

**On the Web: ONE SMALL STEP**

Both resonance contributors to the enolate anion of cyclohexanone must give the same reaction and the same product because they are not separate entities but ways of describing the distribution of electrons within a single species.

**Problem 17.2**    Review the mechanism for the alkylation of the enolate anion of cyclohexanone using the other resonance contributor as your starting point.

Another base that is often used in the formation of enolate ions is the diisopropyl-amide anion, a strong base but a poor nucleophile. For example, the enolate anion generated when 2-benzylcyclohexanone is deprotonated by lithium diisopropylamide reacts in a nucleophilic substitution reaction with methyl iodide. The large, bulky base deprotonates the ketone on the side with the fewer substituents (p. 867).

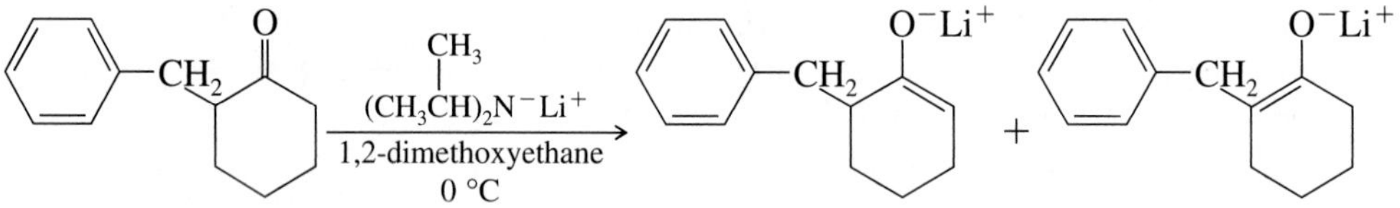

2-benzylcyclohexanone

$(CH_3CH)_2N^-Li^+$
1,2-dimethoxyethane
0 °C

kinetic enolate

$\dfrac{CH_3I \text{ (excess)}}{\begin{array}{c}\text{1,2-dimethoxyethane}\\ 25\ °C\end{array}}$  LiI +

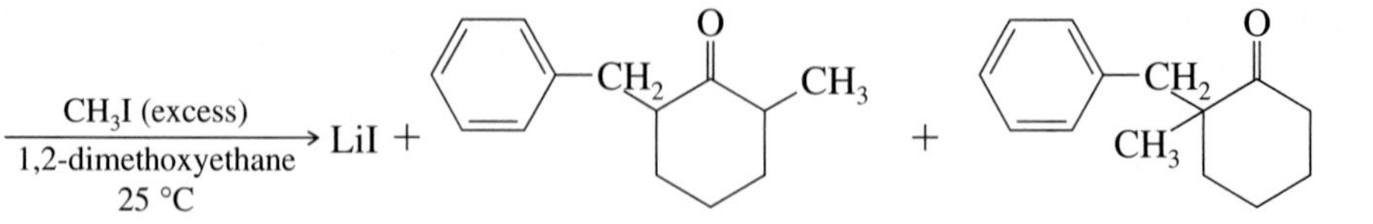

2-benzyl-6-methylcyclohexanone
76%

2-benzyl-2-methylcyclohexanone
6%

The major product is derived from the less highly substituted enolate.

## B. Alkylation of Esters

Enolate anions of esters also can be prepared using lithium diisopropylamide. Methyl butanoate is converted into methyl 2-ethylbutanoate by alkylation of its enolate.

$$CH_3CH_2CH_2\overset{\overset{\displaystyle O}{\|}}{C}OCH_3 \xrightarrow[\substack{\text{tetrahydro-}\\\text{furan}\\-78\,°C}]{(CH_3\overset{\overset{\displaystyle CH_3}{|}}{CH})_2N^-Li^+} CH_3CH_2CH{=}\overset{\overset{\displaystyle O^-Li^+}{|}}{C}OCH_3 \xrightarrow[\substack{\text{hexamethylphosphoric}\\\text{triamide}\\-78\,°C}]{CH_3CH_2I} CH_3CH_2\overset{\overset{\displaystyle O}{\|}}{\underset{\underset{\displaystyle CH_2CH_3}{|}}{C}}HCOCH_3 + LiI$$

methyl butanoate

methyl
2-ethylbutanoate
96%

Lactones undergo similar reactions. For example, the lactone from 4-hydroxy-butanoic acid is substituted at carbon atom 2 of the ring by an allyl group via deprotonation and alkylation.

lactone of
4-hydroxybutanoic
acid

*a γ-lactone*

The fact that lithium diisopropylamide reacts with esters and lactones, as well as ketones, to give products of deprotonation reactions rather than those of nucleophilic substitution reactions at the carbonyl group (p. 628) is further evidence of how low the nucleophilicity of the diisopropylamide anion is.

 **On the Web: ONE SMALL STEP**

**Problem 17.3**    Supply structural formulas for the reagents, intermediates, and products indicated by letters.

(a)

(b)

(c) $CH_3CH_2CH_2\overset{\overset{\displaystyle O}{\|}}{C}OCH_3 \xrightarrow[\substack{\text{tetrahydrofuran}\\-78\,°C}]{D} E \xrightarrow[\substack{\text{hexamethylphosphoric}\\\text{triamide}\\-78\,°C}]{F} CH_3CH_2\overset{\overset{\displaystyle O}{\|}}{\underset{\underset{\displaystyle CH_2OCH_3}{|}}{C}}HCOCH_3$

# 17.3    Reaction of Stabilized Enolate Anions with Alkyl Halides as Electrophiles

## A. Alkylation of Active Methylene Compounds

The enolate anions of compounds containing active methylene groups are formed easily and with high regioselectivity (p. 671). For example, ethyl acetoacetate enolizes regioselectively to one of the two possible enolate ions, the one stabilized by delocalization of charge to two oxygen atoms. The regioselectivity of this enolization allows the ethyl acetoacetate to be alkylated selectively at the active methylene group. The product of the alkylation reaction is a new β-ketoester, ethyl 2-butylacetoacetate.

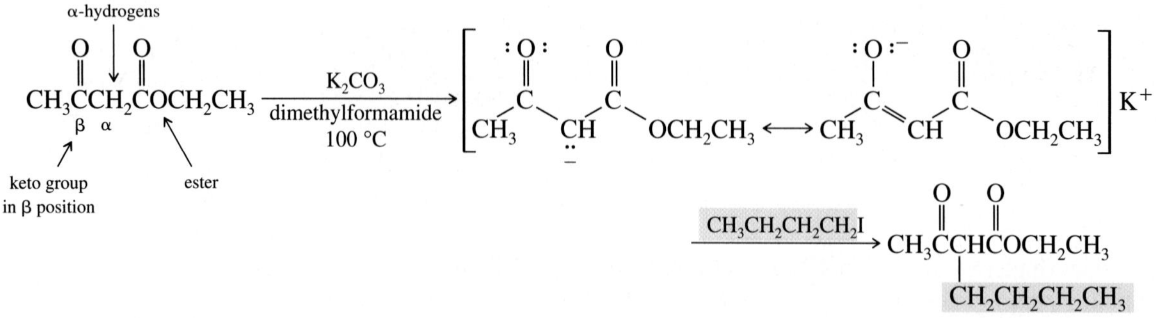

α-hydrogens

$$CH_3CCH_2COCH_2CH_3 \xrightarrow[\text{dimethylformamide}]{\text{K}_2\text{CO}_3 \atop 100\ °\text{C}} \left[ \begin{array}{c} \ddot{\text{:O:}} \quad \text{O} \\ \parallel \\ CH_3 \quad CH \quad OCH_2CH_3 \end{array} \longleftrightarrow \begin{array}{c} \ddot{\text{:O:}}^- \quad \text{O} \\ \parallel \\ CH_3 \quad CH \quad OCH_2CH_3 \end{array} \right] K^+$$

keto group in β position / ester

$$\xrightarrow{CH_3CH_2CH_2CH_2I} CH_3CCHCOCH_2CH_3$$

$$CH_2CH_2CH_2CH_3$$

ethyl 2-butylacetoacetate
99%

Diethyl malonate also has an active methylene group and is a useful starting material for a number of syntheses. The enolate anion displaces bromide ion from *sec*-butyl bromide to give the substituted ester.

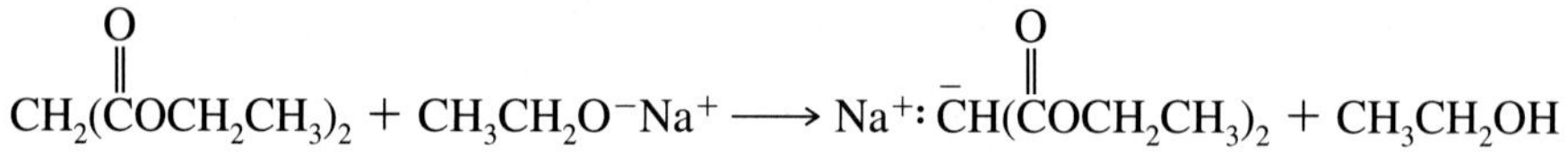

$$CH_2(COCH_2CH_3)_2 + CH_3CH_2O^-Na^+ \longrightarrow Na^+:\overline{C}H(COCH_2CH_3)_2 + CH_3CH_2OH$$

diethyl malonate / sodium ethoxide / sodium salt of the enolate of diethyl malonate, shown as carbanion

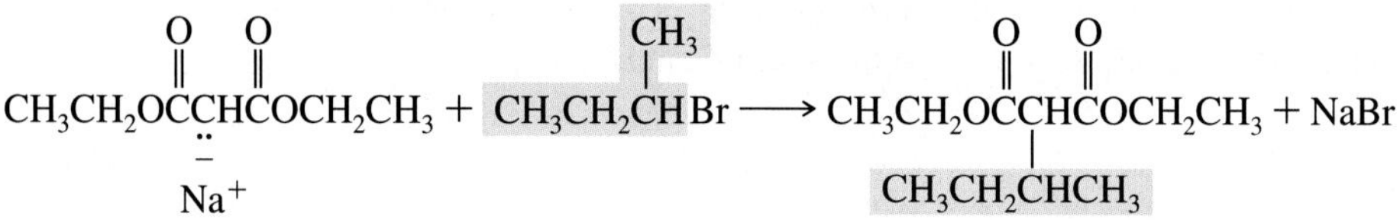

$$CH_3CH_2OCCHCOCH_2CH_3 + CH_3CH_2CHBr \longrightarrow CH_3CH_2OCCHCOCH_2CH_3 + NaBr$$

Na+ / CH_3 / CH_2CH_2CHCH_3

*sec*-butyl bromide / diethyl *sec*-butylmalonate ~80%

A monosubstituted acetoacetic or malonic ester still has a hydrogen atom on the carbon atom between the two carbonyl groups, so another enolization and alkylation are possible. The presence of one alkyl group on that carbon atom makes the remaining hydrogen less acidic, and sometimes a stronger base must be used to remove it. For example, sodium *tert*-butoxide, which is a stronger base than sodium ethoxide, is used to form the enolate from diethyl isopropylmalonate in order to introduce a second alkyl group.

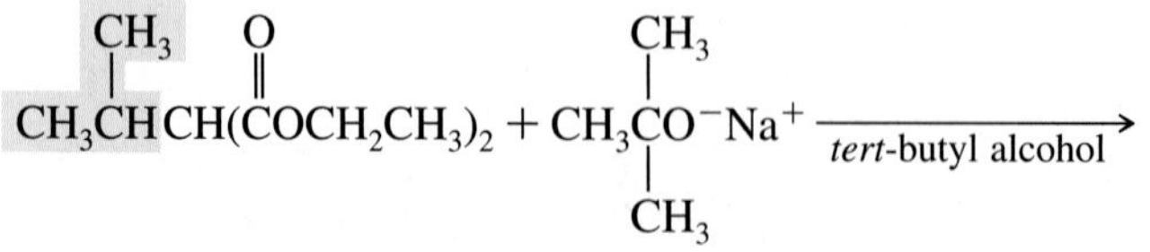

$$CH_3CHCH(COCH_2CH_3)_2 + CH_3CO^-Na^+ \xrightarrow{\text{\textit{tert}-butyl alcohol}}$$

CH_3 / CH_3

diethyl isopropylmalonate / sodium *tert*-butoxide

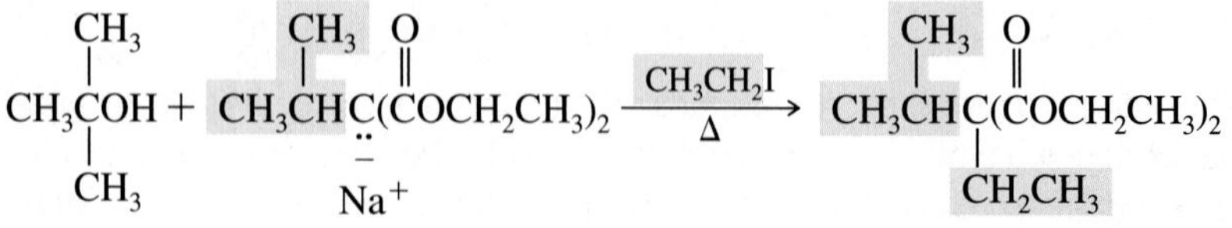

$$CH_3COH + CH_3CHC(COCH_2CH_3)_2 \xrightarrow[\Delta]{CH_3CH_2I} CH_3CHC(COCH_2CH_3)_2$$

CH_3 / CH_3 / Na+ / CH_3 / CH_2CH_3

diethyl ethylisopropylmalonate
65%

**Problem 17.4**    Give structural formulas for all species symbolized by letters in the following equations.

(a) $CH_2(COCH_2CH_3)_2$ $\xrightarrow[\text{ethanol}]{CH_3CH_2ONa}$ A $\xrightarrow{CH_3\overset{\displaystyle Br}{\underset{}{C}}HCH_3}$ B

(b) $CH_3CH_2CH(COCH_2CH_3)_2$ $\xrightarrow[\underset{\text{alcohol}}{tert\text{-butyl}}]{CH_3\overset{\displaystyle CH_3}{\underset{\displaystyle CH_3}{C}}ONa}$ C $\xrightarrow{BrCH_2CH=CH_2}$ D

(c) $CH_3CCH_2COCH_2CH_3$ $\xrightarrow[\text{ethanol}]{CH_3CH_2ONa \text{ (1 molar equivalent)}}$ E $\xrightarrow{ClCH_2CH_2CH_2Br}$ F

(d) $\underset{\displaystyle O}{CH_2CH_2}$ + $NaCH(COCH_2CH_3)_2$ $\longrightarrow$ G

(*Hint:* A review of page 505 may be helpful.)

## B. Decarboxylation of Acids with a β-Carbonyl Group

Carboxylic acids that have a carbonyl group in the β-position are unstable and lose carbon dioxide easily. The instability of such acids makes it possible to use ester groups such as those in ethyl acetoacetate or diethyl malonate to activate α-hydrogen atoms, making their regioselective substitution possible but eliminating the ester once it has performed this necessary function. For example, ethyl acetoacetate can be used to synthesize 2-heptanone in the following sequence of reactions.

β-keto group — $CH_3CCHCOCH_2CH_3$ (ester), $CH_2CH_2CH_2CH_3$

ethyl 2-butylacetoacetate

$\xrightarrow[\substack{H_2O \\ 25\,°C \\ 4\,h}]{5\%\ NaOH}$

$CH_3CCHCO^-Na^+$, $CH_2CH_2CH_2CH_3$

sodium 2-butylacetoacetate

$\xrightarrow{H_2SO_4,\ cold,\ dilute}$

β-keto group — $CH_3CCHCOH$ (acid), $CH_2CH_2CH_2CH_3$

2-butylacetoacetic acid

$\xrightarrow{\Delta}$ $CH_3CCH_2CH_2CH_2CH_2CH_3$ + $CO_2\uparrow$

2-heptanone ~ 60%

Ethyl acetoacetate is selectively alkylated at the active methylene group with *n*-butyl iodide. The product of the alkylation reaction is a β-ketoester. Saponification of the ester group with dilute base followed by careful acidification with cold dilute acid causes an unstable β-ketocarboxylic acid to be formed. This acid loses carbon dioxide on heating. A methyl ketone is the product.

A similar decarboxylation reaction occurs when malonic acid or a substituted malonic acid is heated. Thus diethyl *sec*-butylmalonate, synthesized on page 698, can be saponified and then heated with acid to remove one of the carboxyl groups.

$$CH_3CH_2CHCH(COCH_2CH_3)_2 \xrightarrow[\Delta]{KOH} \underset{\Delta}{H_2O}} CH_3CH_2CHCH(CO^-K^+)_2 \xrightarrow[\Delta, 3\ h]{H_2SO_4} H_2O}$$

diethyl *sec*-butylmalonate          dipotassium *sec*-butylmalonate

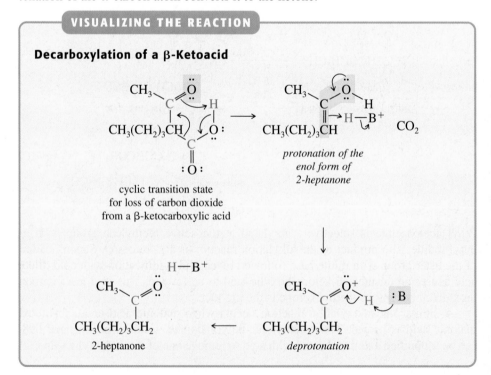

The loss of carbon dioxide does not start until the reaction mixture is acidified and heated; then the evolution of a gas is observed.

The reaction sequence that consists of alkylation of diethyl malonate, hydrolysis of the new diester, and then decarboxylation of the diacid results in the conversion of *sec*-butyl bromide to a carboxylic acid that has two more carbon atoms than the halide. This sequence is a general synthetic method for carboxylic acids. These acids may be regarded as substituted acetic acids. The carboxyl group and the $\alpha$-carbon atom are derived from diethyl malonate, and the rest of the molecule, from the alkyl halide used in the substitution reaction.

Carboxylic acids with carbonyl groups beta to the carboxyl function are unstable because they can react by means of a cyclic transition state. Transfer of a proton, loss of carbon dioxide, and the formation of an enol all occur simultaneously. For example, when carbon dioxide is lost from 2-butylacetoacetic acid, the species formed is the enol form of 2-heptanone. Deprotonation at the oxygen atom and protonation of the $\alpha$-carbon atom converts it to the ketone.

**Problem 17.5**    Give structural formulas for all compounds symbolized by letters in the following equations.

(a) $\overset{O}{\overset{\|}{C}}H_3CCH_2\overset{O}{\overset{\|}{C}}OCH_2CH_3 \xrightarrow[\text{ethanol}]{CH_3CH_2ONa} A \xrightarrow[\Delta]{ClCH_2\overset{O}{\overset{\|}{C}}OCH_2CH_3} B \xrightarrow[\Delta]{H_3O^+} C$

(b) $\text{C}_6\text{H}_5{-}\overset{O}{\overset{\|}{C}}CH_2\overset{O}{\overset{\|}{C}}OCH_2CH_3 \xrightarrow[\text{ethanol}]{CH_3CH_2ONa} D \xrightarrow[\Delta]{BrCH_2C{\equiv}CH} E \xrightarrow[\Delta]{H_3O^+} F$

(c) $CH_2(\overset{O}{\overset{\|}{C}}OCH_2CH_3)_2 \xrightarrow[\text{ethanol}]{CH_3CH_2ONa \ (1 \text{ molar equivalent})} G \xrightarrow{BrCH_2CH_2CH_2Cl} H \xrightarrow[\text{ethanol}]{CH_3CH_2ONa \ (1 \text{ molar equivalent})} I \xrightarrow[\Delta]{H_3O^+} J$

$(C_5H_8O_2)$

## 17.4   Condensation Reactions. Reactions of Enolate Anions with Carbonyl Compounds

### A. The Aldol Condensation

The reaction of nucleophilic carbon atoms with electrophilic carbon atoms to form carbon–carbon bonds is quite general. Reactions of this type form the basis of many synthetic transformations. Reactions of carbanions with carbonyl groups are particularly important in the creation of complex structures. Such reactions have already been discussed in connection with the addition of organometallic reagents to the carbonyl group of aldehydes or ketones (p. 551). Enolate ions react with carbonyl groups to form new carbon–carbon bonds. For example, acetaldehyde reacts with itself in acid or base to give the β-hydroxyaldehyde that is called **aldol** because it is both an **ald**ehyde and an alcoh**ol**.

$$2\ \underset{\substack{\text{acetaldehyde}}}{CH_3\overset{O}{\overset{\|}{C}}H} \xrightarrow[\substack{H_2O \\ 5\,°C \\ 1\ h}]{NaOH} \xrightarrow{H_3O^+} \underset{\substack{\text{3-hydroxybutanal} \\ \text{aldol} \\ 50\%}}{CH_3\overset{OH}{\overset{|}{C}}HCH_2\overset{O}{\overset{\|}{C}}H}$$

The reaction involves the enolization of a molecule of acetaldehyde and then attack by that enolate ion on the carbonyl group of another molecule of acetaldehyde that is present in equilibrium with the enolate ion, because the base that is used is not strong enough to completely deprotonate the aldehyde.

**VISUALIZING THE REACTION**

**The Aldol Condensation**

enolate anion of acetaldehyde

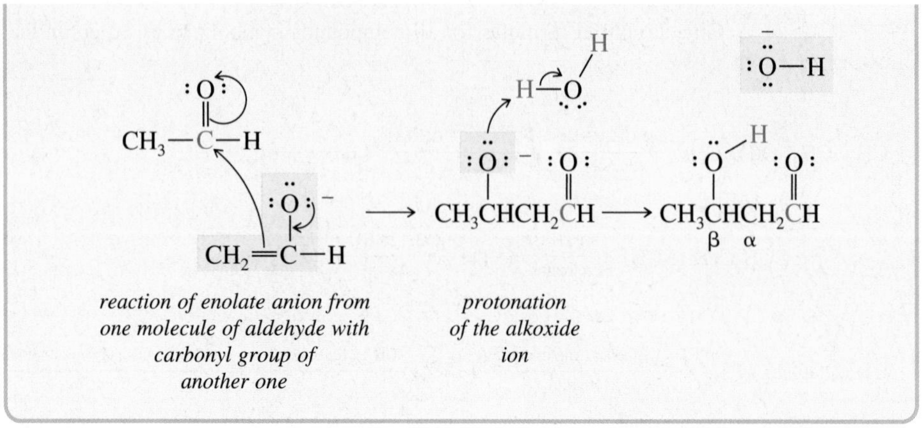

reaction of enolate anion from
one molecule of aldehyde with
carbonyl group of
another one

protonation
of the alkoxide
ion

The resulting alkoxide ion is protonated by the solvent, water, which regenerates the catalyst, hydroxide ion. The product of the reaction shown here is a $\beta$-hydroxy-aldehyde.

A reaction in which the enolate ion of one carbonyl compound reacts with the carbonyl group of another one is called an **aldol condensation.** This type of reaction is called a *condensation* because one larger molecule is created from the union of two smaller ones.

$\beta$-Hydroxyaldehydes are rather unstable and are easily dehydrated to compounds in which the double bond is $\alpha,\beta$ to a carbonyl group. Such a double bond is said to be in conjugation with the carbonyl group. Conjugation makes the double bond particularly stable, even to the point of having water eliminated from the intermediate hydroxy compound under basic conditions. If an aldol condensation is carried out at higher temperatures, the unsaturated compound is the product.

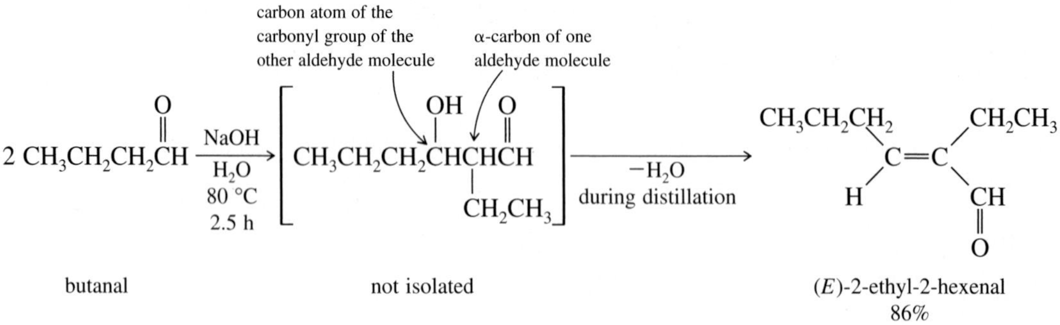

carbon atom of the
carbonyl group of the
other aldehyde molecule

$\alpha$-carbon of one
aldehyde molecule

butanal

not isolated

(E)-2-ethyl-2-hexenal
86%

Note that it is the $\alpha$-carbon atom of one molecule of aldehyde that reacts with the carbonyl group of another molecule of aldehyde. Because acetaldehyde contains no more carbon atoms beyond the $\alpha$-carbon, it gives a straight-chain product. Other aldehydes give a product that is branched at the $\alpha$-carbon atom.

The deyhdration reaction takes place by way of an enolization reaction in a manner similar to the biological conversion of 2-phosphoglyceric acid to 2-phosphoenolpyruvate (p. 679).

**VISUALIZING THE REACTION**

**Loss of Water from an Aldol**

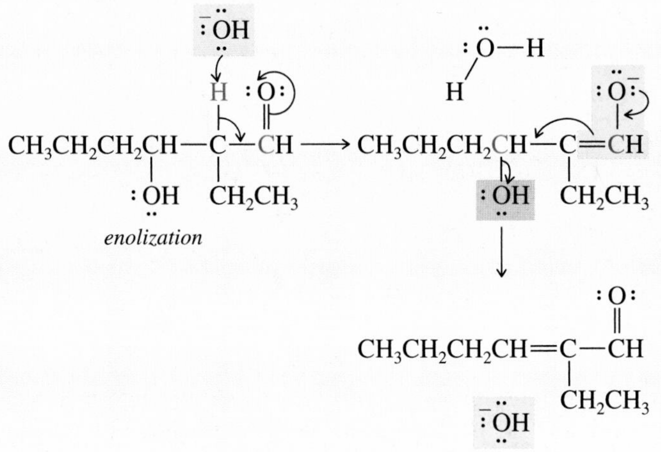

$$CH_3CH_2CH_2CH-C-CH \longrightarrow CH_3CH_2CH_2CH-C=CH$$

*enolization*

$$CH_3CH_2CH_2CH=C-CH$$

The formation of the $\alpha$, $\beta$-unsaturated carbonyl compound is the driving force that allows the hydroxide ion to be the leaving group, something we did not observe for elimination reactions of alcohols (p. 286).

Ketones do not condense with themselves in aldol reactions as readily as aldehydes do. The carbonyl group of a ketone is more hindered and less electrophilic than the carbonyl group of an aldehyde. In the equilibrium between a ketone and its aldol condensation product, the ketone is favored unless the product is continually removed so that it cannot undergo the base-catalyzed retroaldol reaction (p. 719).

The two carbonyl compounds that participate in an aldol condensation do not have to be the same. Aldol reactions between two different compounds are possible. One of the compounds must be a source of enolate anions, and the other must have a carbonyl group for them to attack. For example, 2-pentanone is converted into its enolate by lithium diisopropylamide at low temperatures. Under these conditions, self-condensation of the ketone, which is an unfavorable reaction anyway, does not occur. The enolate anion reacts cleanly with the carbonyl group of butanal in the second step of the reaction.

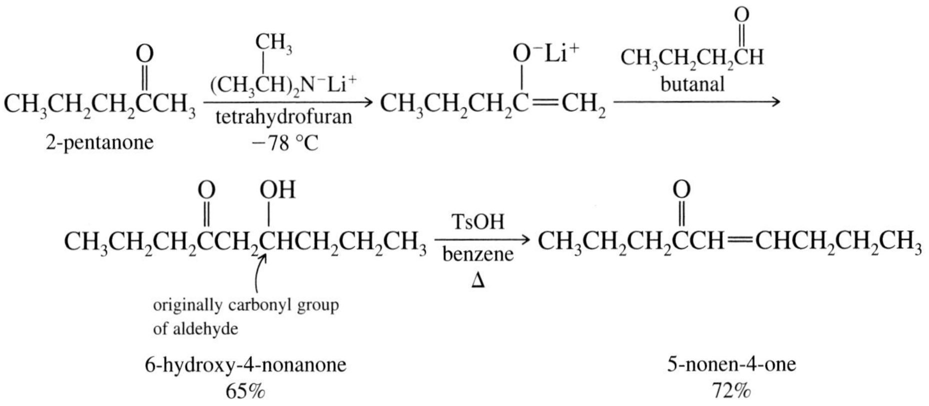

$$CH_3CH_2CH_2CCH_3 \xrightarrow[\substack{\text{tetrahydrofuran} \\ -78\ °C}]{(CH_3CH)_2N^-Li^+} CH_3CH_2CH_2C=CH_2 \xrightarrow{CH_3CH_2CH_2CH \atop \text{butanal}}$$

2-pentanone

$$CH_3CH_2CH_2CCH_2CHCH_2CH_2CH_3 \xrightarrow[\substack{\text{benzene} \\ \Delta}]{\text{TsOH}} CH_3CH_2CH_2CCH=CHCH_2CH_2CH_3$$

*originally carbonyl group of aldehyde*

6-hydroxy-4-nonanone                                    5-nonen-4-one
65%                                                          72%

When one of the participants in an aldol condensation does not have any $\alpha$-hydrogen atoms, the possibility for competing reactions is reduced. One reactant supplies enolate ions, and the other one provides the carbonyl group. Benzaldehyde is an aldehyde that reacts with enolate ions but cannot form one itself. It also reacts with the enolate of acetone to give an $\alpha,\beta$-unsaturated ketone.

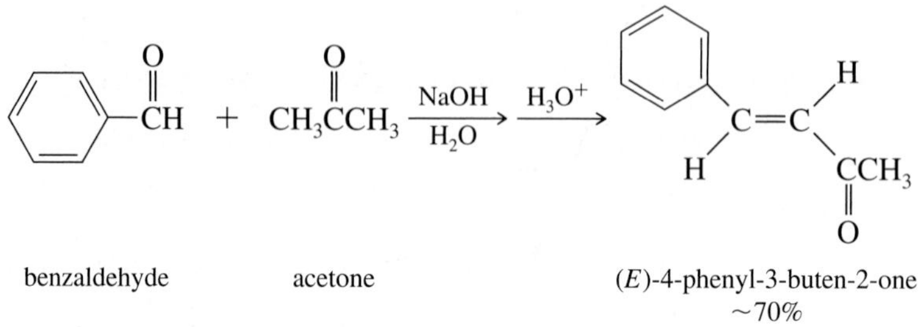

benzaldehyde          acetone                                    (E)-4-phenyl-3-buten-2-one
                                                                              ~70%

In condensation reactions involving aryl aldehydes, the intermediate loses water with great ease because the product of that dehydration has a double bond in conjugation not only with the carbonyl group but also with the aromatic ring. In most cases, the unsaturated compounds obtained from aldol condensations have the E configuration.

Any compound with an $\alpha,\beta$-unsaturated carbonyl function in it is likely to have been synthesized by an aldol condensation. Each product shown in the three preceding equations can be dissected to show the carbonyl compounds from which it was created.

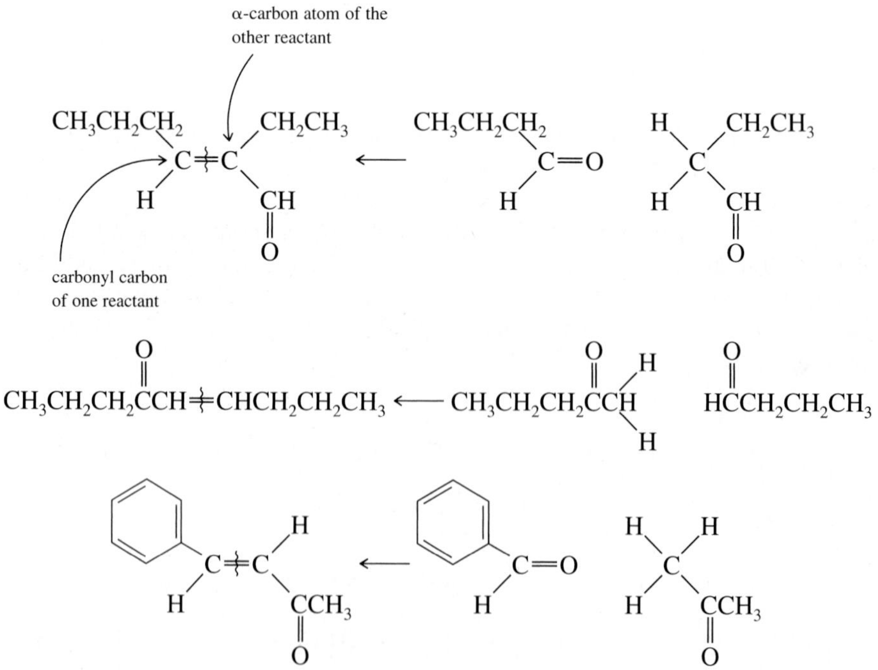

Note that the carbonyl group of one reactant is reconstructed by replacing a double bond to carbon with a double bond to oxygen. Two hydrogen atoms are also put back on the $\alpha$-carbon atom of the other reactant.

**Problem 17.6**   Complete the following equations.

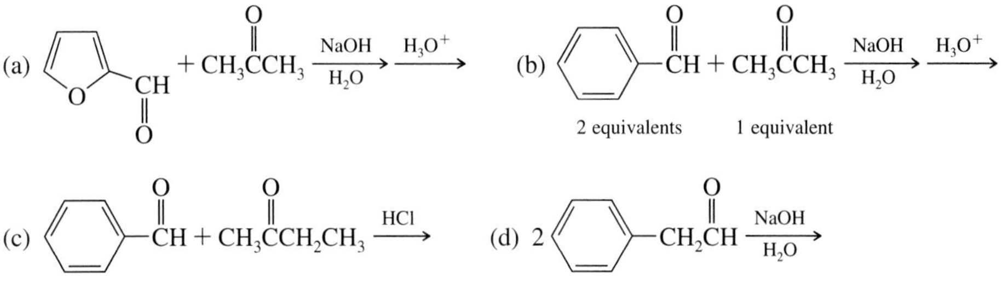

(a) [furan-2-carbaldehyde structure] $+ CH_3CCH_3$ $\xrightarrow[H_2O]{NaOH}$ $\xrightarrow{H_3O^+}$

(b) [benzaldehyde structure] $-CH + CH_3CCH_3$ $\xrightarrow[H_2O]{NaOH}$ $\xrightarrow{H_3O^+}$

2 equivalents     1 equivalent

(c) [benzaldehyde structure] $-CH + CH_3CCH_2CH_3$ $\xrightarrow{HCl}$

(d) 2 [phenylacetaldehyde structure] $-CH_2CH$ $\xrightarrow{NaOH}{H_2O}$

(*Hint:* The more highly stable enol forms.)

**Problem 17.7**   Give the structures of the starting materials that would be required to synthesize the following compounds.

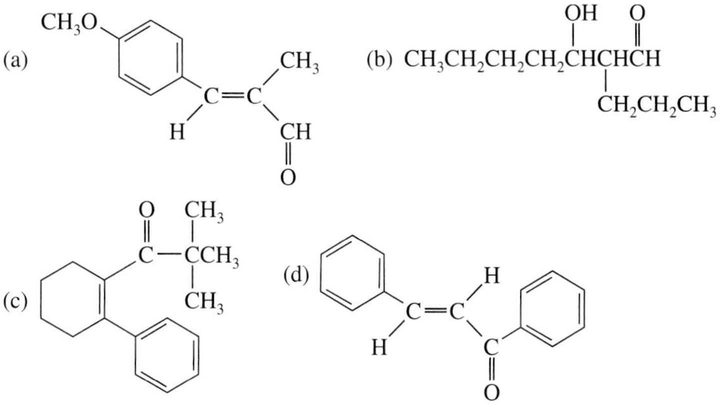

(a) [structure with CH₃O-phenyl, C=C, CH₃, CH=O]

(b) $CH_3CH_2CH_2CH_2\overset{OH}{\underset{|}{CH}}CH\overset{O}{\underset{|}{CH}}$ with $CH_2CH_2CH_3$

(c) [cyclohexene ring with C(=O)-C(CH₃)₂CH₃ and phenyl substituents]

(d) [trans-chalcone structure: phenyl-C=C with H substituents, C=O, phenyl]

## B. The Claisen Condensation

The aldol condensation involves the reaction of an enolate ion with the carbonyl group of an aldehyde or ketone. In a similar reaction, enolate ions from esters react with the carbonyl groups of acid derivatives to form β-ketoesters. Such reactions are **acylation reactions of enolate anions.** In the reaction, the α-carbon atom of the compound giving rise to the enolate ion ends up bonded to an acyl group. For example, the enolate anion of *tert*-butyl 2-methylpropanoate generated by a strong base under conditions that do not allow for equilibration, reacts with benzoyl chloride to give *tert*-butyl 2,2-dimethyl-3-oxo-3-phenylpropanoate.

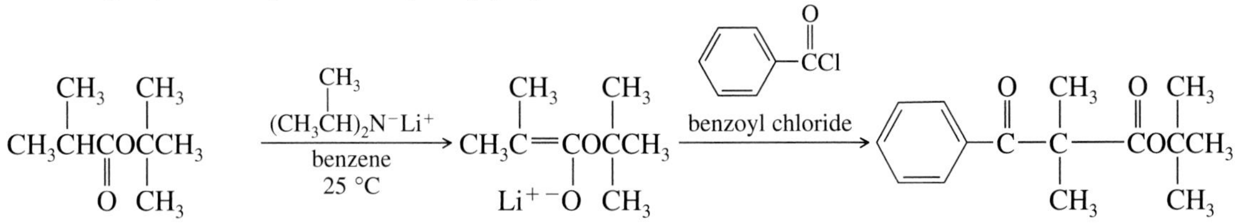

[reaction scheme]

$CH_3CHCOCCH_3$ (tert-butyl 2-methylpropanoate) $\xrightarrow[\text{benzene}\ 25\,°C]{(CH_3CH)_2N^-Li^+}$ $CH_3C=COCCH_3$ (with $Li^+\ ^-O\ CH_3$) $\xrightarrow[\text{benzoyl chloride}]{}$ [product: phenyl-C(=O)-C(CH₃)₂-COCCH₃ structure]

*tert*-butyl 2-methylpropanoate

*tert*-butyl 2,2-dimethyl-3-oxo-
3-phenylpropanoate
78%

An ester is acylated at the carbon alpha to its carbonyl group by this sequence of reactions. The product of an acylation reaction of an enolate anion is always a dicarbonyl compound in which the carbonyl groups are separated by one carbon atom. Such compounds are called **1,3 dicarbonyl compounds.**

When the enolate anion from an ester reacts with the carbonyl group of another ester, the reaction is known as the **Claisen condensation.** The classic example of this type of reaction is the condensation of two molecules of ethyl acetate to give the enolate of ethyl acetoacetate. In a second step, acid is added to the reaction mixture to form the $\beta$-ketoester.

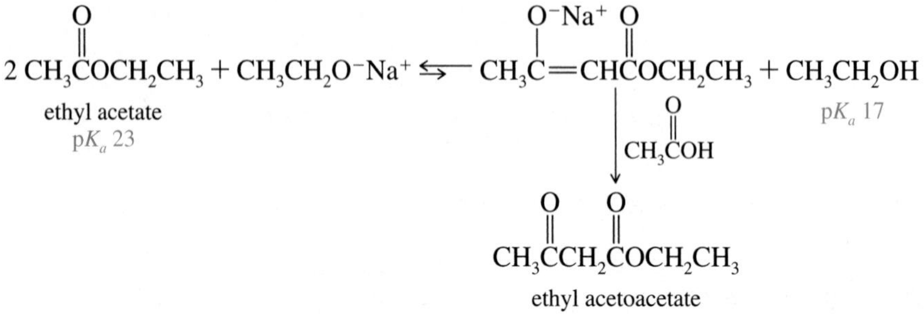

$$2\ CH_3COCH_2CH_3 + CH_3CH_2O^-Na^+ \rightleftharpoons CH_3C=CHCOCH_2CH_3 + CH_3CH_2OH$$

ethyl acetate
$pK_a\ 23$

$pK_a\ 17$

$$CH_3COH$$

$$CH_3CCH_2COCH_2CH_3$$
ethyl acetoacetate

Sodium ethoxide is used to deprotonate the ester in a reaction in which equilibrium lies to the left. This creates a mixture of the enolate ion and the unprotonated ester, which can then serve as the electrophile in the reaction. The alkoxyl group on the ester function serves as a leaving group from the tetrahedral intermediate formed when the carbonyl group is attacked by an enolate ion. The formation of an enolate ion by deprotonation at the active methylene group of the product provides the driving force for the completion of the reaction.

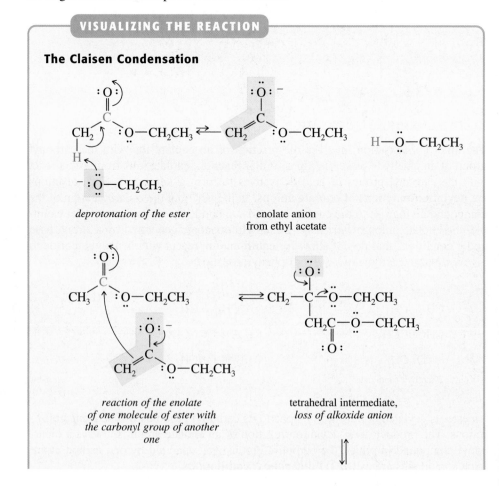

**VISUALIZING THE REACTION**

**The Claisen Condensation**

*deprotonation of the ester*

*enolate anion from ethyl acetate*

*reaction of the enolate of one molecule of ester with the carbonyl group of another one*

*tetrahedral intermediate, loss of alkoxide anion*

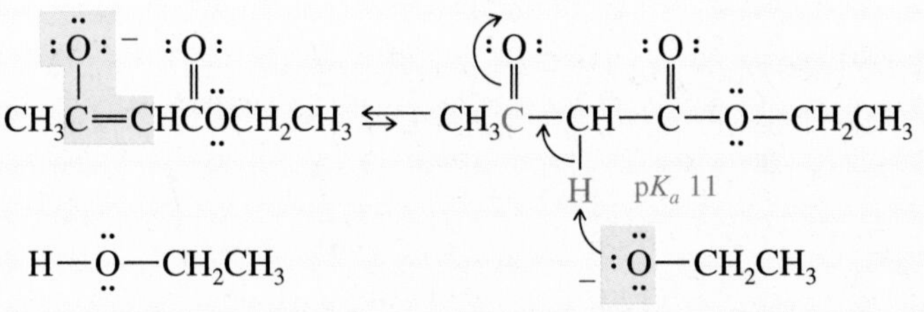

$$
\begin{array}{cc}
\overset{\displaystyle :\ddot{O}:^{-}\ :\ddot{O}:}{\underset{\displaystyle}{CH_3C=CHCOCH_2CH_3}} & \rightleftharpoons\ \ CH_3C\!-\!CH\!-\!C\!-\!\ddot{O}\!-\!CH_2CH_3 \\
\end{array}
$$

enolization of β-ketoester

Note that the reaction of an enolate anion with an ester is reversible. For this reason, the Claisen condensation does not give a good yield of the β-ketoester unless that product is converted to its enolate in the basic reaction mixture. If the β-ketoester is not converted completely to its enolate, it is attacked at the ketone functional group by an alkoxide ion to give the tetrahedral intermediate shown above. The intermediate decomposes to give the enolate anion and the ester. The last stage of the reaction, the reversible protonation and deprotonation of the active methylene group in the β-ketoester, is written as an equilibrium. At equilibrium, the enolate anion of the β-ketoester predominates because the compound with the active methylene group is much more acidic ($pK_a$ 11) than ethanol ($pK_a$ 17). The enolate anion of the ketoester is converted to the desired product by adding an organic acid, such as acetic acid.

Just as aldol reactions are possible between different aldehydes or between an aldehyde and a ketone, Claisen condensations between different esters are possible. If the reactions are run under equilibrium conditions, one of the esters must lack enolizable α-hydrogen atoms, so that the enolate ion comes from only one of the reagents. Esters without enolizable α-hydrogens that are often used in these reactions include ethyl formate, diethyl carbonate, diethyl oxalate, and ethyl benzoate.

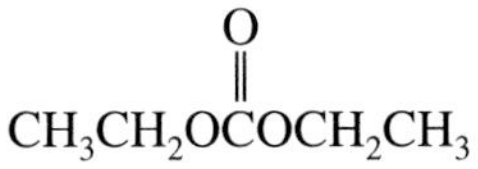

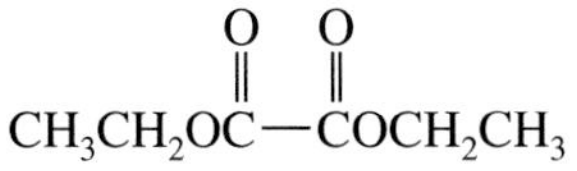

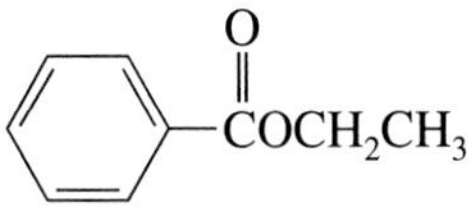

For example, diethyl butanedioate, commonly known as diethyl succinate, reacts with diethyl oxalate to give a triester.

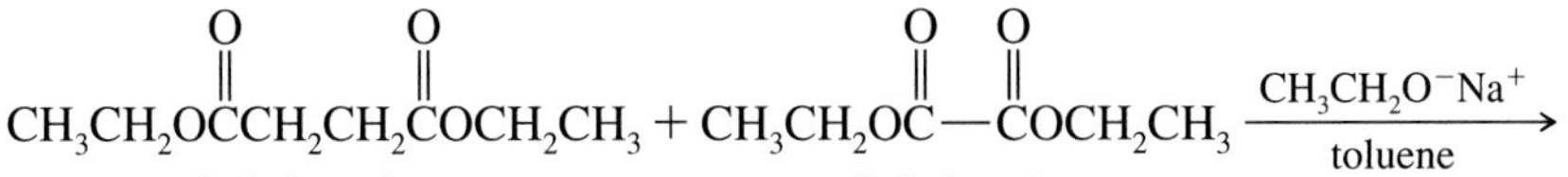

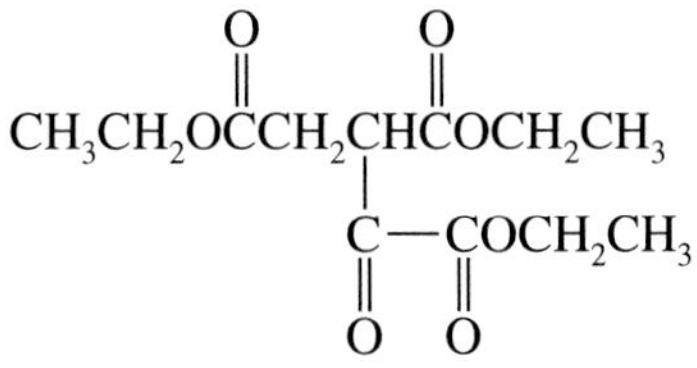

The reaction takes place in two steps. Diethyl succinate is converted to its enolate by base.

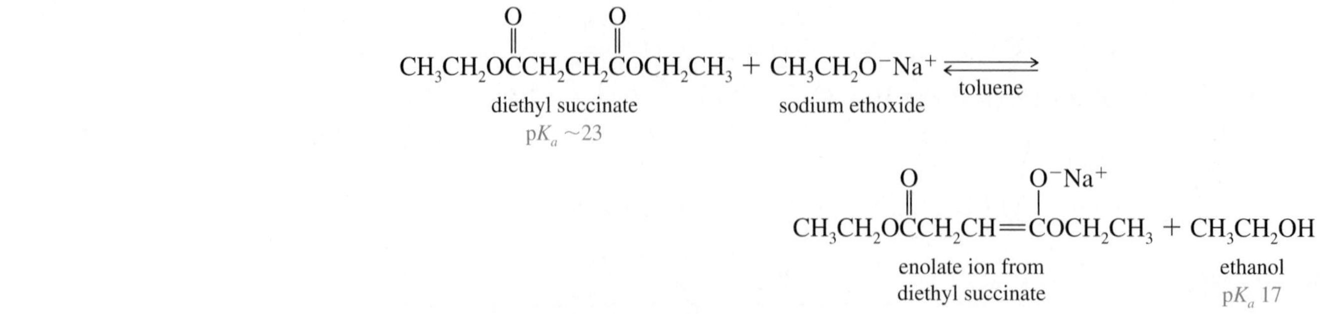

$$CH_3CH_2OCCH_2CH_2COCH_2CH_3 + CH_3CH_2O^-Na^+ \underset{toluene}{\rightleftharpoons}$$

diethyl succinate
p$K_a$ ~23

sodium ethoxide

$$CH_3CH_2OCCH_2CH=COCH_2CH_3 + CH_3CH_2OH$$

enolate ion from
diethyl succinate

ethanol
p$K_a$ 17

In this reaction, the enolate ion of diethyl succinate is in equilibrium with the un-protonated ester. The base, ethoxide ion, is not strong enough to completely depro-tonate the ester. If another ester with enolizable $\alpha$-hydrogens were present in the reaction mixture, it, too, would participate in deprotonation–protonation equilibria. Two enolate ions and two potential electrophiles would be available to react with them and, therefore, four possible products. In this case, however, only one enolate ion is possible, and it reacts with a carbonyl group in diethyl oxalate, displacing ethoxide ion to give the triester product. The carbonyl groups of diethyl oxalate are more electrophilic and less hindered than those of diethyl succinate, so reaction of the enolate ion with the oxalate is favored over reaction with unenolized succinate.

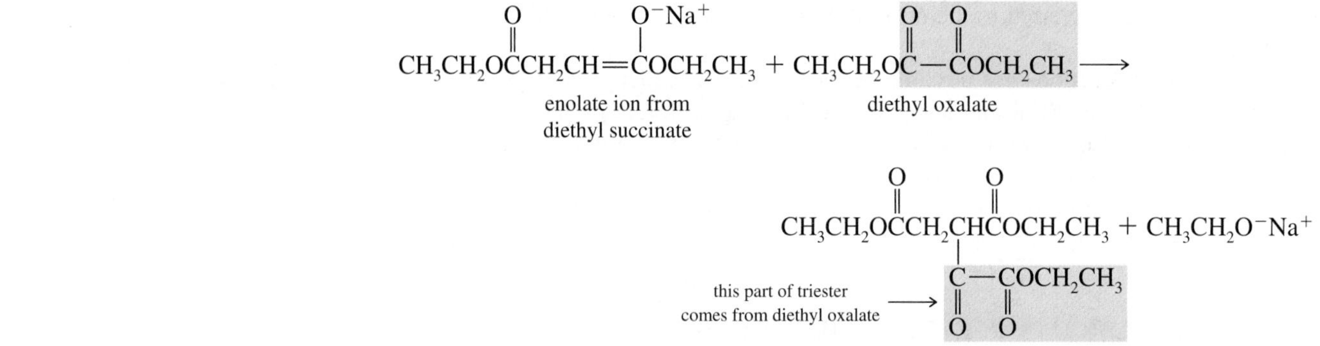

$$CH_3CH_2OCCH_2CH=COCH_2CH_3 + CH_3CH_2OC-COCH_2CH_3 \longrightarrow$$

enolate ion from
diethyl succinate

diethyl oxalate

$$CH_3CH_2OCCH_2CHCOCH_2CH_3 + CH_3CH_2O^-Na^+$$

this part of triester
comes from diethyl oxalate

$\longrightarrow$ C—COCH$_2$CH$_3$

**Problem 17.8**    The triester shown above was synthesized in order to prepare 2-oxopen-tanedioic acid, commonly known as $\alpha$-ketoglutaric acid, an important intermediate in the metabolism of amino acids.

(a)  What is the structure of 2-oxopentanedioic acid?
(b)  How would you get to 2-oxopentanedioic acid from the triester shown above? A review of Section 17.3B will be helpful.

**Problem 17.9**    Ketones and aldehydes can be used as the source of enolate ions that re-act with esters without $\alpha$-hydrogen atoms. Use cyclohexanone as the source of enolate ions and predict what the product of its reaction in the presence of base will be with (a) ethyl for-mate, (b) diethyl oxalate, and (c) diethyl carbonate.

**Problem 17.10**    The products of the reactions in Problem 17.9 are easily enolizable. Draw structural formulas for possible enol forms of the compounds you predicted in Prob-lem 17.9.

**Problem 17.11**    The product from the reaction of cyclohexanone with diethyl oxalate is used to introduce a methyl group onto the substituted cyclohexanone. Write an equa-tion showing how you would do this, and what the product would be.

## C. A Unified Look at Condensation Reactions

Anions from esters also can be used in condensations with aldehydes or ketones. For example, ethyl acetate will condense with benzaldehyde to give an unsaturated carboxylic acid derivative.

benzaldehyde        ethyl acetate                    β-hydroxyester
                                                      intermediate

ethyl (E)-3-phenyl-2-propenoate        (E)-3-phenyl-2-propenoic acid
α, β-unsaturated ester                      trans-cinnamic acid
~ 70%                              α, β-unsaturated acid

In another example, the active methylene group of diethyl malonate condenses with formaldehyde under mild conditions.

diethyl malonate      formaldehyde          diethyl bis(hydroxymethyl)-
                                                  propanedioate

For each of the reactions above, only one of the reactants has an enolizable hydrogen atom, so the reaction goes one way in high yield. Also, note that in both cases the enolate ion reacts much more readily with the carbonyl group of the aldehyde that is present than with the carbonyl group of the ester. In the reaction with benzaldehyde, the intermediate hydroxy compound dehydrates easily to give a conjugated alkene. With formaldehyde, the product retains the β-hydroxy groups.

Clearly, a large variety of reactions are possible between enolate ions and compounds containing carbonyl groups. Many of them are known by the names of their discoverers or developers. But such an array of different names and structures tends to hide the underlying unity and simplicity of this type of reaction. In deciding what kinds of condensation reactions are possible, you have to ask four questions:

1. Is there an enolizable hydrogen atom in one of the reactants? Hydrogens alpha to nitro and cyano groups also count as acidic hydrogen atoms.

2. Is there a carbonyl group that can be attacked by the enolate anion? The carbonyl group of an aldehyde is more reactive (because it is higher in energy) than the carbonyl group of a ketone. The carbonyl group of an ester is the least reactive because it is more resonance-stabilized than either of the other two (p. 597).

3. Is the carbonyl carbon that is being attacked also bonded to a leaving group (usually the alkoxyl group of an ester)? If there is such a group present, it is lost in

going to the product. These types of reactions may be classified as Claisen-type reactions. If there is no good leaving group, the product is a β-hydroxycarbonyl compound or its dehydration product. These reactions are aldol-type reactions.

4. Is there a carbonyl group in the same molecule as an enolizable hydrogen. If a five- or six-membered ring can form, it usually will. We will explore this further in Chapter 21.

**Study Guide**
**Concept Map 17.4**

**On the Web: ONE SMALL STEP**

All you need to bring order to what appears to be a bewildering array of reactions is a sharp eye for these features and a willingness to practice making predictions on a wide variety of examples. The next problem provides such practice.

**Problem 17.12**    Give structures for the intermediates and products designated by the letters.

(a) $C_6H_5-\overset{\overset{\displaystyle O}{\|}}{C}-C_6H_5$ + $\begin{array}{c} CH_2COCH_2CH_3 \\ | \\ CH_2COCH_2CH_3 \end{array}$ $\xrightarrow[\text{alcohol}]{\underset{tert\text{-butyl}}{CH_3COK}}$ A

(b) (cyclooctanone) $\xrightarrow[\underset{\Delta}{\text{benzene}}]{\text{NaH}}$ B $\xrightarrow{CH_3CH_2OCOCH_2CH_3}$ C $\xrightarrow{CH_3COH}$ D

## 17.5  Reactions of Nucleophiles with α,β-Unsaturated Carbonyl Compounds as Electrophiles

### A. The Michael Reaction

Enolate anions react with α,β-unsaturated compounds at the β-carbon atom. For example, the anion generated from diethyl malonate adds to methyl propenoate.

$$CH_2(COCH_2CH_3)_2 \xrightarrow{CH_3CH_2O^-Na^+} Na^+ \text{ }^-:CH(COCH_2CH_3)_2 \xrightarrow[\text{methyl propenoate}]{\overset{\beta\quad\alpha\ \overset{\displaystyle O}{\|}}{CH_2=CHCOCH_3}} (CH_3CH_2OC)_2CHCH_2CH_2COCH_3$$

diethyl malonate          enolate of
                         diethyl malonate

ethyl methyl 2-carboethoxy-
pentanedioate

Alkenes that react with nucleophiles in this way all contain electron-withdrawing groups. In these compounds, the double bond is conjugated with a carbonyl group, a cyano group, a nitro group, or other groups that are capable of delocalizing negative charge. Such alkenes are often called electrophilic alkenes. In them, the electrophilic character of the carbonyl group is transmitted to the β-carbon atom of the α,β-unsaturated carbonyl system.

resonance contributors for an
α,β-unsaturated carbonyl system

The behavior of compounds with this type of structure contrasts with that of ordinary alkenes, which usually react with electrophiles but not with nucleophiles (Chapter 8).

Reactions that proceed by nucleophilic attack at the β-carbon atom of an electrophilic alkene give reactive intermediates that are stabilized by resonance delocalization of negative charge.

**The Michael Reaction**

enolization

attack of nucleophile at the β-carbon
atom of the electrophilic alkene

protonation of the resonance-stabilized intermediate

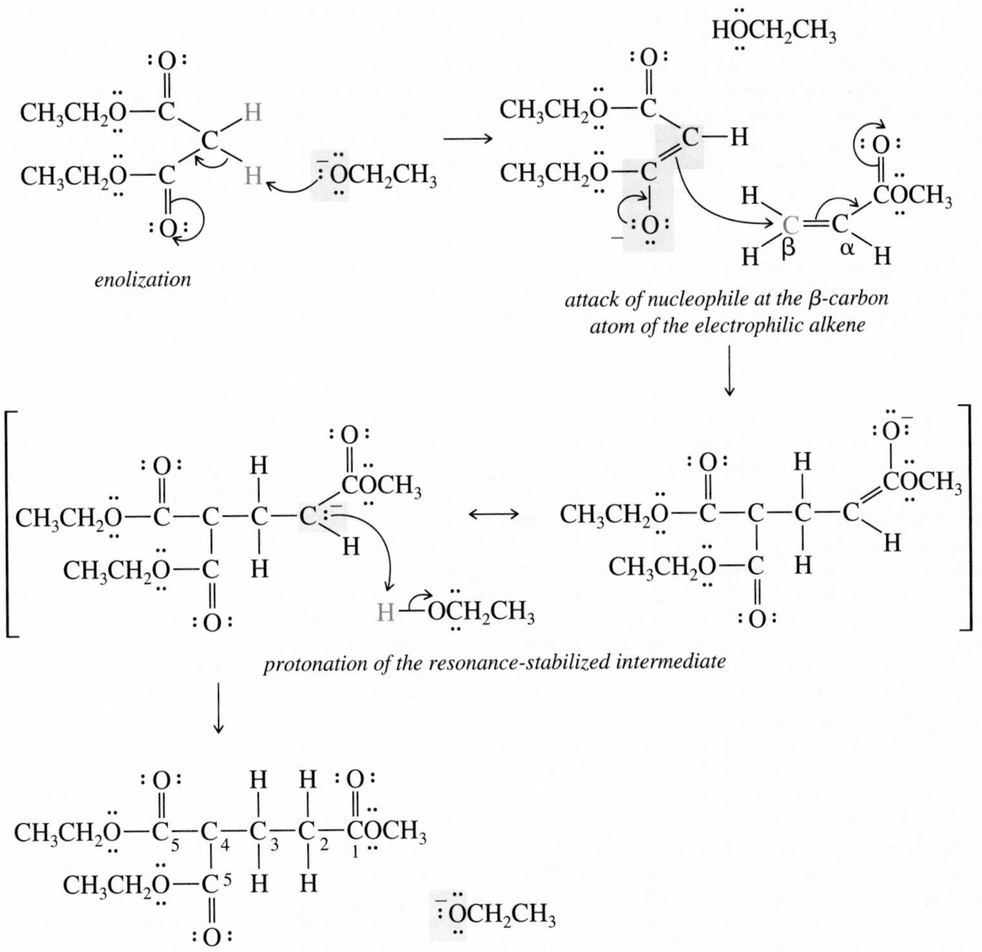

The product of the addition of an enolate anion to an α,β-unsaturated carbonyl compound is a **1,5-dicarbonyl compound.**

In a similar way, both the hydrogen atoms from the active methylene group in ethyl acetoacetate are replaced by alkyl groups through the addition of anions to methyl vinyl ketone.

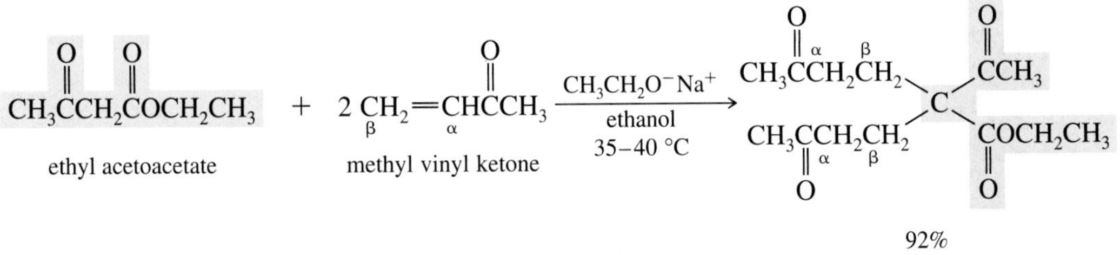

ethyl acetoacetate          methyl vinyl ketone

92%

The carbanion does not have to be derived from a carbonyl compound. Nitro groups also stabilize carbanions (p. 668). For example, 2-nitropropane will add to methyl vinyl ketone in the presence of sodium methoxide.

$$
\underset{\substack{\text{2-nitropropane}}}{\overset{\overset{\displaystyle CH_3}{|}}{CH_3CHNO_2}} + \underset{\substack{\text{methyl vinyl ketone}}}{\underset{\beta\quad\alpha}{CH_2{=}CHCCH_3}} \xrightarrow[\text{diethyl ether}]{CH_3O^-Na^+} \underset{\substack{\text{5-methyl-5-nitro-2-hexanone}\\69\%}}{\overset{\overset{\displaystyle CH_3}{|}}{\underset{\underset{\displaystyle NO_2}{|\ \beta\quad\alpha}}{CH_3CCH_2CH_2CCH_3}}}
$$

A nitro group also can stabilize the carbanion resulting from attack by an anion on an electrophilic alkene. The enolate anion from dimethyl malonate adds to 1-nitro-2-phenylethene with a high yield.

$$
\underset{\substack{\text{dimethyl malonate}}}{\overset{\displaystyle O}{CH_2(COCH_3)_2}} + \underset{\substack{\text{1-nitro-2-phenylethene}}}{\underset{\beta\quad\alpha}{\bigcirc{-}CH{=}CHNO_2}} \xrightarrow[\text{methanol}]{CH_3O^-Na^+} \underset{\substack{\text{methyl 2-carbomethoxy-}\\ \text{3-phenyl-4-nitrobutanoate}\\92\%}}{\underset{\beta\quad\alpha}{(CH_3OC)_2CHCHCH_2NO_2}}
$$

In all these examples, a stabilized carbanion adds to the $\beta$-carbon atom of a double bond that is made electrophilic by an electron-withdrawing substituent.

The reactions described above are known as **Michael reactions,** after Arthur Michael of Tufts University. The compound containing the nucleophilic carbon atom is called the **Michael donor,** and the compound containing the polarized double bond is called the **Michael acceptor.**

⊘ **Study Guide**
**Concept Map 17.5**

**Problem 17.13**  Show the structure and stabilization of the carbanion derived from 2-nitropropane.

**Problem 17.14**  Write structural formulas that rationalize the polarity of the double bond in 1-nitro-2-phenylethene.

**Problem 17.15**  Complete the following equations.

(a) $\underset{\underset{\displaystyle CH_3}{|}}{\overset{\displaystyle O\quad\ O}{CH_3CHCCH_2COCH_2CH_3}} + \overset{\overset{\displaystyle CH_3}{|}}{\underset{\displaystyle O}{CH_3C{-}C{=}CH_2}} \xrightarrow[\text{ethanol}]{KOH}$

(b) $\overset{\displaystyle O}{N{\equiv}CCH_2COCH_3} + \overset{\displaystyle O}{CH_2{=}CHC{-}\bigcirc} \xrightarrow[\text{methanol}]{CH_3ONa}$

(c) $\overset{\displaystyle O}{\bigcirc{-}\bigcirc} \xrightarrow{NaNH_2} \quad CH_2{=}CHC{\equiv}N$

(d) $CH_3NO_2 + \overset{\overset{\displaystyle CH_3}{|}\ \ \overset{\displaystyle O}{}}{CH_3C{=}CHCCH_3} \xrightarrow{CH_3CH_2NHCH_2CH_3}$

(*Hint:* The more highly substituted enolate ion forms.)

## B. 1,2– versus 1,4–Additions to Conjugated Systems

An $\alpha,\beta$-unsaturated carbonyl compound can react with a nucleophile at the carbonyl group to give a product in which the carbonyl function has been modified. This reaction is called a 1,2-addition to the conjugated system, by analogy with the reactions of conjugated dienes (Section 18.3, p. 741). Alternatively, the nucleophile may attack the $\beta$-carbon atom, giving a product in which the carbonyl group is regenerated from an enolate intermediate. The case in which the nucleophile becomes bonded to the $\beta$-carbon atom is known as 1,4-addition or **conjugate addition.**

## 17.31

(a) Enolate ions may have Z or E stereochemistry. Much effort has gone into the study of the stereochemistry of enolate ions and the effect it has on the stereochemistry of the products that are formed in aldol reactions. The trimethylsilyl (TMS, p. 847) ether of 2-hydroxy-2-methyl-3-pentanone gives primarily the Z enolate when deprotonated by lithium diisopropylamide. What is the structure of the enolate?

(b) The enolate reacts with 2-methylpropanal to give an aldol product that has opposite configurations at the two stereocenters that are created in the reaction. What is the structure of the product?

**17.32** Both the enolate ion and the Michael acceptor in a Michael reaction may be a source of stereocenters in the product. The stereochemistry of such reactions was probed using the enolate ion derived from N,N-dimethylpropanamide with 2,2-dimethyl-4-hexen-3-one as the Michael acceptor. Write equations showing the formation of the enolate ion and its reaction with the enone. How many stereocenters are possible? Draw structural formulas and assign configurations to any stereocenters that are created in the reactions.

**17.33** The following reactions were used in the synthesis of an alkaloid. Provide structural formulas for the compounds designated by letters.

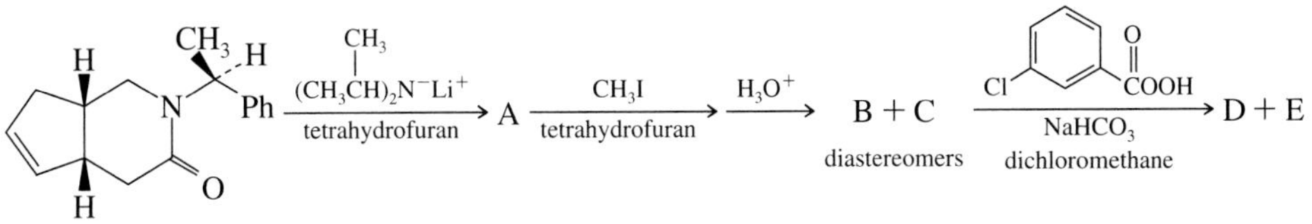

The peroxyacid attacks one face of the bicyclic system preferentially. Which face is more easily reached by the reagent?

**17.34** 1,3-Diketones decompose when heated with base, for example:

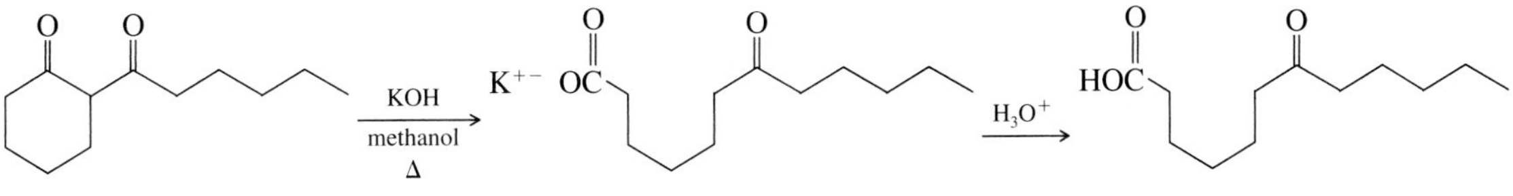

Propose a mechanism for this reaction.

**17.35** The cyclic amino acid proline is often used in the synthesis of chiral compounds. The following reaction was carried out using proline as the catalyst and *trans*-2,5-dimethylpiperazine as the base.

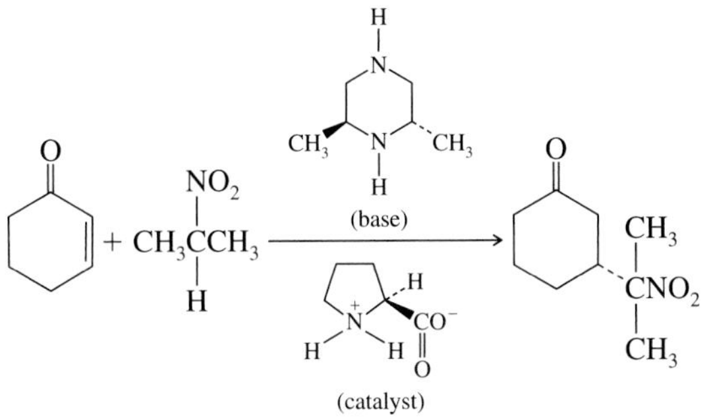

Propose a mechanism for the reaction using the curved-arrow convention and $HB^+$ and B: as acids and bases as needed.

**17.36** The following equation represents part of a synthesis of the batzelladine alkaloids (p. 804), which have a broad range of biological activity, including possible therapy for AIDS.

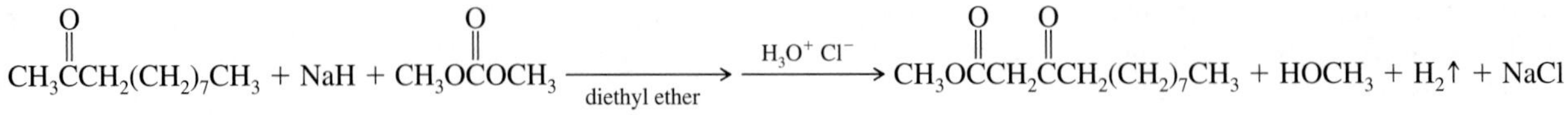

$$CH_3CCH_2(CH_2)_7CH_3 + NaH + CH_3OCOCH_3 \xrightarrow[\text{diethyl ether}]{H_3O^+ \ Cl^-} CH_3OCCH_2CCH_2(CH_2)_7CH_3 + HOCH_3 + H_2\uparrow + NaCl$$

Provide a mechanism for the reaction using the curved-arrow convention.

**17.37** The following reaction was carried out in the synthesis of (−)-horsfiline, an alkaloid. Propose a mechanism for it that rationalizes why the methyl group on the aromatic ring behaves the way it does.

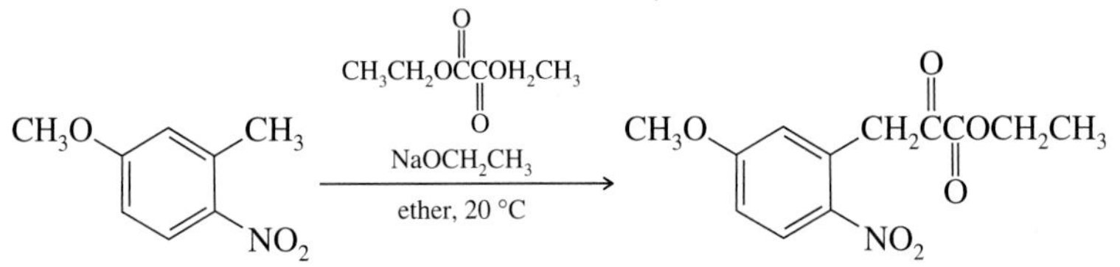

**17.38** Organic chemists have tried to discover whether the metal ion in an enolate salt is primarily associated with a negatively charged oxygen atom or with a negatively charged carbon atom in the ambident nucleophile (p. 695). One of the research tools chemists have used in attempting to answer this question is nuclear magnetic resonance spectroscopy. The proton magnetic resonance spectra of the lithium salt of the enolate ion from 1-phenyl-2-methyl-1-propanone in a variety of solvents (such as benzene and tetrahydrofuran) have two singlets in the region $\delta$ 1.0–2.0. Compare structural formulas for the two resonance contributors for the enolate ion from 1-phenyl-2-methyl-1-propanone and decide what conclusion can be drawn from the proton magnetic resonance data.

**17.39** The following reactions were carried out in the synthesis of large-ring compounds that capture and hold metal ions. Provide structural formulas for Compounds A and B. What structural features of A and B do the infrared absorption bands point to?

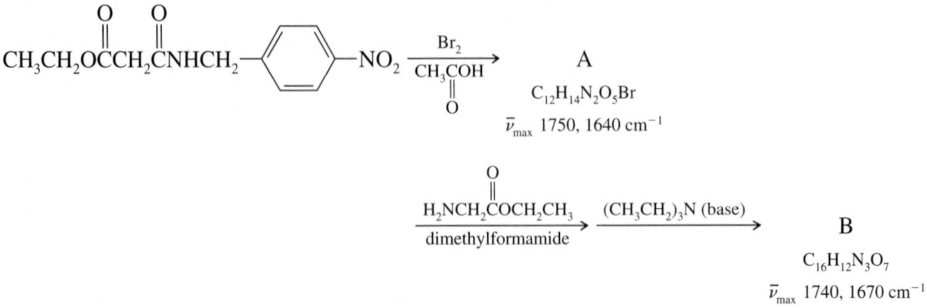

**17.40** Avermectin and its close relative, ivermectin, are antiparasitic agents that are turning out to be very important in the treatment of tropical parasitic diseases. Some steps in a synthesis of avermectin follow. Supply structural formulas for the missing compounds.

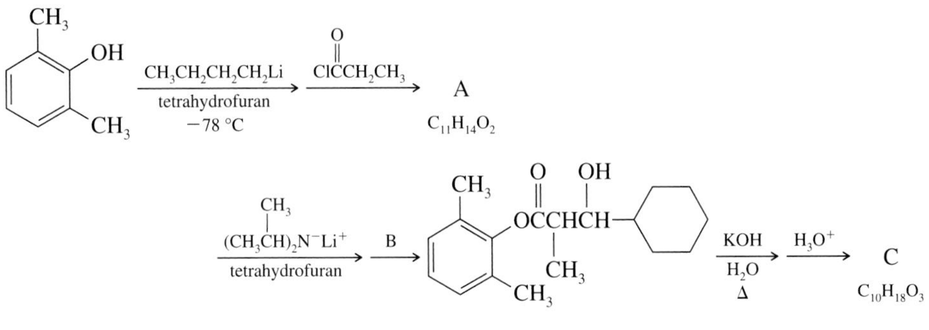

**17.41** A synthesis of (±)-aspidospermidine, a member of a family of alkaloids with important physiologic properties, had the following steps. Supply structural formulas for the reagents and products indicated by letters.

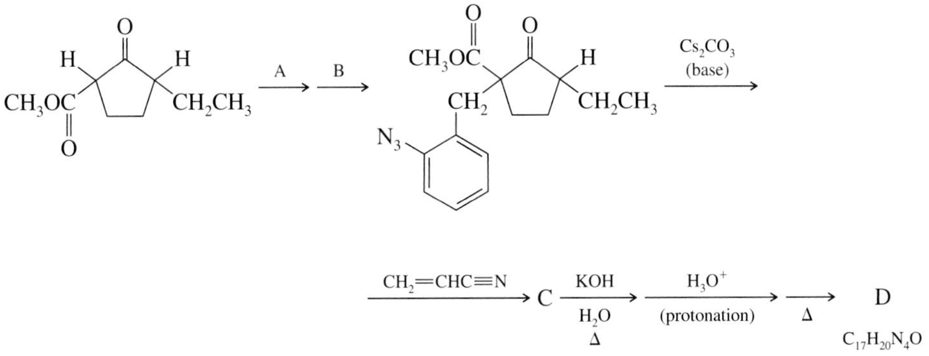

**17.42** The following reaction was observed. Provide a complete stepwise mechanism for the formation of the product from the starting materials.

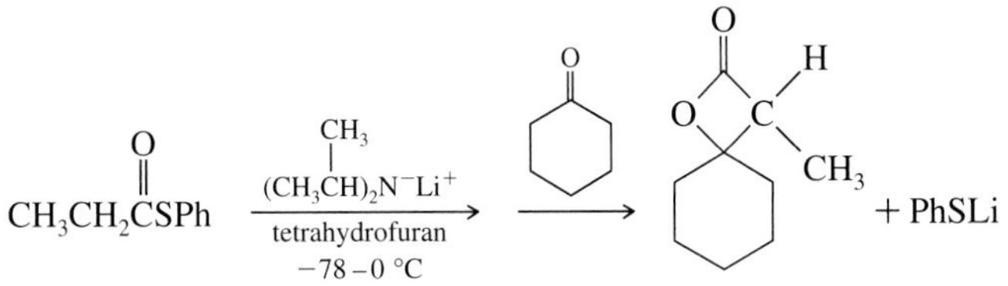

**17.43** An important step in the synthesis of β-amino acids involves the following reaction. Provide a detailed mechanism using the curved-arrow convention for the reaction.

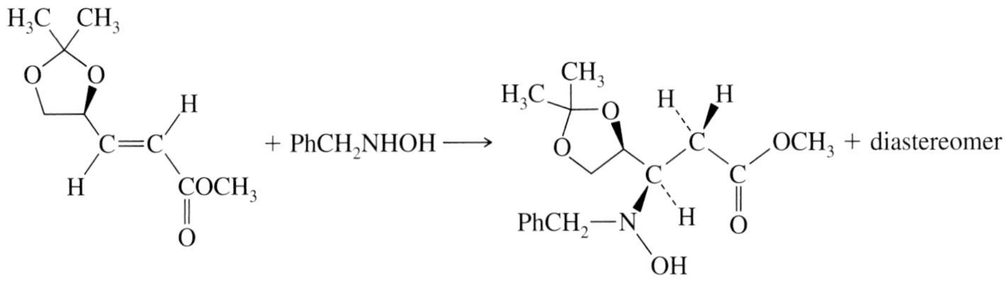

**17.44** (−)-Tanabalin is a natural product isolated from the dried flower of a Brazilian medicinal plant. It is active against the pink bollworm, a pest of cotton plants. The following transformation is part of a synthesis of the compound. Show how you would carry it out, giving structural formulas for any reactive intermediates and products formed along the way.

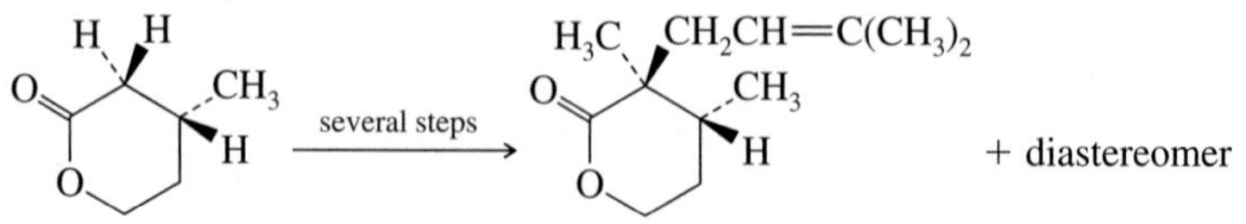

**17.45** Additions to the carbonyl group, 1,2-additions, are in competition with 1,4-additions in reactions of enolate anions with α,β-unsaturated carbonyl compounds. The following reactions were observed.

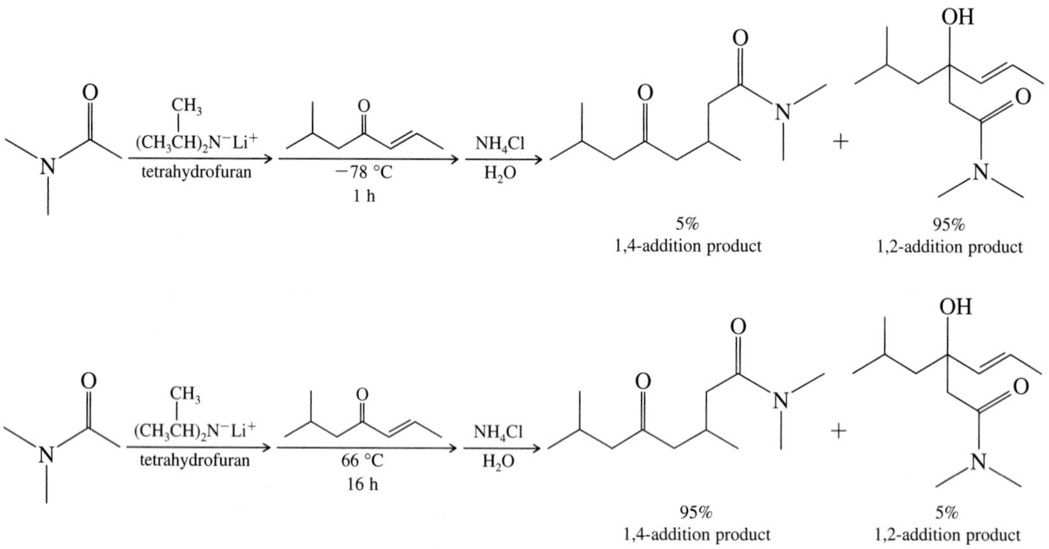

What do these data tell you about the kinetic and the thermodynamic products of the reaction (pp. 133 and 865)? For these results to be obtained, what must be true about the reactions leading to the addition products? Write a mechanism for this process.

**17.46** (R)-Glyceraldehyde (p. 974) is converted in basic or acidic solution to a mixture of fructose and sorbose, shown below. The conversion proceeds faster if dihydroxyacetone (p. 721) is added to the solution. Write equations showing how glyceraldehyde is transformed into the mixture of fructose and sorbose.

fructose                                    sorbose

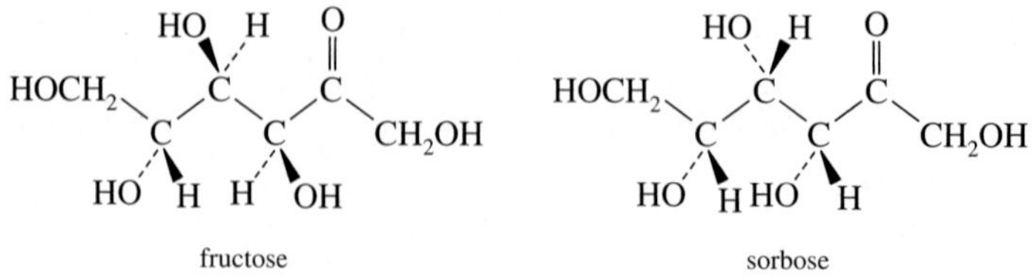

**17.47** Research has shown that $\alpha,\beta$-unsaturated trifluoromethyl ketones are effective inhibitors of glutathione-$S$-transferase, a key enzyme that is involved in the detoxification of foreign chemicals (xenobiotics) that get into the body. The $\alpha,\beta$-unsaturated ketones react readily in a conjugate addition reaction with glutathione, a sulfur-containing peptide. Nuclear magnetic resonance shows the product of the reaction to be an enol as shown below.

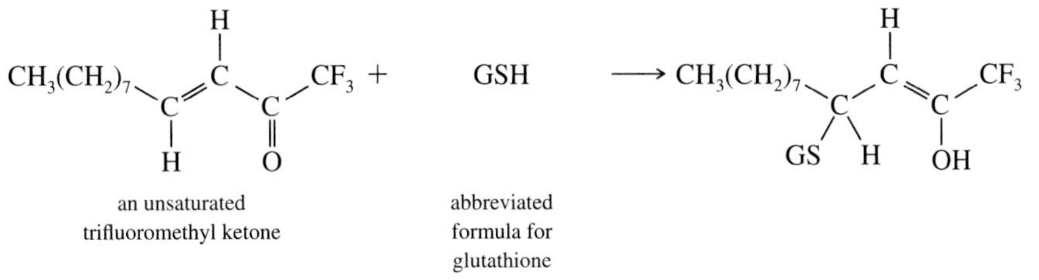

an unsaturated
trifluoromethyl ketone

abbreviated
formula for
glutathione

(a) Propose a mechanism for the reaction shown above using the curved-arrow convention and using $HB^+$ and $B\colon$ to show proton transfers as necessary.
(b) The $\alpha,\beta$-unsaturated trifluoromethyl ketones are more reactive in this reaction than the corresponding compounds without fluorine substitution on the methyl group. Rationalize this observation using structural formulas and words.

**17.48** The following reactions were carried out in the synthesis of compounds related to cylindrospermopsin, a compound produced by blue-green algae and believed responsible for liver infections. Infrared and proton magnetic resonance spectral data are given to help you assign a structure to Compound E, the end point of this synthetic sequence. Provide structural formulas for the compounds designated by letters and an analysis of as much of the spectral data as possible.

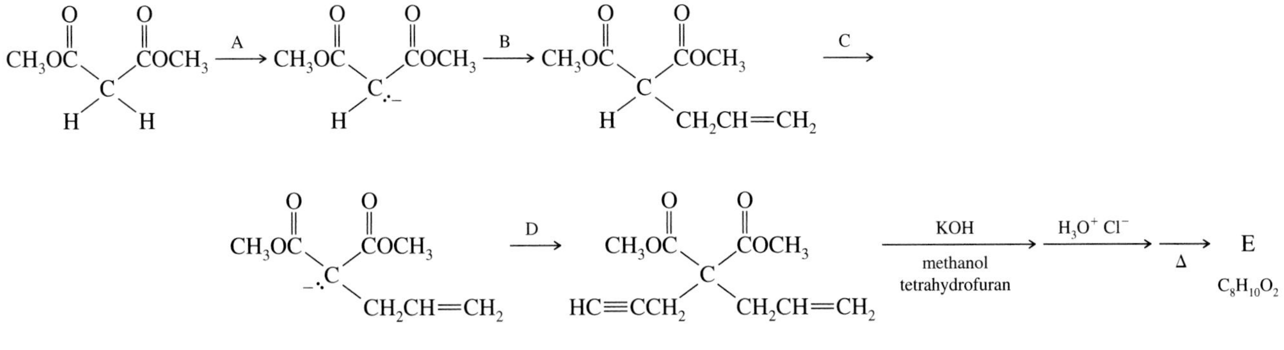

Compound E has bands in its infrared spectrum at 3300, 3000, 2150, 1708, and 1642 cm$^{-1}$. It has peaks in its proton magnetic resonance spectrum at 2.01 (t, $J = 2.7$, 1H), 2.37 − 2.60 (m, 4H), 2.70 (m, 1H), 5.10 (m, 2H), 5.75 (m, 1H), and 10.8 (broad s, 1H) ppm.

**17.49** The proton magnetic resonance spectrum of ethyl ($E$)-3-phenylpropenoate has peaks at $\delta$ 1.32 (3H, t, $J = 7$ Hz), 4.25 (2H, q, $J = 7$ Hz), 6.43 (1H, d, $J = 14$ Hz), 7.35–7.50

(5H, m), 7.68 (1H, d, $J = 14$ Hz). Assign the peaks in the spectrum, and analyze any splitting that is observed. How do you account for the large difference in chemical shift for the two vinylic hydrogen atoms?

**17.50** The $^{13}$C nuclear magnetic resonance spectrum of (E)-2-butenoic acid has peaks at 18.0, 122.6, 147.5, and 172.3 ppm downfield from TMS. Assign these peaks to the carbon atoms in (E)-2-butenoic acid, and explain your reasoning in making these assignments. Table 11.2 (p. 403) may be useful.

# Polyenes

## A Look Ahead

Multiple bonds that are separated from each other by a single bond are said to be conjugated.

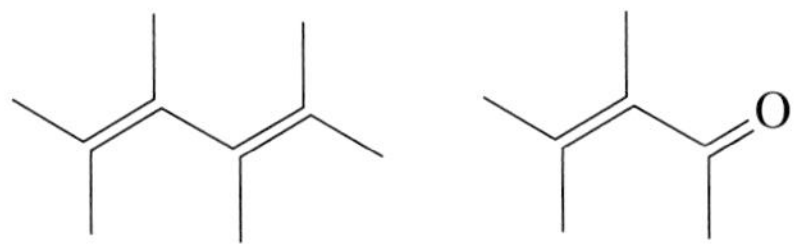

conjugated systems

In Chapter 17 we covered the reactions of compounds in which a carbonyl group is conjugated with a carbon–carbon double bond. This chapter will look at the reactions of conjugated alkenes.

Compounds having conjugated multiple bonds undergo the reactions that are typical of the individual functional groups. They also have special reactivity resulting from the interaction of the multiple bonds. The most important consequence of this is the ability to undergo reactions at the ends of the conjugated system.

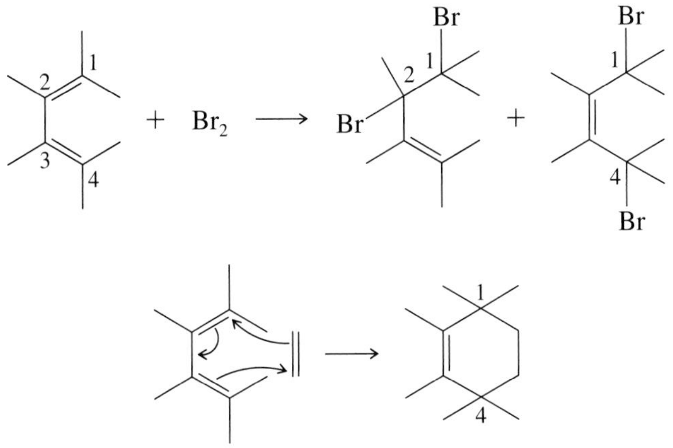

Many biologically important compounds have multiple double bonds or are derived from such compounds. In this chapter, you will see how such complex compounds are built up from simpler units in living organisms.

## 18.1  Isolated, Skipped, Conjugated, and Cumulative Multiple Bonds

The chemistry of carbon–carbon multiple bonds is familiar by this point. Compounds containing double or triple bonds add halogens and acids. They are reduced by hydrogen and oxidized by reagents such as ozone, osmium tetroxide, and peroxyacids. They react with Lewis acids such as diborane to give intermediates that are converted to alcohols. This chemistry is also characteristic of compounds that contain more than one multiple bond, as many examples in preceding chapters have demonstrated. In all those cases, however, the multiple bonds were separated from each other by one or more tetrahedral carbon atoms.

Multiple bonds that are separated from each other by two or more tetrahedral carbon atoms are said to be **isolated multiple bonds.** Some examples of compounds containing isolated multiple bonds are given below with their names.

$$CH_2{=}CHCH_2CH_2CH{=}CH_2 \qquad HC{\equiv}CCH_2CH_2CH_2C{\equiv}CH$$

1,5-hexadiene                         1,6-heptadiyne

$$CH_2{=}CHCH_2CH_2CH{=}CHCH_2CH_2CH{=}CH_2 \qquad HC{\equiv}CCH_2CH_2CH{=}CH_2$$

1,5,9-decatriene                              1-hexen-5-yne

Such compounds are named by indicating the number of multiple bonds in the chain by the suffixes **-adiene** and **-atriene** or **-adiyne** and **-atriyne** and also indicating the position of the multiple bond by giving the number in the chain of the first carbon atom of each multiple bond. When a compound contains both a double bond and a triple bond, the triple bond is named as the suffix.

Multiple bonds that are separated from each other by only one tetrahedral carbon atom are called **skipped multiple bonds.** The polyunsaturated fatty acids (p. 639), for example, contain skipped double bonds. This structural feature makes such compounds especially reactive in ways that are explored in Section 19.6B.

If the multiple bonds are separated from each other by one single bond, the $p$ orbitals on adjacent carbon atoms can interact. The prime example of this kind of interaction occurs in benzene. Sideways overlap of a $p$ orbital from each of the six carbon atoms on the six-membered ring results in delocalization of six electrons over the entire benzene ring (p. 68). Multiple bonds that are separated from each other by one single bond are said to be **conjugated.** The concept of conjugation was first introduced in connection with a carbonyl group separated by a single bond from a double bond or an aromatic ring (pp. 463 and 710). Some other examples of conjugated systems are shown below.

$$CH_2{=}CH{-}CH{=}CH_2 \qquad CH_2{=}CH{-}CH{=}CH{-}CH{=}CH_2$$

1,3-butadiene                         1,3,5-hexatriene

$$HC{\equiv}C{-}C{\equiv}CCH_2CH_3 \qquad HC{\equiv}C{-}CH{=}CHCH_3$$

1,3-hexadiyne                              3-penten-1-yne

$$CH_2{=}CH{-}\overset{\overset{\displaystyle CH_3}{|}}{C}{=}O$$

3-buten-2-one

When at least three adjacent carbon atoms in a molecule are joined by double bonds, the compound is called a **cumulene** or is said to have **cumulative double bonds.** The simplest cumulene is 1,2-propadiene, usually called **allene.** The central carbon atom is *sp*-hybridized and is bonded to the other two carbon atoms of the system by $\pi$ bonds that are at right angles to each other (Figure 18.1).

$$\overset{3}{CH_2}=\overset{2}{C}=\overset{1}{CH_2}$$

1,2-propadiene
allene

**Problem 18.1**   Name the following compounds.

(a) $CH_3C{\equiv}CCH_2C{\equiv}CH$      (b) $CH_3CH{=}CHC{\equiv}CCH_3$      (c) [cyclohexadiene structure]

(d) [cyclooctatetraene structure]   (e) [styrene derivative structure with $C=C$, $H$, $CH_2CH=CH_2$]

(f) $ClCH_2C{\equiv}CC{\equiv}CCH_2CH_3$

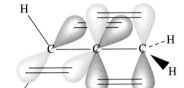

*Figure 18.1*
**Bonding in allene.**

**Problem 18.2**   Write structural formulas for the following compounds.

(a) 5-hexen-2-one      (b) 1,5-hexadiyne      (c) 1,4-pentadiene      (d) 2,5-heptadiyne

(e) (*E*)-4,4-dimethyl-1,6-octadiene      (f) 3,5-hexadien-1-yne

⊘ Study Guide
Concept Map 18.1

## 18.2   1,3-Butadiene

1,3-Butadiene is the simplest compound that contains two double bonds separated by a single bond. Electron diffraction studies have shown 1,3-butadiene to be a planar molecule. The double bonds in 1,3-butadiene are 1.34 Å long, just about the length of the double bond in ethylene; the length of the single bond is 1.48 Å, slightly shorter than a single bond (1.49 Å) adjacent to the double bond in propene and significantly shorter than the single bond in ethane (1.54 Å).

The heat of hydrogenation (p. 282) for 1,3-butadiene (−57.1 kcal/mol) is 4 kcal/mol less than expected for a compound containing two isolated double bonds (~61 kcal/mol). 1,3-Butadiene is thus more stable than expected by that amount. The difference between the stability determined experimentally for a compound (in this case 1,3-butadiene) and that predicted for a hypothetical molecule with isolated double bonds is called the empirical resonance energy (p. 352) for the system.

1,3-Butadiene has a skeleton of four carbon atoms in a chain bonded to six hydrogen atoms by $\sigma$ bonds. The carbon atoms are $sp^2$-hybridized, and a *p* orbital containing one electron is available to each carbon atom in the chain (p. 52). The double bonds in 1,3-butadiene can be pictured as arising from the overlap of pairs of *p* orbitals on adjacent carbon atoms (Figure 18.2, p. 740). But this model for 1,3-butadiene, with double bonds localized between carbon atoms 1 and 2 and carbon atoms 3 and 4, does not explain the planarity of the molecule, the shorter than usual single bond, and the empirical resonance energy determined for the compound.

A better model for the bonding in 1,3-butadiene has been developed using molecular orbital theory (p. 43). In this model, the four atomic *p* orbitals on the carbon atoms combine to give four $\pi$ molecular orbitals, $\psi_1$, $\psi_2$, $\psi_3$, and $\psi_4$. These energy levels were shown in Figure 12.2 (p. 445) and are again in more detail in Figure 25.2 (p. 1084).

The molecular orbitals $\psi_1$ and $\psi_2$ are bonding molecular orbitals, and each has two electrons in it. The other two molecular orbitals are antibonding and do not contain any electrons. According to this picture of 1,3-butadiene, the molecule is planar to allow interaction between the *p* orbitals on carbon atoms 2 and 3. There is some double-bond character between those carbon atoms due to the contribution of the lowest-energy bonding molecular orbital, $\psi_1$ (Figure 18.3, p. 740), and, therefore,

structure of 1,3-butadiene

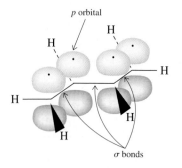

1,3-butadiene showing $\sigma$ bonds
and $p$ orbitals, each orbital
containing one electron

1,3-butadiene shown with
localized double bonds

*Figure 18.2*
An orbital picture of the structure of 1,3-butadiene.

a shortening of the carbon–carbon single bond. 1,3-Butadiene is more stable by 4 kcal/mol than a molecule with two isolated double bonds because of the delocalization of electrons over the whole chain, represented by molecular orbital $\psi_1$.

1,3-Butadiene is shown in Figure 18.2 in an extended conformation in which the two double bonds are trans to each other across the central single bond. This conformation is known as the **s-trans conformation,** the *s* indicating that the stereochemistry refers to the single bond. A higher-energy conformation that is important in the reactions of butadiene is the **s-cis conformation.** Both these conformations exist in equilibrium with each other. In both of them, all four of the carbon atoms and all the hydrogen atoms lie in the same plane, as shown below.

s-trans conformation
of 1,3-butadiene
~ 95%

s-cis conformation
of 1,3-butadiene
5%

steric crowding

**Study Guide**
**Concept Map 18.2**

*Figure 18.3*
The lowest energy $\pi$ bonding molecular orbital, $\psi_1$, in 1,3-butadiene showing delocalization of the electrons.

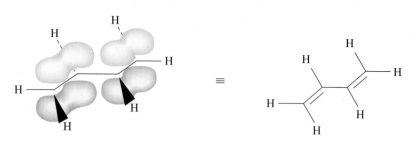

1,3-butadiene showing $\sigma$ bonds
and $p$ orbitals, each orbital
containing one electron

1,3-butadiene showing the
lowest energy $\pi$ bonding
molecular orbital

# 18.3   1,2- and 1,4-Addition of Electrophiles to Conjugated Systems

In Chapter 17 we encountered the concept of 1,4-addition to conjugated systems in the context of the addition of nucleophiles to $\alpha,\beta$-unsaturated carbonyl compounds. Here we see that conjugated dienes act in the same way. For example, 1,3-butadiene undergoes reactions in which electrophilic reagents such as chlorine and hydrogen chloride add to either or both of the double bonds. If 1 equivalent of the electrophilic reagent is used, a mixture of two major products is obtained. The experimental observations about the addition of chlorine to butadiene are given below.

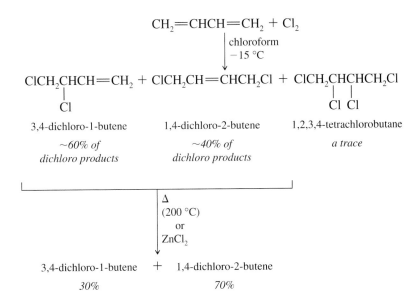

3,4-Dichloro-1-butene is the product from the addition of chlorine to one of the double bonds in 1,3-butadiene. It is called the **1,2-addition product.** It is the product that forms faster, the kinetic product (p. 133). In 1,4-dichloro-2-butene, the chlorine atoms have been attached to the first and fourth carbon atoms of the conjugated system, and the double bond that remains has shifted position in the molecule. This compound is called the **1,4-addition product.**

Experimentally, at low temperatures, 3,4-dichloro-1-butene is formed in larger amounts than is 1,4-dichloro-2-butene. The mixture formed at low temperatures can be equilibrated at high temperatures or in the presence of a Lewis acid, $ZnCl_2$, which aids in the removal of chloride ion. At equilibrium, the mixture contains 70% 1,4-dichloro-2-butene. 1,4-Dichloro-2-butene, with an internal double bond, is thermodynamically more stable than 3,4-dichloro-1-butene.

A conjugated diene system is always attacked at one end by an electrophile, because this gives the most highly stabilized cationic intermediate, as shown on the following page.

**Electrophilic Addition to a Diene**

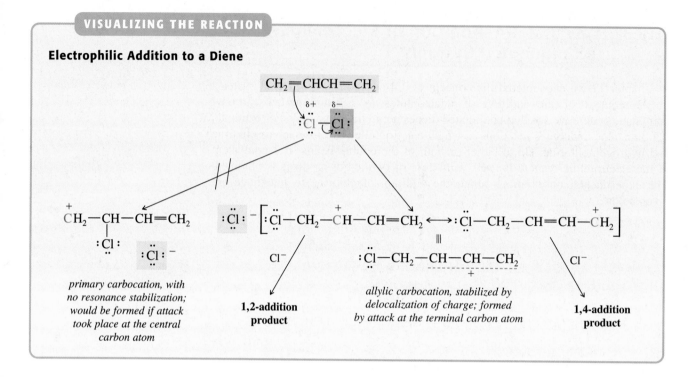

*primary carbocation, with no resonance stabilization; would be formed if attack took place at the central carbon atom*

**1,2-addition product**

*allylic carbocation, stabilized by delocalization of charge; formed by attack at the terminal carbon atom*

**1,4-addition product**

The intermediate formed by addition of a chlorine atom to the end of the conjugated system is an allylic carbocation. An allylic cation is stabilized by delocalization of the positive charge. Addition of a chlorine atom to one of the central carbon atoms in 1,3-butadiene would result in an unstable primary carbocation that had no possibility of resonance stabilization.

An examination of the two contributing resonance structures to the allylic cation shows why 1,2- and 1,4-addition products are formed. The allylic cation has positive charge delocalization to carbon atoms 2 and 4 in the chain. The negatively charged chloride ion can react with the cation at either one of these sites.

**Study Guide**
Concept Map 18.3

**Problem 18.3**    Complete the following equations.

(a)  [cyclohexadiene] $\xrightarrow[\substack{\text{carbon tetrachloride} \\ 0\ °C}]{\text{Br}_2\,(1\ \text{molar equivalent})}$

(b) $CH_2{=}CHCH{=}CH_2 \xrightarrow{\text{HCl (1 molar equivalent)}}$

(c) $CH_2{=}\underset{\underset{\displaystyle CH_3}{|}}{C}{-}\underset{\underset{\displaystyle CH_3}{|}}{C}{=}CH_2 \xrightarrow[\substack{\text{carbon tetrachloride} \\ 0\ °C}]{\text{Br}_2\,(1\ \text{molar equivalent})}$

## 18.4    The Diels–Alder Reaction

### A. Introduction

One of the most important reactions of conjugated dienes is the 1,4-addition of another multiple bond to the conjugated system to give a six-membered ring. A classic example of this reaction is the formation of a substituted cyclohexene from 1,3-butadiene and maleic anhydride.

| | | |
|---|---|---|
| *s*-cis conformation of 1,3-butadiene | maleic anhydride | tetrahydrophthalic anhydride ~95% |
| *diene* | *dienophile* | |

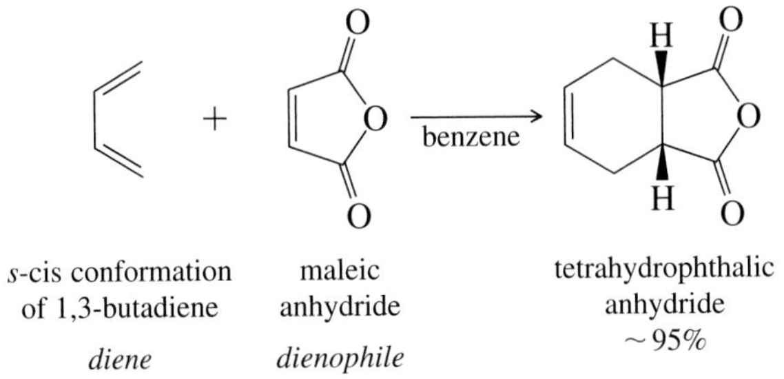

The reaction of 1,3-butadiene with maleic anhydride is an example of the **Diels-Alder reaction,** named after the two German chemists, Otto Diels and Kurt Alder, who recognized the generality of that reaction and jointly received the Nobel Prize in 1950 for their work. This type of addition reaction always has two reactants. One is a conjugated diene, which may have many different types of substituents on it. The other reactant always has a double or a triple bond in it and is known as the **dienophile,** a compound that is attracted to and reacts with the diene. The dienophile may be a simple alkene or part of a diene system. The most reactive dienophiles usually have a carbonyl group or another electron-withdrawing group such as a cyano or nitro group conjugated with a carbon–carbon double bond.

## B. Stereochemistry of the Diels–Alder Reaction

The Diels-Alder reaction involves a redistribution of electrons and bonds. It occurs in one step; no reactive intermediate is formed. Two double bonds disappear, two new single bonds are formed, and a double bond appears between two atoms that formerly shared a single bond.

**VISUALIZING THE REACTION**

### The Diels-Alder Reaction of 1,3-Butadiene and Maleic Anhydride

| | | |
|---|---|---|
| diene in *s*-cis conformation | dienophile, a multiple bond conjugated with electron-withdrawing groups | hydrogen atoms from maleic anhydride retain cis stereochemistry |

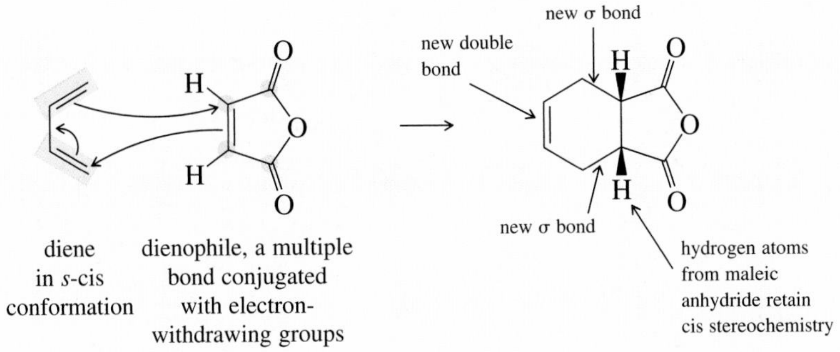

The essential aspects of a Diels-Alder reaction are illustrated by the above reaction. A conjugated diene reacts with a dienophile to give a six-membered ring with a double bond in it. The reaction is thought to proceed via a cyclic transition state. The reaction is highly stereoselective. Substituents on the dienophile as well as

those on the diene retain the stereochemistry that they had relative to each other before the reaction. For example, the hydrogen atoms in maleic anhydride are cis to
each other in the dienophile and cis to each other in the product. The diene, in order
to react, must be in the *s*-cis conformation.

The stereochemistry of the Diels-Alder reaction is further illustrated by the reaction of 2,3-dimethyl-1,3-butadiene with the cis and trans isomers of 1,4-diphenyl-
2-butene-1,4-dione.

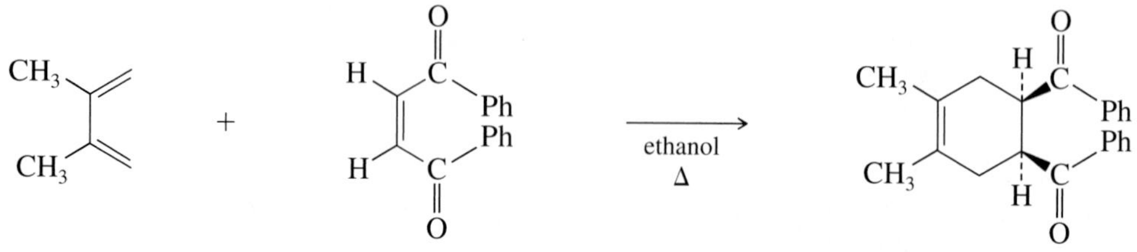

2,3-dimethyl-1,3-butadiene    (*Z*)-1,4-diphenyl-2-butene-1,4-dione      *cis*-1,2-dimethyl-4,5-dibenzoyl-1-cyclohexene
100%

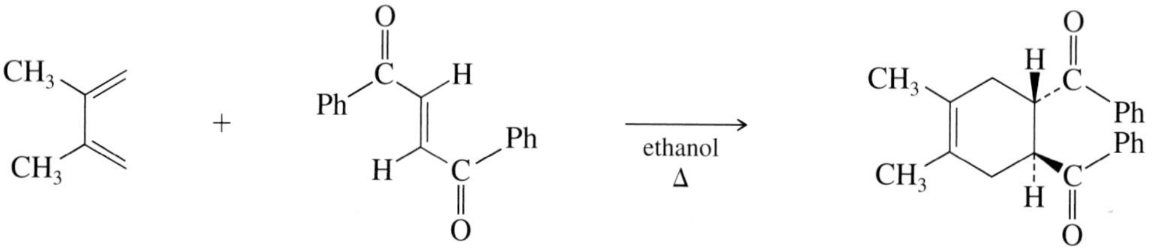

2,3-dimethyl-1,3-butadiene    (*E*)-1,4-diphenyl-2-butene-1,4-dione      *trans*-1,2-dimethyl-4,5-dibenzoyl-1-cyclohexene
100%

The groups that are cis to each other in the dienophile remain cis to each other in
the cyclohexene that forms, whereas those that are trans in the dienophile are trans
in the product.

The dienophile in a Diels-Alder reaction may contain a triple bond. For example, esters of acetylenedicarboxylic acid are highly reactive dienophiles, as demonstrated by the reaction of 1,3-butadiene with dimethyl acetylenedicarboxylate.

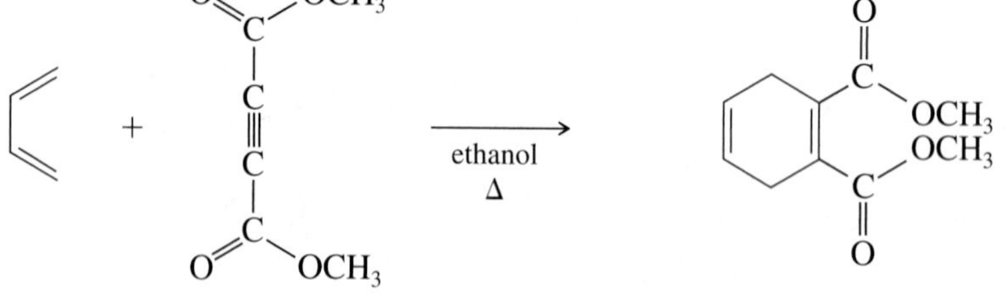

1,3-butadiene         dimethyl              dimethyl 1,4-cyclohexadiene-1,2-dicarboxylate
acetylenedicarboxylate

**Problem 18.4**    Write a mechanism for the Diels-Alder reaction shown above.

## C. Bicyclic Compounds from Diels–Alder Reactions. Endo and Exo Stereochemistry

Cyclic dienes are particularly reactive in Diels-Alder reactions because the two double
bonds are held in an *s*-cis conformation in five- or six-membered rings. (The *s*-trans

conformation is favored for open-chain dienes; see page 740.). For example, cyclopentadiene reacts with maleic anhydride or propenal to give Diels-Alder adducts.

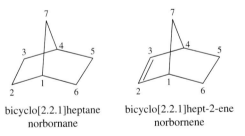

| cyclopentadiene | maleic anhydride | *cis*-norbornene-5,6- |
| diene | dienophile | *endo*-dicarboxylic anhydride |
| | | 98% of addition product |

| cyclopentadiene | 2-propenal | *endo*-bicyclo[2.2.1]hept- |
| diene | dienophile | 5-en-2-carbaldehyde |
| | | 90% |
| | | *racemic mixture* |

Reactions of cyclopentadiene with dienophiles give rise to compounds containing a cyclohexene ring bridged by a single carbon atom. The common name of the saturated hydrocarbon having such a carbon skeleton is norbornane, and the systematic name is bicyclo[2.2.1]heptane.

bicyclo[2.2.1]heptane  
norbornane

bicyclo[2.2.1]hept-2-ene  
norbornene

The systematic name is derived from the way the seven atoms are held together in a structure with two rings, a **bicyclic structure.** The two rings have two carbon atoms in common, numbered 1 and 4 in the structural formula. These positions are called the **bridgehead positions.** Carbon atoms 1 and 4 are tied together by three bridges, two of which have two carbon atoms and one of which has a single carbon atom. The numbers in the brackets, [2.2.1], represent the numbers of atoms in the bridges. The presence of a substituent or a double bond in a bicyclic compound is indicated by using prefixes or suffixes just as for other types of compounds. Thus the compound that results when ethylene adds to cyclopentadiene has the common name norbornene and the systematic name bicyclo[2.2.1]hept-2-ene.

The products of the reactions of cyclopentadiene with dienophiles bring up an interesting question of stereochemistry. The substituents on the bicyclic ring created by the Diels-Alder reaction are oriented away from the carbon bridge on the other face of the six-membered ring. Such an orientation is said to be **endo,** meaning that the substituent projects *into* the cavity on the concave side of the bicyclic ring.

Another orientation in which the substituent extends *out of* the cavity is called the **exo** orientation.

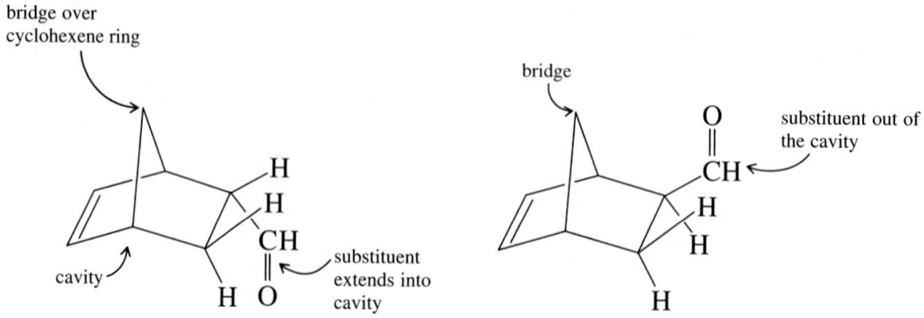

endo orientation
of aldehyde group
on bicyclic ring

exo orientation
of aldehyde group
on bicyclic ring

The reactions of cyclopentadiene with maleic anhydride, with propenal, and with itself give predominantly the endo orientation in the products, a stereochemistry that is quite general for many dienes and dienophiles. This phenomenon is discussed in much greater detail in Section 25.2B, which examines the nature of the interaction between the $\pi$ bonds of the diene and that of the dienophile.

The electronic character of the diene and dienophile have significance with regard to the ease with which a Diels-Alder reaction takes place. Comparing the reaction of cyclopentadiene and maleic anhydride with the reaction of cyclopentadiene and ethylene makes this evident. A mixture of cyclopentadiene and maleic anhydride must be cooled in ice to prevent the reaction from becoming so vigorous that the low-boiling cyclopentadiene (bp 42 °C) is lost. The reaction of cyclopentadiene with ethylene, on the other hand, requires the use of a steel reaction vessel called a bomb, in which the mixture of reagents can be maintained at 800–900 pounds per square inch (psi) at approximately 200 °C for 7 hours.

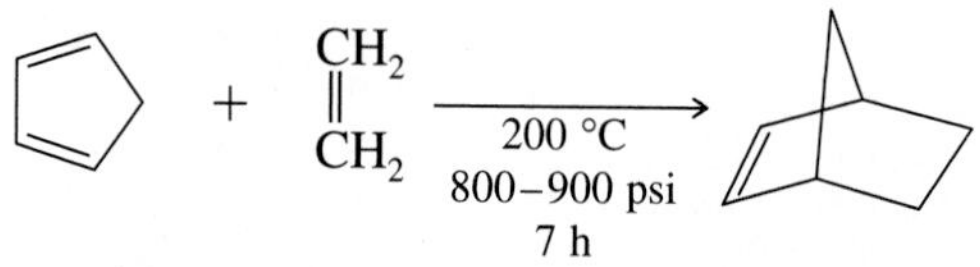

cyclopentadiene   ethylene

norbornene
~60%

Clearly, the electron-withdrawing effect of the conjugated carbonyl groups in maleic anhydride makes that molecule a much better dienophile than ethylene, which has only a double bond.

Electron-rich dienes act as nucleophiles. Dienophiles containing multiple bonds conjugated with electron-withdrawing groups are electrophiles. Classic Diels-Alder reactions occur when these two types of compounds interact. This type of reaction is tremendously useful as a way to synthesize functionalized six-membered rings with high stereoselectivity. The full extent of its usefulness will be explored further in Chapter 21.

Study Guide
Concept Map 18.4

**Problem 18.5**   All the following pairs of reactants undergo Diels-Alder reactions. Complete the equations, showing the stereochemistry of the product when it can be predicted.

(a), 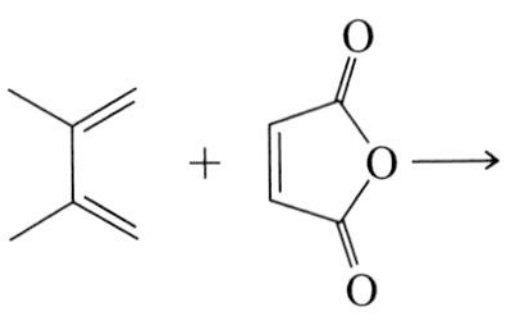 (b), 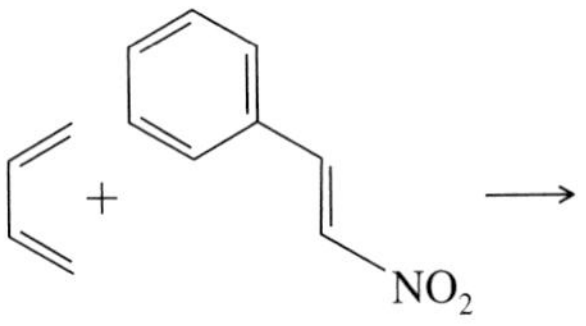 (c), 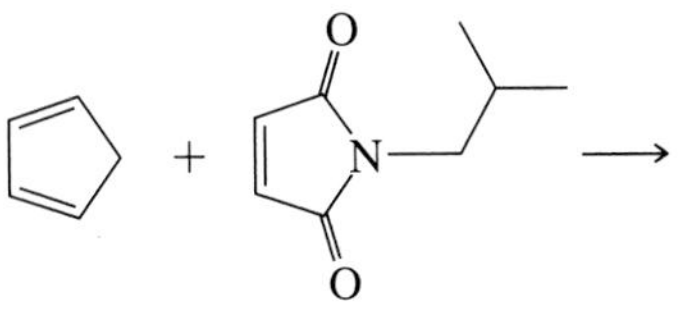 (d), 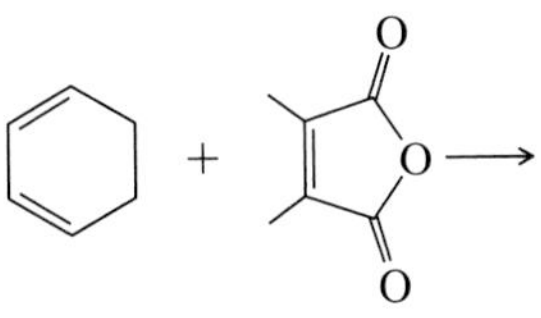 (e), 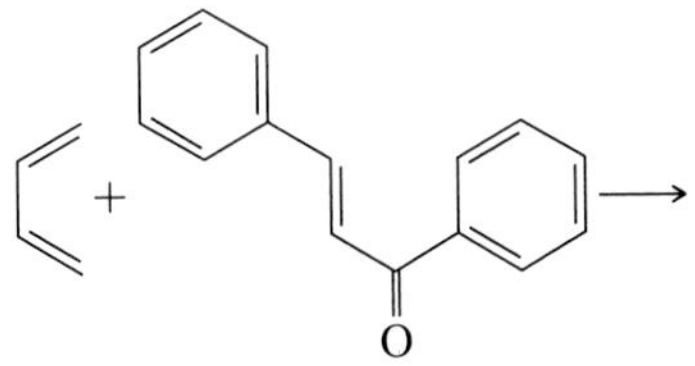 (f) 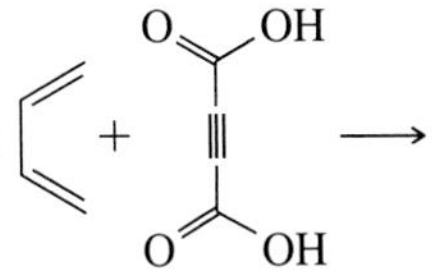 [Diels-Alder reaction schemes]

**Problem 18.6**   Diels-Alder reactions are reversible. Cyclohexene, for example, is broken down into a diene and a dienophile when it is exposed to a red-hot wire in the absence of air. Write an equation showing the products of this reaction.

**Problem 18.7**   What reagents would be required to obtain each Diels-Alder adduct shown below?

(a), 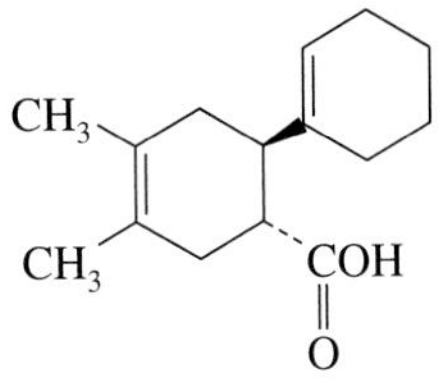 (b), 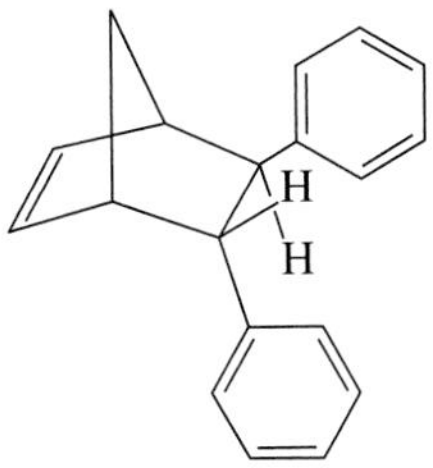 (c), 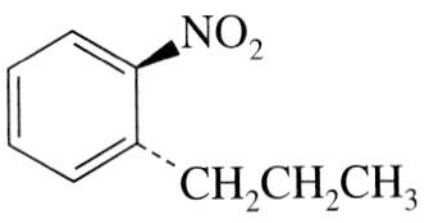 (d), 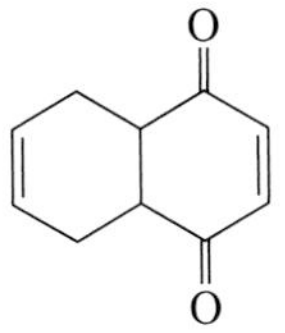 (e), 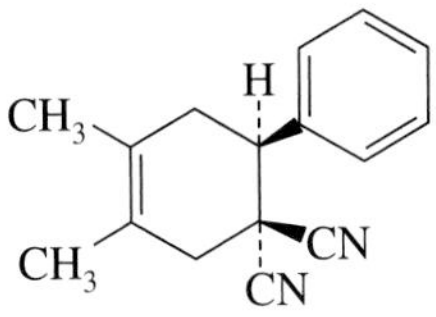 (f) 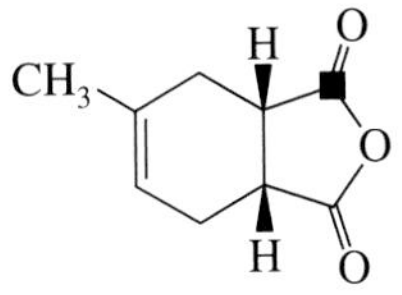 [Diels-Alder adduct structures]

## 18.5  Biologically Interesting Alkenes and Polyenes

### A. Pheromones

Insects communicate with each other, and with their environment, by means of organic chemicals secreted in minute amounts. These substances, known as **pheromones,** have a wide range of structures and include a variety of functional groups.

Because chemists must work with such small amounts of material, the chemistry associated with the isolation of these compounds and the determination of their structures is particularly challenging. Many of the structural determinations were done on only a few milligrams of isolated material. Methods of separation and isolation, as well as spectroscopic techniques, were refined in order to deal with these very small quantities. For example, extraction of the abdominal tips of 500,000 virgin gypsy moth females yielded only 75 mg of the pheromone that attracts males. This quantity of active material had to be separated from approximately 250 g of fatty acids and their esters and 70 g of steroids, mostly cholesterol. This process of purification, which is like looking for the proverbial needle in a haystack, was aided by biological assay methods in which male gypsy moths were exposed to the fractions obtained in various separation steps. The fractions that did not attract the male moths were discarded and those that did were purified further until finally a single component that is highly attractive to the male was isolated.

The sex pheromone for the gypsy moth, which is a devastating pest in hardwood forests, is a chiral oxirane. The dextrorotatory compound, known as disparlure, is a potent attractant for male gypsy moths, but the enantiomer, the levorotatory compound, has no such activity.

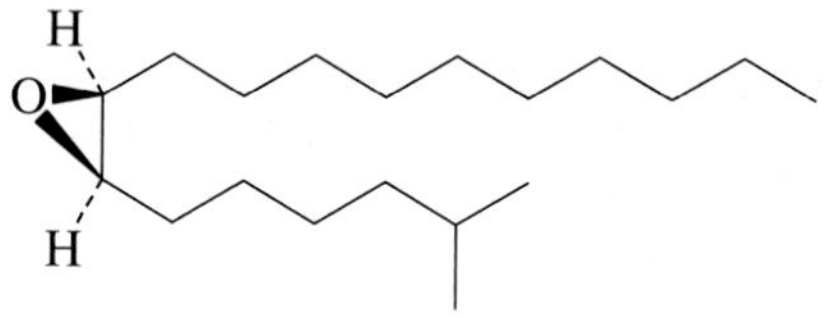

(+)-disparlure

*the sex pheromone of the gypsy moth*

A sex pheromone is usually specific for one species of insect. Understanding how pheromones work might allow scientists to devise ways of controlling certain insect populations without harming the environment. At present, sex pheromones are used as lures in traps that allow foresters to estimate the population of a given insect in a region. For example, a single Mediterranean fruit fly, caught in such a trap in a citrus grove in California, can serve as a warning that a new infestation of the destructive pest is on the way and alert the agricultural experts in that state to be prepared.

The sex pheromones of many insects are long-chain alcohols or esters, with one or more double bonds in them. In some cases, the double bonds are conjugated; in others, they are not. They may have cis or trans stereochemistry. The sex pheromones of the pink bollworm, the silkworm, and the red-banded leaf roller are among those that have been isolated and their structures determined.

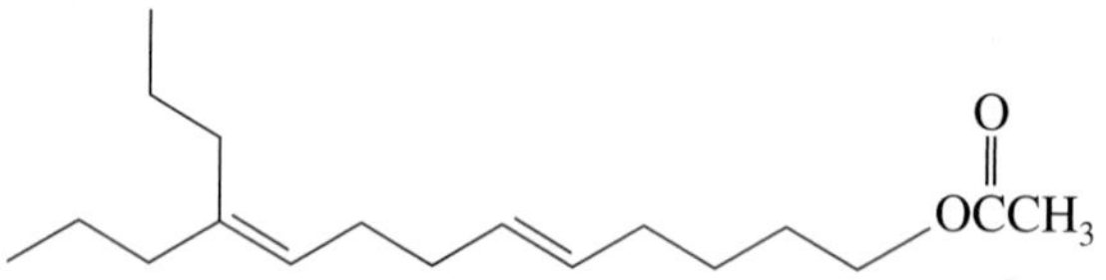

(*E*)-10-propyl-5,9-tridecadienyl acetate

*from the pink bollworm*

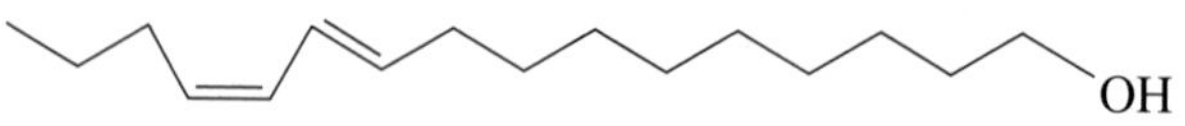

(10*E*,12*Z*)-10,12-hexadecadien-1-ol

*from the silkworm*

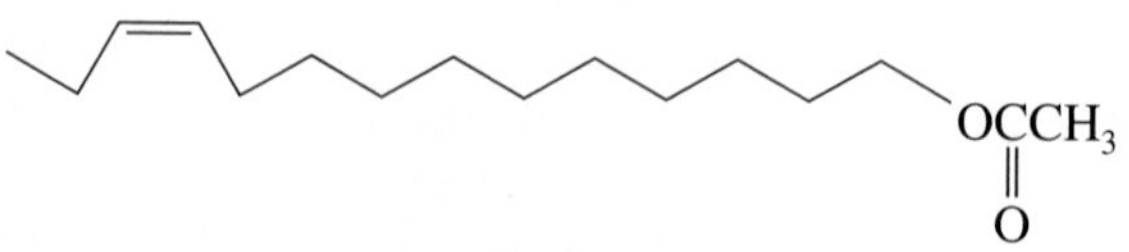

(*Z*)-11-tetradecenyl acetate

*from the red-banded leaf roller*

The insect pheromone released in the greatest concentrations is the alarm pheromone, a chemical substance used to signal a disturbance in an insect colony. The simple ester 3-methylbutyl acetate is the alarm pheromone for the honey bee. Many other insect species use terpenes (p. 750) such as citronellal and limonene to signal alarm.

Even the fact of death is communicated by chemical signals in the insect world. Ants, for example, do not recognize that a freshly frozen ant is dead and treat it quite differently at first than they do after a while. Apparently, bacterial decomposition in the dead insect releases a number of fatty acids, such as oleic and linoleic acids. If mixtures of acids such as these are placed on a small piece of paper and put into the colony, the ants rapidly remove the paper and deposit it on a refuse pile. They do the same thing to a fellow ant once it has been dead for a while.

Organic compounds are used to regulate the relationship between an insect and its environment in many other ways. Plants synthesize compounds that prevent insects from eating them, and insects produce defensive secretions that keep birds and mammals away. Mammals, too, have chemical methods of communication that are just beginning to be recognized as important. Perfume makers, of course, have always understood the importance of scent as a means of communication.

**Problem 18.8**   Disparlure, the sex pheromone of the gypsy moth (p. 748), was synthesized as a racemic mixture by the following sequence of reactions. Give structures for the products or reagents designated by letters.

$$HC\equiv CH \xrightarrow{\text{NaNH}_2} A \xrightarrow{B} CH_3(CH_2)_9C\equiv CH \xrightarrow[\text{diglyme}]{CH_3CH_2CH_2CH_2Li}$$

$$C \xrightarrow{D} CH_3(CH_2)_9C\equiv C(CH_2)_4\overset{\overset{\displaystyle CH_3}{\displaystyle |}}{C}HCH_3 \xrightarrow{E} F \xrightarrow{G} \text{racemic disparlure}$$

$$CH_3\overset{\overset{\displaystyle CH_3}{\displaystyle |}}{C}HCH_2CH_2CH_2CH_2OH \xrightarrow{H} D$$

## B.  Terpenes and Terpenoids. Isoprene and the Isoprene Rule

For centuries, human beings have known that volatile oils with a variety of fragrances and flavors could be isolated from certain plants. These compounds occur in all parts of the plants and are called **essential oils.** They are of great commercial importance in the perfume and flavoring industries, and many also have been used medicinally. The chemical constituents of essential oils have a wide variety of structures; many of them contain rings or one or more double bonds. Some of them are alcohols or ethers; others are ketones or aldehydes.

After many years of carrying out structural determinations on these interesting and challenging compounds, chemists began to detect some patterns. For example, most of these compounds are composed of multiples of 5 carbon atoms. Whole families of compounds were discovered and classified according to their molecular formulas. These natural products, all of which had multiples of 5 carbon atoms in their structures, were called **terpenes.** Compounds with 10 carbon atoms, the **monoterpenes,** those with 15 carbon atoms, the **sesquiterpenes,** and those with 20 carbon atoms, the **diterpenes,** are the chief constituents of essential oils.

Steroids are **triterpenes,** compounds with 30 carbon atoms, or are derived from them. Materials that give plants colors, such as carotene (from carrots), are **tetraterpenes,** containing 40 carbon atoms. Rubber is a polymeric terpenoid (p. 1073).

The structures of a few essential oils as well as some other plant constituents are shown below.

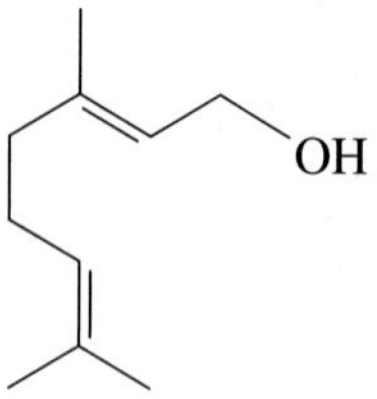

geraniol

*from Turkish geranium oil*

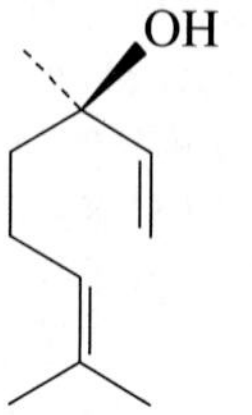

(*R*)-(−)-linaloöl

*(−)-from rose oil*
*(+)-from orange oil*

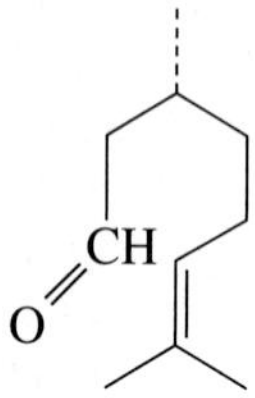

(*R*)-(+)-citronellal

*from citronella oil*

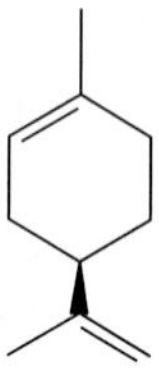

(*R*)-(+)-limonene

*from lemon and orange oils*

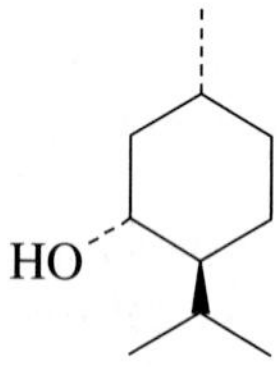

(−)-menthol

*from peppermint oil*

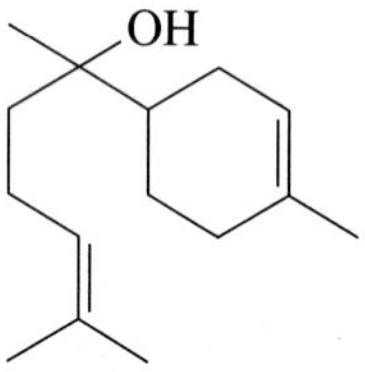

bisabolol

*from chamomile and lavender oils*

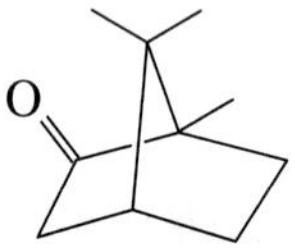

(+)-camphor

*from the trunk of a tree found mostly in Formosa*

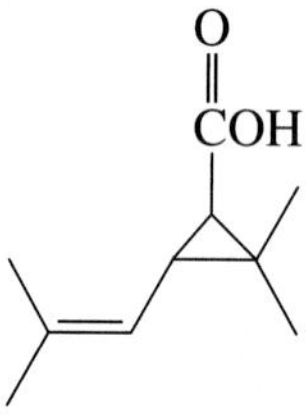

(+)-*trans*-chrysanthemic acid

*from pyrethrin I, an insecticidal constituent of* Chrysanthemum cinerariifolium

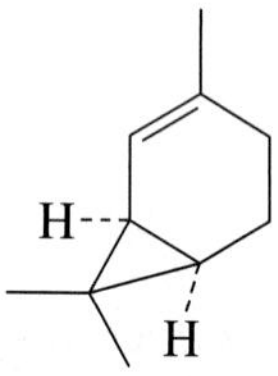

(+)-2-carene

*from various essential oils*

Chrysanthemic acid and 2-carene have historical significance. In 1920, the Yugoslav chemist Leopold Ruzicka, who did his scientific work in Switzerland and later won the Nobel Prize for discoveries in terpene chemistry, finished a proof of the structure of chrysanthemic acid. This compound is a constituent of a pyrethrin, a component of an insecticidal extract from a type of chrysanthemum that grows primarily in East Africa. (Pyrethrins have become important commercially as highly selective and biodegradable insecticides.) In the same year, the structure of 2-carene was determined by the British chemist John L. Simonsen. Ruzicka was struck by the structural resemblance between the two compounds and could see in both of them the elements of a five-carbon structural unit known as **isoprene.** Terpenes are made of isoprene units.

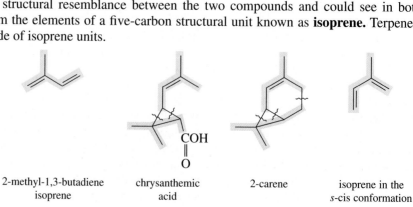

2-methyl-1,3-butadiene
isoprene

chrysanthemic
acid

2-carene

isoprene in the
*s*-cis conformation

In other naturally occurring compounds that were known at that time, especially certain ones that did not contain a ring, another regularity could be seen. Referring to the branched end of the isoprene molecule as its head and the other end as its tail, molecules that contained two or more isoprene units had the head of one isoprene unit attached to the tail of the next one in the chain.

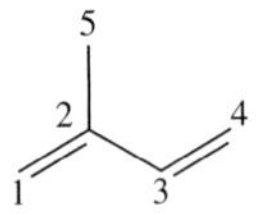

*head of*    *tail of*
*isoprene*    *isoprene*

four-carbon chain with
methyl branch on second carbon

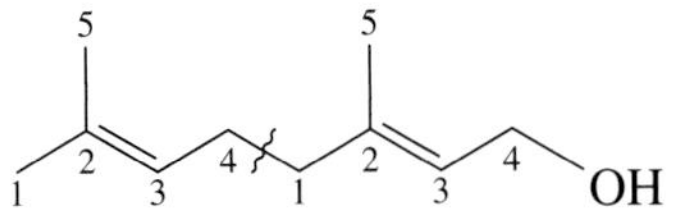

geraniol, a terpene made up of
two isoprene units attached
head to tail

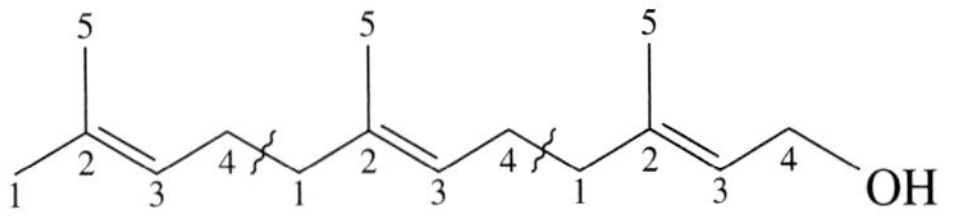

farnesol, scent of lily of the valley,
a sesquiterpene made up of three
isoprene units attached head to tail

Ruzicka hypothesized that these plant constituents were synthesized in nature by a head-to-tail connection of isoprene units. He examined a large number of terpenoid compounds over the years and found that the structures of most of them showed this regularity. Thus he formulated the isoprene rule.

The **isoprene rule** means that given a choice of possible structures for a natural product that contains a multiple of five carbon atoms, chemists will favor one that appears to be made of isoprene units connected in a head-to-tail fashion. The isoprene rule says nothing about whether the compound contains double bonds or functional groups containing oxygen. A wide variety of patterns of rings, unsaturation, and alcohol, carbonyl, and carboxyl groups is present in these natural products. The isoprene rule applies only to the carbon skeleton of the molecule. When rings are present, they indicate that additional points of attachment between various isoprene units have been formed. In those cases, it is still possible to trace the main outline of the chain of isoprene units around the molecule. Some cyclic terpenoid compounds are dissected into isoprene units below.

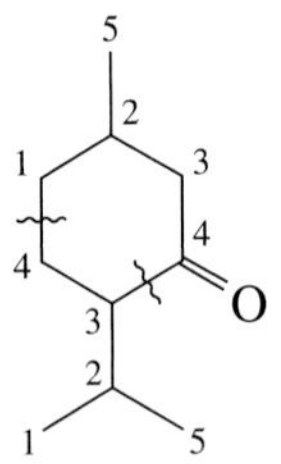

(−)-menthone

*peppermint oils*

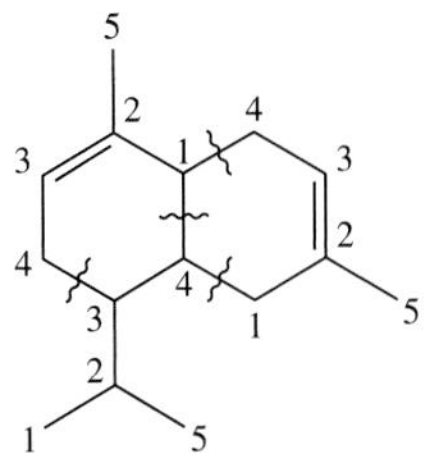

α-cadinene

*oil of citronella*

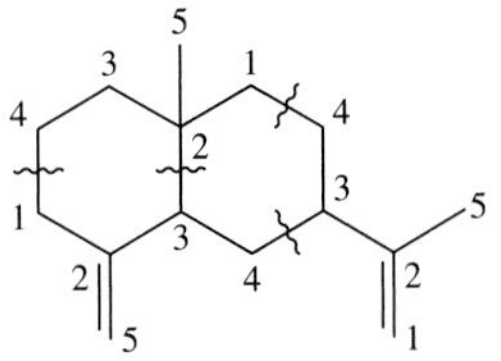

β-selinene

*oil of celery*

In some structures, it is possible to find more than one way to mark off the isoprene units.

As molecules become larger, the final structure is often achieved only after some molecular rearrangements during the biosynthesis. The isoprene rule does not apply to all parts of such structures, though it is still possible to find portions of the molecule that follow the rule.

**Problem 18.9**    Mark off the isoprene units in the following compounds.

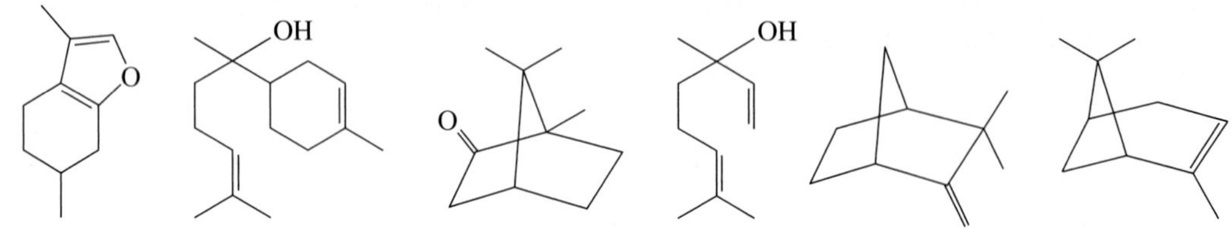

menthofuran          bisabolol          camphor          linaloöl          camphene          α-pinene

## C. Biosynthesis of Terpenes

The pyrophosphate ester of an unsaturated five-carbon alcohol, 3-methyl-3-buten-1-ol, is the structural building block for naturally occurring terpenes. Pyrophosphoric acid is an anhydride of phosphoric acid and seems to be nature's tool for creating good leaving groups (p. 683). 3-Methyl-3-buten-1-yl pyrophosphate, known in the biochemical literature as isopentenyl pyrophosphate, is isomerized enzymatically to 3-methyl-2-buten-1-yl (dimethylallyl) pyrophosphate in a reaction that may be regarded as a protonation at one $sp^2$-hybridized carbon atom and a deprotonation of the incipient carbocation at another site to give the more highly substituted alkene. The participation of an enzyme, a highly specific biological catalyst (see page 1012 for an example of how an enzyme functions), ensures that no high-energy intermediate is formed at any point of the reaction.

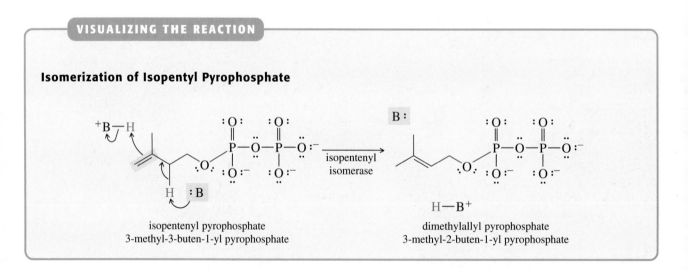

**VISUALIZING THE REACTION**

**Isomerization of Isopentyl Pyrophosphate**

isopentenyl pyrophosphate
3-methyl-3-buten-1-yl pyrophosphate

isopentenyl isomerase

dimethylallyl pyrophosphate
3-methyl-2-buten-1-yl pyrophosphate

Careful experiments have shown that the next step in the biosynthesis of a terpene goes by way of a carbocation intermediate. Ionization of dimethylallyl pyrophosphate gives a resonance-stabilized allylic cation, which is the electrophile that reacts with the double bond of isopentenyl pyrophosphate. The new cation then loses a proton to give a double bond.

---

**VISUALIZING THE REACTION**

**Biosynthesis of a Geranyl Pyrophosphate**

dimethylallyl pyrophosphate

dimethylallyl transferase

isopentenyl pyrophosphate

pyrophosphate anion

geranyl pyrophosphate
the pyrophosphate ester of geraniol

---

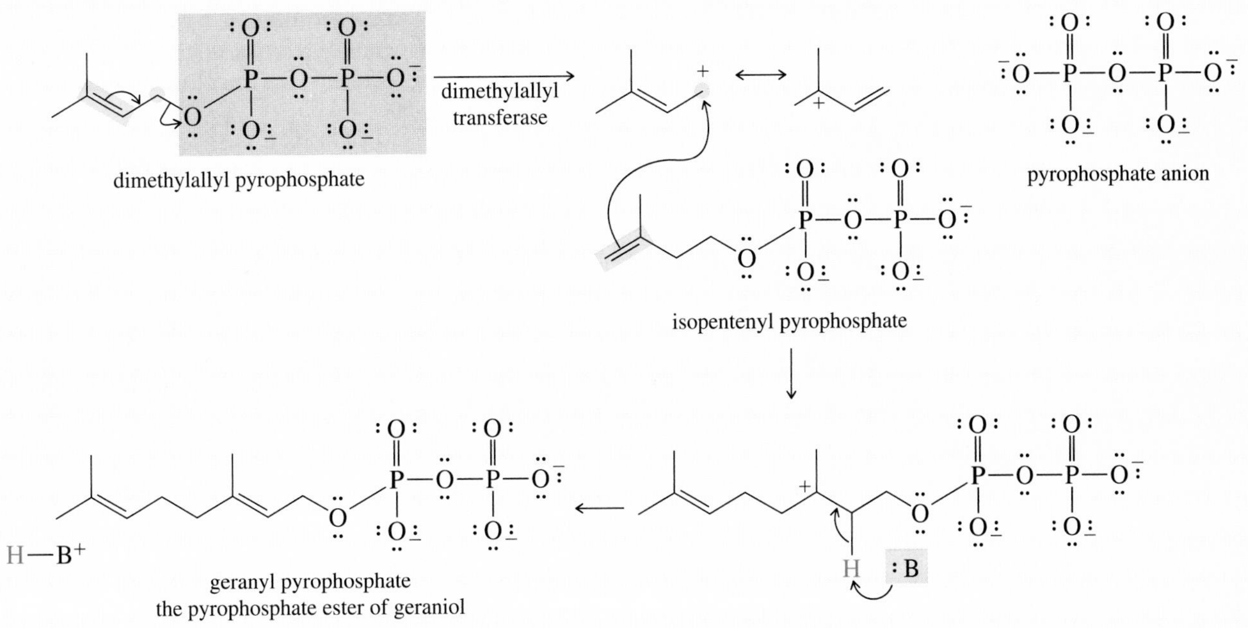

The product is a ten-carbon terpene, geraniol.

Note that the mechanism proposed for terpene synthesis leads in each case to the head-to-tail connection of isoprene units. Five carbon atoms are added, and the end of the chain is left functionalized as a pyrophosphate ester that can serve as a leaving group for further reactions. Further protonations and deprotonations would lead to isomerizations of the positions and of the stereochemistry of double bonds. The oxygen-containing functional groups are modified by oxidation and reduction reactions, and new hydroxyl groups are created by the addition of water to double bonds or by enzymatic oxidation reactions at unactivated sites. Conversely, dehydration of alcohol functions introduces unsaturation into the systems. Biological reduction reactions convert alkenes into alkanes. Carbocations created by protonation of a double bond or loss of water from a protonated alcohol can attack a double bond in another part of the molecule to create rings. The next problem reviews this familiar chemistry in the context of terpene reactions.

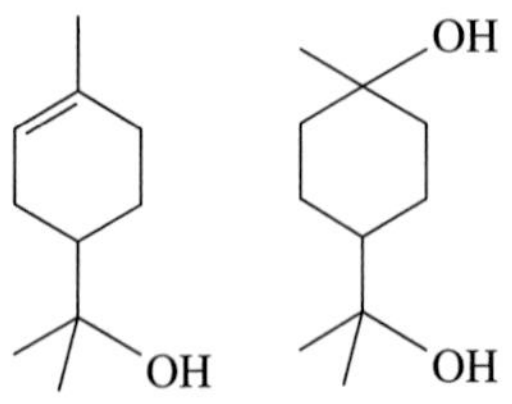

α-terpineol

terpin

### Problem 18.10
In all parts of this problem, you may use $HB^+$ as an acid and B: as a base as necessary. Water is always present as a nucleophile.

(a) Geraniol (p. 750) has a diastereomer called nerol. Nerol is also a natural product but occurs somewhat less abundantly than geraniol. Write a structural formula for nerol, and propose a mechanism for the isomerization of geraniol to nerol.

(b) Treatment of geraniol or nerol with aqueous acid gives rise to α-terpineol and terpin, shown at left. Nerol is converted into these compounds faster than geraniol is. Write mechanisms for the formation of α-terpineol and terpin from nerol.

(c) Limonene, a flavor constituent of citrus fruits (p. 750), is formed from α-terpineol as well as from terpin. Write equations showing how this would take place.

(d) Linaloöl (p. 750) may be considered to arise from either geraniol or nerol. Propose a mechanism for its formation.

### Problem 18.11
Isoprene undergoes a Diels-Alder reaction with itself to give rise to an optically inactive compound known for some time as dipentene. Dipentene was finally identified as the racemic form of a naturally occurring terpene. Write an equation for the reaction and identify the terpene, which appears on page 750.

### Problem 18.12
Show how the sesquiterpene farnesol (p. 751) could be synthesized from geranyl pyrophosphate (p. 753) and isopentenyl pyrophosphate (p. 753).

### Problem 18.13
The experiment that showed that an allylic carbocation is a reactive intermediate in the step bonding isoprene units together in the biosynthesis of terpenes substituted a fluorine atom for a hydrogen atom on geranyl pyrophosphate. Fluorine is approximately the same size as hydrogen but has very different electronic properties. Predict whether the reaction giving rise to fluorofarnesol is faster or slower than with geranyl pyrophosphate itself.

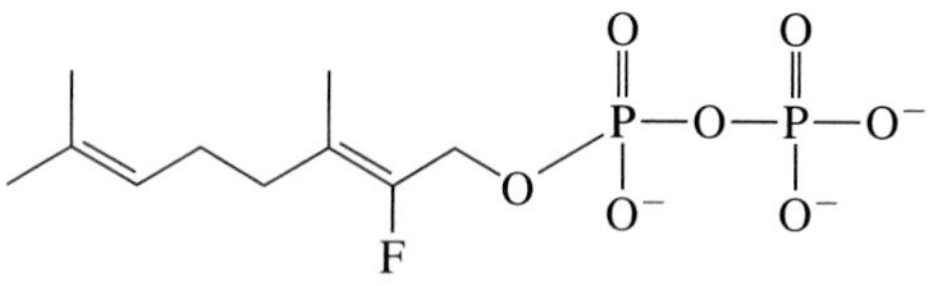

fluorogeranyl pyrophosphate
used to test whether a carbocation is an
intermediate in the biosynthesis of terpenes

### Problem 18.14
Diterpenes, compounds having 20 carbon atoms, are derived from a pyrophosphate that has the following structure.

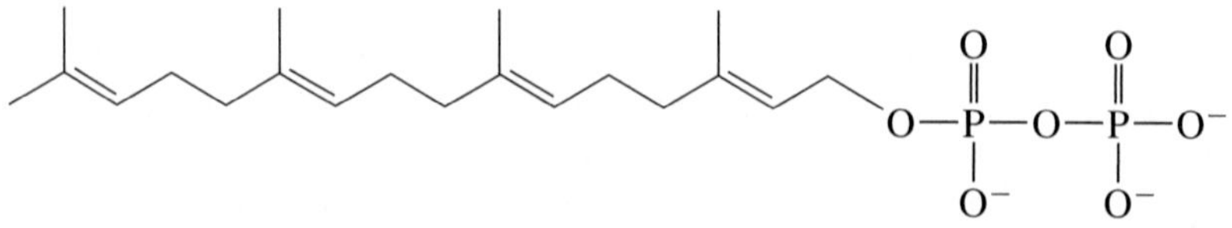

Propose a biosynthesis of this compound, sometimes called geranylgeranyl pyrophosphate.

## D. Steroids

**Steroids** are triterpenoids, a group of compounds produced in plants or animals. They are widespread in nature and have important biological functions. Some hormones and cholesterol (pp. 537 and 757) are steroids. **Terpenoids** are compounds that are derived from terpenes but do not have a multiple of five carbon atoms.

Steroids are characterized by a tetracyclic structure. A typical steroid ring skeleton consists of three six-membered rings and a five-membered ring fused together. Depending on their source and biological function, steroids also may have a variety of functional groups substituted on the skeleton (p. 537).

The rings in a steroid skeleton are designated as the A, B, C, and D rings.

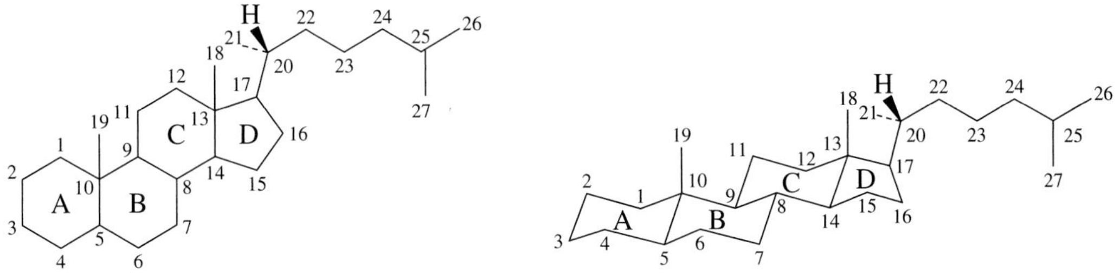

cholestane

The fusion of the four rings gives rigidity to a steroid molecule. The conformational changes (p. 180) that are easy for individual cyclohexane rings are not possible at room temperature for fused rings. Steroids, therefore, have been used for research on the influence of conformation on the course of organic reactions. Sir Derek H. R. Barton of Great Britain received the Nobel Prize in 1969 for recognizing that functional groups could vary in reactivity depending on whether they occupied an axial or an equatorial position on a ring. This way of thinking about stereochemistry, called **conformational analysis,** has greatly influenced the way chemists analyze reactivity.

Squalene, $C_{30}H_{50}$, was first isolated from shark liver oil in 1916. Since that time, research has established its role in the formation of the steroid ring skeleton in living matter. Squalene is a triterpene, but one in which the isoprene rule is violated in one place. Rather than a head-to-tail arrangement of six units of isoprene, there appear to be two farnesyl units that have been connected tail to tail.

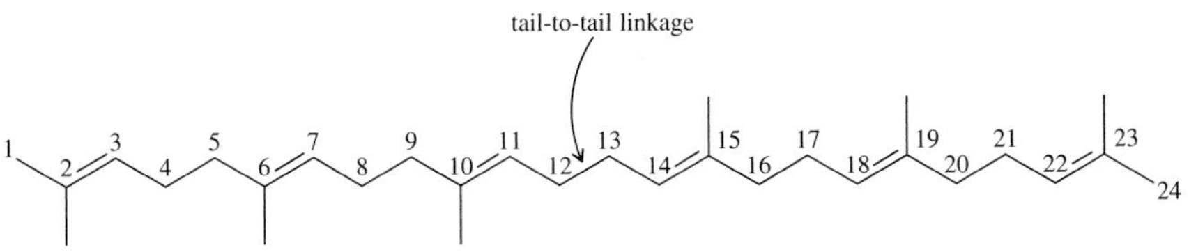

squalene
(6E,10E,14E,18E)-2,6,10,15,19,23-hexamethyl-
2,6,10,14,18,22-tetraicosahexaene

*a triterpene*

A great deal of research, using selective labeling of carbon atoms and hydrogen atoms, has confirmed the tail-to-tail connection. Other evidence indicates that the formation and then the opening of a cyclopropane ring in the middle of the chain occurs as two farnesyl pyrophosphate units interact.

The critical intermediate in the conversion of squalene to the steroid skeleton is an oxirane, squalene-2,3-oxide, which is transformed by enzymes into lanosterol, a steroid alcohol found in wool fat. Similar cyclization reactions take place in the laboratory with Lewis or Brønsted acids as catalysts. The squalene molecule is believed to be folded in a conformation that allows one double bond after another to react as a nucleophile with a cationic center that develops nearby in the molecule. The initial leaving group that starts the process is the oxygen atom of the oxirane ring protonated or complexed in a way that makes it a better leaving group. Oxiranes, as strained small ring compounds, are reactive toward nucleophiles, especially in systems where the oxygen atom develops a positive charge by protonation (p. 502).

---

**VISUALIZING THE REACTION**

**Cyclization of Squalene Oxide**

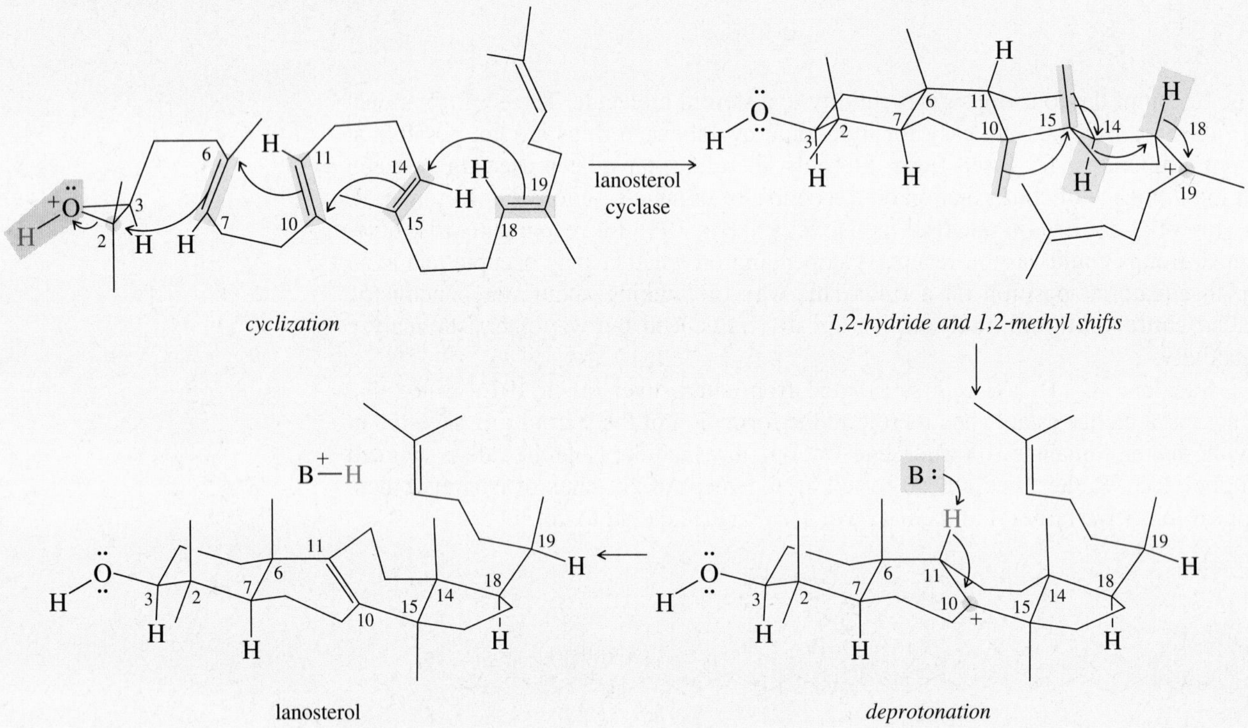

cyclization

lanosterol cyclase

1,2-hydride and 1,2-methyl shifts

deprotonation

lanosterol

---

The whole process, catalyzed by an enzyme, is highly stereoselective. The individual steps of the transformation are familiar as typical carbocation reactions (p. 288). A carbocationic center, as it is generated, behaves as a Lewis acid toward the $\pi$ electrons of an adjacent double bond, which in turn creates a new cationic center, so the cyclization progresses. The initial cyclization product is a cation that differs from lanosterol in the placement of two methyl groups. The double bond is created in the B ring by a series of 1,2-shifts that result in a tertiary cation that undergoes the final deprotonation.

Squalene is the biological precursor of many triterpenoids. Important among these are the steroids, one of which is cholesterol.

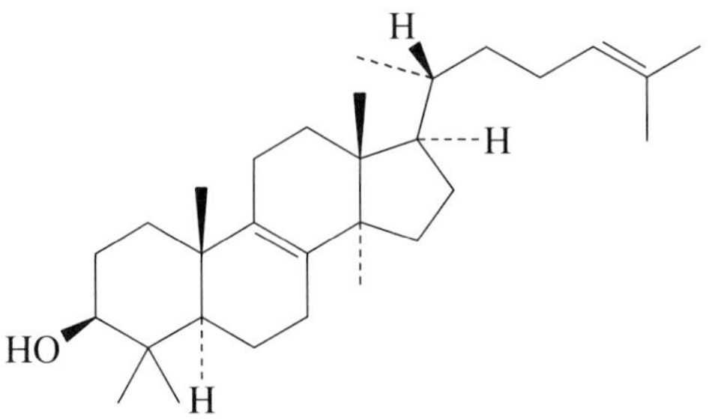

lanosterol 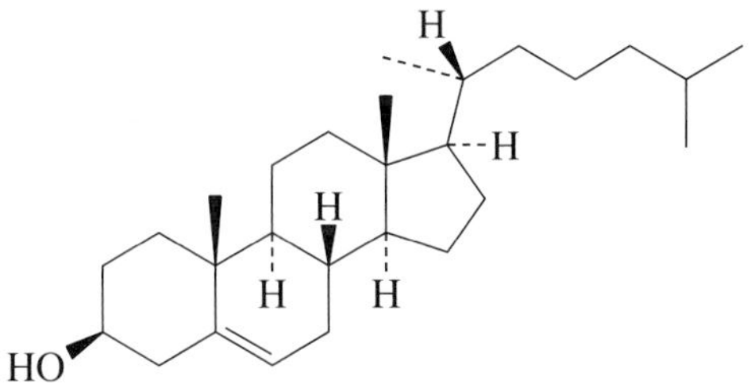          cholesterol

During the biological conversion of lanosterol to cholesterol, three methyl groups are lost, the hydrocarbon chain is reduced to a saturated one, and the position of the double bond in ring B of the steroid skeleton changes. The exact details of how these transformations are carried out are not known. More than one pathway may be involved. That lanosterol is converted to cholesterol by enzymes from the liver has been proved by a number of experiments. For example, when lanosterol, labeled with radioactive $^{14}C$ and carefully purified so that it contains no cholesterol, is incubated with cells from the liver of a rat, cholesterol containing radioactivity is recovered.

## E. Carotenoids. Vitamin A

The **carotenoids** are compounds made up of eight isoprene units and usually containing some rings and a number of conjugated double bonds. The backbone of the chain appears to be constructed by a tail-to-tail union of two geranylgeranyl pyrophosphate units (Problem 18.14), similar to the way squalene is derived from a combination of farnesyl pyrophosphate units. Carotenoids are insoluble in water but soluble in fat and hydrocarbon solvents. All the carotenoids, which are generally yellow to red in color, are widely distributed in nature, in both plants and animals.

Lycopene is the parent carotenoid, related to all other known carotenoids by changes such as reduction of some of the double bonds, cyclization, isomerization of the position of a double bond, and introduction of functional groups containing oxygen.

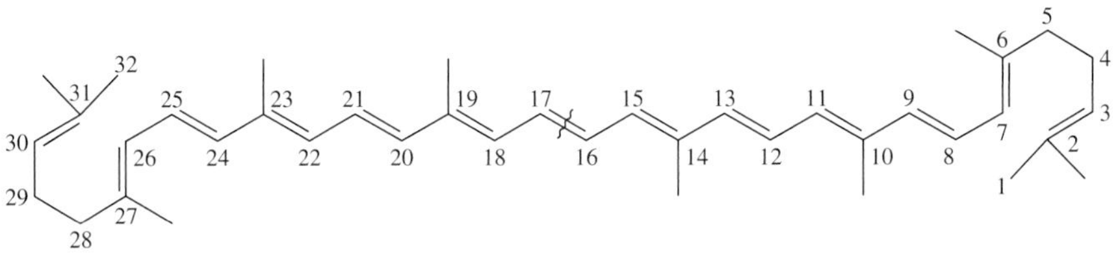

lycopene

*dissected at the point of tail-to-tail
connection of two 20-carbon units*

Note that the double bonds in lycopene are all trans in configuration and that, except for the ones nearest the ends of the molecule, they are all in conjugation. This extended system of conjugation is responsible for the color of these compounds, which are the pigments in many plants (p. 449). Lycopene is abundant in tomatoes, for example.

Carotene, the pigment of carrots, is especially interesting because it is closely related to vitamin A, a diterpenoid essential to growth, to the health of membranes, and to vision. Two isomers of carotene, differing from each other in the position of one of the double bonds, are abundant in carrots. Both of them show strong vitamin A activity in biological tests.

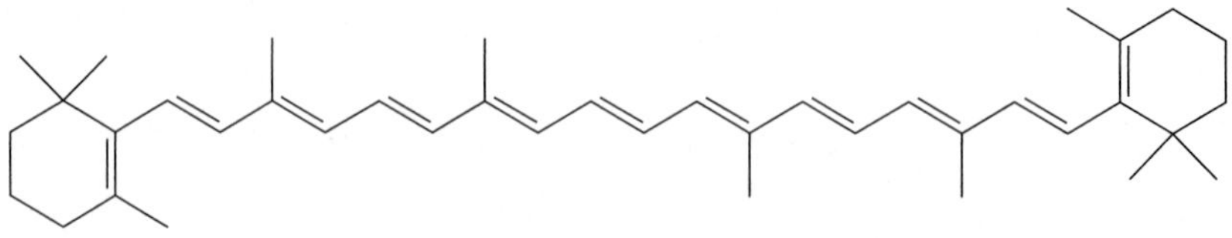

β-carotene

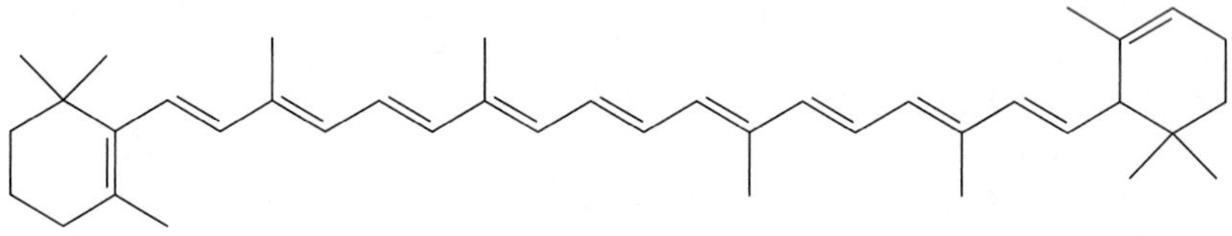

α-carotene

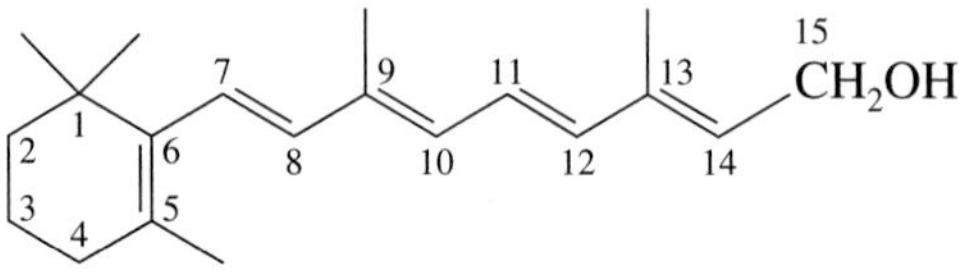

vitamin A
retinol

How the carotenes are derived from lycopene by cyclization of the two ends of the molecule is evident. β-Carotene, in which all the double bonds are in conjugation, is more abundant than α-carotene, in which one of the double bonds (the one in the right-hand ring above) is not conjugated with the rest. A molecule of vitamin A corresponds to half a molecule of carotene and is an alcohol. It is pale yellow, in comparison with the deep red-orange color of crystalline carotene. Animal feeding experiments show that carotenes are transformed into vitamin A in the bodies of most mammals but to differing extents. Lycopene and the carotenes are important nutrients in their own right, serving as antioxidants against the action of oxygen free radicals in the body (p. 794).

Vitamin A, also called **retinol,** is involved in the process of vision as its aldehyde with the cis configuration at the double bond at the 11 position, numbered as is traditional for these systems. This compound, (11Z)-retinal, is bonded with an imine linkage (p. 576) to a protein, opsin, in the visual pigment in the retina of the human eye. The complex molecule that results, called **rhodopsin,** absorbs light energy in the visible region of the spectrum. When light is absorbed, (11Z)-retinal undergoes isomerization to the all-trans configuration of retinal. In this form it can no longer stay bound to the protein and dissociates into (11E)-retinal and opsin, neither of which absorbs light in the visible region of the spectrum. This conversion is known as the bleaching of the visual pigment. (11E)-Retinal is reconverted to the cis form by an enzyme, binds once more to opsin, and the visual cycle begins again. Somehow the change in configuration of the double bond and the resulting dissociation of the visual pigment are translated by the retina into a message that travels through the optic nerve to the brain, to be interpreted there as sight and color vision.

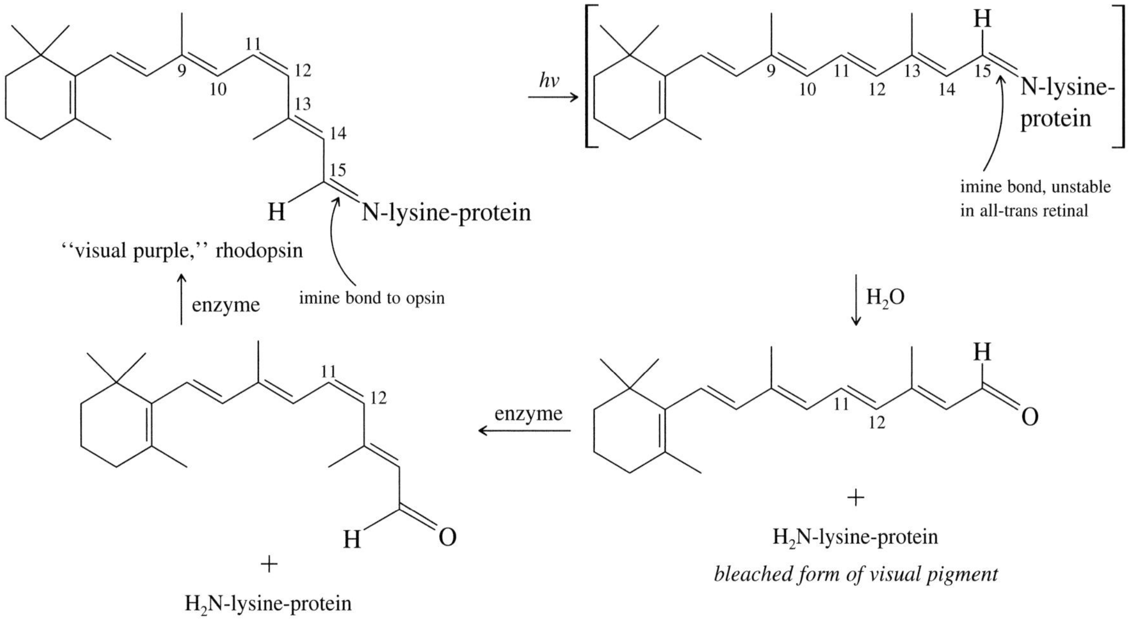

All the details of the changes that take place between the moment when light falls on the retina and the creation of a visual image are not yet understood. We can only wonder at the way nature uses simple chemical reactions in systems of such precise form that one stereochemical change in a small part of a large molecule sets in motion events of such consequence.

The absorption of light by conjugated systems such as vitamin A and how this can lead to isomerization of a double bond was discussed on page 72.

**Problem 18.15**   (a) Vitamin A, as the free alcohol, is sensitive to air. Its esters are more stable. It is found in fish oils as its hexadecanoate (palmitate) ester. What is the structure of this ester?
(b) Vitamin A is often sold as its acetate ester. Write an equation showing how the vitamin could be converted into its acetate ester in the laboratory.

**Problem 18.16**   β-Carotene is converted by monoperoxyphthalic acid into a mixture of a monoepoxide and a diepoxide. Which double bonds in β-carotene would be most vulnerable to oxidation by a peroxyacid? (*Hint:* See page 318.) Write structural formulas for the mono- and diepoxides of β-carotene.

**Table 18.1** **Summary of Reactions of Dienes**

### 1,2- and 1,4-Additions to Dienes

| Diene | Reagent | Intermediate | Product(s) |
|---|---|---|---|

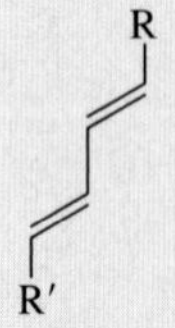

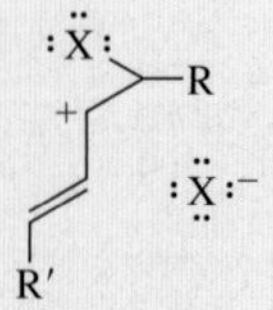

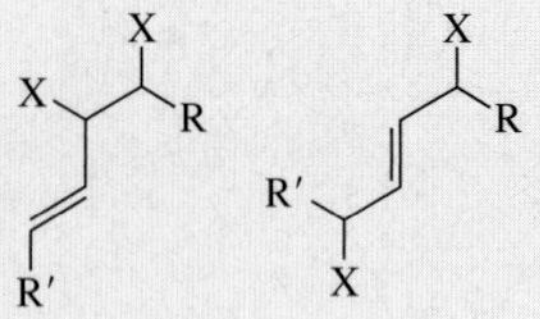

### Diels-Alder Reaction

| Diene | Dienophile | Product |
|---|---|---|

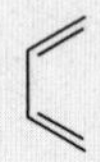

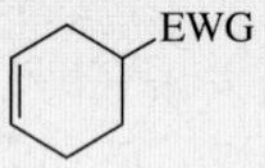

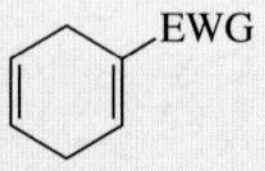

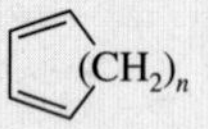

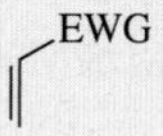

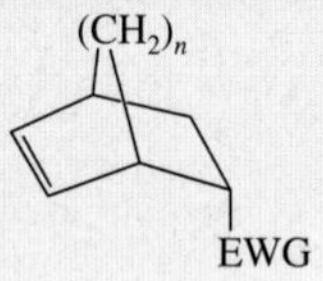

# ADDITIONAL PROBLEMS

**18.17** Give structural formulas for all products symbolized by letters in the following equations. If the stereochemistry of a reaction is known, be sure to show it in your answer.

(a) $A + B \longrightarrow$

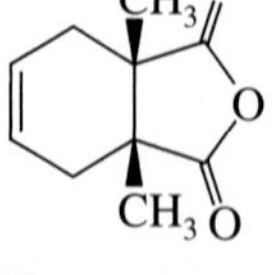

(b)

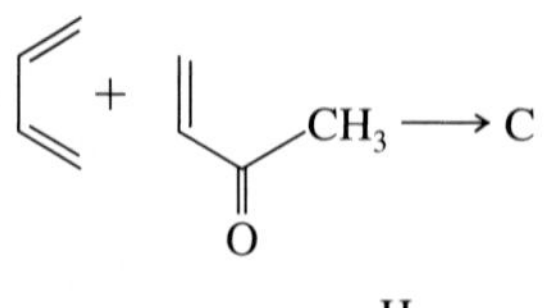

(c) 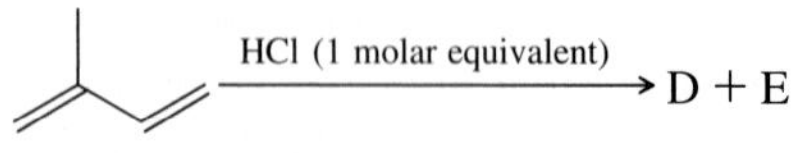 $\xrightarrow{\text{HCl (1 molar equivalent)}} D + E$

(d) $+ F \longrightarrow$

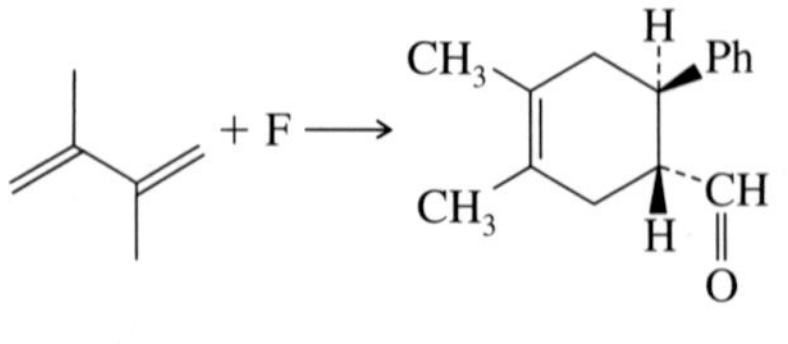

(e) $CH_3C{\equiv}CCH_2C{\equiv}CCH_2CH_3 \xrightarrow[\text{quinoline}]{\substack{H_2 \\ Pd/BaSO_4}} G$

(f)

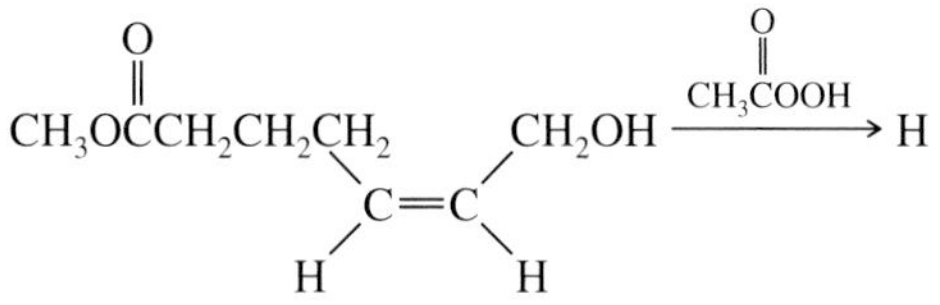

(g)

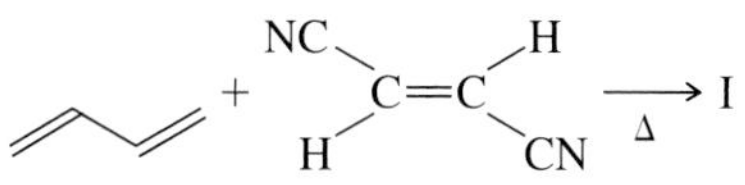

(h)

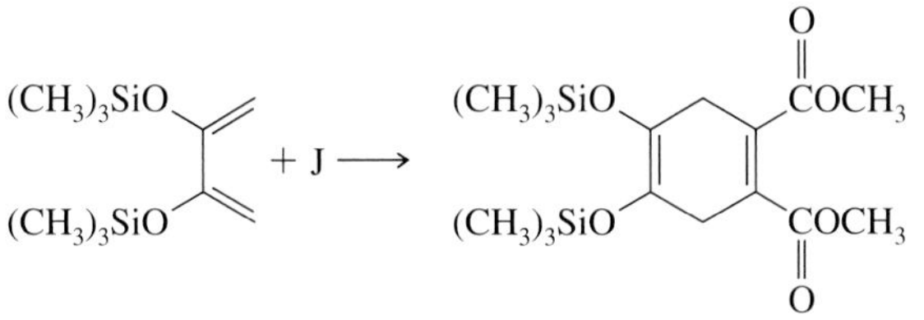

(i)

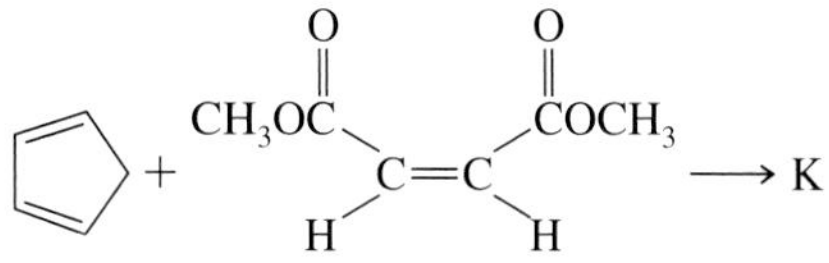

**18.18** Here are some more problems to practice on.

(a)

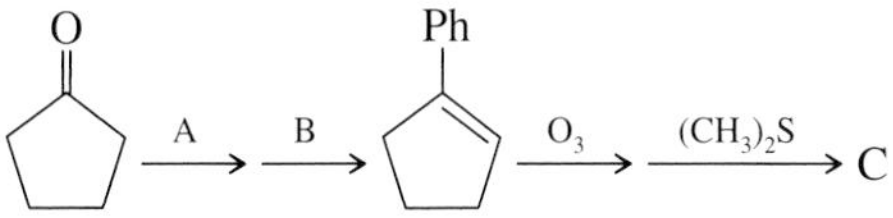

(b)

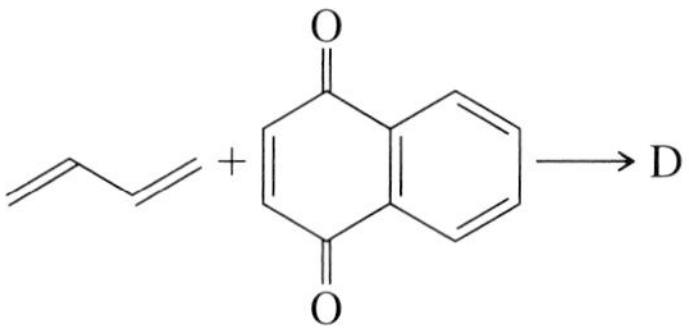

(c)

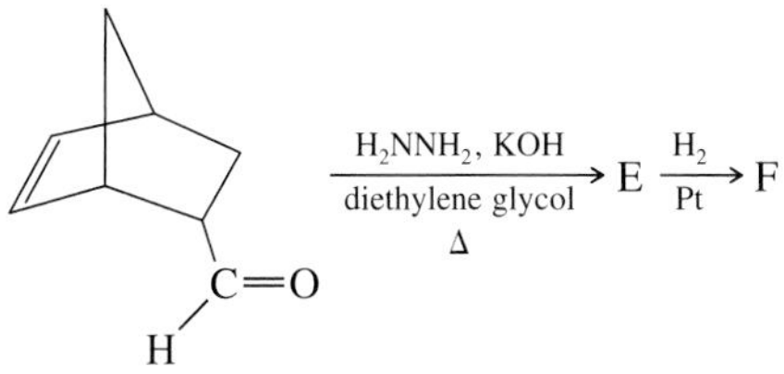

(d)

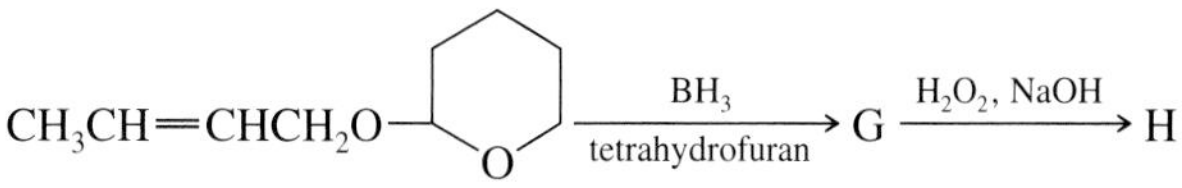

(e)

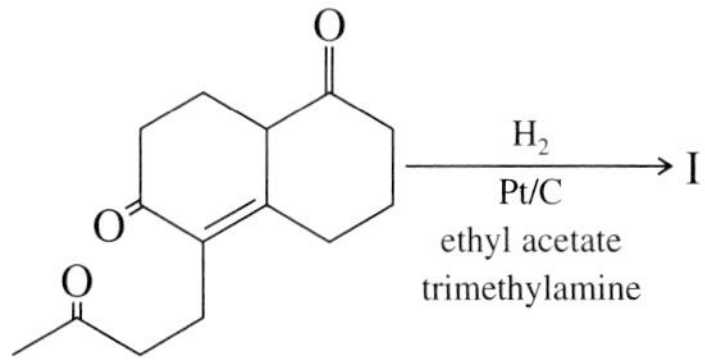

(f)

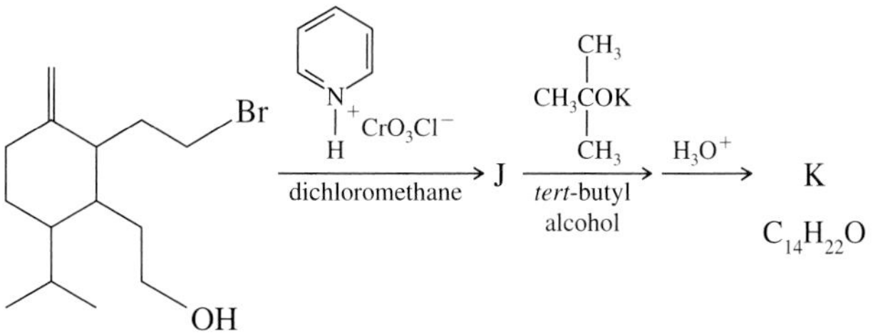

(g)

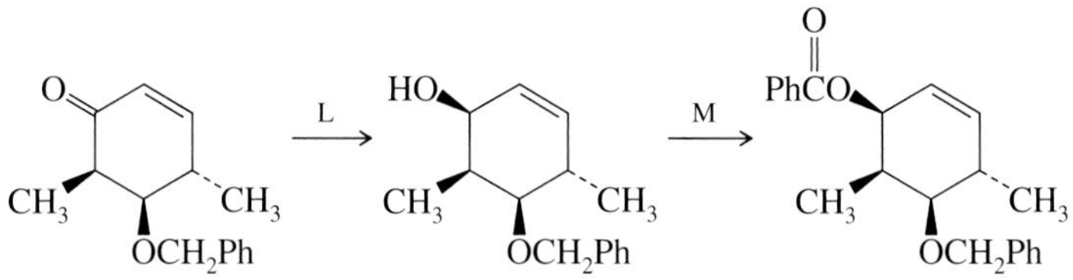

**18.19** Tell what reagents are necessary to carry out the following transformations. More than one step may be required in some cases.

(a)

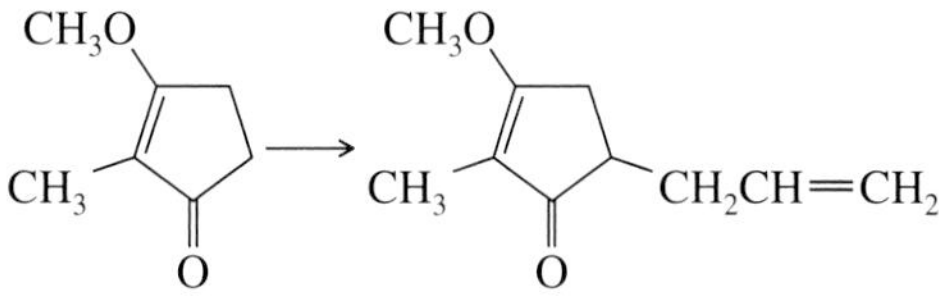

(b)

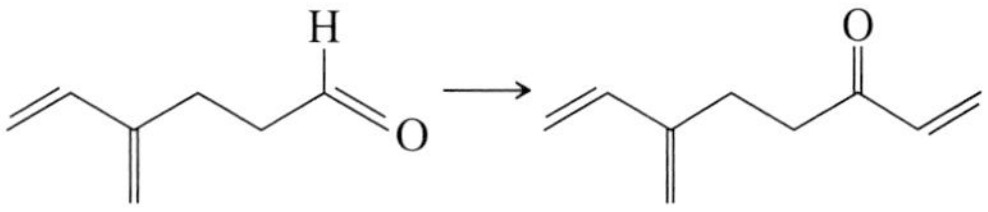

(c)

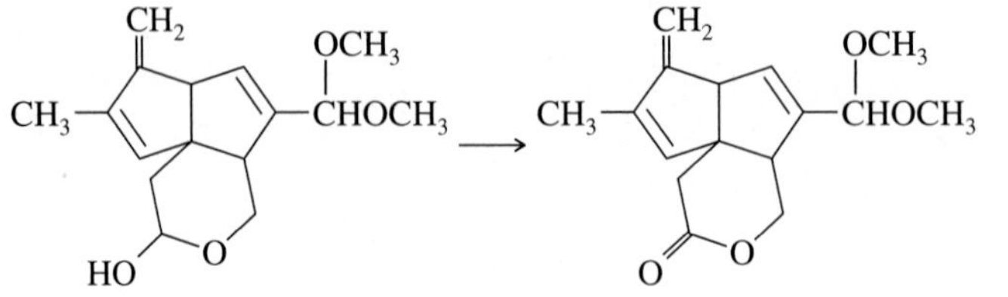

(d)

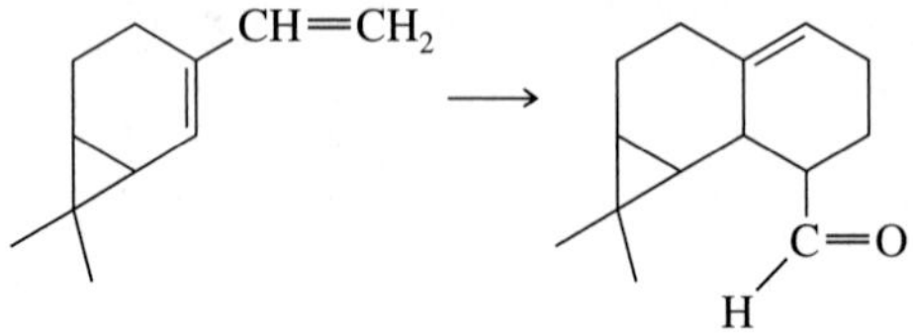

**18.20** The following compounds are synthesized by Diels-Alder reactions. Write equations for their syntheses, showing the diene and dienophile that would have to be used in each case.

(a) 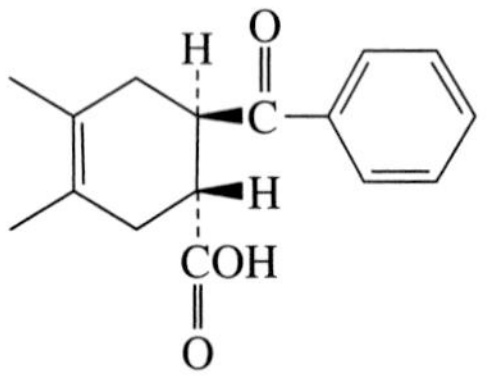    (b) 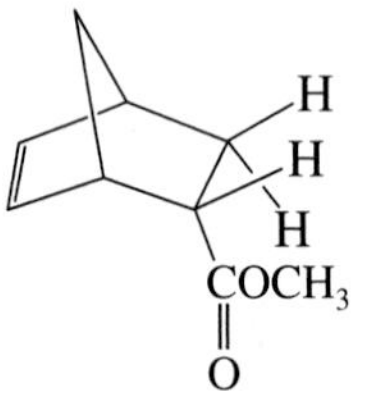    (c)

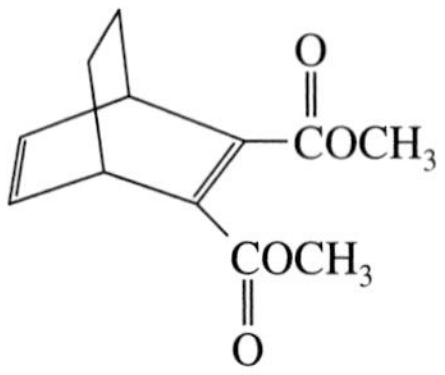

(d) 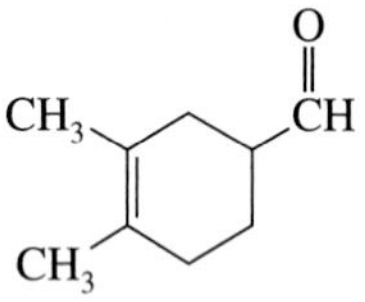    (e) 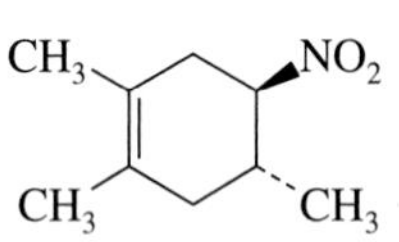    (f)

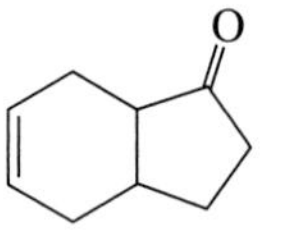

**18.21** The following reactions are part of the synthesis of a monoterpene. Provide structural formulas for the reagents or products designated by the letters.

(a) 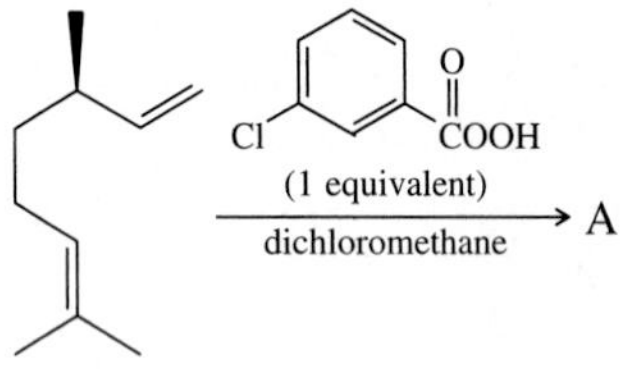    (b)

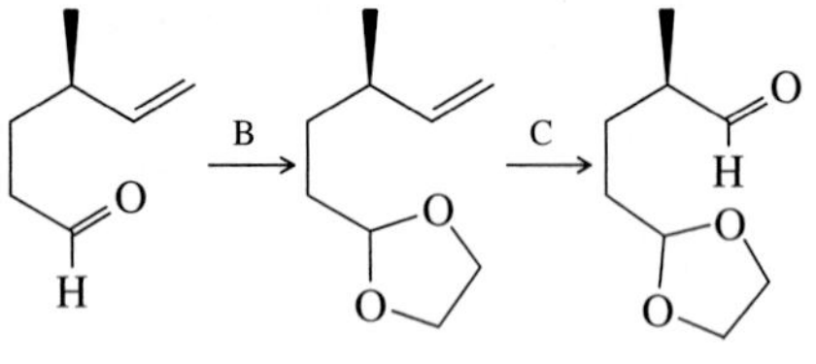

(c) 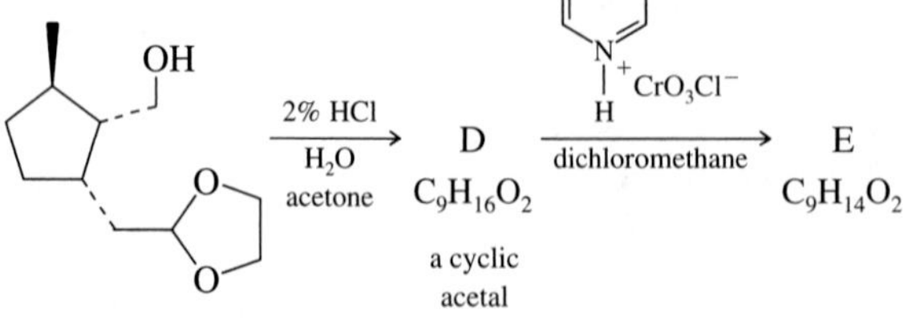    (d) 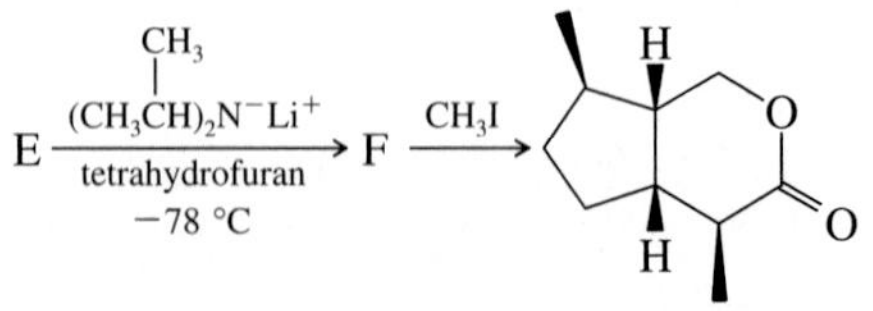 E $\xrightarrow[\text{tetrahydrofuran} \\ -78\ °C]{(CH_3CH)_2N^-Li^+}$ F $\xrightarrow{CH_3I}$

(+)-isoiridomyrmecin

**18.22** 1,3-Butadiene adds to maleic anhydride at a reasonable rate at 100 °C and slowly at room temperature. 2,3-Di-*tert*-butyl-1,3-butadiene does not react with maleic anhydride at all under these conditions. Offer an explanation for these experimental observations.

**18.23** The following reaction has been observed.

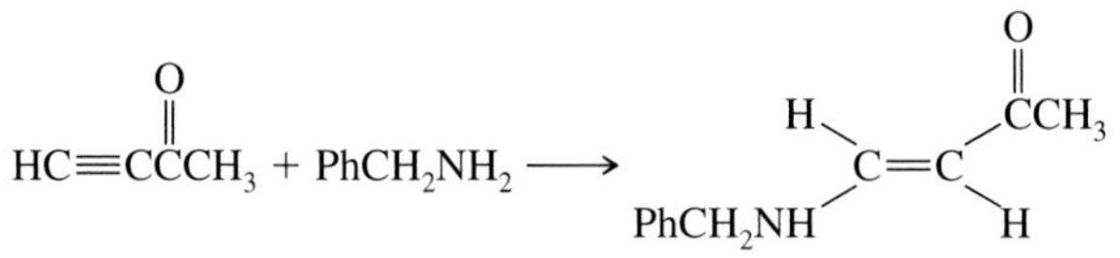
O
HC≡CCCH3 + PhCH2NH2
H
O
CCH3
C=C
PhCH2NH
H

Propose a mechanism for the reaction using the curved-arrow convention. Acids, H—B$^+$, and bases, B:, are available as needed.

**18.24** In order to study the enzyme that catalyzes the cyclization of farnesyl pyrophosphate in biological systems, an inhibitor of the enzyme had to be synthesized stereoselectively. Some of the steps of the synthesis are shown below. Provide structural formulas for the reagents or products designated by letters.

(a)

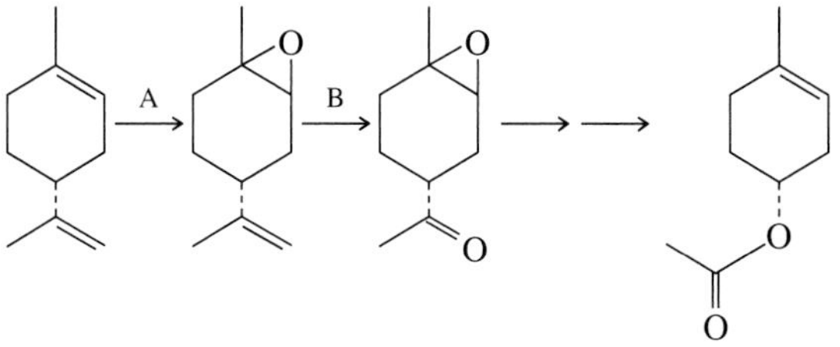
A
B
O
O
O
O
O

(*R*)-(+)-acetate
intermediate

(b)

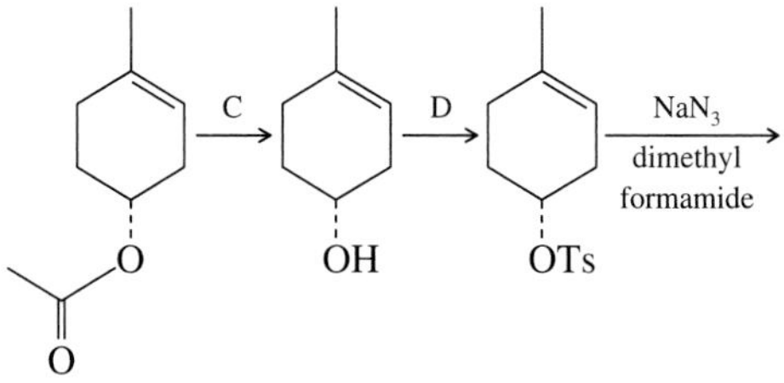
C
D
NaN3
dimethyl formamide
E
O
O
OH
OTs

(c)

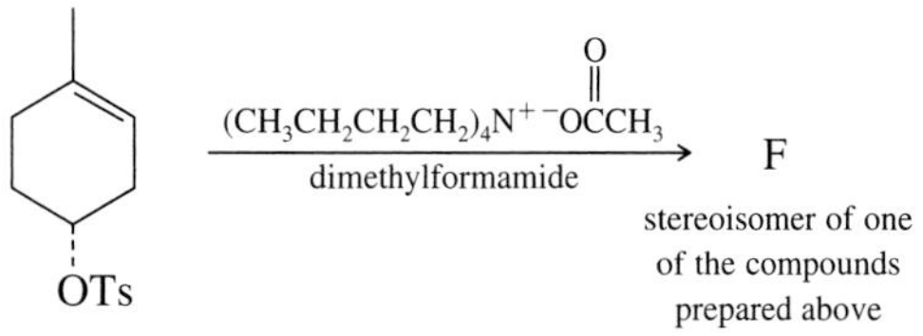
OTs
O
(CH3CH2CH2CH2)4N+ −OCCH3
dimethylformamide
F
stereoisomer of one of the compounds prepared above

**18.25** Predict the products of the following reactions of the terpene chrysanthemic acid and its esters by completing the equations.

(a)

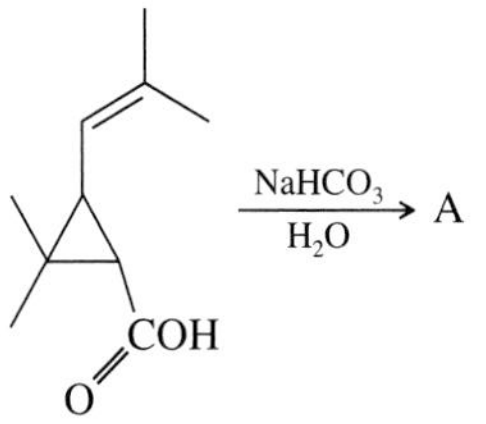
COH
O
NaHCO3
H2O
A

(b)

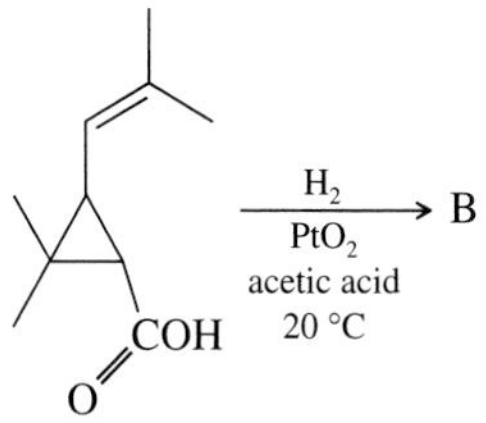
COH
O
H2
PtO2
acetic acid
20 °C
B

(c)

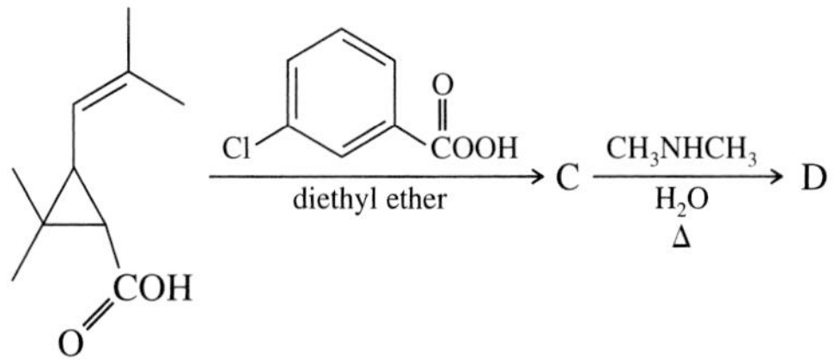
COH
O
Cl
O
COOH
diethyl ether
C
CH3NHCH3
H2O
Δ
D

(d)

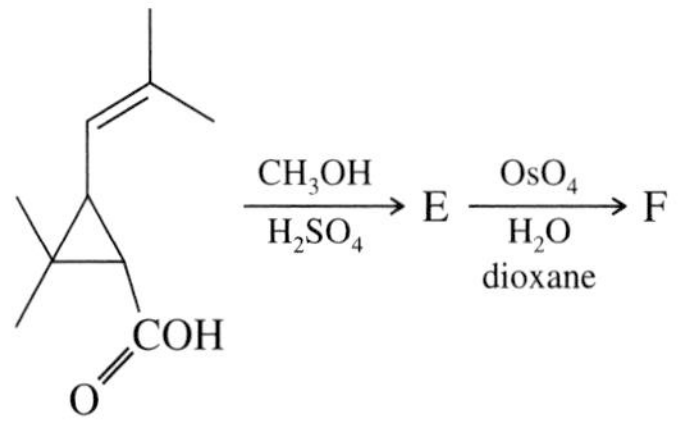
COH
O
CH3OH
H2SO4
E
OsO4
H2O
dioxane
F

(e) $E \xrightarrow{O_3} \xrightarrow[H_2O]{Zn} G$

**18.26**

(a) Bisabolene, $C_{15}H_{24}$, is found widely distributed in nature, especially in myrrh and oil of bergamot. Hydrogenation over platinum in acetic acid gives Compound X, $C_{15}H_{30}$. How many units of unsaturation does bisabolene have? How many of these are rings?

(b) Bisabolene undergoes incomplete hydrogenation in cyclohexane to give Compound Y, $C_{15}H_{28}$. Ozonolysis of Y gives 6-methyl-2-heptanone and 4-methylcyclohexanone. What is the structure of Compound Y?

(c) Ozonolysis of bisabolene gives, among other products, acetone and 4-oxopentanoic acid. Assign a partial structure on the basis of these data. What uncertainty is there about the structure of bisabolene at this point?

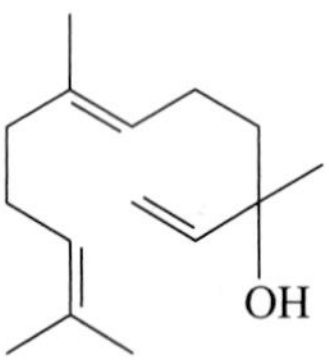

(d) Bisabolene and an alcohol derived from bisabolene are among the products obtained from the acid-catalyzed cyclization of nerolidol, obtained from the flowers of bitter orange. The structural formula for nerolidol is shown at left. Propose a mechanism for an acid-catalyzed cyclization of nerolidol to bisabolene. What does this, in combination with the data given above, suggest about the structure of bisabolene?

(e) What is the structure of the bisabolol also formed in the acid-catalyzed cyclization of nerolidol?

(f) Bisabolene reacts easily with hydrogen chloride to give Compound Z, $C_{15}H_{27}Cl_3$. Compound Z is also formed when the mixture of bisabolene and bisabolol from nerolidol is treated with hydrogen chloride. What is the structure of Compound Z? Write equations showing its formation from bisabolene and bisabolol.

**18.27** The monoepoxide of β-carotene (Problem 18.16, p. 759) rearranges in dilute hydrochloric acid to a compound having the structure shown below. Propose a mechanism for its formation from β-carotene monoepoxide.

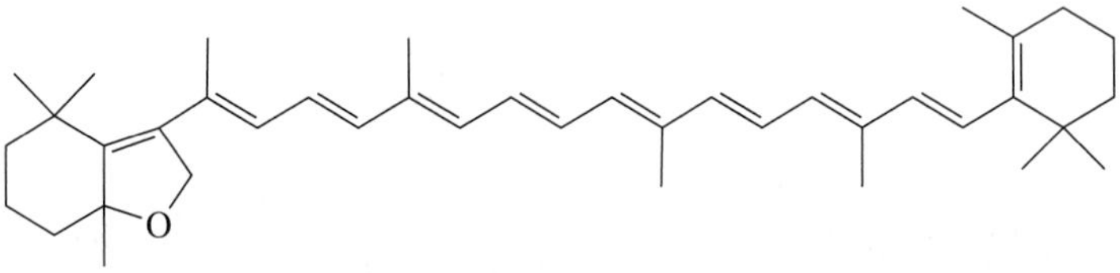

**18.28** Steroids, because of the rigidity of their fused six-membered rings, have been used to study the stereochemistry of many reactions.

(a) Predict which bromonium ion (p. 303) is the intermediate in the addition of bromine to cholesterol (p. 757). What is the stereochemistry of the dibromide that forms? (*Hint:* It may help you to review Section 13.5C.)

(b) Two stereoisomeric bromohydrins derived from a steroid react with base at different rates to give oxiranes (p. 500). Partial structures, A and B, for the two compounds are given below. Predict which compound produces a high yield of oxirane in 30 seconds and which one requires 24 hours to give the same yield. How do you account for the difference in reactivity between the two isomers?

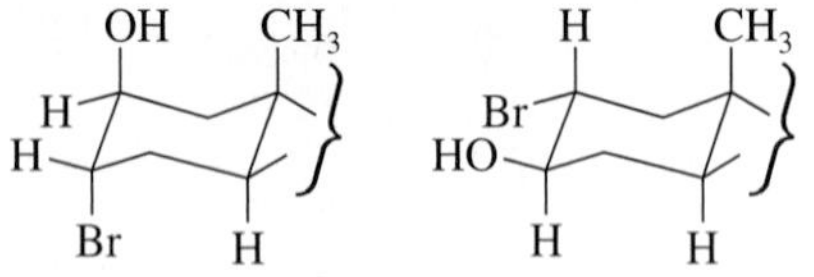

**18.29** Helenalin, a natural product isolated from *Helenium autumnale*, sneezeweed, is of interest because it is cytotoxic. Some steps in a recent synthesis of the compound are shown below. What reagents could be used for these transformations?

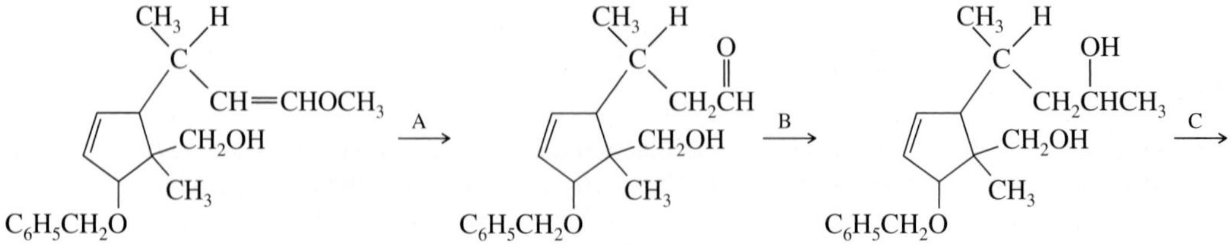

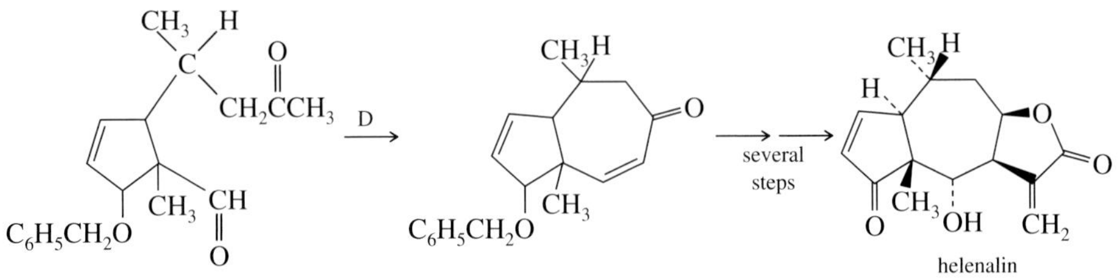

helenalin

(*Hint:* A review of Section 17.4C will be helpful here.)

**18.30** Santonin is a natural product from various species of *Artemisia.* It is used to kill intestinal worms. Use the structure of santonin shown below to decide whether it is a terpene that follows the isoprene rule. If you decide that it does, show clearly the isoprene units in it. If you decide that it does not, show where the rule breaks down.

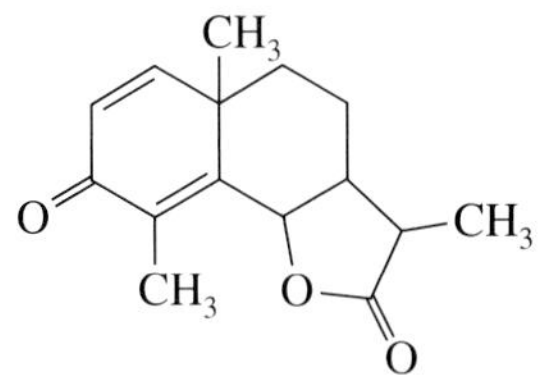

santonin

**18.31** Borjatriol is a compound with anti-inflammatory and antirheumatic properties. Following are some steps in a synthesis of the compound. Supply reagents for the transformations shown.

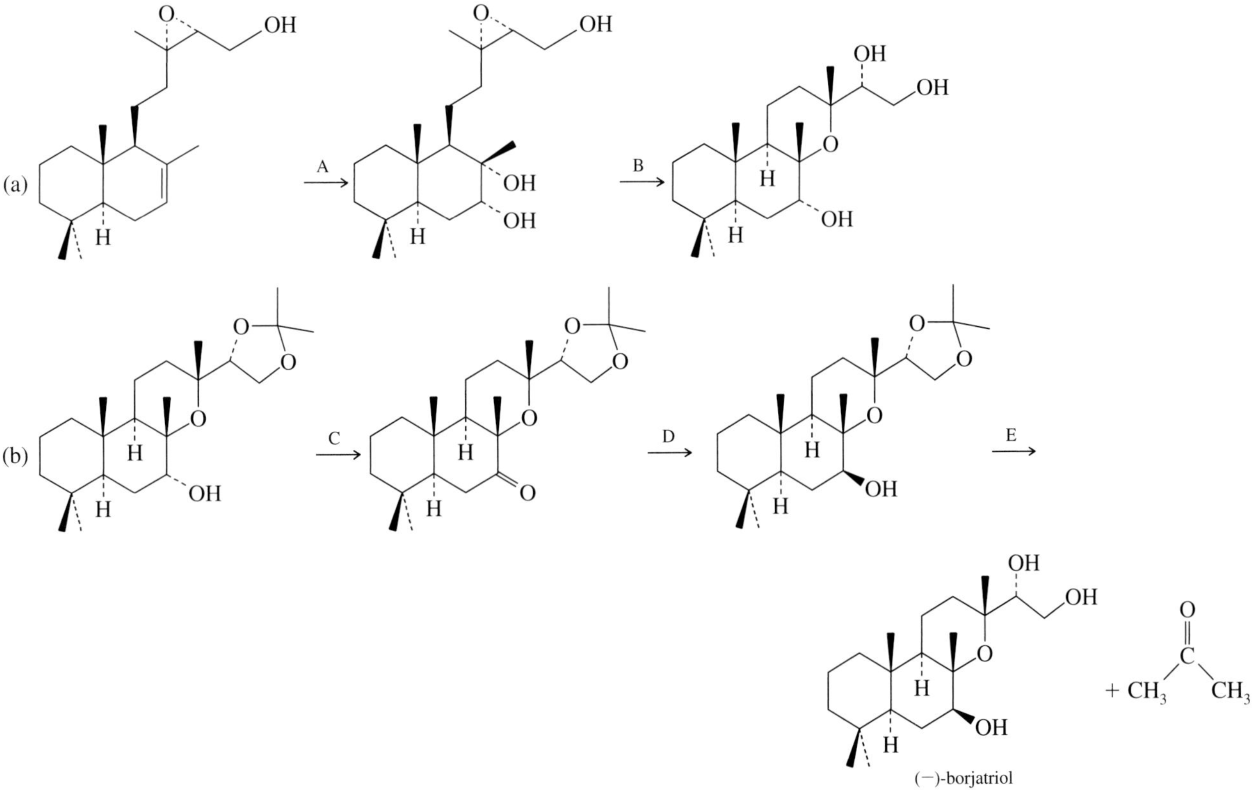

(−)-borjatriol

Draw a structural formula that shows the conformation of (2)-borjatriol. The conformational structure of cholestane (p. 755) will be helpful as a reminder of how fused six-membered rings can be drawn.

**18.32** A natural product, guaiol, is isolated from guaicum resin. When guaiol is heated with sulfur, it is dehydrated and dehydrogenated to a hydrocarbon called guaiazulene.

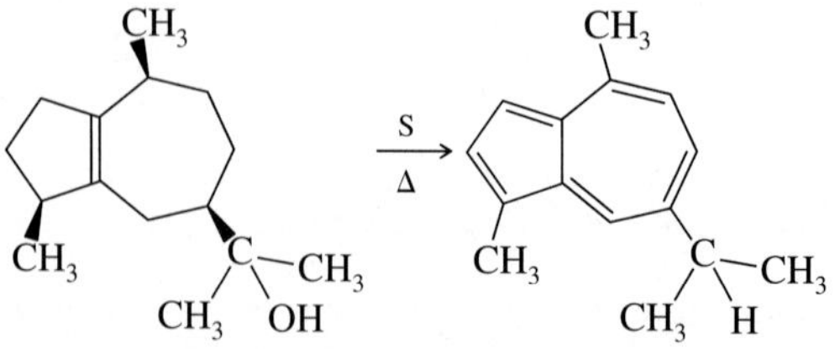

guaicol                              guaiazulene

(a)  Is guaiol a terpene? Justify your answer by showing whether the structure of guaiol was formed in accordance with the isoprene rule.
(b)  How would you rationalize the easy conversion of guaiol to guaiazulene? (*Hint:* A review of Section 10.1B and Problem 10.29 may be helpful.)

**18.33** Isopentenyl pyrophosphate (p. 752) is synthesized biologically from acetyl coenzyme A. Several of the steps at the beginning of this biosynthetic pathway are shown below. Identify the type of reaction that is taking place at each stage, and write a mechanism for it. You may assume that acids, H—B$^+$, and bases, B:, are available as needed.

(a) $CH_3CSCoA + CH_3CSCoA \longrightarrow CH_3CCH_2CSCoA + HSCoA$

(b) $CH_3CCH_2CSCoA + CH_3CSCoA \longrightarrow CoASCCH_2CCH_2CSCoA$
$\qquad\qquad\qquad\qquad\qquad\qquad\qquad\qquad\qquad\qquad OH$

(c) $CoASCCH_2CCH_2CSCoA + H_2O \longrightarrow CoASH + HOCCH_2CCH_2CSCoA$
$\qquad\qquad OH \qquad\qquad\qquad\qquad\qquad\qquad\qquad\qquad\qquad\qquad OH$

**18.34** Lithium in liquid ammonia behaves the way sodium in liquid ammonia does to add hydrogen to alkynes (p. 340, also Problem 13.46). Hydrogen generated by this procedure will also add 1,4- to conjugated double bonds.

When $C_{60}$ (p. 364) is treated with lithium in liquid ammonia and *tert*-butyl alcohol, the dark brown color of the solution of $C_{60}$ disappears and a new compound that is light cream in color is formed. The mass spectrum has a molecular ion at $m/z$ 756. The new compound has broad multiplets in its proton magnetic resonance spectrum at $\delta$ 3.80 and $\delta$ 3.35. The infrared spectrum of $C_{60}$ has no significant bands above 1428.5 cm$^{-1}$; the new compound absorbs at 2925, 2855, 1620, 1450, 1400, and 675 cm$^{-1}$. While $C_{60}$ absorbs in the ultraviolet at 218, 258, 330, and 378 nm, the new compound has absorption only at 222 nm.

Use the spectral data to suggest what happened when $C_{60}$ is treated with lithium in liquid ammonia.

**18.35** The following reactions were carried out in the synthesis of terpenoids. Provide structural formulas for the reagents designated by letters, and show how the spectral data support the structure of the final product.

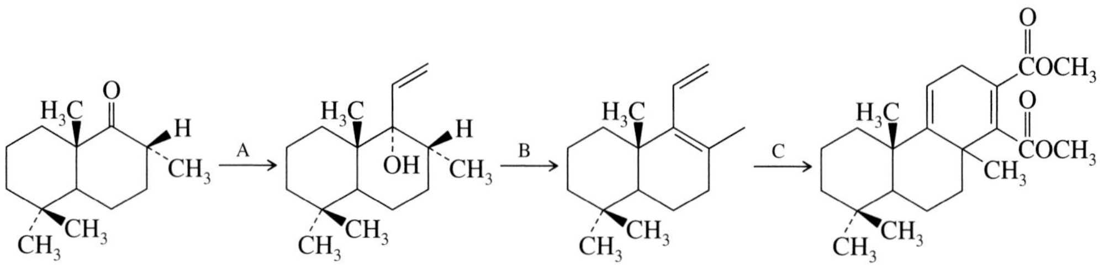

$V_{max}$ 1720, 1660, 1630 cm$^{-1}$
partial $^1$H NMR δ 5.64 (dd, 1H), 3.81 (s, 3H),
3.73 (s, 3H), 3.14 (dd, 1H), 2.77 (dd, 1H),
1.31 (s, 3H), 1.05 (s, 3H), 0.91 (s, 3H), 0.85 (s, 3H)

# 19 Free Radicals

## A Look Ahead

A free radical is a reactive intermediate with an unpaired electron. Free radicals are important in many reactions. Chapter 5 introduced them briefly as the reactive intermediates involved in the halogenation reactions of alkanes (p. 185). The structure of a carbon radical was compared with those of a carbocation and of a carbanion. A radical has an incomplete octet and no charge. Halogen atoms, with seven electrons in the valence shell, may be considered to be radicals.

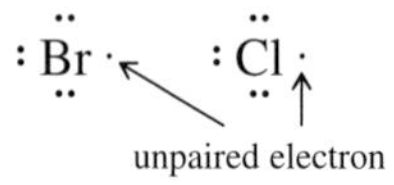

$$: \ddot{Br} \cdot \qquad : \ddot{Cl} \cdot$$

unpaired electron

The oxygen molecule, $O_2$, is a diradical in its ground state. Many important consequences follow from this fact.

A radical behaves as an electrophile that is seeking only a single electron. This electron is often obtained from a sigma bond, usually a carbon–hydrogen bond, in a hydrogen-abstraction reaction.

fishhook symbolizing transfer of single electron

halogen atom with unpaired electron

*bromine atom abstracting a hydrogen atom*

methyl radical

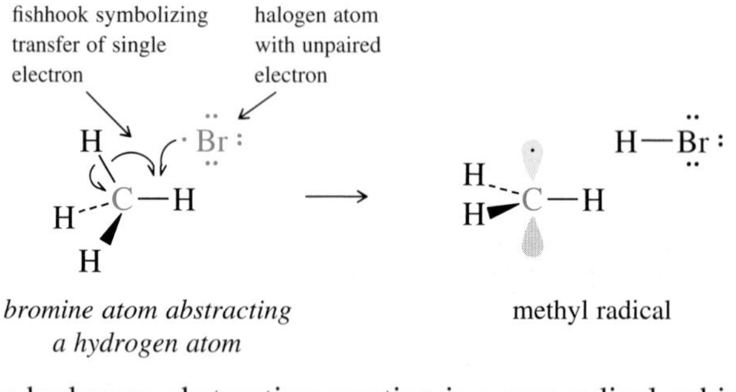

The product of a hydrogen-abstraction reaction is a new radical, which can itself abstract an atom. For example, the methyl radical may abstract a bromine atom from a molecule of bromine, forming bromomethane and a bromine atom.

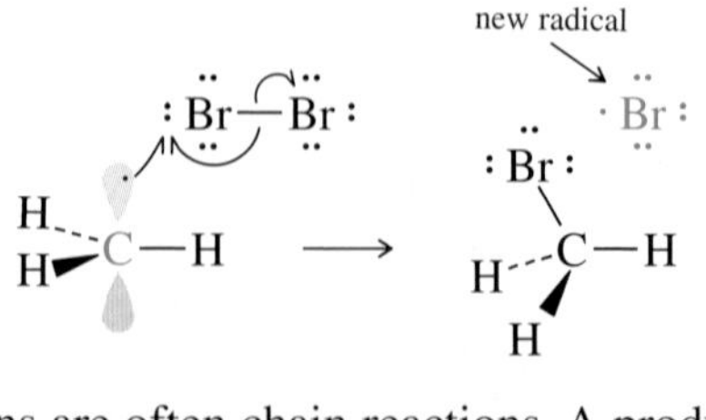

new radical

Free-radical reactions are often chain reactions. A product of one step of the reaction is a reactant in the next step. For example, the bromine atom formed in the second reaction above is a reactant in the first reaction shown. It will react with another molecule of methane to keep the sequence going.

This chapter will discuss the reactions of radicals and will examine some of the practical consequences of their reactivity.

# 19.1 Free-Radical Reactions of Alkanes

## A. Reactions of Chlorine with Alkanes

Chlorine or bromine will react with alkanes in the presence of light to give alkyl halides. For example, methane reacts with chlorine to give a mixture of chlorinated methanes.

$$CH_4 \ + \ Cl_2 \ \xrightarrow{hv} \ CH_3Cl \ + \ CH_2Cl_2 \ + \ CHCl_3 \ + \ CCl_4$$

methane   chlorine     chloro-    dichloro-    trichloro-   tetrachloro-
                        methane    methane      methane      methane

The composition of the product mixture depends on the ratio of starting materials and the temperature, but even when large excesses of the alkane are used, mixtures are formed.

When an alkane with different types of hydrogen atoms is chlorinated, any of the hydrogens will be substituted. For example, when 2-methylbutane is chlorinated at 300 °C, the following mixture is observed.

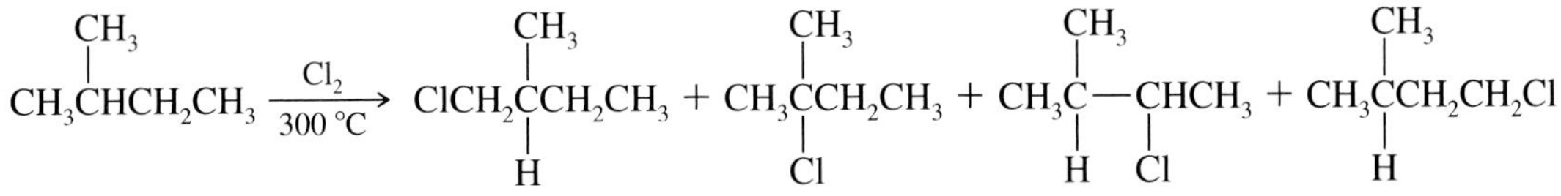

2-methylbutane      1-chloro-2-        2-chloro-2-        3-chloro-3-        4-chloro-2-
                    methylbutane       methylbutane       methylbutane       methylbutane
                    33.5%              22%                28%                16.5%

If the relative numbers of different types of hydrogen atoms are taken into account, these experimental results reveal that a tertiary hydrogen atom is about 4 times more likely to be replaced than a primary one, and a secondary hydrogen is about 2.5 times as reactive as a primary one.

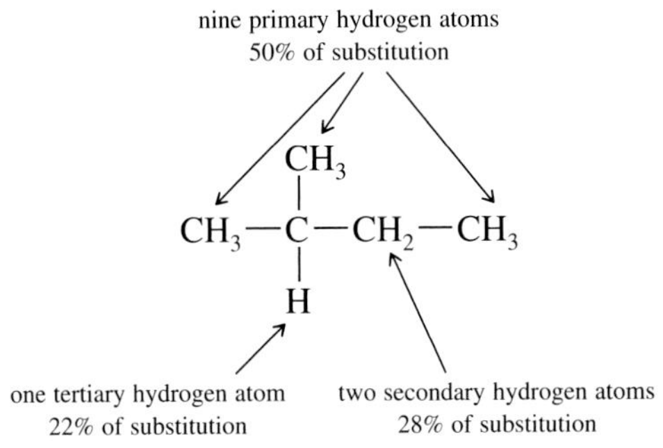

nine primary hydrogen atoms
50% of substitution

one tertiary hydrogen atom     two secondary hydrogen atoms
22% of substitution             28% of substitution

$$\frac{tertiary}{primary} = \frac{22/1}{50/9} = \frac{4}{1}$$

$$\frac{secondary}{primary} = \frac{28/2}{50/9} = \frac{2.5}{1}$$

## B. Free-Radical Chain Reactions

Reactions in which the product of one step is a reactant in the next step are known as **chain reactions.** All chain reactions have three steps: (1) initiation, in which the first reactive intermediate is formed and the chain starts; (2) propagation, in which the chain is continued many times; and (3) termination, in which the chain stops. Research has shown that the reactions of halogens with alkanes proceed via free-radical intermediates. Free radicals are species having an unpaired electron. Highly reactive intermediates, they are involved in oxidation reactions, combustion reactions, and many biological reactions, some of which will be investigated later in this chapter (pp. 786–790).

The halogenation of an alkene is a **free-radical chain reaction.** Chain reactions also occur with other types of reactive intermediates such as anions and cations. Such ionic chain reactions are particularly important in the synthesis of polymers (Sections 24.2C and 24.2D).

On absorbing light energy, the halogen molecule dissociates into two halogen atoms in the initiation step of the chain reaction, leading to the substitution of halogen atoms for hydrogen atoms in alkanes.

**VISUALIZING THE REACTION**

**Initiation**

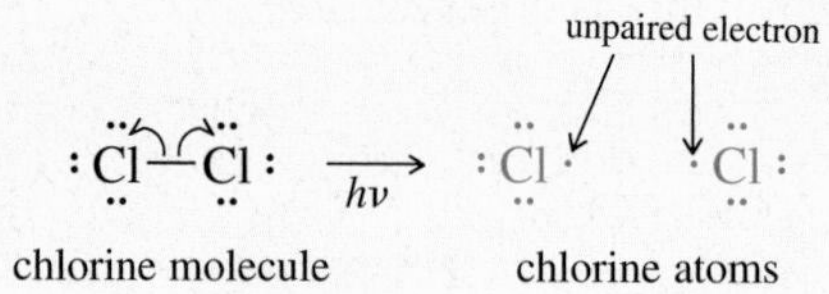

chlorine molecule → chlorine atoms

This bond cleavage is a homolytic cleavage (p. 62) in which one electron of the covalent bond goes to each atom. This cleavage is symbolized using fishhooks (p. 341).

A chlorine atom is highly reactive because of the presence of an unpaired electron in its outermost shell. It is electrophilic, seeking a single electron to complete the octet. It acquires this electron by abstracting a hydrogen atom from the alkane, methane, for example.

**VISUALIZING THE REACTION**

**Propagation**

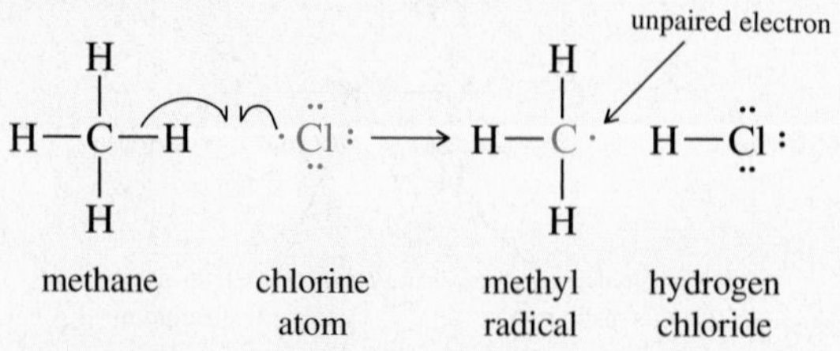

methane    chlorine atom    methyl radical    hydrogen chloride

This reaction gives rise to a new electrophilic species, the methyl radical, which has an unpaired electron. This is a propagation step of the chain reaction. In a second propagation step, the methyl radical abstracts a chlorine atom from a chlorine molecule.

**VISUALIZING THE REACTION**

**Propagation**

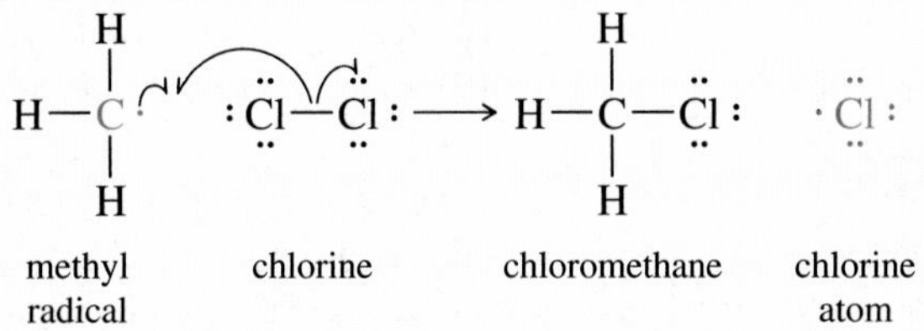

| methyl radical | chlorine | chloromethane | chlorine atom |

The products of this reaction are chloromethane and a chlorine atom, which abstracts a hydrogen atom from another molecule of methane and keeps the chain going. Free radicals are highly reactive intermediates that are involved in oxidation reactions, combustion reactions, and many biological reactions.

The chain reaction shown above has the following steps:

1. **Initiation.** For the free-radical halogenation reaction, formation of halogen atoms by the dissociation of a halogen molecule is the initiation step.

2. **Propagation.** The abstraction of a hydrogen atom by the halogen atom and the abstraction of a halogen atom by the methyl radical are the propagation steps of the free-radical halogenation reaction. For the chlorination of an alkane, about 10,000 propagation steps occur for each initiation step.

3. **Termination** consists of steps that occur when two free radicals happen to collide with each other and form a covalent bond, stopping the chain of reactions. Possible termination steps for the halogenation of an alkane are shown below. In each of these termination reactions, two radicals are destroyed, which stops two chains. However, the concentration of radicals in the reaction mixture is low compared to the concentrations of the other reagents, so the odds greatly favor collision between a radical and a molecule of reagent over collision between two radicals. Propagation steps are therefore much more common than termination steps.

**VISUALIZING THE REACTION**

**Termination**

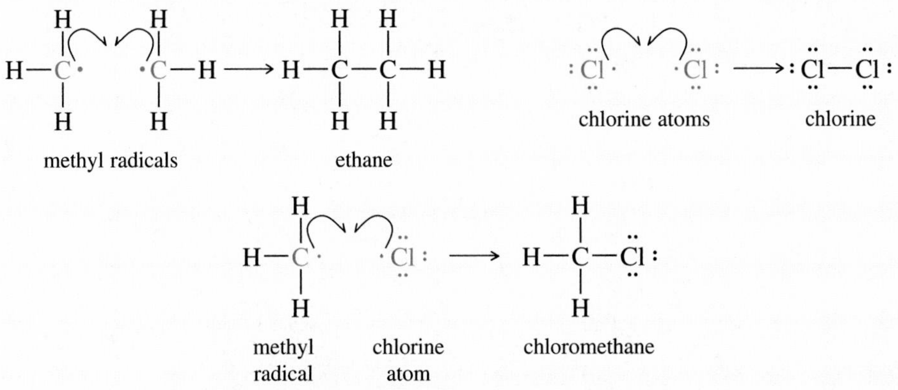

Radicals are formed in initiation steps. The number of radicals stays constant in propagation steps. Radicals are destroyed in termination steps.

Reactions that occur by way of free-radical intermediates are different in many ways from those that go by way of ionic intermediates. Because free-radical intermediates do not have positive or negative charges, their reactions are not seriously affected by solvent polarity. However, the rates of the reactions are affected by substances, called **inhibitors,** that react with free radicals and thus act to terminate chains. The chlorination reaction, for example, is sensitive to the presence of oxygen, which is a diradical (p. 786). No rearrangements are observed for free-radical chain reactions, as they are for reactions that proceed by carbocation intermediates (p. 289). All these facts make it clear that the intermediates involved in free-radical chain reactions are different from those involved in other substitution reactions.

⊘ **Study Guide**
**Concept Map 19.1**

## C. The Selectivity of Chlorination Reactions

The selectivity observed for the chlorination of 2-methylbutane (p. 769) can be explained by an examination of the relative stabilities of the different alkyl radicals that can be intermediates. The relative stabilities of radicals parallel the stabilities of the corresponding carbocations (pp. 129 and 185). A tertiary alkyl radical is more stable than a secondary one, which in turn is more stable than a primary one. For radicals formed from 2-methylbutane, for example, the relative stabilities are as follows:

$$CH_3\overset{\underset{\displaystyle CH_3}{|}}{\underset{.}{C}}CH_2CH_3 > CH_3\overset{\underset{\displaystyle CH_3}{|}}{CH}\underset{.}{C}HCH_3 > \cdot CH_2\overset{\underset{\displaystyle CH_3}{|}}{CH}CH_2CH_3, \quad CH_3\overset{\underset{\displaystyle CH_3}{|}}{CH}CH_2\underset{.}{C}H_2$$

tertiary radical      secondary radical           primary radicals

The ease with which the different types of hydrogen atoms are abstracted reflects the bond dissociation energies for the different carbon–hydrogen bonds (p. 63). An examination of the energy changes that occur during the abstraction of a hydrogen atom by a chlorine atom illustrates the differences.

$$CH_3\overset{\underset{\displaystyle CH_3}{|}}{CH}CH_2CH_2{-}H + \cdot \ddot{C}l\!: \longrightarrow CH_3\overset{\underset{\displaystyle CH_3}{|}}{CH}CH_2\underset{.}{C}H_2 + HCl$$

$DH°$ \qquad 98 kcal/mol \hspace{3cm} primary radical \qquad 103 kcal/mol

$$\Delta H_r = -5 \text{ kcal/mol}$$

$$CH_3\overset{\underset{\displaystyle CH_3}{|}}{\underset{\underset{\displaystyle H}{|}}{CH}}CHCH_3 + \cdot \ddot{C}l\!: \longrightarrow CH_3\overset{\underset{\displaystyle CH_3}{|}}{CH}\underset{.}{C}HCH_3 + HCl$$

$DH°$ \qquad 95 kcal/mol \hspace{3cm} secondary radical \quad 103 kcal/mol

$$\Delta H_r = -8 \text{ kcal/mol}$$

$$CH_3\overset{\underset{\displaystyle CH_3}{|}}{\underset{\underset{\displaystyle H}{|}}{C}}CH_2CH_3 + \cdot \ddot{C}l\!: \longrightarrow CH_3\overset{\underset{\displaystyle CH_3}{|}}{\underset{.}{C}}CH_2CH_3 + HCl$$

$DH°$ \qquad 91 kcal/mol \hspace{3cm} tertiary radical \qquad 103 kcal/mol

$$\Delta H_r = -12 \text{ kcal/mol}$$

In each case, enough energy must be supplied to break a carbon–hydrogen bond, and energy is recovered from the formation of a hydrogen–chlorine bond. The dif-

ference between these two energies is the enthalpy of each reaction. The three preceding reactions involving a chlorine atom are all exothermic, but the amount of energy given off as heat increases going from the abstraction of a primary hydrogen
atom to that of a tertiary one. The hydrogen-abstraction reactions also have
a small energy of activation, ranging from 3.8 kcal/mol for methane to about
1 kcal/mol for the abstraction of primary hydrogens and about 0.7–0.9 kcal/mol for
the abstraction of secondary and tertiary hydrogens. The difference between the
energy of activation for the formation of the primary radical and that for the formation of the tertiary radical is also small. Therefore, the regioselectivity of the
hydrogen-abstraction reaction of chlorine atoms is low.

**Problem 19.1**   Assume that the relative reactivities determined for the reaction of 2-methylbutane with chlorine at
300 °C hold for other alkanes. What kind of product mixture
would result from the chlorination of propane and of 2-
methylpropane under the same conditions?

**Problem 19.2**   The regioselectivity of the chlorination reaction is dependent on temperature. At 600 °C, the relative reactivities for primary, secondary, and tertiary hydrogens are
$1:2.1:2.6$, instead of $1:2.5:4$ as observed at 300 °C. How
would you explain this experimental result?

## D. Reaction of Bromine with Alkanes

Bromination reactions are much more selective than chlorination reactions. For example, when 2-methylpropane is treated with bromine in the presence of light at
127 °C, the product is almost exclusively 2-bromo-2-methylpropane.

|  2-methylpropane  |  2-bromo-2-methylpropane $99 + \%$  |  1-bromo-1-methylpropane trace  |

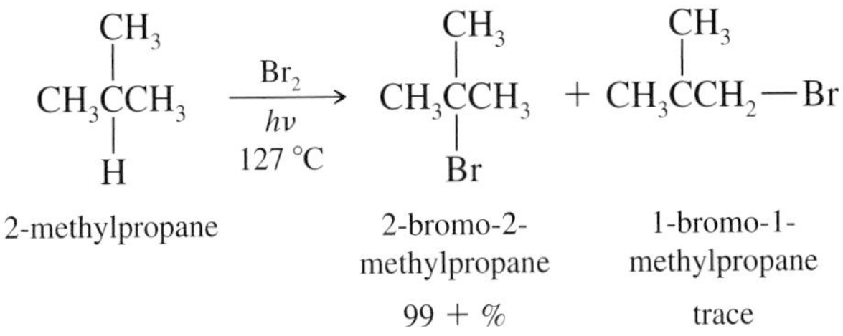

When the above bromination reaction is carried out in a reaction vessel in which butane is also present as a reactant, the relative reactivities for primary, secondary, and
tertiary hydrogen atoms are $1:82:1640$.
   The bond dissociation energies and enthalpies for the abstractions of primary
and tertiary hydrogens by bromine are shown below.

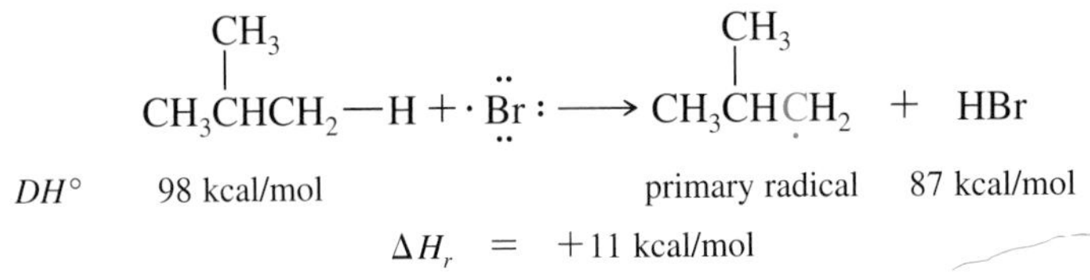

$DH°$   98 kcal/mol     primary radical   87 kcal/mol

$$\Delta H_r = +11 \text{ kcal/mol}$$

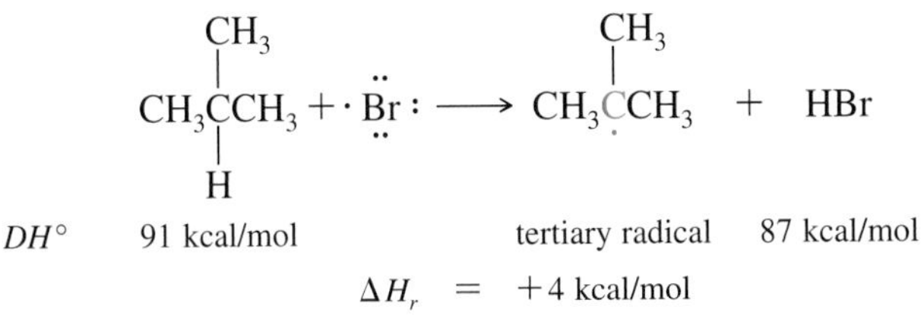

$DH°$   91 kcal/mol     tertiary radical   87 kcal/mol

$$\Delta H_r = +4 \text{ kcal/mol}$$

Both reactions are endothermic. The first reaction has an energy of activation of at
least 11 kcal/mol, and that for the formation of the tertiary radical is at least
4 kcal/mol. More important, the difference between the two is at least 7 kcal/mol,

📌 Study Guide
Concept Map 19.2

leading to a significant difference in the rates of abstraction of primary and tertiary hydrogen atoms. The high regioselectivity of the bromination of alkanes is derived from the differing endothermicity of the hydrogen-abstraction reactions with bromine atoms.

**Problem 19.3**　Neopentane (2,2-dimethylpropane) reacts very slowly with bromine in the presence of light even at relatively high temperatures but reacts easily with chlorine under the same conditions. How would you explain this difference in reactivity?

## E. Free-Radical Chain Reactions of Halocarbons with Ozone

Investigation of the chemistry of the processes that lead to the destruction of ozone, beyond the equilibrium processes described in Section 8.9A, shows that halogen atoms, especially chlorine and bromine atoms, are responsible. Chlorine atoms are generated in the stratosphere by homolytic bond cleavage reactions of chlorofluorocarbons (or CFCs), compounds that have been used in air conditioners and refrigerators for years. When released into the atmosphere, these compounds, which were chosen for their lack of reactivity and low toxicity, survive long enough to reach the stratosphere. There they decompose under the influence of ultraviolet radiation. A typical chlorofluorocarbon is trichlorofluoromethane, also known as Freon-11 or CFC-11. In the stratosphere, trichlorofluoromethane breaks down to give two radicals.

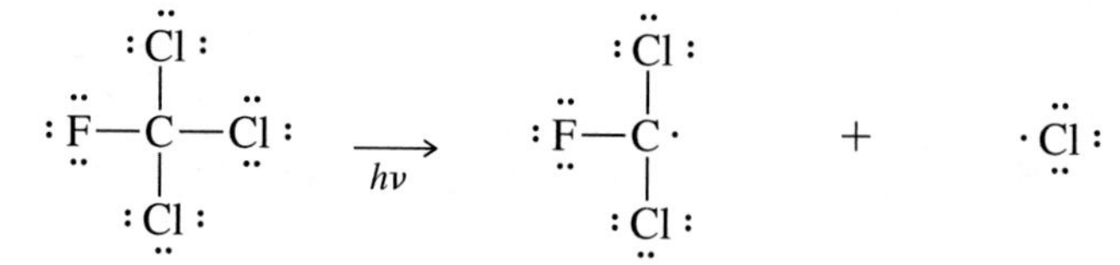

$$:\overset{..}{\underset{..}{Cl}}: \\ | \\ :\overset{..}{\underset{..}{F}}-C-\overset{..}{\underset{..}{Cl}}: \\ | \\ :\overset{..}{\underset{..}{Cl}}: \quad \xrightarrow{h\nu} \quad :\overset{..}{\underset{..}{Cl}}: \\ | \\ :\overset{..}{\underset{..}{F}}-C\cdot \\ | \\ :\overset{..}{\underset{..}{Cl}}: \quad + \quad \cdot\overset{..}{\underset{..}{Cl}}:$$

| trichlorofluoromethane Freon-11 CFC-11 | dichlorofluoromethyl radical | chlorine atom, also a radical with an unpaired electron |

Atomic chlorine has an unpaired electron and is extremely reactive. It abstracts an oxygen atom from ozone to give the chlorine monoxide radical, which plays a major role in the destruction of the ozone layer.

$$:\overset{..}{\underset{..}{Cl}}\cdot \; + \; O_3 \longrightarrow :\overset{..}{\underset{..}{Cl}}-\overset{..}{\underset{..}{O}}\cdot \; + \; O_2$$

| chlorine atom | ozone | chlorine monoxide | oxygen |

Aircraft flying over Antarctica have measured the variations in the concentration of chlorine monoxide and of ozone in regions isolated by the winds that circle the continent. In such zones, an increase in chlorine monoxide correlates with a decrease in ozone.

　　While the formation of a molecule of chlorine monoxide destroys a molecule of ozone, the real damage to the ozone layer comes from free-radical chain reactions in which chlorine atoms are regenerated repeatedly to react with ozone. The cycle that appears to be the most important is shown below. In these equations, M represents a third body, some inert molecule such as nitrogen, that carries away the excess energy generated in reactions. The process starts by the combination of two chlorine monoxide molecules to give a dimer.

$$:\overset{..}{\underset{..}{Cl}}-\overset{..}{\underset{..}{O}}\cdot \; + \; \cdot\overset{..}{\underset{..}{O}}-\overset{..}{\underset{..}{Cl}}: \; + \; M \longrightarrow :\overset{..}{\underset{..}{Cl}}-\overset{..}{\underset{..}{O}}-\overset{..}{\underset{..}{O}}-\overset{..}{\underset{..}{Cl}}: \; + \; M$$

The dimer absorbs ultraviolet radiation and breaks up to give a chlorine atom and a chlorine dioxygen radical.

$$:\!\ddot{C}l\!-\!\ddot{O}\!-\!\ddot{O}\!-\!\ddot{C}l\!:\ \xrightarrow{hv}\ :\!\ddot{C}l\!-\!\ddot{O}\!-\!\ddot{O}\cdot\ +\ \cdot\ddot{C}l\!:$$

The chlorine dioxygen radical loses oxygen to create an oxygen molecule and another chlorine atom.

$$:\!\ddot{C}l\!-\!\ddot{O}\!-\!\ddot{O}\cdot\ +\ M\ \longrightarrow\ :\!\ddot{C}l\cdot\ +\ O_2\ +\ M$$

Each chlorine atom can then attack another ozone molecule. The overall result of the cycle is to convert two ozone molecules into three oxygen molecules, with chlorine atoms serving to catalyze the process. The important point is that a single chlorofluorocarbon decomposing in the stratosphere is capable of catalyzing the destruction of many ozone molecules. Measurements made in 2002 showed that the concentration of chlorofluorocarbons in the stratosphere has peaked and that the ozone hole over Antarctica was smaller that year than it had been earlier (p. 310). It is too early to tell whether this is a sign of long-term recovery of the ozone layer.

Governments all over the world have joined to phase out the use of chlorofluorocarbons. Chemists are creating substitutes for them that will retain their desirable properties but will break down in the lower atmosphere so that they will not survive to reach the ozone layer. Most of the replacement compounds are being designed to contain carbon–hydrogen bonds, which increase the reactivity of the compounds and decrease their lifetime in the atmosphere. One such compound that is now on the market is a hydrofluorocarbon (an HFC), HFC-134a, which has the formula $CF_3CH_2F$.

## 19.2   Free-Radical Substitution of Allylic Hydrogen Atoms

*N*-Bromosuccinimide in a polar solvent, such as a mixture of dimethyl sulfoxide with water, is used as a source of electrophilic bromine, which reacts with alkenes to give halohydrins by way of a bromonium ion intermediate (p. 308). In a nonpolar solvent such as carbon tetrachloride, however, *N*-bromosuccinimide reacts with alkenes that have allylic hydrogen atoms to substitute bromine for one of those hydrogens. For example, cyclohexene reacts with *N*-bromosuccinimide in boiling carbon tetrachloride to give 3-bromocyclohexene.

cyclohexene   *N*-bromosuccinimide          3-bromocyclohexene   succinimide
                   NBS                              60%

This reaction is quite general. *N*-Bromosuccinimide (often abbreviated in equations as NBS) reacts with a variety of alkenes to substitute bromine for an allylic hydrogen atom. The reaction is initiated by peroxides, heat, or light. It is believed that the reaction involves the formation of a bromine atom, which abstracts an allylic hydrogen atom (p. 282) from the alkene.

VISUALIZING THE REACTION

**First Steps in an Allylic Substitution Reaction**

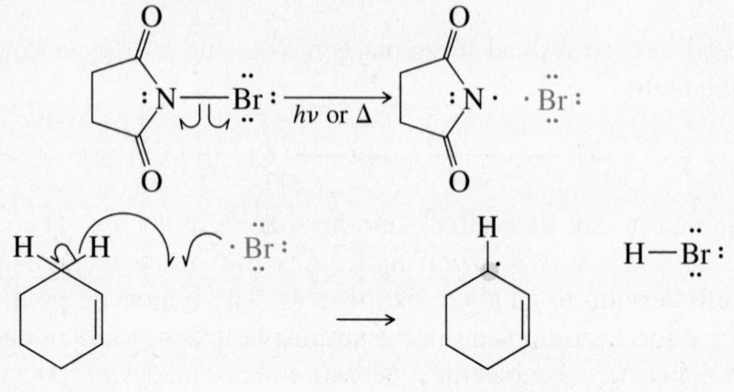

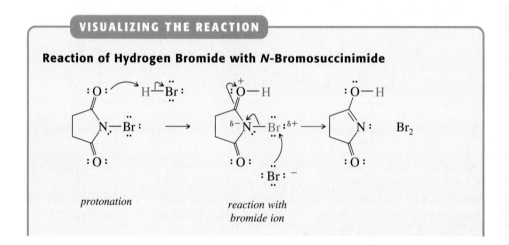

*abstraction of an allylic hydrogen atom*

secondary allylic radical

*most stable intermediate*

The stabilities of various alkyl radicals, which are electron-deficient species, parallel those of carbocations (p. 185). A tertiary alkyl radical is more stable than a secondary one, which in turn is more stable than a primary one. Free radicals stabilized by resonance delocalization of the radical character are the most stable of all and, therefore, the most easily formed. A radical having the electron-deficient carbon adjacent to a double bond, at an allylic position, is stabilized by resonance.

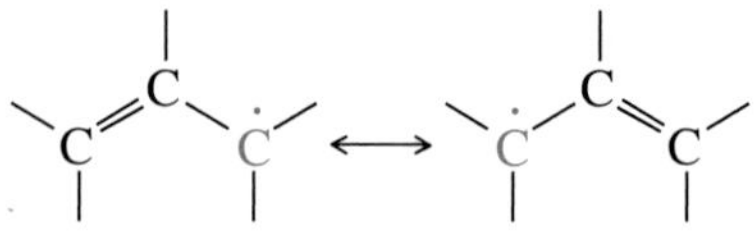

*resonance contributors for an allylic radical*

The major product of the bromination reaction of cyclohexene is derived from the intermediate formed by the abstraction of a hydrogen atom by a bromine atom. The other product of the hydrogen-abstraction reaction is hydrogen bromide, which reacts with *N*-bromosuccinimide to give bromine and succinimide.

VISUALIZING THE REACTION

**Reaction of Hydrogen Bromide with *N*-Bromosuccinimide**

*protonation*

*reaction with bromide ion*

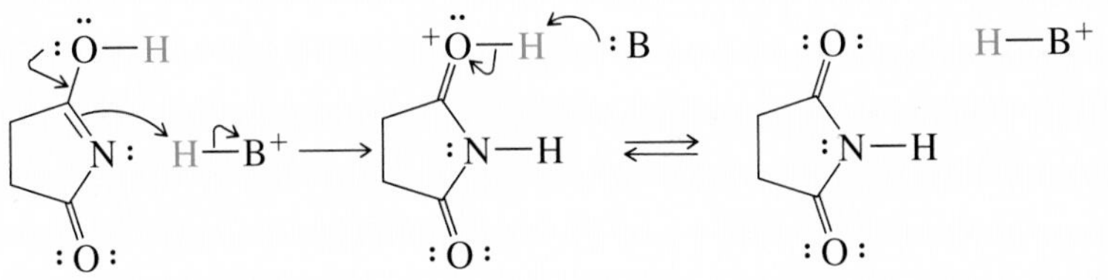

tautomerization                                   succinimide

As a result of this sequence of reactions, hydrogen bromide, which is a product of the hydrogen-abstraction reaction, is converted to bromine. The amount of bromine formed in the reaction mixture is limited by the amount of hydrogen bromide that has been formed in the earlier step of the reaction. The allylic radical formed in the first step of the reaction abstracts a bromine atom from a molecule of bromine to give 3-bromocyclohexene and a bromine atom, which starts the cycle all over again.

---

**VISUALIZING THE REACTION**

**Reaction of an Allylic Radical with Bromine**

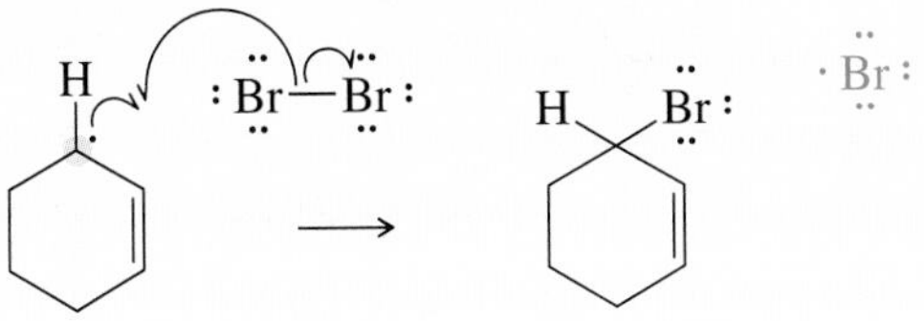

---

When molecular bromine is used as a reagent with alkenes, it adds to the double bond (p. 302). With *N*-bromosuccinimide, addition reactions are not observed. Why does the bromine formed from the reaction of hydrogen bromide with *N*-bromosuccinimide fail to add to the double bond in cyclohexene? One explanation is that a very low concentration of bromine is present in the reaction mixture at any time. The initial addition of bromine to the double bond involves only one of the two bromine atoms in a cyclic bromonium ion intermediate (p. 303). The formation of this intermediate is reversible. If there is no bromide ion nearby, the formation of the dibromide cannot be completed. This theory has been confirmed by running reactions with very low concentrations of molecular bromine, in which case allylic substitution, rather than an addition reaction, is seen. *N*-Bromosuccinimide competes with the bromonium ion for bromide ion, so the allylic substitution reaction is favored.

The mechanism of halogenation reactions that employ *N*-bromosuccinimide has been studied intensively but is not yet completely understood. The ideas presented above offer one picture of what happens under one set of experimental conditions.

**Problem 19.4**　Predict the product(s) of the following reactions.

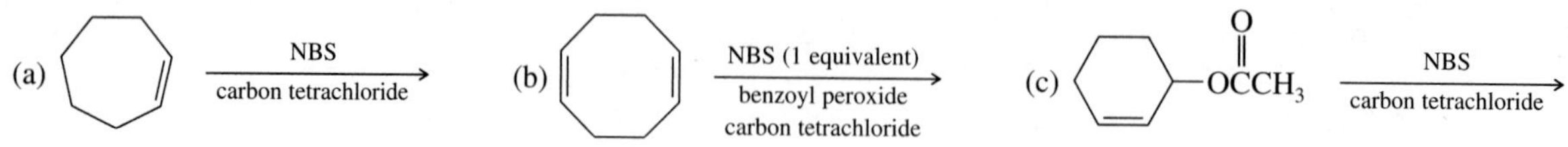

(a) $\xrightarrow[\text{carbon tetrachloride}]{\text{NBS}}$

(b) $\xrightarrow[\substack{\text{benzoyl peroxide} \\ \text{carbon tetrachloride}}]{\text{NBS (1 equivalent)}}$

(c) $\xrightarrow[\text{carbon tetrachloride}]{\text{NBS}}$

**Problem 19.5**　When 2-heptene reacts with *N*-bromosuccinimide, 4-bromo-2-heptene is obtained in 60% yield. What are other possible substitution products of this reaction? Write equations supporting your answer.

## 19.3　Free–Radical Halogenation Reactions at the Benzylic Position

When toluene reacts with bromine in the presence of a Lewis acid, two products are formed: *o*-bromotoluene and *p*-bromotoluene (p. 374).

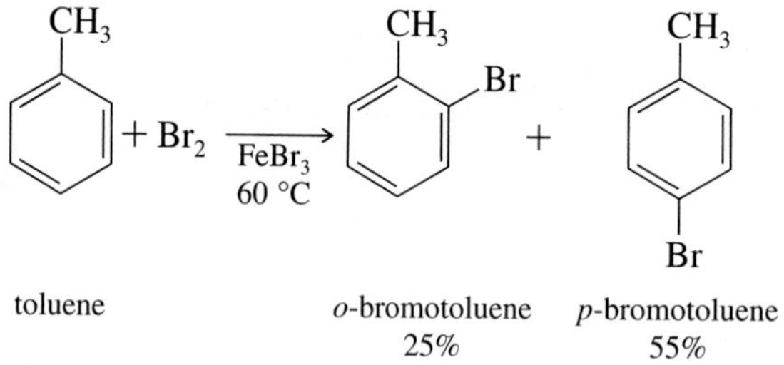

| toluene | | *o*-bromotoluene | *p*-bromotoluene |
|---|---|---|---|
| | | 25% | 55% |

Besides the hydrogen atoms bonded to carbons of the aromatic ring, however, toluene has hydrogen atoms in the methyl group. If bromine and toluene are brought together in the absence of a catalyst, a third substitution product, in which one of the hydrogen atoms of the methyl group has been replaced, is also formed.

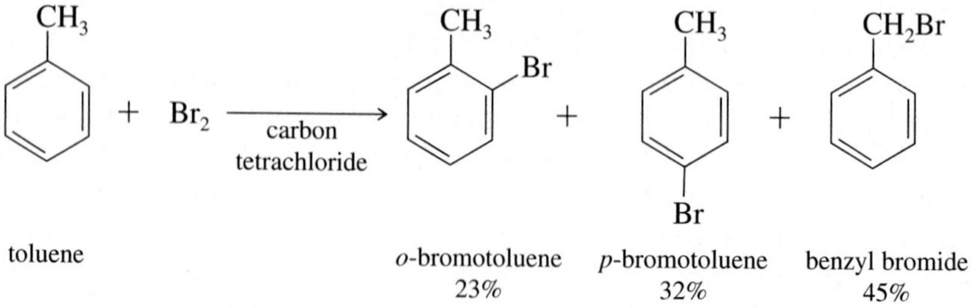

| toluene | | *o*-bromotoluene | *p*-bromotoluene | benzyl bromide |
|---|---|---|---|---|
| | | 23% | 32% | 45% |

The hydrogen atoms of the methyl group of toluene are bonded to a benzylic carbon atom, one adjacent to an aromatic ring. Benzylic hydrogen atoms resemble allylic hydrogen atoms in reactivity (p. 776). They are easily substituted in free-radical reactions because the radical that is formed when a benzylic hydrogen atom is abstracted is stabilized by resonance.

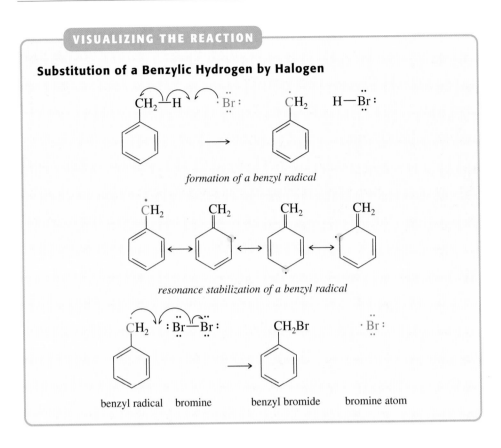

VISUALIZING THE REACTION

**Substitution of a Benzylic Hydrogen by Halogen**

formation of a benzyl radical

resonance stabilization of a benzyl radical

benzyl radical    bromine          benzyl bromide      bromine atom

The deficiency of an electron is delocalized over the aromatic ring, giving rise to a relatively stable species. The radical abstracts a bromine atom from a bromine molecule, forming benzyl bromide and a new bromine atom that can continue the reaction.

Bromine can act both as an electrophile toward the aromatic ring in toluene and as a source of bromine atoms, which give rise to free-radical reactions at the side chain. Free-radical reactions with bromine are more frequent when heat or light is used to cause the dissociation of molecular bromine into two bromine atoms.

*N*-Bromosuccinimide, which is used to substitute bromine for an allylic hydrogen atom (p. 775), has the same kind of reactivity toward benzylic hydrogen atoms. It can be used to substitute selectively at that position, without any competing substitution on the aromatic ring. For example, a benzylic hydrogen atom bonded to the carbon atom between two phenyl rings in diphenylmethane is selectively substituted by *N*-bromosuccinimide.

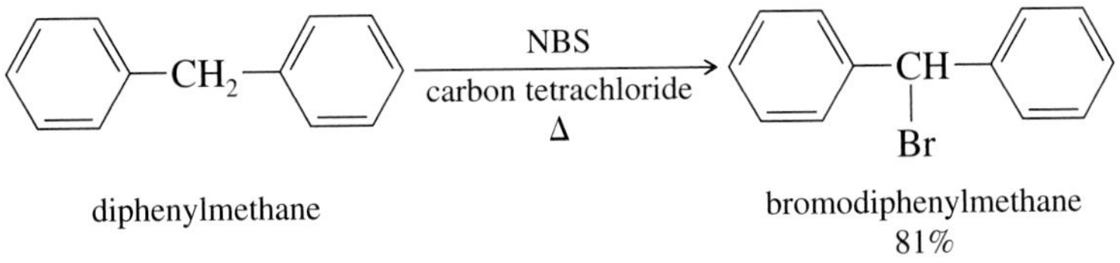

diphenylmethane
$\xrightarrow[\substack{\text{carbon tetrachloride} \\ \Delta}]{\text{NBS}}$
bromodiphenylmethane
81%

The hydrogen atoms of a methyl group on naphthalene are also benzylic in nature and undergo this substitution reaction.

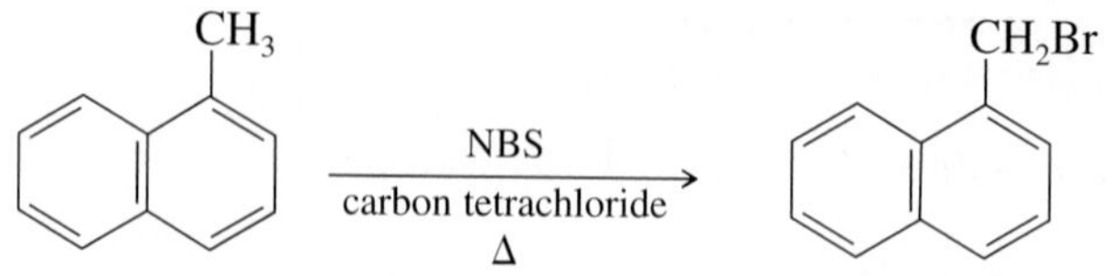

1-methylnaphthalene                    1-(bromomethyl)naphthalene
                                                90%

Note that aromatic hydrogen atoms are not affected by this substitution reaction.

Benzylic radicals, like allylic radicals, are much more stable than alkyl radicals. Transition states leading to the formation of benzylic radicals are of lower energy than the transition states for competing reactions, so substitution reactions are channeled through such intermediates when possible.

**Study Guide**
**Concept Map 19.3**

**Problem 19.6**    Complete the following equations.

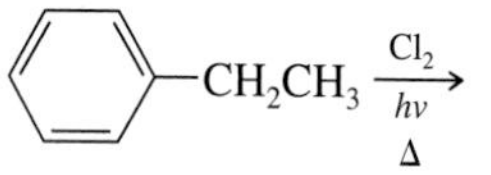

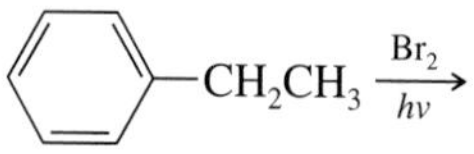

### 19.4    The Triphenylmethyl Radical, a Stable Free Radical

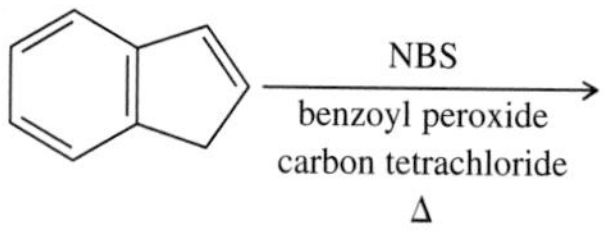

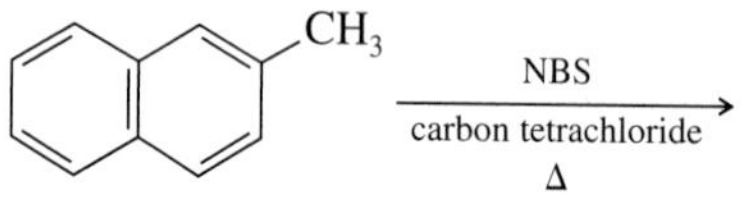

A free radical that is stable enough to be observed as a discrete species was first reported in 1900 by Moses Gomberg of the University of Michigan. He was trying to synthesize hexaphenylethane by the reaction of bromotriphenylmethane with silver. When he ran the reaction under an atmosphere of carbon dioxide, he obtained a white solid that was highly reactive toward oxygen and halogens, including iodine. A solution of the solid in benzene prepared in the absence of air was yellow. The yellow color disappeared when a small amount of air was admitted into the reaction flask but then developed again. Gomberg postulated that the yellow color was due to the formation of the triphenylmethyl radical in equilibrium with hexaphenylethane and that the radical reacted with molecular oxygen, which is a diradical (p. 786), to give a peroxide.

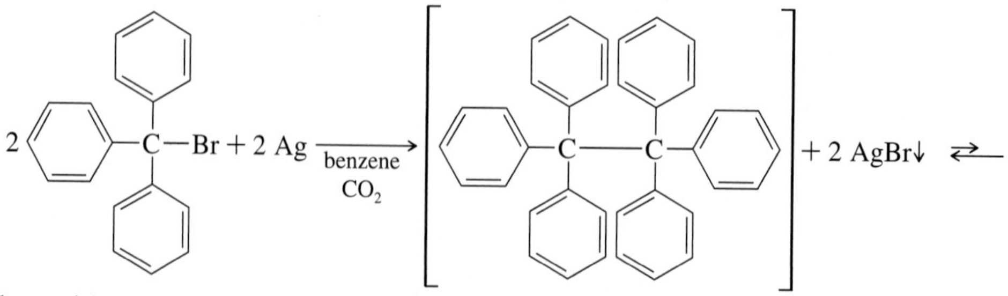

bromotriphenylmethane                    hexaphenylethane postulated
                                              as the product

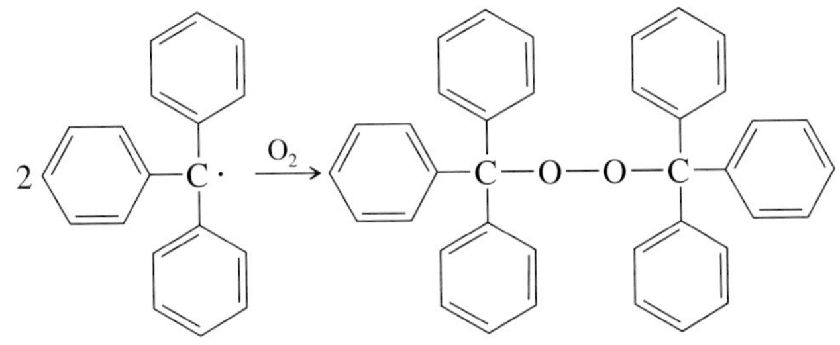

triphenylmethyl
radical,
yellow

peroxide, product
of the reaction
of the radical
with oxygen

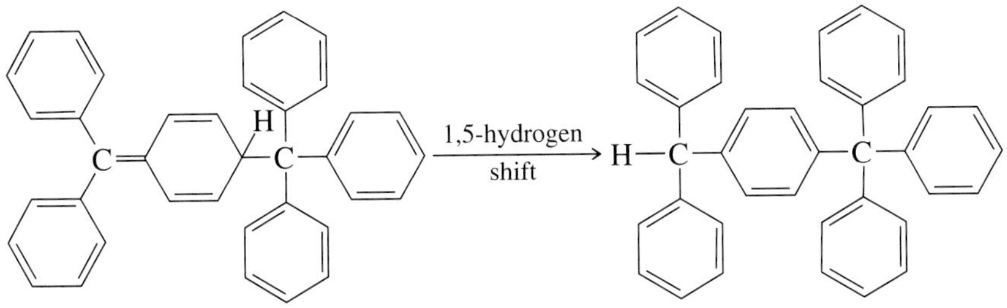

actual product from the
combination of two
triphenylmethyl radicals

rearranged product
usually isolated
from the reaction

Initially, many chemists rejected Gomberg's idea that a free radical could be stable enough to be observed as an intermediate in a reaction, but by 1911 a number of similar radicals had been prepared and characterized. Many years later it was shown that a triphenylmethyl radical adds to the para position of a phenyl ring in a second triphenylmethyl radical to give the product observed and that hexaphenylethane, which would have a highly strained carbon–carbon single bond, does not form.

Radicals that are not only stable, but positively inert, have been synthesized since Gomberg's experiments. One of these is the perchlorotriphenylmethyl radical.

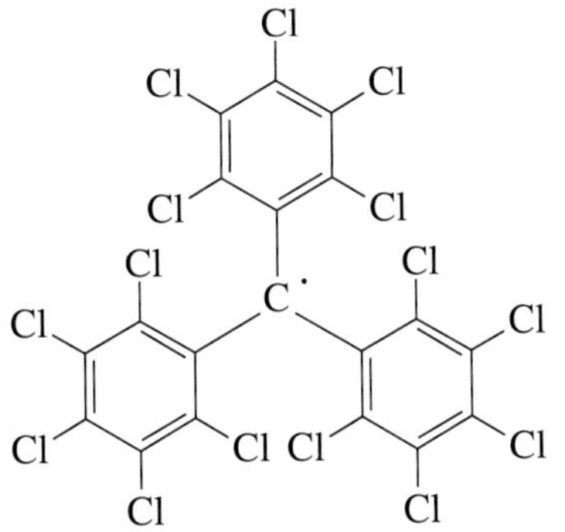

perchlorotriphenylmethyl radical

*an inert free radical*

This radical does not react with oxygen and is unreactive at temperatures up to 300 °C. The unreactivity of the species is attributed to the presence of the large chlorine atoms that shield the electron-deficient carbon atom in the center from contact with reagents.

**Problem 19.7**    Explain the stability of the triphenylmethyl radical.

## 19.5  Free-Radical Addition Reactions of Alkenes

### A. Polymerization Reactions

Industrially, the most important free-radical addition reactions to alkenes are polymerization reactions, which are used to make large quantities of polymeric materials such as plastics and synthetic fibers. The free-radical polymerization of vinyl chloride, for example, gives poly(vinyl chloride), PVC.

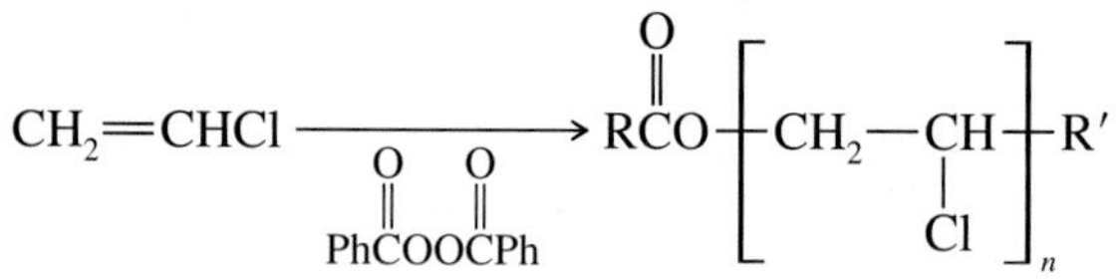

$$CH_2\!=\!CHCl \xrightarrow[\substack{PhCOOCPh\\ \text{initiator}}]{} RCO\!\left[\!CH_2\!-\!\underset{\underset{Cl}{|}}{CH}\!\right]_n\!\!R'$$

poly(vinyl chloride)
polymer

The reaction starts with the homolytic cleavage of the relatively weak oxygen–oxygen single bond in an acyl peroxide such as benzoyl peroxide. The acyloxy radicals that result add to the double bond, creating carbon radicals.

**VISUALIZING THE REACTION**

**A Free-Radical Polymerization Reaction**

benzoyl peroxide          benzoyloxy radical
initiator

initiation

propagation

propagation          dimer

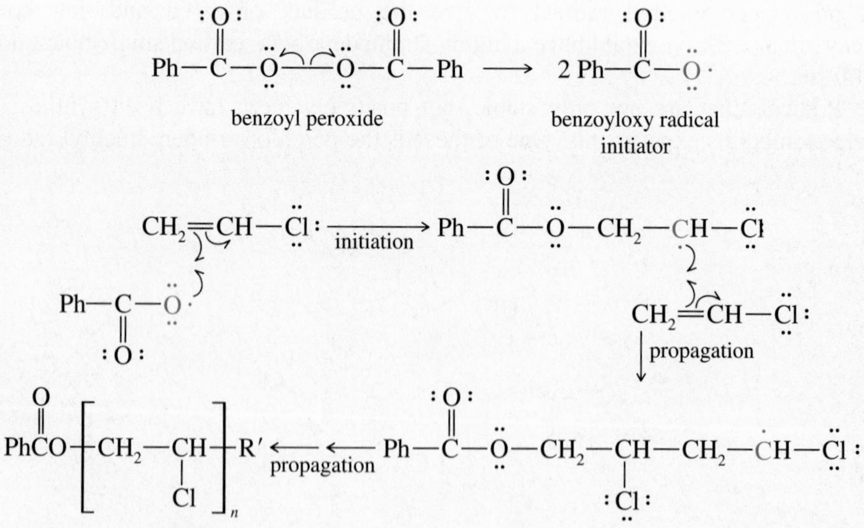

*polymer where R' is
some group that has
resulted in termination
of the chain reaction*

In these reactions, the carbon radical that is the product of one step adds to another molecule of alkene, giving a new carbon radical to carry the chain in a series of propagation steps, until all the monomer is consumed and the chain is terminated (p. 771). Polymerization reactions are favored when a small amount of initiator and a large amount of alkene are mixed together; then all the propagation steps involve additions of radicals to alkenes. Polymerization reactions are discussed in greater detail in Chapter 24.

## B. Other Free-Radical Additions to Alkenes

Many other reagents add to alkenes under conditions that generate free radicals. High temperatures, ultraviolet radiation, gamma-radiation from radioactive sources, peroxides, and oxygen all act as free-radical initiators. Three examples of this type of addition reaction are given below.

$$Cl_3CBr + CH_2{=}CH(CH_2)_5CH_3 \xrightarrow[\substack{CH_3COOCCH_3 \\ h\nu}]{O\ O} Cl_3CCH_2CH(CH_2)_5CH_3$$
$$\overset{|}{Br}$$

bromotrichloromethane    1-octene    3-bromo-1,1,1-trichlorononane
88%

$$CH_3CH_2SH + CH_2{=}\overset{\overset{\displaystyle CH_3}{|}}{C}CH_3 \xrightarrow{\Delta} CH_3CH_2SCH_2\overset{\overset{\displaystyle CH_3}{|}}{C}HCH_3$$

ethanethiol    2-methylpropene    ethyl isobutyl sulfide
94%

$$Cl_3SiH + CH_2{=}CH(CH_2)_5CH_3 \xrightarrow[\substack{CH_3COOCCH_3 \\ \Delta}]{O\ O} Cl_3SiCH_2CH_2(CH_2)_5CH_3$$

trichlorosilane    1-octene    octyltrichlorosilane
99%

In each case, one of the reactants has a bond that is weak enough to break homolytically under the reaction conditions. Products resulting from addition of a carbon, sulfur, or silicon radical to a double bond are obtained in high yields.

**Problem 19.8**    Write a mechanism for each of the three reactions shown above.

**Problem 19.9**    Predict the product(s) of each of the following reactions.

(a) 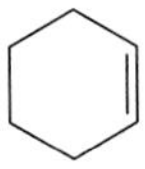  + CH_3SH $\xrightarrow[\substack{hv}]{\text{acetone}}$    (b) Cl_3CBr + CH_2{=}$\overset{\overset{\displaystyle CH_2CH_3}{|}}{C}$CH_2CH_3 $\xrightarrow{\Delta}$

(c) $CH_3(CH_2)_4SiH_3 + CH_2{=}CH(CH_2)_5CH_3 \xrightarrow{\text{peroxides}}$

## C. Intramolecular Free-Radical Addition Reactions

Intramolecular addition of a free radical to a double bond produces a ring. Five-membered rings are usually formed in preference of six-membered rings. For example, when 6-bromo-1-hexene reacts with a tributyltin radical, methylcyclopentane is formed as the major product.

**ONE SMALL STEP**

When the electrophilic addition reaction of hydrogen bromide to propene was first introduced, specific reaction conditions were given (p. 126). Changing the conditions results in a different product. When hydrogen bromide and propene are mixed at −78 °C in the presence of benzoyl peroxide and air, a very rapid reaction takes place; the product is a mixture of 96% *n*-propyl bromide and 4% isopropyl bromide. This change in product composition from Markovnikov to anti-Markovnikov regioselectivity is called the **peroxide effect.**

PROBLEM:   Write an equation for the reaction described above, and identify the electrophile that attacks the double bond.

Hint:   How will HBr react with benzoyl peroxide? The Visualizing the Reaction on page 782 may be helpful.

On the Web: ONE SMALL STEP

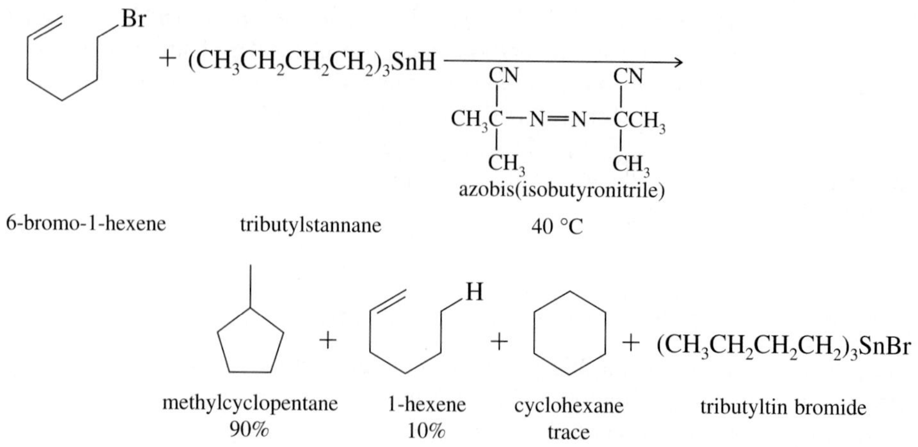

6-bromo-1-hexene + (CH₃CH₂CH₂CH₂)₃SnH

tributylstannane

$$CH_3\overset{\overset{\displaystyle CN}{|}}{\underset{\underset{\displaystyle CH_3}{|}}{C}}-N{=}N-\overset{\overset{\displaystyle CN}{|}}{\underset{\underset{\displaystyle CH_3}{|}}{C}}CH_3$$

azobis(isobutyronitrile)

40 °C

methylcyclopentane    +    1-hexene    +    cyclohexane    +    (CH₃CH₂CH₂CH₂)₃SnBr
90%                          10%             trace                tributyltin bromide

The initiator for the reaction is azobis(isobutyronitrile), widely used because it readily decomposes on heating to give molecular nitrogen and carbon radicals that are stabilized by a substituent cyano group. These carbon radicals abstract hydrogen from tributylstannane (tributyltin hydride).

**VISUALIZING THE REACTION**

**Initiation**

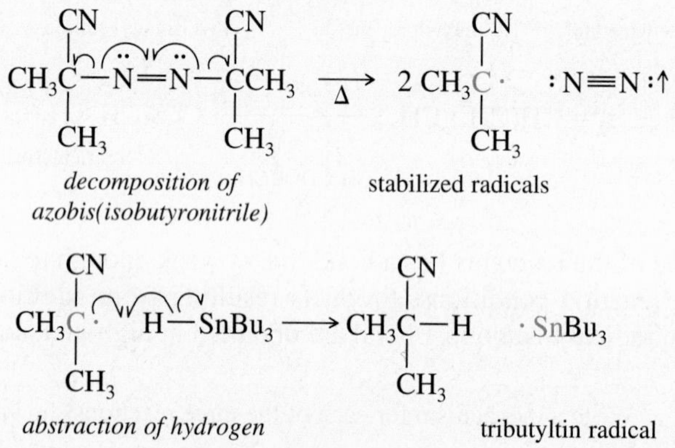

decomposition of azobis(isobutyronitrile)    →    2 CH₃C·    :N≡N:↑

                                                    stabilized radicals

CH₃C· + H—SnBu₃ → CH₃C—H    ·SnBu₃

abstraction of hydrogen                        tributyltin radical

The tin radical then abstracts a halogen atom from the haloalkane, and the resulting carbon radical reacts to give the observed products.

**VISUALIZING THE REACTION**

**Free-Radical Cyclization of a Haloalkene**

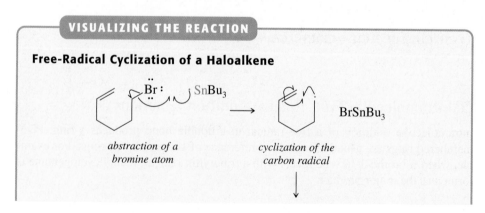

abstraction of a
bromine atom

cyclization of the
carbon radical

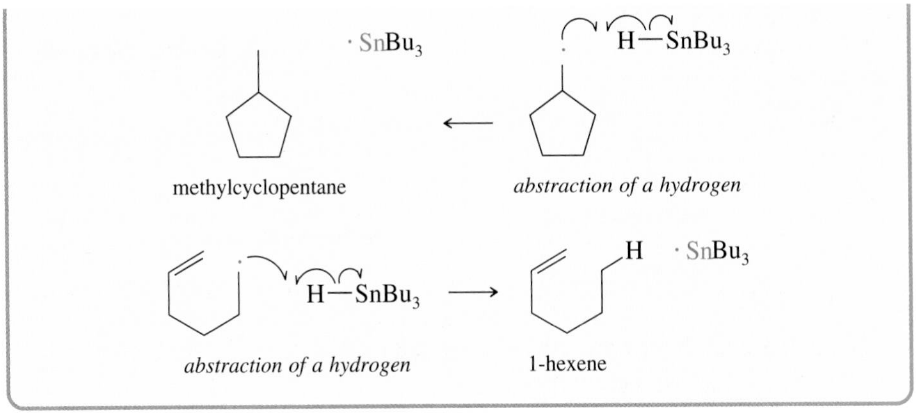

Intramolecular free-radical addition reactions are used in organic syntheses. The following problem concerns an example of such an application.

**Study Guide**
**Concept Map 19.4**

**Problem 19.10** Silphiperfol-6-ene, a natural product, was synthesized from the compound shown below. How would you carry out this transformation?

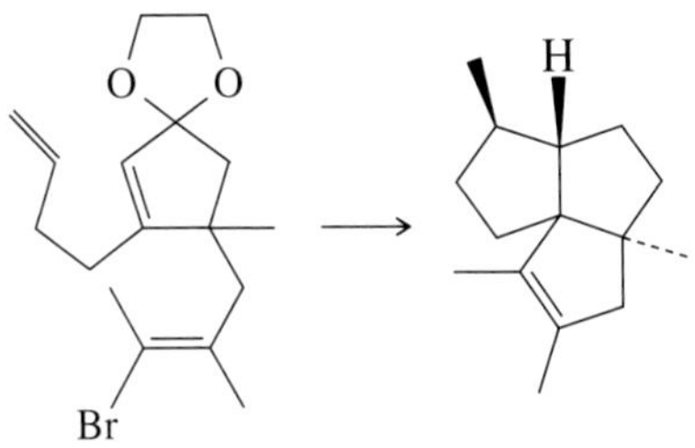

silphiperfol-6-ene

# 19.6 Free-Radical Oxidations with Molecular Oxygen

## A. Autooxidation

When organic compounds are exposed to air, they react slowly with oxygen to give hydroperoxides. This slow oxidation reaction is called **autooxidation.** This reaction is responsible for the slow deterioration in air of foods (pp. 786 and 794), rubber (p. 1073), and paints. The reaction is catalyzed by light, which is why organic reagents are stored in cans or dark-colored bottles. The ease with which a compound undergoes autooxidation is directly related to the ease with which it forms free radicals. For this reason, autooxidation takes place especially easily at allylic or benzylic positions. Cumene, for example, is easily converted to a hydroperoxide.

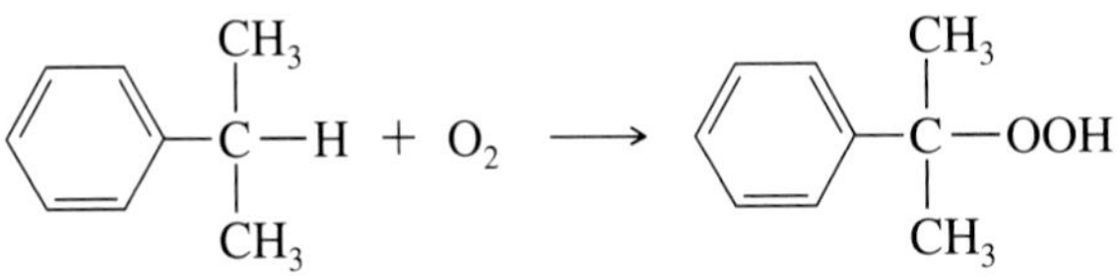

cumene          oxygen          cumene hydroperoxide

The hydrogen atoms bonded to carbons that are adjacent to the oxygen atom in ethers and alcohols are also easily replaced by hydroperoxide groups, shown in the reaction below.

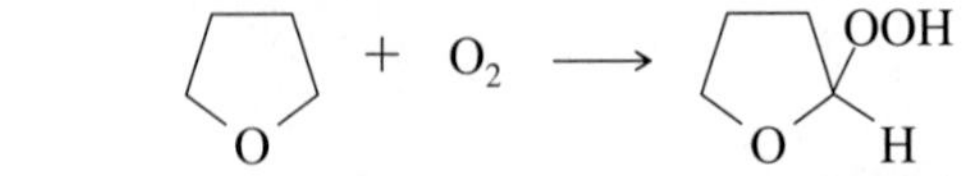

tetrahydrofuran                tetrahydrofuran
                                hydroperoxide

If such solvents have been stored for some time, they almost certainly contain some hydroperoxides. Hydroperoxides are unstable compounds and may decompose violently when heated. For safety, ethers are tested in the laboratory for peroxides and, if necessary, are purified before being used.

The hydrogen atom on the carbonyl group of an aldehyde is also easily abstracted. Aldehydes oxidize in air to carboxylic acids by way of peroxyacids (p. 317). This is so common as to be noticeable in the laboratory, where a bottle of liquid benzaldehyde will develop a crust of solid benzoic acid around its lip once it has been opened and exposed to air.

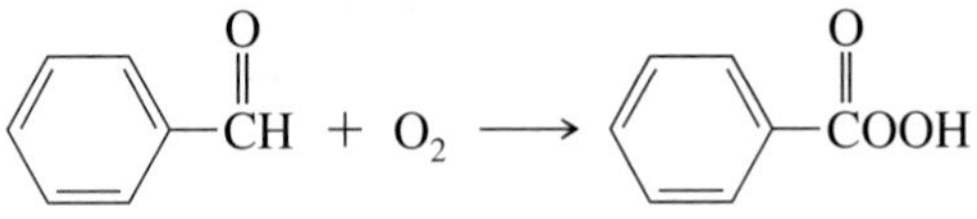

benzaldehyde      oxygen          peroxybenzoic acid

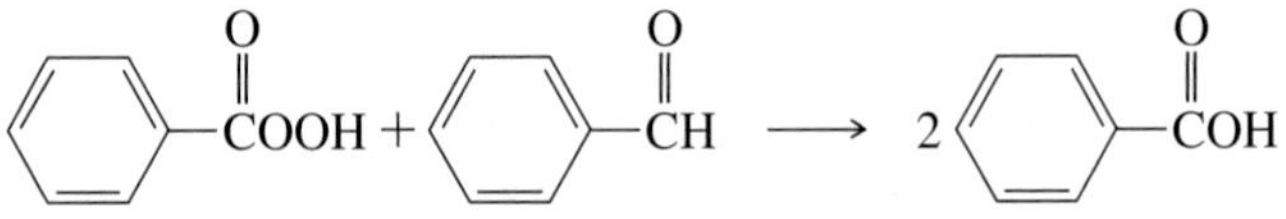

peroxybenzoic acid      benzaldehyde                benzoic acid

In the next section the details of the formation of hydroperoxides are illustrated by the reaction of oxygen with polyunsaturated fatty acids.

## B.  Oxidation of Polyunsaturated Fatty Acids

Polyunsaturated fatty acids (p. 639) are converted by reactions with oxygen to compounds containing oxygen and conjugated double bonds. The formation of such compounds is responsible for the development of rancidity in oils and, therefore, for the spoilage of any food that contains an oil. Polyunsaturated fatty acids, mostly as their esters, are components of the human body. The role of free-radical chemistry in the body and especially its relation to aging are subjects of vigorous debate. Molecular oxygen is a **diradical,** having two unpaired electrons, so the reaction of an unsaturated acid with oxygen involves free-radical intermediates. Oxidation of linoleic acid, catalyzed by an enzyme from soybeans, gives (9Z,11E)-13-hydroperoxy-9,11-octadecadienoic acid.

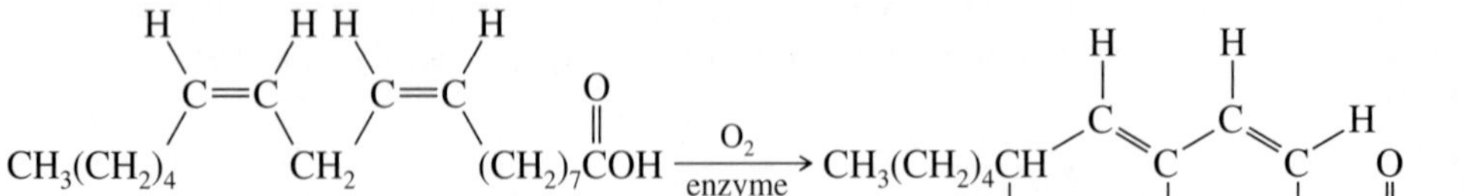

linoleic acid

(9Z,11E)-13-hydroperoxy-9,11-
octadecadienoic acid

Experiments involving isotopic labeling have shown that the oxygen atoms in the product come from oxygen molecules, not from water molecules

**VISUALIZING THE REACTION**

**Formation of a Hydroperoxide**

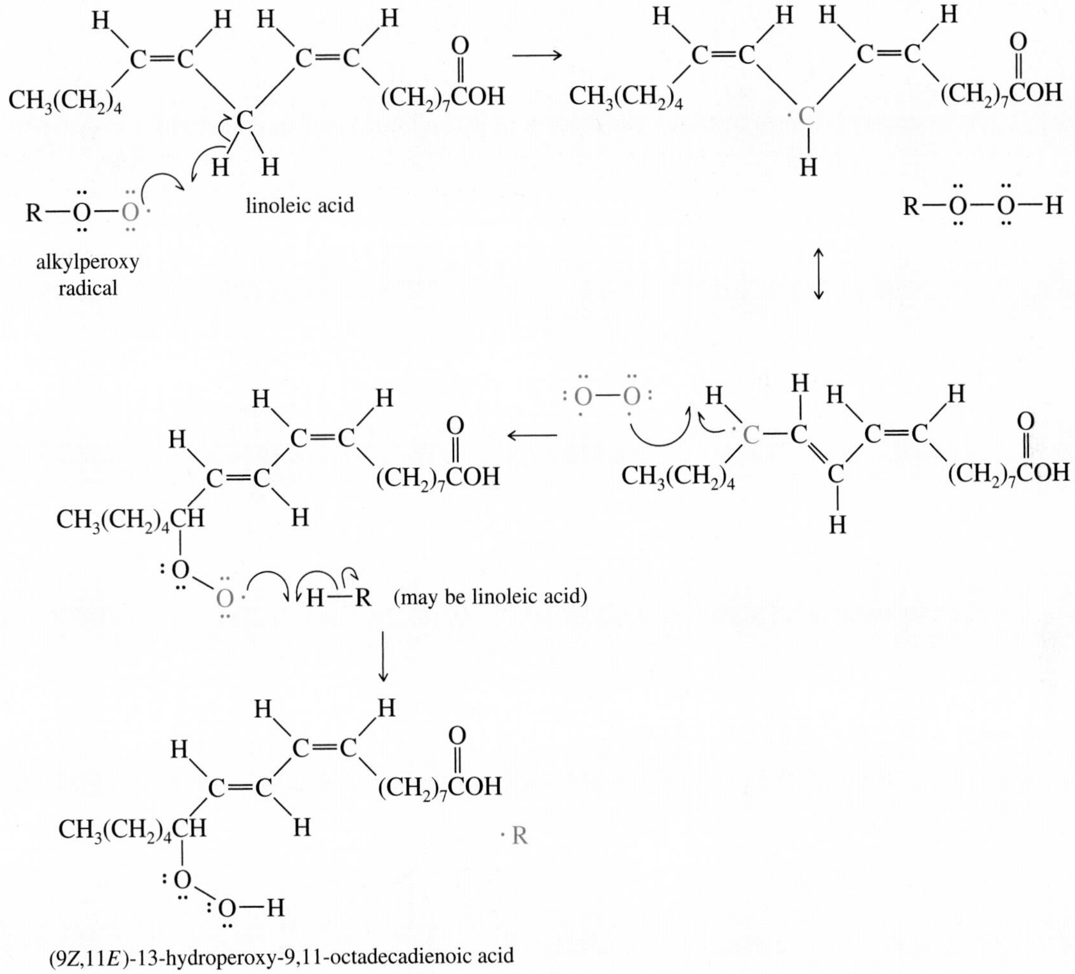

(9Z,11E)-13-hydroperoxy-9,11-octadecadienoic acid

The exact nature of the radical that abstracts the first hydrogen from linoleic acid is not known. It is thought to be some peroxy species. The hydrogen atoms that are allylic to two double bonds are particularly vulnerable to attack by a radical initiator. The carbon radical that forms is allylic and can be stabilized by delocalization of the radical character, giving rise to a conjugated double bond system. The new double bond formed between carbon atoms 11 and 12 is always trans. Combination of the carbon radical with molecular oxygen creates an alkylperoxy radical, which starts the chain reaction going again by abstracting an allylic hydrogen atom. The more double bonds a fatty acid contains, the more sites there are for hydrogen-abstraction reactions, and the greater is the complexity of the products. It is also possible for the alkylperoxy radicals to attack double bonds in other molecules of the unsaturated fatty acid, giving rise to large, complex molecules in which many acid units are held together by oxygen bridges. The characteristic drying of linseed oil on exposure to air, which gives a tough protective coating to paint, is believed to be due to formation of just such a linkage of linolenic acid molecules by oxygen.

**Problem 19.11** When linoleic acid reacts with oxygen in the absence of an enzyme, a 9-hydroperoxyoctadecadienoic acid is formed along with the 13-hydroperoxy isomer shown on page 787. Show how it develops.

## C. Biosynthesis of Prostaglandins

Prostaglandins and related polyunsaturated acids containing twenty carbon atoms constitute a class of compounds that have a wide range of significant biological functions. The name **prostaglandins** comes from the fact that these compounds were first isolated from seminal fluid. Prostaglandins have since been found to be widely distributed in all kinds of body tissues and play roles in reproduction, the nervous system, the intestinal system, blood clotting, and the production of allergic and inflammatory reactions.

Arachidonic acid (p. 639) is the biological precursor of prostaglandin $E_2$. The structures of arachidonic acid, prostaglandin $E_2$, and a key intermediate in the transformation of the former into the latter are shown below.

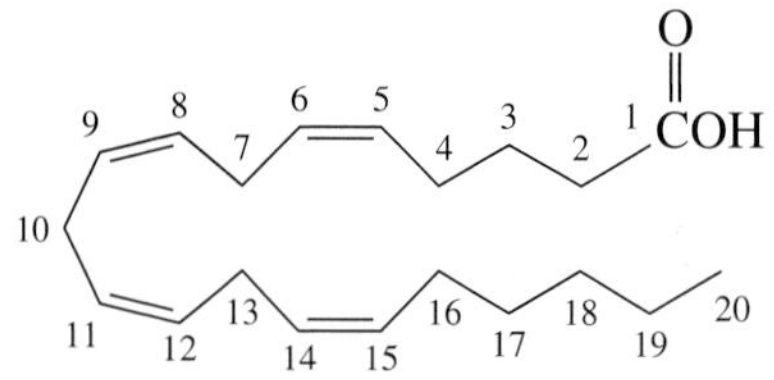

arachidonic acid

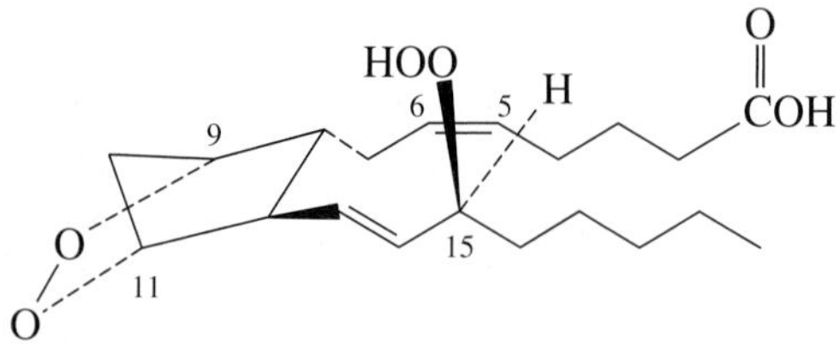

endoperoxide

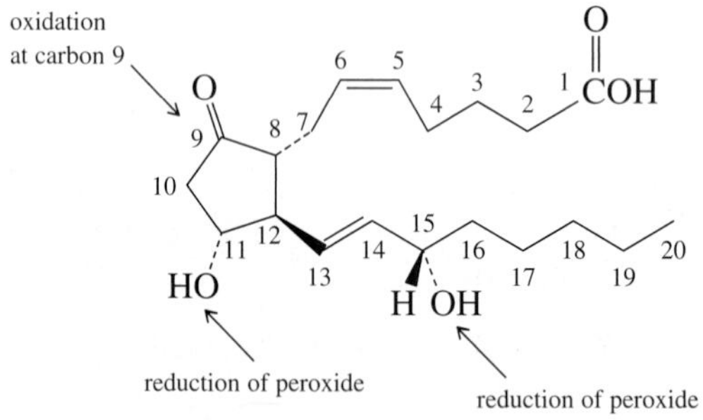

prostaglandin $E_2$

It has been established that two molecules of oxygen are used in the biological transformation of arachidonic acid to prostaglandin $E_2$ and that the oxygen atoms at carbon atoms 9 and 11 of the prostaglandin are derived from the same molecule of oxygen. The enzymatic cyclization of arachidonic acid to prostaglandin involves a number of intermediates and starts with the abstraction of a hydrogen atom at $^{13}C$ to give a free radical. This radical sets off a chain of free-radical reactions, including the formation of a five-membered ring (p. 784) and the addition of oxygen across the ring at carbons 9 and 11 to give the endoperoxide. The endoperoxide is converted to prostaglandin $E_2$ by a series of oxidation and reduction steps. Aspirin and other "anti-inflammatory" medicines work by inhibiting the enzyme that catalyzes the first step of the process by which arachidonic acid is converted to prostaglandins.

The prostaglandin structures shown below are also derived from the endoperoxide intermediate by familiar reactions, such as loss of water from the cyclopentane ring and reduction of ketone functions to alcohols. These structures are not all derived biologically from arachidonic acid, however.

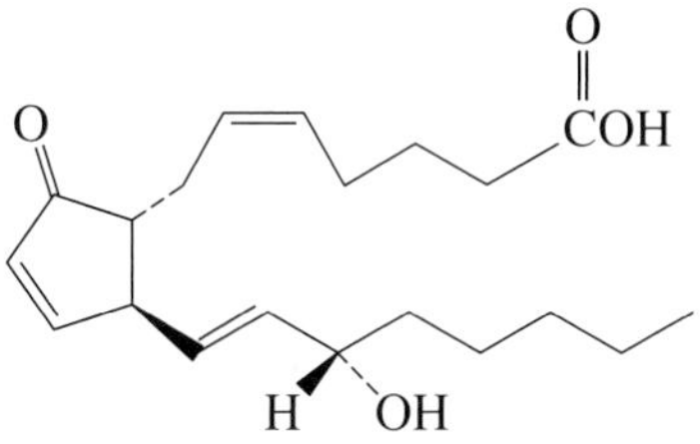

prostaglandin A$_2$

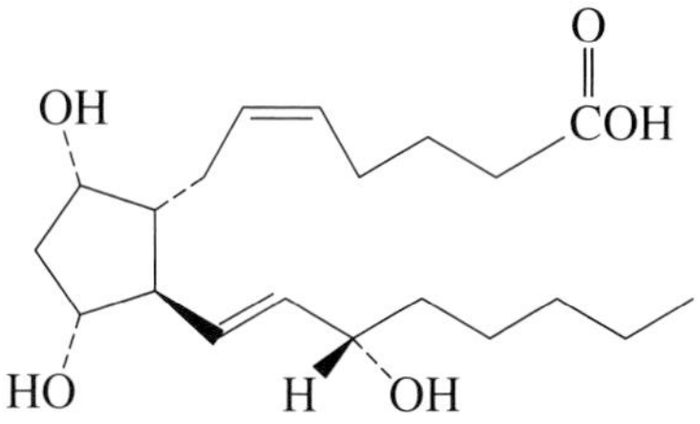

prostaglandin F$_{2\alpha}$

Another series of prostaglandins is derived from 5,9,11,14,17-icosapentaenoic acid, an omega-3 fatty acid, found in fish oils.

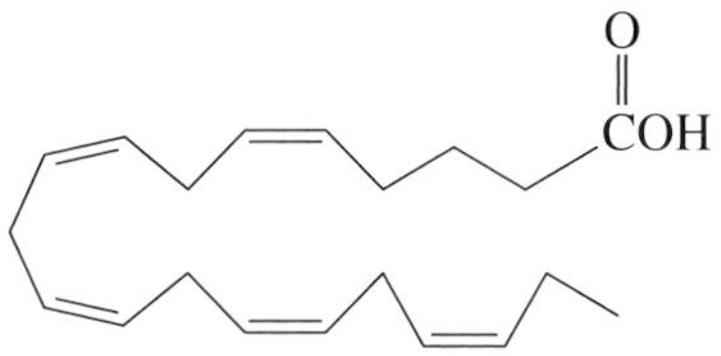

5,8,11,14,17-icosapentaenoic acid
(also known as eicosapentaenoic
acid or EPA)

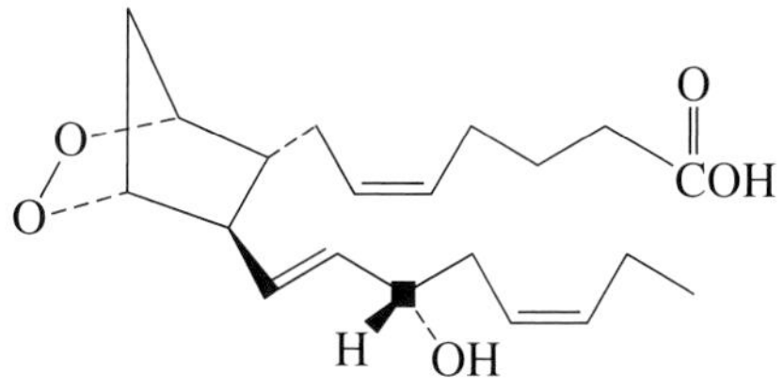

prostaglandin H$_3$
synthesized in the body from EPA

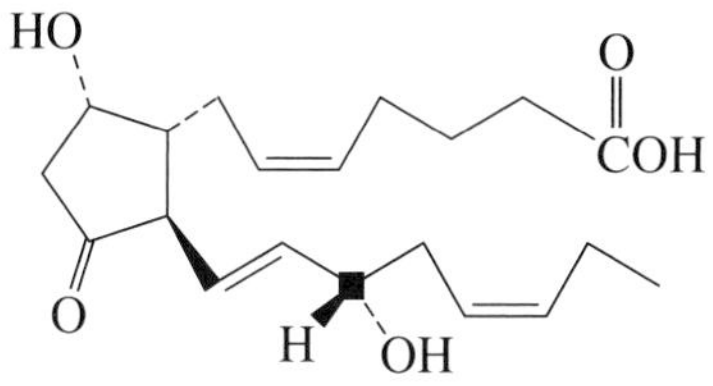

prostaglandin D$_3$
a metabolite of prostaglandin H$_3$

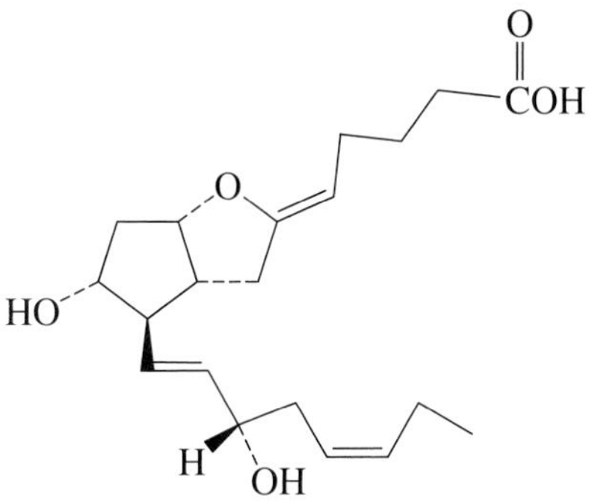

prostaglandin I$_3$
synthesized in the body from EPA

All the prostaglandins derived from EPA inhibit clotting of the blood by preventing the aggregation of blood platelets. Eating oily fish from the ocean appears to protect against heart disease and stroke, probably for this reason.

Another series of compounds that have high biological activity and are also polyunsaturated twenty-carbon acids are the leukotrienes, which despite their name have four double bonds. Of these, leukotriene C is known to be the slow-reacting substance involved in anaphylactic shock, the body's response to a foreign substance that provokes a severe allergic reaction, as in the case of asthma or insect stings. Exactly how leukotriene C is involved is not known. Apparently, cells are stimulated to release it by the arrival of antibodies formed in response to the presence of allergens, substances causing allergies.

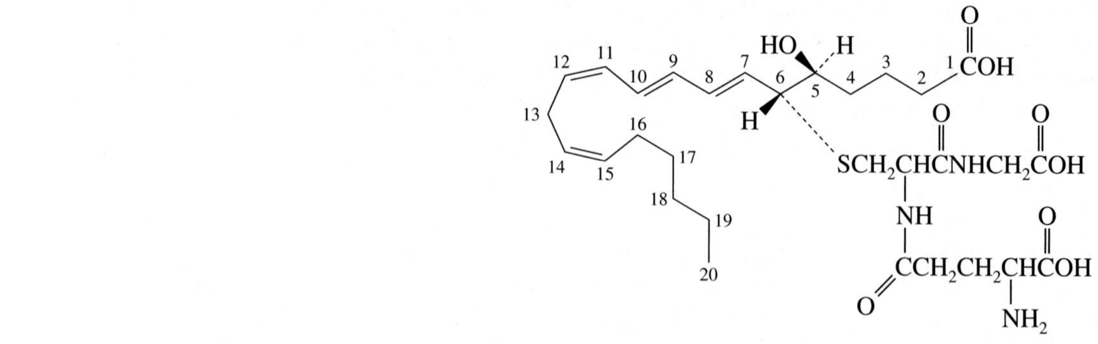

leukotriene C

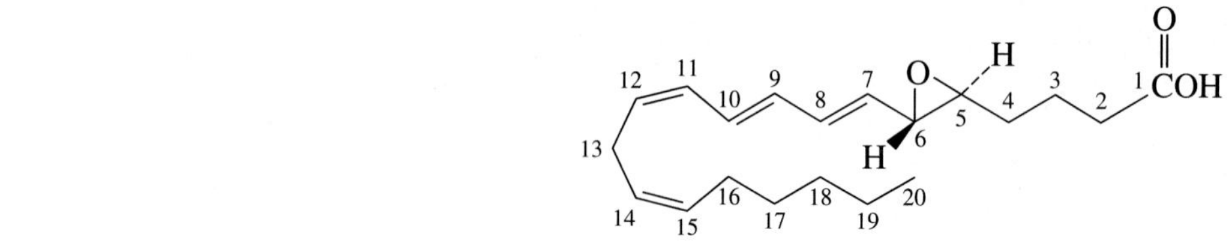

leukotriene A

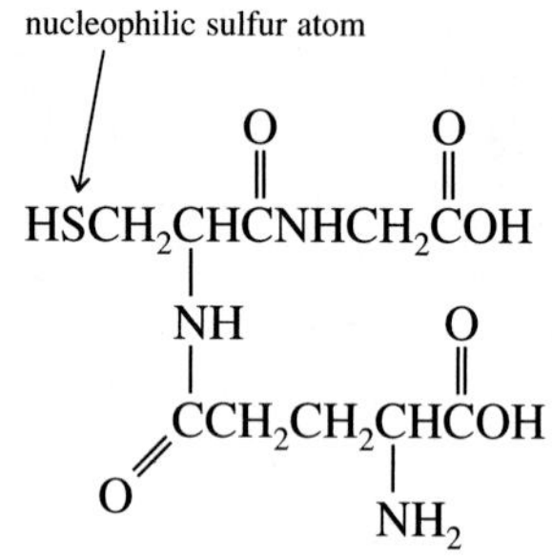

nucleophilic sulfur atom

glutathione

*a peptide containing*
*three amino acids and*
*two peptide linkages*

Both leukotriene A and leukotriene C have an oxygen bonded to carbon atom 5 of the chain. Leukotriene A is an intermediate in the biosynthesis of leukotriene C. The transformation of A to C involves addition of glutathione (a peptide) to the oxirane. Glutathione includes the amino acid cysteine, which has a thiol group. The opening of the oxirane ring in leukotriene A results from attack by the nucleophilic sulfur atom of glutathione (p. 504).

**Problem 19.12**    (8Z,11Z,14Z)-8,11,14-Icosatrienoic acid is converted by an enzyme called soybean lipoxidase and molecular oxygen to (8Z,11Z,13E)-15-hydroperoxy-8,11,13-icosatrienoic acid. Write structural formulas for starting material and product, and propose a mechanism for the conversion.

**Problem 19.13**    Write a mechanism for the reaction of the oxirane ring in leukotriene A with the peptide glutathione, paying close attention to the stereochemistry. The abbreviated structures shown below represent the reactants. In an enzyme system, acids (HB$^+$) and bases (B:) are readily available.

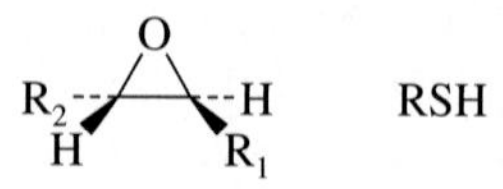

leukotriene A        glutathione

## 19.7  Oxidation of Phenols

### A. Quinones

The readiness of phenols to undergo electrophilic aromatic substitution reactions (p. 366) indicates that the electrons of the aromatic ring must be available to react with electron-deficient species. Oxidizing agents, which, like electrophiles, can be regarded as electron acceptors, react with phenols as well as with aromatic amines. Two types of compounds that are oxidized with particular ease are aromatic compounds that have hydroxyl and/or amino groups ortho or para to each other. These compounds are oxidized to brightly colored carbonyl compounds known as **quinones.**

An important example of this reactivity is the oxidation of 1,4-dihydroxybenzene, commonly known as **hydroquinone,** by silver ion. The reaction is used in developing photographic film, which contains finely divided grains of silver bromide in a gelatin layer. Light activates particles of silver bromide so that they become especially susceptible to reduction to free silver. The development process converts these silver bromide particles to silver metal, which creates a darkening of the film at those points. A basic aqueous solution of hydroquinone is used as the reducing agent. The essential chemistry of the process is shown below.

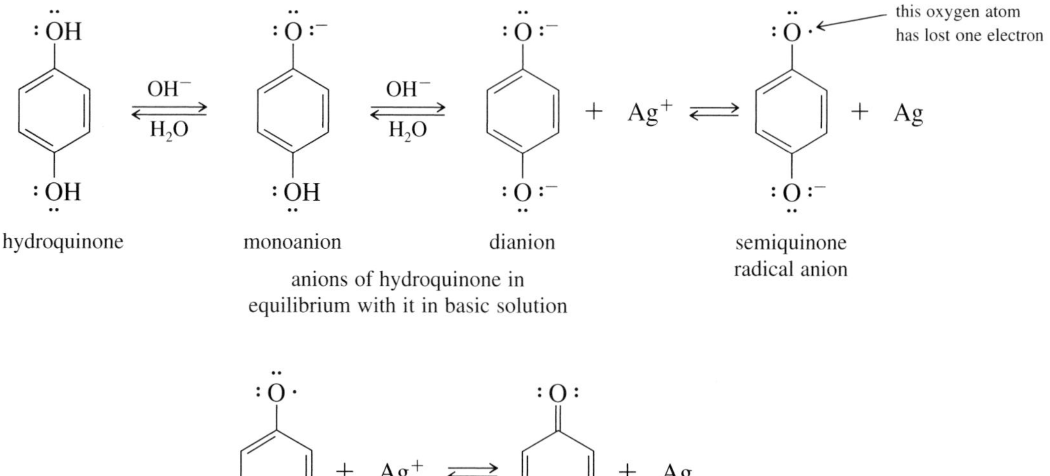

hydroquinone      monoanion      dianion

anions of hydroquinone in
equilibrium with it in basic solution

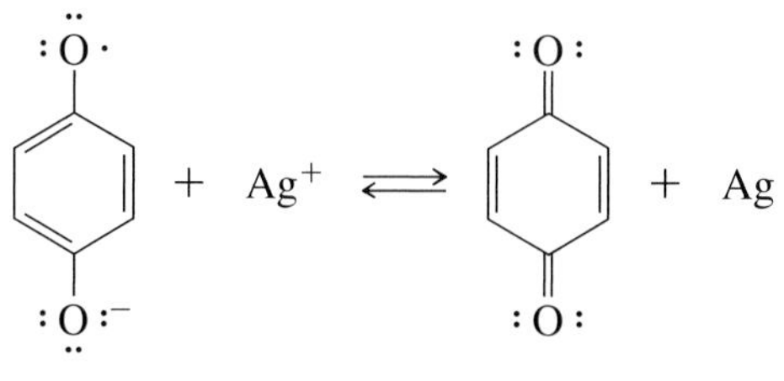

this oxygen atom
has lost one electron

semiquinone
radical anion

semiquinone
radical anion

*p*-benzoquinone

*yellow*

Hydroquinone is converted into its anion by base (p. 655). The anion is oxidized in two steps by silver ion, losing an electron at each stage. Two silver ions gain one electron apiece to become metallic silver. The intermediate in the oxidation is a radical anion known as **semiquinone.** One of its oxygen atoms has an unpaired electron and thus has radical character; the other bears a negative charge. Both the radical character and the negative charge are delocalized to the ring carbon atoms and to the other oxygen atom.

Hydroquinone and $p$-benzoquinone are interconvertible by oxidation and reduction reactions. The metal ion is similarly related to the free metal. The reactions are summarized below as two half reactions.

$$\text{reduction half reaction:} \quad Ag^+ \;+\; e^- \;\rightleftarrows\; Ag$$
$$\underset{\text{silver ion}}{} \qquad\qquad \underset{\text{metallic silver}}{}$$

oxidation half reaction:

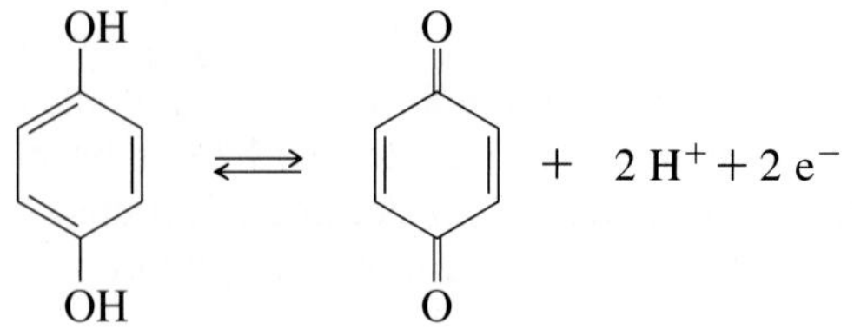

|  |  |
|---|---|
| hydroquinone | $p$-benzoquinone |

It is obvious from an examination of the two half reactions that it will take two silver ions to oxidize one molecule of hydroquinone to quinone.

Hydroquinone, which is colorless, forms a deeply colored complex with quinone. This complex, called **quinhydrone,** is much less soluble than either of its components. A molecule of hydroquinone and a molecule of quinone are held together by the attraction of the electron-deficient quinone ring for the electrons of the electron-rich hydroquinone ring. Such a complex is called a **charge-transfer complex,** indicating that some actual transfer of electronic charge takes place between the rings, so the bond between them has some ionic character.

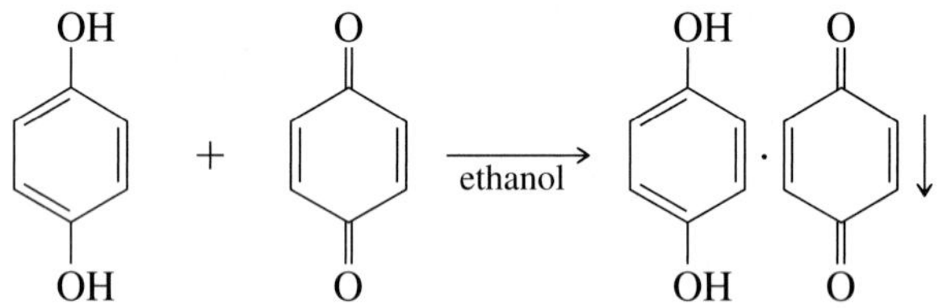

|  |  |  |
|---|---|---|
| hydroquinone | $p$-benzoquinone | quinhydrone |
| *colorless* | *yellow* | *deep green* |

Quinones may have the two carbonyl groups para to each other, as in $p$-benzoquinone, or ortho to each other, as in $o$-benzoquinone.

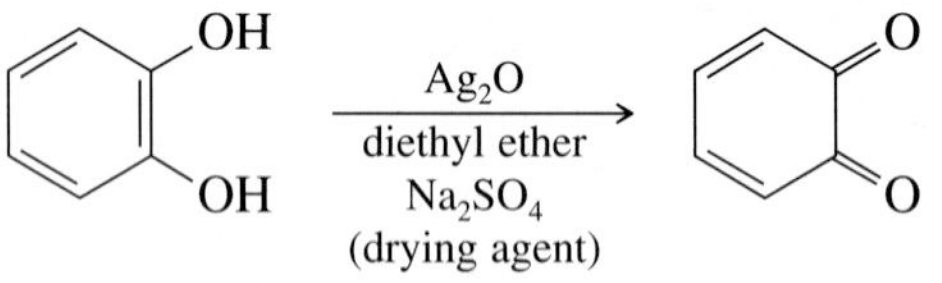

|  |  |
|---|---|
| 1,2-dihydroxybenzene catechol | $o$-benzoquinone |

$o$-Benzoquinone is rather unstable, especially in the presence of moisture, and is prepared with very dry reagents.

All quinones are intensely colored, but ortho quinones are more highly colored than para quinones. The chromophore responsible for the color is the arrangement of conjugated double bonds inside and outside the six-membered ring. The arrangements found in $p$-benzoquinone and $o$-benzoquinone are given the name **quinoid structures** regardless of whether there are oxygen atoms at the ends of the system.

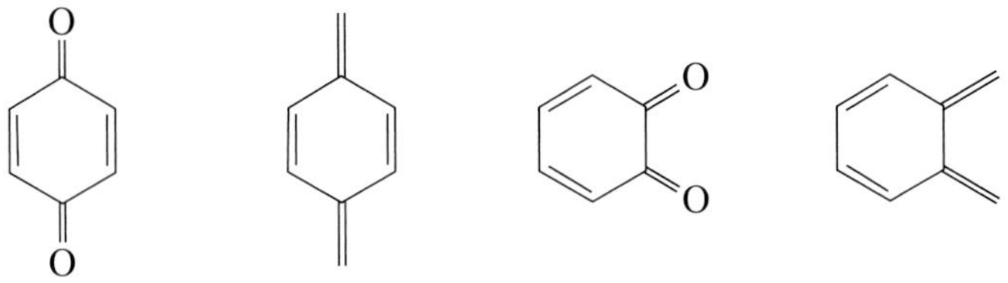

p-benzoquinone        p-quinoid            o-benzoquinone       o-quinoid
                      arrangement of                            arrangement of
                      double bonds                              double bonds

For example, the resonance contributors of polycyclic aromatic hydrocarbons can be analyzed in terms of whether the rings are benzenoid or quinoid in character. Resonance contributors that contain more benzenoid rings than quinoid ones are the major ones for polycyclic aromatic hydrocarbons.

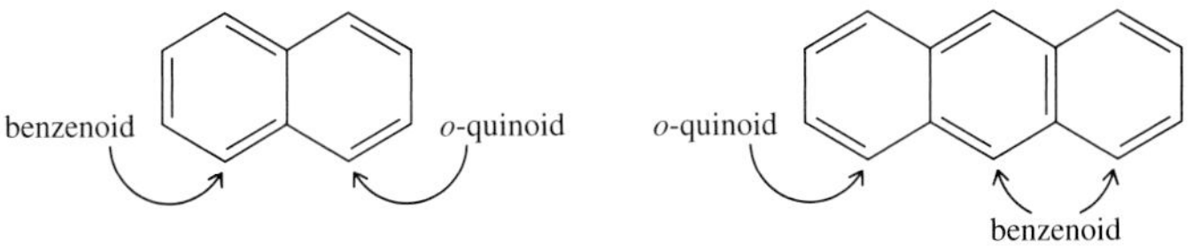

a resonance contributor
of naphthalene with
one benzenoid and
one *o*-quinoid ring

a resonance contributor
of anthracene with one
*o*-quinoid and two
benzenoid rings

Many important natural products are phenols or quinones. A number of the more complex structures, such as vitamin E and vitamin K, are discussed in Section 19.7C (p. 795). Two simple plant products that are *p*-naphthoquinones are 2-hydroxy-*p*-naphthoquinone and 8-hydroxy-*p*-naphthoquinone.

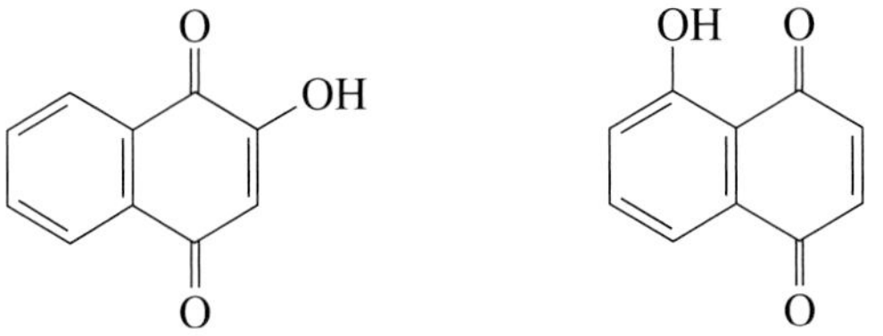

2-hydroxy-*p*-naphthoquinone
lawsone

*pigment in henna*
*orange dye*

8-hydroxy-*p*-naphthoquinone
juglone

*pigment from walnut shells*
*brown dye*

Lawsone is extracted from the leaves of the henna plant and is used by women in the Middle East to dye their hair, their palms, and their fingertips a reddish color. Juglone is the brown stain found in the outer soft shells of walnuts.

Quinones are also important biologically because of the ease with which they are reduced to phenols. Such pairs of compounds serve as mediators of oxidation and reduction reactions in living organisms (p. 795).

**Problem 19.14**   Complete the following equations by first deciding in each case whether the reagent over the arrow is an oxidizing agent (will accept electrons) or a reducing agent (will give up electrons).

(a) 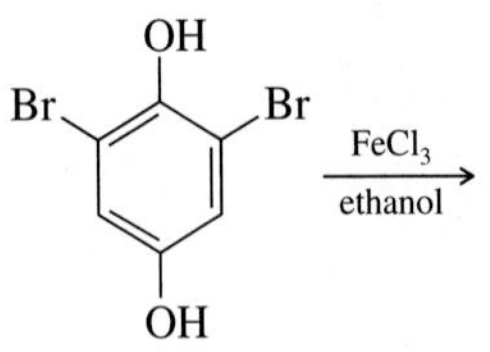 [structure with OH, Br, Br, OH] →(FeCl₃, ethanol)

(b) 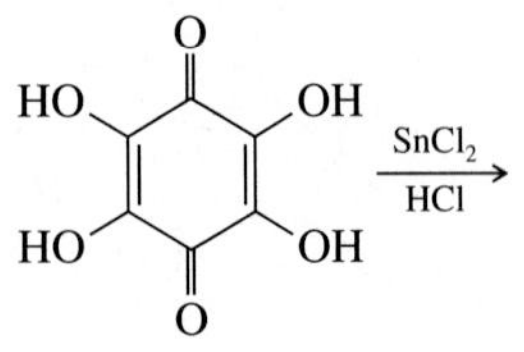 [quinone structure with HO, OH, HO, OH] →(SnCl₂, HCl)

(c) 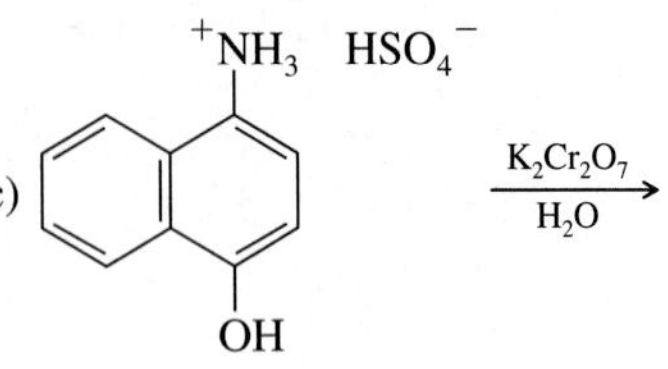 [naphthalene structure with $^+NH_3$ $HSO_4^-$, OH] →(K₂Cr₂O₇, H₂O)

(d) 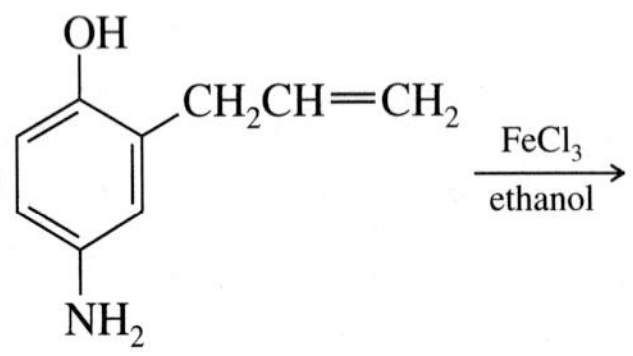 [structure with OH, CH₂CH=CH₂, NH₂] →(FeCl₃, ethanol)

## B.  Phenols as Antioxidants

Substituted phenols are used as **antioxidants** in processed foods. Labels on cereals, cookies, rice products, and many other grocery items frequently list BHA or BHT as an ingredient. The label will sometimes say that the substance was added to retard spoilage or rancidity. BHA stands for *butylated hydroxyanisole* and is a mixture of *tert*-butylmethoxyphenols. BHT is *butylated hydroxytoluene*, or 2,6-di-*tert*-butyl-4-methylphenol.

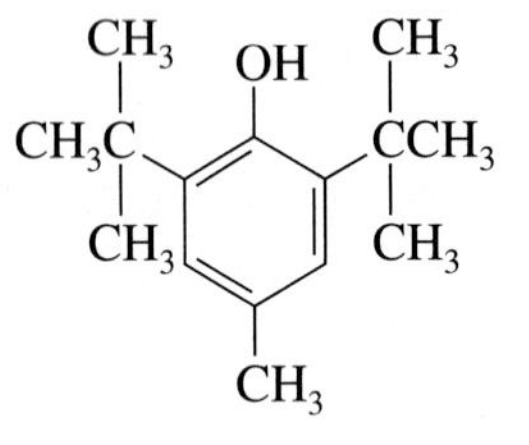

2,6-di-*tert*-butyl-4-methylphenol        2-*tert*-butyl-4-methoxyphenol 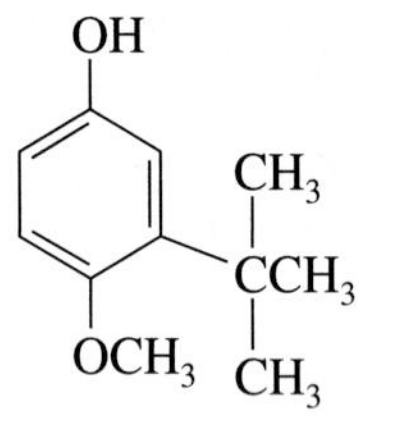        3-*tert*-butyl-4-methoxyphenol

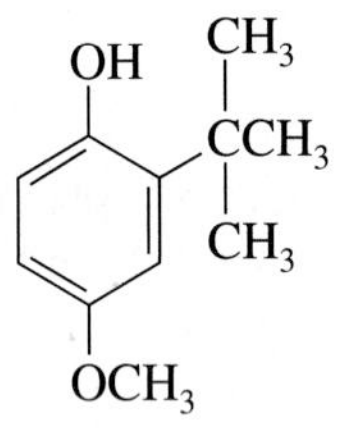

*butylated hydroxytoluene*                *butylated hydroxyanisole*
BHT                                        BHA

*two commercially important antioxidants*

Oxygen from the air reacts with compounds containing double bonds by free-radical chain reactions (p. 786). If these reactions are not stopped, the oils in food products stored at room temperature in warehouses and on supermarket shelves are gradually oxidized and become rancid. Not only do rancid oils taste bad, they are also toxic. As the food-processing industry became more and more centralized, it became necessary for foods to have a long shelf life. Therefore, chemists developed additives that break up the free-radical chain reactions by reacting themselves with any radicals that form. Phenols lose electrons to radicals. BHA and BHT give hindered phenoxy radicals that are unreactive and thus terminate free-radical chain reactions. As a result, they are used to retard free-radical oxidation reactions in food products, among other things. A natural antioxidant found in foods that have a high content of polyunsaturated fatty acids, such as vegetable oils, is vitamin E, discussed in the next section.

---

**Problem 19.15**

(a) BHA is prepared commercially from *p*-methoxyphenol and 2-methylpropene. Suggest how this reaction could be carried out.

(b) A similar reaction is used to make BHT from *p*-methylphenol and 2-methylpropene. Why does the BHA sold commercially consist of a mixture of isomers, whereas the reaction to give BHT has greater regioselectivity?

## C. Vitamin E and Vitamin K

Vitamin E, also called $\alpha$-tocopherol, has phenolic and isoprenoid components. Its structure can be dissected into a trimethylhydroquinone and a chain containing four isoprene units, a phytyl group.

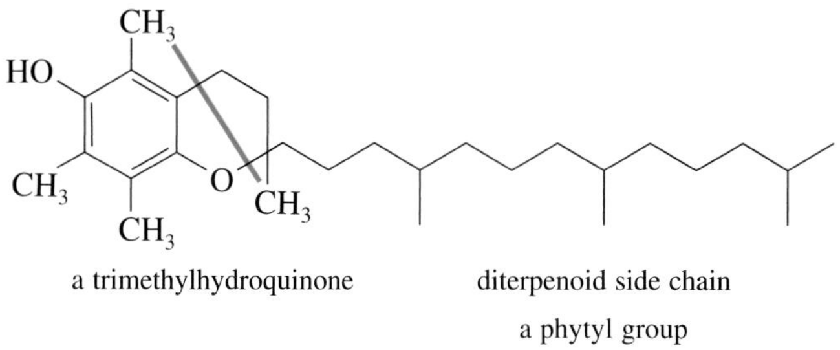

a trimethylhydroquinone          diterpenoid side chain
a phytyl group

$\alpha$-tocopherol, or vitamin E

Vitamin E has been recognized as having an important biological role as a potent antioxidant. It is especially important in preventing the formation of hydroperoxides from polyunsaturated fatty acids (p. 639) and is found most abundantly in seeds that are rich in such fatty acids.

In 1930, researchers discovered that chicks developed hemorrhages when fed a highly artificial diet from which all fatty components had been removed. This observation led to the discovery that a fat-soluble vitamin, named vitamin K, was important in the clotting of blood. Later it was determined that a family of compounds has the activity shown by vitamin K. These compounds have a 2-methyl-1,4-naphthoquinone structure with different isoprenoid side chains attached at carbon 3. Vitamin K is also important in the metabolism of calcium.

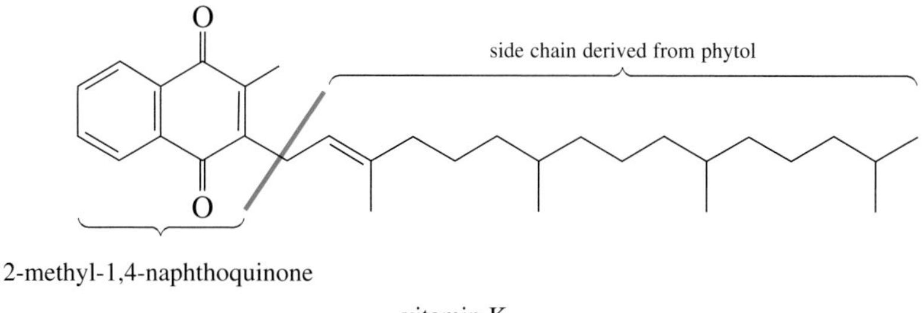

side chain derived from phytol

2-methyl-1,4-naphthoquinone

vitamin $K_1$
phylloquinone

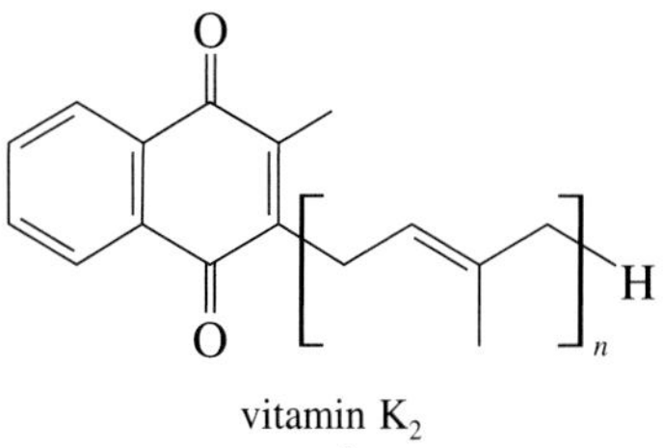

vitamin $K_2$
menaquinones
$n = 6, 7, 8, \text{ or } 9$

Vitamin $K_1$, also known as phylloquinone, has a side chain derived from the diterpene alcohol phytol; the side chain in vitamin E is also derived from phytol. Vitamin $K_2$ is actually a group of compounds, the menaquinones, which have different numbers of isoprene units in the side chain. Vitamin $K_2$ is involved in bacterial **electron transport,** the transfer of electrons from one site to another, which is essential to the occurrence of oxidation and reduction reactions in cells.

The K vitamins are found widely in the leaves of green plants: Vitamin K is also synthesized by bacteria in the lower intestines. Because the compounds are so widespread, it is unlikely that a person would develop a deficiency unless he or she refused to eat any green vegetables at all.

A number of compounds are antagonists of vitamin K; they increase the clotting time of blood. Such compounds are known as **anticoagulants.** For example, salicylic acid, a product of the hydrolysis of aspirin, has this effect. Two more powerful anticoagulants, dicoumarol and Warfarin, are also shown below.

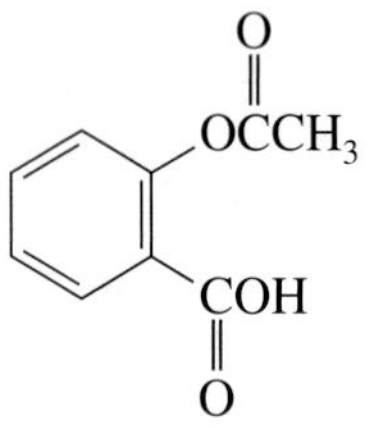

acetylsalicylic acid
aspirin

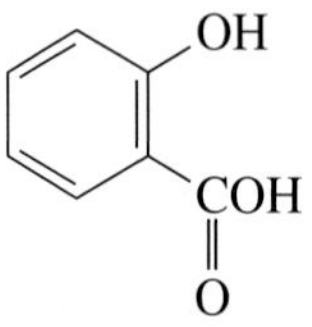

*o*-hydroxybenzoic acid
salicylic acid

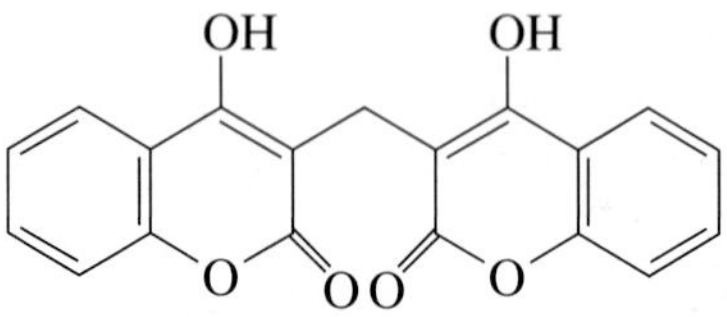

dicoumarol

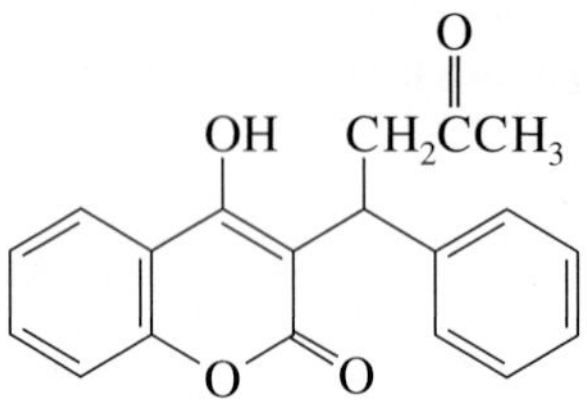

Warfarin

*some compounds that increase blood clotting times*
*anticoagulants*

Dicoumarol is a component of spoiled clover hay that was discovered because animals that ate it started to bleed abnormally. It is used medically to prevent blood clots. Warfarin is a synthetic compound, deliberately designed to be an anticoagulant and used as a rat poison. Note that all the anticoagulants shown have a phenolic or enolic hydroxyl group. Dicoumarol and Warfarin are also lactones of phenols. The phenol lactone system found in these two compounds is also found in coumarin, a fragrant constituent of clover.

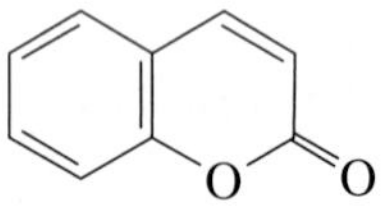

coumarin

*a phenol lactone*

Study Guide
Concept Map 19.5

Coumarin itself does not interfere with the clotting of blood and has been used as a flavoring agent.

**Problem 19.16**    The anticoagulant Warfarin is used in the form of its sodium salt. What is the structure of this compound?

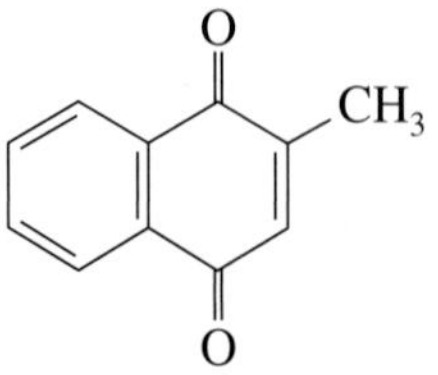

2-methyl-1,4-naphthoquinone

**Problem 19.17**    2-Methyl-1,4-naphthoquinone is a vitamin K analogue. The carbon skeleton of 2-methyl-1,4-naphthoquinone has been synthesized by a Diels-Alder reaction between a diene and a quinone. What diene and quinone would you use? What further steps would be necessary to convert the Diels-Alder product to the naphthoquinone?

## Table 19.1  Summary of Free-Radical Reactions

### Halogenation Reactions

| Compound Reacting | Reagent | Initiation Conditions | Radical Intermediate(s) | Product |
|---|---|---|---|---|
| 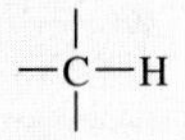 $-\overset{\mid}{\underset{\mid}{C}}-H$  tertiary > secondary > primary > methane  *fastest*　　　*slowest* | $Cl_2$ or $Br_2$ | $h\nu$ or $\Delta$ | 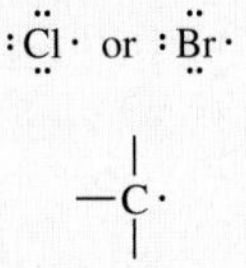 $:\!\overset{..}{\underset{..}{Cl}}\!\cdot$ or $:\!\overset{..}{\underset{..}{Br}}\!\cdot$ ;  $-\overset{\mid}{\underset{\mid}{C}}\cdot$ | 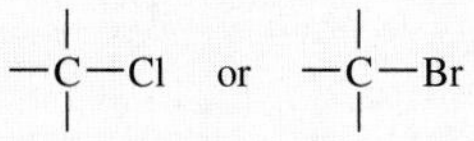 $-\overset{\mid}{\underset{\mid}{C}}-Cl$  or  $-\overset{\mid}{\underset{\mid}{C}}-Br$ |
| 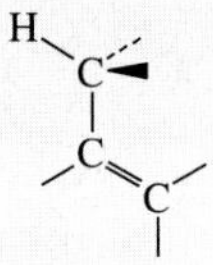 (allylic structure) | 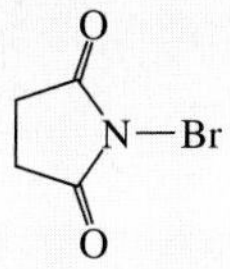 N—Br (succinimide) | $h\nu,\ \Delta$  peroxides | 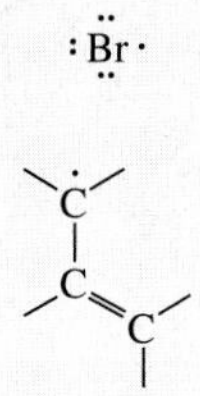 $:\!\overset{..}{\underset{..}{Br}}\!\cdot$ ; (allyl radical) | 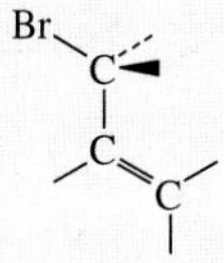 (allylic bromide) |
| 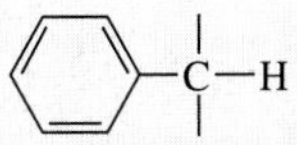 $\text{Ph}-\overset{\mid}{\underset{\mid}{C}}-H$ (benzylic) | 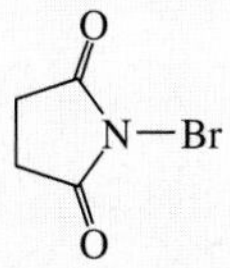 N—Br (succinimide) | $h\nu,\ \Delta$  peroxides | 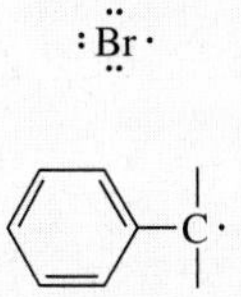 $:\!\overset{..}{\underset{..}{Br}}\!\cdot$ ; $\text{Ph}-\overset{\mid}{\underset{\mid}{C}}\cdot$ | 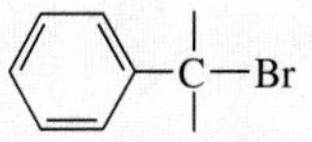 $\text{Ph}-\overset{\mid}{\underset{\mid}{C}}-Br$ |
|  | $Cl_2$ or $Br_2$ | $h\nu$ or $\Delta$ | 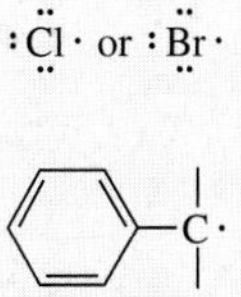 $:\!\overset{..}{\underset{..}{Cl}}\!\cdot$ or $:\!\overset{..}{\underset{..}{Br}}\!\cdot$ ; $\text{Ph}-\overset{\mid}{\underset{\mid}{C}}\cdot$ | 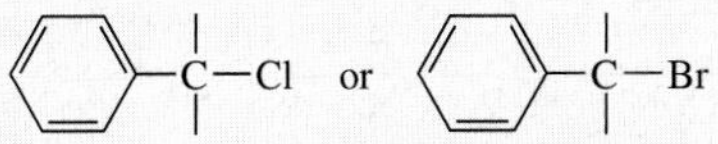 $\text{Ph}-\overset{\mid}{\underset{\mid}{C}}-Cl$  or  $\text{Ph}-\overset{\mid}{\underset{\mid}{C}}-Br$ |

### Addition Reactions to Alkenes

| Alkene | Reagent | Initiation Conditions | Radical Intermediate(s) | Product(s) |
|---|---|---|---|---|
| $R-\overset{R'}{\underset{}{C}}=CH_2$ | $R-\overset{R'}{\underset{}{C}}=CH_2$ | peroxides or 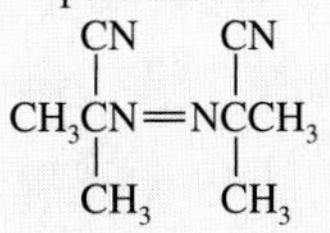  $CH_3\overset{CN}{\underset{CH_3}{C}}N{=}N\overset{CN}{\underset{CH_3}{C}}CH_3$ | 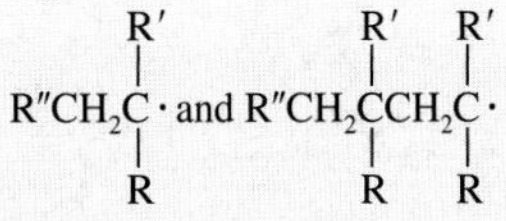 $R''CH_2\overset{R'}{\underset{R}{C}}\cdot$ and $R''CH_2C\overset{R'}{\underset{R}{C}}H_2C\cdot$ | 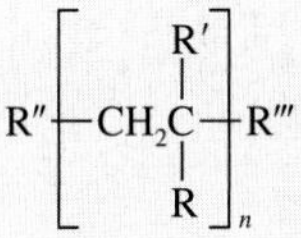 $R''\!\left[\!-CH_2\overset{R'}{\underset{R}{C}}-R'''\right]_n$ |
|  | $Cl_3CBr$ or $CCl_4$ | peroxides, $h\nu$ | 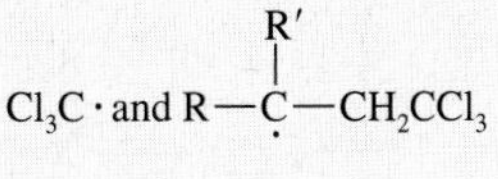 $Cl_3C\cdot$ and $R-\overset{R'}{\underset{\cdot}{C}}-CH_2CCl_3$ | 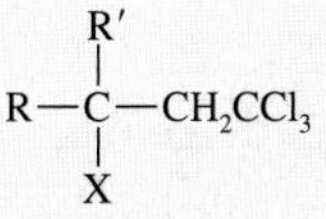 $R-\overset{R'}{\underset{X}{C}}-CH_2CCl_3$  $X = Br$ or $Cl$ |
|  | $R''SH$ | $\Delta$ | 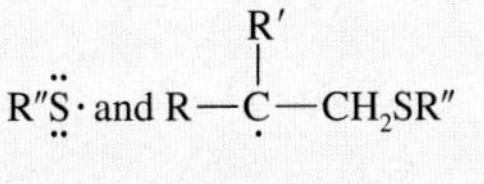 $R''\overset{..}{S}\cdot$ and $R-\overset{R'}{\underset{\cdot}{C}}-CH_2SR''$ | 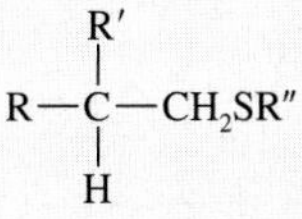 $R-\overset{R'}{\underset{H}{C}}-CH_2SR''$ |
|  | $R''_3SiH$ | peroxides, $\Delta$ | 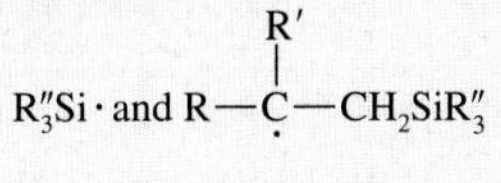 $R''_3Si\cdot$ and $R-\overset{R'}{\underset{\cdot}{C}}-CH_2SiR''_3$ | 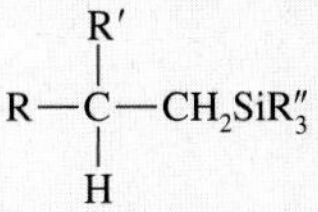 $R-\overset{R'}{\underset{H}{C}}-CH_2SiR''_3$ |

**Table 19.1 (Continued)**

**Addition Reactions to Alkenes**

| Alkene | Reagent | Initiation Conditions | Radical Intermediate(s) | Product(s) |
|---|---|---|---|---|
| 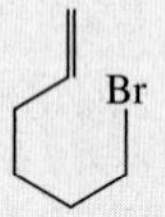 | (intramolecular reaction) | 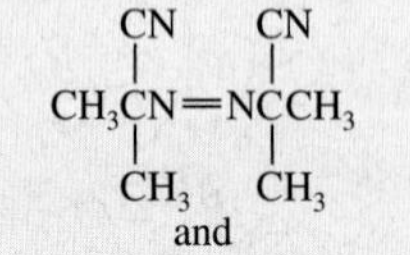 and $(CH_3CH_2CH_2CH_2)_3SnH$ | 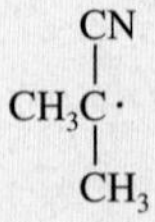 and $(CH_3CH_2CH_2)_3Sn\cdot$ and 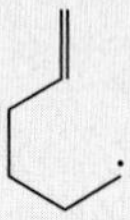 and 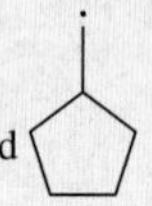 | 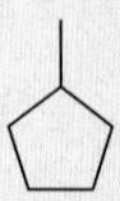 and 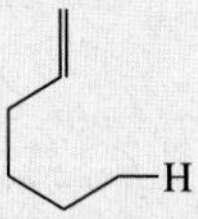 |

**Oxidation Reactions**

| Compound Reacting | Reagent(s) | Radical Intermediate(s) | Product |
|---|---|---|---|
| 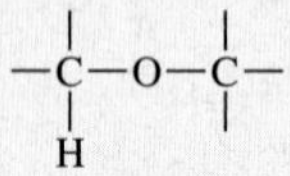 | $O_2$, other R·, metal ion | ROO· and 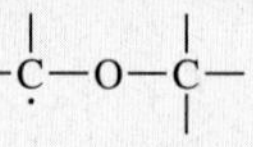 | 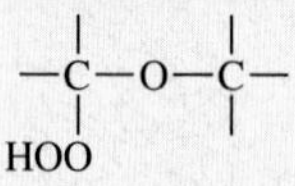 |
| 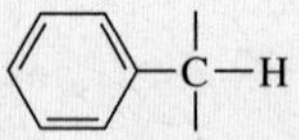 | $O_2$, other R·, metal ion | ROO· and 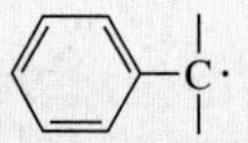 | 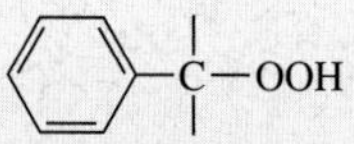 |
| $R\overset{O}{\overset{\|}{C}}H$ | $O_2$ | HOO· and $R\overset{O}{\overset{\|}{C}}\cdot$ | $R\overset{O}{\overset{\|}{C}}OOH \longrightarrow R\overset{O}{\overset{\|}{C}}OH$ |
| 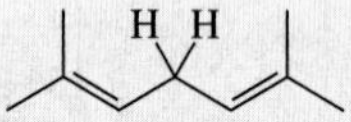 | $O_2$, other R· | ROO· and 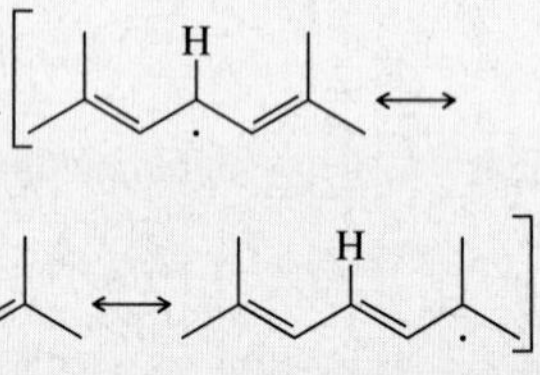 | 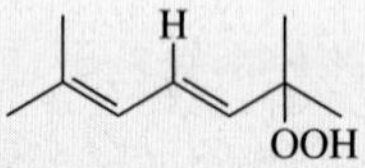 |
| 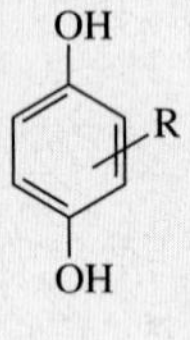 | $Ag^+$, $OH^-$ | 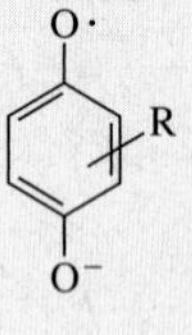 | 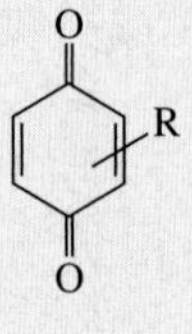 |

## ADDITIONAL PROBLEMS

**19.18** Complete the following equations, showing the product(s) you expect. If the stereochemistry is known for a reaction, make sure your answer shows it.

(a) $CH_3CH=CHCH=CHCH_3$ $\xrightarrow[\text{carbon tetrachloride}]{Br_2 \text{ (1 molar equivalent)}}$

(b) 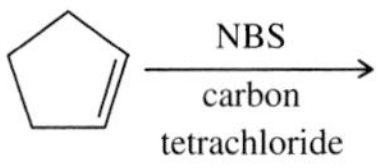 $\xrightarrow[\text{carbon tetrachloride}]{NBS}$

(c) 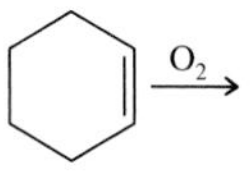 $\xrightarrow{O_2}$

(d) $CH_3CH_2CH_3$ $\xrightarrow[hv]{Cl_2}$

(e) 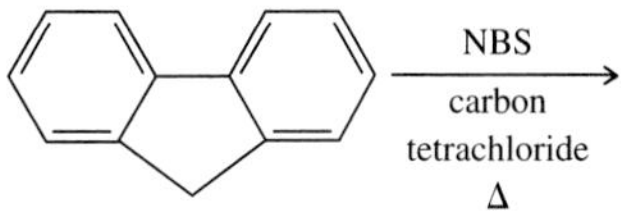 $\xrightarrow[\substack{\text{carbon} \\ \text{tetrachloride} \\ \Delta}]{NBS}$

(f) 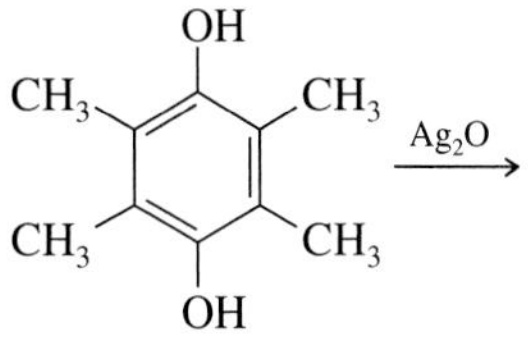 $\xrightarrow{Ag_2O}$

(g) 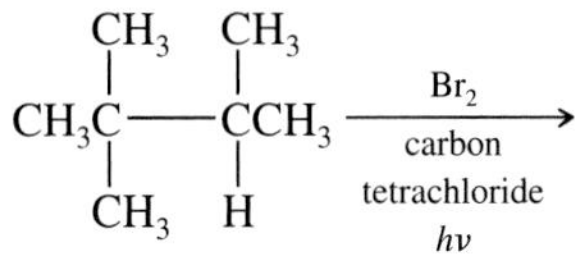 $\xrightarrow[\substack{\text{carbon} \\ \text{tetrachloride} \\ hv}]{Br_2}$

(h) 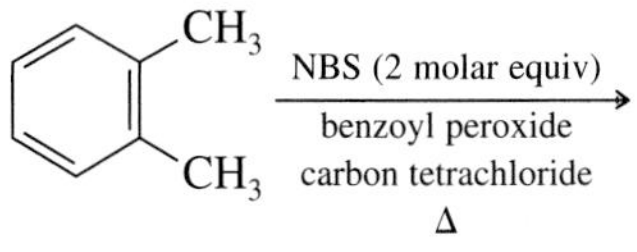 $\xrightarrow[\substack{\text{benzoyl peroxide} \\ \text{carbon tetrachloride} \\ \Delta}]{NBS \text{ (2 molar equiv)}}$

(i) 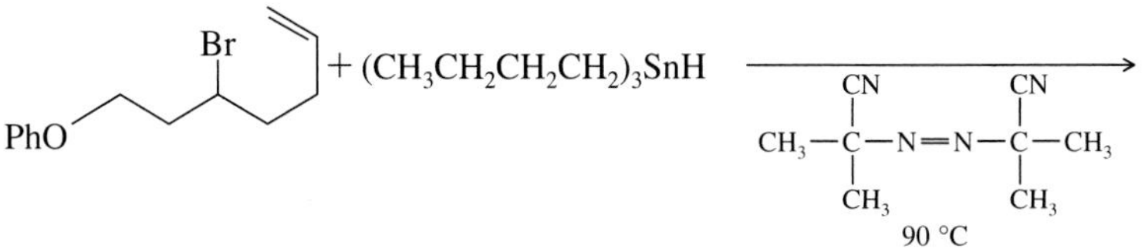 $+ (CH_3CH_2CH_2CH_2)_3SnH$ $\xrightarrow[\substack{\text{(AIBN)} \\ 90\,°C}]{}$

(j) $CH_3CH_2CH_2CH=CH_2 + Cl_3SiH$ $\xrightarrow{280\,°C}$

(k) 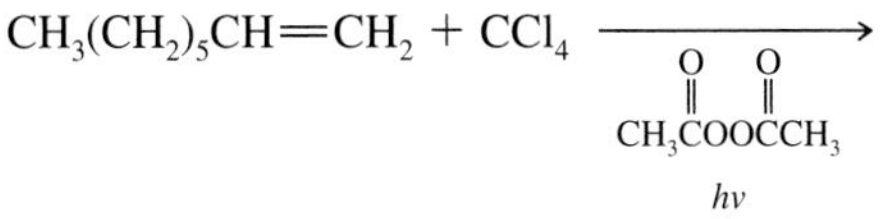 $CH_3(CH_2)_5CH=CH_2 + CCl_4$ $\xrightarrow[hv]{\underset{\text{CH}_3\text{COOCCH}_3}{\overset{O\quad O}{||\quad||}}}$

**19.19** Propose a mechanism that accounts for the formation of both of the products of the following reaction.

major product          minor product

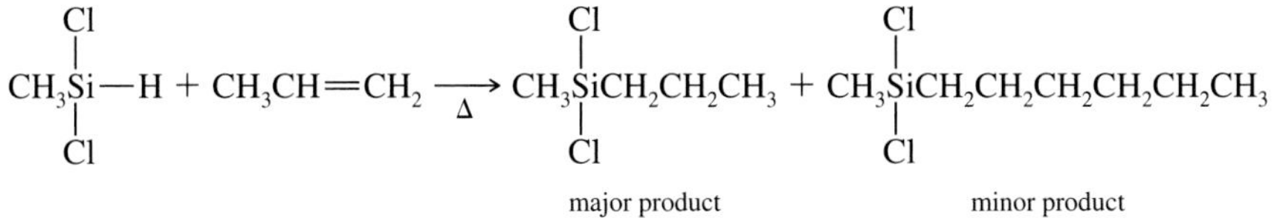

**19.20** The following reactions take place.

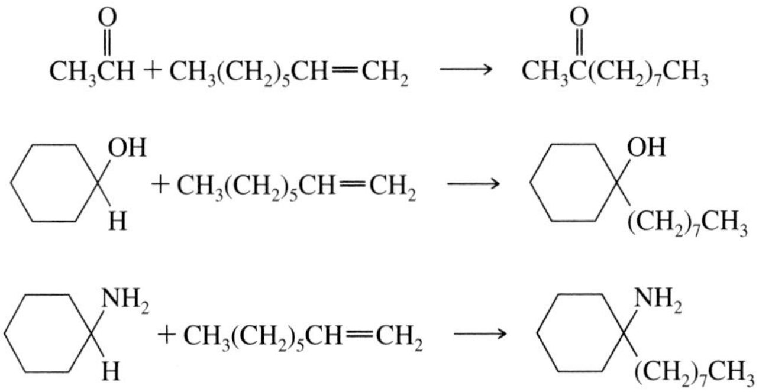

What conditions do you think are necessary for these reactions? Predict whether *tert*-butyl alcohol will react with 1-octene under these conditions.

**19.21** Studies of polymerization reactions involved the following reactions. Provide structural formulas for the reactants or products indicated by letters.

(a)    A  $\xrightarrow[\text{monomer}\quad\underset{\text{PhCOOCPh}}{\overset{O\quad O}{\|\quad\|}}]{}$  B  $\longrightarrow$  poly(methyl methacrylate) product

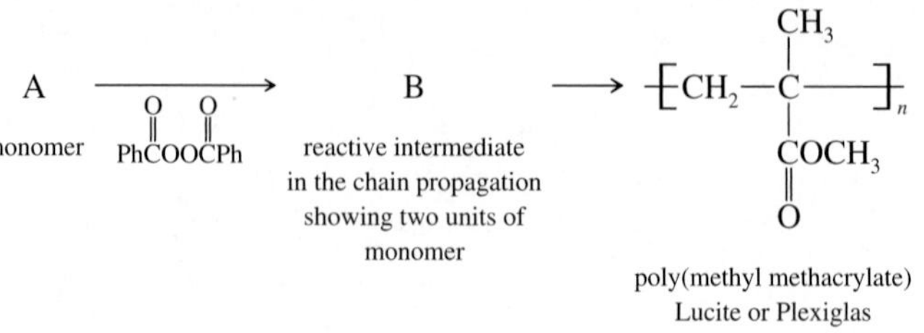

monomer    PhCOOCPh    reactive intermediate in the chain propagation showing two units of monomer

poly(methyl methacrylate)
Lucite or Plexiglas

(b)  styrene-enamine  $\xrightarrow[\underset{\underset{CH_3}{\|}}{CH_3-\overset{CN}{\underset{\underset{CH_3}{\|}}{C}}-N=N-\overset{CN}{\underset{\underset{CH_3}{\|}}{C}}-CH_3}]{}$  C

60 °C

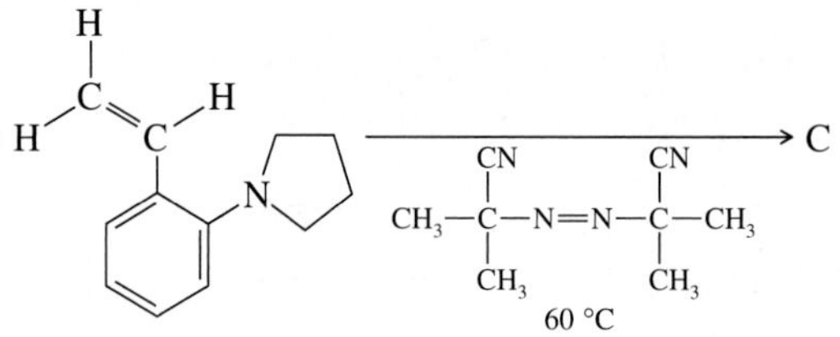

**19.22** How would you carry out each of the following transformations?

(a) diester with iodide and methylene  $\longrightarrow$  bicyclic diester

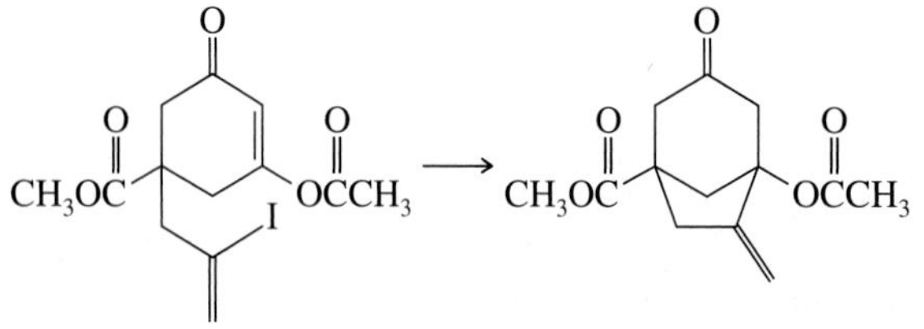

(b)  iodide acetonide  $\longrightarrow$  nitrile acetonide

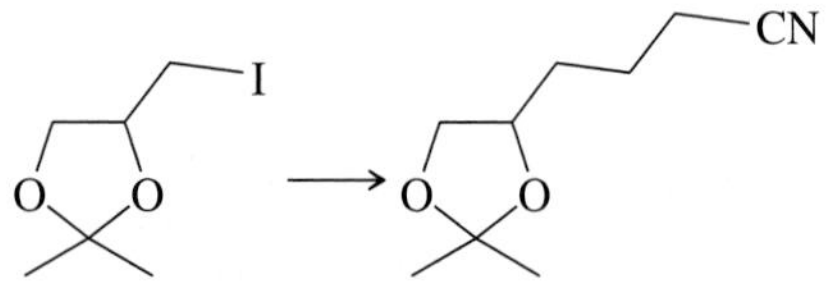

(c)  $CH_3\overset{O}{\underset{\|}{C}}-\overset{CH_3}{\underset{\underset{CH_3}{|}}{C}}-CH_3$  $\longrightarrow$  $CH_2=CHCH_2CH_2CH_2-\overset{O}{\underset{\|}{C}}-\overset{CH_3}{\underset{\underset{CH_3}{|}}{C}}-CH_3$

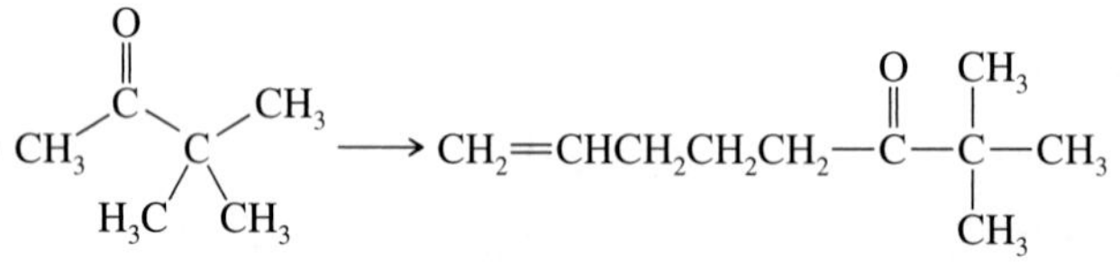

(d)  butadiene  $\longrightarrow$  cyclohexene dicarboxylic acid

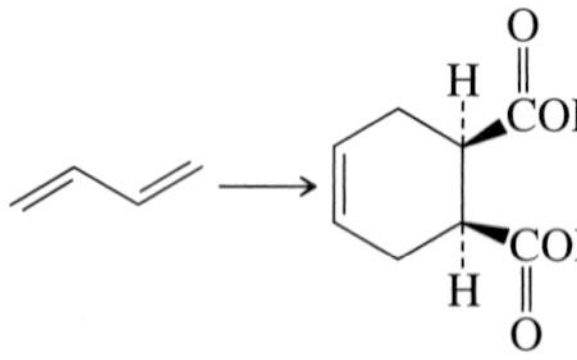

(e)  fluorene  $\longrightarrow$  9-bromo-9-phenylfluorene

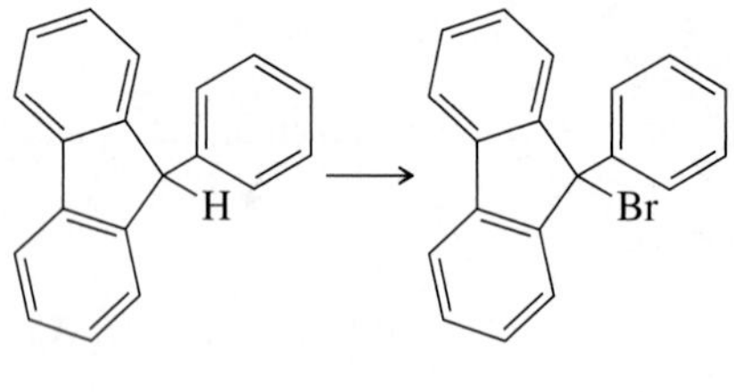

**19.23** The following reactions were used in the synthesis of a prostaglandin. Supply the reagents that are necessary for each transformation.

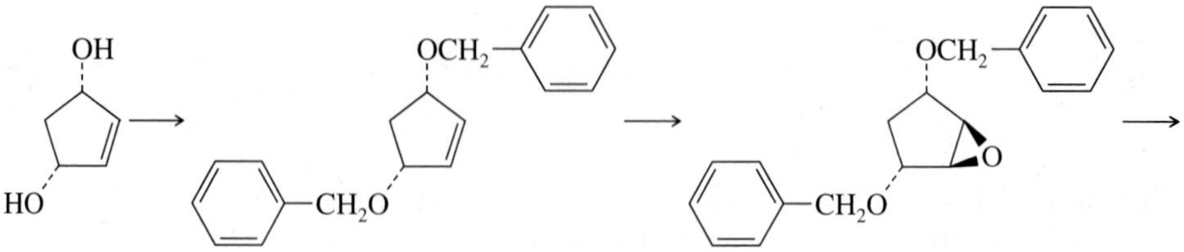

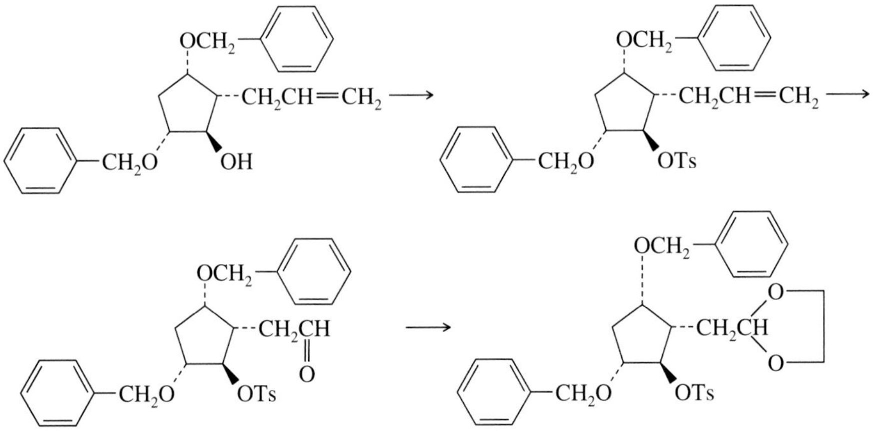

**19.24** When linoleic acid is oxidized in the laboratory using air or pure oxygen, four different hydroperoxides, one of which is the compound formed biologically (p. 787), are the products. Another of the hydroperoxides is the subject of Problem 19.11. These isomers differ in the position of the hydroperoxide group and in the stereochemistry of the double bonds. Write structures for the other two compounds.

**19.25** The species hydrogen trioxide, $HO_3$, has long been postulated to exist in the atmosphere. It has now been produced and detected in a mass spectrometer (p. 467). It was formed by adding one electron to $HO_3^+$, the conjugate acid of ozone.

(a) Draw a Lewis structure for the conjugate acid of ozone.
(b) Draw a Lewis structure for hydrogen trioxide that has no separation of charge and for a resonance contributor with a separation of charge.

**19.26** Some chemistry that is used in the synthesis of compounds for a study of free-radical cyclization reactions is shown below. Supply reagents that will carry out these transformations.

(a)

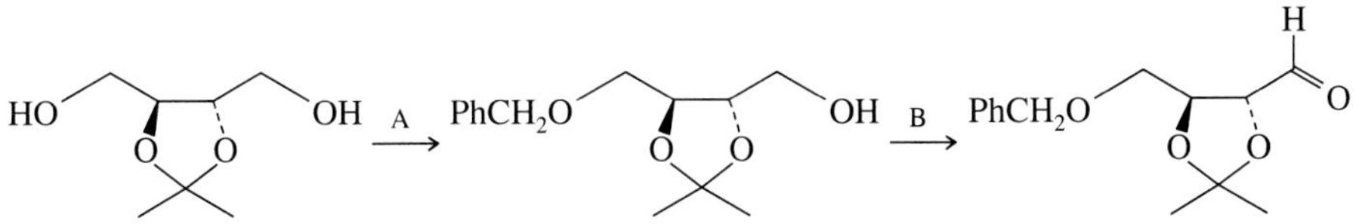

(b)

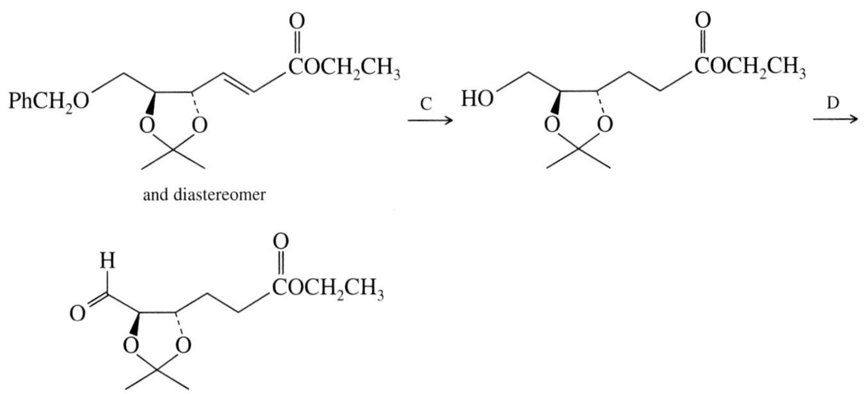

and diastereomer

**19.27** The following compounds were used in studies of the cyclization reactions of radicals.

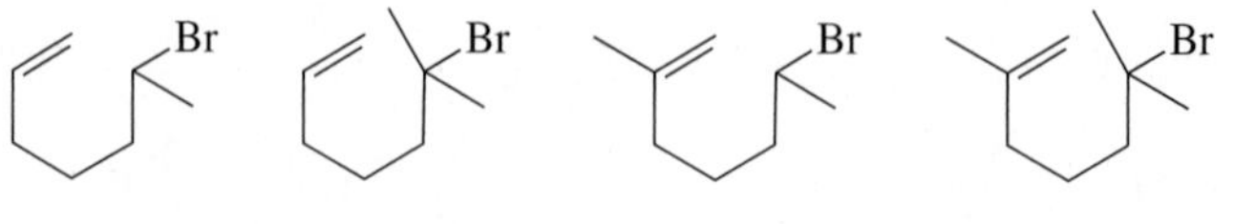

Assuming that 5-bromo-1-pentene and 6-methyl-6-hepten-2-one are available as starting materials, how would you synthesize the bromoalkenes shown above?

**19.28** The starting material used in the synthesis of silphiperfol-6-ene (Problem 19.10, p. 785) was synthesized in the following steps. Supply reagents for each of the transformations.

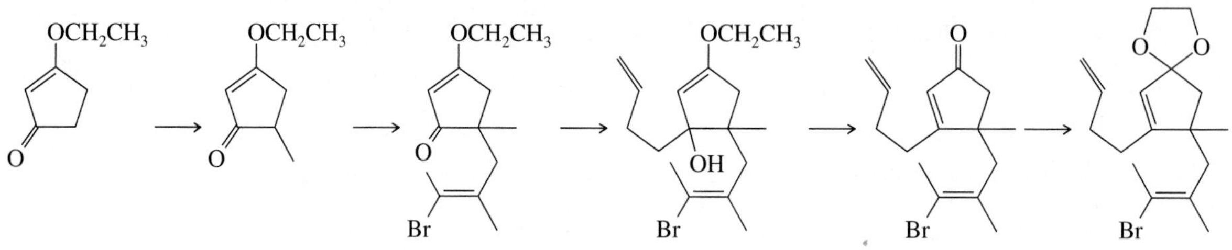

**19.29** The following reactions were part of a synthesis of an inhibitor of an enzyme that is essential to the replication of the HIV-1 virus.

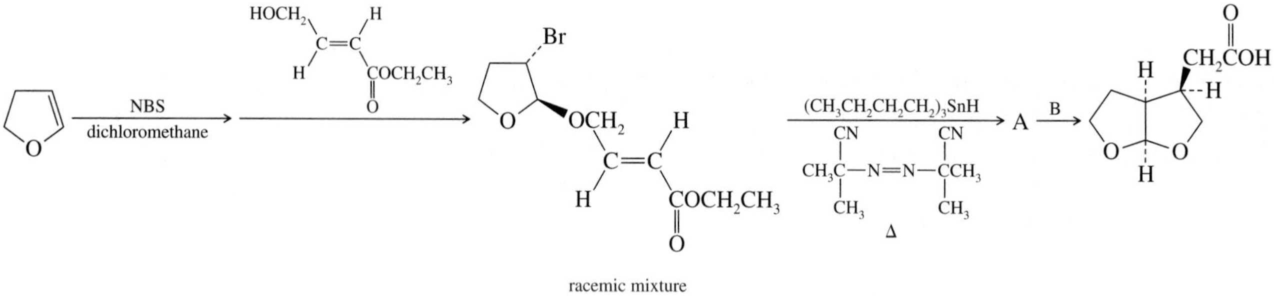

racemic mixture

(a) Provide a mechanism for the first step of the reaction.
(b) Supply structural formulas for Compound A and Reagents B.

**19.30** Cholesteryl acetate is converted in the laboratory into the acetic ester of 7-dehydrocholesterol, a compound that is converted in the skin by sunlight to vitamin $D_3$ (p. 1082). 7-Dehydrocholesterol has a conjugated system in the B ring of the steroid nucleus, with double bonds between carbon atoms 5 and 6 and carbon atoms 7 and 8 (p. 755). How would you convert cholesterol to 7-dehydrocholesteryl acetate? (*Hint:* What is the major route by which double bonds are introduced into molecules?)

**19.31** Natural products isolated from various *Eucalyptus* species show wide antibacterial and anti-HIV activity. Some of the reactions involved in the synthesis of one of these compounds are shown below. Provide structural formulas for the products indicated by letters, and show how spectral data, where given, support your structure.

(a)

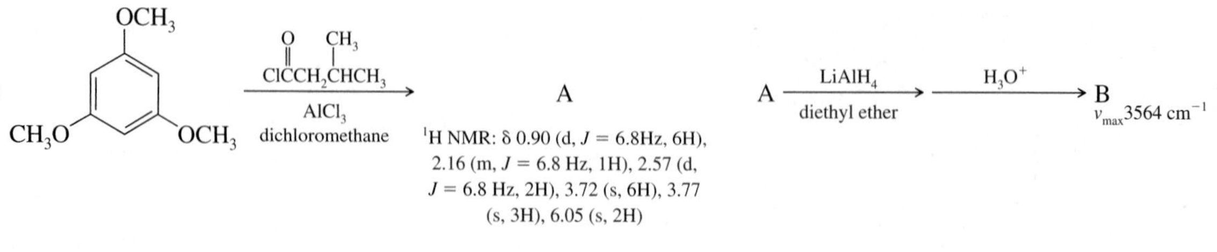

$^1$H NMR: δ 0.90 (d, $J$ = 6.8Hz, 6H), 2.16 (m, $J$ = 6.8 Hz, 1H), 2.57 (d, $J$ = 6.8 Hz, 2H), 3.72 (s, 6H), 3.77 (s, 3H), 6.05 (s, 2H)

(b)

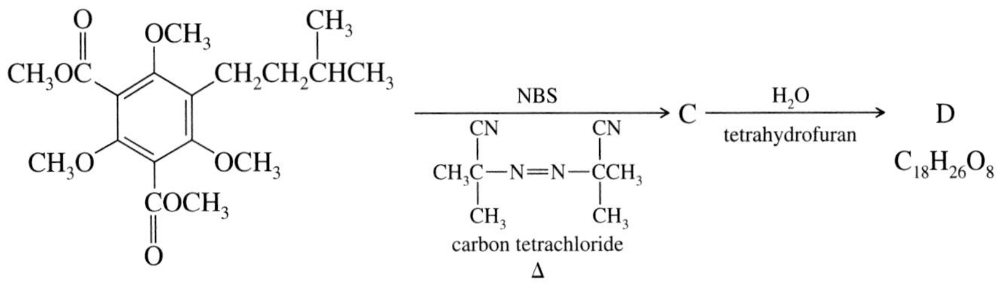

**19.32** Tamoxifen is widely used in the treatment of metastatic breast cancer in humans. It inhibits the growth of tumors by blocking the binding of the female sex hormone estrogen to the estrogen receptors on tumors that are dependent on estrogen for their growth. Synthesis of tamoxifen with radioactive labels helps follow the drug in the body and investigate how it is metabolized. The following reactions were used in the synthesis of tribromotamoxifen, which can be converted to tritium ($^3$H, T)-labeled tamoxifen by treatment with $T_2$ in the presence of a catalyst. Provide structural formulas for reagents or products designated by letters.

(a)

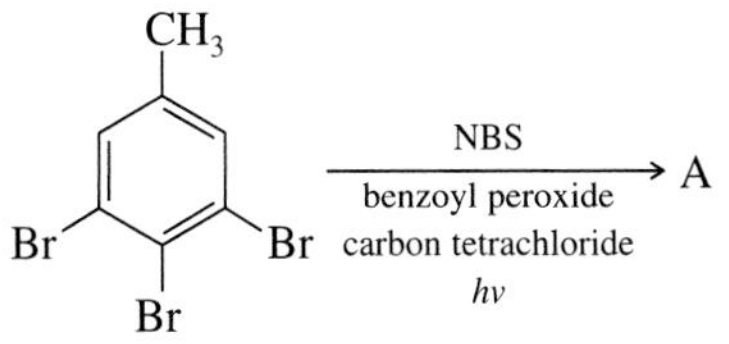

(b)

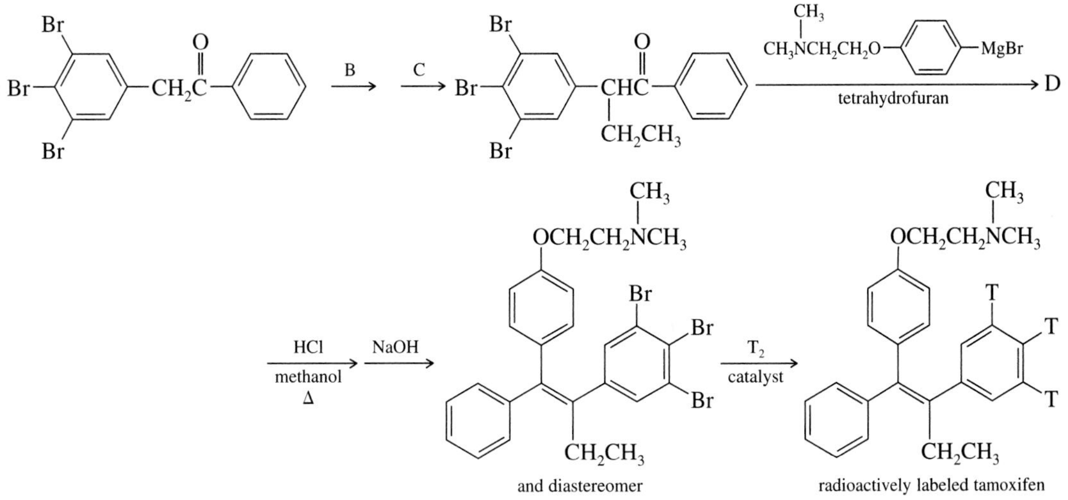

# 20 The Chemistry of Amines

**A Look Ahead**

Earlier chapters presented methods for the preparation of many compounds containing nitrogen. Among these were azides (p. 236), oximes (p. 575), imines (p. 574), amides (p. 628), and nitro compounds (p. 366). This chapter will describe how to convert these types of compounds to amines through reduction reactions.

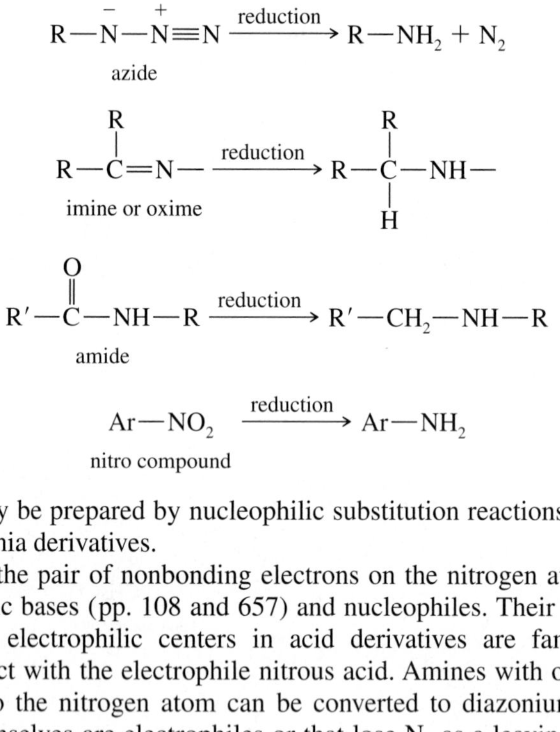

$$R\!-\!\overset{-}{N}\!-\!\overset{+}{N}\!\equiv\!N \xrightarrow{\text{reduction}} R\!-\!NH_2 + N_2$$
azide

$$R\!-\!\underset{\substack{\\ \text{imine or oxime}}}{\overset{\overset{\displaystyle R}{|}}{C}}\!=\!N\!- \xrightarrow{\text{reduction}} R\!-\!\underset{\underset{\displaystyle H}{|}}{\overset{\overset{\displaystyle R}{|}}{C}}\!-\!NH\!-$$

$$R'\!-\!\overset{\overset{\displaystyle O}{\|}}{C}\!-\!NH\!-\!R \xrightarrow{\text{reduction}} R'\!-\!CH_2\!-\!NH\!-\!R$$
amide

$$Ar\!-\!NO_2 \xrightarrow{\text{reduction}} Ar\!-\!NH_2$$
nitro compound

Amines also may be prepared by nucleophilic substitution reactions of ammonia or protected ammonia derivatives.

Because of the pair of nonbonding electrons on the nitrogen atom, amines are important organic bases (pp. 108 and 657) and nucleophiles. Their reactions as nucleophiles with electrophilic centers in acid derivatives are familiar (p. 628). Amines also react with the electrophile nitrous acid. Amines with only one organic group bonded to the nitrogen atom can be converted to diazonium ions, reactive species that themselves are electrophiles or that lose $N_2$ as a leaving group and undergo substitution reactions.

## 20.1 Structure and Natural Occurrence of Amines

**Amines,** compounds in which one or more of the hydrogen atoms of an ammonia molecule have been replaced by an organic group, are found widely in nature. Many plants synthesize complex amines called **alkaloids,** some of which have medicinal or poisonous properties. Morphine, one of the most effective painkillers known, and quinine, an important medication for malaria, are among some of the alkaloids that have been isolated and used by humans. Nicotine, the addictive and stimulating component of tobacco, is also highly toxic and, in fact, can be used as an insecticide.

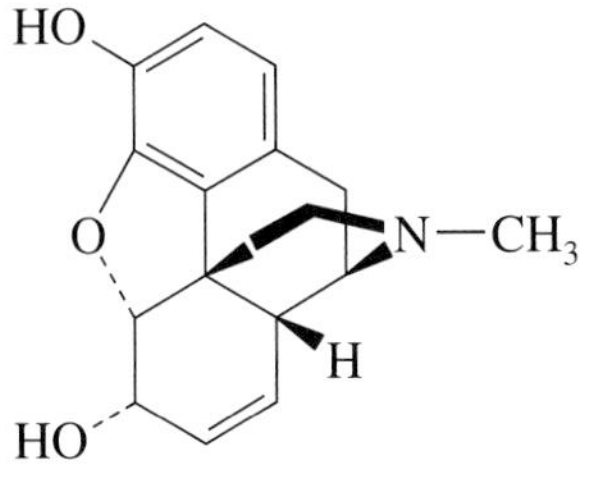

(−)-morphine 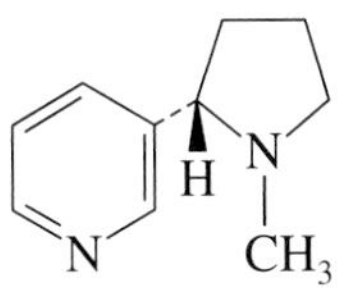          (−)-nicotine 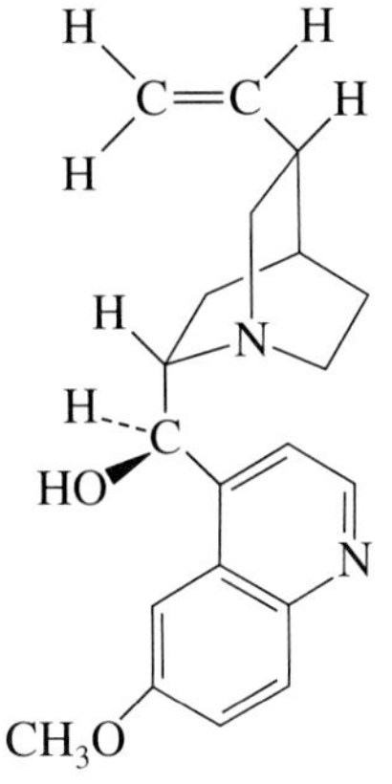          (−)-quinine

*some naturally occurring amines present*
*in plants*

**Proteins,** the complex constituents of living organisms and of the enzyme systems that catalyze chemical processes, are made up of **amino acids.** In amino acids, an amino group is substituted on the second carbon atom of a carboxylic acid. Proteins are formed when amino acids are bonded together by amide bonds, called **peptide linkages** (p. 615), between the amino group of one amino acid and the carboxylic acid group of another.

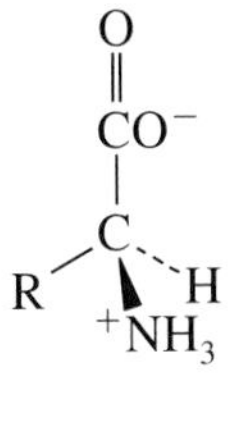

an amino acid 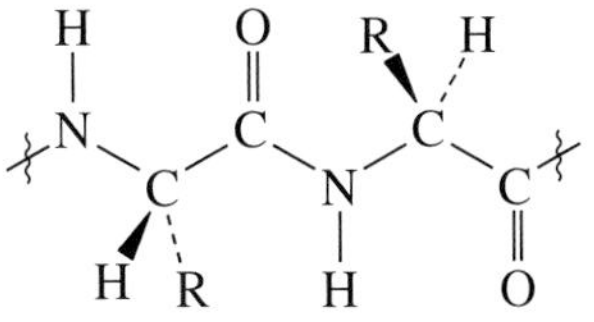          fragment of a protein
molecule showing two
amino acid units

*amino groups, in a free amino acid and in an*
*amide bond in part of a protein chain*

The synthetic methods for preparing amino acids resemble those used for simple alkylamines and are covered in this chapter. The structural features of amino acids that are significant in the chemistry of peptides and proteins are discussed in Chapter 23.

The decomposition of amino acids and proteins in decaying animal matter produces many simple amines such as methylamine, 1,4-butanediamine, and 1,5-pentanediamine. The common names for 1,4-butanediamine and 1,5-pentanediamine are putrescine and cadaverine, respectively. The odor of fish comes from low-molecular-weight amines such as methylamine.

$CH_3NH_2$          $NH_2CH_2CH_2CH_2CH_2NH_2$          $NH_2CH_2CH_2CH_2CH_2CH_2NH_2$

methylamine          1,4-butanediamine          1,5-pentanediamine
putrescine                         cadaverine

*some amines found in decaying animal tissues*

As shown in the structures below, amines are classified as **primary, secondary, or tertiary** depending on the number of organic groups bonded to the nitrogen atom. Amines are also divided into **aryl amines,** in which at least one of the organic substituents is an aryl group, and **alkyl amines,** in which all the substituents are alkyl groups.

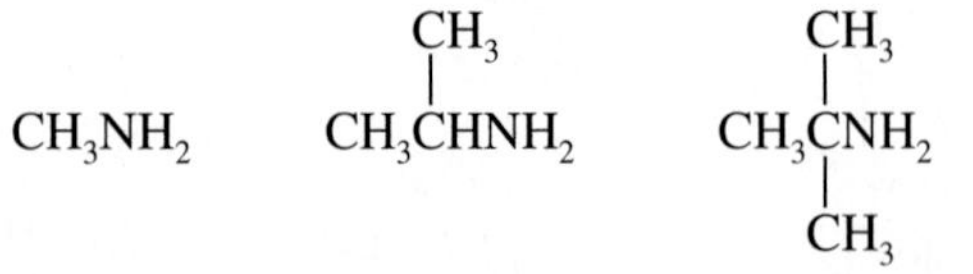

$$CH_3NH_2 \qquad CH_3\overset{\displaystyle CH_3}{\underset{}{CHNH_2}} \qquad CH_3\overset{\displaystyle CH_3}{\underset{\displaystyle CH_3}{CNH_2}}$$

methylamine    isopropylamine    *tert*-butylamine

*a primary*      *a primary*        *a primary*
*alkyl amine*    *alkyl amine*      *alkyl amine*

$$CH_3CH_2NHCH_2CH_3 \qquad CH_3\overset{\displaystyle CH_3}{\underset{}{NCH_3}}$$

diethylamine              trimethylamine

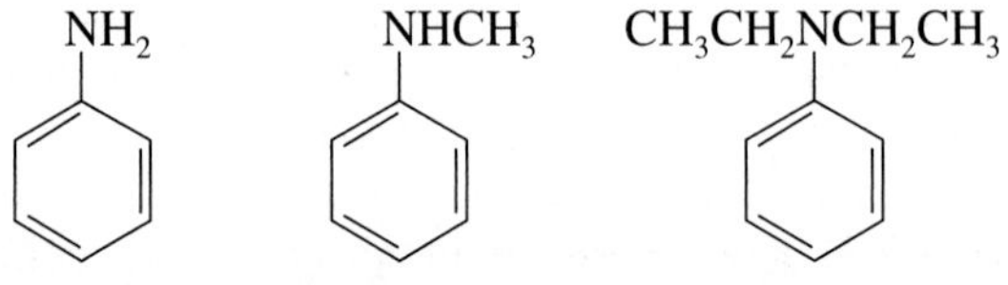

*a secondary*             *a tertiary*
*alkyl amine*             *alkyl amine*

aniline          *N*-methylaniline    *N,N*-diethylaniline

*a primary*      *a secondary*        *a tertiary*
*aryl amine*     *aryl amine*         *aryl amine*

Note that the designations primary, secondary, and tertiary refer to the degree of substitution on the nitrogen atom and not to the carbon atom to which the nitrogen is attached, as in the case of alcohols and alkyl halides.

Primary and secondary amines participate in hydrogen bonding, as both donors and acceptors (p. 34). Nitrogen is not as electronegative as oxygen; therefore, the nitrogen–hydrogen bond is less polar than the oxygen–hydrogen bond, and hydrogen bonding in amines is weaker than it is in alcohols. The boiling points of amines are lower than those of alcohols of similar molecular weight but higher than the boiling points of comparable hydrocarbons and other compounds for which no hydrogen bonding is possible. The trend in boiling points is illustrated with pentane, butylamine, and *n*-butyl alcohol.

$$CH_3CH_2CH_2CH_2CH_3 \qquad CH_3CH_2CH_2CH_2NH_2 \qquad CH_3CH_2CH_2CH_2OH$$

pentane            butylamine          *n*-butyl alcohol
MW 72              MW 73               MW 74
bp 36 °C           bp 78 °C            bp 118 °C

no hydrogen bonding                     strong hydrogen bonding

*the effect of hydrogen bonding on the boiling*
*points of compounds of comparable molecular weight*

Primary, secondary, and tertiary amines in which the organic groups are not too large (p. 33) are soluble in water. The nonbonding electrons of a nitrogen atom are more available to the hydrogen atom of water than are the nonbonding electrons of an oxygen atom. The basicity of amines is related to this phenomenon (pp. 108 and 657).

## 20.2 Nomenclature of Amines

Simple amines are named by combining the name of the alkyl group that is present with the suffix **-amine.** If there are several alkyl groups bonded to the nitrogen atom, they are named in order of increasing complexity. The prefixes **di-** and **tri-** are used to indicate the presence of two or three alkyl groups of the same kind. In IUPAC nomenclature, the ending **amine** is substituted for the final **e** in the name of the alkane, and a number is used to indicate the position of the amino group on the chain or ring.

For more complex amines, the prefix **amino-** is used with a number to indicate the position of the amino group on the hydrocarbon chain. Other substituents that are attached to the nitrogen atom are indicated by an $N$ before the name of the substituent.

$CH_3CH_2NH_2$
ethylamine

$CH_3NHCH_3$
dimethylamine

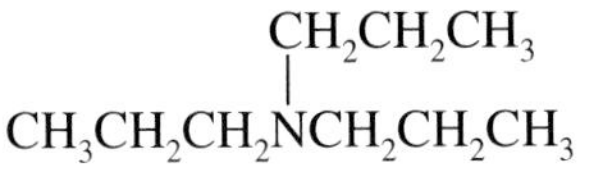

$CH_3CH_2CH_2NCH_2CH_2CH_3$ with $CH_2CH_2CH_3$
tripropylamine

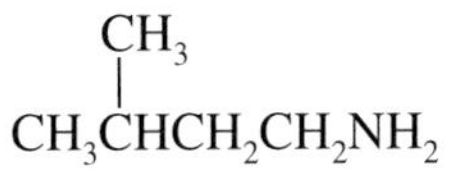

$CH_3CHCH_2CH_2NH_2$ with $CH_3$
3-methyl-1-butylamine

$HOCH_2CH_2CH_2CH_2NH_2$
4-amino-1-butanol

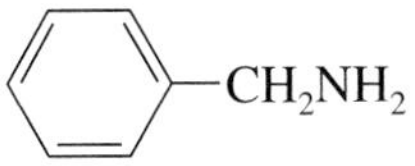

—$CH_2NH_2$
benzylamine

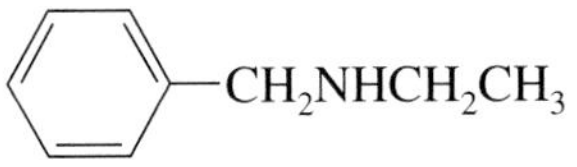

—$CH_2NHCH_2CH_3$
N-ethylbenzylamine

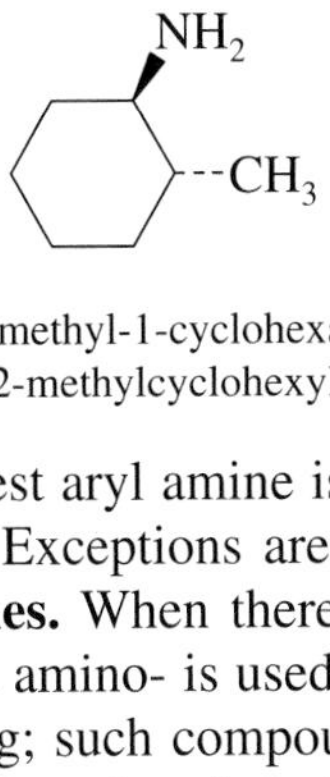

with $NH_2$ and ---$CH_3$
trans-2-methyl-1-cyclohexanamine
trans-(2-methylcyclohexyl)amine

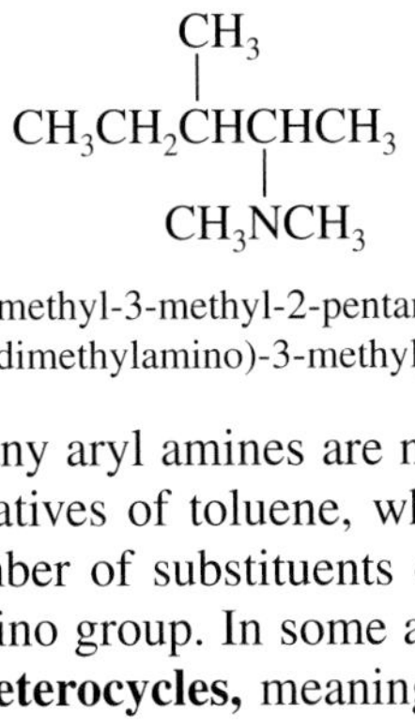

$CH_3CH_2CHCHCH_3$ with $CH_3$ and $CH_3NCH_3$
N,N-dimethyl-3-methyl-2-pentanamine
2-(N,N-dimethylamino)-3-methylpentane

The simplest aryl amine is **aniline,** and many aryl amines are named as substituted anilines. Exceptions are the amino derivatives of toluene, which are usually called **toluidines.** When there are a large number of substituents on the aromatic ring, the prefix amino- is used to locate the amino group. In some amines, nitrogen is part of a ring; such compounds are called **heterocycles,** meaning that they have atoms other than carbon in a ring (p. 360). Many of them have common names, and some that you should recognize are shown below and on the next page.

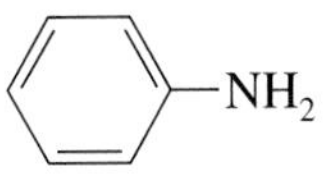

—$NH_2$
aniline

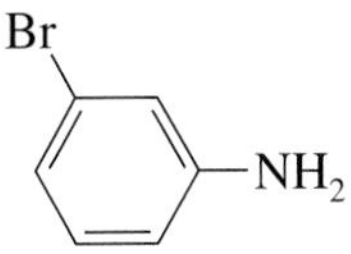

Br, —$NH_2$
m-bromoaniline

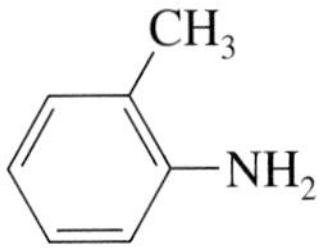

$CH_3$, —$NH_2$
o-toluidine

$CH_3$—, —$NH_2$
p-toluidine

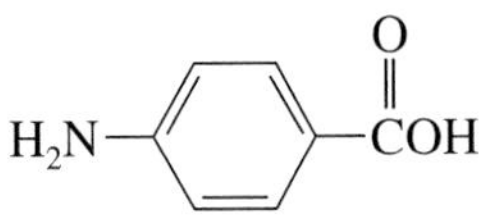

$H_2N$—, —$COH$ (with O)
p-aminobenzoic acid

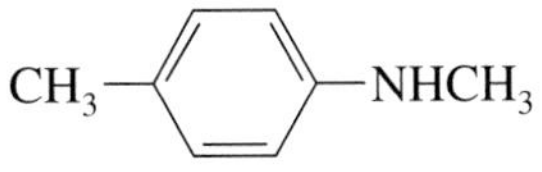

$CH_3$—, —$NHCH_3$
N-methyl-p-toluidine

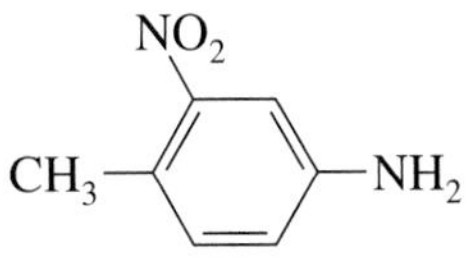

$CH_3$—, —$NH_2$ (with $NO_2$)
4-methyl-3-nitroaniline

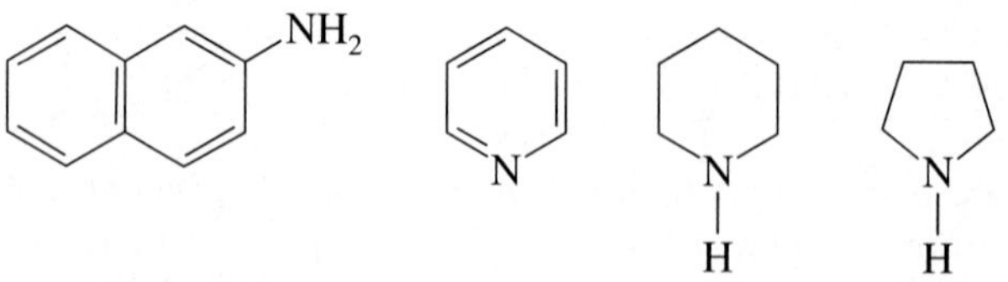

2-naphthylamine
2-aminonaphthalene
β-naphthylamine

pyridine   piperidine   pyrrolidine

heterocyclic amines

The chemistry of heterocyclic amines is discussed in Chapter 22.

**Problem 20.1**   Name the following compounds.

(a) $CH_3CHCH_2NH_2$ (with $CH_3$ substituent)

(b) cyclopentane with $NH_2$ and $CH_2CH_3$

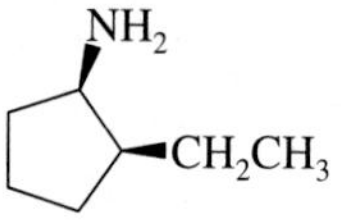

(c) $CH_3CH_2NCH_2CH_2CH_2CH_3$ (with $CH_2CH_3$ substituent)

(d)  cyclopropyl$-CH_2CH_2CH_2NH_2$

(e) 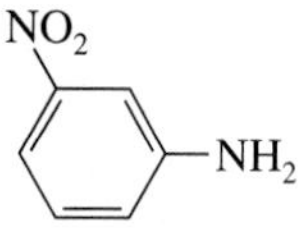 benzene ring with $NO_2$ and $-NH_2$

(f) 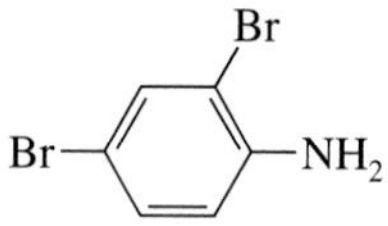 $Br-$ benzene ring with $Br$ and $-NH_2$

## 20.3 Preparation of Amines

### A. Reactions of Ammonia and Amines with Alkyl Halides

Ammonia reacts as a nucleophile with alkyl halides to form amines. The reaction of methyl iodide with ammonia illustrates the course of this nucleophilic substitution reaction.

$$CH_3I \ + \ NH_3 \ \longrightarrow \ CH_3\overset{+}{N}H_3I^-$$

methyl iodide   ammonia      methylammonium
iodide

$$CH_3\overset{+}{N}H_3I^- \ + \ NH_3 \ \rightleftharpoons \ CH_3NH_2 \ + \ \overset{+}{N}H_4I^-$$

methylammonium   ammonia   methylamine   ammonium
iodide                                          iodide

Ammonia first displaces iodide ion from a molecule of the alkyl halide to give methylammonium iodide. The reaction stops here if there is no excess ammonia present. Another molecule of ammonia deprotonates the methylammonium ion and frees methylamine, which is also a nucleophile and reacts further with methyl iodide. Thus dimethylamine, trimethylamine, and tetramethylammonium iodide also may be products of this reaction.

$$CH_3NH_2 \xrightarrow[NH_3]{CH_3I} CH_3NHCH_3 \xrightarrow[NH_3]{CH_3I} CH_3\overset{CH_3}{\underset{}{N}}CH_3 \xrightarrow{CH_3I} CH_3-\overset{CH_3}{\underset{CH_3}{\overset{+}{N}}}-CH_3I^-$$

methylamine      dimethylamine      trimethylamine      tetramethylammonium iodide

It is difficult to prepare pure primary amines using this method.

When this type of reaction is used for synthetic purposes, a large excess of ammonia is used, as illustrated in the preparation of the amino acid glycine from chloroacetic acid. A large excess of ammonia ensures that in random collisions between molecules, most of the time ammonia rather than glycine will encounter and react with chloroacetic acid.

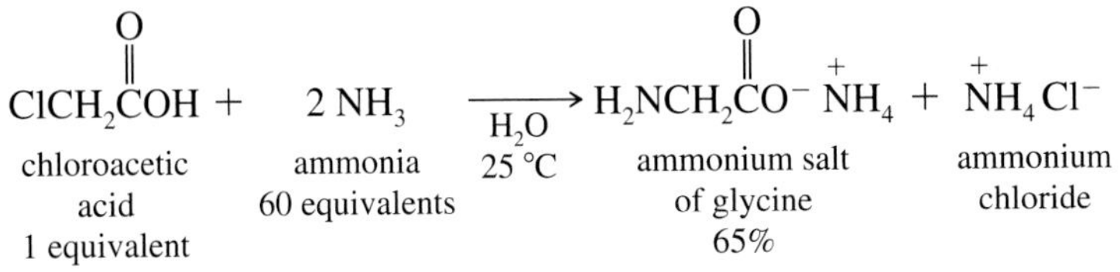

$$ClCH_2\overset{\displaystyle O}{\overset{\|}{C}}OH + 2\,NH_3 \xrightarrow[\substack{H_2O \\ 25\,°C}]{} H_2NCH_2\overset{\displaystyle O}{\overset{\|}{C}}O^-\,\overset{+}{N}H_4 + \overset{+}{N}H_4Cl^-$$

| chloroacetic acid 1 equivalent | ammonia 60 equivalents | ammonium salt of glycine 65% | ammonium chloride |

The same precaution must be used when preparing a secondary amine so as to avoid getting a tertiary amine. For example, an excess of aniline is used in the reaction with benzyl chloride to prevent the formation of N,N-dibenzylaniline.

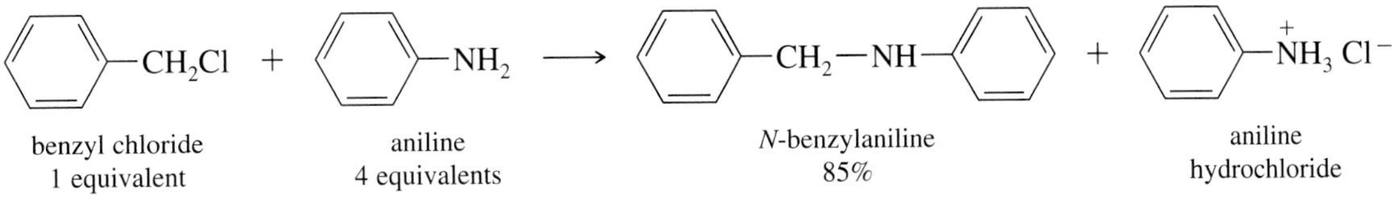

| benzyl chloride 1 equivalent | aniline 4 equivalents | N-benzylaniline 85% | aniline hydrochloride |

**Problem 20.2**  Tell what starting materials would be necessary to synthesize the following compounds. Assume that aniline, benzene, and any other organic compound containing not more than three carbon atoms are available.

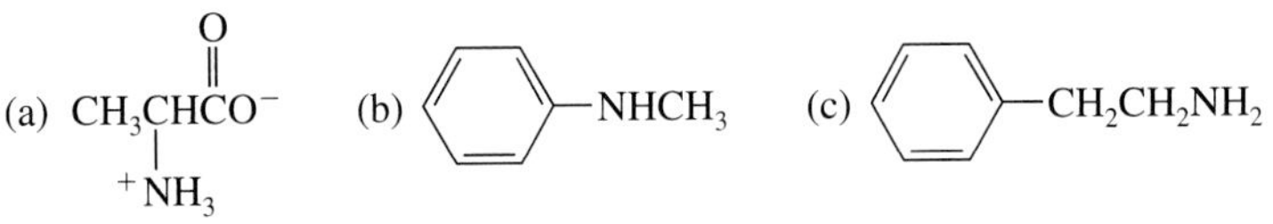

(a) $CH_3\overset{\displaystyle O}{\overset{\|}{C}}HCO^- $  with $\overset{+}{N}H_3$ below CH

(b) ⬡—NHCH₃

(c) ⬡—CH₂CH₂NH₂

## B. Reduction of Azides

The azide ion, $N_3^-$, is a good nucleophile and is used to create carbon–nitrogen bonds (p. 236) in $S_N2$ reactions. Reduction of azides gives primary amines. Hydrogen in the presence of a metallic catalyst or lithium aluminum hydride is the reducing agent used most frequently. For example, (S)-2-octanol is converted into (R)-2-octylamine by the following sequence of reactions.

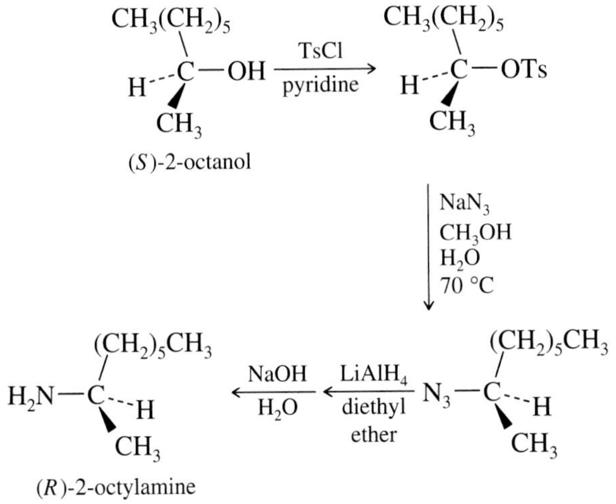

(S)-2-octanol

$$\xrightarrow[\text{pyridine}]{\text{TsCl}}$$

$$\begin{array}{c} NaN_3 \\ CH_3OH \\ H_2O \\ 70\,°C \end{array}$$

$$\xleftarrow[\substack{NaOH \\ H_2O}]{} \xleftarrow[\substack{\text{diethyl} \\ \text{ether}}]{LiAlH_4}$$

(R)-2-octylamine

The displacement of tosylate ion by azide ion is accompanied by inversion of configuration. The azide is reduced to the amine with retention of configuration. The other two nitrogen atoms of the azide group are lost as $N_2$, nitrogen gas.

Azides are also formed when oxiranes undergo a ring-opening reaction with sodium azide. For example, cyclohexene oxide is converted to *trans*-2-aminocyclohexanol in the following way.

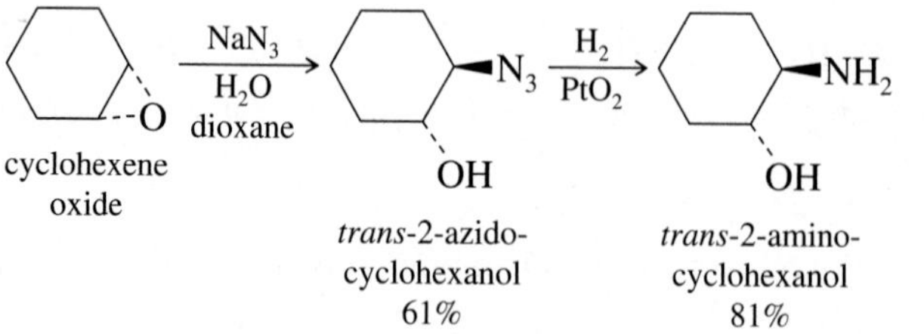

cyclohexene
oxide

*trans*-2-azido-
cyclohexanol
61%

*trans*-2-amino-
cyclohexanol
81%

Note the stereochemistry that results from the opening of the oxirane ring by a nucleophile (p. 505).

## C. Reduction of Carbon–Nitrogen Multiple Bonds

Imines and oximes, compounds containing carbon–nitrogen double bonds, are formed when amines or hydroxylamines react with aldehydes or ketones (pp. 574 and 575). Nitriles, compounds containing carbon–nitrogen triple bonds, are made easily in nucleophilic substitution reactions with cyanide ion. These compounds can all be reduced to amines by metal hydrides or by hydrogen and a catalyst.

The phenylimine of benzaldehyde, which is formed when benzaldehyde and aniline condense, is reduced by sodium borohydride or catalytically to a secondary amine.

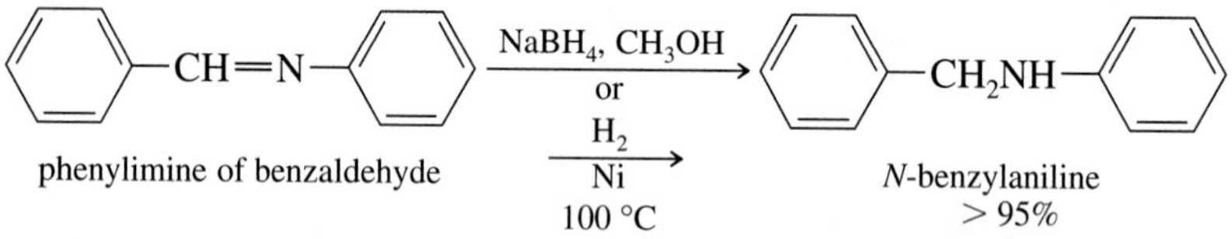

phenylimine of benzaldehyde

*N*-benzylaniline
> 95%

The imine function is like a carbonyl group in its polarization and undergoes hydride reduction in the same way.

---

**VISUALIZING THE REACTION**

**Reduction of an Imine**

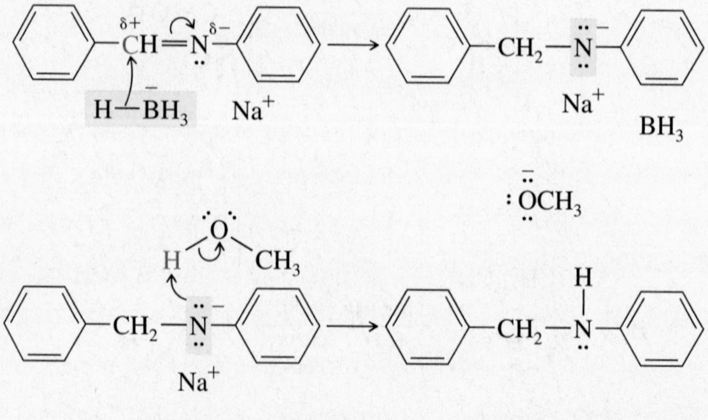

It is not usually necessary to isolate an imine in order to prepare the corresponding amine. The imines generated from aldehydes and ketones with ammonia are highly unstable. In a process known as **reductive amination,** an imine is formed and reduced in the same step by mixing a carbonyl compound and ammonia in the presence of hydrogen gas and a catalyst. Thus acetophenone is converted into $\alpha$-phenylethylamine.

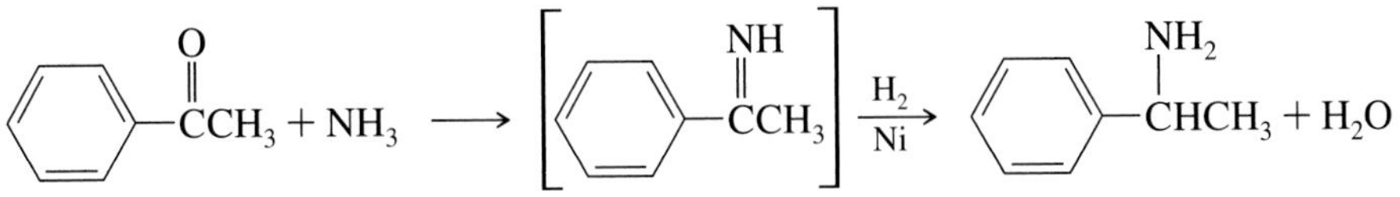

acetophenone          imine          $\alpha$-phenylethylamine
*postulated as intermediate*          64%

## ONE SMALL STEP

The imine function resembles the carbonyl group in its reactivity in many ways.

**PROBLEM:** Predict, by analogy with the carbonyl group, what the product of the following reaction will be.

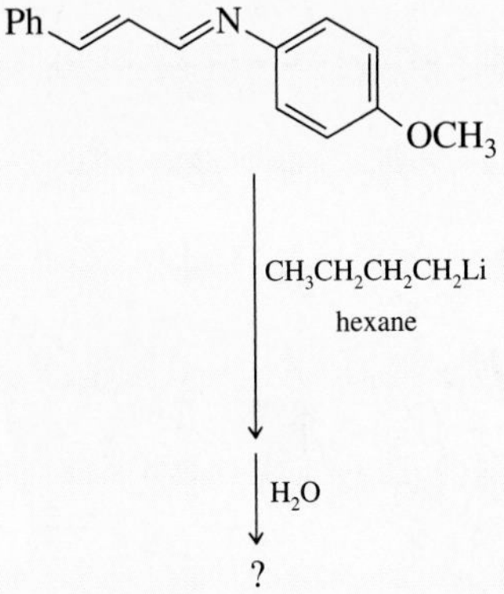

$CH_3CH_2CH_2CH_2Li$
hexane

$H_2O$

?

Amines, as well as ammonia, can be used in reductive amination reactions. Aniline is converted to *N*-isopropylaniline by reaction with acetone in the presence of sodium borohydride.

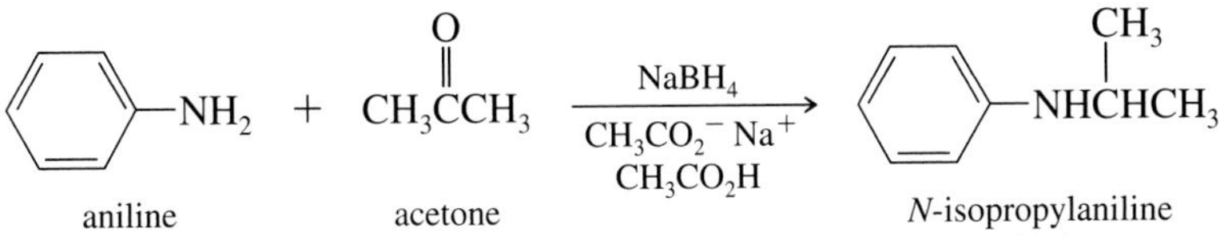

aniline          acetone          *N*-isopropylaniline
91%

Oximes, as well as imines, are reduced by hydrogen to amines. The oxime of pentanal, for example, is converted to pentylamine by catalytic hydrogenation.

$$CH_3CH_2CH_2CH_2CH=NOH \xrightarrow[\text{Ni}]{H_2} CH_3CH_2CH_2CH_2CH_2NH_2 + H_2O$$
oxime of pentanal          100 °C          pentylamine
62%

Nitriles can be reduced to amines by lithium aluminum hydride. Octane nitrile, for example, is reduced to octylamine by this reagent.

$$CH_3(CH_2)_6C\equiv N \xrightarrow[\text{diethyl ether}]{LiAlH_4} \xrightarrow{H_2O} CH_3(CH_2)_6CH_2NH_2$$
octanenitrile          octylamine
90%

Catalytic reduction of nitriles is also possible. Thus butyl bromide is converted to a primary amine containing one more carbon atom by a two-step sequence of nucleophilic substitution and catalytic reduction reactions.

$$CH_3CH_2CH_2CH_2Br + K^+C\equiv N^- \longrightarrow CH_3CH_2CH_2CH_2C\equiv N + K^+Br^- \xrightarrow{H_2}_{Ni}$$
butyl bromide          pentanenitrile

$$CH_3CH_2CH_2CH_2CH_2NH_2$$
pentylamine
90%

Another method for making amines, the reduction of amides, will be discussed on page 858.

**Problem 20.3**   The following reactions were carried out in the synthesis of compounds used to investigate intramolecular hydrogen bonding between amines and carboxylic acids. Provide structural formulas for the compounds designated by letters.

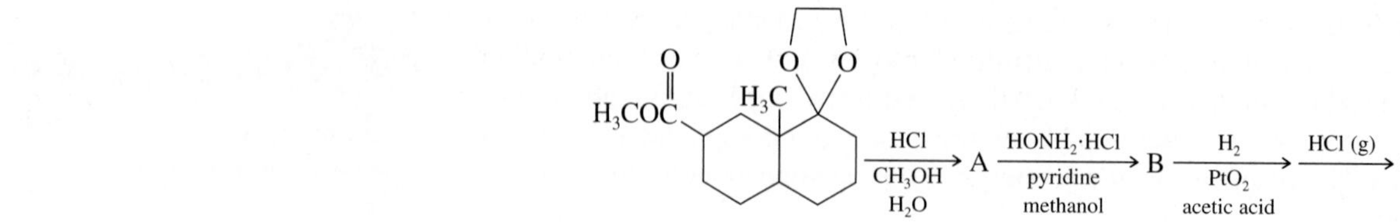

**Problem 20.4** Amino acids often serve as chiral starting materials in syntheses. The methyl ester of the amino acid serine is the starting point for this synthesis. Provide structural formulas for the reagents and products designated by letters.

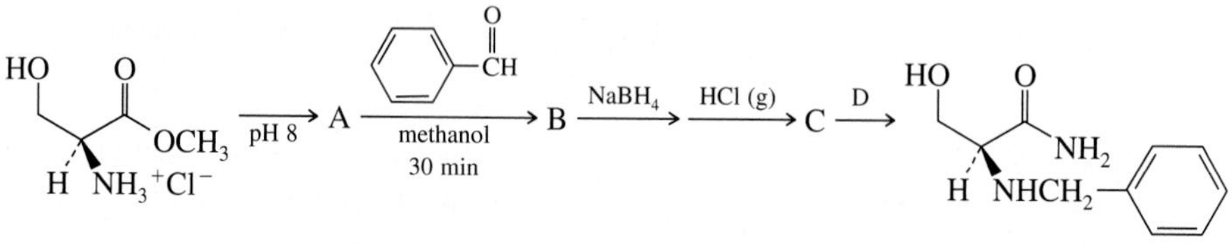

**Problem 20.5** How could each of the following transformations be carried out?

(a)

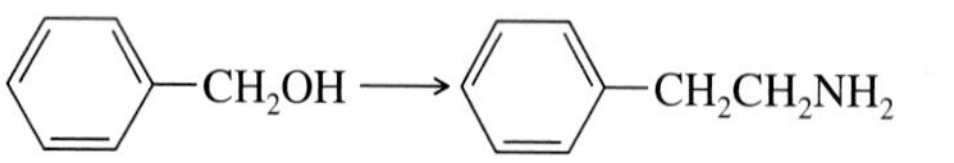

(b) $CH_3CH_2CH_2CH_2OH \longrightarrow CH_3CH_2CH_2CHCH_2CH_3$ with $NH_2$

(c)

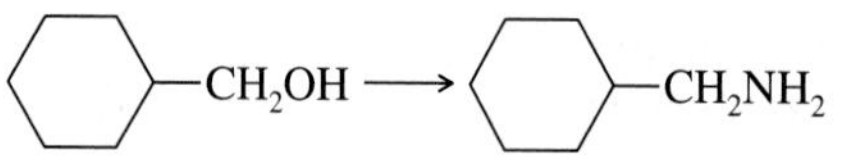

(d) 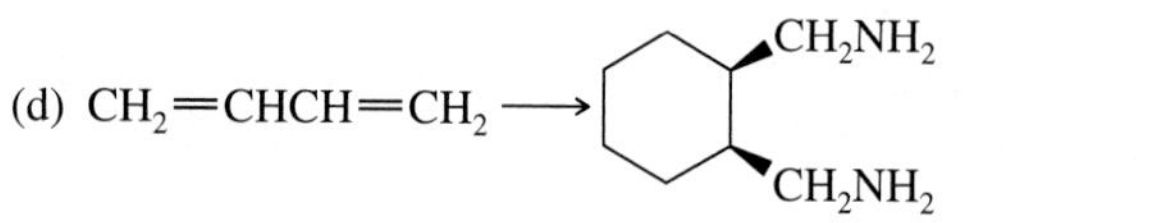 $CH_2{=}CHCH{=}CH_2 \longrightarrow$

## D. Reduction of Nitro Compounds

Aromatic amines are synthesized most conveniently by nitration of an aromatic ring and catalytic reduction of the nitro group to an amino group. Catalytic reduction is part of a synthesis of *p*-aminobenzoic acid from *p*-nitrobenzoic acid.

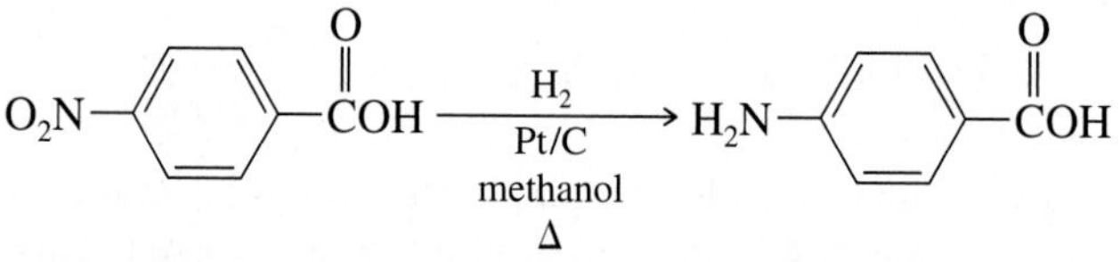

*p*-nitrobenzoic acid       *p*-aminobenzoic acid

Alkyl nitro compounds, synthesized from alkyl halides, are converted to amines in the same way.

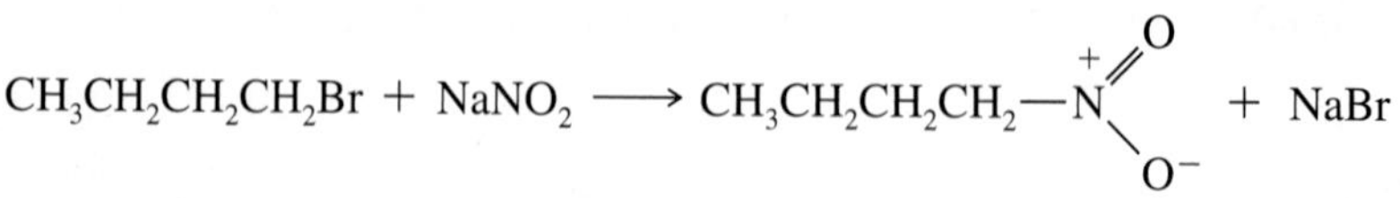

1-bromobutane    sodium nitrite    1-nitrobutane    sodium bromide

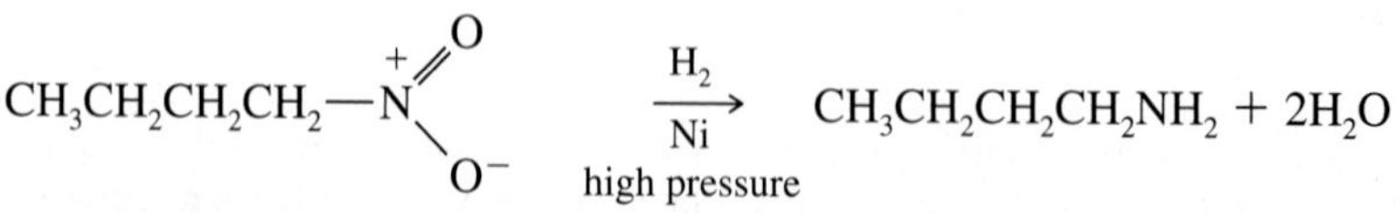

1-nitrobutane      butylamine

Another way to make nitro compounds is by an aldol type condensation of the enolate ion from nitromethane with a carbonyl compound. For example, the following sequence of reactions is used to make an amino alcohol from cyclohexanone and nitromethane.

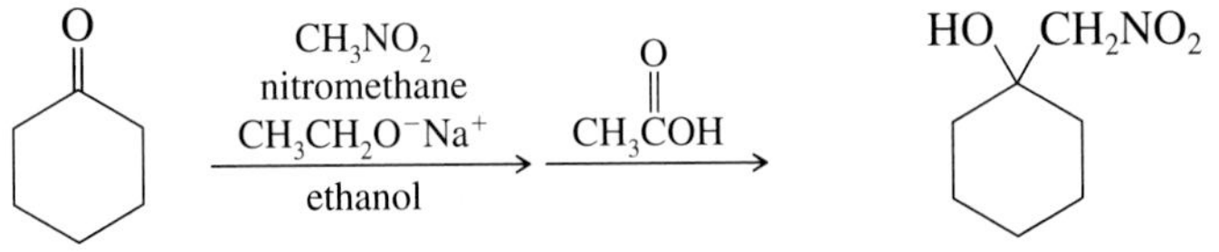

cyclohexanone

aldol condensation product
of cyclohexanone and nitromethane

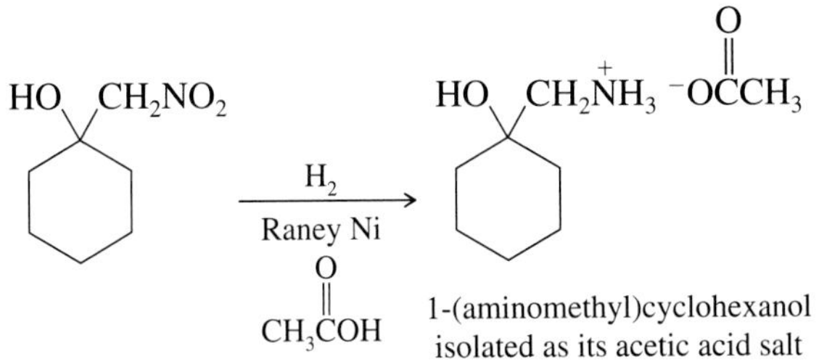

1-(aminomethyl)cyclohexanol
isolated as its acetic acid salt

The reaction is catalyzed by base and involves the addition of the carbanion from nitromethane (p. 668) to the carbonyl group of cyclohexanone to give the enolate anion of the new nitro compound, which is protonated by acetic acid. Reduction of the nitro compound gives the aminoalcohol.

**Problem 20.6**    Write a mechanism for the formation of the aldol condensation product of cyclohexanone and nitromethane.

**Problem 20.7**    Complete the equations below.

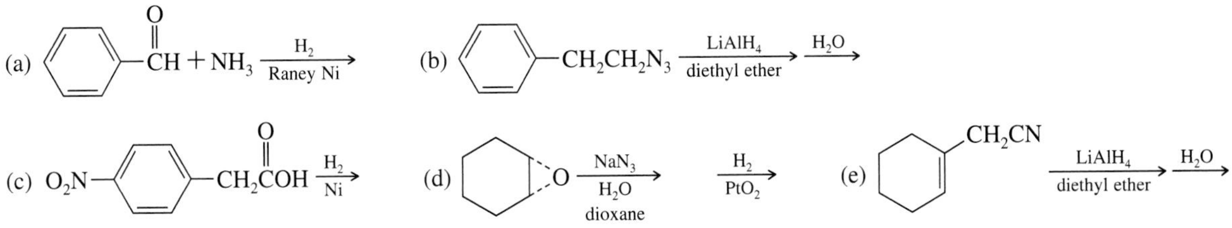

**Problem 20.8**    Show how each of the following transformations could be carried out. Several steps may be required.

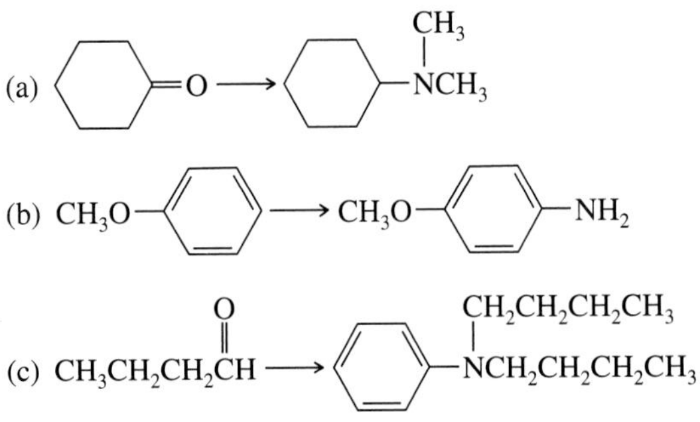

(d) $CH_3CH_2CH_2CH_2Br \longrightarrow CH_3(CH_2)_6CH_2NH_2$ (three different syntheses required)

## 20.4 Biologically Active Amines and Quaternary Nitrogen Compounds

### A. p–Aminobenzoic Acid

*p*-Aminobenzoic acid is a compound with many interesting physiologic properties. It is incorporated by bacteria into folic acid, a vitamin that is essential to bacterial growth.

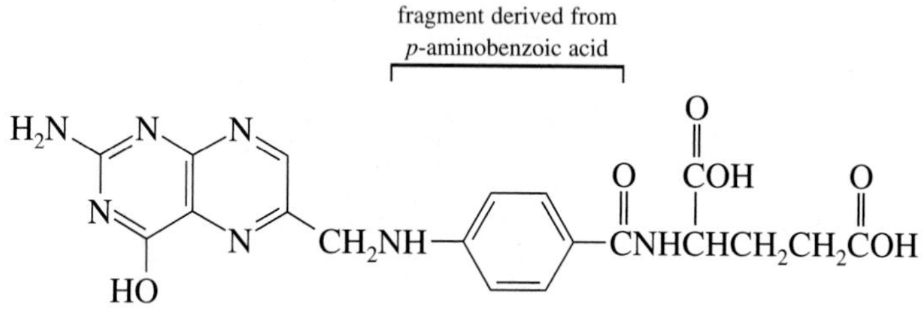

folic acid

The ability of sulfa drugs to inhibit bacterial growth results from the structural resemblance of these compounds, for example, sulfanilamide, to *p*-aminobenzoic acid.

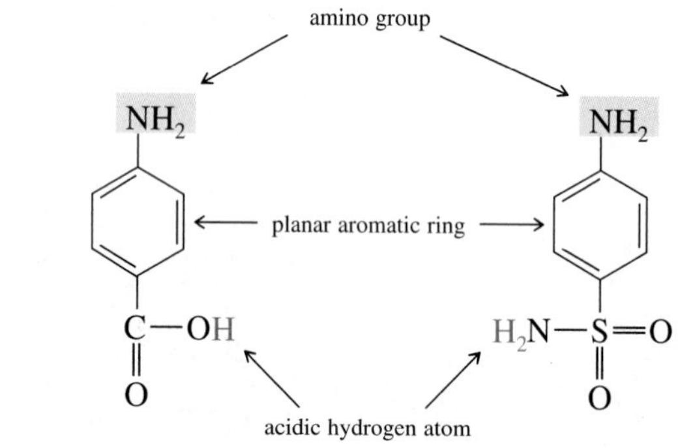

*p*-aminobenzoic acid                    sulfanilimide

Both *p*-aminobenzoic acid and sulfanilamide have the weakly basic amino group on an aromatic ring para to a functional group in which a central atom is doubly bonded to oxygen and an acidic hydrogen atom is bonded to an electronegative atom (oxygen or nitrogen). The sites of basicity and acidity and the spatial arrangement of the functional groups are similar enough in the two molecules that the bacterial enzyme that synthesizes folic acid mistakes the sulfa drug for *p*-aminobenzoic acid. The sulfa drug takes the place of *p*-aminobenzoic acid at the catalytic surface of the enzyme, preventing the synthesis of the essential vitamin and disrupting metabolic processes in the bacteria. Because human beings do not synthesize folic acid, sulfa drugs do not disrupt human metabolism.

Esters of *p*-aminobenzoic acid are local anesthetics. Ethyl *p*-aminobenzoate is commonly known as Benzocaine, and 2-(*N,N*-dimethylamino)ethyl *p*-aminobenzoate is the local anesthetic procaine, widely used in dentistry as its hydrochloride salt, Novocain.

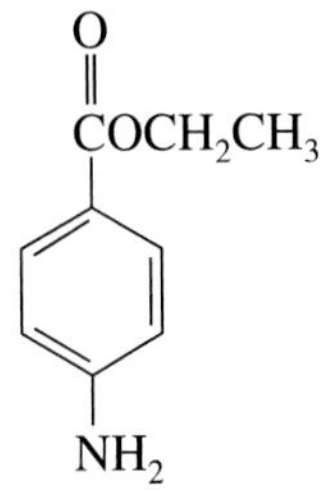

ethyl *p*-aminobenzoate
Benzocaine

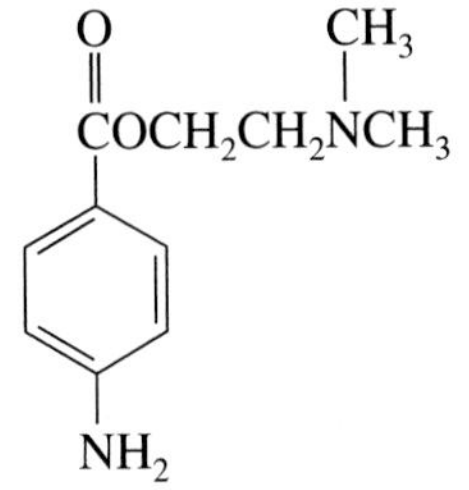

2-(*N*,*N*-dimethylamino)ethyl
*p*-aminobenzoate
procaine

Other derivatives of *p*-aminobenzoic acid are used widely in suntan lotions to absorb ultraviolet radiation and to screen the skin from the more harmful wavelengths of sunlight.

**Problem 20.9**    Write the structural formula for the local anesthetic Novocain, which is procaine hydrochloride and has the molecular formula $C_{11}H_{16}N_2O_2 \cdot HCL$. (*Hint:* Which is the most basic site in procaine, the structure of which is shown above?)

## B. Phenylethylamines

Another biologically important class of amines prepared from nitro compounds are **β-phenylethylamines.** Adrenalin, the hormone secreted by the adrenal glands to mobilize the body to energetic action when either danger or pleasure is anticipated, is a β-phenylethylamine, as is norepinephrine, another amine involved in the transmission of nerve impulses, and mescaline, the hallucinogenic alkaloid of the peyote cactus. These structures appear below.

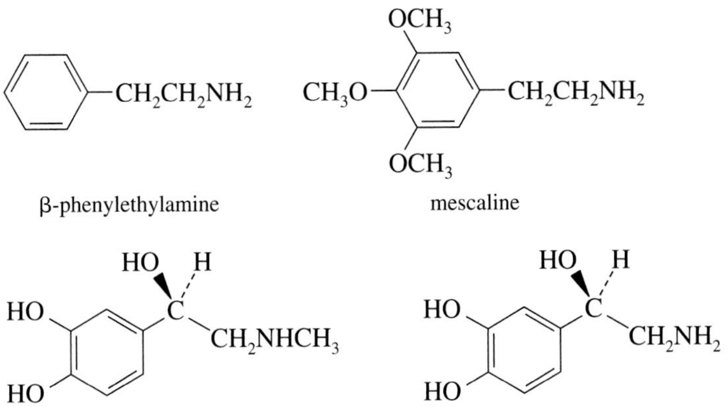

β-phenylethylamine

mescaline

adrenalin
epinephrine

norepinephrine

*biologically important* β-*phenylethylamines*

The ways in which compounds such as adrenalin are generated in the body and how they are deactivated and removed from the body when the momentary need is over have been studied intensively. A large number of compounds that are structurally similar to these hormones have been synthesized so that the relationships between structure and biological activity can be studied. The next section features such a synthesis.

## C. The Art of Solving Problems

**PROBLEM:** One biologically active β-phenylethylamine, β-(3,4-methylenedioxyphenyl)ethylamine, is shown below. How would you synthesize it from 3,4-methylenedioxybenzaldehyde?

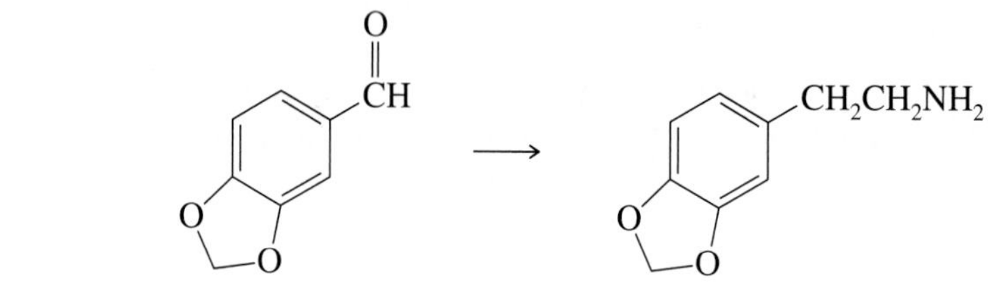

3,4-methylenedioxybenzaldehyde      β-(3,4-methylenedioxyphenyl)ethylamine

### Solution

1. What are the connectivities of the two compounds? How many carbon atoms does each contain? Are there any rings? What are the positions of branches and functional groups on the carbon skeleton?

   The starting material and the product both contain the 3,4-methylenedioxyphenyl group. The starting material has an aldehyde function; the product has an additional carbon atom bearing a nitrogen atom added to the carbon atom that was once a carbonyl group.

2. How do the functional groups change in going from starting material to product? Does the starting material have a good leaving group?

   The carbonyl group of the starting material has added a carbon atom bearing a nitrogen atom.

3. Is it possible to dissect the structures of the starting material and product to see which bonds must be broken and which formed?

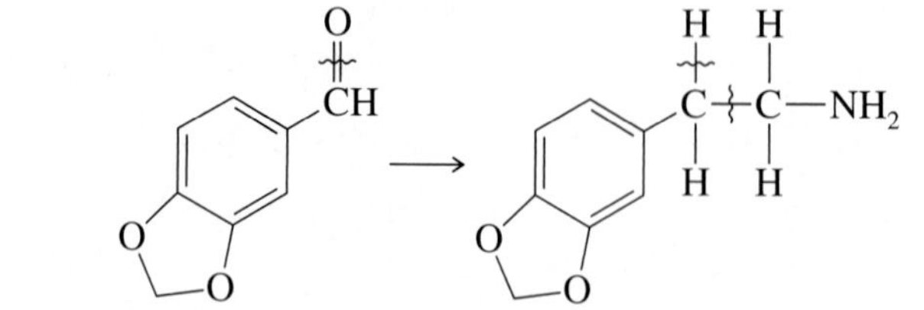

bonds to be broken        bonds to be formed

4. New bonds are created when an electrophile reacts with a nucleophile. Do we recognize any part of the product molecule coming from a good nucleophile or an electrophilic addition?

   The carbonyl group is electrophilic. Loss of the oxygen atom of the carbonyl group suggests an aldol type condensation with loss of water. The carbon atom that is added must be the nucleophile. This is possible only if that carbon bears a nitrogen function that stabilizes negative charge. Two possibilities for nucleophiles that add a carbon atom and a nitrogen atom to the carbonyl group come to mind.

(a) The cyanide ion.

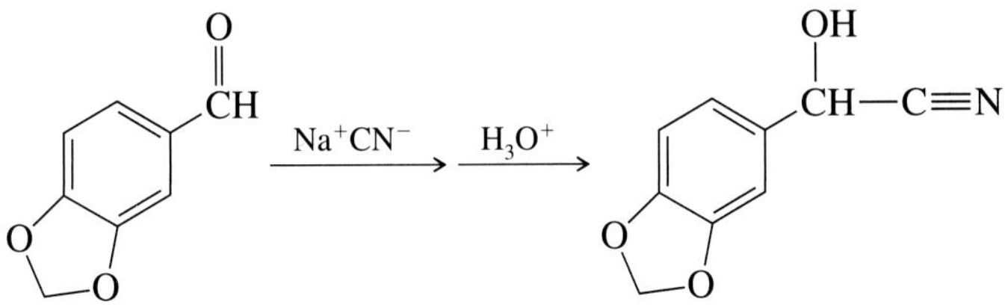

This product does not have the possibility of easily losing the hydroxyl group.

(b) The enolate anion of the nitro group.

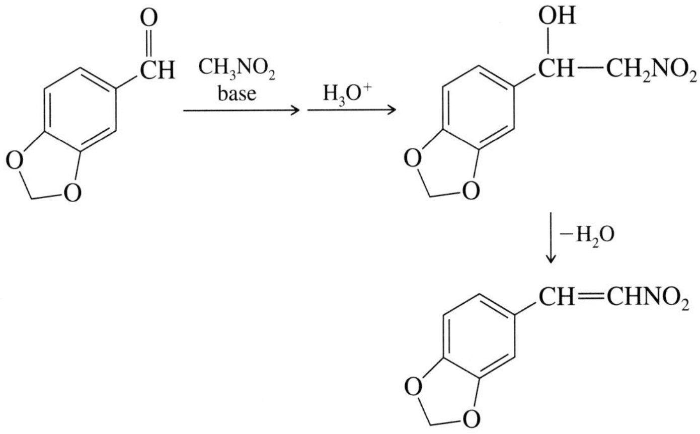

In this case, easy dehydration to give a double bond conjugated with the aromatic ring and the nitro group takes place.

5. What type of compound would be a good precursor to the product?

The unsaturated nitro compound prepared in step 4(b) is a good precursor to the product.

6. After this last step, do we see how to get from starting material to product? If not, we need to analyze the structure obtained in step 5 by applying questions 4 and 5 to it.

The amine we are looking for can be prepared by reduction of the unsaturated nitro compound we made in step 4(b)

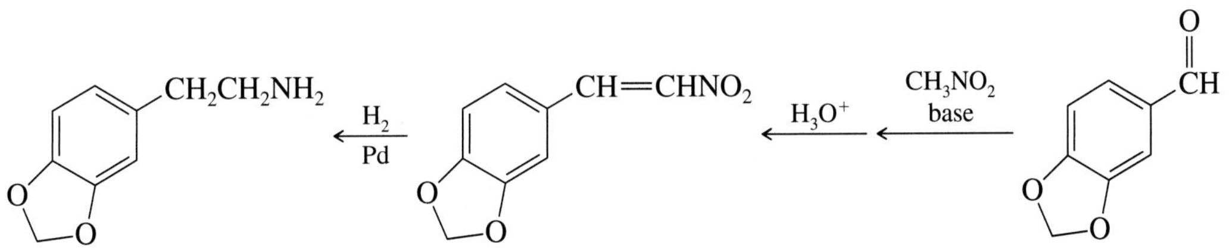

**Study Guide**
Concept Map 20.1

**Problem 20.10**    How would you carry out the following transformation?

$$\text{(furan)—CH=O} \longrightarrow \text{(furan)—C(=CH–H)—C(–CH}_2\text{CH}_2\text{CH}_3\text{)=NO}_2$$

## D. Biologically Active Quaternary Nitrogen Compounds

Choline, the systematic name of which is 2-(trimethylammonium)ethanol, is an important participant in the transmission of nerve impulses. It is also a component of lecithin, a major constituent of cell membranes. Choline is converted in nerve endings to its acetic ester, acetylcholine, by an enzyme with the participation of acetyl coenzyme A (p. 637).

$$\underset{\substack{\text{choline}}}{\overset{\overset{\displaystyle CH_3}{\overset{+|}{CH_3NCH_2CH_2OH}}}{\underset{\displaystyle CH_3}{}}} + \underset{\substack{\text{acetyl}\\ \text{coenzyme A}}}{CoA-\overset{\overset{\displaystyle O}{\|}}{SCCH_3}} \xrightarrow[\substack{\text{choline}\\ \text{acetylase}}]{} \underset{\substack{\text{acetylcholine}}}{\overset{\overset{\displaystyle CH_3}{\overset{+|}{CH_3NCH_2CH_2O}}\overset{\overset{\displaystyle O}{\|}}{CCH_3}}{\underset{\displaystyle CH_3}{}}} + \underset{\substack{\text{coenzyme A}}}{CoA-SH}$$

At the exact instant of the transmission of the nerve impulse, acetylcholine moves across the synapse, the point where a nervous impulse passes from one nerve cell to the other. In motor nerves, this triggers the contraction of the muscle. An enzyme at the receptor site called acetylcholine esterase catalyzes the hydrolysis of acetylcholine back to choline and acetic acid; the nerve impulse stops and the muscle relaxes.

$$\underset{\substack{\text{hydroxyl}\\ \text{function}\\ \text{at enzyme}}}{ROH} + \underset{\substack{\text{acetylcholine}}}{\overset{\overset{\displaystyle CH_3}{\overset{+|}{CH_3NCH_2CH_2O}}\overset{\overset{\displaystyle O}{\|}}{CCH_3}}{\underset{\displaystyle CH_3}{}}} \xrightarrow[\substack{\text{esterase}}]{\text{acetylcholine}} \underset{\substack{\text{choline}}}{\overset{\overset{\displaystyle CH_3}{\overset{+|}{CH_3NCH_2CH_2OH}}}{\underset{\displaystyle CH_3}{}}} + \underset{\substack{\text{acylated}\\ \text{enzyme}}}{\overset{\overset{\displaystyle O}{\|}}{CH_3COR}}$$

Anything that interferes with this alternation of the synthesis and destruction of acetylcholine at the nerve junction creates paralysis. The enzyme acetylcholine esterase is particularly vulnerable. Although the bond that is hydrolyzed at the enzyme active site is the ester linkage, the positively charged quaternary ammonium center is necessary to position acetylcholine so that hydrolysis can take place. The structure of proteins will not be fully explored until Chapter 23, but the exact locations of acidic, basic, charged, or even hydrophobic portions of molecules are crucial to interactions between enzymes and the molecules on which they act.

The importance of the quaternary ammonium group to the action of acetylcholine esterase is made apparent by the fact that a number of compounds that are quaternary ammonium salts have a paralyzing effect. (+)-Tubocurarine, the active component of the curare poison used by natives of South America on the tips of arrows to paralyze their prey, is an example of such a naturally occurring quaternary ammonium compound. A synthetic drug, decamethonium bromide, mimics tubocurarine in the presence of two quaternary ammonium salt centers and the distance between them.

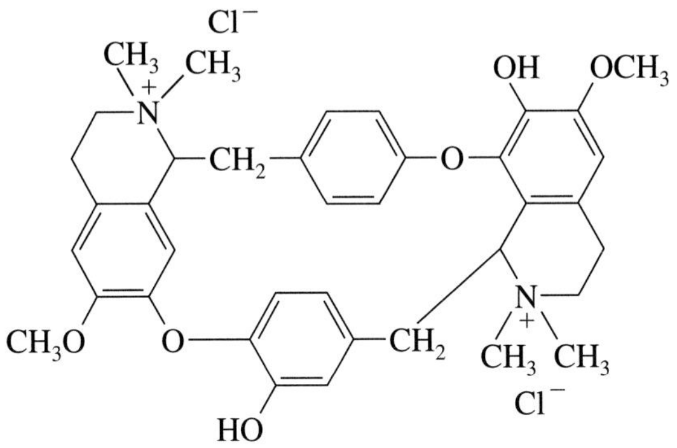

tubocurarine

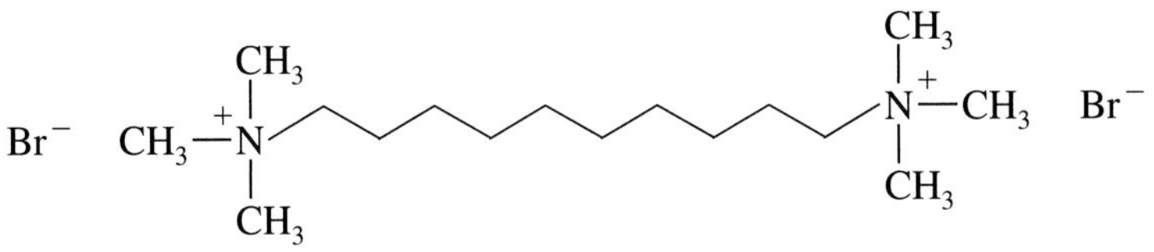

decamethonium bromide

Both tubocurarine and decamethonium bromide are used in surgery as muscle relaxants. They act as inhibitors of acetylcholine esterase, which means that they occupy the active site of the enzyme and interact with it because of the presence of the quaternary ammonium groups. They prevent the enzyme from carrying out its function of destroying acetylcholine and thus interfere with the proper transmission of nerve impulses. An excess of tubocurarine kills by paralyzing the respiratory muscles.

**Problem 20.11**    The antibiotic chloramphenicol is deactivated in the body by an acyl-transfer reaction from acetyl coenzyme A to give first a monoacetylated compound and then the diacetyl product. The structure of chloramphenicol is shown below. Write equations for its reaction with acetyl coenzyme A, predicting which site in the molecule will be acetylated first, and which second.

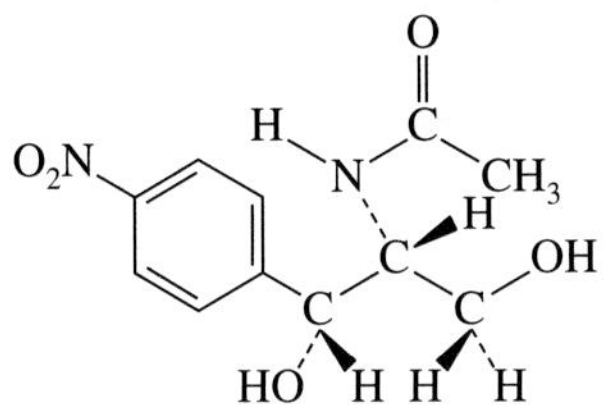

chloramphenicol

Another potent poison is found in the fungus *Amanita muscaria,* commonly called the fly agaric because it was used as a fly poison at one time. The alkaloid muscarine is found in this fungus, along with choline and acetylcholine.

choline residue
incorporated into
muscarine

↓

$$HO \overset{\text{(ring)}}{\underset{CH_3 \quad O}{\bigtriangleup}} CH_2\overset{CH_3}{\underset{CH_3}{\overset{+}{N}}}CH_3 \quad I^-$$

(+)-muscarine

Choline is incorporated into the structure of muscarine, as indicated by the shaded portion of the formula. (+)-Muscarine mimics choline, but (−)-muscarine has very little toxicity, a good example of the high stereoselectivity of enzymatic reactions.

The function of choline as a component of lecithin is quite different from its function in the transmission of nerve impulses. Lecithin is a triglyceride (p. 639) in which one of the primary alcohol groups of glycerol has been esterified by phosphoric acid. The phosphoric acid also forms an ester link to choline. In lecithin, the other two alcohol functions of glycerol have been esterified by fatty acids, usually stearic acid and oleic acid.

$$CH_3(CH_2)_{16}\overset{O}{\overset{\|}{C}}OCH_2$$
$$CH_3(CH_2)_7CH=CH(CH_2)_7\overset{O}{\overset{\|}{C}}O\overset{}{C}H\overset{}{C}H_2-O-\overset{O}{\underset{O^-}{\overset{\|}{P}}}-OCH_2CH_2\overset{CH_3}{\underset{CH_3}{\overset{+}{N}}}CH_3$$

choline esterified
by phosphoric acid

nonpolar end of
the molecule

polar end of
the molecule

choline phosphoglyceride
lecithin

Lecithin is one of a series of compounds known as **phospholipids.** At the pH prevalent in the human body, the phosphoric acid group is negatively charged, and the quaternary ammonium group in the choline residue is positively charged. The molecule of lecithin as a whole is neutral but quite highly charged at one end and hydrocarbon-like and nonpolar at the other. The hydrocarbon portions of phospholipid molecules are believed to interact with each other to form cell membranes consisting of two layers of phospholipids with the charged ends oriented toward both the exterior and interior of the cell (Figure 20.1). The overlapping tails of the phospholipids form a nonpolar region in which polar species are not soluble. The layers are, therefore, not easily penetrated by highly polar molecules. The polar parts of the membrane are placed where they interact with water in the cell and in surrounding fluids. The two layers of lipid molecules can move sideways past each other, giving the membranes flexibility.

**Problem 20.12**    The neurotoxic activity of (+)-muscarine (see above) has made it the subject of much research. Chemists are interested in the synthesis of stereoisomers of muscarine to study the stereochemistry of its reactions with receptors in the nervous system.

(a) Some steps in the synthesis of stereoisomers of muscarine are shown on the next page. Supply structural formulas for the reagents or products designated by letters.

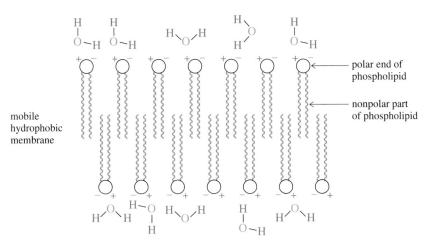

*Figure 20.1*
Model of a cell membrane composed of a phospholipid bilayer.

(b)  One of the final products shown above can be converted into (+)-muscarine. Which one is it? What further reagents would be necessary to make the transformation?

## 20.5   Nitrosation Reactions of Amines

### A. Nitrous Acid

Nitrous acid, $HNO_2$, is an unstable species that exists only as its salts or in solution in equilibrium with a number of other species, depending on the acidity of the solution and the other ions present. In the laboratory, it is generated as needed by treating sodium nitrite with a strong mineral acid, usually hydrochloric acid, at 0–5 °C.

$$Na^+ + NO_2^- + H_3O^+ + Cl^- \xrightarrow[\substack{H_2O \\ 0\,°C}]{} HO{-}N{=}O + Na^+ + Cl^-$$

nitrous acid
$pK_a$ 3.23

In strongly acidic solutions, nitrous acid is protonated; its conjugate acid then loses water to give the nitrosonium ion.

$$HO{-}N{=}O + H_3O^+ \rightleftharpoons \overset{\overset{\displaystyle H}{|}}{\underset{+}{HO}}{-}N{=}O \longrightarrow H_2O + \overset{+}{N}{=}O$$

nitrous acid          conjugate acid          nitrosonium ion
of nitrous acid

The nitrosonium ion is an electrophilic species, like the nitronium ion postulated as the reacting species in electrophilic aromatic substitution reactions (p. 372).

$$:N{\equiv}\overset{+}{O}: \longleftrightarrow :\overset{+}{N}{=}\overset{..}{\underset{..}{O}}$$

*resonance contributors of the nitrosonium ion*

Nitrosonium ion is a weaker electrophile with respect to aromatic rings than nitronium ion is but reacts readily with the nonbonding electrons on nitrogen atoms to initiate reactions that vary with the structure of the amine. Reactions of amines with nitrosonium ions are known as **nitrosation reactions.**

## B. Nitrosation of Amines. *N*–Nitrosamines

The first step of the reaction of nitrous acid with an amine is the formation of an *N*-nitrosamine. In the case of secondary alkylamines, the resulting *N*-nitrosamines are stable and are of great biological interest because they are known to be mutagens and carcinogens. One of the most potent is *N*-nitrosodimethylamine, formed when dimethylamine reacts with nitrous acid.

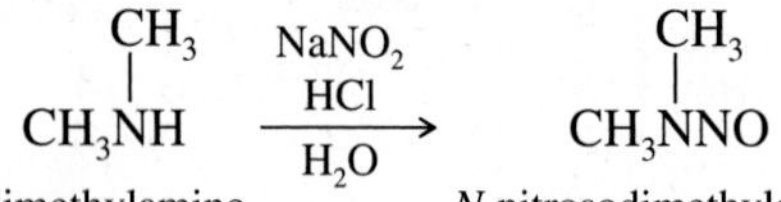

$$\underset{\text{dimethylamine}}{\overset{\overset{\displaystyle CH_3}{|}}{CH_3NH}} \xrightarrow[\substack{H_2O}]{\substack{NaNO_2 \\ HCl}} \underset{N\text{-nitrosodimethylamine}}{\overset{\overset{\displaystyle CH_3}{|}}{CH_3NNO}}$$

The reaction is initiated by attack of the nonbonding electrons of the nitrogen atom of the amine on the electrophilic nitrosonium ion.

---

**VISUALIZING THE REACTION**

**Nitrosation of a Secondary Amine**

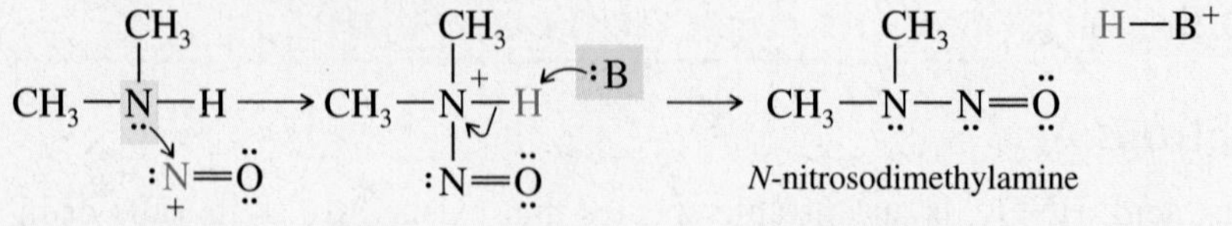

*N*-nitrosodimethylamine

Since the discovery of the carcinogenicity of nitrosamines, extensive research has been conducted on the presence in biological systems of secondary amines capable of forming nitrosamines and on sources of nitrites in food. Hydrochloric acid in gastric juice generates nitrous acid from nitrites that are eaten. The two types of precursors to nitrosamines are found with disturbing frequency. Dimethylamine is found in a number of fish and meat products. Dimethylamine, methylethylamine, and the cyclic secondary amine pyrrolidine (p. 808) are found in tobacco smoke. Sodium nitrite is used as a preservative in meats such as bacon, cold cuts, and frankfurters. Nitrates, which are used widely as fertilizers, are also reduced to nitrites by the body and by some plants, so residues of these nitrates also may contribute to nitrite intake. Concern about the cancer-causing effects of nitrosamines has led to attempts to find ways of preserving foods without using nitrites.

**Problem 20.13**   A secondary amine is soluble in the acidic solution used for a nitrosation reaction. The *N*-nitroso secondary amine that is the product of the reaction usually separates out of the solution as an insoluble oil or precipitate. Why?

If the amine is a primary amine, the *N*-nitrosamine undergoes further reactions, which are important for aryl amines and are discussed in the next section.

## C. Nitrosation of Primary Aryl Amines. Aryl Diazonium Ions

Aniline is converted to benzenediazonium chloride when treated with sodium nitrite and hydrochloric acid at 0 °C. This is called a **diazotization reaction.**

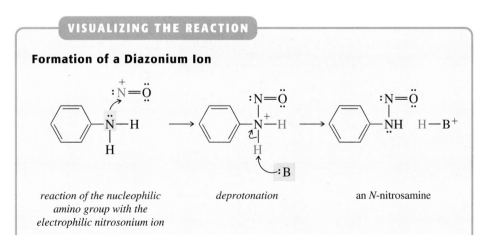

aniline → benzenediazonium chloride
soluble in water
stable at 0 °C

The reaction starts in the same way the nitrosation of a secondary amine does. Deprotonation and protonation then lead to the formation of a **diazonium ion,** which is an unstable species containing a very good leaving group, a nitrogen molecule.

---

**VISUALIZING THE REACTION**

**Formation of a Diazonium Ion**

reaction of the nucleophilic amino group with the electrophilic nitrosonium ion        deprotonation        an *N*-nitrosamine

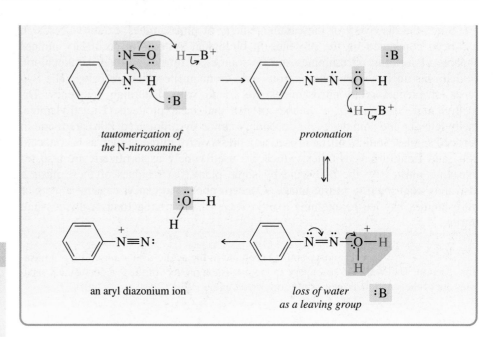

*tautomerization of
the N-nitrosamine*

*protonation*

*an aryl diazonium ion*

*loss of water
as a leaving group*

## ONE SMALL STEP

Primary alkyl amines form diazonium ions in exactly the same way that primary aryl amines do. The only difference is that primary alkyl diazonium ions are unstable and very rapidly lose nitrogen, $N_2$, as a leaving group to give a mixture of substitution and elimination products.

**PROBLEM:** When 1-butanamine is treated with sodium nitrite in aqueous hydrochloric acid at 0 °C, a mixture of 1-butanol, 2-butanol, 1-chlorobutane, 2-chlorobutane, 1-butene, and 2-butene is formed. 1-Butanol (25%) and 1-butene (26%) are the major products. Write mechanisms that account for the formation of all the products starting from the 1-butyldiazonium ion.

**On the Web: ONE SMALL STEP**

Study Guide
Concept Map 20.2

Aryl diazonium ions are stable enough in solution at low temperatures that they can serve as reagents in reactions in which the nitrogen molecule is replaced in a controlled way. These reactions are of major synthetic importance and are discussed in detail in Section 21.7B.

*N*-Methylaniline is a secondary aryl amine and gives an *N*-nitrosamine when treated with nitrous acid.

$$\text{—NHCH}_3 \xrightarrow[\substack{H_2O \\ 0\,°C}]{NaNO_2,\ HCl} \text{—N—N=O}$$

*N*-methylaniline          *N*-methyl-*N*-nitrosoaniline

An electrophilic substitution reaction takes place with the nitrosonium ion acting as the electrophile on tertiary aryl amines that are unsubstituted para to the amino group on the aromatic amine.

$$\text{—NCH}_3 \xrightarrow[\substack{H_2O \\ 0\,°C}]{NaNO_2,\ HCl} O=N\text{—}\text{—NCH}_3$$

*N,N*-dimethylaniline          *N,N*-dimethyl-*p*-nitrosoaniline

**Problem 20.14**    Complete the following equations.

(a) 
$$\text{NH}_2 \xrightarrow[\substack{H_2O \\ 0\,°C}]{NaNO_2,\ HCl}$$

(b) 
$$\text{Br, NH}_2 \xrightarrow[\substack{H_2O \\ -5\,°C}]{NaNO_2,\ HCl}$$

(c) 
$$\text{(piperidine)} \xrightarrow[\substack{H_2O \\ 0\,°C}]{NaNO_2,\ HCl}$$

(d) 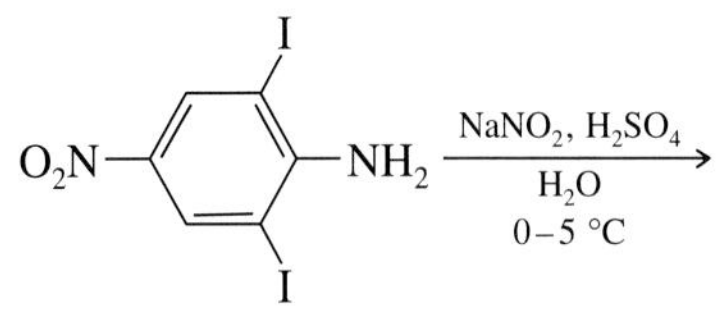 $O_2N$— [benzene ring with two I substituents] —$NH_2$ $\xrightarrow[\substack{H_2O \\ 0-5\,°C}]{NaNO_2,\ H_2SO_4}$

**Problem 20.15**    Synthesis of an enzyme inhibitor for the treatment of asthma involved the following steps. Supply structural formulas for the reagents and products indicated by the letters.

(a) 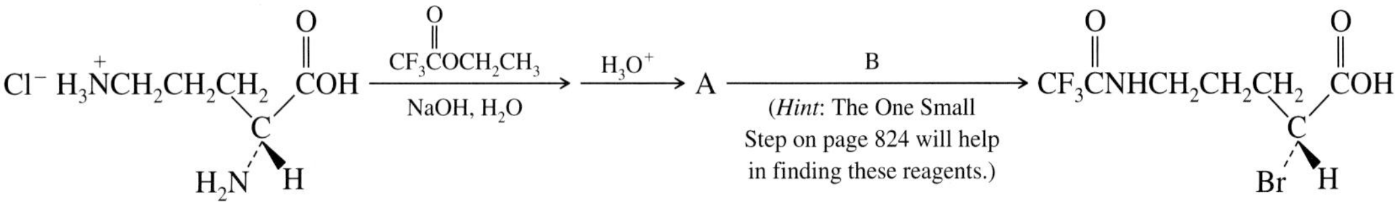 $Cl^-$ $\overset{+}{H_3}NCH_2CH_2CH_2\underset{\underset{H}{\overset{|}{\underset{H_2N}{C}}}}{\overset{O}{\overset{\|}{C}}}OH$ $\xrightarrow[\text{NaOH, } H_2O]{CF_3COCH_2CH_3}$ $\xrightarrow{H_3O^+}$ A $\xrightarrow[\substack{\textit{(Hint: The One Small} \\ \textit{Step on page 824 will help} \\ \textit{in finding these reagents.)}}]{B}$ $CF_3\overset{O}{\overset{\|}{C}}NHCH_2CH_2CH_2\underset{\underset{H}{\overset{|}{\underset{Br}{C}}}}{\overset{O}{\overset{\|}{C}}}OH$

(b) [aromatic ring with CHO (C=O, CH), OCH₃, OCH₃ substituents] $+$ $Cl^-$ $\overset{+}{H_3}N—CH(\overset{O}{\overset{\|}{C}}OCH_3)_2$ $\xrightarrow[\text{methanol}]{(CH_3CH_2)_3N}$ C $\xrightarrow{D}$ [aromatic ring with $CH_2NHCH(\overset{O}{\overset{\|}{C}}OCH_3)_2$, OCH₃, OCH₃ substituents]

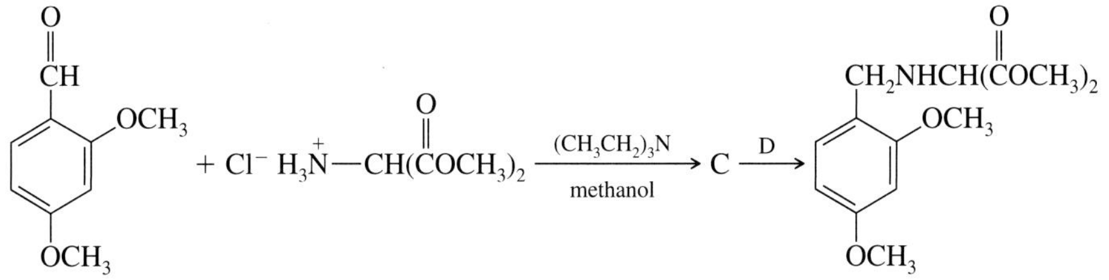

# 20.6  The Diazonium Ion as Electrophile

## A.  Synthesis of Azo Compounds

One reaction of diazonium ions has important practical application in the dye industry. This is the coupling reaction in which the diazonium ion acts as an electrophile, substituting on the activated aromatic ring of either a phenol or an aromatic amine. A typical reaction is that of benzenediazonium chloride and phenol, which is carried out in a weakly basic solution. Electrophilic attack by the benzenediazonium ion takes place at one of the activated positions of the ring, ortho or para to the hydroxyl group.

**VISUALIZING THE REACTION**

**The Diazonium Ion as Electrophile**

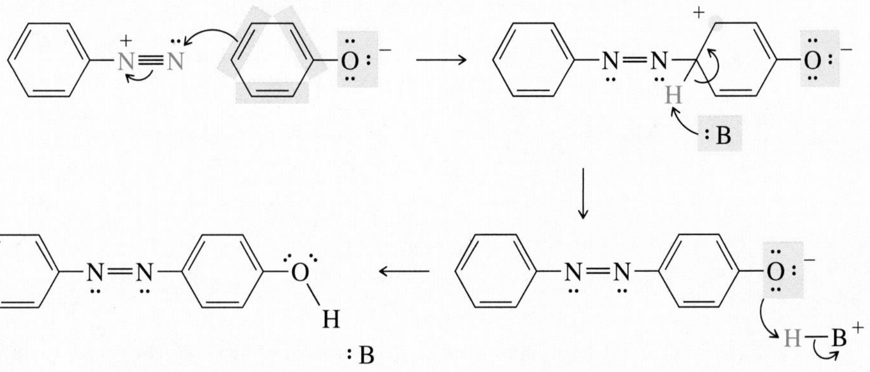

*p*-hydroxyazobenzene
$\lambda_{max}^{ethanol}$ 349 nm ($\epsilon$ 26,300)

The product of the previous reaction is an **azo compound,** containing a nitrogen–nitrogen double bond that can have cis and trans isomers just as a carbon–carbon double bond can. The azo compound with a phenyl group at both ends of the azo linkage is azobenzene, of which both isomers are known.

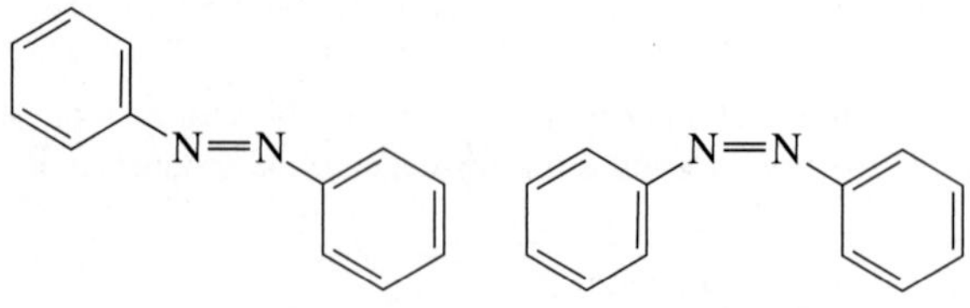

(E)-azobenzene                    (Z)-azobenzene

*stereoisomers of azobenzene*

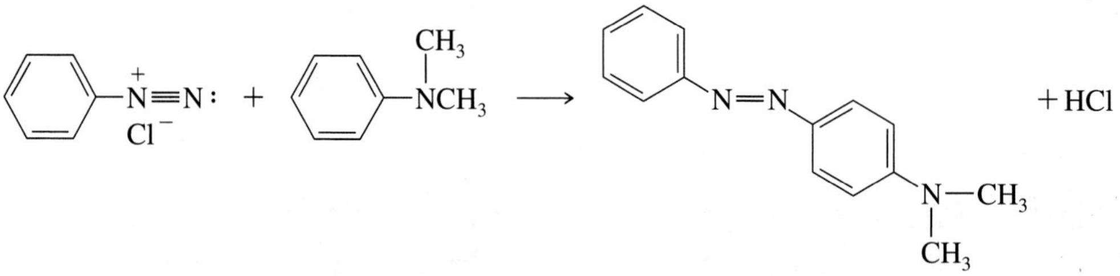

Benzenediazonium chloride also couples with tertiary aromatic amines. For example, it reacts with *N,N*-dimethylaniline to give *p*-dimethylaminoazobenzene, a dye known as butter yellow, which was used to color margarine until it was discovered to be carcinogenic.

benzenediazonium          *N,N*-dimethylaniline          *p*-dimethylaminoazobenzene
chloride                                                            butter yellow

## B. Azo Dyes and Acid–Base Indicators

Compounds in which the azo linkage is between two aromatic rings are highly colored. These compounds have extended conjugated systems, and thus their absorption maxima are in the visible range of the electromagnetic spectrum (p. 449). (E)-Azobenzene, with $\lambda_{max}^{ethanol}$ 318 nm ($\epsilon$ 21,380), absorbs in the near-ultraviolet region and is visibly orange. This phenomenon is the basis for a qualitative analysis test for primary aromatic amines. The formation of red color when diazotized amine solution is added to 2-naphthol indicates the presence of an aromatic diazonium ion and, therefore, of an aromatic primary amine.

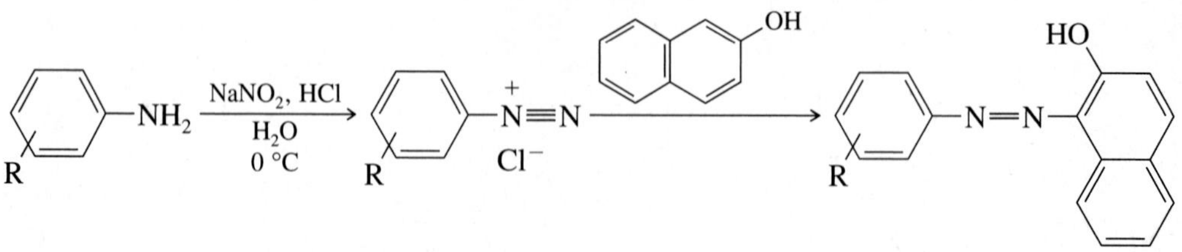

primary aromatic          aryl diazonium          red dye
amine                         ion

Hydroxyl and amino groups, especially if they are ortho or para to the azo bond, intensify the colors of azo compounds. For example, *p*-hydroxyazobenzene has $\lambda_{max}^{ethanol}$ 349 nm ($\epsilon$ 26,300), and *p*-dimethylaminoazobenzene has $\lambda_{max}^{ethanol}$ 408 nm

($\epsilon$ 27,540). Each of these compounds absorbs at a longer wavelength than azobenzene itself does, and in each case, the molar absorptivity, $\epsilon$, is also higher. Azo compounds in which an electron-donating group on one of the aromatic rings is conjugated with an electron-withdrawing group on the other ring have especially deep colors. A good example is the azobenzene in which a nitro group is substituted on one ring para to the azo linkage and a dimethylamino group is substituted on the other ring also para to the azo linkage.

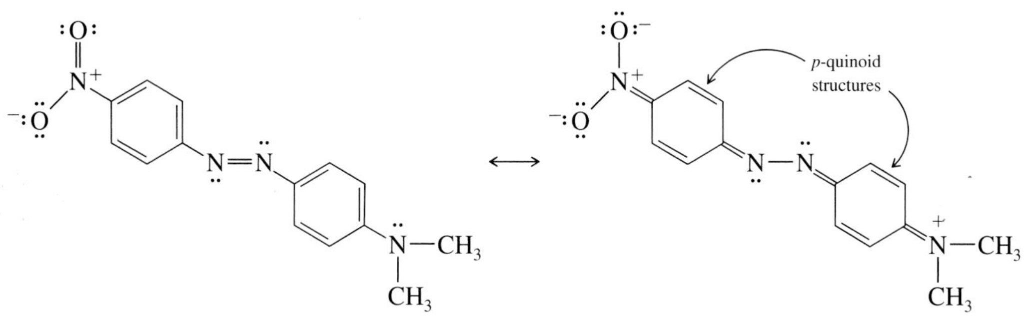

4-dimethylamino-4′-nitroazobenzene
$\lambda_{max}^{ethanol}$ 478 nm ($\epsilon$ 33,110)

One of the resonance contributors of the compound has quinoid structures, a feature that often leads to deep color in compounds (p. 792).

The aromatic rings of azo compounds allow the introduction of a variety of functional groups that can interact chemically with acidic, basic, or polar sites in the fibers used in making cloth or paper. A large number of dyes, tailored to fit the chemical nature of the material to be dyed, have been created by extensions of the reactions shown on the previous page.

Many dye molecules have sulfonic acid groups as substituents, giving them water solubility and enabling them to adhere firmly to fibers such as silk and wool, which are composed chiefly of proteins and have basic functional groups on them. Methyl orange, synthesized from sulfanilic acid and $N,N$-dimethylaniline, is such a dye.

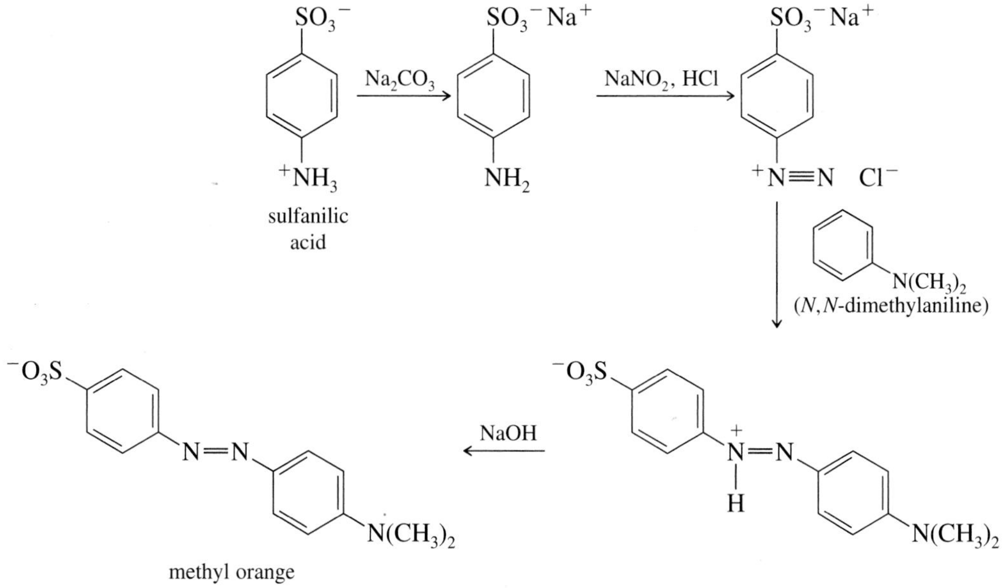

methyl orange

Methyl orange is also used as an acid–base indicator. In dilute solutions with a pH higher than 4.4, it is yellow; $\lambda_{max}$ is 460 nm. When acid is added to the system, methyl orange is protonated, and the resulting dipolar ion predominates at pH values of 3.2 and lower. The protonated form has $\lambda_{max}$ at 520 nm and appears red.

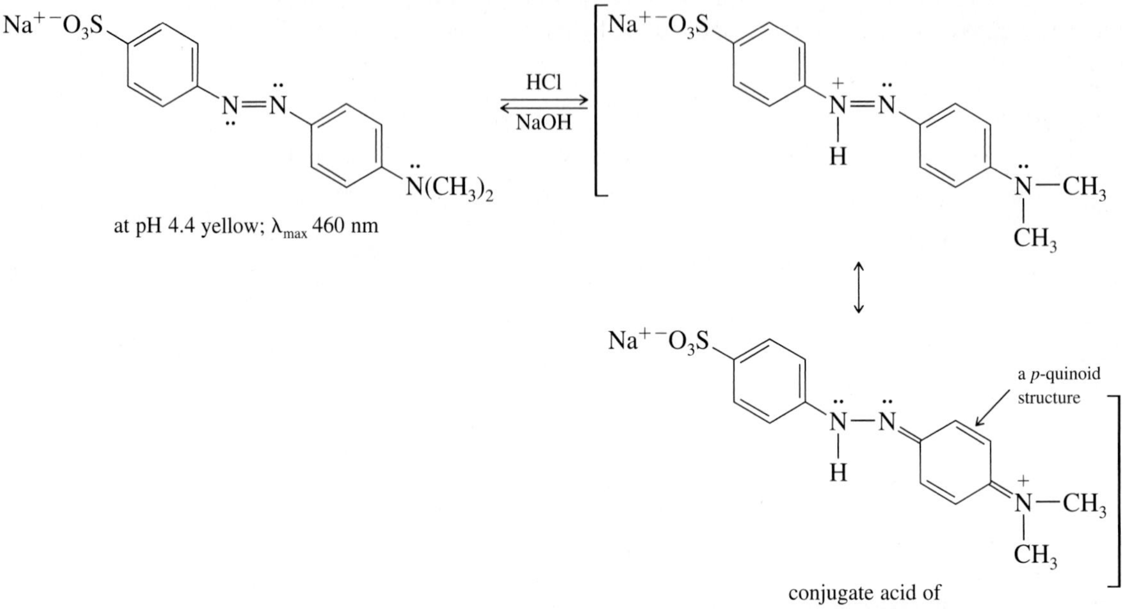

at pH 4.4 yellow; $\lambda_{max}$ 460 nm

a *p*-quinoid structure

conjugate acid of methyl orange, stabilized by delocalization of charge; at pH 3.2 red; $\lambda_{max}$ 520 nm

Methyl orange and its conjugate acid have different chromophores (p. 445) and thus absorb at different wavelengths in the range of visible light. The different colors that the compound shows allow it to be used to detect a change in the acidity of a system around the range of pH at which it is protonated and deprotonated.

*Study Guide*
*Concept Map 20.3*

**Problem 20.16**    There are three nitrogen atoms in methyl orange. Why is it protonated on the particular nitrogen atom shown in the equation above?

**Problem 20.17**    Para red is widely used to dye cotton. The cotton fabric is soaked in one of the components of the dye, and then the diazonium salt derived from the other component is added to the system so that the azo dye forms directly inside the fibers and is trapped there. Para red has the structure shown below. Write structures for the components that you would use to synthesize it.

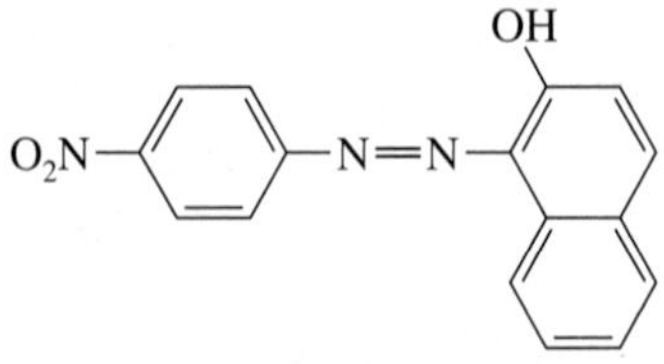

para red

**Problem 20.18**    Extensive conjugation must be present in a molecule of an azo dye for it to have a blue color. Such a blue dye is synthesized by the sequence of reactions shown on page 829. Supply structural formulas for Compounds A and B.

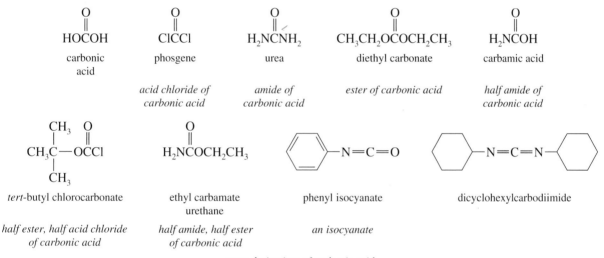

## 20.7 Nitrogen Derivatives of Carbonic Acid

Nitrogen-containing derivatives of carbonic acid are important biologically (urea, for example), industrially (polyurethanes, p. 1061, for example), and in syntheses, especially in the synthesis of peptides (pp. 631 and 901).

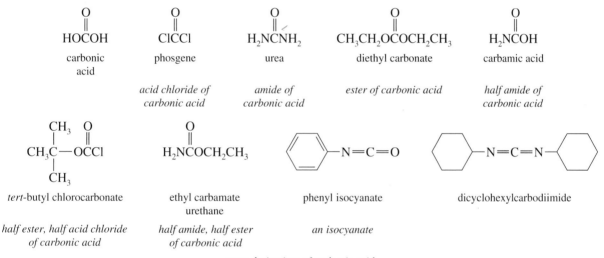

*some derivatives of carbonic acid*

Carbonic acid itself is unstable and decomposes into carbon dioxide and water. Carbamic acid is also unstable, giving carbon dioxide and ammonia. In fact, derivatives of carbonic acid in which only one of the two acid groups has been substituted all lose carbon dioxide with ease. The ones in which both sides are substituted, such as phosgene, urea, and ethyl carbamate, are stable. Phosgene was used in World War I as a poison gas. It reacts with water to give carbon dioxide and hydrogen chloride, which causes fluid to accumulate rapidly in the lungs. Urea is tremendously important in the metabolism of mammals. It is the chief form in which materials from the breakdown of proteins are excreted from the human body; an average man excretes about 30 grams of urea a day.

The unsymmetrical derivatives of carbonic acid retain the reactivity of each type of functional group present. For example, *tert*-butyl chlorocarbonate is an ester and an acid chloride. It reacts readily as an acid chloride with amines to give carbamates, which in turn are easily hydrolyzed to regenerate the amino group, as we saw in Section 15.8B when we used the *tert*-butoxycarbonyl group (boc) as a protecting group in peptide syntheses.

Isocyanates and carbodiimides are reactive compounds to which nucleophiles, such as water, alcohols, and amines, add with ease. 1-Naphthylisocyanate is used, for example, to convert liquid alcohols into solid derivatives that are useful for identifying the alcohols.

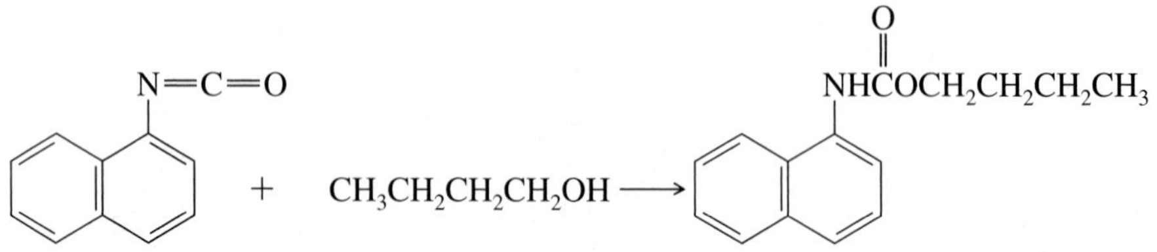
N=C=O
+
CH3CH2CH2CH2OH
NHCOCH2CH2CH2CH3
O

1-naphthylisocyanate — *n*-butyl alcohol, bp 118 °C — butyl *N*-(1-naphthyl)carbamate, butyl naphthylurethane, mp 71 °C

The reaction of polyfunctional isocyanates with polyfunctional alcohols is of considerable industrial importance. Such reactions give polyurethane polymers, which are used to make foam cushions, fibers with elastic qualities, tire treads, and coatings for floors, among many other things (p. 1061).

Dicyclohexylcarbodiimide is useful in promoting the formation of amide bonds. It reacts with carboxylic acids to give intermediates that then react with amines to give amides. The reaction is used in the synthesis of peptides (p. 634).

**Problem 20.19** Complete the following equations.

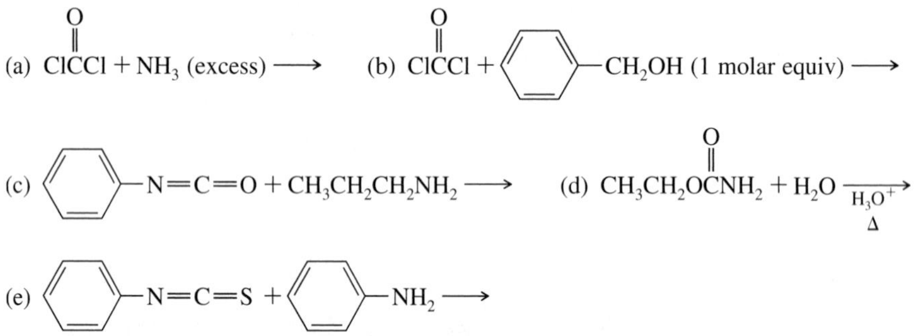
(a) ClCCl + NH3 (excess)
(b) ClCCl + CH2OH (1 molar equiv)
(c) N=C=O + CH3CH2CH2NH2
(d) CH3CH2OCNH2 + H2O
H3O+
Δ
(e) N=C=S + NH2

phenyl isothiocyanate

**Problem 20.20** One of the pesticides developed after the discovery that chlorinated hydrocarbons such as DDT accumulate in the environment is carbaryl. It is a carbamate with the structure below.

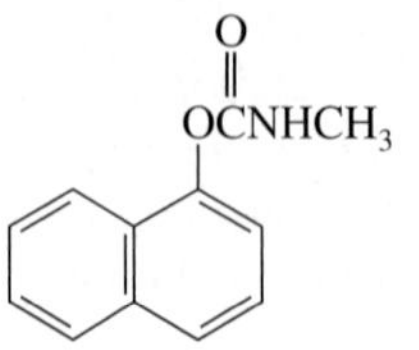
OCNHCH3
O

carbaryl

*an insecticide that is biodegradable*

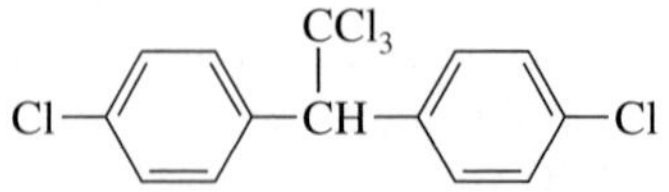
CCl3
Cl
CH
Cl

1,1,1-trichloro-2,2-bis(*p*-chlorophenyl)ethane

DDT

*an insecticide that is not biodegradable*

**On the Web: ONE SMALL STEP**

Write equations showing a mechanism for the degradation of carbaryl by water in the environment. You may assume that the pH of the soil is either below or above 7.

## Table 20.1    Summary of Reactions Used in the Preparation of Amines

### From Alkyl Halides

| Starting Material | Nucleophile | Product of Substitution Reaction | Reagents for Second Step | Final Product |
|---|---|---|---|---|
| RX<br>R = primary or secondary alkyl group | $NH_3$<br>(1 equivalent) | $\overset{+}{R}NH_3X^-$ | | $\overset{+}{R}NH_3X^-$ |
| | $NH_3$<br>(excess) | $RNH_2$ | RX | $RNH_2$<br>$R_2NH$<br>$R_3N$<br>$R_4N^+X^-$ |
| | NaCN | RCN | $H_2$/catalyst<br>or $LiAlH_4$ | $RCH_2NH_2$ |
| | $NaN_3$ | $RN_3$ | $H_2$/catalyst<br>or $LiAlH_4$ | $RNH_2$ |
| | $NaNO_2$ | $RNO_2$ | $H_2$/catalyst | $RNH_2$ |

### From Carbonyl Compounds

| Starting Material | Nucleophile | Product of Substitution Reaction | Reagents for Second Step | Final Product |
|---|---|---|---|---|
| $\overset{O}{\underset{R'CR''}{\parallel}}$ | $RNH_2$ | $\underset{R''}{\overset{R'}{>}}C{=}NR$ | $NaBH_4$, $CH_3OH$<br>or $H_2$/catalyst<br>*(called reductive amination if imine is not isolated and reduction is carried out on mixture of amine and carbonyl compound)* | $\overset{R'}{\underset{}{R''CHNHR}}$ |
| | $H_2NOH$ | $\underset{R''}{\overset{R'}{>}}C{=}NOH$ | $H_2$/catalyst | $\overset{R'}{\underset{}{R''CHNH_2}}$ |

### From Aromatic Compounds

| 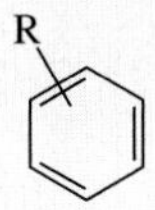 | $HNO_3$<br>$H_2SO_4$ | 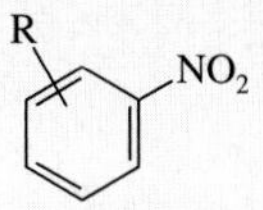 Ar–$NO_2$ | $H_2$/catalyst | 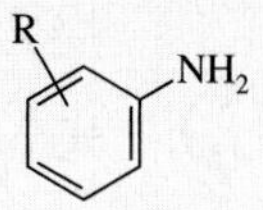 Ar–$NH_2$ |
|---|---|---|---|---|

**Table 20.2** Summary of the Reactions of Amines as Nucleophiles

| Amine | Electrophile | Intermediate | Reagent for Second Step | Product |
|---|---|---|---|---|
| $RNH_2$ | $R'X$ | $R'$ on N: $RNH_2^+ X^-$ | $RNH_2$ | $R'$ on N: $RNH$ |
| $R_2NH$ | $R'X$ | $R'$ on N: $R_2NH^+ X^-$ | $R_2NH$ | $R'$ on N: $R_2N$ |
| $R_3N$ | $R'X$ | — | — | $R_3NR'^+ X^-$ |
| $R_2NH$ | $NO^+$ | — | — | $R_2N{-}NO$ |
| aniline ($NR_2$) | $NO^+$ | — | — | para-nitroso-$NR_2$ (NO at para) |
| R-substituted aniline ($-NH_2$) | $NO^+$ | aryldiazonium ion ($R$—Ar—$N{\equiv}N^+$) | $R'$—Ar—$O^-$; $R'_2N$—Ph; Ph—$NR'_2$ | azo compounds: $R$—Ar—$N{=}N$—Ar—$OH$; $R$—Ar—$N{=}N$—Ar—$NR'_2$ |

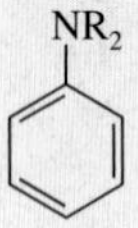

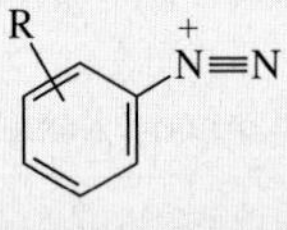

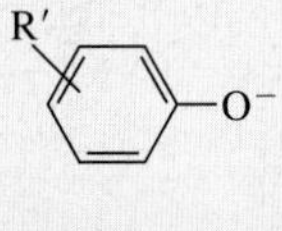

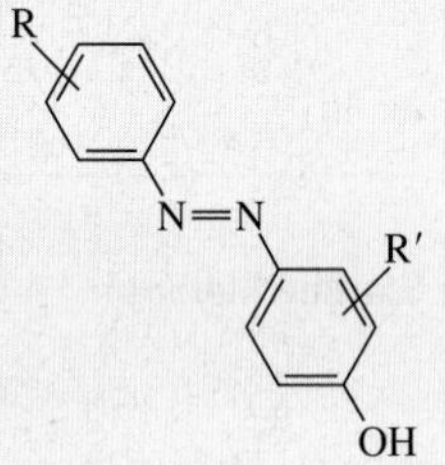

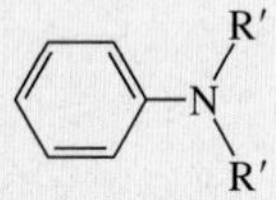

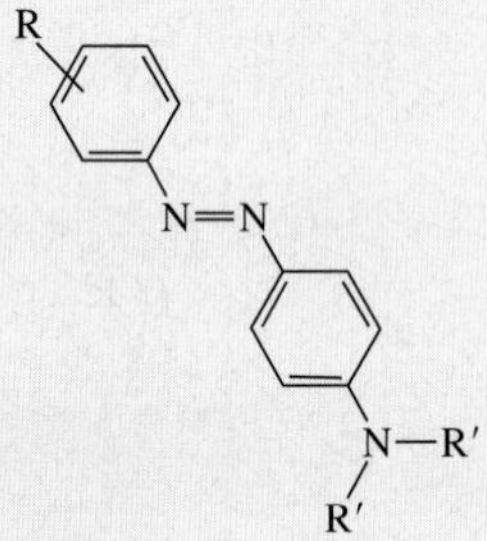

# ADDITIONAL PROBLEMS

**20.21** Name the following compounds.

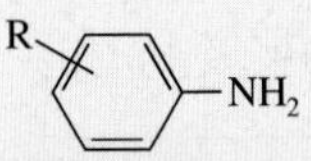

(a) 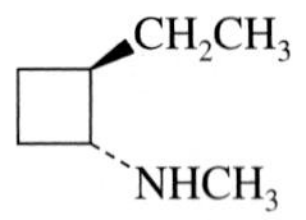 cyclobutane with CH₂CH₃ and NHCH₃ substituents (cis)

(b) 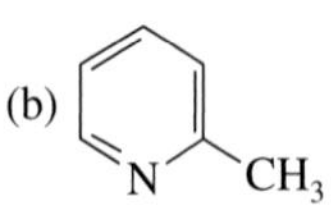 2-methylpyridine

(c) $CH_3CH_2NCH_2CH_3$ with $CH_2CH_3$ on N

(d) 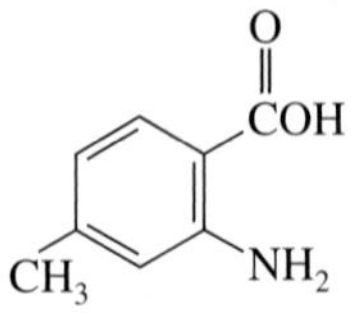 benzoic acid ($-COOH$) with $CH_3$ and $NH_2$ substituents

(e) $H_3NCH_2CH_2CH_2CO^-$ (with $C{=}O$)

(f) [structure: 2,4,6-trichloroaniline — NH₂ with Cl at 2,6 and Cl at 4]

(g) $CH_3CHCH_2CH_2CHCH_3$ with $CH_3$ above and $NH_2$ below

(h) [cyclohexane ring with ---NH₂ and OH]

(i) [structure: N,N-diethyl-4-nitroaniline with $CH_2CH_3$, $NCH_2CH_3$, and $O_2N$]

**20.22** Write structural formulas for the following compounds.

(a) *trans*-2-aminocyclopentanol  (b) *N*-methyl-*N*-propylcyclohexylamine
(c) 3,5-dinitroaniline  (d) *N,N*-dimethyl-*p*-methoxyaniline  (e) *m*-toluidine
(f) 2,5-diaminooctane  (g) 3-hexanamine
(h) 4-amino-2,2-dimethylpentanoic acid

**20.23** Give structural formulas for all compounds designated by letters in the following equations.

(a) [PhCH=N-Ph] $\xrightarrow[\text{diethyl ether}]{\text{LiAlH}_4}$ $\xrightarrow{\text{H}_2\text{O}}$ A

(b) $CH_3\overset{O}{\overset{\|}{C}}CH=\overset{CH_3}{\overset{|}{C}}CH_3$ $\xrightarrow[\text{H}_2\text{O}]{\text{NH}_3}$ B

(c) $CH_3\overset{CH_3}{\overset{|}{C}}CH_2CH_2\overset{O}{\overset{\|}{C}}OCH_3$ with $NO_2$ below $\xrightarrow[\underset{\Delta \quad C_6H_{11}NO}{\text{Ni}}]{\text{H}_2}$ C

(d) $H_2N-$[benzene ring]$-SO_2NH_2$ $\xrightarrow[\text{Fe}]{\text{Br}_2}$ D

(e) $CH_3\overset{O}{\overset{\|}{C}}CH_2CH_2CH_2CH_2CH_3$ $\xrightarrow[\underset{\Delta}{\text{Ni}}]{\text{NH}_3, \text{H}_2}$ E

(f) $CH_3CH_2CH_2NO_2$ $\xrightarrow[\text{NaOH}]{\overset{O}{\overset{\|}{\text{HCH}}}}$ F $\xrightarrow[\underset{\Delta}{\text{Ni}}]{\text{H}_2}$ G $\xrightarrow{\text{Ca(OH)}_2}$ H

**20.24** More practice in recognizing reactions follows.

(a) [cyclohexanone with $C_6H_5$, $C_6H_5$] $\xrightarrow[\underset{\Delta}{\text{pyridine}}]{\overset{+}{\text{HONH}_3\text{Cl}^-}}$ A $\xrightarrow[\text{diethyl ether}]{\text{LiAlH}_4}$ $\xrightarrow{\text{H}_3\text{O}^+}$ B $\xrightarrow[\text{H}_2\text{O}]{\text{NaOH}}$ C

(b) $^-O_3S-$[benzene ring]$-\overset{+}{N}H_3$ $\xrightarrow[\text{H}_2\text{O}]{\text{Na}_2\text{CO}_3}$ D $\xrightarrow[\underset{0-5\,°C}{\text{H}_2\text{O}}]{\text{NaNO}_2, \text{HCl}}$ E

(c) [naphthalene with $N_3$] $\xrightarrow[\text{diethyl ether}]{\text{LiAlH}_4}$ $\xrightarrow{\text{H}_2\text{O}}$ F

(d) [cyclohexanone with CH₃ and two CH₃] $\xrightarrow[\underset{\Delta, \text{ pressure}}{\text{Ni}}]{\text{NH}_3, \text{H}_2}$ G

(e) $CH_3\overset{O}{\overset{\|}{C}}CH_3$ $\xrightarrow[\text{HCl (catalyst)}]{\text{cyclohexyl-NH}_2}$ H $\xrightarrow{\text{NaOH}}$ I

(f) F[CH₂CH₂CH₂]OH $\xrightarrow{J}$ F [CH₂CH₂CH₂]OTs $\xrightarrow{K}$ [pyrrolidine-N-CH₂CH₂-N-CH₂CH₂CH₂-F with CH₂CH₂-(3,4-dichlorophenyl)] F

(g) $PhCH_2O-$[benzene ring with $CH_3O$]$-\overset{H}{\underset{H}{C}}=\overset{}{C}\overset{H}{\underset{CN}{}}$ $\xrightarrow[\underset{\text{chloroform}}{\underset{\text{ethanol}}{\text{PtO}_2}}]{\text{H}_2}$ L

**20.25** The tumors of some breast cancer patients grow in response to estrogen, the female sex hormone. One way of treating such tumors to prevent their growth is to give the patient compounds that react with the estrogen receptors on the tumor cells and thus block the action of estrogen. Much research has gone into other ways of preventing estrogen from reaching tumor cells. Some steps in the synthesis of compounds that block the biosynthesis of estrogen in the organism follow. Provide structural formulas for the reagents and the products designated by letters.

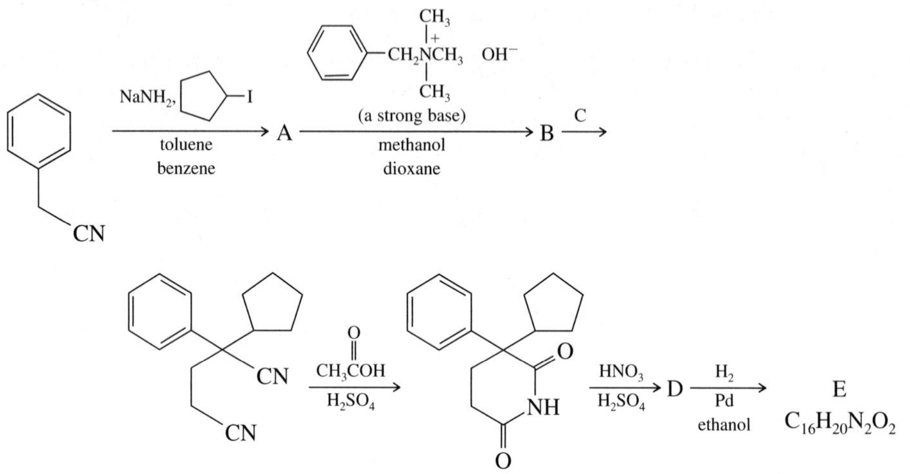

**20.26** The question of how compounds such as morphine and heroin interact with the brain is one that has fascinated chemists. One way to study the question is to synthesize a variety of compounds to study how they interact with the same receptors in the brain. Part of such a synthesis is outlined below. Provide structural formulas for the products designated by letters.

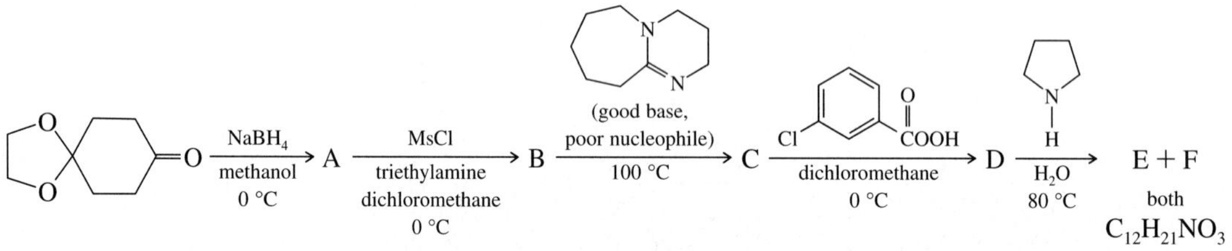

**20.27** Large rings with compatible functional groups on opposite sides of the ring tend to form bridged bicyclic structures. For example,

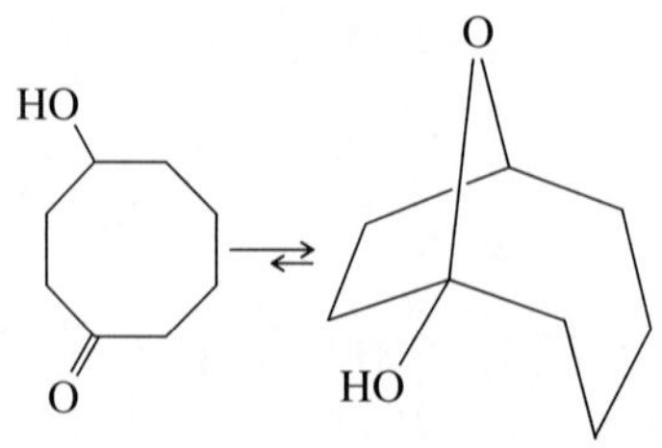

The hydroxy-ketone reacts only slowly with tosyl chloride in pyridine to give a tosylate, which can be displaced by azide ion. Reduction of the azide gives a compound that exists primarily as a bridged structure at temperatures below 0 °C.

Write equations for the reactions described above and propose a structure for the bridged compound that is in equilibrium with another monocyclic form.

**20.28** Synthesis of pharmaceuticals requires the preparation of enantiomerically pure ingredients. Some steps in such a synthesis of (*R*)-fluoxetine, the active component of the antidepressant Prozac, are shown below. Supply structural formulas for the products indicated by letters.

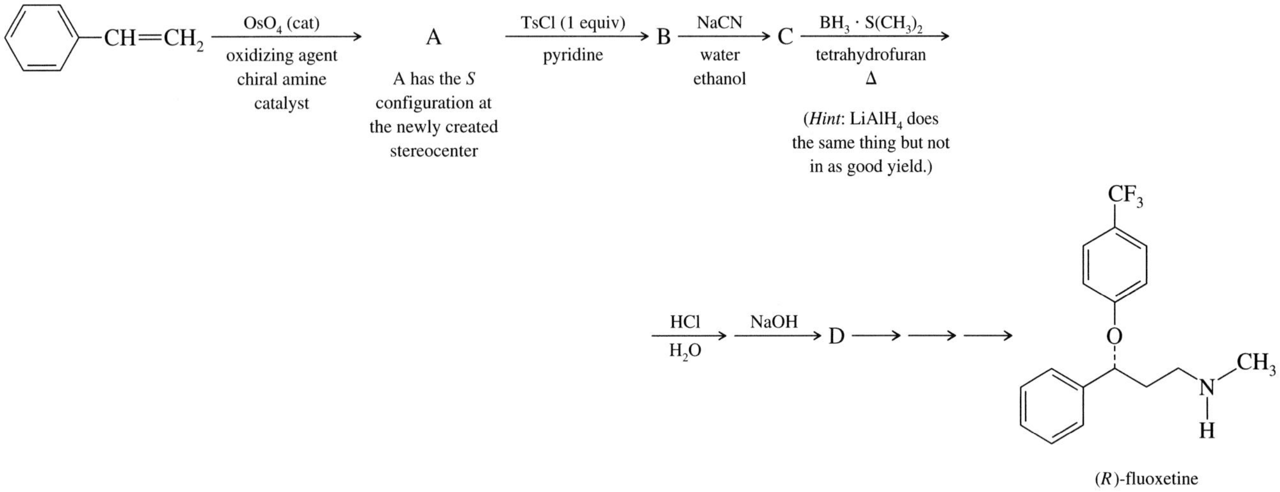

(*R*)-fluoxetine

**20.29** Write a mechanism that explains the following experimental result.

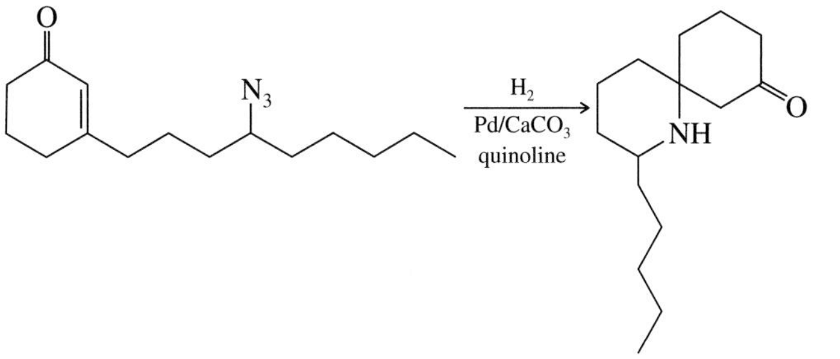

**20.30** One potential method for the treatment of tumors is to covalently bind clusters of boron atoms (such as those found in the compound $B_{10}H_{10}$) to antibodies that target cancer cells. Boron atoms, when bombarded by neutrons, emit $\alpha$-particles that kill cancer cells in their vicinity. The following reactions were carried out in an exploration of the chemistry of carbon compounds covalently bonded to $B_{10}H_{10}$ to see if such compounds could be attached to the proteins of antibodies. Supply the reagents designated by letters for the transformations that are shown. Each letter may stand for more than a single reagent.

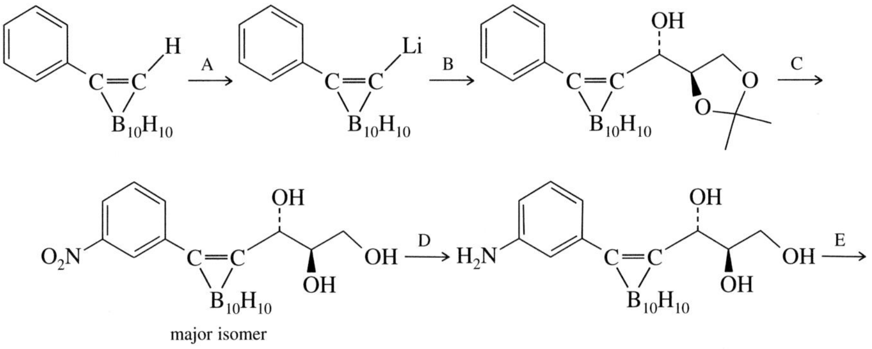

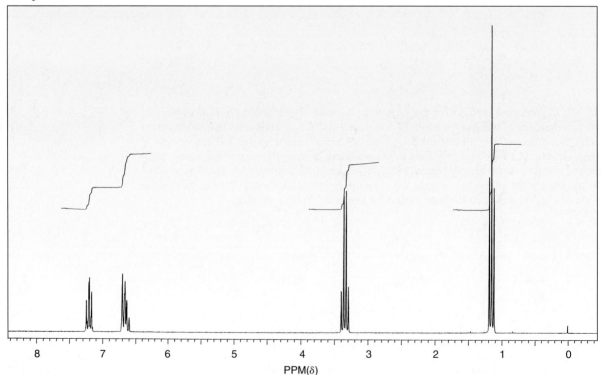

used to model the
amino acid tyrosine

**20.31** Research into compounds aimed at inhibiting the enzyme responsible for the multiplication of the HIV virus used the following reactions. Supply structural formulas for the products indicated by letters, and interpret the spectroscopic data given for Compound B.

$$A \xrightarrow[\text{pyridine}]{\underset{\text{CH}_3\text{CCl}}{\overset{O}{\|}}} \quad A \xrightarrow[\substack{\text{tetrahydrofuran} \\ \text{H}_2\text{O}}]{\text{NaBH}_4} \quad B$$

B
$C_{15}H_{15}NO_2$

$v_{max}$ 1640, 3300, 3410, 3500 cm$^{-1}$
$^1$H NMR: δ 2.15 (s, 3H), 2.32 (br s,
1H, exchangeable with D$_2$O), 5.81
(s, 1H), 7.20–7.49 (m, 9H)

**20.32** Compound A, $C_{10}H_{15}N$, has the proton magnetic resonance spectrum shown in Figure 20.2. Assign a structure to the compound.

Compound A

8    7    6    5    4    3    2    1    0
PPM(δ)

*Figure 20.2*

**20.33** The synthesis of an alkaloid involved the following reactions. Assign structural formulas to the compounds designated by letters, and analyze the spectral data given for them.

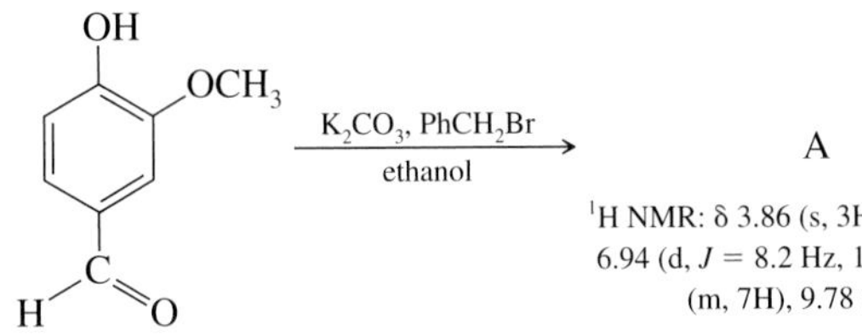

OH
OCH$_3$

$\xrightarrow[\text{ethanol}]{\text{K}_2\text{CO}_3,\ \text{PhCH}_2\text{Br}}$

A

$^1$H NMR: δ 3.86 (s, 3H), 5.17 (s, 2H)
6.94 (d, $J$ = 8.2 Hz, 1H), 7.27–7.43
(m, 7H), 9.78 (s, 1H)

A $\xrightarrow[\text{acetic acid}]{\text{CH}_3\text{NO}_2,\ \text{CH}_3\overset{\text{O}}{\overset{\|}{\text{C}}}\text{O}^-\text{NH}_4^+}$ B

$v_{max}$ (KBr) 1629 cm$^{-1}$
$^1$H NMR: δ 3.91 (s, 3H), 5.20 (s, 2H), 6.89–7.09
(m, 3H), 7.32–7.42 (m, 5H), 7.52 (d, $J$ = 13.4 Hz,
1H), 7.93 (d, $J$ = 13.4 Hz, 1H)

B $\xrightarrow[\substack{\text{ether}\\ \text{tetrahydrofuran}\\ \Delta}]{\text{LiAlH}_4\ (\text{excess})}$ C

$v_{max}$ (NaCl) 3367 cm$^{-1}$
$^1$H NMR: δ 2.32 (br s, 2H), 2.69 (br t, $J$ = 7.0 Hz, 2H),
2.93 (br t, $J$ = 7.0 Hz, 2H), 3.87 (s, 3H), 5.12 (s, 2H),
6.64–6.83 (m, 3H) 7.28–7.45 (m, 5H)

C $\xrightarrow[\substack{\text{K}_2\text{CO}_3\\ \text{acetone}}]{\overset{\text{O}}{\overset{\|}{\text{ClCOCH}_3}}}$ D

$v_{max}$ (KBr) 3345, 1681 cm$^{-1}$
$^1$H NMR: δ 2.72 (br t, $J$ = 7.0 Hz, 2H), 3.39 (br q, $J$ = 7.0 Hz, 2H),
3.64 (s, 3H), 3.87 (s, 3H) 4.85 (br s, 1H), 5.12 (s, 2H),
6.65 (dd, $J$ = 1.8, 8.2 Hz, 1H), 6.73 (d, $J$ = 1.8 Hz, 1H),
6.81 (d, $J$ = 8.2 Hz, 1H), 7.26–7.46 (m, 5H)

**20.34** The $^{13}$C nuclear magnetic resonance spectrum of *N*-ethylacetamide has peaks at 14.6, 22.8, 34.4, and 171.0 ppm. Assign these peaks to the various carbon atoms in the compound. Table 11.2 (p. 403) may be useful.

**20.35** Compound B, C$_3$H$_9$N, has two peaks in its $^{13}$C nuclear magnetic resonance spectrum, at 26.2 and 42.8 ppm. Assign a structure to Compound B.

**20.36** The dye butter yellow, *p*-dimethylaminoazobenzene (p. 826), has $\lambda_{max}$ 408 nm in neutral solution. When acid is added to the solution, two different species are detected spectroscopically. One has $\lambda_{max}$ 320 nm and the other $\lambda_{max}$ 510 nm. Write structural formulas for the species that have chromophores giving rise to these absorption bands.

# 21 Synthesis

**A Look Ahead**

Chemistry, to a large extent, is the science of molecular transformations. Our understanding of structure and reactivity and of reaction mechanisms is put to use in the transformation of one molecular structure into another. From Section 4.4 forward in the text (such as in Section 9.6), in problems, and especially in The Art of Solving Problems sections (pp. 237, 319, and 619, for example), we have learned to look at starting materials and products and to select the reagents that would make such transformations possible. Or we have looked at a desired molecular structure and chosen the starting materials and the reagents that would lead to such a product.

Chemists engage in such synthetic exercises all the time. They may be devising ways to synthesize natural products with useful medicinal properties, especially if the natural product comes from a rare plant or animal. They may be trying to synthesize a compound that is structurally similar to the natural product but with even greater potency or less toxicity. Often chemists use the complex patterns of functional groups and stereochemistry found in such natural products as a challenge to test their skills and the new reagents and techniques they are developing. Chemists also make new materials, plastics, polymers, semiconductors. All these syntheses start with the kinds of analyses that we have been doing in our exercises. In this chapter we will review the kind of thinking that goes into designing syntheses, the thinking backwards known as **retrosynthetic analysis.** We also will extend our knowledge of synthetic reactions and strategies, such as the use of protecting groups, carbon and hydrogen nucleophiles, and Diels-Alder reactions. Finally, we will look at two techniques that are becoming more and more important. The first involves syntheses carried out on the surface of solid supports. The second, combinatorial chemistry, is a way to synthesize large numbers of similar compounds simultaneously to generate what are known as libraries of compounds that can then be screened for useful properties such as biological activity.

## 21.1 Retrosynthetic Analyses

### A. An Introduction

Suppose that we wish to synthesize 1-pentanol from reagents containing three or fewer carbon atoms. We can write equations with the symbol $\Rightarrow$ that chemists use to indicate that they are thinking backwards from a product to the starting material from which it might be made.

$$\text{product} \xrightarrow{\hspace{3cm}} \text{starting material(s)}$$
arrow indicating a<br>retrosynthetic analysis

In the set of retrosynthetic equations shown below, we are summarizing all the ways we know to make a primary alcohol as the first step of our analysis.

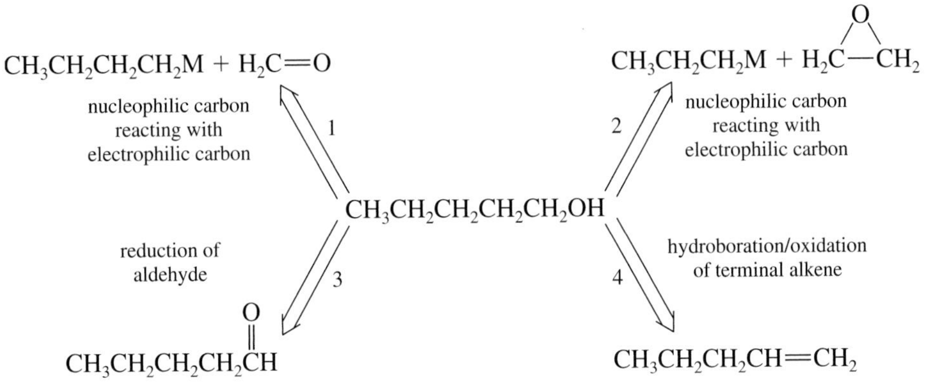

$CH_3CH_2CH_2CH_2M + H_2C{=}O$

nucleophilic carbon
reacting with
electrophilic carbon

1

$CH_3CH_2CH_2M + H_2C{-}CH_2$ (epoxide, O bridging)

nucleophilic carbon
reacting with
electrophilic carbon

2

$CH_3CH_2CH_2CH_2CH_2OH$

reduction of
aldehyde

3

hydroboration/oxidation
of terminal alkene

4

$CH_3CH_2CH_2CH_2\overset{O}{\overset{\|}{C}}H$

$CH_3CH_2CH_2CH{=}CH_2$

Note that at this stage we are not interested in the exact reagents we might use in each reaction but in the general outlines of the types of reactions we might use.

Two of the pathways, 2 and 4, appear to be better than the other two. Pathway 2 has starting material containing two and three carbon atoms, respectively. This method meets the requirement of the problem we have set ourselves. Pathway 4 also can be made to work, but we will have to do a retrosynthetic analysis on 1-pentene to see why this is so.

$$CH_3CH_2CH_2CH{=}CH_2 \xrightarrow[\text{of alkyne}]{\text{reduction}} CH_3CH_2CH_2C{\equiv}CH \xrightarrow[\text{of alkyne}]{\text{preparation}} CH_3CH_2CH_2LG + MC{\equiv}CH$$

The ease with which we can prepare nucleophilic acetylide anions (p. 331) and the selective reduction of an alkyne to an alkene using a poisoned catalyst (p. 339) make the terminal alkene also accessible from reagents containing three or fewer carbon atoms.

Pathway 1 would require several steps to get to an organometallic reagent containing four carbon atoms from a starting material of three or fewer carbon atoms, so we shall ignore that route. And aldehydes are usually prepared from the oxidation of primary alcohols, so pathway 3 is not promising either for this particular case.

$$CH_3CH_2CH_2CH_2\overset{O}{\overset{\|}{C}}H \xrightarrow[\substack{\text{reverse of}\\\text{pathway 3}}]{\substack{\text{oxidation}\\\text{of alcohol}}} CH_3CH_2CH_2CH_2CH_2OH$$

In writing the retrosynthetic equations, we are doing dissections on 1-pentanol and coming up with fragments that would react with one another because of their polarities. In other words, in each case we are looking for a nucleophilic reagent and an electrophilic one. For example, in pathway 2 the important idea is that we have a nucleophilic three-carbon fragment reacting with an electrophilic two-carbon fragment.

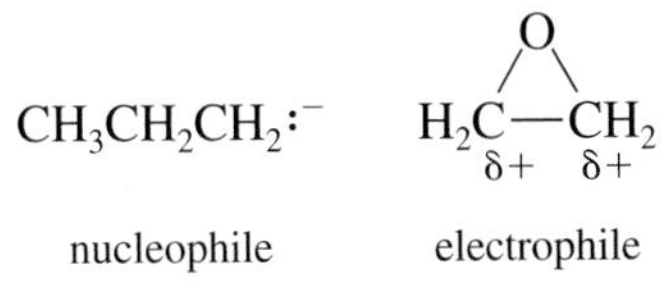

$CH_3CH_2CH_2{:}^-$

nucleophile

$H_2\underset{\delta+}{C}{-}\underset{\delta+}{CH_2}$ (epoxide, O bridging)

electrophile

The carbanion, written as it is above, is not a reagent. We cannot find a bottle of naked carbanions on the shelf in the laboratory. Written this way, it is called a **synthon,** an idealized reactive unit that may be found in a number of different actual reagents. We recognize the synthon as being mechanistically necessary for the reaction to take place.

$$CH_3CH_2CH_2{:}^- \quad H_2\overset{\overset{\displaystyle :\overset{\displaystyle \curvearrowright}{O}:}{|}}{C}{-}CH_2 \longrightarrow CH_3CH_2CH_2CH_2CH_2{-}\overset{..}{\underset{..}{O}}{:}^-$$

nucleophile       electrophile

At this stage we also recognize that to complete the synthesis, we will have to add an acid to protonate the alkoxide ion that results from the reaction of the nucleophile with the electrophile.

$$CH_3CH_2CH_2CH_2CH_2{-}\overset{..}{\underset{..}{O}}{:}^-\overset{\curvearrowright}{\phantom{}}H{-}B^+ \longrightarrow CH_3CH_2CH_2CH_2CH_2{-}\overset{..}{\underset{..}{O}}{-}H$$

$$:B$$

The symbols for the general acid and base that we have been using since early in this book also may be considered synthons. They represent protons or pairs of nonbonding electrons that serve as bases and stand in for the actual reagents we will eventually have to choose.

We are now ready to convert our retrosynthetic analysis to a set of equations for a synthesis. For example, we will now have to choose an actual organometallic reagent and prepare it from an alkyl halide, again working backwards but written as equations now with reagents that are found in the laboratory.

$$CH_3CH_2CH_2CH_2CH_2OH \xleftarrow[H_2O]{HCl} CH_3CH_2CH_2CH_2CH_2O^-\;Li^+$$

$$\uparrow$$

$$CH_3CH_2CH_2Br \xrightarrow{Li} CH_3CH_2CH_2Li + H_2\overset{\overset{\displaystyle O}{\diagup\diagdown}}{C}{-}CH_2$$

**Problem 21.1**    Another nucleophilic reagent may be substituted for the one used in the equation above. Write an equation for its preparation and use it in the synthesis.

The synthons that result from our dissection of the structure of 1-pentene are also a nucleophile and an electrophile. In the retrosynthetic equation shown on page 839, they were shown as

$$CH_3CH_2CH_2LG \qquad MC{\equiv}CH$$

They are not specific reagents but indicate the polarities, electrophile and nucleophile, that we need in order for a reaction to take place. They can be further simplified to synthons.

$$CH_3CH_2CH_2{}^+ \qquad {}^-{:}C{\equiv}CH$$

electrophilic       nucleophilic
synthon            synthon

Neither of these reactive intermediates exists as such. In each case, for the actual synthesis in the laboratory, we must substitute a reagent that has the polarities shown in the synthons. For example, we have already indicated that the electrophilic synthon will be a three-carbon chain with a leaving group on it. Many leaving groups are possible.

$$CH_3CH_2CH_2LG \quad \text{may be} \quad CH_3CH_2CH_2Cl$$
$$CH_3CH_2CH_2Br$$
$$CH_3CH_2CH_2I$$
$$CH_3CH_2CH_2OTs, \text{ etc.}$$

Similarly, the nucleophilic synthon will have to be prepared from acetylene by a deprotonation reaction, but a number of reagents are possible. For example,

$$HC\equiv C:^- \ Li^+ \xleftarrow{\quad CH_3CH_2CH_2CH_2Li \quad} HC\equiv CH \xrightarrow{\quad NaNH_2 \quad} HC\equiv C:^- \ Na^+$$

**Problem 21.2**   Write equations showing how you would prepare 1-pentanol according to pathway 4 on page 839 and the retrosynthetic analysis of 1-pentene on page 840.

**Problem 21.3**   Pathway 1 in the retrosynthetic analysis for 1-pentanol was rejected because of the number of steps that it would take to make the four-carbon organometallic reagent from three or fewer carbon atoms. Provide a retrosynthetic analysis showing how you would prepare that organometallic reagent starting from two- or three-carbon fragments. Convert one of these pathways into a synthesis.

# B. Synthesis of Compounds with More Than One Functional Group

The syntheses examined in detail in Section 21.1A involved the formation of 1-pentanol, a compound with a single functional group in it. Strategies necessary to synthesize compounds with more than one functional group in them or to work with reagents with multiple functional groups are more interesting. For example, suppose that we were to try to synthesize 5-hydroxy-2-pentanone from reagents containing three or fewer carbon atoms. An analysis similar to the one we did for 1-pentanol (p. 839) quickly reveals all kinds of problems.

The first thing we need to recognize is that when there is more than one functional group in a molecule, there may be intramolecular reactions between some of them, especially if three-, five-, or six-membered rings can form (p. 500). In the case of 5-hydroxy-2-pentanone, such an intramolecular interaction does take place, so when we succeed in synthesizing the compound, it will exist mostly as its cyclic hemiketal (p. 559).

|   |   |
|---|---|
| 5-hydroxy-2-pentanone | hemiketal of 5-hydroxy-2-pentanone |

**Problem 21.4**   Write a mechanism showing the formation of the hemiketal of 5-hydroxy-2-pentanone. Acids, $HB^+$, and bases, $B:$, are available as needed.

The formation of the hemiketal from 5-hydroxy-2-pentanone does not interfere with the synthesis itself, so we can proceed with the retrosynthetic analysis.

$$CH_3CCH_2CH_2M + H_2C{=}O$$

nucleophilic carbon
reacting with
electrophilic carbon

$$CH_3CCH_2CH{=}CH_2$$

hydroboration/oxidation
of terminal alkene

$$CH_3CCH_2CH_2CH_2OH$$

oxidation of 2°
alcohol to ketone

reduction of
aldehyde to 1° alcohol

$$CH_3CHCH_2CH_2CH_2OH$$
$$OH$$

$$CH_3CCH_2CH_2CH$$
$$O \qquad O$$

Each one of these reaction pathways is unrealistic. For example, pathway 1 suggests that we can make an organometallic reagent from a compound containing a carbonyl group. This is not possible (unless we are making an enolate anion). The reagent as it forms would be destroyed by reaction with the carbonyl group of another molecule of the same reagent. Similarly, an attempt to add borane to the double bond in the unsaturated carbonyl compound for pathway 2 would result in the reduction of the carbonyl group to the alcohol, with borane behaving very much the way the borohydride anion does (p. 545).

Pathways 3 and 4 require that we be able to oxidize one alcohol function without touching the other or that we be able to reduce one carbonyl group in the presence of another. Both of these would be difficult to do.

In Section 15.8 (p. 629) we encountered for the first time the problem of carrying out syntheses with compounds containing two reactive functional groups when we tried to make peptides from amino acids. At that time we solved the problem by using protecting groups on the amino groups of the amino acids. A similar strategy would work here. For example, to synthesize, 1,4-pentanediol from smaller units, we might do the following dissection.

$$CH_3C{+}CH_2CH_2CH_2OH \Longrightarrow CH_3C{-}H + MCH_2CH_2CH_2OH$$

with OH and H on the left carbon.

electrophile          nucleophile

Immediately we see that we have an organometallic reagent with an acidic hydrogen on the hydroxyl group in it. Such a reagent could not be made; it would be destroyed by the acid (p. 550). However, if the acidic hydrogen had been replaced earlier by a protecting group of some kind, the reaction would be possible. Without being specific about the protecting group (that will be the subject of the next section), a reaction scheme can be written.

$$CH_3CCH_2CH_2CH_2OH \xleftarrow[\text{remove PG}]{} CH_3CCH_2CH_2CH_2OPG \xleftarrow[\substack{\text{oxidation of} \\ \text{2° alcohol} \\ \text{to ketone}}]{} CH_3CCH_2CH_2CH_2OPG$$
$$H$$

protonation

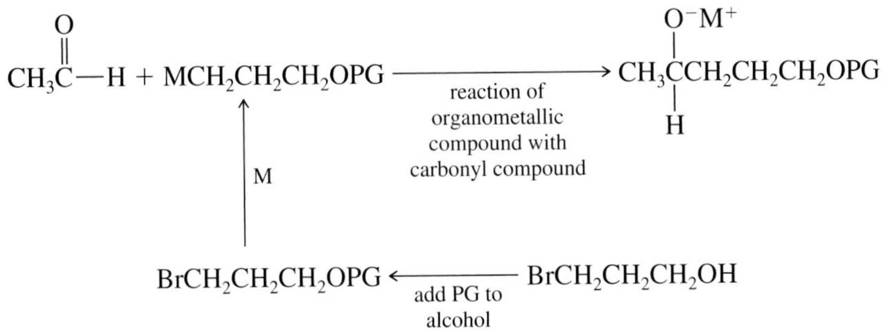

$$\underset{\text{O}}{\overset{\text{O}}{\underset{\|}{\text{CH}_3\text{C}}}}\text{—H} + \text{MCH}_2\text{CH}_2\text{CH}_2\text{OPG} \xrightarrow[\substack{\text{reaction of}\\ \text{organometallic}\\ \text{compound with}\\ \text{carbonyl compound}}]{} \underset{\overset{|}{\text{H}}}{\overset{\text{O}^-\text{M}^+}{\underset{|}{\text{CH}_3\text{CCH}_2\text{CH}_2\text{CH}_2\text{OPG}}}}$$

$$\text{M}$$

$$\text{BrCH}_2\text{CH}_2\text{CH}_2\text{OPG} \xleftarrow[\substack{\text{add PG to}\\ \text{alcohol}}]{} \text{BrCH}_2\text{CH}_2\text{CH}_2\text{OH}$$

Just as was the case with the protecting group for the amino groups in amino acids, the protecting group must be easy to put on, must be resistant to the reagents being used in the rest of the synthesis (in the case above to nucleophiles and oxidizing agents), and must be removed easily when its job is done. The next section will discuss the ways in which this can be done for carbonyl groups and alcohols and expand on what we already know about protecting amines.

**Problem 21.5**     Without worrying about the steps that involve the addition and removal of the protecting group, provide the reagents that you would use to complete the rest of the synthesis shown above. It will be a good idea to stay away from strongly acidic reagents.

## 21.2 Protecting Groups in Synthesis

### A. Acetals and Ketals

Most naturally occurring compounds with interesting biological properties have a variety of functional groups in them. One functional group may be adversely affected by or interfere with a reaction that a chemist wishes to carry out at another one. For example, it is not possible to prepare a Grignard reagent from an alkyl halide containing an alcohol function, as we saw on page 842. And yet it may be useful to make such an organometallic reagent in a complex synthesis (p. 844). To do this, chemists use protecting groups. A protecting group converts a reactive functional group into a different group that is inert to the conditions of some reaction (or reactions) that is to be carried out as part of a synthetic pathway. For example, a hydroxyl group is too acidic to be present while a Grignard reagent is formed. Conversion of an alcohol to an ether would prevent it from reacting with the Grignard reagent. But not just any ether will do, because the protecting group must be easily removable once its job is done so that the original functional group can be restored. Not all ordinary alkyl ethers are easily cleaved (p. 502).

Acetals and ketals are very useful as protecting groups because of the ease with which they are formed and removed. The acetal function is stable to bases (no acidic protons), nucleophiles (no good leaving groups), and reducing agents (no multiple bonds to which hydrogen can add). Acetals can, therefore, be used with a variety of reagents as long as acidic conditions are avoided. The preparation of cyclic ketals from carbonyl compounds was shown in Section 14.7C. Diols in which the two hydroxyl groups are on adjacent carbon atoms (1,2-diols) or in which they are on carbon atoms separated by a single carbon (1,3-diols) also can be protected as cyclic ketals. Acetone is usually used to make the ketal. An example of the protection of a 1,3-diol in order to be able to make a Grignard reagent is shown on the next page.

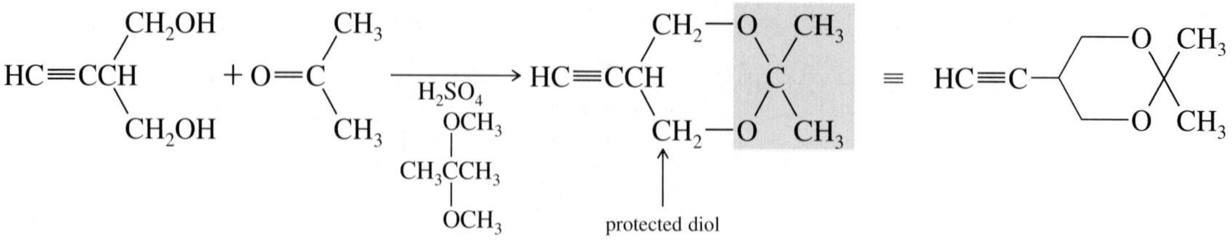

protected diol

This reaction is carried out in the presence of 2,2-dimethoxypropane, the dimethyl ketal of acetone. This ketal serves to remove the water formed as the diol is converted to the cyclic ketal and generates more acetone in the process.

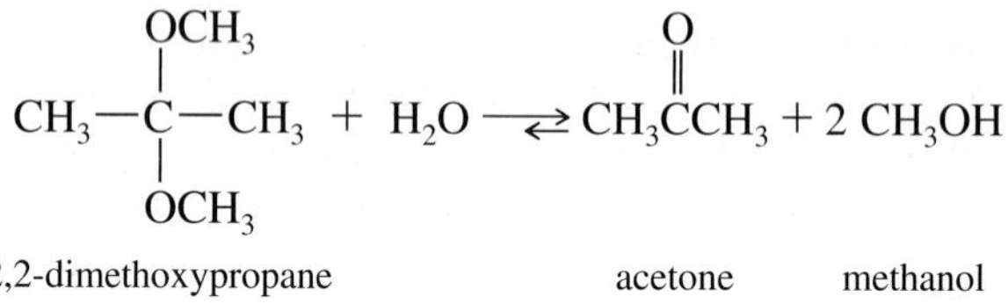

2,2-dimethoxypropane              acetone      methanol

The protected diol is converted into a Grignard reagent, which then reacts with an aldehyde that also contains a protected ketone function.

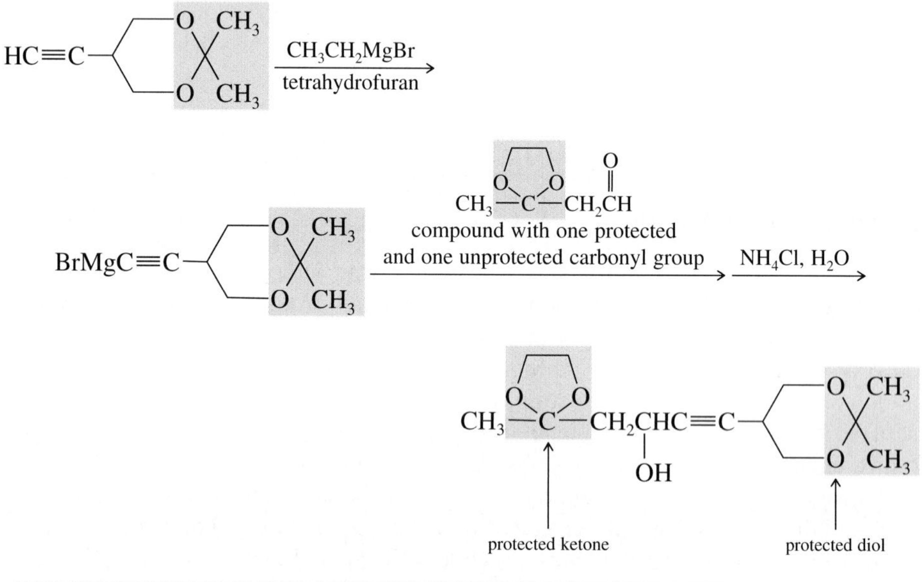

compound with one protected
and one unprotected carbonyl group

protected ketone                protected diol

**Problem 21.6**   The compound synthesized above is converted into the following compound as part of the synthesis of a pseudomonic acid, an antibiotic. How would you carry out this conversion?

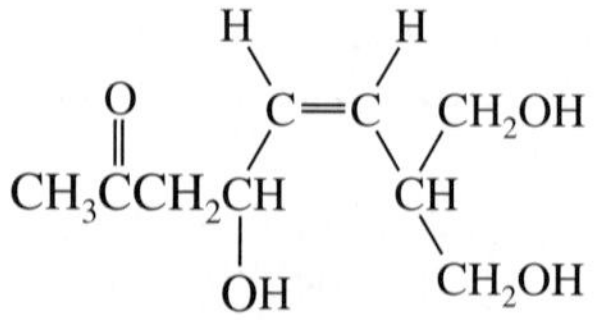

(*Hint:* It will be helpful if you first describe in words the changes that you see between the two structures. The Art of Solving Problems section on page 556 will remind you of the systematic steps you need to take in solving such a problem.)

The cyclic ketals used as protecting groups in the above synthesis serve three functions.

1. One ketal protects two hydroxyl groups so that a Grignard reagent can be made in another part of the molecule.

2. The second ketal protects one carbonyl group in a molecule so that it can be differentiated from a second one.

3. Both ketals can be removed easily so that the two alcohol functions and the ketone function can be restored later in the synthesis (Problem 21.6).

Single alcohol functions also can be protected as acetals. The most common way this is done is by forming what is known as the tetrahydropyranyl ether of the alcohol. In an example that closely parallels the synthesis of 5-hydroxy-2-pentanone in Section 21.1B, the Grignard reagent from 5-chloro-1-pentanol is needed in the synthesis of a constituent of civet, which is a substance isolated from the civet cat and used in making perfumes. The hydroxyl group must be protected before the Grignard reagent can be made. This is done by treating the alcohol with dihydropyran in the presence of *p*-toluenesulfonic acid.

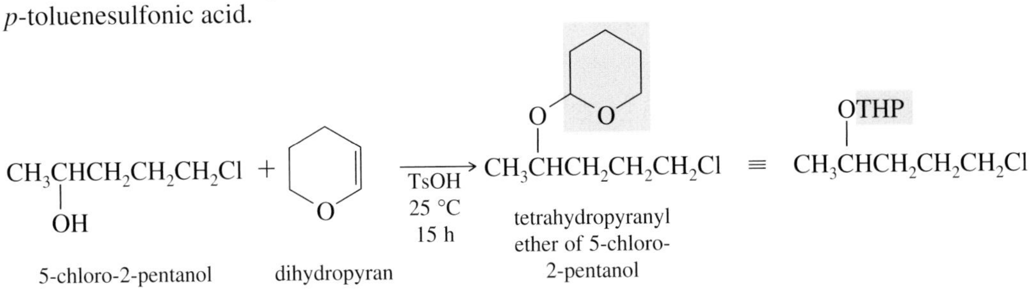

CH₃CHCH₂CH₂CH₂Cl  +  [dihydropyran]  →(TsOH, 25 °C, 15 h)→  CH₃CHCH₂CH₂CH₂Cl  ≡  CH₃CHCH₂CH₂CH₂Cl

$$CH_3CHCH_2CH_2CH_2Cl + \text{(dihydropyran)} \xrightarrow[\substack{25\,°C\\15\,h}]{TsOH} CH_3\overset{\displaystyle |}{C}HCH_2CH_2CH_2Cl \equiv CH_3\overset{\displaystyle \overset{OTHP}{|}}{C}HCH_2CH_2CH_2Cl$$

5-chloro-2-pentanol      dihydropyran      tetrahydropyranyl ether of 5-chloro-2-pentanol

The abbreviation THP is often used to represent the tetrahydropyranyl group in structural formulas, just as Ts is used in structural formulas to represent the *p*-toluenesulfonyl group (p. 259).

How does the tetrahydropyranyl ether form under the conditions of the reaction? Dihydropyran is a cyclic vinyl ether. It is easily protonated to give a stabilized carbocation in which the positive charge can be delocalized to the adjacent oxygen. This cation then reacts with the alcohol, which behaves as a nucleophile.

**VISUALIZING THE REACTION**

**Addition of an Alcohol to Dihydropyran**

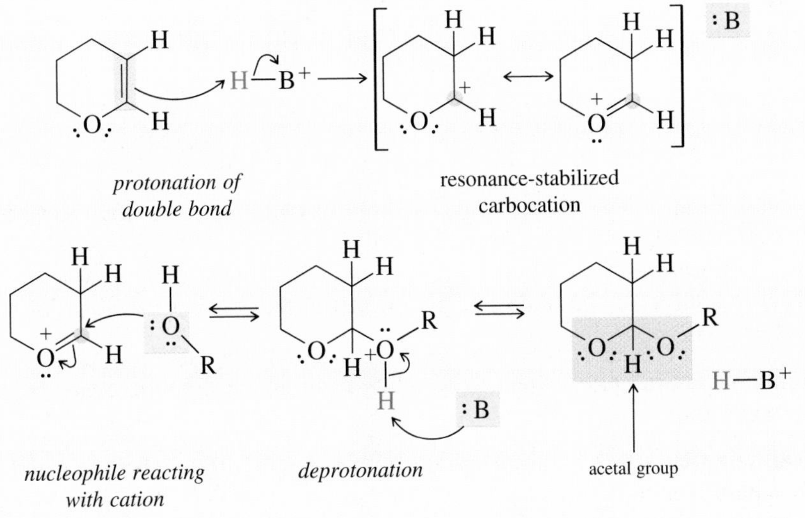

protonation of double bond

resonance-stabilized carbocation

nucleophile reacting with cation

deprotonation

acetal group

The product, though usually called an ether, is really an acetal, with two ether linkages to the same carbon atom. It is easily hydrolyzed with dilute acid, to get back the original alcohol when the need for protection is over. (Such a deprotection step is used later in the synthesis started on page 845.)

The protected 5-chloro-2-pentanol can be converted to a Grignard reagent, which then reacts with cinnamaldehyde (3-phenyl-2-propenal).

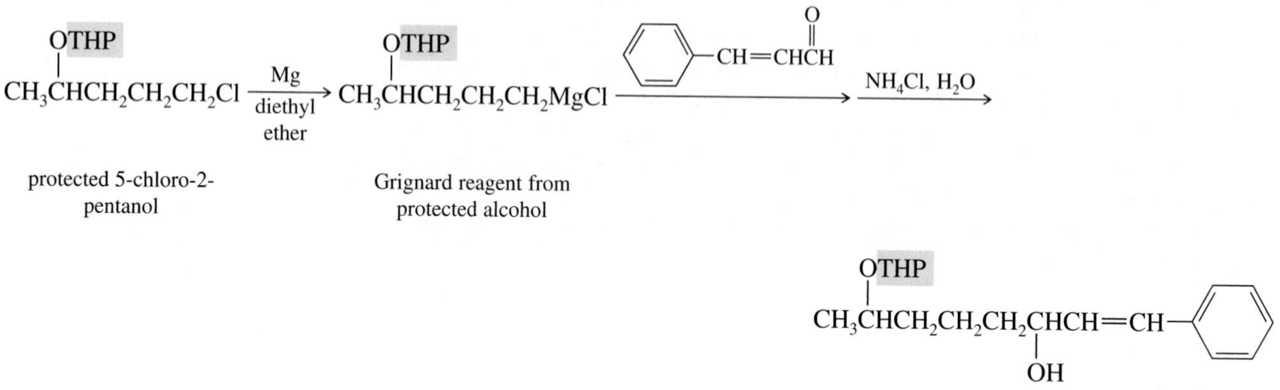

$$\underset{\substack{\text{protected 5-chloro-2-}\\\text{pentanol}}}{\overset{\overset{\text{OTHP}}{|}}{CH_3CHCH_2CH_2CH_2Cl}} \xrightarrow[\text{diethyl}]{\text{Mg}} \underset{\substack{\text{Grignard reagent from}\\\text{protected alcohol}}}{\overset{\overset{\text{OTHP}}{|}}{CH_3CHCH_2CH_2CH_2MgCl}} \xrightarrow{\underset{}{C_6H_5-CH=CHCH=O}} \xrightarrow{NH_4Cl,\ H_2O}$$

$$\overset{\overset{\text{OTHP}}{|}}{CH_3CHCH_2CH_2CH_2\underset{\underset{OH}{|}}{CH}CH=CH-C_6H_5}$$

This product has two hydroxyl groups, one protected and the other one not. The unprotected hydroxyl group is converted into the acetic acid ester (p. 622), and then the other hydroxyl group is deprotected.

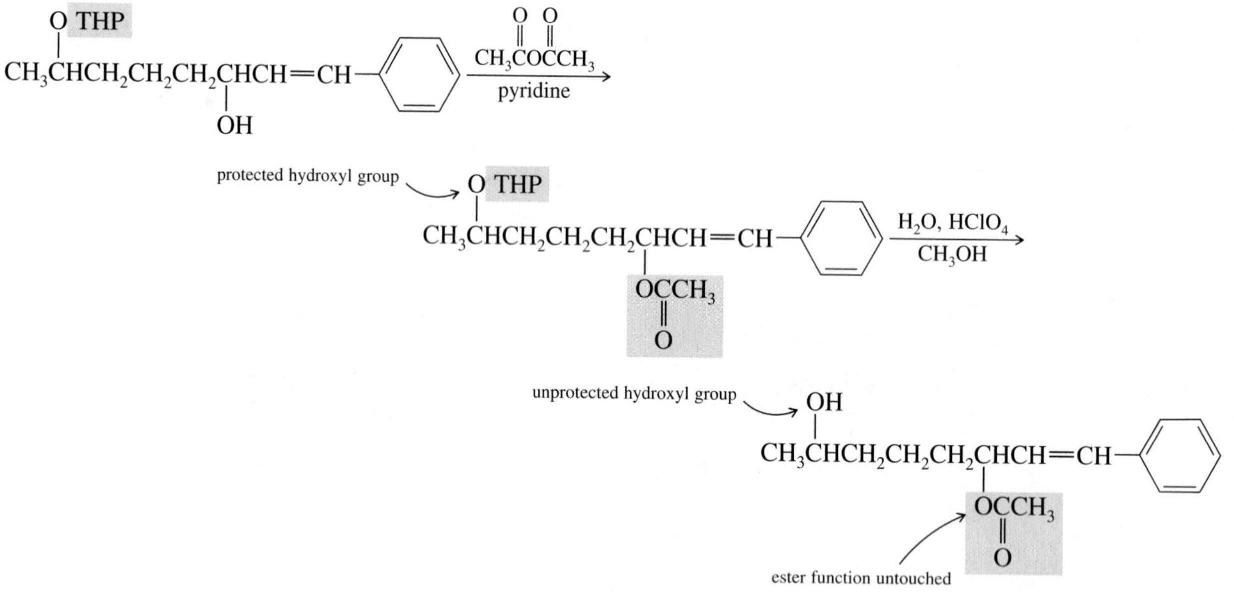

$$\overset{\overset{\text{O THP}}{|}}{CH_3CHCH_2CH_2CH_2\underset{\underset{OH}{|}}{CH}CH=CH-C_6H_5} \xrightarrow[\text{pyridine}]{CH_3COCCH_3}$$

protected hydroxyl group

$$\overset{\overset{\text{O THP}}{|}}{CH_3CHCH_2CH_2CH_2\underset{\underset{\underset{O}{\|}}{OCCH_3}}{|}{CH}CH=CH-C_6H_5} \xrightarrow[CH_3OH]{H_2O,\ HClO_4}$$

unprotected hydroxyl group

$$\overset{\overset{\text{OH}}{|}}{CH_3CHCH_2CH_2CH_2\underset{\underset{\underset{O}{\|}}{OCCH_3}}{|}{CH}CH=CH-C_6H_5}$$

ester function untouched

The acetal group is hydrolyzed under such mild conditions that the ester group is untouched, even though esters generally undergo hydrolysis reactions (p. 616). The acetoxy ester group serves as a leaving group in a later stage of the synthesis.

The protection of the hydroxyl group, therefore, serves three functions in the above synthesis.

1. It allows a Grignard reagent to be made from an alkyl halide containing a hydroxyl group.

2. It differentiates that hydroxyl group from a new hydroxyl group created in the Grignard reaction.

3. It allows the original hydroxyl group to be restored at a later stage of the synthesis.

**Problem 21.7**     Go back to Problem 21.5 and complete the synthesis of 5-hydroxy-2-pentanone by including the protection and deprotection steps of the synthesis.

**Problem 21.8**     An intermediate in the synthesis of a natural product that has antitumor activity is prepared by the following sequence of steps.

$$HC{\equiv}CCH_2OH + \underset{O}{\bigcirc} \xrightarrow{HCl} A \xrightarrow[\substack{\text{dimethyl} \\ \text{sulfoxide}}]{NaH} B \xrightarrow{CH_2{=}CH(CH_2)_8CH_2OTs} C \xrightarrow[\substack{HCl \\ \text{methanol} \ C_{14}H_{24}O}]{H_2O} D$$

Assign structures to Compounds A, B, C, and D.

## B. Ethers

Even some ethers that are not acetals or ketals are easily cleaved. Benzyl ethers are often used as protecting groups for alcohols. The benzylic ether bond is cleaved by hydrogenation reactions. We saw an example of such a cleavage of a bond between oxygen and a benzyl group by hydrogenation in the removal of the carbobenzyloxy protecting group in peptide syntheses (p. 633).

Silyl ethers, in which there is an oxygen–silicon bond, are usually cleaved by fluoride ion. The high bond energy of the silicon–fluorine bond (Table 2.4, p. 64) serves as a thermodynamic driving force for this cleavage. Silyl ethers are prepared by the reaction of an alcohol and a silyl halide in the presence of a base to aid in the deprotonation of the alcohol. For example, the *tert*-butyldimethylsilyl group (abbreviated TBDMS) is used to protect the hydroxyl group of 3-butyn-1-ol so that the terminal alkyne can be selectively deprotonated. The carbanion resulting from that deprotonation adds to acetaldehyde.

hydroxyl group

$$HOCH_2CH_2C{\equiv}CH \ + \ CH_3{-}\underset{\underset{CH_3}{|}}{\overset{\overset{CH_3}{|}}{C}}{-}\underset{\underset{CH_3}{|}}{\overset{\overset{CH_3}{|}}{Si}}{-}Cl \xrightarrow[\substack{\text{(imidazole)}}]{\text{base}} CH_3{-}\underset{\underset{CH_3}{|}}{\overset{\overset{CH_3}{|}}{C}}{-}\underset{\underset{CH_3}{|}}{\overset{\overset{CH_3}{|}}{Si}}{-}OCH_2CH_2C{\equiv}CH$$

protected hydroxyl group

3-butyn-1-ol        *tert*-butyldimethylsilyl chloride
or
*tert*-butylchlorodimethylsilane

*tert*-butyldimethylsilyl ether
of 3-butyn-1-ol

terminal alkyne

$$TBDMSOCH_2CH_2C{\equiv}CH \xrightarrow[\text{tetrahydrofuran}]{CH_3Li} TBDMSOCH_2CH_2C{\equiv}C{:}^-Li^+ \xrightarrow[\text{tetrahydrofuran}]{\overset{\overset{O}{\|}}{CH_3CH}} \xrightarrow{H_2O}$$

*tert*-butyldimethylsilyl ether
of 3-butyn-1-ol

$$TBDMSOCH_2CH_2C{\equiv}CCHCH_3$$
$$\underset{OH}{|}$$

new hydroxyl group

This synthesis is continued, as shown in the reactions on the next page, by hydrogenating the alkyne to a cis alkene, then protecting the secondary alcohol as a tetrahydropyranyl ether, and finally using fluoride ion to deprotect the primary alcohol for further reactions.

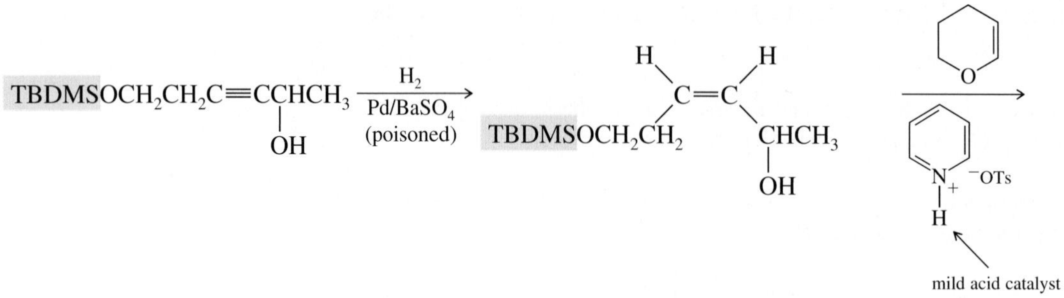

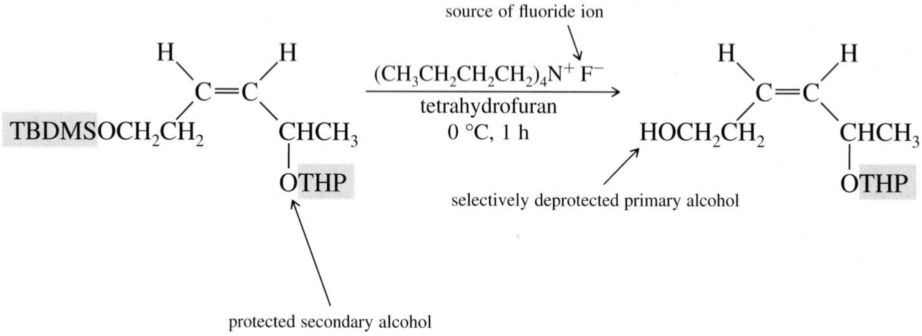

TBDMSOCH₂CH₂C≡CCHCH₃ 
|
OH

$\xrightarrow{\begin{array}{c} H_2 \\ \hline Pd/BaSO_4 \\ (poisoned) \end{array}}$

TBDMSOCH₂CH₂ 

C=C with H, H substituents; CHCH₃ / OH

$\xrightarrow{\text{mild acid catalyst}}$ (dihydropyran; pyridinium ⁺N–H ⁻OTs)

TBDMSOCH₂CH₂

C=C with H, H; CHCH₃ / OTHP

$\xrightarrow[\begin{array}{c} \text{tetrahydrofuran} \\ 0\,°C,\ 1\ h \end{array}]{(CH_3CH_2CH_2CH_2)_4N^+ F^-}$

source of fluoride ion

HOCH₂CH₂

C=C with H, H; CHCH₃ / OTHP

selectively deprotected primary alcohol

protected secondary alcohol

---

**Problem 21.9**    How would you carry out the following transformation, which was part of a synthesis of compounds related to vitamin D?

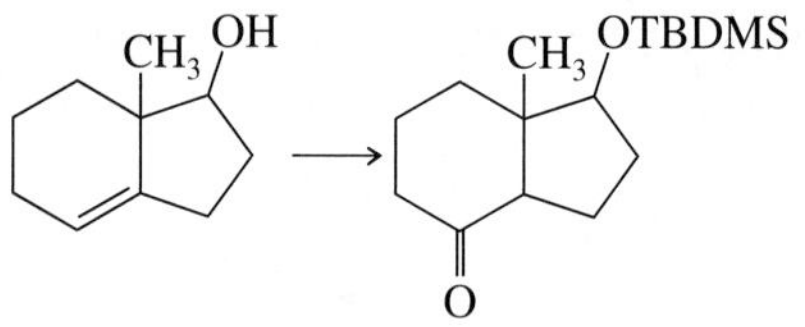

---

**Problem 21.10**    The pheromone of the ambrosia beetle, 6-methyl-5-hepten-2-ol, can be prepared by the sodium borohydride reduction of 6-methyl-5-hepten-2-one (p. 547) but also has been synthesized by another route:

$$
\underset{\substack{| \\ CH_3}}{CH_3C}=CHCH_2CH=CH_2 \longrightarrow A \longrightarrow B \longrightarrow \underset{\substack{| \\ CH_3}}{CH_3C}=CHCH_2CH_2\underset{\substack{| \\ OH}}{CHCH_3}
$$

Supply the structures of Compounds A and B and the reagents necessary to transform 5-methyl-1,4-hexadiene into 6-methyl-5-hepten-2-ol.

---

**Problem 21.11**    A highly mutagenic compound containing five double bonds has been isolated from the feces of people living in industrialized countries. These types of compounds, called fecapentaenes, are suspected of causing colon cancer. Chemists are currently attempting to synthesize fecapentaenes in large enough amounts to test them for carcinogenicity. A portion of a synthesis is shown below. How would you carry out this transformation?

$$
CH_3CH_2CH=CHCH=CHCH \overset{O}{\underset{??}{\longrightarrow}} CH_3CH_2CH=CHCH=CHCHC\equiv CH
$$
|
OTHP

## C. Protecting Groups for Amines

The synthesis of peptides, discussed in Section 15.8B (p. 631), required the use of protecting groups on the amino groups of amino acids to be able to prepare a peptide with the amino acids in the order we desired and not have random reactions between amino groups and carboxylic acid groups. We needed to protect the amino and carboxylic acid groups that we did not want to have react and activate the carboxylic acid group that was to be part of the new peptide bond. The two amine protecting groups we used in Section 15.8B were the *tert*-butoxycarbonyl group (Boc) and benzyloxycarbonyl group (Cbz). These groups are also used as protecting groups in syntheses that do not involve peptides. For example, the following steps are part of the synthesis of a polyamine used to make a complexing agent for metals to be used in connection with radiation therapies.

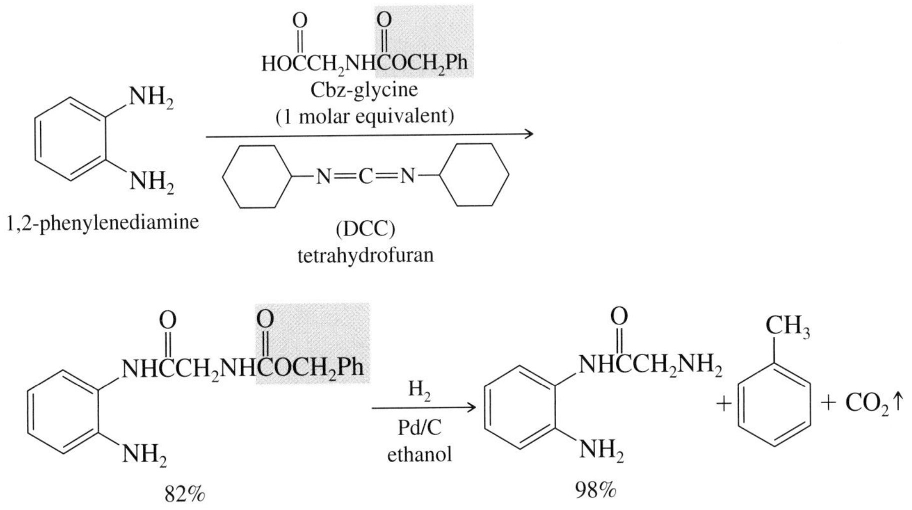

The sequence of reactions shown above reminds us of several things we learned in Section 15.8. The amino acid glycine is protected at the amino group as the benzyloxycarbonyl (Cbz) derivative, making sure that when the carboxylic acid group of the amino acid is activated with dicyclohexylcarbodiimide (DCC, Section 15.8C, p. 634), it will react with one of the amino groups on 1,2-phenylenediamine and not with another molecule of glycine. Once this crucial step is over, the protecting group is easily removed by hydrogenation over a palladium catalyst, a reaction that does not affect the newly formed amide bond.

The Boc and Cbz groups are also easily removed with strong acid, usually trifluoroacetic acid or hydrogen bromide in glacial acetic acid. Chemists are always looking for protecting groups that are resistant to one set of reagents and can be removed by reagents that leave other protecting gtoups untouched. We will see a demonstration of what is possible in this way later in this section. To provide flexibility in syntheses, another protecting group, especially for use in peptide synthesis, the 9-fluorenylmethoxycarbonyl group (Fmoc), has been developed. This group is highly resistant to acid but is easily removed by base. The use of this protecting group is illustrated in the preparation of a dipeptide.

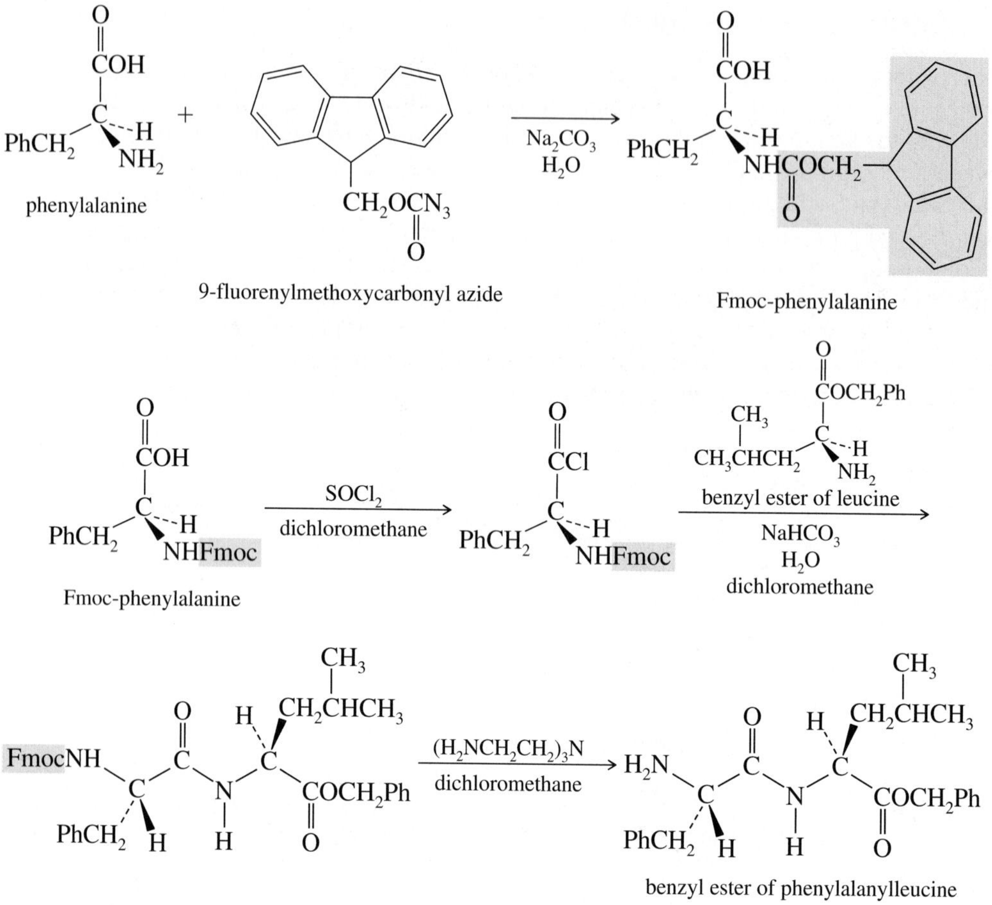

9-fluorenylmethoxycarbonyl azide

Fmoc-phenylalanine

benzyl ester of phenylalanylleucine

Phenylalanine is protected at the amino group as its 9-fluorenylmethoxycarbonyl (Fmoc) derivative. The protecting group is resistant to acid, so it is possible to activate the carboxylic acid group on the amino acid by making acid chloride, with thionyl chloride, a reagent that could not be used with the Boc or Cbz protecting group. The acid chloride of the protected phenylalanine reacts with the amino group of leucine, which is protected at the carboxylic acid group as the benzyl ester. Once the peptide bond is formed, the Fmoc protecting group is removed using an amine, in this case tris(2-aminoethyl)amine. The amino group on the phenylalanine is now exposed for reaction with the activated carboxylic acid group of another amino acid. How the Fmoc group is removed by base is the subject of Problem 21.12.

**Problem 21.12**     The deprotection reaction for the Fmoc-protected phenylalanine portion of the peptide is shown below.

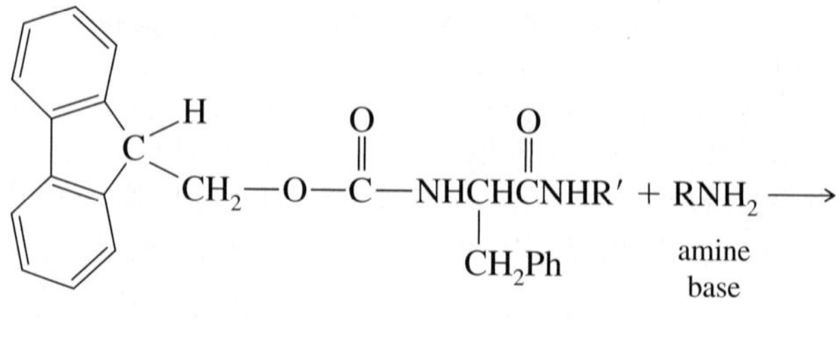

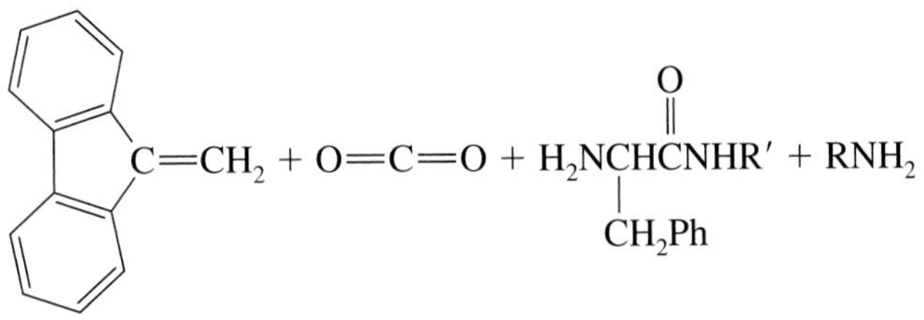

$$\text{(fluorenyl)C}{=}\text{CH}_2 + \text{O}{=}\text{C}{=}\text{O} + \text{H}_2\text{NCHCNHR}' + \text{RNH}_2$$

The deprotection starts with the formation of an anion by removal of the proton that gives rise to the most highly stabilized carbanion. The carbanion, in a second step, ejects carbon dioxide and the amino acid residue, which gets protonated to give the freed phenylalanine shown above.

(a)  Write a mechanism for the reaction using the curved-arrow convention.
(b)  Why is the carbanion intermediate formed so easily?

Ordinary amides can be used as protecting groups if the compound being synthesized does not contain functional groups that would be affected by the relatively harsh conditions under which an amide is hydrolyzed. Amides are often used to protect aromatic amines for reactions with strong electrophiles such as acids or halogens. For example, if we wish to make *p*-nitroaniline from aniline, we need to protect the amino group first because the unprotected amine reacts with the nitric acid to give the anilinium ion in which the positively charged nitrogen is now a meta-director (p. 369).

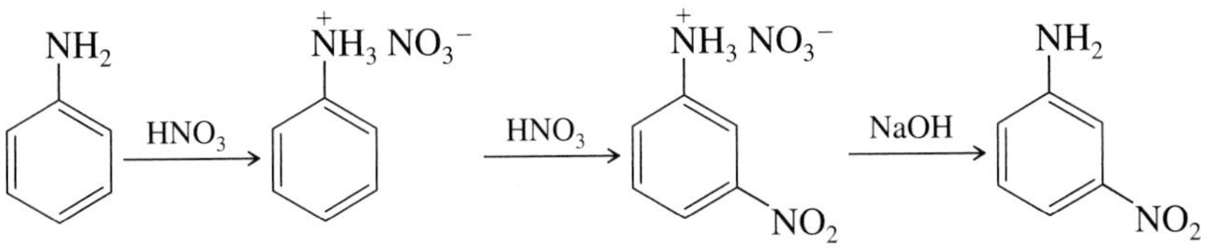

| aniline | anilinium nitrate | anilinium nitrate | *m*-nitroaniline |
|---------|-------------------|-------------------|------------------|

Preparation of the amide reduces the basicity of the amine (Problem 15.4, p. 599) and allows the group to function as an ortho,para-director. For example, the major product from the nitration of benzanilide is *p*-nitrobenzanilide.

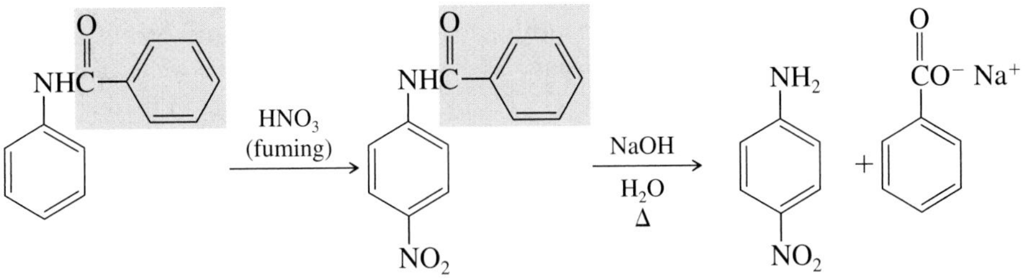

| benzanilide | *p*-nitrobenzanilide | *p*-nitroaniline |
|-------------|----------------------|------------------|

If the acid portion of the amide is substituted with electron-withdrawing groups, the amide becomes easier to hydrolyze. A whole range of ease of hydrolysis is possible by substituting halogens on the acetyl group, for example.

$$\text{CH}_3\text{C}{-}\text{NHR} < \text{ClCH}_2\text{C}{-}\text{NHR} < \text{Cl}_2\text{CHC}{-}\text{NHR} < \text{Cl}_3\text{CC}{-}\text{NHR} < \text{F}_3\text{CC}{-}\text{NHR}$$

hardest to hydrolyze                                        easiest to hydrolyze

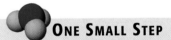

**ONE SMALL STEP**

**PROBLEM:**  When an acetanilide was allowed to stand in ethanol, it was stable for at least 24 hours. A comparable trifluoroacetamide was destroyed so fast that it was not possible to determine the rate at which half the amine was liberated. It took 4.5 hours for half the dichloroacetamide to react with ethanol. Write a mechanism for the reaction of a generic trifluoroacetamide with ethanol, and rationalize the differences in reactivity described above.

The trifluoroacetyl group, because of its ease of hydrolysis in base, is now used widely as a protecting group for amines. A synthesis that illustrates the way the different properties of protecting groups can be used to selectively protect and deprotect amino groups is the preparation of spermidine, differently protected at each amino group.

Spermidine and spermine, a tetraamine, both found widely in nature, have been the subject of much research because they have been implicated in cell proliferation, and synthetic analogs of spermine, especially, have potent antitumor activity.

$$H_2N(CH_2)_3NH(CH_2)_4NH_2 \qquad\qquad H_2N(CH_2)_3NH(CH_2)_4NH(CH_2)_3NH_2$$

*N*-(3-aminopropyl)-1,4-butanediamine          *N,N'*-bis(3-aminopropyl)-1,4-butanediamine
spermidine                                    spermine

To synthesize analogs that are substituted in different ways at the different nitrogen atoms, which are either identical or similar in reactivity, protecting groups are used. Spermidine with a different protecting group at each nitrogen atom has been synthesized by the following sequence of reactions from readily available starting materials. Benzylamine reacts with acrylonitrile in a conjugate addition (p. 712). The nitrile is reduced with hydrogen on a Raney nickel catalyst to an amine. Note that the cyano group is being used as a hidden potential amine in this reaction.

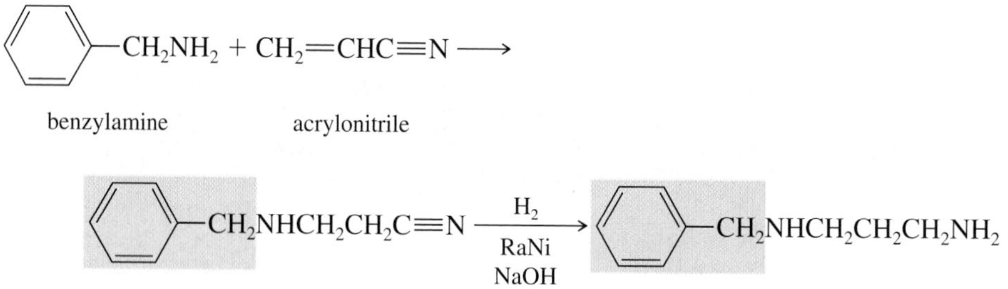

benzylamine                acrylonitrile

*N*-benzylpropane-1,3-diamine

At this stage, two of the amino groups of spermidine have been created, and one of them is already protected by a benzyl group. This is the reason why a nickel catalyst and not a palladium one (p. 849) is used in the hydrogenation step.

Before another amino group is introduced, the primary amine group is protected by the *tert*-butoxycarbonyl group. The reagent used, 2-(*tert*-butoxycarbonyloximino)-2-phenylacetonitrile (Boc-ON) is like an acid anhydride in its reactivity.

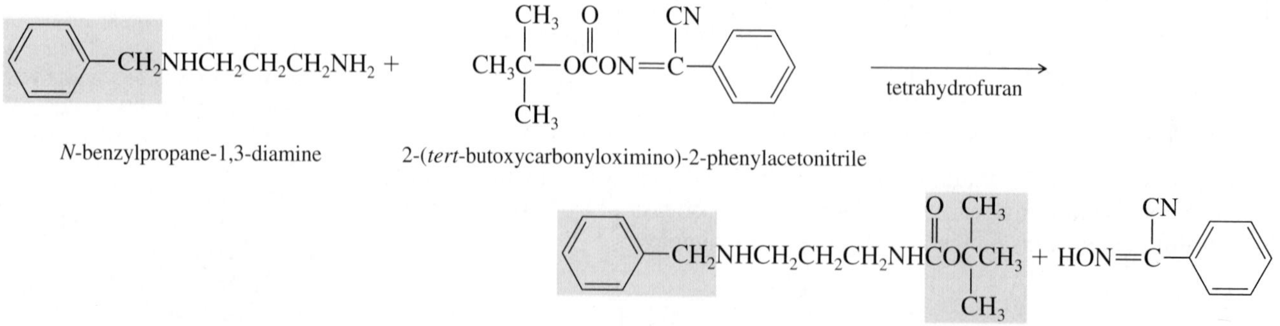

*N*-benzylpropane-1,3-diamine        2-(*tert*-butoxycarbonyloximino)-2-phenylacetonitrile

An $S_N2$ reaction of the secondary amine with 4-chlorobutanenitrile introduces the third amino group, again as a cyano group.

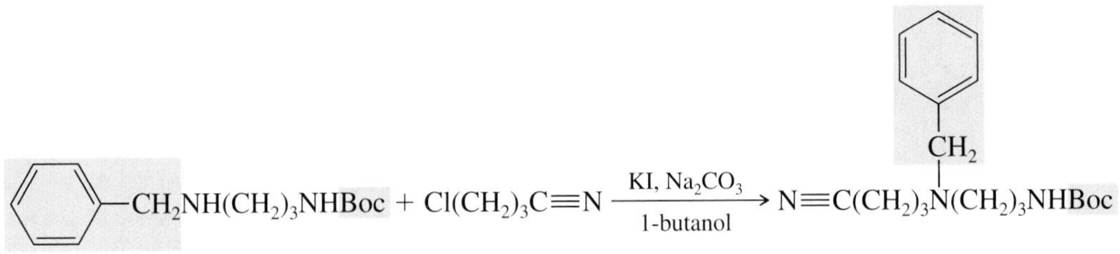

$$\bigcirc\!\!-CH_2NH(CH_2)_3NHBoc \; + \; Cl(CH_2)_3C\!\!\equiv\!\!N \;\; \xrightarrow[\text{1-butanol}]{KI,\, Na_2CO_3} \;\; N\!\!\equiv\!\!C(CH_2)_3\overset{\overset{\displaystyle CH_2}{|}}{N}(CH_2)_3NHBoc$$

Reduction over Raney nickel and protection with the trifluoroacetyl group give spermidine, protected at each amino group with a different protecting group.

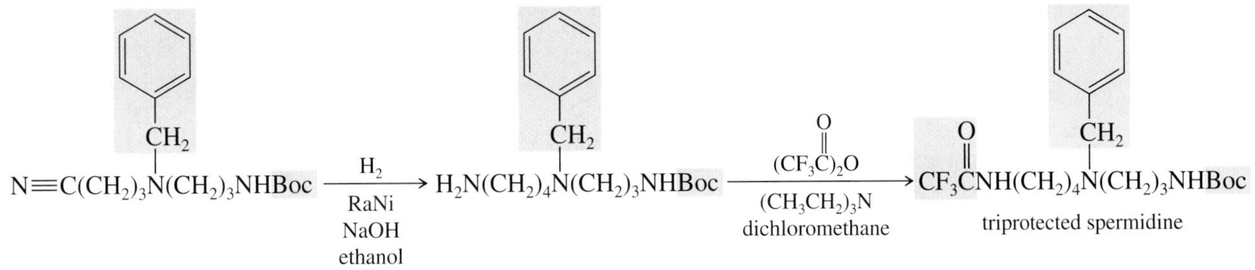

$$N\!\!\equiv\!\!C(CH_2)_3\overset{\overset{\displaystyle CH_2}{|}}{N}(CH_2)_3NHBoc \; \xrightarrow[\substack{RaNi \\ NaOH \\ ethanol}]{H_2} \; H_2N(CH_2)_4\overset{\overset{\displaystyle CH_2}{|}}{N}(CH_2)_3NHBoc \; \xrightarrow[\substack{(CH_3CH_2)_3N \\ dichloromethane}]{(CF_3C)_2O} \; CF_3\overset{O}{\overset{||}{C}}NH(CH_2)_4\overset{\overset{\displaystyle CH_2}{|}}{N}(CH_2)_3NHBoc$$

triprotected spermidine

Each protecting group may be removed in any order and with a reagent that leaves the other two virtually untouched. The benzyl group is removed by hydrogenation over a palladium catalyst (p. 849). The trifluoroacetate group comes off with a relatively weak base, potassium carbonate in water. The Boc group, on the other hand, is removed by a strong acid such as trifluoroacetic acid (p. 632). To prove that this could be done, the researchers removed each protecting group in the order described above and replaced each one with a different acyl group, using acid chlorides as the acyl-transfer reagents. This set of reactions is the subject of the next problem.

**Study Guide**
Concept Map 21.1

**Problem 21.13**    Write equations for the triprotected spermidine shown above, showing the removal of the benzyl protecting group and its replacement with a benzoyl group, then removal of the trifluoroacetyl group and its replacement with an acetyl group, and finally removal of the Boc group and its replacement with a 2,3-dimethoxybenzoyl group.

## 21.3   Oxidation–Reduction Reactions in Functional Group Transformations

### A.  A Look Back

Considerable portions of earlier chapters of this book were devoted to transformations between functional groups involving oxidation–reduction reactions. In Section 13.7A, for example, we examined the concept of oxidation–reduction itself. In Section 13.7C we learned to oxidize primary alcohols to aldehydes and recognized their further oxidation to carboxylic acids.

$$RCH_2OH \; \xrightarrow{\text{oxidation}} \; R\overset{O}{\overset{||}{C}}H \; \xrightarrow{\text{oxidation}} \; R\overset{O}{\overset{||}{C}}OH$$

Secondary alcohols are oxidized to ketones, and tertiary alcohols not at all.

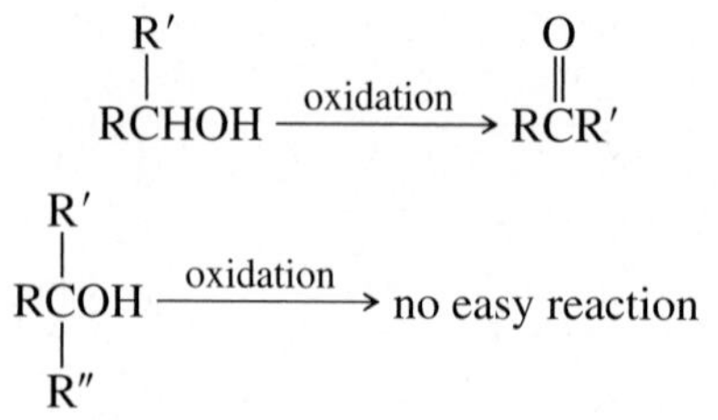

$$\underset{R}{\overset{R'}{\underset{|}{RCHOH}}} \xrightarrow{\text{oxidation}} \overset{O}{\overset{\|}{RCR'}}$$

$$\underset{\underset{R''}{|}}{\overset{R'}{\underset{|}{RCOH}}} \xrightarrow{\text{oxidation}} \text{no easy reaction}$$

The reagents that we use for the selective oxidations that stop at the aldehyde stage for a primary alcohol and also can oxidize secondary alcohols to ketones are pyridinium chlorochromate and Swern's reagent. For example, a diol, protected at the primary alcohol as the benzyl ether (p. 847), is converted to a ketone by the Swern oxidation.

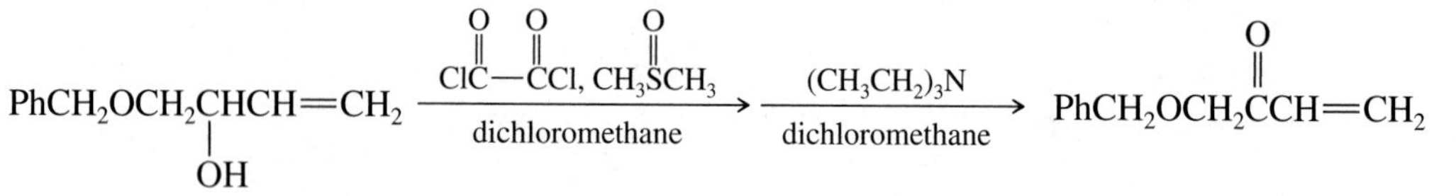

$$\underset{\underset{OH}{|}}{PhCH_2OCH_2CHCH{=}CH_2} \xrightarrow[\text{dichloromethane}]{\overset{O\ O}{\overset{\|\ \|}{ClC-CCl},\ \overset{O}{\overset{\|}{CH_3SCH_3}}}} \xrightarrow[\text{dichloromethane}]{(CH_3CH_2)_3N} \underset{}{PhCH_2OCH_2\overset{O}{\overset{\|}{C}}CH{=}CH_2}$$

1-(phenylmethoxy)-3-buten-2-ol

1-(phenylmethoxy)-3-buten-2-one
68%

Pyridinium chlorochromate is used to oxidize the primary alcohol function in (*E*)-7-hydroxy-6-methyl-5-hepten-2-one to an aldehyde.

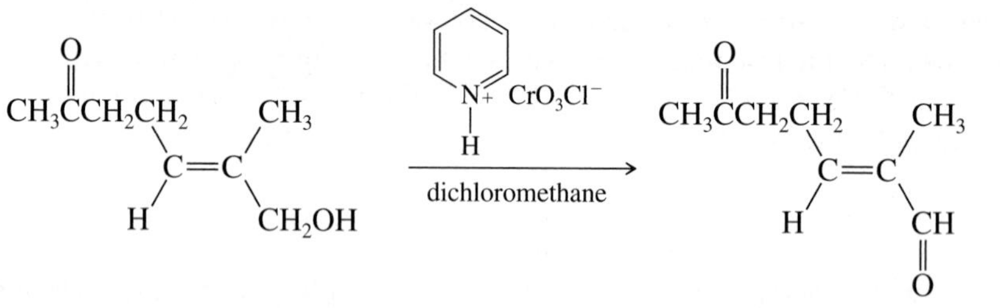

(*E*)-7-hydroxy-6-methyl-5-hepten-2-one

(*E*)-6-oxo-2-methyl-2-heptenal
80%

In Chapter 14 we learned how to go from carbonyl compounds back to alcohols using the metal hydride reagents sodium borohydride and lithium aluminum hydride. For example, an aldehyde group on a cyclopropane ring is reduced to a primary alcohol.

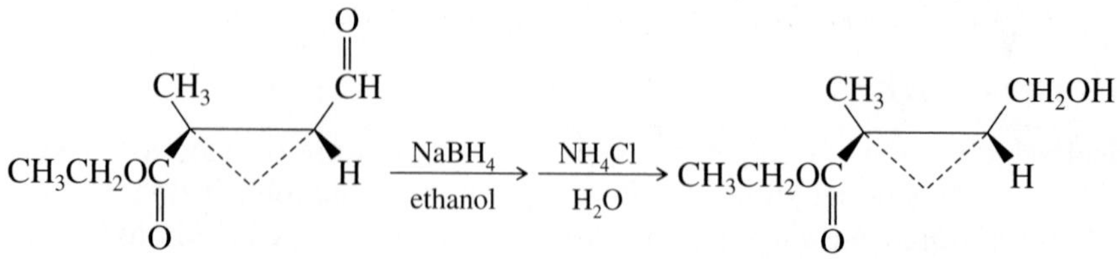

ethyl (1*R*, 2*R*)-2-formyl-1-methylcyclopropanecarboxylate

ethyl (1*R*, 2*R*)-2-hydroxymethyl-1-methylcyclopropanecarboxylate
94%

Note that the carbonyl group in the ester is not reduced by sodium borohydride under these conditions but would be by stronger reducing agents such as lithium aluminum hydride, as we shall see in the next section.

Finally, in Chapter 15 the oxidations of primary alcohols and aldehydes as methods for the preparation of carboxylic acids were discussed. For example, the acid-sensitive cyclic acetal aldehyde shown below is oxidized to the carboxylic acid with silver oxide (p. 608).

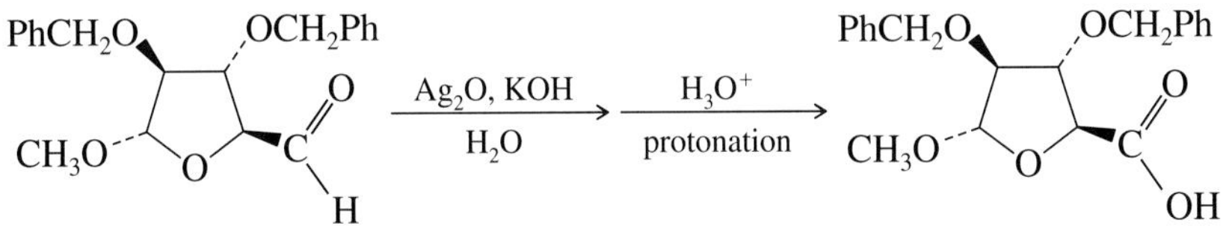

92%

Ozonolysis reactions with oxidative work-ups also give carboxylic acids as products (p. 609). In the case shown below, an aqueous Cr(VI) reagent oxidizes the aldehyde product of ozonolysis to the carboxylic acid.

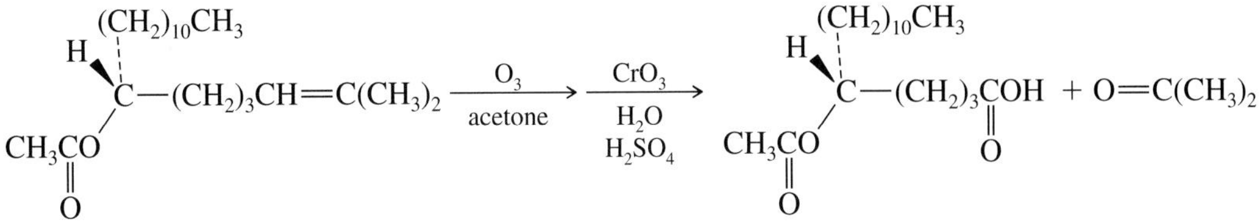

(R)-7-acetoxy-2-methyl-2-octadecene    (R)-5-acetoxyhexadecanoic acid    acetone
83%

What we have not seen so far are reactions that reduce acids and their derivatives. These reactions are important in the conversion of acids and esters to alcohols and of amides as well as nitriles (Section 20.3C) to amines. They are the subject of the next two sections.

## B. Complete Reduction of Acids and Their Derivatives

Carboxylic acids and their derivatives are reduced by metal hydrides, just as aldehydes and ketones are. Sodium borohydride reduces aldehydes and ketones with ease (p. 545) but usually reacts with acid derivatives only with difficulty. Thus it is possible to reduce aldehydes or ketones selectively with sodium borohydride in the presence of an acid or ester if reaction times are short and temperatures are kept low. We saw such an example in the reduction of an aldehyde group in the presence of an ester in ethyl (1R, 2R)-2-formyl-1-methylcyclopropanecarboxylate on page 854.

Lithium aluminum hydride is a more powerful reducing agent and is generally useful in reducing acids and acid derivatives. It reduces acids, esters, amides, and nitriles.

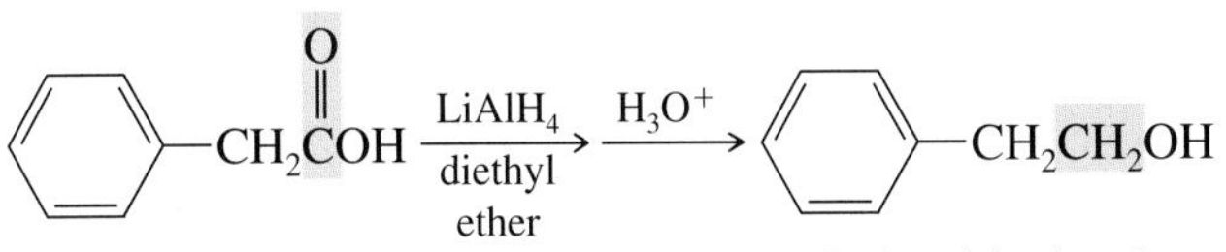

phenylacetic acid                2-phenyl-1-ethanol
                                         92%

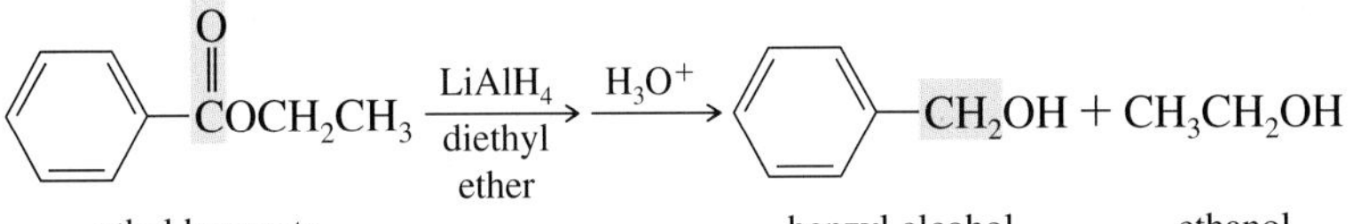

ethyl benzoate          benzyl alcohol          ethanol
                             90%

N-methylacetanilide → N-ethyl-N-methylaniline 91%

tridecanenitrile → tridecylamine 90%

A close examination of these chemical transformations reveals that in each case the carbonyl group (or the carbon atom of the nitrile) has been reduced to a methylene group, —CH$_2$—. A carboxylic acid is reduced to a primary alcohol. In the case of the ethyl ester, clearly, ethanol is formed by protonation of the alkoxyl leaving group. An ester is reduced to two alcohols: a primary alcohol corresponding to the carboxylic acid portion of the molecule and an alcohol corresponding to the alkoxyl group in the ester. Amides and nitriles are reduced to amines.

In the reaction of acid derivatives with lithium aluminum hydride, the first attack is on the electrophilic carbon atom of the carbonyl group by the nucleophilic metal hydride ion in a step analogous to that seen for aldehydes and ketones (p. 546).

## VISUALIZING THE REACTION

### Reduction of an Ester by Lithium Aluminum Hydride

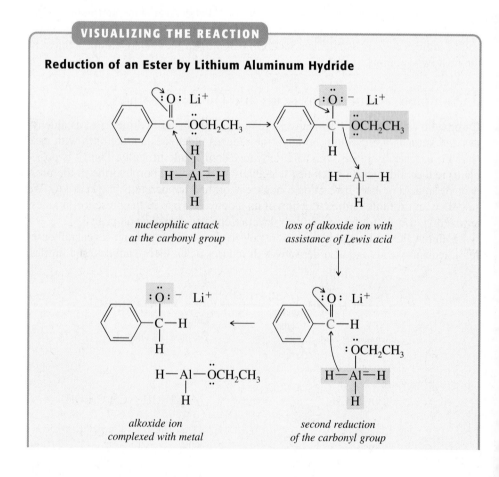

nucleophilic attack at the carbonyl group

loss of alkoxide ion with assistance of Lewis acid

alkoxide ion complexed with metal

second reduction of the carbonyl group

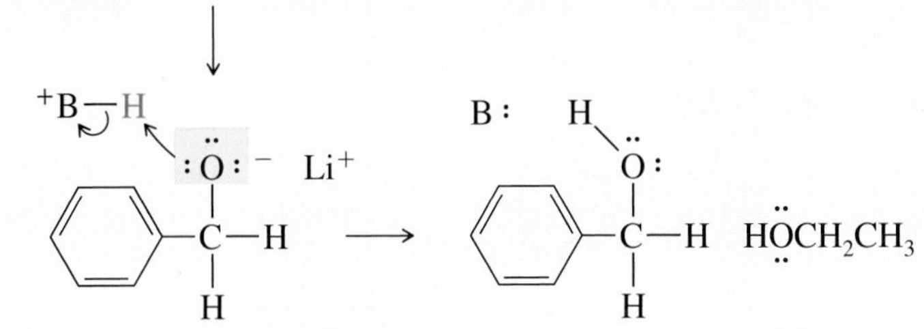

*protonation of alkoxide ion
by acid added in a second step*

The addition of hydride ion to the carbonyl group of the ester gives a tetrahedral intermediate that resembles an acetal in structure and is therefore at the oxidation state of an aldehyde. It loses ethoxide ion with the assistance of the Lewis acid aluminum hydride. The carbon atom of the carbonyl group accepts another hydride ion and is reduced to a primary alcohol. The complexes involving aluminum hydride and alkoxide ions from the ester are sources of hydride ion and will continue to reduce carbonyl groups until all the hydride ions are used up. Cautious addition of a dilute acid protonates the alkoxide ions. This reaction frees the alcohol portion of the ester and reduces the carbonyl group to a primary alcohol. Lithium aluminum hydride is sometimes used to remove acyl groups from valuable alcohols that would be harmed by hydrolysis reactions (as in Problem 21.15, p. 859, for example).

The reductions of carboxylic acids and unsubstituted amides with lithium aluminum hydride do not proceed as smoothly as does the reduction of esters. The hydride ion is a powerful base as well as being a good nucleophile, and it deprotonates the acid or the amide. The resulting salts are often insoluble in the reaction mixture, and their anions are resistant to attack by nucleophiles.

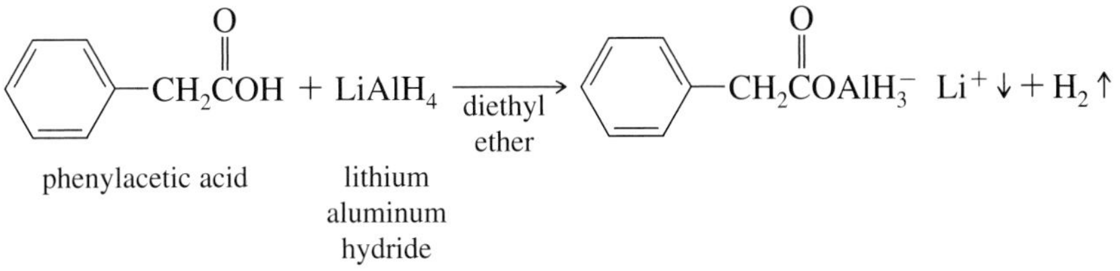

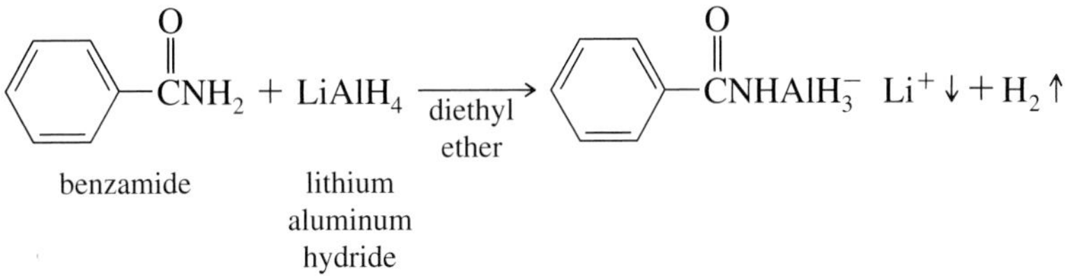

The reduction of an amide by lithium aluminum hydride probably proceeds through an iminium ion intermediate.

**VISUALIZING THE REACTION**

**Reduction of an Amide by Lithium Aluminum Hydride**

*nucleophilic attack at
the carbonyl group*

*alkoxide ion
reacting with
Lewis acid*

*iminium ion
intermediate
being reduced*

*tetrahedral
intermediate*

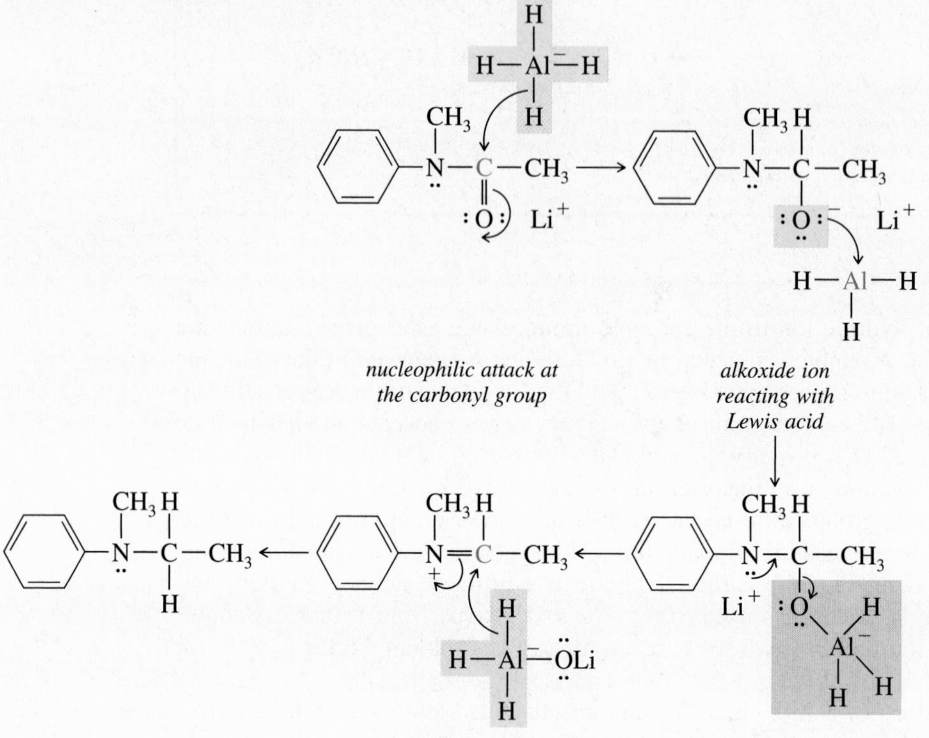

The loss of the oxygen atom of the carbonyl group is thought to take place by the donation of nonbonding electrons from the nitrogen atom of the amide accompanied by the formation of an iminium ion, which is then reduced in another step. The reduction of a nitrile by lithium aluminum hydride follows a mechanism that is similar.

**Problem 21.14**    Complete the following equations.

(a) 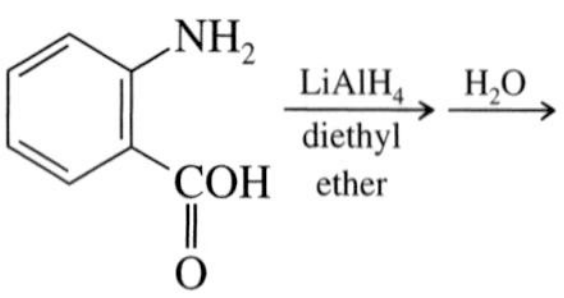 [structure: 2-aminobenzoic acid] $\xrightarrow[\text{diethyl ether}]{\text{LiAlH}_4}$ $\xrightarrow{\text{H}_2\text{O}}$

(b) 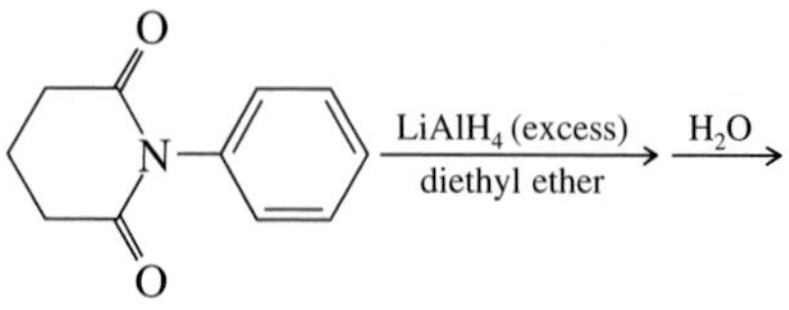 [structure: N-phenyl glutarimide] $\xrightarrow[\text{diethyl ether}]{\text{LiAlH}_4 \text{ (excess)}}$ $\xrightarrow{\text{H}_2\text{O}}$

(c) $CH_3CH{=}CHCH_2CH_2COCH_3$ $\xrightarrow[\text{diethyl ether}]{\text{LiAlH}_4}$ $\xrightarrow{\text{H}_3\text{O}^+}$

(d) 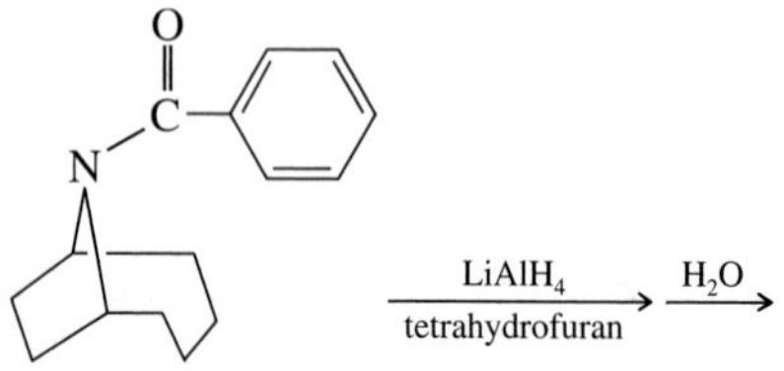 [structure: N-acyl bicyclic amine] $\xrightarrow[\text{tetrahydrofuran}]{\text{LiAlH}_4}$ $\xrightarrow{\text{H}_2\text{O}}$

**Problem 21.15**   The following sequence of reactions is part of a synthesis of compounds that are similar to vitamin D (p. 1095). What are the structures of Compounds A and B?

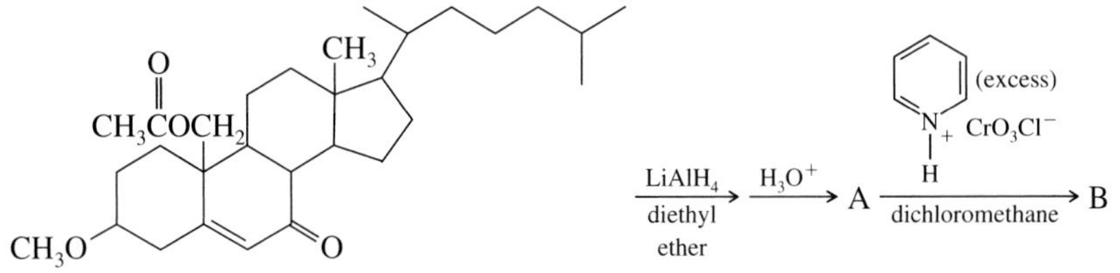

Other reagents that reduce acid derivatives are diisobutylaluminum hydride and the borane–dimethyl sulfide complex. An excess of diisobutylaluminum hydride leads to the same products as would be obtained from reduction by lithium aluminum hydride. For example, butyl hexanoate is reduced to a mixture of 1-hexanol and 1-butanol by 4 equivalents of diisobutylaluminum hydride.

$$CH_3(CH_2)_4\overset{\displaystyle O}{\overset{\|}{C}}OCH_2CH_2CH_2CH_3 + [(CH_3)_2CHCH_2]_2AlH \xrightarrow[\substack{\text{benzene} \\ 45\,°C, 8\,h}]{N_2}$$

butyl hexanoate        diisobutylaluminum
                       hydride
                       4 equivalents

$$\xrightarrow{CH_3OH, H_2O} CH_3(CH_2)_4CH_2OH + CH_3CH_2CH_2CH_2OH$$

                       1-hexanol            1-butanol
                       82%                  58%

The same kinds of results are seen with borane complexed with dimethyl sulfide. For example, ethyl phenylacetate is reduced to 2-phenylethanol by this reagent.

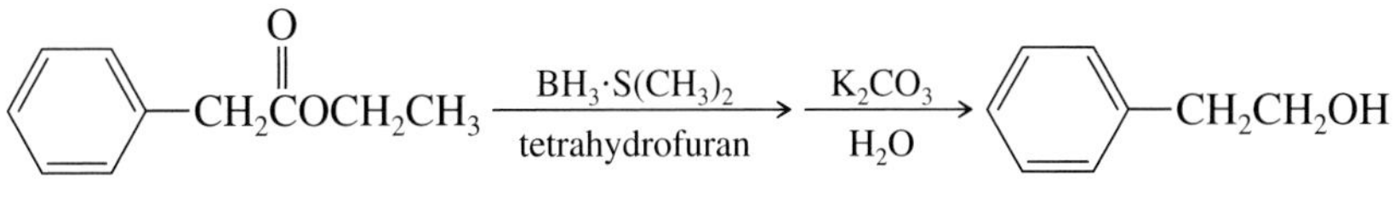

ethyl phenylacetate                              2-phenylethanol
                                                 92%
                                                 +
                                                 $CH_3CH_2OH$

Amides and nitriles are reduced to amines, which complex with the Lewis acid borane, so acid needs to be used in the work-up to break up the complex and liberate the amine. The reduction of *N,N*-dimethylbenzamide to *N,N*-dimethylbenzyl amine illustrates this.

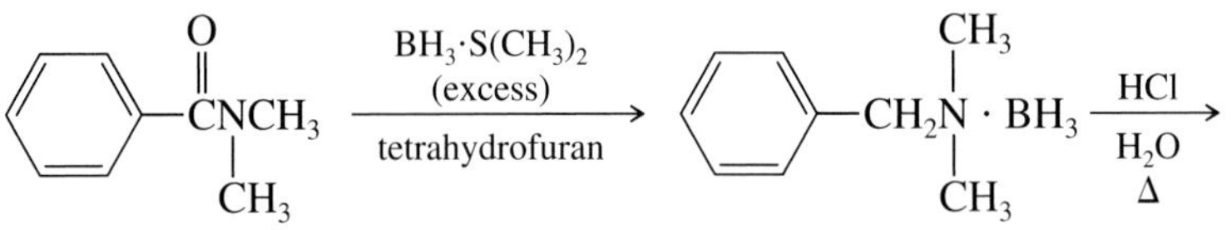

*N,N*-dimethylbenzamide              *N,N*-dimethylbenzamide

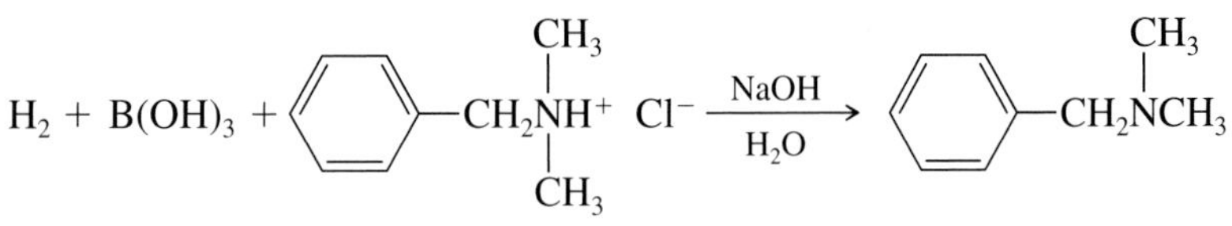

boric acid                                  *N,N*-dimethylbenzyl amine
                                            84%

**Problem 21.16**  Complete the following equations.

(a) PhCOH $\xrightarrow[\substack{\text{benzene} \\ 45\,°C,\,8h}]{\substack{\text{(CH}_3\text{CHCH}_2)_2\text{AlH} \\ \text{(3 equiv)}}}$ $\xrightarrow{\text{CH}_3\text{OH, H}_2\text{O}}$

(b) CH$_3$CH$_2$OCCH$_2$CH$_2$COCH$_2$CH$_3$ $\xrightarrow[\text{tetrahydrofuran}]{\text{BH}_3\cdot\text{S(CH}_3)_2}$ $\xrightarrow[\text{H}_2\text{O}]{\text{K}_2\text{CO}_3}$

(c) CH$_3$C—CN(CH$_3$)$_2$ $\xrightarrow[\text{tetrahydrofuran}]{\text{BH}_3\cdot\text{S(CH}_3)_2}$ $\xrightarrow[\substack{\text{H}_2\text{O} \\ \Delta}]{\text{HCl}}$ $\xrightarrow{\substack{\text{NaOH} \\ \text{H}_2\text{O}}}$

(d) NCCH$_2$CH$_2$CH$_2$CH$_2$CN $\xrightarrow[\text{tetrahydrofuran}]{\text{BH}_3\cdot\text{S(CH}_3)_2}$ $\xrightarrow[\substack{\text{H}_2\text{O} \\ \Delta}]{\text{HCl}}$ $\xrightarrow{\substack{\text{NaOH} \\ \text{H}_2\text{O}}}$

## C. Reduction of Acid Derivatives to Carbonyl Compounds

A close examination of the mechanism of the reduction of an ester to an alcohol by lithium aluminum hydride (p. 856) shows the formation of an aldehyde as an intermediate that gets further reduced to an alcohol. If it were possible to stop the reduction at this stage, the carbonyl compound would be the product. This has been done in two ways. One is to use a limited amount of a reagent that delivers only one reducing hydride ion per equivalent, such as diisobutylaluminum hydride, and keep the temperature low, around −70 °C. Under these conditions, an ester is reduced to an aldehyde in good yield.

CH$_3$(CH$_2$)$_{10}$COCH$_2$CH$_3$ + [(CH$_3$)$_2$CHCH$_2$]$_2$AlH $\xrightarrow[\substack{\text{hexane} \\ -70\,°C}]{}$ CH$_3$(CH$_2$)$_{10}$COCH$_2$CH$_3$ $\xrightarrow{\text{H}_2\text{O}}$

ethyl laurate
ethyl dodecanoate

diisobutylaluminum hydride
1 equivalent

complex from reaction of ester with metal hydride

CH$_3$(CH$_2$)$_{10}$CH + CH$_3$CHCH$_3$ + CH$_3$CH$_2$OH + Al(OH)$_3$↓

dodecanal
88%

isobutane

ethanol

Nitriles are also reduced to aldehydes with high yields when only 1 equivalent of diisobutylaluminum hydride is used.

imine linkage

⬡—CN + [(CH$_3$)$_2$CHCH$_2$]$_2$AlH $\xrightarrow[\text{benzene}]{\text{N}_2}$ ⬡—C=N—Al[CH$_2$CH(CH$_3$)$_2$]$_2$ $\xrightarrow{\text{H}_3\text{O}^+}$ ⬡—C=O

complex from reaction of nitrile with metal hydride

benzaldehyde
90%

The product from the reaction of the nitrile with the metal hydride is an imine (p. 574) complexed with aluminum. When dilute acid is added to the reaction mixture, this complex is hydrolyzed, and a carbonyl group is generated.

The reduction of nitriles to aldehydes is a particularly useful reaction because nitriles can be prepared by S$_N$2 reactions of cyanide ion with primary or secondary alkyl halides or alkyl tosylates. Therefore, an alcohol can be converted by this sequence of reactions into an aldehyde that has one more carbon atom.

**Problem 21.17** Pick an alcohol and convert it into an aldehyde containing one more carbon atom, using the sequence of reactions described above.

The other method for stopping the reduction at the aldehyde stage involves the use of an acid derivative that does not break apart after the attack of the first hydride ion but retains its structure until it is treated with aqueous acid. Such an acid derivative has been created specifically for this purpose, the *N*-methoxy-*N*-methylamides of carboxylic acids. The preparation and reduction of such an amide were used in the preparation of the aldehyde derived from the amino acid leucine, protected as its *tert*-butoxycarbonyl derivative.

*N*-methyl-*O*-methylhydroxyamine hydrochloride

*N*-methylpiperidine

*N*-methyl-*O*-methylhydroxyamine

*N*-methylpiperidine hydrochloride

Boc-leucine

methyl chloroformate

Boc-leucine activated at the carboxyl group

*N*-methoxy-*N*-methylamide of Boc-leucine

Boc-leucinal

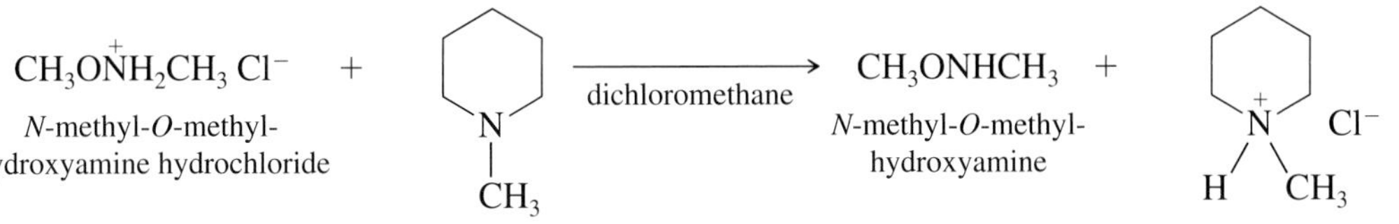

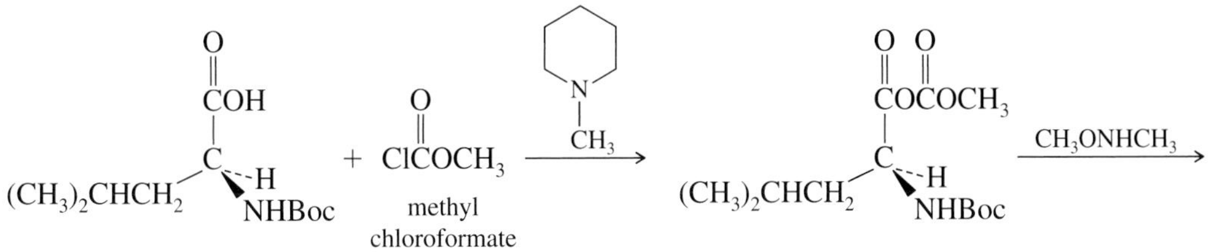

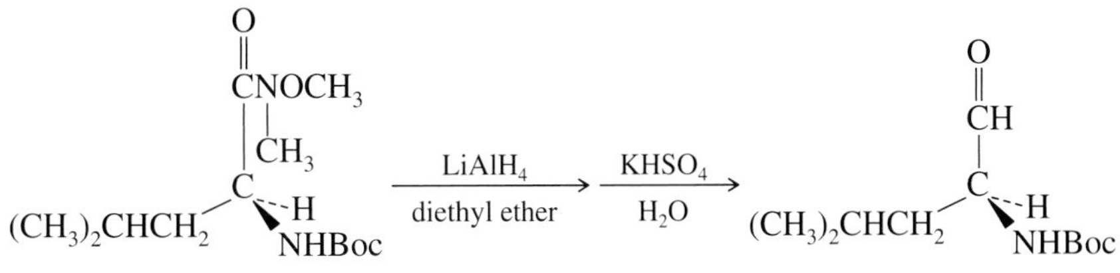

The reason the reduction of this amide stops at the aldehyde stage while others go on to give amines (p. 858) is that the methoxy group on the nitrogen atom is in a position to complex with the lithium ion after the first step of the addition of hydride ion to the carbonyl group.

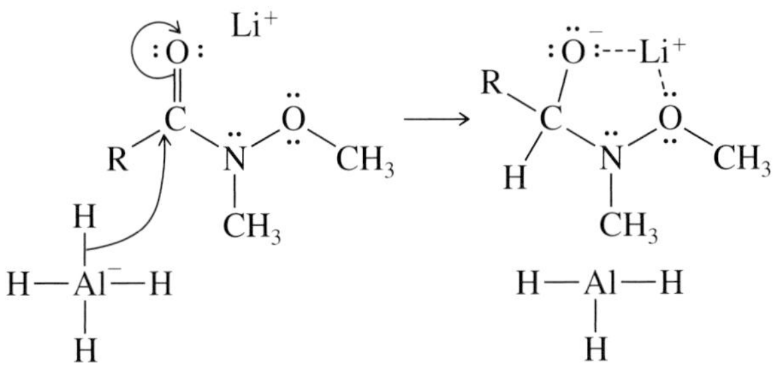

A stable **chelate** (from the Greek word *chele*, meaning "claw") in which the Lewis acid lithium ion is coordinated to two Lewis bases is formed. The bases are the alkoxide ion that results from attack of hydride ion at the carbonyl group and the oxygen atom of the methoxy group. The chelate stabilizes the intermediate to prevent the loss of the carbonyl oxygen and the formation of the iminium ion necessary for further reduction at the original carbonyl group (p. 858). This complex does not break up until acid is added.

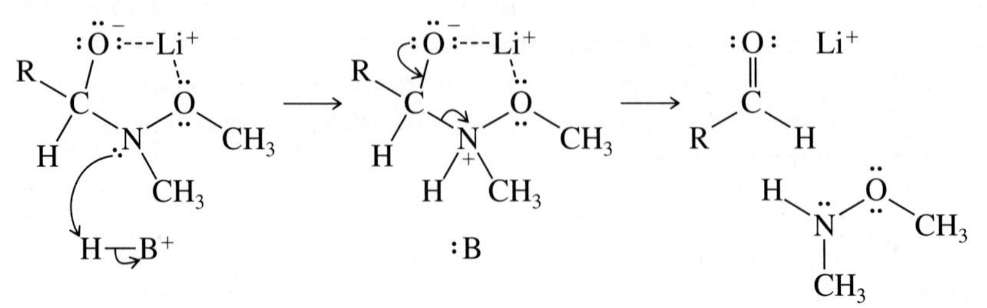

At this point the excess reducing agent has been destroyed, and no further reduction takes place.

*N*-Methoxy-*N*-methylamides are also an exception to the rule that excesses of diisobutylaluminum hydride reduce acid derivatives to the alcohol. The reduction of this acid derivative stops at the aldehyde stage even when excesses of the reducing agent are used for the reasons discussed above. An example of this is the reduction of *N*-methoxy-*N*-methylcyclohexanecarboxamide.

*N*-methoxy-*N*-methyl-cyclohexanecarboxamide

cyclohexanecarbaldehyde 74%

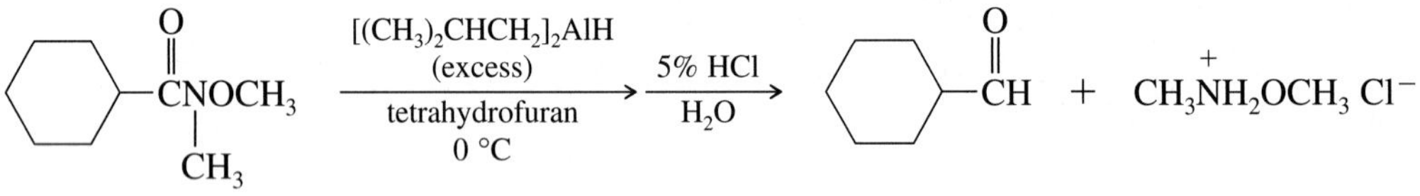

Study Guide Concept Map 21.2

**Problem 21.18** Complete the following equations.

(a) $CH_3CH_2OCH_2(CH_2)_5\overset{O}{\overset{\|}{C}}OCH_2CH_3 \xrightarrow[\text{hexane, } -70\ ^\circ C]{[(CH_3)_2CHCH_2]_2AlH\ (1\ \text{equiv})} \xrightarrow{H_2O}$

(b) 


(c) 


(d)

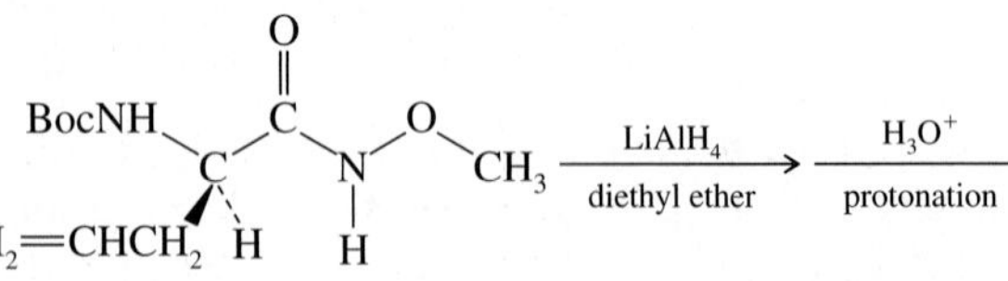

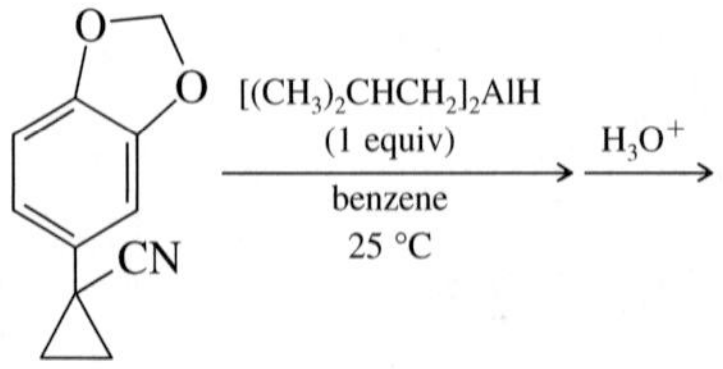

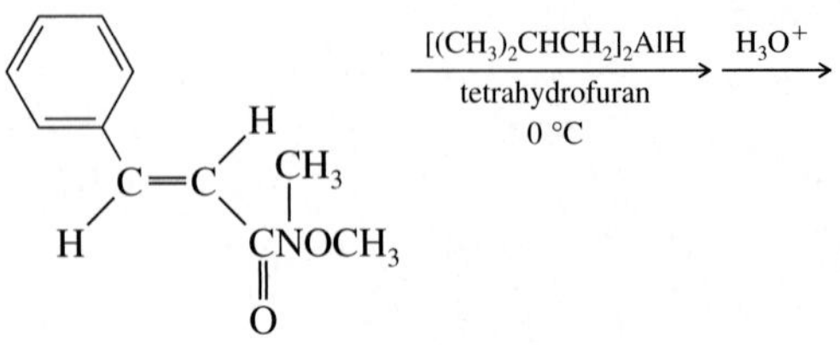

## 21.4   Carbon Nucleophiles Revisited

### A. Organometallic Reagents. Organocuprates

The preparation of and the reactions of organometallic reagents with carbonyl compounds and oxiranes were discussed in Section 14.6. There we saw how the polarity of an alkyl halide, in which the carbon atom is an electrophilic center, is reversed on donation of electrons from reactive metals such as magnesium and lithium to give a carbon atom that is now a nucleophile (p. 550). These nucleophiles can now react with the electrophilic carbon atoms of the carbonyl groups of aldehydes and ketones (p. 551). For example, 3-phenyl-2-propanal (cinnamaldehyde) reacts with methyllithium to give a secondary alcohol.

3-phenyl-2-propenal

and enantiomer
90%
4-phenyl-3-buten-2-ol

Grignard reagents and organolithium compounds are reactive toward carbonyl groups but do not react readily with organic halides to give new carbon–carbon bonds. Copper, however, promotes reactions at electrophilic carbon atoms other than carbonyl carbons. Organocuprate reagents that are particularly useful are prepared by treating an organolithium compound with a copper(I) halide, usually copper(I) iodide.

$$2\ CH_3Li\ +\ CuI\ \xrightarrow[\text{0 °C}]{\text{diethyl ether}}\ (CH_3)_2CuLi\ +\ LiI$$

methyllithium    copper(I)       lithium      lithium
                 iodide          dimethyl-    iodide
                                 cuprate

Just as with the Grignard reagent, the exact structure of the organocuprate reagent is not known; the formula given above represents the stoichiometry observed for the reaction. Two organic groups appear to be associated with the copper atom in a negatively charged species that is a source of nucleophilic carbon atoms. The name lithium dimethylcuprate indicates that copper is associated with the anion in the compound. Primary alkyl halides give reasonably stable organocuprates. Vinyl and aryl organocuprates can also be prepared.

Organocuprates react with organic halides to give compounds with longer carbon chains. For example, lithium dibutylcuprate reacts with 1-bromopentane to give nonane.

$$CH_3CH_2CH_2CH_2CH_2Br\ \xrightarrow[\substack{\text{tetrahydrofuran}\\ \text{25 °C, 1 h}}]{(CH_3CH_2CH_2CH_2)_2CuLi}\ CH_3CH_2CH_2CH_2CH_2CH_2CH_2CH_2CH_3$$

1-bromopentane                                                    nonane
                                                                   98%

Organocuprates may be used to synthesize alkenes or polyenes. The unsaturation in the product may be derived from the organocuprate, from the halide with which it reacts, or from both. 3-Bromo-1-methylcyclohexene, for example, reacts with lithium diisopropenylcuprate to give 3-isopropenyl-1-methylcyclohexene.

3-bromo-1-methylcyclohexene

$$\left(CH_2=\overset{\overset{\displaystyle CH_3}{|}}{C}\right)_2 CuLi$$

$$\xrightarrow[\substack{\text{tetrahydrofuran}\\0\ °C,\ 6\ h}]{}$$

3-isopropenyl-1-methylcyclohexene
75%

The differing reactivities of lithium organocuprates and organolithium reagents are illustrated by the reaction of 4-bromocyclohexanone with lithium diisopropenyl-cuprate.

4-bromocyclohexanone

$$\left(CH_2=\overset{\overset{\displaystyle CH_3}{|}}{C}\right)_2 CuLi$$

$$\xrightarrow[\substack{\text{tetrahydrofuran}\\0\ °C,\ 6\ h}]{}$$

4-isopropenylcyclohexanone
65%

An organolithium reagent would react with the carbonyl group (p. 863) rather than with the alkyl halide function in 4-bromocyclohexanone.

**Problem 21.19**  Complete the following equations.

(a) $CH_3CH_2CH_2CH_2Li + CuI \xrightarrow[\substack{\text{diethyl ether}\\0\ °C}]{}$

(b) $CH_3CH_2CH_2CH_2CH_2I \xrightarrow[\substack{\text{diethyl ether}\\25\ °C}]{(CH_3)_2CuLi}$

(c) $CH_3(CH_2)_6CH_2Cl \xrightarrow[\substack{\text{hexamethylphosphoric}\\\text{triamide}\\25\ °C}]{\left(\substack{CH_3 \\ \diagdown \\ H}C=C\substack{H \\ \diagup \\ H}\right)_2 CuLi}$

(d)  (cyclohexyl)$-I \xrightarrow[\substack{\text{diethyl ether}\\0\ °C}]{(CH_3)_2CuLi}$

Grignard reagents react with $\alpha,\beta$-unsaturated ketones mainly by 1,2-addition.

3,5,5-trimethyl-2-
cyclohexenone

$$\xrightarrow[\text{diethyl ether}]{CH_3MgBr} \quad H_3O^+$$

1,3,5,5-tetramethyl-
2-cyclohexen-1-ol
91%

*major product*
*1,2-addition product*

+

3,3,5,5-tetramethyl-
cyclohexanone
1.5%

*minor product*
*conjugate addition product*

Organocuprates react with $\alpha,\beta$-unsaturated carbonyl compounds at the $\beta$-carbon atom to give high yields of conjugate addition (p. 712) products. For example, 2-cyclohexenone is converted almost quantitatively to 3-methylcyclohexanone by lithium dimethylcuprate; the reaction is run at low temperatures to improve regioselectivity.

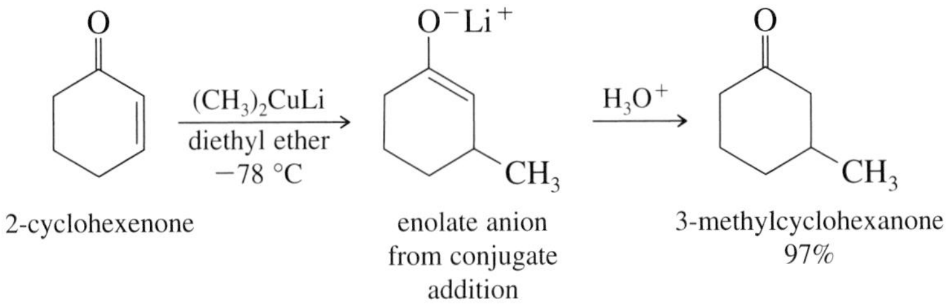

2-cyclohexenone → enolate anion from conjugate addition → 3-methylcyclohexanone 97%

**On the Web: ONE SMALL STEP**

The exact mechanism of the reaction is not known. An excess of the organocuprate is required, and chemists have proposed mechanisms that involve formation of a complex between the organometallic reagent and the carbonyl compound. An enolate anion is formed as an intermediate and can be used in further reactions that form carbon–carbon bonds, for example:

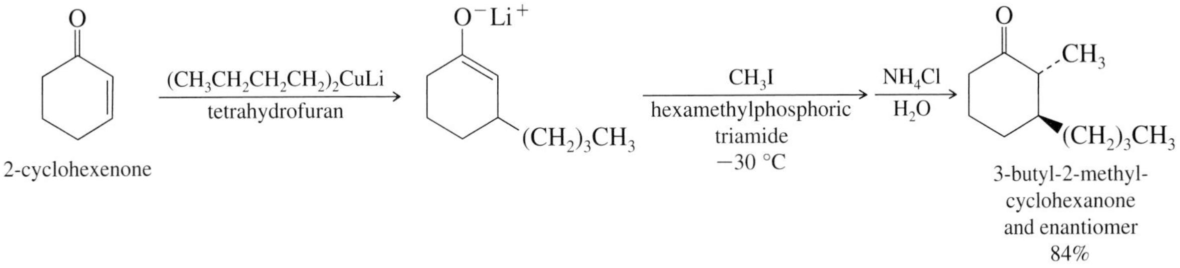

2-cyclohexenone

3-butyl-2-methyl-cyclohexanone and enantiomer 84%

The product that is obtained can be rationalized by postulating nucleophilic attack at the β-carbon atom giving an enolate anion, which is then alkylated by methyl iodide.

Conjugate addition of organocuprate reagents is used widely for creating new carbon skeletons in organic syntheses. We shall see other examples of this in Section 21.5B.

**Study Guide**
Concept Map 17.6

**Problem 21.20**   Give structural formulas for the reagents, intermediates, and products symbolized by letters in the following equations.

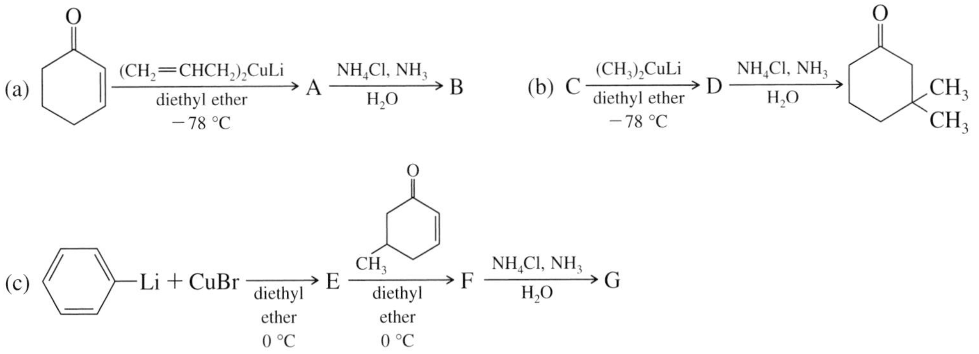

# B. The Regioselectivity of the Enolization Reaction. Thermodynamic Versus Kinetic Enolates

An unsymmetrical carbonyl compound enolizes to give two different enolate anions. For example, when 2-methylcyclohexanone is heated with the base triethylamine in dimethylformamide in the presence of trimethylchlorosilane, a mixture of two trimethylsilyl enol ethers is formed.

2-methylcyclohexanone                                        22%              78%

Trimethylchlorosilane is a reagent that reacts exclusively with the enolate anion at the oxygen atom because a strong silicon–oxygen bond is formed. Therefore, the reagent is used to trap enolate anions as stable trimethylsilyl enol ethers. The composition of the mixture of enol ethers reflects the relative stabilities of the two enolate anions derived from 2-methylcyclohexanone.

more stable                                                less stable
enolate anion,                                             enolate anion,
more highly substituted                                    less highly substituted
double bond                                                double bond

major product                                              minor product
of reaction                                                of reaction

The enolate anion with the more highly substituted double bond is the more stable anion and predominates under conditions that allow equilibrium to be established.

The conditions described for the previous reaction allow an equilibrium to exist between the two enolate anions. The reaction is run at a high temperature. Triethylamine is not a strong base; therefore, the conjugate acid of triethylamine ($pK_a \sim 10$) and unreacted ketone molecules ($pK_a \sim 16$) serve as acids to protonate the enolate anions and interconvert them by way of the ketone. All these factors ensure that the mixture that is observed reflects the relative thermodynamic stabilities of the anions. The more highly substituted enolate anion is said to be the **thermodynamic enolate** of the ketone. Under conditions that allow for equilibrium, free-energy considerations, that is, thermodynamics (p. 95), control what the major product of a reaction will be.

Enolate anions are versatile intermediates in reactions that result in new carbon–carbon bonds being formed (Sections 17.2, 17.3, 17.4, and 17.5, for example). Therefore, it is important that chemists know how to control the compositions of mixtures obtained from unsymmetrical carbonyl compounds. Gilbert Stork at Columbia University and Herbert O. House at the Georgia Institute of Technology have made major contributions by devising methods for regioselective generation of enolate anions. For example, 2-methylcyclohexanone is added slowly to a solution of the very strong base lithium diisopropylamide in 1,2-dimethoxyethane at 0 °C.

The resulting mixture is then treated with trimethylchlorosilane in the presence of triethylamine, and the trimethylsilyl ether of the less highly substituted enol is obtained almost exclusively.

2-methylcyclohexanone

*added dropwise
to cold solution
of base*

major product
formed under
these conditions

99%

major product of
the reaction

1%

The reaction conditions just described do not allow equilibration between the enolate anions. A very strong base, the diisopropylamide anion, is used to deprotonate the ketone. The conjugate acid of this anion, diisopropylamine (p$K_a$ ~36), is not a strong enough acid to protonate the enolate anion that is formed. The other acid in the system, the ketone 2-methylcyclohexanone (p$K_a$ ~16), is never effectively present at the same time as the enolate anion because it is added slowly to the solution of the base. An excess of base is always present, and the ketone is deprotonated completely. The temperature is also kept low. Under these conditions, the major product is the **kinetic enolate,** the enolate that is formed the fastest rather than the one with the greatest thermodynamic stability. Kinetic enolates tend to be the less highly substituted ones. The rate-determining step in an enolization reaction is the breaking of a carbon–hydrogen bond at the $\alpha$-carbon atom of the carbonyl compound. The hydrogen atoms on the less highly substituted $\alpha$-carbon atom are less sterically hindered and react fastest with a strong base. Under conditions that do not allow equilibrium to be attained, rate considerations, that is, kinetics (p. 133), control what the major product of the reaction will be.

The reasons for choosing lithium diisopropylamide as the strong base in the previous reaction are interesting. This reagent is easily prepared by treating diisopropylamine in an ether or hydrocarbon solvent with an alkyllithium, usually *n*-butyllithium. The *n*-butyl anion is a powerful base and deprotonates diisopropylamine completely; the result is a solution of lithium diisopropylamide.

diisopropylamine

*n*-butyllithium

lithium diisopropylamide

butane

Solutions of lithium diisopropylamide in hydrocarbons are stable, but those in ether solvents are not. The base is strong enough to deprotonate ethers, so its ether solutions must be kept cold and used as soon as they are prepared.

The diisopropylamide anion is a strong base but is hindered enough that it is not a good nucleophile. Thus it does not react with alkyl halides or other reagents that react with the enolate anions in subsequent steps. The conjugate acid of the diisopropylamide anion, diisopropylamine, boils at 86 °C and is easily separated from the other products of these reactions. The small lithium ion, with its high concentration of positive charge, coordinates strongly with the oxygen atom and enhances the regioselectivity of the enolization.

⊘ **Study Guide**
**Concept Map 21.3**

**Problem 21.21**    Predict the major product of each of the following reactions.

(a) $CH_3CH_2\overset{\overset{\displaystyle O}{\|}}{C}CH_3$ $\xrightarrow[\underset{\Delta}{\text{dimethylformamide}}]{(CH_3)_3SiCl,\ (CH_3CH_2)_3N}$

(b) $CH_3CH_2\overset{\overset{\displaystyle O}{\|}}{C}CH_3$ $\xrightarrow[\underset{\text{1,2-dimethoxyethane}}{\text{(added to base)}}]{(CH_3CH)_2N^-Li^+\ \overset{CH_3}{|}}$ $\xrightarrow{(CH_3)_3SiCl,\ (CH_3CH_2)_3N}$

## C. Carbanions Stabilized by Phosphorus. Phosphonium Ylides and the Wittig Reaction

Chemists continue to search for new ways of stabilizing carbanions so that they can be generated under conditions that allow them to be used in different types of syntheses. Researchers have found that elements from the third row of the periodic table, such as sulfur and phosphorus, stabilize negatively charged carbon atoms that are adjacent to them in a variety of structures.

A positively charged phosphorus atom is formed when a phosphine, the phosphorus analog of an amine, reacts with an alkyl halide. The phosphonium compound can be deprotonated at the carbon atom adjacent to the phosphorus atom to give a carbanion stabilized by the positively charged phosphorus atom. This sequence of reactions is shown for triphenylphosphine.

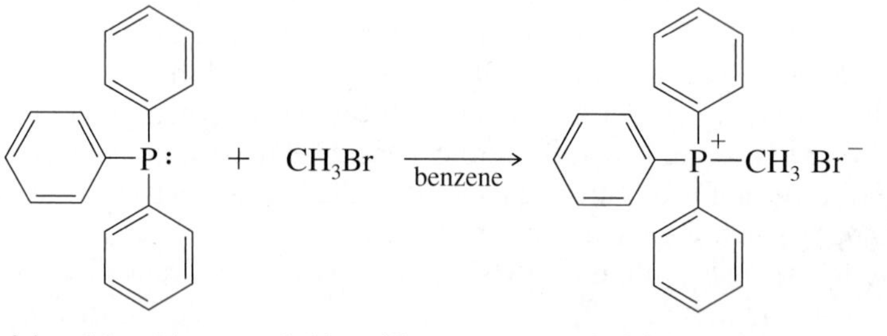

triphenylphosphine    methyl bromide         methyltriphenylphosphonium
                                                              bromide

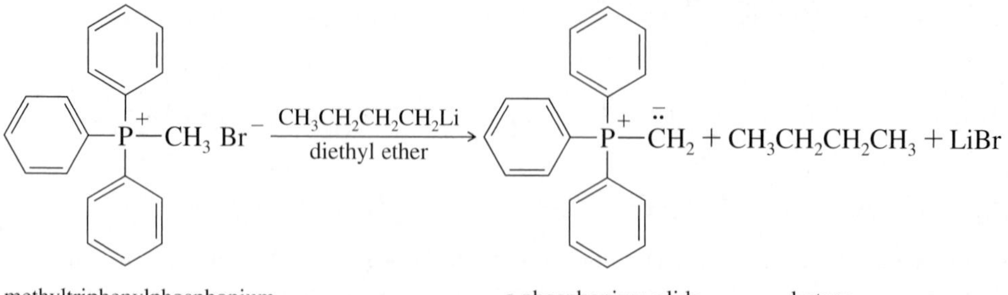

methyltriphenylphosphonium                a phosphonium ylide         butane
           bromide

Such a stabilized carbanion is known as a **phosphonium ylide.** An **ylide** (pronounced "ill · id") is a reactive species having a positively charged atom next to one that is negatively charged. The phosphonium ylide can be written with a double bond symbolizing a donation of electrons from the 2*p* orbital on the carbon atom to the empty 3*d* orbitals of the phosphorus atom.

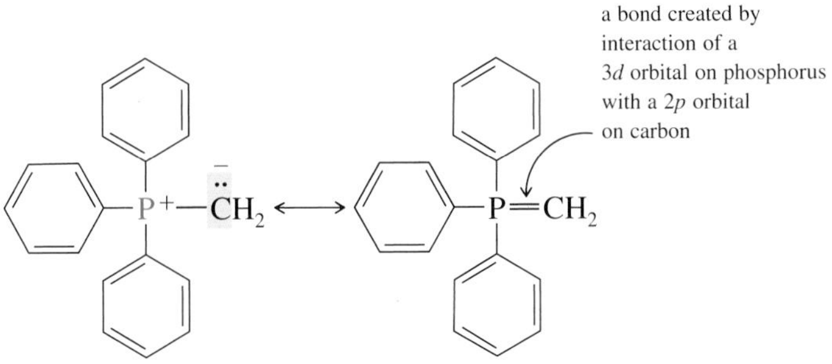


*resonance contributors for a triphenylphosphonium ylide*

In the resonance contributor that has a double bond between phosphorus and carbon, neither atom has a formal charge. There are ten electrons around phosphorus in this resonance contributor, but this is permissible for an element in the third row.

Phosphorus ylides with a wide range of structures can be prepared. A particularly interesting class of ylides comes from $\alpha$-haloesters such as ethyl bromoacetate.

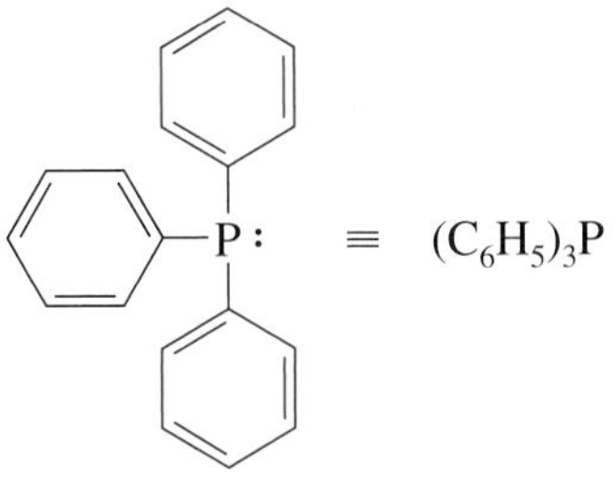


$$\underset{\text{triphenylphosphine}}{(C_6H_5)_3P} + \underset{\text{ethyl bromoacetate}}{BrCH_2\overset{O}{\overset{\|}{C}}OCH_2CH_3} \longrightarrow \underset{\text{phosphonium salt}}{(C_6H_5)_3\overset{+}{P}CH_2\overset{O}{\overset{\|}{C}}OCH_2CH_3\ Br^-} \xrightarrow[H_2O]{NaOH} \underset{\text{phosphonium ylide}}{(C_6H_5)_3\overset{+}{P}-\ddot{\bar{C}}H\overset{O}{\overset{\|}{C}}OCH_2CH_3}$$

A comparison of the above reaction with the one on page 868 indicates that to remove a proton from the phosphonium salt derived from ethyl bromoacetate is much easier than to remove one from methyltriphenylphosphonium bromide. The base used to deprotonate methyltriphenylphosphonium bromide is butyllithium, a strong base. The reaction is carried out in a nonprotic solvent, dry ether. The hydrogens of the methylene group between the positively charged phosphorus atom and the carbonyl group of the ester are much more acidic than the hydrogens of the methyl group in methyltriphenylphosphonium bromide. Aqueous sodium hydroxide is a strong enough base to remove a proton from the phosphonium salt of the ester, and the resulting ylide is stable in water, a medium to be avoided with most other types of ylides. Phosphorus ylides with carbonyl groups on the carbanionic center are generally more stable and less reactive than other types of phosphorus ylides.

Study Guide Concept Map 21.4

Phosphorus ylides react with aldehydes or ketones to introduce a carbon–carbon double bond selectively in place of the carbonyl group. For example, the ylide from methyltriphenylphosphonium bromide reacts with cyclohexanone to give methylenecyclohexane.

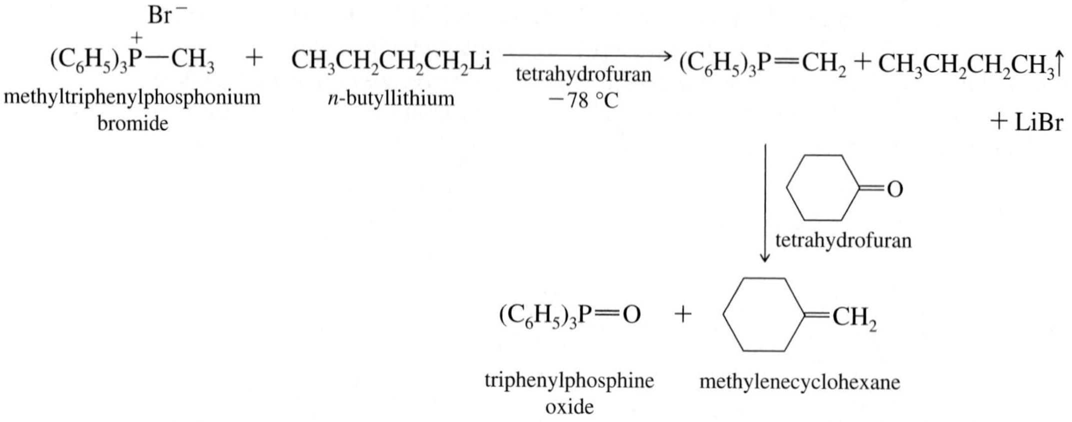

$$
\underset{\substack{\text{methyltriphenylphosphonium}\\ \text{bromide}}}{(C_6H_5)_3\overset{+}{P}-CH_3} \quad + \quad \underset{\text{\textit{n}-butyllithium}}{CH_3CH_2CH_2CH_2Li} \xrightarrow[-78\ °C]{\text{tetrahydrofuran}} (C_6H_5)_3P{=}CH_2 + CH_3CH_2CH_2CH_3{\uparrow}
$$

Br$^-$ (above the P)

$$+ \text{ LiBr}$$

tetrahydrofuran

$$
\underset{\substack{\text{triphenylphosphine}\\ \text{oxide}}}{(C_6H_5)_3P{=}O} \quad + \quad \underset{\text{methylenecyclohexane}}{}{=}CH_2
$$

The base used to deprotonate the phosphonium salt in this reaction is the anion from *n*-butyllithium. The ylide reacts quickly with the ketone to give an alkene and triphenylphosphine oxide. The reaction is believed to start by nucleophilic attack of the ylide on the carbonyl group. The nature of the intermediate in such reactions depends on the conditions used. At low temperatures, an unstable intermediate having a four-membered ring, an **oxaphosphetane,** has been observed using nuclear magnetic resonance spectroscopy. In the presence of lithium bromide, an ionic intermediate stabilized by interaction with lithium and bromide ions also forms. Bond reorganization in either intermediate leads to the products.

---

**VISUALIZING THE REACTION**

**The Wittig Reaction**

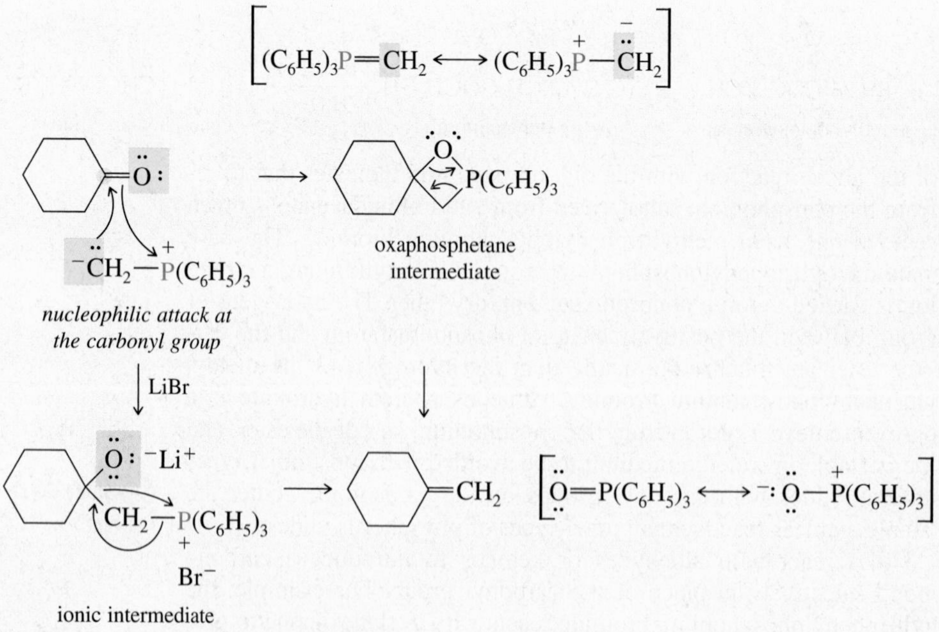

$$
\left[ (C_6H_5)_3P{=}CH_2 \longleftrightarrow (C_6H_5)_3\overset{+}{P}-\overset{..}{\overset{-}{C}}H_2 \right]
$$

oxaphosphetane
intermediate

$^-\ddot{C}H_2 - \overset{+}{P}(C_6H_5)_3$

nucleophilic attack at
the carbonyl group

LiBr

ionic intermediate

$$
=CH_2 \qquad \left[ \ddot{O}{=}P(C_6H_5)_3 \longleftrightarrow {}^-{:}\ddot{O}-\overset{+}{P}(C_6H_5)_3 \right]
$$

Triphenylphosphine oxide is a stable compound, and the formation of the phosphorus−oxygen bond is part of the driving force for the reaction. In the product, the group that can be written as being doubly bonded to the phosphorus atom in the ylide is always doubly bonded to the carbon atom of the former carbonyl group in the aldehyde or ketone that is used in the reaction. For example, benzaldehyde is converted into an $\alpha,\beta$-unsaturated ester by reaction with the phosphonium ylide from ethyl bromoacetate.

phosphonium salt from
ethyl bromoacetate

phosphonium ylide

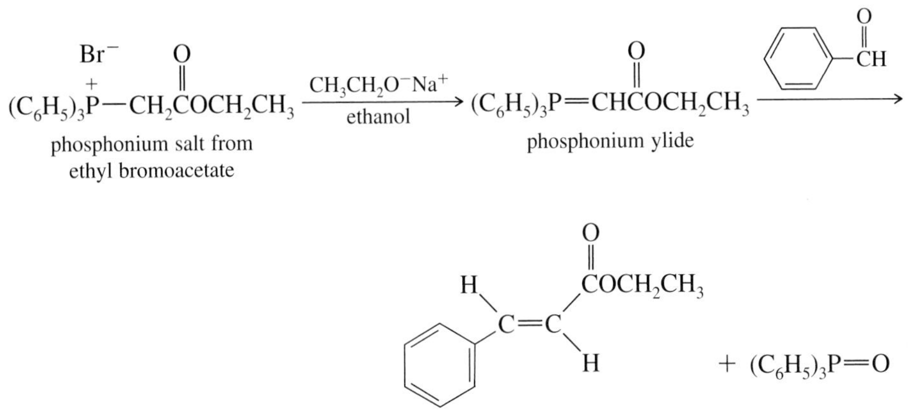

ethyl ($E$)-3-phenyl-2-propenoate
100%

triphenylphosphine
oxide

The product of the above reaction is formed exclusively with the trans configuration around the double bond, but this is not always so. In most cases where stereoisomerism is possible, some of each isomer forms. The composition of the mixture depends on factors such as the solvent used, the presence of inorganic salts in the reaction mixture, and the temperature at which the reaction is carried out.

In both reactions shown in this section, the carbonyl group has been replaced by a double bond to a carbon atom that was the anionic center of a phosphonium ylide. This method of forming carbon−carbon double bonds quite selectively is called the **Wittig reaction** because it was the German chemist Georg Wittig who developed it. Phosphonium ylides are called **Wittig reagents.** The reaction is useful because it introduces double bonds into molecules at specific locations. The Wittig reaction is especially important in preparing compounds such as methylenecyclohexane (p. 870), which has an **exocyclic double bond,** a carbon−carbon double bond that is outside a ring but which shares one carbon atom with the ring. Such a compound is thermodynamically less stable than its isomer in which the double bond is inside the ring and, therefore, difficult to prepare; methods such as the dehydration of alcohols (p. 286) or elimination reactions (p. 264) often give mixtures of products. Because Wittig reactions usually take place at low temperatures and under basic conditions, rearrangements of complex molecules do not generally occur. Note that the Wittig reaction resembles the aldol condensation (p. 702) in its essentials.

⊘ Study Guide
Concept Map 21.5

On the Web: ONE SMALL STEP

**Problem 21.22**   Complete the following reactions.

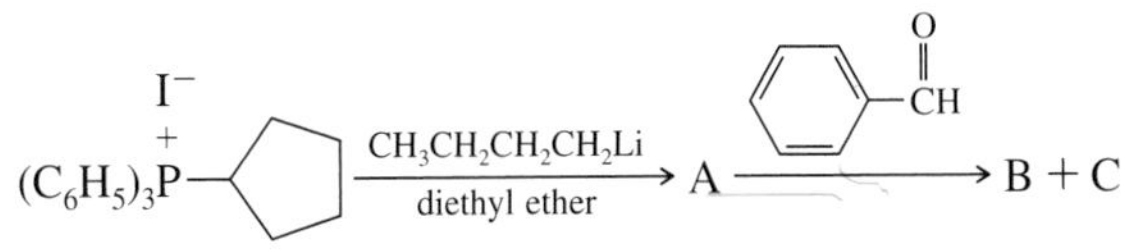

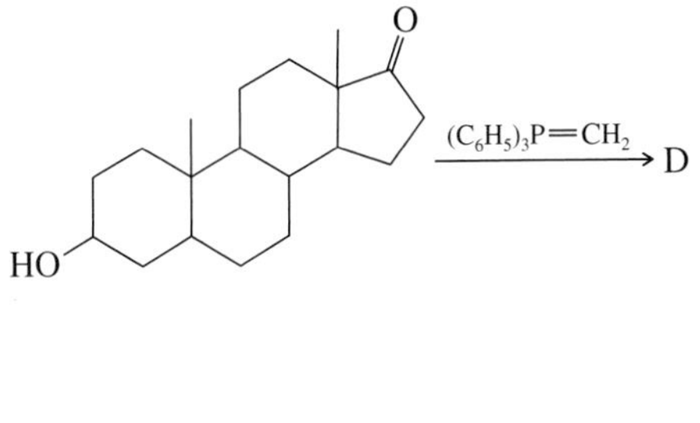

(c) $(C_6H_5)_3\overset{+}{P}CH_2\overset{Br^-}{\underset{}{}}\overset{O}{\overset{\|}{C}}OCH_2CH_3 \xrightarrow[\text{ethanol}]{CH_3CH_2ONa} E \xrightarrow{CH_3CH=CHCH} F + G$

(d) $(C_6H_5)_3\overset{+}{P}CH_2\overset{Br^-}{\underset{}{}}\overset{O}{\overset{\|}{C}}OCH_2CH_3 \xrightarrow[\text{ethanol}]{CH_3CH_2ONa} H \longrightarrow I + J$

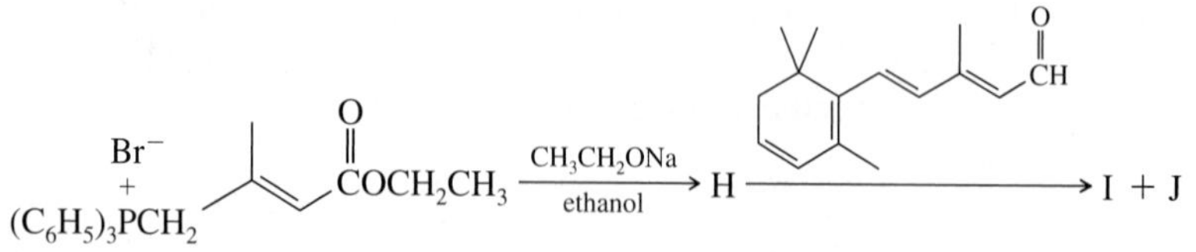

**Problem 21.23** How would you carry out the following transformation?

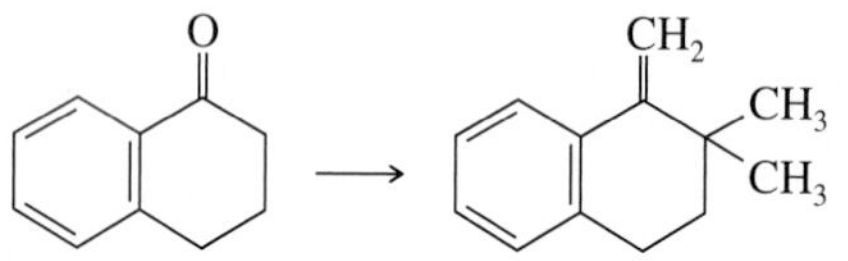

**D. Carbanions Stabilized by Sulfur. Dithiane Anions**

In searching for the synthetic equivalent of a carbonyl group to serve as a nucleo-phile in a nucleophilic substitution reaction, E. J. Corey and Dieter Seebach developed a carbanion stabilized by sulfur, the **1,3-dithiane anion.** The carbon atom of a carbonyl group is an electrophilic center and is attacked by nucle-ophiles. The researchers thought it would be useful to have a synthetic method to introduce a carbonyl group by a nucleophilic substitution reaction initiated by the potential carbonyl group. They chose the cyclic thioacetal of aldehydes, the 1,3-dithiane system, as the source of a carbanionic equivalent of a carbonyl group.

The anion derived from an aldehyde, called an **acyl anion,** usually cannot be prepared directly by removing the hydrogen bonded to the carbonyl carbon. For example, the base butyllithium would add to the carbonyl group of formalde-hyde (p. 553) instead of deprotonating it. Deprotonation of most aldehydes occurs at the $\alpha$-carbon atom (p. 701), not at the carbonyl group. Conversion of the aldehyde to a 1,3-dithiane protects the carbonyl group and increases the acidity of the hydrogen atom to the point that it is removed preferentially by a strong base.

$$\overset{O}{\overset{\|}{R}CH} + CH_3CH_2CH_2CH_2Li \not\longrightarrow \underset{\substack{\text{acyl anion} \\ \text{not } formed \\ \text{in this way}}}{\overset{O}{\overset{\|}{R}C}:^- Li^+} + CH_3CH_2CH_2CH_3$$

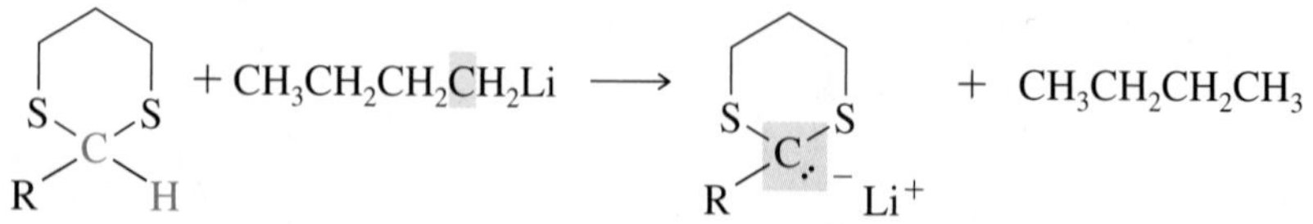

$$\underset{\substack{R \quad H}}{\overset{S \frown S}{\underset{C}{\diagup}}} + CH_3CH_2CH_2CH_2Li \longrightarrow \underset{\substack{R \quad Li^+ \\ \text{acyl anion equivalent}}}{\overset{S \frown S}{\underset{\overset{..}{C}^-}{\diagup}}} + CH_3CH_2CH_2CH_3$$

The electrophilic carbon atom of the carbonyl group is thus converted into a nuc-leophile.

The conversion of formaldehyde to 6-undecanone illustrates the chief features of Corey's synthetic method.

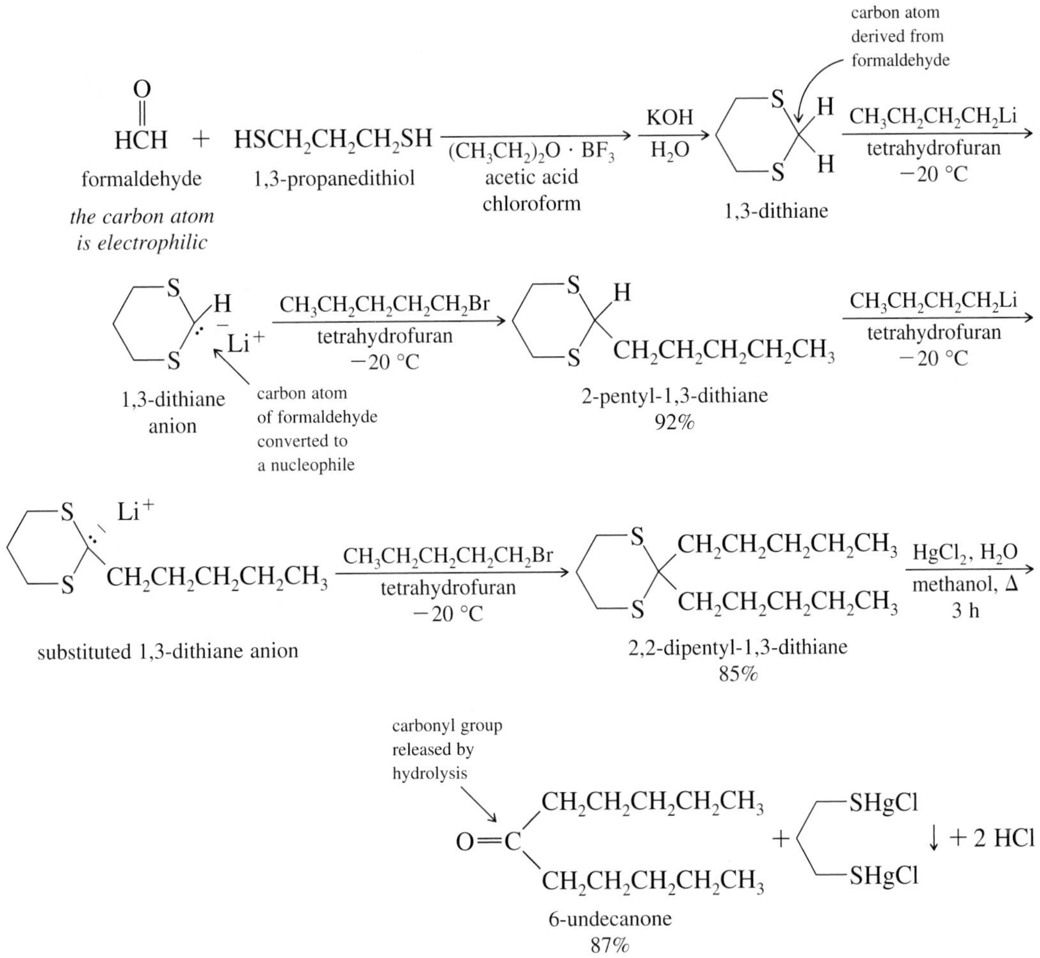

The reaction schemes show:

formaldehyde (HCH with O) + 1,3-propanedithiol (HSCH$_2$CH$_2$CH$_2$SH), with formaldehyde labeled *the carbon atom is electrophilic*, reacting with (CH$_3$CH$_2$)$_2$O · BF$_3$, acetic acid, chloroform, then KOH, H$_2$O to give 1,3-dithiane. The top carbon is labeled *carbon atom derived from formaldehyde*. Then CH$_3$CH$_2$CH$_2$CH$_2$Li, tetrahydrofuran, −20 °C.

1,3-dithiane anion (with Li$^+$), labeled *carbon atom of formaldehyde converted to a nucleophile*, reacts with CH$_3$CH$_2$CH$_2$CH$_2$CH$_2$Br, tetrahydrofuran, −20 °C, to give 2-pentyl-1,3-dithiane, 92%. Then CH$_3$CH$_2$CH$_2$CH$_2$Li, tetrahydrofuran, −20 °C.

substituted 1,3-dithiane anion (with Li$^+$, CH$_2$CH$_2$CH$_2$CH$_2$CH$_3$) reacts with CH$_3$CH$_2$CH$_2$CH$_2$CH$_2$Br, tetrahydrofuran, −20 °C, to give 2,2-dipentyl-1,3-dithiane, 85%. Then HgCl$_2$, H$_2$O, methanol, Δ, 3 h.

6-undecanone, labeled *carbonyl group released by hydrolysis* O=C(CH$_2$CH$_2$CH$_2$CH$_2$CH$_3$)(CH$_2$CH$_2$CH$_2$CH$_2$CH$_3$), 87% + (ring with —SHgCl, —SHgCl) ↓ + 2 HCl

In this sequence of reactions, polarity of the carbonyl group is reversed, resulting in great versatility in the design of syntheses.

The thioacetal (p. 579) is prepared from the aldehyde by treating it with 1,3-propanedithiol in the presence of an acid catalyst. A 1,3-dithiane is deprotonated at the carbon atom between the two sulfur atoms by a strong base, such as *n*-butyllithium, to give a carbanion stabilized by delocalization of charge to the sulfur atoms. The carbanion is a strong nucleophile and reacts with alkyl halides in nucleophilic substitution reactions. In the synthesis shown above, the 1,3-dithiane is deprotonated and alkylated twice. The product, 2,2-dipentyl-1,3-dithiane, contains a carbonyl group hidden in a thioacetal function. The carbonyl group is usually released by hydrolysis in the presence of mercury(II) salts, which serve as Lewis acids in coordinating with the sulfur atoms and help cleave the sulfur–carbon bonds.

1,3-Dithiane anions react with other types of electrophilic carbon atoms, such as those in carbonyl groups. The addition of such a carbanion to a carbonyl compound must be carried out at very low temperatures if the carbonyl compound has enolizable hydrogen atoms. The reaction of the anion from 1,3-dithiane with benzaldehyde, which has no α-hydrogen atoms, goes well.

## ONE SMALL STEP

A disadvantage to using dithianes in synthesis is use of toxic mercury(II) compounds in opening the dithiane ring and freeing the carbonyl group. An alternative method has been developed that uses a large excess of methyl iodide instead.

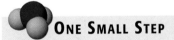

**PROBLEM:** Using

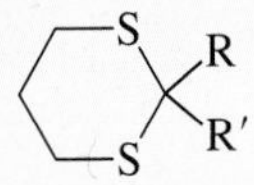

as a generic dithiane, propose a mechanism in which methyl iodide plays a role similar to that of mercury(II) ion in promoting the opening of the dithiane ring. Think in terms of nucleophiles, electrophiles, and leaving groups.

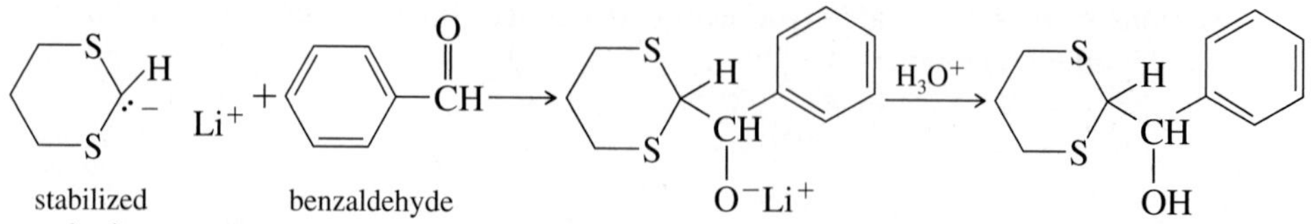

stabilized
carbanion

benzaldehyde

91%

That carbanion also reacts with oxiranes, attacking the less highly substituted carbon atom of unsymmetrical ones (p. 505).

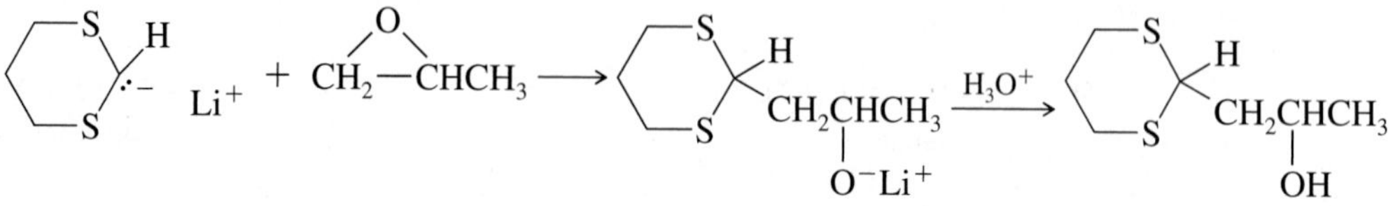

73%

**Study Guide**
**Concept Map 21.6**

The reactions of dithiane anions with carbonyl compounds or oxiranes are routes to potential hydroxycarbonyl compounds.

**Problem 21.24**    Show how each of the following conversions can be carried out.

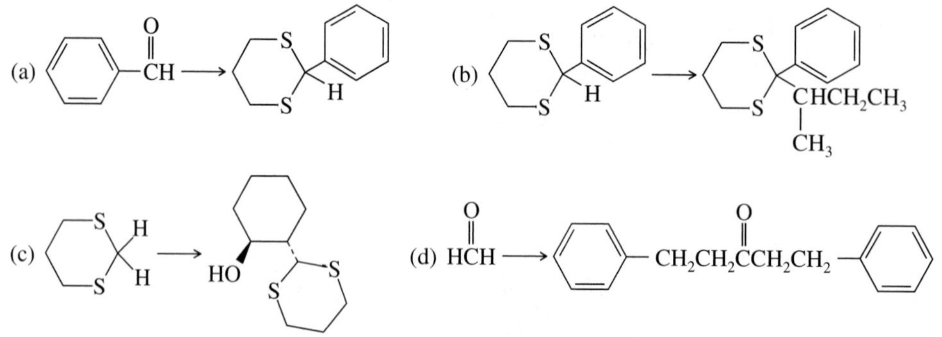

## 21.5   Carbon Nucleophiles in Syntheses

### A. Reactions of Organometallic Reagents with Carboxylic Acids and Their Derivatives

Acid derivatives react with organometallic reagents in acylation reactions. If the carbonyl compound that results from the acylation reaction is protected in some way by the reagent, the reaction stops at that stage. Otherwise, the carbonyl group—aldehyde or ketone—that forms in the reaction reacts further with the organometallic reagent to give an alcohol.

Two equivalents of Grignard reagent react with one equivalent of ester to give an alcohol. A classic application of this reaction is the preparation of triphenylmethanol from methyl benzoate and phenylmagnesium bromide prepared from bromobenzene (p. 550).

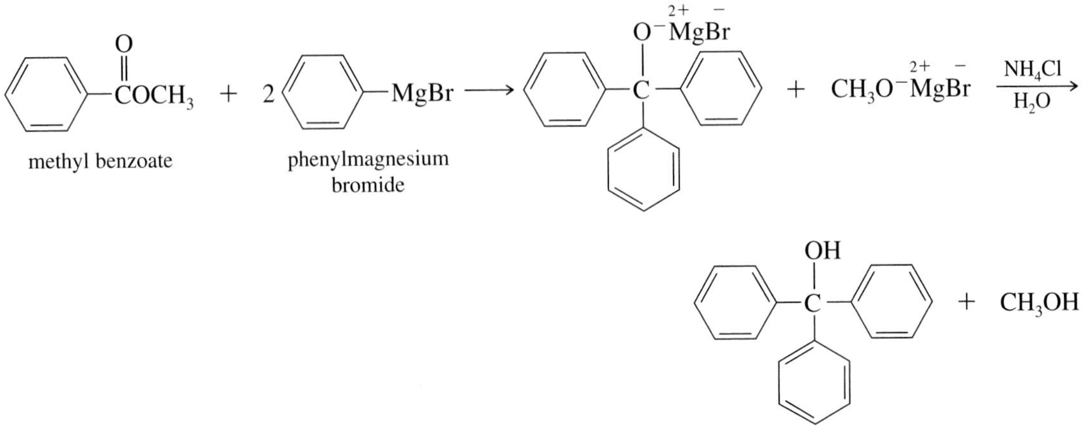

methyl benzoate + phenylmagnesium bromide → ... + triphenylmethanol ~ 90%

The first addition of the Grignard reagent to the carbonyl group of the ester leads to an intermediate that loses methoxide ion, thus regenerating a carbonyl group, which reacts again with the Grignard reagent.

## VISUALIZING THE REACTION

### Reaction of a Grignard Reagent with an Ester

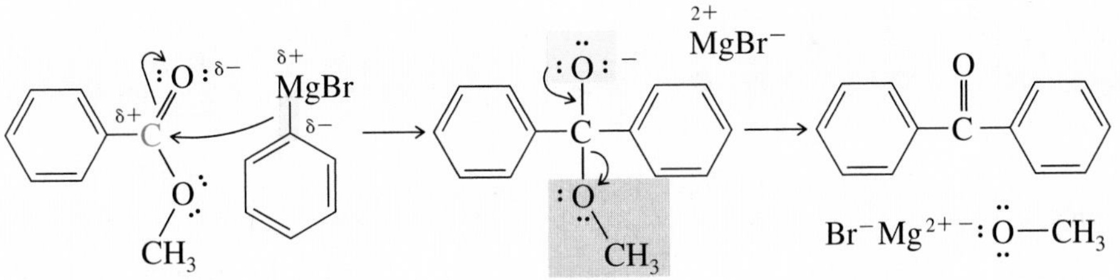

*nucleophilic attack at the carbonyl group*

*loss of an alkoxide ion*

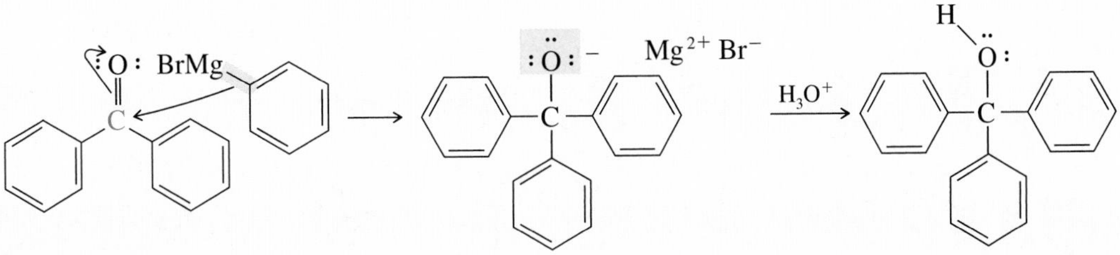

*nucleophilic attack at the new carbonyl group*

alkoxide ion complexed with metal

alcohol after protonation in a second step

The resulting complex contains an alkoxide ion, which is protonated when aqueous acid is added to give the free alcohol. You should be familiar with all these steps from the reactions of aldehydes and ketones with Grignard reagents (p. 552). The reactions of esters and those of aldehydes and ketones differ because of the presence in an ester of the alkoxyl group, which can act as a leaving group.

Reactions of Grignard reagents with most esters give tertiary alcohols in which at least two identical groups are bonded to the tertiary carbon atom. In triphenylmethanol, all three groups are the same because an ester with a benzoyl group was used as a starting material. If a formate ester is used, a secondary alcohol with two identical groups bonded to the secondary carbon atom results, as in the synthesis of 5-nonanol.

$$CH_3CH_2CH_2CH_2Br + Mg \xrightarrow[\text{ether}]{\text{diethyl}} CH_3CH_2CH_2CH_2MgBr$$

n-butyl bromide                    n-butylmagnesium bromide

$$2\ CH_3CH_2CH_2CH_2MgBr + \overset{\overset{\displaystyle O}{\|}}{H}COCH_2CH_3 \xrightarrow[\text{ether}]{\text{diethyl}} CH_3CH_2CH_2CH_2\overset{\overset{\displaystyle O^-MgBr^-\ ^{2+}}{|}}{C}HCH_2CH_2CH_2CH_3 + CH_3CH_2O^-\overset{2+}{M}gBr^-$$

n-butylmagnesium bromide    ethyl formate

$$\downarrow H_3O^+$$

$$CH_3CH_2CH_2CH_2\overset{\overset{\displaystyle}{|}}{C}HCH_2CH_2CH_2CH_3$$
$$\underset{OH}{}$$

5-nonanol
84%

---

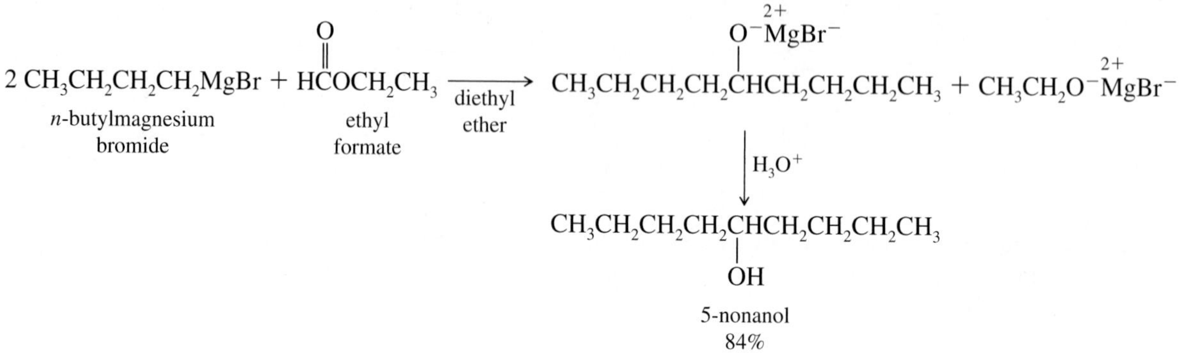

**Problem 21.25**    Suppose you want to synthesize 3-ethyl-3-pentanol with a radioactive $^{14}C$ atom as the tertiary carbon atom. Carbon dioxide labeled with $^{14}C$, $^{14}CO_2$, is readily available. Devise a synthesis for 3-ethyl-3-pentanol-3-$^{14}C$.

**Problem 21.26**    It is possible to use a Grignard reagent to make triphenylmethanol by a different route from the one shown on page 875. Work out another synthesis.

---

N-Methoxy-N-methylamides of carboxylic acids react cleanly with Grignard reagents to give ketones instead of alcohols. For example, N-methoxy-N-methylbenzamide gives acetophenone as the product in greater than 90% yield even when a large excess of Grignard reagent is present.

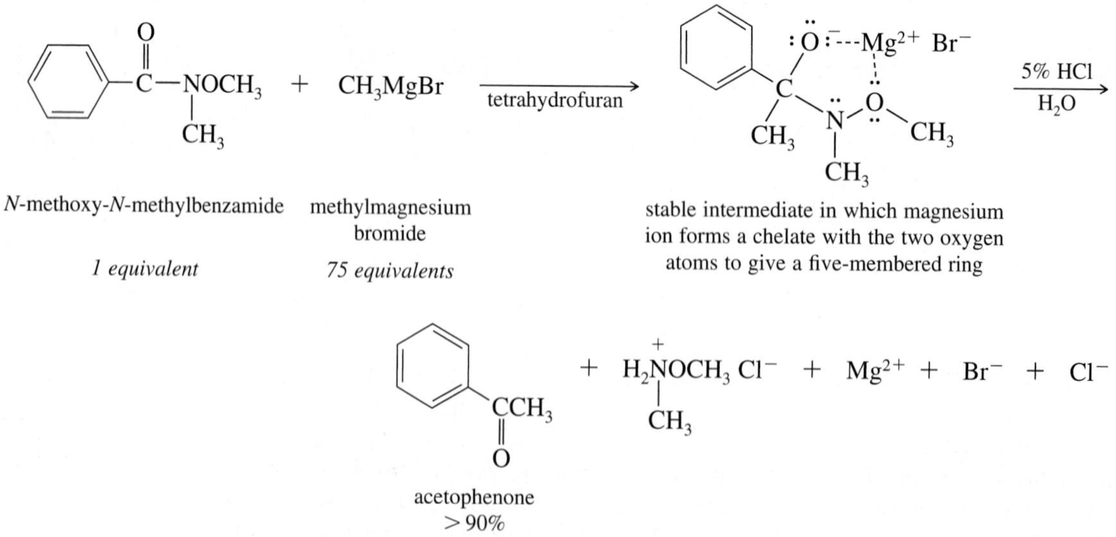

N-methoxy-N-methylbenzamide    methylmagnesium bromide

*1 equivalent*       *75 equivalents*

stable intermediate in which magnesium ion forms a chelate with the two oxygen atoms to give a five-membered ring

$$+\ H_2\overset{+}{N}OCH_3\ Cl^-\ +\ Mg^{2+}\ +\ Br^-\ +\ Cl^-$$
$$\underset{CH_3}{}$$

acetophenone
> 90%

Unlike unsubstituted amides (p$K_a$ ~15), N-methoxy-N-methylamides have no acidic hydrogens on the nitrogen atom and therefore do not destroy the Grignard reagent. Instead, the methoxy group on the nitrogen atom is in a position to complex with the magnesium ion in the first step of the addition of the Grignard reagent to the carbonyl group to form a chelate (p. 862). The chelate stabilizes the intermediate so that it does not decompose to give a ketone, which can react with more Grignard reagent. In addition, the stability of the intermediate is helped by another factor. An amide has a poorer leaving group on it than an ester does. Only when dilute acid is added does the chelate break down to give a ketone. At that point the acid has destroyed the excess Grignard reagent, so no further reaction can take place.

N-Methoxy-N-methylamides react with organolithium reagents to give ketones in the same way as they do with Grignard reagents. The lithium ion can coordinate to two oxygen atoms, just as magnesium ion can, so the reaction is believed to go through a similar intermediate. For example, N-methoxy-N-methylcyclohexanecarboxamide reacts with butyllithium to give butyl cyclohexyl ketone.

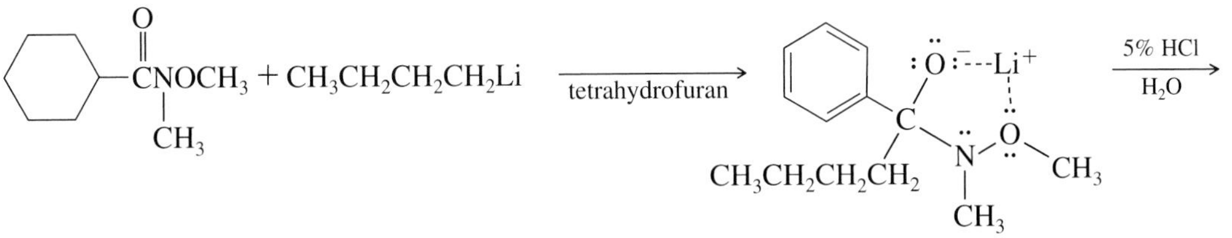

N-methoxy-N-methyl-        butyllithium                          stable intermediate
cyclohexanecarboxamide

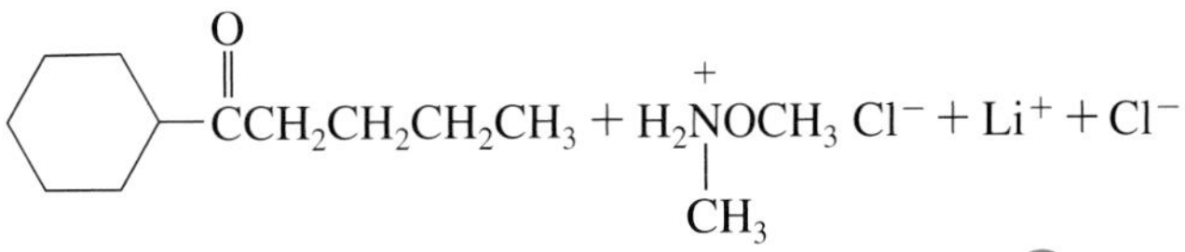

butyl cyclohexyl ketone
86%

Study Guide
Concept Map 21.7

**Problem 21.27**    Predict the major organic products of the following reactions.

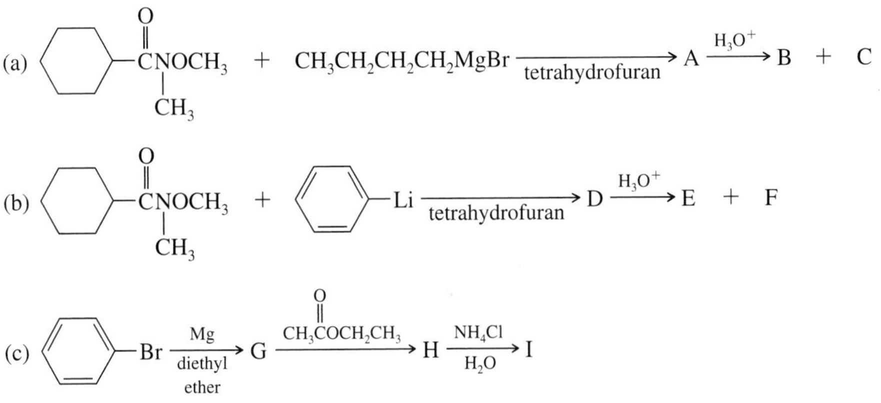

(a) ...

(b) ...

(c) ...

**ONE SMALL STEP**

Grignard reagents react with acid chlorides to give ketones if the Grignard reagent is added at low temperature to a solution of the acid chloride so that the acyl halide is always in excess.

**PROBLEM:** Choose an acid chloride and a Grignard reagent. Write a mechanism showing them reacting to give a ketone. Why is the temperature kept low? Why is the solution of the Grignard reagent added to that of the acyl halide instead of the other way around, as is usually the case?

## B. Reactions of Enolate Anions Revisited

In Chapter 17 we saw how enolate anions could be alkylated (Sections 17.2 and 17.3) and react with the carbonyl groups of aldehydes and ketones (the aldol reaction, Section 17.4A) or with the carbonyl groups of acid derivatives (Claisen-type reactions, Section 17.4B). All these reactions were powerful tools for the creation of carbon–carbon bonds, often leading to compounds with multiple functional groups.

 **On the Web: ONE SMALL STEP**

Sections 17.6B and 17.6C showed how biologically important processes and the biosyntheses of many natural products also involve aldol and Claisen-type reactions.

In this section we will review the reactions of enolate anions as they are used in the synthesis of more complex molecules. In Section 21.4B (p. 865) we discussed the regioselectivity of enolate anion formation. As chemists have gained access to more powerful bases, they have created dianions from active methylene compounds and used them for regioselective reactions of the less highly substituted enolate ion. An example from a synthesis of a component of the glandular secretions of various insects follows.

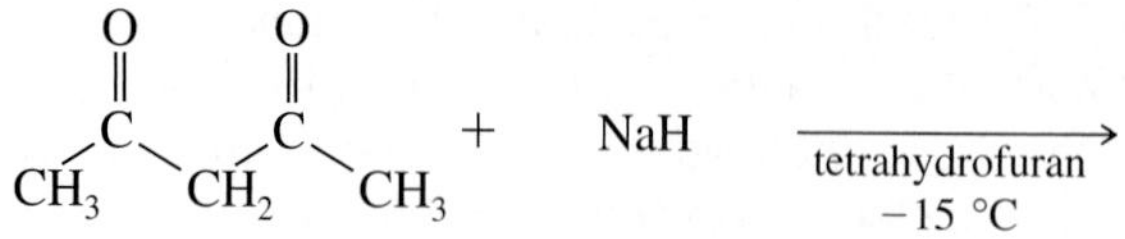

2,4-pentanedione          sodium hydride

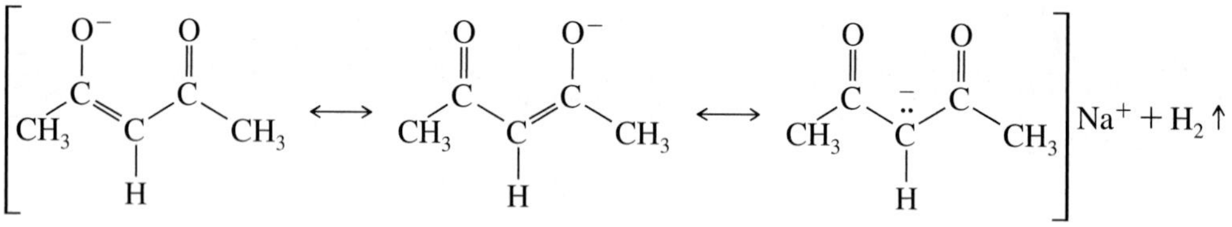

enolate ion from deprotonation
at the active methylene group

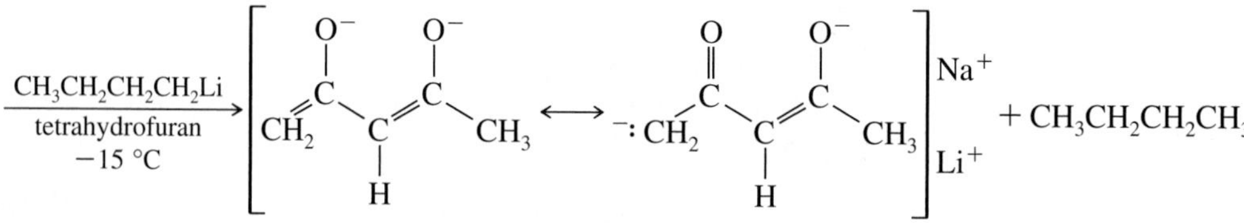

dianion from a second
deprotonation at
the less acidic site

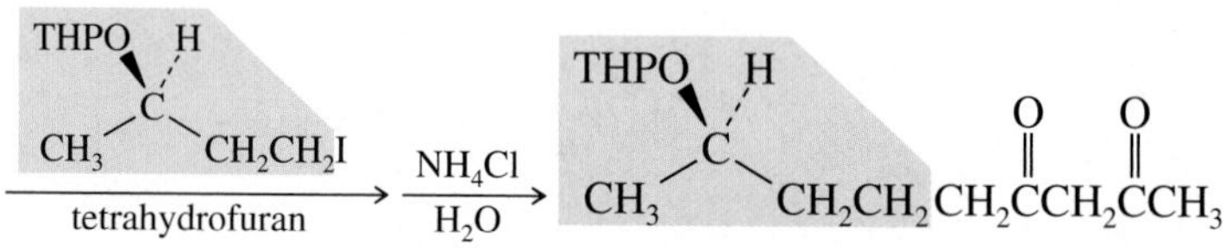

product of alkylation at the
more basic, more nucleophilic site

The two deprotonation reactions occur one after the other to give a dianion that has different basicity, and therefore nucleophilicity, at two sites. The negative charge on the anion derived from deprotonation at the active methylene group is delocalized to two oxygen atoms. The second enolate ion, in which the negative charge is delocalized to only one oxygen atom, is the more basic and more nucleophilic. In the alkylation step, the new carbon–carbon bond is formed to the end carbon of the chain and not to the active methylene group.

**Study Guide**
**Concept Map 17.2**

**Problem 21.28**    Give structural formulas for all species symbolized by letters in the following equation.

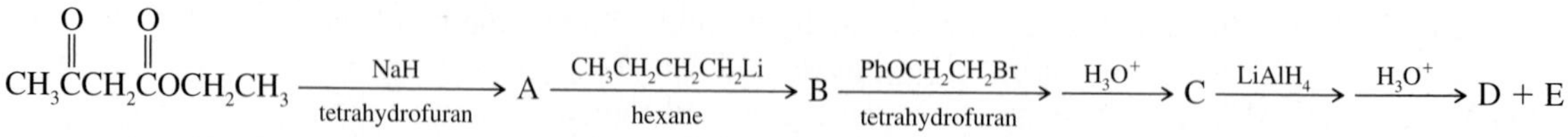

**Problem 21.29**    The chemistry of malonic esters plays an important role in attempts to synthesize the sex pheromones of cockroaches. Some of the steps from a synthesis are shown below. Give structural formulas for Compounds A to E.

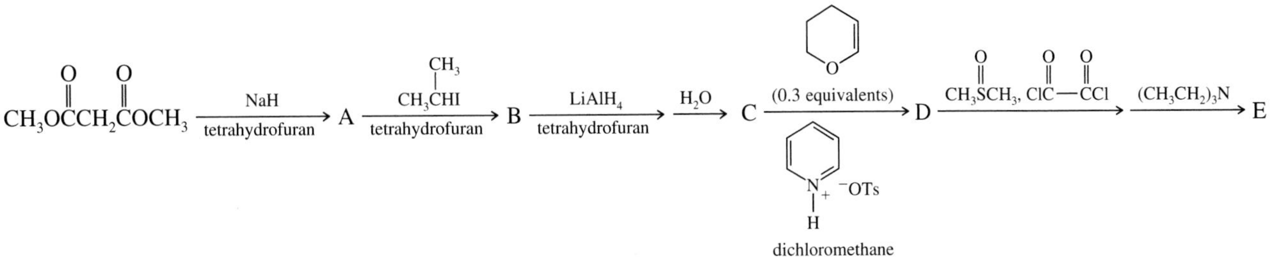

Intramolecular reactions of enolate ions are important in the formation of cyclic compounds. For example, if the alkyl halide function is in the same molecule as the enolate ion, an intramolecular reaction can take place. Intramolecular alkylation reactions of enolate anions have been used to make fused rings. In the example shown below, cyclization of a disubstituted cyclopentane gives two different fused-ring compounds depending on whether the reaction conditions used favor the kinetic enolate or the thermodynamic enolate.

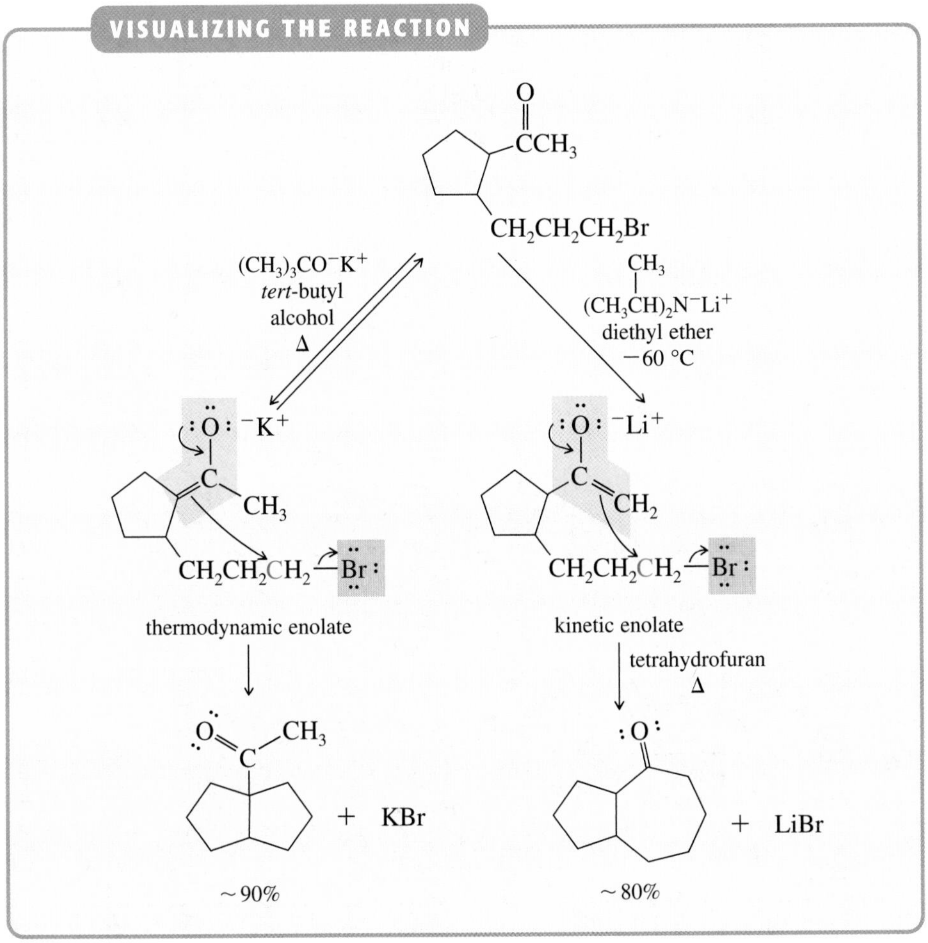

**VISUALIZING THE REACTION**

Similarly, compounds that have two carbonyl groups and an enolizable hydrogen atom next to at least one of them will undergo intramolecular aldol reactions. For example, 2,3-undecanedione, when treated with base, gives 3-methyl-2-pentyl-2-cyclopenten-1-one, also known as dihydrojasmone, a compound with the odor of jasmine.

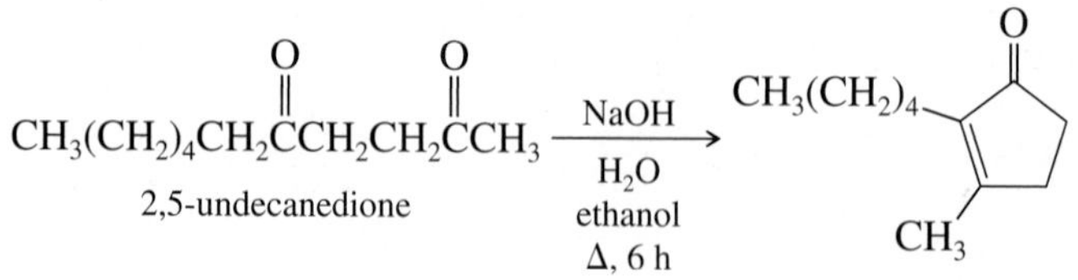

$$CH_3(CH_2)_4CH_2\overset{\overset{\displaystyle O}{\|}}{C}CH_2CH_2\overset{\overset{\displaystyle O}{\|}}{C}CH_3 \xrightarrow[\substack{H_2O \\ ethanol \\ \Delta,\,6\,h}]{NaOH}$$

2,5-undecanedione

3-methyl-pentyl-
2-cyclopenten-1-one
dihydrojasmone
85%

There are four sets of enolizable $\alpha$-hydrogens in 2,5-undecanedione. Enolization of any one of the hydrogen atoms on the two carbons between the carbonyl groups would give enolate ions that would have to form three-membered rings in attacking a carbonyl group. The other two sets of $\alpha$-hydrogens give enolate ions that can reach a carbonyl group to give a five-membered ring. The product that actually forms is the one that has the more highly substituted double bond in the final product.

**VISUALIZING THE REACTION**

**An Intramolecular Aldol Reaction**

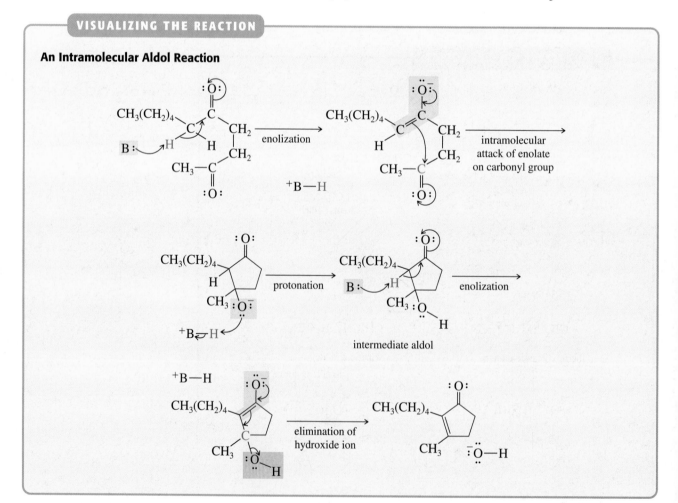

The intermediate aldol product is dehydrated on continued heating with base (p. 702) to give an $\alpha,\beta$-unsaturated ketone.

⊘ **Study Guide**
**Concept Map 17.3**

**Problem 21.30** It has been discovered that it is possible to prepare dianions of carbonyl compounds if very strong bases are used (p. 878). One such dianion derived from a substituted acetoacetic ester was used in another synthesis of dihydrojasmone. The synthesis is outlined below, and some hints are given. Supply structural formulas for the intermediates and products designated by letters.

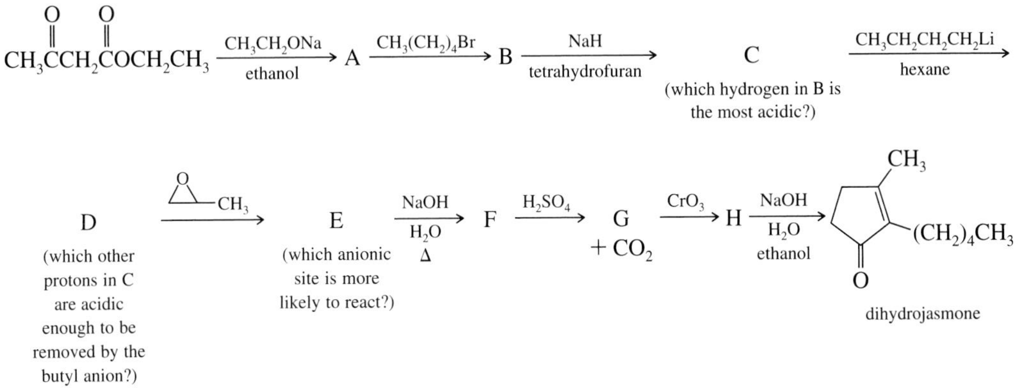

$$\underset{\text{O} \quad\quad \text{O}}{CH_3CCH_2COCH_2CH_3} \xrightarrow[\text{ethanol}]{CH_3CH_2ONa} A \xrightarrow[\text{tetrahydrofuran}]{CH_3(CH_2)_4Br} B \xrightarrow[\text{tetrahydrofuran}]{NaH} C \xrightarrow[\text{hexane}]{CH_3CH_2CH_2CH_2Li}$$

(which hydrogen in B is
the most acidic?)

$$D \xrightarrow{\text{(epoxide)}-CH_3} E \xrightarrow[H_2O]{NaOH} F \xrightarrow{H_2SO_4} G \xrightarrow{CrO_3} H \xrightarrow[\text{ethanol}]{NaOH} \text{dihydrojasmone}$$

D
(which other
protons in C
are acidic
enough to be
removed by the
butyl anion?)

E
(which anionic
site is more
likely to react?)

$\Delta$

$+ CO_2$

dihydrojasmone

What other product is possible in the last step of the synthesis? How do you explain the regioselectivity of the reaction?

A classic intramolecular Claisen condensation takes place if two ester functions are in the same molecule and separated by four or five carbon atoms. Then the carbonyl group of one is in an ideal position to accept the enolate anion created alpha to the other. A cyclic $\beta$-ketoester is the product because of the ease with which five- and six-membered rings form (p. 500). This variation of the Claisen condensation is called the **Dieckmann condensation** and is best illustrated by the formation of ethyl 2-oxocyclopentanecarboxylate.

**VISUALIZING THE REACTION**

**The Dieckmann Condensation**

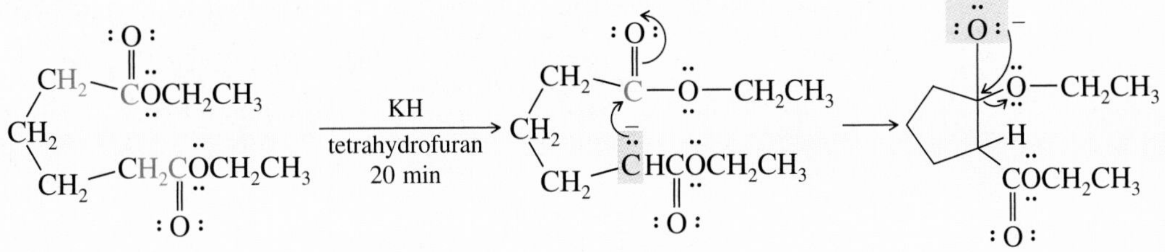

diethyl hexanedioate

enolate anion
attacking carbonyl
group in an
intramolecular
reaction

tetrahedral
intermediate losing
alkoxide anion

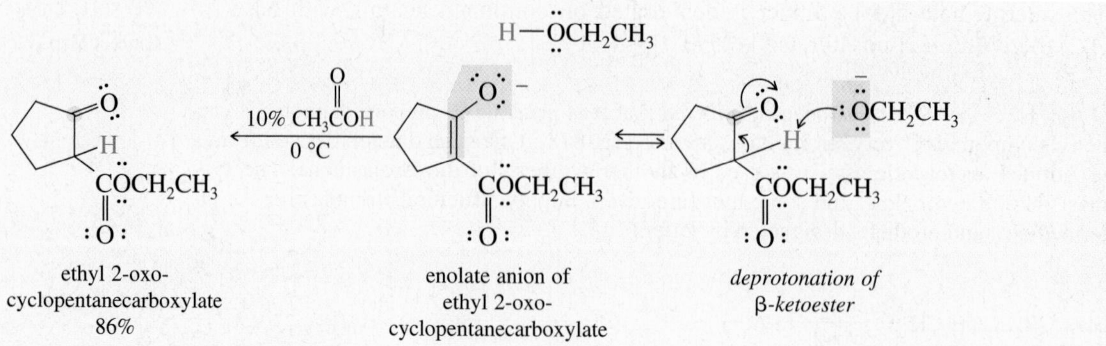

ethyl 2-oxo-
cyclopentanecarboxylate
86%

enolate anion of
ethyl 2-oxo-
cyclopentanecarboxylate

deprotonation of
β-ketoester

Study Guide
Concept Map 17.4

All the factors that affect the Claisen condensation also affect the Dieckmann condensation. The reaction is reversible, and the β-ketoester is stable in basic solution only as an enolate anion.

**Problem 21.31**    (a) Write a mechanism for the reaction shown below. (*Hint:* Reviewing the questions in the Art of Solving Problems section on pages 335–338 may be helpful.)

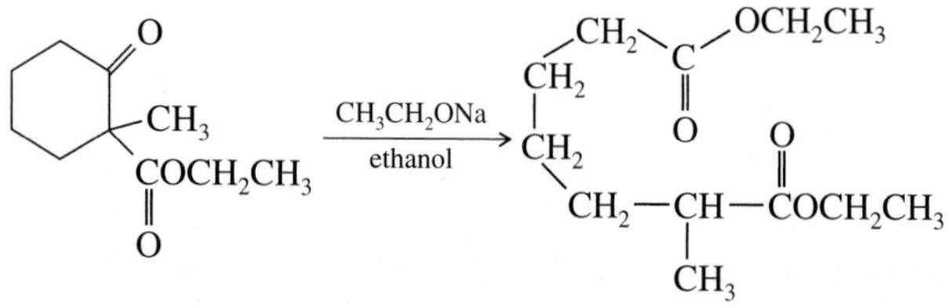

(b) When the diester produced in the above reaction is treated with 1 equivalent of sodium hydride, carefully excluding any alcohol, ethyl 3-methyl-2-oxocyclohexanecarboxylate is formed in 90% yield. Explain why that compound, and not the original ethyl 1-methyl-2-oxocyclohexanecarboxylate, is formed under these conditions.

Conjugate additions of organocuprate reagents to α,β-unsaturated carbonyl systems result in enolate ions that can react in all the ways described earlier, including alkylation of the enolate (p. 865). If the functional group that can react with the enolate is within the same molecule, intramolecular reactions to give cyclic compounds can result here too. For example, an organocuprate reagent made with copper(I) cyanide instead of iodide reacts with an α,β-unsaturated diester to give first conjugate addition and then a Claisen-type reaction to give a cyclic product.

$$2 \, LiCH{=}CH_2 + CuCN \xrightarrow[\text{ether}]{\text{diethyl}} (CH_2{=}CH)_2Cu(CN)Li_2$$

vinyllithium     copper(I)
                 cyanide

diethyl
ether

organocuprate reagent

$(CH_2{=}CH)_2Cu(CN)Li_2 \; +$

dimethyl (*E*)-2-hexenedioate

$\xrightarrow[\text{ether}]{\text{diethyl}} \xrightarrow[\text{H}_2\text{O}]{\text{NH}_4\text{Cl}}$

methyl *trans*-2-oxo-5-vinyl-
cyclopentanecarboxylate
80%

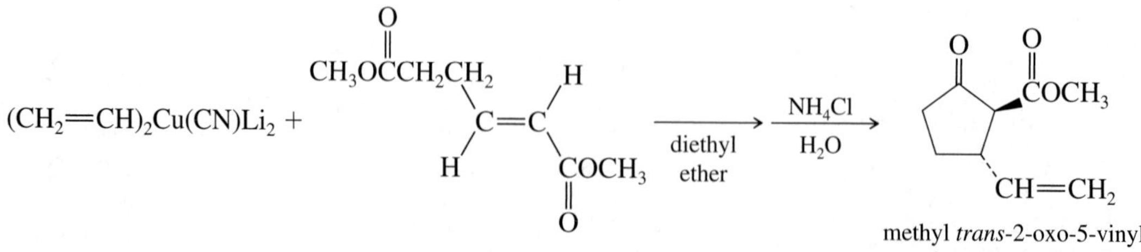

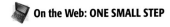

Study Guide
Concept Map 17.6

On the Web: ONE SMALL STEP

**Problem 21.32** Pentalenene is a hydrocarbon related to pentalenolactone, an antibiotic that inhibits the synthesis of nucleic acids in bacterial cells. Part of a synthesis of pentalenene involves the following reaction. Write structural formulas for the compounds indicated by letters.

$$\text{structure} \xrightarrow[\text{tetrahydrofuran}]{(CH_2\!=\!CHCH_2CH_2)_2CuLi} A \xrightarrow{H_3O^+} B$$
$$C_{14}H_{21}ClO$$

## C. Some Case Studies

In this section we will look at some applications of the reactions of carbon nucleophiles to the synthesis of biologically interesting compounds. Our first example comes from studies leading to the synthesis of bafilomycin A, an antibiotic with broad antibacterial and antifungal properties. The compound has chains of carbon atoms bearing oxygen functional groups in precise stereochemical relationships to each other. The researchers, William Roush and his coworkers, chose a simple model compound shown below as a target for investigating the reaction conditions that would lead to the product.

An examination of the compound shows that it is a β-hydroxycarbonyl compound and thus could be synthesized by an aldol condensation.

carbonyl compound
electrophile

enolate anion
nucleophile

ketone

The synthesis of the target compound takes place as expected by treating the ketone with a strong base, the conjugate base of hexamethyldisilazane, $(CH_3)_3SiNHSi(CH_3)_3$, in the presence of the aldehyde.

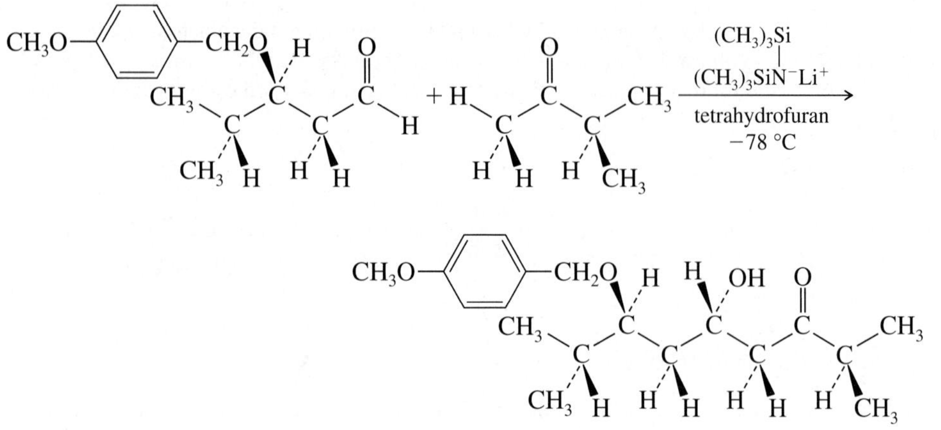

94%
3:1 this stereoisomer

**Problem 21.33**    What commonly used base does the conjugate base of hexamethyldisilazane remind you of?

**Problem 21.34**    Another diastereomer of the product shown above is also formed in this reaction as the minor product. What is it?

In another study, researchers were interested in studying the conformations of the large molecules RNA and DNA (p. 1016) by $^{13}C$ nuclear magnetic resonance (p. 390). In order to do this more easily than relying on the natural low percentage of $^{13}C$ in carbon compounds, they synthesized ribose and deoxyribose (p. 971) labeled with $^{13}C$ at carbon 5. Small molecules such as $CO_2$ and $CH_3I$ enriched with $^{13}C$ are available commercially. The synthesis has to take advantage of such reagents. A target molecule in their synthesis and a retrosynthetic analysis leading back to $^{13}CH_3I$ follows.

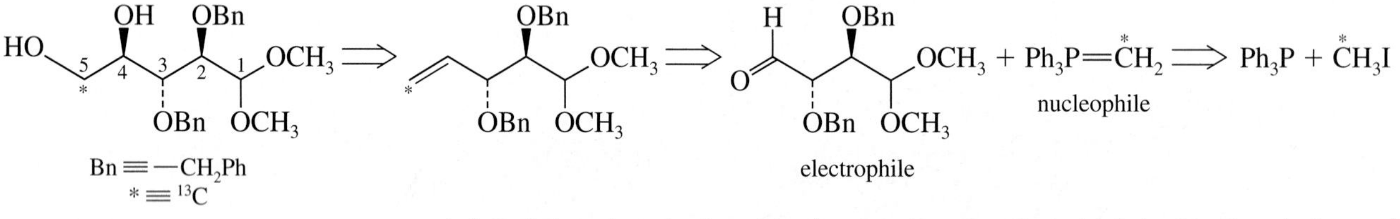

A 1,2-diol such as is found in the target molecule (which is the dimethyl acetal of ribose, protected at two of the hydroxyl groups as the benzyl ethers) can be made by treating an alkene with osmium tetroxide (p. 313). A precise placement of a double bond, especially since we need to put a labeled carbon atom exactly at carbon-5, is best done by a Wittig reaction. The Wittig reagent is prepared from the labeled methyl iodide and triphenylphosphine (p. 868). The synthesis follows.

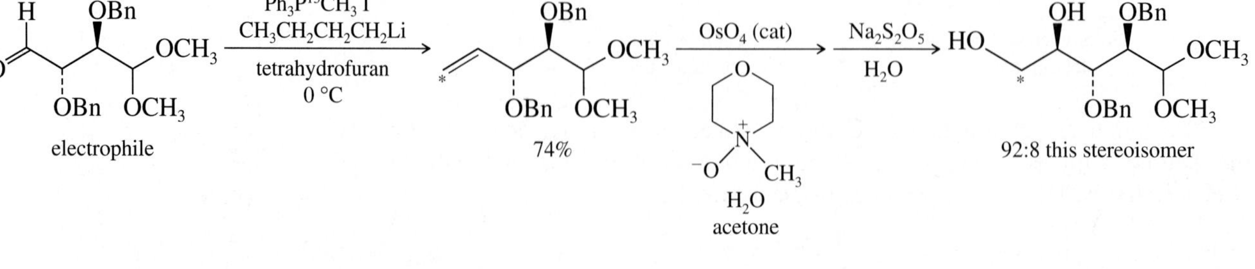

**Problem 21.35** How would you convert the target molecule synthesized above to ribose labeled at carbon-5? See page 971 for the structure of ribose.

Our final example comes from work toward the synthesis of apoptolidin, an antibiotic that is of great interest because of its ability to induce apoptosis, spontaneous cell death, in cells that have been transformed into tumor cells but not in normal cells. In studies toward the synthesis of this antibiotic, K. C. Nicolaou and his colleagues made the following aldehyde, using a chiral oxirane as the source of stereochemistry.

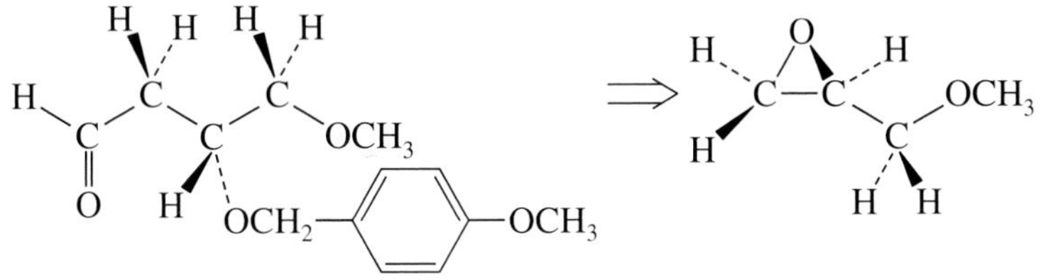

The aldehyde may be seen as containing two hydroxyl groups protected as ethers. One hydroxyl group has been protected with the *p*-methoxybenzyl group, which is easily removable (p. 847). The other oxygen function is a methoxy group, which is much more difficult to convert to a hydroxyl group (p. 502). A dissection of the aldehyde and a comparison with the oxirane shows us that the oxirane contains all the carbon and oxygen atoms (except for those in the benzylic protecting group) other than the carbonyl group.

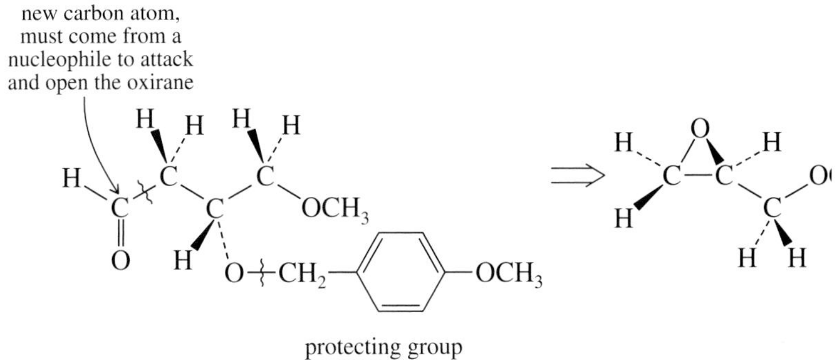


A further consideration reminds us that oxiranes are attacked and opened by nucleophiles (pp. 504 and 554). A carbonyl group is usually an electrophilic center but can be hidden in a nucleophilic dithiane anion, which reacts with oxiranes (p. 874).

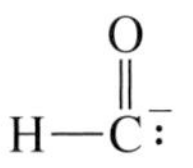
a synthon showing the carbon of the carbonyl group as a nucleophile

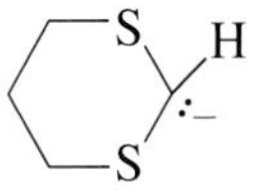
the nucleophilic reagent that masks a carbonyl group

The synthesis is shown below.

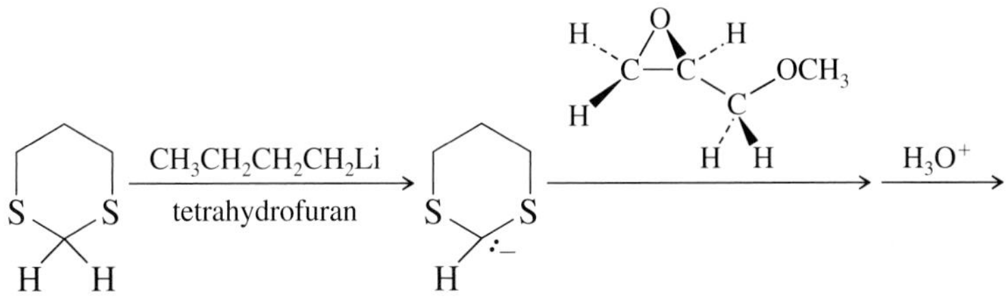

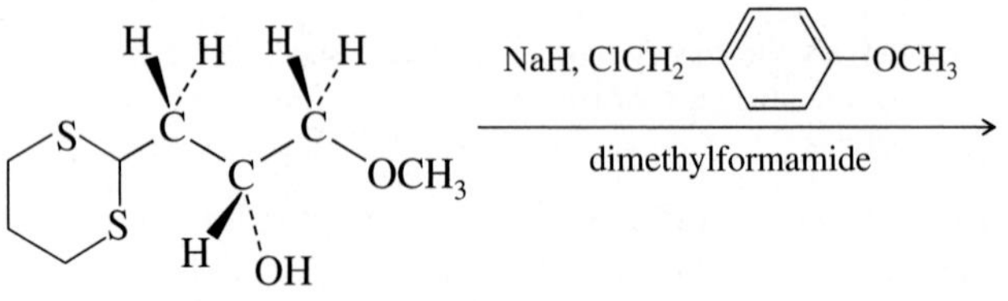

91%

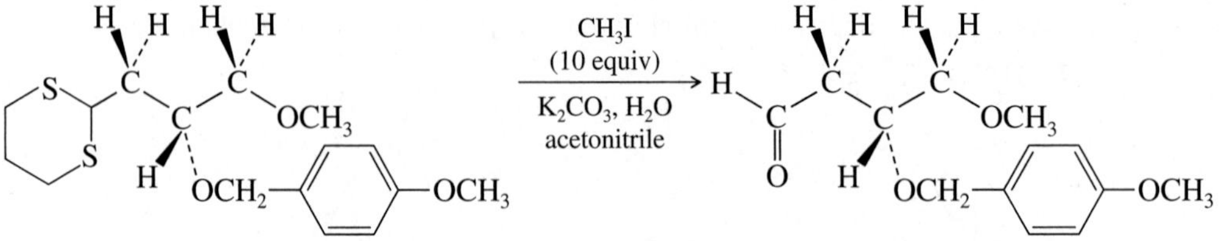

99%                                                                92%

The last step, the removal of the dithiane, uses methyl iodide instead of a mercury salt to react with the sulfur to allow for hydrolysis of the dithiane. How this happens is the subject of the One Small Step on page 873.

**Problem 21.36**    The following differently protected alcohol-ether was synthesized as an intermediate in the preparation of an alkaloid.

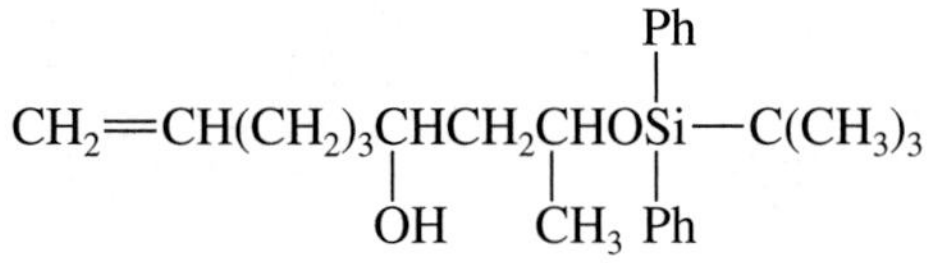

$$CH_2{=}CH(CH_2)_3CHCH_2CHOSi{-}C(CH_3)_3$$
$$\quad\quad\quad\quad\quad\quad | \quad\quad\ | \quad |$$
$$\quad\quad\quad\quad\quad\quad OH \quad CH_3\ Ph$$

with Ph above the Si.

The reagents that were available for the synthesis are

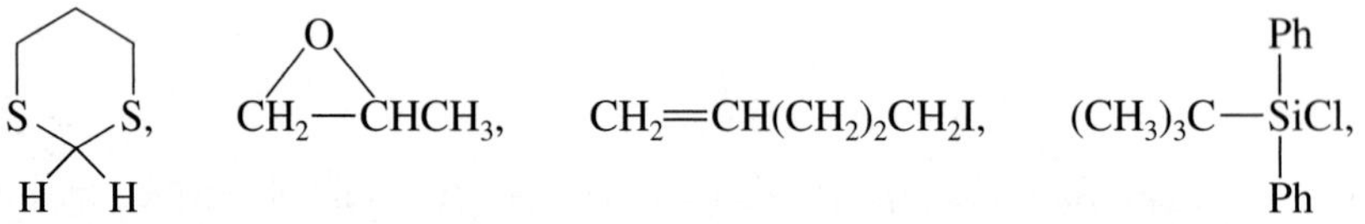

and any other common solvents, acids or bases, and oxidizing agents or reducing agents. How would you prepare the alcohol?

## 21.6    Diels–Alder Reactions

### A. Diels–Alder Reactions of Unsymmetrical Dienes and Dienophiles

The examples of Diels-Alder reactions discussed in Section 18.4 have involved symmetrical dienes or dienophiles, so the question of regioselectivity has not come up. However, a Diels-Alder reaction does have high regioselectivity as well as high stereoselectivity. An example of a reaction between an unsymmetrically substituted diene and dienophile is shown on the next page.

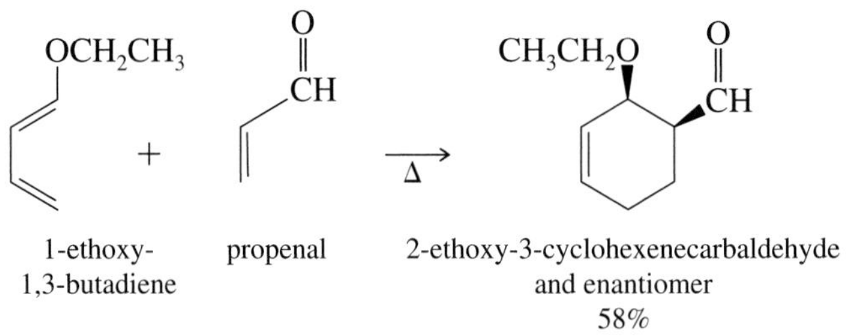

| 1-ethoxy-1,3-butadiene | propenal | 2-ethoxy-3-cyclohexenecarbaldehyde and enantiomer 58% |

The experimental observation above indicates that when a diene substituted on the first carbon atom reacts with an unsymmetrical dienophile, the major product usually has the two substituents of the new ring on adjacent carbons. The stereochemistry shown for the product comes from endo addition (p. 745). More about how to figure this out will come later.

The regioselectivity observed in Diels-Alder reactions can be rationalized for the example shown above by writing resonance contributors for the diene and dienophile and comparing the transition states. The course of the reaction of 1-ethoxy-1,3-butadiene with propenal is quite easy to explain in this way.

---

**VISUALIZING THE REACTION**

**Regioselectivity of the Diels-Alder Reaction**

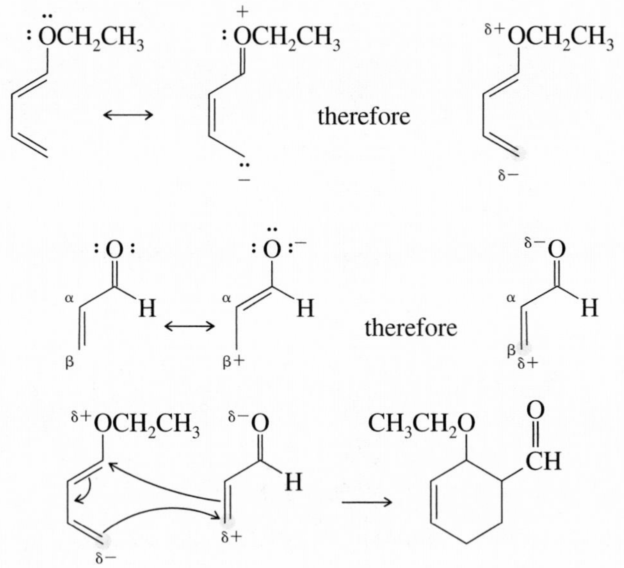

---

Polarization of 1-ethoxy-1,3-butadiene is most likely to increase electron density at carbon 4 of the diene. The $\beta$-carbon atom of the $\alpha,\beta$-unsaturated carbonyl compound, meanwhile, is deficient in electrons. In the transition state, the molecules line up so that the partial charges interact with each other, and the product in which the ethoxyl and aldehyde groups are next to each other is the result.

If the substituent is on the second carbon atom of the diene, most frequently another type of product is obtained.

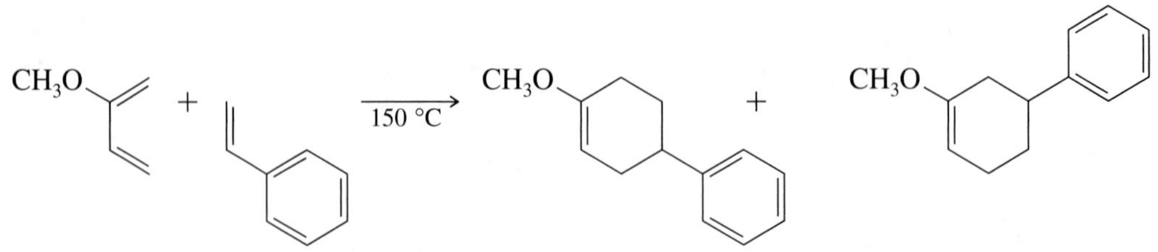

CH₃O ... + ... phenylethene ... $\xrightarrow{150\ °C}$ ... CH₃O ... 1-methoxy-4-phenylcyclohexene ... + ... CH₃O ... 2-methoxy-4-phenylcyclohexene

2-methoxy-1,3-butadiene     phenylethene     1-methoxy-4-phenylcyclohexene    2-methoxy-4-phenylcyclohexene
                                                     12      :              1

In this case, the cyclohexene ring that is formed usually has the substituents located at positions 1 and 4 on the ring. These products are more difficult to rationalize simply. Nevertheless, theoretical considerations outlined in Section 25.2B do make remarkably good predictions that the 1,4-orientation of the substituents should be favored. It is sufficient for you to know that this is the experimental result in most cases.

The addition of 2-methoxy-1,3-butadiene to alkenes is an excellent way to synthesize substituted cyclohexanones. The Diels-Alder products in these cases are enol ethers and are easily hydrolyzed to the corresponding ketones (Problem 14.13, p. 562).

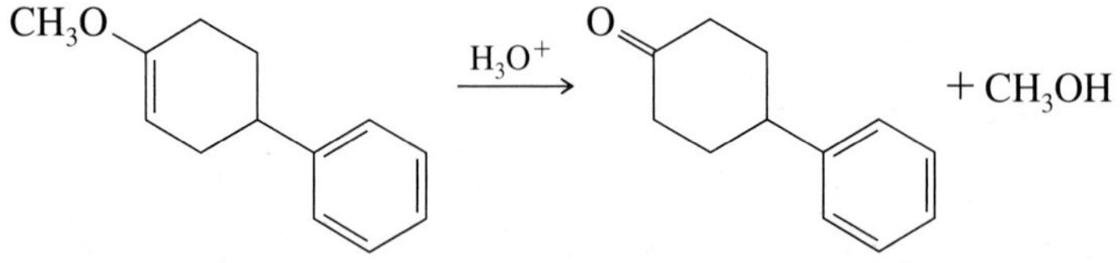

CH₃O ... $\xrightarrow{H_3O^+}$ ... O ... $+\ CH_3OH$

1-methoxy-4-phenylcyclohexene        4-phenylcyclohexanone
                                          88%

A diene that shows particularly good reactivity and regioselectivity was designed by Samuel Danishefsky and is now known as Danishefsky's diene. The compound is 1-methoxy-3-[(trimethylsilyl)oxy]-1,3-butadiene.

OCH₃ ... $\equiv$ ... OCH₃

(CH₃)₃SiO ... TMSO

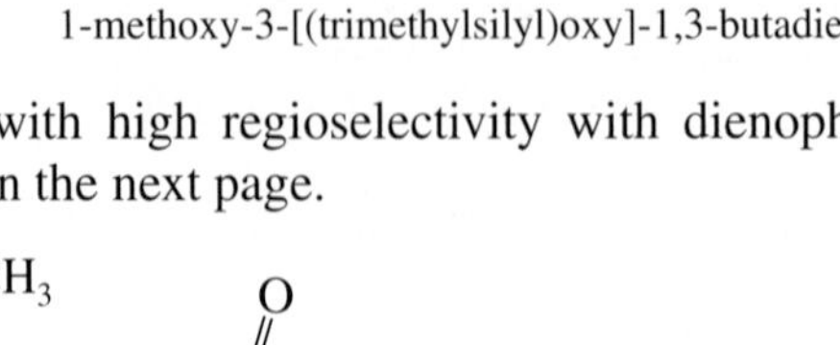

$(CH_3)_3Si\ \equiv\ TMS$

1-methoxy-3-[(trimethylsilyl)oxy]-1,3-butadiene

This diene reacts with high regioselectivity with dienophiles. Two examples are shown below and on the next page.

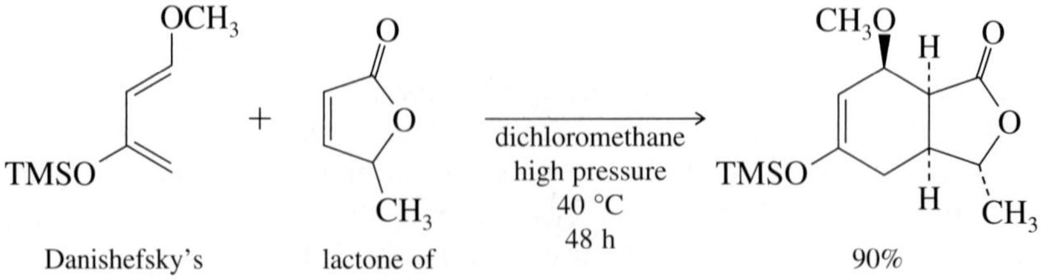

## ONE SMALL STEP

Some dienes, called push-pull dienes, are designed to be polarized so that the one end of the diene is higher in electron density and the other, lower.

PROBLEM: An example of a push-pull diene is shown below with the dienophile with which it reacts. Predict the regioselectivity of the Diels-Alder reaction of these two compounds.

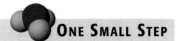

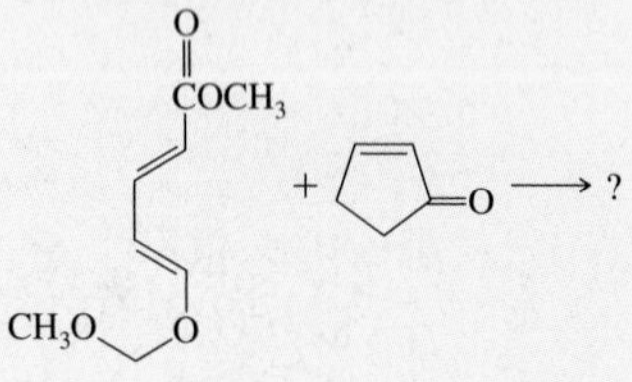

On the Web: ONE SMALL STEP

OCH₃ ... O ... + ... O ... $\xrightarrow[\substack{\text{high pressure}\\ 40\ °C\\ 48\ h}]{\text{dichloromethane}}$ ... CH₃O ... H ... O ... TMSO ... H ... CH₃

TMSO ... CH₃

Danishefsky's diene     lactone of (Z)-4-hydroxy-2-pentenoic acid                  90%

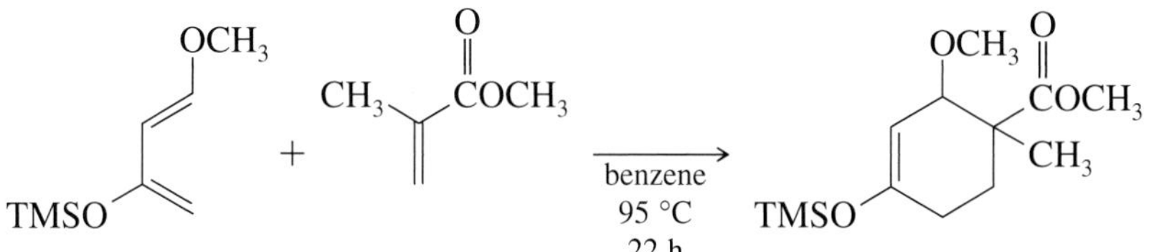


Danishefsky's diene

methyl 2-methylpropenoate

methyl methacrylate

a single product, stereochemistry not determined

In Danishefsky's diene, both oxygen substituents donate electrons toward one end of the conjugated system, polarizing the diene so that it reacts regiospecifically with a dienophile.

**VISUALIZING THE REACTION**

### Polarization of Danishefsky's Diene

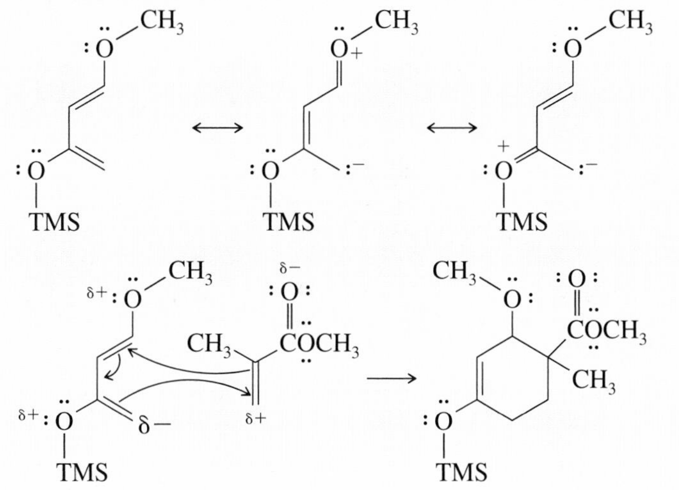


The diene is useful in another way. Products of Diels-Alder reactions with Danishefsky's diene can be converted easily to $\alpha,\beta$-unsaturated ketones by the use of very dilute acid. For example,

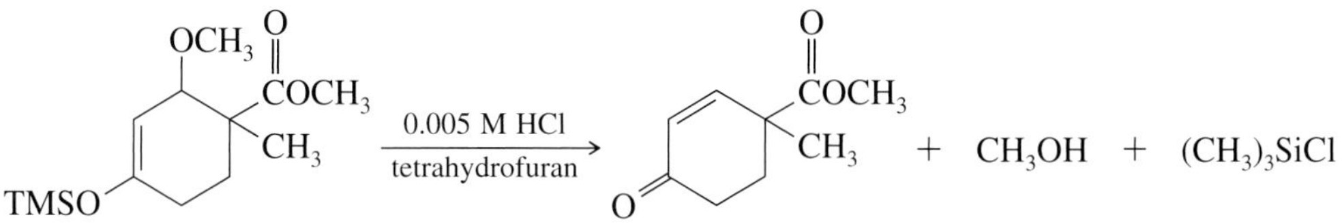


This hydrolysis reaction is similar to the one seen for 1-methoxy-4-phenylcyclohexene on page 888 in which an enol ether is converted to a ketone. The difference is that the presence of the second oxygen function on the ring allows the introduction of a double bond as well as a ketone group. Under the conditions shown above, the reaction is believed to start by protonation of the methoxy group.

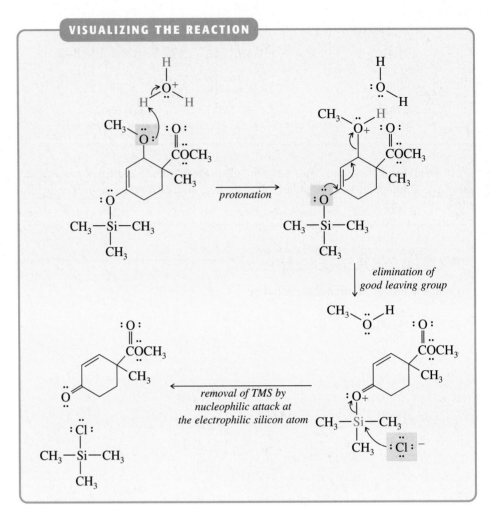

VISUALIZING THE REACTION

⊙ Study Guide
Concept Map 21.8

 **On the Web: ONE SMALL STEP**

---

**Problem 21.37**    Assign structures to the major products designated by letters in the following reactions.

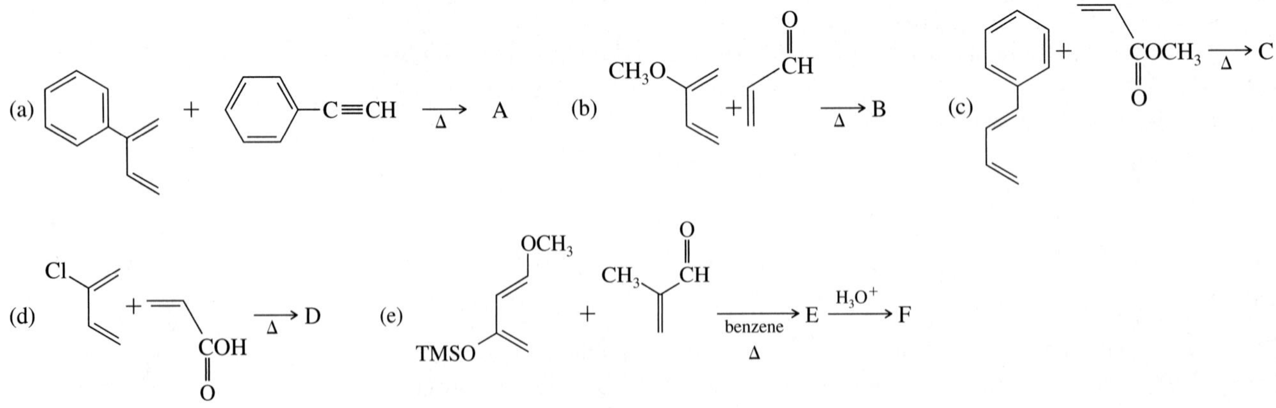

(a) [styrene derivative] + [phenylacetylene] $\xrightarrow{\Delta}$ A    (b) $CH_3O$-diene + [methyl vinyl ketone] $\xrightarrow{\Delta}$ B    (c) [phenylbutadiene] + [methyl vinyl ketone] $\xrightarrow{\Delta}$ C

(d) [chlorodiene] + [acrylic acid] $\xrightarrow{\Delta}$ D    (e) [TMSO/OCH₃ diene] + [dienophile] $\xrightarrow[\Delta]{\text{benzene}}$ E $\xrightarrow{H_3O^+}$ F

---

**Problem 21.38**    What reagents would be required to obtain each Diels-Alder adduct shown on the next page?

(a)

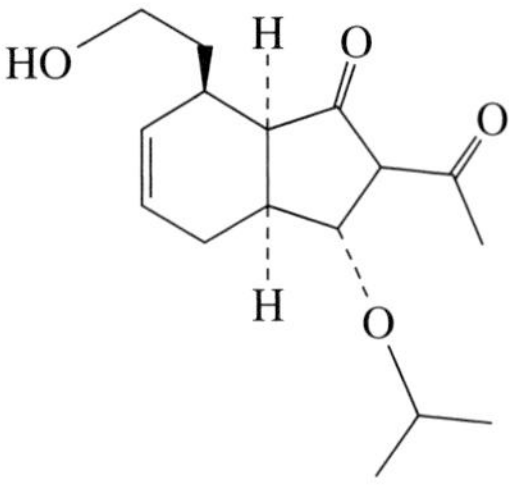

(b)

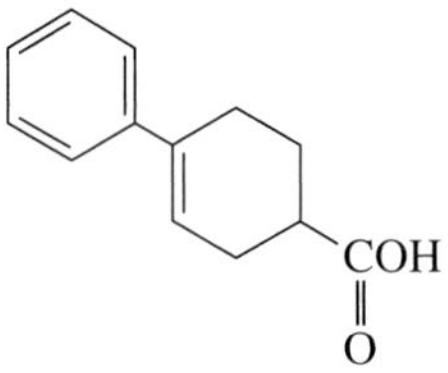

(c)

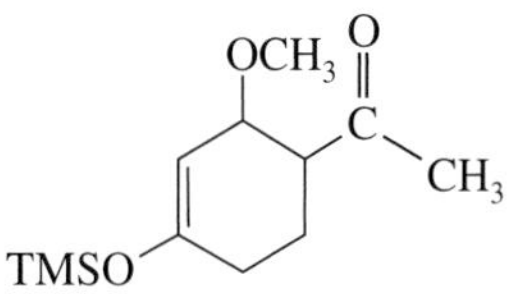

(d)

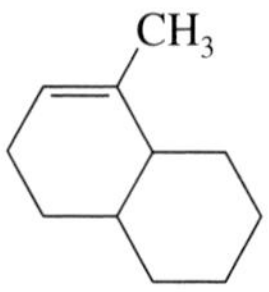

## B. A Review of the Stereochemistry of the Diels–Alder Reaction

In Sections 18.4B and 18.4C we saw that substituents on a dienophile retain their relative stereochemistry in the product of a Diels-Alder reaction. When a cyclic diene reacts with a dienophile with electron-withdrawing groups such as carbonyl, cyano, and nitro groups on it, the major product has this group in the endo orientation; that is, this group is in the cavity created by the bicyclic structure and not the side of the bridge, the exo orientation. For example, cyclopentadiene reacts with methyl propenoate (methyl acrylate) to give mostly the endo bicyclic ester.

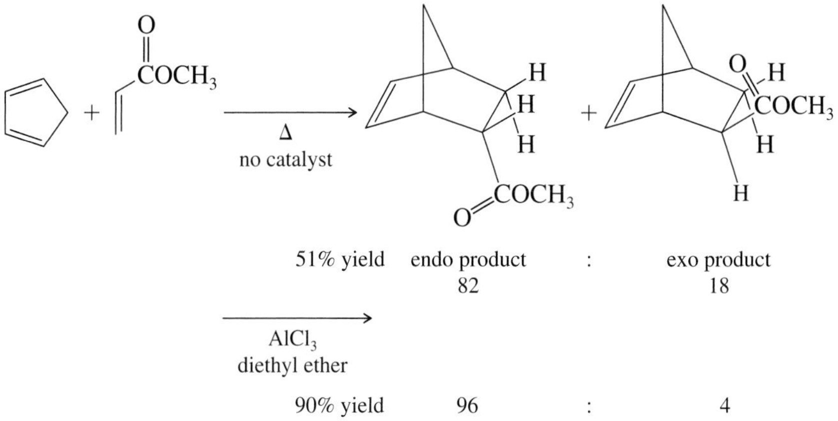

|  | 51% yield | endo product | : | exo product |
|---|---|---|---|---|
|  |  | 82 |  | 18 |

|  | 90% yield | 96 | : | 4 |
|---|---|---|---|---|

AlCl₃ diethyl ether

The preference for the endo stereochemistry arises from the most favorable interactions between the molecular orbitals of the diene and the dienophile in the transition state (see Figure 25.6, p. 1088). This preference (as well as the yield of the products) is improved, as shown above, when a Lewis acid (aluminum chloride in this case) that can complex with the carbonyl group is used.

If we have two substituents on the diene that can give rise to stereochemistry in the product, they, too, retain their stereochemical relationship to each other. For example, (1*E*, 3*E*)-1,4-diphenyl-1,3-butadiene reacts with diethyl acetylenedicarboxylate to give a compound in which the phenyl groups are cis to each other.

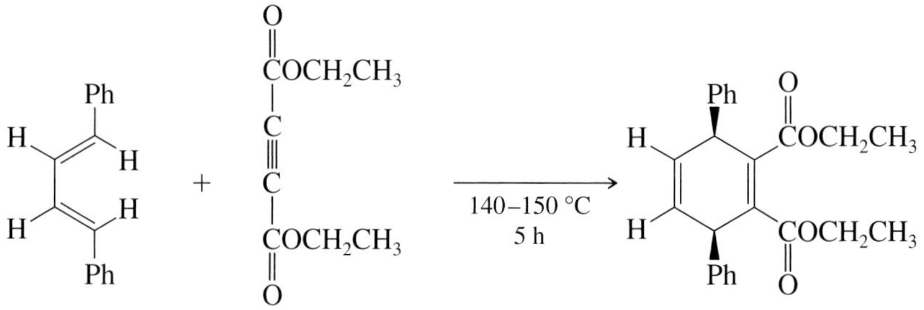

(1*E*, 3*E*)-1,4-diphenyl-1,3-butadiene

diethyl acetylenedicarboxylate

140–150 °C 5 h

ethyl cis-3,6-diphenyl-1,4-cyclohexadiene-1,2-dicarboxylate 90%

In deciding which substituents are going to be on the same side of the newly formed cyclohexene ring, the concept of which substituents on the diene point in and which point out is a useful one. For example, if we compare (1*E*, 3*E*)-1,4-diphenyl-1,3-butadiene to cyclopentadiene, we see the following:

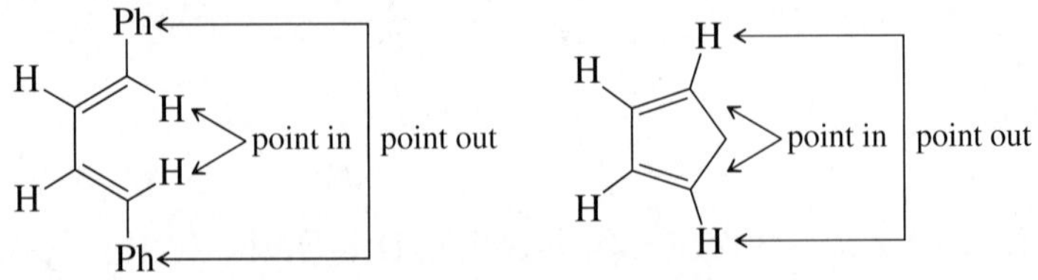

The groups that point out (or those that point in) will end up on the same side of the six-membered ring in the product, as we saw in the reaction of the 1,4-diphenylbutadiene with the acetylenedicarboxylic ester above.

The same kinds of interactions that work to give chiefly endo stereochemistry when a cyclic diene reacts with a dienophile also occur with noncyclic dienes, but the results are not as one-sided. For example, when 1,3-pentadiene reacts with methyl propenoate, a mixture of products is formed.

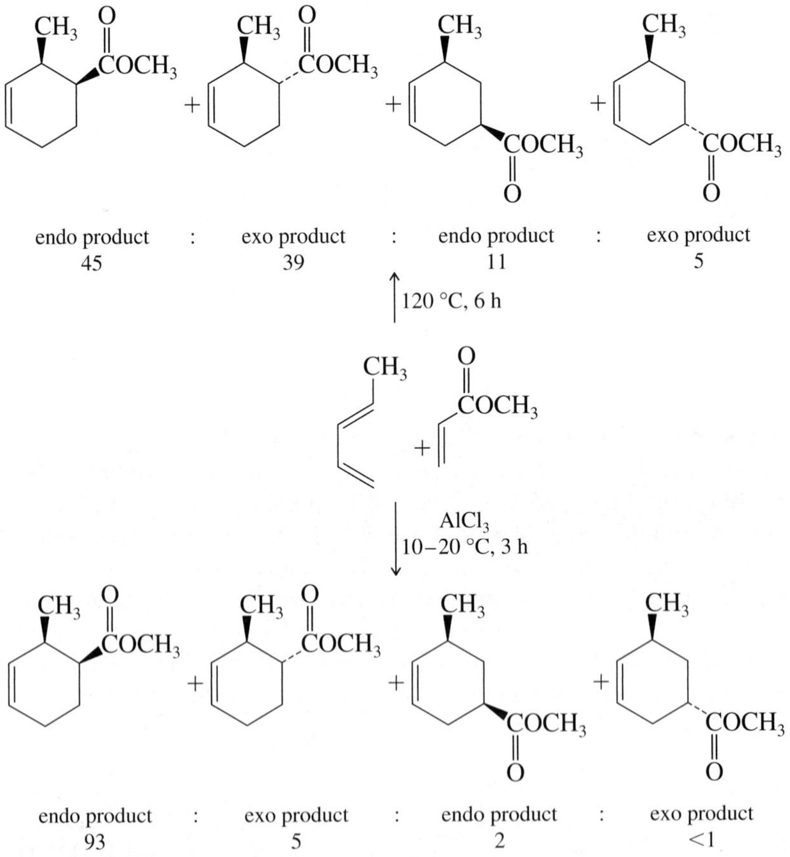

endo product : exo product : endo product : exo product
45 39 11 5

120 °C, 6 h

AlCl₃
10–20 °C, 3 h

endo product : exo product : endo product : exo product
93 5 2 <1

The major products come from the ortho orientation of the substituent on the diene to the substituent on the dienophile (p. 887), but the endo-exo ratio is not high, 45:39. If the reaction is run in the presence of a catalytic amount of Lewis acid (aluminum chloride in this case), the regioselectivity and the stereoselectivity improve greatly. Then most of the product, 93%, is ortho and endo.

How do we figure out that the product with the two substituents cis to each other has come from endo addition when the product is not bicyclic? Again, a look at the substituents that point in and those that point out on the diene is helpful.

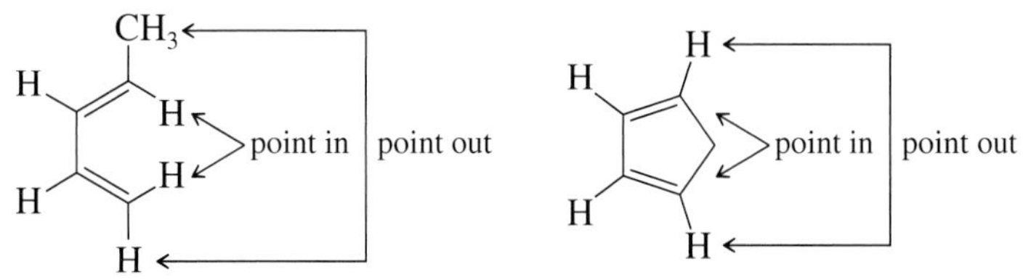

In the bicyclic product, the endo product has the substituent from the dienophile on the side of the bicyclic ring away from the bridge, which comes from the bonds to carbon in cyclopentadiene that point in.

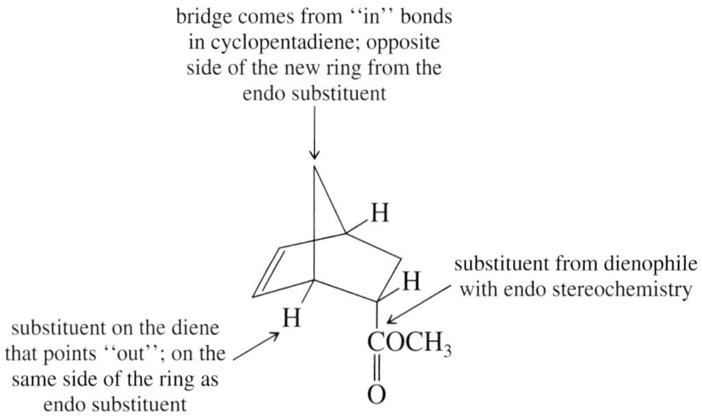

In other words, an "out" substituent on the diene ends up cis to the substituent from the dienophile when endo addition takes place. For the noncyclic diene, the same analysis holds.

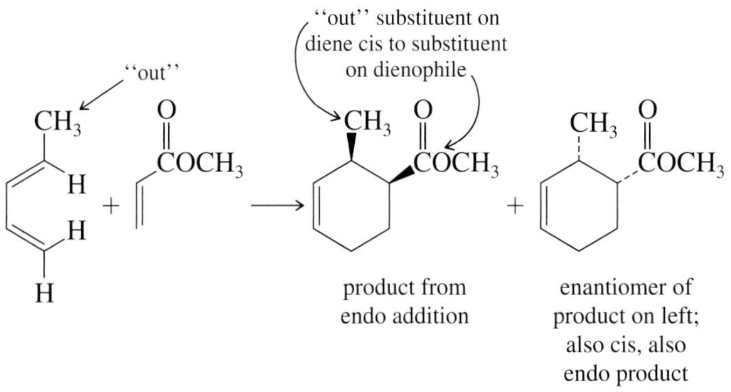

If there are other substituents on the dienophile, predictions about endo-exo ratios become more difficult, and we shall not attempt them.

**Problem 21.39**    The stereochemistry of the product of a Diels-Alder reaction is not always easy to determine. In one case, the decision about the stereochemistry of the reaction was based on the following set of reactions.

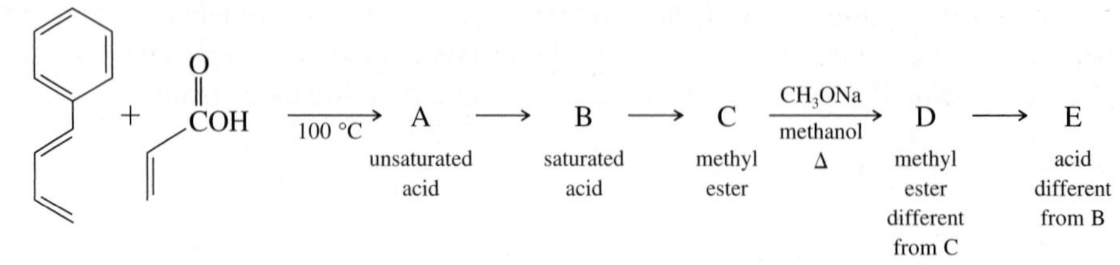

Supply reagents to carry out the transformations where they are missing, and write full structures, including stereochemistry, for Compounds A through E. What do these reactions prove about the stereochemistry of the Diels-Alder addition product, Compound A?

## 21.7 Synthetic Transformations in Aromatic Compounds

### A. Electrophilic Substitution Reactions of Multiply Substituted Aromatic Compounds

In Chapter 10 we determined the effect of a substituent on an aromatic ring on the reactivity of that ring toward further substitution and also on the regioselectivity of the reaction. We divided substituents into ring-activating, *o,p*-directing and ring-deactivating, *m*-directing categories (Table 10.1, p. 369). We will now look at reactions where there is more than one substituent on the ring.

When there are two substituents already on an aromatic ring and both of them act to direct an incoming electrophile to the same position, then predicting where the new substituent will end up is relatively easy. For example, *p*-nitrotoluene is brominated ortho to the methyl group.

4-nitrotoluene — 2-bromo-4-nitrotoluene 90%

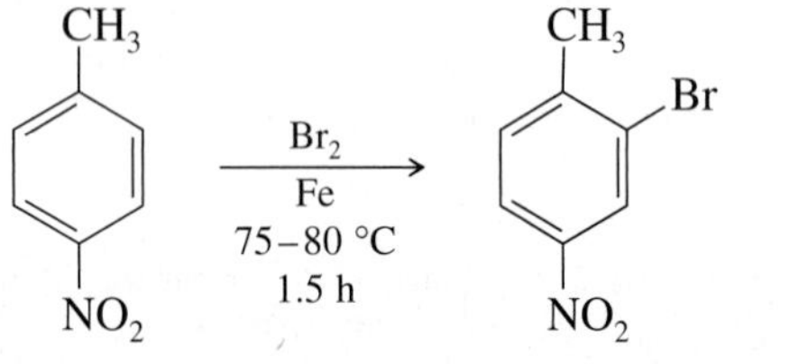

The methyl group is ortho,para-directing, and the nitro group is meta-directing. The bromine substitutes ortho to the methyl group and meta to the nitro group.

When the directing influences of the substituents already on the ring are in conflict, the one that is more strongly activating and ortho,para-directing determines the orientation of the new substituent. For example, 2-fluoromethoxybenzene is nitrated ortho and para to the methoxy group.

2-fluoromethoxybenzene — 2-fluoro-4-nitromethoxybenzene + 2-methoxy-3-nitrofluorobenzene

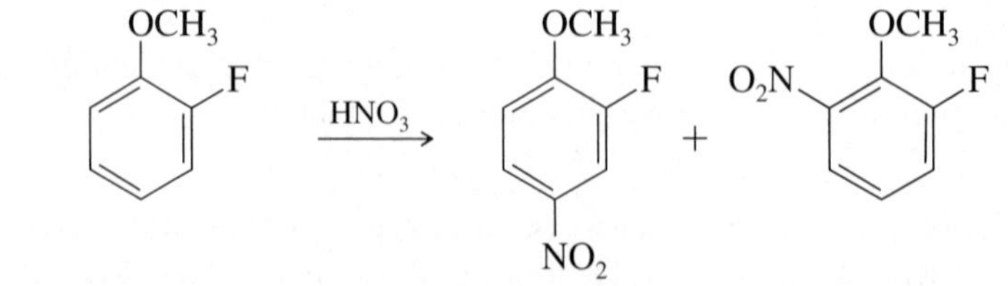

The following gives the relative effectiveness of ring-activating substituents in directing an incoming electrophile to the ortho or para position when two groups are in competition.

$$-NH_2, -OH, -O^- > -OCH_3 \geq -\overset{\overset{\text{O}}{\|}}{O}CR, -NH\overset{\overset{\text{O}}{\|}}{C}R > -X > -CH_3$$

An amino, hydroxy, or alkoxy group is most effective in directing an incoming electrophile ortho or para to itself. The methyl group, which does not have nonbonding electrons to stabilize carbocationic intermediates by resonance, is the least effective. For example, when 2-chlorotoluene is nitrated, the mixture that is formed shows that more substitution has occurred ortho and para to the chlorine than to the methyl group.

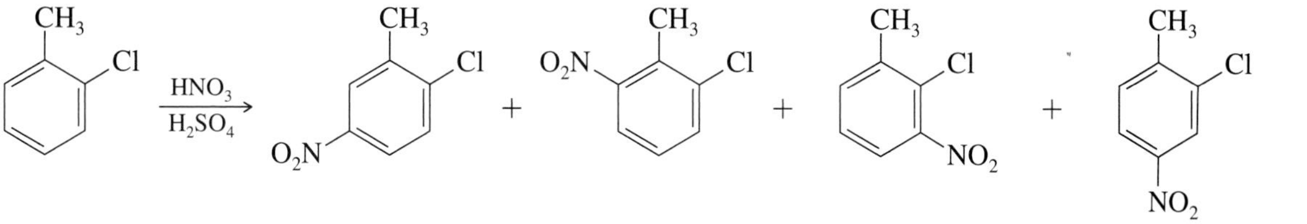

| 2-chlorotoluene | 2-chloro-5-nitrotoluene 43% | 2-chloro-6-nitrotoluene 21% | 2-chloro-3-nitrotoluene 19% | 2-chloro-4-nitrotoluene 17% |

This example also demonstrates the complexity of the mixtures that result when the directing effects of the groups on the ring are more or less equal.

**Problem 21.40**   Predict the major product(s) of the following reactions.

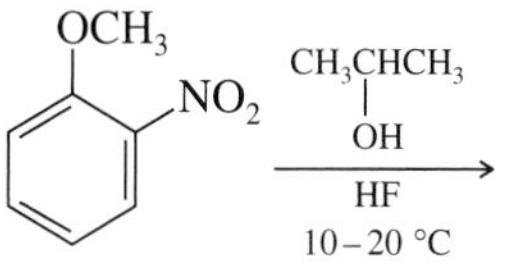

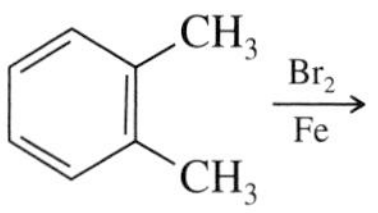

## B. Sandmeyer Reactions

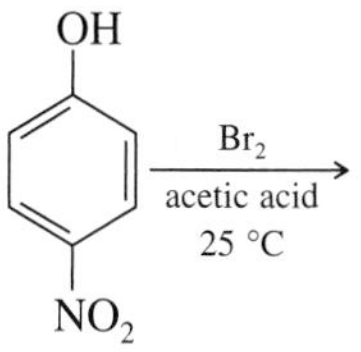

Chemists have discovered that it is possible to have precise placement of substituents such as the halogens, the hydroxyl group, and the cyano group on aromatic rings by preparing diazonium ions from the corresponding primary aryl amines and displacing nitrogen. This is possible because treating aromatic primary amines with nitrous acid gives aryl diazonium ions, which are stable enough in solution at low temperatures (p. 823) to be used as reagents in syntheses. A molecule of nitrogen is the leaving group that is replaced in a variety of substitution reactions on aryl diazonium salts. Some of the reactions require catalysis by copper metal or copper(I) salts and are known as **Sandmeyer reactions.** The exact mechanism of the substitution reactions has been difficult to establish and probably involves radical intermediates in some cases and ionic substitution reactions in others. The important fact to remember is that in each case a primary amine group on an aromatic ring is replaced selectively by a halogen, hydrogen, a cyano, or a hydroxyl group. This type of reaction has wide usefulness.

For example, when toluene is brominated, a mixture of isomers is formed, including o-bromotoluene, p-bromotoluene, and even benzyl bromide when the reaction is carried out in the light (p. 778). If pure o-bromotoluene is needed, the best way to make it is by diazotization of o-toluidine and treatment of the resulting o-toluenediazonium bromide with copper metal or copper(I) bromide.

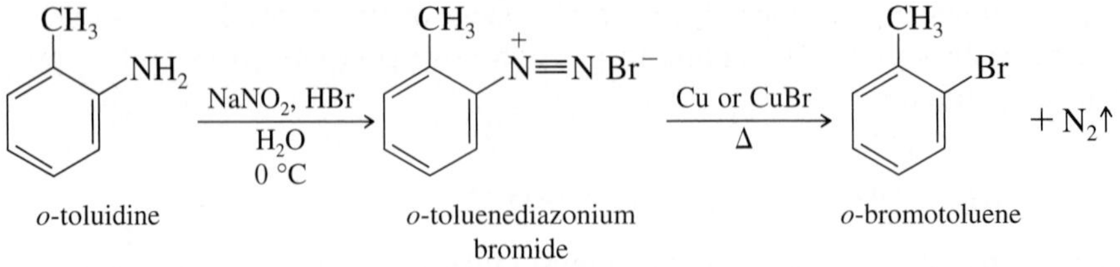

o-toluidine     o-toluenediazonium bromide     o-bromotoluene

Entirely analogous reactions are used to introduce chlorine regioselectively.

Elemental fluorine and iodine cannot be used for electrophilic aromatic substitution reactions. Aromatic fluoro and iodo compounds are prepared from the corresponding amines. Aniline, for example, is converted into fluorobenzene or iodobenzene.

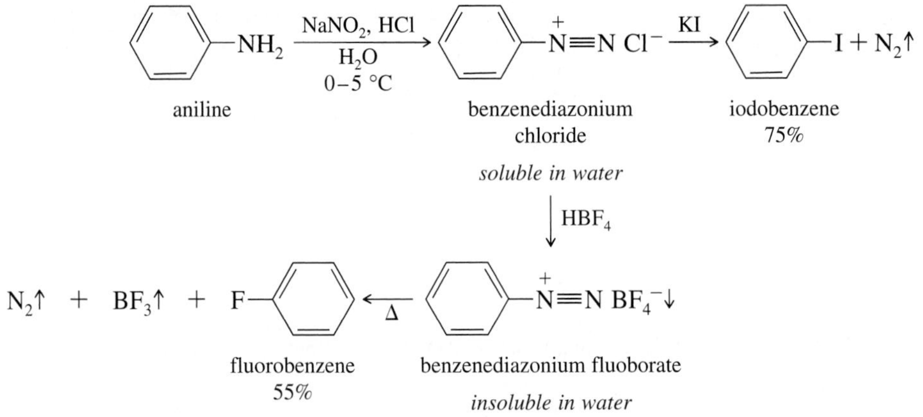

aniline     benzenediazonium chloride     iodobenzene 75%

*soluble in water*

fluorobenzene 55%     benzenediazonium fluoborate

*insoluble in water*

Iodide ion directly replaces the nitrogen in benzenediazonium chloride to give iodobenzene. To make fluorobenzene, it is necessary to replace the chloride ion in benzenediazonium chloride with fluoborate anion, by treating the diazonium salt with fluoboric acid. The resulting diazonium fluoborate is much less soluble in water than the chloride is, and it precipitates. It is isolated, dried, and decomposed by heating to give fluorobenzene, boron trifluoride, and nitrogen.

A cyano group is another group that cannot be introduced directly onto an aromatic ring because aryl halides do not undergo nucleophilic substitution reactions easily (p. 898). Copper(I) cyanide is used to replace nitrogen in an aryl diazonium salt with a cyano group. Once again, the reaction proceeds with complete regioselectivity, as is illustrated by the conversion of *p*-toluidine to *p*-toluonitrile.

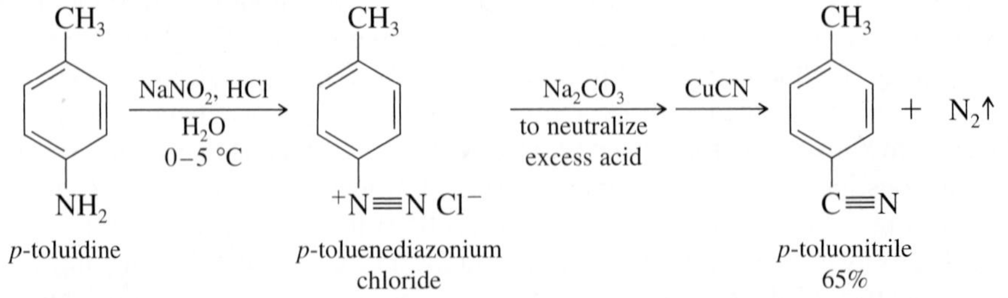

p-toluidine     p-toluenediazonium chloride     p-toluonitrile 65%

In all the reactions described above, phenols are produced as products of a side reaction in which the diazonium ion reacts with water. The replacement of nitrogen by a hydroxyl group can be made the chief reaction by preparing the diazonium salt in an acid with a conjugate base that is a particularly poor nucleophile and then heating the reaction mixture. For example, *m*-nitrophenol, which cannot be prepared by the direct nitration of phenol (p. 367), is made in this way. Note the use of sulfuric acid in the diazotization reaction.

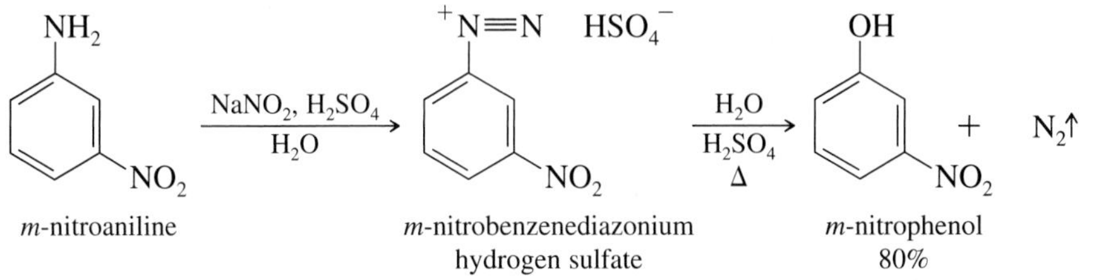

*m*-nitroaniline            *m*-nitrobenzenediazonium            *m*-nitrophenol
                               hydrogen sulfate                      80%

Finally, the amino group can be used to direct electrophilic aromatic substitution on the ring and can then be removed through diazotization and replacement by hydrogen in a reduction reaction. The preparation of 2,4,6-tribromobenzoic acid illustrates this sequence of reactions. Bromination of benzoic acid is expected to give *m*-bromobenzoic acid as the chief product because the carboxyl group is ring-deactivating and meta-directing (p. 368). The amino group in *m*-aminobenzoic acid, however, will activate the ring and direct bromine atoms ortho and para to itself. The amino group can then be removed by diazotization followed by reduction of the diazonium ion with hypophosphorous acid, $H_3PO_2$, a reducing agent.

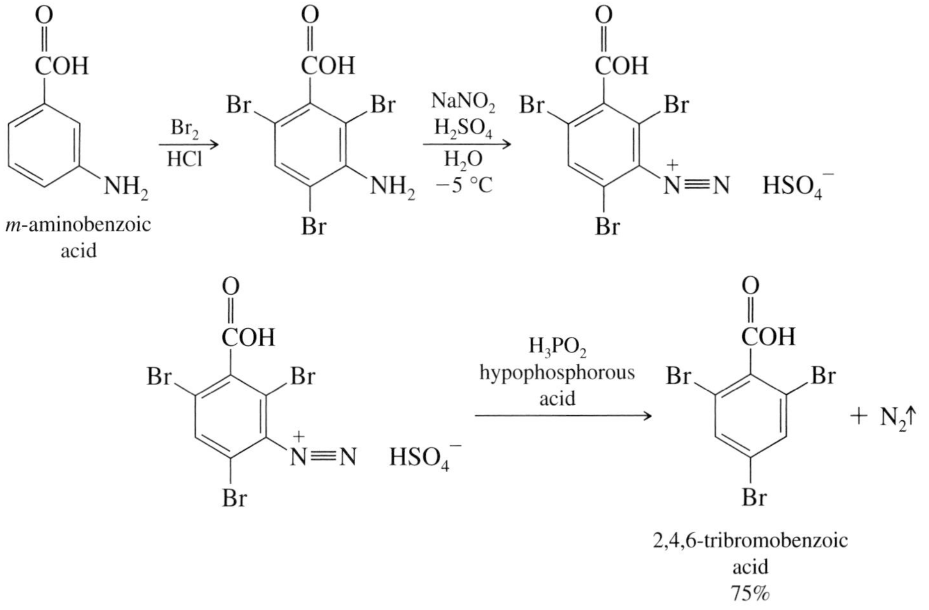

*m*-aminobenzoic
acid

2,4,6-tribromobenzoic
acid
75%

The reactions outlined in the equations in this section represent ways in which functional groups can be positioned regioselectively on aromatic rings.

**Study Guide**
**Concept Map 21.9**

**Problem 21.41**   Devise a synthesis for each of the following six compounds, starting with benzene or toluene.

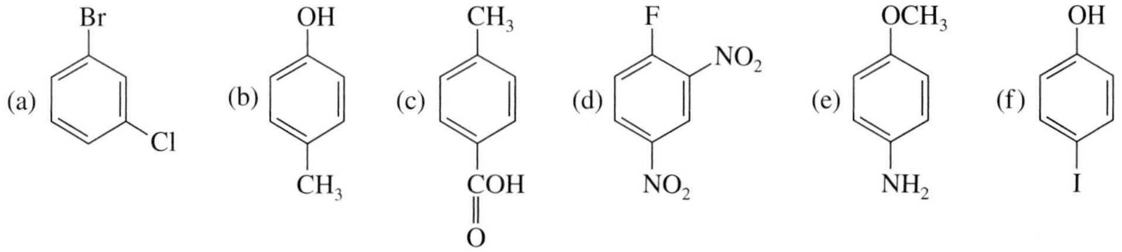

(a)   (b)   (c)   (d)   (e)   (f)

## C. Nucleophilic Aromatic Substitution

Substitution of a nucleophile for a halogen in an aryl halide that does not also have electron-withdrawing groups as substituents is difficult. Such halides are inert to the reagents that react with alkyl halides under the usual conditions. When there are electron-withdrawing groups, especially nitro groups, ortho or para to the halogen, however, nucleophilic substitution takes place with relative ease.

This reaction is of special importance with 2,4-dinitrofluorobenzene, a reagent that reacts very rapidly with amines. It is used extensively in protein chemistry to label free amino groups on a protein or peptide chain (p. 996). The reaction of this reagent with an amino acid, glycine, is illustrated below.

$$\overset{+}{H_3}NCH_2\overset{\overset{\displaystyle O}{\|}}{C}O^- \rightleftharpoons H_2NCH_2\overset{\overset{\displaystyle O}{\|}}{C}OH$$

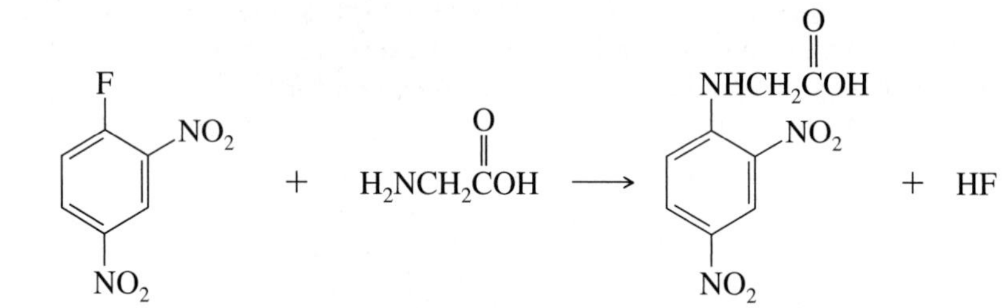

| 2,4-dinitrofluorobenzene | glycine | N-2,4-dinitrophenylglycine |

Although fluoride ion is a poor leaving group in nucleophilic substitution reactions of alkyl halides, it is an excellent one in nucleophilic substitution reactions of aryl halides. In the following reaction, in which the cyclic amine piperidine is used to displace halide ion from a 2,4-dinitrohalobenzene, the fluoro compound reacts 3300 times faster than the iodo compound.

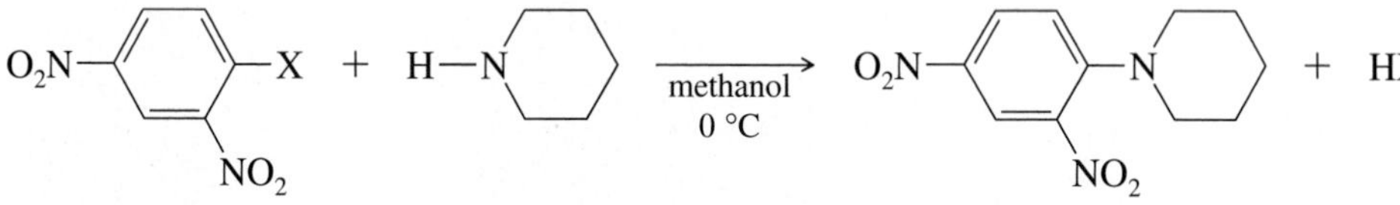

| 2,4-dinitrohalobenzene | piperidine | N-2,4-dinitrophenylpiperidine |

| X | $k_{rel}$ |
| --- | --- |
| F | 3300 |
| I | 1 |

Spectroscopic and kinetic evidence suggest that the reaction takes place by attack of the nucleophile on the aromatic ring at electron-deficient positions ortho and para to the electron-withdrawing nitro groups. An intermediate forms with a tetrahedral carbon atom, like the cation that is formed in electrophilic aromatic substitution. In this case, however, the intermediate is electron-rich rather than electron-deficient and is stabilized by delocalization of the negative charge to the nitro groups.

VISUALIZING THE REACTION

## Nucleophilic Aromatic Substitution. The Rate-Determining Step

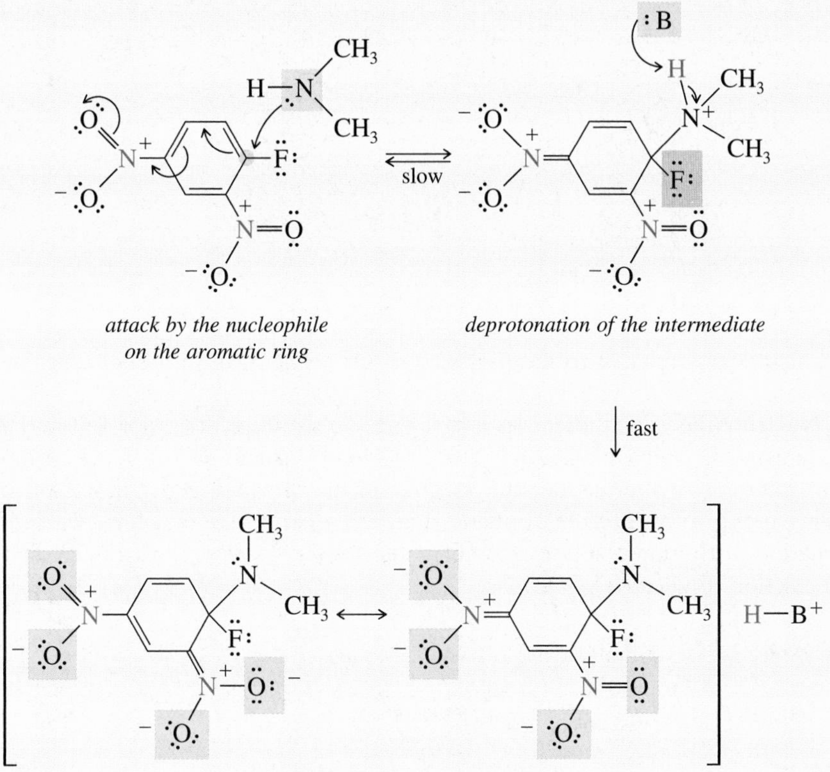

attack by the nucleophile
on the aromatic ring

deprotonation of the intermediate

fast

tetrahedral anionic intermediate
formed from the attack of
nucleophile on aryl halide

The rate-determining step of a nucleophilic aromatic substitution reaction is the formation of the tetrahedral intermediate by nucleophilic attack of the amine at the carbon atom bearing the halogen atom. A fluoro compound reacts much faster than an iodo compound for two reasons. First, fluorine is much more electronegative than iodine. The carbon atom to which the halogen atom is bonded has a much larger positive charge in a fluorobenzene than in an iodobenzene. Second, a fluorine atom is much smaller than an iodine atom and, thus, offers less steric hindrance to an approaching nucleophile that will bond to the carbon atom to give the tetrahedral intermediate. Once the tetrahedral intermediate is formed, the amine loses a proton to a base, and the resulting anion is stabilized by delocalization of charge to the nitro groups. Departure of a fluoride ion reestablishes aromaticity and gives the product.

---

**VISUALIZING THE REACTION**

## Nucleophilic Aromatic Substitution. The Second Step

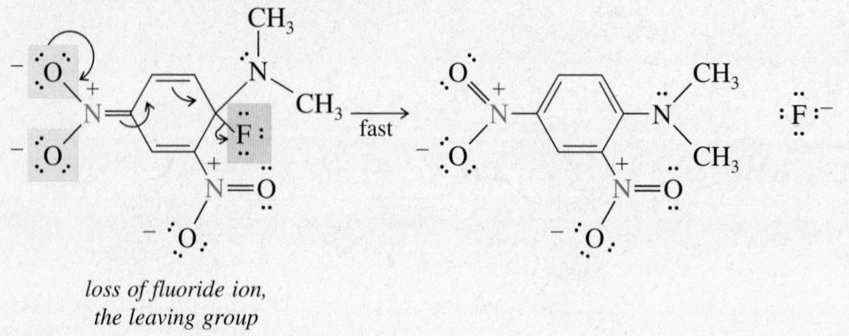

*loss of fluoride ion,
the leaving group*

---

This step is fast relative to the first step of the reaction in protic solvents such as methanol.

Nucleophilic substitution on aryl halides, therefore, does not follow an $S_N2$ mechanism, in which no intermediate is formed. Instead such a reaction has two steps: addition of the nucleophile to give an intermediate and elimination of the leaving group to restore the aromatic ring. The energy diagram for a nucleophilic aromatic substitution reaction resembles that for an electrophilic aromatic substitution reaction (Figure 10.8, p. 371).

⊘ **Study Guide**
**Concept Map 21.10**

**Problem 21.42**   Complete the following equations.

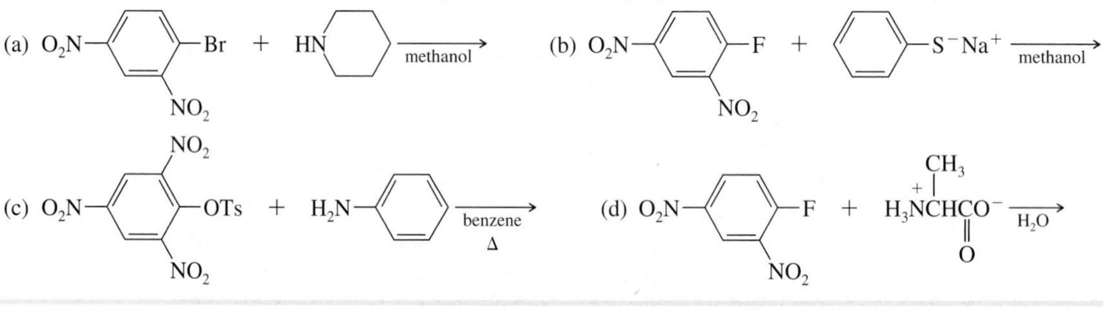

(a) $O_2N$—⟨benzene ring with Br and NO$_2$⟩—Br  +  HN⟨ring⟩ $\xrightarrow{\text{methanol}}$

(b) $O_2N$—⟨benzene ring with F and NO$_2$⟩—F  +  ⟨benzene⟩—$S^-Na^+$ $\xrightarrow{\text{methanol}}$

(c) $O_2N$—⟨benzene ring with NO$_2$, OTs, NO$_2$⟩—OTs  +  $H_2N$—⟨benzene⟩ $\xrightarrow[\Delta]{\text{benzene}}$

(d) $O_2N$—⟨benzene ring with F and NO$_2$⟩—F  +  $H_3\overset{+}{N}\underset{\underset{O}{\|}}{\overset{\overset{CH_3}{|}}{C}}HCO^-$ $\xrightarrow{H_2O}$

# 21.8   Solid–Phase Synthesis

## A. Introduction

The synthesis of a complex molecule usually requires many steps in which reagents are mixed together as they react more or less completely to give often more than one product so that the desired product has to be separated from left-over reagents and undesired products before another step of the reaction can be carried out. The handling of the chemical mixture at each stage leads to losses in material and in time that become significant when many synthetic steps are required. For this reason, chemists try, whenever possible, to design syntheses with the minimum number of steps and with the highest possible yield of the desired product at each step.

The problems associated with multistep syntheses become particularly severe in the synthesis of large molecules such as peptides, in which a number of amino acids have to be strung together in an exact order. We saw some of the difficulties of synthesizing even simple peptides in Section 15.8, where we discussed questions of the protection and activation of the amino and carboxylic acid groups of the amino acids. It was for this reason that a major advance in experimental technique, the use of reagents that are tethered to solid-phase supports so that they can be separated easily from the reagents that surround them in solution, was first developed for the synthesis of peptides. The principles involved there, which we will examine in the next section, are now being used more and more in the synthesis of simple compounds, as we shall see in Section 21.8C.

## B. Solid–Phase Peptide Synthesis

Real progress in the synthesis of larger peptides and proteins began when solid-phase peptide synthesis was introduced in 1963 by Robert B. Merrifield of Rockefeller University. For this achievement, he received the Nobel Prize in Chemistry in 1984. This technique was designed to simplify the isolation and purification of peptides when many steps are necessary for a synthesis. Merrifield reasoned that if the growing peptide could be incorporated into a solid polymer, it could be handled as a solid and purified thoroughly at each stage of the synthesis by repeated washing and filtration steps, minimizing losses of peptides. The rates of the reactions for forming peptide bonds and the degree of completion of the formation of the peptide could be increased greatly by using large excesses of protected amino acids, which could then be washed away from the extended peptide chain on its solid support. Thus it would be possible to add one amino acid at a time in precise order to build up large molecules of high purity in good yield. Ideally, the whole process would be automated so that the multiple cycles necessary to form a peptide bond, purify the extended protected peptide, remove the protecting group, and add a different amino acid to the newly exposed *N*-terminal amino acid could be repeated over and over again with minimal human attention.

It was necessary to establish a solid support system that could be covalently bonded to the *C*-terminal amino acid of the peptide by a linkage that would be stable to the reagents used in making peptide bonds and removing protecting groups. In addition, the support had to be insoluble in the solvents used in the reactions. It had to have structural stability and maintain its particle size so that it would survive the physical manipulations of repeated washings and filtrations. It had to be chemically inert except at the site where the original attachment of the peptide chain was made. Its structure had to allow the solutions of reagents to flow through it so that reactions could take place on a large portion of its surface area.

Much research focused on developing a polymeric material that would meet all these requirements. The material most widely used is a polystyrene that is cross-linked with divinylbenzene (p. 1047). The polystyrene is formed into beads approximately $5 \times 10^{-3}$ cm in diameter. The cross-linking gives the polymer a high molecular weight and very low solubility in organic solvents. The amount of cross-linking is adjusted so that the polymer has some flexibility and the beads swell to double their dry size when they are placed in an organic solvent for the reactions. This allows reactions to take place inside the beads as well as on the surface. Even though the beads are so small that they can hardly be seen by the naked eye, each one has room for the growth of a trillion ($10^{12}$) peptide chains.

The polystyrene resin is converted into a polymer with chloromethyl groups on approximately 22% of the aromatic rings. The *C*-terminal amino acid of the peptide is attached to the polystyrene as a benzyl ester. The amino groups of the amino acids to be added to the chain are protected by *tert*-butoxycarbonyl groups, which can be removed under conditions that do not affect the benzylic ester bond (p. 632). Dicyclohexylcarbodiimide (DCC) is used to activate the carboxylic acid group of the amino acid to be added to the chain (p. 634).

The different steps in a solid-phase synthesis of a tripeptide, glycylalanylvaline, are shown below and on the next page.

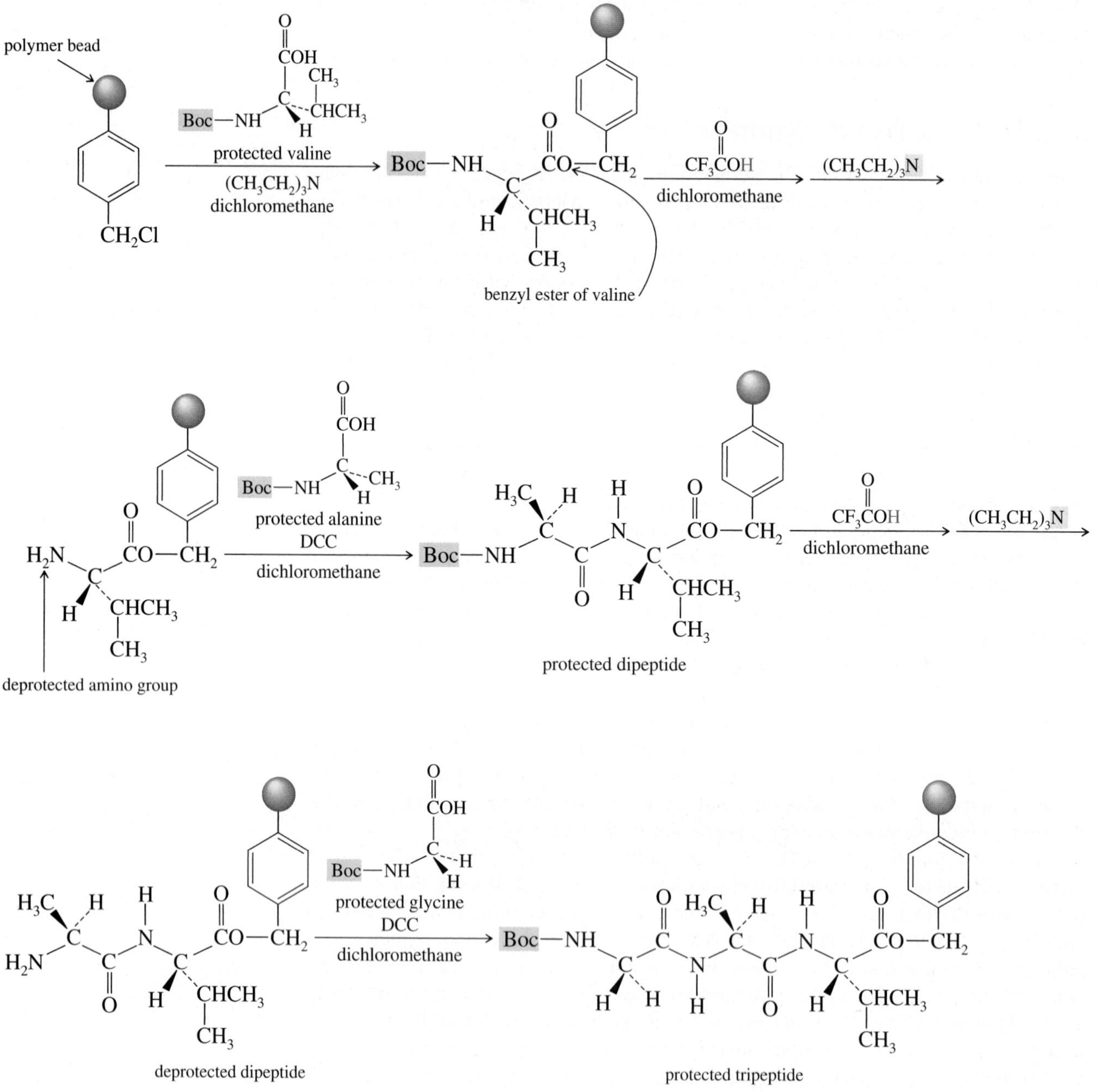

polymer bead

Boc—NH protected valine

(CH$_3$CH$_2$)$_3$N
dichloromethane

benzyl ester of valine

CF$_3$COH
dichloromethane

(CH$_3$CH$_2$)$_3$N

Boc—NH protected alanine
DCC
dichloromethane

deprotected amino group

protected dipeptide

CF$_3$COH
dichloromethane

(CH$_3$CH$_2$)$_3$N

Boc—NH protected glycine
DCC
dichloromethane

deprotected dipeptide

protected tripeptide

The reagents for the peptide synthesis were all chosen so that the reaction sequences are repetitious and, therefore, can be automated. The *tert*-butoxycarbonyl group is removed with trifluoroacetic acid, which leaves behind a protonated amino group. Triethylamine removes the proton to give the free amino group, which reacts with the activated carboxylic acid group from another protected amino acid. Finally, the strong acid hydrogen bromide in trifluoroacetic acid cleaves both the protecting group and the benzylic ester bond. Note that aqueous acids, which would hydrolyze peptide bonds, are avoided at all stages. In addition to the reagents that are shown, there are a number of intermediate stages in which the polymer, with the attached peptide chains, is washed with solvents and filtered to purify it. Altogether, eleven separate steps are necessary in order to add one amino acid to the chain.

**Problem 21.43**   Write equations for a solid-phase synthesis of phenylalanylleucylglycine. The structures of phenylalanine and leucine may be found in Figure 23.3 (p. 983). Give the full structures of the benzylic ester, any protecting groups, and any reagents used to promote the formation of the peptide bond at least once. Show the full structure of the peptide at each stage of the synthesis.

Solid-phase peptide synthesis has been used to prepare a number of naturally occurring proteins, including insulin and the enzyme ribonuclease, which was synthesized in 1969 by Merrifield. Ribonuclease has 124 amino acid residues. Its synthesis required 369 chemical reactions and 11,931 steps in the automated process. The introduction of one amino acid took approximately 4 hours. This is a much more rapid and efficient process than was possible with classic solution-phase chemistry, but it does not begin to approach the efficiency with which proteins are synthesized in the living cell. A bacterial cell completes protein synthesis in seconds and keeps about 3000 different protein syntheses going simultaneously with no confusion of reagents and products.

Since Merrifield's success with the synthesis of peptides on polystyrene supports, many other polymer supports have been developed and used in peptide syntheses, as well as other large- and small-molecule syntheses. One of the most interesting of these technologies is what is known as the Fmoc-polyamide peptide synthesis perfected in the 1980s by British chemists Eric Atherton and Robert C.

Sheppard, which involves the use of a polymer that has amide groups on its backbone and is cross-linked by amide functions. This avoids one of the failings of the Merrifield system, which is that a polar polypeptide (polyamide) chain is growing inside the cavity of the nonpolar, mostly aromatic ring environment of cross-linked polystyrene beads. Traditionally, a relatively nonpolar, non-hydrogen-bonding solvent such as dichloromethane is used in the system. When the peptide chain gets long enough, it may begin to form aggregates between the different peptide chains (remember that there are up to a trillion of them in each tiny bead), interfering with further regular growth of each chain. This problem can be avoided in some cases by using a more polar and hydrogen-bond-accepting solvent such as dimethylformamide (p. 254). Use of a polyamide support with dimethylformamide as the solvent gives the growing peptide chain the polar, hydrogen-bonding environment it needs to prevent aggregation. A fragment of the polyamide polymer is shown below.

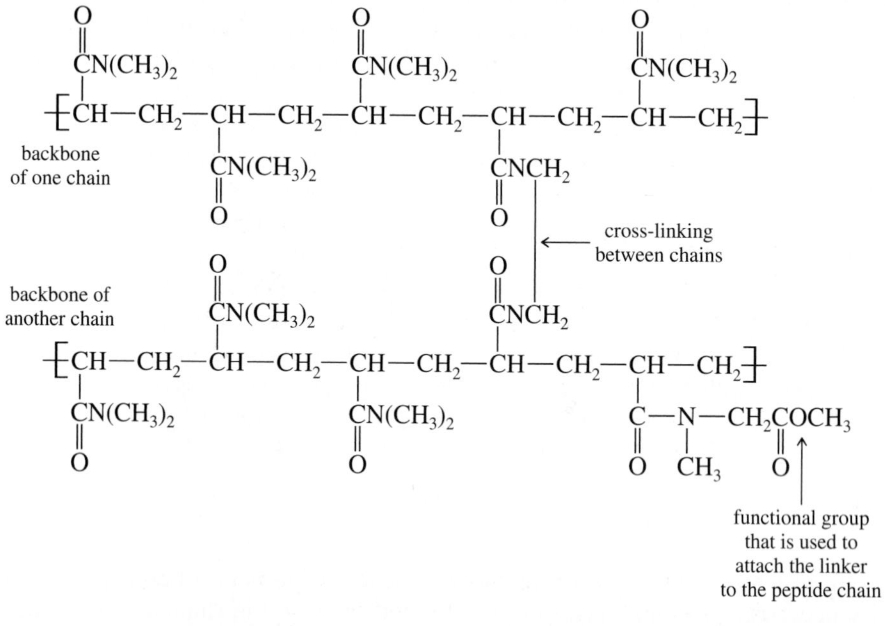

Because the backbone of the polymer contains amide bonds that are reactive toward strong acids or bases, the protecting group most often used in the Merrifield synthesis, the Boc group, cannot be used as a protecting group on the amino acids that are added to the peptide chain. Instead, the 9-fluorenylmethyoxycarbonyl group (Fmoc, p. 849), which can be removed by treatment with an amine, is used. The other change that was made is to avoid the use of dicyclohexylcarbodiimide to activate the carboxylic acid group of the amino acids but instead to add them directly as their active esters. The pentafluorophenyl ester works well in this synthesis. The linker between the polyamide chain and the first amino acid can vary, but the end of the linker is a benzylic alcohol function to which the first amino acid is connected as an ester. This linkage is stable to the amine piperidine, used to remove the Fmoc protecting groups, but is easily cleaved with strong acid. This technique also allows the use of the Boc protecting group on other functional groups on amino acid side chains (see Figures

23.4, 23.5, and 23.6), keeping the Boc groups on until the very last step when the peptide is removed from the polymer support. The application of the Fmoc-polyamide to the synthesis of a fragment of an acyl-carrier protein (p. 722) is shown below and on the next page.

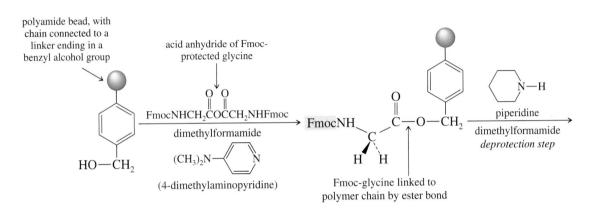

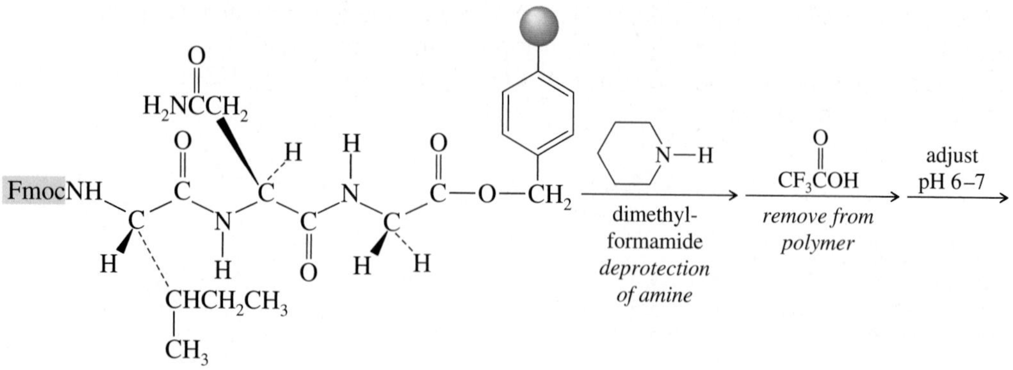

Fmoc-protected tripeptide on the polymer chain

adjust
pH 6–7

$CF_3COH$
*remove from polymer*

dimethyl-
formamide
*deprotection of amine*

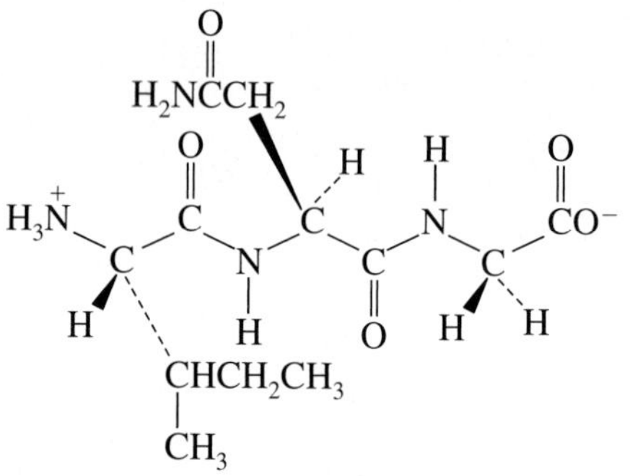

a tripeptide
isoleucylasparagylglycine

The actual synthesis of the peptide fragment goes on to add seven more amino acids, but this portion of the synthesis is sufficient to demonstrate once again the repetitive nature of the process, which allows for automation, and the control, by using protection and deprotection strategies, that allows for the synthesis of the peptide with the order of amino acids exactly as desired.

**Problem 21.44**   The next amino acid to be added to the growing peptide chain shown above was tyrosine, protected at the hydroxyl group on the side chain as the *tert*-butyl ether, which is removable with trifluoroacetic acid. Write equations for the conversion of the tripeptide on the polymer chain to the free tetrapeptide incorporating tyrosine. The protected and activated tyrosine is

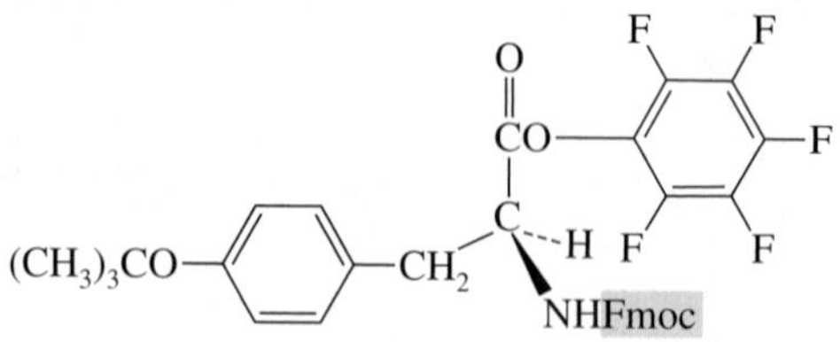

$(CH_3)_3CO-$

NHFmoc

## C. Solid-Phase Organic Synthesis

The techniques that we have seen in the synthesis of peptides are being applied to the synthesis of small molecules also. For example, a reagent such as dicyclohexyl-carbodiimide (p. 634), the urea product of which is sometimes difficult to remove from a reaction mixture, can be anchored to a solid support and then used in large excess to promote an otherwise difficult reaction. An example is the formation of a large-ring lactone by an intramolecular esterification from a long-chain hydroxy acid.

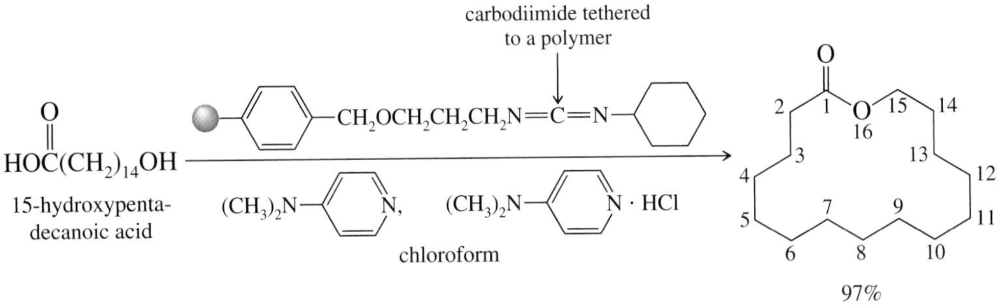

$$HOC(CH_2)_{14}OH$$

15-hydroxypenta-decanoic acid

97%

The product of the reaction can be filtered away from the excess of the carbodiimide and the urea, which also remains tethered to the polymer support. The purification of the lactone is much simplified.

On the other hand, entire syntheses can be carried out on a solid support, and ways are being found to incorporate many reagents onto the surface of a polymer. The following is an example of a couple of steps involving the Wittig reaction (p. 871) in the synthesis of the potent antitumor natural products epothilones A and B.

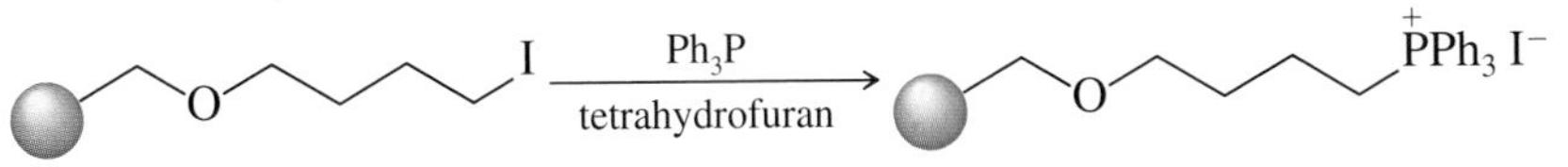

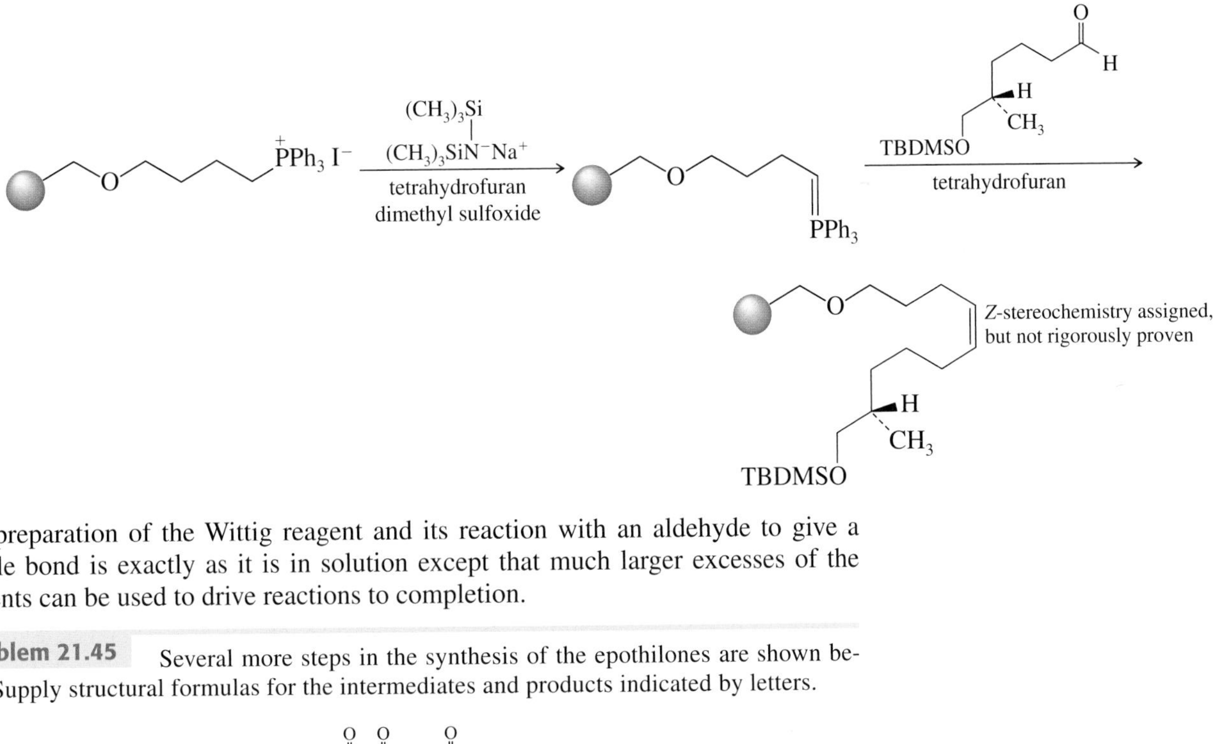

Z-stereochemistry assigned, but not rigorously proven

The preparation of the Wittig reagent and its reaction with an aldehyde to give a double bond is exactly as it is in solution except that much larger excesses of the reagents can be used to drive reactions to completion.

**Problem 21.45**    Several more steps in the synthesis of the epothilones are shown below. Supply structural formulas for the intermediates and products indicated by letters.

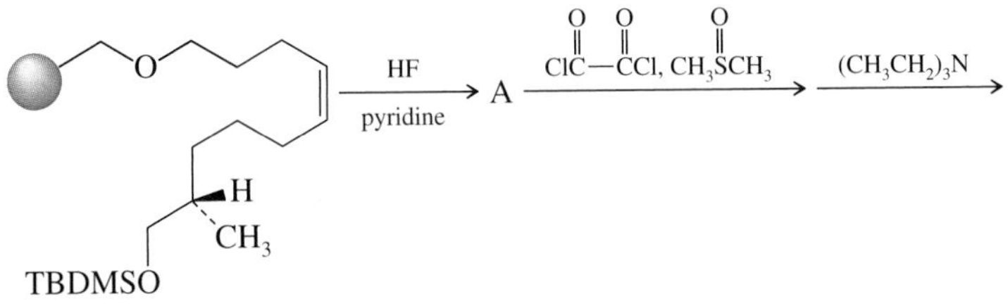

$$B \xrightarrow[\text{tetrahydrofuran}]{(CH_3CH)_2N^-Li^+ \text{ (2.2 equivalents)}} \xrightarrow{\text{protonation}} C$$

a mixture of
diastereomers

## 21.9   Combinatorial Chemistry

### A. Introduction

The success of methods for solid-phase synthesis of simple organic compounds (Section 21.8C) gave rise to another idea. Solid-phase synthesis could be used to create what are called "libraries" of small organic molecules that mimic the variety of structure available in peptides, for example, and that can be screened rapidly for biological activity. A library is a collection of compounds that are diverse in structure but have all been synthesized on a solid support by the application of the same sequence of reactions with different individual reagents at each stage. The library is then removed from the solid support and tested for a particular biological activity. If activity is found in the library as a whole, then subsets of the library can be tested. If, on the other hand, the library shows no activity, then it is not necessary to test the individual compounds in it, saving an enormous amount of time and money in the process of drug discovery, for example.

The kinds of mixtures that are generated by combinatorial chemistry techniques are not unlike the wide variety of natural products synthesized by plants and animals. The process of honing in on the ones that have biological activity is also like the techniques used by chemists as they sort through these multiple natural compounds to find the ones that have the biological activity they are looking for. The process has been described as looking for a needle in a haystack and was described earlier in this book on page 748 for the isolation of the gypsy moth pheromone (+)-disparlure. The difference is that in one case nature creates the haystack; in the case of combinatorial chemistry, it is the chemists who are deliberately making the haystacks!

In the next section we will create a small library, only twenty-seven compounds, of potential antioxidants (p. 794), to see the principles of combinatorial chemistry in action.

### B. Synthesis of a Small Library of Organic Compounds

The purpose of the combinatorial synthesis outlined in this section was to create a library of diversely substituted diols that were tested for antioxidant activity. The general steps of the reaction involved binding a carboxylic acid to Merrifield's resin as an ester, deprotonating it at the $\alpha$-carbon of the ester to give an enolate ion that reacts with an aldehyde or ketone in an aldol-type reaction to give a hydroxy compound, and then removing the product from the resin by reducing the ester group with diisobutylaluminum hydride to create the second hydroxy function. The diversity can arise from the carboxylic acids that are used to bind to the resin and in the aldehydes and ketones that are used at the aldol condensation stage. The actual synthesis used three different carboxylic acids to give three different resin esters. Each one of the esters was then condensed with seven different aldehydes and two differ-

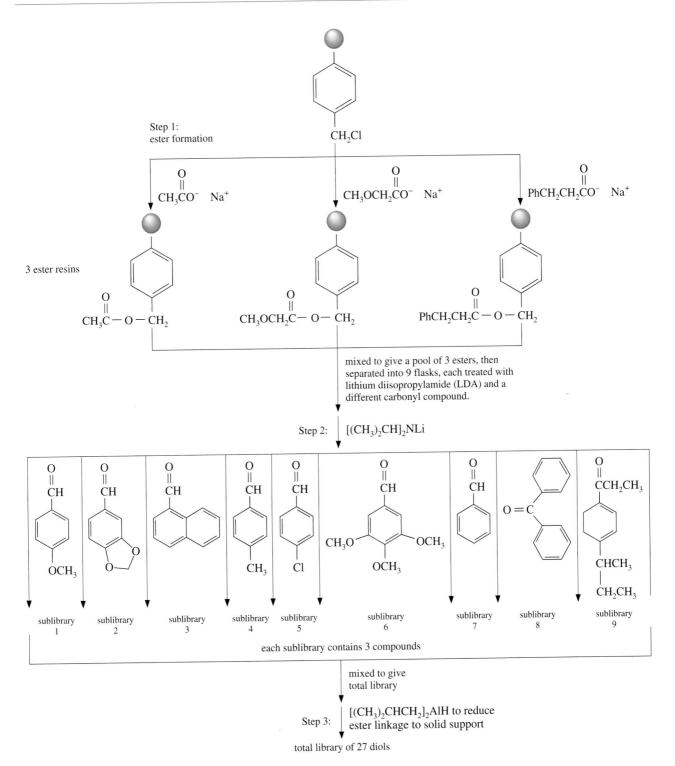

*Figure 21.1*
Outline of the combinatorial synthesis of a library of twenty-seven diols.

ent ketones, giving rise to twenty-seven different compounds. The scheme by which the library was synthesized is outlined in Figure 21.1. The reactions giving rise to sublibrary 1 are shown below and on page 911.

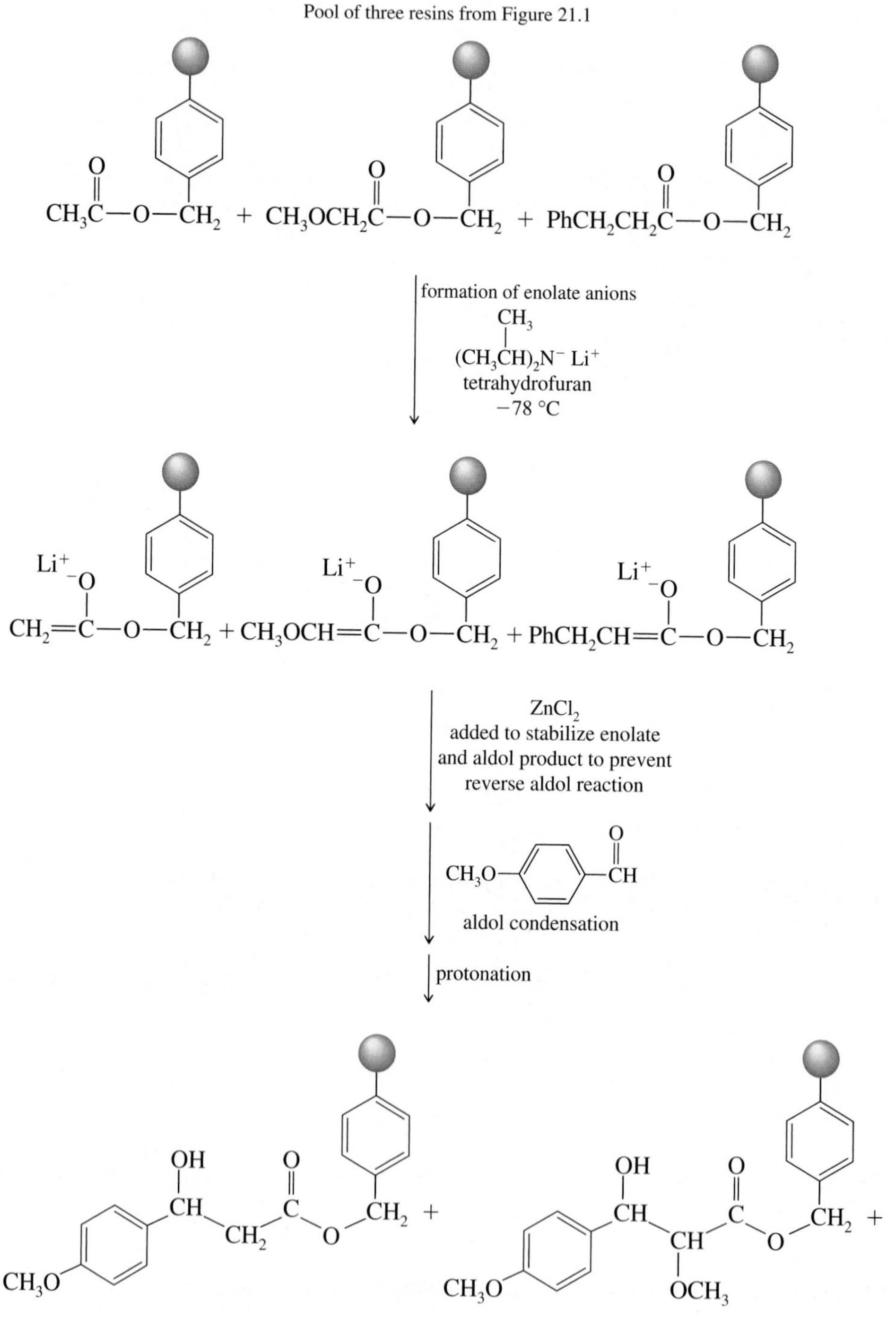

Pool of three resins from Figure 21.1

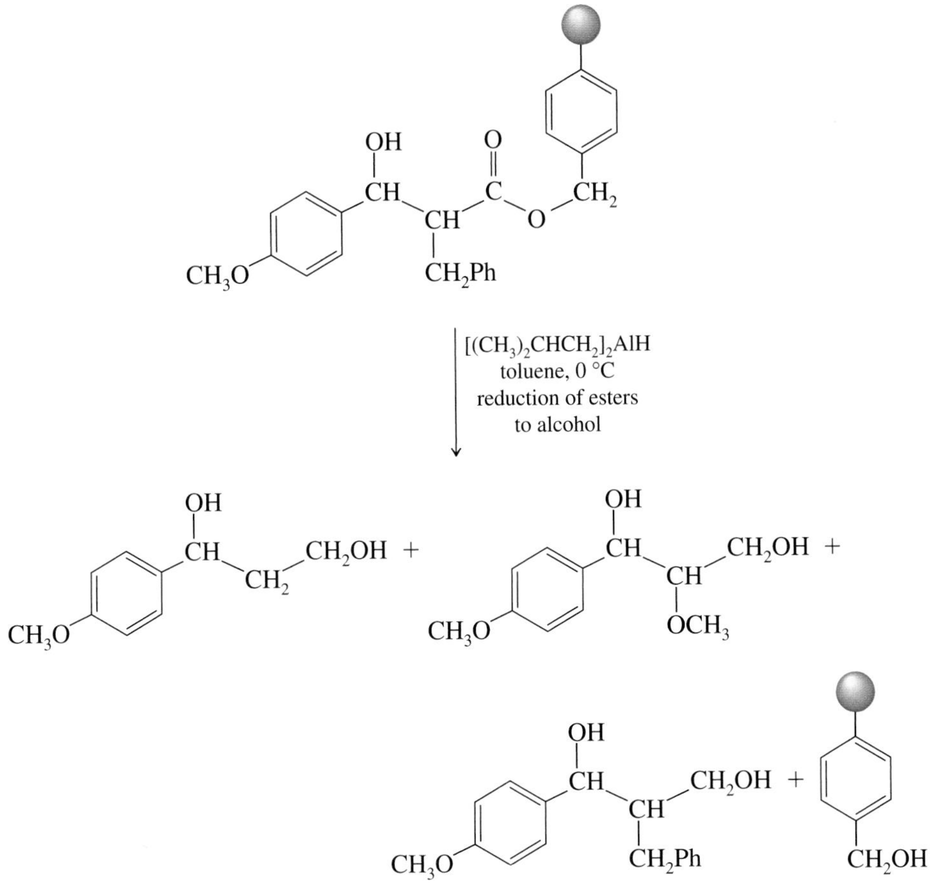

sublibrary 1; three compounds that are derived from the
same aldehyde but different acid components.

Each sublibrary of three compounds, as well as the total library containing all
twenty-seven compounds, was isolated. The total library was tested to see if it sup-
pressed the oxidative activity of linoleic acid hydroperoxide (p. 786), and when it
was shown to have antioxidant activity (p. 794), each sublibrary was tested individually. Sublibrary 6 was shown to have the most antioxidant activity, although some
of the other sublibraries also were active.

The process described above shows how a large number of diverse compounds
can be prepared and screened quickly for a useful property. The technique has now
been automated and reactions carried out simultaneously to give libraries of many
more than twenty-seven compounds, greatly speeding up the process by which
medications or compounds with other desirable properties can be identified.

**Problem 21.46**   Sublibrary 6 of the combinatorial synthesis shown above is the most
active antioxidant pool of the twenty-seven compounds synthesized. Show in detail the reactions that give rise to sublibrary 6 starting from the pool of the three ester resins from
Figure 21.1 (p. 909).

### Table 21.1    Protecting Groups

| Functional Group Protected | Protecting Group | Reagents Used to Make Protecting Group | Reagents Used to Remove Protecting Group |
|---|---|---|---|
| **Alcohol** | | | |
| ROH | 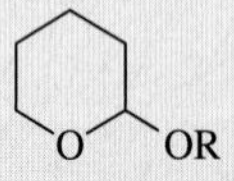 tetrahydropyranyl ether | 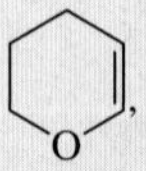 , TsOH | $H_3O^+$ |
| | $-\underset{|}{\overset{|}{Si}}-OR$ silyl ether | $-\underset{|}{\overset{|}{Si}}Cl,$ | $F^-$ |
| | 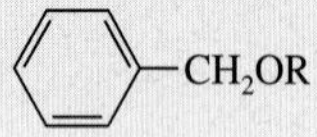 $-CH_2OR$ benzyl ether | NaH, then 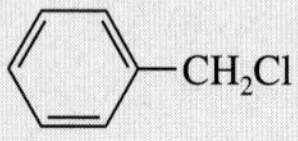 $-CH_2Cl$ | $H_2$ with Pd catalyst |
| **1,2-Diol** | | | |
| | 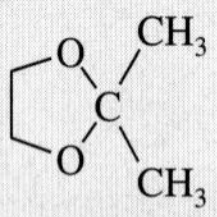 cyclic ketal | 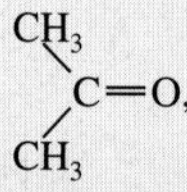 $\underset{CH_3}{\overset{CH_3}{C}}=O$, TsOH | $H_3O^+$ |
| **1,3-Diol** | | | |
| | 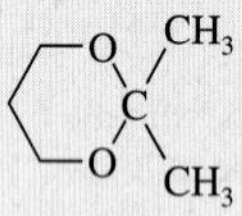 cyclic ketal | 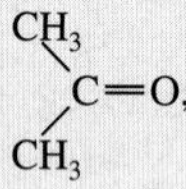 $\underset{CH_3}{\overset{CH_3}{C}}=O$, TsOH | $H_3O^+$ |
| **Aldehyde** | | | |
| $\underset{R}{\overset{O}{\underset{||}{C}}}H$ | $R-\underset{OR}{\overset{OR}{\underset{|}{\overset{|}{C}}}}H$ acetal | ROH (excess), TsOH | $H_3O^+$ |
| | 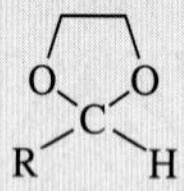 cyclic acetal |  HO    OH, TsOH | $H_3O^+$ |

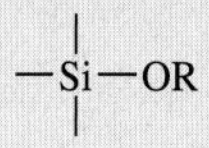

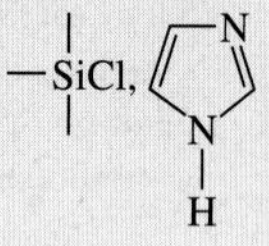

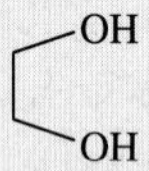

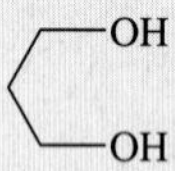

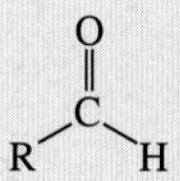

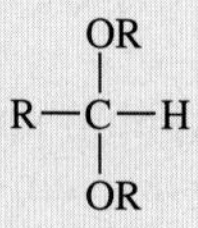

## Table 21.1 (continued)

| Functional Group Protected | Protecting Group | Reagents Used to Make Protecting Group | Reagents Used to Remove Protecting Group |
|---|---|---|---|
| **Ketone** | | | |
| 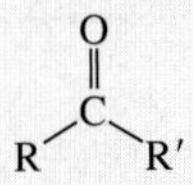 | 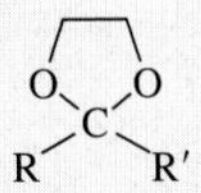 cyclic ketal |  HO⎓OH, TsOH | $H_3O^+$ |
| **Amine** | | | |
| $RNH_2$ | $RNHCOC(CH_3)_3$ *tert*-butoxycarbonyl (Boc) group | $(CH_3)_3COC{-}LG$ LG = Cl, $N_3$, anhydride, etc. | $CF_3COH$ or $HBr, CH_3COH$ |
| $RNH_2$ | 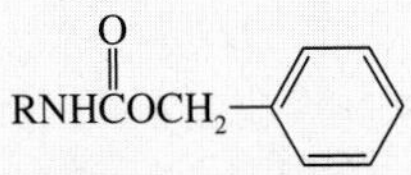 $RNHCOCH_2{-}$ benzyloxycarbonyl (Cbz) group | 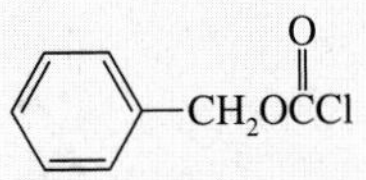 ${-}CH_2OCCl$ | $H_2$ with Pd catalyst, $HBr, CH_3COH$, $CF_3COH$ |
| $RNH_2$ | 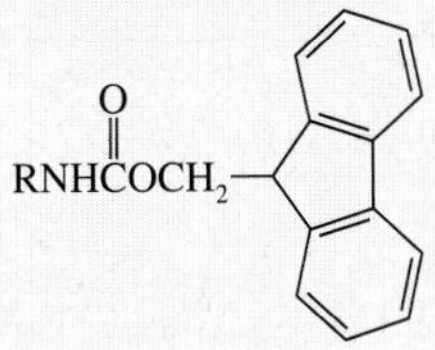 $RNHCOCH_2{-}$ 9-fluorenylmethoxy-carbonyl (Fmoc) group | 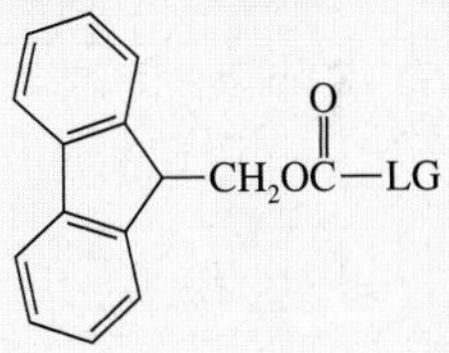 ${-}CH_2OC{-}LG$ LG = Cl, $N_3$, etc. | 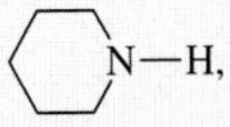 N—H, $R'_2NH$ |
| $RNH_2$ | $RNHCR'$ amide | $R'CCl$ or $R'COCR'$ base | $H_3O^+, \Delta$, or $OH^-, H_2O, \Delta$ |

**Table 21.2** **Reduction of Acid Derivatives with Metal Hydride Reagents**

| Starting Material | Reagent | Intermediate | Reagent in Second Step | Product |
|---|---|---|---|---|
| | | **Complete Reduction** | | |
| $\underset{\text{RCOH}}{\overset{\displaystyle O}{\parallel}}$ | $LiAlH_4$ | $\underset{\underset{\text{H}}{\mid}}{\overset{\overset{\text{O—M}}{\mid}}{RC-H}}$ | $H_3O^+$ | $RCH_2OH$ |
| $\underset{\text{RCNR}'_2}{\overset{\displaystyle O}{\parallel}}$ | $LiAlH_4$ | $\underset{\underset{\text{H}}{\mid}}{RC=NR'_2}$ | $LiAlH_4$ | $RCH_2NR'_2$ |
| $\underset{\text{RCOR}'}{\overset{\displaystyle O}{\parallel}}$ | $LiAlH_4$ | $\underset{\underset{\text{H}}{\mid}}{\overset{\overset{\text{O—M}}{\mid}}{RC-H}} + R'O^-$ | $H_3O^+$ | $RCH_2OH + R'OH$ |
| $RC\equiv N$ | $LiAlH_4$ | $\underset{\underset{\text{H}\quad\text{M}}{\mid\quad\mid}}{\overset{\overset{\text{H}}{\mid}}{RC-N-M}}$ | $H_3O^+$ | $RCH_2NH_2$ |
| $\underset{\text{RCOR}'}{\overset{\displaystyle O}{\parallel}}$ | $[(CH_3)_2CHCH_2]_2AlH$ (excess) | $\underset{\underset{\text{H}}{\mid}}{\overset{\overset{\text{O—M}}{\mid}}{RC-H}} + R'O^-$ | $H_3O^+$ | $RCH_2OH + R'OH$ |
| $\underset{\text{RCOR}'}{\overset{\displaystyle O}{\parallel}}$ | $BH_3 \cdot S(CH_3)_2$ | | $H_2O$ | $RCH_2OH + R'OH$ |
| $\underset{\text{RCNR}'_2}{\overset{\displaystyle O}{\parallel}}$ | $BH_3 \cdot S(CH_3)_2$ | | $HCl/H_2O$, then NaOH | $RCH_2NR'_2$ |
| $RC\equiv N$ | $BH_3 \cdot S(CH_3)_2$ | | $HCl/H_2O$, then NaOH | $RCH_2NH_2$ |

**Table 21.2 (continued)**

| Starting Material | Reagent | Intermediate | Reagent in Second Step | Product |
|---|---|---|---|---|
| **Reduction to Carbonyl Compounds** | | | | |
| RCOR′ (C=O) | $[(CH_3)_2CHCH_2]_2AlH$ 1 equivalent | RC(—OR′)(H)(O—M) | $H_2O$ | RCH + R′OH (C=O) |
| $RC{\equiv}N$ | $[(CH_3)_2CHCH_2]_2AlH$ 1 equivalent | RC(=N—M)(H) | $H_3O^+$ | RC(=O)(H) |
| RCNOCH₃ (C=O, N-CH₃) | $LiAlH_4$ | 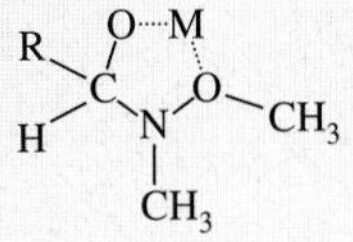 R C(H) ring O···M, N(CH₃)—O—CH₃ | $H_3O^+$ | RC(=O)(H) |
| RCNOCH₃ (C=O, N-CH₃) | $[(CH_3)_2CHCH_2]_2AlH$ | 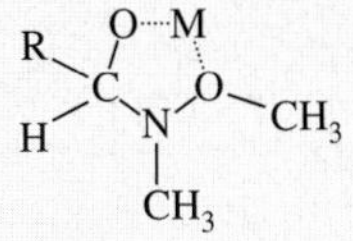 R C(H) ring O···M, N(CH₃)—O—CH₃ | $H_3O^+$ | RC(=O)(H) |

**Table 21.3  Reactions of Organometallic Reagents**

| Starting Material | Reagent | Intermediate | Reagent in Second Step | Product |
|---|---|---|---|---|
| **Grignard and Organolithium Reagents with Acid Derivatives** | | | | |
| RCOR′ (C=O) | R″MgX (twice) | $R-C(O^-Mg^{2+}X^-)(OR′)(R″)$ ; then $R-C(O^-Mg^{2+}X^-)(R″)(R″)$ | $H_3O^+$ | $R-C(OH)(R″)(R″)$ |
| RCNOCH₃ (C=O, N-CH₃) | R′M | R C(R′) ring O···M, N(CH₃)—O—CH₃ | $H_3O^+$ | RCR′ (C=O) |

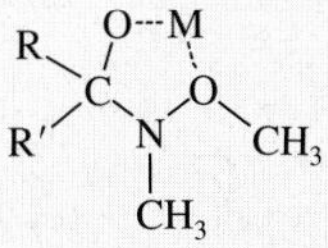

**(continued)**

## Table 21.3 (continued)

| Starting Material | Nucleophile | Intermediate | Reagents for Second Step | Product |
|---|---|---|---|---|
| **Organocuprates** | | | | |
| 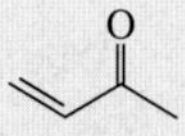 (methyl vinyl ketone) | $R_2CuLi$ | 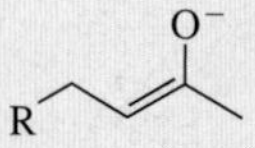 enolate with $O^-$, R | $HB^+$ | 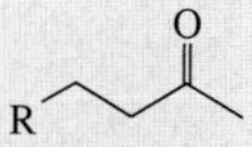 R–CH₂CH₂COCH₃ structure |
| | | | $R'X$ | 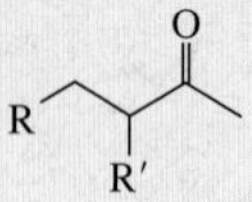 product with R and R' |
| | | | $R'CR''$ (C=O) then $HB^+$ | product with R, R', R'', OH |
| $R'—X$ | | | | $R'—R$ |

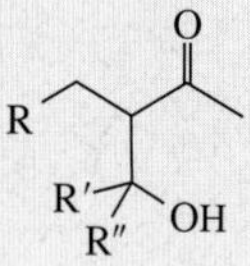

## Table 21.4    Summary of the Preparation and Reactions of Heteroatom-Stabilized Carbanions

| Reactive Intermediate | Reagents That Give Reactive Intermediate | Reagent That Reacts with Ylide | Product |
|---|---|---|---|
| **Wittig Reaction, Phosphonium Ylides** | | | |
| $(C_6H_5)_3P=C$ with R, H | $(C_6H_5)_3P$ $RCH_2X$ BuLi | $R'CH$ (C=O) | $R'CH=CHR$ |
| | | $R'CR''$ (C=O) | $R'R''C=CHR$ |
| $C_6H_5)_3P=C$ with $COCH_2CH_3$, H | $(C_6H_5)_3P$ $XCH_2COCH_2CH_3$ NaOH | $R'CH$ (C=O) | $R'CH=CHCOCH_2CH_3$ |
| | | $R'CR''$ (C=O) | $R'R''C=CHCOCH_2CH_3$ |

**Table 21.4 (continued)**

| Reactive Intermediate | Reagents That Give Reactive Intermediate | Reagent for First Step | Product | Reagents for Second Step | Product |
|---|---|---|---|---|---|
| | | **Dithiane Anions** | | | |
|  |  $BF_3$ then BuLi | R′X |  | $H_2O$ $HgCl_2$ Δ |  |
| | |  |  | $H_2O$ $HgCl_2$ Δ |  |
| | |  |  | $H_2O$ $HgCl_2$ Δ |  |

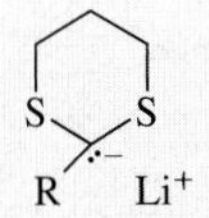

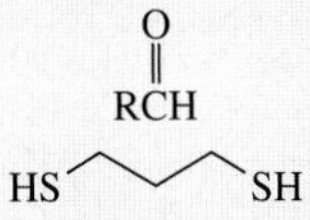

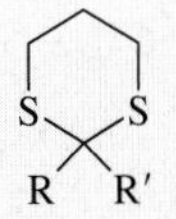

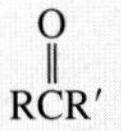

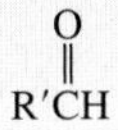

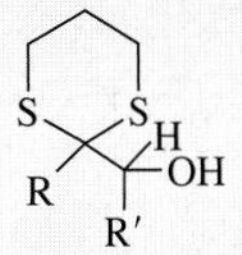

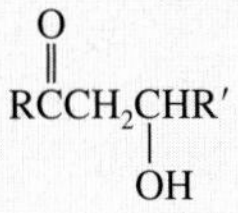

**Table 21.5** **The Diels-Alder Reaction**

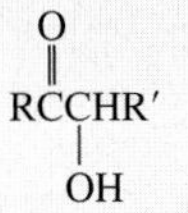

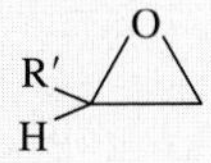

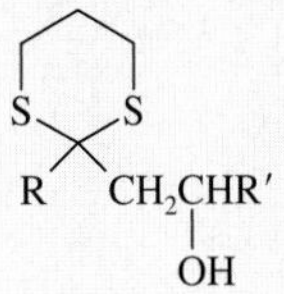

| Diene | Dienophile | Product |
|---|---|---|
|  Z = electron-donating group |  |  |
|  |  |  |

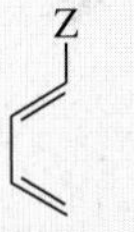

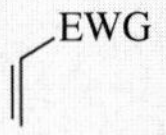

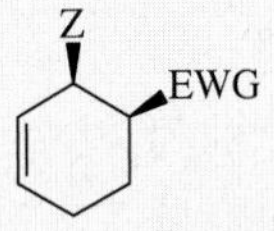

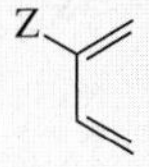

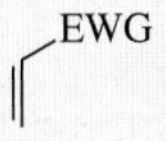

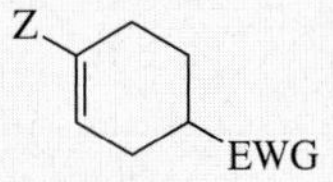

(See also Table 18.1, p. 760.)

## Table 21.6 Synthetic Reactions of Aromatic Compounds

| Ion | Reagent | Product |
|-----|---------|---------|
| **Reactions of Diazonium Ions** | | |
| 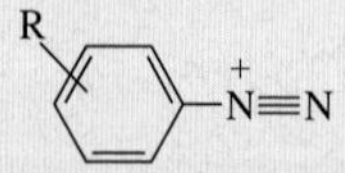 R—⬡—N≡N⁺ | CuCl | 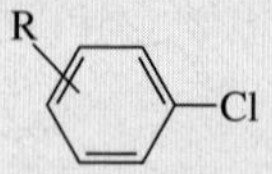 R—⬡—Cl |
| | CuBr | 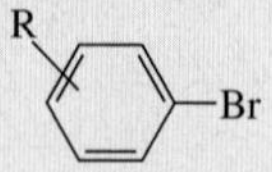 R—⬡—Br |
| | $Na_2CO_3$, CuCN | 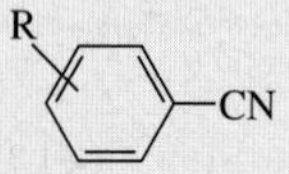 R—⬡—CN |
| | KI | 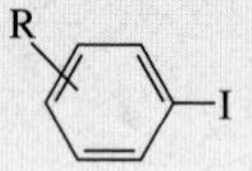 R—⬡—I |
| | $BF_4^-$ | 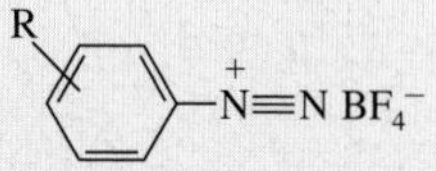 R—⬡—N≡N⁺ $BF_4^-$ |
| 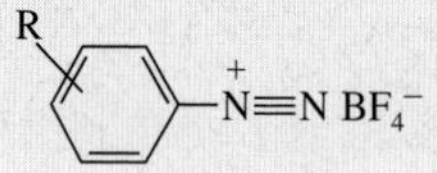 R—⬡—N≡N⁺ $BF_4^-$ | Δ | 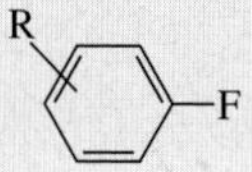 R—⬡—F |
| 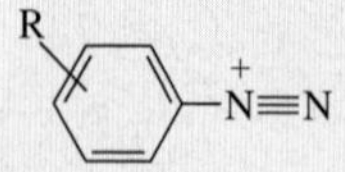 R—⬡—N≡N⁺ | $H_2O$, Δ | 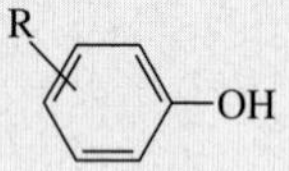 R—⬡—OH |
| | $H_3PO_2$ | 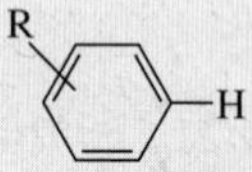 R—⬡—H |

| Aromatic Compound | Nucleophile | Intermediate | Product |
|-------------------|-------------|--------------|---------|
| **Nucleophilic Aromatic Substitution** | | | |
| X, $NO_2$, $NO_2$ (ring) | $RNH_2$ | X, NHR, $NO_2$, $N^+$ intermediate | NHR, $NO_2$, $NO_2$ (ring) |

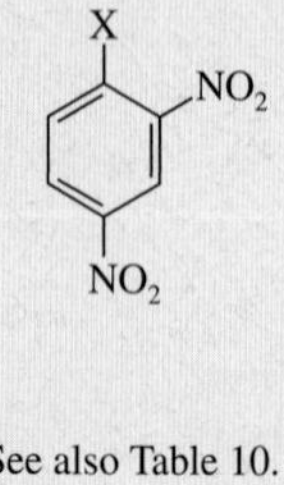

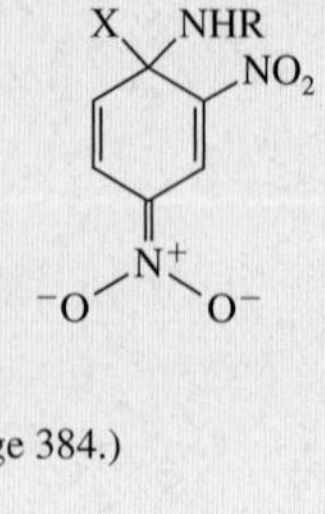

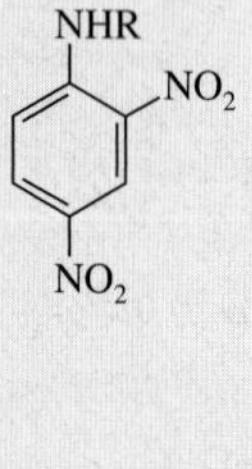

(See also Table 10.1, page 369, and Tables 10.2 and 10.3, page 384.)

## ADDITIONAL PROBLEMS

*Note:* The additional problems in this chapter are grouped together according to the main concepts that are emphasized. Later problems also may incorporate earlier concepts such as the use of protecting groups in syntheses.

### I. Problems Involving the Use of Protecting Groups

**21.47** A laboratory synthesis of chrysomelidial, a compound secreted by the larvae of some beetles to defend themselves from attack, starts in the following way.

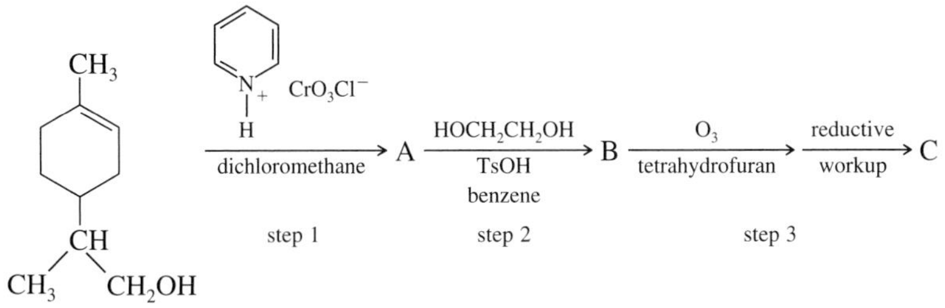

Fill in the structures of Compounds A, B, and C. Why is the second step of this synthesis necessary?

**21.48** During research into methods for synthesizing natural products, the following sequence of reactions was carried out.

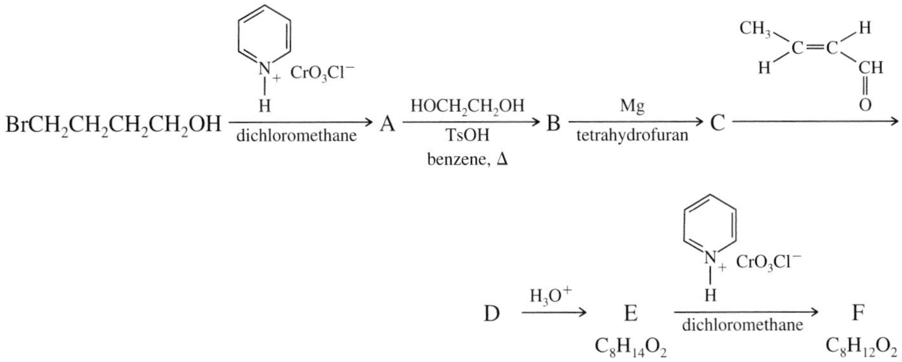

Assign structures to Compounds A, B, C, D, E, and F. (*Hint:* A review of Section 14.7A may be helpful in assigning a structure to E.) How many units of unsaturation does F have? How does the structure you assigned account for all of them?

**21.49** The oxirane derived from 7-methyl-6-octen-3-ol is used as an intermediate in the synthesis of the pheromone of the square-necked grain beetle, an insect that causes great damage to corn.

(a) What is the structure of 7-methyl-6-octen-3-ol, and how would you prepare it from 5-bromo-2-methyl-2-pentene?

(b) Before the double bond is oxidized, the hydroxyl group of 7-methyl-6-octen-3-ol is protected as the tert-butyldimethylsilyl ether. How would you do this?

(c) How would you complete the synthesis of the oxirane? Why was the hydroxyl group protected before the oxirane was made? What type of reaction is possible between an unprotected alcohol and an oxirane ring?

**21.50** The following steps were used in the synthesis of prostaglandins, biologically important fatty acids (p. 788). Assign structures to all intermediates indicated by letters.

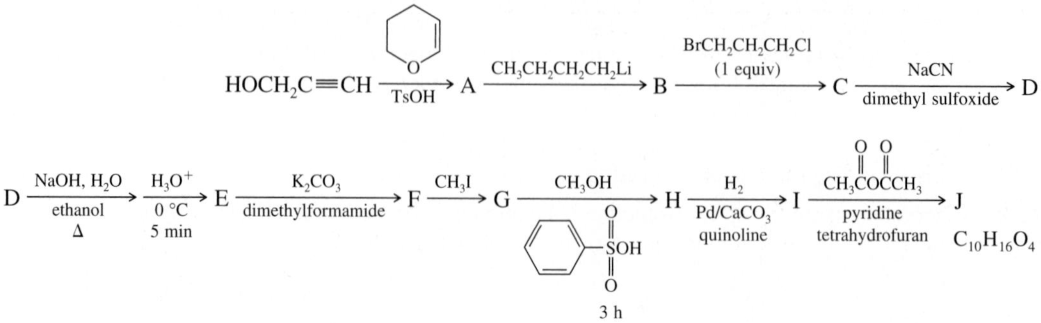

$$HOCH_2C{\equiv}CH \xrightarrow[\text{TsOH}]{} A \xrightarrow{CH_3CH_2CH_2CH_2Li} B \xrightarrow[\text{(1 equiv)}]{BrCH_2CH_2CH_2Cl} C \xrightarrow[\text{dimethyl sulfoxide}]{NaCN} D$$

$$D \xrightarrow[\substack{\text{ethanol} \\ \Delta}]{NaOH, H_2O} \xrightarrow[\substack{0\ ^\circ C \\ 5\ \text{min}}]{H_3O^+} E \xrightarrow[\text{dimethylformamide}]{K_2CO_3} F \xrightarrow{CH_3I} G \xrightarrow[\substack{}]{CH_3OH} H \xrightarrow[\substack{\text{Pd/CaCO}_3 \\ \text{quinoline}}]{H_2} I \xrightarrow[\substack{\text{pyridine} \\ \text{tetrahydrofuran}}]{CH_3COCCH_3} J$$
$$C_{10}H_{16}O_4$$

Compound H has a strong absorption band at 3450 cm$^{-1}$ and another at 1720 cm$^{-1}$ in its infrared spectrum. There are peaks at $\delta$ 1.6–2.0 (2H, m), 2.2–2.6 (5H, m), 3.68 (3H, s), and 4.2–4.3 (2H, m) in its proton magnetic resonance spectrum.

**21.51** In the synthesis of compounds that mimic the properties of peptides, a differentially substituted diamine was prepared using the following reactions. Supply structural formulas for the products indicated by letters.

$$H_2NCH_2CH_2NH_2 \xrightarrow[\substack{\text{1 equiv} \\ \text{NaOH}}]{[(CH_3)_3COC]_2O} A \xrightarrow[\text{acetonitrile}]{} B \xrightarrow[\substack{\text{dimethyl ether} \\ \text{chloroform}}]{HCl} C \xrightarrow[\text{acetonitrile}]{} D$$

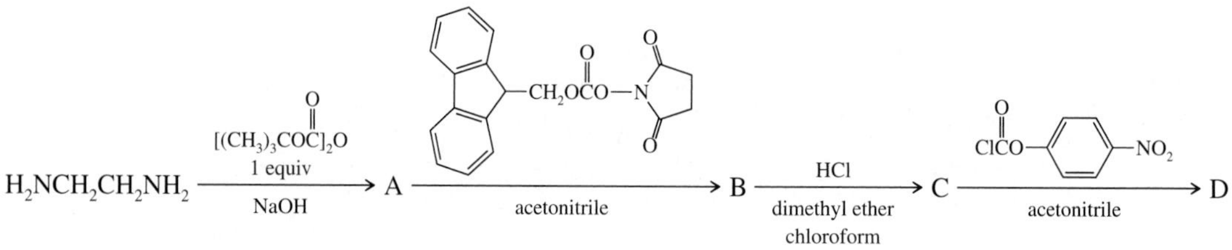

## II. Problems Involving Oxidation–Reduction Reactions

**21.52** Give structural formulas for the organic intermediates and products indicated by letters in the following equations.

(a)

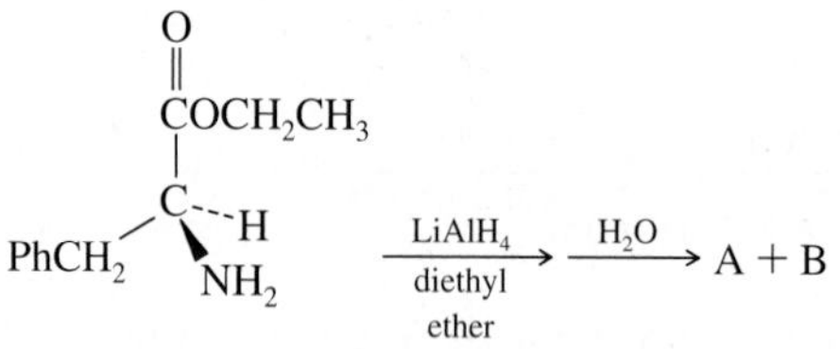

(b)

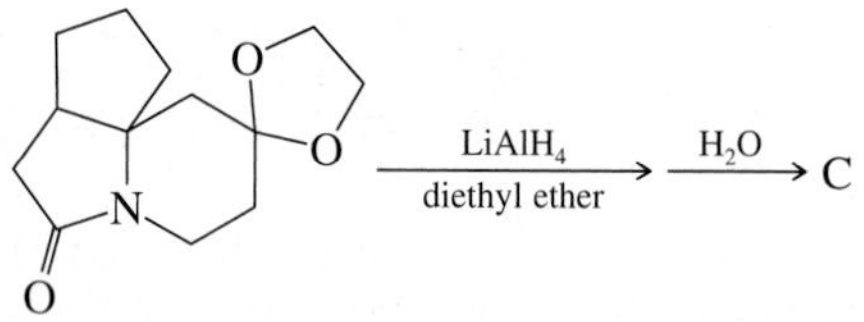

(c)

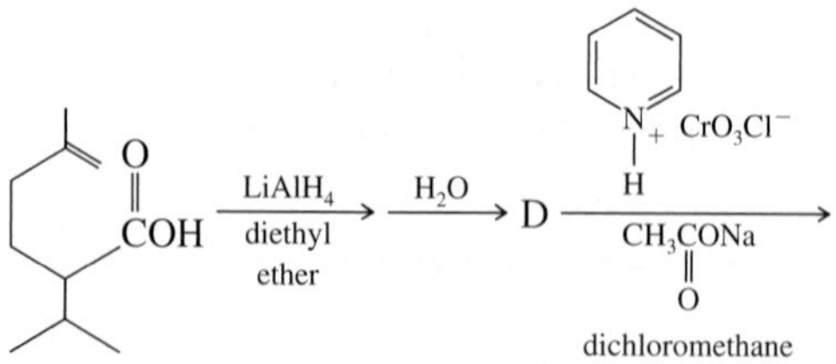

(d)

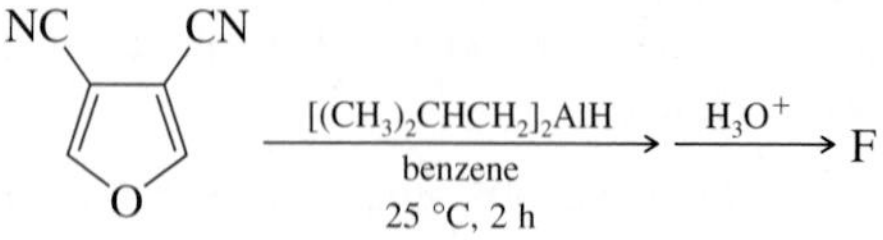

(e)

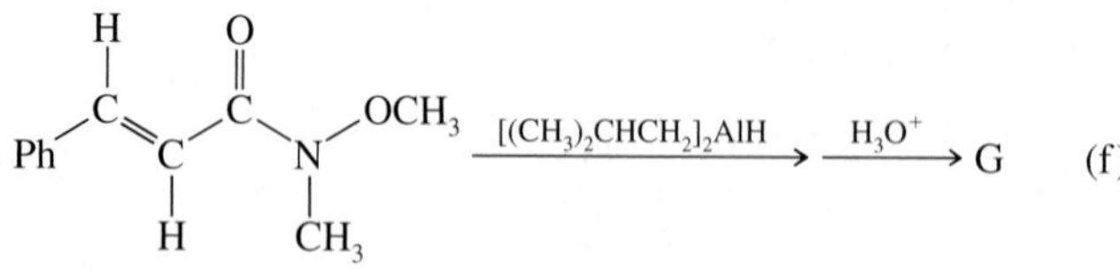

(f)

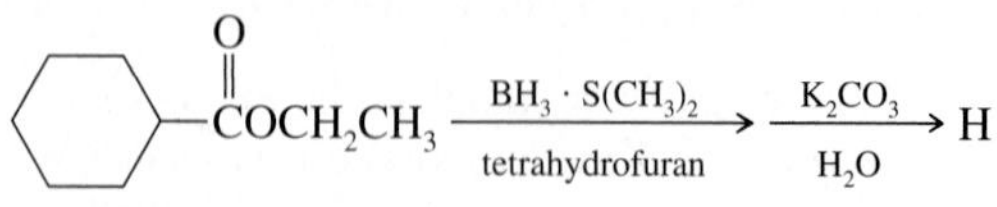

(g) 

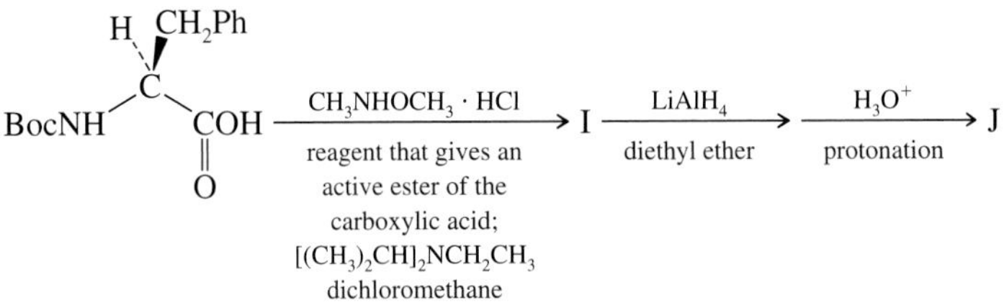

**21.53** A furancarboxylic acid was converted to two esters as shown below. How would you carry out these transformations?

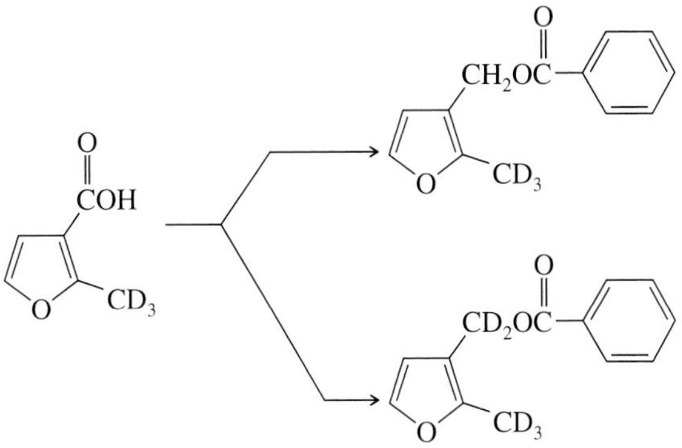

**21.54** A common way to study reaction mechanisms is to prepare compounds that are labeled with isotopes. A source of $^{13}C$ atoms is $^{13}CO_2$. Design a rational synthesis of 2-$^{13}C$-propanoic acid starting with $^{13}CO_2$ and any other reagents that you need.

**21.55** The last three stages in the synthesis of an intermediate in terpene biosynthesis (p. 752) are shown below. Supply structural formulas for the products indicated below.

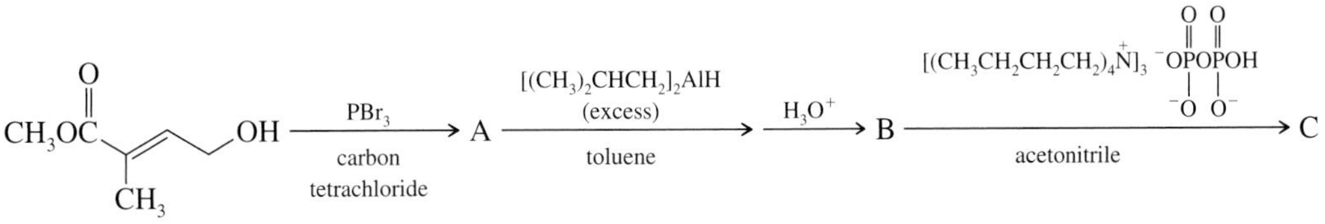

**21.56** A synthesis of certain antibiotics called the pseudomonic acids included the following transformations. Suggest reagents for these reactions.

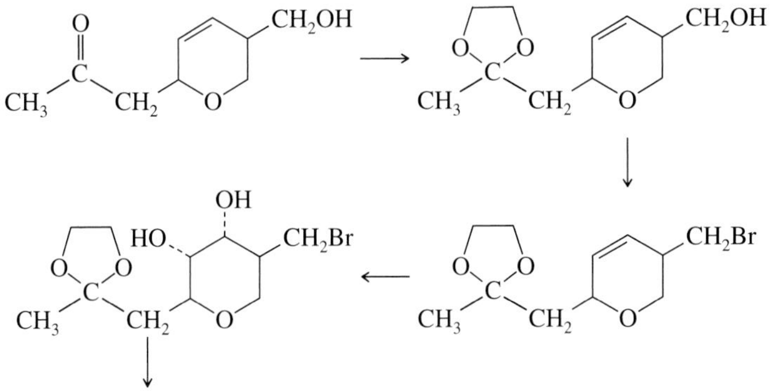

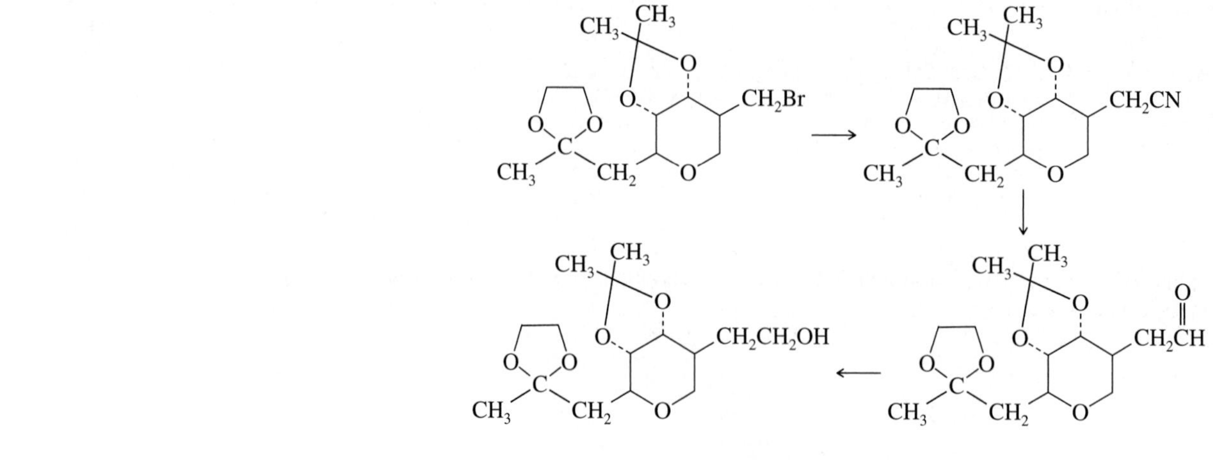

### III. Reactions of Carbon Nucleophiles

**21.57** The following ketoester, which is an intermediate in the synthesis of an inhibitor of an enzyme critical to the survival of cancer cells, was prepared by two routes. Show how you would carry out the synthesis in each case.

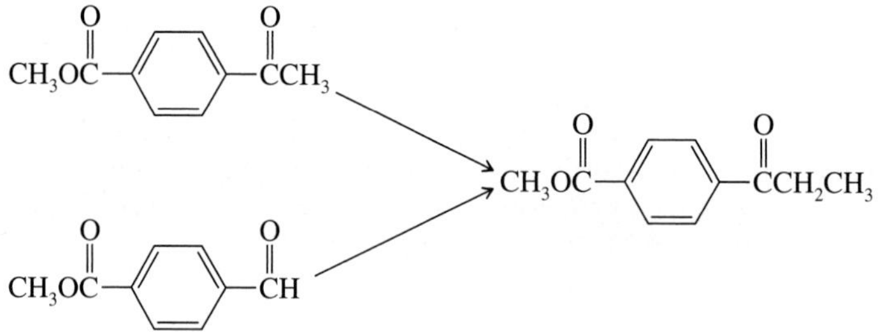

**21.58** Juvenile hormones are substances that block the maturation of the pupae of insects. The juvenile hormone of the giant silkworm moth has the structure shown below.

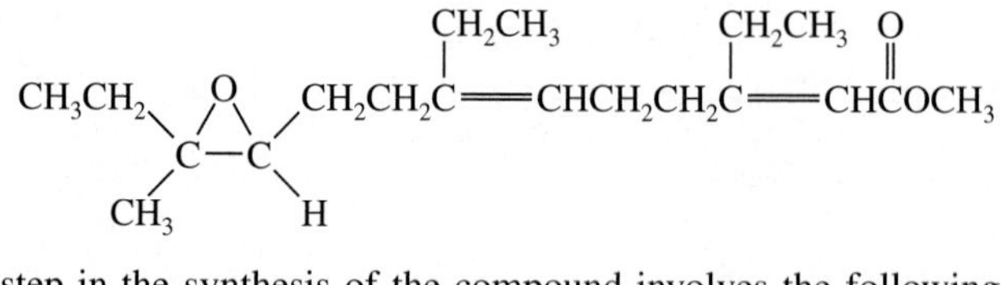

An important step in the synthesis of the compound involves the following transformation. Suggest a reagent for the reaction.

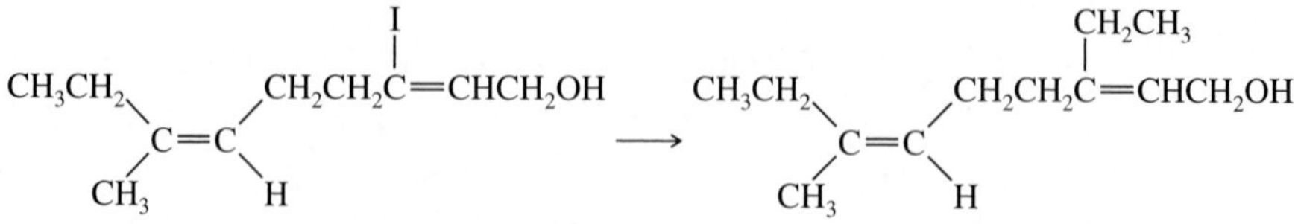

**21.59** A synthesis of philanthotoxin, a neurotoxic poison from the wasp, was explored using the following reactions. Supply structural formulas for the missing products.

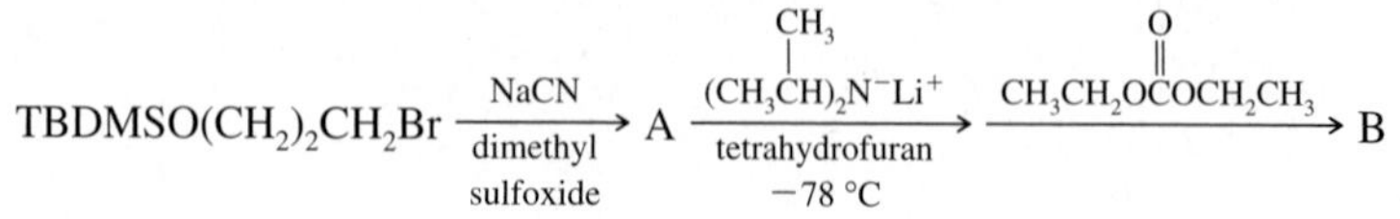

$$\text{TBDMSO(CH}_2)_2\text{CH}_2\text{Br} \xrightarrow[\substack{\text{dimethyl} \\ \text{sulfoxide}}]{\text{NaCN}} \text{A} \xrightarrow[\substack{\text{tetrahydrofuran} \\ -78\,°C}]{(\text{CH}_3\text{CH})_2\text{N}^-\text{Li}^+} \xrightarrow{\text{CH}_3\text{CH}_2\text{OCOCH}_2\text{CH}_3} \text{B}$$

**21.60** A study of how long-chain branched fatty acids are synthesized by marine organisms required acids labeled with radioactive $^{14}C$ at the carbon atom of the carbonyl group. A synthesis of 12-methyltetradecanoic acid was carried out, with 9-bromo-1-nonanol and 2-methylbutanal as the starting materials and $K^{14}CN$ as the source of radioactive carbon. Outline a synthesis that employs the Wittig reaction. The tetrahydropyranyl ether also plays an important role as a protecting group in this synthesis.

**21.61** The cockroach is a major insect pest in many parts of the world, so a great deal of research has been done on ways to control it. Much of this effort concentrates on the synthesis of the sex pheromones of the insects. Some of the steps in such a synthesis were discussed in Problem 21.29 (p. 879). The aldehyde synthesized there is used with the Wittig reagent shown below to make the pheromone. Fill in structural formulas for the reagents, intermediates, and products designated by letters.

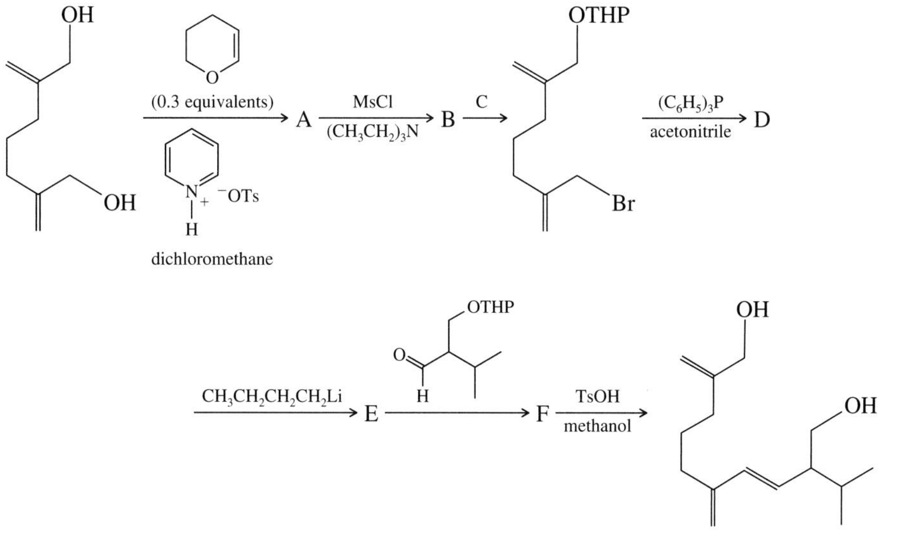

a sex pheromone of the cockroach

**21.62** Some steps in the synthetic studies of calyculins, antitumor compounds isolated from marine sponges, are shown below. Provide structural formulas for the products designated by letters, including stereochemistry when known.

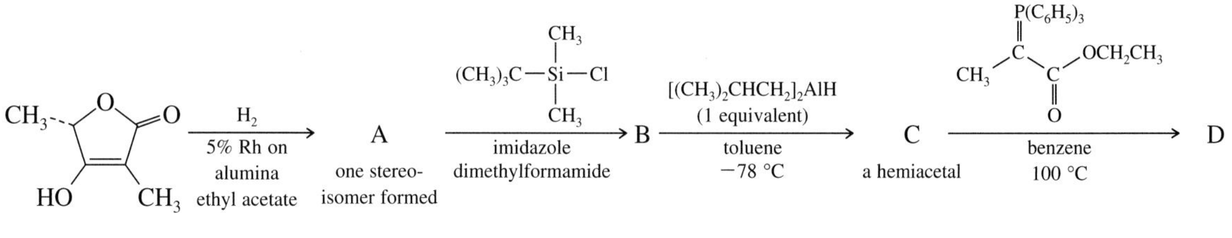

**21.63** The aggregation pheromone of a species of fruit fly was prepared by the following sequence of reactions. Provide structural formulas for the reactive intermediates and products indicated by letters.

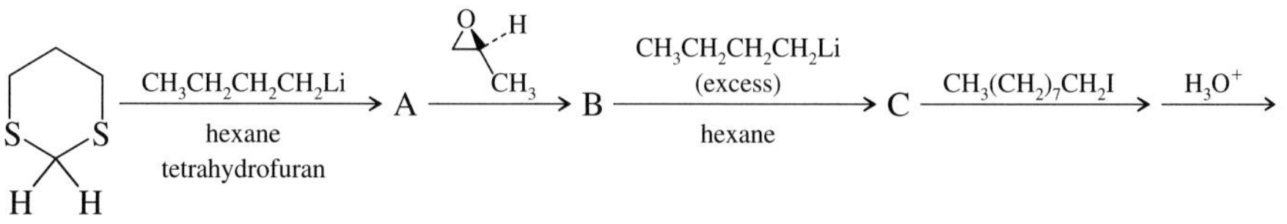

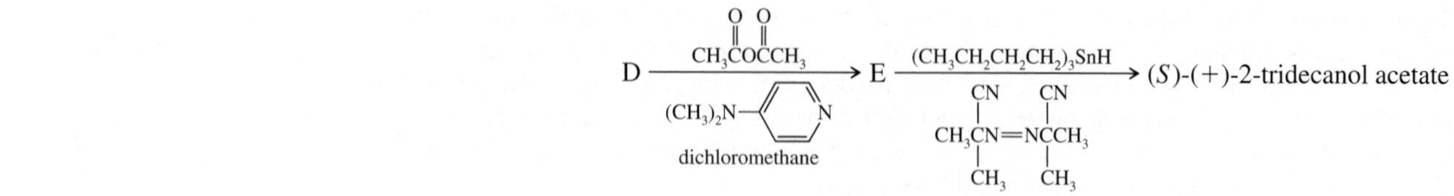

$$D \xrightarrow[\text{dichloromethane}]{\substack{\overset{O}{\overset{\|}{C}}\overset{O}{\overset{\|}{C}} \\ CH_3COCCH_3 \\ (CH_3)_2N\text{—} }} E \xrightarrow{(CH_3CH_2CH_2CH_2)_3SnH} (S)\text{-}(+)\text{-}2\text{-tridecanol acetate}$$

*Hint:* The last step involves the free-radical reduction of C—S bonds in a reaction similar to what takes place with hydrogen over Raney nickel (p. 579).

**21.64** The following transformation is carried out during the synthesis of a natural product. How would you accomplish this transformation?

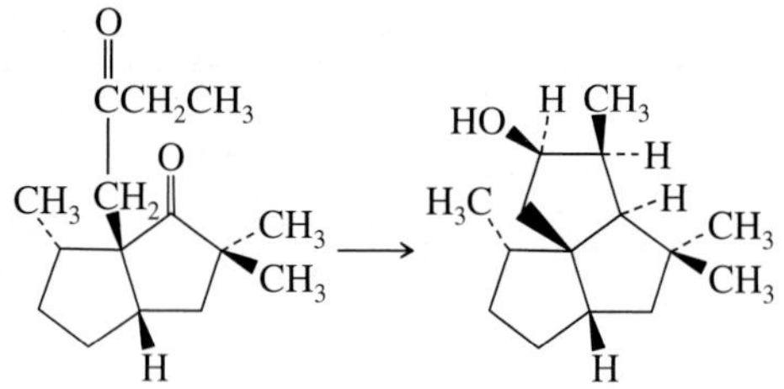

**21.65** A synthesis of pentalenolactone is continued by the conversion of the product ketoacid from Problem 17.20 (p. 717) to a bicyclic diketone. Assign structural formulas to the reagents used in the two steps of this conversion, Compounds A and B in the following equation.

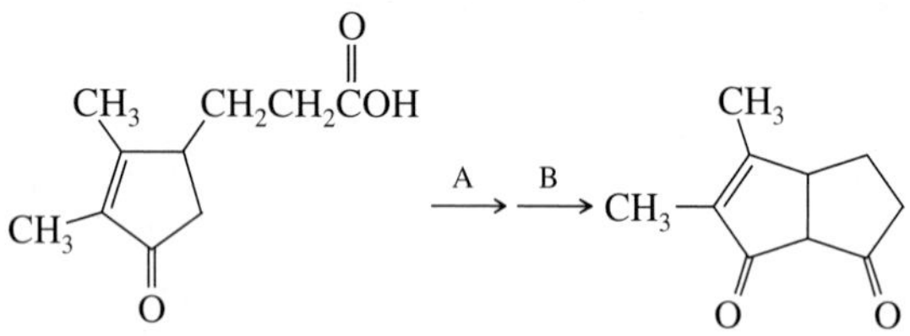

**21.66** In a study of the stereochemistry of the Michael reaction, it became necessary to synthesize some $\alpha,\beta$-unsaturated ketones. The four examples shown below represent four different pathways to such ketones. Show how you would complete the synthesis in each step. Each requires more than one step. (*Note*: These are excellent problems on which to practice your problem-solving skills.)

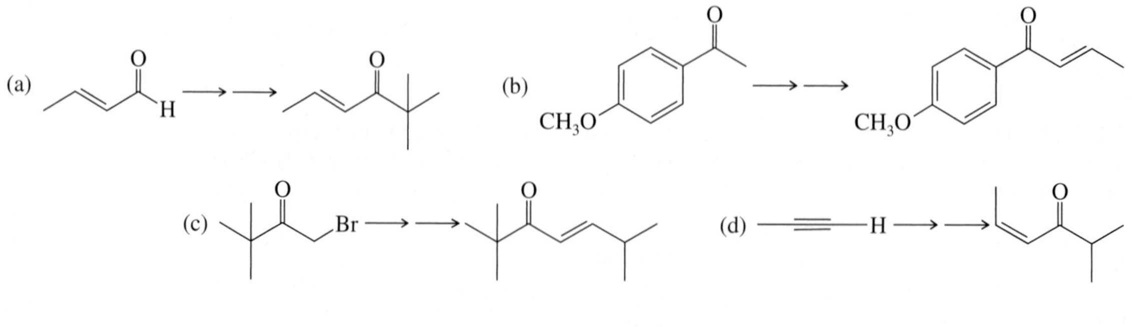

**21.67** The following compounds were prepared as intermediates in the synthesis of an alkaloid-like compound. Supply reagents for all transformations shown.

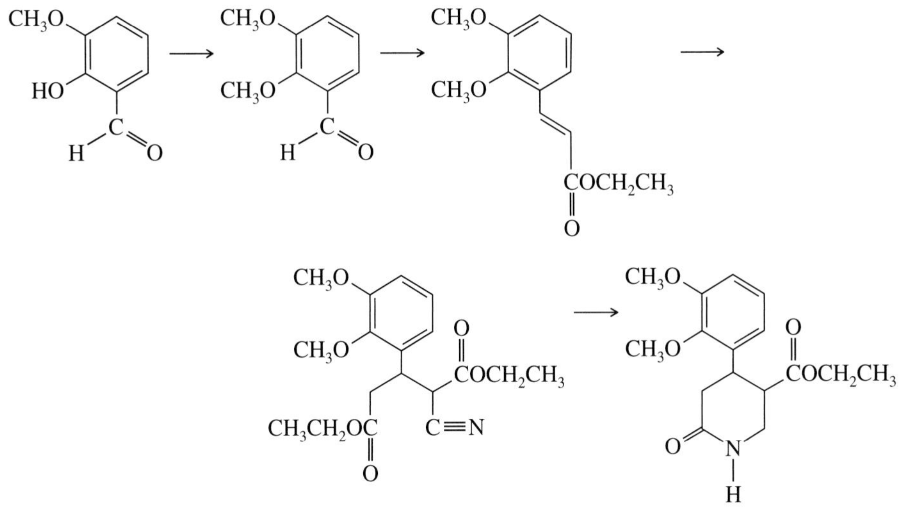

**21.68** Research into the synthesis of (*S*)-dolaphenine, a component of dolastin 10, a natural product isolated from the Indian Ocean sea hare with strong antitumor activity, involved the following steps. Supply structures for the reagents or products indicated by letters.

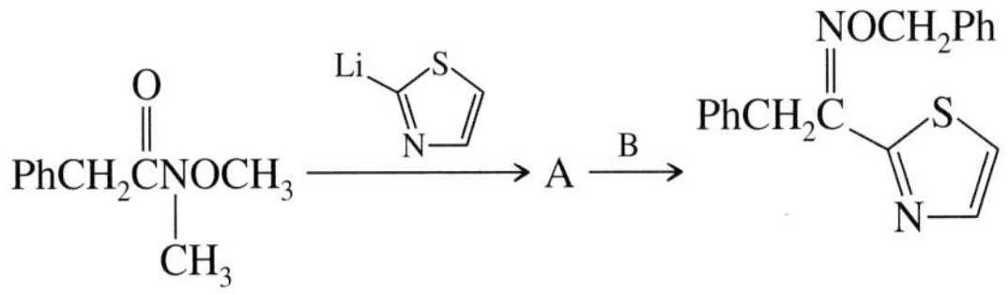

**21.69** The following reactions were carried out in the synthesis of an inhibitor of the protease enzyme used by the HIV virus in replication. Supply structures for the products indicated by letters.

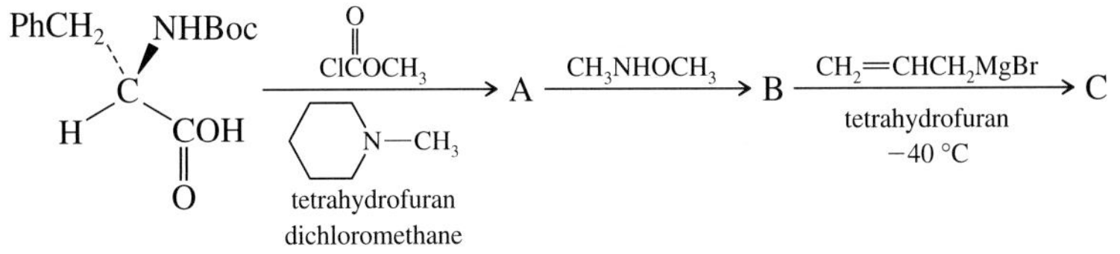

**21.70** The following reaction was carried out in order to synthesize the hydroxyketone shown. The conditions of the reaction are critical in achieving a reasonable yield, 66%, of the desired product. Under these conditions, 33% of unreacted lactone is recovered.

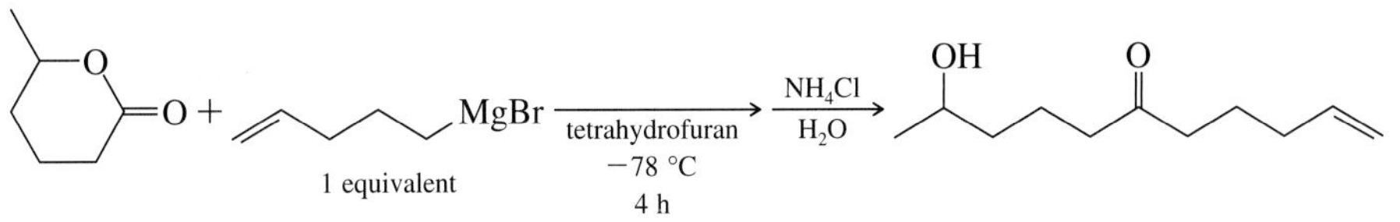

The product has peaks in its $^{13}C$ nuclear magnetic resonance spectrum at $\delta$ 19.67, 22.68, 23.29, 32.95, 38.51, 44.45, 67.32, 115.06, 137.18, and 211.20. Its proton nuclear magnetic resonance spectrum had peaks at $\delta$ 1.0–2.4 (13H, m), 1.2 (3H, d), 3.7 (1H, m), 4.9 (2H, m), and 5.7 (1H, m).

(a) Write a mechanism for the reaction shown above.
(b) Assign as many of the nuclear magnetic resonance peaks as possible to the structure of the product of the reaction.
(c) If the temperature is allowed to rise to 220 °C during the reaction, the desired product is contaminated by a tertiary alcohol. Propose a structure for this side product. How does it arise?

## IV. Diels-Alder Reactions

**21.71** Give structural formulas for the starting material or products symbolized by letters in the following equations. If the stereochemistry of the reaction is known, be sure to show it in your answer.

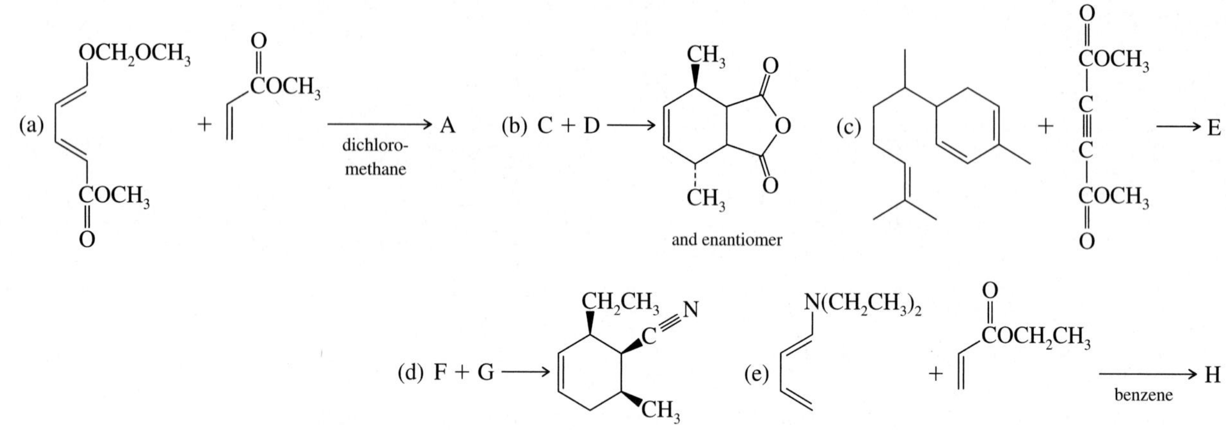

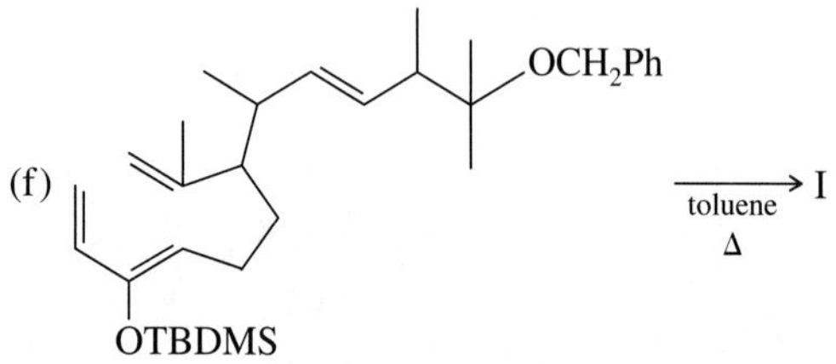

**21.72** Assign structures to the compounds designated by letters shown below. Infrared spectral data are given for the compounds. Assign as many of the bands as you can.

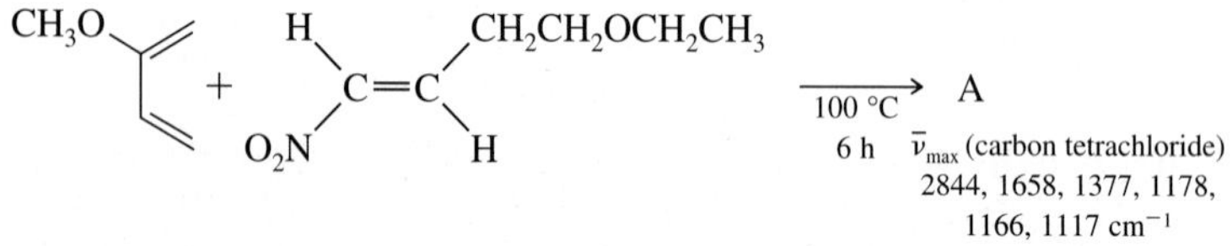

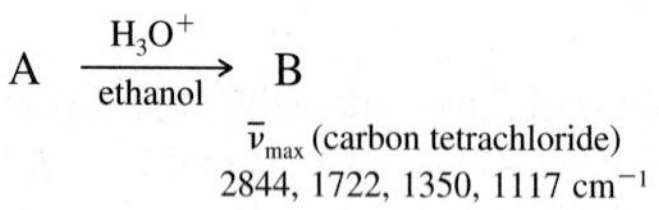

**21.73** Paclitaxel (Taxol), isolated from the bark of the Pacific yew, shows great promise as an antitumor agent, especially against cancers that are difficult to treat, such as ovarian cancer. Chemists are working to devise syntheses of the compound, which has the structure shown below.

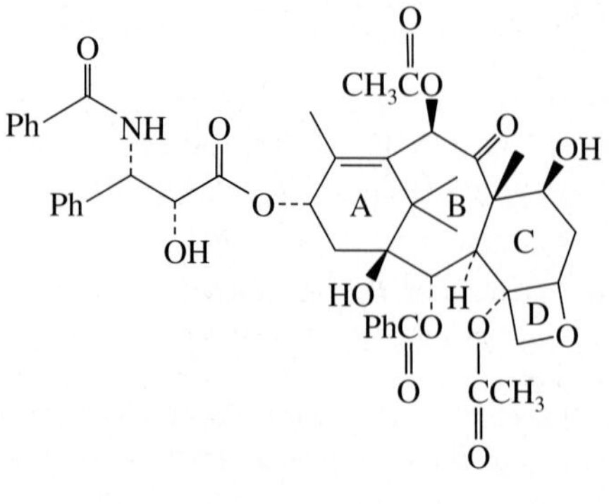

The reactions below were carried out in studies on the synthesis of the A and B ring portions of compounds analogous to paclitaxel. Give structural formulas for the compounds designated by letters.

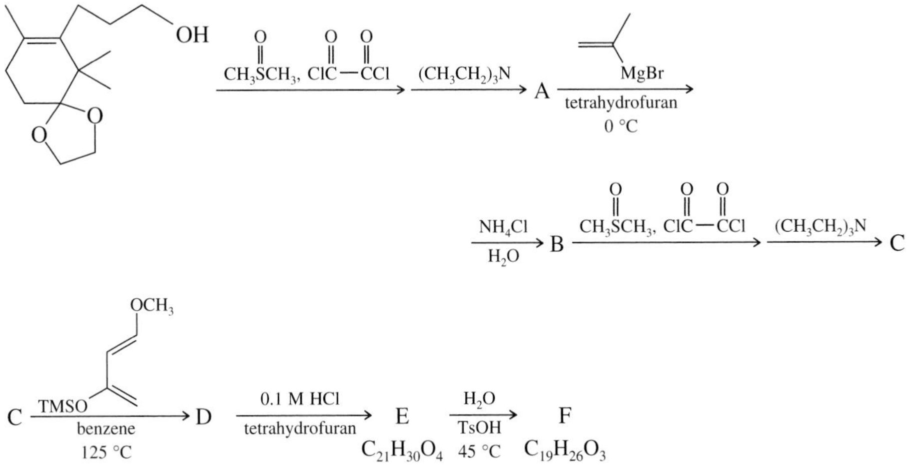

**21.74** Intramolecular Diels-Alder reactions are important in the synthesis of polycyclic compounds. The following reactions were carried out in an investigation of the preparation of systems with fused five- and six-membered rings.

(a) Provide structural formulas for the reagents, intermediates, and products designated by letters.

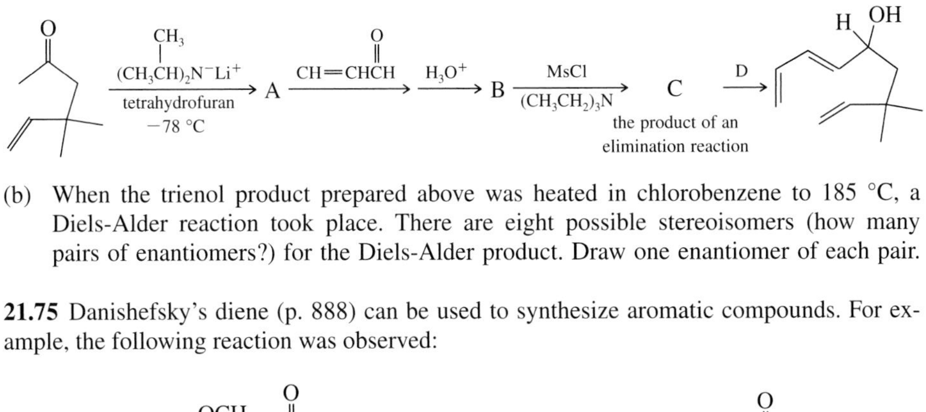

(b) When the trienol product prepared above was heated in chlorobenzene to 185 °C, a Diels-Alder reaction took place. There are eight possible stereoisomers (how many pairs of enantiomers?) for the Diels-Alder product. Draw one enantiomer of each pair.

**21.75** Danishefsky's diene (p. 888) can be used to synthesize aromatic compounds. For example, the following reaction was observed:

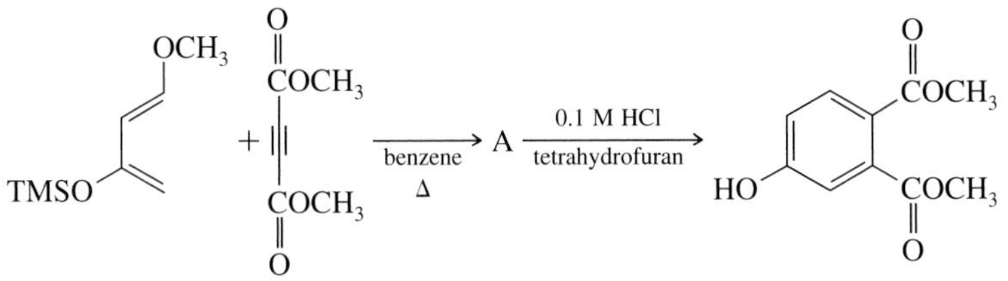

79%

(a) What is the structure of Compound A?
(b) Write a mechanism showing how the product of the hydrolysis of Compound A is converted into the product finally isolated.

## V. Synthetic Transformations of Aromatic Compounds

**21.76** Give structural formulas for the compounds or intermediates designated by letters in the following equations.

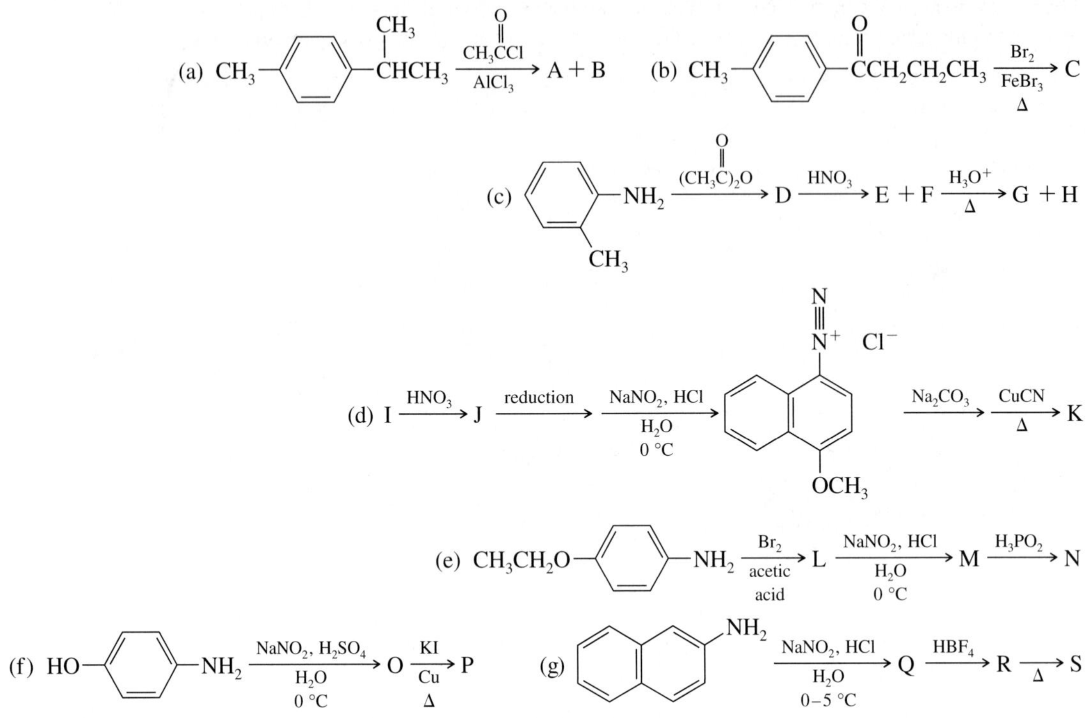

(a) CH₃—⟨benzene⟩—CHCH₃ (with CH₃) $\xrightarrow[\text{AlCl}_3]{\text{CH}_3\text{CCl}}$ A + B    (b) CH₃—⟨benzene⟩—CCH₂CH₂CH₃ $\xrightarrow[\Delta]{\text{Br}_2 \atop \text{FeBr}_3}$ C

(c) ⟨o-toluidine⟩—NH₂ $\xrightarrow{(\text{CH}_3\text{C})_2\text{O}}$ D $\xrightarrow{\text{HNO}_3}$ E + F $\xrightarrow[\Delta]{\text{H}_3\text{O}^+}$ G + H

(d) I $\xrightarrow{\text{HNO}_3}$ J $\xrightarrow{\text{reduction}}$ $\xrightarrow[\substack{\text{H}_2\text{O} \\ 0\,°\text{C}}]{\text{NaNO}_2,\ \text{HCl}}$ ⟨naphthalene diazonium, OCH₃⟩ $\xrightarrow[\Delta]{\text{Na}_2\text{CO}_3 \quad \text{CuCN}}$ K

(e) CH₃CH₂O—⟨benzene⟩—NH₂ $\xrightarrow[\substack{\text{acetic} \\ \text{acid}}]{\text{Br}_2}$ L $\xrightarrow[\substack{\text{H}_2\text{O} \\ 0\,°\text{C}}]{\text{NaNO}_2,\ \text{HCl}}$ M $\xrightarrow{\text{H}_3\text{PO}_2}$ N

(f) HO—⟨benzene⟩—NH₂ $\xrightarrow[\substack{\text{H}_2\text{O} \\ 0\,°\text{C}}]{\text{NaNO}_2,\ \text{H}_2\text{SO}_4}$ O $\xrightarrow[\Delta]{\text{KI} \atop \text{Cu}}$ P     (g) ⟨2-naphthylamine⟩—NH₂ $\xrightarrow[\substack{\text{H}_2\text{O} \\ 0\text{–}5\,°\text{C}}]{\text{NaNO}_2,\ \text{HCl}}$ Q $\xrightarrow[\Delta]{\text{HBF}_4}$ R $\longrightarrow$ S

**21.77** The synthesis of cosalane, a new anti-HIV agent, has the following step:

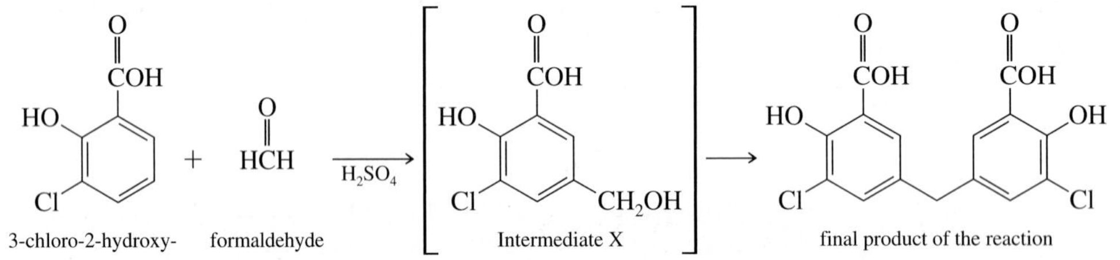

⟨3-chloro-2-hydroxybenzoic acid⟩ + HCH (formaldehyde) $\xrightarrow{\text{H}_2\text{SO}_4}$ [ Intermediate X ] $\longrightarrow$ final product of the reaction

3-chloro-2-hydroxy-      formaldehyde
benzoic acid             1 equivalent
2 equivalents

(a) Using the curved-arrow convention, provide a mechanism for the formation of Intermediate X from 3-chloro-2-hydroxybenzoic acid and formaldehyde in the presence of sulfuric acid.

(b) 3-Chloro-2-hydroxybenzoic acid has three positions that could have been substituted in this reaction with formaldehyde. Show clearly how chemists rationalize the regioselectivity actually observed in the reaction.

(c) Intermediate X is transformed rapidly under the conditions of the reaction to the final product of the reaction. What is the structure of the electrophilic reactive intermediate responsible for this rapid conversion?

**21.78** *p*-Nitrofluorobenzene reacts with a series of nucleophiles in methanol at 25 °C. The relative rates for these reactions are shown below. Explain the experimental observations. (*Hint:* Reviewing Section 7.4 may be helpful.)

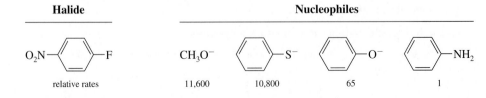

| **Halide** | **Nucleophiles** | | | |
|---|---|---|---|---|
| O₂N—⟨benzene⟩—F | CH₃O⁻ | ⟨C₆H₅⟩—S⁻ | ⟨C₆H₅⟩—O⁻ | ⟨C₆H₅⟩—NH₂ |
| relative rates | 11,600 | 10,800 | 65 | 1 |

## VI. Problems Involving Solid-Phase and Combinatorial Syntheses

**21.79** The following reactions come from research into applications of solid-phase organic syntheses to interesting problems. They demonstrate the range of reactivity possible on solid supports. Supply structures for the reagents, intermediates, or products indicated by letters.

(a) Synthesis of carbocyclic amino acids.

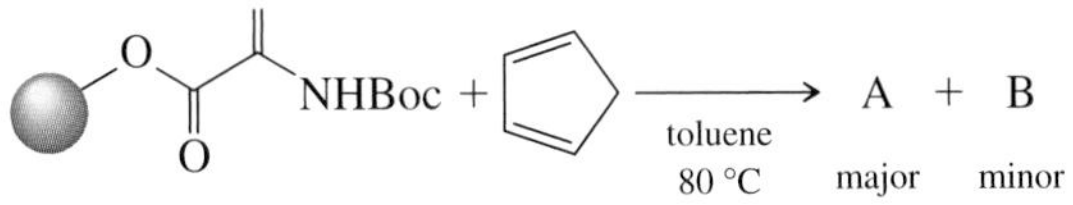

A and B are diastereomers; A has the electronegative group exo.

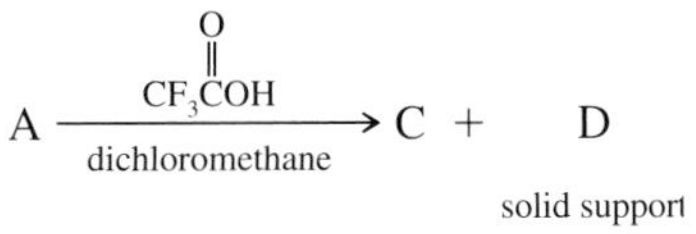

(b) Synthesis of ureas.

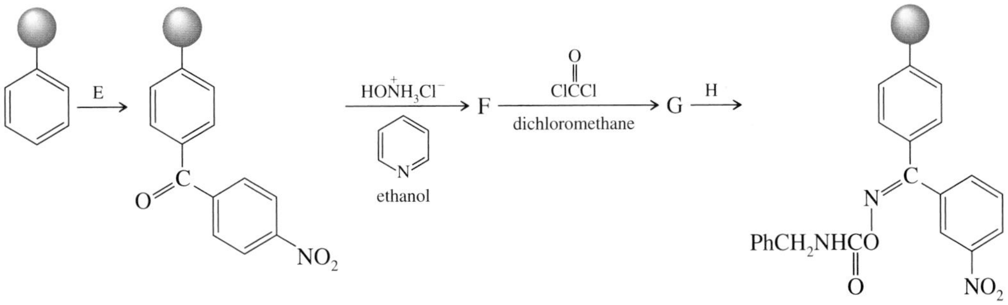

(c) Synthesis of compounds that mimic peptides.

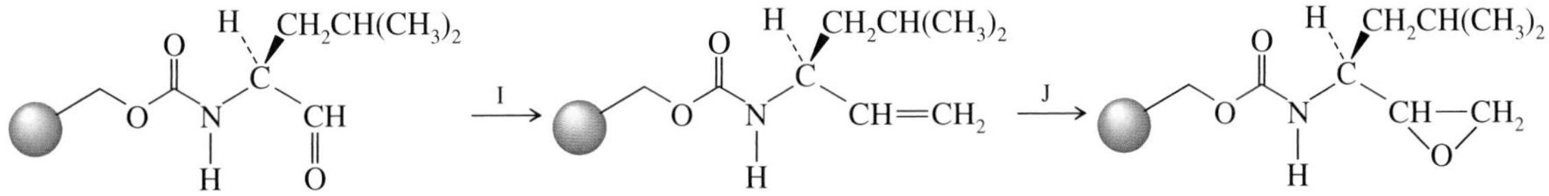

(d) Synthesis of unsaturated acids.

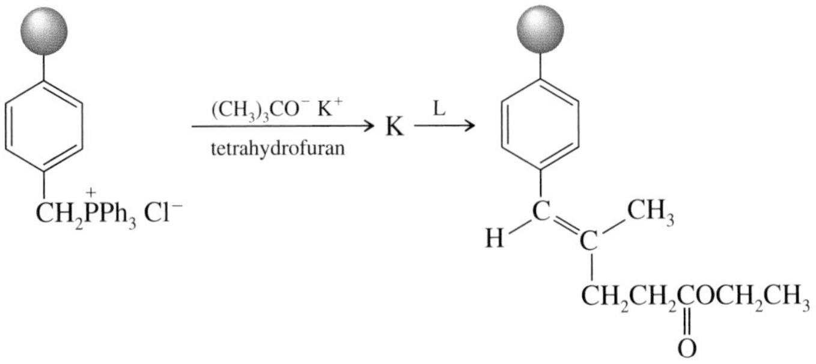

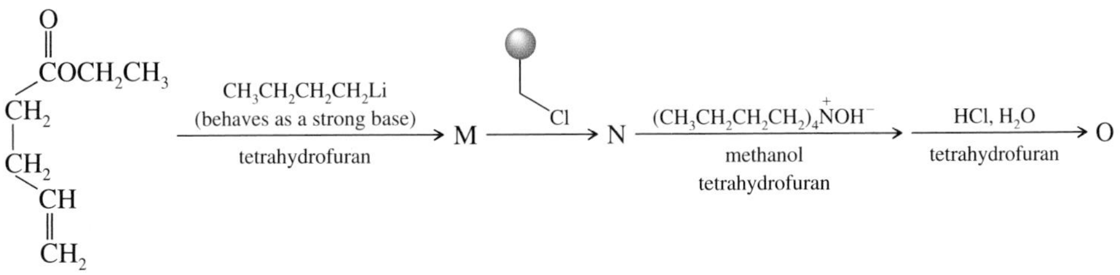

(e) Synthesis of enzyme inhibitors.

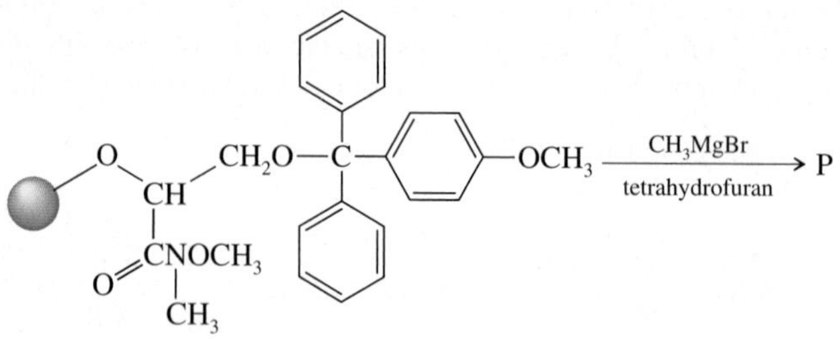

**21.80** Peptide aldehydes have been found to be useful inhibitors of several kinds of proteases, including HIV protease. The following reactions were used in a solid-phase synthesis of a peptide aldehyde that can be easily modified to create a library of such peptides. Provide structural formulas for the reagents or products indicated by letters.

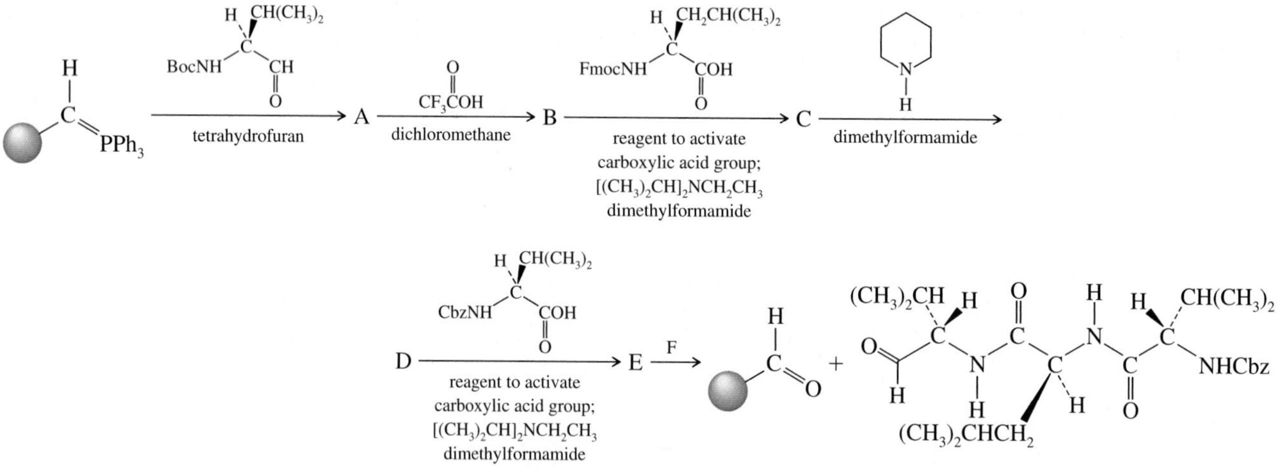

# The Chemistry of Heterocyclic Compounds

## A Look Ahead

Cyclic organic compounds are divided into two large classes: Those with only carbon atoms in their rings are known as **carbocyclic compounds,** and those with atoms of elements other than carbon in their rings are called **heterocyclic compounds.** Heterocyclic compounds occur so widely in nature and are of such importance chemically that any discussion of organic chemistry will not get very far without mentioning them. For example, the heterocyclic compounds pyridine, tetrahydrofuran, and oxirane are already familiar to you.

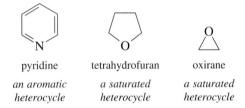

| pyridine | tetrahydrofuran | oxirane |
|---|---|---|
| *an aromatic heterocycle* | *a saturated heterocycle* | *a saturated heterocycle* |

Some heterocyclic compounds, such as pyridine, are aromatic. Others, such as tetrahydrofuran, are not. Aromatic heterocycles differ in reactivity. Some, such as thiophene, are much more reactive toward electrophilic substitution than benzene is. Others, such as pyridine, need extreme conditions to react with electrophiles. This chapter will explore these differences.

Heterocyclic compounds play important medicinal and biochemical roles. The purine and pyrimidine bases that are structural units in RNA and DNA are heterocyclic compounds, as are many drugs such as morphine, heroin, and cocaine. The chemistry of heterocyclic compounds is such an extensive field that this chapter is by necessity a highly selective look at some of it. The chapter emphasizes reactions that will expand and reinforce your understanding of the basic chemical principles that have been developed in the rest of this book.

## 22.1  Nomenclature of Heterocyclic Compounds

Heterocyclic compounds may be classified in a number of ways: by the size of the ring, by the nature and number of **heteroatoms** (nonmetal atoms other than carbon) in the ring, by the degree of unsaturation in the ring, and by whether or not the compound has aromatic character.

Aromatic heterocyclic compounds were included in Section 10.1 to show that nonbonding electrons on heteroatoms could be considered part of an aromatic sextet. The important aromatic heterocyclic systems that contain a single heteroatom are shown below. Nonbonding electrons shown inside a ring are part of the aromatic sextet for that system. Note that some aromatic heterocycles have additional nonbonding electrons on a nitrogen, oxygen, or sulfur atom that do not contribute to the aromaticity of the compound. Some of the chemical consequences of these structural properties are explored in Sections 22.3 and 22.5.

pyridine

quinoline

isoquinoline

pyrrole

thiophene

furan

indole

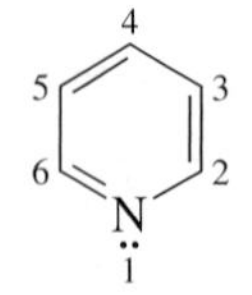

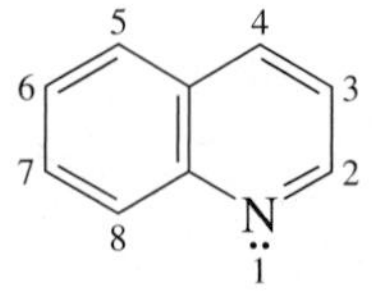

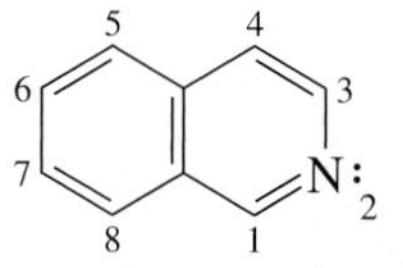

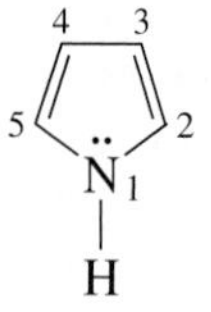

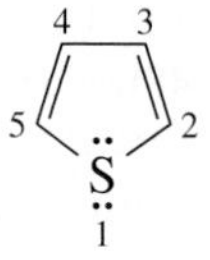

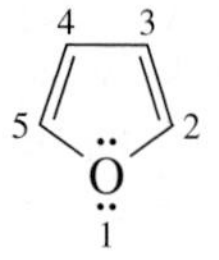

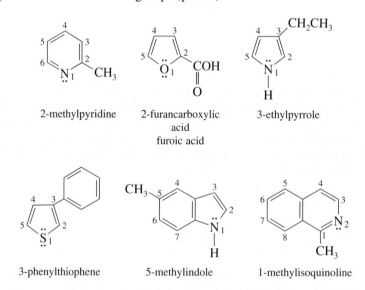

The heteroatom in a heterocycle is given the lowest possible number consistent with an orderly progression around the ring system. In isoquinoline, a carbon atom has a lower number than the nitrogen atom to preserve an orderly sequence around the periphery of the two rings. Substituted heterocyclic compounds are named in the same way as other compounds, by giving the name of the substituent and a number to indicate its position on the ring system. The name of the substituent may appear as either a prefix or a suffix to the name of the heterocycle, depending on the rules for the precedence of functional groups (p. 539).

2-methylpyridine

2-furancarboxylic acid
furoic acid

3-ethylpyrrole

3-phenylthiophene

5-methylindole

1-methylisoquinoline

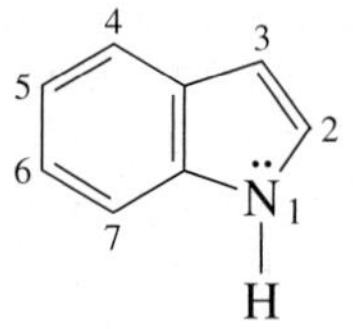

The compounds shown above have only a single heteroatom in their rings. There are other heterocyclic compounds with some aromatic character that have two or more heteroatoms, one of which is nitrogen. When such compounds have a five-membered ring system, their names all end in **azole.** The rest of the name indicates what other heteroatoms are present.

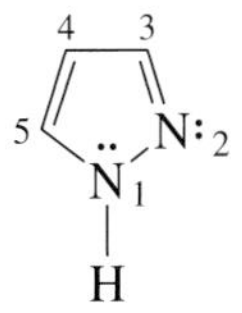

pyrazole 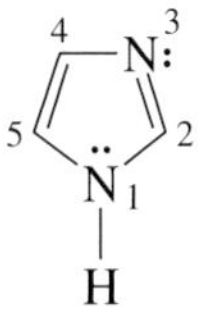   imidazole 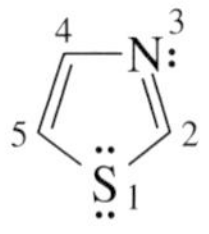   thiazole 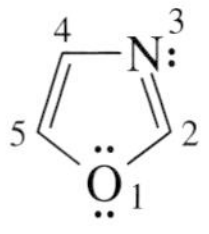   oxazole 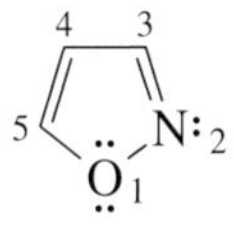   isoxazole

The names **pyrazole** and **imidazole** are given to the two isomeric compounds containing two nitrogen atoms in the ring. The name **thiazole** indicates that the ring has a sulfur atom and a nitrogen atom in it, and **oxazole** indicates the presence of oxygen and nitrogen. That name is reserved for the system in which the two heteroatoms are separated by a carbon atom; the compound in which oxygen and nitrogen are adjacent to each other is known as **isoxazole.**

The six-membered aromatic heterocyclic system with two nitrogens exists in three isomeric forms, the most important of which is **pyrimidine. Purine** is another heterocycle that is biologically important.

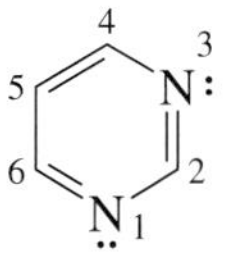

pyrimidine 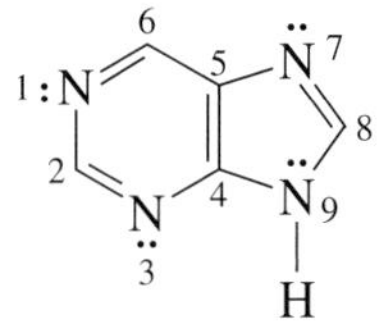   purine

The compounds that have been discussed so far in this section have the maximum degree of unsaturation. More saturated heterocycles also exist; for example, tetrahydrofuran is the fully saturated form of furan. Some other examples of heterocyclic compounds with varying degrees of unsaturation are shown in the structures below.

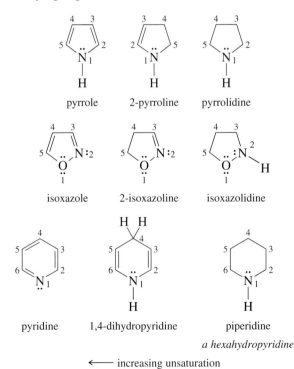

pyrrole   2-pyrroline   pyrrolidine

isoxazole   2-isoxazoline   isoxazolidine

pyridine   1,4-dihydropyridine   piperidine

*a hexahydropyridine*

⟵ increasing unsaturation

Certain heterocycles with four-membered rings are also important. For example, substituted **azetidinones** are components of a number of important antibiotics such as penicillin and cephalosporin.

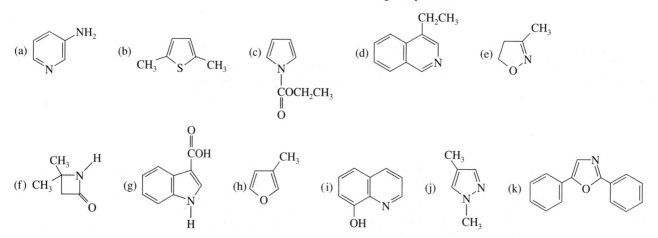

penicillin G                                    cephalosporin C

The reactivity of the carbonyl group in the azetidinone, a strained four-membered amide ring (also called a **β-lactam**), in these antibiotics is responsible for their biological activity. The compounds act as acylating agents and disrupt the synthesis of bacterial membranes.

**Problem 22.1**   Name the following compounds.

(a), (b), (c), (d), (e), (f), (g), (h), (i), (j), (k)

**Problem 22.2**   Write structural formulas for the following compounds.

(a) 4-ethylindole      (b) 6-aminoquinoline      (c) 2-methyl-5-phenylthiophene      (d) 1,4-dimethylisoquinoline

(e) 5-phenylisothiazole      (f) 2,5-dimethylfuran      (g) 3-ethylisoxazole      (h) 5-chloropyrimidine

(i) 2,4-dimethyloxazole      (j) 3-methyltetrahydrofuran      (k) 3,5-dimethylpyrazole      (l) 3-ethyl-2-pyrrolecarboxylic acid

## 22.2   Aromatic Heterocyclic Compounds

Aromatic heterocyclic compounds resemble benzene in that each one can be shown experimentally to have resonance energy. In other words, each is more stable than would be expected of a similar compound with localized double bonds. Resonance energies determined by heats of combustion for different heterocycles, are shown in Table 22.1. Benzene and cyclopentadiene are included for comparison.

None of the aromatic heterocyclic compounds has as much resonance stabilization as benzene. All the five-membered ring heterocycles, however, are much more stable than is expected for a cyclic diene, represented in Table 22.1 by cyclopentadiene. Thiophene, pyrazole, pyridine, pyrrole, and imidazole should have the chemical reactivity of aromatic compounds. Section 22.5 will examine electrophilic aromatic substitution reactions of some heterocycles to see to what extent this prediction is correct.

In Table 22.1, furan stands out as the compound with the least resonance stabilization of the simple heterocycles. The loss of aromaticity is not a large barrier to the reactions of furan. In general, it undergoes addition rather than substitution reactions with much greater ease than do the other heterocyclic compounds (p. 942). The small degree of aromaticity of furan is also illustrated by the relative ease with which it serves as a diene in Diels-Alder reactions. It is much less reactive than cyclopentadiene, which as a cyclic diene reacts readily with all kinds of dienophiles (p. 745), but much more reactive than thiophene or pyrrole. Furan reacts only with highly reactive dienophiles such as dimethyl acetylenedicarboxylate (p. 744).

**Table 22.1**
**Resonance Energies for Some Cyclic Compounds (Determined from Heats of Combustion)**

| Compound | Resonance Energy (kcal/mol) |
|---|---|
| benzene | 36 |
| thiophene | 29 |
| pyrazole | 29 |
| pyridine | 28 |
| pyrrole | 22 |
| imidazole | 22 |
| furan | 16 |
| cyclopentadiene | 3 |

Study Guide
Concept Map 22.1

## 22.3    Reactions of Heterocycles as Acids and Bases

An examination of the structural formulas for aromatic heterocycles raises interesting questions about the acidity and basicity of these compounds. For example, pyrrole, in which the nonbonding electrons on the nitrogen atom are part of the aromatic sextet, has a low basicity. When it accepts a proton, it does so on one of the carbon atoms adjacent to the nitrogen atom. On the other hand, the proton on the nitrogen atom can be removed by hydroxide ion to give the conjugate base of pyrrole. Salts containing the pyrrole anion are easily prepared in this way. The $pK_a$ values for pyrrole and its conjugate acid are compared below with the values for ammonium ion and ammonia, taken from the table of $pK_a$ values inside the front cover of this book.

conjugate acid
of pyrrole
$pK_a$ −3.80

pyrrole
$pK_a$ ~15

conjugate base
of pyrrole

$^+NH_4$  $\underset{\text{acid}}{\overset{\text{base}}{\rightleftharpoons}}$  $:NH_3$  $\underset{\text{acid}}{\overset{\text{base}}{\rightleftharpoons}}$  $:\overset{..}{\overset{..}{N}}H_2$

ammonium ion
conjugate acid
of ammonia
$pK_a$ 9.4

ammonia
$pK_a$ 36

amide anion
conjugate base
of ammonia

The pair of nonbonding electrons on the nitrogen atom in pyrrole is much less available for protonation than the pair on ammonia. The conjugate acid of pyrrole is a strong acid, with $pK_a -3.80$. Pyrrole itself, with $pK_a \sim 15$, is a much stronger acid than ammonia. The conjugate base of pyrrole is still an aromatic species but has an extra pair of nonbonding electrons located in an $sp^2$-hybrid orbital on the nitrogen atom. The negative charge on the amide anion, $NH_2^-$, in contrast, is in an $sp^3$-hybrid orbital, which is less electronegative than an $sp^2$-hybrid orbital. The nonbonding electrons on the pyrrole anion are less available for bonding to an acid than are those on the amide anion (p. 60).

---

**Problem 22.3**    Write resonance contributors for furan, thiophene, and pyrrole. Which resonance contributors are the major ones?

**Problem 22.4**    Why is pyrrole protonated on the carbon atom adjacent to the nitrogen atom rather than on the nitrogen?

---

The reactions of imidazole as an acid or a base are important in many biological systems. Imidazole is a stronger base than pyrrole because the second nitrogen atom has a pair of nonbonding electrons that is not part of the aromatic sextet. Imidazole also has a proton that is lost in base, and therefore, it is an acid. The various species involved in the protonation and deprotonation reactions of imidazole are shown below. The conjugate acid of imidazole is stabilized by delocalization of the charge to both of the nitrogen atoms, giving two equivalent resonance contributors.

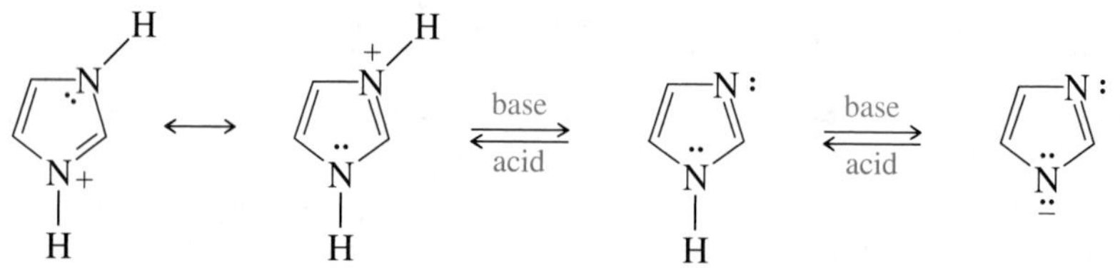

resonance contributors for the
conjugate acid of imidazole
$pK_a$ 6.95

imidazole
$pK_a$ 14.5

conjugate base of imidazole

A comparison of the basicities of the six-membered ring heterocycles pyridine and pyrimidine is also interesting. Pyridine is a weak base but stronger than pyrrole. The conjugate acid has $pK_a$ 5.2. The nonbonding electrons on the nitrogen atom are not part of the aromatic sextet but are present on a nitrogen atom that is $sp^2$-hybridized. It is postulated that the electrons are in an $sp^2$ hybrid orbital rather than an $sp^3$ hybrid orbital, as is the case for ammonia. Thus they are held more closely to the nucleus of the nitrogen atom than is usual for amines because of the greater $s$ character of the orbital.

Introducing a second nitrogen atom into the pyridine ring lowers the basicity of the molecule still further. The conjugate acid of pyrimidine has $pK_a$ 1.3. Nitrogen is more electronegative than carbon. The inductive effect of the second nitrogen atom makes the electrons on the first nitrogen less available for protonation.

Pyridine is the heterocycle that most resembles benzene in its structure and stability. Electrophilic substitution reactions of pyridine are discussed in Section 22.5B, where a closer comparison of the two aromatic compounds is made.

---

**Problem 22.5**    For each of the following sets of bases, discuss the trend observed for the $pK_a$ values given, which are for the conjugate acids of the compounds shown.

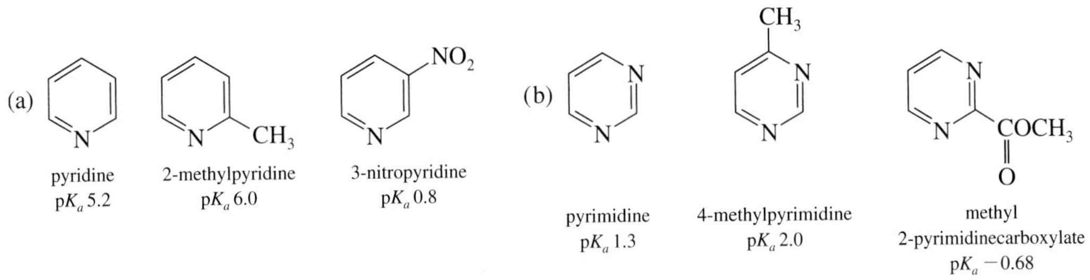

(a)

pyridine
pK$_a$ 5.2

2-methylpyridine
pK$_a$ 6.0

3-nitropyridine
pK$_a$ 0.8

(b)

pyrimidine
pK$_a$ 1.3

4-methylpyrimidine
pK$_a$ 2.0

methyl
2-pyrimidinecarboxylate
pK$_a$ −0.68

## 22.4  Synthesis of Heterocycles by Reactions of Nucleophiles with Carbonyl Compounds

### A. Five-Membered Heterocycles with One Heteroatom

A general way to synthesize heterocyclic compounds is by cyclization of a dicarbonyl compound using a nucleophilic reagent that introduces the desired heteroatom or atoms. An example of such a synthesis is the preparation of 2-methyl-5-phenylpyrrole from 1-phenyl-1,4-pentanedione and ammonia.

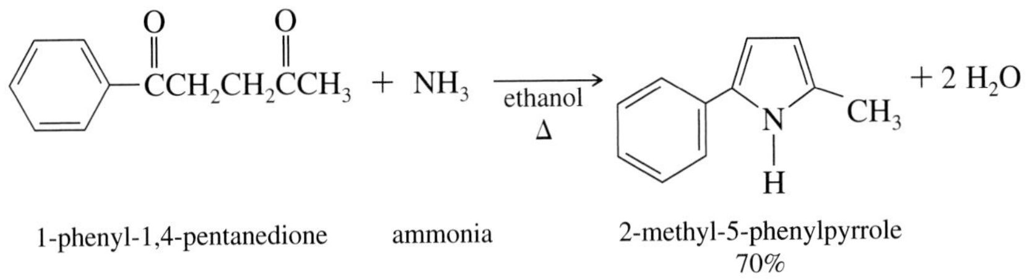

1-phenyl-1,4-pentanedione        ammonia        2-methyl-5-phenylpyrrole
70%

The reaction may be pictured as involving an aminoketal intermediate resulting from nucleophilic attack of ammonia on a carbonyl group. The intermediate undergoes cyclization to a stable five-membered ring and dehydration. The last stage of the mechanism, shown below, represents two sequences of protonation and elimination reactions.

### VISUALIZING THE REACTION

**Formation of a Heterocycle from a 1,4-Dicarbonyl Compound**

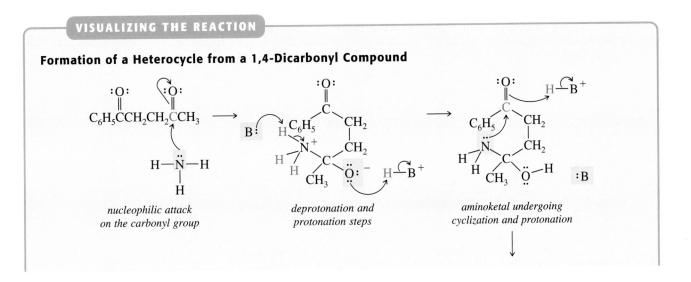

nucleophilic attack
on the carbonyl group

deprotonation and
protonation steps

aminoketal undergoing
cyclization and protonation

*product after a second dehydration step* ← ← *elimination of water* ← *deprotonation and protonation steps*

In this and other mechanisms in this chapter, whenever protonation and deprotonation of the same species occur, the two consecutive steps are shown on a single structure.

1-Phenyl-1,4-pentanedione cyclizes to 2-methyl-5-phenylfuran when it is heated in acid. If there is no nucleophilic reagent present that can supply a heteroatom other than oxygen, the furan ring is formed.

1-phenyl-1,4-pentanedione $\xrightarrow[\Delta]{\text{HCl (conc)}}$ 2-methyl-5-phenylfuran + $H_2O$

In this case, the enol form of one of the ketone functions serves as the nucleophile to form a cyclic hemiketal that then dehydrates.

## ONE SMALL STEP

When a 1,4-dione is treated with phosphorus pentasulfide, which has the molecular formula $P_4S_{10}$, a thiophene is formed. This reaction is believed to go through a thioenol in the same way that the reaction giving rise to a furan goes through an enol.

**PROBLEM:** Predict the product of the reaction of 1-phenyl-1,4-pentanedione with phosphorus pentasulfide, and write a mechanism for the reaction starting with the thioenol.

**Problem 22.6**    Write a mechanism for the formation of 2-methyl-5-phenylfuran from 1-phenyl-1,4-pentanedione according to the description of the reaction given above.

In both these reactions, the driving force is the formation of a stable five-membered ring that has aromaticity. Pyrrole does not undergo reactions that lead to the opening of the ring. Furan, however, may be regarded as a cyclic hemiacetal that has been dehydrated, and it is hydrolyzed back to a dicarbonyl compound easily when heated with dilute acid.

 $CH_3$   $O$   $CH_3$ + $H_2O$ $\xrightarrow[\substack{\text{acetic acid} \\ \Delta}]{H_2SO_4}$ $CH_3CCH_2CH_2CCH_3$

2,5-dimethylfuran                      2,5-hexanedione
                                              86%

**Problem 22.7**    Complete the following equations.

(a) $HCCH_2CH_2CH \xrightarrow{\text{HCl}}$    (b) $CH_3CCH_2CH_2CCH_3 \xrightarrow[\Delta]{CH_3NH_2}$

## B. Five–Membered Heterocycles with Two Heteroatoms

Reagents with two adjacent heteroatoms, such as hydrazine and hydroxylamine, re-act with 1,3-dicarbonyl compounds to give pyrazoles and isoxazoles. For example, 2,4-pentanedione reacts with hydrazine to form 3,5-dimethylpyrazole and with hy-droxylamine to give 3,5-dimethylisoxazole.

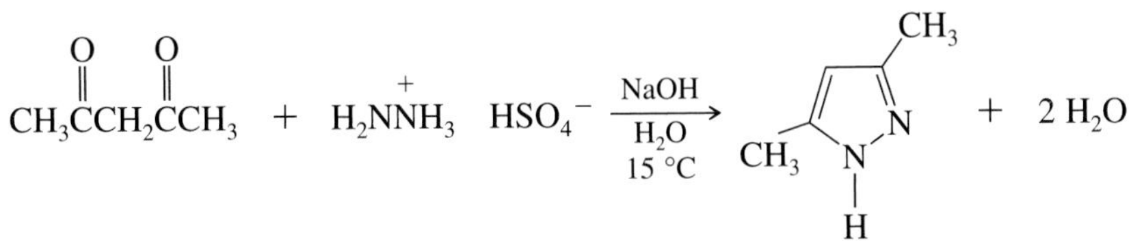

| 2,4-pentanedione | hydrazine sulfate | 3,5-dimethylpyrazole ~ 80% |

| 2,4-pentanedione | hydroxylamine sulfate | 3,5-dimethylisoxazole 84% |

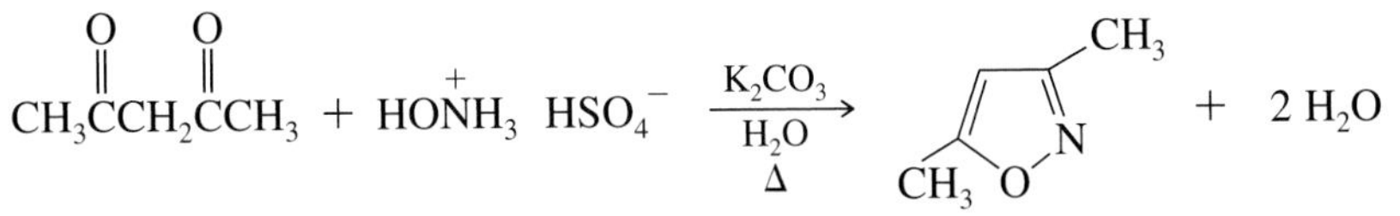

Hydrazine and hydroxylamine are both basic reagents that are most easily stored and handled as their salts. In the presence of bases such as hydroxide or carbonate ions, the free nucleophiles are generated and react with the carbonyl compounds. For example, 2,4-pentanedione reacts with hydroxylamine to give an oxime (p. 574), which cyclizes. The resulting intermediate readily dehydrates to give the aro-matic ring.

**Problem 22.8**   Write mechanisms for the formation of 3,5-dimethylpyrazole and 3,5-di-methyloxazole from 2,4-pentanedione.

Imidazoles are synthesized from two carbonyl compounds that are joined together with nitrogen atoms from ammonia. For example, when a mixture of 1,2-diphenyl-1,2-ethanedione (also called benzil) and benzaldehyde is heated with ammonium acetate in glacial acetic acid, 2,4,5-triphenylimidazole is obtained.

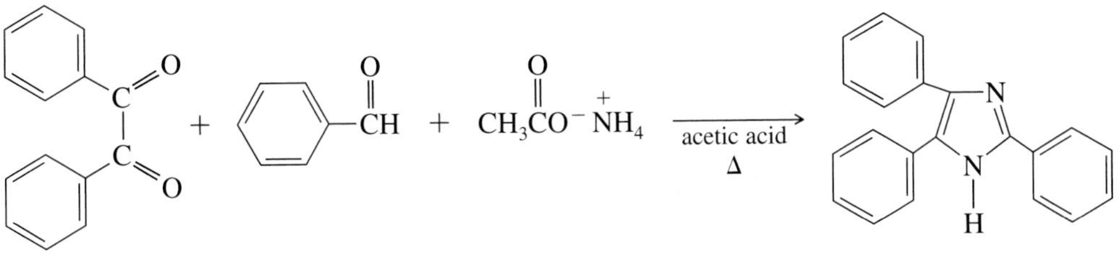

| 1,2-diphenyl-1,2-ethanedione benzil | benzaldehyde | ammonium acetate | 2,4,5-triphenylimidazole 90% |

The three carbon atoms in the imidazole ring are the carbon atoms of the carbonyl groups in the organic reagents; the nitrogen atoms come from ammonia, which is in equilibrium with the ammonium ion from the salt in solution with the weak acid acetic acid.

**Problem 22.9** What is the major product of each of the following reactions?

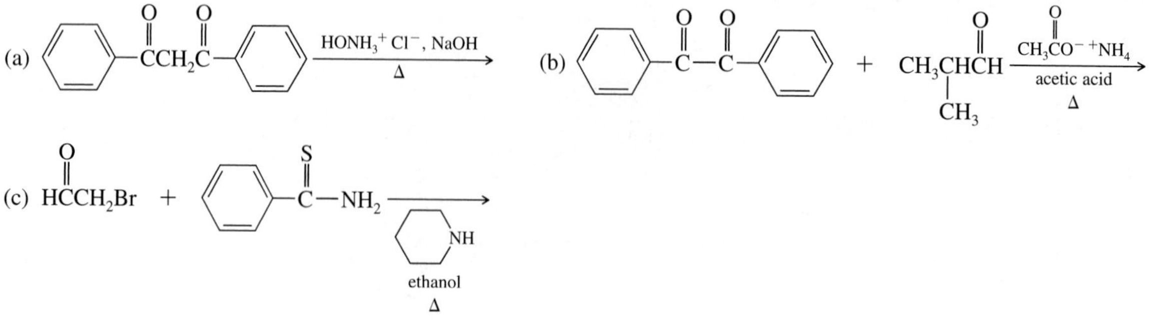

(a) $\xrightarrow[\Delta]{HONH_3^+ Cl^-, NaOH}$

(b) $+ \text{CH}_3\text{CHCH} \xrightarrow[\text{acetic acid}]{\text{CH}_3\text{CO}^- {}^+\text{NH}_4} \atop \Delta$

(c) $\overset{O}{\overset{\|}{\text{HCCH}_2\text{Br}}}$ + $\xrightarrow[\Delta]{\text{ethanol}}$

## C. Six-Membered Heterocycles

Of the six-membered heterocycles, pyridine and various simple substituted pyridines can be obtained conveniently from natural sources. A great deal of work has been done, however, on the synthesis of pyrimidines because of their importance as drugs and as bases found in nucleic acids. The synthesis of barbiturates from derivatives of diethyl malonate and urea (One Small Step, p. 942) is one application of the most general way to create the pyrimidine ring. A 1,3-dicarbonyl compound is condensed with a reagent that is structurally related to urea. The products formed depend on the substituents present on each fragment. Two examples are given below.

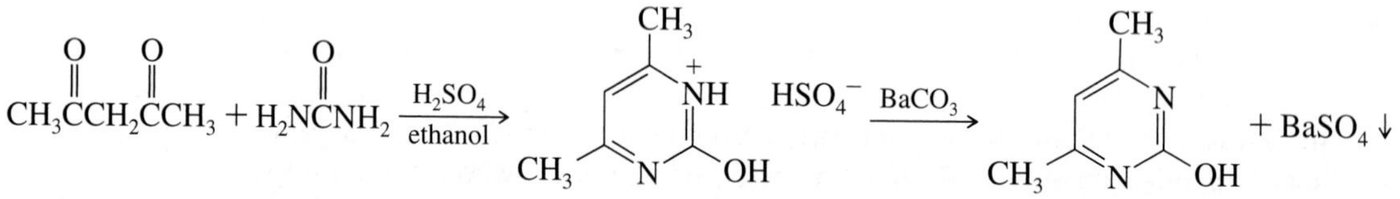

2,4-pentanedione    urea

2-hydroxy-4,6-
dimethylpyrimidine

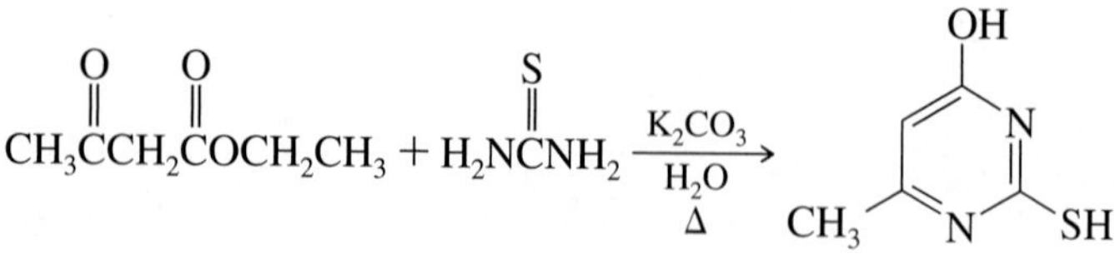

ethyl acetoacetate    thiourea

4-hydroxy-
2-mercapto-
6-methylpyrimidine
95%

In 2,4-pentanedione, the carbonyl groups do not have good leaving groups bonded to them. The reaction with urea proceeds by condensation of the amino groups of urea with the carbonyl groups of the ketone and tautomerization to the aromatic system.

**VISUALIZING THE REACTION**

## Formation of a Pyrimidine from a 1,3-Dicarbonyl Compound and Urea

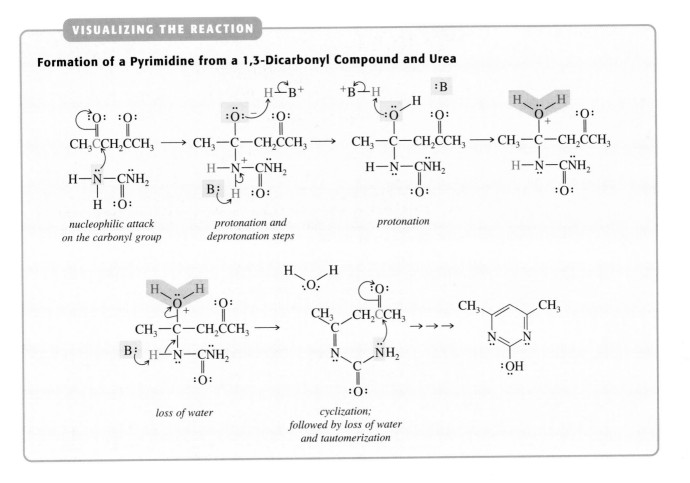

*nucleophilic attack on the carbonyl group*

*protonation and deprotonation steps*

*protonation*

*loss of water*

*cyclization; followed by loss of water and tautomerization*

When an ester group provides one of the carbonyl functions, as in diethyl malonate or ethyl acetoacetate, an alkoxide serves as a leaving group, providing a different pathway for the condensation reaction. The carbonyl group of the ester is retained in the pyrimidine. The carbon atoms that were part of the carbonyl groups of the ester or of urea (or thiourea) are marked by hydroxyl (or thiol) substituents on the fully aromatic form of the ring after tautomerization.

⊘ **Study Guide**
**Concept Map 22.2**

**Problem 22.10**    Complete the following equations.

(a) $CH_3\overset{O}{\overset{\|}{C}}CH_2\overset{O}{\overset{\|}{C}}OCH_2CH_3 + H_2N\overset{O}{\overset{\|}{C}}NH_2 \xrightarrow[\text{ethanol}]{\text{HCl}}$

(b) $CH_3\overset{O}{\overset{\|}{C}}CH_2\overset{O}{\overset{\|}{C}}CH_3 + CH_3ONH\overset{O}{\overset{\|}{C}}NH_2 \xrightarrow[\text{ethanol}]{\text{HCl}}$

(c) 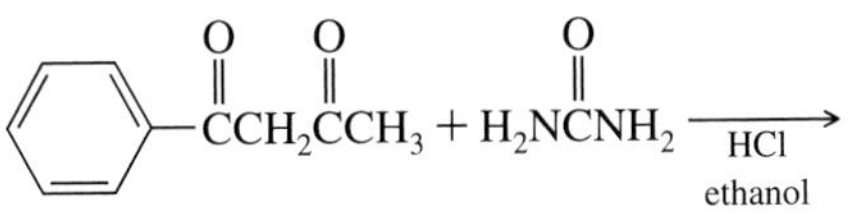 ⬡$-\overset{O}{\overset{\|}{C}}CH_2\overset{O}{\overset{\|}{C}}CH_3 + H_2N\overset{O}{\overset{\|}{C}}NH_2 \xrightarrow[\text{ethanol}]{\text{HCl}}$

## ONE SMALL STEP

Barbituric acid derivatives have the general structure shown below.

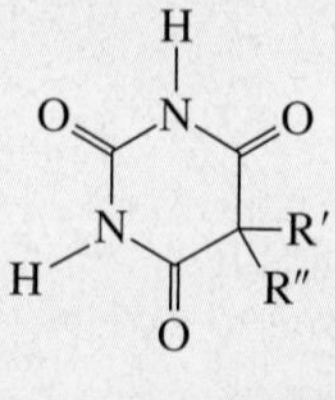

They act as sedatives and hypnotics, that is, sleeping pills. Their physiologic effects depend on the nature of the R′ and R″ groups on the pyrimidine ring. Some take effect quickly; others take longer to begin acting but also have an effect over a longer period of time. Phenobarbital, where R′ is ethyl and R″ is phenyl, also serves as an anticonvulsant.

**PROBLEM:**   Vernonal is the barbituric acid with R′ and R″ both ethyl.
(a) How would you synthesize it from diethyl malonate and urea?
(b) Why are these compounds called "acids?" What is the source of their acidity?

## 22.5   Substitution Reactions of Heterocyclic Compounds

### A. Electrophilic Aromatic Substitution Reactions of Five–Membered Heterocycles

The five-membered aromatic heterocycles are all more reactive toward electrophiles than benzene is. The reactivity of the ring in these compounds resembles that of phenol in the ease with which substitution takes place. For example, thiophene reacts with bromine to give a mixture of bromothiophenes.

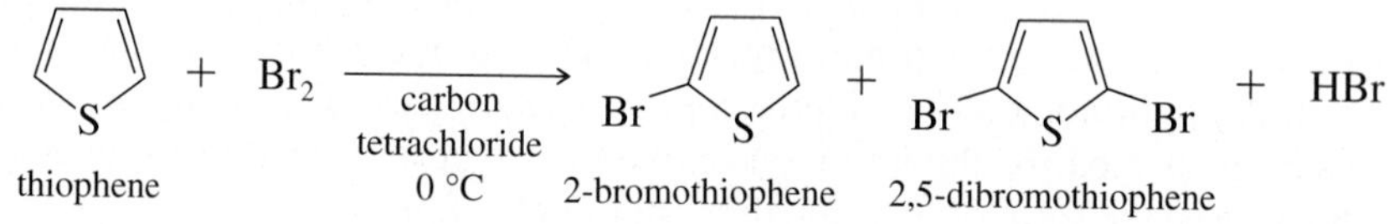

$$\text{thiophene} + Br_2 \xrightarrow[\substack{\text{carbon} \\ \text{tetrachloride} \\ 0\,°C}]{} \text{2-bromothiophene} + \text{2,5-dibromothiophene} + HBr$$

Furan, on the other hand, reacts with bromine by a 1,4-addition reaction, an indication of the relatively low aromaticity of this heterocycle. When the reaction is carried out in methanol, the product that is isolated is formed by solvolysis of the intermediate dibromide.

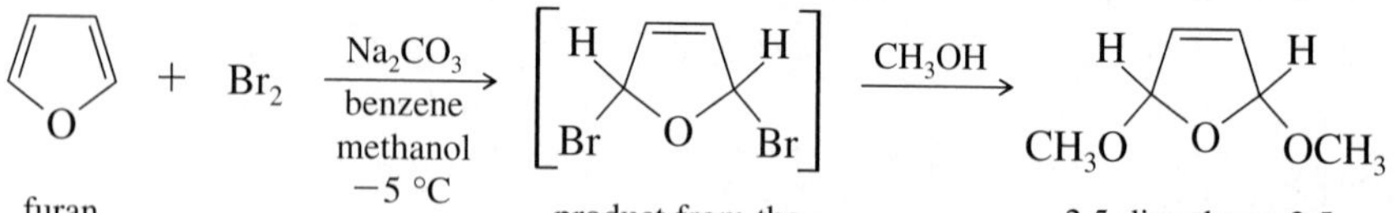

$$\text{furan} + Br_2 \xrightarrow[\substack{\text{benzene} \\ \text{methanol} \\ -5\,°C}]{Na_2CO_3} \left[ \text{product from the 1,4-addition of bromine to furan} \right] \xrightarrow{CH_3OH} \text{2,5-dimethoxy-2,5-dihydrofuran } 75\%$$

Five-membered aromatic heterocycles also undergo nitration reactions. However, because the mixture of nitric acid and sulfuric acid used for the nitration of benzene (p. 368) destroys the heterocycles, a milder nitrating agent prepared by dissolving nitric acid in acetic anhydride is used. Acetic anhydride acts as a dehydrating agent to create nitronium ions from nitric acid. Substitution in thiophene takes place chiefly at one of the carbon atoms adjacent to the heteroatom.

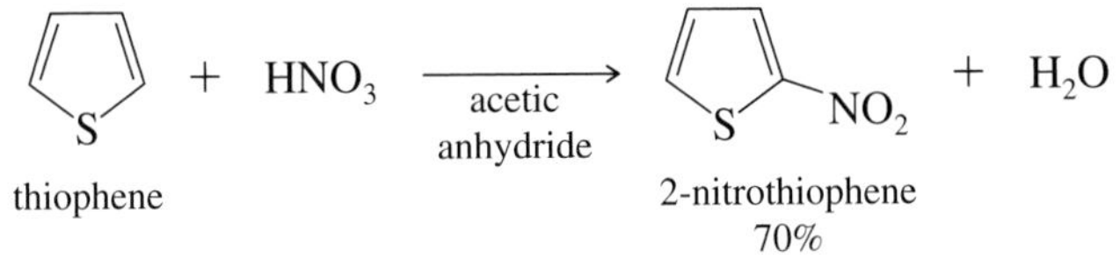

thiophene

2-nitrothiophene
70%

The regioselectivity of the substitution reactions of these heterocycles can be rationalized by the same kind of reasoning that was used to explain the directing effects of substituents on benzene (pp. 373–376). The resonance contributors for the intermediates that would result from attack of the nitronium ion at carbon 2 and at carbon 3 of the thiophene ring are compared below.

**VISUALIZING THE REACTION**

**Regioselectivity of Electrophilic Substitution on Thiophene**

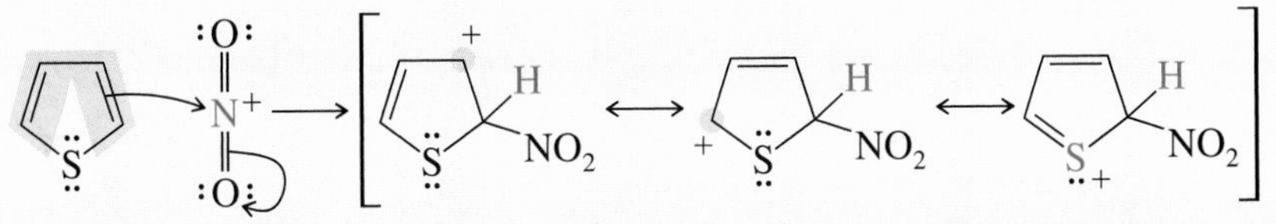

*resonance contributors for the intermediate formed by attack of nitronium ion at carbon 2*

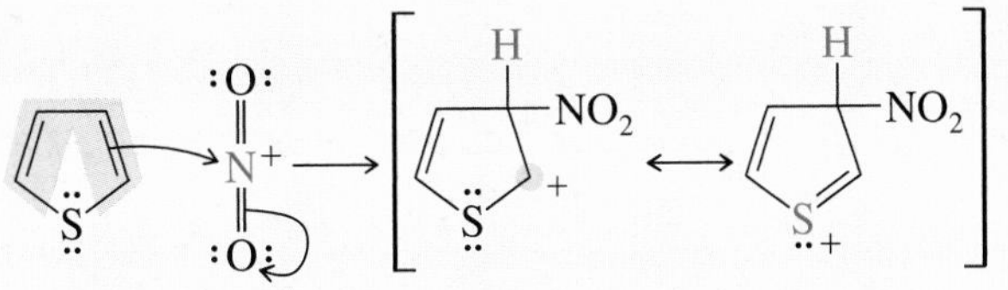

*resonance contributors for the intermediate formed by attack of nitronium ion at carbon 3*

The carbocation formed when nitronium ion attacks carbon 2 of thiophene is more stable than the other one because greater delocalization of charge is possible for it. The reaction thus follows the path leading through that intermediate and the lower-energy transition state corresponding to its formation. The intermediate carbocation loses a proton easily, and the product has the stable aromatic ring.

Substituted aromatic heterocycles usually undergo the reactions typical of the functional groups that are present. For example, in a typical free-radical substitution reaction (p. 778), a hydrogen atom on the methyl group of 3-methylthiophene is replaced by bromine when *N*-bromosuccinimide is used.

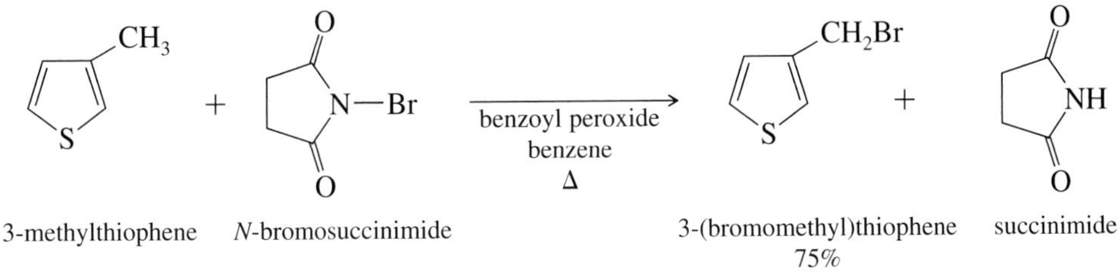

3-methylthiophene    *N*-bromosuccinimide

3-(bromomethyl)thiophene    succinimide
75%

The reactions shown above demonstrate that five-membered aromatic heterocycles behave much like benzene and its derivatives in electrophilic substitution reactions. Substitution occurs preferentially at carbon 2. The presence of the heteroatom makes the heterocyclic compound more reactive than benzene, so some of the reaction conditions must be modified. Substituents on a heterocyclic ring react in ways typical of their functional groups.

**Problem 22.11**  Write structural formulas for all intermediates and products designated by letters in the following equations.

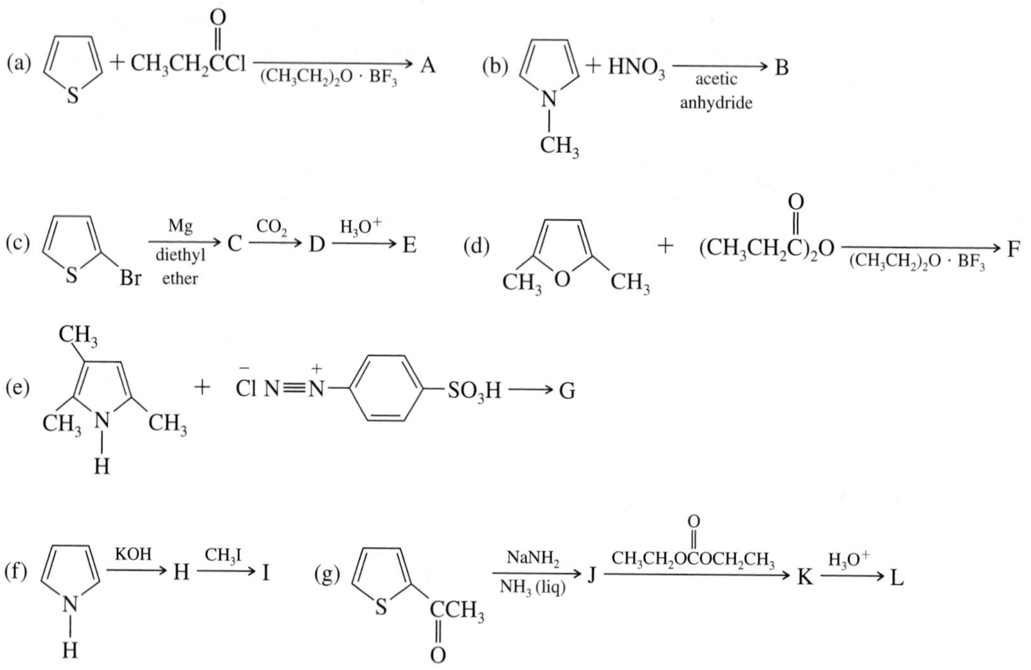

(a) (thiophene) $+ CH_3CH_2\overset{O}{\overset{\|}{C}}Cl \xrightarrow{(CH_3CH_2)_2O \cdot BF_3}$ A

(b) (N-methylpyrrole) $+ HNO_3 \xrightarrow[\text{anhydride}]{\text{acetic}}$ B

(c) (2-bromothiophene) $\xrightarrow[\text{diethyl ether}]{Mg}$ C $\xrightarrow{CO_2}$ D $\xrightarrow{H_3O^+}$ E

(d) (2,5-dimethylfuran) $+ (CH_3CH_2\overset{O}{\overset{\|}{C}})_2O \xrightarrow{(CH_3CH_2)_2O \cdot BF_3}$ F

(e) (2,3,5-trimethylpyrrole) $+ \ \overset{-}{Cl}\ N{\equiv}\overset{+}{N}\ {-}\langle\text{benzene}\rangle{-}SO_3H \longrightarrow$ G

(f) (pyrrole) $\xrightarrow{KOH}$ H $\xrightarrow{CH_3I}$ I

(g) (2-acetylthiophene) $\xrightarrow[\text{NH}_3\text{ (liq)}]{NaNH_2}$ J $\xrightarrow{CH_3CH_2O\overset{O}{\overset{\|}{C}}OCH_2CH_3}$ K $\xrightarrow{H_3O^+}$ L

**Problem 22.12**  When 2-acetyl-1-methylpyrrole is treated with nitric acid in acetic anhydride at 0 °C, two nitration products are obtained. Predict what their structures are by reasoning about the relative stabilities of the intermediates formed. Predict which isomer is the major product.

## B. Aromatic Substitution Reactions of Pyridine

Although the five-membered heterocycles are much more reactive toward electrophilic substitution than benzene is, pyridine is much less reactive than benzene. The conditions that are necessary to bring about substitution on the pyridine ring are often more severe than those required to carry out multiple substitutions on nitrobenzene. An example is the nitration of pyridine, which takes place at 330 °C.

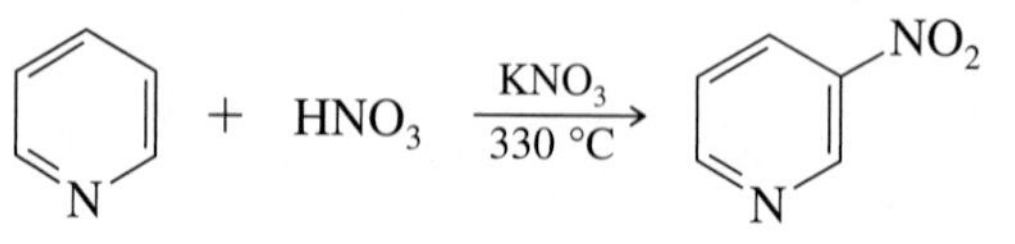

(pyridine) $+ HNO_3 \xrightarrow[\text{330 °C}]{KNO_3}$ (3-nitropyridine)

pyridine

3-nitropyridine
15%

Even at this temperature, only a small fraction of the pyridine molecules are nitrated.

Substitution takes place preferentially at carbon 3 of the pyridine ring. The nitrogen atom in the ring deactivates the positions that are ortho and para to it more than it deactivates the meta positions. In this respect, it has the same effect as a nitro group on benzene. Part of the effect arises because pyridine is protonated or coordinates with Lewis acids under the conditions necessary for most substitution reactions. For example, in the nitration reaction, the species undergoing substitution is the pyridinium ion, not pyridine itself.

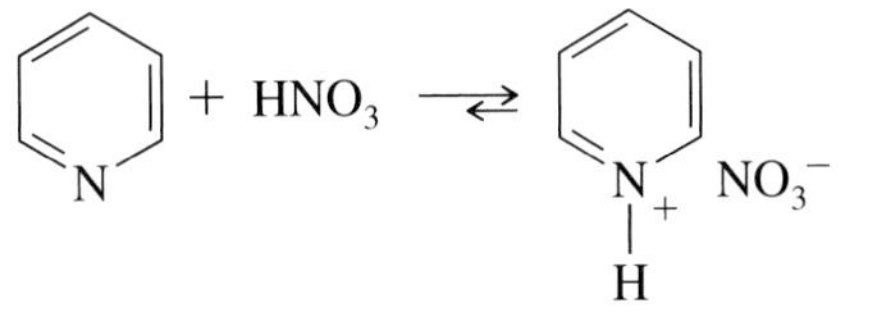

$$\text{pyridine} + HNO_3 \rightleftarrows \text{pyridinium nitrate}$$

pyridine    nitric acid    pyridinium nitrate

The resonance contributors for the intermediates that arise when an electrophile attacks at carbon 2 or 3 of the pyridine ring are shown below. Comparing the resonance contributors shows that the intermediate formed by electrophilic attack at carbon 3 is more stable than the other one. Delocalization of the positive charge in the intermediate from attack at carbon 2 puts the charge on the nitrogen atom, which will already bear a positive charge due to prior protonation or coordination with a Lewis acid. The resonance contributor with the positive charge on nitrogen also shows nitrogen, an element that is more electronegative than carbon, having only a sextet of electrons. The delocalization of charge for the intermediate from attack at carbon 3 does not require such a high-energy contributor.

**VISUALIZING THE REACTION**

**Regioselectivity of Electrophilic Substitution on Pyridine**

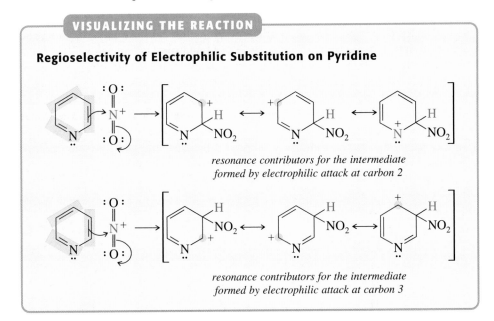

resonance contributors for the intermediate
formed by electrophilic attack at carbon 2

resonance contributors for the intermediate
formed by electrophilic attack at carbon 3

**Problem 22.13**    Write structural formulas for the products represented by letters in the following reactions.

(a) $\xrightarrow[\text{H}_2\text{SO}_4 \ 100\,°C]{\text{KNO}_3}$ A    (b) $\xrightarrow[\text{catalyst}]{\text{H}_2}$ B 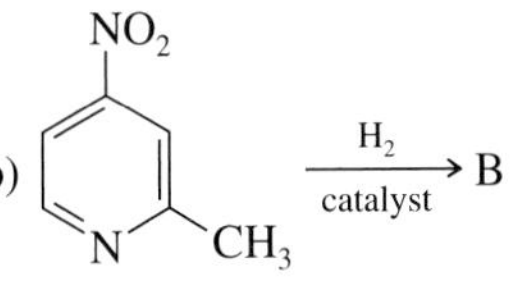    (c) $\xrightarrow{\text{SOCl}_2}$ C $\xrightarrow{\text{AlCl}_3}$ D

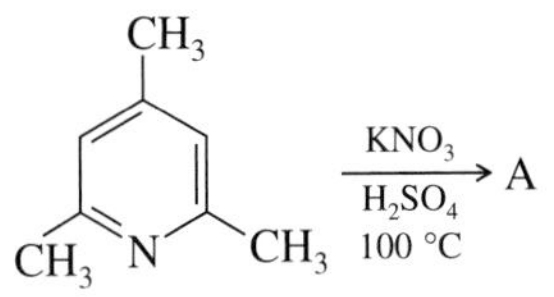

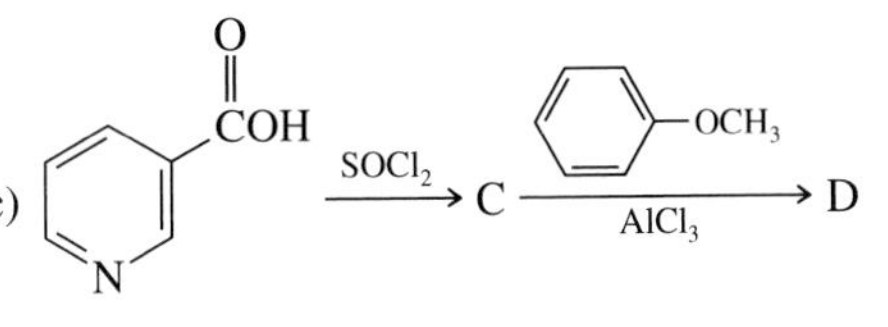

The reactivity of pyridine in electrophilic aromatic substitution reactions is predictable on the basis of its electronic character. Electrophilic substitution occurs predominantly at carbon 3. In most cases, the reactions of substituents on pyridine resemble those of the same substituents on benzene.

**Study Guide**
Concept Map 22.3

**Problem 22.14**  Write structural formulas for all compounds represented by letters in the following equations.

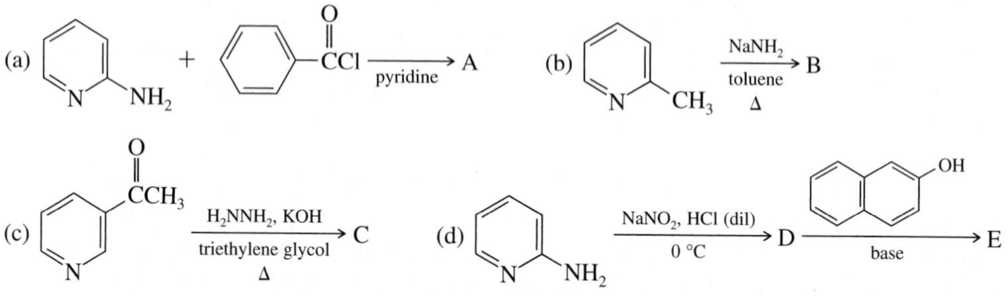

## 22.6 Pyrimidines, Purines, and Pyridines of Biological Significance

### A. Pyrimidines and Purines

The chemistry of pyrimidines and purines is of interest because these heterocyclic rings are found in deoxyribonucleic acids (DNA) and ribonucleic acids (RNA), the complex molecules that transmit genetic information and mediate the synthesis of proteins in cells. The structures and chemical properties of several pyrimidines and purines determine what interactions are possible between different strands of DNA and between molecules of DNA and RNA. These interactions are believed to be largely responsible for the storage and transmission of genetic information in cells. The ways in which these heterocycles influence the structure and function of DNA and RNA will be discussed in greater detail in Sections 23.4 and 23.9.

The pyrimidine and purine bases that are found in DNA and RNA are shown below.

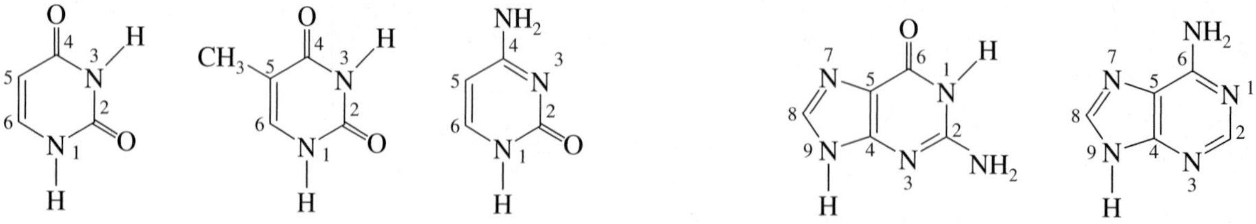

uracil
*found in RNA*

thymine
*found in DNA*

cytosine
*found in both RNA and DNA*

guanine
*found in both RNA and DNA*

adenine
*found in both RNA and DNA*

*pyrimidine bases*

*purine bases*

An examination of the structures of the purines shows that they contain an imidazole ring fused to a pyrimidine ring. The numbering system shown is the one commonly used for these compounds in biologic systems. The tautomeric forms of the bases shown are the ones that are important in water at pH 7, the conditions under which these bases are found in nucleic acids. The question of tautomerism is important because the exact location of hydrogen atoms on oxygen and nitrogen atoms determines the way the bases interact with each other by hydrogen bonding (p. 1018).

A number of other purines occur in nature, including xanthine, hypoxanthine, and uric acid. Caffeine, found in coffee, tea, cola nuts, and cocoa, theobromine from cocoa, and theophylline from tea are methylated xanthines.

hypoxanthine          xanthine          uric acid

caffeine          theobromine          theophylline

*found in tea,          principal alkaloid          small amounts*
*coffee, maté leaves,          of cacao bean,          in tea*
*guarana paste,          also in cola nuts*
*cola nuts          and tea*

Caffeine is a powerful stimulant of the central nervous system. Theophylline is a milder stimulant of the central nervous system and also a relaxant of smooth muscles. Theobromine does not have much activity as a stimulant. Hypoxanthine, xanthine, and uric acid are products of the metabolism of the purine bases adenine and guanine. The disease called gout results from the faulty metabolism and excretion of uric acid.

The purine and pyrimidine bases occupy a central place in the metabolic processes of cells because they are involved in regulating protein synthesis (p. 1020). Their importance has led chemists to design medications that incorporate their ring systems and mimic their structures. Researchers hope that such compounds will disrupt metabolic processes in cancer cells. Many compounds of this type have been synthesized and tested. Two that have been used in the treatment of cancer are 5-fluorouracil and 6-mercaptopurine. 5-Fluorouracil was designed to resemble thymine structurally. It interacts with enzymes that function in the synthesis of RNA and DNA and prevents normal metabolic processes from taking place. 6-Mercaptopurine resembles adenine in structure, except that a nucleophilic sulfur atom replaces a nucleophilic nitrogen atom. This compound also acts in cells by blocking the synthesis of nucleic acids. Cancer cells grow in an uncontrolled way compared with normal cells. The designers of antitumor compounds hoped that the metabolic processes in cancer cells therefore would be more vulnerable to disruptive drugs than those of normal cells. To some extent this is true, but most of the medications that are used in the treatment of cancer are also extremely toxic to normal body cells.

5-fluorouracil

6-mercaptopurine

**Problem 22.15**    Functional groups on pyrimidines and purines show essentially the same reactivity as they do on a benzene ring. Write structural formulas for all intermediates and products designated by letters in the following equations.

(a)

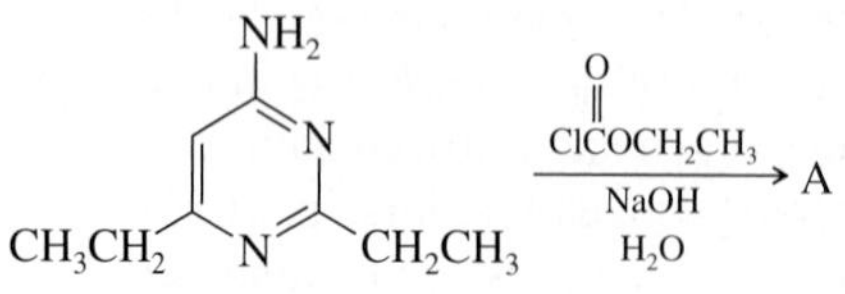

(b)

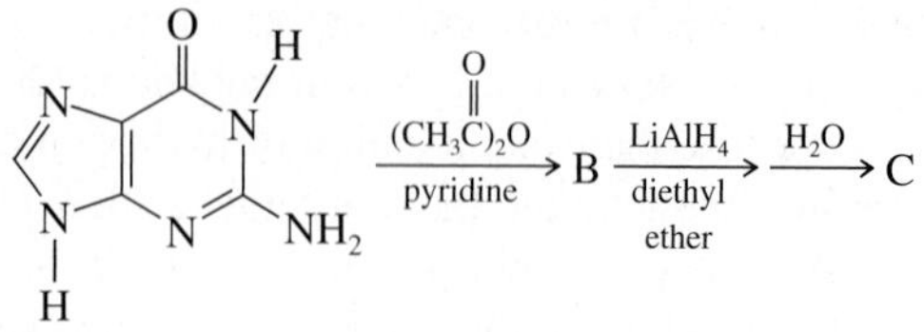

(c)

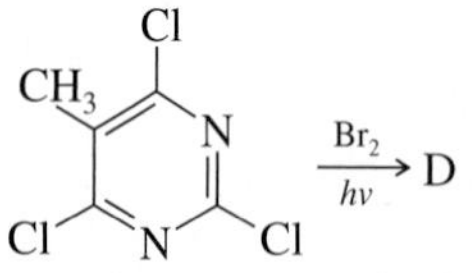

(d)

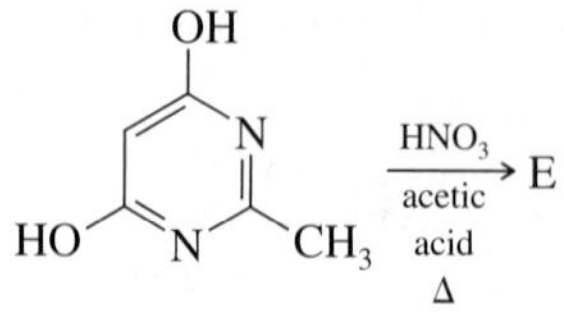

(e)

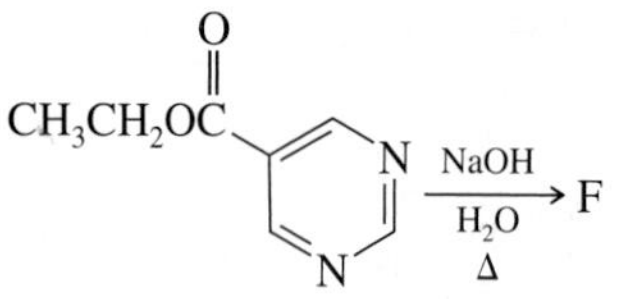

(f)

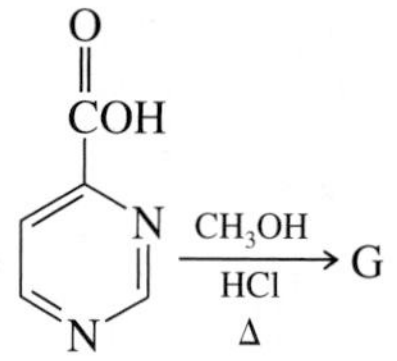

## B. Biological Oxidation – Reduction Reactions

Many biological oxidation–reduction reactions are catalyzed by enzymes that are associated with coenzymes that have nicotinamide as part of their structures. In these coenzymes, nicotinamide, one of the B vitamins, is bonded to ribose (p. 971) at the nitrogen atom of the pyridine ring. The molecule also contains the nucleotide adenosine 5′-phosphate (p. 987).

The oxidation–reduction reactions affect the nicotinamide portion of the molecule, so in equations the structural formula of nicotinamide adenine dinucleotide is abbreviated by using R to symbolize everything that is bonded to the nitrogen atom of the pyridine ring.

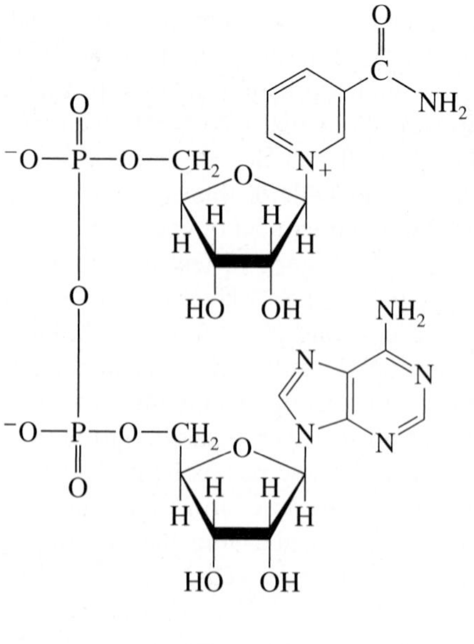

nicotinamide adenine dinucleotide
NAD$^+$

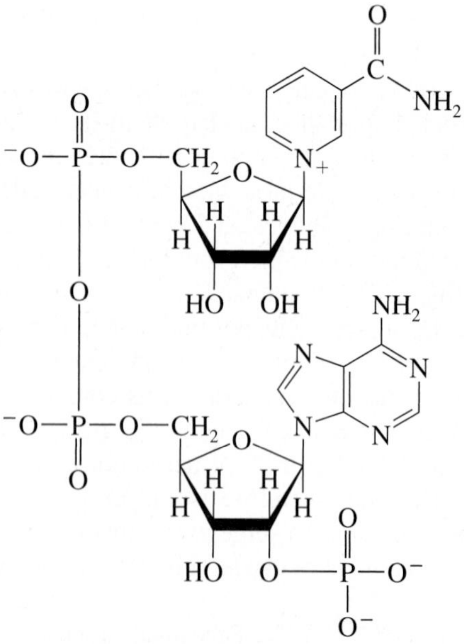

nicotinamide adenine dinucleotide phosphate
NADP$^+$

The presence of a positive charge on the pyridine ring is important. The nitrogen atom of the ring is in the form of a quaternary ammonium ion, and the ring is therefore highly activated toward nucleophilic attack.

The coenzymes $NAD^+$ and $NADP^+$ are associated with a large number of enzymes known as dehydrogenases. A typical reaction catalyzed by a dehydrogenase found in the liver is the oxidation of ethanol to acetaldehyde.

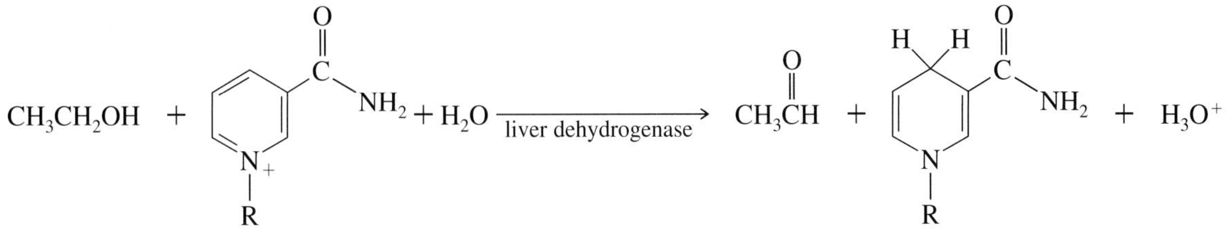


ethanol $NAD^+$ acetaldehyde NADH

In the process, $NAD^+$ is reduced to dihydronicotinamide adenine dinucleotide, abbreviated NADH, and a proton is also transferred to a water molecule.

The reaction starts with the transfer of a hydride ion from the alcohol to $NAD^+$. For example, if ethanol labeled with deuterium at carbon 1 is used, deuterium appears in the reduction product of $NAD^+$.

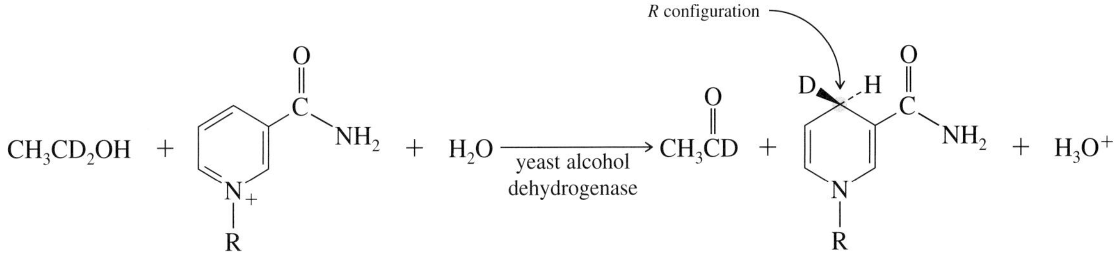


ethanol-1-$d_2$ $NAD^+$ acetaldehyde-1-*d* NADD deuterated reduction product of $NAD^+$

The deuterium atom is transferred with high stereoselectivity to one face of the pyridine ring, so the reduction product with this enzyme and many others always has the *R* configuration at carbon 4. Not all enzymes have the same stereoselectivity. A number of enzymes catalyze the transfer of hydride ion to the other face of the nicotinamide residue.

Dehydrogenases catalyze reduction as well as oxidation reactions. The reduced nicotinamide serves as the reducing agent, again with high stereoselectivity. For example, if acetaldehyde-1-*d*, a product of the reaction shown above, is reduced with NADH, ethanol-1-*d* having the *S* configuration is produced.

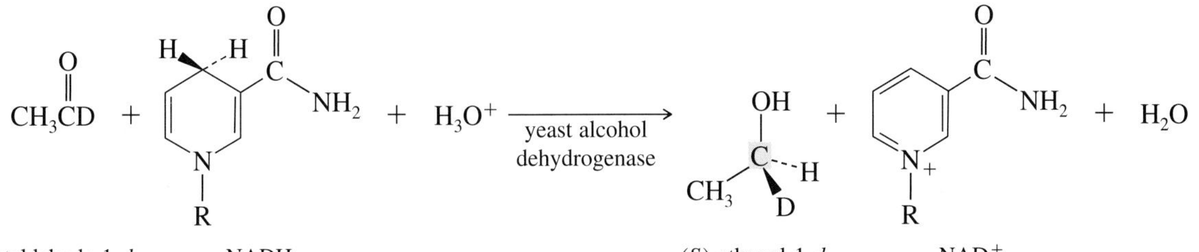


acetaldehyde-1-*d* NADH (*S*)-ethanol-1-*d* $[\alpha]_D^{20}$ −0.28 ° $NAD^+$

If, on the other hand, NADD is used to reduce unlabeled acetaldehyde, the product is (R)-ethanol-1-d.

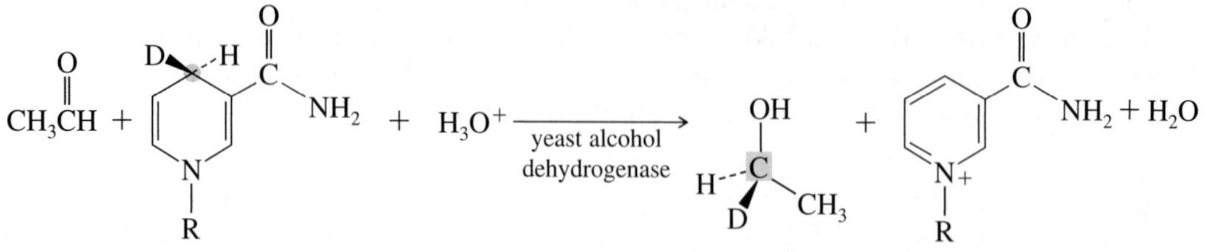

acetaldehyde          NADD                          (R)-ethanol-1-d          NAD⁺

In other words, the enzyme distinguishes between the two faces of the planar carbonyl group. The incoming hydride ion is always attached to the same side of the plane of the carbonyl group and is detached from only one face of the nicotinamide ring.

The reduction of a carbonyl compound by NADH is entirely analogous to reduction by a metal hydride such as sodium borohydride except that the NADH reaction shows stereoselectivity.

**VISUALIZING THE REACTION**

**Reduction of a Carbonyl Group by NADH**

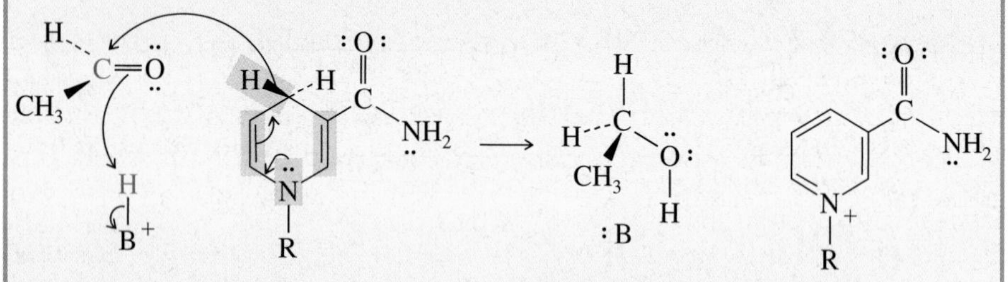

Ethanol that is not labeled with deuterium has no chirality. However, if one of the hydrogen atoms of the methylene group is replaced by another group, such as deuterium, one enantiomer of a pair is formed. Replacement of the other hydrogen atom by deuterium would give the mirror-image isomer of the first compound. The two hydrogen atoms on carbon 1 of ethanol are called **enantiotopic** because enantiomeric compounds are formed by replacing one or the other of them with another group. The two secondary hydrogen atoms on carbon 2 of butane are also enantiotopic. Replacing one of them with a chlorine atom gives (R)-2-chlorobutane, and replacement of the other forms the enantiomeric (S)-2-chlorobutane (p. 191). An achiral reagent does not distinguish between enantiotopic hydrogen atoms, but a chiral one does. Thus, when butane reacts with the achiral reagent, chlorine, there is an equal probability that each one of the two hydrogen atoms on carbon 2 will be replaced, and equal numbers of molecules of (R)- and (S)-2-chlorobutane are formed. When ethanol reacts with NAD⁺ at the active site of the enzyme yeast alcohol dehydrogenase, however, only the hydrogen atom that occupies a certain position in space is transferred to NAD⁺.

This phenomenon is only detectable when the ethanol is labeled with deuterium, but it occurs whether the label is there or not.

**Problem 22.16**    Write equations predicting the products of the reaction of (*R*)-ethanol-1-*d* with NAD⁺ and of (*S*)-ethanol-1-*d* with NAD⁺.

An enzyme can distinguish between the two enantiotopic hydrogen atoms of ethanol because the alcohol molecule fits the active site of the enzyme better in one orientation than it does in the mirror-image orientation. Ethanol fits the active site well in only one of two enantiomeric positions because the active site binds three parts of the molecule. There is no stereoselectivity if the active site interacts with only two parts of the molecule (Figure 22.1).

In a similar way, the two faces of nicotinamide in NAD⁺ may be said to be enantiotopic. Depending on the nature of the enzyme, one face or the other receives the incoming hydride ion stereoselectively. Nicotinamide itself is achiral because it has a plane of symmetry that coincides with the plane of the ring, but the presence of ribose units in the coenzyme makes the molecule as a whole chiral. But even if it were not, the arguments that were made about ethanol could be used to show that

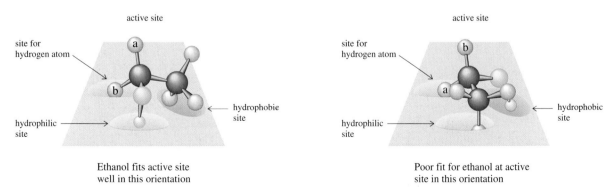

*Figure 22.1*
Two possible ways in which ethanol can interact with the active site of an enzyme.

the active site of an enzyme would interact differently with two enantiomeric orientations of a nicotinamide ring.

**Problem 22.17**    Assume that the active site of a hypothetical enzyme interacts with the nitrogen atom in the pyridine ring of nicotinamide, with the carbonyl group of the amide function, and with carbon 5 of the ring as a hydrophobic site. Prove to yourself that the two faces of an achiral nicotinamide molecule are enantiotopic and thus distinguishable by such an enzyme.

**Problem 22.18**    Inspect the following structural formulas and decide which compounds contain enantiotopic hydrogen atoms.

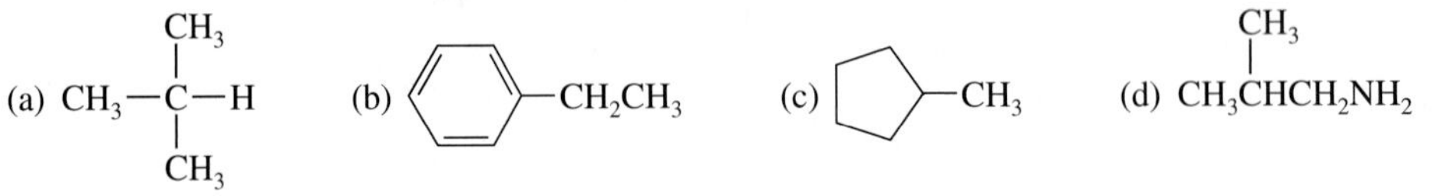

On the Web: ONE SMALL STEP

**Problem 22.19**    $NAD^+$ participates in the oxidation of testosterone, a hormone (p. 537). The enzyme that catalyzes this reaction, testosterone dehydrogenase, has a stereoselectivity that is the opposite of that of yeast alcohol dehydrogenase. Write an equation showing the products that you expect from the reaction. Trace the fate of the hydrogen atoms removed from testosterone.

## 22.7  Five-Membered Heterocycles of Biological Significance

### A. Sulfur Heterocycles

One of the first vitamins to be discovered was thiamine, vitamin $B_1$, which is involved in essential metabolic processes. Thiamine pyrophosphate is found in every cell of the body and functions as a coenzyme in reactions that convert pyruvate ion, one of the products of glycolysis, to acetaldehyde and acetyl coenzyme A (p. 637). A deficiency of thiamine in human beings leads to a disease of the nervous system called beriberi.

In thiamine, a thiazole ring is joined to an aminopyrimidine ring by means of a bridging methylene group.

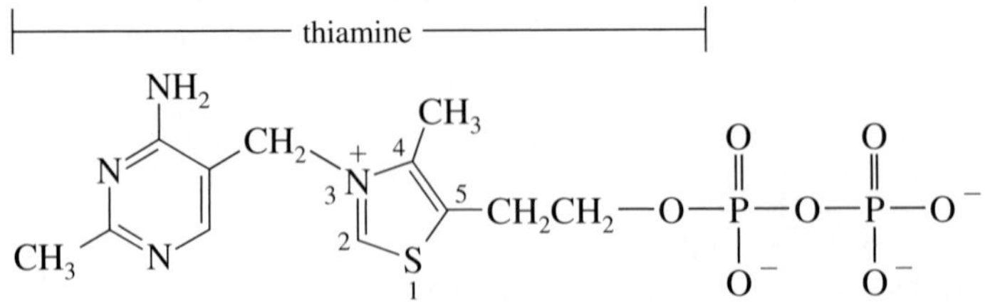

thiamine pyrophosphate

An interesting and important structural feature of the vitamin is the presence of a quaternary nitrogen atom in the thiazole ring. The hydrogen atom on carbon 2 of the thiazolinium ion, the carbon between the positively charged nitrogen atom and the sulfur atom, is extraordinarily acidic. In the 1950s, Ronald Breslow of Columbia University found that such a hydrogen atom in a thiamine analog exchanges for deuterium from deuterium oxide in the absence of either acid or base.

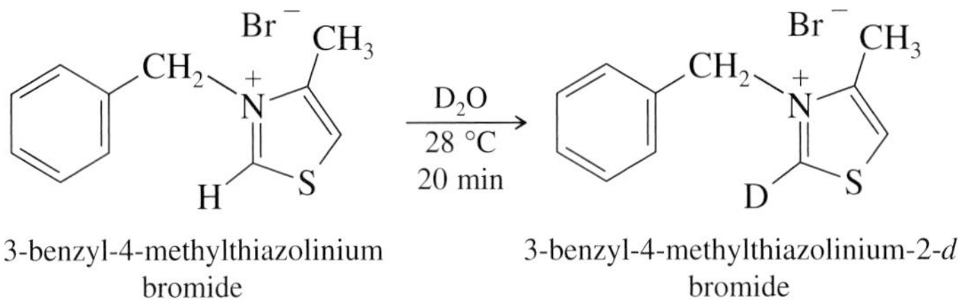

3-benzyl-4-methylthiazolinium
bromide

$\xrightarrow[\substack{28\ °C \\ 20\ \text{min}}]{D_2O}$

3-benzyl-4-methylthiazolinium-2-*d*
bromide

The course of this reaction was followed using nuclear magnetic resonance spectroscopy, and it was found that half the hydrogen atoms at carbon 2 in a sample of the thiazolinium compound are replaced by deuterium atoms in 20 minutes at 28 °C. The rate of the reaction is extraordinarily fast for the breaking of a carbon–hydrogen bond, especially since no strong base is present.

The deuterium-exchange reaction starts with removal of a proton from the thiazolinium ion to give a carbanionic species that is an ylide.

**VISUALIZING THE REACTION**

**The Acidity of the Thiazolinium Ion**

carbanionic
intermediate

*an ylide*

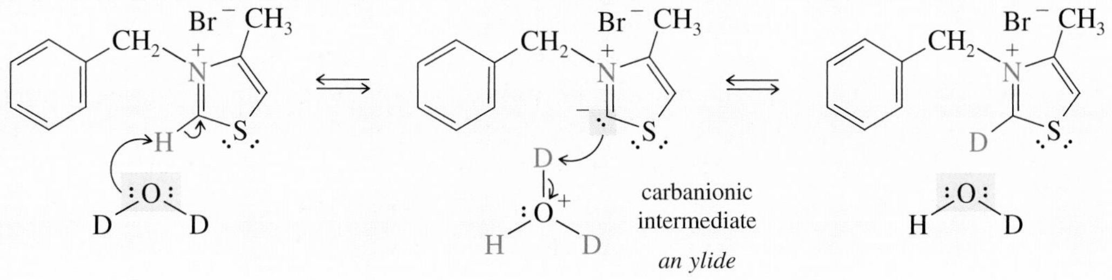

The negative charge is on an $sp^2$-hybridized carbon atom adjacent to a positively charged atom (p. 668), which stabilizes the carbanion. The sulfur atom of the thiazolinium ion also stabilizes the carbanion (p. 872). The combination of these three factors makes it possible for the thiazolinium ring of thiamine to ionize at neutral pH, in other words, under the conditions that exist in cells.

An important reaction that is catalyzed by thiamine pyrophosphate, in the presence of magnesium ions and an enzyme from brewer's yeast known as pyruvate decarboxylase, is the decarboxylation of an $\alpha$-ketoacid to an aldehyde.

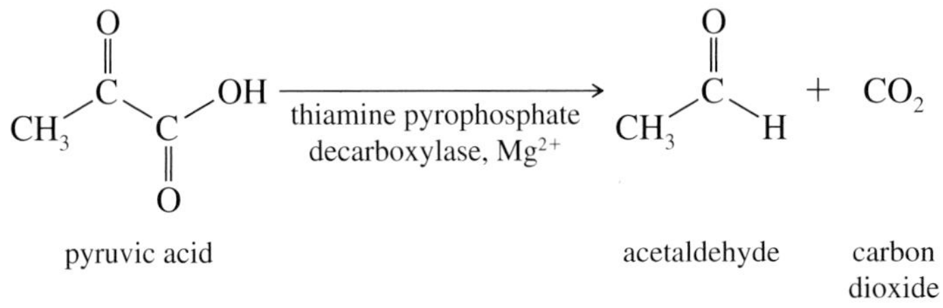

$\xrightarrow[\substack{\text{decarboxylase, Mg}^{2+}}]{\text{thiamine pyrophosphate}}$ $+\ CO_2$

pyruvic acid

acetaldehyde    carbon
dioxide

The reaction occurs by nucleophilic addition of the ylide from thiamine to the ketone function in pyruvic acid, giving an intermediate that loses carbon dioxide easily.

**Problem 22.20**    Draw the structure of the intermediate for the above reaction. Suggest why it decarboxylates easily.

**Problem 22.21**    How is acetaldehyde formed from the intermediate of Problem 22.20? To what class of reactions does this transformation belong?

**Problem 22.22**    Thiamine is reasonably stable in acid. In pure water, it falls apart into a thiazole and a pyrimidine. In strong base, the thiazole ring opens. Suggest mechanisms for these two reactions, shown below.

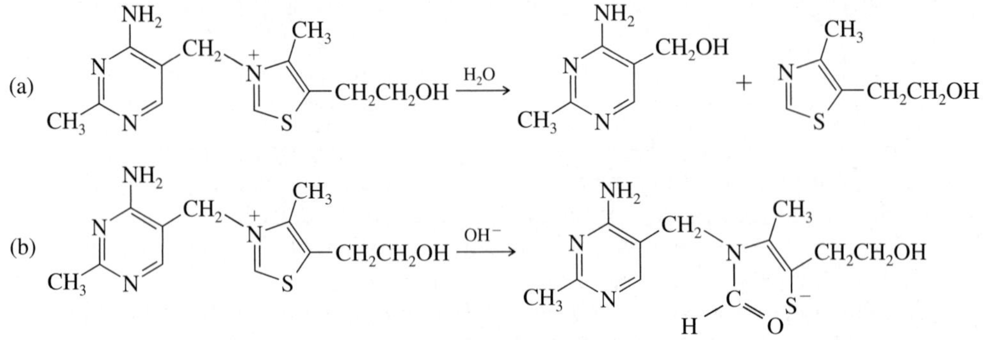

(a)

(b)

## B. Nitrogen Heterocycles. Pyrrole

Pyrrole plays an important role in the chemistry of living organisms. The conversion of light energy from the sun to energy stored in the chemical bonds of the carbohydrates (p. 970) synthesized by green plants is mediated by compounds known as **chlorophylls.** The essential structural feature of chlorophylls is a system of four pyrrole rings held together by bridges, each containing a single carbon atom. This ring system, known as **porphin,** also appears in heme, associated with the proteins hemoglobin (p. 1010) and myoglobin, which are responsible for the transport and storage of oxygen in the body tissues of warm-blooded animals.

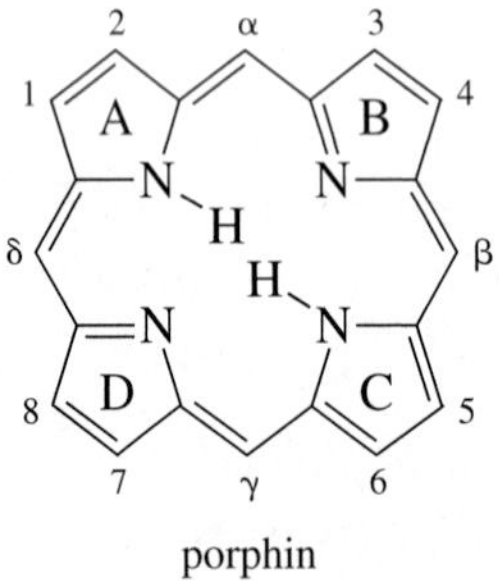

porphin

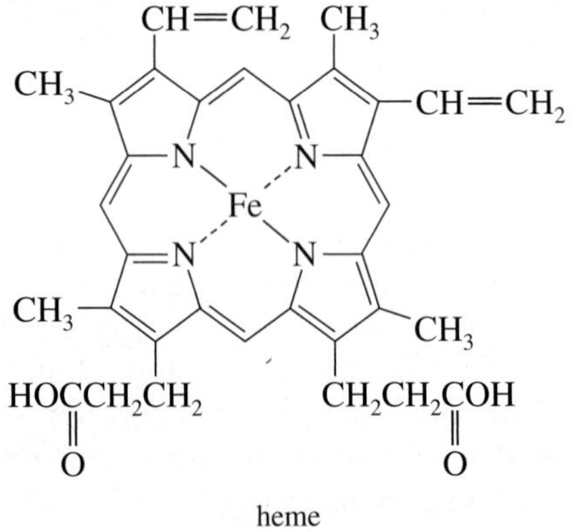

heme

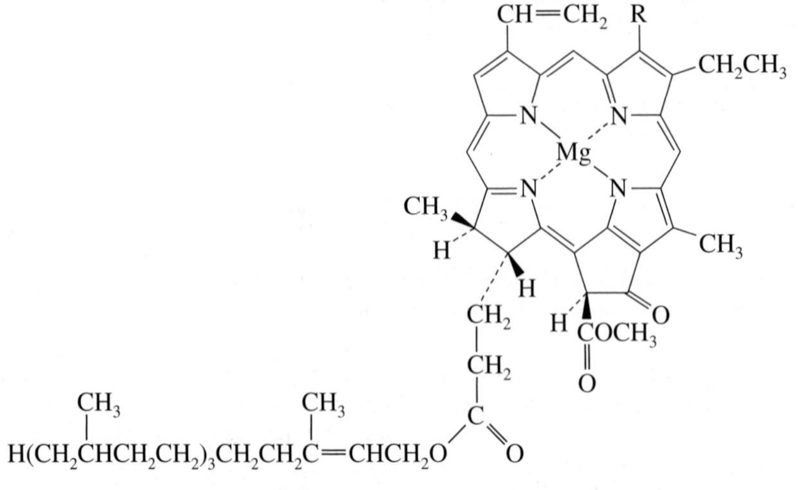

chlorophyll a: R = $CH_3$

chlorophyll b: R = $\overset{\overset{\displaystyle O}{\|}}{CH}$

Porphin, with the four nitrogen atoms of the pyrrole rings pointing toward the center of its large ring system, complexes efficiently with metal ions. In heme, the ion is iron(II); in chlorophylls, it is magnesium(II). These metal ions also may complex with additional ligands above and below the plane of the heterocycle (p. 1010).

The porphin ring systems present in heme and chlorophylls have various substituents on the periphery. Substituted porphins are given the general name of **porphyrins.** The porphyrin in heme has the same degree of unsaturation as porphin does. In the chlorophylls, dihydroporphyrins, with one of the double bonds in the D ring of porphin reduced, are given the special name of **chlorins.** The compounds shown on page 954 are a few of many known porphyrins. The structural complexity means there are wide possibilities for isomerism in these systems.

The porphyrin ring system is very stable and has aromatic character. The extended conjugation present in porphyrins is responsible for the deep colors of these compounds (pp. 445 and 758). Porphin and heme have 22 $\pi$ electrons, but only 18 of these are part of a cyclic array for which resonance contributors can be written. In this respect, a porphyrin resembles [18]annulene (p. 355), having $(4n + 2)$ $\pi$ electrons, where $n$ is 4.

*resonance contributors for porphin drawn in analogy to [18]annulene*

Any attempt to include in the above resonance contributors the four $\pi$ electrons that are outside the thick black lines gives structures that require that the protons on two of the nitrogen atoms be moved. Such a tautomeric transformation does occur in porphyrins. However, two structures that are related to each other by a change in the location of atoms as well as electrons are not resonance contributors.

Porphyrins are synthesized in nature with remarkable ease. The basic unit that combines with itself to give porphyrins substituted in a variety of patterns is a trisubstituted pyrrole called porphobilinogen. This compound is synthesized in living organisms by the condensation of two molecules of 5-amino-4-oxopentanoic acid ($\delta$-aminolevulinic acid). In the presence of the enzyme deaminase (or by simply heating with acid), porphobilinogen is converted into uroporphyrinogen I, which is oxidized by air to the more stable aromatic porphyrin.

5-amino-4-oxopentanoic acid
$\delta$-aminolevulinic acid

porphobilinogen

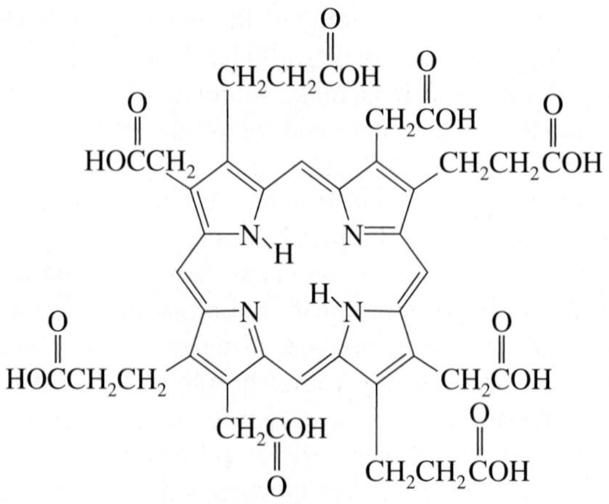

uroporphyrin I

When the two enzymes deaminase and cosynthetase are present, the unsymmetrical porphyrin that is the structural precursor of heme and chlorophyll is formed from porphobilinogen.

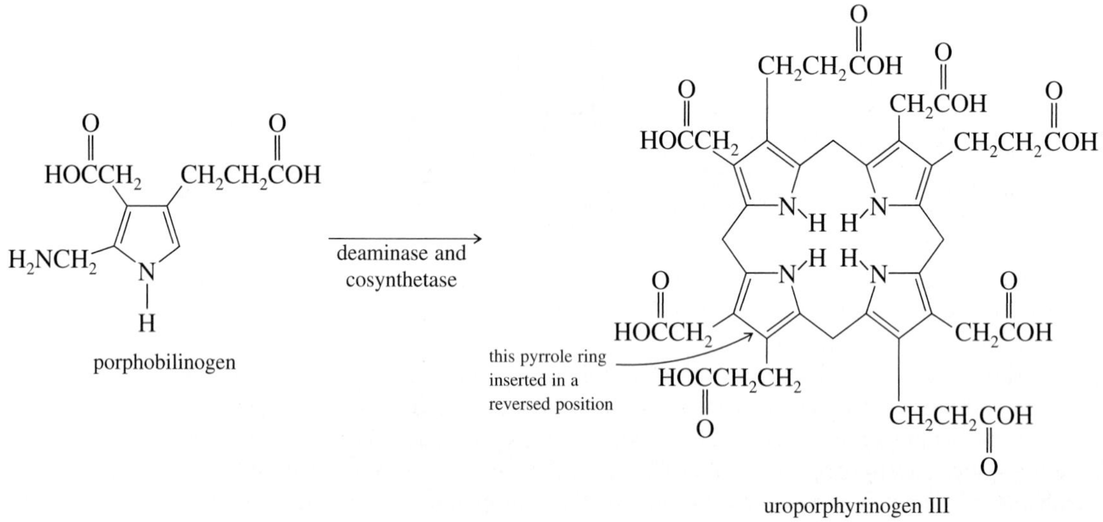

porphobilinogen

deaminase and cosynthetase

this pyrrole ring inserted in a reversed position

uroporphyrinogen III

## 22.8 Alkaloids

### A. Tropane Alkaloids

**Alkaloids,** heterocyclic compounds containing nitrogen, occur most often in the seeds, leaves, and bark of plants. Alkaloids are bases, and many of them have a bitter taste and profound physiologic effects.

A group of alkaloids containing a pyrrolidine ring that is bridged by three carbon atoms between the second and fifth carbons is known as **tropane alkaloids.** To this family belong cocaine, from the leaves of the coca shrub, and atropine, which is the racemic form of (−)-hyoscyamine and is obtained from henbane and the deadly nightshades.

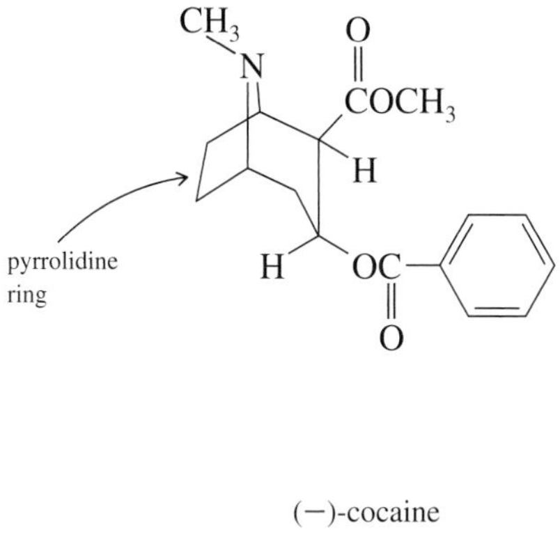

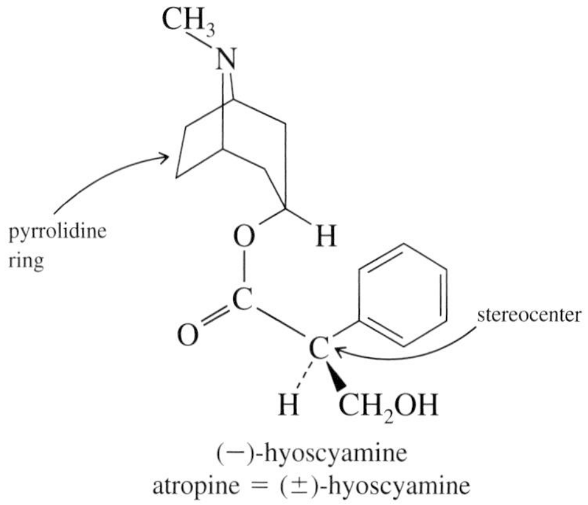

(−)-cocaine

(−)-hyoscyamine
atropine = (±)-hyoscyamine

Cocaine is a stimulant of the central nervous system and a local anesthetic because it blocks the transmission of nerve impulses. The drug is toxic and addictive and disrupts the rhythms of the heart. For this reason, a series of compounds that mimic the action of cocaine as a local anesthetic but lack its more harmful properties has been synthesized. Among them is Novocain (p. 815).

Atropine acts to relax the smooth muscles and thereby ease intestinal and bronchial spasms. Among other medical applications, atropine is used to dilate the pupil of the eye to allow examination of the retina. (−)-Hyoscyamine is the ester of a heterocyclic amino alcohol known as tropine with (S)-(−)-tropic acid, and it is hydrolyzed to these two components in cold water.

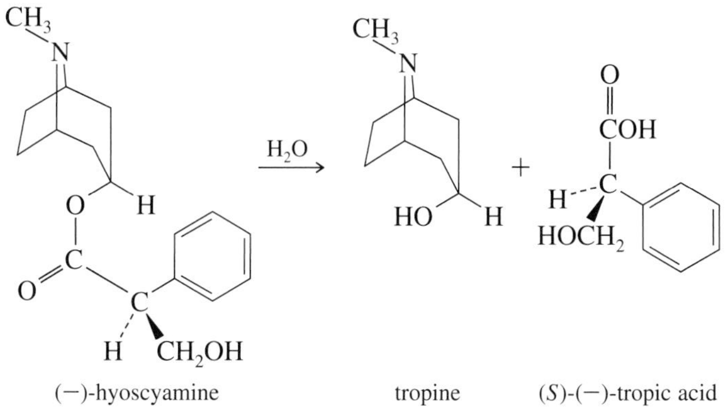

(−)-hyoscyamine          tropine          (S)-(−)-tropic acid

Tropic acid is easily racemized, so (−)-hyoscyamine is converted into optically inactive atropine by warming with base and is hydrolyzed by basic solutions to tropine and racemic tropic acid.

**Problem 22.23** Write an equation outlining a mechanism for converting (−)-hyoscyamine to atropine with a base.

**Problem 22.24** Tropine is oxidized by chromic acid to a ketone, tropinone. When tropinone is reduced, tropine is not formed. Instead, another alcohol, also $C_8H_{15}NO$, called $\psi$-tropine, is obtained. $\psi$-Tropine can be oxidized back to tropinone. Write equations showing what is happening.

Cocaine has essentially the same bicyclic ring structure as atropine, but the pattern of substitution is different. Hydrolysis of (−)-cocaine gives (−)-ecgonine, benzoic acid, and methanol.

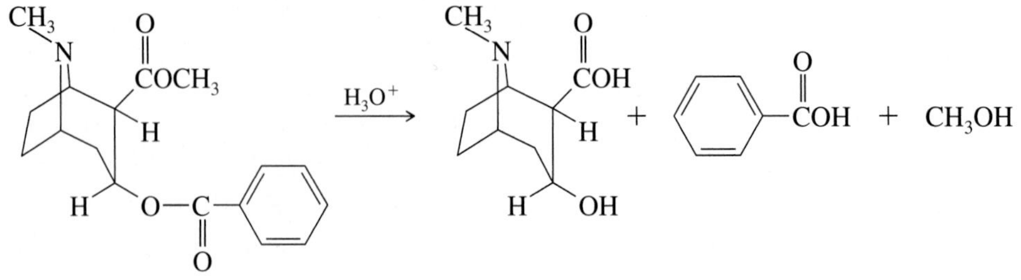

(−)-cocaine (−)-ecgonine benzoic acid methanol

As shown below, the relationship between cocaine and atropine is clear when (−)-ecgonine is oxidized with chromic acid. Tropinone (Problem 22.24) is one of the products obtained, along with other compounds from oxidative cleavage of the ring.

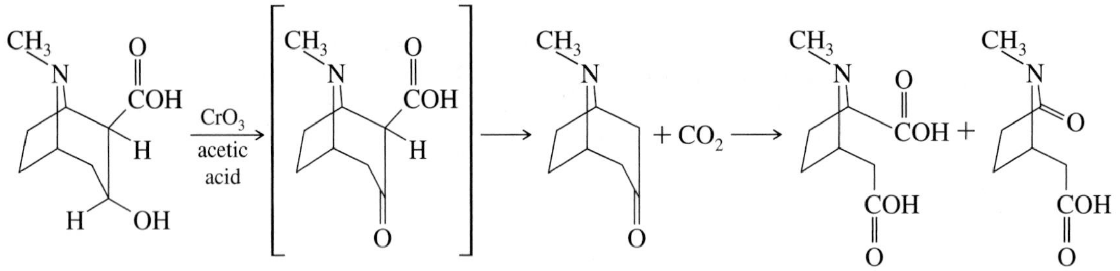

(−)-ecgonine unstable β-keto acid tropinone tropinic acid ecgoninic acid

**Problem 22.25** A base isomeric with ecgonine has been synthesized from tropinone by treating it with hydrocyanic acid followed by hydrolysis of the resulting compound. Propose a structure for this base, known as α-ecgonine, and write equations for the reactions described. Is there any ambiguity about the structure you propose?

## B. Indole Alkaloids

A large and important class of alkaloids is structurally related to the amino acid tryptophan and contains the aromatic indole ring system (pp. 932 and 983). Among the indole alkaloids, one that appears to be of central importance in physiology is 5-hydroxytryptamine, also known as **serotonin.** This compound is widely distributed in nature and stimulates a variety of smooth muscles and nerves. It has an essential function in the central nervous system as a neurotransmitter. Several drugs that interfere with the metabolism of serotonin in the brain because they are structurally similar to it are known to induce mental changes, including symptoms that resemble those of schizophrenia. Other medications used in treatment are known as serotonin uptake inhibitors. Structural formulas for serotonin and three in-dole alkaloids that cause hallucinations, bufotenin, psilocine, and lysergic acid, are given below.

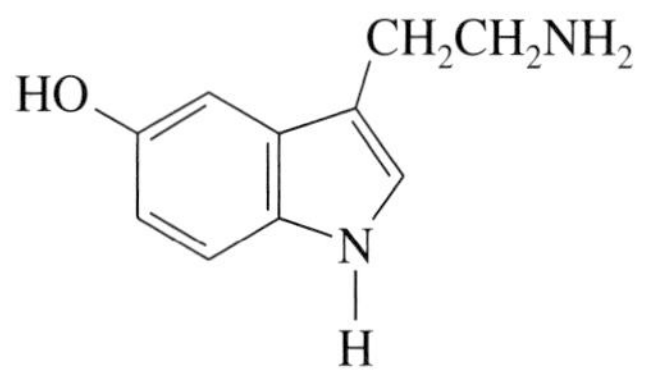

5-hydroxytryptamine
serotonin

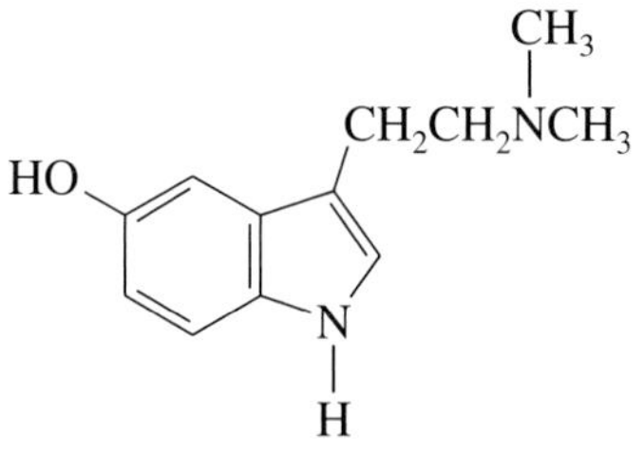

*N,N*-dimethyl-5-hydroxytryptamine
bufotenin

*a psychoactive drug from*
*the cahobe bean*

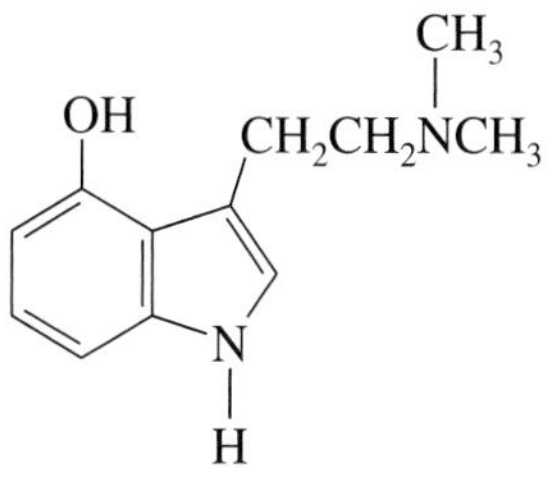

*N,N*-dimethyl-4-hydroxytryptamine
psilocine

*active ingredient of*
*hallucinogenic mushrooms*

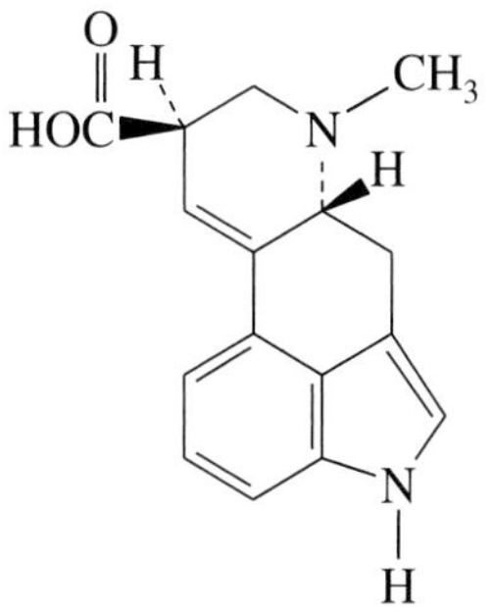

lysergic acid

*an ergot alkaloid*
*from a fungus of rye;*
LSD is the N,N-*diethylamide*
*of this compound*

All the compounds shown have an indole ring substituted at carbon 3 by a two-carbon chain ending in an amino group. Serotonin is a primary amine, and bu-fotenin, psilocine, and lysergic acid are tertiary amines. In lysergic acid, the side chain forms part of two other rings. Serotonin, bufotenin, and psilocine also have phenolic hydroxyl groups on the indole ring. Serotonin is synthesized in mammals from the amino acid tryptophan, by hydroxylation of the aromatic ring (p. 509) and then decarboxylation.

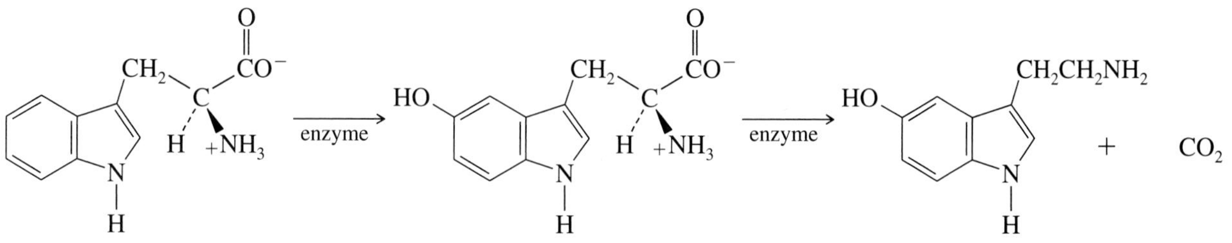

tryptophan

5-hydroxytryptophan

serotonin

**Problem 22.26**    A synthesis of racemic lysergic acid is outlined below. Assign structures to the compounds designated by letters.

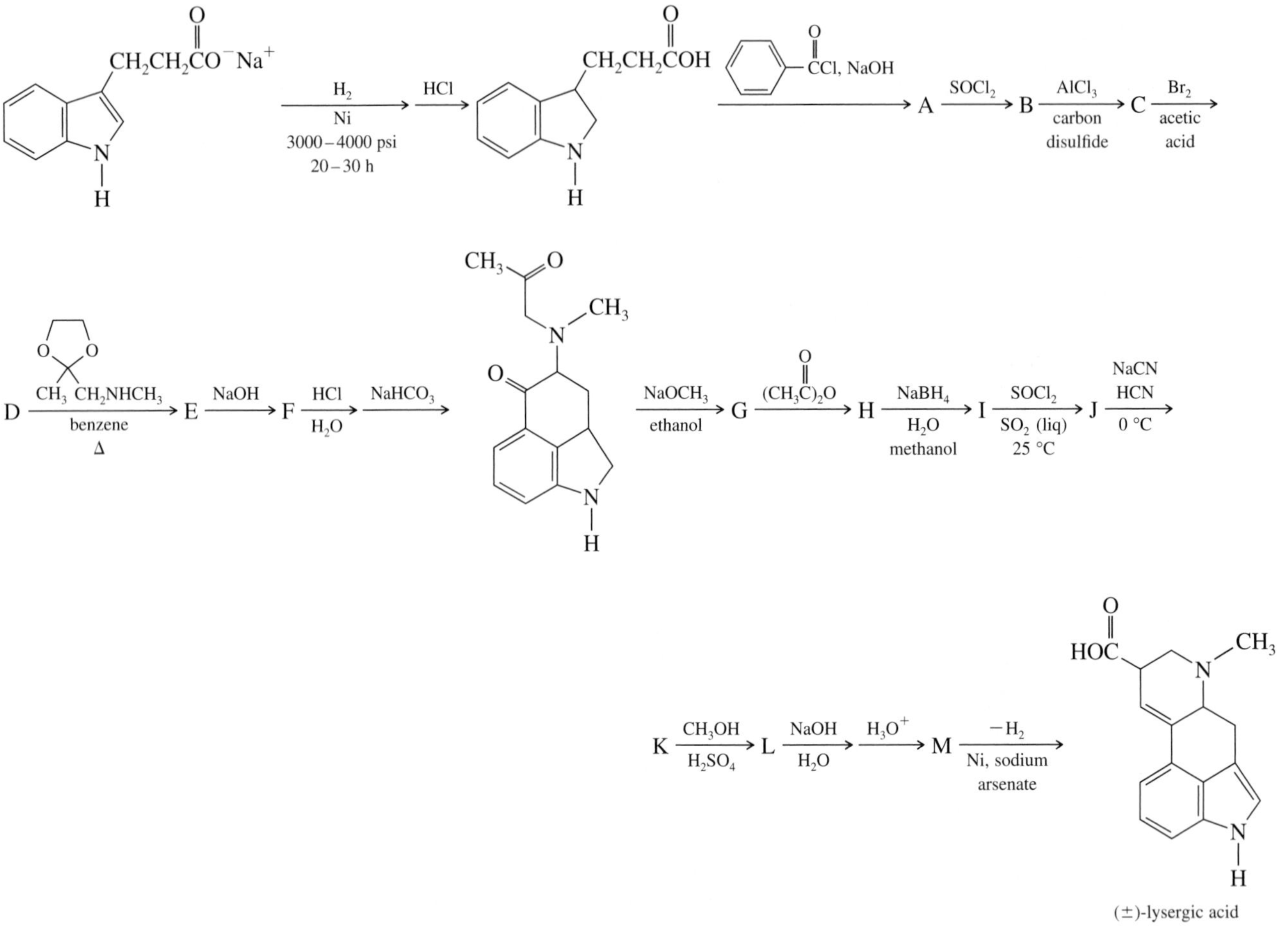

(±)-lysergic acid

## C.  Isoquinoline Alkaloids

The most effective painkiller known is the alkaloid morphine, isolated from the juice obtained from unripe seed pods of the opium poppy, *Papaver somniferum*. Apparently, morphine changes the perception of pain even when the pain itself is not much diminished. For this reason, the drug is valuable in medical practice. Unfortunately, morphine also has two severe disadvantages as a medication: It is addictive, and the body builds up a tolerance to it, so larger and larger doses may be necessary to provide the same relief from pain. The drug also depresses the function of the brain center that controls respiration; large doses of morphine (or of heroin, its synthetic diacetyl derivative) can kill by causing respiratory arrest. Another opium alkaloid is codeine, a monomethyl ether of morphine. Codeine is also a painkiller and is especially useful as a cough suppressant.

morphine codeine heroin

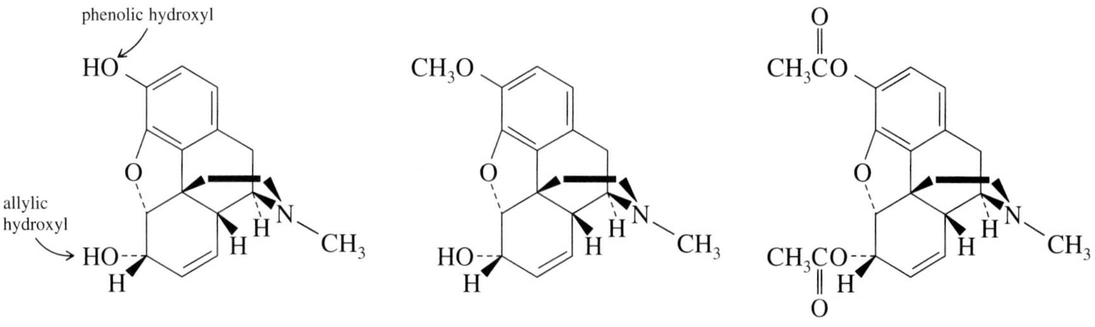

*the opium alkaloids morphine and codeine, and the synthetic derivative heroin*

All these alkaloids have as part of their structure the benzylisoquinoline unit, which can be seen more easily in the structural formula of papaverine, another opium alkaloid.

1-benzylisoquinoline

papaverine

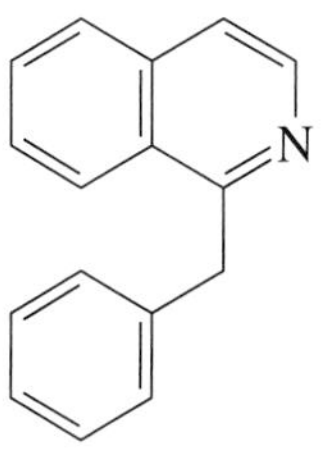

**Problem 22.27** Find the benzylisoquinoline unit in morphine.

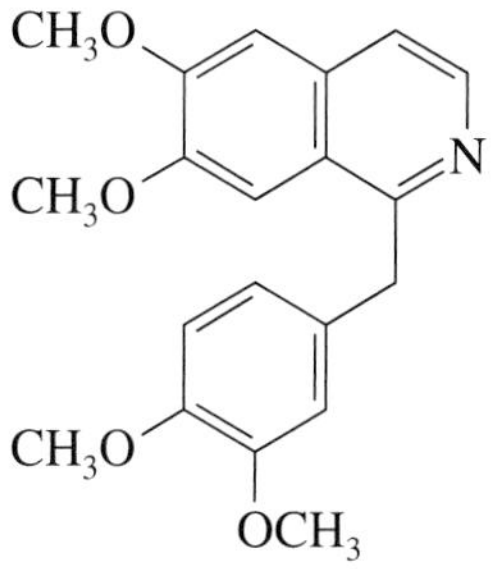

These alkaloids, as well as some simpler ones, are synthesized in plants from the amino acid tyrosine. Tyrosine is first converted by oxidation of the aromatic ring (p. 509) into (3,4-dihydroxyphenyl)alanine (DOPA), a compound that is valuable in treating Parkinson's disease. (3,4-Dihydroxyphenyl)alanine is converted to *β*-(3,4-dihydroxyphenyl)ethylamine by decarboxylation or to (3,4-dihydroxyphenyl)-pyruvic acid by transamination (p. 577).

tyrosine

↓ oxidation of the aromatic ring

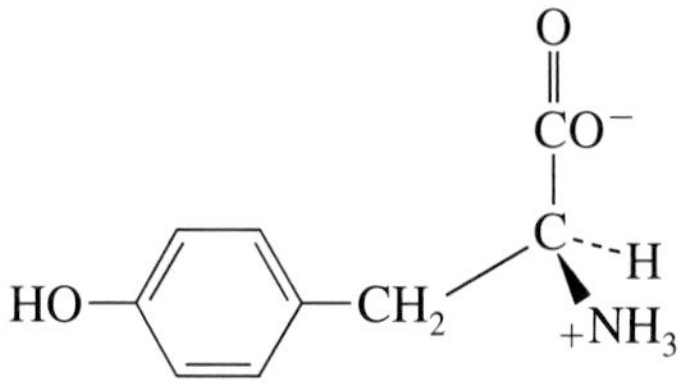

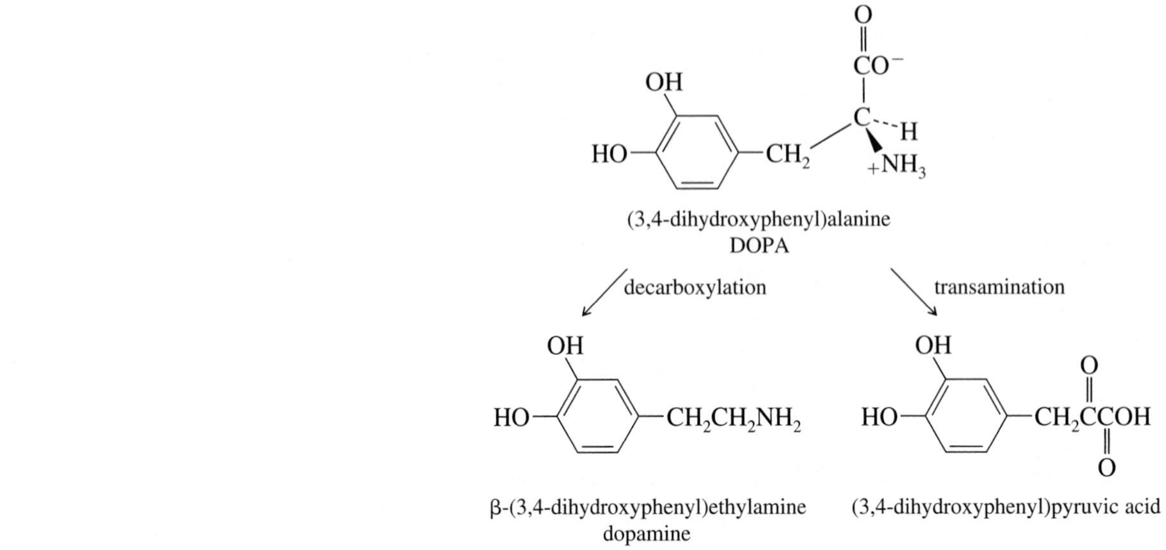

(3,4-dihydroxyphenyl)alanine
DOPA

decarboxylation                                    transamination

β-(3,4-dihydroxyphenyl)ethylamine        (3,4-dihydroxyphenyl)pyruvic acid
dopamine

The decarboxylation product of DOPA, dopamine, is a β-phenylethylamine and is closely related in structure to norepinephrine and adrenalin (p. 815). Dopamine functions as a neurotransmitter. A deficiency of dopamine in the brains of patients with Parkinson's disease is related to the symptoms characteristic of the disease. On the other hand, an overproduction of dopamine appears to be associated with psychological illnesses such as schizophrenia. Medications used for schizophrenia block dopamine receptors in the brain.

**Problem 22.28**    Tubocurarine, the potent curare alkaloid used by South American natives as a poison on the tips of their hunting arrows, is a bis(benzylisoquinoline) alkaloid. Dissect its structural formula, shown on page 819, outlining the benzylisoquinoline units that are present.

**Problem 22.29**    Write structural formulas for the intermediates and products indicated by letters in the following equations.

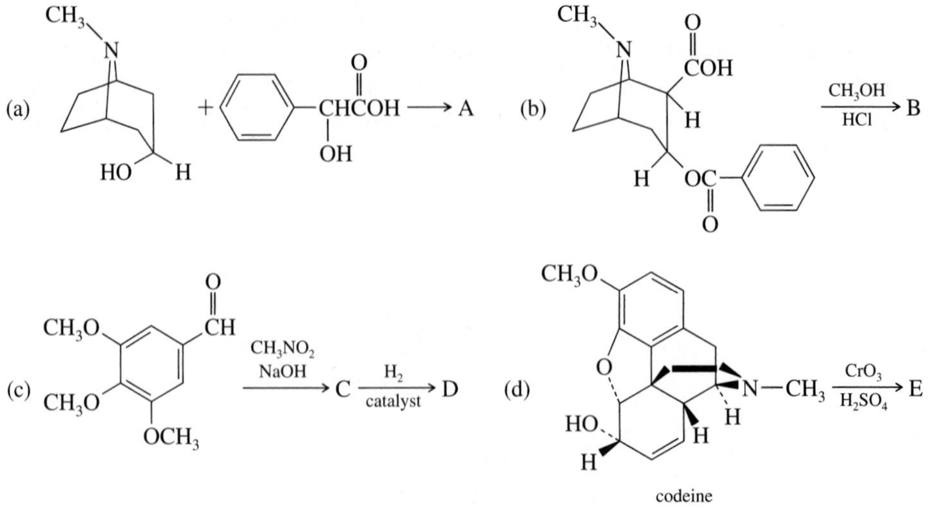

(e)

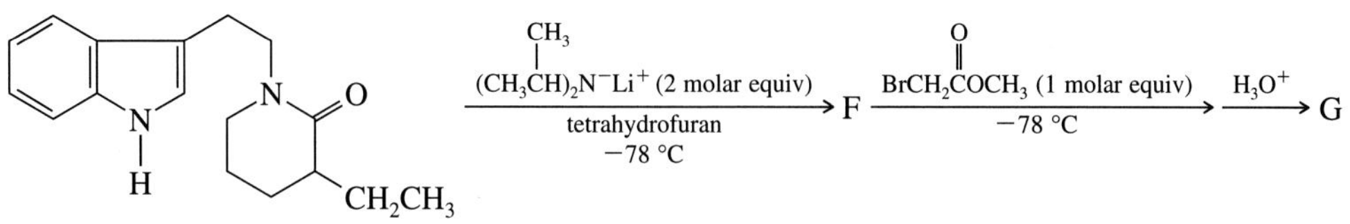

## Table 22.2 Summary of Reactions Used in the Synthesis of Heterocycles from Carbonyl Compounds and Nucleophiles

| Carbonyl Compound | Reagent | Nucleophile or Intermediate | Product |
|---|---|---|---|
| 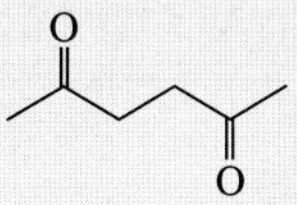 | $NH_3$, $\Delta$ | $NH_3$ | 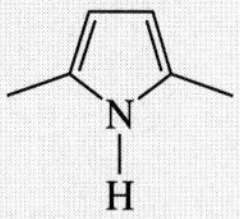 |
| | HCl, $\Delta$ | 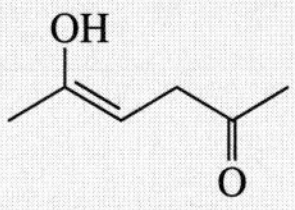 | 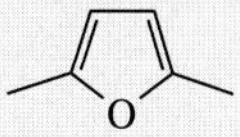 |
| 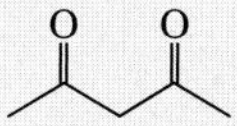 | $H_2NNH_3^+$ $HSO_4^-$ NaOH | $H_2NNH_2$ | 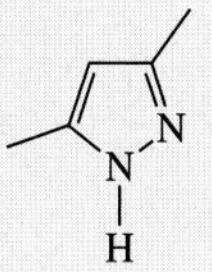 |
| | $HONH_3^+$ $HSO_4^-$ $K_2CO_3$ | $HONH_2$ | 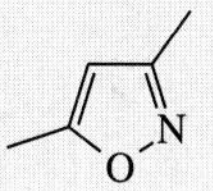 |
| | $H_2N-\overset{O}{\underset{}{C}}-NH_2$ $\Delta$ | $H_2N-\overset{O}{\underset{}{C}}-NH_2$ | |
| | $H_2N-\overset{O}{\underset{}{C}}-NH_2$ $\Delta$ | $H_2N-\overset{O}{\underset{}{C}}-NH_2$ | |
| | $H_2N-\overset{S}{\underset{}{C}}-NH_2$ $\Delta$ | $H_2N-\overset{S}{\underset{}{C}}-NH_2$ | |
| 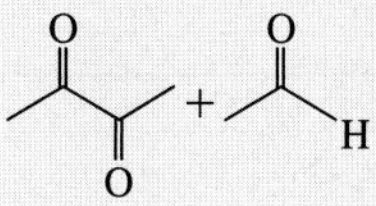 | $CH_3CO^-$ $^+NH_4$ $\Delta$ | $NH_3$ | 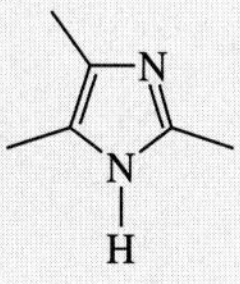 |

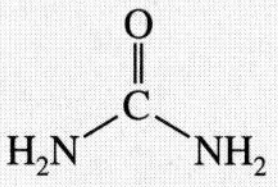

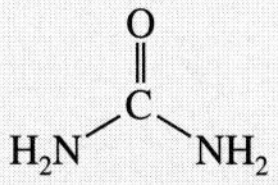

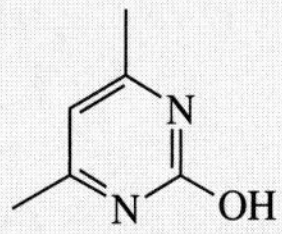

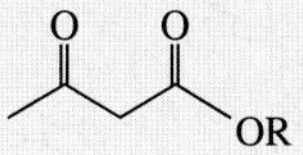

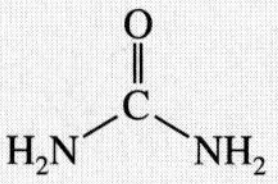

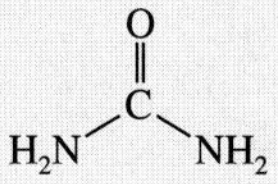

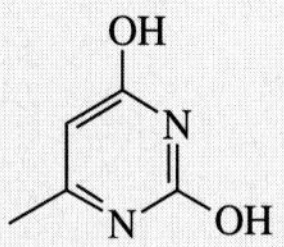

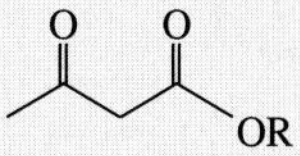

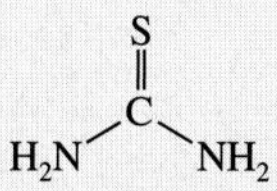

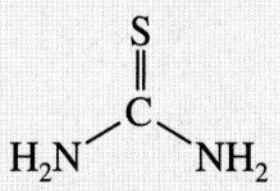

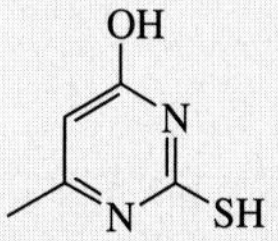

| Table 22.3 | Summary of Electrophilic Aromatic Substitution Reactions of Heterocycles | | |
|---|---|---|---|
| **Heterocycle** | **Reagent** | **Electrophile** | **Product** |
|  (furan, Y = S, NR) | $X_2$ | $X_2$ | 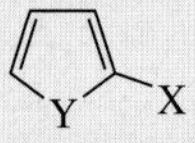 (product with X) |
| | $HNO_3$ $CH_3\overset{O}{\overset{\|}{C}}O\overset{O}{\overset{\|}{C}}CH_3$ | $\overset{+}{N}O_2$ | (product with $NO_2$) |
| 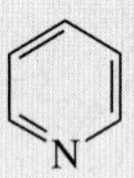 (pyridine) | $HNO_3$, $KNO_3$ 330 °C | $\overset{+}{N}O_2$ | 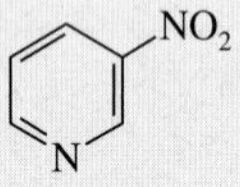 (product with $NO_2$) *low yield* |

# ADDITIONAL PROBLEMS

**22.30** Name the following compounds.

(a) 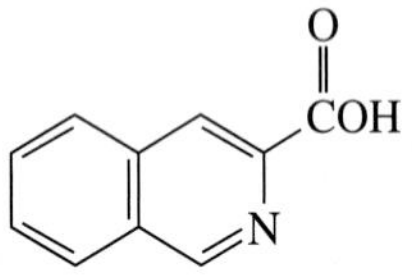 (isoquinoline-3-carboxylic acid structure, COH)

(b) 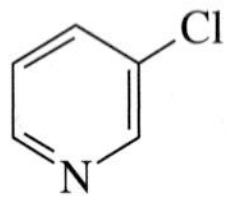 (3-chloropyridine structure, Cl)

(c) 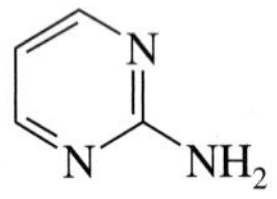 (2-aminopyrimidine structure, NH₂)

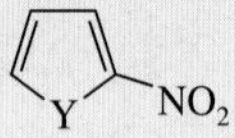

(d) 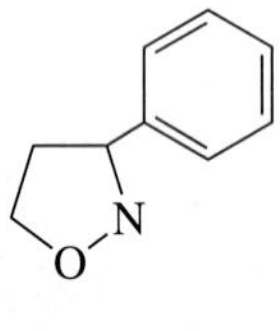 (3-phenyl isoxazoline structure)

(e) 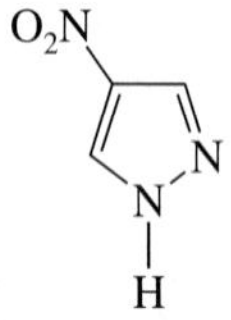 (4-nitropyrazole structure, O₂N)

(f) 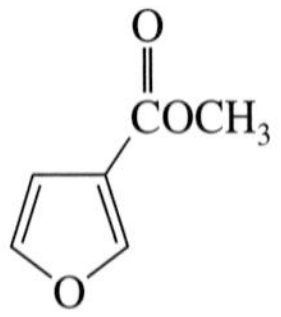 (furan-3-yl methyl ketone, COCH₃)

(g) 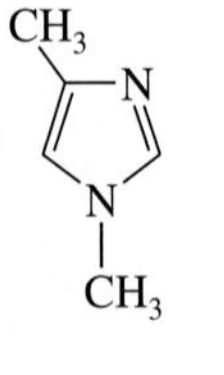 (1,4-dimethylimidazole structure, CH₃)

(h) 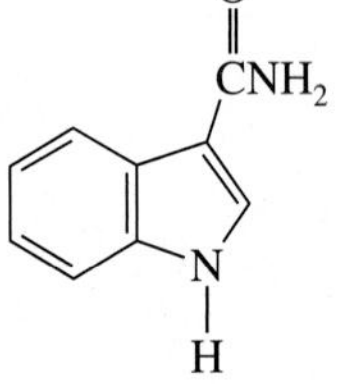 (indole-3-carboxamide structure, CNH₂)

(i) 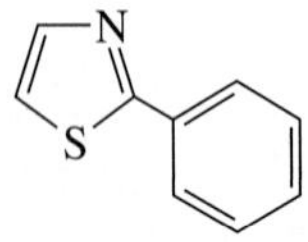 (2-phenylthiazole structure)

(j) 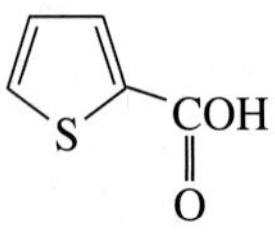 (thiophene-2-carboxylic acid structure, COH)

**22.31** Write structural formulas for all intermediates and products designated by letters in the following equations.

(a) 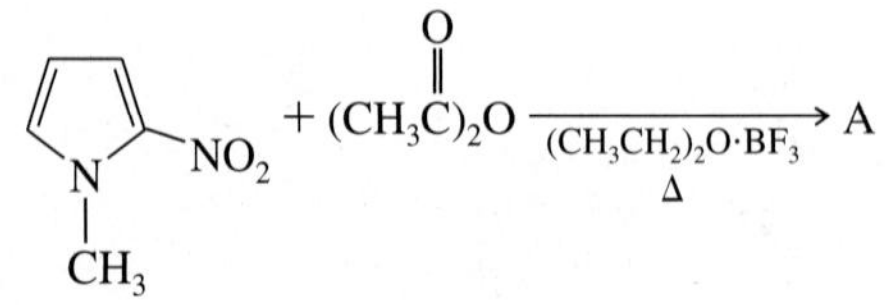 (N-methylpyrrole-2-nitro) $+ (CH_3C)_2O \xrightarrow[\Delta]{(CH_3CH_2)_2O \cdot BF_3}$ A

(b) 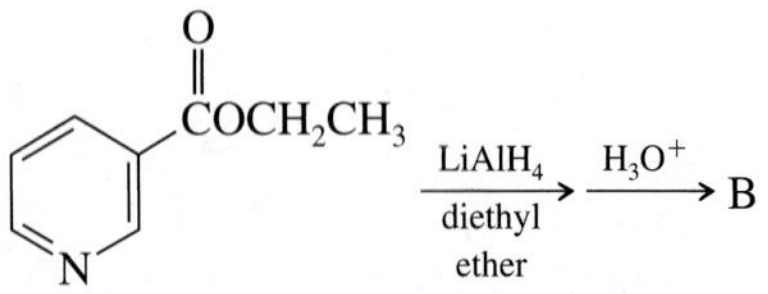 (3-propanoyl pyridine, COCH₂CH₃) $\xrightarrow[\text{diethyl ether}]{LiAlH_4}$ $\xrightarrow{H_3O^+}$ B

(c) $+ H_2 \xrightarrow{Ni} C$

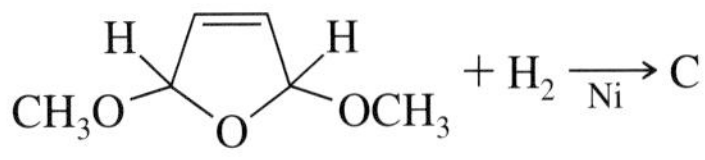

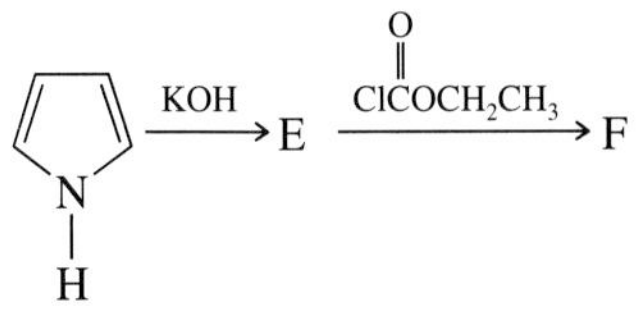

(d) 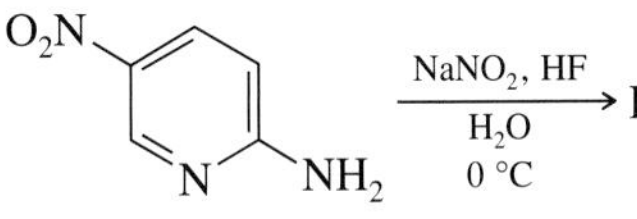 $\xrightarrow[\substack{H_2O \\ 0\ °C}]{NaNO_2,\ HF}$ D

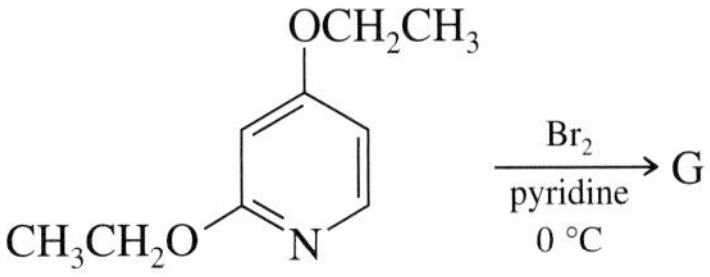

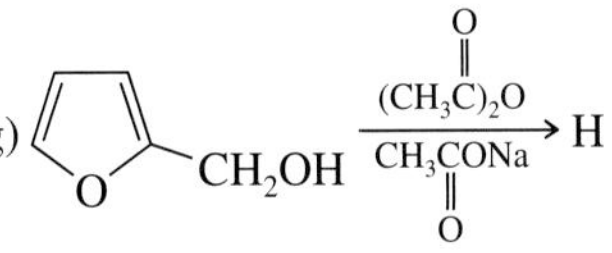

(e) $\xrightarrow{KOH}$ E $\xrightarrow{ClCOCH_2CH_3}$ F

(f) $\xrightarrow[\substack{pyridine \\ 0\ °C}]{Br_2}$ G

(g) $\xrightarrow[CH_3CONa]{(CH_3C)_2O}$ H

(h) $+ CH_3CCH_2CH_2CCH_3 \xrightarrow{HCl}$ I

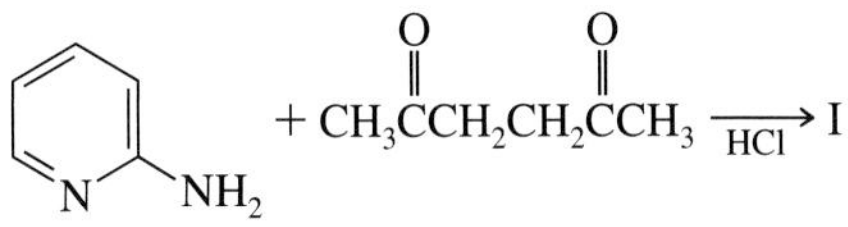

**22.32** More practice in recognizing reactions follows.

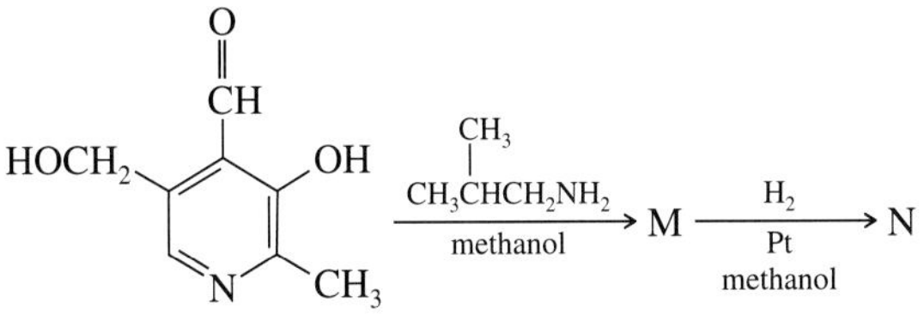

(a) 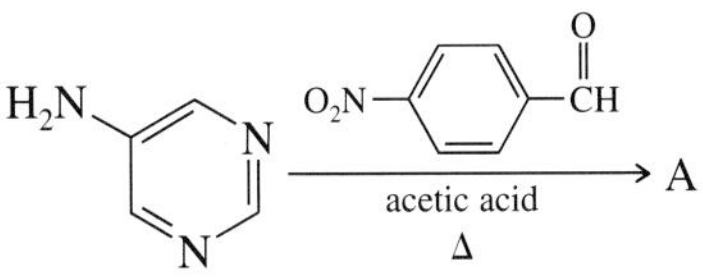 $\xrightarrow[\substack{acetic\ acid \\ \Delta}]{}$ A

(b) 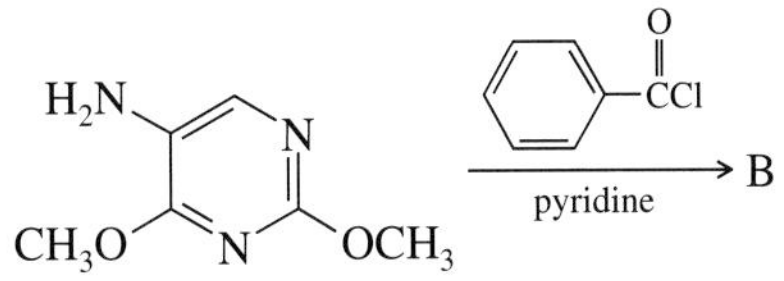 $\xrightarrow{pyridine}$ B

(c) $\xrightarrow{CH_3I}$ C

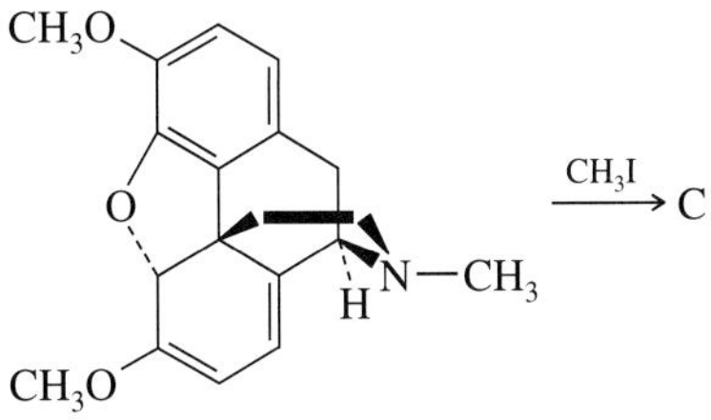

thebaine

(d) 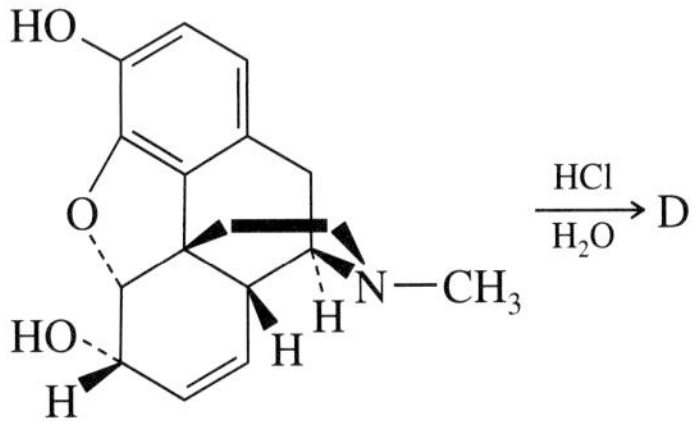 $\xrightarrow[H_2O]{HCl}$ D

morphine

(e) 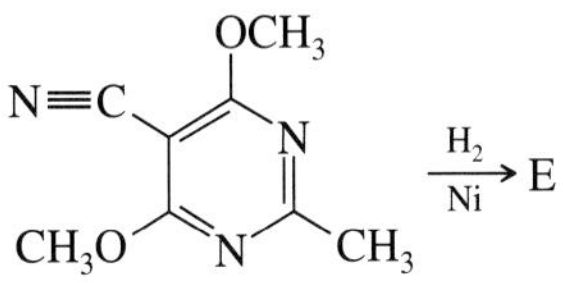 $\xrightarrow{H_2}{Ni}$ E

(f) $\xrightarrow[\substack{H_2O \\ 0\ °C}]{NaNO_2,\ HCl}$ F

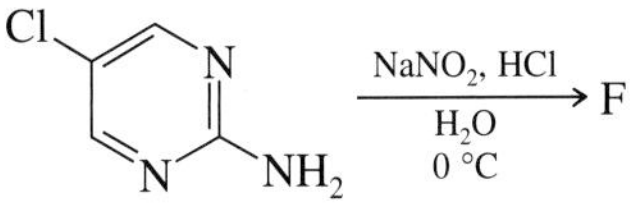

(g) 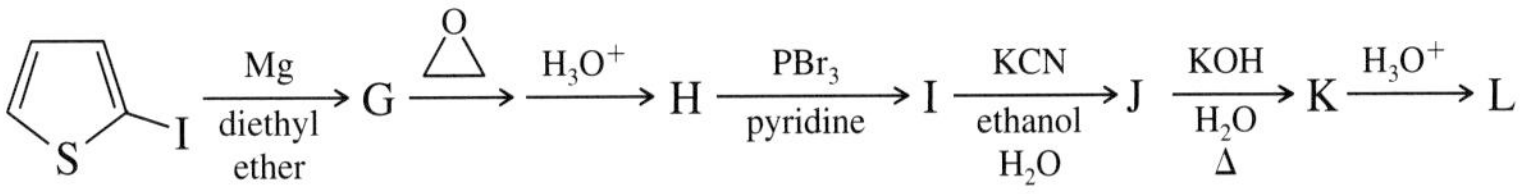 $\xrightarrow[\substack{diethyl \\ ether}]{Mg}$ G $\xrightarrow{}$ $\xrightarrow{H_3O^+}$ H $\xrightarrow[pyridine]{PBr_3}$ I $\xrightarrow[\substack{ethanol \\ H_2O}]{KCN}$ J $\xrightarrow[\substack{H_2O \\ \Delta}]{KOH}$ K $\xrightarrow{H_3O^+}$ L

(h) $\xrightarrow[methanol]{CH_3CHCH_2NH_2}$ M $\xrightarrow[\substack{Pt \\ methanol}]{H_2}$ N

(i) $+$ $\underset{\substack{\\ malic\ acid}}{HOCCH_2CHCOH}$ $\xrightarrow[\substack{dehydrogenase \\ H_2O}]{malic}$ O + P

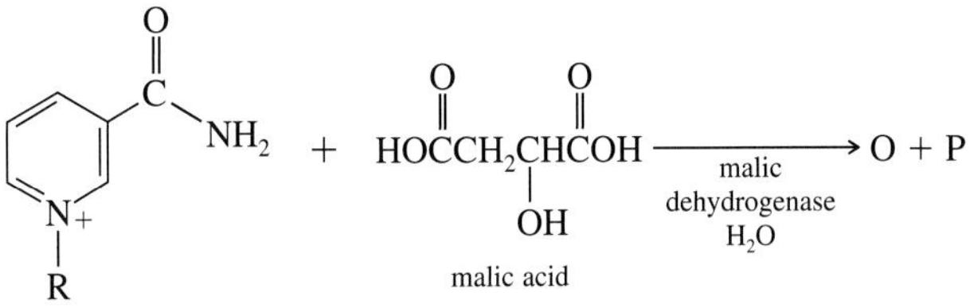

**22.33** Write detailed mechanisms showing how the following transformations could have taken place.

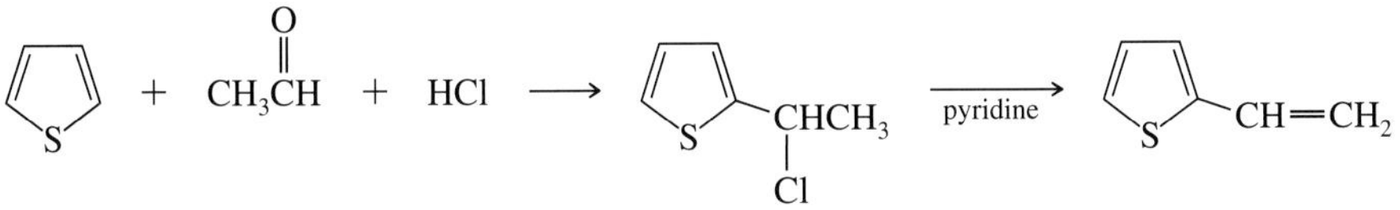

**22.34** The important antibiotics penicillin and cephalosporin include sulfur-containing rings. As chemists look for ways to create better antibiotics that have fewer side effects and new antibiotics to keep ahead of the capacity of microorganisms to develop resistance to medications used, they need to devise ways of creating sulfur heterocycles. Some research into creating the ring system found in penicillins is shown below. Provide structural formulas for the reagents or products designated by letters.

(a)

(b)

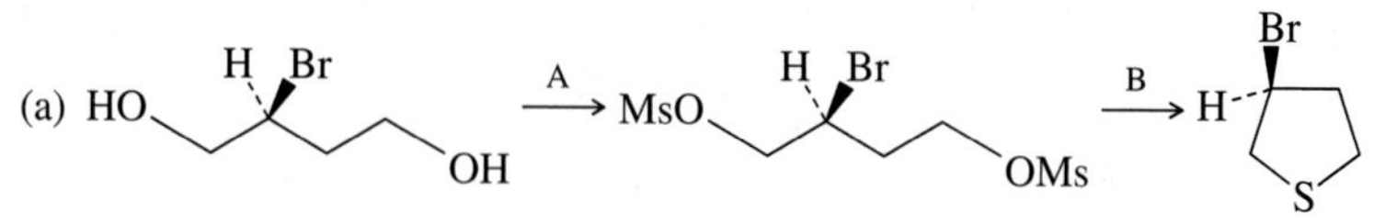

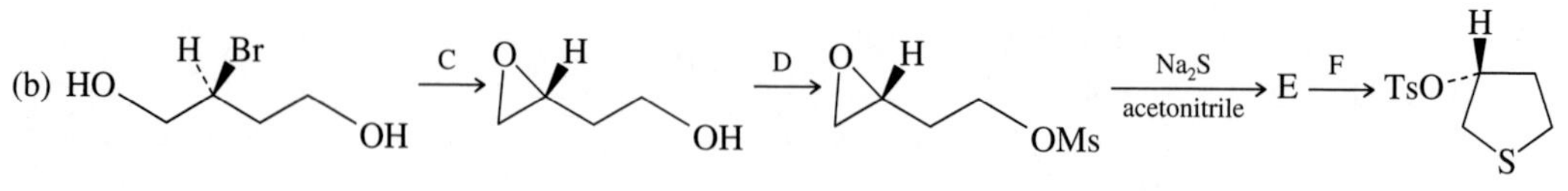

(c) Write a mechanism for the formation of E from the oxirane-mesylate and sodium sulfide.

**22.35** The two enantiomeric ethanol-1-*d* species obtained in the reactions shown on pages 949 and 950 were collected at first in quantities too small to allow their optical rotations to be measured. The fact that they are enantiomers was originally proved by the following sequence of reactions.

$$CH_3CD + NADH \longrightarrow \text{alcohol A} \xrightarrow[\text{pyridine}]{TsCl} B \xrightarrow[\substack{H_2O \\ \Delta}]{NaOH} C$$

C + NAD⁺ ⟶ acetaldehyde + NADD    containing 1
                  containing               molar equivalent
                  no deuterium            of deuterium

A + NAD⁺ ⟶ acetaldehyde + NADH    containing
                  containing               no deuterium
                  deuterium

Write equations showing what is happening in this series of reactions. Be sure to use correct stereochemical representations of all the compounds involved. How do these results prove that the alcohols obtained in the original experiments are enantiomeric?

**22.36** The 9-azabicyclo[4.2.1]nonane system is interesting because it is found in anatoxin-*a*, a toxic metabolite found in algae. (See page 745 for a reminder how bicyclic compounds are named.) The following reactions were carried out in the synthesis of 9-benzyl-9-aza-bicyclo[4.2.1]nonane in a study of routes to compounds similar in structure to anatoxin-*a*. Provide structural formulas for the products designated by letters.

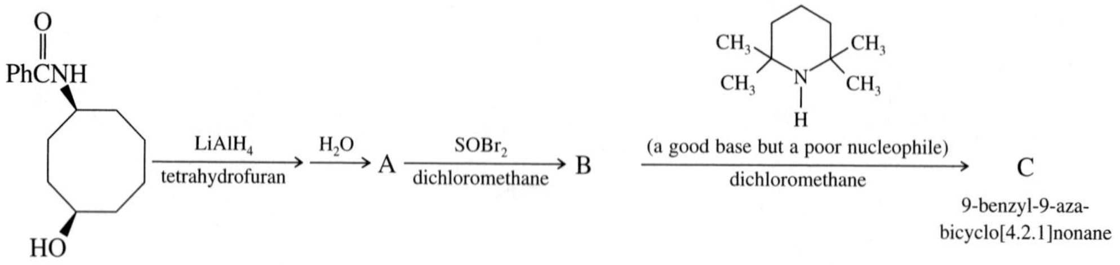

**22.37** Osmium tetroxide forms complexes with amines as it reacts with alkenes. If those amines are chiral, osmium tetroxide will give chiral diols as products of its reactions with alkenes (p. 316). A chiral amine that promotes some stereoselectivity in the reaction of an alkene with osmium tetroxide is the bicyclic diamine synthesized below. Provide structural formulas for the reagents or the products designated by letters. Remember that stereochemistry is important in this synthesis.

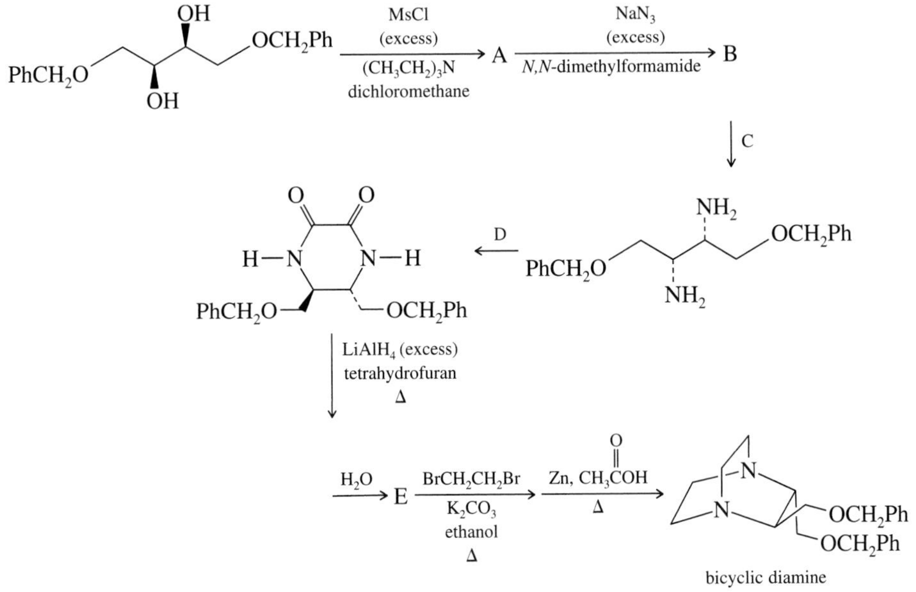

**22.38** Much research is now directed toward finding the cause of Alzheimer's disease and toward finding drugs that may help its victims. The following transformations were carried out in the synthesis of compounds that would bind to the same sites as muscarine does (Problem 20.12). Supply structural formulas for the compounds designated by letters.

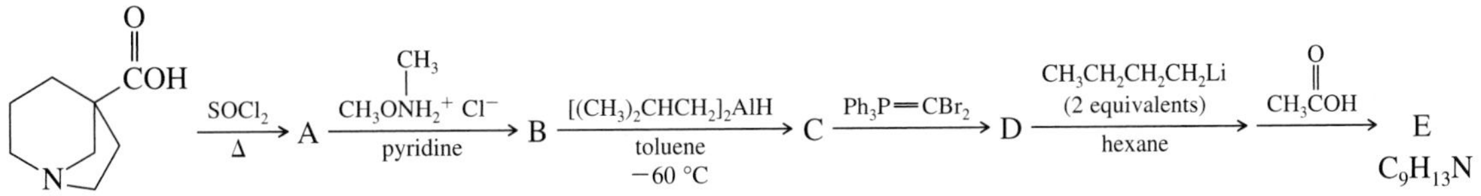

Compound E has a singlet corresponding to one hydrogen at $\delta$ 2.15 in its proton magnetic resonance spectrum. The significant bands in the $^{13}C$ magnetic resonance spectrum of Compound E are at $\delta$ 68.9 and 89.2.

**22.39** Arrange the species shown below in order of increasing acidity, and give your reasons for the order you choose.

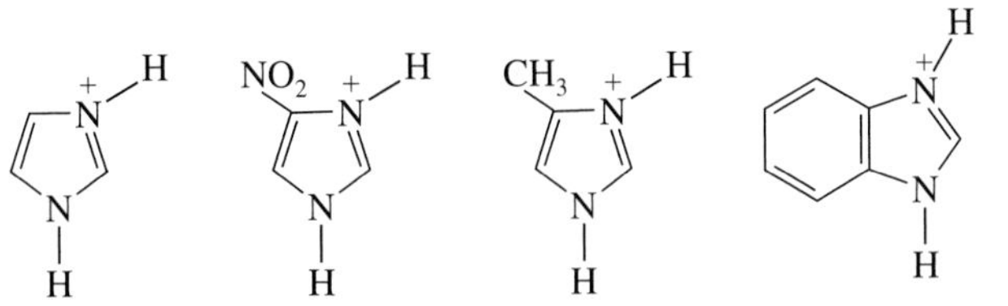

**22.40** Steps in the synthesis of Bao Gong Teng A, the active ingredient in a Chinese herbal remedy that is remarkably effective against glaucoma, are outlined on the next page. Provide structural formulas for the reagents or products designated by letters.

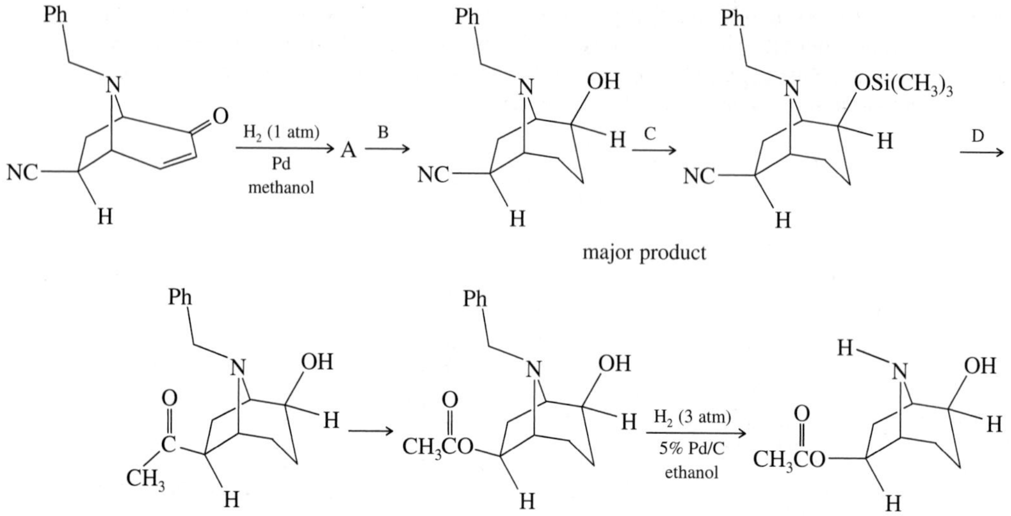

major product

Bao Gong Teng A

**22.41** When phenylhydrazine is heated with 2-butenoic acid, water is evolved and a heterocyclic compound having the molecular formula $C_{10}H_{12}N_2O$ is formed. Propose a structure for this product. (*Hint:* A review of Section 17.5C may be helpful.) When phenylhydrazine is heated with ethyl acetoacetate, another heterocycle, $C_{10}H_{10}N_2O$, is formed. What is the structure of this compound?

**22.42** Hexane and 2-hexanone both have neurotoxicity. This is believed to come from the reaction of a common metabolite of these compounds, 2,5-hexanedione, which reacts with amines to give pyrrole derivatives. Proteins have free amino groups in the amino acid lysine (p. 985). Using Protein —$NH_2$ to represent such a portion of a protein, predict what the product of its reaction with 2,5-hexanedione will be.

**22.43** The amino acid proline often provides a chiral starting material for the synthesis of natural products. Slaframine is a compound found in moldy feeds that make cattle salivate excessively. Studies of the synthesis of slaframine and its analogs used the following reactions. Provide structural formulas for the reagents or the products designated by letters.

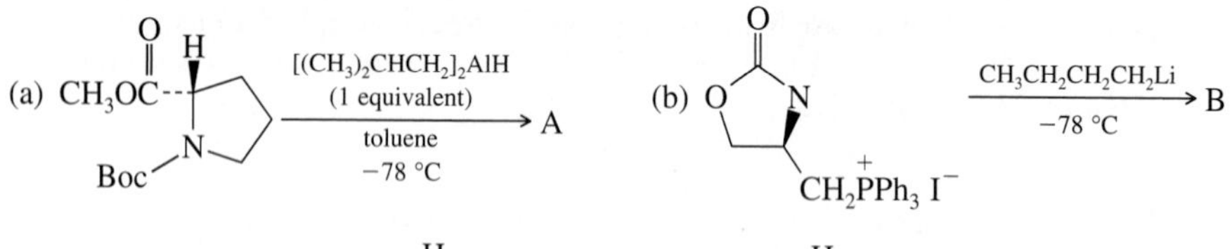

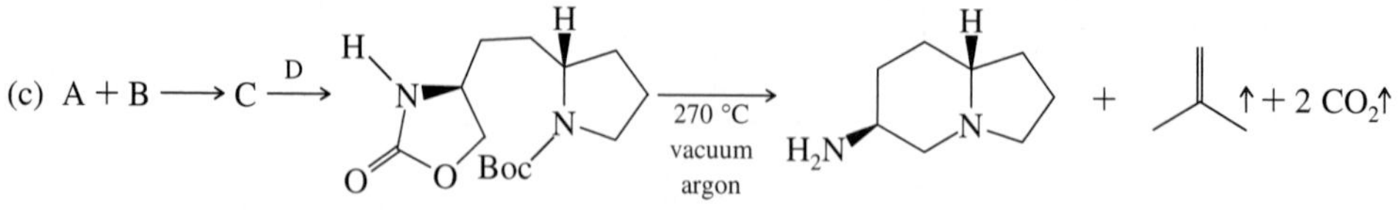

**22.44** Vitamin $B_6$, pyridoxine, has been isolated from rice bran and has the molecular formula $C_8H_{11}NO_3$. Its ultraviolet spectrum changes with pH. At pH 2.1, the spectrum shows $\lambda_{max}^{H_2O}$ 292 ($\epsilon$ 6950); at pH 10.2, $\lambda_{max}^{H_2O}$ 240 ($\epsilon$ 5500) and 315 nm ($\epsilon$ 5800) are observed.

(a)   When the acidic proton in vitamin $B_6$ is replaced by a methyl group, the spectrum of the resulting compound has a single band at 280 nm ($\epsilon$ 5800) that does not change with pH. Given this, what conclusions can you reach about the structure of vitamin $B_6$?

(b) Oxidation of the methyl ether of vitamin $B_6$ with barium permanganate in water at room temperature for 16 hours, followed by filtration of the manganese dioxide that was formed and acidification of the solution with sulfuric acid, gave two products, a lactone and a dicarboxylic acid. Their structures are shown below.

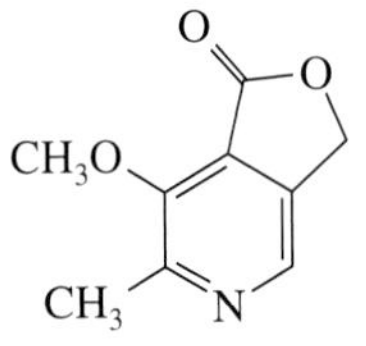

|  |  |
|---|---|
| lactone | dicarboxylic acid |
| from the oxidation of | from the oxidation of |
| the methyl ether | the methyl ether |
| of vitamin $B_6$ | of vitamin $B_6$ |

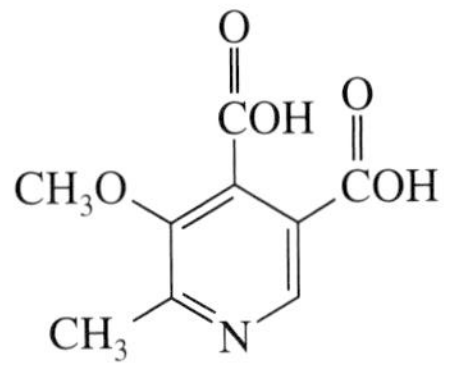

What must be the structure of vitamin $B_6$? Explain why two of the side chains on this aromatic ring were oxidized but the methyl group was untouched. Write equations for a reaction pathway to the lactone.

(c) If you were trying to confirm the structure of vitamin $B_6$ using ultraviolet spectroscopy, what simpler compounds would you use as models? Draw structural formulas for some that would be useful. Under what conditions would you take the ultraviolet spectra of the models?

# 23 Structure and Reactivity in Biological Macromolecules

### A Look Ahead

Among the many kinds of chemical substances found in living organisms, three stand out for the size of their molecules, the variety of their structures, and the relatively few simple structural units from which these large molecules are constructed. One such class of compounds is the carbohydrates. The structures of carbohydrates range from the relatively simple, such as glucose, to the polymeric structures found in starches and cellulose, to glycoproteins and glycolipids in which carbohydrates are bonded to amino acids and fatty acids to create important components of cell membranes and tissues.

Another class of compounds is the proteins. Proteins are one of the primary constituents of living matter. Even in plants, where carbohydrates are more abundant as structural materials, proteins are present in those parts that are responsible for growth and reproduction. About twenty different amino acids are the building blocks of proteins. The number of ways in which these amino acids may be combined in sequence to give proteins is staggeringly large. Molecular weights of proteins range from 5000 to 1 million.

Finally, there are the deoxyribonucleic acids (DNAs) and ribonucleic acids (RNAs), which carry and transmit genetic information in the synthesis of proteins. Each is made up of four nucleotides that contain two purine and two pyrimidine bases.

In this chapter we will learn about the structural units, the sugars, amino acids, and nucleotides, that make up these important biological molecules. We also will learn how these structural units are combined to give the larger molecules. Biological function is closely tied to form in these large molecules, so stereochemistry will be particularly important. Because glucose has several stereocenters, the history of the development of knowledge of stereochemistry is closely related to its chemistry. Fischer projection formulas, another way of representing stereochemistry, will be used in this chapter to clarify stereochemical relationships among carbohydrates. Finally, we will learn about the chemical basis for the functions that these large molecules have in living organisms.

## 23.1 Structures of Monosaccharides. Structural Units of Carbohydrates

**Carbohydrates,** important constituents of both plants and animals, are polyhydroxy aldehydes or ketones or their derivatives. Compounds classified as carbohydrates range from those consisting of a few carbon atoms to gigantic polymeric molecules having molecular weights in the millions.

Carbohydrates that cannot be broken down into simpler units by hydrolysis reactions are known as **monosaccharides.** The most common monosaccharide is glucose, $C_6H_{12}O_6$, the chief form in which carbohydrates are metabolized in the human body. Fructose, $C_6H_{12}O_6$, is known as fruit sugar.

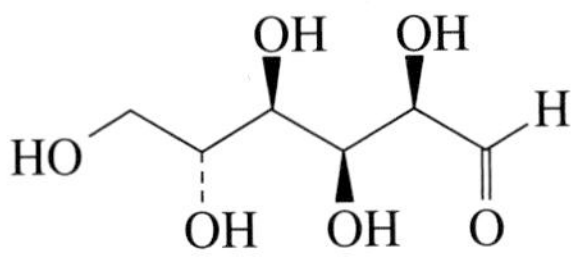

D-glucose
as its open-chain aldehyde form

an aldohexose

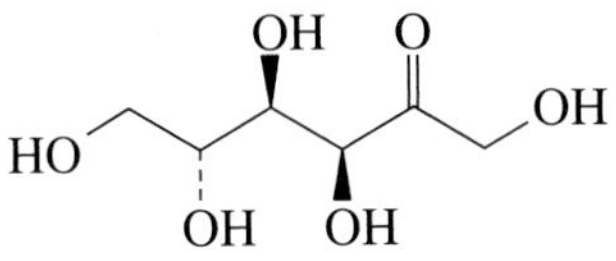

D-fructose
as its open-chain ketone form

a ketohexose

We have already studied how the polyhydroxyaldehyde structure of glucose leads to the formation of cyclic hemiacetals, glucopyranoses (Section 14.8A, p. 563), and acetals, glucopyranosides (Section 14.8B, p. 569). We also have examined how glucose is converted into fructose (Section 16.6A, p. 677) and then how fructose is broken down into three-carbon units in an important reaction in the metabolism of glucose, the retroaldol reaction (Section 17.6B, p. 718).

Ribose, $C_5H_{10}O_5$, and 2-deoxyribose, $C_5H_{10}O_4$, are components of ribonucleic acids (RNAs) and deoxyribonucleic acids (DNAs), respectively, the giant molecules that play an important role in the storage and transmission of genetic information (p. 987).

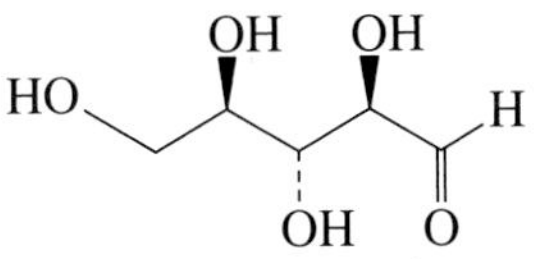

D-ribose
as its open-chain aldehyde form

an aldopentose

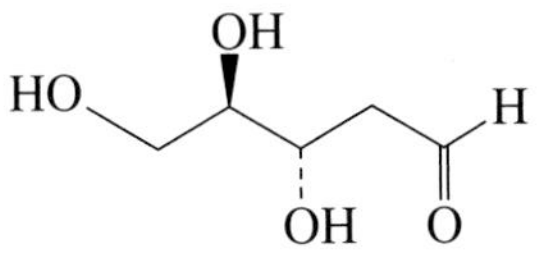

D-2-deoxyribose
as its open-chain aldehyde form

a deoxyaldopentose

Monosaccharides can be further subdivided according to the number of carbon atoms they contain and whether they are aldehydes (**aldoses**) or ketones (**ketoses**). Ribose, with five carbon atoms, is a pentose. It has an aldehyde function, so it is an aldopentose. Glucose is an aldohexose, and fructose is a ketohexose.

**Problem 23.1** Besides pentoses and hexoses, there are trioses, with three carbon atoms, and tetroses, with four carbon atoms. Classify the following monosaccharides according to the number of carbon atoms and the nature of the carbonyl function they contain.

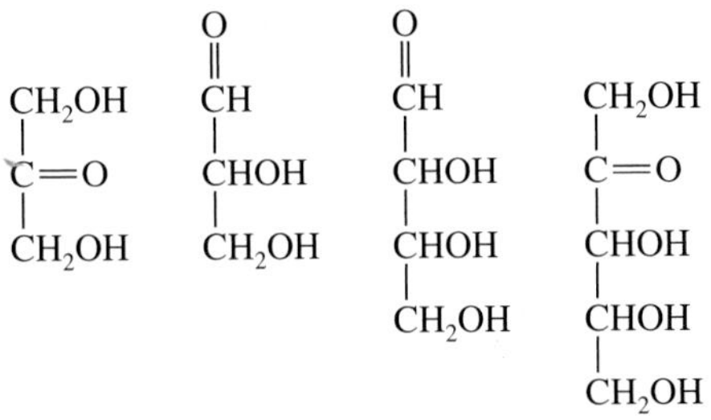

| A | B | C | D |
|---|---|---|---|

Hydroxyl and carbonyl groups are not the only functional groups to appear in monosaccharides. Monosaccharides containing carboxyl groups and amino groups are common structural units in biologically important carbohydrates. Two such monosaccharides are 2-amino-2-deoxy-D-glucose, $C_6H_{13}NO_5$, an amino sugar (earlier called glucosamine), and glucuronic acid, $C_6H_{10}O_7$, a sugar acid.

2-amino-2-deoxy-D-glucose
glucosamine

glucuronic acid

Common table sugar, sucrose, has the molecular formula $C_{12}H_{22}O_{11}$. When it is boiled with water with a trace of acid, as in candy making, it is converted into a mixture of glucose and fructose. Therefore, it is classified as a **disaccharide,** a compound made up of two monosaccharide units.

$$C_{12}H_{22}O_{11} + H_2O \xrightarrow[\Delta]{H_3O^+} C_6H_{12}O_6 + C_6H_{12}O_6$$

sucrose    glucose    fructose

Maltose, another disaccharide, gives two molecules of glucose on hydrolysis.

$$C_{12}H_{22}O_{11} + H_2O \xrightarrow[\Delta]{H_3O^+} 2\ C_6H_{12}O_6$$

maltose    glucose

Similarly, **trisaccharides** and **tetrasaccharides** give three and four saccharide units, respectively, on hydrolysis. Compounds containing relatively few monosaccharide units are called **oligosaccharides.** Saccharides in this molecular weight range are individual, identifiable compounds with definite structures and molecular weights. They are sometimes water soluble and often sweet-tasting. Oligosaccharides have many important physiologic functions. For example, blood groups are determined by glycoproteins—proteins that have oligosaccharides covalently bonded to them—which are found on the surface of red blood cells. Fibrinogen, an important component in blood clotting, is also a glycoprotein, as are immunoglobulins, which are involved in the development of immunity to disease. The chemistry of carbohydrates is thus directly relevant to the process of cell recognition and regulation of cell growth, so important to solving the problem of cancer.

When there are numerous monosaccharide units in molecules, the compounds are defined as **polysaccharides.** Cellulose (p. 1004), the chief structural material of plants, is a high-molecular-weight polysaccharide made up of glucose units. Not all cellulose molecules have the same molecular weight. Instead, any sample contains molecules with a given range of molecular weights. Starch (p. 1002) is the polysaccharide form in which plants store glucose for their energy needs. Animals store glucose as another polysaccharide, glycogen.

## 23.2  Stereochemistry of Monosaccharides

### A. Glyceraldehyde as the Standard for the Assignment of Relative Configurations

The chemistry of carbohydrates has been central to the development of chemists' understanding of stereochemistry and the assignment of absolute configuration (p. 205) to chiral compounds. In 1906, M. Rosanoff, an American chemist, suggested that (+)-glyceraldehyde be assigned a configuration and used as the standard for the configurations of other sugars and, ultimately, other chiral compounds. The configuration that was chosen for (+)-glyceraldehyde was the configuration that came to be called *R* according to the Cahn-Ingold-Prelog Rules many years later, and therefore, (−)-glyceraldehyde was assigned the configuration we now know as *S*.

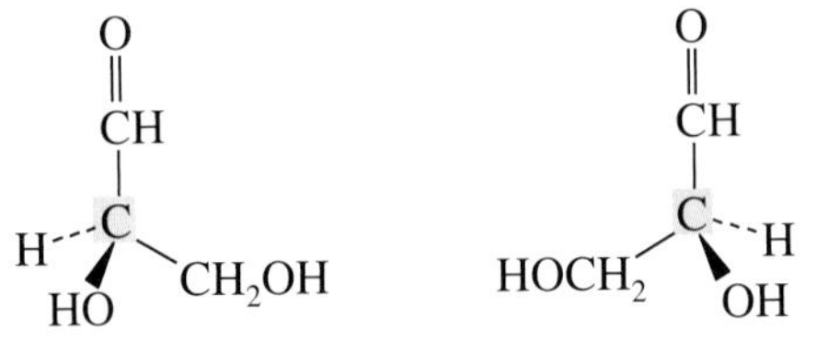

*(R)*-(+)-glyceraldehyde          *(S)*-(−)-glyceraldehyde

*configurations of (+)- and (−)-glyceraldehyde
as assigned in 1906*

This assignment was chosen for (+)-glyceraldehyde because the hydroxyl group and the hydrogen atom at carbon 5 of (+)-glucose had been arbitrarily assigned the corresponding configuration in 1891 by Emil Fischer. Because (+)-glyceraldehyde could be related by chemical transformations to (+)-glucose, the two compounds had to have the same stereochemistry at the stereocenter closest to the primary alcohol group in each molecule. Rosanoff and Fischer, of course, had a 50% chance that their assignments were correct. (+)-Glyceraldehyde had to have either the *R* configuration or the enantiomeric *S* configuration.

Rosanoff's suggestion was widely accepted. Over the years, many compounds were synthesized from or degraded to (+)-glyceraldehyde so that their relative configurations could be established. Each compound that was related to (+)-glyceraldehyde could then serve as a standard for other compounds, until the relative configurations of many compounds were known. The following reaction provides a simple example of the kinds of correlations that can be made. *(R)*-(+)-Glyceraldehyde is oxidized to *(R)*-(−)-glyceric acid (2,3-dihydroxypropanoic acid).

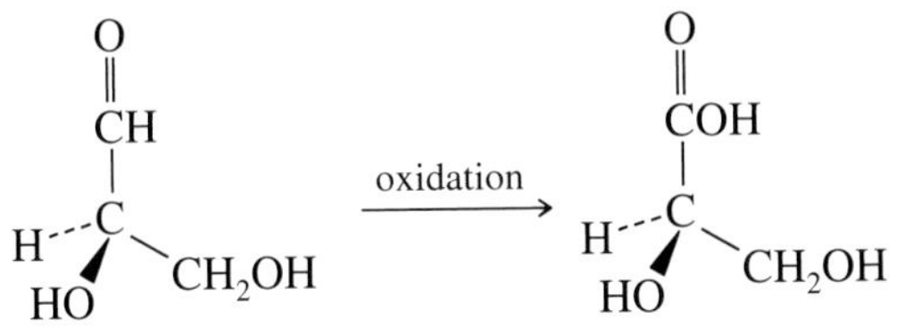

*(R)*-(+)-glyceraldehyde          *(R)*-(−)-glyceric acid

Because this reaction does not break bonds to the stereocenter, the relative configurations of the two compounds must be the same.

O
||
COH
|
H---C
HO   CH₃

(R)-(−)-lactic
acid

**Problem 23.2** (−)-But-1-en-3-ol can be hydrogenated to (−)-2-butanol. Ozonolysis of the butenol, followed by mild oxidation, gives (R)-(−)-lactic acid, shown on the left. Write structures showing the relative configurations of all the compounds mentioned. Give them the proper R or S designations.

## B. Fischer Projections as Two–Dimensional Representations of Chiral Compounds

So far in this book three-dimensional molecules have been represented in two dimensions by perspective formulas (p. 212, for example). Drawing these formulas is not too difficult for compounds containing one or two stereocenters, but it becomes increasingly so for compounds having several stereocenters. Chemists needed a convention for representing three-dimensional molecules in the plane of paper in a consistent and simple way. The great German chemist Emil Fischer introduced such a convention. Fischer called his representations projection formulas because they essentially project a three-dimensional conformation of a molecule onto a two-dimensional surface. These representations are now called **Fischer projection formulas.**

Any two bonds at an $sp^3$-hybridized carbon atom are in a plane at right angles to the plane defined by the other two bonds. The Fischer projections at a given carbon atom represent a rendering in two dimensions of this view of the molecule. Lines intersecting at right angles represent the two planes. The projection formula for (R)-(+)-glyceraldehyde is given in Figure 23.1. In Fischer projection formulas, the longest carbon chain is usually written vertically with the most highly oxidized function at the top.

The conformation that is drawn in a Fischer projection formula of any compound containing more than one stereocenter happens to be the least stable, fully eclipsed conformation of the molecule. Thus Fischer projections do not represent the actual shape of the molecule as it exists in the crystalline state or in solution. They are a convenient way of comparing the configurations of various stereocenters. Fischer's original picture of the structure of glucose is shown on the next page,

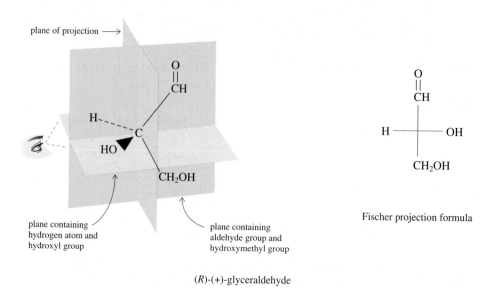

*Figure 23.1*
The planes that intersect at right angles at a tetrahedral carbon atom and the Fischer projection formula that indicates those planes.

plane of projection ⟶

O
||
CH

H---C

HO

CH₂OH

plane containing
hydrogen atom and
hydroxyl group

plane containing
aldehyde group and
hydroxymethyl group

(R)-(+)-glyceraldehyde

O
||
CH

H ——|—— OH

CH₂OH

Fischer projection formula

along with a representation using wedges and dashes as a reminder of what the convention means. The sideways view of the glucose molecule shows that the first and last carbons are in reality quite close to each other. It is instructive to build a molecular model of glucose and explore these relationships. The model also will be useful later in this chapter.

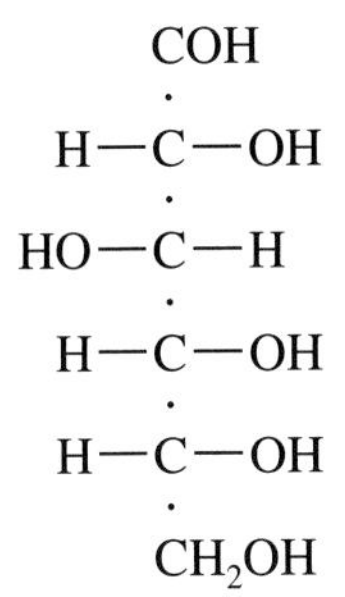

*(+)-glucose as Fischer represented it in his original projection formula*

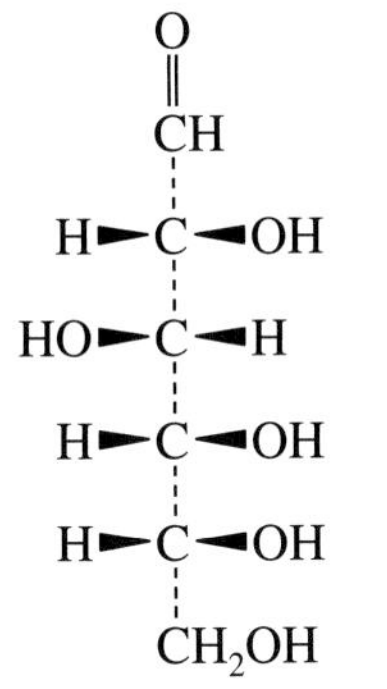

*an indication of which bonds project up and which down at every chiral center viewed individually in (+)-glucose*

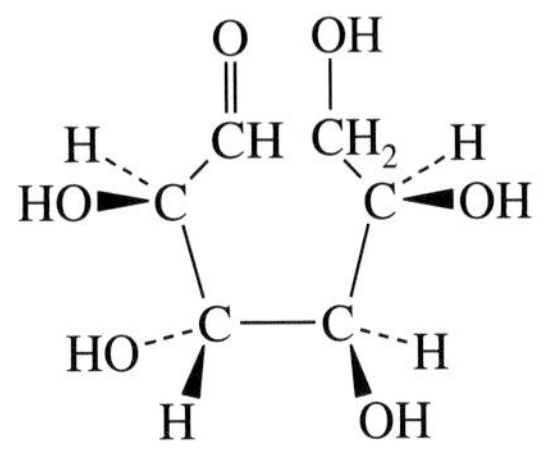

*sideways view*

The original Fischer projections were later changed—the dots were replaced by lines that connected all the groups to the stereocenters. This is the form of Fischer projection currently used in many books. In 1893, Victor Meyer introduced another convention that has been widely used. He suggested that each stereocenter in a Fischer projection be represented as the crossing point of the bonds and that no carbon atom be shown at that point. These conventions are illustrated with various representations of (R)-(+)-glyceraldehyde.

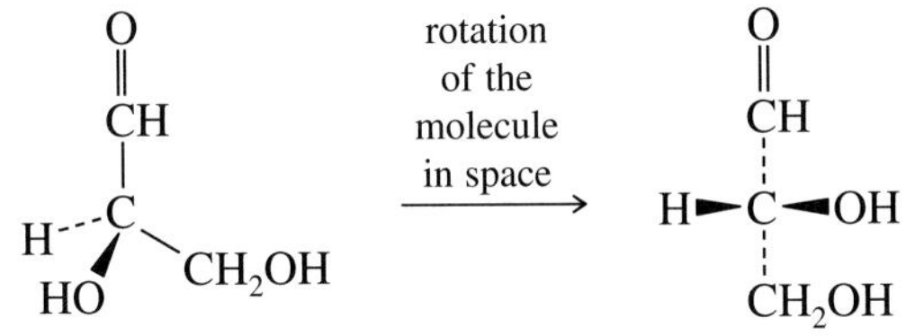

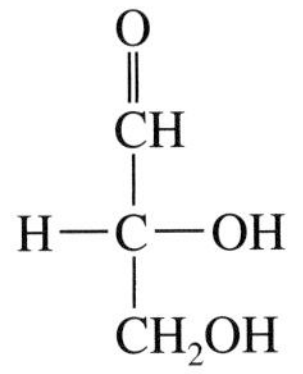

*(R)-(+)-glyceraldehyde Fischer projection formula as often seen*

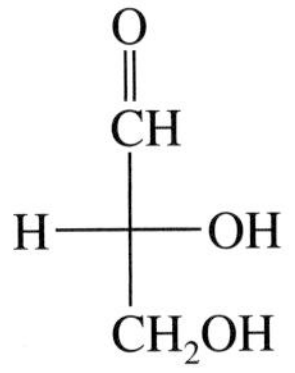

*(R)-(+)-glyceraldehyde Victor Meyer's modification of the Fischer projection formula*

This book will use Meyer's modification exclusively. The other form of the Fischer projection has nothing about it that clearly signals its stereochemical intent. Many times confusion arises because Lewis structures look very similar to this other form of Fischer projections. Lewis structures are not designed to give any stereochemical information. Unless the other form of a Fischer projection is identified as such every time it is used, there is nothing in its appearance to distinguish it from an ordinary representation of a tetrahedral carbon atom with no stereochemical information. The Meyer modification is clearly a different convention, the appearance of which is an immediate reminder that it is showing the specific directions in space for the four groups attached to the crossing point.

Projection formulas impose certain limitations. For example, exchanging the positions of any two substituents at a stereocenter converts the center into its enantiomeric configuration. The most obvious example of this is demonstrated below.

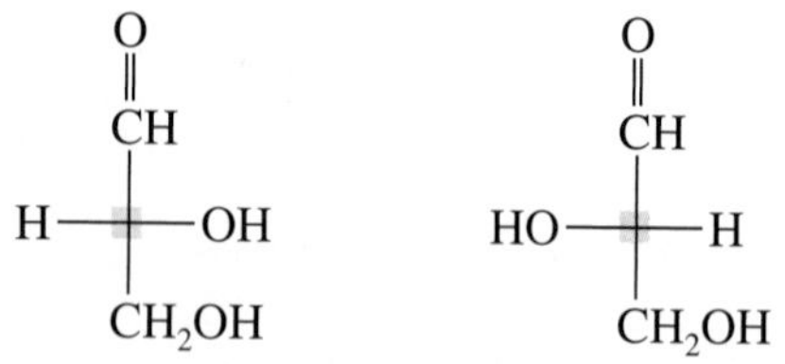

$(R)$-$(+)$-glyceraldehyde     $(S)$-$(-)$-glyceraldehyde

*exchange of two substituents at the stereocenter converts the*
*representation of one enantiomer into that of its mirror-image isomer*

An interchange of the hydrogen atom and the hydroxyl group converts $(R)$-$(+)$-glyceraldehyde to $(S)$-$(-)$-glyceraldehyde. Any interchange of two substituents in a molecule containing a single stereocenter converts the representation to that of the enantiomer. All the structural formulas below are representations of $(S)$-$(-)$-glyceraldehyde arrived at by interchanging two substituents on the projection formula shown above for $(R)$-$(+)$-glyceraldehyde.

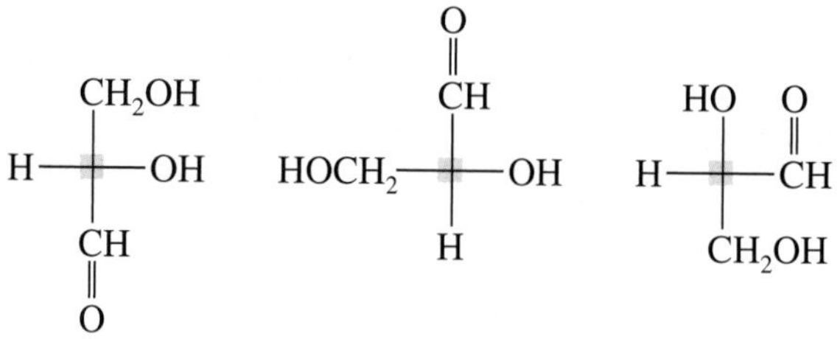

*different Fischer projections of (S)-(−)-glyceraldehyde,*
*all derived from the interchange of two substituents*
*on the Fischer projection of (R)-(+)-glyceraldehyde*

Two of the representations of $(S)$-$(-)$-glyceraldehyde illustrate another point. A Fischer projection formula can be rotated 180° *in* the plane of the paper and retain the same stereochemistry.

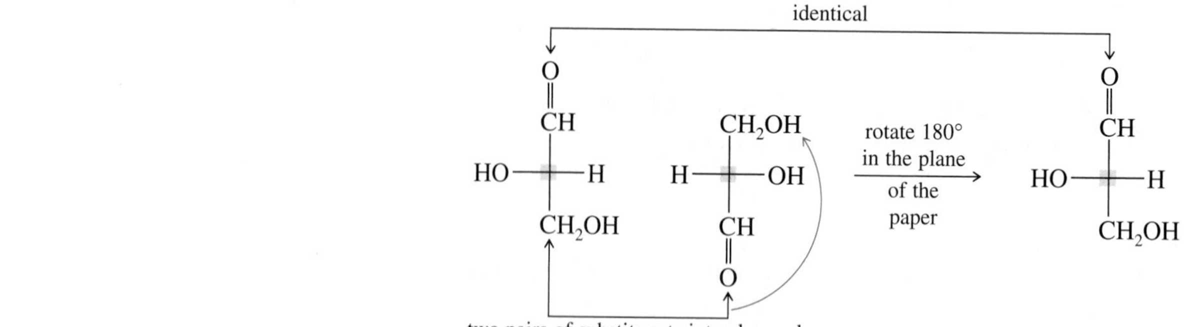

*two pairs of substituents interchanged;*
*molecule retains*
*the same stereochemistry*

The maneuver shown above is the equivalent of interchanging two pairs of substituents at the stereocenter. The aldehyde function was exchanged with the hydroxymethyl group and the hydroxyl group with the hydrogen atom. Such a set of two transformations converts the projection formula into another projection of the *same* compound.

Assignment of configuration to compounds shown as Fischer projections requires care. It is important to remember that the substituents on the vertical lines project back into the page and those on the horizontal lines project out of the page. In assigning configuration, priorities are assigned to each of the substituents at the stereocenter according to the Cahn-Ingold-Prelog Rules (p. 206). If the substituent of lowest priority is on a vertical line, configuration can be assigned by letting the eye travel from the substituent of highest priority to those of second and third priority. A clockwise motion of the eye means that the center has the *R* configuration; a counterclockwise motion of the eye means that the center has the *S* configuration. This is just how configuration is assigned from a perspective formula of a molecule in which the group of lowest priority projects back into the page, as shown at right.

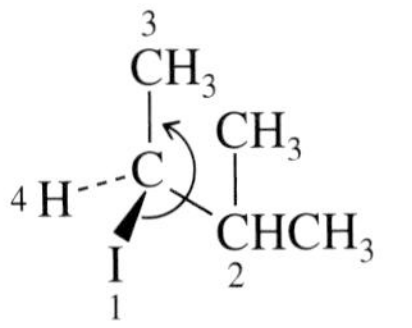

If the Fischer projection is drawn so that the group of lowest priority is on the horizontal line, however, it is important to remember that that substituent is projecting *toward* the viewer, not away from the viewer, as is required for making an assignment of configuration. In such a case, the configuration of the molecule is the opposite of what it appears to be superficially. For example, with the projection on the right, the eye travels counterclockwise from the group of highest priority to that of the second and then the third priority, but because the lowest-priority group is on the horizontal line, though the configuration appears to be *S*, it is in fact *R*. In looking at this Fischer projection to make the assignment, we are looking at the wrong face of the molecule, with the hydrogen atom projecting out of the plane of the page. If we were to view the molecule from behind the page, the eye would have to travel clockwise to go from the hydroxyl group to the aldehyde to the primary alcohol. Thus any time the group of lowest priority is on the horizontal line in a Fischer projection, the correct configuration is the opposite of what it appears to be from looking at the representation as drawn.

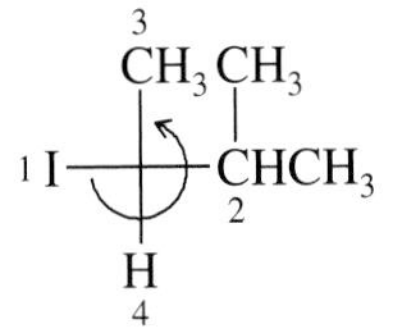

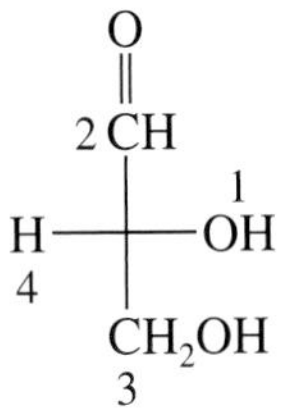

(*S*)-2-iodo-3-methylbutane

(*S*)-2-iodo-3-methylbutane

**Problem 23.3** For each pair of Fischer projection formulas shown below, decide whether they represent the same compound, enantiomers, or diastereomers. Check your conclusions by designating the configuration at each stereocenter as *R* or *S*.

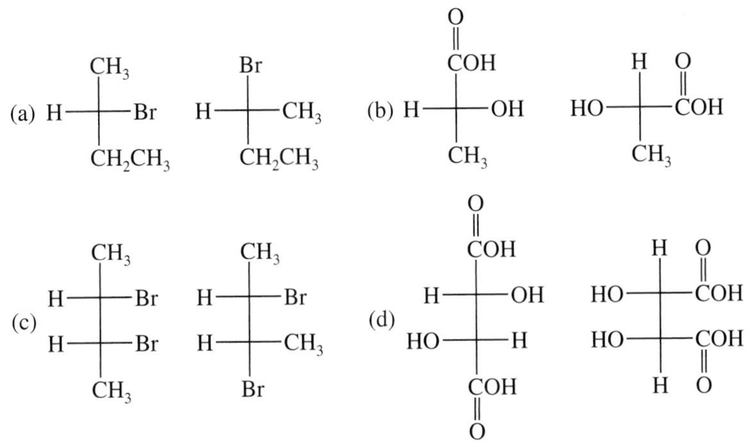

## C. The Designation of Chiral Compounds as D or L

The $R$ and $S$ convention for the designation of configuration was not introduced until the 1950s (p. 206). Before that, chemists used another way of naming relative configurations. When Rosanoff suggested that a particular configuration be assigned to (+)-glyceraldehyde, he also proposed that that arrangement of atoms around the stereocenter be called the D configuration. All compounds having an arrangement of atoms similar to that at the stereocenter of (+)-glyceraldehyde at a comparable carbon atom are members of a D family. Those with the opposite configuration at such a carbon atom belong to an L family. Except for a few cases where relationships are quite easy to see, the system led to many complications and inconsistencies and has been largely abandoned except for carbohydrates and amino acids. This system is still used to some extent, however, especially in the biochemical literature, so you should understand it. Some examples of the D family are given below. The stereocenter that determines the family relationship is shaded in color in each compound.

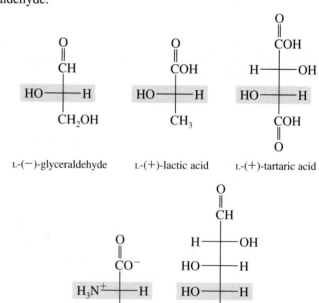

The symbol D has nothing to do with whether the compound is dextrorotatory or levorotatory, just as the designation $R$ or $S$ by itself does not give that information.

Some other compounds are members of the stereochemical family related to L-(−)-glyceraldehyde.

The L designation, again, simply indicates that these compounds all have the same configuration at a certain stereocenter in their molecules.

**Problem 23.4**   Designate as *R* or *S* the configuration at each stereocenter in the compounds of the D and L families given on the preceding page.

## D. Determination of Absolute Configuration

(+)-Tartaric acid is the isomer that occurs most abundantly in nature (p. 204). Bijvoet used it in 1951 for the determination of absolute configuration. This experiment, mentioned on page 210, established the actual orientation in space of the atoms in (+)-tartaric acid.

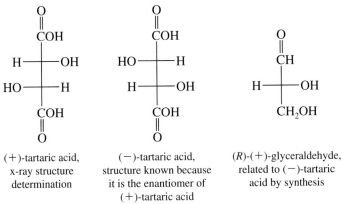

*various representations of the absolute
configuration of (+)-tartaric acid*

The determination of the actual structure of (+)-tartaric acid also established the absolute configuration of its enantiomer, (−)-tartaric acid, which had been related structurally to (+)-glyceraldehyde. This meant that the structure of (+)-glyceraldehyde did, indeed, correspond to the *R* configuration.

|  |  |  |
|---|---|---|
| (+)-tartaric acid, x-ray structure determination | (−)-tartaric acid, structure known because it is the enantiomer of (+)-tartaric acid | (*R*)-(+)-glyceraldehyde, related to (−)-tartaric acid by synthesis |

*absolute configurations for (+)- and (−)-tartaric
acids and for (+)-glyceraldehyde*

With this one experiment, what had been known until that time as the relative configurations of hundreds of compounds suddenly became established as the absolute configurations. Assignments that had been made on the assumption that (+)-glyceraldehyde had the *R* configuration were all correct. The original assignment had been made by Fischer, who had decided arbitrarily that (+)-glucose had a certain configuration at the stereocenter farthest from the aldehyde group. He placed the hydroxyl group at that stereocenter on the right in his projection formula. He knew that his system also would be valid if the structure of glucose was actually the mirror image of the one that he had assigned it, but he had to choose one structure in order to be able to make comparisons with those of other carbohydrates. His selection proved to be correct.

Glucose, in its open-chain form, has four stereocenters and, therefore, $2^4$, or 16, different stereoisomers. The stereoisomers exist as eight pairs of enantiomers. Fischer's assignment of stereochemistry at each of the stereocenters in glucose was the result of painstaking experimental work and logical thought about the implications of the experimental observations. He was given the Nobel Prize for this work in 1902.

**Problem 23.5**    Draw the Fischer projection formula for the enantiomer of D-(+)-glucose (p. 978).

Stereochemistry also has been established for the trioses, tetroses, pentoses, and hexoses related to D- and L-glyceraldehyde. The structural formulas showing the stereochemistry of the members of the D family of aldoses are given in Figure 23.2. The stereochemical relationships shown there form the basis for the system of stereochemical nomenclature recommended for carbohydrates. All the common names shown in Figure 23.2 are accepted names for the particular monosaccharides depicted. Each name also serves as the source of a prefix that can be used to designate

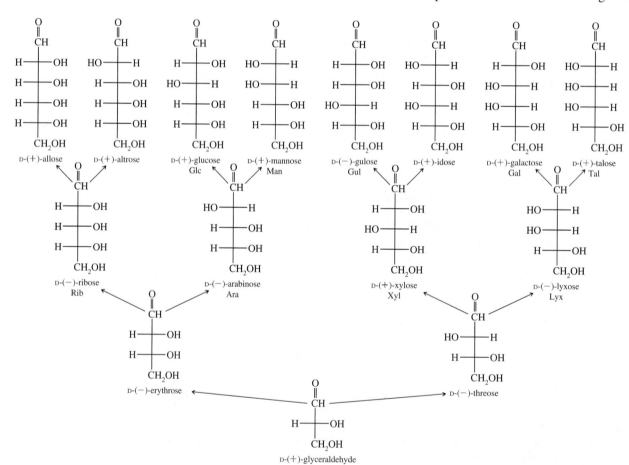

*Figure 23.2*
The family of D-aldoses, with three-letter abbreviations for the most common members.

the relative stereochemistry of stereocenters in other saccharides. One example will show how this is done.

The root of the name of an aldose comes from the number of carbon atoms in the chain. To this root the ending **-ose** is added. The compound is assigned to the D or L family according to the stereochemistry of the stereocenter with the highest number in the chain. The stereochemistry of the whole molecule is then designated by the prefix derived from the names of the monosaccharides given in Figure 23.2 and their enantiomers. According to these rules, the scientific name for glucose is D-*gluco*-hexose. The same *gluco-* prefix is used to designate that arrangement of stereocenters in other molecules no matter whether the sequence is interrupted by atoms that are not stereocenters. An example is the naming of the following eight-carbon saccharide.

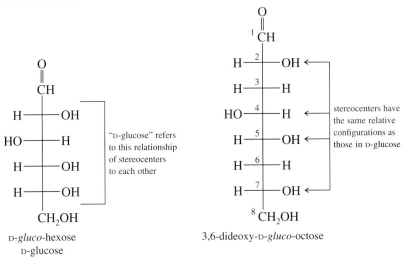

D-*gluco*-hexose
D-glucose

3,6-dideoxy-D-*gluco*-octose

The compound is an aldose with eight carbon atoms, so therefore it is an octose. It has no hydroxyl groups at carbon atoms 3 and 6 of the chain; therefore, it is a deoxysaccharide, in this case a 3,6-dideoxycompound. The stereocenters that do exist in the compound have the same relative configuration as those in D-glucose, so the stereochemistry of the compound is designated by the prefix D-*gluco*-. The full name is 3,6-dideoxy-D-*gluco*-octose.

If a monosaccharide has more than four stereocenters in it, two (or more) prefixes must be used in the name. For example, bacteria and fungi are found to contain a number of heptoses. One is L-*glycero*-D-*manno*-heptose. It has the following structure:

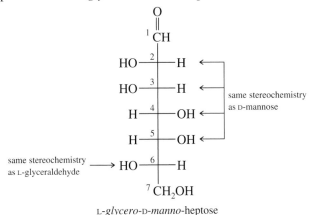

L-*glycero*-D-*manno*-heptose

**ONE SMALL STEP**

A reaction that has been very useful in the determination of the stereochemistry of sugars is oxidation of the sugar with nitric acid. This results in the oxidation of the sugars at both ends of their chains to dicarboxylic acids called aldaric acids. For example, glucose is oxidized by nitric acid to glucaric acid.

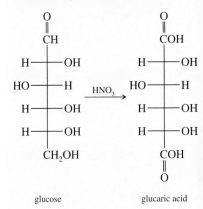

glucose                    glucaric acid

In this reaction, carbon 1 and carbon 6 are converted into the same functional group; thus any symmetry present in the rest of the molecule becomes apparent.

**PROBLEM:** Go to Figure 23.2 and decide which pentoses and which hexoses would give meso, hence optically inactive, compounds on oxidation with nitric acid.

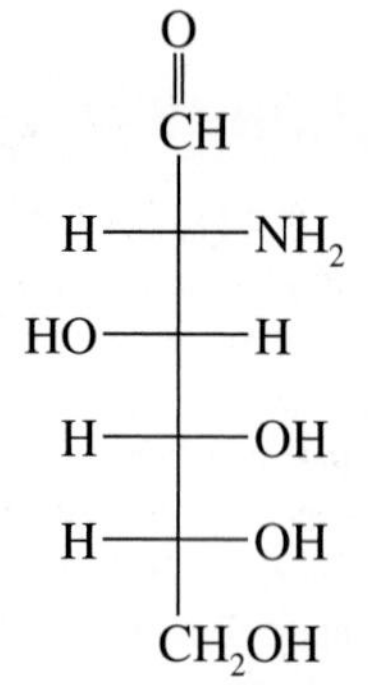

$$
\begin{array}{c}
\text{O} \\
\parallel \\
\text{CH} \\
\text{H}\!-\!\!\!-\!\!\!-\text{NH}_2 \\
\text{HO}\!-\!\!\!-\!\!\!-\text{H} \\
\text{H}\!-\!\!\!-\!\!\!-\text{OH} \\
\text{H}\!-\!\!\!-\!\!\!-\text{OH} \\
\text{CH}_2\text{OH}
\end{array}
$$

2-amino-2-deoxy-D-*gluco*-hexose
2-amino-2-deoxy-D-glucose

The stereocenter with the highest number in the chain has the same configuration as L-glyceraldehyde. The other four stereocenters are the same as those in D-mannose.

In naming saccharides with substituents other than hydroxyl groups at stereocenters, we mentally convert the saccharide into a deoxy compound at that stereocenter, and then the new group is introduced. The name for the amino sugar, 2-amino-2-deoxy-D-*gluco*-hexose is an example.

**Problem 23.6**    The following monosaccharides are important components of bacterial cells. Draw structural formulas for these compounds.

(a) D-*glycero*-D-*galacto*-heptose    (b) 3,6-dideoxy-D-*arabino*-hexose

(c) 4-acetamido-4,6-dideoxy-D-glucose

## E. Haworth Projection Formulas. The Cyclic Hemiacetal Forms of Glucose

In Section 14.8A (p. 563) we explored the interconversion between the open-chain form of glucose, the stereochemistry of which we have been exploring above, and the cyclic hemiacetal forms with which it is in equilibrium. These were chiefly β-D-glucopyranose and α-D-glucopyranose along with small amounts of glucofuranoses (Figure 14.4, p. 567). In Section 14.8 we drew the chair forms of the six-membered pyranose rings and used wedges and dashed lines with the five-membered furanose rings to indicate stereochemistry. Another convention is seen widely, especially in the biochemical literature, in which the planar form of the six-membered ring is used. This way of representing sugars is known as the **Haworth projection formula.** Such formulas are derived easily from the chair form.

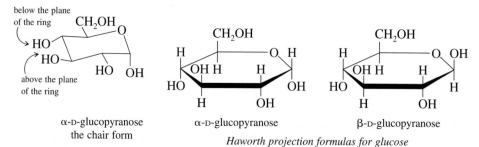

α-D-glucopyranose
the chair form

α-D-glucopyranose

β-D-glucopyranose

*Haworth projection formulas for glucose*

In the Haworth representations, the lower edge of the ring is defined as projecting out of the plane of the paper toward the viewer. Hydroxyl groups, hydrogens, and other substituents are shown as being either above or below the plane of the ring, as shown above.

The same convention applies to five-membered rings. For example, α-D-glucofuranose can be shown as in Figure 14.4 (p. 567) or as its Haworth projection formula.

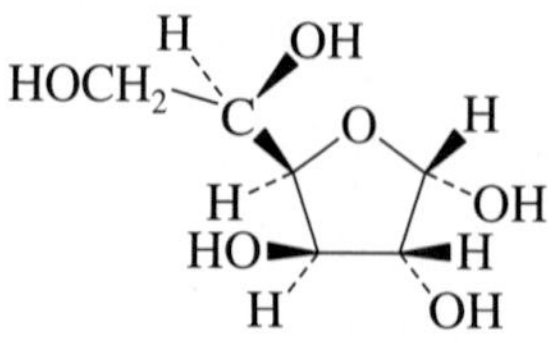

α-D-glucofuranose

*with stereochemistry shown
with wedges and dashed lines*

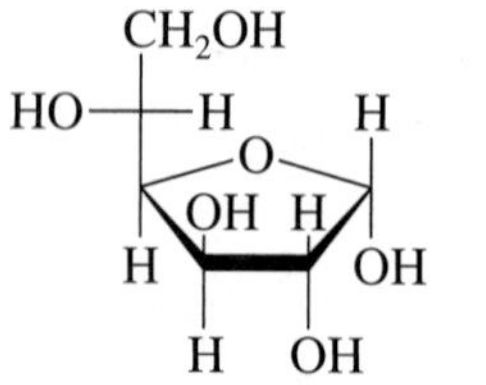

α-D-glucofuranose

*represented by the
Haworth projection formula*

## 23.3 Structure and Properties of Amino Acids. The Structural Units of Proteins

### A. Classification of the Amino Acids Found in Proteins

The amino acids found in proteins derived from animals and higher plants are all $\alpha$-aminocarboxylic acids of a particular stereochemical configuration. The older literature refers to them as L-amino acids, meaning that their configuration is related to that of L-$(-)$-glyceraldehyde (p. 973).

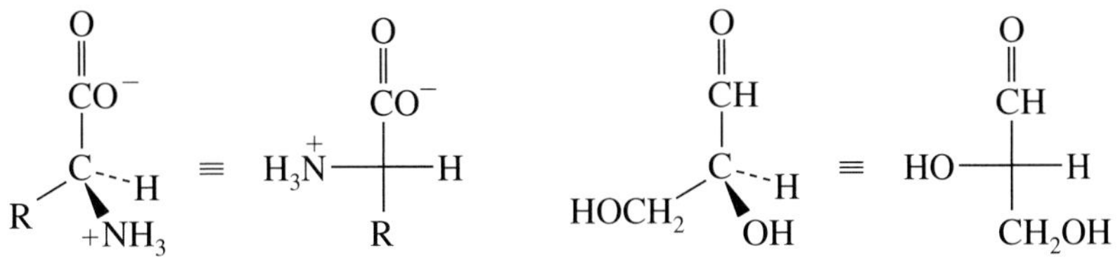

an L-amino acid      L-$(-)$-glyceraldehyde

*configurational relationship between L-amino acids and L-$(-)$-glyceraldehyde*

Their acidity and basicity were discussed in Section 16.2B.

The amino acids that are found in proteins can be classified into four groups on the basis of their properties at physiologic pH. The first group contains the amino acids that have no polar substituents on their side chains. These all have isoelectric points (p. 664) close to 6 and have no net charge at pH 6–7. They are shown in Figure 23.3, along with the abbreviations of their names that will be used in writing formulas for peptides and proteins.

Other amino acids have polar functional groups that can participate in hydrogen bonding or function as nucleophiles in reactions but have no net charge at physiologic

(+)-alanine
Ala

(+)-valine
Val

(−)-leucine
Leu

(+)-isoleucine
Ile

(−)-proline
Pro

(−)-phenylalanine
Phe

(−)-tryptophan
Trp

(−)-methionine
Met

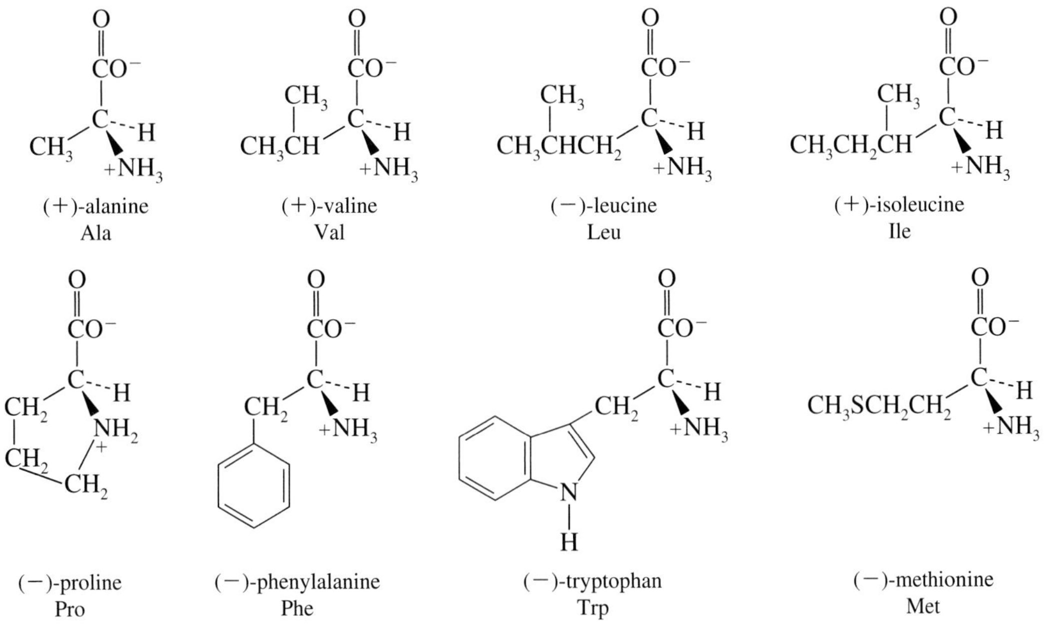

*Figure 23.3*
Amino acids that have nonpolar R groups and no net charge at physiologic pH.

⊘ **Study Guide**
**Concept Maps 23.2, 23.3**

pH (Figure 23.4). Glycine is included among these because it does not have a large, hydrophobic side chain.

Two amino acids have extra carboxylic acid groups on them and therefore have a net negative charge at physiologic pH (Figure 23.5).

Finally, three amino acids have extra basic functional groups and are, therefore, positively charged at the pH of cellular fluids (Figure 23.6).

### Problem 23.7

(a) Would you expect the carboxyl group bonded to carbon 3 of aspartic acid to be more or less acidic than the carboxyl group bonded to carbon 4 of glutamic acid? Explain your answer.

(b) Would the isoelectric point for aspartic acid be higher or lower than that for glutamic acid?

(c) Draw a plot of the relative amounts of the acid–base forms of aspartic acid as a function of pH based on your answers to parts (a) and (b). A review of Section 16.2B will be useful.

### Problem 23.8    The amino acids found in proteins are assigned to the L family according to their stereochemistry at carbon 2. Can they all be assigned the same configuration, $R$ or $S$, at that carbon atom?

### Problem 23.9    Some of the amino acids shown in this section have more than one stereocenter. Pick them out. Assign configuration at the other stereocenters where possible.

### Problem 23.10

(a) Arginine is the most basic of the amino acids. The $pK_a$ of the conjugate acid of the guanidinyl group on its side chain is 12.5, while those for the carboxyl group and the ammonium ion are 2.1 and 8.8, respectively. Why is the guanidinyl group so much more basic than the amino acid group in lysine (p. 985)?

(b) Construct a plot of the relative amounts of the acid–base forms of arginine as a function of pH.

(c) Tryptophan is classified as a nonpolar amino acid even though it has a nitrogen atom in an indole ring (p. 932) as part of its side chain. Why is tryptophan not basic? (*Hint:* A review of Section 10.1B may be helpful.)

## B. Biosynthesis of Amino Acids

Many amino acids are synthesized in the body from ammonia and the degradation products of carbohydrates, such as $\alpha$-ketocarboxylic acids, in the presence of reducing enzymes. The reaction, in fact, closely resembles the reductive amination reactions on page 811. The synthesis of glutamic acid illustrates this reaction.

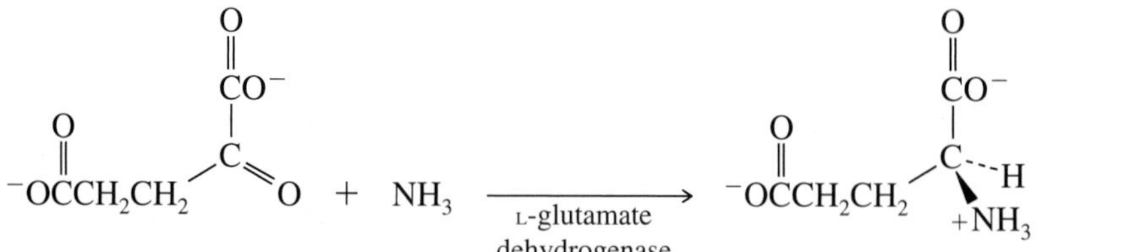

α-ketoglutarate

*an intermediate in the metabolism of glucose to $CO_2$ and $H_2O$*

glutamic acid

The amino group in glutamic acid is transferred to other $\alpha$-ketoacids in the body in another enzymatic reaction. A typical example is shown below.

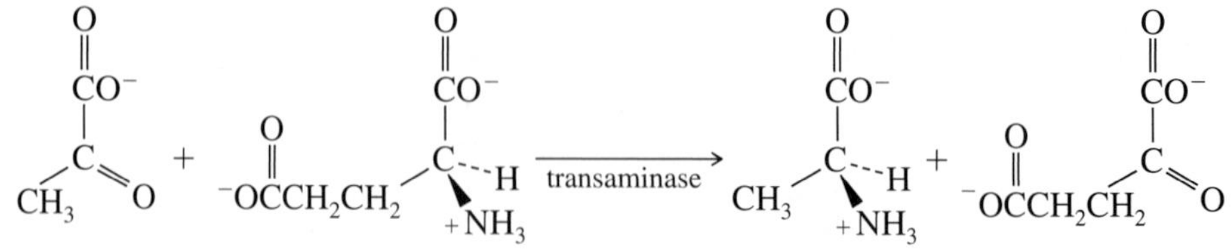

pyruvate anion        glutamate anion        alanine        α-ketoglutarate anion

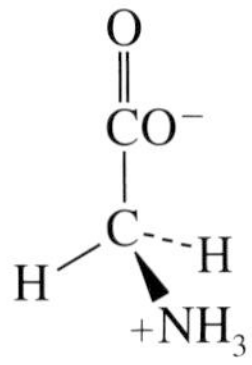

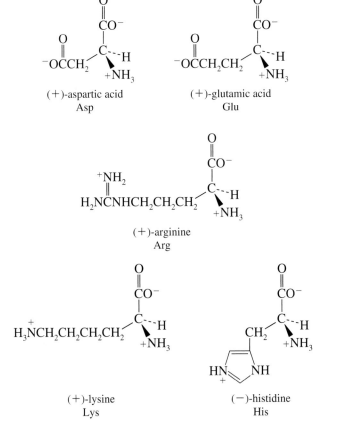

**Figure 23.4**
Amino acids that have polar R groups but no net charge at physiologic pH.

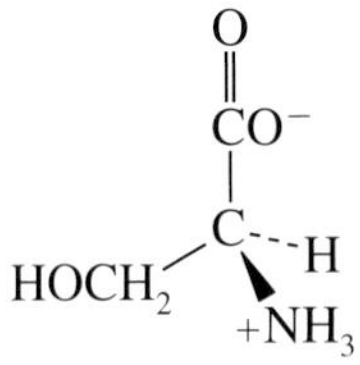

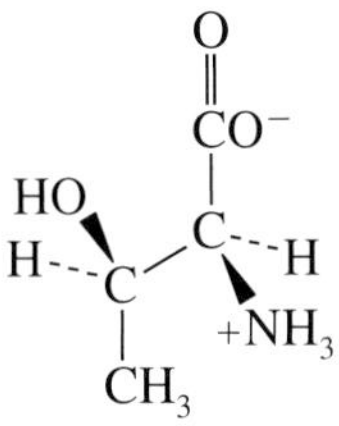

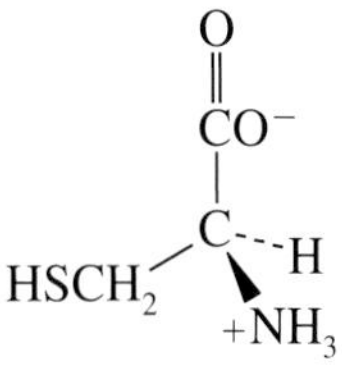

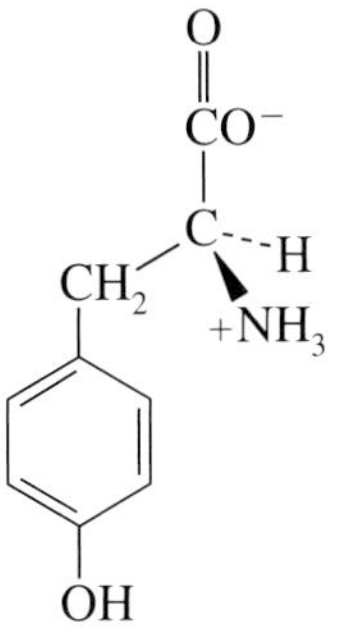

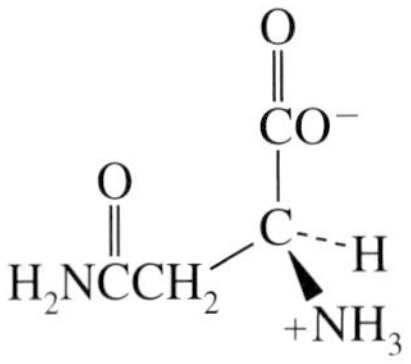

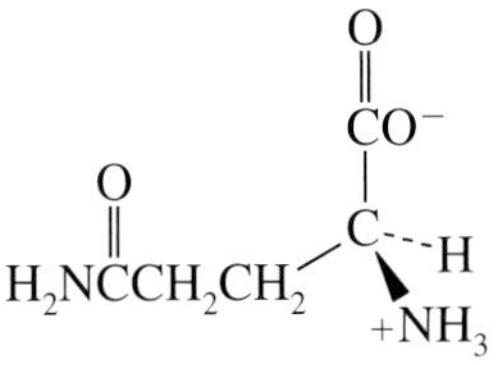

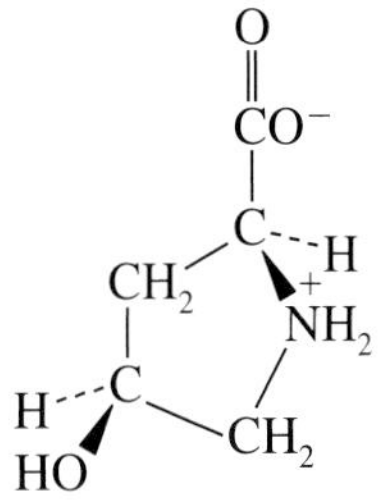

**Figure 23.5**
Amino acids that are negatively charged at physiologic pH.

**Figure 23.6**
Amino acids that are positively charged at physiologic pH.

Such a transformation is called a **transamination** because an amino group is transferred from one species to another. The mechanism of transamination reactions was shown on page 577. A ketoacid is converted into the corresponding amino acid, and glutamic acid becomes $\alpha$-ketoglutaric acid, which then reacts with ammonia to regenerate glutamic acid.

Because the human body cannot synthesize some of the amino acids necessary for protein synthesis, they must be obtained from food. These **essential amino acids** are arginine, histidine, isoleucine, leucine, lysine, methionine, phenylalanine, threonine, tryptophan, and valine.

**Problem 23.11** Amino acids undergo the reactions that are typical of primary alkyl amines and of carboxylic acids. In addition, other functional groups on the side chains can also react in typical ways. For the following equations, show the products that are expected. (*Hint:* For some of the reactions, it is necessary to remember that a small amount of the uncharged form of the amino acid is always in equilibrium with the zwitterionic form.)

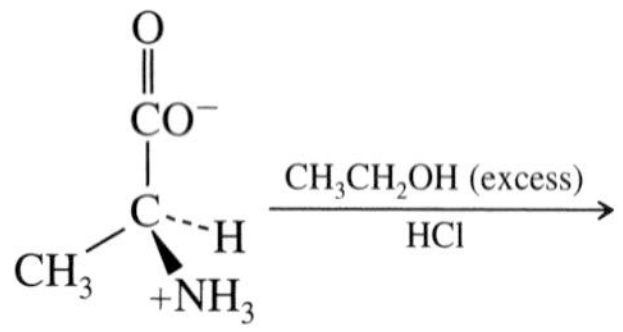

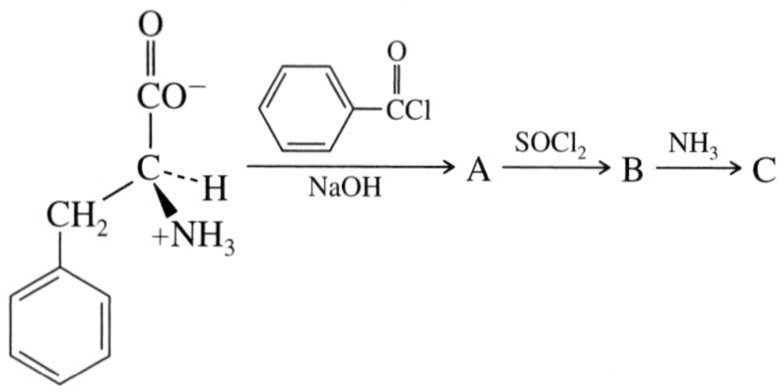

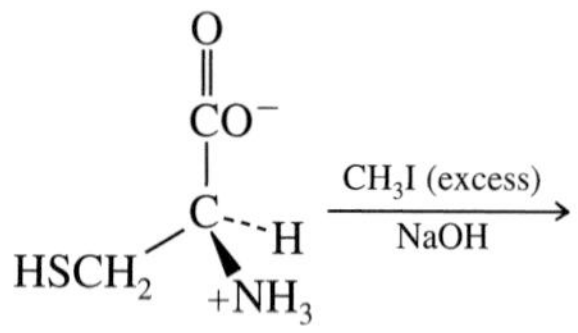

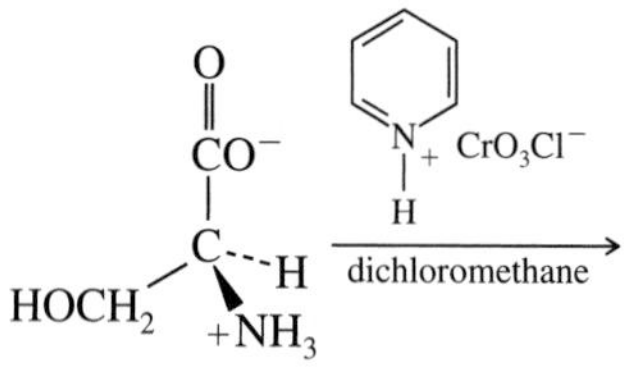

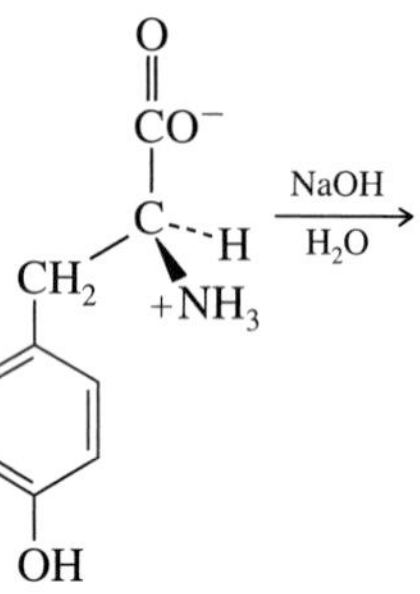

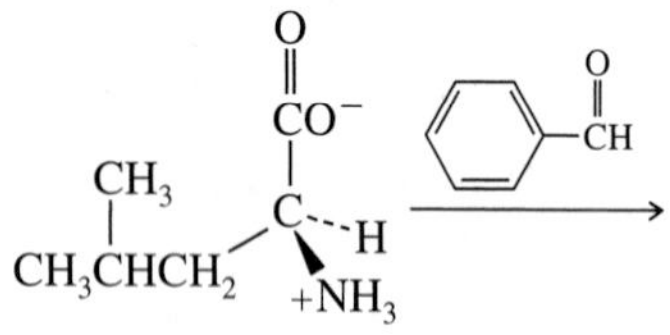

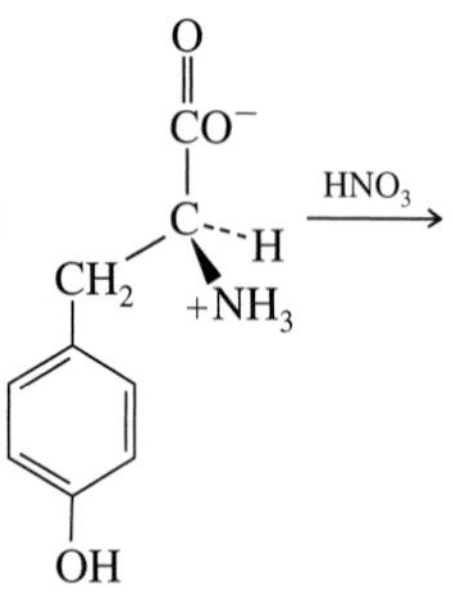

## 23.4 Nucleotides. The Structural Units of DNA and RNA

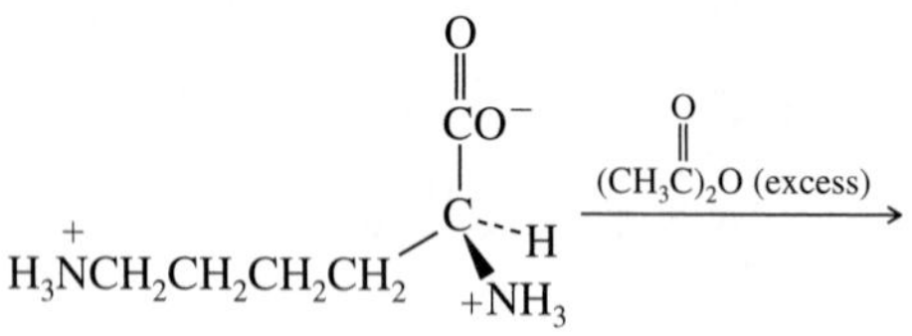

Some of the largest molecules known, often having molecular weights higher than a million, are the nucleic acids found in the nuclei of cells. Nucleic acids have backbones made up of sugar units that are five-membered rings. These rings are held together as the phosphate esters of the hydroxyl groups at carbons 3 and 5. In **ribonucleic acids** (abbreviated RNA), the sugar is D-ribose; in **deoxyribonucleic acids** (abbreviated DNA), it is D-2-deoxyribose.

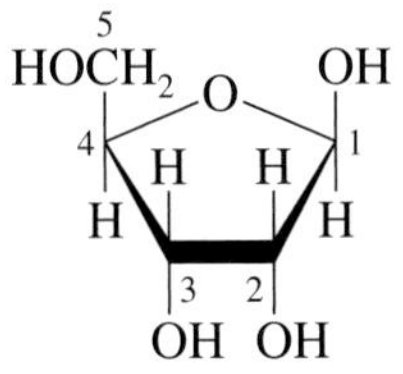

β-D-ribose 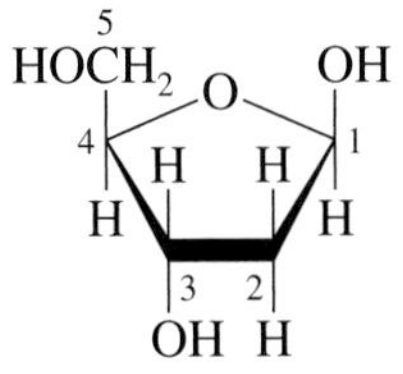          β-D-2-deoxyribose

Each sugar unit is also bonded at carbon 1 to a nitrogen heterocycle, either a purine base (adenine or guanine, p. 946) or a pyrimidine base (thymine, uracil, or cytosine, p. 946), resulting in compounds called **nucleosides.** The phosphate esters of nucleosides are known as **nucleotides.** RNA and DNA are polymers made up of nucleotide units and are therefore also called **polynucleotides.** The structural formulas and names of the nucleotides commonly found in RNA and DNA are shown below.

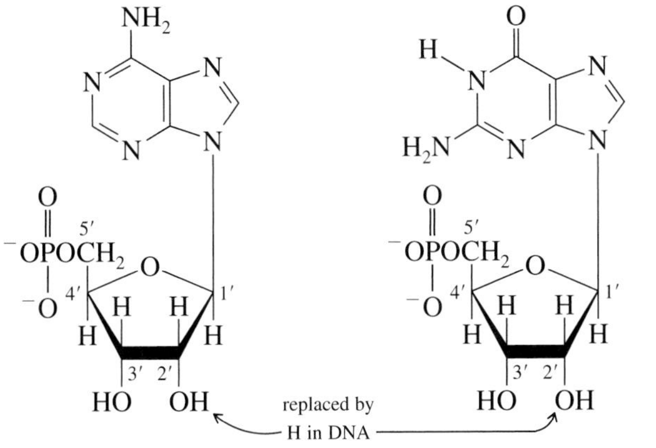

5′-adenylic acid
adenosine 5′-phosphate
5′-AMP

5′-guanylic acid
guanosine 5′-phosphate
5′-GMP

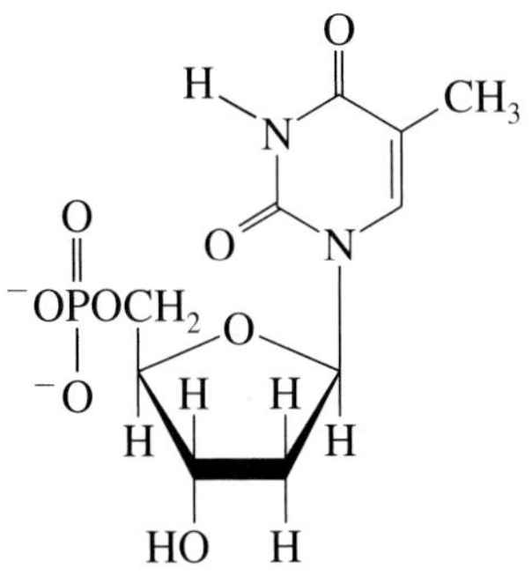

5′-thymidylic acid (DNA)
thymidine 5′-phosphate
5′-TMP

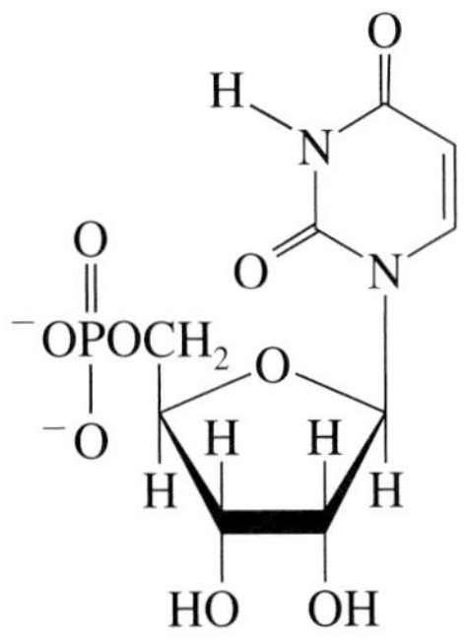

5′-uridylic acid (RNA)
uridine 5′-phosphate
5′-UMP

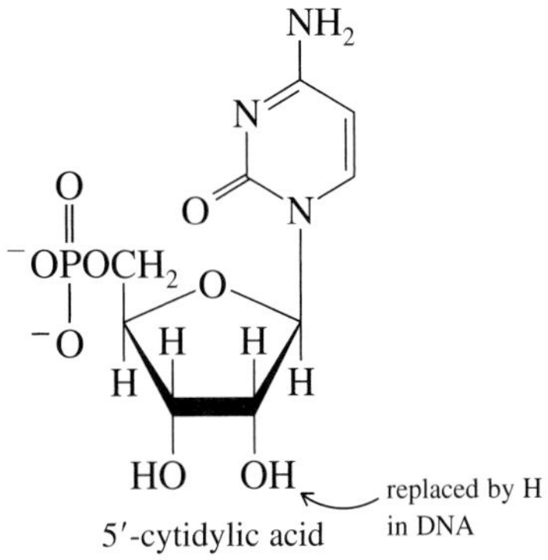

5′-cytidylic acid
cytidine 5′-phosphate
5′-CMP

Except for thymidine 5′-phosphate, which is found only in DNA, and uridine 5′-phosphate, which is found only in RNA, the other three nucleotides come in two forms. The forms found in RNA are shown on page 987. The forms found in DNA, with deoxyribose instead of ribose as the sugar unit, have different names: 5′-deoxyadenylic acid or deoxyadenosine 5′-phosphate, 5′-deoxyguanylic acid or deoxyguanosine 5′-phosphate, and 5′-deoxycytidylic acid or deoxycytidine 5′-phosphate.

We have already encountered the nucleotides adenosine triphosphate (ATP) and adenosine diphosphate (ADP) as important players in the processes of glycolysis, the metabolism of glucose (Section 16.6B, p. 679). Nucleotides are also important components of coenzymes, small molecules usually incorporating vitamins, that are essential to the function of the large proteins that are enzymes. For example, adenosine 3′,5′-diphosphate is a component of coenzyme A, which contains the B vitamin pantothenic acid (p. 637). Adenosine 5′-phosphate is part of the coenzymes NAD⁺ and NADP⁺, which are important in oxidation–reduction reactions in the body (p. 948).

> **Problem 23.12**    Write structural formulas for the following compounds. A review of the structures on page 987 may be helpful.
>
> (a) adenosine 3′,5′-diphosphate    (b) deoxyguanosine 5′-phosphate    (c) cytidine
> (d) thymidine 3′-phosphate    (e) guanosine    (f) uridine 2′-phosphate

## 23.5    Oligosaccharides

### A. Lactose

In Section 14.8B we saw how the cyclic hemiacetal forms of sugars could react with alcohols or phenols to give full acetals known as glycosides (p. 569). The following problem is a reminder and a review of this chemistry.

> **Problem 23.13**    The compound arbutin is a glycoside isolated from the bearberry. It has found use as a diuretic (a medication that increases the production of urine) and an antiseptic for the urinary tract. It can be hydrolyzed by emulsin (p. 989) to glucose and hydroquinone (p. 791). What is the structure of arbutin?

The anomeric carbon atom of one sugar molecule also may react with a hydroxyl group from another sugar molecule to give a disaccharide.

Two disaccharides that occur abundantly in nature are sucrose and lactose. Sucrose is derived from plants and is prepared commercially from sugar beets and sugar cane. Lactose is found in the milk of animals and was known as milk sugar when it was first isolated. Other common disaccharides are prepared by breaking down polysaccharides. For example, maltose is derived from the enzymatic hydrolysis of starch, while the partial hydrolysis of cellulose gives cellobiose.

In disaccharides, two monosaccharide units are held together by a glycosidic linkage. The stereochemistry and the position of the linkage and whether the monosaccharides are in the pyranose or furanose form must be determined in order to establish the structure of a disaccharide. On hydrolysis with dilute acid, lactose gives galactose and glucose and is a substituted glucopyranose.

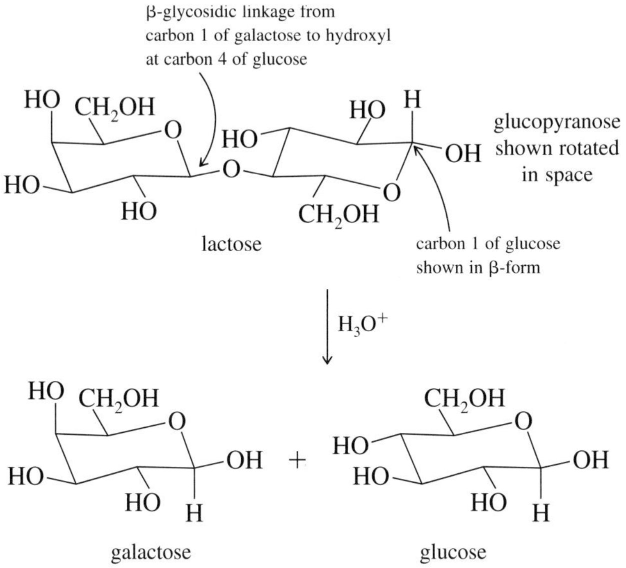

β-glycosidic linkage from carbon 1 of galactose to hydroxyl at carbon 4 of glucose

glucopyranose shown rotated in space

lactose

carbon 1 of glucose shown in β-form

$H_3O^+$

galactose + glucose

The glucose unit is substituted on the oxygen bonded to carbon 4 by a galactopyranose substituent. This substituent is attached at carbon 1 with a β configuration, so it is a β-D-galactopyranosyl group. One systematic name for lactose is therefore 4-O-β-D-galactopyranosyl-β-D-glucopyranose. Another way of indicating the linkage between the galactose and glucose units is to name the compound O-β-D-galactopyranosyl-(1 → 4)-β-D-glucopyranose. This type of nomenclature is useful for more complex saccharides and is commonly abbreviated to β-D-Galp-(1 → 4)-β-D-Glcp, using the three-letter abbreviations for the monosaccharides given in Figure 23.2. The italicized p following the symbols for galactose and glucose means that they are in their pyranose forms.

**Problem 23.14** Is lactose a reducing sugar? See page 608 for a reminder of what a reducing sugar is.

## B. Maltose and Cellobiose

Maltose is 4-O-α-D-glucopyranosyl-D-glucopyranose, and cellobiose is 4-O-β-D-glucopyranosyl-D-glucopyranose. Both are reducing sugars that mutarotate (p. 566). Both are hydrolyzed to two units of glucose. The difference between the two is in the stereochemistry of the glycosidic linkage. Maltose is hydrolyzed by the enzyme maltase, which is specific for α-glycosidic linkages, but emulsin, an enzyme specific for the cleavage of β-glycosidic linkages, is necessary for the hydrolysis of cellobiose.

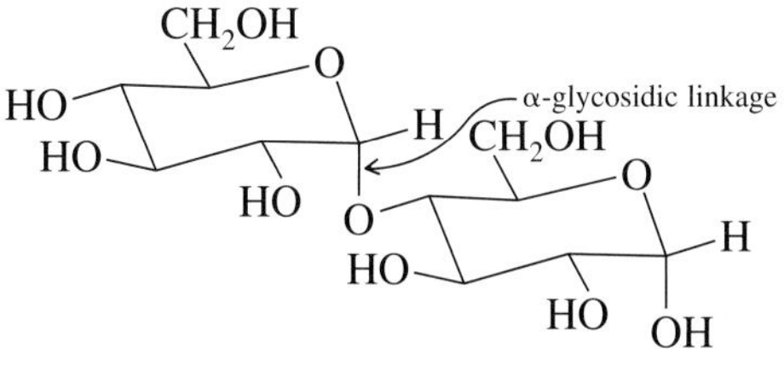

α-glycosidic linkage

β-glycosidic linkage

maltose

*product of the hydrolysis of starch*

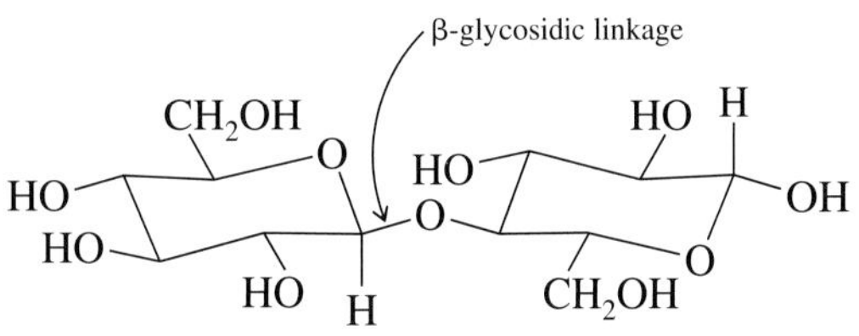

cellobiose

*product of the hydrolysis of cellulose*

These structural differences are important because they suggest that similar differences exist in the polysaccharides starch and cellulose, from which these two disaccharides are derived by hydrolysis reactions.

## C. Sucrose

Sucrose is not a reducing sugar, and it does not exhibit mutarotation. Sucrose itself is dextrorotatory, $[\alpha]_D + 66°$. Hydrolysis of sucrose converts it into glucose, $[\alpha]_D + 52.5°$, and fructose, $[\alpha]_D - 92°$. The resulting mixture is levorotatory. This phenomenon is known as the **inversion of sucrose.** The mixture of sugars that is formed is called **invert sugar.** Invert sugar is sweeter than sucrose chiefly because fructose is about 1.8–2.0 times sweeter than sucrose. Invert sugar is the chief component of honey. The preparation of all sugar syrups and candies involves boiling table sugar with water and a bit of acid such as vinegar, lemon juice, or cream of tartar, which is the monopotassium salt of (+)-tartaric acid. The process described is nothing more than the preparation of invert sugar.

The fact that sucrose is not a reducing sugar means that there is no potential aldehyde or ketone function in the molecule in the hemiacetal form. Carbon 1 of glucose must be bonded to carbon 2 of fructose. Sucrose is hydrolyzed by maltase, so the glucosidic linkage must be $\alpha$ in configuration. The structure and stereochemistry of the fructose portion of the molecule was difficult to establish. An x-ray structure determination in 1947 showed that fructose is present in sucrose as a $\beta$-fructofuranoside.

sucrose
or
β-D-fructofuranosyl-α-D-glucopyranoside
or
*O*-β-D-fructofuranosyl-(2↔1)-α-D-glucopyranoside
or
β-D-Fru*f*-(2↔1)-α-D-Glc*p*

α-glycosidic linkage at carbon 1 of glucose

β-glycosidic linkage at carbon 2 of fructose

---

**Problem 23.15** Gentiobiose is a disaccharide found in a number of natural products. It has the formula $C_{12}H_{22}O_{11}$ and is a reducing sugar. It is hydrolyzed by emulsin to glucose. 2,3,4,6-Tetra-*O*-methyl-D-glucopyranose and 2,3,4-tri-*O*-methyl-D-glucopyranose are obtained when gentiobiose is converted to its fully methylated form (see the One Small Step on page 570 for a reminder of the conversion of sugars to their methyl ethers) and then hydrolyzed. What is the structure of gentiobiose? Give it a systematic name.

---

## D. Cardiac Glycosides

The cardiac glycosides have an effect on the actions of the heart, and many of them are highly toxic. A well-known example of a cardiac glycoside is digitoxin, isolated from the foxglove, *Digitalis purpurea*. This compound reduces the pulse rate, regularizes the rhythm of the heart, and strengthens the heart beat. It consists of a steroid aglycon attached to a trisaccharide that consists of three units of the sugar D-digitoxose.

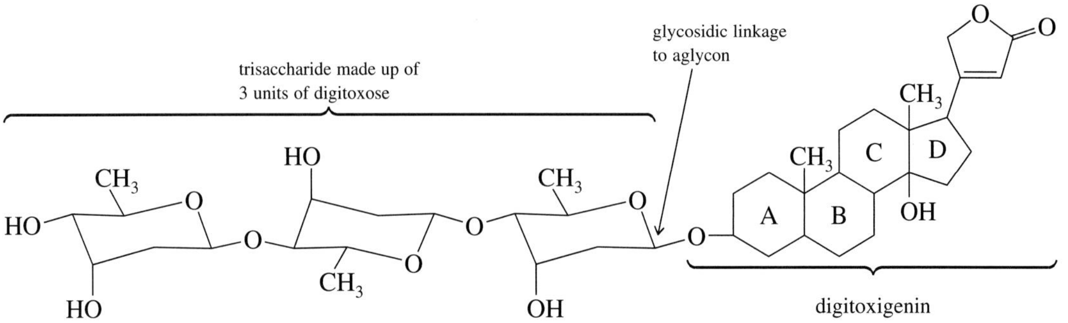

trisaccharide made up of
3 units of digitoxose

glycosidic linkage
to aglycon

digitoxin

digitoxigenin
*aglycon of digitoxin*

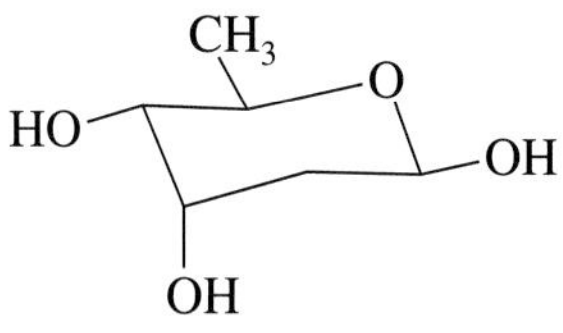

D-digitoxose

Digitoxose is a 2,6-deoxyaldohexose with a configuration resembling that of ribose. The trisaccharide unit has $\beta$-linkages from carbon 1 of one digitoxose unit to the hydroxyl group of carbon 4 of another one. The aglycon in digitoxin is a steroid called digitoxigenin. The five-membered cyclic $\alpha,\beta$-unsaturated lactone unit and the hydroxyl group at the junction of the C and D rings of the steroid are found in other compounds having similar effects on the heart. The sugars of other cardiac glycosides vary, however.

**Problem 23.16**    Uzarin is a cardiac glycoside having the same aglycon as digitoxin does. The aglycon is attached by a $\beta$-glycosidic linkage to cellobiose. What is the structure of uzarin?

## E. Biologically Important Carbohydrate Acids

A biological membrane contains lipids (p. 638), proteins (p. 1007), and carbohydrates in ratios that vary with the source of the membrane. The carbohydrate in the membrane is nearly always covalently bonded to a protein in glycoproteins or with lipids as glycolipids. In glycolipids, carbohydrates are present as esters of fatty acids, or they may form glycosides with hydroxy fatty acids. Lipopolysaccharides are more complex and generally contain polysaccharide and lipid components.

Lipids provide the structural integrity and the barrier to permeability to a cell membrane (p. 820), while the proteins carry out specific functions. The carbohydrates play an important role in cell recognition and, therefore, in the immune system. Bacterial cells are recognized as foreign by the immune system of the host to a large extent because of the presence of carbohydrate chains containing residues that extend through the cell membrane to the surface. An important building block in bacterial cell walls is muramic acid, usually found as its acetyl derivative.

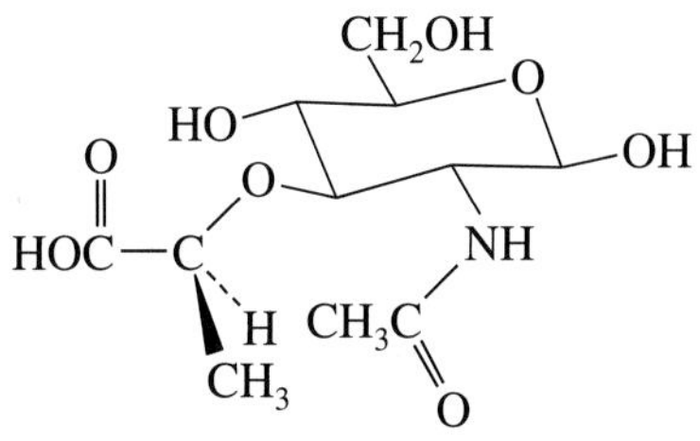

*N*-acetylmuramic acid,
a component of the structural
polymer of bacterial cell walls

*N*-Acetylmuramic acid and 2-acetamido-2-deoxy-D-glucose form repeating units in polysaccharide chains that are linked to each other by short peptide chains (p. 993) containing four amino acids. Such a polymer containing a polysaccharide chain bonded to peptide chains is called a peptidoglycan.

Another important carbohydrate acid is *N*-acetylneuraminic acid, which is one of a group of compounds known as sialic acids. The sialic acids differ in the acyl group present in the amide function.

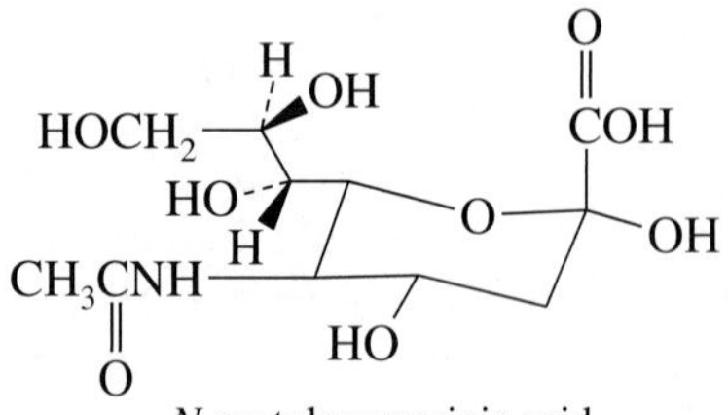

*N*-acetylneuraminic acid
a sialic acid

*N*-Acetylneuraminic acid is found in mucins, secretions from mucous membranes such as those found in the mouth, lungs, stomach, and intestines of higher animals. Mucins are water-soluble glycoproteins of high molecular weight. Their water solutions have high viscosities and serve to lubricate the mouth and esophagus and to form a protective layer over the teeth.

Sialic acids are also found in colostrum, the first milk to come from the breast after the birth of a baby, in immunoglobulins, and in fibrinogen. *N*-Acetylneuraminic acid is also an important component of glycolipids found in the brain known as gangliosides. A typical ganglioside is shown below. An oligosaccharide that includes *N*-acetylneuraminic acid units is bonded by a glycosidic linkage to a unit known as a ceramide.

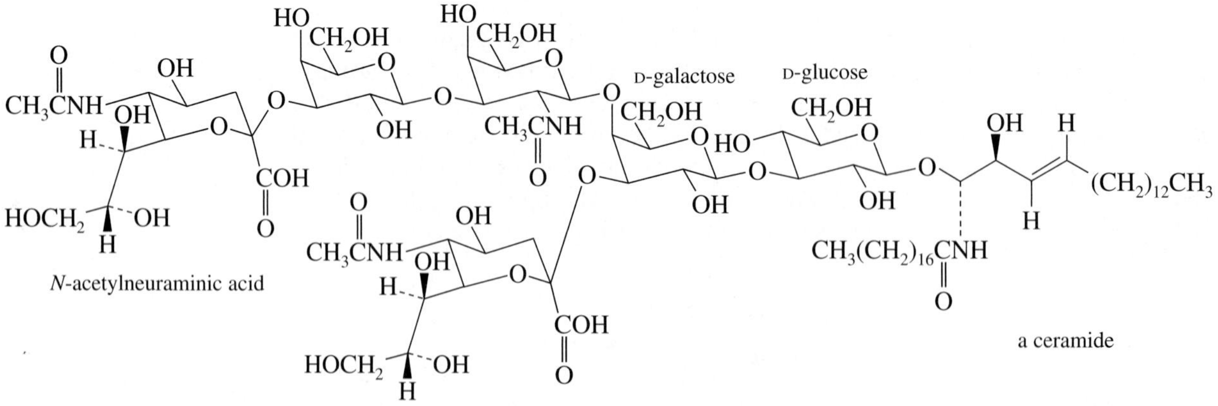

a ganglioside

## 23.6 Peptides

### A. Structures of Peptides. Hydrolysis

In nature, amino acids are linked together by amide bonds, not only in the large molecules known as proteins but also in smaller molecules, many of which have hormonal activity. Such small molecules, containing only a few amino acid residues, are known as peptides. Peptides are familiar to us. We hydrolyzed them to their amino acid constituents (p. 615) and pondered the problems involved in their synthesis from their constituent amino acids (pp. 629–636 and pp. 901–906). In this section we are going to take a close look at peptide structures and some ways in which they are determined.

Peptides are named for the amino acids that constitute them, starting with the amino acid that has the free amino group. The amino acid residue at that end of the peptide chain is known as the ***N*-terminal amino acid,** and the amino acid that has the free carboxylic acid group is called the ***C*-terminal amino acid.** In naming peptides, starting with the *N*-terminal amino acid, each amino acid except the *C*-terminal one is treated as though it were an alkyl substituent on the next unit in the chain. This nomenclature is illustrated for a tetrapeptide.

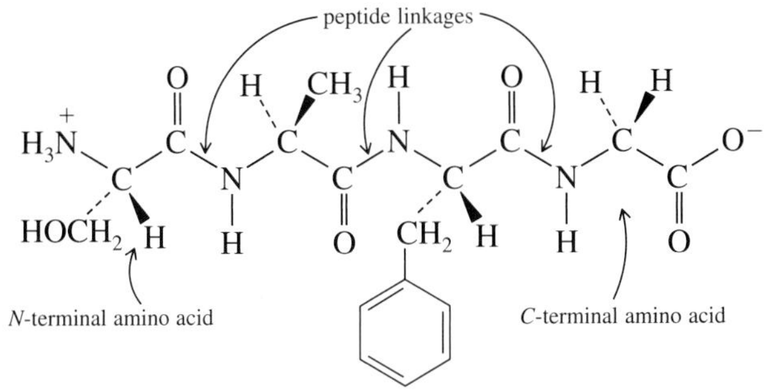


serine alanine phenylalanine glycine

serylalanylphenylalanylglycine

Ser-Ala-Phe-Gly

*N*-terminal amino acid — *C*-terminal amino acid

When written out in full, the name of a peptide containing more than three amino acids is cumbersome. Thus three-letter abbreviations for the names of the amino acids are used to make the names of peptides more manageable. Separating the abbreviations with hyphens indicates that the amino acids are bonded to each other by peptide linkages in the order shown. The first amino acid listed is the *N*-terminal amino acid, and the last one is the *C*-terminal amino acid. This type of notation serves as a substitute for the full structural formula of the peptide. Separating the three-letter abbreviations for the amino acids with commas indicates that the amino acids are present in the peptide but the order of their bonding is not known. For example, complete hydrolysis of the tetrapeptide shown above would yield the information that it contains the amino acids alanine, glycine, phenylalanine, and serine (listed in alphabetical order) but no information on how they are bonded to each other.

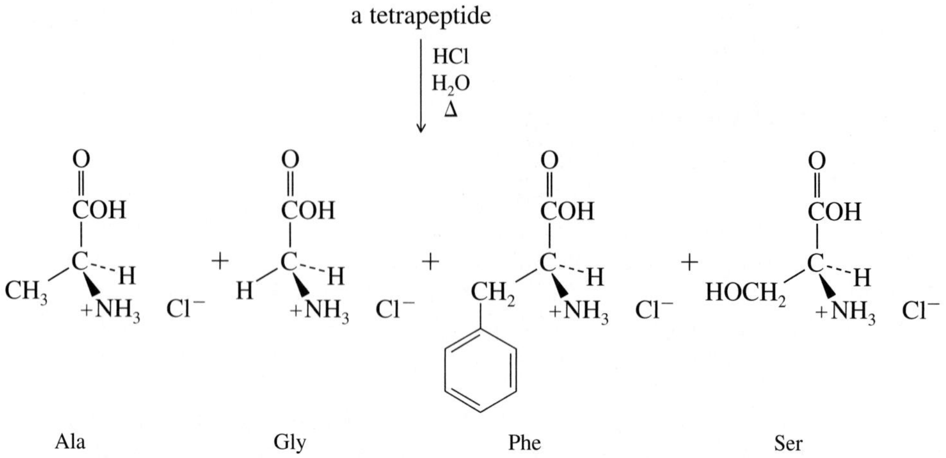

a tetrapeptide

$$\downarrow \begin{array}{c} \text{HCl} \\ \text{H}_2\text{O} \\ \Delta \end{array}$$

Ala        Gly        Phe        Ser

$$\text{a tetrapeptide} \xrightarrow[\substack{\text{H}_2\text{O} \\ \Delta}]{\text{HCl}} \text{Ala, Gly, Phe, Ser as their hydrochloride salts}$$

*no information about the order in which
the amino acids are bonded to each other*

**Problem 23.17** Four amino acids can be combined in various ways to give twenty-four different peptides. Write complete structures for four peptides containing alanine, glycine, phenylalanine, and serine that are different from the one shown above. Show the stereochemistry, and label the *N*-terminal and *C*-terminal amino acids in each one.

**Problem 23.18** Draw the full structure of the peptide Leu-Asn-Lys-Tyr. Show the fragments you would expect to get from its complete hydrolysis with hydrochloric acid.

Peptides have typical isoelectric points that depend on the amino acids present in them. The carboxylic acid groups and amino groups that are part of the peptide linkages no longer have acidic or basic properties, so only the *N*-terminal amino group, the *C*-terminal carboxylic acid group, and any acidic or basic groups that are present in the side chains of the peptide will be ionized. The tetrapeptide Ser-Ala-Phe-Gly (p. 993) has an isoelectric point around 6.0.

**Problem 23.19** For each of the following peptides, predict whether the isoelectric point will be below, above, or approximately at 6. Predict whether each peptide will move to the negative or positive pole of an electric field, or neither, if the pH of the solution is kept at 6.0.

(a) Gly-Phe-Thr-Lys     (b) Tyr-Ala-Val-Asn     (c) Trp-Glu-Leu     (d) Pro-Hypro-Gly

Proteins are cleaved into peptides in the process of digestion by specialized enzymes called **proteases.** It is also possible to degrade proteins into peptides or amino acids by acidic hydrolysis of the amide bonds (p. 615). If a protein is subjected to concentrated hydrochloric acid at 37 °C for several hours, it is degraded into smaller peptide residues. If, on the other hand, it is heated with 20% hydrochloric acid for a period of days, all the amide bonds are cleaved, and the protein is degraded completely into the individual amino acids; in fact, some amino acid molecules are destroyed by this treatment. Tryptophan does not survive vigorous acid hydrolysis, and serine and threonine, which contain hydroxyl groups, are partially destroyed. Asparagine and glutamine, which have amide groups on their side chains, are converted to aspartic acid and glutamic acid by this treatment, and ammonium chloride is formed as another product.

Rapid progress in the determination of the structure of proteins followed the development of chromatographic methods for the analysis of mixtures of peptides or amino acids. Many different techniques are used, and automated equipment can deliver a complete analysis of the amino acid content of a sample in a few hours. Some methods depend on the use of an electric field to separate the amino acids according to the charge they bear at a given pH. In paper chromatography, thin-layer chromatography, or column chromatography, amino acids of differing polarities are distributed differently between two phases, of which one is stationary and the other, mobile. The two phases may be two liquid phases of different polarities or a solid phase and a liquid phase. The more polar amino acids are held back by the more polar phase so that separations take place. At the end of the process, the amino acids are usually treated with ninhydrin, which reacts with amino acids other than proline and 4-hydroxyproline to give a purple dye.

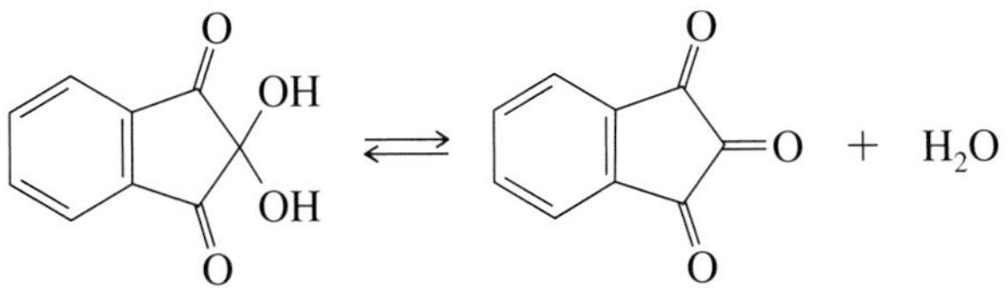

ninhydrin

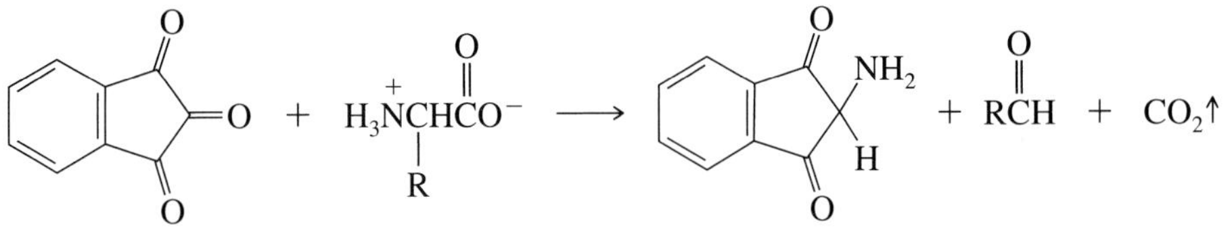

same product from all amino acids except proline and 4-hydroxyproline

different aldehyde from each amino acid

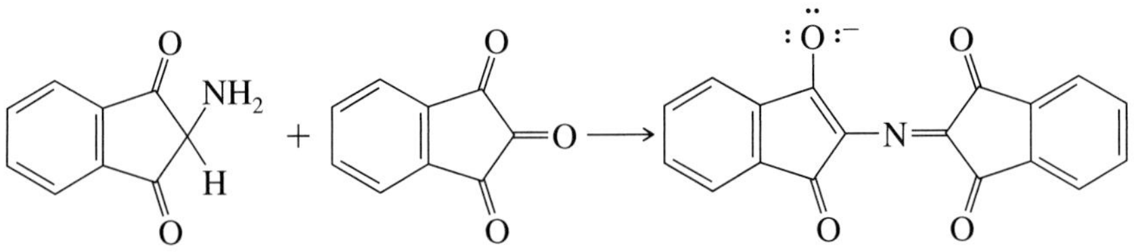

same product from all amino acids except proline and 4-hydroxyproline

from ninhydrin

purple color

If the experimental conditions are carefully standardized, the exact positions at which the dye spots appear on the paper sheet or thin-layer plate can be used to identify the amino acids present. The patterns of spots are sometimes called maps of amino acids. The same kinds of techniques can also be used to separate and identify peptides. Similarly, the absorption of the dye from ninhydrin can be detected spectroscopically and used to plot the appearance of the separated amino acids as they leave the column in column chromatography. Once again, careful standardization of conditions allows chemists to compare the experimental pattern of absorption peaks with known patterns to identify the amino acids present.

Separations may be carried out on the derivatives of amino acids rather than on the amino acids themselves. Automated systems that depend on chromatographic separation of the phenylthiohydantoins (p. 997) derived from the amino acids in a protein are the latest and most rapid analytical tools.

**Problem 23.20** Write a mechanism for the formation of the purple dye from the reaction of excess ninhydrin with the amino derivative of ninhydrin shown on the previous page.

## B. End–Group Analysis of Peptides and Proteins

Chemists have developed a series of reagents that react selectively with the N-terminal amino acid of a polypeptide, transforming it into some derivative that can be separated from all the other amino acids in the chain and identified. One of the most useful of these is 2,4-dinitrofluorobenzene, **Sanger's reagent,** developed by Frederick Sanger during his determination of the structure of insulin. Sanger was given the Nobel Prize in 1958 for being the first to establish the sequence of amino acids in a protein.

2,4-Dinitrofluorobenzene, abbreviated as DNFB, is an aryl halide with nitro groups on the aromatic ring in positions that activate the halogen toward nucleophilic substitution (p. 898). The free amino group of the N-terminal amino acid displaces fluoride ion from the reagent, forming a dinitroarylamino group. Complete hydrolysis of the peptide gives free amino acids from all the residues except the N-terminal one, which turns up labeled with the aryl group. The process is illustrated for the tripeptide Ser-Gly-Val, which is written in the form in which the amino group is not protonated.

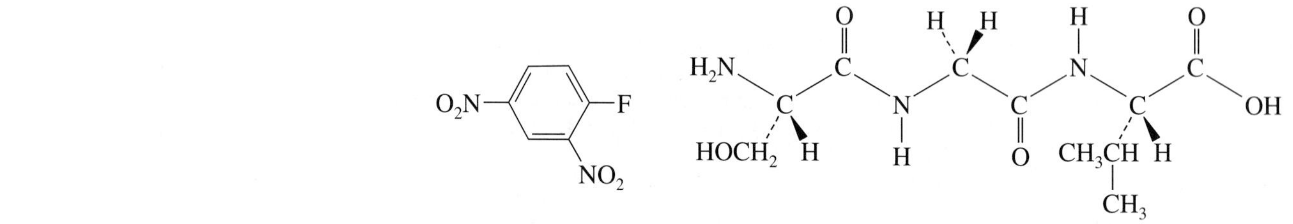

2,4-dinitrofluorobenzene                                          Ser-Gly-Val

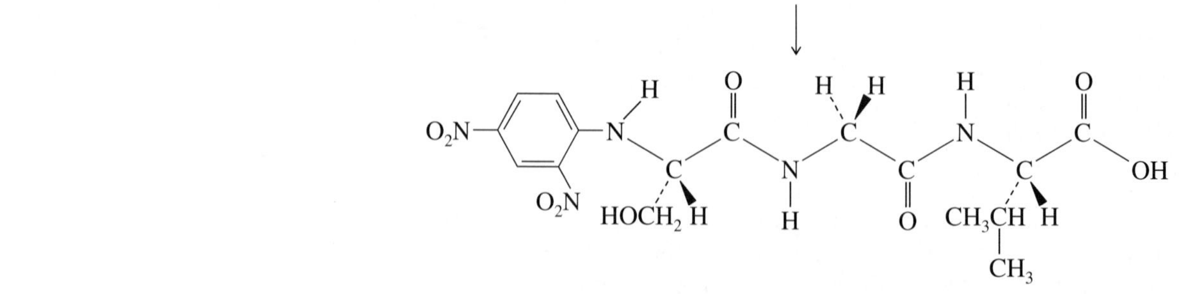

Ser-Val-Gly, labeled at serine with
a 2,4-dinitrophenyl group

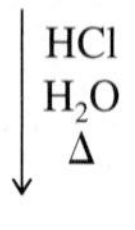

$$HCl$$
$$H_2O$$
$$\Delta$$

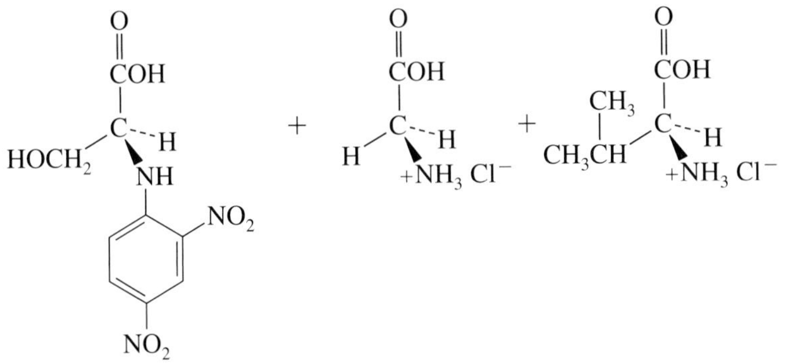

N-2,4-dinitrophenylserine          glycine          valine

*yellow color;*
*behavior unlike that of serine in*
*chromatography or in an electric field*

The presence of the dinitrophenyl group on the nitrogen atom of serine converts the compound from an alkyl amine to an aryl amine and changes its acid–base properties drastically. The derivative of serine behaves differently than serine does with all the techniques used to separate amino acids and is easily identified by comparing it to an authentic sample of N-2,4-dinitrophenylserine.

**Problem 23.21**   2,4-Dinitrofluorobenzene can be used to determine whether lysine is the N-terminal amino acid of a peptide or is somewhere in the middle of the peptide chain. Draw segments of hypothetical peptide chains with lysine in either position, and show with equations how the determination of the position of the lysine molecule could be carried out.

Another reagent used to analyze for N-terminal amino acids is phenyl isothiocyanate. This reagent, which is known as **Edman's reagent,** reacts with the free amino group to give a thiourea (see p. 940) that is then cleaved off in anhydrous acid by the participation of the sulfur atom of the thiourea attacking the carbonyl group of the amino acid at the end of the chain. The reaction proceeds by way of an unstable intermediate that rearranges to give a phenylthiohydantoin derived from the original N-terminal amino acid. The peptide is left with a new N-terminal amino acid that can be subjected to this degradation reaction and the amino acid that is lost becomes part of a substituted phenylthiohydantoin.

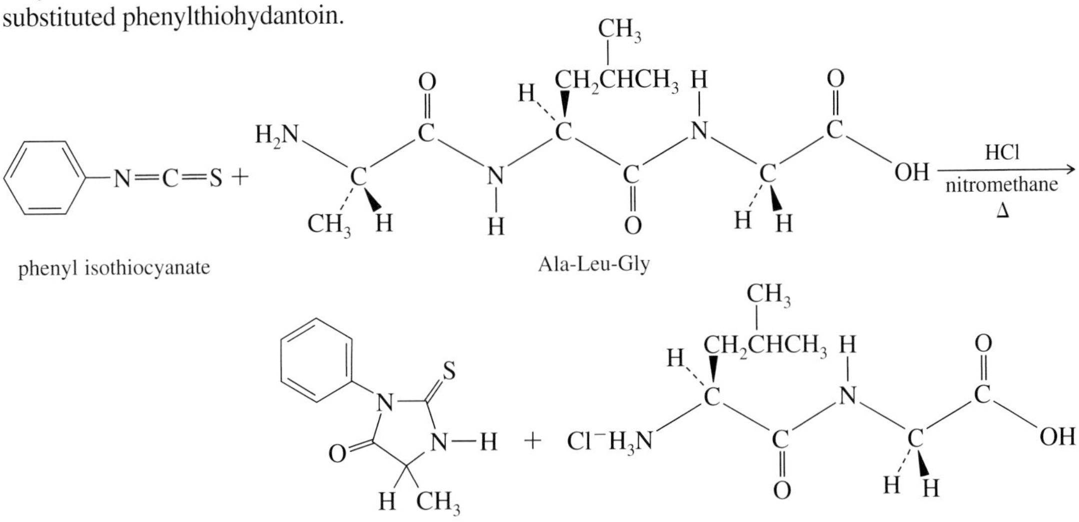

phenyl isothiocyanate          Ala-Leu-Gly

$\xrightarrow[\text{nitromethane}]{\text{HCl}}$  Δ

phenylthiohydantoin          Leu-Gly

*derived from alanine, racemized*

The substituted phenylthiohydantoin can be identified by comparing it with phenyl-thiohydantoins prepared from known amino acids.

In principle, a large polypeptide could be degraded one amino acid at a time and each new *N*-terminal amino acid identified as its phenylthiohydantoin (p. 996). In general, though, the practical problems involved in handling small quantities of material and the side reactions that take place with some of the amino acids limit the number of times the degradation reaction can be used.

The enzyme carboxypeptidase selectively removes the *C*-terminal amino acid of a peptide. This process leaves a smaller peptide with a new *C*-terminal amino acid, which is then also attacked and removed by the enzyme. Therefore, this method must be used in connection with studies of the rate at which different free amino acids appear in the solution.

For example, if the tripeptide Ala-Leu-Gly (shown below) is treated with car-boxypeptidase, the amino acid glycine is liberated into the solution first. After a while, leucine and alanine also will appear.

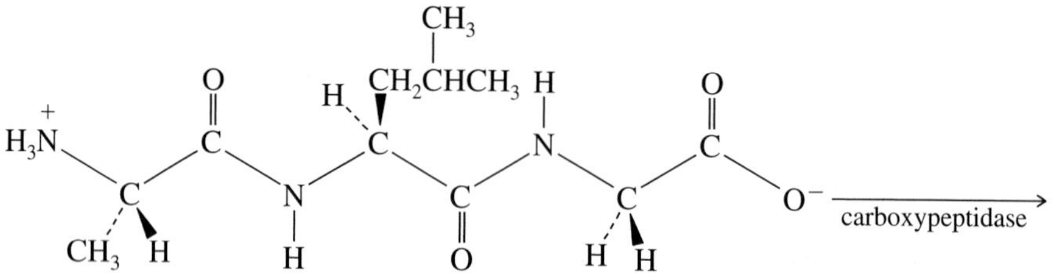

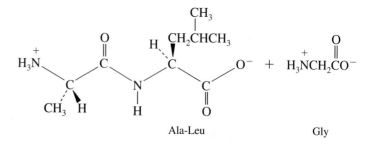

Ala-Leu-Gly

Ala-Leu        Gly

**ONE SMALL STEP**

The reaction described above brings together many mechanistic steps that are familiar to you.

**PROBLEM:**
(a) Write the mechanism of the reaction of Ala-Leu-Gly with phenyl isothiocyanate to give a phenyl-thiourea.
(b) Show the cleavage of the peptide bond by way of nucleophilic attack of sulfur on the carbonyl group of the bond. The unstable intermediate that forms at this stage has the following structure:

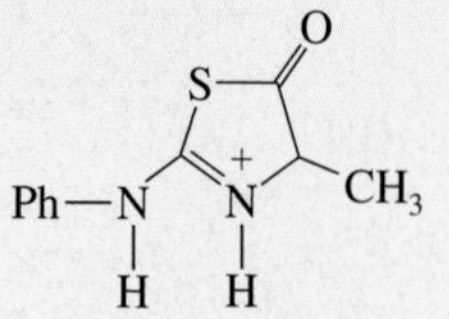

**Problem 23.22**    When the hydrolysis of a peptide by carboxypeptidase is carried out in water labeled with $^{18}O$, the isotopic oxygen appears in the carboxylic acid groups of all the amino acids except the *C*-terminal one, allowing the *C*-terminal amino acid to be identified.

Using Gly-Ala as an example, write a mechanism employing acid and base catalysis and $H_2{}^{18}O$ to show what happens in such a hydrolysis.

End-group analysis of many biologically interesting peptides is complicated by the fact that the *C*-terminal amino acid is in the form of an amide. Also, when glutamic acid is the *N*-terminal amino acid, it often forms a cyclic amide known as py-roglutamic acid, symbolized by a ring added to the three-letter abbreviation Glu. An important hormone secreted by the hypothalamus gland, the thyrotropin-releasing factor, has both these features. The structural formula for thyrotropin-releasing factor is shown on the next page.

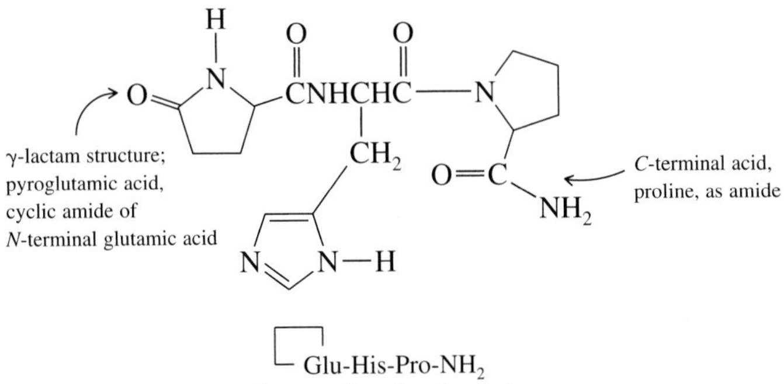

γ-lactam structure; pyroglutamic acid, cyclic amide of N-terminal glutamic acid

C-terminal acid, proline, as amide

Glu-His-Pro-NH₂
thyrotropin-releasing factor

*a tripeptide hormone*

Such peptides do not react with reagents such as Sanger's reagent or Edman's reagent at the *N*-terminal amino acid, nor is the *C*-terminal amino acid cleaved off by carboxypeptidase. Hydrolysis of a molecule of the tripeptide hormone shown would give a molecule of ammonia, as well as the three amino acids glutamic acid, histidine, and proline.

## C. The Degradation of Proteins and Peptides by Enzymes. The Disulfide Bridge

If hydrolysis of a protein is carried out with cold, concentrated hydrochloric acid for a relatively short period of time, a number of peptide fragments are obtained. A protein may be more selectively cleaved at certain points of the peptide chain by enzymes. Trypsin, for example, attacks only the peptide bonds formed by the carboxyl groups of the basic amino acids lysine and arginine. The newly formed peptide fragments have either lysine or arginine as their *C*-terminal amino acid, except for the fragment that includes the original *C*-terminal end of the protein. Another enzyme, chymotrypsin, is less selective. It hydrolyzes the bonds formed by the carboxyl groups of tyrosine, phenylalanine, tryptophan, and methionine. The usefulness of these enzymes in selective cleavages of peptide bonds is illustrated by the determination of the structure of somatostatin, a tetradecapeptide that inhibits the secretion of pituitary growth hormone.

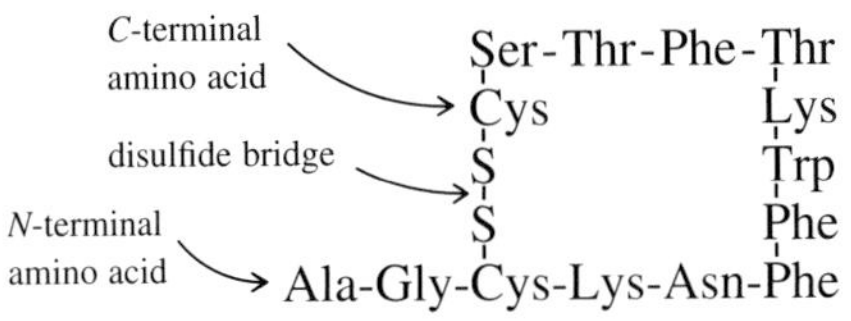

*C*-terminal amino acid

disulfide bridge

*N*-terminal amino acid

Ser-Thr-Phe-Thr
Cys            Lys
 S             Trp
 S             Phe
Ala-Gly-Cys-Lys-Asn-Phe

*the structure of somatostatin, showing the disulfide bridge between two cysteine residues*

The *C*-terminal amino acid of somatostatin and the amino acid third from the *N*-terminal one are cysteine. They are bonded to each other in oxidized form to give a disulfide bridge between the two ends of the peptide. Such bridges are important in stabilizing the tertiary structure of many proteins (p. 1008). The oxidized form of cysteine is known as cystine, a diamino acid with a disulfide bond. Cystine is reduced to give two molecules of cysteine.

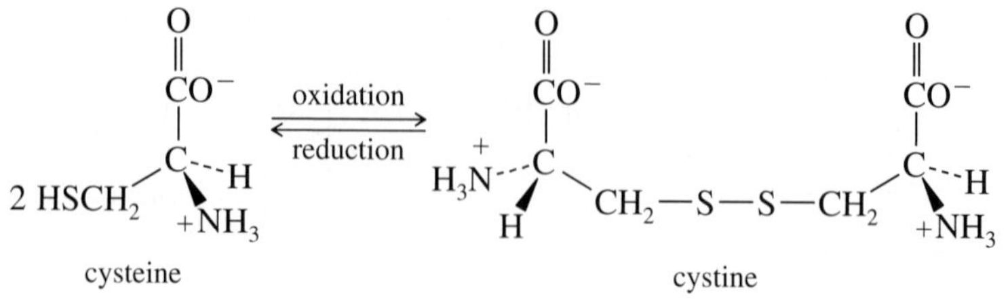

$$2 \; HSCH_2 \underset{+NH_3}{\overset{\overset{\displaystyle O}{\overset{\displaystyle \|}{CO^-}}}{\underset{|}{C}}} \cdots H \quad \underset{\text{reduction}}{\overset{\text{oxidation}}{\rightleftharpoons}} \quad H_3N \overset{+}{\underset{H}{\overset{\overset{\displaystyle O}{\overset{\displaystyle \|}{CO^-}}}{- C}}} CH_2 - S - S - CH_2 \underset{+NH_3}{\overset{\overset{\displaystyle O}{\overset{\displaystyle \|}{CO^-}}}{C}} \cdots H$$

cysteine                                                              cystine

The proof of the structure of somatostatin starts with the reduction of the disulfide linkage with 2-mercaptoethanol, $HSCH_2CH_2OH$; cystine is thereby converted to cysteine. The thiol groups on the cysteine residues are then converted to inert groups by alkylation with iodoacetic acid.

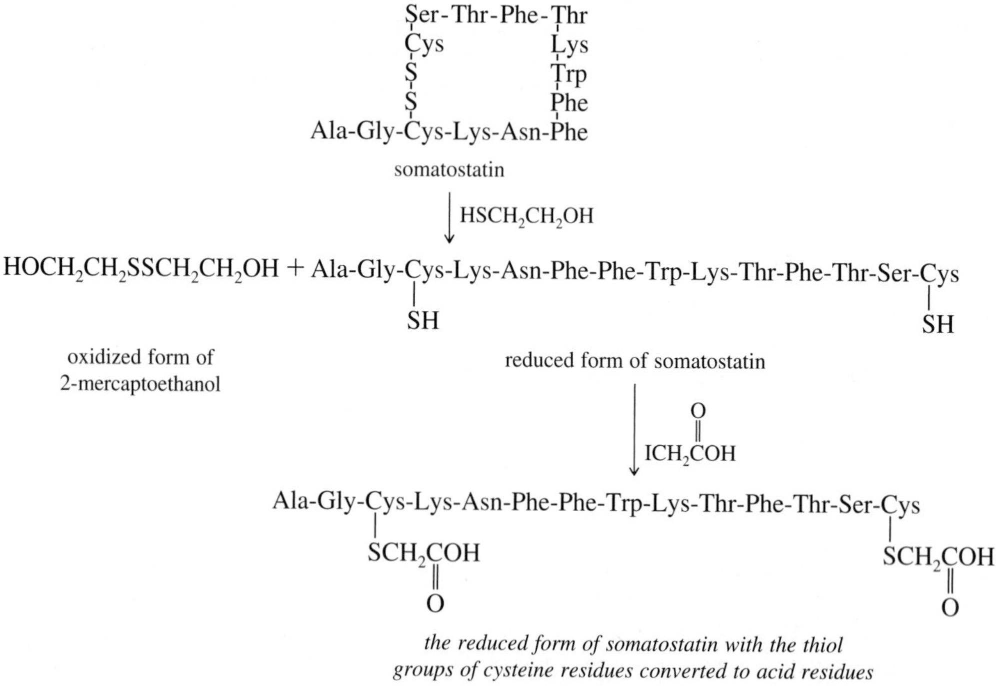

$$
\begin{array}{c}
\text{Ser-Thr-Phe-Thr} \\
\text{Cys} \qquad\qquad \text{Lys} \\
\text{S} \qquad\qquad\quad \text{Trp} \\
\text{S} \qquad\qquad\quad \text{Phe} \\
\text{Ala-Gly-Cys-Lys-Asn-Phe}
\end{array}
$$

somatostatin

$$\downarrow HSCH_2CH_2OH$$

$$HOCH_2CH_2SSCH_2CH_2OH \; + \; \underset{\underset{SH}{|}}{Ala\text{-}Gly\text{-}Cys}\text{-Lys-Asn-Phe-Phe-Trp-Lys-Thr-Phe-Thr-Ser-}\underset{\underset{SH}{|}}{Cys}$$

oxidized form of
2-mercaptoethanol                              reduced form of somatostatin

$$\downarrow \overset{\overset{\displaystyle O}{\overset{\displaystyle \|}{}}}{ICH_2COH}$$

$$Ala\text{-}Gly\text{-}\underset{\underset{\overset{\displaystyle \|}{O}}{SCH_2COH}}{Cys}\text{-Lys-Asn-Phe-Phe-Trp-Lys-Thr-Phe-Thr-Ser-}\underset{\underset{\overset{\displaystyle \|}{O}}{SCH_2COH}}{Cys}$$

*the reduced form of somatostatin with the thiol*
*groups of cysteine residues converted to acid residues*

This step is necessary to prevent the cysteines from being oxidized back to cystine by air.

The peptide somatostatin is cleaved at two points by trypsin.

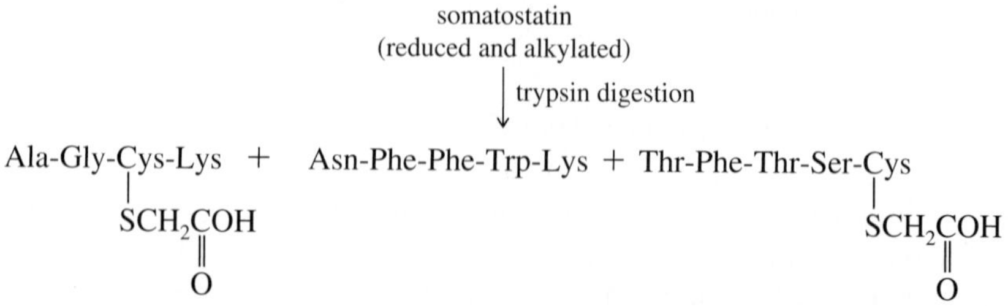

somatostatin
(reduced and alkylated)

$$\downarrow \text{trypsin digestion}$$

$$\underset{\underset{\overset{\displaystyle \|}{O}}{SCH_2COH}}{Ala\text{-}Gly\text{-}Cys}\text{-Lys} \; + \; Asn\text{-Phe-Phe-Trp-Lys} \; + \; Thr\text{-Phe-Thr-Ser-}\underset{\underset{\overset{\displaystyle \|}{O}}{SCH_2COH}}{Cys}$$

Digestion of somatostatin with chymotrypsin gives more fragments than does digestion with trypsin.

somatostatin
(reduced and alkylated)

↓ chymotrypsin digestion

Ala-Gly-Cys-Lys-Asn-Phe + Phe + Trp + Lys-Thr-Phe + Thr-Ser-Cys

| |
SCH₂COH                                                                         SCH₂COH
‖                                                                                  ‖
O                                                                                  O

If the structures of the individual smaller peptides that result from acid or enzymatic hydrolyses can be determined, it is possible to reconstruct the structure of the original larger peptide by piecing together the different overlapping sequences of amino acids found in the fragments. The structures of small peptides are determined by reactions that selectively identify their *N*-terminal and *C*-terminal amino acids by end-group analysis (p. 996). Thus, for example, if a determination is made that the *N*-terminal amino acid in somatostatin is alanine and the *C*-terminal amino acid is cysteine, the fragments obtained from the enzymatic cleavages can be combined to give the full structure of the molecule. The fragments from cleavage by trypsin and those from cleavage by chymotrypsin are lined up below to show how the different points of cleavage provide overlapping information about the order of the amino acids in the chain.

peptides from trypsin cleavage of somatostatin

Ala-Gly-Cys-Lys/Asn-Phe-Phe-Trp-Lys/Thr-Phe-Thr-Ser-Cys

Ala-Gly-Cys-Lys-Asn-Phe/Phe/Trp/Lys-Thr-Phe/Thr-Ser-Cys

peptides from chymotrypsin cleavage of somatostatin

The total structure of somatostatin has also been determined by repeated use of Edman's reagent. The chain was degraded one amino acid at a time starting from the *N*-terminal end, and each amino acid was identified as its phenylthiohydantoin (p. 997).

⊘ **Study Guide**
**Concept Map 23.4**

**Problem 23.23** A heptapeptide is isolated from milk protein that has been degraded by a protease from the bacterium *Bacillus subtilis*. The heptapeptide contains the amino acids arginine, glycine, isoleucine, phenylalanine, proline (two residues), and valine.

(a) Treatment of the heptapeptide with trypsin gives arginine and a hexapeptide. What does this tell you about the structure of the original peptide?
(b) Some of the fragments that were obtained by partial hydrolyses of the heptapeptide are Pro-Phe-Ile, Arg-Gly-Pro, Pro-Phe-Ile-Val, and Pro-Pro. Assign a structure to the heptapeptide.

**Problem 23.24** The hormone arginine vasopressin, which is important in the regulation of water balance in the body, is a nonapeptide with one disulfide bridge. In the proof of its structure, the disulfide bridge was cleaved by oxidation with peroxyformic acid, which oxidizes cysteine residues to cysteic acid. During degradation, the cysteines appear as cysteic acid residues, symbolized below by Cys.

O                                              O
‖                                              ‖
CO⁻           O                              COH
|             ‖                               |
C---H    HCOOH      O      C---H
⁄  ＼    ──────→   ‖     ⁄  ＼
HSCH₂  ⁺NH₃       ⁻OSCH₂  ⁺NH₃
‖
O

cysteine                              cysteic acid

Complete hydrolysis of the oxidized peptide with hydrochloric acid gave 3 equivalents of ammonia and the amino acids Arg, Asp, Cys (2), Glu, Gly, Phe, Pro, and Tyr. 2,4-Dinitro-fluorobenzene reacted with one of the cysteic acid residues. Trypsin digestion gave an octapeptide with arginine as its *C*-terminal amino acid and glycinamide, Gly-NH$_2$. Partial acid hydrolysis gave the following fragments:

Asp-Cys    Phe-Glu    Cys-Tyr-Phe    Glu-Asp-Cys    Phe-Glu-Asp

In addition, a fragment having cysteine as its *N*-terminal amino acid and also containing arginine, glycine, and proline, in an undetermined order, was obtained.

(a)  What is the order of amino acids derived from the degradation experiments?
(b)  What is the full structure for arginine vasopressin?

## 23.7  Polysaccharides

### A. Starch

The two polysaccharides of greatest biological and economic importance are starch and cellulose. Plants store their reserve carbohydrates in the form of starch. Humans get starch from roots, tubers, and seeds. Starch is a high-molecular-weight polysaccharide that can be broken down completely to glucose by acid hydrolysis or, partially, by the enzyme maltase, which establishes the presence of $\alpha$-linkages. Starch can be fractionated into two different kinds of molecules. One is amylose, which has an average molecular weight of over $10^6$. It is a polymer of glucopyranose units attached to each other by $\alpha$-glycosidic linkages between carbon 1 of one glucopyranose and the hydroxyl group on carbon 4 of the next one. The structure of amylose is proved by methylation and hydrolysis, which gives almost exclusively 2,3,6-tri-*O*-methylglucose.

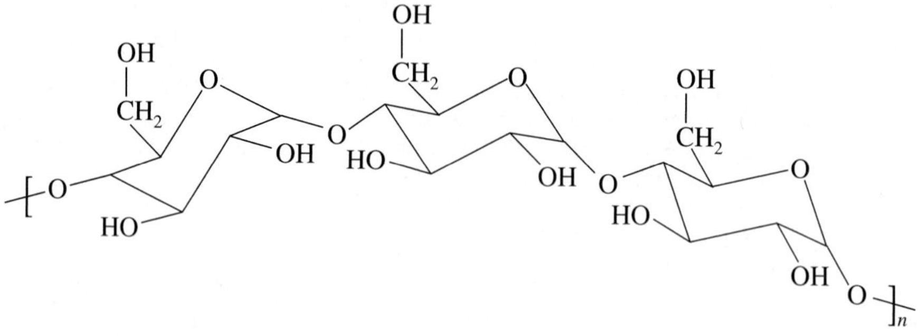

amylose
$n \sim 1000–6000$

The chief fraction of starch is amylopectin, which has a more complicated structure. It is highly polymeric, with as many as a million glucose units in a single molecule. Complete methylation and hydrolysis of amylopectin gives about 3% of 2,3-di-*O*-methylglucose, suggesting that some glucose units are connected to others through the hydroxyl group at carbon 6 as well as through the oxygen at carbons 1 and 4. Enzymatic studies and hydrolysis reactions have shown that amylopectin has a randomly branched structure in which the hydroxyl group at carbon 6 of some glucose units is indeed involved.

HO
|
CH$_2$

main chain of amylopectin
$\alpha$ (1→4) linkage between
glucose units

OH

(1→6) linkage to form
a branch in the chain

HO

CH$_2$

reducing end
of chain

points at which a
1,6-linkage is formed

HO

CH$_2$

OH HO

OH

this glucose would be seen as
2,3-di-*O*-methylglucose after
methylation and hydrolysis

OH

OH

CH$_2$

HO

OH

O

amylopectin, showing
branching of the chain

The branching appears to occur once every 20 to 25 glucose units. The overall structure of amylopectin resembles the branching of a tree.

Glycogen, the form in which animals store carbohydrates, is also composed of glucose units and is similar to amylopectin in structure except that the branching occurs at shorter intervals, with about 12 glucose units in each branch. It is especially abundant in the liver of mammals and also has been isolated from kidneys, brains, and skeletal and cardiac muscles.

In the presence of certain small molecules such as iodine, amylose forms helical structures in which there are about six $\alpha$-D-glucopyranose units per turn. The hydrophilic hydroxyl groups are oriented to the outside to hydrogen-bond with water, while the inside of the helix is relatively hydrophobic. Iodine molecules line up inside the helix, giving rise to a complex that has a deep blue color, the familiar color reaction for iodine in starch. Pure amylopectin does not give the same color.

J. Fraser Stoddart and his colleagues, researching ways to make single-walled carbon nanotubes (SWNT, p. 365) soluble in water, came up with the idea of wrapping them with helical starch molecules. Figure 23.7 (p. 1004) shows, in green, the helical amylose molecule with blue iodine molecules lined up in it and the iodine molecules being displaced by a carbon nanotube depicted as a gray rod. This starch-wrapped nanotube is also the subject of the cover art of this book. Though the helix is amylose, amylopectin molecules in the starch also surround, stabilize, and help to solubilize the nanotubes.

The starch-wrapped individual nanotubes can be separated from clumps of tubes and other soot debris that forms when the nanotubes are made. When purified, they are soluble in water to the extent of about 3 g/L. The individual nanotubes can be recovered by hydrolyzing the starch molecules enzymatically to glucose. The addition of human saliva, which contains $\alpha$-amylase that cleaves the $\alpha$ (1→4) linkage in starch, to a solution of the starch-wrapped nanotubes results in the precipitation of the nanotubes in a few hours.

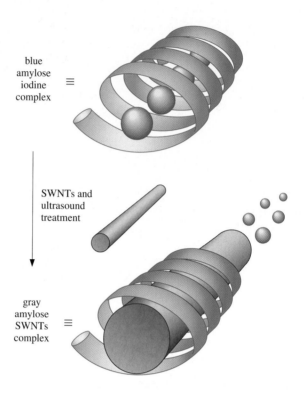

*Figure 23.7*
Schematic representation of the formation of a starch-wrapped carbon nanotube. Ultrasound treatment of an aqueous solution of a blue amylose–iodine complex mixed with carbon nanotubes results in the replacement of iodine molecules by a carbon nanotube to give a gray amylose–nanotube complex.

Adapted with permission from A. Star, D. W. Steurman, J. R. Heath, and J. F. Stoddart, as appeared in *Angewandte Chemie, International Edition* **2002**, *41*, pp. 2508–2512.

## B. Cellulose

Cellulose is the organic substance that is most abundant in nature. It is the structural material of higher plants and is found in all their parts. Wood is about 50% cellulose. Commercially important fibers such as cotton and flax consist almost completely of cellulose.

On hydrolysis, cellulose gives cellobiose and ultimately glucose. This establishes its structure as a linear chain of glucopyranose units attached to each other by $\beta$-glycosidic linkages from carbon 1 of one unit to the hydroxyl group on carbon 4 of another unit. The structure consists of long chains of six-membered rings in the most stable chair conformation, with all the larger substituents in the equatorial positions.

glucopyranose structure rotated in space

cellulose
$n \sim 5000–10,000$

The individual molecules of cellulose are associated with each other in regular structures that have crystalline properties. Approximately 100 to 200 cellulose molecules are held together in each of these larger units. The exact nature of the interactions between the molecules has not been determined. Hydrogen bonding between adjacent strands of molecules does seem to be an important factor in determining the strength and rigidity of cellulose as a structural material.

Cellulose was the first polymeric material that was chemically modified to provide new materials useful to human beings. The polyhydroxy functions in cellulose participate in the typical reactions of alcohols. Thus cellulose treated with acetic anhydride in acetic acid with a little sulfuric acid as a catalyst is converted into its acetate. Exactly how many acetyl groups are added to the molecule depends on the physical state of the cellulose at the beginning of the reaction, the exact reaction conditions, and how the material is treated afterwards. Some of the acetyl groups are hydrolyzed off by exposure to water during processing.

The most useful acetate is cellulose triacetate, which is soluble in mixtures of acetic anhydride and acetic acid or in organic solvents.

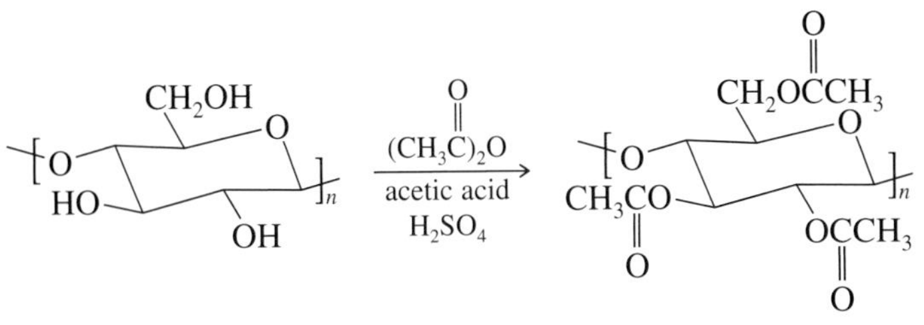

fragment of
cellulose

fragment of
cellulose triacetate

In this compound, all the free hydroxyl groups of cellulose have been converted into ester functions. When a solution of cellulose triacetate is forced through small holes into a solution of dilute acetic acid, the water precipitates it in the form of a continuous thread that can be used to weave fabrics in the textile industry. This product is known as Arnel.

The use of cellulose in a variety of other products is based on similar modifications of structure. The hydroxyl groups are converted to other functional groups. In the process, the cellulose molecule becomes somewhat degraded and much more soluble in organic solvents. In this soluble state, it is manipulated by extruding the solutions as sheets or fine streams into other solutions that precipitate the compounds or reverse the original chemical reactions, regenerating cellulose in a new, more useful form. For example, the hydroxyl groups in cellulose are converted into alkoxide anions by base. These nucleophilic anions add to carbon disulfide to give compounds known as xanthate esters, which are stable as salts in basic solution. The xanthate esters originally form as a mixture at all possible positions on cellulose. The mixture is then allowed to equilibrate to give mostly the xanthate ester at carbon 6, the thermodynamically most stable xanthate ester.

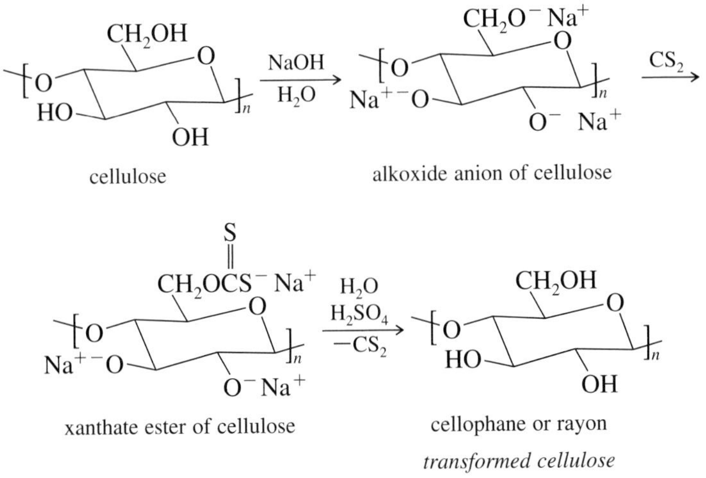

cellulose

alkoxide anion of cellulose

xanthate ester of cellulose

cellophane or rayon

*transformed cellulose*

A basic solution of cellulose xanthate salts can be forced out of spinners into dilute sulfuric acid where they lose carbon disulfide to form rayon threads (viscose rayon). If thin slits are used, sheets of cellophane are formed. Both rayon and cellophane are essentially cellulose in a transformed physical state. The hydrogen bonding between the molecules of cellulose in its original state has been disrupted by the chemical process. When the hydroxyl groups re-form after the reactions are complete, the cellulose molecules have been physically forced by the extrusion process into new conformations, which have new physical properties.

**Problem 23.25**   Cellulose is converted into a nitrate ester by treatment with a mixture of nitric and sulfuric acids. Cellulose trinitrate is called guncotton because of its explosive properties. Write an equation for the formation of cellulose trinitrate.

**Problem 23.26**   Using ROH to represent cellulose, write a mechanism for the formation of cellulose xanthate.

## C. Glycosaminoglycans

Hyaluronic acid is an example of an important group of acidic polysaccharides known as glycosaminoglycans or mucopolysaccharides that are found widely in various animal tissues, especially connective tissues in conjunction with the protein collagen (p. 1009), synovial fluid, which lubricates the joints, and in vitreous humor, the fluid within the eyeball. The disaccharide units that make up hyaluronic acid are D-glucuronic acid and N-acetyl-D-glucosamine.

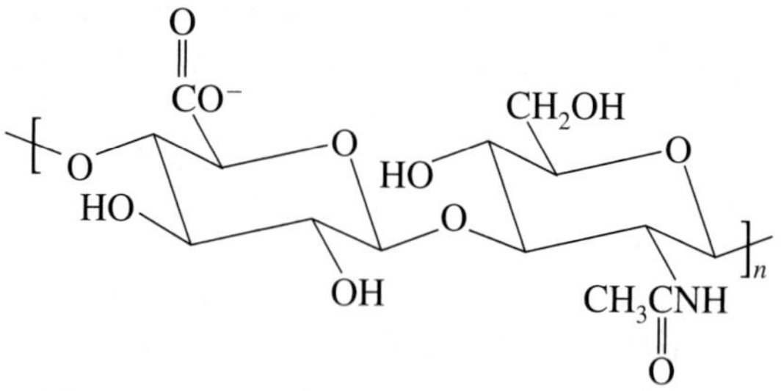

hyaluronic acid

*polysaccharide made up of*
*glucuronic acid and*
*2-acetamido-2-deoxy-D-glucose units*

The chondroitin sulfates are another set of such acidic polysaccharides.

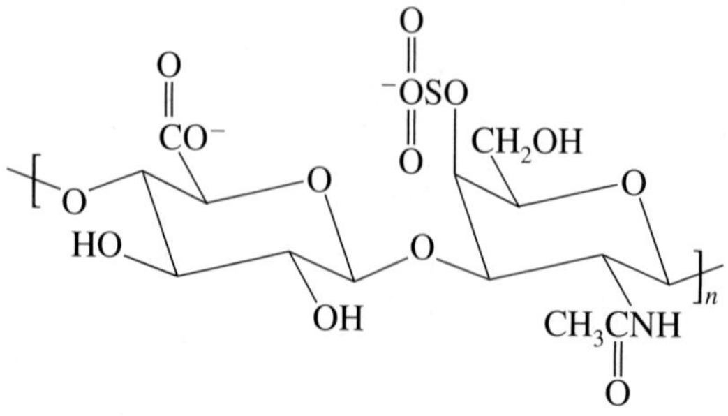

chondroitin-4-sulfate

*polysaccharide made up of*
*D-glucuronic acid and*
*N-acetyl-D-galactosamine-4-sulfate*

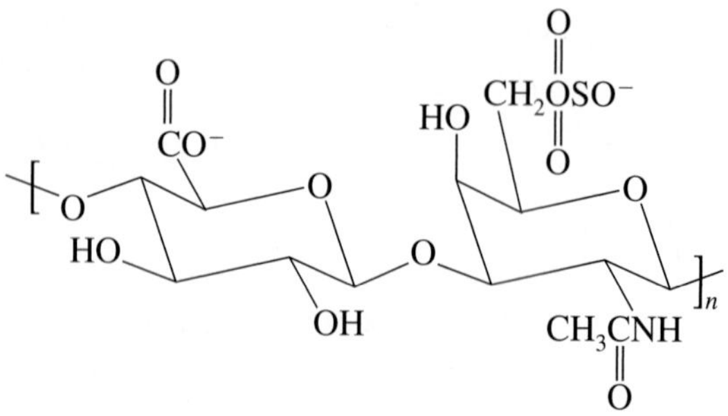

chondroitin-6-sulfate

*polysaccharide made up of*
*D-glucuronic acid and*
*N-acetyl-D-galactosamine-6-sulfate*

These acid polysaccharides have large molecular size, with $n$ up to 25,000. They hydrogen-bond to large amounts of water to give viscous jelly-like solutions. They serve as matrices that hold cells together and lubricate joints. Such compounds are implicated in inflammatory responses of the immune system typical of rheumatoid arthritis. The chondroitins, on the other hand, are sold as nutritional supplements to alleviate the symptoms of osteoarthritis.

**Problem 23.27**  Among the hydrolysis products of chitin, the chief material in the shells of lobsters, is a disaccharide, chitobiose. It has the formula $C_{12}H_{24}N_2O_9$. Further mild hydrolysis of chitobiose gives 2-acetamido-2-deoxy-D-glucose (see page 972 for 2-amino-2-deoxy-D-glucose). Assign a structure to chitobiose, which seems to be structurally analogous to cellobiose.

# 23.8 Proteins

## A. Conformation of Peptides and Secondary Structures of Proteins

Peptides consisting of a number of amino acids assume characteristic conformations. The carbonyl group and the nitrogen and hydrogen atoms around the peptide bond, as well as the two carbon atoms to which the carbonyl and amino groups are bonded, lie in a plane. Resonance interaction between the nitrogen atom and the carbonyl group gives a partial double-bond character to the carbon–nitrogen bond and prevents free rotation.

The oxygen atom of the carbonyl group and the hydrogen atom bonded to the nitrogen atom are trans to each other. The bond lengths and bond angles typical of the peptide group are illustrated in Figure 23.8.

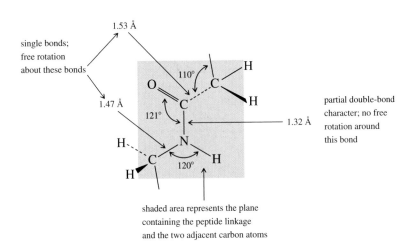

single bonds; free rotation about these bonds

1.53 Å

1.47 Å

110°

121°

120°

1.32 Å

partial double-bond character; no free rotation around this bond

shaded area represents the plane containing the peptide linkage and the two adjacent carbon atoms

*Figure 23.8*
Typical bond lengths and bond angles for the peptide group.

The order of amino acids in the backbone of a polypeptide is known as the **primary structure** of the peptide. The conformation of the peptide chain gives it a **secondary structure.** Further bending and folding of peptide chains creates the **tertiary structure** of proteins. Finally, several polypeptide units may be associated with each other and with other simpler molecules such as sugars, inorganic residues, or coenzymes in what is known as the **quaternary structure** of proteins.

All the amino acids found in naturally occurring peptides and proteins have the same configuration, and this imparts a stereochemical regularity to peptide chains. They assume different conformations depending mainly on the types of amino acids that make up the chain. One common conformation is the **α-helix.** This structure was proposed in 1951 on the basis of x-ray diffraction experiments by L. Pauling and R. Corey. The peptide chain forms a right-handed coil having 3.7 amino acids per full turn. The structure is given rigidity and stability by hydrogen bonding between the hydrogen atom bonded to a nitrogen atom and the carbonyl group of the amino acid four units down the chain at the next turn of the coil. The spacing of the amino acids is such that every peptide linkage is involved in hydrogen bonding (Figure 23.9).

The α-helix is particularly important in the structural proteins, such as the **α-keratins,** which make up the protein part of skin, nails, horns, hair, and feathers. Proteins that form structures such as hair and wool are rich in cystine, the oxidized form of cysteine (p. 1000). In them, several α-helixes are coiled around each other to give multiple strands that are held together by disulfide bridges. Giving a permanent wave, in fact, involves reduction of some of the disulfide linkages so that the hair loses its natural conformation, which is determined genetically. Then the hair is held in a new conformation on curlers and reoxidized to form new disulfide linkages.

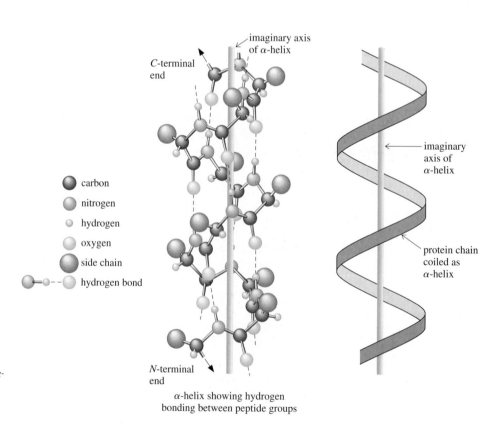

**Figure 23.9**
The α-helix, showing hydrogen bonding between peptide groups and the coiled conformation of the peptide chain.

From Stoker, *General, Organic, and Biological Chemistry.* Copyright © 1998 by Houghton Mifflin Company. Adapted by permission.

- carbon
- nitrogen
- hydrogen
- oxygen
- side chain
- hydrogen bond

imaginary axis of α-helix

*C*-terminal end

imaginary axis of α-helix

protein chain coiled as α-helix

*N*-terminal end

α-helix showing hydrogen bonding between peptide groups

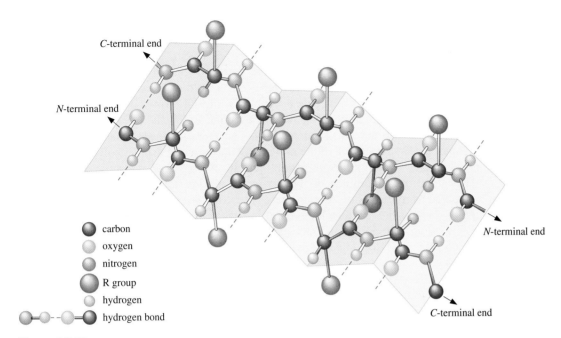

*Figure 23.10*
The pleated sheet conformation of some proteins.
From Stoker, *General, Organic, and Biological Chemistry.* Copyright © 1998 by Houghton Mifflin Company. Adapted by permission.

Not all proteins have $\alpha$-helical structures. The major constituent of silk is a protein called fibroin, which is a **$\beta$-keratin** or **pleated sheet.** The peptide chains are more extended in this conformation and lie side by side, with hydrogen bonding between the chains. In this structure, maximum hydrogen bonding is possible when adjacent polypeptide chains run in opposite directions (Figure 23.10). Such structures are most stable when the side chains of the amino acid residues are small and not charged. Silk fibroin consists mostly of glycine and alanine, for example.

Another important type of structural protein is the **collagens,** which make up connective tissues. When a collagen is boiled in water, gelatin is formed. Collagens are rich in glycine, alanine, and proline and contain 4-hydroxyproline, an amino acid that is seldom found in other proteins. The structure of collagens consists of a triple helix of polypeptide chains, each of which is a left-handed helix. Keratins and collagens are known as **fibrous proteins.**

## B. Tertiary and Quaternary Structures of Proteins

A large class of proteins is the **globular proteins,** in which the tertiary structure of the protein tends to be spherical or ellipsoidal, with much coiling of the polypeptide chain. Many proteins, including all enzymes, have this type of structure.

Parts of globular proteins have the $\alpha$-helix conformation. Other parts are more randomly coiled. Not all side chains of amino acids fit well into the $\alpha$-helix. Proline, in particular, causes a bend in the $\alpha$-helix structure because the nitrogen of the amino group is part of a five-membered ring. Furthermore, proline has no amide hydrogen atom to participate in hydrogen bonding. The side chains of lysine and arginine are positively charged at physiologic pH and repel each other, disrupting the helical structure. Conversely, strong ionic interactions between carboxylate anions on the side chains of glutamic and aspartic acid residues and cationic centers on lysine and arginine residues may be more important than hydrogen bonding in determining conformation. Disulfide bridges between different parts of the peptide chain also may stabilize particular conformations of a protein. The most important factors in the stabilization of a protein structure are the hydrophobic forces, those forces that cause the nonpolar parts of protein molecules to avoid contact with water. In a way, a globular protein molecule resembles a micelle formed by soap (Figure 15.3, p. 641), with the hydrophobic side chains protected on the inside, where they are stabilized by van der Waals interactions. The more polar and charged amino acid residues are on the surface of the protein in contact with the polar solvent water.

Thus the overall structure of a protein in its natural state is a delicate balance determined by a multitude of interactions made possible by the specific order of amino acids in the chain. Among these are attractions between positively and negatively charged sites on the chain, repulsions between sites of similar charge, hydrogen bonding, van der Waals forces acting between hydrocarbon-like side chains on amino acids, and disulfide bridges. The next section looks closely at the structure of an enzyme, chymotrypsin, to see how its function is determined by these complex structural factors. One interesting insight from recent research is that enzymes that perform the same types of functions in different species ranging from fungi to mammals may share as little as 20% of the backbone of their protein structures. In other words, overall three-dimensional structures that can function chemically in a certain way arise in nature from a variety of different building blocks.

A protein retains its structure within a narrow range of pH, temperature, and ionic strength of the solution. Addition of acid, base, metal ions, or urea (which disrupts hydrogen bonding in proteins) to a solution of a protein **denatures** it, that is, changes its structure to a form in which biological activity is lost. An increase in temperature does the same thing—that is what happens when an egg is cooked. Sometimes a denatured protein will return to its original form and recover most of its biological activity if, for example, the optimal pH is restored. In other cases, the damage is irreversible. The particular structure that is native to a given protein is the favored one, and the protein will return to it if possible. The enzyme ribonuclease, for example, has four disulfide bridges between eight cysteine residues, widely spaced in the molecule. If the disulfide bridges are cleaved by reduction, the protein is easily denatured. Addition of urea, for example, causes it to lose all its enzymatic activity. Removal of the urea allows the molecule to regain its activity, indicating that it folds back into the original tertiary structure and even re-forms the same disulfide links by oxidation in air. A random reconnecting of eight cysteine residues by four disulfide bridges could occur in more than 100 different ways, yet only one set of four disulfide bridges is actually formed.

Some proteins such as hemoglobin have quaternary structure. In hemoglobin, four polypeptide chains, each of which has tertiary structure, are associated with each other and with four molecules of heme. Heme is a heterocyclic ring (p. 954) that coordinates with iron(II) ions to perform the oxygen-carrying function in the blood (Figure 23.11). In hemoglobin, the heme units are located in pockets of the protein that are lined with hydrophobic amino acid side chains. This placement protects iron(II) against being oxidized irreversibly by oxygen to iron(III), something that happens easily in ordinary solutions of iron(II) compounds. One ligand to the iron is a histidine residue, which also enables the iron to bind loosely and reversibly with oxygen. Carbon monoxide binds more strongly to hemoglobin than oxygen does. This can lead to carbon monoxide poisoning. Heavy smokers, who inhale large doses of carbon monoxide in tobacco smoke, have about 20% of their hemoglobin bound up in this nonfunctional form.

A protein such as hemoglobin, in which a relatively small molecule (heme) is associated with the protein, is called a **conjugated protein.** The small molecule as-

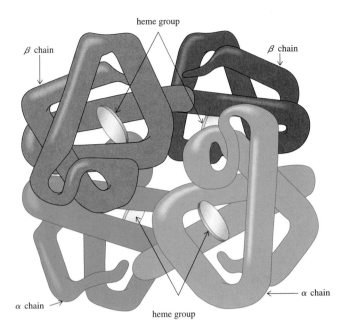

heme group

$\beta$ chain

$\beta$ chain

$\alpha$ chain

$\alpha$ chain

heme group

*Figure 23.11*
Schematic representation of the structure of hemoglobin, showing four polypeptide chains and four heme molecules.

Figure adapted from "The Hemoglobin Molecule." by M.F. Pertuz, *Scientific American,* November 1964, p. 65.

sociated with the protein is known as a **prosthetic group.** Many enzymes are proteins conjugated with coenzymes that are often compounds known to be vitamins.

An issue of great interest to chemists today is the ability to design from scratch a protein that will have a desired secondary structure and predictable biological activity. Much research, some of it involving computer modeling of protein structures, and some of it synthetic, goes into exploring the factors that determine how the individual amino acids along the backbone of a protein ultimately determine its secondary structure. One approach is to synthesize a polypeptide that will have a different spacing between the carbonyl groups and the nitrogen atoms of the amide linkages. Such a change in spacing is expected to change the way in which carbonyl groups and amide hydrogens in different parts of the polypeptide can hydrogen-bond to each other. This change in turn will alter the secondary structure of the peptide.

One way to change spacing is to insert a double bond between the carbonyl group and the $\alpha$-carbon atom of an amino acid and then to create a polypeptide chain from such "vinylogous" amino acids. This approach has been carried out by Stuart L. Schreiber of Harvard University. Just as a typical polypeptide is composed of amino acid units, a vinylogous polypeptide consists of vinylogous amino acids.

a polypeptide

a vinylogous polypeptide

Study Guide
Concept Map 23.5

When synthesized (see Problems 23.28 and 23.52), vinylogous polypeptides give both sheet and helical structures, though with some difference in the number of atoms in the rings formed by the hydrogen bonds.

---

**Problem 23.28**    A vinylogous tyrosine unit was found in the structure of cyclotheon-amide B, a natural product isolated from *Theonella* marine sponges that is a potent inhibitor of the blood-clotting agent thrombin. A synthesis of cyclotheonamide B required the synthesis of a vinylogous tyrosine derivative from tyrosine protected at the amino and hydroxyl groups. The synthesis is outlined below. Supply structural formulas for the reagents or products indicated by the letters.

---

## C. Chymotrypsin, a Look at the Functioning of an Enzyme

Chymotrypsin, one of the digestive enzymes secreted by the pancreas, belongs to a family of enzymes that cleave proteins into smaller peptides. The group, which includes trypsin, is known as the **serine proteases** because the side chain of serine plays an important part in their catalytic activity. Chymotrypsin hydrolyzes the peptide bond at the carboxylic acid group of amino acids that have large hydrophobic side chains, such as phenylalanine, tryptophan, and tyrosine (pp. 1000–1001).

Chymotrypsin is formed in the body when two dipeptide units, consisting of amino acids at positions 14 and 15 and positions 147 and 148, are removed from a precursor molecule, chymotrypsinogen, which has 245 amino acid residues. Chymotrypsin has three polypeptide chains held together by two disulfide bridges. There are three other disulfide bridges in the molecule, stabilizing its conformation. The enzyme has two important regions. The folding of the molecule brings histidine at position 57, aspartic acid at position 102, and serine at position 195 close together in what is known as the **active site** of the enzyme. Near this site is a region lined with hydrophobic groups where the proper portion of the peptide chain is positioned for cleavage. This region is known as the **binding site.** Figure 23.12 is a drawing of chymotrypsin in which each amino acid residue, except those involved in the active site and in the disulfide bridges, is represented only by carbon 2.

The way in which chymotrypsin acts has been determined by many experiments with different types of compounds that are hydrolyzed by the enzyme. Some reagents deactivate the enzyme, and a degradation of the molecule shows which

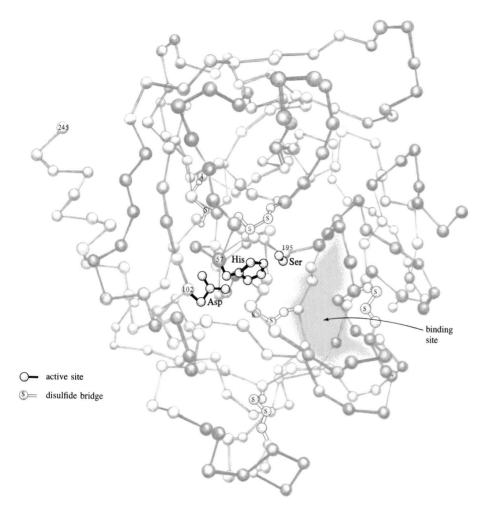

active site

disulfide bridge

*Figure 23.12*
A representation of the chymotrypsin molecule showing the placement of histidine, aspartic acid, and serine at the active site. The hydrophobic pocket that binds the peptide residue during cleavage is below and to the right of the serine at position 195.

Figure adapted from "A Family of Protein-Cutting Proteins," by Robert M. Stroud. Copyright © 1974 by Scientific American, Inc. All rights reserved.

amino acids have been affected. Spectroscopic methods have also been used to follow the course of protonation and deprotonation reactions. The mechanism for the hydrolysis of the peptide bond by the enzyme is an entirely familiar one. The only new aspects are the specific reagents that appear in the roles of acid, base, and nucleophile.

The nucleophile, the hydroxyl group on the serine, attacks the carbonyl group of the peptide bond to give a tetrahedral intermediate (p. 595). Serine is made more nucleophilic by transferring its proton to histidine, which in turn is able to accept that proton because the charge on the ring is stabilized by electrostatic interaction with a carboxylate anion on a conveniently placed aspartic acid residue. The tetrahedral intermediate breaks up after transfer of a proton from histidine to the amide nitrogen atom has created an amine as a leaving group. At this stage, the enzyme is acylated at the serine residue. A transfer of an acyl group from the peptide to the enzyme has taken place. Serine is regenerated by a similar sequence of steps, with water as a participant. The ester bond to the serine hydroxyl group is hydrolyzed, and the peptide fragments diffuse away from the active site, which is then available for cleavage of another peptide bond. A process that requires strong acid or base in water solution and several hours of heating to 100 °C in the laboratory (p. 994) is achieved at pH 6–7, at 37 °C, in a fraction of a second in the body. The mechanisms for the acylation of serine in the active site of chymotrypsin and the regeneration of the active site are shown below and on the next page.

---

**VISUALIZING THE REACTION**

**Acylation of the Active Site of Chymotrypsin**

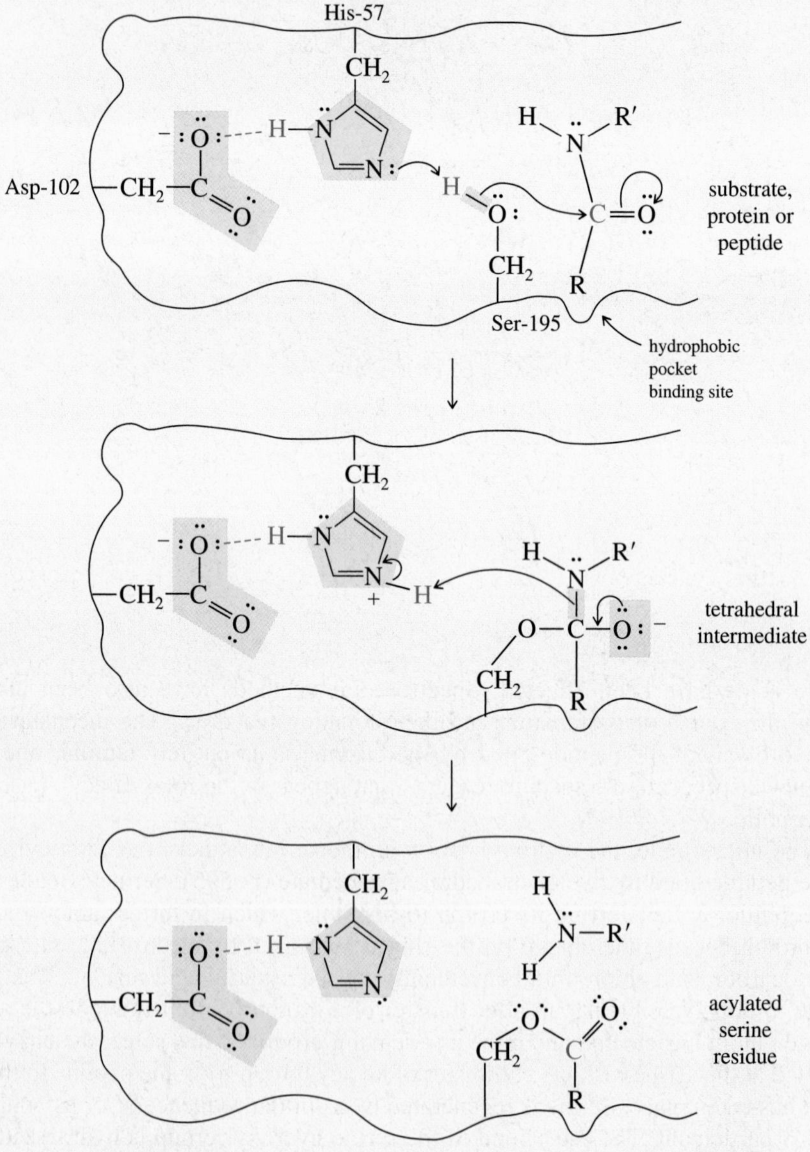

**VISUALIZING THE REACTION**

**Regeneration of the Active Site of Chymotrypsin**

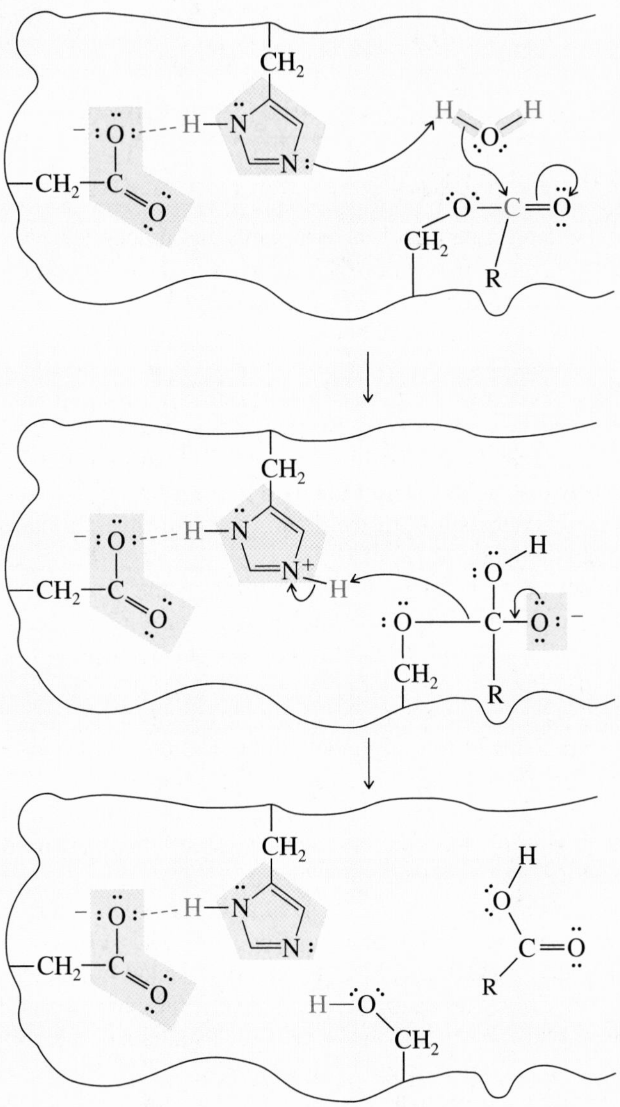

The activity of enzymes is a powerful demonstration of the effect that lowering the energy of the transition state can have on the overall rate of the reaction. The form of an enzyme molecule creates the maximum degree of coordination between the different steps necessary for the cleavage of the peptide bond, ensuring that the transition state for the reaction is achieved with a minimum energy of activation.

Chymotrypsin is only one of the thousands of enzymes that catalyze the chemical processes that support life. The structures of only a few of these enzymes are known well enough for scientists to begin to figure out how they function. Proteins that have similar functions in different organisms have remarkably similar, if not identical, structures. It almost seems as though nature has discovered certain solutions to chemical problems and uses them over and over again with minor modifications to accommodate the unique characteristics of different species.

## 23.9 Polynucleotides: DNA and RNA

### A. The Structure of DNA

The history of the determination of the structures and functions of RNA and DNA is a long one. It starts with the isolation of a material rich in phosphorus from the nuclei of pus cells and from the sperm of salmon by Friedrich Miescher in Germany in 1868. This material was first called nuclein and then nucleic acid. Early in this century, the components of nucleic acids—heterocyclic bases, sugars, and phosphoric acids—were identified. In the 1920s, the structures of the individual nucleotides were determined. Beginning in 1939, the British chemist Sir Alexander Todd investigated the structures of the nucleotides and showed that they are linked in polynucleotides as phosphoric acid esters at the hydroxyl groups of carbons 3 and 5 of the sugar units. For his contribution to the determination of the structures of RNA and DNA, he received the Nobel Prize in 1957.

The phosphate ester linkages between different nucleotides in a fragment of DNA are shown in Figure 23.13. By convention, the oligonucleotides are shown with the 5′ end at the left and the 3′ end at the right. Each unit differs only in the base, thymine (T), guanine (G), adenine (A), or cytosine (C), attached to the deoxyribose at carbon 1. The letters are used as a shorthand to indicate the order in which they are arranged in the oligonucleotide. The one shown in Figure 23.13 is usually abbreviated TGAC, although, strictly speaking, one would have to specify that these were the deoxynucleotides, containing deoxyribose instead of ribose, hence d(TGAC). This is taken for granted when the letters are used to denote the genetic code and not specified each time.

Another large step was taken in the 1950s, when Erwin Chargaff of Columbia University investigated deoxyribonucleic acids from a large number of sources, including viruses, bacteria, molds, yeast, insects, plants, and mammals. He found that nucleic acids from different organisms contain differing amounts of the purine and pyrimidine bases, but in each case the molar ratio of adenine to thymine and of guanine to cytosine in deoxyribonucleic acid is close to 1.00. He reasoned that the constancy of this relationship in such a variety of organisms could not be accidental; it suggested some association of adenine with thymine and of guanine with cytosine.

In the 1950s, other researchers were making progress in developing techniques for determining the structures of complex organic molecules by x-ray diffraction. The patterns that develop on a photographic plate when a crystal is exposed to x rays can be interpreted in terms of the locations of the atoms that diffract the x rays and thus give clues to molecular structure. Dorothy Crowfoot Hodgkins of the United Kingdom was one of the pioneers in this area of research. She received the Nobel Prize in 1964 for her work on the structure of vitamin $B_{12}$. By the 1950s, the technique was advanced enough that it was being used to investigate the structures of proteins and nucleic acids. The best x-ray crystallographic data were obtained by Rosalind Franklin at King's College, London. She worked with the crystalline sodium salt of deoxyribonucleic acid from the thymus gland of a calf.

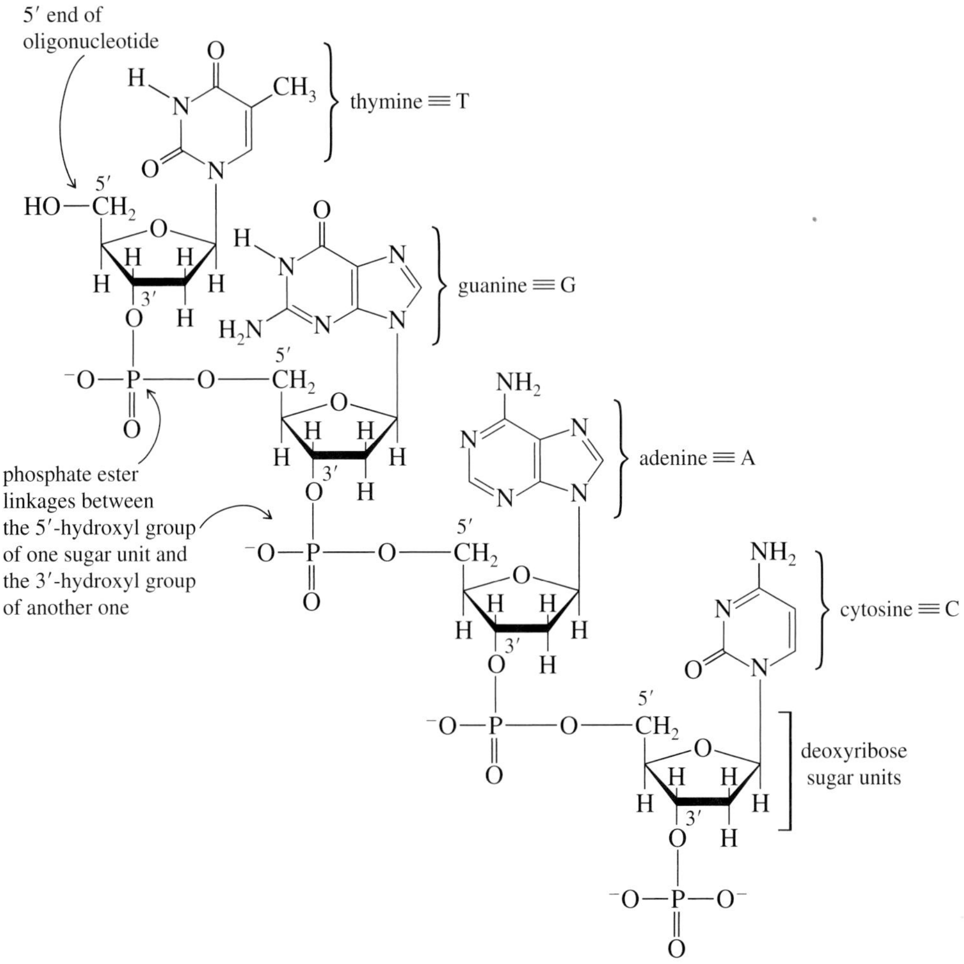

*Figure 23.13*
A fragment of DNA showing the bonding between different nucleotide units.

The diffraction patterns that are seen for this DNA are best explained if it is assumed that the molecule has a helical structure with a distance of 3.4 Å between the different nucleotide units and a diameter of 20 Å.

James Watson and Francis Crick, working in the Cavendish Laboratory at Cambridge University, in England, recognized that a DNA molecule consisting of a single helical strand having these dimensions would not have the density that had already been determined for it. Very shortly after they saw the x-ray data obtained by Franklin, they worked out the idea that in the DNA molecule two strands are twisted around each other in a double helix, arranged so that the more hydrophobic nitrogen bases are inside the helix and the hydrophilic sugar and phosphate groups are on the outside. They proposed, in a paper published in 1953 along with papers by Maurice Wilkins, also at King's College, and Franklin describing the x-ray crystallographic data, that the two strands of the double helix were held together by hydrogen bonding between the bases adenine and thymine and the bases guanine and cytosine. Thus, although there is no regularity to the order in which the bases appear on the backbone of either chain, there is a one-to-one correspondence of the number of adenine units to thymine units and of guanine units to cytosine units for the two strands taken together. Adenine and thymine are said to constitute one **base pair,** and guanine and cytosine make up the other one. The hydrogen bonding responsible for the pairing of nucleotides containing these bases is shown schematically below.

*the hydrogen bonding between nucleotides*
*that is responsible for the pairing of the bases*

The diameter of the double helix is such that a purine base on one strand must be matched with a pyrimidine base on the other. The flat rings of the bases lie parallel to each other in a stack up the axis of the helix. The sugar units to which the bases are attached and the phosphate linkages between the sugars together form a twisting ribbon on the outside of the helix. A polynucleotide has directionality. An unbonded phosphate group is present at carbon 5 of the final sugar unit at one end of the chain, and another is observed at carbon 3 of the analogous unit at the other end of the molecule. The two strands of the double helix are polynucleotide chains, headed in opposite directions. A fragment of a double helix and a view of its overall shape appear in Figure 23.14 to show some of these relationships.

Watson, Crick, and Wilkins received the Nobel Prize in 1962 for their work on the structure of DNA. The model they developed has subsequently led to a tremendous amount of research into the chemistry of polynucleotides and the ways they interact with other constituents of cells. The deoxyribonucleic acid on which the x-ray crystallographic work was done was a highly purified and crystallized sample. Even so, it was observed to have a different structure depending on the humidity of its surroundings. The picture of the molecular structure of DNA, and of the way it may function, that emerges from the x-ray data is a simplified one. It must be modified when the activity of deoxyribonucleic acid in the living cell is considered.

Basic proteins known as **histones** are found with DNA in the nucleus of cells from a calf's thymus gland. Histones are positively charged at pH 7 and are held by ionic and hydrogen bonds to the phosphate groups on the double helix, stabilizing the molecule. A polypeptide chain lies along the groove in the helix structure and is bonded at several points to the polynucleotide strand. The combined polypeptide-polynucleotide structure is known as a **nucleoprotein.** In the nuclei of the cells of a calf's thymus gland, there is also a smaller fraction of acidic proteins. These are also found associated with DNA and with histones. The proteins constitute about 30% of the dry weight of the nucleus of a cell, suggesting that they have an important function there.

DNA has primary, secondary, and tertiary structure. The primary structure of DNA is the order of the nucleotides in the backbone of a single strand. The synthesis of a polynucleotide involves the orderly placement of one nucleotide unit after

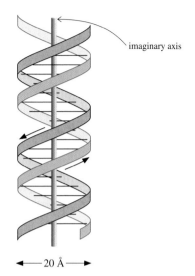

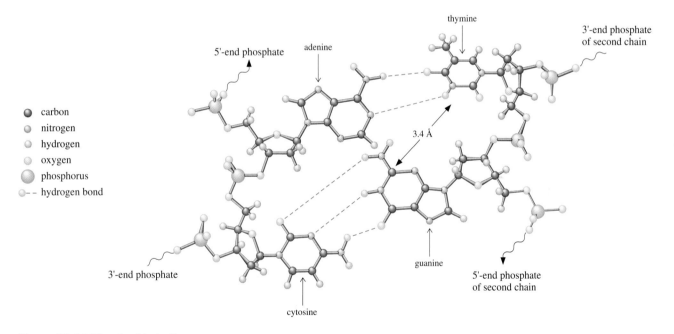

*Figure 23.14*  The double helix.

another in a chain held together by phosphate ester linkages. In the body, this
process is catalyzed by an enzyme, DNA polymerase.

deoxyribonucleotide triphosphates $+$ DNA template

$\downarrow$ DNA polymerase
$Mg^{2+}$

new DNA $+$ pyrophosphate

The secondary structure of DNA is the right-handed helical shape that characterizes a
strand of DNA, and the double helix that forms when the complementary strand is
synthesized. In Figure 23.14 the double helix is shown with its central axis as a
straight line. It is seldom so. The double helix itself curves and twists in a variety of

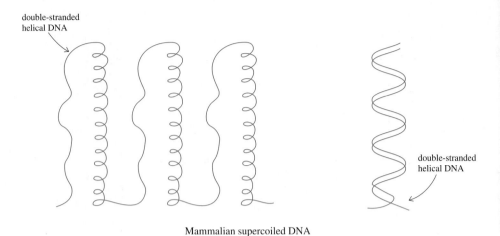

*Figure 23.15*
Supercoiling of a DNA molecule.

Mammalian supercoiled DNA

ways that are termed *supercoiling,* which constitutes the tertiary structure of DNA. Some forms of supercoiling are shown in Figure 23.15. DNA maintains the tertiary structure that is responsible for its biological activity only within a narrow range of pH, temperature, and solvent composition. When conditions are changed too drastically, the DNA is denatured (p. 1010). The most important cause of denaturation is the disruption of the ionic and hydrogen bonds that give the double helix its stability. For example, an increase in pH would result in the loss of protons from the basic proteins (p. 1018) associated with the phosphate groups on the double helix. The ionic bonds that hold them together would be disrupted, and the excess of negative charge on the helix would destabilize it.

**Problem 23.29**    The points in a DNA helix at which adenine is paired with thymine are the first to pull apart when DNA is denatured. In fact, the stability of a particular form of DNA, as indicated by the temperature to which it can be heated before it melts, is directly related to the relative amount of guanine (and cytosine) that it contains. How do you explain this experimental observation? (*Hint:* Another look at the diagram of the pairing of bases on page 1018 may be helpful.)

## B. The Genetic Code

Even before a structure had been assigned to DNA, experimental evidence had indicated that DNA is involved in the storage and transmission of genetic information. Watson and Crick recognized that the pairing of the bases on one strand of the double helix with those on the other strand provided a mechanism by which genetic information could be transmitted. Either strand of a double helix could in principle, if separated from its partner, direct the synthesis of a new strand containing exactly the same purine and pyrimidine bases as were in the missing strand, as shown in Figure 23.16.

Watson and Crick proposed as the "central dogma" of the new science of molecular genetics that the bases are like the letters of an alphabet, that genes are in essence molecules of DNA, and that the order of the bases on the backbone of the polynucleotide chain constitutes a **genetic code.** The code is reproduced when DNA fosters the synthesis of molecules of ribonucleic acid. The order of bases in RNA chains is determined when these bases pair by means of hydrogen bonding with those on a strand of DNA. Ribonucleic acids, in turn, direct the synthesis of proteins.

For a DNA molecule to serve as a template for the synthesis either of other DNA molecules or of RNA molecules, which then direct protein synthesis, the dou-

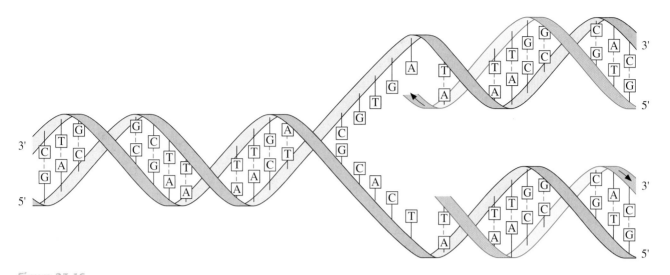

*Figure 23.16*
A schematic representation of the replication of DNA, the conversion of one double helix into two identical double helixes.

ble helix must unwind to some extent. In other words, a disruption of the most stable arrangement of the giant molecule must take place. The histones associated with DNA in multicelled organisms play an important part in the regulation of the process. They must somehow be detached to expose part of the polynucleotide before it can serve as a template. Exactly how this happens and what starts the process in living organisms are not yet understood.

The words that scientists use to describe these processes are interesting. The genes are said to "express" themselves. The process by which genetic information is transferred from DNA to RNA is called "transcription." When the information from RNA is encoded in the structure of a protein, the process is called "translation." All these words are related to language and the relationships between different kinds of languages.

In Section 22.6A (p. 947) we discussed how synthetic purine and pyrimidine bases have been designed to disrupt DNA and RNA synthesis in cancer cells to serve as chemotherapeutic agents. Any reaction that interferes with the correct pairing of bases by hydrogen bonding changes the way in which the genetic code is transmitted. An incorrect transmission of genetic information by DNA may lead to a mutation in the species. Mutations take place all the time in nature, but some compounds, known as mutagens, greatly increase the rate at which mutations occur. Many mutagens are also carcinogens, compounds that increase the probability that a malignant tumor will develop in a living organism.

Another way to interfere with the reproduction of an organism is to prevent the synthesis of its DNA altogether. Much research is now being focused on compounds that prevent the human immunodeficiency virus (HIV), which is believed to be responsible for the development of the acquired immune deficiency syndrome (AIDS), from reproducing itself. Many of the compounds that show promise in clinical trials in slowing the progress of the disease are nucleoside derivatives that lack a hydroxy group at the 3' position of the deoxyribose ring.

The compounds shown above are similar enough in structure to the normal components of DNA that the enzymes that catalyze DNA synthesis incorporate them into the growing DNA chain of the HIV virus through the hydroxy group at

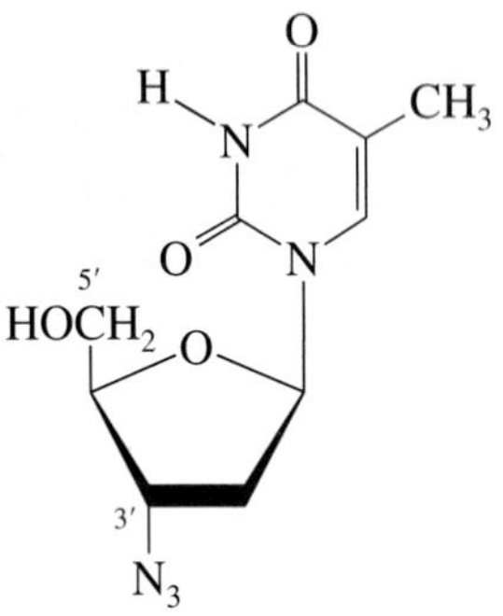

3′-azido-2′-deoxythymidine
AZT

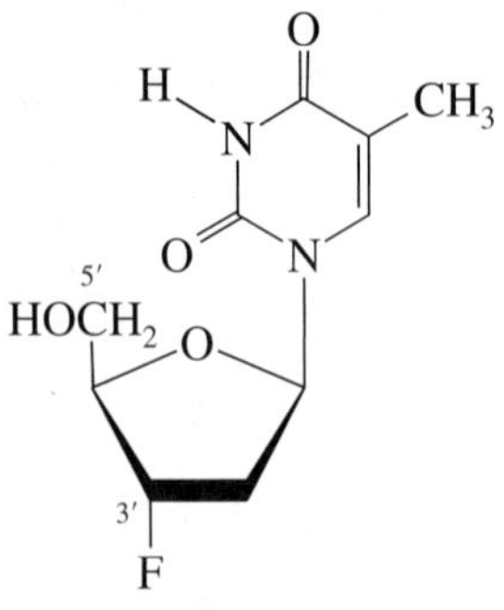

3′-fluoro-2′-deoxythymidine
FLT

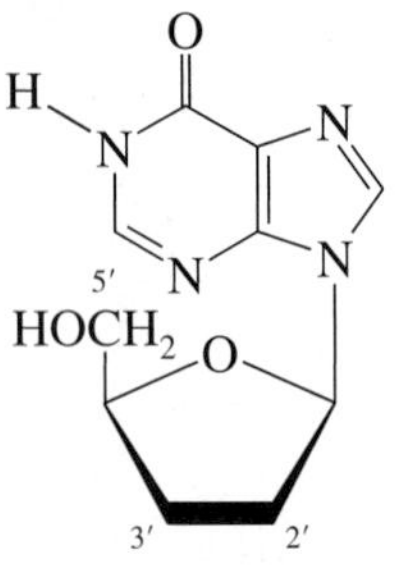

2′,3′-dideoxyinosine
ddI

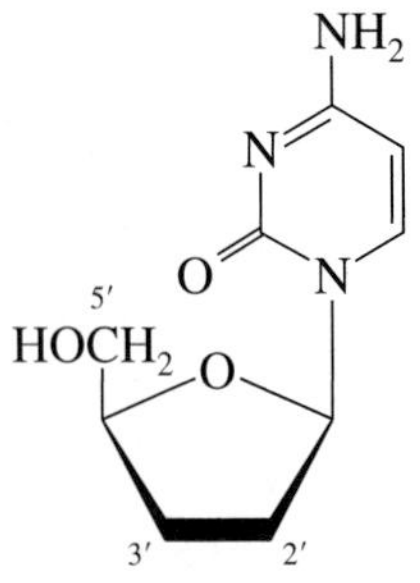

2′,3′-dideoxycytidine
ddC

the 5′ position. For the chain to continue, however, a hydroxy group at the 3′ position would have to form a phosphate ester with the hydroxy group at the 5′ position of another nucleoside unit. Because no hydroxy groups are found at the 3′ position of the compounds shown above, DNA synthesis stops once one of them has been incorporated into the growing chain. This stops the reproduction of that particular virus and slows the development of the disease overall.

**Problem 23.30**     Some of the most potent mutagens are three-membered heterocycles, such as oxiranes or aziridines (three-membered rings containing nitrogen). Write an equation suggesting why such compounds are so destructive to DNA.

**Problem 23.31**     Nitrites (used as food preservatives), which give rise to nitrous acid in the presence of strong acids such as hydrochloric acid (found in the stomach), are believed to be harmful in several ways. On page 822 we saw how nitrous acid gives rise to potent carcinogens, the *N*-nitrosoamines. Another way in which nitrous acid can disrupt cells is by changing the genetic code. Write an equation predicting what would happen if cytidine reacts with nitrous acid. (A review of page 987 may be helpful.)

## C. Ribonucleic Acids

Ribonucleic acid molecules are generally smaller than deoxyribonucleic acid molecules. Their molecular weights range from 30,000 to 2,500,000. A ribonucleic acid does not form a double helix because the hydroxyl group present at carbon 2 of the sugar unit ribose (unlike the hydrogen atom occupying that position in 2-deoxyribose) is too bulky to allow two strands of RNA to come close enough to interact in that way. The RNA molecules exist as single strands that fold and loop with internal base pairing taking place at some points. In RNA, adenosine hydrogen-bonds with uracil, and guanine hydrogen-bonds with cytosine. Because RNA does not have a double helix in which every base on one strand is paired with a corresponding base on the opposite strand, analyses of samples of RNA do not show the regularity in molar ratios of purine bases to pyrimidine bases that is characteristic of DNA. This is an important and striking experimental difference between the two types of polynucleotides.

Experimentally, it has been found that three adjacent bases on one kind of ribonucleic acid, known as **messenger RNA,** are the code for a given amino acid. The first such three-nucleotide code was identified when it was found that polyuridylic acid directed the synthesis of the peptide polyphenylalanine. Thus three uracil bases in a row constitute the code for the amino acid phenylalanine. An amino acid is often coded for by more than one sequence of three nucleotides.

There are also portions of RNA molecules that do not code for anything, as far as is known. Other three-nucleotide sequences serve as punctuation marks, such as a signal to stop the synthesis of a polypeptide. Scientists also have found that some nucleotides are part of a code for two different amino acids. The code sequences for them overlap. The amino acids to be incorporated in a protein chain are transported to the messenger RNA by another type of RNA known as **transfer RNA.** Each amino acid has at least one transfer RNA that is specific for it.

## D.  Synthesis of Oligonucleotides

The synthesis of oligonucleotides presents many of the same challenges that the synthesis of a peptide (p. 629) does. Each nucleotide unit has multiple reactive sites, hydroxyl and amino groups. The nucleotides have to be strung together in a precise order if the compound that results is going to be the carrier of genetic information. Phosphate groups that are not very reactive at one stage of the synthesis have to be activated at another stage in order to create the phosphate ester linkage between two nucleotide units. These phosphate bonds must be between the 5'-hydroxyl group of one nucleotide unit and the 3'-hydroxyl unit of another one.

In a classic experiment in 1968, Hai Gobind Khorana, who shared the Nobel Prize for Medicine that year, demonstrated that oligonucleotides could indeed be synthesized. Just as it was for peptides, syntheses of large oligonucleotides required solid-state syntheses and automation of the steps. Khorana quickly extended his work to synthesis on a solid support. We will look at such a synthesis of a dinucleotide as a reminder of ideas about protecting groups and activating reagents first encountered in Section 15.8.

Khorana started his synthesis with the same resin that Merrifield used for peptide synthesis (p. 901). He modified it by a series of standard organic reactions to create a triphenylmethyl chloride (trityl chloride) substituted on one of the phenyl rings with a *p*-methoxy group (methoxytrityl chloride) attached to the polymer support.

cross-linked
polystyrene resin

resin with methoxytrityl
chloride group on it

**Problem 23.32**    How would you convert the polystyrene resin to a resin carrying the methoxytrityl chloride group?

The hydroxyl group of the 5' end of the oligonucleotide to be synthesized is attached to the polymer support as a trityl ether protecting group. The central carbon atom of the trityl group is a benzylic carbon atom three times over and, therefore, comes off easily as a carbocation. Other protecting groups, those on the 3'-hydroxyl group of nucleotides and those used to protect amino groups on purine and pyrimidine bases, are chosen to be removed by basic reagents such as hydroxide ion, methoxide ion, or ammonia.

The trityl chloride function on the polymer reacts with the 5'-hydroxyl group of thymidine in the presence of base to give the trityl ether.

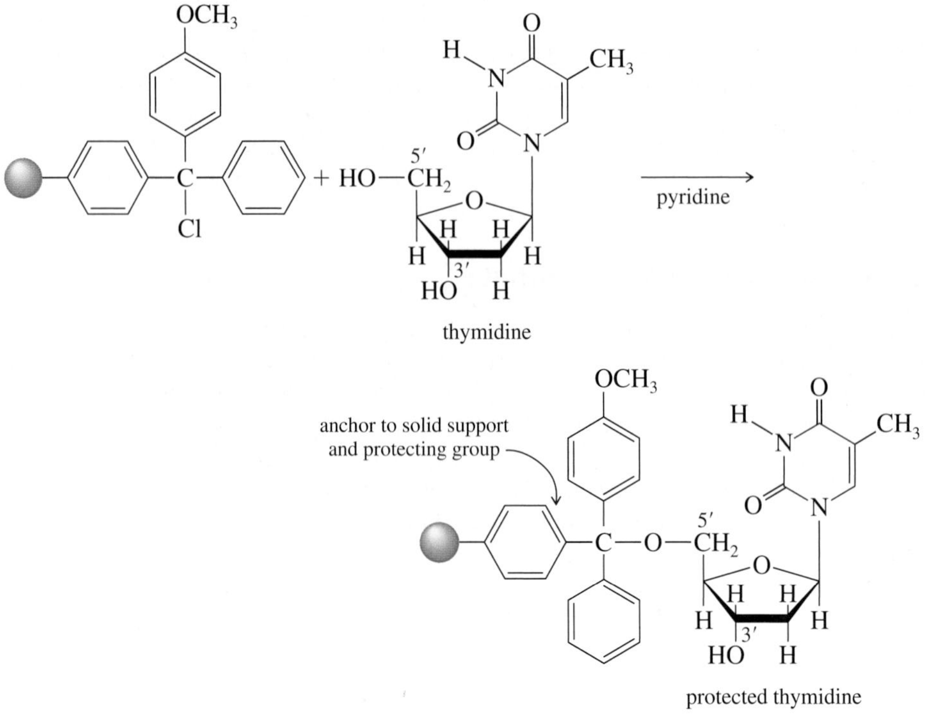

thymidine

anchor to solid support
and protecting group

protected thymidine

The next nucleotide, 5'-deoxycytidylic acid, is protected at both the 3'-hydroxyl group and the amino group with acyl groups. It is added to the polymer-supported thymidine in the presence of mesitylsulfonyl chloride, which activates the phosphoric acid group at the 5'-carbon atom of cytidylic acid. Dicyclohexylcarbodiimide, used to active carboxylic acid groups in peptide syntheses (p. 634), also has been used in forming phosphate ester bonds. In either case, the hydroxyl group on phosphorus is converted into a good leaving group. The resulting dinucleotide is removed from the resin by acid and deprotected at the 3'-hydroxyl group and the amino group by base.

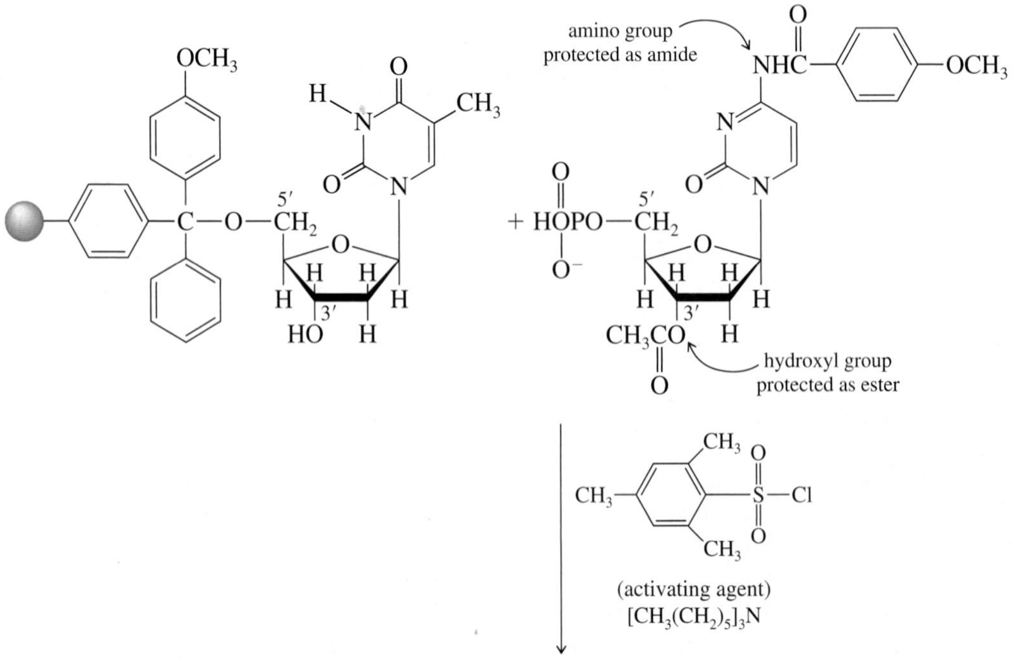

amino group
protected as amide

hydroxyl group
protected as ester

(activating agent)
[CH$_3$(CH$_2$)$_5$]$_3$N

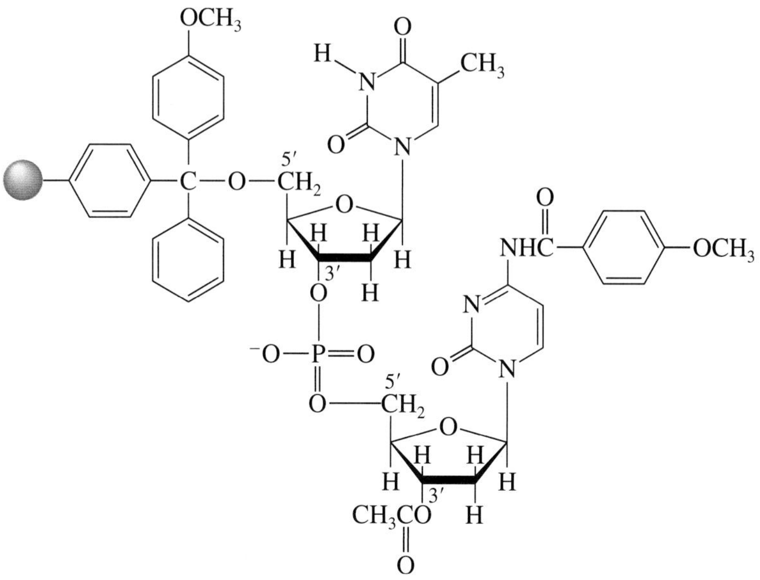

protected dinucleotide on solid support

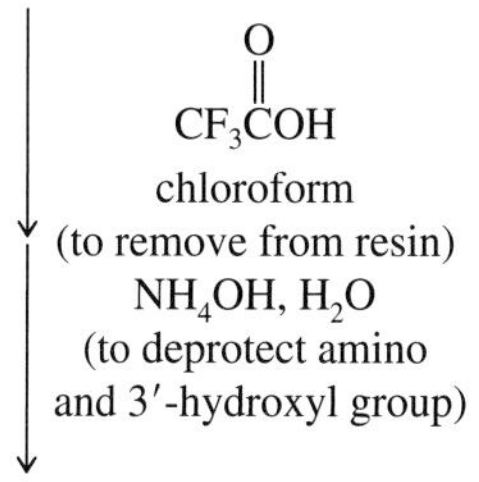

$$CF_3COH$$ (O double bond shown)

chloroform
(to remove from resin)
$NH_4OH$, $H_2O$
(to deprotect amino
and 3'-hydroxyl group)

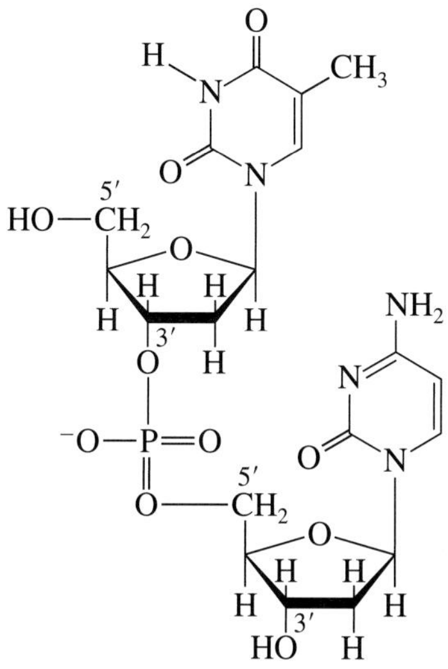

thymidylyl-$(3' \rightarrow 5')$-deoxycytidine
91%

If sodium methoxide in methanol is used as the base in the deprotection step, the ester group is rapidly transesterified to liberate the 3'-hydroxyl group, while the amino group of deoxycytidine remains protected as the amide. In this way it is possible to extend the synthesis by adding successive nucleotides, forming phosphate ester linkages between their phosphate groups and the newly exposed 3'-hydroxyl groups on the chain growing on the solid support. When the oligonucleotide is complete, the chain is removed from the solid by adding acid. These processes have now been extended and refined with different protecting groups for different nucleotides and different support systems that can be used in automated oligonucleotide synthesizers.

In 2002, DNA that codes for the RNA used in replication by the poliovirus was synthesized from its 7741 individual nucleotides according to information from the genome of the virus. When put in the presence of the enzyme RNA polymerase, the virus multiplied. Mice injected with the synthetic virus became infected in the same way as with wild-type virus. The synthetic virus was not as virulent, probably because the researchers had deliberately introduced different oligonucleotides at nineteen sites to make sure that they could distinguish the synthetic from the natural virus.

Viruses fall in an interesting area. They are discrete chemical substances, but they have the ability to invade cells and to use the reproductive systems of those cells to multiply. One can argue whether they are "alive" or not. In any event, the researchers who did this work, Eckard Wimmer and Aniko V. Paul and their colleagues, point to the synthesis of urea by Wöhler (p. 1) as the end of the belief that living organisms made compounds that could not be created in the laboratory. The synthesis of DNA that has the replicating and infectious abilities of the poliovirus in a similar way establishes that the poliovirus is a chemical with the molecular formula $C_{332,652}H_{492,338}N_{98,245}O_{131,196}P_{7,501}S_{2,340}$ that has a life cycle, given the proper environment of enzymes or mice tissues. The medical, social, and ethical implications of our ability to synthesize infectious viruses from chemicals off the shelf, with all the possibilities to either decrease or increase their virulence by changing the genetic code, are subjects of debate and concern.

## E. The Human Genome Project

In February 2001, Francis Collins, representing a publicly funded international group of sixteen major laboratories, and Craig Venter, from Celera Genomics, jointly announced the publication of an almost complete nucleotide sequence map of human DNA. Major efforts to create such a map were started in the mid-1980s in what is known as the Human Genome Project when the technology for determining the order in which nucleotides are bonded together in oligonucleotides had advanced enough to make feasible such an attempt. DNA contains about 3 billion bases and 30,000 to 40,000 genes that code for proteins, far fewer than the approximately 100,000 that had been expected.

In the fifteen years or so between the launching of the Human Genome Project and publication of the maps, major advances in techniques occurred so that it has become easier and cheaper to sequence genomes. In that interval, the genomes of a number of other organisms, including that of a bacterium, *H. influenzae*; a yeast, *S. cerevisiae*; a roundworm, *C. elegans*; a small weed, *A. thaliana*; the fruit fly, *D. melanogaster*; the mouse, *M. musculus*; and the rat, *R. norvegicus*, also have been completed. Many interesting comparisons have already been discovered, including the fact that the sizes of coding regions of genes of different organisms are roughly the same. The human being, with its large and complex body and brain, has approximately twice the number of genes as the roundworm, which has a total of 959 cells, of which about 300 are nerve cells. It appears that human beings do not have more genes but that they use the genes that they do have more creatively.

The human genome shares 223 genes with bacteria. These genes do not exist in the worm or the fly or yeast. Individual human beings differ from each other in their genes by only about 0.1%. We also share most of our genes with the chimpanzee. As the data are further examined, many other such fascinating pieces of information are sure to be forthcoming, raising deep questions about the source of our humanity and our individuality.

While the technical details of how DNA is sequenced are beyond the scope of this book, the general principles behind the work are familiar to us from the kinds of problems that had to be solved in the determination of the structure of a protein (Section

23.6, p. 993). Determining the structure of a protein required that we be able to identify which amino acids are present and the order in which they are bonded to each other. Similarly, to determine the structure of DNA, we have to know which bases are present in the nucleotides and the order in which they are bonded to each other. Protein structure determination involves a combination of chemical labeling and degradation experiments, such as the use of 2,4-dinitrofluorobenzene to determine the *N*-terminal amino acid (p. 996), hydrolysis to peptides and amino acids (p. 994), and sequential Edman's degradation (p. 997). Different enzymes are used to cleave proteins and peptides into smaller fragments at different points in the chain (p. 1000). The total structure is deduced from the information obtained from overlapping sequences of these fragments.

Similar techniques have been developed for the sequencing of oligonucleotides. An example of a complete structure determination is that of the transfer RNA for the amino acid alanine. This relatively small oligonucleotide containing 77 nucleotides was sequenced in the 1960s in the laboratory of Robert W. Holley, who shared the 1968 Nobel Prize in Medicine with Khorana and Marshall Nirenberg for their work on the relationship between the genetic code and protein synthesis. The structure determination took seven years to complete. Enzymes were used to cleave the RNA into smaller fragments, which were separated from each other and further degraded. For example, one enzyme, pancreatic ribonuclease, cleaves the chain to the right of pyrimidine nucleotides. Another enzyme, takadiastase ribonuclease T1, cleaves the chain to the right of purine nucleotides. An enzyme isolated from snake venom, snake venom phosphodiesterase, cleaves off the 3′-terminal nucleotide in sequence, giving rise to a mixture of oligonucleotides that differ from each other by one nucleotide. All these pieces give differing sets of overlapping sequences that can be separated from each other by various chromatographic methods, further analyzed, and then put back together on paper to recreate the sequence of the original oligonucleotide.

The enzymes that cleave DNA are called **endonucleases,** and there are a number of them with different specificities. Chemists also have taken advantage of the different structures and chemical properties of the purine and pyrimidine bases to develop methods to label the individual nucleotides or to cleave the chain at specific bases, to remove every guanine in the chain, for example. It is interesting that even small modifications in the structure of a purine or pyrimidine base destabilize the glycosidic linkage between the base and the sugar and result in easy removal of that base.

The next large area of research that is being opened up by the information now available from the map of the human genome has been called **proteomics,** the task of finding out which genes code for which proteins and what the functions of all those proteins are. For example, each human gene seems to code for about three proteins. How this is controlled is of interest.

Having a map of the human genome also has raised the possibility of altering it. Techniques by which specific parts of genes are changed, one base substituted for another, for example, called **site-specific mutagenesis,** are well developed. Genetic engineering now transfers genetic material from one species to another. Organisms such as corn that has been altered to contain its own pesticides, soybeans that resist herbicides, mice that have certain genes altered or removed so that they better model specific disease processes, and yeast that has been altered to make human insulin are all realities now. A single cell of an adult can now be used to clone the animal to create a biological but younger twin. Such scientific results raise social, ethical, and philosophical questions. Scientists who have worked on the Human Genome Project have been conscious since the beginning of the ethical implications of their work. The public funding for the project has always included money for the study of such issues and the raising of public awareness of them. This kind of science often leads into areas where judgments other than purely scientific ones have to be made.

**Problem 23.33**    Reactions that modify a guanosine residue so that it can be removed from the oligonucleotide strand are shown below. Write a mechanism using the curved-arrow convention for each step.

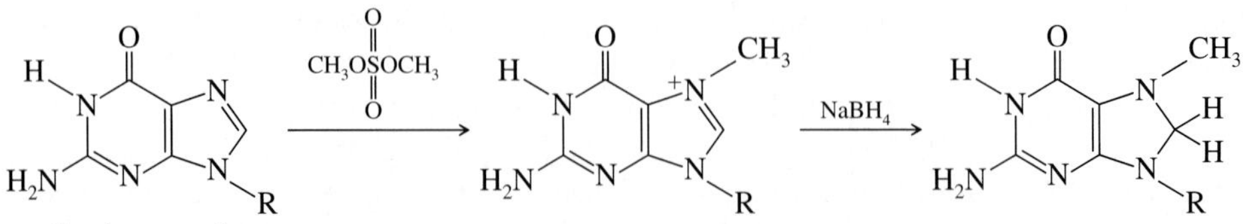

R = the sugar unit

**Problem 23.34**    Once the guanosine unit on the chain is modified as shown in Problem 23.33, it is easily hydrolyzed off the sugar residue, which leaves the anomeric carbon free to react as a carbonyl group with a primary amine such as aniline to give a Schiff base (p. 574). This Schiff base undergoes an E2 reaction to cleave the phosphate ester link between two sugar units. This set of reactions is outlined below. Provide mechanisms for the reactions. You may use HB$^+$ and B : as necessary for protonation and deprotonation reactions. You also may use abbreviated structures for your mechanisms, focusing only on the part of the molecule undergoing reaction.

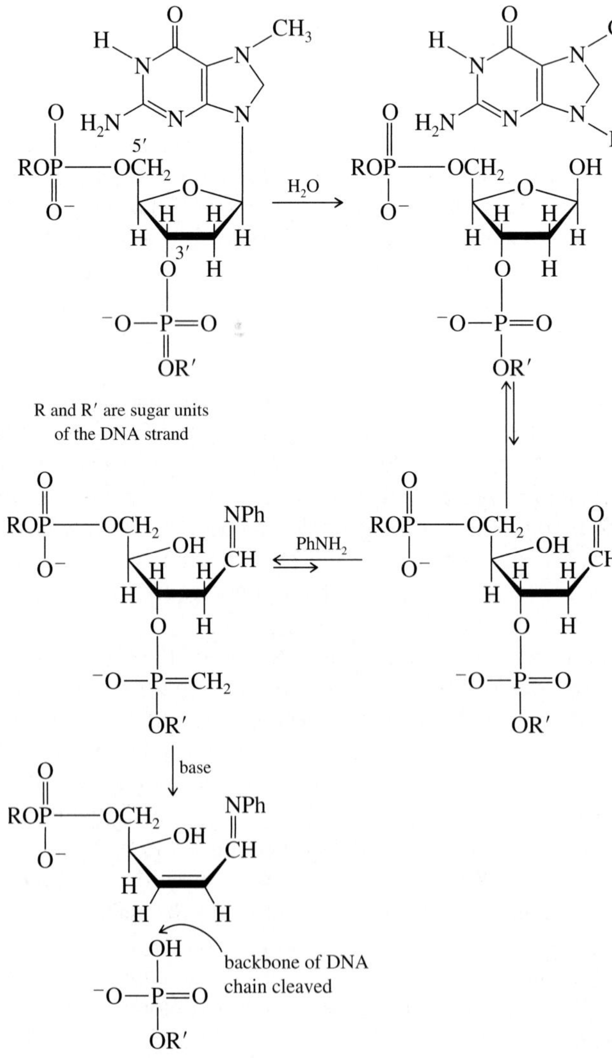

R and R′ are sugar units
of the DNA strand

# ADDITIONAL PROBLEMS

**23.35** Write structures for the following compounds, showing the stereochemistry.

(a) *N*-(α-D-glucopyranosyl)methylamine
(b) methyl 2,3,4,6-tetra-*O*-methyl-β-D-mannopyranoside
(c) D-mannitol        (d) α-D-fructofuranose-6-phosphoric acid
(e) 2-amino-2-deoxy-D-galactose
(f) *O*-α-D-galactopyranosyl- (1 → 6) -*O*-α-D-glucopyranosyl-(1 → 2)-β-D-fructofuranoside
(g) L-*ribo*-D-*manno*-nonose

**23.36** Write full structures for the following compounds.

(a) glycylalanyllysine        (b) ethyl alaninate        (c) *N*-benzyloxycarbonylmethionine
(d) *N*-*tert*-butoxycarbonylvaline        (e) *N*-2,4-dinitrophenylleucine
(f) *p*-nitrophenyl tyrosinate

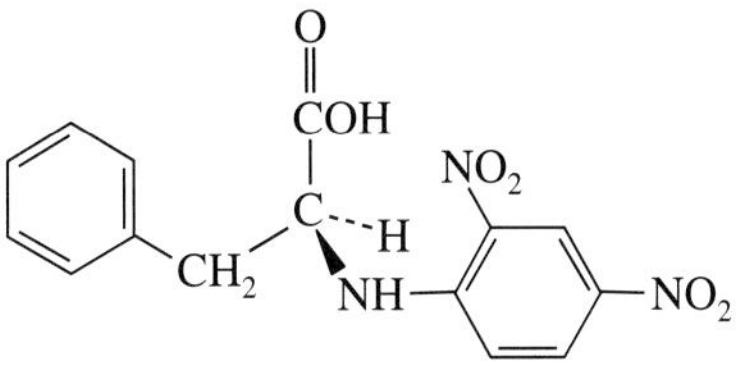

**23.37** Name the following compounds.

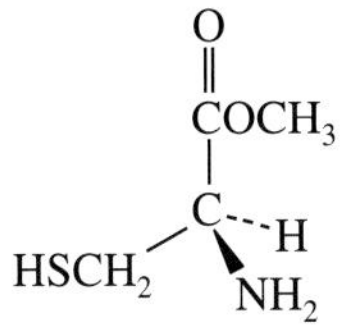

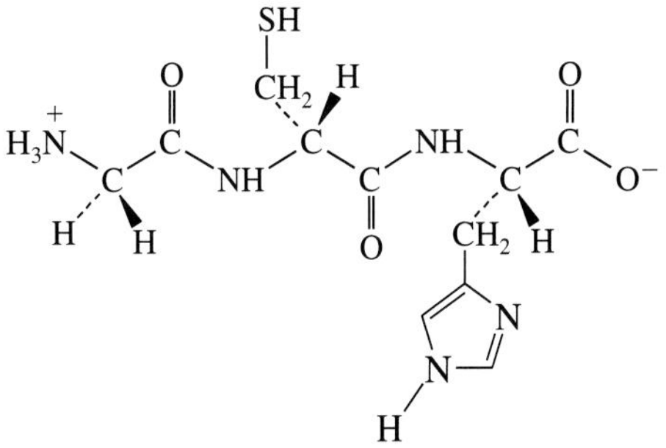

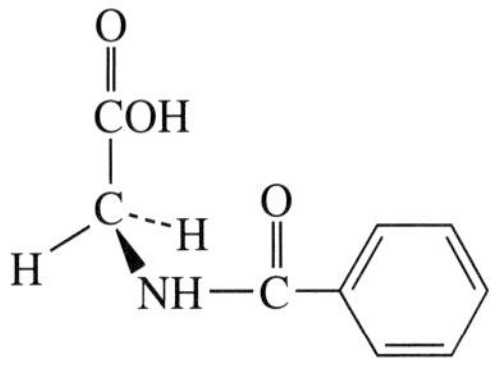

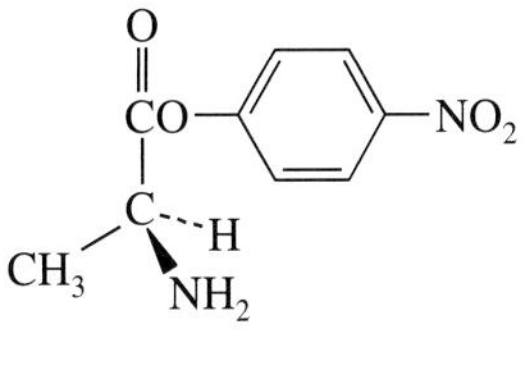

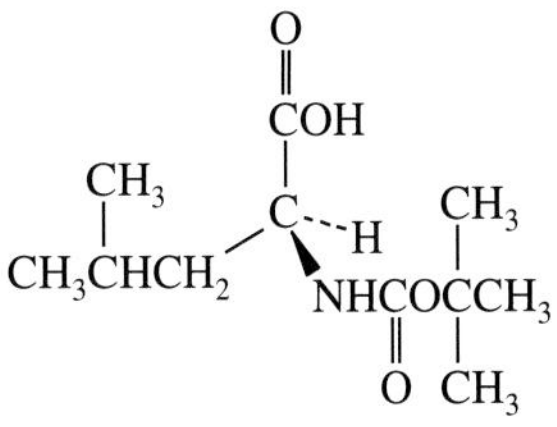

**23.38** Write structural formulas for all compounds indicated by letters. Show stereochemistry whenever it is known.

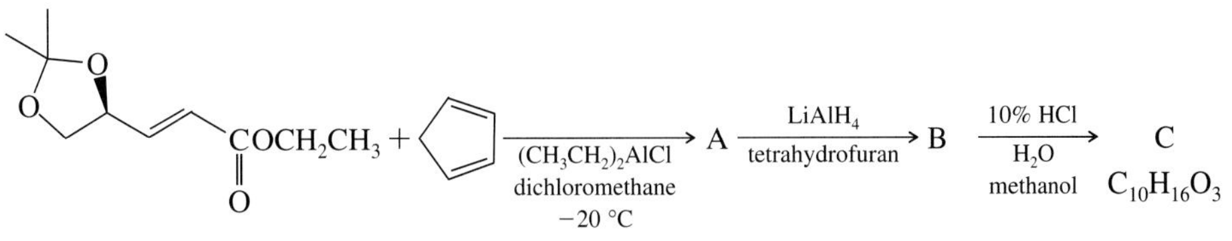

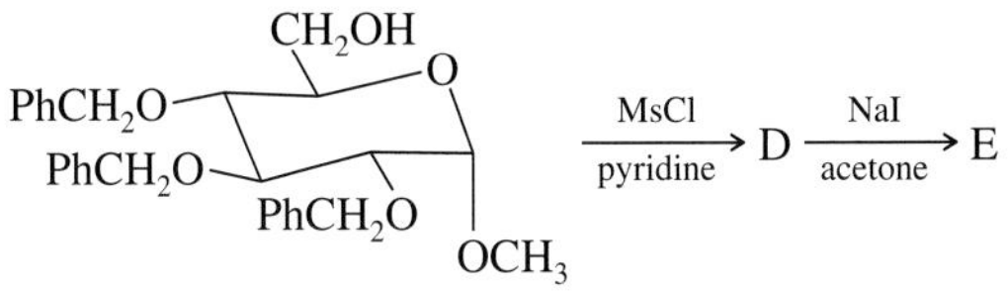

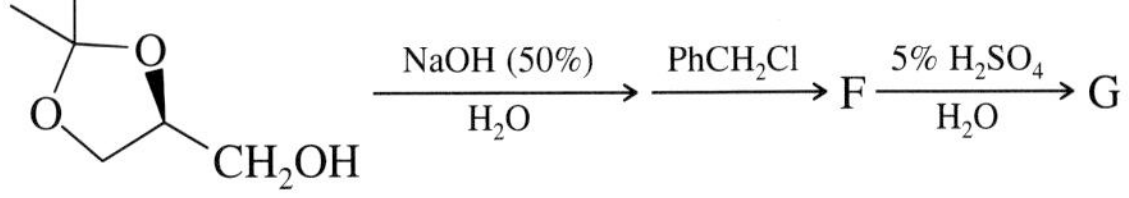

(d) 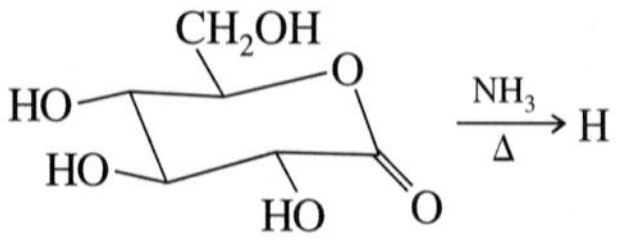 $\xrightarrow[\Delta]{NH_3}$ H

(e) 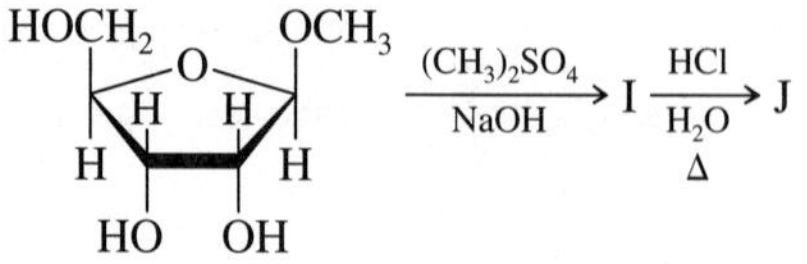 $\xrightarrow[NaOH]{(CH_3)_2SO_4}$ I $\xrightarrow[\substack{H_2O \\ \Delta}]{HCl}$ J

(f) 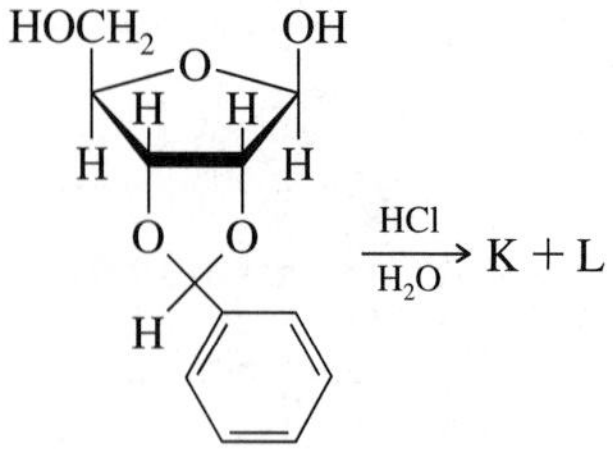 $\xrightarrow[H_2O]{HCl}$ K + L

**23.39** Reagents with which you are familiar in other contexts are used in carbohydrate chemistry. The following reactions give you a chance to practice recognizing some reagents in unfamiliar contexts.

(a) 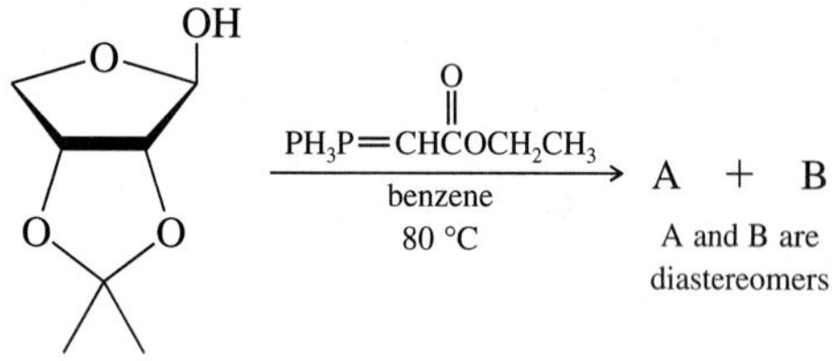 $\xrightarrow[\substack{benzene \\ 80\ °C}]{PH_3P=CHCOCH_2CH_3}$ A + B

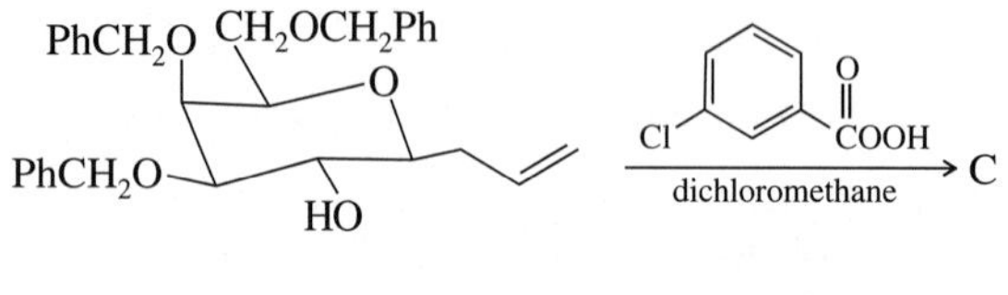

A and B are diastereomers

(b) $\xrightarrow[dichloromethane]{}$ C

(c) $\xrightarrow[\substack{tetrahydrofuran \\ -10\ °C}]{}$ D + E

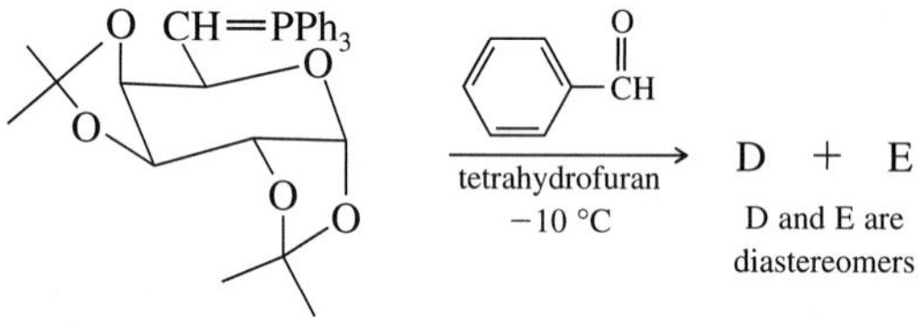

D and E are diastereomers

(d) $\xrightarrow[ethanol]{NaBH_4}$ F $\xrightarrow[\substack{Pd/C \\ ethanol}]{H_2}$ G $\xrightarrow[pyridine]{TsCl}$ H $\xrightarrow{H_3O^+}$ I

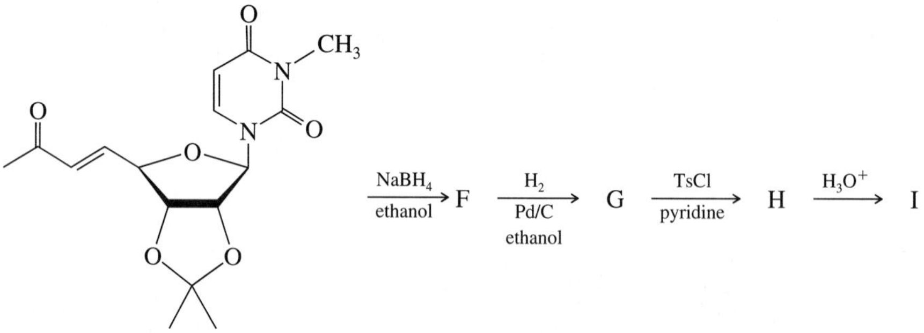

**23.40** Write equations showing the reactions you predict would occur if the amino acid valine were treated with each of the following reagents.

(a) 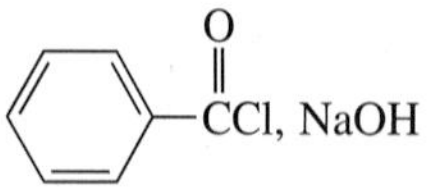 —CCl, NaOH

(b) $CH_3OH$, HCl

(c)

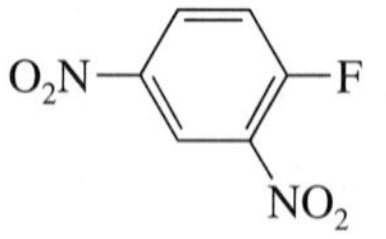

(d) 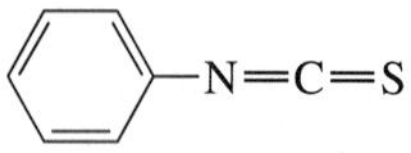 —N=C=S

(e) product of (d), acid, Δ

(f)

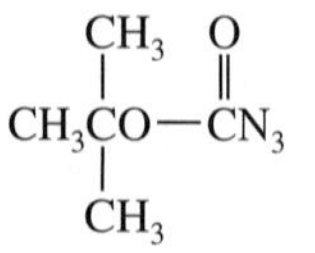

(g) product of (f) +

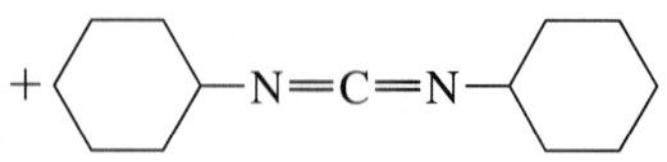

(h) product of (g) +

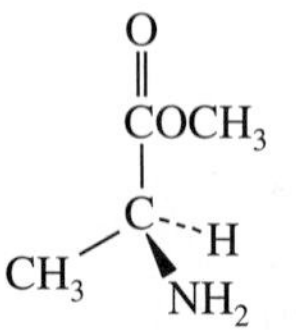

(i)

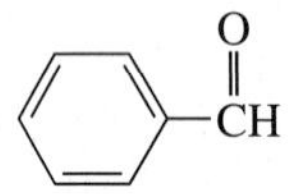

**23.41** Complete the following equations.

(a)

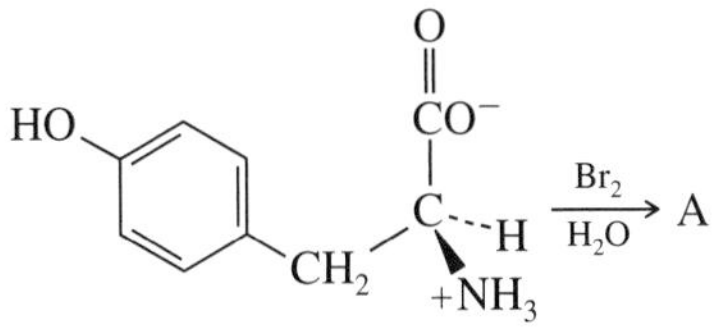

(b)

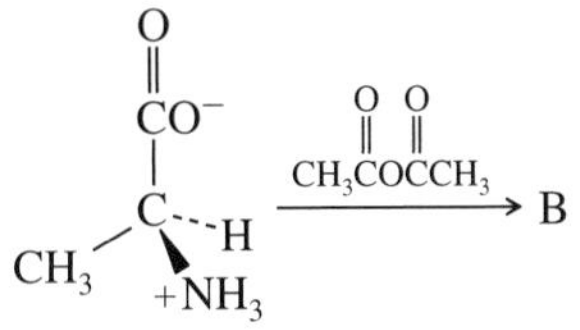

(c)

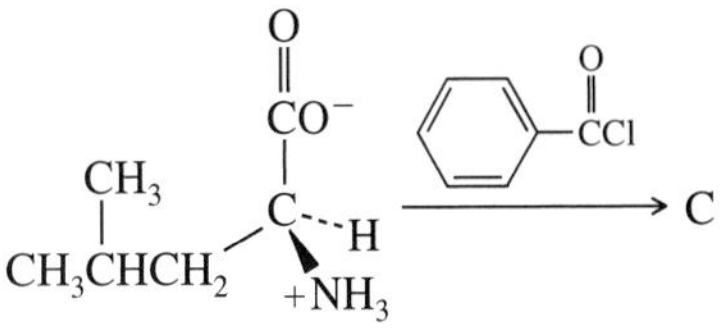

(d)

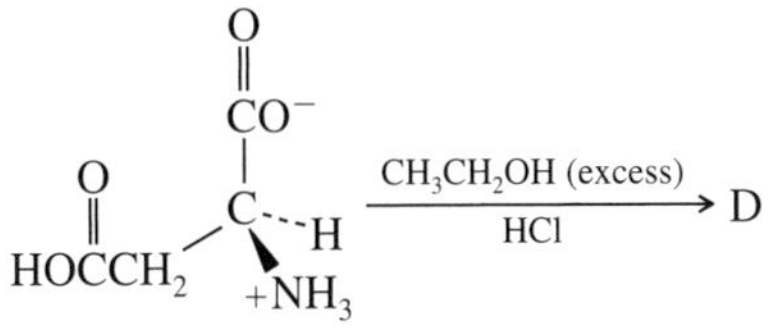

(e)

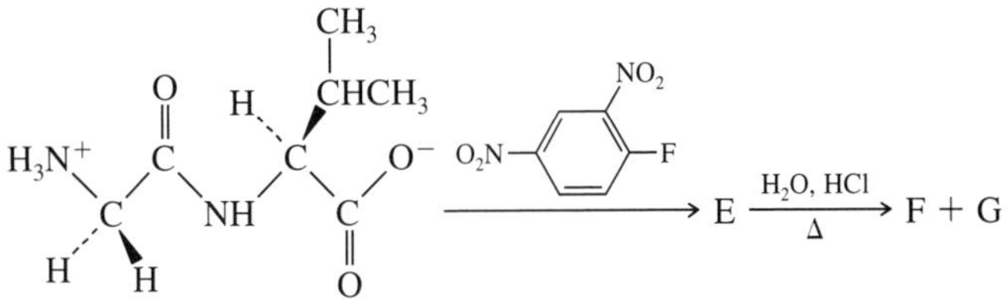

**23.42** More practice in recognizing reactions follows.

(a)

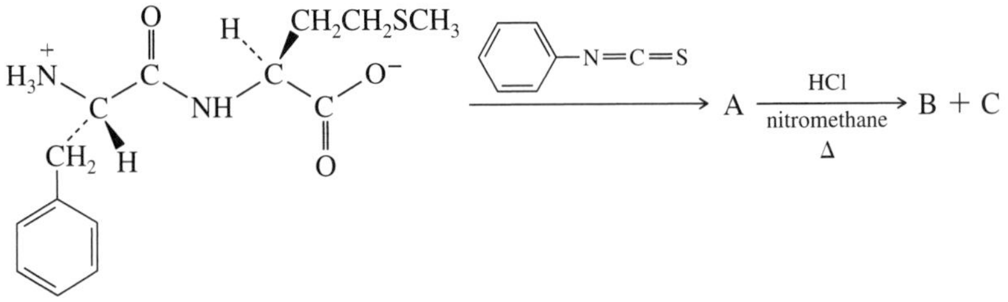

(b)

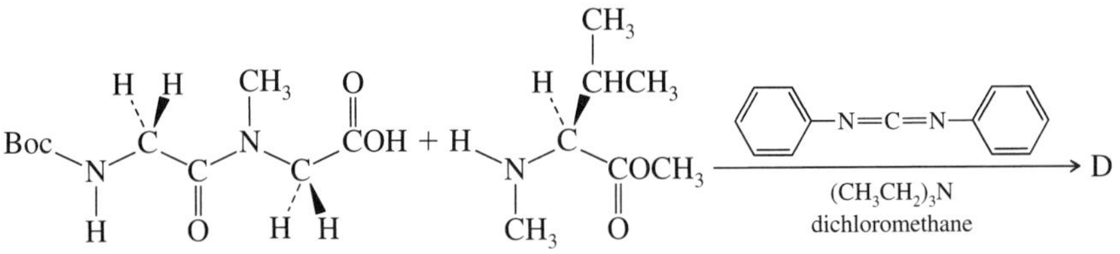

(c)

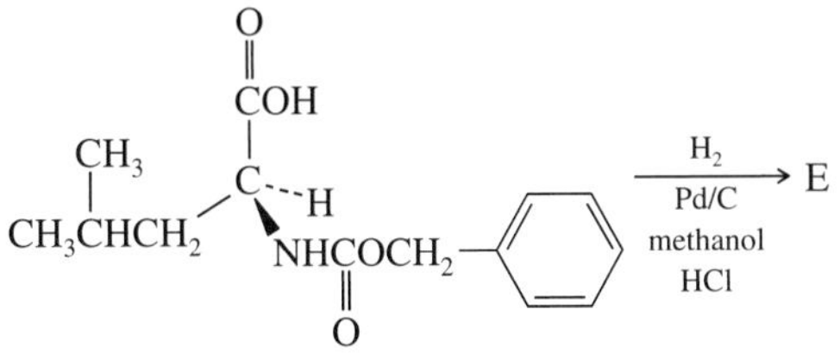

(d)

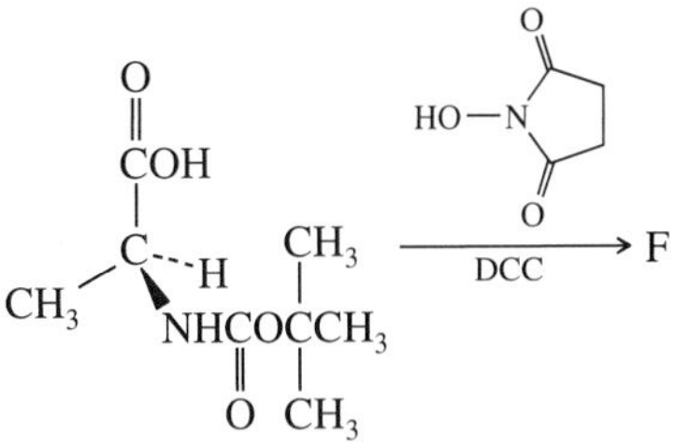

(e)

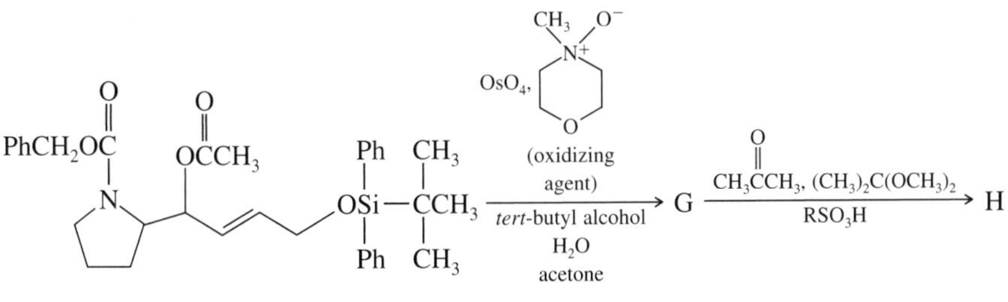

**23.43** Mycosamine is an amino sugar that is a component of many antifungal antibiotics. Its cyclic structure is shown on the next page. Draw a Fischer projection of the open-chain form of the compound.

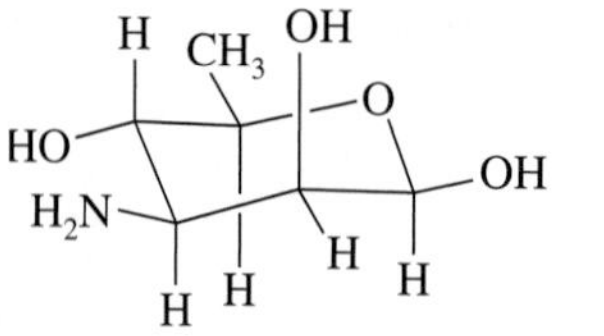

β-mycosamine

**23.44** An antiviral drug, marketed as Vira-A, has the scientific name 9-β-D-arabinofura-nosyladenine. What is the structure of Vira-A? (The structure of adenine is found on page 946.)

**23.45** The β-glucoside of glucose with the potassium salt of gentisic acid is believed to be the compound that controls the leaf-closing response when the plant *Mimosa pudica L.* is touched. The response is triggered by the hydrolysis of the glucoside, shown below, by β-glucosidase. What are the products of this reaction?

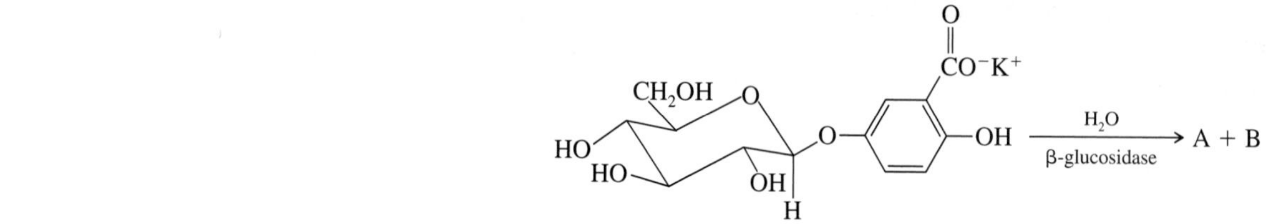

β-glucoside of the
potassium salt of
gentisic acid

**23.46** Sulfur-containing compounds in garlic have been shown to inhibit the development of tumors when given along with carcinogens. Research into similar compounds that inhibit the action of *N*-nitrosodiethylamine on DNA involved the following reaction. Supply structural formulas for the compounds indicated by letters.

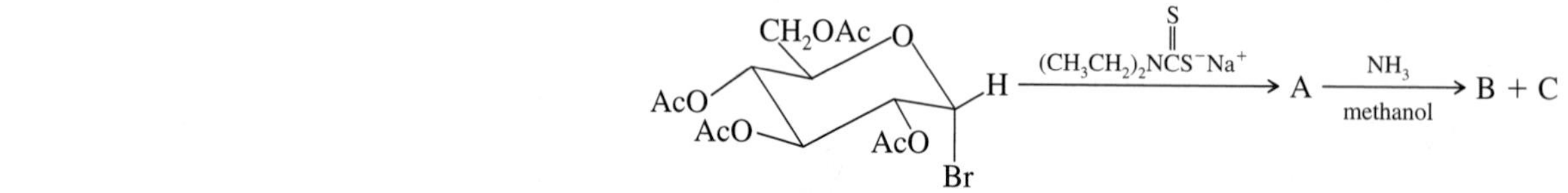

**23.47** A glycoside called gaultherin is isolated from oil of wintergreen. Gaultherin is not a reducing sugar. It can be hydrolyzed by the enzyme primeverosidase into the disaccharide primeverose and the aglycon methyl salicylate. When hydrolyzed with dilute acid, primeverose, a reducing sugar, gives glucose and xylose.

(a) Xylose (p. 980) is present in primeverose in the pyranose form. Draw the structure of β-D-xylopyranose.
(b) Primeverose is reduced by sodium borohydride. The reduction product is hydrolyzed by dilute acid to xylose and sorbitol (the reduction product of glucose). What does this prove about the structure of primeverose?
(c) Primeverose is synthesized from the reaction of α-D-xylopyranosyl bromide 2,3,4-triac-etate with α-D-glucopyranosyl 1,2,3,4-tetraacetate in the presence of pyridine to give primeverose heptaacetate. Primeverose heptaacetate is then converted to primeverose by treatment with sodium methoxide in methanol. Write equations for these reactions, showing complete structures for the reactants, including their stereochemistry.

(d) The linkage between primeverose and the aglycon methyl salicylate is $\beta$. What is the complete structure of gaultherin?

**23.48** Patients who have severe bleeding or burns or who are undergoing surgery need to have huge volumes of blood plasma replaced. It is not always possible to provide as much human blood plasma as necessary, so artificial mixtures called plasma volume extenders have been developed to be used as part of the replacement. For example, a chemical modification of the starch fraction amylopectin has been synthesized and used as a plasma extender. The reaction used in the preparation of this plasma extender is shown below. Predict the structure of the modified polysaccharide formed, assuming that one hydroxyl group has been transformed for each glucose unit.

$$\text{amylopectin} + \text{NaOH} + \underset{\displaystyle CH_2 - CH_2}{\overset{\displaystyle O}{\triangle}} \xrightarrow[H_2O]{} \text{modified polysaccharide}$$

**23.49** As a test case used to develop the protecting groups and activating agents necessary for oligonucleotide synthesis, H. G. Khorana (p. 1023) synthesized thymidylyl-(3'→5')-thymidine. The reagents that he used were thymidine protected at the 5'-hydroxyl group by a trityl (triphenylmethyl) group and thymidine 5'-phosphate protected at the 3'-hydroxyl group as the acetyl ester. He used dicyclohexylcarbodiimide as the activating agent for the condensation. Write equations showing this synthesis. A review of solid-phase synthesis shown in Section 23.9D may be helpful.

**23.50** For each of the following peptides, draw the full structure and show the ionic state in which the compound would exist at pH 1, 6, and 11.

(a) Phe-Val-Asp     (b) Lys-Ala-Gly     (c) Leu-Tyr-Gly-NH$_2$     (d) Met-Gln-Ala

**23.51** Since the discovery in 1975 that there are small peptides in the brain called enkephalins that have analgesic properties, chemists have been synthesizing molecules that are structurally analogous to these compounds. One such synthetic peptide that is 1500 times as active an analgesic agent as a naturally occurring enkephalin has the following structure:

$$\text{Tyr-D-Ala-Gly-Phe-Pro-NH}_2$$

Write the full structure for this peptide, showing correct stereochemistry at every stereocenter in the molecule.

**23.52** The following vinylogous dipeptide was made from Boc-Phe. How would you carry out the synthesis? A review of Problem 23.28 will be helpful.

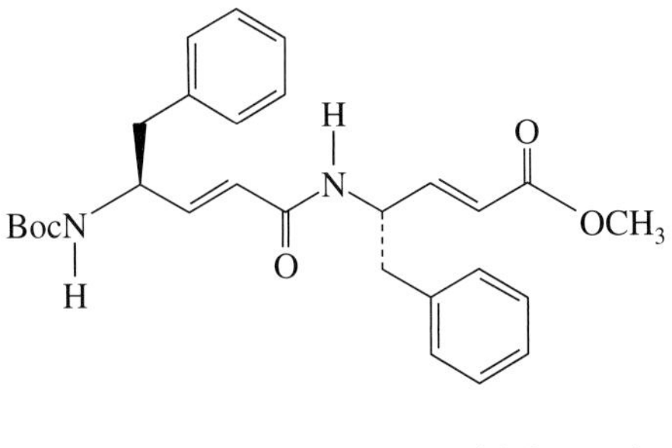

**23.53** Enzymes are used in stereoselective syntheses with increasing frequency. The enzyme Lipase PS-30 at pH 7 in a phosphate buffer selectively hydrolyzes the $S$ enantiomer of ethyl 3-hydroxy-3-phenylpropanoate. The individual enantiomers of the 3-hydroxyphenyl-propanoate are needed as intermediates in the synthesis of important antidepressants such as Prozac.

$$F_3C—\langle\text{ring}\rangle—OCHCH_2CH_2NHCH_3$$

Prozac

Racemic ethyl 3-hydroxy-3-phenylpropanoate can be synthesized from ethyl acetate and benzaldehyde or from the product of the acylation of ethyl acetate with benzoyl chloride.

(a) Show how you would carry out these syntheses.
(b) Show how you would use the enzyme to produce the enantiomerically pure (*S*)- and (*R*)-3-hydroxy-3-phenylpropanoic acids.

**23.54** In the human body, an amino acid is prepared for incorporation into a peptide chain by activation of its carboxyl group through the formation of an acylphosphate linkage. This reaction takes place with ATP (p. 680) to give an acylphosphate group at carbon 5 of the ribose unit and a pyrophosphate anion. Outline a possible mechanism for the activation reaction using any amino acid.

**23.55** The 17-amino acid peptide feline gastrin promotes secretion of HCl in the cells of the lining of the stomach in the cat. Complete hydrolysis of the peptide gives the following array of amino acids in alphabetical order:

$$\text{Ala}_2, \text{Asp}, \text{Gly}_2, \text{Glu}_5, \text{Leu}, \text{Met}, \text{Phe}, \text{Pro}, \text{Trp}_2, \text{Tyr}$$

Digestion of the peptide with chymotrypsin gives the following peptide fragments:

Glu-Gly-Pro-Trp
Gly-Trp
Met-Asp-Phe
Leu-Glu-Glu-Glu- Glu-Ala-Ala-Tyr

The *N*-terminal amino acid is Glu; the *C*-terminal amino acid is Phe.

(a) What are possible structures of feline gastrin?
(b) Write the complete three-dimensional structure of Met-Asp-Phe at physiologic pH, ~6.5. The three p$K_a$ values for Met-Asp-Phe are approximately 2.2, 3.9, and 9.3
(c) What is the structure of Met-Asp-Phe at its isoelectric point, pH 3.1 (p. 664)?

**23.56** An extract of the European mistletoe, *Viscum album,* contains a series of pharmacologically active peptides known as viscotoxins. The amino acid sequence of one of them, viscotoxin A$_2$, was determined using oxidation to break disulfide linkages, followed by digestion of the peptide with trypsin and chymotrypsin. The structures of the peptide fragments that were isolated from the two enzymatic cleavages are given below.

*Peptides from trypsin digestion:* Asn-Ile-Tyr-Asn-Thr-Cys-Arg,
Lys-Ser-Cys-Cys-Pro-Asn-Thr-Thr-Gly-Arg,
Ile-Ile-Ser-Ala-Ser-Thr-Cys-Pro-Ser-Tyr-Pro-Asp-Lys,
Phe-Gly-Gly-Gly-Ser-Arg,
Ser-Cys-Cys-Pro-Asn-Thr-Thr-Gly-Arg.

*Peptides from chymotrypsin digestion:* Asn-Thr-Cys-Arg-Phe,
Gly-Gly-Gly-Ser-Arg-Glu-Val-Cys-Ala-Ser-Leu,
Lys-Ser-Cys-Cys-Pro-Asn-Thr-Thr-Gly-Arg-Asn-Ile-Tyr,
Ser-Gly-Cys-Lys-Ile-Ile-Ser-Ala-Ser-Thr-Cys-Pro-Ser-Tyr-Pro-Asp-Lys.

Viscotoxin A$_2$ has 46 amino acid residues and a molecular weight of 4833. The *N*-terminal and *C*-terminal amino acids are both lysine. What is the order of the amino acids in viscotoxin A$_2$?

**23.57** Peptide P is widely distributed in the human body, especially in the nervous system. It is believed to mediate the body's response to pain. The following data were obtained in a determination of its structure:

1. Vigorous acidic hydrolysis of peptide P gives Arg, Glu (2), Gly, Leu, Lys, Met, Phe (2), Pro (2). Enzymatic hydrolysis gives Arg, Gln (2), Gly, Leu, Lys, Met, Phe (2), Pro (2). Peptide P has eleven amino acid units in all.
2. When peptide P is treated with phenyl isothiocyanate, the phenylthiohydantoins derived from arginine, proline, lysine, and proline can be obtained in that order.
3. Incubation of peptide P with chymotrypsin gives peptides A and B.
4. Peptide A contains Arg, Gln (2), Lys, Phe, Pro (2). Degradation of peptide A with Edman's reagent gives the same phenylthiohydantoins derived from the intact peptide P. Carboxypeptidase releases first Phe and then Gln to the solution.
5. Peptide B reacts with phenyl isothiocyanate to give the phenylthiohydantoins derived from phenylalanine, glycine, and leucine, in that order.
6. Peptide P is a strongly basic peptide with an isoelectric point above 8.9. No amino acid is released to the solution when the peptide is incubated with carboxypeptidase.
7. If peptide P is incubated with 0.03 M HCl at 110 °C for 8–12 hours (a procedure developed to hydrolyze carboxylic acid amide bonds but leave most peptide bonds untouched), the resulting product is peptide C, which reacts with carboxypeptidase. Gly, Met, Leu, Phe appear in the solution. The order in which these amino acids were released was not determined.

(a) What is the *N*-terminal amino acid of peptide P?
(b) What is the *C*-terminal amino acid of peptide P?
(c) What is the sequence of amino acids in peptide A?
(d) What is the sequence of amino acids in peptide B?
(e) What is the complete structure of peptide B?
(f) What is the sequence of amino acids in peptide C?
(g) What is the structure of peptide P?

**23.58** The solid-state synthesis of peptide P (Problem 23.57) was carried out.

(a) Glutamine residues were protected at the α-amino group with *tert*-butoxycarbonyl groups and activated at the carboxylic acid as the *p*-nitrophenyl esters. Write equations, starting with glutamine and showing its conversion to Boc-Gln-*O*-*p*-nitrophenyl, giving the full structure of the protected and activated amino acid.
(b) Lysine was protected at the α-amino group by a *tert*-butoxycarbonyl group and at the ε-amino group by a benzyloxycarbonyl group. Write the full structure of lysine, as it would be used in the peptide synthesis.
(c) The peptide linkages other than those to glutamine were created by activating with dicyclohexylcarbodiimide the carboxylic acid group of the amino acid being added. Show the mechanism for the reaction of Boc-Phe (written out in full) with dicyclohexylcarbodiimide, and the addition of the *N*-terminal end of a peptide fragment (any one) to the activated amino acid to form a new peptide bond.
(d) The Boc groups in this synthesis were removed by the addition of 4 M HCl dissolved in dioxane. Write the equation for removing the protecting group from the peptide fragment synthesized in part c.
(e) It was discovered that methionine esters could not be formed efficiently with the benzylic chloride functions on the polymeric support because the side chain of methionine reacted with the benzylic chloride. What reaction might occur between the side chain of Boc-Met and benzyl chloride?

**23.59** The natural product paclitaxel (Problem 21.73) has an unusual amino acid, *N*-benzoyl-3-phenylisoserine, as its side chain. This side chain has been found to be important to the biological activity of paclitaxel. Chemists have now synthesized the amino acid stereoselectively so that it is available to attach to synthetic paclitaxel or paclitaxel analogs. Steps of the synthesis are shown below. Supply reagents, designated by the letters, that could be used for the transformations shown.

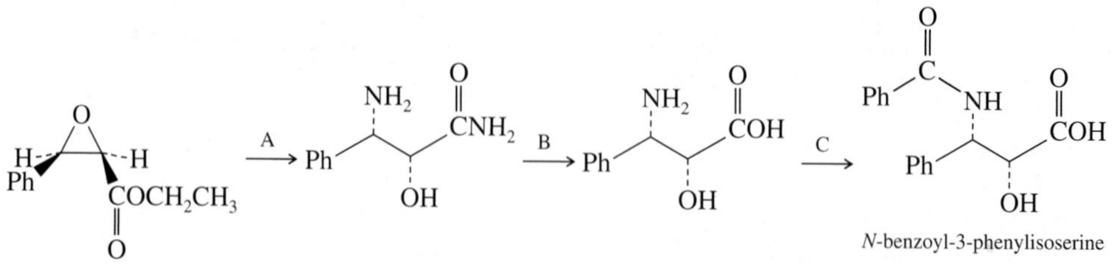

N-benzoyl-3-phenylisoserine

**23.60** The amino acid proline has been used as the starting material for a number of chiral amines. One such synthesis is outlined below. Write structural formulas for the compounds designated by letters.

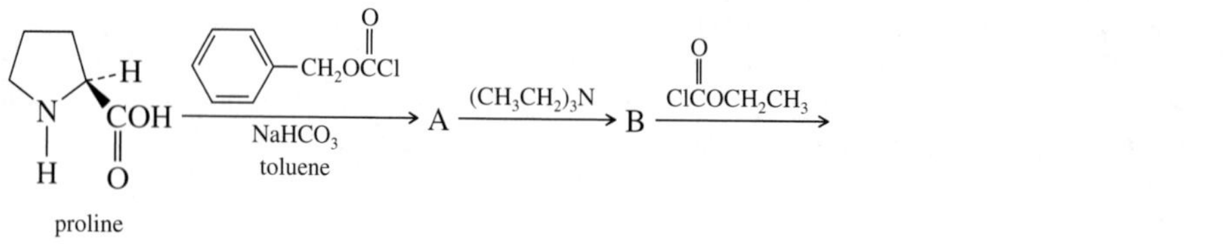

proline

# Macromolecular Chemistry

# 24

## A Look Ahead

**Polymers** are large molecules created by repetitive reactions of simple molecular units. Important in modern industrial society, they are used to make everything from common household utensils to replacement parts for the human body. The small unit that reacts many times to give a polymer is a **monomer** (p. 288). The process by which a monomer is converted into a polymer is called **polymerization.** The growth of a polymer may occur through a steady unit-by-unit addition of monomer molecules to the end of the chain in a **chain-growth process.** Alternatively, several monomer units may combine to give larger fragments, which then react to give the polymer. This is called a **step-growth process.** In some polymers, all the monomer units are the same. Polymers that have more than one kind of repeating unit are called **copolymers.**

The size and stereochemistry of a polymer molecule are important in determining the properties of the material. This chapter will discuss the ways in which polymers are made and how their structures determine their properties.

## 24.1 Introduction to Macromolecules

### A. Macromolecules of Biological Importance

The relationship between the structure and the function of biological macromolecules was the subject of Chapter 23. Many of these biological molecules have great practical and commercial importance. For example, the structural materials of plants, cellulose (p. 1004) and lignin, are giant molecules. The spaces within the long fibers formed by cellulose are filled by lignins, which are complex molecules containing carbon–oxygen and carbon–carbon bonds between adjoining units of phenylpropanes with hydroxy and methoxy substituents on the aromatic rings.

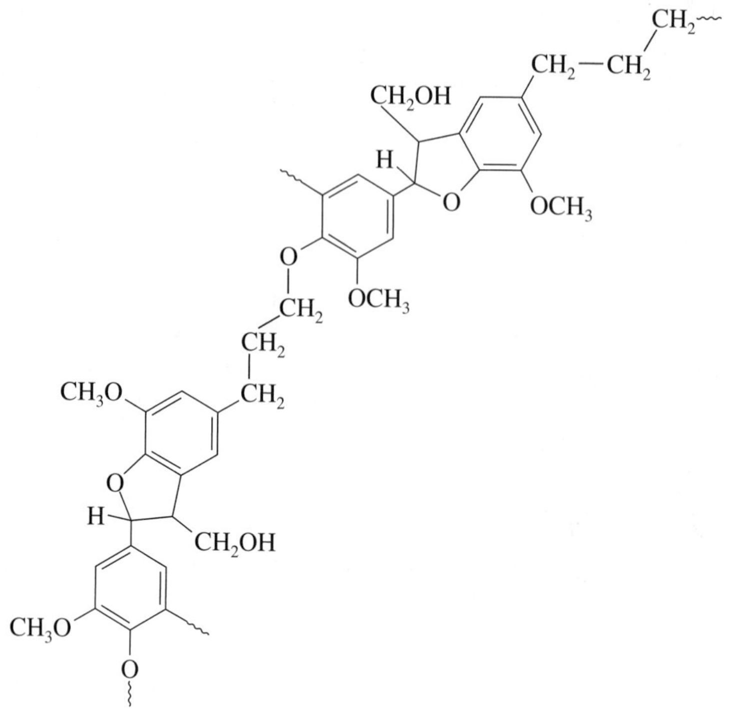

a fragment of lignin

The points at which the phenylpropane units are linked vary. The molecule is not linear but has many points of linkage, so it forms a three-dimensional network. The combination of the long crystalline fibers of cellulose with the network of lignin molecules penetrating them creates the strong, rigid structure of woody plants.

Rubber is another large molecule that owes its useful properties to its molecular size. When rubber is decomposed by heat, isoprene is formed (p. 750). The structure and chemistry of rubber is discussed in Section 24.7.

Proteins, cellulose, lignins, rubber, and the nucleic acids are naturally occurring representatives of a class of organic compounds of biological and, more recently, industrial importance. All these compounds have large molecular size in common. For this reason, they are known as **macromolecules,** or giant molecules. Their molecular structures can be dissected into smaller units. For example, proteins have units of amino acids, cellulose has units of glucose, and rubber has units of isoprene. The small unit that reacts again and again with itself to give the macromolecule is called a **monomer,** and the large molecule that is composed of these units is known as a **polymer.** Isoprene is a monomer, and rubber is a polymer made up of many units of isoprene bonded together.

In some macromolecules, all the monomeric units are the same. Rubber contains only isoprene units. Cellulose is made up only of glucose units. In proteins or nucleic acids, on the other hand, the repeating monomeric units are not all identical. Chapter 23 showed that a large number of different protein molecules are possible because of the different order in which the twenty or so amino acids are incorporated into peptide chains. DNA and RNA are also made up of nucleotides with different structures, arranged in a chain with a varying order. Polymers in which there is more than one type of repeating unit are called **copolymers.** Rubber and cellulose are simple polymers, and proteins and nucleic acids are complex copolymers.

A polymer also may be classified according to the structure of the giant molecule. Rubber, cellulose, proteins, and nucleic acids are all **linear polymers.** The backbone of the macromolecule in each case consists of a long chain of monomer units held together by covalent bonds. The principal linkage between monomers in proteins is the amide bond, called the peptide bond; in cellulose, it is the glycosidic linkage; in DNA and RNA, phosphate ester linkages connect the nucleotides. Isoprene units are held together in rubber by covalent bonds between carbon atoms. In a linear polymer, once the chain is formed, other types of interactions, such as hydrogen bonding in proteins and van der Waals interactions between different parts of the hydrocarbon chain in rubber, contribute to the overall shape of the macromolecule.

Lignins represent another structural class of polymers, those in which there is additional covalent bonding between the monomer units. Not only do the monomers form long chains, but there are also covalent links between adjacent chains. Such polymers are said to be **cross-linked.** These links may be close together, as is the case in lignin, or they may be relatively far apart, as in some synthetic polymers. Such bonding affects the properties of a polymer and its potential uses. Later sections of this chapter will examine the relationship between structure and properties and show how the introduction of cross-linking affects these characteristics and thus changes the uses of a polymer.

Human beings have long made use of natural macromolecular substances. The structural rigidity of wood, which results from the properties of cellulose and lignin, makes it useful in construction. Many macromolecules have fibrous structures that can be spun into threads and then used to weave fabrics. Cellulose fibers in cotton and linen and protein fibers in silk and wool are used in clothing and household furnishings. Modern methods of transportation would not be possible without rubber. Human beings also have discovered ways to modify natural macromolecules to make them more useful. For example, a modification of cellulose led to the invention of paper, an important technological advance. Because only limited modifications could be made on natural molecules without destroying their essential structures, chemists started to think about creating new polymeric materials, starting with small reactive molecules. The remainder of this chapter is the story of their success.

Study Guide
Concept Map 24.1

## B. Macromolecules of Industrial Importance

By the middle of the nineteenth century, organic chemists had obtained as products in some of their experiments high-molecular-weight substances that were usually considered to be evidence of failed reactions. It was not until the early part of this century that chemists started to create polymers deliberately. To do this, they designed reactions that allowed them to control the average molecular weight, and therefore the properties, of the large molecules that were being formed. Hermann Staudinger of Germany, a pioneer in this field, was one of the first to recognize that controlling the conditions of polymerization was essential to the synthesis of useful substances. He received the Nobel Prize in 1953 for his work in this area of chemistry.

At first, chemists in this field tried to imitate nature. For example, the first really successful synthetic fiber was nylon, a polyamide created in the 1930s by the American chemist Wallace Carothers. The structure of nylon resembles a protein in having many amide bonds but is much more regular in its repeating units. Different kinds of nylons are shown on the following page with the structural units that give rise to them.

$$+\!\!\!-\!\!\overset{\displaystyle O}{\overset{\|}{C}}(CH_2)_4\overset{\displaystyle O}{\overset{\|}{C}}NH(CH_2)_6NH-\!\!\!\!+_n \qquad H_2N(CH_2)_6NH_2 \qquad HO\overset{\displaystyle O}{\overset{\|}{C}}(CH_2)_4\overset{\displaystyle O}{\overset{\|}{C}}OH$$

nylon 66, a polyamide of 1,6-hexanediamine and
hexanedioic acid (adipic acid)

$$+\!\!\!-\!\!\overset{\displaystyle O}{\overset{\|}{C}}(CH_2)_{10}\overset{\displaystyle O}{\overset{\|}{C}}NH(CH_2)_6NH-\!\!\!\!+_n \qquad H_2N(CH_2)_6NH_2 \qquad HO\overset{\displaystyle O}{\overset{\|}{C}}(CH_2)_{10}\overset{\displaystyle O}{\overset{\|}{C}}OH$$

nylon 612, a polyamide of 1,6-hexanediamine and
dodecanedioic acid

$$+\!\!\!-\!\!NH(CH_2)_5\overset{\displaystyle O}{\overset{\|}{C}}-\!\!\!\!+_n \qquad H_2N(CH_2)_5\overset{\displaystyle O}{\overset{\|}{C}}OH$$

nylon 6, a polyamide of 6-aminohexanoic acid

$n$ = number of repeating units in the polymer chain;
for nylons that are useful as fibers, $n$ = 50 to 120

The number in the name of a nylon indicates the structure of the polyamide. Nylon 66 is made from an amine and an acid, each having six carbon atoms; nylon 612 contains amine units with six carbon atoms and acid units with 12. Nylons of the appropriate molecular weight, where $n$ is 50 to 120, can be formed into threads and used to make fabrics. In an early practical application, nylon was used to replace silk in women's stockings. It has since been used in a wide variety of fabrics, from carpets to parachutes to clothing of every description. The chemistry of the polymerization processes that give rise to nylons is discussed on pages 1041 and 1059.

Since World War II, with the discovery that useful macromolecular compounds could be synthesized in the laboratory and then produced on a large scale in factories, much research has concentrated on developing new polymers. More industrial organic chemists work in polymer chemistry than in any other field. Objects made of polymers are so pervasive in everyday life that the names, especially the trade names, of polymers have become part of the language. News stories about the health-related and environmental effects of polymers and the monomers that go into their manufacture are common. Structural formulas for some familiar polymers, along with the monomers from which they are made, are given below and on the next page.

$$+\!\!\!-\!\!CH_2CH_2-\!\!\!\!+_n \qquad CH_2\!\!=\!\!CH_2 \qquad\qquad +\!\!\!-\!\!CH_2\overset{\displaystyle Cl}{\overset{|}{C}H}-\!\!\!\!+_n \qquad CH_2\!\!=\!\!CHCl$$

polyethylene          ethylene              poly(vinyl chloride)      vinyl chloride
                                                   PVC

$$+\!\!\!-\!\!CF_2CF_2-\!\!\!\!+_n \qquad CF_2\!\!=\!\!CF_2 \qquad\qquad +\!\!\!-\!\!CH_2\overset{\displaystyle C\equiv N}{\overset{|}{C}H}-\!\!\!\!+_n \qquad CH_2\!\!=\!\!CHC\!\equiv\!N$$

polytetrafluoroethylene    tetrafluoroethylene      a polyacrylonitrile      acrylonitrile
Teflon                                          Orlon, Acrilan

$$+\!\!\!-\!\!OCH_2CH_2O\overset{\displaystyle O}{\overset{\|}{C}}\!\!-\!\!\!\!\bigcirc\!\!\!\!-\!\!\overset{\displaystyle O}{\overset{\|}{C}}-\!\!\!\!+_n \qquad HOCH_2CH_2OH \qquad HO\overset{\displaystyle O}{\overset{\|}{C}}\!\!-\!\!\!\!\bigcirc\!\!\!\!-\!\!\overset{\displaystyle O}{\overset{\|}{C}}OH$$

poly(ethylene terephthalate)          ethylene glycol      terephthalic acid
a polyester
Dacron, Terylene

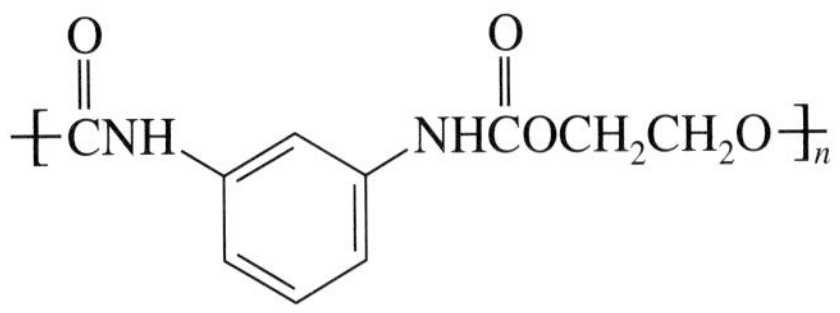

a polyurethane

$HOCH_2CH_2OH$

ethylene glycol

*m*-phenylenediisocyanate
a diisocyanate

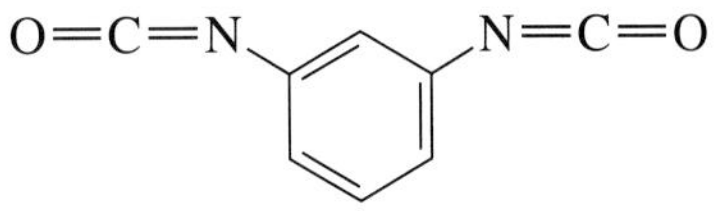

Polymers contain a wide variety of functional groups. These large molecules are made using essentially the same reactions as are applicable to low-molecular-weight compounds. Later sections in this chapter will concentrate on how these reactions generate large molecules having the physical properties that make them useful in practical applications.

**Problem 24.1** Draw structural formulas for the monomers that were used to produce the following polymers.

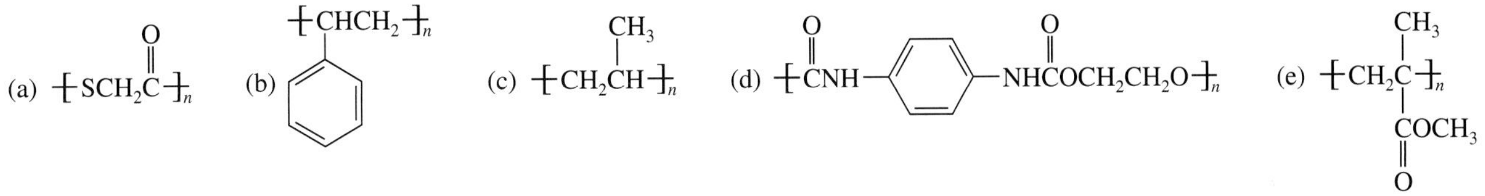

## C. Special Properties of Large Molecules

The literature of polymer chemistry reveals that researchers are intensely interested in the molecular weight of macromolecules. Every synthetic process is judged primarily by the range of molecular weights of the polymers formed. A synthesis of a polymer differs in this way from that of a low-molecular-weight organic compound. All the preparations in earlier chapters of this book have given compounds with well-defined structures. The synthesis of a polymer results in a mixture of compounds having a range of molecular weights. For example, the equation for the preparation of nylon 66 from hexanedioic acid and 1,6-hexanediamine can only indicate an approximate structure for the polymer that is formed.

$$n\ HO\overset{O}{\overset{\|}{C}}(CH_2)_4\overset{O}{\overset{\|}{C}}OH + n\ H_2N(CH_2)_6NH_2 \xrightarrow[\substack{220\ °C \\ \sim200\text{–}250\ \text{psi} \\ 1\text{–}2\ h}]{} \xrightarrow[270\text{–}280\ °C]{} \left[ NH(CH_2)_6NH\overset{O}{\overset{\|}{C}}(CH_2)_4\overset{O}{\overset{\|}{C}} \right]_n + 2n\ H_2O\uparrow$$

hexanedioic acid (adipic acid); 1,6-hexanediamine (hexamethylenediamine); nylon 66

The exact structure of the nylon molecule depends on how many of the repeating units become bonded together before the polymer chain stops growing. The repetitive structure of the polymer is indicated by the subscript *n* outside the brackets that enclose the repeating unit. The larger *n* is, the higher is the molecular weight of the polymer. The molecules of a polymer formed in any reaction mixture do not have identical numbers of repeating units and thus differ in molecular weight. The product is a mixture of similar molecules that differ somewhat in size.

The concern about molecular weights is justified because the physical properties, and thus the practical uses, of a polymer depend heavily on its molecular weight. For example, nylon with a low molecular weight has no useful properties. It is a brittle solid. Only when the molecular weight reaches 10,000 does a nylon begin to show properties that make it useful as a fiber. Nylons that have molecular weights above 100,000 do not make good fibers but have high resistance to heat and high mechanical strength, so they are used in other industrial applications. One such nylon, reinforced with glass fibers, is used instead of steel to make valve covers for automobile engines.

Molecular weight is, of course, a reflection of molecular size. The molecules of a polymer must have a minimum size before the interactions that are important in determining the properties of the substance can take place. These interactions may be between different parts of the same molecule or between adjacent molecules. All the factors discussed in Chapter 23 in connection with the tertiary structure of proteins, such as hydrogen bonding and van der Waals interactions, are also important in determining the shapes of other macromolecules. The nature of the repeating unit in a polymer is, of course, of primary importance in determining which types of interactions are possible. For a nylon, which has a large number of amide linkages, hydrogen bonding may be important. In polyethylene, which is like a gigantic alkane molecule, van der Waals interactions between different parts of the molecule and between adjacent chains are most likely.

The shape of a polymer molecule and the types of interactions that it can have with neighboring molecules are determined by the regularity with which repeating units appear in the chain and by the stereochemistry at points where stereoisomerism is possible. For example, just like a low-molecular-weight alkene, a polymer may contain cis or trans double bonds. Rubber is an example of a polymer with cis double bonds. The trans isomer is a natural product known as gutta percha and has some properties that differ notably from those of rubber (p. 1073). If the repeating units in a polymer are chiral, as is the case for amino acids in proteins and glucose in cellulose, the macromolecule as a whole will exhibit chirality. Even a polymer formed from an achiral monomer such as propene has the possibility of stereoisomerism at the newly created tetrahedral carbon atoms along the backbone of the chain (pp. 1055–1058). In summary, polymers can exist as stereoisomers with differing stereochemistry at double bonds or at tetrahedral carbon atoms.

A given polymer also can exist in a large number of conformations arising from free rotations of the atoms that make up the backbone of the chain. Just as conformational isomerism is possible for butane with four carbon atoms in its chain (p. 166), the molecules in a given sample of a polymer exist in many different and constantly changing conformations. For polymers, too, anti and gauche arrangements of groups on adjacent carbon atoms are preferred over eclipsed conformations. The conformations that are favored and the range of motions possible for different parts of polymer chains are important in determining the properties of a polymer and how it will behave under different conditions.

One of the properties of a polymer that is very important for practical applications is how it behaves at different temperatures. The interactions between the large molecules of a polymer create solids that have a high degree of structural regularity. Linear polyethylene (p. 1056) is a highly crystalline substance with a melting point of approximately 135 °C. The polyethylene produced by some manufacturing processes has branching on the chain and does not form as highly crystalline a solid. Branched polyethylene has a lower melting point, about 120 °C. Some polymers, such as rubber, do not pack together well in regular structures and exist mostly as amorphous solids. Many polymers are partly crystalline, meaning that when the liquid polymer is cooled, parts of it form regions with a high degree of order and other parts solidify before such arrangement takes place. Nylon is such a

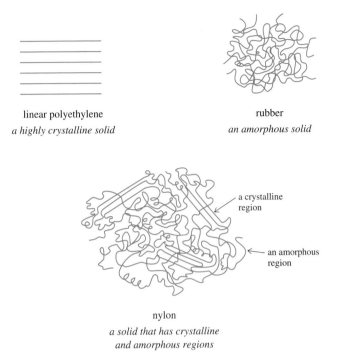

linear polyethylene
*a highly crystalline solid*

rubber
*an amorphous solid*

a crystalline
region

an amorphous
region

nylon
*a solid that has crystalline
and amorphous regions*

*Figure 24.1*
Schematic representations of crystalline, amorphous, and semicrystalline polymers.

partly crystalline polymer. These different kinds of interactions between polymer chains are represented in Figure 24.1.

Whether or not a polymer has crystalline regions determines properties such as its flexibility and mechanical strength. For example, the amorphous and coiled structure of rubber is responsible for its elasticity. Polymers with such elastic qualities are known as **elastomers.** Semicrystalline polymers are rather hard and can be drawn out into strong **fibers.** The process of drawing them out increases the alignment of the polymer chains and thus their crystallinity. Unless a polymer has some crystalline regions, it cannot be formed into fibers; thus an amorphous polymer does not make good fibers. Not all amorphous polymers have elasticity; those that do not are called **plastics** and are used in applications in which they can be molded under heat and pressure into objects such as toys and household goods. Polymers are classified as elastomers, fibers, or plastics according to their properties and their uses.

A polymer can exist in different physical forms at differing temperatures. At low temperatures, it exists as a solid. In the solid state, it may be partly or highly crystalline, or it may be amorphous. A solid that is not crystalline is called a **glass.** Transparency, for example, is typical of a solid that is a glass; a crystalline solid is opaque.

On being warmed to a given temperature, a polymer softens to a more pliable state but does not actually melt. This temperature, the **glass transition temperature, $T_g$,** varies a bit according to the method used to make the determination but is typical for each compound. For rubber, this temperature is $-70\ °C$; for nylon 66, it is $50\ °C$. Thus rubber, as we know it, is in an intermediate state in which it is not a rigid solid but is not a liquid either. If a rubber ball is made very cold by putting it into liquid nitrogen ($-196$ °C, below the $T_g$ of $-70$ °C for rubber), it loses all its bounce and shatters if thrown against a hard surface. This is typical of the brittle character of a glass.

At an even higher temperature, a polymer melts and becomes a liquid. This temperature is the melting point of the polymer, comparable to the melting point of a lower-molecular-weight organic compound. At this temperature, the solid polymer loses its crystalline order. The melting point of raw rubber is 30 °C; for nylon 66, it is 265 °C. Knowledge of these properties of a polymer is clearly important. For example, raw rubber is useful at temperatures between $-70$ °C and $+30$ °C. Raw rubber, therefore, would be useless on a hot day. Similarly, the nylon that is used in valve covers for automobile engines must retain its shape and therefore must remain a solid at high temperatures. This nylon withstands exposure to temperatures of approximately 120 °C for long periods.

⊙ **Study Guide**
**Concept Map 24.2**

## 24.2   Mechanisms of Chain-Growth Polymerization

### A. Chain-Growth Polymerization

How a polymer chain grows is of great importance in determining how long the chain will be, which determines the molecular weight of the polymer. The reactions that give rise to polymers have been classified into two main types. One type is called **chain-growth polymerization.** In this type of polymerization, a reactive intermediate is formed and reacts rapidly with a monomer molecule to give a new reactive intermediate, which in turn reacts with yet another monomer molecule. The monomer is consumed rapidly and always adds onto a chain that gets longer and longer. A new reactive center is created at the end of the polymer chain, so the reaction perpetuates itself until all the monomer molecules have reacted or the reactive intermediate is destroyed by some end reaction. You probably recognize this description as corresponding to the steps of a free-radical chain reaction (pp. 770–772). There is an initiation step, many propagation steps in which each reactive intermediate generates a new reactive intermediate, and finally, a termination step. Such a reaction often involves free radicals, but under the proper conditions, chain reactions with cationic or anionic intermediates are also possible. The important factor in all these reactions is the presence of a reactive intermediate that directs the course of the reaction, making it more probable at one molecular site than at others. In fact, in such a reaction the monomer units are usually incapable of reacting with each other until some reagent is added to create the reactive intermediate. The polymerization of an alkene, such as that of vinyl chloride (p. 782), is usually a chain-growth reaction. The products of chain-growth polymerization reactions are **addition polymers,** so called because they are created by the addition of one molecule of monomer to another one.

### B. Free-Radical Reactions

The polymerization of styrene in the presence of a small amount of benzoyl peroxide is a typical example of a free-radical polymerization reaction.

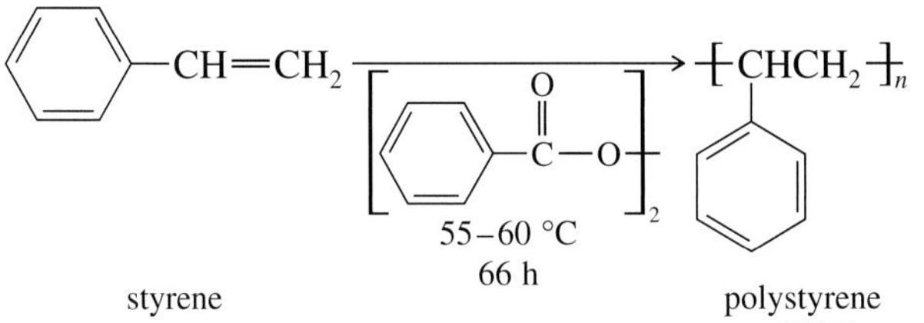

styrene    55–60 °C / 66 h    polystyrene ~100%

The reaction takes place in three stages. First, in the initiation step, benzoyl peroxide dissociates into two benzoyloxy radicals (p. 782). A benzoyloxy radical, an electrophile, then reacts with the $\pi$ electrons of the double bond in styrene to create a new radical, the stable benzylic radical.

**VISUALIZING THE REACTION**

### Initiation Step of a Free-Radical Polymerization

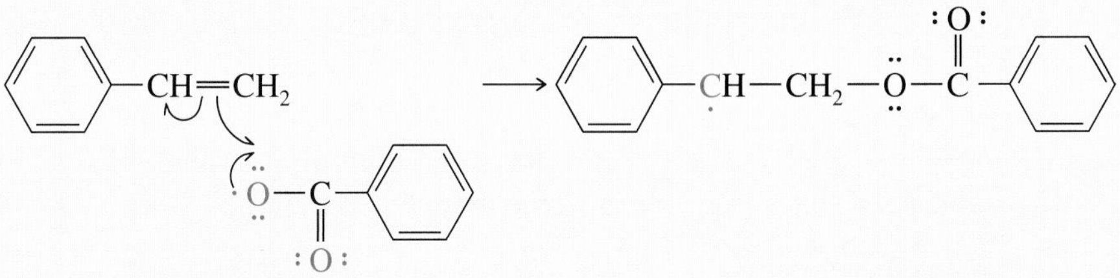

benzoyloxy radical
*an oxy radical*

*a benzylic radical*

The benzylic radical attacks another molecule of styrene, creating yet another radical intermediate.

**VISUALIZING THE REACTION**

### Chain-Propagation Step of a Free-Radical Polymerization

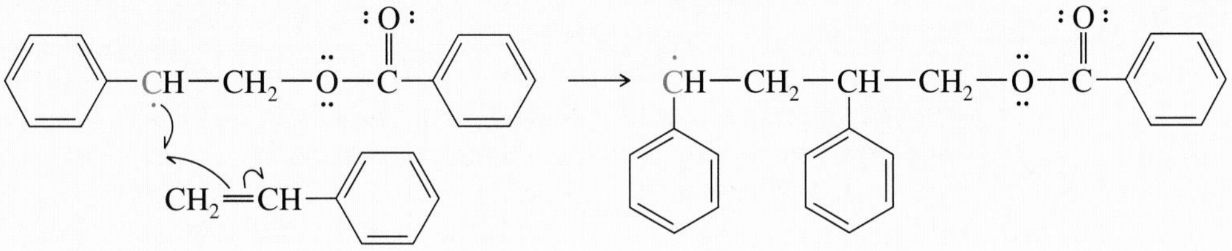

Each new radical continues to react with a molecule of the monomer in a series of chain-propagating steps. The growth of the chain of polystyrene is fast; it has been calculated that approximately 1500 monomer units are added to the chain each second. The reaction continues until all the monomer molecules have reacted or the radical intermediate is destroyed by one of a number of termination reactions. For example, a combination of two radicals will stop the polymerization process. Most polystyrene chains stop growing as a result of the combination of two polystyryl radicals, which gives a species with no reactive sites, called a **"dead" polymer.**

**VISUALIZING THE REACTION**

### Termination Step of a Free-Radical Polymerization

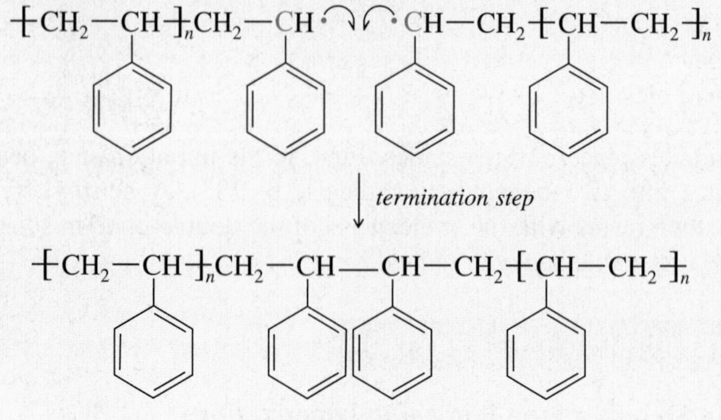

$$\left.\!\!+\!CH_2\!-\!CH\!\right\}_n\!CH_2\!-\!CH \curvearrowright\curvearrowleft CH\!-\!CH_2\!\left\{\!CH\!-\!CH_2\!\right\}_n$$

*termination step*

$$\left.\!\!+\!CH_2\!-\!CH\!\right\}_n\!CH_2\!-\!CH\!-\!CH\!-\!CH_2\!\left\{\!CH\!-\!CH_2\!\right\}_n$$

"dead" polymer

The polymer radical also may abstract a hydrogen atom from another chain, creating a new radical in what is known as a **chain-transfer reaction.**

$$\left.\!\!+\!CH_2\!-\!CH\!\right\}_n\!CH_2\!-\!CH\cdot + \left\{\!CH_2\!-\!CH\!-\!CH_2\!-\!CH\!-\!CH_2\!-\!CH\!\right\}_x \longrightarrow$$

$$\left.\!\!+\!CH_2\!-\!CH\!\right\}_n\!CH_2\!-\!CH_2 + \left\{\!CH_2\!-\!\overset{\cdot}{C}\!-\!CH_2\!-\!CH\!-\!CH_2\!-\!CH\!\right\}_x$$

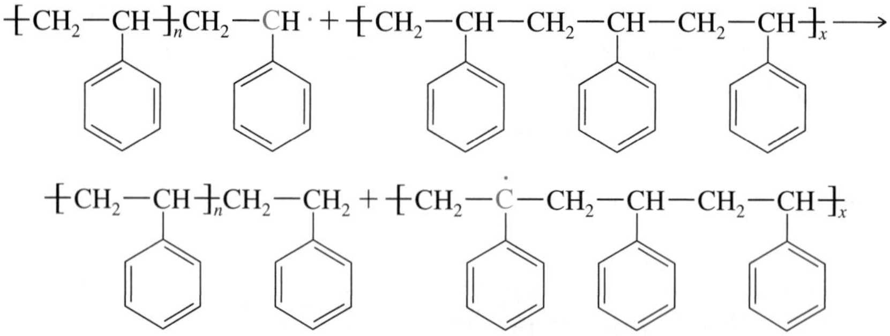

"dead" polymer          *a new radical created by abstraction of a benzylic hydrogen atom from the middle of a polymer chain; a chain-transfer reaction*

This new radical is in the middle of a chain, not at the end. It can react with a monomer unit, giving rise to branching of the chain.

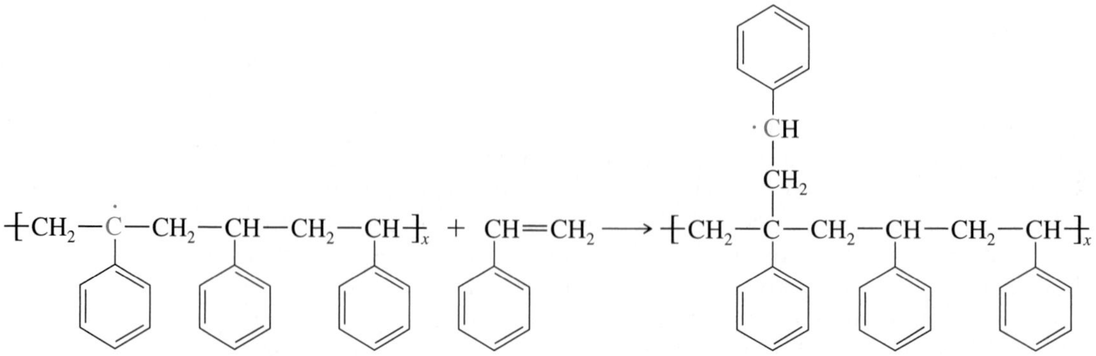

$$\left.\!\!+\!CH_2\!-\!\overset{\cdot}{C}\!-\!CH_2\!-\!CH\!-\!CH_2\!-\!CH\!\right\}_x + CH\!=\!CH_2 \longrightarrow \left\{\!CH_2\!-\!\overset{\overset{\displaystyle CH_2}{\underset{\displaystyle \cdot CH}{|}}}{\underset{\displaystyle |}{C}}\!-\!CH_2\!-\!CH\!-\!CH_2\!-\!CH\!\right\}_x$$

branched chain

Note that in writing the structure of the polymer, the ends of the molecule are left unspecified. The end groups are an insignificant portion of the total chain and are not necessarily identical for every molecule in the sample. The properties of the polymer are determined largely by the main body of the chain. Polystyrene, polymerized simply by adding a small amount of benzoyl peroxide to the monomer, reaches molecular weights of about 2,500,000 and is amorphous.

Polystyrene has many uses. The polymer may be molded into cases for radios and batteries. Toys and all kinds of containers are made from it. If a low-boiling hydrocarbon such as pentane is included in the polystyrene during processing, when the polymer softens as it is heated, the added compound vaporizes and creates bubbles that expand the polystyrene into a rigid but lightweight foam. This foam is used as insulation in construction. Pellets of the foam are supplied to manufacturers, who use pressure and some heat to mold them into ice chests and disposable plastic cups for hot drinks. Egg cartons, which have to be rigid but cushiony at the same time, are made of polystyrene foam. The foam can be molded to the exact shape of an object that needs to be protected during transportation; thus many delicate instruments and bottles of chemicals travel in their own polystyrene foam sheaths.

Many copolymers involving styrene also have been designed to have particular properties. For example, a copolymer of styrene with acrylonitrile is superior to polystyrene in toughness, stability to light, and resistance to chemicals. The polymer, produced by polymerization of a mixture of the two monomers, is believed to have mostly an alternating arrangement of the two units in the chain.

styrene     acrylonitrile         copolymer of styrene
~ 1 equivalent    1 equivalent         and acrylonitrile

The exact arrangement of monomer units in a copolymer depends on the relative reactivities and relative concentrations of the two monomers in the reaction mixture. Styrene-acrylonitrile copolymers have exceptional clarity and strength and have been made into a variety of objects, such as lenses for automobile headlights, household containers, and many devices for medical use, including disposable syringes and parts for artificial kidneys.

Another important copolymer of styrene is the cross-linked one it forms with *p*-divinylbenzene.

styrene     *p*-divinylbenzene

*copolymer of styrene and p-divinylbenzene*
*showing how the p-divinylbenzene forms a*
*cross-link between two polymer chains*

Because *p*-divinylbenzene has two alkene functions, it becomes part of two poly-mer chains, forming a link between them. Depending on the relative amounts of styrene and *p*-divinylbenzene used, the links may be close together or widely separated.

In one practical application, the cross-linked copolymer of styrene and *p*-divinylbenzene, modified by incorporating polar functional groups such as sulfonic acid groups on the aromatic rings, is used as an ion-exchange resin. The whole polymer molecule, as its sodium salt, acts as a gigantic insoluble anion, which can exchange its cations with those in the solution that flows through it. Such resins soften water by exchanging sodium ions for the calcium and magnesium ions in hard water (p. 642) and are used in both water-treatment plants and water-softening units in homes. A styrene-divinylbenzene copolymer is also one of the support sys-tems for automated solid-phase syntheses of peptides, proteins, and oligonu-cleotides (pp. 901 and 1023).

---

**Problem 24.2** For each of the following polymers, show the monomers that would be used to prepare it. Also say whether each one is a copolymer or a simple polymer.

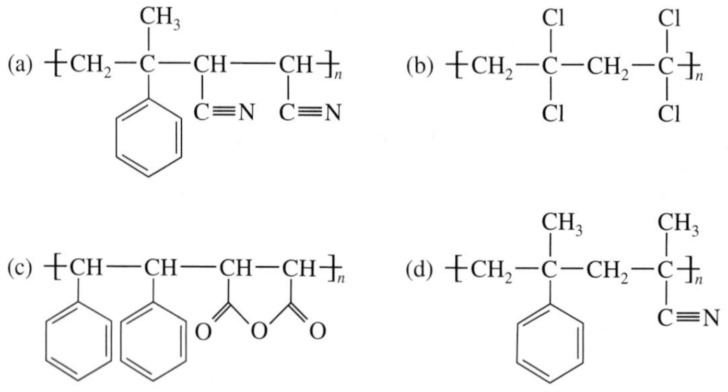

---

**Problem 24.3** The ion-exchange resin used for water softening is made by introducing sulfonic acid groups into the cross-linked copolymer of styrene and divinylbenzene. How would you carry out such a reaction?

---

Free-radical polymerization carried out under the conditions shown above is an uncontrolled process. As chemists have become more interested in designing more precisely the properties of the polymers they make, they have had to create free-radical polymerization methods that they can control better. One of these methods is called **atom-transfer radical polymerization (ATRP).** In such a polymerization, the chain-growth process is controlled by the transfer of an atom, usually a halogen atom, to the radical site on the polymer chain to give the dormant form of the poly-mer. This keeps down the concentration of the radical intermediate and prevents chain termination reactions resulting from the combination of two radicals (p. 771). The dormant polymer can be reactivated by a metal catalyst that removes the halo-gen atom.

A typical atom-transfer radical polymerization of styrene willl serve as an ex-ample of this method. In this reaction styrene is mixed with catalytic amounts of

1-chloro-1-phenylethane, copper(I) chloride, and 2,2′-bipyridine. The reaction starts with the complexing of copper(I) chloride with the bipyridine.

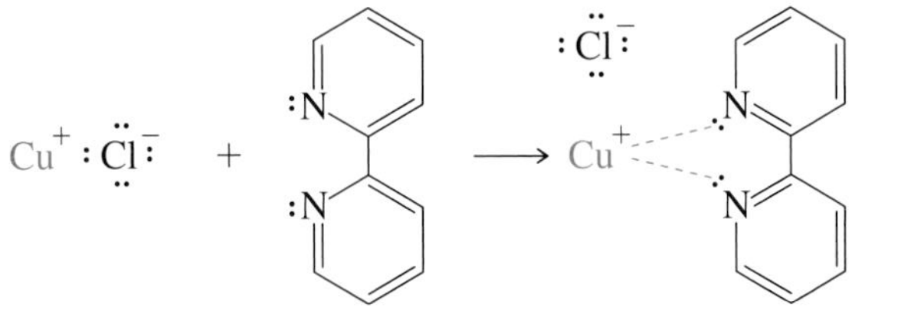

copper(I) chloride    2, 2′-bipyridine

complex of
copper(I) chloride
with bipyridine

Copper is in Group 11 of the periodic table. Copper(II) is its usual stable oxidation state, but as a transition metal, it gains and loses electrons with ease. In the initiation step, the carbon–halogen bond of the alkyl halide is cleaved by the donation of an electron from copper(I). The copper is oxidized to copper(II), the halogen is reduced to halide ion, and an alkyl radical is formed that initiates the polymerization reaction.

## VISUALIZING THE REACTION

### Free-Radical Initiation of Atom-Transfer Radical Polymerization

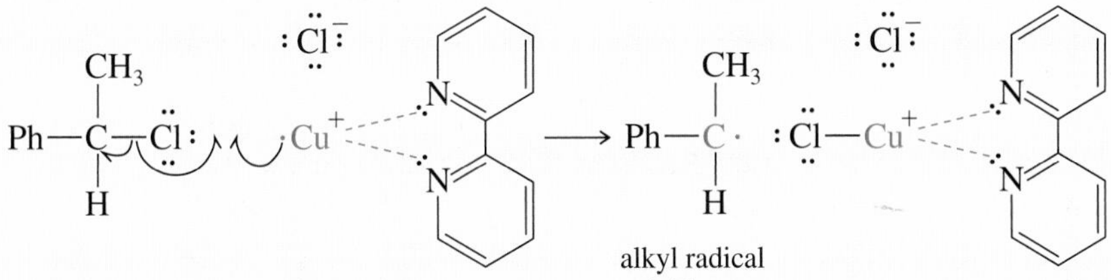

*1-chloro-1-phenylethane transferring
chlorine atom to copper*

alkyl radical
initiator

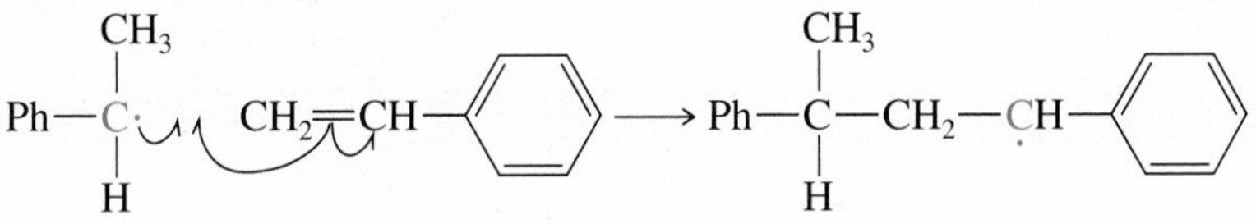

*initiator reacting with styrene*

radical that will propagate the chain

The radical created from the reaction of styene with the initiator reacts with additional molecules of styrene to give a polymeric chain as shown for the chain propagation of free-radical polymerization of styrene on p. 1045. The difference here is that the free radical is in equilibrium with the dormant form of the polymer in which the halogen atom has been transferred back to carbon from copper.

---

**VISUALIZING THE REACTION**

**Atom-Transfer Reaction That Controls the Concentration of Free Radicals**

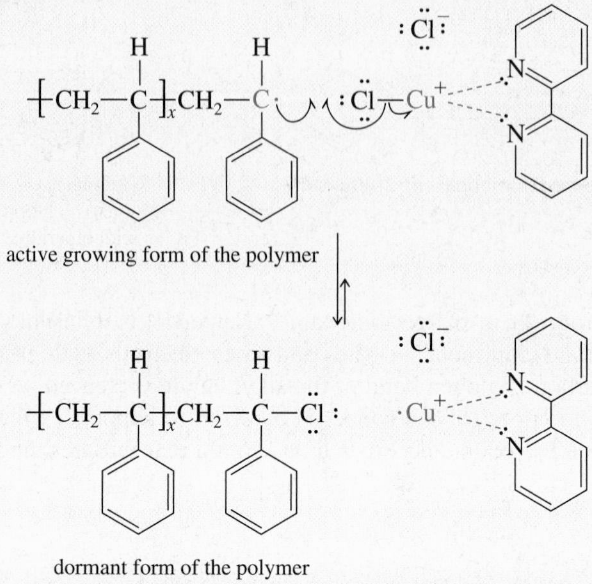

active growing form of the polymer

dormant form of the polymer

---

This controls the polymerization reaction, giving rise to polymers with a much narrower range of molecular weights than achieved in uncontrolled free-radical polymerization reactions.

Atom-transfer radical polymerization is just one of a number of ways that chemists have developed to reversibly control the presence of free radicals in reaction mixtures and, hence, the design of polymer materials in free-radical reactions.

## C. Anionic Reactions

Some polymerization reactions are catalyzed by alkali metals or by organometallic compounds. The reactive species that participates in the growth of the chain in these cases is a carbanion. 2-Phenylpropene, commonly called $\alpha$-methylstyrene, polymerizes by such a reaction.

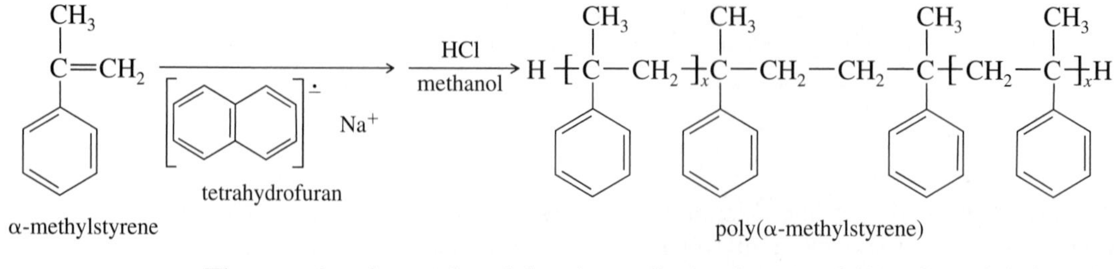

The reaction is catalyzed by the radical-anion (p. 340) of naphthalene in tetrahydrofuran. The radical-anion of naphthalene transfers an electron (and the accompanying negative charge) to $\alpha$-methylstyrene.

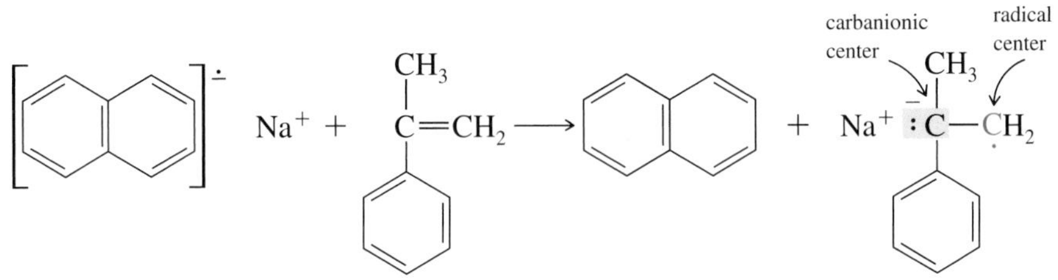

sodium naphthalide;
radical-anion of
naphthalene

α-methylstyrene

naphthalene

sodium salt of
the radical-anion
of α-methylstyrene

Note that one electron has been added to the alkene. For the sake of clarity, this electron and the two electrons of the $\pi$ bond are shown as if they were localized to give a radical center and a carbanionic center in the radical-anion. This is, of course, a highly simplified way of representing the reactive species, but it is useful in rationalizing the course of the reaction, which proceeds by a combination of two radicals to give a dianion.

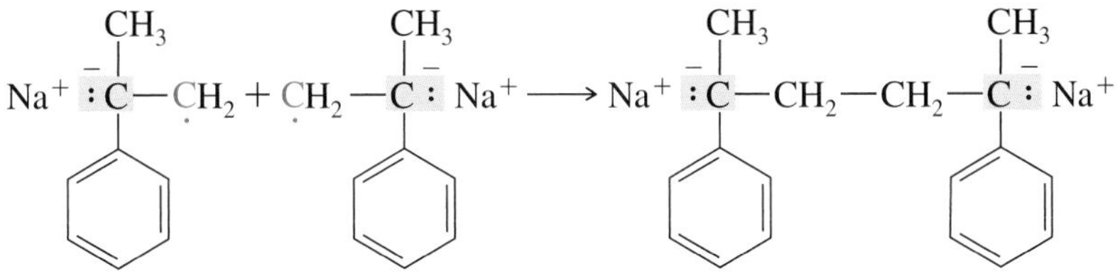

sodium salt of the radical-
anion of α-methylstyrene

dianion
product of the reaction
of 2 radical-anions

The dianion is coordinated with sodium ions. It grows by adding monomer units at either end, creating new anions, which are stabilized by their association with cations.

**VISUALIZING THE REACTION**

**Anionic Polymerization of an Alkene**

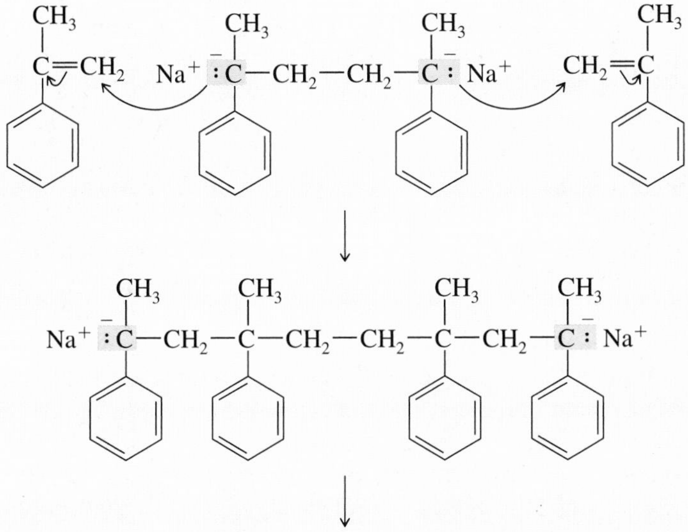

high-molecular-weight dianion (~ 50,000)

HCl
methanol

poly(α-methylstyrene)

The chain grows at both ends until some reagent that reacts with a carbanion is added to the reaction mixture. For example, the anions may be protonated by a dilute solution of hydrochloric acid in methanol.

This type of polymerization reaction was investigated extensively by Michael Szwarc at the State University of New York at Syracuse. He called the systems **living polymers** because the chains remain reactive until some reagent is deliberately added to stop the reaction. Note that two anionic centers have no tendency to react with each other. Thus termination reactions involving the combination of two growing chains, an important reaction for free radicals, do not occur in anionic polymerizations. If the reagents are pure and the reaction mixture is protected from moisture, the polymer chains remain reactive. An anionic polymerization differs from a free-radical reaction in another way. Because all chains start at the same time, at the moment the anionic initiator is added to the reaction mixture, and the intermediates react with monomer at the same rate, polymers of remarkably uniform chain lengths and molecular weights are formed. In comparison, there is a much larger range of molecular weights in any sample of a polymer prepared by free-radical reactions.

Copolymerization of two monomers by an anionic mechanism can be done in such a way that large portions of the polymer chain consist of units of one monomer and other portions contain only units of the second one. Such a polymer is called a **block copolymer,** in contrast to a **random** or **alternating copolymer.** For example, the free-radical polymerization of a mixture of styrene and acrylonitrile gives an alternating copolymer (see p. 1047). It is also possible to create a block copolymer of these two monomers, as is illustrated in Figure 24.2. By controlling the concentration of the initiator and the amount of the first monomer used, the length of the chain in the first dianion can be determined. Addition of a known amount of a second monomer results in the growth of the chain to create a new dianion. Alternating additions of styrene and acrylonitrile give a long chain in which sections consisting of polystyrene are attached to sections that are polyacrylonitrile.

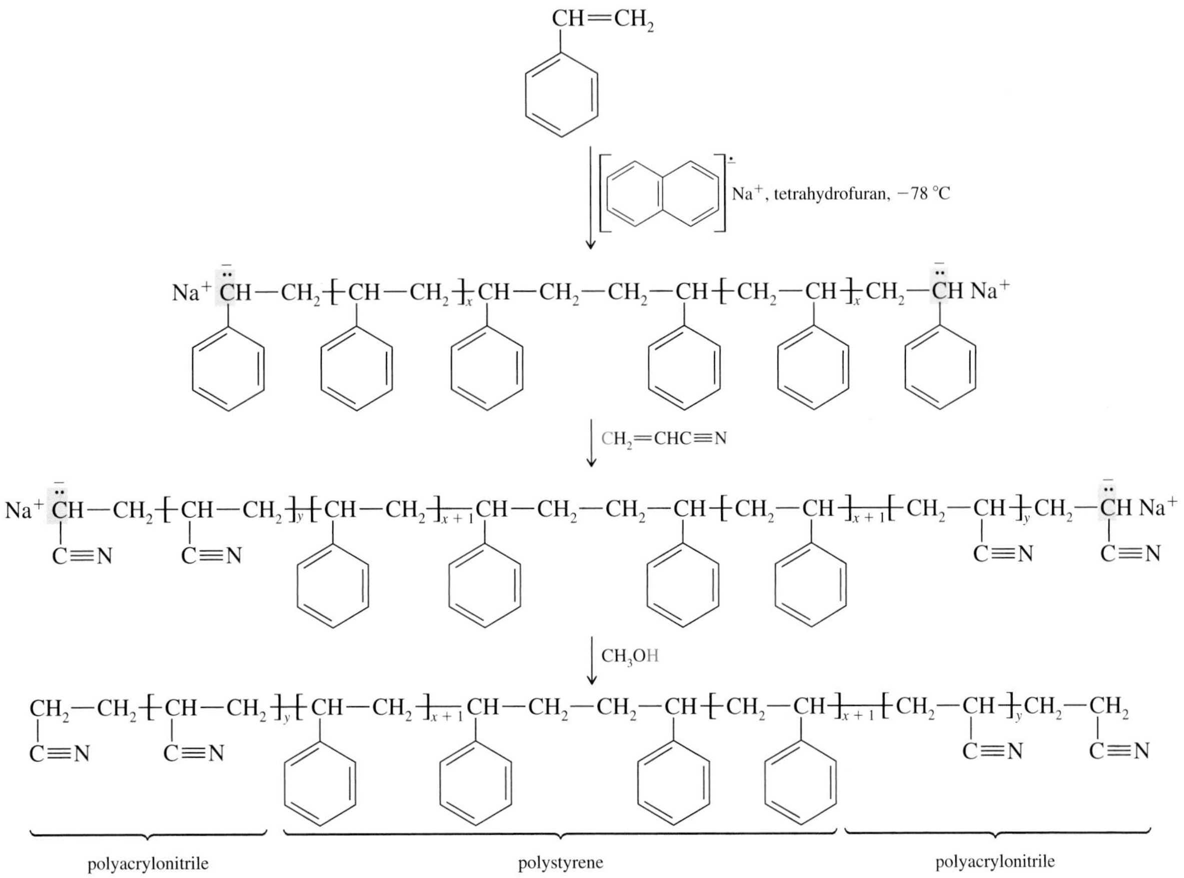

**Figure 24.2**
Formation of a block copolymer by the anionic polymerization of styrene and acrylonitrile.

**Problem 24.4**     Complete the following equations, showing the intermediates that would be responsible for the polymerization reactions.

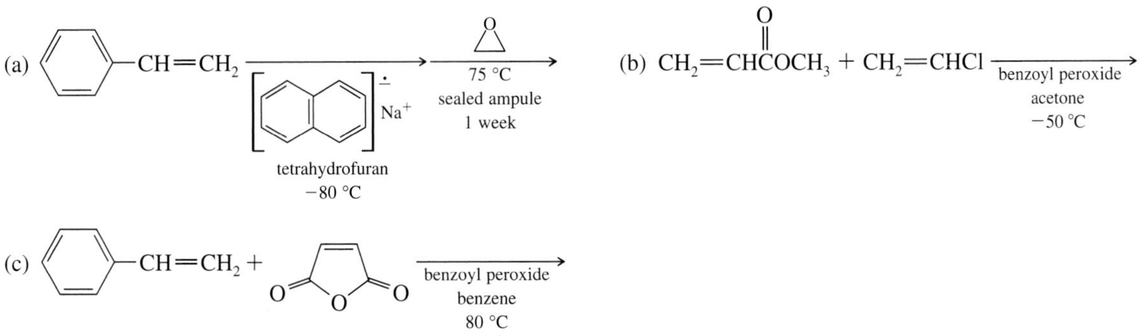

(a)

(b) $CH_2=CHCOCH_3 + CH_2=CHCl$ $\xrightarrow[\substack{\text{acetone} \\ -50\,°C}]{\text{benzoyl peroxide}}$

(c)

**Problem 24.5**     A block copolymer of methyl 2-methylpropenoate (methyl methacrylate) and isopropyl propenoate (isopropyl acrylate) has been prepared. Write equations showing how you would carry out such a preparation.

## D. Cationic Reactions

The acid-catalyzed polymerization of an alkene was explored fairly early in this book (p. 287). Cyclic ethers also undergo polymerization reactions that follow cationic mechanisms. The most interesting of these are the reactions of oxetane and tetrahydrofuran.

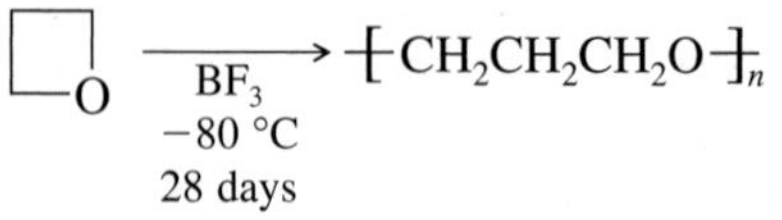

oxetane            *a polyether*
95%

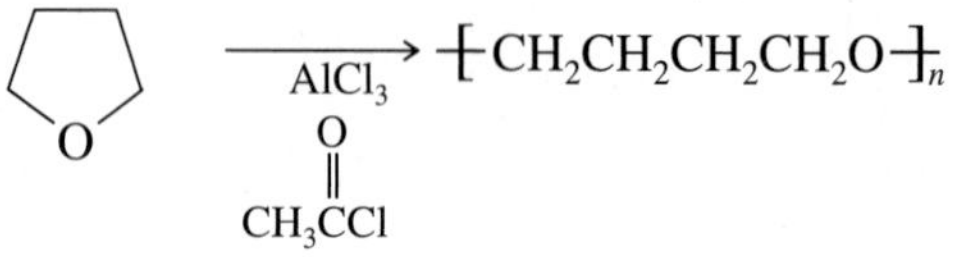

tetrahydrofuran          *a polyether*

In each case, the oxygen atom of the cyclic ether is converted to a good leaving group by protonation or by reaction with a carbocation. The oxygen atoms of other ether molecules then serve as nucleophiles in $S_N2$ reactions and are converted in turn to reactive intermediates that propagate the chain reaction. The sequence is shown below for oxetane.

**VISUALIZING THE REACTION**

**Cationic Polymerization of a Cyclic Ether**

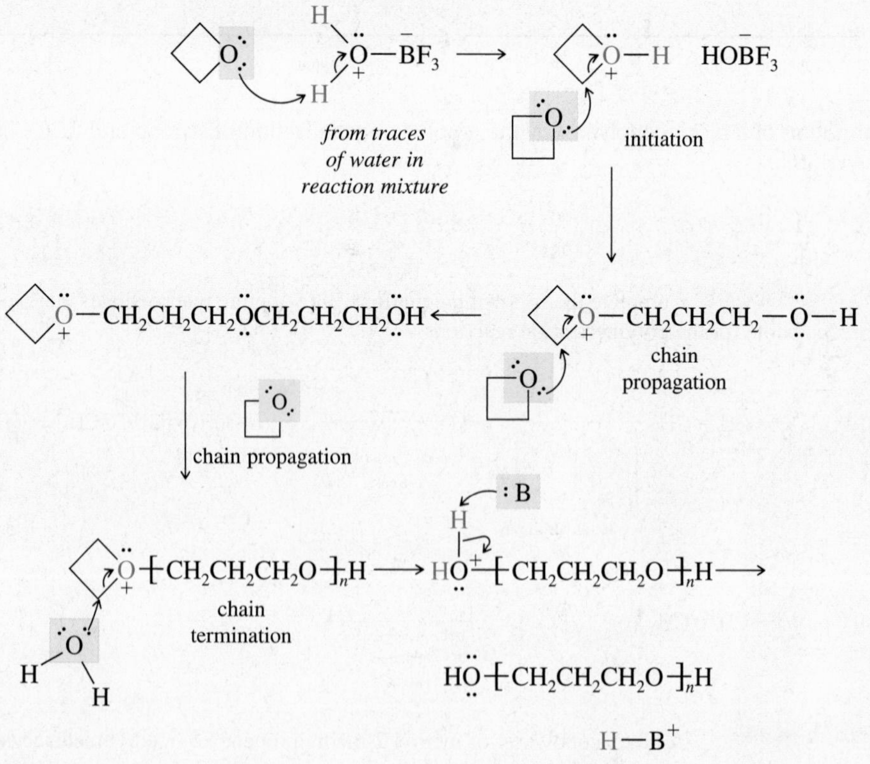

Termination of the reaction occurs when the growing chain reacts with some nucleophile, such as water.

**Problem 24.6** What is the reactive species that initiates the polymerization of tetrahydrofuran in the presence of aluminum chloride and acetyl chloride? Write equations showing the initiation and chain-propagating steps of the reaction.

**Problem 24.7** The following compounds polymerize under the conditions shown. Complete the equations, showing the intermediates that must be involved.

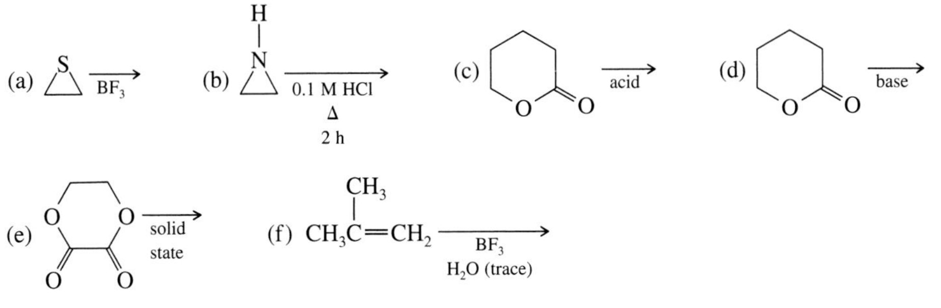

(a) $\overset{S}{\triangle} \xrightarrow{BF_3}$

(b) $\overset{\overset{H}{|}}{\underset{N}{\triangle}} \xrightarrow[\substack{\Delta \\ 2\ h}]{0.1\ M\ HCl}$

(c) [six-membered lactone ring] $\xrightarrow{acid}$

(d) [six-membered lactone ring] $\xrightarrow{base}$

(e) [cyclic dianhydride] $\xrightarrow[state]{solid}$

(f) $CH_3\overset{\overset{\textstyle CH_3}{|}}{C}{=}CH_2 \xrightarrow[H_2O\ (trace)]{BF_3}$

**Problem 24.8** A polymer is synthesized from 1,2-bis(4-hydroxyphenyl)ethane and 1,8-dibromooctane in the presence of base. Draw the structure of a portion of this polymer long enough to identify the repeating unit of the polymer.

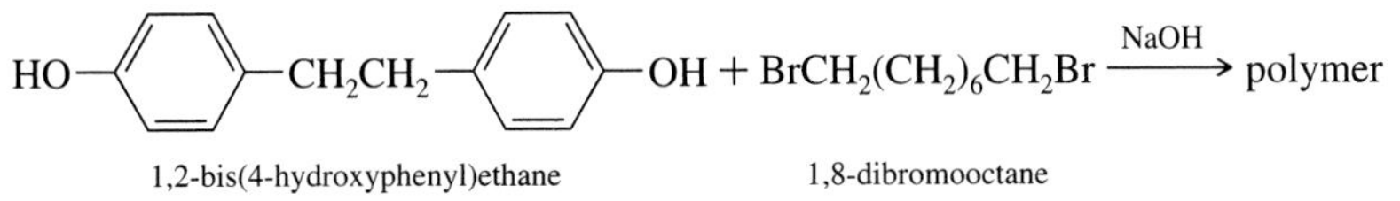

$HO-\bigcirc-CH_2CH_2-\bigcirc-OH + BrCH_2(CH_2)_6CH_2Br \xrightarrow{NaOH} polymer$

1,2-bis(4-hydroxyphenyl)ethane          1,8-dibromooctane

# 24.3 Polymerization Reactions with Controlled Stereochemistry

## A. Stereochemical Regularity in Polymer Structures

The repeating nature of the structure of a polymer chain allows three possible kinds of stereochemical relationships among the substituents on the chain. These are illustrated for polypropylene, formed from propene (propylene).

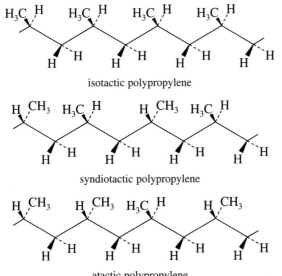

isotactic polypropylene

syndiotactic polypropylene

atactic polypropylene

For the first of these examples, shown with the chain stretched out into an extended form, all the methyl groups are on the same side of the chain. The configuration at each carbon of the chain is the same. Such a polymer is said to be an **isotactic polymer.** In a **syndiotactic polymer,** there is a regular alternation in the configuration at the tetrahedral carbon atoms bearing the substituents. In syndiotactic polypropylene, one methyl group is behind the plane of the paper, the next one in front, and the third one behind again. The third form of polypropylene is an **atactic polymer,** one that has no stereochemical regularity. The arrangement of the methyl groups is completely random.

Whether a polymer is atactic, isotactic, or syndiotactic is important in determining its properties. Of hydrocarbon polymers, only isotactic and syndiotactic ones have the regularity of structure necessary for crystallinity and thus the physical properties required for many applications. Indeed, early attempts to polymerize propylene using radical and cationic initiators resulted in soft polymers that were of no practical use. It was not until the development of new polymerization catalysts in the 1950s made it possible to make isotactic polypropylene (p. 1055) that the polymer was used industrially. Many household objects such as pails, dishpans, and plastic containers are made of polyethylene or polypropylene. Crystalline polypropylene, for example, has a melting point around 170 °C and can therefore be used in objects that will be exposed to boiling water. It can also be made into tough films such as those used in packaging food.

---

**Problem 24.9**    Write structural formulas showing fragments of isotactic, syndiotactic, and atactic polystyrene.

**Problem 24.10**    Why is isotactic or syndiotactic polypropylene more highly crystalline than the atactic polymer?

**Problem 24.11**    A free-radical polymerization reaction of methyl methacrylate (Problem 24.5) produces atactic poly-(methyl methacrylate). Give the structural formula for a fragment of such a polymer chain.

---

## B.  Heterogeneous Catalysis. Ziegler–Natta Catalysts

The polymerization reactions described in the first two sections of this chapter do not give rise to polymers having a high degree of stereochemical regularity. A careful choice of catalysts and solvent systems, however, makes it possible to control the stereochemistry of polymerization for some monomers. In the 1950s, Karl Ziegler of West Germany and Giulio Natta of Italy developed a catalyst system that polymerizes alkenes to give linear polymers with high stereoselectivity. These catalysts, which consist of mixtures of transition metal halides with organometallic compounds, mostly trialkylaluminums, are not soluble in the alkane solvents that are used for polymerizations of alkenes. The reaction mixture is, therefore, a heterogeneous one, and polymerization takes place at the surface of the catalyst. The use of these catalysts has revolutionized the production of alkene polymers. Polyethylene prepared in this way, for example, is almost completely linear and highly crystalline; other methods of polymerization result in a branched product of much lower strength and chemical resistance and thus less usefulness. Ziegler and Natta were jointly awarded the Nobel Prize in 1963 for the work that led to the discovery of these catalysts.

A typical reaction using a Ziegler-Natta catalyst is the preparation of crystalline polystyrene.

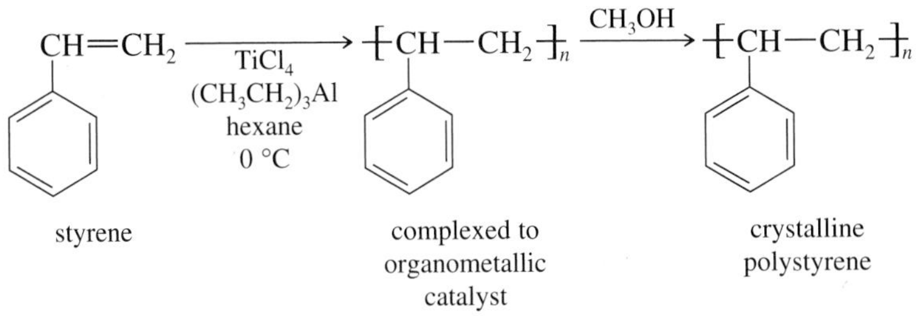

CH=CH2
TiCl4
(CH3CH2)3Al
hexane
0 °C
CH—CH2
CH3OH
CH—CH2
styrene
complexed to organometallic catalyst
crystalline polystyrene


In a similar reaction, ethylene gives a polymer called high-density polyethylene to distinguish it from the softer, more highly branched polymer obtained by other methods.

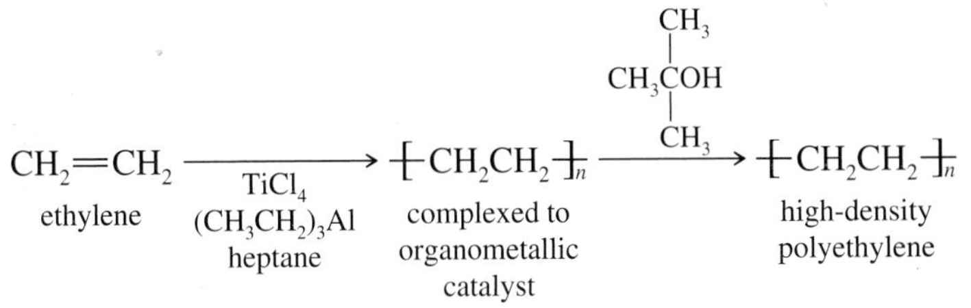

CH2=CH2
ethylene
TiCl4
(CH3CH2)3Al
heptane
CH2CH2
complexed to organometallic catalyst
CH3
CH3COH
CH3
CH2CH2
high-density polyethylene


Titanium tetrachloride is a typical transition metal halide, and triethylaluminum is the kind of organometallic compound used in the catalyst. They react with each other to give an organotitanium compound, which is believed to play the major role in the catalysis.

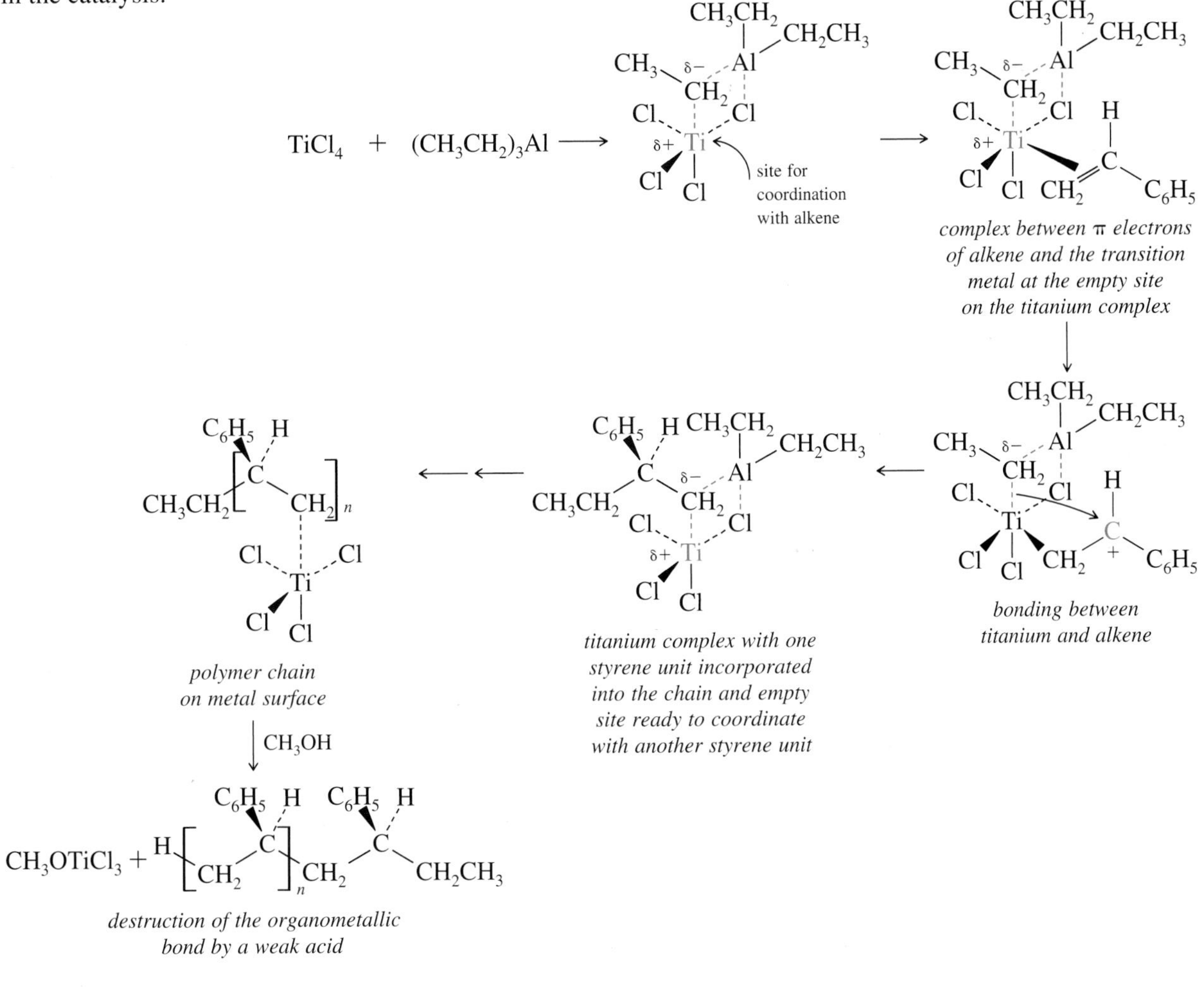

TiCl4 + (CH3CH2)3Al
site for coordination with alkene
complex between π electrons of alkene and the transition metal at the empty site on the titanium complex
bonding between titanium and alkene
titanium complex with one styrene unit incorporated into the chain and empty site ready to coordinate with another styrene unit
polymer chain on metal surface
CH3OH
CH3OTiCl3 +
destruction of the organometallic bond by a weak acid

The reaction is pictured as proceeding through a $\pi$ complex between the transition metal and the monomer. As bonding develops between the metal and the alkene, the metal complex acquires a partial negative charge while the benzylic carbon atom acquires a partial positive charge. At some point, the alkyl group bonded to the metal moves to the developing carbocation with a pair of electrons. The organotitanium compound that is formed is able to repeat this process, successively inserting monomer units between the metal atom and the rest of the growing polymer chain. The reaction takes place in a highly stereoselective way for reasons that are not well understood, partly because the exact nature of the catalyst is not known. A highly simplified representation was used on the previous page. Exactly what the organoaluminum compounds do during the reaction and whether it takes place at a single metal atom or with the cooperation of several are not known. The stereoselectivity is believed to arise from the stereochemistry imposed on the monomer as it is adsorbed onto the catalyst. There is no doubt that the nature of the metal catalyst somehow determines the stereochemistry of the reaction. For example, isoprene can be polymerized to give all *cis*- or all *trans*-1,4-polyisoprene stereoselectively depending on what transition metal is used in the catalyst (pp.1075–1076).

⊘ **Study Guide**
**Concept Map 24.3**

**Problem 24.12**   Complete each of the following equations, showing the structure of the polymer you expect to be produced. Where it is known with certainty, the stereochemical nature of the polymer is indicated. Show this in the structural formula for the polymer. Assume that all copolymers have regular alternating structures.

(a) $CH_3CH_2CHCH=CH_2$ (with $CH_3$ substituent) $\xrightarrow[(CH_3CH_2)_2AlCl]{TiCl_4}$ $\xrightarrow{CH_3CH_2OH}$ isotactic

(b) $CH_2=CHOCH_3$ $\xrightarrow[(CH_3CHCH_2)_3Al \\ heptane]{VCl_3 \\ CH_3}$ $\xrightarrow{CH_3CH_2OH}$ isotactic

(c) $CH_2=CH_2 +$ (butene with two $CH_3$ and two $H$) $\xrightarrow[(CH_3CHCH_2)_2AlCl \\ heptane \\ -30\ ^\circ C]{VCl_3 \\ CH_3}$ $\xrightarrow{CH_3CH_2OH}$ syndiotactic

(d) $CH_2=CCOCH_3$ (with $CH_3$ and $O$ substituents) $\xrightarrow[(CH_3CHCH_2)_3Al \\ heptane]{VCl_3 \\ CH_3}$ $\xrightarrow{CH_3CH_2OH}$

(e) $CH_3CH_2CH=CH_2$ $\xrightarrow[(CH_3CH_2)_2AlCl]{FeCl_3}$ $\xrightarrow{CH_3CH_2OH}$

## 24.4   Step–Growth Polymerization

### A. Polyamides

The preparation of nylon 66 from a mixture of 1,6-hexanediamine and adipic acid (hexanedioic acid) (p. 1041) is an example of a **step-growth polymerization.** In such a reaction, the monomer units contain functional groups that are capable of reacting with each other without the formation of a reactive intermediate. The reactions are generally slower than those of chain-growth polymerizations, and the reaction sites are more random. The formation of a polyamide from a diamine and a diacid is an example of a step-growth reaction.

*First step of polymerization*

$$H_2N(CH_2)_6NH_2 + HOC(CH_2)_4COH \longrightarrow H_2N(CH_2)_6NHC(CH_2)_4COH + H_2O$$

1,6-hexanediamine    hexanedioic acid

monomers

*reactivity of amine function and carboxyl function not much different from reactivity of these functional groups in monomers*

*Second step of polymerization*

$$2 \ H_2N(CH_2)_6NH_2 + 2 \ HOC(CH_2)_4COH + 2 \ H_2N(CH_2)_6NHC(CH_2)_4COH$$

↓

$$H_2N(CH_2)_6NHC(CH_2)_4COH + HOC(CH_2)_4CNH(CH_2)_6NHC(CH_2)_4COH$$

monoamide

diamide from 1 diamine
and 2 diacid units

$$+ \ H_2N(CH_2)_6NHC(CH_2)_4CNH(CH_2)_6NH_2 + 3 \ H_2O$$

diamide from 1 diacid
and 2 diamine units

A closer look at the individual steps of the above reaction reveals the problems associated with step-growth polymerization. The first step of the polymerization gives an amide formed from the diamine and the diacid. The amide retains amino and carboxylic acid groups, and the reactivities of these functional groups are not much different from the reactivities of the same functional groups in the monomers. The next stage is a random reaction of the carboxylic function of the monoamide with the diamine and of the amine function of the monoamide with another molecule of diacid. Instead of a steady and directed growth of a chain, a mixture of molecules results. Large increases in chain length will take place only when the monomer units have been used up and amides containing several units combine with each other. The polymerization process takes place in jumps as well as by unit-by-unit addition to a continuously growing chain. Products of step-growth polymerization reactions are **condensation polymers** formed by condensing together two different types of functional groups with the elimination of a small stable molecule, such as water. Nylon 66 is a typical polyamide.

Polyamides interact with water by hydrogen bonding at the amide bonds. The properties of a polyamide change somewhat, therefore, with changes in humidity. The chains of nylon 66, for example, become more mobile with increasing humidity as the hydrogen bonds between chains are replaced by hydrogen bonds to water molecules. Nylons that are particularly resistant to the absorption of moisture have been produced by making the hydrocarbon-like portions of the molecule larger.

Polyamides containing aromatic rings form highly crystalline structures of exceptional strength. Fibers made from these polyamides, known as aramids, are stiffer than steel at a much lower density. Such polymers are being used to reinforce tires and to make bullet-proof vests and lightweight but very strong cords and cables.

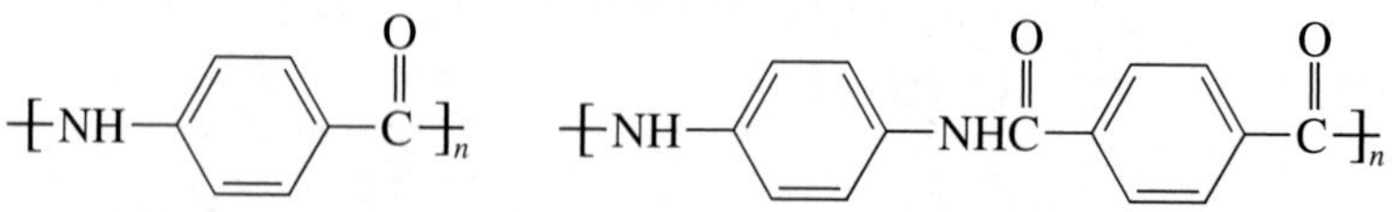

Kevlar

*two polyamides that have high molecular rigidity*

**Problem 24.13**   Polyamides containing aromatic rings (aramids) have very high melting points and glass transition temperatures, which makes it difficult to use them to make articles by processes that use melting. Much research has focused on modifying the structures of aramids to change their properties so that they are easier to use in manufacturing. One such attempt used the following reactions. Draw the structures of the intermediate compounds and of the polymer present at the end of the process.

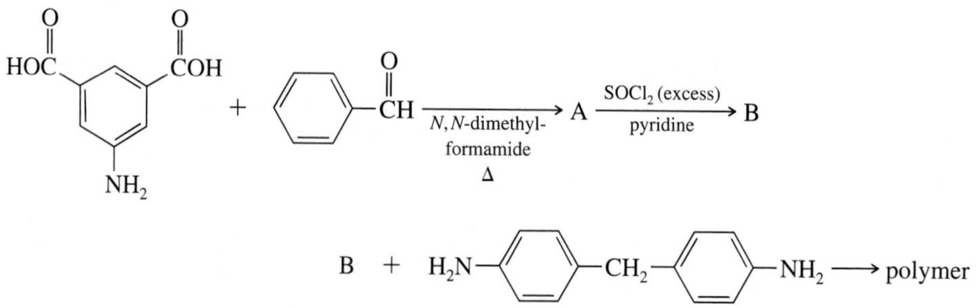

## B. Polyesters

Poly(ethylene terephthalate), a typical polyester, is prepared by two transesterification reactions (p. 626) from dimethyl terephthalate and ethylene glycol. If the lower-boiling alcohol component that is formed at each stage is continuously removed from the reaction mixture, the polymerization is driven to completion. The first stage of this polymerization reaction is shown below.

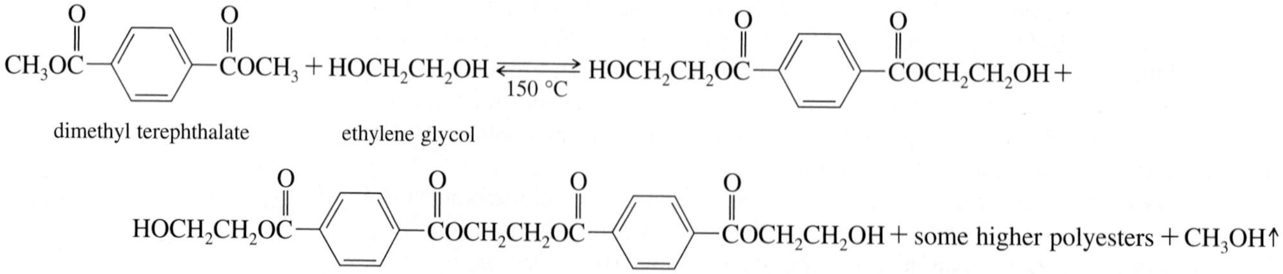

Polyamides and polyesters have been put to many uses. Both are important in the synthetic fibers industry. For example, much cotton clothing contains at least some polyester fiber. Strong, relatively inflexible, but very light polyester films are used in sails for racing yachts and in the wings of human-powered planes.

---

**Problem 24.14**  Two important commercial products are the polycarbonate and polyester having the following structures.

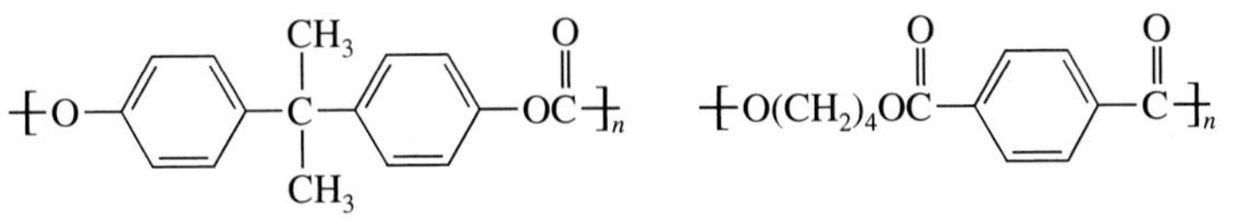

*a polycarbonate*                *a polyester*

Both are prepared by a transesterification reaction (p. 626). Dissect the structure of each polymer, decide what types of reagents are required, and write general equations for its preparation.

---

## C. Polyurethanes

The reaction that forms the backbone of a polyurethane is the addition of an alcohol to an isocyanate (p. 830). For example, ethylene glycol reacts with 4,4'-diphenylmethane diisocyanate to give a polyurethane.

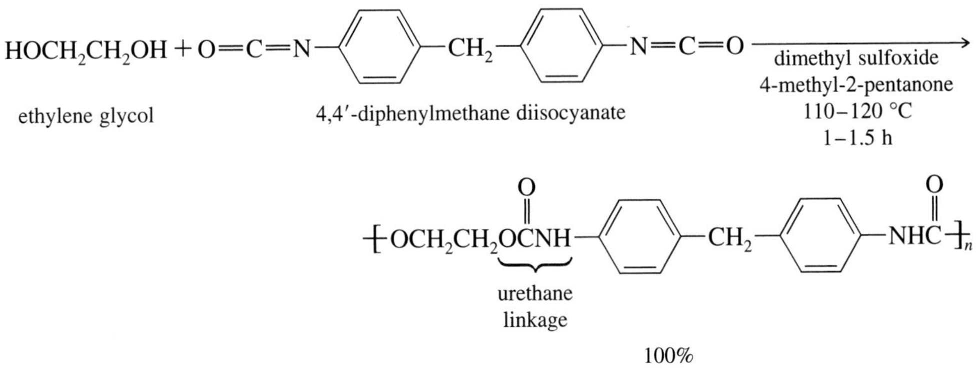

HOCH$_2$CH$_2$OH + O=C=N— ... —CH$_2$— ... —N=C=O $\xrightarrow[\substack{\text{4-methyl-2-pentanone} \\ 110-120\ °C \\ 1-1.5\ h}]{\text{dimethyl sulfoxide}}$

ethylene glycol          4,4'-diphenylmethane diisocyanate

urethane
linkage

100%

This polymerization is carried out in a mixture of dimethyl sulfoxide and a ketone, which keeps the polymer in solution so that the reaction is not stopped by the precipitation of the high-molecular-weight product. Complete polymerization is possible under these conditions.

The chief application of polyurethanes is as foams that are used as cushioning material in furniture, pillows, mattresses, and automobile seats. For polyurethanes to fulfill this function, there must be a moderate amount of cross-linking between polymer chains and some method of creating bubbles in the melted polymer. Compounds that have rubbery properties usually have long, flexible polymer chains made from diol units that are themselves polymers. When cross-linking is desired, an excess of the diisocyanate is used so that some of the polymer chains end in unreacted isocyanate functions. This type of molecule is represented by using a wavy line to join the two isocyanate groups. The polymeric diisocyanate reacts with the urethane linkages in other polymer chains to link them.

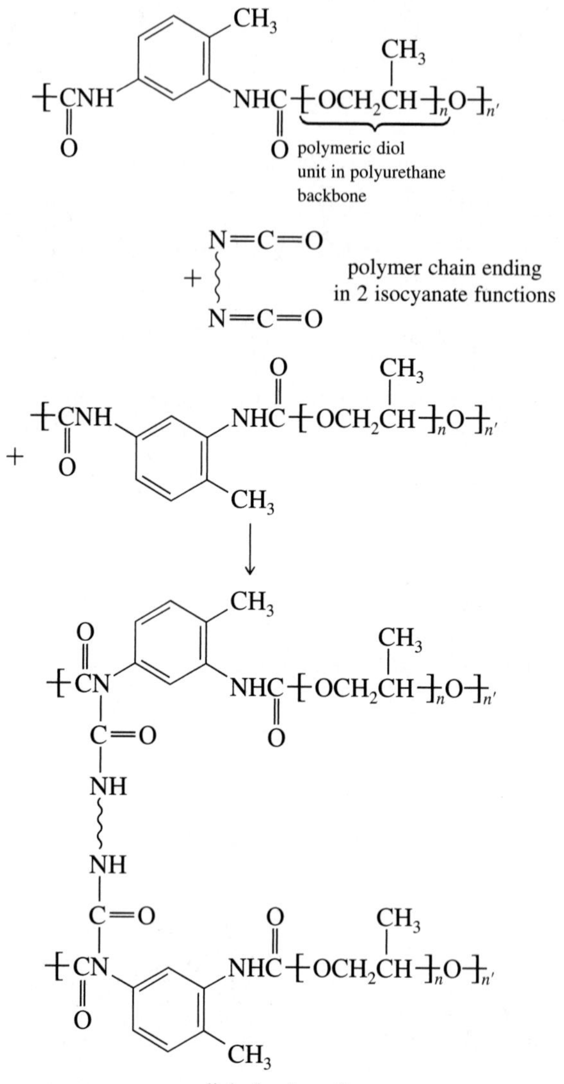

cross-linked polyurethane

If the cross-linking chains are long and flexible, the polyurethane network also will be flexible; that is, the polymer will be an elastomer. If a large number of cross-links are formed and the cross-linking chains are short and rigid, the whole polymer will be hard and inflexible.

An isocyanate reacts with water to give carbon dioxide and an amine. This reaction is one method used to create polyurethane foam (Figure 24.3). A small amount of water is added to the hot polymer at a stage when many isocyanate end groups are present. The carbon dioxide forms bubbles that expand in the hot polyurethane, giving it a foamy texture that it retains when it cools. Meanwhile, the new amine groups that are created in the process react with the remaining excess isocyanate groups to give urea bonds. Toward the end of the polymerization process, no reactive end groups are left, and the structure of the foam is reinforced by further cross-linking reactions.

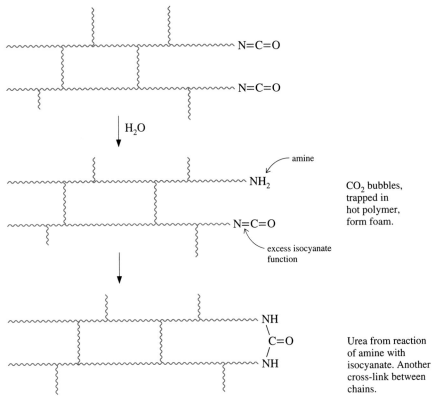

**Figure 24.3**
Reaction of isocyanate end groups in polyurethane to give carbon dioxide and further cross-linking of the polymer.

CO₂ bubbles, trapped in hot polymer, form foam.

Urea from reaction of amine with isocyanate. Another cross-link between chains.

---

**Problem 24.15**   Complete the following equations.

(a) $\underset{\parallel}{\overset{O}{\text{ClCO}}}\text{(CH}_2)_4\underset{\parallel}{\overset{O}{\text{OCCl}}} + \text{H}_2\text{N(CH}_2)_6\text{NH}_2 \xrightarrow[\text{H}_2\text{O}]{\text{NaOH}}$

(b) $\text{O}=\text{C}=\text{N(CH}_2)_6\text{N}=\text{C}=\text{O} + \text{HO(CH}_2)_4\text{OH} \xrightarrow{185-195\ °C}$

(c) image of aromatic diisocyanate $+ \text{HOCH}_2\text{CH}_2\text{OH} \xrightarrow[\substack{\text{4-methyl-2-pentanone} \\ 115\ °C}]{\text{dimethyl sulfoxide}}$

## D. Polymers Produced by Condensation Reactions of Formaldehyde

Polymers produced in reactions of formaldehyde with phenols have been known for over a hundred years. Leo Hendrik Baekeland of the United States was the first chemist to patent such a material at the beginning of this century. The polymer, Bakelite, is a stiff, three-dimensional network with very little solubility in organic solvents and a high resistance to electricity and heat. It is used in a wide variety of household objects and electrical fixtures.

Phenols condense with formaldehyde under either acidic or basic conditions. Under acidic conditions, polymerization gives a network of phenol rings held together by methylene groups at the ortho and para positions. These polymers exhibit a broad range of molecular weights in any one sample.

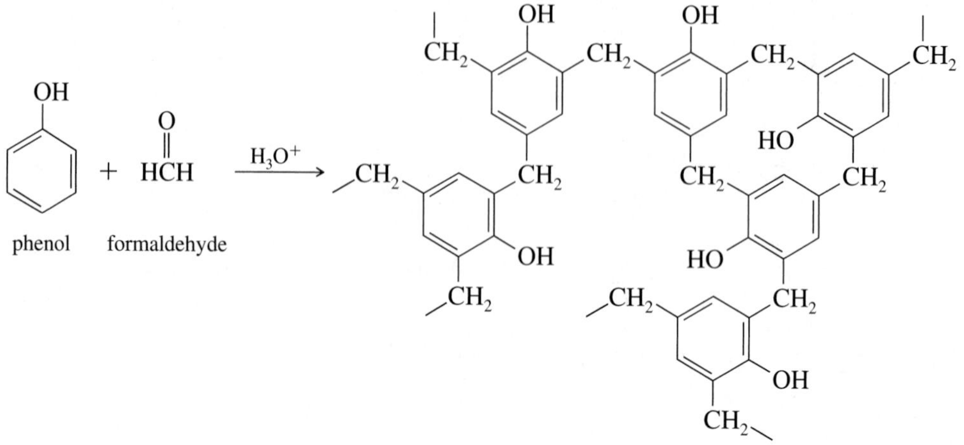

The reaction is thought to start by protonation of the carbonyl group of the aldehyde (or its hydrate) by the acid catalyst to give an electrophile that reacts with the phenol ring, which is highly activated toward electrophilic aromatic substitution at the ortho and para positions (p. 366). The resulting benzylic alcohols are converted into electrophilic cations by protonation and loss of water, and further electrophilic substitution of phenols takes place to create the network.

**Problem 24.16**   Write a mechanism showing how formaldehyde condenses with phenol in the presence of acid. How is the product of that reaction converted into an electrophile that reacts with another unit of phenol?

Commercially important plastics are also obtained when urea reacts with formaldehyde. In such reactions, amino groups add to formaldehyde to give intermediates that have hydroxymethyl groups.

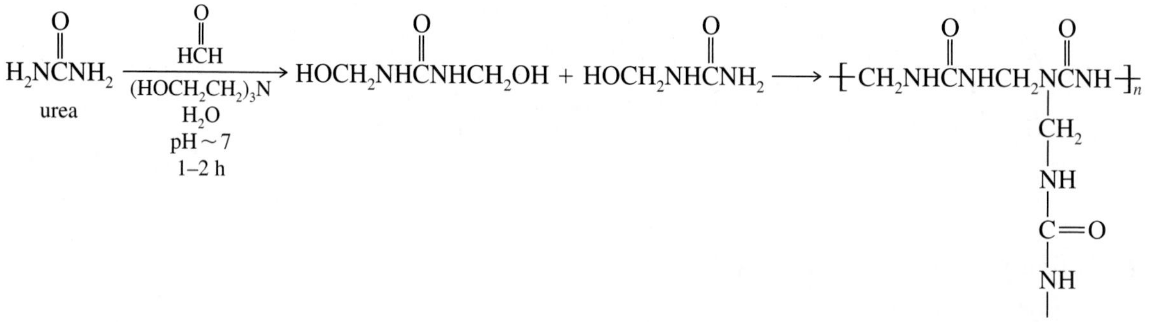

The resins obtained from the kind of reaction shown above have many applications. A small amount of urea-formaldehyde resin applied to a cotton fabric gives it resistance to creasing. In fact, the predominant odor in a large fabric store is often that of formaldehyde. Boards are formed by blending sawdust or wood chips with such a resin and curing the mixture in a hot mold. Packaging paper is made more resistant to moisture by treating the paper pulp with urea-formaldehyde resin. Currently, however, there is some concern that exposure to formaldehyde may be a health hazard.

**Problem 24.17**   Write a detailed mechanism for the reaction of urea with formaldehyde to produce a cross-linked polymer.

**Problem 24.18**   Highly branched polyethers can be synthesized if polyphenols bearing a leaving group at the benzylic position are treated with base. Steps in the synthesis of such a polyether are shown below. Supply a reagent for the first step of the reaction and draw the structural formula for a segment of the product polyether.

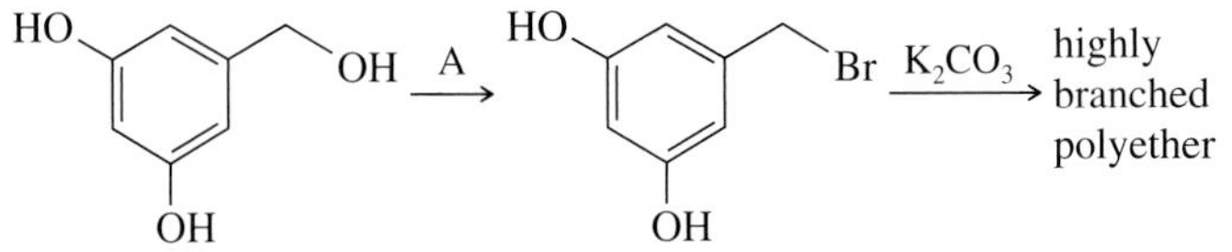

## E. Epoxy Resins

Epoxy resins are in everyday use as adhesives. Such a resin is formed on mixing two solutions: one of a polymer that contains oxirane rings that will form cross-links between polymer chains when treated with a nucleophile and the other containing a polyamino compound, which is the nucleophilic reagent that starts the process and forms some of the cross-links itself. The polymer component contains hydroxyl groups, ether linkages, and oxirane rings. All these functional groups, along with the amino groups that are added during the cross-linking process, hydrogen-bond and coordinate strongly with surfaces such as glass, ceramics, and metal. Thus the epoxy resin glues such surfaces tightly together.

Cross-linking takes place when an amine reacts with the oxirane end groups of the polymer. The linking groups may be amines or the alkoxide ions that are created when the oxirane ring is opened (Figure 24.4). Because the cross-linking takes place very fast, the amine and the linear polymer are mixed just before they are applied to

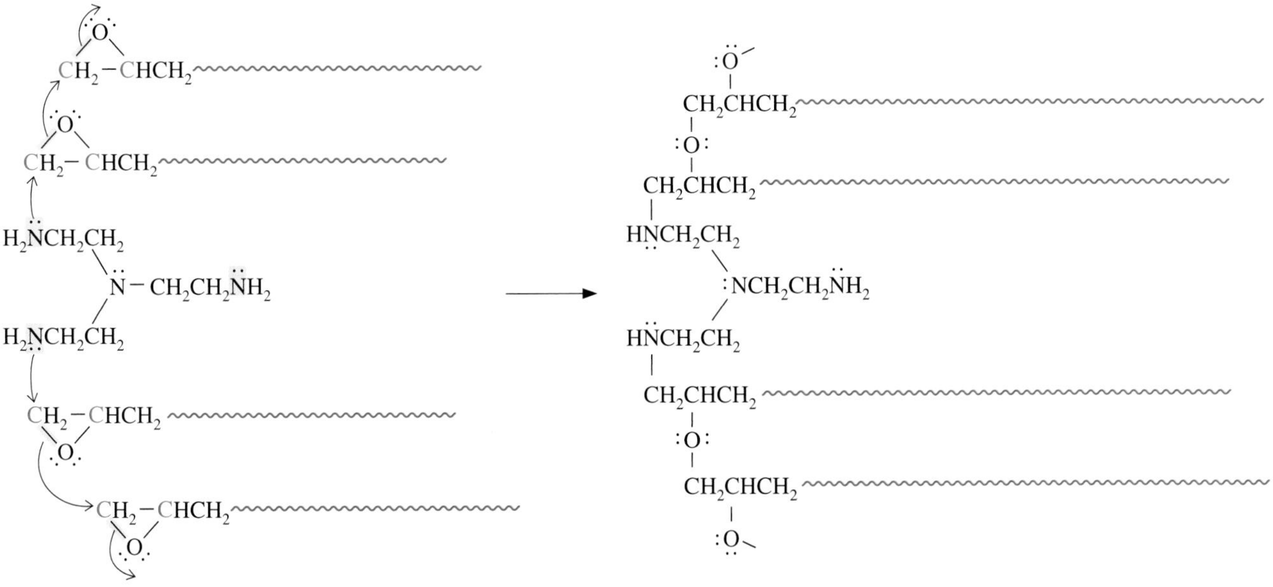

*Figure 24.4*
Cross-linking of an epoxy resin.

the surfaces that they are to bond together. The bonding is so strong that mended objects frequently break elsewhere before they fall apart at the site of repair.

The most common epoxy resin is made from the oxirane derived from 3-chloro-1-propene and 2,2-bis(4′-hydroxyphenyl)propane, shown below.

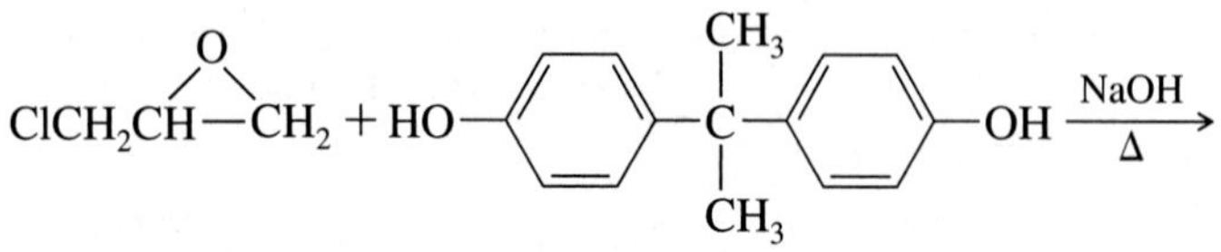

oxirane from
3-chloro-1-propene
epichlorohydrin

2,2-bis(4′-hydroxyphenyl)propane

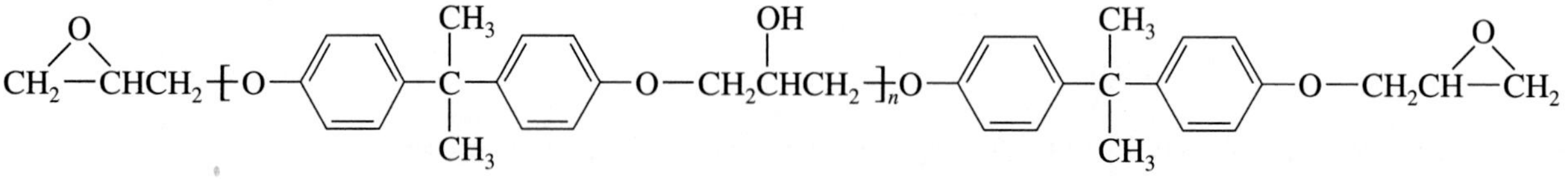

The oxirane is used in excess so that the end groups of the polymer chains are oxiranes.

The oxirane derived from 3-chloro-1-propene has the common name epichlorohydrin and is a toxic compound suspected of being a carcinogen. The reason for its toxicity is the ease with which it reacts with nucleophiles including biological nucleophiles, to link them together. When the oxirane ring is opened by the attack of one nucleophile, the alkoxide ion that is formed displaces chloride ion intramolecularly to give a new oxirane. This ring, in turn, opens in another nucleophilic substitution reaction.

**Study Guide**
**Concept Map 24.4**

---

## VISUALIZING THE REACTION

### Epichlorohydrin as a Linking Agent for Nucleophiles

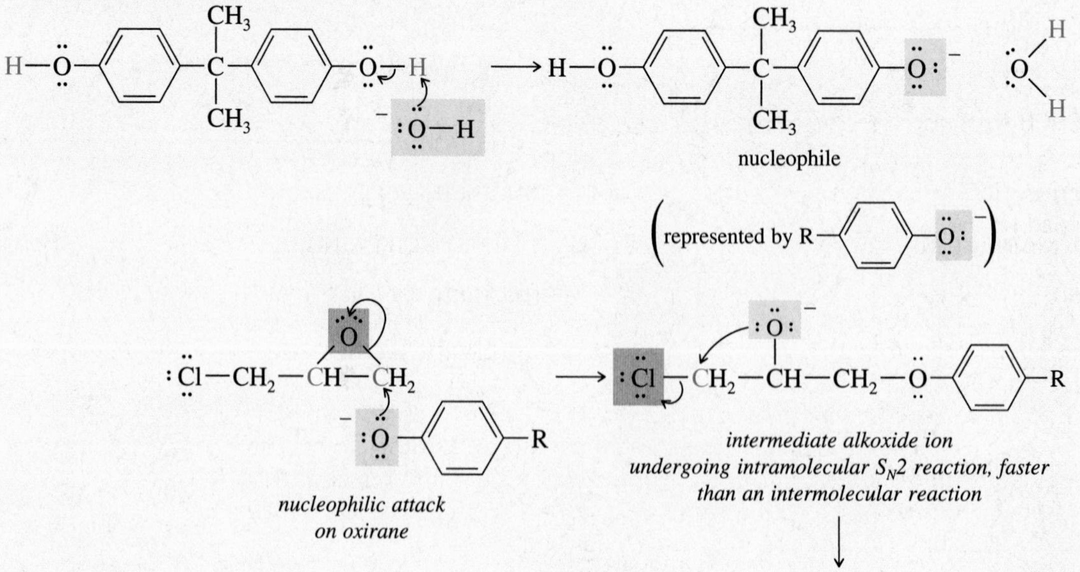

nucleophile

(represented by R—⬡—Ö:⁻)

nucleophilic attack
on oxirane

intermediate alkoxide ion
undergoing intramolecular $S_N2$ reaction, faster
than an intermolecular reaction

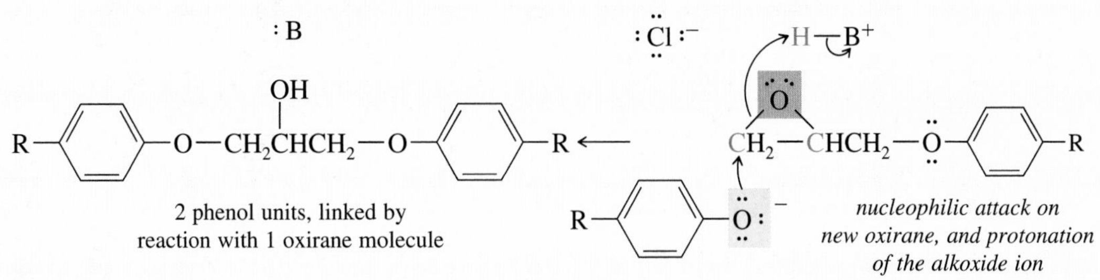

**Problem 24.19**   The diphenol used in the preparation of polycarbonates and epoxy resin has the trivial name bisphenol A because it is synthesized from phenol and acetone. Propose a mechanism for its formation from these reagents.

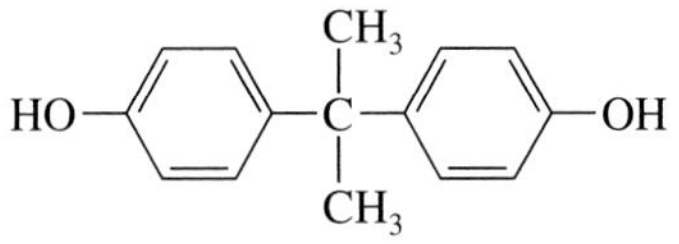

2,2-bis(4′-hydroxyphenyl)propane
bisphenol A

**Problem 24.20**   The commercial synthesis of the oxirane derived from 3-chloropropene starts with propene and involves the use of gaseous chlorine at high temperatures, then chlorine with water at low temperatures, and finally calcium hydroxide. Write equations for these reactions showing how they lead to the formation of 2-(chloromethyl)oxirane. Classify each reaction according to its mechanistic type.

# 24.5  Dendrimers

Over the past 10 years, an increasing amount of research has been done in synthesizing **dendrimers,** macromolecules in which successive layers of branches, like those of a tree, are built onto a central core. A series of repetitive reactions increases the size and the branching until the macromolecule reaches a size and a density of branches that effectively stop further regular growth. At this point, the dendrimer has a three-dimensional shape that depends to some degree on the shape of the original core molecule but usually approximates a sphere. The dendrimer at this stage is a solvent-filled spheroid, which in some ways resembles a cell, except that it is a cell consisting of a single molecule.

Dendrimers differ from the polymers that we discussed earlier in that they have well-defined structures and molecular weights, not the range of molecular weights usual for addition and condensation polymers. Several steps in the synthesis of a polyamidoamine dendrimer are shown on the next page. The core molecule for this dendrimer is ammonia, which adds in a Michael reaction (p. 713) three times to methyl acrylate. The dendrimer starts growing in three directions at once.

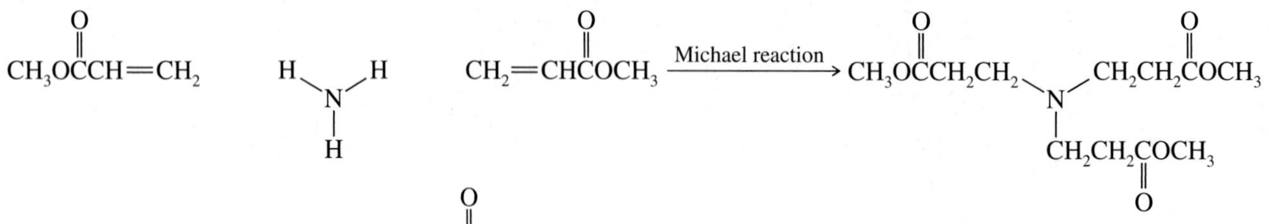

$$CH_3OCCH = CH_2 \qquad H\text{-}N\text{-}H \qquad CH_2 = CHCOCH_3 \xrightarrow{\text{Michael reaction}}$$

beginning of the
dendrimer with
electrophilic groups
on the outside

At this stage the central nitrogen atom is surrounded by electrophilic ester groups, which then react with ethylenediamine in acylation reactions, giving rise to the core that generates the dendrimer.

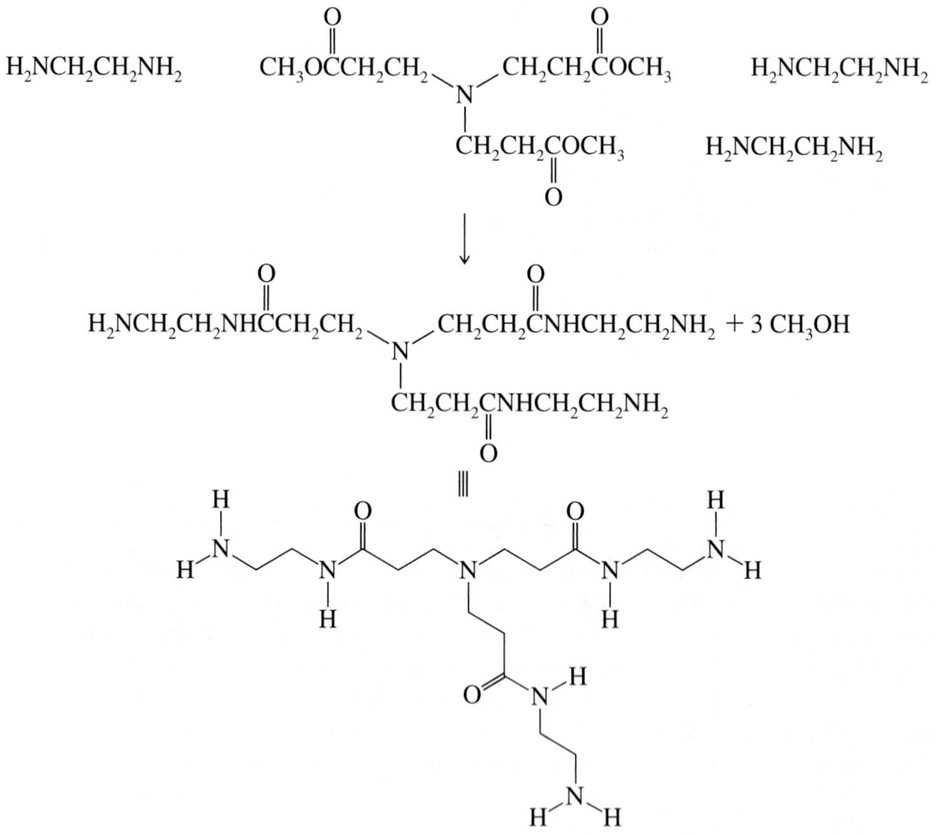

$$H_2NCH_2CH_2NH_2 \qquad \qquad H_2NCH_2CH_2NH_2$$

$$H_2NCH_2CH_2NH_2$$

$$+ 3\ CH_3OH$$

dendrimer core with nucleophilic
groups on the outside

The dendrimer grows from this point by alternate Michael additions to methyl acrylate and reaction with ethylenediamine. The surface of the dendrimer is alternately electrophilic and nucleophilic. After one more cycle, it looks like this:

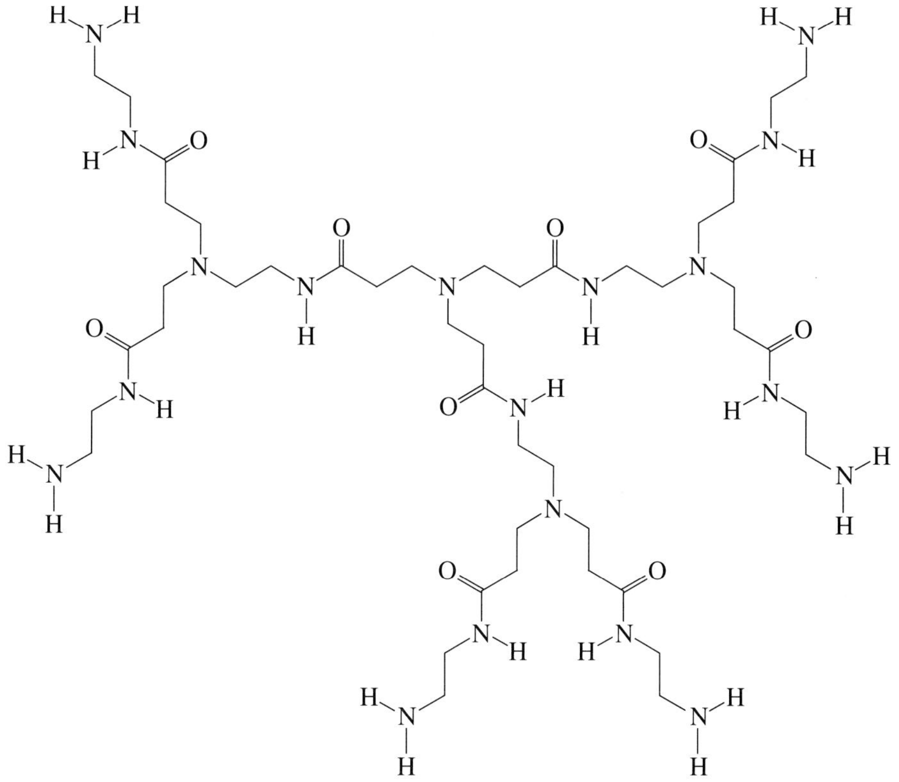

Note that the dendrimer grows geometrically, not linearly. We went from six reactive sites in the core to twelve at the end of what is called the first generation of growth. Figure 24.5 (p. 1070) shows computer-generated three-dimensional representations of the dendrimer at this stage and after two more generations of growth.

The dendrimer increases in size by about 10 Å per generation. It is disklike in shape for the first two generations, then becomes a flattish spheroid with the next two, and becomes a nearly symmetrical spheroid after the fifth generation. The diameter (~58 Å) and shape of the dendrimer after the fifth generation resemble the diameter and shape of the hemoglobin molecule (p. 1011). In many ways dendrimers mimic the properties of biological macromolecules. For example, the polyamidoamine dendrimer discussed above is close enough to the shape and size of the nuclear proteins known as histones (p. 1018) to bind DNA.

The controlled way in which dendrimers can be synthesized opens the way to designing molecules for specific applications in many areas of materials science as well as biological and medical applications. Chemists are exploring these possibilities with a variety of dendritic structures generated from many different functional groups and core molecules.

**Problem 24.21** Another type of dendrimer is synthesized by adding ethylenediamine by a Michael reaction to four equivalents of acrylonitrile (p. 1040). The resulting tetranitrile is reduced catalytically (p. 811) to amino groups, which then add again to acrylonitrile (how many equivalents?). Reduction of the nitrile and addition of more acrylonitrile continue the growth of the dendrimer. Starting from ethylenediamine, grow the dendrimer three times.

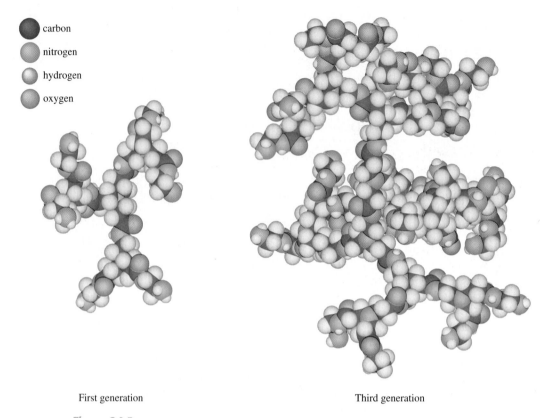

carbon

nitrogen

hydrogen

oxygen

First generation                                    Third generation

*Figure 24.5*
Computer-generated three-dimensional representations of a dendrimer at the end of the first and third generations of growth.

## 24.6  Medical Applications of Macromolecules

Polymers of every description are used in the health care industry. Not only are all kinds of disposable bottles, syringes, and laboratory ware made from different polymers, but replacements for parts of human bodies are being created. Every year, several million artificial parts are implanted into individuals who have suffered some loss as a result of either accident or disease. The development of **biomaterials,** artificial substances that are compatible with human tissues, is an important area of research. In some cases, the implant must become a permanent part of the body and not be rejected by it. Poly(ethylene terephthalate) mesh tubes, for example, are used to replace blood vessels; it is expected that the human tissue will grow into and around the mesh to make the implant part of the body's structure. In other cases, an implant is necessary for a period of time but should eventually be replaced by the body's own tissues. In these cases, a polyester such as poly(lactic acid) can be used so that the implant is absorbed by the body and leaves no permanent residue. This polyester is gradually hydrolyzed to lactic acid by the body and then metabolized to carbon dioxide and water, in the same way as natural lactic acid is.

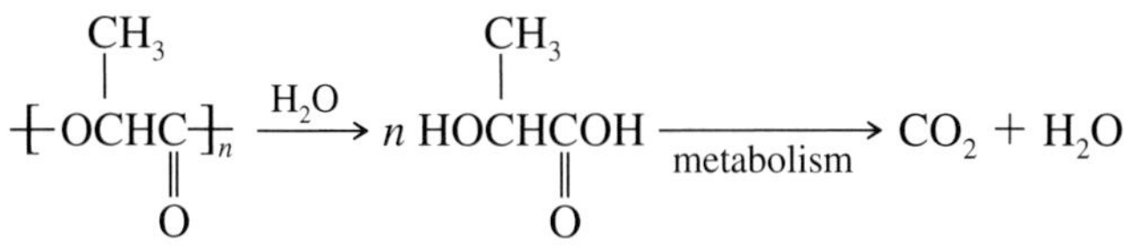

poly(lactic acid)                lactic acid

*a polyester*

Another important problem being attacked by polymer chemists is the use of polymeric materials to control the delivery of drugs to make dosages more convenient, more targeted, and, most important, less toxic. For example, tumor tissues tend to be more leaky and, therefore, allow the entry of larger molecules than do normal tissues. Chemists are working on ways to design molecules, dendrimers (p. 1067) for example, that carry a chemotherapeutic agent into tumor cells, but spare normal cells the toxic effects of the drug.

When a drug is encapsulated in a polymer, a number of factors can determine how it gets out. One advantage of having the drug in such a delivery system would be to control the rate at which the drug is made available so that a slow, steady dosage would be maintained instead of the highs and lows that result when a pill is taken orally. A lot of thought has gone into designing polymer systems that will disintegrate naturally in the body. One system, created by Robert S. Langer and his coworkers, involves a polyanhydride. The rate at which the anhydride linkages hydrolyze and, therefore, the rate at which the polymeric material disintegrates to release the drug embedded in it depend on the structure of a component of the polyanhydride. Anhydrides with longer alkyl chain components are more resistant to hydrolysis than those with short chains. The synthesis of these polyanhydrides are shown below.

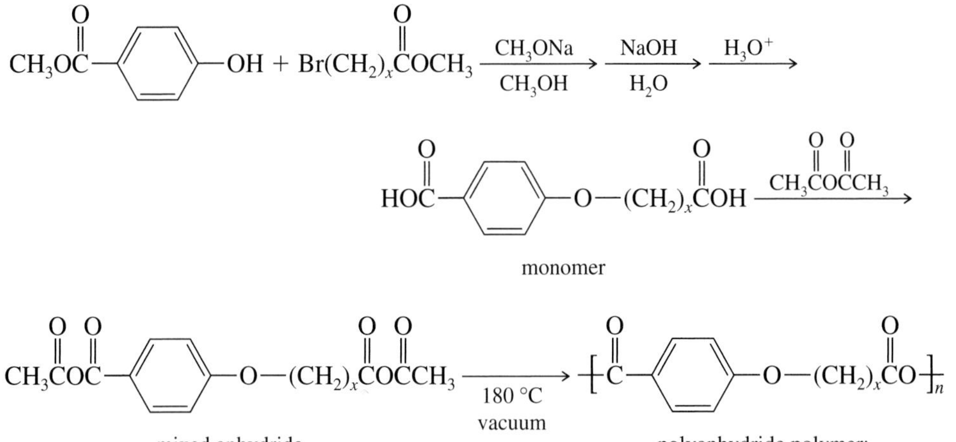

monomer

mixed anhydride

polyanhydride polymer;
$x = 1, 4, 7$
MW up to 44,600

The polymer has a low melting point, which makes it suitable for the processing of heat-sensitive drugs. The polyanhydride in which $x = 7$ is much slower to erode at its surface by hydrolysis than the one with $x = 1$, thus allowing for control of the rate at which such a polymer would release a drug.

**Problem 24.22**   Do you expect polyanhydrides to hydrolyze faster or more slowly than polyesters such as poly(lactic acid)? Rationalize your answer.

Chemists also have started to create polymeric materials that have antibacterial properties so that surfaces and clothing will kill bacteria on contact. Bacteria are killed by reagents containing cationic sites such as ammonium ions. One type of polymer that has been found to be antibacterial contains *N*-alkylated poly(4-vinylpyridine) groups, where the *N*-alkyl chains contain three to eight carbon atoms.

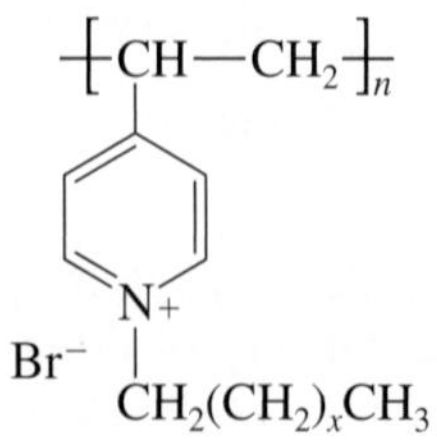

$$x = 1-6; n = \sim 800$$
*N*-alkylated poly(4-vinylpyridine)

These polyvinylpyridines can be attached to surfaces such as glass or to other polymers such as polyethylene and polypropylene, preventing the growth of bacteria on those surfaces.

Other chemists have found a way to attach a group containing ammonium ions to cellulose. The reagent is *N*-hexadecanyl-1,4-diazabicyclo[2.2.2]octane, which reacts with cellulose tosylated at the primary hydroxyl group to give an antibacterial version of cellulose.

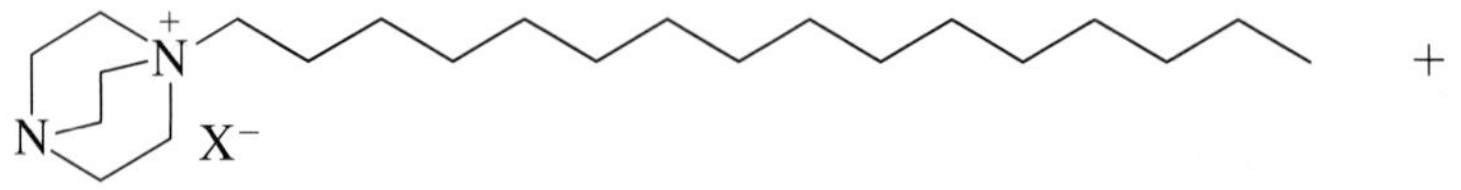

*N*-hexadecanyl-1,4-diazabicyclo[2.2.2]octane

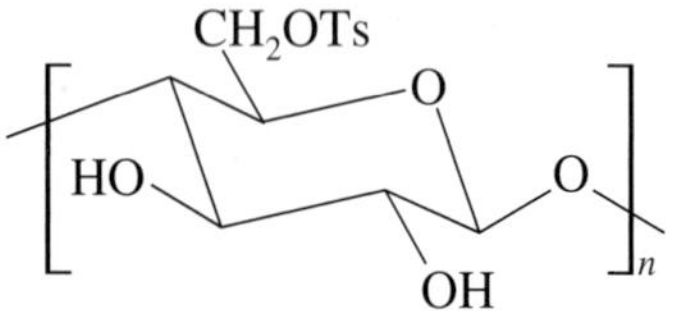

cellulose tosylated at the
primary hydroxyl group

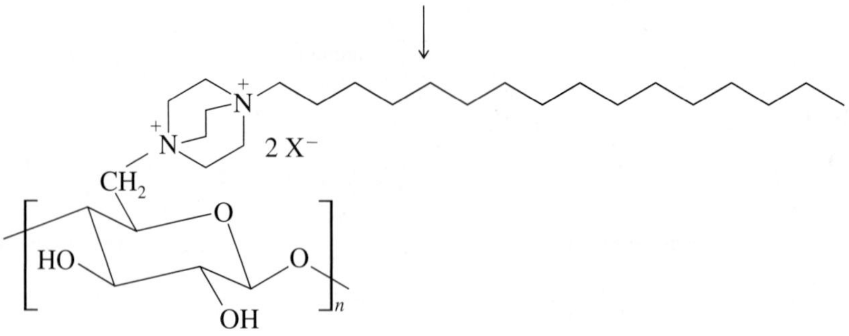

cellulose modified so that it has antibacterial properties

Wood, paper, and cotton, all made of cellulose, can be converted in this way to become antibacterial. Because the ammonium ions are covalently bonded to the cellulose, ordinary washing does not change these properties.

These antibacterial compounds work by disrupting bacterial cell membranes, so it is hoped that the bacteria will not survive long enough to produce offspring that are resistant to the antibacterial action, but chemists are still investigating this. Such resistance is, of course, a major problem with traditional antibiotic.

# 24.7   A Natural Macromolecule, Rubber

## A. The Structure of Rubber and Gutta Percha

Rubber is produced by a number of plants that grow in tropical regions. The one that is most important commercially is a tree, *Hevea brasiliensis,* which was originally found in Brazil but now grows mostly in Southeast Asia. The polymer is contained in a fluid known as latex, which is synthesized by cells under the bark of the tree. Latex is obtained by making cuts in the bark and collecting the liquid that flows out. *Hevea brasiliensis* synthesizes latex at a high rate; a tree may be tapped every other day to harvest rubber.

**Rubber** is a polymer made of isoprene units and containing cis double bonds. An isomeric polymer in which the double bonds are trans is obtained from trees of the genus *Dichopsis,* which are also found in Southeast Asia. This polymeric material, which has properties quite different from those of rubber, is called **gutta percha.**

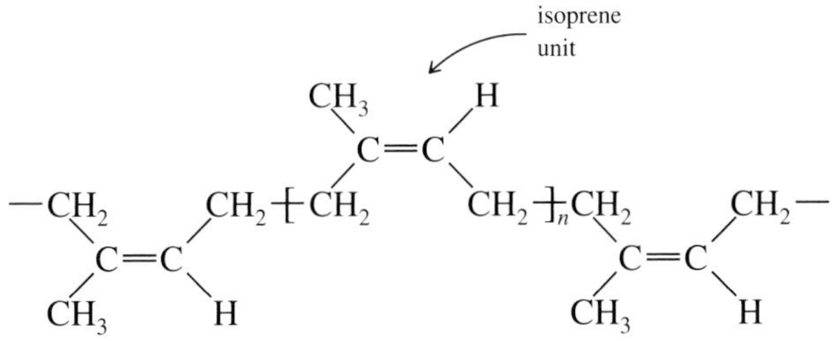

rubber
cis double bonds
$n \sim 1500\text{–}15{,}000$
$MW \sim 100{,}000\text{–}1{,}000{,}000$

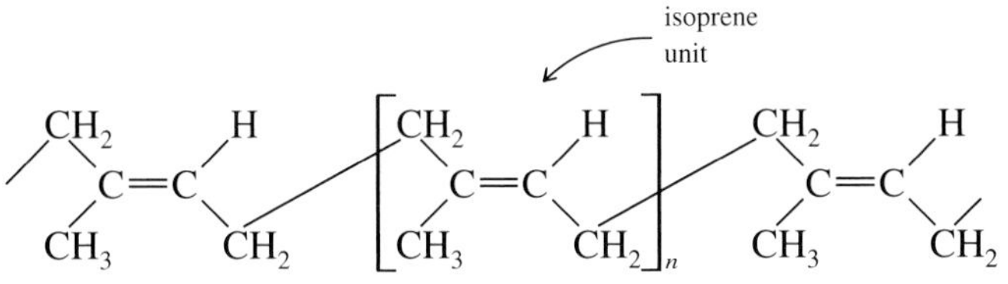

gutta percha
trans double bonds
$n \sim 100$
$MW \sim 7000$

Both rubber and gutta percha are synthesized in plants from isopentenyl pyrophosphate. The reactions that give rise to gutta percha resemble those in the biosynthesis of terpenes such as geraniol and farnesol, which also have trans double bonds (p. 751). The molecular weight of gutta percha is approximately 7000, and that of rubber ranges from 100,000 up. The biosynthesis of rubber is obviously directed by a different type of enzyme than that involved in the production of gutta percha because of the stereoselective formation of cis double bonds in rubber.

The physical properties of rubber and gutta percha are quite different, as should be expected from the different shapes and sizes of their molecules. Rubber has a much more folded structure than gutta percha. The cis arrangement of the double bonds makes it harder for adjacent rubber molecules to fit close to each other in the ordered way that produces a crystalline structure. Thus rubber is highly amorphous. Because of the random coiling of its large molecules, rubber can be stretched easily. When it is stretched, the molecules are forced into a more orderly arrangement, which is unstable for the polymer, so it snaps back into the more random coiled structure when the tension is released. The structure of gutta percha allows the polymer molecules to be packed very close to each other, so this polymer is much more crystalline in its natural state than rubber is. In general, gutta percha is harder and less flexible than rubber. For example, the relatively hard, inflexible covers of golf balls are made mostly of gutta percha. It is also a good electrical insulator and is used in many electrical applications, such as in casings for cables.

Raw rubber is affected by a number of environmental factors such as temperature (p. 1043) and the presence of oxygen. The allylic positions next to the double bonds, for example, are vulnerable to reaction with oxygen as a free radical (pp. 786–788). The double bonds themselves react with ozone, which is always present but is especially abundant in air polluted by exhaust fumes from automobiles. Rubber is affected by organic solvents such as gasoline and oil; it tends to swell and dissolve in such hydrocarbons. Light also breaks down rubber. For these reasons, almost all rubber used industrially is treated to increase its stability; the process is known as **vulcanization.** Vulcanized rubber has greater strength, less stickiness, and greater elasticity than raw rubber. It is less soluble in most solvents and tends to retain its flexibility at lower temperatures.

Charles Goodyear of the United States discovered vulcanization in 1839 when he found that heating rubber with sulfur improved the properties of the rubber. Sulfur reacts irreversibly with rubber to link different chains and parts of the same chain together to give greater stability to the polymer. The chemistry is complicated, and cyclic structures containing sulfur are formed as well. The physical properties of rubber are changed even when only a very small percentage by weight of sulfur is used.

## B. Synthetic Rubbers

Rubber is important to transportation as well as to many other aspects of modern technology. In the world wars, it became clear that industrial states could easily be cut off from the supply of natural rubber from tropical regions, such as Southeast Asia. Therefore, research was started into ways of making synthetic rubber. One of the earliest replacements for natural rubber was neoprene, developed at DuPont in the United States. It is the polymer of 2-chloro-1,3-butadiene, a molecule that resembles isoprene except that a chlorine atom replaces the methyl group on the backbone of the chain, which gives rise to the common name of chloroprene.

A diene may polymerize by 1,2- or 1,4-addition to the conjugated double bonds (p. 741). 2-Chloro-1,3-butadiene polymerizes in a free-radical reaction by both types of addition.

a fragment of poly(chloroprene)
neoprene

The portions of the chain that result from 1,2-addition to the butadiene contain chlorine atoms bonded to carbon atoms that are tertiary and allylic. Heating causes isomerization of such an allylic halide group to the isomer in which the more stable internal double bond is present. Further heating with a metal oxide results in cross-linking of different polymer chains.

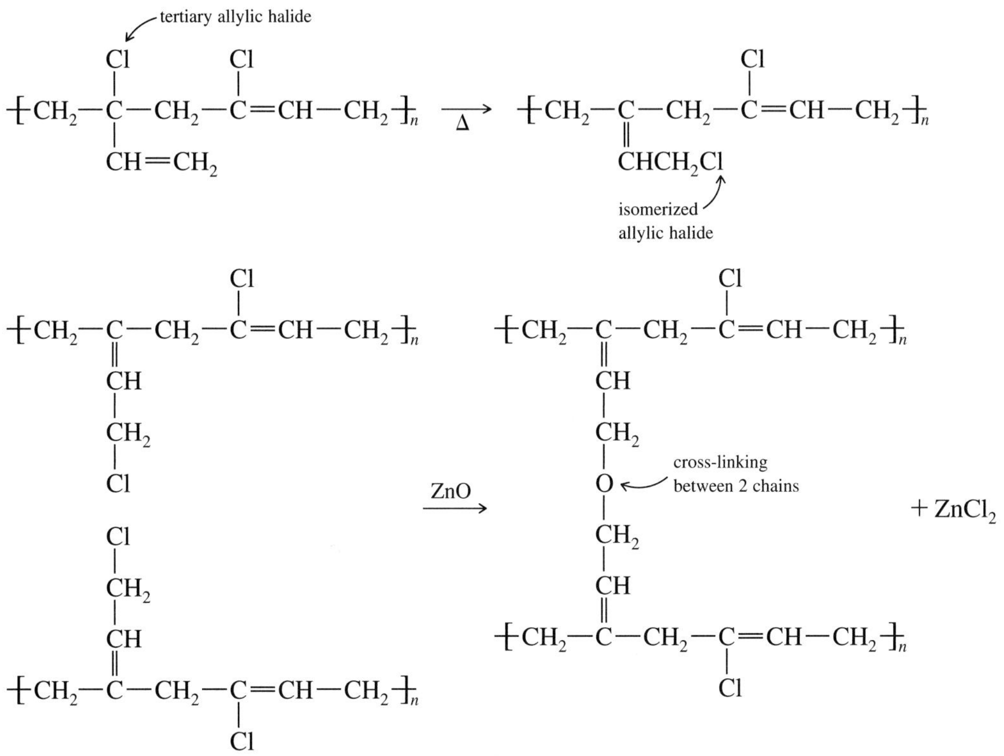

The cross-link is an ether group formed by nucleophilic displacement of the reactive allylic halide substituents on adjacent chains. Note that the unreactive vinylic chlorine atoms are not affected in either the isomerization step or the nucleophilic substitution reaction.

The stereochemistry of the double bond created between the second and third carbon atoms of a butadiene by 1,4-addition depends on the reaction conditions used. Neoprene, formed in emulsion by a free-radical polymerization, has mostly trans double bonds in the chain.

Neoprene is more resistant than natural rubber to oils and solvents and to reaction with ozone. It is tougher and resists wear better than rubber. It is used mostly in applications where its toughness and resistance to oil and grease are important, such as in gaskets, sealing rings, and engine mountings. It is also used in making protective gloves and aprons.

Many attempts had been made to polymerize isoprene to create a synthetic rubber that matched the properties of natural rubber. It was not until Ziegler-Natta catalysts became available in the 1950s that isoprene was polymerized stereoselectively to 1,4-polyisoprene with cis double bonds using one catalyst and 1,4-polyisoprene with trans double bonds using another one.

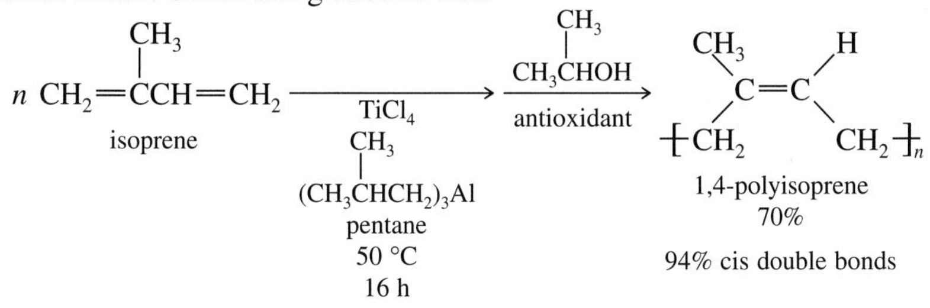

Changing the transition metal used as the primary catalyst from titanium to vanadium changes the stereoselectivity of the polymerization reaction.

This chapter has explored some of the chemical reactions used to transform low-molecular-weight organic compounds into giant molecules that imitate natural molecules in their properties and usefulness. These syntheses are indications of the inventiveness and ingenuity of the human mind when faced with practical problems. Human beings have a great deal of curiosity about the limits of their understanding of natural phenomena. The experimental observations that are accessible to scientists spawn theories, which are followed by new experiments to test the theories. Polymer chemistry is an area where a questioning attitude, a constant asking of "what if?" has been particularly fruitful.

On the other hand, it also must be recognized that polymers have been a mixed blessing. Plastics are useful because they have good mechanical strength and are chemically unreactive. The same properties keep plastics from degrading naturally when they are discarded. Many of them, when they burn, depolymerize to give toxic degradation products; this creates potentially dangerous situations when such materials are used in carpeting and furniture for homes and institutions. The raw materials for polymers come from petroleum; a nonrenewable resource is thus being converted into plastics, many of which are put to only temporary use and are then discarded, creating problems for the solid-waste disposal systems of towns and cities. Each of us contributes to this waste. Perhaps the time has come to ask ourselves which applications of polymers represent a wise use of resources and which are, at best, a small convenience that we could do without.

The materials of the future must come increasingly from renewable sources and be less wasteful both of substance and of energy. Plants are especially versatile in converting the energy of the sun into a wide variety of natural products that can be transformed into other useful substances. Microorganisms, modified by genetic engineering, are already used in fermentation processes that convert readily available plant products such as glucose to both simpler and more complex chemicals. In the future organic chemists will be increasingly challenged to perfect methods for the transformation of common substances isolated from natural sources into other useful products. The stimulating interaction between experimental observation and careful thinking about the implications of the observed phenomenon that is the basis of the science of organic chemistry must continue if human beings are to use the resources of the earth wisely in meeting their needs.

**Problem 24.23**   What products would you expect to get from the reaction of rubber with ozone? Does the process that takes place in the environment end up with a reductive or an oxidative decomposition of the ozonide?

**Problem 24.24** 1,3-Butadiene, when polymerized in the presence of titanium tetrabromide and triethylaluminum, gives a 1,4-polybutadiene with 'mostly cis double bonds. Write an equation for this reaction.

**Problem 24.25** When 1-methoxy-1,3-butadiene is polymerized with a free-radical initiator, a low-molecular-weight, amorphous polymer is obtained. Polymerizing the same diene in the presence of vanadium trichloride and triisobutylaluminum gives a highly crystalline, high-molecular-weight polymer with trans double bonds. Write equations showing the differences in the structures of the two polymers.

**Problem 24.26** A copolymer made of acrylonitrile, butadiene, and styrene has a number of applications because it combines some of the properties of rubber with the toughness of an acrylonitrile-styrene copolymer (p. 1047). Write an equation for the preparation of the acrylonitrile-butadiene-styrene copolymer.

**Table 24.1 Summary of Polymerization Reactions**

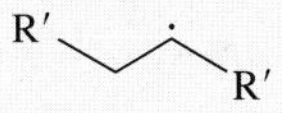

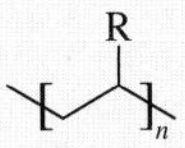

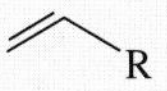

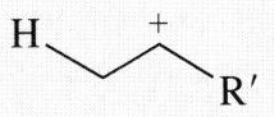

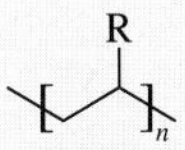

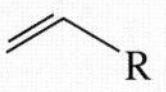

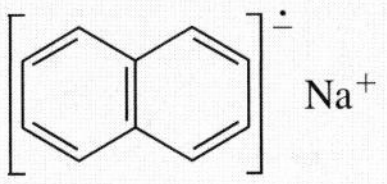

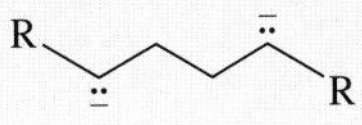

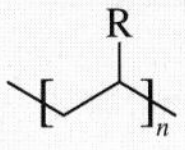

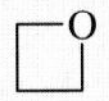

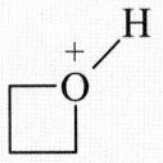

**Examples of Chain-Growth Polymerization**

| Monomer | Initiator | Intermediate | Polymer |
|---|---|---|---|
|  | R′· |  |  |
|  | $HB^+$ |  |  |
|  |  |  |  |
|  | $HB^+$ |  | $\left[ CH_2CH_2CH_2O \right]_n$ |

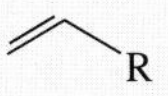

**Examples of Step-Growth Polymerization**

| Monomers | Smallest Intermediate Unit | Polymer |
|---|---|---|
| $HOC(=O)(CH_2)_xC(=O)OH$ + $H_2N(CH_2)_yNH_2$ | $HOC(=O)(CH_2)_xC(=O)N(H)(CH_2)_yNH_2$ | $\left[ C(=O)(CH_2)_xC(=O)N(H)(CH_2)_yN(H) \right]_n$ |
| $HOCH_2CH_2OH$ +  |  |  |
| $HOCH_2CH_2OH$ +  |  |  |

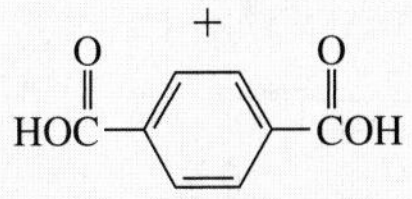

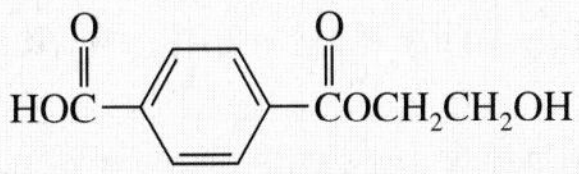

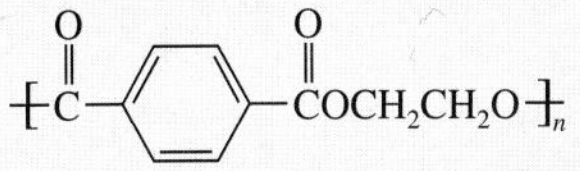

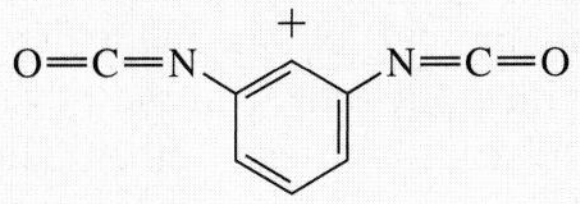

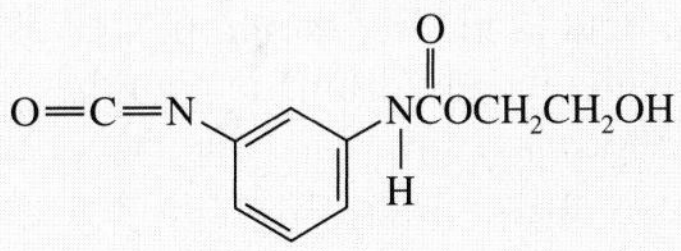

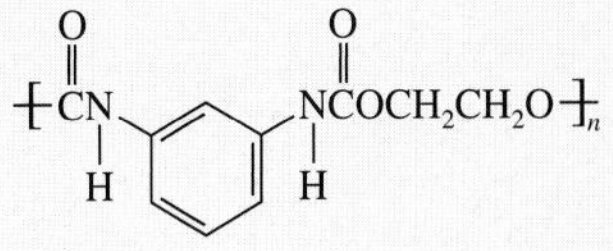

# ADDITIONAL PROBLEMS

**24.27** Predict the structures of the polymers that will result from the following reactions.

(a) 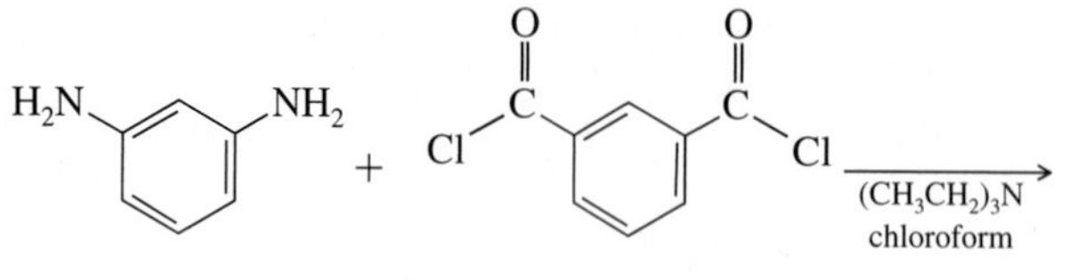 $H_2N$—⬡—$NH_2$ + Cl—C(=O)—⬡—C(=O)—Cl $\xrightarrow[\text{chloroform}]{(CH_3CH_2)_3N}$

(b) $CH_2=CH_2$ + $CH_2=\overset{CH_3}{\underset{\overset{\|}{O}}{\overset{|}{C}}}COCH_3$ $\xrightarrow{\text{peroxide}}$

(c) 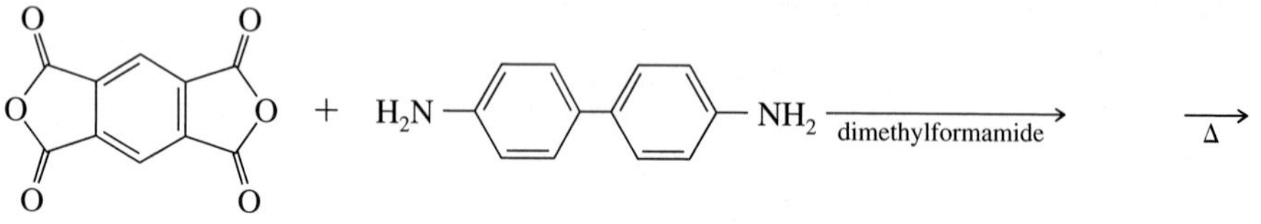 [dianhydride] + $H_2N$—⬡—⬡—$NH_2$ $\xrightarrow{\text{dimethylformamide}}$ $\xrightarrow{\Delta}$

(d)  [thiophene-like ring] $\xrightarrow{(CH_3CH_2)_2O\cdot BF_3}$

(e) $CH_2=\overset{C\equiv N}{\underset{\overset{\|}{O}}{\overset{|}{C}}}COCH_3$ $\xrightarrow[\substack{\text{2-methylpropanenitrile} \\ \Delta}]{\text{free-radical initiator}}$

(f) 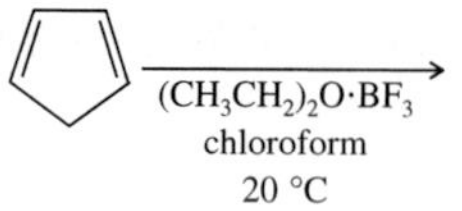 [cyclopentadiene] $\xrightarrow[\substack{\text{chloroform} \\ 20\,°C}]{(CH_3CH_2)_2O\cdot BF_3}$

(Hint: How does 1,3-butadiene polymerize?)

(g) 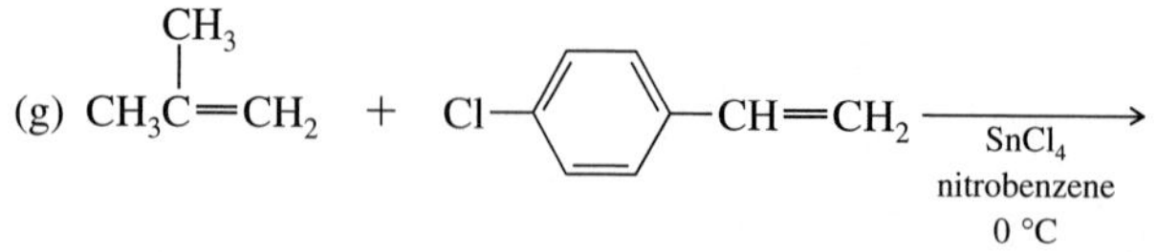 $CH_3\overset{CH_3}{\underset{}{\overset{|}{C}}}=CH_2$ + Cl—⬡—$CH=CH_2$ $\xrightarrow[\substack{\text{nitrobenzene} \\ 0\,°C}]{SnCl_4}$

(h) $CH_2=CHOCH_2\overset{CH_3}{\underset{}{\overset{|}{C}}}HCH_3$ $\xrightarrow{(CH_3CH_2)_2O\cdot BF_3}$

(i) $CH_2=CHC\equiv N$ $\xrightarrow[\substack{CaO \\ \text{dimethylformamide} \\ 20\,°C}]{}$ $\xrightarrow{HCl}$

(j) $CH_2=CHO\overset{CH_3}{\underset{}{\overset{|}{C}}}HCH_3$ $\xrightarrow[\substack{VCl_3 \\ \overset{CH_3}{\underset{}{\overset{|}{}}} \\ (CH_3CHCH_2)_3Al \\ \text{heptane}}]{}$ $\xrightarrow{CH_3CH_2OH}$ isotactic polymer

**24.28** Unsaturated esters having the general formula $CH_2=CHC\overset{O}{\overset{\|}{O}}R$ and acrylonitrile, $CH_2=CHC\equiv N$, are successfully polymerized in chain reactions using free-radical or anionic initiators. Polymerization reactions do not take place when cationic initiators are used. How would you rationalize these experimental observations?

**24.29** The reactivity of an alkene monomer in a free-radical polymerization reaction, as measured by the rate of the propagation step, depends on the vinylic substituent. The following order of reactivity has been observed.

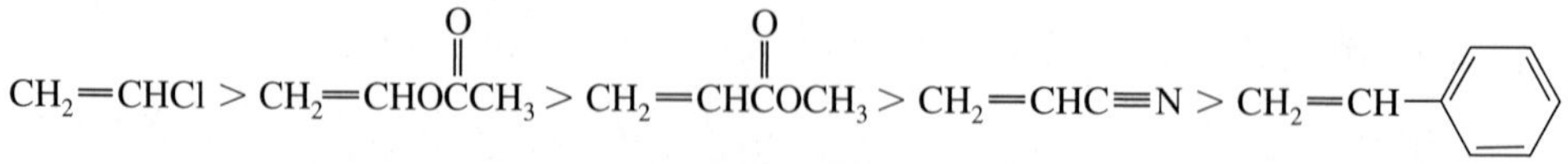

$$CH_2=CHCl > CH_2=CHO\overset{O}{\overset{\|}{}}CCH_3 > CH_2=CHC\overset{O}{\overset{\|}{}}OCH_3 > CH_2=CHC\equiv N > CH_2=CH-⬡$$

How would you rationalize this order of reactivity?

**24.30** In attempts to modify the properties of aromatic polyamides, block copolymers were made of aromatic diacids, aromatic diamines, and 1,3-bis(3-aminopropyl)tetramethyldisiloxane. Two methods were used. In one, all of the reagents were mixed together. In the other, oligomeric polyamides were prepared first and then allowed to react with each other. The equations for the two methods are shown on the next page. How would the polymers obtained by the two methods differ in their structures?

*Method 1:*

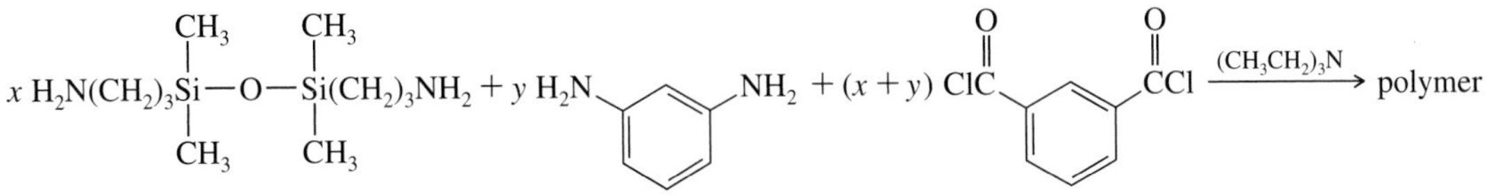

$$x \; H_2N(CH_2)_3\underset{\underset{CH_3}{|}}{\overset{\overset{CH_3}{|}}{Si}}-O-\underset{\underset{CH_3}{|}}{\overset{\overset{CH_3}{|}}{Si}}(CH_2)_3NH_2 + y \; H_2N\text{-}\underbrace{\phantom{xxx}}\text{-}NH_2 + (x+y) \; Cl\overset{O}{\overset{||}{C}}\text{-}\underbrace{\phantom{xxx}}\text{-}\overset{O}{\overset{||}{C}}Cl \xrightarrow{(CH_3CH_2)_3N} \text{polymer}$$

*Method 2:*

$$x \; H_2N(CH_2)_3\underset{\underset{CH_3}{|}}{\overset{\overset{CH_3}{|}}{Si}}-O-\underset{\underset{CH_3}{|}}{\overset{\overset{CH_3}{|}}{Si}}(CH_2)_3NH_2 + (x-1) \; Cl\overset{O}{\overset{||}{C}}\text{-}\underbrace{\phantom{xxx}}\text{-}\overset{O}{\overset{||}{C}}Cl \xrightarrow{(CH_3CH_2)_3N} \text{oligomer A}$$

$$y \; H_2N\text{-}\underbrace{\phantom{xxx}}\text{-}NH_2 + (y+1) \; Cl\overset{O}{\overset{||}{C}}\text{-}\underbrace{\phantom{xxx}}\text{-}\overset{O}{\overset{||}{C}}Cl \xrightarrow{(CH_3CH_2)_3N} \text{oligomer B}$$

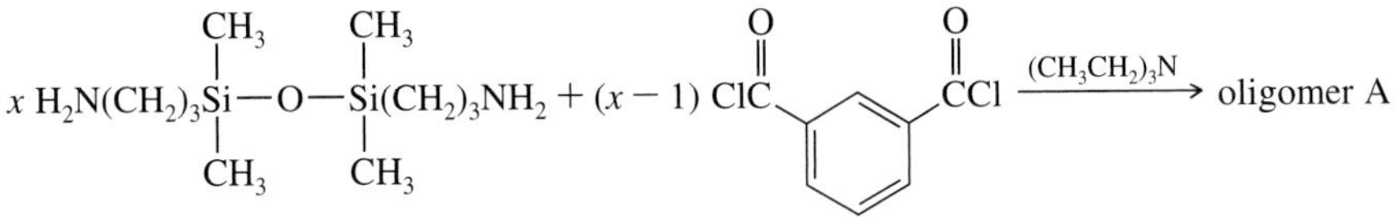

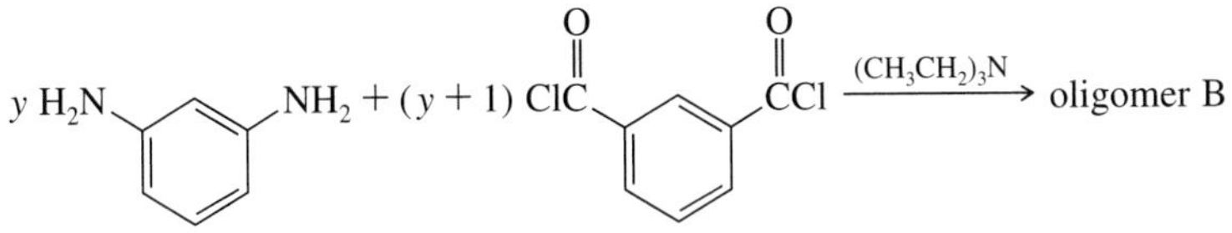

$$\text{oligomer A} + \text{oligomer B} \xrightarrow{(CH_3CH_2)_3N} \text{polymer}$$

**24.31** Ion-exchange resins containing sulfonic acid groups are strongly acidic. Weakly acidic ion-exchange resins have also been made. One of them has the following partial structure. How would you make such a resin?

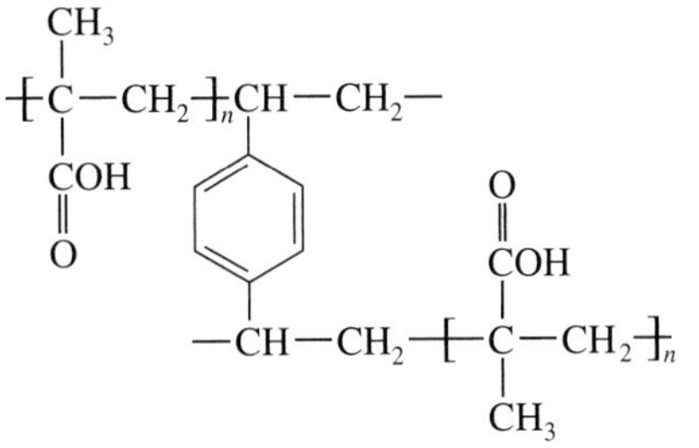

**24.32** In the presence of a free-radical initiator, such as high-energy $\gamma$-radiation, acrylamide polymerizes at the carbon–carbon double bond to give a hydrocarbon chain with amide substituents on it. When a strong base is used to catalyze this polymerization, however, a polyamide that can be hydrolyzed to 3-aminopropanoic acid ($\beta$-alanine) is obtained. Write mechanisms that explain the different products obtained from these two reactions, shown below.

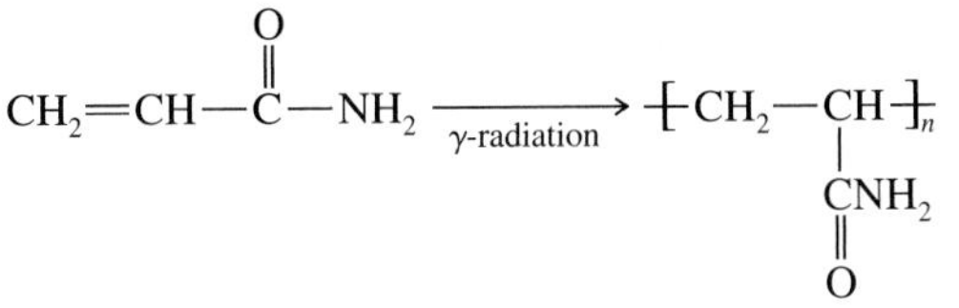

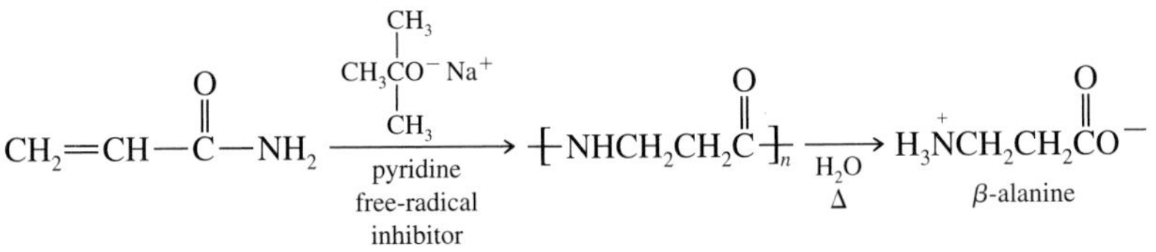

**24.33** Fibers of a polyester, poly(pivalolactone), have a high degree of orientation and recover very quickly from deformation. This is a desirable property in fibers used in carpets;

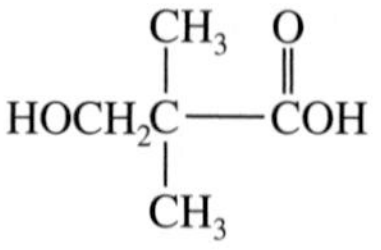

3-hydroxy-2,2-
dimethylpropanoic
acid

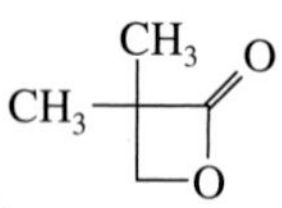

pivalolactone

otherwise, the carpet shows footprints. The polyester is synthesized from either of the monomers shown at left. Suggest a method for synthesizing the polymer from each monomer. For example, what type of catalyst would you use, and what reaction conditions would be necessary for each process?

**24.34** When the polymerization of $(S)$-$(-)$-2-methyloxirane is catalyzed by solid potassium hydroxide, the product is a crystalline, optically active polymer. If the polymerization of the chiral oxirane is catalyzed by iron(III) chloride in ether, however, the resulting polymer is not optically active. Propose mechanisms for the two different types of polymerization that account for the different stereochemistry observed.

**24.35** Under the proper conditions, the carbon–oxygen double bond of a carbonyl group takes part in polymerization reactions; for example, the following two reactions have been observed. Write a mechanism for each one, showing why the polymerization proceeds as it does.

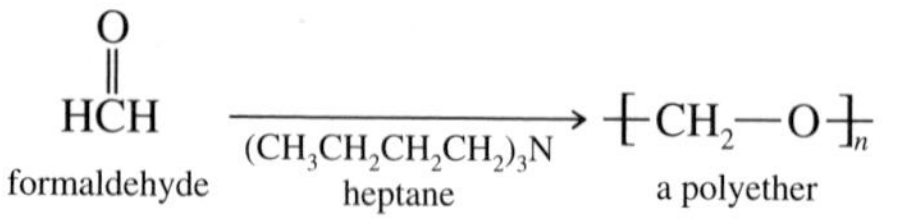

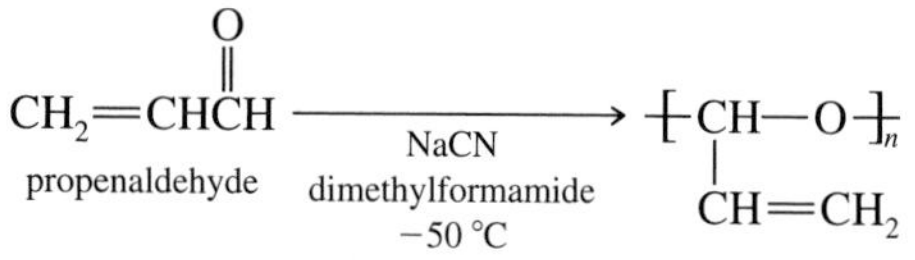

**24.36** Rigid polymers that are useful in fibers can be made by condensation reactions that give fused-ring heterocycles. An example is shown below.

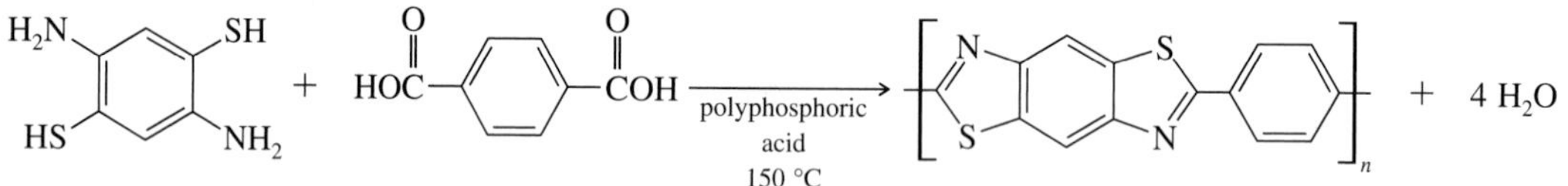

An analogous reaction is shown below. Provide a mechanism for it.

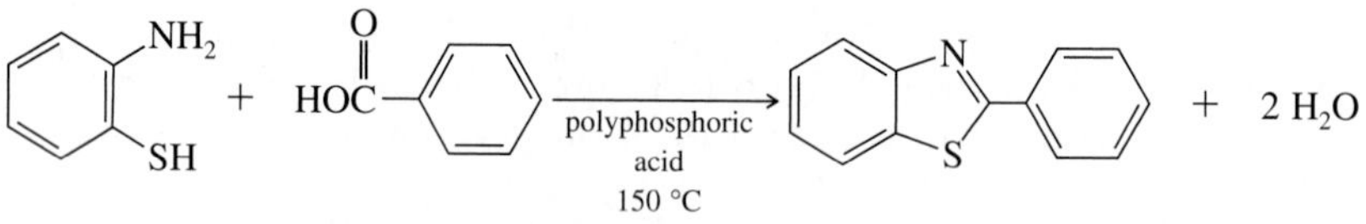

**24.37** Reactions leading to a spherical dendritic structure with 36 hydroxyl groups on its surface use the following reagents. The functional groups in the layer under the hydroxyl groups are amides. Show how the dendrimer can be constructed from these reagents. You may wish to confine yourself to working out one-fourth of the structure in detail.

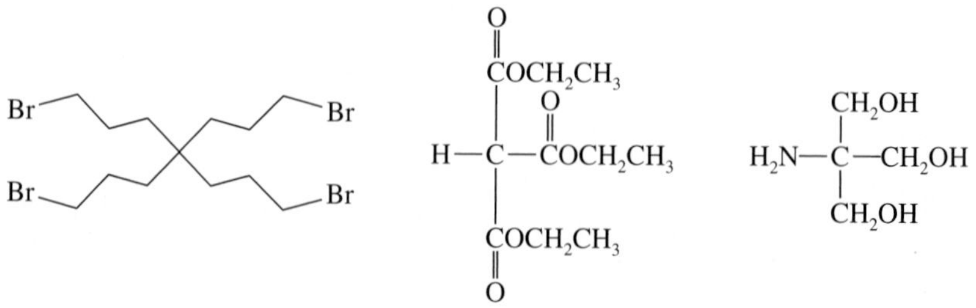

# Concerted Reactions

## A Look Ahead

Concerted reactions take place in one step with high stereoselectivity and without the formation of a reactive intermediate. Bimolecular nucleophilic substitution reactions (p. 240) and Diels-Alder additions of dienes to dienophiles (p. 742) are familiar examples of concerted reactions. This chapter will explore a series of chemical transformations in which rearrangements of the positions of $\sigma$ and $\pi$ bonds take place by way of cyclic transition states. These cycloaddition reactions, electrocyclic reactions, and sigmatropic rearrangements are usually insensitive to reaction conditions such as solvent polarity and catalysis. They are believed to involve only a single step, in which reorganization of the bonding occurs simultaneously.

The reactions that are possible and the resulting stereochemistry depend in many cases on whether the reactions are thermal, meaning that the reactants are in their ground state, or photochemical, meaning that the excited state of a reactant is involved in the reaction. Chemists explain these differences by looking at the symmetry of the molecular orbitals involved in the changes in bonding taking place. An important theory states that the frontier orbitals, the highest-energy molecular orbital occupied by electrons of one reactant and the lowest-energy molecular orbital of the other reactant that is empty, are the ones that determine the course and the stereochemistry of the reaction. This chapter will show how to identify these orbitals and use their properties to make predictions about concerted reactions.

## 25.1   Introduction to Concerted Reactions

### A. Evolution of the Theory of Concerted Reactions

By the 1960s, research mostly into the chemistry and synthesis of natural products had given rise to a series of experimental observations that stimulated a search for theoretical explanations, which in turn would suggest new experiments. An illustration of the close relationship between such practical research and the development of theory in organic chemistry is the work done on vitamin D by Egbert Havinga of the Netherlands. A form of vitamin D is synthesized in the skin on exposure to sunlight.

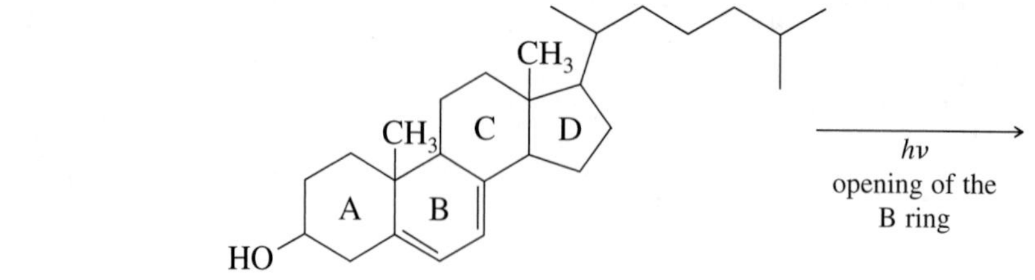

7-dehydrocholesterol

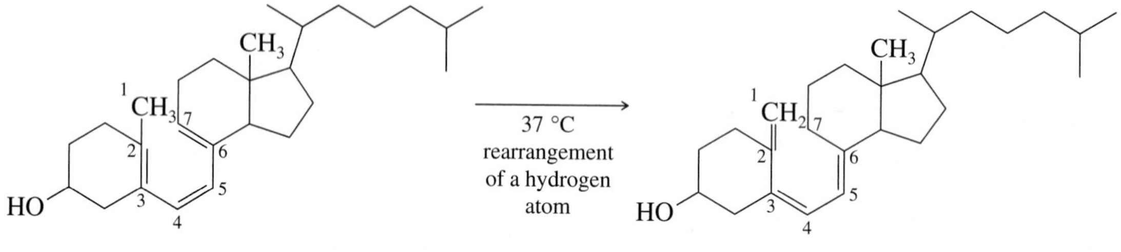

$hv$
opening of the
B ring

37 °C
rearrangement
of a hydrogen
atom

vitamin D$_3$
cholecalciferol

The overall conversion may be classified as a photochemical opening of the B ring of 7-dehydrocholesterol to form a conjugated triene, followed by a rearrangement of a hydrogen atom from the methyl group to the other end of the conjugated system. Recent research has shown that the rearrangement takes place slowly at body temperature, so vitamin D$_3$ continues to be synthesized in the skin for as long as 3 days after exposure to the sun.

The reactions shown above are only two of many similar ones observed for the vitamin D system. The reactions have differing stereochemistry depending on whether they occur photochemically or thermally, as will be seen in Section 25.3B. Havinga's very precise determination of the course of these reactions provided information that led his colleague Luitzen J. Oosterhoff to suggest that the stereochemistry characteristic of these reactions is determined by the symmetry of the molecular orbitals involved in the transformations.

Meanwhile, in the 1960s, Robert B. Woodward at Harvard University was doing research on the synthesis of vitamin B$_{12}$. He observed the precise stereochemistry with which cyclohexadienes were formed from conjugated trienes and then converted back into trienes in thermal and photochemical reactions. Woodward and his colleague Roald Hoffmann, in thinking about these observations, also arrived at the idea that the symmetry of the molecular orbitals that participate in the chemical reaction determines the course of the reaction. In a series of papers published in 1965, they proposed what they called the principle of the **conservation of orbital symmetry** in concerted reactions. In the most general terms, the principle means that in a concerted reaction, the molecular orbitals of the starting materials must be transformed into the molecular orbitals of the products in a smooth, continuous way. This is possible only if the orbitals have similar symmetry. The recognition of the generality and the significance of this way of examining organic reactions was only one of the many contributions made by Woodward, who received the Nobel Prize in 1965 for his work in organic chemistry.

## B. A Review of $\pi$ Molecular Orbitals

The bonding and antibonding $\pi$ orbitals of ethylene (Figure 2.18, p. 54) and the electronic configuration of ethylene in the ground and the excited states are shown

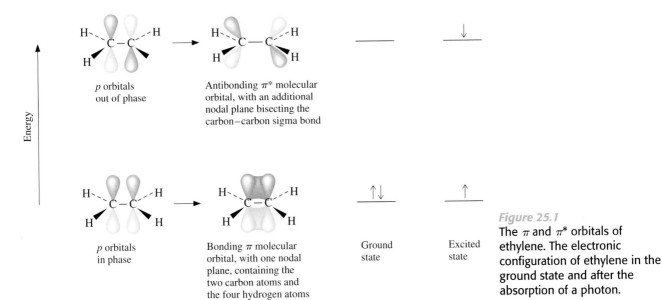

*Figure 25.1*
The $\pi$ and $\pi^*$ orbitals of ethylene. The electronic configuration of ethylene in the ground state and after the absorption of a photon.

in Figure 25.1. The $\pi$ molecular orbital has a nodal plane containing the two carbon atoms and the four hydrogen atoms. The $\pi^*$ orbital has a second nodal plane in addition to the one that includes the $\sigma$-bonded backbone of the molecule and is of higher energy than the bonding $\pi$ orbital. The more nodes there are in a molecular orbital (and in an atomic orbital), the higher is the energy level of the orbital. Ethylene is a stable molecule in the ground state because the two electrons from the two $p$ atomic orbitals are in the bonding $\pi$ orbital. Absorption of ultraviolet radiation by ethylene raises the molecule to the excited state in which one electron has been promoted to the $\pi^*$ orbital (p. 71).

The nodes and the relative energies of the $\pi$ and $\pi^*$ orbitals of ethylene are important because they represent the properties of all other alkenes with one double bond and two $\pi$ electrons. This discussion assumes that substituents on the doubly bonded carbons do not change these properties of the molecular orbitals sufficiently to change the conclusions that can be drawn by looking at the simplified picture.

Just as the molecular orbital picture developed for ethylene can represent all compounds with a single $\pi$ orbital and two electrons in it, a similar molecular orbital picture for 1,3-butadiene can be used to talk about a conjugated diene. The four molecular orbitals, designated as $\psi_1$, $\psi_2$, $\psi_3$, and $\psi_4$, are shown in Figure 25.2 on page 1084. They are constructed by successively introducing additional nodes into the array of $p$ atomic orbitals while retaining the overall symmetry of the system. Note how the energy level of an orbital is raised as the number of nodes in the orbital increases. In 1,3-butadiene, the symmetry of the molecule means that carbon 1 is equivalent to carbon 4 and that carbon 2 is equivalent to carbon 3 (and that carbon 1 is *not* equivalent to carbon 2 and that carbon 3 is *not* equivalent to carbon 4). In constructing molecular orbitals, therefore, if a node is placed between carbon 1 and carbon 2, a node must also appear between carbon 4 and carbon 3.

The molecular orbitals shown in Figure 25.2 will be used to talk about any conjugated diene with four $\pi$ electrons in the system. Once again, the assumption will be made that the substituents on the diene system do not change the nodal properties of the molecular orbitals sufficiently to affect the conclusions that can be drawn.

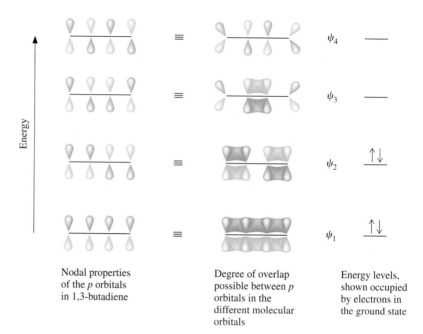

**Figure 25.2**
The four molecular orbitals of butadiene, created from four *p* atomic orbitals.

Nodal properties of the *p* orbitals in 1,3-butadiene

Degree of overlap possible between *p* orbitals in the different molecular orbitals

Energy levels, shown occupied by electrons in the ground state

**Problem 25.1**   How many nodes are there in each of the molecular orbitals, $\psi_1$, $\psi_2$, $\psi_3$, and $\psi_4$, of 1,3-butadiene?

What is the symmetry of a molecular orbital to which Woodward and Hoffmann refer in their principle? Each molecular orbital can be examined for symmetry in ways that are analogous to the method used earlier in this book to decide whether or not molecules are chiral. There are many such ways in which this can be done, but one example will be given to show that in principle each molecular orbital can be classified as symmetrical or antisymmetrical.

The two $\pi$ molecular orbitals shown in Figure 25.1 on page 1083 can be described as symmetrical or antisymmetrical with respect to a plane bisecting the carbon–carbon bond. If you imagine such a plane, you will see that it cuts the $\pi$ orbital into two halves. The left-hand side of the orbital is an exact mirror image of the right-hand side. Thus the $\pi$ orbital is **symmetrical** with respect to this plane. The $\pi^*$ orbital is not. For example, to the left of the plane on the top there is a gray lobe, but to the right of the plane at the top and at the same distance away from the plane there is a colored lobe. In other words, the sign of the wave function is reversed as it goes through the plane. The $\pi^*$ orbital is thus **antisymmetrical** with respect to a plane bisecting the carbon–carbon bond.

**Problem 25.2**   Of the molecular orbitals of 1,3-butadiene (Figure 25.2), decide which are symmetrical and which antisymmetrical with respect to a plane bisecting the bond between carbon 2 and carbon 3.

In a concerted reaction, the symmetry that is present in the reactants is maintained during the course of the reaction and is also present in the product. The Diels-Alder reaction provides a simple demonstration of this principle (Figure 25.3). The reactants, the diene and the dienophile, each have a plane of symmetry. They must approach each other in the transition state in such a way that the same

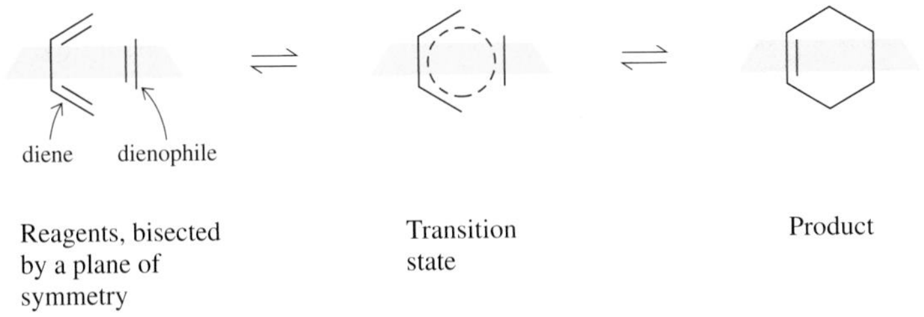

diene     dienophile

Reagents, bisected       Transition            Product
by a plane of            state
symmetry

*Figure 25.3*
A demonstration of the conservation of symmetry in the Diels-Alder reaction.

plane of symmetry is maintained. The product, cyclohexene, also retains the same plane of symmetry.

## C. Interactions Between Molecular Orbitals

Interactions between the molecular orbitals of reactants and transformations of them into the molecular orbitals of products have been described in many different ways. One of the simplest and most useful is to say that a reaction takes place when the highest-energy molecular orbital that contains electrons of one reactant interacts with the lowest-energy molecular orbital that does not contain electrons of the other reactant. This is just like saying that a Lewis base, with electrons to donate, reacts with a Lewis acid, with an empty orbital ready to receive them, or that a nucleophile reacts with an electrophile. The only difference is that the orbitals performing these roles are examined very precisely in order to explain the highly specific chemistry and stereochemistry observed for concerted reactions.

The orbitals that interact have been called the **frontier orbitals** for the reaction by Kenichi Fukui of Japan. His contributions to the development of this way of looking at chemical reactions were recognized in 1981 by the award of the Nobel Prize in chemistry jointly to him and Roald Hoffmann. The frontier orbitals are called the **highest occupied molecular orbital,** abbreviated **HOMO,** and the **lowest unoccupied molecular orbital,** or **LUMO.** For ethylene, for example, the $\pi$ orbital is the HOMO, and the $\pi^*$ orbital is the LUMO in the ground state. When ethylene is raised to its excited state in a photochemical reaction, the $\pi^*$ orbital, with one electron in it, becomes the HOMO (Figure 25.1). In the ground state, $\psi_2$ is the HOMO and $\psi_3$ is the LUMO of 1,3-butadiene (Figure 25.2). These relationships will be taken up again in the next section and applied to reactions actually observed for compounds with molecular orbitals like those of ethylene and butadiene.

Why are these particular orbitals so important in determining the course of a concerted reaction? The electrons in the HOMO of a molecule are like the outer-shell electrons of an atom. They can be removed with the least expenditure of energy because they are already in a higher energy level than any of the other electrons in the molecule. For example, they are the electrons that are lost when the molecule is ionized in the gas phase. Spectroscopic techniques that cause this ionization can be used to determine the energy level of the HOMO for a molecule. The LUMO of a molecule is the orbital to which the electrons can be transferred with the least expenditure of energy.

The higher the energy of the HOMO of a molecule, the more easily electrons can be removed from it. The lower the energy of the LUMO of a molecule, the more easily electrons can be transferred into it. Therefore, the interaction between a molecule with a high HOMO and one with a low LUMO is particularly strong. In general, the smaller the difference in energy between the HOMO of one molecule and the LUMO of another with which it is reacting, the stronger is the interaction between the two molecules.

**Problem 25.3**    Assume that an electron is raised from the HOMO to the LUMO of a molecule of 1,3-butadiene on absorption of energy. Show the electronic configuration after the molecule absorbs ultraviolet radiation, and decide which molecular orbital will be the HOMO and which the LUMO of the diene in the excited state.

## 25.2 Cycloaddition Reactions of Carbon Compounds

### A. Photochemical Dimerization of Alkenes

The usefulness of the theoretical picture of the interactions between molecular orbitals developed in the preceding section is demonstrated by examining the cycloaddition reactions of alkenes. In Figure 25.4, the HOMO of one molecule of (*E*)-2-butene is shown approaching the LUMO of another molecule of (*E*)-2-butene in the ground state, that is, when ultraviolet radiation is not used. An examination of the way in which the two orbitals are approaching each other reveals that it will not be possible for a cyclobutane ring to form in a concerted manner. The bottom right-hand lobe of the HOMO is in phase with the top right-hand lobe of the LUMO and could interact to give a $\sigma$ bond, but the lobes that are facing each other on the left-hand side are out of phase and would only interact in an antibonding way. The second bond necessary to complete the ring would not form.

If one of the molecules of (*E*)-2-butene is raised to the excited state, however, its HOMO will have a different symmetry, as shown in the energy diagram in Figure 25.1 on page 1083. If a sample of (*E*)-2-butene is irradiated with ultraviolet light, the concentration of molecules in the excited state is low; therefore, a molecule in the excited state is most likely to encounter and react with a molecule of (*E*)-2-butene in the ground state. The molecular orbital picture for such an interaction is shown in Figure 25.5. In this case, bonding interactions are possible at both ends of the alkene bonds as the two molecules approach each other face to face. Only one of the two possible stereochemical outcomes for two (*E*)-2-butene molecules reacting in this way has been drawn. The methyl groups on the left-hand carbon atom of each 2-butene are on the same side of the double bond and are cis to each other in the cyclobutane, and the same is true of the methyl groups on the other carbon atom of the reagent. The stereochemistry of the starting materials has been preserved in the product.

The molecular orbital picture thus provides an explanation for the lack of reactivity of 2-butene in the ground state and for the stereoselective dimerization that occurs in the excited state.

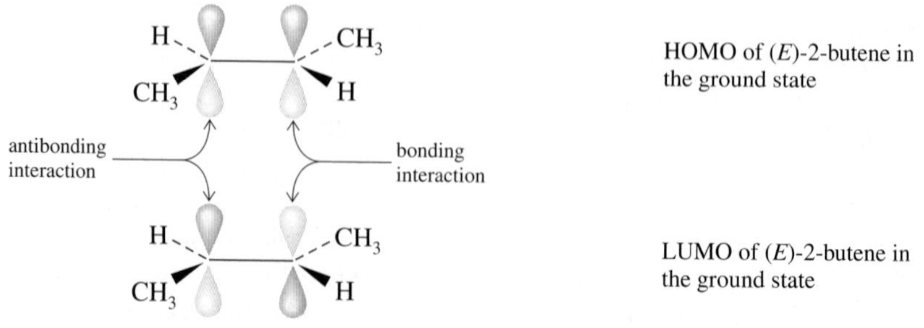

HOMO of (*E*)-2-butene in the ground state

antibonding interaction          bonding interaction

LUMO of (*E*)-2-butene in the ground state

*Figure 25.4*
A schematic representation of the interaction of the HOMO and LUMO of (*E*)-2-butene in the ground state.

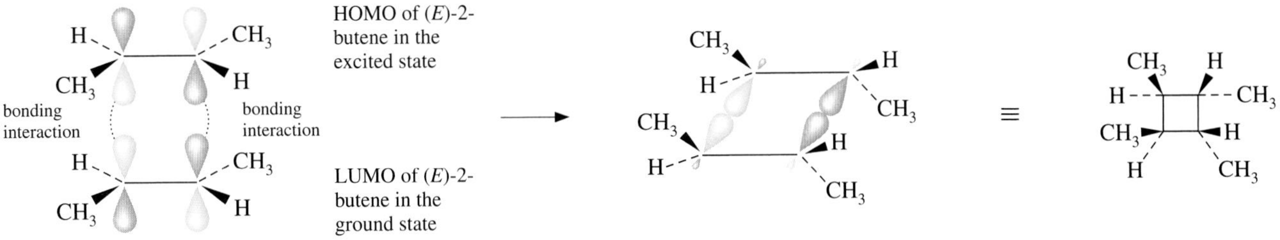

HOMO of (E)-2-butene in the excited state

LUMO of (E)-2-butene in the ground state

bonding interaction

bonding interaction

**Figure 25.5**
A schematic representation of the interaction of the HOMO of (E)-2-butene in the excited state with the LUMO of (E)-2-butene in the ground state.

**Problem 25.4**    Another stereoisomer of the product shown is also formed when (E)-2-butene is irradiated. Draw the molecular orbital picture showing how it arises.

**Problem 25.5**    Draw molecular orbital pictures showing how the two cyclobutanes that are formed when (Z)-2-butene is irradiated with ultraviolet light come into being.

Two alkene functions react photochemically when DNA is damaged by ultraviolet radiation. This reaction is believed to be responsible for most of the harmful effects of such radiation on living organisms. For example, ultraviolet radiation kills bacteria quite efficiently and is used to sterilize equipment. Also, excessive ultraviolet radiation from sunlight (or tanning lamps) can cause skin cancer. Both these effects arise from damage done to DNA so that it can no longer function properly to direct the synthesis of proteins in the cell.

Thymine residues that are adjacent to each other on a strand of DNA can undergo a photochemical [2 + 2] cycloaddition reaction to give a cyclobutane compound known as a thymine dimer.

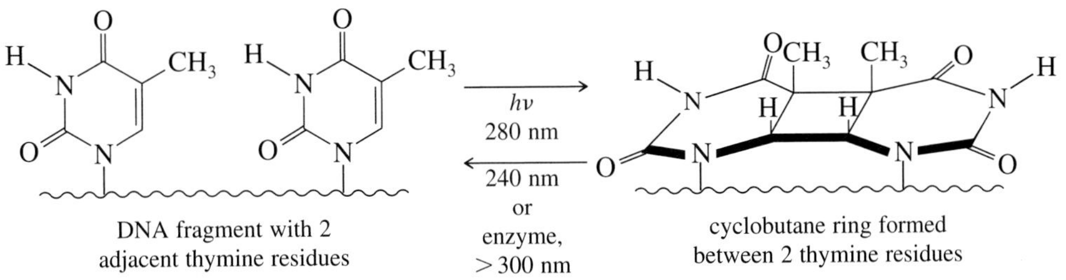

DNA fragment with 2 adjacent thymine residues

$hv$
280 nm
240 nm
or
enzyme,
> 300 nm

cyclobutane ring formed between 2 thymine residues

This photochemical dimerization of thymine disrupts the normal functioning of DNA (p. 1021). Fortunately for creatures that must live under constant exposure to ultraviolet radiation, DNA that is damaged by the formation of dimers can be repaired. The cell has enzymes that repair DNA by cutting out the damaged parts and filling in the missing units. The dimer of thymine also absorbs ultraviolet radiation and undergoes an opening of the cyclobutane ring that converts it back to thymine units. These pathways provide a mechanism for DNA repair, but some damage always remains.

**Problem 25.6**    The chromophore (p. 445) in thymine is an $\alpha,\beta$-unsaturated carbonyl group. In the excited state, such a functional group adds to alkenes or alkynes in a [2 + 2] cycloaddition reaction. Given this information, predict the products of the reactions on the following page.

(a) [structure] + [structure] $\xrightarrow{h\nu}$    (b) [structure] + $CH_3CH_2C{\equiv}CCH_2CH_3 \xrightarrow{h\nu}$

(c) [structure with CH$_3$, CH$_3$, OCH$_2$CH$_3$] + [structure] $\xrightarrow{h\nu}$

**Problem 25.7**    The photochemical [2 + 2] cycloaddition reaction of alkenes has been used to create a cyclobutane ring in several syntheses of grandisol, a sex pheromone of the boll weevil. The cycloaddition products obtained from these syntheses are shown below. Write structures for the two reagents that must have been used in each synthesis. (*Hint:* In part (a), one of them is *not* ethylene.)

(a)  A + B $\longrightarrow$ [structure] $\longrightarrow$ $\longrightarrow$ [structure]

grandisol

(b)  C + D $\longrightarrow$ [structure]    (c)  E + F $\longrightarrow$ [structure]    (d)  G + H $\longrightarrow$ [structure]

**Problem 25.8**    When thymine is irradiated in solution, dimers other than the one shown on page 1087 are produced. What other structures are possible for thymine dimers?

## B.  A Molecular Orbital Picture of the Diels–Alder Reaction

A molecular orbital picture of the Diels-Alder reaction shows the interaction of the HOMO of the diene with the LUMO of the dienophile (Figure 25.6). An examination of the HOMO of the diene shows that the lower lobes of the orbital at carbons 1 and 4 of the conjugated system have the same color as the corresponding upper lobes of the LUMO of the dienophile. Bonding interactions can develop between them if the dienophile approaches the ends of the diene at one face of the molecule. Note that it does not make any difference whether the dienophile is above or below

*Figure 25.6*
A schematic representation of the favorable interaction between the HOMO of a diene and the LUMO of a dienophile.

the diene. Direct overlap of the lobes that are in phase gives rise to two bonding $\sigma$ orbitals and the formation of a six-membered ring.

The substituents at the ends of the diene system and on the doubly bonded carbons of the dienophile retain the stereochemistry they had in the starting compounds because there is no twisting around single bonds during the reaction. The only part of the picture that remains puzzling is how the orbitals between carbons 2 and 3 of the diene system form the $\pi$ bond in the product. At first glance, it looks as if the lobes of the $p$ orbitals that are left over at carbons 2 and 3 will be antibonding. This is so because only two of the orbitals that participate in the reaction are shown in the product; for the sake of simplicity, the others are ignored. As the $\sigma$ bonds form, the interactions between the lobes of the $p$ orbitals at carbons 1 and 2 and those at carbons 3 and 4, which are shown as bonding in the HOMO of the diene, change. The static picture in Figure 25.6 concentrates on only one part of the process and does not tell the whole story.

**Problem 25.9** Draw a molecular orbital picture of the Diels-Alder reaction showing the LUMO of the diene and the HOMO of the dienophile to prove to yourself that the stereochemistry of the reaction and the nature of the transition state are as described on the previous page.

Chemists have used a picture of the interactions of the molecular orbitals of cyclopentadiene and maleic anhydride to rationalize the observed stereochemistry (p. 743) in Diels-Alder reactions. The molecular orbital picture of the cycloaddition of cyclopentadiene with maleic anhydride is shown in Figure 25.7. The LUMO for

cyclopentadiene

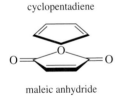

maleic anhydride

cyclopentadiene

maleic anhydride

HOMO of diene

primary interactions leading to the formation of bonds

secondary interactions between LUMO in maleic anhydride and HOMO in cyclopentadiene

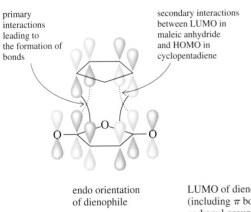

endo orientation of dienophile

LUMO of dienophile (including $\pi$ bonds of carbonyl groups)

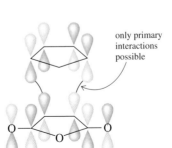

only primary interactions possible

exo orientation of dienophile

*Figure 25.7*
Schematic representation of the interactions between the HOMO of cyclopentadiene and the LUMO of maleic anhydride when the anhydride is endo to the cyclopentadiene ring and when it is exo.

maleic anhydride is shown to include the $\pi$ bonds of the carbonyl groups that are conjugated with the double bond of the diene. The picture drawn is that of the LUMO of a 1,3,5-hexatriene system.

---

**Problem 25.10**    Start with six $p$ orbitals in a row, showing the different signs for their lobes by different kinds of shading, and construct the six $\pi$-type molecular orbitals of 1,3,5-hexatriene. Remember to increase the number of nodes in the orbitals systematically and to maintain the overall symmetry of the system. Show the relative energies for the molecular orbitals, and decide which ones are bonding and which antibonding, where the six $\pi$ electrons will be, and which is the HOMO for the system. The LUMO is already shown in Figure 25.7 on page 1089.

---

Chemists reason in the following way: If maleic anhydride approaches the diene so that the major part of the anhydride molecule lies underneath the cyclopentadiene ring, two kinds of interactions are possible. First, bonding interactions occur between the lobes at the ends of the diene system and the lobes at the double bond in the dienophile. These interactions give rise to the stable six-membered ring in the product. Second, additional weaker interactions are possible between the other lobes of the diene system and those on the carbon atoms of the carbonyl groups. These interactions do not lead to bonding because such a reaction would destroy the stable carbonyl groups and also result in highly strained systems with multiple rings. However, the interactions do stabilize the transition state and lower the energy of activation for the reaction. Thus reaction with the endo orientation of the dienophile is faster than reaction with the exo orientation in which such secondary interactions are not possible (Figure 25.7). As a result, more of the product has the endo stereochemistry.

The regioselectivity observed for reactions of unsymmetrically substituted dienes and dienophiles was discussed in terms of resonance contributors (p. 887). Computer programs are available for calculating the molecular orbitals and other electronic properties of a wide range of molecules. The qualitative results that organic chemists get by drawing curved arrows are confirmed in many cases (and contradicted in a few) by theoretical calculations that describe molecular orbitals more precisely than has been done in this section. For example, all the lobes of the $p$ orbitals making up the molecular orbitals have been shown as being the same size. Actually, the lobes are predicted by theory to be of different sizes, reflecting the different probabilities of finding an electron at any point along the conjugated system. Calculations for compounds such as 1-ethoxybutadiene and propenal (p. 887) confirm that the most important interaction for the HOMO of 1-ethoxybutadiene and the LUMO of propenal is between carbon 4 of the diene system and carbon 3 of propenal, thus determining the orientation observed in the product.

**Study Guide**
**Concept Map 25.1**

## 25.3  Electrocyclic Reactions

### A. Interconversion of Cyclobutenes and Conjugated Dienes

Cyclobutene rings open to conjugated dienes when heated. For example, the ring in *cis*-3,4-dimethylcyclobutene opens smoothly to form (2*Z*,4*E*)-2,4-hexadiene.

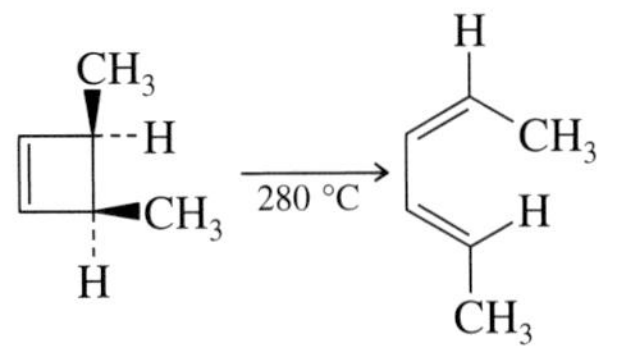

*cis*-3,4-dimethyl-
cyclobutene

(2*Z*,4*E*)-2,4-
hexadiene
> 99% of the
dienes formed

On the other hand, *trans*-3,4-dimethylcyclobutene gives (2*E*,4*E*)-2,4-hexadiene.

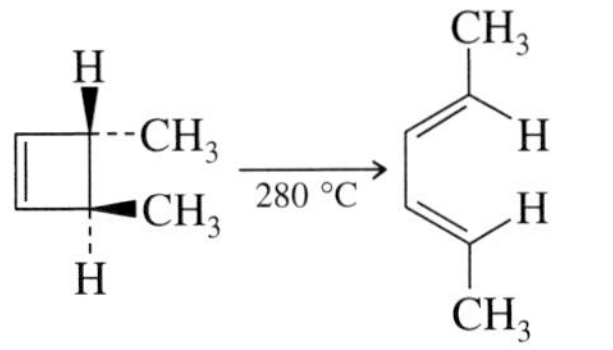

*trans*-3,4-dimethyl-
cyclobutene

(2*E*,4*E*)-2,4-
hexadiene

The ring-opening reaction consists of the breaking of a $\sigma$ bond and the overlap of the lobes of that orbital with those of the adjacent $\pi$ bond so that two new $\pi$ orbitals are formed to accept the four bonding electrons of the diene system. Using the language of frontier orbitals (p. 1085), the reaction may be described as the addition of the HOMO of the $\sigma$ orbital to the LUMO of the $\pi$ bond in the cyclobutene (Figure 25.8).

The HOMO of the $\sigma$ bond is the bonding orbital, shown with two lobes of the $sp^3$ hybrid orbitals on the carbon atoms overlapping. In order for the new $\pi$ bonds to develop, bonding interactions must be possible between the lobes of the orbitals shown for the $\pi$ bond in cyclobutene and the lobes that are freed from the $\sigma$ bond as it breaks. For this to happen, the carbon atoms that are part of the breaking bond must rotate in order to bring the lobes of the $\sigma$ orbital into line with those lobes of similar sign on the other two carbon atoms. As indicated in Figure 25.8, both carbon atoms must rotate in the same direction, either clockwise or counterclockwise, for the new $\pi$ bonds to form smoothly. If you follow the motions of the two methyl groups during the reaction, you will see that they both move in the same direction that the carbon atoms to which they are bonded do, and the product is (2*Z*,4*E*)-2,4-hexadiene. This type of motion, in which the carbon atoms at the ends of the developing polyene system rotate in the same direction, is said to be **conrotatory.** Thus *cis*-3,4-dimethylcyclobutene, when heated, undergoes conrotatory ring opening to (2*Z*,4*E*)-2,4-hexadiene.

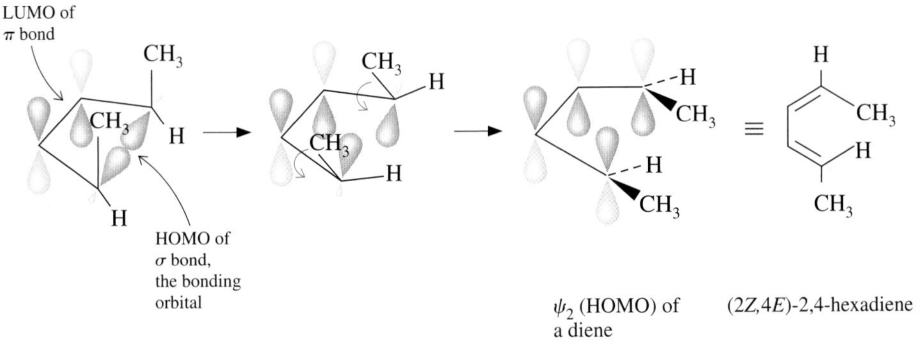

LUMO of
$\pi$ bond

HOMO of
$\sigma$ bond,
the bonding
orbital

$\psi_2$ (HOMO) of
a diene

(2*Z*,4*E*)-2,4-hexadiene

*Figure 25.8*
A schematic representation of the interactions of molecular orbitals that give rise to the stereoselective ring opening of *cis*-3,4-dimethylcyclobutene in a thermal reaction.

**Problem 25.11**    Draw the molecular orbital picture for the thermal opening of the ring in *trans*-3,4-dimethylcyclobutene.

**Problem 25.12**    The following equilibrium has been observed. How do you account for this isomerization? Note that none of the other possible stereoisomers (what are they?) is formed.

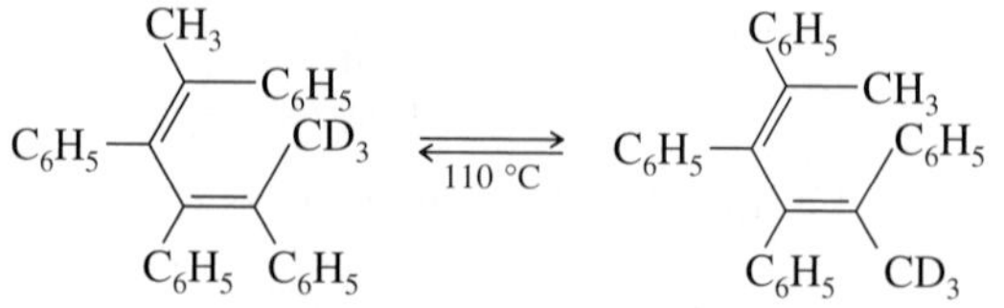

When *cis*-3,4-dimethylcyclobutene undergoes a photochemical ring opening, a mixture of products is obtained, among them (2*E*,4*E*)-2,4-hexadiene.

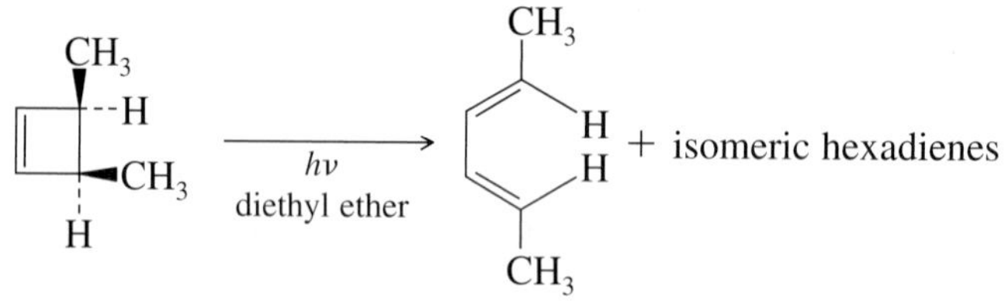

*cis*-3,4-dimethyl-
cyclobutene

(2*E*,4*E*)-2,4-
hexadiene
36%

Practically none of this product was obtained from the thermal ring opening of *cis*-3,4-dimethylcyclobutene (p. 1091). (2*E*,4*E*)-2,4-Hexadiene is believed to develop from the excited state of the cyclobutene. One molecular orbital picture that can be used to rationalize this experimental observation is shown in Figure 25.9.

In the photochemical reaction, the $\pi$ bond of the cyclobutene is assumed to have absorbed ultraviolet energy that has promoted an electron to the $\pi^*$ antibonding orbital. This antibonding orbital then plays the role of the HOMO for that portion of the molecule, and the antibonding $\sigma$ orbital is the LUMO. As is seen in Figure 25.9, the new $\pi$ bonds can form smoothly only if the two carbon atoms of the $\sigma$ bond rotate in opposite directions. One carbon atom must rotate clockwise and the other one counterclockwise so that the lobes being freed from the $\sigma$ bond can overlap with lobes of the same sign from the original $\pi$ bond. When the carbon atoms at the ends of the developing polyene rotate in opposite directions with re-

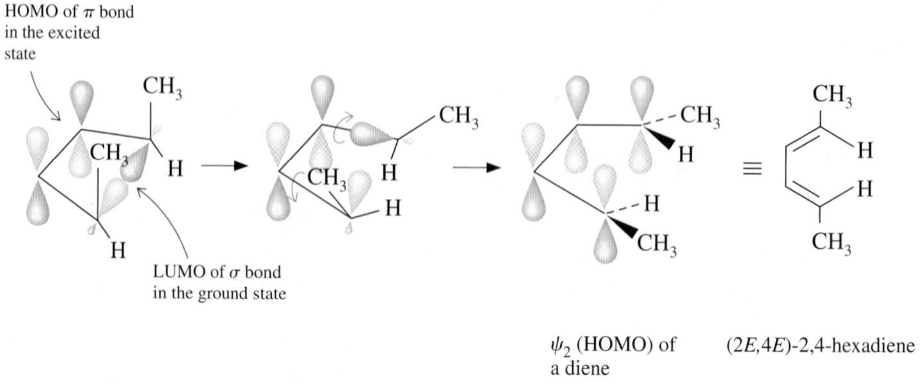

HOMO of $\pi$ bond
in the excited
state

*Figure 25.9*
A schematic representation of the interactions of molecular orbitals that give rise to the stereoselective ring opening of *cis*-3,4-dimethylcyclobutene in a photochemical reaction.

LUMO of $\sigma$ bond
in the ground state

$\psi_2$ (HOMO) of
a diene

(2*E*,4*E*)-2,4-hexadiene

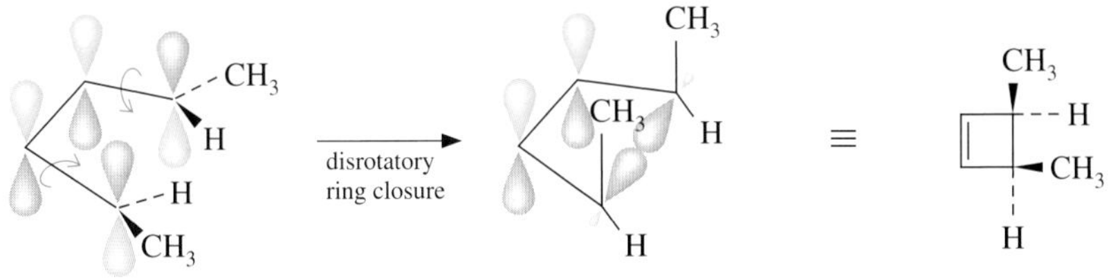

HOMO of the
diene in the
excited state for
(2E,4E)-2,4-hexadiene

cis-3,4-dimethylcyclobutene

**Figure 25.10**
A schematic representation of
the stereochemistry of
photochemical ring closure of
a diene.

spect to each other, their motion is said to be **disrotatory.** The methyl groups move
with the carbon atoms that are rotating, and (2E,4E)-2,4-hexadiene is formed.

(2E,4E)-2,4-Hexadiene undergoes a photochemical ring closure to give cis-3,4-
dimethylcyclobutene.

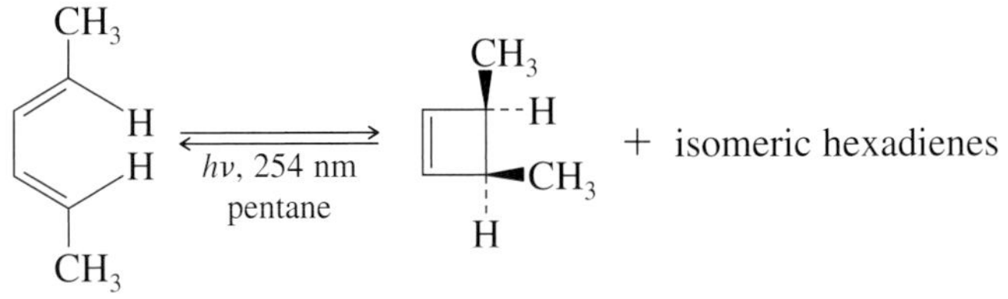

(2E,4E)-2,4-
hexadiene

cis-3,4-dimethyl-
cyclobutene

minor product

The photochemical ring closure is a disrotatory process. The stereochemistry of
a ring closure is determined by the symmetry of the HOMO of the polyene system
undergoing the reaction. The HOMO for the diene in the excited state is $\psi_3$ (Figure
25.2, Problem 25.3). If the end carbon atoms of the diene system are to form a $\sigma$
bond, they must rotate toward each other, one of them clockwise and the other
counterclockwise, to bring the lobes of the same sign into contact so that overlap
and bonding can take place (Figure 25.10).

In summary, cyclobutenes and conjugated dienes interconvert in a conrotatory
way in the ground state. Photochemical interconversions of these systems occur in a
disrotatory fashion. Note that in all these reactions the systems involved either start
or end with four $\pi$ electrons. The difference in the stereochemical course of the
thermal and photochemical reactions can be rationalized by considering the difference in the symmetry of the molecular orbitals of the reacting molecules in the
ground state and in the excited state.

**Problem 25.13** Predict what the product of the reaction shown to the right will be.

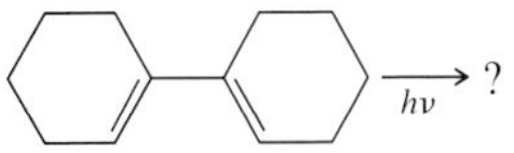

**Problem 25.14** 1,3-Butadiene selectively labeled with deuterium at carbons 1 and 4 has
been prepared. For the isomer shown predict the structures of the products that would be obtained from a thermal and a photochemical cyclization reaction. Use molecular orbital pictures to explain your conclusions.

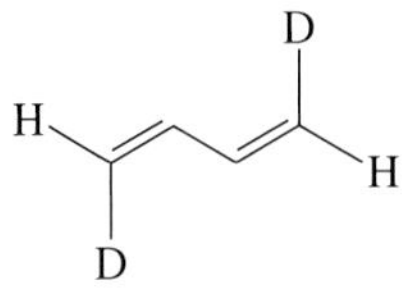

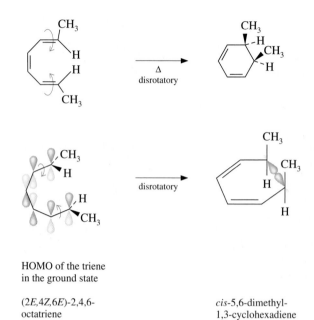

**Figure 25.11**

Schematic representation of the molecular orbital governing the stereochemistry of the thermal ring closure of a triene.

## B. Interconversion of Cyclohexadienes and Trienes

(2E,4Z,6E)-2,4,6-Octatriene undergoes stereoselective ring closure to cis-5,6-dimethyl-1,3-cyclohexadiene. An examination of the stereochemistry of the reactions indicates that the ring is closing in a disrotatory fashion. Thus the thermal ring closure of a conjugated triene takes place with the opposite stereochemistry of that observed for a diene. This experimental observation can be understood by looking at the nature of the molecular orbital undergoing the transformation. The HOMO of (2E,4Z,6E)-2,4,6-octatriene is shown in Figure 25.11.

The HOMO of a triene system has two nodes besides the node in the plane of the molecule (Problem 25.10, p. 1090). The two end carbon atoms of the triene system must rotate toward each other if lobes having the same sign are to overlap to form a bond. The thermal ring closure of a triene is thus disrotatory.

---

**Problem 25.15** Draw the molecular orbitals for (2Z,4Z,6E)-2,4,6-octatriene, and demonstrate to yourself that trans-5,6-dimethyl-1,3-cyclohexadiene is the product of its disrotatory ring closure.

---

When (2E,4Z,6E)-2,4,6-octatriene is irradiated, trans-5,6-dimethyl-1,3-cyclohexadiene is formed, along with stereoisomers of the octatriene. Under the conditions of the reaction, an equilibrium is set up between the cyclohexadiene and the triene.

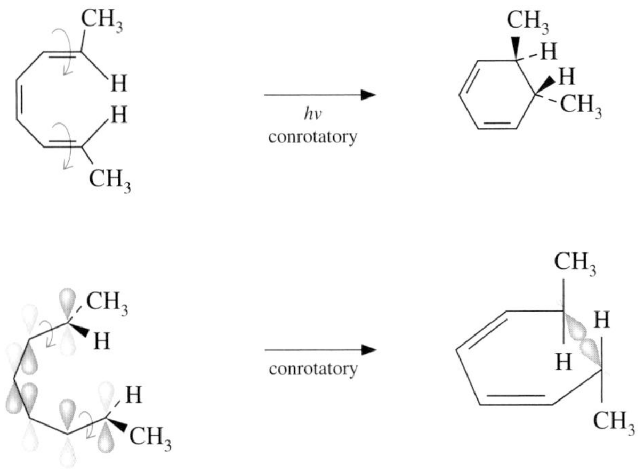

*Figure 25.12*
Schematic representation of the molecular orbitals governing the stereochemistry of the photochemical ring closure of a triene.

HOMO of the triene
in the excited state

(2E,4Z,6E)-2,4,6-
octatriene

*trans*-5,6-dimethyl-
1,3-cyclohexadiene

The photochemical cyclization of a triene is conrotatory (Figure 25.12). When the triene absorbs ultraviolet radiation, an electron is promoted from the HOMO to the LUMO, which then becomes the HOMO of the excited state. It is the symmetry of this orbital that governs the stereochemistry of the photochemical ring closure of the triene. In order to cause the lobes of the same sign on the end carbon atoms of the polyene to overlap, the carbon atoms, and the methyl groups on them, must rotate in the same direction, either clockwise or counterclockwise. The photochemical conrotatory ring closure of (2E,4Z,6E)-2,4,6-octatriene gives *trans*-5,6-dimethyl-1,3-cyclohexadiene; in contrast, the thermal disrotatory ring closure of the triene gives rise to the cis isomer. In the case of the triene-cyclohexadiene transformations, just as for the diene-cyclobutene interconversions, thermal and photochemical reactions proceed with opposite stereochemistry.

**Problem 25.16**    In Figure 25.11, only one of two possible disrotatory ring closures for the (2E,4Z,6E)-2,4,6-octatriene is shown. Similarly, in Figure 25.12, only one kind of conrotatory motion is indicated. Draw molecular orbital pictures to explore the consequences of the other possible disrotatory and conrotatory motions of the end carbon atoms of this polyene system. What is the stereochemical relationship of the new products to those in Figures 25.11 and 25.12?

Photochemical ring-opening reactions of cyclohexadienes are well known and have been most extensively investigated in connection with studies on vitamin D (p. 1081). Vitamin D is a mixture of compounds that are effective in preventing and curing rickets, a disease in which bones become soft and easily deformed. The relationship of the disease to a dietary deficiency that can be remedied by cod liver oil has been recognized for over a hundred years. It was later recognized that certain foods contain compounds that can be transformed into vitamin D by irradiation. These compounds, called **provitamins,** occur widely in both plants and animals. They are all steroids that have a diene system in the B ring. They have no vitamin D activity until that ring is opened on irradiation.

The most widely investigated provitamin D is ergosterol. Ergosterol undergoes a photochemical ring opening to give a compound called precalciferol. All the compounds that are provitamins for vitamin D have the same structure except for variations in the side chain at carbon 17 of the steroid ring. The transformation of ergosterol to precalciferol is a conrotatory opening of the B ring. The stereochemical relationship between the methyl group at carbon 10 and the hydrogen atom at carbon 9 is particularly important, especially for the transformation of precalciferol into vitamin $D_2$ in the next step of the reaction. The photochemical reaction is reversible. The conrotatory opening of the cyclohexadiene ring gives precalciferol in which the methyl group on carbon 10 points toward carbon 9 of the steroid system.

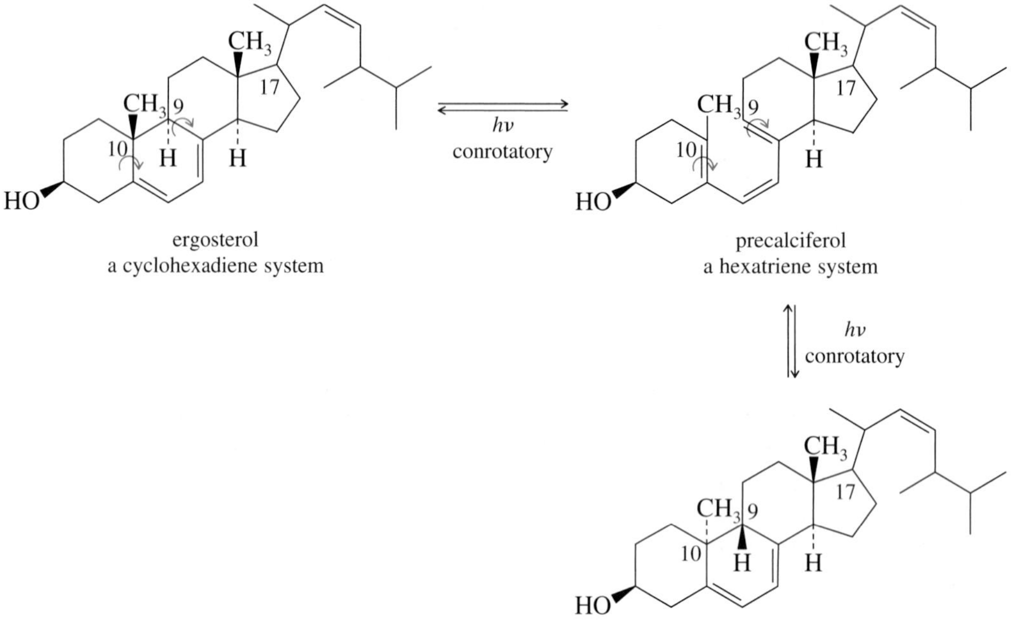

ergosterol
a cyclohexadiene system

precalciferol
a hexatriene system

lumisterol
a cyclohexadiene system

Precalciferol is converted back to ergosterol by a reversal of this motion, but a new cyclohexadiene is formed if ring closure occurs with further rotation of the carbon atoms of the triene in the same direction. This cyclohexadiene is a stereoisomer of ergosterol called lumisterol. In lumisterol, the methyl group at carbon 10 lies below the plane of the steroid ring, and the hydrogen atom at carbon 9 lies above it.

The most important reaction of precalciferol, from a biological viewpoint, is its transformation into calciferol by means of a rearrangement called a **[1,7] sigmatropic rearrangement** (p. 1105).

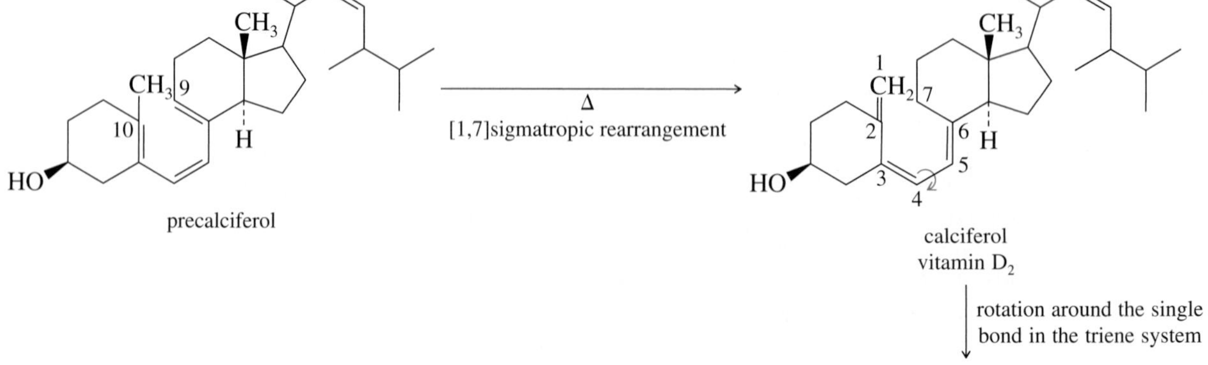

precalciferol

[1,7]sigmatropic rearrangement

calciferol
vitamin $D_2$

rotation around the single bond in the triene system

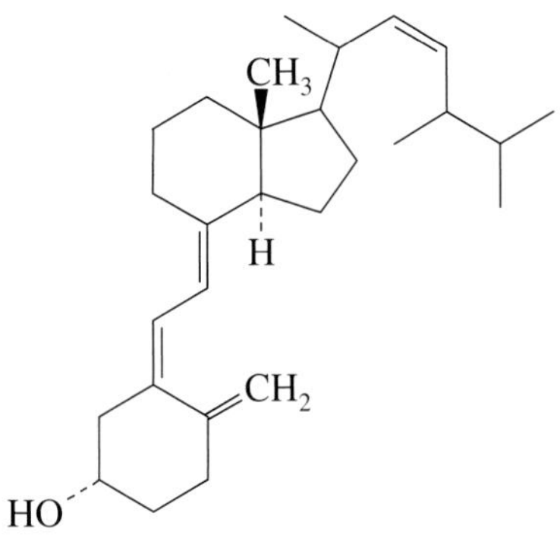

vitamin D$_2$
in its most stable form,
the *s*-trans conformation

The shift of the hydrogen atom results in a change in the position of the double bonds in the triene, allowing the molecule to assume a more stable conformation through rotation around the central single bond of the system. The structure of vitamin D$_2$ is usually drawn with the *s*-trans conformation at that bond.

When precalciferol is heated, two disrotatory ring closures occur. The photochemical and thermal electrocyclic reactions that have been observed, starting with the irradiation of ergosterol, are summarized below.

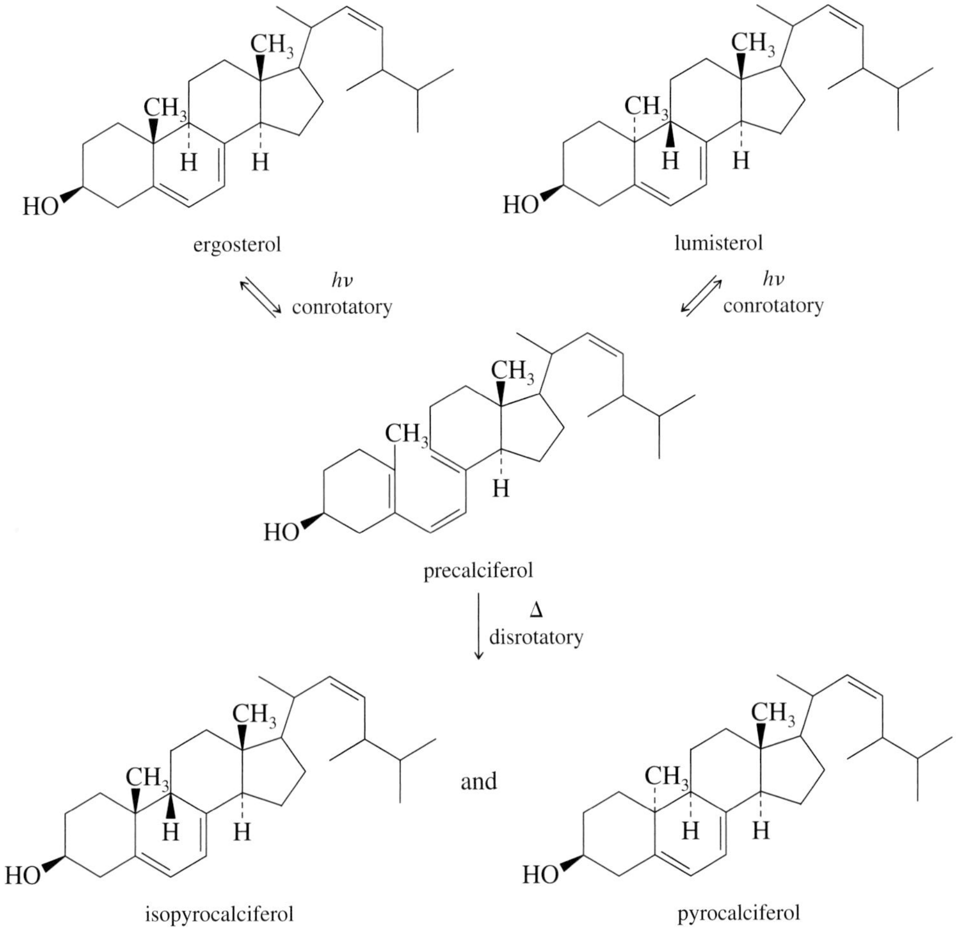

ergosterol

*hv*
conrotatory

lumisterol

*hv*
conrotatory

precalciferol

Δ
disrotatory

isopyrocalciferol

and

pyrocalciferol

**Study Guide**
**Concept Map 25.2**

allo-ocimene

It is easy to understand why chemists were intrigued when they observed reactions in which the stereochemistry depended so precisely on reaction conditions. To explain these observations, they developed the theoretical models discussed in this section.

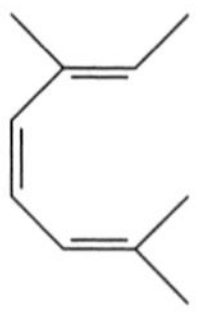

**Problem 25.17**   Using molecular orbital pictures, show how precalciferol is converted thermally into pyrocalciferol and isopyrocalciferol.

**Problem 25.18**   Predict what the products will be when the terpene allo-ocimene shown at left is irradiated.

## C. Some Generalizations about Cycloaddition and Electrocyclic Reactions

A cycloaddition reaction takes place in a concerted manner in the ground state only if $4n + 2$ electrons, where $n = 1, 2, \ldots$, are involved in the transition state. A cycloaddition reaction involving a total of $4n$ electrons participating in the transition state is concerted only in the excited state. These rules apply when the stereochemistry of the reaction has two unsaturated components approaching each other face to face and when bonding occurs on the same side of the molecule at each end of the diene or alkene system. This is the stereochemistry that is shown for the Diels-Alder reaction in Figures 25.6 and 25.7, for example.

These rules, known as the **Woodward-Hoffmann Rules,** indicate how the total number of electrons involved in the transition state of a cycloaddition reaction governs the manner of the cycloaddition. Table 25.1 simply restates the generalizations formulated in the preceding paragraph. A concerted cycloaddition reaction is said to be **allowed** for a system with $4n$ electrons if one of the components of the reaction is raised to the excited state but is **forbidden** if both the components are in the ground state. Forbidden does not mean that no reaction can ever take place. It simply means that if a reaction is observed to occur under these conditions, the theory predicts that it is not a concerted reaction. So far no exceptions have been found.

Note that the generalization refers to the number of electrons and not to the number of orbitals. It is quite easy to remember the rules if you can keep firmly in mind that the Diels-Alder reaction is a $4n + 2$ system in which $n = 1$ and that it is allowed in the ground state. What is allowed in the ground state is forbidden in the excited state, and any change in the number of electrons by two, either up or down, changes the nature of the allowed reaction.

The experimental data that have been given for electrocyclic reactions also lend themselves to generalizations about stereochemistry in terms of the number of electrons involved in the transition state. These Woodward-Hoffmann Rules are

| **Table 25.1** | **Rules for Concerted Cycloaddition Reactions in Which Both Components Retain Stereochemistry** | |
|---|---|---|
| **Total Number of Electrons in the Transition State** | **Concerted Cycloaddition Reaction** | |
| $4n$ | allowed in the excited state | forbidden in the ground state |
| $4n + 2$ | forbidden in the excited state | allowed in the ground state |

**Table 25.2** **Rules for the Stereochemistry of Concerted Electrocyclic Reactions**

| Total Number of Electrons in the Transition State | Stereochemistry of the Allowed Concerted Reaction in the | |
|---|---|---|
| | **Ground State** | **Excited State** |
| 4n | conrotatory | disrotatory |
| 4n + 2 | disrotatory | conrotatory |

summarized in Table 25.2. Once again, these rules do not mean that other types of reactions do not take place, but only that if a reaction is observed to disobey these rules, it is taking place by a different mechanism and is not a concerted reaction.

The two sets of rules given in Tables 25.1 and 25.2 summarize the experimental observations that have been described throughout this chapter. The basis for explaining these observations is molecular orbital theory. The theory allows predictions to be made about systems for which experimental data are not available. One of the predictions based on the rules is that systems containing four electrons on three atoms will add to alkenes in a concerted manner to give five-membered rings. The most fruitful experimental application of this prediction is described in the next section.

⊘ **Study Guide**
**Concept Map 25.3**

**Problem 25.19** A useful synthesis of substituted phenanthrenes is by photochemical cyclization of 1,2-diarylethenes. The reaction gives a dihydrophenanthrene, which is then oxidized easily by oxygen or iodine to the phenanthrene. The reaction is shown below for 1,2-diphenylethene.

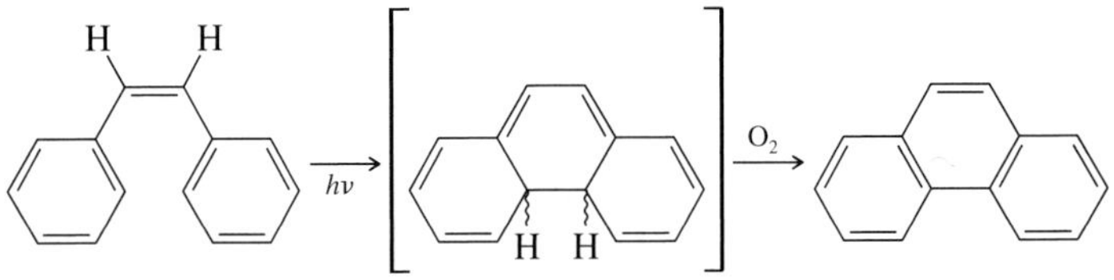

a dihydrophenanthrene
intermediate

How would you classify this reaction according to the Woodward-Hoffmann Rules? What stereochemistry would you expect the dihydrophenanthrene to have?

**Problem 25.20** The Woodward-Hoffmann Rules predict that ionic species also should undergo electrocyclic reactions. For example, the following reaction of a pentadienyl cation has been observed. How would you classify this reaction? What stereochemistry do you expect the product cation to have? (*Hint:* How many electrons are involved in the transition state?)

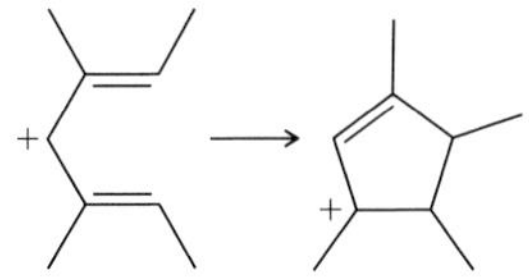

**Problem 25.21** When cyclopentadiene and 3-iodo-2-methyl-1-propene are allowed to react in liquid sulfur dioxide in the presence of silver trifluoroacetate, the products that are formed can be rationalized as arising from the bicyclic tertiary cation shown on the next page.

How would you classify the reaction? What is a likely intermediate? How many electrons are involved in the transition state?

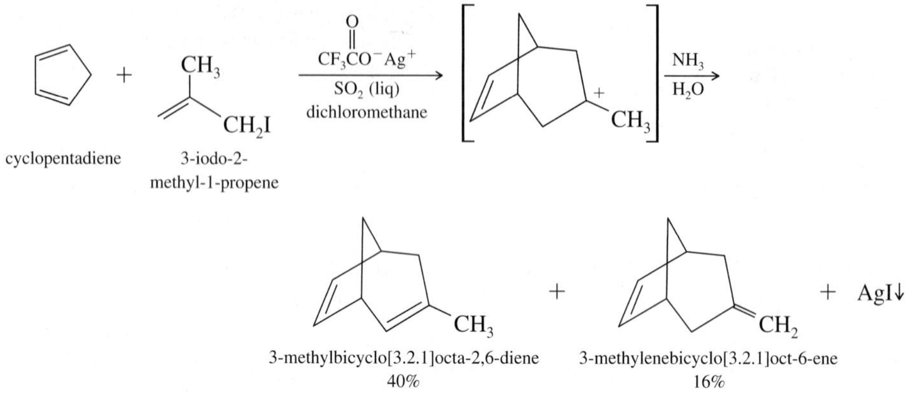
+
$CH_3$
$CH_2I$
O
$CF_3CO^-Ag^+$
$SO_2$ (liq)
dichloromethane
+
$CH_3$
$NH_3$
$H_2O$
cyclopentadiene
3-iodo-2-methyl-1-propene
$CH_3$
+
$CH_2$
+ AgI↓
3-methylbicyclo[3.2.1]octa-2,6-diene
40%
3-methylenebicyclo[3.2.1]oct-6-ene
16%

## 25.4 1,3-Dipolar Additions. Cycloadditions of Compounds Containing Nitrogen and Oxygen

Molecular orbital theory is a powerful tool that can be applied to many different types of reactions. One of the most significant extensions of the theory was made by Rolf Huisgen of Germany, who recognized that certain species containing oxygen or nitrogen could serve as sources of four electrons in [4 + 2] cycloaddition reactions and could therefore be used to create a variety of heterocyclic compounds. These reactions have all the characteristics of concerted reactions, including insensitivity to reaction conditions, such as solvent polarity, and high stereoselectivity.

Some species, though reasonably stable and isolable, have structures that can best be represented as a series of resonance contributors, one of which has a formal positive charge on one atom and a formal negative charge two atoms away. Structures characterized by this distribution of formal charge are called **1,3-dipolar species.** Ozone is a 1,3-dipolar species (p. 310) and so is phenyl azide. All the resonance contributors of phenyl azide have formal charges on various atoms. The 1,3-dipolar character of the compound is represented by the formula shown below.

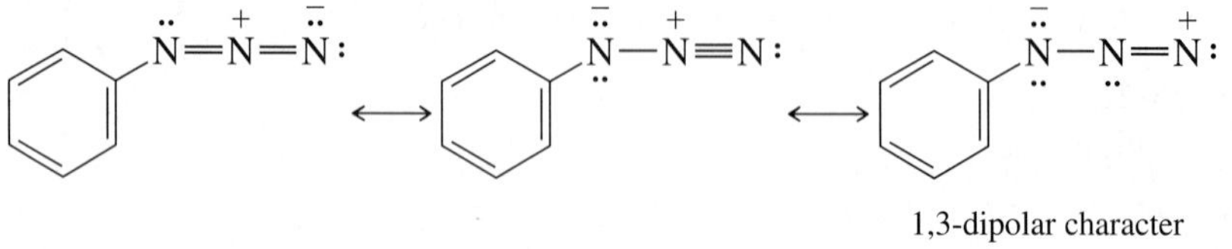
N=N=N:
N—N≡N:
N—N=N:
1,3-dipolar character

*resonance contributors of phenyl azide*

1,3-Dipolar species add to alkenes or alkynes (called **dipolarophiles**) in cycloaddition reactions. The addition of ozone (p. 310) or phenyl azide to an alkene involves the participation of a pair of $\pi$ electrons and a pair of nonbonding electrons from the dipolar species and a pair of $\pi$ electrons from the alkene. The reaction shown on the next page could be classified from an electronic point of view as a [4 + 2] cycloaddition reaction, which is allowed in the ground state. The dipolar species contributes four electrons to the reaction. These electrons are in three $p$ orbitals. Two of the electrons are the electrons of the $\pi$ bond; the other two are nonbonding electrons from an oxygen or a nitrogen atom. All three atoms in the dipolar species participate in the bonding. The product is, therefore, a five-membered ring containing the two doubly bonded carbon atoms of the alkene and three atoms from the dipolar species.

---

**VISUALIZING THE REACTION**

**1,3-Dipolar Addition to a Double Bond**

cyclopentene   phenyl azide        a triazoline

*a 2-electron*   *a 4-electron*   *a [4+2] cycloaddition*
*system*        *system*          *product*

and enantiomer

---

Another type of compound with 1,3-dipolar character is formed when an *N*-substituted hydroxylamine reacts with a carbonyl compound. Such compounds are called **nitrones.** For example, benzaldehyde reacts with hydroxylamine to give an oxime (p. 574). It also reacts with *N*-methylhydroxylamine to give a nitrone.

benzaldehyde   *N*-methylhydroxylamine   *N*-methyl-*C*-phenylnitrone
               hydrochloride              100%

The mechanism for the reaction of a substituted hydroxylamine with a carbonyl compound is similar to that for the formation of the ammonia derivatives of aldehydes and ketones (pp. 575–576). The difference in products comes about from the lack of a second hydrogen atom on the nitrogen atom of the substituted hydroxylamine. The iminium ion intermediate formed in the synthesis of an oxime has a hydrogen atom bonded to the positively charged nitrogen atom. That hydrogen atom is the most acidic one in the molecule and is lost in the last step of the reaction. In the reaction of an *N*-substituted hydroxylamine, the iminium ion intermediate has no hydrogen atom bonded to the quaternary nitrogen atom. In this case, the most acidic hydrogen atom is the one on the hydroxyl group. Removal of that proton by a base gives a nitrone.

A nitrone reacts with alkenes as a 1,3-dipolar compound. For example, it adds to styrene to give a five-membered heterocycle as shown on the next page. A saturated five-membered heterocycle containing adjacent oxygen and nitrogen atoms is known as an isoxazolidine (p. 933). The formation of an isoxazolidine from a nitrone and an alkene is pictured as a [4 + 2] cycloaddition reaction. The nitrone contributes four electrons, two from a $\pi$ bond and two nonbonding electrons from the oxygen atom, to the transition state. The regioselectivity shown for the addition of a nitrone to an alkene is quite general. Terminal alkenes give products in which the substituent is at carbon 5 of the isoxazolidine.

**ONE SMALL STEP**

A road map for the mechanism for the formation of a nitrone is given to the left.

**PROBLEM:** Review the mechanism for the formation of an oxime on pages 575–576. Extend this mechanism to show the stepwise formation of *N*-methyl-*C*-phenylnitrone from benzaldehyde and *N*-methylhydroxylamine.

**Reaction of a Nitrone with an Alkene**

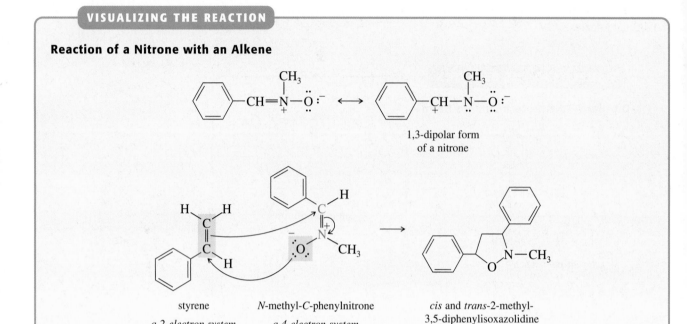

styrene
*a 2-electron system*

*N*-methyl-*C*-phenylnitrone
*a 4-electron system*

1,3-dipolar form
of a nitrone

*cis* and *trans*-2-methyl-
3,5-diphenylisoxazolidine
95%

*racemic*
*a [4+2] cycloaddition product*

It is not necessary to synthesize and isolate a nitrone in order to perform the reaction that produces isoxazolidines. A mixture of a carbonyl compound, an *N*-substituted hydroxylamine, and an alkene heated together will give a high yield of an isoxazolidine. For example, butanal, *N*-phenylhydroxylamine, and styrene react to give 2,5-diphenyl-3-propylisoxazolidine.

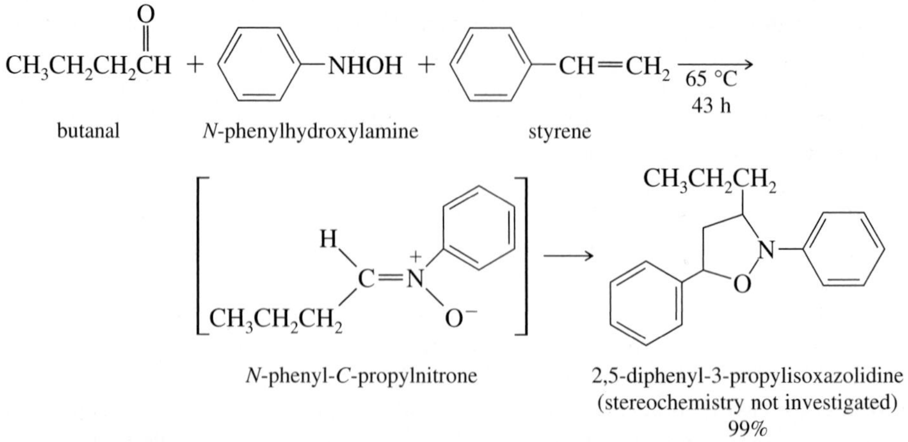

$$CH_3CH_2CH_2\overset{\overset{\displaystyle O}{\|}}{C}H \ + \ \text{⟨⟩}-NHOH \ + \ \text{⟨⟩}-CH=CH_2 \xrightarrow[\text{43 h}]{65\ °C}$$

butanal            *N*-phenylhydroxylamine            styrene

$$\left[ \ \underset{CH_3CH_2CH_2}{\overset{H}{\diagdown}} C \overset{+}{=} N \underset{O^-}{\diagup} \text{⟨⟩} \ \right] \longrightarrow$$

*N*-phenyl-*C*-propylnitrone

2,5-diphenyl-3-propylisoxazolidine
(stereochemistry not investigated)
99%

The reactions described above are only a few examples of the kinds of heterocyclic systems that can be created by the addition of 1,3-dipolar species to alkenes or alkynes.

**Problem 25.22** Predict what the products of the following reactions will be.

(a) 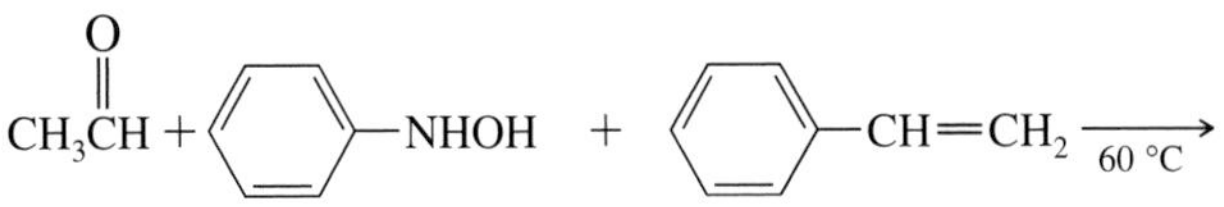 CH₃CH + [phenyl]—NHOH + [phenyl]—CH=CH₂ →60 °C

(b) 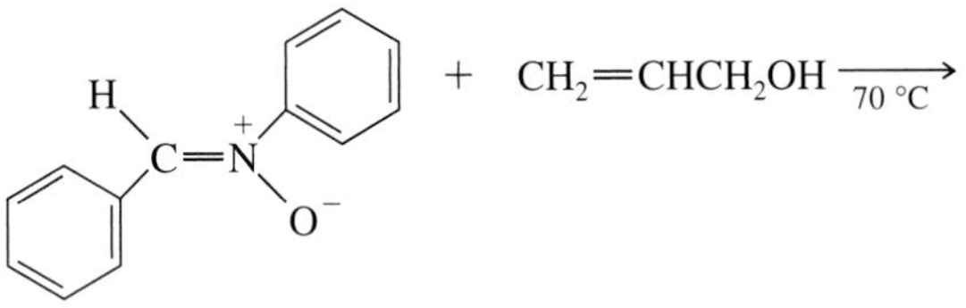
+ CH₂=CHCH₂OH →70 °C

(c) [phenyl]—N=N⁺=N⁻ + CH₃OCC≡CCOCH₃ →

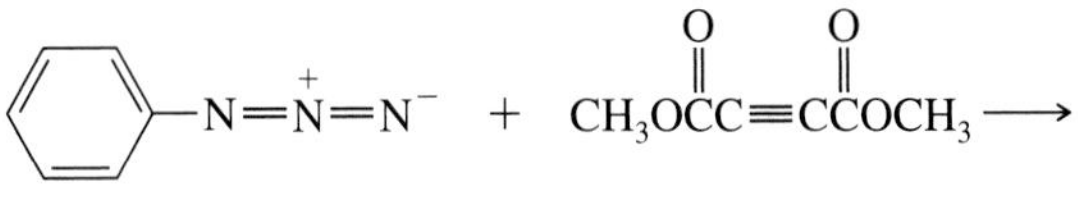

(d) 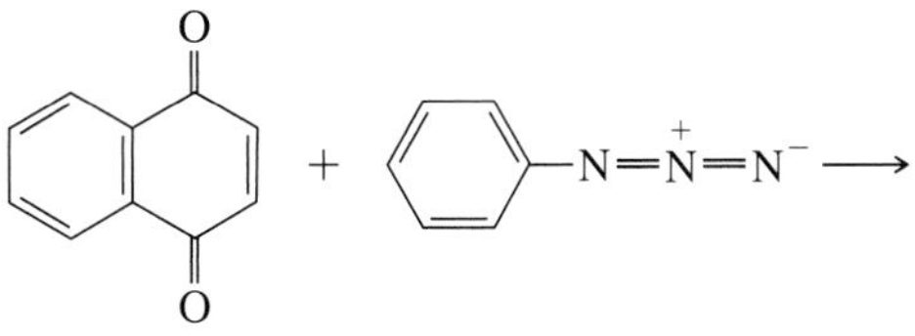 + [phenyl]—N=N⁺=N⁻ →

(e) 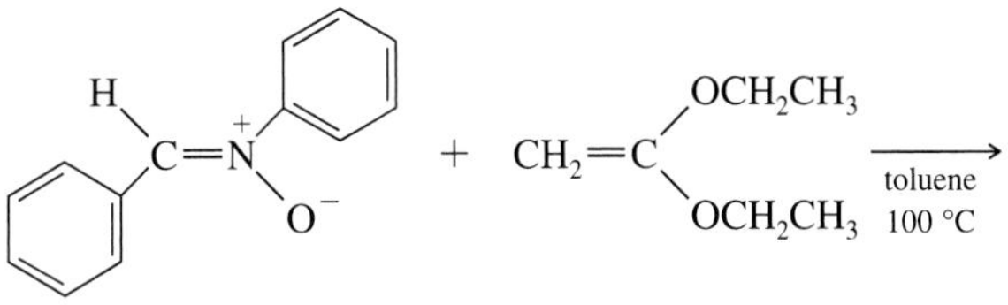
+ CH₂=C(OCH₂CH₃)(OCH₂CH₃) →toluene 100 °C

**Problem 25.23** The following heterocyclic compounds are obtained by 1,3-dipolar addition reactions. Draw the structure of the alkene or alkyne and the 1,3-dipolar species that would be required to synthesize each compound.

(a) 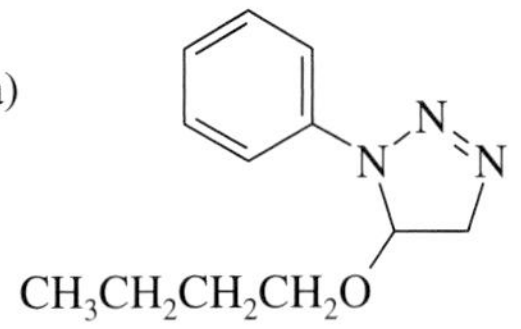
CH₃CH₂CH₂CH₂O

(b)

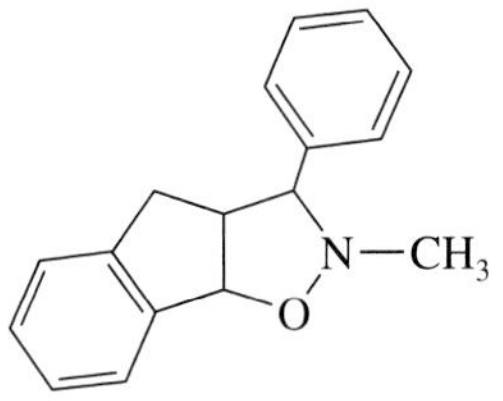

(c)

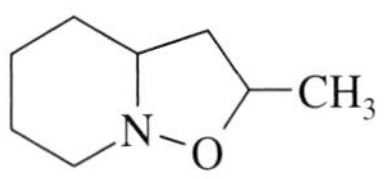

(d)

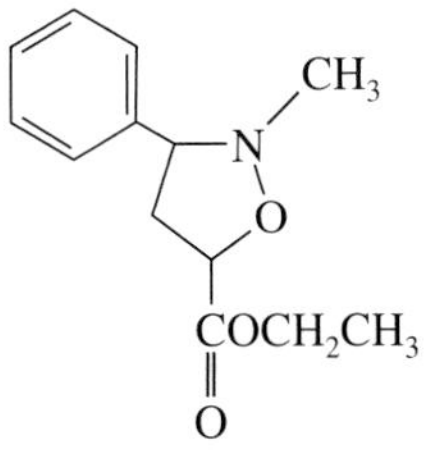

(e)

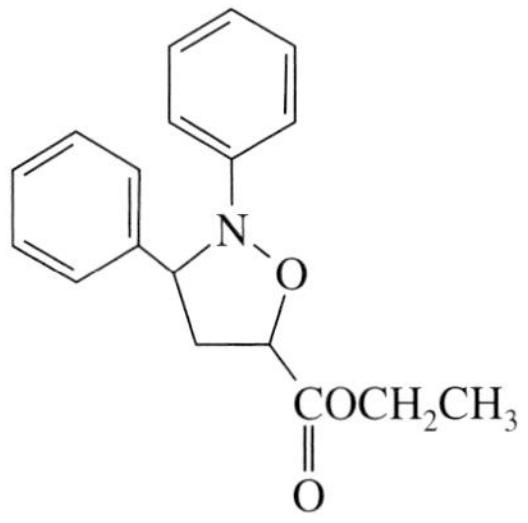

## 25.5 Sigmatropic Rearrangements

### A. Hydrogen Shifts

A sigmatropic rearrangement involves the migration of a σ bond from one position to another one along a polyene chain with a simultaneous shift of π bonds. Such a migration of a hydrogen atom and its σ bond in (Z)-1,3-pentadiene, a [1,5] hydrogen shift, is shown on the next page. The migration occurs in a highly stereoselective way with the hydrogen atom moving along one surface of the polyene system.

(Z)-1,3-pentadiene          (Z)-1,3-pentadiene

The stereoselectivity of the reaction can be explained if a molecular orbital picture is drawn. In Figure 25.13, the process is depicted as the interaction of the HOMO of the carbon–hydrogen $\sigma$ bond at the tetrahedral carbon atom with the LUMO of the diene that is the remainder of the molecule. A bonding interaction can take place between the $s$ orbital of the hydrogen atom and the molecular orbital of the diene system at carbon 5 only if the hydrogen atom moves from the top of carbon 1 to the top of carbon 5. A new diene system with double bonds between carbons 1 and 2 and carbons 3 and 4 forms as the hydrogen atom migrates.

A similar look at the [1,7] sigmatropic rearrangement that occurs when precalciferol is converted to calciferol (p. 1096) suggests an interesting prediction (Figure 25.14). The reaction is pictured as an interaction between the HOMO of the $\sigma$ bond that is migrating and the LUMO of the triene system. If you examine the symmetry of these orbitals, you will see that it is the bottom lobe of the orbital at carbon 7 that is in phase with the $\sigma$ bond at carbon 1. The hydrogen atom must move from the top of carbon 1 to the bottom face of carbon 7 for a bonding interaction to occur between the hydrogen atom and the carbon atom that is becoming tetrahedral. Such a migration is stereochemically possible in this kind of an open-chain triene because one end of the molecule lies above the other end in one conformation of the system.

A more general way of looking at these reactions is to consider the total number of electrons involved in the transition state. The [1,5] hydrogen shift occurs with a reorganization of six electrons, two from the $\sigma$ bond and four in the diene system. In the molecular orbital interactions for the reaction as drawn in Figure 25.13, the

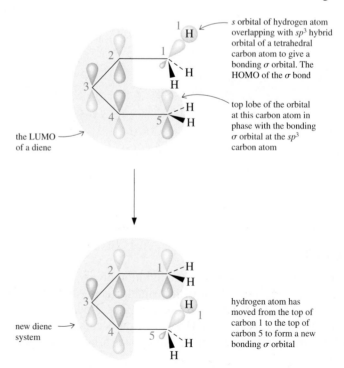

the LUMO of a diene

*s* orbital of hydrogen atom overlapping with $sp^3$ hybrid orbital of a tetrahedral carbon atom to give a bonding $\sigma$ orbital. The HOMO of the $\sigma$ bond

top lobe of the orbital at this carbon atom in phase with the bonding $\sigma$ orbital at the $sp^3$ carbon atom

new diene system

hydrogen atom has moved from the top of carbon 1 to the top of carbon 5 to form a new bonding $\sigma$ orbital

*Figure 25.13*
A molecular orbital picture of the stereochemistry of a [1,5] sigmatropic rearrangement involving hydrogen migration.

reaction has been treated as a [4 + 2] interaction. A further generalization is that all sigmatropic rearrangements involving $4n + 2$ electrons in the transition state take place in a concerted fashion in the ground state with the stereochemistry shown in Figure 25.13. The hydrogen and alkyl shifts seen in the rearrangements of carbocations (p. 289) are concerted [1,2] sigmatropic rearrangements that take place along one face of the molecule. In these cases, only the two electrons of the migrating $\sigma$ bond are involved in the transition state, as the bond shifts to a cationic carbon atom with an empty orbital. Such a case is a rearrangement of a system with $4n + 2$ electrons, where $n = 0$.

Using the same language, a [1,7] hydrogen shift is described as having six electrons from the triene system and two electrons from the $\sigma$ bond involved in the transition state, for a total of eight electrons. Thus such a reaction involves a $4n$ system, with $n = 2$. Rearrangements involving the $4n$ electrons are concerted in the ground state when the stereochemistry is that shown in Figure 25.14, with migration taking place from the top face of one part of the molecule to the bottom face of the other.

Just as in electrocyclic reactions and cycloaddition reactions, in sigmatropic rearrangements, the symmetry of the orbitals involved and the resulting stereochemistry of the reactions change in the excited state. Problem 25.24 on page 1106 provides an opportunity for you to prove this to yourself.

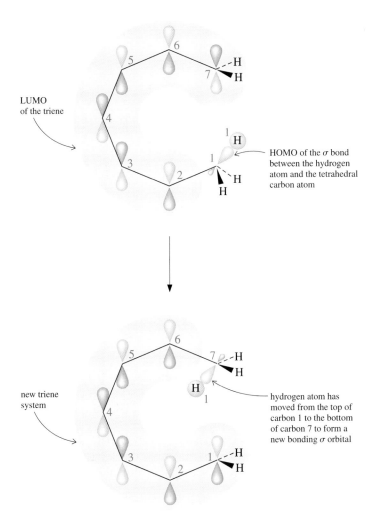

*Figure 25.14*
A molecular orbital picture of the stereochemistry of a [1,7] sigmatropic rearrangement involving hydrogen migration.

**Problem 25.24** When a hexatriene system undergoes a photochemical [1,7] hydrogen shift, the hydrogen atom migrates from the top of carbon 1 to the top of carbon 7. This observation can be rationalized by drawing a molecular orbital picture similar to the one shown in Figure 25.14 but using the HOMO that is appropriate for the excited state of the triene. Prove to yourself that this is so. You may want to refer to Problem 25.10 to see what the HOMO of the excited state of a triene looks like. You must now, of course, use the LUMO of the $\sigma$ bond to the hydrogen atom.

## B. The Cope Rearrangement

Not all sigmatropic rearrangements involve migration of hydrogen atoms. A large and important group of reactions takes place with the migration of a carbon atom and $\sigma$ bond. A 1,2-alkyl shift in a carbocation is an example that you have already learned. A group of such reactions that do not have ionic intermediates is known as the **Cope rearrangement,** for Arthur Cope of the Massachusetts Institute of Technology, who discovered and studied them.

A typical Cope rearrangement is the interconversion of 3-methyl-1,5-hexadiene and (E)-1,5-heptadiene.

3-methyl-1,5-hexadiene          (E)-1,5-heptadiene

When either compound is heated to approximately 300 °C, it is converted into a mixture of the two. The two alkenes are similar to each other in energy, so an equilibrium is established, and the reaction does not go to completion in one direction or the other.

If the diene is chosen so that at least one of the double bonds in the rearranged diene is conjugated with an aryl group or a carbonyl group, then the reaction takes place at lower temperatures and gives good yields. An example of such a reaction is the rearrangement of 3-methyl-4-phenyl-1,5-hexadiene.

3-methyl-4-phenyl-          (1E,5E)-1-phenyl-
1,5-hexadiene          1,5-heptadiene
90%

In the product, (1E,5E)-1-phenyl-1,5-heptadiene, one of the double bonds is conjugated with the aromatic ring, and the other one is an internal double bond. The product diene is more stable than 3-methyl-4-phenyl-1,5-hexadiene, in which both the double bonds are in terminal positions.

A reexamination of the rearrangement of 3-methyl-1,5-hexadiene to (E)-1,5-heptadiene above shows that the reaction involves breaking a $\sigma$ bond at one part of the diene chain and forming a $\sigma$ bond at another, while the $\pi$ bonds in the molecule shift position.

**The Cope Rearrangement**

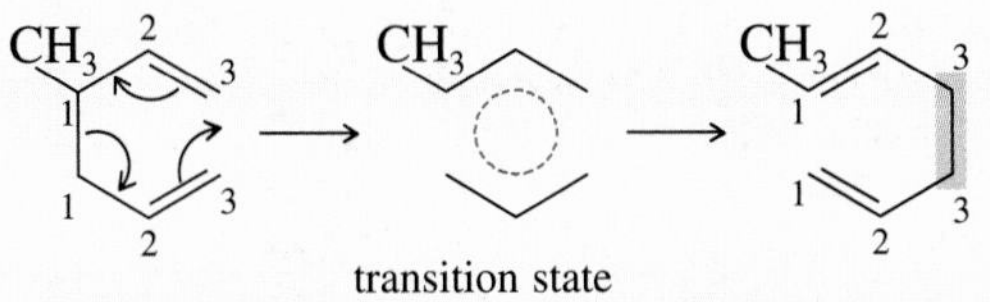

transition state

*a [3,3] sigmatropic rearrangement*

The arrows indicate that the Cope rearrangement has a transition state involving six electrons, two from the $\sigma$ bond that is breaking and four from the two $\pi$ bonds. The hexadiene is numbered with the carbon atoms of the breaking $\sigma$ bond as the first carbon atoms of both halves of the chain. In this way of looking at the molecule, the new $\sigma$ bond forms between the third atoms of each half of the chain. In other words, the $\sigma$ bond has migrated three atoms for the top half of the molecule and also three atoms for the bottom half. For this reason, the reaction is called a [3,3] sigmatropic rearrangement.

**Problem 25.25**    Complete the following equations.

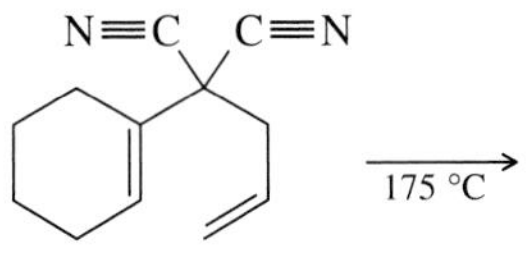

**Problem 25.26**    A synthesis of grandisol, a sex pheromone of the male boll weevil (Problem 25.7, p. 1088), started with a metal-catalyzed dimerization of isoprene to form what the researchers hoped would be a cyclobutane with the structure shown to the right. The product that was isolated when the reaction mixture was allowed to exceed room temperature was 1,5-dimethyl-1,5-cyclooctadiene. How would you rationalize this observation?

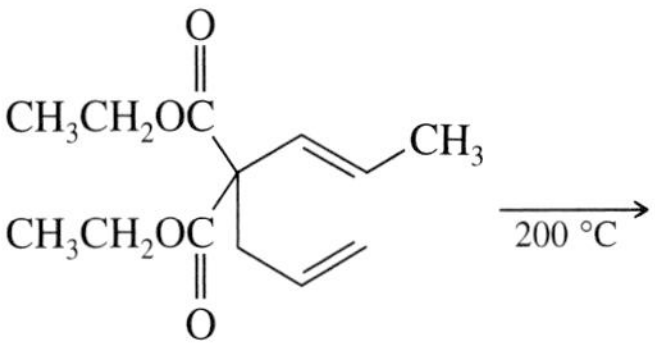

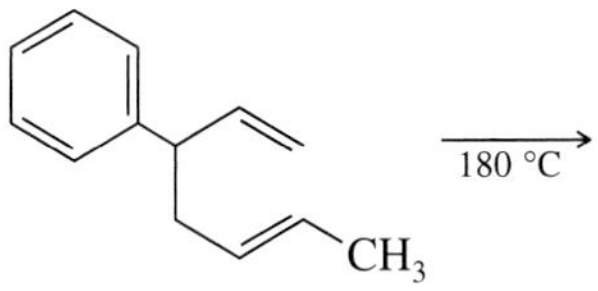

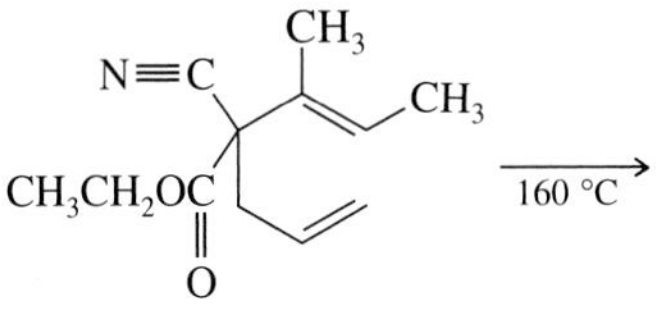

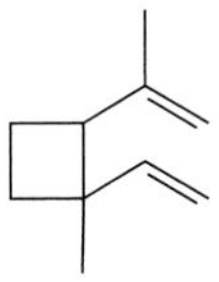

## C. The Claisen Rearrangement

Sigmatropic rearrangements involving the cleavage of a $\sigma$ bond at an oxygen atom are called **Claisen rearrangements.** The same Ludwig Claisen who discovered the condensation reaction of esters to form $\beta$-ketoesters (p. 705) also discovered that the allyl enol ether of ethyl acetoacetate undergoes a rearrangement to an acetoacetic ester substituted by an allyl group at the $\alpha$-carbon atom.

**VISUALIZING THE REACTION**

## Claisen Rearrangement of an Allyl Ether

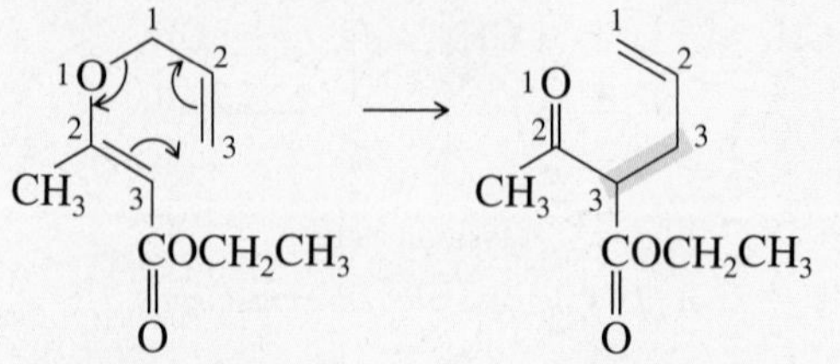

O-allyl ether of
the enol of ethyl
acetoacetate

ethyl 2-allylacetoacetate
85%

*a [3,3] sigmatropic rearrangement*

A $\sigma$ bond is broken between two atoms in this case, an oxygen atom and a carbon atom; another one is formed three atoms away on each portion of the chain. The $\sigma$ bond migrates three atoms away from its starting position for both portions of the chain that participate in the rearrangement; thus this reaction is also called a [3,3] sigmatropic rearrangement. The driving force for the reaction is the formation of the stable carbonyl group in the product.

The Claisen rearrangement is quite general. The rearrangement of the allyl ethers of phenols is another example. For example, allyl phenyl ether, easily prepared from phenol, rearranges to give *o*-allylphenol.

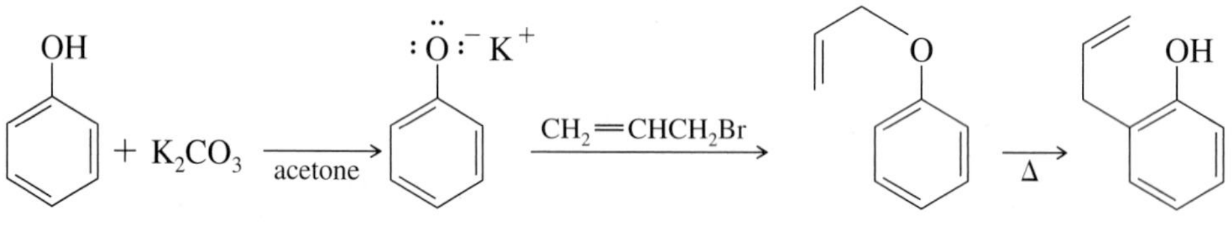

phenol | potassium | potassium | allyl phenyl ether | o-allylphenol
carbonate | phenolate | 86%

The carbon atom attached to the aromatic ring is not the carbon atom that was bonded to the oxygen atom in the ether. If a 2-butenyl instead of an allyl group is present in a phenyl ether, a product phenol has a methyl branch on the side chain.

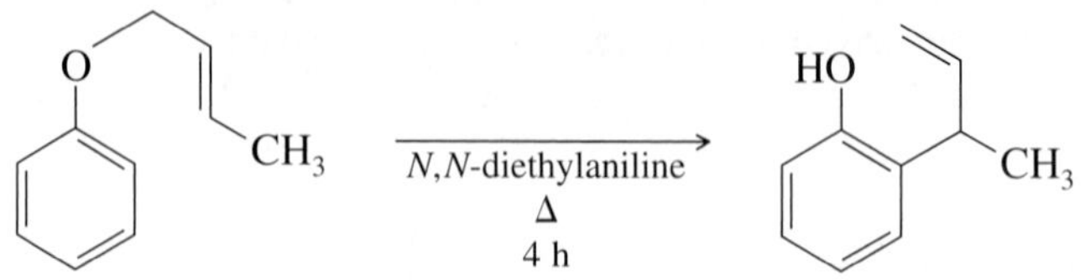

2-butenyl phenyl
ether

2-(1-methyl-2-propenyl)phenol
85%

This experiment and others like it show that attachment between the aromatic ring and the allyl side chain takes place three carbon atoms away from the oxygen atom. The allyl group is not detached from the oxygen atom and reattached to the aromatic ring at the original point of bonding.

The rearrangement goes through a cyclic transition state.

VISUALIZING THE REACTION

**Claisen Rearrangement of an Allyl Aryl Ether**

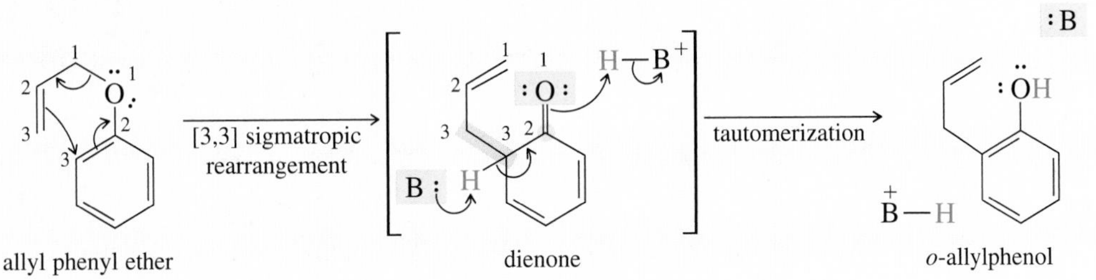

allyl phenyl ether → [3,3] sigmatropic rearrangement → dienone *unstable intermediate* → tautomerization → *o*-allylphenol

Two of the $\pi$ electrons of the aromatic ring participate in the reaction, which involves a total of six electrons in the transition state. The mechanism shown above requires an unstable dienone as an intermediate. Such a dienone is expected to tautomerize rapidly to the stable phenol by loss of a proton at the aromatic ring and protonation of the oxygen atom.

The [1,5] hydrogen shift (p. 1109), the Cope rearrangement (p. 1106), and the Claisen rearrangement (p. 1107) all take place by way of transition states involving six atoms and the reorganization of six electrons. This fact, kept firmly in mind, will help you to recognize these reactions when you see molecular structures that allow for such transformations.

⊘ **Study Guide**
Concept Map 25.4

**Problem 25.27**    Complete the following equations.

(a) $CH_2$=$CHOCH_2CH$=$CH_2 \xrightarrow{255\ °C}$

(b) $\xrightarrow{175\ °C}$

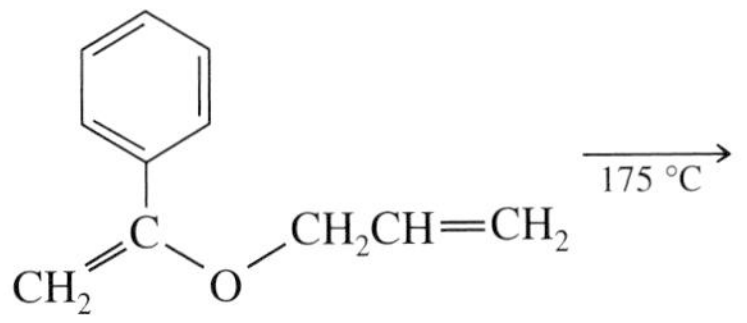

(c) $\xrightarrow{\Delta}$

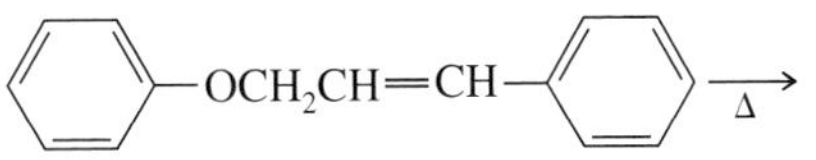

(d) $\xrightarrow{225\ °C}$

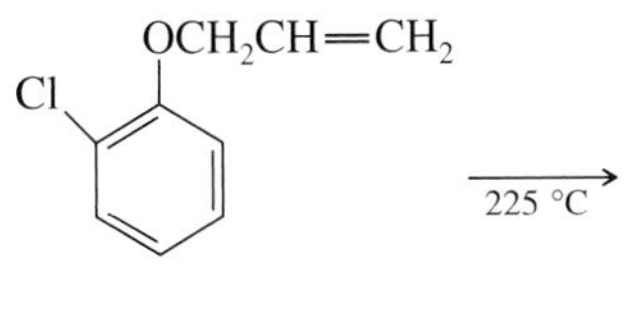

(e) $\xrightarrow[225\ °C]{N,N\text{-dimethylaniline}}$

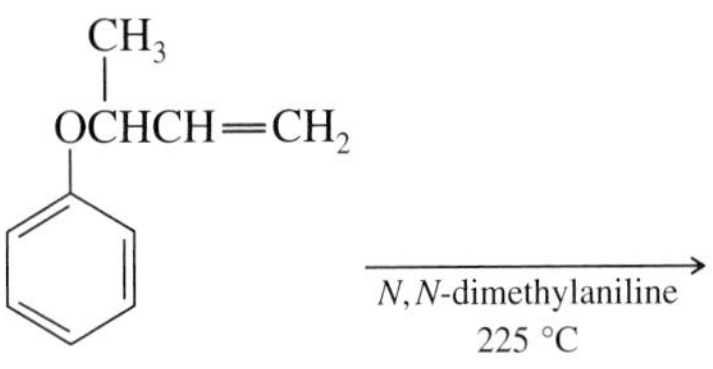

(f) $\xrightarrow{240\ °C}$

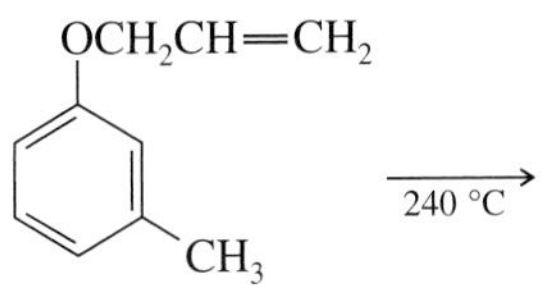

# 25.6  Reactions of Carbenes

**Carbenes** are reactive intermediates containing two single bonds and a pair of nonbonding electrons around a carbon atom. They are neutral species that must acquire another pair of electrons to complete a stable octet. Thus they are electrophiles and add to double bonds to give cyclopropanes. Carbenes are formed when an alkyl halide that cannot undergo elimination to form an alkene reacts with a strong base. For example, tribromomethane reacts with potassium *tert*-butoxide to give dibromomethylene, a carbene.

$$CHBr_3 \;+\; CH_3\overset{\underset{\displaystyle CH_3}{|}}{\underset{\underset{\displaystyle CH_3}{|}}{C}}O^-K^+ \longrightarrow \overset{\underset{\displaystyle Br}{|}}{\underset{\underset{\displaystyle Br}{|}}{C}}: \;+\; CH_3\overset{\underset{\displaystyle CH_3}{|}}{\underset{\underset{\displaystyle CH_3}{|}}{C}}OH + K^+Br^-$$

| tribromomethane bromoform | potassium *tert*-butoxide | dibromomethylene *a carbene* | *tert*-butyl alcohol | potassium bromide |

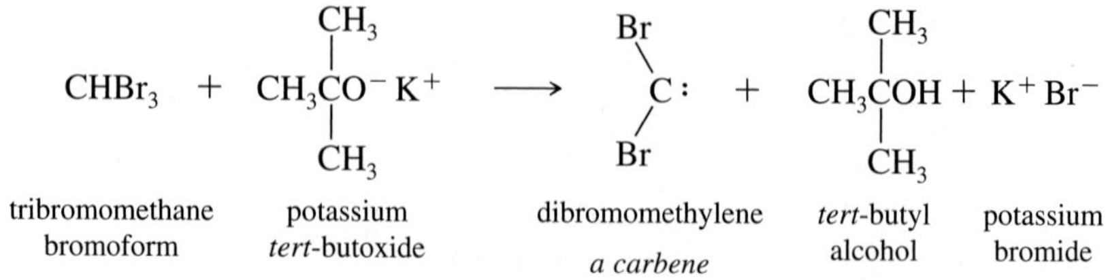

The overall reaction is an $\alpha$-elimination, taking place in two steps.

---

### VISUALIZING THE REACTION

**Formation of a Carbene from a Halide**

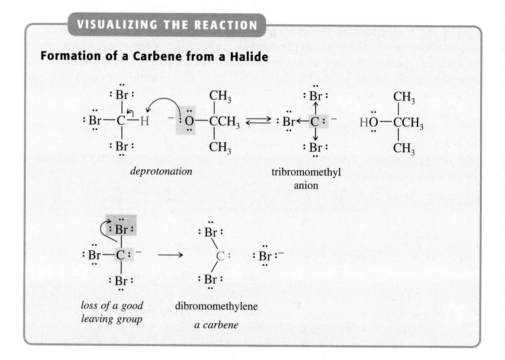

*deprotonation*

tribromomethyl anion

*loss of a good leaving group*

dibromomethylene

*a carbene*

---

The proton and the leaving group are lost from the *same* carbon atom of the alkyl halide rather than from adjacent carbon atoms as in $\beta$-elimination reactions that give rise to alkenes.

The reaction of a carbene such as dibromomethylene with an alkene is stereoselective. The carbene reacts with (*Z*)-2-butene to give *cis*-1,1-dibromo-2,3-dimethylcyclopropane and with (*E*)-2-butene to give the trans isomer.

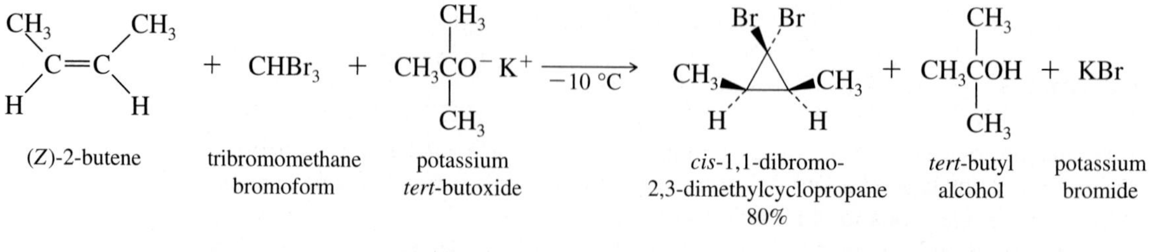

$$\underset{\text{(Z)-2-butene}}{\overset{\underset{\displaystyle H}{|}\;\;\;\underset{\displaystyle H}{|}}{\underset{CH_3}{}C{=}C\underset{CH_3}{}}} + CHBr_3 + CH_3\overset{\underset{\displaystyle CH_3}{|}}{\underset{\underset{\displaystyle CH_3}{|}}{C}}O^-K^+ \xrightarrow{-10\,°C} \underset{\text{cis-1,1-dibromo-}}{} + CH_3\overset{\underset{\displaystyle CH_3}{|}}{\underset{\underset{\displaystyle CH_3}{|}}{C}}OH + KBr$$

| (Z)-2-butene | tribromomethane bromoform | potassium *tert*-butoxide | *cis*-1,1-dibromo-2,3-dimethylcyclopropane 80% | *tert*-butyl alcohol | potassium bromide |

(E)-2-butene     tribromomethane     potassium     *trans*-1,1-dibromo-     *tert*-butyl     potassium
               bromoform     *tert*-butoxide     2,3-dimethylcyclopropane     alcohol     bromide
                                                                 68%

*racemic*

Because the stereochemistry of the alkene is retained in the product cyclopropane, chemists think that the carbene adds to the double bond in a single step, in which both bonds to the doubly bonded carbon atoms of the alkene are formed simultaneously.

## VISUALIZING THE REACTION

### Addition of a Carbene to an Alkene

The number of electrons participating in the reaction, however, presents a problem at first glance. Four electrons are involved in the transition state, two from the $\pi$ bond of the alkene and two from the carbene. This reaction resembles the concerted cycloaddition of two ethylene units, which is forbidden in the ground state (Table 25.1, p. 1098). The difference between the thermal cycloaddition of two alkene units to give cyclobutane and the cycloaddition of a carbene to an alkene lies in the stereochemistry that is possible for the two reactions. The four-carbon atom framework of a cyclobutane molecule requires that the two double bonds approach each other and react in a face-to-face manner (Figure 25.5, p. 1087). Otherwise, an impossibly twisted and strained four-membered ring would form. An examination of the orbitals of a carbene shows that a bonding overlap between the HOMO of a carbene and the LUMO of an alkene (or the other way around) is possible (Figure 25.15, p. 1112).

A carbene has three pairs of electrons around the central carbon atom. It is a trigonal planar carbon atom, with the three pairs of electrons each occupying an $sp^2$ hybrid orbital and an empty $p$ orbital perpendicular to the plane defined by the carbon atom and the two substituents on it. The form of the carbene described above is called its **singlet state,** in which both nonbonding electrons occupy a single orbital and have opposing spins. For some carbenes, another electronic state, in which the nonbonding electrons become unpaired and each one occupies a different orbital, is more stable. Such carbenes are said to be in the **triplet state.** Triplet carbenes, because of their unpaired electrons, behave more like radicals than singlet carbenes do and do not add to double bonds with retention of the stereochemistry of the alkene.

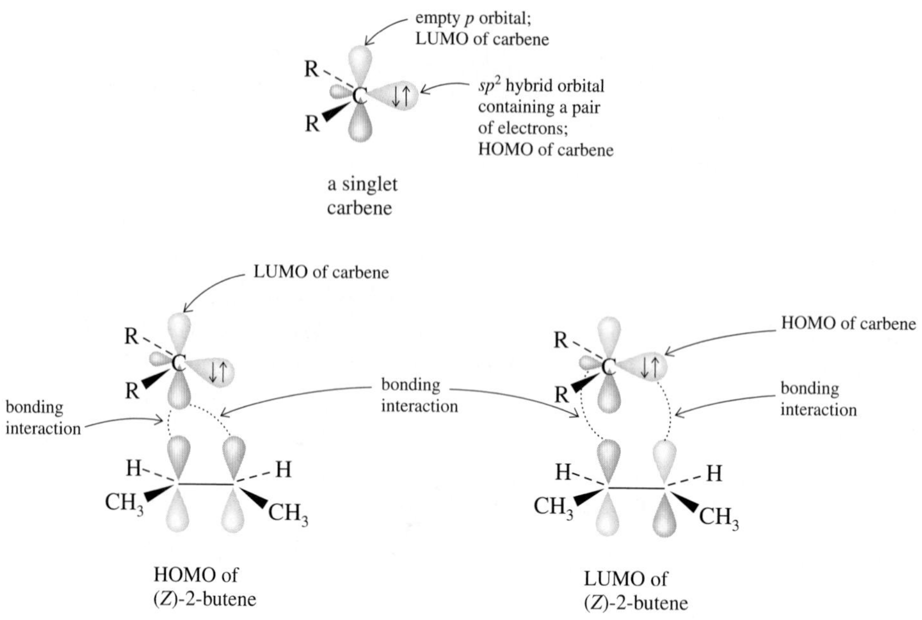

*Figure 25.15*

A representation of the interactions possible between the molecular orbitals of a carbene in its singlet state and an alkene.

The cycloaddition reactions of dibromomethylene are reactions of the singlet state of the carbene. Note that reaction is possible only if the carbene approaches the alkene sideways so that the plane defined by the carbon atom and its two substituents parallels the plane of the alkene. In this orientation, the empty *p* orbital of the carbene is pointing toward the electrons of the $\pi$ bond. The carbene differs from an alkene in having this empty orbital, which makes a sideways approach of the carbene to the $\pi$ bond and reaction to give a three-membered ring possible without the creation of an impossible amount of strain in the transition state.

**Problem 25.28**    Prove to yourself that an approach of a carbene in which its nonbonding electrons are pointing toward the $\pi$ orbital of the alkene cannot lead to a concerted cycloaddition reaction.

Addition of carbenes to double bonds is a useful way to synthesize cyclopropanes. One method often used for this reaction involves an organozinc reagent known as the **Simmons-Smith reagent,** for the two American chemists, Howard E. Simmons and Ronald D. Smith, who developed it. The reagent is prepared by treating metallic zinc with a salt of copper to make a zinc–copper couple, which reacts with diiodomethane to give the active species, usually formulated as $CH_2(ZnI_2)$. The exact structure of the reagent is not known, but it adds a methylene group to a wide variety of alkenes. The preparation of the Simmons-Smith reagent and its reaction with 2-cyclohexenone are shown below and on the next page.

$$CH_2I_2 \ + \ Zn(Cu) \ \xrightarrow{\text{diethyl ether}} \ CH_2(ZnI_2)$$

diiodomethane    zinc–copper
couple

organozinc reagent
Simmons-Smith reagent

$$\text{CH}_2(\text{ZnI}_2) \quad + \quad \underset{\substack{\text{2-cyclohexenone}}}{\text{(2-cyclohexenone)}} \quad \xrightarrow[\substack{\text{ether} \\ \Delta}]{\text{diethyl}} \quad \underset{\substack{\text{bicyclo[4.1.0]heptan-} \\ \text{2-one} \\ 90\%}}{\text{(product)}} \quad + \quad \text{ZnI}_2$$

Simmons-Smith
reagent

A variation of the Simmons-Smith reagent uses ethylzinc iodide with di-iodomethane to create the active species. This reagent often gives better yields of cyclopropanes than the original reagent. For example, styrene can be converted into phenylcyclopropane under these conditions.

$$\underset{\text{ethylzinc iodide}}{\text{CH}_3\text{CH}_2\text{ZnI}} \quad + \quad \underset{\text{diiodomethane}}{\text{CH}_2\text{I}_2} \quad + \quad \underset{\text{styrene}}{\text{(styrene)}-\text{CH}=\text{CH}_2} \quad \xrightarrow[\Delta]{\text{diethyl ether}} \quad \underset{\substack{\text{phenylcyclopropane} \\ 78\% \\ [32\% \text{ with Zn(Cu) reagent}]}}{\text{(phenylcyclopropane)}}$$

⊘ **Study Guide**
**Concept Map 25.5**

**Problem 25.29**    Complete the following equations.

(a) $\text{CH}_2\text{I}_2 \xrightarrow[\substack{\text{diethyl} \\ \text{ether}}]{\text{Zn(Cu)}}$

(b) $\underset{\text{CH}_3}{\overset{\text{CH}_3}{>}}\text{C}=\text{C}\underset{\text{CH}_3}{\overset{\text{CH}_3}{<}} \quad + \text{CHCl}_3 + \text{CH}_3\overset{\text{CH}_3}{\underset{\text{CH}_3}{\text{CO}^-}}\text{K}^+ \xrightarrow{-10\,°\text{C}}$

(c) $\underset{}{\text{(cyclohexene)}} + \text{CH}_2\text{I}_2 \quad + \quad \text{CH}_3\text{CH}_2\text{ZnI} \xrightarrow[\Delta]{\text{diethyl ether}}$

(d) $\text{CH}_2=\text{CHOCH}_2\text{CH}_3 \quad + \quad \text{CHCl}_3 + \text{CH}_3\overset{\text{CH}_3}{\underset{\text{CH}_3}{\text{CO}^-}}\text{K}^+ \xrightarrow{-10\,°\text{C}}$

(e) $\text{CH}_3(\text{CH}_2)_5\text{CH}=\text{CH}_2 \quad + \quad \text{CH}_2\text{I}_2 \xrightarrow[\Delta]{\text{Zn(Cu)}}{\text{diethyl ether}}$

(f) $\underset{}{\text{(cycloheptenol, OH)}} + \text{CH}_2\text{I}_2 \xrightarrow[\Delta]{\text{Zn(Cu)}}{\text{diethyl ether}}$

(g) $\underset{\text{CH}_3\text{CH}_2}{\overset{\text{H}}{>}}\text{C}=\text{C}\underset{\text{H}}{\overset{\text{CH}_2\text{CH}_3}{<}} \quad + \quad \text{CH}_2\text{I}_2 \xrightarrow[\Delta]{\text{Zn(Cu)}}{\text{diethyl ether}}$

In summary, a consideration of the symmetry of molecular orbitals and a classi-fication of a large number of organic reactions according to the number of electrons participating in the transition state have been used by chemists to explain experi-mental phenomena that seem unconnected at first glance. It is important to keep in mind, however, that in each case the theory is a rationalization of the experimental facts and an indication of new directions for experimentation. The concepts de-scribed in this chapter make it easier to correlate and remember facts about known reactions and to predict new reactions. The mere fact that reactions can be described in a certain way is no guarantee that they actually occur by those pathways. Experi-mental and theoretical work is still being done by chemists to improve their under-standing of reactivity, regioselectivity, and stereoselectivity in the transformations of organic compounds.

## Table 25.3    Summary of Concerted Reactions

### Cycloaddition Reactions

| Reactants | Conditions | Product(s) |
|-----------|------------|------------|
| 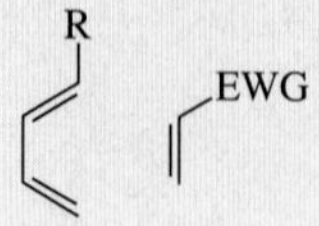 | Δ | 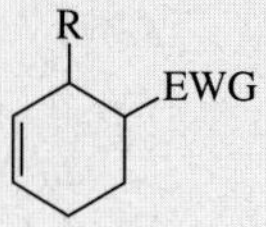 |
| 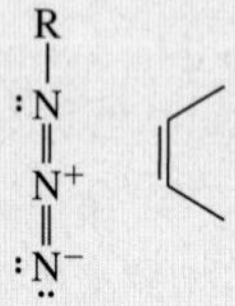 | Δ | 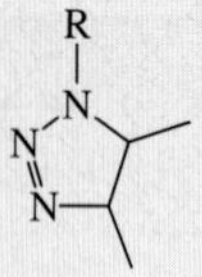 |
| 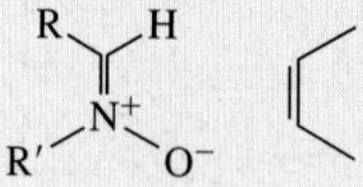 | Δ | 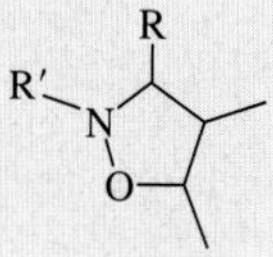 |
| | hv | |

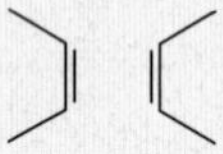

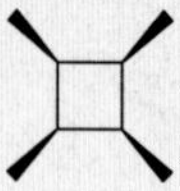

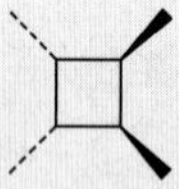

### Electrocyclic Reactions

| Reactant | Conditions | Product |
|----------|------------|---------|
| **4n π Electrons in Transition State** | | |
|  | Δ, conrotatory |  |
| | hv, disrotatory |  |
| **(4n + 2) π Electrons in Transition State** | | |
| 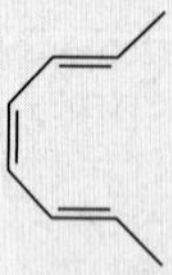 | Δ, disrotatory | 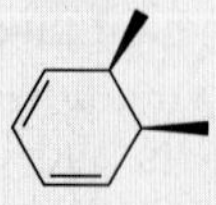 |
| | hv, conrotatory | 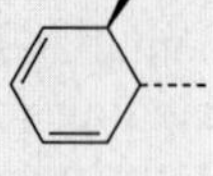 |

(continued)

**Table 25.3** (Continued)

**Sigmatropic Rearrangements**

| Reactant | Conditions | Product |
|---|---|---|
| 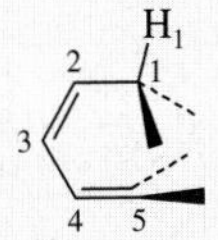 | Δ | 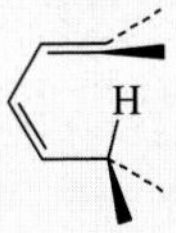 [1,5] |
| 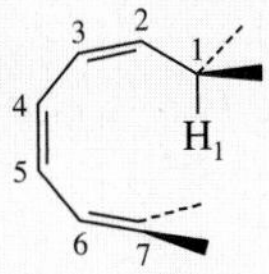 | Δ | 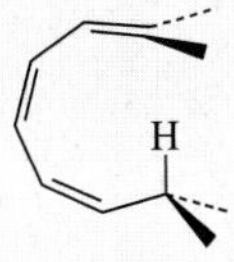 [1,7] |
| | hv | 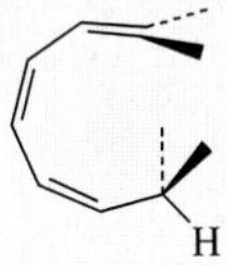 |
| 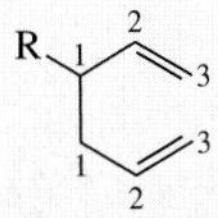 | Δ | 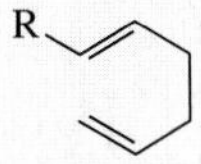 [3,3] |
| 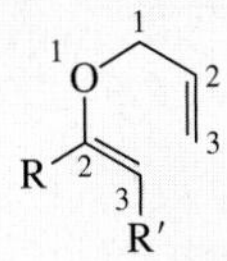 <br>(R and R′ may be part of an aromatic ring.) | Δ | 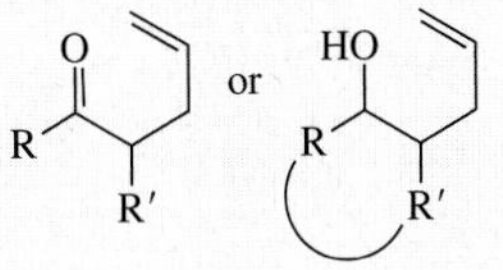 [3,3] |

**Carbene Reactions**

| Reactant | Reaction Conditions | Intermediate | Reagent for Second Step | Product |
|---|---|---|---|---|
| CHX$_3$ | CH$_3$—C(CH$_3$)(CH$_3$)—O$^-$ K$^+$ | X—C:—X |  | 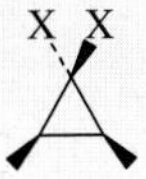 |
| | | | 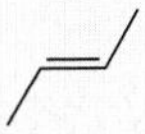 | 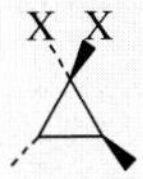 |
| CH$_2$I$_2$ | Zn(Cu) or CH$_3$CH$_2$ZnI | CH$_2$(ZnI$_2$) |  |  |
| | | | 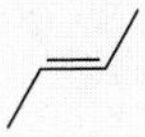 |  |

## ADDITIONAL PROBLEMS

**25.30** Complete the following equations. Be sure to show stereochemistry if it can be predicted.

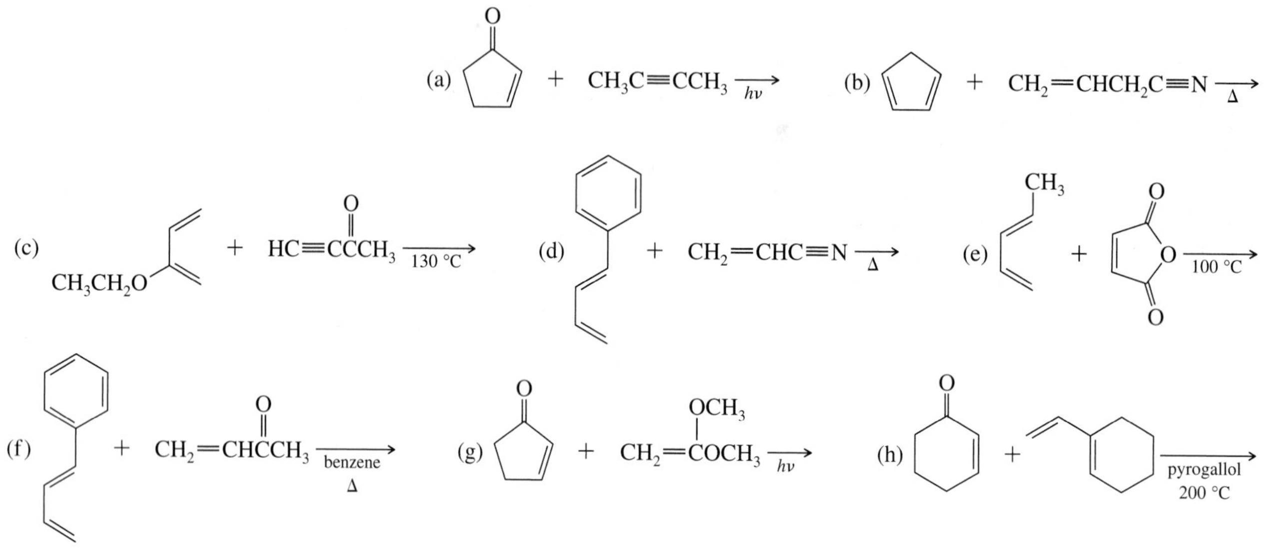

(a)  [cyclopentenone structure]  +  $CH_3C{\equiv}CCH_3$  $\xrightarrow{hv}$          (b)  [cyclopentadiene]  +  $CH_2{=}CHCH_2C{\equiv}N$  $\xrightarrow{\Delta}$

(c)  [1-ethoxybutadiene structure, $CH_3CH_2O$]  +  $HC{\equiv}CCCH_3$ (with O)  $\xrightarrow{130\ °C}$

(d)  [phenyl diene structure]  +  $CH_2{=}CHC{\equiv}N$  $\xrightarrow{\Delta}$

(e)  [methyl-substituted diene, $CH_3$]  +  [maleic anhydride structure]  $\xrightarrow{100\ °C}$

(f)  [phenyl diene structure]  +  $CH_2{=}CHCCH_3$ (with O)  $\xrightarrow[\Delta]{benzene}$

(g)  [cyclopentenone]  +  $CH_2{=}COCH_3$ (with OCH$_3$)  $\xrightarrow{hv}$

(h)  [cyclohexenone]  +  [isopropenyl cyclohexene structure]  $\xrightarrow[200\ °C]{pyrogallol}$

**25.31** Complete the following equations. Show stereochemistry if it is known.

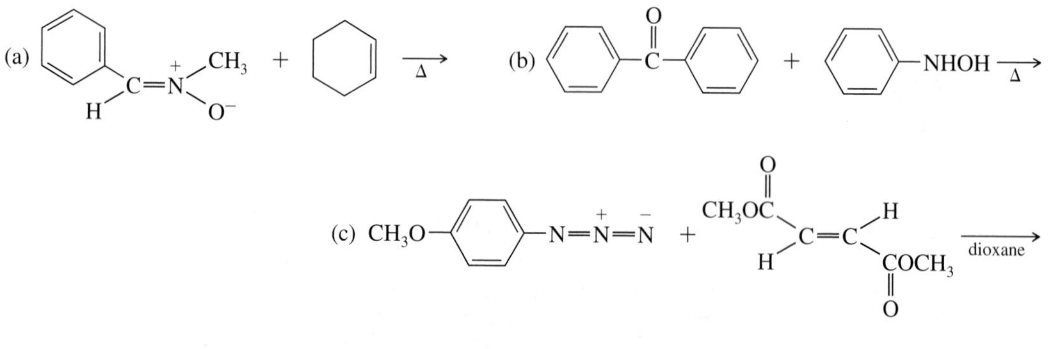

(a)  [phenyl C=N structure with CH$_3$, H, O$^-$]  +  [cyclohexene]  $\xrightarrow{\Delta}$

(b)  [benzophenone structure, diphenyl C=O]  +  [phenyl–NHOH]  $\xrightarrow{\Delta}$

(c)  $CH_3O$—[benzene ring]—$N{=}N{=}N$  +  [fumarate-type structure with $CH_3OC$, O, H, COCH$_3$, O]  $\xrightarrow{dioxane}$

**25.32** Complete the following equations.

(a)  $CH_3C{=}CHCCH_3$ with $OCH_2CH{=}CH_2$ and O  $\xrightarrow{\Delta}$

(b)  [phenyl structure with allyl chains]  $\xrightarrow{180\ °C}$

(c)  $N{\equiv}C$, $N{\equiv}C$ structure with $CH_2CH_3$, $CH_3$, allyl  $\xrightarrow{150\ °C}$

(d)  [phenyl with $OCH_2C{=}CH_2$ and $CH_3$]  $\xrightarrow[200\ °C]{N,N\text{-diethylaniline}}$

(e)  [phenyl with $OCH_2CH{=}CH$—phenyl and $CH_3$]  $\xrightarrow[200\ °C]{N,N\text{-dimethylaniline}}$

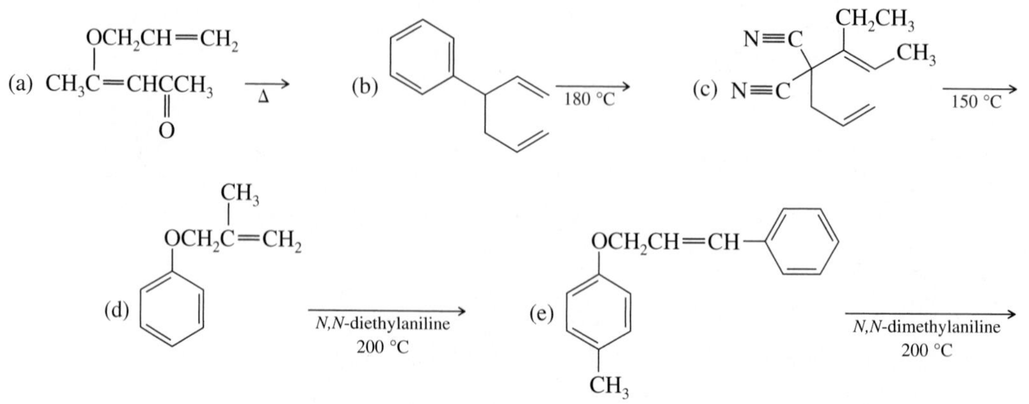

**25.33** Predict what the products of the following reactions will be.

(a)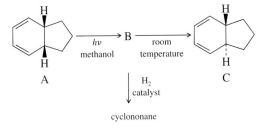

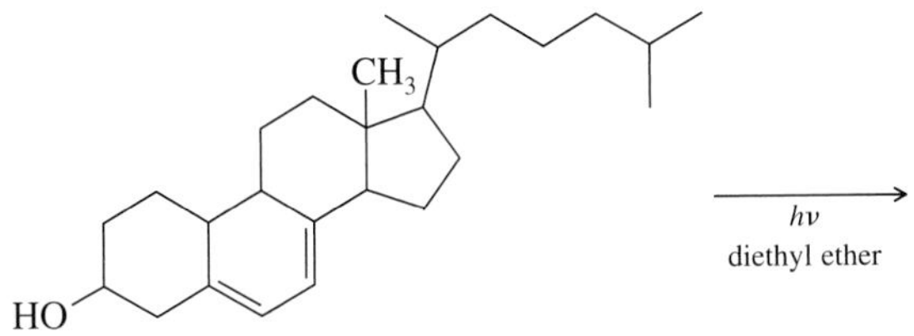

$\xrightarrow[\text{diethyl ether}]{hv}$

7-dehydro-19-norcholesterol
(7-dehydrocholesterol lacking the
methyl group numbered carbon 19)

(b) 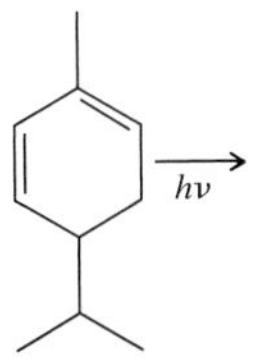 $\xrightarrow{hv}$

(c) 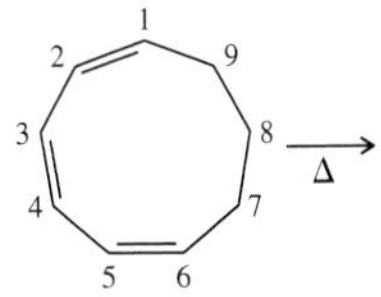 $\xrightarrow{\Delta}$

(d) 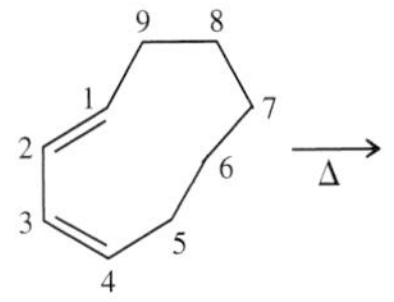 $\xrightarrow{\Delta}$

(e) 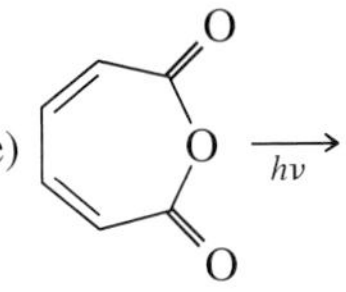 $\xrightarrow{hv}$

(f) 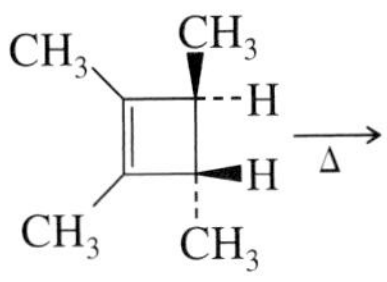 $\xrightarrow{\Delta}$

**25.34** Complete the following equations.

(a)  $\text{cyclopentene-CH}_3$ + $CH_2I_2$ $\xrightarrow[\Delta]{\text{Zn(Cu)}}$ $\xrightarrow{\text{diethyl ether}}$

(b) $\begin{array}{c} CH_3 \quad CH_3 \\ C{=}C \\ H \quad\quad CH_3 \end{array}$ + $CHBr_3$ + $CH_3\overset{CH_3}{\underset{CH_3}{C}}O^-K^+$ $\longrightarrow$

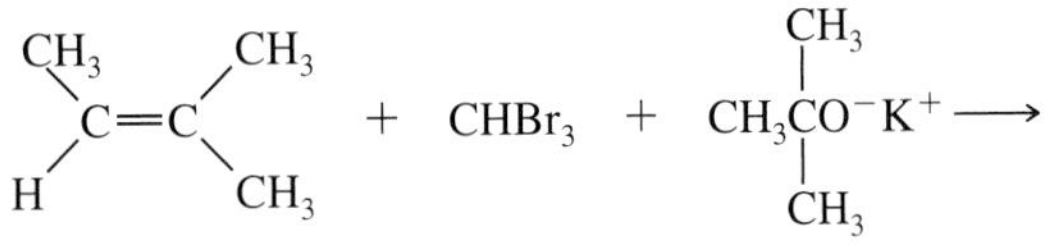

(c) $CH_2{=}CHCCH_3$ ($C{=}O$) + $CH_2I_2$ $\xrightarrow[\Delta]{\text{Zn(Cu)}}$ $\xrightarrow{\text{diethyl ether}}$

(d) 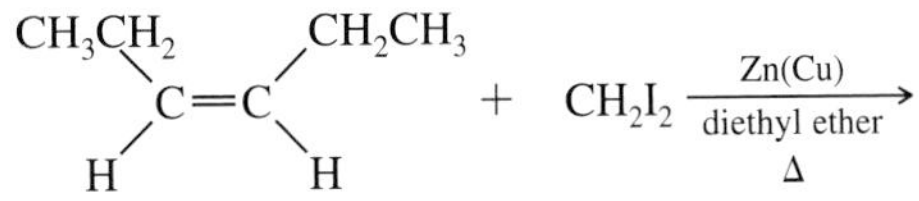 $\begin{array}{c} CH_3CH_2 \quad CH_2CH_3 \\ C{=}C \\ H \quad\quad\quad H \end{array}$ + $CH_2I_2$ $\xrightarrow[\Delta]{\text{Zn(Cu)}}$ $\xrightarrow{\text{diethyl ether}}$

**25.35** When 2-cyclohexenone reacts photochemically with cyclopentene, four isomeric products are formed, Compounds A (68%), B (7%), C (25%), and D (1%). When the mixture is treated over a long period of time with base, Compounds B and C disappear, and the amounts of Compounds A and D increase. Propose possible structures for the compounds. Which of the proposed structures would be easily isomerized by base?

**25.36** The following experimental observations were made:

$A \xrightarrow[\text{methanol}]{hv} B \xrightarrow[\text{temperature}]{\text{room}} C$

$B \xrightarrow[\text{catalyst}]{H_2}$ cyclononane

When *cis*-bicyclo[4.3.0]nona-2,4-diene (Compound A) is irradiated at $-20\ °C$ in methanol, Compound B is obtained. Compound B has $\lambda_{max}$ 290 nm ($\epsilon$ 2050) and bands in the infrared at 1645, 1621, 1598, 975, 960, and 670 cm$^{-1}$. It has two broad multiplets in its nuclear magnetic resonance spectrum, six protons in the region 6.2–4.7 ppm and six protons at 2.6–0.6 ppm. If the temperature is kept below 0 °C, Compound B can be hydrogenated to cyclononane. At room temperature, it is converted to *trans*-bicyclo[4.3.0]nona-2,4-diene (Compound C). How would you explain these observations? What is the structure of Compound B?

**25.37** Cyclopentadiene reacts with (*E*)-1,2-dichloro-1,2-difluoroethylene to give two different products, shown on the following page.

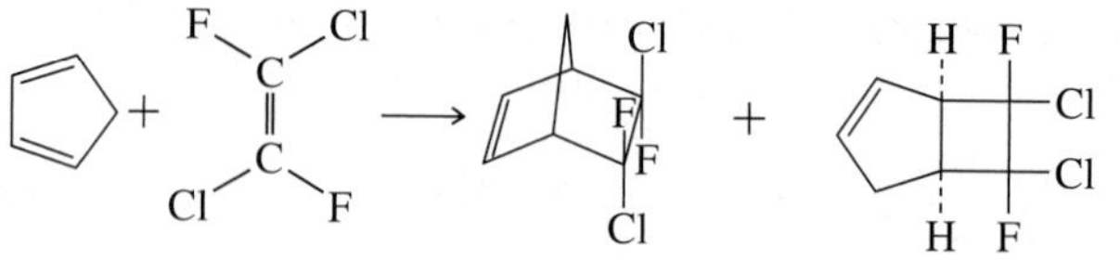

and enantiomer
98%

mixture of 4
possible diastereomers
2%

How do you account for the differing stereoselectivity of the reactions leading to the two types of products?

**25.38** Write a mechanism for the following experimentally observed transformation.

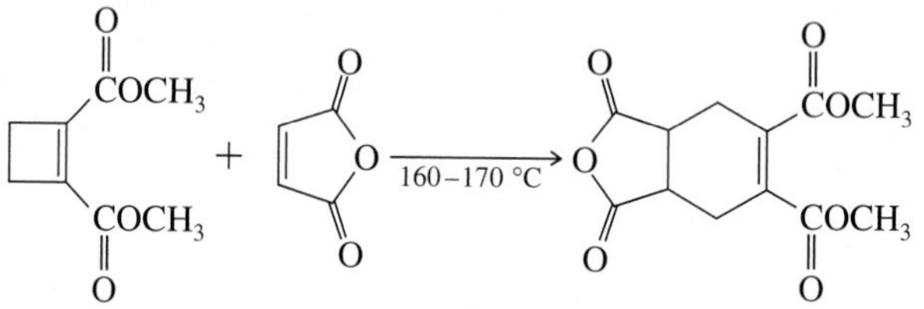

**25.39** The cyclobutene diol shown below gives 3,4-dimethyl-2,5-dione, among other products, on heating. How would you explain the formation of this product?

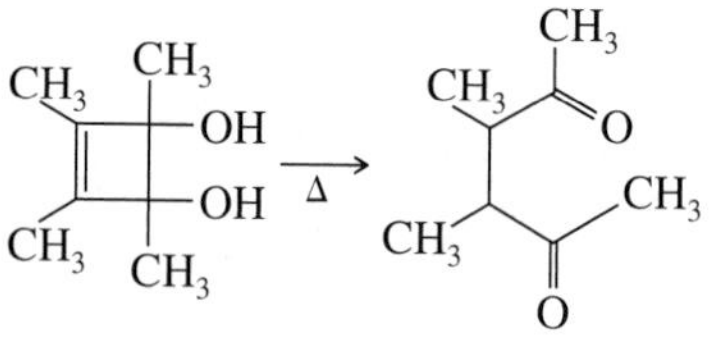

**25.40** A family of small brown seaweeds produces a diterpene with a unique bicyclo[6.1.0] nonane ring system incorporating a cyclopropane ring. A synthesis of the compound had the following two steps. Supply structural formulas for the products.

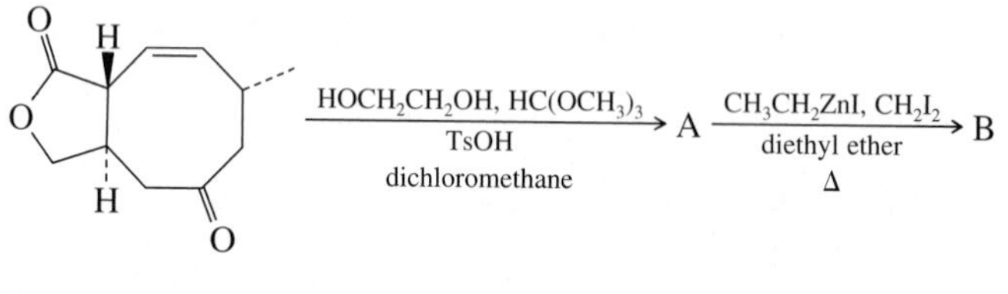

**25.41** An important step in the synthesis of the antibiotic funiculosin involves an electro-cyclic reaction. The steps are outlined below.

(a) Provide structural formulas for the reagents, intermediates, and products designated by letters.

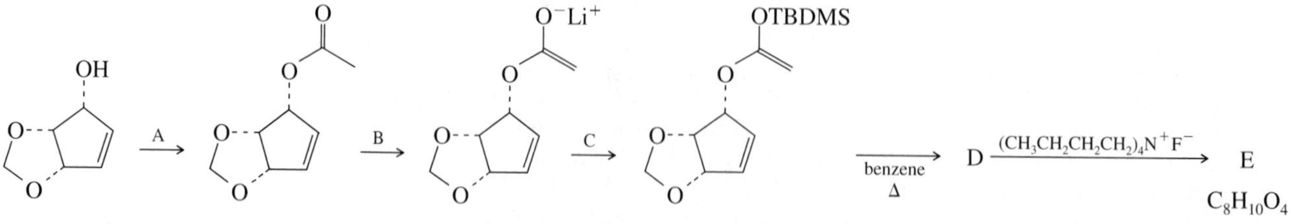

(b) Compound E has peaks in its proton magnetic resonance spectrum at $\delta$ 2.5, 2.6, 3.15, 4.6, 4.7, 5.0, 5.2, 5.7, and 6.0, each representing one hydrogen. A hydrogen that is exchangeable is not reported. The $^{13}C$ magnetic resonance spectrum of Compound E has peaks at δ 33.8 (t), 44.0 (d), 77.8 (d), 84.9 (d), 93.4 (t), 127.4 (d), 138.6 (d), and 177.4 (s). Assign as many of the spectral peaks as possible.

**25.42** The following sequence of reactions was carried out. Assign structures to the compounds designated by letters.

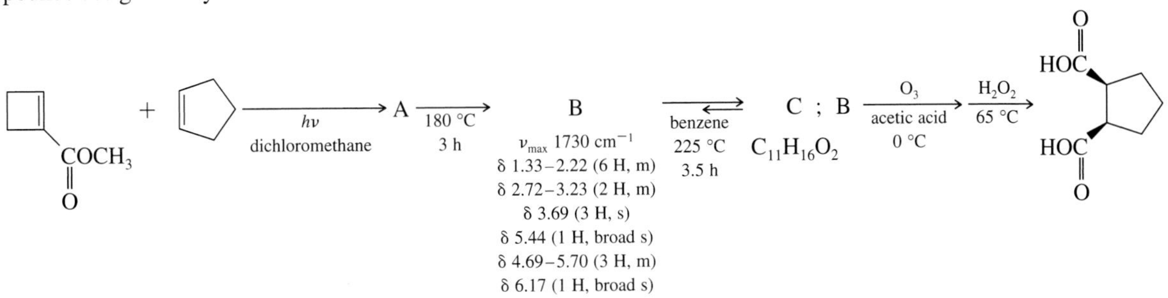

**25.43** Bao Gong Teng A is an alkaloid isolated from a Chinese herb traditionally used to treat fever in humans.

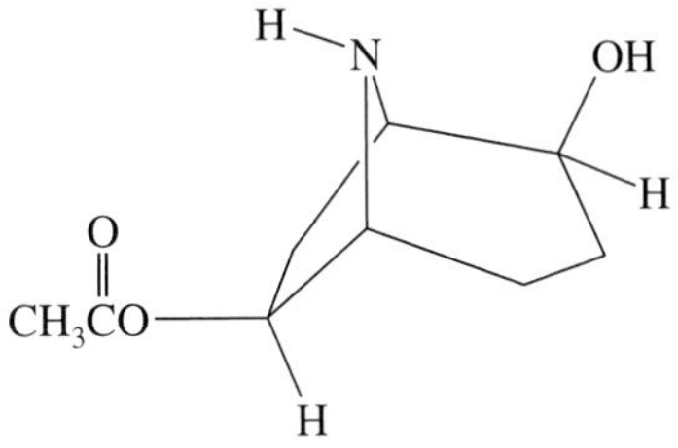

Bao Gong Teng A

It has been found to have fewer side effects than standard medications used to treat glaucoma. The scarcity of the herb requires that a synthesis be developed for the compound (see Problem 22.40). A critical step in one synthesis is a 1,3-dipolar cycloaddition reaction. The dipolar species is prepared as shown below.

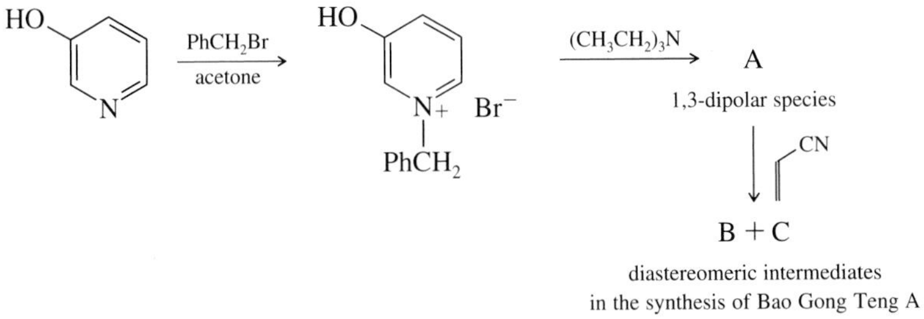

diastereomeric intermediates
in the synthesis of Bao Gong Teng A

(a) What is the structure of the dipolar species A?
(b) What are the structures of B and C, the products of 1,3-dipolar cycloaddition to acrylonitrile? Which one can be converted into Bao Gong Teng A?

**25.44** A critical step in the synthesis of an alkaloid is a photochemical electrocyclic reaction. The compound that undergoes the reaction was synthesized in the following way.

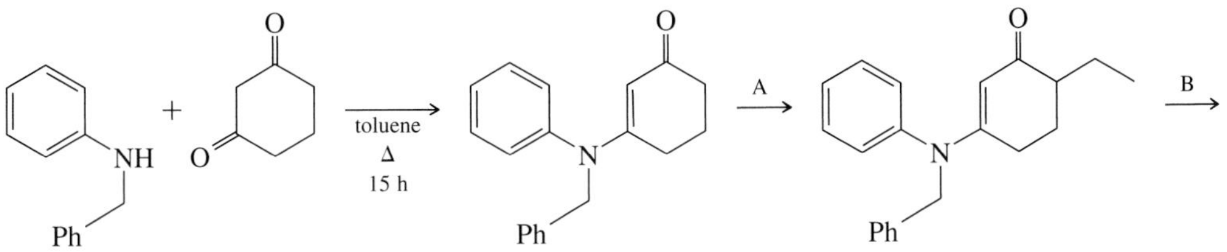

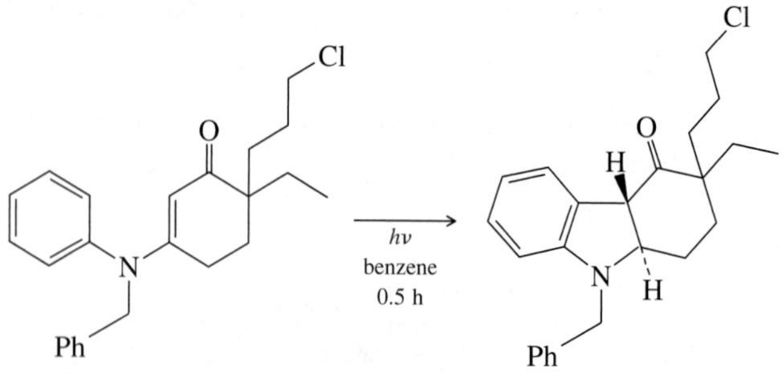

(a) Provide structural formulas for the reagents designated by letters.

(b) The product shown is not the immediate product of the photochemical cyclization reaction but the result of an isomerization. How many electrons are available for the electrocyclic reaction? What is the structure of the intermediate that forms on photolysis?

**25.45** (a) A key step in the synthesis of a natural product is the following. Predict what the product of the reaction will be.

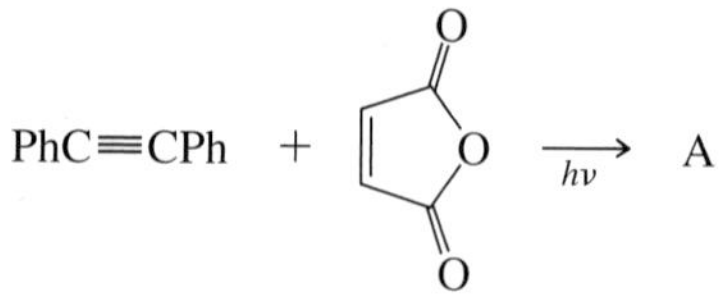

(b) Compound A is further transformed in the following reactions. Provide structural formulas for the reagents or the products designated by letters.

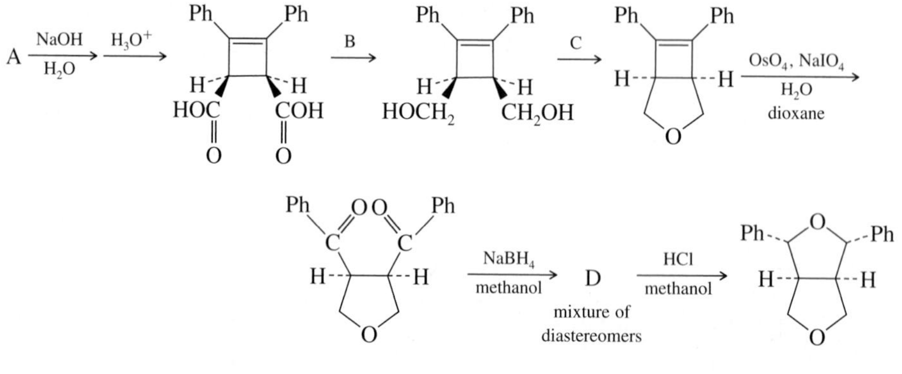

**25.46** Diazirines are compounds containing a nitrogen–nitrogen double bond in a three-membered ring. Such species lose nitrogen easily to give a carbene. The diazirine derived from 2,3,4,6-O-benzyloxyglucose was used to attach a glucose unit to $C_{60}$. Once the protecting groups are removed, the glucose unit would increase the solubility of $C_{60}$ in water, helping studies of the biological properties of the fullerene. Refer back to the structure of buckminsterfullerene, especially the partial structure shown on page 364 , and predict the product of the reaction shown below.

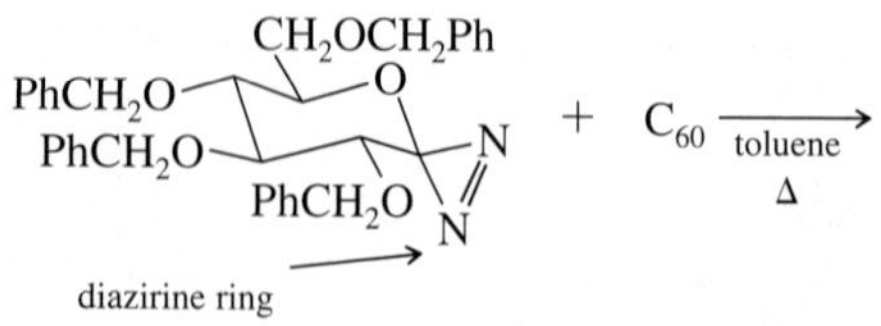

# INDEX

Boldface page numbers indicate text pages on which indexed terms are defined or described. The letter *t* after a page number refers to a table.

Electrons (*Continued*)
delocalization of
in benzene, 68–69
in nitro compounds, 658, 898
nonbonding, **6**
spin of, **42**
valence, 6
Electrophile(s), 115–117, **116**
reaction with aromatic compounds. *See* Substitution reactions, electrophilic aromatic
reaction with multiple bonds. *See* Addition reactions, Alkenes, Alkynes
Electrophilic addition reaction, 126–134, **127**
Electrophilic alkenes, **710**
reactions of, 710–717, 729*t*
Electrophilic center, **235**
Electrophilic reagents, 136*t*
Electrostatic forces, **4**–5
Elimination reactions, **233**–234, **264**–271
bimolecular (E2), 264–271, **267**
mechanism of, 267
stereochemistry of, 269–271
competition with substitution reactions, 233–234, 264–266
dehydrohalogenation, 264–269
α-elimination, mechanism of, 1110
β-elimination, 1094. *See also* Elimination reactions, bimolecular
unimolecular (E1), 247, 264–269, **267**
mechanism of, 247
Empirical resonance energy, **352**
Emulsification, **642**
Emulsin, cleavage of β-glucopyranosides by, 988, 989
Enantiomeric excess, 202–204, **203**
Enantiomers, 191–193, **192**
optical activity of, **193**
separation of. *See* Resolution, of a racemic mixture
Enantiotopic hydrogens, **950**
Endo stereochemistry, 744–747, **745**
Endonucleases, **1027**
Endothermic reaction, **62**
Energy
bond, 62–65
diagram, **123**
for acid–base reactions, 102
for addition reactions to alkenes, 131, 132, 299
for bimolecular substitution reactions, 243, 244
for cleavage of ethers, 506
for conformations of butane, 168
for conformations of ethane, 165
for electrophilic aromatic substitution, 371
for hydrogenation reactions, 299
levels
electronic, 69–72, 395
transitions between, 395, 395*t*, 443–445
energy of transitions between, 395*t*
$\pi \rightarrow \pi^*$, 71, 72
nuclear, 395*t*, 396
transitions between, 395*t*, 396
rotational, 395
energy of transitions between, 395*t*
vibrational, 395
relationship to wavelength, 71–72, 395, 395*t*
standard free energy, 95–96, 101, 120–121
Enkephalins, 1033
Enol ethers, 350, **562**
as products of Diels-Alder reaction, 888–889
trimethylsilyl, 865–866, 888

Enolate anions, **669**–670, 683*t*, 692–736, 877–883. *See also* Carbanions, Enols
as ambident nucleophiles, **695**–696
formation, 669–671, 683*t*, 865–868
regioselectivity of, 699–700, 865–868
kinetic, **867**
as nucleophiles, 674
reactions, 727*t*, 728*t*
with acyl halides, 705
with alkyl halides, 695–697, 697–701, 727*t*
with carbonyl compounds, 701–710
with esters
Claisen condensation, 705–710, 728*t*
Dieckmann condensation, 881–882
with halogens, 693–694, 727*t*
with α,β-unsaturated carbonyl compounds (Michael reaction), 710–712, 727*t*
relative stabilities of, 671, 673
stabilization of by resonance, 673
thermodynamic, **866**
Enolization, **669**–674, 677–683
acid-catalyzed, 670–671
base-initiated, 669–670
reactions, biological importance of, 677–683
in glycolysis, 677–679
Enols, **337**
formation of, 669–671, 683*t*
reactions of
aldol condensation of, mechanism of, 701–702
with halogens, mechanism of, 694
hydrogen exchange, 674–675
tautomerism of, **338**–339
Enols and enolates, 669–671
as intermediates in exchange of protons in carbon acids, 674–677
reactions of, 693–697
alkyl halides as electrophiles, 695–697
halogens as electrophiles, 693–694
a unified look at, 692–693
Entgegen (*E*), designation of stereochemistry at double bonds, **222**
Enthalpy, 62, **96**–97
Entropy, **96**–97
Enzymes, 1
degradation of proteins and peptides by, 992–1002
in the resolution of mixtures of enantiomers, 203
Epichlorohydrin. *See* 3-Chloro-1-propene, oxirane from
Epimerization, **676**
Epimers, **676**
Epinephrine. *See* Adrenalin
Epothilones A and B, synthesis of, 907
Epoxides, **317**–319. *See also* Oxiranes
Epoxy resins, 1065–1067
synthesis of, 1065–1066
uses of, 1065
Equilibria, acid–base, 94–103
Equilibrium, free energy and, 95–97
Equilibrium constant, **94**–96
Equivalence, of groups and atoms, 149–155, 161–164
exploration by chemical substitution, 161–163
as observed with nuclear magnetic resonance spectroscopy, 149–155, 161
carbon spectra for 1,1 and 1,2 dichloroethane, 150, 152
carbon and proton spectra for 1-bromo-2-methylpropane, 162
concept of neighboring atoms, 163–164
proton spectra for 1,1 and 1,2 dichloroethane, 150, 152
Ergosterol
electrocyclic ring opening reaction in, 1096
as precursor to vitamin $D_2$, 1096–1097

Galactose, 980, 989
Gangliosides, 992
Garlic, and carcinogens, 1032
Gasoline, 184
Gauche conformation, **166**–168
Gaultherin, determination of the structure of, 1032–1033
Gelatin, from collagen, 1009
Genetic code, **1020**–1022
   transmission of, importance of base pairing in, 1020, 1021
Genetic engineering, 1027
Gentiobiose, determination of the structure of, 990
Gentisic acid, 1032
Geraniol, 750, 751
   biosynthesis of, 752–753
Geranyl pyrophosphate, in preparation of farnesol, 754
Glass, **1043**
Glass transition temperature, **1043**
Glatial acetic acid, **599**
Gleosporone, synthesis of, 261
Globular proteins, **1009**
Glucaric acid, as product of oxidation of glucose, 981
$\alpha$-D-Glucofuranose, 566, 982. *See also* Glucose
$\beta$-D-Glucofuranose, 566, 982. *See also* Glucose
$\alpha$-D-Glucopyranose, 566, 982. *See also* Glucose
$\beta$-D-Glucopyranose, 566, 982. *See also* Glucose
4-*O*-($\alpha$-D-Glucopyranosyl)-D-glucopyranose. *See* Maltose
4-*O*-($\beta$-D-Glucopyranosyl)-D-glucopyranose. *See* Cellobiose
Glucosamine, 972
Glucose, 2, 537, 971, 975, 978, 980
   conversion to fructose, 677–678
   as cyclic hemiacetal, 564, 982
   determination of structure of, 564–568
   formation of methyl acetal of, 569
   infrared spectrum of, 564
   as monomer in cellulose, 563, 1038
   mutarotation of, **566**–567
   proton magnetic resonance spectra of, 565
   reactions of
      conversion to fructose, mannose, 677–678
      oxidation with
         Benedict's reagent, 608
         bromine, 608–609
         nitric acid, 981
         Tollens reagent, 608
      stereoisomers of, 980
$\alpha$-Glucose, **566**. *See also* Glucose
$\beta$-Glucose, **566**. *See also* Glucose
Glucose-6-phosphate, 682
Glucosides, methyl, formation of, 569
Glucuronic acid, 972, 1006
Glutamic acid, 668–669, 985
   biosynthesis of, 984
   from glutamine, 995
   p$K_a$'s and p$I$ for, 665
Glutamine, 985
   decomposition in acid, 995
Glutaric acid (pentanedioic acid), 602
Glutathione, 790
Glyceraldehyde (2,3-dihydroxypropanal), 57
   configuration of, 210, 973–974, 978–979
      absolute, 979–982
      relative, 973–974
   Fischer projection of, 974–977
   D-fructose and D-sorbose from, 734
   oxidation of, 973
D-Glyceraldehyde 3-phosphate, 678, 720–722
Glyceric acid (2,3-dihydroxypropanoic acid), from glyceraldehyde, 973

Glycine, 113, 985
   synthesis of, 809
Glycogen, 972, 1003
Glycolysis, 677–679, **719**
Glycosaminoglycans, 1006–1007
Glycosides, **569**
   cardiac, **571**, 990–991
   formation of, 569–571
   hydrolysis of, 570
Glycosidic linkage, 988–992
   in cellulose, 1039
5'-GMP. *See* Guanosine 5'-phosphate
Gomberg, Moses, 780
Goodyear, Charles, 1074
Gout, 947
Grain alcohol, **490**. *See also* Ethanol
Grandisol, photochemical reactions in synthesis of, 1088, 1107
Graphite, 363
Grignard, Victor, 550
Grignard reaction, 551–552, **532**. *See also* Grignard reagents, reactions with
Grignard reagents, 549–551, **550**, 874–876, 915*t*
   preparation of, 549–551
   reactions with
      aldehydes and ketones, 551–554, 555*t*
         conjugate addition of, 864
         mechanism of, 552
      alkynes, 551, 552
      carbon dioxide, 609–610
      esters, 874–876
      *N*-methoxy-*N*-methylamides, 876
      oxiranes, 554–556, 555*t*
      water, 550
Ground state, **69**
   for 2-butene, 1086–1087
   cycloaddition reactions allowed in, 1088–1090, 1098*t*, 1098–1100
   for ethylene, 1083
   stereochemistry of electrocyclic reactions in, 1090–1092, 1099*t*
Guaiazulene, 766
Guaiol, 766
Guanine, 946
   base pairing with cytosine, 1017, 1018
Guanosine 5'-phosphate (5'-GMP), 987
5'-Guanylic acid. *See* Guanosine 5'-phosphate
D-Gulose, 980
Gutta percha, 1073
   biosynthesis of, 1073
   properties and uses of, 1073–1074
   stereochemistry of double bonds in, 1042, 1073
   structure of, 1073


Halides. *See* Alkyl halides, Allylic halides, Aryl halides, Vinyl halides
Halogen acids, trends in acidity of, 105–106
Halogenation reactions, 797*t*. *See also individual halogens*
   of alkanes, 769–774
   at allylic positions, 775–778
   of aromatic compounds, 942–944
   at benzylic positions, 778–780
   of carbonyl compounds, 693–694
Halogens, as electrophiles, 302, 369–371
Halohydrins, 302, 306–309
   preparation of, from alkenes and hypohalous acids, 306–309
Halons, 186
Halothane, 12
Hammond Postulate, 124, 244, 373
Hassel, Odd, 180

**1152**

Palmitic acid (hexadecanoic acid), 638, 724
  biosynthesis of, 722–724
  esters of, 640
  melting point of, 638
Pancreatic ribonuclease, used to cleave *t*RNA for alanine, 1027
Pantothenic acid, 637, 722–723
Papaverine, 961
Para position, 360
Para red, synthesis of, 828
Para-directing groups, **368**–369, 369*t*
Parkinson's disease, 961, 962
Pasteur, Louis, 204, 224
Paul, Aniko V., 1026
Pauli exclusion principle, **43**
Pauling, Linus, 19, 26, 1008
Penaresidin A, 530
Penicillin G, 934
Penicillins, 965
*Penicillum glaucum*, 204
1-Pentadecyne, in synthesis of pheromone, 332
1,3-Pentadiene
  sigmatropic rearrangement in, 1103–1104
  ultraviolet spectrum of, 447
Pentadienyl cation, electrocyclic reaction of, 1099
(2*R*,3*S*,4*R*,5*R*)-(+)-2,3,4,5,6-Pentahydroxyhexanal, 563–564
Pentalenene, intermediates in the synthesis of, 589, 883
Pentanal, 538
  oxime of, reduction of, 811
Pentane, 145
  boiling point and/or melting point of, 145, 145t, 806
  condensed formula of, 10
  insolubility in water of, 34
  Lewis structure of, 10
1,5-Pentanediamine (cadaverine), 805
Pentanedioic acid. *See* Glutaric acid
2,4-Pentanedione
  active methylene groups in, 671
  base-catalyzed enolization of, 672
  enol form of, 672
  enolate ion for, 672
  p$K_a$ of, 671, 672, 684
  nuclear magnetic resonance spectrum of, 671
  reaction with hydrazine, 939
  reaction with hydroxylamine, 939
  reaction with urea, 940, 941
Pentanenitrile, 606
  preparation of, 258
  reduction of, 811
Pentanoic acid, 600
  boiling point and melting point of, 600
  solubility in water of, 600
1-Pentanol
  infrared spectrum of, 483
  retrosynthetic analysis for, 838–840
3-Pentanol, reactions with
  hydrogen bromide, 495
  thionyl chloride, 496
2-Pentanone
  condensation with butanal, 703
  mass spectrum of, 471
3-Pentanone, from hydration of 2-pentyne, 335
  mass spectrum of, 471
2-Pentene, reaction with bromine
  mechanism of, 305
  stereochemistry of, 304–305
3-Penten-2-one, ultraviolet spectrum of, 449
Pentose, 971
Pentylamine, from oxime of pentanal, 811

1-Pentyne, 99, 236, 330
  boiling point of, 330
  p$K_a$ of, 236
2-Pentyne, addition of water, 335
Peptide linkage, 615, 805
  bond angles and bond lengths for, 1007
  double bond character of, 1007
  formation of, 902
Peptide P, 1035
Peptides, 615, 993–1002
  cleavage of, 615
  conformation of, 1007–1112
  degradation of, 994–1002
  determination of structure for, 995–1002
  isoelectric point of, 994
  nomenclature of, 993
  synthesis of
    acyl transfer reactions in, 629–636
    polymer support for, 901
    solid phase method, 901–906
    use of protecting groups in, 630–633, 901–903
Perchloroethylene. *See* 1,1,2,2-Tetrachloroethene
Perchlorotriphenylmethyl radical, unreactivity of, 781
Permanganate oxidation
  of alcohols, 607
  of aldehydes, 607
Peroxide, 16
Peroxide effect, 783
Peroxyacids
  as electrophiles, 317
  as oxidizing agents, 317–319
Peroxycarboxylic acids, 317. *See also* Peroxyacids
Peroxyformic acid, preparation of, 317
Perspective formulas, 165, 167
Petroleum
  boiling points of fractions of, 184
  components of, 184
Phenanthrene(s), 360, 362
  bromination of, 362
  photochemical synthesis of, 1099
  reactivity of, 361–362
  resonance contributors of, 361–362
Phenol
  nuclear magnetic resonance spectrum of, 411
  p$K_a$ of, 655, 657
  reactions of
    bromination, 367, 373–374
    nitration, 367
  ultraviolet, major bands of, 451
Phenolate anion
  electrophilic substitution of, 367
  as a nucleophile, 235
  resonance stabilization of, 656
  ultraviolet, major bands of, 451
Phenols, 360, 655
  acidity of, 655–657
  as antioxidants, 794–795
  from diazonium ions, 896
  nuclear magnetic resonance spectra of, 411
  p$K_a$ of, 655, 657
  reactions of
    Claisen rearrangement, of ethers of, 1108–1109
    with diazonium ions, 825
    electrophilic aromatic substitution, 367, 373–374
    oxidation, 791–796
  use in synthesis of polymers, 1063–1064
Phenone suffix, 539
Phenoxy group, 492

Phenyl azide, 1,3-dipolar cycloaddition reactions of, 1100
Phenyl group, 176
Phenyl isocyanate, 829
Phenyl isothiocyanate (Edman's reagent), reaction with *N*-terminal amino acids, 997–998
Phenylacetic acid, reduction of, 855
Phenylacetonitrile
 hydrolysis to amide, 614
 hydrolysis to carboxylic acid, 614
 synthesis of, 614
Phenylalanine, 983
 genetic code for, 1022
1-Phenyl-1-butanone
 infrared spectrum of, 463
 nuclear magnetic resonance spectrum of, 424
4-Phenyl-2-butanone (benzylacetone), nuclear magnetic resonance spectrum of, 593
2-Phenylcyclohexanol, tosylate of, elimination reaction of, 276
4-Phenylcyclohexanone, preparation of, 888
2-Phenylcyclopentanol, tosylate of, stereochemistry of elimination reaction, 271
*m*-Phenylenediisocyanate, in synthesis of polyurethanes, 1041
2-Phenylethanol, nuclear magnetic resonance spectrum of, 534
Phenylethene. *See* Styrene
1-Phenylethyl cation, resonance contributors of, 250
α-Phenylethylamine (1-amino-1-phenylethane)
 from acetophenone, 811
 resolution of racemic mixture of, 227
β-Phenylethylamines, 815–816
 physiological properties of, 815
 synthesis of, 816–817
Phenylethyne, hydrogenation of, 339
Phenylhydrazine, 573, 575
 reaction of, with aldehydes and ketones, 575
Phenylhydrazones, 575
(*S*)-1-Phenyl-2-methyl-1-butanone, 675
 enol form of, 675
 racemization of, 675–676
2-Phenyloxirane, ring opening reactions of, 530
1-Phenyl-1,4-pentanedione
 cyclization to a furan, 938
 reaction with ammonia, mechanism of, 937–938
3-Phenyl-1-propanol, mass spectrum of, 533
3-Phenyl-2-propenal, reduction of, 863
2-Phenylpropene (α-methylstyrene), polymerization of, 1050–1052
2-(Phenylsulfonyl)ethanol, 684
Phenylthiohydantoins, 997
Pheromones, 747–749
 of the ambrosia beetle, 547, 848
 of the boll weevil, 1088, 1107
 of the cockroach, 879, 923
 of the fruit fly, 923–924
 of the gypsy moth (disparlure), 748, 908
 of the honeybee, 749
 of the pink bollworm, 748
 of the red-banded leaf roller, 748
 of the silkworm, 748
 of the square-necked grain beetle, 919
Philanthotoxin, synthesis of, 922
Phosgene, 829
Phosphate ester linkages, 1039
Phosphoenolpyruvate, synthesis of, 679
2-Phosphoglyceric acid, enolization reaction of, 678, 679
Phospholipids, 820
 function of, 820
Phosphonium ylide, 868–869, 916*t*
Phosphoric acid esters
 of adenosine, 262–263, 436–437

as leaving groups, 262–263
Phosphorus tribromide, reaction with alcohols, 497–498
Phosphorus trichloride, in preparation of acid chlorides, 610
Photochemical reactions, 1086–1088, 1092–1098, 1106
Photochemistry, 444
Phthalic acid (1,2-benzenedicarboxylic acid), 602
Phthalic anhydride, 604
Phthalimide, 605
Phylloquinone. *See* Vitamin K$_1$
Physiologic pH, 663
π bonds, 54. *See also* Orbitals, molecular, π
 compounds containing, as bases, 82
 effect on chemical shift of, 407–410
 orbital picture of, 54, 55, 58
Picric acid. *See* 2,4,6-Trinitrophenol
Pinacol. *See* 2,3-Dimethyl-2,3-butanediol
Pinacol-pinacolone rearrangement, 324–325
Pinacolone, 324–325
α-Pinene, 752
Piperidine, 808, 933
 reaction with 2,4-dinitrohalobenzene, 898–899
p$K_a$, 94–95, 100–102. *See also individual compounds or functional groups*
 of carboxylic acids, 103–105, 654–655, 661
 of conjugate acids, of substituted
  amines, 658
  pyridines, 935, 937
  pyrimidines, 936, 937
 effect of solvent on, 103
 in the gas phase, 103
 of substituted phenols, 657
 tables, inside front cover, 105, 107
 use in predicting acid–base reactions, 97–99
Planar molecules, 25
Planck's constant, 395, 444
Plane of symmetry, **194**
Plane-polarized light, 198–**199**
Plastics, **1043**
Platinum, as a catalyst, 298–300
β-Pleated sheet, conformation of proteins, **1009**. *See also* β-Keratin
Poisoned catalyst, **339**
Polarimeter, **199**
Polarity, bond, **26**
Polarizability, **128, 253**
 and nucleophilicity, 252–253
Poliovirus, synthesis of, 1026
Polyacrylonitrile (Orlon), 1, 1040
Polyamides, 1058–1060. *See also* Nylon
 aromatic, 1060, 1078–1079
 hydrogen bonding in, 1059
 preparation of, 1041, 1059
 properties and uses of, 1040, 1042–1044, 1059–1060
Polyanhydrides, synthesis of, 1071
Polychlorinated biphenyls (PCBs), 382–383
Polycyclic aromatic hydrocarbons, **361**–362
Polyenes, 737–767
 biologically interesting, 747–759
 electrocyclic reactions of, 1090–1100
Polyesters. *See* Poly(ethylene terephthalate)
Polyethylene, 1040
 melting point of, dependence on structure, 1042
 use of Ziegler-Natta catalysts in preparation of, 1057
Poly(ethylene terephthalate) (Dacron, polyesters, and Terylene), 1040, 1060–1061
 synthesis of, 1060
 uses of, 1060–1061, 1070
Poly(hexamethylenesebacamide). *See* Nylon
1,4-Polyisoprene, synthesis of, 1075–1076

# Acknowledgments

Infrared spectra from *The Aldrich Library of FT-IR Spectra,* by Charles J. Pouchert. Copyright © 1985 by Aldrich Chemical Company. Reprinted by permission of Aldrich Chemical Company.

Ultraviolet spectra adapted from *UV Atlas of Organic Compounds,* Vols. I and II. Copyright © 1966 by Butterworth and Verlag Chemie. Reprinted by permission of Butterworth and Company, Ltd.

Mass spectra adapted from *Registry of Mass Spectral Data,* Vol. 1, by E. Stenhagen, S. Abrahamsson, and F. W. McLafferty. Copyright © 1974 by John Wiley & Sons, Inc. Reprinted by permission of John Wiley & Sons, Inc.